BIOLOGY

BIOLOGY

SECOND EDITION

NEIL A. CAMPBELL

University of California, Riverside

The Benjamin/Cummings Publishing Company, Inc.

Redwood City, California • Fort Collins, Colorado • Menlo Park, California
Reading, Massachusetts • New York • Don Mills, Ontario
Wokingham, U.K. • Amsterdam • Bonn • Sydney • Singapore
Tokyo • Madrid • San Juan

Sponsoring Editor: Robin J. Williams

Developmental Manager: Jane Reece

Developmental Editor: Deborah Gale

Marketing Manager: Anne Emerson

Production Editor: Karen Gulliver

Production and Art Coordination: Pat Waldo and Deborah Gale/Partners in Publishing

Text and Cover Design: Gary Head

Dummy Artist: Gary Head

Photo Editor: Darcy Lanham

Photo Researchers: Kevin Schafer, Stuart Kenter, Audrey Ross, Carl May, Roberta Spieckerman

Artists: Chris Carothers, Raychel Ciemma, Barbara Cousins, Cecile Duray-Bito, Janet Hayes, Darwen Hennings, Vally Hennings, Georg Klatt, Linda McVay, Kenneth Miller, Fran Milner, Elizabeth Morales, Jackie Osborn, Carla Simmons, Carol Verbeeck, John Waller, Judy Waller

Copyeditor: Carol Dondrea

Proofreaders: Steve Sorensen, Jackie L. Comerford, Susan Scheib-Wixted, Gertrude Sanders

Indexer: Steve Sorensen

Compositor: Graphic Typesetting Service

Film Preparation: Color Response, Inc.

Figure acknowledgments begin on page C-1.

About the cover:

"Dogwood, Yosemite National Park, California, 1938." Photograph by Ansel Adams. Copyright © 1981 by the Trustees of the Ansel Adams Publishing Rights Trust. All rights reserved.

Library of Congress Cataloging-in-Publication Data
Campbell, Neil A., 1946-
    Biology/Neil A. Campbell.—2nd ed.
       p. cm.
    ISBN 0-8053-1800-3
    1. Biology.   I. Title.
QH308.2.C34   1990
574—dc20

BCDEFGHIJ-DO-893210

The Benjamin/Cummings Publishing Company, Inc.
390 Bridge Parkway
Redwood City, CA 94065

To Rochelle and Allison,
with love

About the Author

BIOLOGY is the product of 21 years of teaching experience and 11 years of intensive writing and revision by Dr. Neil A. Campbell. This textbook is a natural outgrowth of Dr. Campbell's broad interest in his science. He earned his M.A. in Zoology from UCLA, where he studied the control of protein synthesis during animal development, and went on to the University of California at Riverside, where he earned a Ph.D. in Biology. Dr. Campbell's research efforts on salt transport in plants and the cellular basis of leaf movements have resulted in publications in *Science*, *The Proceedings of the National Academy of Sciences*, and *Plant Physiology*, among other journals.

In addition to his accomplishments as a research scientist, Dr. Campbell has earned a reputation as an outstanding classroom teacher with a strong commitment to improving undergraduate education. After 10 years of teaching general biology and cell biology at San Bernardino Valley College, he took an academic leave and accepted a faculty position at Cornell University, where he reorganized a two-semester general biology course. After three successful years at Cornell, Dr. Campbell returned to California to reassume his teaching position at San Bernardino Valley College, where in 1986 he received the "Outstanding Professor Award" for excellence in classroom instruction. He frequently returned to Cornell to teach the summer general biology course to advanced placement high school students and Cornell undergraduates on a six-week schedule. In 1988, Dr. Campbell accepted an invitation to teach a one-semester general biology course at Pomona College.

During his many years of teaching general biology—most frequently as the sole lecturer—Dr. Campbell has instructed over 12,000 students. His teaching sensibilities have been honed in both large lecture and small classroom environments and with a diverse group of students. He is currently a Visiting Scholar in the Department of Botany and Plant Science at the University of California at Riverside.

Preface

The first edition of *BIOLOGY* took form in the classroom, shaped in part by advice and encouragement from students who were determined *not* to become casualties of the information explosion. We envisioned a textbook that would explain the ideas of biology clearly and accurately within a context of integrating themes, and would also help students to develop more positive and realistic impressions of what it means to be a scientist. With these dual objectives still in mind, and with the counsel of over a hundred professors and students who reviewed chapters or wrote to me with suggestions, I returned to the writing table to construct a second edition that would perform better than its predecessor. There was, of course, a responsibility to update all chapters to reflect the exciting evolution of biology's many fields. Just as important, however, was the opportunity to refine the presentation of each topic—to make subtle improvements that would elevate the teaching effectiveness of every part of the book. The result, I hope, is a text that will help students succeed in their general biology course and will also serve as a durable reference in their continuing education.

SPECIAL FEATURES

The Themes   The first chapter of *BIOLOGY* articulates a few key themes in biology, and these themes resurface throughout the text to help students synthesize connections in their study of life. The core theme is evolution, which accounts for both the unity and diversity of life. Among the supporting themes is the correlation of structure and function, a relationship that students can apply at all levels of biological organization.

The Interviews   Each of the book's eight units opens with a dialogue between the author and a scientist who has made important contributions to the forthcoming field of study. (A list of these scientists is found on pages xvi–xvii.) These biologists share personal experiences, discuss the cutting edges of their respective fields, and offer career advice to students. The interviews personalize science, portraying it as a social activity of creative men and women rather than an impersonal collection of facts. The interviews also reveal that intellectual debate, the lifeblood of science, abounds in biology.

The Methods Boxes   Learning about important discoveries in biology, students often ask: "How do we know?" To help answer that question, this text includes numerous boxed descriptions, a third of them new in this edition, of methods important in biological research. Many of these boxes describe laboratory and field methods in some experimental context. Some methods boxes will help students understand techniques they are likely to use in the laboratory, such as chromatography and spectroscopy. Other boxes describe methods that have been crucial to progress in biology but that few students will have an opportunity to use. For example, one box explains *RFLP* analysis, the relatively new method that is helping biologists to map the human genome and to solve many other problems in genetics and evolutionary biology. Along with the interviews and descriptions of experiments throughout the text, the methods boxes give students a more realistic view of how scientists think and work.

The Union of Text and Figures   Anyone who has taught general biology over the past decade can attest to the dramatic change in the appearance of textbooks. The trend has been toward more spectacular photographs and more eye-catching art. While this escalation in production standards has certainly produced many beautiful books, there is a risk of distracting students with illustrations that are only marginally relevant to the topic being discussed in the text. Learning complex material requires concentration. The art and photo program of an effective textbook must *support*, not overwhelm, the author's message. For this second edition of *BIOLOGY,* the publisher and I have continued to insist, without compromise, that each figure unite with the text to contribute to a cumulative lesson more effective than the sum of its pedagogical parts. The artists, photo editor, and I began working together during the first draft, embedding the carefully planned figures into the story line of each chapter. In fact, students will find that examining the figures and their captions is one way to preview or review the content of a chapter. The union of words and illustrations is central to the instructional value of this second edition of *BIOLOGY.* Another goal for the art program was to achieve a consistent use of symbols throughout the text. An example of this symbolic logic is the use of a color-coded sunburst to represent ATP wherever this important molecule appears in the book. This subtle consistency will promote quick recognition of repeated structures and processes.

CONTENT AND ORGANIZATION

This text makes no pretense that there is one "correct" way to order the major topics in an introductory biology course. Over the years, I have rearranged my own syllabus in various ways, finding that many different sequences are workable. This book is flexible enough for instructors to adapt its content to a variety of syllabi. The eight major units are self-contained, allowing for rearrangment, and most of the chapters within each unit can be assigned in a different sequence without substantial loss of continuity. For example, instructors who integrate plant and animal physiology can merge chapters from Units Six and Seven to fit their course.

A brief overview of the book's organization will highlight the content of each unit. Specific changes in the content of each chapter of the second edition are too numerous to list in the preface.

Unit One: The Chemistry of Life   While teaching chemistry may be the province of prerequisite or corequisite courses, my colleagues and I have found that many students struggle in their introductory biology course because of inadequate backgrounds in chemistry. Chapters 2–4 are designed to help those students by developing, in carefully paced steps, the concepts of chemistry that are essential for success in biology. This approach makes self-study possible, reducing the need for instructors to spend valuable lecture time on basic chemistry. However, Chapter 5, "Structure and Function of Macromolecules," and Chapter 6, "Introduction to Metabolism," should be read thoroughly even by those students with strong chemistry backgrounds. Among the changes in this unit for the second edition is a new section on the protein-folding problem.

Unit Two: The Cell   Chapters 7–11 emphasize the correlation of structure and function in the study of cells. For example, the importance of membranes in ordering metabolism is highlighted throughout the unit. Extensive reorganization of Chapter 7 in the second edition will make "A Tour of the Cell" easier to follow. Also, This chapter now features much stronger sections on the endoplasmic reticulum and the cytoskeleton. The energy-coupling mechanism known as chemiosmosis is the centerpiece of the chapters on cellular respiration and photosynthesis. New to this edition is consistent use of scale bars on all micrographs as well as abbreviations to indicate the type of microscopy (LM = light microscope; TEM = transmissionelectron microscope; SEM = scanning electron microscope).

Unit Three: The Gene   Chapters 12–19 take a historical approach to genetics, tracing its development from Gregor Mendel to modern recombinant DNA technology and the human genome project. Although certain aspects of molecular biology apply similarly to all organisms, this unit also recognizes the many important differences between prokaryotic and eukaryotic genetics. A separate chapter on human genetics seems, to me, to be an artificial distinction, and thus the extensive coverage of human genetics in Unit 3 emerges topically in each chapter, in close proximity to whatever general principle is being applied. Instructors should note that the molecular basis of development is covered in Chapter 18, while descriptive embryology, morphogenesis, and the cellular aspects of plant and animal development are covered in Chapters 34 and 43, respectively. The genetics unit culminates with an updated chapter on biotechnology.

Unit Four: Mechanisms of Evolution   Evolution is the most fundamental fact of life, and it is the one theme that surfaces in every part of this text. The title of this unit has been changed to better reflect the focus of Chapters 20–23 on *how* life evolves and how biologists study evolution. Among the many changes in this edition are a much clearer discussion of population genetics and a treatment of speciation that better represents the diverse viewpoints among researchers. It is unfortunate that the creationist furor (mentioned nowhere else in this book) has diverted attention from the many legitimate debates among evolutionary biologists. This robust field is alive with controversy about the tempo and mechanisms of evolution, and students deserve to see that not all issues are settled.

Unit Five: The Evolutionary History of Biological Diversity   Chapters 24–30 consider the diversity of life within the context of key evolutionary junctures such as the origin of prokaryotes, the evolution of the eukaryotic cell and diverse protists, the genesis of multicellular life, and the adaptive radiations of prevalent groups of plants, fungi, and animals. Instructors should be aware that viruses are not covered in this unit, but in Chapter 17 of the genetics unit. The relationship between biological and geological history is emphasized throughout Unit Five. The evolutionary theme of this unit contrasts sharply with a "parade of the kingdoms" approach to biological diversity.

Unit Six: Plants: Form and Function   Chapters 31–35 introduce students to the structure and physiology of plants within the evolutionary context of adaptation to terrestrial environments. Flowering plants are stressed because they have been the subjects of most of the research in plant science. The anatomy and life cycles of other plant groups are covered in Chapter 27. A more lucid description of secondary growth, an enriched section on plant–soil interactions, and updated coverage of how plants sense and respond to gravity are just a few of the improvements evident in the second edition. To help reinforce an impression of science as a process of inquiry, many of the experimental approaches that have advanced our understanding of plants are featured in this unit. Students will also learn about important agricultural applications in these chapters. Plants are often underemphasized in introductory biology textbooks. I have tried to help students become more familiar with the lives of plants, not only by improving the chapters devoted exclusively to plants, but also by featuring plants as examples to help teach general biological principles throughout the book.

Unit Seven: Animals: Form and Function The organism–environment interface is the focus of Chapters 36–45, which take a comparative approach in exploring the diverse adaptations that have evolved in the animal kingdom. Humans fit into this comparative format as an important mammalian example. In this second edition, invertebrates and nonmammalian vertebrates are much more prominent than in the first edition. For example, Chapter 40, "Controlling the Internal Environment," compares how different invertebrates and vertebrates solve the problems of osmoregulation, excretion, and thermoregulation in various environments. Chapter 45 has also been thoroughly revised to better represent the diverse sensory adaptations of animals. The progress in immunology over the past few years required that I scrap most of the first edition's version of Chapter 39, "The Immune System," and write what is essentially a new chapter.

Unit Eight: Ecology Chapters 46–50 reflect the synthesis of modern ecology from descriptive natural history and the experimental approaches pioneered by Connell, Paine, and others. The emphasis is on basic ecological study of the factors affecting the distribution and abundance of organisms; but the lessons from these studies are applied—more in this edition than in the first—to the problems of our harmful environmental impact. Math skills are important in ecology, and this unit asks students to think quantitatively about several ideas of ecology. The chapters also attempt to represent the different viewpoints in several of ecology's debates, with the objective of encouraging students to evaluate arguments and evidence critically. The unit and the book close with a chapter on animal behavior, a capstone subject that will help students see the relationships of ecology to other fields of biology, to the other natural sciences, and to the students' general education.

IN-TEXT LEARNING AIDS

Although I have tried to build effective tutelage into every page of *BIOLOGY*, learning aids at the end of each chapter also reinforce the main concepts, vocabulary, and applications. A **Study Outline**, keyed by page number to the major sections of the chapter, summarizes and rephrases essential points. A **Self-Quiz** consisting of multiple-choice questions helps the students measure their comprehension of the chapter contents, but many of these questions also require students to apply material or solve problems. The answers to the Self-Quiz questions are found in Appendix One. **Challenge Questions** ask students to verbalize their intrepretations of concepts; to extrapolate from what they have learned to new situations; to propose hypotheses and experiments of their own; to think critically about ideas; to apply quantitative skills in the context of biological problems; or to consider the implications of the actual research of contemporary biologists. To stimulate students to investigate intriguing topics in greater depth, a list of **Further Readings** completes the learning aids at the end of each chapter. Students will also find a **Glossary** of about 1300 key terms at the end of the book. To assist in still another way, Appendix Three introduces students to the learning tool known as **concept mapping.**

SUPPLEMENTS

The second edition of *BIOLOGY* is supported by supplements that further serve students and instructors:

Student Study Guide by Martha Taylor, Cornell University

Instructor's Guide by Nina Caris, Texas A&M University

Instructors's Test Bank by William Barstow, University of Georgia, and Walter MacDonald, Trenton State College. This test bank is available on Microtest, a microcomputer test generation program for the IBM PC, AC XT, the Apple II family, and the Macintosh computers. (Test bank is available to qualified college and university adopters.)

Laboratory Collection edited by Judith Goodenough, University of Massachusetts, Amherst

Overhead Transparencies A set of 200 color acetates of illustrations and micrographs from *BIOLOGY*, Second Edition.

* * *

The real test of any textbook is how well it helps instructors teach and students learn. I welcome comments from professors and students who use this text. Please address your criticisms and suggestions for improving the next edition directly to me.

NEIL A. CAMPBELL
Department of Botany and Plant Science
University of California
Riverside, California 92521

Acknowledgments

Writing a biology textbook is *not* lonely work. Hundreds of scientists, students, artists, and publishing professionals have helped me revise *BIOLOGY*. Their expertise, creativity, and commitment to science education account for the improvements in this second edition.

More than 120 research specialists and biology instructors strengthened the scientific accuracy and teaching effectiveness of the second edition by reviewing chapters at various stages of revision. Many other professors and their students took the time to volunteer their helpful suggestions by writing directly to me. A number of instructors took part in focus groups, discussing their concerns about introductory biology and helping us to define our objectives for the textbook. Three scientists became even more involved by actually working on some chapters, either making revisions or creating early drafts of new material. These contributors are Lawrence Mitchell of Iowa State University, who collaborated on Unit Five ("The Evolutionary History of Biological Diversity"); Richard Stout, of Montana State University, who helped revise Unit Six ("Plants: Form and Function"); and Gregory Capelli of College of William and Mary, who worked extensively on the revision of Unit Eight ("Ecology"). For their contributions and consultations on the first edition of *BIOLOGY* I would also like to thank Gregory Capelli (College of William and Mary), Wayne Carley (Lamar University), Stanley Faeth (Arizona State University), Jeffrey L. Fox (ASM News), Debra Kirchof-Glazier (Juniata College), William Kritan (UC San Diego), Julie Ann Miller (BioScience), Roger Lederer (California State University, Chico), James Platt (University of Denver), John Ruben (Oregon State University), James W. Valentine (UC Santa Barbara), and James Wittenberger (formerly University of Washington). Martha Taylor of Cornell University improved the end-of-chapter study aids on the second edition. Dr. Taylor also contributed Appendix Three, on "Concept Mapping." Although the final responsibility for errors of fact or judgment rests with me, they are all the fewer because of the efforts of the reviewers, correspondents, focus-group members, and contributors. They worked hard to help me make this book more correct, current, and clear, and I thank them for their participation.

Benjamin/Cummings is fortunate to have a talented scientist in its own house. Jane Reece, who has managed the publisher's development department, also holds a doctorate in microbial genetics. Her combination of publishing and scientific perspectives has been an important factor in the success of many of Benjamin/Cummings books, including the first edition of *BIOLOGY*. I thank Jane for our long partnership and, more specifically, for her excellent work on Unit Three ("The Gene") for this second edition.

One of the greatest pleasures of revising *BIOLOGY* was the opportunity to conduct the new interviews that open the text's eight units. The interviewees for the second edition are George Bartholomew, Jane Goodall, Niles Eldredge, Charles Leblond, Stanley Miller, Ruth Satter, William Schopf, and Nancy Wexler. Their dedication to education adds much to how this book relates to students. The interviews also provided me with insights that influenced my writing.

Three other scientists influenced this book indirectly by contributing so much to the author's personal development as a teacher and scientist. It was Ronald Kroman of California State University at Long Beach who, by his fine example, ignited my desire to teach. He cannot possibly remember me, one of thousands of undergraduates who have passed through his genetics and statistics courses, but I shall never forget him or his teaching artistry. Pius Horner, my colleague for the past twenty years, has sustained my belief in the power of effective teaching. I am also grateful to William Thomson of the University of California, Riverside, who led me to the fun of original, problem-oriented research. These three mentors planted the philosophical seeds of *BIOLOGY*.

The illustration program is such an integral part of *BIOLOGY* that the artists could almost be considered coauthors. More than half of the figures in this second edition are completely new or extensively revised. Raychel Ciemma, Barbara Cousins, Linda McVay, Ken Miller, and Carla Simmons have created illustrations that will enhance the book's reputation, initiated by the fine artists who worked on the first edition, for art that is as pedagogically innovative as it is visually attractive. I thank these talented people for gracing *BIOLOGY*.

My photo editor, Darcy Lanham, led the search for the many striking, instructional photographs that fortify the lessons in *BIOLOGY*. With the skilled help of her photoresearchers—Stuart Kenter, Audrey Ross, and Kevin Schafer—Darcy always found just the right photo to complement the text's discussion of a particular concept. I always look forward to working with Darcy, and value her important role in producing a book where pictures and words work so well together.

Gary Head is responsible for *BIOLOGY's* pleasing and efficient design. I think he is the best in his field, and people who know more than I about book design agree: The first edition won publishing awards for Gary's overall design and for his elegant cover. I thank Gary for pushing this second edition to an even higher standard of functional beauty.

Deborah Gale and Pat Waldo of Partners in Publishing coordinated the production of *BIOLOGY*, transforming typescript, art boards, and photos into a book. That transformation almost seems like magic, but I know the tricks are really experience and hard work, for which I thank Deborah and Pat. Deborah also moonlighted as the developmental editor for the second edition. She always brought humor, intelligence, and a fresh point of view to our development meetings.

Benjamin/Cummings' own production department, including managing editor Karen Gulliver and managers Laura Argento and Glenda Epting, were ultimately responsible for overseeing the entire production process and assuring quality in the manufacturing of the bound book you hold. I am also grateful to Eleanor Brown of Benjamin/Cummings for coordinating the production of the supplements.

Anne Emerson, Rajeev Samantrai, and Erin Connell of the Benjamin/Cummings marketing group keep *BIOLOGY* in touch with the students and professors it serves. I also thank them for announcing the second edition with style and dignity.

The field staff that represents *BIOLOGY* on campuses is my link to the students and professors who use the text. The field representatives tell me when there are compliments or complaints about the book, and they provide prompt, helpful service to college departments. Of course, these field representatives also sell *BIOLOGY*, but they perform this function very successfully without slurring other publishers and their competing books. I commend the field representatives for their professionalism.

When I began the first edition of *BIOLOGY* in 1979, I chose Benjamin/Cummings as the publisher because I believed it was the only company with the vision, convictions, energy, and imaginative personnel required to create a new kind of introductory biology textbook. It took us eight years to put the book we envisioned between hard covers. My first editor, Jim Behnke, nurtured the first edition from the time it was conceived during a meeting in my office at Cornell. Although Jim's meteoric rise as a publisher has separated him from the revision of the book, his publishing values remain evident on every page of the second edition. I thank Jim for his loyalty to *BIOLOGY* and faith in its author. Sally Elliott, the general manager of Benjamin/Cummings, is continuing the company's tradition of responsible publishing.

Robin Williams took over as sponsoring editor of *BIOLOGY* in 1987 and has been my closest partner in the publication of this second edition. Because of Robin's genuine concern for the quality of science education, she has earned the respect of biology instructors on hundreds of campuses. Robin's commitment to those instructors and their students inspired the revision of *BIOLOGY*. I prize Robin's inventiveness, resourcefulness, and sensitivity, and I appreciate all she has done to help me improve this text. I am also grateful to Katie Lilienthal, Robin's hardworking assistant.

Most of all, I thank my family and friends for their support and encouragement. More time to enjoy their company is my greatest reward for finishing this edition of *BIOLOGY*.

FIRST EDITION REVIEWERS

Focus Group Participants

William Barklow, *Framingham State College*

Gary Brusca, *Humboldt State University*

Doug Cheeseman, *De Anza College*

Virginia Fry, *Monterey Peninsula College*

Elizabeth Godrick, *Boston University*

Frank Heppner, *University of Rhode Island*

William Hines, *Foothill College*

Margaret Houk, *Ripon College*

Joseph Levine, *Boston College*

Todd Newbury, *University of California, Santa Cruz*

Cynthia Norton, *University of Maine, Augusta*

Hideo Yonenaka, *San Francisco State University*

Manuscript Reviewers

Richard J. Andren, *Montgomery County Community College*

J. David Archibald, *Yale University*

William Barklow, *Framingham State College*

Wayne Becker, *University of Wisconsin, Madison*

Paulette Bierzychudek, *Pomona College*

Richard Boohar, *University of Nebraska, Omaha*

James L. Botsford, *New Mexico State University*

J. Michael Bowes, *Humboldt State University*

Herbert Bruneau, *Oklahoma State University*

Alan H. Brush, *University of Connecticut*

Edwin Burling, *De Anza College*

Gregory Capelli, *College of William and Mary*

Doug Cheeseman, *De Anza College*

J. John Cohen, *University of Colorado Health Science Center*

John Corliss, *University of Maryland*

Bonnie J. Davis, *San Francisco State University*

Jerry Davis, *University of Wisconsin, La Crosse*

T. Delevoryas, *University of Texas, Austin*

Jean DeSaix, *University of North Carolina*

Betsey Dyer, *Wheaton College*

Robert Eaton, *University of Colorado*

Robert S. Edgar, *University of California, Santa Cruz*

Robert C. Evans, *Rutgers University, Camden*

Lincoln Fairchild, *Ohio State University*

Jerry F. Feldman, *University of California, Santa Cruz*

Milton Fingerman, *Tulane University*

Abraham Flexer, *Manuscript Consultant, Boulder, Colorado*

David Fox, *University of Tennessee, Knoxville*

Virginia Fry, *Monterey Peninsula College*

Alice Fulton, *University of Iowa*

Sara Fultz, *Stanford University*

Berdell Funke, *North Dakota State University*

Arthur W. Galston, *Yale University*

Carl Gans, *University of Michigan*

Patricia Gensel, *University of North Carolina*

Todd Gleeson, *University of Colorado*

Elizabeth Godrick, *Boston University*

A. J. F. Griffiths, *University of British Columbia*

Katherine L. Gross, *Ohio State University*

Gary Gussin, *University of Iowa*

R. Wayne Habermehl, *Montgomery County Community College*

Mac Hadley, *University of Arizona*

Jack P. Hailman, *University of Wisconsin*

Penny Hanchey-Bauer, *Colorado State University*

Laszlo Hanzely, *Northern Illinois University*

George Hechtel, *State University of New York at Stony Brook*

Frank Heppner, *University of Rhode Island*

Ralph Hinegardner, *University of California, Santa Cruz*

Pius F. Horner, *San Bernardino Valley College*

Margaret Houk, *Ripon College*

Ronald R. Hoy, *Cornell University*

Robert J. Huskey, *University of Virginia*

John Jackson, *North Hennepin Community College*

Russell Jones, *University of California, Berkeley*

Alan Journet, *Southeast Missouri State University*

Thomas Kane, *University of Cincinnati*

E. L. Karlstrom, *University of Puget Sound*

Attila O. Klein, *Brandeis University*

Thomas Koppenheffer, *Trinity University*

George Khoury, *National Cancer Institute*

J. A. Lackey, *State University of New York at Oswego*

Kenneth Lang, *Humboldt State University*

Charles Leavell, *Fullerton College*

Joseph Levine, *Boston College*

Bill Lewis, *Shoreline Community College*

James MacMahon, *Utah State University*

Lynn Margulis, *Boston University*

Karl Mattox, *Miami University of Ohio*

Joyce Maxwell, *California State University, Northridge*

Helen Miller, *Oklahoma State University*

John Miller, *University of California, Berkeley*

Kenneth R. Miller, *Brown University*

John E. Minnich, *University of Wisconsin, Milwaukee*

Russell Monson, *University of Colorado*

Frank Moore, *Oregon State University*

Carl Moos, *Veterans Administration Hospital, Albany, New York*

Bette Nicotri, *University of Washington*

Cynthia Norton, *University of Maine, Augusta*

Peter N. Pappas, *County College of Morris*

Crellin Pauling, *San Francisco State University*

Halina Presley, *University of Illinois, Chicago*

Ralph Quatrano, *Oregon State University*

Charles Remington, *Yale University*

Fred Rhoades, *Western Washington State University*

Thomas Rost, *University of California, Davis*

John Ruben, *Oregon State University*

Albert Ruesink, *Indiana University*

Carl Schaefer, *University of Connecticut*

William H. Schlesinger, *Duke University*

Peter Shugarman, *University of Southern California*

Daniel Simberloff, *Florida State University*

Andrew T. Smith, *Arizona State University*

Andrew J. Snope, *Essex Community College*

Barbara Stewart, *Swarthmore College*

John Stolz, *California Institute of Technology*

Daryl Sweeney, *University of Illinois, Urbana-Champaign*

Samuel S. Sweet, *University of California, Santa Barbara*

Roger Thibault, *Bowling Green State University*

John Thornton, *Oklahoma State University*

Maura G. Tyrrell, *Stonehill College*

James W. Valentine, *University of California, Santa Barbara*

Frank Visco, *Orange Coast College*

Susan D. Waaland, *University of Washington*

Peter Wejksnora, *University of Wisconsin, Milwaukee*

Kentwood Wells, *University of Connecticut*

Fred Wilt, *University of California, Berkeley*

Robert T. Woodland, *University of Massachusetts Medical School*

John Zimmerman, *Kansas State University*

SECOND EDITION REVIEWERS

Focus Group Participants

Leigh Auleb, *San Francisco State University*

Steven Barnhart, *Santa Rosa Junior College*

Jane Beiswenger, *University of Wyoming*

Eric Bonde, *University of Colorado, Boulder*

James Dekloe, *University of California, Santa Cruz*

Barbara Finney, *Regis College*

Robert George, *University of Wyoming*

Lorraine Lica, *California State University, Hayward*

Harvey Nichols, *University of Colorado, Boulder*

Gay Ostarello, *Diablo Valley College*

Kay Pauling, *Foothill College*

James Platt, *University of Denver*

Charles Ralph, *Colorado State University*

Manuscript Reviewers

Katherine Anderson, *University of California, Berkeley*

Katherine Baker, *Millersville University*

Steven Barnhart, *Santa Rosa Junior College*

Tom Beatty, *University of British Columbia*

Anne Bekoff, *University of Colorado, Boulder*

Marc Bekoff, *University of Colorado, Boulder*

Adrianne Bendich, *Hoffman-La Roche, Inc.*

Barbara Bentley, *State University of New York, Stony Brook*

Darwin Berg, *University of California, San Diego*

Dorothy Berner, *Temple University*

Robert Blystone, *Trinity University*

Robert Boley, *University of Texas, Arlington*

Erik Bonde, *University of Colorado, Boulder*

Barry Bowman, *University of California, Santa Cruz*

Jerry Brand, *University of Texas, Austin*

James Brenneman, *University of Evansville*

Alan Brush, *University of Connecticut, Storrs*

Meg Burke, *University of North Dakota*

John Bushnell, *University of Colorado*

William Busa, *Johns Hopkins University*

Nina Caris, *Texas A&M University*

Shepley Chen, *University of Illinois, Chicago*

Henry Claman, *University of Colorado Health Science Center*

William Coffman, *University of Pittsburgh*

Stuart Coward, *University of Georgia*

Marianne Dauwalder, *University of Texas, Austin*

Thomas Davis, *University of New Hampshire*

Marvin Druegar, *Syracuse University*

Robert Edgar, *University of California, Santa Cruz*

Betty J. Eidemiller, *Lamar University*

David Evans, *University of Florida*

Sharon Eversman, *Montana State University*

Lynn Fancher, *College of DuPage*

Larry Farrell, *Idaho State University*

Russell Fernald, *University of Oregon*

Norma Fowler, *University of Texas, Austin*

Otto Friesen, *University of Virginia*

Anne Funkhouser, *University of the Pacific*

John Gapter, *University of Northern Colorado*

Reginald Garrett, *University of Virginia*

William Glider, *University of Nebraska*

Elizabeth A. Godrick, *Boston University*

Lynda Goff, *University of California, Santa Cruz*

Paul Goldstein, *University of Texas, El Paso*

Judith Goodenough, *University of Massachusetts, Amherst*

Ester Goudsmit, *Oakland University*

William Grimes, *University of Arizona*

Mark Gromko, *Bowling Green State University*

Richard Harrison, *Cornell University*

H.D. Heath, *California State University, Hayward*

Jean Heitz-Johnson, *University of Wisconsin, Madison*

Frank Heppner, *University of Rhode Island*

David Ho, *University of Missouri*

Carl Hoagstrom, *Ohio Northern University*

James Holland, *Indiana State University, Bloomington*

Laura Hoopes, *Occidental College*

Nancy Hopkins, *Massachusetts Institute of Technology*

Kathy Hornberger, *Widener University*

Alice Jacklet, *State University of New York, Albany*

John Jackson, *North Hennepin Community College*

Robert Kitchen, *University of Wyoming*

Lynn Lamoreux, *Texas A&M University*

Allan Larson, *Washington University*

Robert Leonard, *University of California, Riverside*

Lorraine Lica, *California State University, Hayward*

Harvey Lillywhite, *University of Florida, Gainsville*

Sam Loker, *University of New Mexico*

Jane Lubchenco, *Oregon State University*

Charles Mallery, *University of Miami*

Edith Marsh, *Angelo State University*

Joyce Maxwell, *California State University, Northridge*

Richard McCracken, *Purdue University*

John Merrill, *University of Washington*

Ralph Meyer, *University of Cincinnati*

Roger Milkman, *University of Iowa*

Russell Monson, *University of Colorado, Boulder*

Randy Moore, *Wright State University*

John Neess, *University of Wisconsin, Madison*

Deborah Nickerson, *University of South Florida*

David Norris, *University of Colorado, Boulder*

Brian O'Conner, *University of Massachusetts, Amherst*

Eugene Odum, *University of Georgia*

Stanton Parmeter, *Chemeketa Community College*

Bulah Parker, *North Carolina State University*

Robert Patterson, *San Francisco State University*

Kay Pauling, *Foothill Community College*

Patricia Pearson, *Western Kentucky University*

Jeffrey Pommerville, *Texas A&M University*

Donald Potts, *University of California, Santa Cruz*

David Pratt, *University of California, Davis*

Scott Quackenbush, *Florida International University*

Donna Ritch, *Pennsylvania State University*

Rodney Rogers, *Drake University*

Ted Sargent, *University of Massachusetts, Amherst*

Stephen Sheckler, *Virginia Polytechnic Institute and State University*

David Schimpf, *University of Minnesota, Duluth*

James Shinkle, *Trinity University*

Barbara Shipes, *Hampton University*

Alice Shuttey, *DeKalb Community College*

Erik P. Scully, *Towson State University*

James Sidie, *Ursinus College*

John Smarrelli, *Loyola University*

Stephen Strand, *University of California, Los Angeles*

Cecil Still, *Rutgers University, New Brunswick*

Daryl Sweeney, *University of Illinois, Urbana*

Samuel Tarsitano, *Southwest Texas State University*

David Tauck, *University of Santa Clara*

James Taylor, *University of New Hampshire*

Roger Thibault, *Bowling Green University*

Robert Thornton, *University of California, Davis*

Robert Tuveson, *University of Illinois, Urbana*

Joseph Vanable, *Purdue University*

Theodore Van Bruggen, *University of South Dakota*

Laurie Vitt, *University of California, Los Angeles*

John Waggoner, *Loyola Marymount University*

Dan Walker, *San Jose State University*

Jeffrey Walters, *North Carolina State University*

Margaret Waterman, *University of Pittsburgh*

Terry Webster, *University of Connecticut, Storrs*

Christopher Wills, *University of California, San Diego*

Philip Yant, *University of Michigan*

Uko Zylstra, *Calvin College*

Brief Contents

The Campbell Interviews

DETAILED CONTENTS

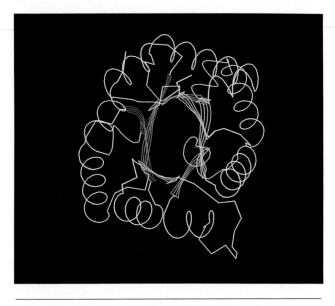

6 Introduction to Metabolism 92

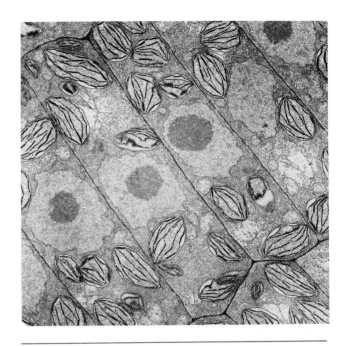

UNIT TWO The Cell 113

7 A Tour of the Cell 116

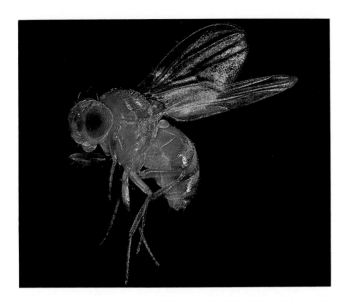

UNIT FIVE   The Evolutionary History of Biological Diversity 505

24 Early Earth and the Origin of Life 510

25 Prokaryotes and the Origins of Metabolic Diversity 522

26 Protists and the Origin of Eukaryotes 540

UNIT SIX   Plants: Form and Function 675

31 Anatomy of a Plant 680

32 Transport in Plants 704

33 Plant Nutrition 721

34 Plant Reproduction 738

UNIT SEVEN   Animals: Form and
              Function 777

Interview: George Bartholomew 777

UNIT EIGHT   Ecology 1045

Interview: Jane Goodall 1045

46 Diverse Environments of the Biosphere: An Introduction to Ecology 1050

BIOLOGY

1 Introduction: Themes in the Study of Life

Biology, the study of life, is rooted in the human spirit. People keep pets, nurture houseplants, invite avian visitors with backyard birdhouses, visit zoos and nature parks, and otherwise express what Harvard biologist E. O. Wilson calls *biophilia*, an innate attraction to life in its diverse forms. Biology is the scientific extension of this human tendency to feel connected to and curious about all forms of life. It is a science for adventurous minds. It takes us, personally or vicariously, into jungles, deserts, seas, and other environments, where a variety of living forms and their physical surroundings are interwoven into complex webs called ecosystems. Studying life leads us into laboratories to examine more closely how living things, which biologists call organisms, work. Biology draws us into the microscopic world of the fundamental units of life known as cells, and into the submicroscopic realm of the molecules that make up those cells. Our intellectual journey also takes us back in time, for biology encompasses not only contemporary life, but also a history of ancestral forms stretching nearly four billion years into the past. The scope of biology is immense. The purpose of this book is to introduce you to this multifaceted science (Figure 1.1).

You are becoming involved with biology during its most exciting era. Armed with new research methods developed in the past few decades, biologists are beginning to unravel some of life's most engaging mysteries. Though stimulating, the information explosion in biology is also intimidating. Most of the biologists who have ever lived are alive today, and they add about 400,000 new research articles to the scientific literature annually. Each of biology's many

(a)

(b)

(c)

(d)

(e)

Figure 1.1
Biologists study life on many different scales. The scope of biology not only reaches across many size scales, from the microscopic and submicroscopic levels of cells and molecules to the global distribution of biological communities, but also stretches back over vast spans of time. (a) Jane Goodall observes and records the social behavior of chimpanzees in their natural community in Tanzania's Gombe National Park. (b) George Bartholomew not only studies animals in their natural settings, but also brings individual organisms to the laboratory for more detailed investigations of the special adaptations that enable these animals to succeed in their environments. Here, Dr. Bartholomew is taking field notes on the behavior of sooty terns on Midway Island. (c) Ruth Satter is interested in the biological clocks that control the daily rhythms of plants, such as the "sleep" movements of species that fold their leaves together at night. This electron microprobe and other powerful instruments extend Dr. Satter's senses to the minute structures within cells. (d) The ability of organisms to reproduce adds a time dimension to biology. Here, Nancy Wexler uses an extensive family tree to trace an inherited disease through the generations of one family in Ecuador. (e) Stanley Miller looks back much farther in time, to life's origin. In a glass apparatus like the one shown in this photograph, Professor Miller first demonstrated that environmental conditions on the lifeless primordial Earth favored the synthesis of some of the organic molecules found in modern cells.

You will learn more about the work of these and other scientists in the interviews that precede each unit of chapters in this text.

subfields is in a continuous state of flux, and it is very difficult for a professional biologist to remain current in more than one narrowly defined specialty. How, then, can beginning biology students hope to keep their heads above water in this deluge of data and discovery? The key is to recognize unifying themes that pervade all of biology—themes that will still apply a decade from now, when much of the specific information presented in any textbook will be obsolete. This chapter enunciates some of the broad, enduring themes in the study of life.

A HIERARCHY OF ORGANIZATION

A basic characteristic of life is a high degree of order. You can see it in the intricate pattern of veins throughout a leaf or in the colorful pattern of a bird's plumage. If you were to scrutinize the vein of a leaf or the feather of a bird under a microscope, you would discover that biological order also exists at levels below what the unaided eye can resolve.

Biological organization is based on a hierarchy of structural levels, with each level building on the levels below it. Atoms, the chemical building blocks of all matter, are ordered into complex biological molecules such as proteins. The molecules of life are arranged into minute structures called organelles, which are in turn the components of cells. Some organisms consist of single cells, but others, including plants and animals, are aggregates of many specialized types of cells. In such multicellular organisms, similar cells are grouped into tissues, and specific arrangements of different tissues form organs. For example, the nervous impulses that coordinate your movements are transmitted along specialized cells called neurons. The nervous tissue within your brain has billions of neurons organized into a communications network of spectacular complexity. The brain, however, is not pure nervous tissue; it is an organ built of many different tissues, including a type called connective tissue that forms the protective covering of the brain. The brain is itself part of the nervous system, which also includes the spinal cord and the many nerves that transmit messages between the spinal cord and other parts of the body. The nervous system is only one of several organ systems characteristic of humans and other complex animals. Another example, shown in Figure 1.2, is the circulatory system.

In the hierarchy of biological organization, there are tiers beyond the individual organism. A population is a localized group of organisms belonging to the same species; populations of species living in the same area make up a biological community; and community interactions that include nonliving features of the environment, such as soil and water, form an ecosystem.

Unfolding biological organization at its many levels, from molecular architecture to ecosystem structure, is fundamental to the study of life. This textbook essentially follows such an organization, beginning by looking at the chemistry of life and ending with ecology. However, we will also see that biological processes transcend this hierarchy, with causes and effects at several organizational levels. For example, when a rattlesnake explodes from its coiled posture and strikes a desert mouse, the predator's coordinated movements result from complex interactions at the molecular, cellular, tissue, and organ levels within the snake. But there are also causes and effects of this behavior that operate on the level of the biological community where the snake and its prey live. That is, the feeding response is triggered when the snake senses and locates the nearby mouse, and at the same time, many such episodes have an important cumulative impact on the population sizes of both the mice and the rattlesnakes. Most biologists specialize in the study of life at a particular level, but they gain broader perspective when they connect their discoveries to processes occurring at lower or higher levels. A narrow focus on a single level of biological organization depreciates the fun and power of biology.

EMERGENT PROPERTIES

With each step upward in the hierarchy of biological order, novel properties emerge that were not present at the simpler levels of organization. These emergent properties result from interactions between components. A molecule such as a protein has attributes not exhibited by any of its component atoms, and a cell is certainly much more than a bag of molecules. If the intricate organization of the human brain is disrupted by a head injury, that organ ceases to function properly even though all of its parts may still be present. And an organism is a living whole greater than the sum of its parts.

Life resists a simple, one-sentence definition because it is associated with numerous emergent properties. Yet, almost any child perceives that a dog or a bug or a tree is alive and a rock is not. We can recognize life without defining it, and we recognize life by what living things do. Figure 1.3 describes some of the properties and processes we associate with the state of being alive.

Recognizing life as a set of emergent properties may seem, at first, to support a doctrine known as vitalism, which views life as a supernatural phenomenon beyond the bounds of physical and chemical laws. However, the concept of emergent properties merely accents the importance of arrangement and applies to inanimate material as well as to life; for instance, diamonds and

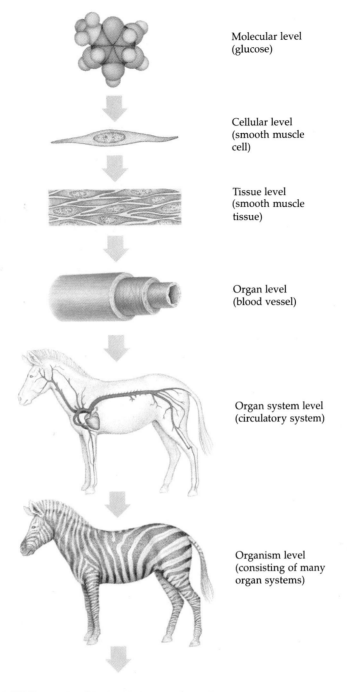

Molecular level
(glucose)

Cellular level
(smooth muscle
cell)

Tissue level
(smooth muscle
tissue)

Organ level
(blood vessel)

Organ system level
(circulatory system)

Organism level
(consisting of many
organ systems)

Higher levels:
(populations,
communities, and
ecosystems)

Figure 1.2
Some integrative levels in the hierarchy of biological order.

graphite have different properties because their carbon atoms are arranged differently. Unique properties of organized matter arise from how parts are arranged and interact, not from supernatural powers. And life is driven not by "vital forces" that defy explanation, but by principles of physics and chemistry extended into a new territory. The emergent properties of life simply reflect a hierarchy of structural organization without counterpart among inanimate objects.

Because the properties of life emerge from complex organization, scientists seeking to understand biological processes confront a dilemma. One horn of the dilemma is that we cannot fully explain a higher level of order by breaking it down into its parts. A dissected animal no longer functions; a cell dismantled to its chemical ingredients is no longer a cell. According to a principle known as holism, disrupting a living system interferes with meaningful explanation of its processes. The other horn of the dilemma is the futility of trying to analyze something as complex as an organism or a cell without taking it apart. Reductionism—reducing complex systems to simpler components that are more manageable to study—has been the most powerful strategy in biology. For example, by studying the molecular structure of a substance called DNA that had been extracted from cells, James Watson and Frances Crick, in 1953, deduced how this molecule could serve as the chemical basis of inheritance. The central role of DNA was better understood, however, when it was possible to study its interactions with other substances in the cell. Biology balances the pragmatic reductionist strategy with the longer-range objective of understanding how the parts of cells and organisms are functionally integrated.

THE CELLULAR BASIS OF LIFE

As the lowest level of structure capable of performing *all* the activities of life, the cell has a very special place in the hierarchy of biological organization. All organisms are composed of cells. They occur singly as a great variety of unicellular organisms, and they occur as the subunits of organs and tissues in plants, animals, and other multicellular organisms. In either case, the cell is life's basic unit of structure and function.

Robert Hooke, an English scientist, first described and named cells in 1665, when he observed a slice of cork (bark from an oak tree) with a microscope that magnified 30 times (30×). Apparently believing that the tiny boxes, or "cells," that he saw were unique to cork, Hooke never realized the significance of his discovery. His contemporary, a Dutchman named Antonie van Leeuwenhoek, discovered organisms we now know to be single-celled. Using grains of sand that he had polished into magnifying glasses as powerful as 300×, Leeuwenhoek discovered a microbial world in droplets of pond water and also observed the blood cells and sperm cells of animals. In 1839, nearly two centuries after the discoveries of Hooke and Leeuwenhoek, cells were finally acknowledged as the ubiquitous units of life by Matthias Schleiden and Theodor Schwann, two German biologists. In a classic case of inductive reasoning—reaching a generalization based on many concurring observations—Schleiden and Schwann summarized their own microscopic studies and those of others by concluding that all living things consist of cells, a concept that forms the basis of what is known as the cell theory. This theory was later expanded to include the idea that all cells come from other cells.

Over the past 30 years, a powerful instrument called the electron microscope has revealed the complex structure of cells. A cell is bounded by a membrane that regulates the passage of materials between the cell and its surroundings. Some cells, including those of plants, have tough walls external to their membranes. Animal cells lack walls. A cell is controlled by its DNA. In the cells of all organisms except bacteria, the DNA is organized along with proteins into structures called chromosomes contained within the nucleus, the largest organelle of most cells. Surrounding the nucleus is the cytoplasm, which contains various organelles that perform most of the cell's functions.

Figure 1.3 ▶
Some properties of life. (a) Organisms are highly ordered, and the other characteristics of life emerge from this complex organization. Order is apparent in this skeleton of a sponge named Venus' flower basket. (b) Organisms reproduce their own kind. Life comes only from life, an axiom known as biogenesis. Here, a roseate spoonbill feeds its offspring. (c) Organisms grow and develop. Heritable programs in the form of DNA direct the pattern of growth and development, producing an organism that is characteristic of its species. Shown here are embryos of a Costa Rican species of frog. Unlike many frogs, this species develops directly from an egg to a legged animal without a tadpole stage. (d) Organisms take in energy and transform it to do many kinds of work, including the maintenance of their ordered state. When the bear eats the fish, it will use the energy stored in the fish's molecules to maintain its own metabolism. (e) Organisms respond to stimuli from their environment. The growth of this strangler fig's roots and stems adjusts to environmental cues, molding the plant to its substratum—in this case, the ruins of a temple in India. (f) Life evolves as a result of the interaction between organisms and their environments. One consequence of evolution is the adaptation of organisms to their environment. The white fur of the arctic fox makes it nearly invisible in the animal's snowy world.

(a)

(b)

(c)

(d)

(e)

(f)

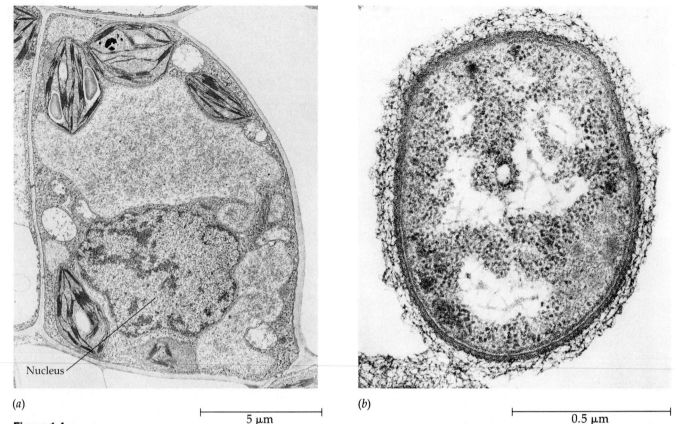

(a)

Nucleus

(b)

| 5 μm | 0.5 μm |

Figure 1.4
Two types of cells, as viewed with the electron microscope. (a) The eukaryotic cell, found in plants, animals, and all other organisms except bacteria, is characterized by an extensive subdivision into many different compartments, or organelles. Only the relatively large organelle known as the nucleus is labeled in this micrograph of a plant cell. Various organelles can also be seen in the cytoplasm outside the nucleus. **(b)** The prokaryotic cell, unique to bacteria, is much simpler, lacking most of the organelles found in eukaryotic cells.

Two major kinds of cells can be distinguished based on structural organization (Figure 1.4). The eukaryotic cell, by far the more complex, is subdivided by internal membranes into many different functional compartments, or organelles, including a nucleus and the various cytoplasmic organelles. The prokaryotic cell is much simpler in organization. Its DNA is not separated from the rest of the cell into a nucleus, and most of the other organelles typical of eukaryotic cells are also lacking. The cells of the microorganisms known as bacteria are prokaryotic. All other forms of life are composed of eukaryotic cells.

Although eukaryotic and prokaryotic cells contrast sharply in structural organization, they have many similarities, especially in their chemical processes. Among eukaryotic cells, there is extensive variation in size, shape, and specific structural features, but certain organelles and processes are remarkably similar from one cell type to another. For the most part, cellular diversity represents variations on common structural and functional motifs.

THE CORRELATION OF STRUCTURE AND FUNCTION

Given a choice of tools, you would not loosen a screw with a hammer or pound a nail with a screwdriver. How a device works is correlated with its structure: Form fits function. Applied to biology, this theme is a guide to the anatomy of life at its many structural levels, from molecules to organisms. Analyzing a biological structure gives us clues about what it does and how it works. Conversely, knowing the function of a structure provides insight about its construction.

The correlation of form and function is apparent in the aerodynamically efficient shape of a bird's wing (Figure 1.5). Beneath these external contours, the skeleton of the bird also has structural qualities that contribute to flight, with bones that have a strong but light honeycombed structure. The flight muscles of a bird are controlled by neurons. Long extensions of the neurons transmit nervous impulses, making these cells

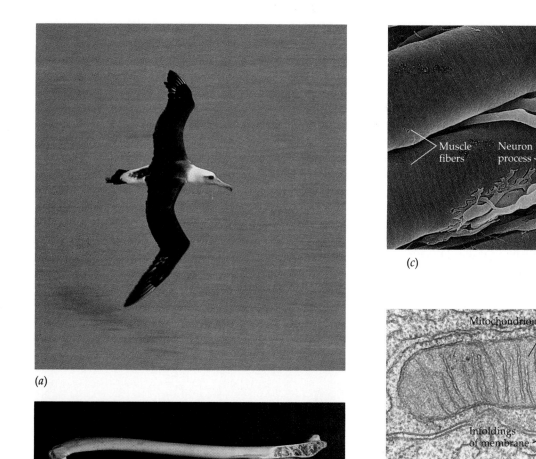

(a)

(c)

Muscle fibers Neuron process

10 μm

(b)

Mitochondrion

Infoldings of membrane

(d)

0.5 μm

Figure 1.5
Form fits function. (a) A bird's build makes flight possible. The correlation of structure and function can apply to the shape of an entire organism, as you can see from this albatross in flight. (b) The principle also applies to the structure of organs and tissues. For example, the honeycombed construction of a bird's bones provides a lightweight skeleton of great strength. (c) The form of a cell fits its specialized function. Nerve cells, or neurons, have long extensions (processes) that transmit nervous impulses—here, to muscle cells (SEM). (d) Functional beauty is also apparent at the subcellular level. This organelle, called a mitochondrion, has an inner membrane that is extensively folded, a structural solution to the problem of packing a relatively large amount of this membrane into a very small container (TEM).

especially well adapted for communication. As an example of functional anatomy at the subcellular level, consider the organelles called mitochondria. They are the sites of cellular respiration, the chemical process that powers the cell by using oxygen to help tap the energy stored in sugar and other food molecules. A mitochondrion is surrounded by an outer membrane, but it also has an inner membrane with many infoldings. Molecules embedded in the inner membrane carry out many of the steps in cellular respiration, and the infoldings pack a large amount of this membrane into a minute container. In exploring life on its different structural levels, we will discover functional beauty at every turn.

THE INTERACTION OF ORGANISMS WITH THEIR ENVIRONMENT

Life does not exist in a vacuum. Each organism interacts continuously with its environment, which includes other organisms as well as nonliving factors. The roots of a tree, for example, absorb water and minerals from the soil, and the leaves take in carbon dioxide from the air. Sunlight absorbed by chlorophyll, the green pigment of leaves, drives the process called photosynthesis, which converts water and carbon dioxide to sugar and oxygen. The tree releases oxygen to the air, and its roots change the soil by breaking up rocks into smaller particles, secreting acid, and absorbing minerals. Both organism and environment are affected by the interaction between them.

The tree also interacts with other life, including soil microorganisms associated with its roots and animals that eat its leaves and fruit. Multifarious interactions between organisms and their environment are interwoven to form the fabric of an ecosystem. The dynamics of any ecosystem include two major processes. One is a cycling of nutrients. For example, minerals acquired by plants will eventually be returned to the soil by microorganisms that decompose leaf litter, dead roots, and other organic debris. The second major process in an ecosystem is a flow of energy from sunlight to photosynthetic life to organisms that feed on plants. The theme of organisms interacting with their environment is essential to understanding life on all levels of organization.

THE INHERITANCE OF BIOLOGICAL INFORMATION

Order implies information; instructions are required to arrange parts or processes in a nonrandom way. Biological instructions are encoded in the molecule known as deoxyribonucleic acid, or DNA. DNA is the substance of genes, the units of inheritance that transmit information from parents to offspring (Figure 1.6).

Each DNA molecule is a long chain made up of four basic chemical building blocks called nucleotides. The way DNA conveys information is analogous to the way we arrange the letters of the alphabet into precise sequences with specific meanings. The word *rat*, for example, conjures up an image of a rodent; but *tar* and *art*, which contain the same letters, mean something quite different. Libraries are filled with books containing information encoded in varying sequences of only 26 letters. We can think of nucleotides as the alphabet of inheritance. Specific sequential arrangements of these four chemical letters encode the precise information in a gene. If the entire library of genes stored within the microscopic nucleus of a single human cell were written in letters the size of those you are now reading, the information would fill more than a hundred books as large as this one. Thus the complex structural organization of an organism is specified by an inherited script conveying an enormous amount of coded information. Inheritance itself is based on a complex mechanism for copying DNA and passing its sequence of chemical letters on to offspring.

All forms of life employ essentially the same genetic code. A particular sequence of nucleotides says the same thing to one organism as it does to another; differences between organisms reflect genetic programs of different nucleotide sequences. The diverse

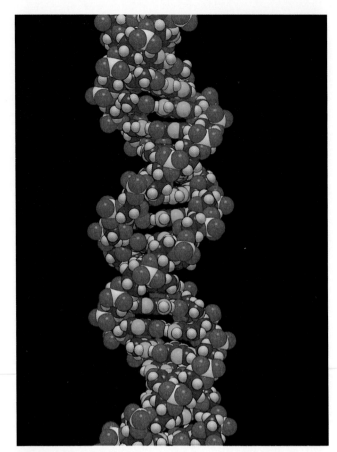

Figure 1.6
The genetic material: DNA. The molecule that conveys biological information from one generation to the next takes the three-dimensional form of a double helix. Along the length of the molecule, information is encoded in specific sequences of nucleotides, the chemical building blocks of DNA.

forms of life are different expressions of a common language for programming biological order.

UNITY IN DIVERSITY

Diversity is a hallmark of life. Biologists have identified and named about 1.7 million species, including over 260,000 plants, almost 50,000 vertebrates (animals with backbones), and more than 750,000 insects; thousands of newly identified species are added to the list each year. Estimates of the total diversity of life range from about 5 million to over 30 million species.

Biological diversity is something to relish and preserve, but it can also be a bit overwhelming. To make the diversity somewhat more comprehensible, people have devised ways of grouping species that are similar. Thus, we may speak of squirrels and pine trees without distinguishing the many different species

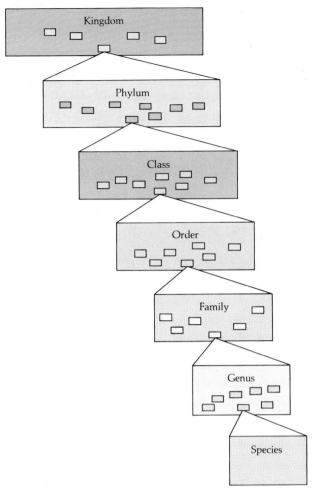

Figure 1.7
Classifying life. The taxonomic scheme classifies species into groups subordinate to more comprehensive groups. Species that are very similar are placed in the same genus, genera are grouped into families, and so on, each level of classification being more comprehensive than those it includes.

belonging to these groups. Taxonomy, the branch of biology concerned with naming and classifying species, groups organisms according to a more formal scheme. The scheme consists of different levels of classification, each more comprehensive than those below it (Figure 1.7). The broadest units of classification are the five kingdoms of life: Monera, Protista, Plantae, Fungi, and Animalia (Figure 1.8).

Kingdom Monera consists of the microscopic organisms known as bacteria. This kingdom is distinguished from the other four kingdoms by cell structure; all organisms included in Monera consist of prokaryotic cells, the simpler of the two major cell types (see Figure 1.4). Organisms with eukaryotic cells are divided among the other four kingdoms.

Kingdom Protista consists mostly of eukaryotic organisms that are unicellular—for example, the microscopic organisms known as protozoa. Also included in this kingdom are certain multicellular forms that seem to be more closely related to single-celled species than to plants, fungi, or animals.

The remaining three kingdoms—Plantae, Fungi, and Animalia—are the major groups of multicellular eukaryotes. Plants are characterized by photosynthesis, the chemical process that converts light energy to the chemical energy of sugar and other foods. Fungi are mostly decomposers that absorb nutrients obtained by breaking down the complex molecules of dead organisms and waste such as leaf litter and feces. Animals obtain food by ingestion, eating and digesting other organisms whole or by the piece. Thus, plants, fungi, and animals are distinguished partly by their contrasting modes of nutrition. Each kingdom, however, has many other unique features, which will be discussed in Unit Five.

If life is so diverse, how can biology have any unifying themes at all? What, for instance, can a mold, a tree, and a human possibly have in common? As it turns out, a great deal! Underlying the diversity of life is a striking unity, especially at the lower levels of organization. We can see it, for example, in the universal genetic code shared by all organisms. Unity is also evident in certain similarities of cell structure (Figure 1.9). Above the cellular level, however, organisms are so variously adapted to their ways of life that describing biological diversity remains an essential goal of biology. But few biologists view taxonomy as mere stamp collecting—a cataloging of seemingly unrelated living objects. In fact, the kinship of all life, though sometimes cryptic, is unmistakable.

EVOLUTION: THE CORE THEME

The history of life is not a story of immutable species individually created on a conservative planet, but is rather a chronicle of a restless Earth billions of years old, inhabited by a changing cast of living forms. Life evolves. Just as an individual has a family history, each species is one tip on a branching tree of life extending back in time through ancestral species more and more remote. Species that are very similar, such as the horse and zebra, share a common ancestor that represents a relatively recent branch point on the tree of life. But through an ancestor that lived much farther back in time, horses and zebras are also related to rabbits, humans, and all other mammals. And mammals, reptiles, birds, and all other vertebrates share a common ancestor even more ancient. Trace evolution back far enough, and there are only the primeval prokaryotes that inhabited the Earth more than three bil-

(a)

50 μm

(b)

0.5 mm

(c)

(d)

(e)

◀ **Figure 1.8**
The five kingdoms. (a) Kingdom Monera includes all prokaryotic organisms. The life forms known as bacteria are placed in this kingdom. This is a mixture of bacteria living in sewage (LM). **(b)** Kingdom Protista consists of unicellular eukaryotes and their relatively simple multicellular relatives.

This is an assortment of protists inhabiting pondwater (LM). **(c)** Kingdom Plantae, entirely multicellular and eukaryotic, is characterized by photosynthesis. **(d)** Kingdom Fungi is defined, in part, by the nutritional mode of its members, organisms that absorb nutrients after

decomposing organic refuse. The species shown here is *Amanita muscaria.* **(e)** Kingdom Animalia consists of multicellular eukaryotes that ingest other organisms. These are rose pelicans in Kenya.

---

lion years ago. All of life is connected. Evolution, the transformation of life on Earth from its earliest beginnings to its apparently unending diversity today, is the one biological theme that ties together all others (Figure 1.10).

Charles Darwin brought biology into focus in 1859 when he published *On The Origin of Species.* Darwin made two major points in his book. First, he argued convincingly from several lines of evidence that contemporary species arose from a succession of ancestors through a process of "descent with modification," his phrase for evolution. So much additional evidence

has accumulated since Darwin's time that nearly all biologists now see evolution as a basic fact of life. (The evidence for evolution is discussed in detail in Chapter 20.) Darwin's second major point was to propose a mechanism of evolution called natural selection.

Darwin synthesized the concept of natural selection from observations that by themselves were neither new nor profound. Others had the pieces of the puzzle, but Darwin saw how they fit together. As evolutionary biologist Stephen Jay Gould has put it, Darwin based natural selection on "two undeniable facts and an inescapable conclusion":

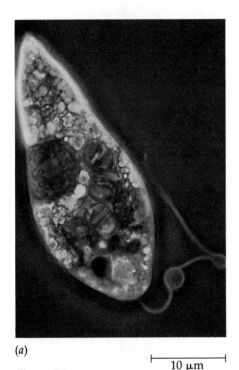

(a)

10 μm

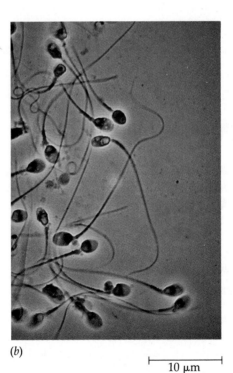

(b)

10 μm

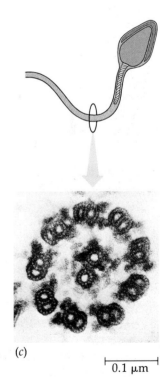

(c)

0.1 μm

Figure 1.9
An example of unity underlying the diversity of life. Eukaryotic organisms as diverse as protozoa (Kingdom Protista) and animals possess flagella, whiplike organelles that propel cells through water. Shown here at approximately the same

magnification are **(a)** a flagellated single-celled organism (*Euglena*) and **(b)** some human sperm cells (LMs). **(c)** Using an electron microscope to compare cross sections of flagella from diverse eukaryotes reveals a common structural organi-

zation. Such striking similarity in complex components contributes to the evidence that organisms as diverse as protozoa and humans are, to some degree, related.

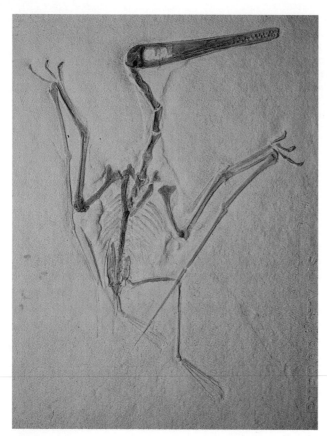

Figure 1.10
The fossil record chronicles evolutionary history.
This giant, flying reptile, called a pterosaur, lived over 135 million years ago. The fossil record tells a story of evolving life on an ever-changing planet.

- Individuals in a population of any species vary in many heritable traits.

- Any population of a species has the potential to produce far more offspring than the environment can possibly support with food, space, and other resources. This overproduction makes a struggle for existence among the variant members of a population inevitable.

- Those individuals with traits best suited to the local environment generally leave a disproportionately large number of offspring, thus increasing the representation of certain heritable variations in the next generation. This selective reproduction is what Darwin called natural selection, and he envisioned it as the cause of evolution.

We see the products of natural selection in the exquisite adaptations of organisms to the special problems of their environments (Figure 1.11). Notice, however, that natural selection does not create adaptations; rather, it increases the frequencies of certain

variations among those that arise randomly in each generation. Adaptation is an editing process, with heritable variations exposed to environmental factors that favor the reproductive success of some individuals over others. The camouflage of the mantids in Figure 1.11 did not result from individuals changing during their lifetimes to look more like their backgrounds and then passing that improvement on to offspring, but by the greater reproductive success of each generation of individuals who were innately better camouflaged than the average mantid.

Darwin argued that by cumulative effects over vast extents of time, natural selection could produce new species from ancestral species. This descent with modification, Darwin speculated, is the source of all biological diversity. It is the one idea that makes sense of both the unity and diversity that can be observed in life. Evolution is the core theme of biology—a theme that will resurface in every unit of this text.

SCIENCE AS A WAY OF KNOWING*

Biology is classified as a natural science. Having introduced unifying themes that apply specifically to the study of life, this chapter closes with a more general description of science.

Like life, science is better understood by observing it than by trying to create a precise definition. The word *science* is derived from a Latin verb meaning "to know." Science emerges from our curiosity about ourselves, the world, and the universe. Striving to understand seems to be one of our basic drives. At the heart of science are people asking questions about nature and believing that those questions are answerable. Scientists tend to be quite passionate in their quest for discovery. Max Perutz, a Nobel Prize–winning biochemist, puts it this way: "A discovery is like falling in love and reaching the top of a mountain after a hard climb all in one, an ecstasy induced not by drugs but by the revelation of a face of nature that no one has seen before."

The process known as the scientific method outlines a series of steps for answering questions, but few scientists adhere rigidly to this prescription. Science is less structured than most people realize. Like other intellectual activities, the best science is a product of minds that are creative, intuitive, and imaginative. Perhaps science is distinguished by its conviction that natural phenomena, including the processes of life,

---

*This section borrows its title from the "Science as a Way of Knowing" project guided by Professor John Moore of the University of California at Riverside and sponsored by the American Society of Zoologists and affiliated organizations. The project has published a series of articles tracing the development of modern biology.

(a)

(b)

(c)

Chapter 1: Introduction: Themes in the Study of Life   **13**

have natural causes and by its obsession with evidence. Scientists are generally skeptics. As you begin each unit of study in this text, you will meet such scientists through personal interviews and come to understand a bit more about how they think and why they enjoy their work.

One of the key ingredients of the scientific process is the hypothesis. Hypotheses are educated hunches that scientists propose as tentative answers to specific questions, or problems. They are tested by experiments or additional observations. Hypotheses are ideas on trial.

We can examine the central role of testable hypotheses in science with an example. Consider the way so many houseplants grow toward the light. Rotate the plants, and their growth will be reoriented until the leaves again face the window. Perhaps you have wondered how plants sense and respond to the direction of light. Asking this same question a hundred years ago, Charles Darwin and his son Francis speculated that the tip of a stem senses the direction of light. To test this hypothesis, they removed tips of grass seedlings and placed them, along with some intact seedlings, on a windowsill. The unaltered seedlings curved normally toward light, but the tipless plants grew straight upward. In another experiment, the Darwins covered the tips of grass seedlings with caps made of tinfoil that blocked light. These plants, apparently unable to sense the direction of light, also grew straight. The results of these experiments were consistent with the hypothesis that it is the tip of a stem that detects the direction of light. Yet, the Darwins observed that the actual bending response of the seedling occurred some distance below the tip. They concluded that the tip of the seedling senses the direction of light and sends some kind of signal down to the responding region, which then curves toward the light as it grows. But what is the nature of this signal? And how does it cause the seedling to curve? As so often happens in science, experiments not only test hypotheses, but also lead to new problems. (We will pursue these particular questions in Chapter 35.)

Notice that the Darwins always included normal, unaltered plants in their experiments. In the vernacular of science, these seedlings were the *controls*. The seedlings that were modified—by removal of tips, for instance—were the *experimentals*. Without controls, we cannot draw any conclusions from the behavior of the experimentals. If tipless seedlings on a windowsill fail to grow toward light, it can be because the day is overcast, or because the room is too warm, or because the grass seedlings are defective, or because of some other unknown factor. Only if we have a control group for comparison can we determine whether removing the tips prevents seedlings from sensing the direction of light. Controls enable us to test the effect of a single variable condition. If the controls bend toward the light and the experimentals do not, it is likely that the difference in behavior is due to the way we have modified the experimentals. We cannot attribute the results to a cloudy day, since it is just as cloudy for both groups of seedlings.

Another key feature of science is its progressive, self-correcting quality. A succession of scientists working on the same problem build on what has been learned earlier. It is also common for scientists to check on the conclusions of others by attempting to repeat observations and experiments. Among contemporary scientists working on the same problem, there is both cooperation and competition. Scientists share information through publications, seminars, meetings, and personal communication. They also subject one another's work to careful scrutiny. Science thrives on intellectual turbulence and new ideas.

Many people associate the word *discovery* with science. Often, what they have in mind is the discovery of new facts. But accumulating facts is not really what science is about; a telephone book is a catalog of facts, but it has little to do with science. It is true that facts, in the form of observations and experimental results, are the prerequisites of science. What really advances a science, however, is a new concept that collectively explains a number of observations that previously seemed to be unrelated. The most exciting ideas in science are those that explain the greatest variety of phenomena. People like Newton, Darwin, and Einstein stand out in the history of science not because they generated a great many facts, but because they synthesized ideas with great explanatory power. Such ideas, broad in scope and supported by a large body of evidence, are known as theories. Compared to a hypothesis, then, a theory is more comprehensive and more widely accepted.

In some ways, biology is the most demanding of all sciences, partly because living systems are so complex and partly because biology is a multidisciplinary science that requires a knowledge of chemistry, physics, and mathematics. Modern biology is the decathlon of natural science. If you are a biology major or a preprofessional student, you have an opportunity to become a versatile scientist. If you are a physical science major or an engineering student, you will discover in the study of life many exciting applications for what you have learned in your other science courses. If you are a nonscience student enrolled in biology as part of a liberal education, you have selected a course in which you can sample many scientific disciplines.

No matter what brings you to biology, you will find the study of life to be challenging and uplifting. The themes introduced in this chapter are intended to provide you with a conceptual framework for fitting together the many things you will learn about biology and to compel you to ask important questions of your own.

The Chemistry of Life

Interview: Stanley Miller

Biology and chemistry are allied sciences. Nowhere is their interdependence more evident than in the study of the origin of life. Chemists and biologists have long speculated about the conditions of primitive Earth and how that environment could have given rise to the precursors of living cells.

In 1953, when Stanley Miller was only 23 years old, he performed an experiment that would attract global attention. In a laboratory at the University of Chicago, Dr. Miller attempted to simulate the chemical dynamics of the primordial Earth in a glass apparatus that he constructed. His landmark experiment was the first to test the

hypothesis that, given the right conditions, the chemical building blocks of life could originate from simpler chemicals. In this interview, Dr. Miller shares with us the events that led to his experiment and the results that brought the world a step closer to understanding how life on earth began.

Today, Dr. Miller is professor of chemistry at the University of California at San Diego where he continues his research on the prebiotic synthesis of organic compounds. In addition to his research, Dr. Miller teaches courses in biochemical evolution, molecular biochemistry, and physical chemistry.

Thinking back to your student days, can you recall how you chose your schools—Berkeley for your undergraduate education and the University of Chicago for graduate school?

I chose Berkeley because it was logical—right near my home in Oakland and it was a good school. It is also costly to go away from home. As for graduate school, the chemistry department at Berkeley didn't allow undergraduates to continue on as graduate students. This is common in chemistry departments. The feeling is that chemistry is so diverse that students should broaden their perspective by attending a different school. I talked to a number of professors about which was the best place to go, and at the time, Chicago seemed the best choice. The real problem was whether or not the stipend that I was offered was adequate. Although shopping around for the biggest fellowship is not a good thing to do, you've got to have enough money to live on.

As a chemical system, what most distinguishes life from nonliving matter?

The essential difference between life and nonlife is replication. There are other differences, but this is the essential one. In addition to replication, there has to be mutation, with the mutations transmitted to the progeny. Thus the origin of life is the origin of replication with mutation. Another way to state this is that the origin of life is the origin of evolution, since reproduction or replication, mutation and selection result in Darwinian evolution.

15

You're best known for your experiments demonstrating that organic molecules could have been produced on the early Earth before there was life. What is an organic molecule?

The old definition, 100 years ago, was a molecule that came from living organisms. It was felt that carbon compounds, except carbon dioxide and carbon monoxide, came only from living organisms and had some sort of vital force in them. But we now know that there is no vital force and organic compounds are just those that contain carbon.

What experiments were historically important in dispelling the myth that organic molecules are products of supernatural vital forces?

The example that's usually cited is Wohler's synthesis of urea in 1828, in which he took ammonium cyanate and converted it to urea, which is a product of animal excretion. But the idea didn't catch on until people started making acetic acid and other organic compounds from starting materials derived from nonliving components.

Prior to your experiments in the early 1950s, what was the prevalent view among biologists about the origin of the first organic molecules?

As far as I can tell, they really didn't think much about it. There was just no rational explanation. What was usually proposed was that life arose by an extremely improbable event. There was

also Oparin's hypothesis, which had been around since 1938. Oparin proposed there was an abundant organic soup, and that the first organisms were derived from these prebiotic organic compounds and ate them rather than making their own.

If organisms didn't make the organic substances present in this primordial soup, then what *was* the source of these molecules?

Oparin proposed that the primitive atmosphere contained the gases methane, ammonia, and hydrogen, and water and that chemical reactions in that primitive atmosphere produced the first organic molecules. That hypothesis had a good deal of appeal, but without the experiments, it was talked about but not very well accepted.

What did you propose to do as an experiment to test the hypothesis?

Harold Urey, my advisor at the University of Chicago, knew that it was more reasonable to make organic compounds from the reduced gases in the proposed primitive atmosphere than in our modern atmosphere. But after some discussion, it was felt that if methane, ammonia, and water were combined and just left to sit, organic compounds would not be made. Energy must be added. And for the source of energy, the choices would be ultraviolet light or electrical discharge (a spark). UV light is difficult to work with, so we settled on an electrical discharge, which was handy around most

chemical labs in the form of a vacuum leak detector. The plan was to simulate the primitive atmosphere in a glass apparatus, energize the mixture of gases with an electrical spark, and see if I made any organic molecules.

When you first proposed the experiment, what was Dr. Urey's initial reaction?

Well, I went up to Urey and told him that I wanted to do this experiment testing his ideas about the primitive atmosphere and the synthesis of organic compounds. He tried to talk me out of it and get me to do another project. When he realized I was determined, he explained that he felt this was very risky and he was responsible for making certain that I could come up with a Ph.D. thesis in two or three years. So we agreed that we'd give it six months or a year, and if nothing came out of it, we would go back to something conventional. But since we obtained encouraging results quickly, there was never any question that we would continue with it.

When Dr. Urey described your proposed experiment at a seminar, a doubtful listener asked, "And what do you expect to get?" Urey replied "Beilstein." What did he mean?

He was referring to *Beilstein's Handbuch der Organischen Chemie*. It used to be a single volume, but by now it has grown to about four hundred volumes. These volumes contain a list of proper-

ties of every organic compound that's ever been reported, a few million by this time. And what he meant was that we would get every possible type of organic compound formed more or less randomly.

When you actually cranked up your apparatus to simulate early Earth conditions, what did you make?

The surprise was that we didn't get Beilstein; we got mainly organic compounds of biological significance. And the amino acids were formed, not in trace quantities, but abundantly! The experiment went beyond our wildest hopes.

What are amino acids?

Amino acids are the building blocks of proteins. Twenty of them occur in proteins, and many thousands more are possible. Since proteins are important to living organisms, making amino acids would be considered very important.

What was the physical appearance of the soup that you made in your discharge apparatus?

The first time I did the experiment, it turned red. Very dramatic! And then after it turned red, it got more yellow and then brown as the sparking went on. The experiment is easily reproducible with the exception of this red color.

After you'd been on your project for about 3½ months, you were ready to publish your first paper reporting your results. Graduate students usually list their advisors as coauthors on their manuscripts, but when you took your paper to Dr. Urey, he suggested that you remove his name from the manuscript. Why was that?

Well, he felt that if his name had been on the paper he would have gotten all the credit and I would have gotten none. So he took his name off. And even at that, it's known as the Miller-Urey experiment. And it's probably the thing he's best known for, with the exception of the discovery of deuterium, for which he won the Nobel Prize, and his work on the atomic bomb.

After the first paper was published in *Science*, Urey expressed concern that the amino acids formed in your discharge apparatus could have been made by bacteria contaminating the glassware or solutions. How did you respond to this concern?

Someone at a seminar had needled him that there were probably bacteria in there contaminating the solution. And

he came back quite concerned, and told me that if I'd made a mistake, it would be better for me to retract than to be proven wrong by someone else. I knew that there were bacteria around, since the solutions were not handled under sterile conditions, but I also knew that they were not synthesizing the compounds—you could see some of the organic material actually being formed and dripping off the electrodes. I decided I was going to put an end to this doubt. I took the whole glass apparatus, filled it up with the gases, sealed it off, and put it in an autoclave for eighteen hours instead of the usual fifteen minutes required to kill bacteria. I then sparked the mixture and obtained the same results I had from the apparatus that had not been sterilized. No one has ever raised the issue of sterility since.

Did other labs rush to repeat your experiments after you published?

It's hard to say. One or two did, because they published papers saying they had obtained similar results. I suspect that others repeated it without reporting it. After a while, many experiments of this type (with variations) were done, and a substantial body of work has been accumulated.

In 1969, a meteorite fell in Murchison, Australia. Chemical analysis of the Murchison meteorite showed that it contained organic molecules. How does this discovery relate to your experiments on the origin of organic molecules?

True, what fell in Murchison, Australia (which is about 300 miles north of Melbourne), was a meteorite that contained organic material. Previously a number of these organic-containing meteorites had fallen and had been sitting in museums. They unquestionably contained organic material, but when they were analyzed for amino acids and other organic compounds, there were two problems. One was that while sitting in the museums they had been contaminated, and the other was that the analytical methods at the time were not adequate to detect and identify the small amounts of organic molecules. But by 1969, when the Murchison meteorite fell, the analytical methods had improved and the meteorite was fresh, so there weren't quite the contamination problems. So, when

the meteorite was analyzed by a team at the NASA Ames Research Center, it turned out that amino acids were present at the parts per million level. These amino acids were very similar to the ones that had been made in the electrical discharge 15 years earlier. They did, however, find some additional amino acids that I hadn't obtained. So I repeated the electrical discharge experiments with a little modification, and I was able to show that all the amino acids in the Murchison meteorite were also obtained in the electrical discharge apparatus.

The significance of this is that we had been doing these prebiotic experiments not knowing whether this really took place, or if it was just a model experiment that bore no relation to reality. The similarity of the Murchison organic compounds and the electrical discharge compounds shows that this kind of synthesis took place on the parent body of the meteorite, probably an asteroid. This makes it plausible, but does not prove, that similar syntheses took place on the primitive Earth.

A few biologists have speculated that the first organic molecules on Earth arrived with extraterrestrial rocks such as the Murchison meteorite. What is your reaction to this idea?

Some organic substances certainly came in this way. There are problems with the survival of organic compounds on meteorites larger than a few meters in diameter, because they are not slowed up by the atmosphere before colliding with the earth. In addition, sugars do not occur in Murchison, so some of the prebiotic synthesis must have occurred on the Earth. And I think that all but a few percent of the synthesis was done on the Earth.

Why doesn't such organic synthesis occur in the present world with our current atmosphere?

Organic synthesis of that sort doesn't take place now because of the presence of oxygen.

Each amino acid occurs in two three-dimensional forms—left- and right-handed versions that mirror one another. In your prebiotic synthesis experiments, both forms are produced, and yet only the left-handed molecules are common in life, especially in proteins. How do you account for this?

Well, that's been an outstanding question for a long time. There really is no good explanation for how this came about. All prebiotic syntheses give equal quantities of the right (D) and left (L) forms. In fact, you can tell whether or not you've got contamination if there's an excess of L-amino acids. That was the reason the amino acid analysis of the Murchison meteorite was convincing—there were equal quantities of the D- and L-forms. Without that, the presence of amino acids would have shouted "contamination." There are a number of rational explanations why living organisms have all L-amino acids. It has to do with selective advantage and the ease of synthesis from building blocks having similar geometry. Although we don't know how this came about, I'm convinced that it happened about the time of the origin of life or shortly thereafter.

But why don't we have both "left-handed" and "right-handed" species?

This is a common question I get asked. An all D-organism is as good as an all L-organism. If life arose only once, then it was by chance that it used L-amino acids, but it could, with equal probability, have used all D-amino acids. If life arose many times, there would have been both D- and L-organisms. In the course of time, one of these would have acquired a selective advantage and would have outgrown all the others.

When you were first proposing your experiments and doing your experiments on prebiotic organic synthesis, were you already thinking of this as a first stage in the origin of life?

Of course. We understood the implications of our discovery; we wouldn't have done the experiment as an exercise in chemical synthesis.

After your provocative paper appeared in the journal *Science*, was there much reaction in the press?

Oh, yes. There was a deluge of requests for interviews and pictures and the like. I was really quite surprised. *Time* magazine had a whole page on the experiment. There was a Gallup poll taken on whether or not people felt life could ever be made in a test tube. The opinion was largely no.

Assuming that mechanisms similar to those at work in your laboratory simulations produced organic molecules on the early Earth, what were some subsequent developments that were important to the origin of life?

The essential process is that of self-replication. Thus we need to find the prebiotic polymers that can contain genetic information. These may be RNA-like molecules, but they could have been quite different.

How early do you think life began on Earth?

Nobody really knows. The Earth is 4.5 billion years old. And the earliest evidence for life is about 3.5 billion. So there's a billion year period in between. There have been various proposals about the temperature of early Earth—that it was cold when it formed or that it was molten. But prebiotic synthesis

did not start until the Earth got down to a low enough temperature for the organic material to be stable—that is, below about a hundred degrees. We're not sure when that happened, but we assume maybe 4 billion years ago. That still leaves hundreds of millions of years in between the origin of suitable conditions on the Earth and the oldest known fossils, and we just don't know when life began during the period.

In what sort of place do you think life began on Earth?

The usual assumption is that it began in the ocean. But you can legitimately propose that some of the processes occurred in different areas. For example, some of the polymerization reactions that made larger organic molecules probably occurred on beaches that had dried out and heated up, and some may have occurred in hot springs. But the oceans form the bulk of the area where organic reactions could take place, and I think most of the chemistry took place there.

One hypothesis is that the first organisms arose in the hot water around volcanic vents on the sea floor. You and a colleague, Jeffrey Bada, have challenged this view in a recent publication. Would you describe your experiments?

The proposal was that life originated in these submarine vents. Water flows out of these vents at 350°C. This water is seawater that flows through the deep sea sediments and is heated by molten rock. The water comes out in a matter of minutes through these black smoker chimneys, and then cools. Living around these vents are some very interesting creatures. There are tube-worms, clams, and the like. These organisms are not living at 350° C, but at around 37° C. And they live off the metabolic products of bacteria in a sort of symbiotic relationship.

The hypothesis is that water and gases start out at 350°C, the amino acids are made, polymerized in the vents, and then somehow the proteins and other polymers are organized into living organisms by the time they get out of the vent. The problem with this hypothesis is that organic compounds cannot be synthesized at 350°C—they would be destroyed. We did an experiment to illustrate this by taking some amino acids and heating them up in a test tube to 350°C. They decomposed before we even got them to that temperature. Furthermore, you can make polymers by heating amino acids, but you have to heat them dry. Well, in a submarine vent, it's obviously not dry. Even if you get these polymers of amino acids, that is not a living organism by any means—it has no genetic material. So on three or four counts the hypothesis is invalid.

If prebiotic synthesis of organic compounds led to the origin of life here on Earth, do you think this is a common process elsewhere in the universe?

I feel it is. But the problem is that we don't know that much about the subsequent steps. Making organic molecules such as amino acids, pyrimidines, purines, and sugars is easy. But how to organize them into the first living organism has not been worked out. So it's very difficult to estimate how frequently this might have occurred. My feeling, though, is that it is very frequent, and that the Earth is not unique in having life.

What work is presently going on in your laboratory?

I'm still studying the synthesis of prebiotic organic compounds, using atmospheres other than methane. There is considerable controversy about the composition of the primitive atmosphere. One of the problems is that methane and ammonia are decomposed by ultraviolet light rapidly. So the question is, could we have had that kind of atmosphere? And if the atmosphere was different, did it still lend itself to organic synthesis?

We have been doing experiments using carbon monoxide and carbon dioxide. You can make organic material with these kinds of atmospheres, but only if there is molecular hydrogen around. It's not that easy to get molecular hydrogen in large amounts into the atmosphere because it tends to escape from the atmosphere into outerspace. Another problem that we have been working on in the laboratory is to try and figure out what primitive RNA was like—RNA is thought to have preceded DNA as the genetic material, but there was some precursor polymer to RNA.

Speaking of DNA, it just occurred to me that your famous paper was published the same year as Watson and Crick published their celebrated paper on the double helix—the structure of DNA, the genetic material.

Yes, their paper was in the April 25 issue of *Nature* and mine was in the May 15 issue of *Science*. I should mention that until then, most biologists thought that the genetic material was protein rather than DNA. That is why the prebiotic synthesis of amino acids was so striking.

Finally, what advice can you offer to students who are just beginning their college science education and are thinking about a career in research?

I think freshmen and sophomores should get a good background in chemistry and biology. Chemistry is very important because if you don't learn it as an undergraduate, you're probably not going to learn it very well as a graduate student. By the time they're juniors or seniors, students should probably do a little project in a research lab. Since biology is going more and more in a molecular direction, a strong background in chemistry and the ability to be efficient in the lab are very important.

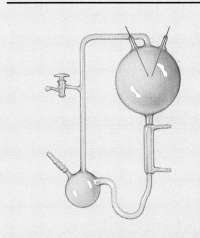

2

Atoms, Molecules, and Chemical Bonds

The phenomenon we call life is the cumulative product of interactions among the many kinds of chemical substances that make up the cells of an organism. As a preliminary step in understanding life, we might dismantle the cell into its chemical components and then study the structure and behavior of these molecules. From there, we could begin to examine how the molecules aggregate and interact. Somewhere in the transition from molecules to cells, we would cross the blurry boundary between nonlife and life. Life emerges from the integrated organization of the whole organism. To understand a biological process, however, we often must reduce it to simpler steps that can be studied at lower levels of organization. This approach, described in Chapter 1 as reductionism, has been very fruitful. Our knowledge of the chemical mechanism of inheritance, for instance, grew from the study of DNA molecules extracted from cells. After we learn how the chemical components involved in a particular biological process work, the next step is to investigate how those parts interact.

One of the themes of this textbook is the organization of life on a hierarchy of structural levels, with additional properties emerging at each successive level (Figure 2.1). The chapters of this unit apply the theme of emergent properties to the lowest levels of biological organization—to the ordering of atoms into molecules, and to the interactions of these molecules within the cell. This chapter introduces general principles of chemistry that we will be able to extend to living matter.

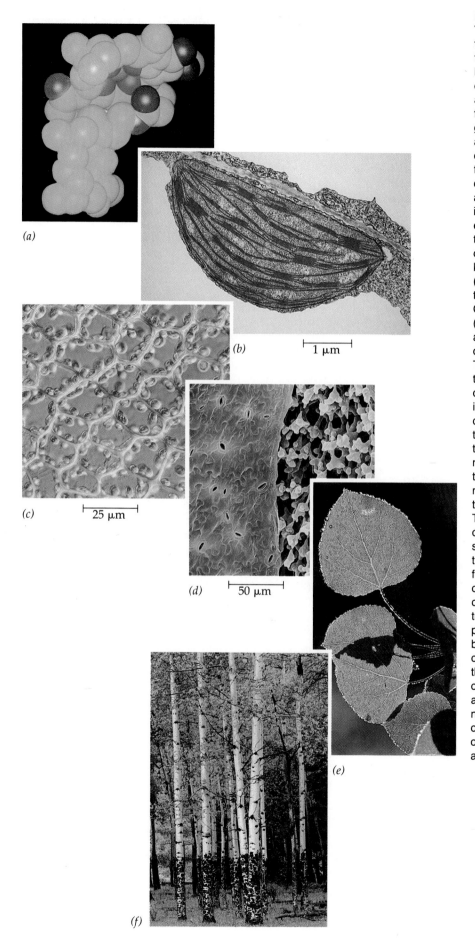

(a)

(b)

1 μm

(c)

25 μm

(d)

50 μm

(e)

(f)

Figure 2.1
The hierarchy of biological organization: a review. This sequence of photos takes us all the way from atoms to a biological community of many interacting species. (**a**) Chlorophyll, represented here by a computer graphic, is a molecule built from more than a hundred atoms. It is the green molecule in the leaves of plants that absorbs sunlight as a source of energy for driving photosynthesis, the manufacture of food in the leaf. The light-absorbing ability of chlorophyll depends on its precise arrangement of component atoms. (**b**) By itself, chlorophyll cannot harness light energy to make food. The process of photosynthesis requires participation of many other molecules organized within the cellular organelle called the chloroplast (TEM). (**c**) Many organelles cooperate in the functioning of the living unit we call a cell. Chloroplasts are evident in these leaf cells (LM). (**d**) In multicellular organisms, cells are usually organized into tissues, groups of similar cells forming a functional unit. The leaf in this micrograph (a photograph taken with a microscope) has been cut obliquely, revealing two different specialized tissues. The honeycomb-like tissue, consisting of photosynthetic cells within the leaf, is called spongy mesophyll. The tissue with the small pores is the epidermis, the "skin" of the plant. The pores in the epidermis allow carbon dioxide, a raw material that is converted to sugar by photosynthesis, to enter the leaf (SEM). (**e**) The leaf, a plant organ, has a specific organization of many different tissues, including mesophyll, epidermis, and the vascular tissue that transports water from the roots to the leaves. No one organ of the plant can survive for long on its own. The organism, in this case an aspen tree, is a whole greater than the sum of its parts. (**f**) These aspens are members of a biological community that includes many other species of organisms. At each of these superimposed levels of biological order, properties emerge from the specific arrangement and interaction of the component parts. This unit of chapters focuses on the lowest levels in the hierarchy of organization—biologically important atoms and the molecules of life they form.

MATTER: ELEMENTS AND COMPOUNDS

Chemistry is the study of matter, and **matter** is defined as anything that takes up space and has mass. (*Note:* Sometimes, we substitute the term *weight* for *mass,* although the two are not equivalent. We can think of mass as the amount of matter an object contains. The weight of an object measures how strongly that mass is pulled by gravity. An astronaut in a space shuttle is weightless, but the astronaut's mass is the same as it would be on earth. However, as long as we are earth-bound, the weight of an object is a measure of its quantity of matter, so for our purposes, we can use the terms interchangeably.)

Matter exists in many diverse forms, each with its own characteristics. Rocks, metals, wood, glass, and you and I are just a few examples of what seems an endless array of matter. The ancient Greek philosophers suggested that the great variety of matter arises from four basic ingredients, or elements. An **element** is a substance that cannot be broken down to other substances by ordinary chemical means. The Greeks imagined the elements of matter to be air, water, fire, and earth—supposedly pure substances that could not be decomposed to other forms of matter. All other substances were thought to be formed by blending various proportions of two or more of the elements. Even though the classical philosophers proposed the wrong elements, their basic idea was correct.

Today, chemists recognize 92 elements occurring in nature—for example, gold, copper, carbon, and oxygen. About a dozen more elements have been made in the laboratory. Each element has a symbol, usually the first letter or two of its name. A few of the symbols are derived from Latin or German names; for instance, the symbol for sodium is Na, from the Latin, *natrium.* Our modern elements fit the Greek definition: Elements cannot be decomposed to other substances by ordinary chemical means.

Two or more elements may be combined in a fixed ratio to produce a **compound.** Table salt, for example, is actually sodium chloride (NaCl), a compound composed of the elements sodium (Na) and chlorine (Cl). Pure sodium is a metal that explodes readily, and pure chlorine is a poisonous gas that was used as a weapon during World War I. Chemically combined, however, sodium and chlorine form an edible compound. This is a simple example of organized matter having emergent properties: A compound has characteristics beyond those of its combined elements.

Elements Essential to Life

About 25 of the 92 natural elements are known to be essential to life, but four of these—carbon (C), oxygen

Symbol	Element	Atomic Number	Percent Wet Weight of Human Body*
O	Oxygen	8	65.0
C	Carbon	6	18.5
H	Hydrogen	1	9.5
N	Nitrogen	7	3.3
Ca	Calcium	20	1.5
P	Phosphorus	15	1.0
K	Potassium	19	0.4
S	Sulfur	16	0.3
Na	Sodium	11	0.2
Cl	Chlorine	17	0.2
Mg	Magnesium	12	0.1

Table 2.1 Naturally occurring elements in the human body

Trace elements (less than 0.01%): boron (B), chromium (Cr), cobalt (Co), copper (Cu), fluorine (F), iodine (I), iron (Fe), manganese (Mn), molybdenum (Mo), selenium (Se), silicon (Si), tin (Sn), vanadium (V), and zinc (Zn).

*Includes water.

(O), hydrogen (H), and nitrogen (N)—make up 96% of living matter. Phosphorus (P), sulfur (S), calcium (Ca), potassium (K), and a few other elements account for most of the remaining 4% of an organism's weight. Table 2.1 lists the elements making up the human body and their amounts. Figure 2.2 illustrates a deficiency of an essential element in plants.

Trace elements are those required by an organism in extremely minute quantities. However, this is not meant to imply that a trace element is a nutrient of marginal importance to the organism. Trace elements are mandatory for good health. Some trace elements, such as iron (Fe), are needed by all forms of life. Other trace elements are required by only certain species. For example, in vertebrates (animals with backbones), the element iodine (I) is a necessary ingredient of a hormone produced by the thyroid gland. A daily intake of only 0.15 milligram (mg) of iodine is adequate for normal activity of the human thyroid, but an iodine deficiency in the diet causes the thyroid gland to grow to abnormal size, producing a deformity called goiter.

THE STRUCTURE AND BEHAVIOR OF ATOMS

The units of matter are called **atoms**. They are so small that it would take about a million of them to stretch across the period printed at the end of this sentence. Each element consists of a certain kind of atom, which

Figure 2.2
Potassium deficiency in plants. Plants require the element potassium, which the roots absorb from the soil. The leaf on the left is from a well-nourished cotton plant. The leaf on the right is from a cotton plant grown in soil deficient in potassium.

is different from the atoms of any other element. We symbolize atoms with the same abbreviation we use for the element made up of those atoms; thus, C stands for both the element carbon and a single atom of it. An atom is the smallest possible amount of an element.

The concept of atoms dates back to ancient Greece. Democritus (4th century BC) envisioned matter as being composed of invisible (and indivisible) particles. As an analogy, he pointed out that the particulate nature of a sandy beach is not apparent until you are close enough to see the individual grains of sand. Democritus reasoned that a grain of sand is made of even smaller particles, too small for us to see. Today, with the help of very powerful microscopes, scientists have actually been able to photograph the atoms only imagined by Democritus over two thousand years ago (Figure 2.3).

Subatomic Particles

Although the atom is the smallest unit having the physical and chemical properties of its element, these tiny bits of matter are composed of even smaller parts called subatomic particles. Physicists have split the atom into more than a hundred types of particles, but only three kinds of particles are stable enough to be of relevance here: **neutrons, protons,** and **electrons**. Neutrons and protons are packed together tightly to form a dense core, or **nucleus,** at the center of the atom. The electrons move about this nucleus at nearly the speed of light (Figure 2.4).

Electrons and protons are electrically charged. Each electron has one unit of negative charge, and each

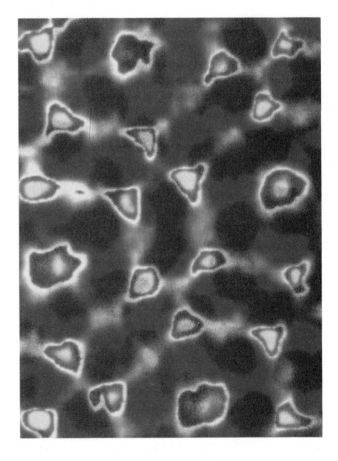

Figure 2.3
Atoms magnified 100 million times. A special type of microscope, called a scanning tunneling microscope, was used to photograph a crystal of the element silicon. Individual silicon atoms (large, yellow/red triangles) on the crystal's surface appear in an orderly array, connected by chemical bonds.

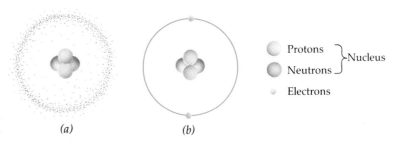

Figure 2.4
Two simplified models of a helium (He) atom. The nucleus consists of two neutrons and two protons. **(a)** The nucleus is surrounded by a cloud of negative charge owing to the rapid movement of two electrons. The probability of electron distribution in an orbital is proportional to the density of the shading. **(b)** The circle indicates the average distance of the electrons from the nucleus, although this distance is not drawn to scale compared to the size of the nucleus. (Our model of an atom will be refined as the chapter progresses.)

proton has one unit of positive charge. A neutron, as its name implies, is electrically neutral. The atomic nucleus is positive because of the presence of protons, and it is the attraction between opposite charges that keeps the rapidly moving electrons in the vicinity of the nucleus.

The neutron and proton are almost identical in mass, each about 1.7×10^{-24} g. Grams and other conventional units are not very useful for describing the mass of objects so minuscule. Thus, for atoms and subatomic particles, scientists use a unit of measurement called the **dalton,** in honor of John Dalton, the English chemist and physicist who helped develop atomic theory around 1800. Neutrons and protons have a mass of almost exactly 1 dalton apiece (actually 1.007 and 1.009, respectively, but close enough to 1 for our purposes). Because the mass of an electron is only about 1/2000 that of a neutron or proton, we can ignore electrons when computing the total mass of an atom (Table 2.2).

Atomic Number and Atomic Weight

Atoms of the various elements differ in their number of subatomic particles. All atoms of a particular element have the same number of protons in their nuclei. This number, which is unique to that element, is referred to as the **atomic number** and is written as a subscript to the left of the symbol for the element. The abbreviation $_2$He, for example, tells us that an atom of the element helium has two protons in its nucleus. Unless otherwise indicated, an atom is neutral in electric charge, which means that its protons must be balanced by an equal number of electrons. Therefore, atomic number tells us the number of protons *and* the number of electrons in a neutral atom.

We can deduce the number of neutrons from a second quantity, the **mass number,** which is the sum of protons plus neutrons in the nucleus of an atom. The mass number is written as a superscript to the left of an element's symbol. For example, we can use this shorthand to write an atom of helium as $_2^4$He. Since the atomic number indicates how many protons there are, we can determine the quantity of neutrons by subtracting atomic number from mass number: A $_2^4$He atom has 2 neutrons. An atom of sodium, $_{11}^{23}$Na, has 11 protons, 11 electrons, and 12 neutrons. The simplest atom is hydrogen, $_1^1$H, which has 1 proton and 1 electron, but no neutrons. A lone proton with a single electron moving about it constitutes a hydrogen atom.

Essentially all of an atom's mass is concentrated in its nucleus, because the contribution of electrons to mass is negligible. Since neutrons and protons each have a mass that is very close to 1 dalton, mass number tells us the approximate mass of the whole atom. The term **atomic weight** is often used to refer to what is technically the total atomic mass. Thus, the atomic weight of helium ($_2^4$He) is 4 daltons (4.003, to be exact).

Isotopes

All atoms of a given element have the same number of protons, but some atoms have more neutrons than other atoms of the same element, and hence weigh more. These different atomic forms are referred to as **isotopes** of the element. In nature, an element occurs as a mixture of its isotopes. For example, consider the three isotopes of the element carbon, which has an atomic number of 6. The most common isotope is carbon-12, $_6^{12}$C, which accounts for about 99% of the carbon in nature. It has six neutrons. Most of the remaining 1% of carbon consists of atoms of the isotope $_6^{13}$C, with seven neutrons. A third isotope, $_6^{14}$C, has eight neutrons; it is present in the environment in minute

Table 2.2 Properties of the subatomic particles		
Particle	**Approximate Weight (Daltons)**	**Charge**
Neutron	1	0
Proton	1	+1
Electron	1/2000	−1

(a)   *(b)*

Figure 2.5
Radiation damage to a forest. (a) In this experiment conducted at the Brookhaven National Laboratory on Long Island, New York, radioactive material was placed in a cannister on the top of a pole to assess the effect of radiation on nearby plants. A circle of dead trees around the radioactive source is obvious. **(b)** This grass survived because it was growing in the "shadow" of a tree that blocked the radiation.

quantities. Notice that all three isotopes of carbon have six protons—otherwise, they would not be carbon.

Both ^{12}C and ^{13}C are stable isotopes, meaning that their nuclei do not have a tendency to lose particles. The isotope ^{14}C, however, is unstable, or radioactive. A **radioactive isotope** is one in which the nucleus decays spontaneously, giving off particles and energy. Electronic instruments can detect this radioactive decay. Each radioactive isotope has a fixed **half-life**, the time it takes for 50% of the radioactive atoms in a sample to decay. The half-life for carbon-14 (^{14}C) is about 5600 years. This means that if a sample has 100 g of carbon-14 to start with, 50 g would remain as ^{14}C after 5600 years, 25 g after another 5600 years, and so on.

Radioactive isotopes have many useful applications in biology, one of which is dating fossils. A live organism takes in carbon-14 and other isotopes in proportions equal to the relative abundance of the isotopes in the environment. When the organism dies, the intake of radioactive isotopes from the environment stops, but the isotopes in its remains continue to decay. The ratio of carbon-14 to carbon-12 in the remains of an organism will be reduced by half each 5600 years. A fossilized tree that has a $^{14}C/^{12}C$ ratio half that of a live tree is about 5600 years old. By decaying at a fixed rate, the dwindling reservoir of ^{14}C acts as a clock, ticking away 5600 years each time its amount is reduced by half. Carbon-14 can be used to date fossils as far back as about 50,000 years; elements with radioactive isotopes having longer half-lives must be used for dating older fossils (Chapter 23).

Radioactive isotopes are also useful as tracers to follow atoms through metabolism, the chemical processes of an organism (see Methods Box, p. 26). Cells use the radioactive atoms as they would nonradioactive isotopes of the same element, but the radioactive tracers can be detected. Radioactive tracers have also become important diagnostic tools in medicine. For example, certain kidney disorders can be diagnosed by injecting small doses of substances containing radioactive isotopes into the blood and then measuring the amount of tracer excreted in the urine.

Although radioactive isotopes are very useful in biological research and medicine, it must be pointed out that radiation from these decaying isotopes poses a hazard to life by damaging cellular molecules. The severity of this threat depends on the type and amount of radiation an organism absorbs (Figure 2.5).

Energy Levels

The simplified models of the atom in Figure 2.4 distort the size of the nucleus relative to the volume of the whole atom. If the nucleus were the size of a golf ball, the electrons would be moving about the nucleus at an average distance of approximately 1 kilometer (km). Atoms are mostly empty space.

When two atoms approach each other, their nuclei never come close enough to interact. Of the three kinds of subatomic particles we have discussed, only electrons are directly involved in the chemical properties of an atom—that is, in the way an atom behaves when it encounters other atoms. This is why isotopes of the same element, which differ only in their neutron number, exhibit the same behavior in chemical processes. The chemical behavior of atoms, then, can be explained by the behavior of electrons.

All electrons are identical in mass and charge, but they vary in the amount of energy they possess. **Energy** is defined as the ability to do work. **Potential energy** is the energy that matter stores because of its position or location. For example, because of its altitude, water in a reservoir on a hill has potential energy. The energy is taken out of storage to do work (turn generators, for example) when the gates of the dam are opened and the water runs downhill. Since potential energy

METHODS: THE USE OF RADIOACTIVE TRACERS IN BIOLOGY

Radioactive isotopes are among the most important tools in biological research. They are used to label certain chemical substances to follow the steps of a metabolic process or to determine the location of the substance within an organism. Organisms do not generally discriminate between radioactive and stable isotopes of the same element; thus, they assimilate and process the labeled substance normally.

The experiments being conducted here were designed to determine how temperature affects the rate at which the genetic material, DNA, replicates in a population of dividing animal cells, and to localize the newly synthesized DNA within the cells. Populations of dividing cells are cultured in an artificial medium that contains, among other things, the chemical ingredients used by cells to make new DNA. One of those ingredients is labeled with ^3H, a radioactive isotope of hydrogen that will be used to trace the incorporation of the ingredient into new DNA.

After a certain amount of time has elapsed, samples of cells grown at various temperatures in the presence of the radioactive tracer are killed, and their DNA is precipitated onto pieces of filter paper. The papers are then placed in vials containing scintillation fluid, which emits flashes of light whenever certain chemicals in the fluid are excited by radiation from the decay of the radioactive tracer in the DNA. The frequency of flashes, proportional to the amount of radioactive material present, is measured in counts per minute by placing the vials in a scintillation counter (**a**). The effect of temperature on the rate of DNA synthesis can be determined by plotting the counts per minute for the various DNA samples against the temperatures at which the cells were grown (**b**).

A technique known as autoradiography can be used to determine the location of the radioactively labeled DNA within the cells. The cells are washed free of any radioactive material that was not incorporated into the DNA. Then they are fixed (preserved), and thin sections of them are placed on glass slides. The sections are covered by a layer of photographic emulsion, and the slides are kept for some time in the dark. Wherever DNA is located in the cells, radiation from the radioactive tracer will expose the photographic emulsion. The emulsion is developed, and the slide is examined with a microscope (**c**). Black grains of the exposed emulsion are superimposed on the nuclei of the cells, the sites of DNA. The nucleus of the cell on the left has been radioactivity labeled.

(a)

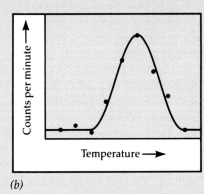

(b)

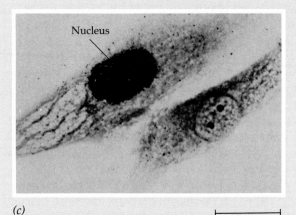

Nucleus

(c)

25 μm

has been expended, the water stores less energy at the bottom of the hill than it did in the reservoir. There is a natural tendency for matter to move to the lowest possible state of potential energy; in this example, water runs downhill. To restore the potential energy of a reservoir, work must be done to elevate the water against gravity.

The electrons of an atom also have potential energy, not because of altitude and gravity, but because of their position in relation to the nucleus. The negatively charged electrons are attracted to the positively charged nucleus; so the more distant the electrons are from the nucleus, the greater their potential energy. Unlike the continuous flow of water downhill, changes in the potential energy of electrons can occur only in steps of fixed amounts. An electron having a certain discrete amount of energy is analogous to a ball on a staircase. The ball can have different amounts of potential energy, depending on which step it is on, but it cannot spend much time between the steps (Figure 2.6).

The different states of potential energy for electrons in an atom are called **energy levels,** or **electron shells.** The first shell is closest to the nucleus, and electrons in this shell have the lowest energy. Electrons in the second shell have more energy, electrons in the third shell more energy still, and so on. An electron can change its shell, but only by absorbing or losing an amount of energy equal to the difference in potential energy between the old shell and the new shell. To move to a shell farther out from the nucleus, the electron must absorb energy. To move to a shell closer in, it must lose energy.

Electron Orbitals

Earlier in this century, the electron shells of an atom were visualized as concentric paths of electrons orbiting the nucleus, something like planets orbiting the sun (see Figure 2.4). The atom is not so simple, however. In fact, we can never know the exact trajectory of an electron. What we can do instead is describe the volume of space in which an electron spends most of its time. The three-dimensional space where an electron is found 90% of the time is called an **orbital.** Unlike a planetary orbit, an orbital is not a defined pathway of movement. The electron orbital is a statistical concept—a volume within which an electron has the greatest probability of being found (Figure 2.7).

No more than two electrons can occupy the same orbital. The first energy shell has a single orbital and so can accommodate a maximum of two electrons. This single orbital, which is spherical in shape, is designated the 1s orbital. The lone electron of a hydrogen atom occupies the 1s orbital, as do the two electrons of a helium atom. Electrons, like all matter, tend to

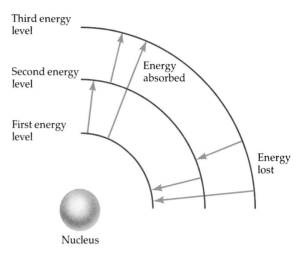

Figure 2.6
Energy levels of electrons. Electrons exist only at fixed levels of potential energy. An electron can move from one level to another only if the energy it gains or loses is exactly equal to the difference in energy between the two levels. Arrows indicate some of the stepwise changes in potential energy that are possible for electrons. Energy levels are also called electron shells.

exist in the lowest available state of potential energy, which they have in the first shell. An atom with more than two electrons must use higher shells, since the first shell is full.

The second electron shell can hold eight electrons, two in each of four orbitals. Electrons in the four different orbitals all have the same energy, but they move in different volumes of space. There is a 2s orbital, spherical in shape like the 1s orbital, but with a slightly greater diameter. The other three orbitals, called p orbitals, are dumbbell-shaped, each oriented at right angles to the other two. They are designated the $2p_x$, $2p_y$, and $2p_z$ orbitals. Besides having an s orbital and three p orbitals, higher electron shells can have additional orbitals with more complex shapes and can therefore accommodate many electrons. However, the outermost energy shell of an atom never has more than four orbitals, the s and three p orbitals, and therefore can never hold more than eight electrons.

Electron Configuration and Chemical Properties

The chemical behavior of an atom is determined by its electron configuration—that is, the distribution of electrons in the atom's electron shells. Beginning with hydrogen, the simplest atom, we can imagine building the atoms of other elements by adding one proton and one electron at a time. Figure 2.8, an abbreviated version of what is called a periodic table, shows this

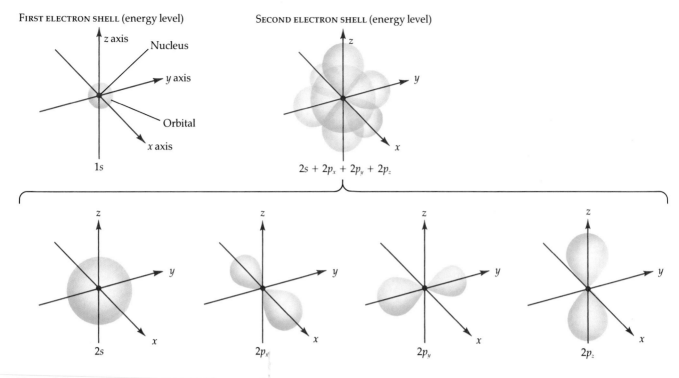

FIRST ELECTRON SHELL (energy level)

z axis
Nucleus
y axis
Orbital
x axis

1s

SECOND ELECTRON SHELL (energy level)

z
y
x

$2s + 2p_x + 2p_y + 2p_z$

z
y
x

2s

z
y
x

$2p_x$

z
y
x

$2p_y$

z
y
x

$2p_z$

Figure 2.7
Electron orbitals. These three-dimensional shapes describe the volumes of space where the constantly moving electrons are most likely to be found. Each electron shell has one spherical *(s)* orbital. The second level and all higher shells also have three dumbbell-shaped *(p)* orbitals, arranged on the imaginary *x, y,* and *z* axes of the atom. The third and higher electron shells can have additional orbitals of more complex shapes.

for the first 18 elements, from hydrogen ($_1$H) to argon ($_{18}$Ar). As electrons are added, you will notice some trends. Lower shells of these elements are filled before electrons occupy higher shells. Helium ($_2$He) has a completed first shell, and beginning with lithium ($_3$Li), electrons start to fill orbitals in the second shell. The first two electrons in the second shell (in beryllium) are both in the *s* orbital. Each of the next three electrons has its own *p* orbital. These *p* electrons are paired as the next three electrons are added. Neon ($_{10}$Ne) has a complete second shell, so the atoms that follow it in the series add electrons to the third shell until its *s* orbital and three *p* orbitals are full, as in argon ($_{18}$Ar). The elements in Figure 2.8 are arranged in three tiers, or periods, corresponding to the sequential filling of the first three electron shells.

The chemical properties of an atom depend on the number of electrons in its outermost shell. We refer to those outer electrons as **valence electrons,** and to the outermost energy shell as the **valence shell.** On the right-hand side of the periodic table are helium, neon, and argon, the only three elements shown that have full valence shells. The fact that a valence shell with eight electrons is complete is referred to as the octet rule.

An atom with a complete valence shell is unreactive; that is, it will not interact with other atoms it encoun- ters. Because of this behavior, such atoms are said to be inert. Each row (period) in Figure 2.8 terminates with an inert element. All other atoms shown in Figure 2.8 have incomplete valence shells, and all are chemically reactive. Atoms with the same number of electrons in their valence shells exhibit similar chemical behavior. For example, fluorine (F) and chlorine (Cl) both have seven valence electrons, and both combine with the element sodium to form compounds.

CHEMICAL BONDS AND MOLECULES

Now that we have looked at the structure of atoms, the next step is to move up in the hierarchy of organization and see how atoms combine to form molecules. Atoms with incomplete valence shells will interact with certain other atoms in such a way that each partner completes its valence shell. Atoms do this by either sharing or completely transferring valence electrons. These interactions usually result in atoms staying close together, held by attractions called **chemical bonds.** A **molecule** consists of two or more atoms held together by chemical bonds. The strongest kinds of chemical bonds are covalent bonds and ionic bonds.

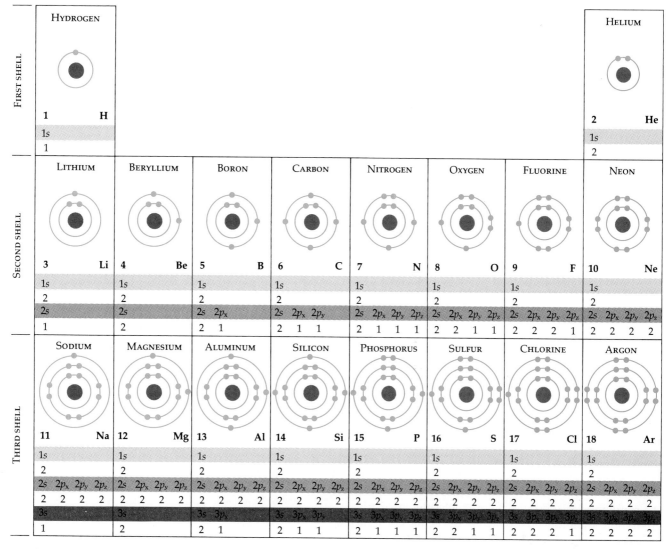

Figure 2.8
Electron configurations of the first 18 elements. The distribution of electrons in orbitals is tabulated, and the total number of electrons in the various shells is symbolized by dots on concentric rings around the nucleus (these rings are only symbols and do not reflect the three-dimensional freedom of electron movement). The elements are arranged in rows, each representing the filling of an electron shell. Those elements with the same number of electrons in their outermost shell behave similarly in chemical reactions.

Covalent Bonds

A **covalent bond** is the sharing of a pair of valence electrons by two atoms. For example, let us see what happens when two hydrogen atoms approach each other. Recall that hydrogen has one valence electron in the first shell, but the shell's capacity is for two electrons. When the two hydrogen atoms come close enough for their 1s orbitals to overlap, they share their electrons. Each hydrogen atom now has two electrons moving through its 1s orbital, and the valence shell is complete. Only when bonded together in this way are the valence shells of both hydrogen atoms full. Thus, the hydrogen molecule consists of two hydrogen atoms held together by a covalent chemical bond. We can abbreviate this molecule by writing H–H, where the line represents a covalent bond—that is, a pair of shared electrons. This type of notation, which represents both atoms and bonding, is called a **structural formula**. We can abbreviate even further by writing H_2, a **molecular formula** that indicates simply that the molecule consists of two atoms of hydrogen (Figure 2.9a).

Oxygen has six electrons in its second electron shell and so needs two more to complete this valence shell. Two oxygen atoms form a molecule by sharing two pairs of valence electrons (Figure 2.9b). The atoms are joined by what is called a **double covalent bond.** The structural formula for this molecule is O=O, and its

Figure 2.9

Covalent bonds. A covalent bond forms when two atoms share a pair of valence electrons. (Molecular formulas are shown at the left, structural formulas at the right.) (**a**) If two unattached hydrogen atoms meet, they will form a single covalent bond by sharing their outer electrons. (**b**) Two oxygen atoms form a molecule by sharing two pairs of valence electrons; the atoms are joined by a double covalent bond. (**c**) Two hydrogen atoms can be joined to one oxygen atom by covalent bonds to produce a molecule of water. (**d**) Four hydrogen atoms satisfy the valence of one carbon atom, forming methane.

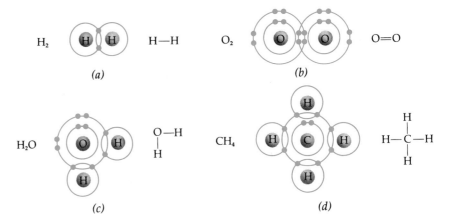

molecular formula is O_2. Nitrogen has five valence electrons, three less than it needs for a complete valence shell. Two nitrogen atoms will join together by a triple covalent bond and share three pairs of valence electrons, forming an N_2, or $N \equiv N$, molecule.

Notice that each atom that shares electrons has a bonding capacity—a certain number of covalent bonds that must be formed for the atom to have a full complement of valence electrons. This bonding capacity is called the atom's **valence**. The valence of hydrogen is 1, of oxygen 2, and of nitrogen 3.

The molecules we have looked at so far—H_2, O_2, and N_2—are not compounds, since each consists of atoms of only a single element. Water is an example of a **compound**, a substance whose molecules are made up of more than one element. The molecular formula of water is H_2O; it takes two atoms of hydrogen to satisfy the valence of one oxygen atom. Figure 2.9c shows the structure of a water molecule. This molecule is so important to life that the next chapter is devoted entirely to its structure and behavior.

Another molecule that is also a compound is methane, a component of natural gas; it has the molecular formula CH_4 (Figure 2.9d). Carbon ($_6C$) has four valence electrons, and thus its bonding capacity, or valence, is 4. It takes four hydrogen atoms, each with a valence of 1, to complement one atom of carbon. To review the valences of the four most abundant elements in life, hydrogen forms one bond, oxygen forms two bonds, nitrogen forms three bonds (usually), and carbon forms four bonds.

Molecular Shapes A molecule has a characteristic size and shape. This is significant because the functions of many molecules in the living cell depend on their geometry.

A molecule such as H_2 or O_2, consisting of two atoms, can only be linear, because the atoms must be side by side to share valence electrons. Molecules consisting of more than two atoms have more complicated shapes.

When an atom forms covalent bonds, the orbitals in the valence shell become rearranged. For atoms with valence electrons in both *s* and *p* orbitals, the single *s* and three *p* orbitals hybridize to form four new orbitals shaped like teardrops extending from the region of the nucleus (Figure 2.10a). These hybrid orbitals are spread as far apart as possible. If we connect the

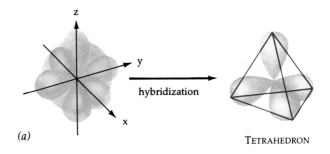

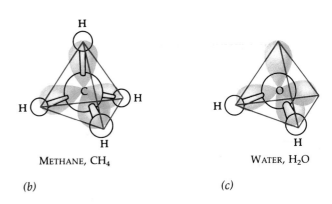

Figure 2.10

Molecular shapes. (**a**) The four teardrop-shaped orbitals of a valence shell involved in covalent bonding form by hybridization of the single *s* and three *p* orbitals. (**b**) The shape of methane. The four hydrogens are at the corners of an imaginary tetrahedron. (**c**) The V shape of a water molecule.

swollen ends of the teardrops with lines, we have the outline of a pyramidal three-dimensional shape called a tetrahedron. Using methane (CH$_4$) as an example, we can see how the geometric arrangement of orbitals determines the shape of a molecule (Figure 2.10b). The nucleus of the carbon atom is at the center of the imaginary tetrahedron, with the four valence orbitals radiating outward. At each of the four corners of the tetrahedron is a hydrogen atom, sharing a pair of electrons in the hybrid orbital with the carbon atom.

The water molecule (H$_2$O) is V-shaped. The four hybridized orbitals in the valence shell of the oxygen atom point to the corners of a tetrahedron, as they do in carbon, but only two of the corners interact with hydrogen atoms (Figure 2.10c).

Nonpolar and Polar Covalent Bonds   The attraction of an atom for the electrons of a covalent bond is called **electronegativity**. The more electronegative an atom, the more strongly it pulls shared electrons toward itself. In a covalent bond between two atoms of the same element, the tug-of-war for common electrons is a standoff; the two atoms are equally electronegative. A covalent bond is said to be **nonpolar** when electrons are shared equally. The covalent bond of hydrogen is nonpolar, as is the double bond of oxygen. The bonds of methane (CH$_4$) are also nonpolar; although the partners are different elements, carbon and hydrogen do not differ substantially in electronegativity. This is not always the case in a compound where covalent bonds join atoms of different elements. If one atom is more electronegative than the other, electrons of the bond will not be shared equally. In such cases, the bond is called a **polar covalent bond**. In a water molecule, the bonds between oxygen and hydrogen are polar. Oxygen is one of the most electronegative of the 92 elements, attracting shared electrons much more strongly than hydrogen does. In a covalent bond between oxygen and hydrogen, the electrons spend more time around the oxygen atom than they do around the hydrogen atom. Since electrons have a negative charge, the unequal sharing of electrons in water causes the oxygen atom to have a slight negative charge and each hydrogen atom a slight positive charge (Figure 2.11).

Ionic Bonds

In some cases, two atoms are so unequal in their attraction for valence electrons that the more electronegative atom strips an electron completely away from its partner. This is what happens when an atom of sodium ($_{11}$Na) encounters an atom of chlorine ($_{17}$Cl) (Figure 2.12). A sodium atom has a total of 11 electrons, with its 1 valence electron in the third electron shell. A chlorine atom has a total of 17 electrons, with

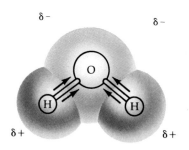

Figure 2.11 H$_2$O
Polar covalent bonds in a water molecule. Oxygen, being much more electronegative than hydrogen, pulls the shared electrons of the bond toward itself. This unequal sharing of electrons gives the oxygen a slight negative charge and the hydrogens a small amount of positive charge. (The symbol δ indicates that the charges are less than full units.)

7 electrons in its valence shell. When these two atoms meet, the lone valence electron of sodium is transferred to the chlorine atom, and both atoms end up with their valence shells complete. (Since sodium no longer has an electron in the third shell, the second shell is now outermost.)

The electron transfer between the two atoms moves one unit of negative charge from sodium to chlorine. Sodium, now with 11 protons but only 10 electrons, has a net electric charge of +1. A charged atom is called an **ion**. When the charge is positive, the ion is specifically called a **cation**. Conversely, the chlorine atom, having gained an extra electron, now has 17 protons and 18 electrons, giving it a net electric charge of −1. It has become a chloride ion—specifically, an **anion**, the term for a negatively charged ion. Because of their opposite charges, cations and anions attract each other in what is called an **ionic bond**.

The combination of sodium with chlorine in sodium chloride is an example of a compound known as a salt—in this case, table salt. Other salts are also ionic. They are often found in nature as crystals of various sizes and shapes, each an aggregate of vast numbers of cations and anions bonded by their electrical attraction and arranged in a three-dimensional lattice. In a crystal of table salt, each sodium ion has six chloride ions as neighbors, and each chloride ion is surrounded by six sodium ions (Figure 2.13). Since crystals can be of any size, there is no fixed number of ions, but sodium (Na$^+$) and chlorine (Cl$^-$) are always present in a one-to-one ratio. A salt crystal does not really consist of molecules in the same sense that a covalent compound does, since a covalently bonded molecule has a definite size and number of atoms. The formula for an ionic compound, such as NaCl, indicates only the ratio of elements in a crystal of the salt.

Not all salts have equal numbers of cations and anions. For example, the ionic compound magnesium chloride (MgCl$_2$) has two chloride ions for each mag-

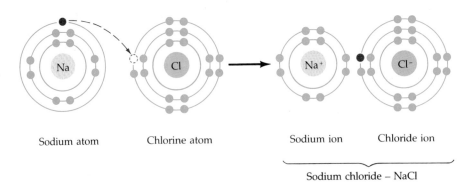

Figure 2.12
Electron transfer and ionic bonding. A valence electron is transferred from sodium (Na) to chlorine (Cl), giving both atoms completed valence shells. The electron transfer leaves the sodium atom with a net charge of $+1$ and the chlorine atom with a net charge of -1. The attraction between the oppositely charged atoms, or ions, is an ionic bond. Ions can bond not only to the atom they reacted with, but to any other ion of opposite charge.

Sodium atom Chlorine atom Sodium ion Chloride ion

Sodium chloride – NaCl

nesium ion. Magnesium ($_{12}$Mg) must lose two outer electrons if the atom is to have a complete valence shell. One magnesium atom can supply valence electrons to two chlorine atoms. After losing two electrons, the magnesium atom is a cation with a new charge of $+2$ (Mg^{2+}).

The term *ion* also applies to entire covalent molecules that are electrically charged. In ammonium chloride (NH_4Cl), for instance, the anion is a single chloride ion (Cl^-), but the cation is ammonium (NH_4^+), a nitrogen atom with four bonded hydrogen atoms. The whole ammonium ion has an electric charge of $+1$ because it is one electron short.

There is no distinct line between covalent bonding and ionic bonding. A nonpolar covalent bond and an ionic bond are opposite extremes in a range of situations where atoms share electrons. In the middle zone is the polar covalent bond, in which electrons are shared, but unequally so. We might think of an ionic bond as a covalent bond that is so polar that one atom has pulled an electron completely away from its less electronegative partner. Indeed, some compounds spend part of their time in a polar covalent state and the rest of their time as ions.

Some Important Weak Bonds

Covalent bonds are strong, meaning that they are relatively hard to break. Ionic bonds in a salt crystal are also very strong, but only when the crystal is dry. Add water, and the salt dissolves; the cations and anions separate from one another and become dispersed throughout the water. In the presence of water, ionic bonds are weak and break easily (you will learn how in the next chapter). Other weak chemical bonds important in biology are hydrogen bonds, van der Waals interactions, and hydrophobic interactions. These weak bonds can form between molecules or between different parts of a single large molecule. Weak bonds play a major role in cellular chemistry, particularly in stabilizing the shapes of large molecules and in helping molecules that interact to recognize one another.

Hydrogen Bonds A **hydrogen bond** occurs when a hydrogen atom covalently bonded to one electronegative atom is also attracted to another electronegative atom (Figure 2.14). In living cells, the electronegative partners involved are usually oxygen or nitrogen atoms. You have seen how the polar covalent bonds of water result in the oxygen atom having partial negative charge and the hydrogen atoms having partial positive charge. A similar situation arises in the ammonia molecule (NH_3), where an electronegative nitrogen atom has a small amount of negative charge because of its pull on the electrons it shares covalently with hydrogen. If a water molecule and an ammonia molecule are close to each other, there will be a weak attraction between the negatively charged nitrogen atom and a positively charged hydrogen atom of the adjacent water molecule. This attraction is a hydrogen bond.

Van der Waals Interactions Even if a molecule's covalent bonds are nonpolar, the molecule may have positively charged and negatively charged regions. This is because the electrons are always in motion and are

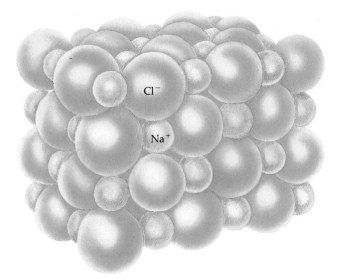

Figure 2.13
A sodium chloride crystal. The sodium ions (Na^+) and chloride ions (Cl^-) are held together by ionic bonds.

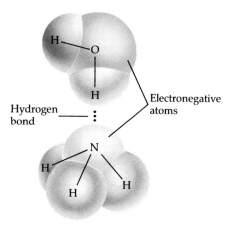

Figure 2.14
A hydrogen bond. A hydrogen atom attached to an electronegative atom by a polar covalent bond is shared with another electronegative atom through a weak electrical attraction. In this figure, a hydrogen bond joins a hydrogen atom of a water molecule (H_2O) with the nitrogen atom of an ammonia molecule (NH_3).

not always symmetrically distributed in the molecule; at any given instant, there are chance accumulations of electrons in one part of the molecule or another. The results are ever-changing "hot spots" of positive and negative charge that enable all atoms and molecules to stick to one another. These **van der Waals interactions** are weak, and they occur only when atoms and molecules are very close together.

Hydrophobic Interactions   Some substances mix readily with water, while others do not. Molecules that do not mix with water, such as those of salad oil, are said to be **hydrophobic** (from the Greek *hydro*, "water," and *phobos*, "fearing"). Placed in water, hydrophobic molecules coalesce to form droplets, which then may join together. This behavior minimizes exposure of the hydrophobic substance to water. What we call a **hydrophobic interaction** between molecules is actually their mutual exclusion of water. Once the molecules are close together, van der Waals attractions reinforce the hydrophobic interaction.

CHEMICAL REACTIONS

The making and breaking of chemical bonds, leading to changes in the composition of matter, are called **chemical reactions**. For example, the electron transfer between magnesium and chlorine that produces the compound magnesium chloride is a chemical reaction:

$$Mg + 2 Cl \longrightarrow MgCl_2$$

When we write a chemical reaction, we use an arrow to indicate the conversion of the starting materials,

called **reactants**, to the **products**. Notice that all atoms of the reactants must be accounted for in the products. Matter is conserved in a chemical reaction: Reactions cannot create or destroy matter, but only rearrange it in various ways. To show water being made by a reaction between hydrogen and oxygen molecules, we can write

$$2 H_2 + O_2 \longrightarrow 2 H_2O$$

The coefficient 2 in front of the H_2 indicates that we are starting with two molecules of hydrogen. Notice that all the atoms on the left side of the arrow also appear on the right side: Matter has been conserved.

Some chemical reactions go to completion; that is, all the reactants are converted to products. But most reactions are reversible, the products of the forward reaction becoming the reactants for the reverse reaction. For example, hydrogen and nitrogen molecules combine to form ammonia, but ammonia can also decompose to regenerate hydrogen and nitrogen:

$$3 H_2 + N_2 \rightleftharpoons 2 NH_3$$

The double arrows indicate that the reaction is reversible.

One of the factors affecting the rate of reaction is the concentration of reactants. The greater the concentration of reactant molecules, the more frequently they collide with one another and have an opportunity to react to form products. The same holds true for the products. As products accumulate, collisions resulting in the reverse reaction become more and more frequent. Eventually, the forward and reverse reactions occur at the same rate, and the relative concentrations of products and reactants remain fixed. The point at which the reactions offset one another exactly is called **chemical equilibrium**. This is a dynamic equilibrium; reactions are still going on, but with no net effect on the concentrations of reactants and products. It is important to realize that equilibrium does not mean that the reactants and products are equal in concentration, but only that their concentrations have stabilized. For the above reaction involving ammonia, equilibrium is reached when ammonia decomposes as rapidly as it forms. In this case, the forward reaction is much more favorable than the reverse reaction; thus, at equilibrium, there will be far more ammonia than hydrogen and nitrogen.

*   *   *

In 1953 when Stanley Miller sparked a mixture of gases and formed molecules characteristic of life, he helped to illuminate the interface between chemistry and biology (see interview at the beginning of this unit). In this chapter, we have followed Dr. Miller's lead by approaching the study of living matter with the principles of chemistry that govern the behavior of atoms and molecules. The theme of emergent properties is manifest, even at these lowest levels of structural organization. The chemical properties of an atom are determined by the number and arrangement of its subatomic particles, particularly its electrons. Additional properties of matter arise when atoms combine to form molecules, and molecules combine to form more complex structures. The next chapter describes how the emergent properties of a particular molecule, water, make life on earth possible.

STUDY OUTLINE

1. A living organism is the product of the numerous chemical interactions within its cells. Understanding the basic principles of chemistry lays the foundation for the study of the phenomenon called life.
2. Organisms are arranged in a hierarchy of structural levels, with novel properties emerging at each level.

Matter: Elements and Compounds (p. 22)

1. The basic ingredients of matter are the elements, substances that cannot be broken down to other types of matter by ordinary chemical means.
2. A compound contains two or more elements in a fixed ratio and often has emergent properties very different from those of its constituent elements.
3. About 25 elements are essential to life. Carbon, oxygen, hydrogen, and nitrogen make up 96% of living matter. The remaining 4% includes trace elements, which are required in minute amounts.

The Structure and Behavior of Atoms (pp. 22–28)

1. An atom is the smallest unit of an element retaining the physical and chemical properties of that element.
2. An atom consists of three types of subatomic particles. Uncharged neutrons and positively charged protons are tightly bound in a nucleus; negatively charged electrons move rapidly about the nucleus.
3. Protons and neutrons each have an atomic mass of 1 dalton. Electrons have a negligible mass.
4. The number of protons in an atom is called the atomic number. Unless otherwise indicated, the number of electrons of an atom is equal to the number of protons, and the atom is electrically neutral.
5. The mass number of an element indicates the sum of the protons and neutrons and approximates the mass of an atom in daltons (also called atomic weight).
6. Most elements consist of two or more isotopes, different in neutron number and mass. Some isotopes are unstable and give off particles and energy as radioactivity. Radioactivity has important uses in science and medicine but can also harm organisms.
7. Electron configuration determines the chemical properties of an atom—its reaction with other atoms.
8. Electrons move within orbitals, three-dimensional spaces located within successive electron shells (energy levels) surrounding the nucleus. Each orbital can hold a maximum of two electrons. Electrons in shells farther from the nucleus have more potential energy than those in shells closer to the nucleus.
9. Chemical properties depend on the number of valence electrons, those in the outermost shell.
10. In atoms with more than one electron shell, the valence shell is complete when it contains eight electrons.

Chemical Bonds and Molecules (pp. 28–33)

1. Chemical bonds form when atoms with incomplete valence shells interact to produce complete valence shells. Molecules consist of two or more atoms held together by chemical bonds.
2. A covalent bond forms when two atoms share a pair of valence electrons. Double covalent bonds result when two atoms share two pairs of electrons.
3. A structural formula shows the atoms and bonds in a molecule. A molecular formula indicates only the number and types of atoms.
4. The valence of an atom reflects the number of covalent bonds needed to complete its valence shell.
5. Molecules have characteristic shapes and sizes that affect how they function in biological systems. Bond angles are often tetrahedral due to the hybridization of s and p orbitals.
6. A nonpolar covalent bond forms when both atoms are equally electronegative. If one atom is more electronegative, a polar covalent bond is formed in which the electron pair is pulled closer to the more electronegative atom.
7. An ionic bond is created when two atoms differ so much in electronegativity that one or more electrons are actually transferred from one atom to the other. The recipient atom becomes a negatively charged anion. The donor atom, called a cation, becomes positively charged. Because of their opposite charges, cations and anions attract each other in an ionic bond.
8. Hydrogen bonds are weaker bonds between a partially positive hydrogen of one polar molecule and the partially negative atom of another polar molecule.
9. Van der Waals interactions occur when transiently positive and negative regions of very close molecules are attracted to each other.
10. Hydrophobic interactions between molecules result in the clumping or coalescing of hydrophobic molecules that exclude water.

Chemical Reactions (pp. 33–34)

1. Chemical reactions break or form chemical bonds to change one form of matter, the reactants, into another form, the products. During a reaction, matter is conserved; the same numbers and kinds of atoms appear in both reactants and products.
2. Most chemical reactions are reversible. Chemical equilibrium is reached when the forward and backward rates are equal and the proportions of reactants and products no longer change.

SELF-QUIZ

1. Which of the following is a *trace* element required by humans?

   a. carbon
   b. oxygen
   c. hydrogen
   d. iodine
   e. calcium

2. Two atoms of the same element *must* have the same number of

   a. neutrons
   b. protons
   c. electrons
   d. neutrons plus protons
   e. protons plus electrons

3. The most common isotope of phosphorus is ^{31}P. Compared to ^{31}P, the radioactive isotope ^{32}P has

   a. a different atomic number
   b. one more neutron
   c. one more proton
   d. one more electron
   e. a different charge

4. What do the four elements most abundant in life—carbon, oxygen, hydrogen, and nitrogen—have in common?

   a. They all have the same number of valence electrons.
   b. Each element exists in only one isotopic form.
   c. They are all relatively light elements, near the top of the periodic table.
   d. They are all about equal in electronegativity.
   e. They are elements produced only by living cells.

5. The atomic number of sulfur is 16. Sulfur combines with hydrogen by covalent bonding to form a compound, hydrogen sulfide. Based on the electron configuration of sulfur, we can predict that the molecular formula of the compound will be

   a. HS
   b. HS_2
   c. H_2S
   d. H_2S_2
   e. H_4S

6. Review the valences of carbon, oxygen, hydrogen, and nitrogen on p. 30, and then determine which of the following molecules is most likely to exist:

   a. O=C—H

   b.
   ```
        H       H
        |       |
    H—C—H—C=O
        |
        H
   ```

   c.
   ```
        H  H
        |  |
   H—O—C—C=O
        |
        H
   ```

   d.
   ```
        O
        |
   H—N=H
   ```

   e. both b and c

7. Which orientation is most likely for two adjacent water molecules?

   a.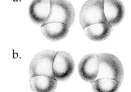

   b.

   c.

   d.

8. Which of these statements is true of *all* anions?

   a. The atom has more electrons than protons.
   b. The atom has more protons than electrons.
   c. The atom has fewer protons than does a neutral atom of the same element.
   d. The atom has more neutrons than protons.
   e. The net charge is −1.

9. What coefficient must be placed in the blank to balance this chemical reaction?

   $$C_6H_{12}O_6 \longrightarrow 2C_2H_6O + \underline{?}\ CO_2$$

   a. 1
   b. 2
   c. 3
   d. 4
   e. 6

10. Which of the following statements correctly describes *any* chemical reaction that has reached equilibrium?

    a. The concentration of products equals the concentration of reactants.
    b. The rate of the forward reaction equals the rate of the reverse reaction.
    c. Both forward and backward reactions have halted.
    d. The reaction is now irreversible
    e. Total conversion of products to reactants has been completed.

CHALLENGE QUESTIONS

1. Recall from Chapter 1 that vitalism is the belief that life possesses supernatural forces that cannot be explained by physical and chemical principles. Explain why the concept of emergent properties does *not* lend credibility to vitalism.

2. A fallen log has come to rest in a place where it is likely to fossilize. Initially, the log contains 64 g of radioactive ^{14}C (half-life = 5600 years). How many years will it take for the amount of ^{14}C in the log to decline to 1 g?

3. The use of radioactive isotopes as tracers in biochemical research is based on the inability of cells to discriminate between the tracers and nonradioactive isotopes. Explain why this ability of radiactive isotopes to infiltrate the chemical processes of the cell also compounds the threat posed by radioactive contaminants in air, soil, and water.

FURTHER READING

1. Atkins, P. W. *Molecules*. New York: Scientific American Library, 1987. Beautifully illustrated tour of the world of atoms and molecules.
2. Baker, J. J. W., and G. E. Allen. *Matter, Energy and Life*. 4th ed. Reading, Mass.: Addison-Wesley, 1981. A paperback primer covering chemical topics essential to biology.
3. Baum, S. J. and C. W. Scaife. *Chemistry: A Life Science Approach*. 3rd ed. New York: Macmillan, 1987.
4. Lewin, R. "Chemistry in the Image of Biology." *Science*, Oct. 30, 1987. How studies of molecular shape led to a Nobel Prize.

3

Water and the Fitness of the Environment

If we could cruise the universe in quest of life, we would do well to search for worlds with water. We might not recognize life on dry planets even if it existed. All organisms familiar to us are made mostly of water and live in a world where water dominates climate and many other features of the environment. Here on Earth, water is the biological medium, the substance that makes possible life as we know it.

As Stanley Miller pointed out in the interview preceding this unit, life probably began in water. And it has been inextricably tied to water ever since. Most cells are surrounded by water; in fact, cells contain from about 70% to 95% water. Earth's surface is also wet, with water covering three-fourths of our planet. Most of this water is in liquid form, enough to cover the United States to a depth of 130 km. Also present on Earth as ice and vapor, water is the only common substance to exist in the natural environment in all three physical states of matter: solid, liquid, and gas (Figure 3.1).

The abundance of water is a major reason Earth is habitable. In a book many biologists consider a classic, *The Fitness of the Environment* (1913), Lawrence Henderson highlights the importance of water to life. While acknowledging that life adapts to its environment through natural selection, Henderson emphasizes that for life to exist at all, the environment must first be a suitable abode. (Indeed, the relationship of organisms to their environments was one of the biological themes introduced in Chapter 1.) We will see throughout this chapter how much water contributes to the fitness of Earth for life.

So common is water that it is easily taken for granted as being nothing special, but water is in fact a most

Figure 3.1
Water: a common oddity. The great abundance of water on Earth tends to obscure its uniqueness. The extraordinary behavior of water is a major factor in the fitness of the environment for life. Here we see one of water's oddities: its presence in three physical states. No other common substance on earth occurs naturally in the solid, liquid, and gaseous states. This cold place is Spitsbergen, an island north of Norway.

exceptional substance with many extraordinary qualities. Following the theme of emergent properties, we can trace water's novel behavior to the structure and interactions of its molecules.

WATER MOLECULES AND HYDROGEN BONDING

Studied in isolation, the water molecule is deceptively simple. As we saw in Chapter 2, its two hydrogen atoms are joined to an oxygen atom by covalent bonds in a lopsided arrangement (Figure 3.2). The four valence orbitals of the oxygen atom point to the corners of a tetrahedron, with two of the corners occupied by the hydrogen atoms. Each of the two remaining orbitals contains a pair of unshared electrons. The covalent bonds between oxygen and hydrogen are polar, with the electronegative oxygen nucleus pulling shared electrons toward itself and away from the hydrogen nuclei.

The water molecule as a whole is electrically neutral, but as a result of the unequal electron distribution in the molecule, the regions where the hydrogen atoms are located have a weak, or partial, positive charge. The opposite side of the molecule has partial negative charge associated with the two unshared orbitals of the oxygen atom. The polarity of its bonds and its asymmetric shape give the water molecule opposite charges on opposite sides; it is a **polar molecule**.

The anomalous properties of water arise from attractions among these polar molecules. The attraction is electrical; a positive hydrogen of one molecule is attracted to the negative oxygen of a nearby mole-

cule. The molecules are thus held together by a hydrogen bond (Figure 3.3). Because the charged regions of the water molecule are associated with the four valence orbitals of oxygen, each molecule can form hydrogen bonds to a maximum of four neighbors. Thus, the extraordinary qualities of water are emergent properties resulting from the hydrogen bonding that orders molecules into a higher level of structural organization.

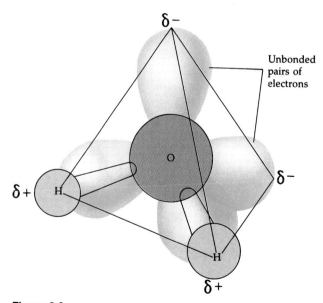

Figure 3.2
The polar water molecule. The unequal sharing of electrons between oxygen and hydrogen, coupled with the lopsided shape of the molecule, results in opposite sides of the molecule having opposite charges.

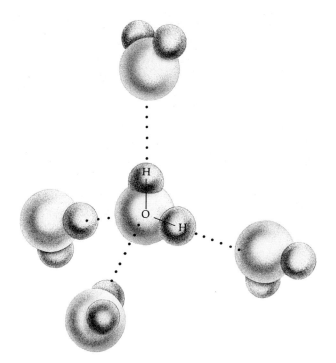

Figure 3.3
Hydrogen bonds between water molecules. The charged regions of a water molecule are attracted to oppositely charged parts of neighboring molecules. Each molecule can hydrogen-bond to a maximum of four partners. At any instant in liquid water at 37°C (human body temperature), about 15% of the molecules are bonded to four partners in short-lived clusters.

SOME EXTRAORDINARY PROPERTIES OF WATER

It is easy to overlook the exceptional behavior of water, since water-based substances are so common in our experience. But compared to other substances, water is indeed unusual. In this section, we will concentrate on some of the qualities of water that contribute to the fitness of the environment.

The Cohesiveness of Liquid Water

Water molecules stick together as a result of hydrogen bonding. When water is in its liquid form, its hydrogen bonds are very fragile, about $1/20$ as strong as covalent bonds. They form, break, and re-form with great frequency. Each hydrogen bond lasts only a few trillionths of a second, but the molecules bond promiscuously to a succession of partners. At any instant, a substantial percentage of all the water molecules are bonded to their neighbors, giving water more structure than most other liquids. Collectively, the hydro-

gen bonds hold the substance together, a phenomenon called **cohesion**.

Cohesion due to hydrogen bonding contributes to the transport of water against gravity in plants. Water reaches leaves through microscopic vessels that extend upward from roots. Water that evaporates from a leaf is replaced by water from the vessels in the veins of the leaf. Hydrogen bonds cause water molecules leaving the veins to tug on molecules farther down in the vessel, and the upward pull is transmitted along the vessel all the way down to the root (Figure 3.4).

Related to cohesion is **surface tension**, a measure of how difficult it is to stretch or break the surface of a liquid. At the interface between water and air is an ordered arrangement of water molecules, hydrogen-bonded to one another and to the water below, making the water behave as though it were coated with an invisible film. Surface tension also causes water on a surface to bead into a spherical shape having the smallest ratio of area to volume, maximizing the number of hydrogen bonds that can form (Figure 3.5a).

Water has a greater surface tension than most other liquids. You can observe the surface tension of water by slightly overfilling a drinking glass; the water will stand above the rim. In a more biological example, the insect known as the water strider distributes its weight over enough area for the animal to walk on water without breaking the surface (Figure 3.5b). It is also because of water's surface tension that we can skip rocks on a pond.

In addition to causing cohesion between water molecules, hydrogen bonds result in **adhesion**, the cling-

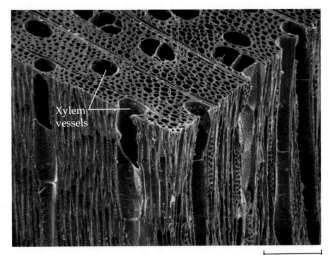

Xylem vessels

100 μm

Figure 3.4
Water transport in plants. Evaporation from leaves pulls water upward from the roots within microscopic conduits, called xylem vessels, in this case located in the trunk of a maple tree. Cohesion due to hydrogen bonding helps hold together the column of water within a vessel (SEM).

(a)

(b)

Figure 3.5
The high surface tension of water. Water has an unusually high surface tension because of the collective strength of its hydrogen bonds. **(a)** Surface tension causes water to bead on a spider web. **(b)** The water strider, though denser than water, can walk on a pond without breaking the surface.

ing of one substance to another. Materials with an affinity for water are said to be **hydrophilic** (from the Greek *hydro*, "water," and *philios*, "loving"). If you have ever tried to separate two glass slides stuck together with a film of water, you can appreciate how tightly water adheres to glass, a hydrophilic substance.

The uptake of water by a porous material that is hydrophilic is called **imbibition**. Wood, paper, and sponges all imbibe water. The germination of a plant seed is triggered by imbibition. When a dry seed is watered, it imbibes water and swells until the coat of the seed ruptures; the seedling then begins to grow.

Water's High Specific Heat

Water stabilizes air temperatures by absorbing heat from air that is warmer and releasing the stored heat to air that is cooler. Water is so effective as a heat bank because of its high specific heat, which means that a slight change in its own temperature is accompanied by the absorption or release of a relatively large amount of heat. To understand this quality of water, we must first look briefly at heat and temperature.

Heat and Temperature   Anything that moves has **kinetic energy**, the energy of motion. Atoms and molecules have kinetic energy because they are always moving, although in no particular direction. The faster a molecule moves, the greater its kinetic energy. **Heat** is the *total* quantity of kinetic energy due to molecular motion in a body of matter. **Temperature** measures the intensity of heat due to the *average* kinetic energy of the molecules. When the average speed of the mole-

cules increases, a thermometer records this as a rise in temperature. Heat and temperature are related, but they are not the same. A swimmer crossing the English Channel has a higher temperature than the water, but the ocean contains far more heat because of its volume.

Whenever two objects of different temperature are brought together, heat passes from the warmer to the cooler body until the two are the same temperature. Molecules in the cooler object speed up at the expense of the kinetic energy of the warmer object. An ice cube cools a drink not by adding coldness to the liquid, but by absorbing heat as the ice melts.

Throughout this book, the **Celsius scale** is used to indicate temperature (Celsius degrees are abbreviated as °C). At sea level, water freezes at 0°C and boils at 100°C (Figure 3.6). Human body temperature is 37°C, and comfortable room temperature is about 20°–25°C.

The unit of heat used in this book is the **calorie** (cal). A calorie is the amount of heat energy it takes to raise the temperature of 1 g of water by 1°C. Conversely, a calorie is also the amount of heat that 1 g of water releases when it cools down by 1°C. (Note: A calorie is equivalent to 4.184 joules, which are metric units of energy.) A **kilocalorie** (kcal), 1000 cal, is the quantity of heat required to raise the temperature of 1 kilogram (kg) of water by 1°C. (The Calorie, with a capital C, a term popularly used to indicate the energy content of foods, is actually a kilocalorie.)

The **specific heat** of a substance is defined as the amount of heat that must be absorbed or lost for 1 g of that substance to change its temperature by 1°C. You already know water's specific heat, because we have *defined* a calorie as the amount of heat that causes water to change its temperature by 1°C. Therefore, the

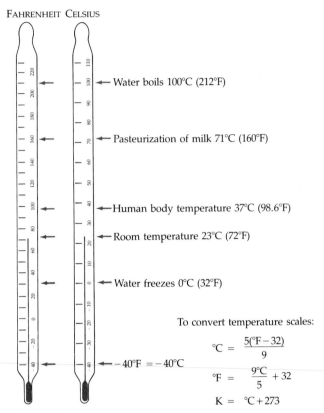

FAHRENHEIT CELSIUS

← Water boils 100°C (212°F)

← Pasteurization of milk 71°C (160°F)

← Human body temperature 37°C (98.6°F)

← Room temperature 23°C (72°F)

← Water freezes 0°C (32°F)

To convert temperature scales:

$$°C = \frac{5(°F - 32)}{9}$$

$$°F = \frac{9°C}{5} + 32$$

$$K = °C + 273$$

← −40°F = −40°C

Figure 3.6
The Celsius thermometer. Water freezes at 0°C and boils at 100°C. Absolute zero, when all molecular motion ceases, is −273°C. Related to the Celsius scale is the Kelvin (K) scale, also called the absolute temperature scale. The degrees are the same size, but the Kelvin scale has its zero point at absolute zero. Thus, 0 K = −273°C, and water freezes at 273 K and boils at 373 K. The Fahrenheit scale is also shown here for comparison.

Table 3.1 Specific heat of some common liquids

Liquid	Specific Heat (cal/g/°C)
Ammonia	1.23
Water	1.00
Ethyl alcohol	0.60
Ether	0.55
Acetone	0.53
Olive oil	0.47
Chloroform	0.23
Mercury	0.03

specific heat of water is 1 calorie per gram per degree Celsius, abbreviated as 1 cal/g/°C. Compared with most other substances, water has an unusually high specific heat (Table 3.1).

How Water Stabilizes Temperature The high specific heat of water means that relative to other materials, water will change its temperature less when it absorbs or loses a given amount of heat. The reason you can burn your fingers by touching the metal handle of a pot on the stove when the water in the pot is still lukewarm is that the specific heat of water is ten times greater than that of iron. In other words, it will take only 0.1 cal to raise the temperature of 1 g of iron 1°C. Specific heat can be thought of as a measure of how well a substance resists changing its temperature when it absorbs or releases heat. Water resists changing its temperature, and when it does change its temperature, it absorbs or loses a relatively large quantity of heat for each degree of change.

We can trace water's high specific heat, like many of its other properties, to hydrogen bonding. Heat must be absorbed to break hydrogen bonds, and heat is released when hydrogen bonds form. A calorie of heat causes a relatively small change in the temperature of water because much of the heat energy is used to disrupt hydrogen bonds before the water molecules can begin moving faster. And when the temperature of water drops slightly, many additional hydrogen bonds form, releasing a considerable amount of energy in the form of heat.

By warming up only a few degrees, a large body of water can absorb and store a huge amount of heat from the sun during daytime and summer. At night and during winter, the gradually cooling water can warm the air. This is the reason coastal areas generally have milder climates than inland regions. The high specific heat of water also makes ocean temperatures quite stable, creating a favorable environment for marine life. Thus, because of its high specific heat, the water that covers most of this planet keeps temperature fluctuations within limits that permit life. Also, since organisms are made primarily of water, they are able to resist changing their own temperatures better than they would be if they were made of a liquid with a lower specific heat.

Water's High Heat of Vaporization

Molecules of any liquid stay close together because they are attracted to one another. Molecules moving fast enough to overcome these attractions can depart from the liquid and enter the air as gas. This transformation from a liquid to a gas is called vaporization or evaporation. Molecules vary in speed; temperature, remember, is the average kinetic energy of molecules. Even at a low temperature, the speediest molecules can escape into the air. Some evaporation occurs

Table 3.2 Heat of vaporization of some common liquids

Liquid	Heat of Vaporization (cal/g)
Water (at 37°C, human body temperature)	576
Water (at 100°C)	540
Ammonia	302
Ethyl alcohol	237
Acetone	125
Mercury	73
Chloroform	59

at any temperature; a glass of water, for example, will eventually evaporate at room temperature. If a liquid is heated, the average kinetic energy of molecules increases and the liquid evaporates more rapidly.

Heat of vaporization is the quantity of heat a liquid must absorb for 1 g of it to be converted from the liquid to the gaseous state. Compared with most other liquids, water has a high heat of vaporization (Table 3.2). To evaporate each gram of water, 540 cal of heat are needed—nearly twice as much heat as needed to vaporize a gram of alcohol or ammonia. Water's high heat of vaporization is another emergent property explained by hydrogen bonds, which restrain the molecules and make their exodus from the liquid state more difficult. A relatively large amount of heat is needed to evaporate water because hydrogen bonds must be broken first. Related to water's high heat of vaporization is its high boiling point, 100°C. It is because

temperatures that high are rare on earth that liquid water is so abundant.

Water's high heat of vaporization helps moderate the climate of the earth. A considerable amount of solar heat absorbed by tropical seas is dissipated during the evaporation of surface water. Then, as moist tropical air circulates poleward, its condensation to form rain releases heat.

As a substance evaporates, the surface of the liquid that remains behind cools down. This **evaporative cooling** occurs because the "hottest" molecules, those with the greatest kinetic energy, are the most likely to leave as gas. It is as if the 100 fastest runners at your college transferred to another school; the average speed of the remaining students would decline.

Evaporative cooling of water contributes to the stability of temperature in lakes and ponds and also provides a mechanism that prevents terrestrial (land-dwelling) organisms from overheating. For example, evaporation of water from the leaves of a plant helps keep the tissues in the leaves from becoming too warm in the sunlight. Evaporation of sweat from human skin cools the surface of the body and helps prevent overheating on a hot day or when excess heat is generated by strenuous activity (Figure 3.7). It is because the concentration of water vapor in the air inhibits evaporation of sweat that high humidity contributes to discomfort on a hot day.

Freezing and Expansion of Water

Ice floats. This observation may seem self-evident, but the reason ice floats is another of water's unusual properties: Water is one of the few substances that is less dense as a solid than it is as a liquid. While other materials contract when they solidify, water expands.

Figure 3.7
Evaporative cooling. Because of water's high heat of vaporization, evaporation of sweat cools the surface of the body.

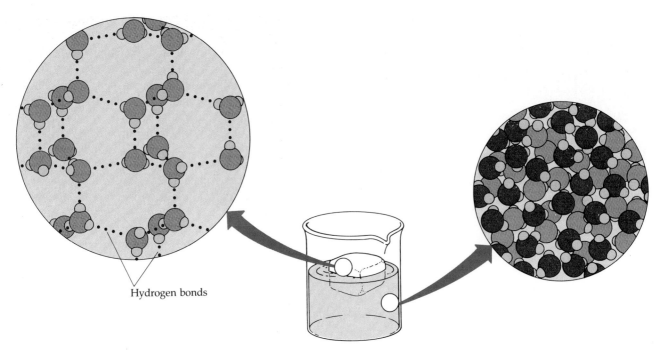

Figure 3.8
The structure of ice. Each molecule is hydrogen-bonded to four neighbors in a three-dimensional crystal with open chan- nels. Because the hydrogen bonds make the crystal spacious, ice has fewer mole- cules than an equal volume of liquid water. In other words, ice is less dense than liq- uid water.

The cause of this exotic behavior is, once again, hydrogen bonding. At temperatures above 4°C, water behaves like other liquids, expanding as it warms and contracting as it cools. Water begins to freeze when its molecules are no longer moving vigorously enough to break hydrogen bonds to neighboring molecules. As the temperature reaches 0°C, the water becomes locked into a crystalline lattice, each water molecule bonded to the maximum of four partners (Figure 3.8). The hydrogen bonds keep the molecules far enough apart from each other to make ice about 10% less dense (have 10% fewer molecules for the same volume) than liquid water at 4°C. When ice absorbs enough heat for its temperature to increase to above 0°C, hydrogen bonds between molecules are disrupted. As the crystal collapses, the ice melts, and molecules are free to slip closer together. Water reaches its greatest density at 4°C and then begins to expand again owing to the increased speed of its molecules. Keep in mind, however, that even liquid water is semistructured because of transient hydrogen bonds. Not until they evaporate are the molecules moving fast enough to completely overcome their attractions for one another.

The expansion of water as it solidifies is an important factor in the fitness of the environment (Figure 3.9). If ice could sink, then eventually all ponds, lakes, and even the oceans would freeze solid, making life as we know it impossible on this planet. During sum- mer, only the upper few inches of the ocean would thaw. Instead, when a deep body of water cools, the floating ice insulates the liquid water below, prevent- ing it from freezing.

The freezing of water and melting of ice also help to make the transitions between seasons less abrupt, enabling organisms to adjust gradually to the chang- ing climate. Again, water releases heat whenever hydrogen bonds form, and absorbs heat whenever hydrogen bonds break. When water solidifies into ice or snow, the heat released warms the surrounding air as hydrogen bonds knit the molecules together into crystals. This helps temper the autumn. During the spring thaw, melting ice absorbs heat as hydrogen bonds are broken, again tempering the change of seasons.

Water as a Versatile Solvent

A sugar cube placed in a glass of water will dissolve, and the glass then contains a uniform mixture of sugar and water. The concentration of dissolved sugar will be the same everywhere in the mixture. A liquid that is a homogeneous mixture of two or more substances is called a **solution**. The dissolving agent of a solution is the **solvent**, and the substance that is dissolved is the **solute**. In this case, water is the solvent and sugar

(a)

(b)

Figure 3.9
Expansion of water during freezing. Ice is less dense than liquid water because hydrogen bonding spaces the molecules relatively far apart in the ice crystal.

(a) The expansion of freezing water uplifts earth in the Alaskan tundra. (b) Floating ice becomes a barrier that protects the liquid water below from the colder air. This

Weddell seal is approaching a hole in the ice sheltering deep water in McMurdo Sound (Antarctica).

is the solute. An **aqueous solution** is one in which water is the solvent.

The medieval alchemists tried to find a universal solvent, one that would dissolve anything. They learned that nothing works better than water. However, water is not a universal solvent; if it were, it could not be stored in any container, and our cells and tissues, being mostly water, would completely dissolve. But water is the most versatile solvent known, a quality we can trace to the polarity of the water molecule.

Suppose, for example, that a crystal of the ionic compound sodium chloride is placed in water (Figure 3.10). At the surface of the crystal, the sodium and chloride ions are exposed to the solvent. The ions and water molecules have an affinity for one another through electrical attraction. The oxygen regions of the water molecules are negatively charged and cling to sodium cations. The hydrogen regions of the water molecules are positively charged and are attracted to chloride anions. Water surrounds the individual ions, separating the sodium from the chloride and shielding the ions from one another. Working inward from the surface of the salt crystal, water eventually dissolves all the ions. This produces a solution of two solutes, sodium and chloride, homogeneously mixed with water, the solvent. Other ionic compounds also dissolve in water. Seawater, for instance, contains a great variety of dissolved ions, as do living cells.

A compound does not need to be ionic to dissolve in water; polar compounds are also water-soluble. For example, ammonia (NH_3) is an electrically neutral molecule, but the covalent bonds between the hydrogen and nitrogen atoms are polar because nitrogen, like oxygen, is very electronegative. The uneven dis-

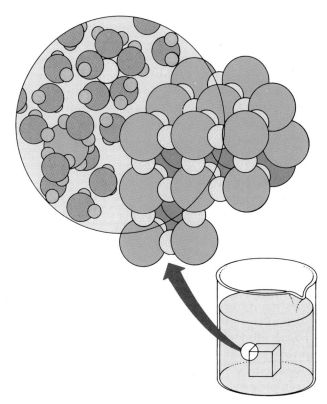

Figure 3.10
A crystal of table salt dissolving in water. The positive hydrogen regions of the polar water molecules are attracted to the chloride (the anions), whereas the negative oxygen regions cling to the sodium (the cations).

Table 3.3 Summary of water's extraordinary properties

Property	Explanation	Example of Benefit to Life
Cohesion and high surface tension	Hydrogen bonds hold molecules together.	Leaves pull water upward from roots in microscopic vessels.
High specific heat	Hydrogen bonds absorb heat when they break and release heat when they form, minimizing temperature changes.	Water stabilizes air and sea temperatures.
High heat of vaporization	Hydrogen bonds must be broken for water to evaporate.	Evaporation of water cools the surfaces of plants and animals.
Expansion upon freezing	Water molecules in an ice crystal are spaced relatively far apart because of hydrogen bonding.	Floating ice insulates the water below and prevents seas and lakes from freezing solid.
Versatility as a solvent	Charged regions of polar water molecules are attracted to ions and polar compounds.	Water is an effective medium for complex chemical reactions in organisms.

tribution of electrons and the asymmetric shape of the ammonia molecule result in opposite sides of the molecule having opposite charge. Therefore, like water itself, ammonia is polar. Through hydrogen bonding, water has an affinity for ammonia and dissolves the compound (see Figure 2.14). Most other polar compounds are also soluble in water. However, water will *not* dissolve nonpolar compounds, neutral molecules that have a symmetric distribution of charge.

Water is the solvent of life. An enormous diversity of solutes are dissolved in the water of biological fluids such as blood, the sap of plants, and the liquid within all cells. The complex solutions characteristic of life are possible only because water is so effective and versatile a solvent. This quality, as well as the other properties of water, are summarized in Table 3.3. We will now look more closely at aqueous solutions.

AQUEOUS SOLUTIONS

Biological chemistry is wet chemistry. Most of the chemical reactions that occur in life involve solutes dissolved in water. For example, oxygen is transported in the bloodstream by bonding to molecules of a protein named hemoglobin. Hemoglobin is dissolved in the water within red blood cells, and oxygen bonds to it only after the oxygen itself becomes dissolved as it is absorbed from air by the lining of the lungs. To understand the chemistry of life, it is important to learn about the properties of aqueous solutions. Here we will address two quantitative aspects of solutions: solute concentration and pH.

Solute Concentration

Suppose we wanted to prepare an aqueous solution of table sugar having a specified concentration of sugar molecules (a certain number of solute molecules in a certain volume of solution). We could begin by counting out the number of sugar molecules we wished to dissolve in water, but counting or weighing individual molecules is not practical. Instead, scientists sometimes measure substances in units called moles (mol). A **mole** is the number of grams of a substance that equals its molecular weight in daltons. Suppose we wanted to weigh out 1 mol of table sugar (sucrose), which has the molecular formula $C_{12}H_{22}O_{11}$. A carbon atom weighs 12 daltons, a hydrogen atom weighs 1 dalton, and an oxygen atom weighs 16 daltons. **Molecular weight** is the sum of the weights of all the atoms in a molecule, and thus the molecular weight of sucrose is 342 daltons. To obtain 1 mol of sucrose, we weigh out 342 g, the molecular weight of sucrose expressed in grams.

The practical advantage of measuring a quantity of chemicals in moles is that a mole of one substance has exactly the same number of molecules as a mole of any other substance. If substance A has a molecular weight of 10 daltons and substance B has a molecular weight of 100 daltons, then 10 g of A will have the same number of molecules as 100 g of B. The number of molecules in a mole, called Avogadro's number, is 6.02×10^{23}. A mole of table sugar contains 6.02×10^{23} sucrose molecules and weighs 342 g. A mole of ethyl alcohol (C_2H_6O) also contains 6.02×10^{23} molecules, but it weighs only 46 g because the molecules are smaller than those of sucrose. The mole concept rescales the weighing of molecules from daltons, used for single molecules, to grams, which are more practical units of weight for the laboratory, and allows scientists to combine substances in fixed ratios of molecules.

Imagine making a liter (L) of solution consisting of 1 mol of sugar dissolved in water. To obtain this concentration, we would weigh out 342 g of sucrose and then gradually add water, while stirring, until the sugar was completely dissolved. We would then add enough additional water to bring the total volume of the solution up to 1 L. At that point, we would have a one-molar (1 M) solution of sucrose. **Molarity** refers to the number of moles of solute per liter of solution, and it is the unit of concentration most often used for aqueous solutions.

Acids, Bases, and pH

Dissociation of Water Molecules Occasionally, a hydrogen atom shared between two water molecules in a hydrogen bond shifts from one molecule to the other (Figure 3.11). When this happens, the hydrogen atom leaves its electron behind, and what is actually transferred is a **hydrogen ion**, a single proton with a charge of $+1$. The water molecule that lost a proton is now a **hydroxide ion** (OH^-), which has a charge of -1. The proton binds to the second water molecule at one of its unshared orbitals, making that molecule a **hydronium ion** (H_3O^+). We can write the chemical reaction this way:

$$H_2O + H_2O \rightleftharpoons \underset{\substack{\text{Hydronium}\\\text{ion}}}{H_3O^+} + \underset{\substack{\text{Hydroxide}\\\text{ion}}}{OH^-}$$

Although this is technically what happens, it is convenient to think of the process as the **dissociation** (separation) of a water molecule into a hydrogen ion (H^+) and a hydroxide ion:

$$H_2O \rightleftharpoons \underset{\substack{\text{Hydrogen}\\\text{ion}}}{H^+} + \underset{\substack{\text{Hydroxide}\\\text{ion}}}{OH^-}$$

As the double arrows indicate, this is a reversible reaction that will reach a state of dynamic equilibrium when water dissociates at the same rate that it is being reformed from H^+ and OH^-. At this equilibrium point, the concentration of water molecules greatly exceeds the concentrations of H^+ and OH^-. In fact, in pure water, only one water molecule in every 554 million is dissociated. Although the dissociation of water is reversible and statistically rare, it is exceedingly important in the chemistry of life. For example, the functioning of molecules within living cells is affected by even slight changes in the concentrations of H^+ and OH^-.

Acids and Bases Since the dissociation of water produces one H^+ for every OH^-, the concentrations of these two ions will be equal in pure water. The concentration of each is $10^{-7}\ M$. This means that there is only one ten-millionth of a mole of hydrogen ions per liter of pure water, and an equal number of hydroxide ions.

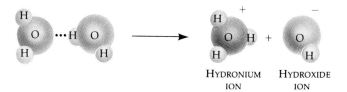

Figure 3.11
Formation of hydronium and hydroxide ions. A proton, the nucleus of a hydrogen atom, shifts from one water molecule to another, forming a hydronium ion and a hydroxide ion.

What would cause an aqueous solution to have an imbalance in its H^+ and OH^- concentrations? When substances called acids dissolve in water, they donate additional hydrogen ions to the solution. An **acid**, according to the definition most biologists use, is a substance that increases the H^+ concentration of a solution. For example, when hydrochloric acid (HCl) is added to water, hydrogen ions dissociate from chloride ions:

$$HCl \longrightarrow H^+ + Cl^-$$

Now there are two sources of H^+ in the solution (dissociation of water is the other), resulting in more H^+ than OH^-. Such a solution is known as an acidic solution.

A substance that reduces the hydrogen ion concentration in a solution is called a **base**. Some bases reduce the H^+ concentration indirectly by dissociating to form hydroxide ions, which then combine with the hydrogen ions to form water. An example of a base that acts this way is sodium hydroxide (NaOH), which in water dissociates into its ions:

$$NaOH \longrightarrow Na^+ + OH^-$$

Other bases reduce H^+ concentration directly by accepting hydrogen ions. Ammonia, for instance, acts as a base by binding a hydrogen ion from the solution, resulting in an ammonium ion:

$$\underset{\text{Ammonia}}{NH_3} + \underset{\substack{\text{Hydrogen}\\\text{ion}}}{H^+} \rightleftharpoons \underset{\text{Ammonium ion}}{NH_4^+}$$

In either case, the base reduces the H^+ concentration, and solutions with a higher concentration of OH^- than H^+ are known as basic solutions. A solution in which the H^+ and OH^- concentrations are equal is said to be **neutral**.

Notice that single arrows were used in the reactions for HCl and NaOH. These compounds dissociate completely when mixed with water. Hydrochloric acid is called a strong acid, and sodium hydroxide is called a strong base because they dissociate so completely. In contrast, ammonia is a weak base. The double arrows in the reaction for ammonia indicate that the binding and release of the hydrogen ion are reversible, and when the reaction reaches equilibrium, there will be a fixed ratio of NH_4^+ to NH_3. There are also weak acids, which dissociate reversibly to release and reaccept hydrogen ions. An example is carbonic acid:

$$H_2CO_3 \rightleftharpoons HCO_3^- + H^+$$

Carbonic acid   Bicarbonate   Hydrogen
                ion   ion

However, the equilibrium so favors the reaction in the left direction that when carbonic acid is added to water, only 1% of the molecules are dissociated at any particular time. Still, that is enough to shift the balance of H^+ and OH^- from neutrality.

The pH Scale   In any solution, the product of the H^+ and the OH^- concentrations is constant at 10^{-14} M. This can be written

$$[H^+][OH^-] = 10^{-14} \, M$$

where brackets indicate molar concentration for the substance enclosed. In a neutral solution at room temperature, $[H^+] = 10^{-7}$ and $[OH^-] = 10^{-7}$, so the product is $10^{-14} \, M$ ($10^{-7} \times 10^{-7}$). If enough acid is added to a solution to increase $[H^+]$ to $10^{-5} \, M$, then $[OH^-]$ will decline by an equivalent amount to 10^{-9} M ($10^{-5} \times 10^{-9} = 10^{-14}$). An acid not only adds hydrogen ions to a solution, but also removes hydroxide ions because of the tendency for H^+ to combine with OH^- to form water. A base has the opposite effect, increasing OH^- concentration, but also reducing H^+ concentration by the formation of water. If enough of a base is added to raise the OH^- concentration to $10^{-4} \, M$, the H^+ concentration will drop to $10^{-10} \, M$. Whenever we know the concentration of either H^+ or OH^- in a solution, we can deduce the concentration of the other ion.

Because the H^+ and OH^- concentrations of solutions can vary by a factor of 100 trillion or more, scientists have developed a way to express this variation more conveniently. This is the purpose of the **pH scale**, which ranges from 0 to 14. The pH scale compresses the range of H^+ and OH^- concentrations by employing a common mathematical device—logarithms. The pH of a solution is defined as the negative logarithm (base 10) of the hydrogen ion concentration expressed in moles per liter:

$$pH = -\log[H^+]$$

For a neutral solution, $[H^+]$ is $10^{-7} \, M$, giving us

$$-\log 10^{-7} = -(-7) = 7$$

Notice that pH declines as H^+ concentration increases. Notice, too, that although the pH scale is based on H^+ concentration, it also implies OH^- concentration. A solution of pH 10 has a hydrogen ion concentration of $10^{-10} \, M$ and a hydroxide ion concentration of $10^{-4} \, M$ (Figure 3.12).

The pH of a neutral solution is 7, the midpoint of the scale. A pH value less than 7 denotes an acidic

solution, and the lower the number, the more acidic the solution. The pH for basic solutions is above 7. Values of pH less than 0 or greater than 14 are rarely encountered, and most biological fluids are within the range pH 6–pH 8. There are a few exceptions, however, including the strongly acidic digestive juice of the human stomach, which has a pH of about 1.5.

It is important to remember that each pH unit represents a tenfold difference in H^+ and OH^- concentrations. It is this mathematical feature that makes the pH scale so compact. A solution of pH 3 is not twice as acidic as a solution of pH 6, but 1000 times more acidic. When the pH of a solution changes slightly, the actual concentrations of H^+ and OH^- in solution change substantially.

Buffers   The internal pH of most living cells is close to 7. Even a slight change in pH can be harmful, because the structures and functions of molecules in the cell are very sensitive to the concentrations of hydrogen and hydroxide ions.

Biological fluids resist changes to their own pH when acids or bases are introduced because of the presence of **buffers**, substances that minimize changes in the concentrations of H^+ and OH^-. Buffers in human blood, for example, normally maintain the blood pH at 7.4. A drop in pH (acidosis) or an increase in pH (alkalosis) is dangerous, and a person cannot survive for more than a few minutes if the blood pH drops to 7 or rises to 7.8. Under normal circumstances, the buffering capacity of the blood prevents such swings in pH.

A buffer works by accepting hydrogen ions from the solution when they are in excess and donating hydrogen ions to the solution when they have been depleted. Most buffers are weak acids or weak bases that combine reversibly with hydrogen ions. One of the most important buffers in human blood and many other biological solutions is carbonic acid (H_2CO_3), which, as we have mentioned, dissociates to yield a bicarbonate ion (HCO_3^-) and a hydrogen ion:

                Response to a
                 rise in pH

$$H_2CO_3 \rightleftharpoons HCO_3^- + H^+$$

H^+ donor   Response to a H^+ acceptor   Hydrogen
(acid)   drop in pH   (base)   ion

The chemical equilibrium between carbonic acid and bicarbonate acts as a pH regulator, the reaction shifting left or right as other processes in the solution add or remove hydrogen ions. If the H^+ concentration in blood begins to fall (that is, if pH rises), more carbonic acid dissociates, replenishing hydrogen ions. But when H^+ concentration in blood begins to rise (pH drops), the bicarbonate ion acts as a base and removes the excess hydrogen ions from solution. Thus, the carbonic acid–bicarbonate buffering system actually consists of an acid and a base in equilibrium with each other. Most other buffers are also acid-base pairs, and

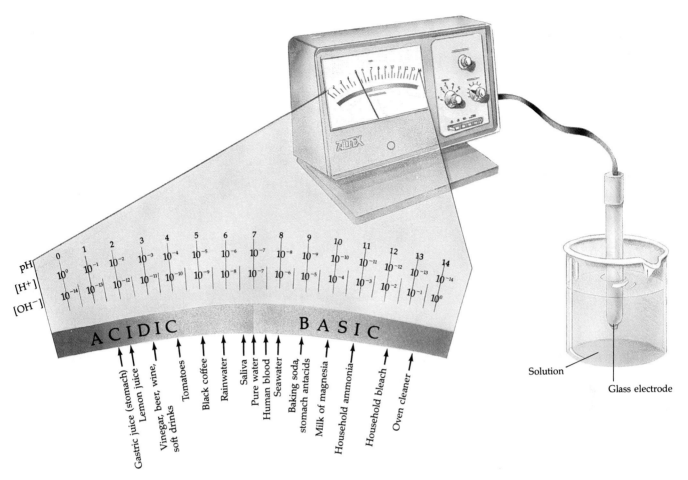

Figure 3.12
The pH of some aqueous solutions. Measurements of pH are usually made with a meter connected to a glass electrode immersed in the solution.

they are most effective at stabilizing pH when their acid and base forms are present in equal concentrations.

Acid Rain: Upsetting the Fitness of the Environment

Acid rain is an environmental problem that has increased public awareness about the sensitivity of life to pH. Uncontaminated rain has a pH of about 5.6, slightly acidic owing to the formation of carbonic acid from carbon dioxide and water. The term *acid rain* applies to rain more strongly acidic than pH 5.6. The questions we must consider are: What causes acid rain? What are its effects on the fitness of the environment? And what can be done to reduce the problem?

Acid rain is caused primarily by the presence in the atmosphere of sulfur oxides and nitrogen oxides, gaseous compounds that react with water in the air to form acids, which fall to the earth with rain or snow. A major source of these oxides is the combustion of fossil fuels by factories and automobiles. Acid rain due to air pollution is as old as the Industrial Revolution, but the problem has escalated and become more widespread in the past two decades. One practice that has

increased the occurrence of acid rain is the construction of taller smokestacks designed to reduce local pollution by dispersing factory exhaust. Unfortunately, prevailing winds simply move the problem, and acid rain falls hundreds or thousands of miles away from industrial centers, often in once pristine regions. In the Adirondack Mountains of upstate New York, the pH of rainfall averages 4.2, about 25 times more acidic than normal rain. Acid rain falls on many other regions, including the Cascade Mountains of the Pacific Northwest and certain parts of Europe. One West Virginia storm dropped rain having a pH of 1.5.

Experiments and observations have confirmed that acid rain is having deleterious effects on both terrestrial and freshwater ecosystems. Acid rain that falls on land lowers the pH of the soil solution, which affects the solubility of minerals. Some mineral nutrients required by plants are washed out of the topsoil, while other minerals, such as aluminum, reach toxic concentrations when acidification increases their solubility. The effects of acid rain on soil chemistry have contributed to the decline of European forests (Figure 3.13). Acid rain has also lowered the pH of lakes and ponds

Figure 3.13
Effects of acid rain on a forest. Rain and snow bearing the acidic products of coal and oil combustion have been blamed for the demise of this forest in Erzgebirge, Czechoslovakia.

in some regions, and the accumulation of certain minerals leached from the soil by acid rain further contaminates freshwater habitats. Many species of fishes, amphibians, and aquatic invertebrates have been adversely affected. More than half the lakes at the higher elevations of the western Adirondacks are now more acidic than pH 5, and fish have completely disappeared from nearly all those lakes. The fish are like the coal mine canary, a warning that something has gone awry in the environment.

It is possible to reduce acid rain through industrial controls and antipollution devices. However, this is a solution that can come only from politicians representing people concerned about their environment.

* * *

In this chapter, we have seen how the emergent properties of water, a substance so common and yet so extraordinary, are crucial to life, and how disturbance of water resources—by a change in the pH of rain, for example—can affect the fitness of the environment. In the next chapter, we will explore another facet of the fitness of the environment: the role of the element carbon in the chemistry of life.

STUDY OUTLINE

1. Water is the medium of life. It is abundant in the environment and makes up a major part of every living cell.
2. The structure of water is responsible for its unusual properties.

Water Molecules and Hydrogen Bonding (p. 37)

1. Water is a polar molecule. In the covalent bonds joining oxygen and hydrogen, oxygen pulls more on the electrons than does hydrogen, making the oxygen partially negative and the hydrogens partially positive.
2. A hydrogen bond is formed when the oxygen of one water molecule is electrically attracted to the hydrogen of an adjacent molecule.
3. Each water molecule can form up to four hydrogen bonds, which are the basis for water's emergent properties.

Some Extraordinary Properties of Water (pp. 38–44)

1. Constant formation and re-formation of hydrogen bonds makes liquid water cohesive. In an example of this behavior, cohesive water is pulled upward in the microscopic vessels of plants.

2. Hydrogen bonding of water molecules on the surface of liquid water exposed to air is responsible for water's surface tension. This property allows water to form droplets and support weight on its surface.
3. Hydrogen bonds are also responsible for adhesion, the strong attraction of water to hydrophilic substances.
4. Adhesion and cohesion are involved in imbibition, an important process in plant growth and development.
5. Heat is the total kinetic energy of the molecules in a body of matter. Temperature is a measure of the average kinetic energy. A calorie is the heat energy required to raise the temperature of 1 g of water by 1°C.
6. Hydrogen bonding gives water a high specific heat; that is, only relatively small temperature changes occur when large amounts of heat are absorbed or released. Heat is absorbed when hydrogen bonds break and is released when hydrogen bonds form, thus minimizing temperature fluctuations to within limits that permit life.
7. Water has a high heat of vaporization; that is, large amounts of heat are required to raise the kinetic energy of liquid water enough to break hydrogen bonds and liberate the

individual molecules to form water vapor. Loss of kinetic energy from the remaining fluid gives water its important property of evaporative cooling.

8. Ice is less dense than liquid water owing to more organized hydrogen bonding that forces the molecules to expand into a characteristic crystalline lattice. In winter, the formation of hydrogen bonds in ice releases heat to temper the season, and the low density of ice allows life to exist in liquid water under the frozen surface of ponds, lakes, and polar seas.

9. Water is an unusually versatile solvent because its polarity attracts it to all charged and polar substances. When ions or molecules of such substances are surrounded by water molecules, they dissolve and are called solutes.

Aqueous Solutions (pp. 44–48)

1. A mole is the number of grams of a substance that equals its molecular weight in daltons. Molarity is a measure of solute concentration in number of moles per liter of solution. The concept of moles allows us to express molecular weight in grams.

2. Water can dissociate into H^+ and OH^-. The reaction is reversible, and at equilibrium, $[H^+] = [OH^-] = 10^{-7}$ M (for pure water at room temperature).

3. The concentration of H^+ is measured in pH units as follows: $pH = -\log[H^+]$. Each pH unit thus represents a tenfold difference in $[H^+]$.

4. Solutes that donate additional H^+ in aqueous solutions raise the ratio of $[H^+]$ to $[OH^-]$ and are called acids.

5. Solutes that donate OH^- or accept H^+ in aqueous solutions lower the ratio of $[H^+]$ to $[OH^-]$ and are called bases.

6. In a neutral solution, $[H^+] = [OH^-] = 10^{-7}$, and pH $= 7$. In an acidic solution, $[H^+]$ is greater than $[OH^-]$, and the pH is less than 7. In a basic solution, $[H^+]$ is less than $[OH^-]$, and the pH is greater than 7.

7. In any solution, the product of $[H^+]$ and $[OH^-]$ is constant and equal to 10^{-14} M.

8. $[H^+]$ critically affects the chemistry of life by influencing the structure and function of biological molecules. The internal pH of most living cells must be kept close to 7.

9. Cell pH would change when acids or bases were formed during metabolism if not for the presence of buffers, which resist changes in pH.

10. A buffer consists of an acid-base pair that combines reversibly with hydrogen ions.

11. If an acid is added to a buffered solution, the excess H^+ combines with the base form of the buffer. If a base is added to a buffered solution, the acid form of the buffer dissociates to replenish H^+ removed by the base.

12. Acid rain occurs when water in the air reacts with sulfur oxides and nitrogen oxides that result from combustion of fossil fuels. The acids that form give rain and snow a pH less than 5.6, sometimes causing serious environmental consequences.

SELF-QUIZ

1. The main thesis of Lawrence Henderson's *The Fitness of the Environment* is
   a. Earth's environment is constant.
   b. It is the physical environment, not life, that has evolved.
   c. The environment of Earth has adapted to life.
   d. Life, as we know it, depends on certain environmental qualities of Earth.
   e. Water and other qualities of Earth's environment exist because they make the planet more suitable for life.

2. Air temperature often increases slightly as clouds begin to drop rain or snow. Which behavior of water is *most directly* responsible for this phenomenon?
   a. water's change in density when it condenses to form a liquid or solid
   b. water's reactions with other atmospheric compounds
   c. release of heat by formation of hydrogen bonds
   d. release of heat by breaking of hydrogen bonds
   e. water's high surface tension

3. For two bodies of matter in contact, heat always flows from
   a. the body with greater heat to the one with less heat
   b. the body of higher temperature to the one of lower temperature
   c. the more dense to the less dense body
   d. the body having more water to the one with less water
   e. the larger to the smaller body

4. A slice of pizza has 500 kilocalories. If we could burn the pizza and use all of the heat to warm a 50-L container of cold water, then what would be the approximate increase in the temperature of the water? (*Note:* A liter of cold water weighs about a kilogram.)
   a. 50° C
   b. 5°C
   c. 10°C
   d. 100°C
   e. 1°C

5. The bonds that are broken when water vaporizes are
   a. ionic bonds
   b. bonds *between* water molecules
   c. bonds between atoms of individual water molecules
   d. covalent bonds
   e. bonds between adjacent hydrogens of water molecules

6. We can be sure that a mole of table sugar and a mole of vitamin C are equal in their
   a. weight in daltons
   b. weight in grams
   c. number of molecules
   d. number of atoms
   e. volume

7. How many grams of acetic acid ($C_2H_4O_2$) would you use to make 10 L of a 0.1 M aqueous solution of the acetic acid? (*Note:* The atomic weights, in daltons, are approximately 12 for carbon, 1 for hydrogen, and 16 for oxygen.)
   a. 10 g

b. 0.1 g

c. 6 g

d. 60 g

e. 0.6 g

8. Acid rain has lowered the pH of a particular lake to 4.0. What is the hydrogen ion concentration of the lake?

a. 4.0 M

b. $10^{-10}\ M$

c. $10^{-4}\ M$

d. $10^4\ M$

e. 4%

9. What is the *hydroxide* ion concentration of the lake described in question 8?

a. $10^{-7}\ M$

b. $10^{-4}\ M$

c. $10^{-10}\ M$

d. $10^{-14}\ M$

e. 10 M

10. Dissolved in water, acetic acid dissociates into an acetate ion and a hydrogen ion:

$$C_2H_4O_2 \rightleftharpoons C_2H_3O_2^- + H^+$$
$$\text{Acetic acid} \qquad \text{Acetate}$$

Assume that this reaction has reached equilibrium, and then the pH of the solution is raised by adding a base. Which substance(s) in the solution will increase in concentration the most as a result of adding the base?

a. acetic acid

b. acetate ion

c. H^+

d. both [acetic acid] and [H^+] will increase equally

e. both [acetate] and [H^+] will increase equally

CHALLENGE QUESTIONS

1. Adhesion and cohesion cooperate to allow water to rise in thin tubes made of hydrophilic material such as glass. This phenomenon is called capillary action (see figure). Water molecules adhering to the glass creep upward, pulling more water along by cohesion. Explain why water rises higher by capillary action in a tube of smaller diameter than in one of wider diameter.

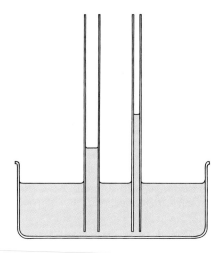

2. Explain how panting helps to regulate a dog's body temperature.

3. Discuss the special political obstacles to reducing acid rain (as compared with environmental issues confined to a more localized region).

FURTHER READING

1. Henderson, L. J. *The Fitness of the Environment.* New York: Macmillan, 1913. A classic book highlighting the importance of water and carbon to life

2. Lehninger, A. L. *Principles of Biochemistry.* New York: Worth, 1981, Chap. 4. This readable, authoritative biochemistry text has an excellent discussion of water.

3. Mohner, V. A. "The Challenge of Acid Rain." *Scientific American,* August, 1988. Analysis of a complex environmental problem.

4. Peterson, I. "Raindrop Oscillations." *Science News,* March 2, 1985. An intriguing article about the properties of water in raindrops.

5. Pimentel, G. C., and A. L. McClellan. *The Hydrogen Bond.* San Francisco: Freeman, 1960. A classic monograph on various aspects of the hydrogen bond.

4

Carbon and Molecular Diversity

Life without carbon is as unimaginable as life without water. Carbon is unparalleled in its ability to form molecules that are large, complex, and diverse—a chemical behavior that casts carbon in a major role in the fitness of Earth's environment for life. A cell is 70%–95% water, but most of the rest of the cell consists of carbon-based compounds. Proteins, DNA, carbohydrates, and other molecules that distinguish living matter from inanimate material are all composed of carbon atoms bonded to one another and to atoms of other elements. Hydrogen, oxygen, nitrogen, sulfur, and phosphorus are other common ingredients of these compounds, but it is carbon that accounts for the endless diversity of organic molecules, some of which are spectacularly complex (Figure 4.1). This chapter focuses on the principles of molecular architecture that make carbon so important to life. The central theme of this unit—emergent properties that arise from the organization of living matter—will be reinforced in our study of the unique molecules assembled from carbon.

THE FOUNDATIONS OF ORGANIC CHEMISTRY

Compounds containing carbon are said to be organic, and the branch of chemistry that specializes in the study of carbon compounds is called **organic chemistry**. Well over two million organic compounds are known, and many more are identified each day. They range from simple molecules to colossal ones with

Figure 4.1
A protein, a complex organic molecule. This mammoth insulin molecule has a few thousand atoms joined by covalent bonds. It is the ability of carbon atoms (shown in white) to bond to multiple partners, including other carbon atoms, that makes such complex molecules possible.

thousands of atoms and molecular weights in excess of 100,000 daltons. The percentages of the major elements of life are quite uniform from individual to individual, and even from species to species (Table 4.1). The atoms of organic molecules, however, can be arranged so many different ways that the uniqueness of each organism is ensured. It is because of carbon's versatility that a limited assortment of atomic building

Table 4.1 Relative abundance of major elements in humans and *Escherichia coli (E. coli)*, a bacterium

Element	Percent Dry Weight*	
	E. coli	Human
Oxygen	20	18
Carbon	50	54
Hydrogen	10	8
Nitrogen	10	9
Phosphorus	4	3
Sulfur	1	0.75

*Dry weight, the weight of an organism if all water were removed, primarily reflects the elemental composition of organic compounds.

blocks, taken in roughly the same proportions, can be used to build organic molecules of inexhaustible variety.

For centuries, humans have used other organisms as sources of valued substances—everything from wine and food to medicines and fabrics. Organic chemistry had its origins in attempts at purifying and improving the yield of these products (Figure 4.2). By the early nineteenth century, chemists had learned to make many simple compounds in the laboratory by combining elements under the right conditions, but artificial synthesis of the complex molecules extracted from living matter seemed hopeless. It was at that time that Jons Jakob Berzelius, a Swedish chemist, first made the distinction between organic compounds, those that apparently could arise only within living organisms, and inorganic compounds, those that were found in the nonliving world. The new discipline of organic chemistry was first built on a foundation of **vitalism**, the belief in a life force outside the jurisdiction of

Figure 4.2
Application of organic chemistry in a nineteenth-century laboratory. Working with his students at Tuskegee Institute in Alabama, George Washington Carver (in bow tie) found over a hundred uses for oils and other organic compounds extracted from peanuts.

physical and chemical laws (see Chapter 1). The dividing line between life and nonlife seemed absolute in the early 1800s.

Chemists began to chip away at the foundation of vitalism when they learned to synthesize organic compounds in their laboratories. In 1828, Friedrich Wöhler, a German chemist who had studied with Berzelius, attempted to make an inorganic salt, ammonium cyanate, by mixing solutions of ammonium (NH_4^+) and cyanate (CNO^-) ions. Wöhler was astonished to find that instead of the expected product, he had made urea, an organic compound present in the urine of animals. Wöhler challenged the vitalists when he wrote: "I must tell you that I can prepare urea without requiring a kidney or an animal, either man or dog." But one of the ingredients used in the synthesis, the cyanate, had been extracted from animal blood, and the vitalists were not swayed by Wöhler's discovery. Then, a few years later, Hermann Kolbe, a student of Wöhler's, made the organic compound acetic acid from inorganic substances that could themselves be prepared directly from pure elements. The foundation of vitalism was shaking. It finally crumbled after several more decades of laboratory synthesis of increasingly complex organic compounds. Stanely Miller, interviewed at the beginning of this unit, extended this work by demonstrating that such abiotic synthesis of organic compounds may have contributed to the origin of life. Researchers also developed new methods for isolating organic compounds from the mixtures found in natural sources (see Methods Box, p. 54).

The pioneers of organic chemistry helped shift the mainstream of biological thought from vitalism to *mechanism*, the belief that all natural phenomena, including the processes of life, are governed by physical and chemical laws. Organic chemistry was redefined as the study of carbon compounds, regardless of their origin. It is true that most naturally occurring organic compounds are products of organisms, and it is also true that these molecules have a diversity and range of complexity unrivaled by inorganic compounds. But the same rules of chemistry apply to inorganic and organic molecules alike. The foundation of organic chemistry is not some "life force," but the unique chemical versatility of the element carbon.

THE VERSATILITY OF CARBON IN MOLECULAR ARCHITECTURE

The key to the chemical characteristics of an atom, as we learned in Chapter 2, is in its configuration of electrons, which determines the kinds and number of bonds it will form with other atoms. Carbon has a total of six electrons, with two in the first electron shell and four in the second shell. Having four valence electrons in a shell that holds eight, carbon has little tendency

Element	Electron Configuration	Valence
Hydrogen		1
Oxygen		2
Nitrogen		3
Carbon		4

Table 4.2 Valences of major elements of organic molecules

to gain or lose electrons and form ionic bonds; it would have to donate or accept four electrons to do so. Instead, a carbon atom completes its valence shell by sharing electrons with other atoms in four covalent bonds. Each carbon atom thus acts as an intersection point from which a molecule branches off in up to four directions. This electron configuration is one facet of carbon's versatility that makes large, complex molecules possible.

The electron configuration of carbon also gives it covalent compatibility with a greater number of different elements than any other type of atom. Table 4.2 reviews the electron configurations and valences of the four major atomic components of organic molecules: carbon and its most frequent partners, oxygen, hydrogen, and nitrogen. You can think of these valences as the rules of covalent bonding in organic chemistry—the building codes that govern the architecture of organic molecules.

In Chapter 2, you learned that when a carbon atom forms covalent bonds, the four electron orbitals in the

Chromatography is used to separate organic compounds from mixtures on the basis of their solubility in one or more solvents passed through a solid support medium, such as paper or columns of powdered material.

Paper Chromatography

A sample of the chemical mixture is applied to paper, and then a solvent is allowed to flow along the paper. The compounds most soluble in the solvent travel the fastest along the paper. For many purposes, a glass plate covered with a gel can be used instead of

paper; this method is called thin-layer chromatography.

Column Chromatography

A powder or resin is packed into a glass column as the support medium. The sample is applied to the top of the column, and then the solvent is passed down the column. The compounds in the sample move down the column at different rates, depending on their individual solubilities in the solvent and how tightly they bind to the support medium.

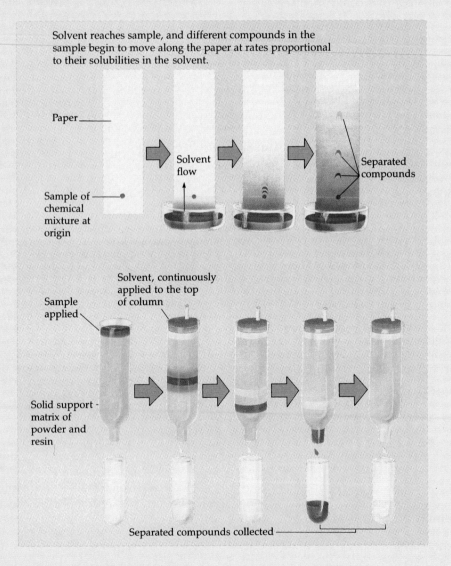

Solvent reaches sample, and different compounds in the sample begin to move along the paper at rates proportional to their solubilities in the solvent.

Paper

Sample of chemical mixture at origin

Solvent flow

Separated compounds

Sample applied

Solvent, continuously applied to the top of column

Solid support matrix of powder and resin

Separated compounds collected

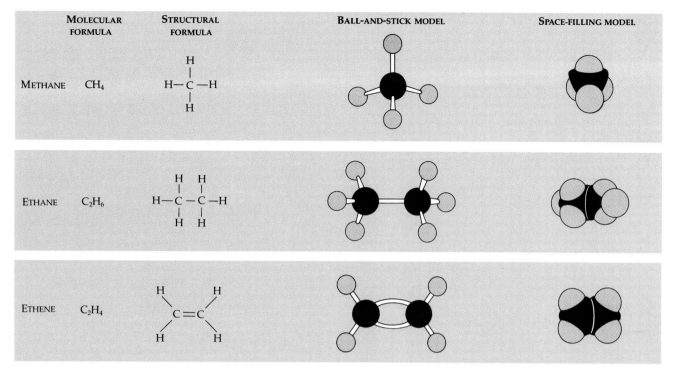

MOLECULAR FORMULA	STRUCTURAL FORMULA	BALL-AND-STICK MODEL	SPACE-FILLING MODEL
METHANE CH_4			
ETHANE C_2H_6			
ETHENE C_2H_4			

Figure 4.3
Shapes of some simple organic molecules. Whenever a carbon atom has four single bonds, the bonds angle toward the corners of an imaginary tetrahedron.

When two carbons are joined by a double bond, all bonds around those atoms are in the same plane. In addition to molecular and structural formulas, two types of

three-dimensional models are shown here. A ball-and-stick model emphasizes bond angles. A space-filling model portrays a molecular shape more accurately.

valence shell become four teardrop-shaped orbitals that angle from the carbon atom toward the corners of a tetrahedron (see Figure 2.10). The bond angles in methane (CH_4) are 109°, and they would be approximately the same in any molecule where carbon has four single bonds. For example, ethane (C_2H_6) is shaped like two tetrahedrons joined at their apexes. It is convenient to write structural formulas as though molecules were flat, but it is important to remember that real molecules have three-dimensional shapes, and the shape of an organic molecule can determine its function in a living cell. Figure 4.3 illustrates the shapes of a few simple organic molecules.

A couple of examples will demonstrate the rules of covalent bonding in organic molecules. In the carbon dioxide molecule (CO_2), a single carbon atom is joined to two atoms of oxygen by double covalent bonds. The structural formula for CO_2 is O=C=O. Remember, each line in a structural formula represents a pair of shared electrons. Notice that the carbon atom in CO_2 is involved in a total of four covalent bonds, two with each oxygen atom. The arrangement completes the valence shells of all atoms in the molecule. Carbon dioxide is such a simple molecule that it is generally considered inorganic, even though it contains carbon. Whether we call carbon dioxide organic or inorganic

is a rather arbitrary distinction, but there is no ambiguity about the importance of CO_2 to the living world. Taken up from the air by plants and incorporated into sugar and other foods during photosynthesis, CO_2 is the source of carbon for all the organic molecules found in organisms.

Another relatively simple molecule is urea, $CO(NH_2)_2$, the organic compound from urine that Wöhler learned to synthesize in the early nineteenth century. The structural formula for urea is

Again, each atom has the correct number of covalent bonds. In this case, one carbon atom is involved in both single and double bonds.

Both urea and carbon dioxide have only one carbon atom, and such simple molecules cannot do justice to carbon's central importance in the chemistry of life. A carbon atom can use one or more of its valence electrons to form covalent bonds to other carbon atoms, making it possible to link the atoms together into chains of seemingly infinite variety.

VARIATION IN CARBON SKELETONS

Carbon chains form the skeletons of organic molecules. The skeletons vary in length and may be straight or branched or even arranged in closed rings (Figure 4.4). Some carbon skeletons have double bonds, which vary in number and location. Such variation in carbon skeletons is one important source of the molecular complexity and diversity that characterize living mat-

ETHANE PROPANE

(a) Carbon skeletons vary in length.

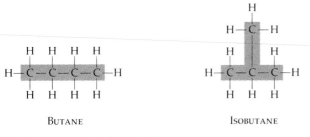

BUTANE ISOBUTANE

(b) Skeletons may be branched or unbranched.

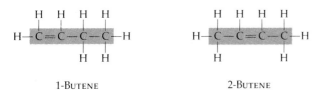

1-BUTENE 2-BUTENE

(c) The skeleton may have double bonds, which can vary in location.

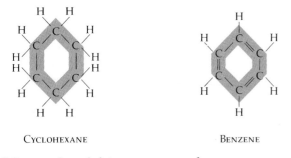

CYCLOHEXANE BENZENE

(d) Some carbon skeletons are arranged in rings.

Figure 4.4
Variations in carbon skeletons. Hydrocarbons, organic molecules consisting only of carbon and hydrogen, are used here to illustrate diversity in the carbon skeletons of organic molecules.

Figure 4.5
Economic and political importance of hydrocarbons. Petroleum is referred to as a fossil fuel because its hydrocarbons are the organic remains of biological material deposited millions of years ago. Dependence on petroleum has amplified international interest in the affairs of the Middle East. These tankers are loading Iranian oil at the Kharg Island terminal.

ter. In addition, atoms of other elements can be bonded to the skeletons at available sites.

All the molecules shown in Figures 4.3 and 4.4 are **hydrocarbons,** organic molecules consisting only of carbon and hydrogen. Atoms of hydrogen are attached to the carbon skeleton wherever electrons are available for covalent bonding. Hydrocarbons are not prevalent in organisms; but as the major components of the fossil fuel petroleum, they have become essential to our standard of living (Figure 4.5). The fossil fuels are accumulations of organic material derived from the remains of organisms living millions of years ago. Here, our interest in hydrocarbons is in the diversity of their carbon skeletons.

Isomers

Variation in the architecture of organic molecules can be seen in **isomers,** compounds that have the same molecular formula but different structures and hence different properties. Compare, for example, the two butanes in Figure 4.4. Both have the molecular formula C_4H_{10}, but they differ in the covalent arrangement of their carbon skeletons. The skeleton is straight in butane, but branched in isobutane. Isomers occur in three types: structural isomers, geometric isomers, and optical isomers (Figure 4.6).

Structural isomers differ in the covalent arrangements of their atoms. The number of possible isomers increases tremendously as carbon skeletons increase in size. There are only two butanes, but there are 18

(a) Structural isomers: variation in covalent arrangement.

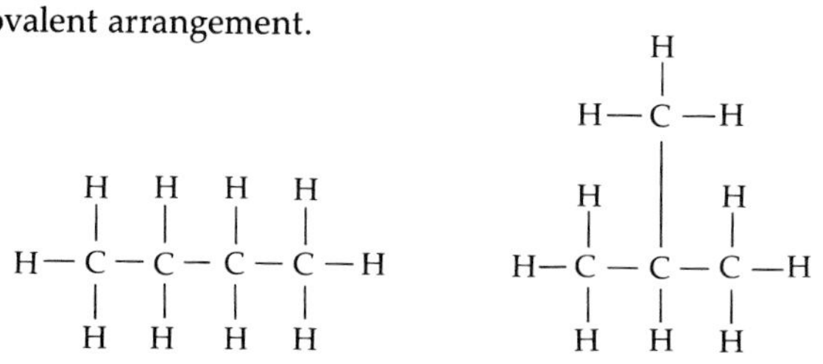

(b) Geometric isomers: variation in arrangement around a double bond.

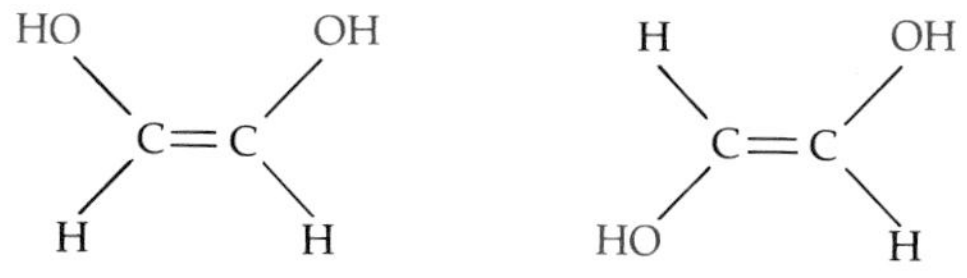

(c) Optical isomers: variation in spatial arrangement around an asymmetric carbon resulting in molecules that are mirror images.

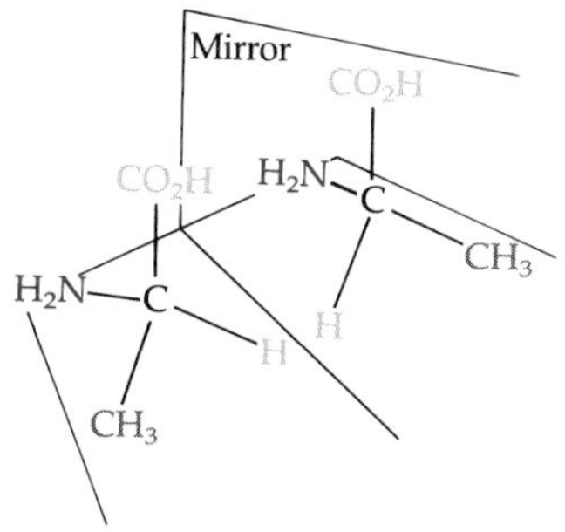

Figure 4.6
Different types of isomers. Compounds with the same molecular formula but different structures, isomers are a source of diversity in organic molecules.

variations of C_8H_{18} and 366,319 possible structural isomers of $C_{20}H_{42}$. Structural isomers may also differ in the location of double bonds.

Geometric isomers of a molecule have all the same covalent partnerships, but they differ in their spatial arrangements. Geometric isomers arise from the inflexibility of double bonds, which, unlike single bonds, will not allow the atoms they join to rotate freely about the axis of the bonds. The subtle difference in shape between geometric isomers can dramatically affect the biological activities of organic molecules. For example, the first step in vision is a light-induced change of rhodopsin, a compound in the eye, from one geometric isomer to another.

Optical isomers are molecules that are mirror images of each other. In the structural formula shown in Figure 4.6c, the middle carbon is called an **asymmetric carbon** because it is attached to four different atoms or groups of atoms. The four groups can be arranged in space about the asymmetric carbon in two different ways that are mirror images. The two molecules are optical isomers of one another. They are, in a way, left- and right-handed versions of the molecule. Cells can tell the difference between the two forms and usually one form is biologically active and its mirror image is not. The concept of optical isomers is important in the pharmaceutical industry, since the two isomers of a drug may not be equally effective. Worse, the inactive isomer may produce harmful side effects. This may have been the case with thalidomide, the sedative that caused many birth defects in the early 1960s. The drug was a mixture of two optical isomers, only one of which has since been demonstrated to cause birth defects in rats. Organisms are sensitive to even the most subtle variations in molecular construction.

FUNCTIONAL GROUPS

The distinctive properties of an organic molecule depend not only on the arrangement of its carbon skeleton, but also on the molecular components attached to that superstructure. Here we will examine certain groups of atoms that are frequently attached to carbon skeletons of organic molecules. These ensembles of atoms are known as **functional groups** because they are the regions of organic molecules most commonly involved in chemical reactions. If we think of hydrocarbons as the simplest organic molecules, we can view functional groups as attachments that replace one or more of the hydrogens bonded to the carbon skeleton of the hydrocarbon.

Each functional group behaves consistently from one organic molecule to another, and the number and arrangement of the groups give each molecule its unique properties. Consider the differences between estradiol and testosterone, respectively female and male sex hormones in humans and other vertebrates (Figure 4.7). Both are steroids, which are organic molecules having a carbon skeleton in the form of four interlocking rings. These sex hormones differ only in the attachment of certain functional groups to the common superstructure. The different actions of these two molecules on many targets throughout the body help produce the contrasting features of females and males. Thus, even our sexuality has its biological basis in the variations of molecular architecture.

The six functional groups most important in the chemistry of life are the hydroxyl, carbonyl, carboxyl, amino, sulfhydryl, and phosphate groups (Table 4.3).

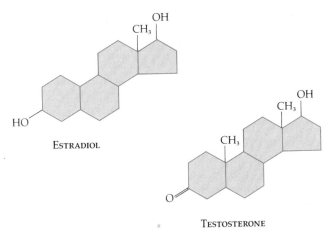

ESTRADIOL

TESTOSTERONE

Figure 4.7
Comparison of functional groups of female (estradiol) and male (testosterone) sex hormones. The two molecules have essentially the same superstructure, but differ in the attachment of functional groups to this common carbon skeleton (which has been simplified here by omitting double bonds and hydrogens of the four fused rings). This subtle variation in molecular architecture is influential in the development of the anatomical and physiological differences between females and males.

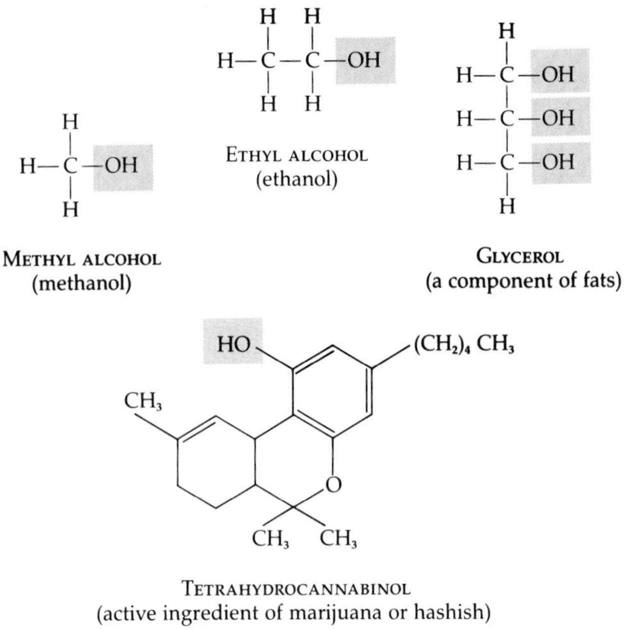

The Hydroxyl Group

In a **hydroxyl group,** a hydrogen atom is bonded to an oxygen atom, which in turn is bonded to the carbon skeleton of the organic molecule. In a structural formula, the hydroxyl group is abbreviated by omission of the covalent bond between the oxygen and hydrogen; it is written as $-OH$ or $HO-$. (This should not be confused with the hydroxide of bases such as sodium hydroxide.) The hydroxyl group is polar as a result of the electronegative oxygen atom drawing electrons toward itself. Consequently, water molecules are attracted to the hydroxyl group, which helps to dissolve organic compounds containing such groups. Sugars, for example, owe their solubility in water to the presence of hydroxyl groups. Organic compounds containing hydroxyl groups are called **alcohols,** and their specific names usually end in *-ol* (Figure 4.8).

The Carbonyl Group

The **carbonyl group** ($-CO$) consists of a carbon atom joined to an oxygen atom by a double bond. If this functional group is on the end of a carbon skeleton, the organic compound is called an **aldehyde** (Figure 4.9). The simplest aldehyde is formaldehyde, commonly used to preserve biological specimens. If the carbonyl group is not on the end of a carbon chain, the compound is called a **ketone.** For this to be the case, the chain must be at least three carbons long, as

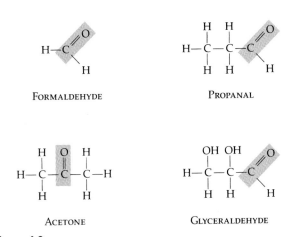

METHYL ALCOHOL
(methanol)

ETHYL ALCOHOL
(ethanol)

GLYCEROL
(a component of fats)

TETRAHYDROCANNABINOL
(active ingredient of marijuana or hashish)

Figure 4.8
Examples of alcohols. The simplest alcohol is methanol, or methyl alcohol, which looks like methane with one hydrogen atom replaced by a hydroxyl group. The two-carbon alcohol ethanol, or ethyl alcohol, is the drug present in alcoholic beverages. Tetrahydrocannabinol is the active ingredient in marijuana. Note that an alcohol may have more than one hydroxyl group attached to its carbon skeleton, as in glycerol.

FORMALDEHYDE

PROPANAL

ACETONE

GLYCERALDEHYDE

Figure 4.9
Examples of aldehydes and ketones. The carbonyl group is terminal on the carbon skeleton of an aldehyde; but in a ketone, such as acetone, the carbonyl group is not located on the end of the chain.

it is in acetone, the simplest ketone. Acetone has different properties from propanal, a three-carbon aldehyde. Molecular diversity in the form of structural isomers arises from variations in the location of a functional group along a carbon skeleton.

The organic compounds known as sugars are characterized by the presence of both carbonyl and hydroxyl

Table 4.3 Functional groups of organic compounds

Functional Group	Formula*	Name of Compounds	Example
Hydroxyl	R—OH	Alcohols	Ethanol
Carbonyl	R—C(=O)H	Aldehydes	Propanal
	R—C(=O)—R	Ketones	Acetone
Carboxyl	R—C(=O)OH	Carboxylic acids	Acetic Acid
Amino	R—N(H)(H)	Amines	Methylamine
Sulfhydryl	R—SH	Thiols	Mercaptoethanol
Phosphate	R—O—P(=O)(O⁻)—O⁻	Organic phosphates	Glycerol phosphate

*The letter R symbolizes the carbon skeleton to which the functional group is attached

groups. Glyceraldehyde, illustrated in Figure 4.9, is the first sugar a plant produces by photosynthesis.

The Carboxyl Group

When an oxygen atom is double-bonded to a carbon atom that is also bonded to a hydroxyl group, the entire assembly of atoms is called a **carboxyl group** ($-COOH$). Compounds containing carboxyl groups

are known as **carboxylic acids**, or organic acids (Figure 4.10). The simplest is the one-carbon compound called formic acid, the stinging substance some ants inject when they bite (Figure 4.11). Acetic acid, which has two carbons, gives vinegar its sour taste. (In general, acids, including carboxylic acids, taste sour.)

Why does a carboxyl group have acidic properties? As we saw in Chapter 3, an acid increases the concentration of hydrogen ions in a solution. A carboxyl group is a source of hydrogen ions, since the covalent bond

Figure 4.10
Examples of carboxylic acids (organic acids). The carboxyl groups are shown here in their undissociated forms.

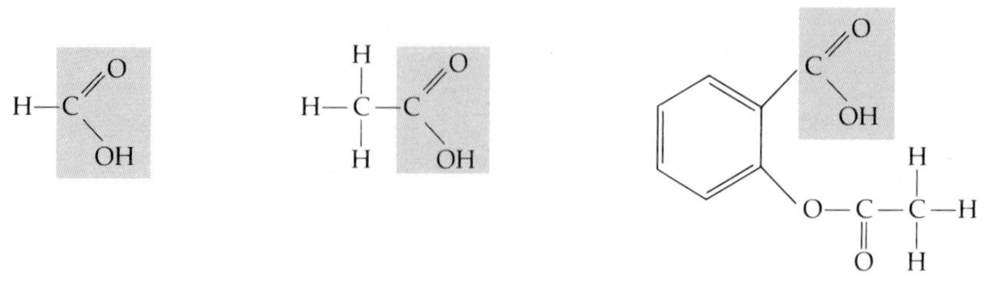

FORMIC ACID

ACETIC ACID

ASPIRIN
(ring portion abbreviated)

Figure 4.11
A minute formic acid factory. Formicine ants, which account for about 10% of all the world's ants, manufacture and secrete formic acid. They use the organic acid for defense and for chemical communication with other members of their population. It is estimated that the 100 trillion formicine ants of the world release about 10 billion kilograms of formic acid into the atmosphere every year.

Dissociation occurs as a result of the teamwork of the two electronegative oxygen atoms of the carboxyl group pulling shared electrons away from hydrogen. If the double-bonded oxygen and the hydroxyl group were attached to separate carbon atoms, the distance would reduce the tendency for dissociation. Here is another example of how emergent properties result from a specific arrangement of building components: Acidic properties arise when an oxygen and a hydroxyl group are attached to the same carbon.

The Amino Group

So far, we have seen organic molecules made up of only carbon, oxygen, and hydrogen. The **amino group** ($-NH_2$) consists of a nitrogen atom bonded to two hydrogen atoms and to the carbon skeleton. Organic compounds with this functional group are called **amines** (Figure 4.12).

The amino group acts as a base; the nitrogen atom has a pair of unshared electrons it can use to bond to a hydrogen ion, which is thus removed from the solution in which the reaction takes place. This reversible process gives the amino group a charge of +1.

The relationship between an organic molecule and the pH of a solution is complicated when a carboxyl group and an amino group are both present in one molecule. Such compounds, named for their two functional groups, are called **amino acids**. The amino acids used in the construction of proteins are among the most biologically important compounds, as you will learn in the next chapter. Figure 4.13 illustrates the effects of pH on amino acids.

between the oxygen and the hydrogen is so polar that there is a tendency for occasional ionization, the hydrogen dissociating reversibly from the molecule as an ion (H^+). In the case of acetic acid, we have

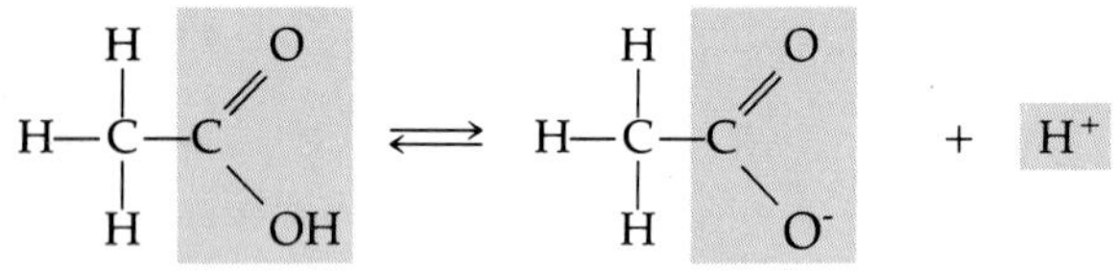

Acetic acid   Acetate ion   Hydrogen ion

Figure 4.12
Examples of amines. The amino groups are shown in their uncharged forms.

METHYLAMINE

GLYCINE (an amino acid)

AMPHETAMINE

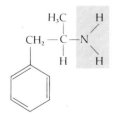

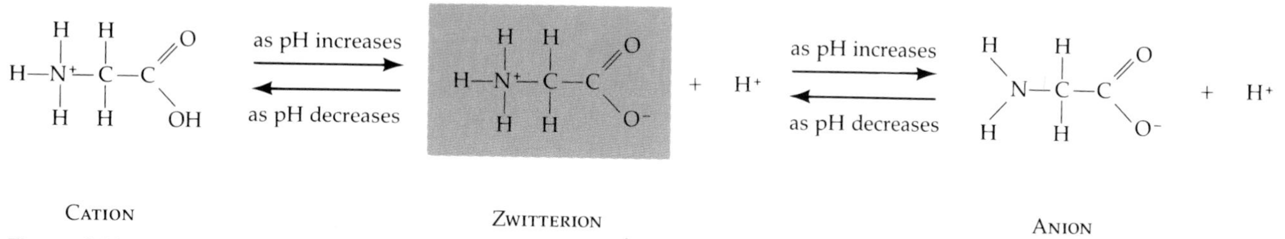

CATION   ZWITTERION   ANION

Figure 4.13
Different ionic forms of glycine, an amino acid. The amino acid can exist in three ionic states, and which one prevails depends on the pH of the solution. The chemical equilibria shift with changes in pH, affecting the electric charge of the amino acid. At cellular pH (about 7), most glycine molecules are neutral, but in the form of a zwitterion, a single molecule having both cationic and anionic groups. Because amino acids can take up and release hydrogen ions reversibly, they are good buffers.

The Sulfhydryl Group

Sulfur is grouped with oxygen in the periodic table; both have six valence electrons and form two covalent bonds. The organic functional group known as the **sulfhydryl group** ($-SH$), which consists of a sulfur atom bonded to an atom of hydrogen, resembles a hydroxyl group in shape (see Table 4.3). Organic compounds containing sulfhydryls are called **thiols** and generally smell like rotten eggs. We shall see in the next chapter how sulfhydryl groups help stabilize the intricate structure of many proteins.

The Phosphate Group

A phosphate ion is the dissociated form of an inorganic acid called phosphoric acid (H_3PO_4). The loss of hydrogen ions by dissociation leaves the phosphate with a negative charge. Organic compounds containing **phosphate groups** have a phosphate ion covalently attached by one of its oxygen atoms to the carbon skeleton (see Table 4.3). In general, organic phosphates store energy that can be passed from one molecule to another by the transfer of a phosphate group. In Chapter 6, you will learn how cells couple the transfer of phosphate groups to the performance of work, such as contraction of muscle cells.

THE ELEMENTS OF LIFE: A REVIEW

Living matter, as you have learned, consists mainly of carbon, oxygen, hydrogen, and nitrogen, with smaller amounts of sulfur and phosphorus. These elements share the characteristic of forming strong covalent bonds, a quality that is essential in the architecture of complex organic molecules. Of all these elements, carbon is the virtuoso of the covalent bond. The chemical behavior of carbon makes it exceptionally versatile as a building block in molecular architecture: It can form four covalent bonds, link together into intricate molecular skeletons, and join with several other elements. The versatility of carbon makes possible the great diversity of organic molecules, each with special properties that emerge from the unique arrangement of its carbon skeleton and the functional groups appended to that skeleton. At the foundation of all biological diversity lies this variation at the molecular level.

Equipped with the basic principles of organic chemistry, you are now ready to move on to the next chapter, where you will discover the specific structures and functions of the most elegant molecules made by living cells: carbohydrates, lipids, proteins, and nucleic acids.

STUDY OUTLINE

1. Carbon is unparalleled in its ability to form the large, complex, and diverse molecules that characterize living matter.

The Foundations of Organic Chemistry (pp. 51–53)

1. Organic chemistry had its origins in vitalism, which held that only living organisms could produce organic compounds because their synthesis required a life force outside physical and chemical laws.

2. The belief in vitalism was later challenged when chemists were able to synthesize organic compounds from inorganic ones. These findings led to the belief that biological phenomena also operate within physical and chemical laws. Organic chemistry was then redefined as the chemistry of carbon compounds.

The Versatility of Carbon in Molecular Architecture (pp. 53–56)

1. A bonding capacity of four contributes to carbon's ability to form complex and diverse molecules.
2. When forming four single covalent bonds, carbon's electron orbitals hybridize into a tetrahedral shape.

Variation in Carbon Skeletons (pp. 56–57)

1. Carbon chains are the skeletons of organic molecules. These skeletons vary in length and shape and possess bonding sites for atoms of other elements.
2. Hydrocarbons, consisting only of carbon and hydrogen, are the simplest organic molecules.
3. Carbon's versatile bonding is the basis for three types of isomers, molecules with the same molecular formula but different structures and thus different properties.
4. Structural isomers differ in the covalent arrangements of atoms and/or the location of double bonds.
5. Geometric isomers differ in spatial arrangement due to the inflexible property of double bonds.
6. Optical isomers are mirror images, right- and left-handed versions of molecules, made possible when an asymmetric carbon attaches to four different atoms or groups of atoms.

Functional Groups (pp. 57–61)

1. Functional groups consist of specific groups of atoms that covalently bond to carbon skeletons and give the overall molecule distinctive chemical properties.
2. The hydroxyl group, found in alcohols, has a polar covalent bond, which makes alcohols soluble in water.
3. The carbonyl group can either be at the end of a carbon skeleton (aldehyde) or within the molecule (ketone). Sugars contain both carbonyl and hydroxyl groups.
4. The carboxyl group is found in carboxylic, or organic, acids. The hydrogen of this group can dissociate to some extent, making the molecule a weak acid.
5. The amino group can accept an H^+, thereby acting as a base. Amino acids, the building blocks of proteins, contain both amino and carboxyl groups.
6. The sulfhydryl group helps stabilize the structure of some proteins.
7. The phosphate group can bond to the carbon skeleton by one of its oxygen atoms and has an important role in cellular energy storage and transfer.

The Elements of Life: A Review (p. 61)

1. Living matter is made mostly of carbon, oxygen, hydrogen, and nitrogen, with some sulfur and phosphorus.
2. Biological diversity has its molecular basis in carbon's ability to form an incredible array of molecules with characteristic shapes and chemical properties.

SELF-QUIZ

1. Organic chemistry is currently defined as
   a. the study of compounds that can be made only by living cells
   b. the study of carbon compounds
   c. the study of vital forces
   d. the study of natural (as opposed to synthetic) compounds
   e. the study of hydrocarbons

2. Column chromatography separates compounds based mainly on their differences in
   a. solubility in the chromatography solvent and affinity for the support medium being used
   b. heats of vaporization
   c. colors
   d. concentrations
   e. degree of radioactivity

3. Which of these hydrocarbons has a double bond in its carbon skeleton?
   a. C_3H_8
   b. C_2H_6
   c. CH_4
   d. C_2H_4
   e. C_2H_2

4. The gasoline consumed by an automobile is a fossil fuel consisting mostly of
   a. aldehydes
   b. amino acids
   c. alcohols
   d. hydrocarbons
   e. thiols

5. Choose the term that correctly describes the relationship between these two sugar molecules

   a. structural isomers
   b. geometric isomers
   c. optical isomers
   d. carbon isotopes

6. Identify the asymmetric carbon in this molecule

7. Which functional group is not present on this molecule?

   a. carboxyl
   b. carbonyl
   c. hydroxyl
   d. amino

8. An organic chemist would classify the molecule in question 7 as a (an)
   a. ketone
   b. aldehyde
   c. hydrocarbon
   d. amino acid
   e. thiol

9. Which functional group is most responsible for some organic molecules behaving as bases?
   a. hydroxyl
   b. carbonyl
   c. carboxyl
   d. amino
   e. phosphate

10. Which of the following molecules would be the strongest acid?

   a.

   b.

   c.

   d.

CHALLENGE QUESTIONS

1. How would you respond to a modern-day vitalist who argues: "Biologists have been unable to discover a physical or chemical basis for a human thought or idea because such abstract aspects of being human *have* no basis in natural processes."

2. Draw an organic molecule having all six functional groups described in this chapter.

3. Lewis Carroll's classic, *Alice in Wonderland*, poses the question: "Is looking-glass milk good to drink?" Respond to this query and justify your answer based on what you have learned about the structure of organic molecules, the importance of isomers, and the biological relevance of molecular shape.

FURTHER READING

1. Asimov, I. *The World of Carbon.* 2d ed. New York: Macmillan, 1962. A primer on the basics of organic chemistry, as told by one of America's most popular science writers.
2. Morrison R. T., and R. N. Boyd. *Organic Chemistry.* 5th ed. Newton, Mass.: Allyn and Bacon, 1987. The most widely used undergraduate organic chemistry text.
3. Ourisson, G., P. Albrecht, and M. Rohmer. "The Microbial Origin of Fossil Fuels." *Scientific American*, August 1984. The biology behind our important energy resources.

5

Structure and Function of Macromolecules

We have applied the concept of emergent properties to our study of water and relatively simple organic molecules, substances central to life and with unique behavior arising from their orderly arrangements of atoms. Another level in the hierarchy of biological organization is attained when cells join together small organic molecules into larger molecules belonging to four classes: carbohydrates, lipids, proteins, and nucleic acids. Many of these cellular molecules are, on the molecular scale, huge. For example, a protein may consist of thousands of atoms covalently connected into a molecular colossus weighing more than 100,000 daltons. Such a molecule is called a **macromolecule**, the term biologists use for the giant molecules of living matter (Figure 5.1).

Considering the size and complexity of macromolecules, it is remarkable that the structures of so many of them have been determined in detail by molecular biologists. Understanding the architecture of a particular macromolecule helps explain how that molecule works in the living cell. In molecular biology, as in the study of life at all levels, form and function are inseparable.

We begin our investigation of the macromolecules of life with a generalization that simplifies their structure: Most macromolecules are polymers.

POLYMERS

Cells make their macromolecules by linking relatively small molecules together end to end, forming chains called polymers (from the Greek *polus*, "many," and

Figure 5.1
Linus Pauling with a model of part of a protein. Pauling discovered several of the basic structural principles for proteins, a major type of macromolecule.

meris, "part"). A **polymer** is a large molecule consisting of many identical or similar subunits strung together, much as a train consists of a chain of cars. The subunits that serve as the building blocks of a polymer are called **monomers**.

Polymers and Molecular Diversity

Fundamental to the diversity of life is variation in the structure of macromolecules. Each cell has thousands of different kinds of macromolecules, many of which vary from one type of cell to another in the same organism. The inherent differences between human siblings reflects variation in polymers, particularly DNA and proteins. Molecular differences between unrelated individuals are more extensive, and from one species to another greater still. The diversity of macromolecules in the living world is vast—one estimate is that about a trillion different proteins actually exist—and the potential variety is essentially infinite.

What is the basis for such diversity in life's polymers? All macromolecules are constructed from a small repertoire of only about 40 to 50 common monomers and some others that occur rarely. (Some small molecules not only function as monomers, but also have other functions of their own.) Building a limitless variety of polymers from such a limited list of monomers

is analogous to constructing hundreds of thousands of words from only 26 letters of the alphabet. The secret is arrangement—variation in the linear sequence in which the subunits are strung together. However, this analogy falls far short of describing the great diversity of macromolecules, since most polymers are much longer than the longest word. Proteins, for example, are built from 20 kinds of amino acids arranged in chains that are typically more than 100 amino acids long.

Polymers account for the molecular uniqueness of an organism. The monomers used to make these polymers, however, are universal. Your proteins and those of a carrot or cow are assembled from the same 20 amino acids, but arranged in polymers varying in sequence. The molecular logic of life is simple but elegant: Small molecules common to all organisms are ordered into distinctive macromolecules.

Making and Breaking Polymers

The classes of macromolecules differ in the nature of their monomers, but the chemical mechanisms that cells use to make and break polymers are basically the same for all macromolecules. Monomers are linked together by a process known as condensation, or **dehydration synthesis.** The net effect of this process is the removal of a water molecule for each monomer added to the chain (although the actual mechanism, as it occurs in cells, is a bit more complicated, involving the formation of "activated" monomers that have phosphate groups attached—this will be discussed in Chapter 6). Whenever two monomers are joined, each contributes part of the H_2O molecule that is removed, one monomer losing a hydroxyl group (OH) and the other losing a hydrogen (H). Both monomers, having lost covalent partners, make new bonds by joining together with a covalent bond (Figure 5.2a). In the dehydration synthesis of a polymer, one water molecule must be removed for every link in the chain of monomers. The cell must expend energy to form these bonds, and the process occurs only with the help of special substances called enzymes, which will be discussed in Chapter 6.

Polymers are disassembled to monomers by **hydrolysis**, a process that is essentially the reverse of dehydration synthesis (Figure 5.2b). The word *hydrolysis* means to break (lyse) with water (hydro). Bonds between monomers are broken by the addition of water molecules, a hydrogen from the water attaching to one monomer, and the hydroxyl joining the adjacent monomer. An example of hydrolysis is the digestion of the food we eat. The bulk of the organic material in our food is in the form of polymers that are much too large to enter our cells. Within the digestive tract, various enzymes attack the polymers, bringing about hydrolysis. The monomers that have been released are then

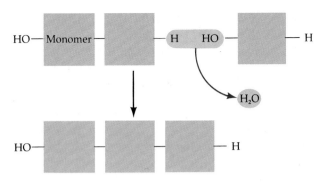

(a) Dehydration synthesis

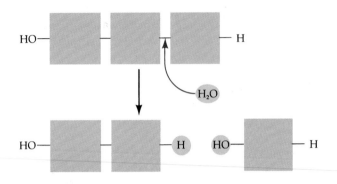

(b) Hydrolysis

Figure 5.2
Synthesis and breakdown of polymers. (a) Monomers are joined by dehydration synthesis, the removal of a water molecule. **(b)** The reverse of this process, hydrolysis, breaks bonds between monomers by adding water molecules.

absorbed into the bloodstream for distribution to all cells of the body.

Having developed an overview of macromolecules as polymers assembled from monomers by dehydration synthesis, we are now prepared to investigate the specific structures and functions of the four major classes of organic compounds found in cells (Table 5.1).

CARBOHYDRATES

Carbohydrates include sugars and their polymers. The simplest carbohydrates are the monosaccharides, or simple sugars. Disaccharides are double sugars, consisting of two monosaccharides joined by dehydration synthesis. Carbohydrates also include macromolecules in the form of polysaccharides, polymers of many sugars.

Monosaccharides

Monosaccharides (from the Greek *mono-*, "single," and *sacchar-*, "sugar") generally have molecular formulas that are some multiple of CH_2O (Figure 5.3). Glucose ($C_6H_{12}O_6$) is the most common monosaccharide, and it is a molecule of central importance in the chemistry of life. In the structure of glucose, we can see all the trademarks of a sugar. A hydroxyl group is attached to each carbon except one, which is double-bonded to an oxygen to form a carbonyl group. Depending on the location of the carbonyl group, a sugar is either an aldehyde or a ketone as well as an alcohol. Glucose, for example, is an aldehyde; fructose, a structural isomer of glucose, is a ketone. (Notice that most names for carbohydrates end in *-ose*.) Another criterion used to classify sugars is the size of the carbon skeleton,

Table 5.1 Major classes of organic compounds in cells		
Class	**Percent Dry Weight of Liver Cell***	**Subunits Joined by Dehydration Synthesis**
Carbohydrates (including polysaccharides)	5	Monosaccharides
Lipids	12	Glycerol and fatty acids (for fats, one lipid subclass)
Proteins	71	Amino acids
Nucleic acids	7	Nucleotides
*The remaining 5% of the dry weight of cells is accounted for by inorganic ions and small organic molecules, including the monomers to make macromolecules.		

ALDEHYDES   KETONES

TRIOSE SUGARS ($C_3H_6O_3$)

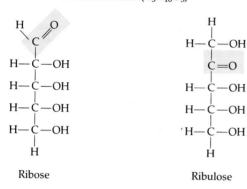

Glyceraldehyde   Dihydroxyacetone

PENTOSE SUGARS ($C_5H_{10}O_5$)

Ribose   Ribulose

HEXOSE SUGARS ($C_6H_{12}O_6$)

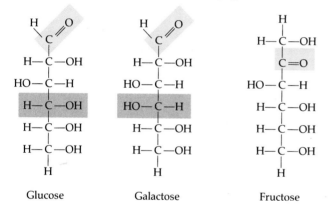

Glucose   Galactose   Fructose

Figure 5.3
Structure and classification of common simple sugars. Sugars may be aldehydes or ketones, depending on the location of the carbonyl group (blue). Sugars are also classified according to the length of their carbon skeletons. A third point of variation is in the spatial arrangement around asymmetric carbons (compare, for example, the gray portions of glucose and galactose).

which ranges from three to seven carbons long. Glucose, fructose, and other sugars having six carbons are called hexoses. Trioses and pentoses are also common. Sugars having four or seven or more carbons

are rarer, often occurring only as intermediates in the synthesis of other sugars.

Still another source of diversity for simple sugars is in the spatial arrangement around asymmetric carbons. (Recall from Chapter 4 that an asymmetric carbon is one attached to four different kinds of covalent partners.) Glucose and galactose, for example, differ only in the placement of parts around one asymmetric carbon. What may seem at first a small difference is significant enough to give the two sugars distinctive shapes, and shape is a major factor by which molecules within cells recognize and interact with one another.

It is convenient to draw glucose as though its carbon skeleton were linear, but this is not an accurate picture. In aqueous solutions, nearly all glucose molecules, as well as most other sugars, form rings (Figure 5.4).

Monosaccharides, particularly glucose, are major nutrients for cells. During photosynthesis, green plants produce glucose from carbon dioxide and water, using sunlight as energy. In the process known as cellular respiration, cells extract the energy stored in the glucose molecules. (Cellular respiration and photosynthesis will be discussed in Chapters 9 and 10, respectively.) Not only are sugar molecules the main fuel for cellular work, but their carbon skeletons serve as raw material for the synthesis of other types of small organic molecules, including amino acids and fatty acids. Sugar molecules that are not immediately used by cells as fuel or as a source of carbon skeletons are generally incorporated as monomers into disaccharides or polysaccharides.

Disaccharides

A **disaccharide**, or double sugar, consists of two monosaccharides joined by a **glycosidic linkage**, the bond formed between two sugar monomers by dehydration synthesis (Figure 5.5). For example, maltose is a disaccharide formed by the linking of two molecules of glucose. Also known as malt sugar, maltose is an important ingredient in the brewing of beer. Lactose, the sugar present in milk, consists of a glucose molecule joined to a galactose molecule. The most prevalent disaccharide is sucrose, better known as table sugar. Its two monomers are glucose and fructose. Plants generally transport carbohydrate from one part of the plant to another in the form of sucrose.

Polysaccharides

Polysaccharides are macromolecules that are polymers of a few hundred to a few thousand monosaccharides linked together. Enzymes join the sugar

Figure 5.4
Ring forms of glucose and fructose. (a)
Chemical equilibrium between the linear structures and rings greatly favors the formation of rings. To form the glucose ring, carbon number one bonds to the oxygen attached to carbon number five. **(b)** In these abbreviated structural formulas for the rings, the thicker edge indicates that you are looking at the ring edge-on.

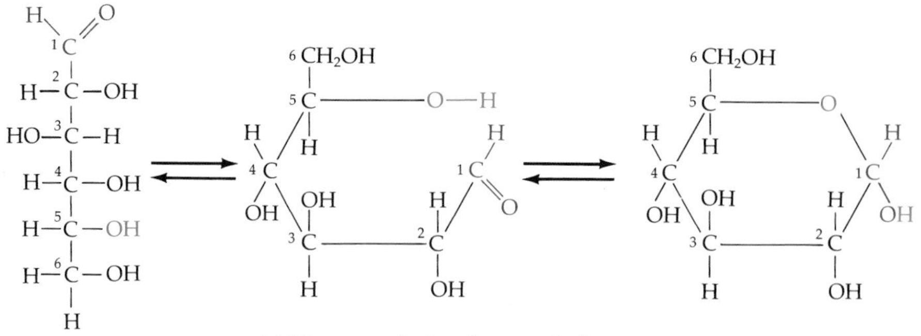

(a) Linear and ring forms of glucose

(b) Abbreviated ring-structures

GLUCOSE

FRUCTOSE

monomers by dehydration synthesis. Some polysaccharides are storage material, hydrolyzed as needed to provide sugar for the cell. Other polysaccharides serve as building material for structures protecting the cells.

Storage Polysaccharides   **Starch**, a storage polysaccharide of plants, is a polymer consisting entirely of glucose (Figure 5.6a). The monomers are joined by 1–4 linkages, like the monomers of maltose, and the angle of these bonds results in the polymer's helical

shape. The simplest form of starch, amylose, is unbranched. Amylopectin, a more complex form of starch, is a branched polymer.

Plants store starch as granules within cellular structures called plastids, including chloroplasts (Figure 5.7a). By synthesizing starch, the plant can stockpile its sugar. The sugar can later be withdrawn from this carbohydrate bank by hydrolysis, at which time the bonds between the glucose monomers are broken. Most animals, including humans, have enzymes that can hydrolyze plant starch, making glucose available as a

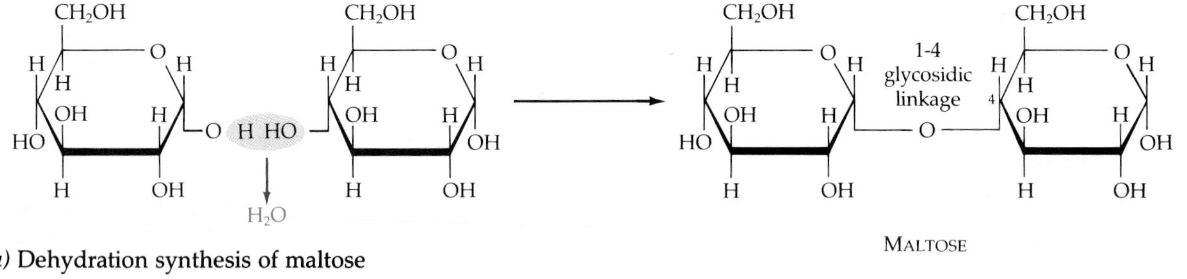

(a) Dehydration synthesis of maltose

MALTOSE

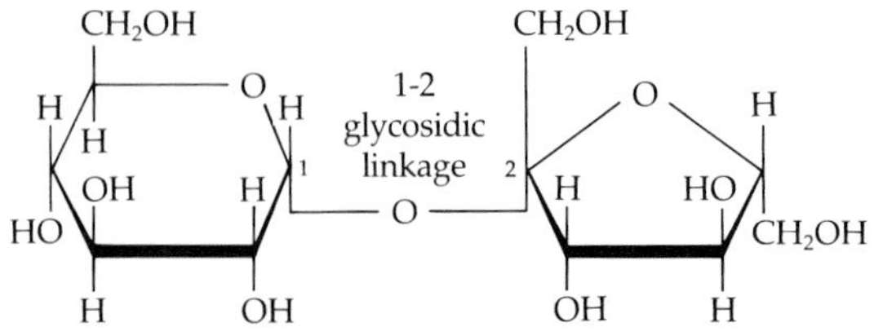

(b) Sucrose

Figure 5.5
Examples of disaccharides. (a) Dehydration synthesis of maltose from two glucose molecules. Notice that the glycosidic link joins the number one carbon of one glucose to the number four carbon of the second glucose. Joining the glucose monomers at different places would result in different disaccharides. **(b)** Sucrose, a disaccharide formed from glucose and fructose.

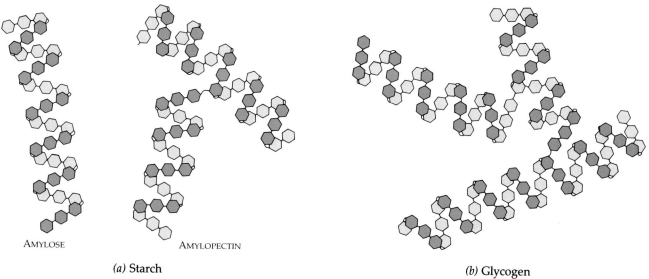

AMYLOSE   AMYLOPECTIN

(a) Starch   *(b)* Glycogen

Figure 5.6
Storage polysaccharides. The examples shown are composed entirely of glucose monomers, which are abbreviated here as hexagons. In these examples, the glucose polymers assume helical shapes. **(a)** Two forms of starch: amylose (unbranched) and amylopectin (branched). **(b)** Glycogen, which is more extensively branched than amylopectin.

nutrient for cells. Potato tubers and grains, which are the fruits of wheat, corn, rice, and other grasses, are the major sources of starch in the human diet.

Animals store a polysaccharide called **glycogen**. Glycogen is a polymer of glucose that is more exten-sively branched than the amylopectin of plants (see Figure 5.6b). Humans and other vertebrates store gly-cogen mainly in the liver and muscle cells, hydrolyz-ing the glycogen to release glucose when demand for sugar peaks (see Figure 5.7b).

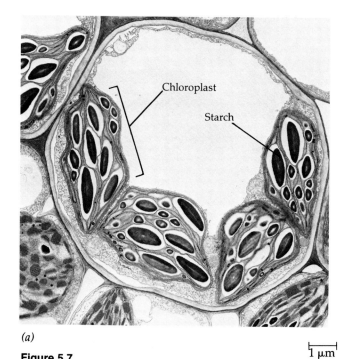

(a)

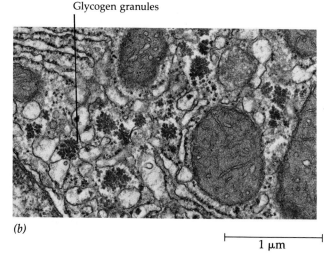

Glycogen granules

(b)

Figure 5.7
Stockpiles of storage polysaccharides.
(a) The dark ovals are granules of starch within the chloroplasts of a plant cell (TEM). (Chloroplasts are one type of plas-tid.) **(b)** Animal cells store the polysaccha-ride glycogen as dense clusters of gran-ules within liver and muscle cells. The electron micrograph shows only a small portion of a liver cell. The object in the lower right is a mitochondrion (TEM).

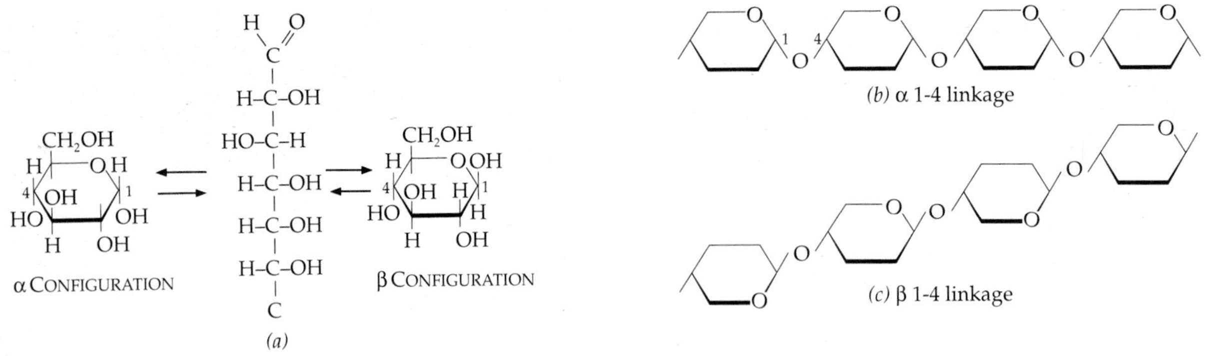

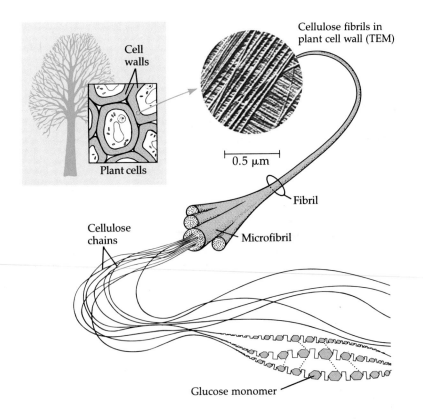

Figure 5.8
Comparison of starch and cellulose structure. (**a**) Glucose forms two interconvertible ring structures, designated α and β, which differ in the placement of the hydroxyl group attached to the number one carbon. (**b**) The α ring form is the monomer for starch. (**c**) Cellulose consists of glucose monomers in the β configuration.

Structural Polysaccharides Structural polysaccharides include cellulose and chitin. **Cellulose**, the most abundant organic compound on the Earth's surface, is a major component of the tough walls that enclose plant cells. Like starch, cellulose is a polymer of glucose, but the glycosidic linkages of these two polymers differ. The difference is based on two possible ring structures for glucose. When the carbon chain of glucose closes to form a ring, the hydroxyl group attached to the number one carbon of the ring is locked into one of two alternative positions, either below or above the plane of the ring. These two ring forms for glucose are called alpha (α) and beta (β), respectively (Figure 5.8a). In starch, all the glucose monomers are α. In contrast, the glucose monomers of cellulose are all in the β configuration (Figure 5.8c). This variation

Figure 5.9
Cellulose and plant cell walls. Cellulose is a linear, unbranched molecule. Parallel cellulose molecules are held together by hydrogen bonds (dotted lines) between hydroxyl groups projecting from both sides of cellulose. A thousand or more cellulose molecules associate to form a microfibril, and several microfibrils intertwine to form a cellulose fibril.

Figure 5.10
Chitin. A polymer of an amino sugar, chitin forms the exoskeleton of arthropods. This cicada is molting, shedding its old exoskeleton and emerging in adult form.

Figure 5.11
Lipids. Fats and oils have the property, common to all lipids, of being insoluble in water. Oils make the feather in this photo repel water, which beads on the surface.

in the geometry of the glycosidic links results in starch and cellulose having different three-dimensional shapes and therefore different properties.

In the cell wall of a plant, many parallel cellulose molecules, held together by hydrogen bonds between hydroxyl groups of the glucose monomers, are arranged as units called microfibrils (Figure 5.9). Several microfibrils intertwined form a cellulose fibril, and several fibrils may in turn be supercoiled. These strong cables are an excellent building material, both for the plant and for humans, who use cellulose-rich wood for lumber.

Enzymes that digest starch are unable to hydrolyze the β linkages of cellulose. In fact, few organisms possess enzymes that can digest cellulose. Humans do not, and the cellulose fibrils in our food pass right on through the digestive tract and are eliminated with the feces. Along the way, the fibrils abrade the wall of the digestive tract and stimulate the lining to secrete mucus, which aids in the smooth passage of food through the tract. Thus, although cellulose is inert to us as a nutrient, it is an important part of a healthful diet. Most fresh fruits, vegetables, and grains are rich in cellulose, or fiber.

Some bacteria and other microorganisms can digest cellulose to glucose. A cow harbors cellulose-digesting bacteria in the rumen, a pouch attached to the stomach. The bacteria hydrolyze the cellulose of hay and grass and convert the glucose molecules to other nutrients, which nourish the cow. Similarly, a termite, unable to digest cellulose for itself, has unicellular

organisms living in its gut that make a meal of wood. Some molds (fungi) can also digest cellulose, serving as decomposers that are crucial in the web of life on Earth (see Chapter 28).

Another important structural polysaccharide is **chitin**, the carbohydrate used by arthropods (insects, spiders, crustaceans, and related animals) to build their exoskeletons (Figure 5.10). An exoskeleton is a hard case external to the soft parts of the animal. Pure chitin is leathery, but it becomes hardened when encrusted with calcium carbonate, a salt. Chitin is also found in many fungi, which use this polysaccharide rather than cellulose as building material for their cell walls. The monomer of chitin is an amino sugar, which has a nitrogen-containing appendage.

LIPIDS

The diverse compounds called **lipids** are grouped together only because they share the one physical property of not mixing with water (Figure 5.11). Some important families of lipids are fats, phospholipids, and steroids.

Fats

Fats are large molecules constructed from two kinds of smaller molecules, glycerol and fatty acids (Figure

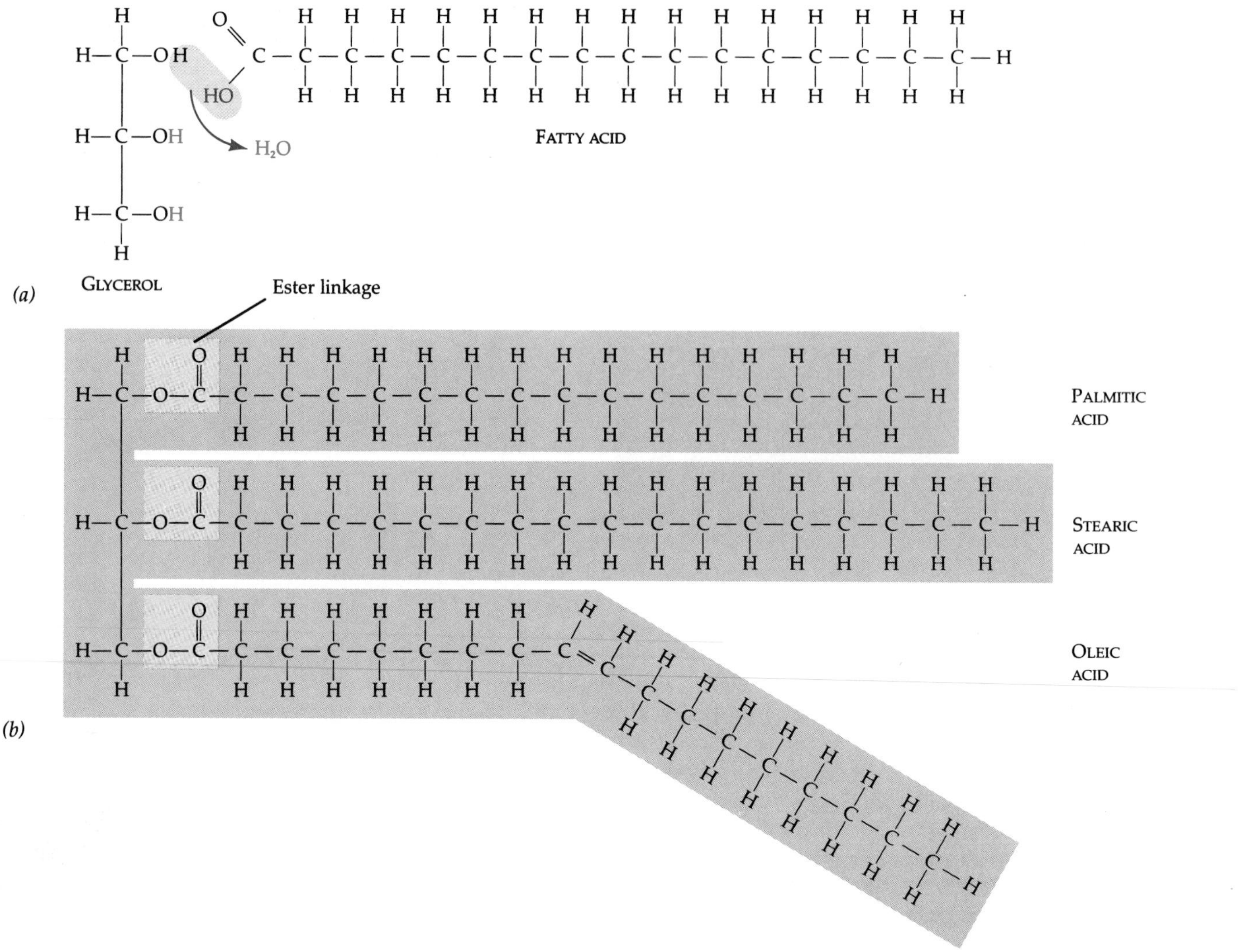

Figure 5.12
Structure of a fat, or triacylglycerol. The molecular building blocks of a fat are one molecule of glycerol and three molecules of fatty acids. Fatty acids vary in length and in the presence and locations of double bonds. *(a)* The fatty acids are joined to glycerol by dehydration synthesis, resulting in *(b)* a fat such as the one shown here. Notice the kink where the double bond is located in oleic acid, an unsaturated fatty acid.

5.12a). Glycerol is an alcohol with three carbons, each bearing a hydroxyl group. A **fatty acid** has a long carbon skeleton, most often 16 or 18 carbon atoms in length (nearly all fatty acids have an even number of carbon atoms). At one end of the fatty acid is a "head" consisting of a carboxyl group, the functional group that gives these molecules the name fatty *acids*. Attached to the carboxyl group is a long hydrocarbon "tail." With their nonpolar C–H bonds, the tails of fatty acids account for fats being insoluble in water (see Chapter 4). The hydrophobic nature of fats causes them to group together and separate from the water surrounding them. You have observed this phenomenon in the separation of vegetable oil from the aqueous vinegar solution in a bottle of salad dressing.

With the help of an enzyme, dehydration synthesis links a fatty acid to glycerol by an ester linkage, the term for a bond between a hydroxyl group and a carboxyl group. Glycerol now has two remaining hydroxyls, and each can also bond to a fatty acid. The product is a fat, or **triacylglycerol**, which consists of three fatty acids linked to one glycerol molecule. All the fatty acids in a fat can be the same, or they can be of two or three different kinds (Figure 5.12b).

Fatty acids vary in length and in the number and locations of double bonds. You have probably heard the terms *saturated fats* and *unsaturated fats* in the context of nutrition. These terms derive from the structure of the hydrocarbon tails of the fatty acids. If there are no double bonds between the carbon atoms composing the tail, then the carbon skeleton is bonded to the maximum number of hydrogen atoms. Such a fatty acid is said to be **saturated** (with hydrogen). An **unsaturated** fatty acid has one or more double bonds, formed

Figure 5.13
Adipose cells. Fat is stored as large droplets within the cells; a droplet takes up most of each cell. Stored fat functions as reserve fuel, protective padding, and thermal insulation (LM).

50 μm

by removal of hydrogen atoms from the carbon skeleton. Wherever a double bond occurs, the fatty acid will have a kink in its shape at that location, as you can see in the oleic acid molecule in Figure 5.12b.

Most fats present in animals are saturated, having fatty acids that lack double bonds. These animal fats solidify at room temperature; bacon grease, lard, and butter are examples. In contrast, plant fats are generally unsaturated. Usually liquid at room temperature, plant fats are referred to as oils—for instance, corn oil, peanut oil, and olive oil. The kinks where the double bonds are located prevent the molecules from packing together closely enough to solidify at room temperature. When you read "hydrogenated vegetable oils" on a food label, it means that unsaturated fats have been synthetically converted to saturated fats by adding hydrogen. Peanut butter, margarine, and many other products are hydrogenated to prevent lipids from separating out in liquid (oil) form.

A diet rich in animal fats is one of several factors that may contribute to the cardiovascular disease known as atherosclerosis. In this condition, deposits called plaques develop on the internal lining of blood vessels, impeding blood flow and reducing the resilience of the vessels (see Chapter 38).

Fat has come to have such negative connotations in our society that you may wonder whether fats serve any useful function. The major function of fats is energy storage. The hydrocarbons of fats are similar to gasoline molecules and just as rich in energy. A gram of fat stores more than twice as much energy as a gram of a polysaccharide such as starch. Because plants are relatively immobile, they can function with bulky energy storage in the form of starch. (Vegetable oils

are generally obtained from seeds, where more compact storage is an asset to the plant.) Animals, on the other hand, must carry their energy baggage with them, so there is an advantage to a more compact reservoir of fuel—fat. Humans and other mammals stock their food reserves in adipose cells, which swell and shrink as fat is deposited and withdrawn from storage (Figure 5.13). In addition to storing energy, adipose tissue also cushions vital organs such as the kidneys, and a layer of fat beneath the skin insulates the body. This subcutaneous (below skin) layer is especially thick in whales, seals, and other marine animals.

Phospholipids

Phospholipids are structurally related to fats, but they have only two fatty acids rather than three (Figure 5.14). The third carbon of glycerol is joined not to a fatty acid, but to a phosphate group, which is negative in charge. Additional small molecules, usually charged or polar, can be linked to the phosphate group to form a variety of phospholipids.

Phospholipids show ambivalent behavior toward water. The tails of the molecule, consisting of hydrocarbons, are hydrophobic and are excluded from water. However, the phosphate group and its attachments form a hydrophilic head that mixes with water.

The structure of phospholipids is well suited to their function as a major constituent of cell membranes. At the surface of a cell, phospholipids are arranged in a bilayer, or double layer (Figure 5.15). The hydrophilic heads of the molecules are on the outside of the bilayer, in contact with the aqueous solutions inside and outside the cell. The hydrophobic tails point toward the interior of the membrane; hydrophobic interactions between the hydrocarbons help to hold the molecules of the membrane together, forming a boundary between the cell and its external environment. (You will learn much more about the structure and function of membranes in Chapter 8.)

Steroids

Steroids are lipids characterized by a carbon skeleton consisting of four fused rings (Figure 5.16). Different steroids vary in the functional groups attached to this ensemble of rings. An important steroid is **cholesterol**, a common component of the membranes of animal cells. Cholesterol is also the precursor from which most other steroids are synthesized. For example, many hormones, including the sex hormones of vertebrates, are steroids modified from cholesterol (see Figure 4.7). Thus, cholesterol has important functions in animals, although a high concentration of cholesterol in the blood may contribute to atherosclerosis.

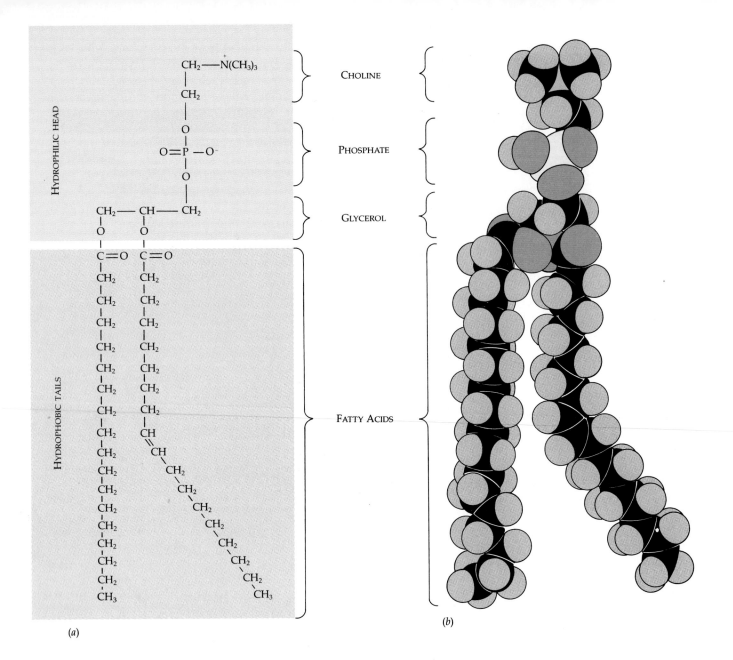

HYDROPHILIC HEAD

CHOLINE

PHOSPHATE

GLYCEROL

FATTY ACIDS

HYDROPHOBIC TAILS

CH_2—$N(CH_3)_3^+$

CH_2

O

O=P—O⁻

O

CH_2—CH—CH_2

(a)

(b)

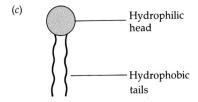

(c)

Hydrophilic head

Hydrophobic tails

Figure 5.14
Structure of a phospholipid. Phospholipid diversity is based on differences in the fatty acids and in the groups attached to the phosphate. This particular phospholipid is named phosphatidyl choline. (**a**) Structural formula, (**b**) Space-filling model. (**c**) Phospholipid symbol that will appear in several chapters of this text.

In addition to fats, phospholipids, and steroids, other families of lipids include waxes and certain pigments in plants and animals.

PROTEINS

The importance of proteins is implied by their name, which comes from the Greek word *proteios*, meaning

"first place." **Proteins** account for more than 50% of the dry weight of most cells, and they are instrumental in almost everything cells do. Proteins are used for structural support, storage, transport of other substances, signaling from one part of the organism to another, movement, defense against foreign substances, and, as enzymes, selective acceleration of chemical reactions in the cell. A human has tens of thousands of different kinds of proteins, each with a

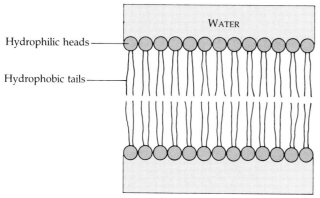

Figure 5.15
Phospholipid bilayer. Such bilayers are the main fabric of biological membranes. The inward-pointing hydrophobic tails of the phospholipids help hold the membrane together; the hydrophilic heads (spheres) of the phospholipids are in contact with water. The diagram shows a cross section through a bilayer.

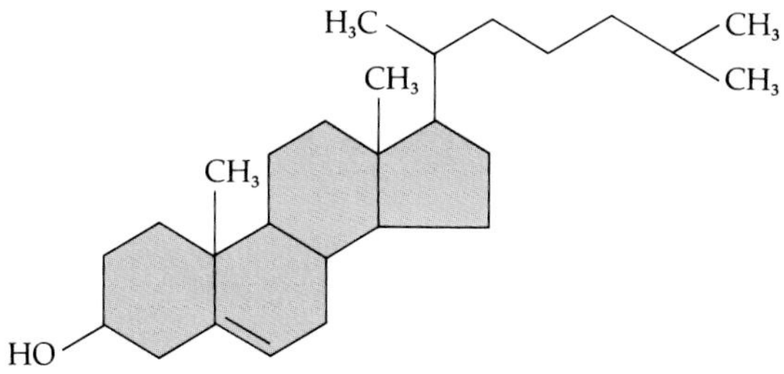

Figure 5.16
A steroid. Cholesterol is the steroid precursor from which other steroids, including the sex hormones, are synthesized. Steroids vary in the functional groups attached to four fused rings (shown in color).

specific structure and function. Table 5.2 summarizes the various types of protein functions.

Proteins are the most structurally sophisticated molecules known. Consistent with their diverse functions, they vary extensively in structure, each type of protein having a unique three-dimensional shape. But as diverse as proteins are, they are all polymers constructed from a universal set of 20 monomers, the amino acids.

Amino Acids

Most amino acids consist of an asymmetric carbon, termed the alpha (α) carbon, bonded to four different covalent partners. Each amino acid has a hydrogen atom, a carboxyl group, and an amino group bonded to the α carbon; the 20 kinds of amino acids that make up proteins differ only in what is attached to the fourth bond of the α carbon (Figure 5.17). This variable part of the amino acid is symbolized by the letter R. The R group, also called the **side chain**, may be as simple as a hydrogen atom, as in the amino acid glycine, or it may be a carbon skeleton with various functional groups

Table 5.2 Survey of protein functions		
Type of Protein	**Function**	**Examples**
Structural proteins	Support	Collagen and elastin provide a fibrous framework in animal connective tissues such as tendons and ligaments. Keratin is the protein of hair, horns, feathers, quills, and other skin appendages of animals.
Storage proteins	Storage of amino acids	Ovalbumin is the protein of egg white, used as an amino acid source for the developing embryo. Casein, the protein of milk, is the major source of amino acids for baby mammals. Plants store proteins in seeds.
Transport proteins	Transport of other substances	Hemoglobin, the iron-containing protein of blood, transports oxygen from the lungs to other parts of the body. Other proteins transport molecules across cell membranes.
Hormonal proteins	Coordination of bodily activities	Insulin, a hormone secreted by the pancreas, helps regulate the concentration of sugar in the blood.
Contractile proteins	Movement	Actin and myosin are responsible for the movement of muscles. Contractile proteins are responsible for the undulations of cilia and flagella, which propel many cells.
Antibodies	Defense	Antibodies combat bacteria, viruses, and other foreign substances in an organism.
Enzymes	Aid in chemical reactions	Enzymes regulate the chemistry of cells by selectively speeding up chemical reactions.

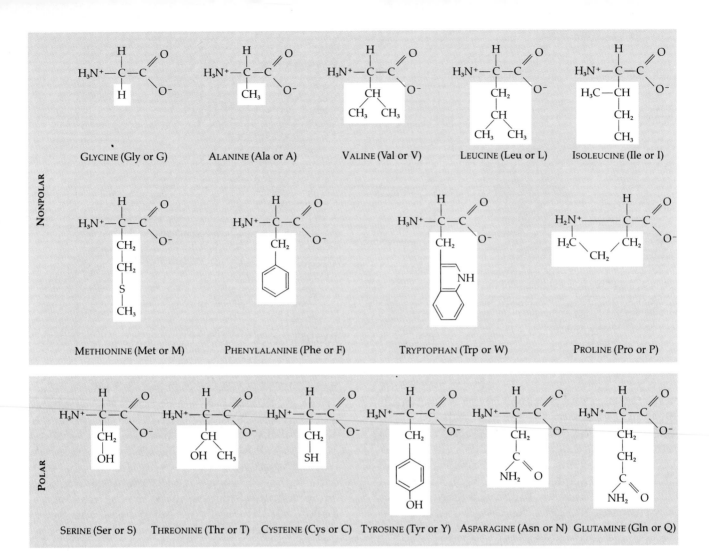

GLYCINE (Gly or G)   ALANINE (Ala or A)   VALINE (Val or V)   LEUCINE (Leu or L)   ISOLEUCINE (Ile or I)

METHIONINE (Met or M)   PHENYLALANINE (Phe or F)   TRYPTOPHAN (Trp or W)   PROLINE (Pro or P)

SERINE (Ser or S)   THREONINE (Thr or T)   CYSTEINE (Cys or C)   TYROSINE (Tyr or Y)   ASPARAGINE (Asn or N)   GLUTAMINE (Gln or Q)

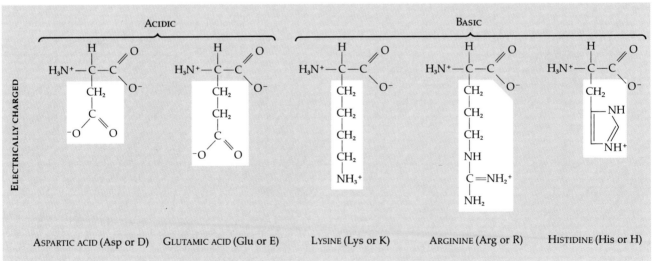

ASPARTIC ACID (Asp or D)   GLUTAMIC ACID (Glu or E)   LYSINE (Lys or K)   ARGININE (Arg or R)   HISTIDINE (His or H)

Figure 5.17
The 20 amino acids. The amino acids are grouped here according to the properties of
their side chains (R groups). The amino acids are shown in their prevailing ionic forms at
pH 7, the pH of the cell. In parentheses are the three-letter abbreviations and the one-
letter symbol for the amino acids.

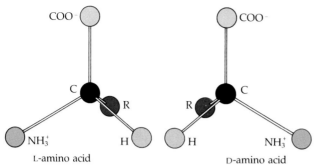

Figure 5.18
Optical isomers of amino acids. The L- and D-amino acids are mirror images of each other. Except for a few rare cases, only the L-amino acids are used by cells to make proteins.

attached, as in glutamine. The physical and chemical properties of the side chain determine the unique characteristics of an amino acid.

In Figure 5.17, the amino acids are grouped according to the properties of their side chains. One group consists of amino acids with nonpolar side chains, which are hydrophobic. Another group comprises amino acids with polar side chains, which are hydrophilic. Acidic amino acids are those with side chains that are generally negative in charge owing to the presence of a carboxyl group, which is usually dissociated at cellular pH. Basic amino acids have amino groups in their side chains, which are generally positive in charge. (Notice that *all* amino acids have carboxyl groups and amino groups bonded to the α carbon, and the terms *acidic* and *basic* in this context refer only to the nature of the side chains.) Because they are ionic, acidic and basic side chains are hydrophilic. Some proteins have rare amino acids, but these are modified from the 20 common amino acids after their incorporation into the protein. (In addition to the 20 amino acids that serve as protein monomers, there are many other amino acids that function on their own in various ways. For example, some of the substances that transmit signals in the brain are amino acids.)

Because the α carbon is asymmetric, each amino acid can exist as two optical isomers, or mirror images (Figure 5.18). The two forms are designated L-amino acid (for the Latin *laevus*, "left") and D-amino acid (for the Latin *dexter*, "right"). When an organic chemist synthesizes amino acids in a test tube, a mixture of the L- and D-isomers is made. However, with very few exceptions, only the L-form of amino acids occurs in proteins (see interview on page 113).

Now that we have examined amino acids, the next step is to see how they are linked together to form polymers.

Polypeptide Chains

When two amino acids are arranged so that the carboxyl group of one is adjacent to the amino group of the other, dehydration synthesis will join the amino acids by a linkage called a **peptide bond** (Figure 5.19). A polymer of many amino acids linked by peptide bonds is referred to as a **polypeptide chain**. At one end of the chain is a free amino group, and at the opposite end is a free carboxyl group. Thus, the chain has a polarity, with an N-terminus (for the nitrogen of the amino group) and a C-terminus (for the carbon of the carboxyl group). All other amino and carboxyl groups of the monomers (except those of side chains) are tied up in the formation of the peptide bonds. The repeating sequence of atoms along the chain (–N–C–C–N–C–C–) is referred to as the polypeptide backbone. Attached to this repetitive backbone are different kinds of appendages, the side chains of the amino acids. Polypeptides range in length from a few monomers to a thousand or more. Each specific type of polypeptide has a unique linear sequence of amino acids.

Protein Conformation

A protein consists of one or more polypeptide chains twisted, wound, and folded upon themselves to form a macromolecule with a definite three-dimensional shape, or **conformation**. A protein's function depends on its unique conformation, which is a consequence of the specific linear sequence of the amino acids that make up the polypeptide chain. Most structural proteins are fibrous in shape, whereas enzymes and many other proteins are generally globular (roughly spherical).

In almost every case, the function of a protein depends on its ability to recognize and bind to some other molecule. For instance, a hormonal protein binds to a cell receptor, an antibody binds to a particular foreign substance that has invaded the body, and an enzyme recognizes and binds to its substrate, the substance the enzyme works on. It is the unique shape of a protein—its conformation—that enables that protein to bind specifically to another molecule.

When a cell synthesizes a polypeptide, the chain folds spontaneously to assume the functional conformation for that protein—its native conformation (Figure 5.20). Amino acids with hydrophobic side chains concentrate at the core of the protein, out of contact with water. Regions of the polypeptide where amino acids with hydrophilic side chains predominate settle at the surface of the protein, in touch with water. Initially, then, the tendency for parts of the polypeptide chain to seek or avoid water is one factor in shaping the protein. As this occurs, the protein's conformation

Figure 5.19
Polypeptide chains. (a) Peptide bonds formed by dehydration synthesis link the carboxyl group of one amino acid to the amino group of the next. **(b)** The polypeptide has a repetitive backbone (gray) with various kinds of appendages, the amino acid side chains (yellow).

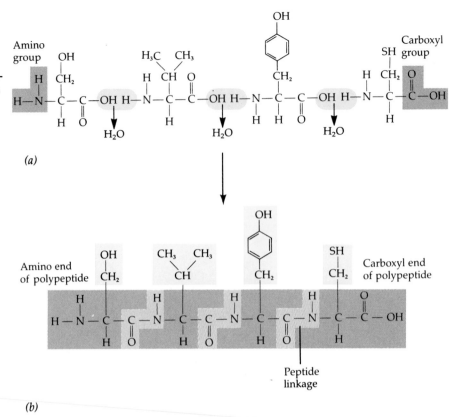

(a)

(b)

is reinforced by a variety of chemical bonds between parts of the chain.

Levels of Protein Structure

In the complex conformation of a protein, we can recognize three superimposed levels of architecture, known as primary, secondary, and tertiary structure. A fourth level, quaternary structure, occurs when a protein consists of two or more polypeptide chains.

Primary Structure   The **primary structure** of a protein is its unique sequence of amino acids. For example, the polypeptide chain of lysozyme, a relatively small enzymatic protein, is 129 amino acids long (Fig-

Figure 5.20
The native conformation of a protein. The polypeptide chain folds spontaneously into a specific shape, which is stabilized by chemical bonds between neighboring regions of the folded protein. The protein shown here is lysozyme, an enzyme. Its structure is simplified by showing only the pattern of folding of the polypeptide backbone. The yellow lines symbolize one type of chemical bond that stabilizes the conformation.

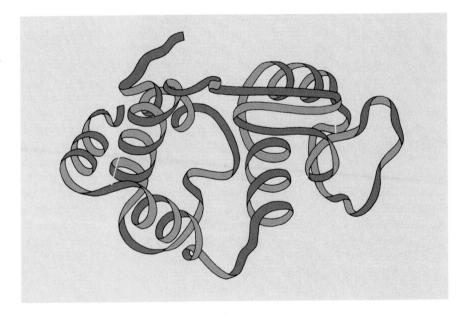

ure 5.21). Each of those 129 positions along the polymer is occupied by a specific one of the 20 amino acids. The primary structure is like the order of letters in a very long word. If left to chance, there would be 20^{129} different ways of arranging amino acids into a polypeptide chain of this length. Actually, the precise primary structure of a protein is determined not by random linking of amino acids, but by inherited genetic information.

Even a slight change in primary structure can affect a protein's conformation and ability to function. For instance, sickle-cell anemia is an inherited blood disorder in which one amino acid is substituted for another in a single position in the primary structure of hemoglobin, the protein that carries oxygen in red blood cells.

Molecular biologists have ascertained the primary structures of hundreds of proteins. The pioneer in this work was Frederick Sanger, who, with his colleagues at Cambridge University in England, determined the sequence of the hormone insulin in the late 1940s and early 50s. Insulin is a small protein consisting of two polypeptide chains joined by covalent bonds. One chain has 30 amino acids, and the other chain is only 21 amino acids long. Even in so small a protein, however, it was a laborious task to determine the precise place of each amino acid. To find out the overall composition of insulin, Sanger hydrolyzed the protein to its component amino acids with acid and then used chromatography (see Chapter 4 Methods Box) to separate the amino acids and measure the quantity of each in insulin. The next stage of the research was to decipher the sequential arrangement of these amino acids in the polypeptide chains. The approach was to use protein-digesting enzymes and other catalysts that break polypeptides at specific places rather than completely hydrolyzing the chain. Treatment with one of these agents would cleave the polypeptide into fragments that could be separated by chromatography. Hydrolysis with another agent would break the polypeptide at different sites, yielding a second group of fragments. Sanger used chemical methods to determine the sequence of amino acids in these small fragments, and then he searched for overlapping regions among the pieces obtained by partial hydrolysis. Consider, for instance, two fragments with the following sequences:

Cys-Ser-Leu-Tyr-Gln-Leu
Tyr-Gln-Leu-Glu-Asn

We can deduce from the overlapping regions that the intact polypeptide contains in its primary structure the segment

Cys-Ser-Leu-Tyr-Gln-Leu-Glu-Asn

Just as we could reconstruct this sentence from a collection of fragments with overlapping sequences of

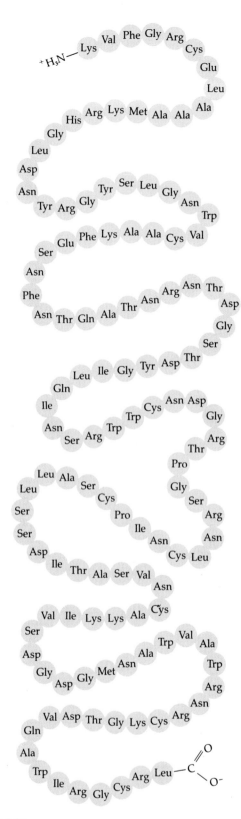

Figure 5.21
Primary structure of a protein. This is the unique amino acid sequence, or primary structure, of the enzyme lysozyme. The names of the amino acids are given as the three-letter abbreviations. (The chain was drawn in this serpentine fashion only so that it would fit on the page. The actual shape of lysozyme is depicted in Figure 5.20.)

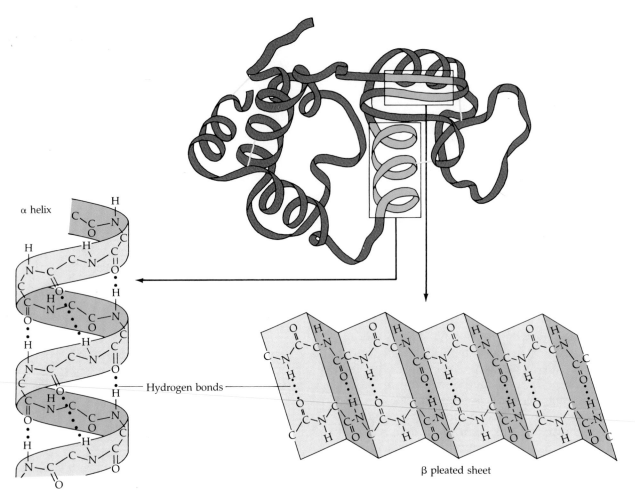

Figure 5.22
Secondary structure. The two types of secondary structure, α helix and β pleated sheet, can both be found in the protein lysozyme. Both patterns depend on hydrogen bonding along the polypeptide chain. The R groups of the amino acids are omitted in these drawings.

letters, Sanger and his co-workers were able, after years of effort, to reconstruct the complete primary structure of insulin. Since then, most of the steps involved in sequencing a polypeptide have been automated. But it was Sanger's analysis of insulin that first demonstrated what is now a fundamental axiom of molecular biology: Each type of protein has a unique primary structure, a precise sequence of amino acids.

Secondary Structure   Most proteins have segments of their polypeptide chain repeatedly coiled or folded in patterns that contribute to the protein's overall conformation. These configurations, called **secondary structure**, are due to hydrogen bonds between functional groups that repeat at regular intervals along the polypeptide backbone (Figure 5.22). Because they are so electronegative, both the oxygen and nitrogen atoms of the backbone have weak negative charge (see Chapter 2). The weakly positive hydrogen atom attached to the nitrogen has an affinity for the oxygen atom of a nearby peptide bond. Individually, these hydrogen

bonds are weak; but repeated many times over a relatively long region of the polypeptide chain, they can support a particular shape for that part of the protein. One such secondary structure is the **alpha (α) helix**, a delicate coil held together by hydrogen bonding between every fourth peptide bond. Linus Pauling and Robert Corey first described the α helix in 1951 while working on protein structure at the California Institute of Technology.

The regions of α helix in the enzyme lysozyme are evident in Figure 5.22, where one α helix is enlarged to show the hydrogen bonds. Lysozyme is fairly typical of a globular protein in having a few stretches of α helix separated by nonhelical regions. In contrast, some fibrous proteins, such as α-keratin, the structural protein of hair, have the α-helix formation over most of their entire length.

The other type of secondary structure is the **beta (β) pleated sheet**, in which the polypeptide chain folds back and forth, or where two regions of the chain parallel one another. Hydrogen bonds hold the struc-

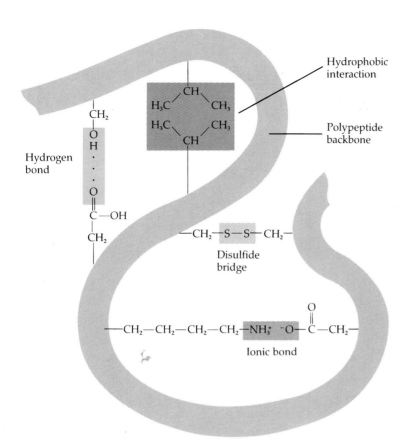

Figure 5.23
Some interactions important in tertiary structure. Hydrogen bonds, ionic bonds, and hydrophobic interactions are weak bonds between side chains that collectively hold the protein in a specific conformation. Much stronger are the disulfide bridges, covalent bonds between the side chains of cysteine pairs that are common in many proteins secreted by cells.

Hydrophobic interaction

Polypeptide backbone

Hydrogen bond

Disulfide bridge

Ionic bond

ture together. The β sheets make up the dense core of many globular proteins, and we can recognize one such region in lysozyme. Also, β sheets dominate some fibrous proteins, including fibroin, the structural protein of silk.

Tertiary Structure Superimposed on the recurrent patterns of secondary structure is a protein's **tertiary structure**, consisting of irregular contortions due to bonding between side chains (R groups) of the various amino acids. (In contrast, remember, secondary structure results from hydrogen bonds formed at regular intervals along the protein's *backbone*.) One type of bonding contributing to tertiary structure has already been mentioned—the hydrophobic interactions between nonpolar side chains in the nonaqueous interior of the protein (Figure 5.23). Hydrogen bonds between certain side chains are also important, as are ionic bonds between positively and negatively charged side chains. These are all weak interactions, but their cumulative effect helps give the protein a stable shape. The conformation of a protein may be reinforced further by strong covalent bonds called **disulfide bridges**. Disulfide bridges form where two cysteine monomers, amino acids with sulfhydryl groups on their side chains, are brought close together by the folding of the protein. The sulfur of one cysteine bonds to the sulfur of a second, and the disulfide bridge rivets parts of the protein together. Disulfide bridges are common in proteins secreted from cells, such as the insulin

secreted from the vertebrate pancreas, but rare in proteins that remain within cells. Note that all of the different kinds of bonds can occur in one protein, as shown hypothetically in Figure 5.23.

Many proteins have a tertiary structure that is modular, with two or more globular regions, called **domains**, connected by relatively flexible regions of the polypeptide chain.

Quaternary Structure As mentioned previously, some proteins consist of two or more polypeptide chains aggregated into one functional macromolecule. Each polypeptide chain is called a **subunit** of the protein. **Quaternary structure** is the structure that results from the relationship between the subunits. For example, collagen is a fibrous protein that has three subunits intertwined into a triple helix (Figure 5.24). Each polypeptide chain, as described earlier, is in the form of a helix. The supercoiled organization of collagen, similar to the design of a rope, gives the long fibers great strength, appropriate to their function as the girders of connective tissue. Hemoglobin is an example of a globular protein with quaternary structure. There are two kinds of polypeptide chains and two of each kind per hemoglobin molecule.

After reducing a protein to its different levels of structure, it is important to remember that it is the overall product, a macromolecule with a unique conformation, that works in a cell. The specific function

Figure 5.24

Quaternary structure of proteins. At this level of structure, two or more polypeptide subunits interact to form a functional protein. **(a)** Collagen is a fibrous protein consisting of three helical polypeptides, which are supercoiled to form a ropelike structure of great strength. **(b)** Hemoglobin is a globular protein with four subunits, two of one kind (α chains) and two of another kind (β chains). (Each subunit has a non-polypeptide component, called heme, with an iron atom that binds oxygen.)

(a) Collagen

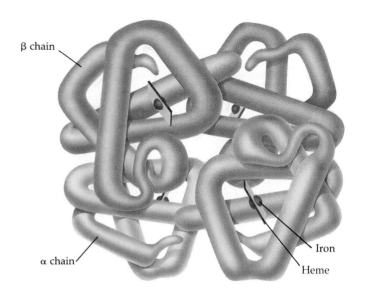

β chain

α chain

Iron

Heme

(b) Hemoglobin

of a protein is an emergent property that arises from the intricate architecture of the molecule (Figure 5.25).

What Determines Conformation?

As long as a protein is in its natural physical and chemical environment, its conformation is quite stable, and the protein will carry out its specific function. As we have seen, a polypeptide chain of given amino acid sequence will spontaneously arrange itself into a three-dimensional shape maintained by the interactions responsible for secondary and tertiary structure. However, if the pH, salt concentration, temperature, or other aspects of its environment are altered, the protein may unravel and lose its native conformation in a process called **denaturation** (Figure 5.26). Misshapen, the denatured protein no longer works.

One way to denature most proteins is to transfer them from an aqueous environment to an organic sol-

PRIMARY STRUCTURE

Ser — Tyr — Ser — Met — Glu — His — Phe — Arg — Trp — Gly — Lys — Pro — Val —

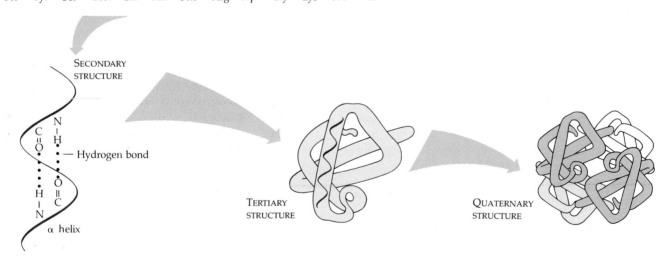

SECONDARY STRUCTURE

— Hydrogen bond

α helix

TERTIARY STRUCTURE

QUATERNARY STRUCTURE

Figure 5.25

Summary of the levels of protein structure. Primary structure is the sequence of covalently joined amino acids in a polypeptide. Secondary structure is the bending and hydrogen bonding of a polypeptide backbone to form α helices and β sheets. Tertiary structure is the overall conformation (shape) of a polypeptide, as reinforced by interactions between the R groups of amino acids. Quaternary structure is the relationship between several polypeptides that make up a protein.

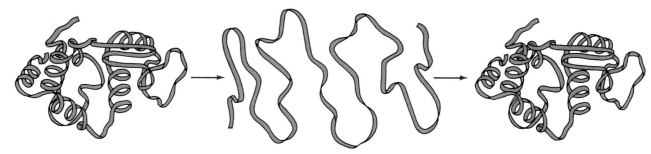

Figure 5.26
Denaturation and renaturation of a protein. High temperature or various chemical treatments will denature a protein, causing it to lose its conformation and ability to function. If the denatured protein remains dissolved, it may renature when the environment is restored to normal.

vent, such as ether or chloroform. The protein turns inside out, the hydrophobic regions changing places with the hydrophilic portions of the protein. Chemical agents that disrupt the hydrogen bonds, ionic bonds, and disulfide bridges that shape a protein also denature the molecule. Denaturation can also result from excessive heat, which agitates the polypeptide chain enough to overpower the weak interactions that stabilize conformation. The white of an egg becomes opaque during cooking because the denatured proteins are insoluble and solidify.

When a protein is denatured without being made to precipitate out of solution, it may re-form to its original shape when returned to its normal environment. We can conclude that the information for building specific shape is intrinsic in the protein's primary structure. The sequence of amino acids determines conformation—where an α helix can form, where β sheets can occur, where disulfide bridges are located, and so on. In the future, there may be computer programs that predict the conformation of any protein of known primary structure. Before such programs can be written, however, much more must be learned about the various interactions that contribute to conformation.

The Protein-Folding Problem

Molecular biologists have determined the amino acid sequences of hundreds of proteins, and the three-dimensional shapes of many of those proteins are also known (see Methods Box). One would think that by correlating the primary structures of many proteins with their conformations, especially with the help of computers, it would be possible to discover the rules of protein folding. Unfortunately, the *protein-folding problem* is not so simple. One complication is that most proteins probably go through several intermediate states on their way to a stable conformation, and looking at the "mature" conformation does not reveal the

stages of folding required to achieve that form. As Thomas Creighton, a British scientist who works on the problem, puts it: "It is like folding the flaps of a cardboard box. To fold them together, you have to put the flaps in a specific order and then distort them." Complicating the problem further, molecular biologists now realize, is the fact that a protein's conformation is much more dynamic than previously imagined. Even after a newly made protein has performed the sequence of folding steps to achieve a stable shape, it is not locked into that particular conformation, but flip-flops between several alternative conformations. Thus, the protein-folding problem is extremely challenging—and important. Once the rules of protein folding are known, it should be possible to design proteins that will carry out specific tasks by making polypeptide chains with appropriate amino acid sequences.

NUCLEIC ACIDS

If primary structure determines the conformation of a protein, then what determines primary structure? As stated earlier, the amino acid sequence is programmed by a unit of inheritance known as a gene. A gene, in turn, is part of a polymer belonging to the class of compounds known as nucleic acids.

Functions of Nucleic Acids: An Overview

There are two types of **nucleic acids: deoxyribonucleic acid (DNA)** and **ribonucleic acid (RNA)**. They are the molecules that enable living organisms to reproduce their complex equipment from one generation to the next. Unique among molecules, DNA can replicate itself, and it is this molecular reproduction that is the basis for the continuity of life.

Three-dimensional structures of biological macromolecules provide important insights into molecular function. Determining the structures of macromolecules as complex as proteins, each made up of a unique combination and arrangement of thousands of atoms, is a formidable task. Early on, the building of molecular models proved a profitable approach, leading Pauling and Corey to the discovery of basic principles of protein structure (see Figure 5.1) and Watson and Crick to the DNA double helix (see Figure 15.8). The cardboard, wood, and wire used in the early days eventually gave way to commercially available plastic model sets, but model building by hand remained extremely laborious.

The data on which a macromolecular model is based come from two kinds of experiments—chemical and physical. Using chemical methods, the primary structure of the macromolecule is determined—that is, the monomers that compose the molecule and how they are joined by covalent bonds. For protein, this information is most conveniently determined as the sequence of amino acids, for a nucleic acid, as the sequence of nucleotides.

From physical methods comes information on the overall shape of the macromolecule. The most powerful physical method is X-ray crystallography. As shown in the diagram, when an X-ray beam passes through a crystal of a particular substance, the regularly spaced atoms of the crystal diffract (deflect) the X-rays into an orderly array. The diffracted X-rays can expose photographic film, producing a pattern of spots. Using complicated mathematical equations, a crystallographer translates the locations and intensities of the spots into information about the coordinates (positions) of atoms in three-dimensional space (Figure a).

As early as the 1950s, crystallographers started using computers to help with the task of solving the equations, but it was decades later before computers came to be used for the actual model building. The development of modern computers and software has made it possible to "build" models much more quickly and accurately than was previously possible. Now the computer does most of the tedious work.

In the photographs in this box (from the Department of Biochemistry at the University of California, Riverside), we follow the development of a computer model for the structure of an enzymatic protein called ribonuclease, whose function involves binding to a nucleic acid molecule. The first step is to crystallize the protein, in this case the protein

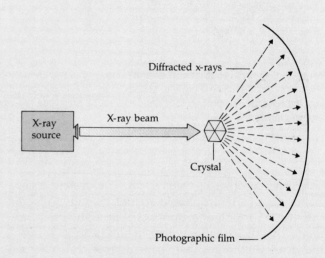

(a) X-ray crystallography

combined with a short strand of nucleic acid (Figure *b*).

Next, X-ray crystallography produces raw data in the form of dot patterns such as the one shown here (Figure *c*).

From such diffraction patterns, computer programs generate electron density maps of successive, cross-sectional slices through the protein (Figure *d*).

When the many electron density maps are all superimposed, the resulting structure reflects the overall shape of the protein. Here we see a low-resolution model such as might be produced at very early stages of analysis (Figure *e*). Before the era of computer graphics, these were constructed from layers of wood.

Combining the information from electron density maps with the known primary structure of the protein, the coordinates of each atom are determined and the graphics software generates a picture showing the position of each atom in the molecule (Figure *f*).

The macromolecule can be represented in a number of ways. Two of the most common are *stick* models, in which lines represent the covalent bonds in the polypeptide backbone, and *space-filling* models, in which each atom is represented by a sphere (Figure *g*).

With the appropriate equipment and software, the computer can move the image around on the screen to simulate the molecule's appearance from various angles—even from within the molecule. Thus, the advantages of being able to manipulate the model by hand are retained and even extended.

(b) Protein crystal

0.5 μm

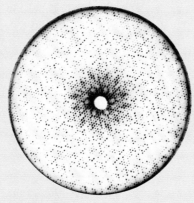

(c) X-ray diffraction pattern from the crystal of a protein

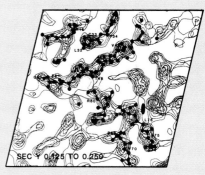

(d) Electron density map

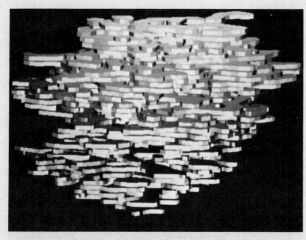

(e) Low-resolution model of a protein

(f) Molecular biologist at a computer monitor

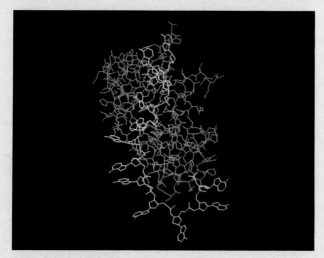

(g) Computer graphic models of the protein ribonuclease (purple) bound to a short strand of nucleic acid (green), shown in stick (left) and space-filling (right) representations

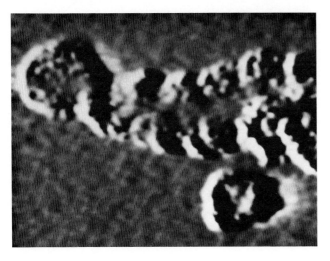

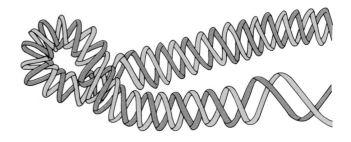

|—————| 3.5 nm

Figure 5.27
DNA, the genetic material. This first photograph ever to show the double-helical structure of DNA was taken with a powerful instrument called a scanning tunneling microscope. The drawing will help you interpret the micrograph, which has a magnification of about a million.

The genetic material that organisms inherit from their parents consists of DNA (Figure 5.27). Each time a cell reproduces itself by dividing, its DNA is copied and passed along from one generation of cells to the next. Written in the structure of DNA are the instructions that program all of the cell's activities. The DNA, however, is not directly involved in running the operations of the cell, any more than computer software by itself can print a bank statement or read one of those parallel-lined product codes on a box of cereal. Just as a printer is needed to print out the statement and a scanner is needed to read the product code, so, too, are proteins needed to enact genetic programs; proteins are the molecular hardware of the cell. It is hemoglobin that carries oxygen in the blood, not the DNA that specifies the structure of hemoglobin.

The other type of nucleic acid, RNA, functions in the actual synthesis of the proteins specified by DNA. We will reserve the details of this chain of command for Chapter 16. Here, we will concentrate on the structure of the nucleic acids.

Nucleotides

Nucleic acids are polymers of monomers called **nucleotides**, which enzymes link together by dehydration synthesis. Each nucleotide is itself composed of three parts: a nitrogenous base, which is joined to a pentose (five-carbon sugar), which in turn is bonded to a phosphate group (Figure 5.28).

There are two families of nitrogenous bases: pyrimidines and purines. A **pyrimidine** is characterized by a six-membered ring made up of carbon and nitrogen atoms. (The nitrogen tends to take up hydrogen ions from the solution, which explains the term "nitrogenous *base*.") The members of the pyrimidine family are cytosine (C), thymine (T), and uracil (U). In the second family, the **purines**, a five-membered ring is fused to the pyrimidine type of ring. The purines are adenine (A) and guanine (G). The specific kinds of pyrimidines and purines differ in the functional groups attached to the rings. Note in Table 5.3 that thymine is found only in DNA, and uracil only in RNA.

The pentose connected to the nitrogenous base is **ribose** in the nucleotides of RNA, and **deoxyribose** in DNA. The only difference between these two sugars is that deoxyribose lacks a hydroxyl group at the number two carbon. So far, we have built a **nucleoside**, a molecule consisting of a nitrogenous base joined to a sugar. To complete the construction of a nucleotide, we attach a phosphate group to the number five carbon of the sugar. The molecule is now a nucleoside monophosphate, better known as a nucleotide.

In addition to serving as the monomers for nucleic acids, nucleotides also perform important functions as individual molecules. One of the most important is adenosine triphosphate, or ATP, which functions in the transfer of chemical energy from one molecule to another in cellular processes (see Chapter 6).

Polynucleotides

A nucleic acid is a **polynucleotide**, a term that emphasizes the way these macromolecules are constructed.

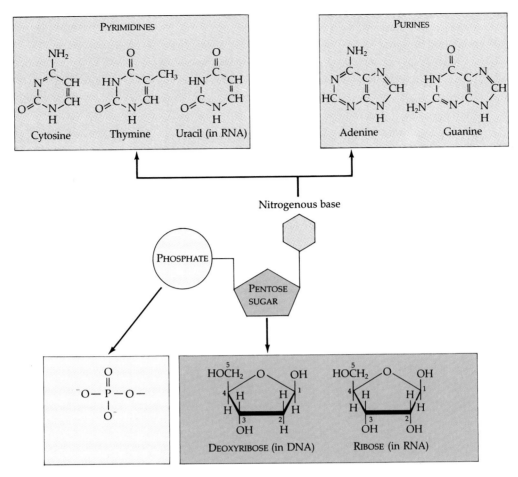

Figure 5.28
The structure of a nucleotide. These monomers of nucleic acids are themselves composed of three smaller molecular building blocks: a nitrogenous base, either a purine or a pyrimidine; a pentose sugar; and a phosphate group.

In a polynucleotide, the monomers are joined by covalent bonds called **phosphodiester linkages** between the phosphate of one nucleotide and the sugar of the next monomer (Figure 5.29). This results in a backbone with a repeating pattern of sugar–phosphate–sugar–phosphate. All along this sugar–phosphate backbone are appendages consisting of the nitrogenous bases. Unlike the regular backbone, the bases are variable.

The Double Helix: An Introduction

The DNA molecules of cells actually consist of two polynucleotide chains spiraled around an imaginary axis to form a **double helix**. (You have probably noticed by now that helices are very common shapes for macromolecules. Starch is a helix, many proteins have regions of α helix, and now we see that DNA has the shape of a double helix.) James Watson and Francis Crick, working at Cambridge University, first proposed the double helix as the three-dimensional structure of DNA in 1953 (Figure 5.30). The two sugar–phosphate backbones are on the outside of the helix, and the nitrogenous bases are paired in the inte-

Table 5.3 Chemical comparison of DNA and RNA		
	DNA	**RNA**
Purines	Adenine (A)	A
	Guanine (G)	G
Pyrimidines	Cytosine (C)	C
	Thymine (T)	Uracil (U)
Pentose	Deoxyribose	Ribose

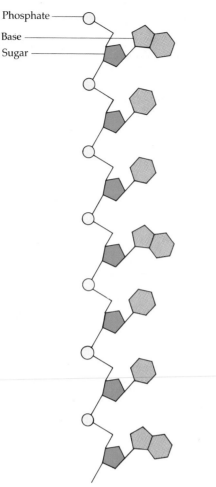

Phosphate

Base

Sugar

Figure 5.29
A polynucleotide. Each nucleotide monomer has its phosphate group bonded to the sugar of the next nucleotide. The polymer has a regular sugar–phosphate backbone with variable appendages, the four kinds of nitrogenous bases.

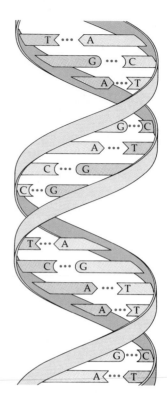

Figure 5.30
The double helix. The DNA molecule is usually double-stranded, with the sugar–phosphate backbone of the polynucleotides on the outside of the helix. In the interior are pairs of nitrogenous bases, holding the two strands together by hydrogen bonds. Hydrogen bonding between the bases is specific: The base adenine (A) can pair only with thymine (T), and guanine (G) can pair only with cytosine (C).

rior of the helix. The two polynucleotide chains, or strands, as they are called, are held together by hydrogen bonds between the paired bases. Each bond is weak, but they have a collective strength like that of the teeth of a zipper. Most DNA molecules are very long, with thousands or even millions of base pairs holding the two chains together. One long DNA molecule represents a large number of genes, each a particular segment along the double helix.

Only certain pairings of bases in the double helix are compatible. Adenine (A) always pairs with thymine (T), and guanine (G) always pairs with cytosine (C). If we were to read the sequence of bases along one strand as we traveled the length of the double helix, we would automatically know the sequence of bases along the other strand. If a stretch of one strand has the base sequence AGGTCCG, then the base-pair-

ing rules tell us that the same stretch of the other strand must have the sequence TCCAGGC. The two strands of the double helix are complementary, each the predictable counterpart of the other. It is this feature of DNA that makes possible the precise copying of genes that is responsible for inheritance. Once again, we see that the properties of life emerge from structural order.

The DNA molecule is so central to life that we will devote Chapters 15 and 16 to its structure and function.

* * *

We have concluded our survey of macromolecules, but not our study of the chemistry of life. Applying the reductionist strategy, we have examined the architecture of molecules without paying much attention to the dynamic interactions between molecules that result in the chemical changes in cells collectively referred to as metabolism. Chapter 6, the concluding chapter of this unit, takes us another step up the hierarchy of biological order by introducing the fundamental principles of metabolism.

STUDY OUTLINE

1. Cells can combine small organic molecules into large macromolecules, forming a higher level in the biological hierarchy of biological order.
2. Carbohydrates, lipids, proteins, and nucleic acids are the four major classes of organic compounds in cells.

Polymers (pp. 64–66)

1. Macromolecules are polymers, chains of identical or similar subunit molecules called monomers. Although there is a limited number of monomers common to all organisms, each organism is unique because of the specific arrangement of these monomers into polymers with distinctive structures and properties.
2. Monomers of all four classes of macromolecules form larger molecules by dehydration synthesis, a chemical reaction in which one monomer donates a hydroxyl group and the other a hydrogen atom, forming a water molecule.
3. Polymers can be disassembled to monomers by the reverse process, called hydrolysis. In this way, large macromolecules in food are digested into monomers small enough to enter our cells.

Carbohydrates (pp. 66–71)

1. Carbohydrates are sugars and their derivatives.
2. Monosaccharides are the simplest carbohydrates, used directly for fuel, converted to other types of organic molecules, or used as monomer units for carbohydrate polymers.
3. All monosaccharides possess a carbon skeleton of three to seven carbons, all but one of which are bonded to a hydroxyl group. The remaining carbon is part of a carbonyl group, and depending on the location of this group, the monosaccharide is either a ketone or an aldehyde.
4. In aqueous solution, most monosaccharides form rings.
5. Disaccharides consist of two monosaccharide monomers connected by a glycosidic bond. The monosaccharides can be the same or different.
6. Polysaccharides may consist of thousands of monosaccharide monomers connected by glycosidic bonds. Starch in plants and glycogen in animals are both storage polymers of glucose. Cellulose is an important structural polysaccharide in the cell walls of plants.

Lipids (pp. 71–74)

1. Lipids make up the most structurally heterogeneous class of macromolecules, but all share the property of being wholly or partly insoluble in water.
2. Fats are high-energy, compact storage molecules also known as triacylglycerols. They are constructed by joining a glycerol molecule to three fatty acids.
3. Fatty acids consist of a carboxyl group and a hydrophobic hydrocarbon tail. The carboxyl group takes part in an ester linkage with the glycerol.
4. Saturated fatty acids have the maximum number of hydrogen atoms because of single bonding between all the carbons. Unsaturated fatty acids (present in oils) have one or more double bonds between the carbons, causing kinks in the molecule and reducing the number of bonding sites for hydrogen atoms.

5. Phospholipids substitute the third fatty acid of a triacylglycerols with a negatively charged phosphate group, which may be joined, in turn, to another small molecule. Such bonding introduces polarity and hence water solubility to one end of the molecule, making phospholipids ideally suited for construction of cell membranes.
6. Steroids, such as cholesterol and the sex hormones, have a carbon skeleton composed of four fused rings, with variation in the number and type of functional groups or atoms attached.

Proteins (pp. 74–83)

1. Proteins are polymers constructed from 20 different amino acids. They are the most complex and versatile macromolecules, with emergent properties arising from their intricate architecture.
2. An amino acid is composed of a central asymmetric carbon singly bonded to a hydrogen atom, a carboxyl group, an amino group, and a variable side chain that confers unique properties on each amino acid.
3. The carboxyl and amino groups of adjacent amino acids link together in a peptide bond, forming long polymers.
4. Protein conformation can be described by three or four superimposed, hierarchical levels. Primary structure is the first level and describes the unique sequence of amino acids.
5. Secondary structure describes how the primary structure is folded into particular, localized configurations, the α helix and β pleated sheet, which result from hydrogen bonding between peptide linkages.
6. Tertiary structure describes the additional, less regular contortions of the molecule caused by the involvement of side groups in hydrophobic interactions, hydrogen bonds, ionic bonds, and covalent bonds called disulfide bridges.
7. Proteins made of more than one polypeptide chain also show a specific arrangement of their constituent subunits in a quaternary level of structure.
8. The function of a protein is an emergent property of its conformation, which is highly sensitive to conditions such as pH, salt concentration, and temperature. Changing these conditions can cause the protein to denature by altering its shape in such a way that it no longer has a biological function.
9. Protein shape is ultimately determined by its primary structure. Molecular biologists are looking for rules that will predict protein folding and final conformation from an amino acid sequence.

Nucleic Acids (pp. 83–88)

1. Nucleic acids are polymers of nucleotides, complex monomers consisting of a pentose (five-carbon sugar) covalently bonded to a phosphate group and to one of five different kinds of nitrogenous bases.
2. In the formation of a polynucleotide, the pentose of one nucleotide joins to the phosphate of another in a phosphodiester linkage, forming a sugar–phosphate backbone from which the nitrogenous bases project.
3. The two kinds of nucleic acids, deoxyribonucleic acid (DNA) and ribonucleic acid (RNA), are named on the

basis of their characteristic pentoses: deoxyribose in DNA, and ribose in RNA.

4. The five nitrogenous bases are members of two families, the purines and pyrimidines, distinctive ring skeletons of carbon and nitrogen with various attached groups.

5. The pyrimidines consist of cytosine (C), thymine (T), and uracil (U); the purines consist of adenine (A) and guanine (G).

6. RNA is a single-stranded nucleotide polymer containing the bases A, G, C, and U.

7. DNA is a helical, double-stranded polymer with bases A, G, C, and T projecting into the interior of the molecule. A always hydrogen-bonds to T, and C to G. Thus, the nucleotide sequence of the two strands is complementary, and one strand can serve as a template for the formation of the other.

8. The property of complementary strands gives DNA its unique ability to replicate itself and provides a mechanism for the continuity of life. Once replicated, specific nucleotide segments of the DNA (genes) program the manufacture of an organism's characteristic proteins. RNA functions in protein synthesis.

SELF-QUIZ

1. Which of these terms is the *most inclusive*?
   a. monosaccharide
   b. disaccharide
   c. starch
   d. carbohydrate
   e. polysaccharide

2. The molecular formula for glucose is $C_6H_{12}O_6$. What would be the molecular formula for a polymer made by linking ten glucose molecules together by dehydration synthesis?
   a. $C_{60}H_{120}O_{60}$
   b. $C_6H_{12}O_6$
   c. $C_{60}H_{102}O_{51}$
   d. $C_{60}H_{100}O_{50}$
   e. $C_{60}H_{111}O_{51}$

3. There are two ring forms of glucose (alpha and beta) because
   a. the two forms are made from two structural isomers of glucose
   b. they arise from different line formulas for glucose
   c. different carbons of the line structure join to form the rings
   d. when the ring closes, one hydroxyl group may be placed on opposite faces of the ring to produce the two forms
   e. one is an aldose and the other is a ketose

4. _____ water molecules are removed for the dehydration synthesis of one fat molecule.
   a. one
   b. two

c. three
d. six
e. cannot answer, because it depends on the size of the fat

5. Which of the following statements concerning *unsaturated* fats is correct?
   a. They are more common in animals than in plants.
   b. They have double bonds in the carbon chains of their fatty acids.
   c. They generally solidify at room temperature.
   d. They contain more hydrogen than do saturated fats having the same number of carbon atoms.
   e. They have fewer fatty acid molecules per fat molecule.

6. Human sex hormones are classified as
   a. proteins
   b. lipids
   c. amino acids
   d. triacylglycerols
   e. carbohydrates

7. For a protein to have a quaternary structure, it *must*
   a. have four domains
   b. consist of two or more polypeptide subunits
   c. consist of four polypeptide subunits
   d. have at least four disulfide bridges
   e. exist in several alternative conformational states

8. What does a protein lose when it denatures?
   a. its primary structure
   b. its three-dimensional shape
   c. its peptide bonds
   d. its sequence of amino acids

9. Which of the following is a complication contributing to the difficulty of the protein-folding problem?
   a. A specific protein has several alternative amino acid sequences.
   b. There are no methods for revealing the three-dimensional shape of a protein.
   c. The same primary structure may yield several equally probable conformations.
   d. It is impossible to determine the precise primary structure of a protein.
   e. One must identify the gene that codes for a particular protein before the protein's folding pattern can be determined.

10. Which of these terms is the *most inclusive*?
    a. nucleoside
    b. nucleotide
    c. nitrogenous base
    d. purine
    e. pyrimidine

CHALLENGE QUESTIONS

1. Compare and contrast starch and cellulose in terms of their structures and functions.

2. Proteins have charged regions due to some amino acid side chains containing carboxyl and amino groups. Explain why the carboxyl and amino groups attached directly to the alpha carbons of amino acids do *not* contribute electrical charges to the proteins (except at the N and C termini).

3. A particular small polypeptide is nine amino acids long. Using three different enzymes to hydrolyze the polypeptide at various sites, the following five fragments are obtained (the N denotes the amino terminus of the chain): Ala-Leu-Asp-Tyr-Val-Leu; Tyr-Val-Leu; N-Gly-Pro-Leu; Asp-Tyr-Val-Leu; N-Gly-Pro-Leu-Ala-Leu. Determine the primary structure of this polypeptide.

FURTHER READING

1. Dickerson, R. E., and I. Geis. *The Structure and Action of Proteins.* Menlo Park, Calif.: Benjamin/Cummings, 1969. A classic, brief text on protein structure, with excellent illustrations.

2. Doolittle, R. F. "Proteins." *Scientific American,* October 1985. Recommended reading for understanding protein structure.

3. Dushesne, L. C. and D. W. Larson. "Cellulose and the Evolution of Plant Life." *Bioscience,* April 1989. Chemistry and natural history of the most abundant organic molecule in the biosphere.

4. Mathews, C. and van Holde, K. *Biochemistry.* Redwood City, CA: Benjamin/Cummings, 1990. Excellent explanations and beautiful art.

5. Sharon, N. "Carbohydrates." *Scientific American,* November 1980. The roles of carbohydrates in organisms.

6. Thompson, E.O.P. "The Insulin Molecule." *Scientific American,* May 1955. The story of how Frederick Sanger and his colleagues first worked out the total chemical structure of a protein.

7. Vogel, S. "The Shape of Proteins." *Discover,* October 1988. How biochemists are redesigning proteins.

8. Weiss, R. "Organic Origami." *Science News,* November 28, 1987. Importance of the protein-folding problem.

6

Introduction to Metabolism

The living cell is a chemical industry in miniature, where thousands of reactions occur within a microscopic space. Metabolism is an emergent property of life that arises from specific interactions between molecules within the orderly environment of the cell. Sugars are converted to amino acids, and vice versa. Small molecules are assembled into polymers, which may later be hydrolyzed as the needs of the cell change. Many cells export chemical products that are used in other parts of the organism. The chemical process known as cellular respiration drives the cellular economy by extracting the energy stored in sugars and other fuels. And always, the entire enterprise of myriad reactions going on in the cell is precisely coordinated. In its complexity, its efficiency, its integration, and its responsiveness to subtle changes, the cell is peerless as a chemical institution.

THE METABOLIC MAP

Metabolism, from the Greek word meaning "change," is the totality of an organism's chemical processes. You can think of a cell's metabolism as an elaborate road map of the thousands of reactions that occur in that cell (Figure 6.1). The reactions are arranged in intricately branched metabolic pathways, which transform molecules by a series of steps. The cell routes matter through the metabolic pathways by means of enzymes, which selectively accelerate each of the steps in the labyrinth of reactions. Analogous to the red, green, and yellow lights that control the flow of traffic and

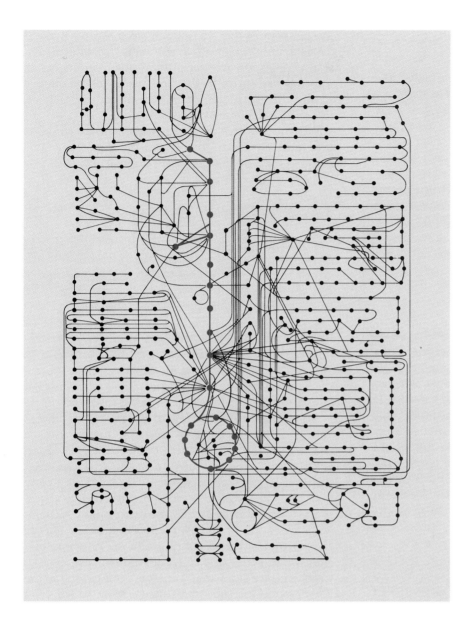

Figure 6.1
The metabolic map. This diagram tracing a few hundred of the reactions that occur in a cell gives an idea of the complexity of cellular metabolism. The dots represent molecules, and the lines represent the chemical reactions that transform them. The reactions proceed in stepwise sequences called metabolic pathways, each step catalyzed by a specific enzyme. The pathway shown in magneta is central to most pathways.

prevent snarls, mechanisms that regulate enzymes balance metabolic supply and demand, averting deficiencies and surpluses of chemicals.

As a whole, metabolism is concerned with managing the material and energy resources of the cell. Some metabolic pathways release energy by breaking down complex molecules to simpler compounds. These degradative processes are called **catabolic pathways**. The main thoroughfare of catabolism is cellular respiration, in which the sugar glucose is broken down to carbon dioxide and water. Energy stored in the glucose molecule becomes available to do the work of the cell. There are also **anabolic pathways**, which consume energy to build complicated molecules from simpler ones. An example of anabolism is the synthesis of a protein from amino acids. Catabolic and ana-

bolic pathways are the downhill and uphill avenues of the metabolic map. The metabolic pathways are interwoven in such a way that energy released from the downhill reactions of catabolism can be transferred to anabolic pathways to drive uphill reactions that require energy.

It is not the objective of this chapter to track specific metabolic pathways. Our focus will be on metabolic mechanisms common to the various pathways. Energy is fundamental to all metabolic processes, making a basic knowledge of energy essential to understanding how the living cell works. Although we will use many physical and mechanical examples to study energy, keep in mind that the same principles apply to organisms. An understanding of energy is as important for students of biology as it is for students of physics.

ENERGY: SOME BASIC PRINCIPLES

The two concepts most basic to science are matter and energy. In Chapter 2, we defined matter as anything that has mass and takes up space. Energy is more abstract. With the help of a microscope, we can witness the whipping of a cell's flagellum, which is matter, but we cannot see the energy that powers this movement. Energy can only be described and measured by how it affects matter. Physicists define **energy** as the capacity to do work—that is, to move matter against an opposing force such as gravity or friction. We can say that energy is the ability to move matter in a direction it would not otherwise move if left alone.

Forms of Energy

Anything that moves possesses a form of energy called **kinetic energy**. It is energy in action—energy actually in the process of doing work. Moving matter does work by transferring its motion to other matter, whether it is a pool player using the motion of the cue stick to push the cue ball, which in turn moves the other balls; water gushing through a dam turning turbines; electrons flowing along a wire and running household appliances; or contraction of your leg muscles pushing bicycle pedals. Heat, or thermal energy, is kinetic energy because of the random movement of molecules. Light also has energy, which is harnessed by green plants to power photosynthesis, the anabolic pathway that builds sugar from carbon dioxide and water.

A resting object not presently performing work may also possess energy, which, remember, is the *capacity* to do work. As we saw in Chapter 2, energy can be stored in the form of **potential energy**, which is energy that matter possesses because of its location or arrangement. Water behind a dam, for instance, stores energy because of its altitude. A subtle form of potential energy, but one particularly relevant to biologists, is chemical energy, which is stored in molecules because of the arrangement of the atoms that are bonded together.

Energy Transformations

Energy can be converted from one form to another (Figure 6.2). Chemical energy can be tapped when a chemical reaction rearranges the atoms of molecules in such a way that potential energy is transformed into kinetic energy. This transformation occurs, for example, in the engine of an automobile when the hydrocarbons of gasoline react explosively with oxygen, releasing the energy that pushes the pistons. Chemical energy also fuels organisms. Cellular respiration and other catabolic pathways release the energy stored

in sugar and other complex molecules and make it available for cellular work. An energy transformation was also responsible for storing potential energy in the fuel molecules in the first place, for plants use photosynthesis to transform light energy to the chemical energy of sugar.

Whether it is an engine or an organism that converts one form of energy to another, the transformations are governed by the same two unbreakable laws.

Two Laws of Thermodynamics

The study of the energy transformations that occur in a collection of matter is called **thermodynamics**. Scientists use the term *system* to denote the collection of matter under study, referring to the rest of the universe as the *surroundings*. A closed system, such as that approximated by liquid in a thermos bottle, is isolated from its surroundings. In an open system, energy can be transferred between the system and its surroundings. Organisms are open systems, absorbing light energy or chemical energy and releasing heat and metabolic waste products to the surroundings.

According to the **first law of thermodynamics**, the energy of the universe is constant. Energy can be transferred and transformed, but it cannot be created or destroyed. The first law is also known as the principle of conservation of energy. The electric company does not make energy, but merely converts it to a form that is convenient to use. By converting light to chemical energy, the green plant is acting as an energy transformer, not an energy producer. When a child climbs the ladder of a slide, she transforms some of the chemical energy that was stored in her food to the kinetic energy of her movements. In turn, climbing upward transforms kinetic energy to the potential energy stored by the girl as she rests at the top of the slide. During the slide down, the stored energy is converted back to kinetic energy. But what has become of the energy when the girl reaches the bottom and comes to a stop? Since energy must be conserved, it must be present *somewhere* in *some* form. The **second law of thermodynamics** can account for this energy.

The second law can be stated many ways. Let us begin with the following idea: Every energy transfer or transformation makes the universe more disordered. There is a quantitative measure of disorder, called **entropy**, whose value increases as disorder increases. Entropy is proportional to randomness, the opposite of order (Figure 6.3). We can now rephrase the second law: Every process increases the entropy of the universe. There is an unstoppable trend toward randomization. The entropy of a particular system may decrease, but the entropy of the system plus its surroundings must increase. If a system becomes more ordered, it is at the expense of the rest of the universe becoming more random.

(a)

(b)

Figure 6.2
Energy transformation.(a) Life is powered by light energy from the sun. The photosynthesis of plants transforms the energy of light to the chemical energy stored in sugar and other organic compounds. Animals eat the plants and transform the potential energy stored in molecules into mechanical movement, heat, and other forms of kinetic energy. Animals that eat animals (or, in this case, drink the cow's milk) are third-hand beneficiaries of photosynthesis. **(b)** Energy stored in food molecules powers the movements of this musician, who transfers the kinetic energy of her movements to the cello strings. In turn, the vibrating strings transfer their energy, via pressure waves in the air, to the eardrums of all who are within hearing range. Finally, through a series of transformations, the energy of vibrating eardrums is converted to signals in the brain that produce perceptions of sound.

Organisms are highly ordered. If life's low entropy seems like a violation of the second law, it is because you are forgetting that organisms are open systems that interact with their surroundings. From the food it eats, an animal obtains starch, proteins, and other complexly arranged molecules and replaces them with carbon dioxide and water, relatively small and simple molecules. Energy enters the animal in the form of ordered molecules rich in potential energy and leaves mainly in the form of heat, the energy of random molecular motion.

In most energy transformations, ordered forms of energy are at least partly converted to heat. Only about 25% of the chemical energy stored in the gasoline of an automobile is transformed into the motion of the car; the remaining 75% is lost from the engine as heat, which dissipates rapidly through the surroundings. In tapping chemical energy to perform work, metabolism also dissipates heat, which can make a room that is crowded with people uncomfortably warm. Conversion of other forms of energy to heat does not violate the first law. Energy has been conserved, because heat is a form of energy, albeit in its most random state. (Now you know what happened to the kinetic energy of the sliding girl mentioned earlier. Friction generated heat, which was dissipated to the surrounding air.) In a sense, heat is a lower grade of energy because it is an uncoordinated movement of molecules that many systems cannot use to do work. A system can use heat to perform work only when there is a temperature difference that results in the heat flowing from where it is warmer to where it is cooler (see Chapter 3). If temperature is uniform throughout a system, as it is in a living cell, then heat is useless energy, except for warming the body.

Combining the first and second laws, we can conclude that the quantity of energy in the universe is constant, but its quality is not. Everything that happens at the expense of organized forms of matter increases the entropy of the universe, either directly or by conversion to heat, which tends to randomize its surroundings.

Now let us see how thermodynamics applies specifically to chemical reactions and metabolism.

CHEMICAL ENERGY: A CLOSER LOOK

When a chemical reaction rearranges the atoms of molecules, old chemical bonds are broken, and new bonds form. Energy must be absorbed by the molecules for their bonds to break, and energy is released from the molecules when bonds form (Figure 6.4). It takes the same amount of energy to break a particular kind of bond as that type of bond releases when it forms. This quantity of energy is called **bond energy**. Table 6.1 lists bond energies for covalent bonds com-

Figure 6.3

Entropy. Entropy measures randomness: The more randomly organized a system, the greater its entropy; the more orderly the system, the lower its entropy. If left alone, any system increases in entropy, losing order. (**a**) Without energy being expended for maintenance and repairs, buildings fall apart. (**b**) Soluble dyes added to water will spread out spontaneously to become randomly distributed throughout the water. (**c**) Cells maintain order at the expense of energy from the surroundings. After a cell, such as this *Paramecium*, dies, its entropy increases (LMs).

(a)

(b)

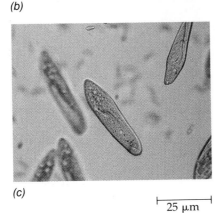
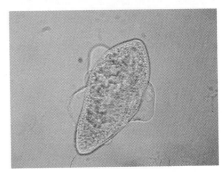

(c)

25 μm   25 μm

mon in organic molecules, expressed in kilocalories per mole of bonds broken or formed (kcal/mol). (The kilocalorie and the mole were defined in Chapter 3.) The bond energies listed are only approximate; they vary somewhat from molecule to molecule because of the influence of neighboring bonds. The stronger the bond, the greater its energy—the harder it is to break, and the more energy it releases when it forms.

Heat of Reaction

As a chemical reaction converts reactants to products, the net release or uptake of energy by the mixture of molecules is equal to the difference between the energy released when bonds form and the energy consumed when bonds break. For example, we can compute the energy change for the reaction that occurs at the burner

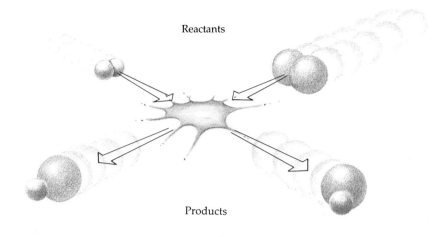

Reactants

Products

Figure 6.4
A chemical reaction. Two different molecules react to form two molecules of product. For the reaction to occur, the reactants must collide with enough energy to break the bonds holding their atoms together. Formation of new bonds releases energy.

of a gas stove, where methane and oxygen react to form carbon dioxide and water:

$$CH_4 + 2\,O_2 \longrightarrow CO_2 + 2\,H_2O$$

The bookkeeping for bond energies goes like this:

Bonds broken (*energy absorbed*)
4 C – H	4 × 99	=	396 kcal
2 O = O	2 × 118	=	236 kcal
			632 kcal

Bonds formed (*energy released*)
2 C = O	2 × 174	=	348 kcal
4 O – H	4 × 111	=	444 kcal
			792 kcal

Net energy released   160 kcal

For each mole of methane burned on the stove, the reaction consumes 632 kcal and releases 792 kcal, for a net release of 160 kcal. The net yield of energy from the reaction is responsible for the heat that radiates from the burner.

The net energy consumed or released when reactants are converted to products is termed the heat of reaction, symbolized by ΔH (read as "delta H"), where Δ stands for a change. For the reaction between methane and oxygen, $\Delta H = -160$ kcal/mol. The minus sign indicates that this much stored energy has been released (lost) from the system of molecules during the reaction. We can think of a molecule as having a heat content, or **enthalpy**, as it is called. Enthalpy is the total potential energy of the molecule. In a chemical reaction, energy must be conserved. If a reaction heats its surroundings, the energy is released at the expense of the enthalpy of the reacting molecules. The heat of reaction, ΔH, is actually the change in enthalpy of the reaction mixture. The reaction of a gas stove is able to release 160 kcal of heat energy because the products of the reaction have 160 kcal less enthalpy than the reactants. A reaction that releases heat—in other words, one with a negative value for ΔH—is said to be exothermic. A reaction that lowers the temperature of the surroundings by absorbing heat is endothermic. An endothermic reaction has a positive value for ΔH, meaning that the products have more enthalpy than the reactants; chemical energy has been upgraded at the expense of energy from the surroundings.

Spontaneous Reactions

A spontaneous process is one that can occur without outside help. Spontaneous does not necessarily mean instantaneous. A spontaneous process may take a very long time to happen; what is important is that it occurs without the addition of external energy. A nonspontaneous process cannot occur on its own; it will happen only if an external energy source intervenes. Water flows downhill spontaneously. Uphill movement of water is nonspontaneous and occurs only when a windmill or some other machine pumps the water against gravity. Notice that the spontaneous process lowers the potential energy of water by decreasing its altitude.

Systems rich in energy, such as a reservoir of water stored at high altitude behind a dam, are intrinsically

Table 6.1 Approximate bond energies	
Bond	**Energy (kcal/mol)**
C—C	83
C—H	99
C—O	84
C=O	174
O—H	111
O=O	118

unstable and tend to change in such a way that their energy decreases. The same trend applies to chemical reactions. Chemical systems rich in energy tend to react in ways that lower their energy. Exothermic reactions are nearly always spontaneous. By contrast, an endothermic reaction increases chemical energy and, like water moving uphill, tends to be nonspontaneous. We can use the sign (positive or negative) of ΔH as a clue to which chemical reactions can and cannot occur on their own. The criterion is fallible, however. Some exothermic processes *are* nonspontaneous, and some endothermic processes *are* spontaneous. For example, water in an open container will evaporate spontaneously, although the energy of the water molecules increases as they absorb heat. In addition to the change in energy, some other factor must be capable of driving spontaneous processes. The additional factor is the trend toward increased entropy, or randomization.

Entropy is symbolized in thermodynamic equations with the letter S. The change in entropy during a process is

$$\Delta S = S_{\text{final state}} - S_{\text{initial state}}$$

A positive ΔS, an increase in entropy (randomness), will contribute to a process being spontaneous. When water evaporates, entropy increases, since water in the gaseous state is less organized than liquid water. This increase in entropy is enough to drive the process despite the increase in energy of the water as it evaporates. On the other hand, there are spontaneous processes that *decrease* the entropy of a system. (Although the second law tells us that the entropy of the system plus its surroundings must increase, we are concerned here with the entropy of only the system.) For example, a highly ordered snowflake forms spontaneously from less organized liquid water when the temperature is below freezing (0°C). In this case, the loss of energy from the cooling water is enough to drive the process despite a decrease in entropy.

From these examples, we can see that neither the energy change nor the entropy change alone can tell us for certain whether a process can occur spontaneously. We need a criterion that combines the effects of these two factors, the tendency for the energy of a system to decrease and for its entropy to increase.

Free Energy

The quantity that combines total energy (enthalpy) and entropy is called **free energy**, represented by the letter G. When some process occurs in a system, the change in free energy (ΔG) is directly related to the change in enthalpy (ΔH), but inversely related to the change in entropy (ΔS):

$$\Delta G = \Delta H - T\Delta S$$

The T stands for the absolute temperature (in Kelvin units, K, equal to °C + 273). A higher temperature amplifies the entropy factor because temperature measures the intensity of random molecular motion, which tends to disrupt order.

Remember, the two tendencies contributing to the spontaneity of a process are a decrease in enthalpy (negative ΔH) and an increase in entropy (positive ΔS). Notice in the given equation that both tendencies reduce the free energy of the system (Figure 6.5). We now have a criterion for spontaneity that combines the effects of changes in both enthalpy and entropy: *In a spontaneous process, the free energy of the system decreases.* (This is another version of the second law of thermodynamics.) A process having a negative free energy change ($\Delta G < 0$) occurs on its own. During a nonspontaneous process, the free energy of the system increases ($\Delta G > 0$).

When the enthalpy and entropy changes oppose each other in their effects on free energy, temperature determines which factor gains the upper hand by its influence on the entropy term (Figure 6.6). Consider

Figure 6.5
How changes in enthalpy and entropy affect free energy. This diagram is based on the relationship $\Delta G = \Delta H - T\Delta S$, where ΔG is the free energy change, ΔH is the enthalpy change, T is the absolute temperature, and ΔS is the entropy change. Shown here are three cases, all of which occur spontaneously because they lead to a decrease in free energy.

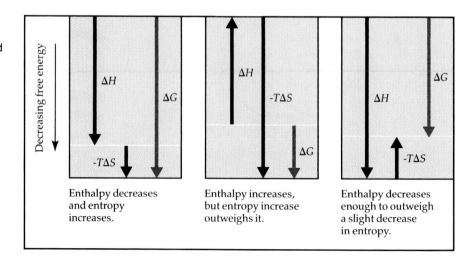

Enthalpy decreases and entropy increases.

Enthalpy increases, but entropy increase outweighs it.

Enthalpy decreases enough to outweigh a slight decrease in entropy.

Figure 6.6
Free energy and the formation of a snowflake. For ice crystals to form, the free energy of the water must decrease, despite a decrease in entropy. Entropy decreases when water freezes because the arrangement of molecules is more ordered in an ice crystal than in liquid water. For water to crystallize spontaneously, its change in free energy, ΔG, or $\Delta H - T\Delta S$, must be negative. Notice that as temperature drops, the impact of the entropy change on ΔG is less. At temperatures of 0°C (273 K) or lower, water loses enough energy in the form of heat for the ΔH factor to offset the entropy factor, and water crystallizes.

the denaturation of a protein (see Chapter 5). When a protein loses its conformation, entropy increases because of the loss of structural order. Denaturation is favored by ΔS. But ΔH opposes denaturation because energy must be absorbed to disrupt hydrogen bonds and other forces before the protein loses its shape. If temperature is increased, at some point (usually above 60°C) the entropy factor wins out and the protein denatures.

The free energy equation is a versatile tool for determining what processes can and cannot occur in nature. But ΔG is more than a criterion for what is feasible; it also tells us how much work a spontaneous process can do. *Free energy is that portion of a system's energy that can be used to do work when temperature is uniform throughout the system, as it is in a living cell.*

Exergonic and Endergonic Reactions

Based on their free energy changes, reactions can be classified as either exergonic (meaning "energy outward") or endergonic (meaning "energy inward"). An

exergonic reaction proceeds with a net release of free energy. Since the chemical mixture loses free energy, ΔG is negative for an exergonic reaction. In other words, exergonic reactions are those that occur spontaneously. The magnitude of ΔG for an exergonic reaction is the maximum amount of work the reaction can perform. We can use cellular respiration as an example:

$$C_6H_{12}O_6 + 6O_2 \longrightarrow 6CO_2 + 6H_2O \quad \Delta G = -686 \text{ kcal/mol}$$

For each mole (180 g) of glucose broken down by respiration, 686 kcal of energy are made available for work. Since energy must be conserved, the chemical products of respiration store 686 kcal less free energy than the reactants. The products are, in a sense, the spent exhaust of a process that tapped most of the free energy stored in the sugar molecules.

An **endergonic reaction** is one that absorbs free energy from its surroundings. Because this kind of reaction stores more free energy in the molecules, ΔG is positive. Such reactions are nonspontaneous, and the magnitude of ΔG is the minimum quantity of work required to drive the reaction. If a chemical process is exergonic in one direction, then the reverse process must be endergonic: You cannot travel downhill in both directions. If $\Delta G = -686$ kcal/mol for respiration, then for photosynthesis, $\Delta G = +686$ kcal/mol. Sugar production in the leaf cells of a plant is steeply endergonic, an uphill process powered by the absorption of light energy from the sun.

Free Energy and Equilibrium

Recall from Chapter 2 that most chemical reactions are reversible and proceed until the forward and backward reactions occur at the same rate. The reaction is then said to be at chemical equilibrium, and there is no further change in the concentration of products or reactants.

Equilibrium and the free energy change (ΔG) for a reaction are related. As a reaction proceeds toward equilibrium, the free energy of the mixture of reactants and products decreases. Free energy increases when a reaction is somehow pushed away from equilibrium. For a reaction at equilibrium, $\Delta G = 0$ because there is no net change in the system. We can think of equilibrium as an energy sink, the bottom of a hill. A process at equilibrium performs no work. A reaction is spontaneous and exergonic when sliding toward equilibrium. To move away from equilibrium is nonspontaneous; it is an endergonic process that can occur only when an outside energy source pushes the reaction "uphill." A key strategy in cellular metabolism is driving endergonic reactions by coupling them to exergonic reactions through an energy shuttle called ATP.

ATP AND CELLULAR WORK

A cell does three main kinds of work:

1. *Mechanical work,* such as the beating of cilia, contraction of muscle cells, flow of cytoplasm within cells, and movement of chromosomes during cellular reproduction

2. *Transport work,* the pumping of substances across membranes

3. *Chemical work,* the pushing of endergonic reactions that would not occur spontaneously, such as the synthesis of polymers from monomers

In nearly every case, the immediate source of energy that drives cellular work is a molecule called adenosine triphosphate, or ATP.

Structure and Hydrolysis of ATP

ATP (adenosine triphosphate) is a nucleoside triphosphate consisting of adenine, bonded to the sugar ribose, which in turn is connected to a chain of three phosphate groups (Figure 6.7a). The only difference between ATP and the adenosine monophosphate found as a monomer in the nucleic acid RNA (see Chapter 5) is the two additional phosphate groups on ATP.

The bonds between the phosphate groups of ATP are unstable and can be broken by hydrolysis. When water hydrolyzes the terminal phosphate bond, a molecule of inorganic phosphate (abbreviated Ⓟ) is removed from ATP, which then becomes adenosine diphosphate, or ADP (Figure 6.8b). The reaction is exergonic and releases 7.3 kcal of energy per mole of ATP hydrolyzed:

$$ATP + H_2O \longrightarrow ADP + Ⓟ \quad \Delta G = -7.3 \text{ kcal/mol}$$

This ΔG figure, technically symbolized $\Delta G°$, is the so-called standard free energy change that is measured in the laboratory under standard conditions of temperature, pH, and certain concentrations of reactants and products. When the reaction occurs under the "nonstandard" conditions of the cell, the actual ΔG is estimated to be -10 to -12 kcal/mol.

Because their hydrolysis releases energy, the phosphate bonds of ATP are referred to as high-energy phosphate bonds, but the term is somewhat misleading. "High-energy bonds" sounds like we are referring to strong bonds, but the phosphate bonds of ATP are actually relatively weak. It is *because* the phosphate bonds are weak, or unstable, that their hydrolysis yields energy. The products of hydrolysis (ADP and Ⓟ) are more stable than ATP. When a system changes in the direction of greater stability—as when a compressed spring relaxes, for instance—the change is generally

(a) **ADENOSINE TRIPHOSPHATE (ATP)**

(b) **ADENOSINE DIPHOSPHATE (ADP)**
+
Ⓟ
INORGANIC PHOSPHATE

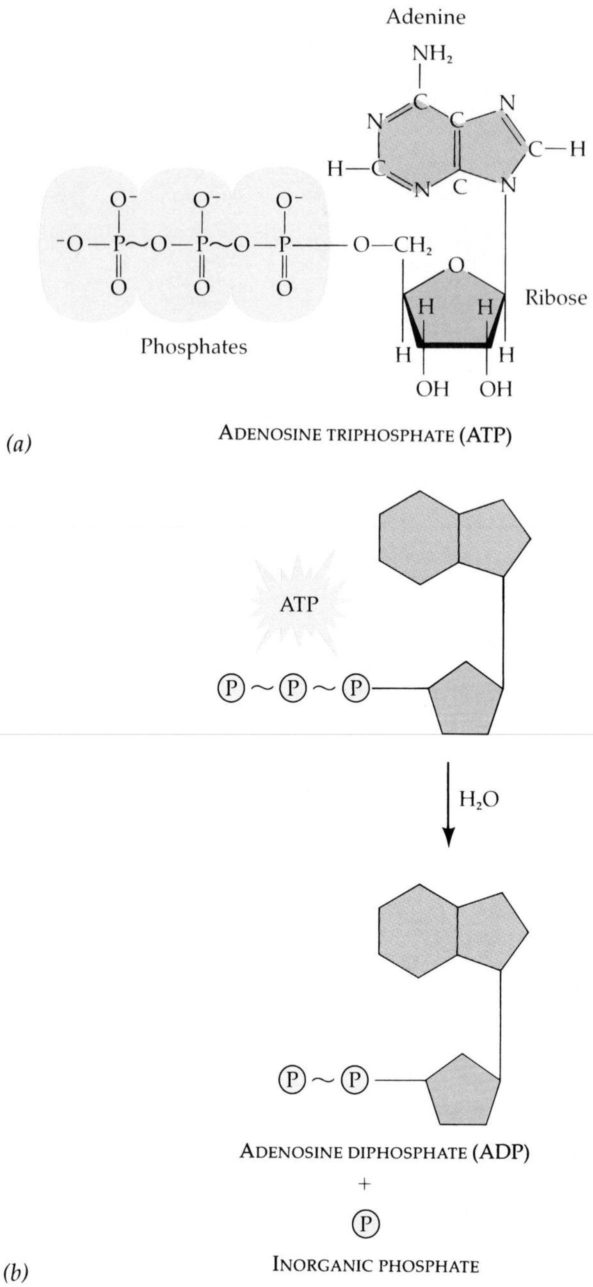

Figure 6.7
ATP. (**a**) The structure of ATP and (**b**) its hydrolysis to yield ADP and inorganic phosphate. In the cell, most hydroxyl groups of phosphates are ionized ($-O^-$).

exergonic. Thus, the release of energy during the hydrolysis of ATP comes from the chemical change to a more stable condition, and not from the phosphate bonds themselves. Why are the phosphate bonds so fragile? Reexamine the ATP molecule in Figure 6.7a, and you can see that all three phosphate groups are negatively charged. These like charges are crowded together and their repulsion contributes to the instability of the phosphate bonds. The triphosphate tail of ATP is the chemical equivalent of a loaded spring.

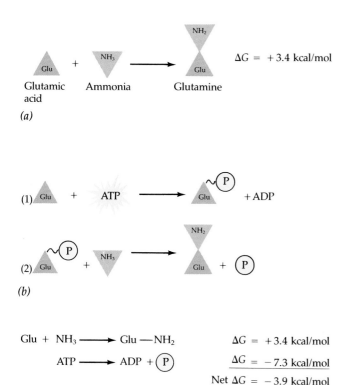

$\Delta G = +3.4$ kcal/mol

Glutamic acid — Ammonia — Glutamine

(a)

(1)

(2)

(b)

Glu + NH₃ ⟶ Glu —NH₂ $\Delta G = +3.4$ kcal/mol

ATP ⟶ ADP + Ⓟ $\Delta G = -7.3$ kcal/mol

Net $\Delta G = -3.9$ kcal/mol

(c)

Figure 6.8
Energy coupling by phosphate transfer. In this example, ATP hydrolysis is used to drive an endergonic reaction, the conversion of the amino acid glutamic acid (Glu) to another amino acid, glutamine (Glu-NH₂). **(a)** Without the help of ATP, the conversion is nonspontaneous. **(b)** As it actually occurs in the cell, the synthesis of glutamine is a two-step reaction driven by ATP. The formation of a phosphorylated intermediate couples the two steps. In the first step, ATP phosphorylates glutamic acid, transferring chemical instability to the amino acid. In the second step, ammonia displaces the phosphate group from the phosphorylated intermediate, forming glutamine. **(c)** We can calculate the free energy change for the overall reaction by dissecting out two reactions for which we already know ΔG. Since the overall process is exergonic (has a negative ΔG), it occurs spontaneously.

How ATP Performs Work

When ATP is hydrolyzed in a test tube, the release of free energy merely heats the surrounding water. In the cell, that would be an inefficient and wasteful use of valuable chemical energy. With the help of specific enzymes, the cell is able to couple the energy of ATP hydrolysis directly to endergonic processes by transferring a phosphate group from ATP to some other molecule, which is then said to be phosphorylated. The key to the coupling is the formation of a **phosphorylated intermediate** that is more reactive (less stable) than the original molecule (Figure 6.8). Nearly all cellular work depends on ATP energizing other molecules by transferring phosphate groups. For

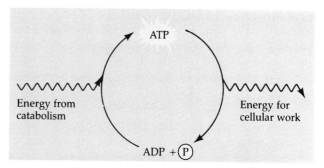

Figure 6.9
The ATP cycle. Energy released by breakdown reactions (catabolism) in the cell is used to phosphorylate ADP, regenerating ATP. Energy stored in ATP drives most cellular work.

instance, ATP powers the movement of muscles by transferring phosphates to contractile proteins.

Regeneration of ATP

An organism at work uses ATP continuously, but ATP is a renewable resource that can be regenerated by the addition of phosphate to ADP (Figure 6.9). The ATP cycle moves at an astonishing pace. A typical cell recycles its entire pool of ATP about once each minute. That turnover represents ten million molecules of ATP consumed and regenerated per second per cell. If ATP could not be regenerated by phosphorylation of ADP, human beings would consume nearly their body weight in ATP each day.

Since a reversible process cannot go downhill both ways, the regeneration of ATP from ADP is endergonic:

$$ADP + Ⓟ \longrightarrow ATP \quad \Delta G = +7.3 \text{ kcal/mol}$$

Catabolic or exergonic pathways, especially cellular respiration, provide the energy to make ATP, an endergonic process. Plants can also use light energy to produce ATP.

Cellular respiration is a stepwise pathway by which enzymes decompose glucose and other complex organic molecules. The process is overwhelmingly exergonic, and the energy it releases drives phosphorylation of ADP to regenerate ATP. In the metaphor of a metabolic map, the ATP cycle is a hub through which energy passes from catabolic to anabolic pathways.

Metabolic Disequilibrium

The chemical reactions of respiration and other catabolic pathways are reversible and would reach equilibrium if they occurred in the isolation of a test tube. Chemical systems at equilibrium have a ΔG of zero and can do no work. In the cell, some of the reversible reactions of respiration are pulled in one direction and kept out of equilibrium. The key to this disequilibrium is that the product of one reaction does not accumulate, but instead becomes a reactant in the next step of the metabolic pathway (Figure 6.10). And what pulls the overall sequence of reactions is the siphoning effect due to the huge free energy difference between glucose at the uphill end of respiration and carbon dioxide and water at the downhill end of the pathway. As long as the cell has a steady supply of glucose or other fuels and is able to expel the CO_2 waste to the surroundings, equilibrium is denied and respiration keeps making ATP.

ENZYMES

Thermodynamics tells us what can and cannot happen, but says nothing about the speed of the processes. A spontaneous chemical reaction may occur so slowly as to be imperceptible. For example, the hydrolysis of sucrose (table sugar) to glucose and fructose is exergonic, occurring spontaneously with a release of free energy ($\Delta G = -7.0$ kcal/mol). Yet, a solution of sugar dissolved in sterile water will sit for years at room temperature with no appreciable hydrolysis. However, if we add a small amount of the enzyme known as sucrase to the solution, then all the sugar may be hydrolyzed within seconds. **Enzymes** are catalysts, chemical agents that change the rate of a reaction without being consumed by the reaction. In the absence of enzymes, chemical traffic through the pathways of the metabolic map would become hopelessly congested. What impedes a spontaneous reaction, and how does an enzyme lower the barrier?

Enzymes and Activation Energy

As we have already seen, before a reaction can occur, energy must be absorbed by the reactants in order to

Figure 6.10
Keeping metabolism away from equilibrium: A hydraulic analogy. (a) The water generates electric energy only while it is falling. Once the levels in the two containers are equal, the turbine ceases to turn and the light goes out. Analogously, the individual steps of respiration, in isolation, would come to equilibrium, and work would cease. **(b)** If there is a series of drops in water level, electric energy can be generated at each drop. Analogously, in respiration, there is a series of drops in free energy between glucose, the starting material, and the metabolic wastes at the end. The overall process never reaches equilibrium as long as the organism lives, because the product of each reaction becomes the reactant for the next, and the metabolic wastes are expelled from the cell.

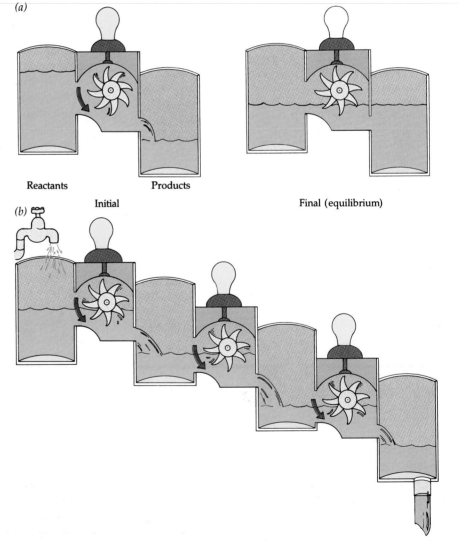

(a)

Reactants   Products

Initial   Final (equilibrium)

(b)

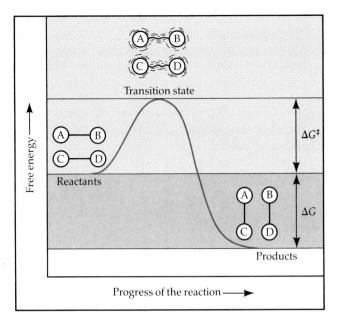

Figure 6.11
Energy profile of a reaction. In this hypothetical reaction, the reactants A—B and C—C must absorb enough energy from the surroundings to surmount the hill of activation energy ($\Delta G^{\ddagger}$) and reach the transition state. The bonds can then break, and as the reaction proceeds, energy is released to the surroundings during the forming of new bonds. This is an exergonic reaction, which has a negative ΔG; the products have less free energy than the reactants.

break bonds. This initial investment of energy to start a reaction is known as the **free energy of activation**, abbreviated $\Delta G^{\ddagger}$. It is usually provided in the form of heat absorbed from the surroundings. If the reaction is exergonic, $\Delta G^{\ddagger}$ will be repaid with dividends as the formation of new bonds releases more energy than was invested in the breaking of old bonds. Figure 6.11 graphs these energy changes for a hypothetical reaction that swaps portions of two reactant molecules:

$$A-B + C-D \longrightarrow A-C + B-D$$

For the bonds of the reactants to break, the molecules must absorb enough energy to become unstable (remember, systems rich in free energy are intrinsically unstable, and unstable systems are reactive). The activation energy is represented by the uphill portion of the graph, with the free energy content of the reactants increasing. The absorption of thermal energy increases the speed of the reactants, so they are colliding more often and more forcefully. Moreover, thermal agitation of the atoms that make up the molecules has made bonds more fragile and more likely to break. At the summit, the reactants are in an unstable condition known as the *transition state;* they are primed,

and the reaction can occur. As the molecules settle into their new bonding arrangements, energy is released to the surroundings. This phase of the reaction corresponds to the downhill portion of the curve, which indicates a loss of free energy by the molecules. The difference in the free energy of the products and reactants is ΔG for the overall reaction, which is negative for an exergonic reaction. Even for an exergonic reaction, which is energetically downhill overall, the barrier of activation energy must be scaled before the reaction can occur.

For some reactions, $\Delta G^{\ddagger}$ is modest enough that even at room temperature there is sufficient thermal energy for the reactants to reach the transition state. In most cases, however, the $\Delta G^{\ddagger}$ barrier is loftier, and the reaction will occur at a noticeable rate only if the material is heated. The spark plugs in an automobile engine heat the gasoline-oxygen mixture so that the molecules reach the transition state and react; only then can there be the explosive release of energy that pushes the pistons. Without a spark, most of the hydrocarbons of gasoline are too stable to react with oxygen.

The barrier of activation energy is essential to life. Proteins, DNA, and other complex molecules of the cell are rich in free energy and have the potential to decompose spontaneously; that is, thermodynamics favors their breakdown. These molecules exist only because at temperatures typical for cells, few molecules can make it over the hump of activation energy. On occasion, however, the barrier for selected reactions must be surmounted, or the cell would be metabolically stagnant. Heat speeds a reaction, but high temperatures kill cells. Organisms must therefore use an alternative—a catalyst.

Enzymes, which are proteins, are the biological catalysts. An enzyme speeds a reaction by lowering the barrier of activation energy so that the precipice of the transition state is within reach even at moderate temperatures (Figure 6.12). An enzyme cannot change the ΔG for a reaction. It cannot make a nonspontaneous reaction spontaneous, an endergonic reaction exergonic. Enzymes can only hasten reactions that would occur eventually anyway, but that is enough for the cell to have a dynamic metabolism. And since enzymes are very selective in the reactions they catalyze, these proteins determine which chemical processes will be going on in the cell at any particular time.

Specificity of Enzymes

The substance an enzyme acts on is referred to as the enzyme's **substrate**. The enzyme binds to its substrate (or substrates, when there are two or more reactants), and while the two are joined, the catalytic action of the enzyme converts the substrate to the product (or

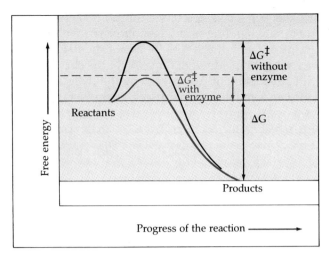

Figure 6.12
Enzymes lower the barrier of activation energy. Without affecting the free energy change (ΔG) for the reaction, an enzyme speeds the reaction by reducing the uphill climb to the transition state. In the graph, the black curve shows the course of the reaction without enzyme, and the magenta curve shows the course of the reaction with enzyme.

products) of the reaction. This process can be generalized in this way:

$$\text{Substrate} \xrightarrow{\text{enzyme}} \text{Product}$$

For example, the enzyme sucrase (many enzyme names end in -ase) breaks the double sugar sucrose into its two monosaccharides, glucose and fructose:

$$\text{Sucrose} \xrightarrow{\text{sucrase}} \text{Glucose} + \text{Fructose}$$

An enzyme can distinguish its substrate from even closely related compounds, such as isomers, so that each type of enzyme catalyzes a particular reaction. For instance, sucrase will act only on sucrose and will reject other disaccharides such as maltose. What accounts for this molecular recognition? Recall that enzymes are proteins, and proteins are macromolecules with unique three-dimensional conformations. The specificity of an enzyme is based on its shape.

Only a restricted region of the enzyme molecule actually binds to the substrate. This receptor, called the **active site**, is typically a pocket or groove on the surface of the protein (Figure 6.13). Usually, the active site is formed by only a few of the enzyme's amino acids, with the rest of the protein molecule providing a framework that reinforces the configuration of the active site.

The specificity of an enzyme is attributed to a compatible fit between the shape of its active site and the shape of the substrate. The active site, however, is not

a rigid receptacle for the substrate. As the substrate enters the active site, it induces the enzyme to change its shape slightly so that the active site fits even more snugly around the substrate. This **induced fit** is like a clasping handshake. The embrace of the substrate by the active site not only tailors fit, but also brings chemical groups of the active site into positions that enhance their ability to work on the substrate and catalyze the chemical reaction.

The Catalytic Cycle of Enzymes

The first step in an enzymatic reaction is the binding of the substrate to the active site to form an enzyme-substrate complex (Figure 6.14). In most cases, the substrate is held in the active site by weak interactions, such as hydrogen bonds and ionic bonds. Side chains (R groups) of a few of the amino acids that make up the active site catalyze the conversion of substrate to product, and the product then departs from the active site. The enzyme is then free to take another substrate molecule into its active site. The entire cycle happens so fast that a single enzyme molecule typically converts about a thousand substrate molecules per second, and some enzymes are much faster. Enzymes, like all catalysts, emerge from the reaction in their original form. Therefore, very small amounts of enzyme can have a huge metabolic impact by functioning over and over again in catalytic cycles.

Enzymes use a variety of mechanisms to lower activation energy and speed up a reaction. In reactions involving two or more reactants, the active site provides a template for the substrates to come together in the proper orientation for a reaction between them to occur. As induced fit causes the active site to clinch the substrates, the enzyme may stress the substrate molecules, stretching and bending critical chemical bonds that must be broken during the reaction. Since ΔG‡ is proportional to the difficulty of breaking bonds, distorting the substrate reduces the amount of thermal energy that must be absorbed to achieve a transition state. The active site may also provide a microenvironment that is conducive to a particular type of reaction. For example, if the active site has a concentration of amino acids with acidic side chains (R groups), the active site may be a pocket of low pH in an otherwise neutral cell. Still another aspect of catalysis is the direct participation of the active site in the chemical reaction. Sometimes this even involves brief covalent bonding between the substrate and a side chain of an amino acid of the enzyme. Subsequent steps of the reaction restore the side chains to their original states, so the active site is the same after the reaction as it was at the beginning.

The rate at which a given amount of enzyme converts substrate to product is partly a function of the initial concentration of substrate; the more substrate

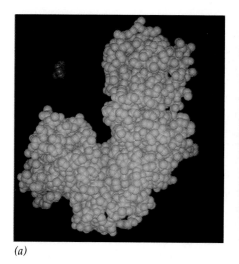

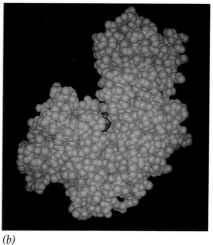

(a)

(b)

Figure 6.13
Induced fit between an enzyme and its substrate. (a) The active site of this enzyme, called hexokinase, can be seen here as a groove on the surface of the protein. **(b)** On entering the active site, the substrate, which is glucose (red), induces a slight change in the shape of the protein that causes the active site to embrace the substrate.

molecules available, the more frequently they wander into the active sites of the enzyme molecules. However, there is a limit to how fast the reaction can be pushed by adding more substrate to a fixed concentration of enzyme. At some point, the concentration of substrate will be high enough that all enzyme molecules have their active sites engaged, and as soon as the product exits an active site, another substrate molecule enters. At this substrate concentration, the enzyme is said to be saturated, and the rate of the reaction is determined by the speed at which the active site can convert substrate to product.

Factors Affecting Enzyme Activity

The activity of an enzyme is affected by general environmental factors, such as temperature and pH, and also by particular chemicals that specifically influence that enzyme.

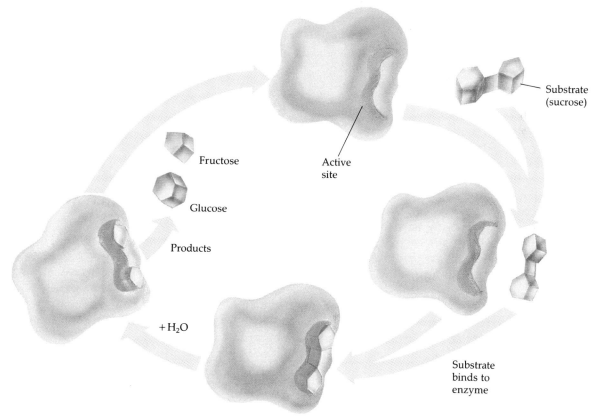

Fructose

Glucose

Products

+ H₂O

Active site

Substrate (sucrose)

Substrate binds to enzyme

Figure 6.14
The catalytic cycle of an enzyme. In this example, the enzyme sucrase catalyzes the hydrolysis of sucrose to glucose and fructose. An enzyme-substrate complex forms when the substrate enters the active site and attaches by weak bonds. The active site changes shape to fit closely around the substrate (induced fit). The substrate is converted to products while in the active site. The enzyme releases the products, and its active site is then available for another molecule of substrate.

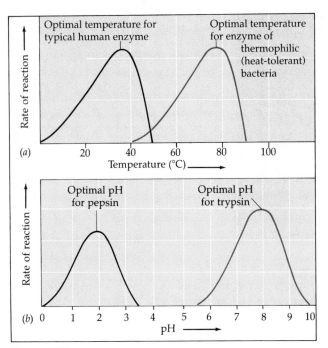

Figure 6.15
How environment affects enzymes. Each enzyme has an optimal temperature and pH that favor the active conformation of the protein molecule.

Environmental Conditions Recall from Chapter 5 that the three-dimensional structures of enzymes and other proteins are sensitive to their environment. Each enzyme has optimal conditions in which it works best, because that environment favors the most active conformation for the enzyme molecule.

Temperature is one environmental factor important in the activity of an enzyme (Figure 6.15a). Up to a point, the velocity of an enzymatic reaction increases with increasing temperatures, partly because substrates collide with active sites more frequently when the molecules move rapidly. But at some point on the temperature scale, the speed of the enzymatic reaction drops sharply with additional increase in temperature. The thermal agitation of the enzyme molecule disrupts the hydrogen bonds, ionic bonds, and other weak interactions that stabilize the active conformation, and the protein molecule denatures. For each type of enzyme, there is an optimal temperature at which reaction rate is fastest. It is the temperature that allows the greatest number of molecular collisions without denaturing the enzyme. Most human enzymes have temperature optima of about 35°–40°C (close to human body temperature). Bacteria that live in hot springs (for example, the hot springs in Yellowstone National Park) have enzymes with temperature optima of 70°C or higher (Figure 6.16).

Another environmental factor that influences the shape of proteins is pH. Just as each enzyme has a temperature optimum, it also has a pH optimum at which it is most active (Figure 6.16b). The pH optima for most enzymes fall in the range of 6–8, but there are exceptions. For example, pepsin, a digestive enzyme in the stomach, works best at a pH of 2. An environment so acidic denatures most enzymes, but the active conformation of pepsin is adapted to the acidic environment of the stomach. In contrast, trypsin, a digestive enzyme residing in the alkaline environment of the intestine, has a pH optimum of 8.

Enzymes are also sensitive to salt concentration. Most enzymes cannot tolerate extremely saline (salty) solutions because the inorganic ions interfere with ionic bonds within the protein molecule. Again, there are exceptions. Some algae and bacteria inhabit pools where the salt concentration is many times greater than seawater; their enzymes and other proteins are active under conditions that would denature the proteins of other organisms.

Cofactors Many enzymes require nonprotein helpers for catalytic activity. These adjuncts, called **cofactors**, may be bound tightly to the active site as permanent residents, or they may bind loosely and reversibly along with the substrate. The cofactors of some enzymes are inorganic, usually metal atoms such as zinc, iron, or copper. If the cofactor is an organic molecule, it is more specifically called a **coenzyme**. Most vitamins are coenzymes or raw materials from which coenzymes are made. Cofactors function in various ways, but in all cases they are necessary for the catalysis to take place.

Enzyme Inhibitors Certain chemicals selectively inhibit the action of specific enzymes. If the inhibitor attaches to the enzyme by covalent bonds, inhibition is usually irreversible. The inactivation is reversible, however, if the inhibitor binds to the enzyme by weak bonds.

Some inhibitors resemble the normal substrate molecule and compete for admission into the active site (Figure 6.17a and b). These mimics, called **competitive inhibitors**, reduce the productivity of enzymes by blocking active sites from the substrate. If the inhibition is reversible, it can be overcome by increasing the concentration of substrate so that as active sites become available, more substrate molecules than inhibitor molecules are around to gain entry to the sites.

Noncompetitive inhibitors impede enzymatic reactions without actually entering active sites (Figure 6.17c). The inhibitor binds to a part of the enzyme separate from the active site, causing the enzyme molecule to change its shape in such a way that the active site is no longer receptive to substrate.

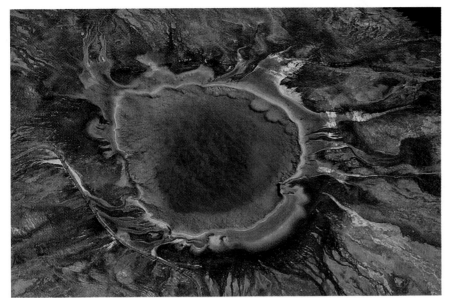

Figure 6.16
Metabolism in a hot spring. Some bacteria thrive in hot springs, where temperatures may top 80°C and the pH may be as low as pH 2. The enzymes of these thermophilic (heat-loving) bacteria not only tolerate these harsh conditions, but actually work best in such environments. In this aerial photo of the Grand Prismatic Pool in Yellowstone National Park, the colonies of bacteria color the edges of the pool. The footpath near the lower left corner will give you some sense of scale.

Enzyme inhibitors may act as metabolic poisons. Some pesticides, including DDT, are inhibitors of key enzymes in the nervous system. Many antibiotics are inhibitors of specific enzymes in bacteria. Penicillin, for instance, blocks the active site of an enzyme that many bacteria use to make their cell walls.

The example of metabolic poisons may give the impression that enzyme inhibition is generally abnormal and harmful. In fact, an essential mechanism in metabolic control is the selective inhibition and activation of enzymes by molecules naturally present in the cell.

Allosteric Regulation   In many cases, the molecules that affect enzyme activity bind to an **allosteric site**, a specific receptor site on some part of the enzyme molecule remote from the active site (allosteric means "another space"). Most enzymes having allosteric sites are proteins constructed from two or more polypeptide chains, or subunits. Each subunit has its own active site, and allosteric sites are usually located where subunits are joined. The entire complex oscillates between two conformational states, one catalytically active and the other inactive. Binding of an activator to an allosteric site stabilizes the conformation that has a functional active site, while an allosteric inhibitor stabilizes the inactive form of the enzyme (Figure 6.18).

The joints between the subunits of an allosteric enzyme articulate in such a way that a conformational change in one subunit is transmitted to all others. Through this interaction of subunits, a single activator or inhibitor molecule that binds to one allosteric site will affect the active sites of all subunits. Induced fit by binding of the substrate is also amplified by a phenomenon called **cooperativity**; a substrate molecule that enters one active site induces a conformational

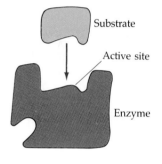

(a) Substrate normally can bind to the active site of an enzyme.

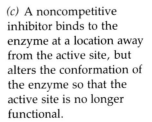

(b) A competitive inhibitor mimics the subtrate and competes for the active site.

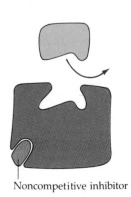

(c) A noncompetitive inhibitor binds to the enzyme at a location away from the active site, but alters the conformation of the enzyme so that the active site is no longer functional.

Figure 6.17
Enzyme inhibition.

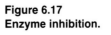
Noncompetitive inhibitor

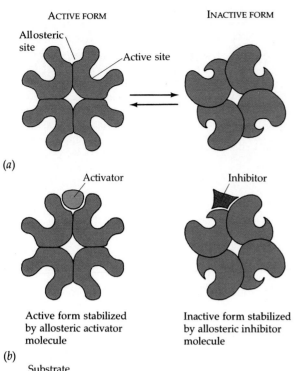

ACTIVE FORM   INACTIVE FORM

Allosteric site

Active site

(a)

Activator   Inhibitor

Active form stabilized by allosteric activator molecule

Inactive form stabilized by allosteric inhibitor molecule

(b)

Substrate

Active form stabilized by substrate molecule

(c)

Figure 6.18
Allosteric regulation of enzyme activity. (a) Most allosteric enzymes are constructed from two or more subunits, each having its own active site. The enzyme oscillates between two conformational states, one active and the other inactive. Remote from the active sites are allosteric sites, specific receptors for regulators of the enzyme, which may be activators or inhibitors. **(b)** Here we see the opposing effects of an allosteric inhibitor and activator on the conformation of an enzyme with four subunits. Binding of a regulator to a single allosteric site affects the conformation of all four subunits of the protein. **(c)** Similarly, through a phenomenon called cooperativity, one substrate molecule can activate all subunits of the enzyme by the mechanism of induced fit. Because substrates, activators, and inhibitors all bind to the enzyme reversibly by weak bonds, the activity of an enzyme at any moment depends on the relative concentrations of all molecules that bind to that enzyme.

change that favors the binding of substrate to the active sites of all other subunits of the protein.

Because allosteric regulators attach to the enzyme by weak bonds, the activity of the enzyme changes from moment to moment in response to fluctuating concentrations of the regulators. In some cases, an inhibitor and an activator are similar enough in shape to compete for the same allosteric site. For example, an enzyme that catalyzes a step in a catabolic pathway such as respiration may have an allosteric site that fits both ATP and ADP. The enzyme is inhibited by ATP and activated by ADP. This control seems logical if we remember that a major function of catabolism is to regenerate ATP from ADP. If ATP production lags behind its use, ADP accumulates and activates key enzymes that speed up catabolism. If the supply of ATP exceeds demand, then catabolism slows down as ATP molecules outnumber ADP molecules in competition for allosteric sites. In this way, allosteric enzymes act as valves that control the rates of key reactions in metabolic pathways.

Let us now see how the regulation of enzymes fits into the overall metabolic program of the cell.

THE CONTROL OF METABOLISM

It would be chemical chaos if all of a cell's metabolic pathways were open simultaneously. Imagine, for example, a substance synthesized by one pathway and broken down by another. If the two pathways were to run at the same time, the cell would be spinning its metabolic wheels. Actually, the operation of each metabolic pathway is tightly regulated. Pathways are switched on and off by controlling catalysis.

Feedback Inhibition

One of the most common modes of metabolic control is **feedback inhibition**, the switching off of a metabolic pathway by its end product, which acts as an inhibitor of an enzyme within the pathway. A specific example of feedback inhibition will reveal the logic of this control mechanism. The cell uses a pathway of five steps to synthesize the amino acid isoleucine from threonine, another amino acid (Figure 6.19). As isoleucine, the end product of the pathway, accumulates, it switches off its own synthesis. This happens because isoleucine is an allosteric inhibitor of the enzyme that catalyzes the very first step of the pathway, the enzyme for which threonine is the substrate. This feedback inhibition prevents the cell from wasting chemical resources to synthesize more isoleucine than is necessary and, by acting on the *first* enzyme in the pathway, also keeps metabolic intermediates from accumulating.

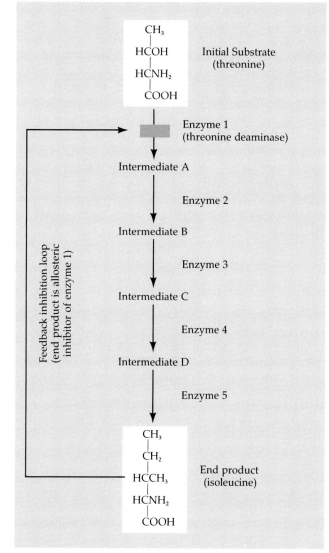

Figure 6.19
Feedback inhibition. Many metabolic pathways are switched off by their end products, which act as allosteric inhibitors of the first enzyme of the pathway. In this case, the amino acid isoleucine inhibits the enzyme that uses threonine as its substrate.

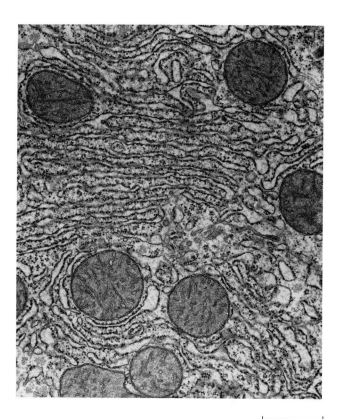

1 μm

Figure 6.20
Structural order and metabolism. Membranes partition a cell into various metabolic compartments, or organelles, each with a corps of enzymes that carry out specific functions (TEM).

Structural Order and Metabolism

The cell is not just a bag of chemicals with thousands of different kinds of enzymes and substrates wandering about randomly. The complex structure of the cell orders metabolic pathways in space and time (Figure 6.20). In some cases, a team of enzymes for several steps of a metabolic pathway are assembled together as a **multienzyme complex**. The arrangement orders the sequence of reactions, as the product from the first enzyme becomes substrate for the adjacent enzyme in the complex, and so on, until the end product is released. Also helping to organize metabolism, some

enzymes have fixed locations within the cell because they are incorporated into the structure of a specific membrane. Even when the enzymes for a metabolic pathway are individually dissolved, they may be highly concentrated along with their substrates within specialized organelles of the cell. Membranes partition the cell into many kinds of compartments, each with its own internal chemical environment and special blend of enzymes. For example, many of the enzymes for cellular respiration reside within organelles called mitochondria. If the cell had the same number of enzymes for respiration, but they were diluted throughout the entire volume of the cell, respiration would be very inefficient.

By examining the structural basis of metabolic order, we have returned to the theme with which this unit of chapters began.

EMERGENT PROPERTIES: A REPRISE

Life, remember, is organized along a hierarchy of structural levels: With each increase in the level of order, new properties emerge in addition to those of the component parts of that level. In these chapters, we have dissected the chemistry of life using the strategy of the reductionist. But we have also flirted with a more integrated view of life as we have seen how properties emerge with increasing order. The peculiar behavior of water, so essential to life on earth, results from interactions of the water molecules, themselves an ordered arrangement of hydrogen and oxygen atoms. We reduced the great complexity and diversity of organic compounds to the chemical characteristics of carbon, but we also saw that the unique properties of organic compounds are related to the specific structural arrangements of carbon skeletons and their appended functional groups. We learned that small organic molecules are often assembled into giant molecules, but we also discovered that a macromolecule does not behave as a simple composite of its mono-mers. For example, the unique form and function of a protein is a consequence of a hierarchy of primary, secondary, and tertiary structure. And now, in this chapter, we have seen that metabolism, that orderly chemistry characteristic of life, is a concerted interplay of thousands of different kinds of molecules and hundreds of metabolic pathways. By completing our overview of metabolism with an introduction to its structural basis in the compartmentalized cell, we have built a bridge to the next unit of this book. In Unit Two, we will study the structure and function of the cell, maintaining our balance between the need to reduce life to a conglomerate of simpler processes and the ultimate satisfaction of viewing those processes in their integrated context.

STUDY OUTLINE

1. Metabolism is the sum of all the chemical reactions occurring in the cells of an organism. Metabolism is complex, efficient, well-integrated, and responsive to subtle changes in conditions.

The Metabolic Map (pp. 92–93)

1. The chemical reactions of metabolism manage the material and energy resources of the cell. Aided by enzymes, metabolism proceeds by steps along interrelated pathways.
2. A specific metabolic pathway is either catabolic or anabolic. Catabolic pathways, such as those of cellular respiration, break complex molecules into simpler compounds, releasing energy in the process. Anabolic pathways build up complex molecules from simpler compounds, requiring energy input usually provided by catabolism.

Energy: Some Basic Principles (pp. 94–95)

1. Energy is the capacity to do work by moving matter against an opposing force.
2. Kinetic energy, the energy of motion, does its work by transferring motion from one body of matter to another.
3. Potential energy is stored energy that results from the specific location or arrangement of matter. Chemical energy is a form of potential energy stored in molecular structure.
4. Energy can be changed from one form to another, governed by the laws of thermodynamics.
5. The first law of thermodynamics, conservation of energy, states that energy cannot be created or destroyed. Thus, the total amount of energy in the universe is constant.
6. The second law of thermodynamics states that every time energy changes form, there is an increase in entropy *(S)*, a measure of disorder, or randomness. Some energy is dissipated as heat, random molecular motion. Whenever matter becomes more ordered it does so at the expense of contributing to the disorder of its surroundings.

Chemical Energy: A Closer Look (pp. 95–99)

1. The amount of energy required to break a particular chemical bond is the same as the amount released when that bond forms. This quantity of energy is called bond energy.
2. Enthalpy *(H)* is the total potential energy of a molecule, as measured by its heat content. The heat of reaction *(ΔH)* is the net change in stored heat content when reactants are converted to products.
3. Reactions with a net release of heat $(-\Delta H)$ are exothermic. Endothermic reactions absorb heat from the surroundings $(+\Delta H)$, reflecting the upgrading of chemical energy in the products at the expense of energy from the surroundings.
4. Exergonic (spontaneous) reactions are those that occur without net input of energy from the surroundings. They proceed only if there is a net decrease in free energy $(-\Delta G)$ of the system, which depends on the magnitudes of the enthalpy and entropy changes and the temperature *(T)*. These relationships are expressed in the formula: $\Delta G = \Delta H - T\Delta S$.
5. Endergonic (nonspontaneous) reactions cannot occur without a supply of energy from the surroundings $(+\Delta G)$.
6. In metabolism, exergonic reactions are used to power endergonic reactions.
7. A reaction approaches equilibrium spontaneously (ΔG is negative). To move a reaction away from its equilibrium, a cell must add free energy.

ATP and Cellular Work (pp. 100–102)

1. ATP (adenosine triphosphate) serves as the main energy shuttle in cells. Hydrolysis of one of its weak phosphate bonds produces ADP (adenosine diphosphate) and inorganic phosphate, in an exergonic reaction that releases free energy.
2. ATP drives endergonic reactions in the cell by the enzymatic transfer of the phosphate group to specific reactants. The phosphorylated intermediates formed are more reactive than the original molecules. In this way, cells can carry out work such as movement, active transport, and anabolism.
3. The regeneration of ATP from ADP and phosphate is an endergonic reaction, driven primarily by cellular respiration and light-driven reactions in photosynthesis.

4. A steady supply of reactants and the removal of end products prevent metabolism from reaching equilibrium.

Enzymes (pp. 102–108)

1. Enzymes are proteins that function as biological catalysts, agents that change the rate of a reaction without being consumed in the reaction.
2. Before a reaction can occur, the reactants must absorb enough energy to break existing bonds. This free energy of activation ($\Delta G^{\ddagger}$) is usually provided in the form of heat absorbed from the surroundings, which causes the reactants to reach an unstable transition state required for the reaction to proceed. Biological macromolecules, with their orderly structure (low entropy), would decompose spontaneously if not for high activation energies.
3. Enzymes allow molecules to react in metabolism by lowering activation energies. This allows bonds to break at the fairly low body temperatures characteristic of most organisms.
4. Each type of enzyme has a uniquely shaped active site, which gives it specificity in combining with its particular substrate molecules.
5. The active site of an enzyme can lower activation energy in a number of ways: by providing a template for substrates to come together in proper orientation; by binding to the substrate in such a way that critical bonds of the substrate are strained; and by providing suitable microenvironments. Some enzymes even participate covalently in a reaction, after which they are chemically restored to their original state.
6. As proteins, enzymes are very sensitive to environmental conditions that influence the weak chemical bonds responsible for their three-dimensional structure. Each enzyme has optimal conditions of temperature, pH, and salt concentration, which determine the most active conformation. Outside these narrow limits, conformation changes and activity decreases.
7. Cofactors are nonprotein ions or molecules required for the function of some enzymes. If the cofactor is organic, it is known as a coenzyme.
8. Enzyme inhibitors are chemicals that selectively inhibit enzyme function, either reversibly through the formation of weak bonds or irreversibly through covalent bonds.
9. A competitive inhibitor is structurally similar to the substrate and can bind to the active site in its place. A noncompetitive inhibitor binds to a place on the enzyme other than the active site, disrupting the shape and function of the active site.
10. Some enzymes change shape when regulator molecules, either activators or inhibitors, bind to specific allosteric receptor sites. Allosteric sites are usually located between the subunits of complex enzymes. Induced fit by the binding of substrate activates other attached subunits in a phenomenon called cooperativity.

The Control of Metabolism (pp. 108–109)

1. One of the most common methods of regulating metabolism is by feedback inhibition, in which the end product of a metabolic pathway inhibits the first enzyme in that pathway. In this way, a cell can conserve resources by producing certain molecules only when they are in low concentration.
2. Some enzymes occur in multienzyme complexes, organized in assembly-line fashion for efficient catalysis of key reaction sequences. Enzymes may also be built into membranes or dissolved in relatively high concentration within specialized cell compartments.

Emergent Properties: A Reprise (p. 109–110)

1. In this unit, we have seen how increasing levels of organization result in the emergence of novel properties different from those of lower levels.

SELF-QUIZ

1. According to the first law of thermodynamics
   a. matter can be neither created nor destroyed
   b. energy is conserved in all processes
   c. all processes increase the entropy of the universe
   d. systems rich in energy are intrinsically unstable
   e. the universe constantly loses energy because of friction

2. For a process to occur spontaneously, which of the following conditions *must* be met?
   a. ΔH must be negative.
   b. ΔS must be positive.
   c. $(-T\Delta S)$ must be negative.
   d. ΔG must be negative.
   e. ΔG must be positive.

3. Which of the following metabolic processes can occur without a *net* influx of energy from some other process?
   a. $ADP + \text{\textcircled{P}} \longrightarrow ATP + H_2O$
   b. $C_6H_{12}O_6 + 6O_2 \longrightarrow 6CO_2 + 6H_2O$
   c. $6CO_2 + 6H_2O \longrightarrow C_6H_{12}O_6 + 6H_2O$
   d. $CO_2 + 2H_2O \longrightarrow CH_4 + 2O_2$
   e. glucose + fructose $\longrightarrow$ sucrose

4. The phosphate bonds of ATP are referred to as high-energy bonds because
   a. their bond energy is relatively high
   b. they are relatively strong bonds
   c. their hydrolysis is exergonic
   d. their formation releases a relatively large amount of energy
   e. their hydrolysis has a relatively high free energy of activation

5. Which molecule binds to the active site of an enzyme?
   a. substrate
   b. allosteric activator
   c. allosteric inhibitor
   d. noncompetitive inhibitor

6. If an enzyme solution is saturated with substrate, then the most effective way to obtain an even faster yield of products would be to
   a. add more of the enzyme

b. heat the solution to 90°C

c. add more substrate

d. add an allosteric inhibitor

e. add a noncompetitive inhibitor

7. An enzyme accelerates a metabolic reaction by

   a. altering the overall free energy change for the reaction

   b. making an endergonic reaction occur spontaneously

   c. lowering the free energy of activation

   d. pushing the reaction away from equilibrium

   e. making the substrate molecule more stable

8. Some bacteria are metabolically active in hot springs because

   a. they are able to maintain an internal temperature much cooler than that of the surrounding water

   b. the high temperatures facilitate active metabolism without need of catalysis

   c. their enzymes have high temperature optima

   d. their enzymes are insensitive to temperature

   e. they use molecules other than proteins as their main catalysts

9. Which metabolic process in bacteria is directly inhibited by the antibiotic penicillin?

   a. cellular respiration

   b. ATP hydrolysis

   c. synthesis of fats

   d. synthesis of chemical components of the cell wall

   e. replication of DNA, the genetic material

10. In the following branched metabolic pathway, a dotted arrow with a minus sign symbolizes inhibition of a metabolic step by an end product

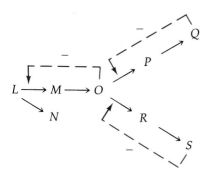

Which reaction would prevail if both Q and S are present in the cell in high concentrations?

a. $L \longrightarrow M$

b. $M \longrightarrow O$

c. $L \longrightarrow N$

d. $O \longrightarrow P$

e. $R \longrightarrow S$

CHALLENGE QUESTIONS

1. The CO_2 produced as a waste product of your cells' metabolism reacts with water to form carbonic acid:

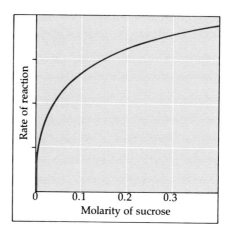

   Use Table 6.1 on p. 97 to compute approximate ΔH for the reaction. To do this, imagine that *all* existing bonds in the reactants are broken, and *all* bonds in the product form anew.

2. Biologists often hear the following type of argument: "Evolutionists claim that the complexity of organisms increased during the history of life. Such evolution of greater biological order contradicts the second law of thermodynamics, which is known to be unbreakable. Therefore, biological evolution is a scientifically invalid concept." How would you respond to this argument?

3. In an experiment, the enzyme sucrase is mixed with various concentrations of its substrate, sucrose. Each test tube begins with a certain sucrose concentration, and all tubes contain the same concentration of the enzyme. The rate of the reaction—conversion of substrate to product—is measured for each of the samples, and the results are plotted on the following graph:

Explain the shape of the curve in the graph.

FURTHER READING

1. Becker, W. M. *The World of the Cell.* Menlo Park, Calif.: Benjamin/Cummings, 1986. Chapters 5 and 6 offer lucid explanation of cellular energetics and enzymes.
2. Harold, F. M. *The Vital Force: A Study of Bioenergetics.* New York: Freeman, 1986. First three chapters provide a challenging but clear introduction to energy and life.
3. Koshland, D. E. "Protein Shape and Biological Control." *Scientific American*, October 1973. How enzymes are regulated.
4. Stryer, L. *Biochemistry.* 3rd ed. San Francisco: Freeman, 1989. Chapters 6 and 11 are introductions to enzymes and metabolism.

The Cell

Interview: Charles Leblond

*Over the mantel in the century-old anat-
omy library at McGill University in Montreal is an imposing portrait of Professor
Charles Leblond, one of Canada's most
famous scientists. Dr. Leblond was born in
France in 1910 and received his M.D.
degree there before moving to Canada,
where he joined the anatomy department at
McGill during the early 1940s. One of the
pioneers of modern cell biology and histology (the study of tissues), Professor Leblond
has spent most of his life studying the
dynamic renewal of the body's cells and
their subcellular components. In this interview, Dr. Leblond traces the historical
development of this work.*

**The 1950s and '60s, when electron
microscopes and other new instruments became available, were golden
decades in the study of cell structure
and function. What was it like to be
involved in cell biology during that
exciting period?**

Well, my own interest was in using
radioactive isotopes to localize key
compounds in cells and tissues. I
mostly did this work at the level of the
simpler light microscopes. But at the
time, we were always thinking, "Ah, if
we could only go a little farther, if we
could only see more details in the
cells." And when we saw the first pictures coming out of Keith Porter's
lab—beautiful pictures taken with the

electron microscope showing all these
details within cells—that was really
fantastic.

**How can we help our students rescale
their spatial thinking to the microscopic level—so they can better understand the cell and its parts?**

This is a very important problem in
teaching biology. In our anatomy and
histology courses here at the medical
school, we try to take the students
through changes in scale in a series of
steps—for example, from the stomach
wall, which you can see with the naked
eye, to the tissue layers and cells of the
wall, which you can see with the light
microscope, to the organelles within
the cells, which you can see only with
the electron microscope. This gradation
makes the scale changes less extreme.
We never show the students pictures
taken with the electron microscope
without also showing them photographs taken with the light microscope
for orientation. It's difficult, but eventually students adjust to the different
scales at which we study life.

**You mentioned that you helped
develop techniques for localizing
chemical compounds within cells and
tissues. How is this done?**

The technique is called radioautography, or more commonly, autoradiography. Radioactive atoms or molecules
are introduced into organisms or cells
via injection. Most radioisotopes
behave in the body in the same way as
nonradioactive isotopes of the same
chemical elements—except researchers

can detect the presence of the radioisotopes, of course. So cells use their normal metabolism to process these radioactive tracers. For example, cells take up radioactively labeled amino acids and incorporate these precursors into proteins. These labeled proteins can then be localized within tissues and cells by placing thin sections of the cells on a glass slide and coating the sections with photographic emulsion. The sections are then placed in the dark for a certain period, and the radiation emanating from radioactive sites in the cells exposes the photographic emulsion. The slide is developed and this preparation, now called a radioautograph, is examined with the microscope. Black silver grains from the photographic emulsion are seen directly over the sites in the cell containing the products made from the labeled precursors. So radioisotopes function in research as spies that report what is going on in the body.

You helped develop this technique?

Well, the first radioautographs were made by Antoine Lacassagne in Paris in the 1920s. He injected radioactive polonium into rabbits, embedded the organs in paraffin, and then placed these paraffin blocks in a dark room with photographic plates kept in contact with the surfaces of the blocks. He used small weights to keep pressure on the plates. He found that the polonium was mostly concentrated in lymph nodes, but the results were very crude. Lacassagne invited me to join his laboratory in Paris in the 1930s, and I began my work on the behavior of iodine in the body.

Using radioautography?

Yes. Well, first, I actually used a Geiger counter to locate radioactive iodine that had been injected into animals. This demonstrated that the iodine was concentrated in the thyroid gland. Then I tried Lacassagne's radioautography—except, instead of using the whole organ in a paraffin block, I cut sections of the block and placed the sections of thyroid on a glass slide. Then I put a photographic plate on the slide. I was able to show that the tracer had been incorporated into the colloidal material within the follicles of the thyroid gland. But the resolution was still poor.

How did you improve the technique?

A few years later, after I was discharged from military service in 1946, I returned to McGill University. A histologist named Leonard Belanger and I then concentrated on improving radioautographs. Instead of using a photographic plate, we found that we could melt the emulsion off film and then paint this emulsion directly onto our radioactively labeled sections of tissues. The improvement was dramatic, and we began to localize tracers in tissues at high magnification with the light microscope. Later, of course, it was possible to examine radioautographs with the electron microscope and study the passage of radioactive labels through the different organelles of cells.

Can you give some examples?

Well, we were able to show how the cells of the thyroid gland produce, store, and secrete the thyroid hormone, which regulates metabolism in the body. In other experiments, we used radioautography to study the metabolic functions of the organelle called the Golgi apparatus. By using chemical precursors labeled with radioisotopes, we learned that the Golgi is where the cell adds sugars to proteins. This turns out to be one of the most important functions of the Golgi. There were many other discoveries. My lab at McGill was like a bee hive—so many researchers using these new techniques that we didn't even have time to publish all our results, although we've managed to publish well over 300 papers since that time.

In one of your articles, you wrote that the use of radioisotopes has added a time dimension to cell biology. Can you explain?

Before radioactive isotopes were first used in this century to trace the chemical dynamics of cells, the prevailing view, especially among anatomists, was that the human body was composed of organs that were stable, with chemical constituents that were permanent for as long as the person lived. There seemed to be some support for this view. For example, in the late 1800s, several German scientists showed that the amount of nitrogen an organism ingests as food protein is equal to the amount of nitrogen leaving the body in feces and urine. The conclusion was that the organs kept their own proteins intact, and simply broke down the proteins from food and got rid of the nitrogenous waste products. The body, in this view, was like a locomotive that simply consumed fuel in the form of food. The organs were like the cogs and wheels of the locomotive, using the food to do work, but not actually incorporating any of the molecules of the food into their own structure. The use of detectable isotopes to trace chemicals in the body overthrew this old view of a stable body with organs that were static in their chemical makeup.

How?

In the late 1930s, Schoenheimer injected animals with amino acids labeled with a heavy isotope of nitrogen. Less than a third of this nitrogen subsequently turned up in the feces and urine. The other two-thirds of the injected radioactive nitrogen was incorporated into the organs of the body. About the same time, Georg Hevesy, who won a Nobel Prize, injected rats with radioactive phosphate and found that this labeled phosphate was incorporated into bone tissue and also turned up in phospholipids and other chemical components of various tissues. These experiments demonstrated the dynamic turnover of substances making up cells, but did not provide precise locations for the movement of these transient chemicals through the cells, tissues and organs. That's where radioautography came in and introduced the time dimension into cell biology.

So some of our organs are constantly changing their cellular parts?

Yes. For example, we were able to show that all of the cells of the stomach lining are replaced every few days. Yet, even as the new cells replace the dead cells that are sloughed from the lining, the histological pattern—the arrangement of cells in the tissue of the stomach lining—is unchanged. Six hundred years ago, Thomas Aquinas realized that the components of the human body are always changing—with age, for instance—but he also pointed out that the identity of the individual is retained. The same can be said of the dynamic states of cells and tissues: There is constant turnover of components, but the same cell types and arrangement of cells in the tissue are continuously regenerated. Methods that give cell biology a time dimension are critical to our understanding of this renewal process.

You mentioned how you began your research in cell biology in France soon after finishing medical college. When did you come to Canada?

Well, before Canada, I went to Yale on a Rockefeller fellowship, where I continued my work on cell biology. I also got involved in a very different type of project. On weekends, I went down to New York, to the American Museum of Natural History, where I became quite friendly with the head of the amphibian section. This was during the depression, and he had 23 WPA workers helping him. I helped find things to keep them busy. Twenty-three people, some of them Ph.D's, and there they were, out of work. Some of them were German immigrants, and so I put them to work translating Konrad Lorenz's book on animal behavior. That was how Lorenz's great work first became widely known in America. Lorenz later received the Nobel Prize.

It was also during my period in the United States that I met and married my wife Gertrude, who is American. Then the war started, and I was in the French Army. For complicated reasons, I was a private, even though I was a medical doctor. I was sent to Morocco, but I managed to get out after six months at the end of 1940. It was then that McGill University offered me a job in their anatomy department. And at the same time, I signed up with the

Free French Army. So just a few years after I began working in Montreal, the Free French sent me to London to collaborate with British scientists on some research involving psychology and the selection of personnel for the war effort. My only connection to psychology was that I had once done some experiments on how injection of certain hormones induced maternal behavior in rats. That didn't seem to

have much to do with personnel selection, but at least I was able to keep intellectually active during the war. After the war, I received a telegram from McGill asking me to come back and I've been here since 1946.

The bulk of your life's work has been accomplished in Canadian universities. Are they very different from their counterparts in the United States?

Not any more. Over the past century, Canadian universities have changed gradually from the British system to an American system. When I arrived at McGill before the war, students still addressed their professors as "sir." By the end of the war, that formality had disappeared.

How do undergraduates prepare for medical education in Canada?

Now that's a little different here in Quebec compared to the United States and even the rest of Canada. After high school, all college students spend their first two years at a junior college, taking general education courses. From there, students can go on to a university and complete their four-year degree. But many premedical students

spend only one year in university studies before entering medical school. During that year, they mostly study chemistry, physics, and biology. In the United States, I think, students graduate from college before entering medicine. About half of our students do the same, which I think has advantages. Students are more mature with another year of undergraduate studies.

After you finished medical school in 1934, why did you decide to concentrate on basic research rather than medicine?

Actually, I wanted to do research before I went to medical school. But my family wouldn't hear of it when I told them I wanted to be a scientist. Then I said, "Well, maybe I'll go into medicine." They were delighted. They thought that by the time I finished medical school, I would forget about science. My small medical school in France wanted me to stay there after graduation and eventually join the medical faculty. But I still had no intentions of practicing medicine. I did practice for about three months. I needed some money. That type of work just didn't appeal to me as much as research did. You see one patient, and then another, and then another, and at the end of a good day of hard work you have seen 25 patients. And then the next day you see 25 other patients. Many doctors like the variety. Myself, I'd like to work with one patient for several days. In any case, I never had any doubt about wanting to be a scientist and having the fun of doing research.

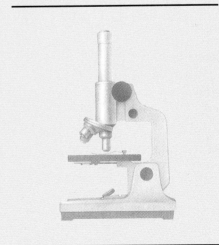

7

A Tour of the Cell

The cell is as fundamental to biology as the atom is to chemistry: All organisms are made of cells. In the hierarchy of biological organization, the cell is the simplest collection of matter that can live. Indeed, there are diverse forms of life existing as single-celled organisms. More complex organisms, including plants and animals, are multicellular; their bodies are cooperatives of many kinds of specialized cells that could not survive for long on their own. However, even when they are arranged into higher levels of organization, such as tissues and organs, cells can be singled out as the organism's basic units of structure and function. It is the contraction of muscle cells that moves your eyes as you read this sentence, and when you turn the page, it will be nerve cells that transmit that decision from your brain to the muscle cells of your hand. Everything an organism does is ultimately happening at the cellular level. This chapter introduces the microscopic world of the cell by surveying its overall geography and describing its various functional components, or organelles.

This text takes a thematic approach to the study of life, and the cell is a microcosmic model of many of the themes introduced in Chapter 1. We shall see that life at the cellular level arises from structural order, reinforcing the theme of emergent properties. For example, the movement of an animal cell depends on an intricate interplay of tubules and filaments within the cell, none of which can contract on its own (Figure 7.1). A related theme is the correlation of structure and function. Because every ordered process is based on an ordered structure, analyzing the anatomy of the cell rewards us with clues about how the cell works.

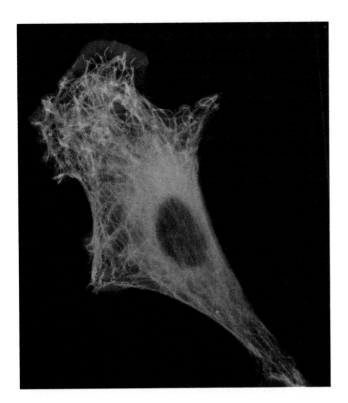

$\vdash\!\!-\!\!-\!\!-\!\!-\dashv$ 10 μm

Figure 7.1
Ordered functions are based on ordered structures.
This animal cell can change its shape and move along a
substratum. Cell motility depends on precise interactions
of tubules and filaments within the cell. Throughout our
study of the cell, we will correlate cellular functions with
structural organization (LM).

Another recurring theme in biology is the interaction
of organisms with their environment. Cells are excitable units that sense and respond to environmental
fluctuations, and, as open systems, they continuously
exchange both materials and energy with their surroundings. Furthermore, as we explore the cell, keep
in mind the one biological theme that unifies all others:
evolution. Though all cells are related to some extent
by their descent from the earliest organisms, they have
been modified in various ways during the long evolutionary history of life on Earth. For example, if one
unicellular organism lives in fresh water and another
inhabits the sea, we should expect these cells to be
somewhat differently equipped as a result of their
divergent adaptations to disparate environments.
Evolution is the basis for the correlations of structure
and function we observe in cells.

Perhaps the greatest mental block to becoming
acquainted with the cell is imagining how something
too small to be seen by the unaided eye can be so
complex. Or, put another way, how can cell biologists,
such as Charles Leblond, whose interview introduces
this unit, possibly dissect so small a package to investigate its inner workings? Before we actually tour the
cell, it will be helpful to learn how cells are studied.

HOW CELLS ARE STUDIED

The evolution of a science often parallels the invention
of instruments that extend human senses to new limits. The discovery and early study of cells progressed
with the invention and improvement of microscopes
in the seventeenth century. Microscopes of various
types are still indispensable tools in the study of cells.

Microscopy

The microscopes first used by Renaissance scientists,
as well as the microscopes you are likely to use in the
laboratory, are all **light microscopes.** Visible light is
passed through the specimen and then through glass
lenses with shapes that refract (bend) the light in such
a way that the image of the specimen is magnified as
it is projected into the eye.

Two important factors in microscopy are magnification and resolving power, or resolution. Magnification is how much larger the object appears compared to its real size. **Resolving power** is a measure
of the clarity of the image; specifically, it is the minimum distance that two points can be separated and
still be distinguished as two separate points. For
example, what appears to the unaided eye as one star
in the sky may be resolved as twin stars with the help
of a telescope.

Just as the resolving power of the human eye is
limited, the resolving power of telescopes and microscopes is limited. Microscopes can be designed to
magnify objects as much as desired, but the light
microscope can never resolve detail finer than about
0.2 μm, which is the size of small bacteria and mitochondria (Figure 7.2). This resolution cannot be
improved upon; it is limited not by flaws in the microscope but by the wavelength of the visible light used
to illuminate the specimen. Light microscopes can
magnify to about 1500 times the size of the actual specimen; greater magnifications are useless, since they
increase blurriness. Most of the improvements in light
microscopy since the beginning of this century have
involved new methods for enhancing contrast, a quality that makes the details that *can* be resolved stand
out better to the eye (Table 7.1, p. 120).

Although cells were discovered by Robert Hooke in
1665, the geography of the cell was largely uncharted
until the past few decades. Most subcellular structures, or **organelles,** are too small to be resolved by
the light microscope. Cell biology began a rapid advance
in the 1950s when investigators began using a powerful new tool, the **electron microscope.** Instead of
using visible light, the electron microscope (EM) focuses
a beam of electrons through the specimen (Figure 7.3).
Resolving power is inversely related to the wavelength of radiation a microscope uses, and electron

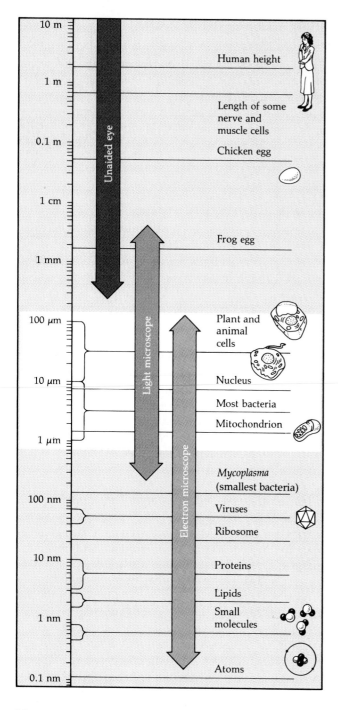

MEASUREMENTS

1 centimeter (cm) = 10^{-2} meter = 0.4 inch
1 millimeter (mm) = 10^{-3} meter
1 micrometer (μm) = 10^{-3} mm
1 nanometer (nm) = 10^{-3} μm

Figure 7.2
The size range of cells. Most cells are between 1 and 100 μm in diameter and therefore are visible only under a microscope. Note that the scale is logarithmic to accommodate the range of sizes shown.

beams have wavelengths much shorter than the wavelengths of visible light. Modern electron microscopes can achieve a resolution of about 0.2 nanometer (nm), a thousandfold improvement over the light microscope. The detailed structure of a cell, as seen with the electron microscope, is referred to as **ultrastructure.**

There are two types of electron microscopes: the **transmission electron microscope (TEM)** and the **scanning electron microscope (SEM).** The TEM aims an electron beam through a thin section of the specimen, similar to the way the light microscope transmits light through a slide. Instead of using glass lenses, which are opaque to electrons, however, the TEM uses electromagnets as lenses to focus and magnify the image by bending the trajectories of the charged electrons. The image is ultimately focused onto a screen for viewing or onto photographic film for a permanent record. The image is a negative, with bright areas corresponding to places of low density in the specimen; through these areas the electrons pass unimpeded. Dark areas in the image correspond to dense areas in the specimen; these areas block or scatter the electrons. Most atoms in cells are small and have little effect on the electron beam. To enhance contrast in the image, very thin sections of preserved cells are stained with heavy atoms of metals, which attach to certain places in the cells and block electrons when the section is viewed in the TEM (Figure 7.4a). Cell biologists use the TEM mainly to study the internal ultrastructure of cells.

The SEM is especially useful for detailed study of the surface of the specimen (Figure 7.4b). The electron beam scans the surface of the sample, which is usually coated with a thin film of gold. The beam excites electrons on the sample surface itself, and these secondary electrons are collected and focused onto a screen, forming an image showing the topography of the specimen. An important attribute of the SEM is its great depth of field, which results in an image that appears three-dimensional.

Electron microscopes have revealed many organelles that are impossible to resolve with the light microscope. But the light microscope offers many advantages, especially for the study of live cells. A disadvantage of electron microscopy is that the chemical and physical methods used to prepare the specimen not only kill cells, but also may introduce artifacts, which are structural features that did not exist in the living cell.

Microscopes of various kinds are the most important tools of **cytology,** the study of cell structure. But simply describing the diverse organelles within the cell reveals little about their function. Modern cell biology developed from an integration of cytology with biochemistry, the study of metabolism and its products. A biochemical approach called cell fractionation has been particularly important in this multidisciplinary synthesis of cell biology.

Figure 7.3
A comparison of light microscopes and transmission electron microscopes.
The details are different, but the principles are similar. In standard light microscopy (bright-field), light from a light source (bottom of the microscope) is focused on a specimen by a glass condenser lens; and the image is subsequently magnified by an objective lens and an ocular lens, for projection on the eye or photographic film. In electron microscopy, a beam of electrons (top of the microscope) is used instead of light, and electromagnets instead of glass lenses (which are opaque to electron beams). The electron beam is focused on the specimen by a condenser lens and the image magnified by an objective lens and projection lens, for projection on a screen or photographic lens.

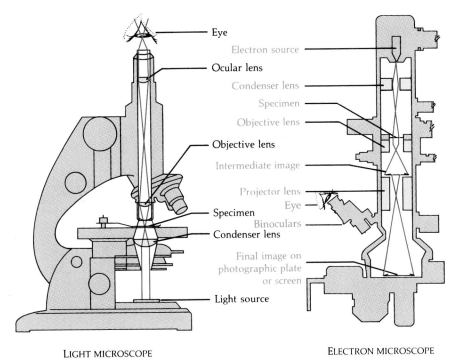

Eye
Electron source
Ocular lens
Condenser lens
Specimen
Objective lens
Objective lens
Intermediate image
Projector lens
Eye
Specimen
Binoculars
Condenser lens
Final image on photographic plate or screen
Light source

LIGHT MICROSCOPE

ELECTRON MICROSCOPE

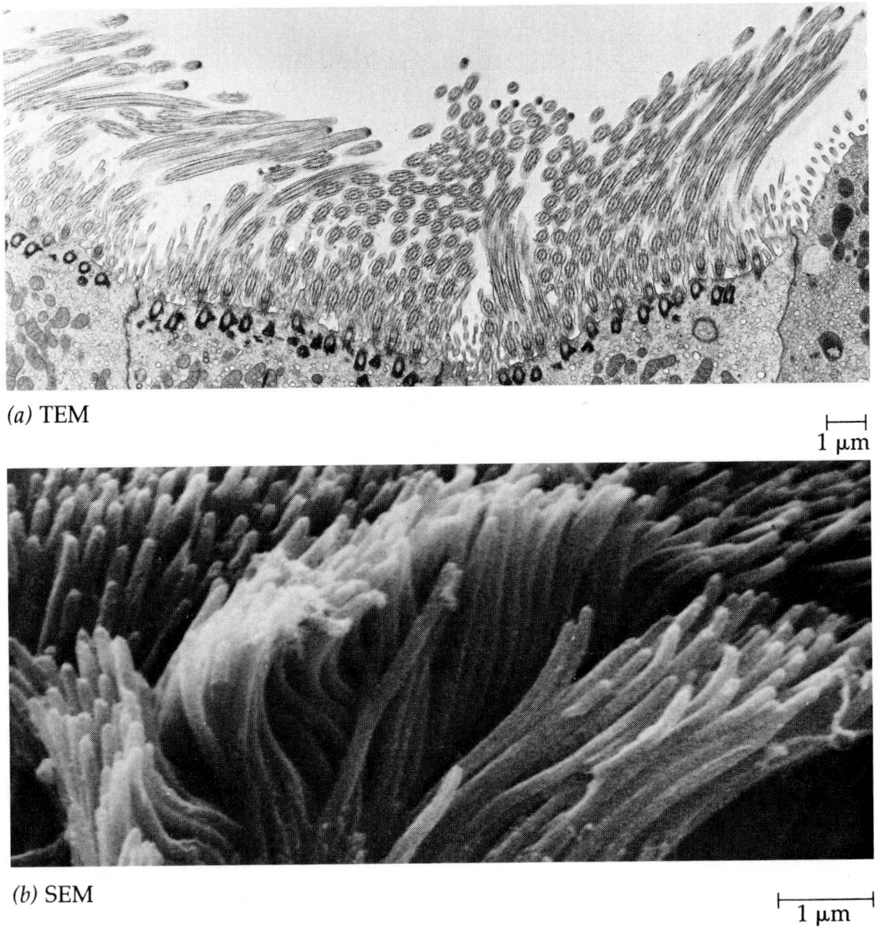

(a) TEM

1 μm

Figure 7.4
Electron micrographs: photographs taken with electron microscopes.
(a) This micrograph taken with a transmission electron microscope (TEM) profiles a thin section of a cell from a rabbit trachea (windpipe), revealing its ultrastructure.
(b) The scanning electron microscope (SEM) produces a three-dimensional image of the surface of the same type of cell and does so with a great depth of field. The projections from the cell surface seen in both micrographs are motile organelles called cilia. Beating of the cilia that line the windpipe helps move inhaled debris upward back toward the pharynx (throat).

(b) SEM

1 μm

Table 7.1 Different types of light microscopy: a comparison

Type of Microscopy	Features	Light Micrographs of Human Cheek Epithelial Cells
Bright-field (unstained specimen)	Passes light directly through specimen; unless cell is naturally pigmented or artificially stained, the image has little contrast	 *(a)*
Bright-field (stained specimen)	Staining with various dyes enhances contrast, but most staining procedures require that the cells be fixed (preserved)	 *(b)*
Dark-field	Passes light through specimen obliquely, and only light scattered by particles can be seen	 *(c)*
Phase contrast	Enhances contrast in unstained cells by amplifying variations in density within the specimen; especially useful for examining living, unpigmented cells	 *(d)*
Differential-interference	Also uses optical modifications to exaggerate differences in density	 *(e)* ⊢——⊣ 50 μm

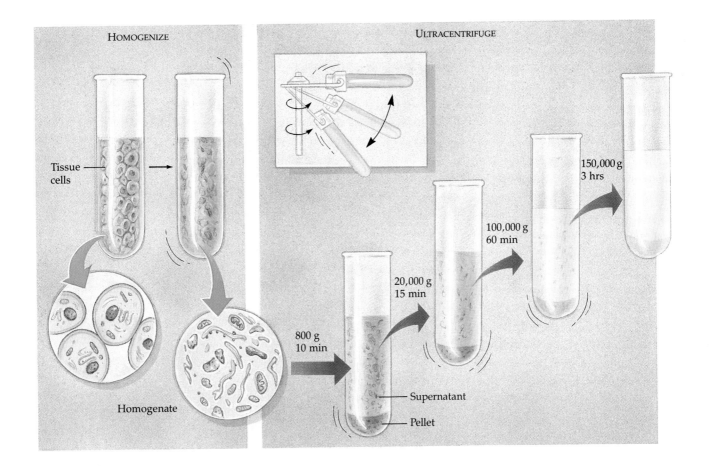

HOMOGENIZE

Tissue cells

Homogenate

ULTRACENTRIFUGE

800 g
10 min

20,000 g
15 min

100,000 g
60 min

150,000 g
3 hrs

Supernatant

Pellet

Figure 7.5
Cell fractionation. Disrupted cells are centrifuged at various speeds and durations to isolate components of different sizes, densities, and shapes. The process begins with homogenization, the disruption of a tissue and its cells with the help of such instruments as pistons, kitchen blenders, or ultrasound devices. The homogenate, a soupy mixture of organelles, bits of membrane, and molecules from the broken cells, is then fractionated by a series of spins in a centrifuge. A slow spin for a short time will cause only the largest components of the homogenate to settle to the bottom of the centrifuge tube as a pellet, which can be resuspended for study. The unpelleted portion, or supernatant, can then be decanted into another tube and centrifuged again, this time at a higher speed. The process is repeated, the speed and/or duration of the centrifugation being increased with each step until very small organelles, or even large molecules, are collected in pellets. By determining which cell fractions are associated with particular metabolic processes, those functions can be tied to certain organelles of the cell. Once a metabolic function has been assigned to a particular cell fraction, the unique enzymes associated with each fraction can serve as "markers" to identify the presence and purity of a particular organelle without researchers having to routinely examine the fraction with an electron microscope.

Cell Fractionation

The objective of **cell fractionation** is to take cells apart, separating the major organelles so that their individual functions can be studied. The instrument used to fractionate cells is the **centrifuge,** a merry-go-round for test tubes that is capable of spinning at various speeds. The fastest machines, called **ultracentrifuges,** can spin at more than 100,000 revolutions per minute (rpm).

Fractionation begins with homogenization, the disruption of cells (Figure 7.5). Vibrating cells with ultrasound, forcing the cells through small spaces, or grinding tissues in ordinary kitchen blenders are among the methods used for homogenization. The usual objective is to break the cells without severely damaging their organelles. The parts of the soupy homogenate are then separated by spinning in a centrifuge, first at a low speed to cause nuclei and other larger particles to settle to the bottom of the test tube, forming a pellet. The supernatant (unsedimented) fraction is then decanted into another tube and centrifuged again at a faster speed, sedimenting organelles that were too small to pellet during the first, slower spin. The process is repeated, increasing the speed with each step, collecting smaller and smaller components of the homogenized cells.

Cell fractionation enables the researcher to prepare specific components of cells in bulk quantity in order to study their composition and metabolism. By fol-

Figure 7.6
The prokaryotic cell. (a) Diagram of a typical rod-shaped bacterium. Lacking the membrane-enclosed organelles typical of a eukaryote, the prokaryotic cell is much simpler in structure. The genetic material (DNA) is coiled up in a region called the nucleoid, which shares the fluid interior of the cell with ribosomes (which synthesize proteins) and a large variety of dissolved molecules. The border of the cell is the plasma membrane, which in some prokaryotes invaginates in places to form structures called mesosomes. Outside the plasma membrane are a fairly rigid cell wall and, often, an outer capsule, usually jellylike. Prokaryotes include the bacteria and the cyanobacteria. **(b)** Electron micrograph of a thin section through a dividing cell of the bacterium *Bacillus cereus*.

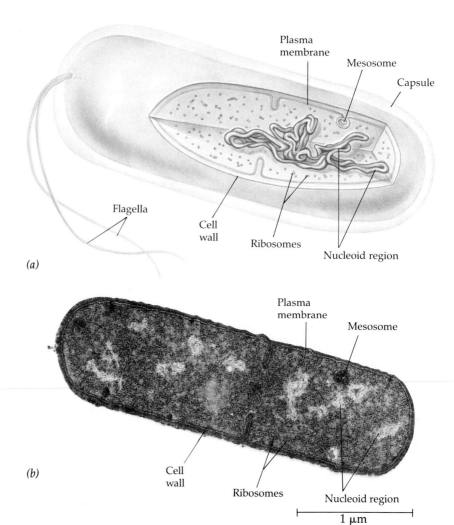

(a)

(b)

1 μm

lowing this approach, biologists have been able to assign various functions of the cell to the different organelles, a task that would be far more difficult if they had to study intact cells. For example, one cellular fraction collected by centrifugation has enzymes that function in the metabolic process known as cellular respiration (see Chapter 6). The prevalent organelles in that fraction match the structures called mitochondria, as visualized in the electron microscope. This discovery helped cell biologists determine that mitochondria are the sites of cellular respiration. And that deduction stimulated further investigation of mitochondrial architecture, which, in turn, helped clarify how cellular respiration works. Cytology and biochemistry complement one another in the correlation of cellular structure and function.

THE GEOGRAPHY OF THE CELL: A PANORAMIC OVERVIEW

Prokaryotic and Eukaryotic Cells

Every contemporary organism is composed of one of two radically different types of cells: **prokaryotic cells** or **eukaryotic cells.** Prokaryotic cells are found only the kingdom Monera, which consists of bacteria and cyanobacteria (formerly called blue-green algae). Protists, plants, fungi, and animals—four of the five kingdoms of life—are all eukaryotes (see Chapter 1). If we were to look for the sharpest division in the diversity of contemporary life, it would be the line separating prokaryotes from eukaryotes.

The two types of cells differ markedly in their internal organization. One difference is denoted by their names. The prokaryotic cell has no true nucleus (the word **prokaryote** is from the Greek *pro,* "before," and *karyon,* "kernel," referring here to the nucleus). Its genetic material is concentrated instead in a region called the nucleoid, but no membrane separates this region from the rest of the cell (Figure 7.6). In contrast, the eukaryotic cell (from the Greek *eu,* "true," and *karyon*) has a true nucleus enclosed by a membranous nuclear envelope. The entire region between the nucleus and the membrane bounding the cell is called the **cytoplasm.** It consists of a semifluid medium called the **cytosol,** in which are suspended organelles of specialized form and function, most of them absent in prokaryotic cells. Thus, the presence or absence of a true nucleus is just one example of the disparity in structural complexity between the two types of cells.

The prokaryotic cell will be described in detail in Chapters 17 and 25, and the possible evolutionary relationships between the two types of cells will be discussed in Chapter 26. Most of the discussion of cell structure that follows in this chapter applies to eukaryotes.

Cell Size

Size is a general feature of cell structure that relates to function. The logistics of carrying out metabolism sets limits on the size range of cells (see Figure 7.2). The smallest cells known are bacteria called mycoplasmas, which have diameters of between 0.1 and 1.0 micrometer (μm). Perhaps these are the smallest packages with enough DNA to program metabolism and enough enzymes and other cellular equipment to carry out the activities necessary for a cell to sustain itself and reproduce. Most bacteria are 1–10 μm in diameter, about ten times larger than the mycoplasmas. At 10–100 μm in diameter, eukaryotic cells are typically ten times larger than bacteria.

Metabolic requirements also impose upper limits on the size that is practical for a single cell. As an object of a particular shape increases in size, its volume grows proportionately more than its surface area. (Area is proportional to a linear dimension squared, whereas volume is proportional to the linear dimension cubed.) If the sides of a box increase by a factor of ten, then the surface area of the box increases a hundredfold, but the volume increases a thousandfold. For objects of the same shape, the smaller the object, the greater its ratio of surface area to volume (Figure 7.7).

At the boundary of every cell, the **plasma membrane** functions as a selective barrier that regulates the chemical composition of the cell by allowing some substances to pass readily between the cell and the external environment, while impeding the entrance or exit of other substances. This membrane must allow for sufficient traffic of oxygen, nutrients, and wastes to service the entire volume of the cell. For each square micrometer of membrane, only so much of a particular substance can cross per second. The need for a surface sufficiently large to accommodate its volume helps to explain the microscopic size of most cells. Another factor limiting cell size is the requirement for a single nucleus to control the entire cytoplasmic volume of the cell. The nucleus can better control a smaller cell, just as a single administrator can manage a smaller factory more efficiently than a larger one.

The Importance of Compartmental Organization

The internal complexity of a eukaryotic cell is another general structural feature correlated with functional requirements. The average eukaryotic cell, with a diameter about ten times greater than the average prokaryotic cell, has a thousand times the volume but only a hundred times the surface area of the prokaryotic cell. Eukaryotic cells compensate for their relatively small ratio of plasma membrane area to cytoplasmic volume by having internal membranes. The area of the plasma membrane alone may be inadequate to satisfy the metabolic needs of the eukaryotic cell. Some of the functions performed by the plasma membrane of a prokaryote are carried out by various internal membranes within the eukaryotic cell. Membranes serve as partitions, dividing the cell into compartments, and also participate directly in much of the cell's metabolism; many enzymes are built right into membranes. In addition, the more extensive compartmental organization of the eukaryotic cell provides many different local environments that facilitate specific metabolic functions. Thus, metabolic processes that are incompatible with each other can go on simultaneously in separate subcellular compartments.

Clearly, membranes of various kinds are fundamental to the complex organization of the cell. In general, biological membranes consist of a double layer of phospholipids and other lipids in which diverse proteins are embedded (see Chapter 5). However, each membrane, whether the plasma membrane or the membrane of a cytoplasmic organelle, has a unique composition of lipids and proteins suited to that membrane's specific functions. For example, many of the enzymes that function in cellular respiration are embedded in the internal membranes of organelles

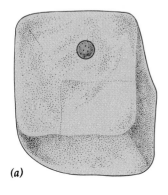

(a)

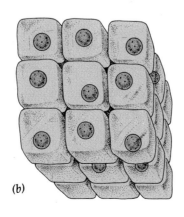

(b)

Figure 7.7
Why are most cells microscopic? (a) A very large cell would have a large volume relative to the surface area that must carry on exchange of materials with the environment to support the cell's metabolism. **(b)** If the same total volume is divided into a number of smaller cells, the surface area will be much greater. Also, the ratio of nucleus to cytoplasm is larger in this case, making control of the cell more manageable.

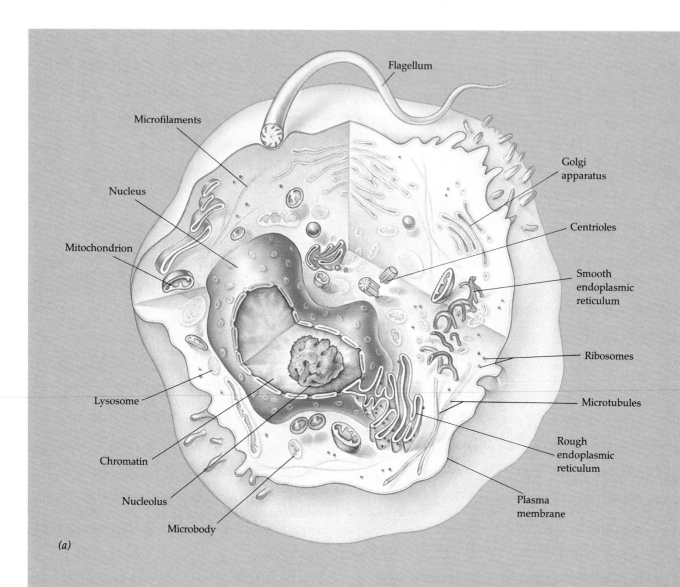

Flagellum

Microfilaments

Nucleus

Mitochondrion

Lysosome

Chromatin

Nucleolus

Microbody

Golgi apparatus

Centrioles

Smooth endoplasmic reticulum

Ribosomes

Microtubules

Rough endoplasmic reticulum

Plasma membrane

(a)

Figure 7.8
An animal cell. (a) The schematic drawing of a generalized animal cell combines the most common structural features found in animal cells. No one cell looks like this. Within the cell are a variety of components collectively called organelles ("little organs"). The most prominent organelle in eukaryotic cells is usually the nucleus, in which most of the cell's inherited genes reside in the form of DNA. The DNA is organized along with proteins into structures called chro-

mosomes; in the nondividing cell, they are not seen as individual structures but as a diffuse material called chromatin. Also present in the nucleus are one or more nucleoli. Nucleoli are involved in the assembly of particles named ribosomes, which function in the synthesis of proteins. The nucleus is bordered by an envelope consisting of two membranes.

Most of the cell's metabolic activities occur in the cytoplasm, the entire region between the nucleus and the plasma

membrane surrounding the cell. The cytoplasm is full of specialized organelles suspended in a semifluid medium called the cytosol. Pervading much of the cytoplasm is the endoplasmic reticulum (ER), a labyrinth of membranes forming flattened sacs and tubes that segregate the contents of the ER from the cytosol. The ER is of two types: rough (studded with ribosomes) and smooth. Many types of proteins are made by ribosomes attached to ER membranes, and the ER

called mitochondria. To a large extent, then, the study of cells is the study of membranes and the functional compartments (organelles) that are sequestered by these diverse membranes. Membranes are so important to how a cell works that they are the subject of the next chapter.

Before going on in this chapter, you should examine the panoramic overviews of eukaryotic cells in Figures 7.8 and 7.9 (p. 126). These figures and their legends introduce the various organelles and provide a map of the cell for the more detailed tour upon which we shall soon embark. Figures 7.8 and 7.9 also contrast

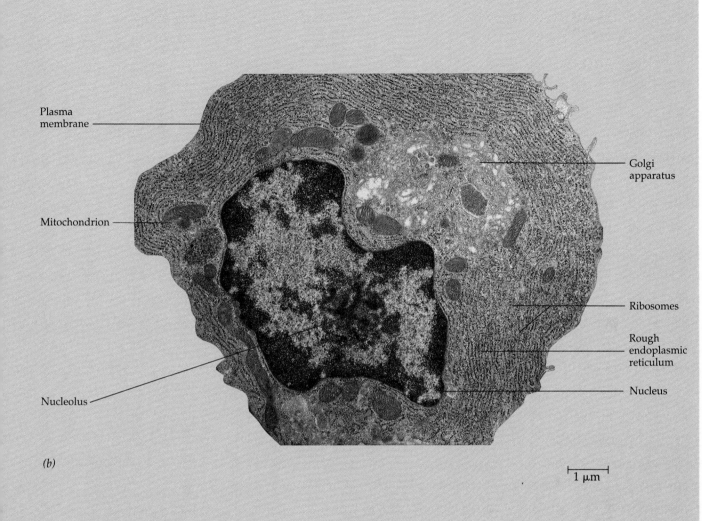

Plasma membrane

Mitochondrion

Nucleolus

Golgi apparatus

Ribosomes

Rough endoplasmic reticulum

Nucleus

(b)

1 μm

also plays a major role in assembling the other membranes of the cell. The Golgi apparatus, another type of membranous organelle in the cytoplasm, consists of stacks of flattened sacs that play an active role in synthesis, refinement, storage, sorting, and secretion of chemical products by the cell.

Other classes of organelles enclosed by membranes are: lysosomes, which contain mixtures of digestive enzymes that hydrolyze macromolecules; micro-bodies, a diverse group of organelles containing specialized enzymes for performing specific metabolic processes; and vacuoles, which have a variety of storage and metabolic functions. The mitochondria are organelles that generate ATP from organic fuel such as sugar in the process called cellular respiration.

Nonmembranous structures seen within the cells include microtubules and microfilaments, which form a framework called the cytoskeleton that reinforces the cell's shape and functions in cell movement. In the drawing we see a flagellum, an organelle of locomotion, which is basically an assembly of microtubules. Also made of microtubules are centrioles, located near the nucleus. These play an important role in cell division. **(b)** The electron micrograph that accompanies the drawing is a thin section of a rat plasma cell, a type of white blood cell.

animal and plant cells. As eukaryotic cells, they have much more in common with each other than either has with *any* prokaryote. There are, however, important differences between plant and animal cells. For example, a plant cell has a relatively thick cell wall external to its outer membrane; animal cells lack walls.

The first stop on our more detailed tour of the cell is one of the membrane-enclosed organelles, the nucleus.

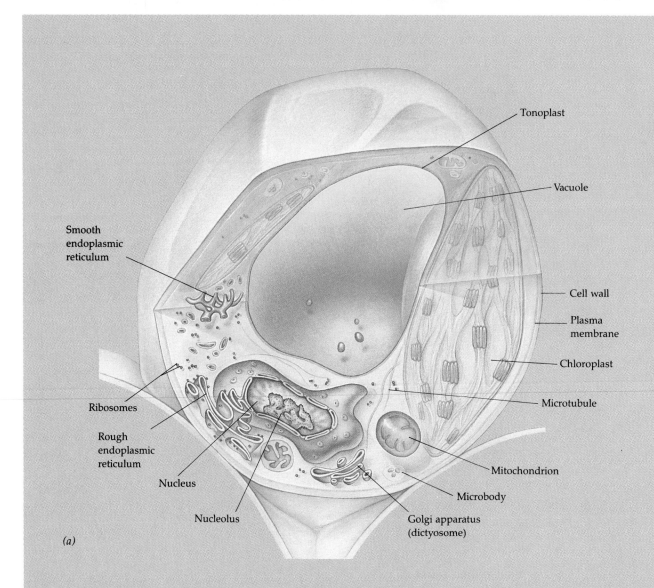

Tonoplast

Vacuole

Cell wall

Plasma membrane

Chloroplast

Microtubule

Mitochondrion

Microbody

Golgi apparatus (dictyosome)

Smooth endoplasmic reticulum

Ribosomes

Rough endoplasmic reticulum

Nucleus

Nucleolus

(a)

Figure 7.9
A plant cell. (a) A schematic drawing of a generalized plant cell reveals the similarities and differences between animal and plant cells. Like the animal cell, the plant cell contains a nucleus, ribosomes, ER, Golgi apparatuses (usually called dictyosomes in plants), mitochondria, microbodies, and microfilaments and microtubules; and it is surrounded by a plasma membrane. However, plant cells also contain organelles called plastids. One important type of plastid is the chloroplast, which carries out photosynthesis, converting sunlight to chemical energy stored in sugar and other organic mole-

THE NUCLEUS

The nucleus, which contains the genes that control the entire cell, is generally the most conspicuous organelle in a eukaryote, averaging about 5 μm in diameter (Figure 7.10 p. 128). The **nuclear envelope** encloses the nucleus, separating its contents from the cytoplasm.

The nuclear envelope is a double membrane. The two membranes, each a phospholipid bilayer with embedded proteins, are separated by a space of about 20–40 nm. Closely associated with the nuclear side of the inner membrane is a layer of protein that helps maintain the shape of the nucleus and may also help maintain the organization of the genetic material. The nuclear envelope is perforated by pores (Figure 7.11 p. 128). At the lip of each pore, the inner and outer membranes of the nuclear envelope are fused. The pore complex seems somehow to regulate the entrance and exit of certain large macromolecules and particles. Traffic between the nucleus and cytoplasm is extensive. For example, the process of making ribosomes requires that ribosomal proteins synthesized in the cytoplasm enter the nucleus and that the assembled ribosomal subunits exit.

Nearly all of the cell's DNA is located in the nucleus, organized along with proteins into **chromosomes.** Unless the cell is in the process of dividing, its chromosomes are too dispersed and entangled to be iden-

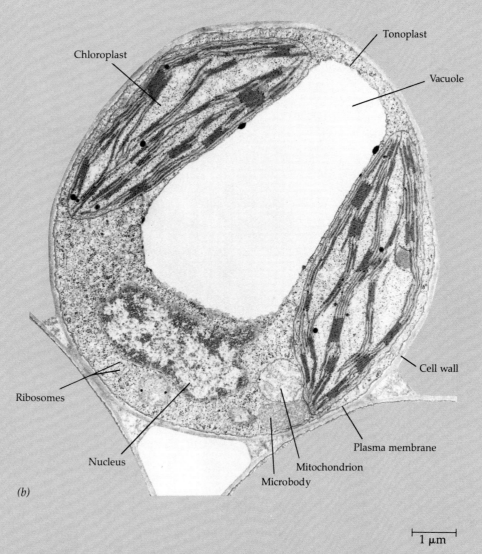

Chloroplast

Tonoplast

Vacuole

Cell wall

Plasma membrane

Mitochondrion

Microbody

Nucleus

Ribosomes

(b)

1 μm

cules. Another prominent organelle in many plant cells is a large central vacuole, which stores chemicals, functions as a lysosome, and, by enlarging, plays a major role in plant growth. Outside the plasma membrane in plants (as well as in fungi and some protists) is a thick cell wall, which helps to maintain the cell's shape and protects the cell from mechanical damage. **(b)** Many of these organelles are visible in the accompanying electron micrograph of a young leaf cell from a coleus.

tified as individual structures. Instead, the aggregate of chromosomes appears through both light microscopes and electron microscopes as a mass of stained material collectively referred to as **chromatin.** Only when the nucleus prepares to divide do the chromosomes condense, becoming thick enough to be discerned as separate structures. Each eukaryotic species has a characteristic number of chromosomes. A human cell, for example, has 46 chromosomes in its nucleus; the exceptions are sex cells—eggs and sperm—which have only 23 chromosomes in humans.

The most visible structure within the nondividing nucleus is the **nucleolus,** which functions in the synthesis of ribosomes (see Figure 7.10). Sometimes, there are two or more nucleoli, the number depending on the species and the stage in the cell's reproductive cycle. The nucleolus is roughly spherical, and in the electron microscope it appears as a mass of densely stained granules and fibers. It consists of **nucleolar organizers,** specialized regions of some chromosomes with multiple copies of genes for ribosome synthesis, along with a considerable amount of RNA and proteins representing ribosomes in various stages of production. An actively growing cell can produce about 10,000 ribosomes per minute.

The nucleus controls protein synthesis in the cytoplasm by sending molecular messengers in the form of RNA (Figure 7.12 p. 129). This **messenger RNA**

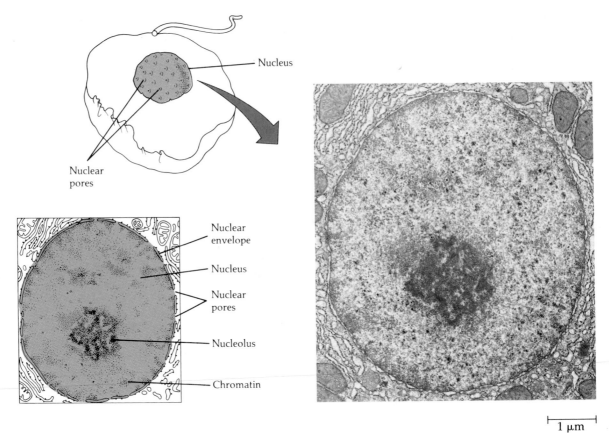

Figure 7.10
The nucleus. The nuclear envelope, which consists of two membranes separated by a narrow space, is perforated with pores. Within the nucleus, the genetic material is in the dispersed form known as chromatin; a darker-staining nucleolus is also visible. The nucleolus is the site of ribosome synthesis (TEM).

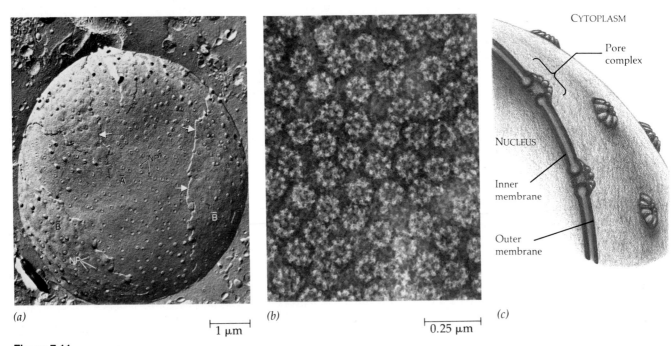

(a)   *(b)*   *(c)*

Figure 7.11
The nuclear envelope. (a) Numerous nuclear pores (NP) through the envelope are evident in this electron micrograph, prepared by a method called freeze-fracture (see p. 158). **(b)** An electron micrograph of the outer surface of the envelope reveals that each pore is bordered by a ring of eight protein particles. **(c)** The pore complexes are thought to regulate the passage of large molecules and molecular complexes into and out of the nucleus.

(mRNA), as it is called, is synthesized in the nucleus according to instructions provided by DNA, after which the mRNA conveys the genetic messages to the cytoplasm via nuclear pores. Once in the cytoplasm, the RNA attaches to ribosomes, and the genetic message is translated into the primary structure of a specific protein. This process is described in detail in Chapter 16.

RIBOSOMES

Ribosomes are the sites where the cell assembles enzymes and all other proteins according to genetic instructions. A bacterial cell may have a few thousand ribosomes, whereas a human liver cell has a few million. Cells that have high rates of protein synthesis have a particularly great number of ribosomes, another example of cell structure fitting function. Cells active in protein synthesis also have prominent nucleoli, which make the ribosomes.

Each ribosome is built from two subunits with the shapes shown in Figure 7.13. In eukaryotes, the ribosomal subunits are constructed in the nucleolus from RNA made in the nucleolus itself, RNA produced elsewhere in the nucleus, and proteins imported from the cytoplasm. The subunits join to form functional ribosomes only after export to the cytoplasm. The ribosomes of prokaryotes are smaller than those of eukaryotes and differ somewhat in their molecular composition. Dissimilarity in a structure so fundamental as the ribosome is an example of the profound dichotomy that separates prokaryotes from eukaryotes. The difference in ribosomes is also medically significant; certain drugs can paralyze prokaryotic ribosomes without inhibiting the ability of eukaryotic ribosomes to make proteins. These drugs, which include tetracycline and streptomycin, are used as antibiotics to combat bacterial infections.

Ribosomes function in two cytoplasmic locales. **Free ribosomes** are suspended in the cytosol, while **bound ribosomes** are attached to the outside of a membranous network called the endoplasmic reticulum (see Figure 7.8). In both cases, the ribosomes usually occur in clusters called polyribosomes, or **polysomes.** A polysome consists of several ribosomes attached to one mRNA molecule, an arrangement that increases the number of polypeptides produced per minute. Most of the proteins made by free ribosomes will function within the cytosol, and free ribosomes are especially abundant in cells growing by the addition of cyto-

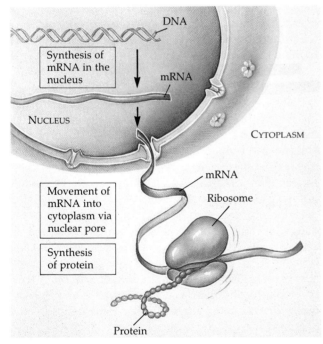

Figure 7.12
Nuclear control of protein synthesis: a diagramatic view. DNA in the nucleus programs protein production in the cytoplasm by producing messenger RNA (mRNA), which travels to the cytoplasm and binds to ribosomes. As a ribosome (much oversized in this drawing) moves along the mRNA (or vice versa), the genetic message conveyed from the nucleus is translated into a protein of specific amino acid sequence.

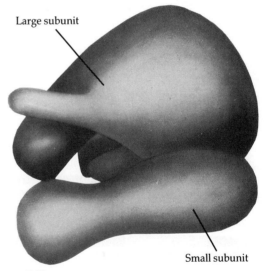

Figure 7.13
Structure of the eukaryotic ribosome. Each ribosome is made up of a large subunit and a small subunit that join together when they attach to messenger RNA in the cytoplasm and begin to make a protein. No membrane encloses the ribosome. The large subunit of a eukaryotic ribosome is constructed in the nucleolus from about 45 different protein molecules and 3 molecules of RNA (specifically called ribosomal RNA). The small subunit has about 33 proteins and 1 molecule of ribosomal RNA. The fully assembled ribosome of a eukaryotic cell measures about 20 nm by 30 nm. The ribosomes of prokaryotes are slightly smaller and differ somewhat in molecular composition.

plasm. Bound ribosomes generally make proteins that are destined either for inclusion into membranes or for export from the cell. Cells that specialize in protein secretion—for instance, the cells of the pancreas and other glands that secrete digestive enzymes—frequently have a high proportion of bound ribosomes. Bound and free ribosomes are structurally identical and interchangeable, and the cell can adjust the relative numbers of each as its metabolism changes.

THE ENDOMEMBRANE SYSTEM

Most cell biologists now consider many of the different membranes of the eukaryotic cell as part of an **endomembrane system.** These membranes are related either through direct physical contact or by the transfer of membrane segments through the movement of tiny vesicles (membrane-enclosed sacs). These relationships, however, do not mean that the various membranes are alike in structure and function. The thickness, molecular composition, and metabolic behavior of a membrane are not fixed, but may be modified several times during the membrane's history. The endomembrane system includes the nuclear envelope, endoplasmic reticulum, Golgi apparatus, lysosomes, microbodies, various kinds of vacuoles, and the plasma membrane (not actually an *endo*membrane in physical location but nevertheless related to the endoplasmic reticulum and other internal membranes). We have already discussed the nuclear envelope. We focus now on the endoplasmic reticulum and the other endomembranes to which it gives rise.

Endoplasmic Reticulum

The **endoplasmic reticulum (ER)** is a membranous labyrinth so extensive that it accounts for more than half of the total membrane in many eukaryotic cells. (*Endoplasmic* means within the cyto*plasm*, and *reticulum* is derived from a Latin word that means "network.") The ER consists of a network of membranous tubules and sacs called **cisternae** (from the Latin *cisterna*, a "box" or "chest"). The ER membrane sequesters its internal compartment, the cisternal space, from the cytosol. And because the ER membrane is continuous with the nuclear envelope, the space between the two membranes of the envelope is contiguous with the cisternal space of the ER (Figure 7.14).

There are two distinct, though confluent, regions of ER that differ in structure and function: **rough ER** and **smooth ER.** Rough ER appears "rough" in the electron microscope because ribosomes stud the cytoplasmic surface of the membrane. Ribosomes are also attached to the cytoplasmic side of the nuclear envelope's outer membrane, which is confluent with rough ER. It is mainly rough ER that manufactures membrane as well as proteins the cell will secrete (secretory proteins). Smooth ER is so named because its cytoplasmic surface lacks ribosomes.

Functions of Smooth ER   Smooth ER of various cell types functions in diverse metabolic processes, including synthesis of lipids, carbohydrate metabolism, and detoxification of drugs and other poisons.

Enzymes of the smooth ER are important to the synthesis of fats, phospholipids, steroids, and other lipids (see Chapter 5). Among the steroids produced by smooth ER are the sex hormones of vertebrates and the various steroid hormones secreted by the adrenal glands. The cells that actually synthesize and secrete these hormones—in the testes and ovaries, for example—are rich in smooth ER, a structural feature that fits the function of these cells.

Liver cells provide one example of the role of smooth ER in carbohydrate metabolism. Liver cells store carbohydrate in the form of glycogen, a polysaccharide (see Figure 5.7 on p. 69). Hydrolysis of glycogen leads to the release of glucose from the liver cells, which is important in the regulation of sugar concentration in the blood. However, the first product of glycogen hydrolysis is glucose–phosphate, an ionic form of the sugar that cannot exit from the cell and enter the blood. It is an enzyme embedded in the membrane of the liver cell's smooth ER that removes the phosphate from the glucose, which can then leave the cell and elevate blood sugar concentration.

Enzymes of the smooth ER help detoxify drugs and other poisons, especially in liver cells. Detoxification usually involves adding hydroxyl groups to drugs, increasing their solubility and making it easier to flush the compounds from the body. The sedative phenobarbital and other barbiturates are examples of drugs metabolized in this manner by smooth ER in liver cells. In fact, barbiturates, alcohol, and many other drugs induce proliferation of smooth ER and its associated detoxification enzymes. This increases tolerance to the drugs, meaning that higher doses are required to achieve a particular effect, such as sedation. Also, because some of the detoxification enzymes have relatively broad action, proliferation of smooth ER in response to one drug can also increase tolerance to other drugs. Barbiturate abuse, for example, may decrease the effectiveness of certain antibiotics and other useful drugs.

Muscle cells exhibit still another specialized function of smooth ER. The ER membrane pumps calcium ions from the cytosol into the cisternal space. When a muscle cell is stimulated by a nervous impulse, the calcium leaks back across the ER membrane into the cytosol, and the calcium triggers contraction of the muscle cell.

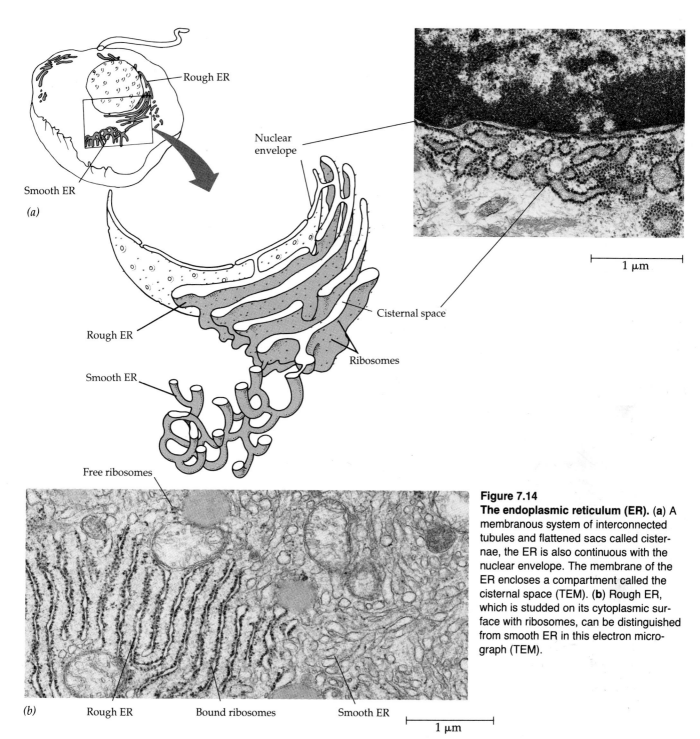

(a)

Rough ER

Smooth ER

Nuclear envelope

Cisternal space

Ribosomes

Rough ER

Smooth ER

Free ribosomes

1 μm

Figure 7.14
The endoplasmic reticulum (ER). (a) A membranous system of interconnected tubules and flattened sacs called cisternae, the ER is also continuous with the nuclear envelope. The membrane of the ER encloses a compartment called the cisternal space (TEM). **(b)** Rough ER, which is studded on its cytoplasmic surface with ribosomes, can be distinguished from smooth ER in this electron micrograph (TEM).

(b)   Rough ER   Bound ribosomes   Smooth ER

1 μm

Rough ER and Protein Synthesis Many types of specialized cells secrete proteins produced by rough ER. For example, white blood cells in humans and other vertebrates secrete antibodies. Proteins destined for secretion are synthesized by ribosomes attached to the rough ER. As a polypeptide chain grows from a bound ribosome, it is threaded through the ER membrane into the cisternal space, possibly through a pore. As it enters the cisternal space, the protein folds into its native conformation. Most secretory proteins are **glycoproteins,** which are proteins covalently bonded to carbohydrates. In the cisternal space, the

carbohydrate is attached to the protein by enzymes built into the ER membrane. The carbohydrate appendage of a glycoprotein is an **oligosaccharide,** a relatively small polymer of sugar units—14 in the case of glycoproteins synthesized by ER.

Once the secretory proteins are formed, the ER membrane keeps them separate from proteins that will remain in the cytosol after being produced by free ribosomes. Secretory proteins depart from the ER wrapped in the membranes of vesicles budded from a specialized region termed **transitional ER.** Such vesicles in transit from one part of the cell to another are

Figure 7.15

The signal hypothesis. During their synthesis, many proteins begin with signal sequences of about 20 amino acids that direct the protein to its proper destination in the cell. (1) One example of signaling is a sequence of mostly hydrophobic amino acids on secretory proteins that enables ribosomes to attach to a receptor site on ER, (2) where synthesis of the protein continues. (3) After the leading end of the new protein is injected into the cisternal space, the signal sequence is clipped off by an enzyme. (4) Released from the ribosome, the polypeptide folds into the conformation of a specific protein.

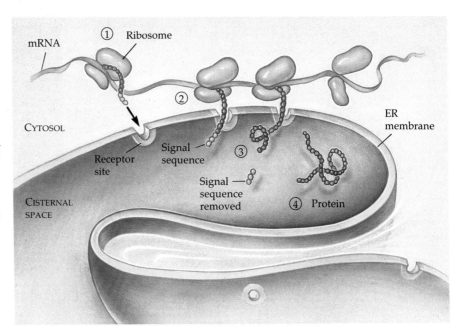

mRNA
① Ribosome
②
ER membrane
CYTOSOL
Receptor site
Signal sequence
③
Signal sequence removed
④ Protein
CISTERNAL SPACE

called **transport vesicles,** and we will soon learn their fate.

Rough ER and Membrane Production   In addition to making secretory proteins, rough ER is a membrane factory that grows in place by adding proteins and phospholipids. As the proteins elongate from the ribosomes, they are inserted into the ER membrane itself. The ER, both rough and smooth, also makes its own membrane phospholipids; enzymes built into the ER membrane assemble the phospholipids from precursors obtained from the cytosol. The ER membrane expands and can be transferred in the form of transport vesicles targeted for other components of the endomembrane system.

The Signal Hypothesis   Recall that free ribosomes and bound ribosomes make proteins destined for different locations. If the ribosomes themselves are identical in structure and can switch their status from "free" to "bound," then what determines whether a ribosome will be free in the cytosol or bound to rough ER at any particular time? The synthesis of all proteins begins in the cytosol when a ribosome starts to translate a messenger RNA molecule. The growing polypeptide chain itself cues the ribosome to either remain in the cytosol or attach to the ER. Secretory proteins are marked by a **signal sequence** of 20 or so amino acids arranged in a specific order (Figure 7.15). This signal sequence, the first part of the polypeptide made, enables the ribosome to attach to a receptor site on the ER membrane (smooth ER lacks these receptors). Synthesis of the protein continues there, and as the growing polypeptide snakes across the membrane into the cisternal space, the signal sequence is removed by

an enzyme. By contrast, if the mRNA molecule lacks a segment that programs synthesis of the ER signal sequence, the ribosome translating that RNA will remain free in the cytosol, where the finished protein will be released.

The use of signal sequences to target secretory proteins to the ER is only one example of what seems to be a general mechanism the cell uses to dispatch proteins to specific sites. Other cases are the transfer of proteins to mitochondria or chloroplasts. These proteins are synthesized by free ribosomes in the cytosol, but have specific signal sequences that are identified by mitochondria or chloroplasts after the newly made proteins are released from ribosomes. As research on the signal hypothesis continues, cell biologists will probably discover additional signal sequences functioning like zip codes, addressing proteins to certain locations in the cell.

The Golgi Apparatus

After leaving the ER, many transport vesicles travel first to the **Golgi apparatus.** We can think of the Golgi as a center of manufacturing, warehousing, sorting and shipping, where products of the ER are modified, stored, and routed to other destinations. Not surprisingly, Golgi apparatuses are especially prevalent in cells specialized for secretion.

The Golgi apparatus consists of flattened membranous sacs, stacked like pita bread (Figure 7.16). Each Golgi stack is called a **dictyosome,** and the Golgi apparatus actually consists of all of the cell's dictyosomes combined. Each of the stacked cisternae of a dictyo-

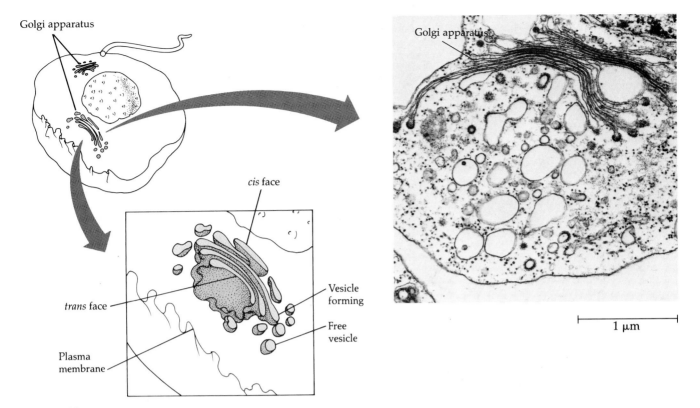

Figure 7.16
The Golgi apparatus. A Golgi apparatus consists of dictyosomes, stacks of flattened membranous sacs. The apparatus receives and dispatches vesicles and the products they contain. Materials received from the ER are modified and stored in the Golgi and eventually shipped to the cell surface or other destinations. Note the vesicles forming at the edges of the stack and also the free vesicles that presumably have just arisen in this way. The dictyosomes of the apparatus have a definite structural and functional polarity, with a *cis* end (also called forming face) that receives vesicles and a *trans* end (also called maturing face) that dispatches vesicles (right, TEM).

some consists of a membrane that separates its internal space from the cytosol. Vesicles concentrated in the vicinity of the Golgi apparatus may be engaged in the transfer of material between the Golgi and other structures.

The dictyosomes of the Golgi apparatus generally have a distinct polarity, with the membranes of cisternae at opposite ends of a stack differing in thickness and molecular composition. The two poles of a Golgi stack are referred to as the ***cis* face** and the ***trans* face;** these act, respectively, as the receiving and shipping departments of the Golgi apparatus. The *cis* face is usually closely associated with smooth ER. In some cases, the smooth ER and the Golgi may physically connect, but it is more likely that transport vesicles move material from the ER to the Golgi. According to this hypothesis, a vesicle that buds from the ER will add its membrane and the contents of its lumen (cavity) to the *cis* face by fusing with a Golgi membrane. The *trans* face gives rise to vesicles, which pinch off from the Golgi apparatus and travel to other sites.

Products of the ER are usually modified during their transit from the *cis* to *trans* poles of the Golgi. Proteins and phospholipids of membranes may be altered. In particular, various Golgi enzymes modify the oligosaccharide portions of glycoproteins. When first added to proteins in the ER, the oligosaccharides of all glycoproteins are identical. The Golgi removes some sugar monomers and substitutes others, producing diverse oligosaccharides. In addition to this finishing work, the Golgi apparatus manufactures certain macromolecules by itself. Many polysaccharides secreted by cells are Golgi products, including hyaluronic acid, a sticky substance that helps glue animal cells together. Golgi products that will be secreted depart from the *trans* faces of Golgi in the lumen of membranous vesicles that eventually fuse with the plasma membrane (see Figure 7.21).

Manufacturing and refining in the Golgi occurs in stages, with different cisternae between the *cis* and *trans* ends of the Golgi stack containing unique teams of enzymes. Products in various stages of processing are probably transferred from one cisterna to the next by vesicles.

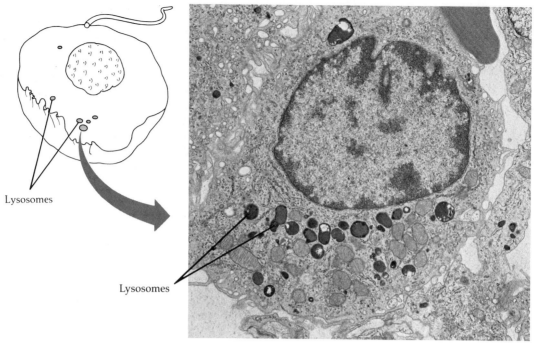

Figure 7.17
Lysosomes. In this white blood cell from a rat, the lysosomes are very dark because of a specific stain that reacts with one of the products of digestion within the lysosome (TEM).

5 μm

Sorting Functions of the Golgi   Before the Golgi apparatus dispatches its products by budding vesicles from the *trans* face, it must sort these products and target them for various parts of the cell. Molecular identification tags, such as phosphate groups and specific oligosaccharides that have been added to the Golgi products, may aid in sorting. And membranous vesicles budded from the Golgi may have external molecules that recognize "docking sites" on the surface of specific organelles. Among the products of the ER and Golgi are lysosomes.

Lysosomes

A **lysosome** is a membrane-enclosed bag of hydrolytic enzymes that the cell uses to digest macromolecules (Figure 7.17). There are lysosomal enzymes that can hydrolyze proteins, polysaccharides, fats, and nucleic acids—all of the major classes of macromolecules. All of these enzymes work best in an acidic environment, at a pH of about 5. The lysosomal membrane maintains this low internal pH by pumping hydrogen ions from the cytosol into the lumen of the lysosome. If the lysosome should break open or leak its contents, the enzymes would not be very active in the neutral environment of the cytosol. However, excessive leakage from a large number of lysosomes can destroy a cell by autodigestion. We see once again how impor-

tant compartmental organization is to the functions of the cell: The lysosome provides a space where the cell can digest macromolecules safely, without the general destruction that would occur if active hydrolytic enzymes roamed at large throughout the cell.

The hydrolytic enzymes and lysosomal membrane are made by rough ER and then transferred to a Golgi apparatus for further processing. Lysosomes probably arise by budding from the *trans* face of the Golgi apparatus. The proteins of the inner surface of the lysosomal membrane and the digestive enzymes themselves are probably spared from self-destruction by having three-dimensional conformations that protect vulnerable bonds from enzymatic attack.

Lysosomes function in intracellular digestion in a variety of circumstances. *Amoeba* and many other protozoa eat by engulfing smaller organisms or other food particles, a process called **phagocytosis** (from the Greek *phagein,* "to eat," and *kytos,* "vessel," referring here to the cell). The particle-containing vacuole formed in this way then fuses with a lysosome, whose enzymes digest the food (Figure 7.18). Some human cells also carry out phagocytosis. Among them are macrophages, cells that help defend the body by eating bacteria and other invaders.

Lysosomes also use their hydrolytic enzymes to recycle the cell's own organic material—a process called autophagy. This occurs when a lysosome engulfs another organelle or a small parcel of cytosol. The

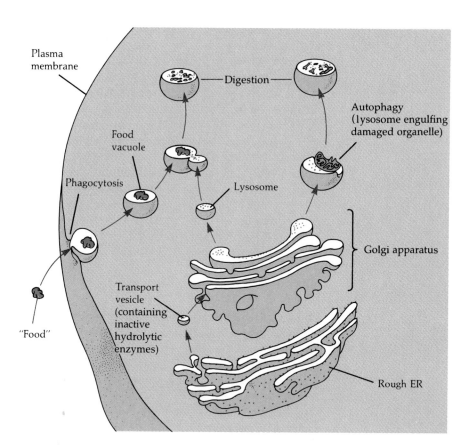

Figure 7.18
Formation and functions of lysosomes.
Lysosomes digest materials taken into the cell, and also recycle materials from intracellular refuse. In phagocytosis, the cell encloses food in a vacuole with a membrane that pinches off internally from the plasma membrane. This food vacuole then fuses with a lysosome, and the hydrolytic enzymes go to work digesting the food. After hydrolysis is completed, simple sugars, amino acids, and other monomers pass across the lysosomal membrane into the cytosol as nutrients for the cell. Lysosomes also play an important role in recycling the molecular ingredients of organelles, a function called autophagy. The ER and Golgi may cooperate in the production of lysosomes, although some lysosomes may bud directly from specialized regions of the ER.

lysosomal enzymes dismantle the ingested material, and the organic monomers are returned to the cytosol for reuse. With the help of lysosomes, the cell continually renews itself. A human liver cell, for example, recycles half of its macromolecules each week.

Programmed destruction of cells by their own lysosomal enzymes is important in the development of many organisms. During the remodeling of a tadpole into a frog, for instance, lysosomes destroy the cells of the tail. And the hands of human embryos are webbed until lysosomes digest the tissue between the fingers.

Lysosomes and Human Disease A variety of inherited disorders called storage diseases affect lysosomal metabolism. A person afflicted with a storage disease is missing one of the hydrolytic enzymes of the lysosome. The lysosomes become engorged with indigestible substrates, which begin to interfere with other cellular functions. In Pompe's disease, for example, the liver is damaged by an accumulation of glycogen due to the absence of a lysosomal enzyme needed to break down the polysaccharide. In Tay-Sachs disease, it is a lipid-digesting enzyme that is missing, and the brain becomes impaired by an accumulation of lipid in the cells. Fortunately, storage diseases are quite rare in the general population. There is encouraging evidence from research laboratories that it may some day be possible to treat storage diseases by injecting the missing enzymes into the blood along with adaptor molecules that target the enzymes for engulfment by cells and fusion with lysosomes.

Microbodies

Cell biologists have discovered a variety of organelles collectively called **microbodies** (Figure 7.19). Bounded by a single membrane, microbodies are compartments specialized for specific metabolic pathways, and each type of microbody has a particular team of enzymes. Nearly all eukaryotic cells have microbodies of one kind or another. Two important families of microbodies are **peroxisomes** and **glyoxysomes**.

Peroxisomes have enzymes that transfer hydrogen from various substrates to oxygen, producing hydrogen peroxide (H_2O_2) as a by-product. These reactions may have many different functions. Some peroxisomes use oxygen to break fats down to smaller molecules that can then be transported to mitochondria as fuel for cellular respiration. Peroxisomes in the liver detoxify alcohol and other harmful compounds by transferring hydrogen from the poisons to oxygen. The H_2O_2 formed by peroxisome metabolism is itself toxic, but the organelle contains an enzyme that converts the H_2O_2 to water. Packaging the enzymes that produce hydrogen peroxide in an enclosure where an enzyme to dispose of the by-product is also concen-

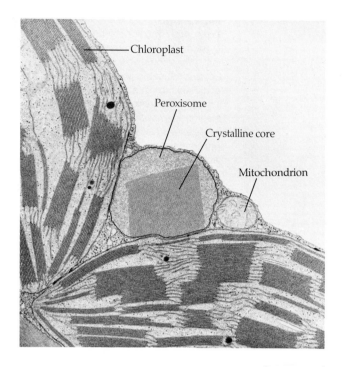

Figure 7.19
Microbodies. Microbodies range in size from about 0.15 to 1.2 μm. They are usually roughly spherical and often have a granular or crystalline core believed to be a dense collection of enzymes. This particular microbody is a peroxisome found in a leaf cell of a plant (TEM).

trated is another example of how the cell's compartmental structure is crucial to its function.

Glyoxysomes are often found in the fat-storing tissues of the germinating seeds of plants. These organelles contain enzymes that initiate the conversion of fats to sugar, a process that makes the energy stored in the oils of the seed available until the seedling is able to produce its own sugar by photosynthesis.

How cells manufacture their microbodies is an unresolved question in cell biology. Rudimentary microbodies probably arise from the ER, but some of the enzymes within a microbody are injected through the membrane of the organelle after being synthesized by free ribosomes in the cytosol.

Vacuoles

The terms *vacuole* and *vesicle* both refer to membrane-enclosed sacs within the cell. The distinction is rather arbitrary: Vacuoles are larger than vesicles. Vacuoles have various functions. **Food vacuoles** formed by phagocytosis have already been mentioned (see Figure 7.18). Many freshwater protists have **contractile vacuoles** that pump excess water out of the cell. Mature plant cells generally contain a large **central vacuole** enclosed by a membrane called the **tonoplast,** which is part of their endomembrane system (Figure 7.20).

The plant cell vacuole is a versatile compartment. It is a place to store organic compounds, including proteins, which are stockpiled in the vacuoles of storage

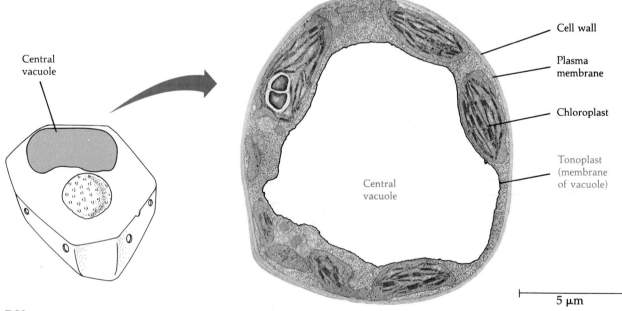

Figure 7.20
The plant cell vacuole. The central vacuole, usually the largest compartment in a plant cell, may fill 80% or more of a mature plant cell. The cytoplasm is generally confined to a narrow zone between the vacuole and the plasma membrane.

The membrane bounding the vacuole is called the tonoplast; it separates the cytosol from the solution inside the vacuole, which is called cell sap. Like all membranes of the cell, the tonoplast is selective in transporting solutes; consequently,

cell sap differs in composition from the cytosol. The vacuole functions in storage, waste disposal, hydrolysis, protection, and growth (right, TEM).

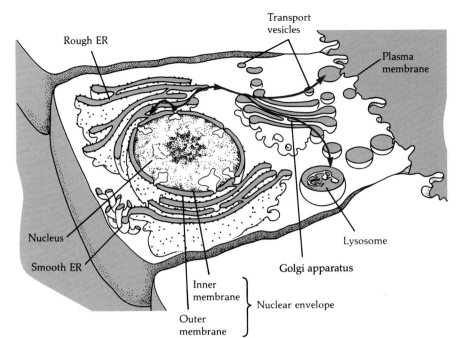

Rough ER

Transport vesicles

Plasma membrane

Nucleus

Smooth ER

Inner membrane

Outer membrane

} Nuclear envelope

Lysosome

Golgi apparatus

Figure 7.21
Relationships among endomembranes. Although each membrane has a unique molecular composition and function, components of the endomembrane system are related through direct physical contact or through the transfer of vesicles. The latter occurs by the pinching off of membrane and its later fusion with the membrane of another organelle or with the plasma membrane. Color is used in this diagram to illustrate that the lumen (cavity) of the ER and its derivatives are topologically related to the *outside* of the cell. That is, a secretory protein or some other product present in the cisternal space of the ER can eventually turn up outside the cell without crossing another membrane.

cells in seeds. The vacuole is also the plant cell's main repository of inorganic ions, such as potassium and chloride. Although plant cells generally lack the specialized lysosomes found in animal cells, the vacuole functions as the plant cell's lysosomal compartment, containing hydrolytic enzymes that digest stored macromolecules and recycle molecular components from organelles. Many plant cells use their vacuoles as disposal sites for metabolic by-products that would be dangerous if they accumulated in the cytoplasm. Some vacuoles are enriched in pigments that color the cells, such as the red and blue pigments of petals that help to attract pollinating insects to flowers. Vacuoles may also help protect the plant against predators by containing compounds that are poisonous or unpalatable to animals. The vacuole has a major role in the growth of plant cells, which elongate as their vacuoles absorb water, enabling the cell to become larger with a minimal investment in new cytoplasm. And because the cytoplasm occupies a thin shell between the plasma membrane and the tonoplast, the ratio of membrane surface to cytoplasmic volume is great, even for a large plant cell.

The large vacuole of a plant cell develops by the coalescence of smaller vacuoles, themselves derived from the endoplasmic reticulum and Golgi apparatus. Through these relationships, the vacuole is an integral part of the endomembrane system.

Relationships of Endomembranes: A Summary

Through fusion of transport vesicles, or in some cases through direct membrane continuity, the components of the endomembrane system are all related (Figure 7.21). The nuclear envelope is an extension of the rough ER, which is also confluent with the smooth ER. Membrane produced by the ER flows in the form of transport vesicles to the Golgi, which in turn pinches off vesicles that give rise to lysosomes, microbodies, and the tonoplast of the plant cell vacuole. Even the cell's outermost membrane, the plasma membrane, grows by the fusion of vesicles born in the ER and Golgi. (Coalescence of vesicles with the plasma membrane also releases cellular products, such as secretory proteins, to the outside of the cell.) As membranes of the system flow from the ER to Golgi and then onto other destinations, their molecular compositions and metabolic functions are modified. The endomembrane system is a complex and dynamic player in the cell's compartmental organization.

Two important organelles that are *not* closely related to the endomembrane system are the mitochondrion and the chloroplast.

ENERGY TRANSDUCERS: MITOCHONDRIA AND CHLOROPLASTS

Mitochondria (singular, *mitochondrion*) and chloroplasts are organelles that convert energy to forms the cell can use for its various kinds of work. **Mitochondria** are the sites of cellular respiration, the elaborate catabolic process that generates ATP by extracting energy from sugars, fats, and other fuels with the help of oxygen. **Chloroplasts,** found only in plants and eukaryotic algae (Kingdom Protista) convert solar energy to chemical energy by absorbing sunlight and using it to drive the synthesis of organic compounds from carbon dioxide and water. The cell, remember,

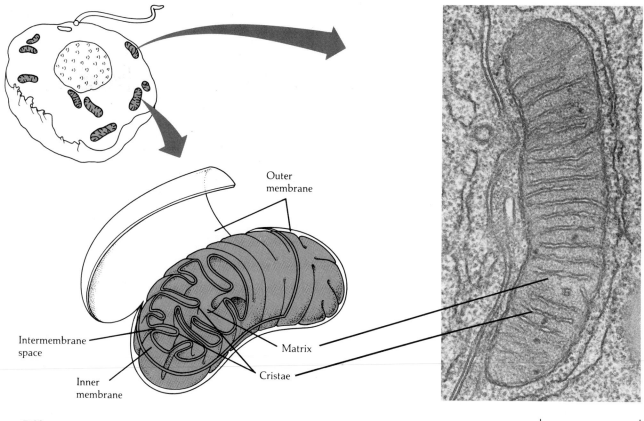

Figure 7.22
The mitochondrion. The double membrane of the mitochondrion is evident in the drawing and micrograph (TEM). The cristae are infoldings of the inner membrane. The three-dimensional drawing emphasizes the relationships between the two membranes and the compartments they bound: the intermembrane space and the mitochondrial matrix.

0.5 μm

is an open system, and both mitochondria and chloroplasts are involved in putting energy acquired from the surroundings to work.

Although mitochondria and chloroplasts are enclosed by membranes, they are not considered part of the endomembrane system. Their membrane proteins are made not by the ER, but by free ribosomes in the cytosol and by ribosomes contained within the mitochondria and chloroplasts themselves. Not only do these organelles have ribosomes, but they also contain a small amount of DNA that programs the synthesis of some of their own proteins (although most of their proteins are made in the cytosol, programmed by messenger RNA sent by nuclear genes). Mitochondria and chloroplasts are semiautonomous organelles that grow and divide to increase their numbers. In Chapters 9 and 10, we will focus on how mitochondria and chloroplasts work. The evolution of these organelles will be considered in Chapter 25. Here, we are concerned mainly with the structure of these energy transducers.

Mitochondria

Mitochondria are found in nearly all eukaryotic cells. In some cases, there is a single, large mitochondrion, but more often, a cell has hundreds or thousands of mitochondria; the number is generally correlated with the metabolic activity of the cell. Mitochondria are about 1–10 μm in length. Time-lapse films of living cells reveal mitochondria moving around, changing their shapes, and dividing in two, unlike the static cylinders seen in electron micrographs of dead cells.

The mitochondrion is enclosed in an envelope of two membranes, each a phospholipid bilayer with a unique collection of embedded proteins (Figure 7.22). The outer membrane is smooth, but the inner membrane is convoluted, with infoldings called **cristae.** The membranes divide the mitochondrion into two internal compartments. The **intermembrane space** is the narrow region between the inner and outer membranes. The outer membrane allows all small molecules to pass, but blocks the passage of proteins and other macromolecules. Consequently, the intermembrane space reflects the composition of the cytosol in its small molecules, but has its own group of enzymes. The **mitochondrial matrix** is the compartment enclosed by the inner membrane. Many of the metabolic steps of cellular respiration occur in the matrix, where several different enzymes are concentrated. Other proteins that function in respiration, including the enzyme that makes ATP, are built into the inner membrane. With its cristae, the inner mitochondrial membrane

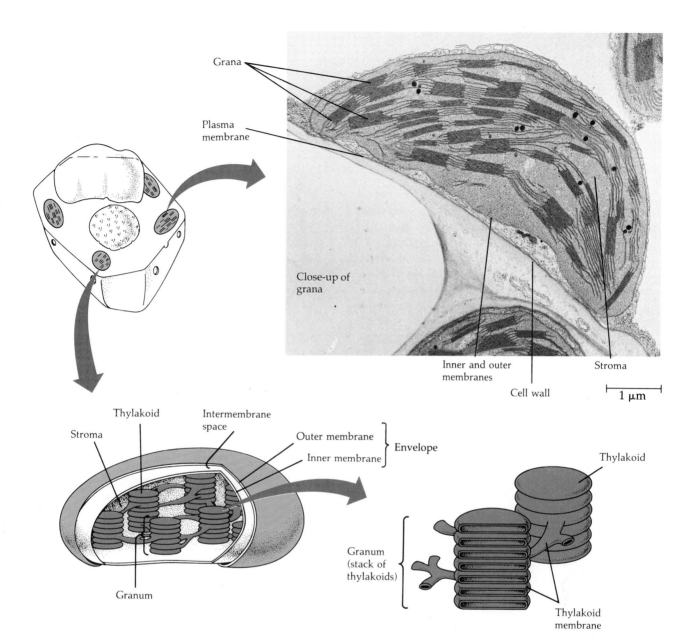

Figure 7.23
The chloroplast. This organelle is the site of photosynthesis. Chloroplasts, like mitochondria, are enclosed by two membranes separated by a narrow intermembrane space. The inner membrane encloses fluid called stroma. The stroma surrounds a third compartment, delineated by its own membrane, the thylakoid membrane. At several locations within the chloroplast, thylakoid sacs are stacked to form structures called grana. Individual thylakoids of one granum are continuous with other grana through extensions that traverse the stroma (upper right, TEM).

has a large surface area that enhances the productivity of cellular respiration. This feature is another example of the correlation of structure and function.

Chloroplasts

The chloroplast is a specialized member of a family of closely related plant organelles called **plastids.** *Amyloplasts* (also called leucoplasts) are colorless plastids that store starch, particularly in roots and tubers.

Chromoplasts are enriched in pigments that give fruits, flowers, and autumn leaves their orange and yellow hues. *Chloroplasts* contain the green pigment chlorophyll along with enzymes and other molecules that function in the photosynthetic production of food. These lens-shaped organelles, measuring about 2 μm by 5 μm, are found in leaves and other green organs of plants and in eukaryotic algae (Figure 7.23).

Chloroplasts, amyloplasts, and chromoplasts can all develop from *proplastids* found in unspecialized cells. As the plant grows, the fate of the proplastids depends on the location of the cells and the environment to which they are exposed. For example, a proplastid will become a chloroplast only if it is exposed to light. Chloroplasts can also arise from the division of previously existing chloroplasts. Under certain conditions, mature plastids are interconvertible. When a fruit ripens, for instance, chloroplasts in its green tissue are transformed to nonphotosynthetic chromoplasts that contribute to the characteristic color indicating that the fruit is ripe.

The contents of a chloroplast are partitioned from the cytosol by an envelope consisting of two membranes separated by a very narrow intermembrane space. Within the chloroplast itself is still another membranous system, which is arranged into flattened sacs called **thylakoids.** In some regions, thylakoids are stacked like poker chips to form structures called **grana** (singular granum). The fluid outside the thylakoids is called the **stroma.** Thus, the thylakoid membrane segregates the interior of the chloroplast into two compartments: the thylakoid space and the stroma. In Chapter 10, you will learn how this compartmental organization enables the chloroplast to convert light energy to chemical energy during photosynthesis.

As with mitochondria, the static and rigid appearance of chloroplasts in electron micrographs belies their dynamic behavior in the living cell. Their shapes are

plastic, and they occasionally pinch in two. They are mobile and move around the cell with mitochondria and other organelles along tracks made up of the tubules and filaments of the cytoskeleton, which will be described next.

THE CYTOSKELETON

Two decades ago, in the early days of electron microscopy, the cell seemed to consist of a variety of organelles suspended or floating in a rather formless, jellylike cytosol. But new techniques in both light microscopy and electron microscopy have revealed a network of fibers throughout the cytoplasm. This mesh has been named the **cytoskeleton** (Figure 7.24).

One function of the cytoskeleton is to give mechanical support to the cell and help maintain its shape. This is especially important for animal cells, which lack walls. Organelles and even cytoplasmic enzymes may be held in place by anchoring to the cytoskeleton. The cytoskeleton also enables a cell to change its shape; like a scaffold, the cytoskeleton can be dismantled in one part of the cell and reassembled in a new location. The cytoskeleton is also associated with motility—movement of the entire cell or movement of organ-

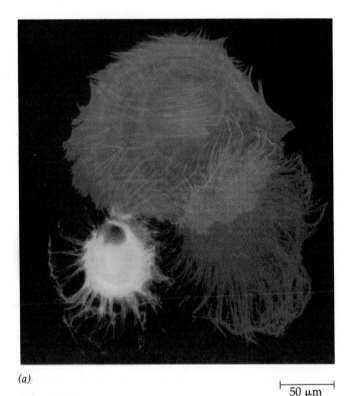

(a)

|⊢——— 50 μm ———⊣|

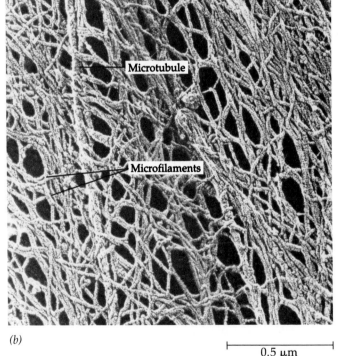

(b)

|⊢——— 0.5 μm ———⊣|

Figure 7.24
The cytoskeleton. (a) The cytoplasm of this fibroblast cell from the skin of a mammal is infiltrated by a fibrous network revealed by a type of light microscopy called fluorescence microscopy. The cytoskeleton gives the cell shape, anchors some organelles and directs the movement of others, and in some cases enables the entire cell to change its shape or move. A special photographic technique was used to rotate images of the three elements of the cytoskeleton so as not to obliterate each other (blue = actin microfilaments; red = microtubules; green = intermediate filaments). **(b)** In this electron micrograph prepared by a method known as deep-etching, microtubules and microfilaments are visible.

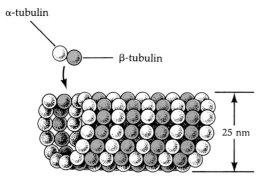

Figure 7.25
Microtubule elongation. The wall of the hollow tube consists of two similar kinds of proteins called tubulins, which are added two by two to one end of a growing microtubule.

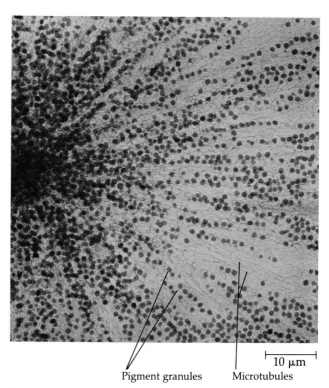

Figure 7.26
Microtubules as tracks. These pigment granules in a skin cell of a fish migrate back and forth along microtubules, changing the coloration of the fish (TEM).

elles within the cell. The fibers of the cytoskeleton are not only the cell's "bones," but also its "muscles." Components of the cytoskeleton wiggle cilia and flagella and enable muscle cells to contract. The cytoskeleton extends the pseudopodia of *Amoeba* and also functions in the streaming of cytoplasm that circulates materials within many large plant cells. Vesicles may travel to their destinations in the cell along "monorails" provided by the cytoskeleton, and contractile components of the cytoskeleton manipulate the plasma membrane to form food vacuoles during phagocytosis.

The cytoskeleton is constructed from at least three types of fibers (Table 7.2 p. 142). **Microtubules** are the thickest of the three types; **microfilaments** (also called actin filaments) are the thinnest. **Intermediate filaments** are a collection of fibers whose diameters fall in a middle range.

Microtubules

Microtubules are straight, hollow rods measuring about 25 nm in diameter and from 200 nm to 25 µm in length. The wall of the hollow tube is constructed from globular proteins called tubulins, of which there are two closely related kinds, named α-tubulin and β-tubulin (Figure 7.25). A new microtubule begins as a flat sheet made of many tubulin units arranged in 13 columns, and this two-dimensional sheet then rolls into a tube. Once formed, the microtubule can elongate by the addition of tubulin proteins to one end of the tubule. The building units are not actually single tubulin molecules but dimers, which are aggregates of two protein molecules. Each dimer consists of one molecule of α-tubulin and one molecule of β-tubulin. Microtubules can be disassembled and their tubulin used to build microtubules elsewhere in the cell.

Microtubules are found in the cytoplasm of all eukaryotic cells. In many cells, they radiate from a **microtubule-organizing center,** which is a mass located near the nucleus. The microtubules that emanate from this center are strong girders that support the cell. In addition to the major framework that radiates from the center, strategically located bundles of microtubules near the plasma membrane reinforce cell shape.

While shaping and supporting the cell, microtubules also serve as tracks along which organelles can move in the cell. For example, microtubules probably help guide secretory vesicles from the Golgi apparatus to the plasma membrane, and Figure 7.26 illustrates a more exotic example involving pigment granules. Microtubules are also involved in the separation of chromosomes when a cell divides (see Figure 11.10 on p. 237).

Within the microtubule-organizing center of an animal cell are two structures called **centrioles** (Figure 7.27). Each centriole is composed of nine sets of triplet microtubules arranged in a ring. When a cell divides, the centrioles replicate. Although centrioles may help to organize microtubule assembly, they are not mandatory for this function in all eukaryotes; microtubule-organizing centers of plant cells lack centrioles altogether.

Figure 7.27

Centrioles. An animal cell has a pair of centrioles within its microtubule-organizing center, a region near the nucleus where the cell's microtubules are initiated. The centrioles, each about 150 nm in diameter, are arranged at right angles to each other, and each is made up of nine sets of three microtubules (TEM).

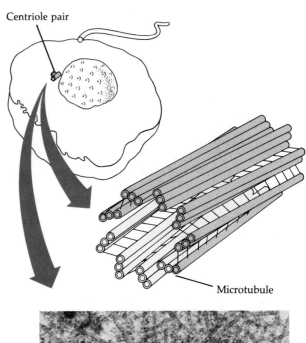

Centriole pair

Microtubule

Cilia and Flagella   A specialized arrangement of microtubules is responsible for the beating of flagella and cilia, locomotive appendages that protrude from some cells (Figure 7.28). Many single-celled organisms (Kingdom Protista) are propelled through water by cilia or flagella, and the sperm of animals, algae, and some plants are flagellated. (To a microscopic vehicle moving through water, the medium is as viscous as liquid asphalt would be to the arms and legs of a human swimmer. Flagellated and ciliated organisms actually crawl or climb through the water.) If cilia or flagella extend from cells that are held in place as part of a tissue layer, then they function to draw fluid over the surface of the tissue. For example, the ciliated lining of the human windpipe sweeps mucus with trapped debris out of the lungs (see Figure 7.4).

Cilia usually occur in large numbers on the cell sur-

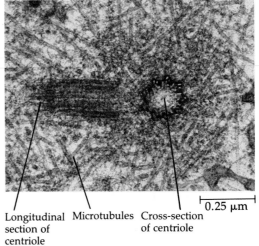

0.25 μm

Longitudinal   Microtubules   Cross-section
section of   of centriole
centriole

Table 7.2 Properties of microtubules, microfilaments, and intermediate filaments

Property	Microtubules	Microfilaments	Intermediate Filaments
Structure	Hollow tubes; wall consists of 13 columns of tubulin proteins	Two intertwined strands of actin	Hollow tubes
Diameter	25 nm with 15-nm lumen	7 nm	8–10 nm
Monomers	α-tubulin β-tubulin	G-actin	Five different proteins depending on cell type
Functions	Cell motility Chromosome movements Organization of cytoplasm Disposition and movement of organelles Maintenance of cell shape	Muscle contraction Cytoplasmic streaming Amoeboid movement Cell division (cleavage furrow formation) Maintenance of cell shape Changes in cell shape	Structural role in cytoskeleton; tension-bearing elements Maintenance of cell shape

SOURCE: Modified from W. M. Becker, *The World of the Cell* (Menlo Park, Calif.: Benjamin/Cummings, 1986), p. 41.

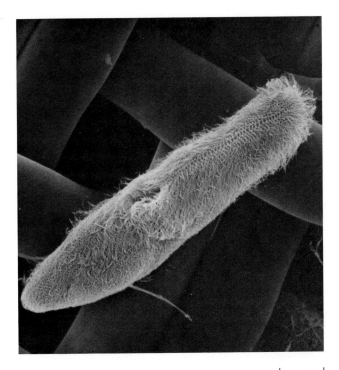

Figure 7.28
Cilia, as viewed with the scanning electron microscope. A dense nap of beating cilia covers this *Paramecium*, a motile protozoa. The cilia beat at a rate of about 40 to 60 strokes per second (500×).

25 μm

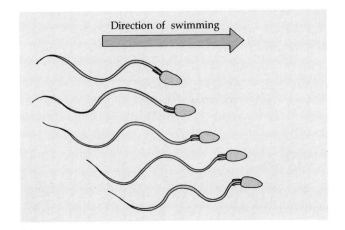

(a)

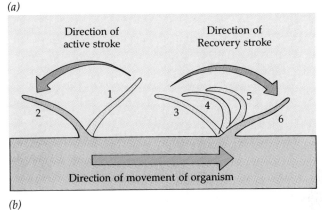

(b)

Figure 7.29
Comparison of the beating of flagella and cilia. (a) A flagellum usually undulates, its snakelike motion driving a cell in the same direction as the axis of the flagellum. Propulsion of a sperm cell is an example of this type of locomotion by a flagellum. **(b)** Cilia usually beat with a back-and-forth motion, alternating active strokes with recovery strokes. This moves the cell, or moves a fluid over the surface of a stationary cell, in a direction perpendicular to the axis of a cilium. For example, the beating of cilia of cells lining the human windpipe helps clean the respiratory system by conveying mucus with trapped debris up the pipe, away from the lungs.

face. They are about 0.25 μm in diameter and about 2–20 μm in length. Flagella are the same diameter as cilia, but at 10–200 μm, they are longer. Also, flagella are usually limited to just one or a few per cell.

Flagella and cilia also differ in their patterns of beating. A flagellum has an undulating motion that generates force in the same axis as the flagellum. In contrast, cilia work more like oars, with a power stroke alternating with a recovery stroke, generating force in a direction perpendicular to the axis of the cilium (Figure 7.29).

Though different in length, number per cell, and beating pattern, cilia and flagella actually share a common ultrastructure. A cilium or flagellum has a core of microtubules ensheathed in an extension of the plasma membrane (Figure 7.30). Nine doublets of microtubules, each doublet a pair of microtubules sharing part of their walls, are arranged in a ring. In the center of the ring are two single microtubules. This arrangement, referred to as the "9 + 2" pattern, is found in nearly all eukaryotic flagella and cilia. (The flagella of motile prokaryotes, which will be discussed in Chapter 25, are entirely different.) The doublets of the outer ring are connected to the center of the cilium

or flagellum by radial spokes that terminate near the central pair of microtubules. Each doublet of the outer ring, as seen in cross sections, also has a pair of sidearms reaching toward the neighboring doublet of microtubules. Numerous pairs of these sidearms are evenly spaced along the length of each doublet.

The microtubule assembly of a cilium or flagellum is anchored in the cell by a **basal body,** which is structurally identical to a centriole. When a cilium or flagellum first begins to grow, the basal body may act as a template for ordering tubulin into microtubules. Once the building of cilia and flagella begins, however, the tubulin subunits are added to the tips of the microtubules, and not to their bases.

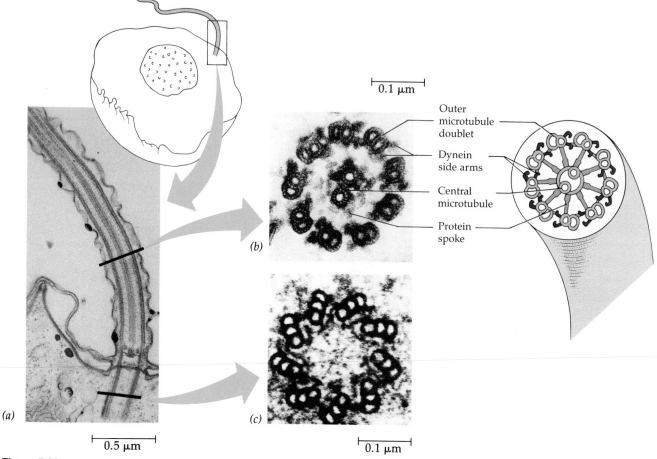

Figure 7.30
Ultrastructure of a eukaryotic flagellum or cilium. (a) In this electron micrograph of a longitudinal section of a cilium, microtubules can be seen running the length of the structure. **(b)** A cross section through the cilium shows the "9 + 2" arrangement of microtubules (TEM). **(c)** The basal body anchoring the cilium or flagellum to the cell has a ring of nine microtubule triplets and is thus structurally identical to a centriole (shown here). The nine doublets of microtubules extend into the basal body, where each doublet joins another microtubule to form the ring of nine triplets. The two central microtubules terminate above the basal body (TEM).

The sidearms extending from each microtubule doublet to the next play a major role in the bending movements of cilia and flagella (see Figure 7.30). The sidearms are made of a very large contractile protein called **dynein**. A dynein sidearm performs a complex cycle of movements caused by changes in the conformation (shape) of the protein, with ATP providing the energy for these changes (Chapter 6). Working something like legs with clawed feet, the sidearms of one doublet attach to the adjacent doublet, swing in such a way that the two doublets slide past one another, release, then reattach to the neighboring doublet a little farther along its length. The cycle is then repeated (Figure 7.31). The mechanics are reminiscent of a cat climbing a tree by attaching its claws, moving its legs, releasing the claws, and grabbing again farther up the tree. In the case of the cilium or flagellum, the grabbing and pulling actions of the dynein sidearms displace one microtubule doublet past its neighbor. In experiments where microtubule doublets are extracted from cilia and then energized with ATP, movements of the dynein sidearms cause one doublet to walk along the surface of another. If this occurred in the intact cilium, the organelle would simply elongate without any lateral movement. The "walking" of the dynein sidearms must be translated into a bending movement of the entire cilium. The function of the radial spokes that anchor the microtubule doublets to the center of the organelle may be to restrain neighboring doublets from sliding past one another very far. Working against this resistance, the grabbing-and-pulling movements of the sidearms distort the microtubules, causing them to bend. In the beating mechanism of cilia and flagella, we see once again that structure fits function.

Microfilaments and Movement

Microfilaments are solid rods about 7 nm in diameter. They are built from molecules of **actin,** a globular pro-

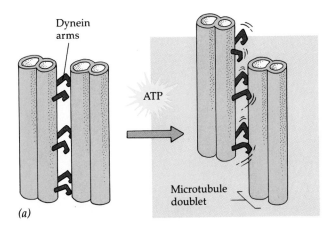

(a)

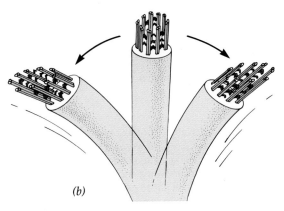

(b)

Figure 7.31
How "dynein walking" moves cilia and flagella. (a) In a solution containing ATP as an energy source, the grabbing and pulling actions of dynein sidearms cause microtubule doublets that have been removed from cilia to slide past one another. (b) In an intact cilium or flagellum, the radial spokes (not shown here—see Figure 7.30) limit linear displacement of microtubules. This restraint causes dynein "walking" to distort the microtubules, which translates into a bending of the whole flagellum or cilium.

tein. The actin molecules are linked into chains, and two of these chains twisted about each other in a helix form the microfilament (Figure 7.32).

Microfilaments are best known for their part in muscle contraction. Thousands of microfilaments made of actin are arranged parallel to one another along the length of a muscle cell, interdigitated with thicker filaments made of a protein called **myosin** (Figure 7.33a).

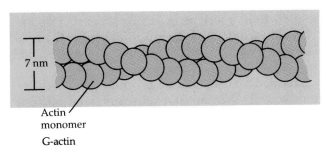

Figure 7.32
Molecular structure of a microfilament. The subunits of the filament are proteins called G-actin ("G" for globular). They join to form long strands of F-actin ("F" for fibrous). A microfilament consists of two F-actin chains wrapped about one another in a helix.

Contraction of the cell results from the actin and myosin filaments sliding past one another, which shortens the cell. The sliding is driven by ATP-powered arms that extend crosswise from the myosin filaments to the actin filaments, much as dynein arms function in cilia and flagella. Muscular contraction is described more thoroughly in Chapter 45.

Although microfilaments are especially concentrated and well ordered in muscle cells, they seem to be present to some extent in all eukaryotic cells. Along with the rest of the cytoskeleton, microfilaments function in support. For example, bundles of microfilaments make up the core of microvilli, delicate projections that increase the surface area of cells specialized for transport of material across the plasma membrane. An example of such specialized cells is the nutrient-absorbing cells that line the human intestine (Figure 7.33b).

In some parts of the cell, microfilaments are associated with myosin in miniature versions of the arrangement found in muscle cells. These actin-myosin aggregates are responsible for localized contractions of cells. For example, when an animal cell divides, it is pinched in two by a contracting belt of microfilaments. In addition, microfilaments function in the elongation and retraction of cellular extensions called

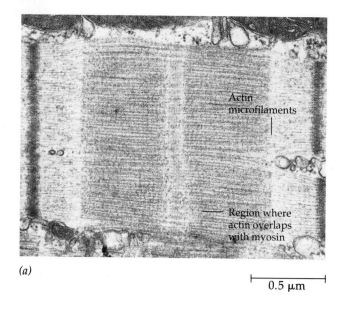

(a)

Actin
microfilaments

Region where
actin overlaps
with myosin

0.5 μm

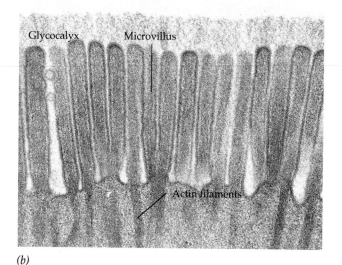

Glycocalyx   Microvillus

Actin filaments

(b)

0.5 μm

Figure 7.33
Microfilaments. (a) This electron micrograph of part of a
muscle cell from a mouse shows actin microfilaments
interdigitated with other filaments made of myosin. **(b)**
The surface area of this intestinal cell is increased by its
many microvilli, each supported by a bundle of microfilaments. A fuzzy glycocalyx (cell coat) covers the cells
(TEM).

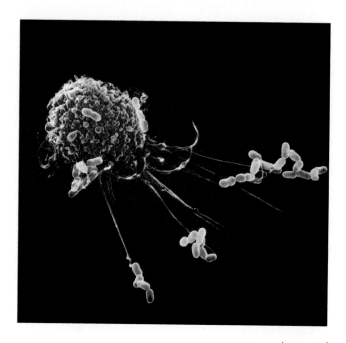

5 μm

Figure 7.34
Amoeboid movement. Named for the locomotion of the
protozoan *Amoeba*, this type of movement is also typical
of many other cells, including certain specialized cells in
humans. An amoeboid cell moves along a substratum by
extending cellular projections called pseudopodia. This
cell is a human macrophage, a phagocytic cell that helps
defend the body by using its pseudopodia to ingest bacteria (shown in green in this artificially colored SEM). The
mechanism of amoeboid movement is not yet completely
understood, but there is evidence that actin microfilaments of the cytoskeleton are essential. The use of vesicles to transfer plasma membrane from the trailing edge
of the cell to the leading tip of a pseudopod may also
contribute to amoeboid movement.

Intermediate Filaments

Intermediate filaments are named for their diameter,
which, at 8–10 nm, is larger than the diameter of
microfilaments but smaller than that of microtubules.
Intermediate filaments actually comprise a diverse class
of cytoskeletal elements, differing in protein composition from one type of cell to another. This heterogeneity contrasts with microtubules and microfilaments, which are consistent in diameter and
composition in all eukaryotic cells.

Intermediate filaments are also more permanent fix-
tures of cells than are microfilaments and microtubules, which are often disassembled and reassembled
in various parts of a cell. Chemical treatments that
remove microfilaments and microtubules from the
cytoplasm leave a web of intermediate filaments that
retains its original shape. Such experiments suggest
that intermediate filaments are especially important
in reinforcing the shape of a cell and fixing the position of certain organelles. For example, the nucleus
commonly sits within a cage made of intermediate

pseudopodia (singular **pseudopodium**) during **amoeboid movement,** although the exact mechanism of this
movement is unknown (Figure 7.34). Plant cells also
have microfilaments, which are involved in **cytoplasmic streaming,** the phenomenon in which the
entire cytoplasm flows around and around the cell in
the space between the vacuole and plasma membrane
(this is especially common in large plant cells). This
movement, also known as cyclosis, speeds the distribution of materials within the cell and from cell to cell.

filaments, fixed in location by branches of the filaments that extend into the cytoplasm. In cases where the shape of the entire cell is correlated with function, intermediate filaments support that shape. For instance, the long extensions (called axons) of nerve cells that transmit impulses are strengthened by one class of intermediate filaments. Specialized for bearing tension, the various kinds of intermediate filaments may function as the superstructure of the entire cytoskeleton.

The entire cytoskeleton must function as an integrated whole, although researchers are not certain about how its different elements are integrated. It has been difficult to obtain an image of the whole cytoskeleton, partly because conventional electron microscopy requires that extremely thin sections of the cell be cut, which makes a three-dimensional assessment difficult. However, high-voltage electron microscopes are now available, with electron beams that can penetrate sections much thicker than those used in regular electron microscopes. Some researchers using these more advanced instruments have produced photographs suggesting that the entire cytoskeleton is physically interconnected by a network, known as a microtrabecular lattice, consisting of short protein fibers that cross-link microtubules, microfilaments, and intermediate filaments. Other investigators argue that the cross-links do not really exist but are artifacts created during sample preparation. No one, however, doubts that the parts of the cytoskeleton are somehow integrated or that the cytoskeleton is a major factor in the overall organization of the cell.

THE CELL SURFACE

Having criss-crossed the interior of the cell to explore various organelles, we complete our tour of the cell by returning to the surface of this microscopic world, where there are additional structures with important functions. Although the plasma membrane is usually regarded as the boundary of the living cell, most cells synthesize and secrete coats of one kind or another that are external to the plasma membrane.

Cell Walls

The **cell wall** is one of the features of plant cells that distinguish them from animal cells. (Table 7.3, p. 150, is a summary of the similarities and differences among plant cells, animal cells, and prokaryotic cells.) Prokaryotes, fungi, and some protists (see Chapter 1) also have cell walls, but we will postpone discussion of them until Unit Five.

Plant cell walls are much thicker than the plasma membrane, ranging from 0.1 μm to several microm-

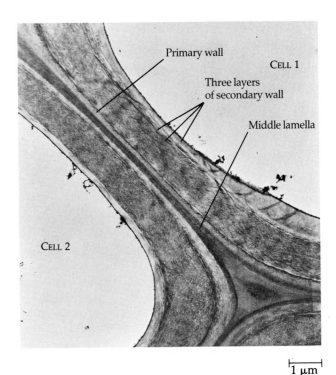

Figure 7.35
Plant cell walls. Young cells first construct thin primary walls, often adding stronger secondary walls to the inside of the primary wall when growth ceases. A sticky middle lamella cements adjacent cells together. Thus, the multi-layered partition between these two plant cells consists of adjoining walls individually secreted by the cells (TEM).

eters. The exact chemical composition of the wall varies from species to species and from one cell type to another in the same plant, but the basic design of the wall is consistent (see Figure 5.9 on p. 70). Fibers made of the polysaccharide cellulose are embedded in a matrix of other polysaccharides and small amounts of protein. This arrangement of strong fibers surrounded by an amorphous matrix is the same basic architectural design found in steel-reinforced concrete and in fiberglass.

A young plant cell first secretes a relatively thin and flexible wall called the **primary cell wall** (Figure 7.35). Between primary walls of adjacent cells is the **middle lamella,** a thin layer rich in sticky polysaccharides called pectins. The middle lamella glues the cells together (pectin is used as a thickening agent to make jams and jellies). The primary wall can stretch as the cell grows. When the cell matures and stops growing, it strengthens its wall. Some cells do this simply by secreting hardening substances into the primary wall. Other plant cells add a **secondary cell wall** between the plasma membrane and the primary wall. The secondary wall, often deposited in several laminated layers, has a strong and durable matrix that affords the cell protection and support. Wood consists mainly of secondary walls. Development of cell walls is discussed in more detail in Chapter 31.

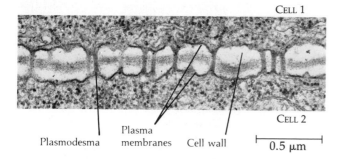

CELL 1

CELL 2

Plasmodesma   Plasma   Cell wall   0.5 μm
membranes

Figure 7.36
Plasmodesmata. These cytoplasmic channels, about
20–40 nm in diameter, pass through perforations in the
adjoining walls between two plant cells (TEM).

Intercellular Junctions

The many cells of an animal or plant must be integrated into one functional organism. Neighboring cells often adhere, interact, and communicate through special patches of direct physical contact.

You might think that the nonliving cell walls of plants would isolate cells from one another. In fact, the walls are perforated with channels called **plasmodesmata** (singular, *plasmodesma*), through which strands of cytoplasm connect the living contents of adjacent cells, unifying most of the plant into one living continuum (Figure 7.36). The plasma membranes of adjacent cells are continuous through a plasmodesma, and the membrane lines the channel. Water and small solutes can pass freely from cell to cell, a transport that is enhanced by cytoplasmic streaming.

In animals, there are three main types of intercellular junctions: **tight junctions, desmosomes,** and **gap junctions.** These are illustrated and described in detail in Figure 7.37. All these junctions are especially common in epithelial tissue, which consists of cells tightly connected into sheets that cover the body and line many of its internal cavities.

*   *   *

From our panoramic view of the cell's overall compartmental organization to our closer inspection of each organelle's architecture, this tour of the cell has provided many opportunities to correlate structure with function. Membranes, which figure so prominently in ordering the functions of cells, come into sharper focus in the next chapter.

The Glycocalyx of Animal Cells

Animal cells lack structured walls, but many have a fuzzy coat called the **glycocalyx,** which is made of sticky oligosaccharides (see Figure 7.33). The glycocalyx strengthens the cell surface and helps glue cells together. The unique molecular structure of the oligosaccharides may contribute to cell-cell recognition by serving as identification tags for specific types of cells. The glycocalyx includes carbohydrates attached by covalent bonds to both the proteins and the lipids of the plasma membrane, as well as glycoproteins secreted by the cell.

The adhesion of animal cells made possible by their sticky cell coats is often augmented by intercellular junctions between the membranes of bordering cells.

Figure 7.37
Intercellular junctions in animals.
(a) Tight junctions. These connections hold cells together tightly enough to block the transport of substances through the intercellular space. At a tight junction, specialized proteins built into the plasma membrane bond directly to similar proteins in the membrane of the adjacent cell, allowing no space between the membranes. Tight junctions usually occur as belts all the way around each cell, forming a continuous barrier to intercellular transport across the layer of cells. These junctions are most often found in epithelial layers that separate two kinds of solutions. In the electron micrograph, tight junctions (TJ) connect three epithelial cells that surround a central Lumen. Microvilli (Mv) project into the lumen, and the cells release

the contents of vesicles called zymogen granules (ZG) into the lumen (TEM).
(b) Desmosomes. At a desmosome, two cells are attached by intercellular filaments that apparently penetrate through the plasma membranes of both cells and cross the space between the cells. In the cytoplasm, just inside the plasma membrane, is a disk of dense material reinforced by intermediate filaments made of the strong structural protein keratin. Desmosomes rivet cells together into strong epithelial sheets, but still permit substances to pass freely through the spaces between the cells (TEM). **(c) Gap junctions.** These connections are specialized for the transfer of material between the cytoplasm of adjacent cells (analogous to the function of plasmodesmata in plants). Gap junctions occur where the plasma mem-

brane has a doughnut-shaped patch of proteins called a connexon that connects across the intercellular space to a connexon embedded in the membrane of the neighboring cell. The connexon proteins protrude from the membranes far enough to leave an intercellular gap of 2–4 nm (hence the name "gap junction"). The pores of the gap junction have diameters of about 1.5 nm, large enough to allow the cells to share inorganic ions, sugars, amino acids, vitamins, and other small molecules. The cells keep their own proteins and other macromolecules, however, for these substances are too large to pass through the pores. The gap junctions illustrated here connect embryonic cells (TEM). Electron micrograph 200,000×.

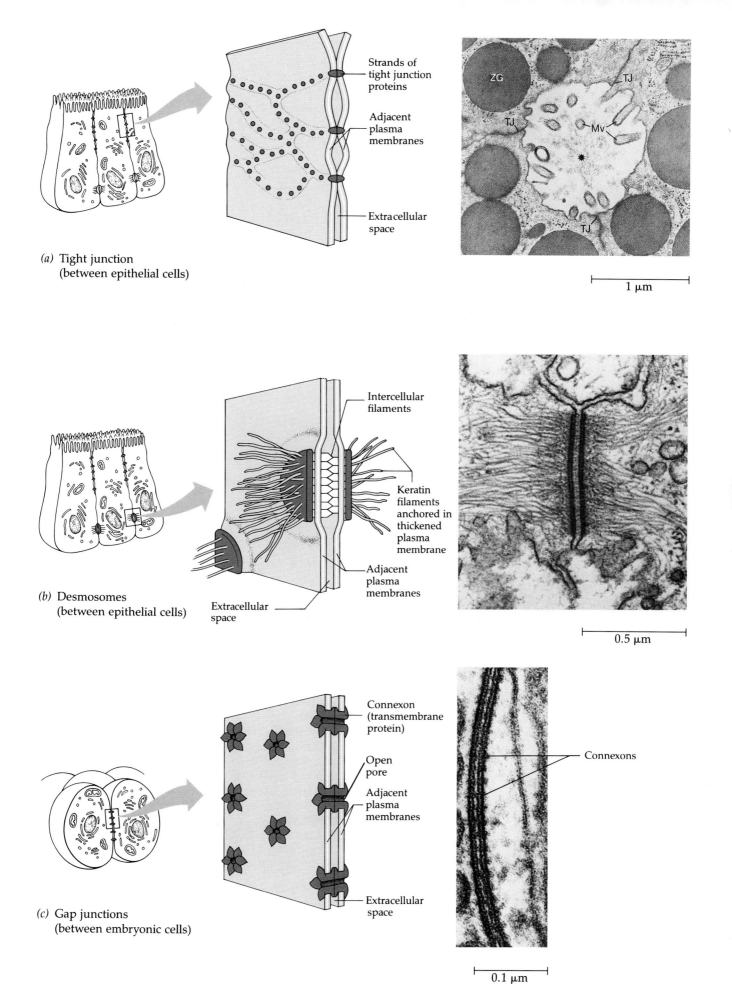

(a) Tight junction
(between epithelial cells)

Strands of tight junction proteins

Adjacent plasma membranes

Extracellular space

1 μm

(b) Desmosomes
(between epithelial cells)

Intercellular filaments

Keratin filaments anchored in thickened plasma membrane

Adjacent plasma membranes

Extracellular space

0.5 μm

(c) Gap junctions
(between embryonic cells)

Connexon (transmembrane protein)

Open pore

Adjacent plasma membranes

Connexons

Extracellular space

0.1 μm

Table 7.3 Structures of prokaryotic and eukaryotic cells

Structure	Prokaryote	Animal Cell	Plant Cell
Plasma membrane	Yes	Yes	Yes
Cell wall	Yes	No	Yes
Nucleus	Lacks nuclear envelope	Bounded by nuclear envelope	Bounded by nuclear envelope
Chromosomes	One continuous DNA molecule	Multiple, consisting of DNA and much protein	Multiple, consisting of DNA and much protein
Ribosomes	Yes (smaller)	Yes	Yes
Endoplasmic reticulum	No	Usually	Usually
Golgi apparatus	No	Yes	Yes
Lysosomes	No	Often	Some vacuoles function as lysosomes
Peroxisomes	No	Often	Often
Glyoxysomes	No	No	Common
Vacuoles	No	Small or none	Usually one large vacuole in mature cell
Mitochondria	No	Yes	Yes
Plastids	No	No	In many cell types (include chloroplasts in photosynthetic cells)
Cilia or flagella	Simple flagella	Complex ("9 + 2" arrangement)	Present on sperm of some plants
Centrioles	No	Yes	No

STUDY OUTLINE

1. The cell is the unit of biological organization. Everything an organism does is a reflection of what is happening at the cellular level, whether that organism is itself a single cell or a complex of many cells.
2. The cell has an intricate structure that is correlated with its impressive array of functions.
3. Cells are open systems, needing to exchange both materials and energy with their environment. They can sense and respond to environmental changes.
4. Evolution is the basis for the correlation of structure and function, a theme encountered repeatedly in our study of the cell.

How Cells Are Studied (pp. 117–122)

1. Most organelles of cells are too small to be viewed with the light microscope. Our current understanding of cell structure and function is due largely to the development of electron microscopy and cell fractionation techniques.
2. Cytology is the study of cell structure. Modern cell biology integrates cytology and biochemistry.

The Geography of the Cell (pp. 122–125)

1. Prokaryotic cells lack true nuclei and membrane-enclosed organelles; bacteria and cyanobacteria in the Kingdom Monera are prokaryotic cells. All other kingdoms have eukaryotic cells with membrane-enclosed nuclei surrounded by cytoplasm in which is suspended a variety of organelles not found in prokaryotic cells.
2. The microscopic size of cells is dictated by their function. Cells require a small volume relative to their surface area to maximize the exchange of materials and energy with the environment. In addition, most cells have only a single nucleus, which must control the entire cytoplasmic volume.
3. Eukaryotic cells are surrounded by a plasma membrane and are partitioned into various compartments by a complex system of internal membranes. These internal membranes provide local environments for specific metabolic processes, some of which are catalyzed by enzymes built into the membranes themselves.
4. All membranes consist of phospholipids and proteins and maintain their integrity by weak chemical bonds. The diversity of membrane function reflects variation in its specific molecular composition.

The Nucleus (pp. 126–129)

1. The trademark of a eukaryotic cell is its distinctive nucleus, enclosed in the nuclear envelope. This envelope is a double membrane that maintains the structure of the nucleus and, through its pores, allows for extensive exchange of

macromolecules between the nucleus and the cytoplasm.

2. The nucleus contains the genetic material, DNA, organized in a characteristic number of chromosomes in each eukaryotic species. Specialized regions of the chromosomes, called nucleolar organizers, form one or more nucleoli, which synthesize ribosomal subunits. These ribosomal subunits, as well as messenger RNA produced at other sites on the chromosomes, pass into the cytoplasm and function together in the synthesis of proteins.

Ribosomes (p. 129–130)

1. Ribosomes are composed of two subunits, which are constructed in the nucleolus from RNA and proteins.

2. Once in the cytoplasm, the ribosomal subunits join with messenger RNA to form functional ribosomes, which carry out protein synthesis while suspended in the cytoplasm (free ribosomes) or attached to the outside of the membranous endoplasmic reticulum (bound ribosomes).

3. Both bound and free ribosomes usually occur in clusters called polysomes, multiple ribosomes attached to a single messenger RNA molecule.

4. Both prokaryotes and eukaryotes possess ribosomes, but those of prokaryotes are smaller and different in molecular composition. This molecular difference is the basis for the selective action of some antibiotics.

The Endomembrane System (pp.130–137)

1. Most of the membranes of a eukaryotic cell are interrelated directly through physical continuity or indirectly through tiny saclike vesicles, pinched-off portions of membrane in transit from one membrane site to another.

2. This interrelated endomembrane system consists of the nuclear envelope, endoplasmic reticulum, Golgi apparatus, lysosomes, microbodies, vacuoles, and plasma membrane.

3. The endoplasmic reticulum (ER) is a network of membrane-enclosed compartments called cisternae. Rough ER, that portion of the ER with bound ribosomes, is continuous with the nuclear envelope and functions in producing cell membrane and manufacturing proteins for secretion. Membrane and secretory proteins can be transferred to other locations in the cell by the budding of transport vesicles from transitional ER.

4. Smooth ER lacks ribosomes and can synthesize steroids, metabolize carbohydrates, store calcium in muscle cells, and detoxify poisons in liver cells.

5. Ribosomes will bind to the ER only if the polypeptides they are manufacturing possess an initial short string of specific amino acids called the signal sequence. The growing polypeptide moves into the ER cisternal space, where the signal sequence is removed by enzymes in preparation for formation of the final secretory product. Other signal sequences can target proteins to other sites in the cell.

6. Proteins without signal sequences cannot induce the binding of their ribosomes to the ER, and such proteins synthesized by free ribosomes remain in the cytosol.

7. The Golgi apparatus consists of dictyosomes, each a stack of membranous sacs that synthesizes various macromolecules, and also modifies, stores, sorts, and exports products of the ER. One side of a dictyosome, the *cis* face, receives secretory proteins from the transitional ER through ER transport vesicles. Once inside, these proteins can be chemically modified and sorted before

release from the *trans* face of the dictyosome in vesicles. The Golgi apparatus manufactures certain macromolecules itself.

8. A lysosome is a membrane-enclosed bag of hydrolytic enzymes originating in the rough ER and processed and released from the Golgi apparatus. Its acidic microenvironment is optimal for the functioning of its enzymes in recycling monomers from cell macromolecules and in digesting substances ingested by phagocytosis.

9. Genetic defects responsible for the absence of one or more lysosomal enzymes cause a variety of specific storage diseases, in which substances that cannot be digested accumulate in the cell and interfere with its function.

10. Microbodies are membrane-enclosed compartments containing enzymes specialized for certain metabolic pathways. One type, the peroxisome, transfers hydrogen from various substrates to oxygen to form hydrogen peroxide, which is subsequently degraded to water. Another type, the glyoxysome, helps convert fat to sugars in germinating seeds.

11. The central vacuole of plant cells is formed by coalescence of smaller vacuoles from the ER and Golgi apparatus. It is an extremely versatile organelle, functioning in storage, breakdown of macromolecules, waste disposal, cell elongation, and protection. The vacuole's membrane is called the tonoplast.

12. The endomembrane system is thus a dynamic assemblage of related components that compartmentalize the cell to facilitate its diverse functions.

Energy Transducers: Mitochondria and Chloroplasts (pp. 137–140)

1. As open systems, cells obtain their energy from their surroundings. Eukaryotic cells accomplish this with mitochondria and chloroplasts, two organelles that act as energy transducers.

2. Both mitochondria and chloroplasts are enclosed in membranes, but they are not considered part of the endomembrane system. Their membrane proteins are not made by the ER, but by free ribosomes and by ribosomes contained within the mitochondria or chloroplasts. These organelles also contain a small amount of DNA that programs the synthesis of some of their proteins.

3. Mitochondria are sites of cellular respiration in eukaryotic cells, capable of using oxygen to convert the chemical energy of sugars, fats, and other molecular fuels into the "high-energy" phosphate bonds of ATP.

4. Mitochondria are compartmentalized by an outer smooth membrane and an inner membrane folded into convolutions called cristae. Many of the metabolic reactions of respiration take place in the space enclosed by the inner membrane called the mitochondrial matrix. Enzymes built into the inner membrane also function in respiration.

5. Chloroplasts, a specialized type of a group of plant organelles called plastids, contain chlorophyll and other pigments, which function in photosynthesis. Chloroplasts are enclosed by two membranes surrounding the fluid stroma in which are embedded the thylakoids. These flattened sacs may be stacked to form grana.

The Cytoskeleton (pp. 140–147)

1. The cytoskeleton is an integrated network of fibers in the cytoplasm that dictates pathways of cell circulation, anchors

organelles, and gives the cell support, shape, and the ability to move. The cytoskeleton is constructed from microtubules, microfilaments, and intermediate fibers.

2. Microtubules are hollow cylinders made of globular proteins called tubulins. In many cells, microtubules radiate out from the microtubule-organizing center, an area near the nucleus that surrounds the centrioles in animal cells. Microtubules shape and support the cell, guide the movement of organelles, and participate in chromosome separation during cell division.

3. Cilia and flagella are motile cellular appendages consisting of a "9 + 2" arrangement of microtubules anchored in the basal body, an organelle structurally identical to a centriole. Movement of cilia and flagella occurs when the sidearms, consisting of the contractile protein dynein and extending from the microtubule doublets, swing and move the doublets past each other.

4. Microfilaments are thinner than microtubules and are solid rods built from actin, a protein. Microfilaments in muscle cells interact with myosin to cause contraction. They also function in cell division and amoeboid movement, cytoplasmic streaming in plants, and support for cellular projections, such as microvilli.

5. In addition to microtubules and microfilaments, many cells have a variety of intermediate filaments that appear to be important in supporting cell shape and fixing various organelles in place.

The Cell Surface (pp. 147–150)

1. The cells of plants, prokaryotes, fungi, and some protists are reinforced by cell walls external to the plasma membrane. Plant cell walls are composed of cellulose fibers embedded in other polysaccharides and protein.

2. Some animal cells possess a glycocalyx, a sticky carbohydrate coat that strengthens the cell surface and helps glue cells together.

3. Various kinds of intercellular junctions help integrate many cells of a plant or animal into one functional organism. Cell-to-cell contact in animals is provided by desmosomes, tight junctions, and gap junctions. Plants have plasmodesmata, cytoplasmic channels that pass through adjoining cell walls.

SELF-QUIZ

1. Electron microscopes have greater resolving power than light microscopes because
   a. electromagnets are more powerful than glass lenses
   b. the specimens used are sliced much thinner, exposing more of a cell's ultrastructure
   c. the specimen is coated with metal atoms or gold to increase contrast
   d. the wavelength of electrons is much shorter than the wavelength of visible light
   e. they can magnify a specimen 1000 times more than a light microscope can

2. From the following, choose the statement that correctly characterizes *bound* ribosomes.
   a. Bound ribosomes are enclosed in their own membrane.

b. Bound ribosomes are structurally different from free ribosomes.
   c. Bound ribosomes generally synthesize membrane proteins and secretory proteins.
   d. The most common location for bound ribosomes is the cytoplasmic surface of the plasma membrane.
   e. Bound ribosomes are concentrated in the cisternal space of rough ER.

3. Which of the following organelles is *least* closely associated with the endomembrane system?
   a. nuclear envelope
   b. chloroplast
   c. Golgi apparatus
   d. plasma membrane
   e. ER

4. If you could follow a small region of membrane as it flowed from organelle to organelle, which of the following sequences would you most likely observe?
   a. Golgi ⟶ lysosome ⟶ ER
   b. tonoplast ⟶ plasma membrane ⟶ nuclear envelope
   c. nuclear envelope ⟶ lysosome ⟶ Golgi
   d. ER ⟶ Golgi ⟶ plasma membrane
   e. ER ⟶ chloroplast ⟶ mitochondrion

5. Which of the following organelles is common to plant *and* animal cells?
   a. chloroplasts
   b. wall made of cellulose
   c. tonoplast
   d. mitochondria
   e. centrioles

6. Which component is present in a prokaryotic cell?
   a. mitochondria
   b. ribosomes
   c. nuclear envelope
   d. chloroplasts
   e. ER

7. A human liver cell measures 20 μm in diameter. This is equivalent to
   a. 200 nm
   b. 2000 nm
   c. 0.02 mm
   d. 0.2 mm
   e. .02 nm

8. Which type of cell would probably provide the best opportunity to study lysosomes?
   a. muscle cell   d. leaf cell of a plant
   b. nerve cell   e. bacterial cell
   c. phagocytic white blood cell

9. Which of the following pairs of structures is *incorrectly* matched?

a. mitochondria–cristae

b. nucleus–chromatin

c. flagella-actin microfilaments

d. chloroplast–grana

e. vacuole–tonoplast

10. For each of the following electron micrographs, name the structure and describe its functions.

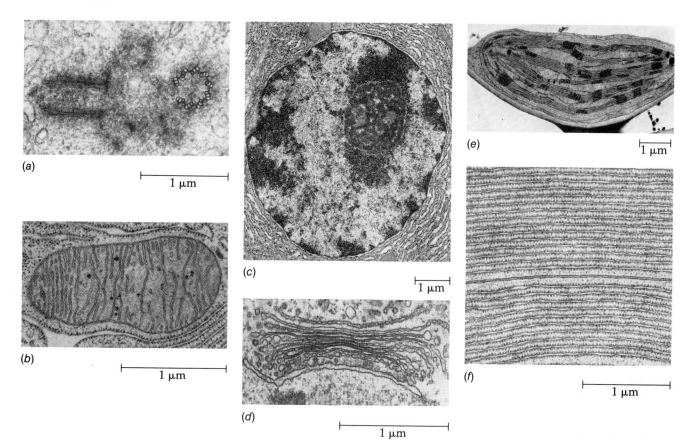

(a) 1 μm

(b) 1 μm

(c) 1 μm

(d) 1 μm

(e) 1 μm

(f) 1 μm

CHALLENGE QUESTIONS

1. An inherited disorder in humans results in the absence of dynein in flagella and cilia. The disease causes respiratory problems and sterility in males. What is the ultrastructural connection between these two symptoms?

2. When very small viruses infect a plant cell by crossing its membrane, the viruses often spread rapidly throughout the entire plant without crossing additional membranes. Explain how this occurs.

3. Vinblastine is a drug that interferes with the assembly of microtubules. It is widely used for chemotherapy in treating cancer patients. Suggest a hypothesis to explain how vinblastine slows tumor growth by inhibiting cell division.

FURTHER READING

1. Alberts, B., D. Bray, J. Lewis, M. Raff, K. Roberts, and J. D. Watson. *Molecular Biology of the Cell.* 2nd ed. New York: Garland, 1989. A popular text; lucidly written, well-illustrated, comprehensive.

2. Becker, W. *The World of the Cell.* Menlo Park, Calif.: Benjamin/ Cummings, 1986. A text of cell biology that is shorter and more accessible than the text by Alberts et al.

3. Bretscher, M. S. "How Animal Cells Move." *Scientific American,* December 1987. The role of vesicles in amoeboid movement.

4. Darnell, J., H. Lodish, and D. Baltimore. *Molecular Cell Biology.* New York: Scientific American Books, 1986. An authoritative survey of modern cell biology.

5. DeDuve, C. *A Guided Tour of the Living Cell.* New York: Scientific American Books, 1986. Beautifully illustrated introduction to the cell presented by the discoverer of lysosomes.

6. Orci, L. J-D Vassalli, and A. Perrelet. "The Insulin Cell." *Scientific American,* September 1988. How cells of the pancreas produce, process, and secrete an important protein.

7. Rothman, J. E. "The Compartmental Organization of the Golgi Apparatus." *Scientific American,* September 1985. A close-up of Golgi structure and function.

8. Symmons, M., A Prescott, and R. Warn. "The Shifting Scaffolds of the Cell." *New Scientist,* February 18, 1989. Dynamics of the cytoskeleton.

8 Membrane Structure and Function

The plasma membrane is the edge of life; it is the boundary that separates the living cell from its nonliving surroundings. A remarkable film only about 8 nm thick, the membrane surrounds the cell and controls the traffic of substances into and out of the cell. Biological membranes are **selectively permeable;** that is, they allow some substances to cross more easily than others. One of the earliest episodes in the evolution of life may have been the origin of a membrane that could maintain a chemical composition within its boundaries that differed from the composition of the solution surrounding it, while still permitting the selective uptake of nutrients and elimination of waste products. This ability of the cell to discriminate in its chemical exchanges with the environment is fundamental to life, and it is the membrane that makes this possible.

The biological membrane is the subject of this chapter. In particular, we will be concerned with how substances cross membranes and how membranes control this traffic. We will concentrate on the plasma membrane, the outermost membrane of the cell. However, the general principles of membrane traffic also apply to the many varieties of internal membranes that partition the eukaryotic cell. As we have seen repeatedly in our study so far, cellular structure fits function. To understand how membranes work, therefore, we begin by examining their architecture.

MODELS OF MEMBRANE STRUCTURE

Lipids and proteins are the staple ingredients of membranes, although carbohydrates may also be present.

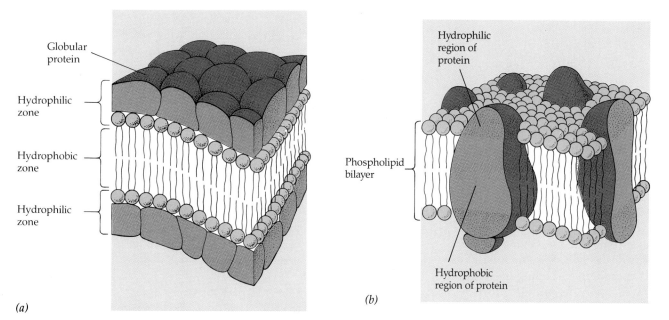

(a)

(b)

Figure 8.3
Evolution of membrane models. (a) The Davson–Danielli model, proposed in 1935, sandwiched the phospholipid bilayer between two protein layers. With later modifications, this model was widely accepted until about 1970. **(b)** The fluid mosaic model has the proteins dispersed and embedded in the phospholipid bilayer, which is in a fluid state. This is our present working model of the membrane.

matrix. Other experiments support the conclusion that the bumps are proteins. Proteins penetrate into the hydrophobic interior of the membrane, which would not be the case if the Davson–Danielli sandwich were correct. The Davson–Danielli model predicts a smooth appearance for a freeze-fractured membrane. The fluid mosaic model, by contrast, predicts the presence of proteins in the interior of membranes.

In science, models are proposed by scientists as ways of organizing and explaining existing information. The replacement of one model of membrane structure with another does not imply that the original model was worthless. The acceptance or rejection of a model depends on how well it fits observations and explains experimental results. A good model also makes predictions that shape future research. Models inspire testing, and few of them survive, without some modification, the experimental assault they invite. New findings may make a model obsolete, but even then it may not be totally scrapped but only revised to account for the new observations. Like its predecessor, which endured for 35 years, the fluid mosaic model may eventually be retailored to fit new observations and experiments; but for now, it is the most acceptable model of membrane structure.

The Fluid Mosaic Model: A Closer Look

What exactly does it mean to describe a membrane as a fluid mosaic? Let us begin with the word *fluid*.

The Fluid Quality of Membranes Membranes are not static, solid sheets of molecules locked rigidly in place. A membrane is held together primarily by hydrophobic attractions, which are weaker than covalent bonds (see Chapter 2). Most of the lipids and some of the proteins can drift about laterally in the plane of the membrane, exchanging places with neighbors (Figure 8.4a). It is very rare, however, for a molecule to flip-flop transversely across the membrane, switching from one phospholipid layer to the other; to do so, the hydrophilic part of the molecule would have to cross the hydrophobic core of the membrane, which is unlikely.

Phospholipids move along the plane of the membrane quite rapidly, averaging about 2 μm—the length of a large bacterial cell—per second. Proteins are much larger than lipids and move more slowly, but there is evidence that some membrane proteins do, in fact, drift. This was elegantly demonstrated by experiments in which a human cell was fused to a mouse cell to form a hybrid cell with a continuous plasma membrane, with each animal species contributing part of the membrane. The membrane proteins of the two species were soon mixed, indicating that the proteins must have drifted in the fluid membrane (see Figure 8.5). Many membrane proteins are unable to move far because they are tethered in place by their attachment to the cytoskeleton.

Membranes remain fluid as temperature decreases until finally, at some critical temperature, the membrane solidifies, much as bacon grease forms lard when

The specimen is frozen at the temperature of liquid nitrogen, and then a cold knife is used to fracture the cells. The knife does not cut cleanly through the frozen cells; instead, it cracks the specimen with the fracture plane following the path of least resistance. The fracture plane often follows the hydrophobic interior of a membrane, splitting the lipid bilayer down the middle into a P (protoplasmic) face and an E (exterior) face. The topography of the fractured surface may be enhanced by etching, the removal of water by sublimation (direct evaporation of frozen water to water vapor). The membrane proteins are not split but go with one or the other of the phospholipid layers.

A fine mist of platinum is sprayed from an angle onto the fractured surface of the cell. There will be "shadows" where the platinum is blocked by elevated regions of the fractured cell. A film of carbon is added to strengthen the platinum coat.

The original specimen is digested away with acids and enzymes, leaving the platinum-carbon film as a replica of the fractured surface. It is this replica, not the membrane itself, that is examined in the electron microscope.

The electron micrographs in the bottom figure have been superimposed on a drawing of a delaminated membrane. Note the protein particles.

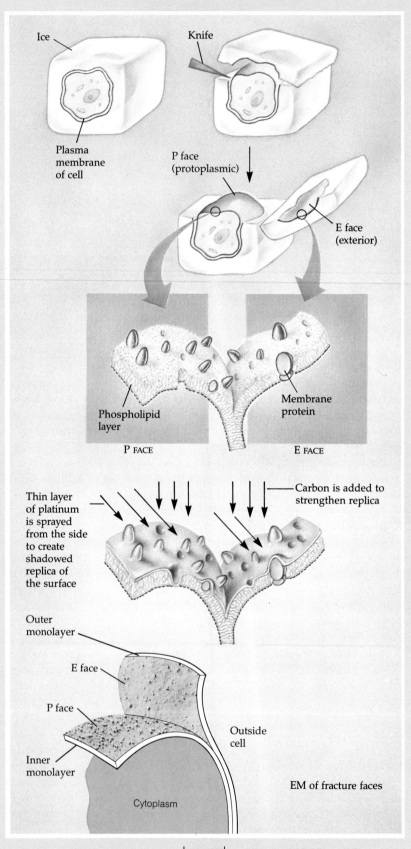

1 μm

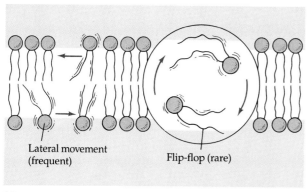

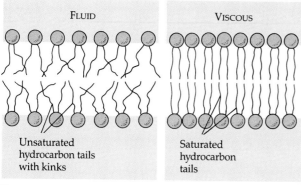

(a)

(b)

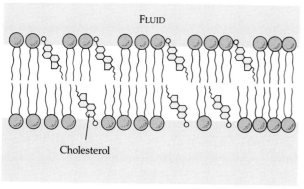

(c)

Figure 8.4
Fluidity of membranes. (a) Lipids are free to move laterally (that is, in two dimensions) in a membrane, but flip-flopping across the membrane (in the third dimension) is rare. (b) Unsaturated hydrocarbon tails of phospholipids have kinks that keep the molecules from packing together, enhancing membrane fluidity. (c) Cholesterol contributes to membrane fluidity by hindering the packing together of phospholipids.

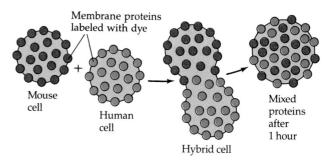

Figure 8.5
Evidence for drifting of membrane proteins. Before the mouse cells and human cells were fused in this experiment, their membrane proteins were specifically labeled with dyes. When the cells were fused, the membrane of the hybrid was at first two-tone; but in less than an hour, the two dyes were completely mixed. The proteins labeled with the dyes had intermingled by migrating along the membrane.

located, unsaturated hydrocarbons do not pack together as closely as saturated hydrocarbons (Figure 8.4b). The steroid cholesterol also enhances membrane fluidity (Figure 8.4c). It is found in the plasma membranes of eukaryotes and is especially abundant in the membranes of animal cells at low temperatures. Cholesterol is itself a lipid, and when wedged between phospholipid molecules, the cholesterol helps to keep the membrane fluid by hindering close packing of the phospholipids.

Membranes must be fluid to work properly. When a membrane solidifies, its permeability changes; also enzymatic proteins in the membrane may become inactive. In the process of renewing its membranes, a cell may alter their lipid composition to some extent as an adjustment to changing temperature. For instance, in many varieties of plants that tolerate extreme cold, such as winter wheat, the percentage of unsaturated phospholipids increases in autumn, an adaptation that keeps the membranes from solidifying during winter. And the membranes of certain hibernating animals, including ground squirrels, are enriched in cholesterol, which helps to keep constant the fluidity of their membranes.

The functioning membrane, then, is a liquid film of proteins dissolved in a lipid bilayer that is normally about as fluid as salad oil.

Membranes as Mosaics of Structure and Function
Now we come to the word mosaic. A membrane is a collage of many different proteins inlaid in the fluid matrix of the lipid bilayer (Figure 8.6). The lipid bilayer is the main fabric of the membrane, but proteins determine most of the specific functions of the mem-

it cools. The temperature at which a membrane solidifies depends on its lipid composition. The membrane remains fluid to a lower temperature if it is rich in phospholipids with unsaturated hydrocarbon tails (see Chapter 5). Because of kinks where double bonds are

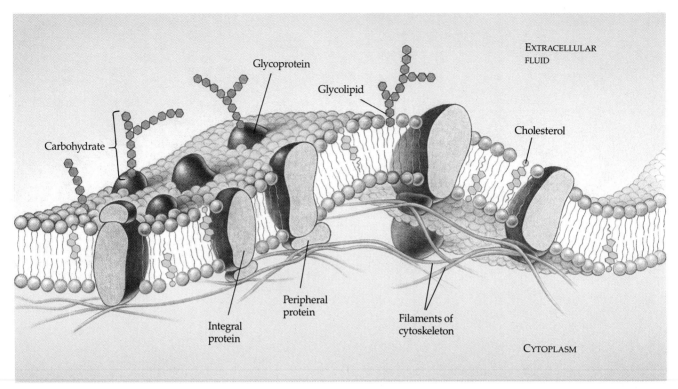

Figure 8.6
Detailed structure of a plasma membrane.

brane (Figure 8.7). The plasma membrane and the membranes of the various organelles each have their unique collections of proteins. More than 50 kinds of proteins have been found in the plasma membrane of red blood cells, for example, and there are probably many more that are just too scarce to be detected.

Notice in Figure 8.6 that there are two major populations of membrane proteins. **Integral proteins** are inserted into the membrane and penetrate far enough for their hydrophobic regions to be surrounded by the hydrocarbon portions of lipids. Some integral proteins may be unilateral, reaching only partway across the membrane. Others—probably most—completely span the membrane. These transmembrane proteins have hydrophobic midsections between hydrophilic ends exposed to the aqueous solutions on both sides of the membrane. **Peripheral proteins** are not embedded in the lipid bilayer at all; they are appendages attached to the surface of the membrane, often to the exposed parts of integral proteins. On the cytoplasmic side of the plasma membrane, some peripheral proteins and their integral protein partners may be held in place by filaments of the cytoskeleton.

Membranes are bifacial, with distinct inside and outside faces. The two lipid layers may differ in specific lipid composition, and each protein has directional orientation in the membrane. The plasma membrane also has carbohydrates, which are restricted to the exterior surface. This asymmetric distribution of proteins, lipids, and carbohydrates is determined as the membrane is being built by the endoplasmic reticulum. Thus, the exterior surface of the plasma membrane is topologically equivalent to the interior surfaces of the ER and the other organelles of the endomembrane system (Figure 8.8).

Membrane Carbohydrates and Cell-Cell Recognition Cell-cell recognition, the ability of a cell to determine if other cells it encounters are alike or different from itself, is crucial in the functioning of an organism. It is important, for example, in the sorting of cells into tissues and organs in an animal embryo. It is also the basis for rejection of foreign cells (including those of transplanted organs) by the immune system, an important line of defense in vertebrate animals (see Chapter 39). The way cells recognize other cells is by keying on surface molecules of the plasma membrane.

Membrane carbohydrates are usually branched oligosaccharides with fewer than 15 sugar units. Some of these oligosaccharides are covalently bonded to a special class of lipids called glycolipids (*glyco*, remember, refers to the presence of carbohydrates). Most membrane carbohydrates, however, are covalently bonded to proteins, which are thereby glycoproteins.

The oligosaccharides on the external side of the plasma membrane vary from species to species, between individuals of the same species, and even

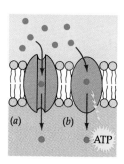

Transport proteins (*a*) A protein that spans the membrane may provide a hydrophilic channel across the membrane that is selective for a particular solute. (*b*) Some transport proteins hydrolyze ATP as an energy source to actively pump substances across the membrane.

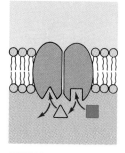

Enzymes A protein built into the membrane may be an enzyme with its active site exposed to substances in the adjacent solution. In some cases, several enzymes are ordered in a membrane as a team that carries out sequential steps of a metabolic pathway.

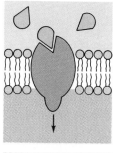

Proteins as receptor sites The portion of a membrane protein exposed to the outside of the cell may have a binding site with a specific shape that complements the shape of a chemical messenger, such as a hormone. If the receptor protein spans the membrane, the external signal may induce a conformational change that activates the portion of the protein facing the cytoplasm to initiate a chain reaction of chemical changes in the cell.

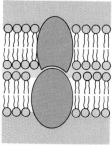

Cell adhesion Membrane proteins of adjacent cells may be hooked together in various kinds of intercellular junctions.

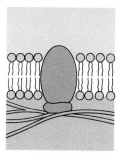

Attachment to the cytoskeleton Actin microfilaments or other elements of the cytoskeleton may be bonded to membrane proteins, a function that is important in maintaining cell shape and in fixing the locations of certain membrane proteins.

Figure 8.7
Some functions of membrane proteins. A single protein may perform some combination of these tasks.

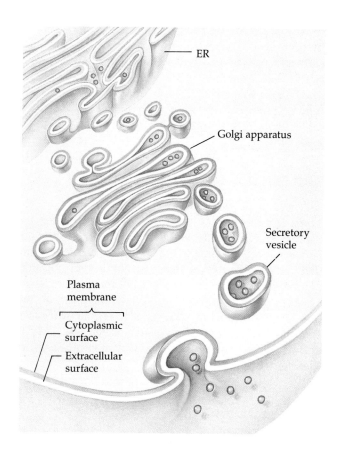

Figure 8.8
Sidedness of the plasma membrane. The membrane is bifacial, with distinct cytoplasmic and extracellular sides. Lipid composition of the two layers may differ, membrane proteins have specific orientations in the membrane, and carbohydrates are restricted to the extracellular side of the membrane. The membrane's bifacial quality is determined when it is first synthesized and modified by the ER and Golgi. This diagram color codes the two leaflets of the membrane to help you see that the side of the membrane facing the lumen (cavity) of the ER, Golgi, and vesicles is topologically equivalent to the extracellular surface of the plasma membrane. The other side of the membrane faces the cytoplasm, from the time the membrane is produced by the ER to the time it is added to the plasma membrane by fusion of a vesicle.

from one cell type to another in a single individual. The diversity of the molecules and their location on the cell's outer fringe are features that make oligosaccharides likely candidates as markers that distinguish one cell from another. Glycoproteins and glycolipids are now major topics of interest in many laboratories studying cell-cell recognition.

The biological membrane is an exquisite example of a supramolecular structure—many molecules ordered into a higher level of organization with emergent properties beyond those of the individual molecules. The remainder of this chapter attends to one of the most important of those properties: the ability of membranes to regulate the transport of key substances across cellular boundaries, a function essential to the cell's existence as an open system. We will see once again that form fits function: The fluid mosaic model of membrane architecture helps to explain how molecules cross membranes.

TRAFFIC OF SMALL MOLECULES

There is a steady traffic of small molecules across the plasma membrane. Consider the chemical exchanges between a human muscle cell and the extracellular fluid that bathes it. Sugars, amino acids, and other nutrients enter the cell, and waste products of metabolism leave. The cell takes in oxygen for cellular respiration and expels carbon dioxide. The cell also regulates its concentrations of inorganic ions, such as Na^+, K^+, Ca^{2+}, and Cl^-, by shuttling them one way or another across the plasma membrane.

Selective Permeability

As mentioned at the beginning of this chapter, biological membranes are selectively permeable. This means that although traffic through the membrane is extensive, substances do not cross the barrier indiscriminately. The cell is able to retain many varieties of small molecules and exclude others. Moreover, substances that move through the membrane do so at different rates.

Permeability of the Lipid Bilayer   The hydrophobic core of the membrane impedes the transport of ions and polar molecules, which are hydrophilic. The ability of various substances to cross this barrier can be tested by measuring the rate of their transport through an artificial phospolipid bilayer (Figure 8.9). Hydrophobic molecules, such as hydrocarbons and oxygen, can dissolve in the membrane and cross it with ease. If two molecules are equally soluble in oil, the smaller of the two will cross the membrane faster. Very small molecules that are polar but uncharged can also pass through the synthetic membrane rapidly. Examples of this are water and carbon dioxide, which apparently are tiny enough to pass between the lipids of the membrane. To larger uncharged polar molecules, such as glucose and other sugars, the lipid bilayer is not very permeable. The lipid membrane is also relatively impermeable to all ions, even small ones such as H^+ and Na^+. A charged atom or molecule and its shell of water (see Chapter 3) find the hydrophobic layer of the membrane difficult to penetrate.

To some extent, we can extrapolate the selective

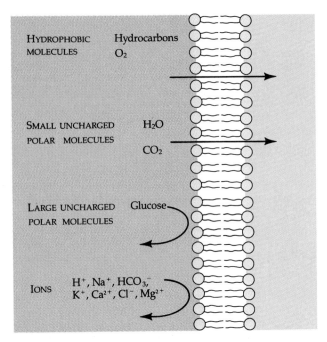

Figure 8.9
Selective permeability of an artificial phospholipid bilayer. Some substances cross the bilayer much more rapidly than others. The rate of transport depends on the size of the substance and its solubility in the hydrocarbon interior of the membrane. The phospholipid bilayer is not very permeable to ions or to polar molecules as large as glucose.

permeability of the artificial membrane to the plasma membrane, whose own lipid bilayer is a differential barrier to molecular traffic. The difference, of course, is that the plasma membrane also has proteins, which greatly affect its permeability.

Transport Proteins   Water, carbon dioxide, and nonpolar molecules, all of which pass through artificial lipid bilayers, also cross the plasma membrane rapidly. But biological membranes, unlike artificial bilayers, are also permeable to specific ions and certain polar molecules of moderate size, including sugars. These hydrophilic substances avoid contact with the lipid bilayer by passing through **transport proteins** that span the membrane (see Figure 8.7). Transport proteins may be classified into three types, depending on how many kinds of molecules or ions they transport at one time, and in what directions (Figure 8.10). In some cases, the transport protein has a channel that certain molecules use as a hydrophilic tunnel through the membrane. Other transport proteins bind to their passengers and physically move them across the membrane. In any case, each transport protein is very specific for the substances it translocates, allowing only a certain molecule or class of closely related molecules to cross the membrane. For example, glucose carried

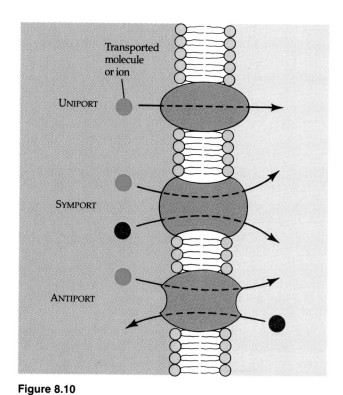

Figure 8.10
Three classes of transport proteins. A *uniport* carries a single solute across the membrane. A *symport* translocates two different solutes simultaneously in the same direction; transport occurs only if *both* solutes bind to the protein. For example, the plasma membranes of many animal cells have symports that transport a sodium ion and a glucose molecule in tandem from the extracellular solution into the cell. If there is no Na$^+$ outside the cell, the symport is unable to transport glucose. An *antiport* exchanges two solutes by transporting one into the cell and the other out of the cell. One antiport, for example, admits sodium ions into the cell, while expelling calcium ions.

to the human liver in blood enters the liver cells rapidly through specific transport proteins inserted in the plasma membrane. The protein is so selective that it even rejects fructose, a structural isomer of glucose.

The selective permeability of a membrane depends on both the discriminating barrier of the lipid bilayer and the specific transport proteins built into the membrane. But what determines the *direction* of traffic across a membrane? At a given time, will a particular substance enter the cell through the membrane, or leave? Part of the answer can be found in the phenomenon known as diffusion.

Diffusion and Passive Transport

Unless the temperature is at absolute zero (0 K), all atoms and molecules have intrinsic kinetic energy called thermal motion, or heat. One result of thermal motion is **diffusion,** the tendency for molecules of any substance to spread out, or diffuse, into the available space. Each molecule moves randomly, and yet diffusion of a population of molecules may be directional. For example, imagine a membrane separating pure water from a solution of a dye dissolved in water, and assume that this membrane is permeable to the dye molecules (Figure 8.11a). Each dye molecule wanders randomly, but there will be a *net* movement of the dye molecules across the membrane to the side that began as pure water simply because there are more dye molecules that can move in that direction. The diffusion—the spreading of the dye across the membrane—will continue until both solutions have equal concentrations of the dye. Once that point is reached, there will be a dynamic equilibrium, with as many dye molecules

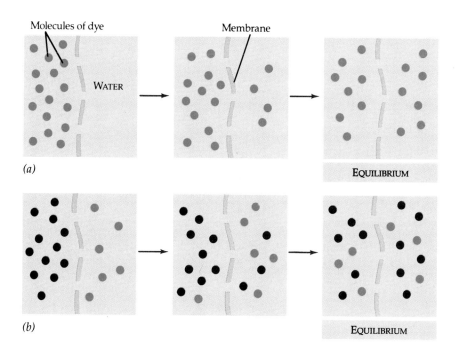

Figure 8.11
Diffusion of a solute. (a) A substance will diffuse from where it is more concentrated to where it is less concentrated. Here, molecules of a dye dissolved in water diffuse across a membrane until dynamic equilibrium is reached. (b) In this case, solutions of two different dyes are separated by a membrane that is permeable to both dyes. Each dye diffuses down its own concentration gradient. There will be a net diffusion of the green dye toward the left, even though the total solute concentration was initially greater on the left side.

moving per second across the membrane one way as in the opposite direction.

We can now state a simple rule of diffusion: In the absence of other forces, a substance will diffuse from where it is more concentrated to where it is less concentrated. Put another way, any substance will diffuse down its **concentration gradient.** No work must be done to get this to occur; diffusion is a spontaneous process because it decreases free energy (see Chapter 6). Remember, in any system there is a tendency for entropy, or disorder, to increase. Diffusion of a solute in water increases entropy by producing a more random mixture than exists when there are localized concentrations of the solute. It is important to note that each substance diffuses down its *own* concentration gradient, unaffected by concentration differences of other substances (Figure 8.11b).

Much of the traffic across the membranes of cells occurs by diffusion. Whenever a substance is more concentrated on one side of a membrane than on the other, there is a tendency for the substance to diffuse across the membrane down its concentration gradient (assuming that the membrane is permeable to that substance). One important example is the uptake of oxygen by a cell performing cellular respiration. Dissolved oxygen diffuses into the cell across the plasma membrane. So long as cellular respiration consumes the O_2 as it enters, diffusion into the cell will continue, because the concentration gradient favors movement in that direction.

The diffusion of a substance across a biological membrane is called **passive transport.** The cell does not have to expend energy for a substance to move across a membrane by passive transport; the molecules are simply diffusing down their concentration gradient, which is a spontaneous process. Remember, however, that membranes are selectively permeable and therefore affect the rates of diffusion of various molecules. One molecule that diffuses freely across membranes is water, a fact that has important consequences for cells.

Osmosis: A Special Case of Passive Transport

In comparing two solutions of unequal solute concentration, the solution with a greater concentration of solutes is said to be **hyperosmotic.** The solution with the lesser solute concentration is **hypoosmotic.** These are relative terms that are meaningful only in a comparative sense. For example, tap water is hyperosmotic to distilled water, but hypoosmotic to seawater. In other words, tap water has a higher concentration of solutes than distilled water, but a lower concentration than seawater. Solutions of equal solute concentration are said to be **isosmotic.**

Picture a U-shaped vessel with a selectively permeable membrane separating two sugar solutions of different concentrations (Figure 8.12). The membrane in this example is permeable to water, but impermeable to sugar. Water will diffuse across the membrane from the more dilute (hypoosmotic) sugar solution into the more concentrated (hyperosmotic) solution (the terms *dilute* and *concentrated* refer to the sugar, not to the water). The concentrations of sugar on opposite sides of the membrane become less different as one solution loses water to the other. Meanwhile, the volume of the more concentrated solution increases and the volume of the more dilute solution decreases as a result of the water transport. The U shape of the vessel makes the changes in volume easy to see.

The diffusion of water across a selectively permeable membrane is a special case of passive transport called **osmosis.** In a sense, the water is diffusing down

Figure 8.12
Osmosis. Two sugar solutions of different concentration are separated by a membrane that is permeable to the solvent (water) but impermeable to the solute (sugar). Water will diffuse from the hypoosmotic solution to the hyperosmotic solution. This diffusion of water across a selectively permeable membrane is called osmosis. Ideally, osmosis would continue until the sugar concentrations on opposite sides of the membrane were equal, and this would probably be the result if the experiment were performed in the weightless conditions of a space shuttle. The two solutions would eventually be isosmotic. Here on Earth, however, the added weight of the rising column of solution into which water moves will eventually force water back across the membrane fast enough to offset water movement due to the remaining difference in sugar concentrations.

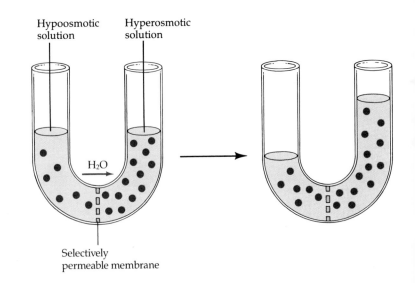

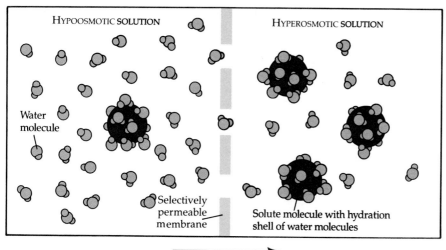

Figure 8.13
The effect of solutes on water mobility. Without reducing the overall concentration of water (the number of water molecules per unit volume), solute molecules *do* reduce the proportion of water molecules that can move freely. Water molecules clustered around solute molecules are not free to diffuse across a membrane. Water will move by osmosis from a hypoosmotic to a hyperosmotic solution because the hypoosmotic solution has a greater concentration of unbound water molecules that can cross the membrane.

Net flow of water

its concentration gradient. The solution with the greater concentration of solute (sugar, in this case) has the lesser concentration of solvent (water). Actually, it is not so simple. For a dilute solution—and most biological fluids are dilute solutions—the presence of solutes does not alter the water concentration significantly. The space taken up by the solute molecules is offset because some of the water becomes more compact as water molecules cluster tightly around molecules of the hydrophilic solutes (see Chapter 3). This bound water is not free to move across the membrane (Figure 8.13). It is not really a difference in water concentration that causes osmosis in the example in Figure 8.12, but a difference in the proportion of unbound water that can cross the membrane. The effect, however, is the same: Water tends to diffuse across a membrane from a hypoosmotic solution to a hyperosmotic solution.

The direction of osmosis is determined only by a difference in *total* solute concentration, not by the nature of the solutes. Water will move from a hypoosmotic to a hyperosmotic solution even if the hypoosmotic solution has more *kinds* of solutes. Seawater, which has a great variety of solutes, will lose water to a solution of sugar that is very concentrated, because the total solute concentration of the seawater is less. Water moves across a membrane separating isosmotic solutions at an equal rate in both directions; that is, there is no net osmotic movement of water between isosmotic solutions.

The tendency for a solution to take up water by osmosis can be measured with instruments called osmometers. One type is an apparatus in which pure water is separated from a solution by a membrane that is permeable to the water but not the solute (Figure 8.14). Ordinarily, osmosis causes the volume of the solution to increase. We can counteract this tendency for the solution to take up water by pushing against

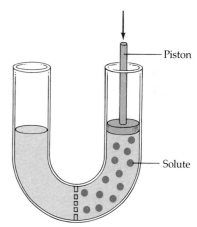

Figure 8.14
Measuring osmotic pressure. In one type of osmometer, a solution is separated from pure water by a membrane that is permeable to water but not to the solute. The amount of pressure that must be applied to the solution to counteract the tendency for water to enter by osmosis is called the osmotic pressure of the solution. The greater the osmotic concentration (total solute concentration), the greater the osmotic pressure of the solution.

the top of the solution with a piston. We can keep the volume of the solution constant by applying just enough pressure with the piston to exactly negate the tendency for water to enter the solution by osmosis. The piston and the solution are thus exerting equal pressure against each other, and this quantity of pressure is said to be the **osmotic pressure** of the solution. (Plant physiologists use another measurement called water potential, which will be introduced in Chapter 32.) Osmotic pressure, then, is a measure of the tendency for a solution to take up water when separated

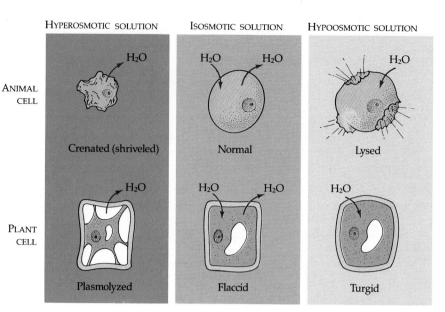

Figure 8.15
Water balance of living cells. How living cells react to changes in the solute concentrations of their environments depends on whether or not they have cell walls. Animal cells do not have cell walls; plant cells do. Unless it has special adaptations to offset the osmotic uptake or loss of water, an animal cell fares best in an isosmotic environment. Plant cells are turgid (firm) and generally healthiest in a hypotonic environment. Under that condition, the tendency for continued uptake of water by the cell is balanced by the elastic wall pushing back on the cell. (Arrows indicate *net* water movement.)

HYPEROSMOTIC SOLUTION   ISOSMOTIC SOLUTION   HYPOOSMOTIC SOLUTION

ANIMAL CELL

Crenated (shriveled)   Normal   Lysed

PLANT CELL

Plasmolyzed   Flaccid   Turgid

from pure water by a selectively permeable membrane. The osmotic pressure of pure water is zero. If our osmometer had pure water on both sides of the membrane, no pressure would have to be applied to the piston. The osmotic pressure of a solution is proportional to its total solute concentration, termed **osmotic concentration.** The greater the osmotic concentration, the greater the osmotic pressure and the greater the tendency for the solution to take up water from a reservoir of pure water. And if two solutions are separated by a membrane, water will pass from the solution with lesser osmotic pressure to the solution with greater osmotic pressure. This is just another way of saying that water moves from a hypoosmotic solution to a hyperosmotic solution.

The movement of water across cell membranes and the balance of water between the cell and its environment are crucial to organisms. Let us now apply to living cells what we have learned about osmosis from artificial systems.

Water Balance of Cells Without Walls   If an animal cell is immersed in an environment that is isosmotic to the cell, there will be no net movement of water across the plasma membrane. Water is flowing across the membrane, but at the same rate in both directions. In an isosmotic environment, the volume of an animal cell is stable (Figure 8.15). Now let us transfer the cell to a solution that is hyperosmotic to the cell—a solution with a greater osmotic concentration and hence greater osmotic pressure than the cell. The cell will lose water to its environment, shrivel (crenate), and probably die. This is one reason why an increase in the salinity (saltiness) of a lake can kill the animals there. However, taking up too much water can be just as hazardous to an animal cell as losing water. If we place the cell in a solution that is hypoosmotic to the

cell, water will enter faster than it leaves, and the cell will swell and pop (lyse) like an overfilled water balloon.

Cells without rigid walls can tolerate neither excessive uptake nor excessive loss of water. The problem of water balance is automatically solved if the cell lives in isosmotic surroundings. Many marine invertebrates are isosmotic to seawater. Terrestrial (land-dwelling) animals generally bathe their cells with an extracellular fluid that is isosmotic to the cells. Animals and other organisms without rigid cell walls that live in environments that are either hyperosmotic or hypoosmotic to their cells must have special adaptations for **osmoregulation,** the control of water balance. For example, the protist *Amoeba* lives in pond water, which is hypoosmotic to the cell. Water continually tends to enter the cell, but *Amoeba* has a plasma membrane that is much less permeable to water than the membranes of most other cells. Also, *Amoeba* is equipped with a contractile vacuole, an organelle that functions as a bilge pump to force water out of the cell as fast as it enters by osmosis. We will examine various adaptations for osmoregulation in Chapter 40.

Water Balance of Cells with Walls   The cells of plants, prokaryotes, fungi, and some protists have walls. Under certain conditions, the wall plays a major role in maintaining water balance between the cell and its external environment. But a wall is of no advantage if the cell is immersed in a hyperosmotic environment. In this case, a plant cell, like an animal cell, will lose water to hyperosmotic surroundings and shrink (see Figure 8.15). As the plant cell shrivels, its plasma membrane pulls away from the wall. This phenomenon, called **plasmolysis,** is usually lethal. The walled cells of bacteria and fungi also plasmolyze in hyperosmotic environments.

It is when a plant cell is in a hypoosmotic solution

that the wall is a factor in water balance. Again like an animal cell, the plant cell swells as water enters by osmosis. However, the elastic wall will expand only so much before it exerts a back pressure on the cell that offsets the tendency for further water uptake from the hypoosmotic surroundings. When the wall pressure exerts a force equal to and opposite to the osmotic pressure of the cell, then water enters and leaves the cell at the same rate, and a dynamic equilibrium is reached. At this point, the cell is **turgid** (very firm). This is the ideal state for most plant cells. Plants that are not woody, such as most house plants, depend on turgid cells for mechanical support. This requires that the cells be hyperosmotic to the solution on the outside of their plasma membranes.

If a plant cell and its surroundings are isosmotic, then there is no net tendency for water to enter, and the cell is **flaccid** (limp). A plant wilts when its cells are flaccid. But as you have probably noticed with your house plants, a wilted plant will perk up again if it is watered so that the solution bathing the cells is once again hypoosmotic enough for the cells to become turgid. If, on the other hand, the wilted plant continues to be deprived of water, then the cells will eventually be surrounded by an increasingly hyperosmotic solution that causes plasmolysis and death.

Facilitated Diffusion

Let us now direct our attention from the transport of water across a membrane to the traffic of specific solutes dissolved in the water. As mentioned earlier, many polar molecules and ions impeded by the lipid bilayer of the membrane diffuse with the help of transport proteins that span the membrane. This phenomenon is called **facilitated diffusion.**

A transport protein has many of the properties of an enzyme (see Chapter 6). The protein is specialized for the solute it transports, just as an enzyme is specific for its substrate. Presumably, the transport protein has a specific binding site akin to the active site of an enzyme. And just as the rate of an enzymatic reaction levels off when the substrate concentration is high enough to keep the active sites of all enzyme molecules occupied, so too can transport proteins be saturated. There are only so many molecules of each type of transport protein built into the plasma membrane, and when they are fully engaged binding and translocating passengers as fast as they can, transport is occurring at a maximum rate. Also like enzymes, transport proteins can be inhibited by molecules that resemble the normal "substrate." This occurs when the imposter competes with the normally transported solute by binding to the transport protein. Unlike enzymes, however, transport proteins do not usually catalyze chemical reactions. Their function is to speed

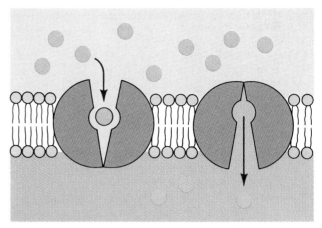

Figure 8.16
One model for facilitated diffusion. The transport protein (violet) alternates between two conformations, moving a solute across the membrane as the shape of the protein changes. The protein can transport the solute in either direction, with the net movement being "down" the concentration gradient of the solute. The process is a passive one, not requiring energy input from the cell itself, but using the potential energy stored in a concentration gradient.

the transport of a molecule across a membrane that would otherwise be impermeable to the substance.

Cell biologists are still trying to learn how membrane proteins facilitate diffusion. It is unlikely that the protein acts as a ferry that picks up its passenger on one side of the membrane and then travels across the width of the membrane to deposit the molecule on the opposite side. It is also improbable that the transport protein functions as a revolving door that spins in the membrane, transporting a molecule with each turn. Both of these models are unattractive because the mechanisms would bring the hydrophilic parts of the protein into contact with the hydrophobic interior of the membrane. A model more consistent with what is known about membrane structure is one in which the protein remains in place in the membrane and helps a molecule across by undergoing a subtle change in shape that translocates the binding site from one side of the membrane to the other (Figure 8.16). The transport protein would wobble between its two states, picking up a molecule in one conformation and depositing the molecule on the opposite side of the membrane in the alternate conformation. The changes in shape could be triggered by the binding and release of the transported molecule.

In certain inherited diseases, specific transport systems are either defective or missing altogether. An example is cystinuria, a human disease characterized by the absence of a transport system that carries cystine and other amino acids into kidney cells. Kidney cells normally reabsorb these amino acids from the urine and return them to the blood, but an individual

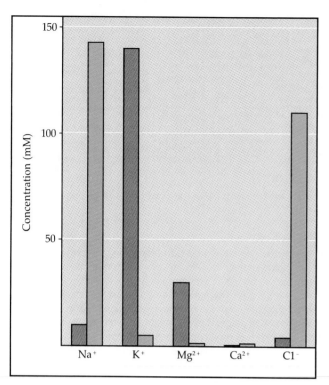

Figure 8.17
Concentrations of various inorganic ions inside (brown) and outside (green) an animal cell. Active transport helps to maintain these steep gradients.

afflicted with cystinuria develops painful kidney stones from amino acids that accumulate and crystallize in the kidneys.

Active Transport

Despite the help of a transport protein, facilitated diffusion is still considered passive transport because the solute is moving down its concentration gradient. Facilitated diffusion speeds the transport of a solute by providing a specific corridor through the membrane, but it does not alter the direction of transport. Some transport proteins, however, can transport solutes against their concentration gradients, across the plasma membrane from the side where they are less concentrated to the side where they are more concentrated. This transport is "uphill" in the sense that it goes against the tendency for substances to diffuse down their concentration gradients. To pump a molecule across a membrane against its gradient, the cell must expend its own energy; therefore, this type of membrane traffic is called **active transport.**

Active transport is a major factor in the ability of a cell to maintain internal concentrations of small molecules that differ from concentrations in the surrounding environment. For example, compared to its surroundings, an animal cell has a much higher concentration of potassium ions and a much lower concentration of sodium ions (Figure 8.17). The plasma membrane helps to maintain these steep gradients by pumping sodium out of the cell and potassium into the cell.

The work of active transport is performed by specific proteins inserted in membranes. These transport proteins share many of the enzymelike properties of the proteins that function in facilitated diffusion, but membrane proteins engaged in active transport must harness cellular energy to pump molecules against concentration gradients. As in other types of cellular work, ATP supplies the energy for most active transport. One way ATP can power active transport is by transferring its terminal phosphate group directly to the transport protein. This may induce the protein to change its conformation in a manner that translocates a solute bound to the protein across the membrane. One transport system that seems to work this way is the **sodium-potassium pump,** which exchanges sodium (Na^+) for potassium (K^+) across the plasma membrane of animal cells (Figure 8.18). We will look further at the function of the sodium-potassium pump in Chapter 44.

The Special Case of Ion Transport

All cells have voltages across their plasma membranes. A voltage is electric potential energy due to opposite charges being separated (see Chapter 6). The cytoplasm of the cell is negative in charge compared to the extracellular fluid because of an unequal distribution of anions and cations across the plasma membrane. Voltage across membranes, called the **membrane potential,** ranges from about -50 to -200 millivolts (mV). (The minus sign indicates that the inside of the cell is negative compared to the outside.)

The membrane potential acts as a battery does and affects the traffic of all charged substances across the membrane, favoring diffusion of cations into the cell and anions out of the cell. Thus two forces drive passive transport of ions across membranes: the concentration gradient of the ion and the effect of the membrane potential on the ion. Figure 8.19 illustrates how these forces interact.

Because of the complication of the membrane potential, it is not correct to say that an ion always diffuses down its concentration gradient. The ion *does* diffuse down its **electrochemical gradient,** a diffusion gradient that combines the influences of the electric force (membrane potential) and the chemical force (concentration gradient). For uncharged solutes, only the concentration gradient is relevant.

Several factors contribute to the membrane potential of a cell. At cellular pH, most proteins and other

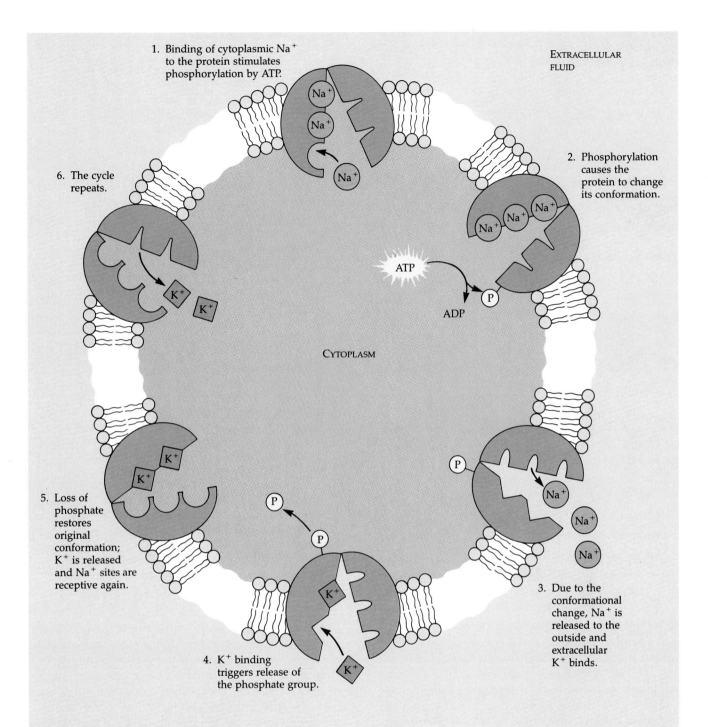

1. Binding of cytoplasmic Na⁺ to the protein stimulates phosphorylation by ATP.

2. Phosphorylation causes the protein to change its conformation.

6. The cycle repeats.

ATP

ADP

P

CYTOPLASM

5. Loss of phosphate restores original conformation; K⁺ is released and Na⁺ sites are receptive again.

3. Due to the conformational change, Na⁺ is released to the outside and extracellular K⁺ binds.

4. K⁺ binding triggers release of the phosphate group.

Figure 8.18
The sodium-potassium pump: a specific case of active transport. This active transport system pumps ions against steep concentration gradients. The pump oscillates between two conformational states in a pumping cycle that translocates three Na⁺ ions out of the cell for every two K⁺ ions pumped into the cell. ATP powers the changes in conformation by phosphorylating the transport protein (that is, by transferring a phosphate group to the protein). The two conformational states differ in the affinity for Na⁺ and K⁺ and in the directional orientation of the ion binding sites. Prior to phosphorylation, the binding sites face the cytoplasm, and only the Na⁺ sites are receptive. Sodium binding induces phosphate transfer from ATP to the pump, triggering the conformational change. In its new conformation, the pump's binding sites face the extracellular side of the membrane, and the protein now has a greater affinity for K⁺ than it does for Na⁺. Potassium binding causes release of the phosphate, and the pump returns to its original conformation. Because the pump also behaves as an enzyme that removes phosphate from ATP, it is sometimes called the Na⁺–K⁺ ATPase. The pump is actually an aggregate of four proteins, treated as one hinged protein in this figure.

Figure 8.19
How membrane potential affects ion transport.

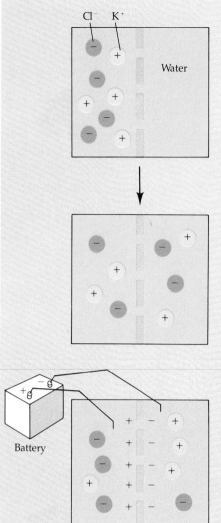

Cl⁻ K⁺

Water

Battery

In this artificial system a membrane separates distilled water from a solution of the salt potassium chloride (KCl). This membrane is not very permeable to the dissolved K^+ and Cl^-, but eventually the ions get through.

If diffusion of the ions is affected only by their concentration gradients, then K^+ and Cl^- will diffuse across the membrane until the concentration of each ion is the same on both sides of the membrane.

Switching on a battery (the membrane potential) makes one side of the membrane negative in charge and the other side positive. The K^+ and Cl^- will redistribute themselves across the membrane by diffusing in response to this new driving force. A dynamic equilibrium will again be reached, but this time the K^+ and Cl^- will each be distributed unequally across the membrane. Since opposite charges attract, K^+ will be more concentrated on the negative side of the membrane, and Cl^- will be more concentrated on the positive side. At the equilibrium point, the tendency for further diffusion due to electric attraction is offset by the tendency of the ions to diffuse down their concentration gradients.

macromolecules are negatively charged. These large anions are trapped within the cell and may make a minor contribution to its membrane potential. A much bigger factor is the selective permeability of the plasma membrane to various ions. Although the membrane is not very permeable to ions in general, some pass faster than others through the membrane. For example, K^+ diffuses across the membrane much faster than Na^+. When the sodium-potassium pumps of animals actively transport Na^+ out of cells and K^+ into cells, the ions tend to diffuse back across the membranes down their electrochemical gradients. Because K^+ leaks out of the cell faster than Na^+ leaks back in, there is a net loss of positive charges from the cell (Figure 8.20a). The sodium-potassium pump also contributes directly to the membrane potential. Notice in Figure 8.18 that the pump does not translocate Na^+ and K^+ on a one-for-one basis, but actually pumps three sodium ions out of the cell for every two potassium ions it pumps into the cell. With each crank of the pump, there is a net transfer of one positive charge

from the cytoplasm to the extracellular fluid. A transport protein that generates voltage across a membrane is called an **electrogenic pump.** The sodium-potassium ATPase seems to be the major electrogenic pump of animal cells. The main electrogenic pump of plants, bacteria, and fungi is a proton pump, which actively transports hydrogen ions (protons) out of the cell. The pumping of H^+ transfers a positive charge from the cytoplasm to the extracellular solution (Figure 8.20b). Proton pumps are also key features of the membranes of mitochondria and chloroplasts.

By generating voltages across membranes, electrogenic pumps store energy that can be tapped for cellular work, including a type of membrane traffic called cotransport.

Cotransport

A single ATP-powered pump that transports a specific solute can indirectly drive the active transport of sev-

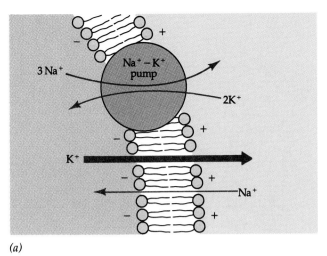

(a)

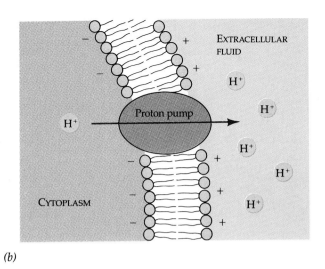

(b)

Figure 8.20
Electrogenic pumps. (a) The sodium-potassium pump (ATPase) of an animal cell pumps three sodium ions out of the cell for every two potassium ions it pumps into the cell. There is a net loss of one positive charge from the cell with each cycle of the pump. Furthermore, the selective permeability of the plasma membrane allows K$^+$ to leak out of the cell down its concentration gradient faster than Na$^+$ leaks in, a difference that contributes to the net charge of the cytoplasm being negative compared to the outside of the cell. **(b)** A proton pump that transports H$^+$ out of the cell helps generate membrane potential in plants, fungi, and bacteria.

eral other solutes in a mechanism known as **cotransport.** By moving a substance across a membrane against its gradient, a membrane pump stores energy. The substance will tend to diffuse back across the membrane down its gradient. Analogous to water that has been pumped uphill and that performs work as it flows back down, a substance that has been pumped across a membrane can do work as it leaks back by diffusion. Another specialized transport protein, separate from the pump, can couple the "downhill" diffusion of this substance to the "uphill" transport of a second substance against its own concentration gradient. For example, a plant cell uses the gradient of hydrogen ions generated by its proton pumps to drive the active transport of amino acids, sugars, and several other nutrients into the cell (Figure 8.21). One specific transport protein enables certain plant cells to accumulate the sugar sucrose. The protein can translocate sucrose into the cell against a concentration gradient, but can do so only if the sucrose travels in the company of a hydrogen ion. The sucrose molecule rides on the coattails of the hydrogen ion, which uses the common transport protein as an avenue to diffuse down the concentration gradient maintained by the proton pump.

TRAFFIC OF LARGE MOLECULES: ENDOCYTOSIS AND EXOCYTOSIS

Water and small solutes enter and leave the cell by passing through the lipid bilayer of the membrane, or

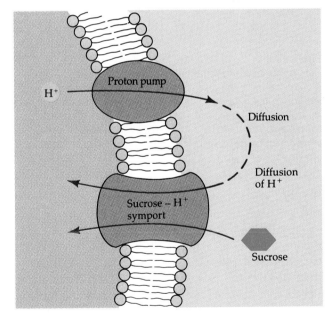

Figure 8.21
Cotransport. An ATP-driven pump stores energy by concentrating a substance (H$^+$, in this case) on one side of the membrane. As the substance leaks back across the membrane through specific transport proteins, other substances are cotransported, through the same proteins, against their own concentration gradients. In this case, the proton pump of the membrane is indirectly driving sucrose accumulation by a plant cell, with the help of a sucrose–H$^+$ symport.

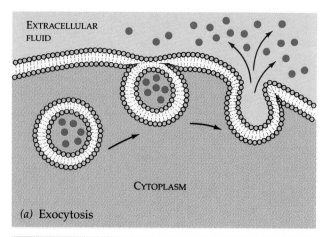

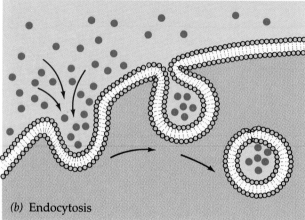

Figure 8.22
Exocytosis and endocytosis. (a) During exocytosis, vesicles fuse with the plasma membrane and dump their contents to the outside of the cell. **(b)** During endocytosis, extracellular substances are incorporated into the cell in vesicles formed by an inward budding of the plasma membrane.

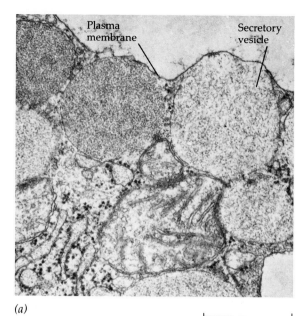

(a)

1 μm

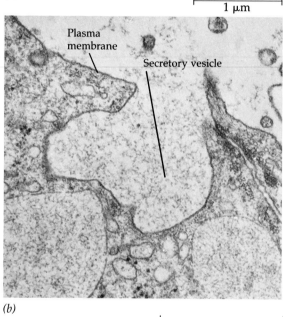

(b)

1 μm

Figure 8.23
Exocytosis in a tear gland. These electron micrographs show portions of cells in the process of secretion. **(a)** A transport vesicle containing secretory products fuses with the plasma membrane and **(b)** releases its contents outside the cell.

they are pumped or carried across the membrane by transport proteins. Large molecules, such as proteins and polysaccharides, generally cross the membrane by a different mechanism. The cell secretes macromolecules by the fusion of vesicles with the plasma membrane, and the cell can take in macromolecules and even particulate matter by forming vesicles derived from the plasma membrane. These processes are called **exocytosis** and **endocytosis,** respectively (Figure 8.22). During exocytosis, a vesicle, usually budded from the endoplasmic reticulum or the Golgi apparatus, migrates to the plasma membrane. The membrane of the vesicle and the plasma membrane come into close apposition, whereupon the lipid molecules of the two bilayers rearrange themselves. The two membranes then fuse to become continuous, and the contents of the vesicle spill to the outside of the cell. The steps are basically reversed during endocytosis. A localized region of the plasma membrane sinks inward to form

a pocket. As the pocket deepens, it pinches into the cytoplasm from the plasma membrane as a vesicle containing material that had been outside the cell.

Many secretory cells use exocytosis to export their products (Figure 8.23). For example, certain cells in the pancreas (a gland) manufacture the hormone insulin and secrete it into the blood by exocytosis. Another example is the neuron, or nerve cell, which uses exocytosis to release chemical signals that stimulate other neurons or muscle cells. When plant cells are making

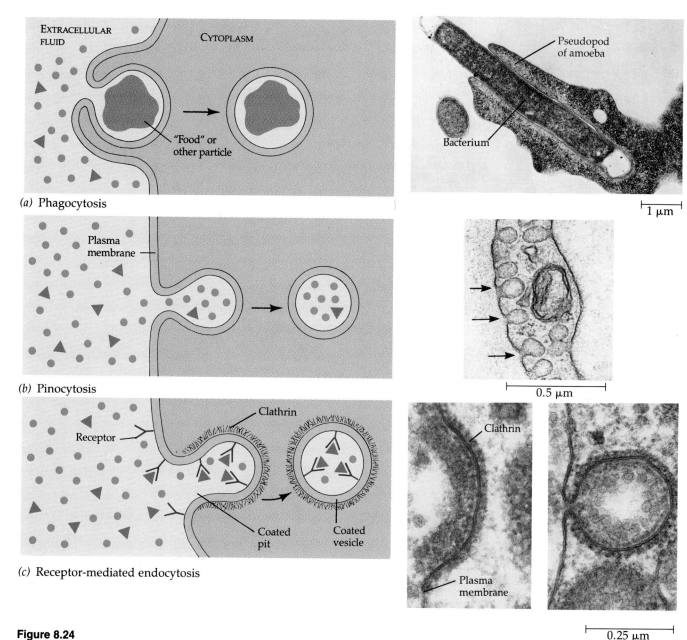

(a) Phagocytosis

Pseudopod of amoeba

Bacterium

1 μm

Plasma membrane

(b) Pinocytosis

0.5 μm

Clathrin

Receptor

Coated pit

Coated vesicle

(c) Receptor-mediated endocytosis

Clathrin

Plasma membrane

0.25 μm

Figure 8.24
Types of endocytosis. (a) In phagocytosis, pseudopodia engulf a particle and package it in a vacuole. The electron micrograph shows an amoeba engulfing a bacterium. **(b)** In pinocytosis, droplets of extracellular fluid are incorporated into the cell in small vesicles. The electron micrograph shows pinocytotic vesicles forming in a cell lining a capillary, a small blood vessel. **(c)** In receptor-mediated endocytosis, coated pits form vesicles when specific molecules bind to receptors on the cell surface. Coated pits are reinforced on their cytoplasmic side by a fibrous protein named clathrin. In the electron micrograph sequence, you can see progressive stages of receptor-mediated endocytosis. After the ingested material is liberated from the vesicle for metabolism, the receptors are recycled to the plasma membrane.

walls, exocytosis delivers carbohydrates from Golgi vesicles to the outside of the cell.

There are three types of endocytosis: **phagocytosis** ("cellular eating"), **pinocytosis** ("cellular drinking"), and **receptor-mediated endocytosis** (Figure 8.24). Phagocytosis was mentioned in Chapter 7. In this process, a cell engulfs a particle by wrapping pseudopodia around it and packaging it within a membrane-enclosed sac large enough to be classified as a vacuole (Figure 8.24a). The particle is digested after the vacuole fuses with a lysosome containing hydrolytic enzymes. In pinocytosis, the cell gulps droplets

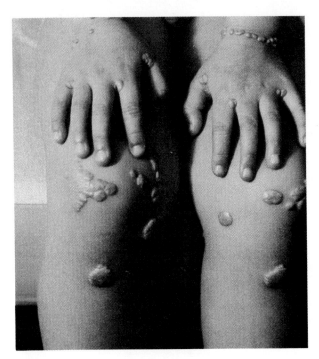

Figure 8.25
An inherited membrane disorder. The yellow deposits of cholesterol and other lipids beneath the skin of this young girl are due to familial hypercholesterolemia (FH), an inherited defect of the plasma membranes of cells. Cholesterol travels in the blood mainly in particles called LDLs (for low-density lipoproteins), each containing about 1500 cholesterol molecules. An LDL also has a large protein molecule that fits a receptor on the membranes of cells. By receptor-mediated endocytosis, cells incorporate LDLs within coated vesicles, and then metabolize the cholesterol or use it in membrane synthesis. A person who has FH lacks LDL receptors or has defective receptors on their membranes, and thus the cholesterol-containing particles accumulate in the blood. This not only results in the cholesterol pockets beneath the skin but, worse, causes early cardiovascular disease by depositing cholesterol on the walls of blood vessels. In this case, a malfunctioning membrane threatens life.

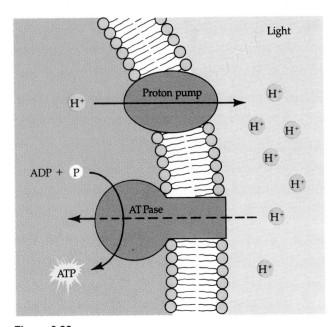

Figure 8.26
A membrane that makes ATP. Purple proteins in the plasma membrane of certain salt-loving (halophilic) bacteria are proton pumps driven by the absorption of light. The protons (H^+) diffuse back through on ATPase, powering the synthesis of ATP. Membranes of mitochondria and chloroplasts also tap the energy stored in H^+ gradients to make ATP.

of extracellular fluid in tiny vesicles (Figure 8.24b). Since any and all solutes dissolved in the droplet are taken into the cell, pinocytosis is unspecific in the substances it transports. In contrast, receptor-mediated endocytosis is very specific. Embedded in the membrane are proteins with specific receptor sites exposed to the extracellular fluid. The receptor proteins are usually clustered in regions of the membrane called **coated pits,** which are lined on their cytoplasmic side by a fuzzy layer consisting of a protein named **clathrin.** The extracellular substances that bind to the receptors are called **ligands,** the general term for any molecule that binds specifically to a receptor site of another molecule. When appropriate ligands bind to the receptor sites, they are carried into the cell by the inward budding of a coated pit to form a **coated vesicle** (Figure 8.24c).

Receptor-mediated endocytosis enables the cell to acquire bulk quantities of specific substances, even though those substances may not be very concentrated in the extracellular fluid. For example, animal cells use the process to take in cholesterol for use in the synthesis of membranes and as a precursor for the synthesis of other steroids. In humans with familial hypercholesterolemia, an inherited disease characterized by a very high level of cholesterol in the blood, the protein receptor sites are missing. Unable to enter the cells, the cholesterol accumulates in the blood, where it contributes to early atherosclerosis (development of fat deposits on the lining of blood vessels) (Figure 8.25).

In spite of all the endocytosis and exocytosis that goes on, the amount of plasma membrane a nongrowing cell has remains relatively constant over the long run. Apparently, the addition of membrane by vesicle fusion during exocytosis offsets the loss of plasma membrane by endocytosis. In fact, these processes go on continually to some extent in most eukaryotic cells. Vesicles not only transport substances between the cell and its surroundings, but also provide a mechanism to rejuvenate or remodel the plasma membrane.

MEMBRANES AND ATP SYNTHESIS

Let us complete our study of membrane traffic by building a bridge to the next two chapters. We have already seen that ATP can power pumps that transport substances across membranes against their gradients. Conversely, the difference in the concentration of a substance on opposite sides of a membrane is an energy source that can drive the synthesis of ATP from ADP and phosphate. The cell can then use the ATP for various kinds of work. In fact, most ATP is made by membranes that tap solute gradients as energy, and the solute is most often H^+. Imagine a proton pump like that of the bacterial membrane, but running in reverse. Instead of *using* ATP to pump H^+, the pump can *make* ATP if a concentration gradient of H^+ already exists. The H^+ diffuses into the cell through the transport protein, which now functions as an ATPase

enzyme that adds a phosphate group to ADP. So long as there is an H^+ gradient, the membrane makes ATP. But if the H^+ gradient is dissipated by passive transport through the ATP-making enzyme, what maintains the gradient? Different membranes use different methods. Perhaps the simplest system is the plasma membrane of the purple bacteria that live in salt ponds (Figure 8.26). The bacteria get their color from a purple protein built into the membrane. The protein functions as a solar-powered H^+ pump. When it absorbs light, it actively transports H^+ out of the cell. Hydrogen ions diffuse back into the cell through an enzyme that makes ATP—an ATPase.

The membranes of mitochondria and chloroplasts also use gradients of H^+ to make ATP. The mitochondrion uses energy from food to generate the H^+ gradient; the chloroplast uses light energy. Membrane traffic is thus essential to both respiration and photosynthesis, the subjects of the next two chapters.

STUDY OUTLINE

1. All cells are separated from their surroundings by the plasma membrane, which controls the traffic of substances into and out of the cell.
2. The plasma membrane has a specialized structure that is well correlated with its function of selective permeability.

Models of Membrane Structure (pp. 154–162)

1. Early in this century, researchers postulated that the membrane phospholipids are arranged in a bilayer with their hydrophobic tails in the interior of the membrane and their hydrophilic heads facing the aqueous compartments inside and outside the cell.
2. The model was modified in 1935 to include layers of proteins on either side of the lipid bilayer. This Davson–Danielli sandwich model seemed consistent with later evidence provided by electron microscopy, which revealed a triple-layered ultrastructure for the plasma membrane.
3. In 1972, S. Singer and G. Nicolson proposed a revised membrane model called the fluid mosaic model. In this model, the membrane is envisioned as a mosaic of dispersed proteins, individually inserted and floating laterally in a fluid bilayer of phospholipids. The fluid mosaic model is consistent with all known properties of cell membranes and with evidence provided by freeze-fracture electron micrographs of delaminated membranes.
4. The fluid quality and distinctive molecular composition of membranes are essential for their dynamic function.
5. Membranes are bifacial, with specific inside and outside faces arising from differences in composition of the two bilayers and/or directional orientation of proteins and any attached carbohydrates.
6. Proteins are either embedded in the lipid bilayer (integral proteins) or occur only on the surface (peripheral proteins).

7. Carbohydrates linked to the proteins and lipids are important for cell-cell recognition.

Traffic of Small Molecules (pp. 162–171)

1. The cell requires an extensive interchange of small nutrient and waste molecules, respiratory gases, and inorganic ions. The plasma membrane regulates the passage of these substances.
2. The selective permeability of the plasma membrane results from its structure. Hydrophobic substances pass through rapidly because of their solubility in the lipid bilayer. Small polar molecules, such as H_2O and C_2O, can also pass through the membrane. Larger polar molecules and ions require specific transport proteins, which provide channels or act as physical vehicles.
3. Diffusion is the spontaneous movement of a substance down its concentration gradient owing to its intrinsic kinetic energy. The diffusion of a substance across a membrane is called passive transport.
4. Osmosis is the diffusion of water across a selectively permeable membrane. Water flows across a membrane from the side with a lesser concentration of solute (hypoosmotic) to the side with the greater solute concentration (hyperosmotic). No net osmosis occurs across membranes separating solutions of equal concentration (isosmotic solutions). The osmotic pressure of a solution is proportional to its solute concentration.
5. Cells lacking walls, the cells of animals and some protists, are either isosmotic with their environments or else have adaptations for osmoregulation.
6. Plants, prokaryotes, fungi, and some protists have an elastic wall around their cells, which keeps the cells from bursting in a hypoosmotic environment. Under such

conditions, these cells are turgid. In isosmotic solutions, such cells are flaccid; in hyperosmotic solutions, they plasmolyze.

7. In facilitated diffusion, transport proteins hasten the movement of certain substances across a membrane down their concentration gradients. The transport protein is specific for its solute.

8. Diffusion, osmosis, and facilitated diffusion are all spontaneous, passive transport processes that do not require energy input from the cell.

9. Active transport, on the other hand, requires energy to pump substances across the membrane against their concentration gradients. The energy, usually in the form of ATP, is harnessed to specific transport proteins that mediate the work. In animal cells, the sodium-potassium pump maintains gradients of these two important ions across cell membranes.

10. Diffusion of uncharged solutes depends only on their concentration gradients. However, ions have both a concentration (chemical) gradient and an electric gradient (voltage). These two forces may be in the same or opposite directions and are combined in an overall force called the electrochemical gradient, which determines the net direction of ionic diffusion.

11. The voltage across the membrane, the membrane potential, depends on an unequal distribution of ions across the plasma membrane of all living cells. Contributing to the potential are the effect of negatively charged proteins and macromolecules, active transport of selected ions, and differential permeability of the membrane to charged substances. Electrogenic pumps, such as the sodium-potassium ATPase and proton pumps, are transport proteins that generate voltage across a membrane.

12. Special membrane proteins can cotransport two solutes, coupling the "downhill" diffusion of one to the "uphill" transport of the other. The diffusion gradient in cotransport usually originates from active transport of that substance by a separate ATP-driven process.

Traffic of Large Molecules: Endocytosis and Exocytosis (pp. 171–174)

1. Large macromolecules leave the cell by exocytosis and enter by endocytosis, two processes involving whole segments of membrane rather than individual membrane molecules.

2. In exocytosis, intracellular vesicles migrate to the plasma membrane, fuse with it, and release their contents.

3. Endocytosis is essentially the reverse of exocytosis and has three subtypes: Phagocytosis is the ingestion of large particles or whole cells; pinocytosis is the intake of tiny droplets of extracellular fluid with all its contained solutes; and receptor-mediated endocytosis is the ingestion of specific substances that bind to receptor proteins located in coated pits on the membrane.

Membranes and ATP Synthesis (p. 175)

1. Specialized membranes can make ATP by tapping energy stored in transmembrane solute gradients. The solute is usually H^+, which diffuses across the membrane through a special transport protein, powering its function as an ATP-synthesizing enzyme.

2. The membranes of mitochondria and chloroplasts make ATP mainly by using the energy of food and light, respectively, to generate H^+ gradients.

SELF-QUIZ

1. Certain cells in the liver, called Kupffer cells, ingest bacteria and various kinds of debris from damaged cells. This function is most likely accomplished by
   a. pinocytosis
   b. phagocytosis
   c. receptor-mediated endocytosis
   d. exocytosis
   e. passive transport

2. According to the fluid mosaic model of membrane structure, proteins of the membrane are
   a. spread in a continuous layer over the inner and outer surfaces of the membrane
   b. confined to the hydrophobic core of the membrane
   c. embedded in a lipid bilayer
   d. randomly oriented in the membrane, with no fixed inside-outside polarity
   e. free to depart from the fluid membrane and dissolve in the surrounding solution

3. When a plasma membrane is split by freeze-fracture into its outer and inner leaflets, the "bumps" visualized in electron micrographs are replicas of
   a. integral proteins
   b. peripheral proteins
   c. phospholipids
   d. cholesterol molecules
   e. clathrin

4. Which of the following factors would tend to increase membrane fluidity?
   a. a greater proportion of unsaturated phospholipids
   b. a lower temperature
   c. a lower cholesterol concentration in the membrane
   d. a greater proportion of relatively large glycolipids compared to lipids having smaller molecular weights
   e. a high membrane potential

5. The sodium-potassium pump is termed electrogenic because
   a. it hydrolyzes ATP
   b. it pumps positive charges out of the cell and negative charges into the cell
   c. it pumps three positive charges out of the cell for every two positive charges it pumps into the cell
   d. it pumps H^+ out of the cell along with Na^+
   e. it pumps electrons into the cell

6. Plant cells are turgid when bathed in a solution that is
   a. hypoosmotic to the cell
   b. hyperosmotic to the cell

c. isosmotic to the cell

d. isosmotic to sea water

e. lower in water concentration than the cell

An artificial cell with an aqueous solution enclosed in a selectively permeable membrane has just been immersed in a beaker containing a different solution.

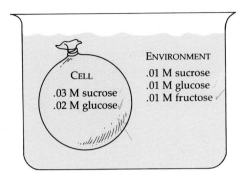

The membrane is permeable to water and to the simple sugars glucose and fructose, but is completely impermeable to the disaccharide sucrose. Now, answer questions 7–10.

7. Which solute(s) will exhibit a net diffusion into the cell?

8. Which solute(s) will exhibit a net diffusion out of the cell?

9. In which direction will there be a net osmotic movement of water?

10. After the cell is placed into the beaker, which of the following changes would occur?

   a. The artificial cell would become more flaccid.

   b. The artificial cell would become more turgid.

   c. The entropy of the system (cell plus surrounding solution) would decrease.

   d. The overall free energy stored in the system would increase.

   e. Both b and d would occur.

CHALLENGE QUESTIONS

1. An experiment is designed to study the mechanism of sucrose uptake by plant cells. Cells are immersed in a sucrose solution, and the pH in this surrounding solution is monitored with a pH meter. The measurements show that sucrose uptake by the plant cells raises the pH of the surrounding solution. The magnitude of the pH change is proportional to the starting concentration of sucrose in the extracellular solution. A metabolic poison that blocks the ability of the cells to regenerate ATP also inhibits the pH change in the surrounding solution. Explain these results.

2. In an adaptation of the preceding experiment, the rates of sucrose uptake from solutions of different sucrose concentrations are compared.

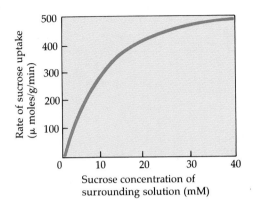

Explain the shape of the curve in terms of what is happening at the membranes of the plant cells.

3. If our cells and body fluids are hyperosmotic to the water of a swimming pool, then why do we not swell and pop when we go for a swim?

FURTHER READING

1. Alberts, B., D. Bray, J. Lewis, M. Raff, K. Roberts, and J. D. Watson. *Molecular Biology of the Cell.* 2d ed. New York: Garland, 1989. Chapter 6 describes the structures and functions of the plasma membrane.

2. Becker, W. M. *The World of the Cell.* Menlo Park, Calif.: Benjamin/Cummings, 1986. Chapters 10 and 13 take the topics of membrane structure and transport a step beyond their coverage here.

3. Bretscher, M. S. "The Molecules of the Cell Membrane." *Scientific American,* October 1985. Superbly depicts shapes and relationships of membrane molecules.

4. Brown, M. S., and J. L. Goldstein. "How LDL Receptors Influence Cholesterol and Atherosclerosis." *Scientific American,* November 1984. Two Nobel Laureates describe their work on a membrane disorder.

5. Kartner, N., and V. Ling. "Multidrug Resistance in Cancer." *Scientific American,* March 1989. Tumor cells have membrane proteins that pump out poisons used for chemotherapy.

6. Slayman, C. L. "Proton Chemistry and the Ubiquity of Proton Pumps." *BioScience,* January 1985. An overview of proton pumps in biology; related articles appear in the same issue.

9

Respiration: How Cells Harvest Chemical Energy

Living is work. A cell organizes small organic molecules into polymers such as proteins and DNA. It pumps substances across membranes. Many cells move or change their shapes. They grow and reproduce. A cell must work just to maintain its complex structure, for order is intrinsically unstable. As we have been emphasizing, the cell is not an autonomous, closed system. To perform their many tasks, cells require transfusions of energy from outside sources (Figure 9.1). We can visualize the process as beginning with the sun as the energy source for plants and other photosynthetic organisms (Figure 9.2). Animals obtain their fuel by eating plants, or by eating other organisms that eat plants. All organisms use the organic molecules in their food not only as energy resources but also as building materials for growth and repair.

How do cells harvest the energy stored in food? With the help of enzymes, the cell systematically degrades complex organic molecules that are rich in potential energy to simpler waste products that have less energy. Some of the energy taken out of chemical storage can be used to do work; the rest is dissipated as heat. As you learned in Chapter 6, metabolic pathways that release stored energy by breaking down complex molecules are called catabolic pathways. One catabolic process, called **fermentation,** is a partial degradation of organic molecules that occurs without the help of oxygen. However, the most prevalent and efficient catabolic pathway is **cellular respiration,** in which oxygen is consumed as a reactant along with the organic fuel. Although very different in mechanism, respiration is in principle similar to the combustion of gas-

Figure 9.1
Obtaining fuel. Organisms are open systems that depend on external energy sources. Animals, such as this desert locust, obtain their energy in chemical form by eating other organisms. After the food is digested and distributed to cells, sugar and other organic molecules are consumed as fuel in the process of cellular respiration. With the help of oxygen, cellular respiration harnesses the energy stored in the fuel molecules for work, such as contraction of muscle cells.

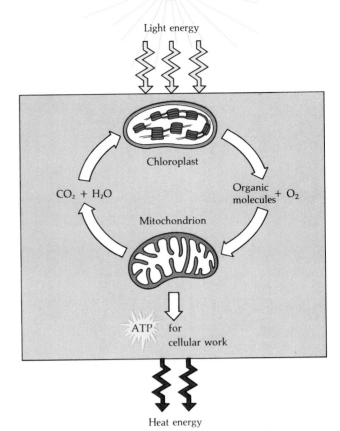

Figure 9.2
Energy flow and chemical recycling in life. The chloroplasts of plants (and photosynthetic protists) absorb light and trap its energy in the chemical bonds of the sugar glucose, which photosynthesis makes from carbon dioxide and water. Oxygen, produced as a by-product of photosynthesis, is released to the air. In mitochondria, present in almost all eukaryotic cells, cellular respiration consumes oxygen in breaking down glucose to carbon dioxide and water, which can be reused in photosynthesis. Mitochondria use the energy that was stored in the glucose to make ATP, which powers nearly all forms of cellular work. Organisms dissipate energy back to the environment in the form of heat. In this and the following chapter we emphasize eukaryotes, but keep in mind that many prokaryotes also carry out respiration and photosynthesis, though these metabolic functions are not compartmentalized into mitochondria and chloroplasts.

oline in an automobile engine after oxygen is mixed with the fuel (hydrocarbons). The main engine of the eukaryotic cell is the mitochondrion, where most of the enzymes and other metabolic gears of respiration are located. The main fuel for respiration is the sugar glucose, and the exhaust is carbon dioxide and water. The overall process can be summarized as follows:

$$C_6H_{12}O_6 + 6\,O_2 \longrightarrow 6\,CO_2 + 6\,H_2O + Energy$$

Glucose   Oxygen   Carbon   Water
dioxide

The catabolic pathway is exergonic, having a standard free energy change of -686 kilocalories per mole of glucose decomposed ($\Delta G = -686$ kcal/mol; remember from Chapter 6 that a negative ΔG indicates that the products of the chemical process store less energy than the reactants). Other carbohydrates, as well as fats, are also common fuels for respiration.

The waste products of respiration, CO_2 and H_2O, are the very substances that chloroplasts use as raw materials for photosynthesis, which in turn makes glucose and returns oxygen to the air. The chemical elements essential to life are recycled. But energy is not. Energy flows into the living world as sunlight and back out as heat. In its inevitable conversion from light to heat, a portion of the energy is temporarily trapped in the ordered structure of organic molecules, until cells use it for work.

Catabolic pathways such as fermentation and cellular respiration do not directly move flagella, pump solutes, polymerize monomers, or perform other cellular work. Catabolism is linked to work by a chemical drive shaft: ATP. In this chapter, you will learn how catabolism, especially respiration, generates the ATP that is expended by the working cell. The processes are complex and challenging to learn. Throughout the chapter, it is important that you keep sight of the objective: to discover how cells use the energy stored in food molecules to make ATP.

HOW CELLS MAKE ATP: AN INTRODUCTION

Recall from Chapter 6 that ATP, short for adenosine triphosphate, is very reactive because the bonds between its three phosphate groups are relatively unstable. These bonds are referred to as high-energy phosphate bonds; their hydrolysis is exergonic, releasing 7.3 kcal for every mole of ATP that loses the terminal phosphate group ($\Delta G = -7.3$ kcal/mol). The products of hydrolysis are ADP (adenosine diphosphate) and inorganic phosphate, which we shall abbreviate with the symbol $\circledP$. The hydrolysis of ATP can be diagrammed this way:

$$\text{ATP} + \text{H}_2\text{O} \longrightarrow \text{ADP} + \circledP + \text{Energy}$$

In some cases, both high-energy phosphate bonds are broken, releasing twice as much energy and leaving adenosine monophosphate as a product. (Figure 6.8 shows the structure of ATP.)

If ATP were to be hydrolyzed directly, energy would be released as heat, which cells cannot use for work. Instead, energy is generally transferred by enzymes that shift a phosphate group with its unstable bond from ATP to some other molecule, which is then said to be phosphorylated (Figure 9.3). Phosphorylation primes the molecule to undergo some kind of change that performs work. We have seen many examples. Solutes are pumped across membranes by transport proteins that change their conformations when they are energized by phosphate groups transferred from ATP (see Chapter 8). Cellular movements are powered by phosphorylation of contractile proteins, such as the dynein arms that slide microtubules past one another in cilia and flagella (see Chapter 7). Chemical work, the endergonic synthesis of the products of anabolic pathways, is also driven by ATP, which increases the reactivity of key substrates by transferring high-energy phosphate groups to them (see Chapter 6). In all these cases, the phosphorylated substance eventually loses its phosphate group as work is performed.

To keep on working, the cell must regenerate its supply of ATP from ADP and inorganic phosphate. Phosphorylation of ADP is exactly as endergonic as hydrolysis is exergonic (ΔG for phosphorylation of ADP is $+7.3$ kcal/mol). The main function of respiration is to provide energy for ATP synthesis.

Figure 9.3
Review of how ATP drives cellular work. The unstable high-energy phosphate bonds of ATP are reactive and can make things happen when they are transferred to other substances in the process known as phosphorylation. ATP drives active transport by phosphorylating membrane proteins, powers movement by phosphorylating contractile proteins, and forces chemical reactions that would otherwise be endergonic by phosphorylating key reactants. All these phosphorylation reactions are catalyzed by enzymes.

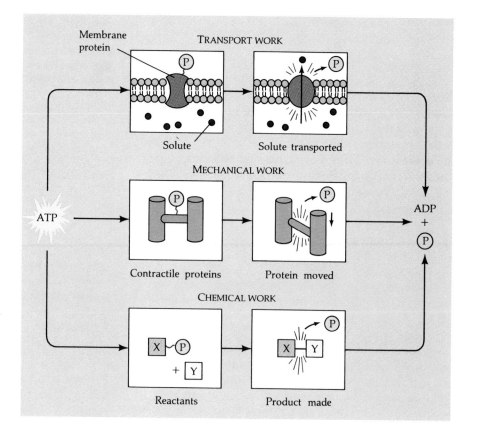

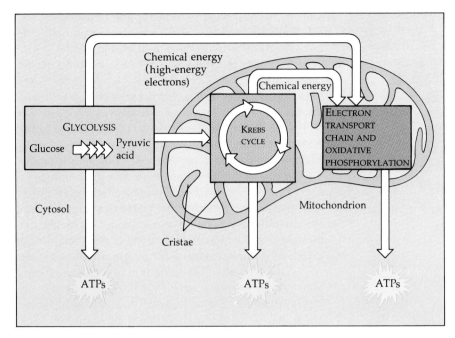

Figure 9.4

An overview of cellular respiration. In a eukaryotic cell, glycolysis occurs outside the mitochondria in the cytosol. The Krebs cycle and the electron transport chain are located inside the mitochondria. During glycolysis, each glucose molecule is broken down to two molecules of a compound called pyruvic acid. The pyruvic acid crosses the double membrane of the mitochondrion to enter the matrix, where the Krebs cycle decomposes it to carbon dioxide. Chemical energy from the Krebs cycle, in the form of energy-rich electrons, is then transferred to the electron transport chain, which is built into the membrane of the cristae. The electron transport chain transforms the chemical energy into a form that can be used to drive oxidative phosphorylation, which accounts for most of the ATP generated by cellular respiration.

An Overview of Cellular Respiration

Respiration is a cumulative function of three metabolic stages:

1. Glycolysis (color coded as light green throughout this chapter)

2. The Krebs cycle (color coded beige)

3. The electron transport chain and **oxidative phosphorylation** (color coded violet)

Glycolysis and the Krebs cycle are catabolic pathways that decompose glucose and other organic fuels. Glycolysis, which occurs in the cytosol, begins the degradation by breaking glucose into two molecules of a compound called pyruvic acid (Figure 9.4). The Krebs cycle, located within the mitochondrion, completes the job by decomposing a derivative of pyruvic acid to carbon dioxide. A limited amount of ATP synthesis is coupled directly to specific steps in glycolysis and the Krebs cycle. However, these two stages function mainly to supply energized electrons to drive oxidative phosphorylation, which accounts for most of the ATP made during respiration. The metabolic machinery for oxidative phosphorylation includes an electron transport chain, a group of molecules built into the inner membrane of the mitochondrion. Electrons removed from food molecules during glycolysis and the Krebs cycle are pulled down the electron transport chain to a lower state of energy by oxygen, a process analogous to objects being pulled downhill by gravity. A complex mechanism completes oxidative phosphorylation by coupling this exergonic slide of electrons to the synthesis of ATP.

Respiration cashes in the large denomination of energy banked in glucose for the small change of ATP, which is more practical for the cell to spend on its work. For each molecule of glucose degraded to carbon dioxide and water by respiration, the cell makes as many as 36 or 38 molecules of ATP, the number depending on the type of cell and its metabolic status. How does the cell couple the exergonic degradation of glucose to the endergonic synthesis of ATP? There are two basic mechanisms: substrate-level phosphorylation and the chemiosmotic mechanism.

Substrate-Level Phosphorylation

Some ATP is made during glycolysis by the direct transfer of phosphate to ADP from organic compounds with high-energy phosphate bonds even more unstable than those of ATP. An example of such a compound is phosphoenolpyruvic acid, or PEP (Figure 9.5). Formed as an intermediate in the breakdown of glucose, PEP is the substrate for an enzyme that brings it together with ADP. When the two molecules convene in the active site of the enzyme, the phosphate group of PEP, with its high-energy bond, is shifted to ADP to form ATP. In this reaction, the last step of glycolysis, the PEP is converted to pyruvic acid. The reaction is energetically feasible because the hydrolysis of PEP has a ΔG that is more negative than the ΔG for phosphorylation of ADP is positive:

PEP + H_2O	⟶ Pyruvic acid + Ⓟ	ΔG =	−14.8 kcal/mol
ADP + Ⓟ	⟶ ATP + H_2O	ΔG =	+7.3 kcal/mol
PEP + ADP	⟶ ATP + Pyruvic acid	ΔG =	−7.5 kcal/mol

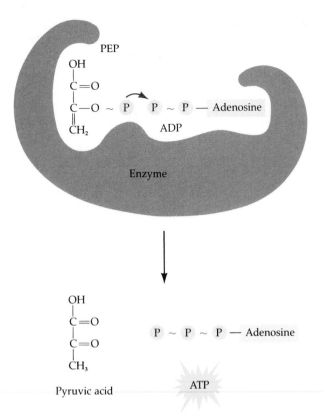

PEP

Enzyme

Pyruvic acid

ATP

Figure 9.5
Substrate-level phosphorylation. Some ATP is made by direct enzymatic transfer of a phosphate group from a substrate to ADP. The reaction is possible because the phosphate bond of the substrate is even more unstable than the phosphate bonds of ATP. The phosphate donor in this case is phosphoenolpyruvic acid (PEP).

This direct coupling of a step in catabolism to ATP synthesis by enzymatic transfer of phosphate from a substrate is called **substrate-level phosphorylation.** It occurs during both fermentation and respiration. Although it may be the simplest way to make ATP, substrate-level phosphorylation accounts for only a small percentage of the ATP generated in most cells. Most cellular ATP is made by oxidative phosphorylation, driven by an energy-coupling mechanism called chemiosmosis.

Chemiosmotic Coupling: The Basic Principle

Oxidative phosphorylation uses the exergonic flow of electrons from food to oxygen to drive the endergonic synthesis of ATP. But how does the mitochondrion actually couple the energy released by the transfer of electrons down the electron transport chain to the work of phosphorylating ADP? In 1961, British biochemist Peter Mitchell postulated a coupling mechanism based on his experiments with bacteria. Mitchell named the process **chemiosmosis** (from the Greek *osmos*, "push"),

a term that emphasizes the coupling between chemical reactions and transport across membranes. (We have used the word *osmosis* in discussing water transport, but here the reference is to the pushing of H^+ across membranes.) Nearly two decades later, after many researchers had confirmed the basic validity of the chemiosmotic theory, Mitchell was awarded the Nobel Prize. The model applies not only to oxidative phosphorylation, but also to ATP synthesis during photosynthesis and many other cases of cellular work.

Membranes have a prominent role in chemiosmosis. Most ATP is made by protein complexes called **ATP synthases,** which are inserted in membranes. These units, which span the membrane, work like ion pumps running in reverse. Instead of consuming ATP to actively transport ions across the membrane, an ATP synthase uses a gradient of ions across the membrane as an energy source to make ATP. Specifically, this occurs when the solution on one side of the membrane has a higher concentration of hydrogen ions than the solution on the opposite side (Figure 9.6a). This concentration difference of H^+ is referred to as the proton gradient, or pH gradient (since pH is a measure of H^+ concentration). The gradient is generated by other proteins in the membrane that translocate H^+ from one side of the membrane to the other. Of course, the ions tend to leak back across the membrane by diffusion, but the lipid bilayer is impermeable to ions and impedes passive transport of H^+, much as a dam prevents the downward flow of water. The proton gradient stores energy. The ATP synthase taps this energy source by allowing the H^+ to diffuse down the gradient. By some mechanism not yet known, the ATP synthase phosphorylates ADP when hydrogen ions pass through the enzyme on their passive return across the membrane.

The proton gradient that drives ATP synthesis is generated either by light, in the case of photosynthesis, or by energy extracted from food molecules, in the case of respiration. In mitochondria, chemiosmosis occurs across the inner membrane (Figure 9.6b). The apparatus for pumping H^+ across this membrane is the electron transport chain, which uses energy released by the breakdown of glucose during glycolysis and the Krebs cycle. The electron transport chain translocates hydrogen ions across the inner membrane from the mitochondrial matrix out to the intermembrane space. This maintains an H^+ reservoir that powers the ATP synthase as the ions flow down their gradient back into the matrix. The folding of the inner membrane into cristae enlarges its surface area, providing space for many electron transport chains and ATP synthase units and thus increasing the ATP output of the mitochondrion. The chemiosmotic model is elegant in its correlation of structure and function.

This brief description of chemiosmosis should give you a general idea of how this complex process works. We will be discussing the individual steps of the process

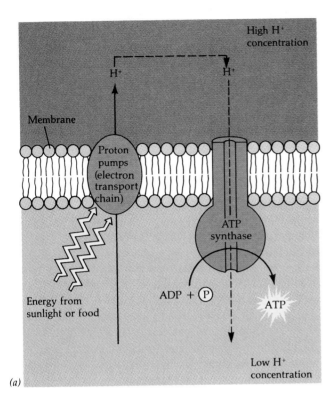

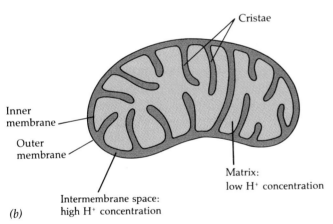

Figure 9.6
Chemiosmotic synthesis of ATP. (a) Energy from food or light powers electron transport chains that translocate hydrogen ions (protons) from one side of a membrane to the other, generating a proton gradient across the membrane. The hydrogen ions diffuse back across the membrane through an ATP synthase, a protein complex that couples the passage of the ions to the phosphorylation of ADP to form ATP. **(b)** In mitochondria, the chemiosmotic machinery is built into the inner membrane, which is made more expansive by its infoldings (cristae). The H^+ concentration is higher in the intermembrane space than in the matrix. Throughout this chapter and the next, brown will be used in illustrations for regions of high H^+ concentration, and beige for regions of low H^+ concentration.

in greater detail throughout the chapter. To begin, we might ask how the electron transport chain harnesses the energy stored in food. To answer this question, we must investigate a chemical process called oxidation.

REDOX REACTIONS IN METABOLISM

In many chemical reactions, there is a transfer of one or more electrons from one reactant to another. These electron transfers are called oxidation-reduction reactions, or **redox reactions** for short. During a redox reaction, the loss of electrons from one substance is called **oxidation,** and the addition of electrons to another substance is known as **reduction.** (This term defies intuition—*adding* electrons is called *reduction.* The term was derived from the electrical effects of adding electrons. When negatively charged electrons are added to a cation, the electrons reduce the amount of positive charge possessed by the cation.) Consider, for example, the reaction between sodium and chlorine to form table salt:

$$
\underset{\text{reduction}}{\overset{\text{oxidation}}{Na \; + \; Cl \longrightarrow Na^+ \; + \; Cl^-}}
$$

Or we could generalize a redox reaction this way:

$$
\underset{\text{reduction}}{\overset{\text{oxidation}}{Xe^- \; + \; Y \longrightarrow X \; + \; Ye^-}}
$$

In the preceding hypothetical reaction, substance X, the electron (e^-) donor, is called the **reducing agent;** it reduces Y. Substance Y, the electron acceptor, is the **oxidizing agent;** it oxidizes X. Since an electron transfer requires both a donor and an acceptor, oxidation and reduction always go together. An electron escapes from one molecule only upon contact with another molecule that has a greater affinity for electrons.

Not all redox reactions involve the complete loss of electrons from the reducing agent; some change the degree of electron sharing in covalent bonds. The reaction between the hydrocarbon methane and oxygen to form carbon dioxide and water, shown in Figure 9.7, is an example of the most common type of redox reaction in energy metabolism. Covalent electrons in methane are shared equally between the bonded atoms because carbon and hydrogen have about the same affinity for valence electrons; they are about equally electronegative (see Chapter 2). But when methane reacts with oxygen, electrons are shifted away from the carbon and hydrogen atoms to their new covalent partner, oxygen, which is very electronegative—it attracts electrons strongly. Methane has thus been oxidized, and oxygen has been reduced. Because

Figure 9.7
Methane combustion as a redox reaction. During the reaction, covalently shared electrons move away from carbon and hydrogen atoms and closer to oxygen, which is very electronegative. The reaction releases energy to the surroundings because the electrons lose potential energy as they move closer to electronegative atoms.

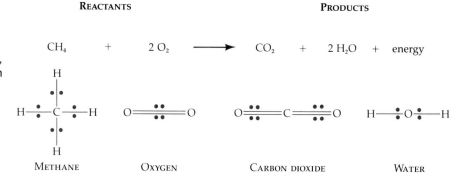

REACTANTS   PRODUCTS

$$CH_4 \quad + \quad 2\,O_2 \quad \longrightarrow \quad CO_2 \quad + \quad 2\,H_2O \quad + \quad energy$$

METHANE   OXYGEN   CARBON DIOXIDE   WATER

oxygen is so electronegative, it is one of the most potent of all oxidizing agents. (Do not let the similarity between the words *oxidation* and *oxygen* mislead you. Oxidation is the partial or complete loss of electrons, whether to oxygen or to some other substance.)

Electrons lose potential energy when they shift toward a more electronegative atom. (Their loss of energy is analogous to a ball rolling downhill.) Therefore, spontaneous redox reactions release energy, most often as heat. The oxidation of methane by oxygen is the main combustion reaction that occurs at the burner of a gas stove. Combustion of gasoline in an automobile engine is also a redox reaction, and the energy released pushes the pistons. But the energy-yielding redox process of greatest interest to us here is respiration: the oxidation of glucose and other fuel molecules in food.

Respiration as a Redox Process

Examine again the summary equation for cellular respiration, but this time think of it as a redox process:

$$\overbrace{C_6H_{12}O_6 \; + \; 6\,O_2 \; \longrightarrow \; 6\,CO_2}^{\text{oxidation}} \; + \; 6\,H_2O$$

As in the combustion of methane or gasoline, the valence electrons of carbon and hydrogen are shifted toward electronegative oxygen atoms. Sugar is oxidized and oxygen is reduced, and the electrons lose potential energy along the way. The energy given up by these electrons is then used by the cells to drive ATP synthesis.

In general, organic molecules that have an abundance of hydrogen are excellent fuels because they are rich in high-energy electrons. The transfer of hydrogen atoms from the organic fuel to oxygen drops the electrons to a lower energy state, which releases energy. Put this way, we can interpret respiration as a redox process that transfers hydrogen from sugar to oxygen, resulting in chemical energy being taken out of storage and made available to the cell.

The main energy foods, carbohydrates and fats, are reservoirs of high-energy electrons associated with hydrogen. Only the barrier of activation energy holds back the flood of electrons to a lower energy state. A mole of glucose, which is about 180 g, gives off 686 kcal of heat when it burns in air. Body temperature is not hot enough for spontaneous combustion. But swallow the glucose in the form of a sugar cube, and when the molecules reach your cells, enzymes will lower the barrier of activation energy, allowing the sugar to be oxidized. In this sense, respiration is like the combustion of methane or gasoline. However, combustion in the cell is more gradual. The wholesale release of energy from a fuel is difficult to harness efficiently for constructive work. The explosion of a gasoline tank cannot drive a car very far. Similarly, cellular respiration does not oxidize glucose in a single explosive step that would transfer all the hydrogen from the fuel to the oxygen at one time. Rather, glucose is broken down gradually during glycolysis and the Krebs cycle in a series of steps, each catabolic step catalyzed by a specific enzyme. At key steps, hydrogen atoms with their high-energy electrons are stripped from the glucose, usually two at a time. They are not transferred directly to oxygen, but are usually passed first to a special electron acceptor named NAD^+.

NAD^+ and the Oxidation of Glucose

Nicotinamide adenine dinucleotide, or NAD^+, is an organic molecule found in all cells. It functions as a coenzyme, which is a nonprotein that is required as an assistant for the catalytic function of certain enzymes (see Chapter 6). NAD^+ belongs to a family of coenzymes that assist enzymes in the transfer of electrons during the redox reactions of metabolism. Another such coenzyme is flavin adenine dinucleotide, or **FAD**, which will reappear later in this chapter.

In most of the redox steps in the catabolism of glucose, enzymes transfer electrons from substrates first to NAD^+, which thus functions as an oxidizing agent. These enzymes are called dehydrogenases because they actually remove a pair of hydrogen atoms from the

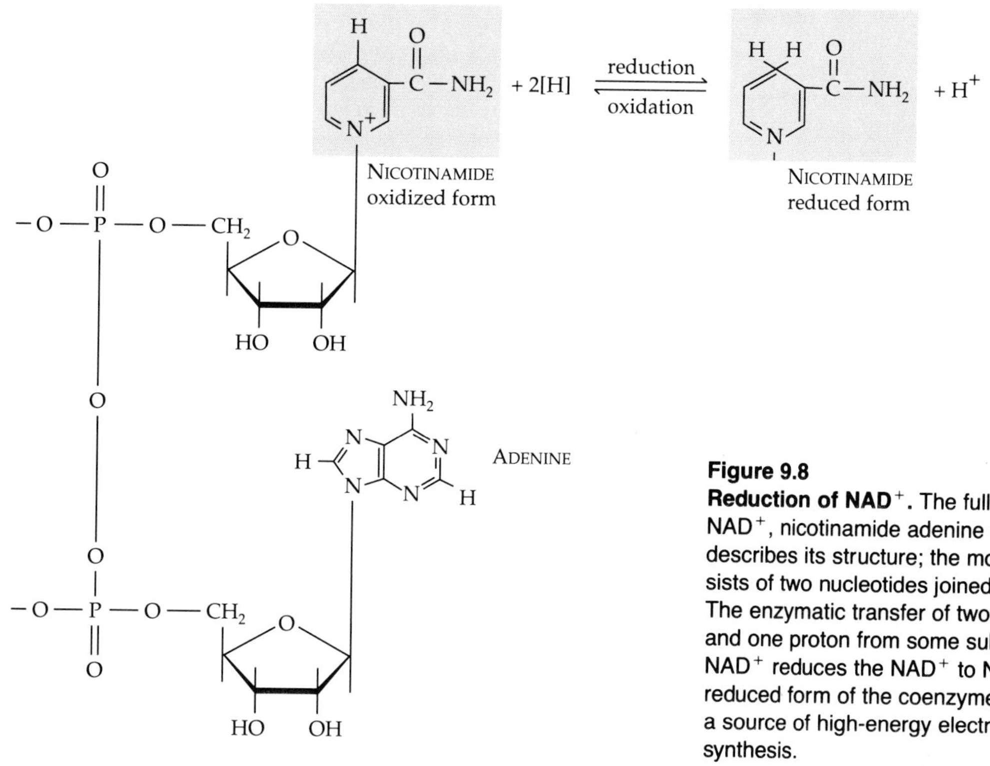

$$\text{NICOTINAMIDE oxidized form} + 2[H] \underset{\text{oxidation}}{\overset{\text{reduction}}{\rightleftharpoons}} \text{NICOTINAMIDE reduced form} + H^+$$

NICOTINAMIDE
oxidized form

ADENINE

NICOTINAMIDE
reduced form

Figure 9.8
Reduction of NAD$^+$. The full name for NAD$^+$, nicotinamide adenine dinucleotide, describes its structure; the molecule consists of two nucleotides joined together. The enzymatic transfer of two electrons and one proton from some substrate to NAD$^+$ reduces the NAD$^+$ to NADH. The reduced form of the coenzyme, NADH, is a source of high-energy electrons for ATP synthesis.

substrate. We can think of this as the removal of two electrons and two protons (the nuclei of hydrogen atoms). The enzyme delivers the *two* electrons along with *one* proton to NAD$^+$ (Figure 9.8). The other proton is released as a hydrogen ion (H$^+$) into the surrounding solution:

$$XH_2 + NAD^+ \xrightarrow{\text{dehydrogenase}} X + NADH + H^+$$

Several dehydrogenase enzymes function in respiration, and X in the preceding equation symbolizes any of the various substrates oxidized by the enzymatic transfer of electrons to NAD$^+$. The reduced form of the coenzyme is **NADH,** an abbreviation that reflects the hydrogen that has been received in the redox reaction. (While the oxidized form, NAD$^+$, had a net positive charge, NADH is electrically neutral. By receiving two negatively charged electrons but only one positively charged proton, NAD$^+$ has had its charge neutralized.) An example of a specific reaction of this type is the oxidation of malic acid, one of the steps to the Krebs cycle:

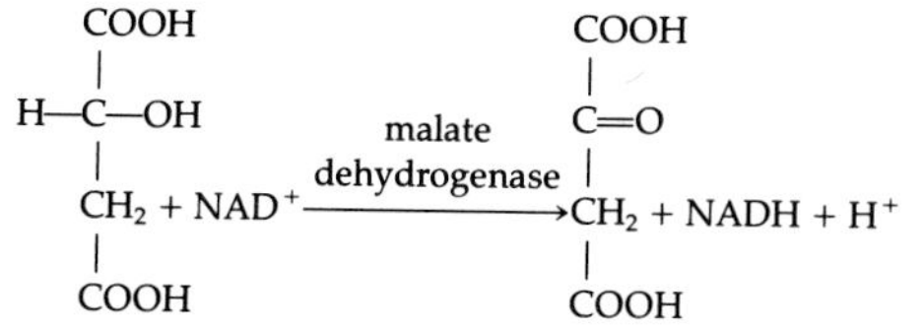

This and many similar redox reactions in glycolysis and the Krebs cycle use NAD$^+$ to trap the energy-rich electrons of food. *These* are the electrons that are subsequently passed down the electron transport chain to oxygen, powering oxidative phosphorylation.

We shall now take a closer look at how glycolysis and the Krebs cycle help generate ATP, directly by substrate-level phosphorylation, and indirectly by passing electrons to the transport chain via NADH.

GLYCOLYSIS

The word *glycolysis* means "splitting of sugar," and that is exactly what happens during this pathway. Glucose, a six-carbon sugar, is split into two three-carbon sugars. These smaller sugars are then oxidized, and their remaining atoms are rearranged to form two molecules of pyruvic acid. During the stepwise conversion of the three-carbon sugars to pyruvic acid, two molecules of NAD$^+$ are reduced to NADH, and there is also a net production of two ATPs by substrate-level phosphorylation. Use Figure 9.9 to follow the ten steps of glycolysis before going on to the next paragraph.

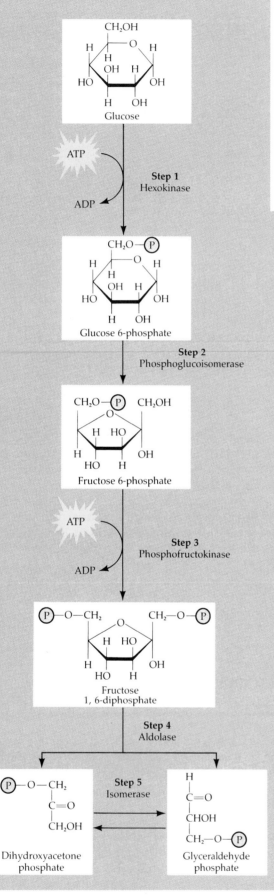

Figure 9.9
The ten steps of glycolysis. Each of the ten steps of glycolysis is catalyzed by a specific enzyme. The enzymes are found in the cytoplasm, dissolved in the cytosol, outside mitochondria. There are two segments to glycolysis. Segment one, consisting of the first five steps, is preparatory; glucose is split in two and altered to compounds that are primed for oxidation. This preparation of glucose actually *consumes* ATP. The payoff comes in the second segment, the last five steps of glycolysis, in which oxidation of the sugar occurs. This second segment generates ATP and NADH. In this diagram tracing glycolysis, coupled arrows (↲) are used to indicate the transfer of a phosphate group or pair of electrons from one reactant to another.

Step 1 Glucose enters the cell and is phosphorylated by the enzyme hexokinase, which transfers a phosphate group from ATP to the number six carbon of the sugar. The product of the reaction is glucose 6-phosphate. The electrical charge of the phosphate group traps the sugar in the cell because of the impermeability of the plasma membrane to ions. Phosphorylation of glucose also makes the molecule more chemically reactive. Although glycolysis is supposed to *produce* ATP, in step 1, ATP is actually consumed— an energy investment that will be repaid with dividends later in glycolysis.

Step 2 Glucose 6-phosphate is rearranged to convert it to its isomer, fructose 6-phosphate. Isomers, remember, have the same number and types of atoms but in different structural arrangements.

Step 3 In this step, still another molecule of ATP is invested in glycolysis. An enzyme transfers a phosphate group from ATP to the sugar, producing fructose 1,6-diphosphate. So far, the ATP ledger shows a debit of −2. With phosphate groups on its opposite ends, the sugar is now ready to be split in half.

Step 4 This is the reaction from which glycolysis gets its name. An enzyme cleaves the sugar molecule into two different three-carbon sugars: glyceraldehyde phosphate and dihydroxyacetone phosphate. These two sugars are isomers of one another.

Step 5 Another enzyme catalyzes the reversible conversion between the two three-carbon sugars, and if left alone in a test tube, the reaction reaches equilibrium. This does not happen in the cell, however, because the next enzyme in glycolysis uses only glyceraldehyde phosphate as its substrate and is unreceptive to dihydroxyacetone phosphate. This pulls the equilibrium between the two three-carbon sugars in the direction of glyceraldehyde phosphate, which is removed as fast as it forms. The net result of steps 4 and 5 is cleavage of a six-carbon sugar into two molecules of glyceraldehyde phosphate; each will progress through the remaining steps of glycolysis.

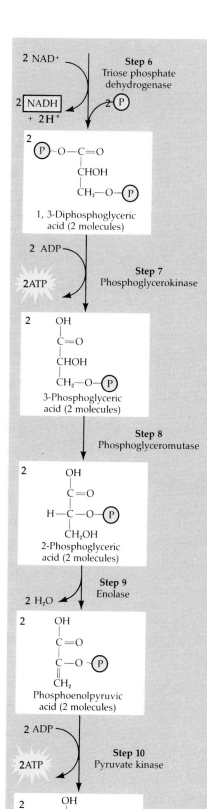

Step 6 An enzyme now catalyzes two sequential reactions while it holds glyceraldehyde phosphate in its active site. First, the sugar is oxidized by the transfer of electrons and H^+ from the number one carbon of the sugar to NAD^+, forming NADH. Here we see in metabolic context the type of redox reaction described on page 185. This reaction is very exergonic ($\Delta G = -10.3$ kcal/mol), and the enzyme capitalizes on this by coupling the reaction to the creation of a high-energy phosphate bond at the number one carbon of the oxidized substrate. The source of the phosphate is inorganic phosphate, which is always present in the cytosol. As products, the enzyme releases NADH and 1,3-diphosphoglyceric acid. Notice in the figure that the new phosphate bond is symbolized with a squiggle (~), which indicates that the bond is at least as energetic as the phosphate bonds of ATP.

Step 7 Finally, glycolysis produces some ATP. The phosphate group with the high-energy bond is transferred from 1,3-diphosphoglyceric acid to ADP. For each glucose molecule that began glycolysis, step 7 produces two molecules of ATP, since every product after the sugar-splitting step (step 4) is doubled. Of course, two ATPs were invested to get sugar ready for splitting. The ATP ledger now stands at zero. By the end of step 7, glucose has been converted to two molecules of 3-phosphoglyceric acid. This compound is not a sugar. The carbonyl group that characterizes a sugar has been oxidized to a carboxyl group, the hallmark of an organic acid. The sugar was oxidized back in step 6, and now the energy made available by that oxidation has been used to make ATP.

Step 8 Next, an enzyme relocates the remaining phosphate group of 3-phosphoglyceric acid to form 2-phosphoglyceric acid. This prepares the substrate for the next reaction.

Step 9 An enzyme forms a double bond in the substrate by extracting a water molecule from 2-phosphoglyceric acid to form phosphoenolpyruvic acid, or PEP. This results in the electrons of the substrate being rearranged in such a way that the remaining phosphate bond becomes very unstable; it has been upgraded to high-energy status.

Step 10 The last reaction of glycolysis produces more ATP by transferring the phosphate group from PEP to ADP. Since this step occurs twice for each glucose molecule, the ATP ledger now shows a net gain of two ATPs. Steps 7 and 10 each produce two ATPs for a total credit of four, but a debt of two ATPs was incurred from steps 1 and 3. Glycolysis has repaid the ATP investment with 100% interest. Additional energy was stored by step 6 in NADH, which can be used to make ATP by oxidative phosphorylation. In the meantime, glucose has been broken down and oxidized to two molecules of pyruvic acid, the compound produced from PEP in step 10.

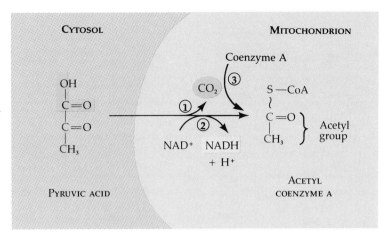

Figure 9.10
Formation of acetyl CoA. (1) The carboxyl group of pyruvic acid, already fully oxidized, is removed as a CO_2 molecule, which diffuses out of the cell. Since pyruvic acid had three carbons, a two-carbon fragment remains. (2) This two-carbon fragment is oxidized while NAD^+ is reduced to NADH. (3) Finally, the oxidized fragment, called an acetyl group, is attached to coenzyme A (CoA). This coenzyme has a sulfur atom, which bonds to the acetyl fragment by a very unstable bond (represented by $\sim$).

After tracing the steps of glycolysis in Figure 9.9, we can now write a summary equation for the overall catabolic pathway:

$$\begin{array}{l} C_6H_{12}O_6 \\ + \ 2\ NAD^+ \\ + \ 2\ ADP + 2\ \textcircled{P} \end{array} \longrightarrow \begin{array}{l} 2\ C_3H_4O_3 \ \text{(pyruvic acid)} \\ + \ 2\ NADH + 2\ H^+ \\ + \ 2\ ATP \\ + \ 2\ H_2O \end{array}$$

For the catabolism of glucose to two molecules of pyruvic acid, the overall free energy change, ΔG, is -140 kcal per mole of glucose oxidized. Most of the energy made available by this exergonic pathway is conserved in the high-energy electrons of NADH and the phosphate bonds of the ATP made by substrate-level phosphorylation. But most of the chemical energy originally stored in glucose still resides in the two molecules of pyruvic acid. If oxygen is present, the pyruvic acid enters the mitochondrion, where the enzymes of the Krebs cycle take over and complete the oxidation of the organic fuel.

THE KREBS CYCLE

As the first stage of cellular respiration, glycolysis releases less than a quarter of the chemical energy stored in glucose. If oxygen is present, cellular respiration can continue to break down pyruvic acid in the mitochondrion. The first order of business when pyruvic acid molecules enter the mitochondrion is to break off the carboxyl group, which has little energy, and attach the remaining two-carbon fragment to a coenzyme to form acetyl coenzyme A, or **acetyl CoA,** the entry compound for the Krebs cycle.

Formation of Acetyl CoA: Linking Glycolysis to the Krebs Cycle

The step that forms acetyl CoA from pyruvic acid is the junction between glycolysis and the Krebs cycle.

This complicated step is not catalyzed by a single enzyme, but by a multienzyme complex having several copies each of three different enzymes assembled together with their various coenzymes into a particle with a diameter of 30 nm, which is larger than a ribosome. Figure 9.10 outlines what happens to pyruvic acid as it is processed by this multienzyme complex. The catalysts remove CO_2 and oxidize the remaining two-carbon fragment to acetic acid, using the extracted electrons to reduce NAD^+ to NADH. Finally, coenzyme A, a sulfur-containing compound, is attached to the acetic acid by a high-energy (unstable) bond. The product of the multienzyme complex, acetyl CoA, is now ready to feed its acetyl group into the Krebs cycle for further oxidation.

How the Krebs Cycle Works

The Krebs cycle is named in honor of Hans Krebs, the German-British scientist who was largely responsible for elucidating the pathway in the 1930s. (The catabolic pathway is also known as the **citric acid cycle,** or **TCA cycle,** for tricarboxylic acid cycle.) The cycle has eight steps, each catalyzed by a specific enzyme in the mitochondrial matrix (Figure 9.11).

For each turn of the Krebs cycle, two carbons enter in the relatively reduced form of the acetyl group, and two different carbons leave in the completely oxidized form of CO_2 (Figure 9.12). Most of the energy made available by the oxidative steps of the cycle is conserved as high-energy electrons in NADH. For each acetyl group that enters the cycle, three molecules of NAD^+ are reduced to NADH. In one oxidative step, electrons are transferred not to NAD^+, but to a different electron acceptor, FAD (see p. 184). The reduced form of this coenzyme, $FADH_2$, donates its electrons to the electron transport chain, as does NADH. There is also a step in the Krebs cycle that forms an ATP molecule directly by substrate-level phosphorylation, similar to the ATP-generating steps of glycolysis. But most of the ATP output of respiration results from oxidative phosphorylation when the NADH and

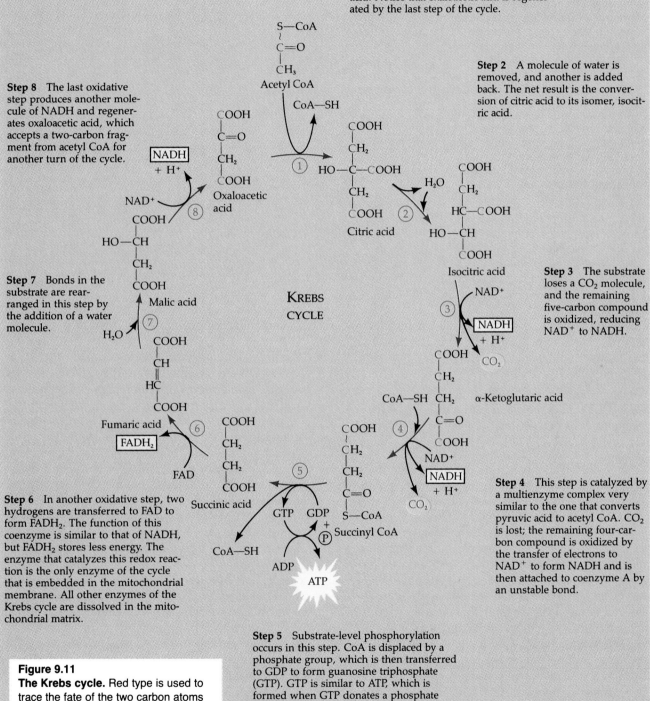

Step 1 Acetyl CoA adds its two-carbon acetyl fragment to oxaloacetic acid, a four-carbon compound. The unstable bond of acetyl CoA is broken as oxaloacetic acid displaces the coenzyme and attaches to the acetyl group. The product is the six-carbon citric acid. CoA is then free to prime another two-carbon fragment derived from pyruvic acid. Notice that oxaloacetic acid is regenerated by the last step of the cycle.

Step 2 A molecule of water is removed, and another is added back. The net result is the conversion of citric acid to its isomer, isocitric acid.

Step 8 The last oxidative step produces another molecule of NADH and regenerates oxaloacetic acid, which accepts a two-carbon fragment from acetyl CoA for another turn of the cycle.

Step 3 The substrate loses a CO_2 molecule, and the remaining five-carbon compound is oxidized, reducing NAD^+ to NADH.

Step 7 Bonds in the substrate are rearranged in this step by the addition of a water molecule.

Step 4 This step is catalyzed by a multienzyme complex very similar to the one that converts pyruvic acid to acetyl CoA. CO_2 is lost; the remaining four-carbon compound is oxidized by the transfer of electrons to NAD^+ to form NADH and is then attached to coenzyme A by an unstable bond.

Step 6 In another oxidative step, two hydrogens are transferred to FAD to form $FADH_2$. The function of this coenzyme is similar to that of NADH, but $FADH_2$ stores less energy. The enzyme that catalyzes this redox reaction is the only enzyme of the cycle that is embedded in the mitochondrial membrane. All other enzymes of the Krebs cycle are dissolved in the mitochondrial matrix.

Step 5 Substrate-level phosphorylation occurs in this step. CoA is displaced by a phosphate group, which is then transferred to GDP to form guanosine triphosphate (GTP). GTP is similar to ATP, which is formed when GTP donates a phosphate group to ADP.

Figure 9.11
The Krebs cycle. Red type is used to trace the fate of the two carbon atoms from pyruvic acid that enter the cycle via acetyl CoA (step 1), and blue type is used to follow the two carbons that are given off as carbon dioxide in steps 3 and 4.

Figure 9.12

Summary of the Krebs cycle. For every turn of the cycle, two carbons enter in the form of the acetyl fragment of acetyl CoA, and two carbons exit the cycle in the form of CO_2. The cycle functions as a metabolic "furnace" for oxidizing acetyl fragments to CO_2. Much of the energy stored in the acetyl group is restocked by the transfer of high-energy electrons to NAD^+ and FAD, forming NADH and $FADH_2$. The reduced coenzymes subsequently shuttle their cargo of high-energy electrons to the electron transport chain, which powers ATP synthesis. In addition to this oxidative phosphorylation, the Krebs cycle also generates 1 ATP per turn by the more direct mechanism of substrate-level phosphorylation. Because each glucose molecule was split in two during glycolysis, it takes two turns of the Krebs cycle to complete the oxidation of each glucose molecule that entered the cell as fuel.

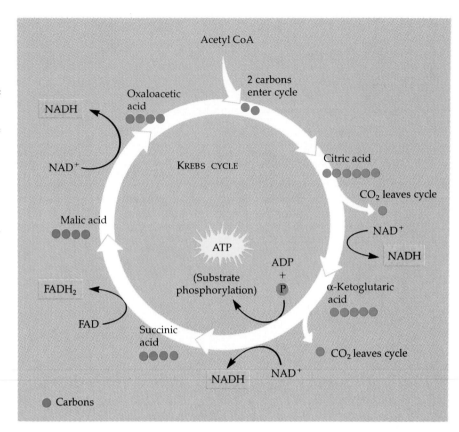

$FADH_2$ produced by the Krebs cycle use the electron transport chain to generate ATP by the chemiosmotic mechanism.

THE ELECTRON TRANSPORT CHAIN AND OXIDATIVE PHOSPHORYLATION

Neither glycolysis nor the Krebs cycle uses oxygen directly. It is the third stage of cellular respiration, consisting of the electron transport chain and oxidative phosphorylation, that requires oxygen. Recall from earlier in this chapter that the electron transport chain is a system of electron carriers embedded in the inner membrane of the mitochondrion. In a series of redox reactions, the chain passes electrons from NADH produced by glycolysis and the Krebs cycle to oxygen. The chain uses the electron flow to pump protons across the inner membrane, storing energy in the form of a proton gradient. The gradient powers an ATP synthase that phosphorylates ADP when protons pass through the enzyme complex on the way back down their gradient. And it is this ATP production driven by the electron transport chain that is called oxidative phosphorylation. The overall function of oxidative phosphorylation is to transform the energy of electrons of glucose into energy stored in the phosphate bonds of ATP.

Electron Transport

Most components of the electron transport chain are proteins, with prosthetic groups (cofactors tightly bound to the proteins) that shift between reduced and oxidized states as they accept and donate electrons.

Figure 9.13 traces the sequence of electron transfers along the electron transport chain. High-energy electrons are transferred from NADH to the first molecule of the electron transport chain, a flavoprotein, so named because it has a prosthetic group called flavin mononucleotide (FMN). In the next redox reaction, flavoprotein reverts back to its oxidized form as it passes electrons to an iron-sulfur protein (Fe·S in Figure 9.13), one of a family of proteins with both iron and sulfur tightly bound as cofactors. The iron-sulfur protein next passes the electrons to a compound called ubiquinone (Q). This electron carrier is the only member of the electron transport chain that is not bound to a protein. Most of the remaining electron carriers between Q and oxygen are proteins called **cytochromes** (Cyt). Their prosthetic group, called a **heme group,** has four organic rings surrounding a single iron atom (Figure 9.14). It is similar to the iron-containing prosthetic group found in hemoglobin, the red protein of blood that transports oxygen. But the iron of cytochromes transfers electrons, not oxygen. The electron transport chain has several types of cytochromes, each a different protein with a heme group. The last cytochrome of the

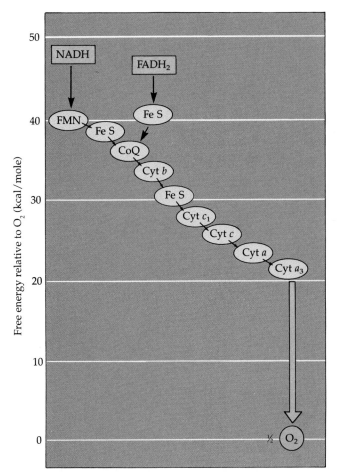

Figure 9.13
The electron transport chain. Each member of the chain oscillates between a reduced state and an oxidized state. A component of the chain becomes reduced when it accepts electrons from its "uphill" neighbor (which has a lower affinity for the electrons). Each member of the chain reverts to its oxidized form as it passes electrons to its "downhill" neighbor (which has a greater affinity for the electrons). At the "bottom" of the chain is oxygen, which is *very* electronegative. The overall energy drop for electrons traveling from NADH to oxygen is 53 kcal/mol, but this fall is broken up into a series of smaller steps by the electron transport chain.

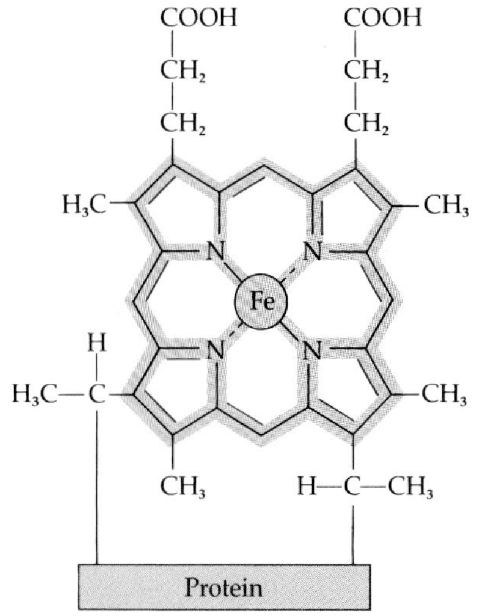

Figure 9.14
The heme group of cytochrome c. A similar, iron (Fe)-containing prosthetic group is joined to each cytochrome protein, as well as to hemoglobin. The various cytochromes differ in their protein components and also have different functional groups attached to the heme.

energy in manageable quantities. As electrons cascade down the chain from carrier to carrier, they lose a small amount of energy with each step until they finally reach oxygen, the terminal electron acceptor. What keeps the electrons moving is that each carrier has a greater affinity for electrons than its "uphill" neighbor in the chain. At the bottom of the chain is oxygen, which has a very great affinity for electrons. High-energy electrons removed from food by NAD^+ fall down the electron transport chain to a far more stable location in the electronegative oxygen atom. Or, put another way, oxygen, the "electron grabber," *pulls* electrons down the chain.

An alternative source of electrons for the transport chain is $FADH_2$, the other reduced product of the Krebs cycle. Notice in Figure 9.13 that $FADH_2$ adds its electrons to the electron transport chain at a lower energy level than NADH. Consequently, the electron transport chain provides about a third less energy for ATP synthesis when the electron donor is $FADH_2$ rather than NADH.

The electron transport chain makes no ATP directly. Its function in oxidative phosphorylation, according to the chemiosmotic model introduced earlier in this chapter, is to generate a proton gradient across the mitochondrial membrane.

Generation of the Proton Gradient

How can the redox reactions that occur along the electron transport chain pump protons across the mitochondrial membrane? Some electron carriers of the

chain, cytochrome a_3, passes its electrons to oxygen, which also picks up a pair of hydrogen ions from the aqueous medium to form water. (An oxygen atom is written in Figure 9.13 as ½ O_2 to emphasize that it is molecular oxygen, O_2, that is reduced by the electron transport chain, and not individual oxygen atoms. For every two NADH molecules, one O_2 molecule is reduced to two molecules of water.)

Electron transfer from NADH to oxygen is very exergonic, having a free energy change of -53 kcal/mol (see Figure 9.13). If the redox reaction took place in a single explosive step, most of that energy released would be dissipated as heat. Instead, the electron transport chain breaks the large free energy drop by the electrons into a series of shorter steps, releasing

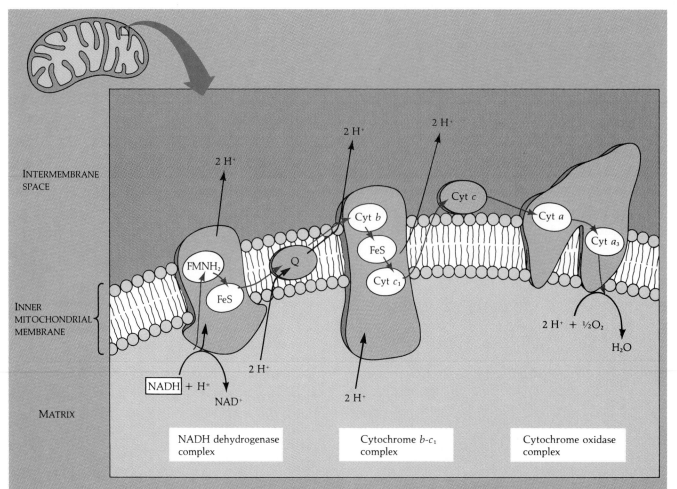

INTERMEMBRANE SPACE

INNER MITOCHONDRIAL MEMBRANE

MATRIX

$2 H^+$

$2 H^+$

$2 H^+$

$2 H^+$

FMNH$_2$

FeS

Q

Cyt b

FeS

Cyt c_1

Cyt c

Cyt a

Cyt a_3

NADH + H$^+$

NAD$^+$

$2 H^+$

$2 H^+$

$2 H^+$ + ½O$_2$

H$_2$O

NADH dehydrogenase complex

Cytochrome b-c_1 complex

Cytochrome oxidase complex

Figure 9.15

Logistics of electron transport: A tentative model. Electron carriers are organized into three complexes. Mobile carriers, ubiquinone (Q) and cytochrome c (Cyt c), transfer electrons between the complexes. Electron flow along the chain causes protons from the matrix side of the membrane to be released to the intermembrane space. The electron path is shown by red arrows and the proton path by black arrows. The exact sites of proton pumping and the total number of protons pumped per electron pair passed down the chain are uncertain. This model suggests three pumping stations. When the NADH dehydrogenase complex is reduced, it also accepts two hydrogen ions from the matrix side of the membrane—one from NADH and one from the solution. When the complex passes the electrons on, it releases two hydrogen ions to the solution in the intermembrane space. The physical mechanism that translocates the protons is unknown, but it may involve conformational changes by proteins in the complex. The complex functions as a proton pump powered by the binding and release of electrons. The second translocation of protons is better understood. The mobile carrier Q accepts electrons from the NADH dehydrogenase complex at a site near the matrix side of the membrane, where Q also picks up two hydrogen ions from the solution. The hydrophobic Q then diffuses across the lipid bilayer and releases its H$^+$ to the intermembrane space as electrons are passed from Q to the cytochrome b-c_1 complex. Thus, Q acts as the second proton pump. As electrons pass through the b-c_1 complex, it may act as the third site of proton pumping, translocating additional H$^+$ from the matrix to the intermembrane space. Cytochrome c then conveys electrons to the cytochrome oxidase complex, which in turn relays the electrons to oxygen. When the transport chain is operating and pumping H$^+$, the solution within the intermembrane space may be one to two pH units lower than the matrix, corresponding to a difference in H$^+$ concentration of 10 to 100 times. (The pH in the intermembrane space is the same as the pH of the cytosol, because the outer membrane of the mitochondrion is permeable to most small solutes, including H$^+$.)

electron transport chain must accept and release H$^+$ (protons) along with electrons. Other carriers transport only electrons. Therefore, electron transfers at certain steps along the chain cause the uptake or release of H$^+$ from the surrounding solution. The electron transport chain generates a proton gradient by accepting hydrogen ions from the matrix side of the inner mitochondrial membrane and releasing them to the intermembrane space on the opposite side of the membrane.

The translocation of protons is based on the spatial organization of the electron transport chain in the membrane. The electron carriers are collected into three complexes (Figure 9.15). Each complex spans the membrane and is present in thousands of copies in a single mitochondrion. Electrons are transferred between the complexes by carriers that are mobile in the membrane. (The complexes may themselves move in the membrane, but only slowly because of their large size. Earlier models of the electron transport chain postulated that electron carriers are arranged in a fixed order in the membrane. This assumption is unnecessary, because the two mobile carriers, Q and cytochrome c, move fast enough to account for the rate of electron transport between the three complexes.) At specific points (perhaps three) along the electron transport chain, protons are pumped from the matrix to the intermembrane space. The electron transport chain thus functions as an energy converter, using the fall of electrons from food molecules to oxygen to store energy in the form of a proton gradient across the inner membrane of the mitochondrion.

The Proton-Motive Force and ATP Synthesis

The potential energy stored in the proton gradient is referred to as the **proton-motive force.** The force has two components: the concentration gradient of hydrogen ions and the voltage across the membrane due to the disparate concentrations of positively charged hydrogen ions on opposite sides of the membrane. In other words, the proton-motive force is an electrochemical gradient, as discussed in Chapter 8.

The proton-motive force tends to drive H^+ across the membrane back into the matrix, but the lipid bilayer of the membrane is not very permeable to H^+. The H^+ must reenter the matrix by passing through ATP synthase, the ATP-synthesizing complex of proteins that spans the mitochondrial membrane (Figure 9.16). The inner membrane, with its many folds (cristae), is studded with multiple copies of the ATP synthase. These function as mills that harness the exergonic passage of H^+ to drive the endergonic phosphorylation of ADP. Chemiosmosis, remember, is this use of a proton-motive force to couple exergonic chemical processes to endergonic ones.

Studying the effects of respiratory poisons has provided important evidence for the chemiosmotic model. A variety of poisons inhibit cellular respiration by disrupting the chemiosmotic mechanism. Some of these deadly poisons block electron flow along the electron transport chain. For example, cyanide blocks passage of electrons from cytochrome a_3 to oxygen. This plugs the electron transport chain, the result being that protons are not pumped and ATP is not made. Another class of poisons, called uncouplers, short-circuit the proton current by making the lipid bilayer of the membrane leaky to hydrogen ions. The electron transport chain works furiously, and oxygen consumption increases, but leakage of H^+ abolishes the proton gradient and no ATP is made. An example of an uncou-

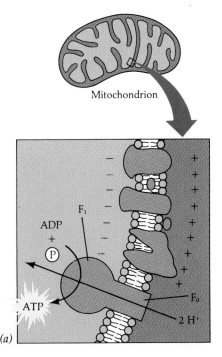

Mitochondrion

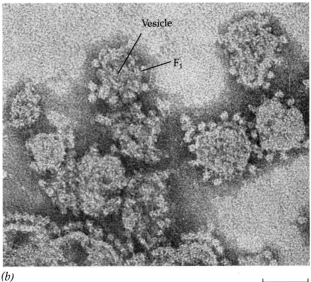

Vesicle

F_1

(a)

(b)

50 nm

Figure 9.16
Structure and function of the mitochondrial ATP synthase. (a) This complex of several proteins has two main parts. One part, named F_0, spans the membrane and functions as a channel specific for diffusion of H^+. Attached to F_0 on the matrix side of the membrane is F_1, a spherical head that catalyzes the phosphorylation of ADP to form ATP. How the enzyme uses the energy of the proton current to attach inorganic phosphate to ADP is still unknown. The hydrogen ions may participate directly in the reaction, or they may induce a conformational change of the ATP synthase that facilitates phosphorylation. **(b)** For this electron micrograph, ultrasonic vibration was used to disrupt mitochondria. Fragments of the inner membrane resealed to form vesicles, but they are inside out; the side of the membrane that originally faced the matrix is now the outer surface of the vesicles. The "lollipops" protruding from the membrane are F_1 portions of the ATP synthase complexes.

pler is the poison dinitrophenol. A third class of poisons, which includes the antibiotic oligomycin, inhibits the ATP synthase directly. As a result, the electron transport chain generates a proton gradient of greater magnitude than normal because the gradient is not dissipated by ATP synthesis. The effects of respiratory poisons have helped to confirm that ATP synthesis is a complex function based on the structural organization of the mitochondrial membrane.

By coupling electron transport to the synthesis of ATP, the proton-motive force functions as the main gear of oxidative phosphorylation; it drives the synthesis of most of the ATP produced during cellular respiration.

The ATP Ledger for Respiration

We can now do some bookkeeping to calculate the net ATP profit when cellular respiration oxidizes a molecule of glucose to six molecules of carbon dioxide. The three main departments of this metabolic enterprise are glycolysis, the Krebs cycle, and the electron transport chain. We first need to tally the few molecules of ATP produced directly by substrate-level phosphorylation during glycolysis and the Krebs cycle. To this we add the many more molecules of ATP generated when chemiosmosis couples electron transport to oxidative phosphorylation. Each NADH that transfers a pair of high-energy electrons from food to the electron transport chain contributes enough to the proton-motive force to generate a maximum of about 3 ATPs. (The average ATP yield per NADH is probably between two and three.) The Krebs cycle also supplies electrons to the electron transport chain via $FADH_2$, but each molecule of this electron carrier is worth a maximum of only 2 ATPs. (Notice again in Figure 9.13 that $FADH_2$ does not donate its electrons to the "top" of the transport chain.) In most eukaryotic cells, this lower ATP yield per electron pair also applies to the NADH produced by glycolysis in the cytosol. The mitochondrial membrane is impermeable to NADH, and thus NADH in the cytosol is segregated from the machinery of oxidative phosphorylation. The high-energy electrons of NADH produced by glycolysis must be shuttled across the membrane to electron acceptors within the mitochondrion. In the most common shuttle system, the electrons are received within the mitochondrion not by NAD^+, but by FAD, and this downgrades the "ATP value" of those electrons. Prokaryotes, of course, do not have to discount the value of NADH because no membrane separates glycolysis from the electron transport chains. In prokaryotes capable of respiration, components of the electron transport chain are built into the plasma membrane, which functions in chemiosmosis in the same way as the inner mitochondrial membrane of eukaryotes.

Figure 9.17, a follow-up to the overview of respiration presented in Figure 9.4, gives a detailed accounting of the ATP yield per glucose molecule oxidized. Subtracting the debit of 2 ATPs incurred during the preparatory steps of glycolysis, and doubling everything after the sugar-splitting step of glycolysis, the bottom line reads 36 ATPs (or 38 in bacteria—see the preceding paragraph).

Our bookkeeping gives only an estimate of the ATP yield from respiration. Before the chemiosmotic model was widely accepted, most biochemists believed that electron transport and oxidative phosphorylation were coupled by direct transfer of phosphate bonds in a manner akin to substrate-level phosphorylation. Such a mechanism would result in a precise yield of ATP for every pair of electrons passed down the electron transport chain. The prevailing view now, however, is that electron transport and ATP synthesis are more loosely linked by a proton gradient. Differences in the leakiness of the mitochondrial membrane to hydrogen ions and use of the proton-motive force to drive other kinds of work, such as the transport of solutes across the membrane, are two of the variables that would affect the ATP yield from respiration.

Since most of the ATP generated by cellular respiration is the work of oxidative phosphorylation, our estimate of 36 ATPs per glucose molecule is contingent on an adequate supply of oxygen to the cell. Without the electronegative oxygen to pull electrons down the transport chain, oxidative phosphorylation ceases. There is, however, a way for many cells to use their organic fuel to generate ATP without the help of oxygen.

FERMENTATION: THE ANAEROBIC ALTERNATIVE

How can food be oxidized without oxygen? Remember, oxidation refers to the loss of electrons to any electron acceptor, not just to oxygen. Glycolysis oxidizes glucose to two molecules of pyruvic acid. The oxidizing agent of glycolysis is NAD^+, *not* oxygen (see step 6 in Figure 9.9). The oxidation of glucose is exergonic, and glycolysis uses some of the energy made available to produce 2 ATPs (net) by substrate-level phosphorylation. If oxygen *is* present, then additional ATP is made by oxidative phosphorylation when NADH passes electrons removed from glucose to the electron transport chain. But glycolysis generates 2 ATPs whether oxygen is present or not—that is, whether conditions are **aerobic** or **anaerobic** (from the Greek *aer*, "air"; the prefix *an* means "without").

The anaerobic catabolism of organic nutrients is called fermentation, as mentioned at the beginning of the chapter. Fermentation can generate ATP by substrate-level phosphorylation as long as there is a sufficient

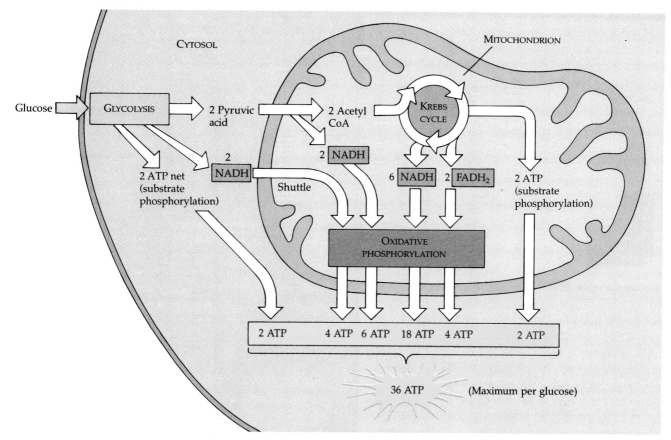

Figure 9.17
Maximum ATP yield for cellular respiration in a eukaryotic cell. For reasons discussed in the accompanying text, the maximum yield of 36 ATPs per glucose molecule is only an estimate.

supply of NAD$^+$ to accept electrons during the oxidation step of glycolysis. Without some mechanism to recycle NAD$^+$ from NADH, glycolysis would soon deplete the cell's pool of NAD$^+$ and shut itself down for lack of an oxidizing agent. Of course, under aerobic conditions, NAD$^+$ would be recycled productively from NADH by transfer of electrons to the electron transport chain. The anaerobic alternative is to transfer electrons from NADH to pyruvic acid, the end product of glycolysis. Fermentation consists of anaerobic glycolysis plus one or two (depending on species) subsequent reactions that regenerate NAD$^+$ by transferring electrons from NADH to pyruvic acid.

There are many types of fermentation; these differ in the waste products formed from pyruvic acid. Two of the most common types are **alcohol fermentation** and **lactic acid fermentation** (Figure 9.18).

In alcohol fermentation, pyruvic acid is converted to ethanol, or ethyl alcohol, in two steps. The first step releases carbon dioxide from the pyruvic acid, which is converted to the two-carbon compound acetaldehyde. In the second step, acetaldehyde is reduced by NADH to ethyl alcohol. This regenerates the supply of NAD$^+$ needed for glycolysis. Alcohol fermen-

tation by yeast, a fungus, is used in brewing (Figure 9.19). Many bacteria also carry out alcohol fermentation under anaerobic conditions.

During lactic acid fermentation, pyruvic acid is reduced directly by NADH to form lactic acid as a waste product, with no release of CO_2. Lactic acid fermentation by certain fungi and bacteria is used in the dairy industry to make cheese and yogurt. Acetone and methyl alcohol are among the by-products of other types of microbial fermentation that are commercially important.

Human muscle cells make ATP by lactic acid fermentation when oxygen is scarce. This occurs during the early stages of strenuous exercise, when sugar catabolism for ATP production outpaces the muscle's supply of oxygen from the blood. The cells switch from aerobic respiration to fermentation. The lactic acid that accumulates as a waste product may cause muscle fatigue, but it is gradually carried away by the blood to the liver. Lactic acid is converted back to pyruvic acid by liver cells, but that process requires oxygen. Thus, fermentation in muscle cells results in an "oxygen debt" that is paid back when you continue to pant after you have stopped exercising.

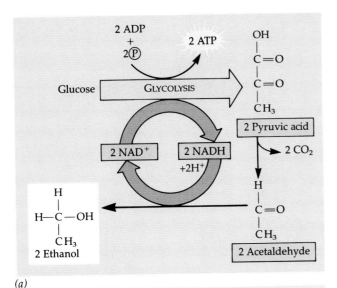

(a)

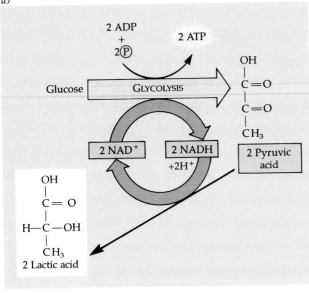

(b)

Figure 9.18
Fermentation. Pyruvic acid, the end product of glycolysis, serves as an electron acceptor for oxidizing NADH back to NAD^+. The NAD^+ can then be reused to oxidize sugar during glycolysis, which yields two net molecules of ATP by substrate-level phosphorylation during fermentation. Two of the common waste products formed from fermentation are (**a**) ethanol and (**b**) lactic acid.

COMPARISON OF AEROBIC AND ANAEROBIC CATABOLISM

There are actually *three* major schemes for harvesting the chemical energy of food molecules: aerobic respiration, and fermentation, which we have already addressed, and a third mechanism called anaerobic respiration. In all three catabolic processes, glucose or other energy-rich substrates are oxidized and their high-energy electrons passed to NAD^+, reducing it to NADH. However, the ultimate fate of these electrons differs for the three catabolic schemes (Table 9.1, p. 198).

Organisms that rely on **aerobic respiration** for ATP, such as plants and animals, are called **strict aerobes** (from the Greek *aer*, "air," and *bios*, "life"), which means that they can survive only in an environment that has oxygen. The final electron acceptor for aerobic respiration, of course, is oxygen, which is reduced by electrons that flow down the electron transport chain from NADH.

Anaerobic respiration occurs only in a few groups of bacteria that live deep in the soil, in stagnant ponds, or in other anaerobic environments. Their metabolism does not require oxygen. In fact, these microorganisms are poisoned by oxygen; they are among the bacteria referred to as **strict anaerobes.** During anaerobic respiration, electrons removed from substrates are passed along an electron transport chain to generate a proton gradient for ATP synthesis. However, the terminal electron acceptor is not oxygen, but some other substance, such as sulfate or nitrate.

The word *respiration* is derived from the Latin term *respirare*, which means "to breathe." Therefore, *aerobic respiration* may seem like a redundant term, and *anaerobic respiration* may seem contradictory. But in recent years, the concept of cellular respiration has been expanded to include all types of catabolism that use electron transport chains to make ATP, whether it is oxygen or some other substance that functions as the final electron acceptor.

Fermentation, as you have learned, makes ATP without the help of an electron transport chain. No oxygen is required, ATP is produced exclusively by substrate-level phosphorylation, and the final electron acceptor is an organic molecule (pyruvic acid or some derivative of pyruvic acid). Among bacteria that are strict anaerobes are those that lack electron transport chains and rely entirely on fermentation for ATP synthesis.

Facultative anaerobes, such as yeasts and many bacteria, can make their ATP either by fermentation or respiration, depending on whether oxygen is available. On the cellular level, our muscle cells behave as facultative anaerobes (see p. 195).

In a facultative anaerobe, pyruvic acid is a fork in the metabolic road that leads to two alternative catabolic routes (Figure 9.20). Under aerobic conditions, pyruvic acid is converted to acetyl CoA, and oxidation continues in the Krebs cycle. Under anaerobic conditions, pyruvic acid is diverted from the Krebs cycle, serving instead as an electron acceptor to recycle NAD^+.

Notice that fermentation not only operates without electron transport chains, but without the Krebs cycle as well. Without oxygen, the energy still stored in

(a)

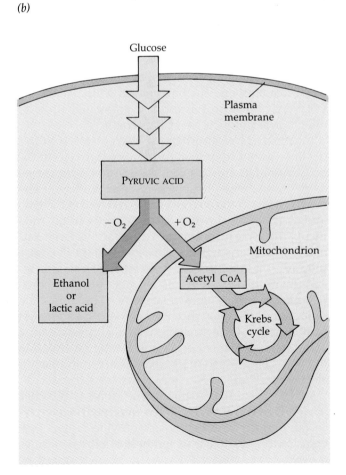

(b)

Figure 9.19
Commercial applications of alcohol fermentation. Wine is prepared by using yeast to convert some of the sugar in fruit juice to alcohol. Yeast must be cultured in the absence of oxygen for fermentation to occur. One-way gas valves allow carbon dioxide to escape from the ferment without letting air in. To make a sparkling wine, such as champagne, the CO_2 is left dissolved in the wine. The basic steps used to make wine from grapes were the same in ancient times (**a**) as they are today (**b**), although modern equipment gives winemakers much more control over the final product. Brewing beer is a similar process, except that the microorganisms are fed a diet of grain and malt sugar rather than fruit juice.

Figure 9.20
Pyruvic acid as a key juncture in catabolism. Glycolysis is common to fermentation and respiration. The end product of glycolysis, pyruvic acid, represents a fork in the catabolic pathways of glucose oxidation. In a cell capable of both respiration and fermentation, pyruvic acid is committed to continue along one of those two pathways, depending on whether or not oxygen is present.

pyruvic acid is unavailable to the cell. Oxygen enables the cell to harvest much more energy from sugar than the cell can tap by fermentation. In fact, for each glucose molecule, respiration yields 18 times more ATP than does fermentation (36 ATPs for respiration compared to the 2 ATPs produced by substrate-level phosphorylation in fermentation).

Glycolysis is the one catabolic pathway common to fermentation and respiration, a similarity with an evolutionary basis.

Evolutionary Significance of Glycolysis

Ancient prokaryotes probably used glycolysis to make ATP long before oxygen was present in the atmo-

Table 9.1 Comparison of catabolic processes

	Aerobic Respiration	Anaerobic Respiration	Fermentation
Growth conditions	Aerobic	Anaerobic	Aerobic or anaerobic
Electron transport chain	Yes	Yes	No
Final hydrogen (electron) acceptor	Free oxygen (O_2)	Usually an inorganic substance (such as NO_3^-, SO_4^{2-}, or CO_3^{2-}), but not free oxygen (O_2)	Organic molecule, such as lactic acid or ethanol
Type of phosphorylation used to build ATP	Mostly oxidative; some substrate-level	Oxidative	Substrate-level

sphere of Earth. The oldest known fossils of bacteria date back over 3.5 billion years. But appreciable quantities of oxygen probably did not begin to accumulate in the atmosphere until about 2.5 billion years ago, when, according to fossil evidence, the cyanobacteria that produce O_2 as a by-product of photosynthesis first evolved. The first prokaryotes must have generated ATP exclusively from glycolysis, which does not require oxygen. Glycolysis is the most widespread metabolic pathway, which suggests that it evolved very early in the history of life. The cytoplasmic location of glycolysis also implies great antiquity; the pathway does not require any of the membrane-enclosed organelles of the eukaryotic cell, which evolved nearly two billion years after the prokaryotic cell. (The evolution of metabolism is covered more thoroughly in Chapters 24 and 25.) Glycolysis is a metabolic heirloom from the earliest cells that continues to function in the fermentation of modern anaerobes and as the first stage in the breakdown of glucose by respiration.

CATABOLISM OF OTHER MOLECULES

Throughout this chapter, we have used glucose as the fuel for cellular respiration. But free glucose molecules are not common in the diets of humans and other animals. We obtain most of our calories in the form of fats, proteins, sucrose and other disaccharides, and starch, a polysaccharide. All of these food molecules can be used by cellular respiration to make ATP (Figure 9.21).

In the digestive tract, starch is hydrolyzed to glucose, which can then be broken down in the cells by glycolysis and the Krebs cycle. Similarly, glycogen, the polysaccharide that humans and many other animals store in their liver and muscle cells, can be hydrolyzed to glucose between meals as fuel for respiration.

Digestion of disaccharides, including sucrose, provides glucose and other monosaccharides that can then be enzymatically coverted to glucose as additional fuel for respiration. Thus, glycolysis can accept a wide range of carbohydrates for catabolism.

Proteins can also be used for fuel, but first they must be digested to their constituent amino acids. Many of the amino acids, of course, are used by the organism to build new proteins. Amino acids present in excess are converted by enzymes to intermediates of glycolysis and the Krebs cycle. The entry point into respiration depends on the structure of the amino acid. Common sites are pyruvic acid, acetyl CoA, and α-ketoglutaric acid, an intermediate of the Krebs cycle. Before amino acids can feed into glycolysis or the Krebs cycle, their amino groups must be removed, a process called deamination. The nitrogenous refuse is excreted from the animal in the form of ammonia, urea, or other waste products (see Chapter 40).

Catabolism can also harvest energy stored in fats obtained either from food or from storage cells in the body. After fats are digested, the glycerol is converted to glyceraldehyde phosphate, an intermediate of glycolysis. Most of the energy of a fat is stored in the fatty acids, which are broken down to two-carbon fragments that enter the Krebs cycle as acetyl CoA. Fats make excellent fuel; they have a lot of calories because they are rich in hydrogen. Remember, electrons associated with hydrogen in organic molecules have high potential energy. A gram of fat oxidized by respiration produces more than twice as much ATP as a gram of carbohydrate. Unfortunately, this also means that a dieter must be patient while consuming fat stored in the body, because so many calories are stockpiled in each gram of fat.

We see, then, that respiration is flexible in the fuels it can oxidize to make ATP. The intermediates of glycolysis and the Krebs cycle also provide carbon skeletons for the cell to use in synthesizing the molecules it needs.

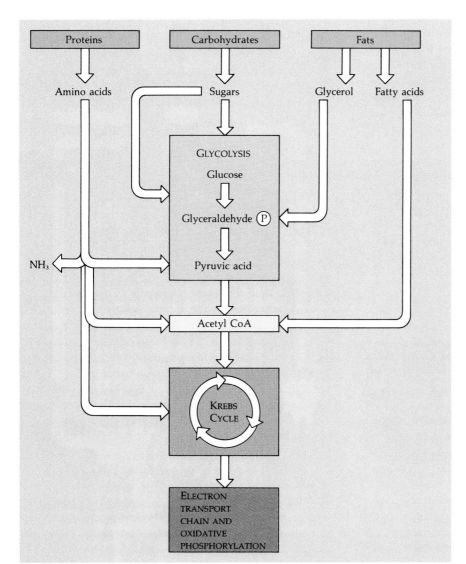

Figure 9.21
Catabolism of various food molecules.
Carbohydrates, fats, and proteins can all be used as fuel for respiration. Monomers of these food molecules enter glycolysis or the Krebs cycle at various points. Glycolysis and the Krebs cycle are catabolic funnels through which high-energy electrons from all kinds of food molecules flow on their exergonic fall to oxygen.

BIOSYNTHESIS

Cells need substance as well as energy. Not all the organic molecules of food are destined to be oxidized as fuel to make ATP. In addition to calories, food must also provide the carbon skeletons that cells require to make their own molecules. Some organic monomers obtained from digestion can be used directly. For example, amino acids from the hydrolysis of proteins in food can be incorporated into the organism's own proteins. Often, however, the body needs specific molecules that are not present as such in food. Compounds formed as intermediates of glycolysis and the Krebs cycle can be diverted into anabolic pathways as precursors from which the cell can synthesize molecules it requires. For example, humans can make about half of the 20 kinds of amino acids by modifying compounds siphoned away from the Krebs cycle. Also,

glucose can be made from pyruvic acid, and fatty acids can be synthesized from acetyl CoA. Of course, these anabolic or biosynthetic pathways do not generate ATP, but consume it instead.

In addition, glycolysis and the Krebs cycle function as metabolic interchanges, enabling our cells to convert some molecules to others as we need them. For instance, carbohydrates and proteins can be converted to fats through intermediates of glycolysis and the Krebs cycle. If we eat more food than we need, we will store fat even if our diet is fat free. Metabolism is remarkably versatile and adaptable.

CONTROL OF RESPIRATION

Basic principles of supply and demand govern the metabolic economy. The cell does not waste energy making more of a particular substance than it needs. If there is a glut of a certain amino acid, for example, the anabolic pathway that synthesizes that amino acid from an intermediate of the Krebs cycle is switched off. The most common mechanism for this control is feedback inhibition; the end product of the anabolic pathway inhibits the enzyme that catalyzes the first step of the pathway (see Chapter 6). This prevents the needless diversion of key metabolic intermediates from uses that are more urgent.

The cell also controls its catabolism. If the cell is working hard and its ATP concentration begins to drop, respiration speeds up. When there is plenty of ATP to meet demand, respiration slows down, sparing valuable organic molecules for other functions. Again, control is based mainly on regulating the activity of enzymes at strategic points in the catabolic pathway. One important switch is phosphofructokinase, the enzyme that catalyzes step 3 of glycolysis (see Figure 9.9). That is the earliest step that commits substrate irreversibly down the glycolytic road. By controlling the rate of this step, the cell can speed up or slow down the entire catabolic process; phosphofructokinase is thus the pacemaker of respiration (Figure 9.22).

An allosteric enzyme with receptor sites for specific inhibitors and activators, phosphofructokinase is inhibited by ATP and stimulated by ADP or AMP. (AMP is what is left of ATP that has donated *two* of its phosphate groups.) The ratio of ATP to ADP and AMP reflects the energy status, or **energy charge,** of the cell, and phosphofructokinase is sensitive to even slight changes in this ratio. As ATP accumulates, inhibition of the enzyme slows down glycolysis. The enzyme becomes active again as cellular work converts ATP to ADP and AMP faster than ATP is being regenerated. Phosphofructokinase is also inhibited by citric acid transported into the cytosol. This control helps to synchronize the rates of glycolysis and the Krebs cycle. If citric acid begins to accumulate, glycolysis slows down, and the supply of acetyl CoA to the Krebs cycle is reduced. If citric acid consumption increases, either because of a demand for more ATP or because anabolic pathways are draining off intermediates of the Krebs cycle, glycolysis accelerates and meets the demand. Metabolic balance is augmented by control of other allosteric enzymes at other key locations in glycolysis and the Krebs cycle. Cells are thrifty, expedient, and responsive in their metabolism.

We have looked at respiration within cells on such a microscopic level that you may have lost the sense of its place in the life of real organisms. But when cells harvest the energy stored in food molecules and use it to make ATP, they are supplying the energy for cellular work that is manifest in all the activities of the whole organism. Without the generation of ATP within your individual cells and, more precisely, within mitochondria, you would not be able to walk, to think, to see—to live.

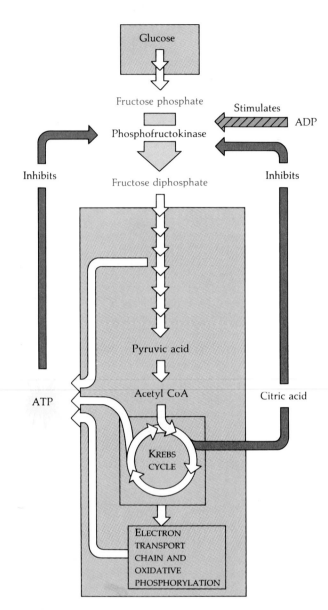

Figure 9.22
Control of cellular respiration. Allosteric enzymes at certain points in the respiratory pathway respond to inhibitors and activators to set the pace of glycolysis and the Krebs cycle. Phosphofructokinase, the enzyme that catalyzes the third step of glycolysis, is one such key enzyme. It is stimulated by ADP but inhibited by ATP and by citric acid.

* * *

We began this chapter by looking at the ultimate energy source for life—the light energy absorbed by the chloroplasts of plants. In the next chapter, you will learn how photosynthesis captures light and converts it to chemical energy.

STUDY OUTLINE

1. The energy that powers most organisms on Earth ultimately comes from the sun. Photosynthetic organisms use sunlight to produce organic molecules from carbon dioxide and water. Animals obtain these organic molecules by eating plants or by eating animals that eat plants.
2. Fermentation, occurring without oxygen, is the partial degradation of organic molecules. Cellular respiration occurs in the mitochondria of eukaryotic cells. Starting with glucose or other organic fuel and using oxygen, respiration yields water, carbon dioxide, and energy in the form of ATP and heat.

How Cells Make ATP: An Introduction (pp. 180–183)

1. The free energy released from exergonic hydrolysis of ATP to ADP + Ⓟ drives essential endergonic processes in cells by transferring unstable phosphate bonds to various substrates, priming them to undergo some change that results in work.
2. To keep on working, a cell must regenerate ATP. Cellular respiration provides the energy to drive the endergonic phosphorylation of ADP.
3. Cellular respiration consists of glycolysis, the Krebs cycle, and the electron transport chain and oxidative phosphorylation. Glycolysis occurs in the cytosol; the other two stages occur inside mitochondria.
4. For each molecule of glucose degraded to carbon dioxide and water in this entire process, about 36 molecules of ATP are made.
5. A small amount of ATP is formed by substrate-level phosphorylation in glycolysis and the Krebs cycle. A phosphate group is directly transferred to ADP from phosphorylated organic compounds produced as intermediates in the catabolism of glucose.
6. Oxidative phosphorylation produces ATP by a chemiosmotic mechanism. During electron transport, some molecules in the inner membrane of the mitochondria generate a proton gradient by translocating H^+ from the mitochondrial matrix to the intermembrane space. The hydrogen ions then diffuse back into the matrix through an enzyme complex called ATP synthase, triggering the phosphorylation of ADP.

Redox Reactions in Metabolism (pp. 183–185)

1. Food molecules store energy in the form of electrons associated mainly with hydrogen atoms. The cell taps this energy through oxidation-reduction, or redox, reactions, in which one substance (the reducing agent) partially or totally shifts electrons to another (the oxidizing agent). The substance receiving the electrons is reduced and the substance losing electrons is oxidized.
2. In cellular respiration, glucose ($C_6H_{12}O_6$) is oxidized to CO_2 and O_2 is reduced to H_2O. Electrons lose potential energy during their transfer from organic compounds to water, and this energy is used to drive ATP synthesis.
3. The initial acceptor of these high-energy electrons removed from organic compounds is usually NAD^+, which functions as a coenzyme. NAD^+ is reduced by gaining electrons and a hydrogen nucleus from the substrate to become NADH.
4. The cell uses NADH to carry the high-energy electrons of food to the electron transport chain.

Glycolysis (pp. 185–188)

1. Glycolysis is the "splitting of sugar" that occurs in the cytosol. It is a metabolic pathway that occurs in every cell.
2. In glycolysis, the six-carbon sugar glucose is oxidized to two three-carbon molecules of pyruvic acid, producing two molecules of ATP by substrate-level phosphorylation and reducing two molecules of NAD^+.
3. In the presence of oxygen, glycolysis functions as the first stage of respiration. Pyruvic acid moves into the mitochondrion, where it is completely oxidized to CO_2 in the Krebs cycle.

The Krebs Cycle (pp. 188–190)

1. The link between glycolysis and the Krebs cycle is the conversion of pyruvic acid to acetyl CoA by a multienzyme complex in the matrix of the mitochondrion.
2. The acetic acid of acetyl CoA joins a four-carbon molecule, oxaloacetic acid, to form the six-carbon citric acid molecule, which is subsequently degraded back to oxaloacetic acid in a series of steps constituting one turn of the cycle. In the process, carbon dioxide is given off, one molecule of ATP is formed by substrate-level phosphorylation, and high-energy electrons are passed to three molecules of NAD^+ and one molecule of FAD, another redox coenzyme.

The Electron Transport Chain and Oxidative Phosphorylation (pp. 190–194)

1. Most of the ATP created from the energy stored in glucose is produced by oxidative phosphorylation when NADH and $FADH_2$ donate their electrons to a system of electron carriers embedded in the mitochondrial cristae.
2. The electron transport chain consists of a series of increasingly electronegative components, starting with a flavoprotein (FMN), progressing through an iron-sulfur protein, then to ubiquinone (Q) and a series of cytochrome proteins with iron-containing heme groups, and finally reaching oxygen, which is very electronegative.
3. The components of the chain receive electrons from NADH and $FADH_2$ and shift between reduced and oxidized states, passing electrons down an energy gradient to oxygen, which then picks up a pair of hydrogen ions and forms water. Two mobile components, Q and cytochrome *c*, transfer electrons between the other electron carriers, which are located in three groups of integrated complexes.
4. The structural order of the carriers causes electron transfers at three steps along the chain to translocate H^+ from the matrix to the intermembrane space, storing energy in an electrochemical gradient known as the proton-motive force. As hydrogen ions diffuse back into the matrix through ATP synthase complexes on the cristae, the exergonic passage of H^+ drives the endergonic phosphorylation of ADP.
5. The effects of various respiratory poisons provide evidence for the chemiosmotic model of ATP synthesis.
6. The complete oxidation of glucose to carbon dioxide during aerobic respiration in eukaryotes produces a net yield of about 36 molecules of ATP, compared to only 2 for incomplete oxidation of glucose during fermentation.

7. The actual yield of ATP during respiration varies, owing to differences in the permeability of the cristae to H^+ and to partial use of the proton gradient to drive active transport of certain solutes across the outer mitochondrial membrane.

Fermentation: The Anaerobic Alternative (pp. 194–196)

1. Glycolysis produces 2 ATPs per sugar molecule by substrate-level phosphorylation, whether under aerobic or anaerobic conditions. Fermentation is the anaerobic catabolism of organic nutrients. It yields the 2 ATPs from glycolysis as long as NAD^+ is regenerated to act as the oxidizing agent.
2. The electrons from NADH are transferred to pyruvic acid, or some derivative of that glycolytic end product. In alcohol fermentation, pyruvic acid is converted to the two-carbon ethyl alcohol and CO_2 is released. Lactic acid fermentation takes place in our muscles during strenuous exercise.

Comparison of Aerobic and Anaerobic Catabolism (pp. 198–200)

1. The ultimate electron acceptor for the electrons removed in the oxidation of glucose differs for three major metabolic schemes. In aerobic respiration, oxygen is the final electron acceptor. In anaerobic respiration, sulfate or nitrate serves as the electron acceptor. Both types of respiration use electron transport chains and oxidative phosphorylation to make ATP. Fermentation uses substrate phosphorylation to make ATP, using neither oxygen nor electron transport chains.
2. Yeast and certain bacteria are facultative anaerobes, capable of making ATP either by aerobic respiration or fermentation, depending on the availability of oxygen.
3. Glycolysis, the breakdown of glucose to pyruvic acid, is a catabolic pathway common to fermentation and respiration. It occurs in the cytoplasm of all organisms and probably evolved in ancient prokaryotes before oxygen was available in the atmosphere.

Catabolism of Other Molecules (p. 198)

1. Fats, proteins, and carbohydrates can all be consumed by cellular respiration to form ATP.
2. Carbohydrates can be enzymatically converted to glucose, as fuel for respiration.
3. The amino acids from protein digestion lose their amino groups and are converted to various intermediates of glycolysis or the Krebs cycle.
4. When fats are used for fuel, the glycerol backbone is converted to an intermediate of glycolysis, and the fatty acids enter the Krebs cycle as acetyl CoA.
5. Thus, glycolysis and the Krebs cycle are catabolic pathways that funnel high-energy electrons from all kinds of food molecules into the electron transport chain, which powers ATP synthesis.

Biosynthesis (p. 199)

1. In addition to providing energy, food must also provide the carbon skeletons needed by cells to make molecules for growth and repair.
2. Organic precursors for anabolism either come directly from digestion or come from glycolysis and the Krebs cycle, which donate intermediates for use in the synthesis of polymers or the conversion of one type of molecule into another.

Control of Respiration (pp. 199–200)

1. The rate of ATP synthesis in respiration is finely controlled by allosteric enzymes at key places in glycolysis and the Krebs cycle.
2. Regulation of the enzymes speeds up or slows down ATP synthesis, based on the presence of specific molecules that signal the moment-to-moment balance between cell catabolism and anabolism.

SELF-QUIZ

1. The *direct* energy source that drives ATP synthesis during oxidative phosphorylation is
   a. oxidation of glucose and other organic compounds
   b. the endergonic flow of electrons down the electron transport chain
   c. the affinity of oxygen for electrons
   d. a difference of H^+ concentration on opposite sides of the inner mitochondrial membrane
   e. the transfer of phosphate from Krebs cycle intermediates to ADP

2. In the reaction:

   $$PEP + NAD^+ \longrightarrow \text{pyruvic acid} + NADH + H^+$$

   the oxidizing agent is
   a. oxygen
   b. NAD^+
   c. NADH
   d. PEP
   e. pyruvic acid

3. Which metabolic pathway is common to both the aerobic and anaerobic catabolism of sugar?
   a. Krebs cycle
   b. electron transport chain
   c. glycolysis
   d. synthesis of acetyl CoA from pyruvic acid
   e. reduction of pyruvic acid to lactic acid

4. In a eukaryotic cell, most of the enzymes of the Krebs cycle are located in the
   a. plasma membrane
   b. cytosol
   c. inner mitochondrial membrane
   d. mitochondrial matrix
   e. intermembrane space

5. The *final* electron acceptor of the electron transport chain that functions in oxidative phosphorylation is
   a. oxygen
   b. water
   c. NAD^+
   d. pyruvic acid
   e. ADP

6. When electrons flow along the electron transport chains of mitochondria, which of the following changes occurs?

a. The pH of the matrix increases.

b. The ATP synthase pumps protons by active transport.

c. The electrons gain free energy.

d. The cytochromes of the chain phosphorylate ADP to form ATP.

e. NAD^+ is oxidized.

7. In the presence of a metabolic poison that specifically inhibits the mitochondrial ATP synthase, you would expect

  a. a decrease in the pH difference across the mitochondrial membrane

  b. an increase in the pH difference across the mitochondrial membrane

  c. increased synthesis of ATP

  d. oxygen consumption to cease

  e. proton pumping by the electron transport chain to cease

8. Most of the ATP made during cellular respiration is generated by

  a. glycolysis

  b. oxidative phosphorylation

  c. substrate-level phosphorylation

  d. direct synthesis of ATP by the Krebs cycle

  e. transfer of phosphate from glucose-phosphate to ADP

9. Which of the following is a true distinction between fermentation and cellular respiration?

  a. Only respiration oxidizes glucose.

  b. NAD^+ is reduced by the electron transport chain only in respiration.

  c. Fermentation, but not respiration, is an example of a catabolic pathway.

  d. Substrate-level phosphorylation is unique to fermentation.

  e. NAD^+ functions as an oxidizing agent only in respiration.

10. The rate of glycolysis is

  a. stimulated by ATP

  b. stimulated by ADP

  c. inhibited by ADP

  d. stimulated by citric acid

  e. stimulated by O_2

CHALLENGE QUESTIONS

1. A century ago, Louis Pasteur, the great French biochemist, investigated the metabolism of yeast, a facultative anaerobe. He observed that the yeast consumed sugar at a much faster rate under anaerobic conditions than it did under aerobic conditions. Explain this "Pasteur effect," as the observation is known.

2. Hibernating bats have a specialized kind of tissue called brown fat, which generates a large amount of heat that quickly raises the bat's body temperature when it comes out of hibernation in the spring. Brown fat contains numerous mitochondria that produce chemiosmotic proton gradients in the usual way but dissipate the gradient without producing ATP. What variation in the property of the cristae do you suppose is necessary for brown fat mitochondria to uncouple respiration from ATP synthesis? How would this adaptation generate heat?

3. In the process of baking bread, an essential ingredient is yeast, which is uniformly distributed throughout the dough during mixing and kneading. Yeast is a facultative anaerobe. How would you expect the metabolism of yeast cells on the surface of the dough mixture to differ from the metabolism of yeast cells in the interior? What causes the bread to rise?

FURTHER READING

1. Becker, W. M. *The World of the Cell.* Menlo Park, Calif.: Benjamin/Cummings, 1986, Chapters 7 and 8. This text is particularly clear and comprehensive in its coverage of how cells harvest energy.

2. Harold, F. M. *The Vital Force: A Study of Bioenergetics.* New York: W. H. Freeman, 1986. The "vital force" is the proton-motive force.

3. Mathews, C. and van Holde, K. *Biochemistry.* Menlo Park, Calif.: Benjamin/Cummings, 1989. Effective diagrams on catabolism.

4. McCarty, R. E. "H^+-ATPases in Oxidative and Photosynthetic Phosphorylation." *BioScience,* January 1985. An article on the structure and function of enzymes involved in ATP synthesis.

5. Stryer, L. *Biochemistry.* 3d ed. San Francisco: Freeman, 1988, Chaps. 11–14.

10

Photosynthesis

L ife on Earth is solar-powered. The chloroplasts of plants capture light energy that has traveled 160 million kilometers from the sun and convert it to chemical energy stored in the bonds of sugar and other organic molecules made from carbon dioxide and water. The process is called photosynthesis, and it is the subject of this chapter.

Cells cannot create their own energy, but as open systems they are able to absorb energy from their surroundings. Photosynthetic cells acquire their energy in the form of light, which they use to make the organic compounds that provide energy to other cells. Hence, photosynthesis nourishes almost all of the living world directly or indirectly.

An organism acquires the organic compounds it uses for energy and carbon skeletons by one of two major modes: **autotrophic** or **heterotrophic nutrition.** At first, the term autotrophic (from the Greek *autos*, "self," and *trophos*, "feed") may seem to contradict the principle that cells are open systems, taking in resources from their environment. Autotrophs are not totally self-sufficient, however. They are merely self-feeders in the sense that they sustain themselves without eating other organisms. Autotrophs make all their own organic molecules from raw materials that are inorganic, and as the sources of organic compounds for other organisms, they are known as the producers in an ecological system. Plants are autotrophs; the only nutrients they require are carbon dioxide from the air, and water and minerals from the soil. Specifically, plants are **photoautotrophs,** organisms that use light as a source of energy to synthesize carbohydrates, lipids, proteins, and other organic substances (Figure 10.1). Photosynthesis also occurs in algae, including certain protists

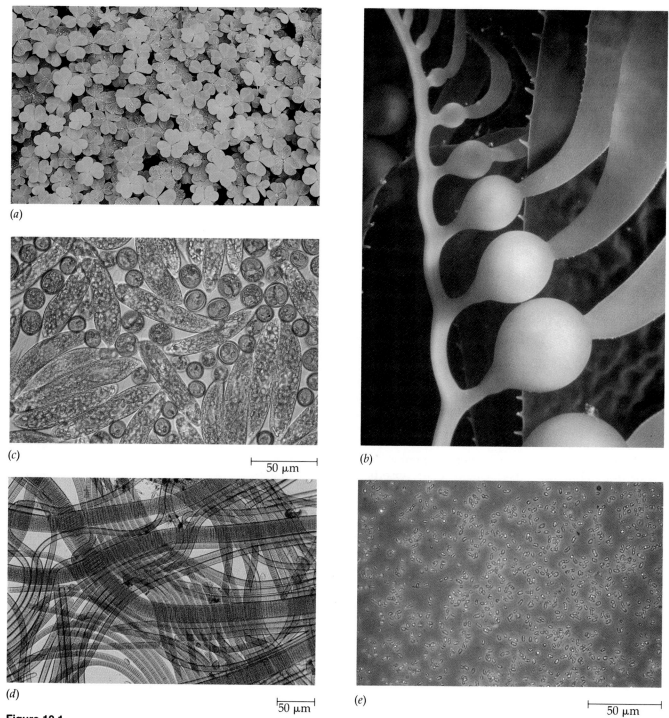

(a)

(c)

| 50 µm |

(b)

(d)

| 50 µm |

(e)

| 50 µm |

Figure 10.1

Photoautotrophs. These organisms use light energy to drive the synthesis of organic molecules from carbon dioxide and (usually) water. They feed not only themselves, but the entire living world.

(a) On land, vascular plants, such as this oxalis, are the predominant producers of food. In oceans, ponds, lakes, and other aquatic environments, photosynthetic organisms include: (b) algae, such as kelp,

(c) some unicellular protists, such as *Euglena,* and certain prokaryotes, including (d) cyanobacteria, and (e) purple sulfur bacteria (c, d, e: LMs).

(see Chapter 1) and some prokaryotes. In this chapter, the emphasis will be on plants. Variations in photosynthesis that occur in algae and bacteria will be discussed in Unit Five. A much rarer form of "self-feeding" is unique to those bacteria that are **chemoautotrophs.** They produce their organic compounds without the help of light, obtaining their energy by oxidizing inorganic substances such as sulfur or ammonia. (We will postpone further discussion of this type of autotrophic nutrition until Chapter 25.)

Heterotrophs obtain their organic material by the second major mode of nutrition. Unable to make their own food, they live on compounds produced by other organisms (note the prefix *heteros*, "other, different"). The most obvious form of this "other-feeding" is when an animal eats plants or other animals. But the process may be more subtle. Some heterotrophs do not kill prey, but instead decompose and feed on organic litter, such as carcasses, feces, and fallen leaves, and thus are known as decomposers. Most fungi and many types of bacteria get their nourishment this way. Almost all heterotrophs, including humans, are completely dependent on plants for food, and also for oxygen, a by-product of photosynthesis. Thus, we can trace the food we eat and the oxygen we breathe to the chloroplast.

CHLOROPLASTS: SITES OF PHOTOSYNTHESIS

All green parts of a plant, including green stems and unripened fruit, have chloroplasts, but the leaves are the major sites of photosynthesis in most plants (Figure 10.2). There are about half a million chloroplasts per square millimeter of leaf surface. The color of the leaf is due to **chlorophyll,** the green pigment located within the chloroplasts. It is the light energy absorbed by chlorophyll that the chloroplast puts to work in the synthesis of food molecules. Chloroplasts are found mainly in the cells of the **mesophyll,** the green tissue in the interior of the leaf. Carbon dioxide enters the leaf, and oxygen exits, by way of microscopic pores called **stomata** (singular, *stoma*). Water absorbed by the roots is delivered to the leaves in veins, or vascular bundles. Leaves also use vascular bundles to export sugar to nonphotosynthetic parts of the plant.

A typical mesophyll cell has about 30–40 chloroplasts, each a lens-shaped organelle measuring about 2–4 μm by 4–7 μm. An envelope of two membranes bounds the stroma, the dense fluid within the chloroplast (Figure 10.3a). An elaborate system of thylakoid membranes segregates the stroma from another compartment, the thylakoid space (see Chapter 7). In some places, thylakoid sacs are layered in dense stacks called grana. Chlorophyll resides in the thylakoid membranes. Thylakoids function in the steps of photosynthesis that initially convert light energy to chemical energy, but the steps that actually use that chemical energy to convert carbon dioxide to sugar occur in the stroma.

Photosynthetic prokaryotes lack chloroplasts, but they do have membranes that function in a manner similar to the thylakoid membranes of chloroplasts. The chlorophyll of photosynthetic bacteria is built into

the plasma membrane or into the membranes of numerous vesicles within the cell (Figure 10.3b). The photosynthetic membranes of cyanobacteria are usually arranged in parallel stacks of flattened sacs, much like the thylakoids of chloroplasts (Figure 10.3c). The photosynthetic membranes of certain photosynthetic bacteria are exceptions to the general rule that prokaryotes do not have membranes physically separate from the plasma membrane.

How do chloroplasts convert light energy to the chemical energy stored in organic molecules? Scientists have tried for centuries to piece together the process by which plants make food. Although some of the steps are still not understood, the overall photosynthetic equation has been known since the early 1800s: In the presence of light, the green parts of plants produce organic material and oxygen from carbon dioxide and water. This simple statement describes what photosynthesis does, but provides no hints as to how the process works. Only in this century have scientists begun to understand how plants trap sunlight and convert it to chemical energy.

HOW PLANTS MAKE FOOD: AN OVERVIEW

Photosynthesis can be summarized with this chemical equation:

$$6\,CO_2 + 12\,H_2O + \text{Light energy} \longrightarrow C_6H_{12}O_6 + 6\,O_2 + 6\,H_2O$$

The carbohydrate $C_6H_{12}O_6$ is glucose, a major product of photosynthesis. Water appears on both sides of the equation because 12 molecules are consumed and 6 molecules are newly formed during photosynthesis. We can simplify the equation by indicating the net consumption of water:

$$6\,CO_2 + 6\,H_2O + \text{Light energy} \longrightarrow C_6H_{12}O_6 + 6\,O_2$$

Writing the equation in this form, we can see that the chemical change during photosynthesis is the reverse of cellular respiration (however, it will soon become apparent that plants do not make food by simply reversing the steps of respiration). Both metabolic processes occur in plant cells.

Now let us rewrite the photosynthetic equation in its simplest possible form:

$$CO_2 + H_2O \longrightarrow CH_2O + O_2$$

Here, CH_2O symbolizes the general formula for a carbohydrate. In other words, we are imagining the synthesis of a sugar molecule one carbon at a time. Six repetitions would produce a glucose molecule. One of the earliest clues to the mechanism of photosynthesis came from the discovery in the 1930s that the

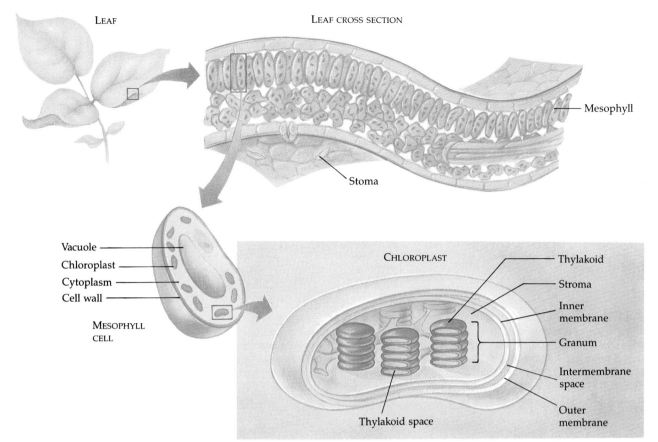

LEAF   LEAF CROSS SECTION

Mesophyll

Stoma

Vacuole
Chloroplast
Cytoplasm
Cell wall

MESOPHYLL
CELL

CHLOROPLAST

Thylakoid
Stroma
Inner membrane
Granum
Intermembrane space
Outer membrane

Thylakoid space

Figure 10.2
The site of photosynthesis in a plant.
Leaves are the major organs of photosynthesis in plants. Chloroplasts are found mainly in the mesophyll, the green tissue in the interior of the leaf. Gas exchange between the mesophyll and the atmosphere occurs through microscopic pores called stomata. The chloroplast is bounded by a double membrane that encloses the stroma, the dense fluid contents of the chloroplast. Membranes of the thylakoid system separate the stroma from the thylakoid space. Thylakoids are concentrated in stacks called grana.

oxygen given off by plants is derived from water and not from carbon dioxide. The chloroplast splits water into hydrogen and oxygen.

The Splitting of Water

An early model of photosynthesis speculated that carbon dioxide is split and then water is added to the carbon:

Step 1: $CO_2 \longrightarrow C + O_2$
Step 2: $C + H_2O \longrightarrow CH_2O$

This model predicted that the O_2 released during photosynthesis came from CO_2. The idea was challenged in the 1930s by C. B. van Niel, then a graduate student at Stanford University. Van Niel was investigating photosynthesis in bacteria, which make their carbohydrate from CO_2, but do not release O_2. Van Niel concluded that, at least in bacteria, CO_2 is not split into carbon and oxygen. One group of bacteria required hydrogen sulfide (H_2S) rather than water for photosynthesis, forming yellow globules of sulfur as a waste product:

$$CO_2 + 2\,H_2S \longrightarrow CH_2O + H_2O + 2\,S$$

Van Niel reasoned that the bacteria split H_2S and used the hydrogen to make sugar. He generalized that all photosynthetic organisms require a hydrogen source, but that the source varies:

Sulfur bacteria: $CO_2 + 2\,H_2S \longrightarrow CH_2O + H_2O + 2\,S$
General: $CO_2 + 2\,H_2X \longrightarrow CH_2O + H_2O + 2\,X$
Plants: $CO_2 + 2\,H_2O \longrightarrow CH_2O + H_2O + O_2$

Thus, van Niel hypothesized that plants split water as a source of hydrogen, releasing oxygen as a by-product.

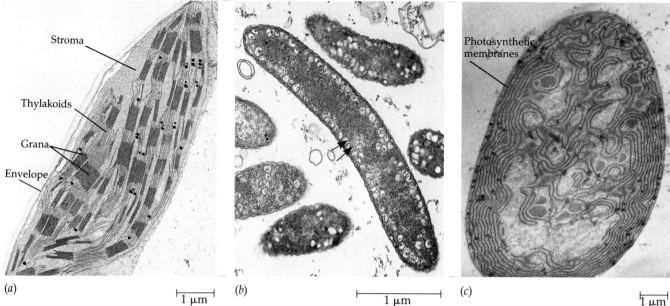

Stroma

Thylakoids

Grana

Envelope

(a)

1 μm

(b)

1 μm

Photosynthetic
membranes

(c)

1 μm

Figure 10.3
Photosynthetic membranes. (a) This electron micrograph of a chloroplast displays the grana, stacks of thylakoid sacs where light energy is converted to chemical energy (TEM). **(b)** In the photosynthetic bacterium *Rhodospirillum rubrum*, chlorophyll is built into the membranes of vesicles (arrows) (TEM). **(c)** Some cyanobacteria resemble chloroplasts because their photosynthetic membranes are stacked, much as they are in the grana of chloroplasts. Shown here is *Anabaena azollae* (TEM).

Nearly 20 years later, scientists confirmed van Niel's hypothesis by using oxygen-18 (^{18}O), a heavy isotope of oxygen, as a tracer to follow the fate of oxygen atoms during photosynthesis (isotopic tracers are explained in Chapter 2). The O_2 that came from plants was ^{18}O *only* if water was the source of the tracer. If the ^{18}O was introduced to the plant in the form of CO_2, the label did not turn up in the O_2. In the following summary of these experiments, red denotes labeled atoms of oxygen:

Experiment 1: $CO_2 + 2\,H_2O \longrightarrow CH_2O + H_2O + O_2$
Experiment 2: $CO_2 + 2\,H_2O \longrightarrow CH_2O + H_2O + O_2$

The most important result of the shuffling of atoms during photosynthesis is the extraction of hydrogen from water and its incorporation into sugar. Remember from Chapter 9 that electrons associated with hydrogen have much more potential energy in organic molecules than they do in water, where the electrons are closer to the electronegative oxygen. And recall that energy is stored in sugar and other food molecules in the form of these high-energy electrons.

Photosynthesis as a Redox Process

Let us briefly contrast photosynthesis with respiration. During respiration, energy is released from sugar when electrons associated with hydrogen are transported by carriers to oxygen, forming water as a by-product. The electrons lose potential energy along the way. The mitochondrion uses the energy made available to synthesize ATP. Photosynthesis, also a redox process, reverses the direction of electron flow. Water is split, and electrons are transferred from the water to carbon dioxide, reducing it to sugar. Electrons cannot travel downhill in both directions; they increase their potential energy when moved from water to sugar. Respiration yields 686 kcal of free energy per mole of glucose oxidized to carbon dioxide, and exactly that much free energy is required to reduce carbon dioxide to glucose; for photosynthesis, $\Delta G = +686$ kcal/mol. It is a redox process that is endergonic, and light provides the energy to boost electrons from water to their high-energy perches in sugar.

The Two Stages of Photosynthesis

The equation for photosynthesis is a deceptively simple summary of a very complex process. Actually, it is not a single process, but two, each with multiple steps. These two stages of photosynthesis are known as the **light reactions** and the **Calvin cycle.**

The light reactions are the steps of photosynthesis that convert solar energy to chemical energy. Light absorbed by chlorophyll drives a transfer of electrons from water to an acceptor named **NADP⁺,** which stands for nicotinamide adenine dinucleotide phosphate. This coenzyme temporarily stores the ener-

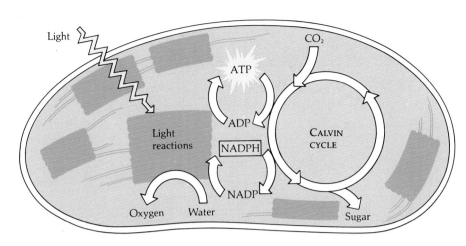

Figure 10.4
Integration of the light reactions and Calvin cycle. The light reactions use solar energy to make ATP and NADPH, which function as chemical energy and reducing power, respectively, in the Calvin cycle. The actual incorporation of CO_2 into organic molecules occurs in the Calvin cycle. Thylakoid membranes, especially those of the grana stacks, are the sites of the light reactions, whereas the Calvin cycle occurs in the stroma.

gized electrons. Water is split in the process, and thus it is the light reactions of photosynthesis that give off O_2 as a by-product. The electron acceptor of the light reactions, $NADP^+$, is first cousin to NAD^+, the coenzyme that functions as an electron carrier in cellular respiration; the two molecules differ only by the presence of an extra phosphate group in the $NADP^+$ molecule. The light reactions use solar power to reduce $NADP^+$ to NADPH by adding a pair of electrons along with a hydrogen nucleus, or H^+. The light reactions also generate ATP by powering the addition of a phosphate group to ADP, a process called **photophosphorylation.** Thus, light energy is initially converted to chemical energy in the form of two compounds: NADPH, a source of energized electrons, and ATP, the versatile energy currency of cells. Note that the light reactions produce no sugar; that happens in the second stage of photosynthesis.

The Calvin cycle, named for M. Calvin, who began to elucidate the steps of the cycle along with his colleagues in the late 1940s, incorporates CO_2 from the air into organic material. This incorporation of carbon into organic compounds is known as **carbon fixation.** The Calvin cycle then reduces the fixed carbon to carbohydrate by the addition of electrons. The reducing power is provided by NADPH, which acquired energized electrons in the light reactions. To convert CO_2 to sugar, the Calvin cycle also requires chemical energy in the form of ATP, also generated by the light reactions. The Calvin cycle makes sugar, but only with the help of the NADPH and ATP produced by the light reactions. The metabolic steps of the Calvin cycle are sometimes referred to as the "dark reactions" because none of the steps requires light *directly*. Nevertheless, the Calvin cycle in most plants occurs during daylight, for only then can the light reactions regenerate the NADPH and ATP spent in the reduction of CO_2 to sugar. In essence, the chloroplast uses light energy to make sugar by coordinating the two stages of photosynthesis.

Figure 10.4 summarizes how the two stages of photosynthesis cooperate. The thylakoids of the chloroplast are the sites of the light reactions, but the Calvin cycle occurs in the stroma. As molecules of $NADP^+$ and ADP bump into the thylakoid membrane, they pick up electrons and phosphate, respectively, and then transfer their high-energy cargo to the Calvin cycle. The two stages of photosynthesis are treated in the figure as "black boxes"— metabolic modules that take in ingredients and crank out products. Our next step toward understanding photosynthesis is a closer look at the inner workings of the first "box," the light reactions.

HOW THE LIGHT REACTIONS CAPTURE SOLAR ENERGY

Chloroplasts are chemical factories powered by the sun. Their thylakoids transform light energy into the chemical energy of ATP and NADPH. To understand this conversion better, it is first necessary to learn some important properties of light.

The Nature of Sunlight

The sun is a giant thermonuclear reactor. The energy it emits comes from fusion reactions much like those that occur in a hydrogen bomb. Four hydrogen atoms fuse to form one helium atom, which has a mass slightly less than the total mass of the hydrogen atoms. The lost mass has been converted to energy, as predicted by Albert Einstein in his famous $E = mc^2$. Each minute, about 120 million tons of solar matter are converted to a colossal amount of energy that radiates out into space. A minuscule fraction of that energy reaches Earth, taking only a few minutes to make the trip (the speed of light is about 300,000 km/sec). Some of the

Figure 10.5

The electromagnetic spectrum. Visible light and other forms of electromagnetic energy radiate through space as waves of various lengths. We perceive different wavelengths of visible light as different colors. Violet and blue have the shortest wavelengths, and orange and red the longest. Ultraviolet (UV) light and infrared (IR) light are just beyond the range of human vision, though some animals can detect these wavelengths. White light is a mixture of wavelengths. By bending light of different wavelengths varying degrees, a prism can sort white light into its component colors. Visible light drives photosynthesis.

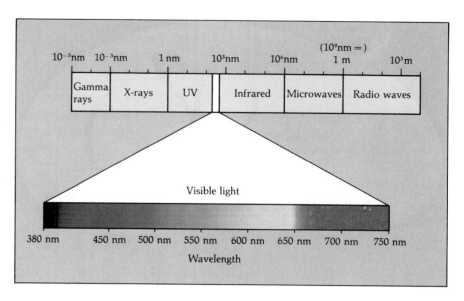

energy that arrives happens to strike the leaves of plants, where it is put to work by chloroplasts.

Sunlight is a form of energy known as **electromagnetic energy,** also called **radiation.** Electromagnetic energy travels in rhythmic waves analogous to those created by dropping a pebble into a puddle of water. Electromagnetic waves, however, are disturbances of electric and magnetic fields rather than disturbances traveling along the surface of water.

The distance between the crests of electromagnetic waves is termed the **wavelength.** Wavelengths range from less than a nanometer (for gamma rays) to more than a kilometer (for radio waves). This entire range of radiation is known as the **electromagnetic spectrum** (Figure 10.5), but the segment most important to life is the narrow band that ranges from about 400 to 700 nm in wavelength. This radiation is known as **visible light** because it is detected as various colors by the human eye.

The theory of light as waves explains most of its properties, but in certain respects, light behaves as though it consists of discrete particles called quanta, or **photons.** Photons may not be tangible objects, but they act that way in that each photon has a fixed quantity of energy. The amount of energy is inversely related to the wavelength of the light; the shorter the wavelength, the greater the energy of each photon of that light. Thus, a photon of violet light packs nearly twice as much energy as a photon of red light.

Although the sun radiates the full spectrum of electromagnetic energy, the atmosphere acts as a selective window, transparent to visible light while screening out a substantial fraction of other radiation. The same part of the spectrum we can see—visible light—is also the radiation that drives photosynthesis. Blue and red, the two wavelengths most effectively absorbed by chlorophyll, are the colors most useful as energy for the light reactions.

Photosynthetic Pigments

As light meets matter, it may be reflected, transmitted, or absorbed (Figure 10.6). Substances that absorb visible light are called **pigments.** Different pigments absorb light of different wavelengths, and the wavelengths that are absorbed disappear. If a pigment absorbs all wavelengths, it appears black. If a pigment is illuminated with white light, the color we see is the color most reflected or transmitted by the pigment. We see green when we look at a leaf because chlorophyll absorbs red and blue light while transmitting and reflecting green light. The ability of a pigment to absorb various wavelengths of light can be quantitatively measured by placing a solution of the pigment in an instrument called a **spectrophotometer** (see Methods Box on p. 212).

The **absorption spectrum** of the type of chlorophyll called **chlorophyll *a*** provides clues to the relative effectiveness of different wavelengths for driving photosynthesis, since light can only perform work in chloroplasts if it is absorbed. Indeed, as previously mentioned, blue and red light work best for photosynthesis, while green is the least effective color. An **action spectrum** profiles the relative performance of the different wavelengths more accurately. An action spectrum is prepared by illuminating chloroplasts with different colors of light and then plotting wavelength against some measure of photosynthetic rate, such as oxygen evolution or carbon dioxide consumption (Figure 10.7).

You can see that the action spectrum for photosynthesis does not exactly match the absorption spectrum of chlorophyll *a*. The absorption spectrum underestimates the effectiveness of certain wavelengths in

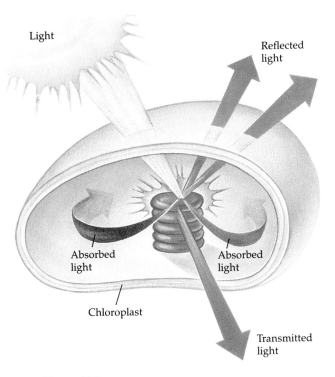

Figure 10.6
Interactions of light with matter. The pigments of chloroplasts absorb blue and red light, the colors most effective in photosynthesis. They reflect or transmit green light, which is why leaves appear green.

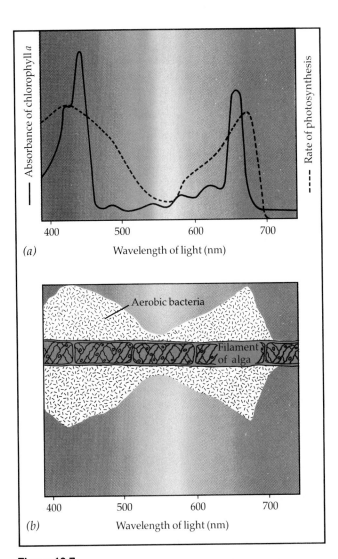

(a) Wavelength of light (nm)

(b) Wavelength of light (nm)

Figure 10.7
Action spectrum for photosynthesis. (a) The dashed line on the graph profiles the effectiveness of different wavelengths of light in driving photosynthesis. Compared to the peaks in the absorption spectrum for chlorophyll *a* (solid line), the peaks in the action spectrum are broader, and the valley is narrower and not as deep. This is partly due to absorption by light by accessory pigments that broaden the spectrum of colors that can be used for photosynthesis. **(b)** An elegant experiment of this type was first performed in 1883 by Thomas Engelmann, a German botanist. He illuminated a filamentous alga with light that had been passed through a prism, thus exposing different segments of the alga to different wavelengths of light. Engelmann used aerobic bacteria, which seek oxygen, to determine which segments of the plant were releasing the most O_2. Bacteria congregated in greatest density around the parts of the alga illuminated with red and blue light.

driving photosynthesis. This is partly because chlorophyll *a* is not the only pigment in chloroplasts important to photosynthesis. Only chlorophyll *a* can participate directly in the light reactions, which convert solar energy to chemical energy. But other pigments can absorb light and transfer the energy to chlorophyll *a*, which then initiates the light reactions. One of these accessory pigments is another form of chlorophyll, chlorophyll *b*. The structure of chlorophyll *b* is almost identical to that of chlorophyll *a* (Figure 10.8), but the slight structural difference between them is enough to give the two pigments slightly different absorption spectra and hence different colors. Chlorophyll *a* is grass-green, whereas chlorophyll *b* is yellow-green. The chloroplast also has a family of accessory pigments called **carotenoids,** which are various shades of yellow and orange. These hydrocarbons are built into the thylakoid membrane along with the two kinds of chlorophyll. Carotenoids can absorb wavelengths of light that chlorophyll cannot, thus broadening the spectrum of colors that can drive photosynthesis. If a photon of sunlight strikes an accessory pigment—a carotenoid or chlorophyll *b*—energy is conveyed to chlorophyll *a*, which then behaves just as though it had absorbed the photon.

A spectrophotometer measures the proportions of light of different wavelengths absorbed and transmitted by a pigment solution. Inside the spectrophotometer, white light is separated into its component colors (wavelengths) by a refracting prism. Then, one by one, the different colors of light are passed through the sample. The transmitted light strikes a photoelectric tube, which converts the light energy to electricity, and the electric current is measured by a galvanometer. Thus, each time the wavelength of light is changed, the meter indicates the

proportion of light transmitted through the sample or, conversely, the proportion of light absorbed. A graph that profiles absorbance at different wavelengths is called an absorption spectrum. As you can see in Figure 10.7a, the absorption spectrum for chlorophyll *a*, the form of chlorophyll most important in photosynthesis, has two peaks, corresponding to wavelengths for blue and red light, the colors chlorophyll *a* absorbs best. The absorption spectrum has a valley in the green region because the pigment transmits that color.

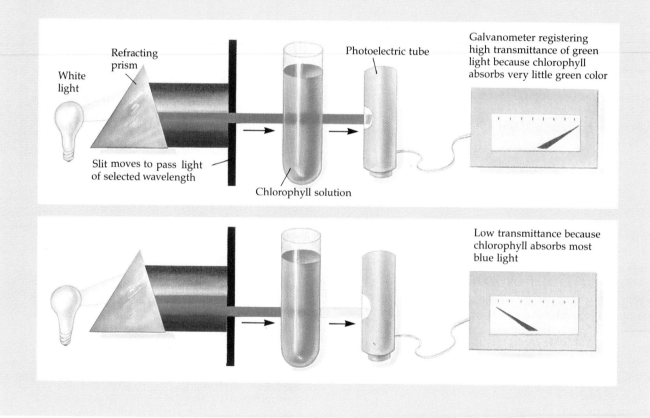

Refracting prism

White light

Photoelectric tube

Galvanometer registering high transmittance of green light because chlorophyll absorbs very little green color

Slit moves to pass light of selected wavelength

Chlorophyll solution

Low transmittance because chlorophyll absorbs most blue light

The Photooxidation of Chlorophyll

What exactly happens when chlorophyll and other pigments absorb photons? The colors corresponding to the wavelengths absorbed disappear from the spectrum of light transmitted and reflected by the pigment, but energy cannot disappear. When a molecule absorbs a photon, one of the molecule's electrons is

elevated to an orbital where it has more potential energy. When the electron is in its normal orbital, the pigment molecule is said to be in its **ground state.** After absorption of a photon boosts an electron to an orbital of greater energy value, the pigment molecule is said to be in an **excited state.** The only photons absorbed are those whose energy is exactly equal to the energy difference between the ground state and the excited state.

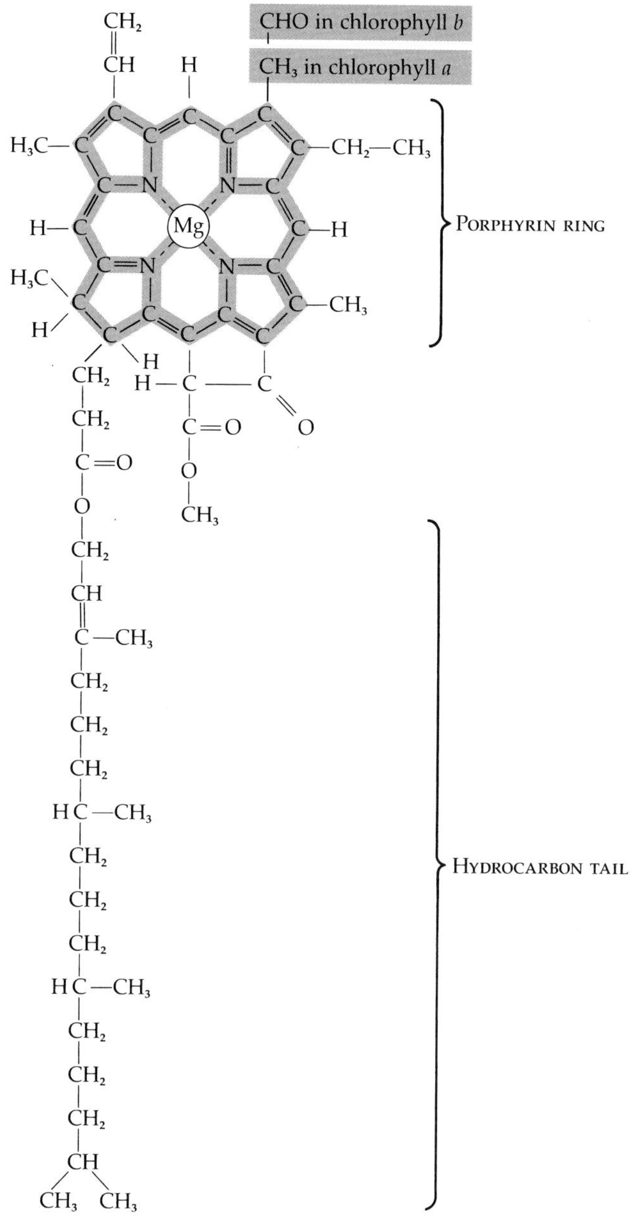

CHO in chlorophyll *b*
CH₃ in chlorophyll *a*

PORPHYRIN RING

HYDROCARBON TAIL

Figure 10.8
The structure of chlorophyll. Chlorophyll *a*, the pigment that participates directly in the light reactions of photosynthesis, has a porphyrin ring assembly with a magnesium atom at its center. Attached to the porphyrin is a hydrophobic tail, which anchors the pigment to the thylakoid membrane. Chlorophyll *b* differs from chlorophyll *a* only in one of the functional groups bonded to the porphyrin.

This is why different pigments, each with a different distribution of electrons, have unique absorption spectra.

The energy of an absorbed photon is converted to the potential energy of an electron raised from the ground state to an excited state. But the electron cannot remain there long; the excited state, like all high-energy states, is unstable. Generally, when pigments absorb light, their excited electrons drop back down to the ground-state orbital in a billionth of a second, releasing their excess energy as heat. This conversion of light energy to heat is what makes the top of an automobile so hot on a sunny day (white cars are coolest because their paint reflects all wavelengths of visible light, although it may absorb ultraviolet and other invisible radiation). Some pigments emit light as well as heat after absorbing photons. The electron jumps to a state of greater energy, and as it falls back to ground state, a photon is given off. This afterglow is called **fluorescence** (Figure 10.9a). The fluorescence has a longer wavelength, and hence less energy, than the light that excited the pigment. This follows from the second law of thermodynamics (see Chapter 6). The difference in energy of the incoming photon and the outgoing photon is dissipated as heat.

If a solution of chlorophyll isolated from chloroplasts is illuminated, it will fluoresce in the red part of the spectrum and also give off heat. But the result of illuminating chlorophyll is quite different when it is in its native environment in the thylakoid membrane of the chloroplast. In the membrane, a nearby molecule, referred to as the **primary electron acceptor,** traps a high-energy electron that has absorbed a photon (Figure 10.9b). A redox reaction occurs, with chlorophyll being photooxidized by the absorption of light energy, and the electron acceptor being reduced. It is because there is no electron acceptor to prevent the electrons of excited chlorophyll from dropping right back to ground state that isolated chlorophyll fluoresces. Defective photosynthetic cells with genetic mutations that disengage chlorophyll from electron acceptors also fluoresce (Figure 10.10). The acceptor molecule functions as a dam that prevents the high-energy electron from plunging immediately back to its ground state in chlorophyll. The solar-powered transfer of electrons from chlorophyll to the primary acceptor is the first step of the light reactions; subsequent steps tap the energy stored in the trapped electrons to power the synthesis of ATP and NADPH.

The Two Photosystems

Chlorophyll *a*, chlorophyll *b*, and the carotenoids are clustered in the thylakoid membrane in assemblies of a few hundred pigment molecules. Of the many chlorophyll *a* molecules in each assembly, only one can trigger the light reactions by donating its excited electron to the primary electron acceptor. The location of the specialized chlorophyll *a* molecule in the pigment assembly is called the **reaction center.** (There is recent evidence that the reaction center has a pair of specialized chlorophyll *a* molecules, but this does not alter the basic concept of a reaction center; we will stick to the hypothesis of a single molecule for simplicity.) The

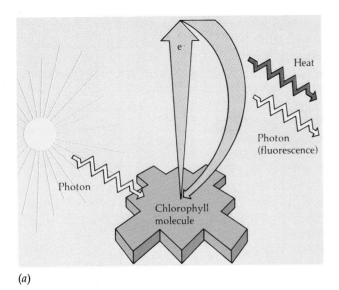

(a)

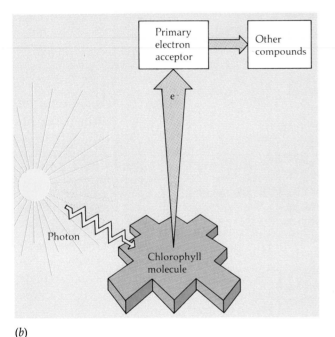

(b)

Figure 10.9
Excitation of chlorophyll. (a) Absorption of a photon causes a transition of the chlorophyll molecule from its ground state to its excited state. The photon boosts an electron to an orbital where it has more potential energy. If isolated chlorophyll is illuminated, its excited electron immediately drops back down to the ground-state orbital, giving off its excess energy as heat and fluorescence (light). **(b)** In the intact chloroplast, illuminated chlorophyll loses its excited electron to a neighboring molecule, the primary electron acceptor, which in turn passes the electron on to other molecules.

other chlorophyll *a* molecules and the molecules of chlorophyll *b* and the carotenoids function collectively as a light-gathering antenna that absorbs photons and passes the energy from molecule to molecule until it

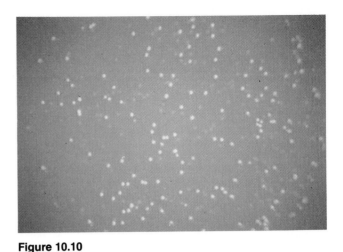

Figure 10.10
Fluorescence of chlorophyll. The dots on this round filter are colonies of *Rhodobacter capsulatus,* a photosynthetic bacterium. Two types of colonies are present: Some colonies consist of normal bacteria, while others are made up of genetically defective bacteria unable to perform the light reactions of photosynthesis. Under visible light, both types of colonies would look alike to the human eye. However, as viewed here, under a light source that enhances fluorescence, the defective colonies can be identified by their brighter glow. Unable to convert light energy to chemical energy, the mutant bacteria are emitting energy they have absorbed but cannot harness for photosynthesis. (Photograph courtesy of Douglas C. Youran, Massachusetts Institute of Technology.)

reaches the reaction center (Figure 10.11). The entire apparatus—the antenna complex along with its reaction-center chlorophyll *a* and the primary electron acceptor—is called a **photosystem.** Photosystems are the light-harvesting units of the thylakoid membrane.

Two types of photosystems have been found in the thylakoid membrane. They are referred to as **photosystem I** and **photosystem II,** in order of their discovery. At the reaction center of photosystem I is a specialized chlorophyll *a* molecule known as **P700,** so named because the light that the pigment absorbs best has a wavelength of 700 nm, which is in the far-red part of the spectrum. At the reaction center of photosystem II is a specialized chlorophyll *a* molecule designated **P680;** its absorption spectrum has a peak at 680 nm, also in the red part of the spectrum. The two pigments, P700 and P680, are actually identical chlorophyll *a* molecules, but they are associated with different proteins, which affects their distribution of electrons and accounts for the slight difference in absorption spectra. In fact, the two reaction-center chlorophylls are no different in structure from the many chlorophyll *a* molecules of the antenna complex. What makes P700 and P680 so special is their particular locations in the thylakoid membrane, where they are bound to specific proteins and are in close proximity to their respective primary electron acceptors. Photosyn-

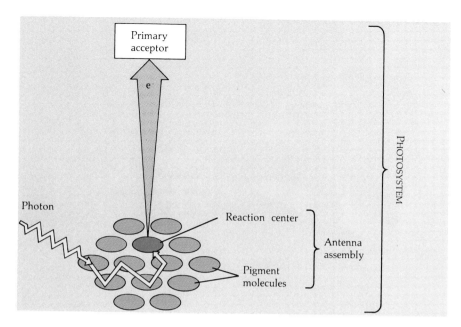

Figure 10.11

How a photosystem harvests light. Photosystems are the light-harvesting units of the thylakoid membrane. The entire apparatus—antenna assembly, reaction-center chlorophyll, and primary electron acceptor—makes up the photosystem. Each photosystem has an antenna of a few hundred pigment molecules, including various chlorophyll and carotenoid molecules. When a photon strikes a pigment molecule, the energy is passed from molecule to molecule until it reaches the reaction center, a specialized chlorophyll *a* molecule (actually, probably a pair of molecules) that is located next to the primary electron acceptor. (The relative distance between the reaction center and the primary acceptor has been exaggerated in this diagram.)

thesis, like any ordered function, is based on ordered structure.

Cyclic Electron Flow

During the light reactions, there are two possible routes for electron flow: cyclic flow and noncyclic flow. **Cyclic electron flow** is the simplest pathway; it involves only photosystem I and generates ATP, but produces no NADPH and evolves no O_2. It is called "cyclic" because excited electrons that depart from chlorophyll *a* at the reaction center eventually return. At some junctures in the round trip, electrons travel in pairs; therefore, we will visualize cyclic electron flow as starting with photosystem I absorbing two photons of light. A series of redox reactions returns the electrons to chlorophyll along an electron transport chain in the thylakoid membrane that is much like the electron transport chain of the mitochondrial membrane (Figure 10.12). With each redox reaction along the chain of electron carriers in the thylakoid membrane, electrons lose potential energy, finally returning to their ground-state orbital in P700. Absorption of another two photons of light by the antenna of pigments excites the reaction center again and sends two more electrons on their way around the circuit.

As the excited electrons give up energy on their way down to P700 along the transport chain, the thylakoid membrane couples the exergonic flow of the electrons to the endergonic reaction of phosphorylating ADP to make ATP. The coupling mechanism is chemiosmosis, as in mitochondria. Certain electron carriers can transport electrons only in the company of hydrogen ions, which are picked up from one side of the thylakoid membrane and then deposited on the opposite side as the electrons move along to the next member

of the transport chain. Electron flow stores energy in the form of a gradient of hydrogen ions across the thylakoid membrane—a proton-motive force. An ATP synthase enzyme very similar to the mitochondrial ATP synthase is built into the thylakoid membrane, and it harnesses the proton-motive force to make ATP. The process is very much like oxidative phosphorylation in the mitochondrion. In chloroplasts, the synthesis of ATP is called photophosphorylation because it is driven by light energy. Specifically, production of ATP during the cyclic pathway of electron flow is referred to as **cyclic photophosphorylation.**

Notice again that cyclic electron flow makes ATP without producing NADPH or releasing oxygen. The other pathway of electron flow in the light reactions generates both ATP and NADPH and also evolves oxygen by splitting water.

Noncyclic Electron Flow

Both photosystems cooperate in **noncyclic electron flow,** in which electrons pass continuously from water to $NADP^+$. As in cyclic electron flow, light excites two electrons from P700, the reaction-center chlorophyll of photosystem I. During noncyclic electron flow, however, the electrons do not return to the reaction-center chlorophyll, but instead are stored as high-energy electrons in NADPH. NADPH will later function as the electron donor when the Calvin cycle reduces carbon dioxide to sugar.

The oxidized chlorophyll itself now becomes a very potent oxidizing agent; its electron "holes" must be filled. This is where photosystem II comes in. It supplies electrons to P700, the reaction-center chlorophyll of photosystem I that has lost electrons to $NADP^+$.

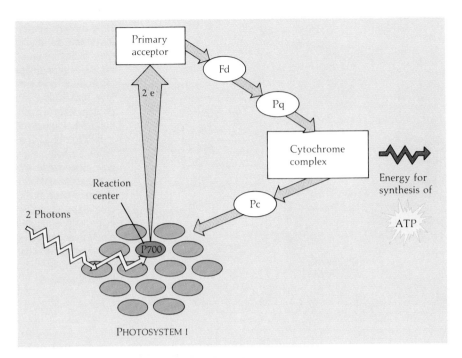

Figure 10.12
Cyclic electron flow. When pigments of the antenna assembly of photosystem I absorb light, the energy reaches P700, the specialized chlorophyll *a* at the reaction center. This boosts electrons of P700 to their excited-state orbital, from which the high-energy electrons are trapped by the primary electron acceptor. The primary acceptor passes electrons to an iron-containing protein named ferrodoxin (Fd), which then transfers the electrons to plastoquinone (Pq), a mobile electron carrier very similar to the ubiquinone of the electron transport chain in mitochondria. Electrons pass next to a complex consisting of two cytochromes. These iron-containing proteins are variations of the cytochromes of mitochondria. The cytochromes of the chloroplast donate their electrons to another protein, named plastocyanin (Pc), which contains copper. The cycle is completed when plastocyanin returns the electrons to P700, the reaction-center chlorophyll. At each step along the transport chain, electrons lose potential energy. Their trip, energetically downhill, is used by the transport chain to pump H^+ across the thylakoid membrane. The proton gradient in turn powers an enzyme that phosphorylates ADP.

When two photons of light are absorbed by pigment molecules of the antenna assembly of photosystem II, the energy reaches P680, the specialized chlorophyll *a* of the reaction center. A pair of electrons is ejected from P680 and trapped by the primary electron acceptor of photosystem II (Figure 10.13). The primary acceptor then transfers the electrons to the same transport chain that functions in cyclic electron flow. The electrons cascade down the chain, losing potential energy as they go, until they reach P700 and fill the vacancies left when photosystem I reduced $NADP^+$. As the electrons slide from photosystem II to photosystem I, the transport chain pumps H^+ across the thylakoid membrane. The proton-motive force can then drive ATP synthesis. When noncyclic electron flow generates ATP, the overall process is termed **noncyclic photophosphorylation.** Notice, however, that the actual mechanism of ATP synthesis is the same as in cyclic photophosphorylation.

So far, noncyclic electron flow has generated NADPH and ATP and has restored electrons to the reaction center of photosystem I. But now P680, the chlorophyll at the reaction center of photosystem II, has electron holes to fill. Oxidized P680 has such a great affinity for electrons that it can acquire them from water, one of the most difficult substances to oxidize. Electron removal splits the water into two hydrogen ions and an oxygen atom, which immediately combines with another oxygen atom to form O_2. This is the water-splitting step of photosynthesis that releases the O_2.

The net result of noncyclic electron flow is the pushing of electrons from water, where they are at a low state of potential energy, to NADPH, where the electrons are stored at a high state of potential energy. For each pair of electrons transferred by the light reactions, four photons of light must be absorbed, two by each photosystem. The two photosystems are connected in series. The electron current flows from water to P680, then to the top of the electron transport chain after being excited by light, on down the chain to P700, where electrons get a second boost from the absorption of two more photons, and finally to $NADP^+$.

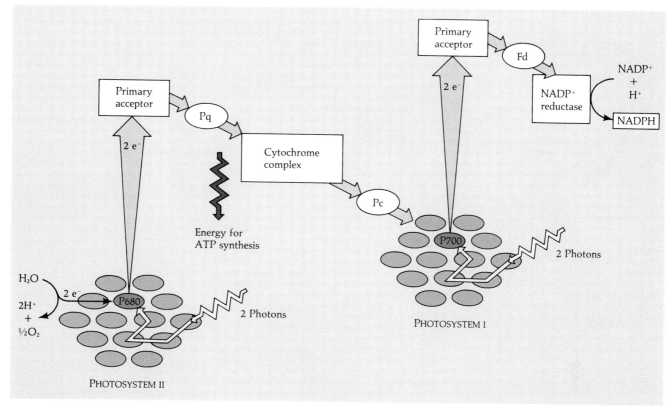

Figure 10.13
Noncyclic electron flow. When both photosystems are illuminated, there is a continuous current of electrons flowing from water to $NADP^+$. Electrons ejected from P680 are replaced by electrons removed from water, a redox process that evolves oxygen. The photoexcited electrons from P680 flow down an electron transport chain to P700, providing energy to generate ATP. Illumination of photosystem I boosts electrons to a high-energy state again, and these electrons are passed with the help of a reductase enzyme to $NADP^+$, reducing it to NADPH. The net products of noncyclic electron flow are ATP, NADPH, and O_2. The flow of electrons is indicated by the orange arrows.

The noncyclic version of the light reactions can be summarized with this equation:

$$\begin{array}{ll} NADP^+ & NADPH \\ ADP + \text{\textcircled{P}} \longrightarrow & ATP \\ H_2O & \frac{1}{2}O_2 + 1H^+ \end{array}$$

The ATP yield for every pair of electrons passed down the transport chain from photosystem II to photosystem I is still unresolved. Figure 10.13 and the preceding equation indicate a yield of one molecule of ATP per electron pair, but the actual number may be greater. For our purposes here, the precise amount of ATP made is not as important as the mechanism by which it is made.

The relationship of cyclic to noncyclic electron flow is evident when we compare Figures 10.12 and 10.13. Cyclic flow is a short circuit; when electrons ejected from P700 reach ferrodoxin, they are shunted back to the chlorophyll molecule rather than being diverted to $NADP^+$. Cyclic electron flow has been observed in chloroplasts in certain experimental conditions, but it is uncertain whether it is important in the light reactions of plants in nature. If it does occur, it would provide a way for chloroplasts to supplement the supply of ATP at times when there is no need for additional NADPH. For the most part, the light reactions take the form of noncyclic electron flow.

Comparison of Chemiosmosis in Chloroplasts and Mitochondria

Chloroplasts and mitochondria generate ATP by the same basic mechanism: chemiosmosis. An electron transport chain assembled in a membrane translocates protons across the membrane as electrons are passed through a series of carriers that are progressively more electronegative. Built into the same membrane is an ATP synthase complex that couples the diffusion of hydrogen ions down their gradient to the phosphorylation of ADP. Some of the electron carriers, including quinones and cytochromes, are very similar in

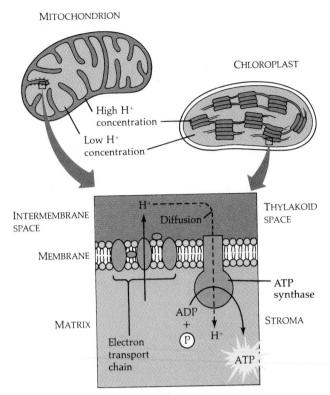

MITOCHONDRION

CHLOROPLAST

High H⁺ concentration

Low H⁺ concentration

INTERMEMBRANE SPACE

MEMBRANE

MATRIX

H⁺

Diffusion

THYLAKOID SPACE

ATP synthase

ADP + Ⓟ

H⁺

STROMA

Electron transport chain

ATP

Figure 10.14
The logistics of chemiosmosis in mitochondria and chloroplasts. The inner membrane of the mitochondrion translocates protons (H⁺) from the matrix into the inter-membrane space (darker brown). ATP is made on the matrix side of the membrane as hydrogen ions diffuse through ATP synthase complexes. In chloroplasts, the thylakoid membrane pumps protons from the stroma into the thylakoid compartment. As the hydrogen ions leak back across the membrane through the ATP synthase, phosophorylation of ADP occurs on the stroma side of the membrane.

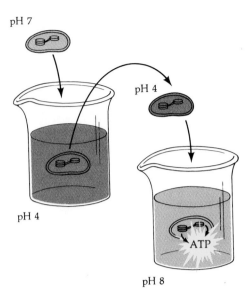

pH 7

pH 4

pH 4

pH 8

ATP

Figure 10.15
An artificially imposed pH gradient drives ATP synthesis. Chloroplasts are first made acidic by soaking them in a solution having a pH of 4. After the thylakoid compartment has reached a pH of 4, the chloroplasts are transferred to a basic solution with a pH of 8. The thylakoid membranes harness the artificially imposed pH gradient between the thylakoid compartment and the stroma to make ATP in the dark.

chloroplasts and mitochondria, and the ATP synthase complexes of the two organelles are also very much alike. But there are also noteworthy differences between oxidative phosphorylation in mitochondria and photophosphorylation in chloroplasts. In mitochondria, the high-energy electrons dropped down the transport chain are extracted by the oxidation of food molecules. Chloroplasts do not need food to make ATP. The photosystems capture light energy and use it to drive electrons to the top of the transport chain. Mitochondria transfer chemical energy from food molecules to ATP. Chloroplasts transform light energy into chemical energy. It is an important difference.

The spatial organization of chemiosmosis in chloroplasts and mitochondria also differs. The inner membrane of the mitochondrion pumps protons from the matrix out to the intermembrane space, which then serves as a reservoir of hydrogen ions that powers the ATP synthase. The thylakoid membrane of the chloroplast pumps protons from the stroma into the thylakoid compartment, which functions as the H⁺ reservoir (Figure 10.14). The membrane makes ATP as the hydrogen ions diffuse from the thylakoid compartment back to the stroma through ATP synthase complexes, whose catalytic heads are on the stroma side of the membrane. Thus, ATP forms in the stroma, where it is used to help drive sugar synthesis during the Calvin cycle.

The proton gradient, or pH gradient, across the thylakoid membrane is substantial. When chloroplasts are illuminated, the pH in the thylakoid compartment drops to about 5, and the pH in the stroma increases to about 8. This gradient of three pH units corresponds to a thousandfold difference in H⁺ concentration. When the lights are turned off, the pH gradient is abolished, but can quickly be restored by turning the lights back on. Such experiments add to the evidence described in Chapter 9 in support of the chemiosmotic model. Perhaps the most compelling evidence for chemiosmosis resulted from experiments performed in the 1960s by André Jagendorf and his co-workers. They induced thylakoids to make ATP in the dark by using artificial means to impose a pH gradient across the membrane (Figure 10.15). Such experiments have demonstrated that the function of the photosystems and the electron transport chain in photophosphorylation is to generate the pH gradient.

Considerably more research is required before the

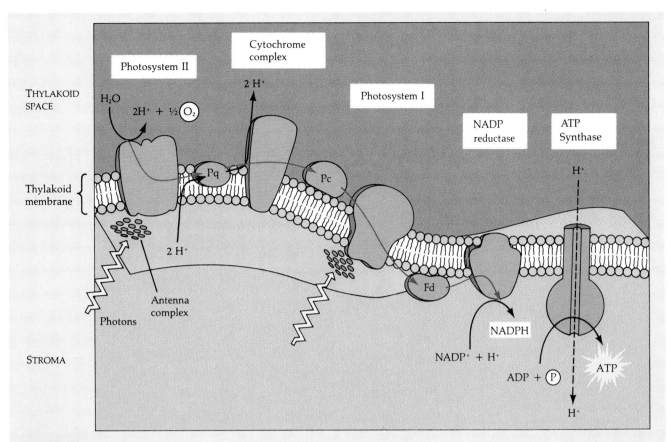

Figure 10.16
Tentative model for organization of the thylakoid membrane. Electron carriers are arranged in such a way that electrons pass from one side of the membrane to the other. As electrons crisscross the membrane, hydrogen ions removed from the stroma are deposited in the thylakoid compartment. There are at least three steps in the light reactions that contribute to the proton gradient: (1) Water is split by photosystem II on the side of the membrane facing the thylakoid compartment; (2) as plastoquinone (Pq), a mobile carrier, transfers electrons to the cytochrome complex, protons are translocated across the membrane; and (3) a hydrogen ion in the stroma is taken up by $NADP^+$ when it is reduced to NADPH. The diffusion of H^+ from the thylakoid space to the stroma (along the H^+ concentration gradient) powers the ATP synthase. This ATP-synthesizing complex of proteins is very similar to the ATP synthase of the mitochondrial membrane. The portion of the chloroplast ATP synthase that spans the membrane and provides a channel for H^+ flow is termed CF_0 (for "coupling factor"). The "head" of the complex, called CF_1, has the active site for ATP synthesis.

precise organization of the thylakoid membrane can be discerned. Figure 10.16 depicts a tentative model based on studies in several laboratories. Notice in the figure that proton pumping by the thylakoid membrane depends on an asymmetric placement of electron carriers that accept and release H^+. Note also that NADPH, like ATP, is produced on the side of the membrane facing the stroma, where sugar is synthesized by the Calvin cycle.

HOW THE CALVIN CYCLE MAKES SUGAR

The Calvin cycle uses the ATP and NADPH generated by the light reactions to reduce carbon dioxide to sugar. The carbon cycle is reminiscent of the Krebs cycle in that a starting material is regenerated after materials enter and leave the cycle. Carbon enters the Calvin cycle in the form of CO_2 and leaves in the form of sugar. The cycle spends ATP as an energy source and consumes NADPH as reducing power for adding high-energy electrons to make the sugar.

The carbohydrate produced directly from the Calvin cycle is actually not glucose, but a three-carbon sugar named **glyceraldehyde 3-phosphate,** referred to in this text as glyceraldehyde phosphate. For the net synthesis of one molecule of this sugar, the cycle must take place three times, fixing three molecules of CO_2 ("carbon fixation," remember, refers to the initial incorporation of CO_2 into organic material). Keep in mind, as we trace the steps of the cycle, that we are following three molecules of CO_2 through the reactions (Figure 10.17).

The Calvin cycle begins when each molecule of CO_2

Figure 10.17

The Calvin cycle. For every three molecules of CO_2 that enter the cycle, the net output is one molecule of glyceraldehyde phosphate, a three-carbon sugar. To fix the three CO_2 molecules, the cycle spends nine molecules of ATP and six molecules of NADPH. The intermediates involved in the conversion of glyceraldehyde phosphate to ribulose bisphosphate are not shown in this simplified version of the cycle.

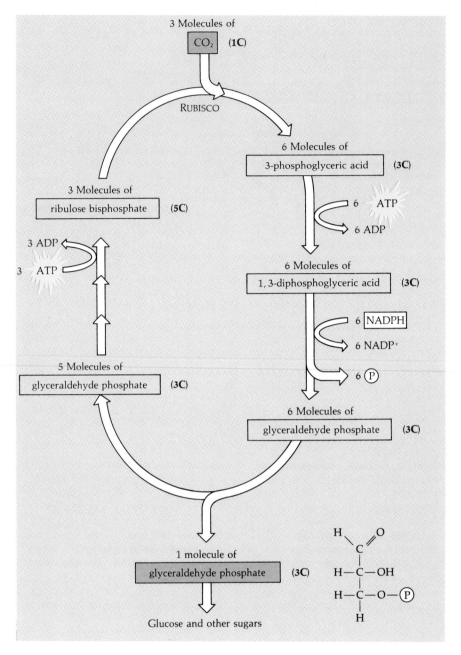

is added to a molecule of ribulose bisphosphate (abbreviated *RuBP*), a five-carbon sugar with a phosphate group on each end. The enzyme that catalyzes this first step is **RuBP carboxylase (rubisco,** for short), the most abundant protein in chloroplasts and perhaps one of the most plentiful proteins on Earth. The product of the reaction is a six-carbon intermediate that is so unstable it immediately splits in half to form two molecules of 3-phosphoglyceric acid. If you refer back to Figure 9.8, you will see that this three-carbon acid is also formed during glycolysis. For every three molecules of CO_2 that enter the Calvin cycle via rubisco, three molecules of RuBP are carboxylated and a total of six molecules of 3-phosphoglyceric acid are made.

In the next step of the cycle, each molecule of 3-phosphoglyceric acid receives an additional phosphate group. An enzyme transfers the phosphate group with its high-energy bond from ATP, forming 1,3-diphosphoglyceric acid as a product. For every three molecules of CO_2 incorporated into the cycle, six molecules of ATP must be used to produce six molecules of 1,3-diphosphoglyceric acid, which are primed by their unstable phosphate bonds for the next step of the cycle. A pair of high-energy electrons donated from NADPH reduces 1,3-diphosphoglyceric acid to glyceraldehyde phosphate. (Helping to drive this redox reaction is the hydrolysis of the phosphate bond transferred to the substrate from ATP in the previous reaction.) Specifically, the electrons from NADPH reduce the carboxyl group of 3-phosphoglyceric acid to the carbonyl group of glyceraldehyde phosphate, which stores more potential energy. Glyceraldehyde phosphate is a sugar—the same three-carbon sugar formed in glycolysis by the splitting of glucose.

We started the Calvin cycle with three molecules of CO_2, and now we have six molecules of glyceraldehyde phosphate. But only one molecule of this three-carbon sugar can be counted as a net gain of carbohydrate. The cycle began with 15 carbons' worth of carbohydrate in the form of three molecules of the five-carbon sugar RuBP. We now have 18 carbons' worth of carbohydrate in the form of six molecules of glyceraldehyde phosphate. One molecule can exit the cycle to be used by the plant cell, but the other five molecules must be recycled to regenerate the three molecules of RuBP. In a complex series of reactions, the carbon skeletons of five molecules of glyceraldehyde phosphate are rearranged by the last steps of the Calvin cycle into three molecules of RuBP. This requires three more molecules of ATP.

For the net synthesis of one glyceraldehyde phosphate molecule by the Calvin cycle, a total of nine molecules of ATP and six molecules of NADPH are consumed. The light reactions regenerate the ATP and NADPH. The glyceraldehyde phosphate spun off from the Calvin cycle becomes the starting material for a variety of metabolic pathways. Although this three-carbon sugar is the carbohydrate directly produced by photosynthesis, two molecules of glyceraldehyde phosphate can be rapidly converted to a molecule of glucose, the six-carbon sugar usually thought of as the photosynthetic product. To make one molecule of glucose, then, the Calvin cycle uses 18 ATPs and 12 molecules of NADPH. The resulting ADP and $NADP^+$ return to the light reactions, where they are again phosphorylated and reduced, respectively. Neither the light reactions nor the Calvin cycle alone can make sugar from CO_2. Photosynthesis is an emergent property of the intact chloroplast that integrates the two stages of photosynthesis.

PHOTORESPIRATION

When the air spaces in a leaf have a much higher concentration of O_2 than CO_2, the active site of rubisco can accept O_2 in place of CO_2. That is, O_2 is a competitive inhibitor of rubisco. The enzyme adds the O_2 to RuBP, and the five-carbon product splits into a three-carbon molecule that remains in the Calvin cycle and a two-carbon compound that leaves the cycle. In fact, the two-carbon compound, named glycolic acid, leaves the chloroplast altogether and enters a peroxisome (see Chapter 7). A metabolic pathway that begins in peroxisomes and is completed in mitochondria breaks

glycolic acid down and releases CO_2. The entire process is termed **photorespiration** because it consumes oxygen, evolves carbon dioxide, and generally occurs only in the light. Unlike the cellular respiration you learned about in Chapter 9, photorespiration generates no ATP. Furthermore, it decreases photosynthetic output by siphoning organic material from the Calvin cycle. According to one hypothesis, photorespiration is evolutionary baggage—a metabolic relic from much earlier times when the atmosphere had less O_2 and more CO_2 than it does today. In that ancient atmosphere, when rubisco first evolved, the inability of the enzyme's active site to distinguish CO_2 from O_2 would have made little difference. This hypothesis goes on to speculate that modern rubisco retains some of its ancestral affinity for O_2, which is so concentrated in the present atmosphere that a certain amount of photorespiration is inevitable.

It is not known if photorespiration is beneficial to plants in any way. It is known that in many types of plants, including some of agricultural importance, such as soybeans, photorespiration drains away as much as 50% of the carbon fixed by the Calvin cycle. As heterotrophs that depend on carbon fixation in chloroplasts for our food, we naturally view photorespiration as wasteful. Indeed, if photorespiration could be reduced in certain plant species without otherwise affecting photosynthetic productivity, crop yields and food supplies would increase.

The environmental conditions that foster photorespiration are hot, dry, bright days. On such days, plants close their stomata, the pores through the leaf surface (see Figure 10.2). This is an adaptation that prevents dehydration by slowing water loss from the leaf. But photosynthesis soon depletes CO_2 and increases O_2 within air spaces of the leaf, and photorespiration commences as rubisco accepts O_2. In certain species, alternate modes of carbon fixation that minimize photorespiration, even in hot, arid climates, have evolved. The two most important of these photosynthetic adaptations are C_4 photosynthesis and CAM.

C_4 PLANTS

Most plants use the Calvin cycle for the initial steps that incorporate CO_2 into organic material. They are called **C_3 plants** because the first stable intermediate formed by carbon fixation is 3-phosphoglyceric acid, a three-carbon compound. Many plant species, however, preface the Calvin cycle with reactions that incorporate CO_2 first into four-carbon compounds; these species are known as **C_4 plants**. Several thousand species in at least 17 plant families use the C_4 pathway. Among the C_4 plants important to agriculture are sugarcane and corn, members of the grass family. The

Figure 10.18

Comparison of C_4 and C_3 leaf anatomy.
The cells where the Calvin cycle occurs are shown in dark green in both drawings. In the leaves of C_3 plants, these are the mesophyll cells of various shapes. In the leaves of C_4 plants, the Calvin cycle occurs in the bundle-sheath cells, which form tight wreaths around the veins. In C_4 plants, preliminary steps that incorporate CO_2 into four-carbon organic compounds take place in the mesophyll cells (light green). Thus, C_4 plants have two types of photosynthetic cells, which differ both structurally and functionally.

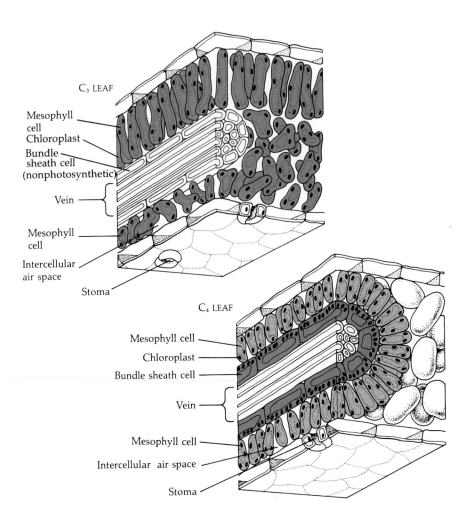

C_3 LEAF

Mesophyll cell
Chloroplast
Bundle sheath cell (nonphotosynthetic)
Vein
Mesophyll cell
Intercellular air space
Stoma

C_4 LEAF

Mesophyll cell
Chloroplast
Bundle sheath cell
Vein
Mesophyll cell
Intercellular air space
Stoma

diagrams of leaf anatomy in Figure 10.18 illustrate how C_4 photosynthesis is spatially ordered in a manner that enhances CO_2 fixation under conditions that cause C_3 plants to lose organic material through photorespiration.

In C_4 plants, there are two distinct types of photosynthetic cells: **bundle-sheath cells** and **mesophyll cells.** Bundle-sheath cells are so named because they are arranged into tightly packed sheaths around the veins of the leaf. Between the bundle sheath and the leaf surface is the more loosely arranged mesophyll. Thylakoids in the chloroplasts of bundle-sheath cells are generally not stacked into grana; the significance of this difference from typical chloroplasts is unknown.

The Calvin cycle is confined to the chloroplasts of the bundle sheath. However, the cycle is preceded by incorporation of CO_2 into organic compounds in the mesophyll (Figure 10.19) The first step is the addition of CO_2 to phosphoenolpyruvic acid (PEP) to form the four-carbon product oxaloacetic acid. (Notice once again that many organic molecules do double duty as intermediates of both photosynthesis and cellular respiration. We encountered both PEP and oxaloacetic acid in Chapter 9.) The enzyme that adds CO_2 to PEP, called **PEP carboxylase,** has a much greater affinity for CO_2 than does RuBP carboxylase (rubisco), the enzyme that incorporates CO_2 into the Calvin cycle. Also, the active site of PEP carboxylase has no affinity for O_2. Therefore, PEP carboxylase can fix CO_2 efficiently when rubisco cannot—when it is hot and dry and CO_2 concentrations in the leaf have fallen and O_2 concentrations have risen because stomata are partially closed. After mesophyll in the C_4 plant fixes CO_2, the cells convert oxaloacetic acid to another four-carbon compound, usually malic acid. The mesophyll cells then export the malic acid to bundle-sheath cells through plasmodesmata (see Chapter 7, Figure 7.36). Within the bundle-sheath cells, malic acid releases CO_2, which is reassimilated into organic material by rubisco and the Calvin cycle. In effect, the mesophyll cells pump CO_2 into the bundle sheath, preventing photorespiration and enhancing sugar production by keeping the CO_2 concentration high enough for rubisco to accept carbon dioxide rather than oxygen. This is especially advantageous in hot regions with intense sunlight, where we are most likely to find C_4 plants. The spatial ordering of photosynthesis in a C_4 plant is an adaptation that clearly illustrates the correlation between structure and function.

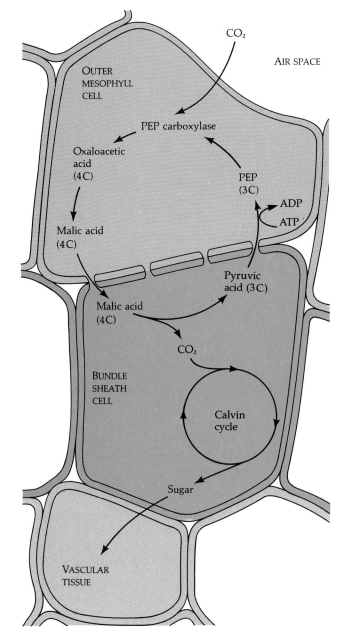

Figure 10.19
The C₄ pathway. Carbon dioxide is initially fixed in mesophyll cells by the enzyme PEP carboxylase. A four-carbon compound, usually malic acid, conveys the CO_2 to bundle-sheath cells, where the CO_2 is transferred to the Calvin cycle. Bundle-sheath cells send the sugar they make into vascular bundles (veins), which transport the sugar throughout the plant. The mesophyll of the C₄ leaf functions as a CO_2 pump that maintains a CO_2 concentration in the bundle sheath high enough for sugar synthesis to continue even on hot, dry days, when C₃ plants are likely to lose organic material via photorespiration.

CAM PLANTS

Still another modification of carbon fixation is found in succulent (water-storing) plants, including "ice plants" and many cacti, which are adapted to very dry climates and live mostly in deserts. These plants open their stomata during the night and close them during the day, just the reverse of how other plants behave. Closing stomata during the day helps desert plants conserve water, but it also prevents CO_2 from entering the leaves. During the night, when their stomata are open, these plants take up CO_2 and incorporate it into a variety of organic acids. This mode of carbon fixation is called crassulacean acid metabolism, or **CAM**, after the plant family Crassulaceae, in which the process was first discovered. The mesophyll cells store the organic acids they make during the night in their vacuoles until morning, when the stomata close. During the daytime, when the light reactions can supply ATP and NADPH for the Calvin cycle, CO_2 is released from the organic acids made the night before to become incorporated into sugar in the chloroplasts. The CAM pathway is similar to the C₄ pathway in that carbon dioxide is first incorporated into organic intermediates before it enters the Calvin cycle. The difference is that in C₄ plants, the initial steps of carbon fixation are separated structurally from the Calvin cycle, whereas in CAM plants the two steps occur at separate times. Keep in mind that CAM, C₄, and C₃ plants all eventually use the Calvin cycle to make sugar from carbon dioxide.

THE FATE OF PHOTOSYNTHETIC PRODUCTS

In this chapter, we have followed photosynthesis from photons to food. The light reactions capture solar energy and use it to make ATP and to transfer electrons from water to $NADP^+$. The Calvin cycle uses the ATP and NADPH to produce sugar from carbon dioxide (Figure 10.20). The energy that entered the chloroplasts as sunlight becomes stored as chemical energy in the bonds of sugar molecules.

The sugar made in the chloroplasts supplies the entire plant with chemical energy and carbon skeletons to synthesize all the major organic molecules of cells. About 50% of the organic material made by photosynthesis, however, is consumed as fuel for cellular respiration in the mitochondria of the plant cells. Sometimes there is a loss of photosynthetic products to photorespiration.

Technically, green cells are the only autotrophic parts of the plant. The rest of the plant depends on organic molecules exported from leaves in vascular bundles.

Figure 10.20
Summary of photosynthesis. This diagram outlines the main reactants and products of photosynthesis as it occurs in the chloroplasts of plant cells. The light reactions convert light energy into the chemical energy of ATP and NADPH. The pigment and protein molecules that carry out the light reactions are found in the thylakoid membranes and include the molecules of two photosystems and an electron transport chain. The light reactions involve the splitting of H_2O with the release of O_2, the source of oxygen in the Earth's atmosphere. The Calvin cycle, which takes place in the stroma of the chloroplast, uses ATP and NADPH to convert CO_2 to carbohydrate (three key intermediates are shown). The by-products ADP, inorganic phosphate, and $NADP^+$ are returned from the Calvin cycle to the light reactions. The entire ordered operation depends on the structural integrity of the chloroplast and its membranes.

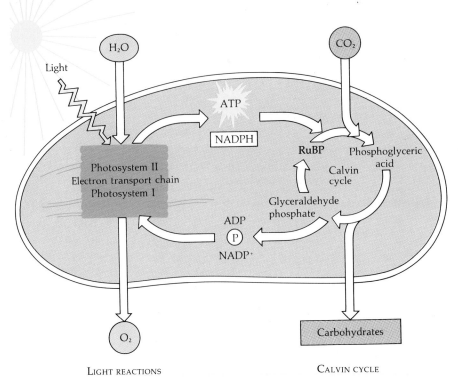

In most plants, carbohydrate is transported out of the leaves in the form of sucrose, a disaccharide. After arriving at nonphotosynthetic cells, the sucrose provides raw material for cellular respiration and a multitude of anabolic pathways. A considerable amount of sugar in the form of glucose is linked together to make the polysaccharide cellulose, especially in plant cells that are still growing and maturing. Cellulose is the most abundant organic molecule in the plant—and probably on the surface of the planet.

Over the 24 hours in a day, most plants manage to make more organic material than they need to use as respiratory fuel and precursors for biosynthesis. They stockpile the extra sugar by synthesizing starch, stor-ing some in the chloroplasts themselves, and some in storage cells of roots, tubers, and fruits. And in accounting for the consumption of the food molecules produced by photosynthesis, let us not forget that most plants lose leaves, roots, stems, fruits, and sometimes their entire bodies to heterotrophs, including humans.

On a global scale, the collective productivity of the minute chloroplasts is prodigious; it is estimated that photosynthesis makes about 160 billion metric tons of carbohydrate per year (a metric ton is 1000 kg—about 1.1 tons). No other chemical process on Earth can match this output. And no process is more important than photosynthesis to the welfare of life on this planet.

STUDY OUTLINE

1. Autotrophs are organisms capable of sustaining themselves without ingesting organic molecules. Photoautotrophs are the producers in ecological systems, using the energy of sunlight to synthesize organic molecules from CO_2 and H_2O. Some bacteria are chemoautotrophs, using inorganic substances rather than sunlight as the source of energy for the formation of their organic molecules.

2. Heterotrophs must ingest other organisms or their by-products to obtain energy and carbon skeletons. They are completely dependent on photosynthesizers to produce food and oxygen to drive aerobic respiration in their mitochondria.

Chloroplasts: Sites of Photosynthesis (p. 206)

1. In autotrophic eukaryotes, photosynthesis occurs inside chloroplasts, organelles enclosing an elaborate system of thylakoid membranes that are layered in places in stack-like grana and separate an outer stroma from an inner thylakoid space.

2. All chloroplasts contain the green pigment chlorophyll, which resides in the thylakoid membranes and absorbs the light energy that initiates photosynthesis.

3. The chloroplasts in plants are especially dense in the cells of the mesophyll, a green tissue in the interior of the leaf, which obtains its CO_2 from and exports its O_2 to the environment through stomata.

4. The conversion of CO_2 to sugar occurs in the stroma. Vascular bundles transport the sugar to other parts of the plant and provide chloroplasts in the leaves with water from the roots.

5. Photosynthetic prokaryotes lack chloroplasts, but their chlorophyll is built into the plasma membrane or vesicle membranes inside the cell.

How Plants Make Food: An Overview (pp. 206–209)

1. The complex process of photosynthesis can be summarized by the following equation:

$$6\,CO_2 + 12\,H_2O + \frac{Light}{energy} \longrightarrow C_6H_{12}O_6 + 6\,O_2 + 6\,H_2O$$

2. The chloroplast splits water into hydrogen and oxygen, incorporating the electrons of hydrogen into the energy-rich bonds of sugar molecules. Photosynthesis is thus an endergonic redox process in which water is oxidized and carbon dioxide is reduced.

3. There are two, linked stages of photosynthesis, the light reactions and the Calvin cycle. The light reactions in the grana produce ATP by photophosphorylation and split water, evolving oxygen and forming NADPH by transferring electrons from water to $NADP^+$.

4. The Calvin cycle occurs in the stroma and uses ATP for energy and NADPH for reducing power to form sugar from CO_2. Although the Calvin cycle, sometimes called the dark reactions, does not require light directly, it usually occurs during the day, when the light reactions are providing ATP and NADPH.

How the Light Reactions Capture Solar Energy (pp. 210–219)

1. Sunlight is a form of electromagnetic energy that travels in waves. The range of wavelengths of this radiation constitutes the electromagnetic spectrum, part of which is detected by us as the colors of visible light.

2. The wavelengths of light are emitted in discrete energy packets called photons, each with a fixed quantity of energy inversely proportional to its wavelength. The two wavelengths most effective in driving photosynthesis are those perceived by the human eye as red and blue.

3. A pigment is a substance that absorbs specific wavelengths of light, determined by using a spectrophotometer to produce an absorption spectrum for that pigment. The action spectrum of photosynthesis and the absorption spectrum of chlorophyll a, however, do not directly correspond. Accessory pigments, chlorophyll b and various carotenoids, have molecular structures that enable them to absorb different wavelengths of light and pass their energy on to chlorophyll a.

4. A pigment goes from a ground state to an excited state when a photon boosts one of its electrons to a higher energy orbital. In isolated pigments, the electron immediately returns to the ground state, releasing the energy as light (fluorescence) and/or heat.

5. The pigments of chloroplasts are built into the thylakoid membrane near molecules known as primary electron acceptors, which trap the high-energy electrons before they return to the ground state in a light-driven redox reaction. Energy stored in these electrons powers the synthesis of ATP and NADPH.

6. Accessory pigments in chloroplasts are clustered in an antenna complex of a few hundred molecules surrounding a molecule of chlorophyll a at the reaction center. Photons absorbed anywhere in the antenna can pass this energy along to energize this chlorophyll a, which then passes its electrons to a nearby primary electron acceptor. The antenna complex, the reaction-center chlorophyll, and the primary electron acceptor make up one of many photosystems, light-harvesting units built into the thylakoid membrane.

7. There are two kinds of photosystems. Photosystem I contains P700, and photosystem II P680, chlorophyll a molecules at the reaction center that have different absorption characteristics due to specific properties of associated proteins.

8. The flow of energized electrons in the light reactions can be cyclic or noncyclic. Cyclic electron flow starts when photosystem I absorbs two photons of light, causing P700 to donate two electrons to an electron transport chain located on the thylakoid membrane. A series of redox reactions returns two electrons to the ground state in P700, generating a proton-motive force across the thylakoid membrane. The passage of hydrogen ions through an ATP synthase enzyme drives the chemiosmotic formation of ATP from ADP.

9. Noncyclic electron flow, the more prevalent pathway in nature, involves both photosystems and produces NADPH and oxygen in addition to ATP. Photons of light boost two electrons from each photosystem to higher energy levels. Electrons from P700 in photosystem I are trapped by $NADP^+$, which stores them in the form of NADPH. Electrons from P680 generate ATP by noncyclic photophosphorylation when they pass from the primary electron acceptor in photosystem II to the same transport chain of photosystem I that functions in cyclic electron flow. Electrons at the end of the chain are accepted by P700, and the electron "holes" in P680 are filled by electrons from water, which is split into hydrogen ions and oxygen. The overall process powers the endergonic transfer of electrons from water to $NADP^+$ and the endergonic phosphorylation of ADP by chemiosmosis.

10. Both chloroplasts and mitochondria use similar ATP synthases to generate ATP by chemiosmotic proton gradients. These gradients are driven by a redox transfer of electrons down a sequence of increasingly electronegative components in a transport chain of a specialized membrane. In the chloroplasts, the thylakoid membrane pumps protons from the stroma into the thylakoid compartment. In the mitochondrion, the cristae pump protons from the matrix to the intermembrane space. The major difference is that light, rather than food energy, powers the process in the chloroplast.

How the Calvin Cycle Makes Sugar (pp. 219–221)

1. The Calvin cycle is organized in a cyclic metabolic pathway in the stroma that combines carbon dioxide with ribulose bisphosphate (RuBP), a five-carbon sugar. Then, using electrons from NADPH and energy from the hydrolysis of ATP, the cycle synthesizes the three-carbon sugar glyceraldehyde phosphate in a series of reactions. Most of the glyceraldehyde phosphate is reused in the

cycle as an intermediate for reconversion to RuBP, but some can exit the cycle and be converted to glucose or other essential organic molecules.

Photorespiration (p. 221)

1. On dry, hot days, plants close their stomata to conserve water, and oxygen from the light reactions builds up. Oxygen is a competitive inhibitor of rubisco, the enzyme that incorporates CO_2 into RuBP. When O_2 substitutes for CO_2 in the active site of rubisco, an intermediate is formed that leaves the cycle to be oxidized to CO_2 and H_2O in the mitochondria. This process, called photorespiration, consumes oxygen, evolves carbon dioxide, produces no ATP, and decreases photosynthetic output.

C_4 Plants (pp. 221–223)

1. Most plants are C_3 plants, named after the number of carbons in the first stable intermediate of the Calvin cycle.

2. C_4 plants are adapted to dry conditions and avert photorespiration by prefacing the Calvin cycle with a series of reactions that incorporate carbon dioxide into four-carbon compounds in specialized mesophyll cells. The final compound, malic acid, is then exported to photosynthetic bundle-sheath cells, where carbon dioxide is released for use in the Calvin cycle.

CAM Plants (p. 223)

1. Some plants use CAM metabolism as an adaptation for carbon fixation in a hot and dry environment. These plants open stomata during the night and incorporate the carbon dioxide that enters into a variety of organic acids that are stored in the vacuoles of mesophyll cells. During the day, the stomata close completely, which conserves water, and the carbon dioxide is released from the organic acids for use in the Calvin cycle while light reactions are supplying ATP and NADPH.

The Fate of Photosynthetic Products (pp. 223–224)

1. Vascular bundles export carbohydrates made in green cells to nonphotosynthetic parts of the plant. Mitochondria degrade about one-half of the carbohydrate made by photosynthesis to form ATP. Much of the remaining carbohydrate is converted to a variety of molecules, including cellulose.

2. Excess organic material is stockpiled as starch and stored in the leaves, roots, tubers, and fruit. Heterotrophs consume much of this organic material.

SELF-QUIZ

1. Chloroplasts can make sugar in the dark if they are provided with a continual supply of
   a. oxygen
   b. presplit water
   c. ATP and NADPH
   d. chlorophyll
   e. ADP and NADP$^+$

2. Which sequence correctly portrays the flow of electrons during photosynthesis?
   a. NADPH $\longrightarrow$ O_2 $\longrightarrow$ CO_2
   b. H_2O $\longrightarrow$ NADPH $\longrightarrow$ Calvin cycle

   c. NADPH $\longrightarrow$ chlorophyll $\longrightarrow$ Calvin cycle
   d. H_2O $\longrightarrow$ photosystem I $\longrightarrow$ photosystem II
   e. NADPH $\longrightarrow$ electron transport chain $\longrightarrow$ O_2

3. Which of the following conclusions does *not* follow from study of the absorption spectrum for chlorophyll *a* and the action spectrum for photosynthesis?
   a. All wavelengths are not equally as effective for photosynthesis.
   b. There must be accessory pigments that broaden the spectrum of light that contributes energy for photosynthesis.
   c. The red and blue areas of the spectrum are most effective in driving photosynthesis.
   d. Chlorophyll owes its color to the absorption of green light.
   e. Chlorophyll *a* has two absorption peaks.

4. Cooperation of the *two* photosystems of the chloroplast is required for
   a. ATP synthesis
   b. reduction of NADP$^+$
   c. cyclic photophosphorylation
   d. photooxidation of the reaction center of photosystem I
   e. generation of a proton-motive force

5. In *mechanism*, photophosphorylation is most similar to
   a. substrate-level phosphorylation
   b. oxidative phosphorylation
   c. the Calvin cycle
   d. carbon fixation
   e. glycolysis

6. In what respect are the photosynthetic adaptations of C_4 plants and CAM plants similar?
   a. In both cases, the stomata normally close during the day.
   b. Both types of plants make their sugar without the Calvin cycle.
   c. In both cases, an enzyme other than rubisco carries out the first step in carbon fixation.
   d. Both types of plants make most of their sugar in the dark.
   e. Neither C_4 plants nor CAM plants have grana in their chloroplasts.

7. Light-driven electron transport in the chloroplast pumps H$^+$
   a. into the stroma
   b. into the intermembrane space between the two layers of the chloroplast envelope
   c. into the thylakoid space
   d. from NADPH to CO_2
   e. out of the chloroplast

8. The stage of photosynthesis that actually produces sugar is

    a. the Calvin cycle

    b. photosystem I

    c. photosystem II

    d. the light reactions

    e. splitting of water

9. On the average, for every CO_2 molecule fixed by photosynthesis, how many molecules of O_2 are released?

    a. 1

    b. 2

    c. 3

    d. 6

    e. 12

10. Which of the following statements is a correct distinction between autotrophs and heterotrophs?

    a. Only heterotrophs need to acquire chemical compounds from the environment.

    b. Cellular respiration is unique to heterotrophs.

    c. Only heterotrophs have mitochondria.

    d. Autotrophs, but not heterotrophs, can nourish themselves beginning with nutrients that are entirely inorganic.

    e. Only heterotrophs require oxygen for their metabolism.

CHALLENGE QUESTIONS

1. Explain how the photosynthetic adaptations of CAM plants and C_4 plants improve photosynthetic yield in hot, arid environments.

2. Compare and contrast the mechanisms of photosynthesis and cellular respiration. How has the chemiosmotic model helped to unify current ideas on the mechanisms of these two processes?

3. The photosynthetic rate of aquatic plants in a test tube can be determined by collecting and measuring the amount of oxygen that gases out of the water. If bicarbonate, the source of CO_2 for aquatic plants, is added to the water, the rate of oxygen evolution increases. If CO_2 is fixed by the Calvin cycle, but oxygen is evolved by the light reactions, then how can an increase in CO_2 supply increase the rate of oxygen evolution?

FURTHER READING

1. Becker, W. M. *The World of the Cell*, Menlo Park, Calif.: Benjamin/Cummings, 1986, Chap. 9.
2. Kemp, P. R., G. L. Cunningham, and H. P. Adams. "Specialization of Mesophyll Structure in C_4 Grasses." *BioScience*, July/August 1983. A good example of correlation between biological structure and function.
3. Mathews, C. and van Holde, K. *Biochemistry*, Menlo Park, Calif.: Benjamin/Cummings, 1989.
4. Percy, R. W., O. Bjorkman, M. M. Caldwell, J. E. Keeley, R. K. Monson, and B. R. Strain. "Carbon Gain by Plants in Natural Environments." *Bioscience*, January 1987. Ecological aspects of photosyntahesis.
5. Prince, R. C. "Redox-driven Proton Gradients." *BioScience*, January 1985. How membranes transform energy.
6. Salisbury, F. B. and C. W. Ross. *Plant Physiology*, 3d ed. Belmont, Calif.: Wadsworth, 1985. The classic upper-division text on plant physiology.
7. Youvan, D. C., and B. L. Marrs. "Molecular Mechanisms of Photosynthesis." *Scientific American*, June 1987.

11

Reproduction of Cells

L ife arises only from life. The ability of organisms to reproduce their kind is the one phenomenon that best distinguishes life from inanimateness. (Though the analogy of making photocopies is frequently mentioned, the reproduction of an organism is actually more analogous to a photocopier making more photocopiers!) This unique capacity to procreate, like all biological functions, has a cellular basis. Rudolf Virchow, a German physician, put it this way in 1855: "Where a cell exists, there must have been a preexisting cell, just as the animal arises only from an animal and the plant only from a plant." He summarized with the axiom, *"Omnis cellula e cellula,"* which means, "All cells from cells." The perpetuation of life is based on the reproduction of cells, or **cell division,** the subject of this chapter.

In some cases, division of one cell to form two reproduces an entire organism, as when a unicellular species, such as *Amoeba,* divides to form duplicate offspring (Figure 11.1a). But cell division also enables a multicellular organism, such as a human, to grow and develop from a single cell—the fertilized egg (Figure 11.1b). Even after the organism is fully grown, cell division continues to function in renewal and repair, replacing cells that die from normal wear and tear or accidents. For example, millions of cells lining your stomach and intestine are abraded away and destroyed every time you digest a meal, but cell division constantly regenerates the lining of the digestive tract.

Reproduction of an ensemble as complex as a cell cannot occur by a mere pinching in half; the cell is not like a soap bubble that simply grows and splits in two. The cell and all its processes, including division, are

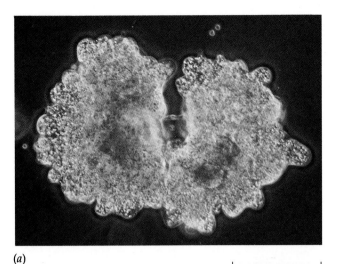

(a)

100 μm

(b)

100 μm

Figure 11.1
Cell division. (a) *Amoeba*, a one-celled eukaryote,
divides to form two cells, each an individual organism
(LM). **(b)** For multicellular organisms, division of embryonic cells is important in growth and development. This
dark-field micrograph shows a sand dollar embryo
shortly after the fertilized egg divided to form two cells.

controlled by DNA, the genetic material. What is most
remarkable about cell division is the fidelity with which
genetic programs are passed along, without dilution,
from one generation of cells to the next. A cell preparing to divide first copies all its genes, allocates them
equally to opposite ends of the cell, and then separates into two daughter cells.*

The objective of this chapter, then, is to help you
learn how cells reproduce to form genetically equiv-

_____
*Although the terms *daughter cells* and *sister chromatids* are traditional and will be used throughout this book, these terms do not
confer a gender on these structures.

alent daughter cells. The focus on *physical* processes
that are visible in the light microscope will seem like
a departure from the preceding three chapters, which
considered submicroscopic, metabolic processes.
However, correlation of structure and function, a
recurrent theme throughout our study of cells, will
prove useful once again as we investigate the mechanism of cell division. Although the emphasis in this
chapter will be on the division of eukaryotic cells, we
will begin with a brief look at prokaryotic cell division.

BACTERIAL REPRODUCTION

Prokaryotes (bacteria and cyanobacteria) reproduce by
a type of cell division called **binary fission,** meaning
literally "to divide in half." Most bacterial genes are
carried on a single chromosome that occurs as a circular DNA molecule. (As we will see, the chromosomes of eukaryotes have a very different organization.) Although bacteria are smaller and simpler than
eukaryotic cells, the problem of replicating their genetic
material in an orderly fashion and distributing the
copies equally to two daughter cells is still formidable.
Consider, for example, the chromosome of the bacterium *Escherichia coli;* when fully stretched out, this
chromosome is about 500 times longer than the length
of the cell. Clearly, such a chromosome must be highly
folded within the cell.

After a bacterial cell copies its chromosome in preparation for fission, the problem of allocating the duplicate chromosomes to daughter cells is solved by
attaching each copy of the chromosome to the plasma
membrane. Then, growth of the membrane between
the two attachment sites separates the two copies of
the chromosome (Figure 11.2). When the bacterium
has reached about twice its initial size, its plasma
membrane pinches inward, and a cell wall forms across
the bacterium between the two chromosomes, dividing the parent cell into two daughters. (Bacterial
reproduction will be discussed further in Chapters 17
and 25.)

EUKARYOTIC CHROMOSOMES AND THEIR DUPLICATION

A dividing cell delivers most of its components in
roughly equal shares to its daughters. However, the
genetic material is divided between the two daughters
with exceptional precision. Each of the tens of thousands of genes in a typical eukaryotic cell is replicated
and allocated so that both daughter cells receive exactly
the same portions and inherit all the genes that were
present in the parent cell. The problem of replicating

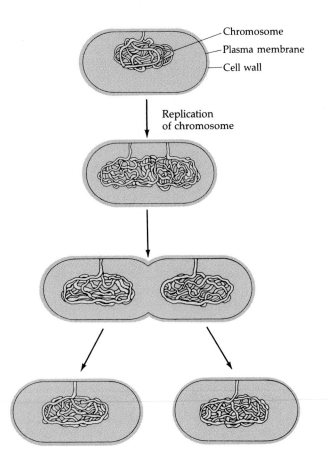

Chromosome

Plasma membrane

Cell wall

Replication
of chromosome

Figure 11.2
Bacterial cell division. In prokaryotic cell division, called binary fission, a membrane attachment mechanism is used to allocate chromosome copies to the two daughter cells. When the bacterial chromosome replicates, it is attached to the plasma membrane; after replication, the duplicate chromosomes attach to the membrane at separate points. Continued growth of the cell gradually separates these attached chromosomes. The two copies are gradually separated by the growth of membrane between them. Eventually, the plasma membrane pinches inward to divide the cell in two as new cell wall material is deposited between the daughter cells.

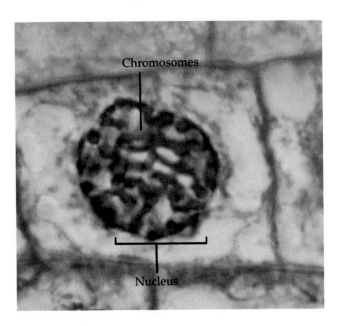

Chromosomes

Nucleus

25 μm

Figure 11.3
Eukaryotic chromosomes. In this light micrograph of an onion cell, the tangles of threadlike chromosomes show up because the cell has been stained with a dye that binds to DNA. This cell is in an early stage of the division process.

and distributing such extensive sets of genes is made somewhat easier by the fact that the genes are grouped into multiple chromosomes—long, threadlike structures found in the nucleus of all eukaryotic cells (Figure 11.3). Well in advance of division, a cell duplicates all of its chromosomes, and therefore copies all genes. (The mechanism by which DNA replicates will be explained in Chapter 15.) Then, during a process called **mitosis** (from the Greek *mitos*, "thread"), the duplicated chromosomes are evenly distributed into two daughter nuclei, one at each end of the cell. Mitosis, the division of the nucleus, is usually followed immediately by **cytokinesis,** the division of the cytoplasm

to form two separate daughter cells, each containing a single nucleus.

Chromosomes get their name (*chromo,* "colored" and *somes,* "bodies"), from their affinity for certain stains used in microscopy. Incorporated into each chromosome is one very long DNA molecule representing thousands of genes. In addition to DNA, the chromosome also has an abundance of proteins that contribute to the structure of the chromosome and help control the activity of the genes. This DNA-protein complex, called **chromatin,** is organized into a long, thin fiber that is compactly folded and coiled to form the chromosome (the details and relevance of this compact packing of chromatin are addressed in Chapter 18).

In most organisms, including humans, all somatic cells (all body cells except sperm or egg cells) have the same number of chromosomes in their nuclei. The egg and sperm cells of such organisms have precisely half the number of chromosomes of any of the somatic cells (see Chapter 12). The chromosome count is characteristic of a given species. For example, goldfish somatic cells have 94 chromosomes, whereas human somatic cells have 46.

When chromosomes replicate, each forms two identical structures called **sister chromatids** (Figure 11.4). During most of mitosis, the sister chromatids remain joined together at a specialized region called the **cen-**

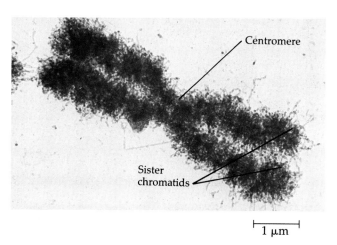

Figure 11.4
A duplicated chromosome. As a cell prepares for mitosis, each chromosome replicates to make two genetically identical sister chromatids attached at their centromeres. The electron micrograph reveals a human chromosome in the duplicated state. The chromosome has a "hairy" appearance because it consists of a very long chromatin fiber, folded and coiled in a compact arrangement. Nucleic acid and protein make up the chromatin fiber (TEM).

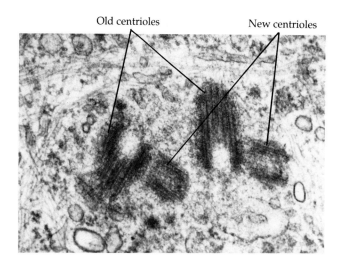

Figure 11.5
Centriole duplication. The two centrioles, located just outside the nucleus, separate and their duplication begins even before the chromosomes start to replicate. In this electron micrograph of a mammalian cell, you can see a new centriole forming at a right angle to each existing centriole. The two pairs of centrioles continue to move farther apart, and eventually each daughter cell gets one centriole pair.

tromere. As mitosis progresses, the sister chromatids separate from each other, with one chromatid from each pair going to a different daughter cell. In a human cell, for instance, each of the two daughter cells receives an identical complement of 46 chromosomes, a complete catalog of genetic instructions.

REPRODUCTION OF CELLULAR ORGANELLES

Besides duplicating their genetic material, most cells must also increase their mass and their supply of essential cytoplasmic organelles before dividing. Among the most important of these organelles are the structures responsible for protein synthesis, the ribosomes, which are made as subunits in the cell's nucleoli. The membranes of rough endoplasmic reticulum (rough ER) must also be expanded by the synthesis of its component proteins and lipids. Some of the membranous material produced along the expanding rough ER is relocated and used for making other components of the endomembrane system, including the plasma membrane and the nuclear envelope (see Chapter 7).

Certain cell organelles, including mitochondria and chloroplasts, contain DNA and reproduce more or less on their own, growing and dividing in two. However,

certain proteins necessary for the functioning of these organelles are generated from instructions carried by the cell's chromosomal DNA. Moreover, almost all the lipids used by these organelles are supplied from the cell's ER. The replication of the organelles' DNA, small circles with little associated protein, occurs in a manner similar to that of bacterial DNA.

The centrioles, a pair of structures composed of microtubules and located just outside the nucleus, also duplicate on their own in a way that is closely coordinated with division of the cell. Before mitosis commences, the two centrioles begin to move farther apart, and a new centriole begins to form at right angles to each existing centriole (Figure 11.5). During mitosis, the two centriole pairs continue to move farther apart until there is a pair at each end of the cell. Thus, each daughter cell will receive one pair of centrioles, which will replicate again before the next cell division.

THE CELL CYCLE

The well-ordered sequence of events between the time a cell divides to form two daughter cells and the time those daughter cells divide again is called the **cell cycle** (Figure 11.6). In brief, it goes like this: A cell roughly doubles its cytoplasm, including the organelles, and

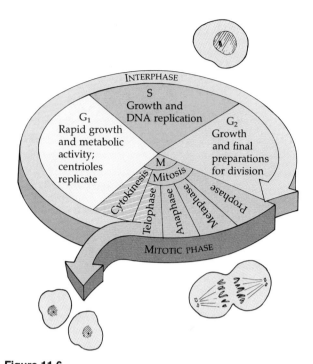

Figure 11.6
The cell cycle. In a dividing cell, the mitotic (M) phase alternates with an interphase, or growth, period. The first part of interphase is called G_1, and it is followed by the S phase, during which the chromosomes replicate; the last part of interphase is called G_2. Next, mitosis divides the nucleus and the duplicated chromosomes within it. Finally, cytokinesis divides the cytoplasm, producing two daughter cells. The stages of mitosis (prophase, metaphase, anaphase, and telophase) are described in Figure 11.8.

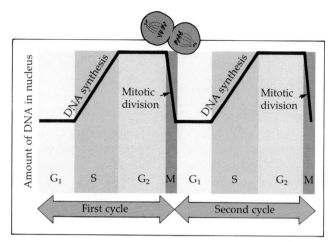

Figure 11.7
Cyclical changes in DNA content of dividing cells. The amount of DNA per cell doubles during the S (synthesis) phase of the cell cycle. At the level of chromosomes, this is when each chromosome duplicates to form two sister chromatids. During M, the mitotic phase, equal parceling of the genetic material to the two daughter cells halves the amount of DNA per cell.

precisely duplicates its DNA. Then the nucleus and its contents, including the duplicated chromosomes, divide (mitosis)—the most dramatic phase of cell life when viewed through the microscope. As mitosis nears completion, cytokinesis begins to divide the cytoplasm between the daughter cells. Each daughter ends up with a single, intact nucleus and some surrounding cytoplasm. Taken together, mitosis and cytokinesis make up the **M phase** (mitotic phase) of the cell cycle. The remainder of the cell cycle, the entire period between successive cell divisions, is called **interphase.**

The duration of cell cycles varies greatly, depending on the cell type and its physiological state. Some cells divide as often as once per hour, while other cell cycles may take more than a day. There are also certain specialized cells in multicellular organisms that, once formed, divide only rarely or not at all; our nerve and muscle cells are examples.

The term *interphase* belies the importance of this phase in the life of the cell. Normally, interphase lasts for at least 90% of the total time required for the cell cycle. When a cell is observed under the microscope during this part of the cycle, it appears to be resting.

This appearance is deceiving, however, because biochemical activity is very high in a cell during interphase. Indeed, most of a cell's growth and metabolic activities occur during this phase.

Although many cell components are made continuously throughout interphase, DNA is synthesized only during a limited period. This part of interphase is called the **S phase** (synthesis phase) (Figure 11.7). The period of interphase before DNA synthesis begins is called the **G_1 phase** (G_1 stands for first gap). And the period of interphase after DNA synthesis occurs but before mitosis begins is called the **G_2 phase** (second gap). New organelles are generally produced throughout interphase, during its G_1, S, and G_2 phases.

During late interphase, the cell has one or more nucleoli, and its nucleus is still surrounded by the nuclear envelope. Just outside the nucleus are two pairs of centrioles, formed by replication of a single pair during interphase (see Figure 11.5). At this stage, the cell's individual chromosomes cannot yet be distinguished under the microscope because they are still in the form of relatively loosely packed chromatin fibers. Soon, however, with the onset of mitosis, there will be visible changes in the chromosomes and other structures.

Phases of Mitosis

Mitosis is unique to eukaryotes and may be an evolutionary adaptation associated with the problem of properly distributing a large amount of genetic material. Although certain details of mitosis may vary from one organism to the next, the overall process is similar in most eukaryotes. The distribution of the parent cell's duplicated genetic material to the two daughter cells is remarkably reliable. Experiments with yeast, for instance, have indicated that an error in chromosome distribution occurs only once in approximately 100,000 cell divisions.

Time-lapse films of living, dividing cells reveal the dynamics of mitosis and cytokinesis as a continuum of changes that divide nucleus and cytoplasm. For purposes of description, however, mitosis is conventionally described as occurring in four stages: **prophase, metaphase, anaphase,** and **telophase.** The details of these stages for a dividing animal cell are provided in Figure 11.8 (pp. 234–235), which you should study before going on to the next section of text.

The Mitotic Spindle

Many of the events of mitosis depend on a structure called the **mitotic spindle,** which appears in the cytoplasm during prophase and consists of fibers made of microtubules and associated proteins (Figure 11.9, p. 236). While the mitotic spindle is being assembled, the microtubules of the cytoskeleton are partially disassembled—probably providing the material used to construct the mitotic spindle. The spindle microtubules, which are aggregates of two proteins called α- and β-tubulin, elongate by incorporating more and more individual protein subunits, usually at one end (see Figure 7.25). In animal cells, the spindle first appears as two **asters,** radial arrays around each centriole pair; plant cells do not have asters. Both animal and plant cells, however, have **polar fibers,** which extend from the two poles of the spindle toward the equator of the cell.

Although the centrioles are the centers of spindle formation in animal cells, these structures are not essential for mitosis. For example, most plant cells lack centrioles, yet they form effective spindles. And in experiments in which the centrioles of animal cells have been destroyed with a laser microbeam, spindles continued to function normally. The true organizing center of the spindle appears to be a cloud of material visible only in the electron microscope. This "amorphous" material, called the **microtubule organizing center (MTOC),** is associated with the centrioles of animal cells, but its mode of action still awaits discovery.

The interaction between the chromosomes and the spindle involves specialized links, called **kinetochores,** that develop at the centromere region. There is one kinetochore on each sister chromatid, and the kinetochores on the two chromatids face opposite directions. A special set of microtubules called **kinetochore fibers** attach to each kinetochore. The kinetochore fibers and polar fibers make up the spindle apparatus. When the spindle first forms during prophase, interactions of the fibers throw the chromosomes into agitated motion. Eventually, the chromosomes become aligned on the **metaphase plate,** the circular plane at the equator of the spindle. Because each chromosome arrives at the metaphase plate with a random orientation relative to the poles, chance determines which chromatid will end up in a particular daughter cell.

The movement of the spindle poles away from each other as a cell elongates during mitosis is a subject of controversy among researchers. Some of the movement may be due to the addition of subunits to the polar fibers and their subsequent elongation. Also, the poles may be pushed apart by a sliding of the interdigitating polar fibers past each other in the region of spindle overlap at the equator. According to this "sliding tubule" hypothesis, elongation of the cell is driven by the hydrolysis of ATP by an enzyme related to dynein, the protein responsible for the sliding of microtubules in eukaryotic cilia and flagella.

How kinetochore microtubules cause the separation and movement of sister chromatids during anaphase is understood even less. Most researchers agree that the force is generated differently from that for pole separation, because blocking dynein's action with special chemical inhibitors does not affect chromosome movement. Recent experiments have also demonstrated that the poleward movement of chromosomes does not require ATP. Thus, a low-energy process, such as dissociation of the kinetochore microtubules into their protein subunits, may play a role in moving the chromosomes. According to one model, the microtubules depolymerize at the ends near the poles of the cell. This hypothesis suggests that the shortening microtubules, with their attached chromosomes, are somehow drawn to the poles as tubulin subunits dissociate from the ends of the microtubules. There is recent experimental evidence, however, that dissociation of the microtubules occurs at their *kinetochore* ends. A model based on this observation suggests that the depolymerization of microtubules at the kinetchore is an exergonic reaction that provides the energy for the kinetochore to creep poleward along the remaining tip of the microtubule, just ahead of the depolymerization (Figure 11.10, p. 237).

Cytokinesis

The events so far described encompass activities relating to mitosis, the division of a cell's nucleus. However, during the later stages of mitosis, at about late anaphase or early telophase, important changes also can be observed in the cytoplasm—changes that result in division of the cell into two. The term for the overall process, *cytokinesis,* means "movement of the cytoplasm."

Figure 11.8

The stages of cell division in an animal cell. The light micrographs show dividing cells from a fish embyro, as seen with the light microscope. The drawings are highly schematic and show details not visible in the micrographs. For the sake of simplicity, only four chromosomes are drawn. In plant cells, centrioles are lacking, and cytokinesis occurs by a different mechanism (see Figures 11.12 and 11.13).

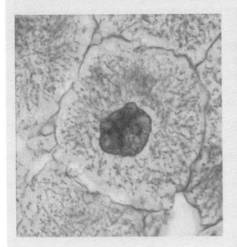

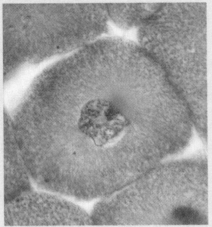

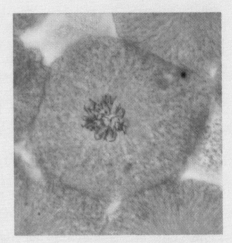

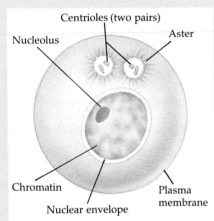

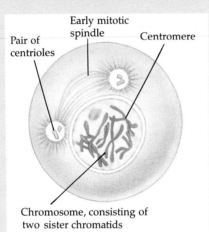

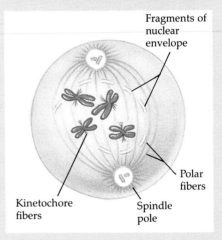

INTERPHASE

During late interphase, the nucleus is well defined and bounded by the nuclear envelope. It contains one or more nucleoli. Just outside the nucleus are two pairs of centrioles, formed earlier by replication from a single pair. In animal cells the two centriole pairs at first lie side by side just outside the nucleus. Around each centriole pair, microtubules form in a radial array, called an *aster* (star). The chromosomes have already duplicated, but at this stage they cannot be distinguished individually because they are still in the form of loosely packed chromatin fibers.

PROPHASE

During prophase, changes occur in both the nucleus and the cytoplasm. In the nucleus, the nucleoli disappear. The chromatin fibers become more tightly coiled and folded into discrete chromosomes observable with a light microscope. Each duplicated chromosome appears as two identical sister chromatids joined at the centromere. In the cytoplasm, the mitotic spindle forms; it is made of microtubules and associated proteins arranged between the two pairs of centrioles. During prophase the centriole pairs move away from each other, apparently propelled along the surface of the nucleus by the lengthening bundles of microtubules between them.

Late in prophase the nuclear envelope fragments. The microtubules of the spindle can now invade the nucleus and interact with the chromosomes, which have become even more condensed. Bundles of microtubules, called *polar fibers*, extend from each pole toward the equator of the cell. Each of the two chromatids of a chromosome now has a specialized structure called the kinetochore, located at the centromere region. Bundles of microtubules called *kinetochore fibers* are attached. These fibers interact with the polar fibers of the spindle, throwing the chromosomes into agitated motion.

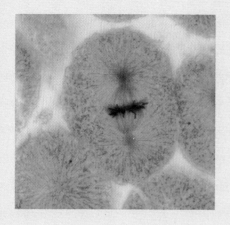

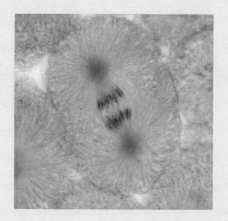

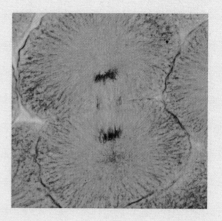

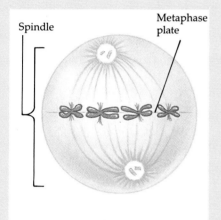

Spindle

Metaphase plate

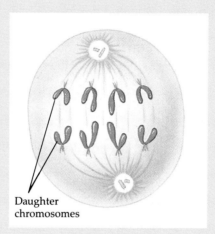

Daughter chromosomes

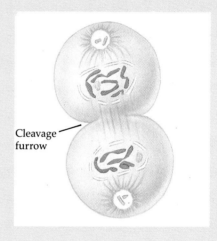

Cleavage furrow

METAPHASE

The centriole pairs are now at opposite ends, or poles, of the cell. The chromosomes convene on the *metaphase plate* (dashed line), the plane that is equidistant between the spindle's two poles. The centromeres of all the chromosomes are aligned with one another at the metaphase plate. The chromosomes lie with their long axes at roughly right angles to the spindle axis. For each chromosome, the kinetochores of the sister chromatids face opposite poles of the cell. Thus, the identical chromatids of each chromosome are attached to kinetochore fibers radiating from opposite ends of the parent cell. The entire apparatus of polar fibers plus kinetochore fibers is called the spindle, for its shape.

ANAPHASE

Anaphase begins when the paired centromeres of each chromosome move apart, liberating the sister chromatids from each other. Each chromatid is now considered a full-fledged chromosome. The spindle apparatus then begins moving the once-joined sisters toward opposite poles of the cell. Because the microtubules of kinetochore fibers are attached to the centromere, the chromosomes move centromere first (their pace is about 1 μm/second). The kinetochore fibers shorten as the chromosomes approach the cell poles. At the same time, the poles of the cell also move farther apart. By the end of anaphase the two poles of the cell have equivalent—and complete—collections of chromosomes.

TELOPHASE AND CYTOKINESIS

At telophase, the polar fibers elongate the cell still more, and the daughter nuclei begin to form at the two poles of the cell where the chromosomes have gathered. Nuclear envelopes are formed from the fragments of the parent cell's nuclear envelope and other portions of the endomembrane system. In further reversal of prophase events, the nucleoli reappear, and the chromatin fiber of each chromosome uncoils. Mitosis, the equal division of one nucleus into two genetically identical nuclei, is now complete. Cytokinesis, the division of the cytoplasm, is usually well under way by this time, so the appearance of two separate daughter cells follows shortly after the end of mitosis. In animal cells, cytokinesis involves formation of a cleavage furrow, which pinches the cell in two.

Figure 11.9
Mitotic spindle and kinetochores. (**a**) An electron micrograph showing the spindle apparatus of a metaphase cell. The two centrioles of each pair are oriented at right angles to each other. The chromosomes are at the metaphase plate, the equator of the spindle. (**b**) The relationship between the chromosomes and the spindle at metaphase. At the centromere region of each chromatid is a structure called a kinetochore, to which are attached microtubules called kinetochore fibers. The kinetochores of sister chromatids face opposite poles of the spindle. The kinetochore fibers interact with the polar fibers to align the chromosomes at the metaphase plate. Polar fibers from the two spindle poles overlap at the equator of the spindle.

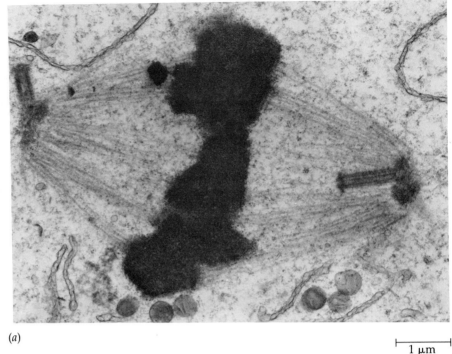

(a)

1 μm

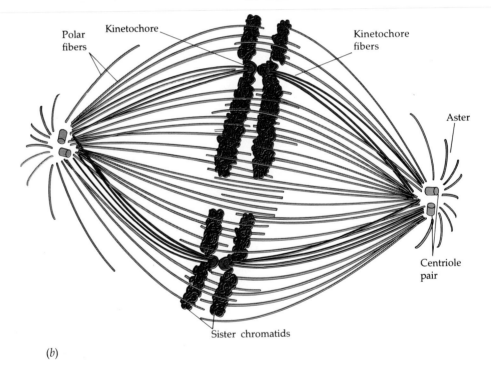

Polar fibers

Kinetochore

Kinetochore fibers

Aster

Centriole pair

Sister chromatids

(b)

In animal cells, cytokinesis occurs by a process known as **cleavage.** The first sign of cleavage is the appearance of a **cleavage furrow,** which begins as a shallow groove in the cell surface near the old metaphase plate (Figure 11.11). On the cytoplasmic side of the furrow is a **contractile ring** of microfilaments made of the protein actin, the same protein that plays a key role in the contraction of muscle as well as many other kinds of cell movement (see Chapter 7). As the dividing cell's ring of microfilaments contracts and its diameter shrinks, the effect is like the pulling of purse strings.

The cleavage furrow deepens until the parent cell is pinched in two. The last bridge between the two daughter cells, containing the remains of the mitotic spindle, finally breaks, leaving two completely separated new cells.

Cytokinesis in plant cells, which have walls, is markedly different. There are no cleavage furrows. Instead, a structure called the **cell plate** forms across the midline of the parent cell where the old metaphase plate was located (Figure 11.12, p. 238). Vesicles derived from the Golgi apparatus coalesce along the middle

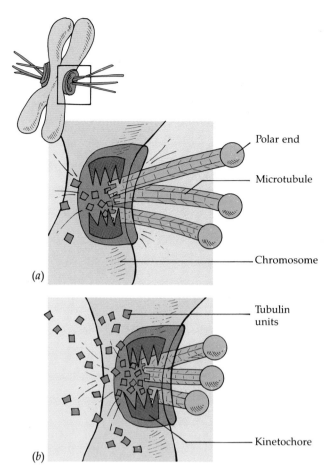

Polar end

Microtubule

Chromosome

(a)

Tubulin
units

Kinetochore

(b)

Figure 11.10
One model for the poleward movement of chromosomes. (a) This hypothesis is based on experimental evidence that kinetochore microtubules depolymerize at their kinetochore ends, not at their polar ends. (b) The kinetochore maintains its grip on the tip of the shortening microtubule, creeping poleward just ahead of the portion of the microtubule that is dissociating into its tubulin subunits. According to this model, depolymerization of the microtubule is an exergonic reaction that provides the energy for the chromosome movement.

of the cell to give rise to the cell plate. The fusion of vesicles forms two membranes, which eventually unite laterally with the existing plasma membrane. This results in the formation of two daughter cells, each with its own plasma membrane. A new cell wall forms between the two membranes of the cell plate. Figure 11.13 (p. 239) shows one complete turn of the cell division cycle for a plant cell. Note the absence of a cleavage furrow and the presence of a cell plate.

In some exceptional cases, mitosis is not followed by cytokinesis. For example, certain slime molds form multinucleated masses called **plasmodia.** These plasmodia may grow to form masses as large as 30 g.

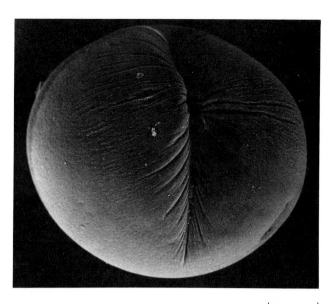

$\vdash$—$100\ \mu m$—$\dashv$

Figure 11.11
Cytokinesis of an animal cell. In this scanning electron micrograph, we see the cleavage furrow of a dividing animal cell in three dimensions. Microfilaments form a ring just inside the plasma membrane at the location of furrowing. These microfilaments are made of contractile proteins, which cause the cleavage furrow to deepen until the cell is pinched in two.

CONTROL OF CELL DIVISION

The timing and rate of cell division in different parts of a plant or animal are critical to normal growth, development, and maintenance. A variety of patterns of cell division is employed by different types of cells. For example, human skin cells divide frequently throughout life, whereas liver cells maintain the ability to divide, but keep it in reserve until an appropriate situation arises—for example, to repair a wound. Some of the most specialized cells, such as nerve cells and muscle cells, do not divide at all in a mature human.

Although many questions about the control of cell division remain unanswered, scientists have learned a great deal by studying the growth of cells in the laboratory. The technique of growing cells outside the organism is often called **tissue culture,** because when first developed, it involved the culture (growth) of tissue fragments that had been removed from an organism. Today, cell biologists usually work with cells dissociated from tissues, in which case the term **cell culture** is more appropriate. Cells are placed into a glass or plastic vessel containing nutrient solution, called a **growth medium.** A typical growth medium for mammalian cells contains glucose, amino acids, vitamins, salts, blood serum, and antibiotics (to prevent bacterial contamination). Given a suitable growth medium and other appropriate conditions, such as a

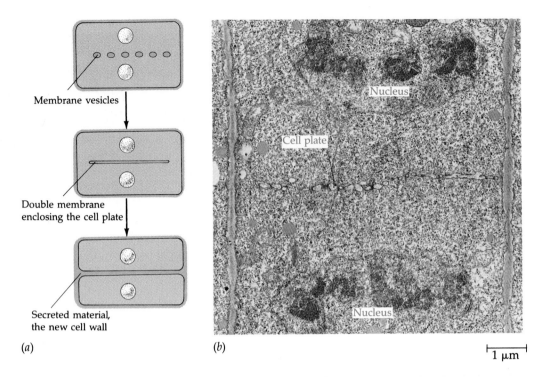

Figure 11.12
Cytokinesis of a plant cell by cell plate formation. The telophase cell has nuclei forming at its two poles. In the meantime, **(a)** membrane vesicles fuse to form a double membrane, which encloses the cell plate at the equator of the cell; and materials are secreted into the space between the membranes to form the new cell wall. **(b)** The electron micrograph shows a cell plate forming in a cell from a soybean root tip.

favorable temperature, many types of cells will grow in an orderly way in culture. The cells usually grow on the inner surface of the culture vessel, which has been prepared to allow cell attachment. Their growth rate will depend on the cell type and on the conditions provided.

Research on cells grown in culture has identified factors that can stimulate or inhibit cell division. For instance, certain hormones and other chemicals are vital for normal cell division. Moreover, cell division can be blocked in various ways—for example, by inhibiting protein synthesis, cutting off essential nutrients, or allowing the cells to become overcrowded. This last condition represents an important observation about normal vertebrate cells. If grown on the inner surface of a container and supplied with a rich growth medium, normal cells divide only until they come in contact with one another—and then cell division stops. This phenomenon is known as **contact inhibition** (Figure 11.14a, p. 240). If some of the cells are removed from the filled surface to create a space, cells bordering the open space begin dividing again until the space is filled.

For most normal cells, the G_1 phase of the cycle—before DNA replication occurs—seems to play a crucial role in controlling cell division. During this phase, a mechanism that so far has eluded researchers some-how determines whether the cell will go through a division cycle or, instead, switch into a nondividing state (sometimes called G_0). If a normal cell is to switch into the nondividing state, it usually stops late in the G_1 phase. However, if it passes a certain stage, called the **restriction point,** it will continue into the S, G_2, and M phases of the cell cycle regardless of external conditions. Recently, one specific gene required for cell division has been identified, and this gene is very similar in organisms as diverse as humans and yeast (a fungus). At this time, however, very little is known about the molecular mechanisms that regulate cell cycles in living organisms.

ABNORMAL CELL DIVISION: CANCER CELLS

Cancer cells do not respond normally to the body's control mechanisms. They divide excessively, invading other tissues. If unchecked, they can kill the whole organism.

By studying cancer cells in culture, researchers have learned that these cells do not heed the normal signals that stop growth. In particular, they do not show contact inhibition when growing in culture (Figure 11.14b).

Figure 11.13
Mitosis of a plant cell. These micrographs show cells of *Tradescantia* (wandering Jew) as viewed by differential-interference-contrast microscopy.

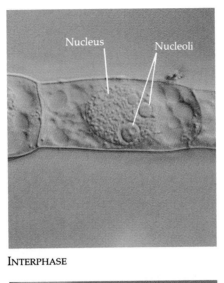

INTERPHASE

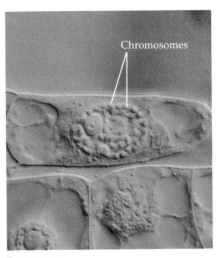

PROPHASE

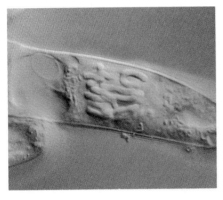

METAPHASE

ANAPHASE

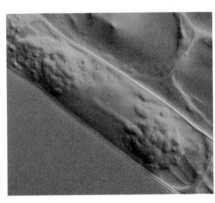

LATE ANAPHASE

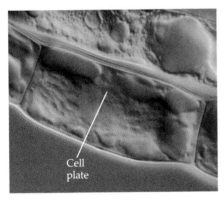

TELOPHASE

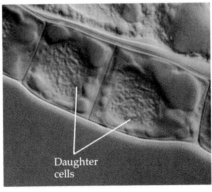

CYTOKINESIS

10 μm

Instead, they continue to multiply even after contacting one another, piling up until the nutrients in the growth medium are exhausted.

There are other important differences between normal cells and cancer cells that reflect derangements of the cell cycle in cancer cells. If and when they stop dividing, cancer cells seem to do so at random points in the cycle, rather than just at the restriction point of G_1. Moreover, in cell culture, cancer cells can go on dividing indefinitely, assuming a continual supply of nutrients, and thus are said to be "immortal." By contrast, nearly all normal mammalian cells growing in culture divide only about 20 to 50 times before they stop dividing, age, and die. Just how the normal controls over mitosis and the cell cycle are disrupted in cancer cells is not known, although it is now clear that genetic changes are ultimately responsible. (The genetic basis of cancer is discussed in Chapter 17.)

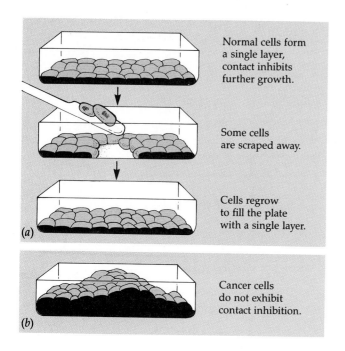

Normal cells form a single layer, contact inhibits further growth.

Some cells are scraped away.

Cells regrow to fill the plate with a single layer.

(a)

Cancer cells do not exhibit contact inhibition.

(b)

Figure 11.14
Contact inhibition. (a) When grown in cell culture, normal cells will multiply only until they come into contact with one another. If some of the cells are removed, the cells at the border will divide until the gap is filled with a single layer of cells. (b) In contrast, cancer cells usually continue to grow even after they come into contact and often pile up on one another. (Individual cells are shown disproportionately large in this figure.)

Perhaps the reason we understand so little about cancer cells is that there are still so many unanswered questions about how normal cells work. The cell, life's basic unit of structure and function, holds enough secrets to engage researchers for well into the future.

*   *   *

With this chapter on the reproduction of cells, we have built a bridge to the next unit of text, which features genes and their role in inheritance.

STUDY OUTLINE

1. The perpetuation of life has its basis in cell division. Unicellular organisms reproduce by cell division, and the development, growth, and repair of multicellular organisms depend on cell division.
2. The ability of DNA to replicate itself is essential to the production of genetically equivalent daughter cells.

Bacterial Reproduction (p. 229)

1. Prokaryotes undergo binary fission, in which the cell splits in two after replication of the single circular chromosome.
2. Daughter cells are formed by the growth and pinching in the plasma membrane between the two chromosomes and the formation of a new cell wall.

Eukaryotic Chromosomes and Their Duplication (pp. 229–231)

1. Eukaryotic cells divide by mitosis, an orderly process of chromosome separation that precedes division of the cytoplasm by cytokinesis.
2. Chromosomes are convenient packages for the replication and distribution of a very large number of genes. All somatic cells have a specific number of chromosomes, which is characteristic for the species and usually twice the number found in egg and sperm cells.
3. Chromosomes are composed of chromatin, a threadlike complex of DNA and protein that becomes tightly coiled and folded during mitosis.
4. When chromosomes replicate, they form identical sister chromatids joined by a centromere. These chromatids separate during mitosis, thereby becoming the chromosomes of the new daughter cells.

Reproduction of Cellular Organelles (p. 231)

1. Dividing cells must reproduce essential organelles in addition to their genetic material.
2. Prior to mitosis, there is increased synthesis of ribosomes, rough ER, and endomembranes, as well as replication of mitochondria, chloroplasts, and centrioles.

The Cell Cycle (pp. 231–237)

1. Mitosis and cytokinesis make up only the M (mitotic) phase of the cell cycle, a sequence of events in the life of dividing cells.
2. Between divisions, cells are in interphase, an active period of growth and metabolism composed of the G_1, S, and G_2 phases. New organelles are produced throughout interphase, but DNA is replicated only during the S (synthesis) phase.
3. The M phase is a dynamic continuum of sequential change, often described by the four stages of prophase, metaphase, anaphase, and telophase.
4. The mitotic spindle is a complex of microtubules that orchestrates chromosome movement during mitosis in eukaryotic cells. During prophase, the spindle forms from the microtubule organizing center, a region near the nucleus that is associated with centrioles in animal cells.
5. The spindle is made up of polar fibers, which extend from the poles of the spindle to the equator, and kinetochore fibers, which link specialized kinetochore structures on the centromeres of sister chromatids to opposite poles of the cell.

6. When the polar fibers contact the kinetochore fibers during prophase, they cause the chromosomes to move toward the metaphase plate, at the equator of the spindle. The cell then elongates and sister chromatids separate and move toward opposite poles by mechanisms that are still being investigated.

7. In most cases, mitosis is followed by cytokinesis, which is characterized by the cleavage furrow in animal cells and the cell plate in plant cells.

Control of Cell Division (pp. 237–238)

1. The rate and timing of cell division are crucial for normal growth, development, and maintenance.

2. Control of cell division has been studied in tissue or cell culture. Factors that stimulate cell division have been identified, such as hormones and other chemicals. Factors that inhibit cell division have also been identified, such as cessation of protein synthesis, depletion of nutrients, and contact inhibition.

3. An unknown mechanism that operates during the G_1 phase determines whether the cell will proceed past the restriction point and through to another cell cycle or stop dividing.

Abnormal Cell Division: Cancer Cells (pp. 238–240)

1. Cancer cells elude normal control mechanisms and divide excessively as a result of an abnormal genetic change. In cell culture, cancer cells do not show contact inhibition and continue to divide indefinitely.

SELF-QUIZ

1. Which process is *not* associated with the reproduction of bacteria?
   a. replication of DNA
   b. binary fission
   c. mitosis
   d. synthesis of new cell wall material
   e. elongation of the parent cell

2. Examining a cell in the microscope, you can see a cell plate beginning to develop across the middle of the cell and nuclei reforming at opposite poles of the cell. This cell is most likely a (an)
   a. animal cell in the process of cytokinesis
   b. plant cell in the process of cytokinesis
   c. animal cell in the S phase of the cell cycle
   d. bacterial cell dividing
   e. plant cell in metaphase

3. DNA replicates during
   a. G_1 phase
   b. S phase
   c. G_2 phase
   d. M phase
   e. cytokinesis

4. If a specialized cell no longer divides, it is generally locked in which stage of the cell cycle?
   a. S
   b. G_1
   c. G_2
   d. M
   e. prophase

5. In a typical cell cycle, cytokinesis generally overlaps in time with which stage?
   a. S phase
   b. prophase
   c. telophase
   d. anaphase
   e. metaphase

6. Chromosomes are in their most extended (least compactly packed) form during:
   a. interphase
   b. prophase
   c. metaphase
   d. anaphase
   e. telophase

7. A particular cell has half as much DNA as some of the other cells in a mitotically active tissue. The cell in question could be in
   a. G_1
   b. G_2
   c. prophase
   d. metaphase
   e. anaphase

8. One difference between a cancer cell and a normal cell is
   a. the cancer cell is unable to synthesize DNA
   b. the cell cycle of the cancer cell is arrested at the S phase
   c. cancer cells continue to divide even when they are tightly packed
   d. cancer cells cannot function properly because they suffer from contact inhibition
   e. cancer cells are always in the M phase of the cell cycle

9. Which event does *not* occur during prophase in an animal cell?
   a. chromosomes condense
   b. spindle begins to form
   c. chromosomes replicate
   d. nuclear envelop disperses
   e. nucleoli disperse

10. In the following light micrograph of dividing cells near the tip of an onion root, identify a cell in: interphase, prophase, metaphase, anaphase, telophase. Describe the major events occurring at each of those stages.

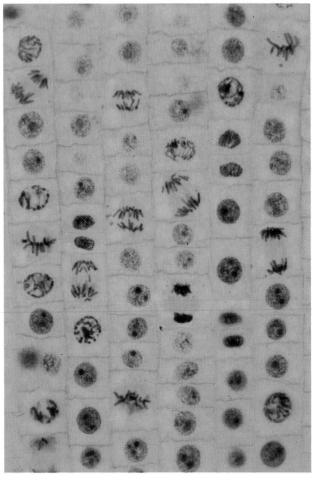

$\overline{}$ 25 μm $\overline{}$

CHALLENGE QUESTIONS

1. When a population of cells is examined with a microscope, the percentage of the cells in the M phase (prophase to telophase) is called the mitotic index. The greater the proportion of cells that are dividing, the higher the mitotic index. In a particular study, cells from a cell culture are spread on a slide, preserved and stained, and then inspected in the microscope. A hundred cells are examined: 9 cells are in prophase; 5 cells are in metaphase; 2 cells are in anaphase; 4 cells are in telophase; the remainder, 80 cells, are in interphase. Now, answer the following questions.

   a. What is the mitotic index for this cell culture?

   b. The average duration for the cell cycle in this culture is known to be 20 hours. What is the duration of interphase? Of metaphase?

   c. Going back to the living culture of these cells, the average quantity of DNA per cell is measured. Of the cells in interphase, 50% contain 10 ng (nanogram = 10^{-9} g) of DNA per cell; 20% contain 20 ng DNA per cell; the remaining 30% of the interphase cells have varying amounts of DNA between 10 and 20 ng. Based on these data, determine the duration of the G_1, S, and G_2 portions of the cell cycle.

2. About a day after a human egg is fertilized by a sperm cell, the zygote (fertilized egg) divides for the first time. The two daughter cells usually stick together and their repeated cell divisions give rise to a multicellular embryo. On rare occasions, however, the two daughter cells formed by the first division of the zygote separate. Each of these cells can go on to form a normal embryo—not a half embryo or otherwise defective embryo. Based on what you have learned in this chapter, explain why these "monozygotic twins" are essentially genetically identical.

3. Cell biologists often need to obtain large populations of cultured cells synchronized with respect to the cell cycle. Hydroxyurea, a chemical that blocks DNA synthesis, can be used in conjunction with deprivation of the amino acids isoleucine or leucine in a stepwise procedure that synchronizes the cycles of the cells. First the cells are arrested in the G_1 phase by deprivation of isoleucine or leucine. Then the amino acid is resupplied, and hydroxyurea is added to the medium. Finally, the hydroxyurea is removed. Explain the rationale for this synchronization process in terms of your knowledge of the cell cycle.

FURTHER READING

1. Becker, W. M. *The World of the Cell.* Menlo Park, Calif.: Benjamin/Cummings, 1986, Chap. 15.
2. Benditt, J. "Genetic Skeleton." *Scientific American,* July, 1988. Short description of a model integrating control of the cytoskeleton and regulation of cell division.
3. Koshland, D. E., T. J. Mitchison, and M. W. Kirschner. "Poleward Chromosome Movement Driven by Microtubule Depolymerization *in vitro.*" *Nature,* February, 1988. One model for how anaphase works.
4. Mazia, D. "The Cell Cycle." *Scientific American,* January 1974. An account of the cell cycle by one of its original elucidators.
5. Prescott, D. M. *Reproduction of Eukaryotic Cells.* New York: Academic Press, 1976. A monograph on cell reproduction.
6. Silberner, J. "Off Switch for Cell Division Found." *Science News,* August 24, 1985. An easily understood summary of the basic findings of Gutowski and her co-workers in Further Reading 2.
7. Sloboda, R. D. "The Role of Microtubules in Cell Structure and Cell Division." *American Scientist,* May/June 1980. A good description of the function of the spindle apparatus.

The Gene

Interview: Nancy Wexler

Many of us know someone whose family has been touched by hereditary disease. Dr. Nancy Wexler's creative detective work is helping to solve some of the mysteries of inherited diseases, especially the fatal genetic disorder called Huntington's disease (HD). Her research blends the high technology of the molecular biology laboratory with the classical approach of tracing genes through family trees.

Dr. Wexler is an associate professor at the College of Physicians and Surgeons, Columbia University, as well as the president of the Hereditary Disease Foundation in Los Angeles, California. In addition to dividing her time between the West and East coasts, Dr. Wexler spends part of her year in South America studying Venezuelan villages where many of the inhabitants are afflicted with Huntington's disease.

The Hereditary Disease Foundation was established by Dr. Wexler's father, Dr. Milton Wexler, after his wife was diagnosed as having Huntington's disease. Nancy Wexler, herself at risk for the disease, has worked tirelessly with scientists and patients all over the world to locate the gene responsible for HD and to learn how it causes the disease.

We joined Dr. Wexler in her offices in Los Angeles, where she works with a small staff to administer the Hereditary Disease Foundation. In this interview, Dr. Wexler shares with us the details of her quest to find the fatal HD gene and discusses her recent appointment to the Human Genome Project, perhaps the most ambitious scientific adventure since the Apollo program.

Dr. Wexler, is it true that you arrived at genetics by a rather unconventional route?

Actually, my only formal education in biology, I confess, is exactly the kind of course for which you're writing this textbook. I went to Radcliffe College as an undergraduate and was required to take Introductory Biology. It was a wonderful class. That's the extent of my formal biology training. My Ph.D is actually in clinical psychology.

Then how did you become involved in genetics?

I started with clinical psychology, but was always very interested in biology and genetics. My mother had a Master's degree in Genetics from Columbia University where I now teach. But I think that I felt (for which I blush) that it was a difficult career for a woman. It was a very stupid feeling. I had a mother who was a geneticist and a father who is a psychoanalyst/clinical psychologist and I chose the psychology route.

Then, in 1968, my father started the Hereditary Disease Foundation, and I began meeting scientists of all disciplines, but particularly geneticists and molecular biologists. These people have really been my teachers since then. They've given me a fantastic education on every napkin you could find in restaurants, bars, and workshops. I also read biology texts and go to lectures to try to learn. I've had great friends who have taught me the subject matter as well as the flavor and excite-

ment of their work. It's been a wonderful education, and, of course, no exams. I consider myself, however, still very much a student. I'm definitely still learning.

Would you tell us more about the Hereditary Disease Foundation?

My father started the Foundation when my mother was diagnosed with Huntington's disease. It was our way of preserving hope. At the Hereditary Disease Foundation we are trying to find a cure not only for Huntington's disease, but for all the hereditary illnesses that afflict people, of which there are many thousands.

The Foundation is not for profit. One hundred percent of all public donations goes directly to science. Not only do we fund grants and postdoctoral fellowships, but we sponsor interdisciplinary workshops that focus on an idea or question, such as how do genes express themselves in the brain? How do you find a gene? We try to excite people about the questions that surround hereditary diseases. We want them thinking about the fundamental issues of biology in terms of human diseases. If we understand what goes wrong in Huntington's disease, then we can apply this understanding more widely to unraveling other genetic diseases.

These workshops have basic scientists and clinical investigators mixed together, representing all varieties of expertise. We feel it is crucial to show the laboratory scientist what the Huntington's gene looks like in action. To do that, we start every workshop by inviting a patient to meet with the participants, allowing them to really see what a gene can do to a person in all

aspects of thinking, feeling, moving, and expressing. The scientists become captivated because the study of genetics is suddenly made more vivid, more than just an intellectual exercise.

These unique workshops are one of the most successful new contributions that the Foundation has made to the study of hereditary diseases. From the workshops come suggestions for grants and postdoctoral fellowships, which we fund. In fact, the idea for using DNA markers to find the Huntington's disease gene was born at a Hereditary Disease Foundation workshop.

What exactly is Huntington's disease?

It's an inherited neurological disorder. It is inherited in what's called an autosomal dominant pattern, which means that the abnormal gene will dominate its normal partner. Each child of a parent with the disease, therefore, has a 50–50 chance of inheriting it. This also means that males and females are affected equally. It usually starts somewhere in mid-life but it can start as early as the age of 2 or as late as the early 80s. It is a slowly progressive disease. The duration can be anywhere from 10 to 20—sometimes even 25 years. The disease is invariably fatal. Unfortunately, the most damaging changes take place fairly early in the disease so that people lose the capacity to work or head a household. The patient becomes impaired relatively early in the illness but lives for a very long time.

What are the symptoms?

The symptoms affect just about everything that makes you human—how you think, move, and feel. It causes

different kinds of uncontrollable, involuntary movements in all parts of the body. It can also cause severe cognitive problems: loss of memory, loss of judgment, loss of the capacity to organize oneself. In almost all standard school tests, patients with Huntington's disease do very poorly. But, they do maintain a social intelligence—an awareness of who they are, where they are, their family and friends, and their social setting. The only trauma in this preservation of faculties is that they recognize the loss of their capacities—their ability to do the kinds of simple things that gave them an identity and some sense of satisfaction and self-worth.

The occurrence of depression in people with Huntington's disease is extremely high, partially because of this deterioration that the patient sees happening, but, also, depression seems to be part of the disease itself. Sometimes people will be hospitalized for depression and never know that they have Huntington's in the family. About 1 in 4 patients with Huntington's disease makes a serious suicide attempt. On the other hand, patients also will maintain an impressive sense of humor. And, if allowed, they can be quite active and involved.

Aspiration pneumonia and infection are the most frequent causes of death. Due to the loss of motor skills, other causes of death include choking, hematomas that are a result of falling, or other accidental forms of death. Patients lose the capacity to swallow and can sometimes die of malnutrition. They also lose the ability to speak. So even though they understand much of what is going on, they can't communicate. It's a difficult, slow decline.

The onset of the disease is so insidious and long that we tend to think of it as a zone of onset rather than a discrete age of onset. There is a perplexing time when neither the person nor the neurologist can be certain if it's Huntington's disease. Sometimes it takes three to five years before you can definitely say, "This is Huntington's disease." People at risk frequently drive themselves nuts questioning whether or not they dropped something because they have the disease or because they're just klutzy. Even if you do a test for the likely presence of the Huntington's disease gene (a new DNA test that was just developed), it's still hard to say definitely when the disease begins. If the test comes out positive—

indicating that the gene is most likely there—you still don't know if the things you're doing are just garden-variety problems or reactions to the traumatic information or the initial manifestations of the disease, or the disease itself. It's a difficult disease, and until very recently, there wasn't any way to tell who was carrying the gene. You just had to wait until the symptoms appeared. People have likened it to a time bomb in a coil of DNA because you just don't know.

How many people have Huntington's disease, and how many people are estimated to be at risk?

The calculation is about 10 patients per 100,000 population, which means that in the United States, we have approximately 25,000 patients and 125,000 people who are at risk. People "at risk" are the siblings and children of diagnosed patients.

Do you think that HD had a single origin as a genetic mutation, or multiple origins?

I don't really know. I guess I'm a little biased because our own work points to a single origin or very few. When we first discovered a genetic marker for the HD gene, our next immediate question was, "How can we get the most disparate families in the world to give blood samples we can test?" In other words, we wanted to see if the disease gene in distant geographical areas and different racial and ethnic groups was also located in the same place, on chromosome 4. Until we actually have the gene, we won't be able to tell if it's the identical defect in the gene worldwide. At the time, there wasn't any particular reason to think there would be only one genetic location since there's heterogeneity (more than one gene for a disorder in different locations) in so many other genetic diseases. We've now tested about 75 different families from Europe, Venezuela, Peru, Japan, and Papua New Guinea, and have found no indication of heterogeneity at all. The Huntington's disease patients in all those countries seem to have their defective gene on the top of chromosome 4. Once we actually find the gene itself, then we can discover if the mutations in these genes from throughout the world are the same. If they are, it may mean that there was one single mutation that spread around the world or that the identical mutation occurred

more than once. We need to know what the actual change is to know how likely it would be to occur once, a few times, or many times.

I hope that it is one gene. To me, that's a very poetic idea. I remember when we were collecting blood samples in China and I was trying to explain to a little girl what we were doing. So I said to her, "You know, your mother has Huntington's and my mother has Huntington's. If we look way back, you and I are cousins." She looked at me in such horror that she could be related to this Western devil! But I thought that it was a great notion.

In the Papua New Guinea case, we were trying to figure out where the disease originated. It turns out that the Boston whalers used to come to nearby islands. The whalers were afraid to go

ashore because the Papua New Guinea people were cannibals. But the Papua New Guineans weren't afraid to go out to the whalers. The books of the whalers were found to say, "Today the natives came aboard, naked and friendly." There's a lot of Huntington's disease in the whole northern New England area, and it is recorded that some of the whalers had Huntington's disease. So, presumably, that's how the gene came to Papua New Guinea. This Huntington's disease gene has probably been spread around the world by a lot of friendly whalers, sailors and other folks going from country to country.

To me, this is the really interesting part about genetic work. It combines everything that you could possibly be interested in. It's anthropology. It's ethnology. It's mystery stories. It's the most fantastic detective story in the entire world.

Now that the gene marker is allowing us to trace Huntington's disease in various populations, the next question for us to address as scientists and a Foundation is, "How do you get to the gene itself and find out what's wrong with it?"

The Huntington's disease gene was initially localized by Drs. James Gusella at Harvard University, P. Michael Conneally at Indiana University, myself, and a number of others working closely together. After the marker was found linked to the gene, the Hereditary Disease Foundation naturally wanted to support research seeking the gene itself. We formed a collaboration of six pioneering scientists in laboratories around the country and in Europe who generously agreed to cooperate with each other instead of competing. This collaboration is unfortunately too rare in science, but it certainly has sped up the work and been extremely productive and fun.

Which brings us to your research. Tell us about your work in Venezuela. How did you find out about these fishing villages in South America where the prevalence of Huntington's disease is so high?

I found out about them in 1972 through a film shown at a meeting of the World Federation of Neurology Research Group on Huntington's Disease. A very smart physician in Venezuela named Dr. Americo Negrette documented the disease in small villages along the shores of Lake Maracaibo, Venezuela, in 1955. At first, he thought that all the villagers were drunk all the time. When it was explained to him that they were sick, he began taking careful pedigrees and searched the world literature to identify the illness. He finally determined that this disease was Huntington's chorea, as it was named in those days (chorea is what the abnormal "dance-like" movements that patients have are called—chorea and choreography have the same root in Greek). Dr. Negrette was very dedicated and he began working with families. He wrote a beautiful monograph in 1962 and then got some other scientists involved. Interest in this disease is inherited, as is the disease itself. His students have carried on his work.

In 1976, the U.S. Congress mandated a Commission for the Control of Huntington's Disease and Its Consequences. We remembered the Venezuelan communities and conducted a workshop with some of the Venezuelan scientists to determine how best to study their extraordinary kindred (a kindred is a large, extended family) with Huntington's disease.

At that workshop, we discussed the idea of finding a person who is a homozygote among the population in Venezuela. A homozygote is a person who inherits *two* copies of the HD gene, one from each parent. The problem with a heterozygote, which is what most patients are, is that you have one normal gene and one HD gene. The normal gene can mask what the defective gene is doing. We thought that if we could just find a homozygote with two HD genes, then the defect would be much more obvious, without the normal gene to confuse things.

So, when we first went to Venezuela in 1979, our original intent was to find a homozygote. Once we began field work in 1981, however, the recombinant DNA revolution had begun and we changed our plans to do genetic linkage studies with these families. We collected extensive pedigrees—who's related to whom—data that were necessary to map the location of the HD gene on a chromosome. We now have a family tree of almost 10,000 people, which is many more than we ever envisioned. The Venezuelan Pedigree has proved spectacular, not only for research on Huntington's disease, but for understanding inheritance in general. The family trees we've collected are perfectly structured for doing genetic linkage studies. They are huge, interconnected families with many children. One man has 29 children by two wives, and the two wives are first cousins. So we not only have these large nuclear families, but they're all related to each other.

The data that we've collected have been used for studying a whole array of genetic diseases. Normal "maps" of the locations of genes and markers on chromosomes can be made by studying the Venezuelan families. These normal maps are then used to study other families with diseases of interest, families that are often too small to do any mapping. The Venezuelan samples have been used to help find the gene for familial Alzheimer's disease, for manic-depressive disease, for two different types of neurofibromatosis, myotonic dystrophy, and others. Other

than having the HD gene on one chromosome, these Venezuelan people have perfectly normal chromosomes. So, if you just want to see how markers and genes are inherited from generation to generation, it's the perfect huge family to study.

Earlier you mentioned that it's now possible to detect the presence of the HD gene through the use of genetic markers. What exactly is a genetic marker?

A marker is a way of holding your place in the chromosome. It's really just what it sounds like; it just sits like a landmark and marks a spot. Consider a situation in which you're looking for one person in the United States—without any boundaries or cities or towns marked off. You know that the person is in the country, but you don't know where. It would be a hopeless task to try to find him or her like that.

So the first order of business is to divide the territory into states. Once that is done, you can say, "This person is in California," and you know where to focus. But California is still a huge state. Where would you go next?

You begin dividing the state into little towns and cities. It turns out that the person you're looking for lives in San Francisco. That's still pretty huge. So then you narrow it down to an area near a ferry terminal. But it turns out that there are three ferry terminals in San Francisco. So you're still in trouble. You must specify that it's the ferry terminal on a particular street.

Those street, city, and state demarcations are all markers. They just tell you where you are relative to the object that you're trying to locate. And that's exactly what a DNA marker is. It's a tiny variant in the DNA itself that has a

specific home in a chromosome and is inherited from generation to generation, exactly like a gene. Sometimes markers *are* genes or are in the middle of genes. You can follow the inheritance of that chromosome by following the marker. Genes near the marker on the chromosome are usually inherited along with the marker.

Now that good markers for the HD gene have been found, a test to determine whether or not a person has the gene is available. What fraction of individuals who know that they are at risk based on family history are opting to be tested?

Well, the fraction has been small, but then the test has been available to only a very few people. Of the people to whom the test has been available, however, only a fraction of those who could use it have actually requested it.

The availability of the test has provoked people to consider in earnest what it would be like to know that they are inevitably going to die, at some unspecified time, of a degenerative disease of the brain. You want to know that you *don't* have the disease, but you don't necessarily want to know that you *do* have the disease. You can't have the opportunity of hearing one answer without the risk of hearing the other. I think that most people who have lived their lives at risk have built certain defenses with respect to having the disease. Knowing, in fact, that you *are* carrying the gene and *will* develop the disease is a very different psychological experience from knowing that there is a *possibility* of developing the disease. Many people have decided not to take the test until there is a treatment. However, there are understandable, pressing reasons for taking the test. People may want to have children free of disease or to clarify the risk for children already born by testing themselves. (The child of a person with a 50% risk has a 25% risk of inheriting HD.) And some people may just want to resolve that ambiguity.

I think that all people should live their lives, to some extent, as though they are in jeopardy because all of us are at risk for something. I think it puts a little edge to your life to be at risk, but to actually know that you *will* have a disease may be more of a push than a person needs. Counseling prior to the test is absolutely essential. People need to know what they're doing when they take this test because you can't erase the results once you've heard them.

For a disease like Huntington's, which has no treatment, the major value of finding the gene is in advancing basic research, not just diagnosis. We now know where the defective gene is. If we can progress from having found the marker that is near the gene to finding the gene itself, then we can find out how the gene works, and what goes wrong, and attempt to fix it. We may be able to develop rational treatments, based on knowing the mistake in the gene. But for now we can only predict, we can't prevent.

Through your work on hereditary disease, you have become involved in the Human Genome Project. Could you tell us about the project?

The Human Genome Project is probably the most ambitious, imaginative, daring effort for humanity to know itself that has ever been attempted. It has extraordinary potential! The Human Genome Project literally sets out to locate every gene on all human chromosomes and to map each one in increasing detail. Eventually, we hope actually to determine the nucleotide sequence for the entire human genome—all of our DNA—all 3 billion base pairs! The real hope is that by learning where these genes are, we will understand what they do. I think this will have profound implications for our understanding of how we function, how we're put together, how diseases originate, how genes interact with each other, and how our genes interact with our environment. At the moment, I think there are fewer than 100 major laboratories around the world working in a concentrated way on the mapping effort.

How long will the project take?

The hope is to have quite a good detailed map in 15 years. In the next five years, we should have a genetic map that will have markers spaced approximately a million base pairs apart. So every million base pairs—rungs on the DNA double helix ladder—will have a marker. And then people will begin filling in the intervals between these markers.

What is your role in the Project.

I think that the Project is fantastic, and I'm overjoyed that they asked me to be a part of it. It's not only excellent science, it's the best human adventure in the world. I'm the chairperson of the Ethics Working Group within the Project. The ethical, social, and legal questions that might be raised as the work is progressing need to be investigated. We cannot afford to wait until after problems arise to consider the ethical ramifications.

Other than the problem of telling someone that he or she has a gene for a disease that we're unable to prevent or cure, what are some of the other ethical, social, and legal issues attached to the Genome Project?

A very big question is the problem of possible discrimination and stigmatization. An insurance company could refuse coverage or an employer could refuse a job if a person was found to carry the gene for a particular disease. People might also be coerced into taking a test that they wouldn't want in order to be considered for a job or an insurance policy. Maybe when the public really understands that each of us has four to ten potentially lethal genes, it will realize that you can't single out any particular patient group as a target for discrimination.

But this issue is complex. The question could be asked, "Do you want people who have genetic susceptibility in positions in which they could have a major impact on other individuals?" For example, do you want an airline pilot with a genetic predisposition toward heart attack? My own feeling is that all of us have genes for something that may be quiescent for a major part of our lives. If you start kicking out everybody who has some genetic susceptibility, then you're going to be in tough shape because there won't be anyone left. It's better to provide excellent medical care and preventive measures or early treatment for problems as they arise.

Another interesting aspect of identifying susceptibility genes is what we are learning about the interaction of genes and the environment. Take, for example, manic depression. A gene was found on the X chromosome in some Israeli families that appears to predispose toward this disease.

The penetrance rate for this gene is not 100%, which means that even though you may have the gene, you might not necessarily develop the disease. Well, what spares some people?

What other influences were brought to bear on the expression of the gene? It could have been another gene or a couple of genes. It could have been the environment. It could have been the fact that you had terrific parents. Or it could have been what you ate. Who knows?

There's a very complicated interaction between our genes and our environment. You're never going to say that the environment has no impact, no matter how sophisticated our genetic information gets. And, in fact, I think we'll have a better appreciation of the environment in light of this new information because I think we'll be able to see better how inheritance and environmental influences mesh.

You actually spend a lot of your time and considerable energy explaining these things to nonscientists. Is educating the general public an important responsibility for scientists?

Oh, yes. Definitely! This is very important when you consider that the Genome Project is going to put a massive amount of information in our hands, and we must choose what to do with it. The myths need to be dispelled because people are going to have to be making policy and medical decisions about how the information is used.

The Genome Project is funded by taxpayers' dollars, so I think that we are obliged to teach people what it is all about. College students are a critical audience because they're the people who will shape the future. They're the people who are going to process this information—whether they're working on it as scientists or not. When people go to college, they begin to make the kinds of decisions that shape their own lives, and eventually the country and the world. If we can just get the students and the public interested in wanting to learn, that education will be reflected in their awareness, their actions, their policies and laws, and in the vitality of the nation. We just have to open our books and get started.

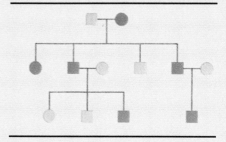

12 Meiosis and Sexual Life Cycles

I n the last unit we discussed how cells manage their material and energy resources by performing various metabolic activities. But cells have another function that is intertwined with their metabolic functions: Cells also process information—information inherited from previous generations.

For thousands of years, people have recognized the fundamental fact of life contained in the adage "Like begets like." Only oaks produce oaks, and only condors can make more condors. Furthermore, offspring generally resemble their parents more than they do less closely related individuals of the same species (Figure 12.1). These observations have been exploited for as long as people have bred domesticated plants and animals. The study of heritable information, called genetics, is the subject of this unit.

THE SCOPE OF GENETICS

Despite the age-old interest in the subject, the mechanisms of heredity eluded biologists until this century, when researchers began to accept the theory that offspring inherit their traits by the transmission of discrete units of information now known as **genes.** The gene concept became less abstract with the discovery that genes consist of the substance DNA. This genetic material can replicate, passing copies of genes along to offspring. In a eukaryotic cell, the DNA is ordered into a manageable number of chromosomes, making it possible to transmit tens of thousands of replicated genes very accurately (see Chapter 11).

Figure 12.1
Two families. Beneath the two sets of parents appear four children in random order; two belong to each set of parents. (All individuals were photographed at about the same age.) Can you match offspring with parents? "Like begets like," but notice that these offspring of sexual reproduction are not carbon copies of their parents and also that siblings are not identical.*

Our study of genetics will encompass three aspects of genes: their transmission, their expression, and their capacity for change. Transmission concerns how genes are organized and how they are passed from one generation to the next. Gene expression involves the way inherited instructions coded by the genes are carried out to produce specific characteristics. Our parents did not, in any literal sense, give us their eyes, hair, or any other traits, but rather endowed us with genetic programs (DNA) for producing specific traits as we developed from fertilized eggs. The third aspect of genes, their capacity for change, is an important link between genetics and evolution. The copying of genes, though remarkably precise, is not completely free of error, and environmental factors can also damage the genetic material. Rare changes in DNA, called mutations, are the ultimate source of new genes, which are

subjected to natural selection during the evolution of a species.

Thus, each organism is the programmed product of genes that are transmitted by orderly mechanisms, expressed by their translation during the development of the organism, and subject to modification. The possession of such genetic programs, a characteristic unique to life, is a central theme of modern biology that will be amplified in this unit.

In this chapter, we begin our study of genetics by focusing on the transmission of chromosomes between generations of organisms that reproduce sexually.

SEXUAL VERSUS ASEXUAL REPRODUCTION

Strictly speaking, "Like begets like" really applies only to organisms that reproduce without sex. In **asexual reproduction,** a single individual is the sole parent and passes on all its genes to its offspring. For exam-

*The boy and girl on either end of the bottom row are children of the parents on the left. The boy and girl in the center of the bottom row are children of the other set of parents.

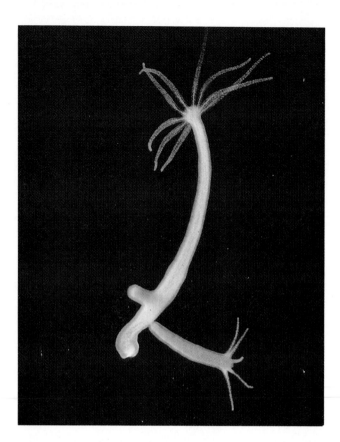

Figure 12.2
Asexual reproduction of *Hydra*. This simple multicellular animal reproduces by budding. The bud, a localized mass of mitotically dividing cells, develops into a small hydra, which detaches from the parent (LM).

1 mm

INTRODUCTION TO SEXUAL LIFE CYCLES: THE HUMAN EXAMPLE

In humans, each **somatic cell**—any cell other than a sperm or egg cell—has 46 chromosomes. Under the microscope, chromosomes can be distinguished from one another by their appearance. They differ in size, position of the centromere, and staining pattern.

Careful examination of a micrograph of the 46 human chromosomes reveals that they can be matched in pairs; that is, there are two of each type. This fact becomes clear when the chromosomes are cut out of the micrograph with scissors and arranged in pairs, starting with the longest chromosomes. The resulting display is called a **karyotype** (see Methods Box, pp. 256–257). The chromosomes that make up a pair—that have the same length, centromere position, and staining pattern—are called **homologous chromosomes,** or homologues. The two chromosomes of each pair carry genes controlling the same inherited traits. For example, if a gene for eye color is located at a particular place, or *locus,* along the length of a certain chromosome, then the homologue of that chromosome will also have a gene specifying eye color at the equivalent locus. Homologous pairs of chromosomes should not be confused with the identical sister chromatids each chromosome has after replication.

There is an important exception to the rule of homologous chromosomes for human somatic cells. It is the two distinct chromosomes referred to as *X* and *Y*. (These chromosomes are in position 23 on the karyotype in the Methods Box.) Human females have a homologous pair of *X* chromosomes, but males have one *X* and one *Y* chromosome, the shortest human chromosome. Because of their role in determining the sex of a person, the *X* and *Y* chromosomes are called **sex chromosomes.** The other chromosomes are called **autosomes.**

The occurrence of pairs of chromosomes in our karyotype is a consequence of our sexual origins. We inherit one member of each chromosome pair from each parent. So the 46 chromosomes in our somatic cells are actually two sets of 23 chromosomes—a maternal set (from one's mother) and a paternal set (from one's father). Cells containing these two sets are called **diploid cells,** and the number of chromosomes in a diploid cell is called the diploid number (abbreviated 2*N*). In humans and most other animals, somatic cells are diploid.

The sperm and the egg are distinct from somatic cells in their chromosome count. Each of these reproductive cells, or **gametes,** has a single set of the 22 autosomes plus a single sex chromosome, either *X* or *Y*. A cell with a single chromosome set is called a **haploid cell.** For humans, the haploid number (abbreviated *N*) is 23. Sexual intercourse allows a haploid

ple, one-celled organisms can reproduce asexually by cell division, in which the genetic material is copied and allocated equally to two daughter cells (see Chapter 11). The offspring are carbon copies of the parent. Some multicellular organisms are also capable of reproducing asexually. *Hydra*, a relative of the jellyfish, can reproduce by budding (Figure 12.2). A new individual begins as a mass of dividing cells growing on the side of the parent. The mass develops into a small hydra, the bud, which eventually detaches from the parent to take up life on its own. Since the cells of the bud were derived by mitosis in the parent, the "chip off the old block" is genetically identical to its parent.

On the other hand, sex-based reproduction usually results in greater variation among the offspring. In **sexual reproduction,** two parents give rise to offspring that have unique combinations of genes inherited from the two parents. Offspring of a sexual union are somewhat different from their parents and siblings (see Figure 12.1). What mechanisms generate this genetic variation among offspring? The key is in the management of chromosomes during the sexual life cycle.

sperm cell from the father to reach and fuse with an egg cell of the mother in the process called **fertilization.** The resulting fertilized egg, or **zygote,** contains the two haploid sets of chromosomes bearing genes representing the maternal and paternal family lines. Together, these 23 chromosome pairs make up the full complement of chromosomes characteristic of a diploid cell.

As a human develops from a zygote to a sexually mature adult, the zygote's genetic endowment is passed on with precision to all somatic cells of the body by the process of mitosis. The sex organs then produce new gametes, which can initiate a new cycle. It has been said that a chicken is just an egg's way of making another egg; in other words, the function of the somatic cells is to allow cells of the **germ line,** the gametes and the cells that give rise to them, to survive and propagate.

Production of gametes requires a special type of cell division. If gametes were made by mitosis, then they would be diploid like the somatic cells. At the next round of fertilization, when two gametes fused, the normal chromosome number of 46 would double to 92, and each subsequent generation would double the number of chromosomes yet again. This does not occur because sexually reproducing organisms carry out a process that halves the chromosome number in the gametes, compensating for the doubling that occurs at fertilization. The special kind of cell division that accomplishes this is called **meiosis.** While mitosis conserves chromosome number, meiosis, which occurs only in our ovaries or testes (gonads), reduces the chromosome number by half. As a result, the human sperm and egg cells that are the products of meiosis have haploid sets of 23 chromosomes. The union of sperm and egg restores the diploid condition, and the human life cycle goes on (Figure 12.3).

In fact, the life cycles of *all* sexually reproducing organisms follow a basic pattern of alternation between the diploid and haploid conditions, even though the details of the life cycles differ. The processes of meiosis and fertilization are the unique trademarks of sexual reproduction.

MEIOTIC CELL DIVISION

Many of the steps of meiosis closely resemble corresponding steps in mitosis. Meiosis, like mitosis, is preceded by the replication of chromosomes. However, this single replication is followed by two consecutive cell divisions, called **meiosis I** and **meiosis II.** These divisions result in four daughter cells (rather than the two daughter cells of mitosis), each with only half as many chromosomes as the parent. The diagrams and text in Figure 12.4 (pp. 252–253) describe

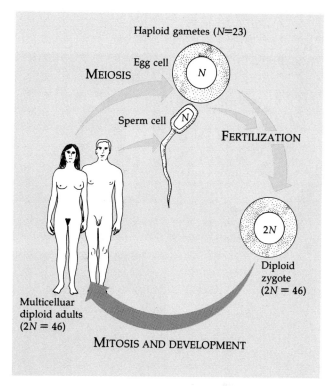

Figure 12.3
The human life cycle. Each generation, the doubling of chromosome number that results from fertilization is offset by the halving of chromosome number that results from meiosis. For humans, the number of chromosomes in a haploid cell is 23 ($N = 23$); the number of chromosomes in the diploid zygote and all somatic cells arising from it is 46 ($2N = 46$).

in some detail the two divisions of meiosis for an animal cell whose diploid number is 4. The micrographs show some of the stages of meiosis for a plant cell. You should examine Figure 12.4 thoroughly before going on to the next section of the chapter.

COMPARISON OF MITOSIS AND MEIOSIS

Let us summarize the key differences between mitosis and meiosis. The chromosome number is reduced by half in meiosis but not in mitosis. The genetic consequences of this difference are important. Whereas mitosis produces diploid daughter cells genetically identical to their parent cell and to each other, meiosis produces haploid cells that differ genetically from their parent cell and from each other.

Figure 12.5, p. 254, compares the key steps in the processes of mitosis and meiosis. Although meiosis involves two nuclear divisions, the events that are unique to meiosis all occur during the first division, meiosis I.

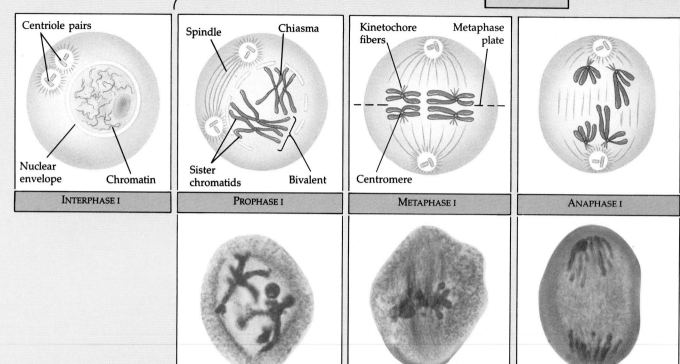

INTERPHASE I   PROPHASE I   METAPHASE I   ANAPHASE I

INTERPHASE I

Meiosis is preceded by an interphase during which each of the chromosomes replicates. This process is similar to the chromosome replication preceding mitosis. For each chromosome, the result is two genetically identical sister chromatids attached at their centromeres. The centriole pairs also replicate.

PROPHASE I

Some important differences between meiosis and mitosis occur in prophase I. Meiotic prophase I lasts longer and is more complex than prophase of mitosis. The chromosomes condense into long thin threads and attach at their ends to the nuclear envelope. In the process called *synapsis*, homologous chromosomes, each made up of two chromatids, come together as pairs into a structure called a *bivalent* (or *tetrad*). Each gene is brought into juxtaposition with its homologous gene on the opposite chromatid. While homologous chromosomes are synapsed, corresponding segments of adjacent nonsister chromatids may exchange by breaking and reattaching to the other chromatid. This event is called *crossing over*, and its genetic significance will be examined later in this chapter. The chromosomes thicken further until they are clearly visible with a light microscope as four separate chromatids, with each pair of sister chromatids linked at their centromeres and nonsister chromatids linked by *chiasmata*, sites of crossing over. The chromosomes then detach from the nuclear envelope.

As prophase I continues, the cell prepares for the division of the nucleus in a manner similar to that observed during mitosis. The centriole pairs move away from each other, and spindle microtubules form between them. The nuclear envelope and nucleoli disperse. Finally, the chromosomes begin their migration to the metaphase plate, midway between the two poles of the spindle apparatus. Prophase I, which can last for days or even longer, typically occupies more than 90% of the time required for meiosis.

METAPHASE I

The chromosome bivalents align themselves on the metaphase plate. In meiosis, both kinetochores of a sister-chromatid pair face the same pole. The centromeres of the homologous chromosome point toward the opposite pole.

ANAPHASE I

As in mitosis, the spindle microtubules interact with the kinetochore fibers and cause the chromosomes to move toward the poles. However, sister chromatids remain attached at their centromeres and move as a single unit toward the same pole. The homologous chromosome moves toward the opposite pole. (In contrast, in mitosis the sister chromatids move toward opposite poles.)

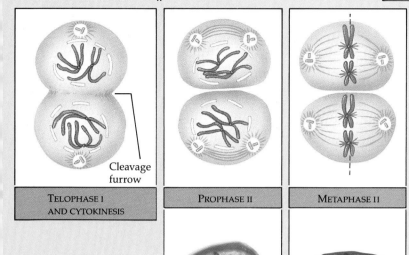

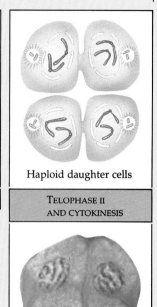

| TELOPHASE I AND CYTOKINESIS | PROPHASE II | METAPHASE II | ANAPHASE II | TELOPHASE II AND CYTOKINESIS |

Cleavage furrow

Haploid daughter cells

25 μm

TELOPHASE I

The spindle apparatus continues to separate the homologous chromosome pairs until the chromosomes reach the poles of the cell. Each pole now has a haploid chromosome set, but each chromosome still has two chromatids. Usually cytokinesis (division of the cytoplasm) occurs simultaneously with telophase I, forming two daughter cells. Cleavage furrows form in animal cells, and cell plates appear in higher plant cells. In some species, nuclear membranes and nucleoli re-form, and there is a period of time, called *interkinesis* (or *interphase II*), before meiosis II. In other species, daughter cells of telophase I begin preparation immediately for the second meiotic division. With or without an interkinesis, there is *no further replication of the genetic material prior to the second division.*

MEIOSIS II

If there has been an interkinesis, the nuclear membrane and nucleoli disperse during **prophase II;** without an interkinesis, this is unnecessary. In either case, during prophase II, a spindle apparatus appears and the chromosomes progress toward its equator, the **metaphase II** plate. The chromosomes align on the metaphase plate in mitosislike fashion, with the kinetochores of sister chromatids of each chromosome pointing toward opposite poles.

In **anaphase II,** the centromeres of sister chromatids finally separate, and the sister chromatids of each pair, now individual chromosomes, move toward opposite poles of the cell.

In **telophase II,** nuclei begin to form at opposite poles of the cell, and cytokinesis occurs. There are now four daughter cells, each with the haploid number of chromosomes.

Figure 12.4
Meiotic cell division. These diagrams show meiotic cell division for an animal cell with a diploid number of 4. The behavior of the chromosomes is emphasized. For information about spindle formation and other nonchromosomal details, see the description of mitosis in Figure 11.8. The photos are light micrographs of some of the meiotic stages as seen in cells from a lily plant.

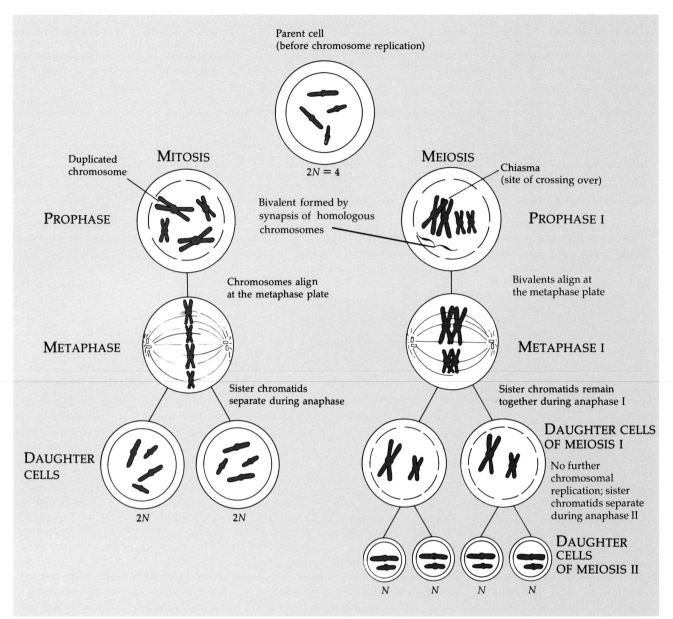

Figure 12.5

Comparison of mitosis with meiosis. In both mitosis and meiosis, the chromosomes are duplicated only once, during the interphase prior to the first prophase. Mitosis entails a single division of the nucleus, usually accompanied by cytokine-sis; the results are two daughter cells genetically identical to the parent cell. Meiosis entails two cell divisions in succession, and the results are four (non-identical) daughter cells, each containing only half the number of chromosomes in the parent cell. The steps of mitosis and meiosis are basically similar, with the events unique to meiosis all taking place during meiosis I. These unique events are called out on the diagram.

1. During prophase I of meiosis, the duplicated chromosomes pair with their homologues, a process called **synapsis.** The four closely associated chromatids are called bivalents, or tetrads. Exchange of genetic material, or **crossing over,** occurs between homologous (nonsister) chromatids; the results of this event are visible in the microscope by the appearance of X-shaped regions called **chiasmata** (singular, *chiasma*). Neither synapsis nor crossing over occurs during mitosis.

2. At metaphase I of meiosis, bivalents consisting of homologous pairs of chromosomes, rather than individual chromosomes, align on the metaphase plate. The homologous pairs are held together by chiasmata.

3. At anaphase I of meiosis, centromeres do not divide and sister chromatids do not separate, as they do in mitosis. Rather, the sister chromatids of each chromosome go to the same pole of the cell. Meiosis I

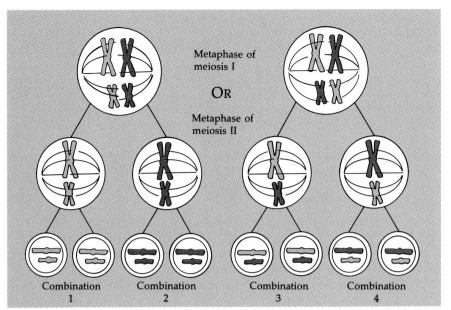

Figure 12.6
Two possible arrangements of chromosomes on the metaphase plate in meiosis I. In this figure, we consider the consequences of meiosis in a hypothetical organism with a diploid chromosome number of 4 ($2N = 4$). The positioning of each homologous pair of chromosomes (bivalent) at metaphase of meiosis I is a matter of chance, like the flip of a coin. The arrangement of chromosomes at metaphase I determines which chromosomes will be packaged together in the haploid daughter cells. For each metaphase I alignment, the chromosomes can be packaged into daughter cells in two different combinations. Thus, a total of four chromosome combinations is possible in the resulting gametes.

Metaphase of meiosis I

OR

Metaphase of meiosis II

Combination 1   Combination 2   Combination 3   Combination 4

separates homologous pairs of chromosomes, not sister chromatids of individual chromosomes.

The second meiotic division, meiosis II, is virtually identical in mechanism to mitosis, separating sister chromatids. Since the chromosomes do not replicate between meiosis I and meiosis II, however, the final outcome of meiosis is a halving of the number of chromosomes per cell.

The process of meiosis is a complicated one, and occasionally errors occur. Such errors can lead to chromosomal abnormalities in gametes and in the diploid individuals arising from those gametes. Chromosomal abnormalities in humans will be discussed in Chapter 14.

SEXUAL SOURCES OF GENETIC VARIATION

In species that reproduce sexually, the sorting out and recombining of chromosomes by meiosis and fertilization are responsible for most of the genetic variation that arises each generation. Variation is important to a population as the raw material on which natural selection works. A closer look at the sexual life cycle will reveal sources of genetic variation.

Independent Assortment of Chromosomes

One way sexual reproduction generates genetic variation is shown in Figure 12.6 above, which colorcodes the chromosomes so that we can keep track of them

as they are transmitted to gametes. At metaphase of meiosis I, each homologous pair of chromosomes, consisting of one maternal and one paternal chromosome, aligns itself on the metaphase plate. The orientation of the homologous pair relative to the two poles of the cell is random; there are two possibilities. Thus, there is a fifty-fifty chance that a particular daughter cell of meiosis I will get the maternal chromosome of a certain homologous pair, and a fifty-fifty chance of receiving the paternal version of that chromosome type. Because each homologous pair of chromosomes is oriented independently of the other pairs at metaphase I, the first meiotic division results in independent assortment of maternal and paternal chromosomes into daughter cells. The gametes produced therefore have all possible combinations of maternal and paternal chromosomes. The number of combinations possible for gametes formed by meiosis starting with two homologous pairs of chromosomes ($2N = 4$, $N = 2$) is four, as shown in Figure 12.6. For the case of $N = 3$, there are eight combinations of chromosomes possible for gametes (Figure 12.7, p. 258). More generally, the number of combinations possible when meiosis packages chromosomes into gametes by independent assortment is 2^N, where N is the haploid number.

In the case of humans, meiosis independently assorts 23 pairs of chromosomes. The number of possible combinations of maternal and paternal chromosomes in the resulting gametes is 2^{23}, which is about eight million. The variations are analogous to the eight million combinations of heads and tails possible for the simultaneous tossing of 23 coins. Thus, each gamete that a human produces contains one of eight million possible assortments of chromosomes inherited from

Karyotypes, ordered displays of an individual's chromosomes, are useful in identifying genetic defects involving microscopically visible abnormalities in the chromosomes. Medical technicians often prepare karyotypes by using lymphocytes, which are a type of white blood cell.

The cells are treated with a drug to stimulate mitosis and grown in culture for several days. They are then treated with another drug to arrest the cell cycle at metaphase, when the chromosomes, each consisting of two joined sister chromatids, are in a state of maximum condensation. (The latter drug, colchicine, destroys the mitotic spindle by binding to tubulin and preventing microtubule formation.) The

drawings that follow outline the further steps in the preparation of a karyotype from lymphocytes.

The micrograph in step 7 shows the karyotype of a normal human male. The 46 chromosomes of a single somatic cell are arranged in 22 homologous pairs, leaving the two sex chromosomes. For a male, these include one X and one Y chromosome. The chromosomes are stained to reveal band patterns, which are helpful in identifying chromosomes and parts of chromosomes. Karyotyping can be used to screen for abnormal numbers of chromosomes or defective chromosomes associated with congenital disorders such as Down syndrome (see Chapter 14).

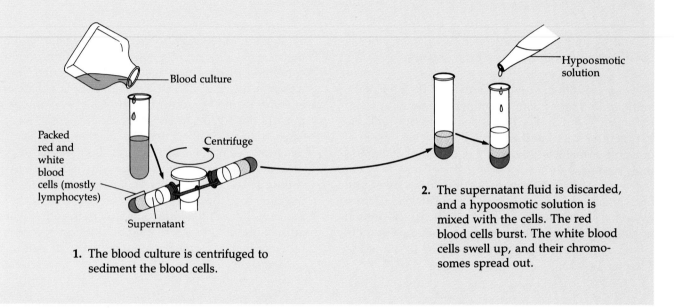

Blood culture

Packed red and white blood cells (mostly lymphocytes)

Centrifuge

Supernatant

1. The blood culture is centrifuged to sediment the blood cells.

Hypoosmotic solution

2. The supernatant fluid is discarded, and a hypoosmotic solution is mixed with the cells. The red blood cells burst. The white blood cells swell up, and their chromosomes spread out.

that individual's mother and father. The significance of this variability is that the maternal and paternal homologues will bear different genetic information at many of their corresponding loci. For example, a paternal chromosome may have a gene that specifies the ability to roll the tongue, while the corresponding locus on the homologous maternal chromosome may have a gene for the lack of tongue-rolling ability. (Tongue-rolling ability actually is controlled by a single genetic locus.)

Crossing Over

The independent assortment of chromosomes is only one way that meiosis contributes to variability. The

meiotic phenomenon of crossing over is also important. Remember that homologous chromosomes come together as pairs during prophase of meiosis I. This pairing, or synapsis, is very precise, with homologues aligning with each other on a gene-by-gene basis. During synapsis, genetic information is exchanged between the paired homologous chromosomes. A segment of a chromatid changes places with the equivalent segment of a homologue. Crossing over is usually a reciprocal process in which identical lengths of chromatids are exchanged between the two homologues. Therefore, there is no loss of genes, but rather an exchange of genes between the two chromosomes.

The exact mechanisms of synapsis and crossing over are not completely understood, but they are known to involve a protein structure called the **synaptonemal**

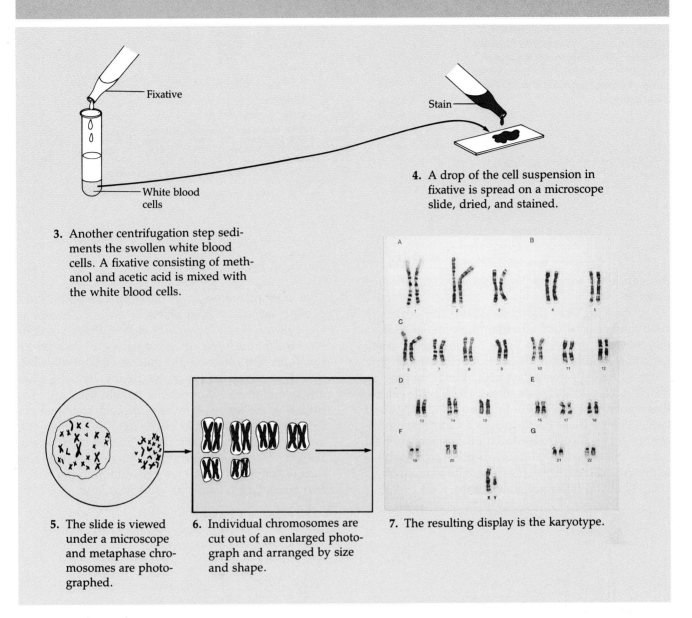

3. Another centrifugation step sediments the swollen white blood cells. A fixative consisting of methanol and acetic acid is mixed with the white blood cells.

Fixative

White blood cells

Stain

4. A drop of the cell suspension in fixative is spread on a microscope slide, dried, and stained.

5. The slide is viewed under a microscope and metaphase chromosomes are photographed.

6. Individual chromosomes are cut out of an enlarged photograph and arranged by size and shape.

7. The resulting display is the karyotype.

complex, which appears in electron micrographs taken in the middle of prophase I (Figure 12.8). In many organisms, the synaptonemal complex looks something like a zipper. On each side of the zipper is attached one homologous chromosome (consisting, at this stage, of two identical sister chromatids). The synaptonemal complex is essential for synapsis and for crossing over between homologous chromosomes. In some micrographs, protein beads are seen scattered along the zipper. These nodules are correlated with the formation of chiasmata later in prophase I. First observed over twenty-five years ago, the synaptonemal complex is an example of a structure for which we do not yet have a detailed functional explanation.

Although no genes are lost in crossing over, new genetic combinations nevertheless result from the exchange of maternal and paternal chromosome segments. In the hypothetical example shown in Figure 12.9, we consider how a crossing-over event affects genes in two loci on the chromosome. (It is actually not known whether the gene loci shown are on the same chromosome.) One gene locus controls tongue-rolling ability. The DNA at this locus on the maternal chromosome dictates this ability, while the DNA at the same locus on the paternal chromosome dictates the lack of this ability. A second gene locus on the same chromosome in our example controls earlobe attachment. The gene on the paternal chromosome is for attached earlobes, while the gene on the maternal chromosome is for free earlobes. Thus, in the absence of crossing over, the gametes could carry either the genes for tongue-rolling ability and free earlobes or

Figure 12.7

Figure 12.7
Possible outcomes of meiosis when
N = 3. For a given gamete, the "choice"
between maternal and paternal chromosomes of any one pair is independent of
how all the other chromosomes have been
distributed. This diagram shows all the
possibilities for the case of three chromosomes. Imagine extending it to 23
chromosomes!

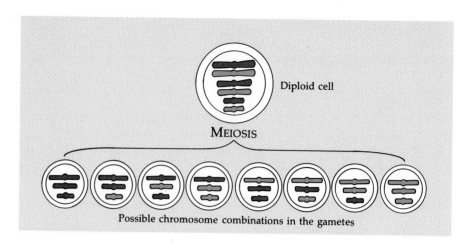

Diploid cell

MEIOSIS

Possible chromosome combinations in the gametes

the genes for tongue-rolling inability and attached
earlobes.

The crossing over shown in Figure 12.9 results in
gene combinations different from those inherited from
the previous generation. This occurrence is called
genetic recombination. In our example, meiosis produces a haploid gamete that carries genes for tongue-
rolling ability and attached earlobes and another gamete
that carries genes for lack of tongue-rolling ability and
free earlobes. These combinations could not have arisen in this particular case without crossing over. Of
course, the recombined segments of chromatids have
many loci besides those shown, and so a single crossover event will influence many genes.

Random Fertilization

The random nature of fertilization compounds the
genetic variability established in meiosis. A human
egg cell, representing one of eight million possibilities, will be fertilized by a single sperm cell, which
can represent one of eight million *different* possibilities. Thus, even without considering crossing over,
any two parents will produce a zygote with any of 64
trillion (8 million × 8 million) diploid combinations
(actually, the exact number is 70,368,744,177,664). It is
no wonder brothers and sisters can be so different.
Looking beyond a single family, we can calculate that
each different set of parents generates another 70 trillion possibilities. Each one of us is unique.

Homologous
chromosomes

Synaptonemal
complex

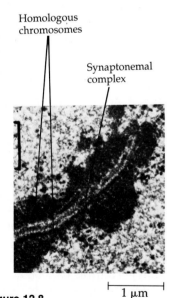

1 μm

Figure 12.8
Synaptonemal complex. Electron micrograph (left) of the synaptonemal complex
between a pair of homologous chromosomes from a lily. The synaptonemal com-

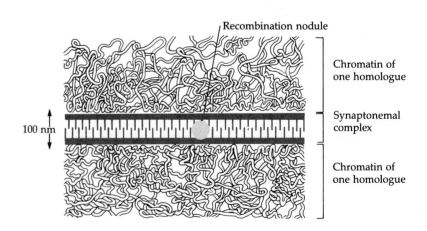

Recombination nodule

Chromatin of
one homologue

100 nm

Synaptonemal
complex

Chromatin of
one homologue

plex is a zipperlike assembly of proteins
that holds homologous chromosomes
tightly together in synapsis, during prophase I of meiosis. It seems to be essen-

tial for crossing over, which may be facilitated by the proteins comprising the so-
called recombination nodules.

Tongue-rolling   Earlobe-attachment
genes   genes

t e ⎤ Paternal   ⎤ Homologous
                    ⎦ chromosome   │ pair of
                    ⎤ Maternal   │ chromosomes
T E ⎦ chromosome   ⎦ in synapsis

↓ Breakage of homologous chromatids

t e

T E

↓ Rejoining of homologous chromatids

t e
                    — Chiasma
T E

↓ Terminalization of chiasma during anaphase I

t e

T E

t e
t E
T e
T E

↓ Anaphase II and telophase

t e Parental type of gamete
t E Recombinant gamete
T e Recombinant gamete
T E Parental type of gamete

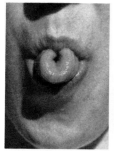
Tongue rolling

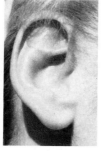
Free earlobes

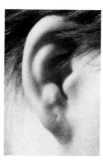
Attached earlobes

Figure 12.9
Genetic recombination as a result of crossing over.
In this hypothetical example, we follow the process of
crossing over between genes for tongue-rolling ability
and earlobe attachment during meiosis. (It is actually not
known if the genes for these two traits are on the same
chromosome.) The following abbreviations are used for
genes: T = tongue-rolling ability, t = lack of tongue-
rolling ability, E = free earlobes, e = attached earlobes.
The photos illustrate tongue rolling and contrast free and
attached earlobes.

So far, we have seen that there are three sources of
genetic variability in a sexually reproducing population of organisms:

• Independent assortment of homologous chromosome pairs during meiosis I

• Recombination between homologous chromosomes
as a result of crossing over in prophase of meiosis I

• Random fertilization of an egg by a sperm

All three mechanisms reshuffle the various genes carried by the individual members of a population and
thus affect how successfully an individual will cope
with its environment. However, as will be discussed
in later chapters, **mutations,** rare changes in the DNA
of genes, are what ultimately create the diversity of
genes in a species. In fact, the very existence of different forms of a gene for a given trait originates in
mutation.

THE VARIETY OF SEXUAL LIFE CYCLES

Although the alternation of meiosis and fertilization
is common to all organisms that reproduce sexually,
the timing of these two events in the life cycle varies,
depending on the species (Figure 12.10). In the human
life cycle, the gametes are the only haploid cells, and
the multicellular organism is diploid. Meiosis occurs
during the production of gametes, which undergo no
further cell division prior to fertilization. Such is the
timing of meiosis and fertilization for most animals.
However, in many fungi and some protists (including
some algae), the only diploid stage is the zygote.
Meiosis occurs immediately after the gametes fuse,
and mitosis then creates a multicellular adult organism that is haploid. Subsequently, gametes are produced from the haploid organism by mitosis rather
than by meiosis.

Plants and some species of algae go through a third
type of life cycle called **alternation of generations.** There
are both diploid and haploid multicellular stages in
this type of life cycle. The multicellular diploid stage
is called the **sporophyte.** Meiosis in the sporophyte
produces haploid cells called **spores.** Unlike a gamete,
a spore gives rise to a multicellular individual without
fusing with another cell. A spore divides mitotically
to generate a multicellular haploid stage called the
gametophyte. The haploid gametophyte makes
gametes by mitosis. Fertilization results in a diploid
zygote, which develops into the next sporophyte generation. Hence, in this type of life cycle, the sporophyte and gametophyte generations take turns repro-

Figure 12.10
Three sexual life cycles differing in the timing of meiosis and fertilization. The common feature of all three cycles is the alternation of these two key events.

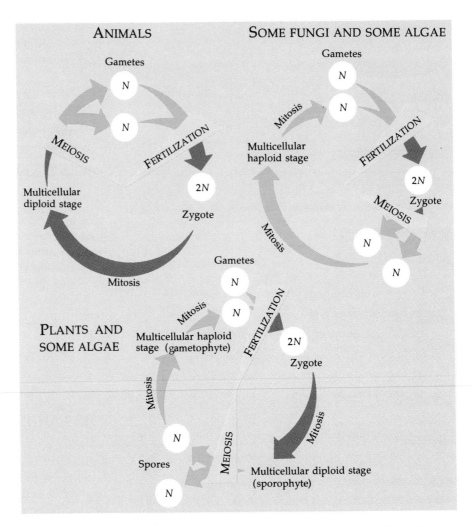

ANIMALS

SOME FUNGI AND SOME ALGAE

Gametes

N

N

MEIOSIS

FERTILIZATION

Multicellular
diploid stage

2N

Zygote

Mitosis

Gametes

N

N

Mitosis

FERTILIZATION

Multicellular
haploid stage

2N

Zygote

MEIOSIS

Mitosis

N

N

PLANTS AND
SOME ALGAE

Gametes

N

Mitosis

N

FERTILIZATION

Multicellular haploid
stage (gametophyte)

2N

Zygote

Mitosis

Mitosis

MEIOSIS

Multicellular diploid stage
(sporophyte)

N

Spores

N

ducing each other. We will examine various kinds of life cycles in more detail in Unit Five.

In this chapter, we have followed the transmission of chromosomes from parents to offspring in sexual life cycles. In the next chapter, we will see how Gregor Mendel discovered the laws of inheritance decades before the role of chromosomes was understood.

STUDY OUTLINE

1. Genetics, traditionally the study of heredity, now encompasses the molecular basis of cellular information.

The Scope of Genetics (pp. 248–249)

1. The genetic material consists of DNA organized into genes, discrete units of information on the chromosomes.
2. Genes are transmitted, expressed as specific characteristics, and subject to mutation.

Sexual Versus Asexual Reproduction (pp. 249–250)

1. In asexual reproduction, a single parent gives rise to genetically identical offspring by mitosis.
2. Sexual reproduction combines two different sets of genes, carried by gametes from two different parents, to produce variable offspring.

Introduction to Sexual Life Cycles: The Human Example (pp. 250–251)

1. Normal human somatic cells contain 46 chromosomes, half inherited from the father and half from the mother.

2. Each of 22 autosomes in the paternal set has a corresponding homologous chromosome in the maternal set. The twenty-third pair, the sex chromosomes, determines whether the person is a female *(XX)* or a male *(XY)*.
3. Single, haploid *(N)* sets of chromosomes in maternal and paternal gametes unite during fertilization to produce a diploid *(2N)*, single-celled zygote. The zygote develops into a multicellular individual by mitosis.
4. At sexual maturity, the gonads produce haploid gametes by meiosis in preparation for a new diploid generation.
5. All sexually reproducing organisms alternate diploid and haploid states through fertilization and meiosis.

Meiotic Cell Division (p. 251)

1. Meiosis is a two-stage process resulting in four daughter cells, each with half the parental chromosome number.

2. The two cell divisions that follow the single replication of chromosomes are called meiosis I and meiosis II.

Comparison of Mitosis and Meiosis (pp. 251–255)

1. Meiosis is distinguished from mitosis by a series of distinctive events that occurs during meiosis I.
2. In prophase I of meiosis, duplicated homologous chromosomes undergo synapsis. This association of two pairs of sister chromatids (the bivalents) allows exchange of genetic material by crossing over of homologous segments. The crossing-over sites are visible as chiasmata.
3. The bivalents align themselves on the metaphase plate, and at anaphase I, homologous pairs (rather than sister chromatids) are pulled toward separate poles, thereby halving the number of chromosomes in the daughter cells.
4. Meiosis II separates the sister chromatids in a process identical to mitosis to form four haploid daughter cells.

Sexual Sources of Genetic Variation (pp. 255–259)

1. Variation in a population of sexually reproducing organisms is due to independent assortment of homologous chromosomes, crossing over, and random fertilization.
2. In independent assortment, the orientation of homologous chromosome pairs relative to the two poles of the cell during metaphase I is random with respect to maternal or paternal origin of the chromosomes. There are 2^N possible combinations of chromosomes in the gametes.
3. The exchange of homologous segments during synapsis generates further variability by genetic recombination.
4. The genetic variation originating in meiosis is compounded by the random nature of fertilization.

The Variety of Sexual Life Cycles (pp. 259–260)

1. Differences in the timing of meiosis with respect to fertilization generate a variety of sexual life cycles. Multicellular organisms may be diploid (as in animals), haploid (as in some fungi), or may alternate generations between haploid and diploid (as in plants).

SELF-QUIZ

1. A human cell containing 22 autosomes and a Y chromosome is probably a
   a. somatic cell of a male   d. sperm cell
   b. zygote   e. unfertilized egg cell
   c. somatic cell of a female

2. Homologous chromosomes segregate during
   a. mitosis   c. meiosis II
   b. meiosis I   d. fertilization

3. Meiosis II is similar to mitosis in that
   a. homologous chromosomes synapse
   b. DNA replicates before the division
   c. the daughter cells are diploid
   d. sister chromatids separate during anaphase
   e. chromosome number is reduced

4. The DNA content of a diploid cell in the G_1 phase of the cell cycle is measured. If this DNA content is X, then the DNA content of the same cell at metaphase of meiosis I would be
   a. $0.25X$ b. $0.5X$ c. X d. $2X$ e. $4X$

5. If we continued to follow the cell lineage from question 4, then the DNA content at metaphase of meiosis II would be
   a. $0.25X$ b. $0.5X$ c. X d. $2X$ e. $4X$

6. Crossing-over most commonly occurs during
   a. prophase I   d. prophase II
   b. anaphase I   e. telophase II
   c. interphase

7. All of the following are sexual sources of genetic variation *except*
   a. crossing-over
   b. mutation
   c. production of offspring from two parents
   d. independent assortment of homologous chromosomes
   e. random fertilization

8. How many different combinations of maternal and paternal chromosomes can be packaged in gametes made by an organism with a diploid number of 8 (2N = 8)?
   a. 2   b. 4   c. 8   d. 16   e. 32

9. The direct product of meiosis in a plant is a
   a. spore   d. sporophyte
   b. gamete   e. gametophyte
   c. zygote

10. Somatic cells of the adult body are haploid in many
    a. vertebrates   c. fungi
    b. invertebrates   d. vascular plants

CHALLENGE QUESTIONS

1. In domestic turkeys, viable offspring are sometimes produced by development of an unfertilized egg cell, in a process called parthenogenesis. Such offspring, like the mother, are diploid. What variation in meiosis could produce a diploid organism without fertilization?

2. Many species can reproduce *either* asexually or sexually. In a favorable, stable environment, these species generally reproduce asexually. It is usually when the environment changes in some way that is unfavorable to the existing populations of these species that the organisms begin to reproduce sexually. Based on your knowledge of natural selection, explain the evolutionary significance of this switch from asexual to sexual reproduction.

FURTHER READING

1. Becker, W. M. *The World of the Cell.* Menlo Park, Calif.: Benjamin/Cummings, 1986, Chap. 16.
2. Pickett-Heaps, J., D. Tippit, and K. Porter. "Rethinking Mitosis." *Cell* 29 (1982): 729–744. An article about the evolution of mitosis and meiosis for the advanced student.
3. Prescott, D. *Cells.* Boston: Jones and Bartlett, 1988. Helpful chapters on mitosis and meiosis.

13 Mendel and the Gene Idea

A person's eyes can be blue, brown, green, gray, or hazel; a person's hair can be different shades of blond, brown, red, or black; a parakeet's feathers can be green, blue, or yellow, with black or gray markings. What causes these biological spectra of colors? That is, what messages are passed from one generation to the next, and what are the rules that govern how messages from two parents interact in the offspring?

Theories of inheritance date back at least to ancient Greece. Aristotle, for instance, suggested that particles called pangenes come together from all parts of the body to form the eggs and semen. This theory, known as *pangenesis*, prevailed into the nineteenth century and was accepted by such important biologists as Jean Baptiste Lamarck and Charles Darwin. With this mechanism of inheritance, changes that occurred in various parts of the body during an organism's life could be passed on to the next generation. Pangenesis has been proved incorrect on several counts: The reproductive cells are not made up of contributions from body cells, and even dramatic changes in body cells do not influence eggs and sperm. But the idea of a particulate basis of heredity has survived, and the basic units of inheritance are now called genes, after the pangenes.

In the seventeenth century, early microscopic observations gave rise to two additional, and contradictory, theories of inheritance. Followers of the Dutch lens maker Anton van Leeuwenhoek thought they could see in human sperm a miniature human being, the *homunculus*. They believed that the mother only serves as an incubator to this tiny creature and that all inherited characteristics come from the father. During the same period, however, the followers of another Dutchman, Regnier de Graaf, subscribed to an opposite and equally exclusive belief. De Graaf was the first to describe the ovarian follicle, the structure in which human egg cells are produced. His followers felt that the egg contains an entire human being in miniature and that the semen's purpose is merely to stimulate its growth. Thus, according to this theory, all inherited characteristics come from the mother.

Not until the early nineteenth century did biologists finally realize that both parents contribute to the characteristics of their offspring. That observation was based on the breeding of ornamental plants. The favored explanation of inheritance then became the "blending" theory, the idea that the hereditary materials contributed by the male and female parents mix to form the offspring, much the way blue and yellow paints blend to make green. According to this theory, a blue parakeet mating with a yellow parakeet would produce green offspring, and once blended, the hereditary material would be as inseparable as are pigments of paint. In the offspring of the green parakeets, the pure blue and yellow feathers would never reappear. And over many generations, a freely mating population of blue and yellow parakeets should reach a uniform green—a simple prediction that is not borne out by reality. The blending theory was also unable to explain certain other common phenomena of inheritance, such as traits that disappear in one generation and then reappear in the next.

Modern genetics had its genesis in an abbey garden, where in the 1860s, an Augustinian monk named Gregor Mendel discovered the fundamental principles of inheritance (Figure 13.1). This chapter describes his theory and its applications.

MENDEL'S MODEL OF INHERITANCE

Mendel's work, which did not influence the general scientific community until years after his death, demonstrated that parents pass on to their offspring discrete heritable factors—genes—which retain their individuality generation after generation. In contrast to the blending theory, heredity is particulate in the Mendelian model.

Mendel's Methods

Mendel discovered patterns of inheritance by breeding garden peas. While studying at the University of Vienna, Mendel was influenced by his physics professor to adopt a quantitative approach to experimentation that contrasted sharply with the nonmathe-

Figure 13.1
Gregor Mendel. Working with pea plants, Mendel discovered the fundamental principles of heredity in the 1860s. He was an Augustinian monk who lived in Brünn, Austria (now Brno, Czechoslovakia).

matical methods that dominated nineteenth-century biology.

Mendel probably chose to work with peas because they are easy to grow and are available in many readily distinguishable varieties. The use of peas also gave Mendel strict control over which plants mated with which. The petals of the pea flower almost completely enclose the female and male parts (carpel and stamens); normally, the plants self-fertilize after pollen grains released from the stamens land on the carpel (Figure 13.2). Mendel could prevent fertilization between different plants, or cross-pollination, by covering the flowers with small bags. When he wished to cross-pollinate his peas, he would remove the immature stamens of a plant before they produced pollen and then dust pollen from another plant onto the emasculated flowers. Whether ensuring self-pollination or executing artificial cross-pollination, Mendel could always be sure of the parentage of new seeds.

Mendel's success was due not only to his quantitative approach, but also to his selection of traits to study. Mendel chose to follow inherited characteristics that differ in a relatively clear-cut manner. For instance, one variety of pea produces green pods, while another produces yellow pods. He recognized this difference in pod color as an either-or, rather than a more-or-less, difference. Including pod color, Mendel selected

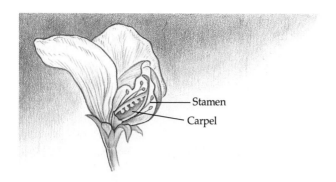

Figure 13.2
Cut-away view of the flower of a garden pea. Male reproductive organs (stamens) and a female reproductive organ (carpel) are both present in the flower of the pea *Pisum sativum.* Pollen, bearing sperm nuclei, forms in the tips (anthers) of the stamens, while eggs are made in the inflated ovary at the lower end of the carpel. In self-fertilization, the pollen's sperm nuclei fertilize eggs of the same plant.

a total of seven characteristics, each occurring in two alternative forms (Figure 13.3). He worked with the plants until he was sure that he had varieties that were **true-breeding;** for instance, he identified a green-pod variety that, when self-fertilized, produced only green-pod offspring.

Now Mendel was ready to see what would happen when he hybridized his true-breeding varieties—for example, when he performed cross-pollination between plants with green pods and plants with yellow pods.

The parental plants of such a cross-fertilization, or **cross,** are referred to as the **P generation** (for parental), and their hybrid offspring are the F_1 **generation** (for first filial). Mendel then allowed the F_1 plants to self-pollinate to produce the next generation, the F_2 **generation.** By following the transmission of well-defined traits such as pod color for several generations, Mendel arrived at two principles of heredity, now known as the law of segregation and the law of independent assortment.

Mendel's Law of Segregation

What would be the pod color of the offspring, or progeny, of a cross between green-pod and yellow-pod parents? Mendel found that pod color was not blended to form chartreuse hybrid peas. Instead, all the F_1 plants had green pods, and the yellow-pod trait had totally and mysteriously disappeared. Was the heritable factor for yellow pods now lost from the plants as a result of hybridization, as the blending theory would predict? If so, then the F_1 plants should only be capable of producing green-pod progeny.

When Mendel allowed the F_1 plants to self-pollinate, however, the yellow-pod trait reappeared in the next generation. Among these F_2 progeny, 428 plants had green pods and 152 plants had yellow pods; that is, there was a ratio of about three plants with green pods to every one with yellow pods (Figure 13.4). Mendel concluded that the heritable factor for yellow pods was not lost in the F_1 plants, but was merely

Figure 13.3
The seven pea characteristics studied by Mendel. Each trait occurs in two alternative forms.

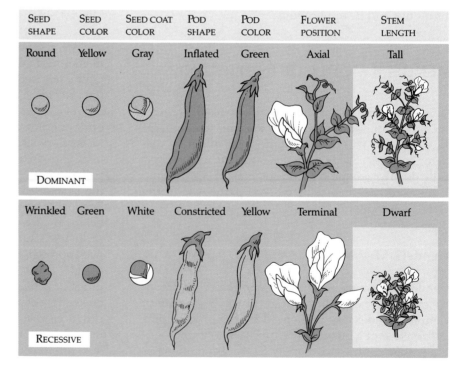

SEED SHAPE	SEED COLOR	SEED COAT COLOR	POD SHAPE	POD COLOR	FLOWER POSITION	STEM LENGTH
Round	Yellow	Gray	Inflated	Green	Axial	Tall
DOMINANT						
Wrinkled	Green	White	Constricted	Yellow	Terminal	Dwarf
RECESSIVE						

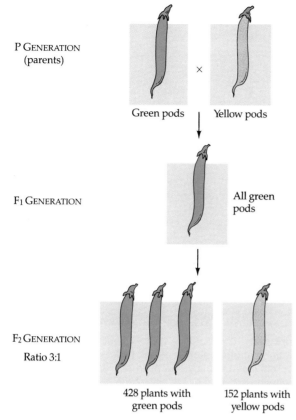

P GENERATION
(parents)

Green pods × Yellow pods

F₁ GENERATION

All green pods

F₂ GENERATION
Ratio 3:1

428 plants with green pods

152 plants with yellow pods

Figure 13.4
The result of a Mendelian cross. Mendel made F₁ hybrids by cross-pollinating between two true-breeding varieties, one with green pods and the other with yellow pods. The F₁ hybrids all had green pods, and when they were allowed to self-pollinate, 75% of the F₂ progeny had green pods while 25% had yellow pods—a 3:1 ratio.

masked by the presence of the green-pod factor, or gene, as we now call such a heritable factor. Based on his observations and conclusions, Mendel formulated his hypothesis of inheritance, which can be broken down into four parts:

1. *There are alternative forms for genes, the units that determine heritable characteristics.* The gene for pod color, for example, exists in two alternative forms, one for green pods and the other for yellow pods: Such alternative forms of genes are now called **alleles.**

2. *For each inherited characteristic, an organism has two alleles, one inherited from each parent.* In Mendel's experiments, one parental variety had a pair of alleles for green pod color, while the other had a pair of alleles for yellow pod color. Mendel's F₁ hybrids had inherited from the parental plants one allele for green pod color and one allele for yellow pod color.

3. *A sperm or egg carries only one allele for each inherited characteristic, because allele pairs separate (segregate) from*

each other during the production of gametes. When a sperm and egg unite during fertilization, both contribute their alleles, thus restoring the gene to the paired condition. In Mendel's experiments, each gamete of a parental plant carried one allele for that plant's pod color, specifying either green or yellow. Cross-pollination resulted in the "mixed" gene combination of the F₁ hybrids.

4. *When the two alleles of a pair are different, one is fully expressed and the other is completely masked. These are called the* **dominant allele** *and the* **recessive allele,** *respectively.* According to this idea, the F₁ hybrids had green pods because the allele for that trait is dominant over the allele for yellow pods, which is recessive.

This hypothesis can explain the 3:1 ratio of progeny plant types that Mendel observed in the F₂ generation. The hypothesis predicts that the F₁ hybrids will produce two classes of gametes. When gene pairs separate, half of the gametes receive a green-pod allele while the other half get a yellow-pod allele. During self-pollination, these two classes of gametes unite randomly. An egg with a green-pod allele has an equal chance of being fertilized by a sperm carrying a green-pod allele or a sperm with a yellow-pod allele. Since the same is true for an egg with a yellow-pod allele, there are a total of four equally likely combinations of sperm and egg. Figure 13.5 illustrates these combinations using a type of diagram called a **Punnett square,** a handy device for predicting the results of a genetic cross.

What will be the physical appearance of these F₂ plants? One-fourth of the plants have two alleles specifying green pod color; clearly, these plants will have green pods. But one-half of the F₂ progeny have inherited one allele for green pods and one allele for yellow pods; like the F₁ plants, these plants will also have green pods, since the allele for green pod color is dominant over the allele for yellow pod color. Finally, one-fourth of the F₂ plants have inherited two alleles specifying yellow pod color and will in fact have yellow pods because there are no dominant alleles present to mask the expression of this recessive trait. Thus, Mendel's model accurately explains the 3:1 ratio that he observed in the F₂ generation.

The pattern of inheritance was the same for each of the seven characteristics that Mendel studied (Table 13.1, p. 267). One parental trait would disappear in the F₁ generation only to reappear in one-fourth of the F₂ progeny. The mechanism underlying this general pattern of inheritance is stated by Mendel's **law of segregation:** *Allele pairs segregate (separate) during gamete formation, and the paired condition is restored by the random fusion of gametes at fertilization.*

Research since Mendel's time has clearly established that the law of segregation applies to all sexually reproducing organisms. For example, breeders

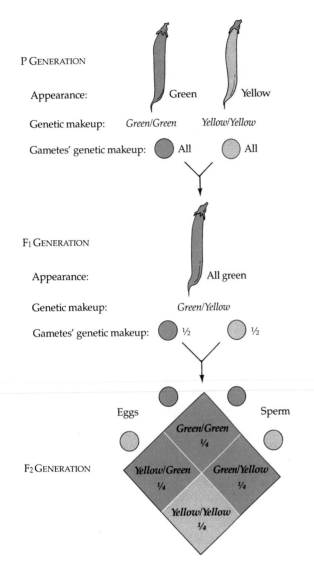

Appearance: Green   Yellow

Genetic makeup: *Green/Green*   *Yellow/Yellow*

Gametes' genetic makeup: All   All

F₁ GENERATION

Appearance: All green

Genetic makeup: *Green/Yellow*

Gametes' genetic makeup: ½   ½

Eggs   Sperm

Green/Green ¼

F₂ GENERATION

Yellow/Green ¼   *Green/Yellow* ¼

Yellow/Yellow ¼

Figure 13.5
Mendel's law of segregation. To explain the results of his cross, Mendel proposed that a plant has a pair of genes for pod color; that the gene pair separates, or segregates, when gametes are formed; and that the paired condition is restored by fertilization. To explain the green pod color of the F₁ hybrids, Mendel hypothesized that the green factor (gene) is dominant over the yellow factor. Segregation in the F₁ plants followed by random fertilization would generate the ratio of three green to one yellow that Mendel observed in the F₂ generation. The diamond at the bottom of the figure is a Punnett square. The circles represent gametes; their colors indicate which pod-color allele each carries (the actual gametes are not colored).

Figure 13.6
Variations in bird color. In nature, budgies are most often light green with a yellow head and black markings, a coloration exhibited by the bird on the left. A variant lacking the yellow pigmentation is called "sky blue" (bird on the right); it has the normal black markings, but has white instead of yellow feathers on its head, and blue instead of green feathers elsewhere. The genetic basis for green and sky-blue feathers is analogous to that for green pods and yellow pods in Mendel's pea plants. In sky-blue birds, both alleles for the yellow-pigment gene are defective; such defective alleles are recessive to the normal yellow-pigment allele. Mendel's law of segregation predicts what will happen if a sky-blue budgie is mated with a true-breeding green budgie: The sky-blue trait will disappear in the first generation of budgie offspring (F₁), but will reappear in the F₂ generation. The ratio of green to sky-blue birds in the F₂ generation will be 3:1. Can you write out the genetic makeup of the F₁ and F₂ generations of birds? Use Figure 13.5 as a model.

of small parrots called budgies observe a striking demonstration of this law in the inheritance of feather color (Figure 13.6).

Some Useful Genetic Vocabulary Geneticists symbolize alleles in various ways. For now, we will indicate a dominant allele with a capital letter (for example, G for green pea pods) and its recessive allele with the lowercase form of the same letter (g for yellow pods) (Figure 13.7).

An organism that has a pair of identical alleles for a characteristic is said to be **homozygous** for that trait. A pea plant with two alleles for green pods (GG) is an example. Allowed to self-pollinate, such homozygotes will produce all green progeny, because all gametes will carry the G allele. Pea plants with yellow pods are homozygous for the recessive allele (gg). They are also true-breeding, producing offspring that resemble the parents in this particular trait. If we cross dominant homozygotes with recessive homozygotes, as in the parental cross of Figure 13.5, all the progeny will

Table 13.1 Results of Mendel's crosses for seven characteristics of peas*

P Cross	F₁ Generation	F₂ Generation	Actual Ratio
Round × wrinkled seeds	All round	5474 round, 1850 wrinkled	2.96:1
Yellow × green seeds	All yellow	6022 yellow, 2001 green	3.01:1
Gray × white seed coats	All gray	705 gray, 224 white	3.15:1
Inflated × constricted pods	All inflated	802 inflated, 229 constricted	2.95:1
Green × yellow pods	All green	428 green, 152 yellow	2.82:1
Axial × terminal flowers	All axial	651 axial, 207 terminal	3.14:1
Tall × dwarf stem	All tall	787 tall, 277 dwarf	2.84:1
All characteristics combined		14,889 dominant, 5010 recessive	2.98:1

*Boldface type indicates the dominant trait.

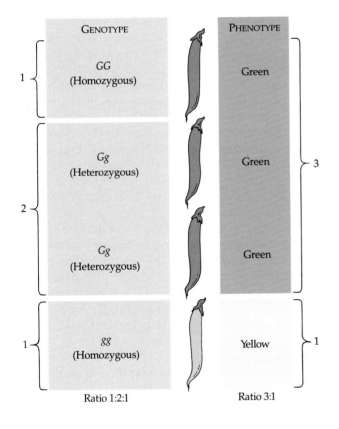

Figure 13.7
Genotype versus phenotype. In this F₂ generation from a cross for pod color, the genotypic ratio is 1*GG*:2*Gg*:1*gg*, and the phenotypic ratio is three green to one yellow. If allowed to self-pollinate to produce an F₃ generation, the dominant homozygotes (*GG*) will be true-breeding for the green-pod trait, and the recessive homozygotes (*gg*) will breed true for the yellow trait. The F₃ progeny of the heterozygotes (*Gg*) will show a 3:1 phenotypic ratio of plants with green pods and plants with yellow pods.

have a mixed allele pair *(Gg)*. Organisms that have two different alleles for a trait are said to be **heterozygous** for that trait. Unlike homozygotes, heterozygotes are not true-breeding, because they produce gametes having one *or* the other of the different alleles. Indeed, we have seen that a *Gg* plant of the F₁ generation will produce both green-pod and yellow-pod progeny when self-pollinated.

Because some alleles are dominant over others, an organism's appearance does not always reflect its genetic composition. Hence, geneticists distinguish between an organism's expressed traits, called its **phenotype,** and its genetic makeup, its **genotype.** In the F₂ generation in Figure 13.7, there is a 3:1 *phenotypic ratio* of plants with green pods to plants with yellow pods. But the *genotypic ratio* of the F₂ generation is 1:2:1 (1*GG* : 2*Gg* : 1*gg*).

The Testcross   Suppose you are presented with a pea plant that has green pods. How can you determine whether it is homozygous or heterozygous? If you cross this pea plant with one having yellow pods, the appearance of the progeny will reveal the genotype of the green-pod parent. The genotype of the yellow-pod parent is known: Because yellow is recessive, the plant must be homozygous. If all the progeny of the cross have green pods, then the other parent is probably homozygous also, since a *GG* × *gg* cross produces nothing but *Gg* progeny. But if both the green and yellow phenotypes appear among the progeny, then the green-pod parent must be heterozygous. The progeny of a *Gg* × *gg* cross will have a 1:1 phenotypic ratio of *Gg* and *gg* (Figure 13.8). This breeding of a recessive homozygote with an organism of unknown genotype, called a **testcross,** was devised by Mendel and continues to be an important tool of geneticists.

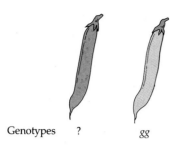

Genotypes ? gg

Two possibilities for the genotype of the green-pod plant:

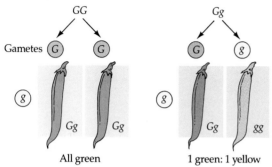

All green   1 green: 1 yellow

Figure 13.8
Mendelian testcross. Frequently, the outward appearance of an organism does not reveal its genotype. But when the organism is bred with one that shows the recessive characteristic (known to reflect a homozygous recessive genotype), the phenotypes of the offspring will indicate the genotype of the parent in question. Such a cross is called a testcross.

Inheritance as a Game of Chance

The segregation of allele pairs during gamete formation and the reconstitution of pairs at fertilization obey the same rules of probability that apply to the tossing of coins, the rolling of dice, or the drawing of cards.

The probability scale ranges from 0 to 1. An event that is certain to occur has a probability of 1, while an event that is certain *not* to occur has a probability of 0. With a two-headed coin, the probability of tossing heads is 1, and the probability of tossing tails is 0. With a normal coin, the chance of tossing heads is $\frac{1}{2}$ and the chance of tossing tails is $\frac{1}{2}$. The probability of rolling the number 3 with a die, which is six-sided, is $\frac{1}{6}$, and the chance of drawing a queen of spades from a full deck of cards is $\frac{1}{52}$. The probabilities of all possible outcomes for an event must add up to 1. With a die, the chance of rolling a number other than 3 is $\frac{5}{6}$. In a deck of cards, the chance of drawing a card other than a queen of spades is $\frac{51}{52}$.

There is an important lesson about probability that we can learn from tossing a coin. For each and every toss of the coin, the probability of heads is $\frac{1}{2}$. The outcome of any particular toss is unaffected by what has happened on previous attempts. We refer to phe-

nomena such as successive coin tosses as *independent events*. It is entirely possible that five successive tosses of an honest coin will produce five successive heads. Before the sixth toss, an observer might predict, "A tail is due to come up because there have already been so many heads." But on the sixth toss, the chance that the outcome will again be heads is still $\frac{1}{2}$. There are two basic laws of probability that will help you both in games of chance and in solving genetic problems.

Rule of Multiplication If two coins are tossed simultaneously, the outcome for each coin is an independent event, unaffected by the other coin. What is the chance that both coins will land heads up? The probability of such a compound event is equal to the product of the separate probabilities of the independent single events. According to this rule of multiplication, the probability that both coins will land heads up is $\frac{1}{2} \times \frac{1}{2} = \frac{1}{4}$. A Mendelian F_1 cross is analogous to this game of chance. With pod color as the heritable characteristic, the genotype of an F_1 plant is *Gg*. What is the probability that a particular F_2 plant will have yellow pods? For this to happen, both the egg *and* the sperm must carry the *g* allele, so we invoke the rule of multiplication. Segregation in the heterozygous plant is like flipping a coin. The probability that an egg will have the *g* allele is $\frac{1}{2}$. The chance that a sperm will have the *g* allele is also $\frac{1}{2}$. Thus, the overall probability that two *g* alleles will come together at fertilization is $\frac{1}{2} \times \frac{1}{2} = \frac{1}{4}$, equivalent to the probability that two independently tossed coins will land heads up (Figure 13.9).

Rule of Addition What is the probability that an F_2 plant will be heterozygous? Notice in Figure 13.9 that there are two ways that F_1 gametes can combine to produce a heterozygous result. The dominant allele can be in the egg and the recessive allele in the sperm, or vice versa. According to the rule of addition, the probability of an event that can occur in two or more alternative ways is the *sum* of the separate probabilities of the different ways. Using this rule, we can calculate the probability of an F_2 heterozygote as $\frac{1}{4} + \frac{1}{4} = \frac{1}{2}$.

The Statistical Nature of Inheritance If we plant a seed from the F_2 generation of Figure 13.5, we cannot predict with absolute certainty that the plant will grow up to produce yellow pods, any more than we can predict with certainty that two tossed coins will both come up heads. What we can say is that there is exactly a $\frac{1}{4}$ chance that the plant will have yellow pods. We can put this another way: Among a large sample of F_2 plants, one-fourth (25%) will have yellow pods. Usually, the larger the sample size, the closer the results will conform to our predictions. The fact that Mendel counted so many progeny from his crosses indicates that he understood this statistical feature of inheri-

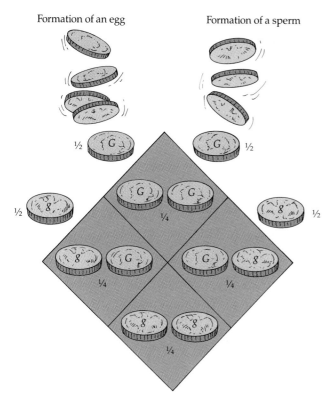

Formation of an egg   Formation of a sperm

½   ½

½   ¼   ½

¼   ¼

¼

Figure 13.9
Segregation and fertilization as chance events. When a heterozygote *(Gg)* forms gametes, segregation of alleles is like the toss of a coin. An egg has a 50% chance of receiving the dominant allele and a 50% chance of receiving the recessive allele. The same odds apply to a sperm cell. Like two separately tossed coins, segregation during sperm and egg formation occurs as independent events. To determine the probability that both the egg and the sperm will carry the dominant allele, we multiply the probabilities of each required event: $\frac{1}{2} \times \frac{1}{2} = \frac{1}{4}$. Similarly, we can predict the probability for any genotype among offspring, as long as we know the genotypes of the parents.

tance. We shall see more evidence of Mendel's keen sense of the rules of chance as we follow the discovery of his second law of inheritance.

Mendel's Law of Independent Assortment

Mendel deduced the law of segregation by performing **monohybrid crosses,** breeding experiments that employ parental varieties differing in a single trait such as pod color. What would happen in a mating of parental varieties differing in two traits—a **dihybrid cross?** For instance, two of the seven characteristics Mendel studied were seed color and seed shape. Seeds may be either yellow or green. They also may be either round or wrinkled. From monohybrid crosses, Mendel knew that the allele for round seeds is dominant

over the allele for wrinkled seeds, and yellow is dominant over green. (This is in contrast to pod color, where green is dominant over yellow.) Are these two traits, seed color and seed shape, transmitted from parents to offspring as a package, or is each trait inherited independently of the other? To find out, Mendel performed a dihybrid cross between homozygous plants having round yellow seeds (genotype *RRYY*) and plants that were homozygous for wrinkled, green seeds *(rryy)* (Figure 13.10). The union of *RY* and *ry* gametes produced hybrids heterozygous for both characteristics *(RrYy)*. All of these F_1 progeny had round yellow seeds, the dominant phenotypes, as we would expect.

The key to learning whether the two traits are transmitted together or independently is to determine what happens when the F_1 plants self-pollinate. Do the F_1 plants transmit their genes in the same combinations in which they were inherited from the P generation? If so, then segregation in the F_1 plants will result in only two classes of gametes: *RY* and *ry*. This hypothesis predicts that the phenotypic ratio of the F_2 generation will be 3:1—three-fourths round yellow seeds and one-fourth wrinkled green seeds (see Figure 13.10a).

The alternative hypothesis is that the two pairs of genes segregate independently of each other. In other words, genes are packaged into gametes in all possible allelic combinations, as long as each gamete has one gene for each trait. In our example, four classes of gametes would be produced in equal quantities: *RY, Ry, rY,* and *ry*. If four classes of sperm are mixed with four classes of eggs, there will be 16 (4 × 4) equally probable ways in which the genes can combine in the F_2 generation, as shown in Figure 13.10b. These combinations make up four phenotypic categories with a ratio of 9:3:3:1 (nine round yellow to three round green to three wrinkled yellow to one wrinkled green). When Mendel did the experiment and categorized ("scored") the F_2 progeny, he obtained a ratio of 315:108:101:32, which is approximately 9:3:3:1. (Notice in Figure 13.10b, however, that there remains a 3:1 phenotypic ratio for each trait: three round to one wrinkled; three yellow to one green. As far as an individual trait is concerned, the segregation behavior is the same as if this were a monohybrid cross.) The experimental results supported the hypothesis that *each allele pair segregates independently during gamete formation*. Mendel tried his seven pea traits in various dihybrid combinations and always observed a 9:3:3:1 phenotypic ratio in the F_2 generation. This behavior of genes during gamete formation is called independent assortment, and the operating principle is now called Mendel's **law of independent assortment.** Figure 13.11 shows how this law applies to a mating of budgies that are heterozygous for two genes.

The rules of probability, as applied to segregation

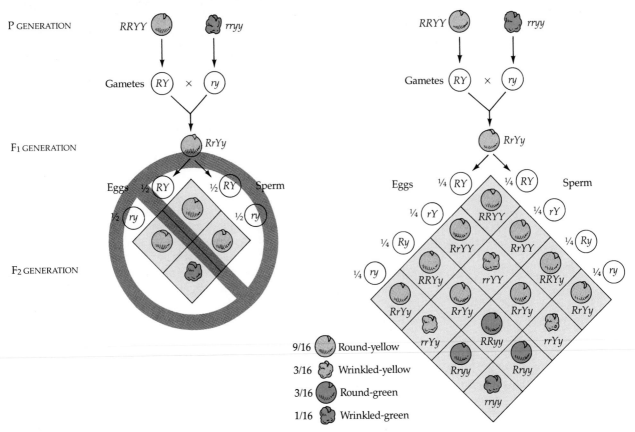

(a) Hypothesis: dependent assortment

P GENERATION   *RRYY* × *rryy*

Gametes   (*RY*) × (*ry*)

F₁ GENERATION   *RrYy*

Eggs ½ (*RY*)   ½ (*RY*) Sperm

½ (*ry*)   ½ (*ry*)

F₂ GENERATION

(b) Hypothesis: independent assortment

RRYY × *rryy*

Gametes   (*RY*) × (*ry*)

RrYy

Eggs ¼ (*RY*)   ¼ (*RY*) Sperm

¼ (*rY*)   ¼ (*rY*)

¼ (*Ry*)   ¼ (*Ry*)

¼ (*ry*)   ¼ (*ry*)

RRYY RrYY RrYY
RrYY RRYy rrYY RRYy RrYy
RrYy RrYy RrYy RrYy
rrYy RRyy rrYy
Rryy Rryy
rryy

9/16 Round-yellow
3/16 Wrinkled-yellow
3/16 Round-green
1/16 Wrinkled-green

Figure 13.10
Comparison of two hypotheses for segregation in a dihybrid cross. A cross between true-breeding parent plants that differ in two traits produces F₁ hybrids that are heterozygous for both traits. (**a**) If the two traits segregate dependently (together), then the F₁ hybrids can only produce the same two classes of gametes that they received from the parents, and the F₂ progeny will show a 3:1 phenotypic ratio. (**b**) If the two traits segregate independently, then four classes of gametes will be produced by the F₁ generation, and there will be a 9:3:3:1 phenotypic ratio in the F₂ generation. Mendel's results supported this latter hypothesis, called independent assortment.

and independent assortment, can solve some rather complex genetics problems. For instance, Mendel crossed pea varieties that differed in three traits, and the results of these trihybrid crosses could be explained by an independent assortment of alleles. Consider a trihybrid cross between two organisms with these genotypes: $AaBbCc \times AaBbCc$. What is the probability that an offspring from this cross will be a recessive homozygote for all three traits ($aabbcc$)? Since each allele pair is sorted independently, we can treat this as three separate monohybrid crosses:

$Aa \times Aa$: Probability for aa offspring $= \frac{1}{4}$

$Bb \times Bb$: Probability for bb offspring $= \frac{1}{4}$

$Cc \times Cc$: Probability for cc offspring $= \frac{1}{4}$

Because the segregation of each allele pair is an independent event, we use the rule of multiplication to calculate the overall probability that the offspring will be $aabbcc$:

$$\tfrac{1}{4}aa \times \tfrac{1}{4}bb \times \tfrac{1}{4}cc = \tfrac{1}{64}$$

You will get a chance to solve genetics problems involving various kinds of hybrid crosses at the end of this chapter.

Mendel's two laws explain inheritance in terms of discrete factors (now called genes) that are passed along, generation after generation, according to simple rules of chance. These principles, discovered in garden peas, are equally valid for figs, flies, fish, birds, and human beings. But the patterns of inheritance described so far in this chapter are simplified generalizations. Experiments on a variety of organisms have revealed more complicated situations. These include situations in which one allele is not completely dominant over the other allele; those in which there are more than two alternative alleles for a trait; and those in which the genotype does not always dictate the phenotype in a rigid manner.

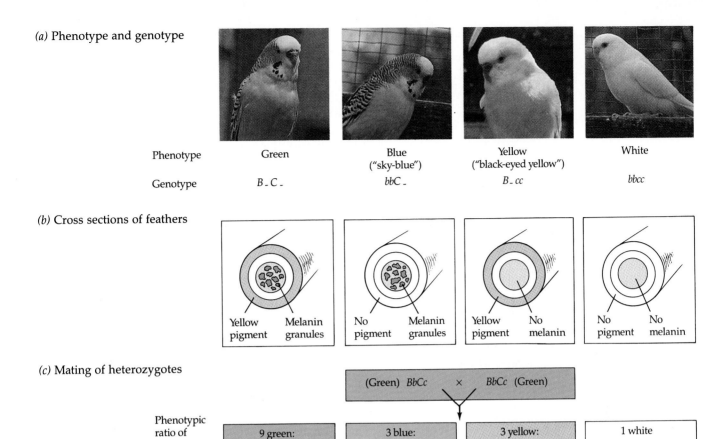

(a) Phenotype and genotype

	Green	Blue ("sky-blue")	Yellow ("black-eyed yellow")	White
Phenotype	Green	Blue ("sky-blue")	Yellow ("black-eyed yellow")	White
Genotype	$B\_C\_$	$bbC\_$	B_cc	$bbcc$

(b) Cross sections of feathers

Yellow pigment — Melanin granules | No pigment — Melanin granules | Yellow pigment — No melanin | No pigment — No melanin

(c) Mating of heterozygotes

(Green) $BbCc$ × $BbCc$ (Green)

Phenotypic ratio of progeny

9 green: | 3 blue: | 3 yellow: | 1 white

Figure 13.11
Independent assortment in the budgie.
The green and sky-blue phenotypes described in Figure 13.6 are actually the result of *two* genes. **(a)** Various combinations of the alleles for these genes give four different phenotypes. The blanks in the genotypes indicate that either the dominant or recessive allele may be present. (The green and sky-blue birds discussed earlier also carried a dominant *C* allele.)

(b) The anatomical and biochemical basis for the four phenotypes is that the *B* and *C* genes are responsible for two feather pigments. The *C* gene controls the production of melanin, the pigment responsible for blue (and black) feather color, and the *B* gene controls the production of a yellow pigment. The recessive alleles for these genes, *c* and *b*, are defective for pigment formation. Birds with both pigments are

green, and birds with neither are white. **(c)** When *BbCc* heterozygotes are mated, progeny result in the phenotypic ratio characteristic of independent assortment among two allele pairs, 9:3:3:1. Can you construct a Punnett square, similar to the one in Figure 13.10b, showing how this ratio comes about?

FROM GENOTYPE TO PHENOTYPE: SOME COMPLICATIONS

Intermediate Inheritance

The F_1 progeny of Mendel's classic pea crosses always looked like one of the two parental varieties because of the complete dominance of one allele over another. But for some traits, there is **intermediate inheritance**, where the F_1 hybrids have an appearance that is somewhere in between the phenotypes of the two parental varieties. For instance, when red snapdragons are crossed with white snapdragons, all the F_1 hybrids have pink flowers (Figure 13.12). This intermediate inheritance is also called **incomplete dominance** and

results from the fact that the heterozygote flowers have only half as much red pigment as the red homozygotes. We should not regard incomplete dominance as evidence of the blending theory, which would predict that the red or white traits could never be retrieved from the pink hybrids. In fact, breeding the F_1 hybrids produces F_2 progeny with a phenotypic ratio of one red to two pink to one white. (Note that when dominance is incomplete, we can distinguish the heterozygotes from the two homozygous varieties, and the genotypic and phenotypic ratios for the F_2 generation are the same—1:2:1.) The segregation of the red and white alleles in the gametes produced by the pink-flowered plants confirms that the genes for flower color are heritable factors that maintain their identity in the hybrids; that is, inheritance is particulate.

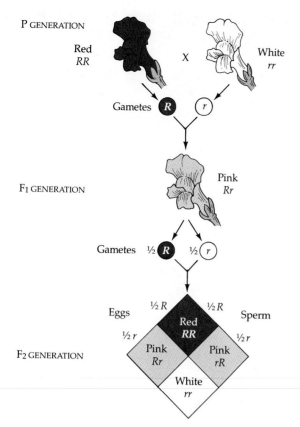

P GENERATION

Red
RR

X

White
rr

Gametes R r

F₁ GENERATION

Pink
Rr

Gametes ½ R ½ r

Eggs ½ R ½ R Sperm

½ r ½ r

Red
RR

Pink
Rr

Pink
rR

White
rr

F₂ GENERATION

Figure 13.12
Incomplete dominance in snapdragon color. When
red snapdragons are crossed with white ones, the F₁
hybrids have pink flowers because the red allele is
incompletely dominant over the white allele. Segregation
of alleles into the gametes of the F₁ plants results in an
F₂ generation with a 1:2:1 ratio for both genotype and
phenotype.

Intermediate inheritance is also observed in humans. One case involves a defective allele responsible for a relatively common inherited disease called *familial hypercholesterolemia (FH)*. About 1 in 500 people is a heterozygote, with one normal allele and one defective allele. These individuals have high levels of cholesterol in their blood, and they are unusually prone to atherosclerosis, the condition in which cholesterol and other materials accumulate in artery walls, eventually blocking the artery and causing a heart attack or a stroke. Heterozygotes have twice the normal blood cholesterol and may have heart attacks by the age of 35. FH is serious enough in heterozygotes that the defective allele is often called dominant, but the severity of the heterozygote disease is only intermediate in comparison to the disease suffered by individuals who are homozygous for the defective allele. Homozygotes (about one in a million people) have six times the normal blood cholesterol and may have heart attacks at the age of 2.

The molecular explanation for FH was worked out in the 1960s. The key gene turns out to be the gene required for synthesis of a cell-surface protein called the *LDL receptor.* The cholesterol that causes atherosclerosis is carried in the blood in particles called low-density lipoproteins (LDLs). Normally the LDL receptors help to keep the level of LDL cholesterol in the blood fairly low by binding LDL for cell uptake (see Chapter 8). But people heterozygous for the FH gene have only half the normal number of LDL receptors, and homozygotes have none, resulting in abnormally high blood LDL levels and high rates of atherosclerosis.

Multiple Alleles

Some genes exist in more than two allelic forms. The ABO blood types in humans are one example of multiple alleles. There are four phenotypes for this trait. A person's blood type may be either A, B, AB, or O (Figure 13.13). These letters refer to two genetically programmed molecules, the A substance and the B substance, which may be found on the surface of red blood cells. A person's blood cells may be coated with one substance or the other (type A or B), with both (type AB), or with neither (type O). Matching compatible blood types is critical for blood transfusions. If the donor's blood has a factor (A or B) that is foreign to the recipient, then specific proteins called antibodies produced by the recipient bind to the foreign molecules and cause the donor's blood cells to agglutinate (clump together). This agglutination can cause the recipient to die.

The four blood types result from various combinations of three different alleles, symbolized as I^A (encoding A), I^B (encoding B), and i (encoding neither A nor B). Every person carries two alleles specifying ABO blood type, one allele inherited from each parent. However, because there are three alleles, there are six possible genotypes, as listed in Figure 13.13. Both the I^A and I^B alleles are dominant to the i allele. Thus, I^AI^A and I^Ai persons have type A blood, and I^BI^B and I^Bi individuals have type B. Recessive homozygotes, ii, have type O blood because neither the A nor B substance is produced. The I^A and I^B alleles are said to be **codominant**, since both will be expressed in the I^AI^B heterozygote, who has type AB blood.

Pleiotropy

A gene can sometimes affect many phenotypic characteristics. The ability of a single gene to have multiple effects is called **pleiotropy.** Many examples are provided by hereditary diseases, such as sickle-cell anemia, in which a single gene (allele pair) causes complex sets of symptoms. A gene can sometimes influence

BLOOD GROUP PHENOTYPE	GENOTYPES	ANTIBODIES PRESENT IN BLOOD SERUM	REACTION WHEN RED BLOOD CELLS FROM GROUPS BELOW ARE ADDED TO SERUM FROM GROUPS LISTED AT LEFT			
			O	A	B	AB
O	ii	Anti-A Anti-B				
A	$I^A I^A$ or $I^A i$	Anti-B				
B	$I^B I^B$ or $I^B i$	Anti-A				
AB	$I^A I^B$	—				

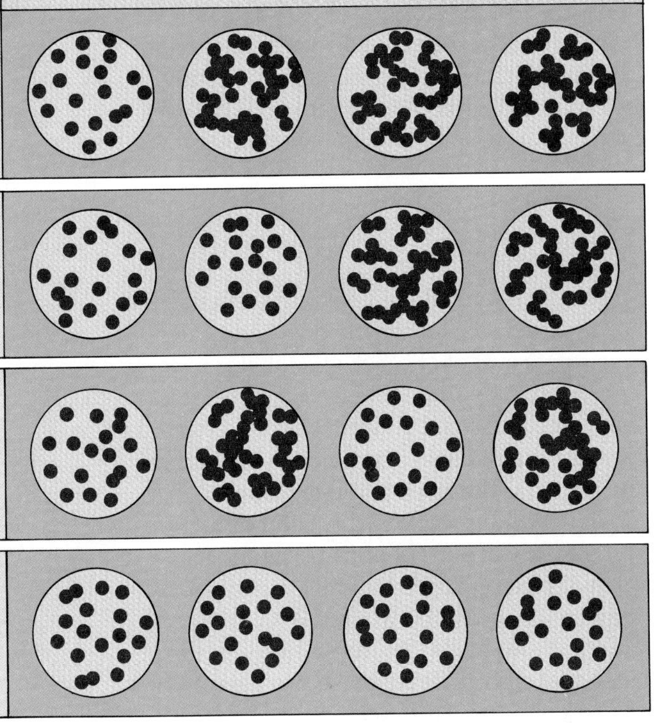

Figure 13.13
Multiple alleles for the ABO blood types. There are three alleles. Because each person carries two alleles, six genotypes are possible. Whenever I^A or I^B is present, the corresponding factor (A or B) is present on the surface of red blood cells. Both of these alleles, which are codominant, are dominant to the i allele, which does not code for any surface factor. A person produces antibodies against foreign blood factors, causing clumping of red blood cells if a transfusion is performed with incompatible blood.

a surprising combination of characteristics. For example, the gene that controls fur pigmentation in Siamese cats also influences the connections between a cat's eyes and its brain. A defective gene causes both abnormal pigmentation and cross-eye. Tigers with abnormal pigmentation also tend to be cross-eyed.

Penetrance and Expressivity

Environmental factors often intervene along the path from genotype to phenotype. The phenotype is the product of a complex interaction between an organism's genetic makeup and its environment. An individual is locked into its inherited genotype, but phenotype may change. If this were not true, physicians would be helpless to relieve the symptoms of inherited diseases, and we would be unable to alter our physiques and athletic skills by exercise and practice. Even identical twins, who are genetic equals, accumulate phenotypic disparities as a result of their different experiences.

In some cases, genotype mandates phenotype in an irrevocable manner. For example, all children who are homozygous for an allele causing Tay-Sachs disease develop symptoms of the disease and die, regardless of their environment and medical treatment. A geneticist would say that this lethal allele has complete, or 100%, penetrance. **Penetrance** is the proportion of individuals who show the phenotype that is expected from their genotype. One example of incomplete penetrance is a type of eye tumor called retinoblastoma, which is due to a dominant allele. Not all individuals who inherit the allele develop the tumor. Furthermore, the severity of the tumor varies among those individuals who show the retinoblastoma phenotype. Therefore, the gene is said to have variable expressivity. **Expressivity** is the degree to which a particular gene is expressed in individuals showing the trait.

Sometimes, an environmental agent can be powerful enough to override inheritance completely. An organism may even display a phenotype characteristic of a genotype other than its own. For example, thalidomide, a tranquilizer taken by hundreds of pregnant women in Europe during the early 1960s, disrupted normal embryonic development so as to mimic the effects of rare mutations that cause a human birth defect known as phocomelia. A baby with this genotype has one or more deformed, flipperlike limbs. Because of thalidomide, babies with normal genotypes were born deformed in this same way. An environmentally produced phenotype that simulates the effects of a particular gene is called a **phenocopy.**

Epistasis

Another twist in the labyrinthine path from genotype to phenotype is **epistasis,** in which one gene interferes with the expression of another gene that is independently inherited. The first gene is said to be epistatic (from *epi,* "on," and *static,* "stand") to the second. For instance, a dominant allele *P* is required for purple flowers in sweet peas. But even if the genotype is *PP* or *Pp,* the flowers will be white if the plant is homozygous for a recessive allele *c* of another gene. Thus, a *PpCc* plant has purple flowers, while a *Ppcc* plant has white flowers. A cross between *PPCC* and *ppcc* plants produces F_1 hybrids having purple flowers (*PpCc*). Independent assortment of these two gene pairs results in an F_2 generation with a 9:7 purple-to-white phenotypic ratio that can be explained by epistasis of the *c* gene over the *P* gene (Figure 13.14). (Do not confuse epistasis with dominance, which is the masking of the effect of one allele by another allele of the same gene). The biochemical basis of epistasis is often quite simple. In the sweet pea case, for example, *P* and *C* are genes that are both needed for the biochemical pathway of purple pigment synthesis.

Polygenic Traits

Mendel studied traits that could be classified on an either-or basis, such as green versus yellow pea pods. There are many traits, however, including human skin color and height, for which an either-or classification is impossible because they are **quantitative traits,** which vary in the population in a continuous way. Although heredity and environment both contribute to the phenotype, continuous variation usually indicates **polygenic inheritance,** an additive effect of two or more genes on a single phenotypic characteristic (the converse of pleiotropy, where a single gene affects many phenotypic traits).

There is evidence, for instance, that skin pigmentation in humans is controlled by at least three separately inherited genes. Let us consider three genes, with the "dark-skin" allele for each gene (*A, B, C*) contributing one "unit" of darkness to the phenotype and being incompletely dominant over the other alleles (*a, b, c*). An *AABBCC* person would be very dark, while

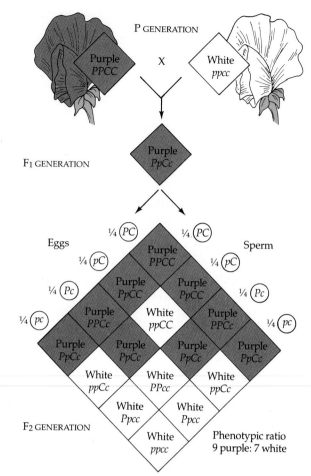

Figure 13.14
Epistasis in sweet peas. A dominant allele, *P,* is required for purple flowers, but this trait is not expressed in plants homozygous for a recessive gene *c.* Hybrids heterozygous for both genes (*PpCc*) have purple flowers. Independent assortment results in an F_2 genotypic ratio that is normal for a dihybrid cross. But the phenotypic ratio is 9:7, rather than the usual 9:3:3:1, because the plants with *cc* genotypes are all white, regardless of the genotype for the other gene pair. When homozygous, the *c* allele masks the *P* allele, and thus is epistatic.

an *aabbcc* individual would be very light. An *AaBbCc* person would have skin of an intermediate shade. Because the alleles have a cumulative effect, the genotypes *AaBbCc* and *AABbcc* would make the same genetic contribution (three "units") to skin darkness. Figure 13.15 shows how this system could result in a bell-shaped curve, called a *normal distribution,* for skin color among the members of a hypothetical population. Environmental factors, such as suntanning, would modulate the skin color phenotype and help to make the graph a smooth curve rather than a stairlike histogram.

We have refined our Mendelian model of heredity by considering several complications, including intermediate inheritance, multiple alleles, pleiotropy, environmental influence, epistasis, and polygenic traits.

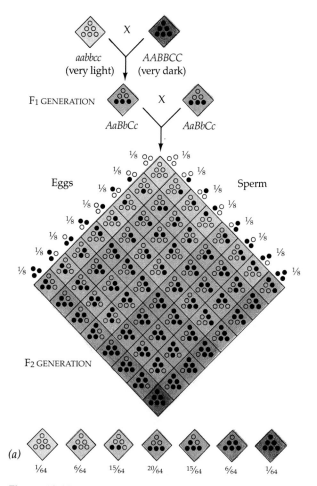

P GENERATION

aabbcc
(very light)

X

AABBCC
(very dark)

F₁ GENERATION

AaBbCc   X   AaBbCc

Eggs   Sperm

F₂ GENERATION

(a)

1/64   6/64   15/64   20/64   15/64   6/64   1/64

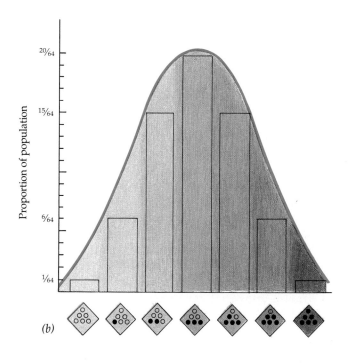

(b)

Figure 13.15
Polygenic inheritance of skin pigmentation, based on a three-gene model.
According to the simple model shown in (**a**), each dominant allele for any one of the three genes contributes one unit of skin darkness to the phenotype. If two persons from the opposite extremes of the pigmentation range mate, their child will have medium skin color owing to heterozygosity for each of the three genes. If this individual mates with another triple heterozygote, independent assortment of the three gene pairs makes possible a wide range of pigmentation for their children. From many such matings, a distribution like the one shown in (**b**) would result. The histogram would be smoothed into a bell-shaped curve by environmental factors. Thus, unlike the color of pea pods, the degree of human skin pigmentation is a quantitative trait that varies continuously in the population.

Do not allow these complications to cloud the basic simplicity of Mendel's laws of segregation and independent assortment. From the abbey garden came data supporting a particulate theory of inheritance, with the particles (genes) being transmitted according to the same rules of chance that govern the tossing of coins. Mendel announced his results in 1865, but it was a long time before other biologists got the message. Mendel's mathematical approach was before its time, and even the few biologists who bothered to read his papers generally regarded the work as unimportant. His teachings disappeared to dusty library stacks, and biologists remained ignorant of the laws of particulate inheritance for another 35 years. Finally, at the turn of the century, 16 years after Mendel's death, his papers were rediscovered. As we shall see further in the next section, Mendel's laws have provided the basis for understanding inheritance in humans.

MENDELIAN INHERITANCE IN HUMANS

Whereas peas are convenient subjects for genetic research, humans are not. The human generation span is about 20 years, and parents produce relatively few offspring (compared to peas and most other species).

Furthermore, well-planned breeding experiments like the ones Mendel performed are impossible (or at least socially unacceptable) with humans. In spite of these difficulties, the study of human genetics continues to advance, powered by the incentive to understand our own inheritance. Recently, tremendous advances have been made using the techniques of cell culture and molecular biology. However, our understanding of Mendelian inheritance in humans is based on more traditional methods, as we shall now discuss.

Human Pedigrees

Unable to manipulate the mating patterns of humans, the geneticist must analyze the results of matings that have already occurred. As much information as possible is collected about a family's history for a particular trait, and this information is assembled into a family tree describing the interrelationships of parents and children across the generations—the family **pedigree**. The pedigree in Figure 13.16 describes the occurrence of a trait called wooly hair for three generations of one family. Caucasian individuals with this trait have curly hair that is fuzzy in texture, superficially similar to the hair of many blacks. Because it is very brittle and breaks at the tips, wooly hair cannot grow long. The trait is due to a dominant allele that we will symbolize by W. Wooly hair is rare in the overall human population; thus, most people are recessive homozygotes (ww). However, the family in the pedigree of Figure 13.16 has a heritage of the wooly-hair phenotype, and we can use Mendel's laws to analyze the pedigree.

By applying Mendel's laws, we can deduce the genotypes of the couple at the top of the pedigree (first generation in this pedigree). We know that the man is ww, for he has the normal phenotype. The woman must be heterozygous (Ww), because three of the six children had normal hair. Had she been homozygous for wooly hair (WW), all children born to the couple would have had wooly hair. This pedigree fits (more perfectly than usual) the 1:1 phenotypic ratio expected from a Mendelian test cross (Ww × ww).

A pedigree not only helps us understand the past; it also helps us predict the future. Suppose that one of the grandsons with wooly hair marries a woman with normal hair (see Figure 13.16, bottom right). They plan to have three children. What is the probability that all three children will have wooly hair? The pedigree tells us that the grandson must have the genotype Ww. Since each of his children is an independent event, genetically speaking, each has a $\frac{1}{2}$ chance of inheriting the wooly allele. Applying the rule of multiplication, the overall probability that all three of the anticipated children will have wooly hair is $\frac{1}{2} \times \frac{1}{2} \times \frac{1}{2} = \frac{1}{8}$. The overall probability that all three children will *not* have

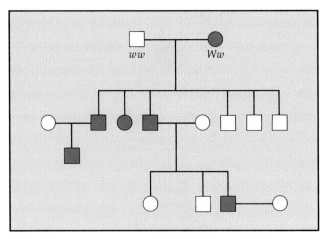

Figure 13.16
A human pedigree. The pedigree, or family tree, for a family in which the unusual dominant trait called wooly hair appears in members of three generations. Colored symbols denote individuals with wooly hair; squares represent males and circles represent females. Horizontal lines indicate matings, with progeny listed below in their order of birth, from left to right. You should be able to fill in the genotypes for all the family members shown on the pedigree.

wooly hair is therefore $\frac{7}{8}$ (since there is a total probability of 1 that the children will have one or the other type of hair).

When the genetic trait to be analyzed can lead to a disabling or lethal disorder, the stakes are much higher than in examining a harmless trait like wooly hair. But for disorders that are inherited as simple Mendelian traits, the same basic techniques for pedigree analysis are applied by geneticists, physicians, and genetic counselors.

Recessively Inherited Disorders

About a thousand genetic disorders are known to be inherited as simple recessive traits. A recessive allele that causes a disorder is usually a defective version of the normal allele. If the organism can get along with one normal allele, the defective allele in heterozygotes will have no noticeable effect.

Recessively inherited disorders range in severity from relatively innocuous handicaps, such as albinism (lack of skin pigmentation), to deadly diseases, such as cystic fibrosis. Because these disorders are caused by recessive alleles, the phenotypes show up only in the homozygous individuals who inherit one recessive allele from each parent. We can symbolize the genotype of such persons as aa, with normal individuals being either AA or Aa. The heterozygotes (Aa), who are phenotypically normal, are called carriers of the disorder, and they may transmit the recessive allele to their offspring.

The vast majority of people afflicted with recessive disorders are born to normal parents who are both carriers. A mating between two carriers corresponds to a Mendelian F_1 cross *(Aa × Aa)*, with the zygote having a $\frac{1}{4}$ chance of inheriting a double dosage of the recessive allele. A normal child from such a cross has a $\frac{2}{3}$ chance of being a carrier. Recessive homozygotes could also result from *Aa × aa* and *aa × aa* matings, but if the disorder is lethal before reproductive age or results in sterility, no *aa* individuals will reproduce. Even if recessive homozygotes are able to reproduce, such individuals will still account for a much smaller percent of the population than heterozygous carriers (for reasons we will discuss in Chapter 21).

In general, a given genetic disorder is not evenly distributed among all racial or cultural groups. These disparities result from the different genetic histories of the world's peoples. We will now examine three examples of such recessively inherited disorders.

The most common lethal genetic disease in the United States is *cystic fibrosis,* which strikes 1 in every 2500 Caucasians but is much rarer in other races. The cystic fibrosis allele, which is recessive, causes excessive secretions of mucus from the pancreas, lungs, and other organs, leading to blockage of the digestive tract, cirrhosis of the liver, pneumonia, and other infections. Untreated, most children with cystic fibrosis die by the time they are four or five years old. A special diet, daily doses of antibiotics to prevent infection, and other palliative treatments can prolong life to adolescence. A small percentage of people with the disease may even live long enough to reproduce, in which case there is a 100% chance that the person will transmit one copy of the cystic fibrosis allele to each of his or her children.

Another lethal disorder inherited as a recessive allele is *Tay-Sachs disease*. The symptoms usually become manifest a few months after birth. The brain cells of a baby with Tay-Sachs disease are unable to metabolize gangliosides, a type of lipid, because a crucial enzyme does not function properly. As the lipids accumulate in the brain, the brain cells gradually cease to function normally. The infant begins to suffer seizures, blindness, and degeneration of motor and mental performance. Inevitably, the child dies within a few years. The allele for this disease is called a late-acting lethal gene; most lethal genes program earlier death, usually during embryonic development. There is a disproportionately high incidence of Tay-Sachs disease among Ashkenazi Jews, Jewish people whose ancestors lived in central Europe. In that population, the frequency of the disease is 1 case in 3600 births, about 100 times greater than the incidence among non-Jews or Mediterranean (Sephardic) Jews.

By far, the most common inherited disease among blacks is *sickle-cell anemia*, which strikes 1 in 500 black children born in the United States. This disease is caused by substitution of a single amino acid in the hemoglobin protein of red blood cells. These abnormal hemoglobin molecules tend to link together and crystallize, especially when the oxygen content of the blood is lower than normal because of high altitude, overexertion, or respiratory ailments. As the hemoglobin crystallizes, the normally disk-shaped red blood cells deform to a sickle shape (Figure 13.17). The life of someone with the disease is punctuated by "sickle-cell crises" when the collapsed, angular cells clog tiny blood vessels, impeding blood flow to body parts. This results in fever and severe pain, especially in the arms and legs, lasting from hours to weeks. The oxygen deprivation caused by blockages of the blood flow results in further sickling of cells, aggravating the problem. Blood transfusions can relieve the symptoms, but there is no cure for this disease, which kills about 100,000 in the world annually.

Individuals who are heterozygous for the sickle-cell

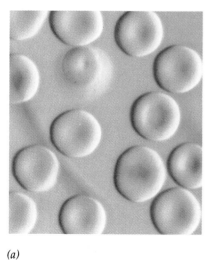 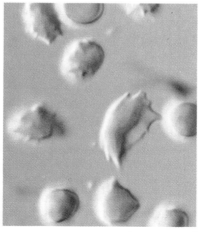

(a)   *(b)* |———— 10 μm ————|

Figure 13.17
Effects of the sickle-cell allele on red blood cells. (a) Normal red blood cells are disk-shaped. The double concave form is possible because the nucleus is absent in mature red cells. **(b)** The jagged shapes of sickled cells cause them to pile up and block small blood vessels, thus starving regions of the body for oxygen and nutrients. Damage may result to one or more organs, including the heart, lungs, brain, digestive organs, kidneys, and spleen. (LMs.)

allele are said to have *sickle-cell trait*. These carriers are usually healthy, although a fraction of the heterozygotes suffer some symptoms of sickle-cell anemia when there is an extended reduction of blood oxygen—at very high altitude, for instance. (The two alleles are codominant at the molecular level; both normal and abnormal hemoglobins are made.) About one in ten American blacks has sickle-cell trait, an unusually high frequency of heterozygotes for an allele with severe detrimental effects in homozygotes. The reason for the prevalence of this allele appears to be that while only individuals who are homozygous for the sickle-cell allele suffer from the disease, heterozygotes have an advantage in certain environments over people who carry no copies of the sickle-cell allele. A single copy of the sickle-cell allele endows the individual with enhanced resistance to malaria. Thus, in tropical Africa, where malaria is common, the sickle-cell allele is both boon and bane. But malaria-resistant heterozygotes far outnumber the unfortunate people who are homozygous for the sickle-cell allele. The relatively high frequency of American blacks with sickle-cell trait is a vestige of their African roots.

Consanguinity   Although it is relatively unlikely that two carriers of the same rare deleterious allele will meet and mate, the probability increases greatly if the man and woman are close relatives (for instance, siblings or first cousins). These matings are called **consanguineous** ("same blood") matings, and they are indicated in pedigrees with double lines. Because persons with recent common ancestors are more likely to carry the same recessive alleles than are unrelated persons, it is more likely that a mating of close blood relatives will produce offspring homozygous for a harmful recessive trait. Such effects can be observed in many types of domesticated and zoo animals that have become inbred.

There is debate among geneticists about the extent to which human consanguinity increases the risk of inherited diseases. Many deleterious mutations have such severe effects that a homozygous embryo spontaneously aborts long before birth. Furthermore, some geneticists argue that consanguinity is just as likely to concentrate favorable alleles as deleterious ones. It must be admitted that some very important people came out of marriages between close relatives. For instance, Cleopatra's mother and father were brother and sister.

Most societies and cultures have laws or taboos forbidding marriages between close relatives. These rules may have evolved out of empirical observation that in most populations, stillbirths and birth defects are more common when parents are closely related. But social and economic factors have also influenced the development of customs and laws against consanguineous marriages, which could concentrate wealth in a few families. In some human populations, such as the Tamils of India, marriage between close relatives (for example, first cousins) has long been the rule. In these cases, there is no observable ill effect of continued inbreeding. And in some zoo populations of endangered species, carefully planned inbreeding is saving small groups of animals from extinction.

Dominantly Inherited Disorders

Although most harmful alleles are recessive, there are several human disorders due to dominant alleles. In medicine, an allele is classified as dominant if a single copy is sufficient to significantly affect phenotype. Thus, physicians classify familial hypercholesterolemia as a dominantly inherited disorder. Another example is *achondroplasia,* a form of dwarfism with an incidence of 1 case among every 10,000 people. A single copy of the defective allele produces the dwarf phenotype (in fact, homozygosity for this dominant allele is lethal, causing spontaneous abortion). Therefore, all persons who are not achondroplasia dwarfs, 99.99% of the population, are homozygous for the recessive allele. The example of achondroplasia should also make it clear that we cannot assume that a dominant allele will be more plentiful than its recessive allele in a population. The relative frequencies of alternative alleles for a gene result mainly from natural selection favoring reproduction by one phenotype over another (see Chapter 21).

Lethal dominant alleles are much less common than lethal recessives. One reason for this difference is that the effects of lethal dominant alleles are not masked in heterozygotes. Many lethal dominant alleles are the result of new changes (mutations) in a gene of the sperm or egg that subsequently kill the developing organism. And if the organism does not survive to reproductive maturity, it will not pass on the new form of the gene. This is in contrast to lethal recessive mutations, which are perpetuated from generation to generation by the reproduction of heterozygous carriers who have normal phenotypes.

However, a lethal dominant allele can escape elimination if it is late-acting, causing death at a relatively advanced age. By the time the symptoms become evident, the afflicted individual may have already transmitted the lethal gene to his or her children. *Huntington's disease* is an example. A degenerative disease of the nervous system, it is caused by a lethal dominant allele that has no obvious phenotypic effect until the individual is about 35 to 45 years old. Recently, medical researchers, using the techniques of modern molecular biology, have developed methods for detecting carriers of the lethal gene before the onset of symptoms, but many individuals at risk prefer not to know their status. Once the deterioration of the nervous system begins, it is irreversible and inevitably lethal. Any child born to a parent who has developed Huntington's disease has a 50% chance of also having the disorder (see the interview on p. 243).

A number of other neurological disorders have recently been shown to result from dominant alleles. These include one type of *Alzheimer's disease*, which leads to degeneration of parts of the brain and associated functions, usually late in life. Also, at least some cases of *manic-depressive disease*, a psychiatric disorder characterized by extreme mood swings, result from the presence of an abnormal dominant allele. Although many cases of both Alzheimer's disease and manic-depressive disease do *not* seem to be hereditary, medical scientists hope that by studying the hereditary cases they will gain insight into the physiological bases of those diseases, which affect millions of Americans.

Genetic Screening and Counseling

A preventive approach to genetic disorders is sometimes possible, since in some cases the risk that a particular genetic disorder will occur can be assessed before a child is conceived or in the early stages of the pregnancy. Many hospitals have genetic counselors who can provide information to prospective parents concerned about a family history for a specific disease.

Let us consider the example of an imaginary couple, John and Carol, who are planning to have their first child and are seeking genetic counseling because of family histories of a lethal disease known to be recessively inherited. John and Carol each had a brother who died of the disorder, so they want to determine the risk of their having a child with the disease. From the information about their brothers, we know that both parents of John and both parents of Carol must have been carriers of the recessive allele. The law of segregation tells us that there is a $\frac{2}{3}$ chance that John is a carrier, and the same is true of Carol. (How do we know that Carol and John are not homozygous recessive?) We can use the rule of multiplication to determine that the probability that *both* John *and* Carol are carriers is $\frac{2}{3} \times \frac{2}{3} = \frac{4}{9}$. Taking these chance factors into consideration, the overall probability of their first-born having the disorder is $\frac{2}{3}$ (the chance that John is a carrier) multiplied by $\frac{2}{3}$ (the chance that Carol is a carrier) multiplied by $\frac{1}{4}$ (the chance of two carriers having a child with the disease), which equals $\frac{1}{9}$.

The availability of new information would allow us to revise our estimate. Suppose that Carol and John decide to take the risk and have a child (after all, there is an $\frac{8}{9}$ chance that their baby will be normal). But their child is born with the disease. We now know that both John and Carol are, in fact, carriers, and we can eliminate two of the factors in our original estimate of risk. If the couple decides to have another child, there is a $\frac{1}{4}$ chance that it will have the disease.

Carrier Recognition Because most children with recessive disorders are born to parents with normal phenotypes, the key to assessing genetic risk for a particular disease is determining if the prospective parents are heterozygous carriers of the recessive trait. For some heritable disorders, there are tests that can distinguish between normal individuals who are dominant homozygotes and normal individuals who are heterozygous, and the number of such tests is increasing. There is, for instance, a biochemical test that can identify carriers of the Tay-Sachs allele. The allele specifying cystic fibrosis can also be detected in carriers, but the test is not simple enough to be practical for mass screening. Nonetheless, the test may be advisable when there is a family history for cystic fibrosis. By contrast, there is a simple, reliable blood test that can identify individuals with the heterozygous genotype for the sickle-cell trait. The cost is only a few cents per test, making mass screening feasible. Detection of many other harmful genes is becoming possible with the help of new methods for directly probing DNA (see Chapter 19).

Fetal Testing Suppose a couple learns that they are both Tay-Sachs carriers, but they decide to have a child anyway. Tests done in conjunction with a technique known as **amniocentesis** (Figure 13.18) can determine between the fourteenth and sixteenth weeks of pregnancy whether the developing fetus has Tay-Sachs disease. To perform this procedure, a physician inserts a needle into the uterus and extracts about 10 milliliters of amniotic fluid, the liquid bathing the fetus. Some genetic disorders can be detected from the presence of certain chemicals in the amniotic fluid itself. Tests for other disorders, including Tay-Sachs disease, are performed on cells grown in the laboratory from the fetal cells that had been sloughed off into the amniotic fluid. These cultured cells can also be used for karyotyping to identify certain chromosomal defects (see Chapter 12 Methods Box).

In a newer technique called **chorionic villi sampling,** the physician suctions off a small amount of fetal tissue from the projections (villi) of the embryonic membrane, or chorion, which forms part of the placenta. Because the cells of the chorionic villi are proliferating rapidly, enough cells are undergoing mitosis to allow karyotyping to be carried out immediately, giving results within 24 hours. The rapidity of this method is an advantage over amniocentesis, in which the cells must be cultured and karyotyping results typically take several weeks. Another advantage of chorionic villi sampling is that it can be performed at only 8 to 10 weeks of pregnancy. However, the risks of the procedure are still being determined.

Other techniques allow a physician to examine a fetus directly for major abnormalities. One such technique is ultrasound, which uses sound waves to pro-

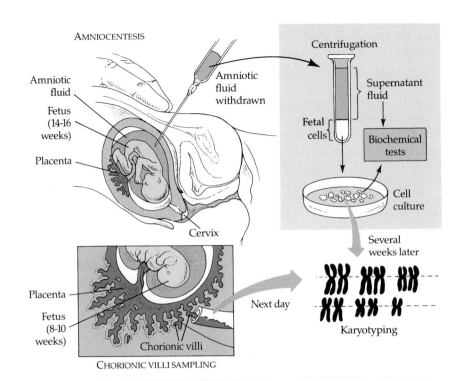

Figure 13.18
Fetal diagnosis. In amniocentesis, ultrasound is used to locate the fetus, and a small amount of amniotic fluid is extracted for testing. Physicians can diagnose some disorders from chemicals in the fluid itself, while other disorders may show up in tests performed on cells cultured from fetal cells present in the fluid. These tests include biochemical tests for the presence of certain enzymes and karyotyping to determine whether the chromosomes of the fetal cells are normal in number and microscopic appearance. In chorionic villi sampling, a physician inserts a narrow tube through the cervix and suctions out a tiny sample of fetal tissue (chorionic villi) from the placenta, the organ that transmits nourishment and wastes between the fetus and the mother. The fetal tissue sample is used for immediate karyotyping.

duce an image of the fetus by a simple noninvasive procedure. This procedure has no known risk to fetus or mother. With another technique, fetoscopy, a needle-thin tube containing a viewing scope and fiber optics (to transmit light) is inserted into the uterus. Fetoscopy allows the physician to examine the fetus for certain anatomical deformities (see Chapter 42).

In about 1% of the cases, amniocentesis or fetoscopy causes complications such as maternal bleeding or fetal death. Thus, these techniques are usually reserved for cases where the risk of a genetic disorder or other type of birth defect is relatively great. If the fetal tests reveal a serious disorder, the parents must choose between terminating the pregnancy or preparing themselves for a defective baby.

Newborn Screening   Some genetic disorders can be detected at birth by simple tests that are now routinely performed in most hospitals in the United States. One of the most significant screening programs is for the recessively inherited disorder called *phenylketonuria (PKU)*. It occurs in about 1 in every 15,000 births in the United States. Children with this disease cannot properly break down the amino acid phenylalanine. This compound and its by-product, phenylpyruvic acid, can accumulate in toxic levels in the blood, causing mental retardation. But if the deficiency is detected in the newborn, retardation can be prevented with a special diet, low in phenylalanine, that allows normal development. Thus, screening newborns for PKU and other phenotypically treatable disorders is vitally important. Unfortunately, very few genetic disorders are treatable at the present time.

*   *   *

In this chapter, you have learned about the Mendelian model of inheritance and its application to human genetics. In the next chapter, you will learn how Mendel's laws have their physical basis in the behavior of chromosomes during sexual life cycles.

STUDY OUTLINE

1. Aristotle maintained that inheritance was based on pangenesis, in which particles called pangenes come together from all parts of the body to form sperm and eggs.
2. Nineteenth-century biologists generally accepted this theory, which served as the basis for Lamarck's theory of inheritance of acquired characteristics.
3. Although the mechanism of pangenesis has been disproved, the particulate idea of inheritance has survived and given rise to the term *gene* (after *pangenes*).
4. Seventeenth-century microscopists, claiming to see complete, miniature human beings in either the sperm or the egg, hotly debated that one or the other parent provided all the offspring's inherited characteristics. The other sex was viewed as necessary only to spark development.
5. Nineteenth-century biologists finally acknowledged the role of both parents, but erroneously accepted a blending theory to explain the mechanism of inheritance.
6. It was not until the 1860s that the fundamental principles

of genetics were discovered in an abbey garden by a monk named Gregor Mendel.

Mendel's Model of Inheritance (pp. 263–270)

1. By breeding garden peas, Mendel demonstrated that parents pass on to their offspring discrete, heritable factors that retain their identity generation after generation. We call these particulate factors genes.

2. Mendel's success is attributed to his quantitative approach and his choice of organism. He worked with seven pea characteristics, each of which occurred in two distinct, alternative forms. His peas were true-breeding, and the morphology of the flowers allowed strict control over the matings.

3. By producing hybrid offspring and allowing them to self-pollinate, Mendel arrived at the law of segregation and the law of independent assortment.

4. A 3:1 ratio of progeny plant types in his F_2 generation showed Mendel that genes have alternative forms (now called alleles) and that each organism inherits one from each parent. These separate (segregate) from each other during gamete formation, so that a sperm or an egg carries only one allele. After fertilization, if the two alleles of the pair are different, one (the dominant allele) is fully expressed in the offspring and the other (the recessive allele) is completely masked.

5. Mendel's law of segregation, which has been found to apply to all sexually reproducing organisms, states that allele pairs segregate during gamete formation and come together again randomly at fertilization.

6. Homozygous individuals have two identical alleles for a given trait and are true-breeding. Heterozygous individuals have two different alleles for a given trait and are not true-breeding, since they produce gametes with one or the other allele.

7. The genetic makeup, or genotype, of an organism cannot be automatically deduced from its appearance, or phenotype, owing to the phenomenon of dominance. Thus, genotypic ratios often differ from phenotypic ratios.

8. The genotype of an organism showing a dominant trait can be determined by breeding it to a recessive homozygote in a so-called testcross.

9. The law of segregation operates according to the rules of probability. According to the rule of multiplication, the probability of a compound event is equal to the product of the separate probabilities of the independent single events. The rule of addition states that the probability of an event that can occur in two or more independent ways is the sum of the separate probabilities of the different ways. The statistical nature of inheritance is such that the larger the sample size, the closer the results will conform to the predictions.

10. Mendel's law of independent assortment, which dictates that different pairs of alleles segregate independently of one another, was based on results of dihybrid crosses that showed that two allele pairs were not inherited as a package.

11. The rules of probability, as applied to segregation and independent assortment, can be used to solve complex genetics problems. However, the solutions are not always straightforward because of complicating variations.

From Genotype to Phenotype: Some Complications(pp. 271–275)

1. Some heterozygous genotypes show the incomplete dominance of intermediate inheritance. Such individuals have an appearance that is intermediate between the phenotypes of the two parents, despite the fact that the genes maintain their identity in the progeny.

2. One human example of intermediate inheritance, familial hypercholesterolemia, is a disease where one or both of the alleles for the LDL receptor is defective. Afflicted individuals are highly prone to atherosclerosis, with the homozygotes more severely affected than the heterozygotes.

3. Some genes have more than two alleles, like the gene for human blood type. Such alleles can be either completely dominant or codominant with respect to each other.

4. Pleiotropy is the ability of a single gene to affect multiple phenotypic traits, some of which may not seem related to each other in the overall functioning of the organism.

5. Various factors may affect how much an individual's phenotype reflects its genotype. Penetrance is the proportion of individuals (up to 100% in the case of complete penetrance) who actually show the phenotype expected of their genotype. Expressivity is the degree to which a particular gene is expressed in individuals showing the trait. And the environment can produce a phenotype that simulates the effects of a particular gene in the phenomenon known as phenocopy.

6. In epistasis, one gene interferes with the expression of another, independently inherited gene.

7. Certain characteristics, such as human skin color, are quantitative traits that vary in a continuous fashion, indicating polygenic inheritance, an additive effect of two or more genes on a single phenotypic characteristic. The environment can often modulate the traits to make the variation occur in a normal distribution.

8. Although these complications do not refute Mendel's laws, they may have contributed to the 35-year lag in the acceptance of segregation, independent assortment, and particulate inheritance by mainstream biologists.

Mendelian Inheritance in Humans (pp. 275–280)

1. Despite intrinsic constraints involved in using humans for genetic research, Mendelian analysis can be applied to advance the understanding of our own inheritance.

2. Family pedigrees of matings that have already occurred can be used to deduce the genotypes of individuals and make predictions about the future. Any predictions are statistical probabilities rather than absolute statements.

3. Certain genetic disorders, such as sickle-cell anemia, Tay-Sachs disease, and cystic fibrosis, are inherited as simple recessive traits from phenotypically normal, heterozygous carriers. Genetic disorders are not evenly distributed among all racial and cultural groups.

4. Consanguineous matings between close relatives often increase the chance that the offspring will be homozygous for a rare deleterious allele. Thus, the incest taboo found in most cultures appears to have a biological basis.

5. Although they are far less common than the recessive type, some human disorders are due to dominant genes. Dominant alleles that are lethal may kill the organism as an embryo or act later, as in Huntington's disease, a fatal

degeneration of the nervous system that is manifested during mid-life. Some cases of Alzheimer's disease and manic-depressive disease are also caused by dominant alleles.

6. Using family histories, genetic counselors aid couples in determining the odds that their children will have genetic defects. For certain diseases, genetic screening for carrier recognition can more accurately define those odds.

7. Once a child is conceived, the techniques of amniocentesis or chorionic villi sampling can help determine whether a suspected genetic disorder is present. Moreover, gross anatomical abnormalities can be detected by ultrasound or fetoscopy.

8. Some genetic disorders can be detected at birth by biochemical tests. If the disorder is amenable to intervention therapy, the individual may be spared the consequences of his genotype.

SELF-QUIZ

1. According to the theory of pangenesis,
   a. the characteristics of the offspring are a blending of parental traits
   b. the sperm contains a miniature human being called the homunculus
   c. eggs and sperm are made in several regions of the body
   d. changes to an organism's body can be passed to the next generation
   e. particulate heritable factors separate in the formation of gametes

2. Discovery of Mendel's law of segregation required the use of
   a. dihybrid parental and F_1 crosses
   b. parental crosses with true-breeding plants followed by crosses with F_1 heterozygous plants
   c. testcrosses
   d. plants showing incomplete dominance
   e. chromosomal analysis

3. A 1:1 phenotypic ratio in a testcross indicates that
   a. the dominant phenotype parent was homozygous
   b. the dominant phenotype parent was heterozygous
   c. the alleles are codominant
   d. intermediate inheritance is involved
   e. both parents were heterozygotes

4. Of 100 individuals born with a certain genetic disorder, 85 die as a result of the disease within ten years. Survivors show a spectrum of clinical symptoms ranging from moderate to severe disability. What would you conclude about this disorder?
   a. It is due to a dominant gene.
   b. It is due to a recessive gene.
   c. The disorder has 85% penetrance and 100% expressivity.
   d. Penetrance is 100%, and expressivity is variable.

e. The disorder is caused by quantitative inheritance of a polygenic trait.

5. With chorionic villi sampling,
   a. a couple is tested to determine if they are carriers of a harmful allele
   b. a fetus is viewed by ultrasound to look for anatomical problems
   c. fetal cells of the placenta are karyotyped
   d. amniotic fluid is biochemically tested and fetal cells are cultured
   e. the probability of a fetus carrying a defective gene is determined

6. A 1:2:1 *phenotypic* ratio in the F_2 generation of a monohybrid cross is an indication of
   a. multiple alleles
   b. intermediate inheritance
   c. complete dominance
   d. pleiotropy
   e. polygenic inheritance

7. Familial hypercholesterolemia is an inherited disease that elevates blood cholesterol level mainly by
   a. increasing absorption of cholesterol from the intestine
   b. increasing cholesterol synthesis in the liver
   c. stimulating conversion of other steroids to cholesterol
   d. reducing cholesterol uptake by cells
   e. excessive decomposition of LDL

8. The main objective of carrier recognition tests is to identify individuals who are
   a. heterozygous for recessively inherited disorders
   b. homozygous for recessively inherited disorders
   c. heterozygous for dominantly inherited disorders
   d. homozygous for dominantly inherited disorders

9. All of the following disorders are due to recessive alleles *except*
   a. Huntington's disease
   b. sickle cell anemia
   c. Tay-Sachs disease
   d. cystic fibrosis
   e. phenylketonuria

10. Which of the following statements about recessively inherited disorders is a correct generalization?
   a. Such disorders are more common in males than in females.
   b. Most copies of the allele that causes a recessive disease are present in individuals who do not have the disease.
   c. Recessive disorders are much rarer than dominantly inherited disorders.
   d. All children born to a parent with the disorder are also afflicted.
   e. Penetrance for a recessive disorder is 50%.

GENETICS PROBLEMS

1. Flower position, stem length, and seed shape were three traits that Mendel chose to study. Each is controlled by an independently assorting gene and has dominant and recessive expression as follows:

Trait	Dominant	Recessive
Flower position	Axial *(A)*	Terminal *(a)*
Stem length	Tall *(L)*	Dwarf *(l)*
Seed shape	Round *(R)*	Wrinkled *(r)*

If a plant that is heterozygous for all three traits were allowed to self-fertilize, what proportion of the offspring would be expected to be:

a. homozygous for the three dominant traits?

b. homozygous for the three recessive traits?

c. heterozygous for the three traits?

d. homozygous for axial and tall, heterozygous for round?

2. A black guinea pig crossed with an albino one gave 12 black progeny. When the albino was crossed with a second black one, 7 blacks and 5 albinos were obtained. What is the best explanation for this genetic situation? Write genotypes for the parents, gametes, and offspring.

3. How many unique gametes can organisms with each of the following genotypes produce: *RrSs, RRss, RrSS, rrss*?

4. In some flowers, a true-breeding, red-flowered strain gives all pink flowers when crossed with a white-flowered strain: *RR* (red) × *rr* (white) → *Rr* (pink). If flower position is inherited as it is in peas (see problem 1), what will be the ratios of genotypes and phenotypes of the generation resulting from the following cross: Axial-red (true-breeding) × terminal-white? What will be the ratios in the F_2 generation?

5. In sesame plants, the one-pod condition *(P)* is dominant to the three-pod condition *(p)*, and normal leaf *(L)* is dominant to wrinkled leaf *(l)*. These traits are inherited independently. Determine the genotypes for the two parents for all possible matings producing the following progeny:

a. 318 one-pod normal, 98 one-pod wrinkled

b. 323 three-pod normal, 106 three-pod wrinkled

c. 401 one-pod normal

d. 150 one-pod normal, 147 one-pod wrinkled, 51 three-pod normal, 48 three-pod wrinkled

e. 223 one-pod normal, 72 one-pod wrinkled, 76 three-pod normal, 27 three-pod wrinkled

6. A man of blood type A marries a woman of blood type B. Their child has blood type O. What are the genotypes of these individuals? What other genotypes, and in what frequencies, would you expect in offspring of this marriage?

7. Phenylketonuria (PKU) is an inherited disease determined by a recessive allele. If a woman and her husband are both carriers, what is the probability that

a. all three of their children will be normal?

b. one *or* more of the three children will have the disease?

c. all three children will be afflicted with the disease?

d. *at least* one child will be normal?

(HINT: Remember that the probabilities of all possible outcomes always add up to 1.)

8. A mouse breeder interested in breeding a heterozygous stock to produce a large number of animals showing intermediate inheritance for a certain trait would expect about how many intermediate progeny out of 1500 offspring?

9. The genotype of F_1 individuals in a tetrahybrid cross is *AaBbCcDd*. Assuming independent assortment of these four genes, what are the probabilities that the F_2 offspring would have the following genotypes?

a. *aabbccdd*

b. *AaBbCcDd*

c. *AABBCCDD*

d. *AaBBccDd*

e. *AaBBCCdd*

10. What is the probability that each of the following pairs of parents will produce the indicated offspring (assume independent assortment of all gene pairs)?

a. *AABBCC × aabbcc → AaBbCc*

b. *AABbCc × AaBbCc → AAbbCC*

c. *AaBbCc × AaBbCc → AaBbCc*

d. *aaBbCC × AABbcc → AaBbCc*

11. Karen and Steve each have a sibling with sickle-cell anemia. Neither Karen, Steve, nor their parents have the disease and neither has been tested to reveal sickle-cell "trait." If this couple has a child, what is the probability that the child will have sickle-cell anemia?

12. Imagine that a newly discovered, recessively inherited disease is only expressed in individuals with type O blood, although the disease and blood type are independently inherited. A normal man with type A blood and a normal woman with type B blood have already had one child with the disease. The woman is now pregnant for a second time. What is the probability that the second child will also have the disease? Assume both parents are heterozygous for the disease gene.

FURTHER READING

1. Diamond, J. "Blood, Genes, and Malaria," *Natural History*, February 1989. Evolutionary history of sickle-cell anemia.
2. Edlen, G. *Genetics Principles: Humans and Social Consequences.* Portola Valley, Calif.: Jones & Bartlett, 1988.
3. Suzuki, D., A. Griffith, J. Miller, and R. Lewontin. *An Introduction to Genetic Analysis.* 4th ed. New York: W. H. Freeman, 1989. A good basic genetics text for undergraduates.

14

The Chromosomal Basis of Inheritance

It was not until the year 1900 that biology finally caught up with Gregor Mendel. At that time, three botanists, working independently, all arrived at the genetic principles of segregation and independent assortment from their own plant-breeding experiments. By searching the literature, however, the German Karl Correns, the Austrian Erich Tschermak von Seysenegg, and the Dutchman Hugo de Vries all found that Mendel had made the same discoveries 35 years before. During the intervening years, biology had grown more experimental and quantitative and thus more receptive to Mendelism. Nevertheless, many biologists remained incredulous about segregation and independent assortment until evidence had mounted that these principles of heredity had a physical basis in the behavior of chromosomes. This chapter integrates and extends what you have learned in the previous two chapters by describing the chromosomal basis for the transmission of genes from parents to offspring.

THE CHROMOSOME THEORY OF INHERITANCE

Cytologists worked out the process of mitosis in 1875 and the process of meiosis in the 1890s. Then, around the turn of the century, cytology and genetics converged as biologists began to see parallels between the behavior of chromosomes and the behavior of Mendel's factors (Figure 14.1). For example, chromosomes and genes are both paired in diploid cells; homolo-

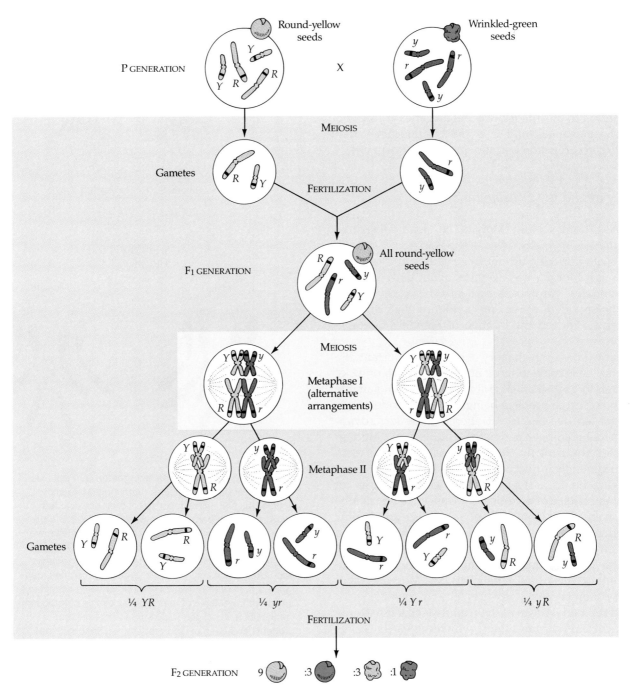

Figure 14.1

The chromosomal basis of Mendel's laws. Here we follow the fates during meiosis and fertilization of two genes that lie on different chromosomes: a gene for seed shape (alleles R and r) and a gene for seed color (alleles Y and y). To understand Mendel's law of segregation, focus on one type of chromosome (for example, the long chromosomes carrying the gene for seed shape). As the homologous F_1 chromosomes separate during the first meiotic division, the two alleles segregate, ending up in different gametes. Random fertilization then leads to F_2 progeny with the ratio of phenotypes that Mendel observed (three round to one wrinkled). To understand Mendel's law of independent assortment, we need to follow both types of chromosome. In the F_1 generation, two alternative, equally likely arrangements of tetrads may occur at metaphase I of meiosis, leading to the production of four genotypes of gametes. The nonhomologous chromosomes (and the genes they carry) have assorted independently. Metaphase I, the crucial stage for setting up both segregation and independent assortment, is highlighted here.

gous chromosomes separate and allele pairs segregate during meiosis; and fertilization restores the paired condition for both chromosomes and genes. Around 1902, Walter S. Sutton, Theodor Boveri, and others independently noted these parallels, and a **chromosome theory of inheritance** began to take form. According to this theory, Mendelian genes are located on chromosomes, and it is the chromosomes that undergo segregation and independent assortment.

Morgan and the *Drosophila* School

It was Thomas Hunt Morgan, an embryologist at Columbia University, who first associated a specific gene with a specific chromosome early in this century. Although Morgan was skeptical about both Mendelism and the chromosome theory, his early experiments provided convincing evidence that chromosomes are indeed the location of Mendel's heritable factors.

Many times in the history of biology, important discoveries have come to those insightful enough or lucky enough to choose an experimental organism suitable for the research problem being tackled. Mendel chose the garden pea because it offered some key advantages for breeding experiments. For his work, Morgan selected the fruit fly, *Drosophila melanogaster,* a common, generally innocuous pest that feeds on the fungi growing on fruit (Figure 14.2a). *Drosophila* was an excellent choice of organism for the genetic questions that were then being asked. Fruit flies are prolific breeders; a single mating will produce hundreds of progeny, and a new generation can be bred every 2 weeks. These characteristics make the fruit fly a convenient organism for genetic studies. Morgan's laboratory soon became known as "the fly room."

Another advantage of the fruit fly is that it has only four pairs of chromosomes, which are easily distinguishable with a microscope (see Figure 14.2b). There are three pairs of autosomes and one pair of sex chromosomes. As in humans, females have a homologous pair of *X* chromosomes and males have one *X* chromosome and one *Y* chromosome.

Location of a Gene   While Mendel could readily obtain different pea varieties, there were no convenient suppliers of fruit fly varieties for Morgan to employ. Indeed, he may have been the first person to want different varieties of this common insect. After a year of breeding flies and looking for variant individuals, Morgan was rewarded with the discovery of a single male fly with white eyes instead of the usual red. Normal phenotypes (those most common in nature), such as red eyes in *Drosophila,* are called **wild type** (Figure 14.3). Traits that are alternatives to the wild type, such as white eyes in *Drosophila,* are called

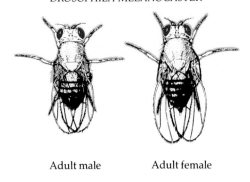

DROSOPHILA MELANOGASTER

Adult male   Adult female

(a)

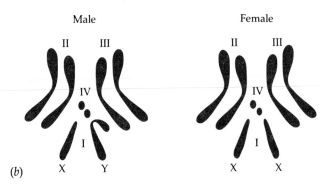

CHROMOSOMES

Male   Female

II   III   II   III

IV   IV

I   I

(b)   X   Y   X   X

Figure 14.2
The fruit fly, *Drosophila melanogaster.* (a) Fruit flies are small, about 3 millimeters (mm) long. Male and female fruit flies differ in certain of their physical features, as shown here. **(b)** The two sexes differ in chromosomal composition. Like humans, the females have two *X* chromosomes and the males have one *X* and one *Y* chromosome. In addition to the sex chromosomes, each *Drosophila* cell has three pairs of autosomes, designated II, III, and IV (which is a pair of very small chromosomes). Because of the relatively simple chromosome composition, short generation time, large number of progeny, and ease of laboratory care, *Drosophila* has long been a favorite subject of genetics researchers.

mutant phenotypes because they are due to alleles assumed to have originated as changes, or **mutations,** in the wild-type gene.

After Morgan discovered his white-eyed male fly, he mated it with a red-eyed female. All the F_1 progeny had red eyes, suggesting that the wild-type allele was dominant over the mutant allele. When Morgan bred the F_1 flies to each other, he observed the classical 3:1 phenotypic ratio among the F_2 progeny. However, there was a surprising result: The white-eye trait showed up only in males. All the F_2 females had red eyes, while half of the males had red eyes and half had white eyes. Somehow, a fly's eye color was linked to its sex.

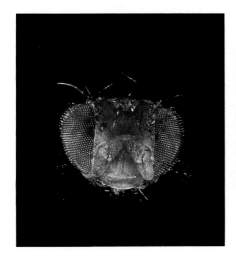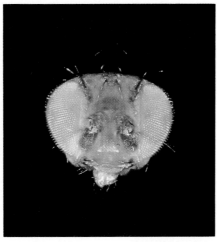

Figure 14.3
Morgan's first mutant. Wild-type *Droso-phila* flies have red eyes (left). Among his flies, Morgan discovered a male with white eyes (right). This variation became the first mutation to be analyzed in *Drosophila*.

├─── 0.5 mm ───┤

From this and other evidence, Morgan deduced that the gene for eye color is located exclusively on the X chromosome; there is no corresponding eye-color site, or locus, on the Y chromosome (Figure 14.4). Thus, females (XX) carry two copies of the gene for this trait, while males (XY) have only one. Since the mutant allele is recessive, a female will have white eyes only if she receives that allele on both X chromosomes—an impossibility for the F_2 females in Morgan's exper-iment. For a male, on the other hand, a single copy of the mutant allele confers white eyes. Since a male has only one X chromosome, there can be no wild-type allele present to mask the recessive allele.

Genes located on a sex chromosome are called **sex-linked genes,** and in fact this term is commonly applied only to genes on the X chromosome or its equivalent. Morgan's evidence that a specific gene is carried on the X chromosome added credibility to the chromo-

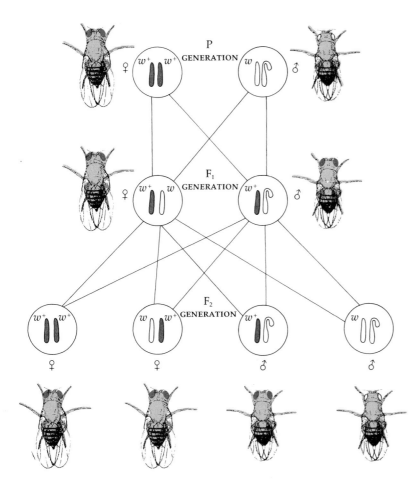

Figure 14.4
Sex-linked inheritance. When Morgan bred his white-eyed male to a wild-type female, all F_1 progeny had red eyes. The F_2 generation showed a typical Mendelian 3:1 ratio of traits, but the recessive trait—white eyes—was linked to sex. All females had red eyes, but half of the males had white eyes. Morgan hypothesized that the gene responsible was located on the X chromosome and there was no corre-sponding site on the Y chromosome. In this figure, the dominant allele (for red eyes) is symbolized w^+, and the recessive allele (for white eyes) is symbolized w. The symbols ♀ and ♂ stand for female and male, respectively.

Figure 14.5
A party in Morgan's fly room. For 30 years Morgan's laboratory at Columbia University in New York catalyzed an explosion in our understanding of inheritance. Here, Thomas Hunt Morgan (back row, far right) and his students in 1919 are celebrating the return of Alfred H. Sturtevant (leaning back in chair) from World War I military service. (The skeleton was just a prop, not an overworked student.)

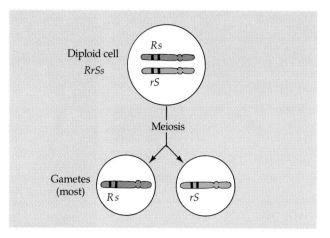

Figure 14.6
Linked genes. Genes located close together on the same chromosome do not follow Mendel's law of independent assortment because they are "linked" together. Thus, meiosis of the diploid cell shown here would yield (mostly) two types of gametes rather than equal numbers of the four types of gametes predicted by Mendel's law. (Actually, the two missing types of gametes, *RS* and *rs*, will be produced, but only in small numbers. See Figure 14.8.)

some theory of inheritance. Realizing the importance of this work, bright students were attracted to Morgan's fly room, and his laboratory dominated genetics research for the next three decades (Figure 14.5). We will see the influence of Morgan and his colleagues as we consider some other important aspects of the chromosomal basis of inheritance.

More Genetic Notation Morgan and other *Drosophila* geneticists employed a genetic notation somewhat different from that which we used in Chapter 13. Rather than using capital letters for dominant alleles and lowercase letters for recessive ones, they symbolized any mutant allele with one or more lowercase letters, such as w for the white-eye allele. The corresponding wild-type allele was then symbolized with a superscript plus sign, as in w^+ for the red-eye allele. A heterozygous female would be labeled w^+w.

Linked Genes: Exceptions to Independent Assortment

The number of genes in a cell is far greater than the number of chromosomes; in fact, each chromosome has hundreds or thousands of genes. Genes located on the same chromosome tend to be inherited together in genetic crosses, because they are part of a single chromosome that is passed along as a unit (Figure 14.6). Genes that tend to be inherited together and do

not follow the law of independent assortment are said to be **linked genes.** (Note that the use of the word *linked* in this way is different from its use in the term *sex-linked*.)

An early demonstration of linked genes came in 1908 from studies of sweet peas, a species closely related to the garden pea that Mendel studied. British biologists William Bateson and Reginald Crundall Punnett (the originator of the Punnett square) observed an anomaly in the inheritance of two traits—pollen color and pollen shape. They mated heterozygous plants, which had pollen with the dominant characteristics: purple color and long shape. The corresponding recessive characteristics are red color and round shape. They found that for each single characteristic, the alleles segregate, in agreement with Mendel's law of segregation. The result was the expected ratios of 3:1 purple to red pollen and 3:1 long to round pollen. However, when the data for the two characteristics were combined, they did not follow the law of independent assortment. Instead of a 9:3:3:1 ratio, a disproportionately large number of plants had either purple long pollen or red round pollen. Bateson and Punnett called this result "partial gametic coupling," but they could not explain how it occurred. Not until several years later did the explanation emerge from the work of T. H. Morgan: The genes for pollen color and shape are on the same chromosome and are usually inherited together. The relatively small number of plants with purple round or red long pollen result from crossing over a phenomenon we will discuss in the next section.

THE CHROMOSOMAL BASIS OF RECOMBINATION

In Chapter 12, we saw that the events of meiosis and random fertilization generate genetic variation among offspring of sexually reproducing organisms. The general term for the production of offspring with new combinations of traits inherited from two parents is **genetic recombination**. Here, we will examine the basis of recombination in more detail.

Recombination of Unlinked Genes: Independent Assortment

Mendel learned from his dihybrid crosses that some offspring have combinations of traits that do not match either parent. For example, in a testcross between a pea plant with round yellow seeds that is heterozygous for both traits and a plant with wrinkled green seeds (homozygous for both recessive traits), half the progeny are unlike either parent (Figure 14.7). The gene loci for these two traits are on separate chromosomes: Seed shape and seed color are unlinked. Notice that one-fourth of the offspring have round yellow seeds, and one-fourth have wrinkled green seeds. Hence, $\frac{1}{4} + \frac{1}{4} = \frac{1}{2}$ of the progeny have the same phenotype as one or the other of the parents; these offspring are called **parental types.** Of the other progeny, one-fourth have round green seeds and one-fourth have wrinkled yellow seeds. Because these offspring have different combinations of seed shape and color than either parent, they are called **recombinants.** When half of all progeny are recombinants, geneticists say that there is a 50% frequency of recombination.

A 50% frequency of recombination is observed for any two genes that are located on different chromosomes. The physical basis of recombination between unlinked genes is the random alignment of homologous chromosomes during metaphase I of meiosis, which leads to the independent assortment of alleles.

Recombination of Linked Genes: Crossing Over

Linked genes do not assort independently because they move together with their common chromosomes through meiosis and fertilization. We would not expect linked genes to recombine into assortments of alleles not found in the parents. But in fact, recombination between linked genes *does* occur. To see how, let us return to Morgan's fly room.

An unexpected observation was made by Morgan and his students when they were studying genes for body color and wing shape. A mutant allele for a body color gene causes black body color rather than the gray

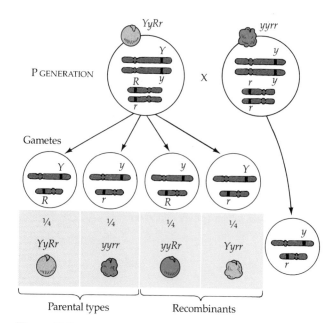

Figure 14.7
Recombination of unlinked genes. The inheritance of two genes carried on different pea chromosomes is analyzed in this testcross. A heterozygous plant with round yellow seeds is mated with a homozygous plant with green wrinkled seeds. The heterozygous parent produces four classes of gametes (see Figure 14.1). The homozygous plant produces only one class of gamete. One-half of the progeny have parental phenotypes; one-half have different trait combinations, or recombinant phenotypes. The recombination is due to the independent assortment of the two traits, which is possible because the genes for the two traits are on different chromosomes.

that is the wild type for *Drosophila.* And a mutant allele for a wing-shape gene causes vestigial wings, stump-like appendages, rather than wild-type wings (Figure 14.8). Both these mutations are recessives carried on autosomes (not on the *X* chromosome like the mutation for white eyes). In the geneticist's symbols, b = black body, b^+ = gray body, vg = vestigial wings, and vg^+ = wild-type wings. Morgan testcrossed flies having both black bodies and vestigial wings ($b\ b\ vg\ vg$) with wild-type flies that were heterozygous for both traits ($b^+\ b\ vg^+\ vg$). If the genes were unlinked, then independent assortment would produce a phenotypic ratio of 1:1:1:1 (one-fourth black body, wild-type wings; one-fourth gray body, normal wings; one-fourth black body, vestigial wings; and one-fourth gray body, vestigial wings). But if the two genes were linked, we might expect to see only the parental phenotypes, in a 1:1 ratio (one-half gray body, wild-type wings and one-half black body, vestigial wings).

Morgan's results fulfilled neither of these expectations. Most of the progeny had parental phenotypes, suggesting linkage between the two genes, but about 17% of the flies were recombinants. Although there

Figure 14.8
Recombination of linked genes. (a) The ratio of phenotypes that result from this dihybrid testcross cannot be explained if the two gene loci are unlinked, nor are the results consistent with complete linkage between the two loci. **(b)** The genetic explanation of the data. The predominance of parental types suggests that the two traits are carried on the same chromosome but that a mechanism exists for occasionally breaking the linkage and exchanging chromosomal segments between homologues. **(c)** The mechanism of crossing over, which occurs during prophase I of meiosis.

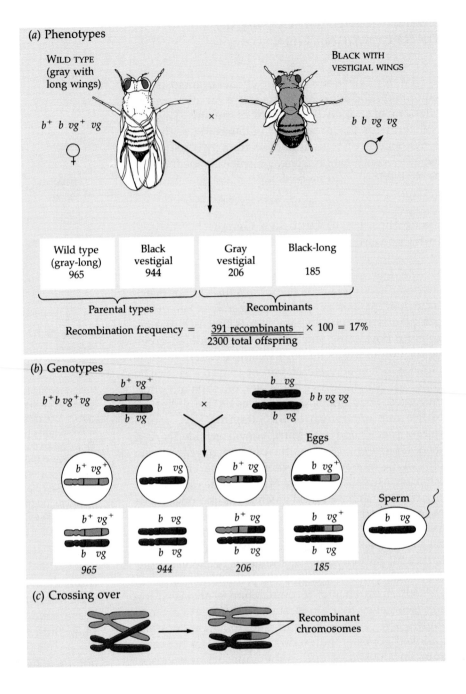

(a) Phenotypes

WILD TYPE (gray with long wings) $b^+ b\ vg^+\ vg$ ♀

×

BLACK WITH VESTIGIAL WINGS $b\ b\ vg\ vg$ ♂

| Wild type (gray-long) 965 | Black vestigial 944 | Gray vestigial 206 | Black-long 185 |

Parental types — Recombinants

Recombination frequency = $\dfrac{391 \text{ recombinants}}{2300 \text{ total offspring}} \times 100 = 17\%$

(b) Genotypes

$b^+ b\ vg^+ vg$ $b^+\ vg^+$ / $b\ vg$ × $b\ vg$ / $b\ vg$ $b\ b\ vg\ vg$

Eggs

$b^+\ vg^+$ $b\ vg$ $b^+\ vg$ $b\ vg^+$

Sperm $b\ vg$

$b^+\ vg^+$ / $b\ vg$ 965
$b\ vg$ / $b\ vg$ 944
$b^+\ vg$ / $b\ vg$ 206
$b\ vg^+$ / $b\ vg$ 185

(c) Crossing over

Recombinant chromosomes

was linkage, it appeared incomplete. Morgan proposed that some mechanism, such as an exchange of segments between homologous chromosomes, must occasionally break the linkage between the two genes. Subsequent experiments have demonstrated that such an exchange, called *crossing over,* accounts for the recombination of linked genes. The physical basis of crossing over was described in Chapter 12 (see Figure 12.9). While homologous chromosomes are paired in synapsis during prophase of meiosis I, nonsister chromatids may break at corresponding points and switch fragments in a reciprocal fashion. Normally, the trade is exactly reciprocal, with the exchanged fragments containing the same genetic loci; otherwise, genes for

certain inherited characters would be lost from some chromosomes and duplicated in others. A crossover between chromatids of homologous chromosomes breaks linkages in the parental chromosomes to form recombinants that may bring together alleles in new combinations. The subsequent events of meiosis distribute the recombinant chromosomes to gametes.

MAPPING CHROMOSOMES

Observations of recombination between genes has allowed scientists to assign, or **map,** genes to particular chromosomes and to specific regions on those chromosomes. It was Morgan's group that first worked out methods for mapping chromosomes.

Maps Based on Crossover Data

As discussed earlier, the mutant genes for black body (*b*) and vestigial wings (*vg*) in *Drosophila* are linked, with a recombination frequency of approximately 17%. Carried on the same chromosome as these two genes is a third gene that has a recessive allele causing "cinnabar eyes," which are a much brighter red than the wild type. The recombination frequency between the cinnabar gene (*cn*) and the *b* locus is 9%. Thus, crossovers between the *b* and *vg* loci are about twice as frequent as crossovers between *b* and *cn* (17% versus 9%). In 1917, Alfred H. Sturtevant, one of Morgan's students, reasoned that different recombination frequencies reflect different distances between genes on a chromosome; that is, if two genes are far apart on a chromosome, there is a higher probability that a crossover event will separate them than if the two genes are close together. If we assume that the probability of crossing over between two genes is directly proportional to the distance between them, then the distance along the *Drosophila* chromosome between *b* and *vg* must be about twice as great as the distance between *b* and *cn*.

Sturtevant began using recombination data to assign genes positions on a map. He defined one **map unit** as the equivalent of a 1% recombination frequency. Thus, the *b* and *cn* loci are separated by 9 map units, while the *b* and *vg* genes are 17 map units apart. But what is the sequence of the three genes? From what we have learned so far, we do not know whether the sequence is *cn–b–vg* or *b–cn–vg*. The frequency of recombination between *cn* and *vg* should reveal the correct sequence of the three genes. The first possible sequence predicts that *cn* and *vg* are about 26 map units (9 + 17) apart, while the second sequence predicts a separation of about 8 map units (17 − 9). Sturtevant found that the frequency of recombination between *vg* and *cn* was 9.5%; he therefore proposed that the genes were arranged along a chromosome in the sequence *b–cn–vg*. This method was soon extended to map the other identified *Drosophila* genes in linear arrays.

Some genes on a chromosome are so far apart from each other that crossovers between them occur very often. The frequency of recombination measured between such genes can have a maximum value of 50%, a result indistinguishable from that for genes on different chromosomes. In fact, the seven characters that Mendel studied in his peas are not all on separate chromosomes, although the pea coincidentally has seven chromosome pairs. Seed color and flower color, for instance, are now known to be on chromosome 1. But they are so far apart on that chromosome that linkage is not observed in genetic crosses. Only for one pair of the genes Mendel studied, plant height and pod shape, do modern biologists observe linkage. Although Mendel observed segregation of alleles for each of these traits, he did not report the results of dihybrid crosses for this particular combination of traits. Genes located far apart on a chromosome are mapped by adding the recombination frequencies from crosses involving each of the distant genes and an intermediate gene.

Using crossover data, Sturtevant and his co-workers were able to cluster the known mutations (and hence the wild-type alleles) of *Drosophila* into four groups of linked genes (Figure 14.9). Because microscopists had found four sets of chromosomes in *Drosophila* cells, this clustering of genes was additional evidence that genes are located on chromosomes. Each chromosome has a linear array of specific gene loci.

Cytological Maps

The information about gene loci determined from crossover data is relative, not absolute. Recombination frequencies tell us the sequence of linked genes and provide clues about comparative distances between them, but this approach discloses neither the actual locations of the genes on the chromosome nor the absolute distances between the genes (in nanometers, for instance). Another technique, **cytological mapping,** pinpoints genes or loci on chromosomes. This method locates a gene by associating a mutant phenotype with the position of a chromosomal defect or other feature that can be seen in the microscope. Cytological maps became more complete as geneticists learned how to induce lesions in chromosomes and identify the resulting chromosomal alterations using the microscope.

One important way to induce chromosomal alterations in laboratory experiments is by using X-irradiation. In the 1920s, Hermann Joseph Muller, another of Morgan's collaborators, discovered that subjecting fruit flies to X-rays increases the frequency of genetic changes. X-rays produce two different classes of mutation. They can cause a change at a single spot within a gene, called a **point mutation,** and they can break chromosomes, causing major alterations called **chromosomal mutations.** Chromosome breaks can result in the loss or rearrangement of segments of chromosomes. These structural alterations can often be seen with the microscope.

The use of X-rays to induce mutations greatly increased the number of mutant varieties of *Drosophila* and other organisms available to geneticists; they no longer were limited to searching for rare, natural

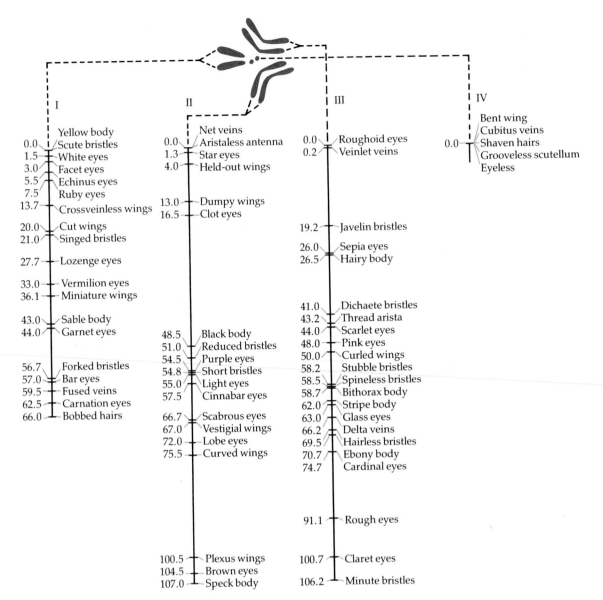

Figure 14.9
A map of *Drosophila* genes. This genetic map of some of the known *Drosophila* genes illustrates that each linkage group corresponds to one chromosome pair. Numbers are map distances determined from crossing-over data. Chromosome I is the X chromosome.

I

0.0	Yellow body / Scute bristles
1.5	White eyes
3.0	Facet eyes
5.5	Echinus eyes
7.5	Ruby eyes
13.7	Crossveinless wings
20.0	Cut wings
21.0	Singed bristles
27.7	Lozenge eyes
33.0	Vermilion eyes
36.1	Miniature wings
43.0	Sable body
44.0	Garnet eyes
56.7	Forked bristles
57.0	Bar eyes
59.5	Fused veins
62.5	Carnation eyes
66.0	Bobbed hairs

II

0.0	Net veins / Aristaless antenna
1.3	Star eyes
4.0	Held-out wings
13.0	Dumpy wings
16.5	Clot eyes
48.5	Black body
51.0	Reduced bristles
54.5	Purple eyes
54.8	Short bristles
55.0	Light eyes
57.5	Cinnabar eyes
66.7	Scabrous eyes
67.0	Vestigial wings
72.0	Lobe eyes
75.5	Curved wings
100.5	Plexus wings
104.5	Brown eyes
107.0	Speck body

III

0.0	Roughoid eyes
0.2	Veinlet veins
19.2	Javelin bristles
26.0	Sepia eyes
26.5	Hairy body
41.0	Dichaete bristles
43.2	Thread arista
44.0	Scarlet eyes
48.0	Pink eyes
50.0	Curled wings
58.2	Stubble bristles
58.5	Spineless bristles
58.7	Bithorax body
62.0	Stripe body
63.0	Glass eyes
66.2	Delta veins
69.5	Hairless bristles
70.7	Ebony body
74.7	Cardinal eyes
91.1	Rough eyes
100.7	Claret eyes
106.2	Minute bristles

IV

0.0	Bent wing / Cubitus veins / Shaven hairs / Grooveless scutellum / Eyeless

10 μm

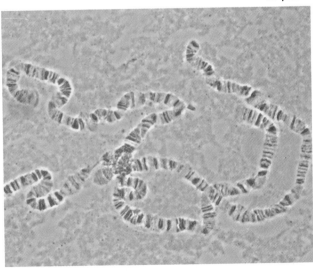

Figure 14.10
Giant chromosomes in the salivary glands of *Drosophila*. The extraordinary thickness of these chromosomes results from the parallel alignment of hundreds of chromatids. When the chromosomes are stained, denser portions retain more stain, giving the banding patterns you see here.

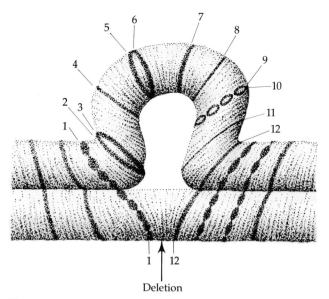

Figure 14.11
A deletion as seen in a polytene chromosome. Homologues are synapsed, but the lower chromosome is missing a segment corresponding to bands 2 through 11. Therefore, the wild-type chromosome buckles out in the region where there are no matching bands on the deficient homologue.

mutants such as Morgan's white-eyed fly. Muller also recognized an alarming implication of his discovery: X-rays and other forms of radiation pose hereditary hazards to people as well as to laboratory animals.

The salivary glands and certain other tissues of *Drosophila* larvae and many other insect larvae contain unusually thick interphase chromosomes, which are very useful to geneticists searching for chromosomal mutations. These giant chromosomes have distinctive banding patterns that can easily be observed with a microscope (Figure 14.10). During the development of a salivary gland, the chromosomes replicate, but the sister chromatids do not separate, and the cells do not divide. This process repeats many times until each

chromosome consists of hundreds of perfectly aligned chromatids, forming structures called **polytene** (many-stranded) **chromosomes.** The parallel alignment of the chromatids aligns differences in density along them, and staining produces patterns of dark and light bands that are characteristic for specific regions of each chromosome. The banding patterns of the polytene chromosomes are further augmented by close pairing of homologues—a rare phenomenon in nonmeiotic cells. Pairing cannot occur in a region where one homologue has a structural alteration, and this makes it possible to identify the location of a chromosomal mutation. For example, where there is a segment missing from one chromosome, the defective chromosome's normal homologue will buckle out in a loop containing the unpaired segment (Figure 14.11).

By matching a specific mutant phenotype, such as vestigial wings, with a localized deformity in polytene chromosome pairing, the chromosomal location of the mutant gene and its wild-type allele can be deduced. By inducing genetic damage with X-rays and other mutagens (agents that cause mutations), screening offspring for mutant phenotypes, and associating those mutants with chromosomal defects seen in the microscope, genetic cartographers have drawn extensive cytological maps of the genes of *Drosophila* and several other organisms.

The spacing of genetic loci derived from crossover data does not match exactly the spacing determined by cytological mapping (Figure 14.12). In fact, a 1% recombination frequency does not correspond exactly to a fixed length of chromosome, because the frequency of crossing over is not the same for all regions of a chromosome.

THE CHROMOSOMAL BASIS OF SEX

We have seen that in *Drosophila*, sex is a phenotypic character determined by inherited chromosomes.

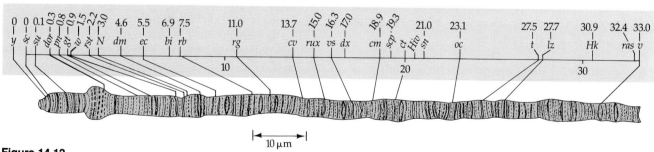

Figure 14.12
Correlation between a genetic map and a cytological map. Shown here are some of the mutations that have been mapped at one end of the X chromosome of *Drosophila melanogaster.* The map prepared from crossover data is shown on the straight line above the polytene chromosome. The physical location of mutations on the chromosome is indicated by connecting lines. The sequences of mutations are the same on the two maps, but the distances between loci vary as a result of different frequencies of crossing over in different regions of the chromosome.

Indeed, in most species, sex is determined by the presence or absence of special chromosomes, but the X-Y mechanism of *Drosophila* is only one of many systems (Figure 14.13).

Systems of Sex Determination

X-Y and X-O Systems Many species, including humans and other mammals, share with *Drosophila* an X-Y mechanism that determines sex at the time of fertilization. As a result of chromosome segregation during meiosis, each haploid gamete contains one sex chromosome along with its haploid set of autosomes. Half of the sperm cells contain an X chromosome and half contain a Y chromosome. Because it produces two kinds of gametes, the male is called the **heterogametic sex.** All of the eggs carry X chromosomes, so the female is called the **homogametic sex.** The sex of an individual is determined by the class of sperm cell, X-bearing or Y-bearing, that fertilizes the egg. Random fertilization results in a 1:1 sex ratio (approximately equal numbers of males and females).

Although both mammal and fruit fly males are *XY,* the mechanisms that determine maleness differ in the two cases. In mammals, it is the presence of the Y chromosome that causes maleness. In *Drosophila,* maleness is determined by the ratio of X chromosomes to autosomes; the Y chromosome is not necessary.

A mechanism similar to the X-Y system in *Drosophila* is the X-O system of grasshoppers, crickets, roaches, and some other insects. Females have two X chromosomes *(XX),* but there is no Y chromosome; males have only a single sex chromosome, the X, and are thus *XO.* As in the X-Y system, the males are the heterogametic sex because they produce two classes of gametes.

Z-W Systems The female is the heterogametic sex in birds, some fishes, and some insects, including butterflies and moths. The sex chromosomes in these species are designated Z and W to avoid confusion with the X-Y system. Male birds are ZZ, while female birds are ZW. Thus, sex is determined by whether the egg carries a Z or a W chromosome.

Haplo-diploidy There are no sex chromosomes in most species of bees and ants, and yet sex determination has a chromosomal basis. Females develop from fertilized eggs and are thus diploid. Males, however, develop from unfertilized eggs—they are fatherless—and remain haploid. The term for such "virgin birth" from an unfertilized egg is **parthenogenesis.**

Sex Determination in Plants Among the species of plants that have two separate sexes, the chromosomal bases of sex determination are often the same as those observed in animals. Plant species that have two separate sexes are called *dioecious* (from the Greek *di* and *oikos,* "two houses"), and male and female flowers are found on separate individuals. Some dioecious species use the X-Y system of sex determination; date palms, spinach, and marijuana are examples. The Z-W system also exists in the plant kingdom; the wild strawberry is one species where females are the heterogametic sex. Often the male and female sex chromosomes in plants look alike and are almost fully homologous, differing only by a single gene.

Organisms Lacking Sex Determination Most plant species and some animals, such as earthworms and garden snails, are *monoecious* (from the Greek, "one house"), meaning that a single individual produces both sperm and eggs. Monoecious animals have both testes and ovaries, and monoecious plants have either one type of flower that has both stamens and carpels (for example, in peas) or two flower types on the same individual, one to produce pollen and the other to produce eggs (for example, in corn—see Chapter 34). In these cases, all haploid or all diploid individuals of a species have the same complement of chromosomes.

SEX-LINKED INHERITANCE

Although certain genes on the sex chromosomes play a role in specifying the sex of an individual, these chromosomes also contain genes for traits unrelated to femaleness or maleness. Because the human X chromosome is much larger than the Y (see the karyotype on p. 257), there are many more X-linked traits than Y-linked traits; and most of the X-linked genes have no homologous loci on the Y chromosome. In humans, the term *sex-linked traits* usually refers to X-linked traits. These traits all follow the same pattern of inheritance that Morgan observed for the white-eye locus in *Drosophila.* Fathers pass X-linked alleles to all their daughters but to none of their sons, since males receive their one X chromosome from their mothers (Figure 14.14). In contrast, mothers can pass sex-linked alleles to both sons and daughters.

If a sex-linked trait is due to a recessive allele, a female will express the phenotype only if she is a homozygote. Because males have only one locus, the terms *homozygous* and *heterozygous* are irrelevant for describing their sex-linked genes (the term *hemizygous* is used in such cases). Any male receiving the mutant allele from his mother will express the trait. For this reason, far more males than females have disorders that are inherited as sex-linked recessives. However, even though the chance of a female's inheriting a double dose of the mutant gene is much less than the probability of a male's inheriting a single dose, there

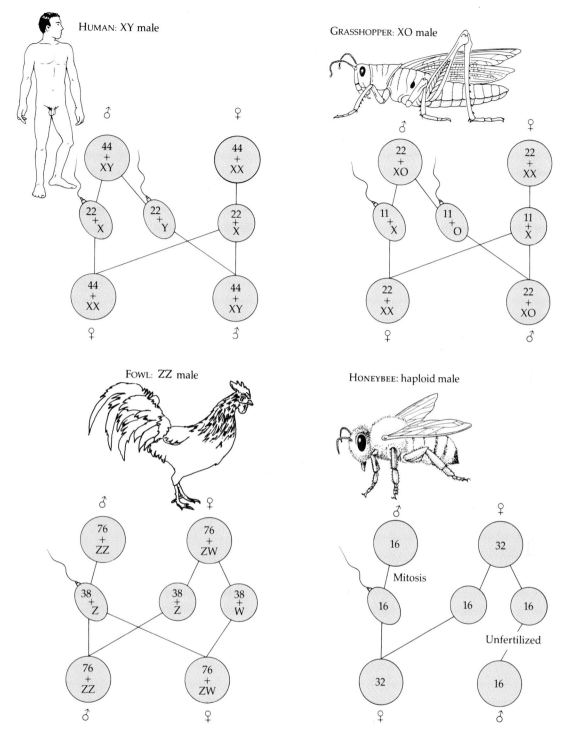

Figure 14.13
Some systems of sex determination. Several different systems by which chromosomes determine sex are found among animals.

are females with sex-linked disorders. For instance, color blindness is a mild disorder inherited as a sex-linked trait. A color-blind daughter may be born to a color-blind father who marries a carrier (see Figure 14.14c). However, because the sex-linked allele for color blindness is rare, the probability that such a man and woman will come together is very low.

Hemophilia is a sex-linked recessive trait with a rich history. Hemophiliacs bleed excessively when injured because they have inherited an abnormal factor

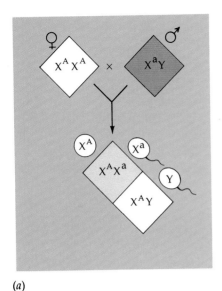

(a)

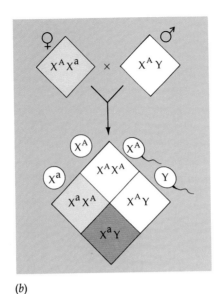

(b)

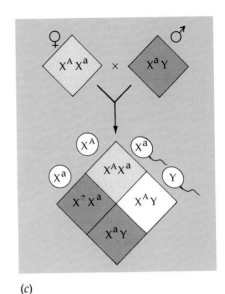

(c)

Figure 14.14

Transmission of sex-linked recessive traits. In this diagram, *X* and *Y* symbolize the sex chromosomes. The superscript *A* represents a dominant allele carried on the *X* chromosome, and the superscript *a* represents a recessive mutant allele. **(a)** A father with the trait will transmit the mutant allele to all daughters, but to no sons. When the mother is a dominant homozy-gote, the daughters will have the normal phenotype, but will be carriers of the muta-tion. **(b)** A carrier who mates with a normal male will pass the mutation to half her sons and half her daughters. The sons with the mutation will have the disease. The daughters who have inherited the mutation in single dosage will have the normal phenotype but will be carriers like their mother. **(c)** If a carrier mates with a male who has the trait, there is a 50% chance that each child born to them will have the trait, regardless of sex. Daugh-ters who do not have the trait will be car-riers, whereas males without the trait will be completely free of the deleterious recessive allele.

involved in blood clotting. The most seriously afflicted individuals may bleed to death after relatively minor skin abrasions, bruises, or cuts. The ancient Hebrews must have had some understanding of the hereditary pattern of hemophilia, for sons born to women having a family history of hemophilia were exempted from circumcision.

A high frequency of sex-linked hemophilia has plagued the royal families of Europe (Figure 14.15). The first hemophiliac in the royal line seems to have been Leopold, son of Queen Victoria (1819–1901) of England. It is likely that the recessive allele for hem-ophilia was introduced to the royal family through a mutation in one of the sex cells of Victoria's mother or father, making Victoria a heterozygote, or carrier, of the deadly allele. Leopold survived to father a daugh-ter who was also a carrier, transmitting hemophilia to one of her sons. Hemophilia was eventually intro-duced to the royal families of Prussia, Russia, and Spain through the marriages of two of Victoria's daughters, Alice and Beatrice, both carriers. The age-old practice of strengthening international alliances by having royalty marry royalty effectively spread hem-ophilia through the royal families of several European kingdoms.

Although a small region of the human *Y* chromo-some may be homologous to a segment of the *X*, most of the genes on the tiny *Y* chromosome appear to have no counterparts on the *X*. These *Y* genes encode traits that are found only in males and that appear in all sons of fathers having the trait.

In 1987, researchers reported evidence that male-ness in humans and other mammals is ultimately determined by a single, "master" gene on the *Y* chro-mosome. The master gene seems to be the *TDF* gene, which is responsible for the production of a protein called *testis-determining factor*. This protein induces testes to develop in embryos. It is not yet known exactly how this gene or its protein product regulates other genes involved in male sexual development.

Gene Dosage Compensation

A normal diploid cell has two copies of each autoso-mal gene and thus a double dose of each autosomal gene's product. But how does an organism compen-sate for the fact that the cells in individuals of one sex may contain two copies of sex-linked genes while the cells in individuals of the other sex contain only one copy? In mammals, including humans, only one *X* chromosome is fully active in most diploid cells. In females, one of the two *X* chromosomes is inactivated during early embryonic development. This explana-tion is called the **Lyon hypothesis,** after British genet-icist Mary F. Lyon. The inactive *X* chromosome con-

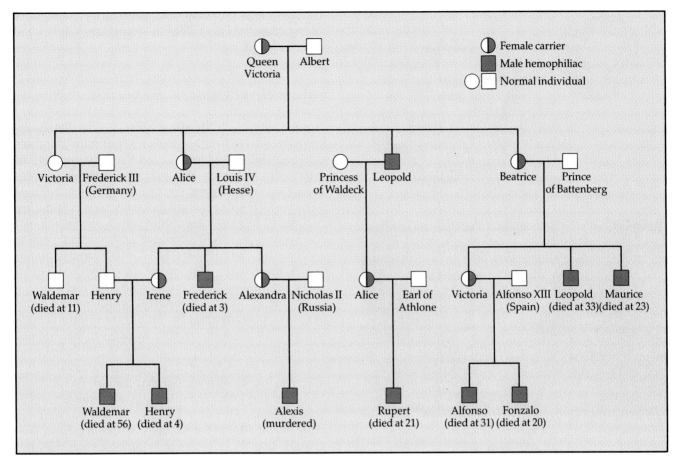

Figure 14.15
Hemophilia in the royal families of Europe. This partial pedigree traces the sex-linked disease from Queen Victoria through three generations of European dynasties. Queen Victoria had nine children, twenty-six grandchildren, and thirty-four great-grandchildren; only the afflicted individuals and their direct ancestors are shown here.

tracts into a dense object, called a **Barr body,** which lies along the inside of the nuclear envelope in cells of females. Most of the genes of the X chromosome that forms the Barr body are not expressed, although small regions of that chromosome remain active.

The "choice" of which of the two Xs will be inactivated occurs randomly and independently in each of the embryonic cells present at the time of X inactivation. As a consequence, females consist of a *mosaic* of two types of cells—those with the active X derived from the father and those with the active X derived from the mother. After an X chromosome is inactivated in a particular cell, all mitotic descendants of that cell have the same inactive X. Therefore, if the female is heterozygous for a sex-linked trait, approximately half of her cells will express one allele, while the others will express the alternate allele. This mosaicism can be seen graphically in the coloration of a calico cat (Figure 14.16). In humans, there is a recessive X-linked mutation that prevents development of sweat glands. A woman who is heterozygous for this

trait will have patches of normal skin and patches of skin lacking sweat glands.

Sex-Limited and Sex-Influenced Traits

The sex chromosomes do not carry all the genes for traits associated primarily with one sex or the other. Some traits, called **sex-limited traits,** appear exclusively in one sex but are determined by autosomal genes present in both sexes. Thus, a man may inherit his beard type from his mother rather than from his father. Although only one sex or the other normally expresses sex-limited genes, both sexes transmit the genes. This is very important in animal breeding. For example, a dairy-cow breeder must consider a bull's family history for milk yield, and a chicken breeder must consider the rooster's family history for egg production.

The penetrance or expressivity of autosomal genes may be sex-dependent. One example of such a **sex-**

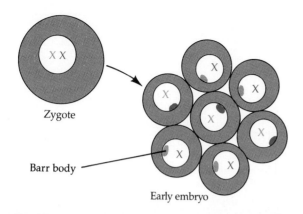

Zygote

Barr body

Early embryo

Figure 14.16
The calico cat. Such cats are females that are heterozygous for an X-linked gene that controls color. One X chromosome has an allele for orange color, while the other X chromosome carries an allele for non-orange color. Random X inactivation in the embryo, followed by mitotic cloning, produces the cat's patchwork coat. Inactive X chromosomes are condensed and located adjacent to the nuclear envelope as Barr bodies.

influenced trait is a form of baldness in which a single copy of the allele results in a bald head for a man, whereas a woman will lose her hair only if she is homozygous for the allele. Usually, it is the different hormonal conditions of the bodies of males and females that dictate which sex-limited and sex-influenced genes will be expressed and how effective they will be.

CHROMOSOMAL ALTERATIONS

Errors in meiosis or mutagens such as radiation can cause major changes in the chromosomes of a cell. As one would predict from the chromosomal basis of inheritance, such alterations can have significant impact on phenotype, often affecting many linked genes at once. These alterations provided evidence for the chromosome theory and continue to be employed in determining the location of genes on chromosomes.

Chromosomal mutations can alter either the number of chromosomes per cell or the structure of individual chromosomes.

Alterations of Chromosome Number

Aneuploidy   Ideally, the meiotic spindle distributes chromosomes to daughter cells without error. But there is an occasional accident, called a **nondisjunction,** in which the members of a pair of homologous chromosomes do not move apart properly during meiosis I or in which sister chromatids fail to separate during meiosis II. In these cases, one gamete receives two of the same type of chromosome and another gamete receives no copy (Figure 14.17). The other chromosomes are usually distributed normally. If either of these aberrant gametes unites with a normal one, the offspring will have an abnormal chromosome number, known as **aneuploidy.** If the chromosome is present in triplicate in the fertilized egg (so that the cell has a total of $2N + 1$ chromosomes), the aneuploid cell is said to be **trisomic** for that chromosome. If a chromosome is missing (so that the cell has $2N - 1$ chromosomes), the aneuploidy is **monosomic** for that chromosome. Mitosis subsequently will transmit the anomaly to all embryonic cells. If the organism survives, it usually has a set of symptoms caused by the abnormal dosage of genes located on the extra or missing chromosome. Down syndrome, for example, results from trisomy for chromosome 21 (see Figure 14.19). (Nondisjunction can also occur in mitosis, but unless it involves cells giving rise to gametes, it is unlikely to have significant consequences.)

Polyploidy   Some mutants have more than two complete chromosome sets. The general term for this chromosomal alteration is **polyploidy,** with specific terms such as **triploidy** ($3N$) and **tetraploidy** ($4N$) used to indicate the number of haploid sets. One way a triploid cell may be produced is by the fertilization of an abnormal diploid egg produced by nondisjunction of all its chromosomes. An example of an accident that would result in tetraploidy is the failure of a $2N$ zygote to divide after replicating its chromosomes. Subsequent mitosis would then produce a $4N$ embryo.

Polyploidy is relatively common in the plant kingdom, and we will see in Chapter 22 that the spontaneous origin of polyploid individuals plays an important role in the evolution of plants. In the animal kingdom, the natural occurrence of polyploids seems to be extremely rare, although polyploidy can be induced experimentally in certain animals, such as frogs and rabbits. In general, polyploids are more normal in appearance than aneuploids. One extra (or missing) chromosome apparently disrupts genetic balance more than having an entire extra set of chromosomes. More

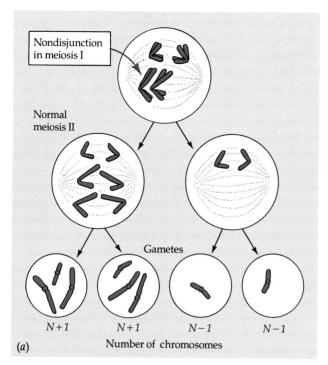

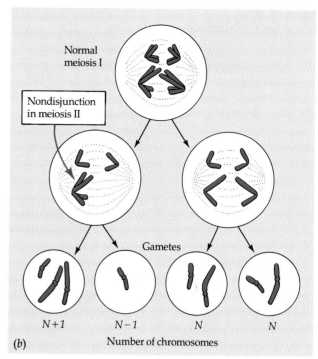

Figure 14.17
Meiotic nondisjunction. (a) Homologues may fail to separate during anaphase of meiosis I, or (b) chromatids may fail to separate during anaphase of meiosis II. Either type of accident will produce gametes with an anomalous chromosome number.

common than complete polyploid animals are mosaic polyploids, animals with patches of polyploid cells. If the sister chromatids for all the chromosomes fail to separate during a *mitotic* division, so that one daughter cell gets all the replicated chromosomes, a tetraploid cell results that can subsequently produce a localized clone of tetraploid cells.

Alterations of Chromosome Structure

Breakage of a chromosome can lead to a variety of rearrangements affecting the genes of that chromosome. Fragments without centromeres are usually lost when the cell divides. The chromosome from which the fragment originated will then have a deficiency, or **deletion**. In some cases, however, the fragment may join to the homologous chromosome, producing a **duplication** there. It also may join a nonhomologous chromosome, an event called a **translocation,** or it may reattach to the original chromosome but in the reverse orientation, a change called an **inversion** (Figure 14.18).

Another source of deletions and duplications is error during crossing over. Although crossovers are normally reciprocal, neighboring chromatids sometimes break at different places, and one partner consequently gives up more genes than it receives. The products of such a nonreciprocal crossover are one

chromosome with a deletion and one chromosome with a duplication.

An organism that inherits a homozygous deletion (or a single *X* chromosome with a deletion, in a male) has a genetic imbalance that is usually lethal. This observation is evidence that most genes are vital to an organism's existence. Duplications and translocations also tend to have deleterious effects. In reciprocal translocations, in which segments are exchanged between chromosomes, and in inversions, the balance of genes is not abnormal—all genes are present in their normal doses. Nevertheless, inversions can alter phenotype because of more subtle **position effects:** A gene's expression can be influenced by its location among neighboring genes.

Chromosomal Alterations in Human Disease

Alterations of chromosome number and structure are associated with a number of serious human disorders. When nondisjunction occurs in meiosis, the result is aneuploidy, an abnormal number of chromosomes in the gamete produced and, later, in the zygote. Although the frequency of aneuploid zygotes may be quite high in humans, most of these chromosomal alterations are so disastrous to development that the embryos are spontaneously (naturally) aborted long before birth.

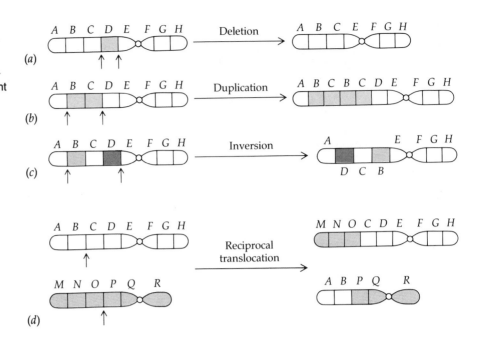

Figure 14.18
Alterations of chromosome structure.
(a) A deletion removes a chromosomal segment. (b) A duplication repeats a segment. (c) An inversion reverses a segment within a chromosome. (d) A translocation moves a segment from one chromosome to another, nonhomologous one.

However, some types of aneuploidy appear to upset the genetic balance less than others, so that individuals are at least occasionally born with a characteristic condition or "syndrome." Genetic diseases caused by aneuploidy can be diagnosed before birth by amniocentesis (see Chapter 13).

Down syndrome is the most common serious birth defect in the United States, affecting approximately 1 out of every 700 children born. Down syndrome is usually the result of aneuploidy: There is an extra chromosome 21, so that each body cell has a total of 47 chromosomes (Figure 14.19a). Although chromosome 21 is the smallest human chromosome, its trisomy severely alters the individual's phenotype. Down syndrome includes characteristic facial features (Figure 14.19b), short stature, heart defects, susceptibility to respiratory infection, and mental retardation. Down syndrome is by far the most common form of severe mental retardation. Furthermore, individuals with Down syndrome are prone to developing leukemia and Alzheimer's disease. (It is probably not a coincidence that genes associated with the latter two diseases have been found to be on chromosome 21.)

Although victims of Down syndrome on the average have a life span much shorter than normal, some individuals with trisomy 21 live to middle age or beyond. Most are sexually underdeveloped and sterile, but a few women with Down syndrome have had children. Since half of these women's eggs have the extra chromosome 21, there is a 50% chance that a woman with Down syndrome will transmit it to each of her children.

Among normal parents, the incidence of Down syndrome in offspring correlates with the age of the mother (Figure 14.20). Down syndrome strikes 0.04%

of children born to women under age 30. The risk climbs to 1.25% for mothers in their early 30s and is even higher for older mothers. Because of this relatively high risk, pregnant women who are over 35 are candidates for amniocentesis in order to check for trisomy 21 and other major chromosomal defects. No one is sure why meiotic nondisjunction increases in frequency as a woman ages, but it may be related to the long time lag between the onset of the first meiotic division, which occurs in the mother's fetal ovaries prior to her birth, and the completion of meiosis, which occurs many years later at the time of ovulation (see Chapter 42).

Far rarer than Down syndrome are several other human diseases caused by autosomal aneuploidy. *Patau syndrome*, caused by trisomy for chromosome 13, is characterized by serious eye, brain, and circulatory defects, as well as harelip and cleft palate. Patau syndrome occurs once in every 5000 live births. Trisomy of chromosome 18 causes the condition known as *Edwards syndrome*, which affects almost every organ system in the body. It occurs about once in 10,000 live births. In both these syndromes, most victims survive less than a year.

Nondisjunction of sex chromosomes produces a variety of aneuploid conditions in humans (Table 14.1). Most sex chromosome aneuploidies appear to upset the genetic balance less than aneuploid conditions involving autosomes. This may be because the Y chromosome carries very few genes and because extra copies of the X chromosome become inactivated as Barr bodies in the somatic cells.

An extra X chromosome in a male, producing XXY, occurs approximately once in every 2,000 live births. Persons with this disorder, called **Klinefelter syn-**

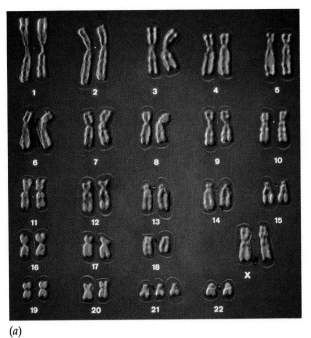

(a)

(b)

Figure 14.19
Down syndrome. (a) The karyotype shows trisomy 21. **(b)** The child with his hand raised has Down syndrome. He is enrolled in a mainstream classroom.

drome, have male sex organs, but the testes are abnormally small and the man is always sterile. The syndrome often leads to breast enlargement and other feminine body contours. The affected individual is usually of normal intelligence. Klinefelter syndrome is also associated with males having more than one additional sex chromosome (*XXYY, XXXY, XXXXY,* and *XXXXXY*). Such individuals are more likely to be mentally retarded than *XXY* individuals.

Human males with a single extra *Y* chromosome (*XYY*) are not characterized by any well-defined syn-

drome, although they tend to be somewhat taller than the average male. Females with trisomy *X* (*XXX*) occur once in approximately 2000 live births. These **metafemales** have limited fertility and may be mentally retarded. Monosomy *X*, called **Turner syndrome,** occurs about once in 10,000 births and is the only known viable human monosomy. Although these *XO* individuals are phenotypically female, their sex organs do not mature at adolescence, and secondary sexual characteristics fail to develop. Such individuals are sterile and of short stature. Turner syndrome patients usually have no mental deficiency.

Table 14.1 Abnormalities of sex chromosome number in humans			
Genotype	**Phenotype**	**Origin of Nondisjunction**	**Frequency in Population**
XO	Turner syndrome (female)	Meiosis in egg or sperm formation	$\frac{1}{10,000}$
XXX	Metafemale	Meiosis in egg formation	$\frac{1}{2000}$
XXY	Klinefelter syndrome (male)	Meiosis in egg or sperm formation	$\frac{1}{2000}$
XYY	Normal male	Meiosis in sperm formation	$\frac{1}{2000}$

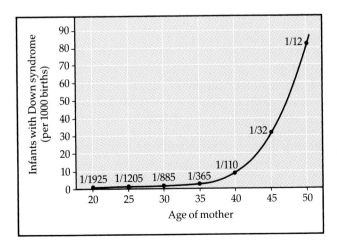

Figure 14.20
Maternal age and Down syndrome. The incidence of Down syndrome increases with maternal age.

Aneuploidy of the sex chromosomes illustrates the basis of sex determination in humans. In general, a single Y chromosome is sufficient to produce "maleness," even in combination with several X chromosomes. The absence of a Y is required for "femaleness."

In addition to aneuploid conditions, there are also structural alterations of chromosomes associated with specific disorders in humans. Many deletions in human chromosomes, even in a heterozygous state, cause severe physical and mental defects. One such chromosomal deficiency is known as *cri du chat* ("cry of the cat") *syndrome*. A human child born with this specific deletion in chromosome 5 is mentally retarded and has a small head with unusual facial features and a cry that sounds like the mewing of a distressed cat. Death usually occurs in infancy or early childhood.

Another type of structural alteration of chromosomes associated with human disorders is the chromosomal translocation, the attachment of a fragment from one chromosome to another, nonhomologous chromosome. Chromosomal translocations have been implicated in certain cancers. One example is *chronic myelogenous leukemia (CML)*. Leukemia is a cancer affecting the cells that give rise to white blood cells, and in the cancerous cells of CML patients a chromosomal translocation has occurred. A portion of chromosome 22 has switched places with a small fragment from a tip of chromosome 9. (How such a switch might cause cancer will be discussed later in Chapter 18.)

A small fraction of individuals with Down syndrome have a chromosomal translocation of a different sort. All the cells of such persons have the normal number of chromosomes, 46. Close inspection of the karyotype, however, shows the presence of part or all of a third chromosome 21 attached to another chromosome by translocation. With two normal chromosomes 21 plus the translocation, some crucial genes must be present in triplicate.

MAPPING HUMAN CHROMOSOMES

Mapping human genes on chromosomes is more difficult than the mapping of *Drosophila* genes. The genes for sex-linked traits can readily be assigned to the X chromosome, of course. Moreover, a particular genetic disorder, such as Down syndrome, may be associated with a microscopically detectable chromosomal mutation in the afflicted individuals, so that the gene in question can be assigned to a particular chromosome. For most cases, however, even the assignment of a gene to a chromosome requires special biological tricks. Since the methods currently used to map human genes depend heavily on new DNA technology, we shall defer further discussion of this topic to Chapter 19.

Figure 14.21
Cytoplasmic inheritance. Variegated leaves may be a result of genes located in the plastids rather than on the nuclear chromosomes of plant cells. The cells of yellow areas contain plastids that are unable to develop into normal green chloroplasts.

EXTRANUCLEAR INHERITANCE

Although there is impressive evidence that eukaryotic genes have specific loci on chromosomes and that the behavior of chromosomes explains Mendel's laws, there are exceptions to the chromosome theory. Not all of a cell's genes are located on nuclear chromosomes, or even in the nucleus. Most of the extranuclear genes are found in cytoplasmic organelles, such as mitochondria and, in plants, plastids, the self-replicating organelles that include chloroplasts. Both mitochondria and plastids replicate and transmit their genes to daughter organelles. These cytoplasmic genes do not display Mendelian inheritance because they are not distributed to offspring according to the same rules that direct the distribution of nuclear chromosomes during meiosis.

Cytoplasmic genes were first observed in plants. In 1909, Karl Correns studied the inheritance of yellow or white patches on the leaves of an otherwise green ornamental plant. He found that the coloration of the offspring was determined only by plants with flowers bearing the seeds and not by plants with flowers producing the pollen. Subsequent research has shown that such variegated (striped or spotted) colorations of leaves are due to differences in pigment production by the plastids. Furthermore, such differences may be controlled by genes located in the plastids (Figure 14.21). In most plants, a zygote receives all of its plastids from the cytoplasm of the egg cell, and none from the sperm. Thus, as the zygote of these plants develops, its pattern of leaf coloration depends only on the maternal cytoplasmic genes.

Maternal inheritance is also the rule for the mitochondrial genes in mammals. The mitochondria are situated in the cytoplasm of a cell, and the egg cell always contributes much more cytoplasm to the zygote than does the sperm. Thus, mitochondria come from the mother. For example, when laboratory rats carrying one type of mitochondrial DNA are mated to rats carrying another type, all the offspring contain only mitochondria of the maternal type. Mitochondrial genes can be used to trace evolutionary relationships, as will be discussed in Chapter 23.

Wherever DNA is located in a cell, its fundamental structure and functioning are the same. In the next chapter, we will discuss the identity and structure of DNA and begin to examine its behavior on a molecular level.

STUDY OUTLINE

1. Thirty-five years after Mendel's discoveries, three botanists independently rediscovered the principles of segregation and independent assortment. This time, the biological community was more receptive to Mendelism, which was shown to have a physical basis in the behavior of chromosomes.

The Chromosome Theory of Inheritance (pp. 284–288)

1. When the details of mitosis and meiosis and the cellular basis of fertilization were revealed in the late nineteenth century, the stage was set for a chromosome theory of inheritance. This theory states that genes are located on chromosomes, which are the actual entities that undergo segregation and independent assortment.
2. Thomas Hunt Morgan was the first to associate a specific gene with a specific chromosome. Like Mendel, he used an organism that had intrinsic advantages for genetic research. The prolific and fast-breeding fruit fly, *Drosophila melanogaster,* has only four pairs of chromosomes.
3. After a year of breeding flies, Morgan obtained a mutant white-eyed male that led him to discover the first known sex-linked gene. This gene for eye color, carried on the X chromosome, gave powerful support to the chromosome theory of inheritance.
4. Each chromosome has hundreds or thousands of genes. Genes near each other on a chromosome are said to be linked and do not obey the law of independent assortment.
5. Morgan used the concept of linkage to explain the anomalous inheritance of pollen color and pollen shape in sweet peas. He explained the rare occurrence of recombinants by the phenomenon of crossing over.

The Chromosomal Basis of Recombination (pp. 289–290)

1. The events of meiosis and random fertilization are responsible for genetic recombination, the production of offspring with new combinations of traits inherited from two parents.
2. Offspring that have the same phenotype as one or the other parent are called parental types. The recombinant offspring, however, have combinations of traits that do not match either of the parents, owing largely to independent assortment of alleles during the first meiotic division. A 50% recombination frequency for two genes usually indicates that the genes are located on separate chromosomes and are thus unlinked.
3. A recombination frequency of less than 50% indicates that the genes are linked but that crossing over has occurred. In this process, homologous chromosomes in synapsis during prophase I of meiosis break at corresponding points and switch reciprocal fragments, thereby creating new combinations of alleles that are subsequently passed on to the gametes.

Mapping Chromosomes (pp. 290–293)

1. Methods have been devised to assign, or map, genes to specific regions on particular chromosomes.
2. Morgan's group was the first to map genes by deducing relative distances between them based on crossover data. Genes that are far apart on a chromosome are more likely to be separated during crossover than are genes that are close together. One map unit is defined as the equivalent of a 1% recombination frequency.
3. Chromosome numbers determined from crossover data are consistent with microscopic evidence, in support of the theory that genes are located on chromosomes.
4. Cytological mapping is a technique that pinpoints the physical locus of a gene by associating a mutant phenotype with a chromosomal defect seen in the microscope.
5. Advances in cytological mapping have been made by irradiating cells with X-rays, which induces observable breakage and rearrangement of chromosomes (chromosomal mutations). X-irradiation also causes tiny changes in genes (point mutations).
6. Giant polytene chromosomes in the salivary glands of *Drosophila* and other insect larvae have distinctive banding patterns and a close pairing of homologues that is rare in nonmeiotic cells. Major chromosomal alterations prevent the proper pairing of homologues, which buckle out and reveal the locations of the defects.
7. Genetic loci determined from crossover data and loci determined by cytological mapping techniques are not completely consistent because the frequency of crossing over varies among different regions of the chromosome.

The Chromosomal Basis of Sex (pp. 293–294)

1. Sex is an inherited phenotypic character usually determined by the presence or absence of special chromosomes, but the exact mechanism for sex determination varies among different species.
2. In humans, other mammals, and *Drosophila,* an X-Y system is operative. The heterogametic XY males apportion either an X or a Y chromosome to their gametes, which combine with gametes containing only X chromosomes

from the *XX* homogametic females. Thus, the sex of the offspring is determined at conception by the sperm cell.

3. Grasshoppers and some other insects have an *X-O* system, in which females possess two *X* chromosomes (*XX*) and males only one (*XO*). Again, the heterogametic males determine the sex of the progeny.

4. In the *Z-W* system found in birds, some fishes, and some insects, the female is the heterogametic sex (*ZW*).

5. Most bees and ants determine sex by haplo-diploidy, without any special sex chromosomes. Diploid females develop from fertilized eggs, and unfertilized eggs give rise to haploid males.

6. Dioecious plants have two separate sexes and often show an *X-Y* or *Z-W* system that is similar to that in animals.

7. Most plants and some animals are monoecious, whereby a single individual produces both sperm and eggs. Such organisms lack sex determination, and all individuals have the same complement of chromosomes.

Sex-Linked Inheritance (pp. 294–298)

1. Certain genes for traits that are unrelated to maleness or femaleness are located on the sex chromosomes. Hemophilia is one of many sex-linked recessive traits whose gene is carried on the *X* chromosome. A few genes that code for male traits are carried on the *Y* chromosome.

2. In mammalian females, only one *X* chromosome per diploid cell is active. The other is condensed into a Barr body located on the inside of the nuclear envelope. According to the Lyon hypothesis, one of the two *X* chromosomes in each cell is randomly inactivated during early embryonic development. Small regions of the inactive chromosome may remain functional.

3. Genes associated with sexual characteristics are not always located on the sex chromosomes. For example, sex-limited traits are generally carried on autosomes present in both sexes, but are expressed exclusively in one sex. And autosomal sex-influenced traits, such as a certain type of baldness, have a sex-dependent penetrance or expressivity that varies with the hormonal status of the individual.

Chromosomal Alterations (pp. 298–302)

1. Mutagens or errors in meiosis can change the number of chromosomes per cell or the structure of individual chromosomes. Such alterations can significantly affect the phenotype, especially if several linked genes or entire chromosomes are affected.

2. Aneuploidy is an abnormal chromosome number. It can arise when a normal parental gamete unites with its counterpart that contains, for example, either two or no copies of a particular chromosome as a result of nondisjunction during meiosis. The resulting zygote will be trisomic or monosomic for the given chromosome.

3. Polyploidy, a condition in which there are more than two entire sets of chromosomes, can occur as a result of complete nondisjunction during gamete formation. Polyploidy is more common in plants, but can also rarely occur in animals, especially in isolated patches of cells.

4. A variety of rearrangements can result from breakage of a chromosome. A lost fragment leaves the original chromosome with a deficiency, or deletion, but may produce

a duplication, translocation, or inversion by reattaching to a chromosome.

5. Deletions and duplications can also result from abnormal, nonreciprocal exchanges between sister chromatids during crossing over.

6. Homozygous or sex-linked deletions are usually lethal; duplications, translocations, and inversions are usually harmful because of gene dosage and position effects.

7. Chromosomal alterations cause a variety of human disorders in individuals whose malfunction does not prevent them from surviving gestation. For example, partial or total trisomy of chromosome 21 is responsible for Down syndrome, a common, serious birth defect that shows increasing incidence with maternal age. Chromosomal translocations are associated with some forms of cancer.

Mapping Human Chromosomes (p. 302)

1. Human genes were first mapped using clues from sex-linked traits and microscopically detectable mutations in chromosomes of individuals afflicted with particular genetic syndromes.

2. Current methods in cell culture and recombinant DNA technology have enhanced our ability map genes.

Extranuclear Inheritance (p. 302)

1. Mitochondria and chloroplasts contain some of their own genes, which are distributed to daughter cells by rules different from those governing the behavior of nuclear genes. Such cytoplasmic genes do not display Mendelian inheritance.

2. In both plants and animals, the zygote receives almost all of its cytoplasm from the egg cell. Such inheritance causes certain aspects of the offspring's phenotype to be dependent solely on maternal cytoplasmic genes.

SELF-QUIZ

1. A key discovery in the early development of the chromosome theory of inheritance was
   a. Mendel's realization that the behavior of chromosomes paralleled the behavior of his "heritable factors" (genes) in peas
   b. the association of specific traits with chromosomes that determine gender in fruit flies
   c. the discovery of the chromosomal basis of Down syndrome
   d. the discovery that DNA is found in chromosomes
   e. the discovery that chromosomes are located in the nucleus

2. Two genes will probably assort independently if
   a. they are very close together on the same chromosome
   b. they are very far apart on the same chromosome
   c. they are on homologous chromosomes
   d. they are both located on the X chromosome
   e. one is dominant and the other is recessive

3. How many linkage groups (groups of linked genes) would be revealed by careful analysis of human genetics?
   a. 1   b. 2   c. 4   d. 23   e. 46

4. An aneuploid person is obviously female, but her cells have *two* Barr bodies. Which aneuploid condition probably accounts for these observations?

   a. XXX   b. XYY   c. XXY   d. XO   e. YYY

5. The genetic event that results in Down syndrome can best be described as

   a. a point mutation

   b. nondisjunction

   c. a chromosomal duplication

   d. a deletion

   e. dosage overcompensation

6. Which of the following gametes would definitely determine a male offspring at fertilization in a *Z-W* system of sex determination?

   a. egg cell containing a *Z*

   b. egg cell containing a *W*

   c. sperm containing a *Z*

   d. sperm containing a *W*

   e. cannot tell unless both gametes are known

7. Determine the order of genes along a chromosome based on the following recombination frequencies: *A—B*, 8 map units; *A—C*, 28 map units; *A—D*, 25 map units; *B—C*, 20 map units; *B—D*, 33 map units.

   a. *A–B–C–D*

   b. *A–C–D–B*

   c. *B–A–C–D*

   d. *D–A–B–C*

   e. *C–D–B–A*

8. According to the Lyon hypothesis,

   a. females are a genetic mosaic due to random nonseparation of chromatids during mitosis

   b. males and females both have equal dosages of most *X*-linked genes because of the formation of Barr bodies

   c. males inherit their sex-linked traits from their mothers

   d. maternal inheritance is a result of the large amount of egg cytoplasm

   e. the calico condition is lethal in male cats

9. In the chromosomal rearrangement known as a duplication,

   a. a fragment of a chromosome may join to its homologous chromosome

   b. homologous chromosomes do not separate in meiosis I

   c. sister chromatids do not separate in meiosis II

   d. a fragment of a chromosome joins to a nonhomologous chromosome

   e. an aneuploid condition results from an extra chromosome

10. Baldness is a sex-influenced trait in that

   a. it is inherited only from fathers

   b. it is inherited only from mothers

   c. it is expressed exclusively in men even though its gene is carried on an autosome

   d. its expressivity varies between men and women

   e. it shows extranuclear inheritance

GENETICS PROBLEMS

1. The normal daughter of a man with hemophilia (a recessive, sex-linked condition) marries a man who is normal for the trait. What is the probability that a daughter will be a hemophiliac? A son? If the couple has four sons, what is the probability that all four will be born with hemophilia?

2. A wild-type fruit fly (heterozygous for gray color and long wings) was mated with a black fly with vestigial wings. The offspring gave the following distribution: wild-type, 778; black-vestigial, 785; black-long, 158; gray-vestigial, 162. What is the recombination frequency between these genes for color and wing type?

3. A space probe discovers a planet inhabited by creatures who reproduce according to the same genetic laws as humans. Three phenotypic characters are height (*T* = tall, *t* = dwarf), head appendages (*A* = antennae, *a* = no antennae), and nose morphology (*S* = upturned snout, *s* = downturned snout). Since the life was not "intelligent," Earth scientists were able to do some controlled breeding experiments, using various heterozygotes in testcrosses. For a tall heterozygote with antennae, the progeny were tall-antennae: 46; dwarf-antennae: 7; dwarf-no antennae: 42; tall-no antennae: 5. For a heterozygote with antennae and an upturned snout, the progeny were antennae-upturned snout: 47; antennae-downturned snout: 2; no antennae-downturned snout: 48; no antennae-upturned snout: 3. Calculate the recombination frequencies for both experiments.

4. Using the information from problem 3, a further testcross was done using a heterozygote for height and nose morphology. The progeny were tall-upturned nose: 40; dwarf-upturned nose: 9; dwarf-downturned nose: 42; tall-downturned nose: 9. Calculate the recombination frequency from these data and then use your answer from problem 3 to determine the correct sequence of the three linked genes.

FURTHER READING

1. Croce, C. M., and G. Klein, "Chromosome Translocation and Human Cancer." *Scientific American*, March 1985. Explains how translocation of chromosomes can activate cancer-causing genes.

2. Marx, J. L. "Evidence Uncovered for a Second Alzheimer's Gene." *Science* 241 (1988): 1432–1433. Alzheimer's disease turns out to be on the same chromosome that is involved in Down syndrome.

3. Miller, J. A. "Common Ground for X, Y Chromosomes." *Science News*, June 15, 1985. Thoughts on the evolution of the sex chromosomes.

4. Patterson, D. "The Causes of Down Syndrome." *Scientific American*, August 1987. How genes thought responsible for the symptoms of Down syndrome are being identified and mapped on chromosome 21.

5. Suzuki, D., A. Griffith, J. Miller, and R. Lewontin. *An Introduction to Genetic Analysis*. 4th ed. New York: W. H. Freeman, 1989.

6. White, R., and J.-M. Lalouel. "Chromosome Mapping with DNA Markers." *Scientific American*, February 1988. A good review of the rationale of classical genetic linkage mapping, as well as a description of the use of more modern methods to map genes associated with human disorders.

15

The Molecular Basis of Inheritance

Deoxyribonucleic acid, the substance of genes, is undoubtedly the most celebrated chemical of our time (Figure 15.1). Better known as DNA, this substance was largely ignored by biologists for nearly a century after its discovery because it seemed far too simple and uniform in structure to serve a very significant purpose. Today, we know that this macromolecule is the genetic material, that Mendel's heritable factors and Morgan's genes on chromosomes are in fact composed of DNA. Chemically speaking, your genetic endowment consists of the DNA inherited from your mother and father.

Of all nature's chemicals, nucleic acids are unique in their ability to direct their own replication. Indeed, the resemblance of offspring to their parents has its molecular basis in the precise replication and transmission of DNA from one generation to the next. In other words, DNA is the substance behind the adage "Like begets like." Traits are not transmitted directly. If you have "your mother's eyes," clearly you did not acquire them in a literal sense. Instead, you inherited genetic programs that are encoded in the chemical language of DNA and reproduced in all cells of the body. It is these DNA programs that direct the development of your eyes, as well as your other biochemical, anatomical, physiological, and, to some extent, behavioral traits.

Today, molecular biologists can make or alter DNA in the laboratory and insert it into a cell, changing the cell's heritable characteristics (see Chapter 19). Earlier in the century, however, no one realized the relationship between DNA and heredity, and identification of the molecules of inheritance then loomed as a major

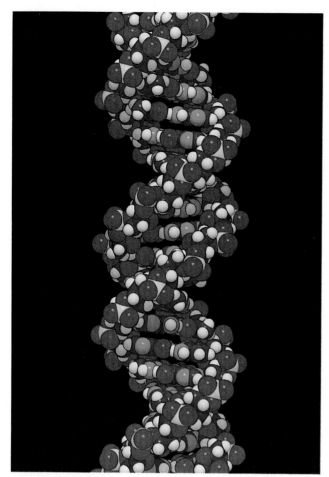

Figure 15.1
DNA. This computer-graphic model of DNA vividly shows the helical structure of the molecule. Each ball represents an atom, with the following color code: blue = C, turquoise = N, white = H, red = O, yellow = P.

challenge to biologists. As with the work of Mendel and Morgan, a key factor in meeting this challenge was the choice of appropriate experimental organisms. Because microscopic organisms—bacteria and the viruses that infect them—are far simpler than peas, fruit flies, or humans, the role of DNA in heredity was first worked out by studying such microbes.

THE SEARCH FOR THE GENETIC MATERIAL

By the 1940s, scientists realized that chromosomes, which were known to carry hereditary information, consisted of two substances, DNA and protein; but most researchers thought it was the protein that was the material of genes. The case for proteins seemed strong, especially since biochemists had identified them as a class of macromolecules with great heterogeneity

and specificity of function, essential requirements for the elusive hereditary material. Moreover, little was known about nucleic acids, whose physical and chemical properties seemed far too monotonous to account for the multitude of specific inherited traits carried by every organism. This view gradually changed as experiments with microorganisms yielded unexpected results.

Evidence That DNA Can Transform Bacteria

The first evidence that the genetic material is a particular chemical was found in 1928. Frederick Griffith, a British medical officer, was studying *Streptococcus pneumoniae*, a bacterium that causes pneumonia in mammals. When Griffith grew colonies of the bacteria in Petri dishes, he could distinguish between two genetic varieties, or strains. One strain produced colonies that appeared smooth, while colonies of the other strain appeared rough. Cells of the smooth strain (abbreviated S) synthesized a polysaccharide that surrounded the cells with a mucous coat, or capsule, that was not formed by the rough cells (R). These alternative phenotypes were inherited: Each strain reproduced its own kind.

When Griffith injected the bacteria into mice, he found that only the S strain was pathogenic (disease-causing). Mice injected with S cells died of pneumonia, while those injected with R cells survived (Figure 15.2). But it was not the polysaccharide of the coat that caused pneumonia, for Griffith found that S cells that had been killed by heat were harmless to the mice.

Then Griffith observed something remarkable. He mixed heat-killed S cells with live R cells and injected the mixture into mice. Although neither the dead S cells nor the live R cells alone were pathogenic, mice injected with the mixture developed pneumonia and died. More startling, Griffith observed live S cells in blood samples taken from the dead mice, although only dead S cells had been injected. Somehow, some of the R cells had acquired from the dead S cells the ability to make polysaccharide coats. Furthermore, this new-found ability was heritable: When Griffith cultured S cells taken from the dead mice, the dividing bacteria produced daughter cells with coats. The phenomenon that Griffith discovered is now called **transformation,** the assimilation of external genetic material by a cell. (This use of the term *transformation* is unrelated to its use in Chapter 11, where it referred to the change of normal eukaryotic cells to a cancerous state.) Although Griffith did not know the chemical nature of the transforming agent, his observations spurred other scientists to search intensively for the elusive genetic material. Moreover, his use of heat to inactivate the S cells hinted that protein might not be the genetic material, because heat denatures most pro-

Figure 15.2

Transformation of bacteria. Griffith discovered that (**a**) the S strain of the bacterium *Streptococcus pneumoniae*, which was protected from the immune system by a capsule, was pathogenic; (**b**) the R strain, a mutant lacking the coat, was non-pathogenic; (**c**) heat-killed S cells were harmless; and (**d**) a mixture of heat-killed S cells and live R cells caused pneumonia and death. Live S bacteria could be retrieved from the dead mice injected with the mixture. Griffith concluded that some chemical from the dead S cells had genetically transformed some of the living R bacteria into S bacteria.

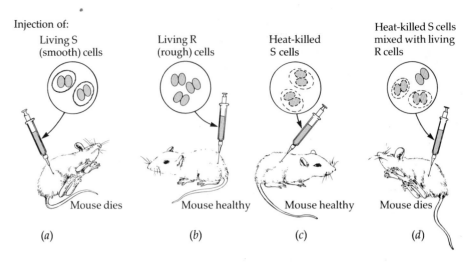

Injection of:

Living S (smooth) cells — Mouse dies — (*a*)

Living R (rough) cells — Mouse healthy — (*b*)

Heat-killed S cells — Mouse healthy — (*c*)

Heat-killed S cells mixed with living R cells — Mouse dies — (*d*)

teins, yet the S-cell genetic material remained able to function in transformation.

For a decade, the American bacteriologist Oswald Avery tried to identify Griffith's transforming agent by purifying various chemicals from the heat-killed S cells and then testing each substance with live R cells to see if it could transform the bacteria. Finally, in 1944, Avery and his colleagues Maclyn McCarty and Colin MacLeod announced that the transforming agent had to be DNA. Their discovery, however, was greeted with considerable skepticism, in part because of the lingering belief that proteins were better candidates for the genetic material. Moreover, many biologists were not convinced that the genes of bacteria would be similar in composition and function to those of more complex organisms. But the major reason for the continued doubt was that so little was known about DNA. No one could imagine how DNA could carry genetic information.

Evidence That Viral DNA Can Program Cells

Additional evidence implicating DNA as the genetic material came from studies of a virus that infects bacteria. Viruses are much simpler than cells. A virus is little more than DNA, or sometimes RNA, enclosed by a protective coat of protein. To reproduce, a virus must infect a cell and take over the cell's metabolic machinery.

Viruses that infect bacteria have been widely used in laboratory research. These viruses are called **bacteriophages** (meaning "bacteria eaters"), or just *phage*, for short. In 1952, Alfred Hershey and Martha Chase discovered that DNA was the genetic material of a phage known as T2. This is one of many phages that infect the bacterium *Escherichia coli (E. coli)*, which normally lives in the intestines of mammals. At that time, biologists already knew that T2, like other viruses, was composed almost entirely of DNA and protein and that the phage could quickly turn an *E. coli* cell into a T2-producing factory that released phages when the cell lysed. Somehow, T2 could reprogram its host cell to produce viruses, but which viral component—protein or DNA—was responsible?

Hershey and Chase answered this question by devising an experiment to determine which substance was transferred from the phage to the *E. coli* during infection (Figure 15.3). They used different radioactive isotopes to tag the molecules of DNA and protein. First, they grew T2 with *E. coli* in the presence of radioactive sulfur. Because protein, but not DNA, contains sulfur, the radioactive atoms were incorporated only into the protein of the phage. Next, in a similar way, the DNA of a separate batch of phage was labeled with atoms of radioactive phosphorus; because nearly all the phage's phosphorus is in its DNA, this procedure left the phage protein unlabeled. Then the protein-labeled and DNA-labeled batches of T2 were each allowed to infect separate samples of nonradioactive *E. coli* cells. Shortly after the onset of infection, the cultures were agitated in a kitchen blender to shake loose any parts of the phages that remained outside the bacterial cells. The mixtures were then spun in a centrifuge, forcing the heavier bacterial cells to form a pellet at the bottom of the centrifuge tubes but allowing the lighter virus particles and other components to remain suspended in the liquid, or supernatant. Radioactivity in the pellet and supernatant was then measured and compared.

Hershey and Chase found that when the bacteria had been infected with the T2 containing labeled pro-

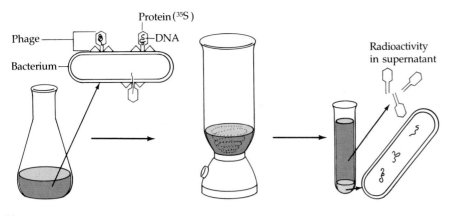

(a)

Phage

Protein (^{35}S)

DNA

Bacterium

Radioactivity
in supernatant

Mix radioactively
labeled phage with
bacteria. The phage
infects the bacterial
cells.

Agitate in a blender to
separate phage outside
the bacteria from the
cells and their
contents.

Centrifuge and mea-
sure the radioactivity
in the pellet and
supernatant.

Figure 15.3
The Hershey-Chase Experiment.
(a) When bacteria were infected with phage T2 whose protein was radioactively labeled (with ^{35}S), the scientists found the radioactivity in the supernatant fraction, which contained the protein coats of the phages. Therefore, the protein did not enter the bacteria. (b) When bacteria were infected with phage whose DNA was radioactively labeled (with ^{32}P), the radioactivity was found in the pellet, indicating that the phage DNA had entered the bacteria. These bacteria could go on to produce progeny phage particles, which contained ^{32}P in their DNA. Hershey and Chase concluded that DNA was the genetic material of this virus.

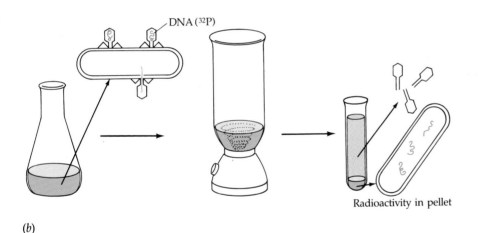

DNA (^{32}P)

Radioactivity in pellet

(b)

teins, most of the radioactivity was found in the supernatant, which contained virus particles (but not bacteria). This suggested that the phage protein did not enter the host cells. But when the bacteria had been infected with T2 phage whose DNA was tagged with radioactive phosphorus, then the pellet of mainly bacterial material contained most of the radioactivity. Moreover, when these bacteria were returned to culture medium, the infection ran its course; the *E. coli* released phages containing radioactive phosphorus.

Hershey and Chase concluded that the DNA of the virus is injected into the host cell, whereas most of the proteins remain outside. More importantly, the injected DNA molecules cause the cells to produce additional viral DNA and proteins—indeed, additional intact viruses—providing powerful evidence that nucleic acids rather than proteins are the hereditary material, at least in viruses.

Additional Evidence That DNA Is the Genetic Material of Cells

Additional circumstantial evidence pointed to DNA as the genetic material in eukaryotes. Prior to mitosis, a eukaryotic cell doubles its DNA content, and during mitosis, this DNA is distributed equally to the two daughter cells. Also, diploid sets of chromosomes have twice as much DNA as the haploid sets found in the gametes of the same organism.

Still more compelling evidence came from the laboratory of biochemist Erwin Chargaff. Recall from Chapter 5 that DNA is a polymer of monomers called nucleotides, each consisting of three components: a nitrogenous base, a pentose sugar called deoxyribose, and a phosphate group (Figure 15.4). The base of each nucleotide can be any one of four different bases: adenine (A), guanine (G), cytosine (C), or thymine

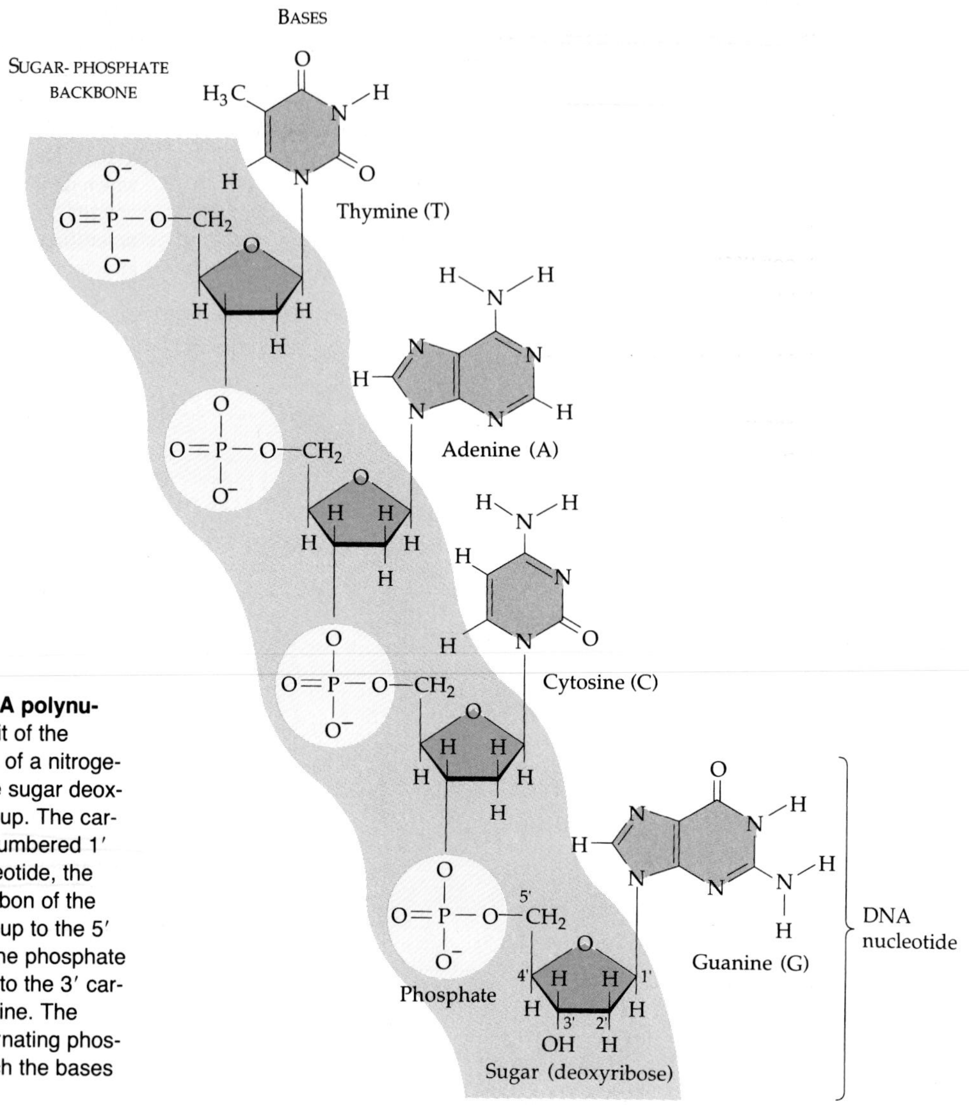

BASES

SUGAR-PHOSPHATE BACKBONE

Thymine (T)

Adenine (A)

Cytosine (C)

Guanine (G)

Phosphate

Sugar (deoxyribose)

DNA nucleotide

Figure 15.4
Chemical structure of a DNA polynu-cleotide. Each nucleotide unit of the polynucleotide chain consists of a nitroge-nous base (A, T, C, or G), the sugar deox-yribose, and a phosphate group. The car-bon atoms of the sugar are numbered 1′ ("one-prime") to 5′. In a nucleotide, the base is attached to the 1′ carbon of the sugar, and the phosphate group to the 5′ carbon. In a polynucleotide, the phosphate of one nucleotide is attached to the 3′ car-bon of the next nucleotide in line. The result is a "backbone" of alternating phos-phates and sugars, from which the bases project.

(T). Adenine and guanine are purines, which are about twice as large as pyrimidines, represented by cytosine and thymine. Using paper chromatography to sepa-rate bases, Chargaff analyzed the composition of DNA from a number of different organisms. In 1947, he reported that DNA composition is species-specific: In the DNA of any one organism, the amounts of the four nitrogenous bases are not all equal, and the ratios of nitrogenous bases vary from one species to another (Table 15.1). Such evidence of molecular diversity, which had been presumed absent from DNA, made DNA a more credible candidate for the genetic material. Fur-thermore, Chargaff found a peculiar regularity in the base ratios. In the DNA of every species he studied, the number of adenine residues approximately equaled the number of thymines, and the number of guanines approximately equaled the number of cytosines. The A = T and G = C equalities, later known as **Char-gaff's rules,** remained unexplained until the discov-ery of the double helix.

DISCOVERY OF THE DOUBLE HELIX

Once most biologists were convinced that DNA was the genetic material, a race was under way to deter-mine how the structure of DNA could account for its role in inheritance. By the beginning of the 1950s, the arrangement of covalent bonds in a nucleic acid poly-mer was well established (Figure 15.4), and the com-petition focused on discovering the three-dimensional structure of DNA. Among the scientists working on the problem were Linus Pauling in California and Maurice Wilkins and Rosalind Franklin in London. First to the finish line, however, were two scientists who were relatively unknown at the time—the Amer-ican James D. Watson and the Englishman Francis Crick.

The brief but celebrated partnership that solved the DNA puzzle began soon after the young Watson jour-neyed to Cambridge University, where Crick was

Table 15.1 Chargaff's data: Nucleotide base compositions of the DNA from various organisms

| Organism | Base Composition (Mole Percent) | | | | A + T |
	A	T	G	C	G + C
Escherichia coli (K12)	26.0	23.9	24.9	25.2	1.00
Streptococcus pneumoniae	29.8	31.6	20.5	18.0	1.59
Mycobacterium tuberculosis	15.1	14.6	34.9	35.4	0.42
Yeast	31.3	32.9	18.7	17.1	1.79
Sea urchin	32.8	32.1	17.7	18.4	1.85
Herring	27.8	27.5	22.2	22.6	1.23
Rat	28.6	28.4	21.4	21.5	1.33
Human	30.9	29.4	19.9	19.8	1.52

Source: E. Chargaff and J. Davidson, eds., *The Nucleic Acids* (New York: Academic Press, 1955).

studying protein structure with a technique called **X-ray crystallography** (see Methods Box in Chapter 5). While visiting the laboratory of Maurice Wilkins at King's College in London, Watson saw an X-ray photograph of DNA, produced by Wilkins's colleague, Rosalind Franklin, that clearly showed the basic shape of DNA to be a helix (Figure 15.5). Moreover, from Watson's recollection of the photograph, he and Crick deduced that the helix had a uniform width of 2 nm, with its purine and pyrimidine bases stacked 0.34 nm apart. The width of the helix suggested that it was made up of two strands—contrary to a three-stranded model that Linus Pauling had recently proposed. The presence of two strands accounts for the now-familiar term **double helix.**

Using molecular models made of wire, Watson and Crick began building scale models of a double helix that would conform to the X-ray measurements and what was then known about the chemistry of DNA. After failing to make a satisfactory model that placed the sugar-phosphate chains on the inside of the molecule, Watson tried putting them on the outside and forcing the nitrogenous bases to swivel to the interior of the double helix. You can imagine this double helix as a rope ladder having rigid rungs, with the ladder twisted into a spiral. The side ropes are the equivalent of the sugar-phosphate backbones, and the rungs represent pairs of nitrogenous bases. Franklin's X-ray data indicated that the helix makes one full turn every 3.4 nm along its length. Because the bases are stacked just 0.34 nm apart, there are ten layers of nucleotide pairs, or rungs on the ladder, in each turn of the helix (Figure 15.6). This arrangement was appealing because it put

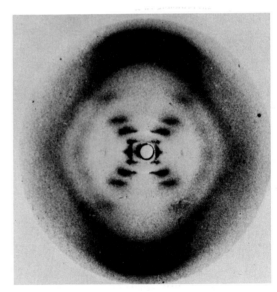

Figure 15.5
X-ray diffraction photograph of DNA. The X shape seen here is a characteristic result from X-ray diffraction of helical objects. This actual photograph suggested to Watson and Crick the double helical structure of DNA. It was made by Rosalind Franklin in the laboratory of Maurice Wilkins.

the more hydrophobic nitrogenous bases in the molecule's interior and thus away from the surrounding aqueous medium. The two sugar-phosphate backbones of the double helix are oriented in opposite directions, an arrangement called **antiparallel.** The sugars on the two strands are upside down with respect to each other (see Figure 15.7).

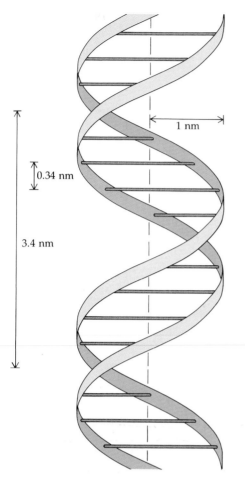

Figure 15.6
The double helix. The ribbons in this diagram represent the sugar-phosphate backbones of the two DNA strands, and the rods are the paired nitrogenous bases. The helix has a radius of 1 nm and completes a 360° turn every 3.4 nm. The base pairs are 0.34 nm apart; there are ten pairs per turn of the helix. The two strands are antiparallel to each other; they run in opposite directions (as is explained further in Figure 15.7). This diagram is similar to the one that appeared in the 1953 paper by Watson and Crick.

1 nm

0.34 nm

3.4 nm

The idea that there is specific pairing of bases was the flash of inspiration that allowed Watson and Crick to solve the DNA structure. If each rung of the DNA ladder contains two nitrogenous bases, and if the bases form specific pairs, then the information on one strand complements that contained along the other. (Such a model immediately suggests the general mechanism for DNA replication, as we shall soon see.)

At first, Watson imagined that the bases pair like with like—for example, A with A and C with C. But that did not fit with the X-ray data, which suggested that the DNA molecule has a uniform cylindrical diameter. It soon became apparent that because of the molecule's 2 nm width, a purine on one strand must always be paired with a pyrimidine on the opposite strand. Moreover, Watson and Crick realized that there must be additional specificity of pairing dictated by the structure of the bases. Each base has chemical side groups that can form hydrogen bonds with its appropriate partner: Adenine can form hydrogen bonds with thymine, and guanine with cytosine (Figure 15.7). In the biologist's shorthand, A pairs with T, and G pairs with C. Although individual hydrogen bonds are weak, their abundance in a DNA molecule (each A–T pair forms two hydrogen bonds, and each G–C pair forms three) can make a significant contribution to the stability of the molecule—another factor that supported this model. Furthermore, van der Waals forces (see Chapter 2) between the stacked bases also contribute to the molecule's stability.

Although the base-pairing rules dictate the side-by-side combinations of nitrogenous bases that form the "rungs" of the double helix, they place no restrictions on the sequence of nucleotides along the length of a DNA strand. Thus, the *sequence* of the four bases can be varied in countless ways, a property that lends itself to the coding of genetic information.

The Watson-Crick model for DNA also explained Chargaff's rules. Wherever one strand of a DNA molecule has an A, the partner strand has a T. And a G in one strand is always paired with a C in the complementary strand. Therefore, in the DNA of any organism, the amount of adenine equals the amount of thymine, and the amount of guanine equals the amount of cytosine.

In April 1953, Watson and Crick surprised the scientific world with a succinct, two-page paper in the British journal *Nature*. The paper reported a new molecular model for DNA—the double helix that has since become the symbol of molecular biology (Figure 15.8).

DNA REPLICATION

The Template Concept

An essential characteristic of the genetic material is its ability to be copied and thus pass on a complete set of inherited instructions, an organism's blueprint, to the next generation. Long before DNA was identified as the genetic material, some theoretical biologists argued that gene duplication was based on something called complementary surfaces. According to this theory, when a gene replicates, a "negative image" is created along the original ("positive") surface, just as clay forms a "negative" shape when it is packed into a mold. The gene's negative image, like the clay, could then function as a template for the synthesis of copies of the original positive image. This theory can also be

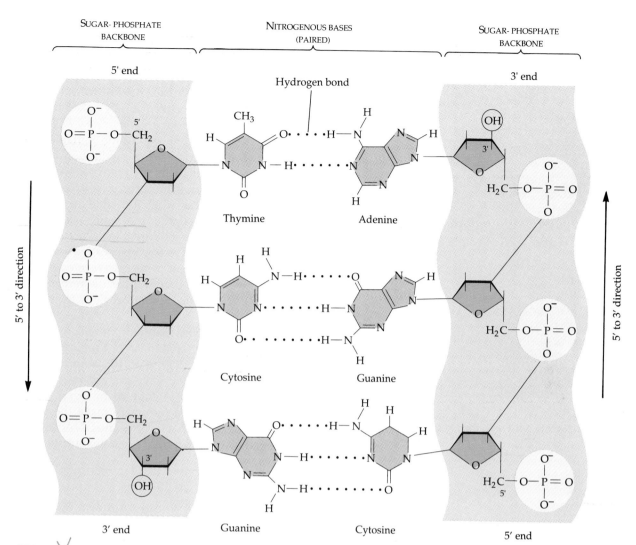

SUGAR- PHOSPHATE BACKBONE | NITROGENOUS BASES (PAIRED) | SUGAR- PHOSPHATE BACKBONE

5' end

3' end

Hydrogen bond

Thymine   Adenine

Cytosine   Guanine

Guanine   Cytosine

3' end   5' end

5' to 3' direction

5' to 3' direction

Figure 15.7

Hydrogen bonding in DNA. In base pairing, adenine and thymine are held together by two hydrogen bonds, while three hydrogen bonds pair guanine with cytosine. Given the locations of the hydrogen-bonding groups on the bases, the DNA helix can maintain a constant diameter only if cytosine always pairs with guanine, and adenine with thymine. In the double helix, the two sugar-phosphate backbones run in opposite directions. Note that the sugars (blue) in the two strands face in opposite directions. For each strand, the end with a phosphate group attached to the number five carbon of the terminal sugar is called the 5' end; the other end, called the 3' end, has a hydroxyl group attached to the number three carbon of the sugar. Because the strands are antiparallel, if one strand is said to have a 5' → 3' orientation (polarity), then the complementary strand has a 3' → 5' orientation. Many of the enzymes that act on DNA in the cell are specific for the 3' or 5' end of a strand.

expressed in terms of photography: A print can be used to make a negative, which can then be used to make copies of the original print. Until 1953, this theory was discounted by many well-known geneticists who championed other models of gene duplication. However, Watson and Crick's model for DNA structure immediately suggested a template mechanism for DNA replication. As they said in the conclusion of their first paper, "It has not escaped our notice that the specific pairing we have postulated immediately suggests a possible copying mechanism for the genetic material."

The Semiconservative Nature of DNA Replication

The logic behind the Watson-Crick proposal for how genes are copied—by specific pairing of complementary bases—is elegantly simple. Cover one of the DNA strands in Figure 15.7 with a piece of paper, and you still can determine its sequence of bases by referring to the unmasked strand and applying the base-pairing rules: A pairs with T, G with C. Watson and Crick predicted that a cell applies the same rules when copying its genes. In a second paper, they suggested

Figure 15.8
Watson and Crick in 1953 with their model of the double helix.

that during replication, the two DNA strands separate, and each strand is used as a template for the assembly of a complementary strand. One at a time, nucleotides line up along the template strand in accordance with the base-pairing rules (Figure 15.9). The nucleotides are enzymatically linked through their sugar-phosphate groups to form new DNA strands.

This model of gene replication remained untested for several years following publication of the DNA structure. The requisite experiments were simple in concept but difficult to perform. Watson and Crick's model predicts that when a double helix reproduces, each of the two daughter molecules will have one old strand derived from the parent molecule and one newly made strand. This **semiconservative model** can be distinguished from a conservative model of replication, in which the parent molecule remains intact (is conserved) and the new molecule is formed entirely from scratch.

In the late 1950s, Matthew Meselson and Franklin W. Stahl demonstrated that DNA replication is indeed semiconservative (Figure 15.10), as predicted by the Watson-Crick model. They grew *E. coli* in a medium containing a heavy isotope of nitrogen, ^{15}N, which was used by the bacterium to make the purine and pyrimidine bases in its DNA. The DNA was then centrifuged in a gradient that separates substances on the

Figure 15.9
Semiconservative replication of DNA. When the two strands of a parent DNA molecule separate, each strand serves as a template for the ordering of nucleotides into new complementary strands. After replication, there will be two double-stranded DNA molecules, each with one old strand (blue) and one new strand (magenta). The sequence of base pairs in each of these two double helices is identical to the sequence of the parent DNA molecule.

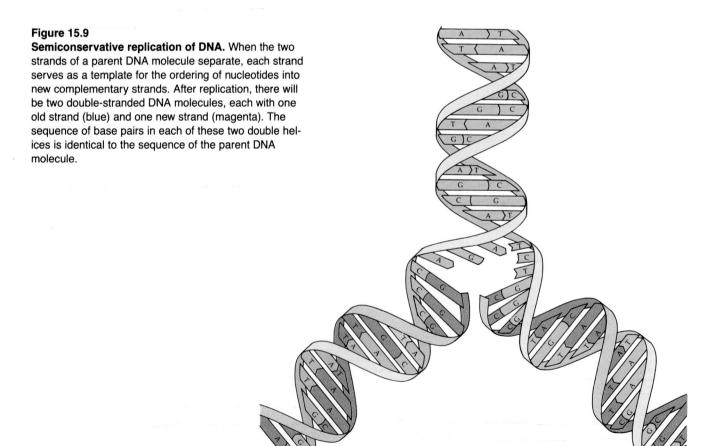

Bacteria are grown in ^{15}N (heavy) medium for many generations, then transferred to ^{14}N (light) medium.

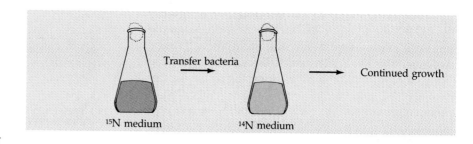

^{15}N medium ^{14}N medium

Transfer bacteria   Continued growth

Labeling of DNA strands predicted by hypothesis of semiconservative replication.

DNA isolated from cells is mixed with CsCl solution and placed in centrifuge.

Solution is centrifuged at very high speed for several days.

Increasing concentration of CsCl toward bottom of tubes is due to its sedimentation under centrifugal force.

DNA molecules move to positions where their density equals that of CsCl solution.

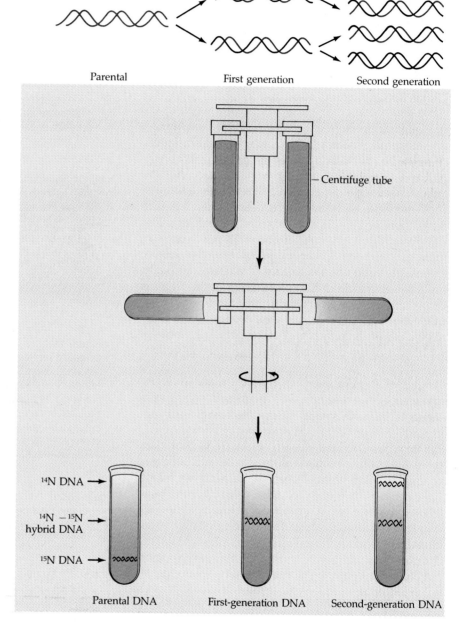

Parental   First generation   Second generation

Centrifuge tube

^{14}N DNA →

$^{14}N - ^{15}N$ hybrid DNA →

^{15}N DNA →

Parental DNA   First-generation DNA   Second-generation DNA

Figure 15.10
The Meselson-Stahl experiment. This experiment demonstrated the semiconservative replication of *E. coli* DNA. It was based on a new technique that used an ultracentrifuge to separate DNA molecules of different densities in a cesium chloride (CsCl) density gradient. The density of the DNA depended on the proportions of ^{14}N (the common, light isotope of nitrogen) and ^{15}N (a heavy isotope) that the DNA contained. Parental strands of DNA were "labeled" with ^{15}N, distinguishing that DNA from newly synthesized DNA containing ^{14}N. Meselson and Stahl observed the density distribution of DNA at different times after the shift of the bacteria from heavy to light medium. The results confirmed predictions based on the semiconservative model for replication.

basis of density, and the DNA was found to be heavier than DNA made when the bacterium is grown in the presence of the more common, and lighter, ^{14}N. The two scientists then transferred the cells grown with ^{15}N to a medium containing an excess of the lighter isotope, ^{14}N. They found that after one generation of bacterial growth, when each DNA molecule had replicated one time, all the DNA was of intermediate density. This result was clearly consistent with the hypothesis of semiconservative replication: The DNA of intermediate density was a hybrid of the parental DNA, which was derived exclusively from the heavier ^{15}N, and of newly added nucleotides, which contained ^{14}N only.

Close-up on Replication

Although the general mechanism of DNA replication is conceptually simple, the actual process involves complex biochemical gymnastics. Some of the complexity is topological, arising from the fact that the helical molecule must untwist as it replicates and must copy its two antiparallel strands simultaneously. The replication process is also extremely rapid. Nucleotides are added at a rate of about 50 per second in mammals and 500 per second in bacteria. Yet, despite its rapidity, replication is amazingly accurate; typically, only about one in a billion nucleotides in DNA is incorrectly paired. In achieving this speed and accuracy, DNA replication involves the cooperation of more than a dozen enzymes and other proteins, some of which are grouped into complexes for efficient action.

Origins of Replication   In a wide variety of organisms, including bacteria and mammals, replication seems to begin at special sites, called **origins of replication** (Figure 15.11). Specific proteins required to initiate replication bind to each origin. Replication then spreads in both directions along the DNA from these central initiation points. A bacterial or viral DNA molecule contains only one replication origin, but the huge DNA molecule of a eukaryotic chromosome has hundreds or thousands of replication origins. Thus, during replication, many different sites along the chromosome are active, forming many replication bubbles that eventually merge to give two continuous, full-sized daughter DNA molecules. The points on a replicating DNA molecule where new strands are growing are called **replication forks** because of their Y shape.

Strand Separation   Involved in the separation of the parental DNA strands are two types of proteins—enzymes called **helicases,** which unwind the helix, and **single-strand binding proteins,** which help keep the separated strands apart. Separating the strands of DNA severely stresses the molecule and puts kinks

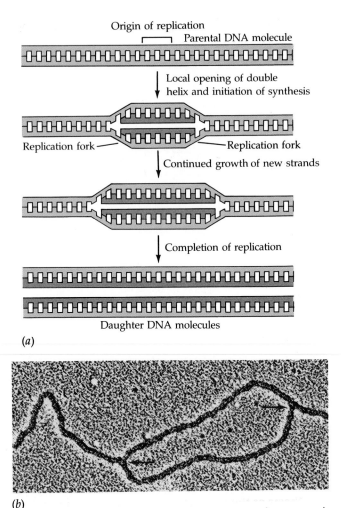

(a)

(b)

0.25 μm

Figure 15.11
Origin of replication. Initiation of DNA replication requires the opening of the DNA double helix at a specific place called the origin of replication. The "bubble" produced then grows in size by expanding in both directions as the new DNA strands are synthesized, until the entire DNA molecule has been replicated. **(a)** The diagram gives an overview of replication for a linear DNA molecule with one origin of replication. Newly made strands are shown in magenta. **(b)** The electron micrograph shows part of a DNA molecule in the process of replication. The arrows point to the two replication forks and also indicate the directions of replication.

into it. Hence, other enzymes, called **topoisomerases,** are required to relieve that stress and untangle the snarls. These enzymes act by breaking and then resealing one strand of the DNA, creating a transient "nick" so that the molecule can rotate freely around a single strand.

Priming   A DNA strand cannot start *de novo*, that is, from scratch. Nucleotides must be added to the end of an already existing chain, called a primer. The primer is not DNA, but a short stretch of ribonucleic acid (RNA), about five nucleotides long, polymerized by

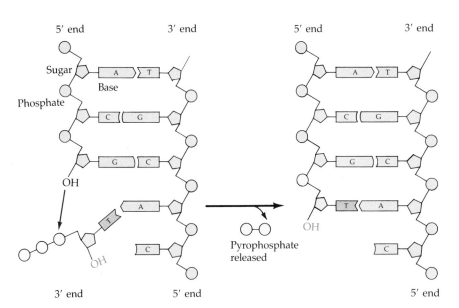

Figure 15.12
Elongation of a nucleic acid strand. A nucleic acid is a polymer of nucleotides joined by links between the 3' carbon of the sugar of one nucleotide and the phosphate group that is attached to the 5' carbon of the next nucleotide. The building blocks for the polymer, nucleoside triphosphates, lose two of their phosphates (as a pyrophosphate molecule) when a nucleotide is added to a growing chain. Because of the specificity of the polymerase enzymes that catalyze such reactions, the polymer always elongates in the 5' → 3' direction. That is, new nucleotides are added to the 3' end of the growing strand.

an enzyme called **primase.** The sequence of bases for this RNA is complementary to the sequence of the portion of the strand that serves as a template for making the primer. (Recall from Chapter 5 that three of the four nucleotides in RNA are the same as in DNA, but RNA contains uracil in place of thymine. Also, the pentose sugar in RNA is ribose rather than deoxyribose.)

Synthesis of the New DNA Strands   Once a primer is in place, enzymes called **DNA polymerases** catalyze the synthesis of a new DNA strand. The old DNA strands, now separated at the origin of replication, are the templates along which nucleotides align themselves according to the base-pairing rules. Then DNA polymerase links the nucleotides to the growing strand, which is made in the 5' → 3' direction. (That is, new nucleotides are joined to the 3' end of the growing strand. See Figure 15.7 to reexamine the 5' and 3' ends of a DNA strand.) The energy to form the new covalent bonds comes from the release of a pyrophosphate group from each incoming nucleotide, which arrives as a triphosphate (Figure 15.12) and thus contains its own store of energy (see Chapter 6).

The simultaneous synthesis of both DNA strands at a replication fork poses a problem, because the two templates are in an antiparallel orientation. One conceivable mechanism for dealing with the synthesis of opposite strands is the presence of two different enzymes to add nucleotides to the growing molecules. That is, one enzyme could be tailored to elongate one strand in the 5' → 3' direction, while the other enzyme could elongate the other strand in the 3' → 5' direction. However, no 3' → 5' DNA polymerase has ever been found, and so the two-enzyme theory is not correct. The new 5' → 3' strand, called the **leading strand,** is synthesized as a single polymer; but the 3' → 5'

strand, called the **lagging strand,** is produced as a series of short segments, each of which is synthesized in the 5' → 3' direction. These segments are called **Okazaki fragments,** after the Japanese scientist who discovered them (Figure 15.13). In bacteria these fragments are each 1000 to 2000 nucleotides long, and in eukaryotes they are 100 to 300 nucleotides in length. Initiating the synthesis of each Okazaki fragment is an RNA primer, similar to the one initiating the lead-

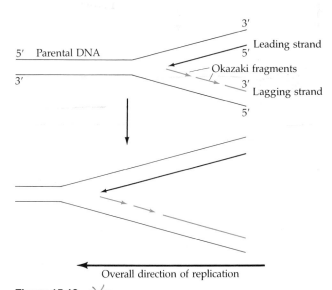

Figure 15.13
Discontinuous synthesis on the lagging strand. DNA polymerases can only elongate strands in the 5' → 3' direction. One new strand, called the leading strand, can therefore elongate continuously in the 5' → 3' direction as the replication fork progresses. But the other new strand, the lagging strand, must grow in an overall 3' → 5' direction by the addition of short segments, Okazaki fragments, that individually grow 5' → 3'.

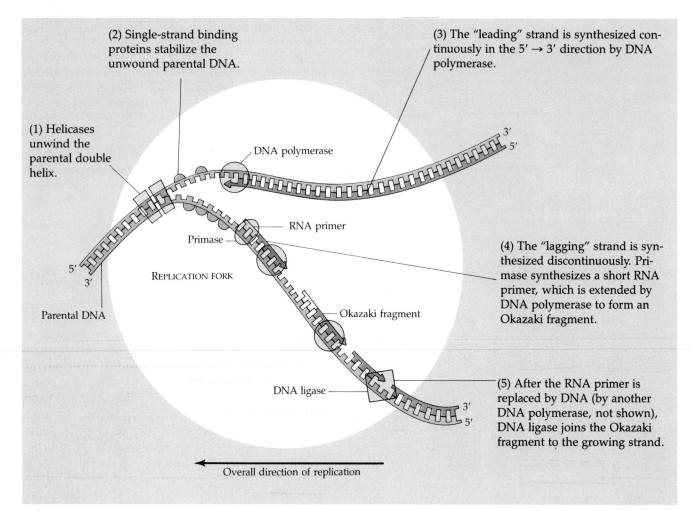

(1) Helicases unwind the parental double helix.

(2) Single-strand binding proteins stabilize the unwound parental DNA.

(3) The "leading" strand is synthesized continuously in the 5' → 3' direction by DNA polymerase.

DNA polymerase

RNA primer

Primase

REPLICATION FORK

5'
3'

Parental DNA

(4) The "lagging" strand is synthesized discontinuously. Primase synthesizes a short RNA primer, which is extended by DNA polymerase to form an Okazaki fragment.

Okazaki fragment

DNA ligase

(5) After the RNA primer is replaced by DNA (by another DNA polymerase, not shown), DNA ligase joins the Okazaki fragment to the growing strand.

3'
5'

Overall direction of replication

Figure 15.14
Events at the DNA replication fork.

ing strand at a replication origin. To produce a continuous DNA strand from the many fragments, two steps are required. First, an enzyme (one of the DNA polymerases, actually) removes the RNA primer and replaces it with DNA. Then a linking enzyme called **DNA ligase** covalently joins the 3' end of each new DNA fragment to the 5' end of the growing chain. Figure 15.14 summarizes the events at the DNA replication fork.

Proofreading   The high degree of accuracy of DNA replication cannot be attributed solely to the specificity of base pairing. Although errors in the completed DNA molecule only amount to one in one billion nucleotides, initial pairing errors between incoming nucleotides and those in the template strand occur at a frequency of about 1 in 10,000. In bacteria, these pairing errors are almost always corrected by the DNA polymerase itself, which checks each nucleotide against its template as soon as it is added to the strand. Upon finding an incorrect nucleotide, the polymerase backs up, removes the incorrect nucleotide, and replaces it

before continuing with 5' → 3' synthesis. This action resembles that of a self-correcting typewriter, which can be directed to backspace to an error and then remove and replace the error before continuing forward. Thus, the DNA polymerases of bacteria are striking examples of enzymes with multiple functions. In eukaryotes, it is not yet known whether DNA polymerase or other enzymes proofread the new strand and remove errors.

DNA REPAIR

Faithful maintenance of the genetic information encoded in DNA requires more than a mechanism for proofreading DNA as a copy is being made; it also requires a way of repairing accidental changes that occur in existing DNA. DNA molecules are constantly subjected to many potentially damaging physical and chemical agents. Reactive chemicals, radioactive emissions, X-rays, and ultraviolet light can change nucleo-

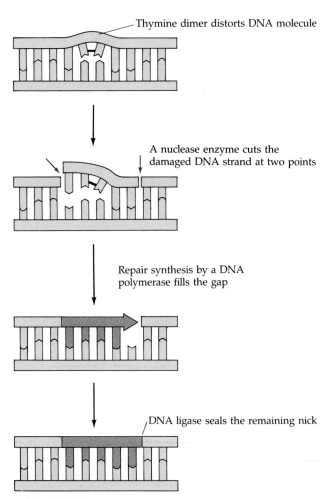

Thymine dimer distorts DNA molecule

A nuclease enzyme cuts the damaged DNA strand at two points

Repair synthesis by a DNA polymerase fills the gap

DNA ligase seals the remaining nick

Figure 15.15
Excision repair of DNA. Damaged DNA can be detected and repaired by a team of enzymes. One type of damage, shown here, is the covalent linking of thymine bases that are adjacent on a DNA strand. Such thymine dimers, induced by ultraviolet radiation, cause the DNA to buckle and will lead to errors during DNA replication. Repair enzymes can excise the damaged region from the DNA and replace it with a normal DNA segment.

tides in ways that can affect encoded genetic information, usually adversely. Fortunately, these changes, or mutations, are usually corrected. Each cell continuously monitors and repairs its genetic material. Biochemists have identified more than 50 different types of DNA repair enzymes. Sometimes, the damage can be directly reversed by an appropriate enzyme. Most often, however, the repair process, like the replication process, takes advantage of the base-paired structure of DNA. For example, in the type of repair called **excision repair,** a segment of the strand containing the damage is cut out by one repair enzyme, and the resulting gap is filled in with nucleotides properly paired with the nucleotides in the undamaged strand.

The enzymes involved in filling the gap are a DNA polymerase and DNA ligase (Figure 15.15).

One function of the DNA repair enzymes in our skin cells is to repair the DNA damage caused by the ultraviolet rays of sunlight. The importance of this function to healthy people is underscored by the skin disease xeroderma pigmentosum, which is caused by a genetic defect in excision repair enzymes. Sunlight readily kills skin cells of people with this disease and invariably causes skin cancers from the damaging effects of ultraviolet radiation.

ALTERNATIVE FORMS OF DNA

The double helix discovered by Watson and Crick was greeted with great elation, largely because of its simplicity and its reaffirmation of a fundamental theme of biology: Structure fits function. The double helix squelched the fears that the structure of genes might be incredibly complicated. It showed that all genes have about the same three-dimensional form and differ from one another only in the number and order of the four nitrogenous bases. Now, however, we know that DNA structure is not as uniform as we once thought. DNA molecules may be linear or circular, and may be single-stranded (in some small viruses). In addition, a DNA molecule may twist upon itself to form a coil called a "supercoil," just like the rubber band that powers a balsa-wood model airplane. (We will discuss some of these variations in later chapters.)

Even linear, double-stranded DNA is not always found in the form corresponding to the Watson-Crick model, although that particular form, called *B-DNA*, is the most common in vivo. Actually, DNA can exist in several other types of helical structures, including *A-DNA, C-DNA,* and *Z-DNA.* In all these forms, the covalent bonds joining the atoms of each strand and the hydrogen bonding between base pairs are basically the same. However, the three-dimensional structures are markedly different, and Z-DNA actually twists in the left-handed direction, opposite to that of B-DNA (Figure 15.16). The biological functions, if any, of the non-B forms of DNA are not yet clearly established.

* * *

In this chapter, we have concentrated on the structure of DNA and how this heritable program is copied. Replication of DNA is the molecular basis for the transmission of the genes we studied in the previous two chapters. However, it is not enough that genes be transmitted; they must also be expressed. How can genes manifest themselves in such phenotypic characteristics as eye color, for instance? In the next chapter, we will examine the molecular basis of gene expression—how the cell translates genetic programs encoded in DNA.

Figure 15.16
Alternative forms of the double helix.
(a) B-DNA has the structure proposed in 1953 by Watson and Crick and is the form DNA usually has in solution. It is a right-handed double helix. The base pairs are approximately perpendicular to the axis of the helix, and there are two noticeable grooves (the major groove and the minor groove) on its surface. (b) Z-DNA is a left-handed double helix. It is longer and thinner than B-DNA, has tilted base pairs, and only one groove. The name derives from the fact that the sugar-phosphate backbone has a zigzag appearance. The Z structure was first proposed only in 1979. There is evidence that it does occur in cells, but probably only in tiny segments.

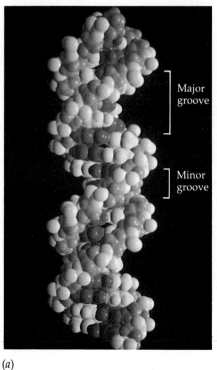
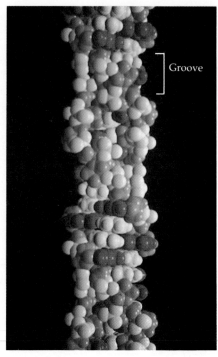

(a)   (b)

STUDY OUTLINE

The Search for the Genetic Material (pp. 307–310)

1. Griffith's experiments on smooth and rough strains of pneumonia bacteria in 1928 revealed the phenomenon of transformation and provided the first evidence that the genetic material was a heat-stable chemical.
2. Although Avery and his colleagues later discovered the transforming agent to be DNA, their findings met with limited acceptance from the scientific community.
3. Further evidence that DNA is the genetic material came from the experiments of Hershey and Chase with radioactively labeled phages and *E. coli*.
4. In 1947, Chargaff reported that DNA composition is species-specific and that there is a one-to-one relationship between adenine and thymine residues and between cytosine and guanine residues. With circumstantial evidence concerning the variation in quantity of DNA during mitosis and meiosis, the stage was set for the serious study of DNA as the molecule of inheritance.

Discovery of the Double Helix (pp. 310–312)

1. Scientists from several laboratories quickly realized that unraveling the three-dimensional structure of DNA was the key to understanding its function.
2. Using wire models, chemical knowledge, and X-ray crystallography, Watson and Crick established the double-helix structure of DNA. Two antiparallel sugar-phosphate chains wind around the outside of the molecule; the nitrogenous bases project into the interior, where they hydrogen-bond in pairs, A with T and G with C.

DNA Replication (pp. 312–318)

1. Meselson and Stahl demonstrated that DNA replication is semiconservative, confirming Watson and Crick's hunch that the parent molecule unwinds and each strand then serves as a template for the synthesis of a new half-molecule according to base-pairing rules.
2. Some impressive complexities underlie the apparent simplicity of DNA replication, with more than a dozen enzymes and other proteins interacting to ensure an extremely rapid and amazingly accurate process.
3. After binding of specific proteins, the initiation of replication begins at special sites called origins of replication, forming Y-shaped replication forks on the molecule.
4. Helicases and single-strand binding proteins, respectively, unwind the parental strands and help keep them apart, aided by topoisomerases, which break and reseal the DNA strands to allow them to "unkink."
5. DNA synthesis must start on the end of a primer, a short segment of RNA whose base sequence is complementary to part of the parental DNA.
6. Using energy from the hydrolysis of nucleotide triphosphate bonds, DNA polymerases catalyze the synthesis of the new DNA strand working in the $5' \rightarrow 3'$ direction.
7. Simultaneous $5' \rightarrow 3'$ synthesis of antiparallel strands at a replication fork yields a continuous leading strand and several short, discontinuous segments of lagging strand called Okazaki fragments. The fragments are later joined together with the help of DNA ligase.
8. Bacterial DNA polymerases also proofread replication errors, correcting any improperly paired nucleotides.

DNA Repair (pp. 318–319)

1. DNA molecules require continuous monitoring and repair because of ongoing assault by mutagens.
2. DNA repair enzymes restore the integrity of the molecule by processes such as excision repair, which take advantage of base-pairing principles.

Alternative Forms of DNA (pp. 319–320)

1. DNA molecules may be linear or circular, may be supercoiled, and may be single-stranded.
2. Double-stranded DNA can manifest itself as B-DNA (the Watson-Crick form), A-DNA, C-DNA, or Z-DNA.

SELF-QUIZ

1. In his work with pneumonia-causing bacteria and mice, Griffith found that

    a. the protein coat from smooth (S) cells was able to transform rough (R) cells

    b. heat-killed S cells were able to cause pneumonia only when they were transformed by the DNA of R cells

    c. some heat-stable chemical from S cells was transferred to R cells to transform them into S cells

    d. the polysaccharide coat of R cells caused pneumonia

    e. bacteriophages injected DNA from S cells into R cells

2. Which of the following experimental techniques was used by Hershey and Chase to establish DNA as the genetic material of phage?

    a. X-ray crystallography

    b. paper chromatography

    c. radioactive isotopes of sulfur and phosphorus and centrifugation

    d. ultracentrifugation through a cesium chloride density gradient

    e. incorporation of ^{15}N into DNA

3. *E. coli* cells grown on ^{15}N medium are transferred to ^{14}N medium and allowed to grow for two generations (two cell replications). DNA extracted from these cells is ultracentrifuged in a cesium chloride density gradient. What density distribution of DNA would you expect in this experiment?

    a. one high and one low density band

    b. one intermediate density band

    c. one high and one intermediate density band

    d. one low and one intermediate density band

    e. one low density band

4. The energy for the polymerization of DNA comes from

    a. the loss of a pyrophosphate from nucleoside triphosphate monomers

    b. the hydrolysis of ATP

    c. the hydrolysis of GTP

    d. DNA ligase

    e. DNA polymerase

5. The elongation of the *leading* strand during DNA syntheses

    a. progresses away from the replication fork

    b. occurs in the $3' \rightarrow 5'$ direction

    c. produces Okazaki fragments

    d. depends on the action of DNA polymerase

    e. requires ligase

6. What is the basis for the difference in the synthesis of the leading and lagging strand of DNA molecules?

    a. The origins of replication occur only at the 5' end of the molecule.

    b. Helicases and single-strand binding proteins work at the 5' end.

    c. DNA polymerase can join new nucleotides only to the 3' end of the growing strand.

    d. DNA ligase works only in the $3' \rightarrow 5'$ direction.

    e. Polymerase can only work on one strand at a time.

7. Not all DNA conforms to the double-helix model because

    a. most DNA may be single-stranded

    b. some DNA may exist as a flat, ladder-shaped molecule

    c. Z-DNA twists in the left-hand direction

    d. some DNA has different covalent bonding patterns in its backbone

    e. base-pairing rules vary

8. In an analysis of the number of different bases in a DNA sample, which result would be consistent with the base-pairing rules?

    a. A = G

    b. A + G = C + T

    c. A + T = G + T

    d. A = C

9. The "primer" required to initiate synthesis of a new DNA strand consists of

    a. RNA

    b. DNA

    c. an Okazaki fragment

    d. a structural protein

    e. a thymine dimer

10. A particular gene measures about 1 μm in length along a double-stranded DNA molecule. What is the approximate number of base pairs in this gene?

    a. 3   b. 10   c. 1000   d. 3000   e. 30,000

CHALLENGE QUESTION

1. Peculiar sequences in DNA allow the molecule to assume interesting shapes. One such sequence, the inverted repeat, consists of a sequence of bases followed on the same strand by its complementary sequence in reverse order. What unusual configuration could occur when the two strands of such a double helix separated from each other? (To learn more about this, see Further Reading 1.)

FURTHER READING

1. Felsenfeld, G. "DNA." *Scientific American,* October 1985. An excellent article on the variability and flexibility of the double helix.
2. Judson, H. F. *The Eighth Day of Creation: Makers of the Revolution in Biology.* New York: Simon and Schuster, 1979. An engaging history of molecular biology.
3. Radman, M., and R. Wagner. "The High Fidelity of DNA Duplication." *Scientific American,* August 1988. Well-illustrated article explaining why so few errors occur when DNA replicates.
4. Watson, J. D. *The Double Helix.* New York: Atheneum, 1968. The brash, controversial best-seller by the codiscoverer of DNA.
5. Watson, J. D., and F. H. C. Crick. "Molecular Structure of Nucleic Acids: A Structure for Deoxynucleic Acids." *Nature* 171 (1953): 737-738. Watson and Crick's classic paper.

16

From Gene to Protein

The DNA inherited by an organism controls the activities of each cell by specifying the synthesis of enzymes and other proteins. A gene does not build a protein directly, but instead dispatches instructions in the form of RNA, which in turn programs protein synthesis. Cells are governed by a chain of command: DNA → RNA → protein. This scheme is known as the *central dogma of molecular biology* (Figure 16.1), a term coined by Francis Crick. This chapter describes the complex steps in the flow of molecular information from gene to protein.

HOW GENES CONTROL METABOLISM

The relationship between genes and proteins was first proposed in 1909, when British physician Archibald Garrod suggested that genes dictate phenotypes through enzymes that catalyze specific chemical processes in the cell. Garrod hypothesized that inherited diseases reflect a patient's inability to make a particular enzyme, and he referred to such diseases as "inborn errors of metabolism." He gave as one example the relatively harmless hereditary condition called alkaptonuria, in which the urine appears very dark red because it contains a chemical, alkapton, that darkens upon exposure to air. Garrod reasoned that normal individuals have an enzyme that breaks down alkapton, whereas alkaptonuric individuals have inherited an inability to make the enzyme that metabolizes alkapton.

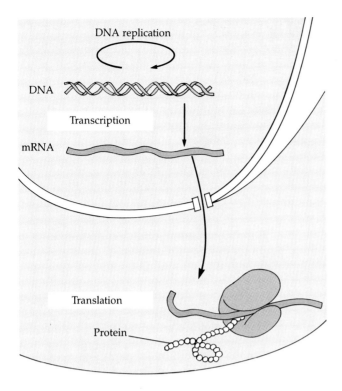

Figure 16.1
The central dogma of molecular biology. Genetic information is stored in the cell as DNA, which serves as the template for its own replication. When genetic information is expressed in a cell, it flows unidirectionally from DNA to mRNA (transcription) and then from mRNA to protein (translation). It is now known that under certain circumstances, a reverse flow of information from RNA to DNA (reverse transcription) can occur (see Chapter 17). The steps in transcription and translation are described in detail later in this chapter.

Evidence That Genes Specify Enzymes

Garrod's hypothesis was ahead of its time, but research conducted several decades later supported the idea that the function of a gene is to dictate production of a specific enzyme. Biochemists accumulated much evidence that cells synthesize and degrade most organic molecules via metabolic pathways, with each step in a sequence catalyzed by a specific enzyme. Such metabolic pathways lead, for instance, to the synthesis of the pigments that give fruit flies their eye color. In the 1930s, George Beadle and Boris Ephrussi speculated that each of the various mutations affecting eye color in *Drosophila* blocks pigment synthesis at a specific step by preventing production of the enzyme that catalyzes that step. However, neither the chemical reactions nor the enzymes that synthesize these pigments were known at the time.

A breakthrough in demonstrating the relationship between genes and enzymes came a few years later, after Beadle and Edward Tatum began to search for mutants of the orange bread mold *Neurospora crassa* that differed from wild type in their nutritional needs. This fungus offers key advantages for studying the links between mutations and abnormal metabolism. Like all organisms that can reproduce sexually, it has both haploid and diploid forms. Since its haploid form is capable of prolific asexual reproduction, the effects of a mutant gene can be readily observed, without being masked by the simultaneous presence of a wild-type allele (as might happen in a diploid organism). After fertilization, the diploid zygote undergoes meiosis followed by a mitotic division within a thin sac called an *ascus,* forming eight haploid cells called *ascospores*. The ascospores can be dissected out and germinated individually on artificial growth medium. (see Methods Box on p. 324).

Wild-type *Neurospora* has modest nutritional needs; it can survive in the laboratory on a moist support medium, called agar, mixed only with inorganic salts, sucrose, and the vitamin biotin. From this **minimal medium,** the mold uses its metabolic pathways to produce all the other molecules it needs. Beadle and Tatum hunted for mutants that could not synthesize certain essential molecules, such as amino acids. Such mutants cannot survive on minimal medium unless the substances they are unable to synthesize are added to the medium. These nutritional mutants are called **auxotrophs** ("increased eaters").

To obtain large numbers of mutants, Beadle and Tatum irradiated *Neurospora* cultures with X-rays. The irradiated fungi were mated to untreated ones and the zygotes allowed to undergo meiosis. Each of the resulting haploid ascospores was transferred to its own vial containing a **complete growth medium,** minimal medium supplemented with all 20 amino acids and some other nutrients. Under these conditions, growth occurred in nearly every vial. Mutants could then be identified by transferring small fragments of the growing fungi to vials containing minimal medium. If the fragment failed to grow, it was considered to be from an auxotrophic mutant. To pinpoint the metabolic defect, Beadle and Tatum took samples from the mutant growing on complete medium and distributed them to several different vials, each containing minimal medium plus a single additional nutrient. The particular supplement that allowed growth indicated the metabolic defect. For example, if the only supplemented vial that supported growth of the mutant was the one fortified with arginine, then one could conclude that the mutant was defective in the pathway that normally synthesizes arginine (see Methods Box).

With further experimentation, a defect could be even more specifically described. For instance, consider three of the steps in the synthesis of arginine, each catalyzed by its own enzyme: A precursor nutrient is converted to ornithine, which is converted to citrulline,

This diagram shows how Beadle and Tatum isolated and identified a mutant of *Neurospora* deficient in arginine synthesis. Key point: The haploid mutant can grow only on medium containing arginine—either "complete" medium (minimal medium supplemented with all the amino acids plus other nutrients) or minimal medium supplemented with arginine only; it cannot grow on medium lacking arginine.

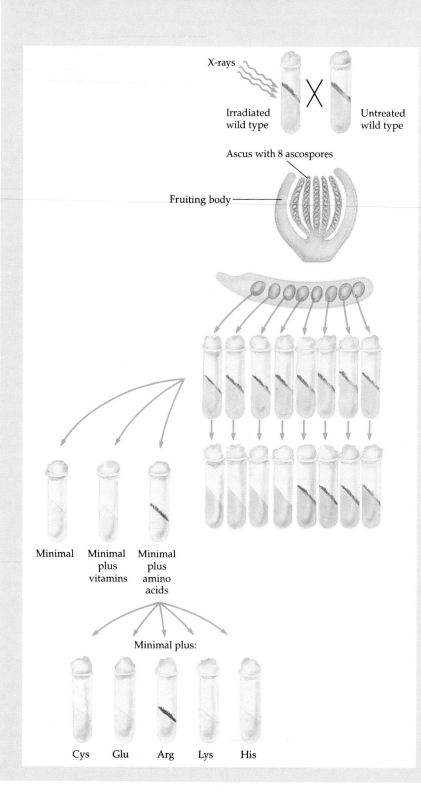

Mutations are induced with X-rays in one culture. Irradiated and untreated cultures are mated.

Zygotes undergo meiosis to produce ascospores.

Individual ascospores are placed on complete medium.

All grow.

Fragments of fungus are transferred to minimal medium. Those that do not grow are mutants.

Fragments of mutant fungi are transferred to minimal medium supplemented with various classes of nutrients. Growth occurs on medium containing amino acids.

Fragments are transferred to tubes containing minimal medium plus one of the amino acids.

Only arginine supports growth. Therefore this mutant is defective in the metabolic pathway that synthesizes arginine.

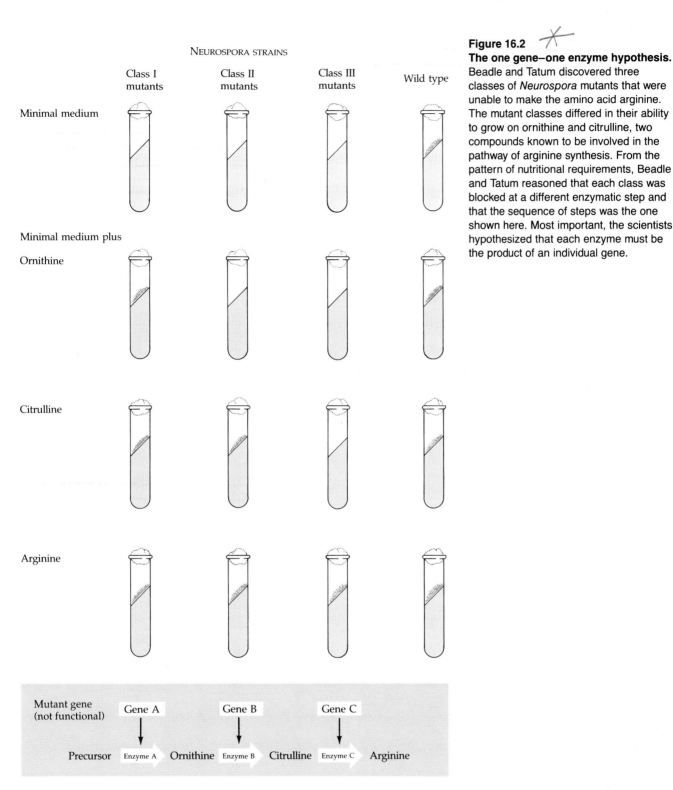

NEUROSPORA STRAINS

	Class I mutants	Class II mutants	Class III mutants	Wild type
Minimal medium				
Minimal medium plus Ornithine				
Citrulline				
Arginine				

Mutant gene (not functional): Gene A, Gene B, Gene C

Precursor → Enzyme A → Ornithine → Enzyme B → Citrulline → Enzyme C → Arginine

Figure 16.2 ✗

The one gene–one enzyme hypothesis. Beadle and Tatum discovered three classes of *Neurospora* mutants that were unable to make the amino acid arginine. The mutant classes differed in their ability to grow on ornithine and citrulline, two compounds known to be involved in the pathway of arginine synthesis. From the pattern of nutritional requirements, Beadle and Tatum reasoned that each class was blocked at a different enzymatic step and that the sequence of steps was the one shown here. Most important, the scientists hypothesized that each enzyme must be the product of an individual gene.

which is converted to arginine (Figure 16.2). Beadle and Tatum could distinguish among the various arginine auxotrophs that they found. Some required arginine, others required either arginine or citrulline, and still others could grow when any of the three compounds—arginine, citrulline, or ornithine—was provided. These three classes of mutants, Beadle and Tatum

reasoned, must be blocked at different steps in the pathway that synthesizes arginine. In fact, the sequence of steps in the pathway can be deduced from the nutritional requirements of the three classes. Beadle and Tatum concluded that each mutant lacked a different enzyme. Assuming that each mutant was defective in a single gene, they formulated what came to be known as the **one gene–one enzyme hypothesis,** which states that the function of a gene is to dictate the production of a specific enzyme.

One Gene–One Polypeptide

As more was learned about proteins, it became necessary to make minor revisions in the one gene–one enzyme hypothesis. With a few very specialized exceptions, all enzymes are proteins, but not all proteins are enzymes. Keratin, the structural protein of animal hair, and the hormone insulin are two examples of nonenzyme proteins. Because proteins that are not enzymes are nevertheless gene products, molecular biologists began to think in terms of one gene–one protein. However, many proteins have been discovered that are constructed from two or more different polypeptide chains, and each subunit may be specified by its own gene. Thus, Beadle and Tatum's axiom has come to be restated as one gene–one polypeptide.

THE LANGUAGES OF MACROMOLECULES

Nucleic acids and proteins are both polymers with specific sequences of monomers that confer information, much as specific sequences of letters confer information in English. In nucleic acids, the monomers are the four types of nucleotides, which differ in their nitrogenous bases. Genes within DNA typically are hundreds or thousands of nucleotides in length, with each gene having a specific sequence of nucleotide bases. Like nucleic acids, proteins each have their monomers arranged in a particular linear order, but the monomers are the 20 amino acids common to all organisms. Thus, these two classes of macromolecules—nucleic acids and proteins—contain information written in two different chemical languages. However, these two languages are related. Encoded in the base sequence along a gene is the information specifying the amino acid sequence of the corresponding protein chain.

In describing how genes dictate the linear arrangement of amino acids, it is customary to employ linguistic terms. The transfer of information from a DNA molecule to an RNA molecule is called **transcription** because the same "language" is used (that of nucleic acids). On the other hand, the transfer of information from an RNA molecule to a polypeptide is called **translation** because there is a change in language (from the language of nucleic acids to that of proteins). Figure 16.1 diagrammed the roles of these processes in the central dogma.

One problem in translation is that there are only 4 nucleotides to specify 20 amino acids. Thus, the genetic code cannot be a language like Chinese, where each written symbol corresponds to a single word. If a one-for-one translation of each nucleotide base into an amino acid occurred, only 4 of the 20 amino acids would be specified. Would a language of two-letter code words suffice? If we read the bases of a gene two at a time, AG, for example, could specify one amino acid, while GT could designate a different amino acid. However, there are only 16 (that is, 4^2) possible arrangements when the 4 bases are taken in doublets—still not enough to code for all 20 amino acids.

Triplets of bases are the smallest units of uniform length that can code for all the amino acids. If the DNA code words consist of triplets, with each arrangement of 3 consecutive bases specifying an amino acid, then there can be 64 (that is, 4^3) possible code words. This is more than enough possibilities to specify the 20 amino acids. In fact, there are enough triplets for there to be synonyms coding for a particular amino acid. Later, we will see how such redundancy in the genetic code is beneficial.

Experiments have verified that the flow of information from gene to protein is based on a triplet code: The genetic instructions for a polypeptide chain are written in the DNA as a series of three-nucleotide words, called **codons** (Figure 16.3).

Figure 16.3
The triplet code. The genetic information in DNA encoding each amino acid in a polypeptide is a series of three nucleotides, called a codon. The order of codons on the DNA corresponds to the order of amino acids in the polypeptide. Only one strand of the DNA codes for a given polypeptide.

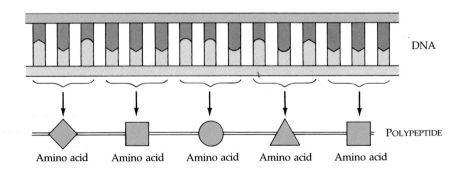

DNA

POLYPEPTIDE

Amino acid   Amino acid   Amino acid   Amino acid   Amino acid

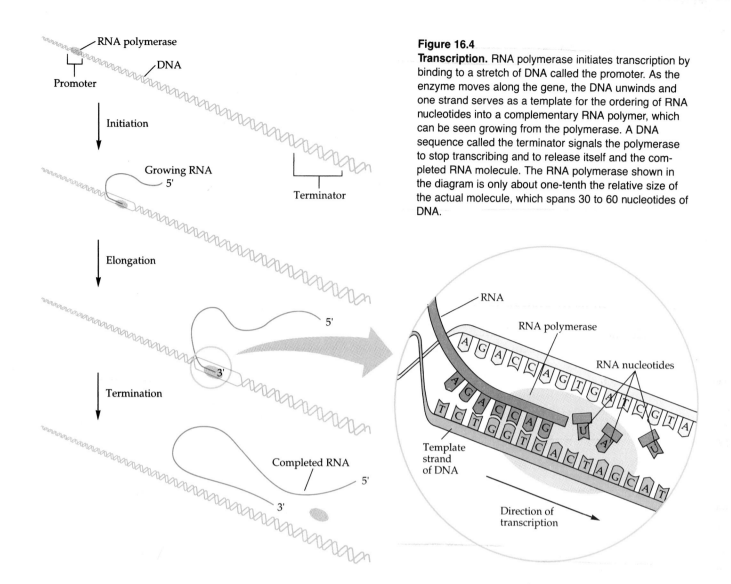

Figure 16.4

Transcription. RNA polymerase initiates transcription by binding to a stretch of DNA called the promoter. As the enzyme moves along the gene, the DNA unwinds and one strand serves as a template for the ordering of RNA nucleotides into a complementary RNA polymer, which can be seen growing from the polymerase. A DNA sequence called the terminator signals the polymerase to stop transcribing and to release itself and the completed RNA molecule. The RNA polymerase shown in the diagram is only about one-tenth the relative size of the actual molecule, which spans 30 to 60 nucleotides of DNA.

TRANSCRIPTION, THE SYNTHESIS OF RNA

As mentioned earlier, the genetic information in DNA is not directly translated but is first transcribed into an intermediary molecule of RNA, appropriately called **messenger RNA (mRNA).** RNA molecules are transcribed from DNA templates by a process that resembles the synthesis of new DNA strands during DNA replication. As with replication (see Chapter 15), the two DNA strands must first separate at the region where transcription will begin, and the new nucleic acid strand grows in the 5′ → 3′ direction. However, during transcription, only one of the DNA strands serves as a template to direct the sequence of nucleotides in the newly forming RNA molecule. The nucleotides that make up the new RNA molecule find their places one at a time along the DNA template strand by forming hydrogen bonds with the nucleotide bases there. The RNA nucleotides follow the same base-pairing rules that govern DNA replication, except that uracil (U), rather than thymine (T), pairs with adenine (A). The RNA nucleotides are linked together by the enzyme **RNA polymerase** (Figure 16.4).

RNA polymerase must be instructed where to start and where to stop the transcribing process. The "start transcribing" signals are specific nucleotide sequences called **promoters,** which are located in the DNA flanking the start end of the gene. RNA polymerase attaches to the promoter and initiates transcription. For any gene, the promoter region signals only one strand to be transcribed. However, transcription of the many genes on a chromosome is not confined to just one of the DNA strands: Some genes are coded by one strand and some by the other.

Elongation of the RNA chain proceeds at a rate of about 60 nucleotides per second. As the RNA is made, it peels away from its DNA template, allowing hydrogen bonds to re-form between the two separated DNA strands. Finally, the RNA polymerase reaches a special sequence of bases in the DNA template called a

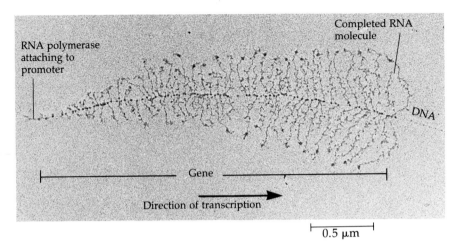

Figure 16.5
Electron micrograph of transcription.
This gene (the horizontal line at the center of this "feather") is being transcribed simultaneously by many molecules of RNA polymerase (the dark dots along the DNA). The newly formed RNA molecules extend perpendicularly from the DNA at the site of each RNA polymerase. The polymerase molecules are transcribing the gene from the left of this photo to the right; thus, the RNA molecules are progressively longer at the right.

In figure: RNA polymerase attaching to promoter — Completed RNA molecule — DNA — Gene — Direction of transcription — 0.5 μm

terminator. This sequence signals termination; the polymerase releases the RNA molecule and departs from the gene.

A single gene can be transcribed simultaneously by several molecules of RNA polymerase, following one another like trucks in a convoy. The growing strands of RNA trail off from each polymerase, with the length of each new strand reflecting how far along the template the enzyme has traveled from the promoter region (Figure 16.5). The congregation of many polymerase molecules on a single gene to produce multiple RNA transcripts helps a cell to produce a particular protein in large amounts.

In prokaryotes, the RNA transcript of protein-coding genes is essentially equivalent to the final form of mRNA; the major peculiarity of prokaryotic mRNA is that a single molecule may be a transcription of several adjacent, related genes (see the discussion of "operons" in Chapter 17). In eukaryotes, there is only one gene per RNA transcript molecule, but, as we shall discuss later in this chapter, the RNA must be modified before it can serve its role in protein synthesis. In eukaryotic cells, the resultant mRNA molecule must cross the nuclear membrane into the cytoplasm, where ribosomes synthesize proteins (Figure 16.6).

TRANSLATION, THE SYNTHESIS OF PROTEIN

In the cytoplasm, the sequence of messenger RNA codons is translated in strict linear order, much as a beginning reader follows the words across a page one

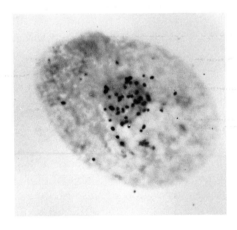

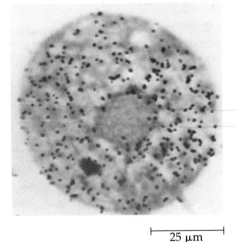

25 μm

Figure 16.6
Movement of messenger RNA from nucleus to cytoplasm. In eukaryotes, messenger RNA carries genetic instructions from the DNA to the nucleus to the cytoplasm, where the equipment for protein synthesis is located. To demonstrate this movement, eukaryotic cells were exposed to radioactive RNA nucleotides and allowed to make RNA from these labeled monomers. The dark spots on these light micrographs show the location of the labeled RNA (see Chapter 2 Methods Box for a description of autoradiography). The cell on the left was killed after 15 minutes of growing in the radioactive medium, and the excess radioactive nucleotides were washed away; the newly synthesized RNA is primarily in the nucleus. The cell on the right was allowed to make labeled RNA for 15 minutes, but was then transferred for 90 minutes to a medium containing no radioactive nucleotides. The labeled RNA, which was made earlier, has moved into the cytoplasm.

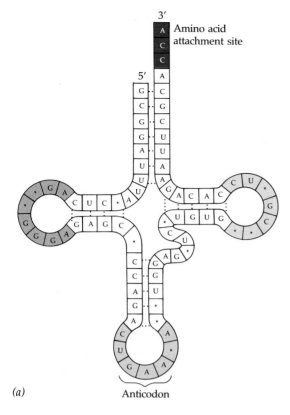

(a)

Anticodon

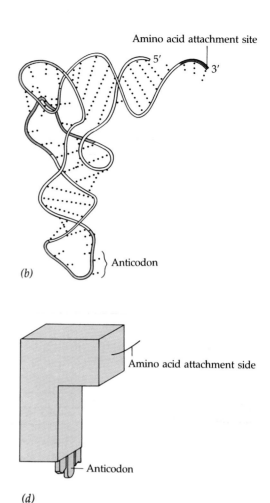

Amino acid attachment site

(b)

Anticodon

Amino acid attachment side

Anticodon

(d)

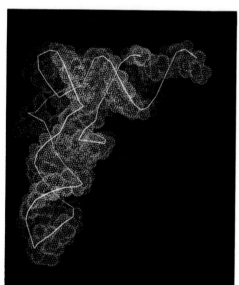

(c)

Figure 16.7
Structure of transfer RNA. **(a)** The two-dimensional structure of a typical tRNA molecule. Unusual bases, indicated by asterisks, are unique to tRNA; they are formed by modification of the usual RNA bases after the tRNA chain is synthesized. Note the four double-stranded regions and three loops characteristic of all tRNAs. At one end of the molecule is the amino acid attachment site, which has the same base sequence for all tRNAs; within the middle loop is the anticodon triplet, which is unique to each tRNA species. **(b)** The three-dimensional structure of a tRNA, shown as a diagram. **(c)** The three-dimensional structure of a tRNA, shown as a computer-graphic. In three dimensions, the molecule is shaped like an L. **(d)** In the figures that follow, tRNA will be represented by the simplified shape shown here.

by one. Among the critical elements in the translation of mRNA is another kind of RNA molecule known as transfer RNA.

Transfer RNA

Transfer RNA (tRNA) functions as an interpreter between the nucleic acid language and the protein language. The codons arranged in sequence along messenger RNA cannot be recognized by the amino acids

themselves. The codon for tryptophan (UGG), for example, is no more attractive to tryptophan than is any other codon. The tRNA molecules are responsible for bringing the appropriate amino acids into alignment to form the new polypeptide. To perform this task, tRNA molecules must carry out two distinct functions: picking up the appropriate amino acids and recognizing the appropriate codons in the mRNA.

The tRNA molecules are made of single-stranded RNA that is only about 80 nucleotides long (Figure 16.7). The single strand of tRNA folds back upon itself to make several double-stranded regions in which short stretches of RNA form hydrogen bonds with other stretches having complementary base sequences. Flattened into one plane to reveal these regions of hydrogen bonding, a tRNA molecule has a cloverleaf shape. The two-dimensional cloverleaf of the typical tRNA molecule twists and folds into a fairly compact three-dimensional structure that is roughly L-shaped. The loop protruding from one end of the L contains a specialized base triplet, called the **anticodon**. The anticodon triplet is complementary to a codon triplet on the mRNA, and it recognizes a particular codon by employing base-pairing rules. From the other end of the L-shaped tRNA molecule protrudes its 3' end, which is the site where an amino acid can attach. Specific enzymes ensure that a tRNA having a particular anticodon attaches to only one kind of amino acid.

Amino Acid Activating Enzymes

The tRNA molecule is like a flashcard with a nucleic acid "word" (anticodon) on one side and a protein "word" (amino acid) on the other. The attachment of the 20 amino acids, each to its particular form of tRNA, is an exacting process and is carried out by enzymes called **amino acid activating enzymes** or, more precisely, **aminoacyl-tRNA synthetases.** There is a whole family of these enzymes, one enzyme for each amino acid. An amino acid activating enzyme has an active site that specifically binds one type of amino acid along with the appropriate tRNA molecule. The enzyme catalyzes the attachment of the amino acid to its tRNA in a two-step process driven by the hydrolysis of ATP (Figure 16.8). The resulting amino acid–tRNA complex is then released from the enzyme and is available to furnish its amino acid to a growing polypeptide chain on a ribosome.

Surprisingly, an amino acid activating enzyme does *not* select its tRNA by recognizing the anticodon; instead it recognizes another part of the tRNA sequence. This recognition is said to be determined by a "second genetic code," to differentiate it from the genetic code that relates codons to amino acids.

Ribosomes

Coordinating the sequential coupling of tRNAs to the series of mRNA codons is the job of the ribosomes. These particles, which can be seen with the electron microscope, are each made up of two subunits (Figure 16.9). (When not involved in protein synthesis, the two subunits are not bound together.) Each ribosomal subunit is an aggregate of proteins and a considerable amount of yet another form of specialized RNA mol-

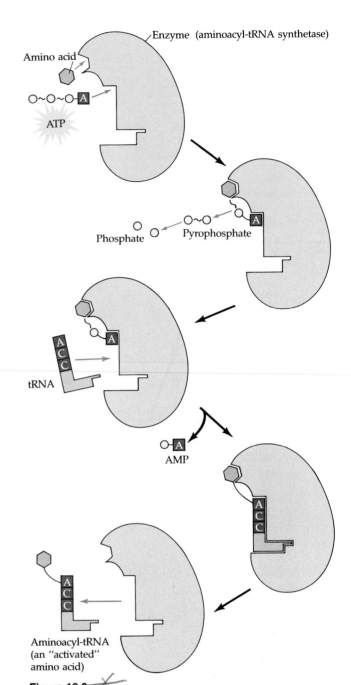

Figure 16.8

Amino acid activating enzymes. Each enzyme mates a specific amino acid to its appropriate tRNA. First, the active site of the enzyme binds the amino acid and an ATP molecule, which loses two phosphate groups and joins to the amino acid as AMP. Next, the appropriate tRNA displaces the AMP from the enzyme's active site and covalently bonds to the amino acid. Finally, the enzyme releases the amino acid-tRNA complex, also known as aminoacyl-tRNA or simply as an "activated" amino acid. The more precise name for the activating enzyme is *aminoacyl-tRNA synthetase.*

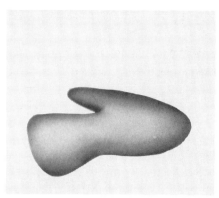

Small subunit (30S)

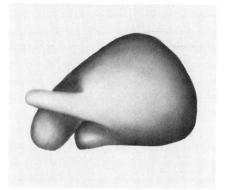

Large subunit (50S)

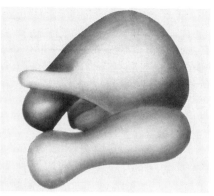

Complete ribosome (70S)

Figure 16.9
Structure of the ribosome. Ribosomes are composed of two subunits. Each has a characteristic shape, as can be seen in the set of drawings of bacterial ribosomes. In prokaryotic cells, the small subunit is made up of 21 proteins and 1 molecule of rRNA, and the large subunit is made up of 34 proteins and 2 molecules of rRNA. The prokaryotic ribosome is referred to as a 70S ribosome, where 70S is its sedimentation constant, a measure of how fast it sediments in the ultracentrifuge; its sub-units are 30S and 50S. The eukaryotic ribosome has a similar structure but is larger (80S), with more molecules of protein and rRNA; its subunits have sedimentation constants of 40S and 60S.

ecule, **ribosomal RNA (rRNA).** About 60% of the weight of each ribosome is rRNA. Because most cells contain thousands of ribosomes, rRNA is an extremely abundant type of RNA.

Ribosomes conduct the specific coupling of tRNA anticodons with mRNA codons during protein synthesis. In addition to a binding site for mRNA, each ribosome has two binding sites for tRNA (Figure 16.10).

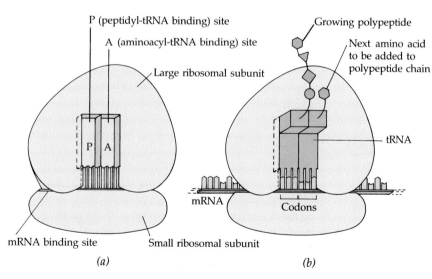

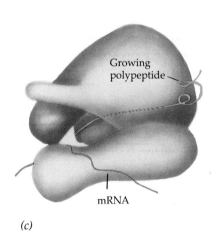

Figure 16.10
Binding sites on ribosomes. (a) In addition to an mRNA binding site, the ribosome has two tRNA binding sites, one called the P site (peptidyl-tRNA binding site) and the other the A site (aminoacyl-tRNA binding site). **(b)** A tRNA fits into each binding site as it base-pairs with the mRNA. The P site holds the tRNA attached to the growing polypeptide chain. The A site holds the tRNA carrying the next amino acid to be added to the chain. The tRNA molecules in the two sites bind to adjacent codons. **(c)** A more realistic rendering of the prokaryotic ribosome and the location of the bound mRNA and growing polypeptide are shown here. The mRNA passes between the two subunits, and the polypeptide exits through a hole in the large subunit.

The **P site** holds the tRNA carrying the growing poly-peptide chain, while the **A site** holds the tRNA car-rying the next amino acid to be added to the chain. Acting like a vise, the ribosome holds the tRNA and mRNA molecules close together while it catalyzes the transfer of an amino acid from its tRNA to the carboxyl end of the growing polypeptide chain.

The Process of Protein Synthesis

The synthesis of a polypeptide chain can be divided into three stages: chain initiation, chain elongation, and chain termination. All three stages require enzymes; initiation and elongation also use up phos-phate bond energy that is provided by molecules of GTP, an energy currency closely related to ATP.

Initiation   Initiation of polypeptide synthesis is a complex process requiring several proteins called **ini-tiation factors** and GTP. The process determines exactly where translation will begin and how the sequence of nucleotides will be read off as base triplets that specify the amino acids. If this grouping of bases into codons, called the **reading frame,** is shifted by one or two nucleotides in either direction, the sequence will pro-duce a completely different sequence of amino acids. The initiation process must bring together the mRNA, the first amino acid attached to its tRNA, and the two subunits of a ribosome.

The first step in initiation is the binding of mRNA and a special initiator tRNA molecule to a small ribo-somal subunit (Figure 16.11). This initiator tRNA mol-ecule carries an amino acid, usually methionine. The start codon, usually AUG, binds to the anticodon of the initiator tRNA. In the case of prokaryotes, we know something about what attracts the mRNA to the ribo-some: In the mRNA on the 5′ side of the start codon is a sequence of nucleotides that constitutes a recog-nition signal for the ribosome; such sequences base-pair to rRNA and help hold the mRNA molecule to the ribosomal subunit, but they are not translated into amino acids.

In the second step of initiation, a large ribosomal subunit binds to the small one, creating a functional ribosome. The initiator tRNA fits into the P site on the ribosome.

The Elongation Cycle   Once initiation is complete, amino acids are added one by one to the initial amino acid. Each addition involves the participation of sev-eral proteins called **elongation factors** and occurs in a three-step cycle (Figure 16.12):

1. In the first step of elongation, the mRNA codon in the A site of the ribosome forms hydrogen bonds with the anticodon of an incoming molecule of tRNA car-

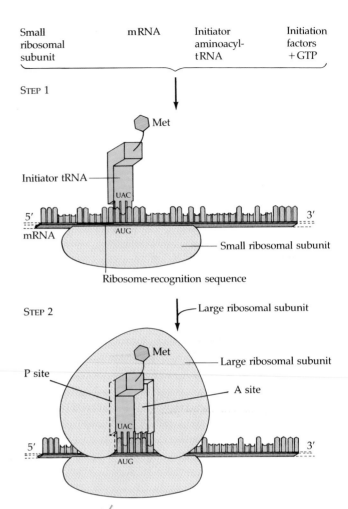

Figure 16.11
The initiation of protein synthesis. (1) In the presence of initiation factors, a small ribosomal subunit binds to a molecule of mRNA. The positioning of the mRNA is sig-naled by a ribosome-recognition sequence on the mRNA that base-pairs with rRNA. At the same time, the initiator methionyl-tRNA, with the anticodon UAC, base-pairs with the initiation codon AUG. (2) The large ribosomal subunit joins the initiation complex. The initiator tRNA is in the P site. The A site is available to the tRNA bearing the next amino acid.

rying its appropriate amino acid. This step requires the hydrolysis of a phosphate bond from GTP.

2. Next, an unusual enzyme that is an integral part of the large ribosomal subunit catalyzes the formation of a peptide bond between the polypeptide in the P site and the newly arrived amino acid in the A site. The enzyme, called **peptidyl transferase,** consists of a collection of ribosomal proteins in association with rRNA. In this step, the polypeptide separates from the tRNA to which it was bound and is transferred to the amino acid carried by the tRNA in the A site.

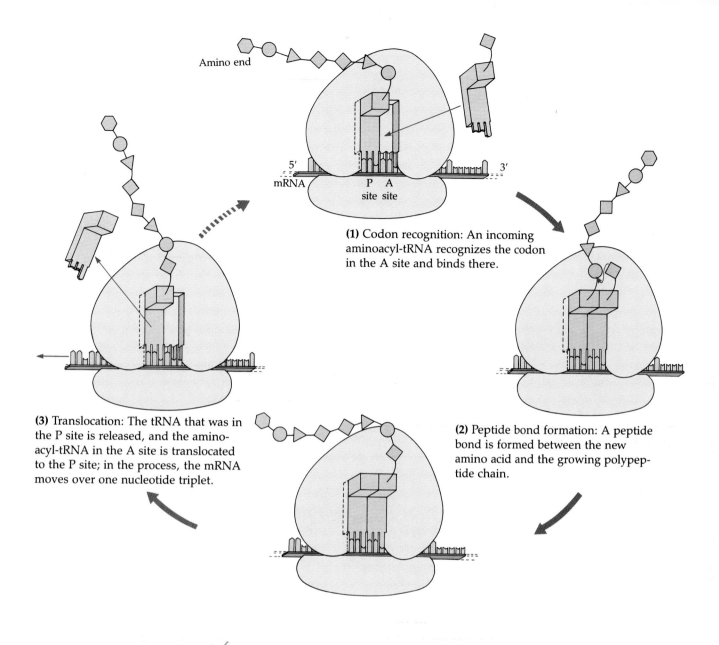

(1) Codon recognition: An incoming aminoacyl-tRNA recognizes the codon in the A site and binds there.

(2) Peptide bond formation: A peptide bond is formed between the new amino acid and the growing polypeptide chain.

(3) Translocation: The tRNA that was in the P site is released, and the aminoacyl-tRNA in the A site is translocated to the P site; in the process, the mRNA moves over one nucleotide triplet.

Figure 16.12
The elongation cycle of protein synthesis.

3. The third step in elongation is called **translocation.** The tRNA in the P site dissociates from the ribosome, and the tRNA in the A site, carrying the growing polypeptide, is translocated to the P site. The codon and anticodon remain hydrogen-bonded, allowing the mRNA and the tRNA to move as a unit. This movement, in turn, brings into the A site the next codon to be translated along the mRNA. The translocation step requires energy, which is provided by hydrolysis of a GTP molecule. The mRNA is moved through the ribosome in the 5' → 3' direction only, much as a ratchet allows a mechanical device to be turned in only one direction.

The elongation cycle takes only about 60 milliseconds and is repeated as each amino acid is added to the chain until the polypeptide is completed.

Termination   The elongation cycle continues until a **termination codon** reaches the A site of the ribosome. These special base triplets—UAA, UAG, and UGA (see Figure 16.14)—do not code for amino acids but instead act as signals to stop translation. A protein called **release factor** binds directly to the termination codon in the A site. The release factor causes the peptidyl transferase to add a water molecule instead of an amino acid to the polypeptide chain. This reaction frees the com-

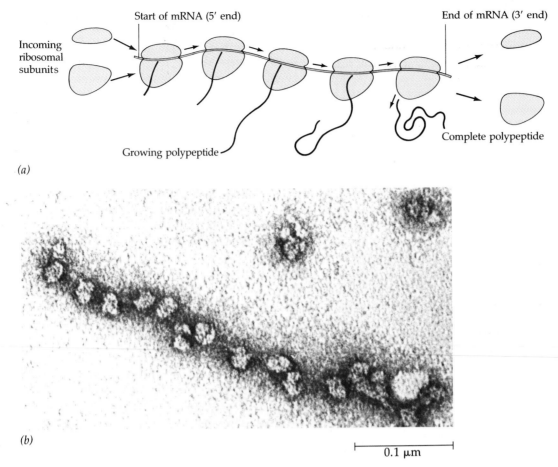

Start of mRNA (5' end)

End of mRNA (3' end)

Incoming ribosomal subunits

Growing polypeptide

Complete polypeptide

(a)

(b)

0.1 μm

Figure 16.13
Polyribosomes. (a) An mRNA molecule is generally translated simultaneously by several ribosomes in clusters called polyribosomes or polysomes. (b) This electron micrograph shows a large polyribosome in a prokaryotic cell.

pleted polypeptide from the tRNA that is in the P site, thereby freeing the polypeptide from the ribosome. The ribosome then separates once again into its small and large subunits.

Creation of a Functional Protein

During and after translation, a polypeptide typically undergoes several changes necessary for its activity in the cell. As the polypeptide spontaneously coils and folds during translation to assume its characteristic three-dimensional shape, disulfide bonds may form between certain cysteine residues. Often, one or more amino acids at the amino terminal of the polypeptide are enzymatically removed, and in some cases, major portions of the polypeptide are removed or the polypeptide is cleaved into several pieces. Certain amino acids may be chemically modified—for example, by the attachment of sugars or phosphate groups. Lastly, several identical or different polypeptides may associate to form a quaternary structure.

Sites of Protein Synthesis

In the prokaryotic cell, which lacks a nucleus, protein synthesis can occur on ribosomes anywhere in the cell. In fact, ribosomes often attach to a growing RNA chain and begin translation even before transcription is complete (see Figure 17.12). In the eukaryotic cell, proteins destined to become part of the cell's membranes or destined to be packaged for exportation from the cell are produced on ribosomes that become bound to the membrane of the endoplasmic reticulum.

As was shown in Figure 7.15, a signal sequence at the beginning ($-NH_2$) end of the growing polypeptide attaches to a special receptor protein in the ER membrane, bringing its mRNA and ribosome along with it. Proteins that are destined to function in a soluble form in the cytoplasm are usually produced on free ribosomes. A single ribosome can make an average-sized polypeptide in less than a minute. Typically, however, a single mRNA is used to make many copies of a polypeptide simultaneously because several ribosomes work on translating the message at the same

Figure 16.14
The genetic code. The three bases in an mRNA codon are designated as the first, second, and third, starting at the 5′ end. Each set of three specifies a particular amino acid, represented here by a three-letter abbreviation (see Figure 5.17). The codon AUG (which specifies the amino acid methionine) is the usual start signal for protein synthesis. The word *stop* indicates the codons that serve as signals to terminate protein synthesis.

First base	Second base: U	Second base: C	Second base: A	Second base: G	Third base
U	UUU Phe	UCU Ser	UAU Tyr	UGU Cys	U
U	UUC Phe	UCC Ser	UAC Tyr	UGC Cys	C
U	UUA Leu	UCA Ser	UAA Stop	UGA Stop	A
U	UUG Leu	UCG Ser	UAG Stop	UGG Trp	G
C	CUU Leu	CCU Pro	CAU His	CGU Arg	U
C	CUC Leu	CCC Pro	CAC His	CGC Arg	C
C	CUA Leu	CCA Pro	CAA Gln	CGA Arg	A
C	CUG Leu	CCG Pro	CAG Gln	CGG Arg	G
A	AUU Ile	ACU Thr	AAU Asn	AGU Ser	U
A	AUC Ile	ACC Thr	AAC Asn	AGC Ser	C
A	AUA Ile	ACA Thr	AAA Lys	AGA Arg	A
A	AUG Met or Start	ACG Thr	AAG Lys	AGG Arg	G
G	GUU Val	GCU Ala	GAU Asp	GGU Gly	U
G	GUC Val	GCC Ala	GAC Asp	GGC Gly	C
G	GUA Val	GCA Ala	GAA Glu	GGA Gly	A
G	GUG Val	GCG Ala	GAG Glu	GGG Gly	G

time (Figure 16.13). Once a ribosome moves past the initiation codon, a second ribosome can attach to it, and thus several ribosomes may trail along the same mRNA. Such clusters, called polyribosomes, can be seen with the electron microscope.

THE GENETIC CODE

Molecular biologists "cracked" the code of life in the early 1960s, when a series of elegant experiments disclosed the amino acid translations of each of the triplet code words of nucleic acids. The first codon was deciphered in 1961 by Marshall Nirenberg. He had synthesized an artificial mRNA by linking together identical RNA nucleotides having uracil as their base. No matter where this message would start or stop, it could contain only one type of triplet codon: UUU. Nirenberg added this "poly U" to a test-tube mixture containing the biochemical ingredients required for protein synthesis. His in vitro system translated the poly U into a polypeptide containing a single amino acid, phenylalanine. Thus, Nirenberg learned that the mRNA codon UUU specifies the amino acid phenyl-

alanine. Soon, the amino acids specified by the codons AAA, GGG, and CCC were also determined.

Although more elaborate techniques were required to decode mixed triplets such as AUA and CGA, all 64 codons were deciphered by the mid-1960s. As you can see in Figure 16.14, 61 of the 64 triplets code for amino acids. As mentioned earlier, the triplet AUG has a dual function: It not only codes for the amino acid methionine, but may also provide a signal for the start of translation from nucleic acid into protein. The remaining three codons do not designate amino acids. Instead, they instruct the ribosomes to stop assembling a polypeptide chain.

Notice in Figure 16.14 that there is redundancy in the code, but no ambiguity. For example, although codons GAA and GAG can both specify glutamic acid (redundancy), neither of them ever represents any other amino acid (no ambiguity). The redundancy in the code, often called "degeneracy," is not altogether random. In many cases, codons that are "synonyms" for a particular amino acid differ only in the third base of the triplet. A possible explanation for this redundancy will be considered in the next section.

The codons in Figure 16.14 are the triplets found in messenger RNA. They bear a straightforward, com-

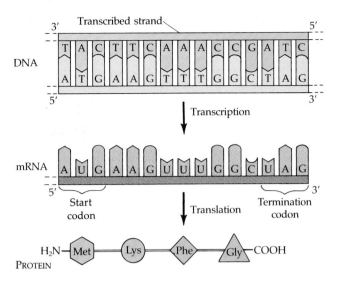

Figure 16.15

Transcription and translation of a DNA sequence.
The genetic message carried by one of the two DNA strands is transcribed into mRNA according to the base-pairing rules. Thus, the mRNA is complementary to one of the DNA strands (and, except for having U instead of T, and ribose instead of deoxyribose), identical to the other DNA strand. The translation apparatus recognizes each triplet of mRNA nucleotides as a codon specifying a particular amino acid to be incorporated into a protein.

plementary relationship to the codons found in DNA. As an exercise in translating the genetic code, consider the small polypeptide in Figure 16.15. It is four amino acids long; therefore, the genetic message for this polypeptide is 12 (that is, 4 × 3) nucleotides in length. Now, let us read the gene as a series of nucleotide triplets. The RNA codon corresponding to the first DNA triplet, TAC, is AUG. In effect, it says, "Place methionine as the first amino acid in the polypeptide." The second DNA triplet, TTC (RNA codon AAG), designates lysine as the second amino acid. We continue reading the triplets until we reach the stop codon

and have completed the entire primary structure of the polypeptide.

Nucleotides are evenly spaced along a gene, with no gaps separating the codons. In general, a particular nucleotide sequence is read in only one reading frame (set of triplet groupings), thereby dictating only a single sequence of amino acids. However, there are exceptions, as in the case of a very small virus in which there are overlapping coding sequences (Figure 16.16).

The Wobble Phenomenon

Although 64 codons can be read from messenger RNA molecules, there are only about 40 distinct types of tRNA molecules. This number is adequate for translation because some tRNA anticodons can pair with two or three different mRNA codons that specify the same amino acid. In an exception to the base-pairing rules called **wobble,** the third nucleotide (5′ end) of the tRNA anticodon can form hydrogen bonds with more than one kind of base in the third position (3′ end) of the codon. For instance, U in the third (wobble) position of a tRNA anticodon can pair with either A or G in the corresponding position of an mRNA codon. In several species of tRNA, the unusual base inosine (I) occupies the third position, from which it can form pairs with either U, C, or A. Thus, for example, the single tRNA species with the anticodon CCI (written 3′ → 5′) will recognize three mRNA codons—GGU, GGC, or GGA—all of which code for the amino acid glycine (refer to Figure 16.14).

Universality of the Genetic Code

Most of the genetic code is shared by all organisms, from the simplest bacteria to humans. The RNA codon CCG, for instance, is translated as proline in all organisms whose genetic code has been examined. In laboratory experiments, bacterial cells can translate the genetic messages obtained from human cells, and

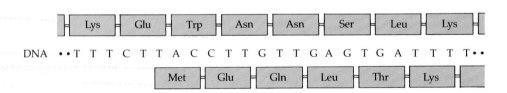

Figure 16.16
Overlapping genes in a virus. This segment of viral DNA is a portion of two of the several overlapping genes in the bacteriophage φX174. The amino acids listed above the DNA are the amino acids that

are encoded by one grouping of the nucleotides and occur in one of the viral proteins. The amino acids listed below the DNA are the amino acids that are encoded by another reading frame and occur in

another viral protein. The genetic material of φX174 is a single strand of DNA. Such overlapping genes have been observed only in a few viruses.

human cells can be made to translate bacterial genes (see Chapter 19). This universality of the genetic vocabulary, a reminder of the kinship among all forms of life, implies that the genetic code was established very early in evolution.

Scientists have recently discovered some exceptions to the universality of the genetic code. In the genes of several single-celled eukaryotes called ciliates (including a *Paramecium* and a *Tetrahymena*), biologists have found a variation from the standard code. In these organisms, the RNA codons UAA and UAG are not stop signals, as usual, but code for the amino acid glutamine. All other known exceptions to the standard genetic code occur within mitochondria, which contain their own DNA for certain genes, as well as the machinery necessary for protein synthesis. Mitochondrial genetic codes vary with the organism. Thus, CUA is read as threonine in yeast mitochondria, whereas in mammalian mitochondria it codes for leucine.

SPLIT GENES AND RNA PROCESSING IN EUKARYOTES

Much of the DNA of most organisms does not code for protein. The noncoding segments of eukaryotic DNA have provided one of the most surprising and important discoveries in recent biological research. In 1977, scientists found that some noncoding sequences fall *within* the boundaries of eukaryotic genes, interrupting segments of DNA that code for the amino acid sequence of a polypeptide. Such unexpected interruptions are puzzling. It is as if unintelligible sequences of letters were randomly interspersed in an otherwise intelligibly written document. Most genes of the more complex eukaryotes are interrupted by long segments of such noncoding regions, which are called intervening sequences or **introns** (Figure 16.17). The coding regions, which introns interrupt, have come to be called **exons**—because they are expressed (translated). Both exons and introns are transcribed into RNA. Then, however, before the RNA leaves the nucleus, the introns are removed from the RNA, and the exons that flanked them are joined to produce an mRNA molecule with a continuous coding sequence. This process is called **RNA splicing** or **RNA processing,** and it is also required for the production of tRNA and rRNA.

The discovery of split genes is an instructive example of two aspects of current research in molecular biology. First, the discovery was made possible by the development of a powerful new set of techniques involving recombinant DNA (see Chapter 19). Second, it was the outcome of experiments with several different kinds of genes carried out simultaneously by several different groups of scientists. With many more

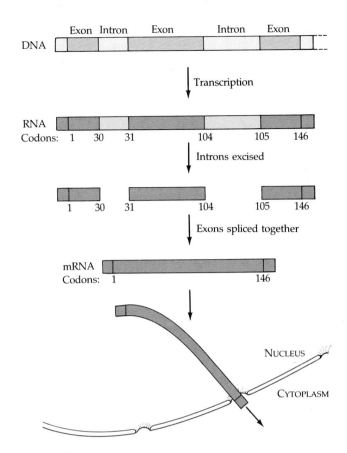

Figure 16.17
Introns in a eukaryotic gene. The gene depicted here codes for β-globin, one of the polypeptides of hemoglobin. β-globin is 146 amino acids long. Its gene has three segments containing coding sequences, exons, that are separated by noncoding introns. The entire gene is transcribed, but before the RNA leaves the nucleus, the introns are excised and the exons of the mRNA are spliced together. The mRNA also contains noncoding sequences at either end (see Figure 16.19). The codons are numbered (starting at the 5' end of the RNA) to show exactly where the introns occur.

scientists at work today than in the past, such simultaneous discoveries are becoming increasingly common.

The details of RNA splicing are still being worked out. Biologists have learned that the signals for RNA splicing are sets of a few nucleotides located at either end of each intron. Particles called **small nuclear ribonucleoproteins,** or **snRNPs** (pronounced "snurps"), play a key role in RNA splicing. As the name implies, these particles are small, are located in the cell nucleus, and are composed of RNA and protein molecules. The RNA in a snRNP particle is called **small nuclear RNA (snRNA),** and it is typically a single molecule about 150 nucleotides in length (about twice as long as a tRNA molecule but much shorter than mRNA). Each

Figure 16.18

The roles of snRNPs and spliceosomes in mRNA splicing. After a eukaryotic gene containing exons and introns is transcribed, the RNA transcript combines with various small nuclear ribonucleoproteins (snRNPs) and other proteins to form a molecular complex called a spliceosome. Within the spliceosome, the RNA of certain snRNPs base-pairs with the ends of each intron, the RNA transcript is cut to release the intron, and the exons are spliced together. The spliceosome then comes apart, releasing mRNA, which now contains only exons.

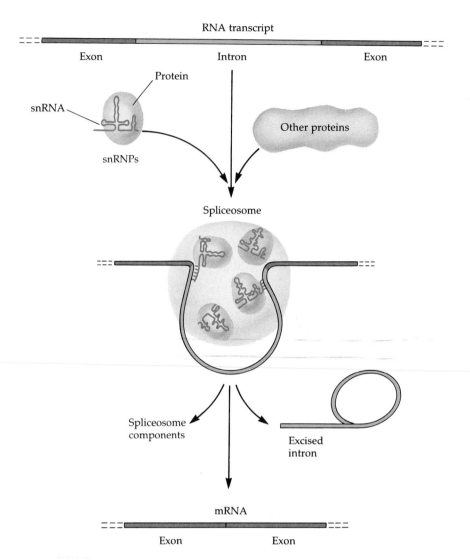

snRNP particle also has seven or more proteins. There appear to be a variety of snRNPs, and their functions have been only partly determined. The several kinds already known to be involved in RNA splicing carry out their roles as components of a larger, more complex assembly called a **spliceosome** (Figure 16.18). The spliceosome interacts with the ends of an RNA intron. It cuts at specific points to release the intron and then directly rejoins the two exons that are now adjacent.

The processing of some other kinds of RNA transcripts—such as those for tRNA and rRNA—occurs in several different ways. However, as with eukaryotic mRNA processing, some kind of RNA is often involved in catalyzing the reactions. One of the more unexpected discoveries in recent years was the discovery that in some cases—for example, in the splicing of rRNA in the ciliated protozoan *Tetrahymena*—the splicing occurs completely without proteins or extra RNA molecules; the intron RNA itself catalyzes the process!

In addition to splicing, eukaryotic mRNA molecules undergo other modifications before they leave the nucleus. The most striking of these is the addition of

nucleotides to the two ends (Figure 16.19). A **cap** consisting of a modified guanosine triphosphate is added to the 5' end, and a stretch of 150 to 200 adenine nucleotides, called **poly A,** is added to the 3' end. Both additions apparently help protect the ends of the mRNA from degradation, and the cap at the 5' end also enhances translation. As shown in Figure 16.19, the cap and the poly-A tail attach to untranslated *leader* and *trailer* sequences at the ends of the mRNA molecule. Thus, we must refine our definition of "exon": An exon at either end of a gene may include a terminal nucleotide sequence that is present in the mRNA but is not translated.

What are the biological functions of introns? One theory is that introns play a regulatory role in the cell. Perhaps intron DNA includes sequences that control gene activity in some way, or perhaps the splicing process itself is a mechanism for regulating the flow of mRNA from nucleus to cytoplasm. There is circumstantial evidence for the latter idea: Spliceosomes, which remain bound to RNA transcripts until processing is complete, are too large to pass through the nuclear

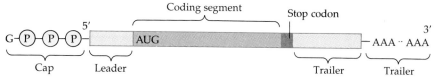

Figure 16.19
Modification of messenger RNA. Before leaving the nucleus as mRNA, eukaryotic RNA transcripts receive a cap of a modified guanosine triphosphate at the 5′ end and a tail of 150 to 200 adenine nucleotides, called poly A, at the 3′ end. Only part of the molecule constitutes the coding segment that is translated into protein. The untranslated "leader" region at the 5′ end contains signals for initiation of translation; the function of the noncoding "trailer" region adjoining the poly A at the 3′ end is not known.

pores. Another hypothesis is that introns play an important part in evolution by facilitating recombination between exons to form a diversity of proteins—that is, by promoting the shuffling of preexisting DNA segments to produce new proteins. According to this hypothesis, an exon generally codes for a polypeptide segment having a specific function, such as a binding site. Such a structural and functional segment of a polypeptide is called a **domain** (see Chapter 5). Because coding regions for a particular protein can be separated by considerable distances along the DNA, the frequency of recombination *within* the gene can be higher than for a continuous coding region. Thus, it is more likely that these distant DNA segments within a single gene will be shuffled by crossovers between homologous chromosomes. In fact, a number of cases are known where exons correspond to polypeptide domains. In several of these cases, there is circumstantial evidence that the genes arose by recombination between exons, followed by mutational changes in the new gene. We will have more to say about the scrambling of exons and the functions of RNA processing in Chapter 18.

From the evolutionary perspective, it is intriguing that as one moves backwards through the eukaryotes from more complex organisms to simpler organisms, the proportion of genes that are interrupted decreases, and the introns that do exist tend to be shorter. In mammals, almost all genes seem to be split, and introns are often much longer than the coding sequences they interrupt. In yeast, at the other end of the spectrum, genes containing introns are rare. The meaning of these differences is a major unanswered question in modern biology.

RNA: A REVIEW

Let us pause at this point to review what we have discussed about the structures and functions of RNA molecules in cells. When you first heard of RNA in Chapter 5, you may have gotten the impression that RNA was merely a sidekick of the master molecule, DNA—that RNA was essential but not very interesting. As described in Chapter 5, the primary structure of RNA differs from that of DNA in only two ways: (1) RNA has the sugar ribose, which has one more oxygen atom than deoxyribose, the DNA sugar; and (2) RNA has the nitrogenous base uracil instead of thymine (uracil and thymine differ by only a $-CH_3$ group). The other structural difference between cellular RNA and DNA is that cellular RNA molecules are single-stranded, rather than double-stranded like DNA.

In this chapter, we have learned that RNA is actually much more varied and interesting than the basic description suggested. First of all, we have seen that cells have several distinct types of RNA (Table 16.1). (And there is already evidence for some additional types of small RNA molecules in both nucleus and cytoplasm.)

Second, we have learned that the lack of a second strand does not mean that RNA lacks secondary or tertiary structure. As we saw in some detail for tRNA, RNA often has regions where the single polynucleotide chain doubles back upon itself to form base-paired double helices (secondary structure), and the entire molecule may fold to take on a characteristic three-dimensional shape (tertiary structure). As with proteins, the shape of an RNA molecule can be important in determining its function. In recent years, researchers have found that secondary and tertiary structure are important in the activities of most types of RNAs, including tRNA, rRNA, snRNA, and other RNA molecules that form parts of certain RNA-processing enzymes.

Third, we have seen that the functions of RNA are more diverse than once thought. In addition to acting as a message for protein synthesis (mRNA), a structural molecule (rRNA), or an adaptor molecule (tRNA), RNA can also act as an enzyme, sometimes with and sometimes without help from proteins. The discovery of this ability has led some scientists to examine more closely the roles of rRNA in ribosomes, and to realize that ribosomes themselves can be regarded as giant enzymes composed of both RNA and proteins. Furthermore, the discovery of RNA enzymatic func-

Table 16.1 Major types of cellular RNA

Type of RNA	Prokaryotic or Eukaryotic Cells	Function
Messenger RNA (mRNA)	Both	Carries information specifying the amino acid sequences of proteins from DNA to ribosomes
Transfer RNA (tRNA)	Both	Serves as adaptor molecule in protein synthesis; translates mRNA nucleotide sequence into protein amino acid sequence
Ribosomal RNA (rRNA)	Both	Plays structural and probably enzymatic roles in ribosomes, the sites of protein synthesis
Small nuclear RNA (snRNA)	Eukaryotic	Plays structural and enzymatic roles in snRNP particles, which help carry out mRNA splicing within spliceosomes

tion, coupled with the discovery that RNA can sometimes serve as a template for DNA synthesis (reverse transcription; see Chapters 17–19), has exciting implications for theories about the origin of life (see Chapter 24).

MUTATIONS AND THEIR EFFECTS ON PROTEINS

Since discovering how genes are translated into proteins, scientists can give a molecular description of heritable changes that arise in an organism—a description that Garrod would have much appreciated. When a child is born with sickle-cell anemia, for instance, the abnormality can be traced back through the difference in a protein to one tiny alteration in a gene (Figure 16.20). A mutation, or change in the nucleotide sequence of DNA, can thus involve large regions of a chromosome or just a single nucleotide pair, as in sickle-cell anemia. When a mutation is limited to about one nucleotide pair, it is called a **point**

mutation. In the remainder of this chapter, we will consider how mutations involving only one or a few nucleotide pairs affect the translation of the genetic code.

Types of Mutations

Mutations within a gene can be divided into two general categories: base-pair substitutions and base-pair insertions or deletions (Figure 16.21, p. 342).

A **base-pair substitution** is the replacement of one nucleotide and its partner from the other DNA strand with another pair of nucleotides. Depending on how a base-pair substitution is translated via the genetic code, base-pair substitutions can result in no change in the protein encoded by the mutated gene, in an insignificant change in that protein, or in a noticeable alteration, which may be crucial to the life of the organism.

It is because of the redundancy of the genetic code that some substitution mutations have no effect. A change in a base pair may transform one codon into another that is translated into the same amino acid. Consider the case of GAA mutated to GAG. (Although the mutation occurs in the DNA, it is convenient to discuss the result in terms of codons in the messenger RNA transcribed from the DNA.) A glutamic acid would still be inserted at the proper location in the protein. In the genetic code, the most common type of redundancy is the equivalence of codons that differ in the third nucleotide. This pattern acts to minimize the effects of base-pair substitutions.

Other changes of a single nucleotide pair may alter an amino acid but have little effect on the protein encoded. The new amino acid may have properties similar to those of the amino acid it replaces or may be in a portion of the protein where the exact sequence of amino acids is not essential to its activity.

However, the base-pair substitutions most frequently studied by geneticists are those that cause a readily detectable change in a protein. The alteration of a single amino acid in a crucial area of a protein—for example, in the active site of an enzyme—will significantly alter protein activity. Occasionally, such a mutation will lead to an improved protein or one with novel capabilities that enhance the success of the mutant organism and its descendants. But much more often, such mutations are detrimental, creating an inactive protein that may cause the death of the cell.

Substitution mutations are usually **missense mutations.** Although the mutations alter the codons, the new codons still code for amino acids and thus make "sense." But what if a point mutation changes a triplet encoding an amino acid into one of the three triplets that are translated as termination signals? Such a change will result in a polypeptide that is prematurely terminated during translation and is therefore shorter than the polypeptide encoded by the normal gene.

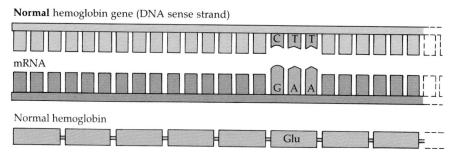

Normal **hemoglobin** gene (DNA sense strand)

mRNA

Normal hemoglobin

Figure 16.20
The molecular basis of sickle-cell anemia. A single base-pair substitution in the sixth triplet of a hemoglobin gene causes a valine to be inserted in sickle-cell hemoglobin, where normal hemoglobin has a glutamic acid. Shown here are the first eight codons of the gene for the β polypeptide chain of hemoglobin, which has a total of 150 amino acids. The hemoglobin molecule has four polypeptides, two α chains and two β chains (see Figure 5.24).

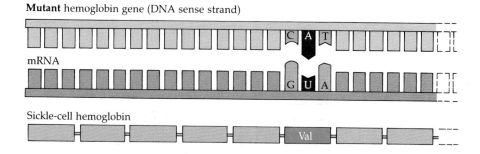

Mutant hemoglobin gene (DNA sense strand)

mRNA

Sickle-cell hemoglobin

Alterations that change an amino acid codon to a stop signal are called **nonsense mutations,** and nearly all nonsense mutations lead to nonfunctional proteins.

Mutations of the second general category involve the **insertion** or **deletion** of one or more nucleotide pairs in a gene. These mutations usually have a more disastrous effect than substitutions on the resulting protein. Because messenger RNA is read as a series of nucleotide triplets during translation, the insertion or deletion of nucleotides may alter the reading frame (triplet grouping) of the genetic message. Such a mutation is called a **frameshift mutation** and will occur whenever the number of nucleotides inserted or deleted is not a multiple of 3. All the nucleotides that are downstream of the deletion or insertion will be improperly grouped into codons, and the result will be extensive missense ending sooner or later in nonsense—premature termination. Unless the frameshift is very near the end of the gene, it will produce a protein that is almost certain to be nonfunctional.

Conditional Mutations

Some mutations are detrimental, or even fatal, to an organism under some conditions but not others. These are called **conditional mutations.** A common type of conditional mutation is a **temperature-sensitive mutation.** The majority of such mutations are detrimental only at elevated temperatures, although some are detrimental only in the cold. Temperature-sensitive mutations are caused by amino acid substitutions that have no obvious effect at one temperature, called the *permissive* temperature, but do change the stability of a protein, making it inactive at another temperature

(usually higher). Temperature-sensitive mutations have been of great value to geneticists because they allow the propagation and study of mutations in genes whose function is essential to the organism. For example, many mutations affecting enzymes essential to protein synthesis are lethal. But if the mutation is temperature-sensitive, it is possible to keep the mutant alive by growing it at the permissive temperature. Then the affected protein can be studied when the temperature is shifted to a nonpermissive one.

Mutagenesis

Mutagenesis, the creation of mutations, can occur in a number of ways. Errors during DNA replication, repair, or recombination can lead to base-pair substitutions, insertions, or deletions. Mutations resulting from such errors are called **spontaneous mutations,** as are other mutations of unknown cause. A number of physical and chemical agents, called **mutagens,** interact with DNA to cause mutations. Table 16.2 lists the main types of mutagens and their effects. The most common physical mutagen in nature and in the laboratory is radiation, and in earlier chapters, we mentioned the mutagenic effects of X-rays and ultraviolet light (see Chapters 13 and 15). Chemical mutagens fall into several categories, including **base analogues,** chemicals that are similar to normal DNA bases but pair incorrectly (Figure 16.22, p. 343).

In addition to physical and chemical mutagens, there are, within the DNA itself, certain DNA segments that can act as mutagens. Scattered throughout the DNA of all cells are segments of DNA that occasionally move out of one DNA site and into another. These mobile

Figure 16.21

Categories and consequences of base-pair mutations. The two major categories of mutations that involve changes in only one or a few base pairs are base-pair substitutions and nucleotide insertions or deletions, which are shown here as they are reflected in mRNA. Base-pair substitutions have a range of effects on the resulting protein, from no effect whatsoever to premature termination of the protein ("stop"). Insertions or deletions usually result in frameshifts (shifting of the translation reading frame) and in extensive missense in the protein, eventually resulting in incorrect termination. The only situation where an insertion or deletion might not be extremely deleterious is if the number of nucleotides added or deleted is a multiple of 3, in which case there would not be a long-term frameshift.

WILD TYPE

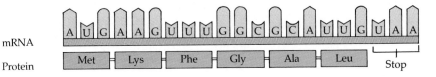

BASE SUBSTITUTION

No effect on amino acid sequence

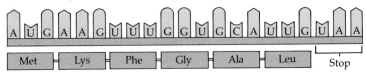

Missense

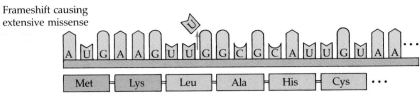

Nonsense

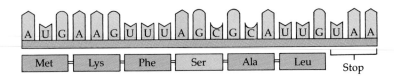

BASE DELETION OR INSERTION

Frameshift causing extensive missense

Frameshift causing immediate nonsense

Insertion or deletion of 3n nucleotides — no extensive frameshift

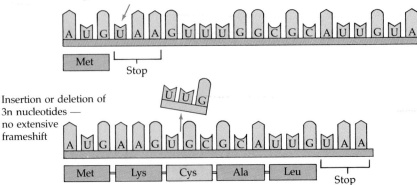
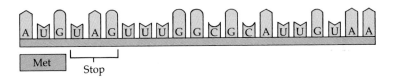

Table 16.2 Categories of mutagens

Mutagen	Effect on DNA	Types of Mutations
Physical agents		
X-rays	Double-strand breaks	Chromosomal rearrangements and deletions
Ultraviolet radiation	Pyrimidine dimers	Base-pair substitutions, insertions, and deletions
Chemical agents		
Base analogues	Substitution for normal bases in DNA synthesis, leading to mispairing	Base-pair substitutions
Reactive chemicals	Addition or deletion of chemical groups to or from normal bases, leading to mispairing during DNA synthesis	Base-pair substitutions
Intercalating chemicals	Insertion of the mutagen between stacked bases in the double helix, leading to small insertions or deletions during DNA synthesis	Frameshift mutations

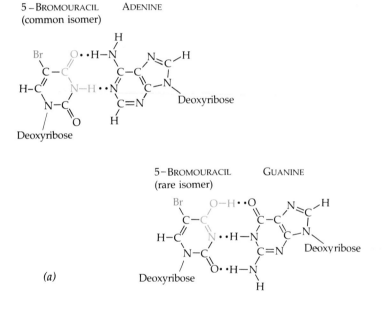

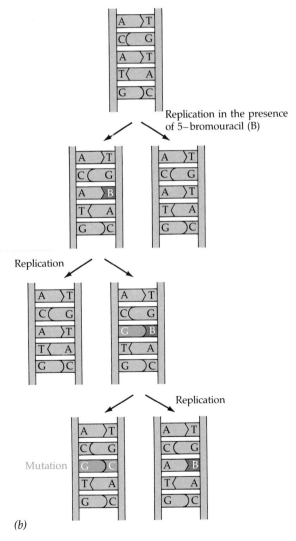

Figure 16.22
Bromouracil, a chemical mutagen that is a base analogue. (a) This mutagen is an analogue of thymine in which the methyl group (CH₃) is replaced by a bromine atom (Br). Its normal isomer pairs with adenine during replication, but sometimes, after incorporation into DNA, the normal isomer shifts to a rare isomeric form that pairs with guanine rather than adenine. **(b)** Here, bromouracil (B) has been incorporated into a DNA strand in place of T. During a replication cycle, the B pairs with G, causing a mutation; now there is a G where there should be an A. During the next replication, the B pairs normally with A, but the damage has already been done. The mutation of A–T to G–C will persist, generation after generation.

Figure 16.23

Overview of transcription and translation. In general, the processes are similar in prokaryotic and eukaryotic cells. The major difference is the occurrence of RNA processing in the eukaryotic nucleus. Eukaryotic RNA processing includes the removal of introns and the addition of a cap and a poly-A tail to the ends of the mRNA.

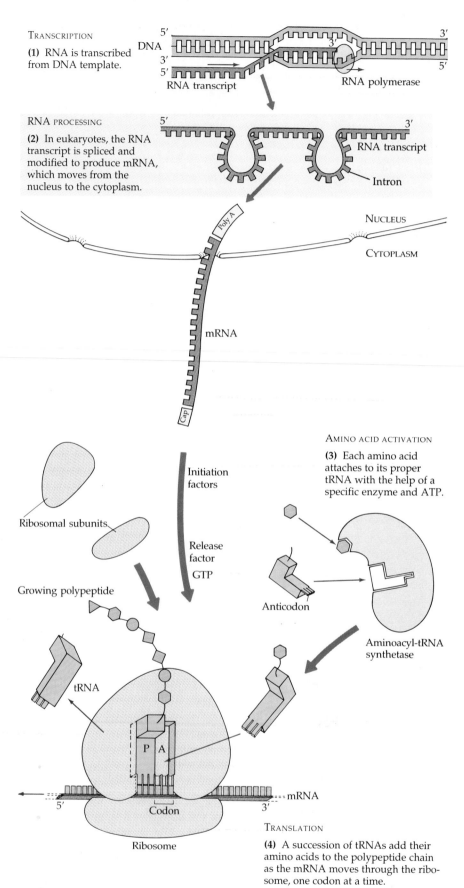

TRANSCRIPTION

(1) RNA is transcribed from DNA template.

DNA

RNA transcript

RNA polymerase

RNA PROCESSING

(2) In eukaryotes, the RNA transcript is spliced and modified to produce mRNA, which moves from the nucleus to the cytoplasm.

RNA transcript

Intron

Poly A

NUCLEUS

CYTOPLASM

mRNA

Cap

AMINO ACID ACTIVATION

(3) Each amino acid attaches to its proper tRNA with the help of a specific enzyme and ATP.

Initiation factors

Ribosomal subunits

Release factor

GTP

Growing polypeptide

Anticodon

Aminoacyl-tRNA synthetase

tRNA

P A

mRNA

5'   3'

Codon

Ribosome

TRANSLATION

(4) A succession of tRNAs add their amino acids to the polypeptide chain as the mRNA moves through the ribosome, one codon at a time.

units, called transposable elements or **transposons,** range in length from a few hundred to tens of thousands of nucleotides. Although transposons have positive functions in the cell (see Chapter 17), these elements can also act as mutagens by disrupting the function of the gene into which they insert. In addition, when transposons move, they can cause deletions of adjacent nucleotides, or they can carry adjacent nucleotides from one site to another.

*   *   *

We have seen in this chapter how genetic information is expressed by its translation into proteins of specific structure and function, which in turn bring about an organism's phenotype. This path from gene to protein is reviewed in Figure 16.23. Genes also produce ribosomal and transfer RNA.

Genes are themselves subject to regulation. The control of gene expression enables a bacterium, for example, to vary the amounts of particular enzymes as the metabolic needs of the cell change. In eukaryotes, the control of gene expression makes it possible for cells with the same DNA to diverge during their development into such different cells as muscle and nerve cells. How gene expression is controlled in eukaryotes is the subject of Chapter 18. The next chapter, Chapter 17, begins our discussion of gene regulation by focusing on the simpler biology of bacteria and viruses.

STUDY OUTLINE

1. DNA controls our inherited characteristics by dispatching instructions via specific RNA molecules, which direct the synthesis of proteins in the cytoplasm.
2. The usual chain of cellular command is from DNA to RNA to protein.

How Genes Control Metabolism (pp. 322–326)

1. The early idea that inherited diseases were a result of "inborn errors of metabolism" was supported by data obtained by Beadle and Tatum on mutant strains of *Neurospora* bread mold. These classic experiments gave rise to the one gene–one enzyme hypothesis.
2. Beadle and Tatum's hypothesis was later modified to one gene–one polypeptide, implying that the function of a gene is to dictate the production of a specific polypeptide chain.

The Languages of Macromolecules (p. 326)

1. Both nucleic acids and proteins are informational polymers assembled from linear sequences of nucleotides and amino acids, respectively.
2. The nucleotide-to-nucleotide transfer of information from DNA to RNA is called transcription, whereas the informational transfer from RNA nucleotides to polypeptide amino acids is called translation.
3. Triplets of the nitrogenous bases in DNA or RNA code for the 20 different amino acids. With 64 total possibilities, there is some redundancy in the genetic code.
4. Experiments have verified that genetic instructions from DNA are written in three-nucleotide units called codons, which indirectly specify particular amino acids.

Transcription, the Synthesis of RNA (pp. 327–328)

1. Transcription produces RNA. In protein synthesis, messenger RNA (mRNA) is an intermediary molecule whose sequence is specified by one or the other of the two DNA strands.
2. RNA synthesis on a DNA template is catalyzed by RNA polymerase. It follows the same base-pairing rules governing DNA replication, except that uracil substitutes for thymine.
3. Promoters, specific nucleotide sequences flanking the start of a gene, signal the initiation of mRNA synthesis. Transcription continues until the RNA polymerase reaches the terminator sequence of nucleotides on the DNA template. As the mRNA peels away, the DNA double helix re-forms.
4. The efficiency of the process is increased by the simultaneous transcription of several mRNA molecules from one gene, utilizing a succession of RNA polymerase molecules.
5. In eukaryotes the transcript must be modified before leaving the nucleus to serve its mRNA function in the cytoplasm.

Translation, the Synthesis of Protein (pp. 328–335)

1. Transfer RNA (tRNA) molecules pick up specific amino acids and line up by means of their anticodon triplets at complementary codon sites on the mRNA molecule, thereby functioning as interpreters between nucleic acid and protein languages.
2. The binding of a specific amino acid to its particular tRNA is a precise, ATP-driven process catalyzed by a family of amino acid activating enzymes.
3. Ribosomes coordinate the coupling of tRNAs to mRNA codons by providing a site for the binding of mRNA and by providing P and A sites for holding adjacent tRNAs during the condensation reactions linking amino acids in the growing polypeptide chain. Each ribosome is composed of two subunits made of aggregates of protein and ribosomal RNA (rRNA).
4. Initiation is the first of three stages of protein synthesis. It is a complex, two-step process that requires energy from GTP and protein initiation factors and which brings together the mRNA, the first amino acid attached to its tRNA, and the two subunits of the ribosome.
5. In the second stage, called the elongation cycle, amino acids are added one by one to the initial amino acid until

the polypeptide chain is completed. This GTP-requiring stage includes the binding of the incoming tRNA to the A site, peptide bond formation, and translocation of the tRNAs and mRNA along the ribosome.

6. Termination, the final stage, occurs when one of three special termination codons reaches the A site of the ribosome, triggering the action of a protein release factor that causes freeing of the polypeptide chain and dissociation of the ribosomal subunits.

7. A protein often undergoes one or more alterations during and after translation that affect its three-dimensional structure and hence its final activity in the cell.

8. In eukaryotic cells, proteins destined for membranes or for export from the cell are synthesized on ribosomes bound to the endoplasmic reticulum, whereas proteins that will remain in the cytosol are manufactured on free ribosomes. Several ribosomes often read a single mRNA, forming polyribosome clusters.

The Genetic Code (pp. 335–337)

1. Artificially constructed mRNA molecules of known nucleotide composition allowed investigators to decipher all 64 codons in the genetic code.

2. Some codons supply start and stop signals, while others serve as synonyms for a particular amino acid.

3. A base in the third position of the anticodon can form pairs with more than one mRNA base in a phenomenon known as wobble, which sometimes involves an unusual base called inosine. The wobble phenomenon accounts for only 40 distinct types of tRNA molecules being able to read 64 different codons.

4. The near universality of the genetic code is a powerful testimony to the evolutionary thread connecting all life. The only variations found in the code so far have been in mitochondria and in some ciliates.

Split Genes and RNA Processing in Eukaryotes (pp. 337–339)

1. Not all the nucleotides in the gene for a protein code for amino acids. In addition to nontranslated sequences at the beginning and end of a gene, most eukaryotic genes are interrupted by long noncoding regions, called introns, that are interspersed among coding regions, known as exons.

2. Processing of eukaryotic RNA involves removing the introns and joining the exons by RNA splicing, a mechanism triggered by sets of nucleotides at either end of the intron.

3. The splicing that produces mRNA is catalyzed by small nuclear ribonucleoproteins (snRNPs), which consist of small nuclear RNA (snRNA) and proteins, and operate within larger assemblies called spliceosomes.

4. In some cases, only RNA is needed to catalyze RNA splicing.

5. Eukaryotic mRNA molecules also receive a modified guanosine triphosphate cap at the 5′ end and a poly-A stretch of nucleotides at the 3′ end, which probably protect the molecule from degradation and enhance translation.

6. Introns are thought to play a regulatory role in the cell and/or facilitate recombination between exons to produce novel proteins. Complex organisms have more and longer introns than simpler organisms.

RNA: A Review (pp. 339–340)

1. The four major types of cellular RNA are mRNA, rRNA, tRNA, and snRNA; each has a different structure and function.

2. RNA can sometimes act as an enzyme, either alone or in association with protein.

Mutations and Their Effects on Proteins (pp. 340–345)

1. Mutations range from changes affecting large sections of a chromosome to point mutations involving changes in a single nucleotide pair.

2. Base-pair substitutions within a gene have a variable effect, depending on whether or not an amino acid is actually altered, and if so, whether the alteration has any effect on the function of the protein. Many substitutions are detrimental, causing missense or nonsense mutations.

3. Base-pair insertions or deletions are almost always disastrous, often resulting in frameshift mutations that disrupt the codon messages downstream of the insertion or deletion.

4. Conditional mutations are harmful to an organism only under certain conditions, such as high or low temperatures in the case of temperature-sensitive mutations.

5. Spontaneous mutations can occur during DNA replication or repair. In addition, various chemical and physical mutagens, and mobile DNA transposons as well, can affect the integrity of the gene.

SELF-QUIZ

1. The creation of an RNA molecule from a section of DNA is known as

   a. transcription

   b. translation

   c. RNA splicing

   d. replication

   e. transposition

2. Which of the following is *not* true of a codon?

   a. It consists of three nucleotides.

   b. It may code for the same amino acid as another codon does.

   c. It never codes for more than one amino acid.

   d. It extends from one end of a tRNA molecule.

   e. It is the basic unit of the genetic code.

3. RNA polymerase

   a. transcribes both DNA strands, but always in a 5′ → 3′ direction

   b. transcribes both introns and exons

   c. creates hydrogen bonds between nucleotides on the DNA strand and their complementary RNA nucleotides

   d. starts transcribing at an AUG triplet on one DNA strand

   e. can produce several polypeptide chains at one time through the creation of polysomes

4. Beadle and Tatum discovered several classes of *Neurospora* mutants that were able to grow on minimal medium

with arginine added. Class I mutants were also able to grow on medium supplemented with either ornithine or citrulline, whereas class II mutants could grow on citrulline medium but not on ornithine medium. The metabolic pathway of arginine synthesis is as follows:

Precursor → ornithine → citrulline → arginine
         A   B   C

From these growth results, they could conclude that

a. one gene codes for one polypeptide chain

b. the genetic code of DNA is a triplet code

c. class I mutants have their mutations later in the nucleotide chain than do class II mutants, and thus have more functional enzymes

d. class I mutants have a nonfunctional enzyme at step A, and class II mutants have a nonfunctional enzyme at step B

e. class I mutants have a nonfunctional enzyme at step B, and class II mutants have a nonfunctional enzyme at step C

5. The bonds between the anticodon of a tRNA molecule and the complementary codon of mRNA are

a. catalyzed by peptidyl transferase

b. formed by the input of energy from ATP

c. hydrogen bonds that form while the codon is in the A site

d. catalyzed by aminoacyl-tRNA synthetase

e. covalent bonds formed with energy from GTP

6. The phenomenon known as wobble refers to

a. the movement of a tRNA from the A to the P site

b. the redundancy of the genetic code

c. the ability of a tRNA to pair with different codons that may differ in the third base

d. the shifting of the reading frame in a deletion or insertion mutation

e. the movement of multiple ribosomes along the same mRNA

7. Which of the following is *not* true of RNA processing?

a. Exons are excised and hydrolyzed before mRNA moves out of the nucleus.

b. The existence of exons and introns may facilitate crossing over between regions of a gene that code for polypeptide domains.

c. Simpler organisms exhibit less RNA processing.

d. RNA splicing may be catalyzed by spliceosomes.

e. An initial RNA transcript is much longer than the final RNA molecule that may leave the nucleus.

8. Using the genetic code in Figure 16.14, identify a pos-

sible sequence of nucleotides in the *DNA* that would code for the polypeptide sequence Phe–Pro–Lys.

a. AAA–GGG–UUU

b. TTC–CCC–AAG

c. TTT–CCA–AAA

d. AAG–GGC–TTC

e. UUU–CCC–AAA

9. Which of the following mutations would be most likely to have a harmful effect on an organism?

a. a base-pair substitution

b. a conditional mutation

c. a single base deletion near the middle of an intron

d. a single base deletion close to the end of the coding sequence

e. a single base insertion near the start of the coding sequence

10. Which component is *not directly* involved in the process known as translation?

a. mRNA

b. DNA

c. tRNA

d. ribosomes

e. GTP

CHALLENGE QUESTION

1. According to one hypothesis, the common ancestors of all cells, called progenotes, had intron-rich DNA, as well as ribosomes to translate the genetic information into proteins using the now virtually universal genetic code. From your understanding of gene splicing, what do you think was the advantage of these introns, and presuming the progenote theory is correct, why might bacteria but not eukaryotes have lost their introns? (See Further Reading 3.)

FURTHER READING

1. Darnell, J. E., Jr. "RNA." *Scientific American*, October 1985. A clear presentation of the role of RNA in protein synthesis and its relationship to DNA.
2. deDuve, C. "The Second Genetic Code." *Nature* 333 (1988): 117–118. Recent experiments concerning how amino acid–activating enzymes recognize the correct tRNA.
3. Gilbert, W. "Genes-in-Pieces Revisited." *Science* 228 (1985):823-824. A case for modular evolution by mix-and-match exons.
4. Moore, P. B. "The Ribosome Returns." *Nature* 331 (1988): 223–227. A refreshingly written overview of ribosome biochemistry and the recent reawakening of interest in these organelles.
5. Steitz, J. A. " 'Snurps.' " *Scientific American*, June 1988. The story of the discovery of snRNPs and their role in the splicing of eukaryotic mRNA.
6. Watson, J. D., N. H. Hopkins, J. W. Roberts, J. A. Steitz, and A. M. Weiner. *Molecular Biology of the Gene.* 4th ed. Menlo Park, Calif.: Benjamin/Cummings, 1987, Chaps. 13–16, 20.

17

The Genetics of Viruses and Bacteria

It was by studying viruses and bacteria that we got our first glimpses of the elegant molecular mechanisms of heredity. Scientists unraveled the threads of genetics using these simplest of biological systems, where biochemistry and genetics are in their most basic, accessible forms.

Discovering the role of DNA in heredity would have been much more difficult had scientists worked solely with peas, fruit flies, or humans. And while most of the molecular principles discovered through research with microbes apply equally well to higher organisms, viruses and bacteria have unique genetic features that make the study of microbial genetics interesting in its own right. These unique mechanisms also have important applications for understanding how viruses and bacteria cause disease. In addition, powerful new techniques for manipulating genes in the laboratory have emerged from the study of the genetic peculiarities of microorganisms—techniques that have had a major impact on both basic research and biotechnology (see Chapter 19).

In this chapter, we will explore the fundamental structures and processes involved in the genetics of viruses and bacteria. We will begin with viruses, particles that are little more than genes in packages (Figure 17.1). Because viruses are essentially genetic entities and not organisms in the usual sense, this book's main discussion of them is located in this chapter.

THE DISCOVERY OF VIRUSES

Microbiologists were able to observe viruses indirectly long before they were actually able to see them. The story of how viruses were discovered begins in 1883

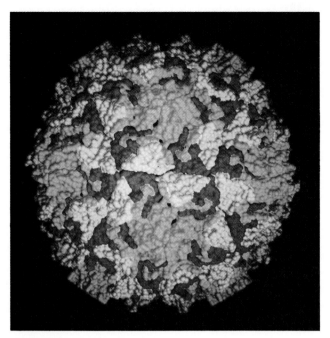

Figure 17.1
Poliovirus. This computer graphic shows a poliovirus, a small virus about 32 nm in diameter. The different colors represent the different proteins that comprise the surface of the virus.

Leaves are mottled by tobacco mosaic disease.

Grind up leaves and filter the sap released.

Spray filtered sap on a healthy plant.

Repeat the procedure.

Figure 17.2
Characteristics of a viral infectious agent. The infectious agent of tobacco mosaic disease is smaller than a bacterium, but still able to reproduce. When sap from infected tobacco leaves (pot A) is passed through a filter designed to remove bacteria, it is still able to produce disease. The infection can be spread by spraying healthy plants with the filtered sap, and once those plants become diseased, spraying other plants with their sap. The pathogen must be reproducing, because its ability to infect is undiluted by several transfers from plant to plant.

with A. Mayer, a German scientist who was seeking the cause of tobacco mosaic disease. This disease stunts the growth of tobacco plants and gives their leaves a mottled, or mosaic, coloration. Mayer discovered that the disease was contagious when he found he could transmit it from plant to plant by spraying sap extracted from diseased leaves onto healthy plants. Mayer searched for a microbe in the infectious sap, but found none. He concluded that the disease was caused by unusually small bacteria that could not be seen with the microscope. This hypothesis was tested a decade later by D. Ivanowsky, a Russian, who passed sap from infected tobacco leaves through a filter designed to remove bacteria. But even after filtering, the sap produced mosaic disease.

Ivanowsky clung to the hypothesis that bacteria caused tobacco mosaic disease. Perhaps, he reasoned, the pathogenic bacteria were so small they could pass through the filter. Or perhaps the bacteria made a filterable toxin that caused the disease. This latter possibility was ruled out in 1897 when Dutch microbiologist Martinus Beijerinck discovered that the infectious agent in the filtered sap could reproduce. Beijerinck sprayed plants with the filtered sap, and after these plants developed mosaic disease, he used their sap to infect more plants, and continued this process through a series of infections (Figure 17.2). The pathogen must have been reproducing, for its ability to cause disease was undiluted after several transfers from plant to plant.

In fact, the pathogen could reproduce only within

the host it infected. Unlike bacteria, the mysterious agent of mosaic disease could not be cultivated on nutrient media in test tubes or Petri dishes. Also, the pathogen was not "killed" by alcohol, which is generally lethal to bacteria. Beijerinck imagined a reproducing particle much smaller and simpler than bacteria. His suspicions were confirmed in 1935 when the American scientist Wendell M. Stanley crystallized the infectious particle, now known as *tobacco mosaic virus (TMV)*. Subsequently, TMV and many other viruses were actually seen with the help of the electron microscope (Figure 17.3a). Let us now examine the architecture of viruses.

VIRAL STRUCTURE

The tiniest viruses are only 20 nm in diameter—smaller than a ribosome. Millions could easily fit on a pinhead, and even the largest viruses can barely be resolved with the light microscope. Stanley's discovery that viruses could be crystallized was exciting and puzzling news. Not even the simplest of cells can aggregate into regular crystals. But if viruses are not cells, then what are they? The virus particle, or virion, is, in its simplest form, just nucleic acid enclosed in a protein shell.

Viral Genomes

We are used to thinking of genes as being made of double-stranded DNA—the conventional double helix; but viruses often defy this convention. Their **genomes** (genetic material) may consist of double-stranded DNA, single-stranded DNA, double-stranded RNA, or single-stranded RNA. Which of these four forms of nucleic acid serves as the genetic material depends on the specific virus. The viral genome is usually organized as a single molecule of nucleic acid that is either linear or circular. The smallest viruses have as few as four genes, while the largest have several hundred.

Capsids and Envelopes

The protein shell that encloses the viral genome is called a **capsid** and may be rod-shaped (more precisely, helical), polyhedral, or complex in structure. Capsids are built from a large number of protein subunits, with the number of different *kinds* of proteins usually small. Tobacco mosaic virus, for example, has a rigid, rod-shaped capsid made from over a thousand molecules of a single type of protein. Adenoviruses, which infect the respiratory tracts of animals, have 252 identical protein molecules arranged into a polyhedral capsid with 20 triangular facets—an icosahedron (Figure 17.3b).

Some viruses have accessory structures that help them to infect their hosts. Flu viruses and many other viruses found in animals have membranous **enve-**lopes cloaking their capsids (Figure 17.3c). This envelope is derived from membrane of the host cell, but in addition to host cell phospholipids and proteins, it also contains proteins and glycoproteins (proteins with carbohydrate covalently attached) of viral origin.

The most complex capsids are found among viruses that infect bacteria (Figure 17.3d). Bacterial viruses are called **bacteriophages** (meaning "bacteria eaters"), or phages. The first phages studied included seven that infect the bacterium *Escherichia coli*. These seven phages were named type 1 (T1), type 2 (T2), type 3 (T3), and so forth. By coincidence, the three "T-even" phages—T2, T4, and T6—turned out to be very similar in structure (Figure 17.4). Their capsids have icosahedral (20-sided) heads that enclose the genetic material. Attached to this head is a protein tailpiece with tail fibers that the phage uses to attach to a bacterium.

THE REPLICATION OF VIRUSES

An isolated virion is an inert particle because viruses lack the metabolic equipment described in Unit Two as essential components of living cells. Viruses are *obligate intracellular parasites*, and as such can express their genes and reproduce only within a living cell. Not surprisingly, the reproduction of a virus differs from that of a cell in many ways. Most striking is the production of hundreds or thousands of progeny in each viral generation. The viral genes use the host cell's enzymes, ribosomes, nutrients, and other resources to make many copies of the viral genome and the viral capsid proteins. As these components accumulate, they are assembled into a large number of virions, which then leave the cell to parasitize new hosts.

Genome Replication

Exactly what happens after viral nucleic acid enters a host cell depends on the particular virus and, in particular, the form of its genome. If the viral genome is double-stranded DNA, then nucleic acid replication will generally resemble that of the cellular genes. However, if the genome is not double-stranded DNA, the replicative process will be somewhat different. As you might expect, the genomes of such viruses usually encode one or more novel enzymes for use in nucleic acid replication. The enzymes of RNA viruses are especially interesting. Most RNA viruses carry a gene for an **RNA replicase,** an enzyme that uses viral RNA as a template to make complementary RNA strands. Some RNA viruses encode an unusual enzyme called **reverse transcriptase,** which uses RNA as a template for *DNA* synthesis; the DNA is later transcribed into both messenger RNA and genomic RNA

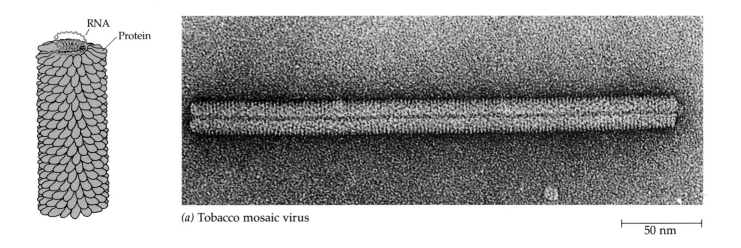

(a) Tobacco mosaic virus

50 nm

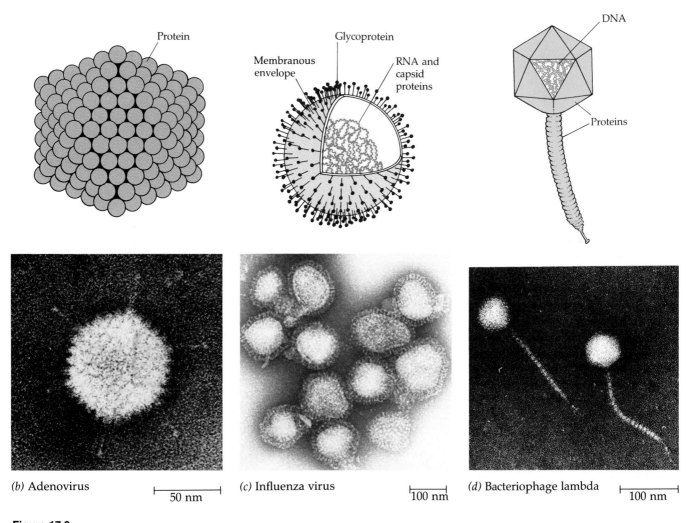

(b) Adenovirus   50 nm

(c) Influenza virus   100 nm

(d) Bacteriophage lambda   100 nm

Figure 17.3
Viral structure. Viruses are made up of nucleic acid (DNA or RNA) enclosed in a protein coat (the capsid) and sometimes further wrapped in a membranous envelope. The individual protein subunits making up the capsid are called capsomeres. Although viruses are diverse in size and shape, there are common structural motifs, most of which appear in the four examples shown here. (**a**) Tobacco mosaic virus has a helical capsid with the overall shape of a rigid rod. (**b**) Adenovirus has a polyhedral capsid. The electron micrograph shows that this virus also has a pro-tein "spike" at each vertex. (**c**) Influenza virus has a flexible helical capsid and an outer membranous envelope studded with glycoprotein spikes. (**d**) Bacteriophage lambda has a complex capsid consisting of a polyhedral head and a flexible rod-shaped tail. (TEMs.)

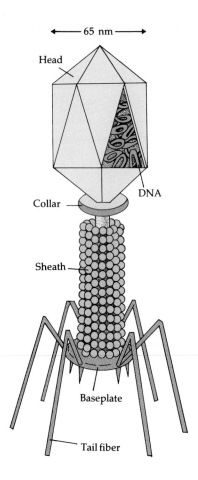

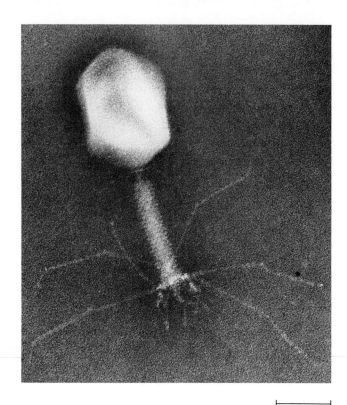

Figure 17.4
Structure of the T-even phage. The genetic material is encased in the head of the phage. The other parts of this complex virus function to inject the DNA into a bacterium. (TEM.)

50 nm

for incorporation into new virions. (These viruses, which include some that cause cancer and AIDS, will be described later in the chapter.) Thus, viral genome replication follows three different overall patterns: DNA→DNA, RNA→RNA, and RNA→DNA→RNA.

Self-Assembly of Virus Particles

The assembly of viral nucleic acid and capsid proteins into new virus particles is a spontaneous process, much like the formation of a functional protein from one or more polypeptide chains. Since the bonds that hold the viral components together are almost always weak bonds (such as hydrogen bonds and van der Waals forces), rather than covalent bonds, enzymes are not usually involved in their formation. The RNA and capsid protein molecules of TMV, for example, can be separated in the laboratory and then reassembled to form perfect TMV virions, simply by mixing the components together again.

Host Specificity

Each type of virus can infect and successfully parasitize only a limited range of host cells, called its **host**

range. Viruses recognize their host cells by a "lock-and-key" fit between proteins on the outside of the virion and specific **receptor sites** on the surface of the cell. (Similar complementary mechanisms are used by enzymes to recognize their substrates and by sperm cells to identify eggs of the same species.) Some viruses have host ranges broad enough to include several species. Swine flu virus, for example, can infect both hogs and humans, and the rabies virus can infect a number of mammalian species, including rodents, dogs, and humans. On the other hand, the host range may be as narrow as a single species or a single type of tissue within a species. For instance, there are a number of phages that can infect only the bacterium *E. coli*, and human cold viruses usually infect only the cells lining the human upper respiratory tract, ignoring other tissues.

BACTERIAL VIRUSES

The bacteriophages have played a special role in molecular biology. They are the best understood of all viruses, although some of them are also among the most complex. These viruses, discovered in 1915, came into the forefront of research in the 1940s when sci-

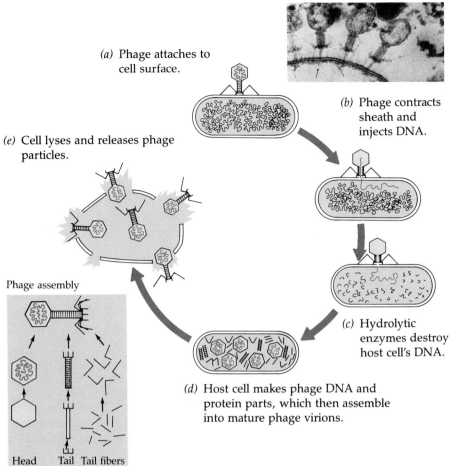

(a) Phage attaches to cell surface.

(b) Phage contracts sheath and injects DNA.

(e) Cell lyses and releases phage particles.

Phage assembly

Head   Tail   Tail fibers

(c) Hydrolytic enzymes destroy host cell's DNA.

(d) Host cell makes phage DNA and protein parts, which then assemble into mature phage virions.

Figure 17.5 The lytic cycle of phage T4.
(a) The T4 virion uses its tail fibers to stick to specific receptor sites on the outer surface of an *E. coli* cell. (b) The sheath of the tail then contracts, thrusting a hollow core through the wall and membrane of the cell. The phage injects its DNA into the cell. (This step can also be seen in the electron micrograph.) (c) The empty capsid of the phage is left as a "ghost" outside the cell. The cell's DNA is hydrolyzed. (d) The cell's metabolic machinery, directed by phage DNA, produces phage proteins, and nucleotides from the cell's degraded DNA are used to make copies of the phage genome. The phage parts come together. Three separate sets of proteins assemble to form phage heads, tails, and tail fibers. (e) The phage then directs production of an enzyme that digests the bacterial cell wall. With a damaged wall, osmotic pressure causes the cell to swell and finally to burst, releasing 100 to 200 phage virions.

entists focused on the seven T phages to determine how phages reproduce within a bacterium. This research helped demonstrate that DNA is the genetic material (see Chapter 15). The study of another phage of *E. coli*, called lambda (λ), led to the discovery that double-stranded DNA viruses can reproduce by two alternative mechanisms, the lytic cycle and the lysogenic cycle.

The Lytic Cycle

Bacteriophages that kill their host cells are said to be **virulent,** and their type of replication cycle is known as the **lytic cycle.** Although they do not actually eat bacteria, phages reproducing by the lytic cycle *lyse* (break open) their host cells and can destroy a bacterial colony in just hours. We will use phage T4 to illustrate the steps of this cycle (Figure 17.5).

The lytic cycle begins when the tail fibers of a T4 virion stick to specific receptor sites on the outer surface of an *E. coli* cell. The sheath of the tail then contracts, thrusting a hollow core through the wall and membrane of the cell. A single phage uses 140 mole-

cules of ATP stored in its tailpiece to pierce the *E. coli* wall and membrane, contract its tail sheath, and, acting like a miniature syringe, inject its DNA into the cell. The empty capsid of the phage is left as a "ghost" outside the cell.

Once infected by the phage DNA, the *E. coli* cell quickly begins to transcribe and translate the viral genes. Phage T4 has about 100 genes, and most of their functions are known. One of the first phage genes translated by the *E. coli* cell codes for an enzyme that chops up the host cell's own DNA. The phage DNA itself is protected, apparently because it contains a modified form of cytosine.

After the host DNA is destroyed, the phage genome gains full control of the cell, inducing its metabolic machinery to produce phage components. Nucleotides salvaged from the cell's degraded DNA are used to make many copies of the phage genome. Three separate sets of capsid proteins are made and assembled into phage tails, tail fibers, and polyhedral heads. The phage completes its subversion of the cell when one of its genes directs production of an enzyme (lysozyme) that digests the bacterial cell wall. With a damaged wall, osmotic pressure causes the cell to swell

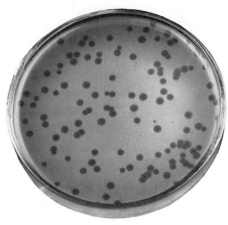

Figure 17.6
Phage plaques. Each plaque (clear spot) in the layer of bacteria growing on this Petri plate results from the lysis of bacteria by the descendants of a single phage virion present in the original sample.

and finally to burst. The lysed bacterium releases 100 to 200 phage particles, which can then initiate another turn of the cycle by infecting other cells nearby.

The entire lytic cycle, from the phage's contact with the cell surface to lysis, takes only 20–30 minutes at 37°C. In that time, a T4 population can increase in size more than a hundredfold, whereas even a fast-growing *E. coli* population can only double within that period. Thus, phage reproduction greatly outpaces growth of a bacterial colony. If we add a single T4 particle to a susceptible culture of *E. coli* growing in a thin layer on a solid medium in a Petri dish, an expanding hole will soon appear in the cloudy bacterial "lawn" (Figure 17.6). The hole, called a **plaque,** results from the lysis of cells by successive generations of phage particles. In fact, we can measure the concentration of phage particles in a liquid sample by mixing part of the sample with a suspension of bacteria, spreading the mixture onto a solid medium, and later counting the number of plaques.

After reading about the lytic cycle, you may wonder why phages haven't exterminated all bacteria. Actually, bacteria are not defenseless. Bacterial mutations can change the receptor sites used by a phage, thus preventing infection. And when phage nucleic acid successfully enters a cell, cellular enzymes may break it down. One important class of degradative enzymes is the **restriction enzymes** (see Chapter 19), which recognize and cut up DNA that is foreign to the cell, including certain phage DNA. Only phages that are somehow resistant to such enzymes can successfully infect a bacterial cell having them. Bacterial hosts and their viral parasites are continually *coevolving*, each changing in response to changes in the other. The most successful bacteria have effective mechanisms for preventing phage entry or reproduction, while the

most successful phages have evolved ways to get around such defenses.

There is still another important reason why bacteria have been spared from extinction due to phage activity. Many phages can check their own destructive tendencies and, instead of lysing their host cells, coexist with them.

The Lysogenic Cycle

Viruses that can reproduce without killing their hosts are called **temperate viruses.** They have two possible modes of reproduction, the lytic cycle and the nonlethal alternative, the **lysogenic cycle.** The details of the lysogenic pathway have been revealed largely through studies of phage λ, which was discovered by Esther Lederberg in 1951. Like the T-even phages, λ has a tail and a polyhedral head with a DNA core, but it has only one tail fiber (see Figure 17.3d).

The life cycle of λ begins when the phage binds to the surface of an *E. coli* cell and injects its DNA (Figure 17.7), much as T4 does. The λ DNA forms a circle and then embarks on one of the two pathways. Either it immediately turns the cell into a virus-producing factory (the lytic cycle) or it inserts by genetic recombination into a specific site in the bacterial chromosome (the lysogenic cycle). Once inserted, the phage genome is referred to as a **prophage.** In the prophage, most of the genes are inactive. However, at least one prophage gene is always active: It codes for a **repressor protein** that keeps most of the other prophage genes switched off. During cell reproduction, the host cell copies the prophage genes along with its cellular DNA and then, upon dividing, passes on both the prophage and cellular DNA to the two daughter cells, which can start the reproductive cycle all over again. A single infected cell can soon give rise to a large population of bacteria carrying prophages.

Rarely, a prophage may leave the bacterial chromosome spontaneously, or departure may be triggered by such environmental conditions as radiation or certain chemicals; the process is the reverse of insertion. Once begun, this excision process usually leads to the phage's lytic cycle. The virions produced may, upon infecting new hosts, use the lytic cycle to reproduce or join bacterial chromosomes as prophages.

Because a host cell carrying a prophage in its chromosome has the potential to lyse and release phages, it is called a **lysogenic cell.** Sometimes, the few prophage genes expressed in a lysogenic cell result in a change in phenotype of the bacterium, a process called **lysogenic conversion.** Some cases of lysogenic conversion have medical significance. For example, the bacteria that cause diphtheria, botulism, and scarlet fever would be harmless were it not for the expression of prophage genes that they carry. Certain prophage

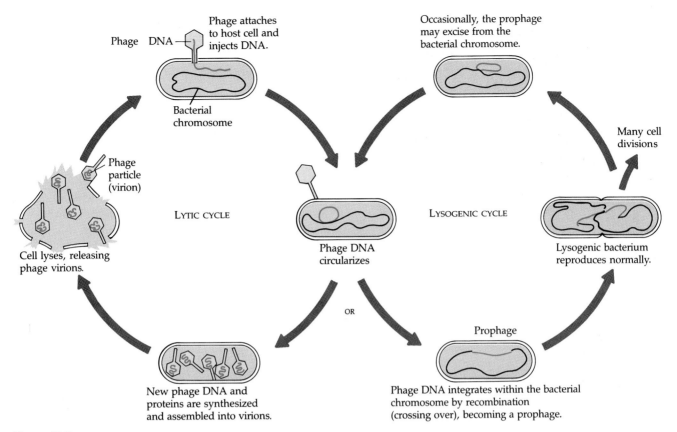

Phage DNA

Phage attaches to host cell and injects DNA.

Phage

Bacterial chromosome

Occasionally, the prophage may excise from the bacterial chromosome.

Phage particle (virion)

LYTIC CYCLE

LYSOGENIC CYCLE

Many cell divisions

Cell lyses, releasing phage virions.

Phage DNA circularizes

Lysogenic bacterium reproduces normally.

OR

Prophage

New phage DNA and proteins are synthesized and assembled into virions.

Phage DNA integrates within the bacterial chromosome by recombination (crossing over), becoming a prophage.

Figure 17.7
The lysogenic and lytic reproductive cycles of phage λ. After entering the bacterial cell, the λ DNA can either integrate into the bacterial chromosome (lysogenic cycle) or immediately initiate the production of a large number of progeny phage particles (lytic cycle). In most cases, the lytic pathway is followed, but once lysogeny occurs, the prophage may be carried in the host cell's chromosome for many generations.

genes direct bacterial production of the toxins that are directly responsible for making people ill.

PLANT VIRUSES AND VIROIDS

Plant viruses are serious agricultural pests that stunt plant growth and diminish crop yields. Most plant viruses discovered so far are RNA viruses. Many of them, including the tobacco mosaic virus, have rod-shaped capsids with capsomeres arranged in a spiral.

There are two major routes by which a plant viral disease may spread. By the first route, called **horizontal transmission,** a plant receives the virus from an external source. Since the invading virus must get past the plant's outer protective layer of cells (the epidermis), the plant becomes more susceptible to viral infections if it has been damaged by wind, chilling, injury, or insects. Insects are a double threat, since they often also act as carriers, or **vectors,** of viruses, transmitting disease from plant to plant. Farmers and

gardeners themselves may transmit plant viruses inadvertently on pruning shears and other tools. The other route of viral infection is **vertical transmission,** in which a plant inherits a viral infection from its parent. Vertical transmission can occur in asexual propagation (for example, by taking cuttings) or in sexual reproduction via infected seeds.

Once a virus enters a plant cell and begins reproducing, virus particles can spread throughout the plant by passing through plasmodesmata, the cytoplasmic connections that penetrate the walls between adjacent plant cells (see Figure 7.36). Agricultural scientists have found no cure for most viral diseases of plants. Therefore, their efforts have largely been focused on reducing the incidence and propagation of such diseases and on breeding genetic varieties of crop plants that are resistant to certain viruses.

As small and simple as viruses are, they dwarf another class of plant pathogens called **viroids.** These are tiny molecules of naked RNA, only several hundred nucleotides in length. Somehow, these RNA molecules can foul up the metabolism of a plant cell and

Figure 17.8
Viroids. This coconut palm plantation in the Philippines has been devastated by a disease called cadang-cadang, which is caused by a viroid.

17.1 (p. 358) is a list of some important families of animal viruses. Like all viruses, those that cause illness in humans and other animals are obligate intracellular parasites that can reproduce only after infecting host cells.

Replication Cycles of Animal Viruses

The replication cycles of animal viruses show many similarities to those of viruses that infect other organisms—and some interesting variations. Consider, for example, the paramyxoviruses, which include the viruses that cause measles and mumps. The paramyxovirus has a genome of single-stranded RNA enclosed in a flexible helical capsid. Enclosing the capsid is a membranous envelope—a feature common to several groups of animal viruses but absent in plant viruses and phages. The envelope helps the virus enter and leave the host cell, which, unlike plant cells and bacteria, lacks a cell wall. When the virus contacts an appropriate cell, glycoprotein "spikes" protruding from the viral envelope attach to receptor sites on the cell's plasma membrane, and the envelope fuses with the plasma membrane (Figure 17.9). This process transports the capsid and its RNA contents into the cytoplasm, where the genome is uncoated. Viral enzymes must be used to replicate the RNA genome and make mRNA (for this group of viruses, the genome is a strand complementary to mRNA), but the host cell's machinery is used for protein synthesis. New capsids assemble around viral genomes and then leave the cell by cloaking themselves in material that buds from the plasma membrane. The virion can then use its envelope to fuse with a new host cell. The entire replicative cycle is called a **productive cycle,** not a lytic cycle, because these viruses can exit by budding from a cell without destroying it.

Not all viral envelopes are derived from plasma membrane. The envelopes of herpesviruses, for example, are derived from the nuclear membrane of the host. The genomes of herpesviruses are double-stranded DNA, and these viruses reproduce within the cell nucleus, using a combination of viral and cellular enzymes to replicate and transcribe their DNA.

While within the nucleus, herpesvirus DNA may be able to integrate into the cell's genome as a **provirus,** similar to a bacterial prophage. The present evidence for integration of herpesvirus DNA is largely clinical: Once acquired, herpes infections (including cold sores and genital sores) tend to recur throughout a person's life. Clearly, the virus somehow remains latent within the body. From time to time, physical stress, such as sunburn, or emotional stress may cause the herpes proviruses to begin a productive cycle, resulting in unpleasant symptoms.

stunt the growth of the whole plant. One viroid disease has killed over ten million coconut palms in the Philippines (Figure 17.8). Another viroid nearly wiped out the chrysanthemum industry in the United States before growers began keeping stock plants in sterile greenhouses. Viroids also threaten potato and tomato crops.

Clues to what viroids do within cells may emerge from an intriguing discovery that relates viroids to certain normal eukaryotic genes, including rRNA genes (among others). The nucleotide sequences of viroid RNA turn out to be extremely similar to the sequences of introns found within those genes—introns that can excise themselves from their RNA transcript without help from enzymatic proteins. This similarity has led some scientists to conclude that viroids originated as "escaped introns." The alternative would be that viroids and the self-splicing introns both evolved from a common ancestor molecule.

In any case, it seems likely that viroids somehow cause errors in the regulatory systems that control the genes of the cell. Indeed, the symptoms that are typically associated with viroid diseases are abnormal development and stunted growth.

ANIMAL VIRUSES

Everyone has suffered from viral infections, whether chicken pox, influenza, or the occasional cold. Table

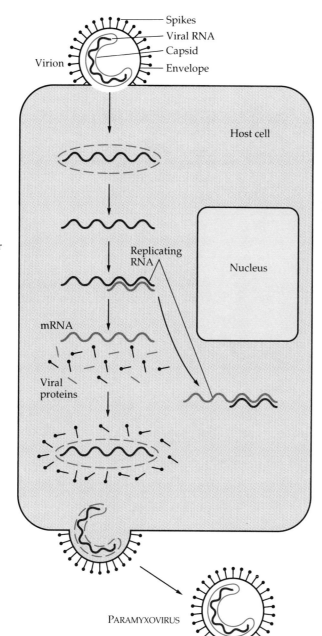

Attachment and entry:
Virion envelope attaches to the cell membrane and fuses with it.

Uncoating:
In the cytoplasm, the capsid is removed from the viral RNA.

Viral RNA and protein synthesis:
The single-stranded RNA genome is the template for the synthesis of complementary strands, which serve both as mRNA and as template for new genome RNA. Viral enzymes catalyze all this RNA synthesis. The cell's protein-synthesizing machinery makes new viral proteins.

Assembly and release:
New capsids assemble around viral RNA and "bud" from the plasma membrane.

Spikes
Viral RNA
Capsid
Envelope
Virion
Host cell
Replicating RNA
Nucleus
mRNA
Viral proteins
PARAMYXOVIRUS

Figure 17.9
Replicative cycle of an enveloped RNA virus. The example depicted here is for a paramyxovirus, such as the virus that causes measles or mumps. The membranous envelope of this virus, derived from the plasma membrane of the previous host cell, facilitates viral infection by fusing with the plasma membrane of the new host cell. (Some other enveloped viruses, as well as all "naked" animal viruses, enter the host cell by endocytosis, rather than by fusion.) The capsid proteins must be removed from the viral genome for it to become active within the cell. Because the viral genome is RNA, viral enzymes are needed to replicate it.

Viral Diseases in Animals

The link between a viral infection and the symptoms it produces is often obscure. Some viruses may damage or kill cells, perhaps by causing the release of hydrolytic enzymes from lysosomes. Infected cells may also produce toxins that lead to certain symptoms, or components of the virions themselves may be toxic. How much damage a virus causes depends partly on the ability of the infected tissue to regenerate by cell division. We usually recover completely from colds because the viruses infect the epithelium of the respiratory tract, and this tissue can efficiently repair itself.

In contrast, the poliovirus attacks nerve cells, which do not divide and cannot be replaced. Polio's damage to such cells, unfortunately, is permanent. Many of the temporary symptoms associated with viral infections, such as fever, aches, and inflammation, may actually result from the body's own efforts at defending itself against the infection.

As you will see in Chapter 39, the immune system is a complex and critical part of the body's natural defense mechanisms and is the basis for the major weapon used by modern medicine to fight viral infections—vaccines. **Vaccines** are harmless variants or derivatives of pathogenic microbes that stimulate the

Table 17.1 Families of animal viruses, grouped by type of nucleic acid

Family	Virion Structure	Diameter (nm)	Examples/Diseases
dsDNA*			
Papovavirus	Naked polyhedral	40–57	Papilloma (human warts, cervical cancer); polyoma (tumors in certain animals); simian virus (tumors)
Adenovirus	Naked polyhedral	70–80	Viruses that cause respiratory disease; some that cause tumors in certain animals
Herpesvirus	Enveloped polyhedral	150–250	Herpes simplex I (cold sores); herpes simplex II (genital); varicella zoster (chicken pox, shingles); Epstein-Barr virus (infectious mononucleosis, Burkitt's lymphoma)
Poxvirus	Enveloped complex	200–350	Variola (smallpox); vaccinia; cowpox
ssDNA			
Parvovirus	Naked polyhedral	18–26	Most depend on coinfection with adenoviruses for growth
ssRNA that can serve as mRNA (+ strand RNA)			
Picornavirus	Naked polyhedral	18–38	Poliovirus; rhinovirus (common cold); enteric viruses
Togavirus	Enveloped polyhedral	40–60	Rubella virus; yellow fever virus; encephalitis viruses (transmitted by insects)
Retrovirus	Enveloped polyhedral; two copies of genome per virion	100–120	RNA tumor viruses (solid tumors and leukemia); AIDS
ssRNA that is a template for mRNA (− strand RNA)			
Rhabdovirus	Enveloped helical	70–180	Rabies
Paramyxovirus	Enveloped helical	150–300	Measles, mumps
Orthomyxovirus	Enveloped helical; RNA in eight segments	80–200	Influenza viruses
dsRNA			
Reovirus	Naked polyhedral; RNA in ten segments	60–80	Diarrhea viruses

*ds = double-stranded; ss = single-stranded.

immune system to mount defenses against the actual pathogen. A vaccine has almost completely eradicated smallpox, which was once a devastating scourge in many parts of the world. There are also effective vaccines against many other viral diseases, including polio, rubella, measles, and mumps. Modern biotechnology has opened up new approaches to the development of antiviral vaccines and to the pharmaceutical production of certain other natural antiviral agents, such as interferons (see Chapter 39).

Although vaccines can prevent illnesses caused by certain viruses, medical technology can do little at this time to cure most viral infections once they occur. The antibiotics that help us recover from bacterial infections are powerless against viruses. Antibiotics kill bacteria by inhibiting enzymes or biosynthetic processes specific to the pathogens, but viruses have few or no enzymes of their own; they use their host's.

A few antiviral drugs have been developed in recent years. Several of these drugs are chemical analogues of purine nucleosides, which interfere with viral nucleic acid synthesis. One such drug is adenine arabinoside (also called Ara-A or vidarabine), which has some effect on a number of human viruses; it acts at concentrations well below those that inhibit the synthesis of host cell nucleic acid. Another such drug is acyclovir, which seems to inhibit herpesvirus DNA synthesis. Two drugs of a different type, amantadine and rimantadine, are effective in the prevention of influenza. They seem to inhibit the influenza virus after it enters a cell, but their mechanism of action remains unknown.

Viruses and Cancer

For many years, scientists have recognized that some viruses can cause cancer in animals; these **tumor viruses**

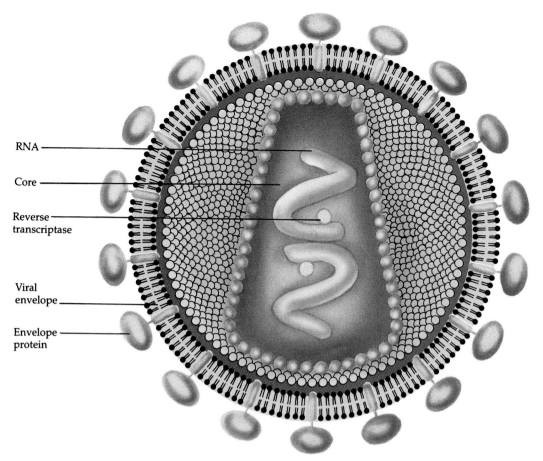

RNA

Core

Reverse transcriptase

Viral envelope

Envelope protein

Figure 17.10

The HIV virion. This diagram shows a cross section through the virion of the virus that causes AIDS. The protein molecules of the viral envelope form knobs that project outward. Within the spherical capsid is a core in the shape of a truncated cone that contains two molecules of RNA, the genome. A molecule of reverse transcriptase is attached to each RNA molecule.

include members of the retrovirus, papovavirus, adenovirus, and herpesvirus groups. Research on these viruses has been facilitated by the development of techniques for growing them in cell cultures. When certain tumor viruses infect animal cells growing in laboratory culture, the cells undergo transformation to a cancerous state. They assume the rounded shapes characteristic of cancer cells and abandon their orderly growth (see Figure 11.14).

In a few cases, there is strong evidence that viruses cause certain types of human cancer. The virus that causes hepatitis B also seems to cause liver cancer in individuals with chronic infections. And the Epstein-Barr virus, the herpesvirus that causes infectious mononucleosis, has been linked to several types of cancer prevalent in parts of Africa, notably Burkitt's lymphoma. Papilloma viruses (of the papovavirus group) have been associated with cancer of the cervix.

Perhaps the most important cancer-causing viruses are **retroviruses,** RNA viruses that reproduce via a DNA intermediate. One retrovirus, called HTLV-I, is known to cause adult T-cell leukemia; and another retrovirus, called HIV, causes AIDS, which weakens the immune system and increases susceptibility to cancer (Figure 17.10; AIDS is discussed in detail in Chapter 39). More generally, research with retroviruses is providing important clues to the mechanisms underlying all forms of cancer.

All tumor viruses transform cells through integration of viral nucleic acid into host cell DNA. (This insertion is permanent; the provirus never excises, as prophages do.) For DNA tumor viruses, the insertion is presumably a straightforward process. Retroviruses, on the other hand, must first carry out reverse transcription to transcribe their genetic material from RNA to DNA (Figure 17.11). Then the viral DNA can be inserted into a cellular chromosome, where it remains as the chromosome replicates in each cell generation.

Scientists have identified a number of viral genes directly involved in triggering cancerous characteristics in cells. To their initial surprise, they have found

Figure 17.11

Replication cycle of a retrovirus. After the single-stranded RNA genome is uncoated, it is not translated but instead serves as a template for the synthesis of double-stranded DNA by the enzyme reverse transcriptase, which was carried in the phage capsid. The DNA is then integrated into the host cell's chromosomal DNA, where it remains as a provirus. Transcription of the proviral DNA may lead to the expression of oncogenes, causing the cell to become cancerous, or to the production of retrovirus virions.

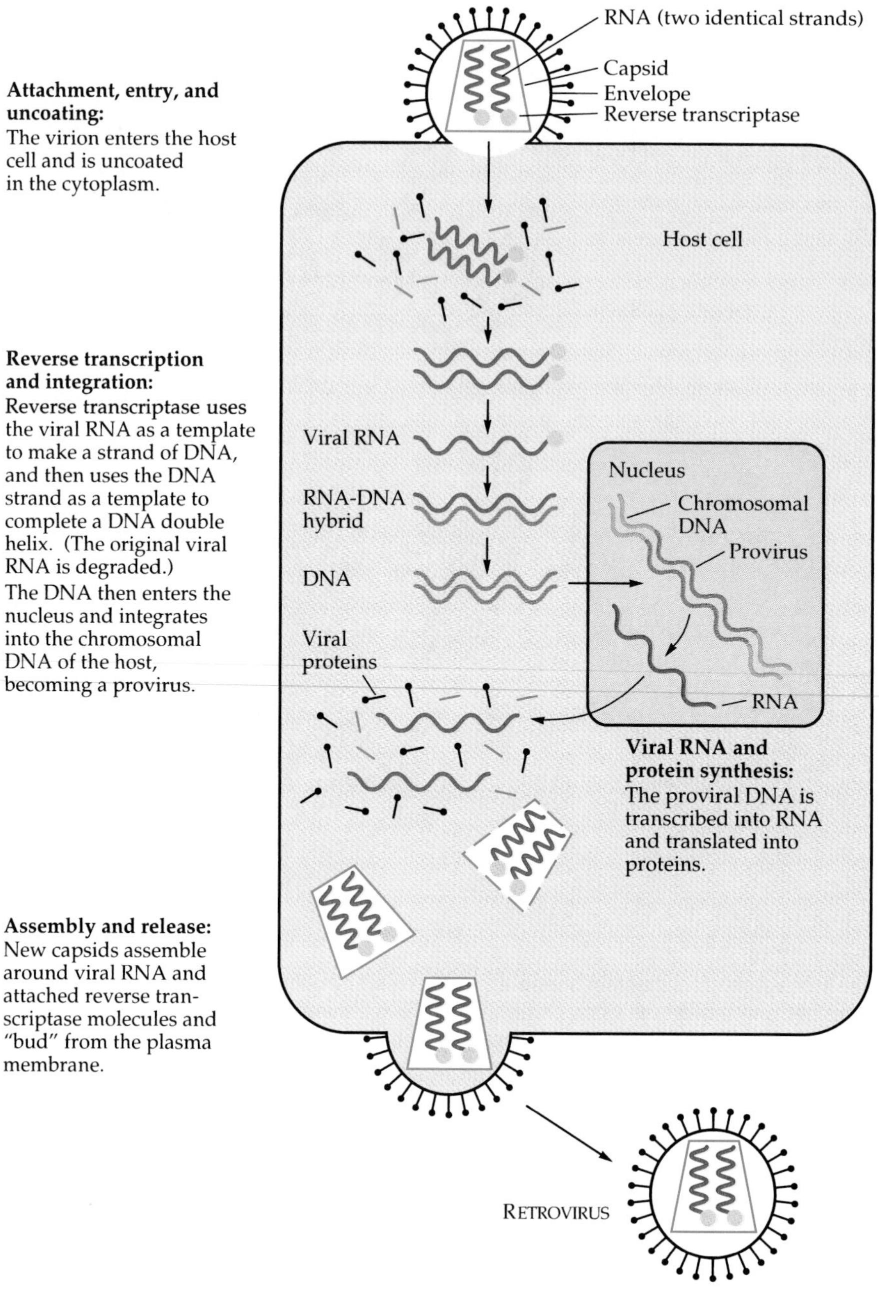

Attachment, entry, and uncoating:
The virion enters the host cell and is uncoated in the cytoplasm.

Reverse transcription and integration:
Reverse transcriptase uses the viral RNA as a template to make a strand of DNA, and then uses the DNA strand as a template to complete a DNA double helix. (The original viral RNA is degraded.) The DNA then enters the nucleus and integrates into the chromosomal DNA of the host, becoming a provirus.

Assembly and release:
New capsids assemble around viral RNA and attached reverse transcriptase molecules and "bud" from the plasma membrane.

RNA (two identical strands)
Capsid
Envelope
Reverse transcriptase

Host cell

Viral RNA

RNA-DNA hybrid

DNA

Viral proteins

Nucleus
Chromosomal DNA
Provirus
RNA

Viral RNA and protein synthesis:
The proviral DNA is transcribed into RNA and translated into proteins.

RETROVIRUS

that many of these genes, called **oncogenes,** are not peculiar to the tumor viruses, but are found in normal cells of many species. The several dozen oncogenes identified to date all seem to code for cellular growth factors or for proteins involved in growth factor action (for example, receptors). In some cases, the tumor virus lacks oncogenes and transforms the cell simply by turning on or increasing the expression of one or more cellular oncogenes. Whatever the mechanism by which a particular virus causes cancer, there is evidence that more than one oncogene must usually be activated to transform a cell to a fully cancerous state. It is likely that most cancer-causing viruses are effective only in combination with other events—and vice versa. Thus, *non*viral cancer-causing agents (carcinogens) probably also act by turning on cellular oncogenes. Under-

standing how cellular gene expression is controlled may be the key to understanding cancer. (Oncogene expression, eukaryotic gene expression in general, and cancer are pursued in Chapter 18.) Furthermore, the occurrence of cellular oncogenes within retroviral genomes provides a clue to the puzzle of how viruses originated.

THE ORIGIN OF VIRUSES

Viruses are in the semantic fog between life and nonlife. Do we think of them as nature's most complex molecules or as the simplest forms of life? Either way, we must bend our usual definitions. An isolated virus

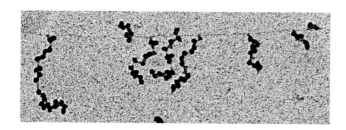

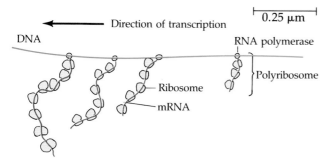

Direction of transcription

0.25 μm

DNA

RNA polymerase

Polyribosome

Ribosome

mRNA

Figure 17.12
Coupled transcription and translation in bacteria. In bacterial cells, where transcription and translation are not segregated by a nuclear envelope, the translation of messenger RNA can begin as soon as the leading end of the mRNA molecule peels away from the DNA template. The electron micrograph shows a strand of *E. coli* DNA being transcribed by RNA polymerase molecules. Attached to each RNA polymerase molecule is a growing strand of mRNA, which is already being translated by ribosomes. The newly synthesized polypeptides are not visible here (TEM).

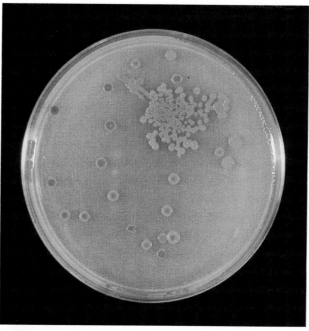

Figure 17.13
Bacterial colonies in the laboratory. A suspension of bacteria (*Serratia marcescens*) was spread on a plate containing a solid nutrient medium. Each bacterial cell reproduced rapidly, in 1 day producing a visible colony composed of millions of cells.

is as static as a rock. Yet, it has a genetic program—a program written in the universal language of life. The program is mutable, and viruses evolve. Although viruses cannot reproduce independently, it is hard to deny their connection to the living world.

How did viruses originate? As we have learned more about the molecular details of viruses, it has become clear that viruses probably evolved from fragments of cellular nucleic acid that acquired very specialized packaging. The genetic material of different families of viruses is much more similar to the genetic material of their hosts than to that of other families of viruses. Indeed, sometimes viral genes are virtually identical to cellular genes, as we have seen in the case of certain retroviruses. The viruses of eukaryotic cells are much more similar in genome structure and function to their cellular hosts than to bacterial viruses, and they have, in fact, served as useful model systems for understanding the control of gene expression in eukaryotes.

Furthermore, viral genomes have marked similarities to certain cellular genetic elements. Those elements include plasmids, which are self-replicating circles of DNA that frequently can transfer from cell to cell, and transposons, which are DNA segments that can move from one spot to another on a chromosome.

Plasmids and transposons will be discussed further as we now examine the transmission and expression of the genome of the simplest type of cell—the prokaryotic cell.

THE BACTERIUM AND ITS GENOME

The relative simplicity of prokaryotic cells is especially apparent in the structures relevant to heredity. Although the average bacterium contains more than 100 times more DNA than a typical virus, it still has only one-thousandth as much DNA as a typical eukaryotic cell. Most of this DNA is found in a single circular molecule, which is called the **bacterial chromosome,** although it is simpler in structure and has fewer associated proteins than a eukaryotic chromosome. Although the area where the bacterial chromosome is located is referred to as the nucleoid region (see Figure 7.6), it is in no way separated from the rest of the cell; in fact, transcription and translation can go on simultaneously (Figure 17.12). Many bacteria also have **plasmids,** smaller rings of DNA that carry accessory genes.

Most bacteria can proliferate rapidly in a favorable environment, whether in nature or in a laboratory (Figure 17.13). Binary fission, described in Chapter 11, is the most common form of reproduction. The bac-

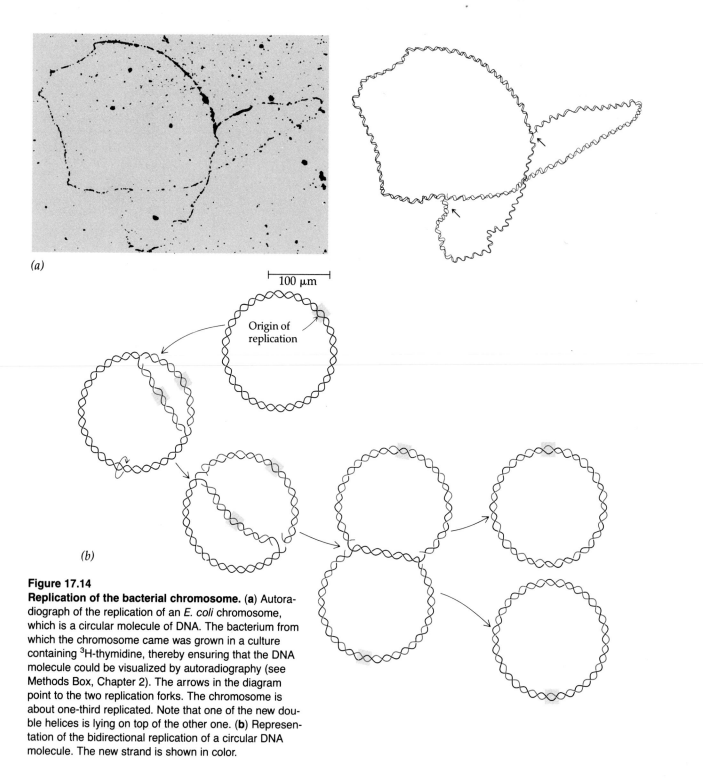

Figure 17.14
Replication of the bacterial chromosome. (a) Autoradiograph of the replication of an *E. coli* chromosome, which is a circular molecule of DNA. The bacterium from which the chromosome came was grown in a culture containing ^{3}H-thymidine, thereby ensuring that the DNA molecule could be visualized by autoradiography (see Methods Box, Chapter 2). The arrows in the diagram point to the two replication forks. The chromosome is about one-third replicated. Note that one of the new double helices is lying on top of the other one. **(b)** Representation of the bidirectional replication of a circular DNA molecule. The new strand is shown in color.

100 μm

Origin of replication

(a)

(b)

terial chromosome has a single origin of replication, and replication proceeds bidirectionally (Figure 17.14). Under optimal conditions, some bacteria can divide as rapidly as once in 20 minutes, a trait that has proved invaluable to geneticists. With plants and animals, it can take weeks or years for the offspring of an experimental mating to reproduce. In contrast, a single bacterial cell, placed on a substrate containing nutrients, can give rise to millions of descendants and form a colony visible to the unaided eye in as little as a day. The color and texture of the colony, its susceptibility

to antibiotics, and the nutrients required for its growth are some easily observable components of a bacterium's phenotype.

TRANSFER AND RECOMBINATION OF BACTERIAL GENES

Until the 1940s, scientists had little hope that bacteria and viruses would make interesting subjects for genetics research. That skepticism was laid to rest,

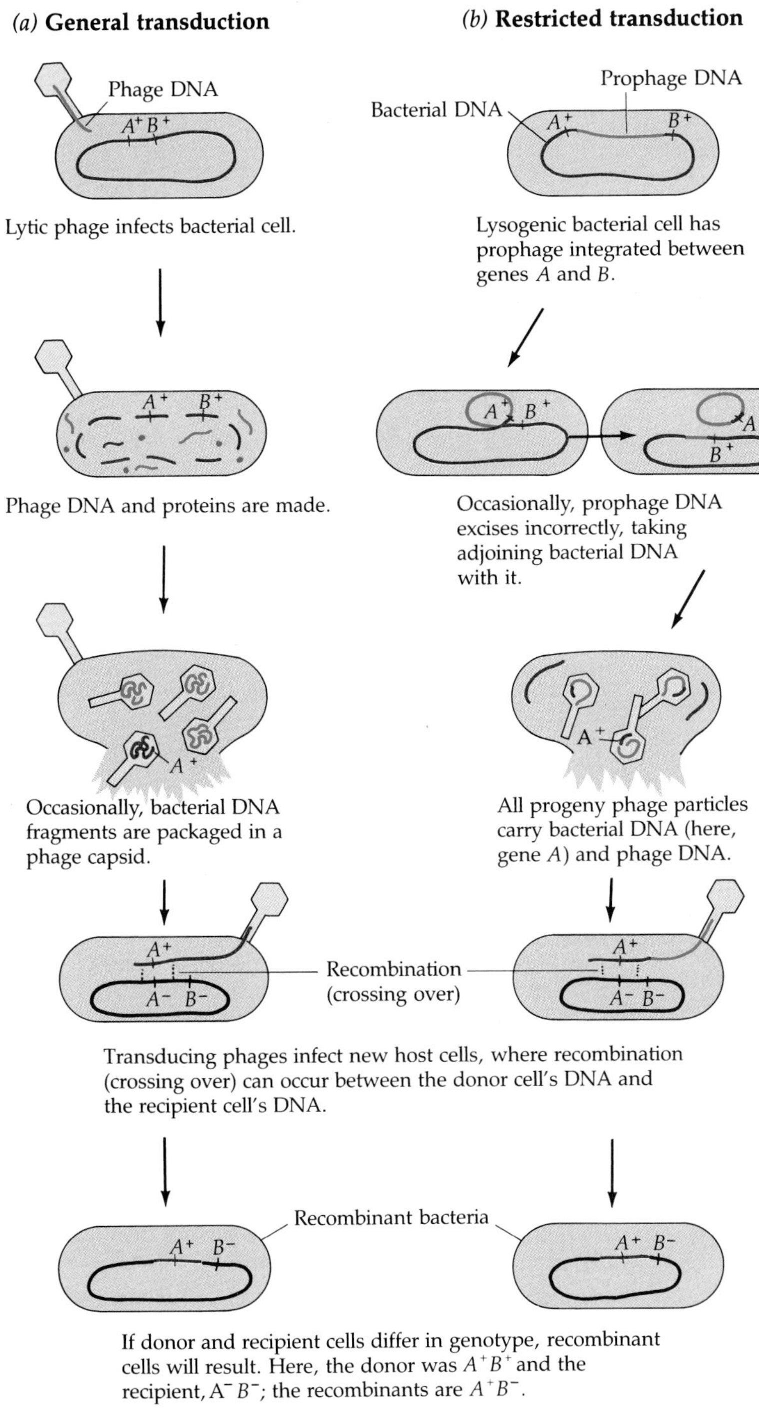

(a) General transduction

(b) Restricted transduction

Figure 17.15

Transduction. Bacterial genes can be carried from one cell to another by a phage. **(a)** In general transduction, random pieces of the host chromosome are packaged into a phage virion. **(b)** In restricted transduction, a prophage excises incorrectly from the host chromosome in such a way that it carries adjacent host genes along with it. In both types of transduction, the transduced DNA later recombines with the genome of the new host cell.

Lytic phage infects bacterial cell.

Lysogenic bacterial cell has prophage integrated between genes *A* and *B*.

Phage DNA and proteins are made.

Occasionally, prophage DNA excises incorrectly, taking adjoining bacterial DNA with it.

Occasionally, bacterial DNA fragments are packaged in a phage capsid.

All progeny phage particles carry bacterial DNA (here, gene *A*) and phage DNA.

Recombination (crossing over)

Transducing phages infect new host cells, where recombination (crossing over) can occur between the donor cell's DNA and the recipient cell's DNA.

Recombinant bacteria

If donor and recipient cells differ in genotype, recombinant cells will result. Here, the donor was A^+B^+ and the recipient, A^-B^-; the recombinants are A^+B^-.

however, when it became apparent just how much could be learned by studying these relatively simple organisms. Although bacteria do not reproduce sexually in the same manner as plants and animals, they do have three mechanisms for transferring genes from one individual organism to another. These mechanisms of gene transfer are called transformation, transduction, and conjugation. In nature, these mechanisms increase the variation in a population; in the laboratory, they make it possible to carry out genetic crosses.

As was described for *Streptococcus pneumoniae* (see Chapter 15), some bacteria can take up segments of naked DNA from the surroundings by the process of **transformation.** After entering the cell, the foreign DNA may be integrated into the bacterial chromosome by recombination, and the progeny of the recipient bacterium will carry a new combination of genes.

Bacteriophages provide the second way that genes of different bacteria can be brought together. This type of process, called **transduction,** has two forms. In **general transduction** (Figure 17.15a), a random piece of host cell DNA, rather than phage DNA, is accidentally packaged within a phage capsid during the lytic cycle of a phage. When that "phage" particle infects a new host cell, the cellular DNA from the donor cell can recombine with the DNA of the recipient. **Restricted transduction** (Figure 17.15b) is carried out only by temperate phages. When a prophage excises from a bacterial chromosome, it may do so incorrectly and take with it some host genes. Again, the progeny phage particle will transport those cellular genes to a new bacterium. Restricted transduction differs from general transduction in that (1) most of the phage genes will also be packaged in the same virion and (2) the bacterial genes transduced will be restricted to those adjacent to the prophage integration site on the bacterial chromosome. Restricted transduction is sometimes called *specialized transduction.*

Until the 1940s, bacteria were believed to be strictly asexual. However, Joshua Lederberg and Edward Tatum then discovered what is sometimes called the "sex life" of bacteria. The bacterial mode of "mating," the third means of genetic transfer, is called **conjugation**—the transfer of genes between two cells that are temporarily joined. In the best-studied example, *E. coli*, a DNA-donating cell uses appendages called *sex pili* to attach to a DNA-receiving cell (Figure 17.16). Then a cytoplasmic bridge forms between the two cells to allow transfer of DNA from one to the other.

The capacity to form sex pili is conferred by a plasmid called the fertility factor or **F factor**. This plasmid carries the genes for pili production and related functions required for transfer of DNA from donor to recipient (Figure 17.17a). Geneticists use the symbol F^+ to denote cells containing the F factor. Because the F factor replicates in synchrony with chromosomal DNA, division of an F^+ cell usually produces two daughter cells that both have the factor and are thus able to donate DNA, a characteristic that is sometimes called "maleness." Cells without the F factor are designated F^- and are sometimes referred to as "female." When an F^+ cell transfers a copy of its F factor to an F^- partner, the recipient becomes F^+. Because the donor cell replicates its F factor during conjugation, it remains F^+.

Occasionally, an F factor inserts into the bacterial chromosome (Figure 17.17b). After integration, the F factor genes (unlike prophage genes) continue to be expressed, and the cell continues to act as a male in conjugation. Now, however, when the F factor is transferred, it carries with it bacterial genes from the chromosome, which can recombine with the genes of the recipient cell (Figure 17.17c). Such cells with integrated F factors are called **Hfr** cells (for "high frequency of recombination"). An Hfr–F^- mating is usually disrupted by random movements of the bacteria before the whole chromosome can be transferred, and most of the donor's genes are usually not transferred.

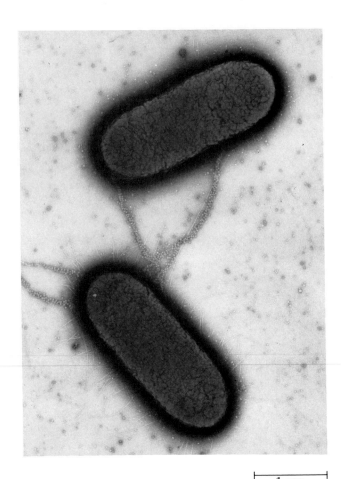

Figure 17.16
A pair of bacteria mating. The DNA-donating (male) *E. coli* cell extends two sex pili to its receptive partner. (The dots covering the sex pili are virions of a phage that attaches specifically to these structures.) Later a cytoplasmic bridge will form, through which DNA will be transferred (TEM).

The recipient cell thus becomes a partial diploid, containing its own chromosome plus a portion of the donor's chromosome. Recombination occurs when a segment of the newly acquired DNA exchanges with the homologous region of the recipient chromosome. Since the chromosome now has a mixture of genes from two bacteria, asexual reproduction of this cell will produce a bacterial colony that is genetically different from both of the cells that mated.

Mapping the Bacterial Chromosome

Hfr bacteria of any given genetic variety, or strain, will always transfer genes in the same sequence during conjugation. The site at which the F factor is incorporated into the chromosome and the factor's orientation in the chromosome determine this sequence of gene transfer. For instance, the bacterial strain in Fig-

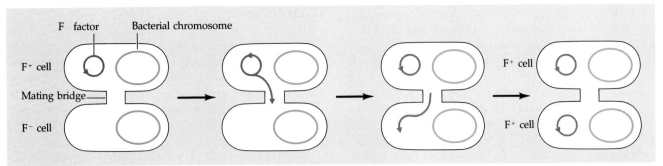

(a) **Conjugation between an F⁺ (male) and an F⁻ (female) bacterium:** Cells that carry an extra-chromosomal fertility factor (F) are called F⁺ cells. They are "male" in that they can transfer the F factor to a "female," F⁻ cell during conjugation. In this way, an F⁻ cell can become F⁺. The F factor replicates as it is transferred so that the donor cell remains F⁺ The replication occurs by a "rolling circle" mechanism, in which one end peels off as a long tail.

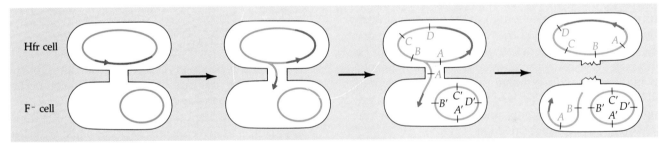

(b) **Conversion of an F⁺ male into an Hfr male by integration of the *F* factor into the chromosome:** An F⁺ cell becomes an Hfr cell if its sex factor integrates into the main bacterial chromosome.

(c) **Conjugation between an Hfr and an F⁻ bacterium:** During conjugation between an Hfr and an F⁻, the integrated F factor of the Hfr cell pulls a copy of the bacterial chromosome along behind its leading end. The F factor always opens up at the same point, and for a particular Hfr strain, the sequence of chromosomal gene transfer is always the same (*A*, *B*, *C*, and *D* represent genes). Usually, the conjugation bridge will break before the entire bacterial chromosome and the tail end of the F factor are transferred.

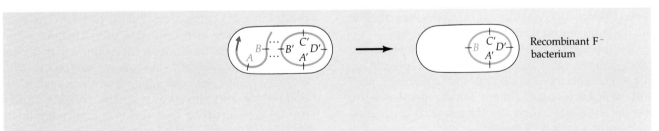

(d) **Recombination between the Hfr chromosome fragment and the F⁻ chromosome:** Crossing over can occur between genes on the fragment of bacterial chromosome transferred from the Hfr cell and the same (homologous) genes on the recipient (F⁻) cell's chromosome. A recombinant F⁻ cell will result. Pieces of DNA ending up outside the bacterial chromosome will eventually be degraded by the cell's enzymes or lost in cell division.

Figure 17.17
Sex and recombination in *E. coli*.

ure 17.17c transfers genes in the sequence A–B–C–D, where each letter stands for a gene on the chromosome. The number of genes the Hfr cell will pass to the F⁻ recipient depends on the duration of conjugation.

By artificially interrupting conjugation, geneticists can determine the rough locations of genes on the bacterial chromosome. The experiment involves mating an Hfr strain and an F⁻ strain with different alleles for the genes of interest. To start the mating, liquid cultures of the two strains are mixed together. At successive time intervals, a sample is taken from the mixture, and mating pairs that have formed are disrupted by agitating the sample in a blender. Genetic analysis of the bacteria indicates which genes were transferred, and from that information, their sequence along the chromosome and the relative distances between them can be determined.

More precise map locations are established as with other organisms, by measuring the frequencies of recombination between genes (see Chapter 14). With bacteria, recombination between the genes of interest can be brought about, as we have seen, by conjugation, transformation, or transduction. A general transducing phage of *E. coli* called P1 is especially useful for such mapping. The length of chromosomal fragment that can be carried by a single P1 virion is very short, so that only genes very close together will be enclosed in the same P1 capsid in transduction. Finally, to map genes with ultimate precision, recombinant DNA techniques can be used to isolate a specific segment of the chromosome, and the exact nucleotide sequence can be determined (see Chapter 19).

Plasmids

The F factor is only one of many small, circular DNA molecules often found in bacterial cells. These plasmids replicate separately from the bacterial chromosome. Those, like the F factor, that can integrate into the bacterial chromosome are called **episomes.**

One interesting class of plasmids, the **R plasmids,** carries genes that code for antibiotic-destroying enzymes. A bacterial strain carrying particular R plasmids will be resistant (hence the designation R) to specific antibiotics, such as tetracycline. Some R plasmids carry as many as seven genes for resistance to different antibiotics. Furthermore, some R plasmids can, like the F factor, mobilize their own transfer to nonresistant cells, even cells of other bacterial species. Hence, R factors carried by pathogenic bacteria can cause severe medical problems by making it difficult to treat a bacterial infection with antibiotics. The formation of plasmids carrying many different combinations of genes is made possible by the existence of transposons.

Transposons

In discussing mutagenesis in Chapter 16, we mentioned that mobile segments of DNA called **transposons** (or transposable elements) serve as natural agents of genetic change. Geneticist Barbara McClintock was able to deduce the existence of such "jumping genes" in maize in the 1940s, but it was not until their discovery in *E. coli* in the late 1960s and the development of recombinant DNA methods (see Chapter 19) that their nature and importance were understood. There is now strong evidence that transposons occur in all organisms.

The DNA of all transposons includes two essential types of nucleotide sequences. The first is a sequence (gene) coding for a *transposase*, an enzyme that catalyzes the cutting and ligating of DNA that occurs in transposition. The second is a pair of sequences called *inverted repeats*, one at each end of the transposon. Inverted repeats are noncoding sequences 20 to 40 nucleotide pairs in length that are virtually identical but are oriented oppositely on the DNA. In the following abbreviated example, note that the inverted repeating sequences are on opposite strands:

$$\frac{...ATCC...\qquad ...GGAT...}{...TAGG...\qquad ...CCTA...}$$

The inverted repeats serve as recognition sites for the transposase enzyme, sites where recombination occurs between transposon and chromosome.

The simplest transposons, consisting only of a transposase gene and inverted repeats at the ends, are called **insertion sequences (IS)** (Figure 17.18a). As they move about on chromosomes, IS elements frequently inactivate genes into which they insert themselves, although occasionally their insertion leads to gene activation or chromosomal rearrangement.

Many transposons, called **complex transposons,** include additional genetic material not connected with the transposition function (Figure 17.18b). The extra DNA may have any sequence of nucleotides. Furthermore, since transposon insertion is not dependent on extensive DNA sequence homology, as is ordinary genetic recombination, transposons can insert themselves into almost any stretch of DNA. Thus, transposons provide a powerful natural mechanism for the movement of genes from one chromosome— or even one species—to another. They have undoubtedly played a role in generating the diversity in gene combination that is the raw material of evolution.

We have already encountered several kinds of genetic elements that contain one or more transposons. These include the F factor and the R plasmids (Figure 17.18c). Moreover, the DNA version of the retrovirus genome is a transposon, which explains how it can insert in a wide variety of places in the cell's genome.

(a) Insertion sequence "IS1"

(b) Complex transposon "Tn 1681"

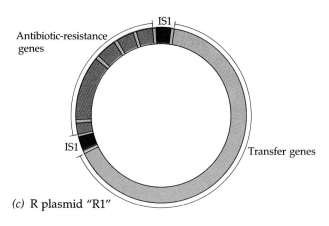

(c) R plasmid "R1"

Figure 17.18
Transposons. **(a)** The simplest type of transposon is the insertion sequence (IS)—a segment of DNA that contains a gene for transposase (the enzyme that catalyzes transposition) and is bounded at each end by nucleotide sequences that are "inverted repeats" (IR). IS1 is one example of an insertion sequence. **(b)** Complex transposons carry other genetic material in addition to transposase genes and inverted repeats. The example shown here, Tn 1681, carries the gene for a bacterial toxin and has complete copies of the insertion sequence IS1 at each end. **(c)** A large transposon makes up a major portion of the plasmid R1, which has two parts. One part contains the genes needed for transfer of the plasmid by conjugation; the other carries genes for resistance to five different antibiotics. The presence of an insertion sequence at each end of the group of antibiotic-resistance genes makes this part of the plasmid a transposon.

The connection between transposons and retroviruses is made more intriguing by strong circumstantial evidence that reverse transcription is part of the mechanism by which certain other eukaryotic transposons move around from one DNA locus to another. (Transposons that move in this way are sometimes called *retrotransposons*.) Furthermore, the genomes of eukaryotes carry numerous examples of nucleotide sequences that appear to be relics of transposition involving reverse transcription. For example, nonfunctional copies of important genes that lack the introns of the functional version are found sprinkled around the human genome; the easiest explanation for such "pseudogenes" is that they arose from reverse

transcription of mRNA (see the discussion of multigene families in Chapter 18).

THE CONTROL OF GENE EXPRESSION IN PROKARYOTES

Metabolic anarchy would overcome a cell that continually expressed its entire genome. Even the simple *E. coli* has about 2500 genes. The genes must be tuned in to the intracellular environment, with genes being switched on and off as conditions change. Think, for instance, of an *E. coli* cell living in the unpredictable environment of a human gut. The bacterium needs the amino acid tryptophan to survive, and it has genes for all the enzymes of a metabolic pathway that synthesizes tryptophan from a precursor substance (Figure 17.19). But what if the milieu bathing the cell is rich in tryptophan that can be absorbed and used by the *E. coli*? The organism would be wasting metabolic energy if it continued to produce this amino acid for itself.

Cells have two main ways of controlling metabolism: by regulating enzyme activity and by regulating enzyme synthesis. As is shown in Figure 17.19, the end product of an anabolic pathway, such as the one that synthesizes tryptophan, may turn off its own production by inhibiting the activity of the first enzyme of the pathway (feedback inhibition). If the *E. coli* continues to enjoy an external source of tryptophan, the cell will respond by ceasing to make the now unneeded enzymes of the tryptophan pathway. Accumulation of tryptophan triggers a specific mechanism for inhibiting the appropriate genes so that they no longer produce messenger RNA. If conditions change so that the need for tryptophan exceeds the supply, the genes that code for the enzymes of the tryptophan pathway must be reactivated. The regulation of gene expression is slower to take effect than feedback inhibition of enzymes, but it is more economical for the cell because it prevents unneeded protein synthesis as well as the synthesis of unneeded small molecules such as tryptophan.

Bacteria—and in particular, *E. coli*—were the first cells for which the mechanisms of gene regulation were worked out, and even today, most of our understanding of such mechanisms at the molecular level is limited to bacteria. Now let us see how some representative bacterial control systems work.

Constitutive Genes and Their Control

The genes for the tryptophan pathway can be switched off, but some of *E. coli*'s genes are continually transcribed. These unregulated genes are said to be **constitutive.** Such genes generally have "housekeeping"

Figure 17.19

Regulation of a metabolic pathway.
Cells can adjust the rates of specific metabolic pathways by regulating both the activity of existing enzymes and the synthesis of new enzyme molecules. The former type of control is useful for immediate, short-term purposes; the latter requires more time to take effect but is more economical for the cell. In the pathway for tryptophan synthesis, the end product of the pathway (tryptophan) can inhibit the activity of the first enzyme in the pathway (feedback inhibition) and/or repress the expression of the genes for all the enzymes needed for the pathway.

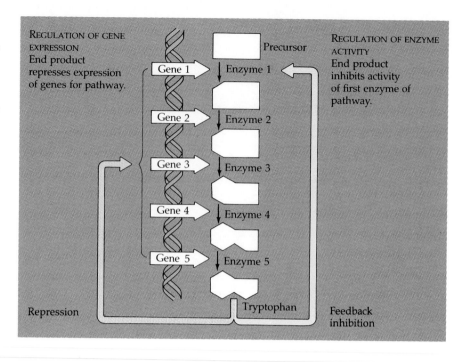

REGULATION OF GENE EXPRESSION
End product represses expression of genes for pathway.

Precursor

Gene 1 — Enzyme 1
Gene 2 — Enzyme 2
Gene 3 — Enzyme 3
Gene 4 — Enzyme 4
Gene 5 — Enzyme 5

REGULATION OF ENZYME ACTIVITY
End product inhibits activity of first enzyme of pathway.

Repression   Tryptophan   Feedback inhibition

functions, producing proteins that are always needed, such as the enzymes for the pathway of glycolysis (see Chapter 9). Although constitutive genes are always active in RNA synthesis, they are not all transcribed at the same rate. Each gene is preceded by a promoter region that serves as a binding site for RNA polymerase (see Figure 16.4). Genes prefaced by "efficient" promoters are transcribed more frequently than genes having less efficient promoter sequences.

These promoter differences are probably important adaptations that have evolved in the *E. coli* genome. Constitutive proteins are not all needed in equal amounts. Perhaps natural selection has maximized metabolic efficiency by favoring "weak" promoter sequences for genes that code for proteins needed in small quantities, and favoring "strong" promoters for genes that specify proteins required in greater abundance. Since these differences are evolutionary adaptations built into the genome, an individual *E. coli* cell is powerless to alter the relative transcriptional rates of its constitutive genes; and although they differ in their mRNA output, all constitutive genes are turned on all of the time.

Regulated genes, however, can be switched on or off as metabolic conditions change. As already mentioned, an accumulation of tryptophan in an *E. coli* cell stops transcription of the genes needed for the pathway of tryptophan synthesis. The mechanism for this regulation was proposed in 1961 by François Jacob and Jacques Monod at the Pasteur Institute in Paris and was based on their research on the *E. coli* genes for lactose catabolism (which will be discussed later).

Operons and Their Control: A Repressible Operon

The genes for the various enzymes of the tryptophan pathway are grouped together as neighbors on the *E. coli* chromosome. Typically, in bacteria and phages, genes with related functions are clustered into units that Jacob and Monod called **operons.** The tryptophan (*trp*) operon, for instance, has five genes, each of which codes for a polypeptide product (Figure 17.20a); such genes are called **structural genes.** A single promoter region serves all the structural genes of an operon. Thus, an RNA polymerase molecule will tend to transcribe the structural genes of an operon on an all-or-none basis, producing a single messenger RNA that carries coding sequences for all the enzymes of a metabolic pathway, such as the one that makes tryptophan. (Such an mRNA molecule is called **polycistronic** because it is a transcript of several genes, sometimes referred to as cistrons.) The cell can translate this message into separate polypeptides because the mRNA is punctuated with codons signaling the termination and initiation of translation (see Chapter 16).

A key advantage to grouping genes into operons is that the expression of these genes can be coordinated. When an *E. coli* cell must make tryptophan, all the enzymes for the metabolic pathway are synthesized at one time. Furthermore, the entire operon, with its several structural genes, can be controlled by a single "on/off switch."

The switch controlling an operon is a segment of

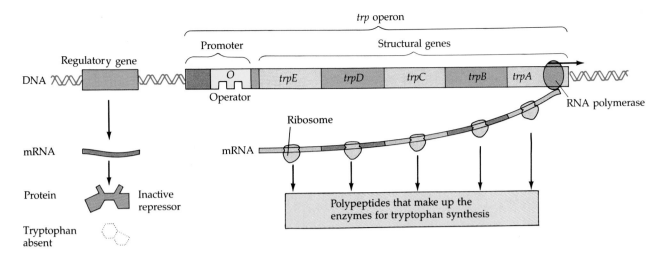

(a)

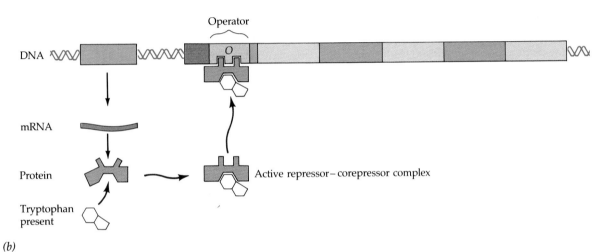

(b)

Figure 17.20
The *trp* operon. In bacteria, genes for the various enzymes of a single metabolic pathway are often grouped into units called operons. Since a single promoter region serves the entire operon, RNA polymerase transcribes the structural genes on an all-or-none basis, and translation of the mRNA produces all of the pathway's different enzymes. Overlapping or within the promoter region is an operator site, where the operon's regulatory protein binds. This figure shows a simplified version of the tryptophan *(trp)* operon, in which the regulatory protein is a repressor. **(a)** In the absence of tryptophan, the repressor molecules encoded by the *trp* regulatory gene are inactive, the *trp* operon is turned on, and the cell produces the enzymes that synthesize tryptophan. **(b)** As tryptophan accumulates in the bacterial cell, it helps to switch off the synthesis of more by activating repressor molecules, which then bind to the operator and block transcription of the structural genes for the *trp* pathway.

DNA called an **operator.** The operator actually overlaps the promoter and sometimes also overlaps the transcription start point for the first structural gene of the operon. Its location and its name both suit the operator's function. Positioned between the promoter and the structural genes, the operator region acts as a red or green light for movement of RNA polymerase onto the promoter and along the structural genes. But what determines whether the operator says "stop" or "go"?

Transcription of the *trp* operon is stopped when a specific protein called a **repressor** binds to the operator region and blocks attachment of RNA polymerase to the promoter (the repressor protein of phage λ does exactly the same thing, by the way, when it blocks transcription of prophage genes). This control is very specific, since each type of repressor protein will recognize and bind only to an operator region having a particular nucleotide sequence. Repressors, like enzymes, have active sites with specific conformations, which discriminate between operators. And the binding of repressor to DNA, like the binding of enzyme to substrate, is reversible.

Some operons are switched *on* by proteins called

activators. Like a repressor, an activator binds to a specific DNA sequence near or within the promoter, but rather than block transcription, it somehow stimulates it.

The genes that code for repressor (and activator) proteins are called **regulatory genes.** The locus of a regulatory gene is often some distance away from the operon it controls. Transcription of the regulatory gene produces messenger RNA that is translated into the regulatory protein. The protein then diffuses until it contacts and binds to the appropriate operator, blocking or activating transcription of the structural genes of the operon.

Each regulatory gene is itself usually constitutive, producing repressor molecules continuously, although at a slow rate. (Apparently, the promoters of regulatory genes are relatively inefficient.) How can an operon ever be expressed if its repressor is always present in the cell? Although repressor molecules can always be found in *E. coli*, they are not always capable of binding to operators and blocking transcription. The ability of a repressor to turn an operon off is determined by the chemical environment within the cell. Specifically, the activity of a repressor depends on whether or not a key metabolite is present in the cell. (A metabolite is any small molecule that is a precursor, intermediate, or end product of some metabolic pathway.) To understand how a metabolite cues a repressor, consider more closely the regulation of the *trp* operon. We have already seen that the amino acid tryptophan blocks the expression of this operon. Tryptophan exerts this negative control by interacting with the operon's repressor protein (Figure 17.20b). The repressor itself is innately inactive because it has a low affinity for the operator. However, in addition to its DNA-binding site, the repressor has an allosteric site (see Chapter 6) that fits tryptophan. If a tryptophan molecule binds to the allosteric site, the conformation of the repressor protein changes from inactive to active and the repressor becomes capable of binding to the operator, switching the *trp* operon off.

Tryptophan functions in this regulatory system as a **corepressor.** Neither the repressor protein nor tryptophan alone can turn off the operon. Only the repressor–corepressor complex can attach to the operator and block synthesis of enzymes for the tryptophan pathway. If the tryptophan concentration in the cell begins to fall below a critical level, the amino acid becomes less likely to be bound to the repressor protein. Once liberated from repression, the *trp* operon commences transcription, and the cell produces enzymes for tryptophan synthesis. This continues until the cell begins to make more tryptophan than it uses, at which time the excess tryptophan switches off the operon by activating the repressor. The structural genes are silenced until there is again a deficiency of tryptophan. In this teeter-totter fashion, tryptophan regulates its own concentration in the cell.

The enzymes of the tryptophan pathway are said to be **repressible enzymes** because their synthesis is inhibited by a metabolite (tryptophan, in this case). *E. coli* also has **inducible enzymes**—enzymes whose synthesis is stimulated, rather than inhibited, by specific metabolites.

An Inducible Operon

In the digestive tract of a person who drinks milk, *E. coli* can absorb the disaccharide lactose, or milk sugar, and use it as a source of energy and carbon atoms for biosynthesis. After the sugar enters the bacterial cell, an enzyme called β-galactosidase cleaves the lactose into its two simpler sugar components, glucose and galactose (see Figure 5.3). Only a few molecules of this enzyme can be found in an *E. coli* cell that has been growing in a lactose-free medium. But if lactose is added to the bacterium's environment, the number of messenger RNA molecules coding for β-galactosidase increases within minutes, and the cell begins making β-galactosidase molecules until there are several thousand of them.

Lactose metabolism is programmed by the *lac* operon, which has three structural genes (Figure 17.21). One of these genes codes for β-galactosidase. A second gene codes for a permease, a transport protein required for accumulation of lactose from the cell's surroundings. The product of the third gene is an enzyme called transacetylase, whose role (if any) in lactose metabolism is unknown. As with the *trp* operon, the *lac* operon has a single promoter and operator that coordinate the expression of all the structural genes. The operon is switched off when the *lac* repressor, the protein product of a specific regulatory gene, binds to the operator.

So far, it sounds as though this system works just like the *trp* operon. But remember, the *trp* repressor was *inactive* until it formed a complex with its corepressor, tryptophan. The *lac* repressor, on the other hand, is innately active, capable of attaching to the *lac* operator without the help of a corepressor. However, the presence of lactose inactivates the repressor by causing it to lose its affinity for the *lac* operator. The operon can now be transcribed, and the cell makes enzymes for lactose metabolism. The metabolite that specifically inactivates a repressor is called an **inducer.** (Actually, it is one of the early metabolites of lactose metabolism that is the true inducer in this system. The few molecules of β-galactosidase in the uninduced cell are sufficient to produce enough of the metabolite to inactivate the repressor. Figure 17.21 abbreviates this process by showing lactose itself as the inducer.)

Now that we have seen how metabolites can repress (*trp* operon) or induce (*lac* operon) the synthesis of specific proteins, we should review what these control systems mean to the *E. coli* cell.

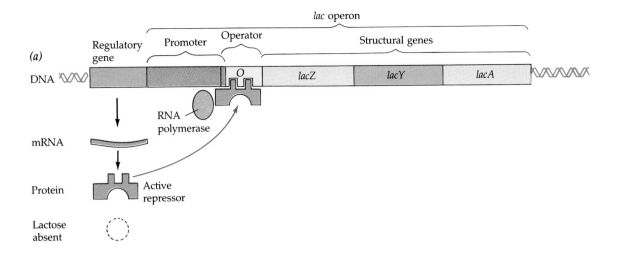

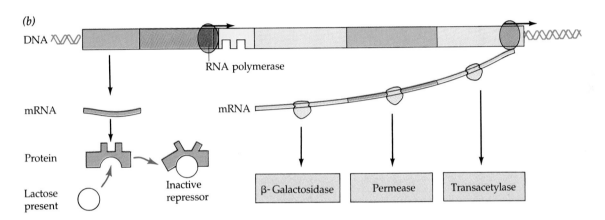

Figure 17.21
The *lac* operon. *E. coli* uses three enzymes to take up and metabolize lactose. The structural genes for these three enzymes are clustered in a single operon, the *lac* operon. One gene, *lacZ*, codes for β-galactosidase, which hydrolyzes lactose and galactose. Another gene, *lacY*, codes for a permease, the membrane protein that transports lactose into the cell. The third gene, *lacA*, codes for an enzyme called transacetylase, whose function in lactose metabolism is unknown. The gene for the *lac* repressor happens to be adjacent to the *lac* operon, an unusual situation. **(a)** The *lac* repressor is innately active, and in the absence of lactose, switches off the operon by binding to the operator region. **(b)** Lactose (more accurately, a derivative of lactose) derepresses the operon by inactivating the repressor protein. Thus, the enzymes for lactose catabolism are induced.

Operon Regulation and the *E. coli* Economy

Repressible enzymes generally function in anabolic pathways, which synthesize organic molecules from precursors, with the pathway's end product switching off enzyme synthesis (Figure 17.22a). This feedback control prevents the expensive overproduction of amino acids or other products. In contrast, inducible enzymes function in catabolic pathways, where nutrient molecules are broken down for energy or carbon atoms, and enzyme synthesis is switched on by the nutrient the pathway uses (Figure 17.22b). This control system prevents the cell from uselessly making enzymes when there are no suitable substrates. Why, for example,

bother to make enzymes that work on milk sugar when there is no milk present?

It is important to see that the main theme of gene regulation is the same for both repressible and inducible enzymes. In both cases, specific repressor proteins control gene expression. In both cases, the repressors can assume two forms: an active form that blocks transcription and an inactive form that allows it. And in both cases, the form of the repressor depends on chemical cues from metabolism, a feature that coordinates gene expression with the changing needs of the cell. The difference between the *trp* repressor and the *lac* repressor is the innate activity of the system. This alone accounts for the opposite characteristics of

(a) Repressible operon

Example: *trp* operon — encodes enzymes for tryptophan synthesis

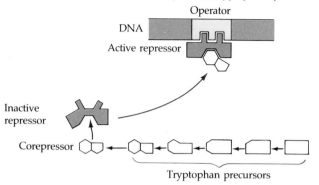

(b) Inducible operon

Example: *lac* operon — encodes enzymes for lactose breakdown

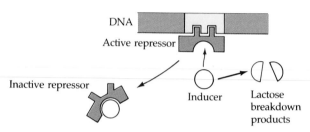

Figure 17.22
Two types of negative control: repressible and inducible operons. In both cases, active repressor proteins bind the operator and switch off structural genes. **(a)** In a repressible system, the repressor is inactive until it associates with a corepressor, usually the end product of an anabolic pathway. **(b)** In an inducible system, the inducer, generally the substrate for a catabolic pathway, derepresses the operon by inhibiting an otherwise active repressor protein.

the *trp* and *lac* operons. Genes for repressible enzymes are switched on until a specific metabolite activates the repressor, whereas genes for inducible enzymes are switched off until a specific metabolite inactivates the repressor.

Both systems are examples of **negative control** in that binding of active repressor to an operator always turns off expression of the structural genes. Lactose does not induce enzyme synthesis by interacting directly with the genome, but by freeing the *lac* operon from the negative effect of the repressor. This derepression involves no positive action on the genome itself. A regulatory system is termed **positive control** only if an activator molecule interacts directly with the genome to turn on transcription.

Catabolite Activator Protein: An Example of Positive Control

For the structural genes of the *lac* operon to be expressed, it is not enough that lactose be present in

the cell. Expression of the *lac* operon also requires an absence of the simple sugar glucose. Even if lactose inactivates the repressor, the operon produces very little mRNA because the *lac* promoter is an inefficient one with a relatively low affinity for RNA polymerase. But transcription is accelerated by a "helper" protein that binds within the promoter region and enhances the ability of the promoter to associate with an RNA polymerase molecule (Figure 17.23). Since this protein, called the **catabolite activator protein (CAP),** stimulates gene expression, it is a positive regulator.

CAP can fasten to the promoter only if glucose is absent from the cell. How does the CAP sense whether glucose is present or absent? When glucose is missing, there is accumulation of **cyclic AMP (cAMP),** a substance derived from ATP. This metabolite binds to the activator protein, and the cAMP–CAP complex can then attach to the *lac* promoter and stimulate transcription. If glucose is added, its metabolism causes the cell's cAMP concentration to decrease. The CAP loses its cAMP and disengages from the promoter; with no activating protein, transcription of the *lac* operon initiates less frequently, even in the presence of lactose.

We have now seen that the dual regulation of the *lac* operon includes negative control by the repressor protein and positive control by the catabolite activator protein. The condition of the repressor (active or inactive) determines whether or not the structural genes can be transcribed. The condition of the CAP (with or without cAMP) determines the rate of RNA synthesis if the operon can be transcribed. The repressor and CAP act like an on/off switch and a volume control, respectively.

How can such a complicated system of negative and positive controls help *E. coli* to economize on its synthesis of RNA and proteins? The CAP has a general effect as an activator of several different operons that program catabolic pathways; the *lac* operon is only one example. By nullifying CAP's effectiveness, glucose causes a general slowdown in the synthesis of enzymes required for the utilization of all catabolites except glucose. *E. coli* preferentially uses glucose, usually available as a nutrient, as its primary carbon and energy source. The enzymes for glucose catabolism are constitutive.

We can view the ability to utilize secondary energy sources, such as lactose, as backup systems enabling a cell deprived of glucose to survive. These emergency systems are relatively inactive in the presence of glucose because CAP does not work when glucose is available. When starved for glucose, the cAMP level rises, and CAP can begin enhancing transcription of operons that program the use of alternate energy sources. Which of these operons is actually transcribed depends on which catabolites are available to the cell. If lactose is present, for instance, the *lac* operon will be switched on via inactivation of the repres-

(a) Lactose present, glucose absent (cAMP level high)

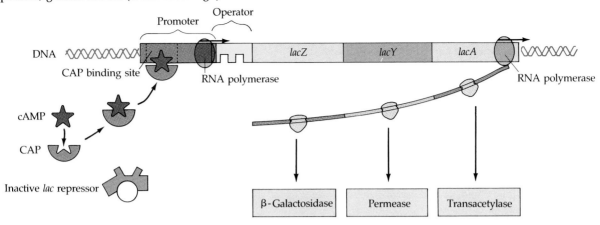

(b) Lactose present, glucose present (cAMP level low)

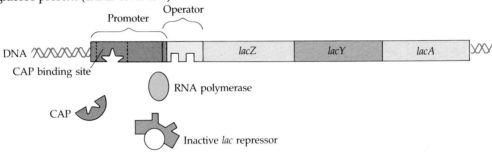

Figure 17.23
Positive control: catabolite activator protein. RNA polymerase has a low affinity for the promoter of the *lac* operon unless helped by catabolite activator protein (CAP), which binds to the DNA. The CAP molecule can attach to the DNA only when associated with cyclic AMP (cAMP), whose concentration in the cell is inversely proportional to the concentration of glucose. (**a**) If glucose is scarce, cAMP activates CAP and the *lac* operon can be efficiently transcribed. (**b**) But when glucose is present, cAMP is scarce and CAP is unable to stimulate transcription.

sor. Dependent on the whimsical eating habits of its human host, *E. coli* has elaborate contingency plans that use chemical stimuli to control gene expression.

*   *   *

This chapter has explored only a sampling of the complex genetic and metabolic regulatory circuits that have been observed in viruses and bacteria. The observations made studying these microbes continue to provide general insights about eukaryotic species too. However, the greater complexity of the eukaryotic genome and eukaryotic cell, as well as the special problems of multicellularity, have led to some mechanisms of gene regulation quite different from those in bacteria, as we will see in the next chapter.

STUDY OUTLINE

1. The molecular mechanisms of heredity were first elucidated in viruses and bacteria, the simplest biological systems. Most, but not all, of the molecular principles are applicable to higher organisms.
2. The unique aspects of microbial genetics made possible a greater understanding of many diseases and the emergence of biotechnology.

The Discovery of Viruses (pp. 348–350)

1. In the nineteenth century, viruses were known only as mysterious agents causing tobacco mosaic disease. These invisible pathogens were found to be extremely small, capable of reproducing within host cells but not on nutrient media, and, unlike bacteria, resistant to destruction by alcohol.
2. It was not until the twentieth century that viruses were finally crystallized and observed under the electron microscope.

Viral Structure (p. 350)

1. Viruses are not cells, but in their simplest form consist only of nucleic acid enclosed in a protein shell.
2. The viral genome may be double- or single-stranded DNA, or single- or double-stranded RNA, depending on the specific virus.

3. The protein shell enclosing the genome is a variously shaped capsid built from many protein subunits. Some animal viruses have a membranous envelope outside the capsid made of both host and viral material. Some bacterial viruses, also called bacteriophages or phages, have the most complex capsids, consisting of complex heads with protein tailpieces.

The Replication of Viruses (pp. 350–352)

1. Viruses are obligate intracellular parasites whose genome uses the enzymes, ribosomes, and small molecules of host cells to synthesize multiple copies of itself and the viral capsid. These constituents self-assemble into virions that leave the cell, ready to infect a new host.
2. Viruses replicate their genome by one of three routes, often using enzymes encoded by their nucleic acid. DNA viruses replicate their DNA much as a cell does. RNA viruses either make complementary RNA strands directly with RNA replicase or use reverse transcriptase to catalyze formation of complementary DNA; the DNA is then transcribed into the viral genome and messenger RNA.
3. The packaging of viral genomes inside capsids is generally a nonenzymatic, spontaneous process resulting from noncovalent interactions between chemical groups within the molecules.
4. Each type of virus has a characteristic host range, determined by specific receptor sites on the surface of host cells.

Bacterial Viruses (pp. 352–355)

1. Bacteriophages have been an important source of knowledge about the nature of DNA and nucleic acid replication.
2. In the lytic cycle of phage replication, injection of a virulent bacteriophage genome into a bacterium programs destruction of host DNA, production of new virions, and lysozyme to digest the bacterial cell wall, which bursts and releases the new virus particles.
3. Bacteria resist viral subversion by producing restriction enzymes that cut up foreign DNA. A continuous coevolution between viruses and their hosts allows perpetuation of the lytic cycle without concomitant destruction of all bacteria.
4. Temperate bacteriophages coexist with their hosts in the lysogenic state by inserting their genome into the bacterial chromosome as a prophage, which codes for a repressor protein that keeps most of its genes inactive. In this innocuous form, the virus can be passed on indefinitely to host daughter cells until at some point it is stimulated to leave the bacterial chromosome and initiate a lytic cycle. The virus progeny subsequently enter into either a lytic cycle or a lysogenic cycle in new host cells.
5. Expression of a few prophage genes in a lysogenic cell can change the phenotype of the bacterium in a process called lysogenic conversion, which may render an otherwise harmless bacterium pathogenic.

Plant Viruses and Viroids (pp. 355–356)

1. Most plant viruses are RNA viruses that seriously compromise plant growth and development.
2. Plant viral diseases may be spread by horizontal transmission, from an external source through a damaged epidermis or via insect vectors; or they may be spread by vertical transmission, the inheritance of viral infections.

3. Plant diseases can also be caused by viroids, tiny molecules of naked RNA that are believed to interfere with plant growth and development by disrupting genetic regulatory systems.

Animal Viruses (pp. 356–360)

1. Animal viruses are often shrouded in an envelope acquired from host cell membrane. Such an envelope allows easy entry and exit through the plasma membrane of host cells.
2. Some viruses, such as the herpesviruses, may integrate into the host chromosome as a latent provirus.
3. Viral infections in animals create a spectrum of effects, depending on the precise mechanism of viral damage and the ability of the infected tissue to regenerate.
4. Vaccines against specific viruses stimulate the immune system to defend the host against an infection. Newly developed antiviral drugs offer promise of halting viral replication once infection has occurred.
5. Viruses have long been implicated in causing cancer, and the study of retroviruses has been especially important in cancer research. Tumor viruses permanently insert viral DNA into host cell DNA, triggering subsequent cancerous changes through their own or host cell oncogenes. Oncogenes code for cellular growth factors or proteins involved in growth factor action.

The Origin of Viruses (pp. 360–361)

1. Although there is debate about whether or not viruses are alive, there is agreement that they evolve.
2. Various lines of evidence point to the origin of viruses as fragments of cellular nucleic acid that came to acquire specialized packaging.

The Bacterium and Its Genome (pp. 361–362)

1. The bacterial chromosome is a circular, relatively simple structure with few associated proteins. Accessory genes are carried on smaller rings of DNA called plasmids.
2. Chromosomal replication in bacteria proceeds bidirectionally from a single origin of replication.

Transfer and Recombination of Bacterial Genes (pp. 362–367)

1. Bacteria have three mechanisms of transferring genes between cells.
2. In the first mechanism, called transformation, naked DNA enters the cell from the surroundings and becomes incorporated into the chromosome by recombination.
3. A second mechanism is general or restricted transduction, in which bacterial DNA is carried from one cell to another by bacteriophages.
4. A third mechanism is conjugation, a primitive kind of "mating" in which an F^+ or Hfr cell transfers DNA to an F^- cell. The transfer is brought about by a plasmid called the F (fertility) factor, which carries genes for the sex pili and other functions needed for mating. In an Hfr cell, the F factor is integrated into the bacterial chromosome, and the Hfr cell will transfer chromosomal DNA along with F-factor DNA in conjugation.
5. Matings between Hfr cells and F^- cells yield partially diploid cells and then recombinant bacteria as a result of crossing over between the newly acquired DNA and the recipient's DNA.

6. Since genes are always transferred in the same sequence during Hfr conjugation, the bacterial chromosome can be mapped by disrupting the process after various durations. Frequencies of recombination are used to refine the map.

7. Episomes are plasmids, such as the F factor, that can integrate into the bacterial chromosome.

8. R plasmids confer resistance to various antibiotics on a bacterium. Their transfer to pathogenic cells poses serious medical problems.

9. Transposons, or "jumping genes," first identified in maize, have been found in many organisms. They generally have at least a transposase gene for cutting and ligating DNA and inverted repeat sequences at each end of the transposon that serve as recognition sites for the transposase.

10. Insertion sequences, the simplest transposons, may affect gene function as they move about.

11. Complex transposons, such as the F factor, R plasmids, and the DNA versions of retrovirus genomes, include additional genetic material not connected with transposition.

The Control of Gene Expression in Prokaryotes (pp. 367–373)

1. Cells control metabolism by regulating enzyme activity or by regulating enzyme synthesis through activation or inactivation of selected genes.

2. Some genes, the constitutive genes, are unregulated, continuously transcribed and translated into proteins that are always needed by the cell. "Weak" and "strong" promoter regions cause transcription of different genes to initiate with different frequencies.

3. In bacteria, regulated genes are often clustered into units called operons, consisting of a single promoter serving adjacent structural genes. A region called the operator, overlapping the promoter, serves as the on/off switch controlling the operon. Binding of a specific repressor protein to the operator shuts off transcription by blocking attachment of RNA polymerase, whereas binding of an activator protein stimulates transcription.

4. Regulatory genes, which code for repressor and activator proteins, are usually constitutive and may be located at some distance from the operon.

5. A repressible operon is switched off in the presence of a key metabolite, usually an end product of a biochemical pathway; the metabolite acts as a corepressor by binding to the normally inactive repressor protein and enhancing its ability to bind to the operator. This action prevents wasteful overproduction of enzymes.

6. Inducible operons are quiescent until activated by a key metabolite. In contrast with the repressible system, binding of the metabolite to the innately active repressor prevents its attachment to the operator, thereby turning on structural genes when necessary.

7. Both repressible and inducible operons are examples of negative control, although the mechanisms and types of metabolic pathways controlled are different.

8. Inducible operons can also involve positive control via a stimulatory activator protein. For example, catabolite activator protein (CAP) stimulates transcription by binding to the promoter and enhancing its ability to associate with RNA polymerase. The promoter-binding ability of CAP is in turn dependent on the presence of cyclic AMP, which accumulates when glucose is scarce.

9. CAP works with repressors to optimize the function of several different operons involved in catabolic backup systems, thereby conferring versatility and economy on the cell.

SELF-QUIZ

1. What aspect of viruses contributed most to the difficulty of their discovery?
   a. their very small size
   b. their inability to reproduce
   c. the unusual nature of some of their RNA genomes
   d. their complex protein coats
   e. their ambiguous state between living and nonliving material

2. Unlike the lytic cycle of a bacteriophage, the productive cycle of an animal virus
   a. involves the incorporation of viral DNA into the host cell's genome
   b. requires the action of reverse transcriptase
   c. involves the entry of the capsid into the cytoplasm of the host cell
   d. usually lyses the host cell on release of new virions
   e. is harmless to the host cell

3. Horizontal transmission of a plant viral disease may involve
   a. the movement of viral particles through plasmodesmata
   b. the inheritance of an infection from a parent plant
   c. the spread of an infection by vegetative propagation
   d. insects as vectors carrying viral particles between plants
   e. a bacteriophage to transport the virus to neighboring plants

4. Restricted transduction occurs when
   a. restriction enzymes cut up viral DNA
   b. reverse transcriptase creates a DNA copy from an RNA template
   c. a prophage includes some bacterial genes when it excises from the bacterial chromosome
   d. naked pieces of double-stranded DNA are picked up by bacterial cells
   e. genetic transfer between an Hfr cell and an F⁻ cell is disrupted, and only part of the donor chromosome moves into the recipient

5. Which of the following is *not* true of transposons?
   a. They have inverted repeat sequences at both ends that serve as a recognition site for reverse transcriptase.
   b. Both F factor and R plasmids are considered to be transposons.

c. They have a nucleotide sequence that codes for transposase, an enzyme that catalyzes the cutting and ligating of DNA necessary for their movement.

d. Barbara McClintock first deduced the existence of these "jumping genes" in maize in the 1940s.

e. Transposons are a mechanism for gene movement from one chromosome to another, and even from one species to another.

6. Inducible operons differ from repressible operons in that

a. inducible operons usually include structural genes that function in catabolic pathways

b. inducible operons use activators and positive control to regulate their transcription

c. inducible operons have an operator that controls binding of RNA polymerase to the promoter region

d. inducible operons are switched off by the binding of a corepressor to a repressor protein

e. the operator of inducible operons is switched off by a repressor protein

7. A mutation that renders the regulatory gene of a repressible operon nonfunctional would result in

a. continuous transcription of the structural genes

b. inhibition of transcription of the structural genes

c. accumulation of large quantities of a substrate for the catabolic pathway controlled by the operon

d. irreversible binding of the repressor to the promoter

e. excessive synthesis of a catabolic activator protein

8. Which of the following information transfers is catalyzed by reverse transcriptase?

a. RNA → RNA

b. DNA → RNA

c. RNA → DNA

d. DNA → DNA

e. RNA → protein

9. A bacterial gene that is perpetually active is said to be

a. inducible

b. constitutive

c. polycistronic

d. facultative

e. repressible

10. Tumor viruses may do all of the following *except*

a. integrate viral nucleic acid into the host genome

b. lyse the host cell in a lysogenic cycle

c. transform tissue culture cells into rounded cells that fail to display contact inhibition

d. introduce oncogenes into the host cell

e. activate the host cell's own oncogenes

CHALLENGE QUESTIONS

1. Scientists have recently worked out the exact three-dimensional molecular structure of the capsid of poliovirus. How might this sort of knowledge be used in devising therapies for viral diseases? (See Further Reading 4.)

2. Bacterial resistance to the antibiotic gentamicin is mediated by a newly discovered resistance gene called *ANT(2″)*. Bacteria of different genera isolated at nine medical centers in the United States and one in Venezuela carried the *ANT(2″)* gene on the same plasmid. What does this evidence suggest about the origin of the gene, its mode of transmission, and the formulation of practical measures for controlling the spread of antibiotic resistance? (See Further Reading 5.)

FURTHER READING

1. Fields, B. N., D. M. Knipe, R. M. Charock, J. L. Melnick, B. Roizman, and R. E. Shope, eds. *Virology.* New York: Raven Press, 1985. An excellent advanced text.

2. Gallo, R. C. "The First Human Retrovirus." *Scientific American,* December 1986. This article discusses the leukemia virus HTLV-1 and also gives an overview of retrovirus biology.

3. Hirsh, M. S., and J. C. Kaplan. "Antiviral Therapy." *Scientific American,* May 1987. Discusses progress and future prospects in the development of drugs to fight viral infections, including AIDS.

4. Hogle, J. M., M. Chow, and D. J. Filman. "The Structure of Poliovirus." *Scientific American,* March 1987. Describes how the detailed structure of this "simple" virus was determined and what it reveals about virus function.

5. O'Brien, T. F., M. d. P. Pla, K. H. Mayer, H. Kishi, F. Gilleece, M. Syvanen, and J. D. Hopkins. "Intercontinental Spread of a New Antibiotic Resistance Gene on an Epidemic Plasmid." *Science* 230 (1985):87–88. An article underscoring a practical problem in the transfer of bacterial genes.

6. Ptashne, M. "How Gene Activators Work." *Scientific American,* January 1989. How regulatory proteins bind to DNA.

7. Varmus, H. "Reverse Transcription." *Scientific American,* September 1987. Discusses the evidence for surprisingly widespread occurrence of reverse transcription in eukaryotic cells.

8. Watson, J. D., N. H. Hopkins, J. W. Roberts, J. A. Steitz, and A. M. Weiner. *Molecular Biology of the Gene,* 4th ed. Menlo Park, Calif.: Benjamin/Cummings, 1987, Chaps. 7, 8, 16, 17, and 24.

9. *What Science Knows About AIDS. Scientific American,* October 1988. This entire issue is devoted to AIDS, with ten articles on various aspects of the disease and the virus that causes it.

Control of Gene Expression and Development in Eukaryotes

18

Eukaryotic cells face the same challenges in regulating their genes as prokaryotic cells, but with added levels of complexity. This complexity is increased even further in multicellular eukaryotes. Like single-celled organisms, the cells of multicellular organisms must continually turn genes on and off in response to signals from their external and internal environments. In addition, however, the control of gene expression is required for **cellular differentiation**—the divergence in structure and function of different types of cells as they become specialized during an organism's development (Figure 18.1).

To appreciate the complexity of controlling gene expression in multicellular organisms, consider the fact that humans have 50,000 to 100,000 genes active in different cells at different times. (Also included in the genomes of humans and other eukaryotes is a large amount of DNA that does not program RNA or protein synthesis.) Any particular human cell synthesizes only a small fraction of the set of human proteins during its lifetime. And at any given time, a cell expresses only about 1% of its genome.

Only fifteen years ago, the mechanisms that control gene expression in eukaryotes were almost entirely unknown. Since then, new research methods have enabled molecular biologists to begin solving some of these previously impenetrable mysteries. Biologists are now using such techniques as gene cloning and nucleic acid sequencing (see Chapter 19) to unveil the organization of eukaryotic genomes. They are beginning to learn how eukaryotic cells alter patterns of protein synthesis as they adapt to both short-term changes

Figure 18.1
Cellular differentiation. As a multicellular
eukaryote develops from a single fertilized
egg cell, cells differentiate (specialize) in
both structure and function. Shown here
are micrographs of a few types of specialized mammalian cells, reproduced at the
same magnification: (**a**) three muscle cells,
each with numerous nuclei (dark discs);
(**b**) a nerve cell (the large cell in the center
with the fibrous extensions; (**c**) sperm
cells; and (**d**) a white blood cell surrounded by smaller red blood cells. (LMs.)
The basis of cellular differentiation is the
turning on (expression) of certain genes
and the turning off of others.

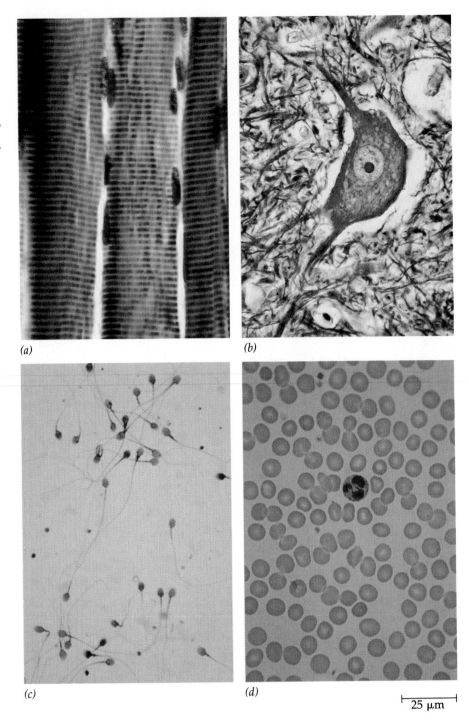

(a)　　　　*(b)*

(c)　　　　*(d)*

$\vdash\!\!-\!\!\dashv$ 25 μm

in their environment and long-term requirements
for specialization within a developing, multicellular
organism. Nevertheless, we are still just at the threshold of understanding the molecular basis of development, and this will be a key area of research for
years to come. In this chapter, we will explore the
general mechanisms of gene regulation in eukaryotes
and their application to molecular and cellular aspects
of development in multicellular organisms.

The control of gene activity in eukaryotes involves
some of the same mechanisms involved in prokaryotic
gene regulation. However, the greater complexity of
chromosome structure and gene organization in
eukaryotes and the greater complexity of eukaryotic
cell structure offer additional opportunities for control. Figure 18.2 provides an overview of the various
levels of control of gene expression in the eukaryotic
cell. Each step boxed in blue offers an opportunity for
regulation. Since the first level of gene control involves
the structure of the chromosomes themselves, we will
begin our study by looking at the arrangement of DNA
within eukaryotic chromosomes.

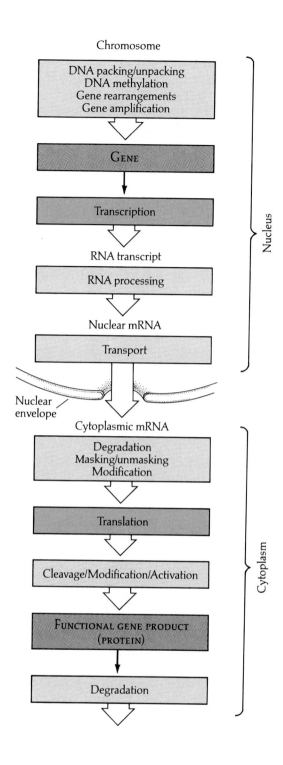

Chromosome

DNA packing/unpacking
DNA methylation
Gene rearrangements
Gene amplification

GENE

Transcription

RNA transcript

RNA processing

Nuclear mRNA

Transport

Nucleus

Nuclear envelope

Cytoplasmic mRNA

Degradation
Masking/unmasking
Modification

Translation

Cleavage/Modification/Activation

FUNCTIONAL GENE PRODUCT
(PROTEIN)

Degradation

Cytoplasm

Figure 18.2
Levels of control of gene expression in eukaryotic cells: an overview. Unlike a prokaryotic cell, a eukaryotic cell has a nuclear envelope that separates transcription and translation in both space and time. This feature offers greater opportunity for posttranscriptional control in the form of RNA processing. In addition, eukaryotes have a greater variety of control opportunities at the gene level (before transcription) and at the protein level (after translation). In this diagram, the processes that offer opportunities for regulation are highlighted by blue boxes.

PACKING OF DNA IN EUKARYOTIC CHROMOSOMES

Eukaryotic chromosomes contain an enormous amount of DNA relative to their size. Each chromosome contains a single uninterrupted DNA double helix, which is thousands of times longer than the diameter of a typical nucleus (Figure 18.3). All of this DNA can fit into the chromosome because of an elaborate, multi-

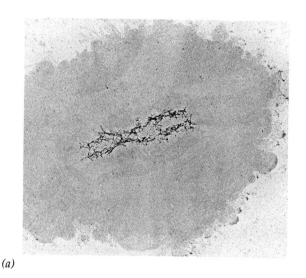

(a)

2 μm

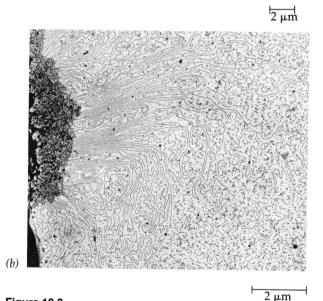

(b)

2 μm

Figure 18.3
Compaction of DNA. A great length of DNA is folded into each chromosome. Each human cell has a total of about 2 m of DNA packed in its 46 chromosomes. **(a)** This electron micrograph shows the DNA spilling out from a single human chromosome, which was isolated during metaphase. The structure of darker, thicker fibers in the center indicates the approximate size of the original metaphase chromosome. **(b)** An enlargement shows loops of DNA coming out from the central structure.

level system of folding, or packing. The degree of DNA compaction in chromosomes can change. For instance, as cells prepare for mitosis, their interphase chromosomes condense greatly to form the short, thick chromosomes that are visible by the end of prophase. The manner in which a region of the DNA molecule is folded may help determine the activity of the genes it contains.

"Beads on a String"

At the first level of DNA packing, proteins attach to the DNA. Eukaryotic cells contain a wide variety of different DNA-binding proteins. The key proteins in folding eukaryotic DNA are the **histones,** which are present in a cell in amounts approximately equal to that of the DNA. Histones are small proteins with a high proportion of positively charged amino acids (lysine and arginine), and they bind tightly to the negatively charged DNA to make up chromatin. There are five types of histone found in most eukaryotic cells.

In electron micrographs, partially unfolded chromatin has the appearance of beads on a string (Figure 18.4b). Each bead is a nucleosome, the basic unit of DNA packing. The **nucleosome** consists of DNA wound around a protein core composed of two copies each of four types of histone. Although nucleosomes appear uniform when examined by electron microscopy, they have subtle chemical variations from cell to cell. For example, there are variations in the extent to which the amino acids have been modified by cellular enzymes. In addition, different nonhistone chromosomal proteins can be associated with different nucleosomes. Nucleosomes may exert control over gene expression by limiting the access of transcription proteins to DNA. The heterogeneity of nucleosomes may also play a role in the control of gene expression.

Higher Levels of DNA Packing

The next order of DNA packing is the **30 nm chromatin fiber** (Figure 18.4c). According to current evidence, it consists of a regular, repeating array of six nucleosomes pulled together by molecules of the histone, called H1, that is not part of the nucleosome core. In at least one type of DNA transcription, the presence of the histone H1 and the intact chromatin structure is necessary for normal control of gene expression. When the structure is disrupted, the gene is transcribed under inappropriate conditions.

The next higher level of organization for eukaryotic DNA is the **looped domain,** which consists of a fold in the 30 nm chromatin fiber (Figure 18.4d). Each loop contains 20,000 to 100,000 base pairs, which some scientists speculate contain genes that are expressed coordinately.

Looped domains themselves coil and fold, further compacting the chromatin (Figure 18.4e). Chromatin that is so highly compacted that it is visible with the light microscope during interphase is called **heterochromatin.** The DNA of heterochromatin is not actively transcribed, whereas **euchromatin,** the much more open form of chromatin, is. The highly condensed DNA of chromosomes at metaphase resembles heterochromatin (Figure 18.4f). Viewed as a whole, Figure 18.4 will give you a sense of how the various levels of folding could enable one chromosome to contain a huge amount of DNA.

THE CONTROL OF GENE EXPRESSION

The Role of DNA Packing and Methylation

Some possible effects on gene expression resulting from nucleosome structure and the higher levels of DNA packing have already been mentioned. The various roles of DNA packing in the control of gene expression are still mostly unknown, but in a few intriguing cases, we have bits of relevant evidence.

One such case is the Barr body, the most striking example of heterochromatin in mammalian cells (see Figure 14.16). A Barr body is an inactive, highly compacted X chromosome, and one of the two X chromosomes in each female somatic cell is in fact a Barr body. The molecular basis for this phenomenon remains unknown, although there is evidence that it may be related to heavy methylation of the Barr body's DNA.

DNA methylation is the addition of methyl groups ($-CH_3$) to bases of DNA after DNA synthesis. The DNA of most plants and animals has methylated bases, usually cytosine. About 5% of the cytosine residues in eukaryotic DNA are methylated. Inactive DNA, such as that of Barr bodies, tends to be highly methylated compared to DNA that is actively transcribed, although there are important exceptions. When molecular biologists look at the same genes in different types of cells (say, from different tissues), they usually find that the genes are more heavily methylated in the cells where they are not expressed. In addition, drugs that inhibit methylation can induce gene reactivation, even in Barr bodies. Thus, DNA methylation may be a cellular mechanism for long-term control of gene expression.

DNA methylation may also influence gene expression by its effect on the structure of the DNA double helix. Methylation of cytosines in certain nucleotide sequences causes the double helix to more readily assume the left-handed Z form (see Figure 15.16). Some molecular biologists have speculated that Z-DNA functions in gene regulation, but what role it plays, if any, is not yet known. Whatever effects DNA structure has on gene expression, they undoubtedly involve control of transcription.

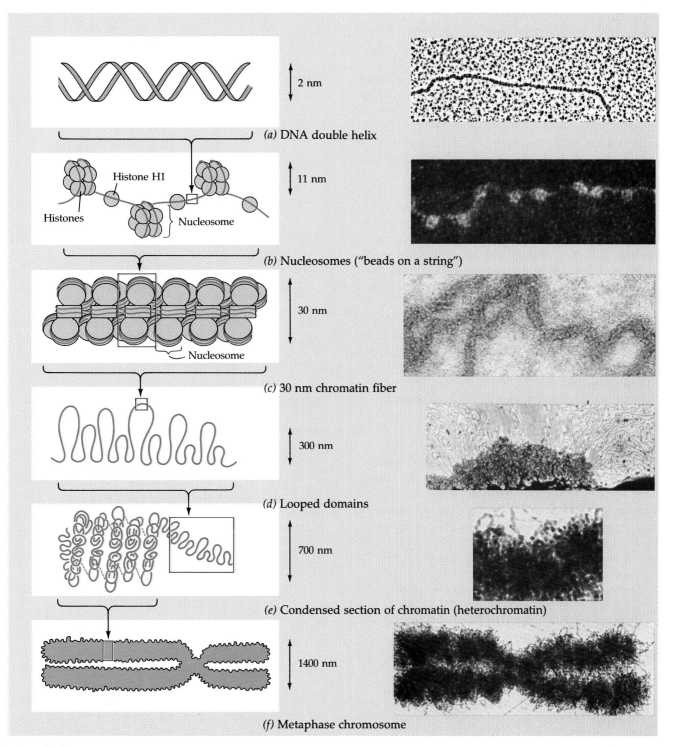

(a) DNA double helix

(b) Nucleosomes ("beads on a string")

(c) 30 nm chromatin fiber

(d) Looped domains

(e) Condensed section of chromatin (heterochromatin)

(f) Metaphase chromosome

Figure 18.4
Overview of chromatin packing. This series of diagrams and electron micrographs shows a current model for the progressive stages of DNA coiling and folding, which culminate in the highly condensed metaphase chromosome. (a) DNA double helix. (b) DNA in association with the five types of histone to form "beads on a string," consisting of nucleosomes in an extended configuration. Each nucleosome has two molecules each of four types of histone. The fifth histone (called H1) binds to the "linker" DNA between the nucleosomes. (c) The 30 nm chromatin fiber, thought to be a tightly wound coil with six nucleosomes per turn. Histone H1 and linker DNA are in the coil's interior. (d) Looped domains.
(e) Condensed section of chromatin. Even in interphase, certain sections of chromosome, called heterochromatin, are condensed in this way. (f) Metaphase chromosome, with all its DNA highly condensed into a compact structure.

Transcriptional Control

Evidence from Chromosome Puffs

Studies of giant (polytene) chromosomes in the salivary gland and other tissues of *Drosophila* larvae have provided evidence that at least some of the control of eukaryotic gene expression occurs at the transcription stage of protein synthesis, as in prokaryotes. As described in Chapter 14, the polytene chromosomes consist of hundreds of parallel chromatids. At characteristic stages in the larva's development, chromosome puffs appear at specific sites on the polytene chromosomes (Figure 18.5). A puff forms when DNA loops out from the chromosome axis, perhaps making the DNA in that region more accessible to the enzyme RNA polymerase. Analysis using autoradiography shows that the puffs indeed correspond to regions of intense RNA synthesis.

The locations of chromosome puffs along a chromosome change as the larva develops. When the larva prepares to molt, some puffs disappear and others form at new sites. The shifting puffs are visual indicators of the selective switching on and off of specific genes during development. The fact that these changes in puffing patterns can be induced by ecdysone, the insect hormone that initiates molting, demonstrates that gene regulation is responsive to specific chemical signals (see Chapter 41).

Molecular Mechanisms of Transcriptional Control

In the DNA of eukaryotes, as in prokaryotes, regulatory proteins interact with specific nucleotide sequences, called regulatory sites. These regulatory proteins can act either as activators or as repressors, and apparently some can even activate one gene and repress another. Because in most eukaryotic cells the

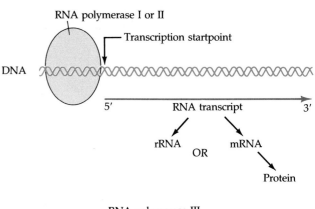

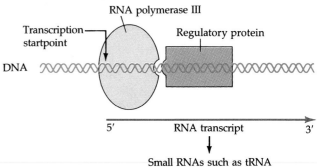

Figure 18.6
Polymerase recognition sites. (a) Eukaryotic RNA polymerases I and II, like bacterial RNA polymerase, recognize specific DNA sequences located upstream from the point where RNA synthesis begins. RNA polymerase I transcribes genes for the large ribosomal RNAs, and RNA polymerase II transcribes genes coding for proteins. **(b)** Eukaryotic RNA polymerase III, however, initially binds to a site within the gene it transcribes, downstream from the start of RNA synthesis. It apparently recognizes a regulatory protein already bound there. RNA polymerase III transcribes the genes for various small RNA molecules, including the tRNAs.

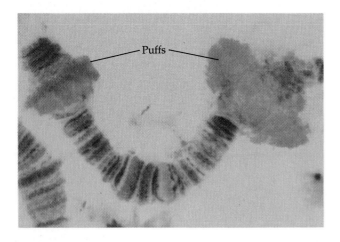

Figure 18.5
Chromosome puffs. This light micrograph shows two puffs in a polytene chromosome of an insect (*Trichosia pubescens*). The locations of puffs along the chromosomes change as a cell develops.

vast majority of the DNA is not expressed, scientists speculate that genes in these cells are usually inactive and that most eukaryotic gene regulatory proteins act to turn *on* transcription of particular genes.

Some sites important for controlling the initiation of transcription have been identified in eukaryotic DNA. These sites are specifically bound by RNA polymerase and thus are promoter regions, like those in prokaryotic DNA. Eukaryotic cells contain three different polymerases for transcribing DNA into RNA, and there are two different kinds of promoters for them (Figure 18.6).

Another class of DNA sequences involved in gene regulation has been discovered in eukaryotic cells and in the viruses that infect them. These so-called **enhancers** boost the activity of any nearby gene several hundredfold. Unlike promoter sequences, an enhancer can be inverted or moved around within a stretch of a few thousand nucleotides and still be effective. Enhancers are sometimes even located within introns of the genes they activate. Enhancer sequences are thought to be recognition sites for proteins that somehow make the DNA in the vicinity more accessible to RNA polymerase molecules.

Posttranscriptional Control

Gene expression is measured in terms of the types and amounts of proteins (or transfer RNA or ribosomal RNA) that a cell makes. Every time a gene is transcribed into a piece of RNA, it is not necessarily expressed. Much happens between transcription and the final creation of a functional eukaryotic protein, and gene expression may be blocked or stimulated at any posttranscriptional step (see Figure 18.2). Unlike a prokaryote, a eukaryotic cell has a nuclear envelope that segregates translation from transcription, a feature that opens new possibilities for controlling gene expression.

RNA Processing   As described in Chapter 16, the eukaryotic cell must "process" its RNA transcripts before they can act as mRNA, tRNA, or rRNA. The RNA segments representing the introns of the genes must be removed and the exons (coding segments) spliced together. Different patterns of splicing can generate different proteins from the same initial RNA transcript, endowing the eukaryotic cell with extra genetic flexibility. In the example shown in Figure 18.7, the splicing apparatus treats a DNA segment as an intron in some situations and as an exon in others.

Processing of messenger RNA also adds nucleotides to both ends of the molecule, modifications that may influence which messenger RNA molecules cross the nuclear membrane into the cytoplasm to reach the cell's protein-making machinery, as well as the fate of an mRNA molecule that does reach the cytoplasm.

Regulation of mRNA Degradation   The lifetime of an mRNA molecule in the cytoplasm is also an important factor in controlling the pattern of protein synthesis in a cell. Prokaryotic messenger RNA molecules have very short lifetimes; they are degraded by enzymes after only a few minutes. This is one reason bacteria can vary their patterns of protein synthesis so quickly in response to environmental changes. In contrast, the messenger molecules of eukaryotes can have lifetimes of hours, even weeks. If two species of mRNA molecules differ in how rapidly they are broken down by enzymes in the cytoplasm, they may differ in how much protein synthesis each directs. A striking example of long-lived mRNA is found in vertebrate red blood cells, which are "factories" for the production of the protein hemoglobin. The mRNAs for hemoglo-

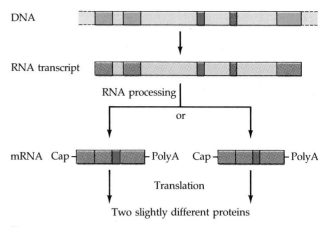

Figure 18.7
Variable RNA processing. The same gene and same RNA transcript can produce two different mRNAs by alternative splicing processes. The two mRNAs are then translated into proteins with slightly different structures and functions. This diagram is based on a gene for a type of protein called troponin T, a regulatory protein of mammalian muscle tissue. The exons that end up in mRNA are indicated with darker colors; lighter colors represent introns. The different forms of the protein appear at different stages of development and in different types of muscle. It is not known what determines which splicing alternative occurs.

bin are unusually stable and are translated repeatedly in the red blood cells of most vertebrate species. (The red blood cells of mammals are an exception: Upon maturation, they lose their nuclei and other organelles, including ribosomes.)

In other cases, eukaryotic mRNA accumulates but translation is delayed until some control signal triggers it. For example, the mRNA used for the active protein synthesis that occurs during the first stage of embryonic development (cleavage) has all been synthesized by the egg cell nucleus prior to fertilization. It is stored in the cytoplasm of the unfertilized egg as "masked messenger"—messenger that is not translated until fertilization. By synthesizing large quantities of specific mRNAs, stocking them in the cytoplasm, and delaying their translation until a signal is given, a developing cell can respond to a stimulus with an explosive burst of synthesis of particular gene products.

Translational and Posttranslational Control   Translation in eukaryotic cells involves many more protein factors, especially initiation factors, than in prokaryotic cells. Thus, there are ample opportunities for control of gene expression at the translation level. For example, in the masked-messenger case, the trigger for the translation of the masked mRNA is the sudden appearance of a necessary initiation factor. Also, the translation of hemoglobin mRNA is controlled at the initiation-factor level by the presence of heme.

The final level of control of gene expression occurs after translation. Often, eukaryotic polypeptides are extensively cleaved to yield smaller, active final products. For example, consider the hormone insulin. It is synthesized as one long polypeptide, which is not active as a hormone. Subsequently, a large center portion is cut away to leave two, shorter chains, which remain linked together by disulfide bridges and which constitute the active hormone (see Figure 41.14). Also operating as control mechanisms in the cell are the selective degradation of particular proteins and the metabolic regulation of the activity of existing enzymes (see Chapter 6).

EUKARYOTIC GENE ORGANIZATION AND ITS EVOLUTION

The arrangement of eukaryotic genes on a chromosome is quite different from that of prokaryotic genes. In prokaryotes, genes that are coordinately controlled are often clustered into an operon; they are adjacent to each other in the DNA molecule and share the regulatory sites located at one end of the cluster. All the genes of the operon are transcribed into a single, polycistronic mRNA molecule and translated together. After prokaryotic operons were described in the 1960s, many molecular biologists expected that operon-like systems would also be the basis for gene regulation in eukaryotes. However, such operons have not been found in eukaryotic cells. Genes coding for the enzymes of a metabolic pathway, for example, are often scattered over different chromosomes in the eukaryotic genome. Even when related genes are located near one another on the same chromosome, each gene has its own promoter and is individually transcribed.

A plausible mechanism that has been suggested for the coordinated control of scattered eukaryotic genes having related functions involves the recognition of a specific nucleotide sequence common to all the genes of a group. There is evidence that such sequences occur adjacent to each of several related genes scattered within the genome (as is the case for a number of bacterial pathways). Such sequences may serve as signals to specific regulatory molecules, saying, in effect, "Transcribe or repress all these genes in synchrony." This mechanism would switch dispersed groups of genes on or off during development.

When "related" genes are clustered in the eukaryotic genome, they do not usually code for enzymes that are sequentially active in a metabolic pathway. Rather, these are genes that are similar in nucleotide sequence and usually in function, and they are thought to be related in evolutionary origin.

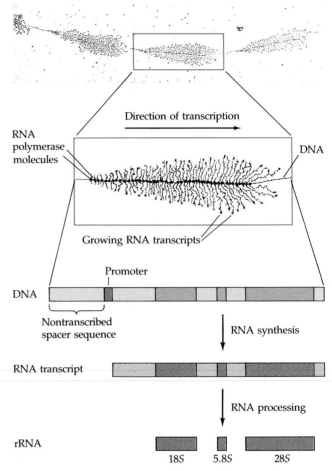

Figure 18.8
Part of a family of identical genes for ribosomal RNA. The electron micrograph (top) shows three of the hundreds of copies of rRNA genes in a salamander genome in the process of being transcribed. Each "feather" corresponds to an rRNA gene being transcribed, from left to right, by about 100 molecules of RNA polymerase; the growing RNA transcripts extend out from the DNA. The genes are arranged in a tandem array, each one separated from the next by a spacer sequence of non-transcribed DNA. The RNA transcripts are processed to yield three kinds of rRNA molecules: 18S, 5.8S, and 28S. (The "S" designations refer to their sedimentation rates in the ultracentrifuge.)

Multigene Families

A **multigene family** is a collection of genes that are similar or identical in sequence and are presumably of common origin. Although the members of a multigene family are often clustered, they may also be dispersed in the genome. With the notable exception of genes for the histone proteins, which occur in multiple copies, the families of *identical* genes consist almost exclusively of genes whose final product is RNA, rather than protein. For families of identical genes, the members are usually clustered. A prominent example is the family of identical genes for the major rRNA molecules (Figure 18.8). These genes are repeated in series

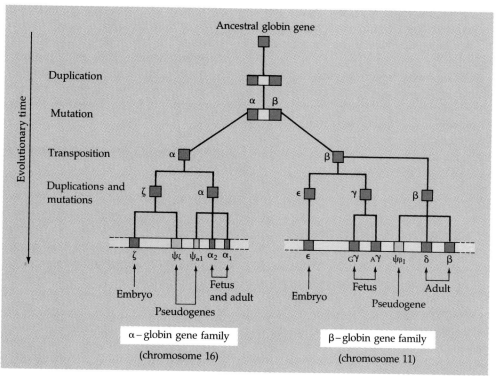

Figure 18.9
Families of related genes for α-globin and β-globin, the hemoglobin subunits. Each gene family consists of a group of similar, but not identical, genes clustered together on a chromosome. In each family, the genes are arranged in order of their expression in development; expression of individual genes is turned on or off in response to the organism's changing environment. At all times in development, func- tional hemoglobin consists of two α-like and two β-like polypeptides. Separating the genes within each family cluster are long stretches of noncoding DNA, which includes pseudogenes, nucleotide sequences very similar to the functional genes. Presumably, the different genes and pseudogenes in each family arose from gene duplication of an original α or β gene followed by mutation. In fact, the original α and β genes themselves undoubtedly arose in much the same way from a common ancestral globin gene. Transposition put the α-globin and β-globin families on different chromosomes, proba- bly early in their evolution. Introns within the individual globin genes are not shown in this diagram.

(tandemly) hundreds to thousands of times in the genomes of higher eukaryotic cells, forming huge tandem arrays of genes that enable the cells to efficiently make the millions of ribosomes needed during active protein synthesis.

The classic examples of multigene families of *nonidentical* genes are the two related families of genes that encode globins, the α and β polypeptide subunits of hemoglobin (Figure 18.9). One family, located on chromosome 16 in humans, encodes various versions of α-globin; the other, on chromosome 11, encodes versions of β-globin. The different versions of each subunit are expressed at different times in development, allowing the hemoglobin to change and function effectively in the changing environment of the organism. The sequences of the various globin genes indicate that the α-like globins and β-like globins evolved from a common ancestral globin.

How do families of genes arise from an ancestor gene? The most likely explanation for families of identical genes is that they arise by repeated gene duplication. Tandem gene duplication seems to be a frequent occurrence, resulting from mistakes made in DNA replication and recombination. In fact, gene duplication can be observed in cells grown in culture under certain conditions.

Families of nonidentical genes probably arise from mutations that accumulate in duplicated genes over evolutionary time. The existence of DNA segments called **pseudogenes** is evidence for the processes of gene duplication and mutation. Pseudogenes have sequences very similar to real (functional) genes but lack the signals (for example, promoters) necessary for gene expression. The globin gene families include several pseudogenes within the stretches of noncoding DNA between the functional genes. The globin pseudogenes also provide evidence that duplicated genes may move about on the genome by a transposition process involving reverse transcription. Two characteristics of the pseudogenes point to a processed-RNA intermediate: They lack introns, and they have poly-A tails—just like mRNA.

Pseudogenes, introns, and the other noncoding DNA regions mentioned so far account for only some of the enormous amount of noncoding DNA in the genomes of higher eukaryotes. In addition, most multicellular eukaryotes have large amounts of DNA consisting of highly repetitive sequences.

Highly Repetitive Sequences

Approximately 10%–25% of the total DNA of complex eukaryotes is made up of short sequences (typically five to ten nucleotides) tandemly repeated thousands of times. The base compositions of such sequences are often sufficiently different from the rest of the cell's DNA to give these DNA segments a different natural density and allow them to be isolated by ultracentrifugation in a cesium chloride density gradient. DNA that can be isolated in this way is called **satellite DNA** because it appears as a "satellite" band separate from the rest of the DNA in the centrifuge tube. In chromosomes, most of the satellite DNA is located at the centromeres. Apparently, this DNA serves a *structural*, rather than genetic, role in the cell, functioning in the replication of the chromosome and the separation of chromatids in mitosis and meiosis.

THE PROGRAM FOR DEVELOPMENT

The goals of developmental biology have wryly been summarized by molecular biologist Sidney Brenner as "how to make a mouse." How does a single fertilized egg cell develop into a complex organism composed of millions of differentiated cells organized into specialized tissues and organs? Discovering the genetic underpinnings for this dramatic transformation is a major challenge of biology today. If we could work out the details of the developmental process for even a single organism (probably *not* a mouse, which is almost as complex as a human!), we might find mechanisms that would be universally applicable to multicellular organisms.

Clearly, regulation of gene expression is a major key to development. As was mentioned earlier, in both eukaryotes and prokaryotes, genes are reversibly activated and deactivated in response to temporary changes in such environmental factors as nutrition, temperature, and light. In multicellular organisms, however, there are also longer-term changes in gene expression that lead to differentiation, the specialization of cells that occurs during development. Each type of differ-

entiated cell makes a characteristic set of proteins employing the appropriate subset of the organism's genes. In this section, we will consider how genes come to be turned on and off during development. It is important to remember, however, that cellular differentiation is not all there is to development. Other aspects of development are covered in Chapter 43.

Determination of Embryonic Cells

The genetic program for development is written in the nucleotide sequences of the DNA in the nucleus of the zygote. The zygote is said to be **totipotent,** meaning that it can give rise to all the different kinds of specialized cells found in the adult organism. Moreover, if a zygote is allowed to divide once, and then an experimenter separates the daughter cells, two normal embryos will develop; that is, each of the zygote's daughter cells is still totipotent (Figure 18.10). Such a separation occasionally occurs spontaneously in two-celled embryos of many organisms (including humans), resulting in identical (monozygotic) twins. Such twins are genetically identical because they are mitotic products of the same zygote.

For most species, embryonic cells remain totipotent through a number of cell divisions (see Figure 18.12). They subsequently lose their developmental versatility so that each cell can no longer develop into all the tissues of the adult organism. This restriction of developmental potential, called **determination,** is a progressive process, with the possible fates of each cell becoming more and more limited as the embryo develops. As a cell's developmental options narrow, differentiation is occurring: The cell is acquiring the molecular and structural features that equip it for its specialized function.

Differentiation

Development can be viewed from a variety of biological perspectives, and scientists of each discipline look for particular characteristics of differentiation. To microscopists, the first signs of specialization are alterations in cellular structure, sometimes subtle differences in cell shape or in the structure of organelles. To biochemists, differentiation is heralded by the appearance of proteins found only in a certain type of cell. The techniques of the molecular biologists reveal an even earlier sign of differentiation: the accumulation of specific messenger RNA molecules.

During differentiation, cells become specialists at making certain proteins. Developing lens cells in vertebrates, for example, synthesize large quantities of crystallins, proteins that aggregate to form transparent fibers that give the lens the ability to transmit and

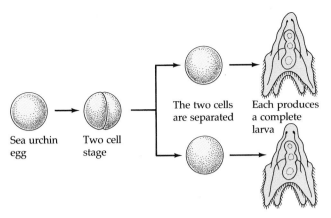

Figure 18.10
Totipotence. After a fertilized sea urchin egg divides once, the two daughter cells can be experimentally isolated. Each will develop into a normal embryo and larva (juvenile form of the animal).

25 μm

Figure 18.11
Differentiation of lens cells. An early step in the differentiation of lens cells is the production of large amounts of special proteins called crystallins. Later, the crystallin-containing cells elongate and lose their nuclei and most other organelles. The tightly packed remnants of these cells from the embryo form the core of the adult lens. This scanning electron micrograph shows the lens cells of a human adult lens stacked like planks in a lumberyard. (From *Tissues and Organs: A Text-Atlas of Scanning Electron Microscopy* by Richard G. Kessel and Randy H. Kardon. W. H. Freeman and Company. Copyright © 1979.)

focus light. Because no other vertebrate cell type makes crystallins, these proteins can be used to follow the progress of differentiation by lens cells. The cue for the immature lens cell to turn on its crystallin genes is contact with the still immature retina. In response to chemical signals from the rudimentary retinal cells, specific messenger RNA molecules coding for crystallin are transcribed and accumulate in the immature lens cell's cytoplasm. Then synthesis of crystallin begins, and the lens cells devote 80% of their capacity for protein synthesis to making this one type of protein. The cells elongate and flatten as they accommodate the crystallin fibers (Figure 18.11).

Genomic Equivalence

If a lens cell makes crystallin, but a blood cell does not, we must conclude that the lens cell expresses a gene that is not expressed in the blood cell. We could account for such differences if cells lost nonessential genes as they differentiated, but most evidence supports the conclusion that all genes are present in all cells of an organism—that all cells have **genomic equivalence.**

The totipotent zygote has a complete library of genes necessary to make all types of specialized cells as an organism develops. What happens to these genes as a cell becomes determined and begins to differentiate? Is differentiation reversible?

Nuclear Transplantation One way to examine whether differentiation is reversible is by replacing the nucleus of a normal egg or zygote with the nucleus from a differentiated cell. If genes are irreversibly inactivated during differentiation, then the trans-

planted nucleus will not be able to guide the development of a normal embryo. The pioneering experiments in nuclear transplantation were carried out by Robert Briggs and Thomas King during the 1950s and were later extended by John Gurdon. These investigators removed or destroyed the nuclei from frog or toad egg cells, which are relatively large, and then transplanted nuclei from embryonic and tadpole cells into the enucleated eggs (Figure 18.12). Many of the embryos containing transplanted nuclei showed substantial developmental progress; a small number even developed into normal tadpoles.

The ability of the transplanted nuclei to support normal development was inversely related to the age of the donor embryos. If the nucleus came from cells of an early embryo, which were relatively undifferentiated, most of the recipient eggs developed into tadpoles. But with nuclei from the differentiated intestinal cells of a tadpole, less than 2% of the eggs developed into normal tadpoles, and most of the embryos

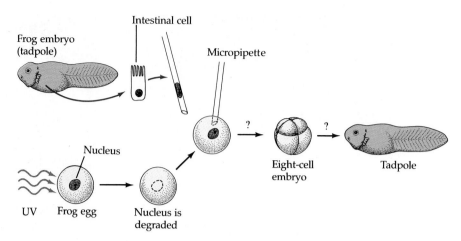

Figure 18.12
Nuclear transplantation. After the egg nucleus is destroyed by ultraviolet (UV) radiation, a nucleus from a more advanced developmental stage is inserted into the egg to test whether nuclei change irreversibly as cells become determined and differentiated. The earlier the developmental stage from which the nucleus comes, the more likely it will support development. Nuclei from very early stages frequently prove to be totipotent, whereas nuclei from late stages (such as a tadpole) rarely are.

failed to make it through even the earliest stages of embryonic development.

Developmental biologists still argue over the meaning of these results, but most agree on two conclusions. First, nuclei *do* change in some way as they become determined and begin to differentiate. Second, this change is not always irreversible, implying that the nucleus of a differentiated cell may have all the genes required for making all other parts of the organism. (Some differences have in fact been found between the genomes of certain cells of embryo and adult; they will be described later in this chapter.)

Development of Plants from Somatic Cells Although the experiment described in Figure 18.12 showed that certain nuclei from the intestinal cells of a tadpole can support development when inserted into eggs, this does not demonstrate that an intestinal cell itself has the potential to form an embryo. Nuclear transplantation in vertebrates involves a trick—the coaxing of the nucleus to return to an embryonic, totipotent state by placing it into the cytoplasmic environment of the egg. No one has yet been able to induce a differentiated cell from a vertebrate to dedifferentiate and then re-form an embryo.

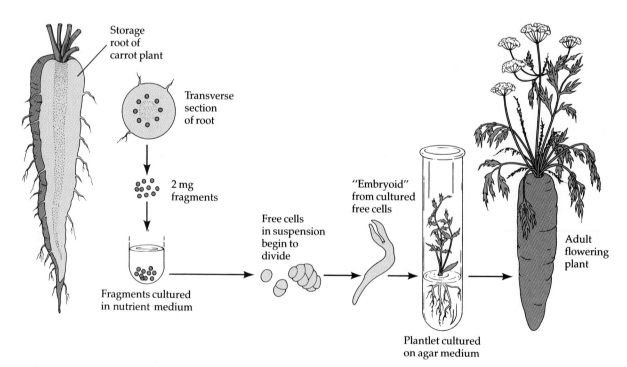

Figure 18.13
Growth of a carrot plant from a differentiated somatic cell. The results of this experiment support the hypothesis that specialized plant cells retain all the genes that were originally present in the zygote. The experiment was first performed in the 1950s by F. C. Steward and his colleagues at Cornell University.

Figure 18.14
Regeneration of a sea star. These invertebrates (Phylum Echinodermata) can replace lost arms. And as long as the central disk of the sea star is present, a single arm can regenerate an entire body, as is shown in this photograph.

This feat is accomplished routinely with differentiated plant cells, however. In the 1950s, carrots were the first plants to be produced from mature somatic cells (Figure 18.13). Now, whole plants are routinely regenerated from single differentiated cells in a variety of plant species. The technique can be used to clone plants, reproducing hundreds or thousands of organisms from the somatic cells of a single plant. This process is being used in conjunction with techniques of genetic engineering in plants (see Chapter 19). The fact that a mature plant cell can dedifferentiate and then give rise to all the plant's specialized cells supports the conclusion from nuclear transplantation that determination and differentiation need not involve irreversible changes in the quality of the genome.

Regeneration   Regeneration is the replacement of parts of an organism that are lost from injury. Many invertebrates have extensive powers of regeneration (Figure 18.14). Although regeneration is less common among vertebrates, there are some striking examples, including the replacement of lost limbs by salamanders and some other amphibians. After the wound from an amputated limb heals, a mass of cells called a *blastema* forms below the skin at the end of the stump. The cells of the blastema resemble the undifferentiated cells of an embryo. As the blastema cells differentiate into muscle, bone, epithelium, and other specialized tissues, the limb begins to re-form.

For decades, developmental biologists debated the source of the blastema cells, with some arguing that the animal must have a supply of embryonic cells to draw on for forming blastemas and regenerating lost parts. More recently, researchers have discovered that the blastema cells are derived from muscle cells, bone cells, and other differentiated cells following amputation of the limb. Somehow, the injury causes these cells to dedifferentiate and form the blastema. As the blastema regenerates the limb, cells are reprogrammed along new developmental pathways.

Differential Gene Expression   Experiments with nuclear transplantation, development of plants from somatic cells, and regeneration are all strong evidence that all the genes of a zygote may be present in an organism's specialized cells, no matter how extensively those cells differ from one another in structure and function. The major mechanism for cell differentiation must therefore be the selective expression of genes (Figure 18.15). The divergence of two cells along different developmental pathways should thus be

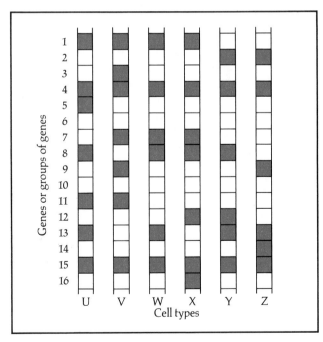

Figure 18.15
Differential gene expression. The numbered segments of DNA symbolize identical genes found in various cell types of a hypothetical organism. The colored genes are those being transcribed and translated. Some genes (such as gene *15*) are expressed in all cell types, while others are expressed in only specific cell types.

viewed as generally due to the differential expression of identical genomes. A muscle cell probably has all the genes necessary to make a nerve cell, but the nerve cell genes are silent in the muscle cell. Let us now examine some of the control mechanisms that may be involved in the differentiation of cells.

GENES THAT CONTROL DEVELOPMENT

In the broadest sense, many (if not most) eukaryotic genes influence development, providing the structural and chemical cues for cell growth and differentiation. The genes of one cell can produce hormones, growth factors, and structural elements that induce a distant or neighboring cell to develop along a certain pathway. In this complex web of interactions, some types of genes stand out as having major developmental impact; they have been identified in situations in which normal development goes awry. In this section, we will look at two such classes of genes: the oncogenes, which are responsible for a cell becoming cancerous, and the homeotic genes, which cause large-scale developmental abnormalities.

Cancer: Cells Out of Control

When normal cells differentiate, control mechanisms limit their growth and division. Among these controls is contact inhibition: As the cells of a developing tissue become crowded, the touching of their surfaces inhibits their movement and reproduction, and somehow, the genes that function in cell division are switched off. But when a normal cell becomes transformed into a cancer cell, it generally escapes from the constraints of contact inhibition and other controls that limit growth and cell division. This phenomenon can be observed in cells growing in the laboratory (see Figure 11.14). As a transformed cell reproduces itself in the body, a mass called a **tumor** forms within the otherwise normal tissue. When grown in culture, tumor cells, unlike normal cells, may continue to divide indefinitely. A striking example of this characteristic can be found in one human cell line that has been reproducing in culture since 1951. (This particular cell line is called HeLa because it originated in a tumor removed from a woman named Henrietta Lacks.)

Cells that multiply excessively but remain at their original site in the body form **benign tumors,** which usually cause few problems and can be completely removed by local surgery. **Malignant tumors** (cancerous tumors), on the other hand, are very harmful. Their cells are abnormal in many ways besides their lack of self-control over cell division. They may have

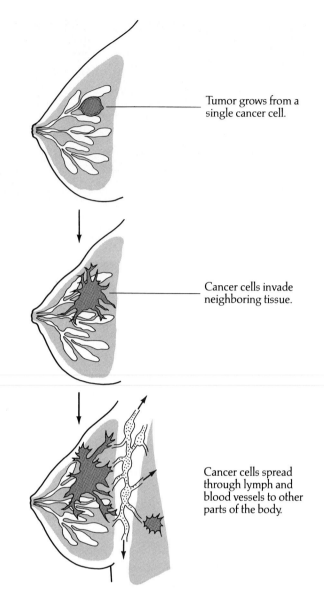

Tumor grows from a single cancer cell.

Cancer cells invade neighboring tissue.

Cancer cells spread through lymph and blood vessels to other parts of the body.

Figure 18.16
Growth and metastasis of a malignant tumor of the breast. The cells of malignant (cancerous) tumors grow in an uncontrolled way and can spread to neighboring tissues and, via the circulatory system, to other parts of the body. The spread of cancer cells beyond their original site in the body is called metastasis.

unusual numbers of chromosomes; many of their metabolic processes may have become deranged; and their cell surfaces are altered.

Abnormal cell surface properties allow cancer cells to escape the normal controls of cell position within the body. Normal cells of a tissue use molecules on their surface to recognize one another and stick together. Cancer cells lose this ability and spread into other tissues surrounding the original tumor (Figure 18.16). They may also separate from the tumor, invade

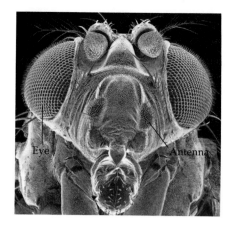 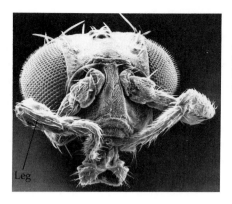

Figure 18.17
Homeotic mutation. These scanning electron micrographs show the heads of two fruit flies. Where small antennae are located in the normal fruit fly (left), one homeotic mutant has legs.

├─────── 0.5 mm ───────┤

blood and lymph vessels of the circulatory system, and then get carried by the circulatory system to new locations where they divide and form additional tumors. The spread of cancer cells beyond their original site is called **metastasis.** If a tumor metastasizes, physicians usually resort to therapy with high-energy radiation and chemicals that are especially detrimental to actively dividing cells.

Cancer can be caused by physical agents such as X-rays and by chemical agents called **carcinogens.** All these agents act by mutating DNA (see Chapter 16). Cancer also can be caused by certain viruses, as described in Chapter 17. Whatever the initial cause of the cancer, the mechanism involves the activation of oncogenes that are either native to the cell or introduced in viral genomes. The normal cellular genes corresponding to oncogenes (often called **proto-oncogenes** when the cell is in a noncancerous state) are thought to be key genes in the control of cell growth and differentiation. By comparing the DNA sequences of known oncogenes with those of known genes for cell growth factors, investigators are beginning to identify these key genes. For example, one oncogene has been found to correspond to a gene that functions as a growth factor in normal wound healing. Thus, studying the causes of cancer is providing insight into the control of normal development, and vice versa.

How might a gene that has an essential function in normal cells cause a cell to run amok? In general, the answer seems to be that the gene is not under its normal controls. There are four mechanisms that can convert a proto-oncogene to an oncogene. They are (1) gene amplification, (2) chromosome translocation, (3) gene transposition, and (4) mutation. Sometimes, oncogenes are present in more copies per cell than is normal (amplification). Frequently, malignant cells contain chromosome translocations, chromosomes that have broken and rejoined, juxtaposing pieces of different chromosmes. In these cases, oncogenes may be found in the new joint region, suggesting that such translocations separate the oncogene from its normal control regions. In gene transposition, the oncogene itself or a regulatory gene may have been transposed to an unusual locus, leading to abnormal expression of the oncogene. In still other cases, the oncogene has a mutated nucleotide sequence, slightly different from that of the corresponding proto-oncogene. Such a change might create a protein that is more active or more resistant to degradation than the normal protein.

Homeotic Genes and Homeoboxes

Genes that control the overall body plan of an animal are currently receiving a great deal of attention from developmental biologists. These **homeotic genes** appear to control the developmental fate of groups of cells. They were first identified in fruit flies, where a mutation in such a gene might produce a fly with such bizarre characteristics as an extra set of wings, extra or missing body segments, or legs growing from the head in place of antennae (Figure 18.17).

Analysis of the DNA sequences of fruit fly homeotic genes revealed a sequence 180 nucleotides long common to all of the genes. This DNA sequence was given the name **homeobox.** Homeoboxes were soon identified in other fruit fly genes involved with development. When scientists looked for homeoboxes in other organisms, they first found them in animals that, like *Drosophila*, had body plans containing repeated segments (see Chapter 36). Some biologists therefore speculated that homeoboxes were involved in segmentation. However, in the past few years, homeobox sequences have been found in virtually every eukaryotic organism examined—from yeast to human. Biologists now think that homeoboxes play a more wideranging role in controlling patterns of development.

Support for a general developmental role for homeoboxes comes from several lines of evidence. Homeobox nucleotide sequences are found within a variety of protein-coding genes, most of which are associated in some way with development. They are translated into peptide sequences of 60 amino acids, called *homeodomains,* which have the property of binding to specific sequences in DNA. Not surprisingly, all the homeodomain-containing proteins examined so far have turned out to be regulatory proteins that activate or repress transcription of other genes by binding to DNA. Presumably they regulate development by coordinating the transcription of batteries of developmental genes, switching them on or off. In early fruit fly embryos, cells in different parts of the embryo show different combinations of active and inactive homeobox genes.

Although much of the research on homeoboxes has been carried out with insects, two recent discoveries illustrate the spectrum of homeobox involvement in vertebrate development. In frogs, a homeobox-containing gene has been identified that apparently controls "posteriorness": When cells expressing this gene are moved to the anterior (front) end of an embryo, a headless frog develops! And in a less sensational but equally important discovery, biologists have recently learned that a homeobox-containing gene is involved in turning on the genes for antibody formation in humans.

The homeodomain amino acid sequence, while varying somewhat from gene to gene and from organism to organism, has been highly conserved in evolution. For example, the homeodomain of the fruit fly protein that controls the phenotype shown in Figure 18.17 differs from one of the frog's homeodomains by only one amino acid out of 60—even though flies and frogs have evolved separately for hundreds of millions of years. Furthermore, eukaryotic homeodomains in general show some similarity to the DNA-binding domains of prokaryotic regulatory proteins. Taken together, the evidence suggests that homeoboxes are all derived from a nucleotide sequence that arose very early in the history of life and that they have continued to be very important in the regulation of gene expression and development throughout the living world.

GENOME MODIFICATION

There are exceptions to the general rule that all cells of an organism have equivalent genomes. Let us examine some specific cases in which physical changes in the genome alter the patterns of protein synthesis during development.

Gene Amplification

The genes that code for ribosomal RNA are repeated many times in the genomes of most eukaryotes. These multiple copies of rRNA genes are built into the genomes of every cell, and they form the core of the nucleolus, where ribosomal subunits are assembled. But one special cell type, the oocyte (developing egg) of amphibians and certain insects, synthesizes a million or more additional copies of the rRNA genes, which exist as extrachromosomal circles of DNA. This selective synthesis of DNA, or **gene amplification,** is a potent way of enhancing expression of the rRNA genes, enabling the oocyte to make tremendous numbers of ribosomes to carry out a burst of protein synthesis once the egg is fertilized.

Selective Gene Loss

In some animal species, whole chromosomes or parts of chromosomes are eliminated from certain cells early in embryonic development. The gall midge, an insect, provides a good example of this **chromosome diminution.** During the first mitotic division after the 16-cell stage, all but two of the cells lose 32 of their initial 40 chromosomes. The two cells that retain the entire genome are the germ cells that later give rise to gametes. The 14 cells that become the somatic cells all keep the same 8 chromosomes. Thus, while chromosome diminution is important in the early divergence of the germ cell line from somatic cells, in at least this case, it does not play a role in the specialization of the somatic cells.

Rearrangements in the Genome

Rearrangements of DNA can activate or inactivate specific genes. One example of such a rearrangement occurs in yeasts, which are single-celled eukaryotes. There, the rearrangement is reversible, so it can provide an alternating pattern of gene activity.

Yeast Mating-Type Genes   Yeasts have two mating types, called α (alpha) and **a,** distinguished by the arrangement of genes at a chromosomal site known as the **mating-type locus.** In an α-type yeast cell, an α gene is present at the mating-type locus, whereas in an **a**-type cell, an *a* gene is present at that site.

Yeasts can be either haploid or diploid; they become diploid by the fusion of two haploid cells that differ in mating type. Haploid yeast cells can change mating type by employing a special mechanism to switch the gene at the mating-type locus. Elsewhere in the same yeast chromosome are "silent" (unexpressed) copies of both the α and *a* genes. When a cell changes mating type, it excises the gene at the mating-type locus and inserts a copy of the other mating-type gene. Silent copies of both mating-type alleles remain available so that the cell's descendants can change mating types again and again. This mechanism is known as the **cassette mechanism** because it involves the removal of one gene from the "play" slot and its replacement with another gene from the genome's two-cassette collection (Figure 18.18).

Immunoglobulin Genes For at least one set of genes in higher organisms, differentiation involves a permanent rearrangement of DNA segments. The mammalian immune system includes white blood cells called B lymphocytes. These cells produce antibodies (also called **immunoglobulins**), which are specific proteins that recognize and help to combat viruses, bacteria, and other invaders. B lymphocytes are very specialized, with each differentiated cell and its descendants producing one specific type of antibody. The human immune system, with its many subpopulations (clones) of such cells, can make millions of different antibody molecules (see Chapter 39).

As a cell of the immune system differentiates into a B lymphocyte, its antibody gene is actually pieced together from several DNA segments that are physically separated in the genome of an embryonic cell. The basic unit of immunoglobulin structure consists of four polypeptide chains held together by disulfide bridges (see Figure 18.19). Each chain has two major parts: a *constant region* that is the same for all antibodies of a particular class and a *variable region* that gives a particular antibody its unique function. In the ge-

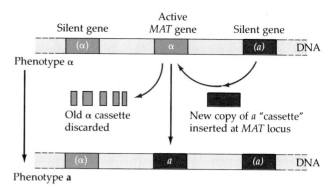

Figure 18.18
Cassette model for mating-type switches in yeast.
The mating type of a yeast cell, **a** or α, is determined by which of two alleles is present at a genetic locus called *MAT*. Wild-type yeasts carry silent (unexpressed) versions of both a and α alleles at loci distant from *MAT* and can move copies of these "cassettes" to *MAT*. Such switching can occur as frequently as once in every cell cycle.

nome of an embryonic cell, the DNA region coding for the constant part of each type of antibody polypeptide is separated by a long stretch of DNA from a region containing a number of variable-coding segments (Figure 18.19). As a B lymphocyte differen-

Figure 18.19

DNA rearrangement in the maturation of an antibody gene. The DNA of antibody genes in undifferentiated cells carries coding segments for a number of different antibody variable (V) regions, for one or more different junction (J) regions, and for one or more different constant (C) regions. During β-lymphocyte differentiation, a long segment of DNA, from the end of one of the V segments to the beginning of one of the J segments, is deleted by recombination. This deletion brings a V segment (in this case V_2) adjacent to a J segment and produces a gene that can be transcribed. The RNA transcript is processed in the usual way to remove introns (and any extra J segments), and the resulting mRNA is translated into one of the polypeptide chains for an antibody molecule. The amino acids coded by the J segment are considered part of the variable region of the polypeptide. This diagram is a simplified representation of one particular kind of antibody polypeptide; in other cases, there are both multiple J segments and multiple C segments in the initial genome.

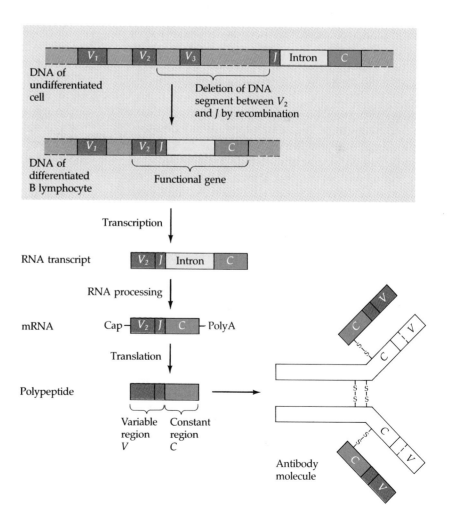

tiates, for each immunoglobulin polypeptide, a specific variable segment of the DNA is connected to a constant segment by the deletion of intervening DNA. The segments join to form the continuous sequence of nucleotides that functions as the gene for that polypeptide chain. Much of the variation in antibodies thus arises from the different combinations of variable and constant regions in the immunoglobulin polypeptides, as well as the different combinations of polypeptides that form the complete antibody molecules. The rearrangement and deletion of DNA involved in the formation of functional antibody genes lead to a striking—and perhaps unique—type of genomic *non*equivalence.

AGING AS A STAGE IN DEVELOPMENT

Organisms continue to change after completing their embryonic development. Obvious examples are growth and maturation (including puberty in humans); metamorphosis from larval to adult forms; wound healing; and regeneration of lost body parts. We should also regard aging and the death that results from it as forms of developmental change in an organism. Like other processes in development, aging may be programmed. In biblical times, the upper limit for the human life span was about "four score and ten," or 90 years. The same is true today. What modern health care has done is increase the numbers that survive long enough to die of old age—not increase the life span.

Genes program the senescence (aging) and death of selected parts of organisms throughout development, sometimes playing a creative role in morphogenesis. Human embryonic hands are webbed, but later, bands of cells in specific locations die, allowing the fingers to separate. Because individuals who inherit a certain mutant allele are born with webbed hands (syndactyly), we can surmise that the function of the normal allele must be to program the death of certain cells when they express this gene during development. Another example of constructive death is the development of a plant's xylem vessels, which do not begin to function in water conduction until they kill themselves with digestive enzymes. In some cases, the senescence and death of whole organs is part of the program of development. When a tadpole becomes a frog, the cells of its tail age and die. The senescence of leaves in the autumn is a particularly beautiful example of programmed death. It is difficult to accept that an entire organism may also be subject to planned obsolescence, but this may be the case.

Cell culture experiments have added credibility to the fatalistic view that aging is bound to happen at a particular stage of development. In these experiments, human embryonic fibroblasts (connective tissue cells) were cultured on an artificial medium in flasks. The cells always ceased to reproduce after about 50 mitotic divisions; they then began to undergo various degenerative changes and die. If a culture was frozen in liquid nitrogen after 20 doublings and thawed later, the cells divided 30 more times and then stopped, even if they had been frozen for years. When similar experiments are performed with other cell types, the number of times they divide differs from fibroblasts, but all normal cell types divide a finite number of times and then begin to age and die.

At least two general hypotheses have been proposed to explain the predictability of aging by organisms and their cells. According to one view, the cumulative effects of mutations and other insults to a cell cause its functional decline. This hypothesis implies that the "timed" death of cells in culture is a statistical phenomenon that reflects the average number of cell divisions required for radiation and other environmental hazards to take their toll, causing cellular function to be impaired. The other hypothesis suggests that aging and death are innate properties of cells, programmed either by the expression of specific genes or by scheduled changes that affect the entire genome, such as declining ability of DNA to replicate or a loss in the effectiveness of DNA repair mechanisms. The two hypotheses need not be mutually exclusive; for instance, an intrinsic crippling of DNA repair would make the cell more susceptible to the accumulative effects of mutagens.

The only cells with a chance for immortality are the germ cells and their derivatives, sperm and eggs. Although somatic cells are dead ends, gametes bridge the generation gap, and their DNA may replicate for millions of years, although changing by mutation and recombination. Viewed in these terms, individuals are but transient custodians for the human germ plasm.

EPIGENESIS VERSUS PREFORMATION

As recently as the eighteenth century, the view prevailed that the egg or sperm contains a preformed, miniature embryo that simply grows during its development. This idea of **preformation** came to include the notion that the miniature embryo must contain a whole series of successively smaller embryos within embryos. According to this version of embryology, the biological history of a species would be analogous to the opening of one of those Russian dolls that comes apart to reveal a family of ever smaller dolls enclosed one within the other. Based on the preformation model, one theologian proposed that Eve, in the Garden of Eden, stored all future humanity within her.

The competing theory of embryology was an idea,

originally proposed 2000 years earlier by Aristotle, that the form of an embryo emerges gradually from a relatively formless egg. This progressive development of form is called **epigenesis.** As microscopy improved during the nineteenth century, biologists could see that embryos took shape gradually, and epigenesis displaced preformation as the favored explanation among embryologists.

In modern biology, the concept of preformation has no merit whatsoever if taken in the strict sense of a tiny person living in an egg or sperm cell. But when interpreted in broader terms, preformation may have something to offer. To be sure, an embryo's form emerges gradually as it develops from a fertilized egg. But in effect, something was preformed in the zygote. The genome of the zygote provides a major part of the plan for making an organism. As the embryo develops, there will be an emergence of inherited traits that is ordered in space and time by mechanisms that selectively control gene expression. (The Methods Box below describes a case in which the unfolding of the genetic plan for development is particularly clear.)

There is another sense in which the zygote has a preformed quality. The nucleus directs the development of a cell, but that nucleus itself is responsive to its surrounding environment, the cytoplasm. Nuclear-cytoplasmic interactions are fundamental to differentiation. The distribution of chemicals and organelles in the zygote will influence later development. As cells move to specific sites during embryonic development, neighboring cells influence the cytoplasmic composition. When selected genes begin to express themselves, this, too, alters the cytoplasm, and the chemical dialogue between nucleus and cytoplasm continues. Which of its genes a differentiated cell expresses is not only a function of its present condition, but also a consequence of the cell's history.

Despite our best efforts to reduce development to a set of understandable steps, the process will always tower as one of the greatest of our many natural wonders. We marvel, justifiably, that complex plants and animals have evolved from unicellular ancestors during the past several hundred million years. But each of us made an analogous journey in only 266 days. As the program of development unfolds, a single egg cell is transformed into a cooperative of specialized cells, tissues, and organs that combine their functions to live as an integrated whole we call an organism.

METHODS: CHOOSING A MODEL ORGANISM FOR STUDYING THE GENETICS OF DEVELOPMENT

When a biologist focuses on a specific problem, choosing an experimental organism well suited for investigating that problem is a key step in a successful research project. Over the years, developmental biologists have worked with a number of experimental organisms, including fruit flies, sea urchins, frogs, chickens, and mice. Each of these organisms has advantages and disadvantages as a model system,

but all of them present difficulties for studies of development on the molecular and cellular levels. So, since the early 1970s, a number of biologists have followed the lead of British molecular biologist Sidney Brenner in turning to a new, simpler model organism, the nematode *Caenorhabditis elegans* (see photo).

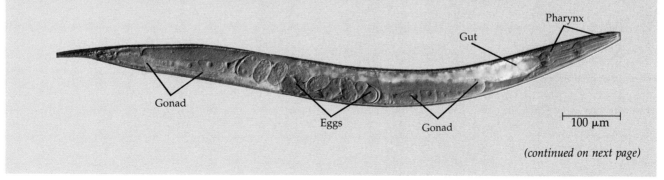

Pharynx

Gut

Gonad

Eggs

Gonad

100 μm

(continued on next page)

C. elegans is a small nematode worm (see Chapter 29) about 1 mm long, with a simple body plan and a small number of cell types. In nature, this animal lives in the soil, but it is easily grown in Petri dishes in the laboratory (it eats bacteria) and has a generation time of only three days. The organism may be either male or hermaphroditic (hermaphrodites produce both sperm and eggs). Genetic studies are facilitated by the fact that the nematode can reproduce either by self-fertilization or by mating with another individual; in either case, hundreds of progeny result. *C. elegans* has about 3000 genes, a relatively small number for a multicellular organism. More than 10% of its genes have already been identified by the techniques of classical and molecular genetics.

Besides being anatomically simple, easy to grow, and well suited for genetic studies, *C. elegans* has two other valuable characteristics for biologists. First, it is transparent at all stages of its life cycle, so that all its individual cells can easily be seen with an appropriately equipped light microscope. Second, unlike most animals (but like certain other invertebrates), the body of the adult always contains exactly the same number of somatic cells, 959, arranged in exactly the same way.

These characteristics made it possible for patient researchers, armed with little more than microscopes and cameras, to determine by direct observation exactly how each cell in the adult arises; the process is the same for every individual. This "cell lineage" information can be represented diagrammatically with a tree-shaped chart (see diagram), which traces every cell division, cell migration, and cell death during the organism's development from zygote to adult. (Each vertical line on the diagram represents a cell, each horizontal line, a cell division.) Although the nematode's highly programmed, invariant development involves little of the cell-cell interaction so important in more complex organisms (see Chapter 43), it provides a model system for asking many very important developmental questions—and for answering such questions in a precise way, using all the resources of modern genetics, biochemistry, and cell biology.

Zygote

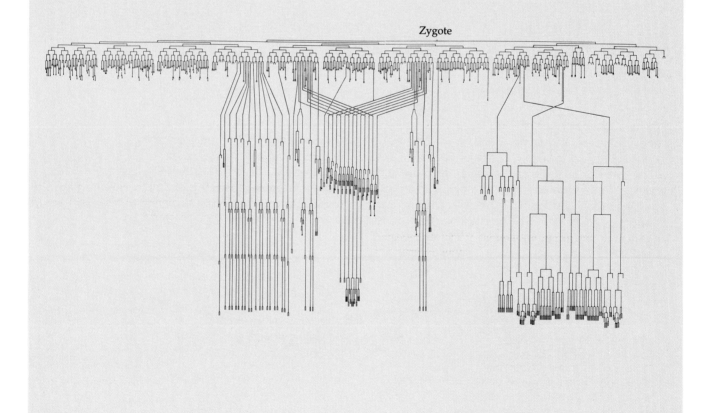

STUDY OUTLINE

1. Cellular differentiation during development depends on the control of gene expression.

Packing of DNA in Eukaryotic Chromosomes (pp. 379–380)

1. The enormous amount of DNA in a eukaryotic chromosome is compacted by a multilevel system of folding.
2. Chromatin is composed of DNA and five types of histone proteins that bind to the DNA and dictate its first level of folding into nucleosomes, basic units of DNA packing.
3. The next order of DNA packing is the 30 nm chromatin fiber, a regular, repeating array of six nucleosomes.
4. The chromatin fiber folds to form the looped domain, the next higher level of DNA organization. Further folding coils the DNA into its most highly compacted, nontranscribed form, called heterochromatin. Active transcription occurs on euchromatin, the more open, unfolded form of the chromosome.

The Control of Gene Expression (pp. 380–384)

1. Methylation of DNA may cause extensive compaction and inactivity of part or all of a chromosome, such as occurs in the Barr bodies of female mammalian cells.
2. The chromosome puffs of polytene chromosomes provide evidence that the control of eukaryotic gene expression can occur at the transcription stage.
3. Transcriptional control is also attested by promoter regions and enhancers, which function as regulatory sites that influence the binding of RNA polymerase.
4. Unlike prokaryotes, eukaryotes have a nuclear envelope that segregates transcription from translation, thereby making posttranscriptional control feasible.
5. In addition, a wide variety of protein factors mediate control of gene expression at the translational level.

Eukaryotic Gene Organization and Its Evolution (pp. 384–386)

1. Unlike prokaryotic operons, eukaryotic genes that need to be transcribed at the same time are often scattered throughout the genome. Coordinated control is believed to be mediated by specific nucleotide sequences common to all the genes of the group.
2. Related genes clustered in the same location tend to be genes of common evolutionary origin, rather than genes for enzymes of the same metabolic pathway.
3. A multigene family is a collection of genes that are similar or identical in nucleotide sequence. They may be clustered or dispersed in the genome.
4. Along with pseudogenes and introns, highly repetitive sequences of five to ten nucleotides tandemly repeated thousands of times make up the enormous reservoir of noncoding DNA in the eukaryotic genome. Highly repetitive sequences include satellite DNA, which is found at centromeres.

The Program for Development (pp. 386–390)

1. Although initially totipotent, individual embryonic cells become progressively restricted in their developmental potential as differentiation proceeds.
2. Genomic equivalence, however obscured by the variable gene expression of differentiated cells, has been demonstrated by nuclear transplantation, development of plants from somatic cells, and regeneration.
3. A major goal of developmental biology is to explain how the divergence of cells along different developmental pathways results from differential expression of identical genomes by specific mechanisms.

Genes That Control Development (pp. 390–392)

1. Although many eukaryotic genes influence cell differentiation, oncogenes and homeotic genes may have a major impact.
2. Oncogenes (called proto-oncogenes when the cell is in the noncancerous state) are thought to be key genes in controlling cellular growth and differentiation. When such genes escape normal control mechanisms, they release cells from the constraints of contact inhibition, allowing the formation of benign or malignant tumors, the latter of which can spread by metastasis. Physical agents, chemical carcinogens, or viruses may activate oncogenes in a variety of ways.
3. Homeotic genes, which control the developmental fate of whole groups of cells, have specific homeobox sequences, which have now been found in a variety of development-related genes in eukaryotes. The genes that contain homeobox sequences appear to code for regulatory proteins, which may activate or deactivate batteries of genes involved in complex morphogenetic events.

Genome Modification (pp. 392–394)

1. There are exceptions to genomic equivalence.
2. The amphibian oocyte can selectively amplify (make extra copies of) the genes for ribosomal RNA.
3. The cells of some species show selective gene loss through chromosome diminution, in which entire chromosomes or parts of chromosomes of certain cells are eliminated.
4. Rearrangements of DNA can activate or inactivate specific genes. In yeast, the alternation of mating types results from a cassette mechanism that switches genes at the mating-type locus. In vertebrates, a permanent rearrangement and selective deletion of DNA segments in differentiating B lymphocytes accounts for antibody diversity.

Aging as a Stage in Development (p. 394)

1. The genetically programmed death of selected cells may play a creative role in morphogenesis.
2. There are two main hypotheses to explain aging. One emphasizes the role of environmental hazards; the other stresses an innate aging program.

Epigenesis Versus Preformation (pp. 394–396)

1. Many early biologists believed in the theory of preformation, which held that a series of preformed, miniature organisms existed inside an egg or sperm.
2. The competing theory of epigenesis maintains that organismal form emerges gradually from a relatively formless egg.
3. Although a literal interpretation of preformation has no basis, the zygote does have a master plan for development in the DNA of its nucleus and a particular cytoplasm that influences nuclear activity.
4. The nematode *Caenorhabditis elegans* is a useful model organism for studying the genetics of development. The lineage of every cell in the adult organism is known.

SELF-QUIZ

1. In a nucleosomes, the DNA is wrapped around
   a. polymerase molecules
   b. ribosomes
   c. histones
   d. the nucleolus
   e. satellite DNA

2. Apparently, your muscle cells are different from your nerve cells mainly because
   a. they express different genes
   b. they contain different genes
   c. they use different genetic codes
   d. they have unique ribosomes
   e. they have different chromosomes

3. Chomosome puffs on the giant chromosomes of *Drosophila* probably represent regions where
   a. genes are inactivated by repressor proteins
   b. hormones are produced
   c. genes have been damaged
   d. genes are especially active in transcription
   e. ribosomes are being synthesized

4. The results of transplantation of frog nuclei into anucleated eggs allow for which conclusion?
   a. Frogs cannot be cloned.
   b. All differentiated cells actually express the same genes.
   c. The later the embryonic stage, the less likely that nuclei from cells at that stage can support the development of an egg into a tadpole.
   d. The nuclei of differentiated cells actually lack some genes found in other cell types.
   e. The differentiated state is unstable.

5. Metastasis is
   a. transformation of a normal cell into a cancer cell
   b. a mutation activates a protooncogene
   c. the spread of cancer cells from their site of origin
   d. loss of contact inhibition
   e. remission of a tumor to a stable condition

6. Enhancers are an example of
   a. a transcriptional control that increases gene expression
   b. a posttranscriptional mechanism for editing mRNA
   c. initiation factors that stimulate translation
   d. posttranslational control activating proteins
   e. a eukaryotic equivalent of the prokaryotic promoter

7. "Masked messenger" refers to
   a. mRNA that is not translated until a control signal is present
   b. the addition of nucleotide sequences to both ends of mRNA before it leaves the nucleus
   c. RNA processing with different splicing patterns that can result in the translation of different proteins
   d. mRNA lacking an initiation codon
   e. unprocessed mRNA trapped in the nucleus

8. Multigene families are
   a. groups of homeotic genes that control key developmental events
   b. pseudogenes that appear to have arisen by reverse transcription of mRNA
   c. equivalent to the operons of prokaryotes
   d. collections of genes whose expression is controlled by the same regulatory proteins
   e. often collections of identical genes, arranged in tandem arrays

9. The formation of blastema cells
   a. is a stage in development of a complete plant from a somatic cell
   b. provides evidence for the irreversible determination
   c. involves the dedifferentiation and redetermination of cells and provides evidence for genomic equivalency
   d. is the result of nuclear transplant experiments in which these cells regain their totipotency
   e. is the result of differentiation, in which the cells express a characteristic subset of their genome

10. Homeotic genes are believed to play a crucial role in controlling development because
    a. they code for homeodomains that can bind to DNA and are part of regulatory proteins
    b. they appear to cause cells to differentiate
    c. they appear to cause cells to become determined
    d. they are identical in all eukaryotic organisms
    e. they have been found in all segmented animals

CHALLENGE QUESTION

1. The amino acid sequences encoded by homeoboxes have been found to be remarkably similar between vertebrates and invertebrates, despite the evolutionary divergence of these animals over 500 million years ago. The similarity implies that all the homeodomains work in the same basic way. Speculate on why evolution has been so conservative with genes that control development.

FURTHER READING

1. Gehring, W. J. "The Molecular Basis of Development." *Scientific American*, October 1985. An excellent article on the function of the homeobox in development.
2. Gilbert, S. F. *Developmental Biology*, 2d ed. Sunderland, Mass.: Sinauer Associates, 1988. An excellent textbook used in upper division developmental biology courses.
3. Holliday, R. "A Different Kind of Inheritance." *Scientific American*, June 1989. Possible role of DNA methylation in development.
4. Ross, J. "The Turnover of Messenger RNA." *Scientific American*, April 1989. What factors control rates of RNA synthesis and degradation?
5. Weinberg, R. A. "Finding the Anti-Oncogene." *Scientific American*, September 1988. Isolation of a gene that helps *prevent* cancer.
6. Weiss, R. "A Genetic Gender Gap." *Science News*, May 20, 1989. The expression of some genes depend on whether they were inherited from the mother or father.

Recombinant DNA Technology

19

Imagine an immense encyclopedia containing the secrets of human life, written in code. We know how to decipher the basic elements of this code, and we have already read a few, small parts of the encyclopedia successfully. But we suspect that deep within its volumes may be found complex new concepts, perhaps beyond our current ability to understand. To complicate matters further, the informational content of the encyclopedia may be buried in a preponderance of gibberish. The characters in which the encyclopedia is written are too small to be seen by even the most powerful microscope, but if printed at the same size as the letters in a standard reference book, the characters would fill 500 volumes. Our goal is to transcribe and translate this entire encyclopedia to better understand the human organism.

Such an encyclopedia actually exists, in the form of the human genome, the DNA that is the blueprint for human life. Only a few years ago, even the task of transcribing the human "encyclopedia"—writing down the complete nucleotide sequence of human DNA—would have seemed unimaginably difficult. But now scientists confidently expect the complete sequence to be determined within the next decade or two, and they are looking with excitement toward the challenge of deciphering its meaning. In the United States, a project to map the human genome is being organized as this paragraph is being written.

The human genome project is made possible by the development of **recombinant DNA technology,** a set of techniques for recombining genes from different sources in vitro and transferring this recombinant DNA into cells, where it may be expressed (Figure 19.1).

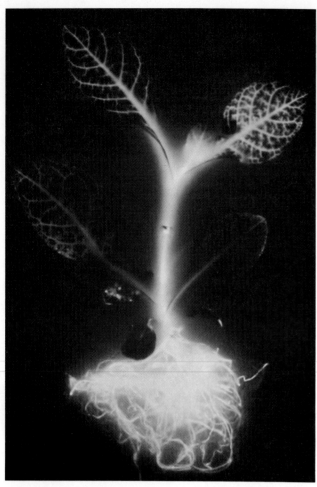

Figure 19.1
Tobacco plant expressing a firefly gene. Recombinant DNA techniques were used to transfer the firefly gene for the enzyme luciferase to a tobacco plant. Luciferase catalyzes the light-producing oxidation of the chemical luciferin. After being watered with a solution of luciferin, the plant glows, indicating that the cells of the plant are actually expressing the firefly gene by synthesizing its product, luciferase.

First developed around 1975 out of basic research in the biochemistry and molecular biology of bacteria, recombinant DNA techniques have given biologists the ability to manipulate genes much more efficiently and precisely than had been possible before. These procedures allow the isolation of specific genes of interest from the rest of a genome and the production of large amounts of these genes and their products. The most important achievements resulting from recombinant DNA technology so far have been in basic eukaryotic molecular biology. Only through the use of these gene-splicing techniques have the detailed structures and functions of eukaryotic genes—the subject of Chapter 18—become open to experimental analysis. The human genome project mentioned above is the most recent and most ambitious extension of this research.

Breakthroughs in molecular biology have launched an industrial revolution in biotechnology. In broad terms, **biotechnology** is the use of living organisms or their components to do practical tasks. Practices that go back centuries, such as the use of microorganisms to make wine and cheese and the selective breeding of livestock and field crops, are aspects of biotechnology. So are the production of antibiotics from microorganisms and the use of even more modern techniques derived from the field of immunology, such as techniques for making monoclonal antibodies (see Chapter 39). Biotechnology based on recombinant DNA is distinct from earlier phases in that it is more precise and systematic. It is also much more powerful, allowing genes to be moved across species barriers, for example. Yet the practical goals remain largely the same—the improvement of human health and food production.

This chapter will examine what the major components of recombinant DNA technology are, how they work, where they are being (or will likely be) applied, and, finally, some of the social and ethical issues that this new biotechnology has raised.

BASIC STRATEGIES OF GENE MANIPULATION

Before 1975, the technology for altering the genes of organisms was severely constrained. The primary strategy was to find desirable mutants and then grow (or breed) them for study or practical use. Sometimes, geneticists relied on nature to produce the mutations. For example, microbiologists looking for new antibiotics combed through vast collections of microbes from soil, sewage, and seawater. In other cases, beginning with Morgan's work with fruit flies, mutagenic radiation or chemicals were used to induce mutations. The screening process—checking the phenotype of each organism for the desired mutant characteristics—was often extremely laborious. Therefore, microbial geneticists, in particular, worked to develop tricks for the **selection** of the desired organisms—that is, methods that would allow only the desired organisms to survive and grow. A simple example of such a method is the inclusion of an antibiotic in the solid growth medium for a strain of bacterium normally sensitive to it; only cells with mutations that confer antibiotic resistance will grow to form colonies on the medium (Figure 19.2).

The selection of mutants is not truly gene manipulation, however, because it does not involve the transfer of the desired gene from one organism to another. Prior to 1975, this was not generally possible except by cumbersome and relatively nonspecific breeding procedures. Only the bacteria and their phages were an exception. As you learned in Chapter 16, genes

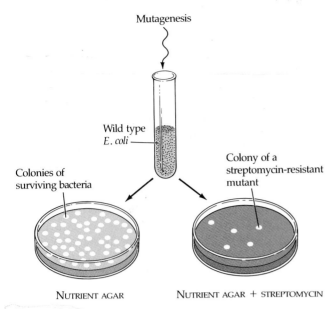

Figure 19.2
Selection for antibiotic-resistant bacterial mutants. In this experiment, radiation or a chemical is used to induce mutations in a culture of streptomycin-sensitive (wild-type) *E. coli*. Then, equal portions of the culture are spread on two types of solid medium—with and without the antibiotic streptomycin. On the plain medium, all cells surviving the mutagen grow; on the medium with streptomycin, only cells that have mutated to streptomycin resistance grow. The streptomycin-containing medium thus "selects" for streptomycin-resistant mutants.

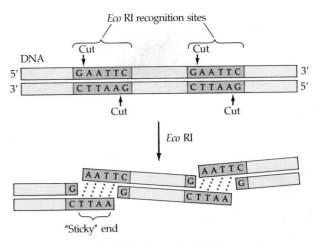

Figure 19.3
Action of a restriction enzyme on DNA. The enzyme *Eco* RI recognizes a six-base-pair sequence and makes staggered cuts in the sugar-phosphate backbone within this sequence. Because of the symmetry of the sequences on the two DNA strands, the result of *Eco* RI action is fragments of DNA with single-stranded, "sticky" ends. Complementary ends will stick to each other by hydrogen bonding, transiently rejoining fragments in their original combinations or in new, recombinant combinations.

can be transferred from one bacterial strain to another, of a different genotype, by the natural biological processes of transformation, conjugation, or transduction by phage. Used as laboratory tools by the geneticist, these processes made possible the detailed molecular study of the structure and functioning of prokaryotic and phage genes. The ease with which these organisms can be propagated and the relatively small size and simplicity of their genomes contributed to the progress made. Eukaryotic organisms were another story. Even though both animal and plant cells could be grown in culture (see Chapters 11 and 34), the detailed workings of their genes remained an apparently impenetrable mystery until the advent of recombinant DNA techniques.

The recombinant DNA technology that is making it possible for biologists to penetrate such mysteries has several key components. These include the biochemical tools that allow the construction of recombinant DNA, methods for purifying DNA molecules and proteins of interest, "vectors" for carrying the recombinant DNA into cells and replicating it, and techniques for determining the nucleotide sequences of DNA molecules. We will begin with a discussion of the special enzymes that allow recombinant DNA to be made in the test tube.

Restriction Enzymes

The major tools of recombinant DNA technology are bacterial enzymes called **restriction enzymes**, which were first discovered in the late 1960s. In nature, these enzymes protect bacteria against intruding DNA from other organisms. They work by cutting up the foreign DNA, a process called **restriction.** Most restriction enzymes are very specific, recognizing short, specific nucleotide sequences in DNA molecules and cutting at specific points within these sequences. The cell protects its own DNA from restriction by adding methyl groups (CH_3) to nucleotide bases within the sequences that would otherwise be recognized by the restriction enzyme. This methylation of DNA, called **modification,** is catalyzed by separate enzymes that recognize the same sequences. There are hundreds of restriction enzymes and about a hundred different specific recognition sequences. As shown in the example in Figure 19.3, recognition sequences are symmetric: The same sequence of four to eight nucleotides is found on both strands, but running in opposite directions. Restriction enzymes cut phosphodiester bonds of both strands, usually in a staggered way, as in the example. The result is a set of double-stranded DNA fragments with single-stranded ends, called "sticky ends." Although not really sticky, these short extensions will form hydrogen-bonded base pairs with complementary single-stranded stretches on other DNA molecules.

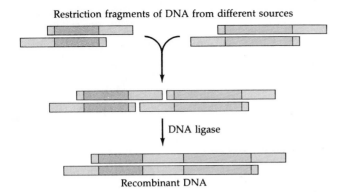

Restriction fragments of DNA from different sources

DNA ligase

Recombinant DNA

Figure 19.4.
Recombinant DNA. After restriction fragments from two different sources come together by base pairing, the enzyme DNA ligase can catalyze the formation of covalent bonds joining their ends. The result is recombinant DNA.

The sticky ends of restriction fragments can be used in the laboratory to join DNA pieces originating from different sources, even from different organisms. Such unions are only temporary, because only a few hydrogen bonds hold them together. They can be made permanent, however, by the enzyme *DNA ligase, which* catalyzes the formation of covalent, phosphodiester bonds (recall from Chapter 15 that DNA ligase is a key enzyme in DNA replication and repair). The biochemical process used by the molecular biologist differs from the natural genetic recombination that goes on within living cells, which does not involve restriction enzymes. But the outcome is similar: the production of recombinant DNA, a DNA molecule carrying a new combination of genes (Figure 19.4).

After a large molecule of DNA is treated with a restriction enzyme, the different "restriction fragments" that result can be separated by **gel electrophoresis.** This valuable method separates macromolecules—either nucleic acids or proteins—on the basis of size and electric charge, by measuring their rate of movement through an electric field (see Methods Box, p. 403). Gel electrophoresis is one of the most valuable tools in molecular biology and is of critical value in many aspects of genetic manipulation and study, as we shall see in this chapter. One use is the identification of particular DNA molecules by the band patterns they yield in gel electrophoresis after digestion with various restriction enzymes. Viral DNA, plasmid DNA, and particular segments of chromosomal DNA can all be identified in this way. Another use of gel electrophoresis is the isolation and purification of individual fragments containing interesting genes, which can be recovered from the gel with full biological activity.

Gene-Cloning Vectors

The recombinant DNA molecules that can be generated by restriction and ligase enzymes are not useful unless they can be made to replicate to produce a large number of copies. Small DNA fragments can be inserted into plasmid DNA from bacteria without impairing its ability to replicate within a bacterial cell. Because isolated plasmids can be introduced into cells by transformation, they can serve as a means—or **vector**—for moving recombinant DNA from test tubes back into cells.

In a typical protocol, many copies of a plasmid are treated with a restriction enzyme so as to cut the DNA ring at a single site. Then pieces of foreign DNA are mixed with the clipped plasmid and joined using DNA ligase into recombinant molecules, each of which is a plasmid. The recombinant plasmids are then introduced into bacterial cells by adding the naked DNA to a bacterial culture; under the right conditions, the bacteria will take up the DNA from solution by the process of transformation (see Chapter 15). Each bacterium, with its recombinant plasmid, is then allowed to reproduce. As the bacterium forms a clone of identical cells, any foreign genes carried by the recombinant DNA are also replicated, a process called **gene cloning** (see Methods Box, p. 404).

Plasmids are not the only means for cloning genes. Bacteriophages are also useful vectors. Special strains of bacteriophage λ (see Chapter 16) have been widely used. In these strains, the middle of the linear genome, containing genes for lysogeny and other nonessential genes, has been deleted, with the cuts made at restriction enzyme cleavage sites. To replace the deleted DNA, restriction fragments of foreign DNA can be inserted. The recombinant phage DNA is then introduced into an *E. coli* cell. Once inside the cell, the phage DNA replicates itself and produces new phage particles, each of which carries the foreign DNA "passenger."

Sources of Genes for Cloning

Where do biologists get the genes they wish to insert into vectors and clone? There are two major sources: DNA isolated directly from an organism and "complementary DNA" made in the laboratory from mRNA templates.

Genomic Libraries The most obvious source of genes is DNA isolated directly from a particular organism. Scientists begin with all the DNA of an organism containing the gene of interest. They use restriction enzymes to snip the DNA into thousands of pieces, each typically a little larger than a gene. Next, all those pieces are inserted into plasmids or

Gel electrophoresis separates macromolecules on the basis of their rate of movement through a gel under the influence of an electric field. A mixture of proteins or nucleic acids is applied near one end of a thin slab of a polymeric gel supported by glass plates and bathed in an aqueous solution. Electrodes are attached to both ends and the current turned on. Each macromolecule then migrates toward the electrode of opposite charge at a rate determined by its charge and size (larger molecules are slowed down more by the gel).

For nucleic acids, which, owing to their phosphate groups, carry negative charges proportionate to their lengths, the rate of migration is a direct measure of molecular size. The gel in the photograph has been treated with a DNA-binding dye that fluoresces pink in ultraviolet light; DNA bands corresponding to DNA restriction fragments can be seen.

For proteins, two main types of gel electrophoresis are used. In one type, the charges due to different amino acid side groups are allowed to influence the rate of migration, along with size. In the other type, the electrophoresis is carried out in the presence of a detergent (sodium dodecyl sulfate) that coats the protein and gives it a roughly uniform negative charge. As a result, size alone determines migration rate, as in the case of DNA.

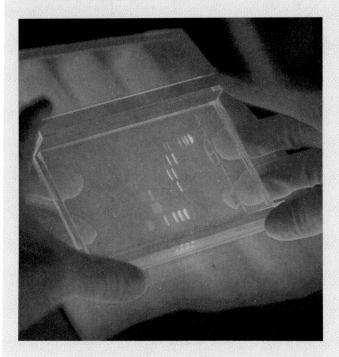

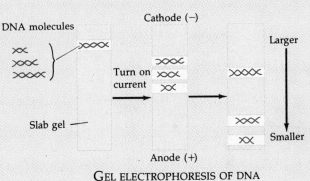

GEL ELECTROPHORESIS OF DNA

viral DNA, and the vectors containing the foreign DNA are introduced into bacteria. Because this procedure involves using a mixture of all the fragments from all of an organism's genetic material, it is often called a "shotgun approach"—an analogy to the barrage of small pellets fired from such weapons. The set of thousands of DNA segments from a genome, each carried by a plasmid or phage, is referred to as a **genomic library** (Figure 19.5).

Complementary DNA One problem with cloning DNA directly from a eukaryotic genome is that genes of interest may be too large to clone easily because they contain long noncoding regions (introns). To avoid this problem, scientists can use **complementary DNA (cDNA)** as their source of eukaryotic DNA. The cDNA is made in the laboratory by using messenger RNA molecules as templates (Figure 19.6). This synthesis of DNA on an RNA template is catalyzed by the enzyme

METHODS: GENE CLONING IN A PLASMID

This diagram shows one method of cloning a gene of interest, using a bacterial plasmid as the vector. The plasmid chosen carries two genes that confer antibiotic resistance on its *E. coli* host cell: *Amp*^{R} (for ampicillin resistance) and *Tet*^{R} (for tetracycline resistance). Furthermore, the plasmid has a single cleav-age site for the restriction enzyme used, and the site lies within the *Tet*^{R} gene. The presence of antibiotic-resistance genes and the location of the restriction site facilitate the identification of bacteria carrying recombinant plasmids, as described below.

The plasmid is isolated from *E. coli*, and "foreign" DNA is isolated from other cells.

A restriction enzyme opens the plasmid at a known cleavage site, disrupting the *Tet*^{R} gene. The same enzyme is used to cut up the foreign DNA.

The sticky ends of the plasmid hydrogen-bond with the complementary sticky ends of a restriction fragment from the foreign DNA. The two DNA molecules are joined covalently by DNA ligase.

The result is recombinant DNA, an *E. coli* plasmid carrying DNA from another source.

The recombinant plasmid is introduced into a bacterial cell by transformation. As the bacterium reproduces, so does the recombinant plasmid. The final result is a bacterial clone (colony) in which the foreign gene has been cloned. Colonies carrying recombinant plasmids can be identified by the fact that their cells are ampicillin-resistant (since they carry the *Amp*^{R} gene) but tetracycline-sensitive (since the *Tet*^{R} gene has been inactivated by the insertion of foreign DNA). Thus, the cells will grow on medium containing ampicillin but not on medium with tetracycline.

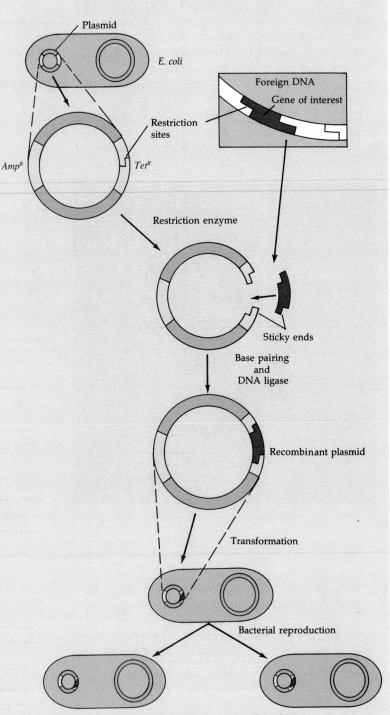

Plasmid

E. coli

Foreign DNA

Gene of interest

Restriction sites

Amp^R

Tet^R

Restriction enzyme

Sticky ends

Base pairing and DNA ligase

Recombinant plasmid

Transformation

Bacterial reproduction

Bacterial clone carrying many copies of the foreign gene

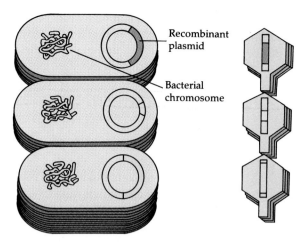

(a) Plasmid cloning vector   *(b)* Phage cloning vector

Figure 19.5
Genomic libraries. A genomic library is a collection of a large number of DNA fragments from a genome. Each fragment, containing about one gene, is carried by a vector, either **(a)** a plasmid within a bacterial cell or **(b)** a phage. The vector is the vehicle for cloning (replicating) the DNA fragment of interest and for conveying it from one cell to another.

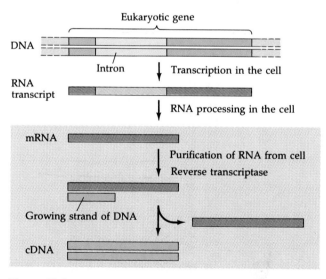

Figure 19.6
Making complementary DNA (cDNA) for a eukaryotic gene. Complementary DNA is DNA made in the laboratory using mRNA as a template and the enzyme reverse transcriptase obtained from retroviruses. Complementary DNA lacks introns and so is smaller than the original gene and easier to clone. It is also more likely to be functional in bacterial cells, which lack the machinery for removing introns from RNA transcripts. To be transcribed, the cDNA must first be joined to an appropriate promoter.

reverse transcriptase, which is obtained from retroviruses (see Figure 17.11). Even if the gene of interest contains introns in its native state, the cDNA does not, because they have been removed from the template mRNA during RNA processing (see Chapter 16). Thus, not only is the cDNA gene more manageable in size than the original gene, but it also has the potential of being translated into protein by bacterial cells, which lack RNA-processing machinery. In order for the cDNA gene to be transcribed in bacteria, it is attached to other DNA containing a promoter and other essential transcription signals.

Because the mRNA molecules for a particular gene cannot usually be isolated from the other mRNA molecules of a cell, the cDNA method also produces genomic libraries (though only partial ones, because no cells transcribe all their genes). However, if the cells used are from a specialized tissue or from a cell culture devoted almost exclusively to making one gene product (protein), a large fraction of the mRNA may be for the gene of interest. For example, most of the mRNA in the precursors of mammalian red blood cells is for the protein hemoglobin.

Finding the Gene

Often, the most difficult challenge in genetic engineering—using either the shotgun or the cDNA approach—is the identification of the bacterial clone containing the gene of interest. If the clones containing the gene of interest actually translate the gene into protein, they can be identified by screening for the presence of the protein. Detection of the protein can be based on either its activity (as with an enzyme) or its structure, using antibodies that combine with it. (Antibodies combine very specifically with particular proteins, as described in Chapter 39.) More often, screening techniques rely on detecting the gene itself, rather than its product. Methods for detecting the gene directly are all based on base pairing between the gene and a complementary sequence on another nucleic acid molecule, either RNA or DNA. For example, when at least part of the nucleotide sequence of the gene is known, or can be guessed from knowledge of the amino acid sequence of the protein, nucleic acid segments can be chemically synthesized and used to identify that gene by hydrogen bonding to complementary sequences. The piece of nucleic acid used to find a gene of interest in this way is called a **probe**, and its location is traced by labeling it with a radioactive isotope (Figure 19.7). Once a bacterial clone carrying the desired gene is identified, the gene of interest can easily be isolated in large amounts and used for further study. Also, the cloned gene itself can be used as a probe to identify similar or identical genes.

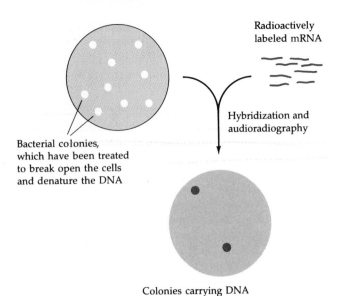

Bacterial colonies, which have been treated to break open the cells and denature the DNA

Radioactively labeled mRNA

Hybridization and audioradiography

Colonies carrying DNA complementary to the mRNA probe

Figure 19.7
Use of a nucleic acid probe to identify a cloned gene of interest. This technique makes use of the fact that nucleic acids of complementary sequence will base-pair (hybridize). Here the probe is mRNA tagged with a radioactive isotope. It is used to find the particular bacterial colonies that carry recombinant plasmids containing the gene of interest.

DNA Synthesis and Sequencing

In some cases, genes for cloning have been chemically synthesized in the laboratory; an example is the artificial genes for the two polypeptide chains of the hormone insulin. The laboratory synthesis of genes was once a tedious process, but machines are now available that can rapidly produce genes several hundred nucleotides long. Still, this approach is currently only practical for short genes whose exact nucleotide sequence is known.

Determining the sequence of a single, simple gene was also once enormously difficult. However, thanks to methods developed during the late 1970s in the United States and England, the sequences of even large genes can now often be figured out in a matter of a few days. Sequencing of DNA molecules was made possible by the availability of restriction enzymes, which cut the very long DNA found in cells and viruses into discrete, reproducible fragments with unique sequences. The two main methods of DNA sequencing use gel electrophoresis to separate strands of DNA differing in length by only a single nucleotide. (For the method used most often today, see Methods Box, p. 407.)

Thousands of DNA sequences are being collected in computer data banks, and they are proving of great value for understanding genes and genetic control elements, as well as for biotechnology. With the help of computers, long sequences can be readily scanned for shorter sequences known to be protein recognition sites or control sequences (for example, promoters) or for similarities to known sequences in other genes or other organisms. Nucleotide sequences can also be automatically translated to give amino acid sequences.

Making the Gene Product

Putting a gene from one organism into another organism is one thing; getting it to function is quite another. The problem of getting bacteria to express eukaryotic genes, even cDNA genes, was expected to be extremely difficult. Even if the lack of RNA-splicing machinery in bacteria was overcome by using cDNA genes, there remained the fact that the signals that control gene transcription and translation and the enzymes that recognize them are different in prokaryotes and eukaryotes, as are the details of protein synthesis. Therefore, it was a pleasant surprise to find that many genes of yeast (a single-celled eukaryote) are expressed in *E. coli*. Furthermore, with the help of various genetic tricks, other genes of yeast, as well as genes of more complex eukaryotes, can also be expressed in bacteria.

When the goal is to make as much of a gene product as possible, such as for commercial purposes, bacteria are usually the organisms of choice. They can be grown rapidly and cheaply in large fermenters (Figure 19.8).

Figure 19.8
Manufacture of a recombinant-DNA gene product.
The bacteria carrying the gene of interest can be grown in large fermenters; the fermenter shown here holds 1500 liters.

METHODS: SEQUENCING OF DNA BY THE SANGER METHOD

Developed by Frederick Sanger, this method and its variations are now the most commonly used techniques for determining the nucleotide sequence of DNA molecules. The Sanger method involves synthesizing in vitro DNA strands complementary to one of the strands of the DNA being sequenced. The method is based on the incorporation of a modified nucleotide (a *dideoxyribonucleotide*, which lacks *two* OH groups) that blocks further DNA synthesis. (See Figures 15.11 and 15.13 for a review of the mechanism of DNA synthesis.) Before beginning the synthesis procedure, the DNA to be sequenced is cut up into restriction fragments. Then the following procedure is carried out with each fragment.

A preparation of one of the strands of the DNA fragment is divided into four portions, and each portion is incubated with all the ingredients needed for the synthesis of complementary strands: a primer (radioactively labeled), DNA polymerase, and the four deoxyribonucleotide triphosphates. In addition, each reaction mixture contains a different *one* of the four nucleotides in the modified, dideoxy (dd) form.

Synthesis of the new strands starts with the primer and continues until a dideoxyribonucleotide is incorporated, at which point synthesis stops. Since the reaction mixture contains both deoxy and dideoxy forms of one nucleotide, the two forms will "compete" for incorporation into the strand. Eventually, a set of radioactive strands of various lengths will be generated. In the diagram, this is shown for only one of the reaction mixtures.

The new DNA strands in each reaction are separated by electrophoresis on a polyacrylamide gel, which can separate strands differing by as little as one nucleotide in length. The sequence of the newly synthesized strands can be read directly from the bands produced in the gel, and from that, the sequence of the original template strand is deduced.

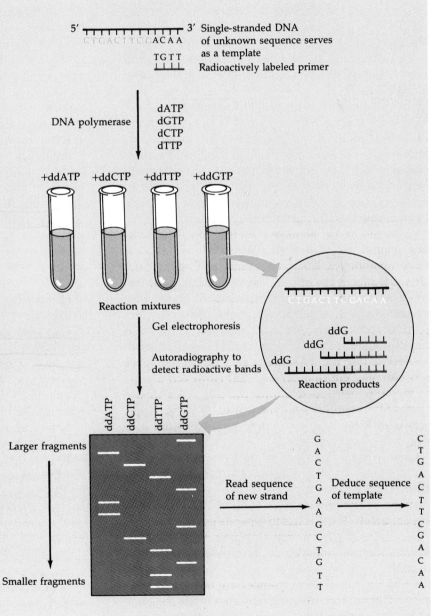

Furthermore, their genomes are relatively simple and easy to manipulate. A number of genetic and biochemical methods can help maximize the expression of a eukaryotic gene in a bacterial cell. They include using a plasmid vector that will produce many copies per cell, changing the promoter controlling the gene's expression to a highly active one, and attaching the eukaryotic gene to the initial portion of a bacterial gene for a protein that is naturally produced in large quantities. (Enzymes are later used to clip off the unwanted bacterial portion of the resulting protein.) Also, bacterial cells can be engineered to secrete a protein as it is made, thereby simplifying the task of purifying it.

Yeast offer some of the advantages of bacteria, such as ease of growth. Although more complex than bacteria, their status as very simple eukaryotes makes them a good choice for some commercial applications and, more important, for basic research into the molecular biology of eukaryotic cells. Information from studies with yeast has exploded in the past few years, leading James Watson to use "Yeasts as the *E. coli* of Eucaryotic Cells" as the title of a chapter in his textbook of molecular genetics (*Molecular Biology of the Gene*, Fourth Edition). To study eukaryotic mechanisms of gene function and regulation, eukaryotic cells as well as eukaryotic genes must be used. The eukaryotic genes of interest are initially cloned in bacteria, but they are later returned to eukaryotic cells to allow gene expression and its study. Reinserting genes into eukaryotic cells is relatively easy with yeast, since yeast can be made to take up DNA by transformation and even have plasmids that facilitate the process.

Even the combination of bacteria and yeast, however, cannot be made to fit every purpose. The cells of more complex eukaryotes carry out certain biochemical processes that are absent in yeast. Thus, cultured animal cells or plant cells need to be used for certain kinds of genetic research and for certain commercial applications. For example, only animal cells have the complex machinery needed to make antibodies, which are highly specialized glycoproteins (see Chapter 39).

APPLICATIONS OF RECOMBINANT DNA TECHNOLOGY

Biological Research

Recombinant DNA technology has triggered research advances in almost all fields of biology by allowing biologists to tackle more specific questions with finer tools. As mentioned earlier, the new techniques have opened up the study of the molecular details of eukaryotic gene structure and function. Once a gene has been cloned, and the clone identified, the gene can be readily produced in large amounts and used as a probe to search out similar DNA segments within the same or other genomes. By virtue of its ability to hydrogen-bond to complementary nucleotide sequences, the DNA probe acts as an agent to locate needles in a haystack. Radioactive isotopes or other kinds of labels tag the probe itself (see Figure 19.7). The beauty of these methods is that they do not depend on whether or not genes are expressed. For the first time, the geneticist can easily study genes directly, without having to infer genotype from phenotype. However, they now are often faced with the opposite problem: determining what a cloned gene does—that is, inferring phenotype from genotype.

The DNA segments fished out by the probe can help the biologist answer important questions. They can provide information about the evolutionary relationships between the gene of interest and other genes within the same organism or within other organisms. They can make possible examination of the natural form of the gene (the probe may have been cDNA) and allow the biologist to look at regulatory sequences and other noncoding sequences adjacent to the gene, or within it. Such information, combined with the results from gene expression experiments, is greatly increasing our understanding of how genes are organized and controlled.

A DNA probe can even be used to map a gene on a eukaryotic chromosome. In the technique called **in situ hybridization,** a radioactive DNA probe is allowed to base-pair (hybridize) with complementary sequences on intact chromosomes on a microscopic slide (*in situ* means "in place"). Autoradiography and chromosome staining are then used to reveal which band on which chromosome the probe has attached to (Figure 19.9).

In addition to allowing the production of large amounts of particular genes, recombinant DNA technology also allows the production of large quantities of proteins that are present naturally in only minute amounts. This is especially important because many molecules crucial for controlling cell metabolism and development are scarce and so cannot be purified and characterized by traditional biochemical methods.

Some researchers, emboldened by the power of these new techniques, are attempting to catalog all the proteins made by particular organs, such as the brain. The task has become feasible now that we are no longer limited to studying only those proteins produced in large enough quantities to detect with antibodies or with assays for enzymatic activities. Instead, complementary DNA from all the messenger RNA molecules present in the cells can be made, cloned in bacteria, and enough protein produced to investigate characteristics and function. Early estimates are that the brain catalog must include tens of thousands of different proteins.

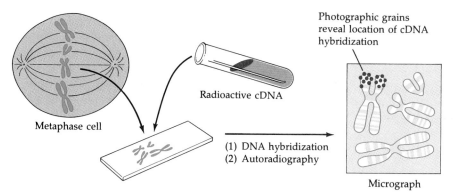

Photographic grains reveal location of cDNA hybridization

Radioactive cDNA

Metaphase cell

(1) DNA hybridization
(2) Autoradiography

Micrograph

Figure 19.9
Mapping a human gene by in situ hybridization of DNA. Radioactively labeled cDNA prepared for the gene of interest is incubated with metaphase chromosomes on a microscope slide. The cDNA binds to complementary sequences on the chromosomes. Its location is detected by autoradiography (the radioactive cDNA exposes a photographic emulsion). For simplicity, only four chromosomes are shown.

The Human Genome Project

The most ambitious research project made possible by modern biotechnology is the project to map the entire human genome. Four complementary approaches are being used:

• *Genetic (linkage) mapping* of the human genome. The initial goal is to locate at least 3000 genetic markers (genes or other identifiable loci on the DNA) spaced evenly throughout the chromosomes. This map will enable researchers to map other genes easily, by testing for genetic linkage to known markers.

• *Physical mapping* of the human genome. This is done by breaking each chromosome into a number of identifiable fragments, and then determining their actual order in the chromosome.

• *Sequencing* the human genome—that is, determining the exact order of the nucleotide pairs of each chromosome. Since a haploid set of human chromosomes contains approximately 3 billion nucleotide pairs, this is potentially the most time-consuming part of the project.

• Carrying out similar analysis of the genomes of other species important in genetic research, such as *E. coli*, yeast, and mouse.

These four approaches will yield data that will come together to give a complete map of the human genome and an understanding of how the human genome compares to those of other organisms. The potential benefits of such information are huge. In the health area, the identification and mapping of the genes responsible for genetic diseases will surely aid in the diagnosis, treatment, and prevention of those conditions. In the area of basic science, detailed knowledge of the genomes of humans and other species will give insight into fundamental questions of genome organization, control of gene expression, cellular growth and differentiation, and evolutionary biology.

Although the approaches just listed will be followed simultaneously, the emphasis in the early years of the project will be on developing a relatively low resolution map based on genetic and physical mapping. This will allow time for the development of more advanced technology for doing the full-scale sequencing in the most rapid and inexpensive way possible. Analysis of the human genome data and comparison with data from other species will proceed throughout all phases of the project, which will involve individual researchers and special research centers across the United States. Other countries are organizing similar efforts.

The methods used to build the human genetic map will combine classical pedigree analysis of large families (see Chapter 13), the use of many of the special molecular techniques described earlier in this chapter, and additional techniques that are still being developed. A significant part of the technological power to achieve the project's goals will undoubtedly come from advances in automation, utilizing the latest electronic technology. Equally important will be advances that are emerging, as such advances have in the past, from basic research in biology and biochemistry. We will now discuss two new techniques that arose in this way and are already being applied for genome mapping and other purposes.

RFLP Analysis   Earlier in this chapter, it was mentioned that the DNA fragments resulting from cutting a particular piece of DNA with a particular restriction enzyme give a characteristic pattern of bands upon gel electrophoresis (see p. 402). Each band corresponds to a DNA restriction fragment of a certain length. Using this technique to examine segments of DNA known to carry different alleles of a given gene, researchers found that DNA of different alleles would sometimes show different band patterns. This result was not surprising, since differences in nucleotide sequence would be expected to result in differences in the number and locations of restriction sites. However, when other segments of DNA were examined in this way—DNA from alleles that give rise to identical phenotypes or DNA from homologous *non*coding regions of a chromosome—the results were somewhat surprising: Differences in restriction fragment patterns were observed much more often than had been expected. Such differences in DNA sequence on homologous chromosomes that result in different patterns of restriction fragment lengths have been dubbed **restriction fragment length polymorphisms (RFLPs).** The Methods Box on pages 412–413 describes how RFLPs are analyzed.

RFLPs (pronounced "riflips") turn out to be scattered abundantly throughout the human genome. Because of this abundance and because RFLPs can be detected whether or not they lead to a difference in the organism's phenotype, they are extremely useful as genetic markers for making linkage maps. They do not have to be located in the coding portion of a gene (exon); they can be located within an intron or indeed in any part of the vast amount of noncoding DNA in the genome (less than 5% of the human genome is thought to code for proteins). Furthermore, a RFLP marker is frequently found in numerous variants in a population (the word *polymorphism* comes from the Greek for "many forms"). The discovery of RFLPs has multiplied manyfold the markers available for mapping the human genome. No longer are human geneticists limited to genetic variations that lead to obvious phenotypic differences (such as genetic diseases) or even to differences in protein products.

Although the current map of RFLPs provides fewer than a thousand markers, it is already proving useful for several purposes. Disease genes are being located by examining known RFLPs for linkage to them. If a mapped RFLP marker is inherited at high frequency along with the diseases, it is probable that the defective gene is located close to the marker on the chromosome. Furthermore, an individual's set of RFLP markers can provide a "genetic fingerprint" that is of forensic use, since the probability that two people (who are not twins) would have the same set of RFLP markers is infinitesimally small.

Polymerase Chain Reaction (PCR)   Another promising new technique, in this case made possible by research on DNA replication, is **polymerase chain reaction (PCR).** This is a technique by which any piece of DNA can be quickly amplified (copied many times) in vitro. The DNA is simply incubated under appropriate conditions with special primers and DNA polymerase molecules (see the description of DNA replication in Chapter 16). Millions of copies of a segment of DNA can be made in a few hours, a much shorter time than the weeks it usually takes to clone a piece of DNA by attaching it to a plasmid or viral genome.

Already PCR is being applied in exciting ways. It has been used to amplify for analysis DNA from a wide variety of sources: fragments of ancient DNA from a 40,000-year-old, frozen woolly mammoth; DNA from tiny amounts of tissue or semen found at the scenes of violent crimes (which can then be "fingerprinted" by RFLP analysis); DNA from single embryonic cells for rapid prenatal diagnosis; and DNA of viral genes from cells infected with such difficult-to-detect viruses as HIV, the virus that causes AIDS.

For the purposes of the human genome project, an important recent accomplishment has been the PCR amplification of DNA from a single human sperm cell. This achievement promises the possibility of doing human linkage mapping without having to find large families for pedigree analysis. As you may recall from Chapter 13, the principle underlying linkage mapping is that the distance between two genes on a chromosome is proportional to the probability of crossing over (recombination) between them. The closer two genetic markers are, the less likely will be recombination between them, and the fewer recombinants will be found among the progeny of a cross. In a human "cross," where the number of progeny is always small, recombination between two very closely linked markers may be virtually undetectable. Thus, the resolution of a human linkage map has been limited by the small sizes of human families. Now, however, by using PCR to amplify DNA from many separate sperm cells produced by one individual, researchers will be able to analyze the immediate products of meiotic recombination—and, theoretically, to analyze as large a sample as they need, even thousands of sperm. They will not have to rely on the chance that the type of recombinant chromosome they seek occurs in an offspring. Thus, they will be able to study the arrangement of genetic markers that are extremely close together.

Medicine

As has been mentioned several times, modern biotechnology is already producing significant benefits to the field of medicine.

Diagnosis of Genetic Diseases   Medical scientists would like to be able to identify individuals afflicted with genetic diseases before the onset of symptoms, preferably before birth, and, better yet, to identify heterozygotic carriers of potentially harmful recessive mutations. Until very recently, the only biochemical tests available were for gene products, such as enzymes, rather than for the genes themselves. Sometimes, however, the cells convenient for testing do not make the product, or a product for testing is not known. Now, gene cloning gives us the possibility of detecting gene mutations directly. Once cloned, the normal gene can be used as a probe to find the corresponding gene in the cells being tested, and the two genes can be compared. One way of doing this uses restriction enzymes and gel electrophoresis: The two segments of DNA are cut up with various restriction enzymes, and the patterns formed by the two sets of restriction fragments in gel electrophoresis are compared. Because a mutation changes the nucleotide sequence, it may also change a restriction enzyme cleavage site and hence, the sizes of the restriction fragments and the band patterns on gels.

Even in cases where the gene has not yet been cloned, it may be possible to diagnose the presence of an abnormal allele with reasonable accuracy if a closely linked RFLP marker is known. Blood samples from relatives of the person at risk must be available for study to determine which variant of the RFLP marker is linked to the abnormal allele in that family, and the RFLP variant(s) linked to the normal allele in that family must be different. If these conditions are met, the RFLP marker variants found in the genome of the person at risk reveal whether the abnormal allele is also likely to be present. Alleles for a number of genetic diseases, including cystic fibrosis and Huntington's disease, can now be detected in this indirect way.

Prospects for Human Gene Therapy   Eventually, genetic engineering may provide means for actually correcting genetic disorders in individuals. For genetic disorders traceable to a single defective gene, it should theoretically be possible to replace or supplement the defective gene with a functional, normal gene using recombinant DNA techniques.

Authorities predict that gene therapy will evolve slowly, starting with efforts to correct somatic cells of individuals with well-defined, life-threatening genetic defects. Since the new gene would presumably be introduced initially into only one or a few somatic cells, the cell type involved must be one that actively reproduces, so that the normal gene would be replicated within the individual. Another requirement would be that the protein product of the normal gene alleviate the mutant phenotype, despite the protein's absence in most tissues of the body. Early experiments will probably be aimed at putting the normal gene for an enzyme a patient lacks into bone marrow cells in vitro and then putting the modified cells back into the patient's body. Even if this procedure works—and bone marrow transplantation, particularly when an individual's own cells are used, is a fairly commonplace medical procedure—it does not guarantee that the presence of the normal enzyme will correct the crucial biochemical defects in the patient.

The possibility of human gene therapy has raised many questions—both technical and ethical. How can the proper genetic control mechanisms be made to operate on the transferred gene? At what stage of biological development is intervention most effective, and when is it too late? Will it be best to introduce the gene

into primitive, undifferentiated embryonic cells or into specific tissues, such as the liver or bone marrow? How can we get the correct gene or gene product to the tissues where it is required? How do we treat developmental diseases, where a gene product may be used for only a brief period during development and then cease to function thereafter?

The most difficult ethical and social question is whether we should try to treat human germ cells in the hope of correcting the defect in future generations. In mice at least, transfer of foreign genes into the germ line is now a routine procedure. For example, scientists have successfully transplanted a human gene for a crucial blood polypeptide (β-globin; see Figure 18.9) into the germ line of mice defective in the corresponding mouse gene. In many of the recipient mice and their descendants, the human gene is active in producing the polypeptide, not only in the appropriate location (red blood cells) but also at the appropriate time in development (at the appropriate fetal stage). So far, getting good expression of foreign genes in live animals has been much less successful in other experimental systems, but it is clear that the technical problems will eventually be overcome. Thus, we may soon have to face the question of whether to intervene in human germ lines.

Some critics have said flatly that tampering with human genes in any way, even to cure individuals afflicted with life-threatening diseases, is wrong. They argue that it will inevitably lead to the practice of eugenics, a deliberate effort to control the distribution of human genes. Other observers see no essential difference between genetic engineering of somatic cells and other by-now conventional medical interventions to save lives.

Vaccines   Recombinant DNA techniques are also being used in the development of vaccines for the prevention of infectious diseases. Especially for the many viral diseases for which there is no effective drug treatment, prevention by vaccination is virtually the only way to fight the disease. Traditional vaccines for viral diseases are of two types: (1) particles of a virulent virus that have been inactivated by chemical or physical means, and (2) active virus particles of an attenuated (nonpathogenic) viral strain. In both cases, the virus particles are similar enough to the active pathogen to trigger an immune response that will protect against it. In the immune response, the animal produces antibodies that will also react very specifically with invading pathogens (see Chapter 39).

There are several ways in which the new biotechnology is being used to modify current vaccines or provide new vaccines against previously recalcitrant diseases. First, recombinant DNA techniques can be used to make large amounts of a specific protein molecule from the protein coat of a particular disease-

METHODS: RFLP ANALYSIS

Solving violent crimes, identifying carriers of genetic diseases, and providing data for mapping the human genome—these are only some of the applications of the powerful method called RFLP analysis. This method is based on naturally occurring, minor differences ("polymorphisms") in DNA sequences, which can be detected because they result in restriction fragment length polymorphisms, or RFLPs.

The principle underlying RFLP analysis is illustrated below, which shows corresponding segments from two homologous chromosomes. The segments differ in sequence by a single base pair. As a consequence of this difference, the segment from chromosome A has one more restriction site (darker color) than is found on the chromosome B segment. So, when these DNA segments are treated with a restriction enzyme, they are cut into fragments that differ in both number (two versus one) and length. Upon gel electrophoresis, the DNA from the two chromosomes shows different patterns of bands.

If a particular RFLP is almost always inherited with a trait of interest in a family line, then we know that this RFLP and the gene for the trait are close together on a common chromosome. Presence of the RFLP in a genome can then serve as a marker indicating that the gene for the trait of interest has probably been inherited as well.

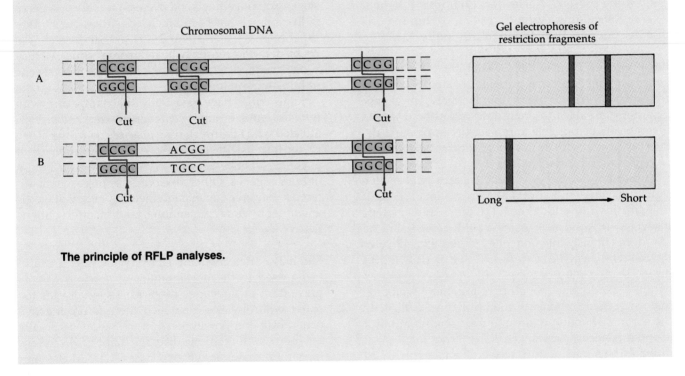

The principle of RFLP analyses.

causing virus, bacterium, or other microbe. If the protein, referred to as a *subunit*, is one that triggers an immune response against the intact pathogen, it can be used as a vaccine. Second, genetic engineering methods can be used to modify the genome of the pathogen directly to attenuate it. Vaccination with a live but attenuated organism is often more effective than a subunit vaccine, because a small amount of material triggers a greater response by the immune system, and pathogens attenuated by gene-splicing techniques may be safer than the natural mutants traditionally used.

Another scheme that uses a live vaccine employs vaccinia, the virus that is the basis of the smallpox vaccine. We can benefit from a wealth of medical experience with smallpox vaccine. With recombinant DNA techniques, the viral genes that induce immunity to smallpox can be replaced with genes that induce immunity to other diseases. In fact, the vaccinia virus can be made to carry the genes required to vaccinate against several diseases simultaneously. Therefore, in the future, a single live vaccinia inoculation may protect people from as many as a dozen diseases.

For practical applications such as medical diagnosis, the problem is to determine which variants of a particular RFLP marker (or set of markers) are carried by the individuals being tested. (In the interview preceding this unit, Nancy Wexler describes how this approach led to a marker correlated with inheritance of the Huntington's disease gene.) There is no need to physically isolate the appropriate chromosome segments before testing. Rather, DNA from the entire genome can be used (typically DNA from white blood cells) and the DNA bands of interest highlighted by a radioactive probe. The probe consists of multiple copies of a radioactively labeled piece of single-stranded DNA that will attach by base-pairing to DNA of the segment being tested (imperfections in base-pairing resulting from the RFLP variations are not extensive enough to interfere with the hybridization). The diagram below shows the details of the procedure, being carried out simultaneously on blood samples from three relatives. The results indicate that individuals I and II carry the same version(s) of the RFLP but that III carries a different version.

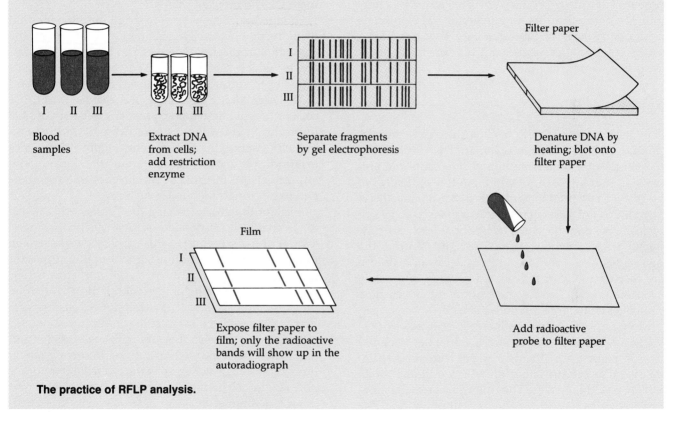

Blood samples

Extract DNA from cells; add restriction enzyme

Separate fragments by gel electrophoresis

Denature DNA by heating; blot onto filter paper

Filter paper

Add radioactive probe to filter paper

Expose filter paper to film; only the radioactive bands will show up in the autoradiograph

Film

The practice of RFLP analysis.

Pharmaceutical Products   Early demonstrations indicated that gene splicing could be used to produce hormones and other mammalian proteins in bacteria. Insulin, growth hormone, and several proteins of the immune system, such as the interferons and interleukins, are among the early examples. Insulin was the first, and human growth hormone the second, polypeptide hormone made by recombinant DNA procedures to be approved for use in treating human patients in the United States.

About two million diabetics in the United States depend on insulin treatment to help control their disease. Before 1982, the principal sources of therapeutic insulin were pig and cattle pancreatic tissues obtained from slaughterhouses. Although the insulin extracted from animals closely resembles the human version, it is not identical and causes adverse reactions in some people. Now there are several ways of producing insulin in genetically engineered bacteria, and the insulin produced is chemically identical to that made in the human pancreas.

Human growth hormone has almost 200 amino acids, making it far larger than insulin. Moreover, growth hormones are more species-specific than insulin, meaning that growth hormone from other animals is generally not an effective growth stimulator if administered to humans. Consequently, children born with hypopituitarism—a syndrome in which the pituitary gland fails to make adequate hormone, causing dwarfism—had to rely on scarce supplies of the hormone obtained from human cadavers. In 1979, a genetically engineered version of this hormone was produced in the laboratory, and after extensive testing, it was approved in 1985 for clinical use in treating children who are pituitary dwarfs. Eventually, the hormone is expected to find uses beyond the treatment of dwarfism—for example, to promote healing in burn victims or in individuals with bone fractures.

In 1989, a biotechnology firm used recombinant DNA methods to produce still another medically important human protein named erythropoietin (EPO). This protein is normally produced by the kidney as a hormone that stimulates the production of red blood cells in the bone marrow. Biotechnology now provides a source of EPO to treat anemia due to a variety of causes, including kidney damage.

Agriculture

Scientists concerned with feeding the Earth's human population hope to use recombinant DNA techniques to gain a better understanding of the plants and animals important to agriculture and, eventually, to improve their productivity. For instance, the direct introduction of particular genetic traits into farm animals is now possible; such *transgenic* animals are produced by the microinjection of foreign DNA into the nuclei of egg cells or early embryos. Farm animals also are receiving new or redesigned vaccines and newly available growth factors and hormones, which now can be made in large quantities cheaply.

More important than an increase in livestock productivity are the predicted effects of the new technology on plant crops. Plants may be made resistant to diseases, to certain harsh growth conditions, or to widely used herbicides. They also may be made more productive by enlarging the agriculturally valuable parts—whether they be roots, leaves, flowers, or stems. The agricultural products themselves may be improved, as in giving a food a balance of amino acids more suitable for the human diet.

Manipulating Plant Genes   In one striking way, plant cells have so far proved much more manipulable than the cells of higher mammals. For many plant species, an adult plant can be regenerated from a single cell growing in tissue culture (see Figure 18.13). This is an important advantage, because many genetic manipulations, such as the introduction of genes from other species, are far easier to perform and assess on single cells than on whole plants. Commercially useful plants

that are readily grown from single somatic cells include asparagus, cabbage, citrus fruits, sunflowers, carrots, alfalfa, millet, tomatoes, potatoes, and tobacco.

Plant molecular biologists have been less successful than researchers working with animal cells in the quest for genetic vectors—that is, means for moving genes freely from one organism (or species) to another. The best-developed vector is a plasmid carried by the bacterium *Agrobacterium tumefaciens*, ordinarily a pathogenic organism that causes tumors called crown galls in the plants it infects. The tumors are induced by the plasmid, called the **Ti plasmid** (tumor inducing), which somehow integrates a segment of its DNA, called T DNA, into the chromosomes of its host plant cells. Several research groups in the United States and Europe have developed ways to eliminate its disease-causing properties while maintaining its ability to move genetic material into infected plants.

Foreign genes have been inserted into this plasmid using recombinant DNA techniques, after which the recombinant plasmid is either put back into *Agrobacterium*, which can then be used to infect plant cells growing in culture, or introduced directly into plant cells. Then, taking advantage of the capacity of those cells to regenerate whole plants, it has been possible to produce plants that contain, express, and pass on to their offspring the foreign gene (Figure 19.10). Once inserted into a plant chromosome, the T DNA and any other DNA inserted in the chromosome are inherited in a normal, Mendelian fashion.

A major drawback to using the Ti plasmid as a vector is that only dicotyledons (plants with two seed leaves) are susceptible to infection by *Agrobacterium*; monocotyledons, including agriculturally important grasses such as corn and wheat, are not. Fortunately, new techniques are helping plant scientists overcome this limitation. Two of these techniques, electroporation and microinjection, were originally developed for use with animal cells; they are both methods for getting naked DNA into cells in culture. In *electroporation*, brief jolts of high-voltage electricity create temporary pores in the cell membrane, through which DNA (and other macromolecules) can enter the cell. In *microinjection*, DNA is injected directly into a cell nucleus by means of a microscopic needle. The combination of such physical techniques with standard recombinant DNA methods is helping scientists to study many aspects of gene expression in plants. These include phenomena of particular importance in plant development, such as the regulation of gene expression by light and by plant hormones.

Even with these new techniques, engineering plant genes is a formidable challenge (Figure 19.11). Although the cloning of plant DNA is straightforward, identifying genes of interest may be very difficult. Furthermore, many agriculturally desirable plant traits, such as crop yield, are polygenic, involving many genes.

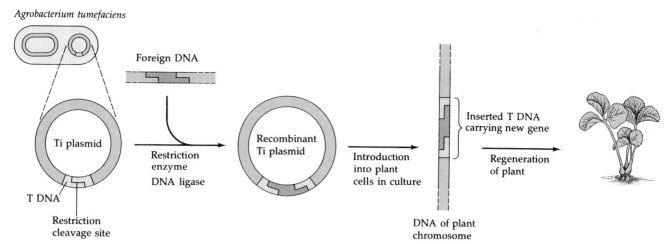

Agrobacterium tumefaciens

Foreign DNA

Ti plasmid

T DNA

Restriction cleavage site

Restriction enzyme
DNA ligase

Recombinant Ti plasmid

Introduction into plant cells in culture

Inserted T DNA carrying new gene

Regeneration of plant

DNA of plant chromosome

Figure 19.10
Using the Ti plasmid as a vector for genetic engineering in plants. The Ti plasmid is isolated from the bacterium *Agrobacterium tumefaciens*, and a fragment of foreign DNA is inserted within its T region by standard recombinant DNA techniques. When the recombinant plasmid is introduced into cultured plant cells, the T DNA integrates into the plant chromosome DNA by an unknown mechanism. As the plant cell divides, each of its descendants receives a copy of the T DNA and any foreign genes it carries. If an entire plant is regenerated, all its cells will carry—and may express—the new genes.

Early Results of Genetic Engineering in Plants

In a few cases where useful traits are determined by single genes, genetic engineering of plants is yielding positive preliminary results. For example, a number of chemical companies have developed strains of crop plants carrying a bacterial gene that makes the plant resistant to the powerful herbicide glyphosate (trade name Roundup). Glyphosate has the advantages of being nontoxic to animals and short-lived in the environment. The glyphosate-resistance trait would make it easier to grow crops while still ensuring that weeds are destroyed.

Crop yields can also be improved by genetic engineering of bacteria that live in association with plants. For example, scientists have transferred a pesticidal gene from one bacterial species into another species that inhabits the soil surrounding plant roots. The gene codes for a toxic protein that is thought to be harmless to mammals but is deadly to crop pests, such as the corn rootworm. The product, if it proves usable, is expected to be safe, not only because the toxin itself is considered safe but also because it is readily degraded in the environment, as is the bacterium that makes it.

Nitrogen Fixation

Perhaps the most exciting potential use of recombinant DNA technology in agriculture involves nitrogen fixation. Nitrogen fixation is the conversion of atmospheric, gaseous nitrogen (N_2), which is useless to plants, into nitrogen-containing compounds that plants can take up from the soil and use for making essential organic molecules, such as amino acids and nucleotides. The availability of appropriate nitrogen compounds is often the limiting factor in plant growth and crop yield, and so the use of nitrogenous fertilizers is a major aspect of modern agriculture; it is also a major expense, since the primary source for such fertilizers is chemical synthesis.

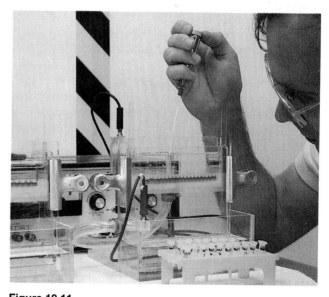

Figure 19.11
Study of plant genes. This scientist is preparing to submit restriction fragments of soybean DNA to electrophoresis on a slab gel (see Methods Box, page 403). This experiment is the first step in determining the nucleotide sequence of the gene of interest.

Recombinant DNA technology offers ways to increase the amount of biological nitrogen fixation carried out by bacteria living in the soil or in association with plants, and perhaps even to engineer plants to fix nitrogen themselves. This topic is discussed further in Chapter 33. The main point here is that it is possible to increase biological nitrogen fixation by gene-splicing techniques, and this could be of tremendous benefit to the world food supply.

SAFETY AND POLICY MATTERS

As soon as scientists realized the potential power of the recombinant DNA techniques, they also worried that there might be dangerous consequences from certain experiments, even experiments that did not involve human cells. The earliest concerns were focused on the possibility that genetic manipulations of microorganisms could create hazardous new pathogens. There were special fears of what might happen if certain tumor cell genes were transferred into microorganisms.

These initial hesitations surfaced in 1973 at a small conference devoted to gene-splicing research. Soon thereafter, scientists began to assess the status of this research and to evaluate more broadly the policy and safety issues surrounding it. The scientists developed an approach that shaped national policy for regulating recombinant DNA technology for at least the next ten years. The idea was that researchers should adhere to a set of voluntary and self-imposed guidelines, whose details would adapt to new data as they were being gathered. The recommendations soon became a formal program administered by the federal agency that funds most biomedical research, the National Institutes of Health (NIH).

The safety measures put into practice were of two types. The first type involved laboratory procedures designed to protect scientists from infection by engineered microorganisms and to prevent the microbes from accidentally leaving the laboratory. These procedures were based on the standard microbiological methods developed over the years to protect scientists and medical personnel who worked with naturally occurring pathogenic microorganisms. The second type of safety measure was biological. The only strains of microorganisms to be used in gene-splicing experiments were strains that had been genetically crippled to ensure that they could not possibly survive outside the laboratory. These special, multiply-mutant strains were quickly developed and experiments undertaken to confirm their safety. In addition, certain types of obviously dangerous experiments were banned.

Meanwhile, a whole new industry was forming, with many of its scientific leaders recruited from academic posts. A good deal of public soul-searching took place over whether it was appropriate for so many university-based biologists to straddle the academic and business worlds. Some critics were especially concerned that safety and ethical questions would be shunted aside in the rush to build profitable business enterprises.

With the worries about potentially dangerous organisms escaping from the laboratory largely assuaged, the focus of concern shifted to organisms that were *designed* to go into the environment—for example, to degrade pollutants or to perform agricultural tasks. The prospects of such products becoming available raised the question of which federal agencies were best suited to evaluate such schemes. The National Institutes of Health (NIH), which had developed a standard procedure for evaluating gene-splicing research proposals, is not a regulatory agency. However, the established regulatory agencies were initially less experienced with the new technology, and a great deal of uncertainty surrounded questions of how to apply established federal regulations to it. In several instances, the federal courts were called in. But the impressive potential of the technology and the specter of other countries taking the lead in applying it to industry, medicine, and agriculture have led federal regulators to build expertise and to work out means to approve more and more practical applications of the techniques. Today, established regulatory agencies such as the Food and Drug Administration, the Environmental Protection Agency, and the U.S. Department of Agriculture—along with the NIH Recombinant DNA Advisory Committee—are actively involved in helping the new biotechnology safely achieve its promise.

STUDY OUTLINE

1. Recombinant DNA technology is a set of techniques for recombining genes from different sources and transferring the recombinant DNA into other cells, where it may be expressed.

2. Such gene manipulation has sparked an explosion of discoveries in molecular biology and an industrial revolution in biotechnology.

Basic Strategies of Gene Manipulation (pp. 400–408)

1. Before the advent in 1975 of recombinant DNA technology, molecular geneticists were largely limited to studying laboriously obtained mutant organisms. Only phage and prokaryotic genes were readily transferable from one cell to another, by use of the natural processes of transformation, conjugation, and transduction.

2. Recombinant DNA technology has now made it possible to study virtually any gene in detail by using one or more of a number of biochemical tools and methods.

3. A variety of bacterial restriction enzymes recognize short, specific nucleotide sequences in DNA and cut the sequences at specific points on both strands to yield a set of double-stranded DNA fragments with single-stranded "sticky ends," which readily form base pairs with complementary single-stranded segments on other DNA molecules. Gel electrophoresis is used to identify, isolate, and purify these individual fragments.

4. Recombinant DNA molecules can be introduced into host cells by means of bacteriophage or plasmid vectors. In either case, gene cloning results when the foreign genes replicate inside the host bacterium.

5. The two major sources of desired genes for insertion into vectors are (a) genomic libraries that contain all the plasmid- or phage-carried DNA segments isolated directly from an organism and (b) complementary DNA synthesized from mRNA templates. The latter source is especially useful for manipulating and expressing intron-rich eukaryotic DNA

6. The presence of a particular gene can be determined directly using radioactively labeled nucleic acid segments of complementary sequence called probes. Detection can also be done indirectly by identifying the protein product of the gene.

7. Machines are now used to synthesize short genes of known nucleotide sequence and to sequence long stretches of DNA, using restriction enzymes and gel electrophoresis as key tools. DNA sequences collected in computer banks allow rapid analysis and automatic translation into amino acid sequences.

8. Bacteria and yeasts have been successfully used to make the gene products of most recombinant DNA. However, the manufacture of certain specialized proteins, such as antibodies, requires the use of cultured animal and plant cells that possess the requisite complex biosynthetic machinery.

Applications of Recombinant DNA Technology (pp. 408–416)

1. Recombinant DNA technology has been a boon for biological research, allowing investigators to answer questions about molecular evolution, probe details of gene organization and control, produce and catalog proteins of interest, and map eukaryotic genes.

2. The human genome project involves linkage mapping, physical mapping, and sequencing of the human genome, as well as similar analysis of the genomes of some simpler species.

3. Restriction fragment length polymorphisms (RFLPs) are differences in DNA sequence on homologous chromosomes that result in different patterns of restriction fragment lengths, which are visualized as bands upon gel electrophoresis. RFLPs can be used as markers for genetic mapping, and for "fingerprinting" DNA for medical or forensic purposes.

4. Polymerase chain reaction (PCR) is a technique for quickly amplifying DNA in vitro.

5. Exciting medical applications of recombinant DNA technology include the large-scale production of previously scarce pharmaceutical products; the design of safer, more effective vaccines; the development of diagnostic tests for detecting mutations that cause genetic disease; and the ultimate prospect of curing and preventing genetic disorders caused by single defective genes. The last application faces a host of formidable technological and ethical challenges that have yet to be resolved.

6. Potential agricultural benefits of genetic engineering include the design of more productive and disease-resistant plants and animals and improvement in food quality.

7. Plant cells are amenable to gene manipulation, using the Ti plasmid from *Agrobacterium* as the principal vector.

8. Preliminary results with engineering of single gene traits in plants have created optimism for increasing crop yields by biotechnology. Increasing biological nitrogen fixation by gene splicing promises to have tremendous potential for increasing the world's food supply.

Safety and Policy Matters (p. 416)

1. When scientists realized that recombinant DNA technology might have some potentially dangerous consequences, they outlined a set of safety measures that have been implemented in laboratories across the country.

2. Several U.S. government agencies are involved in regulating the use of recombinant DNA technology.

SELF-QUIZ

1. Which of the following tools of recombinant DNA technology is incorrectly paired with its use?

   a. restriction enzyme—production of RFLPs

   b. DNA ligase—enzyme that cuts DNA, creating the sticky ends of restriction fragments

   c. DNA polymerase—used in polymerase chain reaction to amplify sections of DNA

   d. reverse transcriptase—production of cDNA from mRNA

   e. electrophoresis–DNA sequencing

2. Which of the following is *not* a technique for introducing recombinant DNA into host cells?

   a. infection by *Agrobacterium*

   b. infection by bacteriophage

   c. transformation using recombinant plasmids

   d. electrophoresis

   e. microinjection

3. Which of the following is *not* true of complementary DNA?

   a. It can be amplified by a polymerase chain reaction.

   b. It can be used to create a complete genomic library.

   c. It is produced from mRNA using reverse transcriptase.

   d. It can be used as a probe to locate a gene of interest.

   e. It eliminates the introns of eukaryotic genes, and thus is more easily introduced into and cloned by bacterial cells.

4. Plants are more readily manipulated by genetic engineering than are animals because

   a. plant genes do not contain introns

   b. more vectors are available for transferring recombinant DNA into plant cells

   c. a somatic plant cell can grow into a complete plant

   d. recombinant genes can be inserted into plant cells by microinjection

   e. plant cells have larger nuclei

5. The human genome project involves all of the following *except*

   a. the location of RFLP markers

   b. the sequencing of the entire nucleotide sequence of the human genome

   c. the physical mapping of the chromosomes

   d. the analysis of the genomes of other species

   e. altering the human genome

6. Recombinant DNA technology has many medical applications. Which of the following has *not* yet been attempted or achieved?

   a. production of hormones for treating diabetes and dwarfism

   b. production of subunits of viruses that may serve as vaccines

   c. introduction of genetically engineered genes into human germ cells

   d. prenatal identification of genetic disease genes

   e. genetic testing for carriers of harmful alleles

7. Which of the following sequences along a double stranded DNA molecule may be recognized as a cutting site for a particular restriction enzyme?

   a. AAGG
      TTCC

   b. AGTC
      TCAG

   c. GGCC
      CCGG

   d. ACCA
      TGGT

   e. AAAA
      TTTT

8. In recombinant DNA methods, the term "vector" refers to

   a. the enzyme that cuts DNA into restriction fragments

   b. the "sticky end" of a DNA fragment

   c. a RFLP marker

   d. a plasmid or other agent used to transfer DNA into a living cell

   e. a DNA probe used to locate in a particular gene

9. The template used to make cDNA is

   a. DNA

   b. mRNA

   c. a plasmid

   d. a DNA probe

   e. a restriction fragment

10. The yeast cell is sometimes referred to as the "*E. coli* of eukaryotes" because

   a. it is an excellent research organism for molecular biology

   b. its habitat, like that of *E. coli*, is the human colon

   c. it is actually more similar to a bacterial cell than it is to a true eukaryotic cell

   d. it lacks a nucleus

   e. it is the simplest known multicellular organism

CHALLENGE QUESTIONS

1. In addition to their use in biological research, DNA probes have important forensic potential. Explain how such technology could be used to identify a rapist. (See Further Reading 4.)

2. In the interview preceding this unit, Nancy Wexler discussed some of the ethical problems associated with the human genome project. Do you think there is a potential in our society for genetic discrimination based on testing for "bad" genes? What policies can you suggest that would prevent abuses of genetic testing? (See also Further Reading 6.)

FURTHER READING

1. Haffie, T. "Mendel to Monctezuma: The End-of-term Review in Genetics." *Bioscience*, April 1989. Sample questions that will help you review many of the topics covered in this unit.
2. Marx, J. L. "Assessing the Risks of Microbial Release." *Science* 237 (1987): 1413–1417. Discussion of safety issues related to the use of genetically engineered microbes in agriculture.
3. Marx, J. L. "Multiplying Genes by Leaps and Bounds." *Science* 240(1988): 1408–1410. A description of the polymerase chain reaction (PCR) technique for making multiple copies of pieces of DNA in vitro, and discussion of its potential applications.
4. Moody, M. D. "DNA Analysis in Forensic Science." *Bioscience*, January 1989. DNA fingerprinting to help solve violent crimes.
5. Watson, J. D., J. Tooze, and D. T. Kurtz. *Recombinant DNA: A Short Course*, 2nd ed. New York: Scientific American Books, 1989. An overview of the principles and applications of recombinant DNA technology.
6. Weiss, R. "Predisposition and Prejudice." *Science News*, January 21, 1989. As molecular biologists develop more tests to detect "bad" genes, will genetic discrimination emerge?
7. White, R., and J.-M. Lalouel. "Chromosome Mapping with DNA Markers." *Scientific American*, February 1988. An excellent article describing in detail how RFLPs act as genetic markers on human chromosomes.

Mechanisms of Evolution

Interview: Niles Eldredge

Dr. Niles Eldredge is a paleontologist at the American Museum of Natural History in New York City, where he is a curator (and chairman) of the Department of Invertebrates. In addition to his research and responsibilities at the museum, Dr. Eldredge has authored eleven books on the subject of evolutionary theory.

Dr. Eldredge is perhaps best known for his work on the tempo of evolution. One view of evolution emphasizes gradualism—the idea that minute evolutionary changes accumulate over vast spans of time to produce evolution on a grander scale. This implies that the fossil record should provide evidence of new species being linked to their

ancestral species though a succession of intermediate fossile forms. In most cases, however, new species appear in the fossil record rather abruptly and then undergo little obvious change for the rest of their existence. Niles Eldredge interprets this feature of the fossil record as evidence that evolution occurs primarily in fits and starts rather than by a smooth continuum of minute changes.

Along with Steven Stanley of Johns Hopkins University and Stephen Jay Gould of Harvard, Dr. Eldredge is one of the principal architects of a theory of evolution known as punctuated equilibrium. Its central point is that most anatomical change is

compressed into bursts of evolution that punctuate longer periods of relative stasis.

In this interview, Dr. Eldredge discusses his work at the museum, his writing, and his thoughts on current issues in evolutionary biology.

Dr. Eldredge, could you tell us about your own education and how your interest in science and paleontology developed?

My interest in natural history and dinosaurs was like that of most kids who grew up in the 1950s. A really important thing happened to me when I went to college at Columbia University in New York City. I took a paleontology course in my junior year, and I became very deeply involved with it. I think I had decided to become an academic almost before I picked a field, and then I got swept away with circumstance, finding a field very quickly for which I developed a real, intense love—invertebrate fossils. There are literally trillions of them out there, so it looked like a wonderful opportunity.

Most academics opt for university careers. How is it that you decided on a museum?

It wasn't really a matter of seeking out a museum environment or museum career. I came up under a program that we still have, part American Museum of Natural History, part Columbia University. At that time, the museum curators were simultaneously full faculty members as opposed to being adjuncts, the way we are now. So it was natural for me to join the faculty at

Columbia at the same time I joined the museum. Of course, there are certain aspects of university life that you don't have here: the extensive teaching, for instance. Here you teach if and when you want. From the point of view of scientific research, the great part about this museum is that it is first and foremost a research institute. Many people have left university environments to come here because of this emphasis on research.

Are there other advantages to a museum appointment besides the extra time you have for research?

It gives you the opportunity for closer contact with the public through exhibitions and educational programs. For example, I like to go on the museum's educational cruises and hang over the ship's rail and have a beer with people and talk about fossils, evolution, and so on, and admit that I don't always know all the answers.

What does it actually mean to be a curator of invertebrates?

The first obligation of a curator is original scientific research. Service to the museum comes next, and in a collection-based department such as ours, that means maintaining the collections. The museum is really a library of natural history items. Curators have a mission to maintain these collections as a sort of sacred public trust. You have to have a commitment to the importance of the work that revolves around the collections before you work in a

museum. Computerizing the collections is another task. We have an on-line collection inventory that people can call from all over the country. The scientific staff is also responsible for the veracity of the content of the exhibits. There are normal administrative duties, teaching at institutions in the vicinity, giving lectures, and so on. And of course, field work takes us out of the museum.

The museum is setting up a laboratory for molecular systematics. How do you think molecular biology will influence modern evolutionary biology?

In several ways. Molecular biology gives us insight into things we can't see with our old technology, and this potential is very exciting. What is even more interesting to me is to see molecular biologists who don't want to be just molecular biologists, but are thinking, "This has to have some evolutionary implications." It is quite obvious already from gene-sequencing data that there are greater similarities in the sequences between organisms that share a relatively recent common ancestor. And comparisons of DNA can be used to assess more distant relationships. The reason I am particularly interested in seeing this work done in our department is that invertebrates were already differentiated into the major phyla about 600 million years ago. I am hoping that the data for highly conserved sequences of DNA will help us analyze the evolutionary relationships between the major groups of invertebrates.

Are there opportunities for undergraduates or new graduates to intern in natural history museums?

We have had a graduate/undergraduate research program; I started here in that program. It is not restricted to summer, though it is used mostly during the summer. In addition, the museum is now giving predoctoral fellowships. It is incumbent upon us to train our successors. We have to become even more dedicated to an educational approach than we have been.

Dr. Eldredge, you have written several books for nonscientists. What moves you to write for the general audience?

One of the first books that I wrote for the general public, in 1982, was *The Monkey Business*. I got drawn into that because I found myself unwittingly being used by the creationism movement. I had given an interview in which I was attempting to say that I thought it would be appropriate for a high school biology teacher getting into evolution to acknowledge that creationist beliefs exist, that students may have such religious beliefs but that they would be expected nonetheless to study what science says about evolution. My position was misrepresented as supporting the teaching of creationism in high schools. I felt embarrassed and used. I got mad and wrote a little article for the New Republic and then later expanded it into a book. As I was doing this, I found out that I had to be able to explain to anybody what the basics of evolution were, and that turned out to not be as easy as I had thought. I also realized that a lot of my colleagues were in the same boat, and I thought that it was incumbent upon all of us to try to address this communication problem.

When you write for the general public, you can write for intelligent people who don't share all the special training in a field and still write things that are very interesting and from which readers can learn. Another side of this is that you can take your message to the intellectual world by spreading it even more widely. Several times in my career I have had colleagues point to a general piece of writing and say, "That is the first time I have ever really understood that point." The midground, as I see it, is the fact that I am always writing for students.

There has been a revival of interest in paleontology in the past ten years or so. How do you account for this?

That is an intriguing question, and I don't have a pat answer. Traditionally, paleontologists have been content to describe the contents, the furniture, of the world and have not been too concerned about cause. They write about *what* has happened and pay less attention to *how* or *why* it happened. In the seventies they started talking about causal relationships rather than simply describing fossils. In addition, they didn't just take a body of biological theory on faith and use it to explain the fossil data; they looked at the fossil data and asked, "What does this tell us about the theory?" This made paleontology more exciting. By the late seventies there were rumblings from the paleontological community about the nature of the evolutionary process, and biologists and geneticists were beginning to take notice. At the same time, creationism was resurfacing, and the creationists were quoting the new theory called "punctuated equilibrium." We "Punc Eckers"—Stephen Jay Gould, Steven Stanley, myself, and a number of other people—seemed to give aid and succor to the enemy by daring to criticize the Darwinian message, when in fact we were turning out to be some of the more vocal, more visible anticreationists. So that brought paleontologists back into view. And then there is the phenomenon of Stephen Jay Gould himself, an exceptionally popular writer for the general audience. More than any other person, he probably symbolizes the revival of paleontology in the minds of people.

You mentioned punctuated equilibrium. Can you describe the theory?

There are really two components to the basic theory, neither one of which was terribly original. First is how we interpret the fundamental observation that there are few good examples of slow, steady, gradual transformation within species in the fossil record through time. Darwin attributed this lack of evidence for gradual transformation to the fact that paleontology was a very young science in his time and the fossil record was incomplete. But a hundred years later there still weren't many satisfying examples of gradual transformation. The theory of punctuated equilibrium takes this observation at face

value and accounts for the rarity of gradual transformation by emphasizing stability in the history of each species. The morphology of a species changes little after that species becomes established.

Second, the theory implies that most anatomical change is compressed into the relatively short time it takes for speciation to occur. Of course, the changes aren't sudden from a genetics point of view; we are talking about 5000 to 50,000 years, ample time for the kinds of changes we are talking about. But it is sudden vis-à-vis the typical long histories in the fossil record, where individual species hardly change at all. So most anatomical change in the fossil record seems to be concentrated, to occur in relatively brief bursts punctuating longer periods of relative stability.

Doesn't this idea that stability prevails in the history of a species challenge conventional Darwinism?

Darwin argued that evolution was an inevitability, and he was basically right. But among many of Darwin's followers the feeling grew that since the environment is always changing, it follows that evolution within species occurs continuously—that no species is going to remain the same. People are now beginning to see that what often happens when the environment changes is that species track suitable habitats. I mean, adaptations do change—there is no question about that and I am sure they change, at least in part, in response to changing environments. But what usually happens is that organisms relocate—even trees, via mechanisms that disperse seeds. That is what happened during the Ice Ages

in North America. There wasn't a tremendous amount of evolutionary change going on, but the geographic ranges of species changed. If it is not possible to relocate to a tolerable environment, then extinction is the next most likely thing. The least likely thing is for a species to become modified in the face of a changing habitat. I think that evolution does basically represent a match between organisms and environment. I just think that it is a tremendously conservative process most of the time.

Then how do new species originate?

I agree with Ernst Mayr and others that speciation is first and foremost the establishment of a new reproductive community from an old one. If a small population is split off from a larger parental species at the periphery of the species range, you are probably looking at organisms that are adapted to the extremes of environmental conditions that the species can tolerate. Populations living out there at the margins of the species' range are already adapted to some extent to a different kind of environment. If that population ever becomes completely isolated and no longer able to interbreed with other populations, adaptation to the local environment can be solidified. By setting up reproductive isolation, what you are really doing is giving a fledgling population a chance to have its own history; it is not going to be sharing genes any more with the parental species—it is sink or swim time. Most probably sink and suffer early extinction. But if an isolated population survives and establishes a new species, most of the anatomical change involved probably occurs relatively rapidly.

Is there evidence for punctuated equilibrium in human evolution?

Humans are really tricky. It is easier to see punctuation in lineages with a lot of adaptive anatomical specialization: Such lineages tend to be rather short-lived in the fossil record, and typically they show a lot of speciation. Hominids are a funny mixture, adaptively speaking. We are very specialized because of our culture, and our big brains are obvious anatomical features that are tied in with our cultural adaptations. On the other hand, we are generalists in how we cope with environments; in fact, our cultural successes sort of enhance our ecological generalness. So the fossil record on hominids is very strange. Some anthropologists claim that statistically there is no significant difference between the brain size of the earliest *Homo erectus* and the latest *erectus*. The picture you get is of a successful species that runs from southern Africa all the way to Java; that is variable, particularly culturally; but that has persisted for about 1.5 million years without any statistically significant change in brain size. Other people don't agree with this interpretation. Human beings are not the best case for punctuated equilibrium.

In your book *Life Pulse*, you wrote that if Darwin's expectation of gradual progressive change does not emerge from the fossil record, the main signal in life's history is extinction. Would you please explain?

I have the feeling now that if extinction hadn't chewed up the ecological fabric of life periodically, nothing much would have happened. You get some background speciation going on, but nothing really dramatic. For example, if you look at the marine forms of the Paleozoic era, you see that during several periods of time the ecosystems are fairly well established. Certain communities of species keep moving around geographically because the environment keeps changing, but pretty much the same species are forming the same sorts of communities over and over again through millions of years. Then all of a sudden the whole community disappears and you get another community with new species. And the greater the precipitating disaster, the greater the difference between the old and new forms. Apparently, there has to be a relatively severe ecosystem collapse before anything truly new accumulates. And that is what evolutionary history really is like.

Some of these upheavals in ecosystems have apparently been global. What could cause such mass extinctions?

Speculation about that is rife. A lot of the older ideas about extinctions were geared toward certain groups of organisms, dinosaurs for instance. They didn't take into account the fact that extinctions are ecosystem-wide and that very often marine, fresh water,

and terrestrial environments are affected simultaneously. More effective theories take account of all these things.

I am well aware of the impact hypothesis—asteroids hitting the Earth. I am prepared to believe that not only do asteroids hit the Earth regularly, but that when they do, they can potentially play tremendous hob with life. On the other hand, extinction seems to be so much a part of the rhythm of things that I can't believe it always has the same extraterrestrial cause. As a general mechanism, I think we have to look at the threshold effects that you get from a linear modification of the climate. A climatic change, which could be as little as a drop of 1°–2°C in mean annual temperature, is going to have a tremendous effect on the distribution of organisms. Some organisms will be able to adjust much better than others. It is a catastrophe only in retrospect: These changes usually take place over tens of thousands if not half a million years. People are concerned about how our own influence on the climate might affect the rest of life. We are clearly worried that something major could happen, particularly with our own negative intervention. So, it is perfectly possible to imagine severe ecological reorganizations occurring without a fireball coming in from space.

Isn't the view that mass extinction is an important force in evolution at odds with some people's impression of evolution as a progressive improvement of life on Earth?

I think our tendency to view evolution as progress stems from Darwin's time, when change could be accepted as a real phenomenon in nature only if it was viewed as improvement and progress. I don't know if we would call a cheetah species that is slightly faster-running than the mid-Pleistocene species progress, but if your criterion is speed, it is improvement. The really complex adaptations—the disguising body forms of walking sticks (insects) are a good example—need a lot of time, a number of speciations to accumulate. You could construe that as a form of progress. But extinction due to an environmental crisis has nothing to do with how well an organism has adapted to its normal living conditions; it is just bad luck who goes and who doesn't go. Dinosaurs are a great exam-

ple. Their extinction doesn't mean that they had come to an evolutionary dead end (although dinosaurs do seem to have been on the decline before they finally disappeared). They lived quite successfully for 150 or 175 million years, and if they hadn't had some environmental bad luck, they might still be here now and, of course, we wouldn't. There is nothing inevitable in the system that human beings would emerge. And that is where the importance of extinction really is—it reshuffles the deck.

Recently a student asked me, "If extinctions have been such an important theme in the history of life, why is everybody so upset about endangered species?" How would you answer that?

I do not see evolution as something that is intrinsically good. We are part of the biota that is now threatened with extinction, and we have every reason to resist. It is easy to take a dispassionate view about other extinctions—the dinosaurs went; they were great but they went—because we weren't here. But we don't want someone down the road saying about us, "They were fine in their time." I think also that the biota is interconnected enough that severe degradations, severe extinctions of other species, will seriously affect human life. There is no question about that. Finally, the conditions that are eliminating other species are also very directly threatening our own existence. That is perhaps even more to the point. For a variety of complex reasons, our future is very much going to be mirrored in the fates of other species around us. So if we care about our own future, we ought to care about the fates of other species.

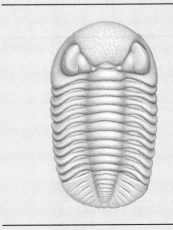

Descent with Modification: A Darwinian View of Life

20

Biology came of age on November 24, 1859, the day Charles Darwin published *On the Origin of Species by Means of Natural Selection*. His book presented the first convincing case for evolution and led the way in the emergence of biology from a bewildering chaos of facts into a cohesive science. In biology, **evolution** refers to the processes that have transformed life on Earth from its earliest beginnings to the seemingly infinite diversity that characterizes it today (Figure 20.1). Darwin addressed the sweeping issues of biology: the great diversity of organisms, their origins and relationships, their similarities and differences, their geographical distribution, and their adaptations to the surrounding environment. None of these aspects of life makes sense except in the context of evolution. Thus, evolution is the most pervasive principle in biology and a thematic thread woven throughout this textbook (see Chapter 1). This unit focuses specifically on the *mechanisms* by which life evolves.

Darwin made two points in *The Origin of Species*. First, he argued from the evidence that species were not specially created in their present forms but had evolved from ancestral species. Second, Darwin proposed a mechanism for evolution, which he termed *natural selection*. The first point can stand on its own, whether or not we accept natural selection as the explanation for how life evolves.

You will find a recurrent theme throughout the chapters of this unit: Evolutionary change is based mainly on the interactions between populations of organisms and their environments. This first chapter defines the Darwinian view of life and traces its his-

Figure 20.1
A small sample of biological diversity.
Shown here are just some of the multitude of species of moths and butterflies in the Lepidoptera collection of the National Museum of Natural History. As diverse as Lepidoptera are, all share a common body plan. A major goal of evolutionary biology is to explain how such diversity arose, while also accounting for characteristics common to different species. Darwin described evolution as descent with modification and attributed the similarities of related species to common ancestry.

torical development. Subsequent chapters discuss how evolutionary theory has been modified in this century.

PRE-DARWINIAN VIEWS

To put the Darwinian view in perspective, we must compare it to earlier ideas about the Earth and its life. The impact of an intellectual revolution such as Darwinism depends as much on timing as on the logic of the theory. The time line in Figure 20.2 will help you place Darwin's ideas in a historical context.

The Origin of Species was truly radical, for not only did it challenge prevailing scientific views, but it also shook the deepest roots of Western culture. Darwin's view of life contrasted sharply with the conventional paradigm of an Earth only a few thousand years old, populated by immutable (unchanging) forms of life that had been individually made by the Creator during the single week in which he formed the entire universe. Darwin's ideas subverted a world view that had been taught for centuries.

The Scale of Life and Natural Theology

A number of classical Greek philosophers believed in the gradual evolution of life. But the philosophers who influenced Western culture most, Plato (427–347 BC) and his student Aristotle (384–322 BC), held opinions that were inimical to any concept of evolution. Plato believed in two worlds: a real world, ideal and eternal, and an illusory world of imperfection that we perceive by the senses. The variations we see in plant and animal populations were to Plato merely imperfect representatives of ideal forms, or essences, and only these perfect forms were real. His philosophy, known as idealism, or essentialism, ruled out evolution, which would be counterproductive in a world where ideal organisms were already perfectly adapted to their environments.

Although Aristotle questioned the Platonic philosophy of dual worlds, his own beliefs also precluded evolution. A careful student of nature, Aristotle recognized that organisms ranged from relatively simple to very complex. He believed that all living forms could be arranged on a scale of increasing complexity, later

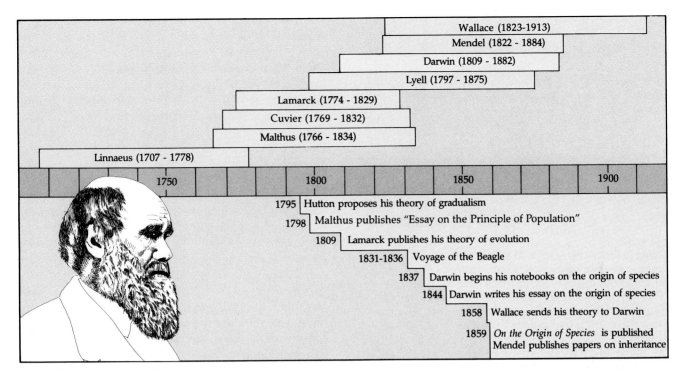

Figure 20.2
Darwinism in historical context.

called the *scala naturae* ("scale of nature"). There were no vacancies and no mobility along this ladder of life; each form had its allotted rung, and every rung was taken. In this view of life, which prevailed for over 2000 years, species are fixed, or permanent, and do not evolve.

Prejudice against evolution was fortified in Judeo-Christian culture by the Old Testament account of creation. The creationist-essentialist dogma that species were individually designed and permanent became firmly embedded in Western thought. There were many evolutionists before Darwin, but none were able to topple the doctrine of fixed species. Even as Darwinism emerged, biology in Europe and America was dominated by natural theology, a philosophy concerned with discovering the Creator's plan by studying his works. Natural theologians saw the adaptations of organisms as evidence that the Creator had designed each and every species for a particular purpose. A major objective of natural theology was to classify species to reveal the steps of the scale of life that God had created.

In the eighteenth century, Carolus Linnaeus (1707–1778), a Swedish physician and botanist, sought order in the diversity of life *ad majorem Dei gloriam*— "for the greater glory of God." Linnaeus was the father of **taxonomy,** the branch of biology concerned with naming and classifying the diverse forms of life. He developed the two-part, or binomial, system of nam-

ing organisms according to genus and species that is still used today. In addition, Linnaeus adopted a filing system for grouping species into a hierarchy of increasingly general categories. For example, similar species are grouped in the same genus, similar genera (plural of genus) are grouped in the same order, and so on. (Taxonomy is discussed in more detail in Chapter 24.)

Linnaeus sought and found order in the diversity of life with his hierarchy of taxonomic categories. But clustering certain species under taxonomic banners implied no evolutionary kinship to Linnaeus, for he believed that species were permanent creations. As a natural theologian, he developed his classification scheme only to reveal God's plan. Or, as Linnaeus himself put it, *Deus creavit, Linnaeus disposuit*—"God creates, Linnaeus arranges." Ironically, a century later, the taxonomic system of Linnaeus would become a focal point in Darwin's arguments for evolution.

Cuvier, Fossils, and Catastrophism

Fossils are relics or impressions of organisms from the past, hermetically sealed in rock (Figure 20.3). Most fossils are found in **sedimentary rocks** that form from the sand and mud that settle to the bottom of seas, lakes, and marshes. New layers of sediment cover older ones and compress them into rock such as sandstone

Figure 20.3
Fossils in sedimentary rock. This assortment of fossils of marine invertebrates was found in sedimentary rock that is over 500 million years old. The fossil record provides unequivocal evidence for the transformation of life on Earth.

Paleontology, the study of fossils, was largely founded by Georges Cuvier (1769–1832), the great French anatomist. Realizing that the history of life is recorded in strata containing fossils, he documented the succession of fossil species in the Paris Basin. He noted that each stratum is characterized by a unique suite of fossil species, and the deeper (older) the stratum, the more dissimilar the flora and fauna are from modern life. Cuvier even understood that extinction had been a common occurrence in the history of life. From stratum to stratum, new species appear and others disappear. Yet, Cuvier was a staunch and effective opponent to the evolutionists of his day. How did he reconcile the dynamic story told by the fossil record with the concept that species are immutable? Cuvier speculated that the boundaries between the fossil strata corresponded in time to catastrophic events such as floods or drought that had destroyed many of the species that had lived at that location at that time. Where there were multiple strata, there had been many catastrophes. This view of Earth history is known as **catastrophism.**

If Cuvier believed that species were fixed, then how did he account for the appearance of species in younger strata that were not present in older rocks? He proposed that the periodic catastrophes that caused mass extinctions were usually confined to local geographical regions. After the extinction of much of the native flora and fauna, the ravaged region would be repopulated by foreign species immigrating from other areas.

and shale. In places where shorelines repeatedly advance and retreat, sedimentary rock will be deposited in many superimposed layers called strata. Later, erosion may scrape or carve through upper (younger) strata and reveal more ancient strata that had been buried (Figure 20.4). The fossil record thus displays graphic and incontrovertible evidence that the Earth has had a succession of flora (plant life) and fauna (animal life).

Figure 20.4
Stratification of sedimentary rock at the Grand Canyon. The Colorado River has cut through 2000 m of sedimentary rock and unearthed many strata of varying color and thickness. Each stratum, or layer, represents a particular period in the history of the Earth and is characterized by a collection of fossils of organisms that lived at that time.

Some of Cuvier's followers had more extreme theories of catastrophism. One theory held that the catastrophes were global and that after each holocaust, God created life anew. Although Cuvier himself left religion out of his writing, his aversion to evolution came through loud and clear. But even as Cuvier was winning his debates against advocates of evolution, a theory of Earth history that would help pave the way for Darwin was gaining popularity among geologists.

Gradualism in Geology

Competing with Cuvier's theory of catastrophism was a very different idea of how geological processes had shaped the crust of the Earth. In 1795, Scottish geologist James Hutton proposed that it was possible to explain the various land forms by looking at mechanisms currently operating in the world. For example, canyons were cut by rivers running down their lengths, and sedimentary rocks with marine fossils were built of particles that had been eroded from the land and carried by rivers to the sea. Hutton explained the state of the earth by applying the principle of **gradualism,** which holds that profound change is the cumulative product of slow but continuous processes.

The leading geologist of Darwin's era, Charles Lyell (1797–1875), embellished Hutton's gradualism into a theory known as **uniformitarianism.** The term refers to Lyell's extreme idea that geological processes are so uniform that their rates and effects must balance out through time. For example, processes that build mountains are eventually balanced by the erosion of mountains. Darwin rejected this extreme version of uniformity in geological processes, but he was strongly influenced by two conclusions that followed directly from the observations of Hutton and Lyell. First, if geological change results from slow, continuous actions rather than sudden events, then the Earth must be very old, certainly much older than the 6000 years assigned by many theologians on the basis of biblical inference. Second, very slow and subtle processes persisting over a great length of time can cause substantial change. Darwin was not the first to apply this principle of gradualism to biological evolution, however.

Lamarck's Theory of Evolution

Toward the end of the eighteenth century, several naturalists suggested that life had evolved along with the Earth. But only one of Darwin's predecessors developed a comprehensive model that attempted to explain how life evolves, and that was Jean Baptiste Lamarck (1744–1829).

Lamarck published his theory of evolution in 1809, the year Charles Darwin was born. Lamarck was in charge of the invertebrate collection at the Natural History Museum in Paris, a position equivalent to the one that Niles Eldredge now holds at the American Museum of Natural History (see interview preceding this chapter). By comparing current species to fossil forms, Lamarck could see what appeared to be several lines of descent, each a chronological series of older to younger fossils leading to a modern species.

Where Aristotle saw one ladder of life, Lamarck saw many, and they were more analogous to escalators. On the ground floor were the microscopic organisms, which Lamarck believed were continually generated spontaneously from inanimate material. At the top of the evolutionary escalators were the most complex plants and animals. Evolution was driven by an innate tendency toward greater and greater complexity, which Lamarck seemed to equate with perfection. As organisms attained perfection, they became better and better adapted to their environments. Thus, Lamarck believed that evolution responded to organisms' *sentiments interieurs,* or "felt needs."

Lamarck is remembered most for the mechanism he proposed to explain how specific adaptations evolve. It entails two related principles. First is use and disuse, the idea that those organs of the body used extensively to cope with the environment become larger and stronger, while those organs that are not used deteriorate. Among the examples Lamarck cited were the blacksmith developing a bigger bicep in the arm that works the hammer and a giraffe stretching its neck to new lengths in pursuit of leaves to eat. Lamarck's second principle of adaptation is the inheritance of acquired characteristics. Lamarck believed that the modifications an organism acquires during its lifetime can be passed along to its offspring. The long neck of the giraffe, Lamarck reasoned, evolved gradually as the cumulative product of a great many generations of ancestors stretching higher and higher. There is, however, no evidence that acquired characteristics can be inherited. Blacksmiths may increase strength and stamina by a lifetime of pounding with a heavy hammer, but these acquired traits do not change genes transmitted by gametes to offspring.

The Lamarckian theory of evolution is ridiculed by some today because of its erroneous assumption that acquired characteristics are inherited; but in Lamarck's era, that concept of inheritance was generally accepted (and, indeed, Darwin could offer no acceptable alternative). To most of Lamarck's contemporaries, however, the mechanism of evolution was an irrelevant issue. In the creationist-essentialist view that still prevailed, species were fixed, and *no* theory of evolution could be taken seriously. Lamarck was vilified, especially by Cuvier, who would have no part of evolution. In retrospect, Lamarck deserves credit for his unorthodox theory, which was quite visionary in many respects: in its claim that evolution is the best explanation for both the fossil record and the current diversity of life; in its emphasis on the great age of the Earth; and in its stress on adaptation to the environment as a primary product of evolution.

ON THE ORIGIN OF DARWINISM

We have set the scene for the Darwinian revolution. Natural theology, with its view of an ordered world where each living form fit its environment perfectly because it had been specially created, still dominated the intellectual climate as the nineteenth century dawned. A few clouds of doubt about the permanence of species were beginning to gather, but no one could have forecast the thundering storm just over the horizon.

Charles Darwin was born in Shrewsbury, in western England, in 1809 (Figure 20.5). Even as a boy, Darwin's consuming interest in nature was evident. When he was not reading nature books, he was in the fields and forests fishing, hunting, and collecting insects. His father, an eminent physician, could see no future for a naturalist and sent Charles to the University of Edinburgh to study medicine. Only 16 years old at the time, Charles found medical school boring and distasteful, although he managed decent grades. He left Edinburgh without a degree and shortly thereafter enrolled at Christ College at Cambridge University, with the intent of becoming a clergyman. At that time in England, most naturalists and other scientists belonged to the clergy, and nearly all saw the world in the context of natural theology. Darwin became the protégé of the Reverend John Henslow, professor of botany at Cambridge. Soon after Darwin received his B.A. degree in 1831, Professor Henslow recommended the young graduate to Captain Robert FitzRoy, who was preparing the survey ship HMS *Beagle* for a voyage around the world.

The Voyage of the Beagle

Darwin was 22 years old when he sailed from England with the *Beagle* in December 1831. The primary mission of the voyage was to chart poorly known stretches of the South American coastline (Figure 20.6). While the crew of the ship surveyed the coast, Darwin spent most of his time on shore, collecting thousands of specimens of the exotic and exceedingly diverse fauna and flora of South America. As the ship worked its way around the continent, Darwin was able to observe the various adaptations of plants and animals that

Figure 20.5
Charles Robert Darwin (1809–1882). By age 31, when he sat for this portrait in 1840, Darwin had already published his journal of the voyage of the *Beagle*, the five-year journey during which Darwin matured as a naturalist. He had already put together the main elements of his theory of evolution by natural selection, though he did not publish *The Origin of Species* until almost 20 years later.

inhabited such diverse environments as the Brazilian jungles, the expansive grasslands of the Argentine pampas, the desolate lands of Tierra del Fuego near Antarctica, and the towering heights of the Andes Mountains.

Their unique adaptations notwithstanding, the fauna and flora of the different regions of the continent all had a definite South American stamp, very distinct from the life forms of Europe. That in itself may not have been surprising. But the plants and animals living in temperate regions of South America were taxonomically closer to species living in tropical regions of that continent than to species in temperate regions of Europe. Furthermore, the South American fossils that Darwin found, though clearly different from modern species, were distinctly South American in their resemblance to the living plants and animals of that continent. Darwin was perplexed by the peculiarities of the geographical distribution of species.

A particularly puzzling case of geographical distribution was the fauna of the Galapagos Islands, which lie on the equator about 900 km west of the South American coast (see Figure 20.6). Most of the animal

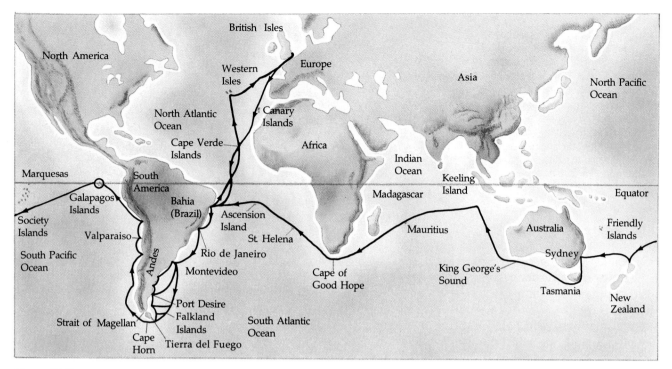

Figure 20.6
The voyage of the *Beagle*. Between 1831 and 1836, the voyage covered 60,000 km.

species on the Galapagos live nowhere else in the world, although they resemble species living on the South American mainland. Among the birds Darwin collected on the Galapagos were 13 types of finches that, although quite similar, seemed to be different species. Some were unique to individual islands, while other species were distributed on two or more islands that were close to each other. However, Darwin was not very careful in arranging his bird collection on an island-by-island basis, apparently because he did not yet appreciate the full significance of the Galapagos fauna and flora.

By the time the *Beagle* sailed from the Galapagos, Darwin had read Lyell's *Principles of Geology*. Lyell's ideas, together with his experiences on the Galapagos, had Darwin doubting the church's position that the Earth was static and had been created only a few thousand years ago. By acknowledging that the Earth was very old and constantly changing, Darwin had taken an important step toward recognizing that life on Earth had also evolved.

Darwin Frames His View of Life

At the time Darwin collected the Galapagos finches, he was not sure whether they were actually different species or merely varieties of a single species. Soon after returning to England in 1836, he learned from

ornithologists (bird specialists) that the finches were indeed separate species. He began to reassess all that he had observed during the voyage of the *Beagle* and in 1837 began the first of several notebooks on the origin of species.

Darwin began to perceive the origin of new species and adaptation as closely related processes. A new species would arise from an ancestral form by the gradual accumulation of adaptations to a different environment. For example, if one species became fragmented into several localized populations isolated in different environments by geographical barriers, the populations would diverge more and more in appearance as each adapted to local conditions, gradually, over many generations, becoming dissimilar enough to be designated separate species. This is apparently what happened to the Galapagos finches. Among the differences between the birds are their beaks, which are adapted to the specific foods available on their home islands (Figure 20.7). Darwin anticipated that explaining how such adaptations arise was essential to understanding evolution.

By the early 1840s, Darwin had worked out the major features of his theory of natural selection as the mechanism of evolution. However, he had not yet published his ideas. He was in poor health, and he rarely left home. Despite his reclusiveness, Darwin was not isolated from the scientific community. Already famous as a naturalist because of the letters and specimens he

Figure 20.7
Galapagos finches. The Galapagos island chain has a total of 13 species of closely related finches, some found on only a single island. The most striking difference among species is in their beaks, which are adapted for specific diets. (**a**) A medium-billed (left) and a large-billed (right) ground finch, both of which have beaks adapted for cracking seeds. (**b**) This tree-dwelling finch uses its beak to hold a stick and probe for food.

(a) *(b)*

sent to England during his voyage on the *Beagle*, Darwin had frequent correspondence and visits from Lyell, Henslow, and other scientists.

In 1844, Darwin wrote a long essay on the origin of species and natural selection. Realizing the significance of this work, he asked his wife to publish the essay should he die before writing a more thorough dissertation on evolution. Evolutionary thinking was emerging in many areas by this time, but Darwin was reluctant to introduce his theory publicly. Apparently, he understood its subversive quality and quite correctly anticipated the stir it would cause. While he procrastinated, he continued to compile evidence in support of his theory. Lyell, not yet convinced of evolution himself, nevertheless admonished Darwin to publish on the subject before someone else came to the same conclusions and published first.

In June 1858, Lyell's prediction came true. Darwin received a letter from Alfred Wallace, a young specimen collector working in the East Indies. The letter was accompanied by a manuscript in which Wallace developed a theory of natural selection essentially identical to Darwin's. Wallace asked Darwin to evaluate the paper and forward it to Lyell if it merited publication. Darwin complied, writing to Lyell: "Your words have come true with a vengeance. . . . I never saw a more striking coincidence . . . so all my originality, whatever it may amount to, will be smashed." That was not to be Darwin's fate, however. Lyell and a colleague presented Wallace's paper along with extracts from Darwin's unpublished 1844 essay to the Linnaean Society of London on July 1, 1858. Darwin quickly finished *The Origin of Species* and published it the next year. Although Wallace wrote up his ideas for publication first, Darwin developed and supported natural selection so much more extensively than Wallace that he is known as the main author of the theory. Darwin's notebooks also prove that he formulated his theory of natural selection fifteen years before reading Wallace's manuscript. Even Wallace felt that Darwin deserved the credit.

Within a decade, Darwin's book and its proponents had convinced the majority of biologists that evolu-

tion was fact. Darwin succeeded where previous evolutionists had failed, partly because science was beginning to shift away from natural theology, but mainly because he convinced his readers with immaculate logic and an avalanche of evidence.

THE CONCEPTS OF DARWINISM

The Principle of Common Descent

In the first edition of *The Origin*, Darwin did not use the word *evolution*, referring instead to **descent with modification,** a term that condensed his view of life. Darwin perceived unity in life, with all organisms related through descent from some unknown prototype that lived in the remote past. As the descendants of that inaugural organism spilled into various habitats over millions of years, they accumulated diverse modifications, or adaptations, that fit them to specific ways of life. In the Darwinian view, the history of life is like a tree, with multiple branching and rebranching from a common trunk all the way to the tips of the living twigs, symbolic of the current diversity of organisms. At each fork of the evolutionary tree is an ancestor common to all lines of evolution branching from that fork. Species that are closely related, such as the domestic cat and the lion, share many characteristics because their lineage of common descent extends to the smallest branches of the tree of life. Most branches of evolution, even some major ones, are dead ends; about 99% of all species that have ever lived are extinct.

Natural Selection and Adaptation

Despite the title of his book, Darwin actually devoted little space to the origin of species, concentrating instead on how populations of individual species become better adapted to their local environments through natural selection (Figure 20.8).

(a)   *(b)*

Figure 20.8
Adaptations that camouflage. (a) The ptarmigan's winter plumage hides it against the snow. **(b)** This sea horse looks so much like its kelp (seaweed) environment that it lures prey into seeming safety.

Ernst Mayr of Harvard University has dissected the logic of Darwin's theory of natural selection into three inferences based on five facts*:

Fact 1: All species have such great potential fertility that their population size would increase exponentially if all individuals that are born would reproduce successfully.

Fact 2: Most populations are normally stable in size, except for seasonal fluctuations.

Fact 3: Natural resources are limited.

Inference 1: Production of more individuals than the environment can support leads to a struggle for existence among individuals of a population, with only a fraction of offspring surviving each generation.

Fact 4: Individuals of a population vary extensively in their characteristics; no two individuals are exactly alike.

Fact 5: Much of this variation is heritable.

Inference 2: Survival in the struggle for existence is not random, but depends in part on the hereditary constitution of the surviving individuals. Those individuals whose inherited characteristics fit them best to their environment are likely to leave more offspring than less fit individuals.

Inference 3: This unequal ability of individuals to survive and reproduce will lead to a gradual change in a population, with favorable characteristics accumulating over the generations.

Natural selection is this differential success in reproduction, and its product is adaptation of organisms to their environment. Even if the advantages of some variations over others are slight, the favorable variations will accumulate in the population after many generations of being disproportionately perpetuated by natural selection.

Thus, natural selection occurs through an interaction between the environment and the variability inherent in any population. Variations arise by chance mechanisms (see Chapter 14), but natural selection is *not* a chance phenomenom. Environmental factors set definite criteria for reproductive success.

A struggle for life is ensured by excessive production of new individuals. Darwin was already aware of the struggle for existence when he read an influential essay on human population that had been written by the Reverend Thomas Malthus in 1798. Malthus contended that much of human suffering—disease, famine, homelessness, and war—were inescapable consequences of the potential for the human population to grow at a much faster rate than increased supplies of food and other resources could keep pace with. The capacity to overproduce seems to be characteristic of

---

*Modified from E. Mayr, *The Growth of Biological Thought: Diversity, Evolution and Inheritance.* (Cambridge, Mass.: Harvard University Press, 1982).

Figure 20.9
Artificial selection. The broccoli, cauliflower, and cabbages shown here, as well as kale and brussel sprouts, have a common ancestor in one species of wild mustard. By selecting different parts of the plant to accentuate, breeders have obtained these divergent results.

all species. Of the many eggs laid, young born, and seeds spread, only a tiny fraction complete their development and leave offspring of their own. The rest are eaten, frozen, starved, diseased, unmated, or unable to reproduce for some other reason.

Variation and overproduction are the two characteristics of populations that make natural selection possible. On the average, the most fit individuals pass their genes on to more offspring than the less fit. The environment screens variations, favoring some over others. Differential reproduction results in the favored traits being disproportionately represented in the next generation. But can selection actually cause substantial change in a population? Darwin found evidence in **artificial selection,** the breeding of domesticated plants and animals. Humans have modified useful species over many generations by selecting individuals with the desired traits as breeding stock. The plants and animals we grow for food bear little resemblance to their wild ancestors (Figure 20.9). The power of selective breeding is especially apparent in our pets, which have been bred more for fancy than utility.

If so much change can be achieved by artificial selection in a relatively short period of time, Darwin reasoned, then natural selection should be capable of considerable modification of species over hundreds or thousands of generations. He postulated that natural selection operating in varying contexts over vast spans of time could account for the entire diversity of life. Darwin did not see life evolving abruptly by quantum leaps, but envisioned instead a gradual accumulation of minute changes. Gradualism is fundamental to the Darwinian view of evolution.

We can now summarize Darwin's view of life: The diverse forms of life have arisen by descent with modification from ancestral species, and the mechanism of modification has been natural selection working continuously over enormous tracts of time.

Some Subtleties of Natural Selection

There are some subtleties to natural selection that require clarification. One is the importance of populations in evolutionary theory. For now, we will define a population as a group of interbreeding individuals belonging to a particular species and sharing a common geographic area. A population is the smallest unit that can evolve. Natural selection involves interactions between individual organisms and their environment, but individuals do not evolve. Evolution can be measured only as change in relative proportions of variations in a population over a succession of generations. Furthermore, natural selection can amplify or diminish only those variations that are heritable. As we have seen, an organism may become modified through its own experiences during its lifetime, and such acquired characteristics may even adapt the organism to its environment, but there is no evidence that acquired characteristics can be inherited. We must distinguish adaptations an organism acquires by its own actions from innate adaptations that evolve in a population over many generations as a result of natural selection.

It must also be emphasized that the specifics of natural selection are regional and timely; environmental factors vary from place to place and from time to time. An adaptation in one situation may be useless or even detrimental in different circumstances. This situational quality of natural selection is evident in a famous case study.

Natural Selection at Work: A Case History

The most cited and extensively documented example of natural selection in action involves the English peppered moth, *Biston betularia*. It is found throughout the English midlands, occurring in two varieties that differ in coloration. The form for which the peppered moth is named is light, with splotches of pigment. The other variety is uniformly dark. Peppered moths feed at night and rest during the daytime, sometimes on trees and rocks encrusted with light-colored lichens. Against this background, light individuals are camouflaged, but the dark moths, being very conspicuous, are easy prey for birds (Figure 20.10). Before the Industrial Revolution, dark peppered moths were very rare, presumably becoming bird food before they could reproduce and pass the genes for darkness on to the next generation. But industrial pollution darkened the

(a)

(b)

Figure 20.10
Natural selection in action: industrial melanism. The English peppered moth, *Biston betularia*, occurs in a light gray variety and a dark variety. In regions where the landscape was darkened when industrial pollution killed lichens, dark moths increased in relative number and light moths nearly disappeared. The two varieties are shown on (**a**) a tree trunk covered with lichens and (**b**) a dark tree trunk lacking lichens due to pollution. In either case, birds find and eat a greater proportion of conspicuous moths than camouflaged individuals (although other factors probably contribute to the comparative success of the two varieties of moths).

landscape of much of the countryside in the late 1800s, mainly by killing lichens that covered rocks and the dark bark of trees. Against this darkened background, light moths stood out, and dark moths were concealed from birds. The frequency of dark individuals in populations of *Biston* began to increase. By the turn of the century, the population in the Manchester region consisted almost entirely of dark moths. This phenomenon, known as industrial melanism, occurred in hundreds of other species of moths in polluted areas.

It is important to realize that industrial melanism is not a case of the inheritance of acquired characteristics. The environment did not *create* favorable characteristics, as in Lamarck's theory, but only acted upon the inherited variations manifest in any population, favoring the survival and reproduction of some individuals over others. Natural selection edits populations. The dark moths were reproductively favored because the newly visible light moths were more commonly eaten by birds and consequently left fewer offspring. Experiments support the hypothesis that natural selection in the form of predation by birds contributed to the shift in the composition of *Biston* populations in industrial regions (although recent research suggests that other factors may also be involved in the relative success of dark and light moths).

The case of *Biston* reinforces the point that natural selection operates in the here and now, tending to adapt organisms to their local environment. Natural selection is utilitarian, picking traits that work best for the present situation. In recent years, the case of the peppered moth has taken a satisfying turn, for much of the pollution has been curbed, enabling some parts of the countryside in industrial areas to return to natural hues. In those places, the light form of *Biston* has made a strong comeback.

THE MODERN SYNTHESIS

Most biologists were convinced of the fact of evolution within a few years after publication of *The Origin of Species*, but Darwin was not nearly so successful in gaining acceptance for natural selection as the mechanism of evolution. A major obstacle was the lack of any theory of genetics that could explain how chance variations arise, while also accounting for the hereditary precision that perpetuates parents' traits in their offspring. Natural selection was based on what seemed to be a paradox: Like begets like . . . but not exactly. Darwin could observe this quirk of inheritance, but he could not explain it. Heredity is such a fundamental aspect of natural selection that Darwin felt compelled to address the problem. For lack of a better explanation, he fell back on a variation of the Lamarckian concept of inheritance of acquired characteristics, which contradicted his own model of evolution. Although Gregor Mendel and Charles Darwin were contemporaries, Mendel's discoveries were unappreciated at the time, and apparently no one noticed that he had elucidated the very principles of inheritance that could have resolved Darwin's paradox and given credibility to natural selection.

Ironically, when Mendel's publication was rediscovered and reassessed at the beginning of this century, many geneticists believed that the laws of inheritance were at odds with Darwin's theory of natural selection. As the raw material for natural selection, Darwin emphasized traits that vary in a continuum in a population, such as the fur length of mammals or the speed with which an animal can flee from a predator. We know today that such quantitative traits are influenced by multiple genetic loci (see Chapter 13 to review polygenic traits and quantitative inheritance). But Mendel, and later the geneticists of the early nineteenth century, recognized only discrete "either-or" traits, such as purple versus white flowers in Mendel's peas, as heritable. Thus, there seemed to be no genetic basis for natural selection to work on the more subtle variations within a population that were central to Darwin's theory. During the 1920s, research focused on mutations, and, as an alternative to Darwin's theory of natural selection, a widely accepted hypothesis held that evolution occurred in rapid leaps as a result of radical changes in phenotype caused by mutations. This idea contrasted sharply with Darwin's view of gradual evolution due to environmental selection acting on continuous variations among individuals of a population. There was also considerable sentiment for *orthogenesis,* the idea that evolution has been a predictable progression to more and more elite forms of life. This notion of goal-oriented evolution, a throwback to Lamarck's "felt needs," opposed Darwin's mechanistic view that evolution simply reflected differential reproductive success extrapolated over many generations.

An important turning point for evolutionary theory was the birth of **population genetics,** which emphasizes the extensive genetic variation within populations and recognizes the importance of quantitative inheritance. With progress in population genetics in the 1930s, Mendelism and Darwinism were reconciled.

A comprehensive theory of evolution that became known as the **modern synthesis,** or neo-Darwinism, was forged in the early 1940s, as the genetic basis of variation and natural selection was worked out. (The genetics of evolution is discussed in the next chapter.) Paleontologists, taxonomists, and biogeographers also contributed to the modern synthesis. The modern synthesis emphasizes the importance of populations as the units of evolution, the essential role of natural selection, and gradualism. Darwin's stamp on the modern synthesis is obvious.

Today, nearly all biologists acknowledge that evolution is a fact. The term *theory* is no longer appropriate except in referring to the various models that attempt to explain *how* life evolves. Lamarck and Darwin had contrasting theories of evolution. Most of Darwin's ideas persist in the modern synthesis, the theory of evolution that has prevailed for the past 50 years. However, many evolutionists, including Niles Eldredge, are now challenging some of the generali-

zations of the modern synthesis. The debate focuses on the tempo of evolution and on the relative importance of evolutionary mechanisms other than natural selection. The study of evolution is more lively and robust than ever, and we will evaluate the current debates in subsequent chapters. Still, it is important to understand that the current questions about how life evolves in no way implies any disagreement over the fact of evolution. Arguing over evolutionary theory is like arguing over different theories of gravity: We know that objects keep right on falling, even as the debate goes on.

EVIDENCE FOR EVOLUTION

Darwin backed the principle of common descent with several lines of evidence. Subsequent discoveries, including those from molecular biology, continue to validate the evolutionary view of life.

Biogeography

It was the geographical distribution of species (**biogeography**) that first suggested common descent to Darwin. Islands have many species of plants and animals that are endemic (native, found nowhere else) and yet closely related to species of the nearest mainland or neighboring island. Some logical questions arise: Why are two islands with similar environments in different parts of the world not populated by closely related species, but inhabited instead by species taxonomically affiliated with the plants and animals of the nearest mainland, where the environment is often quite different? Why are the tropical animals of South America more closely related to species of South American deserts than to species of the African tropics? Why is Australia home to a great diversity of pouched mammals (marsupials) but almost no placental mammals (those in which embryonic development is completed in the uterus)? It is not because Australia is inhospitable to placental mammals; in recent years, humans have introduced rabbits to Australia, and the rabbit population has exploded. Apparently, placental animals are not part of the native fauna of Australia because that continent has been isolated from places where the ancestors of placental mammals lived. The geographical distribution of species makes sense only in the historical context of evolution.

The Fossil Record

The succession of fossil forms is compatible with what is known from other types of evidence about the major

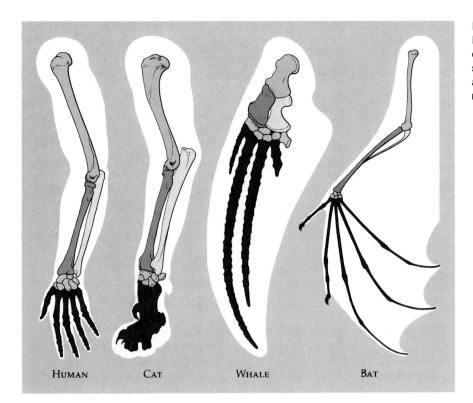

Figure 20.11
Homologous structures. The forelimbs of all mammals are constructed from the same skeletal elements, suggesting that a common ancestral forelimb has been modified for many different functions.

HUMAN   CAT   WHALE   BAT

branches of descent in the tree of life. For instance, evidence from biochemistry, molecular biology, and cell biology places prokaryotes as the ancestors of all life, and indeed the oldest known fossils are prokaryotes. Another example is the chronological appearance of the different classes of vertebrate animals in the fossil record. Fossil fishes predate all other vertebrates, with amphibians next, followed by reptiles, and then mammals and birds. This sequence is consistent with the history of vertebrate descent as revealed by many other types of evidence.

The fossil record was quite fragmentary in Darwin's time, and he was troubled by the dearth of transitional fossils linking modern life to ancestral forms. The record of past life is incomplete even today, although paleontologists continue to find important new fossils, and many of the key links are no longer missing. An example of a transitional fossil is *Archeopteryx*, a bizarre creature linking birds to their reptilian ancestors. In Chapters 22 and 23, we will evaluate a hypothesis that purports to explain why transitional fossils are not more common.

Taxonomy

Ironically, Linnaeus, who apparently believed that species are fixed, provided Darwin with some of the most potent evidence for common descent by recognizing that the great diversity of organisms could be ordered into "groups subordinate to groups" (Darwin's phrase). To Darwin, the natural hierarchy of the Linnaean scheme reflected the branching genealogy of the tree of life, with organisms at the different taxonomic levels related through descent from common ancestors. Classification in itself does not confirm the principle of common descent, but taken with other lines of evidence, the evolutionary significance of taxonomy is unmistakable. For example, genetic analysis reveals that species thought to be closely related on the basis of anatomical features and other criteria are indeed "blood" relatives with a common hereditary background.

Comparative Anatomy

Common descent is evident in anatomical similarities between species grouped in the same taxonomic category. For example, the same skeletal elements make up the forelimbs of cats, bats, whales, humans, and all other mammals, although these appendages have very different functions (Figure 20.11). Surely, the best way to construct the superstructure of a bat's wing is not also the best way to build a whale's flipper. Such anatomical peculiarities make no sense if the structures are uniquely engineered and unrelated. It is more logical, especially in view of corroborative evidence, that the basic similarity of these forelimbs is the consequence of descent of all mammals from a common ancestor. The forelegs, wing, flippers, and arms of different mammals are variations on a common anatomical theme that has been modified for divergent functions. They are **homologous structures,** the term biologists use for structures that are similar because of common ancestry. Comparative anatomy is consistent with all other evidence in testifying that evolution is a remodeling process in which ancestral structures that functioned in one capacity become modified as they take on new functions.

The oddest homologous structures are **vestigial organs,** rudimentary structures of marginal, if any, use to the organism. Vestigial organs are historical remnants of structures that had important functions in ancestors but are no longer essential. For instance, the skeletons of some snakes retain vestiges of the pelvis and leg bones of walking ancestors. (Vestigial organs may seem to support the Lamarckian concept of use and disuse, but they can be explained by natural selection. It would be wasteful to continue providing blood, nutrients, and space to an organ that no longer has a major function. Individuals with reduced versions of those organs would be favored by the environment, and natural selection operating over thousands of generations would tend to phase out obsolete structures.)

Comparative Embryology

Closely related organisms go through similar stages in their embryonic development. For example, all vertebrate embryos go through a stage in which they have paired openings called gill slits on the sides of their throats. Indeed, at this stage of development, similarities between fishes, frogs, snakes, birds, humans, and all other vertebrates are much more apparent than their differences. As development progresses, the various vertebrates diverge more and more, taking on the distinctive characteristics of their classes. In fish, for example, the gill slits develop into gills; but in terrestrial vertebrates, these embryonic structures become modified for other functions, such as the eustachian tubes that connect the middle ear with the throat in humans. The development of organs with diverse functions from the gill slits common to all vertebrate embryos supports the conclusion that all vertebrates have descended from aquatic ancestors with gills.

Inspired by the Darwinian principle of descent with modification, many embryologists in the late nineteenth century proposed the extreme view that "ontogeny recapitulates phylogeny." This notion holds that the embryonic development of an individual organism (**ontogeny**) is a replay of the evolutionary history of the species (**phylogeny**). The theory of recapitulation is an overstatement. What recapitulation does occur is a replay of *embryonic* stages, not a sequence of adultlike stages of ever more advanced vertebrates; although vertebrates share many features of embryonic development, it is not as though a mammal first

goes through a "fish stage," then an "amphibian stage," and so on. Also, because embryonic processes ultimately affect the fitness of the adult organism, they are subject to natural selection. Thus, even relatively early stages of development may become modified in the course of evolution. Nevertheless, ontogeny does provide clues to phylogeny. In particular, comparative embryology can often establish homology among structures that become so altered in later development that their common origin would not be seen by comparing their fully developed forms.

Molecular Biology

The hereditary background of an organism is documented in the DNA that constitutes its genes, and in its proteins, which are products of genes (see Chapter 16). We would expect siblings to have greater similarity in their DNA and proteins than two unrelated individuals of the same species. If the principle of common descent is correct, we should also expect two species judged to be closely related by other criteria to have a greater proportion of their DNA and proteins in common than more distantly related species. And that is the case. To measure the degree of similarity in sequences of nucleotides for DNA extracted from different species, molecular taxonomists use a few different techniques, which will be described in Chapter 23. The closer two species are taxonomically, the greater the percentage of common DNA.

Closely related species also have proteins of similar amino acid sequence, a situation that is consistent with common descent, since proteins are designated by inherited genes (Table 20.1). If two species have librar-

Table 20.1 Phylogenetic relationships and similarities in a polypeptide chain of hemoglobin

Species	Number of Amino Acid Differences in β Chain of Hemoglobin, Compared to Human Hemoglobin (Total Chain Length = 146 Amino Acids)
Human β chain	0
Gorilla	1
Gibbon	2
Rhesus monkey	8
Mouse	27
Gray kangaroo	38
Chicken	45
Frog	67
Lamprey	125
Sea slug (a mollusk)	127

ies of genes and proteins with sequences of monomers that match closely, the sequences must have been copied from a common ancestor. If two long paragraphs were identical except for the substitution of a letter here and there, we would surely attribute them both to a single source.

Darwin's boldest speculation—that *all* forms of life are related to some extent through branching descent from the earliest organisms—has also been substantiated by molecular biology. Even taxonomically remote organisms, such as humans and bacteria, have some proteins in common. An example is cytochrome *c*, the respiratory protein found in all aerobic species (see Chapter 9). Mutations have substituted amino acids at some places in the protein during the long course of evolution, but the cytochrome *c* molecules of all species are clearly akin in structure and function.

A common genetic code is further evidence that all life is related. Evidently, the language of the genetic code has been passed along through all branches of evolution ever since its inception in an early form of life. Molecular biology has thus added the latest chapter to the evidence confirming evolution as the basis for the unity and diversity of life.

* * *

Darwin gave biology a sound scientific basis by attributing the diversity of life to natural causes rather than divine creation. Nevertheless, the products of evolution are elegant, inspiring in their variety and harmony. As Darwin said in the closing pargraph of *The Origin*, "There is grandeur in this view of life. . . ."

STUDY OUTLINE

1. The first convincing case for evolution, *The Origin of Species*, was published by Charles Darwin in 1859.
2. Evolution is the most pervasive principle in biology, referring to all the changes that have transformed life on Earth throughout its history.
3. In *The Origin of Species*, Darwin argued that species evolved from ancestral forms by natural selection.

Pre-Darwinian Views (pp. 424–428)

1. The philosophies of Plato and Aristotle ruled out evolution. In particular, Aristotle envisioned a *scala naturae*, in which fixed species occupied allotted rungs on an increasingly complex scale of life.
2. Prejudice against evolution was fortified by natural theologians, who interpreted the Old Testament account of creation literally. For example, Linnaeus devised a hierarchy for naming and classifying organisms to reveal the specific steps in the divinely created scale of life.
3. Cuvier, a paleontologist and opponent of evolution, used fossils as evidence of catastrophism. He believed that catastrophic extinctions explained the unique sets of fossil species between successive strata and that new forms in younger strata resulted from immigration.
4. Geologists James Hutton and Charles Lyell supported the idea that profound changes in the Earth's surface can result from slow, continuous actions rather than sudden events. Darwin later applied this principle of gradualism to biological evolution.
5. Before Darwin, Jean Baptiste Lamarck proposed a theory of evolution in which increasing complexity and more perfect adaptations result from inheritance of characteristics acquired by organisms interacting with the environment. There is, however, no evidence for inheritance of acquired characteristics.

On the Origin of Darwinism (pp. 428–430)

1. Darwin's view of life apparently began to change when he served as naturalist on an expedition on the HMS *Beagle*.
2. On the voyage, Darwin was impressed by the peculiar geographical distribution and distinctive interrelationships of species, including those of the Galapagos Islands. This experience eventually led him to the idea that new species originate from ancestral forms by the gradual accumulation of adaptations to a new environment.
3. The long-delayed publication of *The Origin of Species* was catalyzed by Alfred Wallace, who independently arrived at the theory of natural selection.

The Concepts of Darwinism (pp. 430–433)

1. Darwin referred to evolution as descent with modification and envisioned all organisms arising from an unknown ancestral prototype, the descendants of which could be symbolized by the diverging branches of an evolutionary tree.
2. Darwin maintained that the mechanism of modification was natural selection working gradually and continuously over long periods of time.
3. Natural selection is based on differential success in reproduction, made possible because of variation in the individuals of any population and the tendency for a population to produce more offspring than the environment can support. The individuals best adapted to the local environment (most fit) leave the most offspring and thereby pass on their adaptive characteristics.

The Modern Synthesis (pp. 433–434)

1. Darwin was unaware of the discoveries of Mendel and could not explain the genetic basis of natural selection.
2. The emergence of population genetics in the 1930s, with its emphasis on quantitative inheritance and genetic variation within populations, reconciled Mendelism with Darwinism.
3. In the early 1940s, the modern synthesis, or neo-Darwinism, provided a comprehensive theory of evolution that reaffirms the essential role of natural selection, stresses gradualism, and acknowledges the importance of populations as the units of evolution.
4. Currently, evolutionary biologists are debating the tempo of evolution and the relative importance of evolutionary mechanisms other than natural selection.

Evidence for Evolution (pp. 434–437)

1. The biogeography of species first suggested common descent to Darwin. He noticed that island species were more closely related to those on the mainland than to those on distant islands with similar environments.
2. The chronological fossil record is compatible with other lines of evidence in support of evolution.
3. The taxonomic hierarchy reflects common descent.
4. Homologous structures testify to an evolutionary remodeling process. The study of embryonic development reveals homologies not apparent in adult species.
5. Closely related species show unmistakable similarities in their DNA and proteins.

SELF-QUIZ

1. The ideas of Hutton and Lyell that Darwin incorporated into his theory concerned
   a. the age of Earth and gradual geological processes producing profound change
   b. extinctions evident in the fossil record
   c. adaptation of species to the environment
   d. a hierarchal classification of organisms
   e. the inheritance of acquired characteristics

2. Which of the following is *not* a fact or inference of natural selection?
   a. There is heritable variation among individuals.
   b. Production of offspring is matched to the abundance of essential resources.
   c. Since only a fraction of offspring survive, there is a struggle for limited resources.
   d. Individuals whose inherited characteristics best fit them to the environment will leave more offspring.
   e. Unequal reproductive success leads to adaptations.

3. In the case of the English peppered moth, which of the following occurred?
   a. Bird predation was probably an important factor in natural selection.
   b. Soot incorporated into the moths resulted in industrial melanism.
   c. Dark moths were unknown before pollution.
   d. Natural selection produced new genes adapted to the darkened landscape.

4. The gill slits of reptile and bird embryos are
   a. vestigial structures
   b. support for "ontogeny recapitulates phylogeny"
   c. homologous structures
   d. used by the embryos to breathe
   e. evidence for the degeneration of unused body parts

5. The best evidence for a common origin of *all* life is
   a. comparative anatomy
   b. comparative embryology
   c. biogeography
   d. molecular biology
   e. the fossil record

6. Perhaps unfairly, Lamarkism is now most associated with
   a. catastrophism
   b. essentialism
   c. creationism
   d. inheritance of acquired characteristics
   e. uniformitarianism

7. Darwin's theory, as presented in *The Origin*, mainly concerned
   a. how new species arise
   b. the origin of life
   c. how adaptations evolve
   d. how extinctions happen
   e. the genetics of evolution

8. Which of these labels best fits Darwin?
   a. Mendelian
   b. gradualist
   c. essentialist
   d. population geneticist
   e. catastrophist

9. The Galapagos Islands are located nearest the
   a. west coast of Africa
   b. west coast of South America
   c. west coast of North America
   d. east coast of Australia
   e. west coast of Greenland

10. Which person is *incorrectly* matched with a term or idea?
    a. Plato-essentialism
    b. Linnaeus-"use-and-disuse"
    c. Malthus-overpopulation
    d. Lyell-uniformitarianism
    e. Aristotle-*scala naturae*

CHALLENGE QUESTIONS

1. Some opponents of evolution have exclaimed, "I just can't believe we came from a chimpanzee!" What is their misconception about evolution?

2. To what extent are humans in a technological society exempt from natural selection? Explain your answer.

FURTHER READING

1. Darwin, C. *The Origin of Species by Means of Natural Selection, or The Preservation of Favored Races in the Struggle for Life.* New York: New American Library, 1963. A modern printing of the historical book that revolutionized biology.
2. Futuyma, E. J. *Evolutionary Biology.* 2d ed. Sunderland, Mass.: Sinauer, 1986. An excellent undergraduate text.
3. Gould, S. J. *Ever Since Darwin: Reflections in Natural History.* New York: Norton, 1977. Essays by a gifted writer.
4. Gould, S. J. "Darwinism Defined: The Difference between Fact and Theory." *Discover,* January 1987. Is the term "scientific creationism" an oxymoron?
5. Stebbins, G. L., and F. J. Ayala. "The Evolution of Darwinism." *Scientific American,* July 1985. The impact of molecular biology on evolution.

How Populations Evolve

21

E volution results from interactions between organisms and their environments. In natural selection, it is individual organisms that are selected, but it is populations that actually evolve. A population is a localized group of individuals belonging to the same species. We can observe the evolution of a population as a change in the prevalence of certain inherited traits over a succession of generations, as in the increasing proportion of darkly pigmented peppered moths in industrial England that we saw in Chapter 20. Darwin understood that a population evolves by the environment's sorting of variations, selecting some over others, but the genetic basis of evolution eluded him. The most important post-Darwinian development in evolutionary biology has been the application of genetics to the theory of natural selection. In this chapter, you will learn about the genetic basis of variation in populations and how natural selection and other mechanisms of evolution alter the genetic constitution of populations.

THE GENETICS OF POPULATIONS

A **species,** simply defined, is a group of populations that have the potential to interbreed in nature (this definition will be examined more critically in Chapter 22). Each species has a geographical range within which individuals are not spread out evenly, but are usually

Figure 21.1
Population distribution. Populations are localized groups of individuals belonging to the same species. **(a)** Here, two dense populations of Douglas fir *(Pseudotsuga memzieii)* are separated by a river bottom where firs are uncommon. The two populations are not totally isolated; interbreeding occurs when wind blows pollen between the populations. Nevertheless, trees are more likely to interbreed with members of the same population than with trees on the other side of the river.
(b) Humans also tend to concentrate in localized populations. In this nighttime satellite view of the United States, you can see the lights of major population centers, or cities. These populations are, of course, not isolated; people move around, and there are low-density suburban and rural communities between cities. But city dwellers are most likely to choose mates who live in the same city—often in the same neighborhood.

(a)

(b)

concentrated in several localized populations. One population may be isolated from others of the same species, exchanging genetic material only rarely. Such isolation is particularly common for populations confined to widely separated islands, unconnected lakes, or mountain ranges separated by lowlands. However, populations are not always isolated, nor do they necessarily have sharp boundaries. One dense population center may blur into another in an intermediate region where members of the species occur, but are less numerous. Although the populations are not isolated, individuals are still concentrated in centers and are more likely to interbreed with members of the same population than with members of other populations. Therefore, individuals near a population center are, on the average, more closely related to one another than to members of other populations (Figure 21.1).

The Gene Pool and Microevolution

The total aggregate of genes in a population at any one time is called the **gene pool.** It consists of all alleles at all gene loci in all individuals of the population; it is the pool of genes from which members of the next generation of that population will draw their alleles. For a diploid species, each locus is represented twice in the genome of an individual, who may be either homozygous or heterozygous for that locus (see Chapter 13). If all members of a population are homozygous for the same allele, then that allele is said to be **fixed** in the gene pool. More often, there are two or more alleles for a gene, each having a relative frequency in the gene pool. For example, in a peppered moth population living in an unpolluted region of England, the allele for light color has a much higher frequency in the population than the allele for dark color. But during the industrialization of England, the allele for dark color increased at the expense of the allele for light color. Evolution is occurring on the smallest scale when the relative frequencies of alleles in a population change over a succession of generations; such change in the gene pool is called **microevolution.** Before we learn about the mechanisms of microevolution, it will be helpful to examine, for comparison, the genetics of a nonevolving population.

The Hardy-Weinberg Theorem

Imagine a plant species with two varieties contrasting in flower color, with red flowers completely dominant to white flowers. If we cross these two varieties in a typical Mendelian monohybrid experiment, 75% of the offspring in the F_2 generation will have red flowers and only 25% will have white flowers (see Chapter 13). Given this numerical prevalence of the dominant variety, it may seem reasonable that, over many generations, the allele for red flowers will become more and more common in a population at the expense of the recessive allele for white flowers. This, in fact, is not the case. In the absence of other factors, the sexual shuffling of genes cannot alter the overall genetic makeup of a population. No matter how many generations alleles are segregated by meiosis and combined by fertilization, the frequencies of alleles in the gene pool will remain constant unless acted upon by other agents. This axiom is known as the **Hardy-Weinberg theorem,** named for the two scientists who derived the principle independently in 1908.

To test the Hardy-Weinberg theorem, let us now imagine an isolated population of our wildflower species (Figure 21.2). We will use the symbols A and a, respectively, for the allele for red flowers (dominant) and the allele for white flowers (recessive). For our simplified problem, these are the only two alleles for this locus in the population. Our imaginary population has 500 plants. Twenty have white flowers because they are homozygous for the recessive allele; their genotype is *aa*. Of the 480 plants with red flowers, 320 are homozygous (*AA*) and 160 are heterozygous (*Aa*). Since these are diploid organisms, there are a total of 1000 genes for flower color in the population. The dominant allele accounts for 800 of these genes (320 × 2 = 640 for *AA* plants, plus 160 × 1 = 160 for *Aa* individuals). Thus, the frequency of the *A* allele in the gene pool of this population is 80%, or 0.8. And since there are only two allelic forms of the gene, we know that the *a* allele has a frequency of 20%, or 0.2.

Now, how will genetic reassortment during sexual reproduction affect the frequencies of the *A* and *a* alleles in the next generation of our wildflower population? We will assume that the union of sperm and ova in

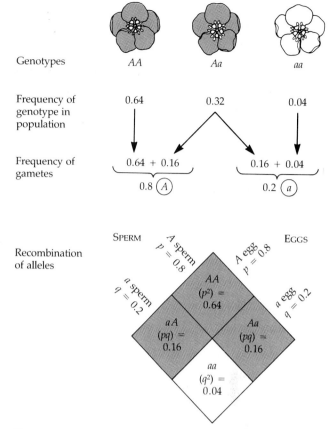

Figure 21.2
The Hardy-Weinberg theorem. The gene pool of a nonevolving population remains constant over the generations. Sexual recombination alone will not alter the relative frequencies of alleles. (p = frequency of A; q = frequency of a.)

the population is completely random—that is, all male-female mating combinations are equally likely. The situation is analogous to mixing all gametes in a sack and then drawing them randomly, two at a time, to determine the genotype for each zygote. Each gamete has one gene for flower color, and the A and a alleles will occur in the same frequencies in which they had occurred in the population that made the gametes. Every time a gamete is drawn from the pool at random, the chance that the gamete will bear an A allele is 0.8, and the chance that an a allele will be present is 0.2.

Using the rule of multiplication (see Chapter 13), we can calculate the frequencies of the three possible genotypes in the next generation of the population. The probability of picking two A alleles from the pool of gametes is 0.64 (that is, 0.8 × 0.8). Thus, about 64% of the plants in the next generation will have the genotype AA. The frequency of aa individuals will be about 4%, or 0.04 (0.2 × 0.2). And 32%, or 0.32, of the plants will be heterozygous—that is, Aa or aA, depending on whether it is the sperm nucleus or ovum that supplies the dominant allele (0.8 × 0.2 = 0.16, and 0.16 × 2 = 0.32).

Regarding the locus for flower color, the genetic makeup in the second generation of the wildflower population will be 0.64 *AA*, 0.32 *Aa*, and 0.04 *aa*. The *AA* individuals account for 64% of all genes for flower color in the population, and heterozygotes account for 32% of the genes, half of which are represented by The *A* allele. The overall frequency of the *A* allele in the population is therefore 0.8 or (0.64 + 0.32/2). The frequency of the *a* allele is 0.2 or (0.04 + 0.32/2). Notice that the alleles are present in the gene pool of the current population in the same frequencies as they were in the previous generation. If we were to repeat the process, segregating alleles to make gametes, and then picking the gametes two at a time to produce the genotypes of still another generation of plants, the frequencies of the alleles and genotypes would remain the same, generation after generation. Thus, the gene pool of the population would be in a state of equilibrium—referred to as **Hardy-Weinberg equilibrium.** (In this example, the wildflower population was at equilibrium initially. If we had started with the population not yet at equilibrium, only a single generation would be required for equilibrium to be attained. You will have a chance to prove this in Challenge Question 2 at the end of the chapter.)

From the specific case of the wildflower population, we can derive a general formula, called the Hardy-Weinberg equation, for calculating the frequencies of alleles and genotypes in populations. We will restrict our analysis to the simplest case of only two alleles, one dominant over the other. However, the Hardy-Weinberg equation can be adapted to situations in which there are three or more alleles for a particular locus and there is no clear-cut dominance.

For a gene locus where only two alleles occur in a population, let us use the letter p to represent the frequency of one allele and the letter q to symbolize the frequency of the other allele (see Figure 21.2). In the imaginary wildflower population, $p = 0.8$ and $q = 0.2$. Note that $p + q = 1$; the combined frequencies of all possible alleles must account for 100% of the genes for that locus in the population. If there are only two alleles and we know the frequency of one, the frequency of the other can be calculated:

$$1 - p = q \quad \text{or} \quad 1 - q = p$$

When gametes combine their alleles to form zygotes, the probability of generating an *AA* genotype is p^2. In the wildflower population, $p = 0.8$, and $p^2 = 0.64$, the probability of an *A* sperm fertilizing an *A* ovum to produce an *AA* zygote. The frequency of individuals homozygous for the other allele (*aa*) is q^2, or $0.2 \times 0.2 = 0.04$ for the wildflower population. Because there are two ways in which an *Aa* genotype can arise, depending on which parent contributes the dominant allele, the frequency of heterozygous individuals in the population is $2pq$. The sum of the frequencies of all possible genotypes in the population must add up to 1:

$$\underset{\substack{\text{Frequency} \\ \text{of } AA}}{p^2} + \underset{\substack{\text{Frequency} \\ \text{of } Aa \text{ and } aA}}{2pq} + \underset{\substack{\text{Frequency} \\ \text{of } aa}}{q^2} = 1$$

The Hardy-Weinberg equation enables us to calculate frequencies of alleles in a gene pool if we know frequencies of genotypes, and vice versa. We can use it to calculate the percentage of the human population that carries the allele for a particular inherited disease. For instance, approximately 1 out of 10,000 babies in the United States is born with phenylketonuria (PKU), a metabolic disorder that, untreated, results in mental retardation and other problems (see Chapter 13). The disease is caused by a recessive allele, and thus the frequency of individuals in the U.S. population born with PKU corresponds to q^2 in the Hardy-Weinberg formula. Given one PKU occurrence per 10,000 births, $q^2 = 0.0001$. Therefore, frequency of the recessive allele for phenylketonuria in the population is $q = \sqrt{0.0001}$, or 0.01. And the frequency of the dominant allele is $p = 1 - q$, or 0.99. The frequency of carriers, heterozygous people who are normal but may pass the PKU allele on to offspring, is

$$2pq = 2 \times 0.99 \times 0.01 = 0.0198$$

About 2% of the U.S. population carries the PKU allele.

How is the Hardy-Weinberg equilibrium relevant to our study of microevolution? It tells us what to expect for a nonevolving population, providing a baseline for comparing actual populations where the gene pools may in fact be changing. The Hardy-Weinberg equilibrium is maintained *only* if *all* five of the following conditions are met:

- The population is very large.

- The population is isolated; that is, there is no migration of individuals into or out of the population.

- There are no net changes in the gene pool due to mutations.

- Mating is random.

- All genotypes are equal in reproductive success.

It is important to realize that Hardy-Weinberg equilibrium describes the genetics of ideal populations that never exist in nature. Let us look now at how real populations evolve.

CAUSES OF MICROEVOLUTION

Five potential agents of microevolution are genetic drift, gene flow, mutation, nonrandom mating, and natural selection (Table 21.1). Each is a deviation from one of

Table 21.1 Causes of microevolution

Mechanism	Action on Gene Pool	Tends to Be Adaptive?
Genetic drift	Random change in small gene pool due to sampling errors in propagation of alleles	No
Gene flow	Change in gene pools due to immigration or emigration of individuals between populations	No
Mutation pressure	Change in allelic frequencies due to net mutation	No
Nonrandom mating	Inbreeding, or selection of mates for specific phenotypes (assortative mating), reduces frequency of heterozygous individuals	Unknown
Natural selection	Differential reproductive success increases frequencies of some alleles and diminishes others	Yes

the five conditions for Hardy-Weinberg equilibrium. Of all the causes of microevolution, only natural selection generally leads to an accumulation of favorable adaptions in a population. The other agents of microevolution are sometimes called non-Darwinian because of their usually nonadaptive nature.

Genetic Drift

Flip a coin a thousand times, and a result of 700 heads and 300 tails would make you very suspicious about that coin. Flip a coin ten times, and an outcome of seven heads and three tails is within reason. The smaller a sample, the greater the chance deviations from an idealized result—an equal number of heads and tails, in the case of a sample of coin tosses. This disproportion of results in a small sample is known as **sampling error,** and it is an important factor in the genetics of small populations of organisms. If a new generation draws its alleles at random, then the larger the sample size, the better it will represent the gene pool of the previous generation. If a population of organisms is small, its existing gene pool may not be accurately represented in the next generation because of sampling error. Chance events can cause the frequencies of alleles in a small population to drift randomly from generation to generation. For example, consider what could happen if the wildflower population discussed earlier consisted of only 25 plants. Assume that 16 of the plants have the genotype AA for flower color, 8

are Aa, and only 1 is aa. Now imagine that three of the plants are accidentally destroyed by a rock slide before they have a chance to reproduce. By chance, all three plants lost from the population could be AA individuals. The event would alter the relative frequencies of the two alleles for flower color in subsequent generations. This is a case of microevolution caused by **genetic drift,** changes in the gene pool of a small population due to chance. Only luck could result in random drift improving adaptation.

Ideally, a population must be infinitely large for genetic drift to be ruled out completely as an agent of evolution. Although that is impossible, many populations are so large that drift may be negligible. However, some populations are small enough for significant genetic drift to occur; chance certainly plays a major role in the microevolution of populations having fewer than 100 or so individuals. The two situations that most often lead to populations small enough for genetic drift to occur are known as the bottleneck effect and the founder effect.

Bottleneck Effect   Disasters such as earthquakes, floods, or fires may reduce the size of a population drastically, killing victims rather unselectively. The result is that the small surviving population is unlikely to be representative of the original population in its genetic makeup—a situation known as the **bottleneck effect.** By chance, certain alleles will be overrepresented among survivors, other alleles will be underrepresented, and some alleles may be eliminated completely (Figure 21.3). The genetic drift that has occurred may continue to affect the population for many generations, until the population is again large enough for random drift to be insignificant.

Genetic drift caused by bottlenecking may have been important in the early evolution of human populations when calamities decimated tribes. The gene pool of each surviving population may have been, just by chance, quite different from that of the larger population that predated the catastrophe.

Bottlenecking reduces the overall genetic variability in a population, since some alleles are lost from the gene pool. One extreme example concerns the population of northern elephant seals, which passed through a bottleneck in the 1890s when hunters reduced the population to about 20 individuals. Since then, the animal has become a protected species, and the population has grown to over 30,000 members. Researchers have examined 24 gene loci in many individuals of the northern elephant seal population, and *no* genetic variation has been found; a single allele has been fixed at each of the 24 loci, probably due in large part to genetic drift. In contrast, genetic variation abounds in populations of the southern elephant seal, which have not been bottlenecked. Bottlenecking may also explain why the South African population of cheetahs displays greater genetic uniformity than do inbred strains

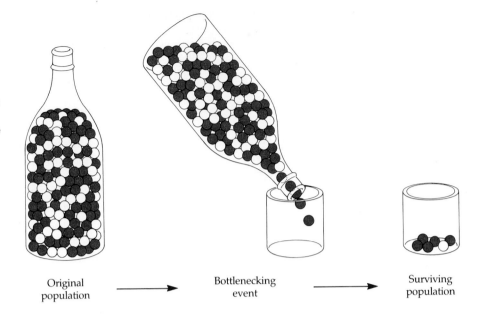

Figure 21.3
The bottleneck effect. A gene pool can drift by chance when the population is drastically reduced by a disaster that kills victims unselectively. In this analogy, a population of red and white marbles in a bottle is reduced in the bottleneck. Notice that the composition of the population in the bottleneck is not representative of the makeup of the larger, original population.

Original population

Bottlenecking event

Surviving population

of laboratory mice. The cheetah population was probably severely reduced during the last ice age about 10,000 years ago and a second time when the animals were hunted to near extinction at the beginning of this century.

Founder Effect Genetic drift is also likely whenever a few individuals colonize an isolated island, lake, or some other habitat new to that species. The smaller the sample size, the less the genetic makeup of the colonists will represent the gene pool of the larger population they left. The most extreme case would be the founding of a new population by one pregnant animal or a single plant seed. If the colony is successful, random drift will continue to affect the frequency of alleles in the gene pool until the population is large enough for sampling errors from generation to generation to be minimal. Genetic drift in a new colony is known as the **founder effect**. The effect undoubtedly contributed to the evolutionary divergence of Darwin's finches after strays from the South American mainland reached the remote Galapagos Islands.

The founder effect is probably responsible for the relatively high frequency of certain inherited disorders among human populations established by a small number of colonists. In 1814, 15 people founded a British colony on Tristan da Cunha, a group of small islands in the Atlantic Ocean midway between Africa and South America. Apparently, one of the colonists carried a recessive allele for retinitis pigmentosa, a progressive form of blindness that afflicts individuals who are homozygous for the allele. Of the 240 descendants who still lived on the island in the late 1960s, 4 had retinitis pigmentosa and at least 9 others were known, based on pedigree analysis, to be carriers. The frequency of this allele is much higher on

Tristan da Cunha than in the populations from which the founders came. Although inherited diseases provide striking examples of the founder effect, this form of genetic drift also alters the frequencies of many alleles in the gene pool that affect more subtle characteristics.

Gene Flow

Hardy-Weinberg equilibrium requires the gene pool to be a closed system, but most populations are not completely isolated. A population may gain or lose alleles by **gene flow,** the migration of fertile individuals, or the transfer of gametes, between populations.

For example, if a wind storm blows pollen to our hypothetical wildflower population from another population of the same species, allelic frequencies may change; perhaps, for example, the outlying population consists entirely of white-flowered individuals. Gene flow tends to reduce between-population differences that have accumulated because of natural selection or genetic drift. If it is extensive enough, gene flow can eventually amalgamate neighboring populations into a single population. As humans began to move about the world more freely, gene flow undoubtedly became an important agent of microevolutionary change in populations that were previously quite isolated.

Mutation

A new mutation that is transmitted in gametes immediately changes the gene pool of a population by substituting one allele for another. For example, a mutation that causes a white-flowered plant in our symbolic wildflower population to produce gametes bearing the dominant allele for red flowers would decrease the frequency of the *a* allele in the population and increase the frequency of the *A* allele. However, mutation by itself does not have much quantitative effect on a large population in a single generation. This is because a mutation at any given gene locus is a very rare event; although mutation rates vary, depending on the species and the gene locus, rates of one mutation per locus per 10^5 to 10^6 gametes are typical. If an allele has a frequency of 0.50 in the gene pool and mutates to another allele at a rate of 10^{-5} mutations per generation, it would take 2000 generations to reduce the frequency of the original allele from 0.50 to 0.49. And the gene pool would be affected even less if the mutation were reversible, as most are. If some new allele produced by mutation increases its frequency in a population, it is not because mutation is generating the allele in abundance, but because individuals carrying the mutant allele are producing a disproportionate number of offspring as a result of natural selection or genetic drift. Over the long run, however, mutation is, in itself, very important to evolution because it is the original source of the genetic variation that serves as raw material for natural selection.

Nonrandom Mating

For Hardy-Weinberg equilibrium to hold, an individual of any genotype must choose its mates at random from the population. But in actuality, individuals usually mate more often with close neighbors than with more distant members of the population, especially in species that do not disperse far. Since other individuals in the same "neighborhood" within a larger population tend to be closely related, mating only with nearby individuals promotes **inbreeding.** The most extreme case of inbreeding is self-fertilization ("selfing"), particularly common in plants.

Inbreeding causes the relative frequencies of genotypes to deviate from that which is expected from Hardy-Weinberg equilibrium. For example, in our imaginary wildflower population, self-pollination would tend to increase the frequencies of homozygous genotypes at the expense of heterozygotes. If *AA* individuals and *aa* individuals "self," then their offspring must also be homozygous. If *Aa* plants "self," however, only half of their offspring will be heterozygous. With each generation, the proportion of heterozygotes decreases and the proportions of dominant and recessive homozygotes increase. Even in less extreme cases of inbreeding without selfing, the decline of heterozygosity occurs, though more slowly. One visible effect of this change in genotypic frequencies is a greater proportion of individuals expressing reces-

sive phenotypes—the frequency of white-flowered individuals in our wildflower population would be greater than the Hardy-Weinberg equation predicts. Regardless of the impact of inbreeding on the ratio of genotypes and phenotypes in the population, however, the values of *p* and *q*, the frequencies of the two alleles, remain the same. It is just that a smaller proportion of recessive alleles are "masked" in heterozygous individuals.

Another type of nonrandom mating is **assortative mating,** in which individuals mate with partners that are like themselves in certain phenotypic characters. For example, snow geese (*Chen hyperborea*) occur in a blue variety and a white variety, with the allele for blue color dominant. The birds mate preferentially with partners of the same color. As with inbreeding, this assortative mating results in fewer heterozygous individuals than we would expect from the Hardy-Weinberg formula (that is, the frequency of heterozygotes is less than $2pq$, where *p* and *q* are the frequencies of the alleles for blue color and white color, respectively). In another case of assortative mating, blister beetles (*Lytta magister*) that live in the Sonoran Desert of Arizona most commonly mate with individuals of the same size (Figure 21.4). To some extent, humans also use size as a criterion for assortative mating—for example, tall women most commonly (but not always) pair with tall men.

Assortative mating can occur without individuals playing any active role in mate selection. This would take place, for instance, in our imaginary wildflower population if one species of insect selectively pollinated red flowers, while another pollinator carried pollen only from white flowers to white flowers.

Note again that nonrandom mating—inbreeding or assortative mating—increases the number of gene loci in the population that are homozygous, but nonrandom mating does not in itself alter the overall frequencies of alleles in a population's gene pool.

Natural Selection

Hardy-Weinberg equilibrium requires that all individuals in a population be equal in their ability to produce viable, fertile offspring. This condition is probably never completely met. Populations of sexually reproducing organisms consist of varied individuals, and on the average, some variants leave more offspring than others. This differential success in reproduction is, of course, natural selection. Selection results in alleles being passed along to the next generation in numbers disproportionate to their relative frequencies in the present generation. For example, in our imaginary wildflower population, plants with red flowers (*AA* or *Aa* genotypes) may for some reason produce more offspring on average than plants having white flowers

Figure 21.4
Assortative mating in blister beetles. These insects (*Lytta magister*) of the Sororan Desert in Arizona pair off, posterior to posterior, according to size. The animals may continue to feed on brittlebush flowers during copulation, which usually lasts several hours. While they are coupled, the male transfers not only sperm, but also nutrients and a chemical called cantharidin, to the female. The cantharidin is a toxic substance used by the insects as a defense against predators—hence the name, blister beetles.

(*aa*)—perhaps white flowers are more visible to herbivorous insects that eat the flowers. This would disturb genetic equilibrium; the frequency of the *A* allele would increase and the frequency of the *a* allele would decline in the gene pool.

Of all agents of microevolution that change the gene pool, only selection is likely to be adaptive. Natural selection accumulates and maintains favorable genotypes in a population. If the environment should change, selection responds by favoring genotypes adapted to the new conditions. But the degree of adaptation can be extended only within the realm of the variability present in the population. Before we examine the process of adaptation by natural selection

more closely, we will direct our attention to the genetic basis of the variation that makes it possible for populations to evolve.

THE GENETIC BASIS OF VARIATION

In what ways do members of a population vary? How extensive is variation? What mechanisms generate and maintain variations in a population? Do all variations function as raw material for selection? These are the questions we will try to answer as we look at the genetic variations so crucial to the process of natural selection.

The Nature and Extent of Variation Within and Between Populations

You have no trouble recognizing your friends in a crowd. Each person has a unique genome, and this is reflected in individual variations of appearance and temperament. Individual variation occurs in populations of all species of sexually reproducing organisms (Figure 21.5). We are very conscious of human diversity; we are less sensitive to individuality in populations of other animals and plants, and the diversity may escape our notice because the variations are subtle. But these slight differences between individuals in a population are the variations Darwin wrote most about as the raw material for natural selection.

Not all of the variation we observe in a population is heritable. Phenotype is the cumulative product of an inherited genotype and a multitude of environmental influences. For example, your campus population probably looks quite a bit different after spring break because many individuals were tanning during their vacation. It is important to remember that only the genetic component of variation can have adaptive impact as a result of natural selection.

Much of heritable variation consists of polygenic traits that vary quantitatively within a population. For example, plant height may vary continuously in our hypothetical wildflower population, from very short individuals to very tall individuals and everything in between. Other traits, such as red versus white flowers, vary categorically, probably because they are determined by a single gene locus with different alleles that produce distinct phenotypes. In such cases, when two or more forms of a Mendelian trait are represented in a population, the contrasting forms are called **morphs**—as in the red-flowered and white-flowered morphs of our wildflower population, for example. A population is said to be **polymorphic** for a trait if two or more morphs are each represented in high enough frequencies to be readily noticeable. (Obviously, this definition is somewhat arbitrary, but a population is

Figure 21.5
Sexual recombination of genes promotes individual variation. Even siblings, such as the puppies of this litter, sired by one male, display individuality due to their unique genomes.

Figure 21.6
Polymorphism. Some populations consist of two or more distinct varieties of individuals. Two forms, or morphs, of the king snake (*Lampropeltis getulus*) that differ markedly in their patterns of coloration coexist in California populations.

not termed polymorphic if it consists almost exclusively of a single morph, with other morphs extremely rare.) Figure 21.6 illustrates a striking example of polymorphism in a population of California king snakes. Polymorphism is extensive in human populations, both in physical traits, such as the presence or absence of freckles, and biochemical traits, such as ABO blood group (for which there are four morphs, *Type A, Type B, Type AB,* and *Type O*—see Chapter 13).

The reservoir of genetic variation in a population is much more extensive than Darwin realized. Much of the genetic variation in a population is invisible, but is manifest in molecular differences that can be detected by biochemical methods. Several laboratories have used electrophoresis, a technique that can separate proteins differing in electric charge (see Methods Box in Chapter 19 on p. 403), to study variations in the protein products of specific gene loci among individuals in a population. Scores of loci have been studied in many different animal species. For example, in populations of the fruit fly *Drosophila*, the gene pool typically has two or more alleles for about 30% of the loci examined, and each fly is heterozygous at about 12% of its loci—that amounts to 700–1200 heterozygous loci per fly. Expressed another way, any two flies in a *Drosophila* population differ in genotype at about 25% of their loci. The extent of genetic variation in human populations, as revealed by electrophoresis, is comparable. And electrophoresis underestimates genetic variation, because proteins produced by different alleles may vary in amino acid composition even though they have

the same overall electric charge. Furthermore, variation in DNA that is not expressed as protein is not detected by this method.

Most species exhibit **geographical variation,** differences *between* populations in their frequencies of alleles. Because at least some environmental factors are likely to be different from one place to another, natural selection can contribute to geographical variation. For example, one population of our now-familiar wildflower species may have a higher frequency of recessive alleles at the flower-color locus than other populations, perhaps because of a local prevalence of pollinators that key on white flowers (recessive homozygotes). Genetic drift can also cause chance variations among different populations. On a more local scale, geographical variation can also occur *within* a population, either because the environment has patchlike variation or because the population is differentiated into subpopulations due to localized inbreeding.

One particular type of geographical variation, called a **cline,** is a graded change in some trait along a geographic transect. In some cases, a cline may represent a graded region of overlap where individuals of neighboring populations are interbreeding. In other cases, a gradation in some environmental variable may produce a cline. For example, average body size of many North American species of mammals increases gradually with increasing latitude. Presumably, the reduced ratio of surface area to volume that accompanies larger size is an adaptation that helps animals living in cold environments conserve body heat. Experimental studies of some clines, such as geographical variation in the height of yarrow plants that grow on the slopes of mountains, confirm the role of genetic variation in

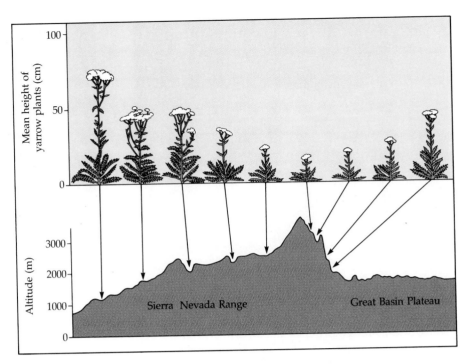

Figure 21.7

A cline. Yarrow plants on the slopes of California's Sierra Nevada mountains gradually decrease in average size at higher and higher elevations. Although the environment affects growth rates directly to some extent, some of the variation has a genetic basis. Researchers established the genetic connection by collecting seeds at different elevations and growing plants in a uniform garden; the average sizes of the plants were correlated with the altitude at which the seeds were collected. (Data from Clausen, Keck, and Hiesey, *Carnegie Institute of Washington Publication 581,* 1948.)

the spatial differences of phenotype (Figure 21.7). A special type of cline, called a **step cline,** is characterized by a gradation of phenotype that occurs over a relatively short distance. For example, in some populations of plants growing in the vicinity of the "tailings" from abandoned mines, there are individuals so tolerant of heavy metals that they can grow right on the mine tailings. Such a high concentration of heavy metals may be toxic to plants of the same species growing just a few meters away.

Sources of Variation

Mutation and sexual recombination (see Chapter 13) are the two processes that generate genetic variation.

Mutation   New alleles originate by mutation (see Chapters 14–16). A mutation affecting any gene locus is an accident that is rare and random. Most mutations occur in somatic cells and die with the individual. Geneticists estimate that in humans an average of only one or two mutations occur in each cell line that produces a gamete, and these are the only mutations that can be passed along to children. A mutation is a shot in the dark. Chance determines where it will strike and how it will alter a gene.

Most point mutations, those affecting a single base in DNA, are probably relatively harmless. Much of the DNA in the eukaryotic genome does not code for protein products, and it is uncertain how a change of a single nucleotide base in this silent DNA will affect the well-being of the organism (see Chapters 16 and 18). Even mutations of structural genes, which do code for proteins, may occur with little or no effect on the

organism, partly because of redundancy in the genetic code (see Chapter 16).

A mutation that alters a protein enough to affect its function is more often harmful than beneficial. Organisms are the refined products of thousands of generations of past selection, and a random change is not likely to improve the genome any more than firing a gunshot blindly through the hood of a car is likely to improve engine performance. On rare occasions, however, a mutant allele may actually fit its bearer to the environment better and enhance the reproductive success of the individual. This is not especially likely in a stable environment, but becomes more probable when the environment is changing and mutations that were once selected against are now favorable under the new conditions. As a result of random mutations, dark peppered moths occurred in populations dominated by the light moths before pollution darkened the English landscape. The mutation became an advantage instead of a liability when the environment changed. Similarly, some mutations that happen to endow house flies with resistance to DDT also reduce growth rate and were deleterious before the pesticide was introduced. A new environmental factor, DDT, tilted the balance in favor of the mutant alleles, and they spread through fly populations by natural selection.

Because chromosomal mutations usually affect many gene loci, they are almost certain to disrupt the development of the organism. But even rearrangements of chromosomes may in rare instances bring benefits. For example, the translocation of a chromosomal piece from one chromosome to another could link alleles that affect the organism in some positive way when they are inherited together as a package. A cluster of genes that have cooperative function and are linked closely on a common chromosome is called a **supergene**. The type of chromosomal mutation known as an inversion may help preserve a supergene by preventing its disruption from crossover with an uninverted segment on the homologous chromosome.

Duplications of chromosome segments, like other chromosomal mutations, are nearly always harmful. But if the repeated segment does not disrupt genetic balance severely, it can persist over the generations and provide an expanded genome with superfluous loci that may eventually take on new functions by mutation while the original genes continue to function at their old locations in the genome. New genes may also arise from existing DNA sequences by the shuffling of exons within the genome, either within a single locus or between loci (see Chapter 18).

Recombination   Although mutations are the source of new genes, they are so infrequent at any one locus that on a generation-to-generation basis, their contribution to genetic variation in a large population is negligible. Members of a population owe nearly all their differences to the unique recombinations of existing alleles each individual draws from the gene pool.

Sex shuffles genes and deals them at random to determine individual genotypes. During meiosis, homologous chromosomes, one inherited from each parent, trade some of their genes by crossing over, and then the homologous chromosomes and the alleles they carry segregate randomly into separate gametes (see Chapter 12). Gametes from one individual vary extensively in their genetic makeup, and each zygote made by a mating pair has a unique assortment of genes resulting from the random union of sperm and ova. And, of course, in a population there are a vast number of possible mating combinations, each bringing together the gametes of individuals that are likely to have different genetic backgrounds. Sexual reproduction recombines old alleles into fresh assortments every generation.

For bacteria and other microorganisms that have very short generation spans, mutation can be an adequate source of genetic variation. Bacteria reproduce asexually by dividing as often as once every 20 minutes; a single cell can potentially give rise to a billion descendants in just 10 hours. A new mutation that happens to be beneficial can increase its frequency in a bacterial population very rapidly. Imagine, for example, exposing a bacterial population to an antibiotic. If a single individual in the population happens to harbor a mutation that renders it resistant to the poison, in just a few hours there may be millions of resistant bacteria, while bacteria sensitive to the antibiotic may have been almost completely eliminated. Bacterial populations can evolve, one mutation at a time, by the explosive asexual expansion of clones favored by the local environment. But even most bacteria increase

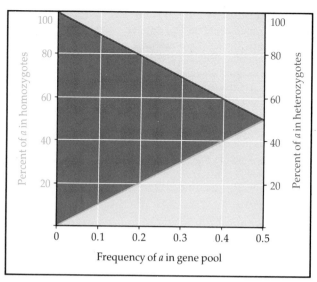

Figure 21.8
How diploidy helps to preserve genetic variation.
Recessive alleles that are selected against in homozygotes persist in a population in heterozygotes. As recessive alleles become rarer, a greater proportion of them are hidden in heterozygotes.

genetic variation by occasionally exchanging and recombining genes through processes that resemble sex (see Chapter 17). Animals and plants depend almost entirely on sexual recombination for the genetic variation that makes adaptation possible.

How Variation Is Preserved

Natural selection tends to consume variation by culling unfavorable genotypes from a population. This trend toward genetic uniformity is opposed by several mechanisms that preserve or restore variation.

Diploidy   The diploid character of most eukaryotes hides a considerable amount of genetic variation from selection in the form of recessive alleles in heterozygotes. Genes that are less favorable than their dominant counterparts, or even harmful in the present environment, can persist in a population through their propagation by heterozygous individuals. This latent variation is exposed to selection only when both parents carry the same recessive allele and combine two copies in one zygote. This happens only rarely if the frequency of the recessive allele is very low. For example, if the frequency of the recessive allele is 0.01 and the frequency of the dominant allele is 0.99, then 99% of the copies of that recessive allele are protected from selection in heterozygotes, and only 1% of the alleles are present in homozygotes (Figure 21.8). And the rarer the recessive allele, the greater the degree of protection afforded by heterozygosity. Heterozygote

Figure 21.9
Balanced polymorphism in a patchy environment. The British land snail, *Cepaea nemoralis*, lives in a patchy environment of wooded areas separated by open fields. There are several morphs, with shells that differ in color and in the presence or absence of stripes. Each form is especially well camouflaged from predation by birds in a particular light or shady area. In this case, polymorphism is apparently associated with a heterogeneous environment.

protection maintains a huge pool of alleles that may not be suitable for present conditions, but could bring new benefits when the environment changes.

Balanced Polymorphism Selection itself may preserve variation at some gene loci. This ability of natural selection to maintain diversity in a population is called **balanced polymorphism.** One of the mechanisms for this preservation of variation is **heterozygote advantage.** If individuals who are heterozygous for a particular locus have greater reproductive success than any type of homozygote, then two or more alleles will be maintained at that locus by natural selection. An interesting case of heterozygote advantage involves the locus in humans for one chain of hemoglobin, the protein of red blood cells that transports oxygen. A specific recessive allele at that locus causes sickle-cell anemia in homozygous individuals. Heterozygotes, however, are resistant to malaria, an important advantage in tropical regions where that disease is a major cause of death. The environment in these regions favors the heterozygotes over both homozygous dominant individuals, who are susceptible to malaria, and homozygous recessive individuals, who are disabled by sickle-cell anemia. The frequency of the sickle-cell allele in Africa is generally highest in areas where the malaria parasite is most common. In some tribes, the recessive allele accounts for 20% of the hemoglobin loci in the gene pool, a very high frequency for a gene that is disastrous in homozygotes. But at this frequency ($q = 0.2$), 32% of the population consists of heterozygotes resistant to malaria ($2pq$), and only 4% of the population suffers from sickle-cell anemia (q^2).

Another example of heterozygote advantage is found in the crossbreeding of crop plants. When corn, for instance, is highly inbred, the number of homozygous gene loci increases, and the corn may gradually become stunted in growth and increasingly sensitive to a variety of diseases. Crossbreeding between two different inbred varieties often produces hybrids that are much more vigorous than either parent stock. This **hybrid vigor** is probably due to two factors: the segregation of deleterious recessives that were homozygous in the inbred varieties, and heterozygote advantage at many loci in the hybrids.

A patchy environment, where natural selection favors different phenotypes in different subregions within a population's geographic boundaries, can also result in balanced polymorphism. For example, in many populations of British land snails (*Cepaea nemoralis*), there are several morphs, each with a coloration that camouflages its shell in a particular patch of the area inhabited by the population (Figure 21.9).

Still another cause of balanced polymorphism is **frequency-dependent selection,** in which the reproductive success of any one morph declines if that phenotypic form becomes too common in the population. A particularly intricate example is a balanced polymorphism that has been observed in populations of *Papilio dardanus*, an African swallowtail butterfly. The males all have similar coloration, but the females occur in several different morphs, each resembling another butterfly species that is noxious to predators (Figure 21.10). *Papilio* females are not noxious, but birds learn to avoid them because they look so much like the distasteful butterflies. This type of protective coloration, called Batesian mimicry, would be less effective if all *Papilio* females copied the same noxious species, because birds would be slow to associate a particular pattern of coloration with bad taste if they encountered good-tasting mimics as often as the noxious models.

Is All Variation Adaptive?

Some of the genetic variations observed in populations are probably trivial in their impact on reproductive success. The diversity of human fingerprints is

Figure 21.10
Frequency-dependent selection. The females of *Papilio dardanus*, an African swallowtail butterfly, occur in several morphs (left column), each a mimic of a different butterfly species that is noxious to predators (right). The advantage of mimicry would be less if any one *Papilio* morph became so common that predators encountered them as often as the distasteful model.

an example of what is called **neutral variation,** which seems to confer no selective advantage for some individuals over others. Much of the protein variation detectable by electrophoresis may represent chemical "fingerprints" that are neutral in their adaptive qualities. For instance, 99 known mutations affect 71 of the 146 amino acids in the β chain of human hemoglobin, one of two kinds of polypeptide chains that make up that protein. Some of those mutations, including the allele for sickle-cell anemia, certainly affect the reproductive potential of the individual. However, according to a **neutral theory** of molecular evolution, many of the variant alleles at this locus and others may convey no selective advantage or disadvantage. The relative frequencies of neutral variations that arise by mutations will not be affected by natural selection; some neutral alleles will increase in the gene pool and others will decrease by the chance effects of genetic drift.

In the context of neutral variation, we should also consider the variation in DNA that does not code for protein. As discussed in Chapter 18, the genomes of most eukaryotes contain large amounts of DNA for which there is no obvious function. Furthermore, there are great variations in the amounts of such DNA in some closely related species. Some scientists have attempted to account for the noncoding DNA by speculating that DNA is inherently "selfish," expanding itself to the limit tolerated by each species. In the extreme version of this view, the DNA of entire genomes exists as a consequence of being self-replicating, rather than because it confers adaptive advantages on the organism. Certainly, particular DNA sequences that have autonomous mechanisms for replication and movement—such as transposons—seem to meet the criteria for "selfish DNA." However, the extent to which such sequences influence the evolution of genomes is unknown.

There is no consensus among evolutionary biologists on how much genetic variation is neutral, or even if any variation can be considered truly neutral. Variations appearing to be neutral may in fact influence reproductive success in ways that are difficult to measure. It is possible to show that a particular allele is detrimental, but it is impossible to demonstrate that an allele brings no benefits at all to an organism. Furthermore, a variation may be neutral in one environment but not in another. We can never know the degree to which genetic variation is neutral. But we can be certain that even if only a fraction of the extensive variation in a gene pool significantly affects the organisms, that is still an enormous resource of raw material for natural selection and the adaptive evolution it causes.

ADAPTIVE EVOLUTION

Adaptive evolution is a blend of chance and sorting—chance in the origin of new genetic variations by mutation and sexual recombination, and sorting in the workings of selection as it favors the propagation of some chance variations over others. From the pool of variations available to it, natural selection orders combinations of genes and fits organisms to their environments.

Fitness

The phrases *struggle for existence* and *survival of the fittest* are loaded with misleading connotations. There are, of course, species in which individuals, usually the males, lock horns or otherwise do combat that determines mating privilege. But direct and violent confrontations between members of a population are rare; success is generally subtle and passive. A barnacle may produce fewer eggs than its neighbors because it is not quite so efficient at collecting food from the water. In a population of moths, certain variants may average more progeny than others because their coloration hides them from predators better. Plants in a wildflower population may differ in reproductive success because some are better able to attract pollinators owing to slight variations in color, shape, or fragrance. Darwinian fitness is measured only by the relative contribution an individual makes to the gene pool of the next generation.

Survival alone does not guarantee reproductive success. Darwinian fitness is zero for a sterile plant or animal, even if it is robust and outlives other members of the population. But, of course, survival is a prerequisite for reproducing, and longevity increases fitness if it results in certain individuals leaving disproportionately high numbers of descendants. Then, again, an individual that matures quickly and becomes fertile at an early age may have a greater reproductive potential than individuals that live longer but mature late. Thus, the components of selection are the many factors that affect both survival and fecundity.

In a more quantitative approach to natural selection, population geneticists define **relative fitness** as the contribution of a genotype to the next generation compared to the contributions of alternative genotypes for the same locus. For example, consider the wildflower population introduced earlier, in which *AA* and *Aa* plants have red flowers and *aa* plants have white flowers. Let us assume that on the average, individuals with red flowers produce more offspring than those with white flowers. The relative fitness of the most fecund variants is set at 1 as a basis for comparison; so in this case, the relative fitness of an *AA* or *Aa* plant is 1. If plants with white flowers average only 80% as many progeny, then their relative fitness is 0.8. The difference between the two fitness values, 0.2 in this case, is termed the **selection coefficient.** It is a relative measure of selection *against* the inferior genotype; the more disadvantageous the allele, the greater the selection coefficient, ranging to a coefficient of 1 for a lethal genotype. Selection coefficients are only statistical estimates, something like a handicapper predicting the order of finish for a horse race that has not yet begun.

The rate at which the frequency of a deleterious allele declines in a population depends on the magnitude of the selection coefficient working against it and on whether the allele is dominant or recessive to the more successful allele (Figure 21.11). Harmful recessives are rarely eliminated completely because of heterozygote protection. Selection acts faster against dominant alleles that are harmful because they are expressed even in heterozygotes. The rate of increase of a beneficial allele is also affected by whether it behaves as a dominant or a recessive. A new recessive mutation spreads very slowly in a population, even if it is very beneficial, because selection cannot act in its favor until the mutation is common enough for two

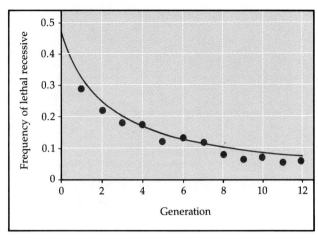

Figure 21.11
Selection against a lethal allele. The points on this graph track the generation-by-generation decline of a lethal recessive allele in a laboratory population of the flour beetle (*Tribolium castaneum*). At the beginning of the experiment, the recessive lethal and the dominant allele were present in the beetle population in equal frequencies ($p = q = 0.5$) The relative fitness of the homozygous dominant genotype is 1, and the fitness of the heterozygous genotype is only slightly less (but not quite 1, because the "normal" allele for this locus is not completely dominant to the lethal recessive). The homozygous recessive genotype has a relative fitness of 0; death results before any offspring are left. Put another way, the homozygous recessive genotype has a selection coefficient of 1 (selection coefficient, a measure of the strength of selection against an inferior genotype, is the difference between the relative fitness values of the most and least successful genotypes). The curve represents the expected decline in the frequency of the lethal allele based on selection theory, and the actual data for the beetle population fits the expected results closely. Note that the rate at which the lethal allele disappears from the population slows as the allele becomes less common. This is because a greater proportion of remaining recessive alleles are present in heterozygotes compared to homozygotes as the allele becomes rarer (see Figure 21.8). (Based on data from P. S. Dawson. *Genetica* 41:147–169, 1970.)

copies to occasionally come together in the same zygote. A new dominant mutation that confers greater advantage than existing alleles can increase in frequency more rapidly because each individual that inherits even a single copy benefits from the allele. The mutant allele responsible for the rapid replacement of light peppered moths by the dark morph in England was dominant. However, most new mutations, whether dominant or recessive, and whether beneficial or harmful, probably disappear from the gene pool early in their history by genetic drift.

(a)   *(b)*

Figure 21.12
Norms of reaction. The effect of a genotype is usually defined by its norm of reaction—a phenotypic range that depends on the environment in which the genotype is expressed. (**a**) The color of hydrangea flowers varies from blue-violet to pink, depending on the acidity of the soil. (**b**) Although genes certainly affect phy-sique, body building shows the extreme to which muscles can be developed through exercise and diet.

What Selection Acts On

An organism exposes its phenotype—its physical traits, metabolism, physiology, and behavior—not its genotype, to the environment. Acting on phenotypes, selection indirectly adapts a population to its environment by increasing or maintaining favorable genotypes in the gene pool.

Throughout this chapter, we have been looking at a wildflower population with alleles for flower color that affect phenotype unambiguously, but the connection between genotype and phenotype is rarely so simple or definite. A genotype may have multiple effects, especially if it influences the development or growth of the organism. This is called pleiotropy (see Chapter 13). The overall fitness of a genotype depends on whether its positive effects outweigh any harmful effects it may have on the reproductive success of the organism. Another complication in the translation of genotype into phenotype arises when many gene loci influence the same characteristic. These are polygenic, or quantitative, traits (see Chapter 13); human height is an example. Individuals in a population do not usually fit into exclusive categories for a polygenic trait, but instead vary continuously over a phenotypic range.

The finished organism acted on by natural selection is an integrated composite of its many phenotypic features, not a collage of individual parts. The fitness of a genotype at any one locus depends on the entire genetic context in which it works. For example, alleles that enhance the growth rate of the trunk and limbs of a tree may be useless or even detrimental in the absence of alleles at other loci that enhance the growth rate of roots required to support the tree. Genes whose functions are related make up a **coadapted gene complex.** Coadapted genes may occur close together on the same chromosome, forming a supergene. The intricate development of any organ, such as the vertebrate eye or the wing of a dragonfly, requires coadaptation of alleles at many gene loci. In the most comprehensive sense, the entire genome of an organism is one coadapted complex.

Phenotype depends on environment as well as on genes (Figure 21.12). A single tree locked into the genome it inherited has leaves that vary in size, shape, and greenness, depending on exposure to wind and sun. For humans, nutrition influences height, exercise

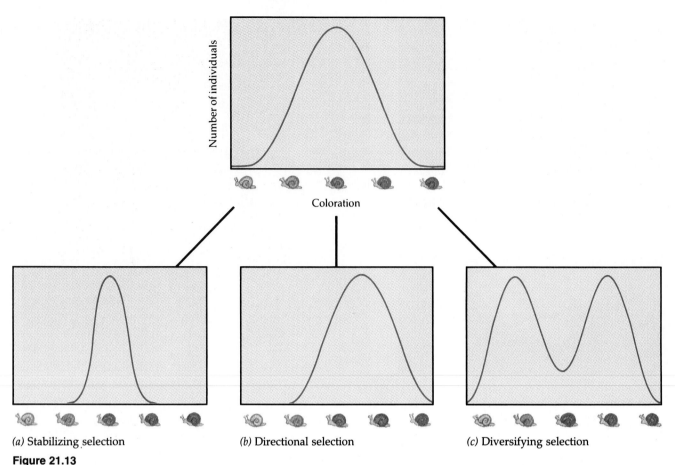

Figure 21.13

Modes of selection. These cases describe the possible fates of a snail population in which there is quantitative variation in coloration. The graphs show how the frequencies of individuals of varying darkness change with time. (a) Stabilizing selection culls extreme variants from the population, in this case eliminating individuals that are unusually light or dark. The trend is toward reduced phenotypic variation and maintenance of the status quo. (b) Directional selection shifts the overall makeup of the population by favoring variants of one extreme. In this case, the trend is toward darker color, perhaps because the landscape has been blackened by volcanic ash. (c) Diversifying selection favors variants of opposite extremes over intermediate individuals, leading to balanced polymorphism. Here, very light and very dark snails have increased their frequencies relative to intermediate variants. Perhaps the snails have recently colonized a patchy habitat where a background of white sand is studded with lava rocks.

(a) Stabilizing selection (b) Directional selection (c) Diversifying selection

alters build, suntanning darkens the skin, and experience increases IQ. Whether it is genes or environment—nature or nurture—that most determines various human characteristics is a very old and hotly contested debate, and we will not attempt to resolve the issue here. We can say, however, that the product of a genotype is generally not a rigidly defined phenotype, but a range of phenotypic possibilities over which there may be variation due to environmental influence. Evolutionists refer to this phenotypic range as the **norm of reaction** for a genotype. There are cases where the norm of reaction has no breadth whatsoever; a given genotype mandates a very specific phenotype. An example is the gene locus that determines a person's blood type, be it type A, B, AB, or O. In contrast, a person's blood count of red cells and white cells varies, depending on such factors as the altitude of one's home, the customary level of physical activity, and the presence of infectious agents. Norms of reaction are broadest perhaps for behavioral traits.

Regardless of these complications in the relationship between genotype and phenotype, the gene pool of a population is certain to feel some impact from selection favoring certain phenotypes over others.

Modes of Natural Selection

Natural selection can affect the frequency of a heritable characteristic in a population in three different ways, depending on which phenotypes in a varying population are favored. These three modes of selection are called stabilizing selection, directional selection, and diversifying selection. These modes can be depicted with graphs that show how the frequencies of different phenotypes change with time (Figure 21.13). This is most meaningful for quantitative traits that depend on many gene loci.

(a)

(b)

(c)

Figure 21.14
Sexual selection and dimorphism. (a) Size. In this family of South American sea lions, the male is markedly larger than the female. **(b)** Ornamentation. The elaborate antlers of male deer may function in defense as well as sexual identification.
(c) Coloration. In most cases of sexual dimorphism, the male is the showier sex, as in this king eider pair.

Stabilizing selection acts against extreme phenotypes and favors the more intermediate variants. The trend is toward reduced phenotypic variation and even greater prevalence of phenotypes best suited to a relatively stable environment. For example, stabilizing selection keeps the majority of human birth weights in the 3–4 kg range. For babies much smaller or larger than this, infant mortality is greater.

Directional selection is most common during periods of environmental change or when members of a species migrate to some new habitat with different environmental conditions. Directional selection shifts the frequency curve for variations in some phenotypic trait in one direction or the other by favoring relatively rare individuals that deviate from the average for that trait. For instance, there is fossil evidence that the average size of black bears in Europe increased with each glacial period of the ice ages, only to decrease again during the warmer interglacial periods.

Diversifying selection (also called disruptive selection) occurs when environmental conditions are varied in a way that favors individuals on both extremes of a phenotypic range over intermediate phenotypes. We can see an example in populations of *Papilio*, the African butterfly discussed earlier. Butterflies with characteristics intermediate between two different noxious models would resemble neither model enough to gain advantage from mimicry. Thus, diversifying

selection can result in balanced polymorphism, such as the multiple mimics that have evolved in *Papilio*.

Sexual Selection

Males and females of many animal species exhibit marked differences in addition to the differences in the reproductive organs that define the sexes. This distinction between the secondary sexual characteristics of males and females is known as **sexual dimorphism.** It is often expressed as a difference in size, with the male usually larger, but it also involves such features as colorful plumage in male birds, manes on male lions, antlers on male deer, and other adornments (Figure 21.14). Notice that in most cases of sexual dimorphism, at least for vertebrates, it is the male that is the showier sex. In some cases, the males with the most impressive masculine features may be the most attractive to females. There are also species in which the secondary sexual structures may be used in direct competition with other males; this is particularly common in species where a single male garners a harem of females. These males may succeed because they defeat smaller, weaker, or less fierce males in combat; more often, however, they are effective in ritual displays that discourage would-be competitors (see Chapter 50).

Darwin was intrigued by **sexual selection,** which he saw as a separate selection process leading to sexual dimorphism. Many secondary sexual features do not seem to be adaptive in the general sense; showy plumage probably does not help male birds cope with their environment and may even attract predators. If such accoutrements give the individual an edge in gaining a mate, however, they will be favored for the most Darwinian of reasons—because they enhance reproductive success. Thus, in many cases, the ultimate evolutionary outcome is a compromise between the two selection forces. In some species, the line between sexual selection and ordinary natural selection blurs because the sexual feature does double duty as an adaptation to the environment. For example, a stag may use his antlers to defend himself against a predator.

Does Evolution Fashion Perfect Organisms?

In a word—no. There are at least four reasons why natural selection cannot breed perfection.

1. As we saw in Chapter 20, each species has a history of descent with modification from a long line of ancestral forms. Evolution does not scrap ancestral anatomy and build each new complex structure from scratch, but co-opts existing structures and adapts them to new situations. For example, the excruciating back problems some humans endure result in part because a skeleton and musculature modified from the anatomy of four-legged ancestors are not fully compatible with upright posture.

2. Adaptations are often compromises. Each organism must do many different things. A seal spends part of its time on rocks; it could probably walk better if it had legs instead of flippers, but it would not swim nearly so well. We owe much of our versatility and athleticism to our prehensile hands and flexible limbs, which also make us prone to sprains, torn ligaments, and dislocations; structural reinforcement has been compromised for agility.

3. Not all evolution is adaptive. Chance probably affects the genetic makeup of populations to a greater extent than was once believed. For instance, when a storm blows insects hundreds of miles over an ocean to an island, the wind does not necessarily pick up the specimens that are best suited to the new environment. And not all alleles fixed by genetic drift in the gene pool of the small founding population are better suited to the environment than alleles that are lost. Similarly, the bottleneck effect can cause nonadaptive or even maladaptive evolution.

4. Natural selection favors only the most fit variations from what is available, which may not be the ideal characteristics. New genes do not arise on demand.

With all these constraints, we cannot expect evolution to result in ideal organisms. Natural selection operates on a "better than" basis. Perhaps the best evidence for evolution is in the subtle imperfections of the organisms it produces.

The Tempo of Microevolution

Natural selection is usually thought of as an agent of change, but it can also act to maintain the status quo. Stabilizing selection probably prevails most of the time, resisting change that may be maladaptive. Evolutionary spurts occur when a population is stressed by a change in the environment, migration to a new place, or a change in the genome. When challenged with a new set of problems, a population either adjusts through natural selection or becomes extinct. The fossil record indicates that extinction is the more common outcome. Those populations that do survive crises often change enough to become new species, as we will see in the next chapter.

STUDY OUTLINE

1. Natural selection acts on individuals, but only populations evolve because evolution is change in the prevalence of inherited traits over a succession of generations.
2. The most important post-Darwinian development in evolutionary biology has been the application of genetics to the theory of natural selection.

The Genetics of Populations (pp. 439–442)

1. A species is a group of populations that have the potential to interbreed in nature. This potential is realized to varying extents, depending on the geographical distribution of breeding individuals.
2. A population is united by its gene pool, the aggregate of all genes in the population. A change in the relative frequencies of alleles in the gene pool over a succession of generations is called microevolution.
3. According to the Hardy-Weinberg theorem, the frequencies of alleles in a population will remain constant if sexual reproduction is the only process that affects the gene pool. The mathematical expression for such a nonevolving population is the Hardy-Weinberg equation, which states that for a two-allele locus, $p^2 + 2pq + q^2 = 1$, where p and q represent the relative frequencies of the dominant and recessive alleles, respectively, p^2 and q^2 the frequencies of the homozygous genotypes, and $2pq$ the frequency of the heterozygous genotype.
4. For Hardy-Weinberg equilibrium to apply, the population must: be very large; be totally isolated; show random

mating; have no *net* mutations; and have equal reproductive success for all individuals. Hardy-Weinberg equilibrium is never met in real populations because conditions in nature violate one or more of these prerequisites.

Causes of Microevolution (pp. 442–446)

1. Genetic drift is the change in gene frequencies observed in small populations due to sampling error or chance events. When large segments of a population are destroyed by disasters (bottleneck effect) or when a small sample of a population colonizes a new habitat (founder effect), the new, small population is unlikely to be representative of the parent population, and genetic drift will continue until the population grows larger.
2. Gene flow is the exchange of alleles between two populations due to migration. It tends to reduce interpopulation differences.
3. Mutation can theoretically affect allele frequencies in a gene pool, but is usually insignificant over the short-term for large populations. Mutation is important in evolution, however, because it generates new variations for natural selection.
4. Nonrandom mating, such as inbreeding or assortative mating, does not ordinarily affect allele frequencies, but does affect the ratio of genotypes in populations.
5. Natural selection, differential success in reproduction, is the only agent of microevolution that tends to cause adaptive change in the gene pool.

The Genetic Basis of Variation (pp. 446–451)

1. Genetic variation in a population includes individual variation in quantitative traits, geographical variation, and polymorphism.
2. Genetic variation is incredibly extensive, but much of it can be detected only at the molecular level.
3. Mutation and sexual recombination produce genetic variation. Mutations are generally harmful, but on rare occasions can be beneficial. Sexual recombination produces zygotes that have unique assortments of genes from two parents.
4. Populations remain variable despite natural selection because of several mechanisms. Diploidy maintains a reservoir of latent variation by hiding recessive alleles from selection in heterozygotes. Balanced polymorphism may maintain variation at some gene loci as a result of heterozygote advantage or frequency-dependent selection.
5. Evolutionary biologists debate what fraction of genetic variation is actually acted on by natural selection and what fraction is neutral in its impact on reproductive success.

Adaptive Evolution (pp. 451–456)

1. Adaptive evolution results from the workings of natural selection on the chance variations due to mutation and sexual recombination.
2. Darwinian fitness is measured only by reproductive success, specifically by the relative contribution of an individual to the gene pool of the next generation. The selection coefficient is the difference between relative fitness values representing the contributions of specific genotypes and is a relative measure of selection against an inferior genotype.
3. Selection maintains favorable genotypes in a population by acting on the phenotype of individual organisms. The

relationship between genotype and phenotype is complicated by pleiotropy, multiple effects of an allele; polygenic traits, in which several genes influence the same characteristic; coadapted gene complexes, the functional association of genes; and the norm of reaction, the range of phenotype possibilities subject to environmental influence.
4. Natural selection can affect the frequency of a phenotype in three different ways: stabilizing selection, which discriminates against extreme phenotypes; directional selection, which favors relatively rare individuals on one end of the phenotypic range; and diversifying selection, which favors individuals at both extremes of a range over intermediate phenotypes.
5. Sexual selection leads to the evolution of secondary sexual characteristics, which give the individual an advantage in mating, but which may or may not help the individual cope with the environment.
6. Evolution by natural selection does not fashion perfect organisms for several reasons: Structures result from modified ancestral anatomy; adaptations are often compromises; the gene pool can be affected by chance as well as adaptation; and natural selection can act only on available variation.
7. Natural selection tends to be stabilizing in a steady environment. However, populations that survive crises may change extensively enough to become new species.

SELF-QUIZ

1. A gene pool consists of
   a. all the genes exposed to natural selection
   b. the total of all alleles present in a population
   c. the entire genome of a reproducing individual
   d. the frequencies of the alleles for a gene locus within a population
   e. all the gametes in a population

2. In a population with two alleles for a particular locus, *B* and *b*, the allele frequency of *B* is 0.7. What would be the frequency of heterozygotes if the population is in Hardy-Weinberg equilibrium?
   a. 0.7
   b. 0.49
   c. 0.21
   d. 0.42
   e. 0.09

3. In a population that is in Hardy-Weinberg equilibrium, 16% of the individuals show the recessive trait. What is the frequency of the dominant allele in the population?
   a. 0.84
   b. 0.36
   c. 0.6
   d. 0.4
   e. 0.48

4. The average length of jackrabbit ears decreases the further north the rabbits live. This variation is an example of

 a. a cline
 b. a step cline
 c. polymorphism
 d. genetic drift
 e. diversifying selection

5. If a particular genotype has a selection coefficient of 0.4,

 a. its relative fitness would be 0.6
 b. individuals with this genotype would leave 40% as many offspring as individuals with the most common genotype leave
 c. the genotype would increase by 40% in each generation
 d. the genotype would gradually disappear from the population
 e. the frequency of the genotype would be 0.4

6. Selection acts directly on

 a. phenotype
 b. genotype
 c. the entire genome
 d. the norm of reaction
 e. the entire gene pool

7. As a mechanism of microevolution, natural selection can be most closely equated with

 a. assortative mating
 b. genetic drift
 c. differential reproductive success
 d. bottlenecking of a population
 e. gene flow

8. Most of the variation we see in coat coloration and pattern in a population of wild mustangs in any generation is probably due to

 a. new mutations that occurred in the preceding generation
 b. sexual recombination of alleles
 c. genetic drift due to the small size of the population
 d. geographic variation within the population
 e. environmental effects

9. In terms of the algebraic symbols used in the Hardy-Weinberg formula (p and q), the most likely effect of assortative mating on the frequencies of two alleles for a locus would be

 a. a decrease in p^2 compared to q^2
 b. an increase in p^2 compared to q^2
 c. an increase in $2pq$ above that expected by the Hardy-Weinberg theorem
 d. a change in p and q, the relative frequencies of the two alleles in the gene pool
 e. a decrease in $2pq$ below the value expected by the Hardy-Weinberg theorem

10. A founder event favors microevolution in the founding population mainly because

 a. mutations are more common in a new environment
 b. a small founding population is subject to extensive sampling error in the composition of its gene pool
 c. the new environment is likely to be patchy, favoring diversifying selection
 d. gene flow increases
 e. members of a small population tend to mate assortatively

CHALLENGE QUESTIONS

1. Some species have been rescued from near extinction by conservationists. What evolutionary problems do such species face as their populations rebound from very small size?

2. Let us return to the wildflowers with which we derived the Hardy-Weinberg theorem (see p. 441). The frequency of A, the dominant allele for red flowers is 0.8, and the frequency of a, the recessive allele for white flowers is 0.2. In our starting population, the frequencies of genotypes do not conform to the Hardy-Weinberg equilibrium: 60% of the plants are AA and 40% of the plants are Aa (at this point, the population has no plants with white flowers). Assuming that all conditions for the Hardy-Weinberg theorem are met, prove that genotypes will reach an equilibrium in the next generation.

FURTHER READING

1. Alcock, J. "Conjugal Chemistry." *Natural History*, April 1986. Adaptive significance of assortative mating in blister beetles.
2. Dawkins, R. *The Blind Watchmaker.* New York: W. W. Norton, 1986. A master of metaphor explains how complexity can arise in the absence of design.
3. Diamond, J. "Founding Fathers and Mothers." *Natural History*, June 1988. Importance of genetic drift in human evolution.
4. Dobzhansky, T., F. J. Ayala, G. L. Stebbins, and J. W. Valentine. *Evolution.* San Francisco: Freeman, 1977. An advanced undergraduate text.
5. Mettler, L. E., T. G. Gregg, and H. E. Schaffer. *Population Genetics and Evolution.* 2d ed. Englewood Cliffs, New Jersey: Prentice Hall, 1988. Concise introduction to the genetics of evolution.
6. Stearns, S. C. "The Evolutionary Significance of Phenotypic Plasticity." *Bioscience*, July/August 1989. One of several articles in this issue emphasizing norms of reaction and their importance in evolution.
7. Stebbins, G. L. *Darwin to DNA, Molecules to Humanity.* San Francisco: Freeman, 1982. An enjoyable, nontechnical book that describes the influence of genetics on evolution.

The Origin of Species

When Darwin saw that the geologically young Galapagos Islands had already become populated with many plants and animals known nowhere else in the world, he realized that he was visiting a place of genesis. Darwin wrote in his diary: "Both in space and time, we seem to be brought somewhat near to that great fact—that mystery of mysteries—the first appearance of new beings on this Earth." The beginning of new forms of life—the origin of species—is at the focal point of evolutionary theory, for it is in new species that biological diversity arises.

The fossil record chronicles two patterns of evolutionary change: anagenesis and cladogenesis (Figure 22.1). **Anagenesis** (from the Greek *ana*, "up," and *genesis*, "origin"), also known as **phyletic evolution,** is the transformation of an unbranched lineage of organisms, sometimes to a state different enough from the ancestral population to justify renaming it as a new species. **Cladogenesis** (from the Greek *clados*, "branch"), also called branching evolution, is the budding of one or more new species from a parent species that continues to exist. Some of the evolutionary lineages that appear to be phyletic based on available fossils may actually be cladogenetic. For example, if species B branched from species A, and species A became extinct a short time (in geological terms) afterward, then the brief coexistence of the two species could not be resolved in the fossil record, and it would appear that A changed to B in a direct, unbranched lineage. Cladogenesis is more important than anagenesis in the history of life, not only because it seems to be the more common pattern, but also because only clado-

patterns of speciation. (a) In anagenesis (phyletic evolution), a single population is transformed enough to be designated a new species. **(b)** Cladogenesis is
branching evolution, in which a new species arises from a small population that
buds from a parent species. Most new
species probably evolve by cladogenesis,
the branching evolution that is the basis
for biological diversity.

(a) Anagenesis   *(b)* Cladogenesis

genesis can promote biological diversity by increasing
the number of species.

In the preceding chapter, we considered the mechanisms of microevolution that change populations, with
natural selection the one agent of change that is adaptive, tending to improve the fit of organisms to their
environment. Evolutionary theory also attempts to
explain biological transformation at a higher level: the
genesis of entirely new species from existing ones.
That the Earth has been inhabited by a changing cast
of living forms is a fact stamped clearly in the fossil
record. Our objective in this chapter is to examine and
evaluate possible mechanisms for the origin of species, or speciation. The first step is to appraise the
assumption that species actually exist in nature as discrete biological units distinct from all others.

THE SPECIES PROBLEM

In 1927, a young biologist named Ernst Mayr led an
expedition into the remote Arafak Mountains of New
Guinea to study the wildlife and collect specimens.
He found a great diversity of birds, identifying 138

separate species on the basis of differences in their
appearance. Mayr was surprised to learn that the local
tribe of Papuan natives, who hunted the animals for
food and feathers, had given names of their own to
137 birds (two birds assigned to separate species by
Mayr are extremely similar, and the Papuans did not
distinguish between them). Although their motives
and training could not have been more different, Mayr
and the indigenous people agreed almost exactly in
their inventory of the local birdlife. There are many
such cases of university-trained taxonomists giving
specific names to organisms in a particular locale, only
to find that the discrete forms they identified correspond to the folk taxonomy of the region. From this
we might conclude that species exist in nature as discrete units, demarcated from other species. Yet devising a formal definition of a species poses a formidable
challenge.

Two Concepts of Species

Species is a Latin word meaning "kind" or "appearance." And indeed we learn to distinguish between
the kinds of plants or animals—between dogs and

(a)

(b)

Figure 22.2
Biological species concept. This view of species emphasizes interfertility rather than physical similarity. **(a)** Members of a single biological species may differ extensively in appearance from one another. **(b)** Conversely, members of two different biological species, such as the eastern meadowlark (top) and the western meadowlark (bottom), may appear very similar.

cats, for instance—from differences in their appearance. Linnaeus, the founder of modern taxonomy, described individual species in terms of their physical form, or morphology, and this is still the method most often used to characterize species. Species defined by their anatomical features are referred to as **morphospecies.** There are certainly pitfalls in applying the morphospecies concept in some situations. It is sometimes difficult, for example, to determine if a set of organisms represents multiple species or a single species with extensive phenotypic variation. Conversely, two populations that are almost indistinguishable by morphological criteria may turn out to be different species based on other criteria. Despite these difficulties, the morphospecies concept, based as it is on observable and measurable anatomy, is usually practical to apply in the field, even on fossils. And most of the species recognized by taxonomists have been designated as separate species based on morphological criteria. From the standpoint of evolutionary theory, however, the morphospecies concept does not address the discontinuity that usually exists between species. An alternative concept is the biological species concept, first enunciated by Mayr in 1942.

A **biological species** is a population or group of populations whose members have the potential to interbreed with one another in nature to produce fertile offspring, but cannot successfully interbreed with members of other populations (Figure 22.2). In other

words, a biological species is the largest unit of population in which gene flow is possible, and which is genetically isolated from other such populations. Put still another way, each species is circumscribed by reproductive barriers that preserve its integrity as a biological package by blocking genetic mixing with other species. Members of a species, said to be conspecific, are united by being reproductively compatible, at least potentially. A businesswoman in Manhattan has little probability of sharing offspring with a dairyman in Outer Mongolia, but if the two should get together, they could have viable babies that develop into fertile adults. All humans belong to the same biological species. In contrast, humans and chimpanzees remain distinct species even where they share territory, because the two species cannot successfully interbreed and produce hybrid offspring.

It is important to remember that biological species are defined by their reproductive isolation from other species in *natural* environments. It is often possible in the laboratory to produce hybrids between two species that do not interbreed in nature.

Limitations of the Biological Species Concept

The biological species concept does not work in all situations. The criterion of interbreeding is useless for organisms that are completely asexual in their repro-

Figure 22.3

A chain species. These four populations of the deer mouse (*Peromyscus maniculatus*) display geographic variation and are designated by some experts as subspecies. Interbreeding between subspecies occurs where their ranges overlap, except for the subspecies *artemisiae* and *nebrascensis*. These two subspecies will not interbreed, yet gene flow between them is possible via the other neighboring populations. The concept of biological species as genetically isolated populations cannot be applied unambiguously to cases such as the deer mouse. Such cases may represent speciation in progress; for example, *artemisae* and *nebrascensis* could legitimately be called different species if, in the future, the populations connecting them became extinct.

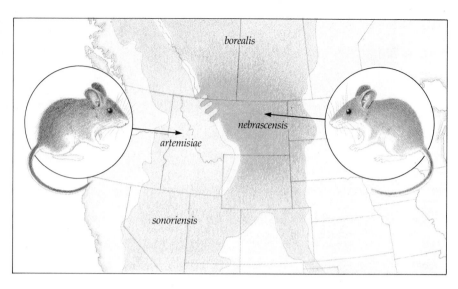

duction, as are all prokaryotes, some protists, and some fungi. There are even plants, including the commercial banana, that are exclusively asexual. Many bacteria do transfer genes on a limited scale by conjugation and other processes, but there is nothing akin to the equal contribution of genetic material from two parents that occurs in sexual reproduction (see Chapter 17). Different lineages of descent give rise to clones, which, genetically speaking, represent single individuals. Asexual organisms can be assigned to species only by the grouping of clones that have the same morphology and biochemical characteristics.

The biological species concept is also inadequate as a criterion for grouping extinct forms of life, the fossils of which must be classified according to the morphospecies concept. Similarly, if two populations are geographically segregated from each other, we do not know if they have the potential to interbreed in nature, even if they are so much alike that they are placed in the same species on morphological grounds.

Even for populations that are sexual, contemporaneous, and geographically contiguous, there are cases where the biological species concept cannot be applied unambiguously. Consider, for example, four populations of the deer mouse (*Peromyscus maniculatus*) found in the Rocky Mountains (Figure 22.3). Such phenotypically distinct populations that are separated geographically are sometimes called subspecies. The deer mouse populations overlap at certain locations, and some interbreeding occurs in these zones of cohabitation; we would therefore consider the populations to belong to the same species according to the biological species concept. The exception is the subspecies *P. m. artemisiae* and the subspecies *P. m. nebrascensis*. Though their ranges overlap, these two populations do not interbreed. Nevertheless, their gene pools are not completely isolated, because each of these populations interbreeds with its other neighboring popu-

lations, and genes could conceivably flow eventually from the *artemisiae* to the *nebrascensis* population. The extent of gene flow by this circuitous route is probably so slight that even the experts disagree on whether or not these two populations should be called subspecies or be assigned to separate species. If the corridors for gene flow were eliminated by extinction of the other two populations, then *artemisiae* and *nebrascensis* could be named separate species without reservation.

Population biologists are discovering more and more cases where the distinction between populations with limited gene flow and full biological species with segregated gene pools blurs. When two populations are completely unable to interbreed where they are in contact, then they are clearly different species, and such sharp species gaps may be true in most cases. But when there is just a trickle of genes between two quite different populations by the formation of fertile hybrids where the populations are in contact, then the biological species concept is difficult to apply. It is as though we are catching populations at different stages in their evolutionary descent from common ancestors. This is to be expected if new species usually arise by the gradual divergence of populations. Additional species concepts have been proposed in recent years to accommodate this dynamic, quantitative aspect of speciation. However, the species problem may have no completely satisfactory resolution; it is unlikely that any single definition of a species can be stretched to cover all cases. These theoretical problems aside, the concept of morphospecies, still the most practical definition of species for taxonomic purposes, and the conceptual definition of biological species usually result in recognizing the same units as species. Ernst Mayr and the Papuan tribe arrived at the same set of species of Arafak birds based on differences in appearance, and it is reproductive isolation that preserves discontinuities at species boundaries.

REPRODUCTIVE BARRIERS

Any factor that impedes two species from producing fertile hybrids contributes to reproductive isolation. No single barrier may be completely impenetrable to gene flow, but most species are genetically sequestered by more than one type of barrier. Here, we are considering only biological barriers to reproduction, which are intrinsic to the organisms. Of course, if two species are geographically segregated, they cannot possibly interbreed, but a geographical barrier is not considered equivalent to reproductive isolation because it is not intrinsic to the organisms themselves. Reproductive isolation prevents populations belonging to different species from interbreeding, even if their ranges overlap.

Clearly, a fly will not mate with a frog or a fern, but what prevents species that are very similar—that is, closely related—from interbreeding? The various reproductive barriers that isolate the gene pools of species can be categorized as prezygotic or postzygotic, depending on whether they function before or after the formation of zygotes, which are fertilized eggs (Table 22.1). **Prezygotic barriers** impede mating between species or hinder fertilization of ova should members of different species attempt to mate. If a sperm cell from one species does fertilize an ovum of another species, then **postzygotic barriers** prevent the hybrid zygote from developing into a viable, fertile adult.

Prezygotic Barriers

Ecological Isolation   Two species that live in different habitats within the same area may encounter each other rarely if at all, even though they are not technically geographically isolated. For example, two species of garter snakes belonging to the genus *Thamnophis* occur in the same areas, but for reasons intrinsic to each species, one lives mainly in water and the other is primarily terrestrial. Ecological isolation also affects parasites, which are generally confined to certain plant or animal host species. Two species of parasites living on different hosts will not have a chance to mate.

Temporal Isolation   Two species that breed during different times of the day, different seasons, or different years cannot mix their gametes. Brown trout and rainbow trout cohabit the same streams, but browns breed in fall and rainbows in spring. Three species of the orchid genus *Dendrobium* living in the same rain forest do not hybridize because they flower on different days. Pollination of each species is limited to a single day because flowers open in the morning and wither that evening. A sudden storm induces all three species to flower, but the number of days that lapse

---

Table 22.1 Reproductive barriers between species

I. *Prezygotic:* Prevent mating or fertilization.
   a. *Ecological isolation:* Populations live in different habitats and do not meet.
   b. *Temporal isolation:* Mating or flowering occur at different seasons or times of day.
   c. *Behavioral isolation:* There is little or no sexual attraction between females and males.
   d. *Mechanical isolation:* Structural differences in genitalia or flowers prevent copulation or pollen transfer.
   e. *Gametic isolation:* Female and male gametes fail to attract each other or are inviable.

II. *Postzygotic:* Prevent the development of viable, fertile adults.
   a. *Hybrid inviability:* Hybrid zygotes fail to develop or fail to reach sexual maturity.
   b. *Hybrid sterility:* Hybrids fail to produce functional gametes.
   c. *Hybrid breakdown:* The offspring of hybrids have reduced viability or fertility.

---

between the stimulus and flowering is eight in one species, nine in another, and ten in the third species; so reproductive isolation is still maintained.

Behavioral Isolation   Special signals that attract mates, as well as elaborate behavior unique to a species, are probably the most important reproductive barriers among closely related animals. Male fireflies of different species signal to females of their kind by blinking their lights in characteristic patterns. The females discriminate among the different signals, responding only to flashes of their own species by flashing back and attracting the males.

Many animals recognize mates of their species by sensing distinctive chemical signals called pheromones. For example, female gypsy moths attract males by emitting a volatile compound. The olfactory (smelling) organs of the males are tuned in to that specific chemical, and males of other moth species do not confuse it with the sex attractant unique to females of their own species.

The eastern and western meadowlarks are almost identical in morphology (see Figure 22.2b) and habitat, and their ranges overlap in the central United States. Yet, they remain two separate species, partly because of the difference in their songs, which enables them to recognize individuals of their own kind. Still another form of behavioral isolation is courtship ritual specific to a species (Figure 22.4).

Mechanical Isolation   Closely related species may attempt to mate, but fail to consummate the act because they are anatomically incompatible. Male dragonflies use a special pair of appendages to clasp females dur-

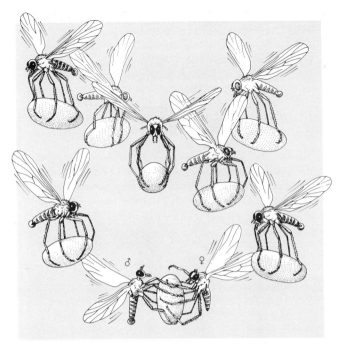

Figure 22.4
Specific courtship behavior. Many animals will not copulate before a courtship ritual unique to that species. For example, male ballooon flies (*Hilara sartor*) fashion exquisite silk balloons, and several males carrying their balloons swarm until they attract a female. When she chooses a partner from the swarm, the male offers his balloon, and the female accepts the gift. Only then will the couple cut away from the swarm and copulate.

ing copulation. When a male tries to mount a female of a different species, he is unsuccessful because his clasping appendages do not fit the female's form well enough to grip securely. Mechanical barriers also contribute to reproductive isolation of flowering plants that are pollinated by insects or other animals. Floral anatomy is often adapted to a specific pollinator that faithfully transfers pollen only among plants of the same species (Figure 22.5).

Gametic Isolation   Even if the gametes of different species should meet, they rarely fuse to form a zygote. For animals whose eggs are fertilized within the female reproductive tract (internal fertilization), the sperm of one species may not be able to survive in the environment of the female reproductive tract of another species. Many aquatic animals release their gametes into the surrounding water, where the eggs are fertilized (external fertilization). Even when two closely related species release their gametes at the same time in the same place, cross-specific fertilization is uncommon. Gamete recognition may be based on the presence of specific molecules on the coats around the egg, which adhere only to complementary molecules on sperm cells of the same species. A similar mechanism of molecular recognition enables a flower to discriminate between pollen of the same species and pollen of different species.

Postzygotic Barriers

Hybrid Inviability   When prezygotic barriers are crossed and hybrid zygotes are formed, genetic incompatibility between the two species may abort development of the hybrid at some embryonic stage.

Figure 22.5
Mechanical isolation of flowering plants. Many plants have exclusive relationships with the animals that carry their pollen. Hummingbirds use their needlelike beaks to reach sugary nectar secreted by glands at the bottom of long floral tubes. As a bird feeds, its head is dusted with pollen, which it then transfers to the next flower it visits. In some cases, the beak of a particular species of hummingbird is just the right length for the floral tube of the plant species it pollinates. The unique floral architectures of plants pollinated by animals impose barriers to the transfer of pollen between species.

Of the several species of frogs belonging to the genus *Rana*, some live in the same regions and habitats, where they may occasionally hybridize. But the hybrids generally do not complete development, and those that do are frail.

Hybrid Sterility   Even if two species mate and produce hybrid offspring that are vigorous, reproductive isolation is intact if the hybrids are sterile because genes cannot flow from one species' gene pool to the other. One cause of this barrier is a failure of meiosis to produce normal gametes in the hybrid if chromosomes of the two parent species differ in number or structure. The most familiar case of a sterile hybrid is the mule, a robust cross between a horse and a donkey; horses and donkeys remain distinct species because, except very rarely, mules cannot backbreed with either parent.

Hybrid Breakdown   In some cases when species cross-mate, the first-generation hybrids are viable *and* fertile, but when these hybrids mate with one another or with either parent species, offspring of the next generation are feeble or sterile. For example, different cotton species can produce fertile hybrids, but breakdown occurs in the next generation when progeny of the hybrids die in their seeds or grow into weak and defective plants.

Introgression

Alleles may occasionally seep through all reproductive barriers and pass between the gene pools of closely related species when fertile hybrids mate successfully with one of the parent species. This transplantation of alleles between species is called **introgression.** For example, corn (*Zea mays*) has some alleles that can be traced to a closely related wild grass called teosinte (*Zea mexicana*). The introgression occurs when the two species hybridize and a fraction of the hybrids manage to cross with corn plants. The transplant of alleles increases the reservoir of genetic variation that can be exploited by breeders trying to produce new corn varieties by artificial selection. But occasional hybridization does not erase the boundary between corn and teosinte. As long as reproductive barriers hold introgression to a trickle, the isolation of the two gene pools is not seriously breached and the two species remain distinct.

If the reproductive barriers we have surveyed form the boundaries around species, then the evolution of these barriers is the key biological event in the origin of new species. Let us now examine situations that make reproductive isolation, and hence speciation, possible.

THE BIOGEOGRAPHY OF SPECIATION

A crucial episode in the origin of a species occurs when the gene pool of a population is severed from other populations of the parent species. With its gene pool isolated, the splinter population can follow its own evolutionary course as changes in allele frequencies caused by selection, genetic drift, and mutations occur undiluted by gene flow from other populations. Speciation episodes can be classified into three modes based on the geographical relationship of a new species to its ancestral species. The initial block to gene flow may be a geographical barrier that physically isolates the population. This mode of speciation is termed **allopatric** (from the Greek *allos,* "other," and the Latin *patria,* "homeland"), and populations segregated by a geographical barrier are known as allopatric populations (Figure 22.6a). A second mode of speciation, called **parapatric speciation** (from the Latin *para,* "near"), may occur at a boundary between two populations where some gene flow occurs, but at a rate too sluggish to overcome divergence of the gene pools of the two neighboring populations (Figure 22.6b). In a third speciation mode, a subpopulation becomes reproductively isolated in the midst of its parent population; this is **sympatric speciation** (from the Greek *syn,* "together"). Populations are said to be sympatric if their ranges overlap (Figure 22.6c).

Allopatric Speciation

Geological processes can cause populations that were once sympatric to become allopatric: A mountain range may emerge and gradually split a population of organisms that can inhabit only lowlands; a creeping glacier may gradually divide a population; a land bridge such as the Isthmus of Panama may form and separate the marine life on either side; or a large lake may subside until there are several smaller lakes with their populations now isolated. Alternatively, a small population may become geographically isolated when individuals from the parent population travel to a new location.

Just how formidable a geographical barrier must be to keep allopatric populations apart depends on the ability of the organisms to disperse due to the mobility of animals or the dispersibility of spores, pollen, and seeds of plants. The Grand Canyon is easily crossed by hawks and many other birds, but it is an impassable barrier to populations of small rodents confined to either the north or south rim of the canyon.

Let us consider an example of allopatric speciation. About 50,000 years ago, during an ice age, what is now the Death Valley region of California and Nevada had a very rainy climate and a system of intercon-

(a)

(b)

(c)

Figure 22.6
Modes of speciation: A biogeographical classification. (a) Allopatric speciation. A population forms a new species while geographically isolated from its parent population. (This symbolic drawing understates the degree of isolation usually required.) (b) Parapatric speciation. Two adjacent populations displaying geographical variation hybridize along a common border, but the populations remain distinct because selection factors unique to their ranges overpower the trickle of alleles flowing across the border. Eventually the gene pools of the two populations may diverge so much that interbreeding can no longer occur. (c) Sympatric speciation. A small population forms a new species in the midst of its parent population. Reproductive isolation is achieved without geographical segregation, usually by genetic changes that result in a reproductive barrier.

Figure 22.7
Pupfish speciation in Death Valley. What is now a desert valley was once a great inland lake that began to dry up about 10,000 years ago. Small isolated springs are all that remain. Some of the springs are home to small animals called pupfishes (*Cyprinodon*), usually with each spring inhabited by a single species found only in that pool. The most logical interpretation is that the different pupfish species descended from a common ancestral population that became fragmented when the lake began to dry up. The isolated populations of pupfishes diverged through genetic drift and natural selection, and have differentiated to the level of true species that cannot interbreed even when brought together in the laboratory.

nected lakes and rivers. A drying trend began about 10,000 years ago, and by 4000 years ago the region had become a desert. Today, all that is left of the network of lakes and rivers is isolated springs scattered in the desert, mostly in deep clefts between rocky walls. The springs vary extensively in water temperature and salinity. Living in many of the springs are tiny fishes called pupfishes, which belong to the genus *Cyprinodon*. Each inhabited spring, often no more than a few meters in diameter, is home to its own species of pupfish adapted to that pool and found nowhere else in the world. The various pupfishes probably descended from a single ancestral species whose range was broken up when the region became arid, cloistering several small populations that diverged in their evolution as they adapted to their home springs. (Figure 22.7).

Conditions Favoring Allopatric Speciation Whenever populations become allopatric, there is a potential for speciation as the isolated gene pools accumulate differences by microevolution that may cause the populations to diverge in phenotype. But an isolated population that is small is more likely than a large population to change substantially enough to become a new species.

The geographical isolation of a small population usually occurs at the fringe of the parent population's range. The splinter population, or peripheral isolate, is a good candidate for speciation for three reasons:

1. The gene pool of the peripheral isolate probably differs from that of the parent population from the outset. Living near the border of the range, the peripheral isolate represents the extremes of any genotypic and phenotypic clines that existed in the original sympatric population. And if the peripheral isolate is small, there will be a founder effect, with chance alone resulting in a gene pool that is not representative of the gene pool of the parent population.

2. Until the peripheral isolate becomes a large population, genetic drift will continue to change its gene pool at random. New mutations or combinations of existing alleles that are neutral in adaptive value may become fixed in the population by chance alone, causing phenotypic divergence from the parent population (see Chapter 21).

3. Evolution caused by selection is likely to take a different direction in the peripheral isolate than in the parent population. Because the peripheral isolate inhabits a frontier, where the environment is somewhat different, the peripheral isolate will probably encounter selection factors that are different from, and generally more severe than, those operating on the parent population.

These factors will cause peripheral isolates to follow an evolutionary course that diverges from that of the parent population so long as the gene pools remain isolated. This does not mean that all peripheral isolates persist long enough or change enough to become new species. Life on the frontier is usually harsh, and most pioneer populations probably become extinct. As evolutionary biologist Stephen Jay Gould puts it: "Status as a peripheral isolate merely gives a lottery ticket to a small population. A population can't win (speciate) without a ticket, but there are very few winning tickets."

There is ample evidence that allopatric speciation is much faster in small populations than in very large ones. The North American sycamore tree and the European sycamore represent large populations that have been allopatric for at least 30 million years, but specimens that are brought together still produce fertile hybrids. Another example can be seen in the similarity of species on either side of the Panamanian land bridge (Figure 22.8). In comparison, the small populations of waifs that managed to reach the Galapagos Islands have given rise to a fauna that has become almost entirely endemic in less than two million years. In fact, evolutionary biologists generally agree that a small population can accumulate enough genetic change to become a new morphospecies in only hundreds to thousands of generations, equivalent to less than a thousand years to a few tens of thousands of years, depending on generation span.

Adaptive Radiation on Island Chains   Flurries of allopatric speciation have occurred on island chains where founding populations that have strayed or

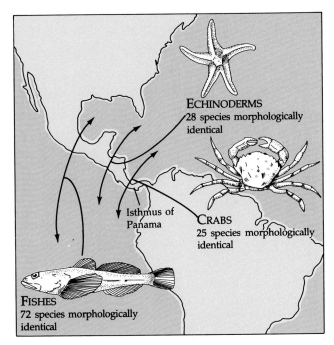

Figure 22.8
Evolutionary stasis of large successful populations. The Panamanian land bridge formed about three million years ago, segregating marine life in the Atlantic Ocean from species in the Pacific. The majority of species on opposite sides of the isthmus have not diverged during their three million years of allopatry, apparently because the two environments are similar and because most large populations evolve very slowly. New alleles and genetic combinations that arise in a large gene pool are swamped by the huge number of existing alleles.

become passively dispersed from their ancestral populations have evolved in isolation. The many endemic species of the Galapagos descended from stragglers that floated, flew, or were blown over the sea from the South American mainland. Consider Darwin's finches. A single dispersal event may have seeded one island with a small population of the ancestral finch, and the peripheral isolate formed a new species for all the reasons already described. Later, a few individuals of this island species may have reached neighboring islands, where geographical isolation permitted more speciation episodes (Figure 22.9). After diverging on the island it invaded, a young species could recolonize the island from which its founding population emigrated and coexist there with its ancestral species or form still another species. Multiple invasions of islands by peripheral isolates of species from neighboring islands would eventually lead to the coexistence of several species on each island. The islands are far enough apart to permit populations to evolve in isolation, but close enough together for occasional dispersion events to occur.

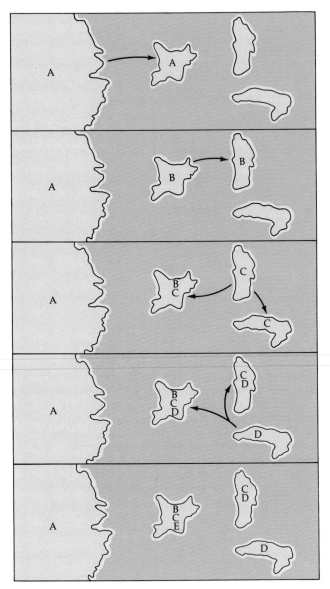

Figure 22.9
Model for adaptive radiation on island chains. One island in this cluster of three is seeded by a small colony founded by waifs of species A blown over from a mainland population. Its gene pool isolated from the ancestral species, the island population evolves into species B as it adapts to its new environment. Storms or other agents of dispersion spread species B to a second island, where the isolated colony evolves into species C. Later, a splinter population from species C recolonizes the first island and cohabits with species B, but reproductive barriers keep the species distinct, each adapted to certain foods and other requirements. A colony of species C may also populate a third island, where it adapts and forms species D. Species D is dispersed to the two islands of its ancestors, finding an unoccupied niche on one island, but forming a new species, E, on the other island. The story could go on and on, with a series of allopatric episodes made possible by the combination of isolation and occasional dispersal.

Thirteen species of finches have evolved on the Galapagos Archipelago, almost certainly from a single ancestral species. Each island now has multiple species, with as many as ten on some islands. In contrast, Cocos Island, about 700 km north of the Galapagos, has only one finch species, apparently derived from an ancestral species that made it to that remote island. Cocos is so isolated that there has been no opportunity for the kind of island hopping that resulted in the many episodes of allopatric speciation on the Galapagos.

The emergence of numerous species from a common ancestor introduced to an environment presenting a diversity of new opportunities and problems is called **adaptive radiation**. Adaptive radiation of Darwin's finches is evident in the many types of beaks specialized for different foods.

The Hawaiian Archipelago is perhaps the world's greatest showcase of evolution. The volcanic islands are about 3500 km from the nearest continent. They become progressively younger to the southeast, terminating with the youngest and largest island, Hawaii, which is less than a million years old and still has active volcanoes. Each island was born naked and was gradually clothed by a fauna and flora derived from strays that rode the ocean currents and winds from distant islands and continents, or from older islands of the archipelago itself (Figure 22.10). The physical diversity of each island, including a range of altitudes and extensive differences in rainfall, provides many environmental opportunities for evolutionary divergence by natural selection. Multiple invasions and allopatric speciations have ignited an explosion of adaptive radiation; most of the thousands of species of plants and animals that now inhabit the islands are found nowhere else in the world (Figure 22.11). In contrast, there are no endemic species on the Florida Keys. Apparently, those islands are so close to the mainland that founding populations are not sequestered long enough for the origin of intrinsic reproductive barriers that block their gene pools from the steady stream of immigrants from the parent populations on the mainland.

Parapatric Speciation

Parapatric populations are those with separate ranges that abut along a common border, which often follows a discontinuity in some important environmental feature. For example, parapatric populations of a plant species that are adapted to different soil conditions may meet along a zone where the soil changes. The gene pools of the two populations on either side of the border would be somewhat different, adapted as they are to disparate conditions, but there would be a limited gene flow between the populations via inter-

Figure 22.10
Long-distance dispersal. Plant seeds cling to this sea bird, which is capable of long flights. This is one mechanism that can disperse terrestrial organisms to isolated islands.

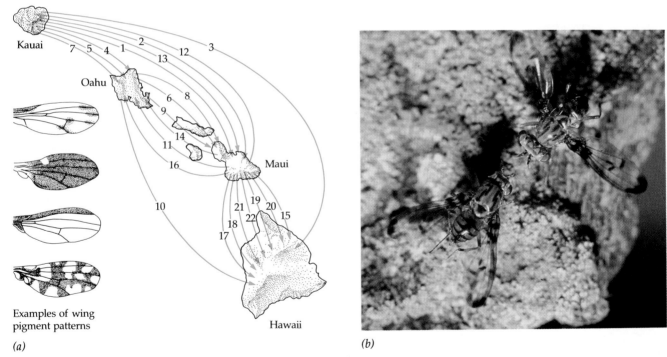

(a)

Examples of wing pigment patterns

(b)

Figure 22.11
Adaptive radiation of picture wing flies on the Hawaiian Islands. (a) Picture wing flies comprise one group of the approximately 500 endemic species of *Drosophila* flies on the islands. Each species has its own picture, which males display to females during courtship. Beginning with a single species of picture wing fly on Kauai, the oldest island, at least 22 interisland journeys by founders have occurred, resulting in a series of allopatric speciations. The numbers on the arrows indicate the probable order of those journeys. Notice that most invasions leading to new species occurred from the older islands in the northwest to the younger islands in the southeast. (b) Sexual selection has played an important role in the speciation of splinter populations of Hawaiian *Drosophila*. Unique courtship rituals, usually dependent on specific morphological features of the male, impose prezygotic barriers between closely related species of the flies. In *D. heteroneura*, shown here as an example, males have "hammerheads," with the eyes set far apart. This trademark may function in species recognition as the male approaches the female during courtship. Also, note the specific "pictures" on the wings of the flies.

breeding at the contact zone. It is conceivable, however, that the two populations could diverge enough to become separate species if the unique selection factors working on their gene pools swamped the trickle of genes flowing across the boundary (see Figure 22.6b).

Because most plants and animals breed with close neighbors, gene flow between parapatric populations may not penetrate far beyond a border. The populations may eventually become reproductively isolated as a result of postzygotic barriers, such as hybrid breakdown at the contact zone; individuals of either population that breed with hybrids would dilute phenotypic characters adapted to conditions on their side of the border.

Many evolutionary biologists believe that parapatric speciation is possible, but no case of speciation by this mode can be cited without reservation. If closely related species are presently parapatric, we cannot be certain that they were in contact during speciation. Parapatric species may have evolved when the populations were allopatric, with contact being established later.

Sympatric Speciation

In sympatric speciation, new species arise within the range of parent populations; reproductive isolation evolves without geographical isolation (see Figure 22.6c). This can occur in a single generation if some genetic change results in a reproductive barrier between the mutants and the parent population. The special genetic events that make sympatric speciation possible may apply to a substantial fraction of plant species. It is less certain how important sympatric speciation is in animal evolution.

Some plant species have their origins in accidents during cell division that result in extra sets of chromosomes. Examine the causes of these mutations in Figure 22.12. An **autopolyploid** results from a single species that doubles its chromosome number to the tetraploid· state. The tetraploids can then fertilize themselves or mate with other tetraploids. However, the mutants cannot interbreed successfully with diploid plants of the original population, because the hybrids would be inviable or sterile. In just one generation, a postzygotic barrier has caused reproductive isolation and interrupted gene flow between a fledgling population of mutants and the parent population that surrounds it—a speciation event that is essentially instantaneous. Sympatric speciation by autopolyploidy was first discovered early in this century by geneticist Hugo De Vries while he was studying the genetics of the evening primrose, *Oenothera lamarckiana,* a diploid species with 14 chromosomes. One day, De Vries noticed an unusual variant that had appeared among his plants, and microscopic inspection revealed that it was a tetraploid with 28 chromosomes. He found that the plant was unable to breed with the diploid primrose, and he named the new species *Oenothera gigas.*

Another type of polyploid species, much more common than autopolyploids, is called an **allopolyploid** (from the Greek *allos,* "other" or "different," referring here to the contribution of two different species to a polyploid hybrid). The potential evolution of an allopolyploid begins when two different species interbreed and combine their chromosomes (see Figure 22.12b). Interspecific hybrids are usually sterile because the haploid set of chromosomes from one species cannot pair during meiosis with the haploid set from the other species. Though infertile, a hybrid may actually be more vigorous than its parents and propagate itself asexually (which many plants can do). At least two mechanisms can transform the sterile hybrids into fertile polyploids. At some instant in the history of the hybrid clone, mitotic nondisjunction affecting the reproductive tissue of an individual may double chromosome number, and then the hybrid will be able to make gametes because each chromosome will have a homologue to synapse with during meiosis. Gametes from this fertile tetraploid could unite and give rise to a new species of interbreeding individuals, reproductively isolated from both parent species.

The second mechanism for the origin of allopolyploid species is probably more common. Meiotic nondisjunction in species A results in a gamete with its chromosome number unreduced—a diploid gamete. This aberrant diploid gamete unites with a normal haploid gamete from species B. The triploid hybrid is sterile, but may propagate asexually. At some point in the history of the asexual clone, meiotic nondisjunction may again produce a gamete with its chromosome number unreduced. Combination of this triploid gamete with a normal haploid gamete from species B results in a fertile hybrid with homologous pairs of chromosomes. As in the first mechanism, the allopolyploid has a chromosome number equal to the sum of the chromosome numbers of the two ancestral species.

Speciation of polyploids, especially allopolyploids, has been very important in plant evolution. Some allopolyploids are especially vigorous, apparently because they combine the best qualities of their two parent species. The conjunction of accidents required to produce these new plants—interspecific hybridization coupled with nondisjunction—has happened often enough that somewhere between 25% and 50% of all plant species are polyploids, including many of the plants we grow for food. The wheat used for bread, *Triticum aestivum,* is an allopolyploid with 42 chromosomes that is believed to have had its origin about 8000 years ago as a spontaneous hybrid of a cultivated wheat having 28 chromosomes and a wild grass hav-

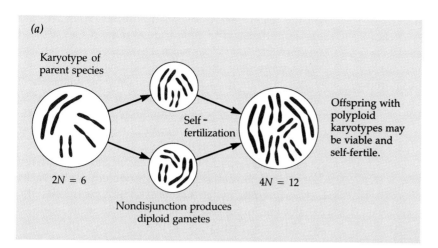

(a)

Karyotype of parent species

Self-fertilization

$2N = 6$

Nondisjunction produces diploid gametes

Offspring with polyploid karyotypes may be viable and self-fertile.

$4N = 12$

Figure 22.12
Sympatric speciation by polyploidy in plants. (a) Autopolyploidy. (b) Two mechanisms of allopolyploidy.

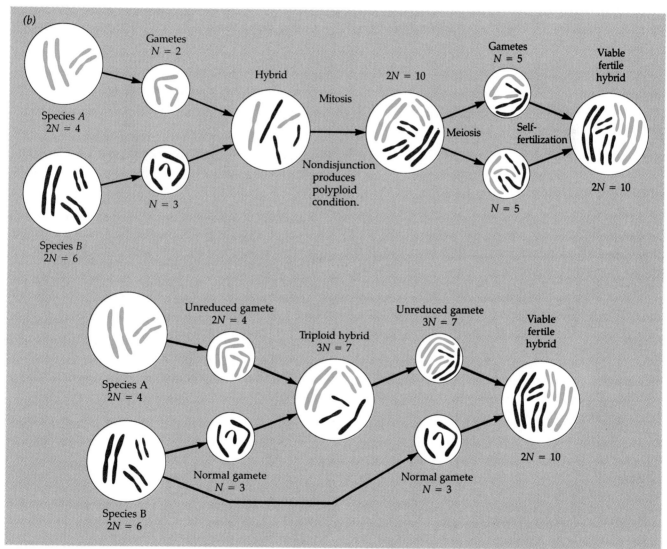

(b)

Species A
$2N = 4$

Gametes
$N = 2$

Species B
$2N = 6$

$N = 3$

Hybrid

Mitosis

Nondisjunction produces polyploid condition.

$2N = 10$

Meiosis

Gametes
$N = 5$

Self-fertilization

$N = 5$

Viable fertile hybrid

$2N = 10$

Species A
$2N = 4$

Unreduced gamete
$2N = 4$

Triploid hybrid
$3N = 7$

Unreduced gamete
$3N = 7$

Viable fertile hybrid

$2N = 10$

Species B
$2N = 6$

Normal gamete
$N = 3$

Normal gamete
$N = 3$

ing 14 chromosomes. Oats, cotton, potatoes, and tobacco are among the other polyploid species of importance to agriculture. Plant geneticists are now hybridizing plants and using chemicals that induce nondisjunction to help create new polyploids with special qualities. For example, there are artificial hybrids that combine the high yield of wheat with the ability of rye to resist disease. (The first attempt at artificial hybridization was not so successful. In 1924, the cabbage was crossed with the radish, resulting in an allopolyploid with the roots of a cabbage and the leaves of a radish.)

With this examination of the chromosomal basis of sympatric speciation in the plant kingdom, we are taking a more genetic approach to the problem of speciation. Let us now see how other genetic mechanisms can actually lead to the reproductive barriers required for the origin of a new species.

GENETIC MECHANISMS OF SPECIATION

Classifying speciation modes as allopatric, parapatric, or sympatric accentuates the biogeographical factors of speciation but does not emphasize the actual genetic mechanisms that lead to reproductive barriers between populations. Many population geneticists, notably Alan Templeton of Washington University in St. Louis, have advocated an alternative classification that groups speciation mechanisms according to genetic criteria rather than geographical scale. The two general categories can be paraphrased as speciation by "divergence" and speciation by "peak shift."

Speciation by Divergence

When two populations adapt to disparate environments, they accumulate differences in the frequencies of genotypes and phenotypes. In the course of this gradual adaptive divergence of two gene pools, reproductive barriers between the two populations may evolve coincidentally, differentiating the populations into two species.

In some cases, it is possible to test if allopatric populations that are closely related have evolved into separate species by attempting to hybridize between representatives of the two populations in the laboratory. In other cases, nature has performed these experiments for us, where two populations previously allopatric have come back into contact at some border between their geographical ranges. Among the possible scenarios when sympatry is restored are two extremes: The two populations hybridize freely when they meet and their gene pools are amalgamated, indicating that evolutionary divergence did not lead to reproductive isolation and speciation; or evolutionary divergence of the two populations during their period of allopatry has resulted in reproductive barriers that maintain the two populations as separate species when they come back into contact.

Between these two clear-cut extremes are intermediate cases where there is limited hybridization between the reunited populations, but the hybrids are less fit than offspring whose parents are from the same population. Such populations, not quite reproductively isolated, are referred to as **semispecies.** According to one hypothesis, the meeting of semispecies should reinforce prezygotic reproductive barriers between the populations. The idea is that any individuals with prezygotic barriers to mating outside their population will generally leave a greater number of fertile offspring than will individuals reproductively isolated from the other population by postzygotic barriers alone (hybrid inviability and hybrid sterility). If this hypothesis is correct, then a speciation episode not quite completed during geographical separation of two populations may be consummated when the populations come back into contact and natural selection favors those individuals least likely to mate outside their own population. For this reinforcement of reproductive isolation to be a common step in the speciation process, postzygotic barriers must generally evolve before prezygotic barriers. Some population geneticists question this basic assumption of the reinforcement hypothesis; prezygotic barriers between populations probably evolve more often during geographical isolation than they do as a consequence of reinforcement after reunion of the populations. Thus, the importance of reinforcement as a common stage of speciation is debated.

A key point in evolution by divergence is that reproductive barriers can arise without being favored directly by natural selection—there is no drive toward speciation for its own sake. Reproductive isolation is usually a secondary consequence of divergence of the two populations as they adapt to their separate environments. Postzygotic barriers may be pleiotropic effects of interspecific differences in genes controlling development (see Chapter 13 to review pleiotropy). For example, in laboratory hybrids between two very similar species of *Drosophila, D. melanogaster* and *D. simulans,* only one of the two sets of genes for synthesis of ribosomal RNA is active. This results in a very low viability for the hybrids. *Prezygotic* barriers can also evolve as by-products of gradual genetic divergence of two populations. For instance, if one population of an insect species adapts to a different host plant than other populations do, then an ecological barrier to interbreeding with the other populations is a side effect.

There *are* cases where reproductive isolation evolves more directly by sexual selection in isolated populations (see Chapter 21 to review sexual selection). For example, the wide head of the *Drosophila heteroneura* male enhances reproductive success with females of the same species, but makes successful courtship with females of other species very unlikely (see Figure 22.11b). However, even when sexual selection results in reproductive barriers, they evolve as adaptations that enhance reproductive success within a single population, not as safeguards against interbreeding with other populations. After all, reproductive barriers usually evolve when populations are allopatric, so they cannot possibly be functioning directly to isolate the gene pools of populations. For this reason, one criticism of the "biological species concept" is its emphasis on reproductive isolating mechanisms. An alternative that is gaining favor is the **recognition concept of species,** which assumes that the reproductive adaptations of a species consist of a set of characteristics that maximize successful mating with members of the same population. Reproductive isolation from other populations would be a spin-off. Whether a species definition stresses "reproductive isolating mechanisms" or "mate-recognition mechanisms" may seem like stressing opposite sides of the same coin. But the recognition concept may help to focus attention on the characteristics that are actually subject to natural selection in an isolated population that is in the process of speciation.

Speciation by Peak Shifts

In the 1930s, Sewell Wright crafted an evolutionary metaphor known as the "adaptive landscape" (Figure 22.13). This symbolic landscape has many **adaptive peaks** separated by valleys. An adaptive peak represents an equilibrium state where the gene pool has allelic frequencies that maximize the average fitness of a population's members. Even in a stable environment, several adaptive peaks for a given population are possible, but natural selection will tend to maintain the population at a single peak. To reach an alternative adaptive peak by some change in the overall gene pool, a population must go through a period corresponding to a valley on the adaptive landscape, where the average fitness of individuals is low. Thus, if some slight change in the frequency of alleles at one or more loci drives a population off an adaptive peak, natural selection will usually push the population back to its original peak. However, if the environment should change, then the adaptive landscape is redefined. A population that survives in this new environment must reach another adaptive peak through microevolution of its gene pool. But this is just another way of looking at speciation by adaptive divergence. What population geneticists call a "peak shift" is triggered *not* by a new physical environment, but by nonadaptive changes in the genetic system.

Peak shifts can be caused by a founder effect or a bottleneck (see Chapter 21). By randomly changing allelic frequencies in the gene pool, genetic drift can knock a small population off its original adaptive peak. If the gene pool is sufficiently destabilized, new adaptive peaks may be within reach. If the population survives, then it will be natural selection that pushes it to a new adaptive peak as the generations pass. Thus, adaptive evolution plays a major role in a peak shift, but it is genetic drift that makes the shift possible. A

Figure 22.13
The adaptive landscape and peak shifts. For any population in a stable environment, there are many alternative states of genetic adaptation symbolized as peaks on an adaptive landscape. The peaks are defined by their unique gene pools, each with allelic frequencies that endow members of the population with a mean fitness that is relatively high. The peaks are separated by valleys representing genetic combinations of relatively low average fitness. Thus, although the adaptive peak occupied by a successful population is only one of several alternative states possible for that population, it is unlikely that the population can reach a different peak by only natural selection if the environment is stable.
If the population is nudged slightly from its adaptive peak—by the introduction of new genes from mutation or gene flow, for example—natural selection will tend to offset this change and push the population back to its original adaptive peak. In a new environment, the adaptive landscape is redefined, and survival of the population may depend on natural selection moving the gene pool to a new adaptive peak. However, population geneticists reserve the term *peak shift* for speciation episodes triggered by nonadaptive changes in the genetic system, such as genetic drift or the origin of a polyploid population. Once a small population is genetically destabilized and dislodged from its original adaptive peak, natural selection may cause a generation-by-generation climb to some new adaptive peak.

peak shift can occur in a bottlenecked population even if its environment is stable, because many adaptive peaks are possible under the same environmental conditions. In the case of a founder effect, a splinter population is not only subject to genetic drift moving the gene pool randomly over the adaptive landscape, but a new set of adaptive peaks now exists. This combination of genetic drift followed by natural selection in a new environment may be responsible for the relatively rapid radiation of island species, such as the Hawaiian *Drosophila*.

The origin of new plant species by the chromosomal accidents that lead to fertile polyploids also depends on a peak shift (see Figure 22.12). The polyploid condition reproductively isolates the fledgling population from its ancestral species. But the polyploid karyotype also represents a novel genetic system that has originated by random processes rather than natural selection. The new karyotype redefines the adaptive land-scape and new peaks are possible. The polyploid species can persist only if natural selection fashions combinations of alleles that move the population to one of these new adaptive peaks.

How Much Genetic Change Is Required for Speciation?

It is not possible to generalize about the "genetic distance" between closely related species. In some cases, reproductive isolation may result from the cumulative divergence of populations at many gene loci. In other cases, changes at only a few loci produce reproductive barriers. For example, two species of Hawaiian *Drosophila*, *D. silverstris* and *D. heteroneura*, differ in alleles at a single gene locus determining head shape, a trait important in mate recognition by these flies (see Figure 22.11b). However, the phenotypic effect of different alleles at this locus is amplified by at least ten other gene loci that interact in an epistatic system (see Chapter 13 for a discussion on the genetic interactions known as epistasis). Changing one gene in a coadapted gene complex can have a substantial impact on the development of the organism. Such examples indicate that massive genetic change involving many loci is not mandatory for speciation. This conclusion is relevant to a debate among evolutionary biologists about the tempo of speciation, an issue we shall now examine.

GRADUAL AND PUNCTUATED INTERPRETATIONS OF SPECIATION

The traditional evolutionary tree that diagrams the descent of species from ancestral forms sprouts branches that diverge gradually, each new species evolving continuously over long spans of time (Figure 22.14a). The theory behind the tree is the extrapolation of the processes of microevolution—changes in the frequencies of alleles in gene pools—to the divergence of species; that is, big changes occur by the accumulation of many small ones. But as Niles Eldredge pointed out in the interview at the beginning of this unit, paleontologists rarely find gradual transitions of fossil forms. Instead, they often observe species appearing as new forms rather suddenly (in geological terms) in a layer of rocks, persisting essentially unchanged for their tenure on Earth and then disappearing from the record of the rocks as suddenly as they appeared. To Darwin, the origin of species was an extension of adaptation by natural selection, with isolated populations from common ancestral stock evolving differences gradually as they adapted to their local environments. But Darwin himself was bewil-dered by the dearth of connecting fossils and wrote: "Although each species must have passed through numerous transitional stages, it is probable that the periods during which each underwent modification, though many and long as measured by years, have been short in comparison with the periods during which each remained in an unchanged condition."

Advocates of a theory known as **punctuated equilibrium** have redrawn the evolutionary tree to represent the fossil evidence for evolution occurring in spurts of relatively rapid change instead of gradual divergence of species (Figure 22.14b). This model, first proposed in 1972 by Eldredge and Harvard's Stephen Jay Gould, depicts species undergoing most of their morphological modification as they first bud from parent species, and then changing little, even as they produce additional species. The theory replaces gradual change with long periods of stasis punctuated by episodes of speciation. A change in the genome, such as occurs in the origin of new polyploid plants, would be one mechanism of sudden speciation. Punctuationalists, as the proponents of this model of jerky evolution are called, point out that allopatric speciation of a splinter population isolated from its parent population by a geographical barrier can also be quite rapid. Remember, genetic drift and selection can cause significant change in just a few hundred to a few thousand generations in the gene pool of a population cloistered in a challenging new environment.

How can speciation in a few thousand generations, which may require several thousand years, be called an abrupt episode? Based on the fossil record, it can be estimated that successful species last for a few million years, on the average. Let us say that a particular species survives for five million years, but most changes in its morphology occurred during the first 50,000 years of its existence. In this case, the speciation episode was compressed into just 1% of the lifetime of the species. On the time scale that can generally be resolved in fossil strata, the species will appear suddenly in rocks of a certain age and then linger with little or no change before becoming extinct. During its formative millennia, the species may have accumulated its modifications gradually, but relative to the overall history of the species, its inception was abrupt.

The scenario of an evolutionary spurt preceding a much longer period of morphological stasis would explain why paleontologists find so few transitional fossils. It also explains why we observe most contemporary species as discrete units clearly separate from other species; according to the theory of punctuated equilibrium, species spend only a small fraction of their lifetimes becoming unique. We do observe some ambiguous cases, such as the populations of deer mice in the Rockies (see p. 462). But even if species spend only an average of 1% of their history diverging from related species, we should expect to encounter some cases of speciation in progress.

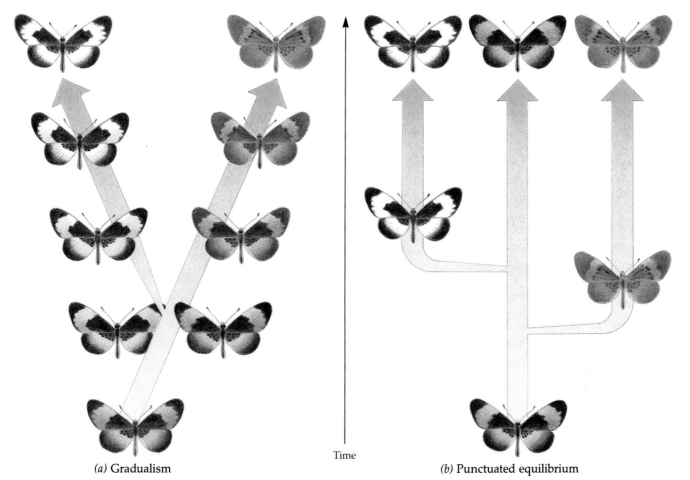

Time

(a) Gradualism

(b) Punctuated equilibrium

Figure 22.14
Two models for the tempo of specia-tion. A new species buds from its parent species as a small isolated population, indicated in this figure by the narow base of the arrow. (**a**) In the gradualist model, species descended from a common ancestor diverge more and more in mor-phology as they acquire unique adapta-tions. (**b**) According to the model known as punctuated equilibrium, a new species changes most as it buds from a parent species, and then changes little for the rest of its existence.

Once it is acknowledged that "sudden" may be many thousands of years on the vast scale of geological time, the debate between punctuationalists and gradualists over the rate of speciation is muted somewhat. The degree to which a species changes after its origin is another issue. If the species is adapted to an environment that stays the same, then natural selection would counter changes in the gene pool (see Chapter 21). And once selection during speciation has honed the genome into a new complex of coadapted genes, conservatism sets in because mutations are likely to impose disharmony on the genome and disrupt the development of the organism. In this view, the tendency for stabilizing selection to hold a population at one adaptive peak results in long periods of stasis.

Some gradualists retort that stasis is an illusion. Many species may continue to change after they come into existence, but in ways that cannot be detected from fossils. By necessity, paleontologists base their theories of descent almost entirely on external anatomy and skeletons. Changes in internal anatomy may go unnoticed, as would modifications in physiology or behavior. And population geneticists point out that many of the effects of microevolution occur at the molecular level without overtly affecting morphology.

Even the claim of long periods of morphological stasis in the history of species is contested. Over the past ten years, Peter Sheldon of Trinity College in Dublin has analyzed about 15,000 fossils of trilobites from shale deposits in Wales (trilobites are extinct arthropods—see Chapter 29). At this study site, the fossil record of trilobites is unusually complete. Paleontologists have arranged these fossils into several evolutionary lineages, and the youngest and oldest fossils of each lineage have been classified as different species because of morphological differences, including a different number of ridges on the tail sections of their shells. However, after Sheldon's extensive study of the fossils, he finds it impossible to draw such species boundaries. In each evolutionary lineage of the trilobites, the average number of tail ridges in the fossil populations changes gradually in the succession of rock layers. Sheldon argues that the requirement for paleontologists to give their fossils species names can artificially lead to a punctuated interpretation of a fossil series that actually changes gradually. What is needed, Sheldon suggests, are many additional exhaustive studies of fossil morphology where specific lineages are well preserved. Only then will we be able to assess the relative importance of gradual and punctuated tempos in the origin of new species. Whatever the outcome, there is no question that the theory of punctuated equilibrium has stimulated research and catalyzed a new interest in paleontology.

In the next chapter, we will see that the debate between gradualists and punctuationalists extends beyond the issue of speciation to major patterns in the history of life.

STUDY OUTLINE

1. New species can arise by transformation of an entire population (anagenesis) or, more commonly, by budding of small isolated populations from an ancestral species (cladogenesis).
2. Cladogenesis, or branching evolution, increases biological diversity.

The Species Problem (pp. 460–462)

1. There are two main concepts for describing species. Morphospecies are defined in terms of their unique anatomy. Biological species are groups of populations that share a distinctive gene pool by having the potential to interbreed.
2. Although the biological species concept is more amenable to evolutionary theory, it cannot apply to exclusively asexual or extinct species. It is questionable in some cases because different degrees of reproductive isolation may accompany populations in the process of speciation.

Reproductive Barriers (pp. 463–465)

1. Any intrinsic factor that prevents two species from producing fertile hybrids is a reproductive barrier that maintains the genetic integrity of a species.
2. Prezygotic barriers prevent cross-specific mating or fertilization. In particular, species that occupy the same geographical area often live in separate habitats (ecological isolation); breed at different times (temporal isolation); possess unique, exclusive mating signals and courtship behaviors (behavioral isolation); and/or have anatomically distinct reproductive organs (mechanical isolation) or incompatible sex cells (gametic isolation).
3. Even if two different species manage to mate, postzygotic barriers usually prevent the interspecific hybrids from developing into adults, breeding with either parent species, or producing viable, fertile offspring.
4. Occasionally, however, fertile hybrids successfully mate with one of the parent species, causing genes to seep through the reproductive barriers. As long as this introgression is limited, the two species remain distinct.

The Biogeography of Speciation (pp. 465–472)

1. The gene pool of a species can diverge in evolution through allopatric, parapatric, or sympatric speciation.
2. Allopatric speciation occurs when a splinter population diverges in evolution from its parent population after becoming geographically isolated.
3. Small splinter populations are better candidates than large ones for allopatric speciation because genetic change can ramify faster through a small gene pool.

4. Geographical isolation usually occurs at the fringe of the parent population's range, forming peripheral isolates. Such isolates usually differ genetically from the parent population at the outset owing to different and generally more severe selection pressures at the boundary of the range. Combined with the founder effect, genetic drift, and more severe selection factors, peripheral isolates follow their own distinctive evolutionary course, a course that leads toward either speciation or extinction.
5. Adaptive radiation is the generation of numerous species from a common ancestor introduced to a diverse environment. Island chains are showcases of adaptive radiation. Beginning with an ancestral species that managed to reach an archipelago, cloistering of populations on separate islands and occasional island hopping by founders result in multiple episodes of allopatric speciation.
6. Parapatric speciation occurs along a boundary between two populations when natural selection in each distinctive range overpowers the limited flow of genes across the border.
7. Sympatric speciation occurs without geographical separation when a segment of the population experiences a mutation that results in instantaneous reproductive isolation. Sympatric speciation is most common in plants, in which the mutation is usually the nondisjunctive doubling of the chromosome number. Autopolyploids are species derived this way from one ancestral species. Allopolyploids are species with multiple sets of chromosomes derived from two different species by nondisjunction before or after hybridization.

Genetic Mechanisms of Speciation (pp. 472–474)

1. Reproductive barriers may arise coincidently as two populations genetically diverge as they adapt to different environments.
2. When two allopatric populations are reunited, they may have developed intrinsic reproductive barriers that maintain them as separate species; they may interbreed freely and reunite their gene pools; or they may have become semispecies, and the hybrids they produce may be less fit. According to a hypothesis for reinforcement of reproductive isolation, postzygotic barriers usually evolve before prezygotic barriers, and the meeting of semispecies may reinforce the development of prezygotic reproductive barriers. Many researchers question this hypothesis.
3. Sexual selection may lead indirectly to reproductive barriers. According to the recognition concept of species,

natural selection would amplify adaptations that enhance reproductive success with members of the same species.

4. According to Wright's evolutionary metaphor, a population's gene pool is perched on an adaptive peak, one of several possible peaks on an adaptive landscape. Natural selection will tend to maintain a population at an adaptive peak that maximizes fitness. Peak shifts, associated with nonadaptive changes in the gene pool, may be initiated by a founder effect or bottleneck and driven by genetic drift. Adaptive evolution can push a destabilized gene pool to a new adaptive peak.

5. Reproductive isolation may result from a change at a single gene locus, especially one in a coadapted gene complex affecting development, or by the cumulative divergence at many gene loci.

Gradual and Punctuated Interpretations of Speciation
(pp. 474–476)

1. The conventional view is that speciation usually occurs gradually by an accumulation of microevolutionary changes in gene pools.

2. According to the theory of punctuated equilibrium, the relatively sudden appearance of species in the fossil record is a real phenomenon rather than a reflection of an incomplete fossil record. In this view, a species changes most when it buds from an ancestral species and then remains fairly static in morphology for the rest of its tenure. Biological history is thus envisioned as long periods of stasis punctuated by episodes of speciation.

3. Additional exhaustive studies of fossil lineages will be necessary to determine the relative importance of gradual and punctuated tempos in the origin of species.

SELF-QUIZ

1. Most of biological diversity has probably arisen by
   a. anagenesis
   b. cladogenesis
   c. phyletic evolution
   d. hybridization
   e. sympatric speciation

2. The largest unit in which gene flow is possible is a
   a. population
   b. species
   c. genus
   d. subspecies
   e. semispecies

3. Some species of *Anopheles* mosquito live in brackish water, some in running fresh water, and others in stagnant water. What type of reproductive barrier is most obviously separating these different species?
   a. ecological isolation
   b. temporal isolation
   c. behavioral isolation
   d. gametic isolation
   e. postzygotic barriers

4. The reproductive barrier that maintains the species boundary between horses and donkeys is
   a. mechanical isolation
   b. gametic isolation
   c. hybrid inviability
   d. hybrid sterility
   e. hybrid breakdown

5. According to advocates of the punctuated equilibrium theory,
   a. natural selection is unimportant as a mechanism of evolution
   b. given enough time, most existing species will branch gradually into new species
   c. a new species accumulates most of its unique features as it comes into existence and changes little for the rest of its duration as a species
   d. most evolution is anagenic
   e. transitional fossils generally link newer species to their parent species.

6. The biological species concept cannot be applied to two putative species that are
   a. sympatric
   b. nearly indistinguishable in morphology
   c. allopatric
   d. capable of forming viable hybrids
   e. exclusively asexual

7. Future cladogenesis of human populations to form new hominid (human) species is probably unlikely because
   a. the environment has stabilized
   b. humans are already perfectly adapted organisms
   c. only one adaptive peak is possible for hominids
   d. most human populations are very large and are incompletely isolated from surrounding populations
   e. human variation is not very extensive

8. According to the reinforcement hypothesis, production of sterile hybrids by the mating of horses and donkeys under natural conditions (without human intervention) would probably tend to
   a. result in the extinction of one of the two species
   b. favor prezygotic barriers between horses and donkeys
   c. gradually reinforce the postzygotic barriers between horses and donkeys
   d. fuse horses and donkeys into a single species
   e. decrease the morphological differences between horses and donkeys

9. Plant species "A" has a diploid number of 12. Plant species "B" has a diploid number of 16. A new species, "C," arises as an allopolyploid from hybridization of "A" and "B." The diploid number of "C" would probably be
   a. 12
   b. 14
   c. 16
   d. 28
   e. 56

10. The speciation episode described in question #9 is most likely a case of
    a. allopatric speciation
    b. sympatric speciation
    c. parapatric speciation
    d. adaptive radiation
    e. anagenic speciation

CHALLENGE QUESTIONS

1. According to one hypothesis, two closely related species should be most distinct from one another where their ranges overlap—that is, in regions of sympatry. This phenomenon is called "character displacement." The hypothesis goes on to propose that character displacement reduces interspecific competition for food, habitat and other resources, and also minimizes hybridization between the species in their region of sympatry. Character displacement has been difficult to demonstrate in nature. A recent experiment has nevertheless addressed the problem by comparing beak sizes of two species of Darwin's ground finches (*Geospiza fortis* and *G. fuliginosa*) in sympatry and allopatry. The experiment controlled for any possible effect on morphology by variation in food supply among locations. Why do you think this control was necessary? After taking this control into account, how would you expect the relative beak sizes of *G. fortis* and *G. fuliginosa* on the Santa Cruz Island, the only island they cohabit, to compare with the relative beak sizes of *G. fortis* on the islands of Daphne Major and *G. fuliginosa* on Los Hermanos if character displacement were operative? (See Further Reading 5.)

2. The study of fossil marine ostracods (small crustaceans) has shown that the frequency, duration, and magnitude of climatic changes have differential effects on evolution. Which climatic pattern do you suppose would have a greater impact on ostracod evolution—rapid climatic oscillations or sustained unidirectional climatic changes? Why? (Check your answer in Further Reading 1.)

FURTHER READING

1. Cronin, T. M. "Speciation and Stasis in Marine Ostracoda: Climatic Modulation of Evolution." *Science*, January 4, 1985. An interesting case study of a mechanism for speciation.
2. Eldredge, N., and S. J. Gould. "Punctuated Equilibria: An Alternative to Phyletic Gradualism." In *Models in Paleobiology*, edited by T.J.M. Schopf. San Francisco: Freeman, Cooper, 1972. The original proposal for punctuated equilibrium.
3. Kaneshiro, K. Y. "Speciation in the Hawaiian *Drosophila*." *Bioscience*, April 1988. The importance of sexual selection in a classic case of adaptive radiation.
4. Sheldon, P. "Making the Most of Evolutionary Diaries." *New Scientist*, January 21, 1988. An exhaustive study of Welsh trilobites supports a gradualistic interpretation of their evolution.
5. Schluter, D., T. D. Price, and P. R. Grant. "Ecological Character Displacement in Darwin's Finches." *Science*, March 1, 1985. A recent investigation of an old observation.

Macroevolution

Macroevolution encompasses the origin of novel designs, such as the feathers and wings of birds and the upright posture of humans; evolutionary trends, such as increasing brain size in mammals; the tremendous diversification of certain groups of organisms following evolutionary breakthroughs, such as the adaptive radiation of flowering plants; and extinction, such as the disappearance of the dinosaurs. Macroevolution is the story of major events in the history of life.

This chapter describes how biologists study macroevolution, discusses theories on the origin of new biological designs, and chronicles a few of the important episodes in the history of life. We will also consider an important question: Is macroevolutionary change primarily a cumulative product of microevolution working gradually over vast spans of time, or is it mostly a product of mechanisms other than the gradual modification of populations by natural selection? In our study of macroevolution, the theme that evolution is a consequence of interactions between organisms and their environments will be extended to environmental and biological change of global proportions.

THE RECORD OF THE ROCKS

Studying the succession of organisms in the fossil record is one method biologists use to reconstruct evolutionary history (Figure 23.1). Scientists who study life of the past are known as paleobiologists, a general name for paleontologists, who are interested mainly

Figure 23.1
The fossil record chronicles the history of life. This paleontologist is recovering parts of a dinosaur skeleton at Utah's Dinosaur National Monument.

in animal fossils, and paleobotanists, who specialize in the fossil record of plants (the prefix *paleo* is from the Greek *palaios*, "old").

Fossils

A fossil (from the Latin *fossilis*, "dug up") is any preserved remnant or impression left by an organism that lived in the past. Fossils that retain organic material are sometimes discovered as thin films pressed between layers of sandstone or shale. Paleobotanists have found leaves millions of years old that are still green with chlorophyll and well enough preserved for their organic composition to be analyzed and the ultrastructure of their cells to be examined with the electron microscope (Figure 23.2a). In rare circumstances, an entire organism, including its soft parts, is fossilized. This can happen only if the individual is interred in a medium that prevents bacteria and fungi from decomposing the corpse (Figure 23.2b).

The organic substances of a dead organism usually decay rapidly, but hard parts of an animal that are rich in minerals, such as the bones and teeth of vertebrates and the shells of many invertebrates, may remain as

fossils (Figure 23.2c). Paleontologists have unearthed nearly complete skeletons of dinosaurs and other forms, but more often, the finds consist of parts of skulls, bone fragments, or teeth. Many of these relics are hardened even more by a process known as petrification. Under the right conditions, minerals dissolved in groundwater seep into the tissues of a dead organism and replace organic material; the plant or animal turns to stone. Bizarre forests of petrified trees can be explored in parts of the southwestern United States that are now desert (Figure 23.2d).

The fossils that paleobiologists find in many of their digs are not the remnants of organisms at all, but replicas cast from molds left when corpses were covered by mud or sand (Figure 23.2e).

Fossils are meaningful in the context of macroevolution only if their vintages relative to other fossils can be determined by methods we shall now discuss.

The Geological Time Scale

Sedimentary rocks are the richest sources of fossils. Sand and silt weathered and eroded from the land are carried by rivers to seas and swamps, where the par-

(a)

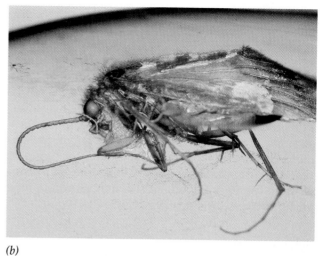

(b)

(c)

(d)

(e)

Figure 23.2
Survey of fossils. (**a**) This leaf is about 40 million years old. Still retaining many of its organic constituents, it is a thin film pressed in rock. (**b**) This insect got stuck in the resin of a tree about 40 million years ago. Since then, the resin has hardened into amber. (**c**) A skull of *Australopithecus africanus,* an ancestor of humans that lived about 2.5 million years ago. (**d**) These petrified trees stood about 190 million years ago in what is now a desert in Arizona. (**e**) Buried organisms, such as these invertebrates called brachiopods, which lived about 375 million years ago, decay and leave empty molds that may be filled by minerals dissolved in water. The cast is preserved if the minerals harden.

ticles settle to the bottom. Deposits pile up and compress the older sediments below into rock, sand into sandstone, and mud into shale. Aquatic organisms, and terrestrial ones swept into the seas and swamps, settle when they die, along with the sediments, and a tiny fraction of them leave fossils.

At any particular location, sedimentation may occur in periods when the sea level changes or lakes and swamps dry up and refill. When a region is submerged, the rate of sedimentation and the types of particles that sediment may vary with time. As a result of these different periods of sedimentation, the rock forms in layers, or strata (see Figure 20.4).

Relative Dating Superimposition of sedimentary rocks tells the relative ages of fossils. The fossils in each layer are a local sampling of the organisms that existed at the time that sediment was deposited. Younger strata are superimposed on top of older ones, and the succession of fossil species is a chronicle of macroevolution that paleobiologists try to read.

The strata at one location can often be correlated with strata at another location by the presence of similar fossils, known as **index fossils.** The best index fossils for correlating strata that are far apart are the shells of sea animals that were widespread. At any one location where a road cut or canyon wall reveals layered rocks, there are likely to be gaps in the sequence. That area may have been above sea level during different periods, and thus no sedimentation occurred. Some of the sedimentary layers that were deposited when the area was submerged may have been scraped away by subsequent periods of erosion. By studying many different sites, geologists have worked out a consistent sequence of geological periods (Table 23.1). These periods can be grouped into four eras. The boundaries between the eras mark major transitions in the forms of life fossilized in the rocks. The periods within each era are further subdivided into finer intervals called epochs (though only the epochs of the current era, the Cenozoic era, are commonly used).

The record of the rocks is a serial that chronicles the *relative* ages of fossils; it tells us the order in which groups of species present in a sequence of strata evolved. However, the series of sedimentary rocks does not tell the *absolute* ages of the embedded fossils. The difference is analogous to peeling the layers of wallpaper from the walls of a very old house that has been inhabited by many owners. You could determine the sequence in which the papers had been applied, but not the year that each layer was added.

Absolute Dating "Absolute" dating does not mean errorless dating, but only that age is given in years instead of relative terms such as *before* or *after* and *early* or *late.* **Radioactive dating** is the method most often

used to determine the ages of rocks and fossils on a scale of absolute time. Fossils contain isotopes of elements that accumulated in the organisms when they were alive. Because each radioactive isotope has a fixed rate of decay, known as its **half-life,** it can be used to date a specimen (see Chapter 2). The half-life is unaffected by temperature, pressure, and other environmental variables. For example, carbon-14, a radioactive isotope, has a half-life of 5600 years, meaning that half of the carbon-14 in a specimen will be gone in 5600 years, half of the remainder will be gone in another 5600 years, and so on, until all of the isotope has decayed. A sample beginning with 8 g of carbon-14 will have 4 g left in 5600 years and 2 g in 11,200 years. Because the half-life of carbon-14 is relatively short, this isotope is reliable for dating fossils less than about 50,000 years old. To date older fossils, paleobiologists must use radioactive isotopes with longer half-lives. For instance, potassium-40, a radioactive isotope with a half-life of 1.3 billion years, can be used to date rocks hundreds of millions of years old and infer the age of fossils embedded in those rocks. Radioactive dating has an error of plus or minus about 10%.

"Clocks" other than radioactive isotopes can be used to date some fossils. Amino acids have either left-handed or right-handed symmetry, designated the L- and D-forms, respectively (see Chapter 5). Organisms synthesize only L-amino acids, which are incorporated into proteins. After an organism dies, however, its population of L-amino acids is slowly converted, resulting in a mixture of L- and D-amino acids. In a fossil the ratio of L- and D-amino acids can be measured. Knowing the rate at which this chemical conversion takes place, we can determine how long the organism has been dead. This clock, unlike radioactive decay, is temperature-sensitive. For fossils found in locations where climate apparently has not changed significantly since the fossils formed, the two kinds of clocks agree closely on the age of the fossils.

The dating of rocks and the fossils they contain has enabled researchers to determine ages for the different geological periods. Notice in Table 23.1 that these periods span unequal intervals of time. The boundaries are not arbitrarily placed, but mark distinct changes in the species composition of sedimentary rocks.

Imperfection of the Fossil Record The discovery of a fossil is the culmination of a sequence of improbable coincidences. First, the organism had to die in the right place at the right time for burial conditions to favor fossilization. Then the rock layer containing the fossil had to escape geological processes that destroy or severely distort rocks, such as erosion, pressure from superimposed strata, or the melting of rocks that occurs at some locations. If the fossil was preserved, there is only a slight chance that a river carving a canyon or some other process will expose the rock

Table 23.1 The geological timetable

Era	Period	Epoch	Age (millions of years)	Some Important Events in the History of Life
Cenozoic	Neogene	Recent		Historic time
			0.01	
		Pleistocene		Ice ages; humans appear
			1.8	
		Pliocene		Apelike ancestors of humans appear
			5	
		Miocene		Continued radiation of mammals and angiosperms
			24	
	Paleogene	Oligocene		Origins of most modern mammalian orders, including apes
			38	
		Eocene		Angiosperm dominance increases; further increase in mammalian diversity
			54	
		Paleocene		Major radiation of mammals, birds, and pollinating insects
			65	
Mesozoic	Cretaceous			Flowering plants (angiosperms) appear; dinosaurs become extinct at end of period
			144	
	Jurassic			Gymnosperms continue as dominant plants; dinosaurs dominant
			213	
	Triassic			Gymnosperms dominate landscape; first dinosaurs, mammals, and birds
			248	
Paleozoic	Permian			Radiation of reptiles; origin of mammal-like reptiles; origins of most modern orders of insects; mass extinction of many marine invertebrates
			286	
	Carboniferous			Extensive forests of vascular plants; first seed plants; origin of reptiles; amphibians dominant
			360	
	Devonian			Diversification of bony fishes; first amphibians and insects
			408	
	Silurian			Diversity of jawless vertebrates; invasion of land by vascular plants and arthropods
			438	
	Ordovician			First vertebrates (jawless fishes); marine algae abundant
			505	
	Cambrian			Origin of most invertebrate phyla; diverse algae
			590	
Precambrian			700	Origin of first animals
			1500	Origin of eukaryotes
			2500	Oxygen-producing photosynthesis
			3500	Oldest definite fossils known (prokaryotes)
			4600	Approximate origin of the Earth

containing the fossil. And then there is only a remote chance that someone will find the fossil, although discovery is more probable for people who are purposefully looking for fossils. No wonder the fossil record is incomplete. A substantial fraction of species that have lived probably left no fossils, most fossils that formed have been destroyed, and only a fraction of existing fossils has been discovered. The fossil record, far from a complete sampling of organisms of the past, is slanted in favor of species that lived a long time, were abundant and widespread, and had shells or hard skeletons.

Darwin devoted an entire chapter in *The Origin of Species* to the imperfection of the fossil record. He was particularly concerned about the rarity of transitional fossils that show a progression of changes as more recent forms of life evolved from their ancestors. Since Darwin's time, many fossils have been discovered that seem to have transitional status, including forms linking fishes to amphibians, amphibians to reptiles, and

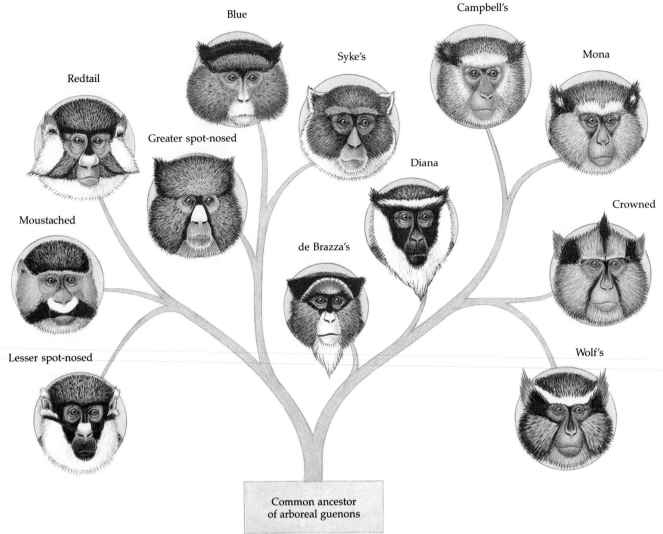

Figure 23.3
A proposed phylogenetic tree for guenon monkeys. A phylogenetic tree is a genealogy depicting probable evolutionary relationships between species and groups of species. This tree traces the evolution of 12 of the 25 known species of guenon monkeys (genus *Cercopithecus*), which inhabit eastern Africa. This particular phylogenetic tree is based on a comparison of blood proteins, but it is compatible with evidence based on facial markings, vocal calls, and biogeographical distribution of the monkeys. Like all evolutionary trees, this one is not gospel, but merely represents the most likely phylogeny based on the available evidence.

reptiles to birds and mammals. The fossil record is much more complete now than it was in Darwin's day, but fossil series showing gradations of change from older to younger species are still quite rare, considering how extensively life has changed over the geological periods. However, if the model of punctuated equilibrium is valid, then most morphological change occurs during relatively short junctures in evolution as new species bud from their ancestors, and we should not expect to observe smooth transitions in a succession of fossils. Niles Eldredge and many other paleontologists interpret discontinuities in the fossil record as evidence that evolution occurs in fits and starts.

The record of the rocks provides an outline of macroevolution, but to fill in the details, biologists must assess evolutionary relationships between modern organisms.

SYSTEMATICS: TRACING PHYLOGENY

The evolutionary history of a species or group of related species is called **phylogeny** (from the Greek *phylon,* "tribe," and *genesis,* "origin"). These genealogies are traditionally diagrammed as phylogenetic trees that trace putative evolutionary relationships (Figure 23.3).

Reconstructing phylogenetic history is part of the scope of **systematics,** the branch of biology concerned with the diversity of life. Systematics encompasses taxonomy, which is the identification and classification of species.

Taxonomy

The system of taxonomy developed by Linnaeus in the eighteenth century had two main features (see Chapter 20). First, it assigned to each species a two-part Latin name, or **binomial.** The first word of the name is the **genus** (plural, genera) to which the species belongs. The second word is the specific epithet (name) of the species. For example, the scientific name for the domestic cat is *Felis cattus.* A given genus may include several similar species, each with its own specific name. For instance, the lynx, *Felis lynx,* belongs to the same genus as the domestic cat. Common names, such as cat, black bear, and mountain lilac, work well in casual communication, but when biologists publish their research, they define the organisms they have studied with scientific names to avoid ambiguity. Many of the scientific names still in use date back to Linnaeus, who assigned binomials to over 11,000 species of plants and animals.

The second major contribution Linnaeus made to taxonomy was adopting a filing system for grouping species into a hierarchy of increasingly general categories. The first step in grouping species is built into binomial nomenclature. Species that are very similar, such as the bobcat and house cat, are placed in the same genus. Grouping species is natural for us, at least in concept. We lump together several trees we know as oaks and distinguish them from several other "types" of trees we call maples. Indeed, oaks and maples belong to separate genera. The Linnaean system formalizes this grouping of species into genera and extends the scheme to progressively broader categories of classification, some of which have been added since the time of Linnaeus. Taxonomists place similar genera in the same **family,** group families into **orders,** orders into **classes,** classes into **phyla** (singular, phylum), and phyla into **kingdoms.** For example, the genus *Felis* is lumped with various species of the genus *Panthera* (lion, tiger, leopard, and jaguar) in the family Felidae, the cat family. This family belongs to the order Carnivora, which also includes the family Canidae (dog family) and a few other allied families. The order Carnivora is grouped with many other orders in the class Mammalia, the mammals. And the class Mammalia is one of several belonging to the phylum Chordata in the kingdom Animalia. Each taxonomic level is more comprehensive than those below. All members of the family Felidae also belong to the order Carnivora and class Mammalia, but not all mammals are cats. Classifying a species by phylum, class, and so on, is analogous to a postal worker sorting mail, first by zip codes and then by streets and house numbers. Appendix Two gives taxonomic classification of major groups of organisms discussed in this text down to class level.

Taxonomy has two main objectives. The first is to sort out closely related organisms and assign them to separate species, describing the diagnostic characteristics that distinguish the species from one another. Related to this function is the naming of newly discovered species. In the Linnaean tradition, the name is a binomial, with the name of the genus to which the species belongs followed by the specific name, or epithet. The second major objective of taxonomy is to order species into the broader taxonomic categories, from genera to kingdoms. In some cases, there are intermediate categories, such as superfamilies (a category between families and orders) or subclasses (between orders and classes). The named taxonomic unit at any level is called a **taxon** (pural, taxa). For example, *Pinus* is a taxon at the genus level, the generic name for the various species of pine trees. Mammalia, a taxon at the class level, includes all the many orders of mammals. Only the genus name and specific epithet are italicized, and all taxa at the genus level or higher (broader) are capitalized. International committees establish rules of nomenclature, which are somewhat different for animals, plants, and bacteria. In Table 23.2, the house cat and the common buttercup are placed in their appropriate taxa as examples of classification.

Of all taxa, only the species actually exists in nature as a biologically cohesive unit, bonded by interbreeding and bounded by reproductive isolation from all other species. In most cases, distinguishing between species can be done objectively if enough is known about their characteristics (remember the story in Chapter 22 of Ernst Mayr and the New Guinea birds), but combining species into higher taxa often involves judgment calls. One taxonomist may value fine distinctions and favor a relatively large number of taxa for each category above species, whereas another may stress unification and propose a minimal number of taxa. For instance, some taxonomists who specialize in the cat family lump all cats except the cheetah into a single genus, *Felis.* Other taxonomists split the same group of species into a genus of small cats (*Felis,* which includes the house cat), a genus of large cats (*Panthera,* which includes the lion), a genus of bobtailed cats (*Lynx*), and other genera.

Ever since Darwin, systematics has had a goal beyond simple organization: to have classification reflect the evolutionary affinities of species. The groups subordinate to groups in the taxonomic hierarchy should represent finer and finer branching of phylogenetic trees (Figure 23.4). A taxon is said to be **monophyletic** if a single ancestor gave rise to all species in that taxon and to no species placed in any other taxon. For example, the group of species in Figure 23.4 represents a monophyletic order of mammals derived from a common carnivorous ancestor. And at the level of a family,

Table 23.2 Classification of domestic cat and common buttercup		
Category	**Domestic Cat**	**Common Buttercup**
Kingdom	Animalia (animals)	Plantae (plants)
Phylum or division	Chordata (chordates)	Anthophyta (flowering plants)
Subphylum	Vertebrata (vertebrates)	—
Class	Mammalia (mammals)	Dicotyledones (dicots—plants with two seed leaves)
Order	Carnivora (carnivores)	Ranunculales
Family	Felidae (cats)	Ranunculaceae (crowfoot family)
Genus Specific name	*Felis cattus* (domestic cat)	*Ranunculus acris* (common buttercup)

Ursidae is a monophyletic group. A taxon is **polyphyletic** if its members are derived from two or more ancestral forms not common to all members. For instance, most plant taxonomists believe that the plant kingdom is a polyphyletic taxon because mosses and vascular plants probably evolved from two different algal ancestors. A **paraphyletic** taxon excludes species that share a common ancestor that gave rise to the species included in the taxon. The reptilian class of vertebrates, for example, excludes birds, which share a reptilian ancestor common to the lizards, crocodiles, snakes, and other animals that *are* classified as reptiles. Ideally, each taxon should be monophyletic, but, for reasons to be explained soon, taxonomists sometimes depart from this ideal.

Sorting Homology from Analogy

A taxonomist classifies species into higher taxa based on the extent of similarities in morphology and other characteristics. Likeness attributed to shared ancestry is called **homology.** The forelimbs of mammals are homologous; that is, the similarity in the intricate skeletal superstructure that supports the limbs has a genealogical basis (see Figure 20.11).

Figure 23.4
Relationship of classification to phylogeny. Our ability to recognize a hierarchy of taxa reflects the branching nature of evolutionary trees. This tree suggests possible genealogical affinities between the taxa subordinate to the order Carnivora, itself a branch of the class Mammalia.

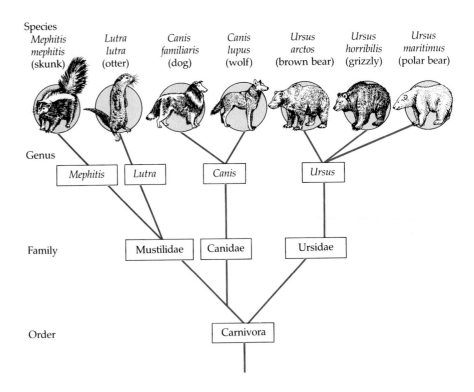

(a)

(b)

Figure 23.5
Convergent evolution and analogous structures. (a)
The marsupial kangaroo of Australia and the placental cary of Argentina are both adapted to similar ecological roles, and convergent evolution has produced analogous equipment—powerful hind legs and large ears—for life in similar environments. However, the marsupial mammals of Australia have evolved in isolation from the placental mammals of other continents for tens of millions of years. Adaptive radiation on Australia has fit marsupials to many of the ecological roles filled by placental mammals on other continents, and the convergence has produced a number of remarkable look-alikes. **(b)** The ocotillo of southwestern North America (left) looks remarkably similar to the allauidia (right) found in Madagascar. The plants are not closely related and owe their resemblance to analogous adaptations that evolved independently in response to similar environmental pressures.

There is a joker in this game of making evolutionary connections by evaluating similarity: Not all likeness is inherited from a common ancestor. Species from different evolutionary branches may come to resemble one another if they have similar ecological roles and natural selection has shaped analogous adaptations. This is called **convergent evolution,** and similarity due to convergence is termed **analogy,** not homology (Figure 23.5). The wings of insects and those of birds, for example, are analogous flight equipment that evolved independently and are built from entirely different structures.

The distinction between homology and analogy is often relative. The wings of birds and bats are modifications of the basic vertebrate forelimb, and on that level, the appendages are homologous. But as wings, they are analogous, having evolved independently from the forelimbs of different flightless ancestors.

To reconstruct evolutionary history, we must sort homology from analogy and build phylogenetic trees on the basis of homologous similarities alone. As a general rule, the greater the amount of homology between two species, the more closely they are related, and this should be reflected in their classification. This guideline is simpler in principle than it is in practice. Adaptation can obscure homologies, and convergence can create misleading analogies. As we saw in Chapter 20, comparing the embryonic development of the features in question can often expose homology that is not apparent in the mature structures.

There is another clue to identifying homology and sorting it from analogy: The more complex two similar structures are, the less likely they have evolved independently. Consider the skulls of a human and a chimpanzee, for example. The skulls are not single bones, but a fusion of many, and the chimp skull and human skull match almost perfectly, bone for bone. It is highly improbable that such complex structures matching in so many details could have separate origins. The multitude of genes required to build these skulls must have been inherited from a common ancestor.

Molecular Systematics

Comparison of information-rich macromolecules—proteins and DNA—has become a powerful taxonomic tool. Sequences of nucleotides in DNA are inherited, and they program corresponding sequences of amino acids in proteins (see Chapter 5). Molecular comparisons go right to the heart of evolutionary relationships.

Protein Comparison Because the primary structures of proteins are genetically determined (see Chapter 5), a close match in the amino acid sequences of two proteins from different species indicates that the genes for those proteins evolved from a common gene present in a shared ancestor. The degree of similarity is evidence of the extent of common genealogy. One advantage of this taxonomic tool is that it is objective and quantitative. A second advantage is that it can be used to assess relationships between groups of organisms that are so phylogenetically distant that morphological similarity is absent.

The amino acid sequence of cytochrome *c*, an ancient protein common to all aerobic organisms, has been determined for a wide variety of species ranging from bacteria to complex plants and animals. The sequences for humans and chimpanzees match perfectly for all 104 positions along the polypeptide chain, and the cytochromes of both species differ from the version found in the rhesus monkey by just one amino acid. All three species belong to the same mammalian order, Primates. Comparing these proteins to the forms found in nonprimates, we find that the differences increase as the species become more taxonomically distant. For instance, human cytochrome c differs from that of the dog by 13 amino acids, from a rattlesnake by 20 amino acids, and from a tuna by 31 amino acids. Phylogenetic trees based on cytochrome *c* are consistent with evidence from comparative anatomy and the fossil record.

DNA Comparison   Comparing the genes or genomes of two species is the most direct measure of common inheritance from shared ancestors. Comparisons can be made by three methods: DNA-DNA hybridization, restriction mapping, and DNA sequencing.

Whole genomes can be compared by **DNA-DNA hybridization,** which measures the extent of hydrogen bonding between single-stranded DNA obtained from two sources. After DNA is extracted, it is heated to separate the complementary strands. Single-stranded DNA from two species is then mixed and cooled to re-form double-stranded DNA. How tightly the DNA of one species can bind to the DNA of the other depends on the degree of similarity, as base pairing between complementary sequences holds the two strands together. The hybrid DNA is again heated to separate the paired strands. The temperature required to do this is correlated with the similarity of the DNA from the two species; the more extensive the pairing, the greater the heat energy required to pull the strands apart. Using the temperature needed to pry apart double-stranded DNA from a single species as a control for complete homology, the temperature at which hybrid DNA separates measures phylogenetic distance.

Evolutionary trees constructed by this technique generally agree with phylogeny as deduced by other methods such as comparative morphology, but DNA-DNA hybridization has the potential to settle some old taxonomic debates. For example, ornithologists (bird specialists) have long disagreed as to whether flamingos are more closely related to storks or geese. Comparison of DNA places the flamingo with storks. There has also been disagreement about whether the giant panda is a true bear or a member of the raccoon family; the evidence from DNA-DNA hybridization groups the giant panda with the bears, but places the red panda in the raccoon family (Figure 23.6).

Although DNA-DNA hybridization can estimate the overall similarity of two genomes, it does not give precise information about the match-up in specific nucleotide sequences of the DNA. An alternative approach is **restriction mapping** of DNA. This method employs the same restriction enzymes used in recombinant DNA technology (see Chapter 19). Each type of restriction enzyme recognizes a specific sequence of a few nucleotides and cleaves DNA wherever such sequences are found in the genome. The DNA fragments obtained after treatment with a restriction enzyme can be separated by electrophoresis (see Methods Box on p. 490) and compared to restriction fragments derived from the DNA of another species. Two samples of DNA with similar maps for the locations of restriction sites will produce similar collections of fragments. In contrast, two genomes that have diverged extensively since their last common ancestor will have a very different distribution of restriction sites, and the DNA will not match closely in the sizes of restriction fragments. Because so many fragments are obtained from the nuclear genome, restriction mapping is more practical for comparing smaller segments of DNA, usually a few thousand nucleotides long. Several laboratories are using restriction maps to compare mitochondrial DNA (mtDNA), which is relatively small. There is the added benefit that mtDNA changes by mutation about ten times faster than does the nuclear genome, which makes it possible to sort out phylogenetic relationships between very closely related species or even between different populations of the same species. For example, a recent comparison of mtDNA from people of several different ethnicities has corroborated the fossil evidence that our species originated in Africa (see Chapter 30).

The most precise method for comparing DNA from two species, and the most tedious, is to actually determine the nucleotide sequences of entire DNA segments that have been cloned by recombinant DNA techniques (see Chapter 19). Such a comparison tells us exactly how much divergence there has been in the evolution of two genes derived from the same ancestral gene. This will become an increasingly powerful taxonomic approach as technology for rapid sequencing of DNA continues to improve. A related approach is the sequencing of ribosomal RNA (rRNA), gene products found in all organisms. Because genes for rRNA change slowly relative to most other DNA, differences in rRNA sequences can be used to trace some of the earliest branching in the tree of life. Comparison

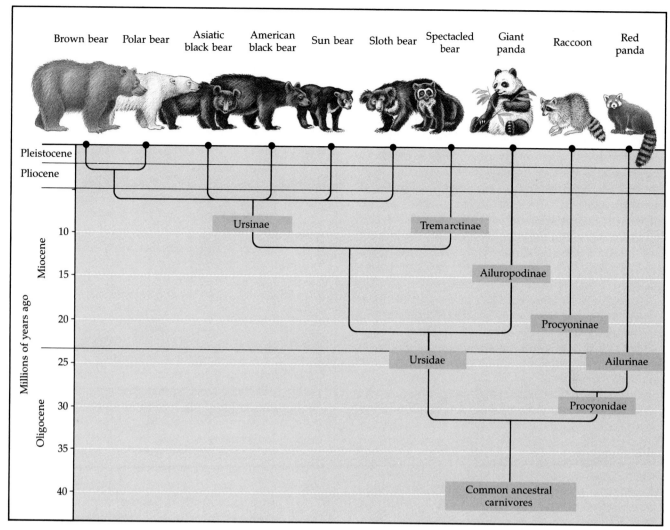

Figure 23.6
A phylogenetic tree based on molecular systematics. Two methods, comparison of blood proteins and comparison of genomes by DNA-DNA hybridization, were used as taxonomic criteria to construct this tree of bears. (The "idae" ending denotes families, while the "inae" ending signifies subfamilies.)

of rRNA sequences has been especially useful in sorting out the phylogenetic relationships among the bacteria (see Chapter 25).

Molecular Clocks Proteins evolve at different rates, but for a given type of protein—cytochrome *c*, for instance—the rate of evolution seems to be quite constant with time. If homologous proteins are compared for taxa that are known from the fossil record to have diverged from common ancestors during certain periods in the past, the number of amino acid substitutions is proportional to the time that has elapsed since the lineages branched apart. The homologous proteins of bats and dolphins are much more alike than those of sharks and tuna, which is consistent with the fossil evidence that sharks and tuna have

been on separate evolutionary lines much longer than bats and dolphins. In this case, molecular divergence has kept better track of the time than superficial changes in body form.

As a clock to date branch points in phylogenetic trees, DNA comparisons are even more promising than protein comparisons. As with protein clocks, the DNA clock may have a reliable beat; dating of phylogenetic branchings based on nucleotide substitutions in DNA generally approximate the dates determined from the fossil record. In many cases, the difference in DNA between two taxa is more closely correlated with how long they have been on separate evolutionary branches than is the degree of morphological difference between the taxa.

Molecular clocks are calibrated by graphing the

METHODS: PRODUCING A PHENOGRAM

A computer has compared five species (A–E) for a large number of characters and has consolidated the data into single numbers that represent overall similarity between pairs of species, with a score of 1.0 signifying a perfect match for all characters measured. For example, species "A" and "B" show extensive correlation (0.9) for the set of characters used to construct this diagram. In step 1, each species is paired with the species it matches most closely, the location of the connecting line on the verticle scale indicating the degree of correlation between the two species. In step 2, each species pair is considered as a unit to be compared to other matched pairs. This forms an even larger unit that can be connected at the appropriate point to less similar species or groups of species (step 3). The result is a phenogram, a tree based on phenotypic similarities and differences.

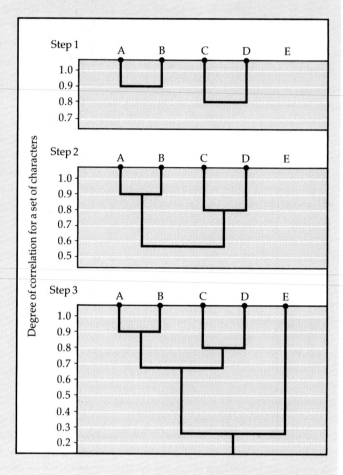

number of amino acid or nucleotide differences against the times for a series of evolutionary branch points known from the fossil record. The graph can then be used to estimate the time of divergence for species when there is no clear fossil evidence for their time of origin from other forms. For example, this was the method used to determine the antiquity of branch points in the phylogenetic tree of bears illustrated in Figure 23.6.

Both the consistent rate of protein change and the rate of DNA divergence imply that there is a significant background of neutral mutations that gradually changes the genome as a whole more than the specific genetic changes associated with adaptation. Evolutionary biologists disagree about the extent of neutral mutations (see Chapter 21). Many evolutionists doubt that neutral evolution is so prevalent, and so they also question the credibility of molecular clocks as tools for

absolute dating of the origin of taxa. There is less skepticism about the value of molecular systematics for determining the relative sequence of branch points in phylogeny. The modern systematist evaluates any available molecular data along with all other taxonomic evidence in order to reconstruct phylogeny.

Schools of Taxonomy

Phylogenetic trees have two significant aspects: the location of branch points along the tree, symbolizing the relative time of origin of different taxa, and the degree of divergence between branches, representing how different two taxa have become since branching from a common ancestor. If taxonomy is to be based on evolutionary history, which property of phylogenetic trees should be given greatest weight in group-

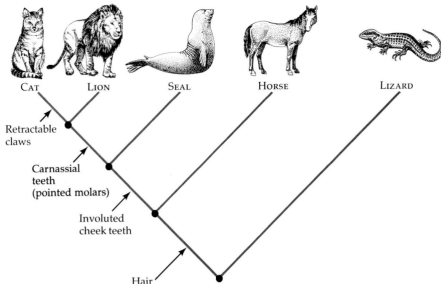

Figure 23.7
A cladogram. Each branch point is defined by apomorphic characters, derived homologies unique to the lineage that arises at that point. The defining characteristics noted here are not the only ones at each branch. Only the sequence of branching is represented in a cladogram, with no consideration of the degree of divergence between branches. The branch points (dots) represent the most recent ancestor common to all species beyond that point. For example, the lion and cat share a common ancestor that lived more recently than did the ancestor that also gave rise to the evolutionary lineage leading to the seal. However, this does *not* mean that the seal itself evolved before lions.

Labels on cladogram: Retractable claws; Carnassial teeth (pointed molars); Involuted cheek teeth; Hair. Species: CAT, LION, SEAL, HORSE, LIZARD.

ing species into taxa? This question has divided taxonomy into three schools of thought: phenetics, cladistics, and classical evolutionary taxonomy.

Phenetics   Endeavoring to make classification less subjective, **phenetics** (from the Greek *phainein*, "to appear"; the term *phenotype* is derived from this same root) makes no phylogenetic assumptions and decides taxonomic affinities entirely on the basis of measurable similarities and differences. As many anatomical characteristics (known as characters) as possible are compared, with no attempt to sort homology from analogy. The data from such comparisons are used to arrange species into dichotomously branched trees called phenograms (see Methods Box). Pheneticists contend that if enough phenotypic characters are examined, then the contribution of analogy to overall similarity will be swamped by the degree of homology. Critics of phenetics argue that overall phenotypic similarity is not a reliable index of phylogenetic proximity. Although a strictly phenetic approach has few proponents today, the methods used in phenetics, especially the emphasis on multiple quantitative comparisons with the help of computers, has had an important impact on taxonomy.

Cladistics   Clades (from the Greek *clados*, "branch") are evolutionary branches. **Cladistics** classifies organisms according to the order in time that branches arise along a phylogenetic tree, excluding the degree of divergence from consideration. The tree takes the form of a cladogram, a series of dichotomous forks. Each branch point is defined by novel homologies unique to the various species on that branch. Let us apply the strategy to five vertebrates: a lizard, a horse, a seal, a lion, and a cat (Figure 23.7). Each species has a mixture of primitive characters that already existed in the common ancestor along with characters that have evolved more recently. The sharing of primitive characters tells us nothing about the pattern of evolutionary branching from a common ancestor. For example, we cannot use the presence of five separate toes to divide these vertebrate species among evolutionary branches. According to fossil evidence, the remote ancestor common to *all* species in our list was five-toed, and hence this homology is termed a primitive character, or **plesiomorphic character.** Seals and horses apparently lost the trait independently, whereas the other species retained the trait. We must seek derived characters, or **apomorphic characters,** which are homologies that evolved after a branch diverged from the phylogenetic tree. Hair and mammary glands are two of the apomorphic characters that define a branching point with the lion, cat, seal, and horse on one limb and the lizard on the other. Now we must determine the branching sequence along the mammalian limb. The lion, cat, and seal share many skeletal modifications not present in the horse, and these are among the apomorphic characters that define the next branch point in our cladogram. The lion and cat branch from the lineage leading to seals at a later point defined by a number of modifications of the skull and teeth. A major difficulty in cladistics is to find characters that are appropriate for each branch point.

The cladistic approach produces some taxonomic surprises. For example, the branch point between birds and crocodiles is more recent than the branch point between crocodiles and the other reptiles. That is, birds and crocodiles share apomorphic characters not present in snakes and lizards, and indeed the fossil record supports the conclusion that birds are closer relatives of crocodiles than are lizards and snakes. In the strictly

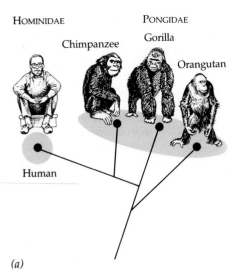

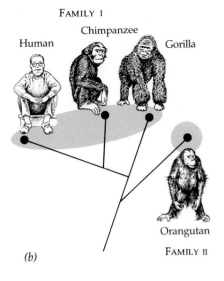

(a)

(b)

Figure 23.8

How should humans and their closest relatives be classified? (a) The classical taxonomic view. On the basis of overall similarities resulting from plesiomorphic (ancestral) characters, apes are grouped into a family separate from humans. The family Pongidae (great apes) is not monophyletic, however, since it stems from an ancestor that also gave rise to a species in another family (humans). (b) The cladistic view. Based on the shared apomorphic (derived) characters that define branching order, chimpanzees and gorillas are more closely related to humans than to orangutans. To a strict cladist, there is no such thing as a phylogenetically pure ape family. Classical evolutionary taxonomists retort that classification of the primates should recognize the extent to which human morphology has diverged from the apes.

cladistic view, there are no such taxa as the class Aves (birds) and the class Reptilia, as we conventionally know them, because birds intrude in the cladogram of the collective of animals we call reptiles. (Some cladists propose retaining the class Reptilia and including birds as a subclass or order.) Birds seem so superficially different because of the extensive morphological remodeling associated with flight that has occurred since birds branched from their reptilian ancestors. Cladistics ignores the extent of morphological divergence between evolutionary branches, information that critics of a strictly cladistic approach argue should be included in a classification scheme.

Classical Evolutionary Taxonomy The classification used in this book and most others is based on **classical evolutionary taxonomy,** an approach that predates phenetics and cladistics and that now attempts to balance the criteria of phenetics and cladistics by considering overall homology along with branching sequence. In cases where this leads to a taxonomic conflict, a subjective judgment is made about which type of information should be given higher priority. For example, classical taxonomists acknowledge that crocodiles have a greater genealogical relationship with birds than with lizards, but opt to combine lizards and crocodiles in a taxon that excludes birds because the ability to fly was an evolutionary breakthrough that placed birds in a major new "adaptive zone." This resulted in adaptive divergence of birds so extensive that classical taxonomists assign birds to their own

class (Aves). A case closer to our taxonomic home is the relationship between humans and their primate cousins, chimpanzees, gorillas, and orangutans (Figure 23.8).

If taxa that split at each branch point in phylogeny diverged in morphology at the same rate, then phenetics and cladistics would give identical classifications; the longer two taxa have evolved separately, the more different they would have become. Life, however, is not so simple. Birds and crocodiles do have a common ancestor, but ever since these two kinds of animals have been on separate evolutionary lines, birds have changed much more than their reptilian relatives. Comparison of proteins and DNA offers the best hope for taxonomic characters that diverge in time from an evolutionary branch point at a reasonably constant rate. Nevertheless, popular classifications will probably continue for many years to reflect morphological similarities and differences. There may be one true phylogeny, but so long as classification compromises between genealogy and the need for a convenient filing system for the diversity of species, there can be no one taxonomy.

MECHANISMS OF MACROEVOLUTION

What processes actually cause the large-scale evolutionary changes that we can trace through paleobiology and taxonomy? How, for instance, do the novel features that define taxa above the species level, such as the flight adaptations of birds, arise? What accounts for evolutionary trends that appear from the fossil record to be progressive, such as the general increase in the size of certain families of reptiles during the age of dinosaurs, or the increase in brain size during human evolution? How have global geological changes affected macroevolution? And how can we explain the major fluctuations in biological diversity that are evident in the fossil record, such as a proliferation of animal diversity at the beginning of the Paleozoic era, or the mass extinctions that have occurred in the past? These are the kinds of questions asked by evolutionary biologists interested in macroevolution.

Origins of Evolutionary Novelties

It is enough to see that a modern animal has feathers to know it is a bird. Feathers are so restricted taxonomically that they can be used as a diagnostic character to distinguish one class of vertebrates from all others (although birds have many other unique adaptations). But for all their structural and functional distinctiveness, the feathers of birds are homologous to the scales of reptiles, and wings are homologous to forelegs. Humans and chimpanzees are very close relatives, and yet their dissimilarities—the differences in posture and brain size, for instance—place the two animals in entirely different ecological roles. What creates the evolutionary novelties that define higher taxa such as families and classes? Put another way, how do new designs for living evolve? One mechanism may be the gradual refinement of existing structures for new functions.

Preaptation   From a retrospective vantage, evolutionists use the term **preaptation** for a structure that evolved in one context and became co-opted for another function. This word does not imply that a structure somehow evolves in anticipation of future use. (For this reason, the term *preaptation*, which only suggests that structures have an evolutionary plasticity that makes alternative functions possible, is gradually replacing an older term, *preadaptation*). Natural selection cannot predict the future and can only improve a structure in the context of its current utility. The light, honeycombed bones of birds (see Figure 1.5 on p. 7) could not have evolved in earthbound reptilian ancestors as an adaptation for upcoming flights. If these honeycombed bones predated flight, as clearly indicated by the fossil record, then they must have had some function on the ground. Birds probably evolved from agile, bipedal dinosaurs that also would have benefited from a light frame. It is possible that wing-like forelimbs, as well as feathers, which increased the surface area of these forelimbs, were also co-opted for flight after functioning in some other capacity, such as

"netting" insects and other small prey chased by small, fleet-footed dinosaurs. The first flight may have been only a glide down from a tree or an extended hop in pursuit of prey or escape from predator. Once flight itself became an advantage, natural selection would have remodeled feathers and wings to better fit their additional function.

We cannot prove that this scenario for the evolution of birds from reptiles is correct, since the known fossil record provides so little history of this transition. However, preaptation offers one explanation for how novel designs can arise gradually through a series of intermediate stages, each of which has some function in the organism's current context. This concept is in the Darwinian tradition of large changes being an accumulation of many small changes crafted by natural selection.

Regulatory Genes and Macroevolution   The evolution of complex structures such as wings and feathers from their antecedents requires so much remodeling that changes are probably involved at many gene loci. In other cases, relatively few changes in the genome can apparently cause major modifications of morphology, such as some of the differences between humans and chimpanzees. How can slight genetic divergence become magnified into major differences between organisms?

The development of an animal depends not only on the structural genes that program the production of proteins, but also on a system of regulatory genes that coordinate the activities of the structural genes, guiding the rate and pattern of development. The slightest alteration of development becomes compounded in its effects on the adult. Let us apply this principle to the body proportions of a plant or animal. Differences in the relative rates of growth of various parts of the body, which is called **allometric growth,** help to shape the organism. Change these relative rates of growth even slightly, and you change the adult form substantially (Figure 23.9). Altering the parameters of allometric growth is one way that relatively small genetic differences can have major morphological impact.

Genetic changes that alter the timing of development can also produce novel organisms. A subtle change in timing that retards the development of some organs compared to others produces a different kind of animal (Figure 23.10). The retention in an adult organism of juvenile features of its evolutionary ancestors is called **paedomorphosis** (from the Greek *paid*, "child," and *morphos*, "form"). Paedomorphosis may have contributed to our own evolution. Humans and chimpanzees are much more alike as fetuses than they are as adults. The fetal skulls, for instance, are the same shape (see Figure 23.8b). Because of different allometric growth patterns, however, juvenile features of the human skull persist into adulthood. In addition, the human brain continues to grow for sev-

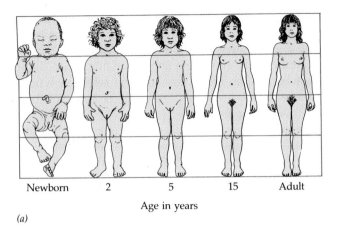

Newborn   2   5   15   Adult

Age in years

(a)

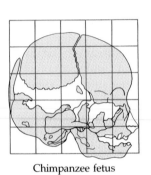
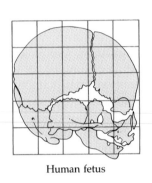

Chimpanzee fetus   Human fetus

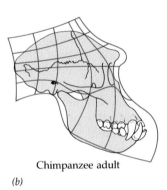
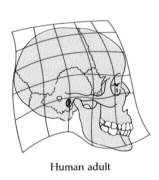

Chimpanzee adult   Human adult

(b)

Figure 23.9
Allometric growth. Differences in the growth rates for some parts of the body compared to others determines body proportions. (**a**) During growth of a human, the arms and legs grow faster than the head and trunk, as can be seen in this conceptualization of different-aged individuals all drawn at the same height. (**b**) The fetal skulls of humans and chimpanzees are the same shape. Allometric growth of the bones transforms the rounded skull of a newborn chimpanzee to the sloping skull characteristic of adult apes. The same allometric pattern occurs in humans, but it is attenuated, and the adult human skull has departed less than the chimpanzee skull from the fetal shape common to primates.

eral years longer than the chimpanzee brain, which can also be interpreted as the prolonging of a juvenile process. The genetic changes responsible for humanness need not be great, but their effects are profound.

Figure 23.10
Paedomorphosis. Some species retain features as adults that were juvenile in ancestors. The salamander in the photo is an axolotl, which grows to full size, becomes sexually mature and reproduces while retaining many larval (tadpole) characteristics.

Because each regulatory gene may influence hundreds of structural genes, there is a potential for some of the evolutionary novelties that define higher taxa to arise much faster than they could by the accumulation of changes in the structural genes themselves. Unfortunately, we still know too little about regulatory genes and the control of development to understand their connection to phylogeny.

What Produces Evolutionary Trends?

Extracting a single evolutionary progression from a fossil record that is likely to be incomplete is as misleading as describing a bush as growing toward a single point by tracing the system of branches that leads from the base of the bush to one particular twig. A case in point is the evolution of the modern horse, which is believed to be a descendant of a much smaller ancestor named *Hyracotherium,* which browsed in the woods of the Eocene epoch about 40 million years ago. In comparison to its ancestor, not only is the modern horse (genus *Equus*) larger, but the number of toes has been reduced from four on each foot to one, and the teeth have become modified for grazing rather than browsing. By selecting certain species from the available fossils, it is possible to arrange a succession of animals intermediate between *Hyracotherium* and modern horses that shows trends toward increased size, reduced number of toes, and grazing dentition (Figure 23.11a). We might interpret this series of fos-

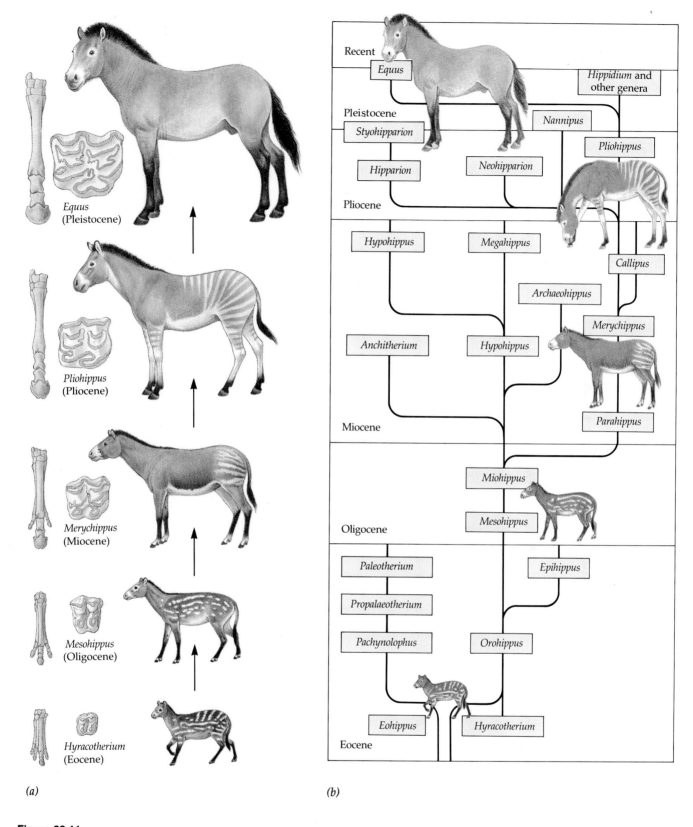

(a)

(b)

Figure 23.11

The branched phylogeny of horses. (a)
One sequence of fossil horses that are
intermediate in form between the modern
horse and its Eocene ancestor, *Hyracoth-* *erium,* suggests phyletic progression with
trends toward larger size, reduced number
of toes, and teeth modified for grazing. **(b)**
A more complete phylogeny reveals that the modern horse is the only surviving
twig of an evolutionary "bush" with many
divergent trends.

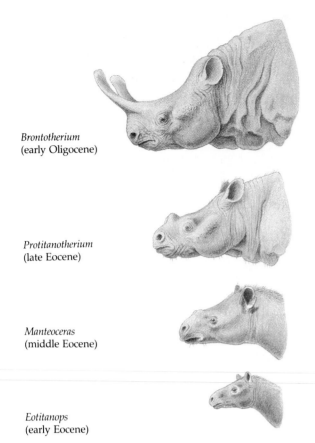

Brontotherium
(early Oligocene)

Protitanotherium
(late Eocene)

Manteoceras
(middle Eocene)

Eotitanops
(early Eocene)

Figure 23.12
Trends in the evolution of titanotheres. In each of the genera of this extinct family of mammals, species became larger and their horns bigger over a period of 45 million years. But no individual species changes significantly in size for its duration in the fossil record. Some paleontologists interpret this to mean that the evolutionary trends were not produced by gradual phyletic evolution of individual populations, but by a series of speciation episodes, with the average size of the species increasing with time.

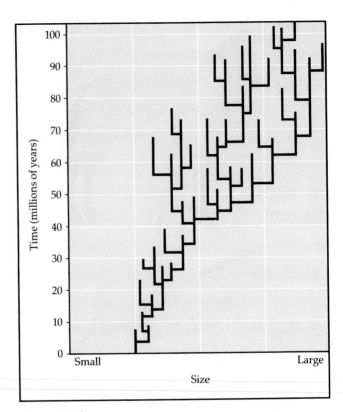

Figure 23.13
The species selection hypothesis of macroevolution.
As an alternative to phyletic evolution, this model explains evolutionary trends as the products of the differential longevity and budding of daughter species. This example shows a trend toward species of larger body size. Species smaller than their parent bud from the evolutionary bush as frequently as species that are larger than their parent. Nevertheless, there is an overall trend because large species live longer and leave the most descendant species, either because of a faster speciation rate, a slower extinction rate, or both.

sils as phyletic evolution, with an unbranched lineage leading directly from *Hyracotherium* to modern horses through a continuum of intermediate stages. If we include all fossil horses known today, however, the illusion of coherent, progressive evolution leading directly to modern horses vanishes. The transition occurs in steps rather than in a smooth gradation of forms; each species appears and disappears in the fossil record without changing noticeably in the interim. *Equus* just happens to be the only surviving twig of a phylogenetic tree that is so branched it is more like a bush (Figure 23.11b). *Hyracotherium* did not become the modern horse by changing gradually, any more than your great grandparents became you. *Equus* descended through a series of speciation episodes that included several adaptive radiations, not all of which led to large, one-toed, grazing horses. Had *Equus* become extinct and some other horse persisted, we might perceive different evolutionary trends.

Evolution *has* produced many trends that seem from the fossil record to be genuine. For example, a family of now-extinct giant mammals called titanotheres, as large as elephants, evolved from a mouse-sized ancestor that lived during the early Cenozoic era (Figure 23.12). There is no evidence of phyletic evolution in the various lineages of titanotheres; there is a succession of progressively larger species, but each species remains the same size for its duration in the fossil record. Punctuated equilibrium seems to be the mode of evolution in this case (see Chapter 22). According to this interpretation of the fossil record, evolutionary trends in most groups of organisms are produced not by a phyletic slurring of forms, but by a staccato of changes occurring in increments during the time that new species branch from ancestral ones.

Branching evolution can produce a trend even if some new species counter the trend. During the Mesozoic era, there was an overall trend in reptilian

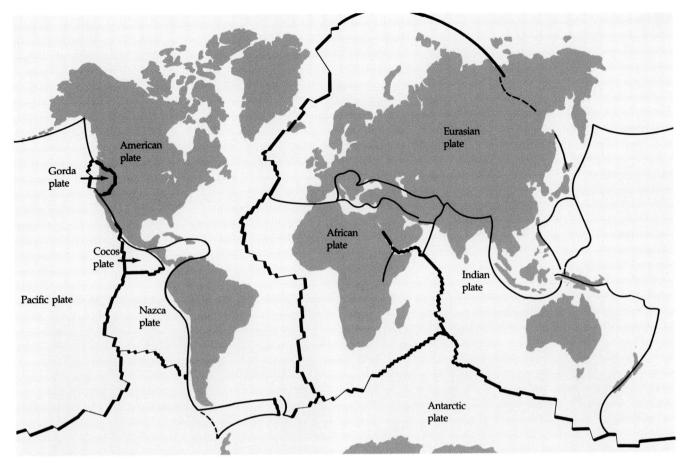

Figure 23.14
Earth's crustal plates. The modern continents are passengers on crustal plates that are swept across the surface of the earth by convection currents of the molten mantle below. This map identifies only the major plates.

evolution toward largeness, a progression that produced the dinosaurs, which became the dominant animals of that era. The trend was sustained even though some new species were smaller than their parent species. In fact, even if the descendant species were smaller as often as they were larger than the species from which they budded, an overall trend would still develop if larger reptiles speciated more often than smaller ones or lasted longer than smaller species before becoming extinct (Figure 23.13). In this view of macroevolution, enunciated by Steven Stanley of Johns Hopkins University, species are analogous to individuals. Speciation is their birth and extinction is their death. New species are their offspring. An evolutionary trend, according to the Stanley model, is produced by **species selection,** which is analogous to the production of a trend within a population by natural selection. The species that live the longest and generate the greatest number of species determine the direction of major evolutionary trends. Differential speciation may play a role in macroevolution similar to the role of differential reproduction in microevolution.

The appearance of an evolutionary trend in some taxa does not imply that there is some intrinsic drive toward a preordained state of being. Evolution is a response to interactions between organisms and their current environments. If conditions change, an evolutionary trend may cease or even reverse itself. The Mesozoic world favored giant reptiles, but by the end of that era, smaller species prevailed.

Continental Drift and Macroevolution

Macroevolution has dimension in space as well as in time. Indeed it was biogeography even more than fossils that first nudged Darwin and Wallace toward an evolutionary view of life. The history of the Earth helps to explain the current geographical distribution of species. For example, the emergence of volcanic islands such as the Galapagos opens new environments for founders that reach the outposts, and adaptive radiation fills many of the available niches with new species. On a global scale, the drifting of continents is the major geographical factor correlated with the spatial distribution of life (Figure 23.14).

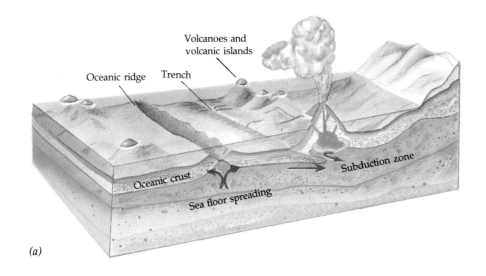

(a)

(b)

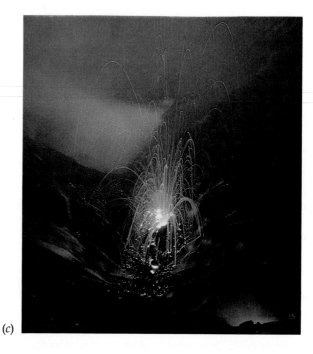

(c)

Figure 23.15
Plate tectonics. (a) At some plate boundaries, such as oceanic ridges, the plates separate, and molten rock wells up in the gap. The rock solidifies and adds crust symmetrically to both plates, a phenomenon called sea floor spreading. There are also areas known as subduction zones, where plates move toward one another, with the denser plate diving below the less dense one and creating a trench. The Marianas Trench of the South Pacific is a subduction zone over 11,000 m deep. The abrasion at subduction zones causes earthquakes and volcanic eruptions. When continents riding on different plates collide, they pile up and build mountains. **(b)** The San Andreas Fault, seen north of Los Angeles in this photo, is the seismically active boundary between the Pacific and American plates. **(c)** A volcano in the New Hebrides erupted in 1984.

The continents are not fixed, but drift about the surface of the Earth as passengers on great plates of crust and upper mantle that float on the molten mantle. Unless two land masses are embedded in the same plate, their positions relative to each other change. For example, North America and Europe are presently drifting apart at a rate of about 2 cm per year. Many important geological phenomena, including mountain building, volcanism, and earthquakes, happen at plate boundaries (Figure 23.15). California's infamous San Andreas fault is part of a border where two plates slide past each other. Mt. St. Helens, the volcano in Washington state that has been so active in recent years, is near another plate boundary.

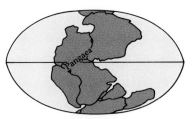

200–250 million years ago

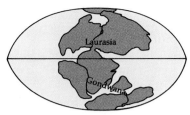

135 million years ago

65 million years ago

Present

Figure 23.16
Continental drift. About 200 to 250 million years ago, all of the Earth's land masses were locked together in a supercontinent named Pangaea. About 180 million years ago, Pangaea began to split into northern and southern land masses, which later separated into the modern continents. The continents continue to drift. India collided with Eurasia just 10 million years ago, forming the Himalayas, the tallest and youngest of Earth's mountain ranges.

Plate movements rearrange geography incessantly, but two chapters in the continuing saga of continental drift must have been especially significant in their influence on life. About 250 million years ago, near the end of the Paleozoic era, plate movements brought all the land masses together into a supercontinent that has been named **Pangaea**, meaning "all land" (Figure 23.16). Imagine some of the possible effects on life. Species that had been evolving in isolation came together and competed. When the land masses coalesced, the total amount of shoreline was reduced, and there is evidence that the ocean basins increased in depth, which drained much of the shallow coastal seas that remained. Then, as now, most marine species inhabited shallow waters, and the formation of Pangaea destroyed a considerable amount of that habitat. It was probably a long traumatic period for terrestrial life as well. The continental interior, which has a harsher climate than coastal regions, increased in area substantially when the land came together. Changing ocean currents also would have affected land life as well as sea life. The formation of Pangaea surely had a tremendous environmental impact that reshaped biological diversity by causing extinctions and providing new opportunities for taxa that survived the crisis.

Another dramatic chapter in the history of continental drift was written about 180 million years ago, during the early Mesozoic era. Pangaea began to break up, and this caused geographical isolation of colossal proportions. As the continents drifted apart, each became a separate evolutionary arena, and the fauna and flora of the different biogeographical realms diverged (see Chapter 48). The pattern of continental separation is the solution to many puzzling cases of geographical distribution, such as the separation of placental and marsupial mammals mentioned earlier.

Punctuations in the History of Biological Diversity

The evolutionary byways from ancient to modern life have not been smooth. Biological diversity has an episodic history, with long, relatively quiescent periods punctuated by briefer intervals when the turnover in species composition was much more extensive. The episodes include explosive adaptive radiations of major taxa as well as mass extinctions.

Major Adaptive Radiations Many taxa have diversified prolifically early in their history after evolving some novel characteristic that opened the window to a new **adaptive zone,** which is a term used to describe a new way of life presenting many opportunities previously unexploited. For example, the development of wings enabled insects to enter an adaptive zone with many new food sources, and adaptive radiation produced hundreds of thousands of variations on the basic insect body plan.

The boundary between the Precambrian era and the Paleozoic era is marked by a large increase in the diversity of sea animals. The oldest animals are creatures found in late Precambrian rocks about 700 million years old (see Table 23.1). These animals, known from fossilized imprints that have been discovered at several sites around the world, were shell-less invertebrates with body plans quite different from those of their Paleozoic successors. Within the first 10 to 20 million years of the Cambrian, the first period of the Paleozoic era, nearly all the animal phyla that exist today evolved, along with many phyla now extinct. One key evolutionary novelty behind this remarkable diversification may have been the origin of shells and skeletons in a few key taxa, an innovation that opened a new adaptive zone by making many new complex designs possible and by rewriting the rules for predator-prey relations.

There are probably empty adaptive zones even today. Flying insects existed at least 100 million years before the flying reptiles and birds that ate the insects evolved. An empty adaptive zone can be exploited only if the appropriate evolutionary novelties arise. Conversely, an evolutionary novelty cannot enable organisms to take advantage of adaptive zones that do not exist or are already occupied. Mammals, with the many unique features characteristic of their class, existed at least 75 million years before their first major adaptive radiation. The rise in mammalian diversity during the early Cenozoic era may have been associated with the ecological void left by the extinction of the dinosaurs. New adaptive radiations have often followed mass extinctions that swept away old tenants of adaptive zones.

Mass Extinctions   A species may become extinct because its habitat has been destroyed or because the environment has changed in a direction unfavorable to the species. If ocean temperatures fall by a few degrees, many species that are otherwise beautifully adapted will perish. Even if physical factors in the environment are stable, the biological factors may change; the environment in which a species lives includes the other organisms that live there, and evolutionary change in one species is likely to have some impact on other species in the community. For example, the evolution of shells by some Cambrian animals may have contributed to the extinction of some shell-less forms.

Thus, extinction is inevitable in a changing world. The average rate of extinction has been between 2.0 and 4.6 families per million years (each family may include many species). However, there have been crises in the history of life when environmental changes on a global scale have been so rapid and disruptive that a majority of species were swept away, perhaps rather indiscriminately. During periods of mass extinctions, the rate of destruction escalates to as high as 19.3 families per million years.

Of the dozen or so mass extinctions chronicled in the fossil record, two have been subjects of the most intense study by paleobiologists. They are known primarily from the decimation of hard-bodied animals of shallow seas, the organisms for which the fossil record is most complete. The Permian extinctions, which define the boundary between the Paleozoic and Mesozoic eras, claimed over 90% of the species of marine animals about 250 million years ago and probably took a tremendous toll on terrestrial life as well. That was about the time the continents merged to form Pangaea, which probably disturbed many habitats and

Figure 23.17
The iridium layer. Located between the Mesozoic and Cenozoic sediments at widely scattered sites on Earth is a layer of clay enriched with the element iridium (the dark band in the photograph; the coin indicates scale). Because iridium is rare on Earth but common in asteroids, Luis Alvarez and his son Walter have concluded that the iridium layer probably resulted from an ancient collision between a large asteroid and Earth, and they have suggested that this collision was a major cause of mass extinctions.

altered climate. Keep in mind, however, that a temporal correlation of two events does not demonstrate a cause-and-effect relationship.

Another emphatic punctuation, the Cretaceous extinction of 65 million years ago, delineates the boundary between the Mesozoic and Cenozoic eras. That debacle doomed more than half of the marine species and exterminated many families of terrestrial plants and animals, including the dinosaurs. The climate was cooling at that time, and it was a period when shallow seas receded from continental lowlands. It is possible that increased volcanic activity contributed to the cooling by releasing materials into the atmosphere that blocked sunlight. There is also evidence that an asteroid or comet struck the Earth while the Cretaceous extinctions were in progress. Separating Mesozoic from Cenozoic sediments is a thin layer of clay enriched in iridium, an element very rare on earth, but common in meteorites and other extraterrestrial debris that occasionally falls to earth (Figure

23.17). Walter and Luis Alvarez and their colleagues at the University of California (Berkeley) studied the anomalous clay and proposed that it is fallout from a huge cloud of dust that billowed into the atmosphere when an asteroid hit Earth. The great cloud would have blocked light and disturbed climate severely for several months. The Alvarez scenario is similar to recent speculations of what a nuclear winter would be like. The evidence for an asteroid collision at the end of the Cretaceous period is credible, but even if it occurred, the coincidence of the impact within a period of mass extinctions does not link the two events as cause and effect. Critics of the asteroid hypothesis point out that the extinctions, while rapid on the scale of geological time, were not *that* abrupt. Many paleontologists and geologists believe that changes in climate due to continental drift and other processes here on Earth are sufficient to account for mass extinctions without looking heavenward for extraterrestrial causes.

Whatever the causes, mass extinctions affect biological diversity profoundly. But there is a creative side to the destruction. The taxa that manage to survive these crises, by their adaptive qualities or by sheer luck, become the stock for new radiations that fill many of the adaptive zones vacated by the extinctions. The world might be a very different place today if a few families of dinosaurs had escaped the Cretaceous extinctions or if *Purgatorius*, the one known primate that lived in the Cretaceous period, had not survived.

IS A NEW SYNTHESIS NECESSARY?

Evolutionary biology has not had a quiet moment since Darwin published *The Origin of Species* a century and a quarter ago. The closest thing to a consensus has been the modern synthesis, the paradigm that has dominated evolutionary theory for the past 50 years (see Chapter 20). Called a synthesis because the ideas were fashioned from several disciplines, including paleontology, biogeography, systematics, and population genetics, it has continued to absorb the discoveries of new fields such as molecular biology. The modern synthesis reaffirmed the Darwinian view of life and updated it by applying principles of genetics. The paradigm is distinctly uniformitarian in its view that large-scale evolutionary changes are the gradual accumulations of many minute changes occurring over vast spans of time. Microevolution, changes in gene frequencies in populations, is extrapolated to explain most macroevolution.

In the view of the modern synthesis, natural selection is the major cause of evolution at all levels. Populations adapt by natural selection, new species arise when isolated populations diverge as different adaptations evolve, and continued divergence due to natural selection differentiates the higher taxa. The modern synthesis recognizes, and in fact first described, how other mechanisms such as genetic drift and chromosomal mutations can cause rapid, nonadaptive evolution. But the major emphasis of the synthesis is on gradualism and natural selection.

A number of evolutionists dissent from the view that the evolution recorded in the fossil record can be explained by extrapolating the processes of microevolution. The debate is partly about the pace of evolution. Many transitions in the fossil record are punctuational, not gradual. Gradualists argue that apparent abruptness in part derives from the imperfection of the fossil record and in part is a semantic issue clouded by the vastness of geological time—Do we call an evolutionary episode that required 10,000 years "sudden" or "gradual"? Punctuationalists counter that the imperfection of the fossil record is not enough to account for the rarity of transitional forms if speciation and the origin of higher taxa were primarily gradual extensions of microevolution. The debate is not just about the tempo of evolution, but also about the degree to which microevolution compounded over time is sufficient to explain macroevolution.

Some evolutionists favor a hierarchical theory that affords mechanisms other than microevolution by natural selection a larger role in macroevolution. In the hierarchical view, most new species begin as small populations isolated from their parent populations by either geographical barriers or genetic accidents such as chromosomal mutations. The small, isolated population can evolve relatively rapidly, its divergence from the parent population due at least as much to genetic drift as to selection. Chance may cause the onset of speciation even before selection has fashioned new adaptations. And some major new adaptations may evolve with minimal genetic change if regulatory genes are involved. Chance also figures prominently in macroevolution. Continental drift and mass extinctions have probably had at least as much effect on the history of biological diversity as gradual adaptation caused by selection operating on gene pools at the population level. Finally, in a hierarchical theory of evolution, most evolutionary trends progress not by phyletic transition due to an accumulation of microevolutionary changes, but by species selection—the differential survival and branching of separate species that change little after they come into existence.

The importance of natural selection is not under fire; the various factions agree that natural selection is the mechanism of adaptation and should therefore be the centerpiece of evolutionary biology. Selection fine-tunes a population to its environment with generation-to-generation changes in the gene pool that are adaptive. And when a new species or higher taxon comes into being, it is natural selection that refines unique adaptations. Although the events that lead to speciation and episodes of macroevolution may have little to do with adaptation, new species only persist long enough to be entered into the fossil record if they have adapted to their environment through natural selection.

Perhaps those who have challenged the orthodoxy of the modern synthesis have erected a straw man. The synthesis has never claimed that evolution is always smooth and gradual or that processes other than changes in gene pools due to selection are unimpor-tant. The questions are not so much about the nature of evolutionary mechanisms as about their relative importance. The modern synthesis may not need rad-ical surgery, but only a simple face-lifting.

Vigorous debate about how life evolves is a healthy sign that evolutionary biology is a robust science that refuses to stagnate in complacency and wallow in dogma. The debates will continue so long as we are curious about our origins and our relationship to the rest of the living world.

STUDY OUTLINE

1. Macroevolution is the origin and history of life above the species level. It involves the origin of evolutionary nov-elties, the study of evolutionary trends, and global epi-sodes of major adaptive radiations and mass extinctions.
2. Although macroevolution takes place on a much larger scale than microevolution, it, too, is a consequence of the interactions of organisms with their environments.

The Record of the Rocks (pp. 479–484)

1. The fossil record provides the historical archives paleo-biologists use to study macroevolution. Sedimentary strata reveal the relative ages of fossils by vertically chronicling specimens from successive geological periods.
2. The absolute ages of fossils in years can be determined by radioactive isotopes or the ratio of amino acid isomers. Absolute dating of sedimentary strata has defined the ages for the different geological periods, each of which corresponds to a major transition in the composition of fossil species.
3. The fossil record is incomplete, owing to improbable chance factors in the formation, endurance, exposure, and discovery of fossil species.

Systematics: Tracing Phylogeny (pp. 484–492)

1. Systematics deals with the diversity of life by tracing the phylogeny, or evolutionary history, of taxa, hierarchical levels of classification ranging from the species level to the kingdom level.
2. Taxonomic affinity and phylogenetic relationships are decided on the basis of homology, structural similarity due to common ancestry. Sometimes, however, two unrelated species possess similar, or analogous, struc-tures as a result of convergent evolution. Similar struc-tures that are complex and share the same embryological origin are usually considered homologous.
3. Molecular systematics is the ultimate level for determin-ing homology. Evolutionary relationships can be revealed by comparing amino acid sequences of proteins and nucleotide sequences of DNA. Molecular evolution may occur at a rate consistent enough to function as a crude clock for determining the relative sequence of branch points in phylogeny.
4. Phylogenetic trees depict both the relative time of origin for different taxa and their degree of divergence from a common ancestor at each branch point. These two aspects of phylogeny have generated three schools of taxonomic thought for ordering species into higher taxa.

5. Phenetics ignores the timing of branch points and clas-sifies solely on the basis of similar anatomical characteristics.
6. Cladistics ignores overall similarity and bases taxonomy on the timing of branch points as decided by the sequence in the origin of apomorphic (shared-derived) characters.
7. Classical evolutionary taxonomy considers overall homology *and* branching sequence.

Mechanisms of Macroevolution (pp. 492–501)

1. An important event in the formation of higher taxa is the appearance of an evolutionary novelty through preap-tation, the gradual modification of an existing structure for a new function, or a change in a regulatory gene that has a major impact on morphology. Regulatory genes can greatly alter adult form by influencing allometric growth or by causing a retention of juvenile features through paedomorphosis.
2. Evolutionary trends are ultimately dictated by environ-mental conditions that affect the survival and reproduc-tive success of organisms.
3. Most evolutionary trends may be the result of species selection rather than gradual phyletic change in an unbranched lineage. Applying the principles of punc-tuated equilibrium, species selection maintains that spe-cies with certain characteristics survive longer and pro-duce more branches than species with other characteristics. Hence, differential speciation may drive macroevolution the way differential reproduction drives microevolution.
4. Continental drift has had a significant impact on the his-tory of life by causing major geographical rearrange-ments affecting biogeography and evolution. The for-mation of the supercontinent Pangaea during the late Paleozoic era and its subsequent breakup during the early Mesozoic era explains many puzzling cases of geograph-ical distribution observed today.
5. Evolutionary history has not been a series of smooth gradations. Long relatively stable quiescent periods have been interrupted by brief intervals of extensive species turnover or mass extinctions.
6. Major radiations may follow the evolution of novel fea-tures that give a new species the opportunity to prolif-erate in a previously unexploited adaptive zone. Mass extinctions have also encouraged adaptive radiation by opening up previously occupied adaptive zones for sur-viving species.

Is a New Synthesis Necessary? (pp. 501–502)

1. The modern synthesis combines a variety of disciplines to explain evolution. It is decidedly Darwinian in its major premise that both micro- and macroevolution are the gradual result of natural selection causing small changes in the gene pool.

2. The hierarchical theory questions the modern synthesis as to the rate of evolution and the degree to which microevolution can lead to macroevolution. It maintains that most morphological change occurs during abrupt speciation events.

3. In the final analysis, the debate among evolutionists is more about the relative importance of evolutionary mechanisms than about their nature. Natural selection, as the mechanism of adaptation, remains central to evolutionary biology.

SELF-QUIZ

1. Index fossils are used by paleobiologists mostly because they provide information about
   a. the absolute age of the rocks containing the fossils
   b. the correlation of the relative ages of rocks in different locations
   c. the origins of major phyla
   d. the causes of mass extinctions
   e. the causes of evolutionary trends

2. If humans and pandas belong to the same class, then they must also belong to the same
   a. order
   b. phylum
   c. family
   d. genus
   e. subclass

3. In the case of comparing birds to other vertebrates, having four appendages is
   a. a plesiomorphic trait
   b. an apomorphic trait
   c. a character useful for distinguishing the birds from other vertebrates
   d. an example of analogy rather than homology
   e. a character useful for sorting the avian (bird) class into orders

4. Advocates of the model known as "species selection" propose that most evolutionary trends result from
   a. the tendency for natural selection to perfect adaptations
   b. stepwise progression of an unbranched lineage, with each step furthering the evolutionary trend
   c. gradual anagenic transformation of a single species
   d. preaptation of species for possible changes in the environment
   e. differences between species in their longevity and/ or rates of speciation

5. The greatest adaptive radiation of the animal kingdom occurred during the
   a. early Precambrian era
   b. late Precambrian era
   c. early Paleozoic era
   d. early Mesozoic era
   e. early Cenozoic era

6. The DNA from two species is compared by the method of restriction mapping. Extensive similarity between the species in the sizes of the collection of DNA fragments that result from treatment with a restriction enzyme indicates that
   a. the genes being compared have the same functions
   b. most sites recognized by the restriction enzyme have equivalent locations in the DNA samples from the two species
   c. the two species normally possess the same restriction enzyme
   d. the DNA fragments that match in size between the two species have identical base sequences
   e. the genomes of the same species are about the same size in their total amount of DNA

7. Extensive adaptive radiations have usually followed in the wake of mass extinctions mainly because
   a. many adaptive zones are vacated
   b. conditions of the physical environment are usually at their most favorable after some crisis has passed
   c. the survivors have superior adaptations that enable them to spill into many environments when conditions improve after the extinction episode
   d. preaptation assures that survivors will radiate to give rise to many new species
   e. given a stable environment, biological diversity tends to increase

8. Which of the following *cannot* be used to help determine the order of phylogenetic branching within a taxon?
   a. the fossil record
   b. restriction mapping of DNA
   c. amino acid comparisons between homologous proteins
   d. comparison of overall similarity
   e. comparison of neutral mutations

9. The evolutionary transformation of a fish's primitive lung into a swim bladder (float) is an example of
   a. convergent evolution
   b. divergent evolution
   c. preaptation
   d. adaptive radiation
   e. paedomorphosis

10. The differences between the modern synthesis and a more hierarchical theory of evolution include all of the following *except*

a. gradualism versus punctuated equilibrium

b. natural selection versus chance as central to adaptation

c. the relative importance of microevolution in macro-evolution

d. phyletic transitions versus species selection

e. the relative role of chance in speciation and biological diversity

CHALLENGE QUESTIONS

1. In the "DNA clock," some nucleotide changes cause amino acid substitutions in the encoded protein (nonsynony-mous changes), and others do not (synonymous changes). In a comparison of rodent and human genes, rodents were found to accumulate synonymous changes two times faster than humans and nonsynonymous substitutions 1.3 times as fast. How do such data complicate the use of molecular clocks in absolute dating? (See Further Reading 1.)

2. Imagine that Pangaea were re-formed today. What macro-evolutionary changes would you expect?

3. Considering that there have been many species on the evolutionary bush that includes *Homo sapiens*, how can a punctuational theory explain an overall trend toward larger brains in human evolution?

FURTHER READING

1. Amato, I. "Tics in the Tocks of Molecular Clocks." *Science News,* January 31, 1987. How reliably can comparisons of DNA reveal the vintage of evolutionary branch points?

2. Avise, J. C. "Nature's Family Archives." *Natural History,* March 1989. Using mtDNA to trace systematics.

3. Eldredge, N. *Life Pulse.* Facts on File, 1987. Eldredge discusses in detail some of the provocative ideas expressed in the interview preceding this unit.

4. Gore, R. "What Caused Earth's Great Dyings?" *National Geographic,* June 1989. Beautifully illustrated article about mass extinctions.

5. Gould, S. J. "The Wheel of Fortune and the Wedge of Progress." *Natural History,* March 1989. The creative role of mass extinctions.

6. Paul, G. S. "Giant Meteor Impacts and Great Eruptions: Dinosaur Killers?" *Bioscience,* March 1989. An intriguing hypothesis critically examined.

7. Simpson, G. G. *Fossils and the History of Life.* New York: Scientific American Library, 1983. Fascinating introduction to paleontology by one of the architects of the modern synthesis.

8. Stanley, S. *Extinction.* New York: Scientific American Library, 1987. More general than its title suggests, a beautiful book discussing several aspects of macroevolution.

The Evolutionary History of Biological Diversity

Interview: William Schopf

What do mountain climbing and rock collecting have to do with biology? Bill Schopf, a geologist in the Department of Earth and Space Sciences at the University of California at Los Angeles (UCLA), has reshaped our understanding of the history of life by studying ancient rocks called stromatolites. Dr. Schopf was among the first to identify Precambrian fossils of bacteria and cyanobacteria in the layers of sediment that make up stromatolites, some of which were formed by the microorganisms over 3.5 billion years ago. His work has revealed much about Earth's earliest known organisms and the primeval environment they inhabited. Dr. Schopf's research has also led to more informed speculation about the origin of life.

Dividing his research time between the field and laboratory, Dr. Schopf has traveled to some of the most remote parts of the world in his search for rocks bearing evidence of life's earliest era. His discoveries and enthusiastic collaboration with other researchers have made him a prominent figure in local and international organizations devoted to the study of early life and evolution. Even with these commitments, Dr. Schopf remains very involved with students, teaching undergraduate courses in paleontology, paleobotany, and the history of life, as well as graduate-level seminars. Professor Schopf views undergraduate science courses as integral parts of a liberal arts education, a value that is apparent in this interview.

We're sitting with you in the offices of the "Center for the Study of Evolution and the Origin of Life." Could you tell us about the Center?

This is an interdisciplinary center on the UCLA campus made up of about 120 people, most of whom are UCLA faculty members, though some are from outside the school. We have representatives from chemistry, biochemistry, microbiology, earth and space sciences, atmospheric sciences, anthropology, biology, and a number of folks from the medical school. Unlike our universities, nature is not compartmentalized into various departments. The natural sciences have no real boundaries; if you want to question one aspect of science, you really must be educated in ancillary disciplines. We come together once a week for dinner and discussion about a broad array of evolutionary problems—the evolution of the universe, the evolution of life, the origin of *Homo sapiens*, and musings about extraterrestrial life and future evolution. When it's my turn to provide dinner, we have Kentucky Fried Chicken.

How did you first become interested in science generally, and then more specifically, in paleobiology?

Well, my father was a paleobotanist and I was always involved with science as a young boy. My father, like many scientists, worked seven days a week. Sometimes he'd have the responsibility of looking after my brother and me;

he'd take the two of us down to the laboratory and assign us tasks to do, which mostly I hated. But I got introduced to the university and to science very early that way. I was always interested in things having to do with evolution.

My professional interests started when I went to Oberlin College in Ohio. Initially I thought I might be a philosophy major, but I maintained my interest in the sciences. When I was a sophomore in college, a professor in a beginning geology course gave a lecture in which he pointed out that essentially nothing was known about the organisms that were around during the Precambrian era—the earliest seven-eighths of earth history. Although he only mentioned this briefly, for some reason it struck me as really interesting. I didn't quite understand why there wasn't a fossil record earlier than the base of the Cambrian period (550 to 600 million years ago). I had a copy of Darwin's *On the Origin of Species*, and that night I started reading it. I found that Darwin devoted a whole lot of space to the fact that there was this missing fossil record. At one point he says that ". . . this matter must remain inexplicable and can be truly urged as an argument against the views here entertained. . . ," that is, his theory of evolution. I thought: Gee whiz, Darwin has got to be right; there

has to be an earlier fossil record. It seemed to me that there ought to be something out there to find. People had been looking for Precambrian fossils since Darwin stated the problem but not much had been written on the subject.

During that same period, I also read a book by George Gaylord Simpson written in 1949, in which he pointed out that the evolutionary distance between man and amoeba ought to be about the same as between an amoeba and the origin of life. And since he thought amoebas might have been around 500 million years ago, he thought that life originated about a billion years ago. I remember thinking that there is no way he could say such a thing, because there were no data. I thought that there ought to be some evidence and so I became very interested in the problem and thought this might be a good thing to study in graduate school.

Bill Chaloner, who is a very distinguished paleobotanist in England, was visiting my father one summer and I remember talking to him about my plans. He said, "Look, you're a nice young boy, don't waste your time on that . . . it won't be successful." He has since told me that he tells this story to his beginning botany students in London to point out that he's not always right. The fact is that I was lucky. I thought that this would be an interesting thing to do. I could have been

absolutely wrong and had no success at all. But I was young and I didn't know any better. It happened that I was fortunate and Darwin was right. There is an early fossil record, and I've had a lot of fun trying to find out about it.

People had been looking for a hundred years for evidence of life before the Cambrian period. What was the key to your discovery?

Well, people were just asking the wrong questions, looking at the wrong rocks, and not really understanding the problem. They had approached the question of the early history of life using the strategy for finding evidence of life in the younger geologic record, namely, tromping across a bunch of rocks out in the field, looking down, and picking up whatever you see. The problem with this method is that the Precambrian era was the age of microscopic life and there are no large body fossils of that sort except at the very end of the Precambrian. The strategy that has paid off really dates back to World War II, when oil companies first started to use the science of palynology. This is the study of fossil plant spores and pollens accomplished by dissolving rocks in mineral acids. A Soviet named Vologdin started applying this technique to Precambrian rocks and had been reporting results in the Soviet literature. But for various reasons, mostly political I think, and also because of language difficulties, the rest of the world, particularly North Americans, didn't read the Soviet literature.

Then came along an economic geologist, Stanley Tyler, at the University of Wisconsin, who in 1953 was doing geological work on the northern shore of Lake Superior in connection with mining companies looking for iron ores. He happened to pick up some rocks, which later turned out to have microscopic fossils from a deposit called the Gunflint Iron Formation. He thought these were probably fossils, but being an economic geologist, not a paleontologist, he wasn't really sure. In 1953, he came to the Geological Society of America meetings, which were held in Boston that year. He had some pictures of these fossils, which he pulled out of his jacket pocket and showed to Elso Barghoorn, a well-known paleobotanist. Barghoorn recognized the beasts and the two of them wrote a paper in 1954. I showed up as a graduate student in Barghoorn's lab at Harvard in 1963 when this thing was just starting to break.

What is the vintage of the oldest known fossils and what do they look like?

The oldest fossils that have been found so far are roughly 3.5 billion years in age—3,500 million years old. They come from Western Australia. They are only a few micrometers in diameter so you cannot see them without a microscope. Normally, you study them with the highest power optical microscope available. They are chains of small cells that look rather like spherical beads on a string; each one of those beads was a living cell. So, they're really colonies of primitive organisms, probably closely related to modern bacteria and, I think, cyanobacteria.

How do you know where to look for these fossils?

Well, thanks to the hard work of field geologists who determine the distribution, nature, and ages of rocks all over the world, scientists like myself know where all of the ancient rocks are. The minute microscopic cells that I look for are very delicate so I want to find not only old rocks, but rocks that have not been heated or subjected to mountain-building processes that could have destroyed the little delicate cells. So, you read the existing literature to find out where the good hunting places are, then you go out, using maps, to collect the rocks, bring them back to the laboratory, and study them.

What are stromatolites?

Stromatolites are mound-shaped objects. Sometimes they're flat, sometimes they're crinkly, sometimes they look like cabbage heads. They vary in size from very small, about the diameter of a pencil, to great big ones that can be as large as a football field. Characteristically, they are made up of layer upon layer of fine rock, each layer about a millimeter in thickness. The layers were formed by, and in fact reflect the presence of, communities of many different types of microscopic organisms. The communities built up these layered masses, often over long periods of time. If the stromatolites are

being formed by organisms in an area where they can become mineralized, particularly with the mineral calcite (calcium carbonate, the stuff of which limestone is made), then the stromatolites can be preserved in the geological record. The stromatolites then represent evidence of diverse communities of microscopic organisms that once lived where the stromatolites are now found.

If the oldest fossils are about 3.5 billion years old, then how much earlier did life first arise on earth?

I'm often asked that question, and I'm afraid that my answer is, I don't know. In fact, no one knows. The two relevant facts here are (1) that the oldest fossils are about 3.5 billion years of age and (2) the planet is about 4.5 billion years of age. So the time span in question is somewhere between 4.5 and 3.5 billion years ago. But within that one billion year period, when did life originate? The answer to that becomes what is called a guess, which is a whole lot different from a scientific statement. All one can do is to extrapolate. We have some knowledge about how fast organisms evolved from 3.5 to 2.5 billion years ago. If that same rate occurred earlier, then how far back might life be traced? The answer, with some sort of a guess, is that it is at least plausible to think that life on Earth might have originated as early as 4 billion or 4.2 billion years ago. That is pretty soon after the origin of the

planet. The interesting corollary to that is that life would seem to be pretty easy to start up. If it happened here relatively rapidly, it might happen elsewhere relatively rapidly, and there might be lots of life throughout the rest of the universe.

What do you think the earth looked like at that time? What type of environment do you think the first cells took form in?

Well, once again, there is a real difference between speculation and fact. We have well-preserved geological sequences of rocks that are 3.5 billion years of age. The rock record tells us that there were liquid water oceans, tides, shallow water environments, and volcanoes. We can paint a reasonably accurate picture of what the Earth looked like at that time.

Unfortunately, only one sequence of rocks is older than 3.5 billion years. It is a sequence from southwestern Greenland that is about 3.8 billion years old. But that set of rocks has been very highly altered by heat and pressure (a thing we call metamorphism). So, it's a little hard to interpret just what they tell us.

One thing that this sequence tells us is that there was liquid water on our planet 3.8 billion years ago. Therefore, the surface of the earth at that time was warmer than 0°C and cooler than 100°C. That's the range at which life exists well today.

There certainly were volcanoes. Remember, the Earth had formed a few hundred million years earlier, and there's a lot of heat in that process. Meteorites smash into the surface of the Earth and get buried; a lot of kinetic energy (heat) is buried with these meteorites. The heat is released by melting rocks, which give off gases. Atmospheric nitrogen and carbon dioxide, as well as the oceans, come from the gases released during these processes.

So, you can image that the early Earth had volcanic cones, water condensing out of the sky, carbon dioxide, some nitrogen, and some argon in small amounts. There were shallow seas and probably deeper seas. But it took a very long time for the Earth to begin to look like it does now with such large continental masses.

organic matter derived from nonbiological processes. The notion is that small organic compounds were synthesized by an electrical discharge, lightning, in a primitive atmosphere without much oxygen. The organisms evolved from that primordial soup and then, in turn, were able to metabolize it. Of course, Stanley Miller demonstrated this in the laboratory and showed that Oparin's idea was right.

In 1976, Oparin, who was by then an elderly man, spent three months here in my lab at UCLA as a visiting scientist. He told us a story that has always impressed me and is worth passing on here. In 1915, at the time he was graduating from high school, Oparin was selected to attend Moscow State University. When visiting the school, he had time to attend only one lecture. Having come from the countryside, he decided to sit in on a botany lecture by a man named Timirjazev, who at that time was the leading Russian botanist of his generation and a leading evolutionist.

In his lecture, Timirjazev told a story about how he had become interested in evolution. When he was a student, Timirjazev had traveled to England in an attempt to meet Charles Darwin. This was about 1870 or 1880, 15 to 20 years after the publication of *Origin of Species*. Darwin was old and often in ill health and unable to receive visitors. Every day Timirjazev went to sit on the front steps outside of Down House (Darwin's home), hoping that he, a young man, would have an opportunity to meet the great Charles Darwin.

Well, after doing that for four to five days, Darwin came out to meet him. The two of them spent two afternoons walking the sand path behind Darwin's home. Darwin told the young Timirjazev what he thought about evolution and about the evolution of plants.

So, Timirjazev goes back to Moscow, becomes a very famous professor, and when he is about 80 years old, meets the young Oparin. He tells Oparin this story, which is the first time that Oparin has heard much about evolution. Oparin gets absolutely enamored with the idea, and he notes that the thing that Darwin didn't deal with was the problem of the origin of life. So, sure enough, Oparin says, by golly, that looks like a good thing for me to worry about!

By 1918, Oparin had written a 15-page manuscript, which he submitted for publication, outlining his theory of the origin of life. Well, it got turned down. In the time of the Czars, the Russian Orthodox church had a set of censors that wouldn't permit publications that were not in line with what the Czar deemed appropriate. Even though this was a year after the Revolution, the censors had not yet been removed from power. So, when Oparin submitted his paper, the censor felt that Oparin's notion about the origin was not consistent with current theological doctrine and rejected it.

Oparin said that was a great thing for him because he spent the next four years on this theory and put together what he always called, affectionately,

his little pamphlet. He resubmitted it in 1921; it was published in 1922 and then finally translated into English in 1938. Oparin's pamphlet formed the framework for all current thinking on the subject.

And now we come to 1976, when Oparin was visiting my lab. He recounted this anecdote to me and about two dozen of my students sitting around having lunch one day.

To me, this entire story shows the remarkable continuity in human science. It illustrates the lineal descent from Darwin to Timirjazev, from Timirjazev to Oparin, and from Oparin to my students. In a hundred and some years of human history, you have this remarkable continuity of the human process of science.

Do you believe that you or your colleagues will ever find fossils of organisms much older than 3.5 billion years old?

The techniques that we're using are good only to detect organisms with cell walls. Since the absolutely first living thing was a lot more like an oil droplet, something that wasn't enclosed by a physically resilient wall, it's well near impossible for us to find that sort of life. Once organisms get cell walls, then we have a chance to find them. And then the only thing that limits us is the available rock record.

How do you think the earliest bacteria nourished themselves?

There is a fair amount of debate on that question. The approach used is to look at the physiology, the metabolism, and the biochemistry of modern organisms (living descendants of the earliest organisms) and try to determine from them what the earliest organisms were like.

A rather famous book, a little pamphlet published in 1922 by a Soviet biochemist named Alexander Oparin, sets the stage for current thought regarding the origin of life. He was one of the first to theorize that the earliest forms of life were probably anaerobic; that is, that they lived in the absence of oxygen. He speculated that they derived energy by metabolizing

Of all your discoveries, what has most surprised you?

I suppose what surprises me most is how often I'm wrong! As a sophomore in college, I intended to trace back the history of life to a billion years and then, based on the geological evidence, investigate the origin of life itself. I thought that if I went back far enough, I'd run out of fossils, and then I could find evidence of the primordial soup that Oparin and Miller postulated. The big surprise is what a very naive notion this was. Who would have thought that for 80% of the history of life, our whole planet was dominated by microscopic organisms? Another big surprise is that the rate of morphological change is so much slower in microbes than in higher organisms. Cyanobacteria, prokaryotes in general, don't evolve very rapidly morphologically. Their big evolutionary changes happen at the biochemical level.

How often do you take field trips to collect rocks? Where do you go and for how long?

Well, I take field trips whenever I get the opportunity. It really depends on the stage of development of a given project. I do a lot of laboratory work too. I'm very lucky to have the opportunity to travel widely because the rocks that I look for don't occur just everywhere. The older the rock is, the greater the chance that it no longer exists. So, as you go back through the geological record, there are fewer and fewer rocks for you to study. It happens that the rocks I look for are in rather interesting places—Western Australia, South Africa, portions of China, Siberia, the Soviet Union.

Is it dangerous?

That depends on where you're working. Western Australia is hot, arid, and fairly desolate. You can get into some mountain climbing predicaments that are a bit scary. I got into a situation in Western Australia on a ledge a couple of feet or so in width. I made the mistake of looking down, and there was about a thousand foot drop directly beneath me. My knees turned to jelly. It was awful. And once a good friend of mine at the University of New Mexico was on a field trip in arctic Canada. They were left off near the summit of a mountain by helicopter and were sup-

posed to hike down. He lost his footing and slid down the side of a huge snowfield. Luckily, two members of the

party below were able to stop him before he went over the edge. He's very experienced, he just lost his footing.

For the most part, it's just dirty, dusty, hard work. But the fact is that you learn a great deal in the field that you simply cannot learn in the laboratory. If someone just gives you a specimen of rock in the lab, you can't tell much about the environment from which it came. If you study the rocks laterally and vertically in the field, you can tell how deep the ocean or the lagoon was in which these rocks were formed. You can see how the environment might have changed over time and you can speculate on what sort of organisms you might expect to find.

What do you do with the rocks once you have them back in the lab?

As mundane as it may sound, it is exceptionally important to make sure that the rocks are labeled properly. Unless you curate a piece of rock properly, it is no good to anyone. You might as well use it as a door stop.

Typically we divide the rock into various fractions. One portion will be ground to a very fine powder, which we'll subject to chemical studies. The chemistry of the organic matter will tell us something about the organisms that were there. Another piece of rock will be dissolved in acid, and the bits of organic matter remaining will be examined under a microscope to see if there are any minute cells preserved. Another portion of that rock will be sliced very thinly and then ground

until it is about the thickness of a piece of paper—thin enough to see through. We will study the slice under a microscope and, if we are fortunate, we will see the fossils of cells that were living at the time.

What is your advice to students who think they would enjoy the detective work of tracing the history of life?

Well, I think I'd like to offer encouragement rather than advice. First of all, it's fun. It's also a lot of work. It's tremendously rewarding to find something that nobody else has ever seen. In my particular area, that happens rather frequently. You might find something that will change a generalization or will change the ideas that go into a beginning biology textbook. Then you feel like you've contributed to the understanding of the world in which we live.

In terms of preparation, I would like to come back to this notion that nature is not compartmentalized. I encourage students at an early stage in their careers to take courses in as many of the related sciences as they can. There is a certain vocabulary associated with any area of science, whether it be physics, geology, mathematics, chemistry, or biology. If you don't learn all these terms early, then you never will. If you never learn the vocabulary, you'll never read the literature. Then you won't understand the framework of that field, and you won't be able to address the questions that involve that field. Only if you educate yourself in several areas of science will you have a chance to make a particular contribution to human knowledge. That's what I'd encourage students to do.

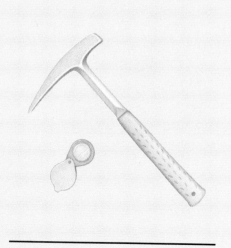

24

Early Earth and the Origin of Life

Life is a continuum extending from the earliest organisms through the various phylogenetic branches to the great variety of forms alive today. This unit of chapters surveys the diversity of contemporary life and traces the evolution of this diversity over 3.5 billion years of history (see Table 23.1).

One recurrent theme in these chapters is the association between biological and geological history. Geological events that alter environments change the courses of biological evolution. The formation and subsequent breakup of the supercontinent Pangaea, for instance, affected the diversity of life tremendously (see Chapter 23). Conversely, life has changed the planet it inhabits, sometimes profoundly (Figure 24.1). For example, the evolution of photosynthetic organisms that released oxygen to air completely altered the atmosphere of the Earth. Much more recently, the emergence of *Homo sapiens* has changed the land, water, and air on a scale and at a rate unprecedented for a single species. The evolution of Earth and its life are inseparable.

These chapters also emphasize key junctures in evolution that have punctuated the history of biological diversity. Earth history and biological history have been episodic, marked by what were in essence revolutions that opened many new ways of life (Figure 24.2).

Historical study of any sort is an inexact discipline, dependent as it is on the preservation, reliability, and interpretation of past records. The fossil record of past life is generally less and less complete the farther into the past we delve. Fortunately, each organism carries

Figure 24.1
The changing Earth and its life. In a scene reminiscent of conditions on the early Earth, violent discharges of lightning and volcanic activity were associated with the birth in 1963 of the island of Surtsey near Iceland in the North Atlantic. Terrestrial life forms began colonizing Surtsey almost immediately after its birth. Since then, the island and its inhabitants have been evolving together, demonstrating on a small scale the inseparable history of the Earth and its life.

traces of its evolutionary history in its molecules, metabolism, and anatomy. As we saw in Unit Four, such traces are clues to the past that augment the fossil record, much as similarities and differences between extant cultures help social scientists understand historical relationships between the cultures. Still, the evolutionary episodes of greatest antiquity are generally the most obscure. This chapter is the most speculative of the unit, for its main subject is the origin of life on a young Earth, and no fossil record of that seminal episode exists. The chapter sets the stage by briefly discussing the origin and early existence of the Earth, in keeping with the theme of the intertwining of geological and biological evolution, and then describes theories of how natural processes on the youthful planet could have created life. The last section of the chapter addresses the various kingdoms of life as a prelude to the survey of biological diversity in the following chapters.

FORMATION OF THE EARTH

The formation of Earth is a fragment of a much bigger story. Earth is one of nine planets orbiting the sun, one of billions of stars in the Milky Way, which is one of millions of galaxies in the universe. The star closest to our sun, Proxima Centauri, is four light years—40 trillion kilometers—away; we see it by the light it emitted four years ago. Some stars are so distant that even if they burned out millions of years ago, we would still see them in the sky tonight; and there are new stars that are invisible because their light has not yet reached Earth. Gazing at stars, we look back in time.

The universe has not always been so spread out. Based on several lines of evidence, most astronomers

now believe that all matter was at one time concentrated in a giant mass that blew apart with a "big bang" sometime between 10 and 20 billion years ago and has been expanding ever since.

Our sun is a second- or third-generation star, born about five billion years ago from the fallout of defunct stars. Compared with the overall universe, the solar system is relatively rich in the heavier elements that formed by fusion from smaller atoms in the crucibles of ancestral stars. Most of the swirling matter in the disk-shaped cloud of dust that formed our solar system condensed in the center as the sun. Peripheral material was left spinning around the infant sun in several concentric rings. The planets, including Earth, formed about 4.6 billion years ago from kernels that used gravity to draw together the dust and ice in their zones.

Most geologists believe that Earth began as a cold world that later melted owing to heat produced by

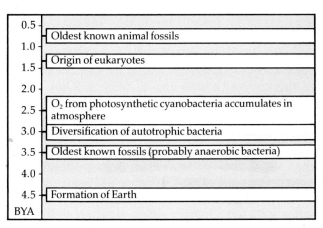

Figure 24.2
Some major episodes in the early history of the Earth and life. (BYA = billion years ago.)

compaction, radioactive decay, and the impact of meteorites. Molten material sorted into layers of varying density. Most of the nickel and iron sank to the center and formed a core. Less dense material became concentrated in a mantle, and the least dense material settled on the surface, eventually solidifying into a thin crust. The present continents are attached to plates of crust that float on the semisolid, flexible mantle (see Chapter 23).

The first atmosphere was probably composed mostly of hot hydrogen gas (H_2), which escaped because the gravity of Earth was not strong enough to hold such small molecules. Volcanoes and other vents through the crust belched gases that formed a new atmosphere. Based on analysis of gases vented by modern volcanoes, scientists have speculated that a second early atmosphere consisted mostly of water vapor (H_2O), carbon monoxide (CO), carbon dioxide (CO_2), nitrogen (N_2), methane (CH_4), and ammonia (NH_3). The first seas formed from torrential rains that began when Earth had cooled enough for water in the atmosphere to condense. In addition to an atmosphere very different from the one we know, lightning, volcanic activity, and ultraviolet radiation were much more intense when Earth was young. On such a world, life began.

THE ANTIQUITY OF LIFE

The metaphor of an evolutionary tree implies that the history of life chronicles an increasing diversity of organisms all descended from the primitive creatures that were the first living things. The apparent absence of fossils of ancestral organisms in Precambrian rocks was incongruous with Charles Darwin's view that complex life evolved from simpler forms, and he wrote in *The Origin of Species*: "To the question why we do not find rich fossiliferous deposits belonging to these assumed earliest periods prior to the Cambrian system, I can give no satisfactory answer. . . . The case at present must remain inexplicable, and may be truly urged as a valid argument against the views here entertained." Some of Darwin's adversaries seized the cue and declared the beginning of the Cambrian period as the time of Genesis and all creation. Only in the past few decades has the discovery of older fossils filled in the Precambrian blank. The Cambrian fauna was preceded by a less diverse collection of animals dating back 700 million years (see Chapter 23). Before then was a succession of microorganisms spanning nearly three billion years (Figure 24.3). For most of that time, only prokaryotes inhabited Earth. One would guess from the relatively simple structure of the prokaryotic cell (compared with the eukaryotic cell) that early organisms were primitive bacteria, and the fossil record now supports that presumption.

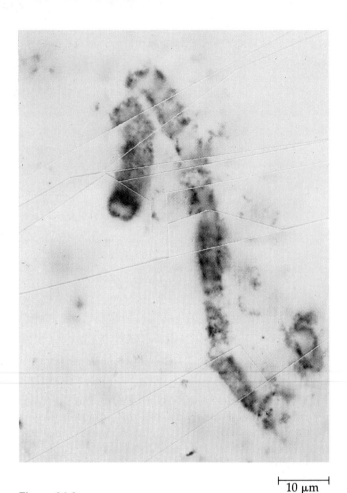

|———————————| 10 μm

Figure 24.3
An early prokaryote. Precambrian fossils escaped notice until about 30 years ago partly because they are microscopic. This filamentous prokaryote, about 3.5 billion years old, was collected in Western Australia.

Prokaryotes originated on Earth within a few hundred million years after the crust cooled and solidified. Though geologists have discovered minerals that crystallized about 4.1 billion years ago and sedimentary rocks that date back 3.8 billion years, no fossils so old have yet been found. However, fossils that appear to be prokaryotes about the size of bacteria *have* been discovered in southern Africa in a rock formation called the Fig Tree Chert, which is 3.4 billion years old. Evidence of even more ancient prokaryotic life has been found in rocks called **stromatolites.** Stromatolites (from the Greek *stroma*, "bed," and *lithos*, "rock") are banded domes of sediment strikingly similar to the layered mats constructed by colonies of bacteria and cyanobacteria living today in very salty marshes. The layers are sediments that stick to the jellylike coats of the motile microbes, which again and again migrate out of one layer of sediment and form a new one above, producing the banded pattern (Figure 24.4). Fossils resembling spherical and filamentous prokaryotes have been found in stromatolites that are 3.5 billion years

(a)

(b)

(c)

Figure 24.4
Bacterial mats and stromatolites. The mats are sedimentary structures produced by colonies of bacteria and cyanobacteria that live, uncropped by predators, in environments inhospitable to most other life. **(a)** Lynn Margulis and Kenneth Nealson, studying the history of life, are collecting bacterial mats in a Baja California lagoon. **(b)** The bands, seen in this section of a mat, are layers of sediment that adhere to the sticky prokaryotes, which produce the succession of layers by migrating. **(c)** Fossilized mats known as stromatolites resemble the layered structures formed by contemporary bacterial colonies. This stromatolite is a Western Australian specimen about 3.5 billion years old. Microfossils, such as the one in Figure 24.3, are present in many stromatolites.

old in western Australia and southern Africa. For now, these are the oldest evidence of life. However, the western Australia fossils appear to be those of photosynthetic organisms, perhaps oxygen producers. If so, it is likely that life had been evolving long before these organisms lived. J. William Schopf, whose interview appears at the beginning of this unit, thinks it possible that the earliest life forms appeared as early as 4 billion years ago.

THE ORIGIN OF LIFE

The question of how life began is more specifically about the genesis of prokaryotes. Sometime between about 4.1 billion years ago, when Earth's crust began to solidify, and 3.5 billion years ago, when the planet was inhabited by bacteria advanced enough to build stromatolites, the first organisms came into being. What was their origin? The great majority of biologists subscribe to the hypothesis that life developed on Earth from nonliving materials that became ordered into molecular aggregates that were eventually capable of self-replication and metabolism. Life cannot arise by spontaneous generation from inanimate material today, so far as we know, but conditions were very different when Earth was only a billion years old. In that ancient environment, the origin of life was evidently possible, and it is likely that at least the early stages of biological inception were inevitable. However, debate abounds about what occurred during these early stages.

According to one hypothesis, the first organisms were products of a chemical evolution in four stages: (1) the abiotic (nonliving) synthesis and accumulation of small organic molecules, or monomers, such as amino acids and nucleotides; (2) the joining of these monomers into polymers, including proteins and nucleic acids; (3) the aggregation of abiotically produced molecules into droplets, called **protobionts,** that had chemical characteristics different from their surroundings; and (4) the origin of heredity (which may have been under way even before the "droplet" stage).

Abiotic Synthesis of Organic Monomers

In the 1920s, A. I. Oparin of Russia and J. B. S. Haldane of England independently postulated that conditions on the primitive Earth favored chemical reactions that synthesized what are now called organic compounds from inorganic precursors present in the early atmosphere and seas. This cannot happen in the modern world, Oparin and Haldane reasoned, because the present atmosphere is rich in oxygen produced by photosynthetic life. The oxidizing atmosphere of today is not conducive to the spontaneous synthesis of complex molecules because the oxygen attacks chemical bonds, extracting electrons. Before oxygen-producing photosynthesis, Earth had a much less oxidizing atmosphere, derived mainly from volcanic vapors. Such a reducing (electron-adding) atmosphere would have enhanced the joining together of simple molecules to form more complex ones. Even with a reducing atmosphere, making organic molecules would require considerable energy, which was probably provided by lightning and the intense ultraviolet radiation that penetrated the primitive atmosphere. The modern atmosphere has a layer of ozone produced from oxygen, and this ozone shield screens out most ultraviolet radiation. Evidence also exists that young suns emit more ultraviolet radiation than older suns. Oparin and Haldane envisioned an ancient world with the necessary chemical conditions and energy resources for the abiotic synthesis of organic molecules.

In 1953, Stanley Miller and Harold Urey tested the Oparin-Haldane hypothesis by creating in the laboratory conditions comparable to those of the early Earth (see interview on p. 15). Their apparatus produced a variety of amino acids and other organic compounds found in living organisms today (Figure 24.5).

The atmosphere in the Miller-Urey model was made up of H_2O, H_2, CH_4 (methane), and NH_3 (ammonia), the gases the researchers in the 1950s believed prevailed in the ancient world. This atmosphere was probably more strongly reducing than the actual atmosphere of the early Earth. The vapors of modern volcanoes include CO, CO_2, and N_2, and it is likely that these gases were present in the ancient atmosphere evolved from volcanoes. Traces of O_2 may even have been present, formed from reactions among other gases as they baked under powerful ultraviolet radiation. Many laboratories have repeated the Miller-Urey

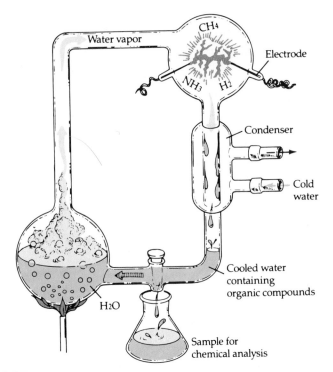

Figure 24.5
Abiotic synthesis of organic molecules in a model system. Stanley Miller and Harold Urey used an apparatus similar to this one to simulate chemical dynamics on the primitive Earth. A warmed flask of water simulated the primeval sea. The "atmosphere" consisted of H_2O, H_2, CH_4, and NH_3. Sparks were discharged in the synthetic atmosphere to mimic lightning. A condenser cooled the atmosphere, raining water and any dissolved compounds back to the miniature sea. As material circulated through the apparatus, the solution in the flask changed from clear to murky brown. After one week, Miller and Urey analyzed the contents of the solution and found a variety of organic compounds, including some of the amino acids that make up the proteins of organisms.

experiment using a variety of recipes for the atmosphere, including a mixture having a very low concentration of O_2. Abiotic synthesis of organic compounds occurred in these modified models, although yields were generally less than in the original experiment. The most relevant characteristic of the early atmosphere seems to be the rarity of the strong oxidizing agent O_2.

Laboratory analogs of the primeval Earth have been used to make all 20 amino acids commonly found in organisms, several sugars, lipids, the purine and pyrimidine bases present in the nucleotides of DNA and RNA, and even ATP (if phosphate is added to the flask). Before there was life, its chemical building blocks may have been accumulating as a natural stage in the chemical evolution of the planet.

Abiotic Synthesis of Polymers

The abiotic synthesis of more complex organic molecules by the joining together of smaller ones also may have been inevitable on the primitive Earth. Organic polymers such as proteins are chains of similar building blocks, or monomers. They are synthesized by dehydration reactions that remove hydrogen and hydroxyl (OH) groups from the monomers, forming a water molecule as a by-product of each new linkage in the polymer (see Chapter 5). In the living cell, specific enzymes catalyze the dehydration reactions. Abiotic synthesis of polymers would have had to occur without the help of these efficient enzymes, and the dilute concentrations of the monomers dissolved in an excess of water would not favor spontaneous dehydration reactions that form more water (see Chapter 2 for a review of chemical equilibrium). Polymerization does occur in laboratory experiments when dilute solutions of organic monomers are dripped onto hot sand, clay, or rock, a process that vaporizes water and concentrates the monomers on the substratum. Using this method, Sidney Fox of the University of Miami has made what he calls *proteinoids*, which are polypeptides produced by abiotic means. It is possible to imagine waves or rain splashing dilute solutions of organic monomers onto fresh lava or other hot rocks on the early Earth and then rinsing proteinoids and other polymers back into the water.

Clay, even cool clay, may have been especially important as a substratum for the polymerization reactions prerequisite to life. Clay concentrates amino acids and other organic monomers from dilute solutions because the monomers bind to charged sites on the clay particles. At some of the binding sites, metal atoms, such as iron and zinc, function as catalysts facilitating the dehydration reactions that link monomers together. Clay with many of these binding sites could have functioned as a lattice that brought monomers close together and then assisted in joining them into polymers. The clay also seems to be able to store energy absorbed from radioactive decay and then discharge this energy at times when the clay changes temperature or degree of hydration. Scientists at NASA's Ames Research Center in California are presently investigating the abiotic synthesis of polypeptides and short nucleic acids on the surface of clay.

Formation of Protobionts

The properties of life emerge from an interaction of molecules organized into higher levels of order (see Chapter 1). Living cells may have been preceded by protobionts—aggregates of abiotically produced molecules not yet capable of precise reproduction but able to maintain an internal chemical environment differ-

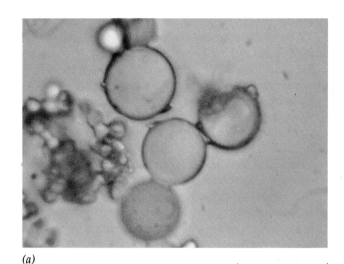

(a)

|——————————| 10 μm

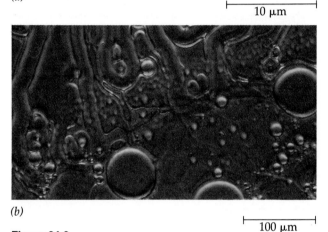

(b)

|——————————| 100 μm

Figure 24.6
Laboratory versions of protobionts, aggregates of organic molecules with some biological properties.
(a) Microspheres are made by cooling solutions of proteinoids, polypeptides created abiotically from amino acids polymerized on hot surfaces. Microspheres grow by absorbing free proteinoids until they reach an unstable size, when they split to form daughter microspheres. Of course, this division lacks the precision of cellular reproduction (LM). (b) Phospholipids, which were probably synthesized abiotically in the early seas, are natural membrane builders. In an aqueous environment, the molecules form films, which, when agitated, break into spheres called liposomes (LM).

ent from the surroundings and exhibiting some of the properties associated with life, including metabolism and excitability.

Laboratory experiments suggest that protobionts could have formed spontaneously from abiotically produced organic compounds. When mixed with cool water, proteinoids self-assemble into tiny droplets called **microspheres** (Figure 24.6a). Coated by a membrane that is selectively permeable, the microspheres undergo osmotic swelling or shrinking when placed in solutions of different salt concentrations. Some microspheres also store energy in the form of a membrane potential, a voltage across the surface (see Chapter 8). The protobionts can discharge the voltage in nervelike fashion; such excitability is characteristic of all life (that is not to say that microspheres are alive but only that they display *some* of the properties of life). Droplets of another type, called **liposomes,** form spontaneously when the organic ingredients include certain lipids, which organize into a molecular bilayer at the surface of the droplet, much like the lipid bilayer of cell membranes (Figure 24.6b). Oparin has made still a different type of protobiont that he calls **coacervates,** which are colloidal droplets that form when a solution of polypeptides, nucleic acids, and polysaccharides is shaken. If enzymes are included among the ingredients, they are incorporated into the coacervates, and the protobionts function as miniature chemical factories that absorb substrates from their surroundings and release the products of the reactions catalyzed by the enzymes (Figure 24.7).

Unlike some laboratory models, protobionts that formed in the ancient seas would not have possessed refined enzymes, which are made in cells according to inherited instructions. Some molecules produced abiotically, however, do have weak catalytic capacities, and there could well have been protobionts that modified the substances they took in across their membranes by a rudimentary metabolism.

The Origin of Genetic Information

Imagine a tidepool, pond, or moist clay on the primeval Earth with a suspension of protobionts varying in chemical composition, permeability, and catalytic capabilities. Those droplets most stable and best able to accumulate organic molecules from the environment would grow and split, distributing their chemical components to the "baby" droplets. Less successful droplets would fall apart or fail to grow and divide. In this way, the environment may have selected in favor of some molecular aggregates and against others. But competition among the various protobionts could not lead to long-range improvement because there was no way to perpetuate success. As prolific droplets grew, split, grew and split again, their unique catalysts and other functional molecules would become increasingly diluted. The chemical aggregates that were the forerunners of cells could not build on the past and evolve until the development of some mechanism for replicating their characteristics—some mechanism of heredity that would pass along not just samples of key molecules but also instructions for making more of those molecules.

A cell stores its genetic information as DNA, transcribes the information into RNA, and then translates the messages into specific enzymes and other proteins (see Chapters 15 and 16). Instructions are transmitted

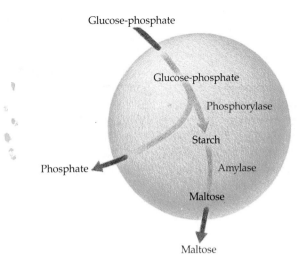

Figure 24.7
Catalysis in protobionts. If enzymes are present in the solution from which molecular droplets self-assemble, the protobionts carry out specific chemical reactions. In this case, the protobiont has incorporated two enzymes that work in tandem to convert glucose-phosphate, a simple sugar, to the dissaccharide maltose. Although such specific and efficient enzymes could not have existed in the prebiotic world, even randomly produced polypeptides have weak catalytic properties.

by the replication of DNA when a cell divides. The DNA → RNA → protein axis of cellular control employs intricate machinery that could not have evolved all at once but must have emerged bit by bit as improvements to much simpler processes. Even before DNA, some primitive mechanism may have existed for aligning amino acids along strands of RNA that could replicate themselves. According to this hypothesis, the first genes were not DNA molecules but short strands of RNA that began self-replicating in the prebiotic world.

Short polymers of ribonucleotides have been produced abiotically in test-tube experiments. If such molecules arose before life, could these RNA strands have self-replicated without the help of elaborate enzymes? The answer seems to be yes, and the evidence is so strong that scientists studying the origin of life now speak of a prebiotic "RNA world." If RNA is added to a test-tube solution containing monomers for making more RNA, sequences about five to ten nucleotides long are copied from the template according to the base-pairing rules (see Chapter 16). If zinc is added as a catalyst, sequences up to 40 nucleotides long are copied with less than 1% error. Recently, Thomas Cech and his co-workers at the University of Colorado at Boulder revolutionized thinking about the evolution of life when they discovered that RNA molecules are important catalysts in modern cells. This disproved the long-held view that only protein enzymes

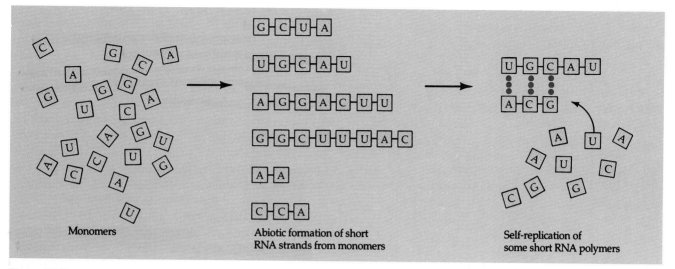

Figure 24.8
Abiotic replication of RNA. The first genes may have been RNA molecules that polymerized abiotically and replicated themselves autocatalytically while bound to clay surfaces.

serve as biological catalysts. Cech and others found that modern cells use RNA catalysts, called **ribozymes,** to do such things as remove introns from RNA (see Chapter 16). Ribozymes also catalyze the synthesis of new RNA, notably ribosomal RNA, tRNA, and mRNA. Thus, RNA is autocatalytic, and in the prebiotic world, long before there were protein enzymes or DNA, RNA molecules may have been fully capable of self-replication (Figure 24.8).

Natural selection on the molecular level works on diverse populations of replicating autocatalytic RNA molecules. Unlike double-stranded DNA, which takes the form of a uniform helix, single-stranded RNA molecules assume a variety of specific three-dimensional shapes mandated by their nucleotide sequences. Each sequence folds into a unique conformation reinforced by hydrogen bonding between regions of the strand having complementary sequences of bases. The molecule thus has a genotype (its nucleotide sequence) and a phenotype (its conformation, which interacts with surrounding molecules in specific ways). In a particular environment, RNA molecules of certain base sequences are more stable and replicate faster with fewer errors than other sequences. Beginning with a diversity of RNA molecules that must compete for monomers to replicate, the sequence best fit to the temperature, salt concentration, and other features of the surrounding solution and having the greatest autocatalytic activity will prevail. Its descendants will be not a single RNA species but a family of closely related sequences (because of copying errors). Selection screens mutations in the original sequence, and occasionally a copying error results in a molecule that

folds into a shape even more stable or more adept at self-replication than the ancestral sequence. Natural selection has been observed operating on RNA populations in test tubes, and perhaps it happened in prebiotic times as well.

The rudiments of RNA-directed protein synthesis may have been in the weak binding of specific amino acids to bases along RNA molecules, which functioned as simple templates holding a few amino acids together long enough for them to be linked, perhaps with zinc or some other metal acting as an adjunctive catalyst. If RNA happened to synthesize a short polypeptide that in turn behaved as an enzyme helping the RNA molecule to replicate, then the early chemical dynamics included molecular cooperation as well as competition (Figure 24.9a).

Thus, the first steps toward the replication and translation of genetic information may have been taken by molecular evolution even before RNA and polypeptides became packaged within membranes. Once primitive genes and their products became confined to membrane-bounded compartments, the protobionts could evolve as units (Figure 24.9b). Molecular cooperation could be refined because components that interacted in ways favorable to the success of the protobiont as a whole were concentrated together in a microscopic volume rather than being spread throughout a puddle or film on the surface of clay. Suppose, for example, that an RNA molecule ordered amino acids into a primitive enzyme that extracted energy from an organic fuel taken up from the surroundings and made the energy available for other reactions within the protobiont, including replication

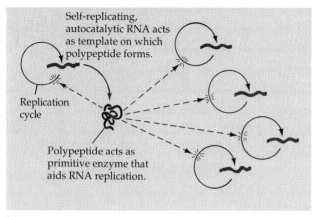

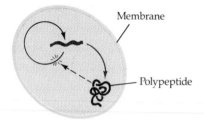

Figure 24.9
The beginnings of molecular cooperation. (a) An RNA strand enhances its own replication if it orders amino acids into a polypeptide that in turn functions as an enzyme helping the RNA strand replicate. In such cooperation may have been the rudiments of the translation of genetic information into protein structure. The benefits of this protein, however, would have been shared by competing RNA molecules not yet segregated within protobiots. **(b)** Once genes were in membrane-enclosed compartments, they would have benefits exclusively from their protein products.

of RNA. Natural selection could favor such a gene only if its diffusible product were kept close by, rather than being shared with competing RNA sequences in its environment.

In this scenario, we have built a hypothetical antecedent of the cell by incorporating genetic information into an aggregate of molecules that selectively accumulates monomers from its surroundings and uses enzymes programmed by genes to make polymers and carry out other chemical reactions. The protobiont grows and splits, distributing copies of its genes to offspring. Even if only one such protobiont arose initially by the abiotic processes that have been described, its descendants would vary because of mutations, errors in the copying of RNA. Evolution in the true Darwinian sense—differential reproductive success of varying individuals—presumably accumulated many refinements to primitive metabolism and inheritance. One trend apparently led to DNA becoming the re-

pository of genetic information. RNA could have provided the template on which DNA nucleotides were assembled. DNA is a much more stable repository for genetic information than RNA, and once DNA appeared, RNA molecules would have begun to take on their modern roles as intermediates in the translation of genetic programs.

Laboratory simulations cannot establish that the kind of chemical evolution that has been described here actually created life on the primitive Earth, but only that some of the key steps *could* have happened. The origin of life remains a matter of scientific speculation, and there are alternative views of how several key processes occurred. For example, was early chemical evolution necessary? It is possible that at least some organic compounds reached the early Earth from space. This idea, advanced by Francis Crick and called **panspermia,** holds that hundreds of thousands of meteorites and comets hitting the early Earth brought with them organic molecules formed in outer space. Extraterrestrial organic compounds, including amino acids, have been found in modern meteorites, and it seems likely that these bodies could have seeded the early Earth with organic compounds (see the interview with Stanley Miller on p. 15). Both panspermia and chemical evolution could have contributed to the pool of organic molecules that formed the earliest life.

However prebiotic chemicals accumulated, polymerized, and eventually reproduced, the leap from an aggregate of molecules that reproduces to even the simplest prokaryotic cell is immense and must have been taken in many smaller evolutionary steps. The point at which we stop calling membrane-enclosed compartments that metabolize and replicate their genetic programs protobionts and begin calling them living cells is as fuzzy as our definitions of life. We do know that prokaryotes were already flourishing at least 3.5 billion years ago, and that all kingdoms of life descended from those ancient prokaryotes.

THE KINGDOMS OF LIFE

In Chapter 23, we looked at taxonomy as a tool scientists use in tracing the evolution of organisms. Now that we have gone backward to the very origins of life on Earth, taxonomy is once again relevant as we attempt to reconstruct evolutionary relationships among the immense diversity of forms that arose from those organisms.

The kingdom is the highest—that is, the most inclusive—taxonomic category. We grow up with the bias that there are only two kingdoms of life—plants and animals—because we live in a macroscopic, terrestrial realm where we rarely encounter organisms that do not fit neatly into a plant–animal dichotomy. The two-

Table 24.1 Comparison of the five kingdoms

	Monera	Protista	Fungi	Plantae	Animalia
Cell Type	Prokaryotic	Eukaryotic	Eukaryotic	Eukaryotic	Eukaryotic
Nutrition	Varied: absorptive, photosynthetic, chemosynthetic, but not ingestive	Absorptive, ingestive, or photosynthetic	Absorptive	Photosynthetic	Ingestive
Oxygen Metabolism	Oxygen poisonous, tolerated, or required, depending on species.	Oxygen required.*	Oxygen required.*	Oxygen required.	Oxygen required.
Reproduction and Development	All can reproduce asexually. Recombination sometimes occurs. No mitosis or meiosis.	All forms can reproduce asexually. In addition, in some forms meiosis and fertilization occur.	Cells are haploid or dikaryotic. Propagation by haploid spores. Fertilization by conjugation. Meiosis.	Fertilization of female by male gamete. Diploid phase develops from embryo. Multicellular haploid phase present ("alternation of generations").	Sperm and egg form a zygote, which forms a diploid blastula and generally then a gastrula.
Life Style	Solitary unicellular, filamentous, colonial, or thinly thready (mycelial). Nonmotile or motile by gliding or by flagella.	Many unicellular aquatic forms. Some are solitary or colonial unicellular, multicellular, and mycelial.	Organisms are often branched into thin threads (mycelial) or secondarily unicellular.	Multicellular, sedentary, most live on land.	Multicellular, typically motile by muscles.
Characteristic Structures or Functions	Flagella composed of flagellin protein. Some produce endospores, sheaths, or fruiting structures. Peptidoglycan walls. No intracellular movement.	All lack embryos and complex cell junctions (e.g., desmosomes). Diverse cell walls and proteinaceous pellicles, some lack walls. Phagocytosis. Intracellular movement. Many bear undulipodia (composed of microtubules in the 9 + 2 pattern composed of tubulin proteins).†	All cells lack undulipodia. Chitinous cell walls. No phagocytosis. Body threads may be divided into segments by perforated cell walls. Extensive intracellular movement.	Extensive tissue differentiation. Cellulosic cell walls. Produce complex secondary compounds (e.g., terpenoids and anthocyanins). Sperm (present in some species) bear undulipodia.	Extensive cellular and tissue differentiation. No cell walls. Phagocytosis. Complex connections between cells (e.g., gap junctions, desmosomes). Sperm and some body cells bear undulipodia.
Examples	Bacteria, cyanobacteria (blue-green algae), actinobacteria (actinomycetes), fruiting and gliding bacteria	Amoebae, flagellates, ciliates, radiolarians, diatoms, slime molds, water molds, seaweeds	Yeasts, molds, puffballs, mushrooms	Mosses, ferns, flowering plants, gymnosperms	Sponges, worms, corals, mollusks, insects, mammals

*Occasional exceptions
†Undulipodia = eukaryote flagella or cilia; distinguished from bacterial flagella by containing tubulin microtubules in a 9 + 2 pattern.
SOURCE: Modified from Kaveski, S., Margulis, L., and Mehos, D. C. *The Science Teacher*, 50 (December 1983).

kingdom scheme also has had a long tradition in formal taxonomy; Linnaeus divided all known forms of life between plant and animal kingdoms. Even with the discovery of the diverse microbial world, the two-kingdom system persisted. Bacteria were placed in the plant kingdom, their rigid cell walls used as justification. Eukaryotic unicellular organisms with chloroplasts were also called plants. Fungi, too, fell under the plant banner, partly because they are sedentary, even though no fungi are photosynthetic and they have little in common structurally with green plants. In the two-kingdom system, unicellular creatures that move and ingest food—protozoa—were called animals. Microbes such as *Euglena* that move but are photosynthetic were claimed by both botanists and zoologists and showed up in the taxonomies of plant *and* animal kingdoms. Schemes with additional kingdoms were proposed, but none became popular with the majority of biologists until Robert H. Whittaker of Cornell University argued effectively for a five-kingdom system in 1969. Lynn Margulis, now at the University of Massachusetts, helped to promote the five-kingdom scheme and has proposed important modifications, some of which this text uses in its classification system. The five kingdoms are Monera, Protista, Plantae, Fungi, and Animalia (Table 24.1).

The five-kingdom system recognizes the two fundamentally different types of cell—prokaryotic and eukaryotic—and sets the prokaryotes apart from all eukaryotes by placing them in their own kingdom, Monera (the name Prokaryotae has also been proposed). The prokaryotes are bacteria, including cyanobacteria, formerly called blue-green algae.

Organisms of the other four kingdoms all consist of cells organized on the eukaryotic plan (see Chapter 7). The kingdoms Plantae, Fungi, and Animalia are multicellular eukaryotes, each kingdom defined by characteristics of structure and life cycle discussed in upcoming chapters (but introduced in Table 24.1). Plants, fungi, and animals also generally differ in their modes of nutrition (the criterion originally used by Whittaker). Plants are autotrophic in nutrition, making their food by photosynthesis. Fungi are heterotrophic organisms that are absorptive in nutrition. Most fungi are decomposers that live embedded in their food source, secreting digestive enzymes and absorbing the small organic molecules that are the products of digestion. Animals live mostly by ingesting food and digesting it within specialized cavities.

We are left with Kingdom Protista (also known as Protoctista) as something of a grab bag containing all eukaryotes that do not fit the definitions of plants, fungi, or animals. Most protists are unicellular forms, but in the version of the five-kingdom scheme used in this text, Protista also includes relatively simple multicellular organisms that are believed to be direct descendants of unicellular protists.

The most natural and unambiguous kingdom is Monera, its boundary defined by prokaryotic organization. Prokaryotes were the first forms of life and the only ones for at least two billion years. They remain tremendously important on the modern Earth. The next chapter addresses the diversity and history of prokaryotic life.

STUDY OUTLINE

1. Biological and geological history are interwoven. Geological events alter environments that affect biological evolution, and organisms in turn change the planet they inhabit.
2. Life has evolved over 3.5 billion years, punctuated by episodes that initiated new ways of life.

Formation of the Earth (pp. 511–512)

1. Our solar system formed from a cloud of matter between 4.6 and 5 billion years ago.
2. The infant Earth was a hot, molten mass that separated into layers of varying density.
3. Life began when the atmosphere had little oxygen and the mixture of gases comprised a reducing atmosphere.

The Antiquity of Life (pp. 512–513)

1. For the first few billion years after the Earth's crust cooled and solidified, only prokaryotes inhabited Earth.
2. The oldest available evidence for life appears in stromatolites containing fossils resembling bacteria dating back 3.5 billion years.

The Origin of Life (pp. 513–518)

1. One hypothesis of the origin of life is based on chemical evolution of protobionts, abiotically produced molecular droplets with distinctive chemical characteristics.
2. Laboratory experiments done under conditions simulating those of the primitive Earth have produced diverse organic molecules from inorganic precursors.
3. Small organic molecules polymerize when they are concentrated on hot sand, rock, or clay.
4. Organic molecules synthesized in the laboratory have spontaneously assembled into a variety of droplets—microspheres, liposomes, and coacervates—with some of the properties associated with life.
5. The first genes may have been abiotically produced RNA, whose base sequence served as a template for both alignment of amino acids in polypeptide synthesis and alignment of complementary nucleotide bases in a primitive form of self-replication.
6. Once genetic information became incorporated inside membrane-bounded compartments, protobionts would have acquired heritability and the ability to evolve as units.

The Kingdoms of Life (pp. 518–520)

1. The five-kingdom system classifies organisms as Monera, Protista, Plantae, Fungi, and Animalia.
2. All prokaryotes belong to kingdom Monera. The remaining four kingdoms, all eukaryotic, are defined by structure, life cycle, and mode of nutrition.

SELF-QUIZ

1. The "big bang" theory applies to the initial formation of
   a. the expanding universe
   b. our solar system
   c. our sun
   d. Earth and the planets
   e. the first protobionts

2. The *main* explanation for the lack of a continuing abiotic origin of life on Earth today is that
   a. there is not sufficient lightning to provide an energy source
   b. our oxidizing atmosphere is not conducive to the spontaneous formation of complex molecules
   c. there is much less visible light reaching Earth to serve as an energy source
   d. there are no molten surfaces on which weak solutions of organic molecules would polymerize
   e. all available niches are filled

3. Stromatolites are
   a. aggregates of abiotically produced organic molecules
   b. meteorites that contain amino acids and may have "seeded" Earth with organic molecules
   c. layers of clay that may have facilitated the polymerization of abiotically produced monomers
   d. a group of ancient prokaryotes
   e. banded domes of sediment that contain the oldest known fossils

4. Which of the following scientists are *incorrectly* paired with their theories or experiments?
   a. Fox—created polypeptides called proteinoids by dripping organic molecules onto hot sand
   b. Cech—discovered RNA catalysts, called ribozymes, that remove introns and catalyze RNA synthesis
   c. Crick—advanced theory of panspermia
   d. Oparin and Haldane—first produced organic compounds by simulating conditions of early Earth
   e. Whittaker—developed the five-kingdom system

5. Formation of microspheres, liposomes, and coacervates all require
   a. RNA
   b. a membrane potential
   c. self-assembly
   d. phospholipids
   e. primitive genes

6. The first genes may have developed when
   a. protobionts grew and divided
   b. short RNA strands self-replicated
   c. DNA molecules produced RNA molecules
   d. polypeptides became catalytic
   e. ribozymes acted as RNA catalysts

7. In the two-kingdom system of classification,
   a. prokaryotes were placed in the plant kingdom
   b. only multicellular organisms were considered animals
   c. the animal kingdom included all heterotrophs
   d. unicellular eukaryotes were placed only in the animal kingdom
   e. prokaryotes and eukaryotes were separated

8. In his laboratory apparatus, Stanley Miller synthesized
   a. proteins
   b. DNA
   c. amino acids
   d. protobionts
   e. proteinoids

9. Which gas was probably *least* abundant in the early atmosphere?
   a. H_2O b. O_2 c. NH_3 d. CO e. CO_2

10. Which of the following steps has *not* yet been accomplished by scientists studying the origin of life?
    a. abiotic synthesis of small RNA polymers
    b. abiotic synthesis of polypeptides
    c. formation of molecular aggregates with selectively permeable membranes
    d. formation of protobionts that use DNA to direct polymerization of amino acids
    e. abiotic synthesis of organic monomers

CHALLENGE QUESTIONS

1. Clifford Matthews of the University of Illinois claims that life on Earth likely originated from proteins that formed from a reaction of hydrogen cyanide polymers with water in the primordial seas. Amino acids, he claims, appeared after the first proteins were formed. Dissenting scientists argue that it is more logical to envision proteins being formed from simple amino acid building blocks than the other way around. Do you feel that this "logical argument" dismantles his hypothesis? Why or why not?

2. Describe the minimum structural, metabolic, and genetic equipment of a post-protobiont that you would consider to be a true primitive cell.

FURTHER READING

1. Amato, I. "RNA Offers Clue to Life's Start." *Science News*, June 17, 1989. The case for autocatalytic RNA as the first genetic material.
2. Cairns-Smith, A. G. "The First Organisms." *Scientific American*, June 1985. A provocative argument for self-replicating minerals as precursors of life.
3. Groves, D. I., Dunlop, J. S. R. "An Early Habitat of Life." *Scientific American*, October 1981. Stromatolites and their bearing on earliest life.
4. Kunzig, R. "Stardust Memories: Kiss of Life." *Discover*, March 1988. The evidence for cosmic origins of organic molecules.
5. Margulis, L., and Schwartz, K. V. *Five Kingdoms: An Illustrated Guide to the Phyla of Life on Earth*, 2d ed. New York: Freeman, 1987. A catalog of the world's living diversity.
6. Whittaker, R. H. "New Concepts of Kingdoms of Organisms." *Science* 163: 150–160, 1969. The proposal for five kingdoms.

25

Prokaryotes and the Origins of Metabolic Diversity

The history of prokaryotic life, or Kingdom Monera, is a success story spanning at least 3.5 billion years. Prokaryotes are commonly called **bacteria** (singular, *bacterium*), a term that encompasses the **cyanobacteria,** photosynthetic organisms formerly known as blue-green algae. The characteristic that defines the kingdom Monera and unifies its otherwise diverse members is the prokaryotic cell, which differs fundamentally from the more complex eukaryotic cells of the other four kingdoms of life. Prokaryotes were the earliest organisms, and they lived and evolved all alone on Earth for two billion years. They have continued to adapt and flourish on a changing Earth, and in turn they have helped to change the Earth.

Except by the criterion of size, prokaryotes still dominate the biosphere, outnumbering all eukaryotes combined. More bacteria inhabit a handful of dirt or the human mouth or skin than the total number of people who have ever lived. Prokaryotes are not only the most numerous organisms by far but also the most pervasive (Figure 25.1). Wherever we find life of any kind, prokaryotes are among the organisms present. Prokaryotic species thrive in habitats too hot, too cold, too salty, too acidic, or too alkaline for any eukaryote. Incomparably bountiful and omnipresent, the kingdom Monera is a dynasty that has endured and expanded through its billions of years of descent from the first cells that were the beginnings of all life.

Prokaryotes are individually microscopic, but their collective impact on the Earth and all of its life is gigantic. We rarely notice these ubiquitous microbes because they are usually invisible to the unaided eye. Illness

PROKARYOTIC FORM AND FUNCTION

Morphology of Prokaryotes

Monera is derived from the Greek word *moneres*, which means "single," referring here to the single-celled nature of the great majority of prokaryotes. Some species exist as aggregates of cells that stick together after dividing; these clusters are not truly multicellular, a term reserved for organisms with a division of labor among specialized cells. However, some cyanobacteria do have a simple multicellular organization of two or three types of specialized cells (see page 531).

We find a diversity of shapes among prokaryotes, the three most common being spherical (cocci), rod shaped (bacilli), and spiral (spirilla) (Figure 25.2). Determining the shapes of bacteria by microscopic examination is an important step in identifying different prokaryotes.

Most prokaryotic cells have diameters in the range of 1–10 μm, compared with 10–100 μm for the majority of eukaryotic cells. Size is not the only difference between the two major types of cells; the prokaryotic cell is unique in the organization of its genome, the relatively simple organization of its cytoplasm, and the structure of its cell wall and other surface components.

The Prokaryotic Genome

Prokaryotes are haploid and on the average have only about 1/1000 as much DNA as a eukaryotic cell. Recall that prokaryotes are named for their lack of true nuclei enclosed by membranes (see Chapter 7). In most prokaryotic cells, the DNA is concentrated as a snarl of fibers in a **nucleoid region** that appears less dense than the surrounding cytoplasm in electron micrographs. The term *genophore* is often used for the bacterial chromosome to distinguish it from eukaryotic chromosomes, which have a very different structure. The snarl of fibers is actually the bacterial chromosome, one double-stranded DNA molecule in the form of a ring. The DNA has very little protein associated with it. This contrasts with the eukaryotic genome, which consists of linear DNA molecules packaged along with proteins into a number of chromosomes characteristic of the species.

In addition to its one major chromosome, the prokaryotic cell may also have much smaller rings of DNA called plasmids, each consisting of only a few genes. In most environments, bacteria can survive without their plasmids because all essential functions are programmed by the chromosome. However, plasmids endow the cell with genes for resistance to antibiotics, for metabolism of unusual nutrients not present in the

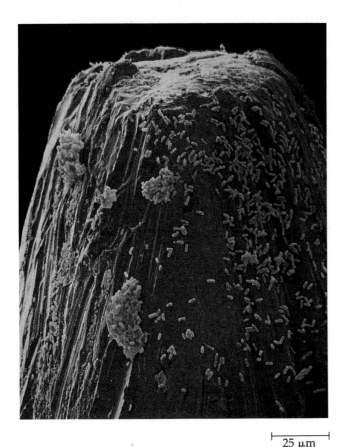

Figure 25.1
The size and ubiquity of bacteria. This scanning electron micrograph (artificially colored) demonstrates the size of bacteria in relation the tip of a pin.

$\vdash\!\!\dashv$ 25 μm

caused by bacterial infections occasionally reminds us that these tiny organisms exist, but the kingdom Monera is no rogues' gallery. Only a minority of prokaryotes cause disease in humans or any other organisms. The great majority of prokaryotic species are benign, and many are essential to all life on Earth. For example, bacteria decompose matter from dead organisms and return vital chemical elements to the environment in the form of inorganic compounds required by plants, which in turn feed animals. If for some reason the entire Kingdom Monera were to suddenly perish, the chemical cycles that sustain life would halt and the other four kingdoms would also be doomed. In contrast, prokaryotic life would undoubtedly persist in the absence of eukaryotes, as it once did for so long.

In this chapter, you will become more familiar with prokaryotes by studying their form and physiology, their diversity, and their ecological significance. The last section of the chapter considers the early history of prokaryotic life, during which all of the major modes of nutrition represented in the five kingdoms evolved.

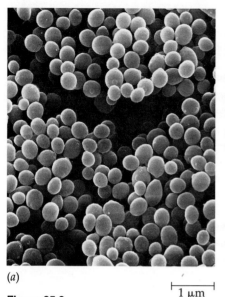
(a)

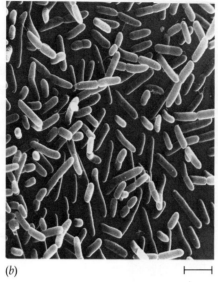
(b)

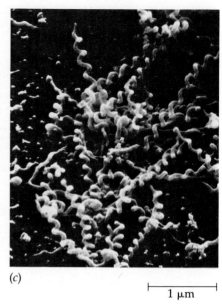
(c)

├─┤ 1 μm

├─┤ 1 μm

├────┤ 1 μm

Figure 25.2
Diversity of form in Kingdom Monera.
(a) Cocci, or spherical bacteria, occur singly or in pairs (diplococcus), in chains of many cells (streptococcus), and in clusters resembling bunches of grapes (staphylo- coccus) (SEM). (b) Rod-shaped bacilli (singular, bacillus) are most commonly solitary, but there are also forms with the rods arranged in chains (SEM). (c) Spiral- shaped bacteria are curved organisms resembling commas (vibrios), helical cells shaped like corkscrews (spirilla), or spiro- chetes, such as these cells (SEM).

normal environment, and for other special contingencies. Plasmids replicate independently of the main chromosome, and many can be readily transferred between partners when bacteria conjugate (see Chapter 17).

Although the broad outlines for replication of DNA and translation of genetic messages into proteins are alike for eukaryotes and prokaryotes, some of the details differ. For example, the prokaryotic ribosome is slightly smaller than the eukaryotic version and differs in its protein and RNA content. The disparity is great enough that selective antibiotics, including tetracycline and chloramphenicol, bind to the ribosomes of prokaryotes and block protein synthesis, while not inhibiting eukaryotic ribosomes.

Membranous Organization

The prokaryotic cell lacks the extensive compartmentalization by internal membranes characteristic of eukaryotes. Various prokaryotes do have specialized membranes to perform many of their metabolic functions, although these membranes may be invaginated regions of the plasma membrane. Figure 25.3 shows some examples.

The Cell Surface

Nearly all prokaryotes have cell walls external to their plasma membranes. The wall maintains the shape of

the cell, affords physical protection, and prevents the cell from bursting in a hypoosmotic environment (see Chapter 8). Like other walled cells, however, prokaryotes plasmolyze and may die in a hyperosmotic medium, which is why heavily salted meat such as jerky can be kept so long without being spoiled by bacteria.

The presence of a cell wall is one reason bacteria were grouped with plants in the old two-kingdom system. But the walls of prokaryotes and plants are analogous rather than homologous; they have a completely different molecular composition. Instead of cellulose, the staple of plant walls, bacterial walls contain a unique material called **peptidoglycan,** which consists of polymers of modified sugars cross-linked by short polypeptides that vary from species to species. The effect is a single, giant, laminated molecule enclosing and protecting the cell. External to this fabric are other substances that also differ from species to species.

Many antibiotics, including penicillins, inhibit the synthesis of cross-links in peptidoglycan and prevent the formation of a functional wall. These drugs are like selective bullets that cripple many species of infectious bacteria without adversely affecting humans and other eukaryotes, which do not make peptidoglycan.

One of the most valuable tools for identifying bacteria is the **Gram stain,** which can be used to separate many bacteria into two groups based on a difference in their cell walls. **Gram-positive** bacteria have the simpler walls, with a relatively large amount of peptidoglycan. The walls of **gram-negative** bacteria have

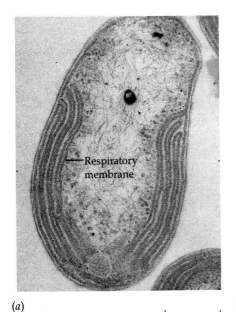

(a)

$\overline{\qquad}$ 0.25 μm

Figure 25.3
Specialized membranes of prokaryotes.
(a) These infoldings of the plasma membrane, reminiscent of the cristae of mitochondria, function in cellular respiration of aerobic bacteria (TEM). (b) Cyanobacteria

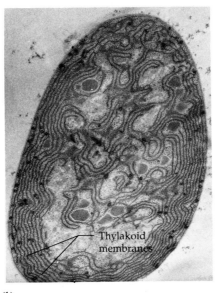

(b)

$\overline{\qquad}$ 1 μm

have thylakoid membranes, much like those in chloroplasts, that function in photosynthesis (TEM). (c) Mesosomes are infoldings of the plasma membrane. Some bacteriologists speculate that these struc-

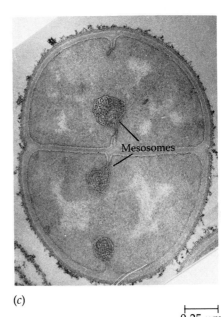

(c)

$\overline{\qquad}$ 0.25 μm

tures function in cell division. Others believe mesosomes are artifacts of sample preparation (TEM).

less peptidoglycan and are more complex in structure. An outer membrane on the gram-negative cell wall contains lipopolysaccharides (see Methods Box, p. 526).

Among pathogenic (from the Greek *pathos*, "suffering" and *gignomai*, "cause"), or disease-causing bacteria, gram-negative species are generally more threatening than gram-positive species. The lipopolysaccharides on the walls of gram-negative bacteria are often toxic, and the outer membrane helps protect the pathogens against the defenses of their hosts. Furthermore, gram-negative bacteria are commonly more resistant than gram-positive species to antibiotics because the outer membrane impedes entry of the drugs.

Many prokaryotes secrete sticky substances that form still another protective layer called a **capsule** outside the cell wall. Capsules provide additional protection and enable the organisms to adhere to a substratum on which they are growing. Gelatinous capsules glue together the cells of many prokaryotes that live as aggregates.

Another way bacteria adhere to one another or to some substratum is by means of surface appendages called **pili** (Figure 25.4). For example, *Neisseria gonorrhoeae*, the pathogen that causes gonorrhea, uses pili to fasten itself to mucous membranes of the host. Some pili are specialized for transferring DNA when bacteria conjugate (see Chapter 17).

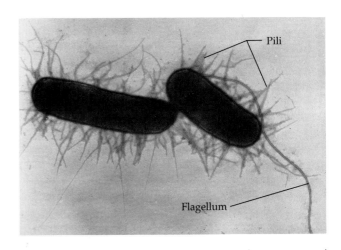

$\overline{\qquad}$ 0.25 μm

Figure 25.4
Pili. Bacteria use these appendages to attach to surfaces or other bacteria. Some pili are specialized for genetic transfer between partners during conjugation (TEM). (See Chapter 17.)

Motility of Prokaryotes

Motile bacteria use one of three mechanisms to move. Some species secrete slime and glide along this substratum. Spirochetes (see Figures 25.2c and 25.13) have **axial filaments,** or bundles of fine fibrils, that spiral

Methods: The Gram Stain

This method, named for Hans Christian Gram, a Danish physician who developed the technique in the late 1800s, distinguishes between two different kinds of bacterial cell walls. Bacteria are stained with a violet dye and iodine, rinsed in alcohol, and then stained again with a red dye. The structure of the cell wall determines the staining response.

Gram-positive bacteria (top), which have cell walls with a large amount of peptidoglycan, retain the violet dye (LM).

Gram-negative bacteria (bottom) have less peptidoglycan, which is located in a periplasmic space between the plasma membrane and an outer membrane. The violet dye used in the Gram stain is easily rinsed from gram-negative bacteria, and the cells then take up the red dye (LM).

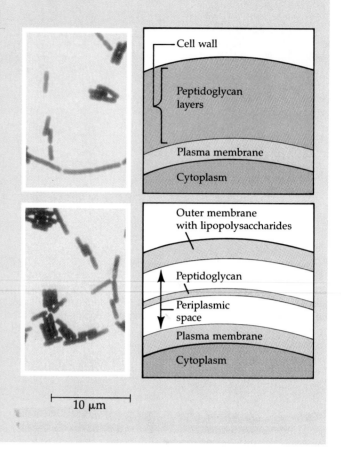

10 μm

around the cell under the outer sheath of the cell wall. The axial filaments cause the cell to rotate and move as a corkscrew (see Figure 25.13). Still other motile bacteria are equipped with flagella, which may be scattered over the entire cell surface or concentrated at one or both ends of the cell. The flagella of prokaryotes and eukaryotes differ entirely in structure and do not work the same way (see Chapter 7 for a review of eukaryotic flagella). The much thinner prokaryotic flagellum is attached to the surface of the cell rather than being a cytoplasmic extension, and it has a unique construction (Figure 25.5).

In an environment that is fairly uniform, flagellated bacteria wander randomly (Figure 25.6). In a heterogeneous environment, however, many bacteria are capable of **taxis**, which is movement oriented toward or away from some stimulus (from the Greek *taxis*, "to arrange or put in order," referring here to oriented movement, as in taxicab). With chemotaxis, for example, bacteria respond to chemical stimuli, perhaps moving toward food or oxygen (a positive chemotaxis) or away from some toxic substance (a negative chemotaxis). Several kinds of receptor molecules that detect

specific substances are located on the surfaces of chemotactic bacteria. Motile prokaryotes that are photosynthetic generally display a positive phototaxis, a behavior that keeps them in the light. There are even bacteria with tiny magnets that help the microbes distinguish up from down (Figure 25.7). Figure 25.8 illustrates the mechanism of taxis in prokaryotes.

Reproduction and Growth of Prokaryotes

Neither mitosis nor meiosis occur in Kingdom Monera; this is another fundamental difference between prokaryotes and eukaryotes. Prokaryotes reproduce asexually by the mode of cell division called binary fission, synthesizing DNA almost continuously. (Binary fission is described in Chapter 17.) A single bacterium in a favorable environment will give rise by repeated divisions to a colony of progeny (Figure 25.9). The term *growth* as applied to bacteria actually refers more to multiplication of cells and population growth than to enlargement of individual cells. The conditions for optimal growth—the best temperature, pH, salt

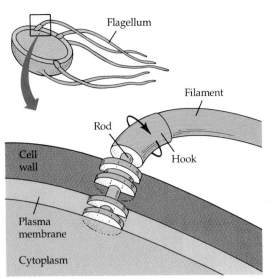

50 nm

Flagellum

Filament

Rod

Cell wall

Hook

Plasma membrane

Cytoplasm

Figure 25.5
How prokaryotic flagella work. The filament of a prokaryotic flagellum is a helix formed by chains of a protein called flagellin. The filament is attached to a hook, which is in turn inserted into a basal apparatus consisting of a system of rings that anchors the flagellum to the wall and plasma membrane while allowing it to turn freely. (The arrangement of rings shown in this electron micrograph and drawing is characteristic of gram-negative bacteria.) The filament is a propeller spun by the basal apparatus, which functions as a rotary motor. The motor is powered by the diffusion of hydrogen ions into the cell after they are pumped out of the cell at the expense of ATP.

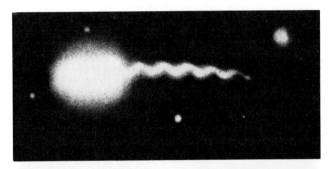

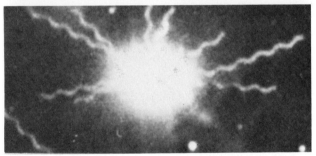

5 μm

Figure 25.6
Running and tumbling. The rotary motor at the base of the flagellum is reversible, spinning either clockwise or counterclockwise. The function of reversal has been studied in *Salmonella typhimurium,* a motile bacillus. When the many flagella that cover the cell spin counterclockwise, they spiral around one another and propel the cell forward, a movement called running (top). When the rotors reverse and spin clockwise, the flagella separate, and their uncoordinated movements cause the cell to stop running and begin tumbling (bottom). These somersaults randomize orientation of the cell, which runs off in a new direction when the rotor again reverses. There are mutant strains of *E. coli* that run continuously because their rotors spin only counterclockwise, and there are also tumbling mutants with flagella that only rotate clockwise.

Magnetite

0.5 μm

Figure 25.7
A bacterium with magnets. This marine species, *Aquaspirillum magnetotacticum,* contains crystals of an iron oxide called magnetite that function as tiny magnets within the cells. North of the equator, the magnetic field lines point downward into the Earth as they point north. Magnetotactic bacteria in the Northern Hemisphere are north seeking, following the magnetic field downward toward sediments rich in food. In the Southern Hemisphere, where the downward field lines aim southward, magnetotactic bacteria are south seeking (TEM).

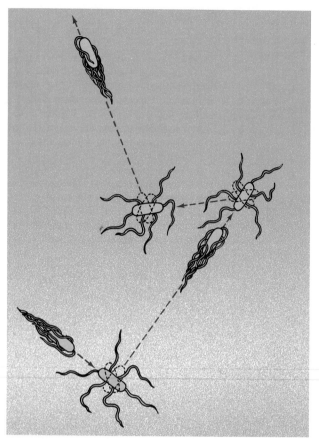

Figure 25.8
Taxis, or directed movement, by a bacterium. Motile prokaryotes move toward some substances and away from others. Bacteria have specific receptor molecules on their membranes that detect the presence of certain chemicals. But how do bacteria sense a concentration that varies over space? They do not seem able to compare concentrations at opposite ends of the cell. Instead, the bacterium uses some form of memory to compare present and past concentrations over short time spans as it moves along a chemical gradient. If the cell senses that it is moving toward an attractant or away from a repellent, its runs between tumbles are relatively long. Movement in the wrong direction causes the organism to tumble more often, shortening the duration of each run. The net effect is drift toward an attractant because the bacteria spend more time running up the gradient than down. It is not yet known how the receptors communicate with the rotary motors, but some chemical messenger may be involved.

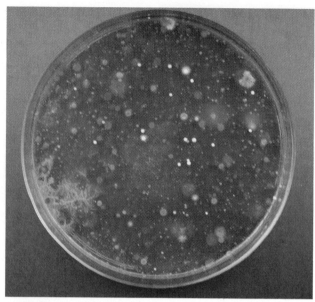

Figure 25.9
Bacterial colonies. Bacteria are grown in the laboratory by culturing them in Petri plates or test tubes that contain liquid or solid media of known composition. The media are sterilized to ensure that no unwanted microbes will grow, and then a sample of bacteria, sometimes just a single cell, is introduced. The plates or tubes are incubated at an appropriate temperature. When the bacteria are grown on solid media, colonies are usually large enough to be visible to the unaided eye after a day or two. Various methods are used for monitoring the growth rates of colonies. The size, shape, texture, and color of a colony provide clues to identification of the bacteria, as do the nutrients and physical conditions required for growth. Shown here are the colonies of several species of bacteria. Microscopic examination of bacteria from the colony is another step in identification.

concentrations, nutrient sources, and so on—vary according to species. Refrigeration retards food spoilage because most bacteria and other microorganisms grow only very slowly at such low temperatures.

The progression for bacterial growth is geometric: One cell divides to form 2, which divide again to produce a total of 4 cells, then 8, 16, and so on, the number doubling with each generation. Most bacteria have generation times in the range of one to three hours, but some species can double every 20 minutes in an optimal environment. If that growth rate were sustained, a single cell would give rise to a colony weighing a million kilograms in just 24 hours. However, growth of bacterial colonies both in the laboratory and nature is usually checked at some point when the cells exhaust some nutrient or the colony poisons itself with an accumulation of metabolic wastes.

The sexual cycle of meiosis and syngamy so important as a source of genetic variation in eukaryotes does not occur in the reproduction of prokaryotes. Genes are occasionally transferred from one bacterium to another by conjugation. Also, viruses, which are discussed in Chapter 17, are important in transferring genes by transduction. These processes, however, involve unilateral passage of a variable amount of DNA—nothing like the meiotic sex of eukaryotes, in which two parents each contribute homologous genomes to a zygote. Mutation is the major source of genetic variation in prokaryotes. Because generation times are measured in minutes or hours, a favorable mutation can be rapidly propagated to a large number of progeny.

Metabolic Diversity

Metabolic diversity is greater in the kingdom Monera than for all eukaryotes combined. Every type of nutrition observed in eukaryotes is represented among prokaryotes, plus some nutritional modes unique to the kingdom Monera. *Nutrition* refers here to how an organism obtains two resources: energy and a source of carbon for synthesizing organic compounds. Species that use light for energy are termed *phototrophs*. *Chemotrophs* obtain their energy from chemicals taken up from the environment. If an organism needs only the inorganic compound CO_2 as a carbon source, it is called an *autotroph*. *Heterotrophs* require at least one organic nutrient—glucose, for instance—as a source of carbon for making other organic compounds. Depending on how they obtain energy and carbon, bacteria can be divided into four major categories:

1. **Photoautotrophs,** including the cyanobacteria and other photosynthetic prokaryotes, harness light energy to drive the synthesis of organic compounds from carbon dioxide.

2. **Photoheterotrophs** can use light to generate ATP, but must obtain their carbon in organic form.

3. **Chemoautotrophs** need only CO_2 as a carbon source, but instead of using light for energy, these bacteria obtain energy by oxidizing inorganic substances such as hydrogen sulfide (H_2S), ammonia (NH_3), ferrous ions (Fe^{2+}), or some other chemical, depending on species.

4. **Chemoheterotrophs** are species that must consume organic molecules both for energy and as a source of carbon.

The majority of bacteria are chemoheterotrophs. This category includes **saprophytes,** which are decomposers that absorb their nutrients from dead organic matter, and **parasites,** which absorb their nutrients from the body fluids of living hosts.

The specific organic nutrients needed for growth vary extensively among chemoheterotrophic bacteria. Some species are very exacting in their requirements; for example, bacteria of the genus *Lactobacillus* will grow well only in a medium containing all 20 amino acids, several vitamins, and other organic compounds. Among species less fastidious in their nutritional needs, *E. coli* can grow on a medium containing glucose as

the only organic ingredient, and the metabolism of the organism is so versatile that many other compounds can substitute for glucose as the sole organic nutrient. There is such a diversity of chemoheterotrophs that almost any organic molecule can serve as food for at least some species. For example, bacteria capable of metabolizing petroleum are used to clean up oil spills. Those few classes of synthetic organic compounds, including some kinds of plastics, that cannot be broken down by any chemoheterotrophs are said to be *nonbiodegradable*.

Another metabolic variation in Kingdom Monera is in the effect that oxygen has on growth (Chapter 9). **Obligate aerobes** use oxygen for cellular respiration and cannot grow without it. **Facultative anaerobes** will use oxygen if it is present but can also grow by fermentation or anaerobic respiration in an oxygen-free environment. **Obligate anaerobes** cannot use oxygen and are poisoned by it. These variations are among the criteria used in classifying the prokaryotes.

The metabolic diversity in modern prokaryotes can be associated with metabolic adaptations of prokaryotes in the past. In the last section of this chapter, we consider how various forms of metabolism developed on the early Earth, an excellent example of the way geological change and biological evolution affected each other. First, let us look at the diverse representatives of the kingdom Monera.

THE DIVERSITY OF PROKARYOTES

The Status of Prokaryotic Taxonomy

The prokaryotic cell is profoundly different from the eukaryotic cell, making Monera the most clearly defined of all kingdoms. But grouping the more than 10,000 known species of prokaryotes into subordinate taxa within the kingdom Monera on the basis of phylogenetic affinities has, until recently, been an exercise in futility. Prokaryotes diversified extensively so long ago that genealogical ties between groups are hazy. Help from the fossil record is very limited because most microfossils found in Precambrian rocks are little more than shadowy outlines of cells. Moneran taxonomy used by practicing microbiologists, especially in medical labs, remains more a classification of convenience to aid in identifying bacteria than an attempt to group according to evolutionary relationships.

Molecular systematics offers the best hope for developing a moneran classification that reflects phylogeny (see Chapter 23). Comparing amino acid sequences of homologous proteins and base sequences of DNA and RNA measures genetic similarity and helps establish the timing of phylogenetic branching along

Figure 25.10

Extreme halophiles. Members of the ancient prokaryotic group known as the archaebacteria, these organisms live in extremely saline waters. The reddish color of these seawater evaporating ponds at the edge of San Francisco Bay results from a dense growth of extreme halophiles that thrive in the ponds when the water reaches a salinity of 15%–20% (before evaporation, the salinity of sea water is about 3%). The ponds are used for commercial salt production and the halophilic bacteria are harmless.

the moneran tree. Comparisons of ribosomal RNAs have been particularly illuminating. One significant discovery from molecular systematics is that prokaryotes split into at least two divergent lineages very early in the history of life. One branch produced the **archaebacteria,** with the only survivors being a few genera of prokaryotes confined to extreme environments that may resemble habitats on the early Earth. A second branch, the **eubacteria,** includes nearly all contemporary prokaryotes. Here we will sample the diversity of the groups archaebacteria and eubacteria. Keep in mind that the organisms discussed and illustrated here are only representatives of the multitude of prokaryotes that inhabit Earth.

Archaebacteria

The name of this group refers to the antiquity of its origin from the earliest prokaryotes (from the Greek *archaio*, "ancient"). The archaebacteria have many distinctive traits, as should be expected of a taxon that has followed a separate evolutionary path for so long. Their cell walls lack peptidoglycan, a component of all walled eubacteria. The plasma membranes of archaebacteria have a lipid composition unlike any other organisms. RNA polymerase and a ribosomal protein of archaebacteria are eukaryote-like and very different from those of eubacteria.

Most archaebacteria live in extreme environments too harsh for other forms of modern life but which may have been prevalent habitats during the early evolution of prokaryotes. For example, archaebacteria have been found living in the hot water at the openings of deep sea vents. Archaebacteria, a lineage nearly as old as life itself, survives as three subgroups: the

methanogens, the extreme halophiles, and the thermoacidophiles.

Methanogens are named for their unique form of energy metabolism, in which H_2 is used to reduce CO_2 to methane (CH_4). Methanogens are among the strictest of anaerobes, poisoned by oxygen. They live in swamps and marshes where other microbes have consumed all the oxygen; the methane that bubbles out at these sites is known as marsh gas. Methanogens are also important decomposers employed for sewage treatment. Some farmers have experimented with using these microbes to convert garbage and dung to methane, a valuable fuel. Other species of methanogens inhabit the anaerobic environment within the guts of animals, playing an important role in the nutrition of cattle, termites, and other herbivores that subsist mainly on a diet of cellulose.

The extreme halophiles live in such saline places as the Great Salt Lake and the Dead Sea (*halophile* means "salt-loving"). Some species merely tolerate salinity, whereas others actually require an environment 10 times saltier than sea water to grow (Figure 25.10). Colonies of halophiles form a pink scum that owes its color to a pigment, bacteriorhodopsin. Bacteriorhodopsin is built into the plasma membrane, where it absorbs light and uses the energy to pump hydrogen ions out of the cell. The gradient of hydrogen ions then drives the synthesis of ATP (see Chapter 8). This is the simplest mechanism of photophosphorylation known, and the halophiles are being studied as model systems of solar energy conversion.

As their name implies, the thermoacidophiles thrive where it is both hot and acidic, a double whammy for nearly all other organisms. The optimal conditions for these archaebacteria are temperatures of 60–80°C and pH between 2 and 4. *Sulfolobus* inhabits hot sulfur

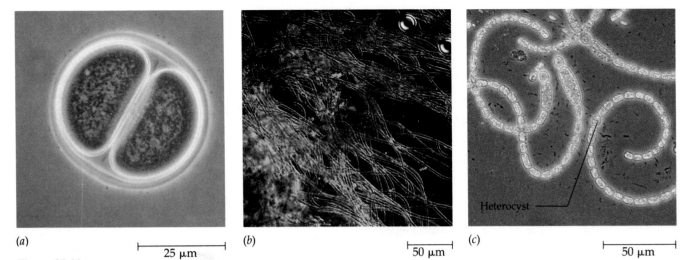

(a)   25 μm
(b)   50 μm
(c)   50 μm

Heterocyst

Figure 25.11
Cyanobacteria. (a) *Chroococcus* (LM). (b) *Oscillatoria*, a filamentous form (LM).
(c) *Nostoc*, another filamentous form. The spherical cells are heterocysts, specialized for
nitrogen fixation (LM).

springs in Yellowstone National Park, obtaining its energy by oxidizing sulfur (see Figure 6.17).

As measured by evolutionary molecular clocks, such as base substitutions in nucleic acids, the archaebacteria and eubacteria did indeed diverge when life was still very young. On the strictly cladistic grounds of classifying organisms only according to the timing of evolutionary branching, some bacterial taxonomists advocate elevating the archaebacteria to the status of a separate kingdom, separate from the eubacteria. The classification scheme used in this text, however, considers the archaebacteria to be a separate branch *within* Monera, a kingdom of life defined by the prokaryotic organization common to archaebacteria and their distant relatives, the eubacteria.

Eubacteria

It would be excessive in a general biology text to present an exhaustive survey of the eubacteria, but a sampling of some of the groups will demonstrate the diversity of these microbes, especially in their mode of nutrition.

Cyanobacteria   The common name for these bacteria, blue-green algae, is a relic of the two-kingdom system. "Alga" (singular) is a generic label used informally for any aquatic organism of a simple plantlike nature. But cyanobacteria are undeniably prokaryotic in organization and in the five-kingdom scheme belong with the other bacteria in the kingdom Monera.

The cells of cyanobacteria are generally larger than other prokaryotes. There are solitary forms as well as colonial and multicellular species, usually filaments of cells embedded in a sticky matrix (Figure 25.11). Flagella are absent, and the motile forms glide. Most cyanobacteria inhabit fresh water, but species also live in seas and damp soils.

Cyanobacteria are photoautotrophs that are very plantlike in their mechanism of photosynthesis. The major pigment in the conversion of light energy to chemical energy is chlorophyll *a*, just as in plants. And cyanobacteria use a tandem of two photosystems to transfer electrons from water to $NADP^+$, releasing O_2 as a waste product, again like plants (see Chapter 10). Of course, prokaryotic cells do not contain chloroplasts, but cyanobacteria do possess thylakoids, sacs bounded by membranes and often arranged in stacks reminiscent of the grana of chloroplasts (see Figure 25.3). Chlorophyll and other components of the light reactions are built into the thylakoid membrane. The blend of chlorophyll and accessory red and blue pigments gives many cyanobacteria a blue-green color that is the basis for the common name of these organisms, though yellow, red, green, brown, and black species also exist, their color depending on the amounts and varieties of the different pigments present.

Some filamentous cyanobacteria have specialized cells called **heterocysts**; these cells contain nitrogenase, an enzyme that reduces atmospheric N_2 to NH_3 (ammonia), which can then be incorporated into amino acids and other nitrogenous organic compounds. The assimilation of nitrogen from N_2 into organic material is called **nitrogen fixation.** The only organisms capable of fixing nitrogen are certain groups of prokaryotes, including the cyanobacteria.

In terms of nutrition, the nitrogen-fixing cyanobacteria are among the most self-sufficient of all organisms. They require only light energy, CO_2, N_2, water, and some minerals to grow.

As the first organisms to release oxygen to the atmo-

Chapter 25: Prokaryotes and the Origins of Metabolic Diversity   **531**

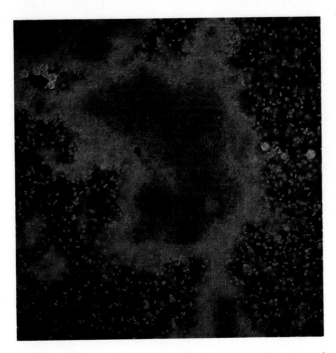

Figure 25.12
Phototrophic bacteria. Colony of *Chromatium*, a purple sulfur bacterium (LM).

| | 100 μm |

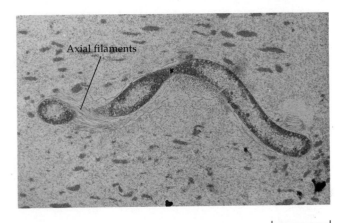

Axial filaments

| | 10 μm |

Figure 25.13
A spirochete. In this transmission electron micrograph of *Cristispira*, you can see the axial filaments wrapped around the cell beneath the outer sheath of the cell wall.

sphere, ancient cyanobacteria played a prominent role in the evolution of Earth and its life, as we shall see in the last section of this chapter.

Phototrophic Bacteria   The green sulfur bacteria and purple sulfur bacteria are photoautotrophs, but they are much less plantlike than cyanobacteria in their photosynthetic equipment (Figure 25.12). They have only one photosystem instead of two and use it to reduce $NADP^+$ with electrons extracted from H_2S, not H_2O. Therefore, these bacteria do not release O_2, and in fact most species can grow only in anaerobic environments such as in pond, lake, and ocean sediments. The chlorophyll of green sulfur bacteria is quite similar to chlorophyll *a*, but purple sulfur bacteria have a pigment called bacteriochlorophyll, which absorbs at longer wavelengths than other chlorophylls.

Pseudomonads   The diverse genus *Pseudomonas* is represented by species in nearly all aquatic and soil habitats. As a group, pseudomonads are the most versatile of all chemoheterotrophs, capable of metabolizing all the usual organic nutrients plus some that no other organisms can use as food. Some soil species decompose pesticides and other synthetic compounds. Pseudomonads, which have been described as microbial weeds, can also be pests. Their ability to feed on minute concentrations of unusual carbon sources enables them to invade hot tubs, drug solutions, and even antiseptic solutions meant to prevent bacterial growth.

Spirochetes   Shaped like helices, spirochetes have a corkscrewlike movement effected by their unique axial filaments (Figure 25.13). This group includes some giants by prokaryotic standards—cells 0.5 mm long, though too thin to be seen without a microscope.

Among the spirochetes are both free-living saprophytes and parasites. The most infamous is *Treponema pallidum*, the pathogen that causes syphilis. It dies quickly in air and relies on contact of moist mucous membranes for transmission between people. Lyme disease is caused by another spirochete, *Borrelia burgdorferi*, which is transmitted by ticks. Antibiotics are very effective against both *Treponema* and *Borrelia*.

Endospore-Forming Bacteria   These bacteria survive periods of harsh conditions by producing endospores, dehydrated cells with thick walls (Figure 25.14). The original cell replicates its chromosome, and one copy becomes surrounded by a durable wall. The outer cell disintegrates, but the endospore it contained is a resistant cell that survives all sorts of trauma, including lack of nutrients and water, extreme heat or cold, and most poisons. Some endospores may remain dormant for centuries. They hydrate and revive to the vegetative state only in hospitable environments.

Boiling water is not hot enough to kill most endospores in a reasonable length of time. To sterilize media, glassware, and utensils in the laboratory, microbiologists use an appliance called an autoclave, a pressure cooker that kills even endospores by heating to temperatures higher than 120°C. The food-canning industry must also take precautions to kill endospores of dangerous bacteria such as *Clostridium botulinum*, which produces the potentially fatal disease botulism.

Enteric Bacteria   These bacteria inhabit the intestinal tracts of animals. Some enteric bacteria are permanent residents that are harmless under normal circumstances. One example that has already been

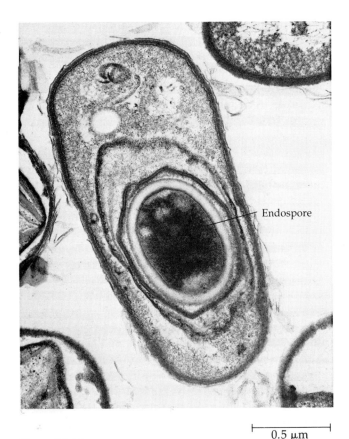

Endospore

0.5 μm

Figure 25.14
An endospore. A nearly mature spore has formed within the mother cell, a bacterium of genus *Bacillus* (TEM).

mentioned is *E. coli*. Detection of this organism in water supplies is a sign of contamination with feces. Pathogenic enterics, such as nearly all members of the genus *Salmonella*, are not normally present in animals. *Salmonella typhi* causes typhoid fever, and several other species of *Salmonella* cause food poisoning.

Rickettsias and Chlamydias Among the smallest bacteria, rickettsias are parasites that can grow only within cells of another organism (Figure 25.15). Nearly all are transmitted to humans by the bites of ticks and insects. Rocky Mountain spotted fever and typhus are two of the diseases caused by rickettsias. Similar to rickettsias are the chlamydias, one of which is the agent of nongonococcal urethritis (NGU), now suspected to be the most common sexually transmitted disease in the United States.

Mycoplasmas Having diameters of only 100–250 nm, the mycoplasmas are even more minute than rickettsias; in fact, they are believed to be the smallest of all cells (Figure 25.16). They are also unique as the only prokaryotes that lack cell walls.

Actinomycetes These bacteria form colonies of branching chains that resemble the filamentous bodies of fungi (Figure 25.17). In fact, the actinomycetes were once mistaken for fungi, which explains their name (from the Greek *aktis*, "ray" and *mykes*, "fungus").

Two actinomycetes of medical importance are a species that causes tuberculosis and another that causes leprosy. But most actinomycetes are free-living organisms among the organic litter of soil, and their chemical secretions are partly responsible for the "earthy" odor of rich soil. Soil bacteria of the genus *Streptomyces* prevent encroachment by other microbes by producing streptomycin and many other antibiotics. Pharmaceutical companies culture various species of *Streptomyces* to produce many commercial antibiotics.

Myxobacteria These gliding bacteria form the most elaborate colonies of all prokaryotes, secreting a slimy substratum on which they glide through soil—hence

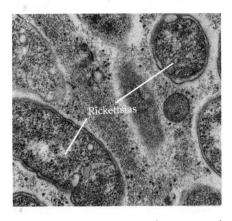

Rickettsias

0.5 μm

Figure 25.15
Rickettsias. A rickettsial infection of tissue cells in a tick (TEM).

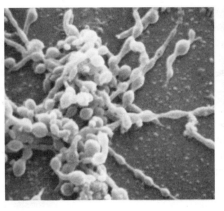

2.5 μm

Figure 25.16
Mycoplasmas. The smallest bacteria, mycoplasmas lack cell walls and inhabit the body fluids of plant and animal hosts. This species, *Mycoplasma pneumoniae*, causes one type of bacterial pneumonia (SEM).

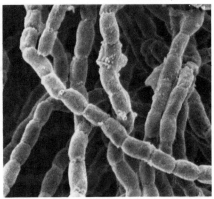

2.5 μm

Figure 25.17
Actinomycetes. *Streptomyces* growing in a compost pile (SEM).

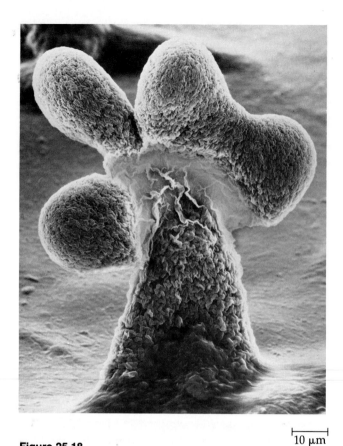

Figure 25.18
Myxobacteria. This mass of cells has formed a fruiting body that will release spores (SEM).

10 μm

the name myxobacteria (from the Greek *myxa*, "mucus"). When the soil dries out or food becomes scarce, the cells congregate into a mass that erects a bulbous stalk called a fruiting body (Figure 25.18). Within the fruiting body, which is often brightly colored and may be as large as a millimeter in diameter, durable spores form by various mechanisms, depending on species. The spores are released, returning to a vegetative state and founding new colonies when conditions are favorable.

THE IMPORTANCE OF PROKARYOTES

One of the themes of this unit is changing life on a changing planet—the interactions of geological history and evolving biological forms. Organisms so pervasive, abundant, and diverse as the prokaryotes have a tremendous impact on the Earth and all of its inhabitants. Here we consider some of these relationships.

Prokaryotes and Chemical Cycles

Not too long ago, in geological terms, the atoms of the organic molecules in our bodies were parts of the inorganic compounds of soil, air, and water, as they will be again. Ongoing life depends on the recycling of chemical elements between the biological and physical components of ecosystems. Prokaryotes are indispensable links in these chemical cycles (which will be discussed in detail in Chapter 49). Along with fungi, bacteria decompose the organic matter of dead organisms and the wastes of live ones, returning the elements to the environment in inorganic forms available for reassimilation by other organisms. If it were not for such **decomposers,** carbon, nitrogen, and other elements essential to life would become locked in the organic molecules of corpses and feces. Prokaryotes are also corridors through which these elements reenter the living world from the reservoirs of air, soil, and water. Autotrophic bacteria fix CO_2, supporting food chains through which organic nutrients pass from the bacteria to bacteria eaters and then on to secondary consumers. Cyanobacteria not only synthesize food but also supplement plants in restoring oxygen to the atmosphere. As previously mentioned, certain prokaryotes are the only organisms capable of fixing nitrogen, stocking the soil and water with nitrogenous minerals plants can use to make proteins. And when plants and the animals that eat them die, it is soil prokaryotes that return the nitrogen to the atmosphere to keep the cycle running. (The nitrogen cycle is discussed in Chapters 33 and 49; see Figure 49.9.) All life on Earth depends on the prokaryotes and their unparalleled metabolic diversity.

Symbiotic Bacteria

Symbiosis, which means "living together," is an ecological relationship between organisms of two different species that are in direct contact. In most cases, the smaller organism, the **symbiont,** lives within or on the larger organism, the **host.** There are three categories of symbiotic relationships: **mutualism, commensalism,** and **parasitism.** In a mutual relationship, both host and symbiont benefit. A commensal relationship is one in which the symbiont receives benefits while neither harming nor helping the host in any significant way. In parasitism, the symbiont, called a parasite in this case, benefits at the expense of the host.

The kingdom Monera is represented extensively in all three types of symbiosis. For example, plants of the legume family (peas, beans, alfalfa, and others) have lumps on their roots called **nodules,** which are home to symbiotic bacteria that fix nitrogen used by the host, while the plant reciprocates with a steady supply of sugar and other organic nutrients. The bacteria inhabiting the human body consist mostly of commensal species, but some species engage in mutualism with their hosts. For instance, fermenting bacteria living in the vagina produce acids that maintain a pH between 4.0 and 4.5, which suppresses the growth of yeast and other potentially harmful microorganisms.

Bacteria and Disease

Bacteria are everywhere, and exposure to pathogenic ones is a certainty. Most of us are well most of the time because our defenses check the growth of harmful bacteria and other pathogens to which we are exposed. Occasionally, the balance shifts in favor of a pathogen, and we become ill. To be pathogenic, a parasite must invade the host, resist internal defenses well enough to begin growing, and then harm the host in some way. It is estimated that about half of all human disease is caused by bacteria.

Bacteria need not be exotic intruders to be pathogenic. Some pathogens are **opportunistic,** meaning they are normal residents of the human body that inflict illness only when defenses have been weakened by such factors as poor nutrition or a recent bout with flu. For example, *Streptococcus pneumoniae* lives in the throats of most healthy people, but this opportunist can multiply and cause pneumonia when the host's defenses are down.

Pasteur, Lister, and other scientists began linking disease to pathogenic microbes in the late 1800s. The first to actually connect certain diseases to specific bacteria was Robert Koch, a German physician who determined the bacteria responsible for anthrax and tuberculosis. His methods established four criteria, now called **Koch's postulates,** which are still the guidelines for medical microbiology. To substantiate a specific pathogen as the cause of a disease, the researcher must (1) find the same pathogen in each diseased individual investigated, (2) isolate the pathogen from a diseased subject and grow the microbe in a pure culture, (3) induce the disease in experimental animals by transferring the pathogen from the culture, and (4) isolate the same pathogen from the experimental animals after the disease develops. The postulates can be applied for most pathogens, but prudent exceptions must be made for some cases. For example, no one has yet been able to culture the bacterium that causes syphilis (*Treponema pallidum*) on artificial media, but the volume of circumstantial evidence associating organism with disease leaves no doubt in this case.

Some bacteria disrupt the physiology of the host by their actual growth and invasion of tissues, but pathogenic bacteria more commonly cause illness by producing toxins. These poisons are of two types: exotoxins and endotoxins. **Exotoxins** are proteins secreted by the bacterial cell. The exotoxin can produce symptoms even without the bacteria actually being present; examples include the botulism toxin produced by *Clostridium botulinum* mentioned on page 532. Exotoxins are among the most potent poisons known; one

gram of botulism toxin would be sufficient to kill a million humans. In contrast to exotoxins, **endotoxins** are not secreted by the pathogens but are instead components of the outer membranes of certain gram-negative bacteria. Endotoxins induce the same general symptoms—fever and aches—regardless of the species of bacteria, whereas exotoxins elicit specific symptoms.

With the discovery in the nineteenth century that "germs" cause disease, people responsible for public health took steps to upgrade hygiene. Sanitation measures played a significant role in reducing infant mortality and extended life expectancy dramatically in developed countries. In the past few decades, medical technology has increased its success at combating bacterial disease with a variety of antibiotics. Bacterial disease has certainly not been conquered, but its decline over the past century, probably due more to public health policies than to "wonder drugs," has so far been the greatest achievement of biomedical research and its application.

Putting Bacteria to Work

Humans have learned many ways of exploiting the diverse metabolic capabilities of prokaryotes, both for scientific research and for practical purposes. Much of what we know about metabolism and molecular biology has been learned in laboratories using bacteria as relatively simple model systems. In fact, *Escherichia coli*, the "white rat" of so many research labs, is the best understood of all organisms. Bacteria are used to digest organic wastes at sewage treatment plants. The chemical industry grows immense cultures of bacteria that produce acetone, butanol, and several other products. Pharmaceutical companies culture bacteria that make vitamins and antibiotics. The food industry uses bacteria to convert milk to yogurt and various kinds of cheese. Recombinant DNA techniques also promise a new era in the commercial importance of prokaryotes.

THE ORIGINS OF METABOLIC DIVERSITY

All forms of nutrition and nearly all metabolic pathways evolved in Kingdom Monera before eukaryotes arose. Reasonable speculations about the early history of prokaryotes and the origins of metabolic diversity are now possible. The hypothetical scenario described here is based on inferences from molecular systematics, comparison of energy metabolism among extant prokaryotes, and evidence from the geological record about conditions on the early Earth.

Nutrition of the Earliest Prokaryotes

The first prokaryotes, which originated at least 3.5 billion years ago, were probably chemoheterotrophs that absorbed free organic compounds generated in the primordial seas by abiotic synthesis (see Chapter 24). The universal role of ATP as an energy currency in all modern organisms implies that prokaryotes became fixed on use of the molecule very early. As the bacteria began to deplete the supply of free ATP, natural selection would have favored cells with enzymes that could regenerate ATP from ADP using energy extracted from other organic nutrients that were still available. The result may have been the step-by-step evolution of glycolysis, a metabolic pathway that breaks organic molecules down to simpler waste products and uses the energy to generate ATP by substrate phosphorylation (see Chapter 9). Glycolysis is the only metabolic pathway common to nearly all modern organisms, suggesting great antiquity.

Further, glycolysis does not require O_2, and indeed the ancient atmosphere had very little of that gas. Fermentation, in which electrons extracted from nutrients during glycolysis are transferred to organic recipients, became a way of life on the anaerobic Earth. The archaebacteria and other obligate anaerobes that live today by fermentation deep in the soil or in stagnant swamps are believed to have forms of nutrition most like that of the original prokaryotes.

The Origin of Electron Transport Chains

Fermenting bacteria produce a variety of waste products, mostly organic acids that are excreted into the surrounding medium. Transmembrane proton pumps may have functioned originally to help early prokaryotes regulate their internal pH in ponds and pools that were becoming acidified by organic wastes (Figure 25.19). But the cell would have to spend a large portion of its ATP to drive the proton pumps. The first electron transport chains may have saved ATP by coupling oxidation of organic acids to the transport of H^+ out of the cell. In some bacteria, electron transport systems efficient enough to extrude more H^+ than necessary for regulating pH evolved, and these cells could use the inward gradient of the ions to reverse the proton pump, which now generated ATP rather than consuming it. This type of energy metabolism, called anaerobic respiration, persists in some modern bacteria, including those that make their ATP by passing electrons down transport chains from organic substrates to sulfur dioxide. But the basic chemiosmotic mechanism of using proton gradients to transfer energy from redox reactions to ATP synthesis is common to cells of all five kingdoms of life.

Stage 1. Proton pumps, driven by ATP, are used to regulate cellular pH by extruding hydrogen ions.

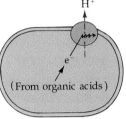

Stage 2. Electron transport chains take over the function of pH regulation by using the oxidation of organic acids to drive proton pumps.

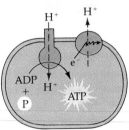

Stage 3. Electron transport chains become efficient enough to generate a gradient of hydrogen ions that can be used to drive ATP synthesis.

Figure 25.19
Evolution of electron transport chains.

The Origin of Photosynthesis

Life confronted its first energy crisis when the supply of free ATP dwindled. It faced its second when the fermenting prokaryotes consumed organic nutrients faster than the compounds could be replaced by abiotic synthesis. An organism that could make its own organic molecules from inorganic precursors would have had a tremendous advantage. Some prokaryotes had pigments and photosystems that used light to drive electrons from hydrogen sulfide (H_2S) to $NADP^+$, generating reducing power that could be used to fix CO_2. They probably coopted components of electron transport chains that had previously functioned in anaerobic respiration and continued to use the chains to power ATP synthesis as well as to provide reducing power. The modern bacteria with nutrition most like the early photosynthetic prokaryotes are believed to be the green sulfur bacteria and purple sulfur bacteria, anaerobes that produce no O_2.

Some of the photosynthetic bacteria were eventually able to use H_2O instead of H_2S or other compounds as a source of electrons and hydrogen for fixing CO_2. It is harder to oxidize water, a problem solved by two photosystems that operated in series to boost electrons from water to $NADP^+$. This capability evolved in the first cyanobacteria. Able to make organic compounds from water and CO_2, they flourished and changed the world by releasing O_2 as a by-product of their photosynthesis.

The Oxygen Revolution and the Origins of Respiration

Cyanobacteria evolved at least 2.5 billion years ago, living along with other bacteria in colonies that built the stromatolites that have been found all over the world. In marine sediments of about that same age are the banded iron formations, red layers rich in iron oxide and valuable as a source of iron ore today. Perhaps the sediments formed over a period when early cyanobacteria released oxygen that reacted with dissolved iron ions, which precipitated as iron oxide. This reaction would have prevented any accumulation of free O_2 for perhaps a few hundred million years until precipitation exhausted the dissolved iron. Only then would the seas become saturated with O_2, which began gassing out and accumulating in the atmosphere. Beginning about two billion years ago, terrestrial rocks rich in iron were rusted red by oxidation with atmospheric O_2.

The gradual change to a more oxidizing atmosphere created a crisis for Precambrian prokaryotes, as oxygen attacks the bonds of organic molecules. The corrosive atmosphere probably caused the extinction of many bacteria unable to cope. Other species survived in habitats that remained anaerobic, where we find their descendants living today as obligate anaerobes. The evolution of antioxidant mechanisms enabled other bacteria to tolerate the rising oxygen levels. Among photosynthetic prokaryotes, some species went a step further than mere oxygen tolerance to actually using the oxidizing power of O_2 to pull electrons from organic molecules down existing transport chains. Thus, aerobic respiration may have originated as modifications of electron transport chains borrowed from photosynthesis. The purple nonsulfur bacteria, photoheterotrophs, still use an electron transport system that is a hybrid of photosynthetic and respiratory equipment. Several other bacterial lineages gave up photosynthesis and reverted to chemoheterotrophic nutrition, their electron transport chains adapted to function exclusively in aerobic respiration.

Thus, on an ancient Earth inhabited only by prokaryotes, all the diverse forms of nutrition and metabolism evolved. Most subsequent evolutionary breakthroughs were structural rather than metabolic. Perhaps the most significant was the origin of eukaryotic cells from prokaryotic ancestors, a juncture in the history of life that is considered in the next chapter.

STUDY OUTLINE

1. All prokaryotes are assigned to the kingdom Monera, which consists of the bacteria, including cyanobacteria.
2. Prokaryotes were the first organisms, predating eukaryotes by two billion years and persisting today as the most numerous and pervasive of all living things.

Prokaryotic Form and Function (pp. 523–529)

1. Prokaryotes are generally single-celled organisms, although some occur as aggregates or simple multicellular forms.
2. The three most common prokaryotic shapes are spherical (cocci), rod-shaped (bacilli), and spiral (spirilla) forms.
3. The prokaryotic genome is haploid, consisting of a single circular DNA molecule in a nucleoid region unbounded by a membrane. Many species also possess smaller separate rings of DNA called plasmids, which code for special metabolic pathways and resistance to antibiotics.
4. Prokaryotic cells are not compartmentalized by endomembranes. However, the plasma membrane may be invaginated to provide internal membrane surface for specialized functions.
5. Nearly all prokaryotes have external cell walls, which protect and shape the cell and prevent osmotic bursting. Cell walls typically contain peptidoglycan, a unique polymer. Gram positive and gram negative bacteria differ in the structure of their walls and other surface layers.
6. Many species secrete sticky substances that form capsules. Some have surface appendages called pili outside the cell wall. Both structures help the cells adhere to one another, and some pili are specialized for conjugation.
7. Prokaryotes either glide on slime secretions or propel themselves by axial filaments or flagella attached to the cell surface. Directional movement, or taxis, occurs in some species in response to chemicals, light, and/or magnetism.
8. Bacteria reproduce asexually by binary fission. The geometric growth of a bacterial colony is usually checked by nutrient exhaustion or accumulation of toxic metabolic wastes.
9. In the absence of meiosis, genetic variation occurs through mutation or gene transfer by bacterial conjugation or viral transduction. Variations are spread rapidly owing to the frenetic pace of binary fission.
10. Prokaryotes are the most metabolically diverse organisms on Earth. Photoautotrophs use light energy and chemoautotrophs use inorganic substances to synthesize their organic compounds from carbon dioxide. Photoheterotrophs require organic molecules for metabolic processes and synthesize ATP using light energy. Most bacteria are chemoheterotrophs, which require organic molecules as a source of both energy and organic carbon.

11. The ability or inability to survive in the presence of oxygen also reflects variation in metabolism. Obligate aerobes require oxygen, obligate anaerobes are poisoned by it, and facultative anaerobes can survive with or without oxygen.

The Diversity of Prokaryotes (pp. 529–534)

1. Because microfossils of ancient prokaryotes are too sketchy to reveal definitive taxonomic relationships, molecular systematics is most often used to unravel moneran phylogeny.
2. Comparisons of selected macromolecules suggest an early split of the prokaryotes into the archaebacteria and the eubacteria.
3. Modern archaebacteria, the methanogens, the extreme halophiles, and the thermoacidophiles, live in harsh environments reminiscent of conditions on the primordial Earth.
4. The eubacteria comprise almost all contemporary prokaryotes and include the following major groups.
5. Cyanobacteria are photoautotrophs whose photosynthetic mechanism is very plantlike. Some species are also capable of nitrogen fixation, which helps recycle atmospheric nitrogen into organic compounds.
6. Phototrophic bacteria consist of the green and purple sulfur bacteria. They use light and hydrogen sulfide to generate energy and obtain electrons to produce organic compounds in anaerobic environments.
7. Pseudomonads are the most versatile group of existing chemoheterotrophs, capable of metabolizing a staggering array of organic and synthetic compounds.
8. Other important groups of chemoheterotrophic eubacteria are the spirochetes; endospore-forming bacteria; enteric bacteria (which inhabit the intestinal tracts of animals); rickettsias and chlamydias; mycoplasmas (the smallest bacteria); actinomycetes (of great medical significance, both positive and negative); and myxobacteria (which form elaborate colonies and move through the soil on a slimy secretion).

The Importance of Prokaryotes (pp. 534–535)

1. Prokaryotes, along with fungi, are decomposers that recycle elements between the biological and physical components of ecosystems. The metabolic versatility of prokaryotes makes them crucial to life on Earth through their ability to decompose organic material, generate oxygen, and fix carbon dioxide and nitrogen into forms other organisms can use.
2. Some prokaryotes live with other species in symbiotic relationships of mutualism, commensalism, and parasitism.
3. Some parasitic prokaryotes are pathogenic, causing disease in the host by invading tissues or poisoning with endotoxins or exotoxins.
4. Bacteria have been put to work in laboratories, sewage treatment plants, and the food and drug industry. One especially exciting development has been the use of prokaryotes in recombinant DNA technology.

The Origins of Metabolic Diversity (pp. 535–537)

1. All forms of nutrition and virtually every kind of metabolic pathway evolved in prokaryotes.
2. The first prokaryotes were likely chemoheterotrophs that absorbed free organic compounds. The central role of ATP and glycolysis in all modern cells suggests that early bacteria used ATP for energy and that glycolysis evolved step by step to regenerate ATP in an environment largely devoid of oxygen.
3. Electron transport chains and chemiosmosis could have evolved from transmembrane pumps that originally served to regulate internal pH in a watery medium increasingly polluted with acidic waste products of fermentation.
4. Early photosynthetic prokaryotes probably used pigments and light-powered photosystems to fix carbon dioxide. The first cyanobacteria began making organic compounds from water and carbon dioxide, releasing free oxygen as a by-product. They drastically changed the ancient atmosphere and affected subsequent biological evolution.
5. The gradual accumulation of oxygen caused the extinction of many prokaryotes, forced others into anaerobic seclusion, and led to the evolution of respiratory mechanisms to either tolerate or capitalize on rising oxygen levels.
6. Ancient prokaryotes thus established virtually all forms of nutrition and metabolism and set the stage for the evolution of eukaryotes.

SELF-QUIZ

1. A prokaryotic genome is different from a eukaryotic genome in that
   a. it has only one-half as much DNA as does a typical eukaryotic genome
   b. it consists of a single-stranded DNA molecule
   c. it has less protein associated with its DNA and is not enclosed in a nuclear envelope
   d. its ribosomes are smaller and chemically distinct
   e. it consists of RNA rather than DNA

2. Photoautotrophs use
   a. light as an energy source and can use water or hydrogen sulfide as a source of electrons for producing organic compounds
   b. light for an energy and oxygen as an electron source
   c. inorganic substances for energy and use CO_2 as a carbon source
   d. light to generate ATP but need organic molecules for a carbon source
   e. light as an energy source and CO_2 to reduce organic nutrients

3. Which of the following statements about bacterial groups is *not* true?
   a. The lipid composition of the plasma membrane found in archaebacteria is different from that of eubacteria.
   b. The archaebacteria and eubacteria probably diverged very early in evolutionary history.
   c. Both archaebacteria and eubacteria have cell walls, but those of archaebacteria lack peptidoglycan.
   d. The three groups of archaebacteria that are living today are anaerobic and pathogenic.
   e. Eubacteria include the cyanobacteria.

4. Home canners pressure cook low-acid vegetables as a precaution primarily against the
   a. mycoplasmas
   b. endospore-forming bacteria
   c. enteric bacteria
   d. pseudomonads
   e. actinomycetes

5. The first prokaryotes were probably
   a. cyanobacteria
   b. chemoheterotrophs that used abiotically-made organic compounds
   c. anaerobic photosynthetic organisms
   d. mycoplasma
   e. thermoacidophiles

6. Banded iron formations in marine sediments indicate that
   a. early cyanobacteria were probably producing oxygen during that time period
   b. the early atmosphere was very reducing
   c. purple sulfur bacteria were the dominant life form in the seas at that time
   d. the pH of the early seas was quite low due to early prokaryotes' excretion of organic acids from fermentation
   e. mats of bacterial colonies were forming stromatolites

7. According to the scenario discussed in this chapter, the first transmembrane proton pumps may have functioned to
   a. produce ATP by the creation of a proton gradient
   b. pass electrons from H_2S to $NADP^+$ in a primitive photosynthesis
   c. regulate internal pH
   d. function in anaerobic respiration
   e. facilitate aerobic metabolism

8. Penicillins function as antibiotics mainly by inhibiting the ability of bacteria to
   a. form spores
   b. replicate DNA
   c. synthesize normal cell walls
   d. produce functional ribosomes
   e. synthesize ATP

9. Coordination of *two* photosystems to carry out plant-like photosynthesis occurs in the
   a. cyanobacteria
   b. purple sulfur bacteria
   c. archaebacteria
   d. actinomycetes
   e. chemoautotrophic bacteria

10. In motile bacteria, the movement of the organism toward a chemical source results from
    a. continuous movement in the direction of increasing concentration of the chemical
    b. longer "runs" when moving toward the chemical source
    c. more frequent "tumbles" when moving toward the chemical source
    d. coming to a complete stop whenever moving in a direction of decreasing concentration of the chemical
    e. controlling the direction of the power stroke taken by the flagella

CHALLENGE QUESTIONS

1. Nitrogen-fixing bacteria either work alone or in conjunction with plants. If you were a scientist investigating the *biochemistry* of nitrogen fixation, would you choose a solitary or symbiotic species for study? Explain your answer.

2. The purple bacterium *Halobacterium halobium* thrives with little competition from other organisms in salt lakes with salt concentrations as high as five times that of seawater. Before chlorophyll-containing green prokaryotes evolved, purple bacteria similar to *Halobacterium* may have been the dominant photosynthesizers. Indeed, the early seas may have been colored purple by the photosynthetic pigment bacteriorhodopsin. Andrew Goldsworthy of Imperial College, London, postulates that modern plants are green and use mainly red and blue components of the spectrum because the ancestors of modern plants evolved in direct competition with purple bacteria. Bacteriorhodopsin absorbs wavelengths of light mainly between 500 and 600 nanometers (nm). Use what you know of the absorption spectrum of chlorophyll pigments to support or refute Goldworthy's hypothesis. (See Further Reading 2.)

3. Professor Lynn Margulis of the University of Massachusetts has suggested that we could probe for extraterrestrial life by simply determining the mixture of gases in atmospheres of planets. If you were to conduct such research, what would you look for? Why?

FURTHER READING

1. Fliermans, C. B., and D. L. Balkwill. "Microbial Life in Deep Terrestrial Subsurfaces." *Bioscience*, June 1989. Bacterial communities thrive 500 meters underground.
2. Goldsworthy, A. "Why Trees are Green." *New Scientist*, December 1987. Why plants have chlorophyll and are green instead of black. An evolutionary explanation for the color of plants.
3. McEvedy, C. "The Bubonic Plague." *Scientific American*, February 1988. A bacterial disease that afflicted humans for over a thousand years.
4. Moser, P. W. "It Must Have Been Something You Ate." *Discover*, February 1987. The role of bacteria in food poisoning.
5. Penny, D. "What Was the First Living Cell?" *Nature*, January 14, 1988. A hypothesis that sulfur bacteria were the first organisms has stirred considerable debate.
6. Postgate, J. "Microbial Happy Families." *New Scientist*, January 21, 1989. Using molecular systematics to unravel bacterial phylogeny.
7. Postgate, J. "A Microbial Way of Death." *New Scientist*, May 20, 1989. Do bacteria age?
8. Rosenthal, E. "Why I Like Lyme Disease." *Discover*, June 1988. A physician describes a painful but curable disease.
9. Shapiro, J. A. "Bacteria as Multicellular Organisms." *Scientific American*, June 1988. Cellular differentiation occurs in some bacteria.
10. Tortora, G. J. et al. *Microbiology*. 3d ed. Redwood City, Calif.: Benjamin/Cummings, 1989. A general text.

26

Protists and the Origin of Eukaryotes

"No more pleasant sight has met my eye than this of so many thousands of living creatures in one small drop of water," wrote Anton von Leeuwenhoek after his discovery of the microbial world more than three centuries ago. It is a world each biology student rediscovers by peering through a microscope into a droplet of pond water filled with diverse creatures of the kingdom Protista (Figure 26.1). Most protists are unicellular, but we also find colonial forms and even some truly multicellular organisms with relatively simple body plans that are included in the kingdom because they are believed to be more closely related to certain unicellular protists than to plants, fungi, or animals.

Protists are eukaryotic, and thus even the simplest are much more complex than the prokaryotes of Kingdom Monera. The first eukaryotes to evolve from prokaryotic ancestors were probably unicellular and would therefore be classified as protists. The very word implies great antiquity (from the Greek *protos*, "first"). It was during the genesis of protists that a true nucleus, mitochondria, chloroplasts, endoplasmic reticulum, Golgi bodies, "9 + 2" flagella and cilia, mitosis and meiosis, and the other structures and processes unique to eukaryotic organization arose. The primal eukaryotes were not only the predecessors of the great variety of modern protists but were also ancestral to plants, fungi, and animals, the other three kingdoms of eukaryotes. Two of the most significant chapters in the history of life—the origin of the eukaryotic cell and the subsequent emergence of multicellular eukaryotes—unfolded during the evolution of protists.

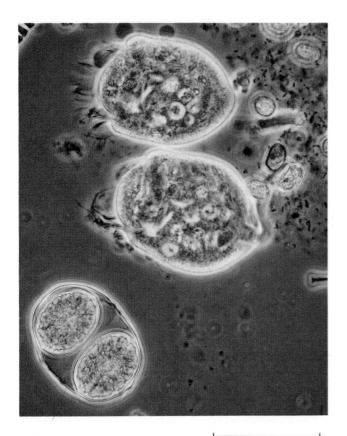

Figure 26.1
Pondwater droplet, home to a variety of protists (LM).

50 μm

This chapter characterizes protists, surveys their diversity, and describes theories on the origins of the eukaryotic cell and multicellular organization.

CHARACTERISTICS OF PROTISTS

Protists vary so extensively in cellular anatomy, ecological roles, and life cycles that few general characteristics can be cited without exceptions.

Protists are found almost anywhere there is water. They are important constituents of **plankton** (from the Greek *planktos,* "wandering"), the communities of organisms, mostly microscopic, that drift passively or swim weakly near the surface of oceans, ponds, and lakes. Among both seawater and freshwater protists are also bottom-dwellers that attach themselves to rocks and other anchorages or creep through the sand and silt. Protists are also common inhabitants of damp soil, leaf litter, and other terrestrial habitats that are sufficiently moist. In addition to free-living protists are the

many symbionts that inhabit the body fluids, tissues, or cells of hosts. These symbiotic relationships span the continuum from mutualism to parasitism, with some parasitic protists being important pathogens.

Nearly all protists are aerobic in their metabolism, using mitochondria for cellular respiration. Some forms are photoautotrophs with chloroplasts, some are heterotrophs that absorb organic molecules or ingest larger food particles, and still others combine photosynthesis *and* eating for their nutrition.

Most members of Kingdom Protista have flagella or cilia at some time in their life cycles. Some biologists argue for calling eukaryotic flagella and cilia **undulipodia** to distinguish them from the very different flagella of prokaryotes (see Chapter 25). Although this text will use the term *flagellum* in both cases, it is important to remember that the prokaryotic and eukaryotic versions of flagella are not homologous structures.

Cell division in the kingdom Protista is perplexingly varied. Mitosis occurs in most phyla of protists, but there are many variations in the process unknown in any other kingdom. All protists can reproduce asexually. Some forms are exclusively asexual; others can also reproduce sexually or at least use the sexual processes of meiosis and syngamy (union of two nuclei) to shuffle genes between two individuals that then go on to reproduce asexually. Many protists persist through periods of harsh conditions by forming resistant **cysts** at some point in their life cycles.

Because most protists are unicellular, they are justifiably considered to be the simplest eukaryotic organisms. But at the *cellular* level, many protists are exceedingly complex; indeed, the kingdom Protista can claim among its members the most elaborate of all cells. We should expect this of organisms that must carry out within the bounds of single cells all the basic functions performed by the collective of specialized cells that make up the bodies of plants and animals. Each unicellular protist is not at all analogous to a single cell from a multicellular organism, but is itself an organism as complete as any whole plant or animal.

THE BOUNDARIES OF KINGDOM PROTISTA

A consensus does not yet exist on which groups of organisms should be included in Kingdom Protista. The problem is that multicellularity apparently evolved several times during the history of protists, giving rise not only to plants, fungi, and animals but also to multicellular organisms that lack the distinctive traits that define those three kingdoms. When Robert H. Whittaker popularized the five-kingdom system in 1969, he assigned unicellular eukaryotes to their own kingdom, Protista. The trend in the past decade has been to expand the boundaries of the kingdom Protista to include some phyla of multicellular organisms classified in earlier renditions of the five-kingdom format as plants or fungi but that seem to have closer relatives among the unicellular organisms. Because the term *protist* had come to connote unicellular life, some advocates of the new classification also recommended changing the name of the kingdom to Protoctista, but the name Protista is still more widely used.

One modification of the five-kingdom scheme moves the multicellular algae, including kelp and other seaweeds, from the plant kingdom to the kingdom Protista. In its expanded form, Kingdom Protista also encompasses phyla of funguslike organisms, which lack important characteristics that define true fungi. This text incorporates these changes, while conceding that all taxonomic boundaries, especially above the species level, are tentative lines drawn by biologists seeking order in the diversity of life. The goal is to draw those lines so that classification is not only convenient but also as phylogenetically relevant as the evidence allows. What may seem to be hair-splitting taxonomic debates are actually dialogues about the history of life.

In morphology and life styles, protists are the most diverse of all organisms. The kingdom Protista, as its limits are defined in this chapter, includes organisms as different as amoebas and giant kelp. In all, more than 60,000 species of extant protists have been described, and about the same number are known from the fossil record. The taxonomy of protists is in a state of flux, and taxonomists do not yet agree on the number or names of phyla. For example, all unicellular organisms with pseudopodia were once grouped in the phylum called Sarcodina. Current classification puts these organisms in at least three phyla—Rhizopoda, Actinopoda, and Foraminifera. Table 26.1 is a partial list of phyla recognized in recent publications by taxonomists, but alternate names are sometimes used in the literature for several of these phyla. Notice in the table that the phyla are grouped into three categories: (1) protozoa (animal-like protists), (2) algal (plantlike) protists, and (3) protists resembling fungi. This is an informal scheme that will make our survey of protistan diversity more convenient, but the categories do *not* reflect evolutionary relationships. A slime mold, for instance, is funguslike only in the sense that a whale is fishlike; the resemblance is due to convergent evolution. Also, many phyla have both photosynthetic and heterotrophic members; there is in the real world no protozoa–algae dichotomy. With these reservations, let us sample the major phyla of protists.

Table 26.1 A partial list of protistan phyla

Phylum	Brief Description
Protozoa (Nutrition Mainly by Ingestion)	
Rhizopoda	Naked and shelled amoebas
Actinopoda	Heliozoans and radiolarians: possess axopodia and usually have siliceous skeletons
Foraminifera	Forams: possess calcareous shells
Apicomplexa	Apicomplexans (formerly called sporozoans): mostly parasitic; have complex life cycles
Zoomastigina	Zoomastigotes: use flagella for motility; mostly unicellular, but some colonial
Ciliophora	Ciliates: use cilia for motility and feeding
Algal Protists (Nutrition Mainly by Photosynthesis)	
Dinoflagellata	Dinoflagellates: two flagella in perpendicular grooves of wall; brownish plastids
Chrysophyta	Golden algae: flagellated; many colonial forms
Bacillariophyta	Diatoms: shells of hydrated silica, with two halves
Euglenophyta	*Euglena* and its relatives: green flagellates lacking walls
Chlorophyta	Green algae: unicellular, colonial, and multicellular forms
Phaeophyta	Brown algae: all multicellular, including giant kelps and other seaweeds
Rhodophyta	Red algae: multicellular, some large enough to be called seaweeds
Protists Resembling Fungi	
Myxomycota	Plasmodial slime molds: feeding stage an amoeboid coenocyte
Acrasiomycota	Cellular slime molds: feeding stage consists of solitary, amoeboid cells; cells gather to form fruiting body that functions in asexual reproduction
Oomycota	Water molds: decomposers or parasites with filamentous bodies resembling fungi, but with walls of cellulose; dispersal by flagellated zygotes or spores

PROTOZOA

Protozoa means "first animals," a misnomer in the context of the five-kingdom system. The term persists to refer informally to protists that live primarily by ingesting food, an animal-like mode of nutrition.

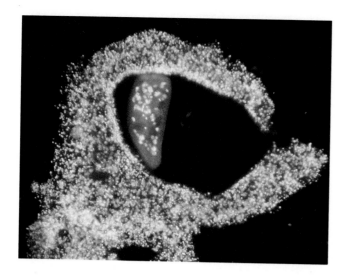

Figure 26.2
Rhizopoda. An amoeba ingests prey (*Paramecium*) by phagocytosis, engulfing the food in a pseudopodium (LM).

50 μm

Rhizopoda

Members of the phylum Rhizopoda, the **amoebas** and their relatives, are all unicellular. With or without shells, they are among the simplest of protists (Figure 26.2). No stages in their life histories are flagellated. Instead, amoebas use cellular extensions called **pseudopodia** (from the Greek *pseudes*, "false," and *pod*, "foot") to move and to feed. You have probably observed this mode of motility in *Amoeba proteus* in the laboratory. Watching a living amoeba, you see one of the most flexible of all cells. Pseudopodia may bulge from virtually anywhere on the cell surface. When an amoeba moves, it extends a pseudopodium and anchors its tip, and then more cytoplasm streams into the pseudopodium (the name *Rhizopoda* means "rootlike feet"). The cytoskeleton consisting of microtubules and microfilaments functions in amoeboid movement. Pseudopodial activity may appear chaotic, but in fact amoebas show directed movement as they creep slowly toward a food source.

Meiosis and sex do not occur in this phylum. The organisms reproduce asexually by various mechanisms of cell division. Spindle fibers form, but the typical stages of mitosis are not apparent in most amoebas. In many genera, for instance, the nuclear envelope persists during cell division.

Amoebas inhabit both freshwater and marine environments and are also abundant in soils. The majority of amoebas are free living, but some are important parasites, including *Entamoeba histolytica*, which causes amoebic dysentery in humans. These organisms spread via contaminated drinking water, food, or eating utensils.

Actinopoda

The name of this phylum, Actinopoda, means "ray feet," a reference to the slender projections called axopodia that radiate from the beautiful protists that compose the phylum (Figure 26.3). Each axopodium is reinforced by a bundle of microtubules, which is covered by a thin layer of cytoplasm. The projections place

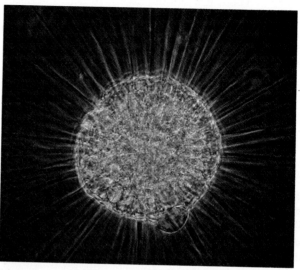

(a)

100 μm

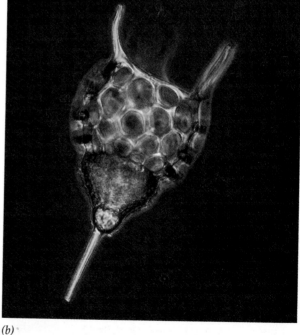

(b)

50 μm

Figure 26.3
Actinopoda. (a) Heliozoans are mainly freshwater protists with stiff axopodia used for feeding (LM).
(b) Radiolarians are mostly marine forms with glassy shells, different in shape for each species (LM).

an extensive area of cellular surface in contact with the surrounding water, help the organisms float, and function in feeding. Smaller protists and other microorganisms stick to the axopodia and are phagocytized by the thin layer of cytoplasm. Cytoplasmic streaming then carries the engulfed prey down to the main part of the cell.

Most heliozoans and radiolarians are components of plankton. Most **heliozoans** ("sun animals") live in fresh water, whereas **radiolarians** are primarily marine. The term *radiolarian* actually applies to several groups of organisms that may not be very closely related, all of which have delicate shells, most commonly made of silicates, the material of glass. After radiolarians die, their shells settle to the sea floor, where they have accumulated as an ooze that is hundreds of meters thick in some locations.

Foraminifera

Exclusively marine, the majority of members of Foraminifera, or **forams**, live in the sand or attach themselves to rocks and algae, but some families are also abundant in plankton. The phylum is named for the porous shells of its members (from the Latin *foramen*, "little hole," and *ferre*, "to bear"). The shells are generally multichambered and consist of organic material hardened with calcium carbonate. Strands of cytoplasm extend through the pores, functioning in swimming, shell formation, and feeding. Many forams also derive nourishment from the photosynthesis of symbiotic algae that live beneath the shells (Figure 26.4).

The shells of forams are important components of marine sediments, including sedimentary rocks that are now land formations, such as the chalky white cliffs of Dover. Foram fossils are excellent markers for correlating the vintages of sedimentary rocks in different parts of the world (see Chapter 23).

Apicomplexa

All members of the phylum Apicomplexa, which were formerly called sporozoans, are parasites of animals, and some cause important human diseases. The parasites disseminate as tiny infectious cells called **sporozoites**. As seen with the electron microscope, one end (the apex) of the sporozoite cell contains a complex of organelles specialized for penetrating host cells and tissues, thus the phylum name *Apicomplexa*. Most apicomplexans have intricate life cycles with both sexual and asexual stages, and these cycles often require two or more different host species for completion. An example is *Plasmodium*, the parasite that causes malaria (Figure 26.5). The incidence of malaria was greatly reduced in the 1960s by the use of insecticides that

100 μm

Figure 26.4
Foraminifera. The calcium carbonate shells of these protists have left an excellent fossil record. This living foram has a snail-like shell. The largest forams grow to diameters of several centimeters (LM).

reduced populations of *Anopheles* mosquitos, which spread the disease, and by drugs that killed the parasites in humans. However, the multiplication of resistant varieties of both the mosquitos and *Plasmodium* species have caused a resurgence of the disease. Each year, more than 200 million people are infected in the tropics, and at least a million die from the disease in Africa alone.

Zoomastigina

The protozoa known as **zoomastigotes** (from the Greek *mastix*, "whip") use whiplike flagella to propel themselves. These members of the phylum Zoomastigina are all heterotrophs that absorb organic molecules from the surrounding medium or engulf prey by phagocytosis. Most live as solitary cells but some form colonies of cells. There are both free-living and symbiotic zoomastigotes. Living within the gut of a termite, for instance, are symbiotic flagellates that digest cellulose in the wood eaten by the host. At the opposite end of the spectrum of symbiotic relations are parasitic zoomastigotes, some of which are pathogenic to humans. Species of *Trypanosoma* cause African sleeping sickness, which is spread by the bite of the tsetse fly (Figure 26.6).

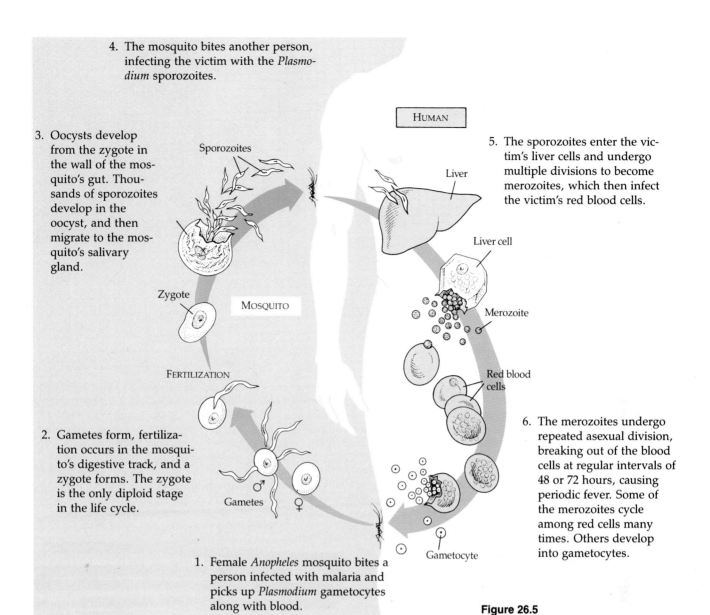

4. The mosquito bites another person, infecting the victim with the *Plasmodium* sporozoites.

3. Oocysts develop from the zygote in the wall of the mosquito's gut. Thousands of sporozoites develop in the oocyst, and then migrate to the mosquito's salivary gland.

Sporozoites

Zygote

MOSQUITO

FERTILIZATION

2. Gametes form, fertilization occurs in the mosquito's digestive track, and a zygote forms. The zygote is the only diploid stage in the life cycle.

Gametes

♂   ♀

1. Female *Anopheles* mosquito bites a person infected with malaria and picks up *Plasmodium* gametocytes along with blood.

HUMAN

Liver

5. The sporozoites enter the victim's liver cells and undergo multiple divisions to become merozoites, which then infect the victim's red blood cells.

Liver cell

Merozoite

Red blood cells

6. The merozoites undergo repeated asexual division, breaking out of the blood cells at regular intervals of 48 or 72 hours, causing periodic fever. Some of the merozoites cycle among red cells many times. Others develop into gametocytes.

Gametocyte

Figure 26.5
Life cycle of *Plasmodium*, the apicomplexan that causes malaria.

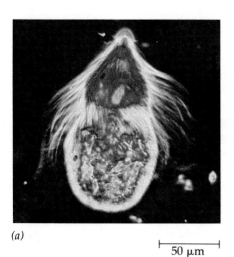

(a)

50 μm

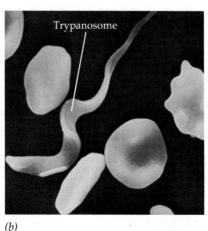

Trypanosome

(b)

5 μm

Figure 26.6
Zoomastigotes. (a) *Trichonympha*, one of several symbiotic flagellates inhabiting the gut of termites. The posterior region of the cell contains wood particles that are being digested (LM). **(b)** *Trypanosoma*, seen here in human blood, is the flagellate that causes African sleeping sickness. The molecular composition of these pathogens' coats changes frequently, preventing immunity from developing in hosts. (Disc-shaped red blood cells surround the elongate trypanosome cell in this scanning electron micrograph.)

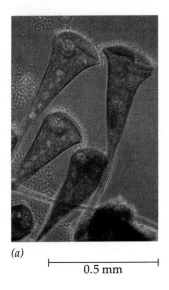

(a)

0.5 mm

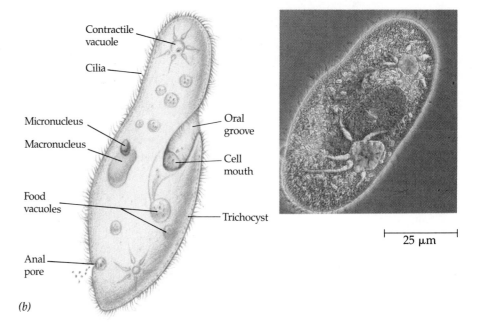

Contractile vacuole

Cilia

Micronucleus

Macronucleus

Food vacuoles

Anal pore

Oral groove

Cell mouth

Trichocyst

(b)

25 μm

Figure 26.7
Ciliates. (a) The beautiful freshwater ciliate *Stentor* moves by means of membranelles formed of many cilia (LM).
(b) *Paramecium*, an example of ciliate
complexity, is covered by countless thousands of cilia. The organism is armed with
an arsenal of trichocysts, tiny darts the cell
discharges, probably for defense. (Some
other ciliates have toxic trichocysts that
disable prey.) *Paramecium* feeds mainly
on bacteria. Cilia that line an indentation
called the oral groove sweep water and
food particles to a cell mouth at the base
of the groove, where the food is engulfed
by phagocytosis. After food vacuoles bud
inward from the mouth, they fuse with lysosomes, and the food is digested as the
vacuoles circulate within the cell by cytoplasmic streaming. Nutrients are transported across the membrane of the vacuole into the cytoplasm, and undigested
wastes are eliminated by exocytosis when
the vacuoles fuse with a specialized region
of the surface that functions as an anal
pore. *Paramecium*, like other freshwater
protists, constantly takes in water by
osmosis from the hypoosmotic environment. Contractile vacuoles accumulate the
excess water, which is expelled from the
cell by powerful contractions of the surrounding cytoplasm (right, LM).

Ciliophora

The diverse protists of the phylum Ciliophora are
characterized by their use of cilia to move and feed.
Most members of Ciliophora, or **ciliates,** live as solitary cells in fresh water. Some ciliates are completely
covered by rows of cilia, whereas others have their
cilia clustered into fewer rows or tufts. The specific
arrangements adapt the ciliates for their diverse life
styles. Some species, for instance, scurry about on
leglike cirri constructed from many cilia bonded
together. Other forms, such as *Stentor,* have rows of
tightly packed cilia that function collectively as locomotor membranelles. Ciliates are probably among the
most complex of all cells (Figure 26.7).

A unique feature of ciliate genetics is the presence
of two types of nuclei, a large **macronucleus** and usually several tiny **micronuclei.** The macronucleus has
50 or more copies of the genome. The genes are not
distributed in typical chromosomes but are instead
packaged into a much larger number of small units,
each with hundreds of copies of just a few genes. The
macronucleus controls the everyday functions of the
cell by synthesizing RNA and is also necessary for
asexual reproduction; ciliates generally reproduce by
binary fission, during which the macronucleus elongates and splits rather than undergoing mitotic division. The micronuclei, of which species of *Paramecium*
have from 1 to as many as 80, do not function in growth,
maintenance, and asexual reproduction of the cell but
are required for sexual processes that generate genetic
variation. The sexual shuffling of genes occurs during
the process known as **conjugation,** diagrammed in
Figure 26.8. In ciliates, sex and reproduction are separate processes.

ALGAL PROTISTS

The phyla grouped here as algal protists consist mainly
of photosynthetic organisms, although some phyla
include heterotrophs as well. The term *alga,* remember, refers to relatively simple aquatic organisms that
are plantlike in the sense that they have chlorophyll.
Except for the prokaryotic cyanobacteria (blue-green
algae), all of the organisms generically called algae
belong to the kingdom Protista according to the version of the five-kingdom scheme used in this text.
Many biologists still prefer to place some multicellular
phyla in the plant kingdom (particularly Chlorophyta,
Rhodophyta, and Phaeophyta—the green algae, red
algae, and brown algae, respectively).

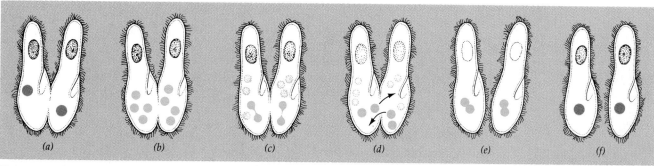

Figure 26.8

Conjugation by *Paramecium caudatum.*
(a) Two individuals of compatible mating strains align side by side and join by forming a conjugation tube. (b) All but one diploid micronucleus in each cell disintegrate, and the one that remains undergoes meiosis to produce four haploid micronuclei. (c) One of these divides by mitosis, and the other three disintegrate. (d) Mates then swap one micronucleus. (e) Syngamy occurs when the micronucleus a cell acquired from its partner fuses with one of its own micronuclei, forming a fresh diploid nucleus with a mixture of chromosomes derived from the two individuals. As partners separate, the newly constituted micronucleus in each divides repeatedly by mitosis. (f) One of the daughter micro- nuclei then replicates its genome many times, without dividing, to form a new macronucleus as the old one breaks down. Notice that throughout this complex process, no reproduction has occurred. But each of the conjugating individuals has been genetically altered by its sexual encounter, and it will go on to produce a unique clone by reproducing asexually.

All photosynthetic protists have chlorophyll *a,* the same pigment found in cyanobacteria and plants. But the accessory pigments—other forms of chlorophyll, carotenoids, and additional pigments—vary among the phyla and are used as characteristics to help classify the algae and assess relationships among phyla. Additional clues to taxonomic affinities are chloroplast structure, the chemistry of cell walls (a few algal phyla lack walls), number and position of flagella, and the form of food stored by the cells (Table 26.2). The following survey of algal protists is incomplete but will give you a sense of the diversity of forms that has evolved.

Dinoflagellata

Dinoflagellates are abundant components of the vast aquatic pastures of phytoplankton, microscopic algae floating near the surface of the sea that provide the foundation of most marine food chains (Figure 26.9). Dinoflagellate blooms, episodes of explosive population growth, cause the red tides that occur on occasion in warm coastal waters. Some dinoflagellates also live as symbionts of animals called cnidarians that build coral reefs; the photosynthetic output of these dinoflagellates is the main food source for reef communities. Other dinoflagellates lack chloroplasts and live as parasites within marine animals. There are even carnivorous species. The existence of both photosynthetic and heterotrophic forms closely related enough to be grouped in the same phylum reinforces the point made earlier in the chapter that the terms *protozoa* and *algae,* although traditional and somewhat useful, have no basis in phylogeny.

Of the several thousand known species, most dinoflagellates are unicellular, but there are some colonial forms. Dinoflagellates have brownish plastids containing chlorophyll *a,* chlorophyll *c,* and a mixture of carotenoid pigments, including one called peridinin that is unique to this phylum. Food is stored in the form of starch. Each dinoflagellate species has a characteristic shape reinforced by plates of cellulose. The beating of two flagella in perpendicular grooves in this "armor" produces a spinning movement for which these organisms are named (from the Greek *dinos,* "whirling").

The structure of the dinoflagellate nucleus and its division during asexual reproduction are unusual. The chromosomes lack histone proteins and are constantly in a condensed state (see Chapter 18). There are no mitotic stages. Although microtubules penetrate the nucleus in some species, they do not attach to the kinetochores of the chromosomes as in mitosis. Instead, the kinetochores are attached to the nuclear envelope and are distributed to daughter cells by splitting of the nucleus.

Chrysophyta

Members of the phylum Chrysophyta (from the Greek *chrysos,* "golden") are named for their color, which results from yellow and brown carotenoid pigments found along with chlorophylls *a* and *c* in the plastids. The cells, which are flagellated, store carbohydrate in the form of laminarin, a polysaccharide with a molecular architecture somewhat different from starch. **Golden algae** live among freshwater plankton in many

Table 26.2 Characteristics of eukaryotic algae

Division	Approximate Number of Species	Photosynthetic Pigments	Carbohydrate Food Reserve	Number and Position of Flagella	Cell Wall Components	Habitat
Dinoflagellata (dinoflagellates)	1,100	Chlorophyll *a*, chlorophyll *c*, carotenoids	Starch	2, lateral	Cellulose	Marine and fresh water
Chrysophyta (golden algae)	850	Chlorophyll *a*, often chlorophyll *c*, carotenoids including fucoxanthin	Laminarin	1 or 2, apical	Pectic compounds with siliceous material	Mostly fresh water
Bacillariophyta (diatoms)	10,000	Chlorophyll *a*, chlorophyll *c*, carotenoids including fucoxanthin	Leucosin	None	Hydrated silica in organic matrix	Fresh water and marine
Euglenophyta (*Euglena* and its relatives)	800	Chlorophyll *a*, chlorophyll *b*, carotenoids	Paramylon	1 to 3, apical	No cell wall	Mostly fresh water
Chlorophyta (green algae)	7,000	Chlorophyll *a*, chlorophyll *b*, carotenoids	Starch	2 or more, apical or subapical	Cellulose	Mostly fresh water, but some marine
Phaeophyta (brown algae)	1,500	Chlorophyll *a*, chlorophyll *c*, carotenoids including fucoxanthin	Laminarin	2, lateral; in reproductive cells only	Cellulose matrix with other polysaccharides	Almost all marine; flourish in cold ocean waters
Rhodophyta (red algae)	4,000	Chlorophyll *a*, carotenoids, phycobilins, chlorophyll *d* in some	Floridean starch	None	Cellulose; pectic materials common	Mostly marine, but some fresh water; many species tropical

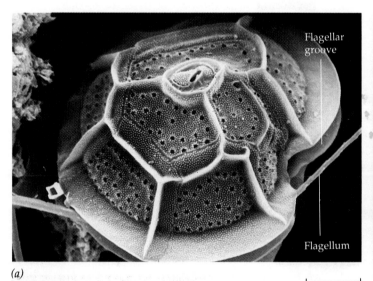

Flagellar groove

Flagellum

(a)

⊢——⊣ 10 μm

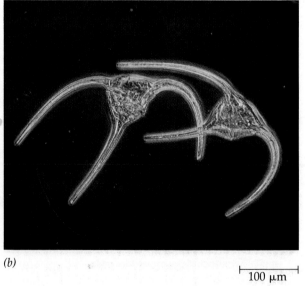

(b)

⊢——⊣ 100 μm

Figure 26.9
Dinoflagellates. These unicellular algae are characterized by a pair of flagella in perpendicular grooves. Beating of the flagella causes the cell to spin as it swims. Each species has a distinctively shaped wall. **(a)** *Gonyaulex tamarensis* (SEM). **(b)** *Ceratium sp.* (LM).

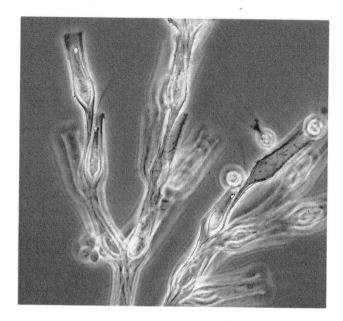

Figure 26.10
A golden alga. *Dinobryon*, a freshwater organism, is one of many colonial forms of this algal phylum (LM).

colonial forms (Figure 26.10). In ponds and lakes that freeze in winter or dry up in summer, golden algae survive by forming resistant cysts, from which active cells emerge when conditions are favorable. Microfossils resembling the ruptured cysts of chrysophytes and other algae have been found in Precambrian rocks.

Bacillariophyta

The members of the phylum Bacillariophyta, or **diatoms,** are yellow or brown in color. They have the same set of photosynthetic pigments as golden algae (Chrysophyta), and the two groups were once assigned to the same phylum. Diatoms, however, have such distinctive cell structure and life cycles that they are now considered to be a separate phylum by most phycologists (specialists in the study of algae). Almost all diatoms are unicellular, although there are some simple colonial forms that are filamentous. Both freshwater and marine plankton are rich in diatoms; a bucket of water scooped from the surface of the sea may have millions of these microscopic algae. The cells store their food reserves in the form of an oil, which also provides buoyancy that keeps the diatoms floating near the surface in the sunlight. Many diatoms are capable of a gliding movement.

Diatoms have unique glasslike walls consisting of hydrated silica embedded in an organic matrix (Figure 26.11). Massive accumulations of fossilized shells are major constituents of the sediments known as diatomaceous earth, which is mined for its quality as a filtering medium and for many other uses.

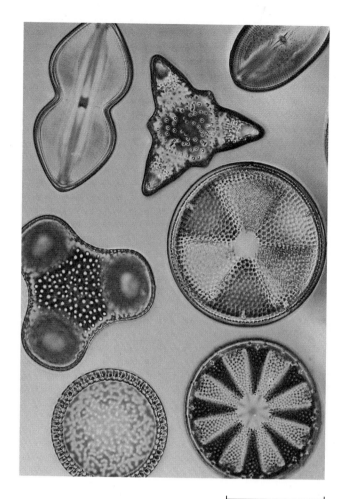

Figure 26.11
Diatoms. The glasslike shells of these algal protists consist of two halves that fit like the bottom and lid of a pillbox. Tiny pores in the ornate shells are avenues for exchange of gases and other substances between the cell and the surrounding medium. The shape of the shell and its pattern of pores are distinctive enough to be used for classifying diatoms (LM).

Euglenophyta

The best-known members of the phylum Euglenophyta belong to the genus *Euglena*, tiny green flagellates among the most common inhabitants of murky pond water (Figure 26.12). In addition to chlorophyll *a*, the grass-green chloroplasts contain the accessory pigment chlorophyll *b*, which is also found in algae of the phylum Chlorophyta (green algae) and in true plants. Euglenophytes store the polysaccharide paramylon.

Euglena is unusually versatile in its nutrition. In the light, it uses its chloroplasts to make a living by photosynthesis, although the organism is not completely autotrophic because it requires minute quantities of vitamin B_{12}. If *Euglena* is placed in the dark, it can live as a heterotroph by ingesting particles of food by phagocytosis. Indeed, some members of the phylum

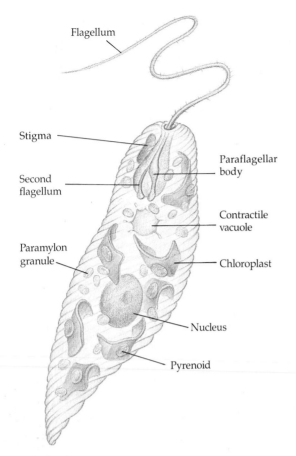

Flagellum

Stigma

Second
flagellum

Paramylon
granule

Paraflagellar
body

Contractile
vacuole

Chloroplast

Nucleus

Pyrenoid

Figure 26.12
Euglena. A motile protist with chloroplasts, this unicellular alga uses its long flagellum for propulsion. Near the
base of the flagellum is a stigma, often called an eyes-
pot, that functions as a pigment shield. Depending on the
position of the organism, the stigma allows light from
only a certain direction to strike a light detector, the paraflagellar body. These structures seem to function in
phototaxis, which is important for these photosynthetic
algae. *Euglena* lacks a cell wall but has a strong, flexible
pellicle made of protein beneath its plasma membrane.
Paramylon granules function in carbohydrate storage.
Pyrenoids within chloroplasts contain enzymatic proteins
active in photosynthesis.

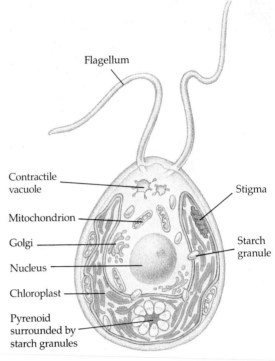

Flagellum

Contractile
vacuole

Mitochondrion

Golgi

Nucleus

Chloroplast

Pyrenoid
surrounded by
starch granules

Stigma

Starch
granule

Figure 26.13
**Structure of *Chlamydomonas*, a unicellular green
alga.** The stigma is a pigment shield that functions as a
light shade. The pyrenoid, as in *Euglena*, contains
enzymes active in photosynthesis.

Euglenophyta lack chloroplasts and depend entirely
on heterotrophism. The euglenophytes remind us once
again that no phylogenetically reliable distinction exists
between protozoa and unicellular algae; all of these
organisms are protists, distinct in many ways from the
other kingdoms of eukaryotes.

Chlorophyta

The phylum Chlorophyta is named for its members'
grass-green chloroplasts (from the Greek, *chloros,*
"green")—which are much like those of plants in

ultrastructure and pigment composition. In fact, most
botanists believe that the ancestors of the plant kingdom were green algae. We examine the evidence for
this relationship in the next chapter; for now, let us
survey the diversity of extant chlorophytes.

More than 7000 species of **green algae** have been
identified. Most live in fresh water, but there are also
many marine species. Various species of unicellular
green algae live as plankton, inhabit damp soil, coat
the surface of snow, or occupy the cells or body cavities of invertebrates as photosynthetic symbionts that
contribute to the food supply of the hosts. Chlorophytes are also among the algae that live symbiotically
with fungi in the mutualistic collectives known as
lichens. The simplest chlorophytes are biflagellated
unicells such as *Chlamydomonas*, which resemble the
gametes of more complex green algae (Figure 26.13).

In addition to unicellular chlorophytes are colonial
species, many of them filamentous forms that contribute to the stringy masses known as pond scum.
There are even some truly multicellular chlorophytes
such as *Ulva*, with bodies large enough and complex
enough that many texts still consider them to be true
plants (Figure 26.14). The similarities to plants are
analogous, not homologous; even the most complex
algae are more closely related to unicellular members

(a)

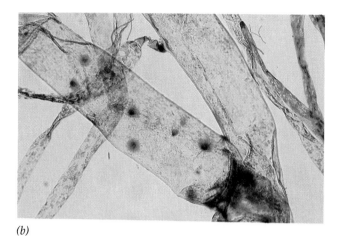

(b)

$\overline{100\,\mu m}$

(c)

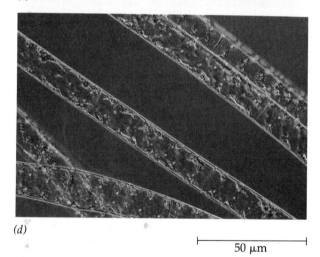

(d)

$\overline{50\,\mu m}$

Figure 26.14

$\overline{50\,\mu m}$

Colonial and multicellular green algae. (a) Species of *Volvox* are colonial chlorophytes that inhabit fresh water. The colony is a hollow ball, with its wall composed of hundreds or thousands of biflagellated cells embedded in a gelatinous matrix. The cells do not share common walls, but usually are connected by strands of cytoplasm. If the cells are isolated, they cannot reproduce. The large colonies seen here will eventually release the small green and red "daughter" colonies within them (LM). (b) Species of *Bryopsis* form branching filaments that lack cross walls and, thus, are multinucleate (LM). (c) *Ulva*, or sea lettuce, is a relatively complex, edible marine alga with a thallus (body) differentiated into leaf-like blades and a rootlike holdfast that anchors the thallus against turbulent tides and waves. (d) *Spirogyra* is a conjugating filamentous alga. Cells of a mating pair are joined by conjugation tubes, through which amoeboid gametes will migrate from the "male" and unite with gametes of the "female." (LM.)

of their phyla than to any true plant. The body of an alga is called a **thallus** (from the Greek *thallos*, "sprout"), and it lacks true roots, stems, and leaves.

Three separate evolutionary trends have probably produced the diverse forms of colonial and multicellular chlorophytes from unicellular, flagellated ancestors. Larger size and greater complexity have evolved by (1) the formation of colonies of individual cells, as seen in species of *Volvox* (see Figure 26.14a); (2) the repeated division of nuclei with no cytoplasmic division, as seen in multinucleate filaments of *Bryopsis* (see Figure 26.14b); and (3) the formation of definite multicellular thalli, as in *Ulva* (see Figure 26.14c).

Most green algae have complex life histories with both sexual and asexual reproductive stages. Nearly all reproduce sexually by way of biflagellated gametes having cup-shaped chloroplasts. The exceptions are the **conjugating algae** such as *Spirogyra* (see Figure 26.14d), which produce amoeboid gametes (the difference is significant enough that many phycologists assign the conjugating algae to a separate phylum).

Let us examine the life cycle of *Chlamydomonas* (Figure 26.15). The mature organism is a single haploid cell. When it reproduces asexually, the cell resorbs its

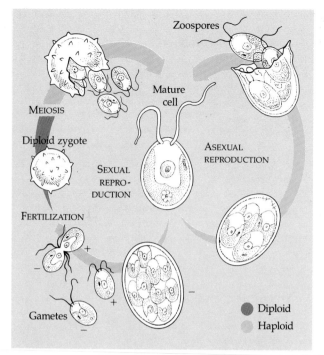

Figure 26.15
Life cycle of *Chlamydomonas*. This green alga exhibits sexual as well as asexual reproduction. The gametes are morphologically identical. Fusion of identical gametes is known as isogamy.

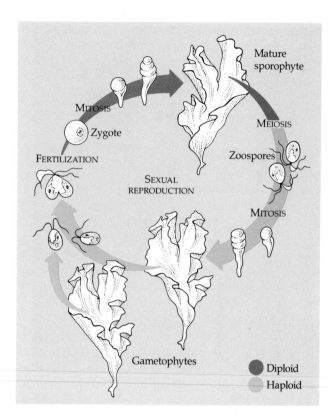

Figure 26.16
Alternation of generations in the life cycle of *Ulva*. The haploid, sexual generation (gametophyte) produces the diploid, asexual generation (sporophyte), and vice versa. In this case, the two generations are isomorphic, or identical in morphology.

flagella and then divides twice by mitosis to form four cells (more in some species). These daughter cells develop flagella and cell walls and then emerge as swimming zoospores from the wall of the parent cell, which had enclosed them. The zoospores grow into mature haploid cells, completing the asexual life cycle. A shortage of nutrients, drying of the pond, or some other stress triggers sexual reproduction. Within the wall of the parent cell, mitosis produces many haploid gametes. After their release, gametes from opposite mating strains (designated + and −) pair off and cling together by the tips of their flagella. The gametes are morphologically indistinguishable, and their fusion is called **isogamy,** which literally means a "marriage of equals." The isogametes fuse slowly and their nuclei unite to make a diploid zygote, which secretes a durable coat that protects the cell against harsh conditions. When the zygote breaks dormancy, meiosis produces four haploid individuals that emerge from the coat and grow into mature cells, completing the sexual life cycle.

Though many of the features of *Chlamydomonas* sex are believed to be primitive, many refinements of the sexual process have evolved among chlorophytes. Some green algae, for instance, produce gametes that differ morphologically from vegetative cells, and in some species the male and female gametes differ in size or morphology. Many species exhibit **oogamy,** a flagellated sperm fertilizing a nonmotile egg.

In the life cycles of some multicellular green algae, haploid and diploid individuals alternate, each producing the other. An example of this phenomenon, called **alternation of generations,** is the life cycle of *Ulva*, the sea lettuce (Figure 26.16). The haploid individual is the **gametophyte** generation, so named because it reproduces sexually by releasing haploid gametes made by mitosis. Fertilization produces a diploid zygote, which develops into the multicellular **sporophyte,** the asexual generation in the cycle. Meiosis in the sporophyte produces haploid cells called zoospores, which are released and divide by mitosis to regenerate the haploid gametophyte generation. In the case of *Ulva*, the two generations are **isomorphic,** meaning they are identical in morphology. In other species, two alternating generations may be morphologically distinct.

Phaeophyta

The largest and most complex protists, members of the phylum Phaeophyta (from the Greek *phaios*, "dusky," "brown"), or **brown algae,** are all multicel-

(a)   *(b)*

Figure 26.17
Brown algae. (a) These giant seaweeds, called kelps, form large underwater "forests,"
common off the U.S. Pacific coast. Kelp beds support a complex community of marine
organisms. **(b)** This close-up shows the holdfasts, stemlike stipe, and leaflike blades of a
seaweed.

lular and most are marine. These algae owe their characteristic brown or olive color to fucoxanthin, a carotenoid pigment in the plastids. In addition to that pigment, brown algae also have chlorophyll *c* and, of course, chlorophyll *a*, the one photosynthetic pigment common to all photosynthetic eukaryotes. The cells store energy in the form of laminarin, also found in golden algae (Chrysophyta). Brown algae and golden algae also share the same set of photosynthetic pigments and have many other biochemical and ultrastructural traits in common; it is likely that brown algae evolved from chrysophytes or that both evolved from a common ancestor.

The algae commonly called seaweeds are either brown algae or red algae (Phylum Rhodophyta). Brown algae are especially common along temperate coasts, where the water is cool. The cell walls of brown algae consist of cellulose and a gelatinous material called algin, which accounts for the slimy and rubbery feel of these seaweeds. The algin cushions the cells of the thallus against the agitation of the waves and also helps prevent the thallus from drying out during low tide.

Beyond the intertidal zone in deeper waters live the giant seaweeds known as kelps (Figure 26.17). They are fastened to the sea floor by holdfasts. At the top of a stemlike stipe, which may be as long as 100 m, are leaflike blades kept floating where it is bright by bladders filled with gas. Sugar produced by photosynthesis in the blades is transported down the stipe in tubular cells similar to those in the vascular tissue of plants. But remember, the morphological complexity of these giant algae and the complex designs of

true plants arose independently from unicellular ancestors; the resemblance is analogy due to convergent evolution.

A variety of life cycles has evolved in the phylum Phaeophyta. Most species alternate generations, with a sexual, haploid gametophyte stage and an asexual, diploid sporophyte stage. In some cases, the two generations are isomorphic. In other cases, such as the life cycle of *Laminaria*, the gametophyte and sporophyte are **heteromorphic,** or morphologically distinct (Figure 26.18). In fact, in some species of brown algae, the two generations are so different in appearance that it takes considerable detective work to reveal the two types of thalli as members of the same species.

Kelps and other seaweeds are very important in the marine economy and are also commercially important as thickeners in food products and cosmetics, fertilizer, animal feed, and other products. The great kelp beds of temperate coastal waters provide habitat and food for a variety of organisms, including many fish caught by humans. The kelps are prodigiously productive. One brown alga, *Macrocystis*, grows to a length of more than 60 m in a single season, the fastest linear growth rate of any organism. Kelp is a renewable resource reaped by special boats that cut and collect the tops of the algae.

Rhodophyta

The majority of members of Rhodophyta, or **red algae,** live in the ocean, but there are also some freshwater

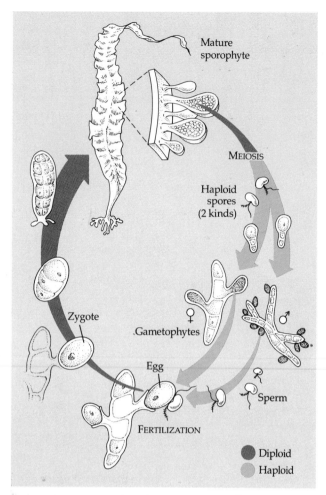

Figure 26.18
Alternation of generations in the life cycle of *Laminaria*. In this case, the sporophyte and gametophyte are heteromorphic, or morphologically different from each other. The large, plantlike sporophyte produces two kinds of spores, one developing into a female gametophyte and the other into a male. Gametophytes are microscopic.

Figure 26.19
A red alga. Coralline algae contribute to the great coral reefs.

and soil species. The red color of the plastids, for which Phylum Rhodophyta is named (from the Greek *rhodos*, "red"), is due to an accessory pigment called phycoerythrin. It belongs to a family of pigments known as phycobiliproteins, found only in red algae and cyanobacteria. Red algae store their food reserves as floridean starch, which is similar to glycogen. The cell walls consist of cellulose and gelatinous materials having physical properties like algin of brown algae.

Red algae are multicellular, and the largest share the title *seaweeds* with the brown algae, although none of the reds are as big as the giant browns. The thalli of most red algae are filamentous, often branched and interwoven in delicate lacy designs. The base of the thallus is usually differentiated as a holdfast. Most species are soft-bodied, but the **coralline algae** are reds with walls encrusted with hard calcium carbonate (Figure 26.19).

Red algae are most abundant in the warm coastal waters of the tropics, whereas cooler seas generally favor brown algae. A species of red algae has recently been discovered living near the Bahamas at a depth of more than 260 m.

Despite the name of the phylum, not all rhodophytes are red, and individuals of the same species can change their pigmentation and optimize photosynthesis at different water depths. Individuals of the same species may be almost black in deep water, bright red at more moderate depths, and greenish when living in very shallow water, owing to less phycoerythrin masking the green of the chlorophyll.

All red algae reproduce sexually. Unlike other algal protists, red algae have no flagellated stages in their life cycles. Gametes rely on water currents to get together. Alternation of generations is common in red algae, but details about reproduction and life cycles are known for very few species.

Coastal people, particularly in the Orient, harvest red algae and other seaweeds for food. Marine algae are rich in iodine and other essential minerals, but much of their organic material consists of unusual polysaccharides that humans cannot digest, which precludes seaweeds from becoming staple foods. Red algae also have commercial uses similar to brown algae.

PROTISTS RESEMBLING FUNGI

Even with five kingdoms instead of only two, the slime molds are a taxonomic enigma. They resemble fungi in appearance and life style, but the similarities are believed to be the result of convergence. In their cellular organization, reproduction, and life cycles, slime molds depart from the true fungi and probably have their closest relatives among the protists. The two phyla

(a)

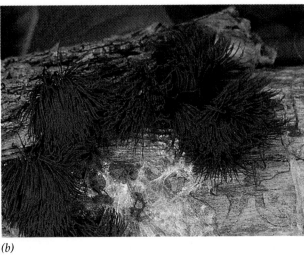

(b)

Figure 26.20
Plasmodial slime molds. (a) The feeding stage is a multinucleate (coenocytic) plasmodium that lives on organic refuse. The plasmodium often takes a weblike form, an adaptation that increases the surface area contacting food, water, and oxygen. Within the fine channels of the plasmodium, cytoplasm first streams one way, then the other in pulsing flows that are beautiful to see with a microscope. The cytoplasmic streaming apparently helps distribute nutrients and oxygen. **(b)** Reproductive structures called fruiting bodies, or sporangia, form when conditions become harsh.

of slime molds described here are now widely regarded as protists, although they are still studied by **mycologists** (biologists who study fungi—from the Greek *mykes,* "fungus"). The classification used in this text also regards the water molds as protists rather than true fungi.

Myxomycota

The phylum Myxomycota comprises **plasmodial slime molds,** which are more attractive than their name (Figure 26.20). Many species are brightly pigmented, usually yellow or orange, but slime molds are not photosynthetic; all are heterotrophs. The feeding stage of the life cycle is an amoeboid mass called a **plasmodium,** which may grow to a diameter of several centimeters. Large as it is, the plasmodium is not multicellular; it is a coenocytic mass, a multinucleated continuum of cytoplasm undivided by membranes or walls. In most species, the nuclei of the plasmodium are diploid. The plasmodium engulfs food particles by phagocytosis as it grows by extending pseudopodia through moist soil, leaf mulch, or rotting logs. If the habitat of a slime mold begins to dry up or there is no food left, the plasmodium ceases growth and differentiates into a stage of the life cycle that functions in sexual reproduction (Figure 26.21).

Acrasiomycota

Members of Acrasiomycota, **the cellular slime molds,** pose a semantic question about what it means to be an individual organism. The feeding stage of the life cycle consists of solitary cells that function individually. When there is no more food, the cells form an aggregate that functions as a unit (Figure 26.22). Although the mass of cells resembles a plasmodial slime mold, the important distinction is that the cells of a cellular slime mold maintain their identity and remain separated by their membranes.

In addition to not being coenocytes, cellular slime molds differ from plasmodial slime molds in other ways. Cellular slime molds are haploid organisms, whereas the diploid condition predominates in the life cycles of most plasmodial slime molds (compare Figures 26.21 and 26.22). Cellular slime molds have fruiting bodies that function in asexual reproduction (Figure 26.23). Also, cellular slime molds have no flagellated stages (except in one recently discovered genus).

Oomycota

Various members of the phylum Oomycota, collectively known as oomycetes, are water molds, white rusts, and downy mildews. Oomycetes resemble fungi in appearance, consisting of finely branched filaments called hyphae, which are coenocytic. Oomycetes and fungi also have similar nutritional modes. Closer inspection, however, suggests that oomycetes are not closely related to the true fungi, and the similarities are analogous, not homologous. Oomycetes have cell walls most commonly made of cellulose, while the walls of true fungi are made of another polysaccharide, chitin. The diploid condition prevails in the life cycles of most oomycetes, but it is the haploid stage

Figure 26.21
Life cycle of a plasmodial slime mold.
The plasmodium rounds into a mound and erects stalked fruiting bodies called sporangia. Within the bulbous tips of the sporangia, meiosis produces haploid spores. The resistant spores germinate to become active haploid cells when conditions are again favorable. These cells are either amoeboid or flagellated, the two forms readily reverting from one to the other. Either two flagellated cells or two amoeboid cells fuse to form diploid zygotes. Repeated division of the nucleus of the zygote by mitosis, without cytoplasmic division, forms a feeding plasmodium and completes the life cycle.

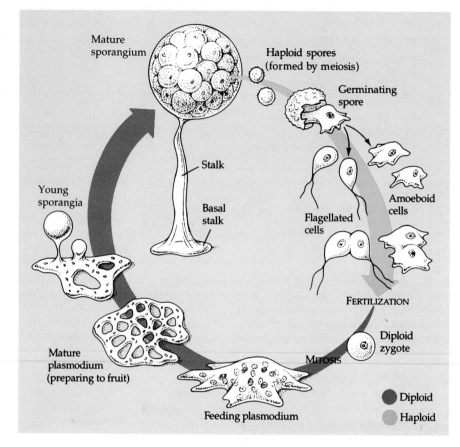

Figure 26.22
Life cycle of a cellular slime mold. The feeding stage of the life cycle consists of solitary cells that engulf bacteria while creeping by amoeboid movement through damp compost. When food is depleted, the amoeboid cells migrate toward an aggregation center where hundreds of the cells congregate in response to a chemical attractant they secrete. The sluglike colony of amoeboid cells may migrate as a unit for a while before settling down and developing stalked fruiting bodies that function in asexual reproduction. A cluster of resistant spores forms at the tip of each fruiting body. After the spores are released and exposed to a favorable environment, amoeboid cells emerge from their protective coats and begin feeding, completing the asexual portion of the life cycle. In the sexual phase in *Dictyostelium*, a pair of haploid amoebas fuse to form a zygote, the only diploid stage in the life cycle. A zygote becomes a giant cell by consuming surrounding haploid amoebas. The giant cell then becomes surrounded by a resistant wall. The encysted giant cell undergoes meiosis, followed by several mitotic divisions. New haploid amoebas are released when the cyst ruptures.

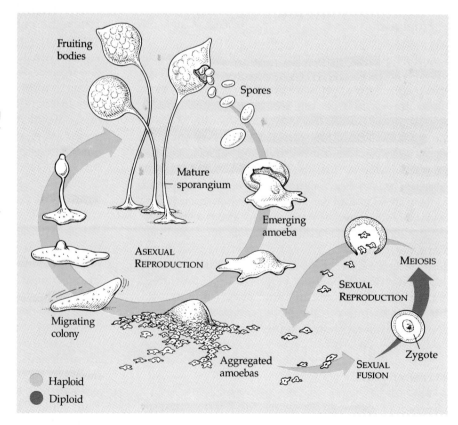

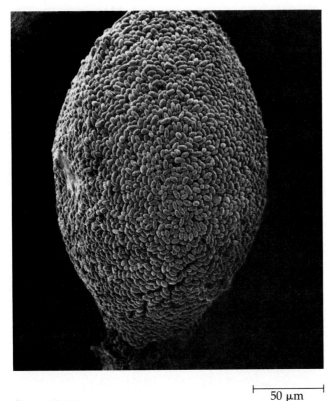

$\vdash$ 50 μm $\dashv$

Figure 26.23
Fruiting structure of a cellular slime mold (SEM).

that dominates the life cycles of fungi. And flagellated cells occur in the life cycles of oomycetes; according to the classification scheme used in this text, true fungi lack flagella.

Oomycota means "egg fungi," a reference to the mode of sexual reproduction in water molds. A relatively large egg cell is fertilized by a smaller sperm cell (Figure 26.24).

Most water molds are saprophytes that grow as cottony masses on dead algae and animals, mainly in fresh water. They are important decomposers in aquatic ecosystems. There are also parasitic water molds, such as those that grow on the skin and gills of fish in ponds or aquariums, but they usually attack only injured tissue. White rusts and downy mildews are close relatives of water molds but generally live on land as parasites of plants. They are dispersed primarily by windblown spores, but they also form flagellated zoospores at some time during their life cycles. Some of the most devastating plant pathogens are oomycetes, including the downy mildew that threatened the French vineyards in the 1870s and the species that causes late potato blight, which contributed to the Irish famine in the 19th century.

With the oomycota, we complete our sampling of the diverse phyla of contemporary protists. We will

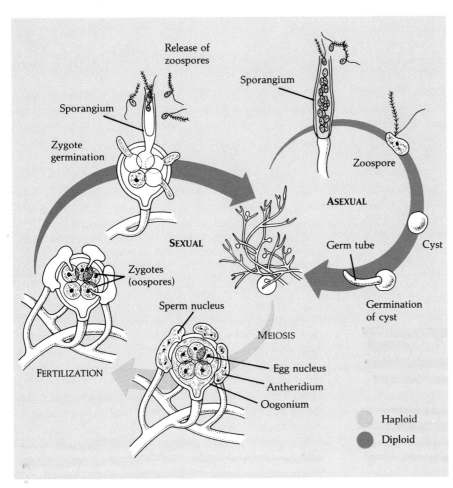

Figure 26.24
Life cycle of a water mold. The egg develops within a structure called an oogonium. The male organs, antheridia, grow like hooks around the oogonium and deposit unflagellated sperm through fertilization tubes that lead to the eggs. The zygotes may develop resistant walls and exist as dormant oospores until conditions are favorable for germination. Upon germination, zygotes give rise to flagellated cells called zoospores. Zoospores are also formed asexually in sporangia.

now step back in time to the origin of protists and other eukaryotes from prokaryotic ancestors.

THE ORIGIN OF EUKARYOTES

The first protists were also the first eukaryotes, and their origin was one of the most important chapters in the history of life. The many differences between prokaryotic and eukaryotic cells represent a distinction greater than that between plant and animal or between any two eukaryotes. Among the most fundamental issues in biology are the questions of when and how the complex eukaryotic cell, with its true nucleus, cytoplasmic organelles, and endomembrane system, evolved from much simpler prokaryotic cells.

The Antiquity of Eukaryotes

The fossil record of prokaryotic life extends back 3.5 billion years (see Chapter 25). Two billion years of prokaryotic history elapsed before the debut of anything resembling eukaryotic cells. The oldest putative fossils of eukaryotes are among Precambrian structures known as **acritarchs** (a Greek derivative meaning "of uncertain origin"). Some acritarchs are about the right size and appearance to be the ruptured coats of cysts similar to those made by certain algal protists today. The oldest are in rocks that have been estimated by radioisotopic dating to be about 1.5 billion years old. Metabolic evidence supports the fossil evidence of a long prokaryotic history before eukaryotes evolved; almost all eukaryotes require oxygen, and the few groups that can live anaerobically probably developed that characteristic secondarily from aerobic ancestors (for instance, zooflagellates within the termite gut have become adapted to that anaerobic environment). The implication is that eukaryotes evolved after cyanobacteria had existed long enough for the oxygen they produced to accumulate in the atmosphere.

Models of Eukaryotic Origins

The two major ideas on the origin of the eukaryotic cell are known as the **autogenous hypothesis** and the **endosymbiotic hypothesis.** According to the autogenous model, eukaryotic cells evolved by the specialization of internal membranes derived originally from the plasma membrane of a prokaryote (Figure 26.25a). The endomembrane system, consisting of nuclear envelope, endoplasmic reticulum, Golgi, and organelles bounded by single membranes, such as lysosomes, are viewed as the differentiated products of invaginated membranes. Mitochondria and chloro-

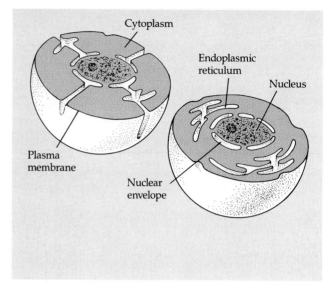

(a)

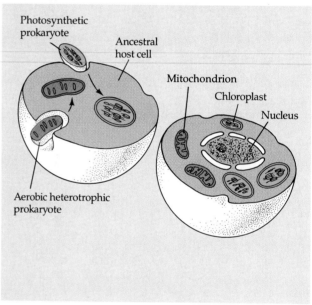

(b)

Figure 26.25
Models of eukaryotic origins. (a) According to the autogenous hypothesis, the complexity of the eukaryotic cell arose by invagination and specialization of the plasma membrane. **(b)** According to the endosymbiotic hypothesis, the eukaryote cell evolved as a consortium of prokaryotes that established symbiotic relationships. The inner membranes of mitochondria and chloroplasts may be remnants of the plasma membranes of early prokaryotes that infected or were engulfed by a larger host cell.

plasts may have acquired their double-membrane status by secondary invagination or more elaborate folding of membranes.

According to the endosymbiotic model, the forerunners of eukaryotic cells were symbiotic consortiums of prokaryotic cells, with certain species, termed endosymbionts, living within larger prokaryotes. Developed most extensively by Lynn Margulis, the endosymbiotic model focuses on the origins of chloroplasts and mitochondria. Chloroplasts are postulated to be descendants of photosynthetic prokaryotes that became endosymbionts within larger cells. The proposed ancestors of mitochondria were endosymbiotic bacteria that were aerobic heterotrophs (Figure 26.25b). Perhaps they first gained entry to the larger cell as undigested prey or internal parasites. By whatever means the relationships began, it is not hard to imagine the symbiosis eventually becoming mutually beneficial. A heterotrophic host could derive nourishment from photosynthetic endosymbionts. And in a world that was becoming increasingly aerobic, a cell that was itself an anaerobe would have benefited from aerobic endosymbionts that turned the oxygen to advantage. As host and endosymbionts became more interdependent, the conglomerate of prokaryotes would gradually be integrated into a single organism, its parts inseparable.

The endosymbiotic hypothesis posits that chloroplasts probably had at least three separate origins, perhaps descending from different prokaryotic endosymbionts. In terms of photosynthetic pigments and chloroplast structure, the phyla of algal protists can be divided into three lineages: a red line, a green line, and a brown line. Chloroplasts of red algae (Rhodophyta) bear a striking resemblance in thylakoid arrangement and pigment composition to cyanobacteria. According to many advocates of the endosymbiotic hypothesis, red algae descended from a prokaryotic syndicate in which the photosynthetic endosymbionts were cyanobacteria. The chloroplasts of green algae are more similar to a photosynthetic prokaryote named *Prochlorothrix*, a grass-green eubacterium with chlorophyll *b* as well as chlorophyll *a*. *Prochlorothrix* is one of the few known prokaryotes with chlorophyll *b*, an accessory pigment also found in green algae (and plants). Perhaps an ancient prokaryote similar to *Prochlorothrix* was the endosymbiotic ancestor of the plastids of chlorophytes. The thylakoids of the brown line generally occur in stacks of three, and chlorophyll *c* and fucoxanthin are present as accessory pigments. Biologists are actively searching for modern prokaryotes to fit the bill as relatives of the possible prokaryotic progenitors of brown plastids.

The feasibility of an endosymbiotic origin of chloroplasts and mitochondria rests partially on the existence of endosymbiotic relationships in the modern world. As another line of evidence, proponents of the endosymbiotic hypothesis cite various similarities between eubacteria and the chloroplasts and mitochondria of eukaryotes. Comparisons of structure and function reveal that chloroplasts and mitochondria are the appropriate size to be descendants of eubacteria. The inner membranes of chloroplasts and mitochondria, perhaps derived from the membranes of endosymbiotic prokaryotes, have several enzymes and transport systems that resemble those found on the plasma membranes of modern prokaryotes. Mitochondria and chloroplasts reproduce by a splitting process reminiscent of binary fission in bacteria. Chloroplasts and mitochondria contain DNA in the form of circular molecules not associated with histones or other proteins, as in prokaryotes. The organelles contain the transfer RNAs, ribosomes, and other equipment needed to transcribe and translate their DNA into proteins. In fact, some of the subunits of the cytochromes and ATPases that function in chloroplasts and mitochondria are known to be made in the organelles themselves. The ribosomes of chloroplasts are more similar in size and biochemical characteristics to prokaryotic ribosomes than to the ribosomes outside the chloroplast in the cytoplasm of the eukaryotic cell. Mitochondrial ribosomes vary extensively from one group of eukaryotes to another, but they are generally more similar to prokaryotic ribosomes than to their counterparts in the eukaryotic cytoplasm.

The limited evidence available so far from molecular systematics also suggests eubacterial origins for chloroplasts and mitochondria. Comparisons of base sequences show that ribosomal RNA of chloroplasts, which is transcribed from genes within the organelles, is more similar to the RNA of certain photosynthetic eubacteria than it is to the RNA of ribosomes in the eukaryotic cytoplasm, which is transcribed from nuclear DNA. Base sequence comparisons also suggest a eubacterial origin for ribosomal RNA of mitochondria, although the similarity is not as close as the one between chloroplasts and eubacteria.

Those skeptical about the endosymbiotic model point out that chloroplasts and mitochondria are not even close to being genetically autonomous. The great majority of proteins in the organelles are made by cytoplasmic ribosomes translating messenger RNA transcribed from nuclear genes. Advocates of the endosymbiotic hypothesis retort that a billion years of coevolution has been sufficient time for the host cell to develop extensive nuclear control over its symbionts, either by the accumulation of mutations or, more likely, by the direct transfer of DNA from the symbionts. In fact, the discovery of transposons (see Chapter 18) has revealed that DNA is suprisingly mobile within the nuclear genome, and there is recent evidence that genes have also jumped between the genomes of organelles and the nucleus.

In considering the origin of eukaryotes, we must point out that the autogenous and endosymbiotic models are not mutually exclusive. Possibly, the nuclear envelope and the rest of the endomembrane system arose by modification of a single cell, whereas chloroplasts and mitochondria may have originated as endosymbionts.

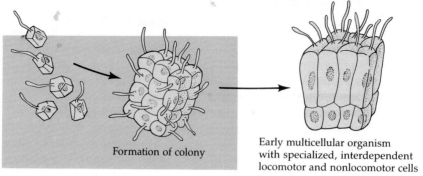

Formation of colony

Early multicellular organism
with specialized, interdependent
locomotor and nonlocomotor cells

Figure 26.26

One hypothesis for the origin of multicellularity. Multicellular organisms may have evolved from colonial protists. This diagram illustrates a possible scenario for the origin of an ancestral multicellular alga, plant, or animal from a colonial aggregate of flagellated cells. In another possible scenario, an initial cell aggregate may have been formed by a mix of flagellated and nonflagellated cells. In any case, increasing cellular specialization and interdependence would have led to multicellularity. (Lacking flagella, fungi probably would have been derived from a nonflagellated ancestor.)

A comprehensive theory of the eukaryotic cell must also account for the evolution of 9 + 2 flagella and cilia. (The more ardent supporters of the endosymbiotic model have proposed that eukaryotic flagella began as spirochetes or other motile bacteria attached to a host cell, but this particular idea does not seem to be widely accepted.) Related to the evolution of the eukaryotic flagellum is the origin of mitosis and meiosis, processes unique to eukaryotes that also employ microtubules. Mitosis made it possible to reproduce the large genomes of the eukaryotic nucleus, and the closely related mechanics of meiosis became an essential process in eukaryotic sex.

THE ORIGINS OF MULTICELLULARITY

The origin of eukaryotic organization could have ignited an explosion of biological diversification. More variations are possible for complex structures than for simpler ones. Unicellular protists, which are organized on the complex eukaryotic plan, are much more diverse in morphology than the simpler prokaryotes. Protists in which multicellular bodies evolved broke through another threshold in structural organization and became the stock for new waves of adaptive radiations. Among the products were the ancestors of plants, fungi, and animals.

Multicellularity undoubtedly evolved many times in the kingdom Protista, and the most widely held view is that links between multicellular organisms and their unicellular ancestors were colonies, or loose aggregates of interconnected cells. Multicellular algae,

plants, fungi, and animals probably descended from several lineages of colonial protists that formed by amalgamations of individual cells. The evolution of multicellularity from colonial aggregates involved increasing cellular specialization and division of labor. Initially, in colonial aggregates ancestral to multicellular algae, plants, and animals, all cells may have been motile with flagella (Figure 26.26). As cells in the colony tended to become increasingly intimate and interdependent, some of the cells may have lost their flagella and become more proficient in performing functions other than locomotion. Another early form of division of labor probably involved the separation of sex cells (gametes) from somatic (nonreproductive) cells. We see this type of specialization and intercellular cooperation today in several colonial protists, such as the green alga *Volvox*. Gametes are specialized for reproduction and they depend on somatic cells while developing. Evolution of the extensive division of labor required to perform all the nonreproductive functions in multicellular organisms as we know them involved many additional steps in somatic cell specialization.

Multicellular life more complex than filamentous algae did not appear until about 700 million years ago during the twilight of the Precambrian era. A variety of animal fossils has been found in late Precambrian strata, and many new forms evolved after the Paleozoic era dawned with the Cambrian period about 570 million years ago. Seaweeds and other complex algae were also abundant in Cambrian oceans and lakes. The land, however, was barren. About 400 million years ago, certain green algae living along the edges of lakes gave rise to primitive plants. In the next chapter, we trace the long evolutionary trek of plants onto land.

STUDY OUTLINE

1. Protists are a diverse group of eukaryotic organisms. Most are unicellular, but colonial and simple multicellular forms also exist.
2. As they evolved from prokaryotes, the earliest protists acquired the structures and processes unique to eukaryotic life. Thus, they are the predecessors not only of their modern protistan descendants, but of the other three eukaryotic kingdoms.

Characteristics of Protists (p. 541)

1. Protists vary enormously in cellular anatomy, ecology, and life cycles. However, there are a few characteristics found throughout the kingdom.
2. Protists are found wherever there is water, living as plankton, submerged bottom-dwellers, or inhabitants of moist soil or body fluids of other organisms.
3. Virtually all engage in aerobic metabolism. They are photoautotrophs, heterotrophs, or both.
4. Most have cilia or flagella during some part of the life cycle.
5. All protists reproduce asexually; some also can reproduce sexually. Many can survive harsh conditions by forming resistant cysts.

The Boundaries of Kingdom Protista (pp. 541–542)

1. Ancient protists gave rise to more than one multicellular line, making their classification complicated and controversial. The kingdom Protista includes multicellular descendants that do not clearly belong to one of the other three eukaryotic kingdoms.
2. Protists exhibit the most diverse spectrum of morphology and life cycles.

Protozoa (pp. 542–546)

1. The simplest protozoa are the rhizopods, unicellular amoebas and their relatives, all of which move by cellular extensions called pseudopodia. Rhizopods inhabit fresh and salt water as well as soil.
2. Actinopods are protozoa with slender, raylike axopodia that help them float and feed. Heliozoan and radiolarian species are components of plankton in freshwater and marine environments, respectively.
3. The marine forams are famous for their lovely porous shells, through which strands of cytoplasm extend for swimming, shell formation, and feeding. Foram fossils serve as useful markers for dating sedimentary rocks.
4. Apicomplexans (formerly called sporozoans) are parasitic protozoa with complex life cycles characterized by both sexual and asexual stages that usually require two or more host species. Apicomplexans have tiny infectious cells or sporozoites. *Plasmodium*, the parasite that causes malaria, is an apicomplexan.
5. Zoomastigina encompasses a varied group of flagellated heterotrophs. One parasitic group causes African sleeping sickness.
6. As their name implies, members of Ciliophora use cilia to move and feed. Ciliates are among the most complex cells, exhibiting an intricate anatomy and unique genetics, with a macronucleus and micronuclei.

Algal Protists (pp. 546–554)

1. Algal protists (algae) are aquatic, plantlike eukaryotes that possess chlorophyll. A few representatives are heterotrophic. Classification characteristics include variations evident in chloroplast structure, accessory pigments, cell walls, flagella, and forms of food storage.
2. Dinoflagellates, abundant in marine plankton, are either photosynthetic or heterotrophic. Most are unicellular, but some forms are colonial. They move in a spinning motion by the beating of flagella. The structure and behavior of the nucleus during cell division is singularly odd.
3. Chrysophytes are the flagellated freshwater golden algae, named for the color of their yellow and brown carotenoid pigments.
4. Bacillariophytes are the diatoms, primarily unicellular organisms with unique glasslike walls of silica.
5. Euglenophyta includes *Euglena* and its relatives, a group that blurs the distinction between plantlike and animal-like characteristics. Most species are photosynthetic, but others are either exclusively heterotrophic or capable of reverting to heterotrophy in the dark.
6. Chlorophytes, the green algae, are the likely ancestors of the plant kingdom. Diverging evolutionary pathways have generated an assortment of unicellular, colonial, multinucleate, and multicellular species that live in a variety of environments. Complicated life histories include both sexual and asexual reproductive stages, including the alternation of generations.
7. Phaeophytes are multicellular, primarily marine, brown algae, commonly called seaweeds. They are the largest and most morphologically complex protists. Phaeophytes are believed to be evolutionarily related to the golden algae, with which they share a number of ultrastructural and biochemical traits. Most species show some type of alternation of generations. Seaweeds, especially kelps, are very important in marine economies.
8. Rhodophytes, the red algae, possess the red accessory pigment phycoerythrin, which may or may not be masked by pigments of other colors. Red algae are multicellular, often lacy protists that store their food as floridean starch and reproduce sexually, usually as part of an alternation of generations. Many species are harvested as food.

Protists Resembling Fungi (pp. 554–558)

1. Slime molds are sufficiently different from the fungi in cellular organization and life cycles to be classified as protists.
2. Myxomycota comprises the plasmodial slime molds, an interesting and often beautiful diploid group that feeds by means of a coenocytic amoeboid plasmodium capable of differentiating into a sexually reproducing structure when moisture or food is scarce. Their heterotrophic mode of nutrition is typical of all slime molds.
3. Acrasiomycetes, the cellular slime molds, include a group of haploid organisms that lead unicellular lives until food is depleted. They then aggregate into a multicellular amoeboid mass that erects asexual fruiting bodies. Sexual reproduction involves the formation of giant cells contained in resistant cysts.

4. Oomycetes, the water molds, are a contested group still regarded by many to be true fungi. However, they have cell walls of cellulose and flagellated stages in their life cycles, traits that don't fit the definition of fungi used in this text. The phylum also includes white rusts and downy mildews, some of which are serious plant pathogens.

The Origin of Eukaryotes (pp. 558–560)

1. The eukaryotic cell made its debut with the advent of the first protist. A fundamental concern in biology is how eukaryotes evolved from prokaryotes.
2. Acritarch fossils, about 1.5 billion years old, are believed to be the first evidence of eukaryotic life.
3. The autogenous hypothesis of the origin of eukaryotes maintains that eukaryotic cells evolved from invaginations of the prokaryotic plasma membrane that subsequently differentiated into the endomembrane system.
4. An alternative, the endosymbiotic model, envisions eukaryotic cells arising as a result of prokaryotes taking up residence inside other prokaryotes. Chloroplasts and mitochondria are thus purported descendants of photosynthetic symbionts and aerobic, heterotrophic symbionts, respectively.

The Origins of Multicellularity (p. 560)

1. Multicellularity evolved many times in Kingdom Protista apparently by the aggregation of individual cells into colonial forms, in which cellular specialization and division of labor developed.

SELF-QUIZ

1. Which of the following is the most accurate general description of the kingdom Protista?
   a. eukaryotic, unicellular organisms that may be photosynthetic or heterotrophic
   b. eukaryotic, heterotrophic and/or photosynthetic, unicellular or simple multicellular organisms, which are different enough from multicellular plants, fungi, or animals to be placed in this diverse group
   c. eukaryotic plankton, which may be flagellated at some point in their life cycle and which reproduce asexually
   d. eukaryotic, photosynthetic or heterotrophic organisms that inhabit moist environments, form resistant cysts, and reproduce with flagellated gametes
   e. relatively simple versions of plants, animals and fungi

2. Which of the following protozoa are incorrectly paired with their description?
   a. rhizopods—naked and shelled amoebas
   b. actinopods—planktonic with slender, raylike axopodia
   c. forams—flagellated heterotrophs, free-living or symbiotic
   d. apicomplexans—parasites with complex life cycles
   e. ciliates—complex, unicellular organisms with macronucleus and micronuclei

3. Mycologists are specialists in the study of
   a. protists
   b. algae
   c. fungi
   d. plants
   e. phylogeny

4. Which of the following algal protists are incorrectly paired with their description?
   a. dinoflagellates—marine plankton, whirling, spinning movement, characteristic shell
   b. chrysophytes—golden algae, fucoxanthin pigment, flagellated, freshwater plankton
   c. bacillariophytes—diatoms, two-piece shells of silica
   d. phaeophytes—multicellular brown algae, seaweeds
   e. rhodophytes—cause red tides, peridinin pigment

5. Plants are believed to have evolved from the
   a. euglenophytes, because they have chlorophyll *a* and *b*
   b. dinoflagellates, because they have similar storage products and cellulose cell walls
   c. chlorophytes, because of similarities in chloroplasts and accessory pigments
   d. phaeophytes, because they have specialized body parts and an alternation of generations
   e. rhodophytes, because they have carotenoids and use floridean starch as a storage product

6. Unlike plasmodial slime molds, cellular slime molds
   a. exhibit phagocytosis
   b. form fruiting bodies
   c. have more than one nucleus per cell
   d. are haploid organisms except for the giant-celled zygote
   e. can move as an amoeboid mass

7. Acritarchs are
   a. symbiotic associations between algae and fungi
   b. eukaryotes that evolved by the autogenous model
   c. the oldest putative eukaryotic fossils
   d. fossils that resemble the ruptured coats of cyanobacteria
   e. diatom shells

8. Which of the following is an *incorrect* statement about the possible endosymbiotic origins of chloroplasts and mitochondria?
   a. They are the appropriate size to be descendants of bacteria.
   b. They contain their own genome and produce all their own proteins.
   c. They contain circular DNA molecules not associated with histones.
   d. Their membranes have enzymes and transport systems that resemble those found in the plasma membranes of prokaryotes.
   e. Their ribosomes are more similar to those of eubacteria than to those of eukaryotes.

9. The endosymbiotic hypothesis maintains that chloroplasts may have had at least three different origins because

   a. the chloroplasts of green plants show three distinct morphologies and pigment compositions

   b. the pigment composition and thylakoid arrangement of cyanobacteria are most like the chloroplasts of green plants

   c. molecular systematics link the red, green, and brown line to three different present-day eubacteria

   ✓d. the phyla of algal protists can be divided into three lineages based on photosynthetic pigments and chloroplast structure

   e. there are three distinct groups of photosynthetic prokaryotes

10. The organism that caused the Irish potato famine is a (an)

   a. actinopod

   b. apicoplexan

   ✓c. oomycete

   d. plasmodial slime mold

   e. cellular slime mold

CHALLENGE QUESTIONS

1. Different genes are expressed at different stages of the life cycle of the malaria-causing apicomplexan *Plasmodium,* which causes different proteins to appear on the outer coat of the infecting cells. Sporozoites are injected by mosquitos and travel in the blood to liver cells, where they continue their life cycle. It was discovered that the sporozoites produce protein coats that are sloughed off and continuously replaced. Host antibodies can attack these proteins by specific complementary binding. How does the continual sloughing and replacing of the protein coat work as an adaptive mechanism to prevent immune destruction of the sporozoite before it gets inside the liver cell where it is protected from blood-borne antibodies? (See Further Reading 1.)

2. Lynn Margulis of the University of Massachusetts has proposed the term *undulipodia* for eukaryotic flagella. Do you feel the distinction in structure between prokaryotic and eukaryotic flagella warrants an addition to biological terminology? Why, or why not? (See Further Reading 3.)

FURTHER READINGS

1. Donelson, J. E., and Turner, M. J. "How the Trypanosome Changes Its Coat." *Scientific American,* February 1985. A clear explanation of how a parasite wriggles out of the immunological clutches of its host.
2. Godson, G. N. "Molecular Approaches to Malaria Vaccines." *Scientific American,* May 1985. A good review of the life cycle of a deadly sporozoan, with emphasis on the biochemical mechanisms it uses to survive.
3. Margulis, L. *Symbiosis in Cell Evolution.* San Francisco: W. H. Freeman and Company, 1981. Extensive arguments for the endosymbiotic theory of eukaryotic origins.
4. Monastersky, R. "Knotty Evolutionary Tree in the Plant World." *Science News,* February 4, 1989. On the endosymbiotic origin of chloroplasts.
5. Saffo, M. B. "New Light on Seaweeds." *Bioscience,* October 1987. The scientific method reasserts itself, as a century-old hypothesis explaining vertical distribution of marine algae turns out to be too simple.
6. Sagan, D., and Margulis, L. "Bacterial Bedfellows." *Natural History,* March 1987. Strong arguments, with modern-day evidence, supporting the hypothesis of the origin of eukaryotes by endosymbiosis.
7. Vidal, G. "The Oldest Eukaryotic Cells." *Scientific American,* February 1984. Evidence that eukaryotes originally evolved in the form of unicellular plankton.
8. Whittaker, R. H., and Margulis, L. "Protist Classification and the Kingdoms of Organisms." *Biosystems* 10, 1978. Some insights on the five-kingdom system.

27

Plants and the Colonization of Land

I t is difficult to picture the continents barren, totally uninhabited by life of any form. But that is how we must imagine Earth for almost the first 90% of the time life has existed. Life was cradled in the seas and ponds, and there it evolved in confinement for three billion years. The long evolutionary pilgrimage onto land finally began about 400 million years ago. Plants led the way, followed by herbivorous animals and their predators. The terrestrial communities founded by green plants transformed the landscape (Figure 27.1).

The evolutionary history of the plant kingdom is a story of increasing adaptation to changing terrestrial conditions. That is the historical context in which this chapter surveys the current diversity of plants and traces their origins. Here we briefly discuss the anatomical characteristics of plants as they are relevant to our survey, reserving detailed treatment of the form and function of flowering plants for Unit Six.

INTRODUCTION TO THE PLANT KINGDOM

General Characteristics of Plants

All plants, as defined in this text, are multicellular eukaryotes that are photosynthetic autotrophs. However, not all organisms with these characteristics are plants; such characteristics also apply to some algae, which we have classified as protists (see Chapter 26).

Figure 27.1
A terrestrial community. Near timberline on Washington's Mt. Rainier, this landscape is colored by flowering plants (angiosperms) and conifers (gymnosperms).

Plants as we are defining them are nearly all terrestrial organisms, although some plants have returned secondarily to water during their evolution. Living on land poses very different problems from living in the water. As plants have adapted to the terrestrial environment, complex bodies with extensive specialization of cells for different functions have evolved. Aerial parts of most plants, such as stems and leaves, are coated with a waxy **cuticle** that helps prevent desiccation, a major problem on land. Gas exchange cannot occur across the waxy surfaces, but carbon dioxide and oxygen diffuse between the interior of leaves and the surrounding air through microscopic pores of the leaf surface called **stomata.** Besides their special adaptations for terrestrial life, land plants share many features with their progenitors, the green algae. For example, their photosynthetic cells contain chloroplasts having the pigments chlorophyll *a*, chlorophyll *b*, and a variety of yellow and orange carotenoids. Plant cells also have walls, and the staple material of their walls is cellulose. Carbohydrate is stored in the form of starch, generally in chloroplasts and other plastids. Mitosis occurs in all plants, proceeding through the typical phases described in Chapter 11.

A Generalized View of Plant Reproduction and Life Cycles

The move onto land paralleled a new mode of reproduction. In contrast to the reproductive style of seaweeds and other algae, gametes now had to be dispersed in a nonaquatic environment, and embryos, like mature body structures, had to be protected against desiccation.

Nearly all plants reproduce sexually, and most are also capable of asexual propagation. Plants produce their gametes within multicellular **gametangia,** organs having protective jackets of sterile (nonreproductive) cells that prevent the delicate gametes from drying out during their development (Figure 27.2). The egg is fertilized within the female organ, where the zygote develops into an embryo that is retained for some time within the jacket of protective cells.

In the life cycles of all plants, an alternation of generations occurs, in which haploid gametophytes and diploid sporophytes take turns producing one another (Figure 27.3; also see Chapter 26). In the life cycles of all extant plants, the sporophyte and gametophyte generations differ in morphology; that is, they are **het-**

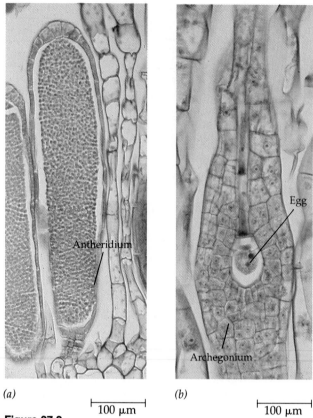

Antheridium

Egg

Archegonium

(a)   (b)

100 μm   100 μm

Figure 27.2
**Jacketed gamete-producing organs as a terrestrial
adaptation.** Gametes and zygotes of plants develop
within gametangia, moist chambers protected by a coat
of sterile cells. Shown here are the gametangia of a
moss. (a) Male gametangia are called antheridia; (b)
female gametangia are called archegonia (LMs).

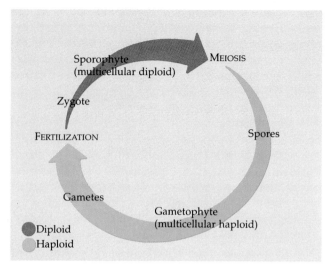

Figure 27.3
Alternation of generations: a generalized scheme.
The life cycles of all plants include a haploid, sexual generation (gametophyte) and a diploid, asexual generation
(sporophyte). The two generations alternate, each producing the other.

eromorphic. In all plants but the bryophytes (mosses
and their relatives), the diploid sporophyte is the more
conspicuous individual. The physiology of plant
reproduction is covered in Chapter 34; in this chapter
we look in some detail at representative plant life cycles.
It is valuable to understand these life cycles for two
reasons. First, they clarify one of the main trends in
plant evolution, toward reduction of the haploid generation and dominance of the diploid. Second, in many
cases, features of the life cycle, such as the replacement of flagellated sperm by pollen, can be interpreted as evolutionary adaptations to a terrestrial
environment.

Some Highlights of Plant Evolution

The fossil record chronicles four major periods of plant
evolution, which are also evident in the diversity of
contemporary plants (Figure 27.4). Each period followed the evolution of structures that opened new
adaptive zones on the land (see Chapter 23).

The first period of evolution was associated with the
origin of plants from aquatic ancestors, probably green

algae, during the Silurian period, which ended about
400 million years ago (see Table 23.1, p. 483). The first
terrestrial adaptations included a cuticle and jacketed
gametangia that protected gametes and embryos. Two
distinct groups of early plants emerged: one group
with specialized tissue known as **vascular** tissue and
one group without vascular tissue. (Vascular tissue
consists of cells joined into tubes that transport water
and nutrients throughout the body of the plant.) The
nonvascular plants were the ancestors of mosses.
During their subsequent evolution, some mosses
acquired tubes that transport water, but these structures are analogous—not homologous—to the conducting tissues of vascular plants.

The early diversification of the vascular plants, the
ancestors of all plants except mosses and their relatives, signaled the beginning of the second major period
of plant evolution. Adaptive radiation produced a
variety of primitive vascular plants during the early
Devonian period.

A third major period of plant evolution began with
the origin of the seed, a structure that advanced the
conquest of land by further protecting plant embryos
from desiccation and other hazards. A **seed** consists
of an embryo packaged along with a store of food
within a protective covering. The first vascular plants
with seeds arose about 360 million years ago, near the
end of the Devonian period. They bore their seeds as
naked structures unenclosed in any specialized chambers. Early seed plants gave rise to many types of
gymnosperms (from the Greek *gymnos*, "naked," and
sperma, "seed"), including the conifers, which are the
pines and other plants with cones. At the same time,
some groups of seedless plants, including ferns, have

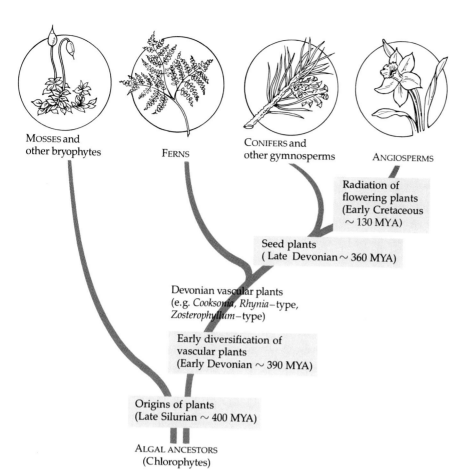

MOSSES and
other bryophytes

FERNS

CONIFERS and
other gymnosperms

ANGIOSPERMS

Radiation of
flowering plants
(Early Cretaceous
~ 130 MYA)

Seed plants
(Late Devonian ~ 360 MYA)

Devonian vascular plants
(e.g. *Cooksonia*, *Rhynia*–type,
Zosterophyllum–type)

Early diversification of
vascular plants
(Early Devonian ~ 390 MYA)

Origins of plants
(Late Silurian ~ 400 MYA)

ALGAL ANCESTORS
(Chlorophytes)

Figure 27.4
Some highlights of plant evolution.
Mosses and vascular plants probably
evolved independently from green algae.
The origins of plants from green algae, the
adaptive radiation of early vascular plants,
the emergence of seed plants, and the origin and diversification of flowering plants
(angiosperms) are four important chapters
in the history of the plant kingdom. (MYA
= million years ago.)

persisted to the present. Great forests of seedless plants and gymnosperms dominated the landscape for more than 200 million years.

The fourth major episode in the evolutionary history of plants was the emergence of flowering plants during the early Cretaceous period, about 130 million years ago. The flower is a complex reproductive structure that bears seeds within protective chambers called ovaries, which contrasts with the bearing of naked seeds by gymnosperms. The great majority of contemporary plants are flowering plants, or **angiosperms** (from the Greek *angion*, "container," referring to the ovary, and *sperma*, "seed").

Classification of Plants

Most botanists prefer to use the term **division** instead of *phylum* for the major taxonomic category under the plant kingdom. Divisions, like phyla, are further subdivided into classes, orders, families, and genera.

The classification scheme used in this text recognizes ten divisions within Kingdom Plantae. An alternative classification divides the plants into just two divisions, Bryophyta (mosses and their relatives) and Tracheophyta (vascular plants), assigning the various vascular plants to lower taxa. In addition to listing the ten divisions, Table 27.1 associates them with the stages

Table 27.1 A classification of plants

	Common Name	Approximate Number of Extant Species
Nonvascular Plants		
Division Bryophyta	Mosses, liverworts, and hornworts	16,000
Vascular Plants		
Seedless Plants		
Division Psilophyta	Whiskferns	10 to 13
Division Lycophyta	Club mosses	1,000
Division Sphenophyta	Horsetails	15
Division Pterophyta	Ferns	12,000
Seed Plants		
■ Gymnosperms		
Division Coniferophyta	Conifers	550
Division Cycadophyta	Cycads	100
Division Ginkgophyta	Ginkgo	1
Division Gnetophyta	Gnetae	70
■ Angiosperms		
Division Anthophyta	Flowering plants	235,000

Figure 27.5
Moss bog. Lacking vascular tissue and rigid supporting tissue, bryophytes are low profile plants most common in damp habitats. The matlike plants of this green carpet are gametophytes, the dominant generation in the life cycles of bryophytes.

3. The cell walls of most green algae, as of plants, are made of cellulose.

4. Green algae, like plants, store their carbohydrate reserves in the form of starch.

5. In some green algae, as in plants, vesicles derived from Golgi form a cell plate that divides the cytoplasm during cytokinesis.

The specific green algae that first colonized land are unknown; the soft thalli of algae left few fossils. The most likely candidates are filamentous chlorophytes, which carpeted the fringes of lakes or salt marshes. During the Silurian period, when the ancestors of plants first established themselves on the beachheads, the continents were relatively flat and they may have been subject to periodic flooding and draining. As water levels changed during the seasons or over longer cycles, natural selection would have favored those algae that could survive through periods when they were not submerged. Some evolutionary lineages accumulated adaptations that made it possible to live permanently above the water line, including waxy cuticles and jacketed reproductive organs, two hallmarks of plants. The evolutionary novelties of the first plants opened an adaptive zone that had never before been occupied. The new frontier was spacious, the bright sunlight was unfiltered by water and algae, the soil was rich in minerals, and, at least at first, there were no herbivores on land.

Bryophytes and vascular plants either had separate origins from green algae or diverged from the same ancestral plant relatively early in their evolution from an algal ancestor.

Division Bryophyta

The division Bryophyta (from the Greek *bryon,* "moss") consists of three classes: the mosses, liverworts, and hornworts. **Bryophytes** inhabit land, but they lack many of the terrestrial adaptations of vascular plants. Damp, shady places are the most common habitats of bryophytes (Figure 27.5).

Bryophytes display the two adaptations that first made the move onto land possible. They are covered by a waxy cuticle that helps the body retain water, and their gametes develop within gametangia, multichambered organs with jackets of sterile cells that keep the gametes moist (see Figure 27.2). The male gametangia, known as **antheridia,** produce flagellated sperm. In each female gametangium, or **archegonium,** one egg is produced. The egg is fertilized within the archegonium, and the zygote develops into an embryo within the protective jacket of the female organ.

Even with their cuticles and protected embryos, mosses are not totally liberated from their ancestral aquatic habitat. First of all, bryophytes need water to

of evolution discussed in the preceding section. Although these broader groupings do not have the status of formal taxonomic categories, they help fit the current diversity of plants into the historical context of a long evolutionary journey onto land.

THE MOVE ONTO LAND

The Case for Green Algae as the Ancestors of Plants

Most botanists agree that plants evolved from green algae (chlorophytes). These protists share the following set of key characteristics with plants:

1. Green algae and plants have, in addition to chlorophyll *a,* the same accessory photosynthetic pigments, including chlorophyll *b* and beta-carotene.

2. The chloroplasts of many green algae have their thylakoid membranes stacked locally into grana like those of plants.

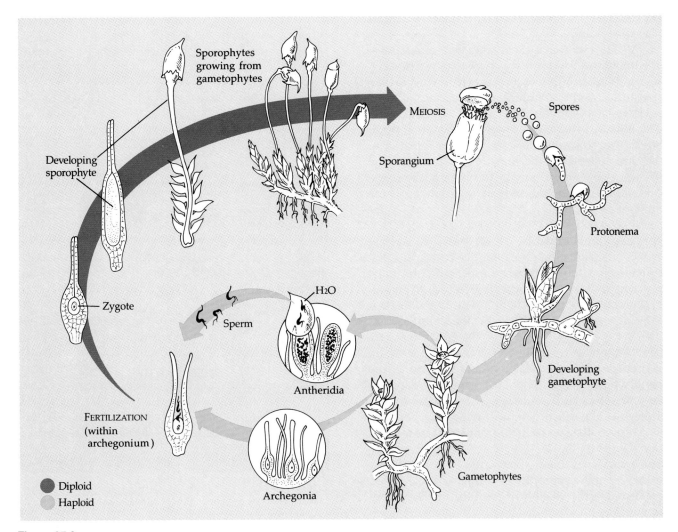

Figure 27.6

Life cycle of a moss. Most species of moss have separate male and female gametophytes, which have antheridia and archegonia, respectively. After a sperm swims through a film of moisture to an archegonium and fertilizes the egg, the diploid zygote divides by mitosis and develops into an embryonic sporophyte within the archegonium. During the next stage of its development, the sporophyte grows a long stalk that emerges from the archegonium, but the base of the sporophyte remains attached to the female gametophyte. At the tip of the stalk is a sporangium, a capsule in which meiosis occurs and haploid spores develop. When the sporangium bursts, the spores scatter. A spore germinates by dividing with mitosis to form a small, green, threadlike protonema resembling a green alga. The haploid protonema continues to grow and differentiates into a new gametophyte, completing the life cycle. The gametophyte is the more conspicuous generation in bryophytes.

reproduce, for their sperm, like those of most green algae, are flagellated and must swim from the antheridium to the archegonium to fertilize the egg (Figure 27.6). For many bryophyte species, a film of rainwater or dew is sufficient for fertilization to occur. In addition, most bryophytes have no vascular tissue to carry water from the soil to the aerial parts of the plant (the exceptions are certain mosses with elongated water-conducting cells). As water moves over the surface of most bryophytes, they must imbibe it like sponges and distribute it throughout the plant by the relatively slow processes of diffusion, capillary action, and cytoplasmic streaming.

Bryophytes lack the woody tissue required to support tall plants on land. Although they may sprawl horizontally as mats over a large surface, bryophytes always have a low profile. Most are only 1–2 cm in height, and even the largest are usually less than 20 cm tall.

Mosses The most familiar bryophytes are **mosses.** A mat of moss actually consists of many plants growing in a tight pack, helping to hold one another up. The mat has a spongy quality that enables it to absorb and retain water. Each plant of the mat grips the substratum with elongate cells or cellular filaments called rhizoids. Most photosynthesis occurs in the upper part of the plant, which has many small stemlike and leaflike appendages. The "stems" and "leaves" of a moss, however, are not homologous with these structures in vascular plants.

In the life cycle of a moss, the two generations, gametophyte and sporophyte, grow together (see Figure 27.6). The sporophyte produces haploid spores in a structure called a **sporangium;** the spores go on to form new gametophytes. The haploid gametophyte is the dominant generation in mosses and other bryophytes. The sporophyte is generally smaller, shorter lived, and depends on the gametophyte for water and nutrients. This contrasts with the life cycles of vascular plants, where the diploid sporophyte is the dominant generation.

Liverworts and Hornworts **Liverworts** are even less conspicuous plants than mosses. The bodies of some are divided into lobes, giving an appearance that must have reminded someone of the lobed liver of an animal (the root *wort* means "herb"). The liverwort class also includes genera with leafy rather than lobed bodies.

The life cycle of a liverwort is much like that of a moss. Within the sporangia of some liverworts are coil-shaped cells, called elaters, that spring out of the capsule when it opens, helping to disperse the spores. Liverworts can also reproduce asexually from little bundles of cells called gemmae, which are bounced out of cups on the surface of the gametophyte by raindrops (Figure 27.7).

Hornworts resemble liverworts but are distinguished by their sporophytes, which are elongated capsules that grow like horns from the matlike gametophyte. The photosynthetic cells of hornworts each have a single large chloroplast rather than the many smaller ones more typical of most plants.

It should be reemphasized that the division Bryophyta is a phylogenetic branch separate from vascular plants, not an evolutionary stepping-stone between algae and vascular plants. The most ancient fossils of mosses are about 350 million years old (late Devonian). By that time, vascular plants were already established on the land. Bryophytes have had a long success—there are more than 16,000 species today— but they probably never dominated much of the landscape. They are elegantly adapted to a limited range of terrestrial habitats. The vascular plants have additional terrestrial adaptations that enabled them to claim much more territory.

Terrestrial Adaptations of Vascular Plants

During their long evolution from aquatic ancestors, vascular plants accumulated many terrestrial adaptations in addition to cuticles and jacketed sex organs. The conquest of land entailed solutions to a new set of problems that aquatic algae did not face (Figure 27.8).

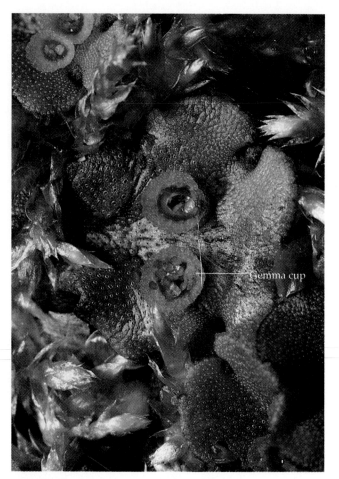

Figure 27.7
Liverworts. The gemmae cups of this leafy species function in asexual reproduction.

The resources a land plant needs to live are spatially segregated. The soil provides water and minerals, but there is no light underground for photosynthesis. The body of a vascular plant is differentiated into a subterranean root system that absorbs water and minerals and an aerial shoot system of stems and leaves that makes food. Roots, which must absorb water across their surface, generally lack the waxy cuticles that limit evaporation of water from stems and leaves.

Regional specialization of the plant body solved one problem but presented new ones. Roots anchor the plant, but for the shoot system to stand up straight in the air, it must have support. This is not a problem in the water: Huge seaweeds need no skeletons because they are essentially weightless in water. An important terrestrial adaptation of vascular plants is **lignin,** a hard material embedded in the cellulose matrix of the walls of cells that function in support. Turgor pressure (see Chapter 8) contributes to the support of small plants, but it is the skeleton of lignified walls that holds up a tree or other large vascular plant.

With increasing specialization of the root system and

Figure 27.8
Comparison of conditions faced by algae and plants.

Medium supportive

Whole alga has direct access to environmental water and minerals

Photosynthesis occurs in most cells

Availability of light often limits photosynthesis

Medium nonsupportive

Aerial parts of plant not in direct contact with water and minerals; tend to lose water to air

Photosynthesis confined to aerial parts of plant

Availability of light less likely to limit photosynthesis

shoot system came the new problem of transporting vital materials between the distant organs. Water and minerals must be conducted upward from the roots to the leaves. Sugar and other organic products of photosynthesis must be distributed from leaves to the roots. These problems are solved by an efficient vascular system that is continuous throughout the plant (Figure 27.9). The two conducting tissues of the vascular system are **xylem** and **phloem.** Tube-shaped cells in the xylem carry water and minerals up from the roots. Functioning xylem cells are actually dead; only their walls remain to provide a system of microscopic water pipes. The walls are generally lignified, and thus xylem functions in support as well as water transport. Phloem is a living tissue of elongated cells arranged into tubes that distribute sugar, amino acids, and other organic nutrients throughout the plant. Transport in plants is discussed further in Chapters 32 and 33.

In some groups of vascular plants, additional adaptations to living on land evolved, including the seed, the replacement of flagellated sperm with pollen as a means for delivering gametes outside water, and the increasing dominance of the diploid sporophyte in the alternation of generations.

The Earliest Vascular Plants

Encased in the sedimentary strata of the late Silurian and early Devonian periods are fossils of a variety of

Figure 27.9
The vascular system of a plant. An extensive network of veins services all parts of this aspen leaf. The vascular tissue includes xylem, specialized for conducting water and dissolved minerals from the roots up to the shoot system, and phloem, specialized for transporting sugar and other organic nutrients.

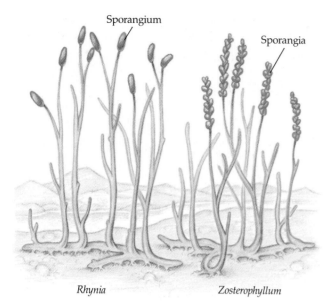

Figure 27.10
Vascular plants of the early Devonian. In form, *Rhynia* resembled *Cooksonia*, with dichotomous branching and terminal sporangia. True roots and leaves were absent. The plant was anchored by a rhizome, a horizontal, underground stem. *Rhynia* grew in dense stands around marshes. The largest species was about 50 cm tall. The several species of *Zosterophyllum* differed from *Rhynia* by bearing clusters of lateral sporangia near the tips of stems, rather than single terminal sporangia.

vascular plants, among the oldest terrestrial organisms known. Many of these petrified plants are beautifully preserved, right down to the microscopic organization of their tissues. The oldest is *Cooksonia*, which has been discovered in late Silurian rocks in both Europe and North America (the two continents were probably joined during the Silurian period). It was a simple plant with dichotomous (repeated "Y") branching. Some of the stems terminated in bulbous sporangia. *Cooksonia* was followed by a diversity of early Devonian species. Two characteristic genera are *Rhynia* and *Zosterophyllum* (Figure 27.10).

Rhynia and *Zosterophyllum* were geographically widespread during the early Devonian. The *Zosterophyllum* type of early Devonian plants is the best candidate as ancestor of the division Lycophyta, one of the nine divisions of vascular plants. Plants resembling *Rhynia* were probably on the evolutionary limb leading to all other divisions of vascular plants. A group of early Devonian plants called trimerophytes were possibly links between plants of the *Rhynia* type and later groups of seedless plants that began to appear in the middle and late Devonian periods.

SEEDLESS VASCULAR PLANTS

The earliest vascular plants were seedless. Four extant divisions of plants have retained this early condition.

Division Psilophyta

This division of relatively simple plants has only two genera, *Psilotum* and *Tmesipteris*. *Psilotum*, the better known, is sometimes called a living fossil because of its relatively simple morphology and its resemblance to early vascular plants (Figure 27.11). Some botanists, however, believe that *Psilotum* is a reduced fern rather than a little-changed remnant of the early Devonian flora.

Psilotum, widespread in the tropics and subtropics, is known in the United States by the common name **whiskfern**. In the diploid sporophyte generation, *Psi-*

Figure 27.11
The sporophyte of *Psilotum*. The scales on these dichotomously branched stems are not true leaves; they lack vascular tissues. The knobs along the stems are sporangia, which release haploid spores that germinate in the soil. Tiny subterranean gametophytes lack chlorophyll and depend for their food on symbiotic soil fungi that decompose organic litter. Flagellated sperm swim through the moist soil from antheridia to archegonia of the gametophytes. The zygote begins its development within the archegonium, and soon a young sporophyte emerges from the gametophyte, which then dies.

lotum has dichotomous branching reminiscent of some of the early vascular plants. True roots and leaves are absent. The subterranean part of the plant consists of a rhizome (horizontal stem) covered with tiny hairs called rhizoids. The upright stems bear emergences, which, unlike true leaves, lack vascular tissue.

Division Lycophyta

The extant **lycopods,** of the division Lycophyta, are bona fide relicts of a far more eminent past. Lycopods first evolved during the Devonian period and became a major part of the landscape during the Carboniferous period, which began about 340 million years ago and lasted until 280 million years ago. By that time, Division Lycophyta split into two evolutionary lines. One group evolved into woody trees that had diameters as large as 2 m and heights of more than 40 m. A second line of lycopods remained small and herbaceous (nonwoody). The giant lycopods thrived in the Carboniferous swamps for millions of years, but became extinct when the swamps began to dry up at the end of that geological period about 280 million years ago. The small lycopods survived, and they are represented today by about a thousand species, most belonging to the genera *Lycopodium* and *Selaginella*. Common names for these plants are club mosses or ground pines, though they are neither mosses nor pines.

Many species of *Lycopodium* are tropical plants that grow on trees as **epiphytes**—plants that use another organism as a substratum but are not parasites. Other species of *Lycopodium* grow close to the ground on forest floors in temperate regions, including the northeastern part of the United States.

The club moss you see in Figure 27.12 is the sporophyte, the diploid generation. The sporangia of *Lycopodium* are borne on **sporophylls,** leaves specialized for reproduction. After their discharge, the spores develop into inconspicuous gametophytes that may live underground for ten years or longer. These tiny haploid plants are, like the gametophytes of *Psilotum*, nonphotosynthetic, and are nurtured by symbiotic fungi. Each gametophyte develops archegonia with eggs and antheridia that make flagellated sperm. After a swimming sperm fertilizes an egg, the diploid zygote gives rise to a new sporophyte.

Lycopodium makes a single type of spore, which develops into a bisexual gametophyte having both female and male sex organs (archegonia and antheridia); it is thus said to be **homosporous.** In **heterosporous** plants, such as the lycopod genus *Selaginella*, the sporophyte makes two kinds of spores. **Megaspores** develop into female gametophytes bearing archegonia, and **microspores** become male gametophytes with antheridia. The gametophytes of het-

Figure 27.12
***Lycopodium*, a club moss.** Club mosses are common inhabitants of forest floors in the northeastern United States. The small plant has a horizontal rhizome that gives rise to roots and vertical branches and has true leaves containing strands of vascular tissue. The sporangia of *Lycopodium* are borne by specialized leaves called *sporophylls*. In some species, such as the one shown here, the sporophylls are clustered at the tips of branches into club-shaped structures called strobili (hence the common name club mosses).

erosporous plants are unisexual, either female or male. We will encounter the homosporous and heterosporous conditions again as we continue our survey of vascular plants.

Division Sphenophyta

Sphenophyta, whose members are commonly called horsetails, is another ancient lineage of seedless plants dating back to the Devonian radiation of early vascular plants (see Figure 27.4). The group reached its zenith during the Carboniferous period, when many species grew up to 15 m tall. All that survives of this division of plants are about 15 species of a single genus, *Equisetum*. *Equisetum* is widely distributed but is most common in the Northern Hemisphere, generally in damp locations such as stream banks (Figure 27.13).

Figure 27.13
***Equisetum* (horsetail).** *Equisetum* has an underground
rhizome from which vertical stems arise. The straight,
hollow stems are jointed, and whorls of small leaves or
branches emerge at the joints. The epidermis, the outer
layer of cells, is embedded with silica, which gives the
plants an abrasive texture. At the tips of some stems of
Equisetum are conelike structures bearing sporangia.
Horsetails are also called "scouring rushes" because
their abrasive stems were often used for scrubbing pots
and pans before modern methods were available.

(a)

(b)

Figure 27.14
Ferns. (a) Beech ferns on a forest floor. **(b)** Unfurling
frond (fiddlehead) of a deer fern.

The conspicuous horsetail plant is the sporophyte
generation. Meiosis occurs in the sporangia, and haploid spores are released. The gametophytes that develop
from these spores are only a few millimeters long, but
they are photosynthetic and free living (that is, not
dependent on the sporophyte for food). Horsetails are
homosporous; the single type of spore gives rise to a
bisexual gametophyte with both antheridia and archegonia. Flagellated sperm fertilize eggs in the archegonia, and young sporophytes later emerge.

Division Pterophyta

From their Devonian beginning, ferns, of the division
Pterophyta, radiated and stood alongside tree lycopods and horsetails in the great forests of the Carboniferous era. Of all seedless plants, ferns are by far the
most extensively represented in the modern flora (Figure 27.14). More than 12,000 species of ferns live today.
They are most diverse in the tropics, but a variety of
species is also found in temperate forests.

The leaves of ferns are generally much larger than
those of lycopods and probably evolved in a different
way. The origin of leaves is currently a subject of much
study. The small leaves of lycopods probably evolved
as emergences from the stem that contained a single
strand of vascular tissue. Leaves with this origin are
called **microphylls.** Each leaf of a fern, termed a **megaphyll,** has a branched system of veins. According to

Figure 27.15
A fern sporophyll. The sori on the underside of the specialized leaf are clusters of sporangia.

one theory, megaphylls evolved by the formation of webbing between many separate branches growing close together.

Most ferns have leaves, commonly called fronds, that are compound, meaning each leaf is divided into several leaflets. The frond grows as its coiled tip, the fiddlehead, unfurls. The leaves may sprout directly from a prostrate stem, as they do in brackens and sword ferns. Large tropical tree ferns, by contrast, have upright stems many meters tall.

The leafy fern plant familiar to us is the sporophyte generation. Some of the leaves are specialized sporophylls with sporangia on their undersides (Figure 27.15). The sporangia of many ferns are arranged in clusters called sori and are equipped with springlike devices that catapult spores several meters. Once airborne, spores can be blown by the wind far from their origin. Figure 27.16 illustrates the life cycle of a fern. With their swimming sperm and fragile gametophytes, the majority of ferns are restricted to relatively damp habitats.

The Coal Forests

The four divisions of plants that we have just surveyed represent the extant lineages of seedless vascular plants that formed vast forests during the Carboniferous period (about 300–350 million years ago) (Figure 27.17).

Seedless plants of the Carboniferous forests left not only living relics but also fossilized fuel in the form of coal. Coal powered the Industrial Revolution, and a resurgence in its use is inevitable as we continue to deplete oil and gas reserves.

Coal formed during several geological periods, but the most extensive beds of coal are found in strata deposited during the Carboniferous period, a time when much of the continents was covered by shallow seas and swamps. Europe and North America, near the equator at that time, were covered by tropical swamp forests. Dead plants did not completely decay in the stagnant waters, and great depths of organic rubble called peat accumulated. The swamps were later covered by the sea, and marine sediments piled on top of the peat. Heat and pressure gradually converted the peat to coal.

Growing along with the seedless plants in the Carboniferous swamps were primitive seed plants. These gymnosperms were not the dominant plants at that time, but they rose to prominence after the swamps began to dry up at the end of the Carboniferous period.

TERRESTRIAL ADAPTATIONS OF SEED PLANTS

Three life-cycle modifications contributed to the success of seed plants as terrestrial organisms:

1. The gametophytes of seed plants became even more reduced than in ferns and other seedless plants. Rather than developing in the soil as an independent generation, the minute gametophytes of seed plants are protected from desiccation by being retained within the moist reproductive tissue of the sporophyte generation.

2. Pollination replaced swimming as the mechanism for delivering sperm nuclei to eggs.

3. The seed evolved. Instead of the zygote developing directly into a young sporophyte that fends for itself, the zygote of a seed plant develops into an embryo that is packaged along with a food supply within a seed coat. This protects the dormant embryo from drought, cold, and other harsh conditions. Seeds also function in overland dispersal; they may be carried far from their parents by wind, water, or animals. In seed plants, the seed has replaced the spore as the stage in the life cycle that disperses the species.

GYMNOSPERMS

Of the two groups of seed plants, gymnosperms appear much earlier in the fossil record, and they lack the

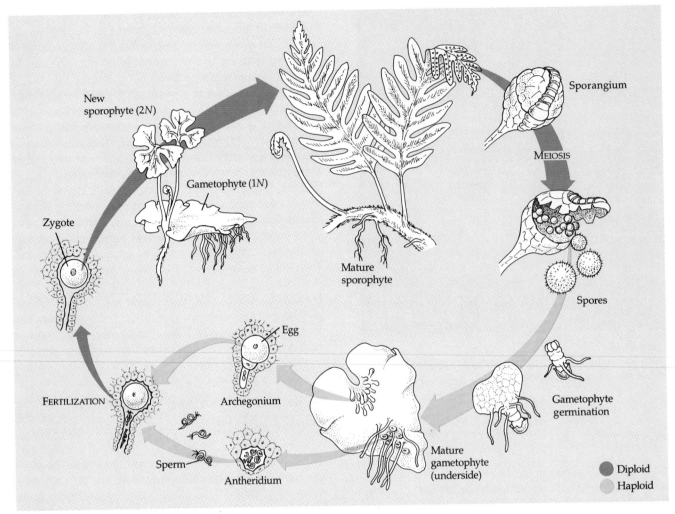

Figure 27.16
Life cycle of a fern. After a fern spore settles in a favorable place, it develops into a small, heart-shaped gametophyte that sustains itself by photosynthesis. Most ferns are homosporous; each gametophyte has both male and female sex organs, but the archegonia and antheridia usually mature at different times, assuring cross-fertilization between gametophytes. Fern sperm, like those of club mosses and horsetails, use flagella to swim through moisture from antheridia to eggs in the archegonia. A sex attractant secreted by archegonia helps direct the sperm. A fertilized egg develops into a new sporophyte, and the young plant grows out from an archegonium of its parent, the gametophyte.

Figure 27.17
A coal-forming forest of the Carboniferous period. Vast stands of seedless vascular plants covered much of what is now the Northern Hemisphere. This painting is based on fossil evidence of the composition of Carboniferous forests.

(a)

(b)

(c)

Figure 27.18
Three divisions of gymnosperms.
(**a**) **Cycadophyta.** Cycads, such as this *Cycas revoluta*, resemble palms and in fact are sometimes called sago palms. But cycads are not true palms, which are flowering plants. The cycad is a gymnosperm that bears naked seeds on the scales of cones. (**b**) **Ginkgophyta.** The ginkgo, also known as the maidenhair tree, has fanlike leaves that turn gold and are deciduous in autumn, an unusual trait for a gymnosperm. The ginkgo is a popular ornamental tree in cities because it can survive air pollution and other environmental insults. (**c**) **Gnetophyta.** Gnetae is a potpourri class of three gymnosperm genera that are probably not closely related. *Gnetum* grows in the tropics as a tree or vine. *Ephedra* is a shrub found in the American deserts. *Welwitschia*, shown here, is a bizarre plant with straplike leaves. It lives only in the deserts of southwestern Africa.

enclosed chambers in which angiosperm seeds develop. There are four divisions of gymnosperms. Three are relatively small: Cycadophyta, Ginkgophyta, and Gnetophyta (Figure 27.18). By far the largest division of gymnosperms is Coniferophyta.

Division Coniferophyta

The name **conifer** (from the Greek *konos,* "cone," and *phero,* "carry") comes from the reproductive structure of these plants, the cone. Pines, firs, spruce, larches, yews, junipers, cedars, cypresses, and redwoods all belong to this division of gymnosperms. Most are large trees. Although there are only about 550 species, conifers dominate vast regions of the Northern Hemisphere, where the growing season is relatively short because of latitude or altitude (Figure 27.19).

Nearly all conifers are evergreens, meaning they retain leaves throughout the year. Even during winter, a limited amount of photosynthesis occurs on sunny days. And when spring comes, conifers already have fully developed leaves that can take advantage of the warmer days.

The needle-shaped leaves of pines and firs are adapted to dry conditions. A thick cuticle covers the leaf, and the stomata are located in pits, further reducing loss of water. The conifer needle, despite its shape, is a megaphyll (see p. 574) as are the leaves of all seed plants.

We get most of our lumber and paper pulp from the wood of conifers. What we call wood is actually an accumulation of lignified xylem tissue, which gives the tree structural support.

Coniferous trees are among the tallest, largest, and oldest living organisms on Earth. Redwoods, found

Figure 27.19
Coniferous forest. As shown in this view of the Peyto Lake region of Banff National Park in Alberta, Canada, conifers form an almost solid band across the northern continents at latitudes where the growing season is short.

only in a narrow coastal strip of Northern California, grow to heights of more than 100 m; only certain eucalyptus trees in Australia are taller. The largest (most massive) organisms alive are the giant sequoias, relatives of redwoods that grow in the Sierra Nevada of California. One, known as the General Sherman tree, has a trunk with a circumference of 26 m. Bristlecone pines, another species of California conifer, are among the oldest organisms alive. One bristlecone, named Methuselah, is more than 4600 years old; it was a young tree when humans invented writing.

Life History of a Pine The pine tree, a representative conifer, is a sporophyte, with its sporangia located on cones. The gametophyte generation develops from haploid spores that are retained within the sporangia. Conifers are heterosporous; male and female gametophytes develop from different types of spores produced by separate cones. Each tree usually has both types of cones. Small pollen cones produce small spores that develop into the male gametophytes. Larger, more complex ovulate cones usually develop on separate branches of the tree and make larger spores that develop into female gametophytes (Figure 27.20). From the time young cones appear on the tree, it takes nearly

three years for a complicated series of events to produce mature seeds. The scales of the ovulate cone then separate and the winged seeds travel on the wind. A seed that lands in a habitable place germinates, its embryo emerging as a pine seedling. Table 27.2, on page 580, puts the complicated life cycle of a conifer into perspective by highlighting some important ways it differs from the life cycles of other plants.

The History of Gymnosperms

Gymnosperms probably descended from a group of Devonian plants called progymnosperms. They were originally seedless plants, but by the end of the Devonian period seeds had evolved. Adaptive radiation during the Carboniferous period and early Permian period produced the various divisions of gymnosperms.

In the history of life, the Permian period was one of great crises. Formation of the supercontinent Pangaea (see Chapter 23) may have been one reason that continental interiors became warmer and drier as the Permian progressed. The flora and fauna of the Earth changed dramatically, as many groups of organisms

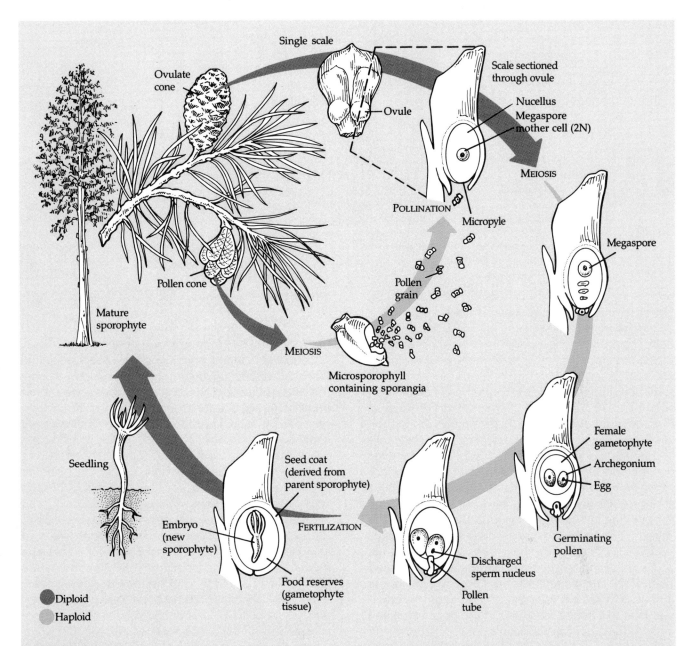

Single scale

Ovulate cone

Ovule

Scale sectioned through ovule

Nucellus

Megaspore mother cell (2N)

POLLINATION

MEIOSIS

Micropyle

Pollen grain

Megaspore

MEIOSIS

Pollen cone

Microsporophyll containing sporangia

Mature sporophyte

Female gametophyte

Archegonium

Egg

Seedling

Germinating pollen

Seed coat (derived from parent sporophyte)

Embryo (new sporophyte)

FERTILIZATION

Discharged sperm nucleus

Pollen tube

Food reserves (gametophyte tissue)

● Diploid

● Haploid

Figure 27.20
Life cycle of a pine, a typical gymnosperm. Trees (sporophytes) of most species bear both pollen cones and ovulate cones. A pollen cone contains hundreds of sporangia held in tiny sporophylls (reproductive leaves). Cells in the sporangia undergo meiosis, giving rise to haploid microspores that develop into pollen grains (immature male gametophytes). An ovulate cone consists of many scales, each with two ovules. Each ovule contains a sporangium, called the nucellus, enclosed in a protective integument with a single opening, the micropyle. During pollination, wind-blown pollen falls on the ovulate cone and is drawn into the ovule through the micro-

pyle. The pollen grain germinates in the ovule, forming a pollen tube that begins to digest its way through the nucellus.

Fertilization usually occurs more than a year after pollination. During that year, a megaspore mother cell in the nucellus undergoes meiosis to produce four haploid cells. One of these cells survives as a megaspore, which divides repeatedly, giving rise to the immature female gametophyte. Two or three archegonia, each with an egg, then develop within the gametophyte. By the time eggs are ready to be fertilized, two sperm nuclei have developed in the male gametophyte and the pollen tube has grown through the nucellus to the female gametophyte.

Fertilization occurs when one of the sperm nuclei, injected into an egg cell by the pollen tube, unites with the egg nucleus. All of the eggs in an ovule may be fertilized, but usually only one zygote develops into an embryo.

The pine embryo, or the new sporophyte, has a rudimentary root and several embryonic leaves called cotyledons. A food supply, consisting of the female gametophyte, surrounds and nourishes the embryo until it is capable of photosynthesis. A pine seed consists of an embryo, its food supply, and a surrounding seed coat derived from the parent tree.

Table 27.2 Comparison of reproduction for some major plant groups

Group	Dominant Stage of Life Cycle	Homosporous or Heterosporous	Mechanism for Combining Gametes
Mosses	Gametophyte	Homosporous	Flagellated sperm swims through film of water to egg
Ferns	Sporophyte	Homosporous	Flagellated sperm
Conifers	Sporophyte	Heterosporous	Sperm nuclei transported in wind-blown pollen
Flowering plants	Sporophyte	Heterosporous	Pollen transferred by wind or animals

disappeared and others emerged as their successors. The changeover was most pronounced in the seas, but terrestrial life was affected as well. In the animal kingdom, amphibians decreased in diversity and were replaced by reptiles, which were better adapted to the arid conditions. Similarly, the lycopods, horsetails, and ferns that dominated the Carboniferous swamps were largely replaced by conifers, which first appeared in the late Carboniferous, and their relatives, the cycads, which were more suited to the drier climate. The world and its life had changed so markedly that geologists use the end of the Permian period as the boundary between the Paleozoic and Mesozoic eras (this boundary was originally defined by the changeover in marine fossils). The Mesozoic is sometimes referred to as the "age of dinosaurs," when giant reptiles were supported by a vegetation consisting mostly of conifers and great palmlike cycads. When the climate changed again at the end of the Mesozoic, becoming cooler, the dinosaurs became extinct. Some of the gymnosperms, particularly conifers, persisted, however, and are still an important part of the Earth's flora.

ANGIOSPERMS

Today, angiosperms, or flowering plants, are by far the most diverse and geographically widespread of all plants. About 235,000 species are known, compared with 550 gymnosperm species. All angiosperms are placed in a single division, Anthophyta (from the Greek *antho*, "flower"). The division is split into two classes: Monocotyledones (monocots) and Dicotyledones (dicots), which differ in several ways that are described in Chapter 31. Examples of monocots are lilies, orchids, yuccas, palms, and grasses, including lawn grasses, sugar cane, and grain crops (corn, wheat, rice, and others). Among the many dicot families are roses, peas,

buttercups, sunflowers, oaks, and maples (Figure 27.21).

Most angiosperms employ insects and other animals as couriers for transferring pollen to female sex organs, which makes pollination less random than the wind-dependent pollination of gymnosperms. Some flowering plants are wind pollinated, but we do not know whether this condition is primitive or evolved secondarily from ancestors that were pollinated by animals.

Vascular tissue also became more refined during angiosperm evolution. The cells that conduct water in conifers are **tracheids,** believed to be a relatively early type of xylem cell (Figure 27.22). The tracheid is an elongated, tapered cell that functions in both mechanical support and movement of water up the plant. In angiosperms, shorter, wider cells called **vessel elements** evolved from tracheids. Vessel elements are arranged end to end to form continuous tubes that are more specialized than tracheids for transporting water but less specialized for support. The xylem of angiosperms is reinforced by a second cell type, the **fiber,** which also evolved from the tracheid. With their thick lignified walls, the xylem fibers are specialized for support. Fibers evolved in conifers, but vessel elements did not.

The refinements in vascular tissue and other structural advances surely contributed to the success of angiosperms, but the greatest factor in the rise of angiosperms was probably the evolution of the flower, a remarkable apparatus that enhances the efficiency of reproduction by attracting and rewarding pollen-carting animals.

The Flower

The **flower** is the reproductive structure of an angiosperm. A flower is a compressed shoot with four whorls

Parallel
venation

(a)

Net
venation

(b)

Figure 27.21
Two classes of angiosperms. (**a**) Monocots, such as
these pink lady's slipper orchids, generally have leaves
with parallel veins. (**b**) The leaves of dicots, such as this
blue violet, usually have netlike venation.

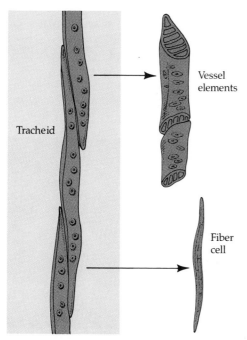

Tracheid

Vessel
elements

Fiber
cell

Figure 27.22
Evolution of xylem cells. In angiosperm evolution,
tracheids gave rise to vessel elements specialized for
conducting water and fiber cells specialized for support.

of modified leaves (Figure 27.23). Starting at the bottom of the flower are the **sepals,** which are usually green. They enclose the flower before it opens (think of a rosebud). Above the sepals are the **petals,** brightly colored in most flowers. They aid in attracting insects and other pollinators. Flowers that are wind pollinated, such as those of many grasses, are generally drab in color. The sepals and petals are sterile floral parts not directly involved in reproduction. Within the ring of petals are the reproductive organs, **stamens** and **carpels.** A stamen consists of a stalk called the **filament** and a terminal sac, the **anther,** where pollen is produced. At the tip of the carpel is a sticky **stigma** that receives pollen. A **style** leads to the **ovary** at the base of the carpel. Protected within the ovary are the ovules, which develop into seeds after fertilization. Remember, the enclosure of seeds within the ovary is one of the features that distinguishes angiosperms from gymnosperms. The carpel probably evolved from a seed-bearing leaf that became rolled into a tube.

Botanists recognize four evolutionary trends in various angiosperm lineages:

1. The number of floral parts has become reduced.

2. Floral parts have become fused. For example, some flowers have compound carpels formed by fusion

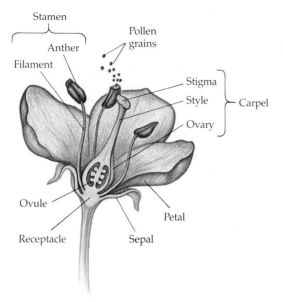

**Figure 27.23
Structure of a flower.**

of several carpels. The general term **pistil** is sometimes used for a single carpel or several fused carpels.

3. Symmetry has changed from radial, in which any cut down its central axis will divide the flower into two equal halves, to bilateral, in which the flower has distinct left and right halves.

4. The ovary has dropped to a position below the petals and sepals, where the ovules are better protected.

Figure 27.24 illustrates these trends. With modification in floral structure, many angiosperms are specialists at using specific animals for pollination, as we will see later in this chapter.

The Fruit

Fruits protect dormant seeds and aid in their dispersal. A **fruit** is a ripened ovary (Figure 27.25). As seeds develop after fertilization, the wall of the ovary thickens. A pea pod is an example of a fruit, with seeds (mature ovules) encased in the ripened ovary. Some fruits, such as apples, incorporate other floral parts along with the ovary. Peas and apples are both simple fruits, meaning they develop from a single ovary. Aggregate fruits, such as raspberries, come from several ovaries that were part of the same flower. A pineapple is an example of a multiple fruit, one that develops from several separate flowers.

Fruits are modified in various ways that help disperse seeds. Some flowering plants, such as dandelions and maples, have seeds within fruits that act as kites or propellers that aid in dispersion by wind. But most angiosperms use animals to carry seeds. Some of these plants have fruits modified as burrs that cling to animal fur (or the clothes of humans). Other angiosperms produce edible fruits. When it eats the fruit, the animal digests the fleshy part, but the tough seeds usually pass unharmed through the digestive tract. Mammals and birds may deposit seeds, along with a fertilizer supply, miles from where the fruit was eaten. The use of animals to tote seeds and pollen has helped angiosperms become the most successful plants on Earth.

(a)

(b)

**Figure 27.24
Primitive and advanced flowers.** (*Primitive* and *advanced* refer only to the closeness of structures to the ancestral condition.) **(a)** The flower of a magnolia, with its radial symmetry and many unfused parts, is considered to be relatively primitive. **(b)** An orchid has many fused parts, has bilateral symmetry, and is specialized for a certain pollinator.

Figure 27.25
A developing fruit. Zucchini squash is a ripened ovary.

Life Cycle of an Angiosperm

Angiosperms are heterosporous. The flower of the sporophyte produces microspores that form male gametophytes and megaspores that produce female gametophytes (Figure 27.26). The immature male gametophytes are **pollen grains,** which develop within the anthers of stamens. Each pollen grain has two haploid nuclei. **Ovules,** which develop in the ovary, contain the female gametophyte, an **embryo sac** with eight haploid nuclei in seven cells (a large central cell has two haploid nuclei). One of the cells is the egg. Development of pollen and embryo sac is described in more detail in Chapter 34.

After its release from the anther, the pollen is carried to the sticky stigma at the tip of a carpel. Although some flowers self-pollinate, most have mechanisms that ensure **cross-pollination,** the transfer of pollen from the flowers of one plant to flowers of another

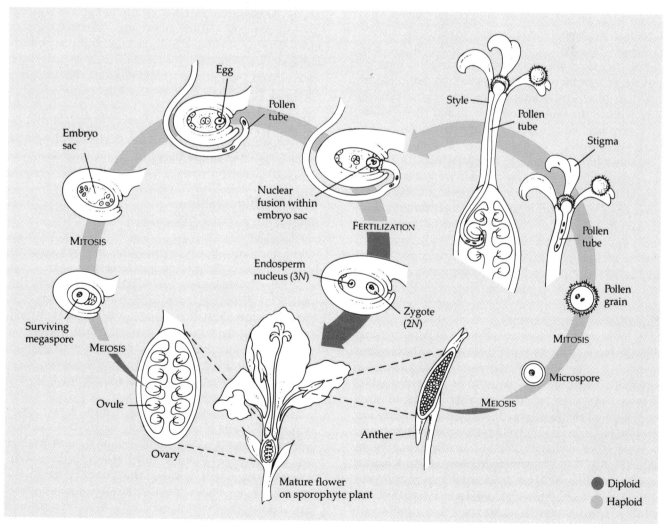

Figure 27.26
Life cycle of an angiosperm. The flower of the sporophyte produces microspores that form male gametophytes (pollen) and megaspores that produce female gametophytes (embryo sacs) within ovules. Pollination brings the gametophytes together in the ovary, fertilization occurs, and zygotes develop into sporophyte embryos that are packaged along with food in seeds.

plant of the same species. For example, stamens and carpels of a single flower may mature at different times, or the organs may be so arranged within the flower that self-pollination is unlikely (see Chapter 34).

The pollen grain, an immature male gametophyte, germinates after it adheres to the stigma of a carpel. The pollen grain extends a tube that grows down the style of the carpel. After it reaches the ovary, the pollen tube penetrates through a pore in the integuments of the ovule and discharges two sperm nuclei into the embryo sac. One sperm nucleus unites with the egg to form a diploid zygote. The other sperm nucleus fuses with two nuclei in the center cell of the embryo sac. This central cell now has a triploid (3N) nucleus. The pollen of conifers, remember, also releases two sperm nuclei, but one disintegrates. In contrast, both sperm nuclei of angiosperm pollen fertilize cells in the embryo sac. This phenomenon, known as **double fertilization,** is unique to angiosperms.

After double fertilization, the ovule matures into a seed. The zygote develops into a sporophyte embryo with a rudimentary root and one or two seed leaves, the **cotyledons** (monocots have one seed leaf and dicots have two). The triploid nucleus in the center of the embryo sac divides repeatedly to give rise to a triploid tissue called **endosperm,** rich in starch and other food reserves. Monocot seeds such as corn store most of their food in the endosperm. Beans and many other dicots restock most of the nutrients into the developing cotyledons.

The seed is a mature ovule, consisting of the embryo, endosperm, and a seed coat derived from the integuments (outer layers of the ovule). In a suitable environment, the seed germinates. The coat ruptures and the embryo emerges as a seedling, using the food stored in the endosperm and cotyledons.

Table 27.2 (p. 580) gives a summary comparison of the life cycles of the various groups of plants.

The Rise of Angiosperms

Charles Darwin called the origin of the angiosperms an "abominable mystery." The mystery endures. The problem is the relatively sudden appearance of angiosperms in the fossil record, with no known transitional links to ancestors. The oldest fossils that are widely accepted as angiosperms are found in rocks of the early Cretaceous period, about 120 million years old. They are sparsely represented among a much greater abundance of ferns and gymnosperms. By the end of the Cretaceous, 65 million years ago, the angiosperms had radiated and become the dominant plants on Earth, as they are today.

Some paleobotanists believe the relative suddenness of the angiosperm appearance in the early Cretaceous is an artifact of an imperfect fossil record. Angiosperms may have originated somewhat earlier

in the highlands or some other location where fossilization was unlikely, and their sudden appearance may be the result of their spread to locations where fossils form more frequently. Another view, in the spirit of the evolutionary theory known as punctuated equilibrium (see Chapter 23), holds that angiosperms actually did evolve and radiate rather abruptly (on the scale of geological time).

The lack of transitional forms in the fossil record obscures the ancestry of flowering plants. Nevertheless, nearly all paleobotanists agree that angiosperms evolved from some group of gymnosperms, perhaps seed ferns (not really ferns, because they had seeds). This ancient group of unspecialized gymnosperms, now extinct, lived in Carboniferous forests and persisted into the Mesozoic era.

Whatever and whenever their origin, the rise to prominence of angiosperms during the Cretaceous is amply documented in the fossil record. The Cretaceous was another crisis period when many old groups of organisms were replaced by new ones. Cooler climates may have contributed to the changeover. Again, the frequency of extinctions was greatest in the seas, but significant changes in terrestrial fauna and flora also occurred. The dinosaurs disappeared, as did many of the cycads and conifers that had thrived during the Mesozoic era. They were replaced by mammals and flowering plants. The change in fossils during the Cretaceous, especially in marine sediments, is so extreme that geologists use the end of that period as the boundary between the Mesozoic and Cenozoic eras.

Relationships Between Angiosperms and Animals

Ever since they followed plants onto the land, animals have influenced the evolution of terrestrial plants, and vice versa. The fact that animals must eat affects the natural selection of both animals and plants. For instance, with animals crawling and foraging for food on the forest floor, there must have been selection pressure in favor of plants that kept their spores and gametophytes up in the treetops, rather than dropping these critical structures to hungry animals on the ground. This in turn may have been a selection factor in the evolution of flying insects. On the other hand, as plants with flowers and fruits evolved, some herbivores became beneficial to the plants by carrying the pollen and seeds of plants they used as food. Certain animals became specialists at these tasks, feeding on specific plants. Natural selection reinforced these interactions, for they improved the reproductive success of both partners. The plant got pollinated and the animal got fed. The mutual evolutionary influence between two species is termed **coevolution** (this definition will be refined in Chapter 48).

Coevolution of angiosperms and their pollinators is

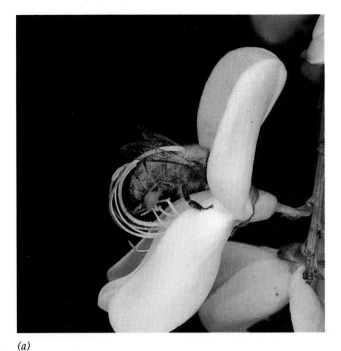

(a)

(b)

Figure 27.27
Relationships of angiosperms and their pollinators.
(**a**) This scotch broom has a tripping mechanism that dusts pollen onto the back of a visiting bee. (**b**) Some flowers have nectaries at the bottom of long tubes. This butterfly (*Heliconius erato*) has a slender proboscis (coiled in this photo) that reaches the nectary. Such mechanisms often result in exclusive relationships between angiosperms and their pollinators.
(**c**) Wahlberg's epauleted bat (*Epomophorus wahlbergi*) feeds on the baobab flower. Its body collects and distributes the plant's pollen. The baobab, like many plants that depend on bats, has blossoms that are lightly colored, large, and scented, and are therefore easily found by nocturnal feeders.

(c)

largely responsible for the diversity of flowers. Many flowers are pollinated by a specific animal—say, a particular type of bee, beetle, bird, or bat. These exclusive relationships ensure that the plant's pollen will not be wasted by being carried to the flower of a different species. At the same time, the pollinator has a monopoly on a food source. The color and fragrance of a flower is usually keyed to its pollinator's senses of sight and smell. Flowers pollinated by bees, for instance, often have nectar guides, markings that help direct the bee as it taxies to the nectaries, glands that secrete the sugary solution bees eat (Figure 27.27). On its way to or from the nectar, the bee becomes dusted with pollen. The markings on some bee-pollinated flowers are invisible to humans but are apparently vivid to bees, whose eyes are sensitive to ultraviolet light. Flowers pollinated by birds are usually red, to which bird eyes are especially sensitive. The shape of the flower may also specify a particular pollinator. Flowers pollinated by hummingbirds, for instance, have their

nectaries located deep in a floral tube where only the long, thin tongue of the hummingbird is likely to reach.

Relationships between angiosperms and animals are also evident in the edible fruits of angiosperms. Fruits that are not yet ripe are usually green, hard, and distasteful (at least to us). This helps the plant retain its fruit until the seeds are mature and ready for dispersal. As it ripens, the fruit becomes softer and its sugar content increases. Many fruits also become fragrant and brightly colored, advertising their ripeness to animals. One of the most common colors for ripe fruit is red, which insects cannot see very well. Thus, most of the fruit is saved for the birds and mammals that disperse the seeds. Again, we see that one of the keys to angiosperm success has been interaction with animals.

Angiosperms and Agriculture   Flowering plants provide nearly all our food. All of our fruit and vegetable crops are angiosperms. Corn, rice, wheat, and the other grains are grass fruits. The endosperm of the grain seeds is the main food source for most of the people of the world and their domesticated animals. We also grow angiosperms for fiber, drugs, perfumes, and decoration.

Like other animals, early humans probably collected wild seeds and fruits. Agriculture was gradually invented as humans began sowing seeds and cultivating plants to have a more dependable food source. As they domesticated certain plants, humans began to intervene in plant evolution by selective breeding designed to improve the quantity and quality of the foods the crops produced. We have developed a very special relationship with the plants we cultivate. We water them and fertilize them, try to protect them from insects, and plant their seeds. Many of these plants are so genetically removed from their origins that they probably could not survive in the wild. Agriculture is a unique case of an evolutionary relationship between plants and animals. And it is a precarious relationship for the billions of people now living on Earth, because cultivated crops are very vulnerable to natural and human-caused disasters. In an attempt to protect the world's food supply, many countries try to maintain seed banks of agriculturally important plants. The U.S. National Seed Storage Laboratory at Fort Collins, Colorado, maintains nearly 250,000 varieties representing about 1300 crop species.

*   *   *

In tracking 400 million years of plant evolution, from the descendants of green algae that moved onto shore to the flowering plants that now dominate most landscapes, we have seen once again that the organisms of today are best understood by their history. This chapter has also reinforced the theme of the connectedness of biology and geology. Transitions such as changes in climate certainly influenced plant evolution, but plants have also changed the course of geology by altering soil and transforming the land in many other ways. The next chapter considers the history and environmental impact of the fungi, another kingdom of organisms that spread onto land with the plants.

STUDY OUTLINE

1. Since the colonization of land by plants 400 million years ago, plant evolution has been interwoven with major changes in terrestrial conditions.

Introduction to the Plant Kingdom (pp. 564–568)

1. All plants are photosynthetic multicellular eukaryotes. Plants have cellulose in their cell walls and store their carbohydrates as starch. Plant cells possess chlorophyll *a*, chlorophyll *b*, and a variety of carotenoid pigments. Stomata and the cuticle of stems and leaves are two important adaptations to a terrestrial life style. Mitosis and meiosis occur in all plants.

2. As plants colonized the land, jacketed sex organs, called gametangia, evolved and protected gametes and embryos from desiccation.

3. All modern plants show a heteromorphic alternation of generations with distinctive haploid gametophyte and diploid sporophyte forms. The sporophyte is more conspicuous in all but the bryophytes.

4. Important innovations in plant structure catalyzed four major periods of plant evolution. First, terrestrial adaptations allowed plants to make the transition from water to the land about 400 million years ago. The second major period occurred with the emergence of vascular tissue in one of the two groups of plants that evolved from the earliest terrestrial plants. Third, the origin of the seed about 360 million years ago allowed embryos to leave the parent plant encased in a resistant coat complete with a food supply. The earliest seed plants were gymnosperms, which produce naked seeds. The fourth major episode occurred 130 million years ago with the evolution of the flower, a specialized reproductive structure that produces seeds enclosed within an ovary. Today the angiosperms or flowering plants are the most diverse members of the plant kingdom.

5. In plant taxonomy, the division is the largest category under the kingdom level.

The Move onto Land (pp. 568–572)

1. The green algae are considered the ancestors of plants because representatives of both possess chlorophyll *a* and identical accessory pigments in similar chloroplasts, have cell walls made of cellulose, store starch, and form cell plates during cytokinesis.

2. The early separation of plants into vascular and nonvascular species could have been the result of divergence from a common ancestral stock or separate origins from different groups of algae.

3. The nonvascular division Bryophyta consists of the mosses, liverworts, and hornworts. Bryophytes have a waxy cuticle and gametangia that protect the gametes and the embryo on land, but they still require a moist habitat for fertilization and for imbibing water in the absence of vascular tissue. Lack of woody tissue dictates their short stature.

4. Mosses grow tightly together, clinging to the soil with hairlike rhizoids. The small sporophyte grows out of the leafy gametophyte, on which it depends for water and nutrients.

5. Liverworts and hornworts are inconspicuous bryophytes with distinctive reproductive structures.

6. Vascular plants absorb water and minerals from the soil through roots and produce food in aerial parts held erect by turgor pressure or incorporation of lignin into the cell walls. Distant parts of the plant are interconnected by the vascular tissue, with water and minerals transported by xylem and organic nutrients by phloem.

7. Over evolutionary time, there was a trend toward increasing dominance of the diploid sporophyte in the life cycles of vascular plants.

8. The earliest known vascular plant belonged to the Silurian genus *Cooksonia*, which gave rise to several genera, two of which may have evolved into the extant divisions of vascular plants.

Seedless Vascular Plants (pp. 572–575)

1. There are four divisions of seedless plants.
2. The division Psilophyta is a small group with simple morphology to which the widespread genus *Psilotum* (whiskferns) belongs.
3. The division Lycophyta consists of the lycopods or club mosses, small, herbaceous survivors of a dominant ancient division that once included large treelike forms. Modern species use specialized leaves called sporophylls to produce either a single type of spore or megaspores and microspores, which develop into female and male gametophytes, respectively.
4. The horsetails belong to a single surviving genus of the division Sphenophyta. The conspicuous sporophyte produces tiny bisexual gametophytes from a single type of spore.
5. The division Pterophyta consists of the ferns, the most species-rich group of living seedless plants. The fronds of the sporophyte generation form sporangia that produce spores that germinate into small gametophytes.
6. The ancient seedless plants of the Carboniferous forests left a legacy of fossilized fuel in the form of coal.

Terrestrial Adaptations of Seed Plants (p. 575)

1. The success of seed plants on land may be attributed to three developments: (1) further reduction of the gametophyte and its retention within the sporophyte; (2) replacement of swimming sperm with pollination; and (3) development of the seed which functions in protection and dispersal.

Gymnosperms (pp. 575–580)

1. Gymnosperms were the first of the two groups of seed plants to appear in the fossil record.
2. Coniferophyta comprises the largest of the four gymnosperm divisions. Almost all conifers are evergreens with needle-shaped leaves. Conifers are among the tallest, largest, and oldest living organisms.
3. The cone is the distinguishing characteristic of conifers. In the pine, two different types of cones produce male and female gametophytes on the sporophyte tree. Pollen grains released as immature male gametophytes land on female cones housing immature female gametophytes inside complex ovules. After a period of gametophyte maturation, fertilization occurs. The zygote develops into an embryo, which is packaged into a winged seed that disperses by the wind from exposed scales of the mature ovulate cone.
4. The divisions of gymnosperms arose during the Carboniferous and early Permian periods. They became dominant during the Mesozoic.

Angiosperms (pp. 580–586)

1. Angiosperms, flowering plants, belong to the division Anthophyta, which contains the most diverse and widespread members of the plant kingdom. Angiosperms are divided into monocots and dicots.
2. The use of insect and animal vectors for pollination and the origin of refined vessel elements for transport and support contributed to the rise of the angiosperms, but the greatest contribution to their success was probably the evolution of the flower, which greatly improved reproductive efficiency.
3. The flower is a reproductive structure housing stamens and carpels within sterile sepals and petals. Evolutionary trends have been the reduction and fusion of floral parts, a change from radial to bilateral symmetry, and the settling of the ovary with its ovules into a more protected site below the sepals and petals.
4. Fruits form from the ripened ovaries of one or more flowers, with or without associated floral parts. They protect dormant seeds and are modified in various ways that aid in seed dispersal.
5. Pollen grains are immature male gametophytes which form in the anthers of stamens and germinate on the sticky stigma of the carpel. A pollen tube grows down to the ovary, where one sperm nucleus fertilizes the egg and another nucleus combines with two female haploid nuclei to make a triploid endosperm that functions in food storage. This double fertilization, unique to angiosperms, produces a seed containing the embryonic sporophyte complete with a primordial root and one or two cotyledon seed leaves, surrounded by the endosperm and seed coat.
6. Most paleobotanists agree that angiosperms arose from gymnosperms. By the end of the Cretaceous, 65 million years ago, angiosperms had become the dominant plants on Earth.
7. Coevolution of plants and pollinating animals has contributed to the reproductive success of the angiosperms and has influenced the color, fragrance, and shape of flowers and the production of nectar and edible fruits.
8. A special case of plant and animal relationship is the human invention of agriculture.

SELF-QUIZ

1. Which of the following is *not* evidence that plants evolved from chlorophytes?
   a. The chloroplasts of both groups are similar in structure, with thylakoid membranes stacked into grana.
   b. Both groups use chlorophyll *a* and *b* and the same accessory pigments.
   c. Organisms of both groups have jacketed reproductive organs called gametangia.
   d. Cellulose is the major component of their cell walls, and starch is used as a storage product.
   e. Some members of Chlorophyta and all green plants have a similar formation of cell plates during cytokinesis.

2. Most bryophytes require a damp habitat because
   a. they do not have a cuticle
   b. they lack gametangia to protect their developing embryo
   c. their sporophyte generation, which lacks vascular tissue, is dependent on the gametophyte
   d. the flagellated sperm must swim to fertilize the egg
   e. they are only a few centimeters tall

3. Which of the following is *not* characteristic of all divisions of vascular plants?

a. the development of seeds

b. an alternation of generations

c. differentiation into roots, stems, and leaves

d. xylem and phloem for transporting materials between roots and leaves

e. addition of lignin to cell walls to provide aerial support

4. A heterosporous plant is one that

a. produces a gametophyte that bears both sex organs

b. produces microspores and megaspores giving rise to to separate male and female gametophytes

c. is a seedless vascular plant

d. produces two kinds of spores, one asexually by mitosis and one type by meiosis

e. reproduces only sexually

5. During the Carboniferous period, the dominant plants, which later formed the great coal beds, were mainly

a. the giant lycopods, horsetails, and ferns

b. the gymnosperms

c. the tree lycopods and cycads

d. the treelike early vascular seed plants

e. the bryophytes that dominated early swamps

6. The male gametophyte of an angiosperm consists of

a. an anther

b. a sac containing eight haploid nuclei

c. a microspore

d. a germinated pollen grain

e. a microscopic, nonphotosynthetic thallus

7. A fruit is most commonly

a. a ripened ovary

b. a thickened style

c. an enlarged ovule

d. an enlarged aggregate of several flowers

e. a mature female gametophyte

8. Important terrestrial adaptations that evolved *exclusively* in seed plants include all of the following *except*

a. pollination by wind or animal instead of swimming sperm

b. transport of water through vascular tissue

c. retention of the gametophyte plant within the sporophyte

d. dispersal of new plants by seeds

e. protection and nourishment of the embryo within the seed

9. Of the following, the most recent ancestors of vascular plants were probably

a. mosses

b. cyanobacteria

c. green algae

d. red algae

e. horsetails

10. A land plant produces flagellated sperm and the dominant generation is diploid. The plant is most likely a

a. fern

b. moss

c. conifer

d. chlorophyte

e. dicot

CHALLENGE QUESTIONS

1. Your roommate cries that his favorite fern has caught "some dreadful disease" after he notices brown spots on the bottoms of a few of the fronds. What is the likely explanation of these spots?

2. What evidence from anatomy, life cycles, and the fossil record suggests that bryophytes and vascular plants are not closely related?

3. Plant evolution has been characterized by compromise. For example, branching patterns that allow growth mainly in the vertical direction are more efficient at gathering light because they can reach beyond the shadows of other plants. However, tall plants must bear the mechanical stresses associated with their height and have woody stems that increase in girth as the plant becomes taller. What trade-off might limit the bushiness of plants, considering the advantage that more extensively branched plants have in outcompeting neighboring plants by shading them? (See Further Reading 5.)

FURTHER READING

1. Fearnside, P. M. "Extractive Reserves in Brazilian Amazonia." *Bioscience*, June 1989. Economics of the tropical rain forest.

2. Hartmann, E. et al. "How Mosses Detect Light Direction." *BioScience*, May 1984. An enlightening article on bryophyte phototropism.

3. Heyler, D., and C. M. Poplin. "The Fossils of Montceau-les Mines." *Scientific American*, September 1988. Going back in time to a Carboniferous swamp.

4. Milot, C. "Blueprint for Conserving Plant Diversity." *Bioscience*, June 1989. The importance of genetic diversity in endangered plant species.

5. Niklas, K. J. "Computer-simulated Plant Evolution." *Scientific American*, March 1986. An ingenious account of how to recreate trends in plant evolution with a desktop computer.

6. Raven, P. H., Evert, R. F., and Eichhorn, S. *Biology of Plants*, 4th ed. New York: Worth, 1986. A good, solid general botany textbook.

7. Rosenthal, G. A. "The Chemical Defenses of Higher Plants." *Scientific American*, January 1986. A fascinating and illuminating article about how plants fight back.

8. Wickelgren, I. "Plants Poised at Extinction's Edge." *Science News*, December 1988. A brief article describing the impending extinction of nearly 700 plant species in the United States.

Fungi

28

The words *fungus* and *mold* may evoke some unpleasant images. Fungi spoil food, rot timbers, attack plants, lead to famines, and afflict humans with athlete's foot and worse maladies. On the other hand, ecosystems would collapse without fungi to decompose dead organisms, fallen leaves, feces, and other organic materials, thus recycling vital chemical elements back to the environment in forms other organisms can assimilate (Figure 28.1). Humans have also found many uses for fungi. We eat them (mushrooms, for instance), culture them to produce antibiotics and other drugs, and add them to dough to make bread rise. Perhaps it is unavoidable that we will judge the various fungi, as we do all organisms, according to our self-interest. Whatever those subjective perceptions, as objects of study fungi are fascinating. They are a form of life so distinctive that in the five-kingdom system of taxonomy they have been accorded their own kingdom.

This chapter is an introduction to fungi. The chapter characterizes the members of the kingdom Fungi, surveys their diversity, discusses their ecological and commercial impact, and considers speculations about their phylogeny. As we did with the plant kingdom, we will look in some detail at life cycles, partly for what they tell us about the evolutionary adaptations of fungi.

CHARACTERISTICS OF FUNGI

Fungi are eukaryotes, and nearly all are multicellular. Unicellular fungi, such as yeasts, are believed to have

(a)

(b)

Figure 28.1
Fungi as decomposers. Fungi secrete enzymes that break down organic molecules; the fungi then absorb these smaller nutrients. **(a)** Growth of the mold *Aspergillus* is common on strawberries and many other fruits. **(b)** These mushrooms are the reproductive structures of a fungus, *Amanita panthering*, that is living on the compost of a forest floor.

evolved secondarily from multicellular forms. First studied by botanists, fungi were grouped with plants in the two-kingdom system. Now it is clear that fungi are not primitive or degenerate plants lacking chlorophyll but unique organisms that generally differ from other eukaryotes in style of nutrition, structural organization, and reproduction.

Nutrition

All fungi are heterotrophs. In contrast to animals—heterotrophs that, for the most part, ingest food—fungi acquire their nutrients by **absorption.** In this mode of nutrition, small organic molecules are absorbed from the surrounding medium. A fungus digests food outside its body by secreting into the food acids and powerful hydrolytic enzymes that decompose complex molecules to the simpler compounds that the fungus can absorb and use.

In their absorptive nutrition, fungi are specialized as saprophytes, parasites, or mutualistic symbionts. Saprophytes absorb nutrients from nonliving organic

material, such as animal corpses, fallen logs, or the wastes of live organisms; saprophytes are decomposers. Parasitic fungi absorb nutrients from the body fluids of living hosts. Many of these fungi, such as certain species infecting human lungs, are pathogenic. Some fungal species can cross the line between saprophyte and parasite; for example, the fungus that causes Dutch elm disease attacks the living tree, eventually kills it, and then continues to live off the wood as a saprophyte. Mutualistic fungi also absorb nutrients from a host, but they reciprocate with functions beneficial to the host in some way, such as aiding a plant in the uptake of minerals from the soil.

Structure

The basic body plan of fungi is actually a netlike mass of filaments called **hyphae.** Hyphae begin as tubular extensions of spores and branch repeatedly to form the network, which is known as a **mycelium** (Figure 28.2a). Even the common mushroom, solid though it appears, is actually a mass of tightly packed hyphae.

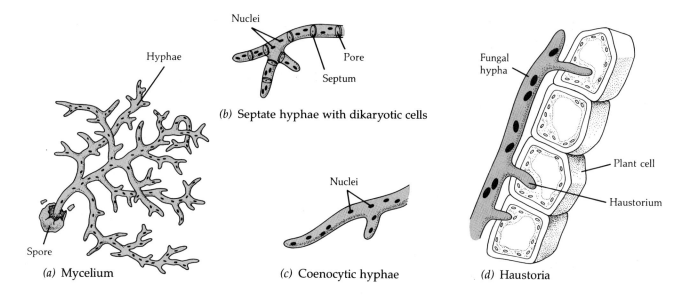

Hyphae

Nuclei

Pore

Septum

(b) Septate hyphae with dikaryotic cells

Nuclei

Fungal hypha

Plant cell

Haustorium

Spore

(a) Mycelium

(c) Coenocytic hyphae

(d) Haustoria

Figure 28.2
Significant features of fungal anatomy.

And the mushroom is just the tip of the iceberg, an aerial reproductive structure attached to a comparatively huge, underground mycelium.

Most fungal hyphae are divided into cells by cross walls, or **septa.** The septa generally have pores large enough to allow ribosomes, mitochondria, and even nuclei to flow from cell to cell (Figure 28.2b). The cell walls of fungi differ from the cellulose walls of plants. Most fungi build their walls mainly of chitin, a strong but flexible nitrogen-containing polysaccharide similar to the chitin found in the external skeletons of insects and other arthropods. (This is an example at the molecular level of a similarity that is analogous rather than homologous.) Some fungi are **aseptate;** that is, their hyphae are not divided into cells by cross walls. These so-called **coenocytic** fungi consist of a continuous cytoplasmic mass with hundreds or thousands of nuclei (Figure 28.2c). The coenocytic condition results from the repeated division of nuclei without cytoplasmic division.

The filamentous structure of the mycelium provides an extensive surface area that suits the absorptive nutritional style of fungi. Parasitic fungi usually have some of their hyphae modified as **haustoria,** nutrient-absorbing threads that penetrate the tissues of the host (Figure 28.2d).

The fungal mycelium grows rapidly, adding as much as a kilometer of hyphae each day as it ramifies within a food source. Growth so fast is possible because proteins and other materials synthesized by the entire mycelium are channeled by cytoplasmic streaming to the tips of the extending hyphae. The fungus concentrates its energy and resources on adding hyphal length rather than girth, another growth pattern adapted to the absorptive life style. Most fungi are nonmotile

organisms; they cannot run, swim, or fly in search of food or mates. But the mycelium makes up for the lack of mobility by swiftly extending the tips of its hyphae into new territory.

The nuclei of fungi divide by mitosis, but in a manner different from most other eukaryotes. The nuclear envelope remains intact and a spindle forms within the nucleus, which divides only after the spindle has separated the chromosomes. The nuclei of fungal mycelia are haploid. A unique condition of some septate fungi is the presence of two separate haploid nuclei in each cell. Such mycelia are called **dikaryons.** In some cases, the two nuclei are genetically dissimilar, usually because they are derived from different parents during sexual reproduction. The condition has some of the advantages of diploidy; one haploid genome may be able to compensate for deleterious mutations in the other nucleus, and vice versa.

According to the classification scheme used in this text, none of the fungi has flagellated stages in its life cycle. (This feature excludes the water molds—Phylum Oomycota—from Kingdom Fungi; this text classifies water molds as protists.)

Reproduction

Fungi reproduce by releasing spores that are produced either sexually or asexually. For many fungi, sex is a contingency mode of reproduction that occurs when there has been some unfavorable change in the environment. When conditions are habitable and stable, fungi generally clone themselves by producing enormous numbers of spores asexually. Carried by wind or water, the spores germinate if they land in a moist place where there is food. Spores thus function in dispersal and account for the wide geographic distribution of many species of fungi. The airborne spores of fungi have been found circling the globe at altitudes greater than 100 miles.

The haploid condition prevails in the life cycles of most fungi, with diploid nuclei forming only as a transient stage in sexual reproduction. After a diploid zygote forms by syngamy, it divides by meiosis to restore the haploid condition characteristic of the nuclei of fungal mycelia. This contrasts to the sexual life cycles of vascular plants and animals, in which the diploid condition is dominant.

Within this general pattern, the details of sexual reproduction vary among the groups of fungi, but often involve some form of conjugation. Hyphae of opposite mating strains join, and haploid nuclei from the two parents come together in common cells. Syngamy may follow conjugation immediately, the two haploid nuclei of each hybrid cell joining to produce the diploid condition. Alternatively, the two nuclei in each cell may remain separate for some time, each dividing independently while the hybrid portion of the mycelium formed by conjugation grows as a dikaryon. Later, syngamy occurs in cells at the tips of dikaryotic hyphae. Whether syngamy follows conjugation directly or is delayed, diploid spores are formed, and they must divide by meiosis as a prerequisite to forming new haploid mycelia. Because sexual recombination has occurred, these offspring will be genetically distinct from one another and from both parents.

DIVERSITY OF FUNGI

More than 100,000 species of fungi are known, and mycologists describe another thousand or so each year. The taxonomic scheme used in this chapter classifies the species into four divisions, based primarily on variations in sexual reproduction, plus lichens, a unique symbiotic association of fungi and algae (Table 28.1). Use of the botanical term *division* instead of *phylum* is a vestige of the two-kingdom system. Most of the names of the divisions are derived from some sexual structure that characterizes that group of fungi.

Division Zygomycota

Zygomycetes are mostly terrestrial fungi that live in soil or on decaying plant and animal material. Their hyphae are coenocytic, with many haploid nuclei. Asexual spores, usually wind dispersed, develop in **sporangia** at the tips of aerial hyphae. Sexual reproduction involves the formation of resistant bodies called zygosporangia that can remain dormant when the

Table 28.1 Divisions of fungi

Division	Common Name	Number of Species
Zygomycota	Zygomycetes	600
Ascomycota	Sac fungi	30,000
Basidiomycota	Club fungi	25,000
Deuteromycota	Imperfect fungi	25,000
Lichens (symbiotic associations of algae and fungi)	Lichens	25,000

environment is too harsh for growth of the fungus (it is for these sexual structures that Division Zygomycota is named).

One zygomycete you may have encountered is black bread mold, *Rhizopus stolonifer,* still an occasional household pest despite the addition of preservatives to most processed foods (Figure 28.3). Horizontal hyphae spread out over the food and rhizoids penetrate it, absorbing nutrients and anchoring the fungus. In the asexual phase, bulbous black sporangia develop at the tips of upright hyphae. Within each sporan-

Spores

10 μm

Figure 28.3
Black bread mold, a zygomycete. Growth of *Rhizopus* and other bread molds gives moldy bread its furry appearance. Stalked, spore-forming structures called sporangia grow from the mycelia of these fungi. The top of the sporangium shown in this scanning electron micrograph was originally a spore-forming sac. The sporangial sac has opened and is releasing spores.

Figure 28.4
***Pilobolus*, a zygomycete, aiming its sporangia.** The spores released will eventually be eaten along with grass by grazing animals and dispersed in their dung.

gium, hundreds of haploid spores develop and are dispersed through the air. Spores that happen to land on moist food germinate, growing into new mycelia.

Air currents are not a very precise way to disperse spores, but *Rhizopus* releases the tiny cells in great numbers. Though they drift aimlessly, enough land in hospitable places. Some zygomycetes, however, are actually able to aim their spores. One is *Pilobolus*, a fungus that decomposes animal dung. *Pilobolus* bends its spore-bearing hyphae toward bright light, where grass is likely to be growing (Figure 28.4). The fungus

then shoots its sporangia like cannonballs, and they stick to the grass. Grazing animals eat the sporangia and scatter the spores in feces.

Rhizopus and other zygomycetes usually reproduce asexually, but if they begin to starve, freeze, or desiccate, they may reproduce sexually. Sex also offers the advantages of genetic recombination and a dormant stage (the zygosporangium—Figure 28.5).

Division Ascomycota

Ascomycetes, sac fungi, range in complexity from unicellular yeasts to elaborate cup fungi (Figure 28.6). Ascomycota gets its name from **asci** (the Greek *ascus* means "little sac"), sacs of sexually produced spores. The most complex ascomycetes have many asci packed together into "fruiting" structures called **ascocarps** (the cups of a cup fungus). The hyphae of multicellular ascomycetes are septate.

Ascomycetes lack sporangia but reproduce asexually by producing chains of spores at the tips of specialized hyphae. These spores, called **conidia** (from the Greek word for dust), are wind dispersed in most species. Sexual reproduction of ascomycetes is complex (Figure 28.7).

Notice that the ascomycetes produce two kinds of haploid spores, conidia and ascospores. Conidia are asexual spores that clone the fungus. The **ascospores** of a mycelium are genetically heterogeneous, their genomes being the products of sexual recombination between two parents. The eight ascospores of each ascus are lined up in a row in the order in which they formed from a single zygote. This arrangement provides geneticists with a unique opportunity to study

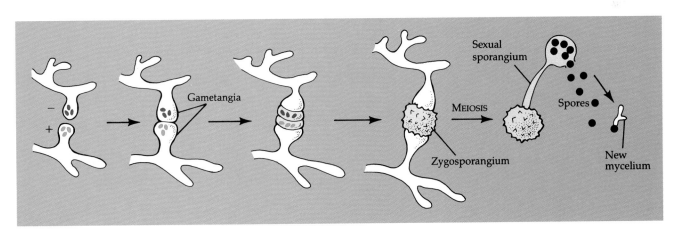

Figure 28.5
Sexual reproduction in *Rhizopus*. Neighboring mycelia of opposite mating strains (designated + and −) form hyphal extensions called gametangia, which fuse. Fused gametangia act as sex organs. The walls between the gametangia dissolve and the fused gametangia become a rough, thick-walled structure called a zygosporangium. The zygosporangium, containing one or more cells, each with haploid nuclei of the + and − strain, survives in a dormant state long after its parent mycelia have died. When conditions are favorable, haploid nuclei within the zygosporangium fuse, meiosis ensues, and the zygosporangium breaks dormancy. The resulting haploid sporangium forms spores that develop into new mycelia.

(a)

Figure 28.6
Ascomycota. Ascomycetes range from parasites that cause serious plant diseases such as Dutch elm disease to gastronomical delicacies such as truffles and morels. (**a**) Scarlet cup (*Sarcoscypha coccinea*), a common cup fungus that lives on rotting wood. (**b**) *Morchella deliciosa* (a morel).

(b)

Figure 28.7
Life cycle of a cup fungus. Haploid mycelia of opposite mating strains become intertwined and fuse. One acts as a "female," producing a structure called an ascogonium, which receives a number of haploid nuclei from the antheridium of the "male." The ascogonium then has a pool of nuclei from both parents, but these haploid nuclei do not fuse at this time. The ascogonium grows hyphae with cells that are dikaryotic. In terminal cells of the dikaryotic hyphae, syngamy (nuclear fusion) finally occurs. The diploid nucleus then divides by meiosis, yielding four haploid nuclei, each of which divides once more by mitosis. The tip of the hypha is now an ascus, with eight haploid nuclei. Next, the nuclei form walls and become partitioned into ascospores, which germinate and begin the cycle again. Ascomycetes can also reproduce asexually by producing airborne spores called conidia.

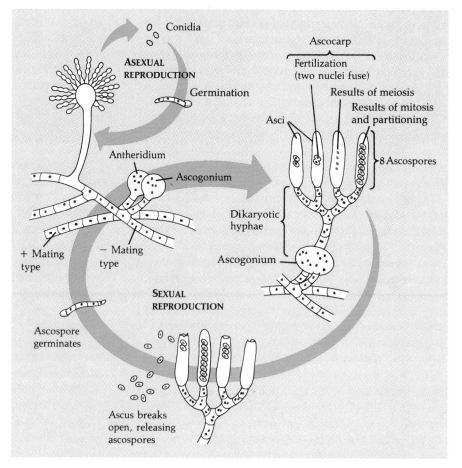

(a)

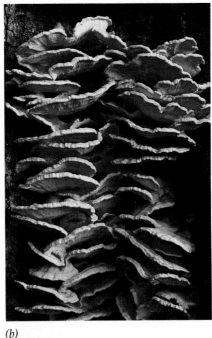

(b)

(c)

(d)

Figure 28.8
Basidiomycota. Basidiomycetes are the club fungi.
(**a**) Miniature waxy cap (*Hygrophorus*). (**b**) A shelf fungus growing on the trunk of a tree. Shelf fungi are important decomposers of wood. (**c**) Common puffball discharging a cloud of spores. (**d**) Corn smut (*Ustilago maydis*). Smuts and rusts, which are common on grain crops, cause tremendous economic losses each year.

genetic recombination. Genetic differences between mycelia grown from ascospores taken from the same ascus reflect crossing-over and independent assortment of chromosomes during meiosis of a single cell.

The division Ascomycota includes the unicellular yeasts. This may seem odd, but many yeasts produce the equivalent of an ascus during sexual reproduction. After two haploid yeasts fuse, the diploid nucleus undergoes meiosis, and the four daughter cells (eight in species where meiosis is followed by a mitotic division) remain enclosed for a while within the original wall of the diploid cell. This sac of sexually produced spores is the ascus. Even asexual reproduction of yeasts, a process called budding, resembles the way conidia form on the multicellular ascomycetes.

Division Ascomycota includes important decomposers that secrete enzymes capable of breaking down

such tough materials as the lignin of dead plants and the collagen of dead animals. There are also symbiotic ascomycetes, some mutualistic and others parasitic. The majority of fungi that live symbiotically with algae as lichens are ascomycetes. Many ascomycetes are parasitic on plants and animals. Among those causing serious plant diseases are powdery mildews that infest cereal grains and the highly destructive species *Ceratocystis ulmi*, which causes Dutch elm disease.

Division Basidiomycota

The mushrooms, shelf fungi, puffballs, and stinkhorns are all classified in the division Basidiomycota (Figure 28.8). The name derives from the **basidium** (Latin for "little pedestal"), which is a transient dip-

Figure 28.9
Sexual reproduction of a club fungus.
When mycelia of opposite mating strains grow together, their monokaryotic hyphae fuse, but the haploid nuclei remain separate, as in ascomycetes. The hybrid mycelium gives rise to dikaryotic hyphae, compact masses of which make up the mushrooms, or basidiocarps. In the cells at the tips of the dikaryotic hyphae, the haploid nuclei of the two parents fuse, forming a transient diploid stage. The cell swells to form a club-shaped basidium, and the diploid nucleus undergoes meiosis, yielding four haploid nuclei. The basidium then grows four appendages, and one haploid nucleus enters each appendage and develops into a basidiospore. Thus, four basidiospores are attached to the basidium. Dispersed by the wind, the basidiospores germinate under appropriate conditions, developing into mycelia, to begin the cycle again.

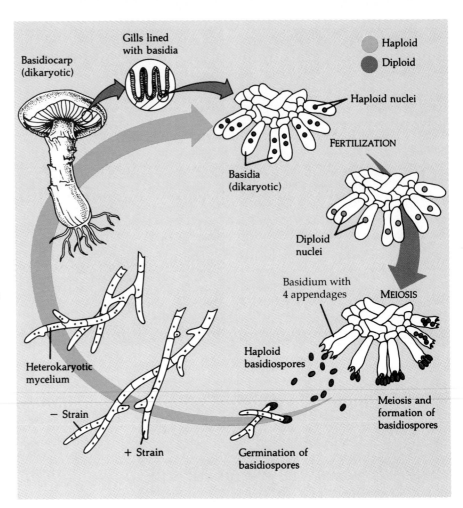

loid stage in the organism's life cycle. The clublike shape of the basidium also gives rise to the common name *club fungus*. The common characteristic of this division is a distinctive mode of sexual reproduction (Figure 28.9).

The mushroom we see poking out from the ground is a reproductive structure supported by a large subterranean mycelium of septate hyphae that decompose organic matter in the soil and absorb the nutrients. Syngamy and meiosis occur in the mushroom, or **basidiocarp.** The "gills" on the underside of the cap are lined with basidia, each of which bears four basidiospores. As the basidiospores drop from the gills, they are dispersed by the wind. A typical mushroom releases more than ten billion spores, and giant puffballs produce spores by the trillions.

By concentrating growth in the hyphae of mushrooms, the mycelium can erect the fruiting structures in just a few hours. A ring of mushrooms, popularly called a fairy ring, may appear on a lawn overnight (Figure 28.10). Although the grass in the center of the ring is normal, you may notice after a few days that the grass near the ring is stunted and the grass just outside the garland of mushrooms is especially lush.

As the underground mycelium grows outward, its center portion and the mushrooms above it die because the mycelium has consumed all the available nutrients. Thus, the living mycelium is an expanding ring that produces mushrooms above it. The grass beneath the mushrooms is stunted because it cannot compete for minerals with the active mycelium. But the advancing mycelium secretes digestive agents ahead of it that decompose the organic matter in the soil, producing a lush growth of grass that absorbs the minerals that become available. The fairy ring slowly increases in diameter as the mycelium advances at a rate of about 30 cm per year. Some giant fairy rings may be centuries old.

Division Deuteromycota

We have seen that the fungi of each division have characteristic sexual structures for which the division is named. But thousands of fungal species have no sex life, as far as we know. Either these fungi reproduce only asexually or their sexual phases have not yet been observed by mycologists. All of these fungi are grouped

Figure 28.10
A fairy ring. The legendary explanation of these circles of fungi is that mushrooms spring up where fairies have danced in a ring on moonlit nights. Afterward, the tired fairies sit down on the mushrooms, but some of the mushrooms are sat on by toads. The mushrooms the fairies choose are edible by humans, but the "toadstools" are poisonous. Mycologists offer a different explanation for fairy rings (see text).

together, for convenience, in the division Deuteromycota (meaning second or "other" fungi). They are also sometimes called **imperfect fungi,** from the botanical use of the term *perfect* to refer to the sexual stage in a life cycle. Whenever a sexual stage is discovered, the fungus is reclassified into the appropriate division. Many imperfect fungi, including *Penicillium,* the source of the antibiotic penicillin, are probably related to ascomycetes, as suggested by their asexual formation of conidia. Among the more unusual deuteromycetes are predaceous fungi in soil that trap and kill small animals, especially roundworms called nematodes (Figure 28.11).

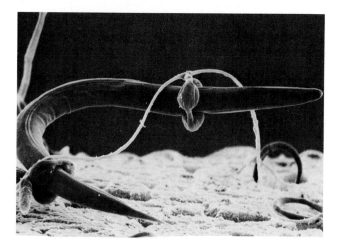

25 μm

Figure 28.11
Arthrobotrys, **a predaceous deuteromycete.** Portions of the hyphae are modified as hoops that constrict around nematodes, small worms, in a fraction of a second. The fungus then penetrates its prey with hyphae and digests the meal.

Lichens

Lichens often resemble mosses or other simple plants growing on rocks, tree trunks, the sides of buildings, and other substrata (Figure 28.12). But lichens are not plants at all, nor are they technically even individual organisms. Lichens are highly integrated symbiotic associations of millions of algal cells tangled in a lattice of fungal hyphae. The fungal component is most commonly an ascomycete, and the algae are usually unicellular or filamentous green algae (chlorophytes) or blue-green algae (cyanobacteria). Although lichens differ in the details of their architecture, most are variations on a common theme (Figure 28.13).

Lichens can reproduce asexually as symbiotic units, either from fragments or by dispersing tiny airborne starters called **soredia,** each with a clump of algae embedded in hyphae. The alga and fungus also reproduce independently, either sexually or asexually, and new lichens may later form where the two organisms came together.

The nature of the symbiosis in lichens has long been debated. The alga definitely provides the fungus with food, but it is not known if the alga receives any reciprocal benefits. Researchers have speculated that the fungal mycelium helps retain moisture and provides the alga with water, minerals, and protection. An argument for the symbiosis being mutual is the ability of the lichen to survive in habitats that are inhospitable to either organism alone. However, no contributions of the fungus to the growth of the alga have ever been demonstrated. Furthermore, the fungus actually kills some of its algal associates, though not as fast as the alga replenishes its numbers by reproduction. Some students of lichen biology believe the

Figure 28.12
Crustose lichens. These and other lichens are symbiotic associations of a fungus and an alga. Some of the algae in these associations can live independently of the fungi, but a lichen fungus does not grow independently of its algal symbiont.

symbiosis is more of controlled parasitism than a mutualistic partnership.

In any case, the merger of fungus and alga is so complete that lichens are actually given genus and specific names, as though they were single organisms. The convention is to name them for their fungal component. Naturalists categorize lichens according to overall appearance as being foliose (leafy), fruticose (shrubby), or crustose (crusty). Many lichens are vividly colored, owing to the photosynthetic pigments of the alga and the fruiting structures of the fungus, and their pigments make valuable dyes.

Lichens are rugged, able to survive on bare rocks and other forbidding substrata. They are important pioneers on new land, such as volcanic flows. Growth and secretion by the lichens break down the rock and make it possible for a succession of plants to grow. Some lichens tolerate severe cold. In the arctic tundra, great herds of caribou and reindeer graze on carpets of reindeer moss, which are actually lichens. Lichens can also survive desiccation. When it is foggy or rainy, the lichen may absorb more than ten times its weight in water. Photosynthesis occurs when the water content of the lichen is 65%–90%. In dry air, the lichens rapidly dehydrate and photosynthesis stops. Thus, in arid climates, lichens grow very slowly, perhaps in spurts of less than a millimeter per year. Some lichens may be thousands of years old, rivaling the oldest plants as the elder organisms on Earth.

As tough as lichens are, they do not stand up very well against air pollution. Lichens get most of their minerals from air in the form of dust or compounds

Figure 28.13
Anatomy of a lichen. The upper and lower surfaces are protective layers constructed from tightly packed fungal hyphae. The upper layer of the mycelium also forms the reproductive structures of the fungus. Just beneath the upper surface are the algae, enmeshed in the net of hyphae. The middle of a lichen generally consists of loosely woven hyphae of the fungus.

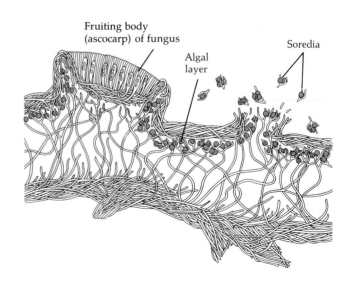

Fruiting body (ascocarp) of fungus

Algal layer

Soredia

dissolved in rain. This mode of mineral uptake makes lichens particularly sensitive to sulfur dioxide and other aerial poisons. The death of lichens in an area may be a warning that air quality is deteriorating.

ECOLOGICAL AND COMMERCIAL IMPORTANCE OF FUNGI

Fungi as Decomposers

Fungi and bacteria are the principal decomposers that keep ecosystems stocked with the inorganic nutrients essential for plant growth. Without decomposers, carbon, nitrogen, and other elements would accumulate in corpses and organic wastes, no longer available as raw material for new generations of life. Imagine what would become of a forest if its decomposers rested for even a few years. Leaves, logs, feces, and dead animals would pile up on the forest floor. Plants and the animals they feed would starve because elements taken from the soil would not be returned. Gradually, the forest would die. This hypothetical forest is only a symbol of the fate that all ecosystems would meet without decomposers. In reality, an endless recycling of chemical elements occurs between organisms and their nonliving surroundings. The metabolism of saprophytic fungi and bacteria decomposes such complex molecules as polysaccharides and proteins to carbon dioxide, nitrate, and other simple inorganic compounds that plants can assimilate as raw materials for photosynthesis. The air is so loaded with spores that as soon as a leaf falls or an insect dies, it is covered with fungal spores and is soon infiltrated by saprotrophic hyphae.

Mycorrhizae

Mycorrhizae are mutualistic associations of plant roots and fungi. The word *mycorrhizae* means "fungus roots." The fungal hyphae of the mycorrhizae greatly increase the absorptive surface of the roots, and the fungus exchanges minerals accumulated from the soil for organic nutrients synthesized by the plant. About 90% of tree species have mycorrhizae, as do the majority of smaller vascular plants.

About 90% of the fungi of tree mycorrhizae are basidiomycetes; the remainder are zygomycetes and ascomycetes. The underground mycelia of about half the species of mushrooms are associated with the roots of trees.

Mycorrhizae are enormously important in natural ecosystems and agriculture. Plants generally do not grow as well if they are deprived of their mycorrhizae. The relationships between plants and their root fungi

are specific. When trees that normally live in the forest are grown from seeds in greenhouses and are then transplanted to grasslands, they often die or grow poorly. But a few spadesful of forest soil mixed with the dirt around the tree is enough to inoculate the roots with mutualistic fungi and improve the growth of the tree. If we knew more about the natural microbiology of soil—about the specific relationships between plants and their mutualistic fungi and bacteria—we could probably reduce the amount of artificial fertilizer that is added to soil.

The structure and physiology of mycorrhizae is described in more detail in Chapter 33.

Commercial Uses of Fungi

Most of us have eaten mushrooms, though it rarely occurs to us that it is a fungus we are ingesting. Mushrooms (basidiomycetes) are cultivated commercially on compost in the dark. Edible mushrooms also grow in fields, forests, and backyards, but so do poisonous ones. Mycologists make no distinction between mushrooms and toadstools, referring to the structures as mushrooms, edible or not. There are no simple rules to help the novice distinguish edible from deadly mushrooms. Only experts in mushroom taxonomy should dare to collect mushrooms for eating.

Mushrooms are not the only fungi people eat. The turquoise streaks in blue cheese and Roquefort are mycelia of certain species of *Penicillium* (Deuteromycota). Perhaps the fungi most prized by gourmets are truffles, the fruiting bodies of mycelia that are mycorrhizal on the roots of trees. Their complex flavor is variously described as nutty, musky, cheesy, or all three. Truffle hunters traditionally used pigs to locate the underground fungi. The pig has a keen sense of smell, and the truffle emits a chemical that resembles the musk secreted by pigs as sex attractants. Today, dogs are used more commonly than pigs for locating truffles because the dogs do not try to eat the truffles when they find them.

The yeast *Saccharomyces cerevisiae* is the most important domesticated fungus. The tiny yeast cells, available as many strains of baker's yeast and brewer's yeast, are very active metabolically. When bakers add yeast to dough, the metabolism of the cells produces carbon dioxide gas that makes bread rise. *Saccharomyces* is a facultative anaerobe that ferments sugars to alcohol when forced to live without oxygen. Brewers add yeast to water and grains to make beer, and winemakers use the native yeast growing on grape skins to convert grape juice to wine.

The pharmaceutical industry grows fungi to produce antibiotics, chemicals produced by one organism that kill or inhibit growth of another organism. Many fungi secrete antibiotics as weapons against bacteria that may be attacking the fungus or competing with it for food. In 1928, Alexander Fleming, a British bacteriologist, accidently contaminated one of his bacterial cultures with a strain of the fungus *Penicillium*. After observing that the fungus killed his bacteria, Fleming proposed that the *Penicillium* was secreting an antibacterial chemical. A decade later, Howard Florey of Oxford University purified penicillin and began promoting its use as an antibiotic to treat bacterial infections in humans. Penicillin and the many other antibiotics subsequently discovered revolutionized the treatment of bacterial disease.

Fungi as Spoilers

We may applaud fungi that decompose forest litter or dung, but it is a different story when molds attack our bowls of fruit or our shower curtains. A wood-digesting saprophyte does not distinguish between a fallen oak limb and the oak planks of a ship. During the revolutionary war, the British lost more ships to dry rot than to enemy attack. Soldiers stationed in the tropics during World War II watched their tents, clothing, boots, and binoculars be destroyed by molds. Some fungi can even decompose certain plastics. The best way to protect materials from mold is to keep them as dry as possible.

All of us have lost food to molds. When a fungus starts decomposing our bread, fruits, vegetables, and other groceries, it is competing with us for food. As we have seen, fungal spores are everywhere, but if we eat fresh foods soon enough, the fungi do not have a chance to grow.

Pathogenic Fungi

Many fungi are pathogens. Among the diseases that fungi cause in humans are athlete's foot, ringworm, yeast infections of the vagina, and lung infections that may be fatal.

Plants are particularly susceptible to fungal diseases. For example, an ascomycete caused Dutch elm disease, which has drastically changed the landscape of the northeastern United States. The fungus was accidently introduced to the United States on logs that were sent from Europe to help pay World War I debts. Carried from tree to tree by bark beetles, the fungus is on its way to completely eliminating the American elm.

Some of the fungi that attack food crops are toxic to humans. One ascomycete forms purple structures called ergots on rye (Figure 28.14). If diseased rye is inadvertently milled into flour and consumed, poisons from the ergots cause gangrene, nervous spasms, burning sensations, hallucinations, and temporary

Figure 28.14
Ergots on rye. The ascomycete *Claviceps purpurea* forms dark purple growths (ergots) on seed heads of rye and certain other grasses. Ergots, which are the overwintering bodies of the fungus, contain toxins poisonous to humans and domestic animals.

insanity. One epidemic in 944 A.D. killed more than 40,000 people. During the Middle Ages, the disease (ergotism) became known as St. Anthony's fire because many of its victims were cared for by a Catholic nursing order dedicated to Saint Anthony. One of the hallucinogens that has been isolated from ergots is lysergic acid, the raw material from which LSD is made. Toxins extracted from fungi often have medical uses when administered in weak doses. For example, an ergot compound is helpful in treating high blood pressure and stopping maternal bleeding after childbirth.

EVOLUTION OF FUNGI

The origin of fungi is a mystery. They do not seem to be closely related to either plants or animals, and probably arose directly from protistan ancestors.

The phylogeny of fungi after their origin from protists is also uncertain. In one possible scenario, based on comparisons of morphology and life cycles of the fungal divisions, Basidiomycota evolved from Ascomycota, itself descendant from Zygomycota. Among possible candidates as the protistan ancestors of Zygomycota are red algae and conjugating green algae.

All major groups of fungi had evolved by the end of the Carboniferous period, about 300 million years ago. Fossil traces resembling fungi are found in Precambrian rocks dating back 900 million years, but they are probably oomycetes (water molds, which this text classifies as protists) or even algae. The oldest undisputed fossils of fungi are in Ordovician strata about 450–500 million years old. Plants and fungi moved from water to land together; fossils of the first vascular plants that colonized land during the late Silurian period 400 million years ago have petrified mycorrhizae. From their inception, terrestrial communities have depended on fungi.

STUDY OUTLINE

1. The fungi constitute a kingdom of organisms that have tremendous impact, both positive and negative, on other organisms.

Characteristics of Fungi (pp. 589–592)

1. The fungi are a eukaryotic, primarily multicellular group.
2. All fungi are heterotrophs, acquiring their nutrients by absorption. They digest food outside their bodies by secretion of acids and enzymes. There are saprophytic decomposers, parasitic species, and mutualistic forms.
3. The fungal body plan consists of mycelia, netlike masses of branched hyphae ideally suited for absorption. Parasitic fungi penetrate their hosts with specialized hyphae called haustoria.
4. The majority of fungi construct their cell walls from the polysaccharide chitin.
5. Although aseptate (coenocytic) forms occur, most fungi have their hyphae partitioned into cells by septa, with large pores allowing cell-to-cell continuity.
6. Fungi begin as spores that sprout tubular extensions of hyphae, branching rapidly outward in all directions.
7. Most fungi are haploid and generate new cells by a variation of mitosis. Some septate species are dikaryotic, possessing two separate haploid nuclei per cell. Sometimes the nuclei are genetically dissimilar as a result of sex.
8. Fungi produce astounding numbers of airborne or aquatic spores. Most spores are asexually cloned until unfavorable conditions trigger sexual reproduction. If sex occurs, the haploid condition is usually quickly restored by meiosis following syngamy.
9. Sex in fungi usually involves various forms of conjugation between hyphae of opposite mating strains.

Diversity of Fungi (pp. 592–598)

1. The division Zygomycota, fungi that live in soil or decaying organic matter, includes the familiar black bread mold. Zygomycetes have coenocytic hyphae with asexual spores that develop in aerial sporangia. The division is named for its sexually produced zygosporangia, which are tough, dormant structures capable of persisting through unfavorable conditions.

2. The sac fungi of Division Ascomycota produce two kinds of haploid spores. Chains of genetically identical conidia are produced asexually. Sexually generated, genetically heterogeneous ascospores are packaged into characteristic sacs, or asci. Ascomycetes include yeasts.
3. The division Basidiomycota, or club fungi, contains the mushrooms, shelf fungi, puffballs, and stinkhorns. Basidiomycetes reproduce sexually. The basidium, a club-like transient diploid stage, generates haploid basidiospores within basidiocarps, the mushrooms.
4. The imperfect fungi, Division Deuteromycota, are a motley group characterized by the lack of an observed sexual phase in the life cycle. Imperfect fungi include species that produce penicillin, as well as predaceous members that trap and kill small animals in soil.
5. Lichens are such highly integrated symbiotic associations of algae and fungi that they are classified as single organisms and as fungi by convention. The fungal component is usually an ascomycete and the algal component is generally a green alga or cyanobacterium. Reproduction can be by a collective asexual effort or as a result of secondary association of sexual or asexual progeny produced independently by the alga and the fungus. Lichens are rugged organisms that set the stage for plant growth by slowly breaking down bare rocks. Scientists still do not know the exact nature of the symbiosis in lichens.

Ecological and Commercial Importance of Fungi (pp. 599–600)

1. Without fungi as decomposers, we would be surrounded by the organic debris of dead organisms and deprived of the essential recycling of chemical elements between the biological and nonbiological world.
2. Natural ecosystems and agriculture benefit enormously from mycorrhizae, mutualistic associations of fungi and the roots of plants.
3. Beneficial fungi are eaten as food or used to raise bread, brew beer, and make wine, cheese, and antibiotics.
4. Fungi also decompose food and useful natural and artificial objects.
5. Some fungi cause disease, plaguing humans with a variety of ills from athlete's foot to fatal lung infections. Plants are especially vulnerable to fungal infections.

Evolution of Fungi (p. 601)

1. The origin of fungi is unknown, but taxonomists generally agree that this enigmatic kingdom arose directly from ancestral protists rather than from plants or animals.

2. Subsequent phylogeny is also far from clear-cut, but it is certain that all major groups of fungi had evolved by the end of the Carboniferous period and that fungi accompanied plants in the move from water to land.

SELF-QUIZ

1. Which of the following is *not* descriptive of the kingdom Fungi?

   a. absorptive form of nutrition

   b. alternation of generations, although haploid state is dominant

   c. eukaryotic heterotrophs

   d. most with body form of hyphae

   e. flagellated cells do not occur during the life cycle

2. The pores in septa are

   a. regions of high rates of absorption of dissolved organic molecules through the cell walls

   b. the openings through which spores are expelled

   c. a means for proteins and materials from other cells to move to the rapidly growing tips of hyphae

   d. involved in the unusual mitosis found in many fungi

   e. only found in coenocytic species

3. Which of the following cells or structures are associated with *asexual* reproduction?

   a. ascospores   d. zygosporangia

   b. basidiospores   e. antheridia

   c. conidia

4. Sporangia on erect hyphae that produce asexual spores are characteristic of

   a. Ascomycota   d. Zygomycota

   b. Basidiomycota   e. lichens

   c. Deuteromycota

5. Members of Deuteromycota

   a. have no known asexual stage

   b. include yeasts

   c. include fungi that reproduce by a form of conjugation

   d. include members that appear to be related to the ascomycetes because of their conidia

   e. are probably not true fungi.

6. Among members of Basidiomycota,

   a. hyphae fuse to grow into a dikaryotic mycelium

   b. spores line up in a sac after they are formed by meiosis

   c. the vast majority of spores formed are asexual

   d. are the most important commercial sources of antibiotics

   e. no sexual stage has been found

7. Mycorrhizae are

   a. asexual reproductive structures formed by lichens

   b. thin hyphae that grow directly into host tissues

   c. the mycelium that forms fairy rings

   d. compact, dikaryotic hyphae that form a basidiocarp

   e. mutualistic associations between plant roots and fungi

8. Parasitic fungi have specialized hyphae called

   a. haustoria   d. soredia

   b. ascogonia   e. ergots

   c. conidia

9. The fungus responsible for Dutch elm disease is a (an)

   a. zygomycete   d. deuteromycete

   b. ascomycete   e. oomycete

   c. basidiomycete

10. The photosynthetic symbiont of a lichen is most commonly a (an)

    a. moss   d. ascomycete

    b. green alga   e. small vascular plant

    c. red alga

CHALLENGE QUESTIONS

1. In what way might use of a wide-spectrum fungicide that kills all fungal species have a harmful effect on vegetation, such as a forest?

2. The symbiotic nature of lichens has been hotly debated by mycologists. Current investigations of lichens have shown that an alga commonly present in lichens will excrete only small amounts of a sugar alcohol when it is grown in an isolated culture, but will use up to 90% of the carbon it fixes in photosynthesis to produce the same compound when paired with a fungus as a lichen. Does this finding lend more weight to the mutualistic or the parasitic argument for lichen symbiosis? (Check your opinion against Further Reading 2.)

3. Zygomycetes fuse haploid nuclei to form zygotes only to restore the haploid state again by meiosis before growth of new mycelia. What does this formation of a transient diploid stage accomplish?

FURTHER READING

1. Abelson, P. H. "Plant-fungal Symbiosis." *Science* 229:617, 1985. A general article presenting hopeful prospects for efficient reforestation by infecting trees with symbiotic fungi.

2. Ahmadjian, V. "The Nature of Lichens." *Natural History*, March 1982. A beautifully illustrated article challenging the idea that the algal-fungal association is a blissful copartnership.

3. Alexopoulos, C. J. and C. W. Mims. *Introduction to Mycology*. 3d ed. New York: Wiley, 1979. A general text.

4. Dusheck, J. "Fungus Degrades Toxic Chemicals." *Science News*, June 22, 1985. The versatile metabolism of fungi appears to have exciting potential for environmental cleanup.

5. Raloff, J. "Fungi Feel Their Way to Feast." *Science News*, April 1987. How the bean rust fungus finds and infects leaf stomata.

6. Rudolph, E. D. "Lichens in US Introductory Botany Textbooks, 1836–1986." *Bioscience*, June 1988.

Invertebrates and the Origin of Animal Diversity

29

A nimal life began in Precambrian seas with the evolution of multicellular forms that lived by eating other organisms. It was a way of life that opened many new adaptive zones, and led to a series of evolutionary radiations that populated the seas, lakes, and eventually the land with animals of dazzling diversity. More than a million extant species of animals are known, and at least as many will probably be identified by future generations of zoologists. Based largely on anatomical and embryological criteria, animals are grouped into about 35 phyla, the exact number depending on the extent to which groups are lumped together in taxonomic categories. We are most familiar with the vertebrates, a single subphylum of backboned animals within the phylum Chordata. All other animals, about 95% of the species, lack backbones and are collectively called **invertebrates** (Figure 29.1). They are the main subject of this chapter. The chapter also describes general characteristics of animals, discusses possible relationships among the phyla, and addresses theories on the origin and early radiation of animals.

CHARACTERISTICS OF ANIMALS

All animals are multicellular heterotrophic eukaryotes. Their mode of nutrition, termed **ingestion,** is one characteristic that distinguishes the kingdom Animalia from the other two kingdoms of multicellular eukaryotes. In contrast to the autotrophic nutrition of plants and the absorptive nutrition of fungi, most ani-

Figure 29.1
The fauna of a tidepool. Living as we do on land, where vertebrates are dominant animals, we have a biased sense of animal diversity. Tidepools and coral reefs reveal a multitude of invertebrate phyla that live in the sea.

mals nourish themselves by eating other organisms or detritus (decomposing organic material), either whole or by the piece. Parasitic animals that absorb organic molecules directly across their outer body surfaces probably evolved secondarily from ancestors that lived by ingestion.

Several features in addition to ingestion are shared by most animals. In contrast to plants, animals store their carbohydrate reserves as glycogen. Their cells also lack walls, and the intercellular junctions described in Chapter 7—desmosomes, gap junctions, and tight junctions—are found only in animals. Animals have highly differentiated body cells, and most species have cells organized into tissues, tissues organized into organs, and organs teamed into organ systems specialized for such functions as digestion, internal transport, gas exchange, movement, coordination, excretion, and reproduction. The animal way of life requires dynamic behavior; muscles, and nerves that control them, are unique to animals.

Reproduction in the animal kingdom is typically sexual, with the diploid stage usually dominating the life cycle. In most species, a small flagellated sperm fertilizes a larger, nonmotile egg to form a diploid zygote. The zygote then undergoes **cleavage,** a succession of mitotic cell divisions. During the development of most animals, cleavage leads to the formation of a multicellular stage called a **blastula,** which often takes the form of a hollow ball (variation in the formation and developmental fate of the blastula is discussed in Chapter 43). Some animals develop directly through transient stages of maturation into adults, but the life cycles of many include larval stages. The **larva** is a free-living, sexually immature form. It is morphologically distinct from the adult stage, usually eats different food, and may even have a different habitat than the adult, as in the case of a frog tadpole. Animal larvae eventually undergo **metamorphosis,** a resurgence of development that transforms the animal into a sexually mature adult.

Animals inhabit nearly all environments of the biosphere. The seas, where the first animals probably arose, are still home to the greatest number of animal phyla. The freshwater fauna is extensive, but not nearly as rich in diversity as the marine fauna.

Terrestrial habitats pose special problems for animals as they do for plants (see Chapter 27), and few animal phyla have made successful evolutionary treks onto land. Earthworms (Phylum Annelida) and land snails (Phylum Mollusca) are confined to moist soil and vegetation. Only the vertebrates and arthropods, including insects and spiders, are represented by a great diversity of species adapted to various terrestrial environments.

CLUES TO ANIMAL PHYLOGENY

Our understanding of the evolutionary relationships among the animal phyla is generally hazy. Because of the dearth of transitional fossils linking the phyla to their predecessors (see Chapter 23), zoologists attempting to reconstruct animal phylogeny depend mostly on clues from comparative anatomy and embryology.

Major Branches of the Animal Kingdom

Although lively debate continues, many zoologists maintain that the modern animal phyla were all derived

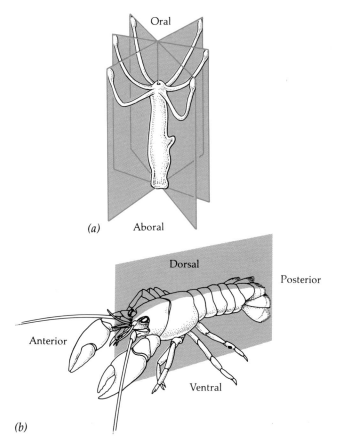

(a) Oral

Aboral

(b) Dorsal

Posterior

Anterior

Ventral

Figure 29.2
Body symmetry. (a) The parts of a radial animal, such as a hydra, are arranged like spokes of a wheel that radiate from the center. Any imaginary slice through the central axis divides the animal into mirror images. (b) A bilateral animal has a left and right side, and only one imaginary cut will divide the animal into mirror-image halves.

traveling animal that is usually first to encounter food, danger, and other stimuli. A head end is an adaptation for movement such as crawling, burrowing, and swimming. The symmetry of an animal appears to fit its life style. Many radial animals are sessile forms (attached to a substratum) or plankton (drifting or weakly swimming aquatic forms), and their symmetry equips them to meet the environment equally well from all sides. More active animals are generally bilateral.

Symmetry alone is not a foolproof criterion for assigning an animal phylum to Branch Radiata or Branch Bilateria. The radial symmetry of some animals has apparently evolved secondarily from a bilateral condition as an adaptation to a more sedentary life style. For example, sea stars (Phylum Echinodermata) are radially symmetrical, but their embryonic development and internal anatomy place them in the branch Bilateria rather than in the branch Radiata.

Branches Radiata and Bilateria probably diverged very early in the history of animal life; indeed, there is some evidence that jellyfishes and their relatives had different protistan ancestors than the eumetazoa of Branch Bilateria. The remainder of our discussion of phylogeny focuses on animals of Branch Bilateria.

Development and Body Plan

Early in the development of nearly all animals of the branch Bilateria, the embryo becomes triple layered. As a general rule, these concentric layers, termed the **germ layers,** form the various tissues and organs of the body as development progresses. **Ectoderm,** covering the surface of the embryo, gives rise to the outer covering of the animal and, in some phyla, to the central nervous system. **Endoderm,** the innermost germ layer, lines the primitive gut, or **archenteron,** and gives rise to the lining of the digestive tract and its outpocketings such as the liver and lungs of vertebrates. Between ectoderm and endoderm is the **mesoderm,** the germ layer that forms the muscles and most other organs between the gut and outer covering of the animal.

Animals with solid bodies—that is, without a cavity between the gut and outer body wall, are referred to as the **acoelomates** (from the Greek *a,* "without," and *koilos,* "a hollow"). This group comprises flatworms (Phylum Platyhelminthes) and animals of a few other phyla (Figure 29.3a). (Sponges and jellyfishes also lack body cavities, but we will restrict the term *acoelomate* to Branch Bilateria.) The other phyla of the branch Bilateria have tube-within-a-tube body plans, with a fluid-filled cavity separating the digestive tract from the outer body wall. If the cavity is not completely lined by mesoderm, it is termed a **pseudocoelom.** Animals with this body plan, such as rotifers (Phylum

from one group of protists. Sponges (Phylum Porifera), however, have unique development and an anatomical simplicity that separates them from all other animal phyla. They are classified in **Subkingdom Parazoa** (which means "beside the animals"). Other animal phyla are grouped in **Subkingdom Eumetazoa.**

Subkingdom Eumetazoa has been divided into two major branches, partly on the basis of body symmetry. **Branch Radiata** consists of the hydras, jellyfishes, and their relatives, which have **radial symmetry** (Figure 29.2a). A radial animal has a top and bottom, or an oral and an aboral side, but no front and back and no left and right. **Bilateria,** the other major branch of eumetazoan evolution, led to animals with **bilateral** (two-sided) **symmetry** (Figure 29.2b). A bilateral animal has not only a top (**dorsal** side) and bottom (**ventral** side) but also a head (**anterior**) end and tail (**posterior**) end and a left and right side.

Associated with bilateral symmetry is **cephalization,** an evolutionary trend toward concentration of sensory equipment on the anterior end, the end of a

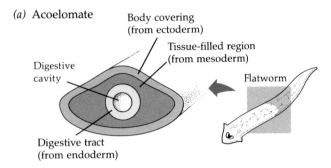

(a) Acoelomate

Digestive cavity

Body covering (from ectoderm)

Tissue-filled region (from mesoderm)

Flatworm

Digestive tract (from endoderm)

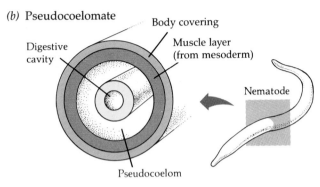

(b) Pseudocoelomate

Body covering

Muscle layer (from mesoderm)

Digestive cavity

Nematode

Pseudocoelom

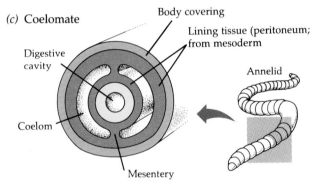

(c) Coelomate

Body covering

Lining tissue (peritoneum; from mesoderm

Digestive cavity

Annelid

Coelom

Mesentery

Figure 29.3
Body plans of the Bilateria. The various organ systems of the animal develop from the three germ layers that form in the embryo. **(a)** Acoelomates lack a body cavity between the gut and outer body wall. **(b)** Pseudocoelomates have a body cavity only partially lined by mesoderm. **(c)** A true coelom, a body cavity completely lined by mesoderm, is characteristic of the coelomate phyla.

Rotifera), roundworms (Phylum Nematoda), and a few other phyla, are called **pseudocoelomates** (Figure 29.3b). **Coelomates** are animals with a true **coelom,** a body cavity completely lined by mesoderm. The inner and outer layers of mesoderm that surround the cavity connect dorsally and ventrally to form mesenteries, tissues that suspend the internal organs in the fluid-filled coelom (Figure 29.3c).

A body cavity has many functions. Its fluid cushions the suspended organs, helping to prevent internal injury. The cavity also enables the internal organs to grow and move independently of the outer body wall. If it were not for your coelom, every beat of your heart or ripple of your intestine could deform your body surface, and exercise would distort the shapes of the internal organs. In soft-bodied coelomates such as earthworms, the noncompressible fluid of the body cavity functions as a hydrostatic skeleton against which muscles can work. Though they may have first evolved as adaptations for burrowing by soft-bodied animals, coeloms evolved independently at least twice.

The Protostome–Deuterostome Dichotomy

The coelomate phyla can be divided into two distinct evolutionary lines. Mollusks, annelids, and arthropods represent one of these lines and are collectively called **protostomes.** Echinoderms and chordates, collectively called **deuterostomes,** represent the other line. Protostomes and deuterostomes are distinguished by several fundamental differences in their development. Differences are evident as early as the cleavage divisions that transform the zygote into a ball of cells. Many protostomes undergo **spiral cleavage,** in which planes of cell division are diagonal to the vertical axis of the embryo. As seen in the eight-cell stage resulting from spiral cleavage, small cells lie in the grooves between larger, underlying cells (Figure 29.4a). Furthermore, the so-called determinate cleavage of some protostomes rigidly casts the developmental fate of each embryonic cell very early. A cell isolated at the four-cell stage from a protostome such as a snail forms an inviable embryo that lacks parts.

In contrast to the protostome pattern, the zygote of many deuterostomes undergoes radial cleavage, during which the cleavage planes are either parallel or perpendicular to the vertical axis of the egg; as seen in the eight-cell stage, the cells are aligned, one directly above the other. Deuterostomes are further characterized by **indeterminate cleavage,** meaning that each cell produced by early cleavage divisions retains the capacity to develop into a complete embryo. If the cells of a sea star embryo, for example, are separated at the four-cell stage, each will go on to form a normal larva. It is the indeterminate cleavage of the human zygote that makes identical twins possible.

Another difference between protostomes and deuterostomes is apparent later in development. In stages following the blastula, the rudimentary gut of an embryo forms as a blind pouch (the archenteron), which has a single opening to the outside known as the **blastopore** (Figure 29.4b). A second opening forms later at the opposite end of the archenteron to produce a digestive tube with a mouth and anus. The mouth of a typical protostome develops from the first opening, the blastopore, and it is for this characteristic that the protostome line is named (from the Greek *protos,* "first," and *stoma,* "mouth"). By contrast, the mouth of a deu-

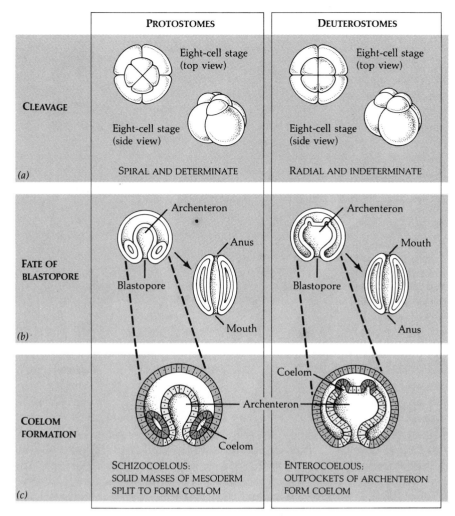

PROTOSTOMES	DEUTEROSTOMES

CLEAVAGE

Eight-cell stage (top view)

Eight-cell stage (side view)

(a) SPIRAL AND DETERMINATE

Eight-cell stage (top view)

Eight-cell stage (side view)

RADIAL AND INDETERMINATE

FATE OF BLASTOPORE

Archenteron

Anus

Blastopore

(b) Mouth

Archenteron

Mouth

Blastopore

Anus

COELOM FORMATION

Archenteron

Coelom

(c) SCHIZOCOELOUS: SOLID MASSES OF MESODERM SPLIT TO FORM COELOM

Coelom

Archenteron

ENTEROCOELOUS: OUTPOCKETS OF ARCHENTERON FORM COELOM

Figure 29.4
Comparison of early development in protostomes and deuterostomes.
(a) Protostomes have spiral, determinate cleavage; deuterostomes have radial, indeterminate cleavage. (b) The stages diagrammed here, called gastrulas, form after the blastula stage. The blastopore of the gastrula forms the mouth in protostomes; the mouth forms from a secondary opening in the deuterostomes.
(c) Coelom formation also occurs in the gastrula stage. Protostomes are called schizocoels because their coelom forms from splits in mesoderm. Deuterostomes are called enterocoels because their coelom forms from mesodermal outpocketings of the archenteron (blue = ectoderm, yellow = endoderm, red = mesoderm).

terostome (from the Greek *deuteros,* "second") is derived from a secondary opening, and the blastopore becomes the anus. Thus, the anterior-posterior axis of a deuterostome is 180° reversed from that of a protostome, a fundamental difference that helps to justify splitting these two groups of animals phylogenetically.

A third fundamental difference between protostomes and deuterostomes is in the development of the coelom (Figure 29.4c). As the archenteron forms in a protostome, the coelom begins as splits within what were initially solid masses of mesoderm; this is called **schizocoelous** development (from the Greek *schizo,* "split"). Development of the body cavities of deuterostomes is termed **enterocoelous:** The mesoderm arises as lateral outpocketings of the archenteron with hollows that become the coelomic cavities.

Keeping in mind that any phylogenetic grouping is tentative, we can now see Kingdom Animalia as arranged on the evolutionary tree shown in Figure 29.5. In the following pages, we survey some of the most important phyla of invertebrate animals, which are summarized in Table 29.1. Although we consider some aspects of function, such as reproduction and

nutrition, as they are characteristic of particular groups, the main discussion of animal physiology is left for Unit Seven.

PARAZOA (PHYLUM PORIFERA)

Sponges, of the phylum Porifera, are sessile animals that appear so sedate to the human eye that the ancient Greeks believed them to be plants (Figure 29.6). Sponges range in height from about 1 cm to 2 m. Of the 5,000 or so species of sponges, only about a hundred live in fresh water; the rest are marine. The body of a simple sponge resembles a sac perforated with holes (*porifera* means "pore bearers"). Water is drawn through the pores into a central cavity, the **spongocoel,** and then flows out of the sponge through a larger opening called the **osculum.** (This process is described in more detail in Figure 29.7.) More complex sponges have folded body walls and branched spongocoels. Sponges are filter-feeders that collect food particles from the water that streams through the porous body.

Figure 29.5
Tentative phylogenetic tree of the animal kingdom. Not all phyla are represented. Some zoologists include acoelomates and pseudocoelomates with the protostomes because of similarities in embryonic development, but this text uses the term *protostome* only as a subgroup of coelomate animals. A few phyla of coelomates known collectively as the lophophorate animals are placed in neither the protostome nor the deuterostome branch because the relationship of these phyla to other coelomates has not yet been resolved.

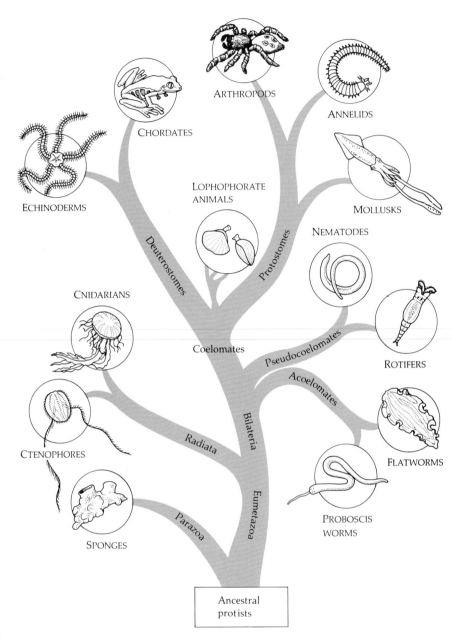

ARTHROPODS

ANNELIDS

CHORDATES

MOLLUSKS

LOPHOPHORATE ANIMALS

NEMATODES

Deuterostomes

Protostomes

ECHINODERMS

CNIDARIANS

Pseudocoelomates

ROTIFERS

Coelomates

Acoelomates

Radiata

Bilateria

FLATWORMS

CTENOPHORES

Eumetazoa

Parazoa

PROBOSCIS WORMS

SPONGES

Ancestral protists

Figure 29.6
Sponges. Sessile animals without specialized organs and tissues, sponges filter food from water pumped through their porous bodies. It has been estimated that a sponge must filter about 1 ton of water to grow by 1 oz. The diverse species of sponges vary in shape and color, some brightly pigmented by symbiotic algae.

Table 29.1 Characteristics of some animal phyla

Phylum	Symmetry	Cleavage	Body Cavity	Digestive Tract	Circulatory System
Cnidaria	Radial	Determinate	None	Gastrovascular cavity	Absent
Ctenophora	Radial	Determinate	None	Gastrovascular cavity	Absent
Platyhelminthes	Bilateral	Determinate	None	Gastrovascular cavity	Absent
Nemertea	Bilateral	Determinate	None	Complete; with mouth from blastopore	Closed; no heart
Rotifera	Bilateral	Determinate	Pseudocoelom	Complete; with mouth from blastopore	Absent
Nematoda	Bilateral	Determinate	Pseudocoelom	Complete; with mouth from blastopore	Absent
Annelida	Bilateral	Determinate	Coelom	Complete; with mouth from blastopore	Closed or open
Mollusca	Bilateral	Determinate	Reduced coelom and hemocoel	Complete; with mouth from blastopore	Open except in cephalopods
Arthropoda	Bilateral	Determinate	Hemocoel	Complete; with mouth from blastopore	Open
Lophophorate phyla	Bilateral	Determinate	Coelom	Complete; development mixes protostome and deuterostome characteristics	Absent in most
Echinodermata	Secondarily radial	Indeterminate	Coelom	Complete; with anus from blastopore	Open or absent
Chordata	Bilateral	Indeterminate	Coelom	Complete; with anus from blastopore	Closed, except in tunicates

Wandering through a gelatinous layer called the **mesohyl** between the two layers of the sponge body wall are cells called **amoebocytes,** named for their use of pseudopodia. Amoebocytes have many functions. They take up food from feeding cells called choanocytes, digest it, and carry nutrients to the epidermal cells. Amoebocytes also form tough skeletal fibers within the mesohyl. In some groups of sponges, these fibers are sharp spicules made from calcium carbonate or silicate; other sponges produce more flexible fibers composed of a protein called spongin. Variation in the chemical composition of the skeleton is one way taxonomists group sponges into the different classes of Phylum Porifera.

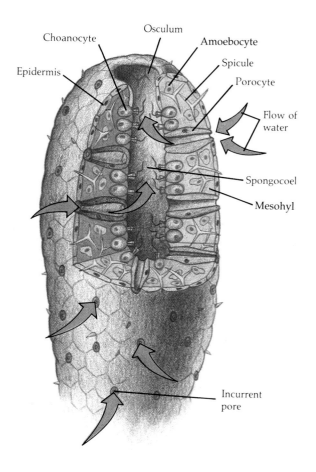

Figure 29.7
Anatomy of a sponge. The wall of a sponge has two layers of cells separated by a gelatinous matrix, the mesohyl (middle matter). The outer layer consists of tightly packed epidermal cells. The incurrent pores are channels through porocytes, cells shaped like elongated donuts that span the body wall. The spongocoel is lined mainly by choanocytes, or collar cells, each with a flagellum ringed by a collar of fingerlike projections. The beating flagella pump water through a sponge, and the collars help trap food particles that the choanocytes ingest by phagocytosis.

Most sponges are **hermaphrodites** (from the Greek names for the male god *Hermes* and the female goddess *Aphrodite*), meaning that each individual functions as both male and female in sexual reproduction by producing sperm *and* eggs. Amoebocytes differentiate into both types of gametes within the mesohyl. Sperm are released into the spongocoel and are carried out of the sponge by the water current. Cross-fertilization results from some of the sperm being drawn into neighboring individuals. Fertilization occurs in the mesohyl, where the zygote develops into a larva covered with flagella. The swimming larvae disperse after their release into the spongocoel and their exit through the osculum. A tiny fraction of the larvae survive to settle on a suitable substratum and begin the sessile existence characteristic of sponges. In some cases, the larva turns inside out during metamorphosis, and the flagellated cells then face the interior.

Sponges are capable of extensive **regeneration,** the replacement of lost parts. They use this power not only for repair but also to reproduce asexually from fragments broken off a parent sponge.

Sponges are among the least complex of all animals. They lack organs, and the layers of loose federations of cells are not really tissues because the cells are relatively unspecialized. Sponges have no nerves or muscles, but the individual cells can sense and react to changes in the environment. Under certain conditions, the cells around the pores and osculum contract to close the openings. The ancestors of Phylum Porifera were possibly choanoflagellates, collared protists that resemble the choanocytes of sponges. Some choanoflagellates form colonies, and the first sponges may have evolved from similar colonial protists (see Figure 29.42).

RADIATA

Phylum Cnidaria

More than 10,000 species of cnidarians exist, mostly marine. The phylum (formerly called Coelenterata) includes hydras, jellyfishes, sea anemones, and corals. These animals have radial symmetry and relatively simple anatomy.

The basic body plan of a cnidarian is a sac with a central digestive compartment, the **gastrovascular cavity,** and a single opening that functions as both mouth and anus. This basic body plan has two variations—the sessile **polyp** and the floating **medusa** (Figure 29.8). An example of the polyp form is *Hydra.* The animals we generally call jellyfishes are medusas. Polyps are cylindrical forms that adhere by the aboral end of a body stalk to the substratum and wave their tentacles, waiting for prey. A medusa is a flattened,

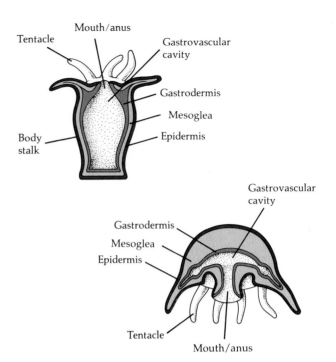

Figure 29.8
Polyp and medusa forms of cnidaria. The body wall of a polyp (top) or medusa (bottom) has two layers of cells, an outer layer of epidermis specialized for protection and an inner layer of gastrodermis for digestion. After the animal ingests food, the gastrodermis secretes digestive enzymes into the gastrovascular cavity. The gastrodermal cells engulf small pieces of the partially digested food by phagocytosis, and digestion is completed within the cells in food vacuoles. Flagella on the gastrodermal cells keep the contents of the gastrovascular cavity agitated and help to distribute nutrients. Sandwiched between the epidermis and gastrodermis is a gelatinous layer of mesoglea. In some polyps, such as hydra, the mesoglea is quite thin and noncellular, but in others, such as sea anemones, it is thick and cellular. (Strictly speaking, cellular "mesoglea" is called mesenchyme, but most zoologists call the middle layer in all cnidarians mesoglea for the sake of simplicity.) In many medusas, the mesoglea is thick and jellylike—thus their name jellyfish.

mouth-down version of the polyp. It moves freely in the water by a combination of passive drifting and weak contractions of the bell. The serpentine tentacles of the jellyfish dangle from the oral surface, which points downward. Some cnidarians exist only as polyps, others only as medusas, and still others pass sequentially through both medusa and polyp stages in their life cycles.

Cnidarians are carnivores that use tentacles arranged in a ring around the mouth to capture prey and push the food into the gastrovascular cavity where digestion begins. The undigested remains are egested through the mouth/anus. The tentacles are armed with batteries of **cnidocytes,** cells of the epidermis that function in defense and capture of prey (Figure 29.9).

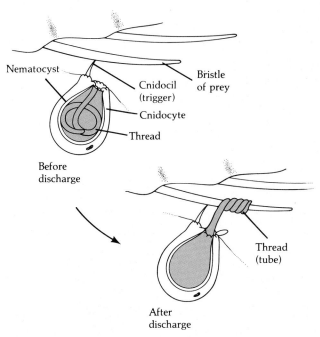

Nematocyst

Cnidocil (trigger)

Bristle of prey

Cnidocyte

Thread

Before discharge

Thread (tube)

After discharge

Figure 29.9
Cnidocytes of cnidaria. Each cnidocyte contains a stinging capsule, the nematocyst, which itself contains a threadlike weapon that waits in an inside-out state. When a trigger on the cnidocyte is stimulated by touch or by certain chemicals, the thread everts and shoots out from the nematocyst. Some of the threads are long and entangle the tiny appendages of small animals that bump haplessly into the tentacles, whereas other threads puncture the prey and inject a poison that paralyzes the victim.

Cnidocytes, which contain stinging capsules called nematocysts, give Phylum Cnidaria its name (from the Greek *cnide*, "nettle").

Muscles and nerves occur in their simplest forms in cnidarians. Cells of the epidermis and gastrodermis have bundles of microfilaments arranged into contractile fibers (see Chapter 7). The gastrovascular cavity, with its noncompressible water, acts as a hydrostatic skeleton against which the contractile cells can work. When the animal closes its mouth, the volume of the cavity is fixed, and contraction of selected cells causes the animal to change shape. The slow movements are coordinated by a nerve net. Cnidarians have no brain, and cnidarian behavior seems to be completely rigid; no one has yet trained a jellyfish.

Phylum Cnidaria is divided into three classes: **Hydrozoa, Scyphozoa,** and **Anthozoa** (Figure 29.10).

Hydra, one of the few cnidarian genera found in fresh water, is an unusual member of the class Hydrozoa in that it exists only in the polyp form. When environmental conditions are favorable, *Hydra* reproduces asexually by budding, the formation of outgrowths that pinch off from the parent to live independently. When conditions deteriorate, *Hydra*

reproduces sexually, forming resistant zygotes that remain dormant until the environment improves. Most hydrozoans alternate polyp and medusa forms, as in the life cycle of *Obelia,* a colonial animal (Figure 29.11). The polyp stage is the more conspicuous animal, a characteristic of the class Hydrozoa.

In Class Scyphozoa, it is generally the medusa rather than the polyp that prevails in the life cycle. The medusas of most species live among the plankton as the animals commonly called jellyfishes. Most coastal scyphozoans go through a small polyp stage during their life cycle, but jellyfishes that live in the open ocean have generally eliminated the sessile polyp.

Sea anemones and corals belong to the class Anthozoa, which means "flower animals." They occur only as polyps. Coral animals live as solitary or colonial forms and secrete hard external skeletons of calcium carbonate. Each polyp generation builds on the skeletal remains of earlier generations to construct "rocks" having shape and color characteristic of the species; it is these skeletons that we call coral.

Phylum Ctenophora

Comb jellies, of the phylum Ctenophora, resemble small cnidarian medusas, but the relationship between ctenophores and cnidarians is uncertain (Figure 29.12). There are only about one hundred species of comb jellies, all of which are marine. The transparent animals range in diameter from about 1 to 10 cm. Ctenophorans ("comb-bearers") are named for their eight rows of cilia; they are the largest animals to use cilia for locomotion. A sensory organ with calcareous particles that settle to the low point acts as an orientation cue, and nerves running from the sensory organ to the combs of cilia coordinate movement. A comb jelly has a pair of long, retractable tentacles, but cnidocytes have been found on only one species.

ACOELOMATES

Phylum Platyhelminthes

Flatworms, of the phylum Platyhelminthes, are ribbonlike animals with soft bodies. Flatworms lack a body cavity and have bilateral symmetry. It is logical to compare them with cnidarians because flatworms are the simplest in organization of all members of the branch Bilateria. Flatworms differ from cnidarians by having several distinct organs and organ systems and by developing from three-layered embryos having mesoderm. However, the gut of a typical flatworm, like that of a cnidarian, is a gastrovascular cavity with only one opening. There are over 15,000 species of

(a)

(c)

(b)

(d)

Figure 29.10
Representatives of the cnidarian classes. (a) The common freshwater cnidarian *Hydra* is a member of Class Hydrozoa. (b) Polyps of a colonial species belonging to the class Hydrozoa. (c) A jellyfish of Class Scyphozoa. The medusa is the conspicuous stage of the scyphozoan life cycle. The largest scyphozoan species have tentacles over 30 m long dangling from umbrellas 2–3 m in diameter. (d) Sea anemones and other members of Class Anthozoa exist only as polyps. (e) A colony of coral polyps (Class Anthozoa). Many corals harbor symbiotic algae that contribute to the food supply of the polyps. Coral reefs, which provide habitats for an enormous variety of invertebrates and fishes, are restricted to warm, shallow seas. This is a star coral.

(e)

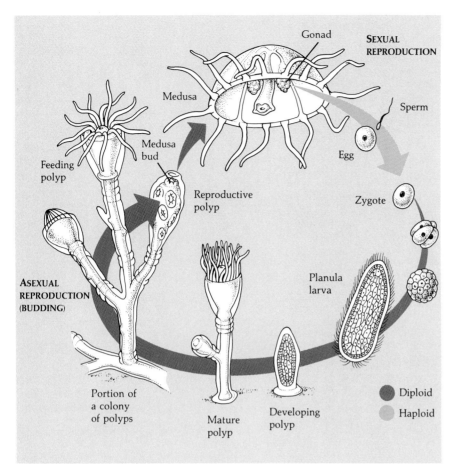

Figure 29.11
Life cycle of the hydrozoan *Obelia*. The polyp reproduces asexually by budding to form a colony of interconnected polyps. Some polyps, equipped with tentacles, are specialized for feeding. Other polyps, specialized for reproduction, lack tentacles and produce tiny medusas by asexual budding. The medusas swim off, grow, and reproduce sexually. The zygote develops into a solid, ciliated larva called the planula. The planula eventually settles and develops into a new polyp. Notice that the polyp stage is asexual and the medusa stage is sexual, and these two stages alternate, one producing the other. But do not confuse this with the alternation of generations that occurs in the plant kingdom. Both polyp and medusa are diploid organisms, whereas one of the plant generations is haploid.

flatworms divided into three classes: **Turbellaria** (turbellarians; mostly free-living flatworms), **Trematoda** (flukes), and **Cestoda** (tapeworms).

Class Turbellaria   Turbellarians are nearly all free living (nonparasitic) and mostly marine (Figure 29.13), but species called **planarians** abound in ponds and streams. Planarians are carnivores that subdue smaller animals with a secretion of mucus, but they also feed on carrion (dead animals). The anatomy of a planarian is described in Figure 29.14.

Planarians and other flatworms lack organs specialized for gas exchange and circulation; the flat shape of the body places all cells close to the surrounding

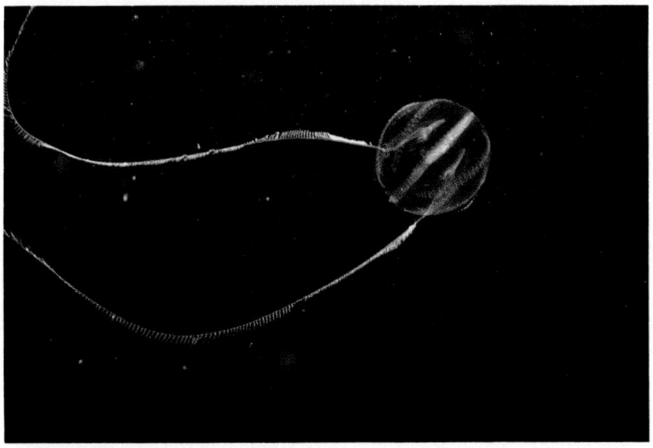

Figure 29.12
A ctenophoran, or comb jelly. This planktonic marine animal is named for its eight combs of cilia, used for locomotion. Ctenophorans and cnidarians are radiate animals, but many zoologists believe these groups have little else in common.

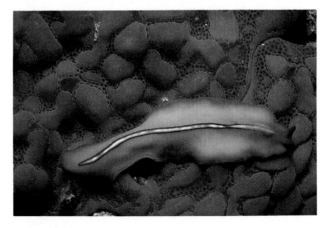

Figure 29.13
A flatworm. Class Turbellaria consists mainly of free-living marine flatworms.

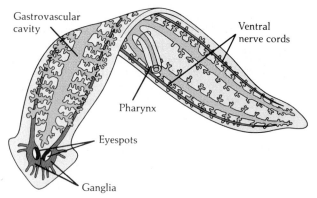

Figure 29.14
Anatomy of a planarian. The mouth is at the tip of a muscular pharynx that extends from the middle of the ventral side of the animal. Digestive juices are spilled onto the prey and the pharynx sucks small pieces of the food into the gastrovascular cavity, where digestion continues. Digestion is completed within the cells lining the gastrovascular cavity, which has three branches, each ramified to provide an extensive surface area. Undigested wastes are egested through the mouth. Located at the anterior end of the worm, near the main sources of sensory input, are a pair of ganglia, dense clusters of nerve cells—a rudimentary brain. From the ganglia, a pair of ventral nerve cords runs the length of the body. With cross nerves connecting the two cords, the planarian nervous system is ladderlike.

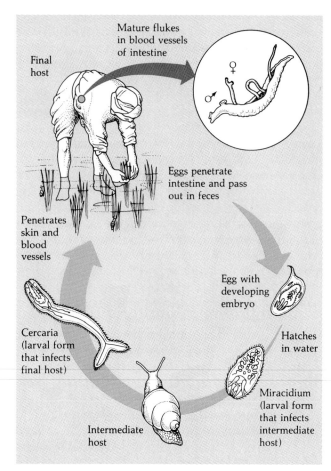

Figure 29.15
Life cycle of a blood fluke (*Schistosoma mansoni*). Workers in irrigated fields that are contaminated with human feces may be exposed to larvae of blood flukes that have escaped from their intermediate host, snails. The mature female fluke fits into a groove running the length of the larger male's body.

water, and ramification of the gastrovascular cavity distributes food throughout the animal. Nitrogenous waste in the form of ammonia diffuses directly from the cells into the surrounding water. Flatworms also have a relatively simple excretory apparatus that functions mainly to maintain osmotic balance between the animal and its surroundings. This system consists of ciliated cells known as flame cells that waft fluid through branched ducts opening to the outside (see Chapter 40 for more details).

Planarians move by using cilia on the ventral epidermis to glide along a film of mucus. Some of the worms also use their muscles to swim through water with an undulating motion.

A planarian has a head with a pair of eyespots that detect light and lateral flaps called auricles that function mainly for smell. The planarian nervous system is more complex than the nerve nets of cnidarians. With a rudimentary brain, planarians can learn to modify their responses to stimuli.

Planarians can reproduce asexually through regeneration: The parent constricts in the middle, and each half regenerates the missing end. Sexual reproduction also occurs. Although planarians are hermaphrodites, copulating mates cross-fertilize.

Class Trematoda Flukes, of the class Trematoda, live as parasites in other animals. Many have suckers they

use to attach to internal organs of the host, and a tough covering helps protect the parasite. Gonads nearly fill the interior of a mature fluke.

Flukes generally have life cycles complicated by an alternation of sexual and asexual stages and a requirement for an intermediate host where larvae develop before infecting the final host, the species in which the adult fluke lives. For example, flukes that parasitize humans spend parts of their life cycles in snails (Figure 29.15). The 200 million people around the world who are infected with blood flukes (*Schistosoma*) suffer body pains, anemia, and dysentery.

Class Cestoda Tapeworms, of the class Cestoda, are parasitic flatworms that as adults live mostly in vertebrates, including humans. The tapeworm head, or scolex, is armed with suckers and often menacing hooks that lock the worm to the intestinal lining of the host. Posterior to the scolex is a long ribbon of units called

Figure 29.22
A bivalve. This scallop has many eyes peering out between the two halves of the hinged shell.

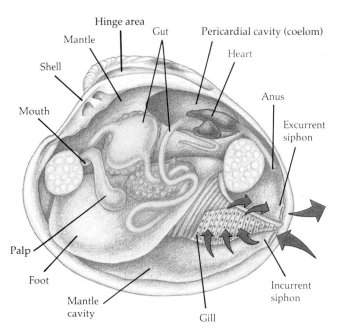

Labels: Hinge area, Mantle, Gut, Pericardial cavity (coelom), Shell, Heart, Mouth, Anus, Excurrent siphon, Palp, Foot, Incurrent siphon, Mantle cavity, Gill

Figure 29.23
Anatomy of a clam. The left half of the bivalve shell has been removed. Food particles suspended in water that enters through the incurrent siphon are collected by the gills and passed to the mouth via elongate flaps, called palps.

shells divided into two halves (Figure 29.22). The two parts of the shell are hinged at the mid-dorsal line, and powerful adductor muscles draw the two halves tightly together to protect the soft-bodied animal. Bivalves have no heads. When the shell is open, the bivalve may extend its hatchet-shaped foot for motility or to anchor itself.

The mantle cavity of a bivalve contains gills that are used for feeding as well as gas exchange (Figure 29.23). Most bivalves are filter-feeders that trap fine food particles in mucus that coats the gills and then use cilia to convey the particles to the mouth. Water flows into the gill chamber through an incurrent siphon, passes over the gills, and then exits the mantle cavity through an excurrent siphon. Bivalves have no radula.

Being filter-feeders, most bivalves lead rather sedentary lives. Clams can pull themselves into the sand or mud, using the muscular foot for an anchor.

Sessile mussels secrete strong threads that tether to rocks, docks, boats, and the shells of other animals. In addition to digging, scallops can also skitter along the sea floor by flapping their shells, rather like the mechanical false teeth sold in novelty shops.

Class Cephalopoda Unlike the sluggish gastropods and the sedentary bivalves, cephalopods are built for speed, an adaptation that fits their carnivorous diet (Figure 29.24). Squid and octopuses use beaklike jaws

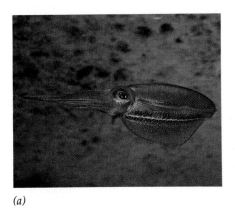

(a)

(b)

(c)

Figure 29.24
Cephalopods. (a) The squid is a speedy carnivore with a beaklike jaw and well-developed eyes. **(b)** The octopus is believed to be one of the most intelligent invertebrates. **(c)** The chambered nautilus is the only shelled cephalopod alive today.

to crush their prey. The mouth is at the center of the foot, which is drawn out into several long tentacles (*cephalopod* means "head foot"). A mantle covers the visceral mass, but the shell is either reduced and internal (squid) or missing altogether (octopuses).

A squid darts about, usually backward, by drawing water into its mantle cavity and then firing a jetstream of water through the excurrent siphon that points anteriorly. The animal steers by pointing the siphon in different directions. Most species of squid are less than 75 cm long, but there are also giant squid, the largest of all invertebrates. The record specimen with documented measurements was 17 m long and weighed about 2 tons.

Rather than swimming as squid do in the open seas, most octopuses live on the sea floor where they scurry about in search of crabs and other food.

Cephalopods are the only mollusks with **closed circulatory systems,** meaning that their blood is always contained in vessels. Cephalopods also have well-developed nervous systems with complex brains. The ability to learn and behave in a complex manner is probably more critical to a fast-moving predator than to sedentary filter-feeders such as clams. Squid and octopuses also have well-developed sense organs.

The ancestors of octopuses and squid were probably shelled mollusks that took up a predaceous life style, the loss of the shell occurring in later evolution. Shelled cephalopods called **ammonites,** many of them very large, were the dominant invertebrate predators of the seas for hundreds of millions of years until their disappearance during the mass extinctions at the end of the Cretaceous period. One shelled cephalopod, the chambered nautilus, survives today (see Figure 29.24c).

Phylum Annelida

Annelids have segmented bodies that give them a ringed appearance (*annelida* means "little rings"). There are about 10,000 annelid species, ranging in length from less than a millimeter to the 3 m length of a giant Australian earthworm. Annelids live in the sea, most freshwater habitats, and damp soil. We can describe the anatomy of annelids in terms of a well-known member of the phylum, the earthworm (Figure 29.25).

The coelom of the earthworm is partitioned by septa, but the digestive tract and longitudinal blood vessels penetrate the septa and run the length of the animal without segmentation (the major vessels have segmental branches). The digestive system has several specialized regions—the pharynx, the esophagus, the crop, the gizzard, and the intestine. The complex closed circulatory system consists of a network of vessels containing blood with oxygen-carrying hemoglobin.

Dorsal and ventral vessels are connected by segmental pairs of vessels. The dorsal vessel and five pairs of vessels that circle the esophagus of an earthworm function as hearts that pump blood through the circulatory system.

In each segment of the worm is a pair of metanephridia with ciliated funnels, called nephrostomes, that remove wastes from the blood and coelomic fluid. The metanephridia lead to exterior pores, through which the metabolic wastes are discharged. Tiny blood vessels are abundant in the skin, which the earthworm uses as its respiratory organ to obtain oxygen. Because gas exchange must occur across a film of moisture, an earthworm suffocates if its skin dries.

The brain of an earthworm is a pair of cerebral ganglia above and in front of the pharynx. A ring of nerves around the pharynx connects the brain to subpharyngeal ganglia, from which a pair of fused nerve cords run posteriorly. All along these ventral nerve cords are segmental ganglia, also fused.

Earthworms are hermaphrodites, but they cross-fertilize. Two earthworms copulate by joining anteriorly so that each deposits sperm near the female opening of its partner. A few days later, the fertilized eggs are packaged in a protective cocoon fashioned by a special organ, the clitellum. Earthworms can also reproduce asexually by fragmentation followed by regeneration.

Some aquatic annelids swim in pursuit of food, but most are bottom-dwellers that burrow in the sand and silt; earthworms, of course, are burrowers. Many annelids creep along by coordinating two sets of muscles, one longitudinal and the other circular. These muscles work against the noncompressible coelomic fluid, a hydrostatic skeleton. The muscles can alter the shape of each segment individually because the coelom is segregated into separate compartments. When the circular muscles of a segment contract, that segment becomes thinner and elongates. Contraction of the longitudinal muscles causes the segment to shorten and thicken. The worm probes forward as alternating contractions of circular and longitudinal muscles progress along the segments like waves. Segmentation may have first functioned as an adaptation for this type of movement.

Phylum Annelida is divided into three classes (Figure 29.26): **Oligochaeta** (earthworms and their relatives), **Polychaeta** (polychaetes), and **Hirudinea** (leeches).

Class Oligochaeta This class of segmented worms includes the earthworms and a variety of aquatic species. An earthworm eats its way through the soil, extracting nutrients as the soil passes through the digestive tube. Undigested material, mixed with mucus secreted into the digestive tract, is egested as casts through the anus. Farmers value earthworms because

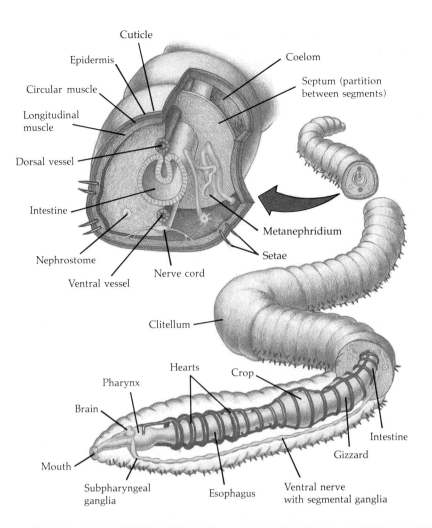

Cuticle

Epidermis

Circular muscle

Longitudinal muscle

Dorsal vessel

Intestine

Nephrostome

Ventral vessel

Nerve cord

Coelom

Septum (partition between segments)

Metanephridium

Setae

Clitellum

Hearts

Crop

Pharynx

Brain

Mouth

Intestine

Gizzard

Subpharyngeal ganglia

Esophagus

Ventral nerve with segmental ganglia

Figure 29.25
Anatomy of an earthworm. Annelids are segmented both externally and internally. Many of the internal structures are repeated in tandem, segment by segment. Externally each segment has four pairs of setae, bristles that provide traction for burrowing.

(a)

(b)

(c)

Figure 29.26
Annelids, or segmented worms.
(a) Most annelids of Class Polychaeta are marine worms. Each segment has a pair of lateral flaps that function as gills for exchange of respiratory gases with the surrounding water. (b) Fanworms are tube-dwelling polychaetes that use their feathery headdresses to filter food from the seawater. This species is known as a Christmas-tree worm. (c) Leeches (Class Hirudinea) are free-living carnivores or parasites that suck blood from other animals. This species inhabits the Malaysian rain forest.

the animals till the earth and the castings improve the texture of the soil. Charles Darwin estimated that 1 acre of British farmland had about 50,000 earthworms that produced 18 tons of castings per year.

Class Polychaeta   Each segment of a polychaete has a pair of paddlelike or ridgelike structures called parapodia ("almost feet"). These parapodia, richly vascularized, function as gills that provide an extended area of skin for gas exchange (Figure 29.26a). Parapodia are also instrumental in locomotion. Each parapodium has several setae made of the polysaccharide chitin (*polychaete* means "many setae"). The setae provide traction when the worm crawls.

Most polychaetes are marine. A few adult forms drift and swim among the plankton, some crawl on the sea floor, and many others live in tubes, which the worms make by mixing mucus with bits of sand and broken shells. The tube-dwellers include the brightly colored fanworms, which trap microscopic food particles in feathery filters that wave from the opening of the tube (Figure 29.26b).

Class Hirudinea   The majority of leeches inhabit fresh water, but there are also land leeches that move through moist vegetation. Many leeches feed on snails and other small invertebrates, and others are blood-sucking parasites that feed by attaching temporarily to other animals, including humans (Figure 29.26c). Leeches range in length from about 1 to 30 cm. Some use bladelike jaws to slit the skin of the host, whereas others secrete enzymes that digest a hole through the skin. The host is usually oblivious to this attack because

the leech secretes an anesthetic. After making the incision, the leech secretes another chemical, hirudin, which keeps the blood of the host from coagulating. The parasite then sucks as much blood as it can hold, often more than ten times its own weight. After this gorging, a leech can last for months without another meal. Until this century, leeches were frequently used by physicians for bloodletting. Leeches are still used for treating bruised tissues and for stimulating circulation of blood to fingers or toes that have been sewn back to hands or feet after accidents.

Phylum Arthropoda

It is estimated that the arthropod population of the world, including crustaceans, spiders, and insects, numbers about a billion billion (10^{18}) individuals. Nearly a million arthropod species have been described, mostly insects. In fact, two out of every three organisms known are arthropods, and the phylum is represented in nearly all habitats of the biosphere. On the criteria of species diversity, distribution, and sheer numbers, Arthropoda must be regarded as the most successful phylum of animals ever to live.

Characteristics of Arthropods   Arthropods are equipped with jointed appendages, for which the phylum is named (from the Greek *arthron*, "joint," and *podus*, "foot"). The appendages are variously modified for walking, feeding, sensory reception, copulation, and defense (Figure 29.27). The body of an arthropod is completely covered by the **cuticle**, an **exoskeleton**

Figure 29.27
External anatomy of a lobster, an arthropod. In this dorsal view of the animal, many of the characteristic features of arthropods are apparent, including the jointed exoskeleton, sensory antennae and eyes, and multiple appendages modified for different functions. Arthropods are segmented, but this characteristic is pronounced only in the abdominal portion of the lobster.

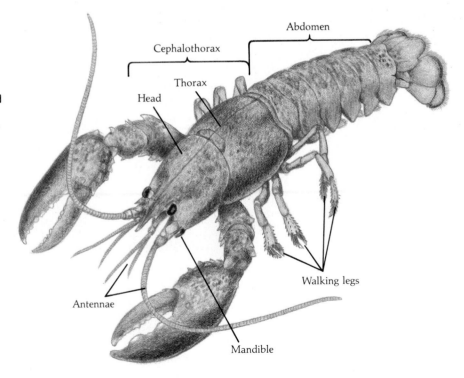

Figure 29.28
***Peripatus*, the walking worm (Phylum Onychophora).**
Peripatus is distinctly segmented and has excretory
organs, musculature, and certain other features that are
annelid-like. *Peripatus* resembles arthropods in its respiratory and circulatory systems, its cuticle made of chitin,
and its jaws modified from appendages.

Figure 29.29
A fossil arthropod. Trilobites were prevalent arthropods
throughout the Paleozoic era. About 4000 trilobite species have been described from fossils.

(external skeleton) constructed from layers of protein
and chitin. The cuticle can be modified into thick, hard
armor over some parts of the body or be paper-thin
and flexible in other locations, such as the joints of
the appendages. The exoskeleton protects the animal
and provides points of attachment for the muscles that
move the appendages. To grow, an arthropod must
occasionally shed its old exoskeleton and secrete a larger
one, a process called **molting.**

Arthropods tune in to their environment with well-
developed sensory organs, including eyes, olfactory
receptors for smell, and antennae for touch and smell.
Cephalization is extensive, with most sensory organs
concentrated on the anterior end of the animal.

Arthropods have open circulatory systems in which
fluid called hemolymph (the term "blood" is reserved
for fluid in a closed circulatory system) is propelled
by a heart. Hemolymph leaves the heart via short
arteries and passes into spaces called sinuses surrounding the tissues and organs. Hemolymph reenters the arthropod heart through pores that are usually equipped with valves. Collectively, the body sinuses
are called the hemocoel, which is not part of the coelom. In most arthropods, the coelom that forms in the
embryo becomes much reduced as development progresses, and the hemocoel becomes the main body
cavity in adults.

A variety of organs specialized for gas exchange have
evolved in arthropods. Most aquatic species have gills
with feathery extensions that place an extensive surface area in contact with the surrounding water. Ter-

restrial arthropods generally have internal surfaces
specialized for gas exchange. Most insects, for instance,
have tracheal systems, branched air ducts leading into
the interior from pores in the cuticle.

Arthropod Evolution   Arthropods are segmented
animals that probably evolved from annelids or from
a segmented protostome that was a common ancestor
of annelids and arthropods. Some zoologists have
speculated that the parapodia of ancient polychaetes
were the forerunners of arthropod appendages. Perhaps early arthropods resembled onychophorans, a
separate phylum of animals that look like walking
worms (Figure 29.28). The multiple appendages of
onychophorans are unjointed, but there are enough
Cambrian fossils of jointed-legged animals that
resemble segmented worms to support other evidence of an evolutionary link between Annelida and
Arthropoda.

Among the early arthropods were the **trilobites**
(Figure 29.29). They were common denizens of the
shallow seas throughout the Paleozoic era but disappeared with the great Permian extinctions that closed
that era about 280 million years ago. Trilobites had
pronounced segmentation, and their appendages
showed little variation from segment to segment. As
arthropods continued to evolve, the segments tended
to fuse and become fewer in number and the appendages to become specialized for a variety of functions
(compare the trilobites with the lobster in Figure 29.27).

The trilobites were outlasted by the **eurypterids**, or
sea scorpions. These marine predators, up to 3 m in
length, belonged to a subphylum of arthropods called
chelicerates. The body of a chelicerate is divided into
an anterior cephalothorax and a posterior abdomen.
The appendages are more specialized than those of

Figure 29.30
Horseshoe crabs. These "living fossils," which have changed little in hundreds of millions of years, have survived from a rich diversity of chelicerates that once filled the seas. Horseshoe crabs are common on the Atlantic and Gulf coasts of the United States.

trilobites, and the most anterior appendages are modified as either pincers or fangs. Chelicerates (from the Greek *cheilos*, "lips," and *cheir*, "arm") are named for these feeding appendages, the **chelicerae.** Most of the marine chelicerates, including the eurypterids, are extinct; one survivor is the horseshoe crab (Figure 29.30). The bulk of modern chelicerates are found on land in the form of Class Arachnida, which includes the spiders, scorpions, ticks, and mites.

Aside from the chelicerates, another major line of arthropod evolution produced several groups, including the crustaceans, insects, centipedes, and millipedes. Rather than having clawlike chelicerae, these arthropods have jawlike **mandibles.** They are also distinguished from chelicerates in having one or two pairs of sensory **antennae** and a pair of **compound eyes** (multifaceted eyes with many separate focusing elements). Chelicerates lack antennae and many have simple eyes (eyes with a single lens). Crustaceans constitute a subphylum of arthropods, whereas insects, centipedes, and millipedes are classes in another subphylum, Uniramia. Uniramians have one pair of antennae and uniramous (unbranched) appendages; crustaceans have two pairs of antennae and typically biramous (branched) appendages. Crustaceans are primarily aquatic and are believed to have evolved in the ocean. Uniramians, on the other hand, are believed to have evolved on land.

The move onto land by chelicerates and uniramians was made possible in part by the cuticle, or exoskeleton. When the exoskeleton first evolved in the seas, its main functions were probably protection and anchorage for muscles, but it was a preaptation that helped certain arthropods live on land by solving the problems of water loss and support. (Recall from Chapter 23 that a preaptation is a structure that evolves

and functions in one environmental context but can perform additional functions when placed in some new environment.) The arthropod cuticle is relatively impermeable to water, helping to prevent desiccation. The firm exoskeleton also solved the problem of support when arthropods left the buoyancy of water. Uniramians and chelicerates both spread onto land during the early Devonian period, following the colonization by plants. The oldest fossil evidence of terrestrial animals is burrows of millipede-like arthropods about 450 million years old. Fossilized arachnids almost as old have also been found.

Of the extant arthropods, the five groups with the greatest number of species are the chelicerate class **Arachnida,** the subphylum (and its one class) **Crustacea,** and the uniramian classes **Chilopoda** (centipedes), **Diplopoda** (millipedes), and **Insecta.**

Class Arachnida   Scorpions, spiders, ticks, and mites are grouped in the class Arachnida (Figure 29.31). These chelicerates have a cephalothorax with six pairs of appendages: the chelicerae, a pair of appendages called pedipalps that usually function in feeding, and four pairs of walking legs. Spiders use their fanglike chelicerae, equipped with poison glands, to attack prey. As the chelicerae and pedipalps masticate the prey, the spider spills digestive juices onto the torn tissues. The food softens, and the spider sucks up the liquid meal. In most spiders, gas exchange is carried out by **book lungs,** stacked plates contained in an internal chamber. Figure 29.32 illustrates the anatomy of a spider in more detail.

A unique adaptation of many spiders is catching flying insects by stringing webs of silk, a protein produced as a liquid by special abdominal glands. The silk is spun by organs called spinnerets into fibers that solidify when contacting air. Each spider engineers a style of web characteristic of its species and constructs the web perfectly on the first try. This complex behavior is apparently inherited. Besides building their webs from silk, various spiders use these fibers in other ways—as droplines for rapid escape, as coats that cover eggs, and even as gift wrapping for food that certain male spiders offer females during courtship.

Class Crustacea   While arachnids and insects thrived on land, crustaceans, for the most part, remained in the seas and ponds where they are now represented by about 40,000 species (Figure 29.33, p. 626).

The multiple appendages of crustaceans are extensively specialized. Lobsters and crayfish, for instance, have a tool kit of 20 pairs of appendages (see Figure 29.27). Crustaceans are the only arthropods with two pairs of antennae. Three or more pairs of appendages are modified as mouthparts, including the hard mandibles. Walking legs are present on the thorax, and, unlike insects, crustaceans have appendages on the abdomen. A lost appendage can be regenerated.

(a)

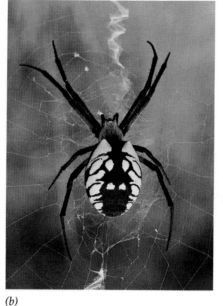

(b)

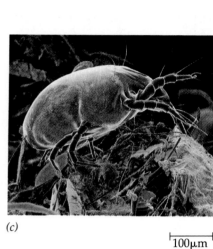

(c)

⊢100µm⊣

Figure 29.31
Arachnids. (a) Scorpions, which hunt by night, were among the first terrestrial carnivores, preying on vegetarian arthropods that fed on the early land plants. The pedipalps of scorpions are pincers specialized for defense and capture of food. The tip of the tail bears a poisonous sting. **(b)** Spiders are generally most active during the daytime, when they hunt for prey or trap insects in webs. **(c)** This house-dust mite, magnified by a scanning electron microscope (artificially colored micrograph), is an ubiquitous scavenger in human dwellings. Unlike some mites that carry disease-causing bacteria, dust mites are harmless except to people who are allergic to them.

Small crustaceans exchange gases across thin areas of the cuticle, but larger forms have gills. The circulatory system is open, with a heart pumping hemolymph through arteries into sinuses that bathe the organs. Crustaceans excrete nitrogenous wastes by diffusion through thin areas of the cuticle, but a pair of glands regulates the salt balance of the hemolymph.

Sexes are separate in most crustaceans. In the case of the lobster, the male uses a specialized pair of appendages to transfer sperm to the reproductive pore of the female during copulation. Most aquatic crustaceans go through a swimming larval stage.

Lobsters, crayfish, crabs, and shrimp are all relatively large crustaceans called decapods. The exoskeleton, or cuticle, is hardened by calcium carbonate; the portion that covers the dorsal side of the cephalothorax forms a shield called the carapace. Most decapods are marine. Crayfish, however, live in fresh water, and some tropical crabs live on land.

The isopods are mostly small marine crustaceans,

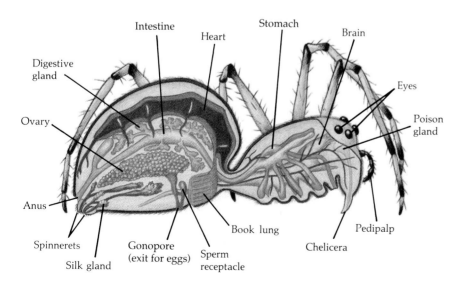

Figure 29.32
Anatomy of a spider, an arachnid.

Labels: Intestine, Heart, Stomach, Brain, Digestive gland, Eyes, Poison gland, Ovary, Anus, Spinnerets, Silk gland, Gonopore (exit for eggs), Sperm receptacle, Book lung, Chelicera, Pedipalp

(a)

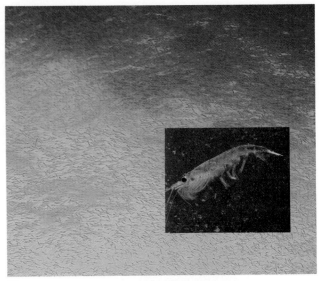

(b)

Figure 29.33
Crustaceans. (a) Crabs, lobsters, crayfish, and shrimp are the most familiar members of Class Crustacea. This is a red crab. (b) Tiny planktonic crustaceans known as krill are consumed by the ton by whales and other large suspension feeders. The inset is a magnification of one of the many krill seen as tiny flecks in the main photograph. (c) Barnacles are sessile crustaceans with a shell (exoskeleton) hardened by calcium salts. Note the jointed appendages projecting from the shell. These are used to capture small plankton and organic particles suspended in the water.

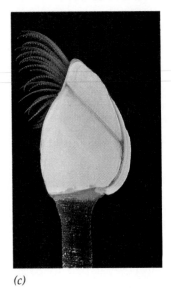

(c)

but this group also includes sow bugs and pill bugs, familiar land animals. These terrestrial crustaceans live mostly in moist soil and other damp places.

Another group of small crustaceans, the copepods, are among the most numerous of all animals. They are important members of the plankton communities that are the foundation of marine and freshwater food chains. Plankton also includes larvae of many larger crustaceans.

Barnacles are sessile crustaceans with parts of their cuticles hardened into shells by salts. They feed by using their feet to paddle food toward the mouth (see Figure 29.33c).

Class Diplopoda and Class Chilopoda The evolutionary link between arthropods and annelids is evident in the distinct segmentation of these two classes (Figure 29.34).

Millipedes are wormlike, with a large number of walking legs (two pairs per segment), though fewer than the thousand their name implies. They eat decaying leaves and other plant matter. Millipedes were probably among the earliest animals on land, living

on mosses and primitive vascular plants. The fossil record suggests that predaceous scorpions moved onto land along with the vegetarian millipedes.

Centipedes are terrestrial carnivores. The head has a pair of antennae and three pairs of appendages modified as mouthparts, including the jawlike mandibles. Each segment of the trunk region has one pair of walking legs. Centipedes use poison claws on the most anterior trunk segment to paralyze prey and for defense.

Class Insecta In species diversity, the insects outnumber all other forms of life combined. They live in almost every terrestrial habitat and in fresh water, and flying insects fill the air. Insects are rare, though not absent, in the seas, where crustaceans are the dominant arthropods. Class Insecta is divided into about 26 orders, some of which are described in Table 29.2, pp. 628–629. **Entomology,** the study of insects, is a

(a)

(b)

Figure 29.34
Diplopods (millipedes) and Chilopods (centipedes). **(a)** Millipedes feed on decaying plant matter. **(b)** The house centipede (*Scutigera coleoptrata*), a fast-moving carnivore, feeds on insects, including cockroaches, and other small invertebrates.

vast field having many subspecialties such as physiology, ecology, and taxonomy. But here we can only examine the general characteristics of this class of animals.

The oldest insect fossils date back to the Devonian period, which began about 400 million years ago. But it was during the Carboniferous and Permian periods when flight evolved and the first explosion in insect diversity occurred. Another major radiation of insect species took place during the Cretaceous period, when insects diversified along with the flowering plants they ate and pollinated (see Chapter 27). Flight is obviously one key to the great success of insects. A flyer can escape predators, find food and mates, and disperse to new habitats much faster than animals that must crawl about on the ground.

Many insects have one or two pairs of wings that emerge from the dorsal side of the thorax (Figure 29.35, p. 630). Since the wings are extensions of the cuticle rather than true appendages, insects are able to fly without sacrificing any walking legs. (By contrast, the flying vertebrates—birds and bats—have one of their two pairs of walking legs modified for wings and as a result are generally quite clumsy on the ground.) Insect wings may have first evolved as cuticular extensions that helped the insect body absorb heat, only later becoming organs for flight. Insects beat their wings, often at speeds of several hundred cycles per second, by using muscles to warp the shape of the entire cuticle covering the thorax. As the wings flap, they change angles so that they produce lift on both the up and down strokes.

Dragonflies, with two coordinated pairs of wings, were among the first insects to fly. Several insect orders that evolved later than dragonflies have modified flight equipment. Bees and wasps, for instance, hook their

wings together, which move as a single pair. Butterflies get a similar result by overlapping their anterior and posterior wings. In beetles, it is the posterior pair of wings that is used for flight, with the anterior pair modified as covers that protect the wings when the beetle is on the ground and burrowing.

The grasshopper illustrates the anatomy of insects (Figure 29.36, p. 630). The insect body has three regions: a head, a thorax, and an abdomen. Segmentation is apparent along the thorax and abdomen, but the head segments are fused tightly. On the insect head are one pair of antennae and a pair of compound eyes. Several pairs of appendages modified for chewing (in grasshoppers) or for lapping, piercing, and sucking (in certain other insects) form the mouthparts; the insect mouth is a very busy place when all its dexterous parts go to work on a piece of food. The thorax of the insect bears three pairs of walking legs.

The internal anatomy of an insect consists of several complex organ systems. The digestive tract is a tube pinched into several regions, each with its own function in breakdown of food and absorption of nutrients. Like other arthropods, an insect has an open circulatory system, with a heart pumping hemolymph into a large artery that then conveys the hemolymph into the sinuses of the hemocoel. Metabolic wastes are removed from the hemolymph by unique excretory organs called **Malpighian tubules,** which are outpocketings of the gut. Gas exchange in insects is accomplished by a **tracheal system** of branched, chitin-lined tubes that infiltrate the body and carry oxygen to nearly every cell. The tracheal system opens to the outside of the body through spiracles, which can open or close to regulate air flow and limit water loss.

The insect nervous system consists of a pair of ven-

Table 29.2 Some major orders of insects

Order	Approximate Number of Species	Main Characteristics	Examples	
Isoptera	2,000	Two pairs of wings, but some stages are wingless; chewing mouthparts; social; division of labor for reproduction, work, defense; incomplete metamorphosis	Termites	
Lepidoptera	140,000	Two pairs of wings; hairy bodies; long coiled tongue for sucking; complete metamorphosis	Butterflies, moths	
Odonata	5,000	Two pairs of wings; biting mouthparts; incomplete metamorphosis	Damselflies, dragonflies	
Orthoptera	30,000	Two pairs of horny, membranous wings; biting and chewing mouthparts in adults; incomplete metamorphosis	Crickets, roaches, grasshoppers, mantids	
Siphonaptera	1,200	Small, wingless, laterally compressed; piercing and sucking mouthparts; jumping legs; complete metamorphosis	Fleas	
Trichoptera	7,000	Two pairs of hairy wings; lapping mouthparts; complete metamorphosis; aquatic larvae build movable cases of sand and gravel bound together by secreted silk	Caddis flies	

Table 29.2 *(Continued)*

Order	Approximate Number of Species	Main Characteristics	Examples	
Anoplura	2,400	Wingless; sucking mouthparts; small with flattened body, reduced eyes; legs with clawlike tarsi for clinging to skin; incomplete metamorphosis; very host-specific	Sucking lice	
Coleoptera	500,000	Two pairs of horny, membranous wings; heavy, armored exoskeleton; biting and chewing mouthparts; complete metamorphosis	Beetles, weevils	
Dermaptera	1,000	Two pairs of leathery, membranous wings; biting mouthparts; large pincers in males; incomplete metamorphosis	Earwigs	
Diptera	80,000	One pair of wings and halteres; sucking, piercing, lapping mouthparts; complete metamorphosis	Flies, mosquitos	
Hemiptera	55,000	Two pairs of horny, membranous wings; piercing, sucking mouthparts; incomplete metamorphosis	True bugs: assassin bug, bedbug, chinch bug	
Hymenoptera	90,000	Two pairs of membranous wings; head mobile; well-developed eyes; chewing and sucking mouthparts; stinging; complete metamorphosis; many species social	Ants, bees, wasps	

Figure 29.35
Insect flight. Insect wings are not modified appendages but are extensions of the cuticle that are flapped by flight muscles that bend the cuticle of the thorax. Here, a lacewing takes off from a leaf.

tral nerve cords with several segmental ganglia. The two cords meet in the head, where the ganglia of several anterior segments are fused into a dorsal brain close to the antennae, eyes, and other sense organs concentrated on the head. Insects are capable of complex behavior, though this behavior seems to be largely innate. Even the intricate social behavior of some bees and ants is apparently inherited. We discuss insect digestion, circulation, excretion, gas exchange, nervous control, and behavior in more detail in Unit Seven.

Many insects undergo metamorphosis in their development. In the **incomplete metamorphosis** of grasshoppers and some other orders, the young resemble adults but are smaller and have different body proportions. The animal goes through a series of molts, each time looking more like an adult, until it reaches full size. Insects with **complete metamorphosis** have larval stages, known by such names as maggot, grub, or caterpillar, that look entirely different from the adult stage (Figure 29.37). The main job of the larva is to eat and grow. The primary function of the adult is to find a mate and reproduce. Mates come together and recognize each other as members of the same species by advertising with bright colors (butterflies), sound (crickets), or odors (moths). After mating, a female lays her eggs on an appropriate food source where the larvae can begin eating as soon as they hatch.

Reproduction in insects is usually sexual, with separate male and female animals. Fertilization is usually internal. In most species, sperm are deposited directly into the female's vagina at the time of copulation, though in some species the male deposits a sperm

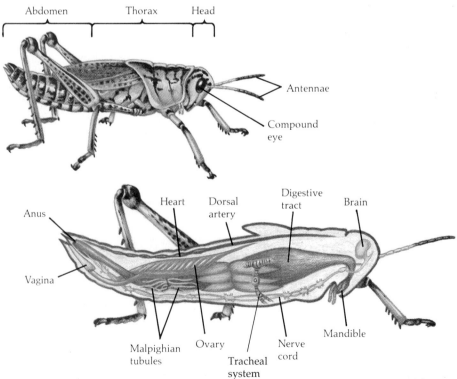

Figure 29.36
Anatomy of a grasshopper, an insect.

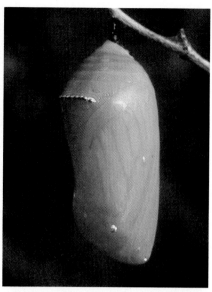

Figure 29.37
Metamorphosis of a butterfly. The larva (caterpillar) spends its time eating and growing, molting as it grows. After several molts, the larva encases itself in a cocoon and becomes a pupa. Within the pupa, the larval tissues are broken down and the adult is built by the division and differentiation of cells that were quiescent in the larva. Finally, the adult emerges from the cocoon. Fluid is pumped into veins of the wings, and then the fluid is withdrawn to leave the hardened veins as struts supporting the wings. Now the insect flies off and reproduces, consuming the calories stored by the feeding larva.

packet ouside the female, who then picks it up. Inside the female, a structure called the spermatheca stores the sperm—usually enough to fertilize more than one batch of eggs. Though most insect species produce a multitude of eggs, some flies produce live offspring, usually a single young at a time. Many insects mate only once in a lifetime.

Animals so numerous, diverse, and widespread as insects are bound to have impact on the lives of all other terrestrial organisms, including humans. On the one hand, we depend on insects to pollinate many of our crops and orchards. On the other hand, insects are vectors for many diseases, including malaria and African sleeping sickness. Furthermore, insects compete with humans for food. In parts of Africa, for instance, insects claim about 75% of the crops. Trying to minimize their losses, farmers in the United States spend billions of dollars each year on pesticides, spraying crops with massive doses of some of the

deadliest poisons ever invented. Try as they may, not even humans have challenged the preeminence of insects and their arthropod kin.

THE LOPHOPHORATE ANIMALS

In the division of coelomate animals into protostomes and deuterostomes, three phyla—**Phoronida, Bryozoa,** and **Brachiopoda**—are difficult to assign. These enigmatic phyla are collectively called the **lophophorate animals,** a reference to the most distinctive structure they share—the **lophophore** (Figure 29.38). A feeding organ, the lophophore is a horseshoe-shaped or circular fold of the body wall bearing ciliated tentacles that surround the mouth at the anterior end of the animal. The cilia draw water toward the mouth between the tentacles, which help trap food particles

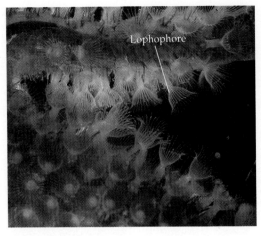

(a)   (b)

Figure 29.38
The lophophorate animals. The most distinctive characteristic of this group is the lophophore, an organ that functions in filter feeding. **(a)** Bryozoans are colonial lophophorates with hard exoskeletons. **(b)** Brachiopods are lophophorates with a bivalve shell.

for these filter-feeders. The common occurrence of this complex apparatus in the lophophorate animals suggests that the three phyla are related. However, other similarities, such as a U-shaped digestive tract and absence of a distinct head, are adaptations to a sessile existence that may have evolved convergently.

In their embryonic development, the lophophorate animals as a group are neither full-fledged protostomes nor deuterostomes, and zoologists are divided on the evolutionary affinities of these phyla. According to one hypothesis, the lophophorate animals are more closely related to deuterostomes, but the two groups branched from a common coelomate ancestor before all of the developmental features characteristic of deuterostomes evolved.

Phoronids are marine worms ranging from 1 mm to 50 cm in length. They live buried in the sand within tubes made of chitin, extending their lophophore from the opening of the tube and withdrawing the feeding organ into the tube when threatened. There are only about ten species of phoronid worms in two genera.

Bryozoans are tiny animals living in colonies that superficially resemble mosses (*Bryozoa* means "moss animals"). In most species, the colony is encased in a hard exoskeleton with pores through which the lophophores of the animals extend (Figure 29.38a). Of the 5000 species of bryozoans, most live in the sea where they are among the most widespread and numerous sessile animals. Several species are important reef builders.

Brachiopods, or lamp shells, superficially resemble clams and other bivalve mollusks, but the two halves of the brachiopod shell are dorsal and ventral to the

animal rather than lateral, as in clams (Figure 29.38b). A brachiopod lives attached to its substratum by a stalk, opening its shell slightly to allow water to flow through the lophophore. All brachiopods are marine. The living brachiopods are remnants of a much richer past; only about 280 extant species are known, but there are 30,000 species of Paleozoic and Mesozoic fossils. A tie to the past is *Lingula*, a living brachiopod genus that has changed little in 400 million years.

DEUTEROSTOMES

At first glance, Phylum Echinodermata, which includes the starfish, may seem to have little in common with Phylum Chordata, which includes birds and mammals, but animals in both of these phyla share features characteristic of deuterostomes: radial and indeterminate cleavage, development of the coelom from outpocketings of the archenteron, and formation of the mouth at the end of the embryo opposite the blastopore.

Phylum Echinodermata

Sea stars and most other **echinoderms** (from the Greek *echin*, "spiny," and *derma*, "skin") are sessile or sedentary animals with radial symmetry (Figure 29.39). The internal and external parts of the animal radiate from the center, often as five spokes. A thin skin covers an endoskeleton of hard calcareous plates. Most echi-

(a)

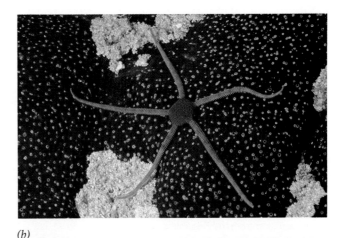

(b)

(c)

(d)

Figure 29.39
Echinoderms. Exclusively marine, the mostly sedentary or slow-moving echinoderms are characterized by radial symmetry, bony endoskeletons, spiny skins, and hydraulic tube feet that function in movement, feeding, and, in some echinoderms, gas exchange. (**a**) Sea star (Class Asteroidea) on coral. (**b**) Brittle star (Class Ophiuroidea). (**c**) Sea urchin (Class Echinoidea). (**d**) Sea lily (Class Crinoidea). (**e**) Sea cucumber (Class Holothuroidea).

(e)

noderms are prickly, owing to bumps and spines of various functions. Unique to echinoderms is the **water vascular system,** a network of hydraulic canals that branch into extensions called tube feet that function in locomotion, feeding, and gas exchange.

Sexual reproduction of echinoderms usually involves separate male and female individuals that release their gametes into the seawater. The radial adults develop by metamorphosis from bilateral larvae.

The 6000 or so echinoderms, all marine, are divided into five classes: **Asteroidea** (sea stars), **Ophiuroidea** (brittle stars), **Echinoidea** (sea urchins and sand dollars), **Crinoidea** (sea lilies), and **Holothuroidea** (sea cucumbers).

Sea stars have five arms (sometimes more) radiating from a central disc (Figure 29.40). The undersurfaces of the arms bear tube feet. When the ampulla attached to each tube foot contracts, the fluid it expels causes the tube foot to extend. After a suction cup on the end of each tube foot grips the substratum, muscles in the wall of the tube foot contract and shorten the foot. The sea star coordinates its tube feet to adhere firmly

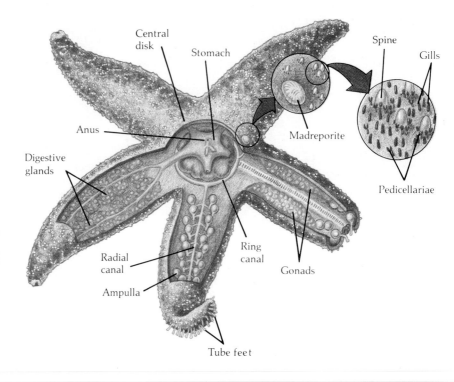

Figure 29.40

Anatomy of a sea star. The surface of a sea star is covered by spines and tiny pincers called pedicellariae that keep the surface free of debris and help defend against small animals. The skin is also studded with small gills for gas exchange. Internal organs are suspended by mesenteries in a well-developed coelom. A short digestive tract runs from the mouth on the bottom of the central disc to the anus on the top of the disc. Digestive glands secrete digestive juices and aid in the absorption and storage of nutrients. The central disc has a nerve ring and nerve cords radiating from the ring into the arms. The water vascular system consists of a ring canal in the central disc and five radial canals, each running the length of an arm in a groove. The system connects to the outside by way of the madreporite. Branching from each radial canal are hundreds of tube feet filled with fluid continuous with the rest of the water vascular system. Attached to each tube foot, and instrumental in its functioning, is a water bulb called the ampulla.

to rocks or to creep along slowly as the tube feet extend, grip, contract, release, extend, and grip again. Sea stars also use their tube feet to prey on clams and oysters. The arms of the sea star embrace the closed bivalve, hanging on tightly by the tube feet. Eventually, the muscles of the mollusk begin to fatigue and the shell opens slightly. The sea star then turns its stomach inside-out, everting it through its mouth into the cracked door of the bivalve. The digestive tract of the sea star secretes juices that begin digesting the soft body of the mollusk within its shell.

Sea stars and other echinoderms have strong powers of regeneration. Shell fishermen have occasionally hacked up sea stars to limit losses of clams and oysters, only to find that the sea star population increased because the fragments regenerated their lost parts.

Brittle stars have smaller central discs than sea stars, and the arms are comparatively longer. Lacking tube feet with suckers, brittle stars move by a serpentine lashing of the prehensile arms. Feeding mechanisms vary among the different species.

Sea urchins and sand dollars have no arms, but they do have five rows of tube feet that can be used to move slowly. These echinoderms also use muscles for pivoting their long spines to aid in movement. The mouth is ringed by complex jawlike structures that are used to eat seaweeds and other food. Sea urchins are roughly spherical in shape, whereas sand dollars are flattened in the oral-aboral axis.

Sea lilies are sessile echinoderms that live attached to the substratum by stalks. The arms circle the mouth,

which points upward. Crinoidea is an ancient class that has been very conservative in its evolution; fossilized sea lilies some 500 million years old could pass for contemporary members of the class.

On casual inspection, sea cucumbers do not look much like other echinoderms. They lack spines and the hard endoskeleton is much reduced. Sea cucumbers are elongated in the oral-aboral axis, giving them their name but further disguising their relationship to sea stars and sea urchins. Closer examination, however, reveals five rows of tube feet, part of the water vascular system found only in echinoderms.

Phylum Chordata

This phylum, our own, consists of two subphyla of invertebrate animals plus the subphylum Vertebrata, the animals with backbones. Grouping the chordates with echinoderms as deuterostomes on the basis of similarities in early embryonic development is not meant to imply that one phylum evolved from the other. Chordates and echinoderms have existed as distinct phyla for at least a half billion years; if the developmental similarities stem from shared ancestry, the evolutionary paths of the two phyla must have diverged very early in the history of animal life. We trace the phylogeny of chordates in the next chapter, concentrating on the history of vertebrates. This chapter concludes by considering some more general questions about early diversification of the animal kingdom.

THE ORIGINS OF ANIMAL DIVERSITY

Ever since the Darwinian revolution, zoologists have speculated about the origins of animals from unicellular ancestors. One hypothesis, called the coenocytic hypothesis, holds that animals arose from a multinucleate protist (perhaps a ciliate) that became subdivided by membranes into many cells. According to this hypothesis, early animals may have resembled acoel worms, a group of flatworms lacking digestive cavities (here, acoel refers to the absence of a gut). Acoels, like other flatworms, are bilaterally symmetrical. They have a ciliated epidermis and a pharynx leading into a solid coenocytic mass that takes in food by phagocytosis and digests within food vacuoles. Because of problems in deriving the sponges and radiate cnidarians from a bilateral flatworm, most zoologists adhere to a second idea, the colonial hypothesis, which postulates that ancestral animals were heterotrophic colonial flagellates (see Figure 26.26). Such ancestral organisms would have been free-swimming, hollow spheres of heterotrophic cells with an anterior-posterior orientation and distinct reproductive and somatic cells. An increasing number of zoologists think that a group of extant protists known as choanoflagellates are similar to the animal ancestors (Figure 29.41).

Most proponents of the colonial hypothesis believe that all animals, including sponges, the radially symmetrical cnidarians, and the bilateral phyla, evolved from hollow flagellate colonies. But several innovations must have evolved in these colonies prior to the origin of the modern animal phyla. Cell layers, a fundamental feature of animals, may have developed as cells proliferated into the center of the ancestral hollow colonies. This could have produced a protoanimal composed of an outer layer of flagellated locomotor cells and an inner mass of digestive and reproductive cells. Such a hypothetical organism, called a planuloid, would have consisted of a solid mass of cells and may have resembled the planula larva of certain cnidarians (Figure 29.42a). An alternative to the planuloid/cell proliferation model proposes that cell layers formed by invagination of the hollow colonial flagellates. This would have produced a two-layered protoanimal similar to the gastrula stage in animal development (Figure 29.42b). Yet another hypothesis holds that a contemporary animal named *Trichoplax adhaerens* resembles the form of the earliest animals (Figure 29.43).

All or none of these hypotheses may approach what actually happened in animal evolution, and the question of what protoanimals looked like remains unresolved because no transitional fossils between unicellular forms and eumetazoa have been discovered. In fact, the first animals, diverse and relatively complex, appear in the fossil record without discernible ties to

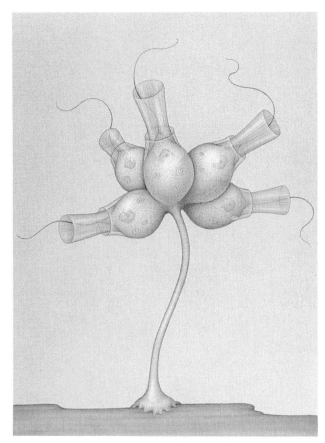

Figure 29.41
Colonial choanoflagellates. Some zoologists consider these protists, which resemble the collar cells (choanocytes) of sponges, to be similar to the ancestors of the animal kingdom. Electron microscopy has shown that mitochondria in choanoflagellates differ markedly from those of other protists, but closely resemble those of animal cells.

their protistan predecessors. As a phylogenetic puzzle, the early diversification of animals is as tough a problem as the origin of animals from protists.

The oldest known fauna lived from about 700 million years ago to about 590 million years ago, a span of Earth history at the end of the Precambrian era that has been named the **Ediacaran period** (for the Ediacara Hills of Australia, where these fossils were first studied extensively). A few fossils of this same period have also been found in Africa and southern China. The Ediacaran fauna has been nicknamed the worm–jellyfish stage of animal evolution; most of the creatures were soft-bodied forms that resembled, at least superficially, cnidarian medusas, colonial cnidarians called sea pens, and annelids.

A second radiation produced a much more diverse fauna as the Paleozoic era dawned with the Cambrian period about 590 million years ago. The increase in animal diversity may have been set off by the explo-

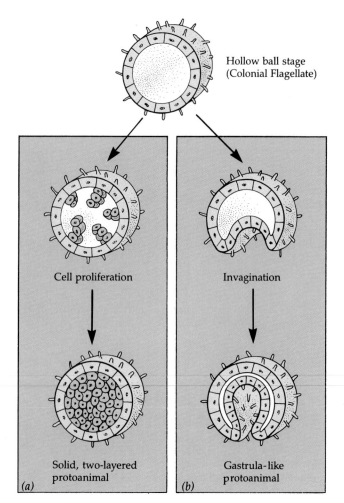

Hollow ball stage (Colonial Flagellate)

Cell proliferation

Invagination

Solid, two-layered protoanimal

Gastrula-like protoanimal

(a)

(b)

Figure 29.42
Hypothetical models for the evolution of protoanimals from hollow colonial flagellates. (a) Formation of a layered planuloid ancestor by internal cell proliferation. (b) An alternate model leading to a gastrula-like organism via invagination.

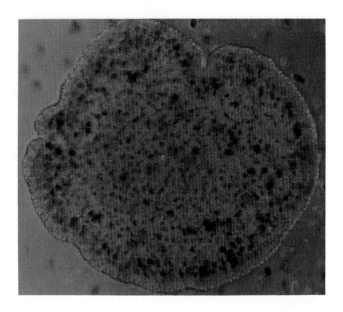

⊢ 100 μm ⊣

Figure 29.43
***Trichoplax adhaerens* (Phylum Placozoa).** *Trichoplax,* the simplest known animal, is a tiny, flattened marine creature with a ciliated epidermis covering a solid core of relatively unspecialized cells. (This is a dorsal view.) The animal has no gut, but as it crawls over a food particle, it hunches its back to form a temporary digestive cavity around the food; a similar habit may have formed the first gut cavity (LM).

sive evolution of shells and hard skeletons, which opened new adaptive zones and revolutionized predator–prey relations. Nearly all modern phyla are represented among the Cambrian fossils, along with many phyla that subsequently disappeared. Figure 29.44 compares the fossil of a Cambrian animal with a fossil representing the Ediacaran fauna.

The Cambrian debut of so many kinds of complex animals is one of the great mysteries of biological history. There is speculation that this highly diverse fauna resulted from environmental changes that occurred during the transition from the Precambrian to the Cambrian. Breakup of continents that occurred during that time would have expanded tropical shorelines, causing an increase in the nutrient content of the seas. Discovery of the Ediacaran fauna has reduced the Cambrian explosion in the minds of some paleo-

biologists to a mere flare-up in a diversification of body plans that began with Precambrian worms and jellyfish. Adolf Seilacher, a paleontologist at the University of Tübingen in Germany, has a different interpretation of the Ediacaran fossils. He argues that the Precambrian animals have been erroneously associated with cnidarians, annelids, and other modern phyla. Seilacher sees a coherent Ediacaran fauna discontinuous with the Cambrian fauna that followed. In his view, the two groups of animals represent different solutions to the problems associated with increased size. The bodies of most Ediacaran animals were flat and foliate (leafy) or quilted, and these finely branched designs exposed most cells to the surrounding environment. In contrast, most modern animals and their Cambrian ancestors are thicker and have expansive internal surfaces for exchange with the environment—the lining of the digestive tract, for instance. Seilacher speculates that the Ediacaran fossils may represent an "experiment" in the evolution of body plans separate from the Cambrian radiation. If he is correct, then the Cambrian radiation was not simply a continuation of phylogenetic branching that began during the Ediacaran period but represents a new wave of diverse designs that evolved in the aftermath of a mass extinction. Perhaps the survivors that served as the ancestral stock for the new radiation were burrow-

(a)

(b)

Figure 29.44
Fossils of early animals. (a) This sandstone imprint of an animal from the Ediacaran Hills of southern Australia is approximately 650 million years old. Although this radially symmetrical animal resembles a jellyfish, the evolutionary relationships between the Ediacaran fauna and the Cambrian animals that followed are still uncertain. **(b)** Adaptive radiation during the Cambrian period, which began about 590 million years ago, produced nearly all known phyla of animals. This fossil of a polychaete worm was collected from the Burgess Shale of British Columbia, one of the richest Cambrian beds.

ing animals of rounded or cylindrical shape that coexisted with the thin, foliated animals that dominated the Ediacaran seascape. Debate on the nature of the Cambrian radiation is one contest in the more general intellectual scrimmage about the tempo of evolution (see Chapter 23).

Explosive revolution or gradual transition, the Cambrian radiation produced nearly all the basic designs we observe in modern animals. In the last half billion years, animal evolution mainly generated new variations on old body plans. One set of variations on the deuterostome plan led to the chordates and eventually to the vertebrates, the subjects of the next chapter.

STUDY OUTLINE

1. Animal life began with the Precambrian evolution of multicellular marine forms that lived by eating other organisms.
2. Subsequent evolution has generated a diversity of animals, grouped into about 35 phyla. About 95% of animal species are invertebrates (lacking backbones).

Characteristics of Animals (pp. 603–604)

1. All animals are multicellular eukaryotes distinguished by a specific type of heterotrophy called ingestion.
2. Animal cells lack walls and store carbohydrate reserves as glycogen. In most animals, cells are successively organized into tissues, organs, and organ systems.
3. Animal reproduction is primarily sexual, and the life cycle is dominated by the diploid stage, gametes generally being the only haploid cells. Asexual budding or regeneration occurs in some species.
4. In sexual reproduction, fertilization of an egg by a flagellated sperm initiates cleavage in the zygote and the formation of a hollow ball of cells called the blastula.

5. Many animals go through metamorphosis, a second stage of development that transforms a sexually immature larva into a morphologically distinct sexual adult.
6. Muscles and nerves, which control active behavior, are unique to animals.

Clues to Animal Phylogeny (pp. 604–607)

1. Taxonomists rely mainly on comparative anatomy and embryology to reconstruct animal phylogeny.
2. Sponges are classified in the subkingdom Parazoa. Other animal phyla are grouped in the subkingdom Eumetazoa.
3. Members of Eumetazoa diverged early into two major branches, the Radiata and the Bilateria. Members of Branch Radiata are the jellyfish and their relatives, sedentary and planktonic forms with radial symmetry. Members of Branch Bilateria are characterized by bilateral symmetry and cephalization.
4. Animals of Branch Bilateria develop from embryos constructed of three concentric primary germ layers: an inner endoderm, an outer ectoderm, and a middle mesoderm.

5. The body plan of Bilateria members is either solid, as in the acoelomates, or with the digestive tube separated from the outer body wall by a cavity. In pseudocoelomates, the cavity is incompletely lined by embryonic mesoderm. Coelomates have a true coelom, a body cavity completely lined by mesoderm.

6. Based on features of embryonic development, coelomate phyla are divided into two main groups: the protostomes, comprising the annelids, mollusks, and arthropods; and the deuterostomes, consisting of the echinoderms and chordates.

Parazoa (Phylum Porifera) (pp. 607–610)

1. The least complex animals are the sponges, which lack tissues and organs such as muscles or nerves.

2. Sponges filter feed by drawing water through pores into a central spongocoel and out an osculum.

3. Most sponges are hermaphroditic, and all are capable of regeneration, which functions in both body repair and asexual reproduction.

Radiata (pp. 610–611)

1. Phylum Cnidaria consists of primarily marine carnivores possessing tentacles armed with stinging cnidocytes that aid in defense and capture of prey. The simple, radial body exists as a sessile polyp or a floating medusa (or both stages in some cases) organized around a central gastrovascular cavity with a single opening for both mouth and anus.

2. Phylum Cnidaria is divided into the following three classes: Class Hydrozoa usually alternates polyp and medusa forms, although the polyp is more conspicuous and the only stage in the genus *Hydra*; the jellyfishes belong to Class Scyphozoa, in which the medusa is the prevalent form of the life cycle; Class Anthozoa contains the sea anemones and corals, which occur only as polyps. Corals secrete external skeletons of calcium carbonate that persist as elaborate rocklike formations.

3. Phylum Ctenophora, the comb jellies, are transparent medusas whose locomotion depends on eight rows of cilia.

Acoelomates (pp. 611–615)

1. The flatworms, Phylum Platyhelminthes, are the simplest members of Bilateria. Ribbonlike animals with a single opening to their gastrovascular cavities, the phylum has the following three classes: Class Turbellaria is made up of mostly free-living, primarily marine species. Planarians have a rudimentary brain and a head containing a pair of eyespots and lateral auricles for olfaction; Class Trematoda, the flukes, live as parasites in animals. Most alternate sexual and asexual stages, requiring an intermediate host; Class Cestoda consists of the tapeworms, parasitic flatworms with a headlike scolex connected to a ribbon of detachable units of sex organs called proglottids.

2. The proboscis worms of Phylum Nemertea are named for the retractable tube they use for search, defense, and prey capture. This group has a simple circulatory system and a digestive tube with both mouth and anus.

Pseudocoelomates (pp. 615–616)

1. Phylum Rotifera is made up of tiny animals possessing a crown of cilia that draws food into the mouth.

2. The roundworms of Phylum Nematoda are among the most numerous animals in both species and individuals. Nematodes have tapered ends and exclusively sexual reproduction. A well-known species is the parasite that causes trichinosis.

Protostomes (pp. 616–631)

1. Phylum Mollusca includes a diverse spectrum of soft-bodied species possessing various modifications of a muscular foot, a visceral mass, and an overlying mantle, which in many species secretes a shell of calcium carbonate. In addition, many mollusks have a radula, a rasping organ used in feeding.

2. Four of the seven classes of Phylum Mollusca are the following: Class Polyplacophora is made up of the chitons, oval-shaped marine animals that cling to exposed rocks and are encased in an armor of dorsal plates; the largest molluscan class is Gastropoda, the snails and their relatives. This group is often protected by single, spiraled shells, although unshelled slugs also occur. Embryonic torsion of the body is a distinctive characteristic; the clams and their relatives of Class Bivalvia have hinged shells divided into two halves. Bivalves are sedentary, headless filter-feeders that use gills for both gas exchange and feeding; Class Cephalopoda includes squid and octopuses, speedy carnivores with beaklike jaws surrounded by tentacles of the modified foot. The shell is either reduced or absent in most genera, and, consistent with their active life styles, the group has a closed circulatory system and sense organs integrated with a well-developed nervous system and brain.

3. Phylum Annelida is a group characterized by body segmentation. Their progressive, wavelike locomotion results from alternating contractions of circular and longitudinal muscles against a fluid-filled, compartmentalized coelom. Annelids have well developed regulatory and nervous systems. Many are hermaphrodites but cross-fertilize. The phylum is divided into the following three classes: Class Oligochaeta includes the earthworm and various aquatic species; representatives of Class Polychaeta possess vascularized, paddlelike parapodia that function as gills and aid in locomotion; Class Hirudinea consists of the leeches, some of which are blood-sucking parasites.

4. Phylum Arthropoda has more known species than all other phyla combined. The arthropods have bodies with various modified, jointed appendages and an exoskeleton that must be shed during molting to allow growth. Arthropods also possess well-developed sense organs, an open circulatory system, and a variety of specializations for gas exchange.

5. The segmentation of arthropods suggests evolution from annelids. Chelicerates are arthropods with pincer or fanglike feeding appendages. A separate line of evolution produced arthropods with jawlike mandibles. The subphylum Crustacea contains primarily aquatic organisms with two pairs of antennae and branched appendages. The subphylum Uniramia contains the insects, centipedes, and millipedes, which have one pair of antennae and unbranched (uniramous) appendages. The five principal classes of Phylum Arthropoda follow.

6. Class Arachnida includes the spiders, ticks, scorpions, and mites, extant chelicerates with simple eyes, two pairs of feeding appendages, and four pairs of walking legs.

7. Lobsters and crayfish are members of the class Crustacea. Among a number of multiple appendages are two pairs of antennae and appendages modified for pinching, chewing, locomotion, and copulation. The circulatory system is open, and gas exchange is through gills or thin areas of the cuticle that also function in excretion.

8. Classes Diplopoda and Chilopoda, the millipedes and centipedes, are animals whose segmentation attests to annelid ancestry. The vegetarian millipedes have two walking legs per segment, whereas the carnivorous centipedes have one pair per segment and are armed with poison claws.

9. In terms of species diversity, members of the class Insecta outnumber all other forms of life combined and have exploited virtually every habitat on Earth. Part of their success is due to the evolution of flight and their relationship with flowering plants.

   The head, which contains one pair of antennae, a pair of compound eyes, and variously modified mouthparts, is connected to a segmented thorax and abdomen. The thorax bears three pairs of walking legs and often one or two pairs of wings. Insects have a complex digestive tract, an open circulatory system, a tracheal respiratory system, unique excretory Malpighian tubules, and a well-developed nervous system that permits complex, though largely innate, behavior. In many species young go through complete or incomplete metamorphosis.

The Lophophorate Animals (pp. 631–632)

1. The phyla Phoronida, Bryozoa, and Brachiopoda evade simple classification as protostomes or deuterostomes. They are grouped together on the basis of their lophophores, horseshoe-shaped, filter-feeding organs bearing ciliated tentacles.

2. These sessile animals, which lack a distinct head and possess U-shaped digestive tracts, may have shared a common coelomate ancestor with the deuterostomes.

Deuterostomes (pp. 632–634)

1. Sea stars and their relatives comprise five classes of the marine phylum Echinodermata, radially symmetrical animals with a unique water vascular system ending in tube feet used for locomotion and feeding. A thin, bumpy or spiny skin covers a calcareous endoskeleton.

2. Phylum Chordata, discussed in detail in Chapter 30, is made up of two subphyla of invertebrates plus a vertebrate subphylum, to which humans belong.

The Origins of Animal Diversity (pp. 635–637)

1. Sponges almost certainly evolved from colonial flagellates, but because transitional fossils are lacking, ideas on eumetazoan ancestry have been based on inferences drawn from the simple anatomy of certain modern organisms. Different hypotheses have variously envisioned protoanimals as colonial flagellates, planuloids, plakulas, or ciliated coenocytic forms that later became subdivided by membranes.

2. The oldest known fauna consisted of soft-bodied animals that lived during the Ediacaran period. During the ensuing Cambrian period, a much more diverse fauna evolved, which included many species with hard shells and skeletons and nearly all the modern phyla.

3. A debate is currently taking place about whether the Ediacaran fauna was mainly an evolutionary dead end or the sparks for the Cambrian explosion. In any event, it is certain that all the basic designs observed in modern animals arose during the Cambrian radiation.

SELF-QUIZ

1. The subkingdom Parazoa consists of
   a. the radially symmetrical sponges, jellyfishes, and their relatives
   b. the only animals that do not feed by ingestion
   c. animals with protostome-type development
   d. colonial choanoflagellates
   e. animals lacking tissues and organs

2. As a group, acoelomates are characterized by
   a. gastrovascular cavities
   b. a body cavity called a hemocoel
   c. deuterostome development
   d. a coelom that is not completely lined with mesoderm
   e. a solid body without a cavity surrounding internal organs

3. Which of the following is *not* descriptive of deuterostomes?
   a. radial cleavage
   b. determinate cleavage
   c. enterocoelous formation of the body cavity
   d. blastopore develops into anus
   e. includes the echinoderms and chordates

4. The branches Radiata and Bilateria of the eumetazoa both exhibit
   a. cephalization
   b. bilateral symmetry of larval forms
   c. dominance of diploid life stage
   d. a complete digestive tract with separate mouth and anus
   e. three germ layers in embryonic development

5. A land snail, a clam, and an octopus all share
   a. a mantle
   b. a radula
   c. gills
   d. embryonic torsion
   e. distinct cephalization

6. Which of the following is *not* a characteristic of the phylum Annelida?
   a. hydrostatic skeleton
   b. segmentation
   c. metanephridia

d. pseudocoelom

e. closed circulatory system

7. In arthropods, the possession of a cuticle usually dictates

a. an aquatic environment

b. molting

c. metamorphosis

d. external fertilization

e. gills or tracheal system for gas exchange

8. Which of the following is *not* true of the chelicerates?

a. They have biramous appendages.

b. Their body is divided into a cephalothorax and an abdomen.

c. The horseshoe crab is one surviving marine member.

d. They include ticks, scorpions, and spiders.

e. Their anterior appendages are modified as pincers or fangs.

9. Which of the following combinations of phyla and characteristics is *incorrect*?

a. Nemertea—proboscis worms, complete digestive tract

b. Nematoda—round worms, pseudocoelomate

c. Cnidaria—radial symmetry, polyp and medusa body forms

d. Platyhelminthes—flatworms, gastrovascular cavity, acoelomate

e. Porifera—gastrovascular cavity, mouth from blastopore

10. Among Cambrian fossils of animals, paleontologists find

a. the oldest known animal fossils

b. mostly soft-bodied, foliate animals

c. early representatives of nearly all animal phyla that exist today

d. evidence for the origin of animals from a coenocytic ciliate resembling an acoel worm

e. transitional fossils revealing the phylogenetic affinities of most animal phyla

CHALLENGE QUESTIONS

1. Pick a representative acoelomate, pseudocoelomate, and coelomate and explain at least one salient embryological difference among your three animals.

2. In the late 1970s, J. Kingsolver and M. Koehl proposed that insect wings were evolutionary derivatives of thermoregulatory structures that projected laterally from the body. The scientists maintained that natural selection worked first on the heat-exchange benefits of the "pro-towings" and later on flight. Experimental data indicated that thermoregulatory effects occurred for wing lengths up to 1 cm, but wings had to be at least 1.5 cm long before aerodynamic effects became significant. What problem did this finding pose for the model, and how might it be conceptually resolved?

3. Professor Lynn Margulis of the University of Massachusetts has suggested that observing an explosion of animal diversity in Cambrian strata is like viewing Earth from a satellite over a long period of time and noticing the emergence of cities only after they are large enough to be evident at that distance. What do you think Dr. Margulis was saying about the Cambrian "explosion"?

FURTHER READING

1. Barnes, R. D. *Invertebrate Zoology*. 5th ed. Philadelphia: Saunders, 1987. A comprehensive text.
2. Daly, H. V., J. T. Doyen, P. R. Ehrlich. *Introduction to Insect Biology and Diversity*. New York: McGraw-Hill, 1978. An excellent introduction to the insects.
3. Goreau, T. F., N. I. Goreau, T. J. Goreau. "Corals and Coral Reefs." *Scientific American*, August 1979. An informative article on the biology and ecology of some beautiful reef-building cnidarians.
4. Gould, S. J. "Play It Again, Life." *Natural History*, February 1986. The role of chance in the Cambrian explosion.
5. Jackson, R. R. "A Web-building Jumping Spider." *Scientific American*, September 1985. A fascinating account of adaptations in an unusual jumping spider that preys on other spiders.
6. Kingsolver, J. G. "Butterfly Engineering." *Scientific American*, August 1985. An ingenious approach to studying butterfly design, based on observations of how the insects drink nectar, fly, and bask in the sun.
7. McMenamin, M. A. S. "The Emergence of Animals." *Scientific American*, May 1987. Possible explanations for the diversification of faunas after the Precambrian.
8. Mitchell, L. G., J. A. Mutchmor, W. D. Dolphin. *Zoology*. Menlo Park, Calif.: Benjamin/Cummings, 1988. An introductory text.
9. Polis, G. A. "The Unkindest Sting of All." *Natural History*, July 1989. The feeding behavior of scorpions.

The Vertebrate Genealogy

Most of us are curious about our genealogies. On the personal level, we wonder about our family ancestry. As biology students, we are interested in retracing human ancestry on the scale of evolutionary history. The questions we ask are basically the same in both cases: What were our ancestors like? How are we related to other animals? Who are our closest relatives? In this chapter, we trace the phylogeny of the vertebrates, the group that includes humans and their closest relatives. Mammals, birds, reptiles, amphibians, and the various classes of fishes are all classified as vertebrates because they have a backbone and many other features in common (Figure 30.1). Following the plan of the other chapters in this unit, we discuss some aspects of vertebrate anatomy and physiology as they pertain to our evolutionary theme. Detailed discussion of animal form and function is reserved for Unit Seven. Our first step in tracking the vertebrate genealogy is to determine where vertebrates fit in the animal kingdom.

PHYLUM CHORDATA

The vertebrates make up one subphylum within the phylum Chordata. **Chordates** also include two subphyla of animals without backbones, the **cephalochordates** and the **urochordates.**

Chordate Characteristics

Based on certain similarities in early embryonic development, chordates are grouped as deuterostomes along

Figure 30.1
A vertebrate skeleton. Vertebrates are named for their backbone, which consists of a series of vertebrae. This hallmark is apparent in this skeleton of a dolphin. The vertebrates make up one subphylum within the phylum Chordata.

with the echinoderms and a few smaller phyla (see Chapter 29). Although chordates vary widely in appearance, they are distinguished as a phylum by the presence of four anatomical structures in the embryo stage (Figure 30.2):

1. Notochord. All chordate embryos have a notochord, which is a longitudinal, flexible rod located between the gut and the nerve cord. The notochord is composed of large, fluid-filled cells encased in fairly stiff, fibrous tissue. It extends through most of the length of the animal as a relatively simple skeleton. It is for this structure that chordates are named. In some invertebrate chordates and primitive vertebrates, the notochord persists to support the adult, but in most vertebrates a more complex, jointed skeleton develops and the adult retains only remnants of the embryonic notochord—as gelatinous material of the discs between vertebrae of humans, for example.

2. Dorsal, hollow nerve cord. The nerve cord of a chordate embryo develops from a plate of dorsal ectoderm that rolls into a tube located dorsal to the notochord. The result is a dorsal, hollow nerve cord unique

to chordates. Other animal phyla have solid nerve cords, usually ventrally located. The chordate nerve cord forms the central nervous system—the brain and spinal cord.

3. Pharyngeal slits. The lumen of the digestive tube of nearly all chordate embryos opens to the outside through several pairs of slits located on the flanks of the pharynx, the region of the digestive tube just posterior to the mouth. The pharyngeal slits probably functioned in early chordates as devices for filter-feeding but became modified for gas exchange and other functions during vertebrate evolution.

4. Postanal tail. Most chordates have a tail extending beyond the anus. By contrast, most nonchordates have a digestive tract that extends nearly the whole length of the body. The chordate tail contains skeletal elements and muscles and provides much of the propulsive force in many aquatic species.

Chordates Without Backbones

Subphylum Cephalochordata Known as lancelets because of their bladelike shape, cephalochordates closely resemble the idealized chordate in Figure 30.2. The notochord, dorsal nerve cord, numerous gill slits, and postanal tail are all prominent (Figure 30.3). A tiny marine animal only a few centimeters long, the lancelet wiggles backward into the sand, leaving only its anterior end exposed. The animal feeds by using a mucous net secreted across the pharyngeal slits to filter tiny food particles from seawater drawn into the mouth by ciliary pumping. The water exits through the slits, and the trapped food passes down the digestive tube.

The lancelet frequently leaves its burrow and swims to a new location. Though a feeble swimmer, the lancelet displays, in rudimentary form, the method of swimming that fishes use. Coordinated contraction of muscle segments serially arranged like rows of chevrons (<<<<) along the sides of the notochord

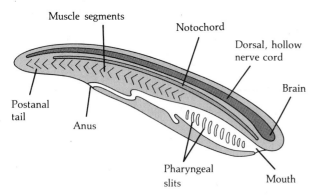

Figure 30.2
Chordate characteristics. This "stripped-down" version of the basic chordate plan possesses the four trademarks of the phylum: a notochord, a dorsal hollow nerve cord, pharyngeal slits, and a postanal tail.

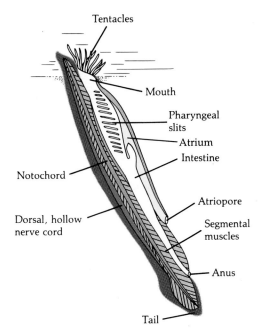

Figure 30.3
Subphylum Cephalochordata: The lancelet *Branchiostoma* (formerly called *Amphioxus*). This small invertebrate displays all four chordate characteristics. The pharyngeal slits are used for filter-feeding. Water passes into the pharynx and through slits into a chamber called the atrium that vents to the outside via the atriopore. Food particles trapped by a mucous net are swept by cilia into the digestive tract. The muscle segments produce the sinusoidal swimming of these animals.

flexes the notochord from side to side in a sinusoidal (~) pattern. This serial musculature makes evident the segmentation of a lancelet. The muscle segments develop from blocks of mesoderm called **somites** that are arranged along each side of the notochord of a chordate embryo. Analogous to the segmentation of annelids, the segmentation of chordates evolved independently.

Subphylum Urochordata Urochordates are commonly called **tunicates**. Most tunicates are sessile marine animals that adhere to rocks, docks, and boats. Some species are colonial. Seawater enters the animal through an incurrent siphon, passes through the slits of the dilated pharynx into a chamber called the atrium, and exits through an excurrent siphon (Figure 30.4a). The food filtered from this water current by a mucous net is passed by cilia into the intestine. The anus empties into the excurrent siphon. The entire animal is cloaked in a tunic made of a cellulose-like carbohydrate called tunicin. Because they shoot a jet of water through the excurrent siphon when molested, tunicates are also called sea squirts.

The adult tunicate scarcely resembles a chordate. It displays no trace of a notochord, nor is there a nerve cord or tail. Only the pharyngeal slits suggest a link to other chordates. But all four chordate trademarks are manifest in the larval form of some groups of tunicates (Figure 30.4b). The larva swims until it attaches by its head to a surface and undergoes metamorphosis, during which most of its chordate characteristics disappear.

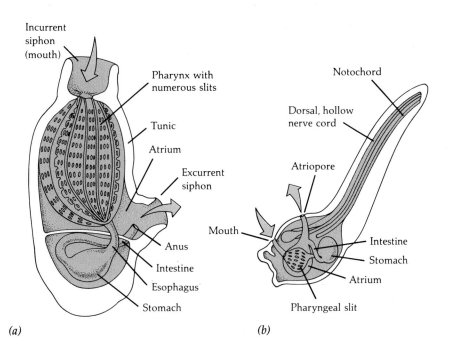

(a)   (b)

Figure 30.4
Anatomy of a urochordate. **(a)** The adult tunicate is a sessile animal commonly arranged in a U shape. Prominent pharyngeal slits function in filter-feeding, but the other chordate characteristics are not obvious in the adult. **(b)** The chordate characteristics are evident in the tunicate larva, which is a free-swimming filter-feeder. Some urochordates lack the sessile "adult" stage, existing as free-swimming, larvalike forms throughout life.

THE ORIGIN OF VERTEBRATES

Phylum Chordata, like most animal phyla, makes its first appearance in the fossil record in Cambrian rocks (see Table 23.1 to review the geological timetable). Paleontologists have found fossilized invertebrates resembling cephalochordates in the Burgess Shale of British Columbia, which is about 550 million years old, about 50 million years older than the oldest known vertebrates. The record of the rocks is too incomplete for us to retrace the origin of the earliest vertebrates from invertebrate ancestors; we can only speculate about this evolution, based mainly on comparative anatomy and embryology.

Most zoologists believe that vertebrate ancestors were filter-feeders with all four of the fundamental chordate characteristics. Protovertebrates may have resembled lancelets, but as specialized bottom-dwelling burrowers, cephalochordates probably represent an offshoot of the evolutionary lineage that gave rise to the vertebrates. Most zoologists consider it more likely that protovertebrates were derived from a urochordate-like ancestor, a simpler, less specialized creature than a lancelet, perhaps similar to the free-swimming, planktonic larvae of tunicates (see Figure 30.4b). Such an ancestor would have been a filter-feeder, weakly swimming and drifting about in ocean currents.

The major milestones in protovertebrate evolution probably occurred early in the Cambrian period. During that time, the phenomenon called **paedogenesis,** the precocious attainment of sexual maturity in a larva, may have had a major impact on vertebrate evolution. Zoologists postulate that some early urochordate-like larval forms became sexually mature and reproduced before undergoing metamorphosis to the sessile tunicate stage. If reproducing larvae were very successful, natural selection may have reinforced the absence of metamorphosis, and a life cycle similar to that of vertebrates may have evolved. Paedogenesis (from the Greek *paid,* "child," and the Latin *genus,* "birth") may be caused by mutations in regulatory genes (see Chapter 23). Perhaps reminiscent of ancestral events, some living species of urochordates exist only as free-swimming larvalike forms.

Eventually, in the vertebrate lineage, some of the reproducing, larvalike forms may have assumed a more active existence, perhaps developing segmental muscles and stronger skeletal support in their tails, becoming more powerful swimmers. This would have allowed them to forage actively in search of more nutritious and abundant food resources. Actively foraging animals must deal with navigation in currents and avoidance of predators. In this context, natural selection may have resulted in the evolution of a protovertebrate with a distinct head equipped with a brain and acute sensory organs.

VERTEBRATE CHARACTERISTICS

Vertebrates retain chordate features while adding other specializations. Cephalization, the development of sensory structures (eyes, nose, inner ears) and a highly specialized brain to process sensory input, is perhaps the most fundamental difference between invertebrate and vertebrate chordates. In addition, most vertebrates possess a column of serially arranged skeletal units, the **vertebrae,** that enclose the nerve cord (see Figure 30.1). In most adult vertebrates, the vertebral column is the most obvious sign of segmentation. The brain, the enlarged anterior end of the nerve cord, is also encased in a skeletal structure, the skull. Together, the skull and vertebral column make up much of the axial (central) skeleton of the vertebrate. Other axial elements, present in many vertebrates, are the ribs and breastbone. Most vertebrates also have an appendicular skeleton supporting the two pairs of appendages (fins or limbs) of the body (see Chapter 45). The vertebrate endoskeleton may be made of either hard bone or flexible cartilage or of some combination of these two materials. Though the skeleton consists mostly of a nonliving matrix, living cells that secrete the matrix are present. The living endoskeleton of a vertebrate can grow with the animal, unlike the nonliving exoskeleton of arthropods, which must be periodically shed by molting.

Vertebrates have closed circulatory systems, with blood pumped by a ventral heart through arteries to microscopic capillaries that reach nearly every cell of the body. After flowing through the capillaries, blood returns to the heart in veins. The vertebrate heart is divided into two to four chambers. Oxygen is transported from respiratory organs to other parts of the body by red blood cells containing the respiratory pigment hemoglobin. The blood is oxygenated as it flows through the skin or highly vascularized membranes lining gills or lungs. Waste products are removed from the blood as it passes through excretory structures collected into compact organs, the kidneys.

Male and female sexes are separated in most vertebrates. Reproduction is nearly always sexual, though in some species in most vertebrate classes, eggs develop without fertilization, a mode of reproduction called parthenogenesis. Fertilization may be either external or internal, depending on species.

There are seven extant classes of the subphylum Vertebrata (Table 30.1). Three of these are commonly called fishes: Agnatha (jawless vertebrates), Chondrichthyes (sharks and rays), and Osteichthyes (bony fishes). The other four classes—Amphibia (frogs and salamanders), Reptilia (reptiles), Aves (birds), and Mammalia (mammals)—are collectively called **tetrapods** (from the Greek *tetra,* "four," and *pod,* "foot") because most have two pairs of limbs that support

Table 30.1 The vertebrate classes

Class	Main Characteristics	Examples
Agnatha	Jawless vertebrates: cartilaginous skeleton; notochord persists throughout life; marine and fresh water; living species lack paired appendages	Lampreys, hagfishes
Chondrichthyes	Cartilaginous fishes: cartilaginous skeleton; jaws; notochord replaced by vertebrae in adult; respiration through gills; internal fertilization, may lay eggs or bear live young; acute senses, including lateral line system	Sharks, skates, rays
Osteichthyes	Bony fishes: bony skeletons and jaws; most species have external fertilization and lay large numbers of eggs; respiration through gills except in a few species of lungfishes; many have a swim bladder; marine and fresh water	Bass, trout, perch, tuna
Amphibia	Aquatic larval stage metamorphosing into terrestrial adult (many species); may lay eggs or bear live young; respiration through lungs and/or skin.	Salamanders, newts, frogs, toads
Reptilia	Terrestrial tetrapods with scaly skin: respiration via lungs; lay amniotic shelled eggs or bear live young	Snakes, lizards, turtles, crocodiles
Aves	Tetrapods with feathers: forelimbs modified to form wings; respiration through lungs; endothermic; internal fertilization; shelled amniotic eggs; acute vision	Owls, sparrows, penguins, eagles
Mammalia	Tetrapods with young nourished from mammary glands of females: respiratory system with diaphragm to ventilate lungs; endothermic; most bear live young; possess hair	Monotremes (echidnas, platypuses); marsupials (opossums, kangaroos); placental mammals (humans, dogs, horses, whales, rodents, elephants, cattle, bats)

them on land. In the following pages we survey the diversity of organisms in each of these vertebrate classes. The survey also includes a brief description of an extinct class of early jawed vertebrates called placoderms.

CLASS AGNATHA

The oldest vertebrate fossils are diverse jawless creatures called **agnathans** ("without jaws"), including fishlike animals called **ostracoderms** that were encased in an armor of bony plates. Traces of agnathans are found in Cambrian strata, but most date back to the Ordovician and Silurian period, about 400–500 million years ago. The fossil record provides few clues directly linking agnathans to invertebrate ancestors, nor is the record complete enough to explain the tremendous evolutionary radiation of agnathan forms that had occurred by the time of the late Silurian.

Early agnathans were generally small, less than 50 cm in length. Many lacked paired fins and apparently were bottom-dwellers that wiggled along stream beds or the sea floor, but there were some active, midwater forms with paired fins as well. All had mouths with circular or slitlike openings that lacked jaws. Most agnathans were probably mud-suckers or filter-feeders that used their mouths as intakes for sediments or suspended organic debris that was passed through the gill slits where food was trapped. Thus, the pharyngeal apparatus retained the primitive filter-feeding function, although gills in agnathans were probably also the major sites of oxygen–carbon dioxide exchange.

The ostracoderms and most other agnathan groups declined and finally disappeared during the Devonian period, but jawless vertebrates are alive today in the form of lampreys and hagfishes. In common with many extinct agnathans, lampreys and hagfishes lack paired appendages. In contrast to many ostracoderms, however, they have no external armor. The eel-shaped sea lamprey feeds by clamping its round mouth onto the flank of a live fish, uses a rasping tongue to penetrate the skin of its prey, and then feeds on the prey's blood (Figure 30.5). Sea lampreys live as larvae for years in freshwater streams and then migrate to the sea or lakes as they mature into adults. The larva is a filter-feeder that looks very much like the lancelet, a cephalochordate. Some species of lampreys feed only as larvae. Following several years as filter-feeders in streams, they attain sexual maturity, reproduce, and die within a few days.

Hagfishes superficially resemble lampreys, but they are mainly scavengers rather than blood-suckers or filter-feeders and their mouthparts are not adapted for rasping. Some species feed on sick or dead fish, whereas other hagfishes eat marine worms. Hagfishes lack a larval stage and live entirely in salt water.

Figure 30.5
The sea lamprey, an agnathan (jawless vertebrate).
Acting like both a predator and a parasite, this lamprey uses its rasping mouth to bore a hole in the side of a fish, living on the blood of its host.

CLASS PLACODERMI

During the late Silurian and early Devonian periods, the agnathans were largely replaced by a diversity of armored fishes called **placoderms** (Figure 30.6). The largest were more than 10 m long, but most were less than a meter in length. In contrast to many ostracoderms, placoderms had paired fins, which greatly enhanced their swimming ability. Placoderms also had jaws, making their mouth more than just a fixed orifice for scooping sediments. With paired fins and hinged jaws, placoderms did not have to rely on filtering bottom sediments to obtain nutrients. Many species were predators, capable of chasing prey and biting off chunks of food.

The hinged jaws of vertebrates evolved by modification of the skeletal rods that had previously supported the anterior pharyngeal (gill) slits (Figure 30.7). The remaining gill slits, no longer required for filter-feeding, remained as the major sites of respiratory gas exchange with the external environment.

The origin of vertebrate jaws illustrates a general feature of evolutionary change: New adaptations usually evolve by the modification of existing structures. Hinged jaws also evolved in arthropods (see Chapter 29), but these had a totally different origin than vertebrate jaws. Arthropod jaws are modified appendages that work from side to side rather than up and down like vertebrate jaws. As a mechanism of adaptation, evolution is limited by the raw material with which it must work; evolution is more of a remodeling process than a creative one.

The Devonian period (about 350–400 million years ago) is known as the age of fishes, as the placoderms

Figure 30.6
A placoderm. An extinct class of vertebrates, placoderms were armored fishes with jaws. They were important predators in the Devonian seas. This is a dorsal view of a museum model of *Lunapsis heroidi*.

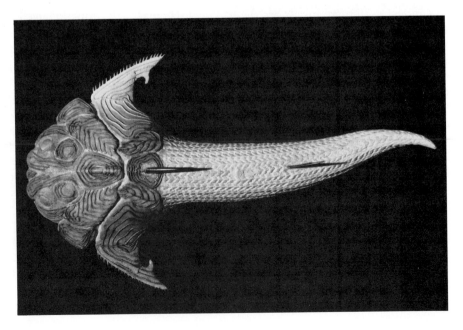

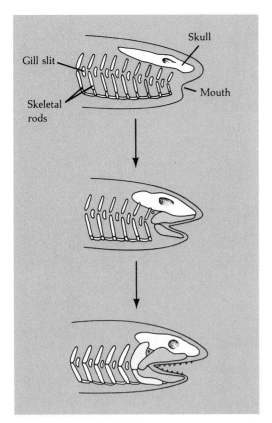

Figure 30.7
Evolution of vertebrate jaws. The skeleton of the jaws and their supports evolved from two pairs of skeletal rods located between gill slits that were near the mouth. Pairs of rods anterior to those that formed the jaws were either lost or incorporated into the jaws.

and another group of jawed fishes called acanthodians (placed in a separate class) radiated and many new forms evolved in both fresh and salt water. Placoderms and acanthodians dwindled and disappeared almost completely by the beginning of the Carboniferous period, 350 million years ago. During the Devonian or earlier, however, ancestors of placoderms and acanthodians may have given rise to sharks (Class Chondrichthyes). Another group of large, jawed predators—the bony fishes (Class Osteichthyes)—diverged from the common ancestor some 425–450 million years ago. Sharks and bony fishes are the reigning vertebrates over the watery two-thirds of the surface of Earth.

CLASS CHONDRICHTHYES

Sharks and their relatives are called cartilaginous fishes because they have somewhat flexible skeletons made of cartilage rather than bone. Jaws and paired fins, which first appeared in the placoderms, are well developed in the cartilaginous fishes. The most widespread and diverse members of Class Chondrichthyes are the sharks and rays (Figure 30.8).

Most sharks have streamlined bodies and are swift swimmers, but they do not maneuver very well. Their stiff caudal (tail) and dorsal fins are used more like keels than rudders, and the paired pectoral (fore) and pelvic (hind) fins function like submarine vanes for lift in the water. Though it gains some buoyancy by storing a large amount of oil in its huge liver, a shark is still more dense than water, and it sinks if it stops swimming. Continual swimming also ensures that

(a)

Figure 30.8
Cartilaginous fishes. (a) Fast swimmers with acute senses and powerful jaws, sharks, such as this blacktip reef shark, are well adapted to their predatory way of life. Note the paired pectoral and pelvic fins. **(b)** Most rays are flat bottom-dwellers that crush mollusks and crustaceans for food.

(b)

water will flow into the mouth and out through the gills, where gas exchange occurs. Some sharks, however, take breaks from swimming and sleep on the sea floor or in underwater caves.

The largest sharks filter-feed on plankton (an example is the whale shark, which filters a million liters of water per hour), but most sharks are carnivores that swallow their prey whole or use their powerful jaws and sharp teeth to tear flesh from animals too large to swallow in one piece. Shark teeth evolved from the jagged scales that cover the abrasive skin. The digestive tract of many sharks is proportionately shorter than the digestive tube of many other vertebrates. Within the shark intestine is a spiral valve, a corkscrew-shaped ridge that increases surface area and prolongs the passage of food along the short digestive tract.

Acute senses are adaptations that go along with the restless, carnivorous life style of sharks. Sharks have sharp vision but cannot distinguish colors. The nostrils of sharks, like those of most fishes, open into dead-end cups, and they can be used only for olfaction, not for breathing. Along with eyes and nostrils, the shark head also has a pair of regions in the skin that can detect the electric fields generated by muscle contractions of nearby fish and other animals. Running the length of each flank of the shark is the **lateral line system,** a row of microscopic organs sensitive to changes in the surrounding water pressure, enabling the shark to detect minor vibrations (see Chapter 45). Sharks can also hear by sensing percussions with a pair of auditory organs. Sharks and other fishes have no eardrums, structures that terrestrial vertebrates use to transmit sound waves traveling through air into the ear toward the auditory organs. Sound reaches the shark through water, and the animal can use its entire body to transmit the sound to the hearing organs of the inner ear.

Shark eggs are fertilized internally. The male has a pair of claspers on its pelvic fins that transfer sperm into the reproductive tract of the female. Some species of sharks are **oviparous,** laying eggs that hatch outside the mother's body. These sharks release their eggs after encasing them in protective coats. Other species are **ovoviviparous;** that is, they retain the fertilized eggs in the oviduct. Nourished by the egg yolk, the embryos develop into young that are born after hatching within the uterus. A few species are **viviparous**—the young develop within the uterus, nourished until they are born by nutrients received from the mother's blood through a placenta. The reproductive tract of the shark empties along with the excretory system and digestive tract into a common chamber, the **cloaca,** which exits through a single vent.

Although rays are closely related to sharks, they have adopted a very different life style. Most rays are flattened bottom-dwellers that feed by using their jaws to crush mollusks and crustaceans. The tail of many rays is whiplike and, in some species, bears venomous barbs.

CLASS OSTEICHTHYES

Of all vertebrate classes, bony fishes, of the class Osteichthyes, are the most numerous, both in individuals and in species (about 30,000). Fishes of this class are abundant in the seas and in nearly every freshwater habitat. Bony fishes range in size from about 1 cm to more than 6 m. Most of the fishes familiar to us are osteichthyans (Figure 30.9).

In contrast to the cartilaginous fishes, the skeleton of most bony fishes is reinforced by a hard matrix of calcium phosphate. The skin is often covered by flattened bony scales that differ in structure from the toothlike scales of sharks. Typical of fishes, glands in the skin of a bony fish secrete a mucus that gives the animal its characteristic sliminess, an adaptation that reduces drag during swimming. In common with sharks, bony fishes have a lateral line system clearly evident as a row of tiny pits in the skin on either side of the body.

Bony fishes breathe by drawing water over four or five pairs of gills that are located in chambers covered by a protective flap, the **operculum.** Water is drawn into the mouth, through the pharynx, and out between the gills by movement of the operculum and contraction of muscles contained within the gill chambers. This enables a bony fish to breath while stationary; sharks lack operculi and must move to pass water over the gills.

Another adaptation of bony fishes not found in sharks is the **swim bladder,** an air sac that helps control the buoyancy of the fish (Figure 30.10). Transfer of gases between the swim bladder and the blood varies the inflation of the bladder and adjusts the density of the fish. Many bony fishes, in contrast to sharks, can rest almost motionless in the water.

Bony fishes are generally maneuverable swimmers, their flexible fins better for steering and propulsion than the stiffer fins of sharks. The fastest bony fishes, which can swim in short bursts of up to 80 km per hour, have the same basic body shape as a shark. In fact, this body shape, termed *fusiform* (tapering on both ends), is common to all fast fishes and aquatic mammals such as seals and whales. Water is about a thousand times more dense than air, and thus the slightest bump that causes drag is even more impeding to a fish than to a bird. Regardless of their different origins, we should expect speedy fishes and marine mammals to have similar streamlined shapes, because the laws of hydrodynamics are universal. This is one more example of convergent evolution.

(a)

(b)

Figure 30.9
Bony fishes. The species depicted here are ray-finned fishes, one of three groups of bony fishes that had evolved by the end of the Devonian period. **(a)** Longnose gar. **(b)** Porcupine fish.

Details in reproduction of bony fishes vary extensively. Most species are oviparous, reproducing by external fertilization after the female sheds large numbers of small eggs. However, internal fertilization and birthing characterize other species. Some bony fishes display complex mating rituals (see Chapter 50).

Both cartilaginous and bony fishes diversified extensively during the Devonian and Carboniferous, but whereas sharks arose in the sea, bony fishes probably had their origin in fresh water. The swim bladder was modified from lungs that had been used to augment the gills for gas exchange, perhaps in stagnant swamps with low oxygen content. By the end of the Devonian, three distinct subclasses of bony fishes had evolved: the **ray-finned fishes** (Subclass Actinopterygii), the **lobe-finned fishes** (Subclass Crossopterygii), and the **lungfishes** (Subclass Dipnoi).

Nearly all the families of fishes familiar to us are ray-fins. The various species of bass, trout, perch, tuna, and herring are examples. The fins, supported mainly by long flexible rays, have become modified for maneuvering, defense, and other functions. Ray-finned fishes spread from fresh water to the seas during their long history. Adaptations that solve the osmotic problems of this move to salt water are discussed in Chapter 40.

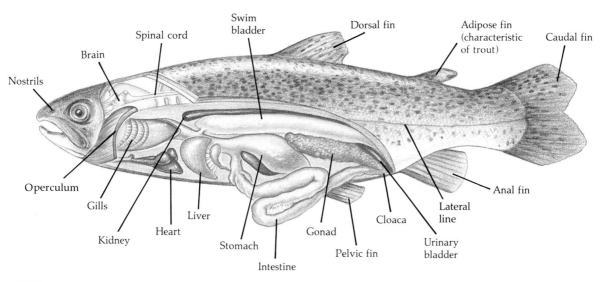

Figure 30.10
Anatomy of a trout, a representative bony fish.

Figure 30.11
Latimeria, a lobe-finned fish. The only living representative of the Subclass Crossopterygii, *Latimeria* lives in the deep sea off Madagascar.

Numerous species of ray-finned fishes returned to fresh water at some point in their evolution. Some of these, including salmon and sea-run trout, replay their evolutionary round-trip from fresh water to seawater back to fresh water during their life cycle.

In contrast to the ray-fins, most lobe-finned fishes and lungfishes remained in fresh water and continued to use their lungs to aid the gills in breathing. Lobe-fins and some lungfishes had fleshy, muscular pectoral and pelvic fins that were supported by extensions of the bony skeleton. Many were large, apparently bottom-dwelling forms that may have used their paired fins as aids to "walking" on the substrate under water. Some may also have been able to waddle occasionally on land.

Three genera of lungfishes live today in the Southern Hemisphere. They generally live in stagnant ponds and swamps, surfacing to gulp air into lungs connected to the pharynx of the digestive tract. When ponds shrink during the dry season, lungfishes can burrow in the mud and aestivate, which means to wait in a state of torpor.

Lobe-finned fishes are represented today by only one known species, the coelacanth (*Latimeria*). Although most Devonian lobe-fins were probably freshwater animals with lungs, the coelacanth is a lungless species belonging to a lineage that entered the seas at some point in its evolution (Figure 30.11).

Few of us encounter lungfishes or lobe-fins, and today these animals are far less numerous than the ray-fins. However, the lobe-finned fishes of the Devonian are of great importance in the vertebrate genealogy, for they gave rise to amphibians (Figure 30.12).

CLASS AMPHIBIA

The amphibians were the first vertebrates on land. Today the class is represented by a total of about 4000 species of frogs, salamanders, and caecilians (wormlike creatures that live in tropical forests).

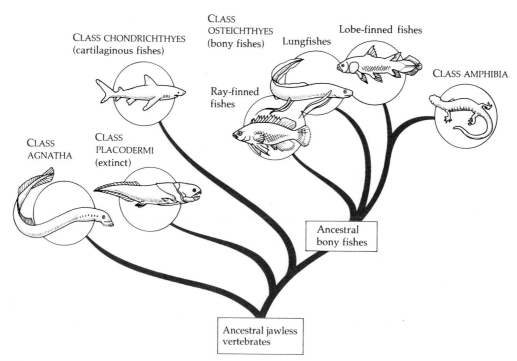

Figure 30.12
An evolutionary tree of fishes and amphibians.

Early Amphibians

As is the case today, parts of the Devonian world were subject to cycles of drought, followed by heavy rainfall and then drought again. In such areas, some of the freshwater fishes apparently adapted to the unreliable conditions, and lobe-finned species with lungs prevailed. The lobe-fins are of particular interest because the skeletal structure of their fins suggest that these paired appendages could have assisted in movement on land. Some of the fossil lobe-fins, including a creature named *Eusthenopteron*, exhibited many other anatomical similarities to the earliest amphibians, making them good candidates as the ancestors of tetrapods (Figure 30.13a).

The oldest amphibian fossils date back to late Devonian times, about 350 million years ago (Figure 30.13b). New adaptive zones opened to these first vertebrates on land, with a cornucopia of food previously unexploited by backboned animals. From the start, amphibians were predators that ate insects and other invertebrates that preceded them onto land. Adaptive radiation of the earliest amphibians gave rise to a diversity of new forms. Many Carboniferous species superficially resembled reptiles. Some reached 4 m in length. Amphibians were the only vertebrates on land in late Devonian and early Carboniferous times, making the name Age of Amphibians appropriate for the Carboniferous period. Amphibians began to decline during the late Carboniferous period. As the Mesozoic era dawned with the Triassic period, about 230 million years ago, most survivors of the amphibian lineage resembled modern species.

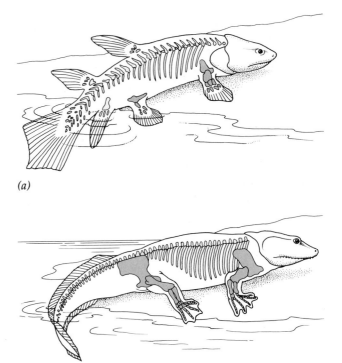

(a)

(b)

Figure 30.13
The emergence of amphibians. (a) The ancestors of tetrapods were lobe-finned fishes, such as *Eusthenopteron*, that had muscular fins with extensions of the skeleton that provided some support for the animal on land. **(b)** This reconstruction, based on Devonian fossils, depicts an early amphibian.

Modern Amphibians

There are three extant orders of amphibians (Figure 30.14): **urodeles** ("tailed ones"—salamanders), **anurans** ("tail-less ones"—frogs, including toads), and **apodans** ("legless ones"—caecilians).

There are only about 400 species of urodeles, some of which are entirely aquatic whereas others live on land as adults or throughout life. Most salamanders that live on land walk with a side-to-side bending of the body that may resemble the swagger of the early amphibians. Aquatic salamanders swim sinusoidally or walk along the bottom of streams or ponds.

Anurans, numbering nearly 3500 species, are more specialized than urodeles for moving on land. Lacking a tail, adult frogs use their powerful hindlegs to hop along the terrain. A frog nabs insects by flicking out its long sticky tongue, which is attached to the front of the mouth. Frogs display a great variety of adaptations that help them avoid being eaten by larger predators. In common with other amphibians, many exhibit color patterns that camouflage, and their skin

glands secrete distasteful, or even poisonous, mucus (see Figure 30.14b).

Apodans, the caecilians (about 150 species), are legless, nearly blind, and superficially resemble earthworms. Caecilians inhabit tropical areas where most species burrow in moist forest soil; a few South American apodans live in freshwater ponds and streams.

Amphibian means "two lives," a reference to the metamorphosis of many frogs (Figure 30.15). The tadpole, the larval stage of a frog, is usually an aquatic herbivore with internal gills, a lateral line system resembling that of fishes, and a long finned tail. The tadpole lacks legs and swims by undulating like its fishlike ancestors. During the metamorphosis that leads to the "second life," legs develop and the gills and lateral line system disappear. The young tetrapod with air-breathing lungs, a pair of external eardrums, and a digestive system capable of digesting animal protein crawls onto shore and begins life as a terrestrial hunter. In spite of the name *amphibian*, however, many members of the class, including some frogs, do not go through the aquatic tadpole stage, and many do not live a dualistic—aquatic and terrestrial—life. There

(a)

(b)

(c)

Figure 30.14
Amphibian orders. (a) Urodeles (salamanders) retain their tails as adults. Some are entirely aquatic, but others live on land. This northern red salamander is a member of a large family (Plethodontidae) whose members have lost their lungs during their evolution and rely entirely on their skin and moist mouth surfaces for gas exchange. **(b)** Anurans, such as this blue poison arrow frog, lack tails as adults. Poison arrow frogs inhabit tropical forests and are being depleted in numbers as the forests are being destroyed; their skin glands secrete deadly nerve toxins used by Central and South American natives to coat arrow tips. **(c)** Apodans, such as this Sri Lankan caecilian, are legless, mainly burrowing amphibians.

(a)

(b)

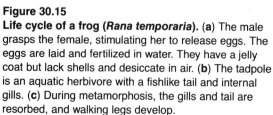

(c)

Figure 30.15
Life cycle of a frog (*Rana temporaria*). (a) The male grasps the female, stimulating her to release eggs. The eggs are laid and fertilized in water. They have a jelly coat but lack shells and desiccate in air. **(b)** The tadpole is an aquatic herbivore with a fishlike tail and internal gills. **(c)** During metamorphosis, the gills and tail are resorbed, and walking legs develop.

are strictly aquatic and strictly terrestrial species in all three extant orders. Moreover, salamander and caecilian larvae look much like adults, and typically both the larvae and adults are carnivorous. Paedogenesis is common among some groups of salamanders; the mudpuppy (*Necturus*), for instance, retains gills and other larval features when sexually mature.

Most amphibians maintain close ties with water and are most abundant in damp habitats such as swamps and rain forests. Even those frogs that are adapted to drier habitats spend much of their time in burrows or under moist leaves where the humidity is high. Adult amphibians either have small, rather inefficient lungs or lack lungs altogether. Most species rely heavily on their skin to carry out gas exchange with the environment, and amphibians that live on land must cope with the problem of keeping their skin moist to allow gases to diffuse in and out. Many species also use moist surfaces of the mouth for gas exchange.

The amphibian egg has no shell, and it dehydrates quickly in dry air. Fertilization is external, with the male grasping the female and spilling his sperm over the eggs as the female sheds them (see Figure 30.15a). Oviparous amphibians generally lay their egs in ponds or swamps or at least in moist environments. Some species lay vast numbers of eggs, and mortality is high. Desert frogs often breed explosively in temporary pools. In contrast are species displaying various types of parental care; in these cases, the number of eggs laid is small. Depending on the species, either males or females may incubate eggs on their back, in the mouth, or even in the stomach. Certain tropical tree frogs stir their egg masses into moist foamy nests that resist drying. There are also live-bearing amphibians (ovoviviparous and even some viviparous species) that retain the eggs in the female reproductive tract, where embryos can develop without drying out.

Amphibians, particularly anurans, exhibit complex and diverse social behavior, especially during their breeding seasons. Frogs are usually quiet creatures, but many species fill the air with their mating calls during the breeding season. Males may vocalize in defense of breeding territory or to attract females. In some terrestrial species, migrations to specific breeding sites may involve vocal communication, celestial navigation, or chemical signaling.

CLASS REPTILIA

Reptiles, a diverse group with a wide array of extinct lineages, are represented today by about 7000 species of lizards, snakes, turtles, and crocodilians. Reptiles have several adaptations for terrestrial living not generally found in amphibians.

Figure 30.16
Terrestrial adaptations of reptiles. The shelled egg and a waterproof skin were two features that enabled reptiles to succeed on land. This species is bullsnake (*Pituophis melanoleucus*).

Reptilian Characteristics

Scales containing the protein keratin waterproof the skin of a reptile, helping to prevent dehydration in dry air. Keratinized skin is the vertebrate analogue of the chitinized cuticle of insects and the waxy cuticle of land plants. Because they cannot breathe through their dry skin, most reptiles obtain all of their oxygen with lungs. Many turtles also use the moist surfaces of their cloaca for gas exchange.

Although many viviparous reptiles exist, most species lay eggs on land; parchmentlike shells prevent desiccation (Figure 30.16). The embryo develops in the fluid of an amniotic sac within the egg. The evolution of the **amniote egg,** a shelled egg with a self-contained "pond" of amniotic fluid, enabled vertebrates to complete their life cycles on land and sever their last ties with their aquatic origins. (The seed played a similar role in the evolution of land plants; see Chapter 27.) Fertilization in reptiles must occur internally, before the shell is secreted as the egg passes through the reproductive tract of the female.

Reptiles are sometimes labeled "cold-blooded" animals because they do not use their metabolism to control body temperature. But reptiles *do* regulate body temperature by using behavioral adaptations. Many lizards can raise their internal temperature to 35°C or more by basking in the sun when the air is cool and seeking shade when the air is too warm. Since they absorb external heat rather than generating much of their own, reptiles are said to be **ectothermic,** a term more appropriate than "cold-blooded." (Control of body temperature is discussed in more detail in Chapter 40.) By heating directly with solar energy rather than through the metabolic breakdown of food, a reptile can survive on less than 10% of the calories required by a mammal of equivalent size. Having relatively modest food requirements and being adapted to arid conditions, many reptiles thrive in deserts.

During the Mesozoic era, reptiles were far more widespread, numerous, and diverse than they are today. Let us look briefly at this period, known as the Age of Reptiles.

The Age of Reptiles

Origin and Early Evolutionary Radiation of Reptiles The oldest reptilian fossils are found in rocks from the upper Carboniferous period, about 300 million years old. Their ancestors were among the Devonian amphibians. By the end of the Carboniferous period, there was a diversity of reptilian forms known as **cotylosaurs,** or "**stem reptiles.**" These mostly small, lizardlike insect eaters were the ancestral stock from which the various reptilian orders arose. In two great waves of adaptive radiation, reptiles became the dominant terrestrial vertebrates in a dynasty that lasted more than 200 million years.

The first major reptilian radiation occurred during the Permian period, giving rise to terrestrial predators called **synapsids** among other groups (Figure 30.17). One lineage of these reptiles was progenitor of the **therapsids,** which are sometimes called mammal-like

reptiles. Indeed mammals are believed to have evolved from the therapsid line. The early therapsids were large, dog-sized predators, but a variety of other forms evolved as the Permian progressed. Some of the stem reptiles returned to the water, giving rise to the plesiosaurs and ichthyosaurs, which had seallike or fishlike shapes. The **thecodonts** were another group that descended from the stem reptiles during the reptilian radiation of the Permian period. They are of special interest because they were the ancestors of dinosaurs, crocodilians, and birds. Both groups of ancient reptiles—synapsids and thecodonts—survived the mass extinctions of the Permian period (see Chapter 23), and in fact carnivorous and herbivorous therapsids continued as the dominant terrestrial vertebrates until well into the Triassic period.

Dinosaurs and Their Relatives The second great reptilian radiation was underway by late Triassic times (a little more than 200 million years ago), when several lineages arose from thecodont stock (see Figure 30.17). Descendants of the thecodonts included two groups of reptiles—the dinosaurs, which lived on land, and the pterosaurs, or flying reptiles. These groups were the dominant vertebrates on Earth for millions of years. Dinosaurs, an extremely diverse group varying in body shape, size, and habitat, included the largest animals ever to inhabit land. Fossils of gigantic dinosaurs that

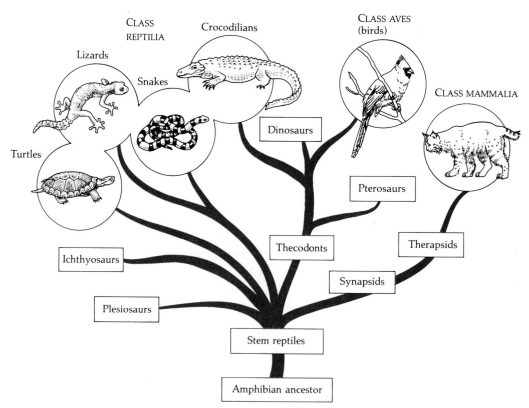

Figure 30.17
Phylogeny of reptiles and their descendants.

probably weighed nearly 100 tons have recently been unearthed in New Mexico. Pterosaurs had wings formed from a membrane of skin stretched from the body wall, along the forelimb, to the tip of an elongated finger. Stiff fibers provided support for the skin of the wing.

Contradicting the long-standing view that dinosaurs were slow, sluggish creatures, there is increasing evidence that many dinosaurs were agile, fast moving, and endothermic. (**Endothermy** is the ability to keep the body warm through metabolism.) Likewise, reevaluation of pterosaur fossils strongly indicates that many of these reptiles were active, endothermic flyers, rather than inefficient ectothermic gliders. Despite considerable anatomical evidence supporting these hypotheses, some experts remain skeptical. The Mesozoic climate was relatively warm and consistent, and behavioral adaptations such as basking may have been sufficient for maintaining a suitable body temperature, especially for the land-dwelling dinosaurs. Also, large dinosaurs had low surface-to-volume ratios that reduced the effects of daily fluctuations in air temperature on the internal temperature of the animal.

The Cretaceous Crisis   During the Cretaceous, the last period of the Mesozoic era, the climate became cooler and more variable. It was another crisis in the history of life—another period of mass extinctions. A quarter of the families of marine invertebrates disappeared, and the Cretaceous crisis was the curtain call for the dinosaurs. Nearly all of them were gone by about the close of the Mesozoic era, some 65 million years ago (fossils of some dinosaurs that survived into the early Cenozoic have recently been discovered). What caused the demise of the dinosaurs? There is evidence that the decline of the dinosaurs occurred over five to ten million years. But even this is rather sudden on the scale of geological time for a group so diverse and dominant to become extinct. Disappearance of the dinosaurs, one of the most engaging mysteries in the history of life, is discussed in Chapter 23. The Age of Reptiles ended when the dinosaurs disappeared, but a few reptilian groups survived the crisis and are still successful today.

Reptiles of Today

The three largest and most diverse extant orders of reptiles are **Squamata** (lizards and snakes), **Chelonia** (turtles), and **Crocodilia** (alligators and crocodiles).

Lizards are by far the most numerous and diverse reptiles alive today (Figure 30.18a). Most are relatively small; perhaps they were able to survive the Cretaceous "crunch" by nesting in crevices and decreasing their activity during cold periods, a practice many modern lizards use.

Snakes are apparently descendants of lizards that adopted a burrowing life style (Figure 30.18b). Today most snakes live above ground, but they have retained the limbless condition. Vestigial pelvic and limb bones in primitive snakes such as boas, however, are evidence that snakes evolved from reptiles with legs.

Snakes are carnivorous, and a number of adaptations aid them in hunting prey. They have acute chemical sensors, and though they lack eardrums, snakes are sensitive to ground vibrations, which helps them detect movements of prey. Heat-detecting organs between the eyes and nostrils of pit vipers, including rattlesnakes, are sensitive to minute temperature changes, enabling these night hunters to locate warm animals. Poisonous snakes inject their toxin through a pair of sharp, hollow teeth. The flicking tongue is not poisonous but helps transmit odors toward olfactory organs on the roof of the mouth. Loosely articulated jaws enable most snakes to swallow prey larger than the diameter of the snake itself.

Turtles evolved from stem reptiles during the Mesozoic era and have scarcely changed since (Figure 30.18c). The usually hard shell, an adaptation that protects against predators, has certainly contributed to this long success. Those turtles that returned to water during their evolution must crawl ashore to lay their eggs on land.

Crocodiles and alligators are among the largest living reptiles (some turtles are heavier) (Figure 30.18d). They spend most of their time in water, breathing air through their upturned nostrils. Crocodiles and alligators are confined to the warm regions of Africa, China, Indonesia, India, Australia, South America, and the southeastern United States, where alligators are making a strong comeback after spending years on the endangered species list. The crocodilians evolved from thecodonts and are thus the living reptiles most closely related to the dinosaurs. However, the only modern animals that seem to have descended from a group of dinosaurs are not reptiles at all, but birds (see Figure 30.17).

CLASS AVES

Birds (Class Aves) evolved from dinosaurs during the great reptilian radiation of the Mesozoic era (see Figure 30.17). Amniote eggs and scales on the legs are just two of the reptilian remnants we see in birds. But birds look quite different from lizards and other reptiles because of their distinctive flight equipment (Figure 30.19).

Characteristics of Birds

Almost every part of a typical bird's anatomy is modified in some way that enhances flight. The bones are

(a)

(b)

(c)

Figure 30.18
Extant reptiles. (a) Lizards, such as this collared lizard, are the most numerous and diverse of the extant reptiles. **(b)** Snakes may have evolved from lizards that adapted to a burrowing lifestyle. This is an annulated boa. **(c)** Turtles have changed little since their evolution from stem reptiles early in the Mesozoic era. This species is a yellow-bellied turtle. **(d)** Crocodiles and alligators are the reptiles most closely related to dinosaurs, having evolved from the same ancestral stock (thecodonts). An alligator is shown here.

(d)

honeycombed—strong but light. The skeleton of a frigate bird, for instance, has a wingspan of more than 2 m but weighs only about 4 oz. Another adaptation reducing the weight of birds is the absence of some organs. Females, for instance, have only one ovary. Also, birds are toothless, an adaptation that trims the weight of the head. Food is not chewed in the mouth

but ground in the gizzard, a digestive organ near the stomach (crocodiles also have gizzards, as did some dinosaurs). The bird's beak, made of keratin, has proven to be very malleable during avian evolution, taking on a great variety of shapes suitable for different diets.

Flying requires great expenditure of energy from an active metabolism. Birds are endothermic; they use

Figure 30.19
Flight. The special adaptations associated with flight are superimposed on reptilian traits that have persisted since birds evolved from dinosaurs. Birds can reach great speeds and cover enormous distances. The fastest bird is the swift, which can fly 170 km per hour. The bird that travels farthest in its annual migration is the arctic tern, which flies round trip between the North Pole and South Pole each year. The drinking bird shown here is a swallow.

their own metabolic heat to maintain a warm, constant body temperature, aided by insulation provided by feathers and a layer of fat (Figure 30.20). An efficient circulatory system with a four-chambered heart that segregates oxygenated blood from oxygen-poor blood supports the high metabolic rate of the bird's cells. The efficient lungs have tiny tubes leading to and from elastic air sacs that help dissipate heat and contribute to the trimming of density.

For safe flight, senses, especially vision, must be acute. Birds have excellent eyes, perhaps the best of all vertebrates. The visual areas of the brain are well developed, as are the motor areas; flight also requires fine coordination.

With brains proportionately larger than those of reptiles and amphibians, birds generally display very complex behavior. Avian behavior is particularly intricate during breeding season when birds engage in elaborate rituals of courtship. Since eggs are shelled when laid, fertilization must be internal. The act of copulation is somewhat awkward because the male of most bird species has no penis. He must climb atop the female's back and then twist her tail so the mates' vents, the opening to their cloacas, can come together. After eggs are laid, the avian embryo must be kept warm through brooding by the mother, father, or both, depending on the species.

A bird's most obvious adaptation for flight is its wings. Bird wings are airfoils that illustrate the same principles of aerodynamics as the wings of an airplane. Providing power for flight, birds flap their wings by contractions of large pectoral (breast) muscles anchored to a keel on the sternum (breastbone). Some birds, such as hawks, have wings adapted for soaring on air currents and flap their wings only occasionally, whereas others, including hummingbirds, must flap

Figure 30.20
Emperor penguins in Antarctica. Birds are endotherms, using insulation provided by feathers and fat to slow the dissipation of body heat.

Figure 30.21
Structure of feathers. The feather consists of a central hollow shaft, the rachis, from which radiate the vanes. The vanes are made up of barbs, in turn bearing even smaller branches called barbules. Birds have contour feathers and downy feathers. Contour feathers are the stiff ones that shape the wings and body. Their barbules have hooks that cling to barbules on neighboring barbs. When a bird preens, it runs the length of a feather through the beak, engaging the hooks and uniting the barbs into a precisely shaped vane. Downy feathers lack hooks, and the free-form arrangement of barbs produces a fluffiness that provides excellent insulation because of the trapped air.

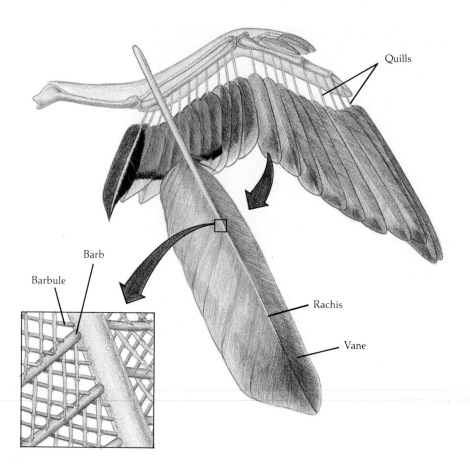

Quills

Barb

Barbule

Rachis

Vane

continuously to stay aloft. In either case, it is the shape and arrangement of the feathers that form the wing into an airfoil.

In being both extremely light and strong, feathers are among the most remarkable of vertebrate adaptations (Figure 30.21). Feathers are made of keratin, the same protein that forms our hair and fingernails and the scales of reptiles. In fact, feathers evolved from reptilian scales. Feathers may have functioned first as insulation during the evolution of endothermy, only later being coopted as flight equipment. Besides providing support and wing shape, feathers can be manipulated to control air movements around the wing.

The Origin of Birds

For many zoologists, the presence of feathers is enough to classify an animal as a bird. And certainly, if we want to trace the ancestry of birds, we must search for the oldest fossils with feathers. Fossils of an ancient bird named *Archaeopteryx lithographica* have been found in Bavarian limestone in Germany, dating back some 150 million years into the Jurassic period. Unlike modern birds, but similar to reptiles, *Archaeopteryx* had clawed forelimbs, teeth, and a long tail containing vertebrae. Indeed, if it were not for the preservation of

its feathers, *Archaeopteryx* would be regarded as a member of a diverse group of dinosaurs called the **theropods.** In common with birds, many dinosaurs, including some theropods, were probably endothermic, built nests, and cared for their young. In fact, some zoologists argue that modern birds are living dinosaurs and should be classified as reptiles.

Archaeopteryx is not considered the ancestor of modern birds, and paleontologists place it on a sidebranch of the avian lineage. Nonetheless, *Archaeopteryx* probably was derived from ancestral forms that also gave rise to modern birds. Its skeletal anatomy indicates that it was a weak flyer, prehaps mainly a tree-dwelling glider, and this may have been the earliest mode of flying in the bird lineage. A fossil recently discovered in Spain seems to represent a later stage of avian evolution. Dating from about 125 million years ago, this form had a tail similar to that of modern birds and forelimb bones that would have allowed stronger, more active flight than that of *Archaeopteryx*.

The evolution of flight required radical alteration in body form, but flight provides many benefits. It allows aerial reconnaissance that enhances hunting and scavenging; it allows birds to exploit flying insects as an abundant, highly nutritious food resource; it also provides ready escape from land-bound predators and allows some birds to migrate great distances to utilize different food resources and seasonal breeding areas.

Figure 30.22
Ratite birds. Flightless birds, or ratites, such as this emu of Australia, are completely land-bound. Having reduced wings or lacking wings altogether, ratites walk or run on strong hind limbs. Flightless birds of today probably evolved from a group of flying birds.

Modern Birds

There are about 8600 extant species of birds classified in about 28 orders. Flight ability is typical of birds, but there are several flightless species, including the ostrich, kiwi, and emu (Figure 30.22). Flightless birds are collectively called ratites (from the Latin word for "flat-bottomed") because their breastbone lacks a keel. Ratite birds also lack the large breast muscles that attach to the sternal keel and provide flight power in flying birds.

In contrast to the ratites, other birds are said to be carinate, because they have a carina, or sternal keel supporting their large breast muscles. The demands of flight have rendered the body form of carinate birds similar to each other, yet experienced bird watchers can distinguish many species by their body profile. Carinate birds also exhibit great variety in feather colors, beak and foot shape, behavior, and flight ability (Figure 30.23). Nearly 60% of the living bird species belong in one carinate order called the passeriforms, or the perching birds. These are the familiar jays, swallows, sparrows, warblers, and many others (Figure 30.23d). Among the most unusual carinate birds are the penguins, which do not fly, but use their powerful breast muscles in swimming (see Figure 30.20).

CLASS MAMMALIA

As mammals ourselves, we naturally have a special interest in this class of vertebrates. Let us examine some of the features we share with all other mammals.

Mammalian Characteristics

Mammals have hair, a characteristic as diagnostic as the feathers of birds. Hair, like feathers, is made of keratin, though it did not evolve from the scales of reptilian ancestors. Also like feathers, hair insulates to help the animal maintain a warm and constant body temperature; mammals are endothermic. Their active metabolism is supported by an efficient respiratory system that uses a sheet of muscle called the **diaphragm** to help ventilate the lungs. The four-chambered heart of a mammal prevents the mixing of oxygen-rich blood with oxygen-poor blood.

Mammary glands that produce milk are as distinctively mammalian as hair (Figure 30.24). All mammalian mothers nourish their babies with milk, a balanced diet rich in fats, sugars, proteins, minerals, and vitamins.

Most mammals are born rather than hatched. Fertilization is internal, and the egg develops into an embryo within the uterus of the female reproductive tract. In placental mammals, the lining of the mother's uterus and membranes arising from the embryo collectively form a **placenta,** where nutrients diffuse into the embryo's blood.

Mammals have larger brains than other vertebrates of equivalent size, and seem to be the most capable learners. The relatively long duration of parental care extends the time for parents to teach important skills to their impressionable progeny.

Differentiation of teeth is another important mammalian trait. Whereas the teeth of reptiles are generally conical and uniform in size, the teeth of mammals come in a variety of sizes and shapes adapted for chewing many kinds of foods. Our own dentition, for example, comprises a mixture of knifelike teeth (incisors) modified for shearing and grinding teeth (molars) specialized for crushing.

Evolution of Mammals

Mammals evolved from reptilian stock even earlier than birds. The oldest fossils believed to be mammalian date back 190 million years (Triassic period). The ancestors of mammals were among the mammal-like reptiles known as therapsids (see Figure 30.17). The therapsids disappeared during the reign of dinosaurs, but their mammalian offshoots coexisted with the

(a)

(b)

(c)

(d)

Figure 30.23
Carinate birds. (a) In common with many species, the harlequin duck, which inhabits mountain streams and coastal areas of the Pacific Northwest, exhibits pronounced color differences between the sexes.
(b) Most birds exhibit specific courtship and mating behavior. The western grebe virtually runs on the water surface while courting. **(c)** Hummingbirds can hover while they feed on nectar because they have short, rigid wings that can turn like propellers in virtually any direction.

(d) The western tanager is a member of the order Passeriformes. Passeriforms are called perching birds because the toes of their feet can lock around a tree branch, allowing the bird to rest in place for long periods of time.

dinosaurs throughout the Mesozoic era. Mesozoic mammals were very small—about the size of shrews—and most probably ate insects. A variety of evidence, such as the size of the eye sockets, suggests that these tiny mammals were nocturnal.

The extinction of the dinosaurs at the close of the Mesozoic era reopened many adaptive zones for new animal forms, and mammals underwent an extensive adaptive radiation that filled the void. The flora was also changing. Flowering plants became abundant in the Cretaceous and replaced gymnosperms as the dominant land plants in most locations. (Some scientists speculate that this change in vegetation contributed to the decline of the dinosaurs, which may have been fixed on gymnosperms for food.) The transformations of the Cretaceous marked the end of the Mesozoic era. As the Cenozoic era dawned, mammals continued to diversify. That diversity is represented today in the three major groups: **monotremes** (egg-laying mammals), **marsupials** (mammals with pouches), and **placental mammals** (Table 30.2, pp. 662–663).

Monotremes

The monotremes, the platypuses and the echidnas (spiny anteaters), are the only living mammals that lay eggs. The egg, which is reptilian in structure and development, contains enough yolk to nourish the developing embryo. Most mammals maintain a temperature of 36°–39°C, but the internal temperature of a platypus is only about 30°C. Monotremes have hair and milk, two of the most important trademarks of Mammalia. On the belly of a monotreme mother are specialized glands that secrete milk. After hatching, the baby sucks the milk from the fur of the mother,

Figure 30.24
Mammalian trademarks. The presence of hair and mammary glands that nourish the young with milk are distinctively mammalian traits.

who has no nipples. The mixture of ancestral reptilian traits and derived traits of mammals suggests that monotremes descended from a very early branch in the mammalian genealogy. Today monotremes are found only in Australia and New Guinea.

Marsupials

Opossums, kangaroos, and koalas are examples of marsupials, mammals that complete their embryonic development in a maternal pouch called a **marsupium.** The egg contains a moderate amount of yolk that feeds the embryo as it begins its development within the mother's reproductive tract. Marsupials are born very early in their development. A red kangaroo, for instance, is about the size of a honeybee at its birth, just 33 days after fertilization. Its hind legs are merely buds, but the forelimbs are strong enough to crawl from the exit of the reproductive tract to the mother's pouch, a journey lasting a few minutes. In the pouch, the neonate fixes its mouth to a teat and completes its development while nursing.

In Australia marsupials have radiated and filled niches occupied by placental mammals in other parts of the world. As we might expect, convergent evolution has resulted in a diversity of marsupials that resemble their placental counterparts who have similar ecological roles (see Chapter 23). The opossums of North and South America are the only extant marsupials outside the Australian region, though South America had a diverse marsupial fauna throughout the Tertiary period. The distribution of modern and fossil marsupials begins to make sense in the context

of plate tectonics and continental drift. According to recent fossil evidence, marsupials probably originated in what is now North America, spreading southward when the land masses were still joined. After the breakup of Pangaea, South America and Australia became island continents, and their marsupials diversified in isolation from the placental mammals that began an adaptive radiation on the northern continents. Australia has not been in contact with another continent since early in the Cenozoic era, about 65 million years ago. The South American fauna, meanwhile, has not remained cloistered. Placental mammals reached South America throughout the Cenozoic era. The most important migrations occurred about 12 million years ago and then again about 3 million years ago, when North and South America joined at the Panamanian isthmus, and extensive two-way traffic of animals took place over the land bridge. The distribution of mammals is another example of the interplay of biological and geological evolution.

Placental Mammals

Placental mammals complete their embryonic development within the uterus, joined to the mother by the placenta (Figure 30.25, p. 664). Adaptive radiation during late Cretaceous and early Tertiary periods (about 70–45 million years ago) produced the orders of placental mammals we recognize today (see Table 30.2). The phylogenetic relationship between marsupials and placental mammals is somewhat obscure, but scientists believe they are more closely related to each other than either is to the monotremes. Fossil evidence indi-

Table 30.2 Major orders of mammals

	Order	Major Characteristics	Examples	
Monotremes	Monotremata	Lay eggs; have no nipples; suck milk from fur of mother	Platypuses, echidnas	
Marsupial mammals	Marsupialia	Embryonic development completed in marsupial pouch	Kangaroos, opossums, koalas	
Placental mammals	Artiodactyla	Possess hooves with an even number of toes on each foot; herbivorous	Sheep, pigs, cattle, deer, giraffes	
	Carnivora	Carnivorous; possess sharp, pointed canine teeth and molars for shearing	Dogs, wolves, bears, cats, weasels, otters, seals, walruses	
	Cetacea	Marine forms with fish-shaped bodies, paddlelike forelimbs and no hindlimbs; thick layer of insulating blubber	Whales, dolphins, porpoises	
	Chiroptera	Adapted for flying; possess a broad skinfold that extends from elongated fingers to body and legs	Bats	
	Edentata	Have reduced or no teeth	Sloths, anteaters, armadillos	

Table 30.2 *(Continued)*

Order	Major Characteristics	Examples	
Insectivora	Insect-eating mammals	Moles, shrews, hedgehogs	
Lagomorpha	Possess chisel-like incisors, hind legs longer than front legs and adapted for jumping	Rabbits, hares, pikas	
Perissodactyla	Possess hooves with an odd number of toes on each foot; herbivorous	Horses, zebras, tapirs, rhinoceroses	
Primates	Opposable thumb; forward-facing eyes; well-developed cerebral cortex; omnivorous	Lemurs, monkeys, apes, humans	
Proboscidea	Have a long, muscular trunk; thick, loose skin; upper incisors elongated as tusks	Elephants	
Rodentia	Possess chisel-like continuously growing incisor teeth	Squirrels, beavers, rats, porcupines, mice	
Sirenia	Aquatic herbivores; possess finlike forelimbs and no hindlimbs	Sea cows (manatees)	

Figure 30.25
Birth of a placental mammal. Placental mammals develop within the uterus of the mother, nurtured by the flow of maternal blood through the dense capillary network of the placenta. The placenta is the reddish portion of the afterbirth clinging to this newborn zebra. The large silvery membrane is the amniotic sac, which contained the developing embryo in a bath of protective amniotic fluid.

cates placentals and marsupials may have diverged from a common ancestor about 80–100 million years ago.

The evolutionary relationships among the many orders of placental mammals are, for the most part, also unsettled. Most mammalogists now favor a tentative genealogy that recognizes at least four main evolutionary lines of placental mammals.

One branch of placental mammals consists of those in the orders Chiroptera (bats) and Insectivora (shrews), which resemble early mammals. Bats, whose forelimbs are modified as wings, probably evolved from insectivores that fed on flying insects. In addition to the insect eaters, some bat species feed on fruit, whereas others bite mammals and lap up their blood. Most bats are nocturnal.

A lineage of medium-sized herbivores underwent a spectacular adaptive radiation during the Tertiary period, leading to such modern orders as lagomorphs (rabbits and their relatives); perissodactyls (odd-toed ungulates, including horses and rhinoceroses; ungulates are mammals that walk on the tips of toes); artiodactyls (even-toed ungulates, including deer and swine); sirenians (sea cows); proboscideans (elephants); and cetaceans (porpoises and whales). Porpoises and some whales feed on fish and squid, but the largest whales are filter-feeders that trap huge quantities of planktonic crustaceans in grilles called baleens suspended from the roof of the whale's mouth. Some blue whales are more than 30 m long and weigh as much as 25 elephants, making them the largest animals ever to live. Cetaceans seem to be very intelligent and social, using sound to communicate. Whales also produce high-pitched sounds that function in sonar navigation.

A third thrust in the evolution of placental mammals produced the order Carnivora, which includes the cats, dogs, raccoons, skunks, and the pinnipeds (seals, sea lions, and walruses). The Carnivora probably first appeared during the early Cenozoic era. Seals and their relatives apparently evolved from middle Cenozoic carnivores that became adapted for swimming.

The most extensive adaptive radiation of placental mammals produced the primate-rodent complex. The order Rodentia includes rats, squirrels, and beavers. The order Primates comprises monkeys, apes, and humans.

THE HUMAN ANCESTRY

We have now tracked the vertebrate genealogy to the mammalian order that includes *Homo sapiens* and its closest kin. We are primates, and to learn what that means we must trace our ancestry back to the trees, where some of our most treasured traits had their beginnings as adaptations to arboreal existence.

Evolutionary Trends in Primates

The first primates were probably small arboreal mammals, and dental structure suggests that primates descended from insectivores late in the Cretaceous period. A fossil species called *Purgatorius unio*, found in Montana in beds of the Cretaceous/Tertiary boundary, is considered by many authorities to be the oldest known primate. Thus, by the end of the Mesozoic era, some 65 million years ago, our order was already defined by characteristics that had been shaped, through natural selection, by the demands of living in the trees.

For example, primates have limber shoulder joints, which make it possible to **brachiate** (swing from one hold to another). The dexterous hands of primates can hang onto branches and manipulate food. Claws have been replaced by nails in many species, and the fingers are very sensitive. The eyes of primates are close together on the front of the face. The overlapping fields of vision of the two eyes enhance depth perception, an obvious attribute when brachiating. Excellent eye-hand coordination is also important for arboreal maneuvering.

Figure 30.26
The slender loris, a prosimian.

Parental care would seem to be essential for young animals in the trees. Mammals devote more energy to caring for their young than most other vertebrates, and primates are among the most devoted parents of all mammals. Most primates have single births and nurture their offspring for a long time. Though humans do not live in trees, we retain in modified form many of the traits that evolved there.

Modern Primates

Two commonly recognized subgroups of primates are **prosimians** ("premonkeys") and **anthropoids** (monkeys, apes, and humans). Prosimians probably resemble early arboreal primates. The lemurs of Madagascar and the lorises, pottos, and tarsiers that live in tropical Africa and southern Asia are examples of prosimians (Figure 30.26).

The first anthropoids to appear in the fossil record are monkeylike primates that probably evolved from prosimian stock about 40 million years ago in Africa and possibly even earlier in Asia. Africa and South America had already drifted apart, and it is not clear if ancestors of New World monkeys reached South America by rafting on logs or other debris from Africa or by migration southward from North America. What is certain is that New World monkeys and Old World monkeys have been evolving along separate pathways for many millions of years (Figure 30.27). All New World monkeys are arboreal, whereas Old World monkeys include ground-dwelling as well as arboreal species. Most monkeys of both groups are diurnal (active during the day) and usually live in bands held together by social behavior.

There are four genera of apes, shown in Figure 30.28: *Hylobates* (gibbons), *Pongo* (orangutans), *Gorilla* (gorillas), and *Pan* (chimpanzees). Modern apes are confined exclusively to tropical regions of the Old World. With the exception of gibbons, modern apes are larger than monkeys, with relatively long legs and short arms

Figure 30.27
Comparison of New World monkeys and Old World monkeys. (a) New World monkeys, such as spider monkeys, squirrel monkeys, and capuchins, have prehensile tails and nostrils that open to the sides. This is a Mexican spider monkey. **(b)** Old World monkeys lack prehensile tails, and their nostrils open downward. The tough seat pad is unique to the Old World group, which includes macaques, mandrills, baboons, and rhesus monkeys. This is a family of baboons.

(a)

(b)

Figure 30.28

Apes. (a) Gibbons have long arms and are among the most acrobatic of all primates. These Asian primates are also the only monogamous apes. **(b)** The orangutan is a shy and solitary ape that lives in the rain forests of Sumatra and Borneo. Orangutans spend most of their time in trees, but they do venture onto the forest floor occasionally. **(c)** The gorilla is the largest ape, with some males almost 2 m tall and weighing about 200 kg. These gentle herbivores are confined to Africa, where they usually live in small groups of about 10 to 20 individuals. **(d)** The chimpanzee lives in tropical Africa. Chimpanzees feed and sleep in trees, but also spend a great deal of time on the ground. Chimpanzees are intelligent, communicative, and social. Some biochemical evidence indicates that chimpanzees are more closely related to humans than they are to other apes.

(a)

(b)

(c)

(d)

and no tails. Although all the apes are capable of brachiation, only gibbons and orangutans preserve a primarily arboreal existence. Social organization varies among the genera of apes; gorillas and chimpanzees are highly social. Apes have larger brains than monkeys and their behavior is consequently more adaptable.

The Emergence of Humankind

Paleoanthropology, the study of human origins and evolution, has a checkered history. Until about 20 years ago, researchers, perhaps paternalistic about their

discoveries, often gave new names to fossil forms that were undoubtedly the same species as fossils found by competing scientists. Elaborate theories have often been based on a few teeth or a fragment of jawbone. During the early part of this century, baseless speculations spawned many misconceptions about human evolution that still persist in the minds of much of the general population, long after these myths have been overthrown by fossil discoveries.

Some Popular Misconceptions Let us first dispose of the myth that our ancestors resembled chimpanzees or any other modern apes. Chimpanzees and

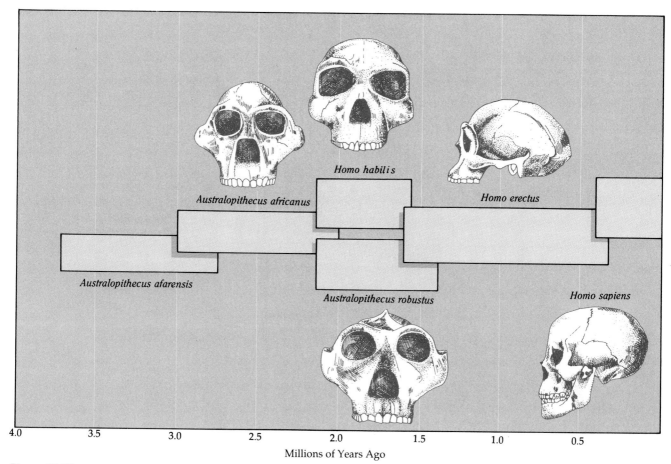

Figure 30.29
Timeline of some hominid species. Several times in the history of human evolution, two or more hominids coexisted.

humans represent two divergent branches of the anthropoid tree that evolved from a common, less-specialized ancestor.

Another misconception envisions human evolution as a ladder with a series of steps leading directly from an ancestral anthropoid to *Homo sapiens.* This is often illustrated as a parade of fossil hominids (members of the human family) becoming progressively more modern as they march across the page. If human evolution is a parade, then it is a disorderly one, with many splinter groups having traveled down dead ends. At times in hominid history, several different human species coexisted (Figure 30.29). Human phylogeny is more like a multibranched bush than a ladder, our species being the tip of the only twig that still lives. If the punctuated mode of evolution holds for humans, most change occurred as new hominid species came into existence, not by phyletic change within an unbranched hominid lineage.

One more myth we must bury is the notion that various human characteristics, such as upright posture and an enlarged brain, evolved in unison. A popular image is of early humans as half-stooped, half-witted cave dwellers. Different features evolved at different rates—a phenomenon known as **mosaic evolution**—with erect posture, or bipedalism, leading the way. Our pedigree includes ancestors who walked upright but had brains as small as an ape's.

After dismissing some of the folklore on human evolution, however, we must admit that present understanding of our ancestry remains somewhat muddled.

Early Anthropoids   The oldest known fossils of apes have been assigned to the genus *Aegyptopithecus,* the "dawn ape." This anthropoid was a cat-sized tree-dweller that lived about 35 million years ago. During the Miocene epoch, which began about 25 million years ago, descendants of the first apes diversified and spread into Eurasia. About 20 million years ago, the Indian plate collided with Asia and thrust up the Himalayan range. The climate became drier and the forests of what is now Africa and Asia contracted. Some of the Miocene apes may have made an evolutionary descent from the trees and begun living on the edge of the forest and foraging for food on the adjacent savanna

Figure 30.30
The antiquity of upright posture.

(a) "Lucy" (*Australopithecus afarensis*) and her kind lived about 2.9 to 3.4 million years ago. They walked upright but retained many apelike traits, including curved toes. (b) The Laetoli footprints, over 3.5 million years old, confirm that upright posture evolved quite early in hominid history.

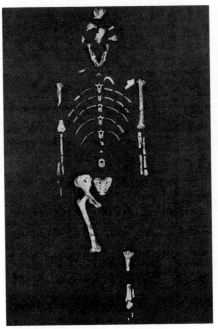

(a)   *(b)*

(grassy plains dotted with trees). One of these anthropoids may have been the ancestor of humans. Attention has focused on *Ramapithecus*, an anthropoid known from Indian and African fossils 8 to 14 million years old. However, most anthropologists now believe that humans and apes diverged in their evolutionary paths less than 8 million years ago and probably only 4 to 5 million years ago, implying that *Ramapithecus* was part of the anthropoid heritage common to humans and modern apes.

Australopithecines: The First Humans In 1924, British anthropologist Raymond Dart announced that a fossilized skull discovered in a South African quarry was the remains of an early human. He named his "ape-man" *Australopithecus africanus* ("Southern Ape"). With the discovery of more fossils, it became clear that *Australopithecus* was a legitimate hominid that walked fully erect and had humanlike hands and teeth. However, the brain of *Australopithecus* was only about one-third the size of a modern human's. *Australopithecus* foraged on the African savannas for nearly two million years, beginning about three million years ago. Two forms existed, one slender and the other more robust; some paleoanthropologists assign these two forms to different species.

In 1974, in the Afar region of Ethiopia, an *Australopithecus* skeleton that was 40% complete was discovered (Figure 30.30). "Lucy," as the fossil was named, was petite—only about a meter tall, with a head about the size of a softball. Lucy and similar fossils have been considered sufficiently different from *Australopithecus africanus* to be named a separate species, *Australopithecus afarensis* (for the Afar region). The place-

ment of *A. afarensis* on the hominid bush has stirred considerable debate. Lucy was first estimated to be about 3.5 million years old, making her species the most ancient of the australopithecine group. According to one tentative genealogy, *A. afarensis* was the common ancestor of two hominid branches, one leading to *A. africanus* and the other to *Homo*. Estimates of Lucy's age have recently been revised to as young as 2.9 million years, and some anthropologists believe that *A. afarensis* was part of an australopithecine branch that had already diverged from the lineage leading to *Homo*, our genus. In any case, the discovery of Lucy confirms that upright posture evolved relatively early in hominid history. This important conclusion is supported by fossilized footprints more than 3.5 million years old that were discovered in Laetoli, Tanzania.

Australopithecus walked erect for more than a million years with no substantial enlargement of the brain. Perhaps an upright posture freed the hands for gathering food or caring for babies. The making of tools came much later. As evolutionary biologist Stephen Jay Gould puts it: "Mankind stood up first and got smart later."

Enlargement of the human brain is first evident in fossils dating back to the latter part of australopithecine times, about two million years ago. Skulls have been found with brain capacities of about 650 cubic centimeters (cc), compared with 500 cc for *Australopithecus africanus*. Simple stone tools are sometimes found with the larger-brained fossils. Some paleoanthropologists believe the advances are great enough to place these fossils in the genus *Homo*, naming it *Homo habilis* ("handy man"). Other scientists argue that the handy man was just a variant of *Australo-*

pithecus and that australopithecines were the first hominids to use and make tools. Regardless of the taxonomic debate, the message of the fossils is clear: After walking upright for more than a million years, hominids were finally beginning to use their brains and hands to fashion tools.

Homo habilis coexisted for nearly a million years with the smaller-brained *Australopithecus*. The two may not have competed directly, although both may have scavenged for food, hunted animals, and gathered fruits and vegetables. According to one theory of human origins, *Australopithecus africanus* and *Homo habilis* were two distinct lines of hominids, neither evolving from the other. If this scenario is correct, then *A. africanus* was an evolutionary dead-end, but *H. habilis* was on the path to modern humans, leading first to *Homo erectus*, which later gave rise to *Homo sapiens*.

Homo Erectus and Homo Sapiens   The first hominid to migrate out of Africa into Europe and Asia was *Homo erectus* ("upright man"). The fossils known as Java Man and Peking Man are examples of this species. *Homo erectus* was taller than *H. habilis* and had a larger brain capacity. Fossils of *H. erectus* range in age from 1.5 million years to about 300,000 years. During that period, the brain continued to enlarge, eventually reaching capacities of more than 1000 cc.

To survive in the colder climates of the north, humans had to live by their wits. *Homo erectus* resided in huts or caves, built fires, clothed themselves in animal skins, and designed stone tools that were more elaborate than the tools of *Homo habilis*. In anatomical and physiological adaptations, *H. erectus* was poorly equipped for life outside the tropics but made up for the deficiencies with intelligence and social cooperation. Deterioration of the climate during the ice ages (Pleistocene epoch) provided a testing ground for the human brain, and *Homo sapiens* emerged.

The oldest fossils classified as *Homo sapiens* are the Neanderthals (first found in the Neander Valley of Germany), who lived from about 130,000 years ago to about 30,000 years ago. Compared with us, Neanderthals had slightly heavier brow ridges and less pronounced chins, but their brains were slightly larger than ours. They were skilled toolmakers, and they participated in burials and other rituals that required abstract thought.

The oldest fully modern fossils of *Homo sapiens*, discovered in a cave in Israel, date back some 90,000 years. Fragmentary fossils of modern humans of similar age have been found in South Africa. All human fossils less than 30,000 years old are completely modern, but Neanderthals may have overlapped modern humans in Eurasia by some 30,000 years. Debate continues concerning the fate of the Neanderthals and their genes, but a growing body of evidence derived from molecular genetics indicates that these archaic humans became extinct without contributing to the modern human gene pool. Recent studies of DNA found in the mitochondria of cells obtained from diverse human groups indicate that today's human population is very uniform for this DNA (mitochondrial DNA is discussed in Chapter 14). This genetic uniformity leads some researchers to conclude that the entire gene pool of the current human population was derived from a very small number of individuals. Furthermore, because mitochondrial DNA is passed from one generation to the next only in eggs, not in sperm, mitochondrial DNA in modern humans has been inherited entirely from females. As a result, some authorities speculate that the modern human gene pool is traceable to a single female that lived between 200,000 and 400,000 years ago, probably in Africa. Perhaps derived from this female after a bottlenecking event or a founding event (see Chapter 21), a small population of modern *Homo sapiens* may have spread from Africa and eventually populated the Old World. As this occurred, archaic humans, including the Neanderthal lineage, may have died out.

The hypothesis that modern humans had a single African origin has stimulated a lively debate. Stressing the need for more genetic data, skeptics of the idea maintain that modern *Homo sapiens* had a multiregional origin. Some authorities also contest the idea that the Neanderthals made no contribution to the modern human gene pool and postulate that the distinctive traits of Neanderthals faded away by interbreeding with modern humans. This debate and the melding of molecular and fossil studies will fuel a significant expansion of our knowledge of human origins in the next few years.

Cultural Evolution: A New Force in the History of Life   An erect stance was the most radical anatomical change in our evolution; it required major remodeling of the foot, pelvis, and vertebral column. Enlargement of the brain was a secondary alteration made possible by prolonging the growth period of the skull and its contents. The brains of mammalian fetuses grow rapidly, but the growth usually slows down and stops not long after birth. The primate brain continues to grow after birth, and the period of growth is longer for a human than for any other primate. The extended period of human development also lengthens the time parents care for their offspring, which contributes to the child's learning to repeat the experiences of earlier generations. This is the basis of culture—the transmission of accumulated knowledge over the generations. The major means of this transmission is language, written and spoken.

The first stage in our cultural evolution began with nomads who hunted and gathered food on the African grasslands. They made tools, organized communal activities, and divided labor. The second stage came with the development of agriculture in Eurasia and the Americas about 10,000–15,000 years ago. Along with agriculture came permanent settlements and the first cities. The third major stage in our cultural evolution was the Industrial Revolution, which began in the eighteenth century. Since then, new technology and human population have escalated exponentially; the lives of some people spanned the flight of the Wright brothers and Neil Armstrong's moon walk. Through all of this cultural evolution, from hunting/gathering to high-tech societies and an overpopulated world, we have not changed biologically in any significant way. We are probably no more intelligent than our forebears in Africa and in Eurasian caves. The same toolmaker who chipped away at stones now fashions microchips. The know-how to build skyscrapers, computers, and spaceships is stored not in our genes but in the cumulative product of hundreds of generations of human experience, passed along by parents, teachers, and books.

Evolution of the human brain may have been anatomically simpler than acquiring an upright stance, but the consequences of cerebral growth have been enormous. Cultural evolution made *Homo sapiens* a new force in the history of life—a species that could defy its physical limitations and shortcut biological evolution. We do not have to wait to adapt to an environment through natural selection; we simply change the environment to meet our needs. We are the most numerous and widespread of all large animals, and everywhere we go, we bring environmental change. There is nothing new about environmental change. As we have seen in this unit, the history of life has been the history of biological evolution on a changing planet. But it is unlikely that change has ever been so rapid as in the age of humans. Cultural evolution outpaces biological evolution by orders of magnitude. We may be changing the world faster than many species can adapt; the rate of extinctions in this century is 50 times greater than the average for the past 100,000 years.

This overwhelming rate of species extinction is mainly a result of habitat destruction and chemical pollution, both functions of human cultural changes and overpopulation. Simply feeding, clothing, and housing the 5 billion plus people now in existence imposes an enormous strain on the Earth's capacity to sustain life. If all of these people suddenly assumed the high standard of living enjoyed by many people in developed nations, it is likely that the earth's support systems would be overwhelmed. Already, for example, current rates of fossil fuel consumption, mainly by developed nations, are so great that waste carbon dioxide may be causing the temperature of the atmosphere to increase enough to alter world climates (see Chapter 49). Today, not just individual species, but entire ecosystems, the global atmosphere, and the oceans, are seriously threatened. Tropical rain forests, which play a vital role in the maintenance of atmospheric gas balance and in moderating global weather, are being cut down at a startling rate. Scientists have hardly begun to study these ecosystems, and many species in them may become extinct before they are even discovered. Of the many crises in the history of life, the impact of one species, *Homo sapiens*, is the latest and potentially the most devastating.

STUDY OUTLINE

1. Vertebrates, the backboned animals, are a subphylum of chordates that comprises the mammals, birds, reptiles, amphibians, and various classes of fishes.

Phylum Chordata (pp. 641–643)

1. Chordates are a diverse group of deuterostomes, classified together by virtue of shared structures: a notochord, a dorsal hollow nerve cord, pharyngeal slits, and a post-anal tail. These structures become variously modified in adult chordates of different species.
2. Subphylum Cephalochordata is an invertebrate group best known for the lancelet, an exemplary chordate. Subphylum Urochordata, also invertebrate, includes the marine, filter-feeding tunicates, or sea squirts.

The Origin of Vertebrates (p. 644)

1. The first vertebrates may have arisen during the early Cambrian period from a filter-feeding cephalochordate-like ancestor.

2. It is considered more likely that vertebrates arose from an ancestor resembling a tunicate larva that became sexually mature before metamorphosis, a phenomenon called paedogenesis.

Vertebrate Characteristics (pp. 644–645)

1. One hallmark of the vertebrates is cephalization, the development of highly specialized brains capable of processing extensive sensory input.
2. In addition, most vertebrates possess a column of vertebrae that encloses the nerve cord. The skull and vertebral column are part of a bony or cartilaginous endoskeleton.
3. The vertebrate heart pumps blood containing hemoglobin through a closed circulatory system to gills, skin, or lungs, where it is oxygenated. Wastes in the blood are removed by kidneys.
4. Reproduction is almost always sexual, with separate sexes. Fertilization can be external or internal.

5. Of the seven extant classes of vertebrates, three are fishes; members of the other four classes generally have two pairs of limbs and are called tetrapods.

Class Agnatha (p. 645)

1. The oldest vertebrate fossils are jawless agnathans, a group with round or slitlike mouths, which include the armored ostracoderms. Today, Class Agnatha is represented only by lampreys and hagfishes.

Class Placodermi (pp. 646–647)

1. Placoderms, now extinct, were Devonian armored fishes that had paired fins and hinged, biting jaws that evolved from skeletal supports of gills.

Class Chondrichthyes (pp. 647–648)

1. Sharks and rays, the most widespread members of Class Chondrichthyes, have paired fins, cartilaginous skeletons, and biting jaws characteristic of the group.
2. Primarily carnivorous, the streamlined sharks have keen senses for vision, olfaction, hearing, and electroreception. A specialized lateral line system of microscopic organs sensitive to changes in water pressure enables sharks to perceive minor vibrations.
3. The teeth of sharks are evolutionarily derived from jagged skin scales.
4. Sharks possess a spiral valve in the intestines, which increases surface area and prolongs the passage of food down the unusually short intestine.
5. Sharks have internal fertilization, and development can be oviparous, ovoviviparous, or viviparous. The eggs or young exit the female's body through the cloaca, an opening common to the excretory, digestive, and reproductive tracts.
6. Rays are flattened bottom-dwellers with whiplike tails.

Class Osteichthyes (pp. 648–650)

1. Osteichthyes is the most species-rich vertebrate class. The bony fishes have skeletons of calcium phosphate and a slimy, low-friction skin generally covered by flattened bony scales.
2. Unlike sharks, which must move to breathe, bony fishes can breathe while stationary by moving a flaplike operculum that helps draw water through the mouth and into the gill chamber.
3. Bony fishes can adjust their density and thus control their buoyancy by means of a unique swim bladder. A fusiform body shape and flexible fins are further adaptations for aquatic life.
4. Although external fertilization and ovipary are common, bony fishes exhibit great reproductive variation.
5. The ancestor of bony fishes was likely a freshwater species, which by the end of the Devonian had spawned three distinct subclasses: ray-finned fishes, lobe-finned fishes, and lungfishes.
6. Most fish familiar to us are ray-fins. Supported by long, flexible rays, their fins are modified for various functions. During their history, various species of ray-fins have invaded the sea and returned again to fresh water.
7. Most lobe-finned fishes remained in fresh water and used lungs to aid their gills in breathing. Devonian lobe-finned fishes gave rise to amphibians.

Class Amphibia (pp. 650–653)

1. Ancient lobe-finned fishes with lungs were equipped to survive the cycles of drought and rain during Devonian times. The skeletal structure of their fins allowed mobility on land by some of these fishes.
2. As the first terrestrial vertebrates, amphibians radiated during early Carboniferous times. They later declined, and the survivors at the dawn of the Mesozoic era largely resembled modern forms.
3. The life styles and life cycles of modern amphibians attest to their aquatic heritage. A moist skin complements the lungs in gas exchange. Oviparous forms lay their unshelled eggs in wet environments, though ovoviviparous and viviparous species also exist. Most frogs and their relatives undergo metamorphosis of an aquatic larval stage into a terrestrial adult.
4. The salamanders make up the tailed urodele order, an aquatic and terrestrial group that generally develops directly without metamorphosis; paedogenesis occurs in some.
5. The tail-less, hopping anuran frogs catch prey with sticky tongues. Skins specialized for camouflage or production of distasteful mucus or poison help them avoid predators. Fertilization is external and metamorphosis is common.
6. The apodans, or caecilians, are limbless, virtually blind aquatic or small burrowing species.

Class Reptilia (pp. 653–655)

1. Reptiles, a diverse group represented today by lizards, snakes, turtles, and crocodilians, have numerous terrestrial adaptations not present in amphibians.
2. Reptiles have lungs, are covered with waterproof scales, and have a shelled amniote egg that permits development in a dry environment.
3. Fertilization is internal for both viviparous and oviparous species.
4. Reptiles are ectothermic, regulating body temperature through behavioral modifications that either increase or decrease absorption of external heat.
5. The Age of Reptiles traces its origin to the Carboniferous cotylosaurs or "stem reptiles," lizardlike insectivores that evolved from Devonian amphibians.
6. The first major reptilian radiation produced the Permian synapsids, a group of terrestrial predators that gave rise to the mammal-like therapsids. Other descendants of stem reptiles, the thecodonts, survived the mass extinctions of the Permian period with the synapsids to become the stock for the evolution of dinosaurs, crocodilians, and birds.
7. The second great reptilian radiation produced flying pterosaurs and terrestrial dinosaurs from Triassic thecodonts, thereby introducing the largest terrestrial animals that ever lived.
8. Controversy exists about whether dinosaurs were endothermic, but there is no doubt that during the cooler, more climatically unstable Cretaceous period, something dramatic caused the abrupt demise of these magnificent reptiles.
9. The major reptilian orders that survived the Cretaceous crisis are the Squamata, Chelonia, and Crocodilia.

10. Squamata comprises the lizards and snakes. Lizards include the most numerous and diverse reptiles today. Ancestral burrowing species gave rise to snakes. A limbless group with little or no trace of the original pelvic and pectoral bones, snakes are carnivores with loosely articulated jaws. They home in on prey through olfaction or detection of heat. Poisonous snakes inject toxins through modified hollow teeth.

11. Chelonians, the turtles, generally hard-shelled reptiles that lay their eggs on land, have changed little since their evolution from stem reptiles.

12. Crocodilians, the crocodiles and alligators, are close relatives of the dinosaurs. Crocodilians are restricted to warm regions of the world.

Class Aves (pp. 655–659)

1. The amniote eggs and scaled legs of birds give testimony to their reptilian heritage; birds descended from dinosaurs.

2. Almost every part of avian anatomy enhances flight.

3. A body of low density is the result of honeycombed bones and lack of certain bilateral organs.

4. Variously shaped beaks take in foods that are ground in a gizzard.

5. A heart that completely separates oxygen-poor from oxygen-rich blood and a highly efficient lung and air sac system make possible the active, endothermic metabolism required for flight.

6. Flight requires fine coordination and acute senses. Vision has reached its peak development in birds. Birds also have relatively large brains and display complex behavior, especially when breeding.

7. Fertilization is internal, and eggs are laid and kept warm by brooding until hatching.

8. Only birds have feathers, which shape wings into airfoils that provide both propulsion and lift. Pectoral muscles anchored to a specialized keel on the sternum power the wings.

9. Feathers evolved from reptilian scales and likely first functioned as insulation during the development of endothermy.

10. *Archaeopteryx*, an early bird, was a feathered dinosaur with wings that evolved from the entire forelimb.

11. The ratites include the flightless species, such as the ostrich and kiwi, which lack a keeled breastbone. Carinate birds have a sternal keel to which attach the large flight muscles.

Class Mammalia (pp. 659–664)

1. Hair and mammary glands are the two diagnostic characteristics of mammals. Like feathers, hair is made of the protein keratin and helps insulate the body.

2. An active endothermic metabolism is supported by a four-chambered heart and respiratory system ventilated by a muscular diaphragm.

3. Fertilization is internal, and most mammals give birth to young after development in the uterus.

4. Of all vertebrates, mammals are endowed with the largest brains and the capability for learning. Relatively long periods of parental care provide the opportunity for parental teaching.

5. Mammals are characterized by highly differentiated teeth, which are adapted for chewing many kinds of foods.

6. Small, insect-eating mammals arose from the therapsids and coexisted with the dinosaurs throughout the Mesozoic era. Extinction of the dinosaurs reopened many adaptive zones and catalyzed an extensive mammalian radiation that resulted in the modern monotremes, marsupials, and placental mammals.

7. The monotremes are egg-laying mammals, today represented only by the platypuses and the echidnas found in Australia and New Guinea. After hatching, the young suck milk from the fur of the mother.

8. The marsupials include opossums, kangaroos, and the koalas, animals whose young complete their embryonic development inside a maternal pouch, the marsupium, attached to a teat from which they get their nourishment. Marsupials isolated in Australia show convergent evolution with placentals in other parts of the world.

9. The most widespread and diverse modern mammals are the placentals, a group whose young complete their embryonic development attached to a placenta inside the mother's uterus. Placental mammals and marsupials may have shared a common ancestor 80–100 million years ago. Subsequently, placental mammals branched out into several main evolutionary lines.

The Human Ancestry (pp. 664–671)

1. The first primates were probably small arboreal animals, which descended from insectivores in the late Cretaceous period. Life in trees demanded limber shoulder joints; dexterous, sensitive hands and fingers; close-set eyes with overlapping fields of vision for depth perception; excellent eye-hand coordination; and extended parental care, all of which extant primates retain.

2. Two subgroups of modern primates are the prosimians—lemurs and their relatives—and the anthropoids. Anthropoids probably evolved from prosimian stock and diverged early into New World and Old World monkeys that maintained separate evolutionary pathways. Modern apes are confined to the Old World.

3. The earliest known ape was *Aegyptopithecus*, a cat-sized tree-dweller that gave rise to several forms. One of these, *Ramapithecus*, was likely ancestor of modern apes and humans.

4. The first humanlike fossil to be discovered was *Australopithecus africanus*, a hominid with a small brain that walked erect and had human-type hands and teeth. Subsequent discovery of "Lucy," a petite fossil classified as *Australopithecus afarensis*, has confirmed the early evolution of upright posture in the hominids.

5. Enlargement of the brain and the appearance of simple stone tools occurred about two million years ago, as indicated by fossils of *Homo habilis*, a species that coexisted with the smaller-brained *Australopithecus* and led to the evolution of *Homo erectus*.

6. *Homo erectus* was the first hominid to venture out of the tropics and into colder climates. Intelligence and social cooperation contributed to survival, especially during the ensuing ice ages. *Homo sapiens*, our species, emerged.

7. The heavy-browed Neanderthals, the earliest examples of *Homo sapiens*, possessed brain sizes even larger than

our own. They were contemporaries of the first fully modern humans.

8. Although the erect stance was the most radical anatomical change in our evolution, enlargement of the brain and its extended period of development, which required long periods of parental care, gave rise to language and the resultant far-reaching consequences of culture. The first stage of cultural evolution, which began in Africa with wandering hunters and gatherers, progressed to the development of agriculture in Eurasia and the Americas about 15,000 years ago. The third stage of cultural evolution is the Industrial Revolution, which today continues as accelerating technological change. The exploding human population, so capable of altering environments, now threatens Earth's ecosystems.

SELF-QUIZ

1. Which of the following is *not* a general characteristic of the phylum Chordata?
   a. dorsal, hollow nerve cord
   b. vertebral column
   c. notochord
   d. pharyngeal slits
   e. postanal tail

2. Which of the following groups is entirely extinct?
   a. cephalochordates
   b. agnathans
   c. placoderms
   d. lobe-fin fishes
   e. ratite birds

3. Which of the following is *not* an adaptation that enhances flight in birds?
   a. honeycombed bones
   b. keeled sternum for attachment of flight muscles
   c. external fertilization
   d. large pectoral muscles
   e. relatively large brains for visual processing and motor coordination

4. The amniote egg first evolved in
   a. bony fishes
   b. amphibians
   c. reptiles
   d. birds
   e. mammals

5. Mammals and birds share all of the following characteristics *except*
   a. endothermy
   b. descent from reptiles
   c. four-chambered heart
   d. teeth specialized for diverse diets
   e. the ability of some species to fly

6. If you were to observe a monkey in a zoo, which characteristic would indicate a New World origin for that monkey species?
   a. distinct "seat pads"
   b. eyes close together on the front of the skull
   c. use of the tail to hang from a tree limb
   d. occasional bipedal walking
   e. downward orientation of the nostrils

7. Only an animal species with a diaphragm can be expected to have
   a. hair
   b. feathers
   c. scales
   d. lungs
   e. moist skin

8. Unlike placental mammals, both monotremes and marsupials
   a. lack nipples
   b. have some embryonic development outside the mother's uterus
   c. lay eggs
   d. are found in Australia and Africa
   e. include only insectivores and herbivores

9. Mosaic evolution refers to the observation that
   a. reptiles had two major periods of adaptive radiation
   b. preexisting structures are often coopted for new functions
   c. different anatomical features unique to a species evolve at different rates
   d. some fish that evolved in fresh water and moved to the seas have returned to fresh water
   e. groups that evolve in separated geographical areas often develop superficially similar animal forms that fill the same niches

10. The possibility that the Neanderthals did not interbreed extensively with modern humans is supported by evidence that shows
    a. the two groups are anatomically very different
    b. the two groups did not coexist at the same time—Neanderthals died out before modern humans first appeared
    c. the two groups have been shown to have very different gene pools
    d. the uniformity of human mitochondrial DNA may point to the origin of all modern humans from a very small splinter population
    e. fossil evidence shows that modern humans evolved in Africa, whereas Neanderthals lived in Eurasia

CHALLENGE QUESTIONS

1. Discuss similarities and differences among the eggs of fishes, amphibians, reptiles, birds, and monotreme

mammals as they relate to the life cycles of these various classes of vertebrates.

2. Field biologists that capture sharks for tagging and study must "kick start" them after placing them back in the water by repeatedly propelling them forward a short distance. From what you have learned in this chapter, why do you think this procedure is necessary and what does it accomplish?

3. Compare and contrast bats and birds, two flying vertebrates, in as many ways as you can.

4. Throughout chordate evolution, many existing structures have been coopted for new functions. Describe several such cases of preaptation.

FURTHER READING

1. Carroll, R. C. *Vertebrate Paleontology and Evolution.* New York: W. H. Freeman, 1987. Comprehensive, authoritative, yet accessible text.
2. del Pino, E. M. "Marsupial Frogs." *Scientific American,* May 1989. Some tree frogs incubate eggs in a pouch rather than laying them in water.
3. Diamond, J. "The Great Leap Forward." *Discover,* May 1989. Entertaining speculations on human evolution.
4. French, A. R. "The Patterns of Mammalian Hibernation." *American Scientist,* November–December 1988. Which mammals hibernate and why.
5. Fricke, H. "Coelacanths. The Fish That Time Forgot." *National Geographic,* June 1988. Historical and modern studies of this "living fossil," with photographs taken from a deep-sea submersible.
6. Griffiths, M. "The Platypus." *Scientific American,* May 1988. Biology of one of the three extant species of egg-laying mammals.
7. Lovejoy, C. O. "Evolution of Human Walking." *Scientific American,* November 1988. Evidence that human bipedalism dates back at least 3 million years to "Lucy" and her australopithecine relatives.
8. Morell, V. "Announcing the Birth of a Heresy." *Discover,* March 1987. Fossil finds in the western United States indicating that dinosaurs were endothermic and cared for their young.
9. Padian, K. "The Flight of Pterosaurs." *Natural History,* December 1988. The first flying vertebrates.
10. Simons, T., S. K. Sherrod, M. W. Collopy, and M. A. Jenkins. "Restoring the Bald Eagle." *American Scientist,* May–June 1988. Encouraging results of attempts to conserve a threatened species.
11. Stringer, C. B., and P. Andrews. "Genetic and Fossil Evidence for the Origin of Modern Humans." *Science,* March 11, 1988. The case for our genesis in Africa from a small, splinter population of *Homo.*
12. Weaver, K. F. "The Search for Our Ancestors." *National Geographic,* November 1985. An artfully illustrated essay on human evolution and how it is studied.

Plants: Form and Function

Interview: Ruth Satter

Ruth Satter is a plant physiologist whose fascination with plants began with an early interest in gardening. After graduate work at the University of Connecticut, Dr. Satter was a research scientist for several years at Yale University. She is now a professor in residence in the Department of Molecular and Cell Biology at the University of Connecticut at Storrs. Dr. Satter's research focuses on rhythmic functions of plants, such as the daily leaf movements exhibited by many species. Her imaginative experiments have made her a leader in the field of chronobiology, the study of how organisms measure time and control rhythms.

This interview took place at the 1988 annual meeting of plant physiologists in Reno, Nevada. Between meeting with colleagues, helping her graduate students with their poster sessions, and attending presentations, Professor Satter shared the story of her extraordinary career.

(**Note**: As this book was going to press, Ruth Satter passed away after a long, intrepid struggle with leukemia. We will treasure our memories of this remarkable woman who enriched so many lives.)

Ruth, why have you and so many other scientists converged here in Reno, Nevada, this week?

We are here for the annual meeting of the American Society of Plant Physiology. It's an organization of about four thousand faculty members, graduate students, postdoctoral fellows, and undergraduates who are interested in plant research. We meet once a year to learn about the research of our colleagues and to discuss our own research.

How do scientists actually report their work at a meeting like this?

They report their work either by an oral presentation of about 15 minutes or on a display board (of about 4 feet by 4 feet) on which the scientist describes the purpose, the processes, the results, and the significance of his or her research. Individuals attending the meeting then have the opportunity to examine the poster and discuss it with the person whose work it illustrates.

Let's back up and trace the path by which you've come to attend this conference today. How did you begin your college career? Did you major in biology?

I didn't take any biology or botany in college. In fact, I had a double major in math and physics at Barnard College in New York City. After I graduated in 1944, I worked at Bell Laboratories doing research on radar development during the latter part of World War II.

When I was a graduate student, Beatrice Sweeney, a pioneer in the field of circadian rhythms, gave a seminar at the University of Connecticut where she discussed her research on bioluminescence in certain algae. She showed us data indicating that bioluminescence only occurs at certain times of day, and that if one kept the algae in a constant environment—constant light and constant temperature—these periods of bioluminescence occurred approximately 24 hours apart. I found it fascinating that there was some sort of an internal oscillator or clock that was controlling bioluminescence, and I knew then that at some time in my life I wanted to study circadian rhythms.

Following your work with Bell Labs, did you go to graduate school right away?

No, after that I married and had children. At that time, child care was inadequate, and opportunities for women to work while they took care of their children were scarce. We had four children and I was out of formal academic life for about fifteen years.

Can you recall what induced your return to school for graduate work?

It was something that I had always wanted to do. During my child-raising years, I continued to pursue my intellectual interests. I read books on physics and kept up with new developments. I also became very interested in horticulture. I read books and articles about plant growth and conducted experiments in my own garden. I took courses in horticulture at the New York Botanical Gardens and taught horticulture for an adult education program in Hartford every winter for five or six years.

Your interest in botanical research developed from your home gardening?

Yes, my grandmother stimulated my interest when I was a child. She was an immigrant and had no formal education, but she just had a great curiosity about the way plants grew. She was a natural experimentalist, and her gardening consisted of taking dried lima beans purchased to make soup and saving some of them to plant out in the yard. Or taking potatoes that had sprouted, and cutting out the sprouts and planting them. I worked at these

experiments with her and was fascinated by them. These experiences led to my switch from math and physics to the life sciences when I started graduate school at the University of Connecticut.

Was it difficult to change the focus in your life from raising a family to attending graduate school? And, what was it like going back to school when you were probably a bit older than most of your peers?

In response to the first part of your question: No, the change was not difficult—going to graduate school was a real privilege and something that I had looked forward to. It added a whole new dimension to my life. Yes, I was much older than most of my peers, and at that time some of the younger students were not particularly friendly toward the older students. Returning to school was not an accepted thing to do at that time. Of course, I did find some friends, but I was really there because I wanted to learn.

Considering that research support and job opportunities are more ample for animal biologists than for plant scientists, do you ever regret specializing in plant science as opposed to animal physiology?

Never. I have always been so interested in plants. It would never have occurred to me to switch to the animal sciences.

At what point did you become interested in circadian rhythms and the biological clocks?

So it was the impact of a guest speaker that directed your future research?

Yes. After I received my Ph.D. degree, I had a postdoctoral fellowship at Yale, where I studied leaf movements. Initially, I looked at the effect of light on the movements, but I was most interested in the fact that one had to apply the light stimulus at a certain time of day to get the effect in which we were interested.

Could you describe what you mean by leaf movements?

There are certain plants that have leaves that are horizontal in the daytime and vertical at night. Plants that exhibit these sleep movements are called nyctinastic plants.

How do we know that these sleep movements are not just a response of the plant detecting light and darkness—how can we distinguish between the plant simply responding to an environmental cue as opposed to actually keeping track of time for itself?

To make this distinction, one has to maintain light, temperature, and humidity constant for a long period of time and see whether the movements persist. If they persist in constant conditions, one can assume the movements are probably controlled by an internal oscillator. To be officially called a circadian rhythm, the plant must repeat its movement cycle at approximately, although rarely exactly, 24-hour intervals in a constant environment.

Presumably, this internal clock can be adjusted by changes in the environment—by the seasonal changes in the timing of sunrise and sunset, for instance. How does the clock interact with the environment?

This is a very important question because if the clock did not interact with the environment, it would fall out of synchrony with solar day and wouldn't function usefully. Let's use the analogy of a watch that runs fast, for instance: You can either take such a watch back to the store and have it fixed so that it keeps proper time, or you can set the hand of the clock back each day so that you repeat part of the cycle. The latter method is the method that organisms use to adjust their rhythms to solar light. Every day at the transition between light and darkness, at dawn and at dusk, the environmental cue resets the clock.

Do you have any ideas on the mechanism by which the clock works?

Unfortunately, we really don't know how the clock works, but if we understood how the clock is reset, we would know a great deal about the clock mechanism. We are now looking at the complex series of events between perception of the light signal and resetting of the clock. We are approaching this in several different ways. First, we are looking at biochemical changes. After the light signal is detected, there are certain changes in biochemical compounds in membranes of cells. Our hypothesis is that these biochemical changes lead to changes in the amount of calcium within the cell and in the activation of enzymes called kinases that phosphorylate proteins, and that changes in calcium and phosphorylation change the properties of membrane transport proteins.

How does the plant actually move its leaves? Does it have something like muscles?

At the base of each leaf or leaflet is an organ called a pulvinus, which looks like a horizontal cylinder in the daytime but curves to form an inverted U at night. The changes in curvature of the pulvinus cause the leaf to move. The blade doesn't change; it's just the pulvinus that curves.

What causes the pulvinus to curve?

When the pulvinus is horizontal in the daytime, cells on the underside are swollen and those on the upper side are shrunken. In the nighttime, when the pulvinus curves, cells on the upper side swell and the under cells shrink. Changes in the size of the cells depend upon changes in the amount of potassium, chloride, and other solutes within the cell. When the cell takes up potassium, chloride, and other ions, water moves in by osmosis and the cells swell. When the cells lose these ions, water moves out of the cell through osmosis and the cell shrinks.

Do you have any speculations as to the possible function within the plants of these sleep movements?

Charles Darwin suggested that sleep movements help plants conserve heat at night, but this is not the whole answer. Perhaps someone in the young generation of biologists just beginning to study biology will answer this question.

Are there any connections between the circadian rhythms in your plants and those in humans and other animals?

The basic principles that govern rhythms in plants, animals, and microorganisms are very similar, but we don't now whether or not the clocks that generate the rhythms function in the same way. Most people don't realize that our first evidence for biological rhythms came from studies in plants. In the year 1729, a scientist named DeMairan conducted a very simple experiment. He knew of certain plants with leaves that are horizontal in the day and vertical at night. It had been generally assumed that light and dark caused the movements. DeMairan put one of the plants on his windowsill and one in his basement where the light was very dim but of constant

intensity; perhaps he used a candle in those days. He found that the movements persisted in the plant that was in constant conditions as well as in the one that was on the windowsill. From these results he predicted that there had to be an internal oscillator in plants, and he speculated that animals also have internal oscillators. It took 200 more years, however, before there was any solid evidence for rhythms in vertebrates. We now know that all eukaryotic organisms probably have clocks of some kind.

Do zoologists who work on circadian rhythms and botanists who study rhythms in plants now have much interaction?

Yes, they meet together every year or two at special chronobiology conferences where they read about each other's research and discuss ideas. They learn from one another.

How do you define chronobiology?

Chronobiology indicates time measurement by an organism.

You describe your work as basic research in the sense that you are investigating a basic biological process rather than searching for any kind of specific applications. How is basic botanical research related to agriculture?

We need to understand fundamental botanical processes in order to make improvements in agriculture. In my own research, for instance, although our study of time-keeping processes seems very theoretical, it has a practical application. A large number of processes in many plants occur at close to the same date each year. These include the time at which a plant switches from vegetative growth to flowering, and the time at which certain trees begin processes that will eventually lead them to become dormant in the winter. The plant has to have some environmental cue to synchronize its growth with changes in the external environment; a change in the length of the dark and light periods is the most reliable cue. For instance, spinach is a plant that grows vegetatively in the spring and flowers during the long days of summer. By knowing this, one can manipulate day length artificially and grow several crops during the year instead of just growing one. Understanding how the plant measures time should have additional practical benefits, possibly by genetic engineering to alter the plant's photoperiodic requirements.

Students read a lot today, even in the newspaper, about genetic engineering and possible applications. What promise does genetic engineering hold for agriculture?

At this particular time, genetic engineering is helping us to understand a lot of basic processes. This includes learning more about how the photosynthetic machinery works, why some plants survive periods of unusually low temperature whereas other plants don't, why some plants are more tolerant of high temperature than others, and why some plants are resistant to pathogens such as viruses, bacteria, and fungi while others are not. We are beginning to discover this information through genetic engineering techniques. There certainly will be many practical applications in the future.

With all of the present problems of regional famine, do you see any special opportunities and responsibilities that plant scientists in developed countries have toward alleviating these problems?

We have a very great responsibility to do just that. We have the technical ability, the educational facilities, and the capital, and we certainly should use our scientific resources as wisely as we can to help prevent famine in other countries. This would include both bringing scientists from other countries

here for training and sending some of our people to their countries. It also includes focusing research on nutritionally important crops like rice, beans, and corn.

This all requires money. Could you tell us about research grants and how one goes about getting them?

The application process for a research grant is highly competitive. It requires a description of the proposed research, the experimental plan to be followed, and an indication of the potential application of the research. The proposal then is sent to a funding agency composed of a panel of experts who determine which proposals to fund. The two largest funding agencies are the National Institutes of Health (NIH) and the National Science Foundation (NSF). Both are supported by the United States government. Plant research is also funded by the Department of Agriculture.

Let's imagine a young professor who is just out of graduate school or has just completed postdoctoral work and is starting a faculty position at a university with a major research program. What kind of a situation would he or she actually walk into in terms of facilities and how would the young professor go about getting started?

A young professor would have a problem that he or she is interested in addressing and would think of the important questions posed by that problem. The professor would hope that the university would help initially by providing some equipment and money for supplies. Basic research would be conducted to indicate whether the methods and the project as a whole are feasible. Then the professor would write a grant proposal and submit it to one of the agencies that supplies funding. The long-term survival of each research program depends upon obtaining outside funding.

Are you provided with not only lab space but space in which to grow your plants?

Yes, my university provides us with both greenhouse space and space for growth under controlled conditions. If I were working on crop plants, I would also be given study fields.

How do you obtain the exotic plants with which you work, such as *Samanea*?

I first obtained the seed for *Samanea* while vacationing with my husband in the Virgin Islands. We looked at the various plants with moving leaves and found one that had a very large pulvinus. I asked the person who owned the tree if I could have some seeds. After that, I found out that the plants grow near the botany building at the University of Hawaii. A fine friend sends me seeds each year.

Training graduate students and postdoctoral students is a very important function of the university professor. About how many students of various kinds have you had in your labs over the years, and how do you select them?

Close to two dozen. I like to include undergraduate students too, most of whom initially find me. They usually come looking for jobs—sometimes to earn money by washing glassware or doing other routine tasks. When they are in the lab, I can determine whether or not they are reliable and are truly interested in our work. If so, I encourage them by including them in our research. Some have done publishable research. Two undergraduates presented posters at regional meetings of the Plant Physiology Society this year. One of them has been the co-author on two papers. Another is in the process of writing a paper now.

Could you describe the social environment in a lab that includes you as the principal investigator, graduate students, one or two postdoctorates, and an undergraduate student?

There is a lot of interaction and sharing. We try to have lab meetings once a week. People discuss their research. They give each other ideas and they stimulate each other. The professor should provide a laboratory environment that encourages students to learn and ask questions, to think critically, and to challenge dogma. When a student criticizes an idea, I feel that he or she is growing up. The students must understand that research involves a lot of discipline and hard work. They should also find it stimulating, exciting, and fun. They should learn that they can challenge me and that I learn from them as they learn from me.

What qualities do you look for when choosing students or postdoctoral researchers for your lab?

Intelligence. Dependability. Diligence. Dedication. It is very important that the student be enthusiastic about our work. And hopefully he or she has a real passion for learning. I look for people who will contribute to the enjoyment of researching and learning. It is important that the person be cooperative, honest, and willing to admit to making mistakes. If a student tries to hide mistakes, I really don't want to keep him or her.

You mentioned earlier that you were very much influenced by a seminar given by Beatrice Sweeney, a woman working in the area of circadian rhythms. Today though, there are still far fewer women in science than there are men. Do you have any recommendations as to how we can attract more women to careers in science?

The gender gap is much smaller today that it was a decade ago. But, there are a few basic problems. A generation or two ago it was very difficult for a woman to have a career in science. Opportunities were very limited, and science in many institutions was a male enclave. Barriers are breaking down, but it will take a while before the break is complete. Consequently, there are not enough older women in science to act as role models. I think that the situation will continue to improve, resulting in a more comfortable environment for women. Also, a career in science is very demanding, particularly if one is at an academic institution where one is required to teach, keep up with modern research, read journals, and understand how the field is progressing. Science is changing rapidly and staying abreast of these changes is time consuming. Many young women are concerned that it would be difficult to find time to raise children and lead a normal family life. To women who want a career in the sciences, I would say, you *can* have it all. It takes a lot of dedication and possibly a readjustment of priorities at certain times of your life. You will have years during which you may not be able to concentrate on your science as much as you had hoped, but you can make up for that at some later time. Don't give up easily. If you are interested in science, pursue your dream. Universities can help by providing child care at the work place and flexible work schedules.

How early do you think students need to develop the confidence that they can succeed in science?

The earlier the better, of course. If they develop it as children, they are better off, but they can develop it in college or in graduate school.

What advice or recommendations would you give to a college freshman who is just starting a curriculum in biology, or even more generally, in science?

If you find an aspect of science that interests you particularly, identify the faculty members at your university who are working in this area. Go to see them, visit their laboratories, express your interest, and ask if you can do any work in the laboratory, even if it just involves washing glassware. While there, you will come into contact with people working in the lab and will have the opportunity to talk with them about their work. Express an interest and an enthusiasm. I think that relatively few faculty members will turn down an enthusiastic student. In fact, they will encourage you to become more involved.

31 Anatomy of a Plant

Botany, the study of plants, is one of the oldest and most practical sciences, having its origins in the identification of plants that are edible, poisonous, or medicinal. The same pragmatism still compels us to learn about plants; we depend on plants for lumber, fabric, paper, landscaping and home decorations, medicine, and, most important, for food. In the preceding interview, Professor Ruth Satter discussed the relationship between basic botanical research, including her own work on how plants keep time, and the potential agricultural applications so vital to our welfare. Botany is a human concern (Figure 31.1).

The subject of this unit is the green plant—its form and function. The chapters focus on the angiosperms, or flowering plants. The structure and physiology of algae, bryophytes (mosses and their relatives), and gymnosperms, as well as the evolutionary relationships of these plants to the angiosperms, were addressed in Unit Five. Angiosperms are characterized by flowers and fruits, specialized structures that function in reproduction and dispersal of seeds. With about 275,000 known species, angiosperms are by far the most diverse and widespread of all plant divisions. The flowering plants are split into two classes: **monocots,** which have one cotyledon, or embryonic seed leaf, and **dicots,** which have two cotyledons. Monocots and dicots also have several other structural differences (Figure 31.2). Humans have domesticated and selectively bred only a tiny fraction of angiosperm species, but these cultivated plants supply nearly all of our food, either directly as grains, vegetables, and fruits or indirectly as animal fodder.

Figure 31.4
Root hairs of a radish seedling. Growing by the thousands just behind the tip of each root, the hairs cling tightly to soil particles and increase the surface area for the absorption of water and minerals by the roots.

Although the entire root system helps anchor a plant, most absorption of water and minerals occurs near the root tips, where vast numbers of tiny **root hairs** increase the surface area of the root tremendously (Figure 31.4).

In addition to roots that extend from the base of the shoot, some plants have roots arising above ground from stems or even from leaves. Such roots are said to be **adventitious,** the term used for any plant part that grows in an unusual location (from the Latin *adventicius,* "not belonging to"). The adventitious roots of some plants, including corn, function as props that help support stems.

The Shoot System

Stems and leaves are the two organs that make up the shoot system. Flowers, the angiosperm organs specialized for reproduction, are constructed mainly from modified leaves. Discussion of the structure and function of flowers is deferred until Chapter 34.

Stems   Along the stem, **nodes,** the points at which leaves are attached, alternate with **internodes,** the stem segments between nodes (Figure 31.5; see also Figure

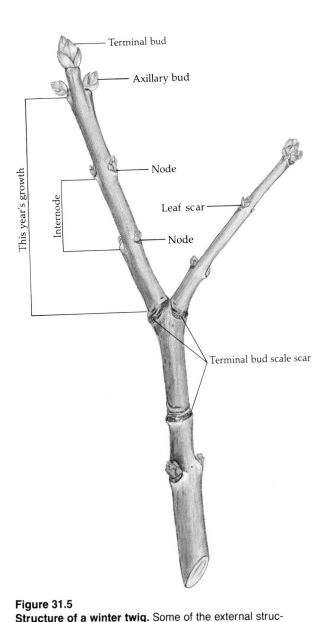

Figure 31.5
Structure of a winter twig. Some of the external structures of a woody stem can be observed by inspecting a twig of a deciduous tree that has lost its leaves for the winter. At the tip of this lilac twig is the dormant terminal bud, enclosed by scales that protect its embryonic tissues. In spring the bud will shed its scales and begin growing, producing a series of nodes and internodes. Farther down the twig are whorls of scars left by the scales that enclosed the terminal bud during the previous winter. Thus, the region of the twig between the terminal bud and the first ring of bud scales was produced during the past spring and summer, and the region between these bud scars and the next whorl of scars was formed during the growing season of the previous year. The number of whorls of bud scars indicates the age of a twig. Along each growth segment, nodes are marked by scars left when leaves fell during autumn. Above each leaf scar is either an axillary bud or a branch twig produced by previous growth of the axillary bud.

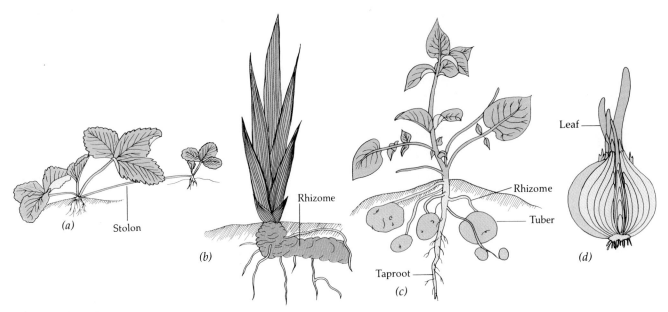

Figure 31.6
Modified stems. (a) Stolons, shown here on a strawberry plant, grow on the surface of the ground. **(b)** Rhizomes, like the one on this iris plant, are horizontal stems that grow below ground. **(c)** Tubers are swollen ends of rhizomes specialized for storing food. The "eyes" arranged in a spiral pattern around the potato are clusters of buds that mark the nodes. **(d)** Bulbs are vertical, underground stems. You can see the many layers of modified leaves attached to the stem by slicing an onion bulb lengthwise.

31.3). In the angle formed by each leaf and the stem is an **axillary bud,** itself an embryonic shoot. Most axillary buds are dormant, however; growth of a shoot is usually concentrated at the tip, where there is a **terminal bud** with developing leaves and a compact series of nodes and internodes. Presence of the terminal bud is partly responsible for inhibiting the growth of axillary buds, a phenomenon called **apical dominance.** Under certain conditions, axillary buds begin growing. Some develop into shoots bearing flowers, and others become vegetative (leaf-bearing) branches complete with their own terminal buds, leaves, and axillary buds. Sometimes it is possible to stimulate growth of axillary buds by removing the terminal bud. This is the rationale for pruning and "pinching back" of houseplants.

Modified stems of some plants are often mistaken for roots. Stolons are horizontal stems that grow along the surface of the ground (Figure 31.6). The "runners" of Bermuda grass and strawberry plants are examples. Rhizomes, such as those of irises, are horizontal stems that grow underground. Some rhizomes end in enlarged tubers where food is stored, as in white potatoes. Bulbs, such as those of tulips or onions, are vertical, underground shoots with leaves modified for food storage.

Leaves Leaves are the main photosynthetic organs of most plants (green stems also perform photosynthesis). Leaves vary extensively in form, but they generally consist of a flattened blade and a stalk, the **petiole,** which joins the leaf to a node of the stem (see Figure 31.3). Grasses and most other monocots lack petioles; instead, the base of the leaf forms a sheath that envelops the stem.

The leaves of monocots and dicots differ in how their major veins, or vascular bundles, are arranged. Most monocots have parallel major veins that run the length of the leaf blade. In contrast, dicot leaves generally have a multibranched network of major veins. All leaves have numerous minor cross veins. Vascular arrangement, leaf shape, and leaf pattern on the stem are characteristics used by plant taxonomists to help identify or classify plants (Figure 31.7).

Figure 31.7 ▶
Classification of leaves. Leaves are arranged on the stem in a variety of patterns. If each node has a pair of leaves 180° apart, the leaves are said to be opposite. The leaf pattern is alternate when each node has a single leaf and the leaves of adjacent nodes point in opposite directions. If a node has three or more leaves attached, the arrangement is termed whorled. Botanists refer to a leaf as being simple if it has a single, undivided blade. If the blade is divided into several leaflets, then the leaf is compound. You can distinguish a compound leaf from a stem with several closely spaced leaves that are simple by examining the locations of axillary buds. Each leaf has only one axillary bud. There will be no bud at the base of a leaflet of a compound leaf but a bud where the petiole of the compound leaf joins the stem. Leaves also vary in shape, in the contour of their margins, and in their pattern of veins.

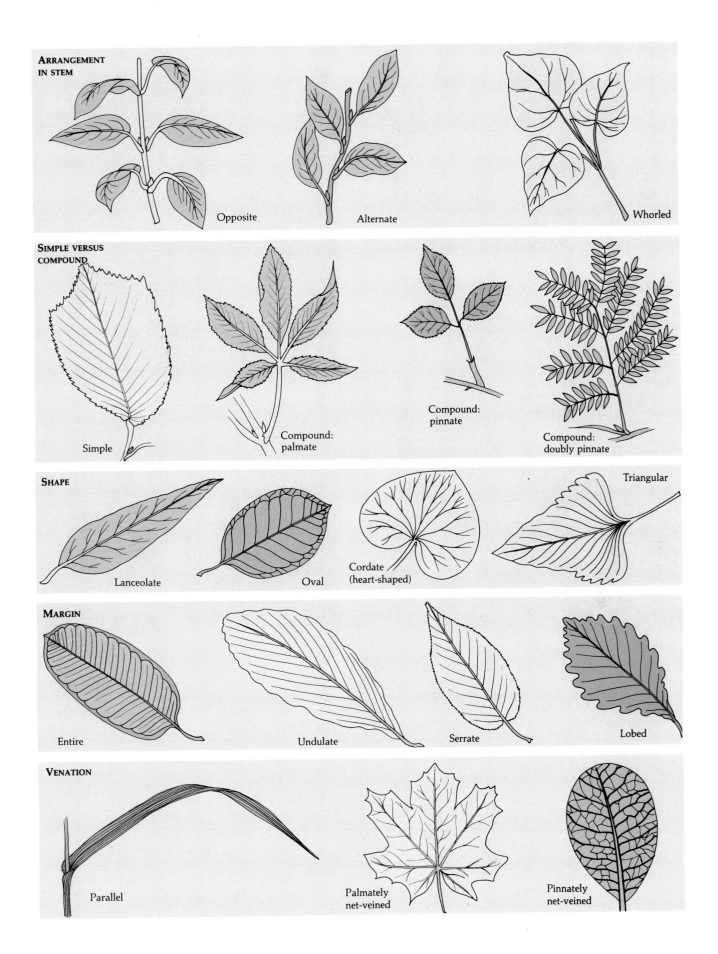

ARRANGEMENT IN STEM

Opposite

Alternate

Whorled

SIMPLE VERSUS COMPOUND

Simple

Compound: palmate

Compound: pinnate

Compound: doubly pinnate

SHAPE

Lanceolate

Oval

Cordate (heart-shaped)

Triangular

MARGIN

Entire

Undulate

Serrate

Lobed

VENATION

Parallel

Palmately net-veined

Pinnately net-veined

(a)

(c)

(b)

(d)

Figure 31.8
Modification of leaves. (a) The tendrils used by this cucumber to cling to supports are modified leaflets. **(b)** The spines of cacti, such as this prickly pear, are actually leaves, and photosynthesis is carried out mainly by the fleshy green stems. **(c)** Many succulents, such as ice plant, have leaves modified for storing water. **(d)** In many plants, brightly colored leaves help attract pollinators to the flower. The red "petals" of the poinsettia are actually leaves.

Although most leaves are specialized for photosynthesis, some plants have leaves that have become adapted for other functions (Figure 31.8).

So far we have examined the structural organization of the whole plant as we see it with the unaided eye. With this overview, we can now begin to dissect the plant and explore its microscopic organization.

PLANT CELLS AND TISSUES

Cell structure was described in detail in Chapter 7. Figure 31.9 reviews major features of plant cells.

How Plant Cells Grow

Although cell division sustains plant growth, the actual increase in the size of roots and stems is mostly due to cell growth, defined as an irreversible increase in cell size. The mechanism of cell growth is one important difference between plants and animals. Animal cells grow mainly by synthesizing organic molecules and adding more cytoplasm. Although plant cells do synthesize additional cytoplasm, most of the growth of a plant cell occurs when the vacuole and cytoplasm take in more water as the cell wall becomes temporarily more plastic (Figure 31.10). The wall of the swelling

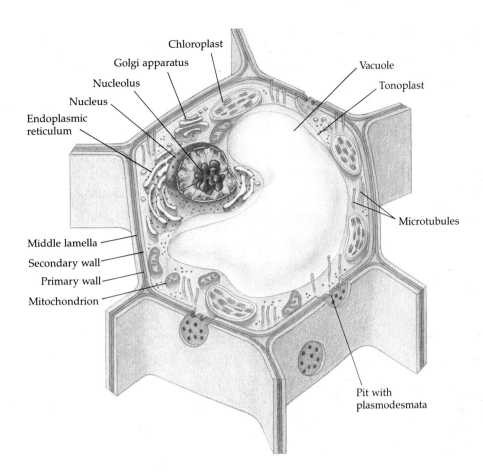

Chloroplast
Golgi apparatus
Nucleolus
Nucleus
Endoplasmic reticulum

Vacuole
Tonoplast

Microtubules

Middle lamella
Secondary wall
Primary wall
Mitochondrion

Pit with plasmodesmata

Figure 31.9
Review of plant cell structure. A plant cell consists of a protoplast enclosed in a cell wall. The protoplast is bounded by the plasma membrane. Outside the plasma membrane is the primary cell wall and in some plants a secondary cell wall, constructed from cellulose fibers and other components added between the plasma membrane and primary wall. Between the primary walls of adjacent plant cells is the middle lamella, a sticky layer that cements the cells together. The protoplasts of neighboring cells are generally connected by plasmodesmata, cytoplasmic channels that pass through pores in the walls (see Figure 7.35 in Chapter 7). The plasmodesmata may be concentrated in areas called pits, where the distance between adjacent protoplasts is narrowed because secondary walls are absent. When mature, most living plant cells have a large central vacuole that occupies as much as 90% of the volume of the protoplast. A membrane called the tonoplast separates the contents of the vacuole from the thin layer of cytoplasm, in which the mitochondria, plastids, and other organelles are located. Within the vacuole is the cell sap, a complex aqueous solution that helps the vacuole play an important role in maintaining the turgor, or firmness, of the cell (see Chapter 8).

cell extends, but usually the orientation of cellulose fibers favors growth in one direction. For example, cells of the stem may elongate to 20 times their original length, with little increase in width. There is evidence that microtubules beneath the plasma membrane help determine the orientation of cellulose fibers, but how this process is controlled is one of many unanswered questions about plant growth. As a plant cell stops growing, additional material is applied to the inside of the existing wall (outside the plasma membrane), making the wall more rigid.

The uptake of water by the vacuole accounts for about 90% of the enlargement of cells. A plant can therefore grow rapidly and economically because a small amount of new cytoplasm can go a long way. Bamboo shoots, for instance, may elongate more than 2 m per week. Rapid extension of shoots and roots increases exposure to light and soil, an important adaptation to the immobile life style of plants.

Types of Plant Cells

As a plant grows and develops, its cells become specialized for a variety of functions. This section surveys some of the most important types of plant cells (Figure 31.11, p. 689). Each of these cell types may be found in any of the three plant organs—roots, stems, or leaves. As you consider each type of cell, notice the way its structure is appropriate to its function. Also notice that the cell wall is often a major factor determining the structure and function of each cell type.

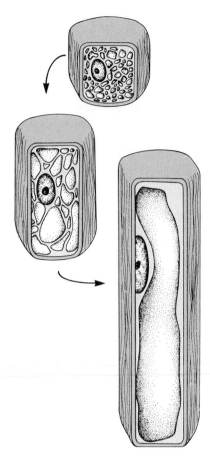

Figure 31.10
Growth of a plant cell. Most of the increase in size of a plant cell is due to accumulation of water by the vacuole. A young cell has many small vacuoles that coalesce to form the central vacuole. The wall of the growing cell stretches, but the cell maintains the thickness of its wall by secreting molecules for wall enlargement. The orientation of cellulose fibrils in the wall determines the plane of cell growth; in this illustration, the cell is growing in length without increasing much in diameter, a typical growth pattern for cells in stems and roots that are elongating.

Parenchyma Cells   Because they are structurally unspecialized, **parenchyma cells** are often depicted as "typical" plant cells (Figure 31.11a). Most parenchyma cells lack secondary walls, and their primary walls remain thin and flexible at maturity. The protoplast generally has a large central vacuole.

Parenchyma cells carry on most of the metabolism of the plant, synthesizing and storing various organic products. For example, photosynthesis occurs within the chloroplasts of parenchyma cells in the leaf. Some parenchyma cells in stems and roots have colorless plastids that store starch. The flesh of most fruit is composed mostly of parenchyma cells.

Developing plant cells of all types usually have the generalized structure of parenchyma cells before specializing further in structure and function. Mature parenchyma cells themselves do not generally undergo cell division, but most retain the ability to divide and differentiate into other types of plant cells under special conditions—during the repair and replacement of organs after injury to the plant, for instance. It is even possible in the laboratory to regenerate an entire plant from a single parenchyma cell.

Collenchyma Cells   Like parenchyma, **collenchyma cells** have protoplasts and usually lack secondary walls. Collenchyma cells have thicker primary walls than parenchyma, though the walls are unevenly thickened (Figure 31.11b). Usually grouped in strands or cylinders, collenchyma cells help support young parts of the plant. Young stems, for instance, often have a cylinder of collenchyma just below their surface. Because they lack secondary walls and the hardening agent lignin is absent in their primary walls, collenchyma cells provide support without restraining growth. They elongate with the stems and leaves they support.

Sclerenchyma Cells   **Sclerenchyma cells** also function as supporting elements in the plant, but with thick secondary walls strengthened by lignin, sclerenchyma cells are much more rigid than collenchyma (Figure 31.11c and d). Mature sclerenchyma cells cannot elongate, and they occur in regions of the plant that have stopped growing in length. So specialized are sclerenchyma cells for support that many lack protoplasts at functional maturity, the stage in a cell's development when it is fully specialized for its function. Thus, at functional maturity a sclerenchyma cell may actually be dead, its rigid wall serving as scaffolding to support the plant.

The two forms of sclerenchyma cells are fibers and sclereids. Long, slender, and tapered, **fibers** usually occur in bundles. Some plant fibers are commercially important, such as hemp fibers used to make rope and flax fibers that are woven into linen. **Sclereids** are shorter than fibers and irregular in shape. Nutshells and seed coats owe their hardness to sclereids, and sclereids scattered among the soft parenchyma tissue of a pear give that fruit its gritty texture.

Water-Conducting Cells   The water-conducting elements of xylem are elongated cells of two types: **tracheids** and **vessel elements** (Figure 31.12). Both types of cells are dead at functional maturity, but secondary walls are produced before the protoplast dies. In parts of the plant that are still elongating, the secondary walls are deposited unevenly in spiral or ring patterns that enable the cells to stretch like springs. Tracheids and vessel elements that form in parts of the plant that are no longer elongating usually have secondary walls with **pits,** thinner regions where only primary walls are present (Figure 31.13). A tracheid or vessel element completes its differentiation when its proto-

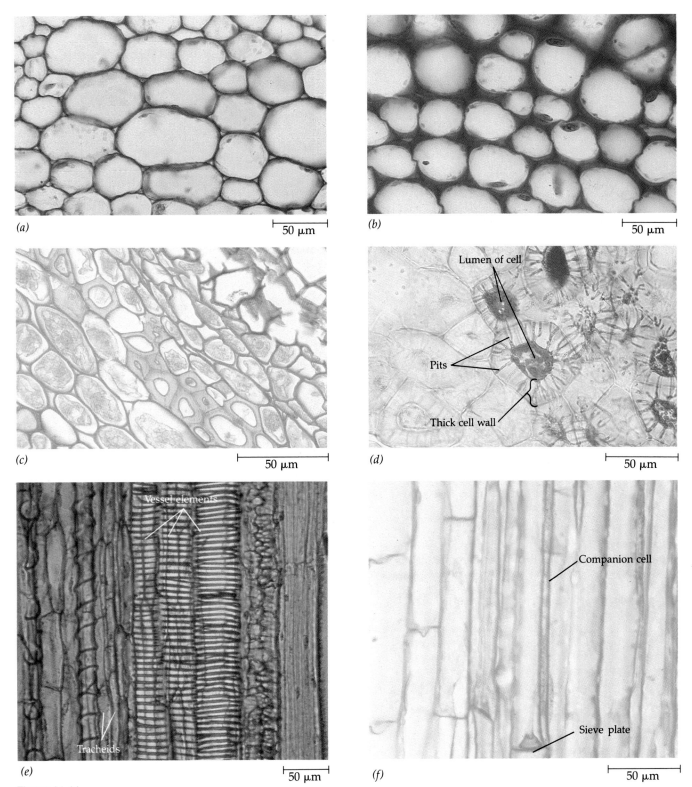

(a) ⟶ 50 μm

(b) ⟶ 50 μm

(c) ⟶ 50 μm

(d)
Lumen of cell
Pits
Thick cell wall
⟶ 50 μm

(e)
Vessel elements
Tracheids
⟶ 50 μm

(f)
Companion cell
Sieve plate
⟶ 50 μm

Figure 31.11
Types of plant cells. (All cells are shown in cross section except (e) and (f), which are shown in longitudinal section.) (**a**) Parenchyma cells are relatively unspecialized and usually lack secondary walls. These cells carry on most of the plant's metabolic functions. (**b**) Collenchyma cells have unevenly thickened primary walls, and provide support to parts of the plant that are still growing. (**c**) Sclerenchyma cells, spe-cialized for support, have secondary walls hardened with lignin and may be dead (lacking protoplasts) at functional maturity. Shown here are fiber cells, elongated sclerenchyma cells. (**d**) Sclereids are irregularly shaped sclerenchyma cells with very thick, lignified walls. (**e**) The water-conducting cells of xylem include tapered tracheids and vessel elements (the cells that look like springs) arranged end to end to form vessels. Both cell types have secondary walls and are dead at functional maturity. (**f**) The food-conducting cells of phloem are sieve-tube members, which are arranged end to end with perforated walls (sieve plates) between them. The cells are living at functional maturity, but lack nuclei. Alongside each sieve-tube member is a nucleated companion cell. (All LMs.)

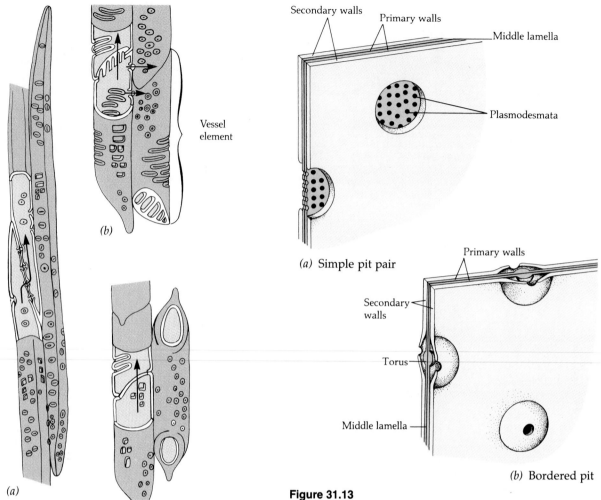

Figure 31.12
Water-conducting cells of the xylem. Arrows indicate the flow of water. **(a)** Tracheids are spindle-shaped cells with pits, through which water flows from cell to cell. **(b)** Vessel elements are individual cells linked together end to end to form long tubes called xylem vessels. Water streams from element to element through end walls that are perforated. Water can also migrate laterally between neighboring vessels through pits. **(c)** Resistance to water flow in some vessels is lowered by the complete absence of walls between the vessel elements.

(b)

Vessel element

(a)

(c)

(a) Simple pit pair

Secondary walls
Primary walls
Middle lamella
Plasmodesmata

Primary walls
Secondary walls
Torus
Middle lamella

(b) Bordered pit

Figure 31.13
Structure of pits. **(a)** At a simple pit pair, adjacent cells have aligned depressions where secondary walls are absent. Passing through the primary walls and middle lamella are numerous plasmodesmata connecting the protoplasts of the two cells. **(b)** At a bordered pit, extensions of the secondary wall arch over a partition consisting of the primary walls and middle lamella. In the tracheids of conifers, the partition has a thickened region called the torus. When hydrostatic pressure is greater in one cell than in its neighbor, the torus is pushed against the pit aperture of the cell with the lower pressure. This prevents flow of water or gases through the pit.

plast disintegrates, leaving behind a nonliving conduit through which water can flow.

Tracheids are long, thin cells with tapered ends. Water moves from cell to cell mainly through pits. Because the secondary walls are hardened with lignin, tracheids function in support as well as water transport.

Compared with tracheids, vessel elements are generally wider, shorter, thinner walled, and less tapered. The end walls of vessel elements are perforated, and thus water can flow freely through long xylem vessels consisting of chains of vessel elements.

Of the two types of water-conducting cells, tracheids evolved first. Most gymnosperms have only tracheids. Angiosperms generally have both tracheids and vessel elements; the latter probably evolved from tracheids in ancient flowering plants. (See Chapter 27.) Vessel elements are generally considered to be more efficient water conductors than tracheids.

Sieve-Tube Members Sucrose, other organic compounds, and some mineral ions are transported within phloem through tubes formed by chains of cells called **sieve-tube members** (see Figure 31.11f). In contrast to the water-conducting cells of xylem, sieve-tube members are alive at functional maturity, although their protoplasts lack such organelles as the nucleus, ribosomes, and a distinct vacuole. The end walls between sieve-tube members, called **sieve plates,** have pores that presumably facilitate the flow of fluid from cell to cell along the sieve tube. Found in sieve-tube members of dicots (and some monocots) are long strands of P-protein ("P" for *phloem*). The function of these strands has not yet been resolved. Another abundant material in sieve-tube members is callose, a polysaccharide that probably plays some role in formation of the sieve plate.

Alongside each sieve-tube member is at least one **companion cell** (Figure 31.11f), which is connected to the sieve-tube member by numerous plasmodesmata. The nucleus and ribosomes of the companion cell may serve not only that cell but also the adjacent sieve-tube member, which has no nucleus or ribosomes of its own.

Plant Tissues

The cells of the plant are grouped into tissues. **Simple tissues** consist of a single cell type, such as parenchyma, collenchyma, or sclerenchyma (Figure 31.14a). Thus, we can speak of a parenchyma cell, but we can also refer to parenchyma tissue, an association of many parenchyma cells with a common function.

A plant tissue composed of more than one type of cell is called a **complex tissue** (Figure 31.14b). The two vascular tissues, xylem and phloem, are complex tissues. Xylem generally has sclerenchyma cells (fibers) and parenchyma cells in addition to the water-conducting tracheids and vessel elements. The fibers add support, and the parenchyma cells function in storage and localized transport of solutes, such as pumping salts into xylem vessels. Phloem also contains fibers, sclereids, and parenchyma cells, as well as sieve-tube members and companion cells.

Tissue Systems

Still a higher level of structure is the organization of the plant into three tissue systems: the dermal, vascular, and ground tissue systems. Each tissue system is continuous throughout the plant body, although the specific characteristics of the tissues and their spatial relationships to one another vary in different organs of the plant (Figure 31.15).

The **dermal tissue system,** or **epidermis,** is generally a single layer of tightly packed cells that covers and protects all young parts of the plant—the "skin" of the plant. In addition to the general function of protection, the epidermis has more specialized characteristics consistent with the function of the partic-

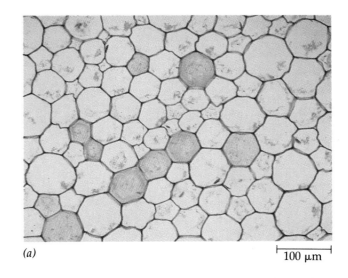

(a)   ⊢——— 100 μm ———⊣

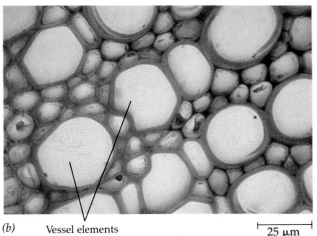

(b)   Vessel elements   ⊢——— 25 μm ———⊣

Figure 31.14
Simple and complex tissues. (a) A simple tissue, such as this parenchyma, consists of a single type of cell (LM). **(b)** Xylem is a complex tissue composed of more than one cell type, often including tracheids, vessel elements, fibers, and parenchyma (LM).

ular organ it covers. For example, the root hairs so important in the absorption of water and minerals are extensions of epidermal cells near the tips of roots. The epidermis of leaves and some stems secretes a waxy coating called the **cuticle** that helps the aerial parts of the plant retain water, an important adaptation to living on land. Roots, which must absorb water from the soil, generally lack cuticles.

The continuum of xylem and phloem throughout the plant forms the **vascular tissue system,** which functions in transport and support. The specific organization of vascular tissue in stems and roots is discussed in the next section.

The **ground tissue system** makes up the bulk of a young plant, filling the space between the dermal and vascular tissue systems. Ground tissue is predominately parenchyma, but collenchyma and scleren-

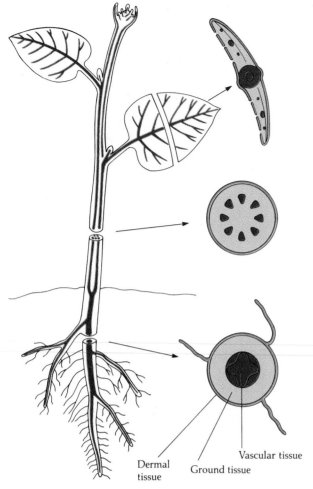

Figure 31.15
The three tissue systems. The dermal system, or epidermis, is a single layer of cells that covers the entire body of a young plant. The vascular tissue system is also continuous throughout the plant, but is arranged differently in each of the organs. The ground tissue system is located between the dermal tissue and vascular tissue in each plant organ.

Dermal tissue

Ground tissue

Vascular tissue

chyma are also commonly present. Among the diverse functions of ground tissue are photosynthesis, storage, and support.

Examination of how a plant grows will help us to understand how the tissue systems are organized in the different plant organs.

PRIMARY GROWTH

Indeterminate Growth

Most plants continue to grow as long as they live, a condition known as **indeterminate growth.** Plants have this capability because they possess tissues called **meristems** that remain embryonic as long as the plant lives. Most animals, in contrast, are characterized by **determinate growth;** that is, they cease growing after reaching a certain size. While whole plants may show indeterminate growth, certain plant organs, such as leaves or flower parts, exhibit determinate growth.

Indeterminate growth does not imply immortality. Some plants have finite life spans that are genetically programmed; such plants have a fixed longevity even when grown in constant, favorable conditions. Other plants have lifespans that are environmentally determined; that is, if the plants are grown under controlled temperature and light conditions and are protected from disease, they may live much longer than they typically do in natural environments. Plants known as **annuals** complete their life cycles—from germination through flowering and seed production to death—in a single year or growing season. A great diversity of wildflowers are annuals, as are the most important food crops, including the cereal grains and legumes. A plant is called a **biennial** if its life generally spans two years. Flowering usually occurs during the second year, after a year of vegetative growth. Beets and carrots are biennials (their lifespans are environmentally controlled), but we rarely leave them in the ground long enough to see them flower. Plants that live many years, including trees, shrubs, and some grasses, are known as **perennials.** Some of the buffalo grass of the North American plains is believed to have been growing for 10,000 years from seeds that sprouted at the close of the last ice age. Perhaps some perennials have the potential for immortality, if only they could escape disease, accidents such as floods and wildfires, and encroachment by humans.

As mentioned above, indeterminate growth in plants is possible because of meristems. Meristematic cells are unspecialized and divide to generate new cells near the growing points of the plant. **Apical meristems,** located at the tips of roots and in the buds of shoots, supply cells for the plant to grow in length. Growth initiated by the apical meristems is called **primary growth** (Figure 31.16). It produces the primary tissues, which are organized as the three tissue systems discussed earlier.

Primary Growth of Roots

Primary growth pushes roots through the soil. The root tip is covered by a thimblelike **root cap,** which protects the delicate meristem as the root elongates through the abrasive soil. The cap also secretes a polysaccharide slime that lubricates the soil ahead of the growing root.

Growth in length of a root is concentrated near its tip, where three zones of cells at successive stages of primary growth are located. From the root tip upward,

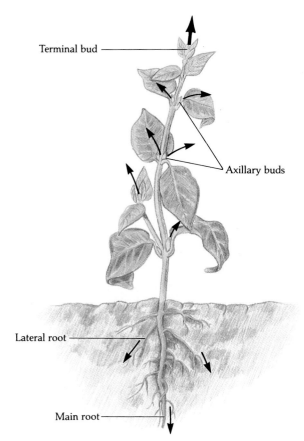

Figure 31.16
Primary growth of the plant. Shoots and roots grow in length mainly at their tips (arrows), where apical meristems produce new cells.

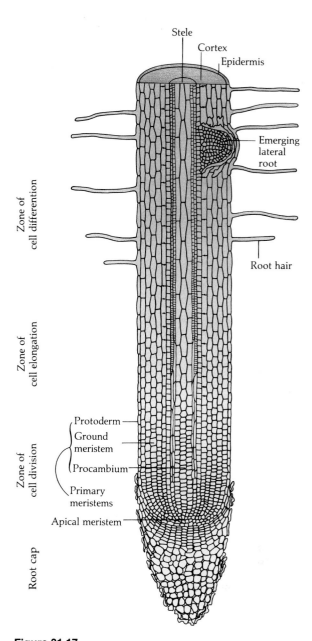

Figure 31.17
Primary growth of a root. Mitosis is concentrated in the zone of cell division, where the apical meristem and its products, the three primary meristems, are located. Most growth of cells takes place in the zone of elongation. Cells become functionally mature in the zone of differentiation. This zone is sometimes known as the root hair zone, as root hairs extend from epidermal cells there.

they are the **zone of cell division**, the **zone of cell elongation**, and the **zone of cell differentiation.** These regions grade together, with no sharp boundaries (Figure 31.17).

At the heart of the zone of cell division is the apical meristem, which produces the cells for primary growth and also replaces cells of the root cap that are abraded away. Just above the apical meristem, its products form three concentric cylinders of cells that continue to divide for some time. These are the primary meristems, the **protoderm, procambium,** and **ground meristem,** which will produce the three tissue systems of the root.

The zone of cell division blends into the zone of elongation. Here the cells elongate to more than ten times their original length. Although the meristem provides the new cells for growth, the elongation of cells is mainly responsible for pushing the root tip, including the meristem, ahead. The meristem sustains growth by continuously adding cells to the youngest end of the zone of elongation (Figure 31.18).

Even before they finish elongating, the cells of the root begin to specialize in structure and function where the zone of elongation grades into the zone of differentiation. In this latter zone of the root, the three tissue systems produced by primary growth complete their differentiation.

Primary Tissues of Roots   The protoderm, the outermost primary meristem, gives rise to the epidermis, a single layer of cells covering the root. Water and

Figure 31.18
Apical meristem of a root. "Hot spots" of cell division have been located here by providing a growing root with radioactive monomers of DNA, and then using autora-diography (see Chapter 2) to identify regions of active DNA synthesis. Most of the radioactive isotope has been accumulated by nuclei of dividing cells of the apical mer-istem. At the core of the apical meristem is a region of undividing cells called the quiescent center (LM).

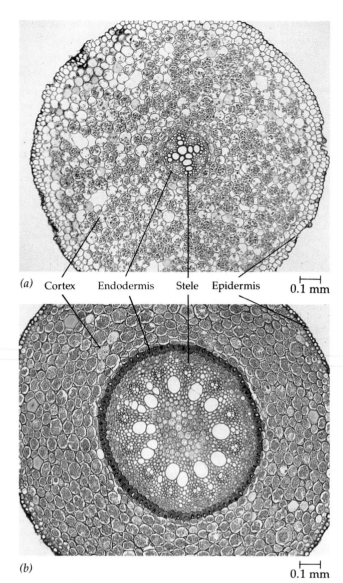

(a) Cortex Endodermis Stele Epidermis |⟶| 0.1 mm

(b)
|⟶| 0.1 mm

Figure 31.19
Organization of primary tissues in young roots.
(a) Cross section of a dicot root (*Ranunculus*) (LM). (b) Root of *Smilax*, a monocot. In contrast to the dicot root, the stele has a pith of parenchyma surrounded by the rest of the vascular tissue (LM).

minerals that enter the plant from the soil must cross the epidermis. The root hairs enhance this process by greatly increasing the surface area of epidermal cells.

The procambium forms a central vascular cylinder, or **stele**, where xylem and phloem develop (Figure 31.19). The specific arrangement of the two vascular tissues varies. In most dicots, the xylem cells radiate from the center of the stele in two or more spokes, with phloem developing in the wedges between the spokes. The stele of a monocot generally has a central core of parenchyma cells, often called the **pith**, which is ringed by vascular tissue with the same alternating pattern of xylem and phloem as in dicots.

Between the protoderm and procambium is the ground meristem, which gives rise to the ground tis-sue system. The ground tissue fills the **cortex**, the region of the root between the stele and epidermis. Mostly parenchyma, the ground tissue of the root stores food, and the membranes of the cells are active in the uptake of minerals that enter the root with the soil solution. The innermost layer of the cortex is the **endodermis**, a cylinder one cell thick that forms the boundary between the cortex and stele. The endodermis func-tions as a selective barrier that regulates the passage of substances from the soil solution into the vascular tissue of the stele. (The structure and function of endodermis is discussed in detail in Chapter 32.)

An established root may sprout **lateral roots**, also called **secondary roots**, which arise from the outer-most layer of the stele (Figure 31.20). Just inside the endodermis is the **pericycle**, a layer of cells that may become meristematic and begin dividing again. Orig-inating as a clump of cells formed by mitosis in the pericycle, the lateral root elongates and pushes through

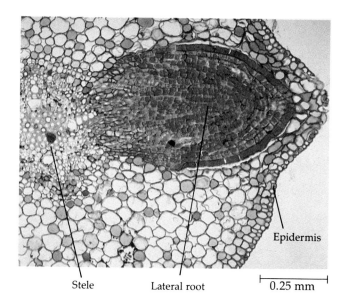

Figure 31.20
Formation of lateral roots. In this cross section of a root, a lateral root emerges from the pericycle, the outermost layer of the stele (LM).

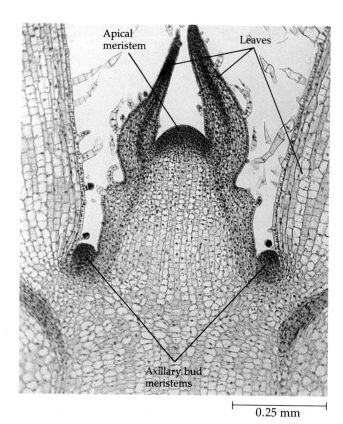

Figure 31.21
The terminal bud and primary growth of a shoot.
Leaf primordia arise from the flanks of the apical dome. The apical meristem gives rise to protoderm, ground meristem, and procambium, which in turn develop into the three tissue systems. Shown here is the shoot tip of *Coleus* (LM).

the cortex until it emerges from the primary root. The stele of the lateral root retains its connection with the stele of the primary root, making the vascular tissue continuous throughout the root system.

Primary Growth of Shoots

The apical meristem of a shoot is a dome-shaped mass of dividing cells at the tip of the terminal bud (Figure 31.21). As in the root, the apical meristem of the shoot tip forms the primary meristems—protoderm, procambium, and ground meristem—which will differentiate into the three tissue systems. Leaves arise as tiny bulges called leaf primordia on the flanks of the apical dome. Axillary buds develop from islands of meristematic cells left by the apical meristem at the bases of the leaf primordia.

Most of the growth in length of the shoot of a dicot is concentrated near the tip, where the cells derived from the apical meristem elongate. The shoots of many monocots, however, grow at each node along the stem because meristematic cells at the nodes continue to divide.

Primary Tissues of Stems Vascular tissue runs the length of a stem in several strands called **vascular bundles** (Figure 31.22). This contrasts with the root, where the vascular tissue forms a single stele consisting of the entire united set of vascular bundles. At the transition zone where the shoot grades into the root, the vascular bundles converge to join the root stele.

Each vascular bundle of the stem is surrounded by ground tissue. In most dicots, the vascular bundles are arranged in a ring, with a pith to the inside of the ring and a cortex external to the ring. Both pith and cortex are part of the ground tissue system. The vascular bundles have their xylem facing the pith and their phloem facing the cortex side. The pith and cortex are connected by pith rays, thin layers of ground tissue between the vascular bundles. In the stems of most monocots, the vascular bundles are arranged throughout the ground tissue rather than in a ring; thus it is not possible to distinguish pith and cortex regions.

The ground tissue of the stem is mostly parenchyma, but many stems are strengthened by collenchyma located just beneath the epidermis.

The protoderm of the terminal bud gives rise to the epidermis that covers stems and leaves as part of the continuous dermal tissue system (see Figure 31.15).

Tissue Organization of Leaves The leaf is cloaked by its epidermis, with cells tightly interlocked like pieces

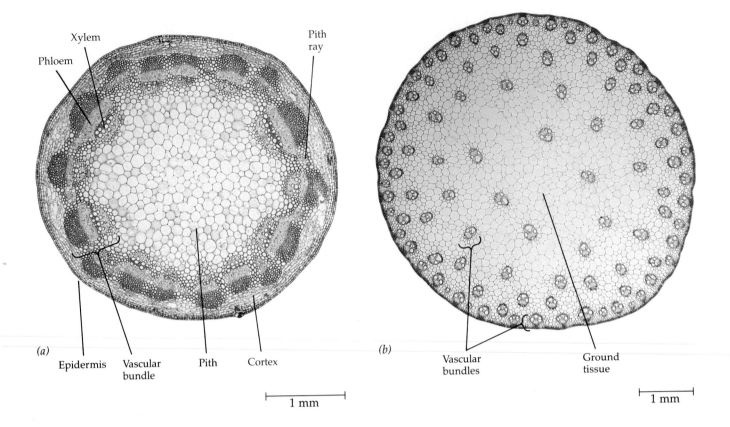

Xylem

Phloem

Pith ray

(a)

Epidermis

Vascular bundle

Pith

Cortex

(b)

Vascular bundles

Ground tissue

1 mm

1 mm

Figure 31.22
Organization of primary tissues in young stems. (a) Dicot stem (sunflower) with vascular bundles arranged in a ring. The ground tissue system consists of an outer cortex and an inner pith, surrounded by vascular bundles. Between the bundles are pith rays (LM). (b) Monocot stem (corn) with vascular bundles arranged in a complex manner throughout the ground tissue (LM).

Figure 31.23
Anatomy of a leaf.

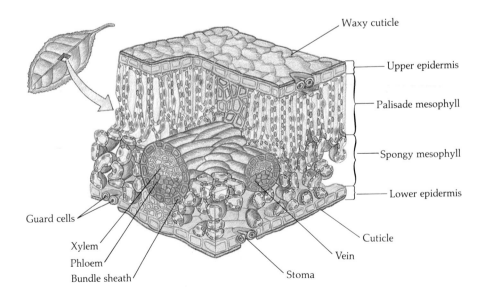

Waxy cuticle

Upper epidermis

Palisade mesophyll

Spongy mesophyll

Lower epidermis

Cuticle

Vein

Stoma

Bundle sheath

Phloem

Xylem

Guard cells

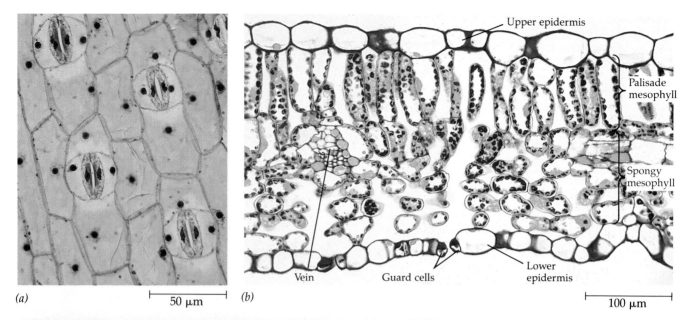

(a)   *(b)*

50 μm   100 μm

Upper epidermis

Palisade mesophyll

Spongy mesophyll

Vein   Guard cells   Lower epidermis

(c)

Figure 31.24
Leaf architecture. (a) This surface view of a *Tradescantia* leaf shows the cells of the epidermis and stomata, with their guard cells (LM). **(b)** Palisade and spongy regions of mesophyll are present within the leaf of a lilac, a dicot (LM). **(c)** An extensive network of venation ramifies throughout the leaf.

of a puzzle (Figure 31.23). This epidermis, like our own skin, is a first line of protection against physical damage and pathogenic organisms. Its waxy cuticle makes the epidermis also a barrier to the loss of water from the plant. The epidermal barrier is interrupted only by the **stomata,** tiny pores flanked by specialized epidermal cells called **guard cells** (Figure 31.24a). Each stoma is actually a gap between a pair of guard cells. The stomata allow gas exchange between the surrounding air and the photosynthetic cells inside the leaf, but they are also the major avenues for the loss of water from the plant by evaporation, a process called **transpiration.** The role of the guard cells in regulating gas exchange and transpiration is discussed in Chapter 32.

The ground tissue of a leaf, sandwiched between the upper and lower epidermis, is called **mesophyll** (from the Greek *mesos,* "middle," and *phyll,* "leaf"). It consists mainly of parenchyma cells equipped with chloroplasts and specialized for photosynthesis. The leaves of many dicots have two distinct regions of mesophyll (Figure 31.24b). On the upper half of the leaf is the palisade mesophyll, made up of cells that are columnar in shape. Below the palisade layer is the spongy mesophyll, which gets its name from the labyrinth of air spaces through which carbon dioxide and oxygen circulate around the irregularly shaped cells. The air spaces are particularly large in the vicinity of stomata, where gas exchange with the outside air occurs.

Figure 31.25
Palms, trees without secondary growth.
Only primary growth occurs in palms and most other monocots. The tall trunk of the palm is uniformly thick because most growth in girth occurs when the primary tissues develop from an apical meristem that trails downward for some distance below treetop. The base of the trunk fans out because of layers of adventitious roots that help support the tree.

The vascular tissue of the leaf is continuous with the xylem and phloem of the stem by a **leaf trace,** a strand of vascular tissue that branches at a node from a vascular bundle in the stem. The leaf trace continues in the petiole as a vein, which then branches repeatedly within the blade of the leaf to ramify throughout the mesophyll. This brings xylem and phloem into intimate contact with the photosynthetic tissue (Figure 31.24c). The vascular infrastructure also functions as a skeleton that supports the parenchyma of the leaf. Small veins of the leaf are ringed by bundle sheaths of tightly packed parenchyma cells. At their extremities, veins end with single tracheids that deliver water transported all the way up from the roots within uninterrupted chains of xylem cells.

SECONDARY GROWTH

The stems and roots of most vascular plants grow in girth as well as length. This increase in diameter is called **secondary growth.** After the apical meristem produces the primary tissues in a young region of a stem or root, two new layers of meristematic tissue develop. These two **lateral meristems** are the **vascular cambium** and the **cork cambium.** Secondary growth, which produces the secondary tissues described below, results from cell division in these two layers of cells. The formation of secondary tissues not only increases the thickness of a stem or a root, but also changes its structure.

Secondary growth does not occur in most mono-cots, which increase in size solely by primary growth (Figure 31.25). Lateral meristems and secondary growth do occur, however, in a few groups of monocots such as Yucca trees, which grow in parts of the deserts of the southwestern United States.

Secondary Growth of Stems

Vascular Cambium In the stem of a plant capable of secondary growth, cells between the primary xylem and primary phloem of each vascular bundle become meristematic. This meristem eventually forms the vascular cambium, which adds new vascular tissue to the stem (Figure 31.26). The cells of the vascular cambium divide radially, producing new cells both *internal* to itself, which differentiate into **secondary xylem,** and cells *external* to itself, which differentiate into **secondary phloem.** The vascular cambium spreads into the parenchyma tissue between the vascular bundles, eventually producing a continuous cylinder of meristematic tissue surrounding the xylem and pith. Some cells in the vascular cambium produce radial files called **rays,** which consist mostly of parenchyma cells (Figure 31.27). The rays mainly function in the lateral transport of water and nutrients and in the storage of such substances as starch.

As secondary growth continues over the years, layer upon layer of secondary xylem accumulates to produce what we call wood. Wood consists mainly of tracheids, vessel elements (in dicots), and fibers. These cells, dead at functional maturity, have thick, lignified walls that give wood its hardness and strength. In temperate regions of the Earth, secondary growth in

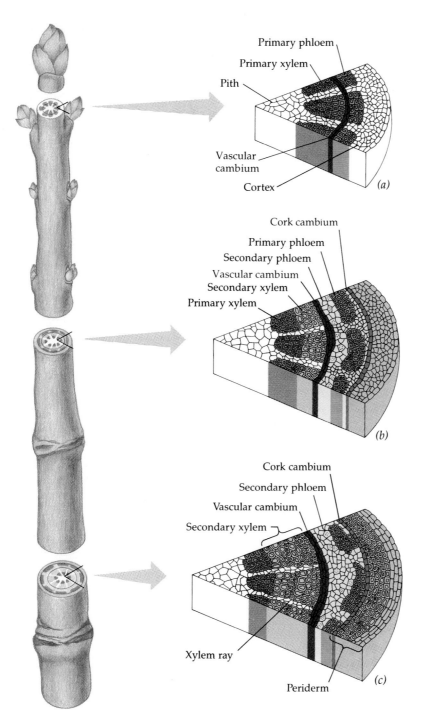

Figure 31.26
Secondary growth of stems. A stem grows in diameter by cell division in two lateral meristems, the vascular cambium and the cork cambium. The vascular cambium arises initially from the procambium, which lies between the primary xylem and the primary phloem. Vascular cambium produces secondary xylem to its inside and secondary phloem to its outside. Each year a new layer, or growth ring, of secondary xylem is added outside the secondary xylem produced in previous years. Wood consists of this accumulation of xylem. The cork cambium forms a new surface tissue, the periderm, that replaces the epidermis produced by primary growth. Cork cambium produces cork cells to its exterior. When functionally mature, the cork cells are dead and have waxy walls. Each year, as the outer layers of cork are sloughed away, the cork cambium forms in deeper and deeper tissues. These cross sections through progressively older regions of a twig show the organization of tissues at the completion of primary growth (top), during the first season of secondary growth (middle), and during the second year of secondary growth (bottom).

perennial plants is interrupted each year when the vascular cambium becomes dormant during winter. When secondary growth resumes in the spring, the first secondary xylem cells to develop usually have relatively large diameters and thin walls compared to the secondary xylem cells that are produced later in the summer. Thus, it is usually possible to distinguish spring wood from summer wood (see Figure 31.27). The annual growth rings that are evident in a cross section of most tree trunks result from this yearly activity of the vascular cambium: cambium dormancy,

spring wood production, and summer wood production. The boundary between one year's growth and the next is usually quite conspicuous (see Figure 31.27), and this sometimes allows us to estimate the age of a tree by counting its annual rings.

The secondary phloem, external to the vascular cambium, does not accumulate over the years as the secondary xylem does. As a tree grows in girth, the older (outermost) secondary phloem, and all tissues external to it, develop into bark, which eventually splits and sloughs off the tree trunk.

Figure 31.27

The woody dicot stem. Several years of secondary growth are apparent in this cross section of a stem from *Tilia*, the American linden (basswood). At the boundary of one season's growth to the next, note the spring wood and summer wood. Ray cells grow horizontally, extending through the secondary xylem. The bark consists of all tissues external to the vascular cambium (LM).

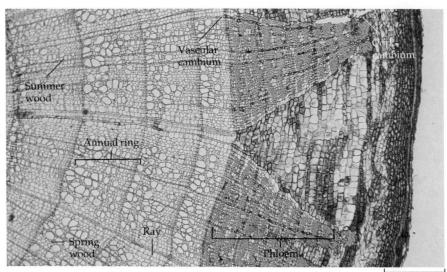

0.25 mm

Cork Cambium During secondary growth, the epidermis produced by primary growth splits, dries, and falls off the stem. It is replaced by new protective tissues produced by the cork cambium, a cylinder of meristematic tissue that first forms in the outer cortex of the stem (see Figure 31.27). To its outside, the cork cambium produces cells that differentiate into cork cells, which are dead at functional maturity and have walls impregnated with a waxy substance called suberin. To its inside, the cork cambium may produce a parenchyma tissue called phelloderm. Together, the cork cambium, cork, and phelloderm make up the **periderm,** the protective coat that replaces the epidermis. As continued secondary growth splits the periderm, it is replaced again and again by new cork cambium that forms deeper and deeper in the cortex. Eventually, no cortex is left, and the cork cambium then develops from parenchyma cells in the secondary phloem. Only the youngest secondary phloem, which is internal to the cork cambium, functions in sugar transport. The older secondary phloem, outside the cork cambium, dies and helps protect the stem until it is sloughed off as part of the bark during later seasons of secondary growth. The term *bark* refers to all tissues external to the vascular cambium. Spongy regions in the bark called **lenticels** make it possible for living cells within the trunk to exchange gases with the outside air for cellular respiration.

The result of many years of secondary stem growth can be seen by examining an old tree trunk in cross section (Figure 31.28). In many old trees, we can distinguish two zones of wood: the **heartwood** and **sapwood.** The darker heartwood of the trunk consists of older layers of secondary xylem. These cells are nonfunctional for water transport because they are clogged

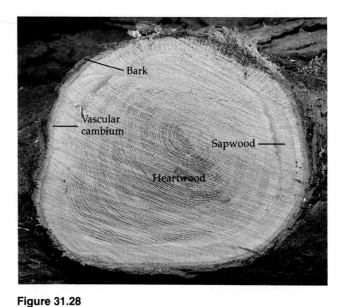

Figure 31.28

Anatomy of tree trunk. The vascular cambium is the dividing line between the wood—the secondary xylem, older and older extending toward the center of the trunk—and the bark—everything external to the vascular cambium. Notice that the bulk of a large tree is dead tissue. The only living tissues in the trunk and large branches are the youngest secondary phloem, differentiating xylem cells, xylem rays, vascular cambium, and cork cambium. The older, darker secondary xylem in the center of the trunk is called heartwood. Its cells are clogged with resins and other by-products of the plant's metabolism. The sapwood is the younger, outer secondary xylem, which still conducts water.

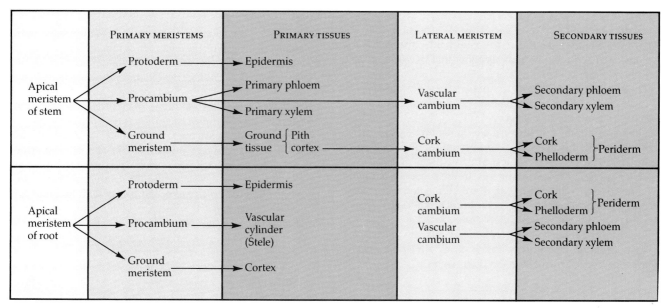

Figure 31.29
Summary of primary and secondary growth of a woody dicot.

with resins and other metabolic by-products. The lighter-colored sapwood consists mainly of younger secondary xylem, vascular cambium, young secondary phloem, and cork cambium. Unlike the heartwood, the vascular tissue of the sapwood conducts water and nutrients.

Secondary Growth of Roots

The two lateral meristems, vascular cambium and cork cambium, function also in secondary growth of roots. The vascular cambium, first located between the xylem and phloem of the stele, produces secondary xylem to its inside and secondary phloem to its outside. As the stele grows in diameter, the cortex and epidermis are split and shed. A cork cambium forms from the pericycle of the stele and produces the periderm, which becomes the secondary dermal tissue.

Over the years, the root becomes more woody, and annual rings are usually evident in the secondary xylem. The tissues external to the vascular cambium form a thick, tough bark. After extensive secondary growth, old stems and old roots are quite similar.

Figure 31.29 summarizes the development of primary and secondary tissues in roots and stems.

* * *

In dissecting the plant to examine its parts, as we have done in this chapter, we must remember that the whole plant functions as an integrated organism, not as an aggregate of independent cells, tissues, or organs. In the following chapters, you will learn more about how materials are transported within the plant, how plants obtain nutrients, how plants reproduce and develop, and how the various functions of the plant are coordinated. Your understanding of the working plant will be enhanced by remembering that structure fits function and that interactions with the environment affect the anatomy and physiology of plants.

STUDY OUTLINE

1. Angiosperms (flowering plants) are the most diverse and widespread of plant divisions. Based on differences in anatomy, the angiosperms can be divided into two classes: monocots and dicots.

The Parts of a Flowering Plant (pp. 682–686)

1. The differentiation of the plant body into an underground root system and an aerial shoot system is an adaptation to terrestrial life. It provides water and nutrients from the soil to a photosynthetic apparatus requiring light and carbon dioxide.

2. Vascular tissues integrate the parts of the plant body. Water and minerals move up from the roots in the xylem, and synthesized or stored food molecules travel to nonphotosynthetic parts in the phloem.

3. The structure of roots is adapted to anchor the plant,

absorb and conduct water and minerals, and store food. Tiny root hairs near the root tips enhance absorption. Some plants have adventitious roots that may function to help support stems.

4. The shoot system consists of stems, leaves, and flowers, the latter of which are organs of reproduction constructed from modified leaves.

5. Leaves are attached by their petioles to the nodes of stems. The growth of axillary buds, embryonic shoots formed in the angle between the leaf and the stem, is kept in check partly by apical dominance in the presence of terminal buds. Axillary buds that are stimulated to grow may become flowers or vegetative branches. Stolons, rhizomes, and bulbs are modified stems.

6. Leaves, the primary photosynthetic organs, show extensive variation. Monocots differ from dicots in the arrangement of major veins in their leaves.

Plant Cells and Tissues (pp. 686–692)

1. Plant cells grow mainly by taking up water into the vacuole as the cell wall loosens and extends. As growth stops, the cell wall is fortified with additional material. This rapid, economical extension of shoots and roots increases the immobile plant's exposure to light, minerals, and water.

2. Parenchyma cells are the least specialized plant cells, serving general metabolic, synthetic, and storage functions. They retain the ability to divide and differentiate into other cell types under certain conditions

4. Collenchyma cells often occur in strands or cylinders that support young parts of the plant shoot without restraining growth. They lack secondary walls and have protoplasts, which allows them to elongate with growing stems and leaves.

5. Sclerenchyma cells have thick, lignified secondary walls and lack protoplasts; thus, at maturity they are unable to elongate. Both fiber and sclereid types occur, acting as scaffolding for the plant.

6. Water-conducting xylem tissue is composed of elongated tracheid and vessel element cells that are dead at functional maturity. Tracheids are long, thin, tapered cells with lignified secondary walls that function in support and permit water flow through pits. The wider, shorter, and thinner-walled vessel elements have perforated ends through which water flows freely.

7. Sieve-tube members are living cells that form phloem tubes for transport of sucrose and other organic nutrients. These cells lack nuclei and ribosomes and possess porous end walls called sieve plates. Each sieve-tube member is connected to one or more companion cells through plasmodesmata.

8. Plant cells are grouped into simple tissues consisting of a single cell type or into complex tissues, such as xylem and phloem, made of two or more cell types.

9. Tissues, in turn, are arranged into three continuous systems. The dermal tissue system, or epidermis, is an external layer of tightly packed cells that functions in protection, formation of root hairs, and secretion of a waxy cuticle that helps the plant retain water. The vascular tissue system provides transport and support. The predominantly parenchymous ground tissue system functions in organic synthesis, storage, and support.

Primary Growth (pp. 692–698)

1. Because they possess permanently embryonic meristems, plants, unlike animals, show indeterminate growth.

2. Apical meristems at root tips and shoot buds initiate primary growth and the formation of the three tissue systems.

3. Root tips, protected by the root caps, grow by activity of cells in the successive zones of cell division, cell elongation, and cell differentiation.

4. Just above the apical meristem in the zone of cell division are the three primary meristems of the root. The protoderm gives rise to the epidermis; the procambium forms the central vascular stele; and the ground meristem produces the ground tissue of the cortex with its inner cylinder of endodermis. Subsequent lateral roots arise from the pericycle of the stele.

5. The primary growth of shoots comes from the dome-shaped apical meristem at the top of the terminal bud. Leaves arise from the sides of the apical dome from leaf primordia, and axillary buds from residual islands of meristematic cells at the bases of leaf primordia.

6. In contrast to the single stele of the root, the vascular tissue of stems runs in vascular bundles surrounded by ground tissue in characteristic patterns that differ between monocots and dicots. Ground tissue is mostly parenchyma, but it may be strengthened by collenchyma just beneath the epidermis. Protoderm gives rise to the epidermis.

7. Leaves are covered with a waxy epidermis. Pairs of guard cells flank openings called stomata, through which gas exchange and transpiration occur. Between the upper and lower epidermis, the ground tissue, or mesophyll, consists mainly of parenchyma cells equipped with chloroplasts for photosynthesis. A strand of vascular tissue called the leaf trace connects the veins of the leaf with the vascular tissue of the stem.

Secondary Growth (pp. 698–701)

1. The increase in girth of stems and roots is due to secondary production of new cells by the vascular cambium and the cork cambium, two lateral meristems.

2. The vascular cambium, a continuous cylinder of meristematic cells arising from parenchyma cells located between the xylem and phloem of each vascular bundle, produces secondary xylem internally and secondary phloem externally. The secondary xylem, known as wood, is traversed by rays of parenchyma cells, produced as living channels functioning in lateral transport. In temperate regions, seasonal wood production cycles result in annual growth rings. The external secondary phloem eventually splits and sloughs off during growth.

3. The cork cambium, a meristematic cylinder in the outer cortex of the stem, produces waxy cork cells externally and phelloderm internally. The cork cambium, cork, and phelloderm make up the periderm, which replaces the epidermis that sloughs off during secondary growth. Secondary phloem gives rise to new cork cambium after the original cortex is shed.

4. In roots, the vascular cambium arises between the xylem and phloem of the stele and functions similarly to that in stems. Cork cambium, produced from the pericycle of the stele, forms the periderm that replaces cortex and epidermis. Old roots resemble old stems, even in possessing annual growth rings.

1. Most absorption of water and dissolved minerals occurs through

   a. adventitious roots

   b. stolons

   c. taproots

   d. root caps

   e. root hairs

2. Which of the following is *not* a correctly stated difference between monocots and dicots?

   a. parallel veins in monocots; branching, netlike venation in dicot leaves

   b. vascular bundles scattered in monocot stem; central vascular stele in dicot stem

   c. flower parts in threes in monocots; flower parts in multiples of four or fives in dicots

   d. usually only primary growth in monocots; secondary growth in many dicots

   e. one cotyledon in monocots; two cotyledons in dicot seeds

3. The lateral roots of a young dicot originate from the

   a. pericycle of the taproot

   b. endodermis of fibrous roots

   c. meristematic cells of the protoderm

   d. vascular cambium

   e. root cortex

4. A sieve-tube member would likely lose its nucleus in which zone of growth in a root?

   a. zone of cell division

   b. zone of cell elongation

   c. zone of cell differentiation

   d. zone of cell proliferation

   e. none of the above; the functionally mature cell retains its nucleus

5. Tracheids of the primary plant body originate from the

   a. protoderm

   b. procambium

   c. ground meristem

   d. xylem rays

   e. cork cambium

6. Unlike primary growth, secondary growth in both roots and stems

   a. is indeterminate

   b. produces xylem and phloem

   c. involves vascular and cork cambiums

   d. results in a rapid increase in root or stem length

   e. is a function of meristematic tissue

7. Which of the following is *not* part of an older tree's bark?

   a. cork

   b. cork cambium

   c. phelloderm

   d. secondary xylem

   e. secondary phloem

8. When a leaf trace extends from a vascular bundle in a dicot stem, what would be the arrangement of vascular tissues in the veins of the leaf?

   a. The xylem would be on top and the phloem on the bottom.

   b. The phloem would be on top and the xylem on the bottom.

   c. The xylem would encircle the phloem.

   d. The phloem would encircle the xylem.

   e. There is no way to determine this.

9. Which of the following cell types or structures is incorrectly paired with its meristematic tissue?

   a. epidermis—protoderm

   b. stele—procambium

   c. cortex—ground meristem

   d. secondary phloem—cork cambium

   e. three primary meristems—apical meristem

10. One important difference between the anatomy and physiology of roots and leaves is that

    a. unlike roots, the xylem vessels in leaves normally transport sap in both directions

    b. a waxy cuticle covers most leaves but is absent in roots

    c. leaves, but not roots, have ground tissue consisting mainly of parenchyma cells

    d. phloem always conducts sap away from leaves and toward roots

    e. the root has an epidermis, while the leaf has only an endodermis

CHALLENGE QUESTIONS

1. If you were to live for the next several years in a treehouse built on the large, lower branches of a tree, would you gain altitude as the tree grew? Explain your answer.

2. Considering the difference between how plants and animals grow, explain why most fruits and vegetables have fewer calories than meat per gram of fresh weight.

3. Examine leaves of several plants and classify them according to the criteria introduced in Figure 31.7.

FURTHER READING

1. "The Trees Told Him So." *Science News*, September 7, 1985. Growth rings may give clues to volcanic eruptions.
2. Bolz, D. M. "A World of Leaves: Familiar Forms and Surprising Twists." *Smithsonian*, April 1985. A delightful article on the adaptations of leaves, featuring exquisite photographs.
3. Galston, A. W., P. J. Davies, and R. L. Satter. *The Life of the Green Plant*. 3d ed. Englewood Cliffs, NJ: Prentice Hall, 1980. A compact introduction to plant form and function.
4. Raven, P. H., R. F. Evert, and S. E. Eichorn. *Biology of Plants*. 4th ed. New York: Worth, 1986. A profusely illustrated, comprehensive text on plants.

32

Transport in Plants

The aquatic ancestors of plants were completely surrounded by water and dissolved minerals, and every cell had direct access to those resources. Differentiation of the plant body into roots and leaf-bearing shoots is one adaptation that paralleled the colonization of land, but with this change came the problem of transporting substances between the separated organs (Figure 32.1). Sap flows throughout the plant in a vascular system structurally specialized for its function in internal transport. Water and minerals absorbed by roots are drawn upward in xylem vessels to shoots. Sugar produced by photosynthesis is exported from leaves to other parts of the plant via the sieve tubes of phloem. The whole plant depends on this interorgan commerce to integrate the activities of its specialized parts.

Transport in plants occurs on three levels: (1) the uptake of water and solutes by individual cells, (2) the short-distance transport of substances from cell to cell, and (3) the long-distance transport of sap in xylem and phloem. In this chapter, you will learn how these transport processes combine to move water and solutes within the plant. (See Chapter 8 for a review of the general principles of active and passive transport.)

ABSORPTION OF WATER AND MINERALS BY ROOTS

Water and mineral salts enter the plant through the epidermis of roots, cross the root cortex, pass into the stele (vascular cylinder), and then flow up xylem vessels to the shoot system.

Figure 32.1
The problem of long-distance transport in plants. These redwoods illustrate the spatial separation of leaves and roots and the resulting problem of long-distance transport. The problem is solved by vascular tissues specialized in structure and function for transporting materials between the parts of the plant.

The soil solution soaks into the hydrophilic walls of epidermal cells and their extensions, the root hairs that adhere to soil particles. Walls of adjacent plant cells touch, and soil solution wetting the epidermis can seep freely along the matrix of walls into the root cortex, just as water soaks along a paper towel. This exposes all of the parenchyma cells in the cortex to the soil solution, providing a much greater surface area of plasma membrane for the uptake of water and minerals into cytoplasm than the surface area of the epidermis alone.

Active Accumulation of Mineral Ions

An essential mineral present in the soil solution in high concentration may enter cells of the root cortex by diffusing across plasma membranes down a concentration gradient, but plants do not rely on diffusion alone to absorb most minerals. The soil solution is usually very dilute, and roots can accumulate essential minerals to concentrations that are hundreds of times greater than the concentrations of these minerals in

soil. This is accomplished by active transport, the pumping of mineral ions across membranes against gradients. Certain mineral ions enter the cells of the root through specific carrier proteins embedded in the plasma membrane. One carrier transports potassium (K^+), and it is able to distinguish between this ion and the similar ion sodium (Na^+), much as the active site of an enzyme can recognize its substrate. This specificity in active transport enables the plant to selectively extract K^+, an essential nutrient, from a soil solution in which Na^+ might be more concentrated than K^+.

Actively absorbing minerals against gradients is work that requires plant cells to expend energy in the form of ATP. Much experimental evidence shows that plants use cellular respiration to power the accumulation of minerals. For instance, mineral uptake can be depressed by removing oxygen, adding poisons that inhibit respiration, or lowering the temperature. In some cases, carrier proteins responsible for active transport may be powered directly with ATP generated by respiration. There is evidence, for instance, that K^+ uptake by the cell requires the hydrolysis of ATP.

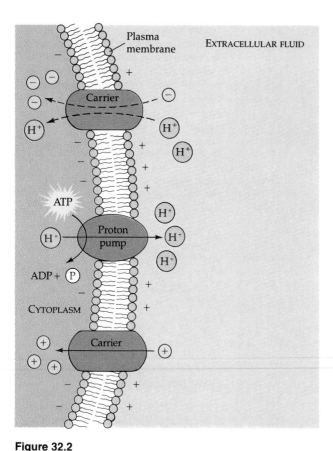

Figure 32.2
Active absorption of minerals. Proton (H⁺) pumps of
plant cell membranes can drive the accumulation of various cations and anions indirectly by generating a membrane potential (voltage) and a proton gradient, making
the inside of the cell negative and higher in pH than the
outside. Cations have a tendency to enter the cell due to
the membrane potential, but usually diffuse through a
specific carrier. Anions (such as NO_3^-) may enter the cell
in association with protons via carriers that are specialized for cotransporting the two ions together into the cell
(see Chapter 8).

One membrane pump that may affect the transport
of many different minerals simultaneously is the **proton pump,** a general feature of plant cell membranes
that functions in a variety of processes. The proton
pump consumes ATP to force hydrogen ions (H⁺) out
of the cell, a process that generates a voltage across
the plasma membrane because positive charge is being
translocated from the cytoplasm to the surrounding
solution (Figure 32.2). The resulting voltage, or membrane potential (with the inside of the cell more negative in charge than the outside) and the pH gradient
(with a higher proton concentration on the outside)
together help to drive selective cation and anion transport across the plasma membrane. Scientists have much
more to learn about how the proton pump works and
how its activity is coupled to the transport of other
substances before its role in mineral uptake can be
fully understood.

Entry of Water and Minerals into Xylem

Two routes are available for the transit of water and
minerals across the root cortex from the epidermis to
the stele, where xylem is located (Figure 32.3). One
pathway is the **symplast,** the living continuum of cytoplasm connected by plasmodesmata between cells
(Figure 32.4). Ions and water absorbed by parenchyma
cells of the root cortex may move through the symplast all the way into the stele through plasmodesmata
of the endodermis, the innermost layer of the cortex
at its boundary with the stele. However, much of the
water and some of the ions that soak into the root
remain in the walls of the cortex cells and move toward
the stele along an extracellular pathway called the apo-
plast. At the endodermis, this extracellular avenue is
blocked. Each cell of the endodermis has in its wall a
belt made of a waxy material, suberin. This ring of
wax around the cell, called the **Casparian strip,** is
impervious to water and dissolved minerals, thus
blocking the apoplast at the boundary between the
cortex and stele. Water and minerals that make it to
the endodermis along the apoplast can enter the stele
only by crossing the plasma membrane of an endodermal cell and passing through plasmodesmata.

The endodermis, with its Casparian strip, ensures
that no substance can reach the vascular tissue of the
stele without having to cross at least one membrane.
The Casparian strip also keeps ions accumulated in
the vascular tissue from leaking back out of the roots
and into the soil. Since transport across membranes
is selective, the endodermis is important in the ability
of the root to preferentially transport certain minerals
from the soil into the xylem and up to the shoot. The
key to this function is the structure of the endodermis
and its strategic location in the root.

Within the stele are **transfer cells,** which are specialized for the transfer of solutes between the symplast and apoplast. The water-conducting tracheids
and vessels of the xylem lack protoplasts, and thus
the lumen of these cells, as well as their walls, are part
of apoplast. Water and minerals, however, enter the
stele via the symplast. A transfer cell selectively pumps
minerals out of its cytoplasm and into its wall; water
and minerals can then flow freely into tracheids and
vessel elements. Transfer cells have numerous
ingrowths of their walls and associated infoldings of
their membranes, which increases the surface areas
of the cells and enhances the transfer of substances.
Mitochondria are abundant in transfer cells, which
require much energy in the form of ATP to drive the
active transport of solutes. Transfer cells are not only
found in xylem but are also present in phloem, where
there also is extensive traffic of solutes between symplast and apoplast.

The water and minerals we have tracked from the
soil to the xylem can now be transported upward as
xylem sap to the shoot system.

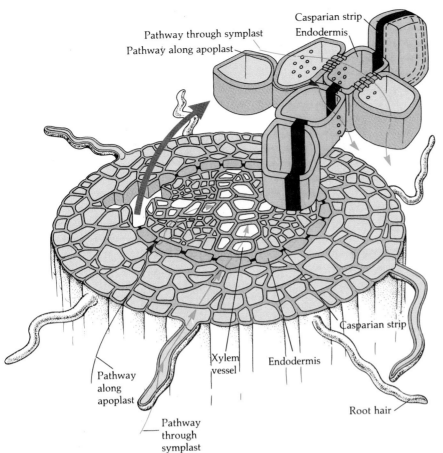

Figure 32.3
Pathways of mineral transport in roots. Minerals are absorbed with the soil solution by the root surface, especially by root hairs, which are extensions of epidermal cells. To reach the vascular cylinder (stele), minerals and water must cross the root cortex. The solution may move either along the apoplast—the matrix of cell walls—or through the symplast—the living continuum of protoplasts connected by plasmodesmata. At the endodermis, any minerals remaining in the apoplast are blocked from entering the vascular cylinder by the Casparian strip and must cross the selective membrane of an endodermal cell before being transported up into the shoot system within xylem.

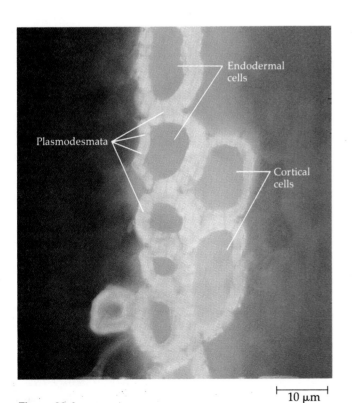

Figure 32.4
Plasmodesmata of the root endodermis. Cell walls are stained with a fluorescent dye. Plasmodesmata, cytoplasmic channels through cell walls, link adjacent cells into a living continuum called the symplast (LM).

ASCENT OF XYLEM SAP

Plants lose an astonishing amount of water by **transpiration,** the evaporation of water from leaves and other aerial parts of the plant. An average-sized maple tree, for instance, loses more than 200 liters of water per hour during the summer. Unless the transpired water is replaced by water transported up from the roots in xylem, leaves wilt and eventually die. The flow of xylem sap upward in the plant also brings mineral nutrients to the shoot system.

Xylem sap rises against gravity, without the help of any mechanical pump, to reach heights of more than 100 m in the tallest trees. To understand how the plant accomplishes this feat, we must examine the physical forces that cause water to move.

Water Potential

In predicting which direction water will flow from one part of a plant to another, the most useful measurement is the quantity known as **water potential,** abbreviated by the Greek letter *psi* (ψ). The most important thing for us to know about water potential is that water always flows passively from a location with a greater

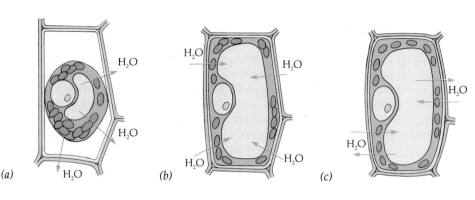

Figure 32.5
Water relations of plant cells. (a) In a hyperosmotic environment, the cell has a greater water potential than its surroundings. The cell loses water and plasmolyzes. (b) In a hypoosmotic environment, the cell initially has a lower water potential than its surroundings. There is a net uptake of water by osmosis, causing the cell to become turgid. (c) When this tendency for water to enter is offset by the back pressure of the elastic wall, water potentials are equal for cell and surroundings.

water potential to a location where the water potential is less. Water potential, then, measures the tendency for water to leave one place in favor of another place. Plant physiologists have traditionally expressed water potential in units of pressure. In the past, the unit most commonly used for water potential was the bar, the pressure required to push a column of water upward about 10 m. (A bar is about the same as one atmosphere of pressure. You may be familiar with this measurement on a *barometer*.) Today, most plant physiologists prefer **megapascals** (MPa) as pressure units. One MPa is equal to 10 bars.

For purposes of comparison, the water potential of pure water in a container open to the atmosphere is defined as zero ($\psi = 0$ MPa). The two factors that are most important in changing the water potential are the addition of solutes and the existence of a pressure, either positive or negative. The addition of solutes lowers the water potential. If ψ is 0 MPa for pure water, then any solution has a negative water potential. For instance, a 0.1 molar solution of any solute has a water potential of -0.23 MPa. If a solution is separated from pure water by a selectively permeable membrane, water will flow by osmosis into the solution, moving from where water potential is greater (0) to where it is less (negative). The tendency for the solution to take up water can be measured in an osmometer (See Chapter 8 for a review of osmosis and for an example of how one type of osmometer works.)

Contrary to the effect of solutes, pressure increases water potential, as we can see with an osmometer. The flow of water into an osmometer containing a solution can be prevented by using a piston to push down just hard enough to keep the solution from rising in the tube (see Figure 8.14). For a 0.1 M solution ($\psi = -0.23$ MPa), we must apply a pressure of $+0.23$ MPa. The solution now has a water potential of 0, because the pressure exactly counteracts the effect of the solutes. Since ψ on both sides of the membrane is now equal, there will be no net movement of water.

Let us now apply what we have learned about water potential to the uptake and loss of water by plant cells

(Figure 32.5). First, imagine the cell bathed in a solution more concentrated (hyperosmotic) than the cell itself. Since the external solution has the lower water potential, water will leave the cell by osmosis and the cell will plasmolyze, meaning that the protoplast pulls away from the wall. Now let us place a plant cell in pure water ($\psi = 0$). The cell has a lower water potential because of the presence of solutes, and water enters the cell by osmosis. The cell begins to swell and push against the wall to produce a turgor pressure. The partially elastic wall pushes back against the turgid cell. When this wall pressure is great enough to offset the tendency for water to enter because of the solutes in the cell, then ψ for the cell and its environment are equal. A dynamic equilibrium has been reached, and there is no further net movement of water, although a brisk, equal exchange of water across the membrane continues.

Differences in water potential not only govern uptake and loss of water by individual cells but, as we will now see, also account for long-distance transport of water within the plant.

Root Pressure

The stele of a root behaves something like an osmometer. At night, when transpiration is very low or zero, the root cells are still expending energy to pump mineral ions into the xylem. The endodermis surrounding the stele of the root helps to prevent the leakage of these ions back out of the stele. Accumulation of minerals in the stele lowers its water potential, and water flows in, forcing fluid up the xylem. This upward push within the stele is called **root pressure.**

Root pressure causes **guttation,** the exudation of water droplets that can be seen in the morning on tips of grass blades or the leaf margins of some herbaceous dicots (Figure 32.6). During the night, when the rate of transpiration is low, the roots of some plants keep accumulating minerals, and root pressure pushes xylem

Figure 32.6
Guttation. Although root pressure can force water droplets from some small plants under certain conditions, it plays only a minor role in the ascent of xylem sap in most cases.

sap into the shoot system. More water enters leaves than is transpired, and the excess is forced through specialized structures called hydathodes, which function as escape valves.

In most plants, root pressure is not the major mechanism driving the ascent of xylem sap. At most, root pressure can force water upward only a few meters, and many plants, including some of the tallest trees, generate no root pressure at all. Even in most small plants that display guttation, root pressure cannot keep pace with transpiration after sunrise. For the most part, xylem sap is not pushed from below by root pressure but pulled upward by the leaves themselves.

The Transpiration-Cohesion-Adhesion Theory

If we want to move an object upward, it is intuitive that we can either push it from below or pull it from above. However, it is not so intuitive that something as seemingly fluid as water could be pulled up a pipe. Nevertheless, that is what happens in xylem vessels of plants. According to the most widely accepted explanation for the ascent of xylem sap, transpiration provides the pull, and the cohesion of water due to hydrogen bonding transmits the upward pull along the entire length of the xylem to the roots.

Transpirational Pull   Stomata, the microscopic pores through the surface of a leaf, lead to a labyrinth of internal air spaces that expose the mesophyll cells to the carbon dioxide they need for photosynthesis (Figure 32.7). In contact with the moist walls of the cells, the air in the leaf is saturated with water vapor. Damper air has a greater water potential than drier air, and thus on most days gaseous water within the leaf diffuses through stomata to the drier air outside. The water vapor lost from the air spaces within the leaf is replaced by evaporation of water from mesophyll cells that border the air spaces. This lowers the water potential of these cells, causing them to take up water from neighboring cells that have a higher water potential because they are more distant from the air spaces. Cells still farther from the site of evaporation have even higher water potentials, and thus water will move from cell to cell toward the air spaces. At the "uphill" end of this stream of water are the cells that front on tracheids, and these cells replace their water from the xylem sap. From xylem to air spaces to the outside of the leaf, water moves along a gradient of decreasing water potential maintained by transpiration. There is a transpirational pull on the fluid within xylem vessels.

Cohesion and Adhesion of Water   The transpirational pull on xylem sap is transmitted all the way from the leaves to the root tips and even into the soil solution (Figure 32.8). It is the cohesion of water, due to hydrogen bonds (see Chapter 3), that makes it possible to pull a column of sap from above without the water separating. Water molecules exiting from tracheids in the leaf tug on adjacent molecules, and this pull is relayed, molecule by molecule, down the entire column of water in the xylem. Also helping to fight gravity is the strong adhesion of water molecules (again by hydrogen bonds) to the hydrophilic walls of the xylem cells. The very small diameter of tracheids and xylem elements contributes to the importance of this factor in overcoming the downward pull of gravity. Indeed, the adhesion of water molecules to the walls of the xylem conduits is just as important as the cohesion of water in the ascent of xylem sap.

The upward pull on the cohesive sap creates tension within the xylem. Tension is the opposite of pressure—it is, in effect, a negative pressure. Pressure will cause an elastic pipe to swell, but tension will pull the walls of the pipe inward. (You can actually measure a decrease in the diameter of a tree trunk on a warm day when transpiration pull puts the xylem under tension). As we know, pressure increases water potential and tends to force water outward. Tension, a negative pressure, *decreases* water potential, favoring the uptake of water. Transpirational pull puts the xylem under tension all the way down to the root tips, even in the tallest trees. This tension lowers water potential in the

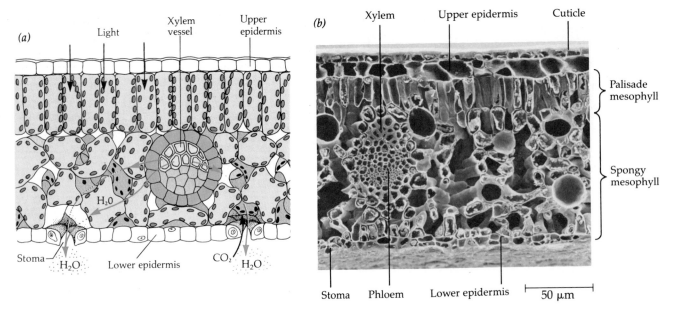

(a) Light Xylem vessel Upper epidermis

H₂O

Stoma H₂O Lower epidermis CO₂ H₂O

(b) Xylem Upper epidermis Cuticle

Palisade mesophyll

Spongy mesophyll

Stoma Phloem Lower epidermis 50 μm

Figure 32.7
Structural aspects of transpirational pull. (a) Water in the leaf moves down a gradient of water potential from xylem vessels to the air outside the leaf, and this pulls on the xylem sap. Carbon dioxide dif- fuses through the stomata into the honey- combed interior of the leaf. (b) A scanning electron micrograph of a leaf cross section reveals the extensive surface area of the spongy mesophyll. This enhances CO_2 uptake but also results in rapid evapora- tion of water, which transpires mainly through stomata that must remain open for photosynthesis to occur.

Figure 32.8
Ascent of water in a tree. Hydrogen bonding forms an unbroken chain of water molecules extending from leaves all the way to the soil. The force that drives the ascent of xylem sap is a gradient of water potential (Ψ).

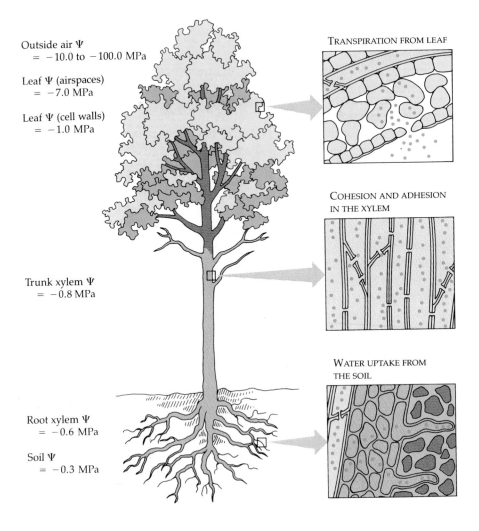

Outside air Ψ = −10.0 to −100.0 MPa

Leaf Ψ (airspaces) = −7.0 MPa

Leaf Ψ (cell walls) = −1.0 MPa

TRANSPIRATION FROM LEAF

COHESION AND ADHESION IN THE XYLEM

Trunk xylem Ψ = −0.8 MPa

Root xylem Ψ = −0.6 MPa

Soil Ψ = −0.3 MPa

WATER UPTAKE FROM THE SOIL

root xylem to such an extent that water flows passively from the soil, across the root cortex, and into the stele.

Transpirational pull can extend down to the roots only through an unbroken chain of water molecules. Cavitation, the formation of a water vapor pocket in a xylem vessel, breaks the chain. If xylem sap freezes in winter, cavitation may occur. Small plants can use root pressure to refill xylem vessels in spring, but in trees root pressure cannot push water to the top, so a vessel with a water vapor pocket can never function as a water pipe again. However, the transpiration stream can detour around the water vapor pocket through pits between adjacent xylem vessels, and secondary growth adds a layer of new xylem vessels each year. In some angiosperm trees, including oaks and elms, only the youngest, outermost growth ring of xylem transports water, with the older xylem functioning to support the tree.

The Driving Force: A Gradient of Water Potential

The transpiration-cohesion-adhesion mechanism that causes the ascent of xylem sap is an excellent example of the application of physical principles to biological problems. To summarize the theory, water flows passively through the plant, from soil to air, along a gradient of decreasing water potential (see Figure 32.8). It is a physical process, relying only on solar-powered evaporation to maintain a low water potential in leaves and on the cohesion and adhesion of water molecules to hold microscopic water columns together. Plant physiologists estimate that leaves at the top of a very tall tree must have water potentials at least as low as -3.0 MPa to draw xylem sap strongly enough to overcome gravity and resistance; negative water potentials of that magnitude have actually been measured in leaves removed from the tops of trees (see Methods Box, p. 712). The transpiration-cohesion-adhesion theory is the best answer we have to a very old question: How does water reach treetops?

THE CONTROL OF TRANSPIRATION

Because they do not recirculate much of their water, plants generally require more water than animals. A leaf may transpire its weight in water each day. Leaves are kept from wilting by a transpiration stream in xylem vessels that flows as fast as 75 cm per minute, about the speed of the tip of a second hand sweeping around a wall clock. The tremendous requirement for water by a plant is part of the cost of making food by photosynthesis.

The Photosynthesis-Transpiration Compromise

To make food, a plant must spread its leaves to the sun and obtain CO_2 from the air. Carbon dioxide dif-fuses into the leaf, and oxygen produced as a by-product of photosynthesis diffuses out of the leaf through the stomata (see Figure 32.7). The stomata lead to the honeycomb of air spaces through which CO_2 diffuses to the photosynthetic cells of the mesophyll. The internal surface area of the leaf may be 10–30 times greater than the external surface area we see when we look at the leaf. This structural feature of leaves amplifies photosynthesis by increasing exposure to CO_2, but at the same time it increases the surface area for the evaporation of water, which exits the plant freely through open stomata. About 90% of the water a plant loses escapes through stomata, though these pores account for only 1%–2% of the external leaf surface. The waxy cuticle limits water loss through the remaining surface of the leaf. The stomata of most plants are more concentrated on the lower surfaces of leaves, reducing transpiration because the bottom of the leaf receives less sunlight than the top surface.

One way to evaluate how efficiently a plant uses water is to determine its transpiration-to-photosynthesis ratio, the amount of water lost per gram of CO_2 assimilated into organic material by photosynthesis. A common ratio is 600, meaning the plant transpires 600 g of water for each gram of CO_2 that becomes incorporated into organic material. However, corn and other plants that assimilate atmospheric CO_2 by the C_4 pathway have transpiration-to-photosynthesis ratios of 300 or less. With the same concentration of CO_2 within the air spaces of the leaf, C_4 plants can assimilate that CO_2 into carbohydrates at a greater rate than C_3 plants. Since water loss is the trade-off for enabling CO_2 to diffuse into the leaf, the photosynthetic return for each gram of water sacrificed is greater for plants that can assimilate CO_2 at greater rates.

By now you may have the impression that transpiration is a "necessary evil" of assimilating CO_2, but transpiration does have some benefits. The transpiration stream assists in the transfer of minerals and other substances from the roots to the shoot and leaves. Transpiration also results in evaporative cooling, which can lower the temperature of a leaf by as much as $10°$–$15°C$ compared with the surrounding air. This prevents the leaf from reaching temperatures that could inhibit the enzymes that catalyze the photosynthetic reactions, as well as other enzymes involved in the leaf's metabolic processes. Desert succulents, which have low rates of transpiration, can tolerate high leaf temperatures; in this case, the loss of water due to transpiration is a greater threat than overheating.

As long as leaves can pull water from the soil fast enough to replace what they lose, transpiration is no problem. However, when transpiration exceeds delivery of water by xylem, as when the soil begins to dry out, the leaf begins to wilt as its cells lose turgor pressure. The potential rate of transpiration will be greatest on a day that is sunny, warm, dry, and windy, because these are the climatic factors that increase the

When a branch is cut, the very negative water potential in the leaves draws water in xylem away from the cut surface. The plant is placed in a container called a pressure bomb, and external pressure is increased until it is just great enough to counteract the negative water potential in leaves and push water back to the cut surface. The amount of pressure that must be applied is equivalent to the average water potential of the leaves. This type of experiment has confirmed that water potentials at the tops of the tallest trees are negative enough to pull water up from the soil.

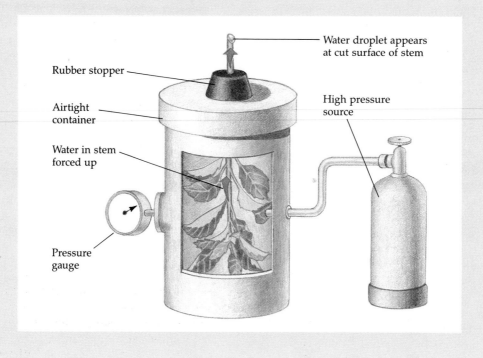

- Rubber stopper
- Airtight container
- Water in stem forced up
- Pressure gauge
- Water droplet appears at cut surface of stem
- High pressure source

evaporation of water. Plants are not helpless against the elements, however, for they are capable of adjusting to their environment. In the photosynthesis-transpiration compromise, mechanisms that regulate the size of the stomatal openings strike a balance.

How Stomata Open and Close

Each stoma is flanked by a pair of guard cells, which are kidney-shaped in dicots and dumbbell-shaped in most grasses. The guard cells are suspended by subsidiary cells over an air chamber, which leads to the maze of air spaces within the leaf.

Guard cells control the diameter of the stoma by changing shape, which widens or narrows the gap between the two cells. (Figure 32.9). When guard cells take in water by osmosis, they become more turgid and swell. In most dicots, the cell walls of guard cells are not uniformly thick, and the cellulose microfibrils are oriented in a radial manner (see Figure 32.9). These structural adaptations cause the guard cells to buckle outward when they are turgid, increasing the size of the gap between the cells. When the cells lose water and become flaccid, they sag together and close the space between them. The dumbbell-shaped guard cells of grasses (monocots) are thin-walled on the ends of the cell and thick-walled along the middle. As these guard cells become turgid, the ends swell and produce a gap between the cells' mid-regions, which remain narrow. The changes in turgor pressure that open and close stomata result primarily from the reversible uptake and loss of potassium ions (K^+) by the guard cells. Stomata open when guard cells actively accumulate

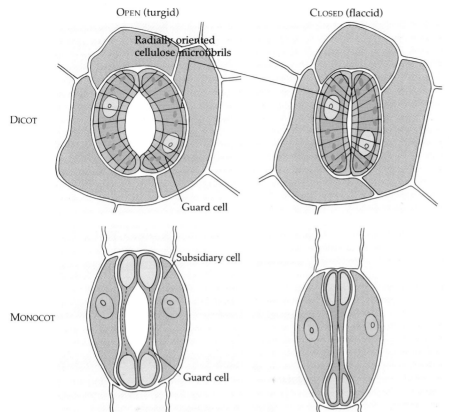

OPEN (turgid)   CLOSED (flaccid)

Radially oriented
cellulose microfibrils

DICOT

Guard cell

MONOCOT

Subsidiary cell

Guard cell

Figure 32.9
Control of stomatal aperture. Guard
cells of a dicot and a grass are shown in
their flaccid and turgid states. Guard cells
respond to a complex set of signals, including environmental factors and cues
within the plant itself. The uptake or loss of
water causes the guard cells to change
their shape and widen or narrow the gap
between them, a response that affects
the rate of photosynthesis and controls
transpiration.

K^+ from subsidiary cells or neighboring epidermal cells (Figure 32.10). This uptake of solute causes the water potential to become more negative within the guard cells, and the cells become more turgid as water enters by osmosis. The increase in positive charge due to the influx of K^+ ions is offset by the uptake of chloride (Cl^-), by the pumping out of the cell of hydrogen ions released from organic acids, and by the negative charges of these organic acids after they lose their hydrogen ions. Stomatal closure parallels an exodus of K^+ from guard cells, which leads to an osmotic loss of water. The K^+ fluxes across the guard cell membrane are probably coupled to the generation of membrane potentials by proton pumps.

In general, stomata are open during daytime and closed at night (Figure 32.11). This prevents the plant from needlessly losing water when it is too dark for photosynthesis. At least three cues contribute to stomatal opening at dawn. First, light itself stimulates guard cells to accumulate potassium and become turgid. This response is probably triggered by illumination of a blue-light receptor that has been discovered on the tonoplast surrounding the vacuoles of guard cells. Activation of these blue-light receptors some-

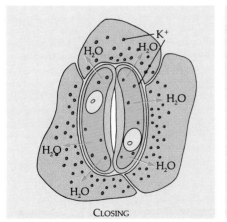

CLOSING

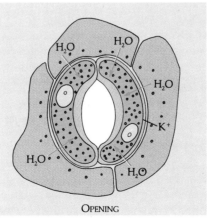

OPENING

Figure 32.10
Role of potassium in guard cell function. Stomata open when guard cells accumulate potassium (red dots), take in water in response to the increase in solute concentration, and become turgid.

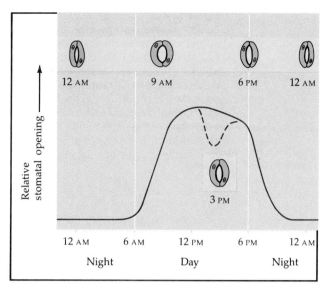

Figure 32.11
Rhythmic opening and closing of a stoma. An internal clock as well as a variety of environmental cues control the turgor changes that cause guard cells to change shape. During the hottest part of the day, the stomata of some plant species may close if transpiration exceeds water uptake (dotted line). This closure may be caused by an increase in abscisic acid produced by the leaf in response to water deficiency. Stomatal closure during the day helps reduce water loss, but also lowers the photosynthetic rate.

how stimulates the activity of ATP-powered proton pumps in the plasma membranes of the guard cells, which, in turn, promotes the uptake of K^+. Light may also stimulate stomatal opening by driving photosynthesis in guard cell chloroplasts, making ATP available for the active transport of K^+ and H^+ ions. A second factor causing stomata to open is depletion of CO_2 within air spaces of the leaf, which occurs when photosynthesis begins in the mesophyll. A plant can be tricked into opening its stomata at night by placing it in a chamber devoid of CO_2. A third cue in stomatal opening is an internal clock located in the guard cells. Even if you keep a plant in a dark closet, stomata will continue their daily rhythm of opening and closing (see Figure 32.11). As Dr. Satter mentioned in the interview at the beginning of this unit, all eukaryotic organisms have internal clocks that somehow keep track of time and regulate cyclical processes. Cycles that have intervals of approximately 24 hours are called **circadian rhythms.** You will learn more about circadian rhythms and the biological clocks that control them in Chapter 35.

Environmental stress of various kinds can cause stomata to close during the daytime. When the plant is suffering a water deficiency, guard cells may lose turgor. In addition, a hormone called abscisic acid, which is produced in the mesophyll cells in response to water

deficiency, signals guard cells to close stomata. This response reduces further wilting, but it also slows down photosynthesis, which is one reason droughts reduce crop yields. High temperatures also induce stomatal closure, probably by stimulating cellular respiration and increasing CO_2 concentration within the air spaces of the leaf. High temperature and excessive transpiration may combine to cause stomata to close briefly during midday (see Figure 32.11). Thus, guard cells arbitrate the photosynthesis-transpiration compromise on a moment-to-moment basis by integrating a variety of internal and external stimuli.

Leaf Adaptations That Reduce Transpiration

Plants adapted to arid climates are called **xerophytes.** They have leaves that are modified in various ways to reduce the rate of transpiration. Many xerophytes have small, thick leaves, an adaptation that limits water loss by reducing surface area relative to volume. A thick cuticle gives some of these leaves a leathery consistency. The stomata are concentrated on the lower leaf surface, and they are often located in cryptlike depressions that shelter the pores from the dry wind (Figure 32.12). During the driest months, some desert plants shed their leaves. Others, such as cacti, subsist on water the plant stores in its fleshy stems during the rainy season (these modified stems are the photosynthetic organs of cacti; the leaves are the spines).

One of the most elegant adaptations to an arid habitat is found in succulent plants of the family Crassulaceae. These plants, and a few members of other families, including pineapples, assimilate their CO_2 by an alternative photosynthetic pathway known as CAM, for crassulacean acid metabolism (see Chapter 10). Mesophyll cells in a CAM plant have enzymes that can assimilate CO_2 into organic acids during the night. During the daytime, the organic acids are broken down to release CO_2 in the same cells, and sugars are synthesized by the conventional (C_3) photosynthetic pathway. Since the leaf takes in its CO_2 at night, the stomata can close during the day when transpiration would be most severe. The circadian rhythm for stomatal opening in CAM plants is out of phase with other plants by about 12 hours. In stomatal behavior, both short-term physiological adjustments and evolutionary adaptations are manifest.

TRANSPORT IN PHLOEM

The transpiration stream within xylem generally flows in the wrong direction to function in exporting sugar from leaves to other parts of the plant. A second vascular tissue, the phloem, transports the organic prod-

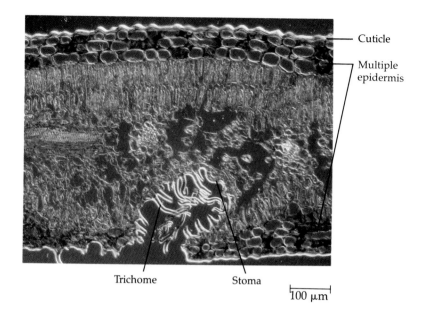

— Cuticle

— Multiple
epidermis

Trichome   Stoma

|‾‾‾‾‾‾‾‾|
100 μm

Figure 32.12
Structural adaptations of leaves to arid conditions. The stomata of this oleander, a xerophyte, are recessed in crypts equipped with hairs called trichomes. This structural adaptation protects the stomata from the hot, dry wind. The upper epidermis has several layers of cells, another adaptation that reduces water loss (LM).

ucts of photosynthesis throughout the plant. The transport of food in the plant is called **translocation.**

Phloem is a complex tissue consisting of parenchyma cells, fiber cells, sieve-tube members, and companion cells. Sugar is transported by the sieve-tube members, which are arranged end to end to form long sieve tubes (Figure 32.13). Between the members are sieve plates, porous cross walls that probably impose little resistance to the flow of sap along the sieve tube.

Phloem sap differs markedly in composition from xylem sap. By far the prevalent solute in phloem sap is sugar, primarily the disaccharide sucrose. The sucrose concentration may be as high as 30% by weight, giving the sap a syrupy thickness. Phloem sap may also contain minerals, amino acids, and hormones in transit from one part of the plant to another.

Source-to-Sink Transport

In contrast to the unidirectional transport of xylem sap, from roots to leaves, the direction that phloem sap travels is more variable. The one generalization that holds is that sieve tubes carry food from a sugar source to a sugar sink. A **source** is a plant organ in which sugar is being produced either by photosynthesis or breakdown of starch. Leaves, obviously, are usually sources. A **sink** is an organ that consumes or stores sugar. Growing roots, shoot tips, and fruits are sugar sinks supplied by phloem, as are nongreen stems and trunks. A storage organ, such as a tuber or bulb, may be either a source or a sink depending on the season. When the storage organ is stockpiling carbohydrates during the summer, it is a sugar sink. After breaking dormancy in early spring, however, the storage organ becomes a source as its starch is broken

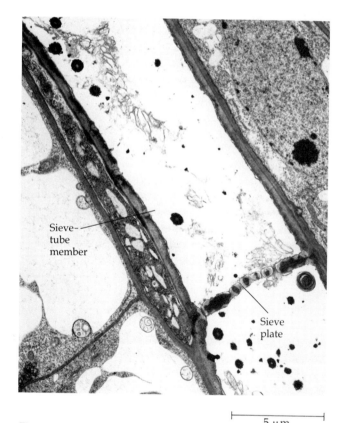

Sieve-tube member

• Sieve plate

|‾‾‾‾‾‾‾‾|
5 μm

Figure 32.13
Sieve tubes and translocation. Sap flows between the sieve-tube members through porous sieve plates (TEM).

down to sugar, which is carried away in the phloem to the growing buds of the shoot system.

Other solutes may be transported to sinks along with sugar. For example, minerals that reach leaves in xylem may later be transferred in the phloem to developing fruit.

Figure 32.14
Loading of sucrose into phloem. According to the hypothesis illustrated here, sucrose (red dots) produced in the mesophyll is transported into the apoplast of a small vein. The companion cell of a sieve-tube member actively accumulates the sugar, which can pass into the sieve-tube member through the plasmodesmata that connect it to its companion cell. The active loading of sucrose is probably driven by proton pumps. This model envisions the proton pump generating a gradient of hydrogen ions and the H^+ then reentering the companion cell through a common carrier that cotransports sucrose into the cell along with the protons. Supporting this hypothesis are measurements of pH changes in solutions bathing phloem tissue. When sucrose is added to the bathing medium, the pH of the solution increases (H^+ concentration decreases). The extent of the pH change depends on the amount of sucrose added. One interpretation of these results is that hydrogen ions are moving from the external solution into the cells, but can do so only in the company of sucrose molecules, thus allowing for the selective accumulation of sucrose from the apoplast.

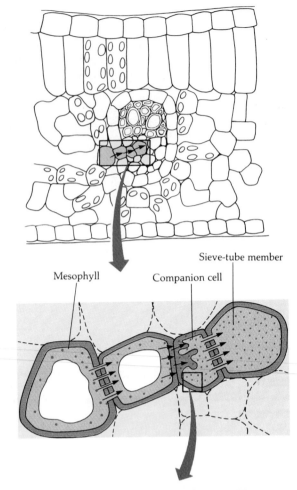

Mesophyll · Sieve-tube member · Companion cell

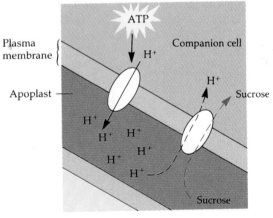

ATP

Plasma membrane

Companion cell

H^+

H^+

Apoplast

Sucrose

H^+

H^+ H^+

H^+ H^+

H^+

Sucrose

A sugar sink is usually supplied by the sources nearest to it. The upper leaves on a branch may send sugar to the growing shoot tip, whereas the lower leaves export sugar to roots. A growing fruit requires so much food that it may monopolize the sugar sources all around it. One sieve tube in a vascular bundle may carry phloem sap in one direction, while a different tube in the same bundle flows in the opposite direction. For each sieve tube, the direction of transport depends only on the locations of the source and sink connected by that tube, and the direction may change with the season or developmental stage of the plant.

Phloem Loading and Unloading

At the source end of a sieve tube, sucrose is loaded into the phloem, probably by active transport. Experiments using radioactive tracers to follow movements of carbohydrate within the leaf suggest that sugar moves through the mesophyll in the symplast—that is, through plasmodesmata connecting the mesophyll cells. Where mesophyll borders small veins, sucrose is apparently transported out of the cells and into the apoplast (walls). Sucrose is then actively accumulated by sieve-tube members or more likely by the companion cells that are connected by plasmodesmata to the tube members. (Some companion cells have the structure of transfer cells, with convoluted plasma membranes that increase surface area for solute absorption.) The loading of sucrose into the sieve tubes requires metabolic energy and is believed to be driven indirectly by the proton pumps that are common components of plant membranes (Figure 32.14). Sieve-tube members in the leaf accumulate sucrose to concentrations that are two to three times higher than in the surrounding mesophyll.

Our knowledge about the mechanism for phloem loading just described comes largely from research on sugar beets and corn. In other plants, particularly members of the squash family, sugars and other nutrients may travel from the mesophyll cells to the sieve-tube members mainly through the symplast. The sieve-tube members in these plants may have sucrose concentrations similar to that of the surrounding mesophyll. Thus, the active loading of sucrose may play a relatively minor role in phloem loading in such plants.

Downstream at the sink end of the sieve tube, phloem unloads its sucrose. The unloading of sugar has not been studied as extensively as loading, but sucrose may be pumped out of sieve tubes at the sink end by active transport.

Pressure Flow

Phloem sap flows from source to sink at rates as great as 1 m per hour, which is much too fast to be accounted for by either diffusion or cytoplasmic streaming. There is now a consensus among plant physiologists that phloem sap moves by a pressure-flow mechanism (Figure 32.15). Phloem loading results in a high solute concentration at the source end of a sieve tube, which lowers the water potential and causes water to flow into the tube. Hydrostatic pressure develops within the sieve tube, and the pressure is greatest at the source end of the tube. At the sink end, the pressure is relieved by the loss of water, owing to water potential being lowered outside the sieve tube by the exodus of sucrose. The building of pressure at one end of the tube and reduction of that pressure at the opposite end causes water to flow from source to sink, carrying the sugar along. Water is recycled back from sink to source by xylem vessels.

The pressure-flow model explains why phloem sap always flows from a sugar source to a sugar sink, regardless of their locations in the plant. However, the model is somewhat difficult to test because most experimental procedures disrupt the structure and function of the sieve tubes. Some of the most interesting studies have taken advantage of natural phloem probes—aphids that feed on phloem sap (Figure 32.16). For now, the pressure-flow model is the most satisfactory explanation for the flow of sap in phloem.

*   *   *

Plant physiologists still have much to learn about the mechanisms of transport in the elaborate vascular system of xylem and phloem. William Harvey, the great seventeenth-century physiologist, speculated that plants and animals have similar circulatory systems. The idea was abandoned after careful dissection failed to turn up a heart in plants. We are only now beginning to understand how the plant keeps sap flowing through its veins without the help of moving parts.

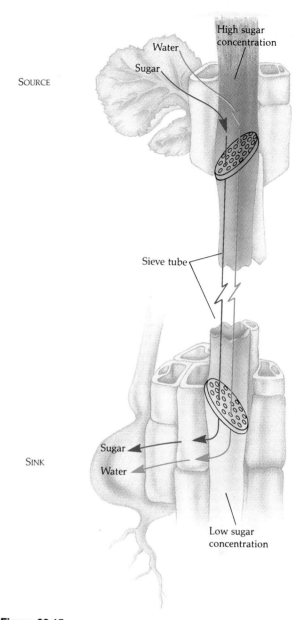

Figure 32.15
Pressure flow in a sieve tube. Loading of sugar into the tube at the source reduces the water potential inside the tube, and increases it outside the tube. Water flowing into the tube from a less negative to a more negative water potential generates a hydrostatic pressure that causes the sap to flow along the tube. The gradient of pressure in the tube is reinforced by the unloading of sugar and the consequent loss of water at the sink. Xylem recycles water from sink to source.

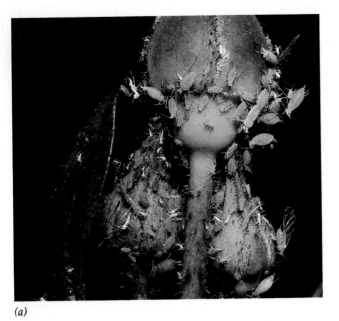

(a)

Figure 32.16
Tapping phloem sap with an aphid. (a) Aphids feeding
on rose buds. **(b)** The insect inserts a modified mouth-
part called a stylet into the plant and probes until the tip
of this hypodermic-like organ penetrates a single sieve-
tube member. The pressure within the sieve tube force-
feeds the aphid, swelling it to several times its original
size. While the aphid is feeding, it can be anesthetized
and severed from its stylet, which then serves the
researcher as a miniature tap that exudes phloem sap
for hours. The closer the stylet to a sugar source, the
faster sap flows out and the greater its sugar concentration. This is what we would expect if pressure is generated at the source end by an inward pumping of sugar.

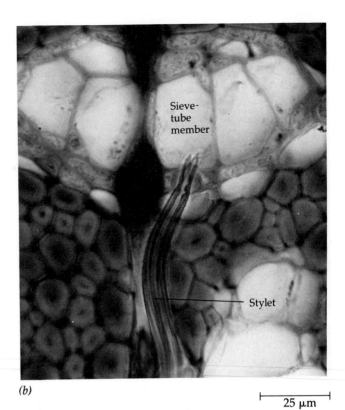

(b)

25 μm

STUDY OUTLINE

1. The life of a terrestrial plant depends on an integrated
   system of transport between the roots and the leaf-bearing shoots.
2. Transport occurs at the level of the individual cell, between
   neighboring cells, and over long distances.

Absorption of Water and Minerals by Roots (pp. 704–707)

1. Water and dissolved solutes gain access to the roots
   through epidermal cells and their root hairs. The hydrophilic cell walls, in contact with those of the root cortex,
   provide pathways for the soil solution across the cortex
   to the endodermis.
2. Minerals enter the root cortex by diffusion or active transport. In the latter case, transport proteins on the plasma
   membranes selectively absorb certain ions, powered by
   ATP, either directly or indirectly by means of gradients
   produced by the proton pump.
3. Water and minerals enter xylem by one of two routes. In
   the symplastic pathway, the substances are absorbed into
   parenchymal cells and progress directly to the stele via
   plasmodesmata. The second route is the extracellular
   apoplastic pathway, in which solutions travel along cell

walls until they reach an impervious Casparian strip at
the endodermal boundary to the stele. Minerals traveling
in the apoplast can only enter the stele by crossing the
selectively permeable membrane of the endodermis.
4. Transfer cells within the stele have physical and functional specializations that enhance their transfer of solutes between symplast and apoplast.

Ascent of Xylem Sap (pp. 707–711)

1. The upward flow of xylem sap supplies minerals to the
   shoots and replaces water lost by transpiration.
2. The net movement of water, both with respect to individual cells and over long distances, is determined by
   water potential, the tendency for water to leave one place
   and move to another. Water will always flow passively
   from a location with greater water potential to a region
   with lower water potential. Increased concentration of
   solutes lowers water potential, and increased pressure
   increases it.
3. Osmotic flow of water into the stele causes the upward
   push of root pressure that forces water up the xylem.
   This is responsible for guttation, the exudation of water

droplets. However, root pressure alone cannot sustain the flow of water through the plant.

4. The transpiration-cohesion-adhesion theory best explains the main method of transporting xylem sap. Evaporative water loss during transpiration lowers the water potential of cells in the leaf compared to the root. With the help of cohesion of water molecules due to hydrogen bonds, transpiration pulls the water upward and creates a tension, or negative pressure, that is transmitted throughout even the tallest plants. The adhesion of water molecules to the walls of the tracheids and vessel elements also contributes to this mechanism.

The Control of Transpiration (pp. 711–714)

1. Plants must strike a balance between gas exchange and transpirational water loss associated with the presence of stomata.

2. The transpiration-to-photosynthesis ratio is a convenient way to determine how efficiently a plant uses water. Since water loss is the trade-off for allowing carbon dioxide to diffuse into the leaf, plants with lower ratios are generally at an advantage in arid habitats.

3. Transpiration is important for evaporative cooling and for delivering minerals to leaves.

4. Plants strike a balance between photosynthesis and transpiration by regulating the size of stomatal openings through changes in turgor pressure of guard cells. These physical changes are due to fluxes in potassium ions that are probably coupled to the generation of membrane potentials by proton pumps in guard cell membranes.

5. Guard cells usually open at dawn, due to carbon dioxide depletion, an inherent circadian rhythm, and ion movements triggered by light-detecting pigments.

6. Water deficiency and high temperatures can cause stomata to close during the daytime.

7. Xerophytic plants in dry habitats have leaves with morphological or physiological adaptations that reduce transpiration.

Transport in Phloem (pp. 714–718)

1. Translocation is the process of transporting photosynthetically produced food throughout the body of the plant in the phloem. In addition to sucrose, phloem sap contains minerals, amino acids, and hormones.

2. Unlike the unidirectional transport of xylem sap, the sieve tubes of phloem carry food and other solutes from a sugar source to a sugar sink. A source is an organ that produces sugar by photosynthesis or breakdown of starch, whereas a sink consumes or stores sugar.

3. Sources and sinks vary between and within various parts of the plant. However, for any given sieve tube, the direction of transport depends only on the relative locations of the source and sink.

4. Mesophyll cells surrounding small veins in the leaf may secrete synthesized sugar from the symplast into the apoplast, where it is loaded into sieve tubes, probably by active transport involving proton pumps. In some plants, sugar may travel from the mesophyll to the sieve tubes mainly through the symplast. Unloading at the end of the sieve tube may also require active transport.

5. Phloem sap is believed to move by a pressure-flow mechanism. The high solute concentration produced at the source end of a sieve tube during sucrose loading generates a hydrostatic pressure that is relieved at the sink end by the loss of water accompanying the unloading of sucrose.

SELF-QUIZ

1. Water and minerals that cross the cortex through the symplast move
   a. up a water potential gradient
   b. extracellularly along the matrix of cell walls
   c. from cell to cell through plasmodesmata
   d. by active transport
   e. by pressure flow

2. In order to enter the stele, water and minerals must pass through
   a. the apoplast
   b. the Casparian strip
   c. endodermal cells
   d. transfer cells
   e. xylem vessels

3. Stomata open when guard cells
   a. sense an increase in CO_2 in the air spaces of the leaf
   b. flop open due to a decrease in turgor pressure
   c. become more turgid due to an influx of K^+, followed by the osmotic flow of water
   d. reverse their circadian rhythm
   e. accumulate water by active transport

4. Which of the following is *not* part of the transpiration-cohesion-adhesion theory of the ascent of xylem sap?
   a. the evaporation of water from the mesophyll cells, which initiates a pull of water molecules from neighboring cells and eventually from the tracheids
   b. the transfer of this pull from one water molecule to the next due to the cohesion caused by hydrogen bonds
   c. the hydrophilic walls of the narrow tracheids and vessels that help to maintain the column of water against the force of gravity
   d. the use of proton pumps to create membrane potentials that facilitate the accumulation of mineral ions
   e. the higher water potential of the roots compared to the water potential of the leaves

5. The movement of sap from a sugar source to a sugar sink
   a. occurs through the apoplast of sieve-tube members
   b. may translocate sugars from the breakdown of stored starch up to developing shoots
   c. may be due to a pressure-flow mechanism caused by lower water potential in the source than in the sink
   d. is primarily a passive process
   e. results mainly from diffusion

6. Guttation in plants results from
   a. root pressure

b. transpirational pull

c. pressure flow in phloem

d. injury to the plant

e. condensation of atmospheric water

7. Water potential is generally most negative in

a. the soil solution

b. mesophyll cell walls

c. xylem sap of roots

d. xylem sap of leaves

e. the cytoplasm of root hairs

8. The productivity of a crop begins to decline when the leaves begin to wilt mainly because

a. the chlorophyll of wilting leaves decomposes

b. flaccid mesophyll cells are incapable of photosynthesis

c. stomata close, preventing CO_2 from entering the leaf

d. photolysis, the water-splitting step of photosynthesis, cannot occur when there is a water deficiency

e. accumulation of CO_2 in the leaf inhibits enzymes required for photosynthesis

9. Imagine cutting a live twig from a tree and examining the cut surface with a magnifying glass. You locate the vascular tissue and observe a growing droplet of fluid exuding from the cut surface. This fluid is probably

a. phloem sap

b. xylem sap

c. guttation fluid

d. fluid of the transpiration stream

e. cell sap from the broken vacuoles of cells

10. Which structure or compartment is *not* part of the plant's apoplast?

a. the lumen of a xylem vessel

b. the lumen of a sieve tube

c. the cell wall of a mesophyll cell

d. the cell walls of transfer cells

e. the cell walls of root hairs

CHALLENGE QUESTIONS

1. Wind ventilates the leaf surface, increases transpiration, and replenishes the supply of carbon dioxide for photosynthesis. Stomata close when the carbon dioxide concentration increases in air by sensing changes in carbon dioxide concentration inside the leaf. How do you think stomatal response to carbon dioxide is related to wind speed, and what might be the adaptive value of this relationship? (See Further Reading 2, pp. 158–159.)

2. Research has shown that stomatal opening is triggered by absorption of blue light, which stimulates plant growth by a light-sensing apparatus distinct from that involved in photosynthesis. Blue light turns on a pump that removes H^+ from guard cells and creates an electrical gradient across the membranes. Explain how this pumping of H^+ out of guard cells would affect K^+ and H_2O transport and the stomatal aperture. (See Further Reading 3.)

3. Describe the environmental conditions that would minimize the transpiration–photosynthesis ratio for a C_3 plant.

FURTHER READING

1. Greulach, V. *Plant Structure and Function*. 2d ed. New York: Macmillan, 1983. A book that stresses the form and function theme of this unit.
2. Mansfield, T. A., and W. J. Davies. "Mechanisms for Leaf Control of Gas Exchange." *BioScience*, March 1985. An excellent article on the various functions of stomata.
3. Miller, J. A. "Plant 'Sight' from Pores and Pumps." *Science News*, November 30, 1985. Current knowledge about how stomatal function may increase crop yields.
4. Salisbury, F. B., and C. W. Ross. *Plant Physiology*. 3rd ed. Belmont, CA: Wadsworth, 1985. A rigorous text with excellent chapters on transport.
5. Schulze, E.-D., R. H. Robichaux, J. Grace, P. W. Rundel, and J. R. Ehleringer. "Plant Water Balance." *Bioscience*, January 1987. Evolutionary adaptations that minimize water loss.
6. Wickelgren, I. "Plant Ion-Pump Gene Cloned, Sequenced." *Science News*, March 4, 1989. Exciting research on the proton pump that drives so much transport work in plants.

Plant Nutrition

33

The ability of plants to assimilate nutrients from the environment and synthesize organic material is not only essential to the survival of the plants but also to humans and all other animals, consumers dependent on the productivity of plants. Because the quality of plant nutrition determines the quality of our own nutrition, much agricultural research focuses on the nutritional requirements and mechanisms of crops (Figure 33.1).

All organisms, remember, are open systems that continuously exchange energy and materials with their environment. As photosynthetic autotrophs, plants can make all of their own organic compounds but must obtain inorganic raw materials to do so. For a plant to survive and grow, it must be nourished with carbon dioxide, water, and a variety of minerals in the form of inorganic ions present in the soil. (The plant must also take in oxygen for cellular respiration, although during the daytime the plant releases more O_2 from photosynthesis than it consumes for respiration.) Roots and shoots are anatomically adapted for absorbing essential nutrients from the soil and air, demonstrating the correlation of structure and function (Figure 33.2). Photosynthesis and other metabolic pathways in plant cells convert the "diet" of inorganic nutrients to carbohydrates, proteins, and other classes of organic compounds, which in turn provide nutrition for animals. The physiology of photosynthesis is covered in Chapter 10; you may want to review that chapter at some point during your study of plant nutrition.

In this chapter, you will learn how a plant obtains and uses its nutrients. The chapter emphasizes how interactions between plants and their environments affect the nutritional status of the plants.

Figure 33.1

Plant nutrition determines human nutrition. Compare these two corn crops, one (left) growing on soil rich in the mineral nitrogen, which is essential to plants, and the other growing on soil deficient in nitrogen. Even if a crop that has been poorly nourished can be harvested, it yields less food and its specific mineral deficiencies are passed on to livestock or human consumers.

NUTRITIONAL REQUIREMENTS OF PLANTS

Chemical Composition of Plants

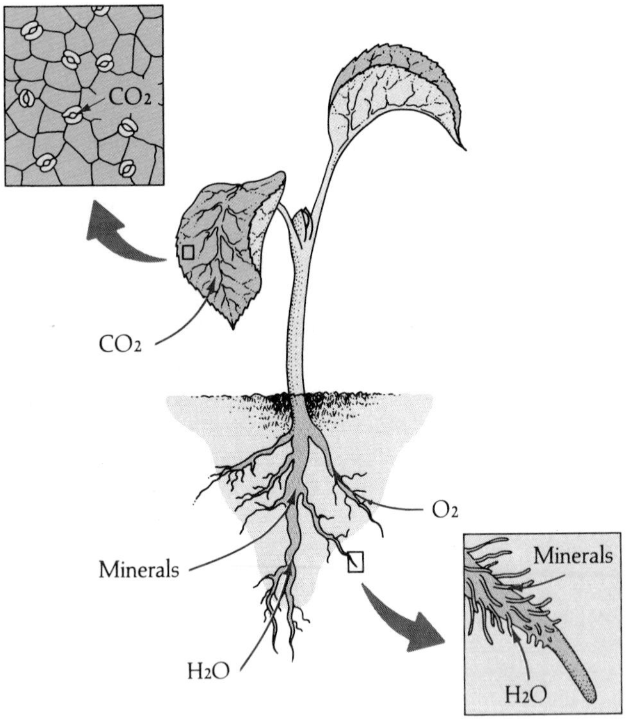

Watch a large plant grow from a tiny seed and you cannot help wondering where all the mass comes from. Aristotle thought soil provided the substance for plant growth, since plants seemed to spring from the earth. Leaves, he believed, functioned only to shade the developing fruit. In the seventeenth century, a Belgian physician named Jean-Baptiste van Helmont performed an experiment to find out if plants grew by absorbing soil. He planted a willow seedling in a pot that contained 90.9 kg of soil. After five years, the willow had grown into a tree weighing 76.8 kg, but only 56.8 g of soil had disappeared from the pot. Van Helmont concluded that the willow had grown mainly from the water he had added regularly. A century later, Stephen Hales, an English physiologist, postulated that plants were nourished mostly by air.

As it turns out, none of the early ideas about plant nutrition is entirely incorrect. Plants *do* extract minerals that are essential nutrients from the soil, but these minerals make only a small contribution to the overall mass of the plant, as van Helmont discovered. About 80%–85% of a herbaceous (nonwoody) plant is water, and plants grow mainly by accumulating water in the central vacuoles of their cells (see Chapter 31). Furthermore, water can truly be considered a nutrient since it supplies most of the hydrogen and some of the oxygen that is incorporated into organic compounds by photosynthesis. However, only a small fraction of the water that enters a plant contributes atoms to organic molecules. Generally, more than 90% of the water absorbed by plants is lost by transpira-

Figure 33.2

Uptake of nutrients by a plant. Roots absorb water and minerals from the soil, with root hairs greatly increasing the area of the epidermal surface that functions in this absorption. Carbon dioxide, the source of carbon for photosynthesis, diffuses into leaves from the surrounding air through stomata. (Plants also need O_2 for cellular respiration, although the plant is a net producer of O_2.) From these inorganic nutrients, the plant can produce all of its own organic material.

CO_2

CO_2

Minerals

O_2

Minerals

H_2O

H_2O

The roots of plants are bathed in solutions of various minerals dissolved in known concentrations. The water must be aerated to provide the roots with oxygen for cellular respiration. A particular mineral, such as potassium, can be omitted from the culture medium to test whether it is essential to the plants.

If the element deleted from the mineral solution is an essential nutrient, then the incomplete medium will cause plants to become abnormal in appearance compared with controls grown on a complete mineral medium. The most common symptoms of a mineral deficiency are stunted growth and discolored leaves.

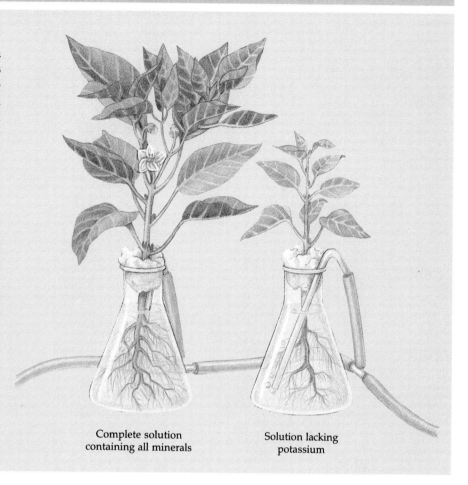

Complete solution containing all minerals

Solution lacking potassium

tion, and the water that is retained by the plant actually functions as a solvent, makes cell elongation possible, and serves to maintain the form of soft tissue by keeping cells turgid. By weight, the bulk of the organic material of a plant is derived from the CO_2 that is assimilated from the atmosphere.

We can measure water content by comparing the weight of plant material before and after it is dried. We can then analyze the chemical composition of the dry residue. Organic substances account for about 95% of the dry weight, with inorganic minerals making up the remaining 5%. Most of the organic material is carbohydrate, including the cellulose of cell walls. Thus, carbon, oxygen, and hydrogen, the ingredients of carbohydrates, are the most abundant elements in the dry weight of a plant. Because some organic molecules contain nitrogen, sulfur, or phosphorus, these elements are also relatively abundant in plants.

More than 50 chemical elements have been identified among the minerals present in plants, but it is unlikely that all of these elements are essential. Roots are able to absorb minerals somewhat selectively, en-abling the plant to accumulate essential elements that may be present in the soil in very minute quantities. To a certain extent, however, the minerals in a plant reflect the composition of the soil in which the plant is growing. Plants growing on mine tailings, for instance, may contain gold or silver. Studying the chemical composition of plants provides clues about their nutritional requirements, but we must distinguish elements that are essential from those that are merely present in the plant.

Essential Nutrients

A particular chemical element is considered to be an essential nutrient if it is required for a plant to grow from a seed and complete the life cycle, producing another generation of seeds. A method known as **hydroponic culture** can be used to determine which of the mineral elements are actually essential nutrients (see Methods Box). Such studies have helped to identify 16 elements that are essential nutrients in all plants

Table 33.1 Essential nutrients in plants

Element	Form Available to Plants	% of Dry Weight	Major Functions	Some Symptoms of Deficiency
Macronutrients				
Oxygen	O_2, H_2O	45.0	Major component of plant's organic compounds	*
Carbon	CO_2	45.0	Major component of plant's organic compounds	*
Hydrogen	H_2O	6.0	Major component of plant's organic compounds	*
Nitrogen	NO_3^-, NH_4^+	1.5	Component of nucleic acids, proteins, hormones, coenzymes, etc.	Stunted growth, chlorosis (yellowing) of leaves; affects whole plant
Potassium	K^+	1.0	Cofactor functional in protein synthesis; osmosis, operation of stomata	Chlorosis, necrosis (spots of dead tissue), weak stems and roots; older leaves most affected
Calcium	Ca^{2+}	0.5	Important in formation and stability of cell walls; maintenance of membrane structure and permeability; activates some enzymes	Death of shoot and root tips; young leaves and shoots most affected
Magnesium	Mg^{2+}	0.2	Component of chlorophyll; activates many enzymes	Chlorosis of leaves; older leaves most affected
Phosphorus	$H_2PO_4^-$, HPO_4^-	0.2	Component of nucleic acids, phospholipids, ATP, several coenzymes	Stunted growth, plant dark green; affects entire plant
Sulfur	SO_4^-	0.1	Component of proteins, coenzymes	Chlorosis, with veins remaining dark and tissue between light; affects young leaves
Micronutrients				
Chlorine	Cl^-	0.010	Activates photosynthetic elements; functions in water balance	Wilted leaves, stunted roots, chlorosis, necrosis
Iron	Fe^{3+}, Fe^{2+}	0.010	Component of cytochromes; may activate some enzymes	Chlorosis of tissue between veins, stems short and slender; affects young leaves
Boron	$H_2BO_3^-$	0.002	Uncertain; may be involved in carbohydrate transport and nucleic acid synthesis	Death of stem and root apical meristem, leaves twisted; young tissue most affected
Manganese	Mn^{2+}	0.0050	Active in formation of amino acids; activates some enzymes	Chlorosis of young leaves, with smallest veins remaining green; necrosis between veins
Zinc	Zn^{2+}	0.0020	Active in formation of chlorophyll; activates some enzymes	Reduced leaf size, shortened internodes, chlorosis, spotted leaves; older leaves most affected
Copper	Cu^+, Cu^{2+}	0.0006	Component of many redox and lignin-biosynthetic enzymes	Young leaves dark green, twisted, wilted; tip remains alive
Molybdenum	$MO_4^=$	0.00001	Essential for nitrogen fixation; cofactor functional in nitrate reduction	Chlorosis, twisting, death of young leaves

*Rarely limiting enough as nutrients to cause specific symptoms.

and a few other elements that are essential to certain groups of plants (Table 33.1). Most research has involved crop plants; little is known about the specific nutritional needs of uncultivated plants, even some of the most commercially important conifers.

Macronutrients Elements required by plants in relatively large amounts are called **macronutrients.** There are nine macronutrients in all, including the six major ingredients of organic compounds: carbon, oxygen, hydrogen, nitrogen, sulfur, and phosphorus. The other three macronutrients are calcium, potassium, and magnesium. Calcium is found in cell walls, where it combines with pectins in the middle lamella to form a glue that holds plant cells together. Free (soluble) calcium helps regulate the selective permeability of membranes and has other important regulatory functions in the plant. Potassium is the major solute for osmotic regulation in plants. For instance, when a plant cell elongates, it accumulates K^+ in its central vacuole, which causes water to enter the cell by osmosis. Potassium is also important as an activator of several enzymes. Magnesium is a basic component of chlorophyll, the photosynthetic pigment. This mineral is also a cofactor required for the activity of several enzymes (recall from Chapter 6 that a cofactor is a substance that cooperates with an enzyme in catalysis).

Micronutrients   Elements that plants need in very small amounts are called **micronutrients.** The seven micronutrients are iron, chlorine, copper, manganese, zinc, molybdenum, and boron. These elements function in the plant mainly as cofactors of enzymatic reactions. Iron, for example, is a metallic component of cytochromes, the proteins that function in the electron transport chains of chloroplasts and mitochondria. It is because micronutrients generally serve catalytic roles that plants need only minute quantities of these elements. The requirement for molybdenum, for example, is so modest that there is only 1 atom of this rare element for every 16 million atoms of hydrogen in dried plant material. Yet a deficiency of molybdenum or any other micronutrient can weaken or kill a plant.

The optimal concentrations of micronutrients vary for different plant species, and some species need additional elements that are not required by others. For instance, sodium is an essential nutrient for plants that use the C_4 pathway to fix carbon during photosynthesis.

Additional micronutrients may exist that have not yet been identified. The hydroponic method determines if elements are essential by testing the effect of their deletion from the mineral solution, but even the slightest traces of a micronutrient present in the seed or as a contaminant of the glassware, water, or other chemicals used in the hydroponic culture may satisfy the needs of the plant. Only recently, for instance, have methods for purifying water become thorough enough for chlorine to be identified as an essential nutrient.

Mineral Deficiencies

The symptoms of a mineral deficiency depend partly on the function of that nutrient in the plant (see Table 33.1). A magnesium deficiency, for example, reduces the amount of chlorophyll that can be synthesized and

Figure 33.3
Magnesium deficiency in tomato plants. Yellowing of leaves (chlorosis) is caused by inability to synthesize chlorophyll, which contains magnesium.

causes yellowing of the leaves, or **chlorosis** (Figure 33.3). Since calcium is required for the synthesis of new cell walls, a deficiency of this mineral retards growth of roots and shoots because the cells in the apical meristems cannot complete their division without making new cell walls. In some cases, the relationship between a mineral deficiency and its symptoms is less direct. For instance, iron deficiency can cause chlorosis even though chlorophyll contains no iron because this metal is required as a cofactor in one of the steps of chlorophyll synthesis.

The symptoms of a mineral deficiency depend not only on the role of that nutrient in the plant but also on its mobility within the plant. If a nutrient moves about freely from one part of the plant to another, symptoms of a deficiency will show up first in older organs, because young, growing tissues have more "drawing power" than old tissues for nutrients that are in short supply. A plant starved for magnesium, for example, will show signs of chlorosis first in its older leaves. Magnesium, which is relatively mobile in the plant, is shunted preferentially to young leaves. On the other hand, a deficiency of a nutrient that is relatively immobile within a plant will affect young parts of the plant first. Older tissues may have adequate amounts of the mineral, which they are able to retain during periods of short supply. A deficiency of iron, which does not move freely in the plant, will cause yellowing of young leaves before we can see any effect on older leaves.

The symptoms of a mineral deficiency are often distinctive enough for a plant physiologist or farmer to diagnose its cause. One way to confirm the diagnosis of a specific deficiency is to analyze the mineral content of the plant and soil. Deficiencies of nitrogen, potassium, and phosphorus are the most common problems; shortages of micronutrients are less common and tend to be geographically localized because of differences in soil composition. The amount of a micronutrient needed to correct a deficiency is usually quite small. For example, to treat a zinc deficiency in fruit trees, it is often enough to hammer a few zinc nails into each tree trunk. Moderation is important because overdoses of some micronutrients can be toxic to plants.

One way to ensure optimal mineral nutrition is to grow plants hydroponically on nutrient solutions that can be precisely regulated (Figure 33.4). Hydroponics is currently practiced commercially, but only on a limited scale because it is a very expensive way to grow food compared with growing crops on soil.

Before we turn to the ways that plants assimilate their nutrients, let us consider one of the major environmental factors in plant nutrition, the soil.

SOIL

The texture and chemical composition of soil are major factors determining what kinds of plants can grow well in a particular location, be it a natural ecosystem or an agricultural region. (Climate, of course, is another important factor.) Plants that grow naturally in a certain type of soil are adapted to its mineral content and texture, and are able to absorb water and extract essential nutrients from that soil. In interacting with the soil that supports their growth, plants in turn affect the characteristics of the soil.

Texture and Composition of Soils

Soil has its origin in the weathering of solid rock. Water that seeps into crevices and freezes in winter fractures the rock, and acids dissolved in the water also help to break down the rock. Once organisms are able to invade the rock, they accelerate the decomposition. Lichens, fungi, bacteria, mosses, and plant roots all secrete acids, and the expansion of roots growing in fissures cracks rocks and pebbles into smaller pieces. The eventual result of all this activity is **topsoil**, a mixture of decomposed rock of varying texture, living organisms, and humus, a residue of partially decayed organic material. The topsoil and other distinct soil layers, or **horizons**, are often visible in vertical profile where there is a roadcut (Figure 33.5).

The texture of a topsoil depends on the size of its particles, which are classified in a range going from coarse sand to microscopic clay particles (Table 33.2).

Figure 33.4
Hydroponic farming. In this apparatus, a nutrient solution flows over the roots of lettuce growing on a slat. Perhaps astronauts living in a space station will one day grow their vegetables hydroponically, but because of the expense, it is unlikely that this type of farming will soon relieve hunger here on Earth.

The most fertile soils are usually **loams,** made up of a mixture of sand, silt, and clay. Loamy soils have enough fine particles to provide a large surface area for retaining water and minerals but enough coarse particles to provide air spaces containing oxygen that can be used by roots for cellular respiration. If soil does not drain adequately, roots suffocate because the air spaces are replaced by water; the roots may also be attacked by molds favored by the soaked soil. These are common hazards for houseplants that are over-

Table 33.2 Sizes of soil particles	
Type of Particles	**Diameter of Particles**
Coarse sand	0.2–2 mm
Sand	20–200 μm
Silt	2–20 μm
Clay	Less than 2 μm

Figure 33.5
Soil horizons. This roadcut exposes a vertical profile of three soil layers, or horizons. The A horizon is the topsoil. It is a mixture of decomposed rock of varying textures, living organisms, and decaying organic matter. Topsoil is subject to extensive physical and chemical weathering. The B horizon contains much less organic matter than the A horizon and is less weathered. Clay particles and minerals leached from the A horizon by water may accumulate in the B horizon. The C horizon serves as the parent material for the upper layers of soil. This horizon is mainly composed of partially weathered rock.

watered in pots that do not drain. Some plants, however, are adapted to waterlogged soil. For example, mangroves and many other plants that inhabit swamps and marshes have some of their roots modified as hollow tubes that grow upward and function as snorkels bringing down oxygen from the air. Variation in the degree of drainage required by plants is one of the many ways soil conditions help determine what type of vegetation prevails in a particular location.

Soil is home to an astonishing number and variety of organisms. A teaspoon of soil has about five billion bacteria that cohabit with various fungi, algae and other protists, insects, earthworms, nematodes, and the roots of plants. The activities of all of these organisms affect the physical and chemical properties of the soil. Earthworms, for instance, aerate the soil by their burrowing and add mucus that holds fine soil particles together. The metabolism of bacteria alters the mineral composition of the soil. Plant roots extract water and minerals but also affect soil pH and reinforce the soil against erosion.

Humus, the decomposing organic material formed by the action of bacteria and fungi on dead organisms feces, fallen leaves, and other organic refuse, prevents clay from packing together and builds a crumbly soil that retains water but is still porous enough for good aeration of roots. Humus is also a reservoir of mineral nutrients that are returned gradually to the soil as microorganisms decompose the organic matter.

Availability of Soil Water and Minerals

After a heavy rainfall, water drains away from the larger spaces of the soil, but smaller spaces retain water because of its attraction for the soil particles. Some of this water is imbibed so tightly by the hydrophilic soil

particles that it cannot be extracted by plants. In the tiniest spaces of the soil, however, is a film of water bound less tightly to the particles; this water is generally available to the plant (Figure 33.6). It is not pure water but in fact a soil solution containing dissolved minerals. The soil solution can diffuse into the root via the apoplast up to the endodermis (see Chapter 32).

Many minerals in soil, especially those that are positively charged, such as potassium (K^+), calcium

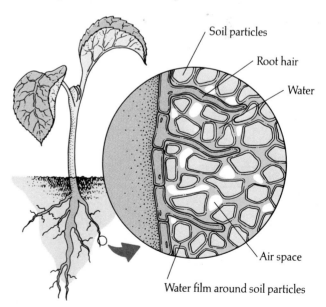

Figure 33.6
Availability of soil water. A plant cannot extract all of the water in the soil because some of it is so tightly held by hydrophilic soil particles. Water retained in tiny spaces in the soil, bound less tightly to soil particles, can be imbibed by the root, especially by the epidermis, which includes the root hairs.

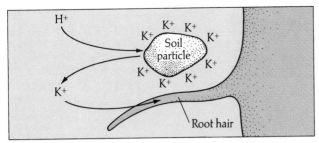

Figure 33.7
Cation exchange in soil. Hydrogen ions in the soil solution help make certain nutrients available to plants by displacing positively charged minerals that were bound tightly to the surface of fine soil particles. Secretion of hydrogen ions from roots facilitates cation exchange.

(Ca^{2+}), and magnesium (Mg^{2+}), adhere by electrical attraction to the negatively charged surfaces of clay particles. The presence of clay in a soil helps prevent the leaching (draining away) of mineral nutrients during heavy rain or irrigation because the finely divided particles provide so much surface area for binding minerals. Minerals that are negatively charged, such as nitrate (NO_3^-), phosphate ($H_2PO_4^-$), and sulfate (SO_4^{2-}), are usually not bound tightly to soil particles and thus tend to leach away more quickly. On the other hand, clay particles must release their bound minerals to the soil solution for roots to absorb the nutrients. Positively charged minerals are made available to the plant when hydrogen ions in the soil displace the mineral ions from the clay particles. This process, called **cation exchange**, is stimulated by the roots themselves, which release acids that add hydrogen ions to the soil solution (Figure 33.7).

As roots absorb water and minerals, the supply of these resources is depleted in the immediate surroundings of the existing roots. However, rapid growth of roots, with surface areas expanded by root hairs, brings the plant into contact with a larger and larger volume of soil and thus a new supply of nutrients.

Soil Management

It may take centuries for a soil to become fertile through decomposition and accumulation of organic material, but human mismanagement can reduce that fertility within a few years. Good soil management, on the other hand, can preserve fertility, resulting in sustained agricultural productivity.

To understand soil management, we must begin with the premise that agriculture is unnatural. In forests, grasslands, and other natural ecosystems, mineral nutrients are usually recycled by the decomposition of dead organic material in the soil. In contrast, when farmers harvest a crop, or a gardener uses a grass catcher on the lawn mower, essential elements are diverted from the chemical cycles going on in that location. In general, agriculture depletes the mineral content of the soil. To grow a ton of wheat grain, the soil gives up 18.2 kg of nitrogen, 3.6 kg of phosphorus, and 4.1 kg of potassium. Each year the fertility of the soil diminishes, unless fertilizers are applied to replace the lost minerals. Many crops also use far more water than the natural vegetation that once grew on that land, forcing farmers to irrigate. Prudent fertilization, thoughtful irrigation, and prevention of erosion are three of the most important aspects of soil management.

Fertilizers Prehistoric farmers may have started fertilizing their fields after noticing that grass grew faster and greener where animals had defecated. The Romans used manure to fertilize their crops, and North American Indians buried fish along with seeds when they planted corn. Today in developed nations, most farmers use commercially produced fertilizers containing minerals that are either mined or prepared by industrial processes. These fertilizers are usually enriched in nitrogen, phosphorus, and potassium, the three mineral elements that are most commonly deficient in farm soils. Commercial fertilizers are labeled with a three-number code that indicates the mineral content. A fertilizer marked "10-12-8," for instance, is 10% nitrogen (as ammonium or nitrate), 12% phosphorus (as phosphoric acid), and 8% potassium (as the mineral potash).

Manure, fishmeal, and compost are referred to as "organic" fertilizers because they are of biological origin and contain organic material that is in the process of decomposing. However, before the elements in compost can be of any use to plants, the organic material must be decomposed to the inorganic nutrients that roots can absorb. In the end, the minerals a plant extracts from the soil are in the same form whether they came from organic fertilizer or from a chemical factory. Compost releases minerals gradually, however, whereas the minerals in commercial fertilizers are available immediately, but may not be retained by the soil for long. Excess minerals not taken up by the plants are wasted because they may be rapidly leached from the soil by rainwater or irrigation. As a side effect, this mineral runoff may enter the groundwater and eventually pollute streams and lakes. These problems can sometimes be reduced by spraying a commercial fertilizer solution directly onto the leaves of the plants. Although the leaves are covered by the cuticle, enough minerals may enter the leaves to nourish the plants. Commercial fertilizers have sometimes been abused, but without their use, agricultural productivity would be unacceptably reduced. Agricultural researchers are

(a)

(b)

Figure 33.8
Irrigation. (a) Flood irrigation. After a field is flooded, much of the water evaporates, leaving the salts that were dissolved in the irrigation water behind to accumulate in the soil. (b) Drip irrigation. Instead of flooding the field or filling ditches with water, perforated pipes drip the water slowly into the soil close to the plant roots. Drip irrigation reduces the loss of water from evaporation and drainage. Both photos were taken in citrus groves in a Southern California desert.

attempting to develop ways to reduce fertilizer inputs while maintaining crop yields.

To fertilize judiciously, the farmer must pay close attention to the pH of the soil. Acidity not only affects cation exchange but also influences the chemical form of all minerals. Even though an essential element may be abundant in the soil, plants may be starving for that element because it is bound too tightly to clay or it is in a chemical form the plant cannot absorb. Managing the pH of soil is touchy; a change in hydrogen ion concentration may make one mineral more available to the plant while causing another mineral to become less available. At pH 8, for instance, the plant can absorb calcium, but iron is almost completely unavailable. The pH of the soil should be matched to the specific mineral needs of the crop. If the soil is too alkaline, sulfate can be added to lower the pH. Soil that is too acidic can be adjusted by "liming" (adding calcium carbonate or calcium hydroxide).

Irrigation   Even more than mineral deficiencies, the availability of water most often limits the growth of plants. Irrigation can transform a desert into a garden, but farming in arid regions is a huge drain on water resources. Many of the rivers in the southwestern United States have been reduced to trickles by the diversion of water for irrigation (quenching the thirst of growing cities adds to the problem). Another problem is that irrigation in an arid region can gradually make the soil so salty that it becomes completely infertile (Figure 33.8a). As the world population continues to grow, more and more acres of arid land will have to be cultivated. New methods of irrigation may reduce the risks of running out of water or losing farmland to salinization (salt accumulation). For instance, drip irrigation is now used to water many of the crops and orchards in Israel (Figure 33.8b). In another approach to solving some of the problems of dryland farming, plant breeders are working to develop varieties of plants that require less water or that can tolerate more salinity.

Erosion   Thousands of acres of farmland are lost to water and wind erosion each year in the United States alone (Figure 33.9). Certain precautions can help reduce these losses. Rows of trees dividing fields make effective windbreaks, and terracing a hillside can prevent the topsoil from washing away in a heavy rain. Crops such as alfalfa and wheat provide good ground cover and protect the soil better than corn and other crops that are usually planted in rows.

If managed properly, soil is a renewable resource on which farmers can grow food for generations to come.

Figure 33.9
Erosion in a corn field. A field that has just been tilled is especially vulnerable to erosion because it is barren. Farmers till before planting, not only to aerate the soil, but also to plow under weeds and the stubble left from the previous crop. A method known as minimal tillage farming can reduce erosion. The land is tilled less often, and instead herbicides are used to kill weeds.

NITROGEN ASSIMILATION BY PLANTS

Of all mineral elements, nitrogen is the one that most often limits the growth of plants and the yields of crops. Plants require nitrogen as an ingredient of pro-teins, nucleic acids, and other important organic molecules. It is ironic that plants sometimes suffer nitrogen deficiencies, for the atmosphere is nearly 80% nitrogen. This atmospheric nitrogen, however, is gaseous N_2, and plants cannot use nitrogen in that form. For plants to absorb nitrogen, it must first be converted to ammonium (NH_4^+) or nitrate (NO_3^-). Though these minerals are present in the soil, they are not derived from the decomposition of the parent rock, but are produced from atmospheric N_2, primarily by the metabolism of certain bacteria living in the soil, and by microbes decomposing humus. The cycling of nitrogen in ecosystems is described in detail in Chapter 49. Here we are concerned only with the steps of the cycle that lead directly to nitrogen assimilation by plants (Figure 33.10).

Nitrogen Fixation

Living in the soil are several genera of bacteria possessing **nitrogenase**, an enzyme that reduces N_2 by adding hydrogen ions and electrons to form ammonia (NH_3):

$$N_2 + 6\,H^+ + 6e^- \xrightarrow{\text{nitrogenase}} 2\,NH_3$$

This process, called **nitrogen fixation**, is very expensive in energy, costing the bacteria at least 12 ATPs for each ammonia molecule synthesized. Nitrogen-fixing bacteria are most abundant in soils rich in organic material, which provides fuel for cellular respiration.

In the soil solution, ammonia picks up another hydrogen ion to form ammonium (NH_4^+), which plants

Figure 33.10
Assimilation of nitrogen by plants. Soil bacteria form ammonium by fixing atmospheric N_2 (nitrogen-fixing bacteria) and by decomposing organic material (ammonifying bacteria). Plants absorb mainly nitrate from the soil. The nitrate is transported to the leaves, where it is reduced to ammonium, which is then used to make amino acids and other nitrogen-containing organic compounds.

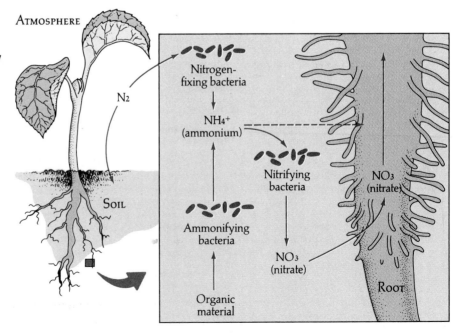

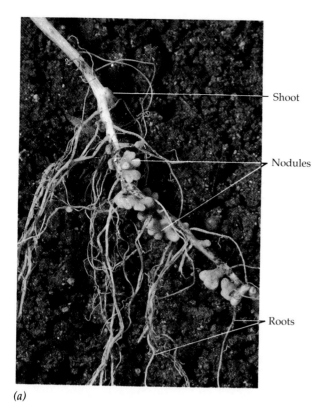

(a)

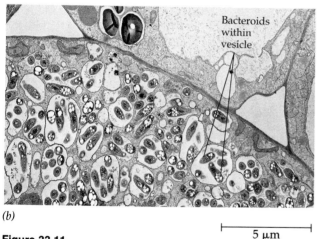

(b)

Figure 33.11
Root nodules on a legume. (a) The nodules of this pea root contain symbiotic bacteria that fix nitrogen and obtain photosynthetic products supplied from the plant. (b) In this electron micrograph, a cell from a root nodule of soybean is filled with bacteroids. The adjoining cell remains uninfected.

5 μm

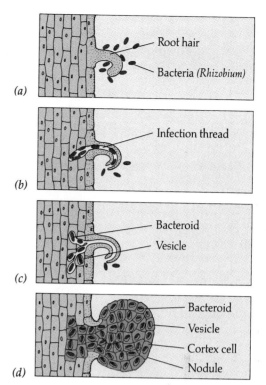

(a)

(b)

(c)

(d)

Figure 33.12
Development of a root nodule. (a) Root hairs and *Rhizobium* of the appropriate species are believed to recognize one another by molecular identification tags on their surfaces. Root hairs will curl to prepare for infection by the rhizobial bacteria. (b) Bacteria enter the root through an "infection thread" that elongates and carries the bacteria into the cortex of the root. (c) Enclosed within vesicles in the host cells of the root cortex, the bacteria, or bacteroids as they are now called, produce a chemical that induces their host cells to grow and divide. (d) These dividing cells form a mass, the nodule, containing bacteroids that cooperate with the plant to fix atmospheric nitrogen.

can absorb. However, plants acquire their nitrogen mainly in the form of nitrate (NO_3^-), which is produced in the soil by bacteria that oxidize ammonium (Figure 33.10). After nitrate is absorbed by roots, most of it is transported in the xylem to the leaves. Most of the nitrate is reduced to ammonium in the leaves by nitrate reductase and other enzymes. The plant can then use the ammonium to synthesize amino acids.

Symbiotic Nitrogen Fixation

Plants of the legume family, including peas, beans, soybeans, peanuts, alfalfa, and clover, have a built-in source of fixed nitrogen (Figure 33.11). Their roots may have swellings called **nodules** composed of plant cells that contain nitrogen-fixing bacteria of the genus *Rhizobium* (which means "root living"). Inside the nodule, the *Rhizobium* assume a form called **bacteroids.** Each legume is associated with a particular species of *Rhizobium*. Figure 33.12 describes the steps in the development of root nodules.

The symbiotic relationship between a legume and its nitrogen-fixing bacteria is mutual, with both partners benefiting. The bacteria supply the legume with fixed nitrogen, and the plant provides the bacteria with carbohydrates and other organic compounds. The exquisite coevolution of partners is evident in their cooperative synthesis of a molecule named **leghe-moglobin,** with the plant and the bacteria each making part of the molecule. Leghemoglobin is an iron-containing protein that, like the hemoglobin of our red blood cells, binds to oxygen (the *leg* is for legume). Leghemoglobin releases oxygen for the intense respiration required to produce ATP for nitrogen fixation. More important, however, leghemoglobin keeps the concentration of free oxygen low in root nodules, an important function because the enzyme nitrogenase is inhibited by oxygen.

Most of the ammonium produced by symbiotic nitrogen fixation is used by the nodules to make amino acids, which are then transported to the shoot and leaves via the xylem. When conditions are favorable, root nodules fix so much nitrogen that they actually secrete the excess ammonium, which increases the fertility of the soil for nonlegumes. This is the basis for crop rotation. One year a nonlegume such as corn is planted, and the following year alfalfa or some other legume is planted to restore the concentration of fixed nitrogen in the soil. Instead of being harvested, the legume crop is often plowed under so that it will decompose and add even more fixed nitrogen to the soil. To ensure that the legume encounters its specific *Rhizobium*, the seeds are soaked in a culture of the bacteria or dusted with bacterial spores before sowing.

A few groups of nonlegumes, including alders and certain tropical grasses, host nitrogen-fixing actinomycetes (see Chapter 25). Rice, a crop of great commercial importance, benefits indirectly from symbiotic nitrogen fixation. Rice farmers culture a water fern called *Azolla* in their paddies. The fern has symbiotic cyanobacteria that fix atmospheric nitrogen and increase the fertility of the rice paddy. The growing rice shades and kills the *Azolla*, and decomposition of this organic material adds even more nitrogenous minerals to the paddy.

Improving the Protein Yield of Crops

The ability of plants to incorporate fixed nitrogen into proteins and other organic substances has a major impact on human welfare; the most common form of malnutrition in humans is protein deficiency. Either by choice or by economic necessity, the majority of people in the world have a predominantly vegetarian diet, and thus, particularly in the developing and underdeveloped countries, depend mainly on plants for protein. Unfortunately, many plants have a low protein content, and the proteins that are present may be deficient in one or more of the amino acids that humans need from their diet. Improving the quality and quantity of proteins in crops is a major goal of agricultural research.

Plant breeding has resulted in new varieties of corn, wheat, and rice that are enriched in protein. However, many of these "super" varieties have an extraordinary demand for nitrogen, which is usually supplied in the form of commercial fertilizer. The industrial production of ammonia and nitrate from atmospheric nitrogen is, like biological nitrogen fixation, very expensive in energy costs, and a chemical factory making fertilizer consumes large quantities of fossil fuels. Generally, the countries that most need high-protein crops are the ones least able to afford the fuel bill.

Improving the productivity of symbiotic nitrogen fixation is another strategy with the potential to increase protein yields of crops. Normally, the accumulation of fixed nitrogen in the nodules of legumes switches off the bacterial genes that code for nitrogenase and other enzymes involved in nitrogen fixation. Microbiologists have isolated mutant strains of *Rhizobium* that keep on making these enzymes even after fixed nitrogen accumulates. Farmers may someday grow plants infected with the mutant bacteria, a breakthrough that would increase the protein content of the legume crop and also add more fixed nitrogen to the soil. Geneticists are also working to improve the efficiency of symbiotic nitrogen fixation. The total food yield of a legume crop is often relatively low because so much of the carbohydrate produced in photosynthesis is consumed as energy for nitrogen fixation. Selecting for legume and *Rhizobium* varieties that can fix nitrogen at a lower cost in photosynthetic energy could be a boon to human nutrition.

Genetic engineering promises additional improvements in the protein yields of crops. Molecular biologists have already succeeded in transferring some of the genes required for nitrogen fixation from *Rhizobium* into other bacteria, and it may be possible through genetic engineering to create varieties of *Rhizobium* that can infect nonlegumes. Transplanting the genes for nitrogen fixation directly into the genomes of plants is also a possibility, using the plasmids of bacteria that cause a plant tumor called crown gall as vehicles for the gene transfer (see Chapter 19). This research has a long way to go from the laboratory to the field, but the prospect of wheat, potatoes, and other nonlegumes fixing their own nitrogen is a strong incentive for the work to continue.

SOME NUTRITIONAL ADAPTATIONS OF PLANTS

Adaptation arising from interactions of plants and their environment is one of the major themes of this unit, and we complete this chapter by examining some of the modifications for nutrition that have evolved among plants. Some plants parasitize other plants, others prey

(a)

Figure 33.13
Parasitic plants. (a) Mistletoe growing on an oak. **(b)** Dodder growing on a California pickleweed. **(c)** In this cross section of a host stem supporting dodder, a haustorium of the parasite can be seen tapping the host plant's vascular tissue for nutrients (LM).

on animals, and most use fungi to absorb minerals. The correlation of structure and function is manifest in the specialized anatomical equipment of these plants.

Parasitic Plants

The mistletoe we find tacked above doorways during the holiday season lives in nature as a parasite on oaks and other trees (Figure 33.13). Mistletoe is photosynthetic, but it supplements its nutrition by using projections called haustoria to siphon sap from the vascular tissue of the host tree. (These structures are analogous, not homologous, to the haustoria of parasitic fungi.) Some parasitic plants, such as dodder, have given up photosynthesis entirely, drawing all of their nutrients from other plants.

Plants called **epiphytes** (from the Greek *epi*, "upon," and *phyton*, "plant") are sometimes mistaken for parasites. An epiphyte is a plant that nourishes itself but grows on the surface of another plant, usually on the branches or trunks of tropical trees. An epiphyte is anchored to its living substratum by roots, but it absorbs water and minerals mostly from rain that falls on the leaves (Figure 33.14, p. 734). Examples of epiphytes are staghorn ferns, Spanish moss (actually an angiosperm), and many species of orchids and bromeliads.

Carnivorous Plants

Living in acid bogs and other habitats where soil conditions are poor, especially in nitrogen, are plants that fortify themselves by occasionally eating meat. These carnivorous plants make their own carbohydrates from photosynthesis, but they obtain some of their nitrogen and minerals by killing and digesting insects. Various kinds of insect traps have evolved by the modification of leaves (Figure 33.15). The traps are usually equipped with glands that secrete digestive juices.

Figure 33.14
Epiphytes. Though they are anchored by their roots to another plant, epiphytes, such as this bromeliad growing on a tree in the Florida Everglades, are not parasitic in their nutrition. Epiphytes absorb their own minerals and water from rain that collects on their leaves and make their own organic molecules by photosynthesis.

(a)

(b)

(c)

Figure 33.15
Carnivorous plants. (a) The Venus flytrap consists of two lobes (modified leaves) that close together rapidly enough to capture an insect. An insect that enters the trap touches sensory hairs, initiating an electrical impulse that triggers closure of the trap. Movement of the trap is essentially a very rapid growth response, with cells in the outer region of each lobe accumulating water and enlarging. This changes the shape of the lobes, bringing their margins together. Glands in the trap then secrete digestive enzymes, and nutrients are later absorbed by the leaves that make up the trap. In spite of its name, the flytrap catches fewer flies than it does ants and grasshoppers. **(b)** Sundews capture their prey in a sticky secretion that makes the trap like flypaper. Hairs on the trap bend over the insect and restrain it, and the entire leaf cups around the prey. The hairs then secrete digestive enzymes. **(c)** Pitcher plants use a pitfall to capture insects. Insects slip into a long funnel containing water in the bottom. After the insect drowns, it is digested by enzymes secreted into the water.

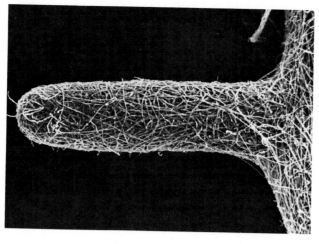

Figure 33.16
Mycorrhizae. Symbiotic associations between fungi and roots, mycorrhizae enhance the absorption of minerals. This scanning electron micrograph of fungal hyphae on a small root of a eucalyptus reveals the intimate association between the symbiotic partners of mycorrhizae.

100 μm

Mycorrhizae

Many plants have roots modified as **mycorrhizae** (from the Greek *mykos*, "fungus," and *rhiza*, "root"), which are actually symbiotic associations between the roots and fungi. The fungal partners of mycorrhizae are discussed in Chapter 28. The fungus either forms a sheath around the root or penetrates the root tissue, greatly increasing the surface area of the root through outward extensions of the hyphae (Figure 33.16). The fungus is proficient at absorbing minerals, especially phosphate, and it also may secrete acid that increases the solubility of some minerals and converts them to forms that are more readily used by the plant. The symbiosis is mutual; experiments have demonstrated that minerals taken up by the fungus are transferred to the plant and that photosynthetic products from the plant nourish the fungus. In addition to enhancing mineral nutrition, mycorrhizae also absorb water. The fungi may also help to protect the plant against certain soil pathogens.

Mycorrhizae are common adaptations of plants growing in poor soils, but almost all plants are capable of forming mycorrhizae if they are exposed to the appropriate species of fungi. As plant physiologists learn more about the development and function of mycorrhizae, important agricultural applications are likely. For example, inoculation of pine seeds with spores of mycorrhizal fungi promotes the formation of mycorrhizae by the seedlings. Pine seedlings so infected grow more vigorously than those trees without the fungal association.

Fossil evidence indicates that the earliest plants on land had mycorrhizae (see Chapter 27). This symbiosis of plants and fungi enabled roots to extract from soil enough minerals to supply the entire plant and thus may have contributed to the colonization of land. Mycorrhizae are an excellent example of the two biological themes of adaptation to the environment and the correlation of structure and function.

STUDY OUTLINE

1. As photosynthetic autotrophs, plants produce all their own organic compounds from nutrients consisting of carbon dioxide, water, and minerals.

Nutritional Requirements of Plants (pp. 722–726)

1. Water, the most abundant compound in most plants, is an essential nutrient in photosynthesis, as well as a solvent involved in cell elongation and growth.

2. Atmospheric carbon dioxide is incorporated into a plant's organic material, most of which is carbohydrate.

3. More than 50 inorganic minerals have been identified in plants. Such substances are selectively absorbed by roots, but also reflect the composition of the surrounding soil. Hydroponic culture has determined which of these minerals are actually essential for plant growth.

4. Plants require nine elements, the macronutrients, in fairly

large amounts. Carbon, oxygen, hydrogen, nitrogen, sulfur, and phosphorus are six elements required for the formation of organic compounds. Calcium, potassium, and magnesium, the remaining three essential macronutrients, serve a variety of essential regulatory functions.

5. Seven micronutrients are required in extremely small quantities: chlorine, iron, boron, manganese, zinc, copper, and molybdenum. These elements function mainly as cofactors in enzymatic reactions. Different plants vary in their optimal concentrations and specific requirements of micronutrients.

6. Mineral deficiencies reflect the composition of the soil and cause an array of symptoms that depend on the role of the nutrient and its mobility within the plant.

Soil (pp. 726–730)

1. Soil is one of the most important environmental factors interacting with plants.

2. Soil texture depends on the size of particles in topsoil. The most fertile soils are generally loams, which contain both fine particles that retain adequate water and minerals and coarse particles that provide adequate drainage.

3. An incredible number and variety of living organisms inhabit the soil. Some of these organisms are involved in producing humus, decomposing organic matter that improves the texture and mineral content of soil.

4. Fine particles of clay in soil are negatively charged and attract water and positively charged solutes. By releasing acids, roots obtain these solutes in a process called cation exchange. Constant expansion of the root system allows a continuing supply of mineral nutrients.

5. Unlike natural ecosystems, agriculture depletes the mineral content of soil, taxes water reserves, and encourages erosion. Proper soil management is essential to avoid squandering a precious resource that often takes centuries to acquire.

6. Organic fertilizers of biological origin have been used since antiquity to maintain or improve the quality of soil. Modern, commercial fertilizers contain minerals in pure form, sidestepping the degradative processes required by organic fertilizers. Today's farmers must use fertilizers wisely, considering such factors as method of application and pH of the soil.

7. Although irrigation can dramatically increase crop yields, it drains water reserves and contributes to a detrimental salt accumulation in the soil, factors that may be ameliorated through innovative irrigation methods and plant breeding.

8. The erection of natural windbreaks, the terracing of hillsides, and the planting of ground cover can greatly decrease wind and water erosion and help make soil a renewable resource.

Nitrogen Assimilation by Plants (pp. 730–732)

1. Plant growth and crop yields are often limited by nitrogen, an essential mineral ingredient of proteins and nucleic acids.

2. Although the atmosphere is rich in nitrogen, plants require the assistance of bacteria living in soil and humus to supply them with the forms of nitrogen they need. Such bacteria possess nitrogenase, an enzyme that converts atmospheric nitrogen into ammonia by an energy-expensive process called nitrogen fixation. The ammonia is then converted in the soil to nitrate or ammonium, which is absorbed by the plant.

3. The roots of legumes, alders, and some tropical grasses have nodular swellings that house nitrogen-fixing bacteria, which have coevolved with the plants into a mutualistic symbiotic relationship. Excess ammonium is secreted into the soil, where it becomes available to other plant species.

4. Improving the quality and quantity of protein in crops can help alleviate the problem of human malnutrition due to deficiencies in essential amino acids. Agricultural research in developing protein-rich plant breeds and enhancing nitrogen-fixing capabilities through selection and gene-splicing may yield practical benefits for a hungry world.

Some Nutritional Adaptations of Plants (pp. 732–735)

1. The modifications for nutrition that have evolved among plants illustrate the interactions of plants and their environment. The correlation between structure and function is clear in the anatomy of parasitic plants, carnivorous plants, and mycorrhizae.

2. Parasitic plants either supplement their photosynthetic nutrition or give up photosynthesis entirely by tapping into the vascular tissues of host plants. Epiphytes, often mistaken for parasites, are actually free-living organisms that use other plants simply for a substratum.

3. The chronically poor soil conditions of acid bogs are associated with the evolution of carnivorous plants. By a variety of methods, these plants obtain nitrogen and minerals by killing and digesting insects.

4. Mycorrhizae, mutualistic associations between roots and fungi, help the plant by enhancing mineral nutrition, water absorption, and pathogen resistance. The earliest land plants had mycorrhizae.

SELF-QUIZ

1. Most of the mass of organic material of a plant comes from
   a. water
   b. carbon dioxide
   c. soil minerals
   d. atmospheric oxygen
   e. nitrogen

2. Micronutrients are needed in very small amounts because
   a. most of them are mobile in the plant
   b. most function as cofactors in enzymes
   c. most are supplied in large enough quantities in seeds
   d. they play only a minor role in the health of the plant
   e. only the growing regions of the plants require them

3. Which of the following is a correct statement about nitrogen fixation?
   a. Plants convert atmospheric nitrogen to ammonia.
   b. Ammonia is converted to nitrates, which is the form of nitrogen most easily absorbed by plants.

c. Bacteria housed in mycorrhizae are capable of producing ammonium.

d. Mutant strains of *Rhizobium* are able to secrete excess nitrogen into the soil.

e. The enzyme nitrogenase reduces N_2 to form ammonia.

4. Which of the following is *not* a true comparison of organic or chemical fertilizers?

a. Organic fertilizers supply nutrients to plants more slowly.

b. Commercial fertilizers are generally more expensive to produce than organic fertilizers.

c. Chemical fertilizers are more likely to leach out of the soil than are organic fertilizers.

d. Organic fertilizers provide already formed organic molecules to plants, whereas commercial fertilizers provide only inorganic nutrients.

e. Commercial fertilizers are usually enriched in nitrogen, phosphorus, and potassium.

5. Mycorrhizae are

a. nodules containing nitrogen-fixing bacteria found on the roots of legumes

b. cellular extensions of parasitic plants that tap into the host's vascular tissue

c. plants that use other plants as a substratum, but do not take nutrients from the host

d. symbiotic associations between roots and fungi

e. root hairs resembling the hyphae of fungi

6. Which of the following nutrients is incorrectly paired with its function in a plant?

a. calcium—formation of cell walls, combines with pectin to glue cells together

b. magnesium—constituent of chlorophyll, cofactor for important enzymes

c. iron—component of chlorophyll

d. potassium—important in osmotic regulations, cofactor

e. phosphorus—component of ATP, phospholipids, nucleic acids

7. Most of the water taken up by the plant is

a. split during photosynthesis as a source of electrons and hydrogen

b. lost by transpiration through stomata

c. absorbed by cells during their elongation

d. returned to the soil by osmosis from the roots

e. incorporated directly into organic material

8. A mineral deficiency is likely to affect older leaves more than younger leaves if

a. the mineral is a micronutrient

b. the mineral is very mobile within the plant

c. the mineral is required for chlorophyll synthesis

d. the deficiency persists for a long period of time

e. the older leaves are in direct sunlight

9. Carnivorous adaptations of plants mainly compensate for soil that has a relatively low content of

a. potassium

b. nitrogen

c. calcium

d. water

e. phosphate

10. Based on our retrospective view, von Helmont's famous experiment on the growth of a willow tree mainly demonstrated that

a. the tree increased in mass mainly by absorbing water

b. the increase in the mass of the tree could not be accounted for by the consumption of soil

c. most of the increase in the mass of the tree was due to the uptake of O_2

d. soil simply provides physical support for the tree without providing nutrients

e. it should be possible to cultivate trees without soil

CHALLENGE QUESTIONS

1. An acre of corn actually yields more total protein than an acre of soybeans. Explain this.

2. Acid rain not only harms aquatic life, but can also affect the growth of plants growing on soil. Explain this in terms of mineral nutrition.

3. Having read this chapter, articulate some of the problems the world faces for future food production and outline some potential and time-tested remedies.

FURTHER READING

1. Brill, W. J. "Agricultural Microbiology." *Scientific American*, September 1981. Manipulating microorganisms that live with plants may benefit agriculture.
2. Brill, W. J. "Biological Nitrogen Fixation." *Scientific American*, March 1977. A clear description of an essential and complex process.
3. Epstein, E. "Rhizostats: Controlling the Ionic Environment of Roots." *BioScience*, November 1984. Computer technology has applications in plant nutrition research.
4. Gibbons, B. "Do We Treat Our Soil Like Dirt?" *National Geographic*, September 1984. Some thought-provoking perspectives on soil management.
5. Mohlenbrock, R. H. "Croatan National Forest, North Carolina." *Natural History*, June 1985. A glimpse of some insectivorous plants in the southeastern United States.
6. Pimental, D., and numerous coauthors. "World Agriculture and Soil Erosion." *Bioscience*, April 1987. Quantitative assessment of a serious threat to world food production.
7. Tangley, L. "Crop Productivity Revisited." *BioScience*, March 1986. An article supporting the emphasis on basic research in botany.

34

Plant Reproduction

Sexual Reproduction of Flowering Plants

Asexual Reproduction

It has been said that an oak is an acorn's way of making more acorns. Indeed, in the Darwinian view of life, the fitness of an organism is measured only by its ability to replace itself with healthy, fertile progeny. Consider the century plant. It lives for decades without flowering, and then one spring, the century plant grows a floral stalk that may be as tall as a telephone pole (Figure 34.1). That season the plant produces seeds and then withers and dies, its food reserves, minerals, and water spent in the formation of its massive bloom. Although not all flowering plants are so completely consumed as the century plant in leaving offspring, most of their other functions can be interpreted, in the broadest Darwinian sense, as mechanisms contributing to propagation.

Modifications in reproduction were key adaptations that enabled plants descended from aquatic ancestors to spread into a variety of terrestrial habitats. During the life cycles of most algae, flagellated sperm swim to eggs, and the offspring develop in the water without protection. In the conifers and angiosperms, the plants currently most widespread on land, pollen carried by wind or animals has replaced flagellated sperm as a means of bringing gametes together, and zygotes develop into embryos protected within seeds. Many plants can also reproduce without sex by mechanisms that promote propagation of successful individuals in specific environments. This chapter focuses on the reproduction of flowering plants, both sexual and asexual (the life cycles of other plant and algal groups are covered in Chapters 26 and 27). The emphasis is on the mechanics of reproduction and development of the offspring. The complex interactions of hormones and environmental cues that control these events are discussed in the next chapter.

Figure 34.1
Floral stalk of the century plant. After decades of vegetative growth, the century plant *Agave* produces its once-in-a-lifetime bloom and consumes all of its food reserves in the act of reproducing. Flowers, specialized reproductive structures, have contributed to making angiosperms the most diverse and widespread plants on Earth.

SEXUAL REPRODUCTION OF FLOWERING PLANTS

Figure 34.2 provides an overview of the angiosperm life cycle, which was introduced from an evolutionary perspective in Chapter 27. Here we examine more closely the structure and function of the reproductive equipment, looking first at the flower as a whole before turning to the details of gamete development, pollination, and fertilization.

The Flower

The angiosperm organ specialized for sexual reproduction is the flower, which is actually a compressed shoot with four whorls of modified leaves separated by very short internodes (Figure 34.3). These four floral parts, in their order from outside to inside of the flower, are the sepals, petals, stamens, and carpels. Usually green, the **sepals** of most flowers have retained a more leaflike appearance than the other floral parts. Sepals enclose and protect a floral bud before it opens. **Petals,** generally more brightly colored than sepals, advertise the flower to insects and other pollinators. Sepals and petals are sterile floral parts; that is, no gametes (sex cells) develop within them. Stamens and carpels, in contrast, are the fertile parts of flowers.

Each stamen consists of a stalk, the **filament,** and a terminal structure called the **anther.** Within the anther are usually four chambers, where the pollen grains develop. Pollen grains contain male gametes in the form of sperm nuclei.

A **carpel** has a slender neck, the **style,** leading to an **ovary** located at the base of the carpel. Developing within the ovary are one or more **ovules,** multicellular structures that each contain an egg, the female gamete. At the tip of the carpel is a sticky **stigma,** which serves as a landing platform for pollen brought from other flowers by wind or animals. Many flowers have two or more carpels fused to form an ovary having more than one chamber, or locule, where ovules develop. The term **pistil** is sometimes used for the "female" parts of the flower, whether they are separate carpels or formed from fused carpels.

A flower with all four kinds of structures—sepals, petals, stamens, and carpels—is said to be a **complete flower.** An **incomplete flower** is missing one or more of these parts. A flower that has both stamens and carpels is referred to as **perfect,** although it may be incomplete by lacking sepals or petals. If the missing parts of an incomplete flower are either stamens or carpels, then the flower is termed **imperfect.** An imperfect flower is either staminate (having stamens but no carpels) or carpellate (having carpels but no stamens). A plant species is said to be **monoecious** (from the Greek, meaning "one house") if it has both staminate flowers and carpellate flowers on the same individual plant (Figure 34.4a). A **dioecious** ("two houses") species has staminate flowers and carpellate flowers on separate plants, analogous to the exclusive possession of testes and ovaries by male and female animals (Figure 34.4b).

Alternation of Generations: A Review

As you may recall from Chapter 27, the life cycles of plants are characterized by an alternation of haploid (*N*) and diploid (2*N*) generations. The diploid plant, called the **sporophyte,** produces haploid spores by meiosis in specialized chambers called sporangia. A spore divides by mitosis to become a multicellular male or female **gametophyte,** the haploid generation, which produces gametes by mitosis. Union of haploid gametes

Figure 34.2
Review of the angiosperm life cycle.
Within the ovary of a flower, the egg of an ovule is fertilized by a sperm nucleus released from a pollen tube. After fertilization, the ovule matures into a seed containing the embryo, and the ovary develops into a fruit, which aids in dispersal of the seed. In a suitable habitat, the seed germinates, its embryo developing into a seedling.

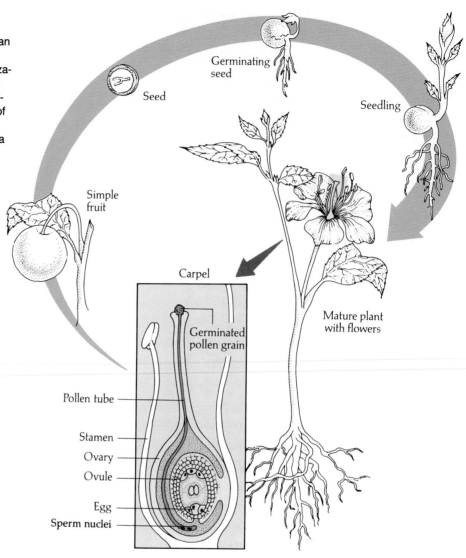

Figure 34.3
Structure of an idealized flower. Among the quarter of a million angiosperm species are many variations on this "basic flower plan"; there is really no such thing as a typical flower. Some of the trends in the evolution of flowers are discussed in Chapter 27.

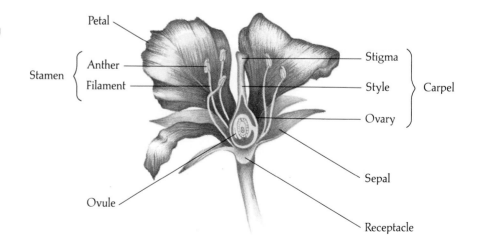

(a)

(b)

Figure 34.4
Monoecious and dioecious plants.
These terms apply to plants having imperfect flowers, those that lack either stamens or carpels. (a) Corn and other monoecious species have separate staminate flowers and carpellate flowers on the same individual plant. The "tassles" are the staminate flowers of corn, and the "ears" are derived from the carpellate flowers. (b) Dioecious species, such as these *Cannabis* plants, bear carpellate flowers (left) and staminate flowers (right) on separate individuals.

during fertilization results in a diploid zygote, which divides by mitosis and develops into a new sporophyte.

In angiosperms, the diploid sporophyte is the dominant generation in the sense that it is the conspicuous plant we see. The haploid gametophytes are reduced in size and develop while still dependent on their sporophyte parents. The modified leaves that make up the flower, including stamens and carpels, are part of the sporophyte.

It is convenient to speak of stamens and carpels as male and female sex organs, respectively, but this is technically incorrect. These structures produce spores, not gametes. However, the spores are retained within the flower, where they develop into tiny gametophytes that are totally dependent on the much larger sporophyte. A pollen grain is a male gametophyte not yet mature. The female gametophyte is a structure called an embryo sac that develops within an ovule. It is these gametophytes that produce sperm nuclei and eggs, the gametes that unite during fertilization to form the zygote. Thus, it is the gametophytes, not the sporophyte organs that house them, that are sexual. We can skirt this technicality by saying that the processes of sex—production of gametes and their fusion to form zygotes—take place *within* flowers.

In the following sections, we look in some detail at the development of gametophytes, pollination, and fertilization. As you read these sections, refer to Figure 34.5, which illustrates the different processes and their relationships. (There are many variations in details in these processes, depending on species.)

Development of Pollen

Within the sporangial chambers of an anther, diploid cells called microsporocytes undergo meiosis, each forming four haploid **microspores.** In each microspore, the haploid nucleus eventually divides once by mitosis to give rise to a generative nucleus and a tube nucleus. The wall of the microspore thickens and becomes sculptured into an elaborate pattern unique to the plant species. This pollen grain, with its two nuclei, is the male gametophyte in an immature form (Figure 34.6).

Pollen grains are extremely durable. Their tough coats are chemically different from other plant cell walls and are resistant to biodegradation. Fossilized pollen has provided many important clues to the evolutionary history of flowering plants.

About the time the anther opens to release pollen or after the pollen lands on a stigma, depending on species, the generative nucleus divides to form two **sperm nuclei,** which function as male gametes. The pollen grain is now ready to germinate as the male gametophyte on the stigma at the tip of a carpel.

Figure 34.5
Gametophyte development and fertilization. Pollen develops within anthers, and ovules form in ovaries. After a pollen grain is carried by wind or an animal to the stigma, a long pollen tube begins growing down the style toward the ovary. The tube discharges two sperm nuclei into the ovule. One nucleus fertilizes the egg to form the zygote, and the other combines with two polar nuclei to form a triploid cell that will develop into a nutritive tissue called endosperm.

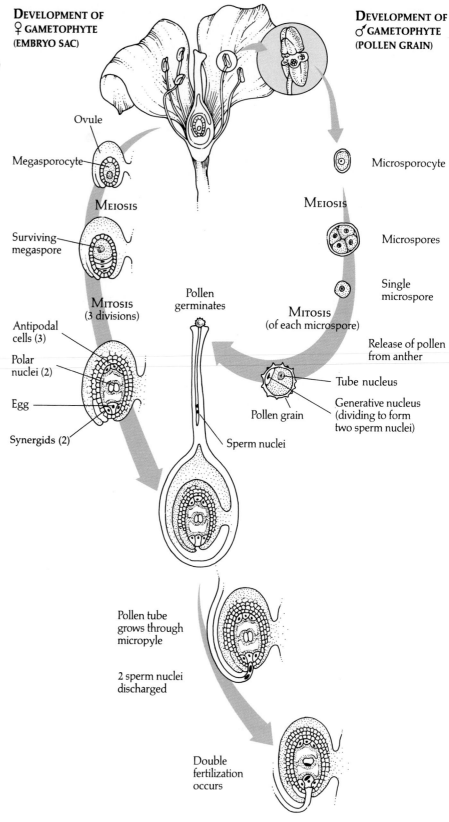

DEVELOPMENT OF
♀ GAMETOPHYTE
(EMBRYO SAC)

DEVELOPMENT OF
♂ GAMETOPHYTE
(POLLEN GRAIN)

Ovule

Megasporocyte

Microsporocyte

MEIOSIS

MEIOSIS

Surviving
megaspore

Microspores

Single
microspore

MITOSIS
(3 divisions)

Pollen
germinates

MITOSIS
(of each microspore)

Antipodal
cells (3)

Release of pollen
from anther

Polar
nuclei (2)

Tube nucleus

Egg

Generative nucleus
(dividing to form
two sperm nuclei)

Pollen grain

Synergids (2)

Sperm nuclei

Pollen tube
grows through
micropyle

2 sperm nuclei
discharged

Double
fertilization
occurs

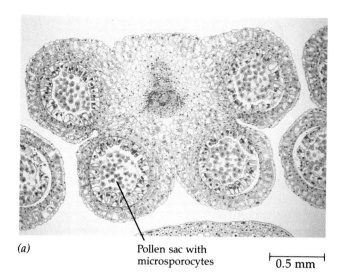

(a) Pollen sac with microsporocytes

0.5 mm

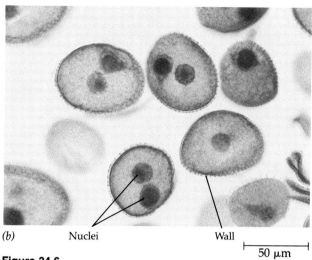

(b) Nuclei   Wall

50 μm

Figure 34.6
Pollen development. (a) This transverse section of an immature lily anther shows four pollen sacs (LM). **(b)** At the time of its release from the anther, a pollen grain generally has two nuclei, a generative nucleus and a tube nucleus (LM). **(c)** Species diversity in the sculpturing of pollen walls is displayed in this light micrograph.

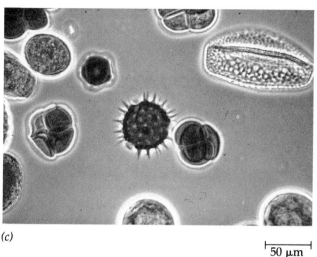

(c)

50 μm

Development of Ovules

Ovules, which contain the female gametophytes, form within the chambers of the ovary. One cell in the sporangium of each ovule, the megasporocyte, grows and then goes through meiosis to produce four haploid **megaspores.** In most angiosperms, only one of these survives. This megaspore continues to grow, and its nucleus divides by mitosis three times, resulting in one large cell with eight haploid nuclei. Membranes then partition this mass into a multicellular structure called the **embryo sac,** which is the female gametophyte (Figure 34.7). At one end of the embryo sac are two cells called synergids that flank the egg cell. At the opposite end are three antipodal cells. The other two nuclei, called polar nuclei, are not separated by membranes but share the cytoplasm of the large central cell of the embryo sac. The ovule consists of the embryo sac, or female gametophyte, along with its integuments, protective layers of sporophyte tissue around the embryo sac. At the end of the embryo sac containing the egg is an opening through the integ-

uments called the micropyle. Thus, in angiosperms, the female gametophyte (embryo sac) usually consists of seven cells (with eight nuclei). The female gametophytes in angiosperms are more reduced than those in gymnosperms, which consist of a multicellular structure with several archegonia (see Chapter 27).

Pollination and Fertilization

For the egg to be fertilized, the male and female gametophytes must meet and unite their gametes. The first step is **pollination,** the placement of pollen onto the stigma of a carpel. Some plants, including grasses and many trees, use wind as a pollinating agent. They compensate for this random dispersal by releasing enormous quantities of the tiny grains. At certain times of the year, the air is loaded with pollen dust, as anyone plagued with allergies to pollen can attest. Many angiosperms, however, do not rely on the aimless wind to carry pollen, but instead have relationships with animals that transfer pollen directly between flowers.

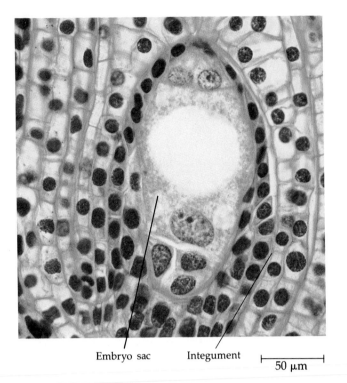

Embryo sac   Integument   ⊢——————⊣
                                       50 μm

Figure 34.7
The embryo sac of a lily ovule. An ovule consists of
the embryo sac, which is the female gametophyte, along
with its surrounding integuments (LM).

The coevolution of angiosperms and their pollinators
was discussed in Chapter 27.

Some flowers self-pollinate, but the majority of
angiosperms have mechanisms that make it difficult
or impossible for a flower to pollinate itself. In some
cases, the stamens and carpels of the flower mature
at different times. Many flowers that are pollinated by
animals are structurally arranged in such a way that
it is unlikely the pollinator could transfer pollen from
anthers to the stigma of the same flower. Other flowers are self-incompatible; if a pollen grain from an anther
happens to land on a stigma of the same flower, a
biochemical block prevents the pollen from completing its development and fertilizing an egg. Dioecious
plants, of course, cannot self-pollinate because they
are unisexual, being either staminate or carpellate. The
various mechanisms that prevent self-pollination contribute to genetic variety by ensuring that sperm and
eggs come from different parents.

After adhering to the sticky stigma, the pollen grain
germinates, growing a tube that extends down between
the cells of the style toward the ovary. Directed by a
chemical attractant, usually calcium, the tip of the pollen tube enters the ovary, probes through the micropyle, and discharges its two sperm nuclei within the
embryo sac (see Figure 34.5). One sperm fertilizes the
egg to form the zygote, and the other combines with
the two polar nuclei to form a triploid (3N) nucleus in

the center of the large central cell of the embryo sac,
which will give rise to the endosperm. This union
between two sperm and two cells in the embryo sac
of angiosperms, termed **double fertilization,** is unique
in all of life. After fertilization, the ovule begins developing into a seed.

The Seed

Endosperm   Endosperm development usually begins
before embryo development. After double fertilization, the triploid nucleus of the ovule's central cell
divides to form a multinucleate "supercell" having a
milky consistency. This mass, the **endosperm,** becomes
multicellular and more solid when cytokinesis forms
membranes and walls between the nuclei.

The endosperm is rich in nutrients, which it provides to the developing embryo. It also stockpiles
nutrients that can be used later when the seed germinates and the embryo grows into a seedling. In many
dicots, the food reserves of the endosperm are restocked in the cotyledons before the seed completes
its development, and consequently the mature seed
lacks endosperm.

Development of the Embryo   The first mitotic division of the zygote is transverse, splitting the egg into
a basal cell and a terminal cell (Figure 34.8). Only the
terminal cell goes on to form the embryo. The basal
cell continues to divide transversely to produce a thread
of cells called the suspensor, which will anchor the
embryo and transfer nutrients to it from the parent
plant. Meanwhile, the terminal cell divides several
times to form a spherical proembryo attached to the
suspensor. The cotyledons, or seed leaves, begin to
form as bumps on the proembryo. A dicot, with its
two cotyledons, will be heart-shaped at this stage.
Only one cotyledon develops in monocots.

Soon after the rudimentary cotyledons appear, the
embryo elongates during what is called the torpedo
stage. Cradled between the cotyledons is the apex of
the embryonic shoot, with its apical meristem. At the
opposite end of the embryo's axis, where the suspensor attaches, is the apex of the embryonic root, also
with a meristem. After the seed germinates, the apical
meristems at the tips of shoot and root will sustain
primary growth as long as the plant lives. The three
primary meristems—protoderm, ground meristem,
and procambium—are also present in the embryo.

Notice that the root–shoot polarity of the embryo is
determined with the very first division of the zygote.
In the egg, organelles and chemicals were distributed
heterogeneously throughout the cytoplasm; division
of the zygote segregates different cytoplasmic components among the embryonic cells. Although these
cells have equivalent nuclei, they will differ in cytoplasmic composition and in their locations within the

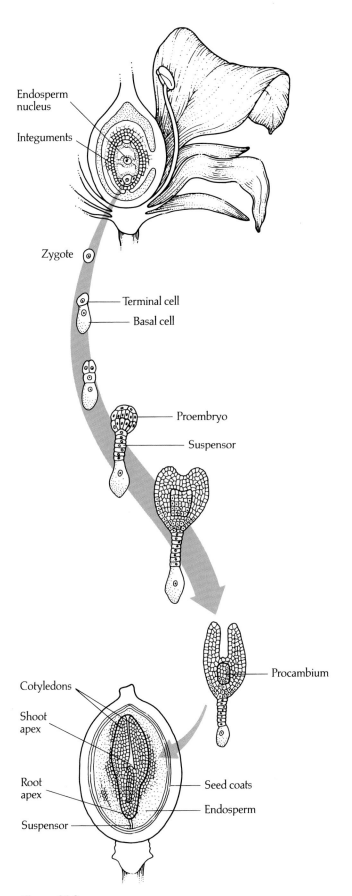

Figure 34.8
Development of a dicot plant embryo. By the time the ovule becomes a mature seed, the zygote has given rise to an embryonic plant with rudimentary organs.

Labels (top to bottom, left side):
Endosperm nucleus
Integuments
Zygote
Terminal cell
Basal cell
Proembryo
Suspensor
Procambium
Cotyledons
Shoot apex
Root apex
Suspensor
Seed coats
Endosperm

embryonic mass. Perhaps the cytoplasmic environment and position of an embryonic cell help to determine its developmental fate by providing cues to the genome within the nucleus. The means by which the cytoplasmic and extracellular environments influence the selective switching on and off of genes during cellular differentiation remains one of the most engaging mysteries in biology (see Chapter 18).

Structure of the Mature Seed During the last stages of its maturation, the seed dehydrates until its water content is only about 5%–15% of its weight. By now, the embryo has ceased growing and will remain quiescent until the seed germinates. The embryo is surrounded by its enlarged cotyledons or by endosperm or by both. The embryo and its food supply are enclosed by a **seed coat** formed from the integuments of the ovule, the progenitor of the seed.

We can take a closer look at one type of dicot seed by splitting open the seed of a common bean (Figure 34.9a). At this stage, the embryo is an elongate structure, the embryonic axis, attached to fleshy cotyledons that absorbed food from the endosperm when the seed developed. Below the point at which the cotyledons are attached, the embryonic axis is called the **hypocotyl** (from the Greek *hypo*, "under"). The hypocotyl terminates in the **radicle,** or embryonic root. The portion of the embryonic axis above the cotyledons is the **epicotyl** (from the Greek *epi*, "on" or "over"). At its tip is the plumule, consisting of the shoot tip with a pair of miniature leaves.

The seeds of some dicots, such as castor beans, retain their food supply in the endosperm and have cotyledons that are very thin (Figure 34.9b). The cotyledons will absorb nutrients from the endosperm and transfer them to the embryo when the seed germinates.

A corn kernel is an example of a monocot seed (Figure 34.9c). The single cotyledon, also called the scutellum, is very thin but has a large surface area, all the better to absorb nutrients from the endosperm during germination. The embryo is enclosed by a sheath consisting of a **coleorhiza,** which covers the root, and a **coleoptile,** which cloaks the embryonic shoot.

While the seeds develop from ovules, the ovary of the flower becomes transformed into a fruit, which protects the enclosed seeds and aids in their dispersal.

Development of the Fruit

A **fruit** is a ripened ovary. In some angiosperms, other floral parts also contribute to what we call a fruit in grocery store vernacular. The fleshy part of an apple, for instance, is derived mainly from the fusion of flower parts located at the base of the flower, but only the core is a true fruit—a ripened ovary.

The fruit begins to develop after pollination triggers hormonal changes that cause the ovary to grow tre-

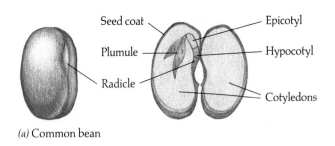

(a) Common bean

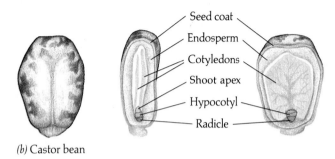

(b) Castor bean

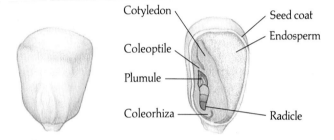

(c) Corn

Figure 34.9
Seed structure. **(a)** The fleshy cotyledons of the common garden bean store food that was absorbed from the endosperm when the seed developed. **(b)** The castor bean has membranous cotyledons that will absorb food from the endosperm when the seed germinates. **(c)** Corn, a monocot, has only one cotyledon. The rudimentary shoot is sheathed in a coleoptile.

(a)

(b)

(c)

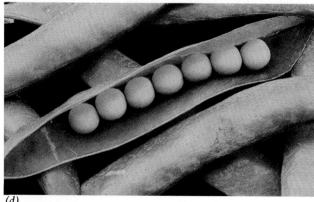

(d)

Figure 34.10
Fruit development. In this sequence of photos, the flower of a pea plant is transformed to the fruit as the wall of the ovary thickens to become the pod.

mendously. Its wall thickens, becoming the **pericarp** of the fruit (Figure 34.10). This transformation of the flower, called fruit set, parallels the development of the seeds. If a flower has not been pollinated, fruit usually does not set, and the flower withers and falls away. Some plants (called parthenocarpic plants) do produce fruit without fertilization of their ovules. An example of such a seedless fruit is the banana. Although sterile, banana pollen can apparently initiate fruit set.

Fruits are classified as one of several types, depending on their origin (Figure 34.11). A fruit derived from a single ovary is called a **simple fruit.** A simple fruit may be fleshy, such as a cherry, or dry, such as a soy-

(a)

(b)

(c)

(d)

Figure 34.11
Types of fruit. (a) A cherry is a simple, fleshy fruit derived from a single ovary. (b) The soybean is a simple, dry fruit. (c) The strawberry is an aggregate fruit derived from a flower with several carpels. (d) A multiple fruit, the pineapple develops from several separate flowers.

bean pod. An **aggregate fruit,** such as a strawberry, results from a single flower that has several separate carpels. A **multiple fruit** develops from an **inflorescence,** a group of separate flowers tightly clustered together. When the walls of the many ovaries start to thicken, they fuse together and become incorporated into one fruit. The pineapple is an example of a multiple fruit.

The fruit usually ripens about the time the seeds it contains are completing their development. For a dry fruit, ripening is little more than senescence (aging) of the fruit tissues. Ripening of a fleshy fruit is more elaborate, its steps guided by complex interactions of hormones, discussed in the next chapter. The "pulp" of the fruit becomes softer as a result of enzymes digesting components of the cell walls. There is usually a color change from green to some other color such as red, orange, or yellow. The fruit becomes

sweeter as organic acids or starch molecules are converted to sugar, which may reach a concentration of as much as 20% in a ripe fruit. Unripe fruit is sour to the taste because of the high acid concentration. The ripening of some fruits parallels a rise in cellular respiration that peaks at a stage called the climacteric, when the ripening processes are most intense. Not long after the climacteric, many fruits begin to spoil.

In addition to protecting seeds, fruits help disperse the seeds to a location some distance from the parent plants (Figure 34.12). Both wind and animals aid in seed dispersal.

By selectively breeding plants, humans have capitalized on the production of edible fruits. The apples, oranges, and other fruits we gather in grocery stores are amplified versions of much smaller, natural varieties that functioned as enticements to animals who spread seeds. However, the staple foods for humans

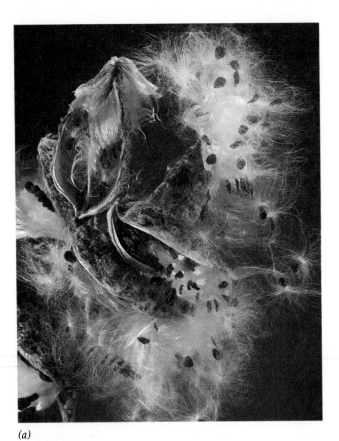

(a)

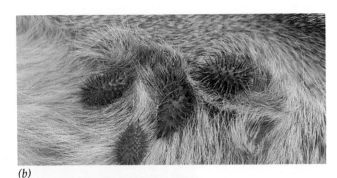

(b)

(c)

Figure 34.12
Adaptations of fruits for seed dispersal.
(a) The fruit of the milkweed aids in dispersal of the seed by wind. (b) Some fruits are modified as burrs that enable seeds to hitchhike on animals. These are cocklebur fruits attached to fur of a donkey. (c) The seeds of edible fruits are dispersed by animals. After a bird or mammal eats the fruit, the indigestible seeds pass through the digestive tract until they are sown, together with a little fertilizer, when the animal defecates.

are the wind-dispersed fruits of grasses, which are harvested while still on the parent plant. The cereal grains of wheat, rice, corn, and other grasses are easily mistaken for seeds, but they are actually fruits with dry pericarps that adhere tightly to the seed coats.

Seed Germination

To many people, the germination of a seed symbolizes the beginning of life, but in fact the seed already contains a miniature plant, complete with an embryonic root and shoot. At germination, the plant does not begin life but rather resumes the growth and development that was temporarily suspended when the seed matured and its embryo became quiescent. Some seeds germinate as soon as they are in a suitable environment. Other seeds are dormant and will not germinate, even if sown in a proper place, until a specific environmental cue causes them to break dormancy.

Seed Dormancy Evolution of the seed was one of the most important factors in the adaptation of plants to the special problems of living and reproducing on land (see Chapter 27). Compared with lakes and seas, terrestrial habitats are generally less stable, with environmental factors such as temperature and water availability fluctuating. Seed dormancy is an adaptation that increases the chances that germination will occur at a time and place most advantageous to success of the seedling. The conditions for breaking dormancy vary in ways that seem to be adaptive. Many desert plants, for instance, will not germinate unless there has been a substantial rainfall. If these seeds were to germinate after a modest drizzle, the soil might soon be too dry to support the seedlings. Where natural fires are common, such as the chaparral regions of California, natural selection has favored seeds that require intense heat to break dormancy. Seedlings will be most abundant after a brush fire has cleared away old, competing vegetation. Plants that live where win-

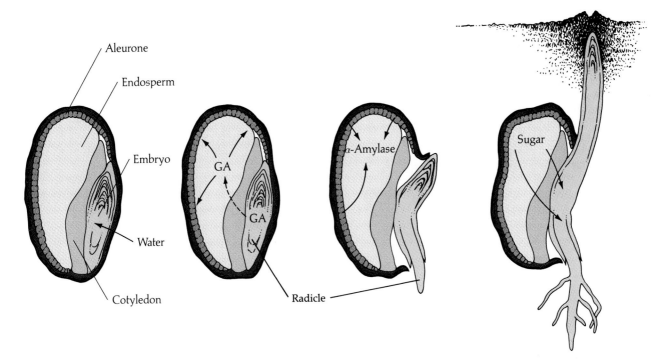

Figure 34.13

Mobilization of nutrients during germination of a cereal. After water is imbibed, the embryo releases a hormone (GA, or gibberellic acid) as a signal to the aleurone, which then synthesizes and secretes α-amylase and other digestive enzymes that hydrolyze stored foods in the endosperm, producing small, soluble molecules. Nutrients absorbed from the endosperm by the cotyledon are consumed during growth of the embryo into a seedling.

ters are harsh often have seeds that germinate only after an extended exposure to cold; seeds sown during summer or fall do not germinate until the following spring. Very small seeds, such as those of some lettuce varieties, require light for germination and will break dormancy only if they are buried shallow enough for the seedlings to poke through the soil surface. Some seeds have coats that must be scarified (abraded) before they can germinate. In some cases, tumbling over rocks in a stream scarifies the seeds. Other seeds must have their coats weakened by chemical attack as they pass through an animal's digestive tract. Such requirements also help ensure that seeds will be carried some distance before germinating.

The length of time a dormant seed remains viable and capable of germinating varies from a few days to decades or even longer, depending on species and environmental conditions. Most seeds are durable enough to last a year or two until conditions are favorable for germinating. Thus, the soil has a pool of ungerminated seeds that may have accumulated for several years. This is one reason vegetation can come back so rapidly after a fire, drought, flood, or some other catastrophe.

From Seed to Seedling   The first step in germination of many seeds is **imbibition,** the absorption of water by the dry seed. Hydration causes the seed to expand and rupture its coat, and also triggers a series of metabolic changes in the embryo that cause it to resume growth. Enzymes begin digesting the storage materials of the endosperm or cotyledons, and the nutrients are shunted to the growing regions of the embryo. This mobilization of food reserves has been studied most extensively in the seeds of barley and other cereals, so we will use a cereal as an example of the process (Figure 34.13). Soon after water is imbibed, the aleurone, the thin outer layer of the endosperm, begins making α-amylase and other enzymes that digest the starch stored in the endosperm. (A similar enzyme is found in our saliva, which helps us digest bread and other foods made from the starchy endosperm of ungerminated cereal grains.) If the embryo is dissected out of the seed before water is added, no α-amylase is produced, suggesting that the embryo sends some kind of messenger to the aleurone to initiate enzyme production. This chemical signal has been identified as gibberellic acid (GA), one of the hormones discussed in more detail in the following chapter.

The first organ to emerge from the germinating seed is the radicle, the embryonic root (Figure 34.14). Next, the shoot tip must break through the soil surface. In garden beans and many other dicots, a hook forms in the hypocotyl and growth pushes the hook above

Figure 34.14

Germination. The radicle, the root of the embryo, emerges from the seed first. Then the shoot breaks the soil surface by one of the following mechanisms: **(a)** In beans, straightening of a hook in the hypocotyl pulls the shoot and cotyledons from the soil. **(b)** In peas, the hook is above the cotyledons on the epicotyl, and the cotyledons remain in the ground. **(c)** In corn and other monocots, the shoot grows straight up through the tube of the coleoptile.

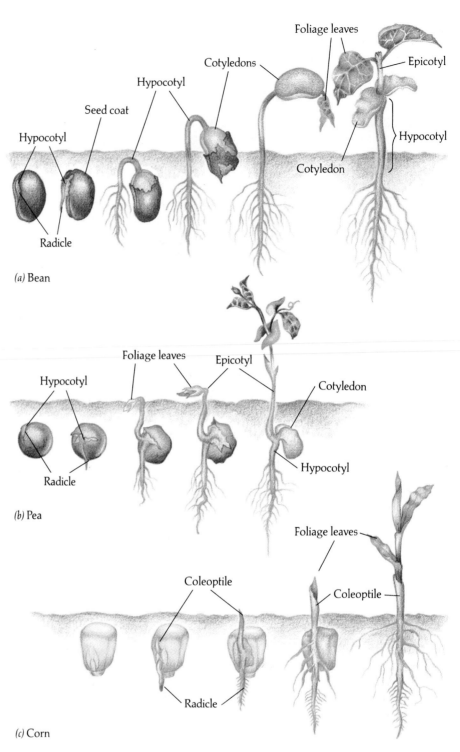

(a) Bean

(b) Pea

(c) Corn

ground. Stimulated by light, the hypocotyl straightens, raising the cotyledons and epicotyl. Thus, the delicate shoot apex and bulky cotyledons are pulled above ground, rather than being pushed tip first through the abrasive soil. The elevated epicotyl now spreads its plumule, and the first foliage leaves expand, become green, and begin making food by photosynthesis. The cotyledons shrivel and fall away from the

seedling, their food reserves having been consumed by the germinating embryo.

Light seems to be the main cue that tells the seedling it has broken ground. The hypocotyl of a common garden bean will continue to elongate and push its hook upward until it is out of darkness. Only when the seedling senses light will the hook straighten and the epicotyl begin to elongate. We can trick a bean seedling into behaving as though it is still buried by germinating the seed in darkness, a treatment called **etiolation.** The etiolated seedling extends an exaggerated hypocotyl with a hook at its tip, and the foliage leaves fail to green. After it exhausts its food reserves, the spindly seedling stops growing and dies.

Peas, although in the same family as beans, have a different style of germinating. A hook forms in the epicotyl rather than hypocotyl, and the shoot tip is lifted gently out of the soil by elongation of the epicotyl and straightening of the hook. Pea cotyledons, unlike those of beans, remain behind in the ground.

Monocots, such as corn, use yet a different method for breaking ground when they germinate. The coleoptile, the sheath enclosing and protecting the embryonic shoot, pushes upward through the soil and into the air. The shoot tip then grows straight up through the tunnel provided by the tubular coleoptile.

Germination of a plant seed, like the birth or hatching of an animal, is a critical stage in the life cycle. The tough seed gives rise to a fragile seedling that will be confronted with predators, parasites, wind, and other hazards. Natural selection has ample opportunity to screen the many different genetic combinations that arise from sexual reproduction. In the wild, only a small fraction of seedlings endure long enough to become parents themselves. However, production of enormous numbers of seeds and fruits compensates for the odds against individual success. This, of course, is usually very expensive in terms of the resources consumed in flowering and fruiting. Asexual reproduction, generally simpler and less hazardous for offspring than sexual reproduction, is an alternative means of plant propagation.

ASEXUAL REPRODUCTION

Imagine some of your fingers separating from your body, taking up life on their own, and eventually developing into carbon copies of yourself. This would be an asexual reproduction, the offspring derived from a single parent without genetic recombination occurring. The result would be a clone, a population of asexually produced, genetically identical organisms. In reality, of course, we cannot reproduce in this way (though some animals *are* capable of asexual reproduction). Many plant species do clone themselves by asexual reproduction, also called **vegetative reproduction.**

Natural Mechanisms of Vegetative Reproduction

Vegetative reproduction is an extension of the capacity of plants for indeterminate growth (see Chapter 31).

Plants, remember, have meristematic tissues of dividing, undifferentiated cells that can sustain or renew growth indefinitely. In addition, parenchyma cells throughout the plant can divide and differentiate into the various types of specialized cells, enabling plants to regenerate lost parts. Detached fragments of some plants can develop into whole offspring; a severed stem, for instance, may develop adventitious roots that reestablish the plant.

Fragmentation, the separation of a parent plant into parts that re-form whole plants, is one of the most common modes of vegetative reproduction (Figure 34.15). An entirely different mechanism of asexual reproduction has evolved in dandelions and some other plants. The dandelion produces seeds without its flowers being fertilized. A diploid cell in the ovule gives rise to the embryo, and the ovules mature into seeds, which are dispersed by windblown fruit. Though the dandelion clones itself by this asexual process, it also has the advantage of seed dispersal, an adaptation usually reserved for sexual reproduction of plants. This asexual production of seeds is called **apomixis.**

Vegetative Propagation in Agriculture

With the objective of improving crops, orchards, and ornamental plants, humans have devised various methods for vegetative propagation. Most of these are based on the ability of plants to form adventitious roots or shoots.

Clones from Cuttings Most houseplants, woody ornamentals, and orchard trees are asexually propagated from cuttings. In some cases, shoot or stem cuttings are used. At the cut end of the shoot, a mass of dividing, undifferentiated cells called a **callus** forms, and then adventitious roots develop from the callus. Some plants, including African violets, can be propagated from single leaves rather than stems. For other plants, the cuttings are taken from specialized storage stems. For example, a potato can be cut up into several pieces, each with an "eye" that regenerates a whole plant.

In a modification of vegetative propagation from cuttings, a twig or bud from one plant can be grafted onto a plant of a closely related species or a different variety of the same species (Figure 34.16). Grafting makes it possible to combine the best qualities of different species or varieties into a single plant. The graft is usually done when the plant is very young. The plant that provides the root system is called the stock; the twig grafted onto the stock is referred to as the scion. For example, scions from French varieties of vines that produce superior wine grapes are grafted onto root stock of American varieties, which are more resistant to certain diseases. The quality of the fruit, determined by the genes of the scion, is not dimin-

(a)

(b)

(c)

(d)

(e)

Figure 34.15

Natural mechanisms of vegetative reproduction. (a) Each "spider" of a spider plant can grow adventitious roots and become an independent plant after fragmenting from the parent. (b) A storage organ such as the garlic bulb can give rise to several plants by fragmenting into cloves. (c) Grass, such as this dune grass in Australia, can propagate asexually by sprouting shoots and roots from runners (stolons). A small patch of grass can spread in this way until it covers an acre or more of surface. The distinction between a sprawling superplant and a clone of smaller plants is only a matter of fragmentation of the parent when its connecting stolons are somehow broken. (d) In some species of dicots the root system of a single parent gives rise to many adventitious shoots that become separate shoot systems. The result is a clone formed by asexual reproduction from one parent. The oldest of all known plant clones, this ring of creosote bushes propagated asexually from a single plant, is believed to be at least 12,000 years old. (e) Some aspen groves, such as those shown here, are actually clones of thousands of trees descended by asexual reproduction from the root system of one parent. Notice that genetic differences among the clones result in different timing in the development of fall color and loss of leaves.

Figure 34.16
Grafting. The scion, a cutting from an English walnut, has been grafted onto root stock of a black walnut. The graft is still apparent in this mature tree.

Figure 34.17
Test-tube cloning. This apple plantlet was grown from a meristem cultured on agar supplemented with mineral nutrients and plant hormones. A "super" variety of tree can be cloned to produce progeny in large numbers by this method of propagation.

ished by the genetic makeup of the stock. In some cases of grafting, however, the stock can alter the characteristics of the shoot system that develops from the scion. For example, dwarf fruit trees are made by grafting normal twigs onto dwarf stock varieties that retard the vegetative growth of the shoot system. Since seeds are produced by the part of the plant derived from the scion, they would give rise to plants of the scion species if planted.

Test-Tube Cloning and Related Techniques In a new kind of botanical alchemy, test-tube methods are now being used to create and clone novel plant varieties. It is possible to grow whole plants by culturing small explants (pieces of tissue cut from the parent), or even single parenchyma cells, on an artificial medium containing nutrients and hormones (Figure 34.17). The cultured cells divide to form an undifferentiated callus. When the hormonal balance is manipulated in the culture medium, the callus can sprout shoots and roots with fully differentiated cells. The test-tube plantlets can then be transferred to soil, where they continue their growth. A single plant can be cloned into thousands of copies by subdividing calluses as they grow. This method is used for propagating orchids and also

for cloning pine trees that deposit wood at unusually fast rates. Plant tissue culture also facilitates genetic engineering of plants (see Chapter 19). Most techniques for the introduction of foreign genes into plants require the use of single plant cells or small pieces of plant tissue as the starting material. Test-tube culture makes it possible to regenerate genetically altered plants from a single plant cell into which the foreign DNA has been incorporated.

A technique known as **protoplast fusion** is being coupled with tissue culture methods to actually invent new plant varieties that can be cloned. Before they are cultured, the protoplasts can be screened for mutations that may improve the agricultural value of the plant. It is also possible in some cases to fuse two protoplasts from different plant species that would otherwise be sexually incompatible and then culture the hybrid protoplasts. A protoplast, remember, is the living part of a plant cell bounded by the plasma membrane; it is a plant cell without the cell wall. Beginning with a piece of tissue, it is possible to produce a mixture of protoplasts by using enzymes to digest away the walls of the plant cells (Figure 34.18). Each of the

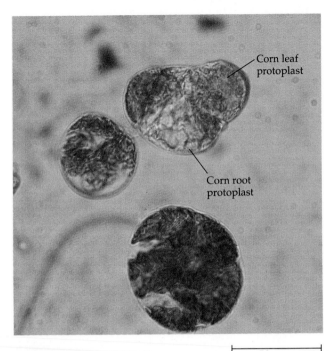

Figure 34.18
Protoplast fusion. Protoplasts, plant cells that have had their walls removed, may fuse to form hybrid cells, sometimes crossing species boundaries. The hybrid protoplasts re-form walls and can be cloned by tissue culture methods. In the case shown here, however, leaf and root protoplasts of a single species, corn, have been fused (LM).

50 μm

many protoplasts can regenerate its wall and form a callus by cell division.

As already mentioned, direct genetic manipulation of cultured cells and protoplasts by recombinant DNA techniques is the latest twist in creating and cloning new plant varieties. The potential agricultural impact of genetic engineering is discussed in Chapter 19.

The Benefits and Risks of Monoculture By conscious effort, genetic variability in many crops and orchards has been virtually eliminated. For grains and other crops that are grown from seed, plant breeders have selected for varieties that self-pollinate to prevent dilution of desirable traits. Such plants in a field grow at the same rate, fruit on all of the trees of an orchard ripen in unison, and yields at harvest time are dependable. Whenever possible, vegetative propagation is used to clone exceptional plants and make farming more reliable. There is no doubt that **mono-**culture, the cultivation of large areas of land with a single plant variety, has helped farmers feed the world. But modern farms are very fragile ecosystems: Where there is little genetic variability, there is also little adaptability. A clone is a kind of superorganism; in terms of natural selection, it is one genetic individual. What is good for one is good for all, and what is bad for one threatens the entire clone. Some plant scientists fear that a disease could devastate thousands of acres of grain or some other important monoculture. The risk must be taken seriously. Plant breeders have responded by maintaining "gene banks" where seeds of many different plant varieties that can be used to breed new hybrids are stored.

A Comparison of Sexual and Asexual Reproduction in Plants

The dilemma of monoculture reflects on the more general issue of the relative advantages of sexual versus asexual reproduction in the wild. Many plants are capable of both modes of reproduction, and each offers advantages in certain situations. Sex generates variation in a population, an asset when the environment changes (see Chapter 21). An additional benefit of sexual reproduction in plants is the seed, a stage in the life cycle that can disperse to new locations and can also wait until hostile environmental conditions have improved. On the other hand, a plant well suited to a stable environment can use asexual reproduction to clone many copies of itself. Moreover, the progeny of vegetative reproduction, usually mature fragments of the parent plant, are not as frail as the seedlings produced by sexual reproduction. A sprawling clone of prairie grass may cover an area so thoroughly that seedlings have little chance of competing. But in the soil is a pool of seeds, waiting in the wings for some cue to germinate. After a fire, drought, or some other disturbance clears patches of the turf, seedlings can finally get a foothold when conditions improve. The seedlings are unequal in their traits, for their genomes are products of the sexual recombination of genes. A new competition ensues, and from this struggle for existence certain plants excel and propagate themselves asexually. Both modes of reproduction, sexual and asexual, have had featured roles in the adaptation of plant populations to their environments.

STUDY OUTLINE

1. The use of pollen to convey sperm nuclei and evolution of the seed as a resistant stage are two of the important terrestrial adaptations of angiosperms and conifers. In many cases, asexual reproduction serves as an alternative to sexual reproduction.

Sexual Reproduction of Flowering Plants (pp. 739–751)

1. The flower is an angiosperm apparatus of modified leaves specialized for reproduction.

2. Floral parts consist of sterile sepals and petals and fertile stamens and carpels. The presence or absence of these parts allows classification of flowers as complete or incomplete, perfect or imperfect, and plants as monoecious or dioecious.

3. Stamens consist of a stalk, the filament, and a terminal sac, the anther. Pollen grains containing sperm nuclei develop inside the sporangia of the anther.

4. Female gametes are housed in one or more ovules inside the ovary of a carpel. The style leads to the ovary from the stigma, a sticky tip that receives pollen.

5. In alternation of generations in angiosperms, the dominant stage is the diploid sporophyte. The spores develop inside the flower into tiny, haploid gametophytes—the male pollen grain and the female embryo sac that depend on the sporophyte for nutrition. These gametophytes produce sperm nuclei and eggs, which unite inside the ovule to form the zygote.

6. Pollen grains begin development as microspores, meiotic products of microsporocytes inside the anther. Each pollen grain has a generative nucleus and a tube nucleus and is covered by a thickened, sculptured coat. In maturation of the pollen grain, the generative nucleus divides into two sperm nuclei, which function as the male gametes.

7. A single megasporocyte in each ovule of the ovary divides by meiosis into four haploid megaspores, only one of which survives. Mitosis of this megaspore produces the embryo sac, the female gametophyte consisting of eight nuclei in seven cells surrounded by integuments of sporophyte tissue. At the end of the ovule containing the egg cell is an opening between the integuments called the micropyle.

8. Fertilization is preceded by pollination, the placement of pollen by wind or animal agents onto the stigma of the carpel. Although some species are self-pollinating, the majority have morphological or developmental mechanisms that ensure cross-pollination and thus genetic variability.

9. The pollen grain germinates, growing a pollen tube that extends down the style and gains access to the embryo sac through the micropyle. Two sperm nuclei are released and effect a unique double fertilization that forms a diploid zygote and a triploid endosperm, triggering the development of the seed.

10. The embryo develops from the terminal cell, one of two products of the first mitotic division following fertilization. The embryo develops one or two cotyledons and elongates. Apical meristems for the future root and shoot take form, and protoderm, ground meristem, and procambium also appear. The varying cytoplasmic environments and positions that develop during early cell division may be responsible for the developmental fate of the cell and the groups of cells that arise from it.

11. Mitotic division of the triploid endosperm gives rise to a multicellular, nutrient-rich mass that feeds the developing embryo and later the young seedling.

12. The developing seed gradually dehydrates, and embryonic growth is suspended. The integuments of the ovule form a seed coat around the embryo and its food supply.

13. A fruit is a ripened ovary that protects the enclosed seeds and aids in their dispersal via wind or animals. Hormonal changes initiated by pollination cause a transformation of the flower known as fruit set, in which the ovarian wall thickens and becomes the pericarp of the fruit. Although fruit set parallels development of seeds in most species, parthenocarpic plants produce fruit without fertilization.

14. Fruits can be classified as simple, aggregate, or multiple, depending on their origin.

15. Ripening of fruit usually coincides with maturation of the seeds and can involve simple aging of the fruit tissue or complex, hormonally orchestrated changes in color, texture, and sugar content.

16. Germination resumes the growth and development of the miniature plant that was temporarily suspended at an embryonic stage during the maturation of the seed. Specific environmental cues are often required to initiate germination.

17. The evolution of the seed allowed plants to firmly establish themselves in unstable terrestrial environments. Conditions that break dormancy are finely tuned by selection to the environment of the species. Ungerminated seeds may remain viable for years until conditions are suitable.

18. Germination begins when seeds absorb water by imbibition, which frees the expanded seed from its coat and sets off a series of metabolic changes that trigger the resumption of growth.

19. The embryonic root, or radicle, is the first structure to emerge from the germinating seed. Next the embryonic shoot breaks through the soil surface. In many dicot species a hypocotyl hook pulls the shoot above ground. In some monocots (grasses) the shoot grows up through the coleoptile that pushes through the soil.

Asexual Reproduction (pp. 751–754)

1. Asexual or vegetative reproduction, the production of genetically identical offspring from a single parent, is a simpler alternative to sexual reproduction that can be accomplished in a variety of natural and artificial ways.

2. Fragmentation of a parent plant into parts that re-form whole plants demonstrates the versatility and latent potential of meristematic and parenchymal tissues. Another natural means of asexual reproduction occurs in the asexual production of seeds by apomixis in certain plants.

3. Isolated leaves, pieces of specialized storage stems, and the cut end of severed shoots in many plants can form an undifferentiated callus from which adventitious roots emerge. Grafting a twig or bud of one species onto a closely related species has generated a host of commercially important plants.

4. Laboratory methods can clone large numbers of desired plant varieties by culturing small explants or single parenchyma cells and subdividing callus tissue. In addition, new plant varieties can be created and cloned by coupling protoplast fusion or recombinant DNA techniques with tissue culture methods.

5. Although monoculture of desirable clones has greatly increased agricultural productivity, the danger is that the absence of genetic variability may prove disastrous in the face of uncertain future conditions. Gene banks contain-

ing seeds of different varieties are maintained as a potential source for generating genetic variation if the need arises.

6. Both sexual and asexual modes of reproduction have been important in the adaptation of plants to their environment by offering them relative advantages in different environmental situations.

SELF-QUIZ

1. Which of the following would definitely be an imperfect flower? A flower that
   a. is also incomplete
   b. has no sepals
   c. is found on a monoecious plant
   d. is staminate
   e. cannot self-pollinate

2. Which of the following structures is incorrectly paired with the proper life-cycle generation?
   a. anther—sporophyte
   b. petals and sepals—sporophyte
   c. ovary—gametophyte
   d. pollen grain—gametophyte
   e. proembryo—sporophyte

3. In double fertilization, both of the sperm nuclei
   a. enter the embryo sac
   b. fertilize the egg cell
   c. fuse with the polar cell to produce the triploid endosperm
   d. are produced by the tube nucleus, and enter the ovule through the micropyle
   e. fertilize two egg cells

4. The basal cell in a zygote
   a. develops into the root of the embryo
   b. forms the suspensor that anchors the embryo and transfers nutrients
   c. results from the fertilization of polar and sperm nuclei and develops into the endosperm
   d. divides to form the two cotyledons of the proembryo
   e. eventually develops into the shoot apex

5. A fruit is a
   a. ripened ovary
   b. ripened ovule
   c. seed plus its integuments
   d. fused carpel
   e. enlarged embryo sac

6. Which of these statements about the hypocotyl hook is *not* true?
   a. It develops below the attachment of the cotyledons.
   b. It is the first structure to emerge from a dicot seed.
   c. It precedes the cotyledons and shoot apex up through the soil.

d. It straightens when exposed to light.
   e. It fails to straighten in an etiolated seedling.

7. Which of these conditions is needed by almost all seeds to break dormancy?
   a. exposure to light
   b. imbibition
   c. abrasion of the seed coat
   d. exposure to cold temperatures
   e. covering of fertile soil

8. Which of the following is *not* an example of asexual propagation?
   a. fragmentation—the separation of plant parts that develop into whole plants
   b. grafting—the attachment of a scion onto a stock, or plant that provides a root system.
   c. apomixis—the asexual production of seeds
   d. test-tube culturing of divided calluses
   e. production of fruit without fertilization in parthenocarpic plants

9. Which of the following pairs of terms represents a correct match?
   a. ovule-egg
   b. pollen grain-sperm
   c. seed-zygote
   d. embryo sac-female gametophyte
   e. endosperm-integuments

10. Which of the following structures is unique to the seed of a monocot?
    a. coleoptile
    b. radicle
    c. seed coat
    d. endosperm
    e. cotyledon

CHALLENGE QUESTIONS

1. What are the advantages of propagating house plants by asexual reproduction?
2. Distinguish between pollination and fertilization.
3. Explain why it is technically incorrect to refer to stamens and carpels as the sex organs of an angiosperm.

FURTHER READING

1. Christopher, T. "Seeds in Space." *Omni*, October 1984. A new twist for an old adaptation.
2. Johri, B. M. *Embryology of Angiosperms.* New York: Springer-Verlag, 1984. An in-depth survey of sexual reproduction in angiosperms.
3. Tilton, V. R., and S. H. Russell. "Applications of In Vitro Pollination/Fertilization Technology." *BioScience,* April 1984. An article describing how some natural barriers to pollination and fertilization can be overcome in developing new plant varieties.
4. Weiss, R. "Blazing Blossoms." *Science News,* June 24, 1989. Some flowers that resemble carrion generate considerable heat, which helps attract beetles and other pollinators.

Control Systems in Plants

35

At every stage in the life of a plant, sensitivity to the environment and coordination of responses are evident. One part of a plant can send signals to other parts. For example, the terminal bud at the apex of a shoot is able to suppress the growth of axillary buds that may be many meters away. Plants keep track of the time of day and also the time of year. They can sense gravity and the direction of light and respond to these stimuli in ways that seem to us to be completely appropriate. It is tempting to explain this response in terms of such human qualities as desire, need, wisdom, and decisiveness or to imagine that a plant behaves a certain way to accomplish a particular result. The way of science, however, is to search for a mechanism—a link between cause and effect. A houseplant orients its leaves toward a window, not because of conscious choice or because it is trying to find more light, but because cells on the darker side of stems and petioles grow faster than cells on the brighter side, causing the organs to curve toward the light (Figure 35.1). Natural selection has favored mechanisms of plant response that enhance reproductive success, but this implies no purposeful planning on the part of the plant.

As you learn in this chapter about the control mechanisms at work in plants, you will come to appreciate plants as elaborate, integrated organisms, able to respond adaptively to what goes on around them. Our emphasis is on the control systems that enable individual plants to adjust to their surroundings, but these control systems are themselves adaptations that evolved over countless generations of plants interacting with their environments.

Figure 35.1
Plants sense and respond to their environment. The shoots of this *Impatiens* plant bend toward light, a response called phototropism.

For the most part, plants and animals respond to environmental stimuli by very different means. Animals, being mobile, respond mainly by behavioral mechanisms, moving toward positive stimuli and away from negative stimuli. Rooted to one location for life, a plant generally responds to environmental cues by adjusting its pattern of growth and development. Because the program for development of the plant remains somewhat plastic, plants of the same species vary in body form much more than animals of the same species; all lions have four limbs and approximately the same body proportions, but oak trees are less regular in their number of limbs and their shapes.

The growth and development of a plant is controlled by complex interactions of environmental factors and internal signals. Some of the most important signals are chemical messengers called hormones.

THE SEARCH FOR A PLANT HORMONE

The word **hormone** is derived from a Greek verb meaning "to excite." Found in all multicellular organisms, hormones are chemical signals that coordinate the parts of the organism. As first defined by animal physiologists, a hormone is a compound that is produced by one part of the body and is then translocated to other parts of the body, where it triggers responses in target cells and tissues.

The concept of chemical messengers in plants emerged from a series of classic experiments on how stems respond to light. A houseplant on a windowsill grows toward light. If you rotate the plant, it will soon reorient its growth until its leaves again face the window. The growth of a shoot toward light is called positive **phototropism** (growth away from light is negative phototropism). In a forest or other natural ecosystem where plants may be crowded, phototropism directs growing seedlings toward the sunlight they need for photosynthesis. What is the mechanism for this adaptive response? Much of what is known about phototropism has been learned from studies of grass seedlings, particularly oats. The shoot of a grass seedling is enclosed in a sheath called the coleoptile, which grows straight upward if the seedling is kept in the dark or if it is illuminated uniformly from all sides. If the growing coleoptile is illuminated from one side, it will curve toward the light. This response results from differential growth of cells on opposite sides of the coleoptile; the cells on the darker side elongate faster than cells on the brighter side (Figure 35.2).

Some of the earliest experiments on phototropism were conducted in the late nineteenth century by Charles Darwin and his son, Francis. They discovered that a grass seedling could bend toward light only if the tip of the coleoptile was present (Figure 35.3). If the tip was removed, the coleoptile would not curve. The seedling would also fail to grow toward light if the tip was covered with an opaque cap; neither a transparent cap over the tip nor an opaque shield placed farther down the coleoptile prevented the phototropic response. It was the tip of the coleoptile, the Darwins concluded, that was responsible for sensing light. However, the actual growth response, the curvature of the coleoptile, occurred some distance below the tip. Charles and Francis Darwin speculated that some signal must have been transmitted downward from the tip to the elongating region of the coleoptile. A few decades later, Peter Boysen-Jensen of Denmark

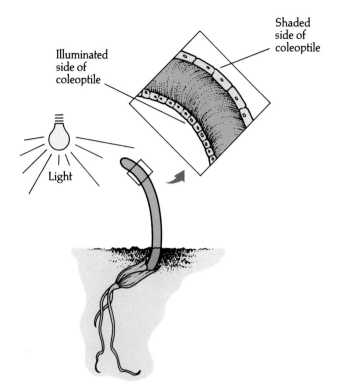

Figure 35.2
Phototropism. The coleoptile of an oat seedling bends toward light because cells on the darker side of the organ elongate faster than cells on the brighter side.

demonstrated that the signal was a mobile substance of some kind. He separated the tip from the remainder of the coleoptile by a block of gelatin, which would prevent cellular contact but through which chemicals could diffuse. These seedlings behaved normally, bending toward light. However, if the tip was segregated from the lower coleoptile by an impermeable barrier of mica, no phototropic response occurred.

In 1926, F. W. Went, then a young plant physiologist in Holland, extracted the chemical messenger for phototropism by modifying the experiments of Boysen-Jensen (Figure 35.4). Went removed the coleoptile tip and placed it on a block of agar, a gelatinous material extracted from red algae. The chemical messenger from the tip, Went reasoned, should diffuse into the agar, and the agar block should then be able to substitute for the coleoptile tip. Went placed the agar blocks on decapitated coleoptiles that were kept in the dark. A block that was centered on top of the coleoptile caused the stem to grow straight upward. However, if the block was placed off center, then the coleoptile began to bend away from the side with the agar block, as though growing toward light. Went concluded that the agar block contained a chemical produced in the coleoptile tip, that this chemical stimulated growth as it passed down the coleoptile, and that a coleoptile curved toward light because of a higher concentration of the growth-promoting chemical on the darker side

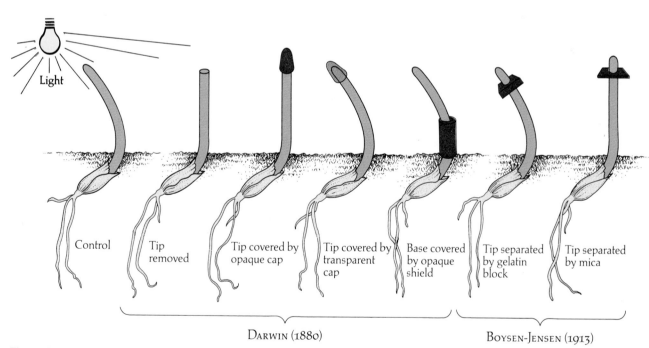

Control | Tip removed | Tip covered by opaque cap | Tip covered by transparent cap | Base covered by opaque shield | Tip separated by gelatin block | Tip separated by mica

DARWIN (1880)   BOYSEN-JENSEN (1913)

Figure 35.3
Early experiments on phototropism.
Only the tip of the coleoptile can sense the direction of light, but the bending response occurs some distance below the tip. A signal of some kind must travel downward from the tip. The signal can pass through a permeable barrier (gelatin block) but not through a solid barrier (mica), suggesting that the signal for phototropism is a mobile chemical.

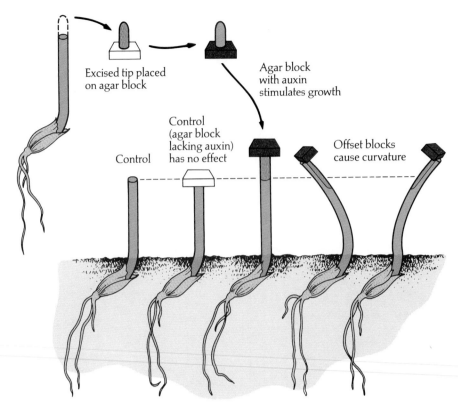

Figure 35.4
The Went experiments. Some chemical (indicated by color) that can pass into an agar block from a coleoptile tip stimulates elongation of the coleoptile when the block is substituted for a tip. If the block is placed off center on the top of a decapitated coleoptile kept in the dark, the organ bends as if responding to illumination from the side. The chemical is the hormone auxin, which stimulates elongation of cells in the shoot.

Excised tip placed on agar block

Agar block with auxin stimulates growth

Control

Control (agar block lacking auxin) has no effect

Offset blocks cause curvature

of the coleoptile. For this chemical messenger, or hormone, Went chose the name auxin (from the Greek *auxein*, "to increase"). Auxin was later purified and its structure determined by Kenneth Thimann and his colleagues at the California Institute of Technology. Other plant hormones were subsequently discovered.

FUNCTIONS OF PLANT HORMONES

So far, five classes of plant hormones have been positively identified. They are **auxin, cytokinins** (actually a class of related chemicals), **gibberellins** (also a class of similar chemicals), **abscisic acid,** and **ethylene.** In general, these hormones control plant growth and development by affecting the division, elongation, and differentiation of cells. Each hormone has a multiplicity of effects, depending on its site of action, the developmental stage of the plant, and the concentration of the hormone.

Plant hormones are produced in very small concentrations, but a minute amount of hormone can have a profound effect on the growth and development of a plant organ. This implies that the hormonal signal must be amplified in some way. A hormone may act by altering the expression of genes, by affecting the activity of existing enzymes, or by changing properties of membranes. Any of these actions could cause a major redirection in the metabolism and develop-

ment of a cell responding to a small number of hormone molecules.

Reaction to a hormone usually depends not so much on the absolute amount of that hormone as on its relative concentration compared with other hormones. It is hormonal balance, rather than hormones acting in isolation, that may control the growth and development of the plant. These interactions will become apparent in the following survey of hormone function.

Auxin

The term *auxin* is used to describe any chemical substance that promotes elongation of young developing stems or coleoptiles. The natural auxin that has been extracted from plants is a compound named indoleacetic acid, or IAA (Table 35.1). In addition to this natural auxin, several other compounds, including some synthetic ones, have auxin activity. Although auxin affects several aspects of plant development, its most important function may be to stimulate the elongation of cells in young developing shoots.

Auxin and Cell Elongation   The apical meristem of a shoot is a major site of auxin synthesis. As auxin from the shoot apex moves down to the region of cell elongation (see Chapter 31), the hormone stimulates growth of the cells. Auxin has this effect only over a

Table 35.1 Functions of plant hormones

Hormone	Major Functions	Where Produced or Found in Plant
Auxin (such as IAA)	Stimulates stem elongation, root growth, differentiation and branching, apical dominance, development of fruit; instrumental in phototropism and gravitropism	Endosperm and embryo of seed; meristems of apical buds and young leaves
Cytokinins (such as kinetin)	Affect root growth and differentiation; stimulate cell division and growth, germination, and flowering; delay senescence	Synthesized in roots and transported to other organs
Gibberellins (such as GA_1)	Promote seed and bud germination, stem elongation, leaf growth; stimulate flowering and development of fruit; affect root growth and differentiation	Meristems of apical buds, roots, and young leaves; embryo
Abscisic acid	Inhibits growth; closes stomata during water stress; counteracts breaking of dormancy	Leaves, stems, green fruit
Ethylene	Promotes fruit ripening; opposes or reduces some auxin effects; promotes or inhibits growth and development of roots, leaves, flowers, depending on species	Tissues of ripening fruits, nodes of stems, senescent leaves

certain concentration range; at higher concentrations, auxin may inhibit cell elongation (Figure 35.5). This is probably due to a high level of auxin inducing synthesis of another hormone, ethylene, which generally acts as an inhibitor of plant growth due to cell elongation. Auxin from the shoot apex also reaches the root, whose cells are more sensitive than stem cells to the hormone. An auxin concentration too low to stimulate stem cells will cause root cells to elongate. On the other hand, an auxin concentration high enough to stimulate elongation of stem cells is in the concentration range that inhibits elongation of root cells. The effects of auxin on cell elongation reinforce the point that the same chemical messenger may have different effects at different concentrations in one target cell and that a given concentration of the hormone may have varying effects on different target cells.

Polar Transport of Auxin The speed at which auxin is transported down the stem from the shoot apex is about 10 mm per hour—much too fast for diffusion, although slower than translocation in phloem. Auxin seems to be transported directly through parenchyma tissue, from one cell to the next. This transport is unidirectional, or polar, moving only from shoot tip to base. Polar transport of auxin has nothing to do with

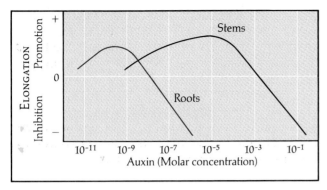

Figure 35.5
Effect of auxin concentration on cell elongation.
Notice that a hormone concentration high enough to inhibit root growth is optimal for stimulating stem growth.

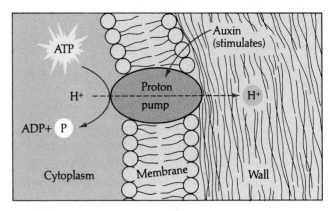

Figure 35.6
The acid-growth hypothesis. There is evidence that auxin's initial effect on cell elongation is due to stimulation of a proton pump, which causes acidification of the wall. According to the acid-growth hypothesis, acidification of the wall activates enzymes that break cross-links between microfibrils of the wall. This loosens the wall, allowing elongation of the cell owing to its turgor pressure.

gravity, for auxin travels upward in experiments where a stem or coleoptile segment is placed upside down. Polar auxin transport is an active process requiring metabolic energy. Experiments suggest that IAA is actively transported down a stem or coleoptile by auxin carriers that are located on the basal ends of cells but are absent on the apical ends.

The Acid-Growth Hypothesis One way auxin initiates cell elongation is by loosening the cell wall (Figure 35.6). According to a hypothesis currently being experimentally tested, auxin loosens walls indirectly by stimulating proton pumps. As the cell secretes hydrogen ions, the wall becomes more acidic, and pH-sensitive enzymes in the wall are activated. According to the acid-growth hypothesis, the activated enzymes break cross-links between cellulose microfibrils, making the wall less rigid. With the fabric of the retaining wall loosened, the turgor pressure of the cell exceeds wall pressure, causing the cell to elongate by the uptake of water. If auxin is applied to young stems or coleoptile segments, it can induce wall loosening and cell elongation within minutes. For sustained growth after this initial elongation, however, cells must make more wall material and cytoplasm.

Other Effects of Auxin Auxin has several effects in addition to stimulating cell elongation for primary growth. It affects secondary growth by inducing cell division in the vascular cambium, and it influences the differentiation of secondary xylem. Auxin also promotes the formation of adventitious roots at the cut base of a stem, an effect obtained in horticulture by dipping cuttings in rooting media containing synthetic auxins. Auxin produced by leaf primordia is responsible for the development of leaf traces, the strands of vascular tissue that will connect to vascular bundles in the stem. Developing seeds also synthe-

size auxin, which promotes growth of fruit in many plants. Synthetic auxins sprayed on tomato vines induce parthenocarpic fruit development (see Chapter 34), eliminating the need for pollination and making it possible to grow seedless tomatoes by substituting for auxin that would normally be synthesized by seeds.

One of the most widely used herbicides, or weed killers, is 2,4-D, a synthetic auxin that disrupts the normal balance of plant growth. Dicots are more sensitive than monocots to this herbicide, and thus 2,4-D can be used to selectively remove dandelions and other "broad leaf" weeds from a lawn or grain field.

Cytokinins

Cytokinins were discovered during trial-and-error attempts to find chemical additives that would enhance growth and development of plant cells in tissue culture. In the 1940s, Johannes van Overbeek, working at the Cold Spring Harbor Laboratory in New York, found he could stimulate the growth of plant embryos by adding coconut milk, the liquid endosperm of the giant coconut seed, to his culture medium. A decade later, Folke Skoog and Carlos O. Miller, at the University of Wisconsin, induced tobacco cells being grown in culture to divide by adding degraded samples of DNA. The active ingredients of both additives, coconut milk and deteriorated DNA, turned out to be modified forms of adenine, one of the components of nucleic acids. These growth regulators were named cytokinins because they stimulate cytokinesis, or cell division. A variety of cytokinins from plant sources have now been identified, and several synthetic compounds with cytokinin activity have also been produced.

Control of Cell Division and Differentiation Cytokinins are produced in actively growing tissues, particularly in roots, embryos, and fruits. Cytokinins produced in the root reach their target tissues by moving up the plant in the xylem sap. Acting in concert with auxin, cytokinins stimulate cell division and influence the pathway of differentiation.

The effects of cytokinins on cells growing in tissue culture provide clues about how this class of hormones may function in an intact plant. When an explant, such as a piece of parenchyma tissue from a stem, is cultured in the absence of cytokinins, the cells grow very large but they do not divide. If cytokinins alone are added to the culture, they have no effect. If cytokinins are added along with auxin, however, the cells divide. Furthermore, the ratio of cytokinin to auxin controls the differentiation of the cells. When the concentration of the two hormones is about equal, the mass of cells continues to grow, but it remains an undifferentiated callus. If there is more cytokinin than auxin, shoot buds develop from the callus. If auxin is more concentrated than cytokinin, roots form. It is remarkable that gene expression can be controlled so extensively by manipulating the concentration of just two chemical signals.

The ability of these hormones to trigger cell division and influence differentiation could result from the fact that cytokinins stimulate RNA and protein synthesis. These newly synthesized proteins may be involved in cell division. By what mechanisms do cytokinins promote biosynthesis of RNA and proteins? Can these hormones selectively switch on production of proteins needed for cell division or for development of buds? We have still much to learn about how cytokinins and other hormones act at the molecular level.

Control of Apical Dominance We can see another interaction of cytokinins and auxin in the control of apical dominance, the ability of the terminal bud to suppress development of axillary buds (see Chapter 31). In this case, the two hormones are antagonistic. Auxin transported down the shoot from the terminal bud restrains axillary buds from growing, causing a shoot to lengthen at the expense of lateral branching. If the terminal bud is removed, the plant may become bushier. However, cytokinins entering the shoot system from roots counter the action of auxin by signaling axillary buds to begin growing. Auxin cannot suppress growth of these buds once it has begun. Lower buds on a shoot usually break dormancy before buds closer to the apex, reflecting the relative distance away from auxin and cytokinin sources.

The check-and-balance control of lateral branching by auxin and cytokinins may be one way the plant coordinates the growth of its shoot and root systems. As roots become more extensive, the increased level of cytokinins would signal the shoot system to form more branches. The two hormones reverse their roles in the development of lateral roots; auxin stimulates and cytokinins inhibit the growth of branch roots.

Both auxin and cytokinins may regulate the growth of axillary buds indirectly by changing the concentration of still another hormone, ethylene. The levels of different nutrients in a bud may also affect its response to hormones. Most plant physiologists once believed that the control of apical dominance was an exclusive function of auxin from the terminal bud. The situation is obviously far more complex.

Cytokinins as Anti-aging Hormones Cytokinins can retard the aging of some plant organs, perhaps by inhibiting protein breakdown, by stimulating RNA and protein synthesis, and by mobilizing nutrients from surrounding tissues. If leaves removed from a plant are dipped in a cytokinin solution, they stay green much longer than they otherwise would. In intact plants, cytokinins may inhibit leaf senescence by promoting stomatal opening. This would increase the flow of water and nutrients to the leaf, which might help to prolong its life. Because of their anti-aging effect, cytokinin sprays are used to keep cut flowers fresh. They can also prolong the shelf life of fruits and vegetables after harvest, although this latter application has not yet been approved by the U.S. Food and Drug Administration.

Gibberellins

A century ago, rice farmers in Asia noticed some outrageously tall seedlings growing in their paddies. Before these rice seedlings could mature and flower, they grew so tall and spindly that they toppled over. In Japan, this aberration in growth pattern became known as *bakanae,* or "foolish seedling disease." In 1926, E. Kurosawa, a Japanese plant scientist, discovered that the disease was caused by a fungus of the genus *Gibberella.* By the 1930s, Japanese scientists had determined that the fungus produced hyperelongation of rice stems by secreting a chemical, which was given the name gibberellin. Western scientists finally learned of gibberellin after World War II. In the past 30 years, scientists have identified more than 70 different gibberellins, many of them occurring naturally in plants. All of the gibberellins are subtle variations on a common molecular theme, but some forms are much more active than others in the plant. Foolish rice seedlings, it seems, suffer an overdose of growth regulators normally found in plants in lower concentrations. Gibberellins have a variety of effects in plants.

A technique that determines the concentration of a chemical by measuring the response of living material is called a **bioassay**. In this example, a plant's quantitative response to gibberellin is used to measure concentration of the hormone. A sample of unknown concentration is applied to dwarf pea plants, and after a certain amount of time, their height is compared with dwarfs treated with a range of known gibberellin concentrations. Using the degree of coleoptile curvature in phototropism to determine auxin concentration is another example of a bioassay.

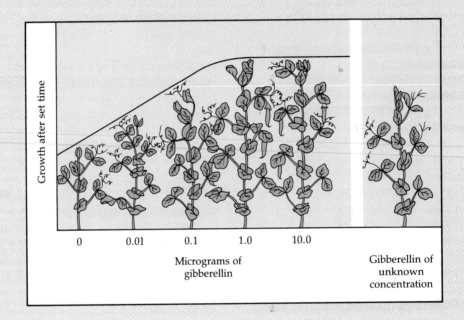

Stem Elongation　　Roots and young leaves near the shoot apex are major sites of gibberellin production. Gibberellins stimulate growth in both the leaves and the stem, but they have little effect on root growth. In a growing stem, gibberellins and auxin must be acting simultaneously in some synergistic manner we do not yet understand.

Enhancement of stem elongation by gibberellins can be seen by applying the hormones to certain dwarf varieties of plants. Dwarf pea plants, for instance, grow to normal height if treated with gibberellins. The extent of the growth of dwarf plants is generally correlated with the concentration of the added hormone (see Methods Box). If gibberellins are applied to plants of normal size, there is often no response. Apparently, these plants are already producing their own optimal dose of the hormone. Gibberellins may be at low levels or absent in dwarf varieties of some species or the target cells may be less responsive to the hormones.

A specific case in which gibberellins cause rapid elongation of stems is **bolting**, the growth of a floral stalk. In their nonflowering stage, some plants develop a rosette form; that is, they are low to the ground with very short internodes. The plant switches to reproductive growth when a surge of gibberellins induces stems to elongate rapidly, which elevates flowers developing from buds at the tips of the stems. Cabbage plants have been selectively bred to maintain a rosette form, but a gibberellin treatment will produce bolting (Figure 35.7).

Gibberellins and Fruit Growth　　Fruit development is another case in which we can observe dual control by auxin and gibberellins. In some plants, both hormones must be present for fruit to set. Sprays containing a combination of auxin and gibberellins will induce parthenocarpic fruit development. The most important commercial application of gibberellins is in the spraying of Thompson seedless grapes. The hormones cause the grapes to grow larger and farther apart.

Figure 35.7
Bolting of cabbage induced by gibberellin treatment.
As a result of selective breeding, cabbage shoots normally do not bolt unless artificially treated with large doses of gibberellin.

Gibberellins and Germination   Many seeds have a high concentration of gibberellins, particularly in the embryo. After water is imbibed, release of gibberellins from the embryo signals the seeds to break dormancy and germinate. Many seeds that require special environmental conditions to germinate, such as exposure to light or cold temperatures, will break dormancy if they are treated with a gibberellin solution. In nature, gibberellins in the seed are probably the link between the environmental cue and the metabolic processes that renew growth of the embryo.

Gibberellins trigger germination of cereal grains by stimulating the synthesis of digestive enzymes such as α-amylase that mobilize stored nutrients (see Chapter 34). Even before these enzymes appear, gibberellin stimulates the synthesis of messenger RNA coding for α-amylase. Here is a case in which a hormone controls

development by affecting gene expression, although the hormone molecule may not be acting directly on the genome. Many responses to hormones are probably mediated by second messengers within the cell that relay the hormonal signal to the nucleus and other sites of action. Calcium ions (Ca^{2+}) may act as second messengers in the gibberellin-induced secretion of α-amylase in germinating grains. We shall see that Ca^{2+} may also be involved with auxin in root gravitropism.

Gibberellins also function to break dormancy in the resumption of growth by apical buds in spring. In both seed dormancy and bud dormancy, gibberellins act antagonistically to another hormone, abscisic acid, which generally inhibits plant growth.

Abscisic Acid

The hormones we have studied so far—auxin, cytokinins, and gibberellins—usually stimulate plant growth (there are exceptions, as in the inhibition of root growth by a high concentration of auxin). By contrast, there are times in the life of a plant when it is adaptive to slow down growth and assume a dormant state. The hormone abscisic acid (ABA), produced in the bud itself, slows growth and directs leaf primordia to develop into the scales that will protect the dormant bud during winter. The hormone also inhibits cell division in the vascular cambium. Thus, ABA helps prepare the plant for winter by suspending both primary and secondary growth. Abscisic acid was named when it was believed that the hormone caused the abscission (from the Latin *ab*, "loss," and *caedere*, "cut") of leaves from deciduous trees during autumn, but no clear-cut role for ABA in abscission has been demonstrated.

Another stage in the life of a plant when it is advantageous to suspend growth is the onset of seed dormancy, and again it may be abscisic acid that acts as the growth inhibitor. The seed will germinate when ABA is overcome by its inactivation or removal or by increased activity of gibberellins. The seeds of some desert plants break dormancy when a heavy rain washes ABA out of the seed. Other seeds require light or some other stimulus to trigger degradation of abscisic acid. In most cases, the ratio of ABA to gibberellins determines whether the seed will remain dormant or germinate. (As noted in Chapter 34, many cases of seed dormancy may not involve ABA.) Similarly, dormancy of terminal buds is controlled more by a balance of growth regulators than by their absolute concentrations. In apple trees, for instance, the concentration of ABA is actually higher in growing buds than in dormant buds, but an excess of gibberellins overpowers the inhibitory hormone.

In addition to its role as a growth inhibitor, abscisic acid acts as a "stress hormone," helping the plant cope with adverse conditions. For example, when the plant is water-stressed (dehydrated), ABA accumulates in leaves and causes stomata to close, reducing transpiration and preventing further water loss. In one variety of tomato that is deficient in ABA and suffers chronic wilting, experimental addition of abscisic acid closes stomata by triggering potassium loss from the guard cells. In this response, we see how a small amount of hormone has its impact amplified by acting on a membrane.

Ethylene

Early in this century, citrus was ripened by "curing" the fruit in sheds equipped with kerosene stoves. Fruit growers believed it was the heat that ripened the fruit, but newer, cleaner-burning stoves did not work. Plant physiologists learned later that ripening in the sheds was actually due to ethylene, a gaseous by-product of kerosene combustion. Now it is known that plants produce their own ethylene as a hormone that elicits a variety of responses in addition to fruit ripening. Unique among plant hormones because it is a gas, ethylene diffuses through the plant in the air spaces between cells. In some cases, ethylene acts to inhibit cell elongation. Many of the inhibitory effects once attributed to auxin are now believed to be due to synthesis of ethylene induced by a high concentration of auxin. It is probably ethylene, for example, that inhibits root elongation and development of axillary buds in the presence of an excess of auxin. In addition to its role as a growth inhibitor, ethylene is also associated with a variety of aging processes in plants.

Senescence in Plants   Aging, or **senescence**, is defined as a progression of irreversible change that leads eventually to death. A normal part of plant development, senescence may occur at the level of individual cells, entire organs, or the whole plant. Tracheids, fiber cells, and cork cells age and die before assuming their specialized functions. Autumn leaves and withering flower petals are examples of senescent organs. Plants that are annuals age and die soon after flowering. Ethylene probably has important functions in all of these cases of senescence, but the aging processes where the effects of the hormone have been studied most extensively are fruit ripening and leaf abscission.

Fruit Ripening   Several changes in structure and metabolism accompany the ripening of an ovary into a fruit (see Chapter 34). Some of these changes, including the degradation of cell walls, which softens the fruit, and the decrease in chlorophyll content, which causes loss of greenness, can be regarded as aging processes. Ethylene initiates or hastens these deteri-

orative changes and also causes some ripened fruits to drop from the plant. A chain reaction occurs during ripening as ethylene triggers senescence and then the aging cells release more ethylene. Because ethylene is a gas, the signal to ripen even spreads from fruit to fruit: One bad apple really does spoil the lot. If you pick or buy green fruit, you may be able to speed ripening by storing the fruit in a plastic bag so that ethylene gas will accumulate. On a commercial scale, many kinds of fruit are ripened in huge storage containers into which ethylene gas is piped—a modern variation on the old curing shed. In other cases, measures are taken to retard ripening caused by natural ethylene. Apples, for instance, are stored in bins flushed with carbon dioxide. Circulating the air prevents ethylene from accumulating, and carbon dioxide somehow inhibits the action of whatever ethylene has not been flushed away. Stored in this way, apples picked in autumn can be shipped to grocery stores the following summer.

Leaf Abscission   The loss of leaves each autumn is an adaptation that keeps deciduous trees from desiccating during winter when the roots cannot absorb water from the frozen ground.

Before leaves are abscised, many of their essential elements are shunted to storage tissues in the stem. These nutrients are recycled back to developing leaves the following spring. The autumn leaf stops making new chlorophyll and loses its greenness. The fall colors are a combination of new pigments made during autumn and pigments that were already present in the leaf but concealed by the dark green chlorophyll. The oranges and yellows are carotenoids and other pigments in chloroplasts (see Chapter 10). The crimson reds and purples are anthocyanins, pigments located in vacuoles and made in fall. Also in the vacuoles are brown tannins.

When an autumn leaf falls, the break point is an abscission zone located near the base of the petiole (Figure 35.8). The small parenchyma cells of the abscission layer have very thin walls, and there are no fiber cells around the vascular tissue. The abscission layer is further weakened when enzymes hydrolyze polysaccharides in the cell walls. Finally, the weight of the leaf, with the help of wind, causes a separation within the abscission layer. Even before the leaf falls, a layer of cork forms a protective scar on the twig's side of the abscission layer, preventing invasion of the plant by pathogens.

The environmental stimuli for leaf abscission are the shortening days and cooler temperatures of autumn. City trees growing near streetlights may retain their leaves longer than usual. The mechanics of abscission are controlled by a change in the balance of ethylene and auxin. An aging leaf produces less and less auxin, and this drop in concentration initiates changes in the

Twig | Protective layer | Abscission layer | 0.5 mm | Petiole

Figure 35.8
Abscission layer at the base of a petiole. Abscission is controlled by a change in the balance of ethylene and auxin. The abscission layer can be seen in this light micrograph as a vertical band at the base of the petiole. After the leaf falls, a protective layer helps prevent pathogens from invading the plant.

abscission layer. Then cells in the abscission layer produce ethylene, which induces synthesis of the enzymes that digest walls. This causes cells on the stem side of the zone to swell, owing to their reduced wall pressure, and press on the weakened layer. A decreased supply of cytokinins moving up to leaves from the roots may also contribute to the onset of senescence.

Some Unanswered Questions About Plant Hormones

There are so many more questions than answers about the internal chemical signals of plants that some plant physiologists have argued it is premature to even call these growth regulators hormones, a term implying that the chemicals function as messages from one part of the organism to another. For instance, A. Trewavas of the University of Edinburgh points out that whereas experimental application of hormones to various parts of the plant causes certain responses, natural increases in the concentration of hormones that would be expected in organs responding to an influx of the chemicals from some other part of the plant are generally not observed. For example, an increase in the auxin concentration in the zone of elongation of growing stems has not been detected. It has been suggested that responses to growth regulators may be due more to changes in the *sensitivity* of local cells to regulators already present than to the arrival of these compounds from other parts of the plant. Although plant physiologists debate the nature of chemical control within the plant, they do agree that auxin, cytokinins, gibberellins, ABA, and ethylene, whether true hormones or not, have a profound influence on growth and development.

Future research will surely bring the discovery of new hormones and of additional functions for hormones that are already known. Discovering the ultimate site of a hormone's action will help researchers understand how one triggering effect can have multiple results. There is also much more to learn about the manipulation of hormonal balance by stimuli from the environment. How, for instance, do shorter days or colder temperatures induce the complex hormonal changes that control leaf abscission? In our exploration of the control systems that help plants respond to their surroundings, we now shift focus to various kinds of plant movements that are useful for studying how plants sense and react to environmental stimuli.

PLANT MOVEMENTS

To the casual observer, most plants do not appear to be very dynamic, but time-lapse photography reveals they are capable of quite precise movements. The two types of plant movements are tropisms and turgor movements.

Tropisms

It is important to remember that the environment has a great influence on the body shape of plants. **Tropisms** (from the Greek *tropos*, "turn") are growth responses that result in curvatures of whole plant organs toward or away from stimuli. The mechanism for a tropism is a differential rate of elongation of cells on opposite sides of the organ. Three of the stimuli that induce tropisms, and a consequent change of body shape, are light (phototropism), gravity (gravitropism), and touch (thigmotropism).

Phototropism We have already examined the hormonal basis for the bending of shoots toward light. Cells on the darker side of a stem, remember, elongate

Figure 35.9
Gravitropism. The 3-day-old corn seedling on the left was germinated so that the shoot would grow straight up and the root straight down. The seedling on the right was germinated the same way, but on day three it was turned on its side so that the shoot and root were horizontal. The photograph was taken 20 hours later, after the shoot had turned back upward (negative gravitropism) and the root had turned down (positive gravitropism).

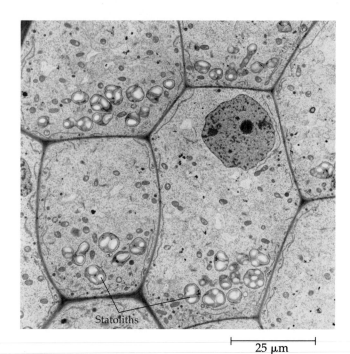

25 µm

Figure 35.10
Statoliths in root cap cells. This electron micrograph of a longitudinal section through a corn root cap shows statoliths, filled with large starch grains, resting on the lower side of several cells. This may be the way roots detect gravity. According to one hypothesis, statoliths resting against the plasma membrane may affect the transport of chemical signals that control the direction of root growth.

faster than cells on the brighter side because of an asymmetric distribution of auxin moving down from the shoot tip. But how does this unequal distribution of auxin come about? Experiments have demonstrated that illuminating a coleoptile from one side causes auxin to migrate laterally across the tip from the bright side to the dark side. It is not yet known how light induces this redistribution of auxin. More is known about the nature of the photoreceptor in the shoot tip, which is most sensitive to blue light and is believed to be a yellow pigment related to the vitamin riboflavin. The same receptor may be involved in stomatal opening and some of the other responses of plants to light.

Gravitropism   Place a seedling on its side and it will adjust its growth so that the shoot bends upward and the root curves downward (Figure 35.9). In their responses to gravity, roots display positive gravitropism and shoots exhibit negative **gravitropism.** Plants apparently tell up from down by the settling of **statoliths,** specialized plastids containing dense starch grains, to the low points of cells (Figure 35.10). This is analogous to the way certain animal organs, including our own inner ears, monitor body position. In roots, the statoliths occur in some root cap cells, and, in primary shoots, they typically occur in parenchyma

cells adjacent to the vascular bundles. Somehow, the sensation of the aggregating statoliths causes certain hormones to accumulate on the low side of the plant organ. In stems, cells on the lower side elongate faster than cells on the upper side because of a higher concentration of auxin and gibberellins. Roots apparently curve down because IAA, which inhibits root elongation at relatively moderate concentrations, accumulates on the lower side. It is unknown how the location of statoliths causes lateral movements of hormones; since the hormones are dissolved, they are themselves unresponsive to gravity, and must be moved downward by some transport mechanism.

It is now clear that Ca^{2+} plays an essential role in root gravitropism. Like auxin, Ca^{2+} accumulates on the lower side of horizontal roots. Although we do not yet know the primary role of Ca^{2+} in gravitropism, it may be involved in the transport of IAA. Since the statoliths have a significant amount of Ca^{2+} associated with them, their redistribution within the cell may cause localized increases in Ca^{2+} at the plasma membrane, which, in turn, may affect IAA transport.

Thigmotropism   Most vines and other climbing plants have tendrils that coil around supports (Figure 35.11). These grasping organs usually grow straight

Figure 35.11
Coiling response of a green bean.

Mechanical stimulation can also cause a much more general response. One experiment has shown that rubbing stems with a stick a few times results in plants that are shorter and thicker than controls that are not so molested. In the wild, wind can cause a similar stunting of growth, enabling the plant to hold its ground against strong gusts. A tree growing on a windy mountain ridge, for instance, will usually have a shorter, stockier trunk than a tree of the same species growing in a more sheltered location. This decrease in stem length with concomitant thickening caused by mechanical perturbation is referred to as **thigmomorphogenesis.** It usually results from an increased production of ethylene in response to chronic mechanical stimulation.

Turgor Movements

In addition to the relatively long-lasting changes in body shape resulting from tropisms, plants are also capable of reversible movements caused by changes in the turgor pressure of specialized cells in response to stimuli.

until they touch something; the contact stimulates a coiling response caused by differential growth of cells on opposite sides of the tendril. This directional growth in response to touch is called **thigmotropism** (from the Greek *thigma,* "touch").

Rapid Leaf Movements When the compound leaf of *Mimosa,* the sensitive plant, is touched, it collapses and its leaflets fold together (Figure 35.12). This response, which takes only a second or two, results from a rapid loss of turgor by cells within pulvini, specialized motor organs located at the joints of the leaf. There is evidence that the motor cells suddenly become flaccid after stimulation because they lose potassium, which causes water to leave the cells by osmosis. It takes about ten minutes for the cells to regain their turgor and restore the natural form of the

(a)

(b)

Figure 35.12
The sensitive plant (*Mimosa pudica*). (a) In the unstimulated plant, leaflets are spread apart. **(b)** Within a second or two after a mechanical stimulus, the leaflets have folded together, and the entire leaf has dropped.

leaf. The function of the sensitive plant's behavior invites speculation. Perhaps by folding its leaves and reducing its surface area when jostled by strong winds, the plant conserves water. Or maybe the rapid response of the sensitive plant discourages herbivores. Indeed, the collapse of leaves does expose thorns on the stem.

A remarkable feature of rapid leaf movements is the transmission of the stimulus through the plant. If one leaf on a sensitive plant is touched with a hot needle, first that leaf collapses, then the adjacent leaf responds, then the next leaf along the stem, and so on until all of the plant's leaves are drooping. From the point of stimulation, the message to respond travels wavelike through the plant at a speed of about a centimeter per second. Chemical messengers probably have a role in this transmission, but also an electrical impulse can be detected by attaching electrodes to the plant. These impulses, called **action potentials,** resemble nervous messages in animals, though the action potentials of plants are thousands of times slower than those of animals. Action potentials, which have been discovered in many species of algae and plants, may be widely used as a form of internal communication.

Sleep Movements   Bean plants and many other members of the legume family lower their leaves in the evening and then raise them to a horizontal position in the morning (Figure 35.13). These **sleep movements** are powered by daily changes in the turgor pressure of motor cells in pulvini similar to those of the sensitive plant. When leaves are horizontal, cells on one side of the pulvinus are turgid, whereas cells on the opposite side are flaccid. This is reversed at night when the leaves close to their "sleeping" position. Paralleling the opposing changes in volume of the motor cells is a massive migration of potassium ions from one side of the pulvinus to the other. Apparently, the potassium is an osmotic agent that leads to the reversible uptake and loss of water by the motor cells. In this respect, the mechanism of sleep movements is similar to stomatal opening and closing. Sleep movements are only one example of the many responses that depend on the ability of plants to keep track of time.

CIRCADIAN RHYTHMS AND THE BIOLOGICAL CLOCK

Your pulse, blood pressure, temperature, rate of cell division, blood cell count, alertness, urine composition, metabolic rate, sex drive, and responsiveness to drugs all fluctuate with the time of day. Some insects are more vulnerable to insecticides in the afternoon than in the morning. Certain fungi produce spores for several hours the same time each day. Unicellular algae that glow in the dark switch on their bioluminescence

Figure 35.13
Sleep movements of the bean. *Top*, leaf position at noon. *Bottom*, leaf position at midnight. The movements are caused by reversible changes in the turgor of cells on opposing sides of the pulvini, swollen regions of the petiole.

like clockwork. As Ruth Satter told us in this unit's interview, plants also display rhythmic behavior—the sleep movements of legumes and the opening and closing of stomata are examples. All of these rhythmic phenomena and many others are controlled by biological clocks, internal oscillators that keep accurate time. Biological clocks seem to be ubiquitous features of eukaryotic organisms, and, as Dr. Satter mentioned in her interview, our first evidence for biological rhythms came from studies of plants.

A physiological cycle with a frequency of about 24 hours is called a **circadian rhythm** (from the Latin *circa*, "approximately," and *dies*, "day"). Are these rhythms truly prompted by an internal clock, or are they merely daily responses to some environmental cycle, such as the rotation of the Earth? Circadian rhythms persist even when the organism is sheltered from environmental cues. A bean plant, for example, will continue its sleep movements even if kept in constant light or constant darkness; the leaves are not simply responding to sunrise and sunset. Organisms, including humans, continue their rhythms when placed in the deepest mine shafts or orbited in satellites. All research so far indicates that the oscillator for circadian rhythms is endogenous (internal). This clock, however, is entrained (set) to a period of precisely 24 hours by daily signals from the environment. If an organism is kept in a constant environment, circadian rhythms deviate from a 24-hour period (a "period" is the duration of one cycle). These **free-running periods,** as they are called, vary from about 21 to 27 hours, depending on the particular rhythmic response. The sleep movements of bean plants, for instance, have a period of 26 hours when kept under the free-running conditions of constant darkness. The observation that various circadian rhythms have different free-running periods is further evidence that these rhythms are not simply responses to cosmic cues.

Deviation of the free-running period from exactly 24 hours does not mean that biological clocks drift erratically. The clocks are still keeping perfect time, but they are not synchronized with the outside world. The light-dark cycle due to the Earth's rotation is the most common factor that entrains biological clocks. If the timing of these cues changes, it takes a few days for the clock to be reset. Thus, a plant kept for several days in the dark will be out of phase with plants growing in a normal environment where the sun rises and sets each day. The same thing happens when we cross several time zones in an airplane; when we reach our destination, the clocks on the wall are not synchronized with our internal clocks. All eukaryotes are probably prone to jet lag.

Most biologists now agree that organisms possess built-in clocks, but the nature of the internal oscillator is still a mystery. Where is the clock, and how does it work? In attempting to answer this question, we must take care to differentiate between the oscillator and the rhythmic processes it controls. The sweeping leaves of sleep movements are the "hands" of the biological clock, but these movements are not the essence of the clockwork itself. If the leaves of a bean plant are restrained for several hours so they cannot move, they will, on release, rush to the position appropriate for the time of day. We can interfere with a biological rhythm, but the clock goes right on ticking off the time. Most scientists who study circadian rhythms place the clock at the cellular level, either in membranes or in the machinery for protein synthesis.

Whatever the mechanism for circadian rhythms, their periods are, in most cases, affected little by temperature. The rates of metabolic processes are generally sensitive to temperature changes, but the clock compensates somehow. A clock that speeds up or slows down with a rise and fall of temperature would be an unreliable timepiece.

PHOTOPERIODISM

Seasonal events are important in the life cycles of most plants. Seed germination, flowering, and the onset and breaking of bud dormancy are examples of stages in plant development that usually occur at specific times of the year. The environmental stimulus plants most often use to detect the time of year is the photoperiod, the relative lengths of night and day. A physiological response to day length, such as flowering, is called **photoperiodism.**

Photoperiodic Control of Flowering

One of the earliest clues to how plants detect the progression of seasons came from an unusual variety of tobacco studied by W. W. Garner and H. A. Allard in 1920. This variety, named Maryland Mammoth, grew exceptionally tall but failed to flower during summer when normal tobacco plants flowered. Maryland Mammoth finally bloomed in a greenhouse in December. After trying to induce earlier flowering by varying temperature, moisture, and mineral nutrition, Garner and Allard learned that it was the shortening days of winter that stimulated Maryland Mammoth to flower. If the plants were kept in light-tight boxes so that lamps could be used to manipulate durations of "day" and "night," flowering would occur only if the day length was 14 hours or shorter. The Maryland Mammoths did not flower during summer because, at Maryland's latitude, the days were too long during that season.

Garner and Allard termed Maryland Mammoth a **short-day plant,** because it apparently required a light period *shorter* than a critical length to flower. Chrysanthemums, poinsettias, and some soybean varieties are a few of the other short-day plants, which generally flower in late summer, fall, or winter. Another group of plants dependent on photoperiod will flower only when the light period is *longer* than a certain number of hours. These **long-day plants** generally flower in late spring or early summer. Spinach, for example, flowers when days are 14 hours or longer. Radish, lettuce, iris, and many cereal varieties are also long-day plants. Flowering in a third group, **day-neutral plants,** is unaffected by photoperiod. Tomatoes, garden peas, rice, and dandelions are examples of day-neutral plants.

Critical Night Length　In the 1940s, it was discovered that night length, not day length, actually controls flowering and other responses to photoperiod. Researchers worked with cocklebur, a short-day plant that flowers only when days are 16 hours or less in length (and nights, therefore, at least 8 hours long). If the daytime portion of the photoperiod is broken by a brief exposure to darkness, there is no effect on flowering (Figure 35.14). However, if the nighttime part of the photoperiod is interrupted by even a few minutes of dim light, the plants will not flower. Cocklebur requires at least 8 hours of *continuous* darkness to flower. Short-day plants are really long-night plants, but the older term is embedded firmly in the jargon of plant physiology. Long-day plants are actually short-night plants; grown on photoperiods of long nights

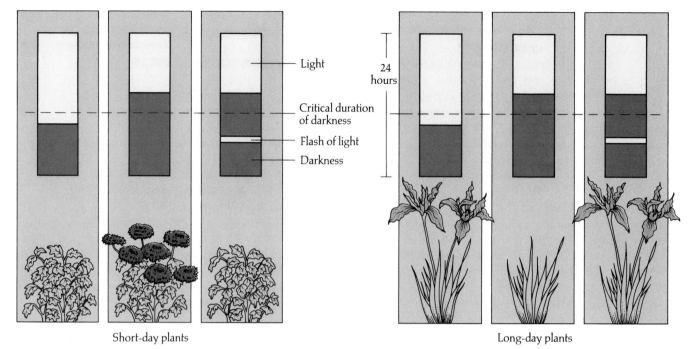

Figure 35.14

Photoperiodic control of flowering. A short-day plant, really a long-night plant, flowers when night exceeds a critical dark period. A flash of light interrupting the dark period prevents flowering. A long-day plant, actually a short-night plant, flowers only if the night is shorter than a critical dark period. The night can be artificially shortened with a flash of light.

Labels in figure: Light · Critical duration of darkness · Flash of light · Darkness · 24 hours · Short-day plants · Long-day plants

that would not normally induce flowering, long-day plants will flower if the period of continuous darkness is shortened by a few minutes of light at any point. Thus, photoperiodic responses depend on a critical night length. Short-day plants will flower if the duration of night is *longer* than the critical length (8 hours for cocklebur), and long-day plants flower when the night is *shorter* than the critical length. The floriculture (flower-growing) industry has applied this knowledge so that we can have flowers out of season. Chrysanthemums, for instance, are short-day plants that normally bloom in fall, but their blooming can be stalled until Mother's Day by punctuating each long night with a flash of light, thus turning a long night into two short nights.

Notice that we distinguish long-day from short-day plants *not* by an absolute night length but by whether the critical night length sets a maximum (long-day plants) or minimum (short-day plants) number of hours of uninterrupted darkness required for flowering.

Some plants bloom after a single exposure to the photoperiod required for flowering. Other species need several successive days of the appropriate photoperiod. Still other plants will respond to photoperiod only if they have been previously exposed to some other environmental stimulus, such as a period of cold temperatures. Winter wheat, for example, will not flower unless it has been exposed to several weeks of temperatures below 10°C. This requirement for pretreatment with cold before flowering is called **vernaliza-**

tion. After winter wheat is vernalized, a photoperiod with long days (short nights) will induce flowering.

Evidence for a Flowering Hormone Buds produce flowers, but the photoperiod is detected by leaves. To induce a short-day plant or long-day plant to flower, it is enough in many species to expose a single leaf to the appropriate photoperiod. Indeed, if only one leaf is left attached to the plant, photoperiod is detected and floral buds are induced. If all leaves are removed, however, the plant is blind to photoperiod. Apparently, some message to flower is transported from leaves to buds. Most plant physiologists believe this message is a hormone, which has been named florigen (Figure 35.15). The signal to flower that travels from leaves to buds appears to be the same for short-day and long-day plants, although the two groups of plants differ in the photoperiodic conditions required for leaves to send this signal.

The evidence for a flowering hormone is compelling, but so far florigen has not been identified. It is possible that the flowering impulse is not a single chemical but a specific mixture of several hormones or even the absence of inhibitors.

Phytochrome

The discovery that night length is the critical factor controlling seasonal responses of the plant poses

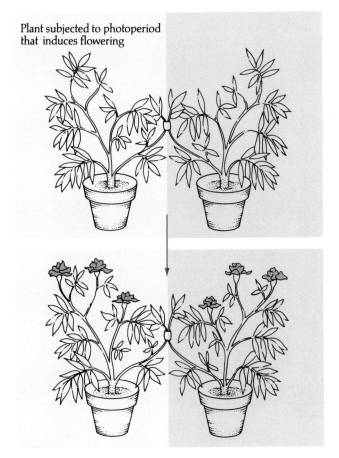

Plant subjected to photoperiod that induces flowering

Figure 35.15
Evidence for a hormone that induces flowering. Evidence for a flowering hormone comes from grafting experiments. If a plant that has been induced to flower by photoperiod is connected to a plant that has not been induced, both plants flower. This works in some cases even if one is a short-day plant and the other is a long-day plant.

another question: How does the plant measure the length of darkness in a photoperiod? A pigment named **phytochrome** is part of the answer. It was discovered as a result of studies on how different colors of light affect photoperiodic control of flowering, seed germination, and other processes.

Red light having a wavelength of 660 nm is the light most effective in interrupting night length. A short-day plant kept under conditions of critical night length fails to flower if a brief exposure to red light breaks the dark period. Conversely, a flash of red light during the dark period will induce a long-day plant to flower even if the total night length exceeds the critical number of hours (long-day plants, remember, require nights shorter than a critical length). The effect of the red flash is to shorten the plant's perception of night length.

The shortening of night length by red light can be negated by a subsequent flash of light having a wavelength of about 730 nm. This wavelength is in the far-

red part of the spectrum, just barely visible to humans. If red light (R) given during the dark period is followed by far-red (FR) light, the plant perceives no interruption in night length. A short-day plant will not flower if a night of critical length is broken by an R flash, but the plant *will* flower if it receives two flashes, first R and then FR. Reversing this sequence to FR–R prevents flowering. Each wavelength of light cancels the effect of the other. No matter how many flashes of light are given, the wavelength of only the last one affects the plant's measurement of night length. Thus, a succession of flashes with the sequence R–FR–R–FR–R prevents short-day plants from flowering, but flowering occurs if the sequence is R–FR–R–FR–R–FR. As expected, the opposite behavior occurs in long-day plants (Figure 35.16).

The photoreceptor responsible for the reversible effects of red and far-red light is phytochrome, a protein containing a chromophore (the light-absorbing portion). Phytochrome alternates between two forms that differ only slightly in structure, but one absorbs red light and the other absorbs far-red light (Figure 35.17). The two variations of phytochrome—P_r (red absorbing) and P_{fr} (far-red absorbing)—are said to be photoreversible. The $P_r \rightleftarrows P_{fr}$ interconversion acts as a switching mechanism controlling various events in the life of the plant.

One function of the phytochrome system is to tell the plant that light is present. Plants synthesize phytochrome as P_r, and if they are kept in the dark, the pigment remains in that form. Then if the phytochrome is illuminated in sunlight, most of the P_r is converted to P_{fr} because sunlight is more enriched in red than far-red light. The appearance of P_{fr} is one way plants detect sunlight, and P_{fr} triggers many of the responses of plants to light, such as the germination of many seeds that require light to break dormancy. Although the cellular mechanism of phytochrome action is not completely known at the present time, the expression of a number of nuclear genes has been shown to be controlled by phytochrome.

Role of the Biological Clock in Photoperiodism

In darkness, the pool of P_{fr} molecules gradually reverts by biochemical processes to P_r; this reversion occurs each day after sunset. Then at sunrise, the P_{fr} level suddenly increases again by rapid photochemical conversion from P_r. Phytochrome conversion marks the beginning and end of the dark segment of a photoperiod. Does the plant use these signals to measure night length, the environmental variable that controls flowering and other responses to photoperiod? Perhaps the gradual conversion of P_{fr} to P_r in the dark is a chemical hourglass that gauges the passage of night.

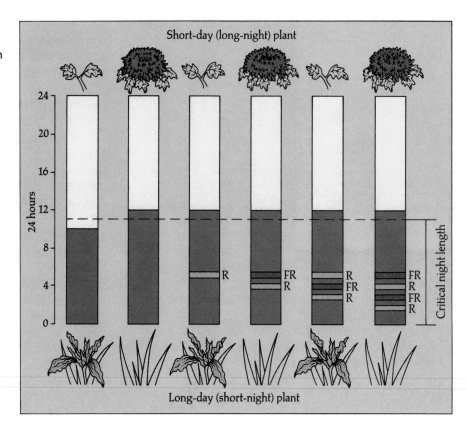

Figure 35.16
Reversible effects of red and far-red light on photoperiodic response. A flash of red light shortens the dark period. A subsequent flash of far-red light cancels the effect of the red flash.

However, the conversion is usually completed within a few hours after sunset. If the plant used the disappearance of P_{fr} to measure the length of the dark period, it would lose track of time during the middle of the night. Furthermore, temperature affects the rate of conversion of P_{fr} to P_r. Night length is measured not by phytochrome but by the biological clock. The role of phytochrome in photoperiodism may be to synchronize the clock to the environment by telling it when the sun sets and rises.

If the photoperiodic requirement for flowering is met, the clock sounds some sort of alarm that causes leaves to send a flowering stimulus (perhaps florigen) to buds. Night length is measured very accurately; some short-day plants will not flower if night is even one minute shorter than the critical length. Some plant species always flower on the same day each year. According to the hypothesis described here, plants tell the season of year by using the clock, apparently entrained with the help of phytochrome, to keep track of photoperiod.

Photoperiodism is a fitting finale to this chapter, for it integrates what we have learned about internal and external stimuli controlling plant growth and development. Genes, hormones, an endogenous clock, photoreceptors, and the tilt and rotation of the Earth are all intertwined in the control of flowering and other seasonal events. In the title of this chapter, "Control Systems in Plants," the term *systems* should be underscored. It is a word that connotes complex interactions of many components. Through a systems approach to plant physiology, we are beginning to understand how plants work. Progress must continue, not only to satisfy our curiosity, but also because basic research in plant physiology is the think tank of modern agriculture. We cannot predict which discoveries, no matter how esoteric they may seem at the time, will lead to agricultural applications that will enable farmers to grow more food. To recall the words of Ruth Satter in this unit's interview, support of botanical research and agricultural development may be a social responsibility.

Figure 35.17
Interconversion of phytochrome. The two forms of the phytochrome molecule are designated P_r (red absorbing) and P_{fr} (far-red absorbing). Absorption of light of the appropriate wavelength changes phytochrome from one form to the other. Red light converts P_r to P_{fr}, and far-red light converts P_{fr} to P_r.

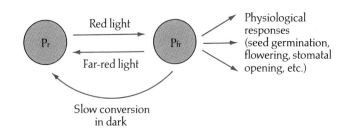

STUDY OUTLINE

1. Plants adjust their growth and development in response to environmental cues.

The Search for a Plant Hormone (pp. 758–760)

1. Hormones are traveling chemical messengers that coordinate functions between distant parts of an organism.
2. Experiments on phototropism led to identification of the first plant hormone, named auxin.

Functions of Plant Hormones (pp. 760–767)

1. Five classes of hormones control plant growth and development by multiple effects on division, elongation, and differentiation of cells. The site of action, the developmental stage of the plant, and the concentration of the hormone all affect reaction to a hormone.
2. Hormone effects may be due more to a balance between hormones than to action of any one hormone.
3. Auxins stimulate elongation in selected cells.
4. Produced primarily in the apical meristem of the shoot, auxin stimulates cell elongation in different target tissues, depending on its concentration. High concentrations may inhibit growth through induction of ethylene synthesis.
5. The acid-growth hypothesis maintains that auxin causes stimulation of proton pumps that acidify the cell wall region. This, in turn, activates enzymes that break cross-links between cellulose microfibrils in the cell wall.
6. Auxin also affects secondary growth and differentiation, initiates adventitious root formation, induces formation of leaf traces, and promotes fruit growth.
7. Cytokinins, a second class of plant hormones, are adenine derivatives that stimulate cell division, or cytokinesis.
8. Actively growing tissues, such as roots, embryos, and fruits, are rich in cytokinins, which, in concert with auxin, stimulate cell division and influence differentiation. Subtle changes in the cytokinin/auxin ratio have different effects on plant development.
9. Cytokinins and auxin also work together to control apical dominance in a complex interaction that likely involves ethylene and specific nutrient levels. These interactions help coordinate the growth of root and shoot systems.
10. Cytokinins retard the aging process in some plant organs by prolonging protein synthesis and mobilizing nutrients.
11. Gibberellins produced in roots and young leaves stimulate growth in leaves and stems. In conjunction with auxin, gibberellins increase elongation in stems.
12. Gibberellins and auxin also stimulate fruit set.
13. Gibberellins signal seeds to germinate by stimulating the synthesis of key enzymes involved in the mobilization of seed storage material.
14. Abscisic acid slows plant growth and favors the dormant state by inducing development of bud scales, inhibiting cell division in the vascular cambium, and suspending growth in buds and seeds. Whether buds and seeds break dormancy depends on the ratio of abscisic acid to gibberellins. Abscisic acid is also a "stress hormone," helping plants cope with adverse conditions.
15. Ethylene, a gaseous hormone, diffuses through the plant in intercellular air spaces. It inhibits root growth and development of axillary buds in the presence of high auxin concentrations. Ethylene is also implicated in several aspects of senescence of plant cells and parts.
16. Fruit ripening was one of the earliest observed functions of ethylene.
17. Leaf fall, or abscission, is a complex process triggered in autumn by short days and cool temperatures. The mechanics of abscission involve decreased auxin and cytokinin levels and increased ethylene production.

Plant Movements (pp. 767–770)

1. Tropisms are growth responses that result in curvature of entire plant organs toward or away from stimuli.
2. Light causes phototropic responses by inducing redistribution of auxin.
3. Gravitropism is a response mediated by statoliths, which may affect IAA transport.
4. Thigmotropism leads to an adaptive coiling of tendrils on touching a support. Thickening of stems by chronic, strong winds is an example of thigmorphogenesis.
5. Turgor movements are rapid, reversible responses to stimuli caused by potassium-mediated changes in turgor pressure of specialized cells.
6. Rapid leaf movements in *Mimosa* result from turgor changes in pulvini at the joints of each leaf. A signal is transmitted from the point of stimulation as a nervelike action potential.
7. Sleep movements of legumes are also powered by changes in turgor pressure in pulvini.

Circadian Rhythms and the Biological Clock (pp. 770–771)

1. Sleep movements and other rhythmic behavior of plants are controlled by biological clocks, internal oscillators present in all eukaryotes that keep track of time.
2. Circadian rhythms are physiological cycles that have a frequency of about 24 hours. Absence of environmental cues leads to free-running periods, in which the rhythms may deviate by a few hours from a 24-hour period but are still precise within their own cycles. The light-dark cycle probably entrains biological clocks.

Photoperiodism (pp. 771–774)

1. Photoperiodism, the response to relative lengths of night and day, helps regulate stages of plant development.
2. Photoperiodic responses actually depend on a critical night length, which sets a minimum (short-day plants) or maximum (long-day plants) number of hours of uninterrupted darkness required for flowering. Flowering in day-neutral plants is unaffected by photoperiod.
3. A signal from the leaves, postulated by many to be a hormone called florigen, causes flowering of buds.
4. Phytochrome, a pigment that exists in two photoreversible states, is one factor that signals sunrise and sunset. Actual night length is measured by the biological clock, which uses the interconversion of phytochrome to mark the beginning and end of the dark segment of each day.

SELF-QUIZ

1. Which of the following plant hormones is incorrectly paired with its function?

a. auxin—promotes stem growth through cell elongation

b. cytokinin—initiates senescence

c. gibberellin—stimulates seed and bud germination

d. abscisic acid—promotes seed and bud dormancy

e. ethylene—inhibits cell elongation

2. Spraying some plants with a combination of auxin and gibberellins

   a. promotes formation of parthenocarpic fruit

   b. kills broad-leaved dicot plants

   c. prevents senescence

   d. promotes fruit ripening

   e. is used to treat dwarfism in plants

3. Which of the following is *not* part of the acid-growth hypothesis?

   a. Auxin stimulates proton pumps in cell membranes.

   b. Lowered pH activates enzymes in the cell walls.

   c. pH-sensitive enzymes break cross-links between cellulose microfibrils.

   d. Auxin-activated proton pumps stimulate cell division in meristems.

   e. The turgor pressure of the cell exceeds the restraining pressure of the loosened cell wall, and the cell takes up water and elongates.

4. Statoliths are involved in

   a. sleep movements

   b. leaf abscisions

   c. gravitropism

   d. phototropism

   e. thigmotropism

5. The hypothetical plant hormone florigen could be released prematurely in a long-day plant experimentally exposed to flashes of

   a. far-red light during the night

   b. red light during the night

   c. red light followed by far-red light during the night

   d. far-red light during the day

   e. red light during the day

6. Which of the following is *not* true of free-running periods?

   a. They may vary in length from 24 hours.

   b. They are of a set period for each rhythmic response.

   c. Their lengths are generally sensitive to temperature.

   d. They provide evidence that the biological clock is endogenous.

   e. They can be measured when an organism is kept in a constant environment.

7. The phytochrome system helps to entrain the biological clock by indicating to a plant that light is present when

   a. P_r is rapidly converted to P_{fr}.

   b. P_{fr} is slowly converted to P_r.

c. P_r and P_{fr} are equal in concentration.

d. Red light is absorbed by P_{fr}.

e. Photosynthetic production of ATP powers phytochrome conversion.

8. Bolting of floral shoots is triggered by a high concentration of

   a. auxin

   b. gibberellins

   c. cytokinins

   d. florigen

   e. ethylene

9. If a long-day plant has a critical night length of 9 hours, then which of the following 24-hour cycles would *prevent* flowering?

   a. 16 hours light/8 hours dark

   b. 14 hours light/10 hours dark

   c. 15.5 hours light/8.5 hours dark

   d. 4 hours light/8 hours dark/4 hours light/8 hours dark

   e. 8 hours light/8 hours dark/light flash/8 hours dark

10. In the control of flowering, the organs that monitor photoperiod are the

    a. floral buds

    b. lateral buds

    c. leaves

    d. roots

    e. shoot apex

CHALLENGE QUESTIONS

1. Discuss how day length, phytochrome, the biological clock, gibberellins, and abscisic acid may interrelate in the germination of seeds planted just below the soil surface.

2. Suggest possible functions of sleep movements.

3. Explain how it is possible that a short-day plant and a long-day plant growing in the same location could flower on the same day of the year.

FURTHER READING

1. Cleland, C. F. "The Flowering Enigma." *BioScience*, April 1978. The search for florigen.

2. Evans, M. L., R. Moore, and K-H Hasenstein. "How Roots Respond to Gravity." *Scientific American*, December 1986.

3. Mandoli, D. F., and W. R. Briggs. "Fiber Optics in Plants." *Scientific American*, August 1984. Illuminating information on how "light pipes" coordinate plant physiology.

4. Mores, P. B., and N-H Chua. "Light Switches and Plant Genes." *Scientific American*, April 1988. A link between environment and gene expression.

5. Sisler, E. C., and S. F. Yang. "Ethylene, the Gaseous Plant Hormone." *BioScience*, April 1984. Information on the biosynthesis, effects, and mechanism of action of an important plant hormone.

Animals: Form and Function

Interview: George Bartholomew

George Bartholomew helped to define a new field of biology. Starting his career as a physiologist, he worked with a variety of animals: elephant seals, kangaroo rats, pocket mice, insects, and many others. In his research he frequently found himself making comparisons between animals, observing the different ways in which organisms meet the challenges of their environment and how their limits of tolerance for environmental stresses determine where they can live. Dr. Bartholomew's work brought into focus the young scientific field now referred to as physiological ecology.

In addition to his research accomplishments, Dr. Bartholomew has a reputation for training an impressive number of out-

standing graduate and postgraduate students. During his 40 years on the faculty at the University of California, Los Angeles, scores of students have worked with him in his laboratory. Based on his thorough research and consistently important publications, Dr. Bartholomew has earned membership in the prestigious National Academy of Sciences. By inspiring excellent scientific work in others, he has also earned admiration as the model scientist of nearly all working physiological ecologists today.

In this interview, George Bartholomew discusses his scientific philosophy, the case for the organism–environment interface as the focal point of physiology, and the importance of natural history in modern biology.

You must have had more than the average child's curosity about animals. How did that interest develop?

When I was a young boy, I spent a lot of time at the Field Museum of Natural History in Chicago, so it probably started there. In any event, by the time I was in junior high school, I knew I was going to be a zoologist. When I was a little older, I thought it would be fascinating to go to distant parts of the world and study strange animals. It wasn't until I was an undergraduate at the University of California at Berkeley that I realized that to spend my life studying animals in the modern world, I would have to get a Ph.D. So I went on to graduate school at Harvard. I got my job here at UCLA directly after I got my Ph.D. degree in 1947 and have been here ever since.

You're a member of the National Academy of Sciences. Can you say a little about the academy and how it selects its members?

The official business of the academy is to advise the government on matters of policy relating to science and technology. It operates through the National Research Council, which appoints committees usually chaired by a member of the National Academy, though nonmembers are on the committees, also. It was founded, in the time of Abraham Lincoln, as an advisory group to the executive branch; but now it works for the government generally. It also is a place where scientific policy is formulated. Sometimes it seems to be primarily a place where people get honors.

A lot of energy goes into deciding who's going to get elected. I don't know of any people in the academy who are not first-rate scientists. But there are a very large number of equally good scientists who are not in. It depends on many, many things.

Your work has helped define the field of physiological ecology. What exactly is this field of research?

In zoology, physiological ecology is an examination of those aspects of physiology that relate to the animal as an organism and to its interactions with the environment. So the obvious things would be energetics, nutrition, water economy—where there are exchanges between organism and environment. By extension, this has come to include a good bit of behavior—for instance, migration and the environmental timing of reproduction.

Within that broad field, what are some of the specific problems that have captured your interest?

A long time ago I decided that I would try to study things as basic as possible. The basic things that every animal has to do are expend energy to keep itself going, reproduce, process water, and be a member of a social group. So all of my research, however diverse it may appear, relates to social behavior, energetics, water economy, or reproductive timing. Each of these has its own literature in a separate field, but they're all very closely related. If you are studying organisms, you cannot afford to isolate the standard fields of specialization. The animal doesn't distinguish

between its behavior and its physiology, or its ecology and its morphology; it's an integrated system that does all these things. The specialties are a convenience for human beings so they don't have to know so much. In my own case, since I'm first and foremost a naturalist interested in animals, I would rather let the animal tell me what questions to ask.

I've studied many, many different animals, ranging from sea lions to insects. Energetics is my usual point of departure in physiology. For instance, I've studied the energetics of flight in hummingbirds, which meant moving into aeronautics and fluid dynamics and such things. That ties in very closely with pollination by hummingbirds. And as far as plants are concerned, Sphinx moths occupy the same general pollinating role, so the first thing you know, you end up worrying about flight in birds *and* in insects.

One of the themes of our textbook is the correlation of structure and function. Can you give an example of how this basic integrating principle has been important in your work?

In my own work I really don't distinguish between structure and function. For an animal to do something, he's got to have the machinery to do it with. This machinery is part morphological, part physiological, and part behavioral. Structure and function are not merely close together, they are essentially the same thing. I'll give you a simple, direct example. When I first decided to do some work on insects, I wanted to do a field project that was physiological and also simple enough

that I would have some chance of doing it as a novice insect student. I decided to study the temperature of flying moths in Costa Rica from sea level, which is humid tropics, to the top of the mountains (an altitude of 12,000 feet), where it freezes at night. Bernd Heinrich and I started out capturing all different kinds of moths at different altitudes. As we got to measuring the body temperature, which is a physiological attribute of the animal, we realized very quickly that the ones that were hot had small wings and heavy bodies. The ones that were not so hot had great big wings and small bodies. So, even before we analyzed our data, we found out that body temperature was closely correlated with wing loading—the ratio of wing area to the mass of the animal. This is a morphological attribute. An insect with small wings can't fly unless he's hot because his muscles won't contract fast enough to generate the force for lifting. However, an insect with very big wings that will flap very slowly doesn't have to be warm—his muscles work just fine at a lower temperature. We ended up producing a rather elaborate study of the interplay between morphology and physiology in terms of wing size. So structure and function are inseparable.

You've pointed out that evolution by natural selection adds a dimension to biology that makes this science fundamentally different from physics and chemistry.

I'm convinced that this is so although I have come to realize that, if you are taking a broad view, physics and chemistry have evolved as the universe has evolved. But when you introduce natural selection into the process—with reproduction, with genetics—it throws into very sharp relief the fact that chance and history determine what organisms do and become and what they are. In that situation, you really have to measure something and study it very carefully because the results cannot be predicted from logic. Once you make a measurement, an observation, then you can rationalize it and make it fit logically. But you can't predict it beforehand. For instance, I would never have predicted, given what I know about vertebrates, that small insects would be endothermic ("warm-blooded"). That's because there are no small endothermic vertebrates; all the little ones, really little ones, are ectothermic ("cold-blooded"). It never occurred to me, until I began studying insects, that maybe some little ones could produce enough heat to warm up. In fact, it turns out that there are lots of tiny insects that are endothermic when they are active. Once I discovered it by measuring them, a whole new world was opened. How can they be warm blooded when they are so small? What are the factors that favor it? What mechanisms allow it? What's the machinery? So then you have the morphological, the physiological, and the evolutionary all intertwined.

In the realm of biological research, how would you describe creativity?

A creative scientist has to be a good scientist. The ideas have to be innovative—a new way of looking at the world—and have to lead to general conclusions concerning widespread phenomena. Innovativeness and generality, I would think, are the criteria of very good science. The problem becomes how to be innovative. Luck has a great deal to do with it, of course. My own approach to innovation is to take a nonstandard animal, see what it wants to tell me, and then go measure it. The consequence is that you turn up all kinds of things that are interesting. Some of them are important, some are just fun. Using the animal and its evolutionary history as your guide will, I think, let you do innovative research.

What did you mean when you wrote that expertise is creatively stifling?

Expertise means that at some point in your life you have put boundaries on what you know. You can become an expert on only some little bit of the world. And if you are going to become the world expert on this phenomenon you are closing your eyes to the diversity of the rest of the world. In that sense, I think it's stifling. If it fits your temperament to become an expert, to look in ever greater detail, and if you have a good mind, you can find out cosmically important things by looking at small issues. It's not that specialization is evil. Every working scientist is a specialist. It's just that if you become an expert in one little thing you don't see where you are.

Can you give a specific example of the kinds of problems that arise from biology having become fragmented into so many subspecialties?

Because the literature is so big, nobody can follow it all, even with computerized support. You really can't. There are whole important traditions of thought that somebody like myself knows nothing of, and I'm a part of a tradition that another kind of biologist will know nothing of. If I were concerned with DNA and cloning, I couldn't possibly master that and also be competent as an ecologically oriented population geneticist. There is an unbroken series of connections from DNA to population biology, yet there are people working at each end who don't even know each other's vocabulary.

You've advocated an enlarged role for natural history in modern biology. How do you define natural history? What are its main questions and how can natural history improve physiological research?

The traditional questions of natural history have been: What organism is it? Where does it live? How does it live? How did it get to do things the way it does? If you dress these questions up into scientific language, the questions become: What is the systematic status of this animal, its evolutionary history, its functional physiology and morphology, and its ecology? All of these fields I've just mentioned are subsets of natural history. The subset means that you're looking at only part of the animal's function. In this sea of information we are all swimming in, if we keep going back to the natural history of the organism, we are offered an integrated theme to make our limited findings meaningful.

You've argued that traditional physiological approaches that sometimes ignore considerations of natural history have mistreated some important topics. Can you give an example?

The one that's nearest and dearest to my heart is hibernation, which I have studied at intervals all of my professional career. The most striking thing about hibernation is that mammals cool off, they reset their thermostat. Thus, hibernation is often considered to be a special case of thermoregulation and most students of mammals have stud-

ied hibernation as a way to look at mammalian thermoregulation. That's because they are used to thinking in terms of thermoregulation as a discipline. In fact, hibernation is one of many devices for scheduling energy expenditure. If you look at it that way, hibernation then becomes a special case of all other sorts of energy schedulings, from circadian rhythms to seasonal cycles, to fat deposition, to sleep. And it involves not just mammals, but all animals that schedule energy transactions. The traditional study of hibernation has blinded people who came into it from biochemistry or neurophysiology to what they were really studying.

You were mentioning earlier that you advise students not to become overspecialized too early in their career. But you've also suggested that young scientists should develop an effective scientific orientation early in their careers. What do you mean by that?

Specializing too early is like making any decision early in life that will affect your later life: You're closing many doors. You should keep the doors open as long as possible. By effective scientific orientation, I mean that that even though you avoid overspecialization, you do not avoid the rigor that goes with trying to be a sound scientist. In contemporary terms this means that you have the necessary physical, chemical and statistical background in addition to your biology that will let you go in any one of a number of directions depending on where a research problem goes.

If you look over the history of science, aside from perhaps a few towering figures who can do it by brute power, most of us do it by serendipity. We work hard and long on some things and some things fall in our lap. And unless you have an open mind and sufficient breadth in your approach to exploit the things that fall in your lap, you miss them.

I have made an issue of this in my own career. It is very important to do your own science—not to be a science administrator—and to be open to whatever happens. This hands-on approach means that when something interesting comes up, you see it and can change your problem, redesign your experiment, instantly. But if your data are gathered by technicians and something unusual comes up, the chances are you won't even know about it. I'll give you an example from my own experience. When I first came to UCLA, I thought I would study desert birds, but I was interested in mammals as well. I caught some little pocket mice and brought them into the lab, mostly just because they were cute and easy to take care of. I had them right in front of me above my desk where I would see them every day. When I came in in the mornings, I noticed they looked pretty sluggish. I thought, they must need water or something. When I took their temperature, it turned out to be 20° instead of 37°, which is what all self-respecting mammals are supposed to be. It turned out that this was quite commonplace. These animals turned off; they would drop their body temperature. In this way, I stumbled onto heterothermic mammals, and I think that I was the first one who documented it as a routine phenomenon in this family (Heteromyidae) of mammals. It happened because I was watching the animals. If I had a technician go in and stir them up in the morning and feed them and so on, I would never have seen them except in their active state, would never have stumbled on an interesting problem.

I think most of the fun in science is the unusual and unexpected. If you knew what was going to happen with such precision that you could successfully predict it, you wouldn't have to make any measurements. And if you are just a research administrator, no matter how clever you are, and how much money you get, and how many assistants you have, you are missing a lot of the fun. You're also missing a chance to find some new things. You may find what you're looking for, but you won't find the related things that may be even more important.

You view science as a social enterprise. Can you explain what you mean by this?

The ideas of science are not things that come out of one's head. Scientific thought is a historical phenomenon and it's an extended phenomenon spatially. A scientist cannot operate in a vacuum. He can operate only in association with other scientists and as a part of the stream of time. Each of us stands, of course, on the shoulders of

the generation who went before. Science is a part of human culture and, to be effective in it, you have to participate, at least to the extent of taking advantage of other people's ideas, insights, methods, and machinery.

You have a remarkable history of training graduate students who go on to productive and distinguished careers. Do you have some ideas about what makes you such an effective mentor?

The main thing is to get good students. I have been very, very choosy. Each of the graduate students I accepted became an important person to me and I gave them all loving attention. I worked in the lab continuously, so I was with them every day. Anything I knew, they soon knew. When it came to writing the dissertation, they wrote the first draft and I would tear it to bits, absolutely tear it to bits, because I felt the most important thing I could do was to make sure that they were good scientific writers by the time they finished. I never published or put my name on a dissertation that was a student's own. So I gave them the immediate rewards. The combination of hands-on attention and aggressive editing of the dissertation is the only thing I can think of, really, aside from getting good students.

With each student I've entered into a contract that after a year I will tell you whether I will accept you permanently. The first year is provisional. If you don't like me, if you and I don't strike it off, if I don't think you have it, or if your approach is inappropriate for my style, I will do my best to get you

another advisor or another niche someplace, either here at UCLA or at another school.

One other thing that was terribly important was a weekly meeting to which my graduate students came. It was a continuing seminar that ran year after year. The students participated in it from the beginning to the end of their graduate studies, so that there were always beginning graduate students and advanced graduate students and a couple of post-docs with me in the seminar and we all worked together. We would report on our research. We'd read books and articles together, and discuss them. This kept me alive and up to date because I was drawing on the library resources of eight or ten people, continuously. This let me communicate to them everything I knew about science, and it let the senior grad students teach the junior ones, and let the post-docs put in different points of view that came from other institutions.

You've described creativity as a form of youthful play. How can we keep playing as we get older? How can middle-aged and older professors benefit from their students?

Associate with them. I started saying, when I was about 45, that I'm intellectually dead without my students. I know I would have bogged down into teaching the same stuff year after year and doing the same kind of research. But with students, you can't. You know, they're all different. They all want to do different things. Related to that, I always let the students choose their own problems. I never assign problems, which means that you have people working on all kinds of things in the lab. And that's one way you retain your youthful point of view. I think the student is the center of the research enterprise.

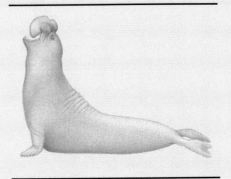

Introduction to Animal Structure and Function

36

Plants and animals confront many of the same problems. For example, those that live in terrestrial environments must meet the architectural challenge of supporting themselves against gravity, and almost all multicellular organisms need internal transport systems to distribute vital materials throughout the body. We can even find superficial similarities in the solutions to some of these problems. For instance, rigid structural materials that provide body support and networks of tubes that transport body fluids have evolved in both the plant and animal kingdoms. But on closer examination, we see that the equipment in plants and animals is very different in its structure and its workings. The multicellular body plans of plants and animals evolved independently from different unicellular ancestors with entirely different modes of nutrition. However, when we try to understand how *any* organism works, the same two themes apply. One is the correlation of structure and function (Figure 36.1). The second theme is the capacity of life to adjust to its environment on two temporal scales: over the short term by physiological responses and over the long term by adaptation due to natural selection (see Chapter 31). As George Bartholomew pointed out in the preceding interview, it is essential to consider organisms in their environmental context. We will follow his lead in this unit of chapters addressing the form and function of animals.

Many students taking a general biology course have a healthy preoccupation with human anatomy and physiology. In the chapters that follow, you will indeed learn much about how your body works. But the scope of this unit is not limited to humans or even to ver-

Figure 36.1
Form fits function. The gills of this salmon are finely subdivided, providing a large surface area for exchange of oxygen and carbon dioxide between the surrounding water and blood flowing through vessels within the gills. We will correlate structure with function throughout our study of animal anatomy and physiology.

tebrates. By following a comparative approach we can see how physiological problems have been solved by animals of diverse evolutionary history and varying complexity, from protozoa to vertebrates. (Although protozoa are actually protists, we include them in this unit to illustrate unicellular analogues of certain processes that occur at higher levels of organization in animals.)

The objective of this chapter is to orient you to the body plans of animals. The use of the terms *plan* and *design* in no way implies that animal forms are products of a conscious invention. The design of an animal—its shape, symmetry, internal organization, and so on—results from a pattern of development programmed by the animal's genome, itself the product of millions of years of natural selection.

LEVELS OF STRUCTURAL ORGANIZATION

Life is characterized by a hierarchy of structural order (see Chapter 1). Atoms are built into molecules, mol-

ecules are assembled into supramolecular structures such as membranes, ribosomes, and chromosomes, and these structures are organized into the living units we call cells. The cell holds a special place in the hierarchy of life because it is the lowest level of organization that can live as an organism. Protozoa, for example, have organelles that are specialized to perform particular jobs, enabling them to digest food, move about, sense environmental change, excrete waste products, and reproduce—all within the confines of a single cell. Protozoa represent the cellular level of organization, the simplest level possible for an organism. Multicellular organisms, including animals, have specialized cells grouped into tissues, the next higher level of structure and function. In most animals, various tissues are combined into functional units called organs, and in the most complex animals, organs cooperate in teams called organ systems. For example, our own digestive system consists of a stomach, a small and a large intestine, a gallbladder, and several other organs, each a composite of different types of tissues.

Animal Tissues

Tissues are groups of cells with a common structure and function. A tissue may be held together by sticky substances that coat the cells (see Chapter 7), or the cells may be woven together in a fabric of extracellular fibers. Indeed, the term *tissue* is from a Latin word meaning "weave."

Histologists, biologists who specialize in the study of animal tissues, classify tissues into four main categories: epithelial tissue, connective tissue, muscle tissue, and nervous tissue. These are present to some extent in all but the simplest animals, but the following survey emphasizes the tissues as they appear in vertebrates.

Epithelial Tissue Occurring in sheets of tightly packed cells, **epithelial tissue** covers the outside of the body and lines organs and cavities within the body. The cells of an epithelium are closely joined, with little material between them. In many epithelia, the cells are riveted together by tight junctions like those described in Chapter 7. This tight packing is consistent with the function of an epithelium as a barrier protecting against mechanical injury, invading microorganisms, and fluid loss. The free surface of the epithelium is exposed to air or fluid, whereas the cells at the base of the barrier are attached to a **basement membrane.**

Two criteria for classifying epithelia are the number of cell layers and the shape of the cells on the free surface (Figure 36.2). A **simple epithelium** has a single layer of cells, whereas a **stratified epithelium** has multiple tiers of cells. A pseudostratified epithelium

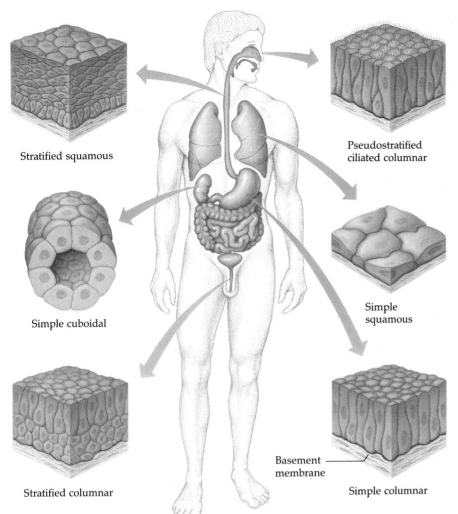

Figure 36.2
Structure and function of epithelial tissue. The structure of an epithelium fits its function. For instance, simple squamous epithelia, which are relatively leaky, are specialized for exchange of materials by diffusion. We find these epithelia in such places as the linings of blood vessels and air sacs of the lungs. A stratified squamous epithelium regenerates rapidly by cell division near the basement membrane. The new cells are pushed to the free surface as replacements for cells that are continually sloughed off. This type of epithelium is commonly found on surfaces subject to abrasion, such as the outer skin and linings of the esophagus, anus, and vagina. Columnar epithelia, having cells with relatively large cytoplasmic volumes, are often located where secretion or active absorption of substances are important functions. For example, the stomach and intestines are lined by a columnar epithelium. Cuboidal cells also commonly function in secretion and make up the epithelia of kidney tubules and many glands, including the thyroid gland and salivary glands.

Labels on figure:
Stratified squamous
Simple cuboidal
Stratified columnar
Pseudostratified ciliated columnar
Simple squamous
Basement membrane
Simple columnar

is single-layered, but appears stratified because the cells vary in length. The shape of the cells at the free surface of an epithelium may be **cuboidal** (like dice), **columnar** (like bricks on end), or **squamous** (like flat floor tiles). Combining the features of cell shape and number of layers, we get such terms as *simple cuboidal epithelium* and *stratified squamous epithelium.*

In addition to protecting the organs they line, some epithelia are specialized for absorbing or secreting chemical solutions. For example, the epithelial cells that line the lumen (cavity) of the small intestine absorb nutrients. Another kind of epithelium, called **mucous membrane,** secretes a slimy solution that lubricates a surface and keeps it moist. The digestive tract and air tubes of the respiratory system are lined with mucous membranes. The free epithelial surfaces of some mucous membranes have beating cilia that move the film of mucus along the surface. The ciliated epithelium of our respiratory tubes helps keep our lungs clean by trapping dust, pollen, and other particles and sweeping them back up the trachea (windpipe).

Connective Tissue Connective tissue functions mainly to bind and support other tissues. In contrast to epithelia with their tightly packed cells, connective tissues have a sparse population of cells scattered through an extracellular **matrix.** This nonliving matrix generally consists of a web of fibers embedded in a homogeneous ground substance that may be liquid, jellylike, or solid.

The most widespread connective tissue in the vertebrate body is **loose connective tissue** (Figure 36.3). It is used as binding to attach epithelia to underlying tissues and as packing material to hold organs in place. This type of connective tissue gets its name from the loose weave of its fibers. These fibers, which are made of protein, are of three kinds: collagenous fibers, elastic fibers, and reticular fibers.

Collagenous fibers are made of collagen, perhaps the most abundant protein in the animal kingdom. A collagenous fiber is a bundle of several fibrils, each a ropelike coil of three collagen molecules (Figure 36.4). Collagenous fibers have great tensile strength, which

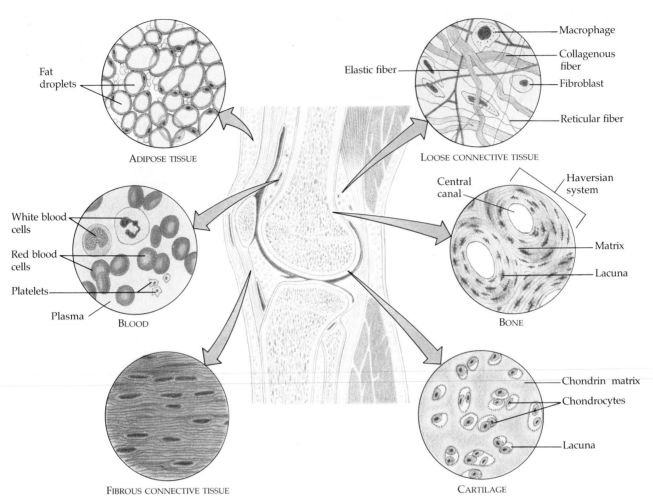

Figure 36.3
Some representative types of connective tissue. The area pictured is the region around the knee joint.

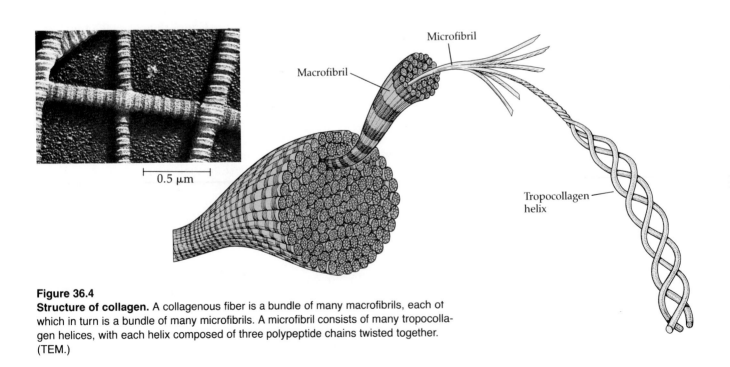

0.5 μm

Figure 36.4
Structure of collagen. A collagenous fiber is a bundle of many macrofibrils, each of which in turn is a bundle of many microfibrils. A microfibril consists of many tropocollagen helices, with each helix composed of three polypeptide chains twisted together. (TEM.)

means that they do not tear easily when pulled lengthwise. If you pinch and pull some skin on the back of your hand, it is mainly collagen that keeps you from tearing your flesh from the bone.

Elastic fibers are long threads made of a protein called elastin. As their name implies, these fibers are elastic, unlike collagenous fibers, which resist stretching. Elastic fibers endow loose connective tissue with a resilience that complements the tensile strength of collagenous fibers. When you pinch the back of your hand and then let go, the rubbery elastic fibers quickly restore your skin to its original shape. You will lose some of this resilience as you get older because collagenous fibers become cross-linked as we age, reducing the elasticity in our connective tissue.

Reticular fibers are branched and form a tightly woven fabric that joins connective tissue to adjacent tissues.

Among the cells scattered in the fibrous mesh of loose connective tissue, two types predominate: fibroblasts and macrophages. **Fibroblasts** secrete the protein ingredients of the extracellular fibers. **Macrophages** are amoeboid cells that roam the maze of fibers, engulfing bacteria and the debris of dead cells by phagocytosis (see Chapter 8). They are weapons in an elaborate arsenal of defense you will learn more about in Chapter 39.

Adipose tissue is a specialized form of loose connective tissue that stores fat in adipose cells distributed throughout its matrix. Adipose tissue pads and insulates the body and stores fuel molecules. Each adipose cell contains a large fat droplet that swells when fat is stored and shrinks when fat is used by the body as fuel. Inheritance is an important factor in obesity, but there is also some evidence that the number of fat cells in our connective tissue is determined partly by the amount of fat we stored when we were babies. This is discouraging news for dieting adults who were overweight at an early age, since it suggests that they have more fat cells than people who may have put on a few pounds in later life.

Fibrous connective tissue is dense, owing to its enrichment in collagenous fibers. The fibers are organized into parallel bundles, an arrangement that maximizes tensile strength. We find this type of connective tissue in the **tendons,** which attach muscles to bones, and in the **ligaments,** which join bones together at joints.

Cartilage has an abundance of collagenous fibers embedded in a rubbery ground substance called **chondrin,** which is a protein-carbohydrate complex. The chondrin is secreted by **chondrocytes,** cells confined to scattered spaces called lacunae in the ground substance. The composite of collagenous fibers and chondrin makes cartilage a strong yet somewhat flexible support material. The skeleton of a shark is made of cartilage. Other vertebrates, including humans, have

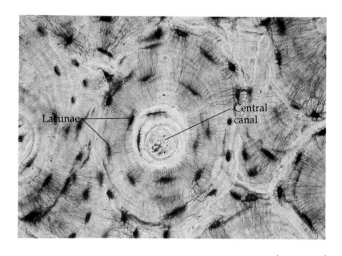

$100 \, \mu m$

Figure 36.5
Histology of bone. The compact bone of mammals consists of Haversian systems, each with concentric layers of mineralized matrix surrounding a central canal. The collagen and calcium phosphate of the matrix are secreted by osteocytes, cells confined to spaces called lacunae. (LM.)

cartilagenous skeletons during the embryo stage, but the skeleton hardens into bone as the embryo matures. We nevertheless retain cartilage as flexible support in certain locations, such as the nose, the ears, the rings that reinforce the windpipe, the discs that act as cushions between our vertebrae, and the caps on the ends of some bones.

The skeletons that support the bodies of most vertebrates are made of **bone,** a mineralized connective tissue (Figure 36.5). Cells called **osteocytes** deposit a matrix of collagen, but also release calcium phosphate, which hardens within the matrix into needles of the mineral hydroxyapatite. The combination of hard mineral and flexible collagen makes bone harder than cartilage without being brittle. The microscopic structure of hard mammalian bone consists of repeating units called **Haversian systems.** Each system has concentric layers of the mineralized matrix called lamellae, which are deposited around a central canal containing blood vessels and nerves that service the bone. Osteocytes are located in lacunae, spaces surrounded by the hard matrix and connected to each other by extensions of the cells called canaliculi. In long bones, such as the femur (shank) of your thigh, only the hard outer region is compact bone built from Haversian systems. The interior is a spongy bone tissue also known as marrow. Blood cells are manufactured in the red marrow located near the ends of long bones. (Bones and skeletons will be examined in more detail in Chapter 45.)

Although blood functions differently from other connective tissues, it does meet the criterion of having an extensive extracellular matrix. In this case, the matrix

Figure 36.6

Types of vertebrate muscle. Skeletal muscle consists of bundles of long cells called fibers; each fiber in turn is a bundle of strands called myofibrils. The myofibrils are a linear array of sarcomeres, the basic contractile units of the muscle. Skeletal muscle is said to be striated because the alignment of sarcomere subunits in adjacent myofibrils forms light and dark bands. Cardiac muscle, also striated, has contractile properties similar to those of skeletal muscle. Unlike skeletal muscle, however, cardiac muscle fibers branch and interconnect via intercalated discs, which help to synchronize the heartbeat. Visceral muscle, or smooth muscle, consists of spindle-shaped cells lacking cross striations.

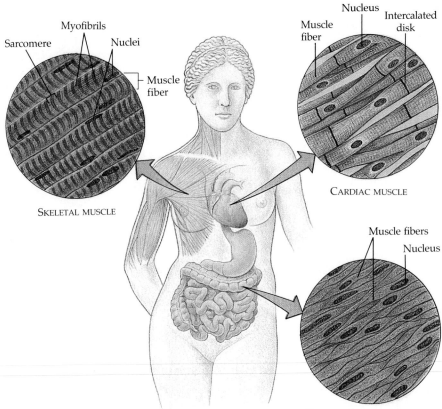

SKELETAL MUSCLE

CARDIAC MUSCLE

VISCERAL (SMOOTH) MUSCLE

is a liquid called plasma, consisting of water, salts, and a variety of dissolved proteins. Suspended in the plasma are three classes of blood cells: erythrocytes (red blood cells), leukocytes (white blood cells), and platelets. Red cells carry oxygen, white cells function in defense, and platelets are involved in the clotting of blood. The composition and functions of blood are discussed in detail in Chapters 38 and 39.

Muscle Tissue Muscle tissue is composed of long, excitable cells capable of considerable contraction. Arranged in parallel within the cytoplasm of muscle cells are large numbers of microfilaments made of the contractile proteins actin and myosin (see Chapter 7). Consistent with a high priority on movement, muscle is the most abundant tissue in most animals. It accounts for about two-thirds of the bulk of a well-conditioned human.

In the vertebrate body, we find three types of muscle tissue: skeletal muscle, cardiac muscle, and visceral muscle (Figure 36.6). Attached to bones by tendons, **skeletal muscle** is generally responsible for the voluntary movements of the body. Adults have a fixed number of muscle cells; weight lifting and other methods of building muscle do not increase the number of cells, but simply enlarge those already present. Skeletal muscle is also called striated muscle because the arrangement of overlapping filaments gives the cells a striped (striated) appearance in the microscope.

Cardiac muscle forms the contractile wall of the heart. It is striated like skeletal muscle, but cardiac cells are branched. The ends of cells are joined by structures called intercalated discs, which relay the impulse to contract from cell to cell during a heartbeat.

Visceral muscle, also called smooth muscle because it lacks cross striations, is found in the walls of the digestive tract, bladder, arteries, and other internal organs. The cells are spindle-shaped. They contract more slowly than skeletal muscles, but they can sustain their contracted condition for a longer period of time.

Skeletal and visceral muscles are controlled by different kinds of nerves. Skeletal muscles are termed "voluntary" because an animal can generally contract them at will. Visceral muscles are involuntary; they are not generally subject to conscious control. You can decide to raise your hand, but you are usually unaware when visceral muscles churn your stomach or constrict your arteries. You will learn more about the control and contraction of muscles in Chapter 45.

Nervous Tissue Nervous tissue senses stimuli and transmits signals from one part of the animal to another. The functional unit of nervous tissue is the **neuron,** or nerve cell, which is uniquely specialized to conduct an impulse (Figure 36.7). It consists of a cell body and two or more appendages called nerve processes, which may be as long as a meter in humans. Dendrites are

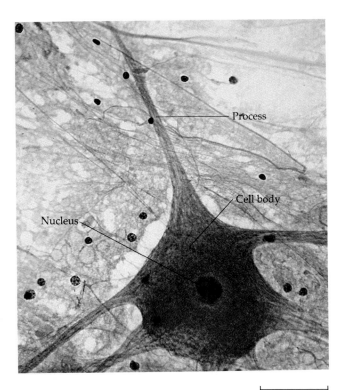

Process

Cell body

Nucleus

50 μm

Figure 36.7
Neuron structure. This nerve cell from the spinal cord has a large cell body with multiple processes that transmit impulses. (LM.)

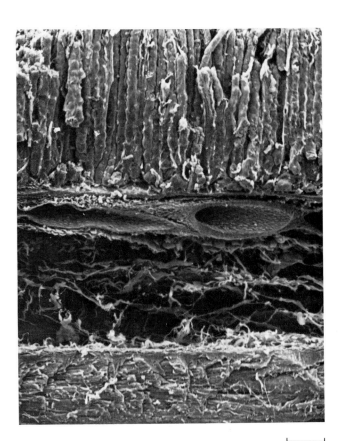

100 μm

Figure 36.8
Tissue layers of the stomach, a digestive organ. Each organ of the digestive system has a wall with four main tissue layers (top to bottom in this scanning electron micrograph): the mucosa, submucosa, muscularis, and adventitia. The mucosa surrounds the lumen of the tract with an epithelial layer on a layer of connective tissue. The submucosa is a matrix of connective tissue that contains blood vessels, lymph vessels, and nerves. The muscularis has an inner layer of circular muscles and an outer layer of longitudinal muscles. The adventitia is the outer layer of connective tissue. (SEM.) (From *Tissues and Organs: A Text-Atlas of Scanning Electron Microscopy* by Richard G. Kessel & Randy H. Kardon. W. H. Freeman & Company. Copyright © 1979.)

processes that conduct impulses toward the cell body, and axons transmit impulses away from the cell body. We postpone a detailed discussion of the structure and function of neurons until Chapter 44.

Organs and Organ Systems

In all but the simplest animals (sponges and cnidarians), different tissues are organized into the specialized centers of function we call **organs.** Some organs have their tissues arranged in layers. The stomach, for example, has four major layers (Figure 36.8). The lumen is lined by a thick epithelium that secretes mucus and digestive juices. Outside this layer is a zone of connective tissue followed by a thick layer of smooth muscle. The entire stomach is encapsulated by another layer of connective tissue. A laminated organization is also apparent in the dermis, the outer skin of the body.

Many of the organs of vertebrates are suspended by sheets of connective tissue called **mesenteries** into body cavities filled with fluid. Mammals have an upper **thoracic cavity** separated from a lower **abdominal cavity** by a sheet of muscle called the diaphragm. The embryonic origin of the body cavity (coelom) and mesenteries was described in Chapter 30.

There is a level of organization still higher than organs. The major functions of vertebrates and most invertebrate phyla are carried out by groups of **organ systems,** each consisting of several organs. Examples are the digestive, circulatory, and respiratory systems. Each has specific functions, but the efforts of all systems must be coordinated for the animal to survive. For instance, nutrients absorbed from the digestive tract are distributed throughout the body by the circulatory system. But the heart that pumps blood through the circulatory system depends on nutrients absorbed by the digestive tract and also on oxygen

Figure 36.9
Contact with the environment. (a) In a single-celled organism, the entire surface area contacts the environment. Because of its small size, the cell has a large surface area relative to its volume through which to exchange materials with the external world. **(b)** Each cell in the bilayered *Hydra* directly contacts the environment and exchanges materials with it.

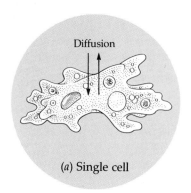

(*a*) Single cell

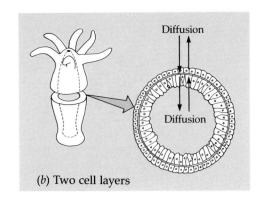

(*b*) Two cell layers

obtained from the air by the respiratory system. The organism, be it a protozoan or an assembly of organ systems, is a living whole greater than the sum of its parts.

SIZE, SHAPE, AND THE EXTERNAL ENVIRONMENT

An animal's size, shape, and symmetry are fundamental features of form and function that affect how the animal interacts with its environment. For instance, to take in oxygen and other materials and get rid of waste products, the animal must be arranged so every cell is bathed in an aqueous medium. Only in such an environment can dissolved substances diffuse across the plasma membrane. Because it is so small, a protozoan living in aqueous surroundings has a sufficient surface area of plasma membrane to service its entire volume of cytoplasm (Figure 36.9a). Geometry imposes limits on the size that is practical for a single cell. A large cell has less surface relative to its volume than a smaller cell of the same shape (see Chapter 7). This is one reason why nearly all cells are microscopic. A multicellular animal is composed of microscopic cells, each with its own plasma membrane to function as a loading and unloading platform for a modest volume of cytoplasm. But this only works if all the cells of the animal have access to a suitable aqueous environment. The freshwater invertebrate *Hydra*, built on the sac plan, has a body wall only two cell layers thick (Figure 36.9b). Since its body cavity opens to the exterior, both outer and inner layers of cells are bathed in water. A flat body shape is another way to maximize exposure to the surrounding medium. For instance, tapeworms may be several meters long, but they are paper thin, and most cells are bathed in the intestinal fluid of its vertebrate host.

Two-layered sacs and planar shapes are designs that put a large surface area in contact with the environment, but these simple forms do not allow much complexity in internal organization. Most animals are bulkier, with outer surfaces that are relatively small compared with the animal's volume. To take an extreme comparison, the surface-to-volume ratio of a whale is millions of times smaller than that of a protozoan, and yet every cell in the whale must be bathed in fluid and have access to oxygen, nutrients, and other resources.

Most complex animals have internal surfaces specialized for exchanging materials with the environment (Figure 36.10). Extensive folding or ramification (branching) gives these moist internal membranes expansive surface areas. The internal membrane of our lungs, for example, is specialized for absorbing oxygen and getting rid of carbon dioxide; it lines millions of microscopic air chambers with a total surface area of about 100 m², the approximate area of a tennis court. The internal surface of the vertebrate kidney, consisting of about a million microscopic tubules, is specialized to filter waste products out of the blood. Still another surface, the lining of the digestive tract, absorbs nutrients. The human intestine is about 6 m long, and its wavy lining has a surface area roughly the size of a baseball diamond (Figure 36.10b). Materials are shuttled between all these exchange surfaces by the circulatory system.

Although the logistical problems of exchange with the environment are much more complicated for an animal with a compact form than for one with all of its cells exposed directly to its surroundings, a compact form has some distinct benefits. Since the animal's external surface need not be bathed in water, it is possible for the animal to live on land. Also, since the immediate environment for the cells is the internal body fluid, the animal can control the quality of the solution that bathes its cells.

THE INTERNAL ENVIRONMENT

More than a century ago, the French physiologist Claude Bernard made the distinction between the

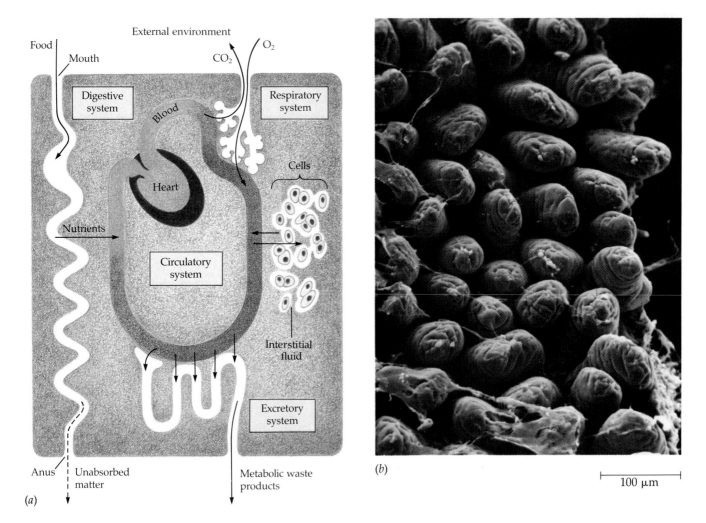

(a)

(b)

100 μm

Figure 36.10
Internal surfaces of complex animals.
(a) Rather than using the outer body surface, most complex animals use specialized surfaces for exchanging materials with the environment. Usually internal, but connected to the environment via openings on the body surface, these exchange surfaces are large in area due to a branched or folded arrangement. A circulatory system transports substances among the various internal surfaces and other parts of the body. (b) A large surface specialized for absorption. The lining of the mammalian small intestine has numerous projections called villi, giving the organ a very large surface area that absorbs nutrients (SEM).

external environment surrounding an animal and the internal environment in which the cells of the animal actually live. The internal environment of vertebrates is the **interstitial fluid.** This fluid, which fills the spaces between our cells, exchanges nutrients and wastes with blood contained in microscopic vessels called capillaries. Bernard also recognized the power of many animals to maintain relatively constant conditions in their internal environment, even when the external environment changes. A pond-dwelling *Hydra* is powerless to affect the temperature of the fluid that soaks its cells, but a human can maintain its "internal pond" at a constant temperature of 37°C. We can also control the pH of our blood and interstitial fluid to within a tenth of a pH unit of 7.4, and we regulate the amount of sugar in our blood so that it does not fluctuate for

long from a concentration of 0.1%. There are times, of course, during the development of an animal when major changes in the internal environment are programmed to occur. For example, the balance of hormones in the blood of a human is altered radically during puberty. Still, the stability of the internal environment is remarkable.

Today, Bernard's "constant internal milieu" is incorporated into the concept of **homeostasis,** which means "steady state." One of the main objectives of modern physiology is to learn how animals maintain homeostasis. Most of the mechanisms of homeostasis that have been discovered are based on **negative feedback.** We can see an analogy in the thermostat that controls the temperature of a room. When room temperature falls below a **set point,** say 20°C (68°F), a thermometer activates a switch that turns on the heater. When the sensor detects that the temperature has risen above the set point, the heating equipment is switched off. This type of control circuit is called negative feedback because a change in the variable being monitored triggers a response that counteracts the initial fluctuation. Owing to the lag time between sensation and response, the variable drifts slightly above and below the set point, but the fluctuations are moderate. Negative feedback mechanisms prevent small changes from becoming too large.

Our own body temperature is kept close to a set point of 37°C by the cooperation of several negative feedback circuits. One of these involves sweating as a means to cool the body. A part of the brain called the hypothalamus monitors the temperature of the blood. If this thermostat detects a rise in body temperature above the set point, it sends nervous impulses directing sweat glands to increase their production of sweat, which lowers body temperature by evaporative cooling (see Chapter 3). When body temperature drops below the set point, the thermostat in the hypothal-

amus stops sending the signals to the glands. We shall see in the chapters that follow how animals use other negative feedback circuits to control several other characteristics of the internal environment.

We can also note physiological examples of **positive feedback,** where a change in some variable triggers mechanisms that amplify rather than reverse the change. During childbirth, for instance, the pressure of the baby's head against sensors near the opening of the uterus stimulates uterine contractions, which cause greater pressure against the uterine opening, which heightens the contractions, causing still greater pressure. Positive feedback brings childbirth to completion, a very different sort of process from maintaining a physiological steady state.

* * *

Having learned some general principles of animal body form and physiology, we are now ready to compare how diverse animals perform such functions as digestion, circulation, gas exchange, excretion of wastes, reproduction, and coordination.

STUDY OUTLINE

1. In our study of animals, we will apply two central themes of biology: the correlation between structure and function, and the capacity to adapt to the environment by both physiological adjustments and natural selection.

Levels of Structural Organization (pp. 782–788)

1. The organization of animals emerges from the grouping of specialized cells into tissues, tissues into organs, and organs into organ systems.
2. Epithelial tissue, present in all animals, covers the outside of the body and lines internal organs and cavities. Consistent with their barrier function, epithelial cells are tightly packed and rest on a basement membrane. Epithelia are described according to the number of cell layers (simple or stratified) and the shape of the surface cells (cuboidal, columnar, or squamous).
3. Epithelia may be specialized for absorption and secretion. The mucus secreted by the mucous membranes lining the gut and respiratory system lubricates and moistens these surfaces.
4. Connective tissues bind and support other tissues. Unlike epithelia, the cells of connective tissue are sparsely scattered through a nonliving extracellular matrix of fibers and ground substance.
5. Loose connective tissue, used as binding and packing material, consists of fibroblasts and macrophages interspersed among collagenous, elastic, and reticular fibers. Adipose (fat) tissue is a specialized type of loose connec-

tive tissue. Fibrous connective tissue, found in tendons and ligaments, is made of dense, parallel bundles of collagenous fibers. Cartilage, bone, and blood are also classified as connective tissues. Cartilage is a strong yet flexible support material consisting of collagenous fibers and a rubbery ground substance secreted by chondrocytes. The osteocytes of bone, which secrete calcium phosphate, are embedded in repeating units called Haversian systems.
6. Muscle tissue, composed of long, excitable cells containing parallel microfilaments of contractile proteins, makes up most of the body of mobile animals. The muscle tissue of vertebrates is subdivided into skeletal, cardiac, and visceral muscles, which differ in shape, striation, and nervous control.
7. Neurons are the functional units of nervous tissue. The sensory and transmission function of nervous tissue is reflected in the anatomy of the neuron, which is composed of a cell body with extensions called dendrites and axons that transmit impulses.
8. In all but the simplest animals, tissues are organized into organs, functional units with complex structures. Many organs of vertebrates are suspended by mesenteries inside fluid-filled body cavities, such as the thoracic and abdominal cavities of mammals. Each organ system, or group of organs, has its specific function, but the activities of all systems are coordinated in the successful functioning of the whole organism.

Size, Shape, and the External Environment (p. 788)

1. Each cell of a multicellular animal must have access to an aqueous environment, a fact that dictates body shape. Simple two-layered sacs and flat, planar shapes maximize exposure to the surrounding medium. More complex, compact body plans have highly folded, moist internal surfaces specialized for exchanging substances with the environment. A circulatory system carries these substances among the parts of an animal.

The Internal Environment (pp. 788–790)

1. The internal environment surrounding the cells is usually very different from the external environment surrounding the entire animal. In addition, the internal environment is carefully controlled and regulated, a fact that was first appreciated by the French physiologist Claude Bernard in his concept of the "constant internal milieu."

2. The internal environment is maintained within narrow limits by steady-state, homeostatic mechanisms based on feedback loops.

SELF-QUIZ

1. A histologist would describe the layered, flat, scaly epithelium of human skin as

   a. simple squamous

   b. stratified squamous

   c. stratified columnar

   d. pseudostratified cuboidal

   e. ciliated columnar

2. Which of the following structures or materials is *incorrectly* matched with a tissue?

   a. Haversian system–bone

   b. platelets–blood

   c. fibroblasts–skeletal muscle

   d. chondrin–cartilage

   e. basement membrane–epithelium

3. Which structures contribute most to the tensile strength of loose connective tissues?

   a. elastic fibers

   b. myofibrils

   c. collagenous fibers

   d. chondrin

   e. reticular fibers

4. The involuntary muscles that cause the wavelike contractions pushing food along our intestine are

   a. striated muscles

   b. cardiac muscles

   c. skeletal muscles

   d. smooth muscles

   e. intercalated muscles

5. Which of the following is *not* considered to be a tissue?

   a. cartilage

   b. mucous membrane lining the stomach

   c. blood

   d. brain

   e. cardiac muscle

6. The membranes that suspend vertebrate organs in the body cavity are called

   a. visceral muscle

   b. loose connective tissue

   c. coelomic membranes

   d. ligaments

   e. mesenteries

7. Compared to a smaller cell, a larger cell of the same shape has

   a. less surface area

   b. less surface area per unit of volume

   c. the same surface to volume ratio

   d. a lesser average distance between its mitochondria and the external source of oxygen

   e. a smaller cytoplasm to nucleus ratio

8. Which of the following vertebrate organ systems does *not* open directly to the external environment?

   a. digestive system

   b. circulatory system

   c. excretory system

   d. respiratory system

   e. reproductive system

9. Most of our cells are surrounded by

   a. blood

   b. fluid equivalent to seawater in salt composition

   c. interstitial fluid

   d. pure water

   e. air

10. Which of the following physiological responses is an example of *positive* feedback.

    a. An increase in the concentration of glucose in the blood stimulates the pancreas to secrete insulin, a hormone that lowers blood glucose concentration.

    b. A high concentration of carbon dioxide in the blood causes deeper, more rapid breathing, which expels carbon dioxide.

    c. Stimulation of a nerve cell causes sodium ions to leak into the cell, and the sodium influx triggers the inward leaking of even more sodium.

    d. The body's production of red blood cells, which transport oxygen from lungs to other organs, is stimulated by a low concentration of oxygen.

    e. The pituitary gland secretes a hormone called TSH, which stimulates the thyroid gland to secrete another hormone called thyroxine. A high concentration of thyroxine suppresses the pituitary's secretion of TSH.

CHALLENGE QUESTIONS

1. Identify each of the following tissues (LMs). Be as specific as possible.

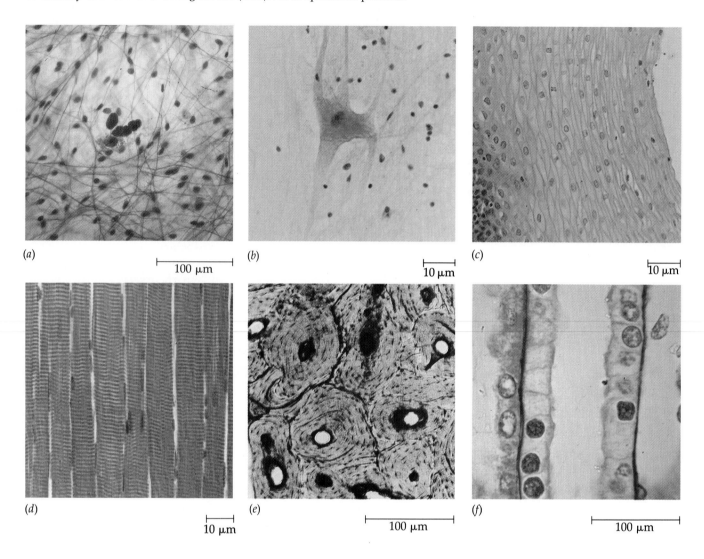

(a) 100 μm

(b) 10 μm

(c) 10 μm

(d) 10 μm

(e) 100 μm

(f) 100 μm

2. Red blood cells pick up oxygen as they travel through the capillaries (microscopic blood vessels) of the lungs, and then deposit the oxygen as they pass through capillaries in other organs. Considering this function, suggest an advantage for our blood being populated by enormous numbers of very small red blood cells rather than fewer large cells. (Assume that the *total* volume of red blood cells is the same in both cases.)

3. Choose three vertebrate tissues and describe how their structures fit their functions.

FURTHER READING

1. Caplan, A. "Cartilage." *Scientific American,* October 1984. Structure and function of an important tissue.
2. Eckert, R., and D. Randall. *Animal Physiology: Mechanisms and Adaptations,* 3rd ed. San Francisco: W. H. Freeman, 1988. A comparative physiology text applying basic principles of structure and function.
3. Flannery, M. "Small, Medium or Large: Why Is Size So Important?" *The American Biology Teacher.* February 1989. How spatial relations constrain evolution.
4. Marieb, E. *Human Anatomy and Physiology.* Redwood City, CA: Benjamin/Cummings, 1989. Basic text, beautifully illustrated.
5. Nilsson, L., and J. Lindberg. *Behold Man.* Boston: Little, Brown, 1974. A visual introduction to human tissues from both a scientific and artistic perspective.
6. Schmidt-Nielsen, K. *Animal Physiology: Adaptation and Environment,* 3d ed. New York: Cambridge University Press, 1983. Comparative physiology text emphasizing homeostasis.
7. Sochurek, H., and P. Miller. "Medicine's New Vision." *National Geographic,* January 1987. Remarkable methods for photographing the human interior.

Animal Nutrition

37

Every mealtime is a reminder that we are animals, dependent on a regular supply of food derived from other organisms. As heterotrophs, animals are unable to live on inorganic nutrients alone and rely on organic compounds in their food for energy and raw materials for growth and repair (Figure 37.1). The ability of an animal to feed itself figures prominently in reproductive success, and natural selection has produced many fascinating nutritional adaptations during the long evolution of the animal kingdom. In this chapter, we shall compare and contrast the diverse mechanisms by which animals obtain and process their food.

FEEDING MECHANISMS

Most animals are holotrophs, organisms that ingest other organisms—dead or alive, whole or by the piece (among the exceptions are certain parasitic animals, such as tapeworms, which absorb organic molecules directly across their outer body surface). Diets vary. **Herbivores,** including gorillas, cows, and fish that graze on algae, eat plants (the Latin *vorare* means "to eat"). **Carnivores,** such as sharks, hawks, spiders, and the python of Figure 37.1, ingest other animals. **Omnivores** (from the Latin *omnis,* "all") consume both meat and plant material. Cockroaches, crows, raccoons, and humans, who began as hunters and gatherers, are examples of omnivores.

The mechanisms by which animals obtain their food also vary. Many aquatic animals are **filter-feeders** that

Figure 37.1
Ingestion. Most animals nourish themselves by eating other organisms, often whole. This rock python is beginning to ingest a gazelle.

sift small food particles from the water. Clams and oysters, for example, use a film of mucus on their gills to trap tiny morsels, which are then swept along to the mouth by beating cilia. Baleen whales, the largest animals ever to live, are also filter-feeders. They swim with their mouths agape, straining millions of small animals from huge volumes of water forced through screenlike plates attached to their jaws (Figure 37.2).

Substrate-feeders live in or on their food source, eating their way through the food. Examples are leaf miners, which are larvae of various insects that tunnel through the soft mesophyll in the interior of leaves (Figure 37.3). Earthworms are also substrate-feeders or, more specifically, **deposit-feeders.** Eating its way through the dirt, the earthworm salvages detritus, partially decayed organic material ingested along with soil.

Fluid-feeders make their living by sucking nutrient-rich fluids from a living host (Figure 37.4). Aphids, for example, tap the phloem sap of plants, and leeches

Figure 37.2
Filter-feeding: baleen whales. (a) Gray whales and other baleen whales use screenlike plates suspended from the upper jaw to sift small crustaceans called krill from enormous volumes of water. The whale opens its mouth and fills an expandable oral pouch with water, and then closes the mouth and contracts the pouch. This forces water out of the mouth through the comblike baleen, leaving a mouthful of trapped food. **(b)** A behavioral adaptation enables the humpback whale to concentrate krill before filtering. The whale sets a "bubble net" by blowing a ring of bubbles as it swims in an upward spiral from a depth of about 15 m. Krill and other small animals near the surface tend to swim away from the bubbles in an avoidance reaction, causing them to concentrate in the center of the bubble net. After herding the krill, the whale harvests its catch by swimming through the center of the ring with its mouth agape.

(a)

(b)

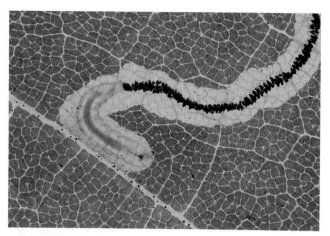

Figure 37.3
Substrate-feeding: a leaf miner. Eating its way through the soft mesophyll of an oak leaf, this caterpillar, the larva of a moth, leaves a trail of feces in its wake.

Figure 37.4
Fluid-feeding: the mosquito. This parasite has impaled its human host with a hypodermic-like mouth part and has filled its abdomen with a blood meal.

suck blood from animals. Since these particular fluid-feeders harm their hosts, they are considered parasites. Hummingbirds and bees, on the other hand, perform a service for their hosts, as they transfer pollen when they visit flowers to suck nectar.

Most animals, rather than filtering food from water, eating their way through the substrate, or sucking fluids, ingest relatively large pieces of food. They use such diverse utensils as tentacles, pincers, claws, poisonous fangs, and jaws and teeth to kill their prey or to tear off pieces of meat or vegetation. Humans, of course, feed in this manner.

DIGESTION: A COMPARATIVE INTRODUCTION

Regardless of what it eats or how it feeds, an animal must digest its food. **Digestion** is the process of breaking food down into molecules small enough for the body to absorb. The bulk of the organic material in food consists of proteins, fats, and carbohydrates in the form of starch and other polysaccharides. Although these are suitable raw materials, animals cannot use them directly, for two reasons. First, these molecules are too large to pass through membranes and enter the cells of the animal. Second, the macromolecules that make up an animal are not identical to those of its food. In building their macromolecules, however, all organisms use common monomers. For instance, soybeans, cattle, and people all assemble their proteins from the same 20 kinds of amino acids (see Chapter 5). Digestion cleaves macromolecules into their

component monomers. Polysaccharides and disaccharides are split into simple sugars, fats are digested to glycerol and fatty acids, proteins are broken down to amino acids, and nucleic acids are cleaved into nucleotides. The digestion of macromolecules is catalyzed by a special class of enzymes.

Enzymatic Hydrolysis

Recall from Chapter 5 that when a cell makes a macromolecule by linking together monomers, it does so by removing a molecule of water for each new covalent bond formed. Digestion reverses this process by breaking bonds with the enzymatic addition of water. (Actually, a hydrogen is added to one side of the bond and a hydroxyl group is added to the other side—the net effect is the addition of H_2O.) This splitting process is called **hydrolysis** (lysing with water). There are hydrolytic enzymes that digest each of the classes of macromolecules found in food. This chemical digestion is usually preceded by mechanical fragmentation of the food—by chewing, for instance. Breaking food into smaller pieces increases the surface area exposed to digestive juices containing hydrolytic enzymes.

An animal must digest its food in some type of specialized compartment where enzymes can attack the food molecules without damaging the animal's own cells. After the food is digested, amino acids, simple sugars, and other small molecules are absorbed—that is, they cross the membrane of the digestive compartment and enter the cells of the animal. Undigested material is then eliminated from the digestive compartment.

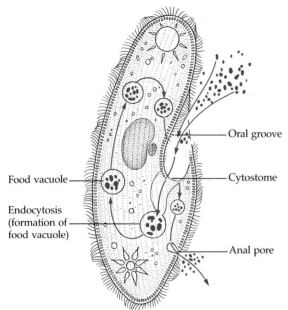

Oral groove

Food vacuole

Cytostome

Endocytosis
(formation of
food vacuole)

Anal pore

Figure 37.5
Intracellular digestion in *Paramecium*. *Paramecium*
has a specialized feeding structure called the oral
groove, which leads to the cell's "mouth" (cytostome).
Cilia that line the groove draw water and suspended food
particles toward the mouth, where the food is packaged
by endocytosis into a food vacuole that functions as a
miniature digestive compartment. Cytoplasmic streaming
carries food vacuoles around the cell, while hydrolytic
enzymes are secreted into the vacuoles. As molecules in
the food are digested, the nutrients (sugars, amino acids,
and other small molecules) are transported across the
membrane of the vacuole into the cytoplasm. Later, the
vacuole fuses with an anal pore, a specialized region of
the plasma membrane where undigested material can be
eliminated by exocytosis.

Intracellular Digestion in Food Vacuoles

The simplest digestive compartments are food vacuoles, organelles where a single cell can digest its food
without the hydrolytic enzymes mixing with the cell's
own cytoplasm. Protozoa digest their meals in food
vacuoles, usually after engulfing the food by endocytosis (phagocytosis or pinocytosis—see Chapter 8).
This digestive mechanism is termed **intracellular
digestion** (Figure 37.5). Sponges are among the true
animals that digest their food entirely by the intracellular mechanism (Figure 37.6). In most animals,
however, at least some hydrolysis occurs by **extracellular digestion** within compartments that are continuous via passages with the outside of the body.

Digestion in Gastrovascular Cavities

Some of the simplest animals have digestive sacs with
single openings. These pouches, called **gastrovascular cavities,** function both in digestion and distribution of nutrients throughout the body (hence, the *vascular* part of the term). The cnidarian *Hydra* provides
a good example of how a gastrovascular cavity works
(Figure 37.7). *Hydra* is a carnivore that stings its prey
with specialized organelles called nematocysts and then
uses prehensile tentacles to stuff the food into a mouth
that can stretch to accommodate an animal larger than
the *Hydra* itself (see Chapter 29). With the food now
in the gastrovascular cavity, specialized cells of the
gastrodermis, the tissue layer that lines the cavity,
secrete digestive enzymes that fragment the soft tissues of the prey into tiny pieces. Beating flagella on
the gastrodermal cells keep the small food particles
from settling and distribute them throughout the cavity. The food particles are taken into gastrodermal cells
by phagocytosis, and food molecules are then hydrolyzed within food vacuoles. Extracellular digestion
within the gastrovascular cavity initiates breakdown
of the food, but most of the actual hydrolysis of
macromolecules is accomplished by intracellular
digestion. After *Hydra* has digested its meal, undigested materials remaining in the gastrovascular cavity, such as the exoskeletons of crustaceans, are eliminated through the single opening, which functions
in the dual role of mouth and anus.

If the gastrodermal cells are capable of phagocytosis
and digestion, then what is the function of an extracellular cavity? Phagocytosis is limited to the ingestion
of microscopic food. Extracellular cavities for digestion are adaptations enabling animals to devour larger
prey.

Most flatworms, such as cnidarians, have gastrovascular cavities with single openings (see Figure 37.7b).
The pharynx, a muscular tube that can be everted
through the mouth on the ventral side of a planarian,
penetrates the prey and releases a digestive juice. The
partly digested food is sucked into the gastrovascular
cavity, which has three main branches and many secondary ramifications that increase the surface area of
the cavity extensively. As in *Hydra*, a planarian begins
digesting its food within the gastrovascular cavity, but
digestion is continued within cells that take up food
particles by phagocytosis.

Digestion in Alimentary Canals

In contrast to the saclike anatomy of cnidarians and
flatworms, more complex animals, including nematodes, annelids, mollusks, arthropods, echinoderms,
and chordates, have digestive tubes running between
two openings, a mouth and an anus (Figure 37.8).

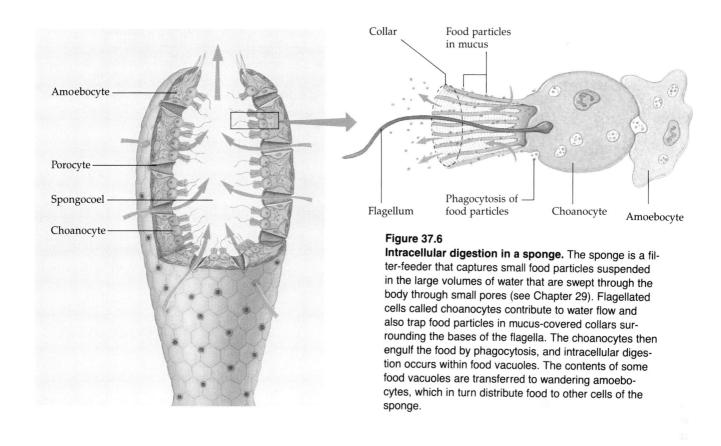

Figure 37.6
Intracellular digestion in a sponge. The sponge is a filter-feeder that captures small food particles suspended in the large volumes of water that are swept through the body through small pores (see Chapter 29). Flagellated cells called choanocytes contribute to water flow and also trap food particles in mucus-covered collars surrounding the bases of the flagella. The choanocytes then engulf the food by phagocytosis, and intracellular digestion occurs within food vacuoles. The contents of some food vacuoles are transferred to wandering amoebocytes, which in turn distribute food to other cells of the sponge.

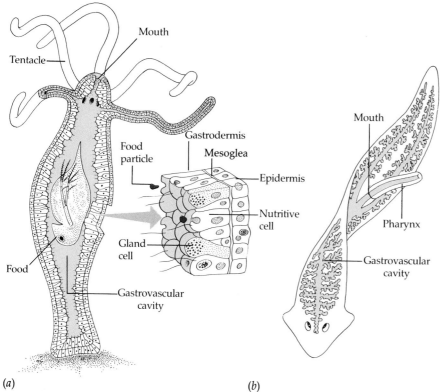

(a)   *(b)*

Figure 37.7
Gastrovascular cavities. (a) *Hydra.* The outer epidermis has protective and sensory functions, whereas the inner gastrodermis is specialized for digestion. The mesoglea, a jellylike layer, separates the two layers of cells. Enzymes released from gland cells into the gastrovascular cavity initiate digestion, which is completed intracellularly after small food particles are taken into nutritive cells by phagocytosis. Undigested waste is egested through the mouth/anus, the single opening of the gastrovascular cavity. **(b)** *Planaria.* The gastrovascular cavity of a planarian digests food in a fashion similar to that of *Hydra.* Its extensive branching increases the surface area available for digestion and transports nutrients to all parts of the animal.

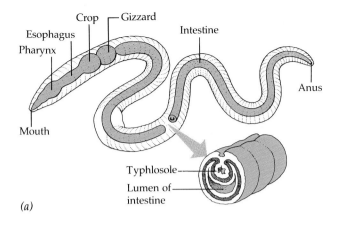

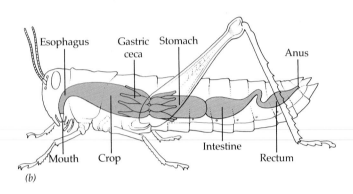

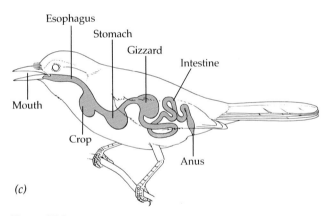

Figure 37.8
Alimentary canals. (a) The digestive tract of the earthworm consists of five organs. A muscular pharynx sucks food in through the mouth. Food passes through the esophagus and is stored and moistened in the crop. The muscular gizzard, which contains small bits of sand and gravel, pulverizes the food. Digestion and absorption occur in the intestine, which has a dorsal fold, the typhlosole, that increases the surface area for nutrient absorption. Finally, undigested material is expelled through the anus. **(b)** The grasshopper is a herbivore. As in the earthworm, there is a crop where food is moistened and stored, but the grasshopper also has a stomach, where most digestion occurs. Gastric ceca, pouches extending from the stomach, transfer nutrients to the grasshopper's hemolymph (blood). Undigested material is eliminated along with metabolic wastes via the intestine and rectum. **(c)** A bird has three separate chambers—the crop, stomach, and gizzard—where food is pulverized and churned before passing into the intestine.

These tubes are called **complete digestive tracts** or **alimentary canals.** Since food moves along the canal in one direction, the tube can be organized into specialized regions that carry out digestion and absorption of nutrients in a stepwise fashion. Some regions of the alimentary canal have a similar function in most species, although the tubes are adapted in various ways to specific diets. Food ingested through the mouth and pharynx passes through an esophagus that leads—depending on the species—to a crop, gizzard, or stomach. Crops and stomachs are usually organs that store the food, whereas gizzards actually grind the food. The pulverized meal next enters the long intestine, where digestive enzymes hydrolyze the food molecules and nutrients are absorbed across the lining of the tube into the blood. Undigested wastes are egested through the **anus.**

THE MAMMALIAN DIGESTIVE SYSTEM

The mammalian digestive system consists of the alimentary canal along with various accessory glands that secrete digestive juices into the canal through ducts. From the esophagus to the large intestine, the mammalian digestive tract has a four-layered wall (see Figure 36.8 in Chapter 36). The lumen (cavity) is lined by a mucous membrane, the mucosa. Next comes a layer made up mostly of connective tissue, followed by a layer of smooth muscle. The outermost layer is a sheath of connective tissue attached to the membrane of the body cavity. **Peristalsis,** rhythmic waves of contraction by the smooth muscles, pushes the food along the tract. At some of the junctions between specialized segments of the tube, the muscular layer is modified into ringlike valves called **sphincters,** which close off the tube like purse strings to regulate the passage of material between chambers of the canal. In addition to the alimentary canal itself, the mammalian digestive system also includes accessory glands that deliver digestive juices to the canal through ducts. These accessory organs are three pairs of **salivary glands,** the **pancreas,** and the **liver** with its storage organ, the **gallbladder.** The accessory glands originate in the embryo as outpocketings of the primordial gut; the ducts are remnants of the connections of the glands to the alimentary canal.

Using the human as an example, we can now follow a meal through the alimentary canal (Figure 37.9), seeing in more detail what happens to the food in each of the processing stations along the way.

Oral Cavity

Both physical and chemical digestion of foods begin in the mouth. During chewing, teeth of various shapes cut, smash, and grind food, making the food easier

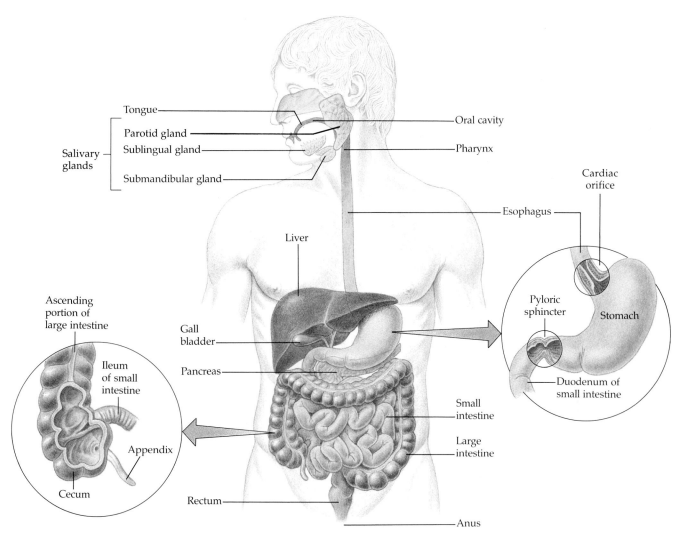

Figure 37.9
The human digestive system. Food enters the system through the mouth, where it is chewed for some seconds before being swallowed. Food takes 5–10 seconds to pass down the esopha- gus and into the stomach, where it spends 2–6 hours being partially digested. Final digestion and nutrient absorption occur in the small intestine over a period of 5–6 hours. In 12–24 hours, any undigested material passes through the large intes- tine, and feces are expelled through the anus.

to swallow and increasing its surface area. The presence of food in the oral cavity also triggers a nervous reflex that causes the salivary glands to deliver saliva through ducts to the oral cavity. Even before food is actually in the mouth, salivation may occur in anticipation because of learned associations between eating and the time of day, cooking odors, or other stimuli.

In humans, more than a liter of saliva is secreted into the oral cavity each day. Dissolved in the saliva is a slippery glycoprotein (carbohydrate-protein complex) called mucin, which protects the soft lining of the mouth from abrasion and lubricates the food for easier swallowing. Saliva contains buffers that help prevent dental cavities by neutralizing acid in the mouth. Also, antibacterial agents in saliva kill many of the bacteria that enter the mouth with food. Finally, **salivary amylase,** a digestive enzyme that hydrolyzes starch, is present in saliva. Salivary amylase can break only every other bond in the polysaccharide, and hence the smallest product of this digestion is the double sugar maltose. Food is not usually in the mouth long enough for amylase to do more than split starch into smaller polysaccharides; perhaps a major function of salivary amylase is to prevent an accumulation of sticky starch between teeth.

The tongue is used not only to taste, but also to manipulate the food during chewing and then to help shape the food into a ball called a **bolus.** Food is swallowed when the tongue pushes the bolus to the very back of the oral cavity and into the pharynx.

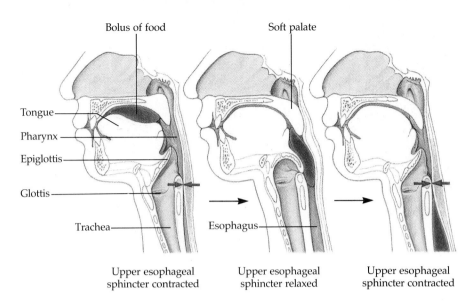

Figure 37.10
Swallowing reflex. The swallowing reflex is triggered when a bolus of food reaches the pharynx (left). The upper esophageal sphincter (blue arrows), which is ordinarily contracted, relaxes and allows the bolus to enter the esophagus (middle). The larynx, the upper part of the respiratory tract, moves upward and tips the epiglottis over the glottis, which prevents food from entering the trachea. After the food has entered the esophagus, the larynx moves downward and opens the breathing passage (right).

Labels on figure: Bolus of food · Soft palate · Tongue · Pharynx · Epiglottis · Glottis · Trachea · Esophagus

Upper esophageal sphincter contracted | Upper esophageal sphincter relaxed | Upper esophageal sphincter contracted

Pharynx

The region we commonly call our throat is the **pharynx,** an intersection that leads to both the esophagus and the windpipe (trachea). When we swallow, the top of the windpipe moves so that its opening is blocked by a cartilaginous flap called the **epiglottis** (Figure 37.10). You can see this motion in the bobbing of the "Adam's apple" during swallowing. The epiglottis and also the vocal cords in the opening of the windpipe guard the respiratory system against the entry of food or fluids. The swallowing mechanism normally always ensures that a bolus will be guided into the entrance of the esophagus.

Esophagus

The esophagus is the segment of the alimentary canal that conducts food from the pharynx down to the stomach. Peristalsis, the wavelike contractions of the smooth muscles lining the digestive tract, squeezes a bolus along the narrow esophagus (Figure 37.11). Only the muscles at the very top of the esophagus are striated (voluntary). Thus, swallowing is initiated voluntarily, but then the involuntary waves of contraction by the smooth muscles take over.

Stomach

The J-shaped stomach is located on the left side of the abdominal cavity, just below the diaphragm. Because this large organ can store an entire meal, we do not need to eat constantly. With a very elastic wall and accordionlike folds called rugae, the stomach can stretch to accommodate about 2 L of food and fluid.

The epithelium that lines the lumen of the stomach secretes **gastric juice,** a digestive fluid that mixes with the food. With its high concentration of hydrochloric acid, the gastric juice has a pH of about 2, acidic enough to dissolve nails. One function of the acid is to dismantle the tissues in food by disrupting the intercellular glue that binds cells together in meat and plant material. The acid also kills most bacteria that are swallowed with the food.

Also present in gastric juice is **pepsin,** an enzyme that hydrolyzes proteins. Hydrolysis is incomplete, however, since pepsin can only break peptide bonds adjacent to specific amino acids. The result of pepsin's

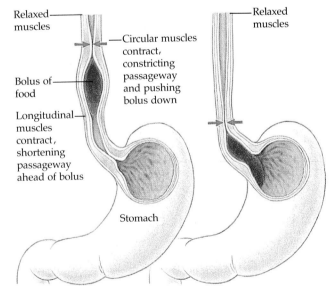

Labels on figure: Relaxed muscles · Circular muscles contract, constricting passageway and pushing bolus down · Bolus of food · Longitudinal muscles contract, shortening passageway ahead of bolus · Stomach · Relaxed muscles

Figure 37.11
Peristalsis. Waves of muscular contraction move the bolus of food down the esophagus to the stomach.

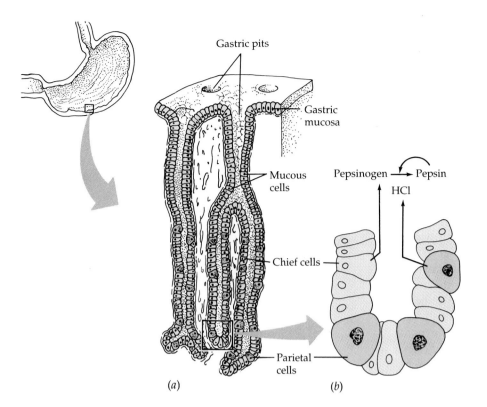

Figure 37.12
Secretion of gastric juice in the stomach. (a) The mucosa that lines the stomach consists of simple columnar epithelium organized into tubular gastric glands. The gastric glands have three types of secretory cells. The cells that line the openings of the glands secrete mucus that helps protect the stomach from digesting itself. Deeper in the gland are parietal cells, which secrete hydrochloric acid, and chief cells, which secrete pepsinogen. (b) Within the gastric pits, hydrochloric acid converts the inactive pepsinogen to pepsin, which, in an example of positive feedback, activates more pepsin.

action, then, is to cleave proteins into smaller polypeptides. Pepsin is one of the few enzymes that works best in a strongly acidic environment. Indeed, the low pH of gastric juice denatures the proteins in food, increasing exposure of their peptide bonds to the pepsin. Even the salivary amylase that is swallowed ceases to work soon after reaching the stomach and is digested along with the food proteins.

What prevents pepsin from destroying the cells of the stomach wall that produce this hydrolytic enzyme? Pepsin is synthesized and secreted in an inactive form called **pepsinogen** (Figure 37.12). Hydrochloric acid in the gastric juice converts pepsinogen to active pepsin by removing a short segment of the protein's polypeptide chain, an alteration that exposes the active site of pepsin. Since the acid and pepsinogen are secreted by different kinds of cells, the two ingredients do not mix until their release into the lumen of the stomach. Once some of the pepsinogen is activated by acid, a chain reaction occurs because pepsin itself can activate additional molecules of pepsinogen. This domino effect is an example of positive feedback. We shall see other cases where protein-digesting enzymes are secreted in inactive forms, generally called **zymogens.**

A coating of mucus secreted by the epithelial cells helps protect the stomach lining from being digested by the pepsin and acid in gastric juice. Still, the epithelium is constantly eroded, and mitosis must generate enough cells to completely replace the stomach lining every 3 days. When pepsin and acid destroy the lining faster than it can regenerate, lesions called gastric ulcers occur.

Gastric secretion is controlled by a combination of nervous impulses and hormones. When we see, smell, or taste food, a nervous message from the brain to the stomach initiates secretion of gastric juice. Then certain substances in the food stimulate the stomach wall to release a hormone called **gastrin** into the circulatory system. As the gastrin gradually recirculates in the bloodstream back to the stomach wall, the hormone stimulates further secretion of gastric juice. Thus, an initial burst of gastric secretion at mealtime is followed by a sustained secretion that continues to add gastric juice to the food for some time. If the pH of the stomach contents becomes too low, the acid inhibits release of the hormone gastrin, decreasing secretion of gastric juice; this is an example of negative feedback. Each day, the stomach wall secretes about 3 L of gastric juice.

About every 20 seconds, the stomach contents are mixed by the churning action of smooth muscles. When an empty stomach churns, hunger pangs are felt. (Sensations of hunger are also associated with centers in the brain that monitor the nutritional status of the blood.) Sometimes the stomach rumbles loudly enough to be heard without the help of a stethoscope. What begins in the stomach as a meal mixed with gastric juice soon becomes a nutrient broth known as **acid chyme.**

Much of the time, the stomach is closed off at either end (see Figure 37.9). The opening from the stomach to the esophagus normally dilates only when a bolus driven by peristalsis arrives. Occasional backflow of acid chyme into the lower end of the esophagus causes "heartburn." If backflow is a persistent problem, an ulcer may develop in the esophagus.) The cardiac sphincter opens intermittently with each wave of peristalsis that delivers a bolus of food. At the bottom of the stomach is the **pyloric sphincter,** which helps regulate the passage of chyme from the stomach into the small intestine. A squirt at a time, it takes about 2–6 hours after a meal for the stomach to empty.

Small Intestine

Although limited digestion of starch takes place in the oral cavity and partial digestion of proteins by pepsin in the stomach, most enzymatic hydrolysis of the macromolecules in food occurs in the small intestine. With a length of more than 6 m, the small intestine is the longest section of the alimentary canal (its name is based on its small diameter, compared with the diameter of the large intestine). It is not only the major organ of digestion but also the part of the digestive system responsible for the absorption of most nutrients into the blood.

Regulation of Digestive Secretions Several other organs contribute to digestion in the small intestine by producing, storing, and secreting various digestive juices.

One of these organs is the pancreas, which produces an assortment of hydrolytic enzymes as well as an alkaline solution rich in bicarbonate. The bicarbonate acts as a buffer to offset the acidity of chyme from the stomach.

Another accessory organ is the liver, which performs a wide variety of important functions in the body, including the production of **bile,** a mixture of substances that is stored in another accessory organ, the gallbladder, until needed. Bile contains no digestive enzymes, but it does contain bile salts, which aid in the digestion and absorption of fats. Bile also contains pigments that are by-products of red blood cell destruction in the liver; these bile pigments are eliminated from the body with the feces.

The first 25 cm or so of the small intestine is called the **duodenum,** and it is here that chyme seeping in from the stomach mixes with digestive juices secreted by the pancreas, liver and gallbladder, and gland cells of the intestinal wall itself.

At least four regulatory hormones help to ensure that digestive secretions are present only when they are needed (Figure 37.13). We have already seen that gastrin is released from the stomach lining in response to the presence of food. The acidic pH of the chyme that enters the duodenum stimulates the intestinal wall to release a second hormone, **secretin.** This hormone signals the pancreas to release bicarbonate, which neutralizes the acid chyme. **Cholecystokinin (CCK),**

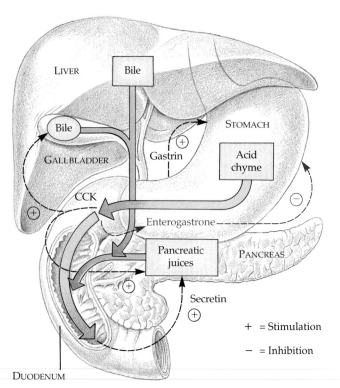

Figure 37.13
Control of digestion. Gastrin, the release of which is triggered by events associated with introduction of food into the stomach (presence of peptides, stomach distension, previous gastrin secretion), stimulates production of gastric juices. Most digestion of food occurs in the duodenum. The acid chyme seeps in from the stomach and is first neutralized. Secretin mediates this neutralization by stimulating the release of sodium bicarbonate by the pancreas. The presence of amino acids or fatty acids in the duodenum triggers the release of cholecystokinin (CCK), which stimulates the release of digestive enzymes by the pancreas and bile by the gallbladder. A fourth hormone, enterogastrone, slows digestion by inhibiting stomach peristalsis and acid secretion when acid chyme rich in fats (which require additional digestion time) enters the duodenum.

another hormone produced by cells in the lining of the duodenum, causes the gallbladder to contract and release bile into the small intestine; CCK also triggers the release of pancreatic enzymes. The chyme, particularly if it is rich in fats, also causes the duodenum to release **enterogastrone,** a hormone that inhibits peristalsis in the stomach and thus slows down the entry of food into the small intestine.

Digestion in the Small Intestine Let us now see how enzymes from the pancreas and intestinal wall combine to digest macromolecules.

The digestion of starch that was initiated by salivary amylase in the oral cavity is continued in the small intestine (Table 37.1). A pancreatic amylase hydro-

Table 37.1 Enzymatic digestion

	Carbohydrate Digestion	Protein Digestion	Fat Digestion
Oral Cavity	Starch $\xrightarrow[\text{AMYLASE}]{\text{SALIVARY}}$ Maltose		
Stomach		Protein $\xrightarrow{\text{PEPSIN}}$ Smaller polypeptides	
Lumen of Small Intestine	Undigested polysaccharides $\xrightarrow[\text{AMYLASE}]{\text{PANCREATIC}}$ Maltose	Polypeptides $\xrightarrow[\text{CHYMOTRYPSIN}]{\text{TRYPSIN}}$ Smaller Polypeptides Small polypeptides $\xrightarrow[\text{CARBOXYPEPTIDASE}]{\text{AMINOPEPTIDASE}}$ Amino acids	Fat globule $\xrightarrow[\text{SALTS}]{\text{BILE}}$ Emulsified fat Fats $\xrightarrow{\text{LIPASE}}$ Glycerol Fatty acids Glycerides
Brush Border of Small Intestine	Disaccharides $\xrightarrow[\substack{\text{SUCRASE}\\\text{LACTASE}\\\text{ETC.}}]{\text{MALTASE}}$ Monosaccharides	Dipeptides $\xrightarrow{\text{DIPEPTIDASE}}$ Amino acids	

lyzes starch into maltose, a disaccharide. Digestion is completed by the enzyme maltase, which splits the maltose and releases two molecules of the simple sugar glucose. Maltase is one of a family of **disaccharidases,** each specific for the hydrolysis of a different disaccharide. Sucrase, for instance, hydrolyzes table sugar (sucrose), and lactase digests milk sugar (lactose). (In general, adults have much less lactase than children. In some populations, such as certain tribes in Africa, lactase is missing altogether. If these people were to drink milk in quantity, they would develop cramps and diarrhea.) The disaccharidases are not dissolved within the lumen of the intestine, but are actually built into the membranes and coating (glycocalyx) of the intestinal epithelium (the "brush border" in Table 37.1). Thus, the terminal steps in carbohydrate digestion occur at the site of sugar absorption.

Pepsin in the stomach primes proteins for digestion by breaking them into smaller pieces, and then a team of enzymes in the small intestine completely dismantles polypeptides into their component amino acids. **Trypsin** and **chymotrypsin** are specific for peptide bonds adjacent to certain amino acids, and thus, like pepsin, break polypeptides into shorter chains. **Carboxypeptidase** splits off one amino acid at a time, beginning at the end of the polypeptide that has a free carboxyl group. **Aminopeptidase** works in the opposite direction. Notice that either aminopeptidase or carboxypeptidase alone could completely digest a protein. But teamwork between these enzymes and the trypsin and chymotrypsin that attack the interior of the protein speeds up hydrolysis tremendously. Protein digestion is further hastened by various **dipeptidases** attached to the intestinal lining that digest fragments only two or three amino acids long.

The protein-digesting enzymes, including trypsin, chymotrypsin, and carboxypeptidase, are secreted as inactive zymogens by the pancreas. An intestinal enzyme named **enterokinase** triggers activation of these enzymes within the lumen of the small intestine (Figure 37.14).

In a cooperative hydrolytic assault similar to the digestion of proteins, a team of enzymes called **nucleases** hydrolyzes DNA and RNA present in food.

Nearly all the fat in a meal reaches the small intestine completely undigested. Hydrolysis of fats is a special problem because fat molecules are insoluble in water. Bile salts secreted into the duodenum coat tiny fat droplets and keep them from coalescing, a process called **emulsification.** Because the droplets are small, there is a large surface area of fat exposed to **lipase,** an enzyme that hydrolyzes the fat molecules.

Thus, the macromolecules from food are completely hydrolyzed to their component monomers as peristalsis moves the mixture of chyme and digestive juices along the small intestine. Most digestion is completed early in this journey, while the chyme is still in the duodenum. The remaining regions of the small intes-

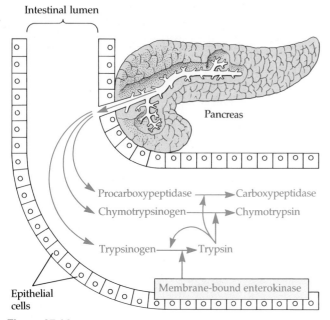

Intestinal lumen

Pancreas

Procarboxypeptidase ⟶ Carboxypeptidase

Chymotrypsinogen ⟶ Chymotrypsin

Trypsinogen ⟶ Trypsin

Membrane-bound enterokinase

Epithelial cells

Figure 37.14
Activation of zymogens in the small intestine. The pancreas secretes protein-digesting enzymes in an inactive form, which are activated in the duodenum by enterokinase. Enterokinase, a proteolytic enzyme bound to the intestinal membrane, converts trypsinogen to trypsin by cleaving off part of the precursor peptide. The trypsin, itself a proteolytic enzyme, then activates procarboxypeptidase and chymotrypsinogen.

tine, the **jejunum** and **ileum,** are specialized for the absorption of nutrients.

Absorption and Distribution of Nutrients Topologically speaking, the lumen of the alimentary canal is outside the body. To enter the body, nutrients that accumulate in the lumen when food is digested must cross the lining of the digestive tract. A limited number of nutrients are absorbed in the stomach and large intestine, but most absorption occurs in the small intestine.

The lining of the small intestine has wrinkles upon wrinkles, giving it a huge surface area of 600 m^2, about the size of a baseball diamond. Large folds are decorated with fingerlike projections called **villi,** and each of the epithelial cells of a villus has many microscopic appendages, the **microvilli,** which are exposed to the lumen of the intestine. Architecturally, the epithelium of the small intestine is well suited for its task of absorbing nutrients (Figure 37.15).

Only two single layers of epithelial cells separate nutrients in the lumen of the intestine from the bloodstream. Penetrating the hollow core of each villus is a net of microscopic blood vessels called capillaries. Also extending into the core of the villus is a tiny lymph vessel called a **lacteal.** (In addition to a circulatory

system that carries blood, vertebrates have an auxiliary system of vessels—the lymphatic system—which carries a clear fluid called lymph. Lymph drains into the circulatory system where the two systems connect near the left shoulder.) Nutrients are absorbed across the epithelium and then across the single-celled wall of the capillaries or lacteals. In some cases, the transport is passive. The simple sugar fructose, for example, is apparently absorbed by diffusion down its concentration gradient from the lumen of the intestine into the epithelial cells and then out of the epithelial cells into capillaries. Other nutrients, including amino acids, vitamins, and glucose and several other simple sugars, are pumped against gradients by the epithelial membranes. The absorption of some nutrients seems to be coupled to the active transport of sodium across the membranes of the epithelial cells. The membrane pumps sodium out of the cell and into the lumen, and the passive reentry of the ions is harnessed by cotransport to drive the uptake of nutrients (see Chapter 8).

Amino acids and sugars pass through the epithelium, enter capillaries, and are carried away from the intestine by the bloodstream. After glycerol and fatty acids are absorbed by epithelial cells, they are recombined within those cells to form fats again. The fats are then coated with special proteins to make tiny globules called **chylomicrons,** which are transported by exocytosis out of the epithelial cell and into a lacteal. Some fat molecules, however, are bound to specialized proteins and are transported as **lipoproteins** into capillaries.

The capillaries that drain nutrients away from the villi all converge into a single vessel, the **hepatic portal vein,** which leads directly to the liver. The rate of flow in this large vein is about 1 L per minute. This express route ensures that the liver, which has the metabolic versatility to interconvert various organic molecules, has first access to nutrients absorbed after a meal is digested. The blood that leaves the liver may have a very different balance of nutrients from the blood that entered via the hepatic portal vein. For example, blood exiting from the liver usually has a glucose concentration of very close to 0.1%, regardless of the carbohydrate content of a meal. From the liver, blood travels to the heart, which pumps the blood and the nutrients it contains to all parts of the body.

Large Intestine

The large intestine, or colon, is connected to the small intestine at a T-shaped junction, where a sphincter acts as a valve controlling the movement of material. One arm of the T is a blind pouch called the **cecum** (see Figure 37.9). Compared to many other mammals, humans have a relatively small cecum with a fingerlike

An animal's metabolic rate is the number of calories consumed per unit of time. The apparatus shown here helps estimate the metabolic rate of a human by measuring oxygen consumed by cellular respiration, the oxidation of food (see Chapter 9). For every liter of O_2 consumed, respiration liberates about 4.83 kcal of energy from food molecules. If, for example, the subject uses 16 L of oxygen per hour, then the metabolic rate is 77.28 kcal/h (16 L/h × 4.83 kcal/L). Measured at rest on an empty stomach (digestion requires energy), this value is the subject's basal metabolic rate, or BMR. For comparisons, BMRs are standardized as kilocalories per hour per kilogram of body weight, or per square meter of body surface area. A person's age, sex, genetic makeup, and many other factors affect BMR. And, of course, actual metabolic rate usually exceeds BMR, depending on the level of physical activity. For example, the metabolic rate of this man is being monitored as he works out on a treadmill.

Similar methods can be used to monitor the metabolic rates of other animals. The ghost crab in the photo below is on a treadmill enclosed in a plastic chamber. Air of known O_2 concentration flows through the chamber and the metabolic rate of the crab is calculated from the difference between the amount of O_2 entering and leaving the chamber.

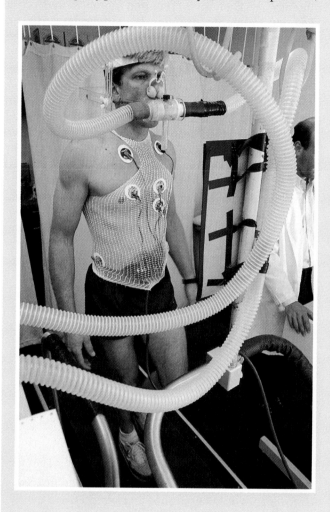

you were to carry a 50-lb suitcase around with you all day. How heavy we can be before we are considered unhealthy is a disputed matter. Charts of "ideal weights" for people of various heights have been liberalized in the past few years, based on statistics from insurance companies showing that "pleasantly plump" does not necessarily mean increased health risk. Nevertheless, the billions of dollars spent each year on diet books and diet aids indicate that a substantial portion of the population would like to lose weight because of concern with health and appearance.

Unfortunately, there are no magic methods for slimming, and some of the fad diets—"eat all you want as long as it's carbohydrate," for instance—can actually be dangerous. Effective weight watching is a matter of caloric bookkeeping. If we take in more calories than we need, the excess will be stored as fat. If we expend more calories than we take in, we balance the deficit by consuming body fat. Our weight is stable when caloric demand is matched by caloric supply from our diet. To lose weight, we must eat less, exercise more, or both.

Food for Fabrication

Heterotrophs cannot make organic molecules from raw materials that are entirely inorganic. To synthesize the molecules it needs to grow and replenish itself, the animal must obtain organic precursors from its food. Given a source of organic carbon (such as sugar) and a source of organic nitrogen (such as amino acids from the digestion of protein), the animal can fabricate a great variety of organic molecules by using enzymes to rearrange the molecular skeletons of the precursors acquired from food. For instance, a single type of amino acid can supply nitrogen for the synthesis of several other types of amino acids that may not be present in the food. Also, we have seen that animals can synthesize fats from carbohydrates. In vertebrates, the conversion of nutrients from one type of organic molecule to another occurs mainly in the liver.

So we see that although animals depend on food as a source of organic carbon and nitrogen, they exhibit considerable versatility in the biosynthesis of their own organic molecules. However, there are some substances the animal cannot fabricate from any precursors.

Essential Nutrients

In addition to providing fuel and raw materials for biosynthesis, an animal's diet must also supply certain substances in preassembled form. Chemicals an animal requires but cannot make are called essential nutrients. They vary from species to species, depending on the biosynthetic capabilities of the animal. A particular molecule can be an essential nutrient for one animal and yet be nonessential in the diet of another animal—one that can produce the molecule for itself from some precursor.

An animal whose diet is missing one or more essential nutrients is said to be **malnourished** (the term *undernourished*, remember, refers to caloric deficiency). For example, cattle and other herbivores may suffer mineral deficiencies if they graze on plants grown in soil that itself is deficient in key minerals. Malnutrition is much more common than undernutrition in human populations, and it is even possible for an overnourished individual to be malnourished.

There are four classes of essential nutrients: (1) essential amino acids, (2) essential fatty acids, (3) vitamins, and (4) minerals.

Essential Amino Acids Of the 20 kinds of amino acids required to make proteins, about half can be synthesized by most animals as long as their diet includes organic nitrogen. The remaining amino acids must be obtained from food in prefabricated form. Eight amino acids are essential in the adult human diet (a ninth is essential for infants). Do not let the term *essential* mislead you; an animal requires all 20 amino acids. Essential amino acids are those that must be present in the diet.

A diet that lacks one or more of the essential amino acids results in a form of malnutrition known generally as protein deficiency. The most common type of malnutrition among humans, protein deficiency is concentrated in geographical regions where there is a great gap between food supply and population size. The victims are usually children, who, if they survive infancy, are likely to be retarded in mental and physical development. In Africa, the syndrome is named **kwashiorkor,** which means "rejected one"—a reference to the onset of the disease when a child is weaned from its mother's milk and placed on a starchy diet after a sibling is born. The problem of protein deficiency in some less-developed countries has been compounded by a trend away from breast-feeding altogether, an unfortunate dietary change that has been catalyzed by an aggressive marketing campaign by companies that make baby formula. Impoverished mothers often "stretch" the expensive formula by diluting it with water to such an extent that the protein content is insufficient. Also, uninformed consumers pay too little attention to hygiene when mixing and storing formula. In contrast, nursing not only provides a balanced meal, but also transfers temporary immunity to several infectious diseases from mother to child.

There is, of course, an economic component to protein deficiency. The most reliable sources of essential amino acids are meat and animal by-products (for example, eggs and cheese), which are relatively expensive. The proteins in animal products are complete, which means that they provide all the essential amino acids in their proper proportions. Most plant proteins, on the other hand, are incomplete, since they are deficient in one or more essential amino acids. Zein, for example, the principal protein in corn, is deficient in the amino acid lysine. People forced by economic necessity to obtain nearly all their calories from corn would begin showing symptoms of protein deficiency. The result would be similar for people on monotonous diets of rice, wheat, potatoes, or any other single staple. A vegetarian, however, can obtain sufficient quantities of all essential amino acids (Figure 37.19). The key is to select a combination of plant foods that complement one another; two plants with incomplete proteins may not be deficient in the same amino acids. Beans, for example, supply the lysine that is missing in corn, and whereas beans are deficient in methionine, this amino acid is present in corn. Thus, a meal of beans and corn can provide all the essential amino acids. The combination of vegetables must be consumed at the same meal; the body cannot store amino acids, and hence a deficiency of a single essential amino acid retards protein synthesis and limits the

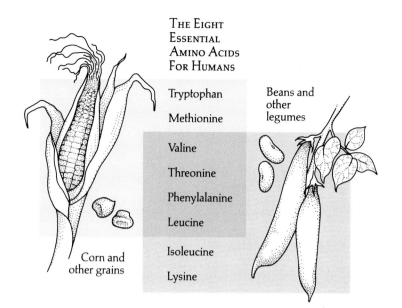

THE EIGHT
ESSENTIAL
AMINO ACIDS
FOR HUMANS

Tryptophan

Methionine

Valine

Threonine

Phenylalanine

Leucine

Isoleucine

Lysine

Corn and
other grains

Beans and
other
legumes

Figure 37.19
Essential amino acids from a vegetarian diet. Corn has little isoleucine and lysine. Beans have ample isoleucine and lysine, but little tryptophan and methionine. A person can obtain all eight essential amino acids by consuming a meal of corn and beans.

use of other amino acids. Most cultures have by trial and error developed balanced diets that prevent protein deficiency.

Essential Fatty Acids   Animals are able to synthesize most of the fatty acids they need, but they are unable to make certain unsaturated fatty acids (fatty acids having double bonds; see Chapter 5). In humans, for example, an unsaturated fatty acid named linoleic acid must be present in the diet in prefabricated form. This essential fatty acid is required to make some of the phospholipids found in membranes. Most diets furnish ample quantities of essential fatty acids, and thus deficiencies are rare.

Vitamins   Vitamins are organic molecules required in the diet in amounts that are quite small compared with the relatively large quantities of essential amino acids and fatty acids the animal needs. Minute doses of vitamins suffice because most of these molecules serve as coenzymes or parts of coenzymes and thus have catalytic functions (see Chapter 6).

Although requirements for vitamins are modest, these molecules are absolutely essential in a healthful diet. Deficiencies can cause severe syndromes. Indeed, the first vitamin to be identified, thiamine (vitamin B_1), was discovered as a result of the search for the cause of a mysterious disease called beriberi. Its symptoms include loss of appetite, fatigue, and nervous disorders. The syndrome was first described when it struck soldiers and prisoners in the Dutch East Indies in the nineteenth century. The dietary staple for these men was polished rice, which had the hulls (seed coats and other outer layers) removed to increase storage life. Not only did the men who were fed this diet develop beriberi, but so did the chickens that ate the table scraps. It was found that beriberi could be prevented in both the men and the chickens by supplementing their diets with unpolished rice. Later, the active ingredient of rice hulls was isolated. Since it belongs to the chemical family known as amines, the compound was named vitamine (a vital amine). The "e" was later dropped from the end, and the term has persisted, even though many of the vitamins subsequently discovered are not amines.

So far, 13 vitamins essential to humans have been identified (Table 37.2). The compounds are grouped into two categories: water-soluble vitamins and fat-soluble vitamins. Water-soluble vitamins include the B complex, which consists of several compounds that generally function as coenzymes in key metabolic processes. Vitamin C (ascorbic acid) is also water-soluble. Ascorbic acid is required for the production of connective tissue. Excesses of water-soluble vitamins are excreted with the urine, and moderate overdoses of these vitamins are probably harmless. Recently, a nervous disorder linked to excessive intake of vitamin B_6 has been described, but only in people who have been taking at least a thousand times the daily requirement of the vitamin.

The fat-soluble vitamins are A, D, E, and K. Vitamin A is incorporated into visual pigments of the eye. Vitamin D aids in calcium absorption and bone formation. The functions of vitamin E are not yet understood, but it seems to protect the phospholipids in membranes from oxidation. Vitamin K is required for blood clotting. Excesses of fat-soluble vitamins are not excreted, but instead are deposited in body fat, so overdoses may result in accumulation of these compounds to toxic levels.

The subject of vitamin dosage has aroused heated debate. One faction believes that it is enough to fill recommended daily allowances (RDAs), which are nutrient intake recommendations for healthy people,

Table 37.2 Vitamin requirements of humans

Vitamin	RDA* (Milligrams)	Dietary Sources	Major Body Functions	Possible Outcomes of Deficiency
Water-Soluble				
Vitamin B$_1$ (thiamine)	1.5	Pork, organ meats, whole grains, legumes	Coenzyme in the removal of carbon dioxide	Beriberi (peripheral nerve changes, edema, heart failure)
Vitamin B$_2$ (riboflavin)	1.8	Widely distributed in foods	Constituent of two coenzymes involved in energy metabolism (FAD and FMN)	Reddened lips, cracks at corner of mouth (cheilosis), lesions of eyes
Niacin	20	Liver, lean meats, grains, legumes	Constituent of two coenzymes involved in oxidation-reduction reactions (NAD$^+$ and NADP$^+$)	Pellagra (skin and gastrointestinal lesions; nervous, mental disorders)
Vitamin B$_6$ (pyridoxine)	2	Meats, vegetables, whole grain cereals	Coenzyme involved in amino acid metabolism	Irritability, convulsions, muscular twitching, kidney stones
Pantothenic acid	5–10	Widely distributed in foods	Constituent of coenzyme A, which plays a central role in metabolism	Fatigue, sleep disturbances, impaired coordination, nausea (rare in humans)
Folacin (folic acid)	0.4	Legumes, green vegetables, whole wheat products	Coenzyme in carbon transfer in nucleic acid and amino acid metabolism	Anemia, gastrointestinal disturbances, diarrhea, red tongue
Vitamin B$_{12}$	0.003	Muscle meats, eggs, dairy products	Coenzyme in carbon transfer in nucleic acid metabolism; maturation of red blood cells	Pernicious anemia, neurological disorders
Biotin	Unknown	Legumes, vegetables, meats	Coenzyme in fat synthesis, amino acid metabolism, glycogen formation	Fatigue, depression, nausea, dermatitis, muscular pains
Vitamin C (ascorbic acid)	45	Citrus fruits, tomatoes, green peppers, salad greens	Maintains intercellular matrix of cartilage, bone, and dentin; important in collagen synthesis	Scurvy (degeneration of skin, teeth, blood vessels; epithelial hemorrhages)
Fat-Soluble				
Vitamin A (retinol)	1	Provitamin A in green vegetables; retinol in milk, butter, cheese, margarine	Constituent of rhodopsin (visual pigment); maintenance of epithelial tissues	Keratinization of ocular tissue, night blindness, permanent blindness
Vitamin D	0.01	Cod liver oil, eggs, dairy products, margarine	Promotes bone growth, mineralization; increases calcium absorption	Rickets (bone deformities) in children; osteomalacia in adults
Vitamin E (tocopherol)	15	Seeds, green leafy vegetables, margarine	Functions as an antioxidant to prevent cell membrane damage	Possibly anemia; never observed in humans
Vitamin K (phylloquinone)	0.03	Green leafy vegetables; small amount in cereals, fruits, and meats	Important in blood clotting (involved in formation of active prothrombin)	Deficiencies associated with severe bleeding, internal hemorrhages

*Recommended daily allowance, for an adult in good health.
Source: *The Requirements of Human Nutrition*, by Nevin S. Scrimshaw and Vernon R. Young. Copyright 1976 by Scientific American, Inc. All rights reserved.

deliberately set to be higher than most people's actual needs. Others argue that RDAs are set too low for some vitamins and that we should be thinking in terms of *optimal* requirements. The dust has not yet settled, particularly in the debates over optimal doses of vitamins C and E. About all that can be said with any certainty at this time is that people who eat a balanced diet are unlikely to develop symptoms of vitamin deficiency.

Research involving experimental animals is of limited value in assessing the vitamin needs of humans. A compound that is a vitamin for one species will not

Table 37.3 Mineral requirements of humans

Mineral	Amount in Adult Body (Grams)	RDA* (Milligrams)	Dietary Sources	Major Body Functions	Possible Outcomes of Deficiency
Calcium	1500	800	Milk, cheese, dark-green vegetables, dried legumes	Bone and tooth formation; blood clotting; nerve transmission	Stunted growth; rickets, osteoporosis; convulsions
Phosphorus	860	800	Milk, cheese, meat, poultry, grains	Bone and tooth formation; acid-base balance; ATP formation	Weakness, demineralization of bone, loss of calcium
Sulfur	300	(Provided by sulfur amino acids)	Sulfur amino acids (methionine and cystine) in dietary proteins	Constituent of tissue compounds, cartilage, and tendon	Related to intake and deficiency of sulfur amino acids
Potassium	180	2,500	Meats, milk, many fruits	Acid-base balance; body water balance; nerve function	Muscular weakness; paralysis
Chlorine	74	2,000	Common salt	Formation of gastric juice; acid-base balance	Muscle cramps; mental apathy; reduced appetite
Sodium	64	2,500	Common salt	Acid-base balance; body water balance; nerve function	Muscle cramps; mental apathy; reduced appetite
Magnesium	25	350	Whole grains, green leafy vegetables	Activates enzymes; involved in protein synthesis	Growth failure; behavioral disturbances; weakness; spasms
Iron	4.5	10	Eggs, lean meats, legumes, whole grains, green leafy vegetables	Constituent of hemoglobin and enzymes involved in energy metabolism	Iron-deficiency anemia, reduced resistance to infection
Fluorine	2.6	2	Drinking water, tea, seafood	May be important in maintenance of bone structure	Higher frequency of tooth decay
Zinc	2	15	Widely distributed in foods	Constituent of enzymes involved in digestion	Growth failure; small sex glands
Copper	0.1	2	Meats, drinking water	Constituent of enzymes associated with iron metabolism	Anemia; bone changes (rare in humans)
Manganese	0.02	3	Whole grain cereals, egg yolks, green vegetables	Activates several enzymes, including one required for urea production	None reported for humans
Iodine	0.011	0.14	Marine fish and shellfish, dairy products	Constituent of thyroid hormones	Goiter (enlarged thyroid)
Cobalt	0.0015	(Required as vitamin B_{12})	Organ and muscle meats, milk	Constituent of vitamin B_{12}	None reported for humans

*Recommended daily allowance, for an adult in good health
Source: *The Requirements of Human Nutrition*, by Nevin S. Scrimshaw and Vernon R. Young. Copyright 1976 by Scientific American, Inc. All rights reserved

be essential in the diet of a second species that can synthesize the compound for itself. By definition, a substance is not a vitamin if it can be made by the animal. Also, symbiotic microorganisms living within the gut of an animal may produce vitamins that supplement or substitute for vitamins in food. Rabbits, for example, require no vitamin C in their diet because their flora of intestinal bacteria produces ample quantities of the compound.

Minerals Minerals are inorganic nutrients, usually required in very small amounts (Table 37.3). As with vitamins, mineral requirements vary with animal species. Humans and other vertebrates require relatively

Figure 37.20
A natural salt lick. Many herbivores cannot obtain enough sodium and chloride from plants but supplement their diet by visiting natural salt licks, as this female bighorn sheep and her lambs are doing here.

Chapter 9) and of hemoglobin, the oxygen-binding protein of red blood cells. Magnesium, manganese, zinc, and cobalt are cofactors built into the structure of certain enzymes; magnesium, for example, is present in enzymes that split ATP. Vertebrates need iodine to make thyroxine, a thyroid hormone that regulates metabolic rate. Sodium, potassium, and chlorine are important in nerve function and also have a major influence on osmotic balance between cells and the interstitial fluid. Herbivores, such as deer and cattle, often seem to crave salt; vegetation generally contains a very low concentration of sodium chloride. Salt licks, either natural or placed on the range by ranchers, attract large numbers of herbivores (Figure 37.20). Most humans, on the other hand, ingest far more salt than they need. In the United States, the average person consumes enough salt to provide about 20 times the required amount of sodium.

A healthful diet, then, must supply enough calories to satisfy energy needs, carbon skeletons and organic nitrogen for biosynthesis, and ample quantities of the essential nutrients.

* * *

Although the focus of this chapter has been on nutrition and digestion, nerves, blood vessels, and hormones have all been part of the discussion. No organ system is a solo act. The animal is an integrated whole, a concert of organ systems working together. In the next chapter, you will learn about the parts played by the circulatory and respiratory systems.

large quantities of calcium and phosphorus for the construction and maintenance of bone. Calcium is also necessary for the normal functioning of nerves and muscles, and phosphorus is also an ingredient of ATP and nucleic acids. Iron is a component of the cytochromes that function in cellular respiration (see

STUDY OUTLINE

Feeding Mechanisms (pp. 793–795)

1. All animals are heterotrophs and must obtain their nutrients by eating plants, animals, or both. This variation in diets allows classification of animals into herbivores, carnivores, and omnivores, respectively.
2. Animals may obtain nutrients by filter-, deposit-, fluid-, or substrate-feeding, but most animals are holotrophs.
3. Animal nutrition involves digestion and absorption. Digestion enzymatically breaks down the macromolecules of food into their component monomers inside compartments separated from living protoplasm. The monomers are then absorbed across the membranes of the digestive compartment. Any undigested material is eliminated.

Digestion: A Comparative Introduction (pp. 795–798)

1. Digestion breaks down food particles into the various monomers of organic polymers. This hydrolysis is catalyzed by specific enzymes.
2. In a process known as intracellular digestion, large food molecules may be engulfed by endocytosis and digested within food vacuoles. Protozoans and sponges use this type of digestion exclusively. In the extracellular diges-

tion of most animals, hydrolysis occurs in a separate compartment and smaller food molecules are then absorbed into the cells.
3. The gastrovascular cavities of the simplest animals have a single opening through which food enters and undigested wastes pass. *Hydra* initiates digestion inside such a cavity, but accomplishes most hydrolysis intracellularly after the particles are small enough to enter the gastrodermal cells. A similar process occurs in flatworms, which suck food into a branched digestive sac through a muscular pharynx.
4. More complex animals have digestive tracts, or alimentary canals, which move food through a one-way tube with specialized regions for digestion and absorption that are variously modified in different species. The basic pathway leads sequentially from the mouth, pharynx, and esophagus into a crop, gizzard, or stomach, and then through intestines to the anus.

The Mammalian Digestive System (pp. 798–806)

1. The mammalian digestive tract has a four-layered wall from the esophagus to the large intestine. The smooth muscle layer propels food along the tract by peristalsis

and regulates its passage through strategic points by means of sphincters.

2. Mammals have accessory glands that add digestive secretions to the tract through ducts. These include the salivary glands, pancreas, liver, and gallbladder.

3. Digestion begins in the oral cavity, where teeth chew food into smaller particles that are exposed to a starch-digesting enzyme in saliva called salivary amylase. Saliva also contains buffers, antibacterial agents, and mucin for lubricating the food. The tongue is adapted for tasting and manipulating food into a bolus for swallowing.

4. The pharynx is the intersection leading to the trachea and the esophagus. Food is usually prevented from entering the trachea by the epiglottis.

5. The esophagus conducts food from the pharynx to the stomach by involuntary peristaltic waves.

6. The stomach stores food and secretes gastric juice, which converts a meal to acid chyme. Gastric juice includes HCl and the proteolytic enzyme pepsin. A thick coating of mucus and a continual replacement of epithelial cells prevent the stomach from digesting itself. Nervous impulses, the hormone gastrin, and sphincters regulate gastric motility and secretion. Food is reduced in the stomach to an acid chyme that intermittently enters the small intestine.

7. Most of digestion and virtually all of absorption occur in the small intestine, the longest segment of the alimentary canal.

8. The pancreas, liver, and gallbladder empty by ducts into the duodenum, the first part of the small intestine. Regulatory hormones, such as secretin and cholecystokinin, modulate the activity of the accessory glands.

9. Carbohydrate digestion, begun in the mouth, continues in the duodenum in the presence of pancreatic amylase in the lumen and disaccharidases built into the plasma membranes of intestinal epithelial cells.

10. Trypsin, chymotrypsin, carboxypeptidase, and aminopeptidase, pancreatic enzymes that work together to hydrolyze the polypeptide fragments from the stomach, are secreted into the small intestine as inactive zymogens that are subsequently activated by enterokinase, an intestinal enzyme. Peptidases on the intestinal cells finish protein digestion by hydrolyzing small peptide fragments.

11. Fats are broken up into smaller droplets during emulsification by bile salts in the intestinal lumen, thereby allowing maximum exposure to the fat-digesting enzyme, lipase.

12. Most digestion is completed in the duodenum, and the remaining regions of the small intestine, the jejunum and ileum, are involved in absorption.

13. The anatomy of the small intestine is well correlated with its absorptive function. The large folds of the lining have fingerlike villi, whose cells have microscopic microvilli, all greatly increasing the surface area. In the core of each villus are networks of capillaries and lacteals that take up and distribute the products of absorption.

14. Nutrients are absorbed by passive diffusion or by active transport. Amino acids and sugars pass directly into the capillaries, but glycerol and certain fatty acids are reassembled into fats and coated with protein to form tiny chylomicrons inside intestinal cells, from which they are transported by exocytosis into lacteals that eventually empty into the circulation. Some lipids are attached to carrier proteins and enter the capillaries directly as lipoproteins.

15. The nutrient-laden blood from the villi capillaries travels through the large hepatic portal vein to the liver, where organic molecules are interconverted and the nutrient content of the blood is regulated.

16. A sphincter and a cecum with its attached appendix mark the junction between the small intestine and the large intestine, or colon. The colon aids the small intestine in reabsorbing water and houses bacteria, some of which synthesize vitamin K. The feces, which pass through the rectum and out the anus, contain undigested parts of food, cellulose, bile pigments, salts excreted by the colon, and a large proportion of intestinal bacteria.

Some Adaptations of Vertebrate Digestive Systems (pp. 806–807)

1. Dentition is correlated with diet: Herbivores have broad, grinding molars; carnivores have sharp, ripping canines.

2. Herbivores generally have longer alimentary canals, reflecting the longer time needed to digest vegetation. Many herbivorous mammals have special fermentation chambers in the stomach, cecum, or intestines, in which symbiotic microorganisms digest cellulose.

Nutritional Requirements (pp. 807–814)

1. A proper diet must supply fuel for cellular respiration, organic raw materials for synthesis of macromolecules, and essential nutrients.

2. Even at complete rest, an animal requires calories to stay alive (basal metabolic rate). Among vertebrates, the endothermic birds and mammals have a higher basal metabolic rate than the ectothermic reptiles, amphibians, and fishes. In most animal groups, metabolic rate is inversely related to body size.

3. Animals store excess calories in the form of glycogen in the liver and muscles and as fat in adipose tissue. Fat can be synthesized from excess proteins or carbohydrates in the liver before distribution to the adipose tissue. Undernourished animals have diets deficient in calories.

4. Essential nutrients must be supplied in preassembled form because the body lacks the biosynthetic machinery for their synthesis. Malnourished animals are missing one or more of the essential nutrients.

5. Essential amino acids are those the animal cannot make from nitrogen-containing precursors. If one or more essential amino acids are not supplied in the diet, protein deficiencies develop, such as kwashiorkor in humans.

6. Animals can synthesize most necessary fatty acids; there are a few essential unsaturated fatty acids.

7. Vitamins are organic molecules that serve as coenzymes or parts of coenzymes and thus are required in small amounts. Water-soluble vitamins include the B complex, most of which serve as coenzymes in key metabolic processes, and vitamin C. A, D, E, and K are fat-soluble vitamins.

8. Minerals are inorganic nutrients, such as calcium, iron, sodium, and zinc, that are required in varying amounts, depending on their role in physiology and metabolism.

1. In *function*, a paramecium's food vacuole is most analogous to our
   a. mouth   d. liver
   b. small intestine   e. anus
   c. esophagus

2. Which of the following animals lack alimentary canals (complete digestive systems)?
   a. earthworms   d. fishes
   b. jellyfish   e. birds
   c. insects

3. Our oral cavity, with its dentition, is most *functionally* analogous to an earthworm's
   a. intestine   d. stomach
   b. pharynx   e. anus
   c. gizzard

4. Which of the following enzymes has the lowest pH optimum?
   a. salivary amylase   d. pancreatic amylase
   b. trypsin   e. pancreatic lipase
   c. pepsin

5. After surgical removal of an infected gallbladder, a person must be especially careful to restrict his or her dietary intake of
   a. starch   d. fat
   b. protein   e. water
   c. sugar

6. Trypsinogen, a pancreatic zymogen secreted into the duodenum, can be activated by
   a. chymotrypsin   d. trypsin
   b. enterogastrone e. pepsin
   c. secretin

7. Carnassial teeth, which are pointed molars, would most likely be part of the dentition of a
   a. human   d. rabbit
   b. cow   e. hawk
   c. lion

8. What do the typhlosole of an earthworm, the spiral valve of a shark, and the villi of a mammal all have in common?
   a. All are adaptations for the efficient digestion and absorption of meat.
   b. They are all adaptations of the stomach.
   c. They are all microscopic structures.
   d. They all increase the absorptive surface area of intestinal epithelium.
   e. They are all homologous structures.

9. *Basal* metabolic rate, expressed on a "per gram" basis, would be greatest for a
   a. cheetah
   b. chipmunk
   c. shark
   d. whale
   e. human

10. If you were to sprint a hundred meters a few hours after lunch, which stored fuel would you probably tap?
    a. muscle proteins
    b. muscle glycogen
    c. fat stored in the liver
    d. fats stored in adipose tissue
    e. blood proteins

CHALLENGE QUESTIONS

1. Trace a bacon, lettuce, and tomato sandwich through the human alimentary canal, describing what happens to the food in each region of the tract.

2. Suggest your own hypothesis to explain the inverse relationship between body size and metabolic rate per gram of tissue. How would you test your hypothesis?

3. Famine plagues certain parts of the world today. Some people argue that distributing food more equitably to the various countries would reduce starvation, at least for a while. Others counter that it is erroneous and ultimately harmful to perceive starvation as a global problem, because the causes and long-term solutions are usually regional. Evaluate the biological, political, and ethical facets of this debate.

FURTHER READING

1. Davenport, H. W. "Why the Stomach Does Not Digest Itself." *Scientific American*, January 1972. Anatomy and physiology of a tough organ.
2. Christian, J. L. and J. L. Gregor. *Nutrition for Living*, 2 ed. Menlo Park, Calif.: Benjamin/Cummings, 1988. Detailed discussion of nutrition related to life circumstances.
3. Heinrich, B. "The Raven's Feast." *Natural History*, February 1989. Feeding behavior of a scavenger.
4. Hume, I. "Reading the Entrails of Evolution." *New Scientist*, April 15, 1989. Convergent evolution of digestive systems of marsupials and placental mammals.
5. Jennings, J. B. *Feeding, Digestion, and Assimilation in Animals.* New York: St. Martin's, 1973. A comparative approach to feeding mechanisms and digestion in vertebrates.
6. Monmaney, T. "Vitamins: Much Ado About Milligrams." *Science 86*, January/February 1986. Recent input on the raging debate over vitamin requirements.
7. Moog, F. "The Lining of the Small Intestine." *Scientific American*, November 1981. A look at the site of absorption.
8. Stiling, P. D. "Eating a Thin Line." *Natural History*, February 1988. The natural history of leaf miners, insect larvae that eat their way through the soft tissue of leaves.

Circulation and Gas Exchange

E very organism must exchange materials with its environment. We have seen that this chemical commerce ultimately occurs at the cellular level, with substances passing across the plasma membrane between the cell and its immediate surroundings. Since substances can permeate the membrane only if they are dissolved in water, every living cell must be bathed by an aqueous environment, which provides oxygen, nutrients, and other resources the cell needs and which also serves as a disposal site for carbon dioxide and other metabolic waste products that diffuse out of the cell. For a protozoan living in an aquatic habitat, the environment is the surrounding pond water or seawater, and chemical exchange is accomplished simply by diffusion or active transport across the plasma membrane. Because the protozoan is so small, its external surface is sufficient in area to service the entire volume of the organism. The same strategy works for the simplest multicellular animals, which have body plans that expose every cell to the surroundings (Chapter 36).

The length of time it takes for a substance to diffuse from one place to another is proportional to the square of the distance the chemical must travel. For example, if it takes 1 second for a given quantity of glucose to diffuse 100 μm, it will take 100 seconds for the same quantity to move 1 mm and about 3 years for that quantity to diffuse 1 m! Clearly, diffusion is quite inadequate for transporting chemicals over macroscopic distances in animals—for example, for moving oxygen from the lungs to the brain in humans.

All but the simplest animals have special systems for internal transport of body fluids. Chemicals are

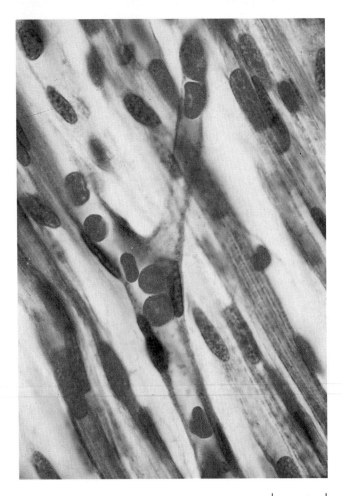

|← 10 μm →|

Figure 38.1
Capillary traffic. As red blood cells pass single file through these microscopic capillaries in a human, they release oxygen, which then diffuses out of the capillary, through the interstitial fluid, and into the cells of the tissue. Specialized systems for internal transport, such as the vertebrate circulatory system, convey materials very rapidly over relatively long distances within animals—transporting oxygen from the lungs to brain, for instance. (LM.)

transferred between the body fluid (blood or interstitial fluid) and the environment across the thin epithelia of organs specialized for gas exchange, nutrient absorption, or waste expulsion. In our lungs, for example, oxygen from the air we inhale diffuses across a thin epithelium and into the blood, while carbon dioxide diffuses in the opposite direction down its own concentration gradient. The circulatory system then carries the oxygen-rich blood away to all parts of the body. As the blood streams through our tissues within microscopic vessels called capillaries, chemicals are transported between the blood and the interstitial fluid that directly bathes our cells (Figure 38.1). No substance has to diffuse far to enter or leave a cell. And because our "pond"—blood and the interstitial fluid it services—is internal, it is possible to control

the chemical and physical properties of the milieu in which our cells live. By circulating the blood frequently through organs such as the liver and kidneys, where its contents of nutrients and wastes can be regulated, the circulatory system plays a central role in homeostasis (see Chapter 36).

In this chapter, you will learn about mechanisms of internal transport in animals. You will also learn about one of the most important cases of chemical transfer between animals and their environment: the exchange of the respiratory gases oxygen and carbon dioxide. We begin by surveying some of the transport systems that have evolved among the animal phyla.

INTERNAL TRANSPORT IN INVERTEBRATES

Gastrovascular Cavities

The saclike body plan of *Hydra* and other cnidarians makes any specialized system for internal transport unnecessary. A body wall only two cells thick encloses a central gastrovascular cavity, which serves the dual functions of digestion and distribution of substances throughout the body (see Chapter 37). The fluid inside the cavity is continuous with the water outside through a single opening; thus, both the inner and outer layers of tissue are bathed by the milieu. Thin strands of the gastrovascular cavity extend into the tentacles of *Hydra*, and some jellyfishes have even more elaborate gastrovascular cavities (Figure 38.2). Some of the cells lining the cavity have beating flagella that stir the contents, helping to distribute materials throughout the animal. Since digestion begins in the cavity, only the cells of the inner layer have direct access to nutrients, but the nutrients have only a short distance to diffuse to the cells of the outer layer.

Planarians and other flatworms also have gastrovascular cavities that exchange materials with the environment through a single opening (see Figure 37.7). The flat shape of the body and the ramification of the gastrovascular cavity throughout the animal ensure that all cells are bathed by a suitable milieu.

Open and Closed Circulatory Systems

A gastrovascular cavity is inadequate for internal transport within animals possessing many layers of cells, especially if the animals live out of water. In insects and other arthropods and most mollusks, blood bathes the internal organs directly (Figure 38.3a). This arrangement is called an **open circulatory system.** There is no distinction between blood and interstitial fluid, and the general body fluid is more correctly termed **hemolymph.** Chemical exchange between the fluid and

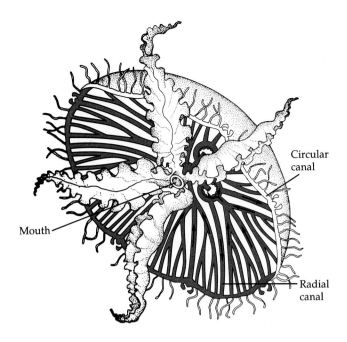

Figure 38.2
Transport in the jellyfish *Aurelia.* The mouth leads to an elaborate gastrovascular cavity (shown in color) that has branches radiating to and from a circular canal. Ciliated cells lining the cavity circulate fluid in the directions indicated by the arrows. The animal is viewed here from its underside.

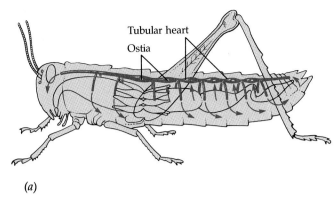

(a)

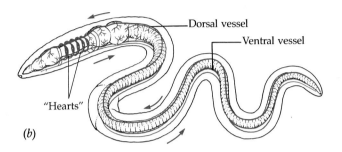

(b)

Figure 38.3
Open and closed circulatory systems of invertebrates. (a) Grasshoppers and other arthropods have an open circulatory system, in which the hemolymph has direct contact with the body tissues. Circulation is driven by the pumping of tubular hearts and by movement of the animal. Hemolymph enters the hearts through pores called ostia when the hearts relax, and it is pumped out when the hearts contract. (b) In the closed circulatory system of the earthworm, blood is confined to vessels. The dorsal and ventral vessels carry blood anteriorly and posteriorly, respectively, and are connected by large vessels that loop around the digestive tract. The dorsal vessel functions as the main heart. Five pairs of the connecting vessels act as auxiliary hearts.

body cells occurs as the hemolymph oozes through **sinuses,** which are spaces surrounding the organs. Hemolymph is "circulated" in a limited fashion by body movements that squeeze the sinuses and by contraction of a heart, usually part of a dorsal vessel. Contraction of the heart pumps hemolymph through vessels, which open into the interconnected system of sinuses. When the heart relaxes, it draws hemolymph in through pores called ostia, which are equipped with valves that close when the heart contracts.

An earthworm (Phylum Annelida) has a **closed circulatory system,** meaning that blood is confined to vessels (Figure 38.3b). There are two major vessels, one dorsal and one ventral, from which branch smaller vessels that supply blood to the various organs. The dorsal vessel functions as the main heart, pumping blood forward by waves of peristalsis. Near the anterior end of the worm, pairs of vessels loop around the digestive tract, connecting the dorsal and ventral vessels. Five pairs of these vessels function as auxiliary hearts by pulsating. The blood exchanges materials with the interstitial fluid, which bathes the cells. Vertebrates and some mollusks (squids and octopuses) also have closed circulatory systems.

Blood percolates through an open circulatory system more slowly than it flows through a closed system. Since most classes of animals depend on their circulatory system to transport oxygen for cellular respiration, we might expect to find open systems only in animals that move sluggishly. And yet, flying insects, among the most active of all animals, have open circulatory systems. Insects, however, do not use blood to carry oxygen long distances. Oxygen infiltrates the insect body through microscopic air ducts called tracheae, which we will discuss in more detail later in this chapter.

CIRCULATION IN VERTEBRATES

Internal transport is accomplished in humans and other vertebrates by a closed circulatory system, also called the **cardiovascular system.** The components of the cardiovascular system are the heart, blood vessels,

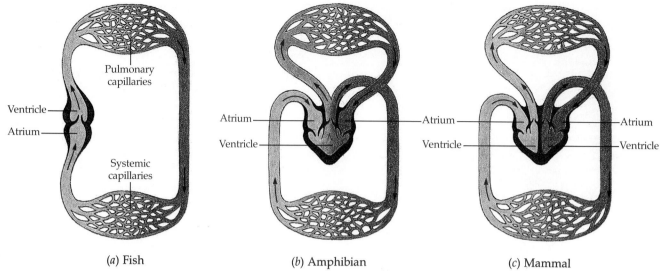

(a) Fish

(b) Amphibian

(c) Mammal

Figure 38.4
Circulatory schemes of vertebrates.
Red is used to symbolize oxygen-rich blood, and blue represents oxygen-poor blood. (**a**) The fish has a two-chambered heart and a single circuit of blood flow. (**b**) The amphibian has a three-chambered heart and two circuits of blood flow, pulmonary and systemic. This double circulation delivers blood to systemic organs under high pressure. In the single ventricle, there is some mixing of oxygen-rich with oxygen-poor blood. (**c**) Mammals and birds have four-chambered hearts and double circulation. Oxygen-rich blood is kept completely segregated from oxygen-poor blood within the heart.

and blood. The heart consists of one or more **atria,** the chambers that receive blood returning to the heart, and one or more **ventricles,** the chambers that pump blood out of the heart. Arteries, veins, and capillaries are the three kinds of blood vessels, which in the human body have been estimated to extend a total distance of 100,000 km. **Arteries** carry blood away from the heart to organs throughout the body. Within these organs, arteries branch into **arterioles,** tiny vessels that give rise to the capillaries. The **capillaries** form networks of microscopic vessels that infiltrate each tissue. It is across the thin walls of capillaries that chemicals are exchanged between the blood and the interstitial fluid surrounding the cells. At their "downstream" end, capillaries rejoin to form **venules,** and these small vessels converge into veins. The **veins** return blood to the heart. Notice that arteries and veins are distinguished by the *direction* in which they carry blood, and not by the quality of the blood they contain. Not all arteries carry oxygenated blood, and not all veins carry blood depleted of oxygen. But all arteries do carry blood from the heart to capillaries, and only veins return blood to the heart from capillaries. We will now examine the routes of blood flow in various classes of vertebrates.

Vertebrate Circulatory Schemes

Various adaptations of the general circulatory scheme just described have evolved in the different vertebrate classes.

A fish has a two-chambered heart, with one atrium and one ventricle (Figure 38.4a). Blood pumped from the ventricle travels first to the gills, where the blood picks up oxygen and disposes of carbon dioxide across the walls of capillaries. The gill capillaries reconvene to form a vessel that carries the oxygenated blood to capillary beds in all other parts of the body. Blood then returns in veins to the atrium of the heart. Notice that in a fish, blood must pass through *two* capillary beds during each circuit, one in the gills and a second one in some other organ. When blood flows through a capillary bed, blood pressure, the hydrostatic pressure that pushes blood through vessels, drops substantially (for reasons that will be explained shortly). Therefore, oxygenated blood leaving the gills flows to other organs in the fish quite slowly. The process is aided by the whole-body movements during swimming.

Frogs and other amphibians have three-chambered hearts, with two atria and one ventricle (Figure 38.4b). The ventricle pumps blood into a forked artery that directs the blood through two circuits: the **pulmonary circuit** and the **systemic circuit.** The pulmonary circuit leads to the lungs and skin, where the blood picks up oxygen as it flows through capillaries. The oxygenated blood returns to the left atrium of the heart, and then most of it is pumped into the systemic circuit. The systemic circuit carries blood to all organs except the lungs and then returns the blood to the right atrium in veins. This scheme, called **double circulation,** ensures a vigorous flow of blood to the brain, muscles, and other organs because the blood is pumped a second time after it loses pressure in the capillary beds of the lungs. This is distinctly different from the single circulation in the fish, where blood flows directly from the respiratory organs (gills) to other organs under reduced pressure.

In the single ventricle of the frog, there is some mixing of oxygen-rich blood that has returned from the lungs with oxygen-poor blood that has returned from the rest of the body. However, a ridge within the ventricle diverts most of the oxygenated blood from the left atrium into the systemic circuit and most of the deoxygenated blood from the right atrium into the pulmonary circuit. In reptiles, there is even less mixing of oxygen-rich with oxygen-poor blood. Although the reptilian heart is three-chambered, the single ventricle is partially divided by a septum (wall). One order of reptiles, the crocodiles, has a complete septum that divides the ventricle into two chambers.

The four-chambered heart of a bird or mammal has two atria and two completely separated ventricles (Figure 38.4c). There is double circulation, as in amphibians and reptiles, but the heart keeps oxygen-rich blood fully segregated from oxygen-poor blood. The left side of the heart handles only oxygenated blood, and the right side receives and pumps only deoxygenated blood. With no mixing of the two kinds of blood, and with a double circulation that restores pressure after blood has passed through the lung capillaries, delivery of oxygen to all parts of the body for cellular respiration is enhanced. As endotherms, which use heat released from metabolism to warm the body, birds and mammals require more oxygen per gram of body weight than other vertebrates of equal size. A more detailed diagram of blood flow through the mammalian circulatory system is shown in Figure 38.5.

We will now see how the heart actually works. Although the process is described with reference to humans, all mammalian hearts work in essentially the same way.

The Heart

The human heart is a cone-shaped organ about the size of a clenched fist, located just beneath the breastbone (sternum). It is enclosed in a sac having a two-layered wall; a lubricating fluid fills the space between the two membranes, enabling them to slide past each other as the heart pulsates. The wall of the heart itself consists mostly of cardiac muscle tissue (see Chapter 36). The atria have relatively thin walls and function as collection chambers for blood returning to the heart, pumping blood only the short distance to the ventricles. The ventricles have thicker walls and are much more powerful than the atria—especially the left ventricle, which must pump blood through the systemic circuit (see Figure 38.5).

The Heart Cycle The sequence of events during each heartbeat is referred to as the **heart cycle** (Figure 38.6). The cycle has two alternating phases: systole and diastole. During **systole,** the heart muscle contracts and the chambers pump blood. (Systole actually refers to only the contraction of the ventricles, but we shall include atrial contraction in systole.) During **diastole,** the ventricles are filling. In an average human at rest, the entire heart cycle takes about 0.8 second (giving a pulse of about 75 beats per minute). Systole and diastole are generally equal in duration, lasting about 0.4 second each. During the first 0.1 second of systole, the atria contract, squeezing blood into the ventricles. Then, in a slow but powerful contraction, the ventricles pump blood into the arteries during the remaining 0.3 second of systole. Notice that seven-eighths of the time—all but the first 0.1 second of the heart cycle—the atria are relaxed and filling with blood returning in veins. Throughout diastole, the ventricles are also filling, since blood can flow into them freely from the atria when the heart is relaxed. In fact, ventricles hold about 70% of their capacity by the end of diastole, and atrial contraction at the beginning of systole only finishes the job of filling them with blood.

Heart Valves and Heart Sounds Four valves in the heart prevent backflow of blood when the ventricles contract (see Figure 38.5 inset). Between each atrium and ventricle is an **atrioventricular valve. Semilunar valves** are located at the two exits of the heart, where the aorta leaves the left ventricle and the pulmonary artery leaves the right ventricle. The valves consist of flaps of connective tissue anchored by strong fibers that prevent the valves from turning inside out. Hydrostatic pressure generated by the powerful contraction of the ventricles forces the atrioventricular valves closed, keeping blood from flowing back into the atria. The blood is pumped out into the arteries through the semilunar valves, which are forced open by ventricular contraction. The elastic walls of the arteries expand and then snap back. Along with relaxation of the ventricles, recoil of the arteries closes the semilunar valves and prevents blood from flowing back into the ventricles.

The heart sounds we can hear with a stethoscope are caused by the vigorous closing of the valves. (You can even hear them without a stethoscope by pressing your ear against the sternum of a friend.) The sound pattern is "lub-dupp, lub-dupp, lub-dupp," the first tone a lower pitch than the second. The first heart sound ("lub") is created by the forceful contraction of the ventricles and the closing of the atrioventricular valves. The second sound ("dupp") is the closing of the semilunar valves.

A defect in one or more of the valves causes a condition known as a heart murmur, which may be detectable as a hissing sound when a stream of blood squirts backward through a valve. Some people are

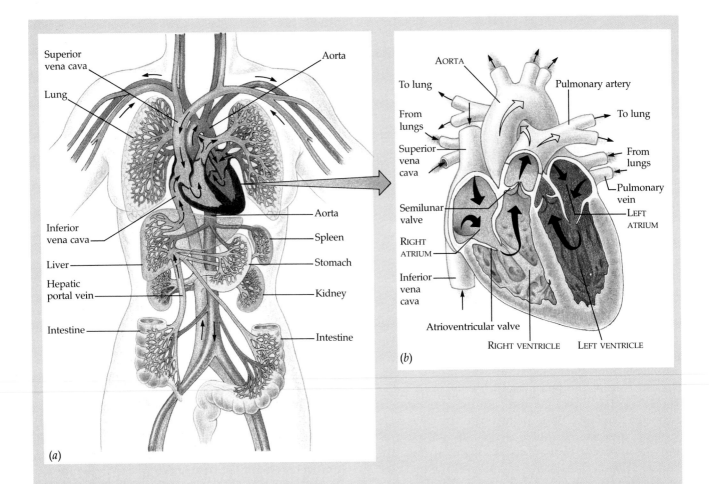

Figure 38.5

The human circulatory system. Beginning our tour with the systemic circuit, oxygenated blood is pumped by the left ventricle into the aorta; this artery is the largest blood vessel in the body, in humans having a diameter about the size of a quarter. The aorta arches over the heart and gives rise to arteries leading throughout the body. The first branches from the aorta are the coronary arteries, which supply blood to the heart muscle itself. Then come branches leading to the head and forelimbs. The aorta continues in a posterior direction, supplying blood to abdominal organs and the hind limbs. Within each organ, the arteries branch into arterioles, which in turn give rise to capillaries, where the blood gives up its oxygen and receives the carbon dioxide produced by cellular respiration. Capillaries rejoin to form venules, which lead to veins. The systemic veins return deoxygenated blood to the heart. Blood from the head, neck, and forelimbs is channeled into a large vein called the anterior ("superior," in humans) vena cava. A posterior ("inferior," in humans) vena cava drains blood from the trunk and hind limbs. The two venae cavae empty their blood into the right atrium, completing the systemic circuit.

The function of the pulmonary circuit is to oxygenate blood by passing it through the lungs. Blood that has returned to the heart via the systemic circuit is pumped by the right ventricle into the pulmonary artery. (Notice here that we have an artery carrying deoxygenated blood.) The pulmonary artery forks, one branch going to each lung. As the blood flows through capillaries in the lungs, it takes up oxygen and unloads carbon dioxide. The pulmonary circuit is completed when oxygenated blood returns to the heart in pulmonary veins, which empty into the left atrium. This blood is ready to be pumped again through the systemic circuit. The inset shows the route that blood follows through the human heart.

born with heart murmurs, while others have their valves damaged by infection (from rheumatic fever, for instance). Most heart murmurs do not reduce efficiency of blood flow enough to warrant surgery. More serious murmurs may be corrected by replacing the damaged valves with artificial ones or with human valves taken from a cadaver.

Heart Rate and Cardiac Output   Heart rate, or **pulse,** is the number of heartbeats per minute. You can easily measure your own heart rate by counting the pulsations of arteries in your wrist or neck; each heart cycle, the contractions of the ventricles during systole pump blood with such force that the elastic arteries stretch from the pressure. For an average human at rest, the

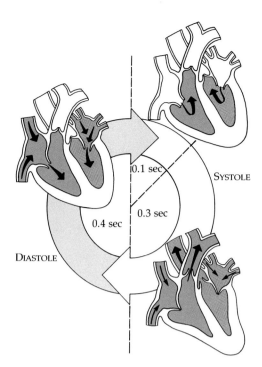

Figure 38.6
Heart cycle. Each heart cycle consists of two phases of roughly equal duration, systole and diastole. During the first 0.1 second of the cycle, the atria contract and force blood into the ventricles. The ventricles contract for about 0.3 second and push the blood out of the heart, completing systole. During diastole, the heart relaxes and blood flows into the atria and ventricles. Blood also flows into the atria during the second part of systole.

pulse is about 65 to 75 beats per minute, although individuals who exercise regularly often have slower resting pulses than those who are less fit. Your own pulse will vary, depending on your level of activity and other factors.

In comparing the heart rates of different mammals, we see an inverse relationship between size and pulse. An elephant, for instance, has a pulse of only 25 beats per minute, while the heart of a tiny shrew races at about 600 beats per minute. To understand the significance of this difference, it will help to recall from Chapter 37 that metabolic rate per gram of tissue is proportionately greater for smaller mammals than for larger ones. A rapid pulse is one adaptation that enhances the delivery of oxygen for cellular respiration.

The volume of blood per minute that the left ventricle pumps into the systemic circuit is called **cardiac output.** This volume depends on two factors: heart rate (pulse) and **stroke volume,** the amount of blood pumped by the left ventricle each time it contracts. The average stroke volume for a human is about 75 ml per beat. A person with this stroke volume and a resting pulse of 70 beats per minute has a cardiac output of 5.25 L/min. This is about equivalent to the total

volume of blood in the human body. Cardiac output can increase about fivefold during heavy exercise.

Excitation and Control of the Heart The cells of cardiac muscle are self-excitable, or myogenic; they can contract without any signal from the nervous system. A heart removed from a frog and placed in a beaker of saline solution will continue to beat for an hour or more. Even individual cardiac muscle cells removed from the heart and viewed with a microscope can be seen to pulsate, but they do so at irregular intervals. Although the cells of cardiac muscle have an intrinsic ability to contract, they must be coordinated with one another and the rhythm of the contractions must somehow be controlled. The rate of contraction is set by a specialized region of the heart called the **sinoatrial (SA) node,** or **pacemaker.** The SA node is located in the wall of the right atrium, near the point where the anterior vena cava enters the heart (Figure 38.7). It is composed of specialized muscle tissue that combines characteristics of both muscle and nerve. Nodal tissue contracts like muscle, but in so doing, it generates electrical impulses much like those found in nervous tissue. Each time the SA node contracts, it initiates a wave of excitation that travels through the wall of the heart. The impulse spreads rapidly, and the two atria contract in unison. (Cardiac muscle cells are electrically coupled by the intercalated discs between adjacent cells; see Chapter 36.) At the bottom of the wall separating the two atria is another patch of nodal tissue, the **atrioventricular (AV) node.** When the wave of excitation reaches the AV node, it is delayed for about 0.1 second and then relayed down to the ventricles. The delay ensures that the atria will contract first and empty completely before the ventricles contract. The impulses that travel through cardiac muscle during the heart cycle produce electrical currents that are conducted through body fluids to the body surface, where the currents can be detected by electrodes placed on the skin and recorded as an **electrocardiogram** (EKG or ECG).

Contraction of the SA node sets the tempo for the entire heart, but the pacemaker itself is controlled by a variety of cues. Two sets of nerves oppose each other in adjusting heart rate; one set speeds up the pacemaker, and the other set slows it down. At any given time, heart rate is a compromise regulated by the opposing actions of these two sets of nerves. The pacemaker is also controlled by hormones secreted into the blood by glands. For example, epinephrine, the "fight-or-flight" hormone from the adrenal glands, increases heart rate (see Chapter 41). Body temperature is another factor that affects the pacemaker. A temperature increase of only 1°C increases the heart rate by about 10 to 20 beats per minute. This is the reason your pulse increases substantially when you have a fever. Exercise also increases the heart rate,

Figure 38.7
Excitation and conduction in the heart.
The impulse originating in the sinoatrial node spreads quickly to the left atrium and produces a unified contraction of the atria. When the impulse from the SA node reaches the atrioventricular node, it travels slowly through thin transition fibers, creating a 0.1-second delay. The signal is then conducted quickly down the AV bundle to the bundle branches and on to the Purkinje fibers, which terminate on ventricular cells. When the signal reaches the ventricular cells, it generates a strong contraction. Hence, the SA node drives the contraction of the entire heart. If the SA node malfunctions, perhaps because it has been damaged by a heart attack, the heart cycle may become erratic. In many cases, an artificial pacemaker can be implanted to take over the function of pacing the heart.

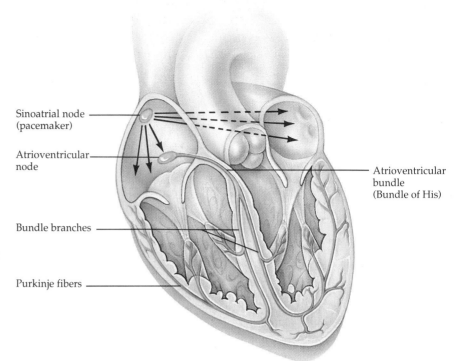

Sinoatrial node (pacemaker)

Atrioventricular node

Atrioventricular bundle (Bundle of His)

Bundle branches

Purkinje fibers

partly because the increased load that the returning venous blood places on the heart stimulates the SA node. This adaptation enables the circulatory system to provide the additional oxygen needed by muscles hard at work.

Blood Flow

Blood Vessel Structure   The wall of an artery or vein has three layers (Figure 38.8). On the outside is a zone of connective tissue with elastic fibers that enable the

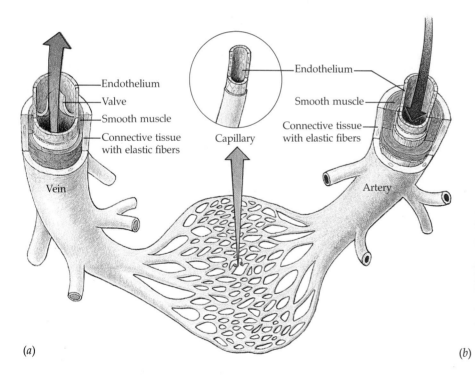

Endothelium
Valve
Smooth muscle
Connective tissue with elastic fibers
Vein

Endothelium
Smooth muscle
Connective tissue with elastic fibers
Capillary
Artery

(a)

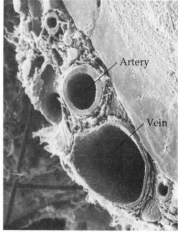

Artery

Vein

(b)

100 μm

Figure 38.8
Structure of blood vessels. (a) The wall of an artery or vein has three layers: an inner layer of endothelium, a middle layer of smooth muscle and elastic fibers, and an outer layer of connective tissue with elastic fibers. A capillary wall consists of a single layer of endothelium. **(b)** In this scanning electron micrograph, an artery can be seen next to a thinner-walled vein.

vessel to stretch and recoil. The middle layer consists of smooth muscle and more elastic fibers. This layer is especially thick in the arteries, which must be stronger and more elastic than veins. (The walls of major arteries are so thick that they must themselves be supplied by blood vessels.) Blood vessels are lined by the **endothelium,** a simple squamous epithelium (see Chapter 36). The endothelium, along with its basement membrane and a thin ring of connective tissue, forms the inner layer of a blood vessel. Capillaries lack the outer layers, and their very thin walls consist only of the endothelium.

Blood Flow Velocity   Blood does not flow through the circulatory system at a uniform speed. After it is pumped into the aorta by the left ventricle, it initially travels at a velocity of about 2 m/sec, but by the time it reaches the capillaries, it is flowing much more slowly. To understand why the blood decelerates, we need to consider the *law of continuity,* a rule that governs the flow of fluids through pipes. If a pipe changes diameter over its length, a fluid will stream through narrower segments of the pipe faster than it flows through wider segments. Since the *volume* of flow per second must be constant through the entire pipe, the fluid must flow faster as the cross-sectional area of the pipe narrows. For instance, compare the velocity of water squirted by a hose with and without a nozzle. Based on the law of continuity, it may at first seem as though blood should travel faster through capillaries than through arteries, since the diameter of capillaries is much smaller. However, it is the *total* cross-sectional area of the pipes delivering the fluid that determines flow rate. Although an individual capillary is very narrow, each artery gives rise to such an enormous number of capillaries that the *total* diameter of the conduits is actually much greater in capillary beds than in any other part of the circulatory system. Furthermore, resistance to blood flow is much greater in capillary beds than in arteries. Thus, blood slows down substantially as it enters capillaries from arteries, but then it speeds up again as it passes along to the veins—a result of the reduction in total cross-sectional area (Figure 38.9). Capillaries are the only vessels with walls thin enough to permit the transfer of substances between the blood and interstitial fluid, and the leisurely flow of blood through these tiny vessels enhances this chemical exchange.

Blood Pressure   Hydrostatic pressure is the force that moves fluids through pipes. (Perhaps you have experienced the annoyance of a slow-running faucet when there is inadequate water pressure.) The hydrostatic force that blood exerts against the wall of a vessel is called **blood pressure** (see Methods Box, p. 826). This pressure is much greater in arteries than in veins and is greatest in arteries during systole, when the heart

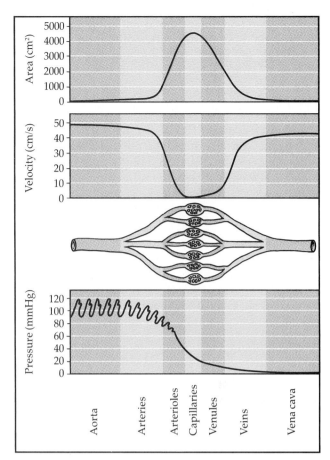

Figure 38.9
Velocity of blood flow and blood pressure. Blood slows down as it flows through a capillary bed owing to the large, total cross-sectional area of the numerous microscopic vessels. Resistance to flow through arterioles and capillaries reduces blood pressure and eliminates the pressure peaks caused by systole.

contracts. When you take your pulse by placing your fingers on your wrist, you can actually feel an artery bulge with each heartbeat. The surge of pressure is partly due to the narrow openings of arterioles impeding the exit of blood from the arteries. Thus, when the heart contracts, blood enters the arteries faster than it can leave, and the vessels stretch from the pressure. The elastic walls of the arteries snap back during diastole, but the heart contracts again before enough blood has flowed into the arterioles to completely relieve pressure in the arteries. This impedance by the arterioles is called **peripheral resistance.** As a consequence of the elastic arteries working against peripheral resistance, there is a blood pressure even during diastole, driving blood into arterioles and capillaries continuously.

Blood pressure is determined partly by cardiac output and partly by the degree of peripheral resistance to blood flow due to the arterioles, the bottlenecks of

METHODS: MEASUREMENT OF BLOOD PRESSURE

(a) Blood pressure is recorded as two numbers separated by a slash; the higher number is the blood pressure during systole, and the lower number is the diastolic pressure. A typical blood pressure for a 20-year-old is 120/70. The units for these numbers are millimeters of mercury (mm Hg); a blood pressure of 120 is a force that can support a column of mercury 120 mm high. (b) A sphygmomanometer, an inflatable cuff attached to a pressure gauge, measures blood pressure in an artery. The cuff is wrapped around the upper arm and inflated until the pressure closes the artery so that no blood flows past the cuff. A stethoscope is used to listen for sounds of blood flow

below the cuff to verify that the artery is closed. (c) The cuff is gradually deflated until blood begins to flow into the forearm and sounds due to blood pulsing into the artery below the cuff can be heard with the stethoscope, which occurs when the blood pressure is greater than the pressure exerted by the cuff. The pressure at this point is the systolic pressure, the high pressure exerted by the ventricles contracting. (d) The cuff is loosened further until blood flows freely through the artery, and the sounds below the cuff disappear. The pressure at this point is the diastolic pressure, the residual pressure between heart contractions.

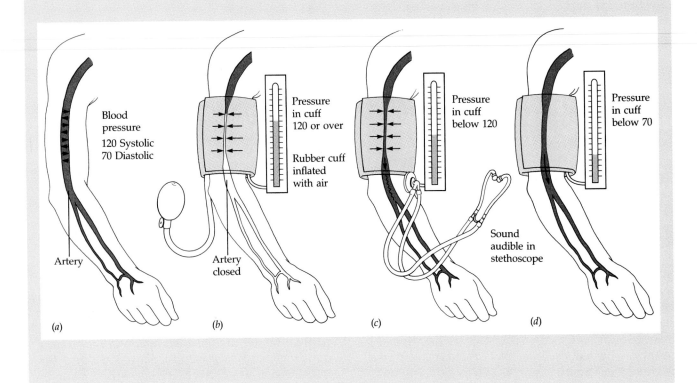

(a) Blood pressure 120 Systolic 70 Diastolic — Artery

(b) Pressure in cuff 120 or over — Rubber cuff inflated with air — Artery closed

(c) Pressure in cuff below 120 — Sound audible in stethoscope

(d) Pressure in cuff below 70

the circulatory system. Contraction of smooth muscles in the walls of the arterioles constricts the tiny vessels, increases resistance, and therefore increases blood pressure in the arteries. When the muscles relax, the arterioles dilate, and pressure in the arteries falls. These muscles are controlled by nerves, hormones, and other signals. Stress, both physical and emotional, can raise blood pressure by triggering neural and hormonal responses that constrict the blood vessels.

By the time blood reaches the veins, its pressure has dropped to near zero. This is because the blood

encounters so much resistance as it passes through the millions of tiny arterioles and capillaries that the force from the pumping heart can no longer propel the blood in the veins. How, then, does blood return to the heart, especially when it must travel from the lower extremities against gravity? The answer is that veins are sandwiched in between muscles, and whenever we move, our skeletal muscles pinch our veins and squeeze blood through the vessels (Figure 38.10). Within large veins are flaps of tissue that function as one-way valves, allowing the blood to flow only in the

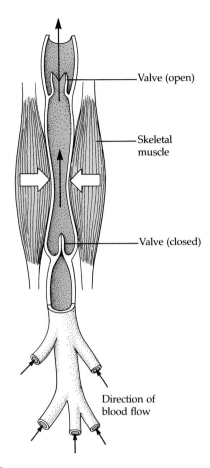

Valve (open)

Skeletal muscle

Valve (closed)

Direction of blood flow

Figure 38.10
Blood flow in veins. Contracting muscles squeeze the veins, in which flaps of tissue act as one-way valves that keep blood moving toward the heart and prevent back-flow. Muscular activity during exercise increases this rate of blood flow. If we sit or stand too long, the lack of mus-cular activity causes our feet to swell with stranded fluid unable to return to the heart. Hairdressers, assembly-line workers, and other people in occupations that require standing for long periods are prone to varicose veins, which occur when valves collapse from the unrelenting downward pull of gravity on the blood.

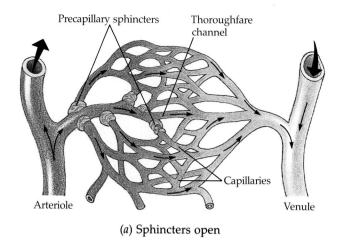

Precapillary sphincters   Thoroughfare channel

Capillaries

Arteriole   Venule

(a) Sphincters open

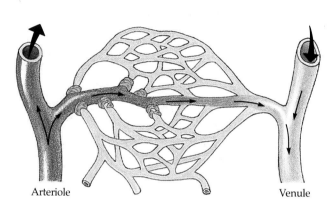

Arteriole   Venule

(b) Sphincters closed

Figure 38.11
Microcirculation. Blood flows from arterioles to venules through thoroughfare channels, vessels that are always open. **(a)** When the precapillary sphincters are open, blood also flows into the capillaries. **(b)** Blood flow through capillaries is shut down when their sphincters close.

direction of the heart. (Such valves are not found in arteries, where blood pressure keeps the blood flow-ing in the right direction.) Muscular activity during exercise increases the rate at which blood returns to the heart, and the increased load stimulates the heart to speed up. Breathing also helps return blood to the heart; when we inhale, the change in pressure within the thoracic (chest) cavity causes the vena cava and other large veins near the heart to expand and fill.

Microcirculation and Blood Distribution   Microcir-culation refers to the flow of blood between arterioles and venules through capillary nets (Figure 38.11). Some blood streams directly from arterioles to venules through **thoroughfare channels,** which are always open. True capillaries branch off from a thoroughfare channel. Passage of blood into the capillary is regu-lated by a sphincter, a ring of smooth muscle at the capillary entrance.

The distribution of blood to the various organs is controlled by the dilation and constriction of arterioles and by the precapillary sphincters. At any given time, only about 5%–10% of the body's capillaries have blood flowing through them. Because each tissue has so many capillaries, however, every part of the body is sup-plied with blood at all times. The supply varies locally as blood is diverted from one destination to another. After a meal, for instance, arterioles in the wall of the digestive tract dilate, capillary sphincters open, and the digestive tract receives a larger share of blood. During strenuous exercise, blood is diverted from the digestive tract and supplied more generously to skel-etal muscles. (This is one reason why heavy exercise immediately after a big meal may cause indigestion.) Most of the time, the organs most heavily perfused

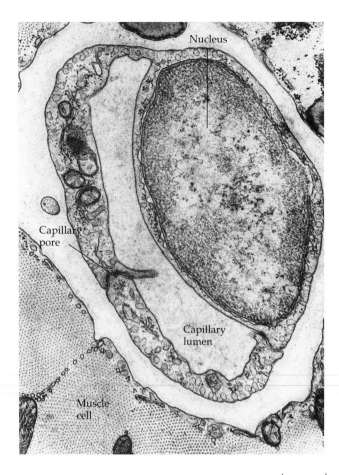

Figure 38.12
Capillary wall. Overlapping endothelial cells enclose the lumen of the capillary, as seen in this electron micrograph. The spaces between the cells function as capillary pores. Substances cross the capillary wall by diffusion, by bulk transport in vesicles (arrows), and by pressure-driven filtration through the clefts between cells.

0.5 μm

with blood are the brain, kidneys, liver, and the heart itself.

Capillary Exchange

Now we come to the most important business of the circulatory system: the transfer of substances between the blood and the interstitial fluid that bathes the cells. This exchange takes place across the thin walls of the capillaries. The capillary wall, remember, is the endothelium, a single layer of flattened cells that overlap at their edges (Figure 38.12).

Some materials may be carried in bulk across an endothelial cell in vesicles that form by endocytosis on one side of the cell and then release their contents by exocytosis on the opposite side; other substances simply diffuse between the blood and the interstitial fluid. Small molecules diffuse down their concentration gradients across the membranes of the endothelial cells. Diffusion can also occur through the clefts between adjoining cells. Water and small solutes, such as sugars, salts, oxygen, and urea, move freely through these capillary pores, but proteins dissolved in the blood and blood cells are too large to pass readily through the endothelium.

In addition to the passive diffusion of substances out of the blood, fluid is pushed through the leaky endothelium by hydrostatic pressure (blood pressure) within the capillary. Fluid flows out of a capillary at the upstream end near an arteriole, but reenters downstream near a venule. The mechanism behind this cycling of material between the blood and interstitial fluid is illustrated in Figure 38.13. About 99% of the fluid that leaves the blood at the arterial end of a capillary bed reenters from the interstitial fluid at the venous end, and the remaining 1% of the fluid lost from capillaries is eventually returned to the blood by the vessels of the lymphatic system.

The Lymphatic System

Although capillaries lose only about 1% of the volume of fluid they carry, so much blood passes through the capillaries that the cumulative loss of fluid adds up to about 3 L per day. There is also some leakage of blood proteins, even though the capillary wall is not very permeable to these large molecules. The lost fluid and proteins return to the blood via the **lymphatic system** (covered in more detail in Chapter 39). Fluid enters this system by diffusing into tiny lymph capillaries that are intermingled among capillaries of the true circulatory system. Once inside the lymphatic system, the fluid is called **lymph;** its composition is about the same as that of interstitial fluid. The lymphatic system drains into the circulatory system at two locations near the shoulders.

Whenever interstitial fluid accumulates rather than being returned to blood by the lymphatic system, tissues and body cavities become bloated, a condition known as edema. One type of severe, localized edema is elephantiasis, which is caused by parasitic worms that block the lymph vessels. Another cause of edema is severe dietary protein deficiency. When starved for amino acids, the body consumes its own blood proteins. This reduces the osmotic pressure of the blood, causing interstitial fluid to accumulate in body tissues rather than being drawn back into capillaries. A child suffering from protein deficiency may have a bloated belly because of all the fluid that collects in the body cavity.

Lymph vessels, like veins, have valves that prevent back flow of fluid toward the capillaries. And like veins, lymph vessels depend mainly on the movement of skeletal muscles to squeeze fluid along. Rhythmic

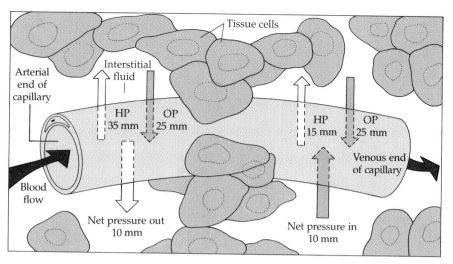

Figure 38.13

Movement of fluid in capillaries. Fluid flows out of a capillary at the upstream end near an arteriole and reenters a capillary downstream near a venule. The direction of fluid movement at any point along the capillary depends on the difference between two opposing forces: hydrostatic pressure (HP) and osmotic pressure (OP). The hydrostatic pressure, or blood pressure, tends to force fluid out of the capillary. The osmotic pressure is a tendency for water to enter the capillary owing to the relatively high solute concentration of the blood. At the arterial end of the capillary, the hydrostatic pressure forcing fluid outward exceeds the osmotic pressure drawing water inward, resulting in a net exodus of fluid from the capillary. Since the endothelium is selectively permeable, it filters the fluid so that water and small solutes leave the capillary while most proteins remain behind in the blood. As blood continues along the capillary, the hydrostatic pressure decreases as a result of resistance and loss of fluid volume. At the downstream end of the capillary, the tendency for fluid to exit due to hydrostatic pressure is overpowered by the tendency for fluid to enter due to osmotic pressure.

contractions of the vessel walls also help draw fluid into lymphatic capillaries.

Along a lymph vessel are specialized swellings called **lymph nodes.** By filtering the lymph and attacking viruses and bacteria, the nodes play an important role in the body's defense. Inside a lymph node is a honeycomb of connective tissue whose spaces are filled by white blood cells specialized for defense. When the body is fighting an infection, these cells multiply rapidly, and the lymph nodes become swollen and tender (which is why your physician checks for swollen nodes in your neck).

The lymphatic system, then, helps to defend the body against infection, as well as maintaining the fluid level and protein concentration of the blood. In addition, as mentioned in Chapter 37, lymph capillaries penetrate the villi of the small intestine and absorb fats; thus, the lymphatic system also has the job of transporting fats from the digestive tract to the circulatory system.

BLOOD

We now shift our focus from the structure and function of blood vessels to the composition of the blood itself. Vertebrate blood is considered a connective tissue with several types of cells suspended in a liquid matrix called **plasma.** The average human body contains about 4–6 L of blood. If a blood sample is taken, the cells can be separated from the plasma by spinning the **whole blood** in a centrifuge. The cells, or **formed elements,** which occupy about 45% of the volume of blood, settle to the bottom of the centrifuge tube to form a dense red pellet. Above this pellet is the transparent, straw-colored plasma (Figure 38.14).

Plasma

Blood plasma consists of an extensive variety of solutes dissolved in water, which accounts for about 90% of the plasma. Among these solutes are inorganic salts, sometimes referred to as blood **electrolytes,** which are present in the plasma in the form of dissolved ions. The combined concentration of these ions is an important factor in maintaining osmotic balance between the blood and interstitial fluid. Some of the ions also help buffer the blood, which has a pH of 7.4 in humans. And the ability of muscles and nerves to function normally depends on the concentration of key ions in the interstitial fluid, which reflects their concentration in plasma. The kidney maintains plasma electrolytes at precise concentrations, an example of homeostasis.

Another important class of solutes is the plasma proteins, which have a number of functions. Collectively they act as buffers to help maintain constant pH,

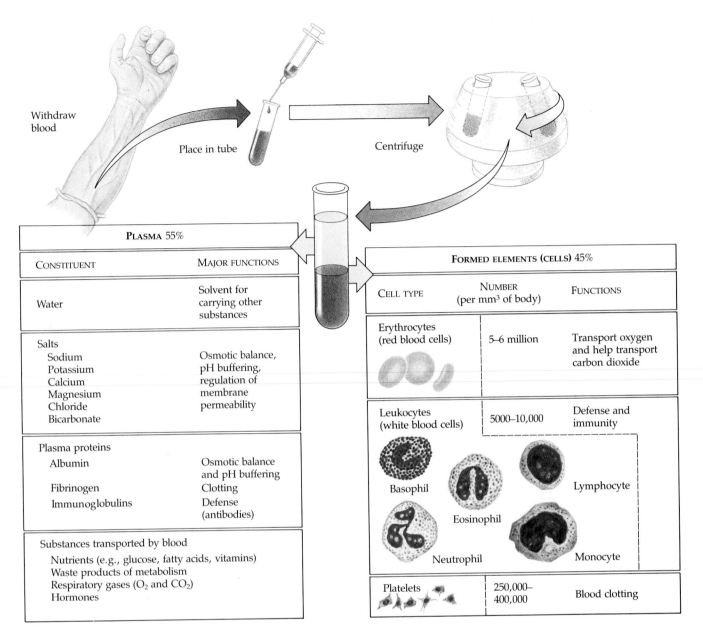

Figure 38.14
The composition of blood.

help determine the osmotic strength of blood, and contribute to its viscosity (thickness). The various types of plasma proteins also have specific functions. Some serve as escorts for lipids, which are insoluble in water and can travel in blood only when bound to proteins. Another class of proteins, the immunoglobulins, are the antibodies that help combat viruses and other foreign agents that invade the body (see Chapter 39). And some of the plasma proteins, called fibrinogens, are clotting factors that help plug leaks when blood vessels are injured. Blood plasma that has had these clotting factors removed is called serum.

Plasma also contains various substances in transit from one part of the body to another, including nutrients, metabolic waste products, respiratory gases, and hormones. Blood plasma and interstitial fluid are similar in composition, except that plasma has a much higher protein concentration than interstitial fluid (capillary walls, remember, are not very permeable to proteins).

Blood Cells

Dispersed throughout blood plasma are three classes of cells: red blood cells, which transport oxygen; white blood cells, which function in defense; and platelets, which are involved in blood clotting.

Red Blood Cells   Red cells, or **erythrocytes,** are by far the most numerous blood cells. Each cubic millimeter of human blood contains about five million red cells, and there are about 25 trillion of these tiny cells in the body's 5 L of blood.

The structure of the red blood cell is another excellent example of structure fitting function. A human erythrocyte is a biconcave disk, flatter in the center than at its edges. Mammalian erythrocytes lack nuclei, an unusual characteristic for living cells (the other vertebrate classes have nucleated erythrocytes). Moreover, all red blood cells lack mitochondria and generate their ATP exclusively by anaerobic metabolism. The major function of erythrocytes is to carry oxygen, and they would not be very efficient if their own metabolism were aerobic and the oxygen they carried was consumed in transit. The small size of erythrocytes also suits their function. For oxygen to be transported, it must diffuse across the plasma membranes of the red blood cells. The smaller the cells, the greater the total area of plasma membrane in a given volume of blood. The biconcave shape of the erythrocyte also adds to its surface area.

As small as a red cell is, it contains about 250 million molecules of **hemoglobin,** a protein containing iron. As red cells pass through the capillary beds of lungs, gills, or other respiratory organs, oxygen diffuses into the erythrocytes and hemoglobin binds the oxygen. This process is reversed in the capillaries of the systemic circuit, with the hemoglobin unloading its cargo of oxygen. (Oxygen loading and unloading are described in more detail later in this chapter.)

Erythrocytes are formed in the red marrow of bones, particularly the ribs, vertebrae, breastbone, and pelvis. Within the marrow are **stem cells** that can develop into any type of blood cell. Red cell production is controlled by a negative feedback mechanism that is sensitive to the amount of oxygen reaching the tissues via the blood. If the tissues are not receiving enough oxygen, the kidney secretes a hormone called **erythropoietin,** which stimulates production of erythrocytes in the bone marrow (see Chapter 19). If blood is delivering more oxygen than the tissues can use, the level of erythropoietin is reduced and erythrocyte production slows. On the average, erythrocytes circulate for about 3 to 4 months before being destroyed by phagocytic cells located mainly in the liver. The hemoglobin is digested and the amino acids are incorporated into other proteins made in the liver. Much of the iron of the hemoglobin is cycled back to bone marrow, where it is reused in erythrocyte production.

White Blood Cells   White cells, or **leukocytes,** fight infections. Of the five types of leukocytes (see Figure 38.14), some are phagocytes, which eat bacteria and debris from our own dead cells. One type of white cell, the lymphocyte, gives rise to the cells that produce antibodies, the plasma proteins that react against foreign substances. The leukocytes we see in blood are in transit. White cells actually spend most of their time outside the circulatory system, patrolling through interstitial fluid, where most of the battles against pathogens are waged. There are also great numbers of white cells, especially lymphocytes, in lymph nodes and other parts of the lymphatic system (see Chapter 39).

Leukocytes arise in bone marrow from the stem cells that can also differentiate into erythrocytes. Some lymphocytes mature, after leaving the marrow, in the spleen, thymus, tonsils, adenoids, and lymph nodes, all of which are called lymphoid organs. Normally, a cubic millimeter of human blood has 5000 to 10,000 leukocytes, but the number increases whenever the body is fighting an infection.

Platelets   Platelets are not really cells at all, but chips of cells about 2 to 3 μm in diameter. They have no nuclei and originate as pinched-off cytoplasmic fragments of large cells in the bone marrow. Platelets then enter the blood and function in the important process of blood clotting.

Blood Clotting

Each of us suffers cuts and scrapes from time to time, and yet we do not bleed to death because blood contains a self-sealing material that plugs leaks in our vessels. The sealant is always present in our blood, but in an inactive form called **fibrinogen.** A clot forms only when this plasma protein is converted to its active form, **fibrin,** which aggregates into threads that form the fabric of the clot. The clotting mechanism usually begins with release of clotting factors from platelets and involves a complex chain of reactions that ultimately transforms fibrinogen to fibrin (Figure 38.15). More than a dozen clotting factors have been discovered, and the mechanism is still not fully understood. An inherited defect in any step of the clotting process causes **hemophilia,** a disease characterized by excessive bleeding from even minor cuts and bruises.

Anticlotting factors in the blood normally prevent spontaneous clotting in the absence of injury. Sometimes, however, platelets clump and fibrin coagulates within a blood vessel, blocking the flow of blood. Such a clot is called a **thrombus.** These clots are more likely to form in individuals with cardiovascular disease.

CARDIOVASCULAR DISEASE

More than half of all deaths in the United States are caused by **cardiovascular disease**—diseases of the heart

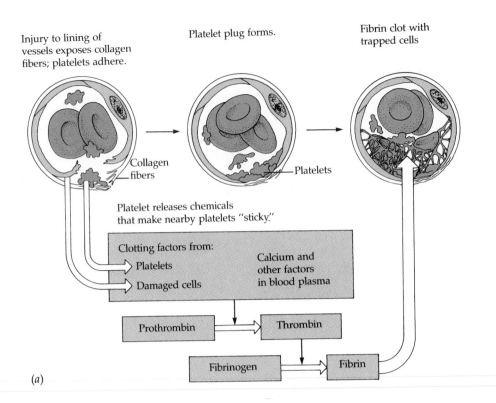

Injury to lining of vessels exposes collagen fibers; platelets adhere.

Platelet plug forms.

Fibrin clot with trapped cells

Collagen fibers

Platelets

Platelet releases chemicals that make nearby platelets "sticky."

Clotting factors from:

Platelets

Damaged cells

Calcium and other factors in blood plasma

Prothrombin → Thrombin

Fibrinogen → Fibrin

(a)

Figure 38.15

Blood clotting. (a) The clotting process begins when the endothelium of a vessel is damaged and connective tissue in the wall of the vessel is exposed to blood. Platelets adhere to collagen fibers in the connective tissue and release a substance that makes nearby platelets sticky. The platelets clump together to form a plug that provides emergency protection against blood loss. Although the plug is adequate to seal minor lesions in the endothelium, it is reinforced by a clot of fibrin when damage to the vessel is more severe. Clotting factors released from the clumped platelets or damaged cells mix with clotting factors in the plasma to form an activator, which converts a plasma protein called prothrombin to its active form, thrombin. Calcium and vitamin K are among the plasma factors required for this step. Thrombin itself is an enzyme that catalyzes the final step of the clotting process, the conversion of fibrinogen to fibrin. Thus, in a cascade of reactions, injury activates prothrombin, which then activates fibrinogen. The threads of fibrin become interwoven into a patch that traps blood cells and seals the injured vessel until the wound is healed by regeneration of the vessel wall. (b) Scanning electron micrograph (artificially colored) of red blood cells trapped in a clot of fibrin.

(b)

10 μm

and blood vessels. Most often, the final blow from cardiovascular disease is either a heart attack or a stroke. These disasters are often associated with a thrombus, a blood clot that clogs a key artery. If the thrombus blocks one of the coronary arteries that supply blood to the cardiac muscle, a heart attack occurs. A thrombus that causes a heart attack may form in a coronary artery itself, or it may develop elsewhere in the circulatory system and reach a coronary artery via the bloodstream. Such a moving clot is called an **embolus.**

The embolus is swept along until it becomes lodged in an artery too small for the clot to pass. Since the heart cannot obtain oxygen or nutrients from the blood within its chambers, the cardiac muscle tissue downstream from the obstruction dies. If the damage is located where it interrupts the conduction of electrical impulses through the cardiac muscle, the heart may begin beating erratically (arrhythmia) or stop altogether. (Still, the victim may survive if heartbeat is restored by cardiopulmonary resuscitation—CPR—or some other emergency procedure within a few minutes of the attack.) Similarly, many strokes are associated with a thrombus or embolus that clogs an artery in the brain. The brain tissue supplied by that artery dies. The effects of the stroke and the individual's chance of survival depend on the extent and location of the damaged tissue.

The suddenness of a heart attack or stroke belies the fact that the arteries of most victims had become gradually impaired by a chronic disease known as **atherosclerosis** (Figure 38.16). Atherosclerosis greatly increases the risk of a blood clot plugging an artery. During the course of this cardiovascular disease, growths called **plaques** develop on the inner walls of the arteries and narrow the bore of the vessels. A plaque forms when lipids infiltrate an abnormally thick matrix of smooth muscle. In some cases, the plaques even become hardened by calcium deposits, resulting in a form of atherosclerosis called **arteriosclerosis,** or "hardening of the arteries." An embolus is more likely to become trapped in a vessel that has been narrowed by plaques. Furthermore, plaques are common sites of thrombus formation. Healthy arteries have smooth linings. The rougher lining of an artery affected by atherosclerosis seems to encourage the adhesion of platelets, which triggers the clotting process.

As atherosclerosis progresses, arteries become more and more clogged by plaque, and the threat of heart attack or stroke becomes much greater. Sometimes, there are warnings. For example, if a coronary artery is partially blocked by atherosclerosis, a person may feel occasional chest pains, a condition known as angina pectoris. The pain is a signal that part of the heart is not receiving a sufficient supply of oxygen, and it is most likely to occur when the heart is laboring hard because of physical or emotional stress. However, many people with atherosclerosis are completely unaware of their disease until catastrophe strikes.

Hypertension (high blood pressure) promotes atherosclerosis and increases the risk of heart attacks and strokes (and, conversely, atherosclerosis tends to increase blood pressure by narrowing the bore of the vessels and reducing their elasticity). According to one hypothesis, the chronic punishment to the lining of arteries by hypertension damages the endothelium and initiates plaque formation. Acting alone or in lethal combination, hypertension and atherosclerosis, the two most common cardiovascular diseases, lead to the majority of deaths in the United States and other developed nations. Hypertension is sometimes called the "silent killer" because a person with the disease may experience no symptoms until a stroke or some other tragedy occurs. Fortunately, hypertension is simple to diagnose and can usually be controlled by drugs, diet, exercise, or a combination of these treatments. A diastolic pressure above 90 may be cause for

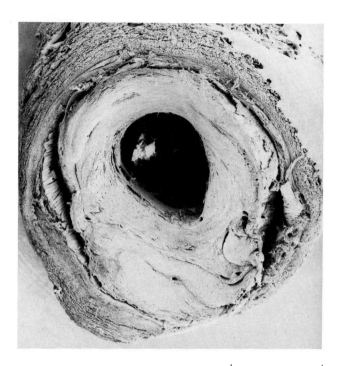

50 μm

Figure 38.16
Atherosclerosis. Atherosclerosis is a disease that narrows the lumen of arteries by growths (plaques) on the inner arterial walls. A plaque forms when lipids such as cholesterol infiltrate a matrix of smooth muscle that proliferates abnormally. In some cases, the plaques become hardened by calcium deposits, resulting in a form of atherosclerosis called arteriosclerosis, or hardening of the arteries. This scanning electron micrograph shows a human blood vessel with its lumen almost completely occluded by a plaque of lipids and fibrous connective tissue.

concern, and living with extreme hypertension—say, 200/120—is courting disaster.

To some extent, the tendency for hypertension and atherosclerosis is inherited, making certain people more predisposed than others to cardiovascular disease. We cannot do much about our genes, but the health of our cardiovascular system is not completely out of our hands. Smoking, lack of exercise, and a diet rich in animal fats and cholesterol are among the factors that have been correlated with an increased risk of cardiovascular disease.

An abnormally high concentration of cholesterol in blood plasma is one of the most important correlates of potential atherosclerosis. Cholesterol travels in the blood mainly in the form of **low-density lipoproteins (LDLs),** plasma particles consisting of thousands of cholesterol molecules and other lipids bound to a protein (see Chapters 8 and 13). Cells of the liver and other organs remove LDLs from the blood when the particles bind to membrane receptors and enter the cells by endocytosis. Certain genetic diseases reduce the number of LDL receptors, raising plasma cholesterol levels and accelerating atherosclerosis. Such genetically impaired individuals, however, account for only a small fraction of those who have heart attacks. What role does high blood cholesterol play in cardiovascular disease of the *general* population? In contrast to LDLs, another form of cholesterol carriers, called **HDLs** (for **high-density lipoproteins**) actually *reduces* deposition of cholesterol in arterial plaques. Many researchers now believe that the ratio of LDLs to HDLs is more reliable than total plasma cholesterol as an indicator of impending cardiovascular disease. Exercise tends to increase HDL concentration, while smoking has the opposite effect on the LDL–HDL ratio.

Let us end the discussion of cardiovascular disease with some good news: Over the past 15 years, the death rate from cardiovascular disease in the United States has declined by more than 25%. Health scientists are not sure of the reasons. So far, heart transplants, artificial hearts, bypass surgery, and other radical methods for treating heart disease have not made any statistically significant contribution to the substantial decrease in deaths from heart attacks. On the other hand, diagnosis and treatment of hypertension may be preventing a large number of heart attacks and strokes, and improved methods of intensive care for cardiovascular patients may be increasing the chances of surviving heart attacks and strokes once they occur. Also, many Americans are now more conscious of their health; as a group, we are smoking less, exercising more, and watching our diets. Education is potent medicine.

GAS EXCHANGE

A major function of circulatory systems is to transport oxygen and carbon dioxide between respiratory organs and other parts of the body. We now focus on the actual exchange of these gases between animals and their environments.

General Problems of Gas Exchange

Animals require a continuous supply of oxygen (O_2) for cellular respiration (see Chapter 9), and they must expel carbon dioxide (CO_2), the waste product of this process. It is important not to confuse gas exchange, the traffic of O_2 and CO_2 between the animal and its environment, with the metabolic process of cellular respiration. Gas exchange supports cellular respiration by supplying oxygen and removing carbon dioxide.

The Earth's main reservoir of oxygen is the atmosphere, which is about 21% O_2. Oceans, lakes, and other bodies of water also contain oxygen in the form of dissolved O_2. The source of oxygen, called the **res-piratory medium,** is air for a terrestrial animal and water for an aquatic one. The portion of an animal's surface where gas exchange with the respiratory medium occurs is called the **respiratory surface.** Oxygen and carbon dioxide cannot bubble across membranes, but can only diffuse through membranes if they are first dissolved in the water that coats the respiratory surface. Thus, the respiratory surface must be moist for both aquatic and terrestrial animals, and it must also be large enough to provide O_2 and expel CO_2 for the entire body. A variety of solutions to this problem has evolved, depending mainly on the size of the animal and whether it lives in water or on land. The most common solution is specialization of a localized region of the body surface as an efficient respiratory surface to supply oxygen for the entire animal.

General Structure and Function of Respiratory Organs

The respiratory surface of a lung, gill, or other respiratory organ is a thin, moist epithelium, usually with a rich blood supply (Figure 38.17). Only this single layer of cells separates the respiratory medium, whether air or water, from the blood or capillaries.

Some animals use their entire outer skin as a respiratory organ. An earthworm, for example, exchanges gases by diffusion across the general body surface. Just below the skin is a dense net of capillaries. Since the respiratory surface must be moist, earthworms and other "skin breathers," including some amphibians, must live in water or damp places.

Most animals that use their general body surface as a respiratory organ are relatively small and have a long, thin (wormlike) shape or flat shape with a high ratio of surface to volume. For most other animals, the general body surface lacks sufficient area to exchange gases for the whole body. The solution is a localized region of the body surface that is extensively folded or branched, enlarging the area of the respiratory surface and enhancing the efficiency of gas exchange. The expanded respiratory surface of most aquatic animals is external and bathed by either fresh water or seawater. These localized extensions of the body surface are called **gills.** Gills are generally unsuitable for an animal living on land because an expansive surface of wet membrane exposed to air would soon desiccate the animal by evaporation. Most terrestrial animals have their respiratory surfaces invaginated into the body, opening to the atmosphere only through narrow tubes. This arrangement reduces evaporative water loss from the moist respiratory epithelium. The **lungs** of terrestrial vertebrates and the **tracheae** of insects are two variations of this plan. We will now examine more closely the three most common respiratory organs: gills, tracheae, and lungs.

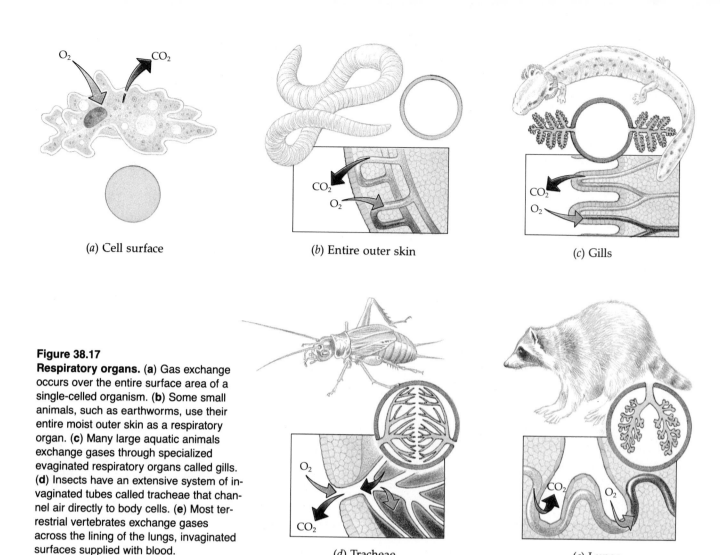

Figure 38.17
Respiratory organs. (a) Gas exchange occurs over the entire surface area of a single-celled organism. (b) Some small animals, such as earthworms, use their entire moist outer skin as a respiratory organ. (c) Many large aquatic animals exchange gases through specialized evaginated respiratory organs called gills. (d) Insects have an extensive system of invaginated tubes called tracheae that channel air directly to body cells. (e) Most terrestrial vertebrates exchange gases across the lining of the lungs, invaginated surfaces supplied with blood.

(a) Cell surface

(b) Entire outer skin

(c) Gills

(d) Tracheae

(e) Lungs

Gills: Respiratory Adaptations of Aquatic Animals

Gills are evaginations (outfoldings) of the body surface specialized for gas exchange (Figure 38.18). Some gills have simple shapes. Those of sea stars and other echinoderms, for instance, are mere bumps that dot the skin. The gills of many segmented marine worms (Phylum Annelida) are flaps that extend from the sides of each segment of the animal. Rather than being distributed over the entire body, the complex gills of many other animals are restricted to a local region of body surface where the skin is finely dissected to form a feathery respiratory surface having a large area. Although only a limited region of the body is devoted to gas exchange, the gills may have a total surface area that is much greater than that of the rest of the body surface. We find such finely divided gills in most mollusks, crustaceans (Phylum Arthropoda), fishes, and some amphibians (some salamanders and the tad-

poles of frogs). Since they are external, the delicate gills are vulnerable to physical damage and attack from other organisms, and in most cases, they are sheltered by a protective cover of some kind. A flap called the operculum covers the gills of bony fishes, for instance.

As a respiratory medium, water has both advantages and disadvantages. On the positive side, there is no problem keeping the respiratory surface wet, since the gills are completely surrounded by the aqueous milieu in which the animal lives. But the oxygen concentration in water is much lower than in air; and the warmer and saltier the water, the less dissolved oxygen it holds. Thus, gills must be very efficient to obtain enough oxygen from water. One process that helps is **ventilation,** a term that refers to any method of increasing contact between the respiratory medium (air or water) and the respiratory surface (lungs or gills). For example, crayfish and lobsters ventilate by using tiny appendages modified as paddles to beat a current of water over the gills. The gills of a bony

Figure 38.18
Invertebrate gills. (a) The gills of a sea star are evaginations of the coelom. Although they contact the external environment directly, they are protected by bony spines. **(b)** Polychaetes (Phylum Annelida) have a pair of gills on each segment. **(c)** Cilia move water through the fingerlike gills of a clam. **(d)** Crayfish and other crustaceans have feathery gills beneath a thoracic exoskeleton. Modified appendages sweep water over the gills.

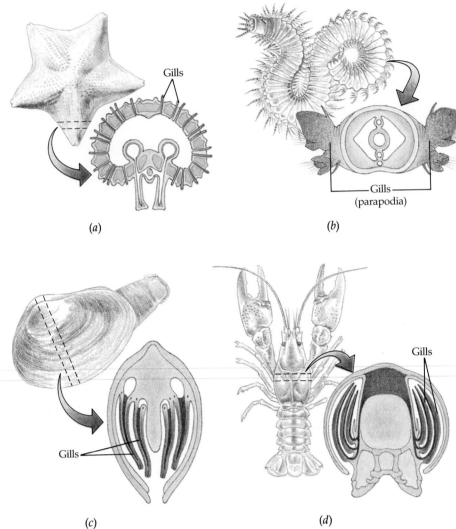

(a)

(b)

(c)

(d)

fish are ventilated continuously by a current of water that enters the mouth, passes through slits in the pharynx, flows over the gills, and exits at the back of the operculum (Figure 38.19). If it were not for ventilation, water around the gills would soon stagnate, becoming depleted in oxygen and saturated with carbon dioxide. Ventilation brings a fresh supply of oxygen and removes carbon dioxide expelled by the gills. Because water is much denser and contains much less oxygen per unit volume than air, a fish must expend a considerable amount of energy to ventilate its gills.

The arrangement of capillaries in the gills of a fish also enhances gas exchange. Blood flows opposite to the direction in which water passes over the gills. This pattern makes it possible for oxygen to be transferred to the blood by a very efficient process called **countercurrent exchange** (Figures 38.19 and 38.20). As blood flows through the capillary, it becomes more and more loaded with oxygen, but at the same time, it is encountering water that is more and more concentrated in oxygen because the water is just beginning its passage over the gills. This means that along the entire length of the capillary, there is a diffusion gradient favoring

the transfer of oxygen from the water to the blood. So efficient is this countercurrent exchange mechanism that the gill is able to remove more than 80% of the oxygen dissolved in the water passing over the respiratory surface. The mechanism of countercurrent exchange is also important in temperature regulation and several other physiological processes, as we shall see in other chapters.

Tracheae: Respiratory Adaptations of Insects

As a respiratory medium, air poses a different set of problems than water. Air has many advantages, not the least of which is a much higher concentration of oxygen. Also, since O_2 and CO_2 diffuse much faster in air than in water, respiratory surfaces exposed to air do not have to be ventilated as thoroughly as gills. As the respiratory surface removes oxygen from the air and expels carbon dioxide, diffusion rapidly brings more oxygen to the surface and carries the carbon dioxide away. When a terrestrial animal does venti-

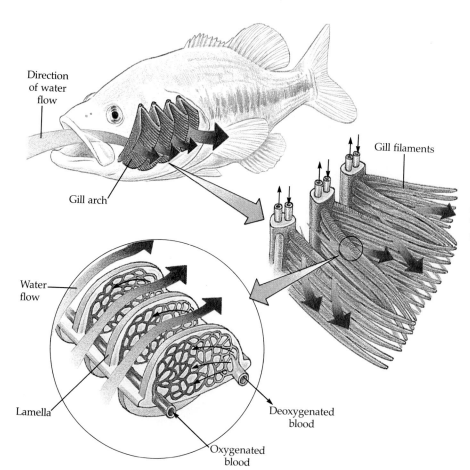

Direction of water flow

Gill arch

Gill filaments

Water flow

Lamella

Deoxygenated blood

Oxygenated blood

Figure 38.19
Physiology of fish gills. Fish continuously pump water through the mouth and over the gill arches, using coordinated movements of the jaws and operculum (gill cover) for this ventilation. Each arch has two rows of gill filaments, and surface area is increased even more by tiny lamellae, which protrude from the filaments (inset). Blood flowing through capillaries within the lamellae picks up oxygen from the water. Notice that water flows over the lamellae in a direction opposite to the blood flow, an arrangement called a countercurrent, which enhances oxygen transfer (see Figure 38.20).

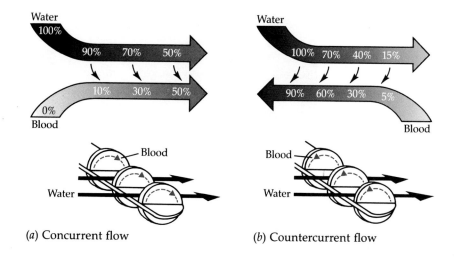

(a) Concurrent flow

(b) Countercurrent flow

Figure 38.20
Countercurrent exchange. The vascular arrangement of fish gills is a marvelous adaptation that maximizes oxygen transfer from the water to the blood. (**a**) If blood flowed through capillaries of the lamellae in the same direction as water flowing over the lamellae, gills could, at best, pick up only 50% of the oxygen dissolved in the water. As O_2 diffused from the water into the blood, the concentration gradient would become less and less steep until blood and water equilibrated at the same O_2 concentration. (**b**) With countercurrent flow, the blood continues to pick up oxygen from the water all along the length of the capillary. As the blood flows along the vessel and becomes more and more loaded with oxygen, it passes by water that has given up less of its O_2. Over the length of the capillary, more than 80% of the O_2 dissolved in the water is transferred by diffusion into the blood.

Figure 38.21
Tracheal systems. Insects and some other terrestrial arthropods (for example, some spiders) have a respiratory system that delivers air directly to body cells. **(a)** Tracheae branch repeatedly and extend to all parts of the body. Enlarged sections of tracheae form air sacs near organs that require a large supply of oxygen. **(b)** Air enters the respiratory system through openings called spiracles, which are located on both sides of the body. Air passes through tracheae and the smaller tracheoles until they terminate on the plasma membrane of individual cells. The liquid in the tracheole endings regulates the amount of air that contacts the cells. When the oxygen requirement increases, some liquid is withdrawn, increasing the surface area of air in contact with cells. **(c)** In this light micrograph of tracheae in a cockroach, you can see rings of chitin that reinforce the air tubes and keep them from collapsing.

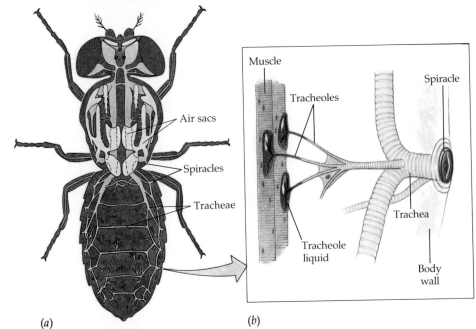

(a)

(b)

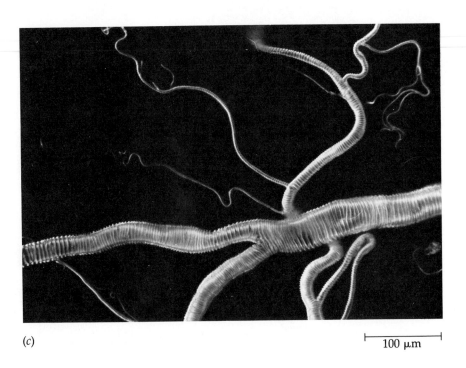

(c)

$\vdash\!\!\!\!\!\!-\!\!\!\!\!\!\dashv$ 100 μm

late, less energy is expended because air is so much easier to move than water. But offsetting these advantages of air as a respiratory medium is a problem: The respiratory surface, which must be large and moist, continuously loses water to the air by evaporation. The solution is a respiratory surface invaginated (infolded) into the body, rather than evaginated like gills. One variation of this plan is the tracheal system of insects.

Tracheae are tiny air tubes that ramify throughout the insect body (Figure 38.21). The finest branches extend to the surface of nearly every cell, where gas is exchanged by diffusion across the moist epithelium that lines the terminal ends of tracheal system. The entrances to the tracheal system are **spiracles,** minute pores distributed over the surface of the insect's body. Small insects rely on diffusion to bring O_2 from the air into the tracheal system and carry CO_2 out of the system. Some larger insects ventilate their tracheal systems with rhythmic body movements that compress and expand the air tubes like bellows.

It is because the tracheal system exposes all cells directly to the respiratory medium and because oxygen diffuses so quickly in air that insects do not need

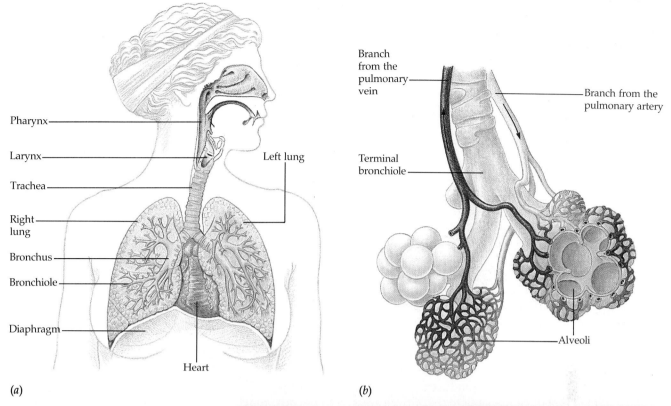

Pharynx

Larynx

Trachea

Right lung

Bronchus

Bronchiole

Diaphragm

Left lung

Heart

(a)

Branch from the pulmonary vein

Terminal bronchiole

Branch from the pulmonary artery

Alveoli

(b)

Figure 38.22
Mammalian respiratory system. (a) The organs of the respiratory system. Air is conveyed down the trachea and bronchi to the tiniest bronchioles, which dead-end as lobed, microscopic air sacs called alveoli. **(b)** Structure of alveoli. Oxygen within the air sac dissolves in the moist film coating the respiratory surface and diffuses across this thin epithelium into the capillaries that surround each alveolus. **(c)** Scanning electron micrograph of alveoli.

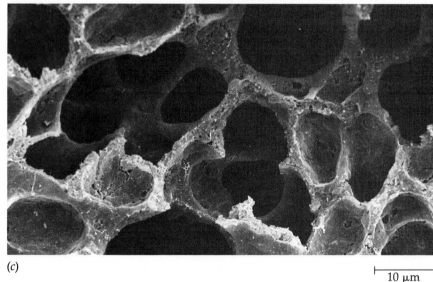

(c)

10 μm

to use their circulatory systems to transport oxygen and carbon dioxide. This is a major reason why an open circulatory system, which moves blood much more slowly than a closed system, suffices for even the most active flying insects (see p. 819).

Lungs: Respiratory Adaptations of Terrestrial Vertebrates

In contrast to the respiratory channels that pervade the entire insect body, lungs are invaginations of the body surface that are restricted to one location. Since the respiratory surface of a lung is not in direct contact with all other parts of the body, the gap must be bridged by the circulatory system, which transports oxygen from the lungs to the rest of the body. Lungs are heavily vascularized with a dense net of capillaries located just beneath the epithelium that forms the respiratory surface. This solution to the problem of gas exchange has evolved as a vascularized mantle in land snails, as book lungs in spiders (see Chapter 29), and as lungs in the terrestrial vertebrates—most amphibians, reptiles, birds, and mammals (Figure 38.22). The lungs of most frogs are balloonlike, with the respiratory sur-

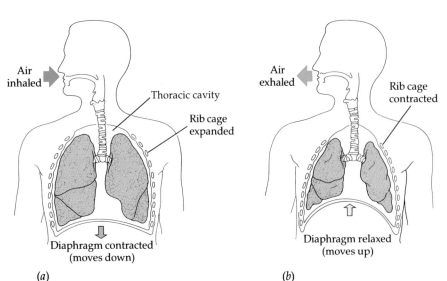

Air inhaled

Thoracic cavity

Rib cage expanded

Diaphragm contracted (moves down)

(a)

Air exhaled

Rib cage contracted

Diaphragm relaxed (moves up)

(b)

Figure 38.23
Negative pressure breathing. Mammals breathe by changing the air pressure within their lungs relative to the pressure of the outside atmosphere. They do this by changing the volume of their thoracic cavity with the rib cage and/or diaphragm. **(a)** To inhale, a mammal increases the volume of the lungs by expanding the rib cage and contracting the diaphragm. Air pressure in the lungs falls below that of the atmosphere, and air rushes into the lungs. **(b)** Exhalation occurs when the rib cage contracts and/or the diaphragm relaxes.

face limited to the outer surface area of the lungs. This is not a large area, but frogs also obtain some of their oxygen by diffusion across the moist outer skin of the body. In contrast, the lungs of mammals have a spongy texture and are honeycombed with epithelium having a total surface area much greater than the outer surface area of the lung itself. The total respiratory surface in a human is about equivalent to the surface area of a tennis court. This entire respiratory surface is coated by a thin film of moisture in which the oxygen in the lung's air spaces dissolves before diffusing across the epithelium and into the blood. Because the passageways connecting the air pockets in the lungs to the outside air are narrow, loss of water by evaporation from the respiratory surface is minimized.

Anatomy of the Mammalian Respiratory System
The lungs of a mammal are located in the thoracic (chest) cavity, enclosed by a double-walled sac (see Figure 38.22). The inner layer of the sac adheres tightly to the outside of the lungs, and the outer layer is attached to the wall of the chest cavity. The two layers are separated by a thin space filled with fluid. Because of surface tension, the two layers behave like two plates of glass stuck together by a film of water. The layers can slide smoothly past each other, but they cannot be pulled apart easily.

Air is conveyed to the lungs by a system of branching ducts. Air enters this system through the nostrils and is then filtered by hairs, warmed, humidified, and sampled for odors as it flows through a maze of spaces in the nasal cavity. This cavity leads to the pharynx, an intersection where the paths for air and food cross. When food is swallowed, the opening of the windpipe (glottis) is pushed against the epiglottis, and food is diverted down the esophagus to the stomach (see Figure 37.11). The rest of the time, the glottis is open, and we can breathe. The glottis leads to the **larynx** (also called the Adam's apple), which has a wall rein-

forced with cartilage. Humans and many other mammals use the larynx as a voice box. As air is exhaled through the chamber, a pair of **vocal cords** vibrates and produces sounds. Pitch is controlled by changing the tension of the cords.

From the larynx, air passes into the **trachea,** or windpipe. Rings of cartilage maintain the shape of the trachea, much as metal rings keep the hose of a vacuum cleaner from collapsing. The trachea forks into two **bronchi** (singular, **bronchus**), one leading to each lung. Within the lung, the bronchus branches repeatedly into finer and finer tubes called **bronchioles.** The entire system of air ducts has the appearance of an inverted tree, the trunk being the trachea. The epithelium lining the major branches of this respiratory tree is covered by cilia and a thin film of mucus. The mucus traps dust, pollen, and other contaminants, and the beating cilia move the mucus upward to the pharynx, where it can be swallowed into the esophagus. This process helps cleanse the respiratory system.

At their tips, the tiniest bronchioles dead-end as multilobed air sacs called **alveoli** (singular, **alveolus**). The thin epithelium of the millions of alveoli in the lung serves as the respiratory surface. Oxygen in the air conveyed to the alveoli by the respiratory tree dissolves in the moist film and diffuses across the epithelium and into the web of capillaries that embraces each alveolus. Carbon dioxide diffuses from the capillaries, across the epithelium of the alveolus, and into the air space.

Ventilating the Lungs Vertebrates ventilate their lungs by **breathing,** the alternate inhalation and exhalation of air. Ventilation maintains a maximal oxygen concentration and minimal carbon dioxide concentration within the alveoli.

A frog inhales by pushing air down its windpipe. The animal first lowers the floor of its mouth, enlarging the oral cavity and drawing air in through the open

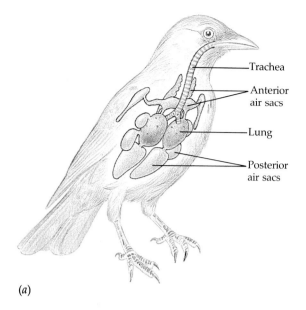

(a)

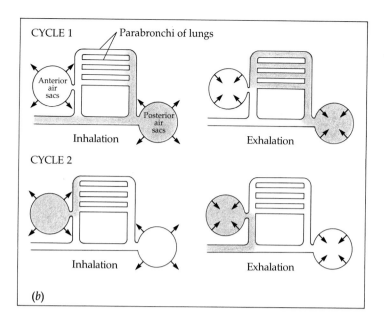

CYCLE 1 — Parabronchi of lungs

Anterior air sacs — Posterior air sacs

Inhalation — Exhalation

CYCLE 2

Inhalation — Exhalation

(b)

Figure 38.24

Avian respiratory system. (a) A bird possesses air sacs in addition to lungs. The air sacs function in ventilating the lungs, where gas exchange occurs. **(b)** In this sequence of diagrams, we follow one "breath" of fresh air (color) through the respiratory system of a bird. Two cycles of inhalation and exhalation are required for the air to pass all the way through the system and out of the bird. During the first inhalation, most of the air bypasses the lungs and enters the posterior air sacs. That air passes through the lungs during exhalation and is drawn into the anterior air sacs during the next inhalation; at the same time, the posterior sacs draw in another breath of fresh air. Together, the anterior and posterior air sacs function as bellows that keep air flowing through the air ducts of the lungs continuously and in one direction (posterior to anterior).

nostrils. Then, with the nostrils and mouth closed, the frog raises the floor of its mouth, which forces air down the trachea. Elastic recoil of the lungs, together with their compression by the muscular body wall, forces air back out of the lungs during exhalation. Oxygen is also absorbed across the lining of the mouth, which is ventilated by fluttering of the throat.

In contrast to frogs, mammals ventilate their lungs by **negative pressure breathing,** which works like a suction pump rather than a pressure pump: The forces are generated near the branches of the respiratory tree, rather than at the trunk, so air is pulled down into the lungs rather than being pushed in from above (Figure 38.23). This is accomplished by changing the volume of the thoracic cavity, which houses the lungs. One way to do this is to use the rib muscles to pull the ribs upward from their normal position, which expands the rib cage. Movement of the lungs is coupled to movement of the rib cage by the surface tension of the fluid in the thin space between the two layers of the sac that encloses the lungs. When the rib cage expands, so do the lungs. Because this movement increases the lung volume, air pressure within the alveoli is reduced to less than atmospheric pressure. Because air always flows from a region of higher pressure to a region of lower pressure, air rushes through the nostrils and down the respiratory tree to the alveoli. When the rib muscles relax, the lungs are compressed, and the increase in air pressure within the alveoli forces air up the respiratory tree and out through the nostrils. This type of breathing usually occurs only during vigorous exercise. For more shallow breathing, when we are at rest, we use the diaphragm rather than the rib muscles. The **diaphragm** is a thin sheet of muscle that forms the bottom wall of the thoracic cavity. When relaxed, the diaphragm is dome-shaped. Contraction of the diaphragm takes up the slack and causes it to flatten, which enlarges the thoracic cavity. In this way, contraction of the diaphragm lowers pressure in the lungs and causes inhalation. Air is exhaled when the diaphragm relaxes to its dome shape.

The volume of air an animal inhales and exhales with each breath is called **tidal volume.** It averages about 500 ml in humans. The maximum volume of air that can be inhaled and exhaled during forced breathing is called **vital capacity,** which is about 4000–5000 ml for a college-age male (somewhat less for a female). Among other factors, the vital capacity depends on the resilience (springiness) of the lungs. The lungs actually hold more air than the vital capacity, but since it is impossible to completely collapse the alveoli, a **residual volume** of air remains in the lungs even after we forcefully blow out as much air as we can. As lungs lose their resilience as a result of aging or disease (such as emphysema), residual volume increases at the expense of vital capacity.

The most complex ventilation occurs in birds (Figure 38.24). Besides their lungs, birds have eight or nine air sacs that penetrate the abdomen, neck, and even the wings. These air sacs trim the density of the bird and also serve as sinks for the considerable amount of heat dissipated by the metabolism of flight muscles. The entire system, lungs and air sacs, is ventilated when the bird inhales and exhales. Air flows through the interconnected system in a circuit that passes through the lungs in one direction only, regardless of whether the bird is inhaling or exhaling. Alveoli, which are dead ends, would not be suitable in such a system. Instead, the lungs of a bird have tiny channels called **parabronchi,** through which air can flow continuously in one direction. The air sacs do not function directly in gas exchange, but act as bellows that keep air flowing through the lungs.

Control of Breathing   Although we can hold our breath voluntarily a short while or consciously breathe faster and deeper, normally our breathing is controlled by automatic mechanisms. The control center is located in the medulla, the stem of the brain that tapers into the spinal cord. We inhale when the nerves in this **breathing center** send impulses to the rib muscles or diaphragm, stimulating the muscles to contract. These nerves fire rhythmically, about 10 to 14 times per minute when we are at rest. A negative feedback loop from the lungs to the respiratory center prevents us from overexpanding our lungs when we take a deep breath. As inhalation deepens, stretch sensors in the lung tissue send nervous impulses that inhibit the breathing center.

In addition to receiving input from the nervous system, the breathing center also monitors the pH of the blood, which begins to drop slightly as the amount of carbon dioxide in the blood increases (carbon dioxide reacts with water to form carbonic acid, which lowers pH). When the center senses a rise in the CO_2 level (a drop in pH), the tempo and depth of breathing are increased. This occurs when we exercise. Oxygen has little affect on the breathing center. Oxygen sensors in key arteries send alarm signals to the breathing center when the oxygen level is too low, but this occurs only when the deficiency is severe—at very high altitudes, for instance. Fortunately, a rise in carbon dioxide concentration is usually a good indication of a drop in oxygen concentration, because CO_2 is produced by the same process that consumes O_2—cellular respiration. It is possible, however, to trick the breathing center by hyperventilating. Excessively deep, rapid breathing purges the blood of so much CO_2 that the breathing center temporarily ceases to send impulses to the rib muscles and diaphragm. Breathing stops until the CO_2 level increases enough to switch the breathing center back on.

The breathing center, then, responds to a variety of neural and chemical signals, adjusting the rate and depth of breathing to meet the changing demands of the body. Control of breathing is only effective, however, if it is coordinated with control of the circulatory system. During exercise, for instance, heart output is matched to the increased breathing rate, which enhances O_2 supply and CO_2 removal as blood flows through the lungs.

Loading and Unloading of Oxygen and Carbon Dioxide   To understand how gases are exchanged at various locations around the body, recall that substances diffuse down their concentration gradients. The concentration of gases, whether in air or dissolved in water, is measured as **partial pressure.** At sea level, the atmosphere exerts a total pressure of 760 mm Hg. This is a downward force equivalent to that exerted by a column of mercury 760 mm high. Since the atmosphere is 21% oxygen, the partial pressure of oxygen is 0.21×760, or about 160 mm Hg. (This is the portion of atmospheric pressure contributed by oxygen—hence the term *partial pressure.*) The partial pressure of carbon dioxide at sea level is only 0.23 mm Hg. These partial pressures are abbreviated as P_{O_2} and P_{CO_2}. When water is exposed to air, the amount of any gas that dissolves in the water is proportional to its partial pressure in the air and its solubility in water. An equilibrium is eventually reached when the gas molecules enter and leave the solution at the same rate. At this point, the gas is said to have the same partial pressure in the solution as it does in the air. Thus, the P_{O_2} in a glass of water exposed to air is 160 mm Hg, and the P_{CO_2} is 0.23 mm Hg. A gas will always diffuse from a region of higher partial pressure to a region of lower partial pressure.

Blood arriving at a lung via the pulmonary artery has a lower P_{O_2} and a higher P_{CO_2} than the air in the alveoli (Figure 38.25). As blood enters a capillary net around an alveolus, carbon dioxide diffuses from the blood to the air within the alveolus. Oxygen in the air dissolves in the fluid that coats the epithelium and diffuses across the surface and into a capillary. By the time the blood leaves the lungs in the pulmonary veins, its P_{O_2} has been raised and its P_{CO_2} has been lowered. After returning to the heart, this blood is pumped through the systemic circuit. In the systemic capillaries, gradients of partial pressure favor the diffusion of oxygen out of the blood and carbon dioxide into the blood. This is because cellular respiration rapidly depletes the oxygen content of interstitial fluid and adds carbon dioxide to the fluid (again, by diffusion). After the blood unloads oxygen and loads carbon dioxide, it is returned to the heart by systemic veins. The blood is then pumped to the lungs again, where it exchanges gases with air in the alveoli.

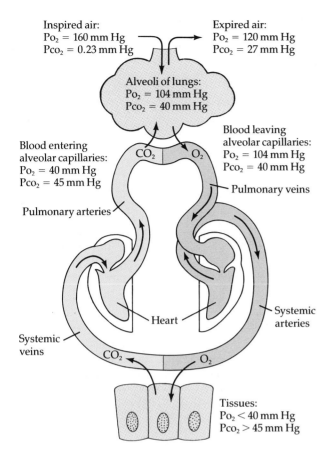

Inspired air:
$P_{O_2} = 160$ mm Hg
$P_{CO_2} = 0.23$ mm Hg

Expired air:
$P_{O_2} = 120$ mm Hg
$P_{CO_2} = 27$ mm Hg

Alveoli of lungs:
$P_{O_2} = 104$ mm Hg
$P_{CO_2} = 40$ mm Hg

CO_2 O_2

Blood entering
alveolar capillaries:
$P_{O_2} = 40$ mm Hg
$P_{CO_2} = 45$ mm Hg

Blood leaving
alveolar capillaries:
$P_{O_2} = 104$ mm Hg
$P_{CO_2} = 40$ mm Hg

Pulmonary veins

Pulmonary arteries

Systemic
arteries

Heart

Systemic
veins

CO_2 O_2

Tissues:
$P_{O_2} < 40$ mm Hg
$P_{CO_2} > 45$ mm Hg

Figure 38.25
Loading and unloading of respiratory gases. Partial pressures of O_2 and CO_2 (symbolized by P_{O_2} and P_{CO_2}, respectively) are proportional to concentrations, and thus each gas diffuses from a region of higher partial pressure to a region of lower partial pressure.

Respiratory Pigments and Oxygen Transport

Because oxygen is not very soluble in water, very little oxygen is transported in blood in the form of dissolved O_2. In most animals, oxygen is carried by **respiratory pigments** in the blood. These pigments are proteins that owe their color to metal atoms built into the molecules. The respiratory pigment of almost all vertebrates is hemoglobin, the iron-containing protein located in red blood cells. It is the iron of hemoglobin that actually binds to oxygen. In some animals, such as earthworms, the hemoglobin is dissolved directly in the blood plasma rather than being located in the cells. (Presumably, one advantage of packaging the hemoglobin in cells is that the blood can have a high concentration of this protein without increasing the osmotic pressure of the plasma.) In addition to hemoglobin, several other proteins that carry oxygen are found in the blood of various invertebrates. One of these, called **hemocyanin,** has copper rather than iron as its oxygen-binding component, coloring the

blood blue. Common in arthropods and many mollusks, hemocyanin is always dissolved directly in plasma.

Hemoglobin consists of four subunits, each with a cofactor called a heme group that has an iron atom at its center; thus, each hemoglobin molecule can carry four molecules of O_2 (see Figure 5.24). To function as an oxygen vehicle, hemoglobin must bind the gas reversibly, loading oxygen in the lungs or gills and unloading it in other parts of the body. In this loading and unloading, there is cooperation between the four subunits of the hemoglobin molecule (see Chapter 6 to review the concept of cooperativity in allosteric proteins). The binding of oxygen to one subunit induces the remaining subunits to change their shape slightly so that their affinity for oxygen increases. The hesitant loading of the first O_2 molecule results in the rapid loading of three more. And when one subunit unloads its oxygen, the other three quickly follow the lead as a conformational change lowers their affinity for oxygen. This mechanism of cooperative oxygen binding and release is evident in the **dissociation curve** for hemoglobin (Figure 38.26). Over the range of oxygen concentrations where the dissociation curve has a steep slope, even a slight change in concentration will cause hemoglobin to load or unload a substantial amount of oxygen. Notice that the steep part of the curve corresponds to the range of oxygen concentrations found in body tissues. When cells in a particular location begin working harder—during exercise, for instance—oxygen concentration dips in the vicinity as the O_2 is consumed in cellular respiration. Because of the effect of subunit cooperativity, a slight drop in O_2 concentration is enough to cause a relatively large increase in the amount of oxygen that the blood unloads.

As with all proteins, hemoglobin's conformation is sensitive to a variety of environmental factors. For example, a drop in pH lowers the affinity of hemoglobin for O_2, an effect called the *Bohr shift* (Figure 38.26b). Because CO_2 reacts with water to form carbonic acid, an active tissue will lower the pH of its surroundings and induce hemoglobin to give up more of its oxygen, which can be used for cellular respiration. Hemoglobin is a remarkable molecule, its structure well suited for its role of transporting oxygen from regions of supply to regions of demand.

Carbon Dioxide Transport   About 7% of the carbon dioxide released by respiring cells is transported as dissolved CO_2 in blood plasma. Another 23% binds to the multiple amino groups of hemoglobin. However, most carbon dioxide, about 70%, is transported in the blood in the form of bicarbonate ions. Carbon dioxide expelled by respiring cells diffuses into the blood plasma and then into the red blood cells, where the CO_2 is converted to bicarbonate. Carbon dioxide first reacts with water to form carbonic acid, which

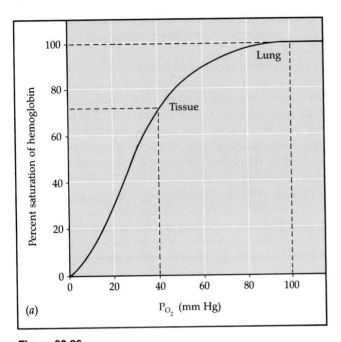

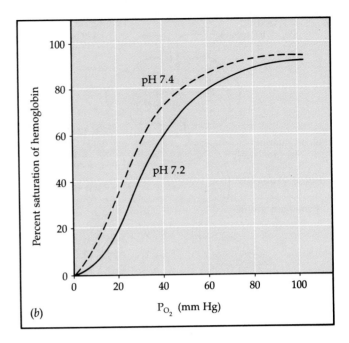

Figure 38.26

Oxygen dissociation curves for hemoglobin. (a) This is the dissociation curve for hemoglobin at 38°C and pH 7.4. The curve shows the relative amounts of oxygen bound to hemoglobin when the pigment is exposed to solutions varying in their partial pressure of dissolved oxygen. At a P_{O_2} of 100 mm, typical in the lungs, hemoglobin is about 98% saturated with oxygen. At a P_{O_2} of 40 mm, common in the vicinity of respiring tissues, hemoglobin is only about 70% saturated—that is, it gives up about 30% of its oxygen. **(b)** Hydrogen ions affect the conformation of hemoglobin, and thus a drop in pH shifts the oxygen-dissociation curve toward the right. Notice that at an equivalent P_{O_2}, say 40 mm, hemoglobin gives up more oxygen at pH 7.2 than it does at pH 7.4, the normal pH of human blood. This occurs in very active tissues because the CO_2 produced by respiration reacts with water to form carbonic acid, thus lowering the pH. Hemoglobin then releases more O_2, which supports a high level of cellular respiration.

then dissociates into a hydrogen ion and a bicarbonate ion. Most of the hydrogen ions produced as a byproduct of this reaction attach to various sites on hemoglobin and other proteins. In this way, hemoglobin acts as a buffer, and CO_2 transport has only a small effect on the pH of blood (venous blood has a pH of 7.34, versus 7.4 for arterial blood). As blood flows through the lungs, the process is reversed. Diffusion of CO_2 out of the blood shifts the chemical equilibrium within red cells in favor of the conversion of bicarbonate to CO_2. The transport of carbon dioxide is illustrated and described in detail in Figure 38.27.

Adaptations of Diving Mammals　We can further our appreciation of physiological diversity by examining some of the special adaptations that make it possible for air-breathing mammals, such as certain species of seals, whales, and dolphins, to make long underwater dives in search of food. One diving mammal that has been studied extensively is the Weddell seal (*Leptonychotes weddelli*), a large Antarctic predator (adults weigh about 400 kg). Weddell seals catch large cod and other deep-water fish by plunging to depths of 200 to 500 m during dives that typically last about 20 minutes. They occasionally submerge for more than

an hour, perhaps to escape predators or explore new routes beneath the ice (Figure 38.28).

One diving adaptation of the Weddell seal is its ability to store oxygen. Compared to humans, the seal contains about twice as much oxygen per kilogram of body weight, mostly in the blood and muscles. About 36% of our total oxygen is in our lungs, and 51% is in our blood. In contrast, the Weddell seal holds only about 5% of its oxygen in its relatively small lungs, stocking 70% in the blood. This is possible partly because the seal has about twice the volume of blood per kilogram of body weight as a human. Another adaptation is the seal's huge spleen, which can store about 24 L of blood. The spleen probably contracts after a dive begins, fortifying the blood with additional erythrocytes loaded with oxygen. Diving mammals also have a higher concentration of an oxygen-storing protein called **myoglobin** in their muscles than most other mammals; the Weddell seal can stow about 25% of its oxygen in muscle, compared to only 13% in humans. Thus, the Weddell seal owes part of its diving virtuosity to blood and muscles enriched with oxygen.

Diving mammals not only begin an underwater trip with a relatively large reservoir of oxygen, they also

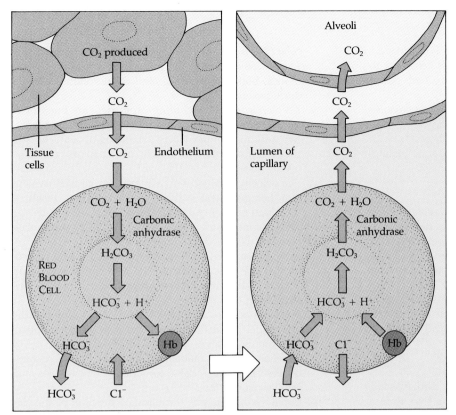

Figure 38.27
Carbon dioxide transport in the blood.
Carbon dioxide produced by body tissues diffuses into the plasma and then into the erythrocytes. In the erythrocytes, the enzyme carbonic anhydrase catalyzes a reaction of CO_2 and H_2O to form carbonic acid, which dissociates into a hydrogen ion (H^+) and a bicarbonate ion. Bicarbonate diffuses out of the erythrocyte into the blood. Electrical balance of the cell is maintained by uptake of chloride, which, like the bicarbonate, is an anion. The hydrogen ions from carbonic acid, which have the potential to acidify the blood, are mostly bound to hemoglobin (Hb) molecules within the erythrocytes. The reversibility of the carbonic acid–bicarbonate conversion also helps buffer the blood, releasing or removing H^+, depending on pH (see Chapter 3). The processes that occur in the tissue capillaries are reversed in the lungs. CO_2 is reconstituted, diffuses out of the blood and into the lungs, and is expelled in an exhaled breath. About 70% of the carbon dioxide in blood is transported in the form of bicarbonate. The remainder travels as dissolved CO_2 or CO_2 bound to the amino groups of proteins.

Figure 38.28
The Weddell seal, a diving mammal. Adaptations of the circulatory and respiratory systems enable these Antarctic seals to travel under water for more than an hour.

have adaptations that conserve that oxygen. A diving reflex slows the pulse, and an overall reduction in oxygen consumption parallels the lower cardiac output. Regulatory mechanisms affecting peripheral resistance route most blood to the brain, spinal cord, eyes, adrenal glands, and placenta (in pregnant seals). Blood supply to the muscles is restricted, and it is shut off altogether during the longest dives. Thus, during dives of more than about 20 minutes, the muscles deplete the oxygen stored in their myoglobin and then derive their ATP from fermentation (see Chapter 9).

Response to environmental pressures—over the short term by physiological adjustments and over the long term by natural selection—is a central theme in our study of organisms that is showcased by the unusual adaptations of diving mammals.

*   *   *

We have seen throughout this chapter that the circulatory and respiratory systems cooperate extensively. Still another activity of the body that depends on the circulatory system is defense, which is discussed in the next chapter.

STUDY OUTLINE

1. The exchange of oxygen, nutrients, carbon dioxide, and metabolic wastes between an organism's cells and the environment must occur across fluid-bathed plasma membranes.
2. In all but the simplest animals, special systems are required for transporting substances within the body.
3. Chemical transfer of gases, nutrients, and wastes occurs across thin, specialized epithelia that separate blood or interstitial fluid from the environment.

Internal Transport in Invertebrates (pp. 812–813)

1. Cnidarians and flatworms have gastrovascular cavities that function for circulation as well as digestion.
2. Arthropods and most mollusks have open circulatory systems, in which tissues are bathed directly in hemolymph pumped by a heart into an interconnected system of sinuses.
3. Annelids have closed circulatory systems, with blood confined to vessels, some of which pulsate and function as hearts. Vertebrates and some mollusks also have closed circulatory systems.

Circulation in Vertebrates (pp. 813–822)

1. In vertebrates, blood flows in a closed cardiovascular system consisting of blood vessels and a two- to four-chambered heart. The heart has one or two atria, which receive blood from veins, and one or two ventricles, which pump blood into arteries. Arteries branch into arterioles, which give rise to capillaries—the sites of chemical exchange between blood and interstitial fluid. Capillaries rejoin into venules that converge into veins.
2. In fishes, a two-chambered heart pumps blood to gills for oxygenation and the blood then travels to other capillary beds of the body before returning to the heart.
3. Amphibians and most reptiles have a three-chambered heart in which the single ventricle pumps blood to both lungs and body in the pulmonary and systemic circuits. These two circuits return blood to separate atria. This double circulation repumps blood returning from the capillary beds of the respiratory organ, thus ensuring a strong flow of blood to the rest of the body.
4. Birds and mammals, both endotherms, have four-chambered hearts that keep oxygen-rich and oxygen-poor blood completely separated.
5. The heart cycle is the sequence of events during each heartbeat consisting of a period of contraction, called systole, and a period of relaxation, called diastole.
6. Atrioventricular valves between each atrium and ventricle and semilunar valves at the exit of each ventricle dictate a one-way flow of blood through the heart.
7. Together with the stroke volume, heart rate (pulse) determines cardiac output, the volume of blood pumped into the systemic circulation per minute by the left ventricle.
8. The intrinsic contraction of the cardiac muscle is coordinated by a conduction system originating in the sino-atrial (SA) node (pacemaker) of the right atrium. The pacemaker initiates a wave of contraction that spreads to both atria, hesitates momentarily at the atrioventri-

cular (AV) node, and then progresses to both ventricles. The pacemaker is itself influenced by nerves, hormones, body temperature, and atrial volume changes during exercise.
9. All blood vessels, including capillaries, are lined by a single layer of endothelium. Arteries and veins have two additional outer layers composed of characteristic proportions of smooth muscle, elastic fibers, and connective tissue.
10. The velocity of blood flow varies in the circulatory system, being slowest in the capillary beds as a result of their high resistance and large total cross-sectional area. This leisurely flow enhances exchange of substances between the blood and interstitial fluid.
11. Blood pressure, the hydrostatic force that the blood exerts against the wall of a vessel, is determined by cardiac output and peripheral resistance due to variable constriction of the arterioles.
12. Muscular activity and pressure changes during breathing propel blood back to the heart in veins equipped with one-way valves.
13. The moment-to-moment blood supply to different organs is determined by variable constriction of arterioles and capillary sphincters.
14. Capillary exchange is the ultimate function of the circulatory system. Substances traverse the endothelium in endocytotic-exocytotic vesicles, by diffusion, or dissolved in fluids forced out by blood pressure at the arterial end of the capillary. Exuded fluid reenters the circulation directly at the venous end of the capillary or indirectly through the lymphatic system.

Blood (pp. 822–825)

1. Whole blood consists of cells (formed elements) suspended in a liquid matrix called plasma.
2. Plasma is a complex aqueous solution of inorganic electrolytes, proteins, nutrients, metabolic waste products, respiratory gases, and hormones.
3. Plasma proteins influence blood pH, osmotic pressure, and viscosity and function in lipid transport, immunity (the immunoglobulins, or antibodies), and blood clotting (fibrinogens).
4. Red blood cells, or erythrocytes, transport oxygen, a function reflected in their small size, biconcave shape, anaerobic metabolism, and hemoglobin content.
5. Five types of white blood cells, or leukocytes, function in defense by phagocytizing bacteria and debris or by producing antibodies.
6. Platelets are fragments of cells produced in the bone marrow that function in blood clotting, a cascade of complex reactions that converts the plasma protein fibrinogen into fibrin.

Cardiovascular Disease (pp. 825–827)

1. A leading cause of death in the United States is cardiovascular disease, a deterioration of the heart and blood vessels. Gradual plaque buildup during atherosclerosis or arteriosclerosis narrows the diameter of vessels and encourages the formation of thrombi or emboli that can

cause heart attack or stroke by blocking strategic vessels in the heart and brain.

2. Both hypertension and a high plasma cholesterol level (high LDL-to-HDL ratio) correlate with an increased risk of cardiovascular disease.

Gas Exchange (pp. 827–838)

1. Metabolism demands a constant supply of oxygen and removal of carbon dioxide. Animals require large, moist respiratory surfaces for adequate diffusion of respiratory gases between their cells and the respiratory medium.

2. A gill is any localized evagination of the external body surface specialized for gas exchange. The efficiency of gas exchange in some gills, including those of fishes, is increased by ventilation and countercurrent flow of blood and water.

3. The tracheae of insects are tiny branching tubes that penetrate the entire body, bringing oxygen directly to cells.

4. Terrestrial vertebrates, land snails, and spiders have internal lungs restricted to one location.

5. Mammalian lungs are enclosed in a double-walled sac whose layers adhere to one another, to the lungs, and to the chest cavity. Air inhaled through the nostrils passes through the pharynx and glottis, between the vocal cords of the larynx, into the trachea, bronchi, and bronchioles, and into the dead-end alveoli, where gas exchange occurs.

6. Lungs must be ventilated by breathing. Mammals ventilate with negative pressure by contracting and relaxing rib muscles and a diaphragm, which changes the volume and hence the pressure of the thoracic cavity and lungs relative to the atmosphere.

7. Tidal volume, the amount of air normally inhaled or exhaled, is less than the vital capacity of the lungs. Even after forceful exhalation, a residual volume of air remains in the alveoli.

8. Birds have complex respiratory systems, with one-way, highly efficient ventilation of the lungs made possible by a system of air sacs, which also decrease density and dissipate heat, important adaptations for flight.

9. Breathing is regulated automatically by complex control systems involving a breathing center in the medulla of the brain stem. Rhythmic nervous impulses, blood pH changes reflecting carbon dioxide concentrations, and oxygen sensors work to adjust the rate and depth of breathing to match the metabolic demands of the body.

10. Oxygen and carbon dioxide diffuse from where their partial pressures are higher to where they are lower. In the lungs, oxygen enters the blood and carbon dioxide leaves. The blood then enters the heart through the pulmonary vein and is pumped through the systemic circuit where the diffusion process is reversed in the metabolic environment of the tissues.

11. Respiratory pigments increase the amount of oxygen that blood can carry. Arthropods and many mollusks use copper-containing hemocyanin, whereas vertebrates and some invertebrates use iron-containing hemoglobin.

12. Hemoglobin molecules, carried inside red blood cells, have four iron-containing subunits, each one capable of binding a molecule of oxygen. The binding of the first oxygen molecule increases the affinity of the other subunits for oxygen. The dissociation curve for hemoglobin is steepest in the range of oxygen concentrations found in body tissues. A drop in pH reduces hemoglobin's affinity for oxygen, thus providing more oxygen to active tissues.

13. Most carbon dioxide generated during metabolism is transported in the form of bicarbonate ions, which result from the dissociation of carbonic acid formed in the red blood cells from the chemical union of carbon dioxide and water. Hydrogen ions from the dissociation are bound to hemoglobin and other proteins, serving to buffer the blood. The entire process is reversed when blood enters the lungs, allowing free carbon dioxide to diffuse into the environment.

14. Diving mammals have a larger volume of blood, a higher concentration of myoglobin in their muscles, and a diving reflex that slows heart rate and reroutes blood flow away from muscles and noncritical organs.

SELF-QUIZ

1. Which of the following respiratory systems is *not* closely associated with a blood supply?

   a. vertebrate lungs

   b. fish gills

   c. tracheal systems of insects

   d. outer skin of earthworm

   e. parapodia of a polychaete worm

2. Blood returning to the mammalian heart in a pulmonary vein will drain *first* into the

   a. vena cava

   b. left atrium

   c. right atrium

   d. left ventricle

   e. right ventricle

3. Pulse is a direct measure of

   a. blood pressure

   b. stroke volume

   c. cardiac output

   d. heart rate

   e. breathing rate

4. The respiratory medium flows unidirectionally over the respiratory surface in the gas-exchange systems of

   a. mammals

   b. frogs

   c. birds

   d. insects

   e. starfish

5. In negative pressure breathing, inhalation results from
   a. forcing air from the throat down into the lungs
   b. contracting the diaphragm
   c. relaxing the muscles of the rib cage
   d. using muscles of the lungs to expand the alveoli
   e. contracting the abdominal muscles

6. The maximum volume of air you can forcefully exhale after taking the deepest possible breath is called
   a. tidal volume
   b. residual volume
   c. vital capacity
   d. total respiratory volume
   e. alveolar volume

7. A decrease in the pH of human blood would
   a. decrease breathing rate
   b. increase heart rate
   c. decrease the amount of O_2 unloaded from hemoglobin
   d. decrease cardiac output
   e. decrease CO_2 binding to hemoglobin

8. Compared to the interstitial fluid that bathes active muscle cells, blood reaching that tissue in arteries has a
   a. higher P_{O_2}
   b. higher P_{CO_2}
   c. greater bicarbonate concentration
   d. lower pH
   e. greater osmotic pressure

9. Which of the following reactions prevails in red blood cells traveling through pulmonary capillaries? (Hb = hemoglobin)
   a. $Hb + 4 O_2 \rightarrow Hb(O_2)_4$
   b. $Hb(O_2)_4 \rightarrow Hb + 4 O_2$
   c. $CO_2 + H_2O \rightarrow H_2CO_3$
   d. $H_2CO_3 \rightarrow H^+ + HCO_3^-$
   e. $Hb + (CO_2)_4 \rightarrow Hb(CO_2)_4$

10. Compared to a human, a diving mammal of equal size has
    a. less blood, an adaptation that helps to conserve oxygen
    b. larger lungs
    c. a larger spleen
    d. less oxygen stored in muscles
    e. less oxygen stored in blood

CHALLENGE QUESTIONS

1. Because they support their weight against gravity and repeatedly accelerate limbs from standing starts, terrestrial vertebrates consume more energy in locomotion than do fishes swimming through water. That is, it takes more calories per gram of animal to move a meter on land than it does to move a meter in water (assuming, of course, that the animal is in its natural habitat—either land or water). How does this disparity fit in with the evolution of the vertebrate cardiovascular system?

2. The hemoglobin of a human fetus differs from adult hemoglobin. Compare the dissociation curves of the two hemoglobins in the following graph, and then explain the physiological significance of the difference.

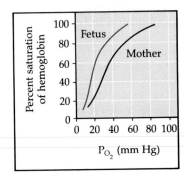

3. Describe some adaptations you think would be helpful for a mammal living at very high altitude, where the air is "thin" (relatively low P_{O_2}).

FURTHER READING

1. Cantin, M., and J. Genest. "The Heart as an Endocrine Gland." *Scientific American*, February 1986. Reports how in addition to its pumping function, the heart produces a hormone that regulates blood pressure and blood volume.
2. Dickerson, R. E., and I. Geis. *Hemoglobin: Structure, Function, Evolution and Pathology*. Menlo Park, Calif.: Benjamin/Cummings, 1983. A comprehensive treatment of the vertebrate respiratory pigment.
3. Lillywhite, H. B. "Snakes, Blood Circulation, and Gravity." *Scientific American*, December 1988. How does a snake slithering up a tree get blood to its brain?
4. Randall, D. J., W. W. Burggren, A. P. Farrell, and M. S. Haswell. *The Evolution of Air Breathing in Vertebrates*. New York: Cambridge University Press, 1981. A book on adaptations of form and function in terrestrial respiration.
5. Schmidt-Nielsen, K. "How Birds Breathe." *Scientific American*, December 1971. A lucid explanation of bird respiration by the expert in the field.
6. Vines, G. "Diet, Drugs, and Heart Disease." *New Scientist*, February 25, 1989. How to lower blood cholesterol.
7. Zapol, W. M. "Diving Adaptations of the Weddell Seal." *Scientific American*, June 1987. Lab and field studies of a living diving machine.

The Immune System

39

T he **immune system** is a security system that helps protect the animal's body from intruders. Martial imagery is often used to describe the immune system's activities. The body is likened to a citadel under constant attack from without and in danger of treacherous assaults from within, while the immune system is the armed force ready to mobilize for defense at a moment's notice. Such images may be simplistic, but they vividly convey the role of this system in the vertebrate body. Innumerable bacteria and viruses are everywhere—in the air, on surfaces throughout the environment, and in food and water. Within the body, insurgents, such as early-stage cancer cells, must be kept under control (Figure 39.1). In this chapter, you will learn how the defense system of the body counters these threats.

NONSPECIFIC DEFENSE MECHANISMS

The immune response is very specific, recognizing a certain offender, such as a strain of virus, and then producing large numbers of cells specialized for combating that particular invader. This specific defense backs up several mechanisms of resistance termed *nonspecific* because they employ defensive cells and chemicals that do not distinguish one infectious agent from another. The nonspecific defenses, which help prevent the entry and spread of harmful microorganisms, include the skin and mucous membranes, defensive white blood cells such as phagocytes and natural killer cells, the inflammatory response, and antimicrobial proteins.

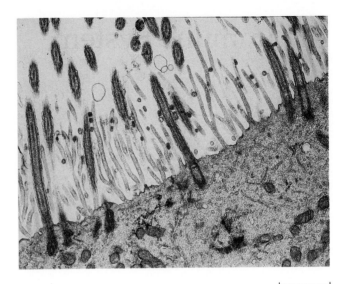

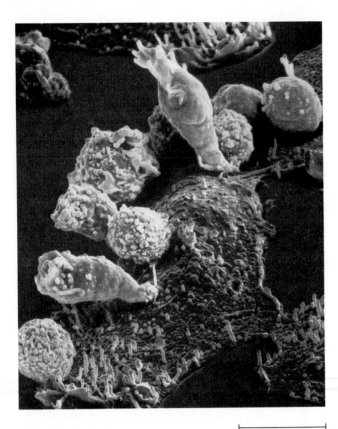

Figure 39.2
The ciliary escalator. Epithelia and their secretions keep many pathogens outside the body's cells and internal fluids. In this electron micrograph, influenza viruses (black dots) can be seen among the cilia of the epithelium lining a human trachea (windpipe). Trapped in the mucus secreted by the epithelium, viruses and other harmful particles are swept away from the lungs by the cilia.

⊢————⊣ 1 μm

⊢————⊣ 5 μm

Figure 39.1
Cells of the immune system attack a cancer cell. An animal's health is threatened not only from the outside by pathogenic microorganisms, but also from within by the body's own cells that go awry, such as cancer cells. The immune system helps defend against both threats. In this scanning electron micrograph (artificially colored), a troop of cells called cytotoxic T cells (white) attacks a tumor cell.

The Skin and Mucous Membranes

The first line of the body's defense system consists of physical and chemical obstacles to the entry of intruders. The skin's outer layer is a toughened barrier of cells through which most bacteria and viruses cannot penetrate. This physical barrier is reinforced by chemical defenses. In humans, for example, secretions from oil and sweat glands give the skin a pH of 3–5, which is acidic enough to discourage many microorganisms from taking up residence there. Bacteria that make up the normal flora of the skin are adapted to this acidic environment and may contribute to defense by releasing acids and other metabolic wastes that inhibit the growth of pathogens. In addition, perspiration contains an enzyme called **lysozyme,** which attacks the cell walls of many bacteria. Lysozyme is also present in tears and saliva, helping to protect the eyes from infection and defending against bacteria present in food and water. Most bacteria that are still alive when food is swallowed are killed by stomach acid (gastric

juice). Another potential route of pathogen entry that must be guarded is the respiratory tract. The mesh of tiny hairs in the nostrils filters inhaled particles that may carry microorganisms. Organisms that penetrate this defense are trapped in the mucus that coats the epithelium of the respiratory tract. Cilia keep this mucus and its trapped particles moving out of the lungs (Figure 39.2).

Phagocytes and Natural Killer Cells

Microorganisms that manage to penetrate the skin and mucous membranes and enter the underlying connective tissue are soon confronted by amoeboid cells capable of phagocytosis (see Chapter 8). Chief among the phagocytes are **macrophages** ("big eaters"), the majority of which wander through interstitial fluid engulfing bacteria, viruses, and the debris of damaged cells (Figure 39.3). There are also fixed macrophages, which reside permanently in most organs, including the brain and lungs. Macrophages develop from a class of white blood cells called monocytes, which migrate from capillaries into the interstitial fluid. **Neutrophils,** another class of white blood cells, also become phagocytic in infected tissue.

Natural killer cells do not attack microorganisms directly, but rather destroy the body's own infected cells, especially cells harboring viruses, which can

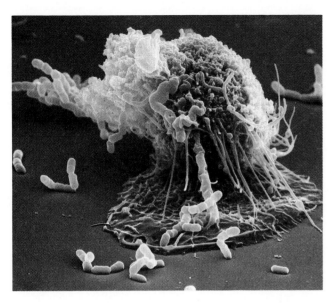

Figure 39.3
Phagocytosis by a macrophage. Using multiple pseudopodia to snare its prey, this macrophage, photographed in the scanning electron microscope, is dining on a rod-shaped bacterium.

⊢————⊣ 5 μm

reproduce only within host cells. The natural killers also assault aberrant cells that could form tumors. The mode of destruction is not phagocytosis, but an attack on the membrane of the target cell, which causes that cell to lyse (break open).

The Inflammatory Response

Damage to tissue by physical injury, such as a scratch, or by microorganisms triggers an **inflammatory response** (which means "setting on fire"). Small blood vessels in the vicinity of an injury dilate and become leakier. This increased blood supply and permeability of capillaries produces the redness, heat, and swelling associated with infection. Most important to actual defense, the increased flow of blood to the site of injury and the leakiness of the capillaries enhance the migration of phagocytic white blood cells from the blood to the interstitial fluid, where these cells begin engulfing microorganisms and cleaning up the debris. Neutrophils arrive first, followed by monocytes that develop into macrophages. The macrophages not only devour pathogens, but also clean up the remains of tissue cells killed by the injurious agent and remove the litter of neutrophils that have themselves died after eliminating many microorganisms. The pus that often accumulates at the site of infection consists mostly of dead cells and fluid that leaked from the capillaries during the inflammatory response.

Clotting proteins present in blood plasma also pass into the interstitial fluid during inflammation, begin-

ning the repair process and sealing off the infected region. Thus, the inflammatory response not only disposes of infectious agents and dead cells, but also helps prevent the spread of the infection to surrounding tissue.

Various chemical signals initiate and mediate the inflammatory response (Figure 39.4). Injured cells release an alarm signal called **histamine,** which induces dilation of neighboring blood vessels and increases the leakiness of capillaries. Other substances discharged from damaged cells and white blood cells also increase blood flow to the injured locale. Still other chemical signals attract amoeboid phagocytes to the site of infection. And injured cells also emit a compound called *leukocytosis-inducing factor,* which is carried by the blood as a signal that stimulates release of neutrophils from bone marrow—a call for reinforcements. This is one reason the number of white cells in blood may increase severalfold within a few hours after inflammation begins.

The inflammatory responses described so far are localized, as in infections caused by splinters, for example. But the body's reaction to an infection may also be more widespread, or systemic. One systemic reaction is an increase in the number of circulating white blood cells. Another response is fever. Toxins produced by the pathogens may trigger the fever, but certain white blood cells also release compounds called **pyrogens,** which set the body's thermostat to a higher temperature. A very high fever is dangerous, but a moderate fever contributes to defense by stimulating phagocytosis and inhibiting growth of many kinds of microorganisms.

Antimicrobial Proteins

A variety of proteins functions in nonspecific defense by attacking microorganisms directly or by impeding their reproduction. Particularly important are the interferons and complement proteins.

Interferons   Interferon was first identified in 1957, when it was discovered that virus-infected cells produce a substance that helps other cells resist the virus—the substance "interferes" with the virus. Several types of interferons, each produced by a particular type of cell, have since been identified. Although interferons are proteins made only in small quantities by the body, recombinant DNA technology has made it possible to produce large quantities of these molecules (see Chapter 19). By the early 1980s, they were being tested clinically for their effectiveness in treating viral infections and cancer. The results of those early tests were mixed, with interferons getting passing but not outstanding marks across the board. One early but persistent problem has been that interferon treatment has often given patients flulike symptoms, including chills and fever.

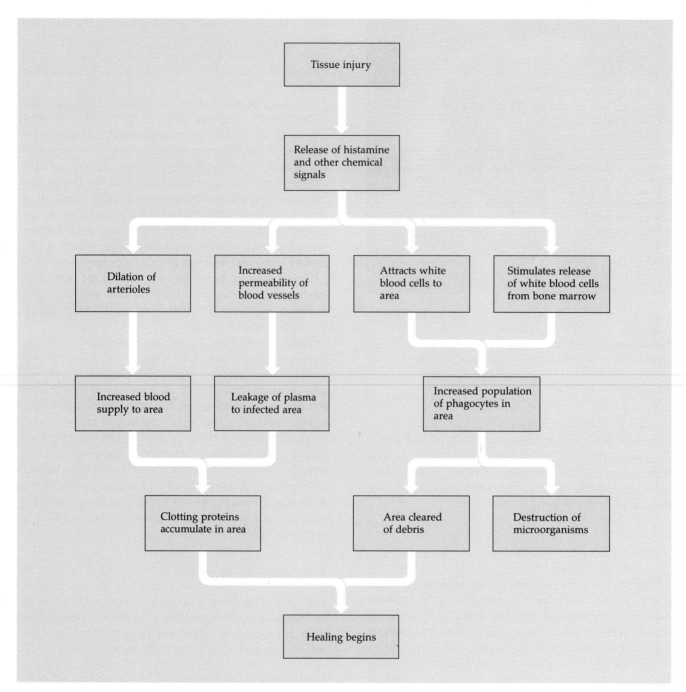

Figure 39.4
The inflammatory response. Triggered and mediated by the release of chemical signals from damaged cells, the inflammatory response combats local infection, helps prevent spread of the infection to surrounding tissue, and initiates the healing process.

One type of interferon is believed to fight viral infections in the following way. When viruses first contact mammalian cells, the viruses turn on a set of interferon genes, causing the cells to produce interferon. Interferon cannot save an infected cell, but it diffuses to neighboring cells, where the interferon stimulates production of other proteins that inhibit viral replication in those cells. Thus, an infected cell can help protect uninfected cells. The defense is not virus-specific; interferon produced in response to one viral strain confers resistance to unrelated viruses. However, interferons *are* host-specific; for example, the interferons native to mice are impotent when injected into another mammal.

Scientists are uncertain how interferon may act against cancer. They speculate that because some tumors may be of viral origin, interferon might slow down the production of viral proteins necessary for the tumor cells' growth. Also, at least one type of interferon mobilizes natural killer cells, which may help keep tumors in check. Moreover, interferon changes some of the properties of malignant cell membranes, making them less likely to metastasize (spread into different regions of the body).

Interferons are also among the chemical signals that mediate the inflammatory response by activating macrophages. And, as one of the many chemical coordinators tying together the defense network, interferons, themselves considered nonspecific defenses because they are effective against a variety of infections, stimulate the specific defense known as the immune response.

Complement   A group of at least 20 proteins, **complement** is a defense system named for its cooperation (complementation) with other defense mechanisms. The complement proteins circulate in the blood in inactive form. The system is activated either by the onset of the immune response or by chemical markers on the surfaces of microorganisms. Some of the activated proteins amplify the inflammatory response by stimulating histamine release from cells and by attracting phagocytes. Complement proteins also coat the invading microorganisms and roughen their surfaces, which facilitates phagocytosis by macrophages. This is called **opsonization** (which means "making tasty"). Still other complement proteins team up to form a **membrane attack complex,** which inserts in the membrane of a foreign cell, causing the pathogen to lyse (Figure 39.5).

In its multiple interactions with other branches of the body's defense network, complement illustrates that resistance to infection, like all biological processes, is an emergent property that arises from complex organization. In the next section we will see many examples of links between the nonspecific defenses, such as phagocytes and complement, and the immune system, which responds with weapons specifically targeted for certain invaders.

SPECIFIC DEFENSE MECHANISMS: THE IMMUNE RESPONSE

The immune system is a recognition system that distinguishes "self" (the body's own molecules) from "nonself" (foreign molecules). When the immune system detects a foreign substance, called an **antigen,** it responds with a proliferation of cells that either attack the invader directly or produce specific defensive proteins called **antibodies,** which help to counter the antigen in various ways. Among antigens triggering an immune response are certain molecules on the sur-

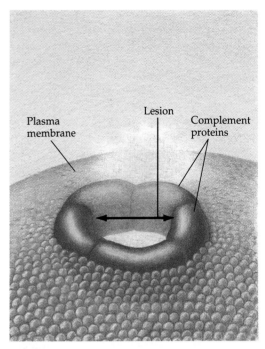

Figure 39.5
Lysis of foreign cells by complement. Activated complement forms a membrane attack complex that creates a lethal hole in the membrane of a pathogen.

faces of viruses, bacteria, and other pathogenic organisms, as well as biochemical markers on the cell surfaces of organs transplanted from one individual to another. In contrast to the nonspecific defenses, which are always ready to fight a variety of infections, the immune response must be primed by the presence of the antigen, and the defensive cells and antibodies produced against that antigen are ineffective against any other foreign substance.

In addition to its specificity, another remarkable feature of the immune system is its ability to "remember" antigens it has encountered before and to react against them more promptly on second and subsequent exposures. The term **immunity,** in fact, originally referred to protection conferred on a host organism by earlier exposure to certain viral and bacterial pathogens. The basic phenomenon of immunity was recognized at least 2500 years ago by Thucydides of Athens, who described how those sick and dying during an epidemic of plague were cared for by those who had recovered, "for no one was ever attacked a second time." Similarly, if you had mumps and chicken pox as a child, then your immune system "remembers" the biochemical signatures of the viruses that cause these diseases. You are immune to reinfection because your body will recognize and destroy the viruses before they can produce symptoms of illness. Thus, the immune response, unlike the nonspecific defenses, is adaptive; exposure to a particular foreign agent enhances future response to that same invader.

Immunity can also be achieved by the process known as **vaccination,** in which the immune system is presented with an inactive or attenuated (weakened) form of the pathogen—one that cannot itself cause the disease but closely resembles the form that can. This inactive agent initiates a long-term capability of the immune system to respond quickly to the real infective agent. Vaccination has been a particularly effective way of combating many viruses, including those responsible for diseases such as polio, mumps, measles, and smallpox.

Whether antigens enter the body naturally (if you catch the flu) or artificially (if you get a flu shot), the immunity thus achieved is called **active immunity** because the body is stimulated to produce antibodies in its own defense. It is also possible to acquire **passive immunity.** For example, when antibodies cross the placenta from mother to fetus, or when a traveler gets a shot containing antibodies to a variety of pathogens, the fetus or the traveler has passively acquired antibodies that provide temporary immunity. It is temporary because the individual's own immune system has not been stimulated by the antigens, and the injected antibodies will circulate in the bloodstream for only a few weeks or months.

Duality of the Immune System

The immune system has two different but overlapping components. The first is known as **humoral immunity** because it results in the production of antibodies, which circulate as soluble proteins in blood plasma and lymph—fluids formerly called "humors." The second component is known as **cell-mediated immunity** because its defensive activities are carried out by highly specialized cells that circulate in the blood and lymphoid tissue. The humoral system defends primarily against free bacteria and viruses present in body fluids, whereas the cell-mediated system works against bacteria and viruses that have already infected host cells and also against fungi and protozoa. The cell-mediated immune system also reacts against tissue transplants (which the body recognizes as "foreign") and is thought to be important in protecting the body from its own cells if they become cancerous. The two responses of the immune system—humoral and cell-mediated responses—are mounted by two different kinds of cells.

Cells of the Immune System

The cells responsible for the immune response are white blood cells called **lymphocytes** (see Figure 38.14). Although all lymphocytes look much the same in the light microscope, there are actually two distinct populations of these cells, each responsible for one of the two components of immunity. The B lymphocytes, or **B cells,** secrete antibodies and thus function in humoral immunity. The T lymphocytes, or **T cells,** function in cell-mediated immunity, including attacks on infected or defective cells.

All blood cells, including lymphocytes, originate from common stem cells in the red bone marrow. Initially, all lymphocytes are alike, but subsequently they specialize as T or B cells, depending on where they continue their maturation (Figure 39.6). Lymphocytes that migrate from the bone marrow to the thymus, a gland in the upper chest region, develop into T cells (the "T" stands for "thymus"). Lymphocytes that remain in the bone marrow and continue their maturation there become B cells (the "B" actually stands for "bursa of Fabricius," a digestive organ unique to birds where these lymphocytes were first discovered, but it may help you remember the site of maturation for these cells if you equate "B" with "bone").

As T cells and B cells mature in the thymus and bone marrow, respectively, they develop **immunocompetence.** This means that each cell becomes competent at recognizing one specific antigen and mounting an immune response against that particular antigen. Immunocompetence involves the synthesis of specific receptor proteins on the surface of the lymphocyte; each cell has receptors capable of binding to one type of antigen. The receptors on B cells are actually bound copies of the very type of antibody that is subsequently secreted during an immune response. The T-cell receptors are not antibodies, but probably recognize foreign molecules by complementary fit between the antigen and the binding site of the receptor, much as antibodies recognize specific antigens. Wearing its receptors, each immunocompetent T cell and B cell is tuned into a specific antigenic threat. One cell may recognize the mumps virus, while another detects a particular species of bacterium. The T- and B-cell populations actually consist of hundreds of thousands of subpopulations, each made up of cells with unique antigen receptors on their surfaces. Each lymphocyte becomes rigidly programmed to recognize and respond to a specific antigen *before* the cell actually encounters that antigen. In fact, that particular antigen may *never* invade the body of the animal. An important feature of the immune system is its preparedness for an almost unlimited variety of potential infections.

Immunocompetent T and B cells next migrate to the lymph nodes and spleen (Figure 39.7). Lymph nodes are organs arranged along the extensive system of lymph vessels, which return fluid from the interstitial spaces to the circulatory system (see Chapter 38). Most pathogens that penetrate the outer defenses provided by skin and mucous membranes eventually turn up in lymphatic fluid, since the capillaries of the lymphatic system drain all the tissues of the body. Thus,

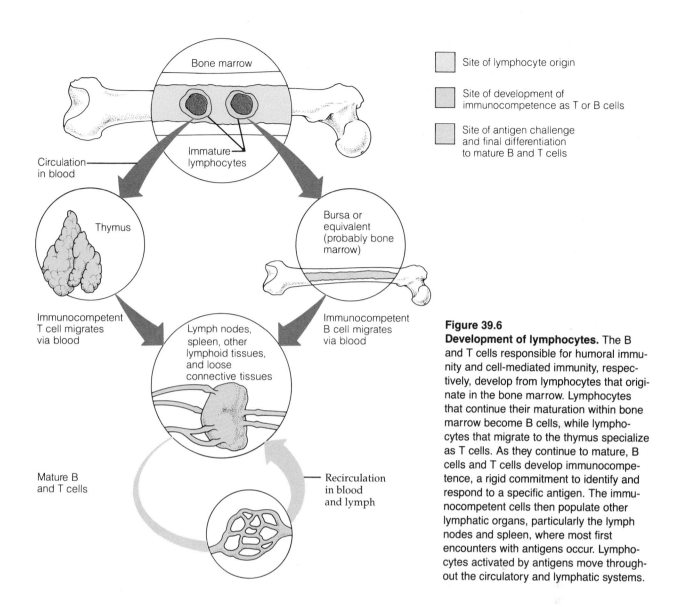

Figure 39.6
Development of lymphocytes. The B and T cells responsible for humoral immunity and cell-mediated immunity, respectively, develop from lymphocytes that originate in the bone marrow. Lymphocytes that continue their maturation within bone marrow become B cells, while lymphocytes that migrate to the thymus specialize as T cells. As they continue to mature, B cells and T cells develop immunocompetence, a rigid commitment to identify and respond to a specific antigen. The immunocompetent cells then populate other lymphatic organs, particularly the lymph nodes and spleen, where most first encounters with antigens occur. Lymphocytes activated by antigens move throughout the circulatory and lymphatic systems.

Legend (top right):
- Site of lymphocyte origin
- Site of development of immunocompetence as T or B cells
- Site of antigen challenge and final differentiation to mature B and T cells

Diagram labels:
- Bone marrow
- Immature lymphocytes
- Circulation in blood
- Thymus
- Bursa or equivalent (probably bone marrow)
- Immunocompetent T cell migrates via blood
- Immunocompetent B cell migrates via blood
- Lymph nodes, spleen, other lymphoid tissues, and loose connective tissues
- Mature B and T cells
- Recirculation in blood and lymph

immunocompetent lymphocytes waiting in the lymph nodes are strategically located to encounter antigens. Macrophages also populate the lymph nodes, ingesting pathogens that are swept into the nodes by the flow of lymphatic fluid. The macrophages play a key role in triggering the activation of lymphocytes, as we will soon see. As lymphocytes and macrophages mount their attack on pathogens, the lymph nodes may swell, a common symptom of infection. Once activated by antigens, lymphocytes multiply by cell division and develop into **effector cells,** the derivatives of B and T cells that are actually equipped to counteract the intruders by various effector mechanisms. The effector cells are mobilized, moving from lymph node to lymph node via the circulatory and lymphatic vessels, fighting the infectious agent throughout the body. Before we examine the two arms of this defense—humoral and cell-mediated immunity—in more detail, we will take closer looks at the nature of antigens, at how antigens provoke the cloning of specific lympho-

cytes, and at the cellular basis of immunological memory.

Antigens

As mentioned earlier, the immune system can recognize and distinguish thousands (probably millions) of different antigens, macromolecules that do not belong to the host organism. (The word *antigen* is actually a contraction of "*anti*body *gen*erating," a reference to the foreign agent's provoking the immune system to react.) In general, antigens are large molecules having molecular weights of 10,000 daltons or more. The vast majority are proteins or large polysaccharides that are often outer components of invading microorganisms: the coats of viruses, the capsules and cell walls of bacteria, and the surfaces of many other types of cells. Foreign molecules associated with blood cells from other persons or species, or with transplanted tissues and

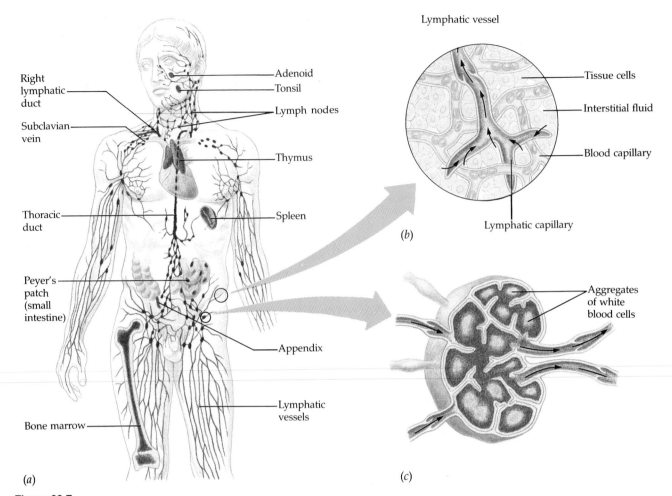

Figure 39.7
Human lymphatic system. (a) The lymphatic system returns fluid from the interstitial spaces to the circulatory system (see Chapter 38). It consists of lymphatic vessels with nodules (nodes) densely populated by lymphocytes and macrophages. The system also includes various satellite organs that cooperate with the vessels and nodes in defense against pathogens.

The thymus, adenoids, tonsils, spleen, Peyer's patch, appendix, and bone marrow, where white blood cells originate, are all important lymphatic organs. **(b)** Some of the interstitial fluid that bathes cells is taken up by lymphatic capillaries. The fluid, now called lymph, flows through the system of vessels, eventually returning to the circulatory system where two large

lymphatic vessels, the right lymphatic duct and the thoracic duct, drain into veins near the shoulders. **(c)** Along the way, lymph must pass through numerous lymph nodes, where any pathogens present in the lymph encounter macrophages and immunocompetent lymphocytes.

organs, can also provoke an immune response. In any case, notice that it is not an entire cell or invading organism that is identified as foreign, but biochemical markers that cover the invader. In fact, antibodies do not generally recognize the antigen as a whole molecule. Rather, antibodies identify localized regions on the surface of an antigen, called **antigenic determinants** (Figure 39.8). Most antigens have many different determinants on their surface, some of which are more effective than others at provoking an immune response. Since different determinants are recognized by different antibodies, a single antigen molecule may stimulate the immune system to make several distinct antibodies against it.

Clonal Selection

It is not the case that each type of antigen evokes a specific immune response by programming generic lymphocytes to focus their defensive activities against that particular antigen. As we have seen, the ability of a lymphocyte to recognize and respond to one particular antigenic determinant is rigidly predetermined before an encounter with that foreign chemical group ever takes place. And, recall, the development of this immunocompetence in lymphatic organs involves the placement of specific antigen receptors on the surface of each lymphocyte. Each B cell, for example, has thousands of identical antibody molecules embedded

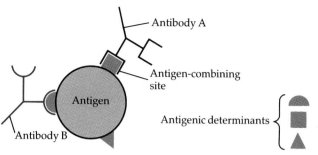

Figure 39.8
Antigenic determinants. Antibodies bind to antigenic determinants on the surface of antigens. Since an antigen generally possesses several different determinants, different antibodies can bind to the same complex antigen.

in its membrane, and this commits the cell to identify and react to a single type of antigenic determinant. The versatility of the immune system—its ability to defend against an almost infinite variety of antigens—is not a function of flexible cells that can change their antigenic targets on demand, but rather depends on the presence in the lymphocyte population of a great diversity of cells with different receptor specificities. (The genetic basis for this protein diversity was discussed in Chapter 18.) Thus, an antigen introduced

to the body selectively activates a tiny fraction of the quiescent lymphocytes, which grow and divide to form a clone of effector cells. This mechanism of immunological specificity is called **clonal selection** (Figure 39.9). Since most antigens have many determinants, several different clones of B and T cells may mount an immune response against the same antigen.

In addition to accounting for the specificity of the immune response, clonal selection is also important in the ability of the immune system to "remember" antigens it has encountered before.

Immunological Memory

Selective proliferation of lymphocytes to form clones of effector cells keyed to a newly introduced antigen constitutes the **primary immune response.** This primary response is characterized by a lag period of several days, since it takes that long for lymphocytes selected by the antigen to proliferate and differentiate into effector B and T cells ready to actually counteract the antigen. In the case of B cells, the lag period is evident in the delayed appearance of antibodies in the blood plasma after an antigen has been introduced to the body for the first time (Figure 39.10). If the body is exposed to the same antigen at some later time,

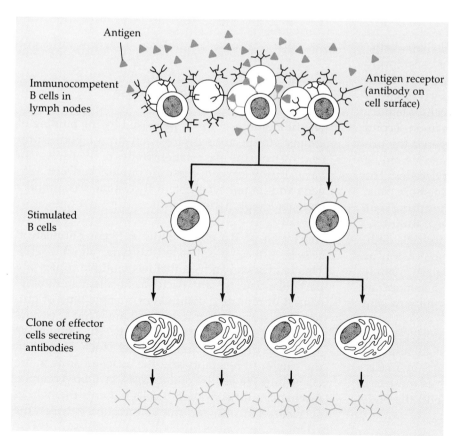

Figure 39.9
Clonal selection. Lymph nodes are populated by diverse lymphocytes with different antigen receptors on their surfaces. With multiple copies of a single type of receptor protruding from its plasma membrane, each lymphocyte is committed to recognizing and reacting to one kind of antigenic determinant. Thus, the specificity of a lymphocyte is predetermined before the cell ever encounters an antigen. When an antigen does invade the body fluids, it selectively activates, by binding to the receptors, the tiny fraction of lymphocytes keyed to that antigen's determinants. Primed by this interaction with the antigen, the selected lymphocytes begin to grow and divide, producing a clone of effector cells specialized for defending against the very antigen that triggered the response. This diagram represents the cloning of B cells to form effector cells that fight the infection by secreting specific antibodies, but the specificity of the cell-mediated response by T cells is also based on clonal selection.

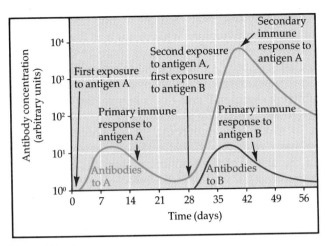

Figure 39.10
Immunological memory. In this example, the first exposure to antigen A produces a primary immune response that is characterized by a lag and a relatively small response. A second exposure to antigen A at day 28 produces a faster and greater secondary immune response. The specificity of the secondary immune response is demonstrated by injecting antigen B, which elicits a primary rather than a secondary response. Although this graph traces the humoral response, as indicated by antibody concentration in blood plasma, immunological memory is also a feature of the cell-mediated response.

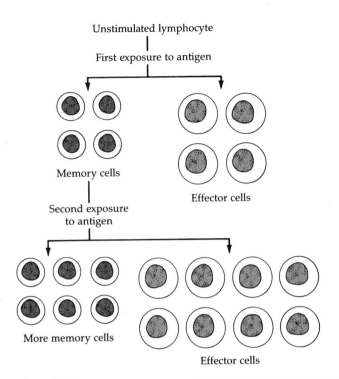

Figure 39.11
Cellular bases of immunological memory. First exposure to an antigen stimulates a lymphocyte, which produces both memory and effector cells. The effector cells are responsible for the primary immune response. Another exposure to the same antigen reactivates memory cells, which, already primed by the first exposure to the antigen, quickly produce more memory cells and large numbers of effector cells that mediate the secondary immune response.

there is a **secondary immune response,** which is faster, more effective, and more prolonged than the primary response.

The secondary response is based on **memory cells,** which are produced along with effector cells during the primary response (Figure 39.11). Only the effector cells combat the antigen during the primary response—by secreting antibodies, in the case of B cells. Memory cells give rise to additional effector cells upon second exposure to the same antigen that triggered the primary response. Effector cells usually survive only a few days, but memory cells may live for decades. The secondary response is mounted when memory cells, already sensitized by one exposure to the appropriate antigen, proliferate rapidly to produce additional memory cells and a new clone of effector cells dedicated to countering that antigen. In some cases, the production of memory cells during a primary immune response confers lifetime immunity for the animal, as in such human childhood diseases as mumps and chicken pox.

Clonal selection and the role of memory cells are features of immunity common to the humoral response by B cells and to the cell-mediated response by T cells. Let us now see in more detail how each of these two arms of the immune system helps defend the body against pathogens.

The Humoral Immune Response

The humoral response is provoked by the binding of antigens to specific receptors, actually antibodies protruding from the plasma membranes of B cells. As you have already learned, these encounters between antigens and immunocompetent B cells most commonly occur in lymph nodes.

Activation of B Cells   Some antigens stimulate B cells by a mechanism called **capping** (Figure 39.12). This requires an antigen that has multiple copies of a particular determinant. Bacterial flagella, for example, consist of repeating subunits of a single type of protein. The serial antigenic determinants bind to several antibodies on the surface of the B cell, pulling the antibodies closer together. The B cell is activated when the cluster of receptors and the attached antigen are taken into the cell by endocytosis. A more common mechanism for activating B cells requires the help of specialized T cells, and this cooperation between the

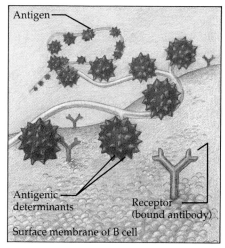

Figure 39.12

The capping hypothesis. Polyvalent antigens, those having multiple copies of identical determinants, can bind to several receptors, specific antibodies projecting from the surface of the immunocompetent B cell. According to the capping hypothesis, the polyvalent antigen pulls the receptors together to form a cap, which is taken into the cell by endocytosis. This may be one way that some antigens activate B cells.

cell-mediated and humoral arms of immunity will be discussed in a later section.

We have seen that each B cell is coated with a single type of receptor representing the specific antibody that cell is capable of producing, and thus the presence of a particular antigen selectively activates only those cells programmed to mount an immune response against that antigen. The activated cells begin to grow and multiply, generating a large clone of cells primed for response to a specific antigen. This clone includes memory cells, but most of the cells differentiate to become **plasma cells,** the effector cells of humoral immunity (Figure 39.13). Although B cells themselves secrete very few antibodies, the plasma cells that develop from B cells may secrete as many as 2000 antibody molecules per second for the 4- or 5-day lifetime of the cell. These discharged antibodies, identical in antigen-specificity to the receptors on the cell surface, circulate in the blood and lymphatic fluid, binding to antigens and contributing to their destruction.

Antibody Structure Antibodies constitute a class of proteins called **immunoglobulins,** abbreviated **Ig.** Every antibody molecule has the ability to recognize and bind to an antigen as well as an effector mechanism to assist in the destruction and elimination of that antigen. These two functions are reflected in antibody structure: Every antibody molecule consists of discrete regions that carry out one of the two functions. Since different antibodies must be able to recognize different antigenic determinants, the regions that bind antigens differ distinctly from one antibody to another and are called **variable (V) regions.** In contrast, only a few effector mechanisms are involved in antigen elimination, so there are only a few different kinds of regions involved in these functions. These are **constant (C) regions.**

A typical antibody molecule consists of two pairs of polypeptide chains—two short identical **light (L) chains** and two longer identical **heavy (H) chains.** The chains are joined by disulfide bridges and noncovalent associations to form a Y-shaped molecule (Figure 39.14). Both H and L chains have variable sections located at the ends of the two arms of the Y, where they form an area called the **antigen-binding site.** There are thus two binding sites per antibody molecule, one at the tip of each of the antibody's two arms. The dimensions and contours of the binding sites are determined by the unique amino acid composition of the V regions. The association between an antibody and an antigenic determinant resembles the association between an enzyme and its substrate: Several weak bonds form between contiguous chemical groups on the respective molecules. However, in contrast to the pocketlike active site of an enzyme, which changes its conformation to embrace a substrate, the binding site of an antibody is a relatively flat, rigid surface.

The rest of the antibody molecule, the part consisting of constant regions of the H and L chains, determines the antibody's effector function. There are five types of constant regions, and hence five major classes of antibodies in mammals: IgM, IgG, IgA, IgD, and IgE (Table 39.1, p. 862). Each class plays a different role in the immune response; for example, IgG attacks

Figure 39.13

Figure 39.13
The humoral immune response. An antigen selects immunocompetent B cells having receptors that fit the antigen's determinants. Cloning of these activated B cells during the primary response produces antibody-secreting plasma cells and memory cells, which persist after the first infection and which mount a secondary response if the same antigen challenges the body again.

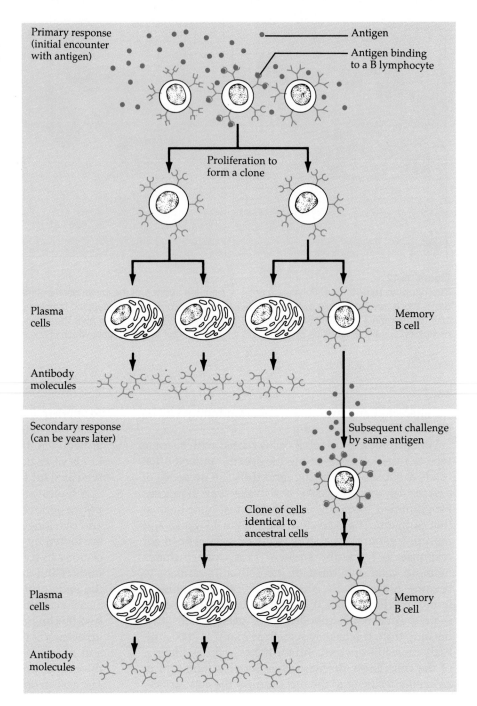

bacteria and viruses, whereas IgE triggers allergic reactions. Within each class, of course, are a great variety of specific antibodies characterized by their unique antigen-binding sites.

Humoral Effector Mechanisms Antibodies do not generally possess the power to destroy antigen-bearing invaders directly. Instead, the soluble antibodies "tag" foreign molecules and cells for destruction by a variety of effector mechanisms (Figure 39.15). Each mechanism is triggered by the selective binding of antibodies to antigens to form **antigen–antibody complexes.** The simplest effector mechanism is neutrali-

zation, in which the binding of antibodies simply blocks the harmful chemical groups of viruses and bacterial toxins. Phagocytes then dispose of the neutralized antigen–antibody complexes. Another effector mechanism is the agglutination of bacteria and other foreign cells by antibodies; because each antibody has two or more binding sites (depending on the Ig class), antibodies can cross-link invading cells to form clumps that are easily engulfed by phagocytes. A similar mechanism is precipitation, the cross-linking of soluble antigen molecules (rather than cells) to form immobile precipitates that are captured by phagocytes. One of the most important effector mechanisms

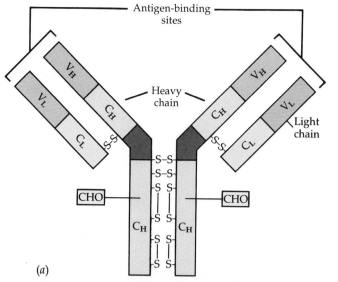

(a)

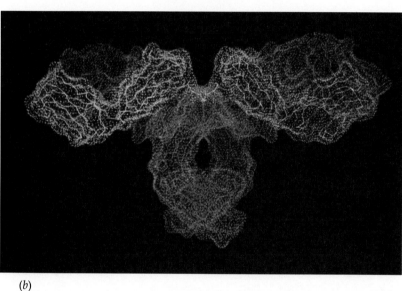

(b)

Figure 39.14
Structure of an antibody molecule.
(a) An antibody is a Y-shaped protein consisting of two identical heavy chains and two identical light chains bound together by disulfide bridges (S–S). Carbohydrate groups (CHO) that may help determine the molecule's effector functions are attached to the heavy chains. Most of the molecule is constant (C), being the same in all antibodies of the same major class. The antigen-binding sites lie within the variable regions (V) at the ends of the two arms. (b) Computer graphic image of an antibody.

of the humoral response is the activation of the complement system by antigen–antibody complexes. Recall from earlier in the chapter that various complement proteins incite the inflammatory response, cause foreign cells to lyse by forming a membrane attack complex, and opsonize antigens for easier consumption by phagocytes.

In these multiple effector mechanisms of humoral immunity are many examples of how the various components of defense, both specific and nonspecific mechanisms, cooperate. Most pathogens are actually destroyed by phagocytes and complement, which are nonspecific, but the pathogens are identified as foreign to the body by the specific binding of antibodies to antigens.

Monoclonal Antibodies   Because of their specificity and affinity for antigens, antibodies have been widely used in biological research and in clinical testing to detect molecules of interest. Until recently, however, it was very difficult to isolate antibodies of a single

kind (that is, antibodies specific for a single antigenic determinant). The usual procedure was to make antibodies by injecting a small sample of the antigen to be detected into a rabbit, mouse, or goat. The animal would then mount an immune response that produced antibodies against the antigen. The problem was that an animal produces more than one type of antibody against a single antigen because the antigen carries many different antigenic determinants. Moreover, to make any reasonable amount of antibody and to ensure a steady supply of it, more than one animal had to be used as a source.

Such problems sharply limited the use of antibodies until the late 1970s, when a technique for making **monoclonal antibodies** was developed. The term *monoclonal* means that all the cells producing such antibodies are descendants of a single cell; thus, they all produce identical antibody molecules. The ability to make such uniform antibodies has spawned a new industry. A typical application has been in clinical tests, where monoclonal antibodies provide a reliable reagent

Table 39.1 The five classes of immunoglobulins

IgM
(pentamer)

IgM is produced during the early response to an invading microorganism. It is the largest immunoglobulin, altogether containing five Y-shaped units (monomers) of two light and two heavy polypeptides each. The units are held together by a component called a J chain. The relatively large size of IgM restricts it to the bloodstream, where these antibodies combat infection by agglutinating antigens and activating complement.

IgG
(monomer)

IgG molecules are the most abundant of circulating antibodies. They consist of a single Y-shaped monomer and can traverse capillary walls rather readily; IgG also crosses the placenta to carry some of the mother's immune protection to the developing fetus. High concentrations of antigen must be present before a particular IgG is made, which usually occurs when an infection is well established. IgG cooperates in the destruction of foreign cells by activating complement.

IgA
(monomer
or dimer)

IgA is found in body secretions, including saliva, sweat, and tears, and along the walls of the intestines. It is also the major antibody of colostrum, the initial secretion from a mother's breasts after birth, and of milk. IgA occurs as a monomer or as double-unit aggregates of the Y-shaped protein molecule. IgA molecules tend to be arranged along the surface of body cells and to combine there with antigens, such as those on a bacterium, thus preventing the foreign substance from directly attaching to the body cell. The invading substance can then be swept out of the body together with the IgA molecule.

IgD (monomer)

IgD molecules, which occur as monomers, are almost always bound to the plasma membranes of B cells. They are probably the antigen receptors of immunocompetent B cells.

IgE
(monomer)

Less is known about the IgE immunoglobulin. IgE is associated with some of the body's allergic responses, and its levels are elevated in individuals known to have allergies. The constant regions of IgE molecules can bind tightly to mast cells, a type of connective tissue cell that releases histamines as part of the allergic response.

---

for assaying medically important antigens, such as bacteria that cause sexually transmitted diseases or the hormones indicative of pregnancy. A monoclonal antibody is also being used in bone marrow transplants to destroy those cells in the donor's bone marrow thought to attack the recipient's cells. It is also possible to couple a monoclonal antibody to a toxin, and then use the specificity of the antibody to target the toxin for a particular type of cell. Such "guided missiles" may soon be useful as chemotherapy to treat cancer and other diseases.

The technology that makes monoclonal antibodies possible is the fusion of two cells to form a hybrid cell called a **hybridoma,** which is then cultured on an artificial medium to produce a mass of cells (see Methods Box, p. 864). The two types of cells from which a hybridoma is made have distinct abilities, neither of which can be exploited alone. One cell type is derived from a tumor known as a myeloma. These tumor cells grow readily and indefinitely in culture dishes in the

laboratory, unlike many normal cells, which do not survive under such conditions. The other type of cell used in the fusion is a normal antibody-producing lymphocyte, obtained from the spleen of an animal that has been exposed to the antigenic determinant of interest. Fusing the two kinds of cells enables researchers to produce a hybridoma that makes a single type of antibody but is also able to grow successfully and indefinitely in culture.

Cell-Mediated Immunity

Many pathogens, including all viruses, are obligate intracellular parasites that can reproduce only within the cells of the host. The humoral immune response helps the defense network identify and destroy free pathogens, but it is the cell-mediated arm of immunity that battles pathogens that have already entered cells.

In contrast to B cells, T cells, the main agents of cell-mediated immunity, cannot be activated by free anti-

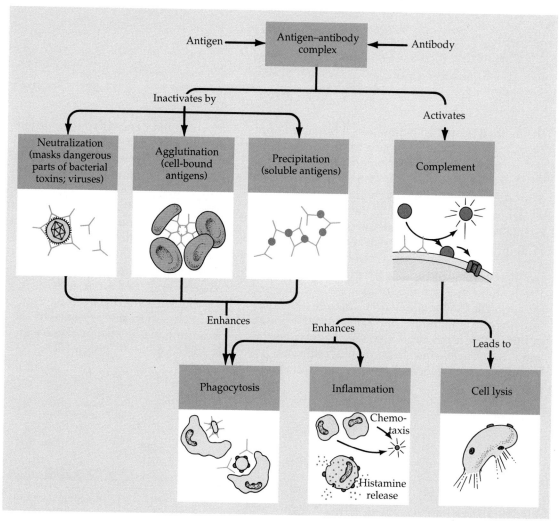

Figure 39.15
Effector mechanisms of humoral immunity. Formation of antigen–antibody complexes tags foreign cells and molecules for destruction by complement and phagocytes.

gens present in body fluids. Immunocompetent T cells can respond only to antigenic determinants displayed on the surfaces of the body's own cells. The ability to recognize these bound antigens is based on the presence of **T-cell receptors,** specific proteins embedded in the plasma membranes of T cells.

T-cell Receptors and Histocompatibility Restriction
A T-cell receptor actually recognizes a combination of the antigen along with one of the body's own "self" markers (Figure 39.16). "Self" is signaled by the **major histocompatibility complex (MHC),** a group of proteins, unique to the individual, that is present on the surface of cells. There are two classes of MHC markers, termed MHC I and MHC II. **MHC** I markers are found on all nucleated cells of the body. **MHC** II markers—found only on macrophages, B cells, and some T cells—enable the cells of the immune system to recognize one another in interactions.

Some cells can process antigens and display them on the cell surface along with an MHC protein. Among these **antigen-presenting cells (APCs)** are macrophages, which ingest foreign objects and then attach portions of the antigens to MHC proteins on the cell surface. It is neither the antigenic determinant alone nor the MHC protein that the T-cell receptor "sees," but rather a "self–nonself" complex of the two molecules together. The MHC protein that presents the antigen is shaped something like a hammock that nestles the antigen. An MHC molecule can associate with a variety of antigens. The T-cell receptors, however, are very specific. The receptors of each cell can bind only to an MHC protein that is itself associated with a specific antigenic determinant. This constraint on a T cell's responsiveness, which is acquired during the development of immunocompetence in the thymus, is called **histocompatibility restriction.**

The MHC–antigen complex is like a red flag to T cells, calling them into action against cells infected by the pathogen represented by that particular antigen.

METHODS: PRODUCTION OF MONOCLONAL ANTIBODIES WITH HYBRIDOMAS

The method described here makes it possible to produce large quantities of pure antibodies reactive against one antigenic determinant. The application illustrated here is the preparation of monoclonal antibodies to detect human chorionic gonadotropin (HCG), a hormone present in the blood or urine of pregnant women. A mouse injected with an antigenic determinant characteristic of HCG will begin to produce antibodies against the antigen. The spleen can then be removed and its B lymphocytes isolated. The lymphocytes are mixed with mutant myeloma (tumor) cells having a mutation that prevents them from surviving without a particular nutrient present in their growth medium. Some of the lymphocytes and myeloma cells fuse to become hybridomas. The hybridoma cells are identified when all cells are grown in a medium deficient in the nutrient needed by the mutant myeloma cells. Any unfused myeloma cells die, but the hybridoma cells live because the lymphocyte DNA supplies the necessary nutrient gene and its product. Each hybridoma clone is tested for production of antibodies that react with the specific antigenic determinant. Clones that test positive are isolated and cultured for large-scale production of the antibody.

Monoclonal antibodies against HCG are the reagents now used for very sensitive pregnancy tests; a positive reaction between the antibodies and blood or urine shows the presence of the HCG antigen, indicating that the woman who supplied the sample is pregnant.

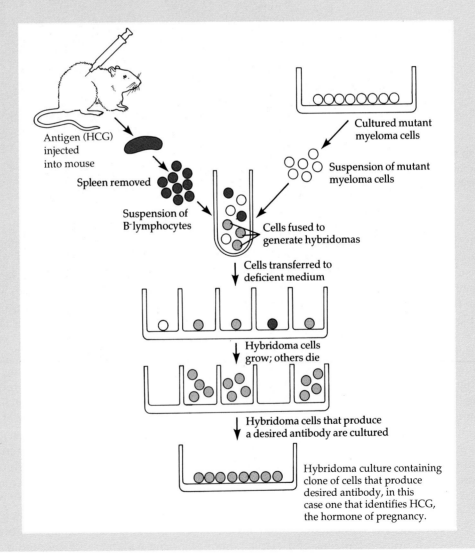

Hybridoma culture containing clone of cells that produce desired antibody, in this case one that identifies HCG, the hormone of pregnancy.

Figure 39.16

A T-cell receptor. (a) T cells are activated when their receptors recognize a unique combination of "self" and "nonself" molecules on the surfaces of the body's own cells. The "self" component is a surface glycoprotein belonging to the major histocompatibility complex (MHC), a kind of molecular fingerprint unique to the cell surfaces of each individual. Antigen-presenting cells (APCs), including macrophages, display portions of antigens along with MHC proteins, and it is this "self–nonself" combination that is actually identified by a T-cell receptor. Each T cell is restricted by its specific type of receptor to identifying a single antigenic determinant presented in the context of the MHC protein. This "double keying" of the T cell is analogous to the opening of a safety deposit box, which can be done only by using the banker's key along with your specific key. (b) An MHC I protein is shaped something like a hammock. The antigen presented by an infected cell probably nestles in the groove of the MHC protein, between two ridges consisting of helical portions of the protein. It is the overall shape of the MHC–antigen complex that is recognized by the T-cell receptor.

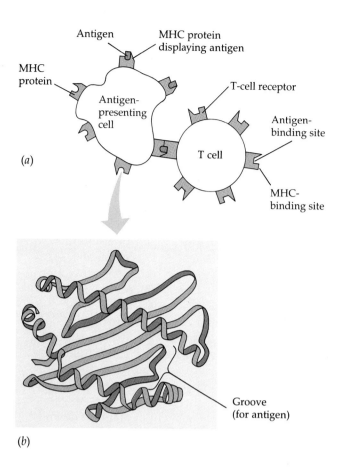

The situation stimulates T cells that have the appropriate receptors to proliferate to form a clone of cells specialized for fighting a particular pathogen. The activated products include memory cells and effector cells called **cytotoxic T cells,** which actually attack infected cells. But there are also two types of T cells that have regulatory functions in immunity—**helper T cells** and **suppressor T cells.** Let us first examine the helpers, which play such an important role in mobilizing the entire immune response.

Helper T Cells   A key episode in defense is an interaction between antigen-presenting macrophages and helper T cells (Figure 39.17). Macrophages that have

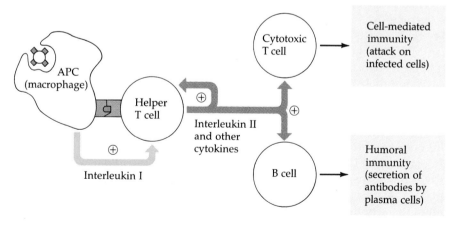

Figure 39.17

The central role of helper T cells. Helper cells mobilize both humoral and cell-mediated arms of the immune response. The T-cell receptor of the helper recognizes the MHC–antigen complex displayed on the surface of an antigen-presenting cell, usually a macrophage. The macrophage secrets interleukin I, a chemical signal that contributes to the activation of the helper T cells. An activated T cell grows and divides, producing a clone of helper T cells, all with receptors keyed to MHC markers in combination with the specific antigen that triggered the response. The helper cells discharge a second chemical signal, interleukin II, which amplifies the cell-mediated response by stimulating proliferation and activity of other helper T cells, all specific for the same antigenic determinant. Interleukin II also helps activate cytotoxic T cells and B cells.

ingested a pathogen or foreign molecule process the antigens in some manner and then transfer the processed antigens back to the cell surface, where the antigens are displayed along with MHC proteins. As these antigen-presenting cells meet lymphocytes in lymph nodes, the APCs are recognized by those helper T cells bearing complementary receptors. Binding of a helper T cell to a macrophage causes that macrophage to release a chemical signal called **interleukin I,** which in turn stimulates growth and cell division of the T cell. Interleukin I is an example of a **cytokine,** the general term for a chemical that is secreted by one cell as a regulator of neighboring cells. The activated T cells also release a cytokine, called **interleukin II.** In an example of positive feedback (see Chapter 36), interleukin II stimulates helper T cells to grow and divide even more rapidly. This cytokine also amplifies the proliferation of cytotoxic T cells.

In addition to their important function in mobilizing cell-mediated immunity, helper T cells also figure prominently in the ability of B cells to mount a humoral response against antigens. You read earlier in the chapter that immunocompetent B cells can be activated directly by antigens that cause capping—that is, antigens with a serial arrangement of identical determinants. Such foreign molecules are called **T-independent antigens.** More common are **T-dependent antigens,** those that can activate only B cells that have also been stimulated by interleukin II and other cytokines released from helper T cells. This double activation by antigens and helper cells results in a clone of plasma cells, which secrete antibodies against free antigens. In addition to arming the humoral defense, of course, helper T cells have bolstered cell-mediated immunity by stimulating cytotoxic T cells.

Cytotoxic T Cells As the effectors of cell-mediated immunity, cytotoxic cells are the only T cells that actually kill other cells. They identify their targets by the fit of their specific receptor to an MHC–antigen complex. Cells displaying fragments of antigens have probably been infected by viruses or other microorganisms bearing those antigens. A cytotoxic T cell with a receptor specific for that particular antigenic determinant, combined with an MHC protein, can dock on the surface of the infected cell. The T cell then kills the target cell by releasing **perforin,** a protein that inserts in the plasma membrane of the target and forms a lesion that causes the infected cell to lyse (Figure 39.18). This effector mechanism resembles the actions of complement and natural killer cells, but those defensive forces, remember, are nonspecific; they do not distinguish one type of infection from another. In contrast, cytotoxic T cells, as part of the specific defense, recognize and destroy only a cell displaying antigenic determinants that represent a particular infectious agent. In addition to killing infected cells, cytotoxic T cells prob-

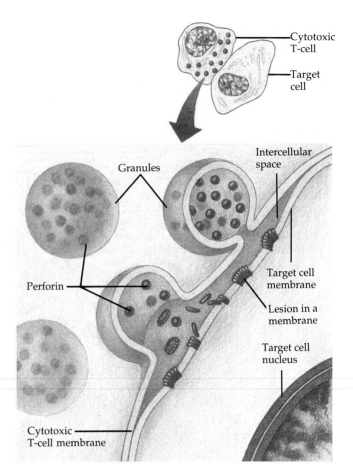

Figure 39.18
How cytotoxic T cells kill their targets. Recognizing a specific antigenic determinant displayed on the membrane of the target cell along with an MHC protein, the cytotoxic cell docks on the surface of the target and discharges a protein called perforin. The perforin molecules insert in the membrane of the target cell and form a lesion, causing the cell to lyse.

ably attack cancer cells and the foreign human cells introduced by tissue grafts and organ transplants (see Figure 39.1).

Suppressor T Cells Suppressor T cells release cytokines that inhibit the activity of other T cells as well as B cells. This probably occurs late in an immune response, after the effector cells have eliminated the threatening agent. Thus, suppressor T cells terminate immune activities that are no longer required. Of course, the primary response has left a population of memory cells that are ready to mount a rapid, vigorous response should the same infectious agent enter the body again.

Now that we have examined the mechanisms of humoral and cell-mediated responses to foreign cells and molecules, let us take a closer look at how the body actually distinguishes self from nonself.

SELF VERSUS NONSELF

So far, we have emphasized the role of the immune system in recognizing foreign substances and fighting infections. Implicit in the immune system's ability to selectively battle foreign molecules and cells, however, is its capacity to recognize the body's own molecules as native—that is, to distinguish "self" from "nonself." The mechanism for chemical recognition involves molecular identification tags on the surface of cells. As immunocompetence develops in the fetus, those lymphocytes displaying receptors keyed to chemical markers present in the body at that time are apparently destroyed, leaving only those lymphocytes capable of mounting immune responses against foreign molecules. Thus, the ability of the immune system to distinguish self from nonself is established by the time of birth.

Perhaps it is not surprising that the immune system can distinguish the animal's own cells from a virus or bacterium, but the ability of the system to discriminate self from nonself is so sophisticated that immunological war is waged against cells from other individuals of the same animal species. The chemical markers that determine blood groups and the proteins of the histocompatibility complex are among the molecules most important in immune responses against cells of another individual.

Blood Groups

Blood group is one important facet of a person's chemical identity. The genetic basis for ABO blood groups has already been discussed, in Chapter 13. Individuals of group A have the A antigen on the surface of their red blood cells; group B individuals have the B antigen; group AB individuals have both A and B antigens; and group O individuals have neither A nor B antigens. These surface molecules are antigenic only in the sense that they could be foreign if injected into someone else. An individual belonging to the B group does not produce antibodies against the B-type antigen, but does produce antibodies against A (see Figure 13.13). Thus, if such an individual were to receive a blood transfusion containing the A antigen, the antibodies present in the type B recipient would cause the donated blood to agglutinate; that is, antibodies and blood cells would form clumps in the bloodstream and in the body's organs—a life-threatening reaction. AB individuals can receive blood from all donors because they carry both antigens on their blood cells and do not make either anti-A or anti-B antibodies. Group O individuals, on the other hand, make both anti-A and anti-B antibodies; hence, the only blood they can receive is group O. However, group O indi-

viduals are known as universal donors because their blood cells carry neither A nor B antigens, and thus their blood cells will not cause an adverse reaction in individuals who produce the anti-A and anti-B antibodies.

Another blood group antigen, the **Rh factor,** is most notorious in cases where antibodies produced by the immune system of a pregnant woman react with the blood of her developing fetus. The situation arises if the mother is Rh-negative (she lacks the Rh determinant on her own red blood cells), but her unborn child has inherited the factor from the father and is thus Rh-positive. The mother develops antibodies against the Rh factor when small amounts of fetal blood cross the placenta and come in contact with her lymphocytes, usually late in pregnancy or during delivery of the baby. Typically, her response to this first exposure is mild and without medical consequences for the baby. The real danger occurs in subsequent pregnancies, when the mother's immune response against the Rh factor has already been primed (because of immunological memory) and her antibodies can cross the placenta during the final weeks of gestation to destroy the red blood cells of the fetus. As a precaution, the mother may be injected with anti-Rh antibodies after delivering her *first* Rh-positive baby. The antibodies destroy the Rh-positive red cells before the mother becomes sensitized and develops immunological memory for the Rh antigen.

The Major Histocompatibility Complex

Especially important in the ability of the immune system to distinguish native cells from those of another individual is the major histocompatibility complex (MHC), which was introduced earlier in this chapter in connection with cell-mediated immunity. The MHC consists of a set of several cell-surface molecules encoded by a family of genes. Because there are about 50 or more alleles for each gene and several MHC genes (at least 20), it is virtually impossible for any two people to have matching sets of MHC markers on their cells (unless the people are identical twins). Thus, the major histocompatibility complex is a biochemical fingerprint unique to each individual.

The MHC complicates tissue grafts and organ transplants. Foreign MHC molecules are antigenic to the immune system, causing T cells to mount a cell-mediated reponse against the donated tissue or organ. The "rejection" of the transplanted organ is really an attack on the foreign cells by cytotoxic T cells. To minimize rejection, attempts are made to match the MHC of organ donor and recipient as closely as possible, but a perfect match is possible only between identical twins. Various drugs are therefore used to suppress the immune response when most transplants are performed, but these drugs are unselective and compromise the ability of the immune system to fight infections. One drug, cyclosporine, has the advantage of suppressing only cell-mediated immunity, without crippling humoral immunity.

Note that the body's disastrous reaction against an incompatible blood transfusion or a transplanted organ is not a disorder of the immune system, but a normal action taken by a healthy immune system exposed to foreign chemicals. However, as with any complex system, the immune apparatus does malfunction in some individuals, resulting in a variety of serious disorders.

DISORDERS OF THE IMMUNE SYSTEM

Autoimmunity

Sometimes, the immune system goes awry and turns against self, leading to a variety of **autoimmune diseases.** In systemic lupus erythematosus, for example, people develop immune reactions against components of their own cells, particularly nucleic acids. One result is inflammation of joints, producing arthritis-like symptoms. People with the disease, predominantly young women, must take immunosuppressive drugs and anti-inflammatory agents, both of which tend to damage the kidneys, as does the disease itself. Other diseases that probably involve autoimmunity are rheumatoid arthritis, rheumatic fever, and juvenile diabetes.

Autoimmune disorders are not well understood, but they may involve several mechanisms. In some cases, antibodies or effector T cells that respond to a foreign invader may cross-react with the individual's own tissues. For example, some of the damage caused by rheumatic fever occurs when antibodies produced in response to a streptococcal infection coincidentally react against the muscle tissue of the heart. In other cases, viruses, drugs, or genetic mutations may alter the surface components of some cells enough for them to be identified as foreign by the immune system. Still another cause of autoimmunity may be the unmasking of body substances that were previously hidden from the immune system. For example, the cornea normally has no contact with the blood. But if injury to the cornea exposes certain molecules, which are then perceived as foreign to the immune system, then an autoimmune reaction may cause the cornea of the eye to become opaque. In any case, the results of mutiny in the immune system are likely to be devastating.

Immunological Aspects of Cancer

Cancer is really a variety of diseases affecting many different cell types in the body. Regardless of their multiple causes and origins, however, all cancers involve changes in normal body cells, and some of these changes take place along the outer membrane surfaces of the cells. Such changes can result in the immune system identifying the cancer cell as a foreign intruder. Some antigenic determinants are widely distributed on cancer cells; others are peculiar to a particular cell type.

Individuals with deficiencies of the immune system are often far more susceptible to cancer than individuals with healthy immune systems, an observation that suggests that this system may play an important role in fighting cancer. Studies with animals have also linked a deficient immune system to an increased incidence of malignant disease. Such research has led scientists to speculate that the immune system plays a watchdog role against cancer. This theory, called the **immune surveillance theory,** suggests that the immune system is constantly finding and destroying cancer cells in very early stages of development, protecting the body from a nearly constant threat of malignancy. Why this surveillance system sometimes fails, allowing tumors to develop, is still largely a mystery, although several plausible mechanisms have been identified in laboratory studies. Formation of a tumor may involve immunological escape by the cancer cells, either because they shed the surface molecules that marked them as foreign, or because they secrete chemicals that suppress the entire immune system.

The implication that the immune system may help combat cancer is the basis for some chemotherapy, employing drugs that stimulate the immune response. So far, research on the immunological factors in cancer has done little more than reveal just how much remains to be learned, both about cancer and about the complex components of the immune system.

Allergy

Allergies are hypersensitivities of the body's defense system to certain environmental antigens, referred to in the case of allergies as **allergens.** Allergic reactions are typically very rapid and show extraordinary sensitivity to minute amounts of allergen. Allergic reactions can occur in many parts of the body, including the nasal passages and bronchi, the gastrointestinal tract, and the skin.

Antibodies of the IgE family participate in allergic reactions. These immunoglobulins are found as free antibodies in the blood and as bound receptors on the surfaces of mast cells, noncirculating cells found in connective tissue. When these attached IgE molecules bind to antigens, the mast cells respond with a process called **degranulation,** in which the mast cells release a flood of histamine and other inflammatory mediators (Figure 39.19). Remember, histamine causes dilation and leakiness of small blood vessels during an

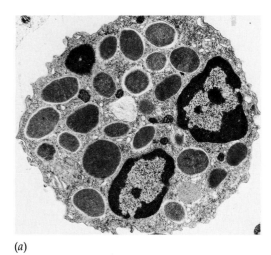

(a)

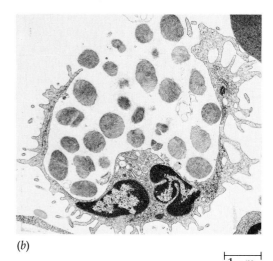

(b)

1 μm

Figure 39.19
Degranulation and allergies. (a) An intact mast cell has numerous cytoplasmic granules. **(b)** When an allergen binds to IgE molecules that are attached to the cell surface, the cell releases the granules' contents, including histamine. This triggers an inflammatory response associated with most of the symptoms of an allergy (TEM).

inflammatory response. Histamine and other compounds released in response to allergens causes some of the best-recognized symptoms of allergy: sneezing and nasal irritation, itchiness of the skin, and tearing of the eyes. Antihistamines are drugs that interfere with the action of histamine.

One of the most serious problems related to acute allergic responses is **anaphylaxis,** a severe and life-threatening response by the immune system to an allergen such as the venom injected by a bee sting. Mast cells can release their alarm substance so suddenly that the abrupt dilation of peripheral blood vessels causes a precipitous drop in blood pressure, leading to a potentially fatal shock reaction. Anaphylaxis and other severe allergic reactions can be counteracted with injections of the hormone epinephrine.

Immunodeficiency

A defect in any one of the many components of the immune system compromises defense. Individuals lacking or deficient in T cells are susceptible to viruses, certain bacterial infections, and cancer, whereas individuals lacking B cells are prone to extracellular infections, caused mainly by bacteria. The most serious disorder of the immune system is a rare congenital disease known as severe combined immunodeficiency (SCID), in which both T and B cells are absent or inactive. Individuals with this deficiency are extremely sensitive to even minor infections, and without successful bone marrow transplants, their only chance for survival is to lead an isolated existence behind protective barriers that keep out all infectious agents. The patient contacts only filtered air, sterilized food, and disinfected objects. Although bone marrow transplants can help some afflicted individuals, such operations are not always successful. Sometimes the transplanted cells do not survive and multiply in the new host, and sometimes they actually mount an immune attack against the host ("graft-versus-host response"), threatening the recipient's life.

There are a variety of other immunodeficiency diseases. For instance, certain cancers, such as Hodgkin's disease, affect the lymph node cells and can thus lead to depression of the immune system. Immunodeficiencies can also result from certain drug treatments, such as certain chemotherapy and radiation therapy used for treating cancer patients. As mentioned earlier, organ transplants must be followed by immunosuppressive therapy, which makes the patient more susceptible to infectious diseases.

AIDS

Acquired immune deficiency syndrome (AIDS) is an immune disorder characterized by a marked reduction in the number of helper T cells that activate other T and B lymphocytes. The immune system is weakened by this reduction, and in some cases people with AIDS cannot form antibodies in response to newly presented test antigens. Most individuals with full-blown AIDS die within 3 years of being diagnosed, usually from opportunistic infections or cancers that normally present no threat to a healthy immune system. A rare form of pneumonia caused by a protozoan, severe diarrhea, and various infections caused by viruses and fungi are among the diseases that accompany full-blown AIDS. The incidence of some rare cancers, including a cancer of blood vessel walls called Kaposi's sarcoma, is also much higher among AIDS patients than in the general population. In addition to full-blown AIDS, there is a transitional stage of the disease known as **AIDS-related complex (ARC),** which manifests itself in swollen lymph nodes, fever, night sweats, and weight loss. It is probable that most individuals with ARC will eventually develop full-blown AIDS.

AIDS was first recognized in 1981 when the Centers for Disease Control (CDC) in Atlanta began investigating reports of Kaposi's sarcoma occurring in homosexual men. The CDC then identified several high-risk groups, including sexually active homosexual and bisexual males, intravenous drug users, recipients of blood products, and sexual partners of people in these high-risk groups. The CDC also established that the routes of transmission involved intimate sexual contact, contaminated needles, and blood-to-blood contact (by a transfusion, for instance, or by transmission from mother to fetus via the placental blood supply). The number of AIDS cases has been doubling every year, with about 100,000 cases having been diagnosed in the United States by the end of 1989. About 60% of those individuals have died.

In 1984, scientists working in the United States and France independently identified the infectious agent responsible for AIDS, a virus now known as **HIV,** for **human immunodeficiency virus** (Figure 39.20; see also Figure 17.10). It infects and eventually kills helper T cells, delivering a crippling blow to the entire immune system. A glycoprotein named gp 120 on the surface of HIV binds specifically to a receptor called **CD4** on the surface of a helper T cell, and the virus then enters the cell. HIV also infects macrophages, including those in the brain; this may account for the high incidence of neurological disorders among AIDS victims. It is the presence of infected cells in blood and semen that spreads the virus from one individual to another. HIV has also been found at low levels in saliva and tears, but there are no known cases of transmission via these fluids. Indeed, there is substantial evidence that AIDS is very difficult to transmit by casual contact.

Those who have been exposed to HIV have circulating antibodies to the virus, and testing for the presence of these antibodies is the most common method for identifying infected individuals. This test is used to screen blood donations, for example. As a result of such blood screening, it has been learned that only a fraction of individuals who have been exposed to the virus presently display symptoms of AIDS or ARC. This is because the virus has a long incubation time of several months to several years before it manifests itself.

At this time, AIDS remains incurable. Some antiviral drugs, such as AZT, may extend the lifetimes of some patients considerably, but no drug yet tested rids the body of HIV. Many new drugs are useful for treating some of the opportunistic infections common in AIDS patients, but again, these drugs do not cure AIDS. Vaccines to protect uninfected individuals, and perhaps prevent asymptomatic infections from progressing to AIDS, are now being tested. The problem is that HIV apparently evolves very rapidly, and so a vaccine effective against one strain of the virus may

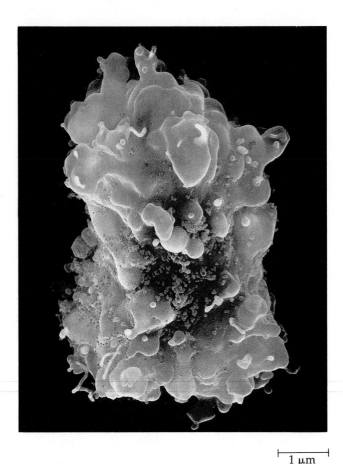

Figure 39.20
AIDS viruses attack a T cell. HIV viruses, artificially colored blue in this scanning electron micrograph, surround a T lymphocyte. The virus invades helper T cells, killing them and seriously weakening the immune system of the infected person.

1 μm

offer little protection against another strain. So far, educating people about unsafe sex and the risks of sharing needles to inject drugs offers the best hope for controlling the AIDS epidemic. Perhaps it is because so many people still erroneously think of AIDS as a disease of homosexuals and drug addicts that much of the information about high-risk behavior has so far fallen on deaf ears; the rate of new infections is increasing significantly among heterosexuals who do not inject drugs. Anyone who has sex with a person whose entire sexual history for the past decade is unknown risks exposure to HIV. Condoms minimize but do not completely eliminate the risks.

Stress and Immunity

Two thousand years ago, Galen, a Greek physician, recorded that people suffering from depression were more likely than others to develop cancer. At one time or another, most of us have been objects of the admonishment: "Don't get run down. It'll lower your resistance and you'll get sick." Indeed, there is growing evidence that physical and emotional stress can compromise immunity. Hormones secreted by the adrenal gland during stress affect the numbers of white blood cells and may suppress the immune system in other ways. People who report poor marital quality have a lower ratio of helper T cells to suppressor T cells than do those who are satisfied with their personal relationships. In one study, a group of students was examined just after a vacation and then again during final exams, and it was found that defense systems were impaired during exam week—natural killer cells were less effective and less interferon was produced.

A network of nerve fibers penetrates deep into lymphoid tissue, including the thymus, and receptors for chemical signals secreted by nerve cells have been discovered on the surfaces of lymphocytes. This suggests a direct link between the nervous and immune systems. Some of these chemical signals, secreted by nerve cells when we are relaxed and happy, may actually enhance immunity. These and other observations and speculations have led physiologists to take a serious look at how general health and state of mind affect immunity, giving birth to a new field, psychoneuroimmunology.

DEFENSE IN INVERTEBRATES

This chapter has emphasized the nonspecific and specific defenses as they occur in vertebrates, because relatively little is known about how invertebrates defend against pathogens that have penetrated the skin and other outer barriers. However, experiments have established that one fundamental facet of defense, the ability to distinguish "self" from "nonself," is well developed in invertebrates. For example, if the cells of two different sponges of the same species are mixed, the cells from each individual sort themselves out and aggregate, excluding cells from the other individual. In many invertebrates, amoeboid cells called coelomocytes identify and destroy foreign materials.

Studies of tissue grafting in earthworms have shown a memory response in the defense systems of these annelids. When a portion of body wall is grafted from one worm to another, the recipient's coelomocytes attack the foreign tissue. If donor and recipient are from the same population, then the grafted tissue survives for about eight months before it is completely rejected. However, if a worm receives a graft from a donor taken from a distant location, the graft is rejected in just two weeks. If a second graft from that same donor to the same recipient is attempted, it takes coelomocytes less than a week to eliminate the foreign tissue. Additional research on the defense systems of invertebrates may help biologists understand how the vertebrate immune system evolved.

* * *

The immune response is one of many adaptive processes that enable animals to adjust to the adversities of the environment. The next chapter describes several other processes that help maintain favorable conditions within animals as they cope with varying and sometimes harsh external environments.

STUDY OUTLINE

Nonspecific Defense Mechanisms (pp. 849–853)

1. Nonspecific defense mechanisms include physical and chemical barriers, phagocytes, inflammatory responses, and antimicrobial proteins.
2. The first line of defense against attack by bacteria and viruses is the physical barrier of the skin and mucous membranes, aided by lysozyme in sweat, tears, and saliva. Gastric juices kill bacteria that reach the stomach, and the ciliated mucous membranes of the respiratory tract trap and expel particles.
3. Macrophages are phagocytes that devour bacteria, viruses, and cellular debris. Natural killer cells attack the body's own infected or aberrant cells.
4. An inflammatory response is triggered by tissue damage. Injured cells release histamine, a chemical signal that dilates blood vessels and increases capillary permeability, allowing large numbers of phagocytic white blood cells to enter the interstitial fluid.
5. Interferons are proteins produced by virus-infected cells that diffuse to neighboring cells and inhibit those neighboring cells from making viral proteins.

6. Complement consists of proteins that circulate in the blood and are activated by an immune response or by contact with microorganisms. These proteins may amplify the inflammatory response, coat microbes to facilitate phagocytosis (opsonization), or lyse pathogens.

Specific Defense Mechanisms: The Immune Response (pp. 853–866)

1. The immune system distinguishes between "self" and "nonself" cell surface markers, and responds to specific antigens, or foreign molecules, by proliferating cells that either attack the invader or produce antibodies against the antigen.
2. Immunity is the capability of the immune response to "remember" an antigen and mount a rapid, more effective assault upon subsequent exposure to that pathogen. Vaccination primes the immune system's memory to protect against future exposure to disease organisms. In active immunity, the body has produced antibodies against a pathogen; passive immunity provides temporary protection, with antibodies supplied by an injection or passage through the placenta.

3. Humoral immunity involves circulating antibodies that defend against free bacteria and viruses. Cell-mediated immunity is carried out by specialized cells that target infected body cells.

4. Lymphocytes function in both humoral and cell-mediated immunity. B cells produce antibodies; T cells attack infected or defective cells. Developing lymphocytes that remain in the bone marrow differentiate into B cells; those that migrate to the thymus develop into T cells. As both types of cells mature, they develop immuno-competence—the ability to recognize one specific antigen. This programming of receptor proteins occurs before the cells have met their specific antigen.

5. Immunocompetent lymphocytes and macrophages await antigens in lymph nodes. Encountering their specific antigen, lymphocytes multiply into effector cells that move out to fight an infection.

6. Antigens are usually proteins or polysaccharides that are part of the outer covering of bacteria or viruses, or the cell membrane of foreign cells. Antibodies recognize localized regions of an antigen, called antigenic determinants.

7. Clonal selection occurs when a lymphocyte is activated by the binding of its specific antigenic determinant and proliferates to produce a clone of effector cells, all specific for that particular antigen.

8. In the clonal selection of T and B cells, long-lived memory cells are produced along with effector cells. The primary immune response has a lag period of several days. A secondary immune response is faster and more effective because the memory cells clone rapidly upon subsequent antigen exposure.

9. In humoral immunity, preprogrammed B cells exposed to their antigens are activated to produce clones of plasma cells and some memory cells. Plasma cells secrete antibodies, which circulate in blood and lymph and bind to antigens.

10. Antibodies, proteins called immunoglobulins (Ig), are Y-shaped molecules consisting of a pair of identical short, light polypeptide chains and a pair of longer, heavy chains. The variable regions of the chains at the ends of the Y's arms form antigen-binding sites. The constant regions of the H and L chains, which interact with an effector mechanism to destroy the antigen, fall into five major classes of antibodies, each with its own role in the immune system.

11. The selective binding of antibodies to antigens can trigger one or more humoral effector mechanisms. The antigen–antibody complex may neutralize toxic chemical groups, agglutinate bacteria, or precipitate antigen molecules—all of which are then engulfed by phagocytes, and activate the complement system.

12. Monoclonal antibodies, identical antibodies used in research and clinical testing, are produced from clones of hybridomas. These hybrid cells are produced by fusing tumor cells with specific lymphocytes.

13. Cell-mediated immunity is triggered when T-cell receptors on immunocompetent T cells recognize a complex of antigenic determinant and "self" markers (MHC) displayed by infected cells.

14. The binding of a helper T cell to an antigen-presenting macrophage causes the macrophage to release interleukin I, which stimulates division of the T cell. Activated T cells release another cytokine, called interleukin II, which amplifies the proliferation of both helper T and cytotoxic T cells. Cytotoxic T cells kill infected cells by releasing perforin, a protein that lyses the target cell's membrane. The interleukin II and other cytokines from the helper T cells also help activate B cells.

15. Late in an immune response, suppressor T cells release cytokines that inhibit the action of other T and B cells.

Self versus Nonself (pp. 867–868)

1. During fetal development, clones of lymphocytes capable of reacting against one's own chemical markers are apparently destroyed.

2. The antigens on red blood cells determine whether a person has A, B, AB, or O type blood. Individuals have antibodies for the antigens not displayed on their own blood cells, a necessary consideration in blood transfusions. The Rh factor is another blood-group antigen that may create difficulties when an Rh-negative mother carries successive Rh-positive fetuses.

3. Each individual has a unique major histocompatibility complex, a biochemical fingerprint coded for by several genes, each with numerous alleles in the population. Foreign MHC molecules on tissue grafts or organ transplants cause the immune system to reject the transplant.

Disorders of the Immune System (pp. 868–871)

1. Autoimmune diseases, such as lupus and rheumatoid arthritis, occur when an individual's immune system turns against its own body cells.

2. The immune system can identify cancer cells as nonself and, according to the immune surveillance theory, destroy them before they gain a foothold.

3. Allergies, hypersensitivities to environmental antigens called allergens, can range from minor irritation of the respiratory system to life-threatening anaphylaxis. When antigens bind with IgE antibodies, which are attached to mast cells, the mast cells release inflammatory compounds.

4. Immunodeficiency disorders occur when a component of the immune system is defective or suppressed by certain cancers or drug treatments.

5. AIDS, acquired immune deficiency syndrome, is a fatal immune system disorder caused by a virus (HIV) that attacks and kills helper T cells and macrophages, leaving the immune system defenseless against opportunistic infections or cancers. AIDS and ARC, a less severe stage of the disease, are transmitted by intimate sexual contact, contaminated needles, and other blood-to-blood contact.

6. The new field of psychoneuroimmunology examines the connection between stress and the immune system.

Defense in Invertebrates (p. 871)

1. Invertebrates have the ability to distinguish between self and nonself. Amoeboid cells called coelomocytes can identify and destroy foreign substances. Experiments in earthworms show that their defense systems can both reject and form a memory response to tissue grafts.

SELF-QUIZ

1. Which of the following molecules is *incorrectly* paired with its source?
   a. lysozyme—gastric juices
   b. histamine—injured cells

c. interferons—virus-infected cells

d. immunoglobulins—plasma cells

e. interleukin I—macrophage

2. Which of the following is *not* characteristic of the early stages of a localized inflammatory response?

a. increased permeability of capillaries

b. attack by cytotoxic T cells

c. release of clotting proteins

d. release of histamine

e. dilation of blood vessels

3. Immunocompetence refers to

a. the ability of the immune system to distinguish "self" from "nonself"

b. the differentiation, depending on location, of lymphocytes into B or T cells

c. the development of receptors on B and T cells that are specific for one particular antigenic determinant

d. the ability of helper T cells to recognize the major histocompatibility complex

e. the ability of the immune system to react to a second antigen exposure with a faster and more effective response

4. The major difference between humoral immunity and cell-mediated immunity is that

a. humoral immunity is nonspecific, whereas cell-mediated immunity is specific for particular antigens

b. the agents of humoral immunity are carried in the bloodstream, whereas the cells of the latter are concentrated in lymph nodes

c. humoral immunity cannot function independently; it is always activated by cell-mediated immunity

d. humoral immunity acts against free-floating antigens, whereas cell-mediated immunity works against pathogens that have entered body cells

e. humoral immunity, but not cell-mediated immunity, displays immunological memory

5. Monoclonal antibodies are

a. produced by clones formed from memory cells

b. used to produce large quantities of interferon

c. produced by cultures of hybridoma cells

d. produced by clones of T cells fused with tumor cells

e. produced by recombinant DNA methods

6. Antigenic determinants bind to which portions of an antibody?

a. variable regions

b. constant regions

c. only light chains

d. only heavy chains

e. the effector region

d. only heavy chains

e. the effector region

7. Which of the following molecules is *incorrectly* paired with its action?

a. interleukin I—stimulates division of helper T cells

b. interleukin II—increases proliferations of helper T and cytotoxic T cells

c. interferon—helps neighboring cells to resist viral infection

d. histamine—fights allergic reactions

e. lysozyme—attacks bacterial cell walls

8. Which of the following cells is *incorrectly* paired with its function?

a. plasma cell—produces antibodies

b. helper T cell—lyses foreign cells

c. memory cell—rapidly proliferates into clones of effector cells when it encounters antigen

d. macrophage—engulfs bacteria and viruses

e. cytotoxic T cell—releases perforin that lyses infected cell

9. Which blood transfusion would agglutinate blood?

a. A donor → A recipient   d. O donor → A recipient

b. A donor → O recipient   e. O donor → AB recipient

c. A donor → AB recipient

10. The HIV virus compromises the immune system mainly by infecting

a. cytotoxic T cells   d. plasma cells

b. helper T cells   e. B cells

c. suppressor T cells

CHALLENGE QUESTIONS

1. Assuming that immunological memory is intact, how can you explain people getting colds or flu year after year?

2. Explain why the passive immunity conferred from mother to child is only temporary.

3. List all possible combinations of ABO blood groups for donor and recipient in blood transfusions, indicating which transfusions would induce an immune response by the recipient.

FURTHER READING

1. Burton, D., P. Artymiuk, and G. Ford. "Death by Antibody." *New Scientist*, April 22, 1989. How antibodies trigger destruction of foreign cells.
2. Marrack, P., and J. Kappler. "The T Cell and Its Receptor." *Scientific American*, February 1986.
3. Montgomery, G. "The Human Mouse." *Discover*, February 1989. An AIDS researcher transplants the human immune system into mice.
4. Patlak, M. "The Fickle Virus." *Discover*, February 1989. The problems of developing an AIDS vaccine.
5. Robbins, A., and P. Freeman. "Obstacles to Developing Vaccines for the Third World." *Scientific American*, November 1988. Who will pay for immunizing the children of developing countries.
6. Tonegawa, S. "The Molecules of the Immune System." *Scientific American*, October 1985. An excellent overview of the molecules and battle plans involved in immunological defense.
7. "What Science Knows About AIDS." *Scientific American*, October 1988. A single-topic issue.
8. Young, J. D.-E., and Z. A. Cohn. "How Killer Cells Kill." *Scientific American*, January 1988.

40

Controlling the Internal Environment

M ost animals can survive fluctuations in the external environment more extreme than any of their individual cells could tolerate (Figure 40.1). A goldfish can endure water as acidic as pH 3 or as alkaline as pH 10 for an hour or more, but the cells within the fish will die if their internal pH drifts only slightly. A salmon migrates from fresh water to seawater and then back again, encountering changes in salinity that would inflict osmotic shock if they were experienced by cells within the body of the animal. During the course of a day, a human may be exposed to substantial changes in ambient temperature, but will die if *internal* body temperature fluctuates more than a few degrees about a mean of 37°C. In all these examples, the animals survive changes in their external environment by maintaining their internal environment within ranges that can be tolerated by their cells, a condition known as homeostasis (see Chapter 36). Homeostasis relates directly to one of the major themes in our study of animals, the ability of organisms to cope with environmental change, over the short term by physiological compensation and over the long term by adaptation based on natural selection. (These two levels of adjustment to the environment are, of course, related. Homeostatic mechanisms can best be interpreted as adaptations that have evolved in populations facing certain environmental problems.) A second important theme in biology, the correlation of structure and function, is also evident in the form and physiology of the tissues and organs responsible for homeostasis.

In all but the simplest animals (sponges, cnidarians, and flatworms), most body cells are not in direct con-

tact with the external environment, but are bathed instead by an internal body fluid. This internal "pond" is typically either hemolymph, as in insects and other animals with open circulatory systems, or interstitial fluid serviced by blood, as in vertebrates and other animals with closed circulatory systems. (Note that animals with closed circulatory systems have *three* internal fluid compartments: the intracellular compartment consisting of the cytosol of cells, and two extracellular compartments: the blood plasma and interstitial fluid.) Changes in these body fluids are tempered by a variety of regulatory systems (Figure 40.2), usually involving feedback mechanisms, as discussed in Chapter 36. The problems and solutions of controlling the internal environment vary, depending on the phylogenetic history of the animal and the environment in which the species of animal has evolved. This chapter takes a comparative approach

to homeostasis, concentrating on how animals control solute balance and the gain and loss of water (**osmoregulation**), how they get rid of the nitrogen-containing waste products of metabolism such as urea (**excretion**), and how they maintain internal temperature within a tolerable range (**thermoregulation**).

OSMOREGULATION

Think about the ways that water enters and leaves your body. We acquire most of our water in our food and drink, and obtain a smaller amount by oxidative metabolism ("metabolic water" is produced by cellular respiration when electrons and hydrogen are added to oxygen—see Chapter 9). We lose water by urinating and defecating, and by evaporative loss due to sweat-

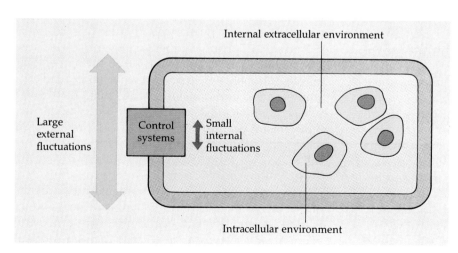

Figure 40.2
Homeostasis. Control systems cushion the internal environment from the impact of fluctuations in the external environment.

ing and breathing. For aquatic animals, evaporation is unimportant, but these animals experience the uptake and loss of water across the body surface by osmosis. Even if an animal is protected by a covering that impedes water loss or gain, specialized epithelia that must be exposed to the environment in order to exchange gases (such as gills, lungs, and tracheae) cannot be waterproof.

Whether an animal inhabits land, fresh water, or saltwater, one general problem occurs: The cells of the animal cannot survive a net water gain or loss. Although water continuously enters and leaves an animal cell across the plasma membrane, uptake and loss must balance. Animal cells swell and burst if there is a net uptake of water or shrivel and die if there is a net loss of water (see Chapter 8). We shall now see that there are two basic solutions to this problem of water balance.

Osmoconformers and Osmoregulators

Recall from Chapter 8 that osmosis is the movement of water across a selectively permeable membrane. It occurs whenever two solutions separated by the membrane differ in total solute concentration, or **osmolarity** (total solute concentration expressed as molarity, or moles of solute per liter of solution; see Chapter 3). The units for osmolarity used in this chapter are milliosmoles per liter (mosm/L). This unit is equivalent to a total solute concentration of 10^{-3} M. For example, the osmolarity of human blood is about 300 mosm/L, while seawater commonly has an osmolarity of about 500 mosm/L. Two solutions are said to be isosmotic if they are equal in osmolarity. There is no *net* osmosis between isosmotic solutions. When two solutions differ in osmolarity, the one with the greater concentration of solutes is referred to as hyperosmotic and the more dilute solution as hypoosmotic. Water will flow by osmosis across a membrane from a hypoosmotic solution to a hyperosmotic one.

One way for a saltwater animal to balance water exchange with the environment is to be isosmotic with its aqueous surroundings. Such animals, which do not actively adjust their internal osmolarity, are known as **osmoconformers**. Animals that are not isosmotic with their surroundings, called **osmoregulators**, must either discharge excess water if they live in a hypoosmotic environment or continuously take in water to offset osmotic loss if they inhabit a hyperosmotic environment. All freshwater animals and many marine animals are osmoregulators that maintain an internal osmolarity that differs from the surrounding water. Humans and other terrestrial animals, also osmoregulators, must compensate for water loss. With these general approaches to water balance in mind, we can now survey some specific examples of osmoregulatory adaptations to various environments.

Problems of Osmoregulation in Different Environments

Marine Animals   Animals first evolved in the sea, which remains the most common environment for the majority of phyla. Most marine invertebrates are osmoconformers, with body fluids isosmotic to the surrounding seawater. However, these animals differ from seawater in their concentrations of specific salts. The difference is usually slight, but in some cases it is substantial. Thus, an animal that conforms to the osmolarity of its surroundings may still regulate its internal composition of ions.

The hagfish, a jawless vertebrate (Class Agnatha), is isosmotic with the surrounding seawater, but most marine vertebrates osmoregulate. Sharks and most other cartilaginous fishes (Class Chondrichthyes) maintain internal salt concentrations that are relatively low compared with that of seawater mainly by the use of rectal glands that pump salt out of the body through the anus. However, the shark has an osmolarity close to that of seawater, a characteristic attributed to the retention of a large amount of organic solute in the form of urea, a nitrogenous (nitrogen-containing) waste product other animals excrete rather than accumulate. (Shark meat must be soaked in fresh water before it is eaten to wash out this high concentration of urea.) Sharks produce and retain another organic compound, trimethylamine oxide (TMAO), which helps to protect proteins from the damaging effects of urea. As a result of the high concentration of organic solutes in their body fluids, sharks are actually slightly hyperosmotic to seawater. They do not drink, and they balance the osmotic uptake of water by copius urination. Among other osmoregulators that maintain body fluids hyperosmotic to seawater as a result of accumulating urea are the crab-eating frogs of Southeast Asia, the only saltwater amphibians.

Bony fishes (Class Osteichthyes) evolved from ancestors that entered freshwater habitats. In their subsequent evolution, many groups of bony fishes became marine, but internally remained more similar to fresh water in osmolarity. Marine bony fishes constantly lose water by osmosis to their hyperosmotic surroundings. They compensate by drinking large amounts of seawater and using the epithelium of their gills to pump the excess salt out of the body. Similarly, many marine birds, including sea gulls, have nasal salt glands that rid the animals of much of the salt they obtain by drinking seawater. Marine reptiles, including the marine iguana and sea turtles, also have salt glands that function in osmoregulation.

Freshwater Animals   The osmoregulatory problems of freshwater animals are the opposite to those of marine animals. Freshwater animals are constantly taking in water by osmosis because the osmolarity of their internal fluids is much higher than that of their

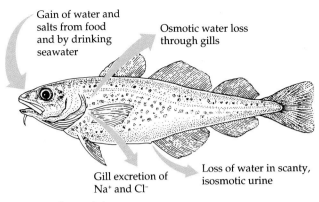

Gain of water and
salts from food
and by drinking
seawater

Osmotic water loss
through gills

Gill excretion of
Na⁺ and Cl⁻

Loss of water in scanty,
isosmotic urine

(a) Marine bony fish

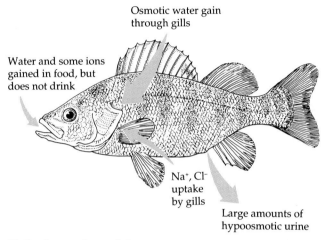

Osmotic water gain
through gills

Water and some ions
gained in food, but
does not drink

Na⁺, Cl⁻
uptake
by gills

Large amounts of
hypoosmotic urine

(b) Freshwater bony fish

Figure 40.3
Comparison of osmoregulation in marine and fresh-water bony fishes. (a) A marine fish, such as this cod, is hypoosmotic to the surrounding seawater and must cope with a constant loss of water by osmosis, mainly across the epithelium of the gills. The animal compensates for the fluid loss by drinking large amounts of seawater and then pumping salt out of the body across the gill epithelium. Urine is scanty and isosmotic to the body fluids of the fish. **(b)** The opposite problem is faced by freshwater fishes, such as this perch, which constantly gain water from their hypoosmotic surroundings by osmosis. This water uptake is balanced by copious excretion of urine that is hypoosmotic to the body fluids of the fish. Although the urine is dilute, the animal still loses important salts during excretion, and compensates by the active uptake of ions across the epithelium of the gills.

immediate environment. Freshwater protozoa such as *Amoeba* and *Paramecium* have contractile vacuoles that function as bilge pumps (see Figure 26.7). Many freshwater animals, including fishes, bail out water by excreting large amounts of very dilute urine. As they excrete water, however, they experience a small net loss of salts. Their salt content is replenished by eating plants and animals that have much higher salt con-

centrations than the water and, at least in some fish, by the gills' active accumulation of sodium and chloride ions from the external environment into the bloodstream (Figure 40.3).

Euryhaline Animals   Most animals, whether osmoconformers or osmoregulators, cannot tolerate substantial changes in external osmolarity. Such animals are said to be **stenohaline** (from the Greek *stenos*, "narrow"; haline refers to salt). However, some animals, called **euryhaline** animals (from the Greek *eurys*, "broad"), do survive radical fluctuations of osmolarity in the surrounding water, either by conforming to the changes or by regulating their internal osmolarity within a narrow range even as the external osmolarity changes. Among euryhaline animals is a diversity of invertebrates and fishes that inhabit the brackish water of estuaries, where salinity changes with each rainfall and daily tides. Salmon and other fishes that migrate between seawater and fresh water are also euryhaline. Their adaptations include an ability to alter osmoregulatory mechanisms. While in the ocean, salmon drink seawater and excrete excess salt from the gills, osmoregulating like other marine fishes. With their transition to fresh water, salmon cease drinking and the epithelia of their gills are modified for the accumulation of salt from the dilute environment.

Anhydrobiosis   Dehydration dooms most animals, but some aquatic invertebrates living in ponds and films of water around soil particles can lose almost all their body water and survive in a dormant state when their habitats dry up. This remarkable adaptation is called **anhydrobiosis** ("life without water"), or cryptobiosis ("hidden life"). Among the most striking examples are the tardigrades, or water bears, tiny invertebrates less than a millimeter long (Figure 40.4). In their active, hydrated state, these animals contain about 85% water by weight, but can dehydrate to less than 2% water and survive in an inactive state, dry as dust, for a decade or more. Just add water, and within minutes the rehydrated tardigrades are moving about and feeding.

One mechanism that makes anhydrobiosis possible is the animal's production of a large amount of disaccharides, particularly trehalose, a double sugar consisting of two glucose units. With their multiple hydroxyl groups capable of forming hydrogen bonds, the sugars apparently replace water associated with membranes and proteins, protecting these cellular structures and molecules from extreme distortion during dehydration.

Terrestrial Animals   Unfortunately, few truly terrestrial animals are capable of anhydrobiosis. Humans, for example, die if they lose about 12% of their body water. The threat of desiccation is perhaps the most important problem confronting terrestrial life, both plants and animals. The severity of this problem may

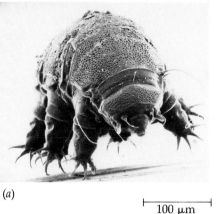

(a)

(b)

├─ 100 μm ─┤

Figure 40.4
Anhydrobiosis (cryptobiosis). (a) Tardigrades, or water bears, inhabit ponds and films of water in soil and on mosses and lichens. **(b)** When their habitat completely evaporates, these invertebrates can lose more than 95% of their body water and survive for decades in this dehydrated, inactive state. If the animals are rehydrated, they begin walking around again within minutes (SEM).

be one reason why only two groups of animals, arthropods and vertebrates, have colonized the land with great success (although other phyla have some representatives on land, most of their species are aquatic).

What *are* some of the evolutionary adaptations that have made it possible for animals, which consist mostly of water, to survive on land? Much as a waxy cuticle contributes to the success of plants on land (see Chapter 27), most terrestrial animals are covered by relatively impervious surfaces that help prevent dehydration. Examples are the waxy layers of the exoskeletons of insects, the shells of land snails, and the multiple layers of dead, keratinized skin cells covering most terrestrial vertebrates (see Chapter 30). Still, most terrestrial animals lose a considerable amount of water that must be replenished by drinking and eating moist foods. Behavioral adaptations, such as nervous and hormonal mechanisms that control thirst, are important osmoregulatory mechanisms in land-dwelling animals (these mechanisms are discussed later in the chapter). Many terrestrial animals, especially in the deserts, are nocturnal, another important behavioral adaptation that reduces dehydration. The kidneys and other excretory organs of terrestrial animals often exhibit adaptations that help conserve water (and these adaptations will also be highlighted later in the chapter). Some mammals are so well adapted to minimizing water loss that they can survive in deserts without drinking. For example, kangaroo rats conserve water so frugally that they can compensate for water loss by the production of metabolic water and the intake of very small quantities of water in their food (Figure 40.5).

Transport Epithelia and Osmoregulation

Although the problems of water balance in environments as diverse as saltwater, fresh water, and land are very different, the solutions in osmoregulators are mostly variations on a common theme: the use of specialized epithelia, called **transport epithelia**, to regu-

Figure 40.5
Water balance in kangaroo rats. Kangaroo rats, which live in the American southwestern desert, survive without drinking any water by regulating body fluids very efficiently. The animal obtains 90% of its water as a product of metabolic reactions and the remaining 10% from free water present in the food it eats. It also reduces evaporative water loss by staying inside an underground burrow during the day and reduces respiratory water loss by condensing moisture in its nasal passages. It excretes little water because its kidneys are specialized to concentrate urine. The feces are also relatively dry.

late the transport of salt (and hence water, which follows solute movement by osmosis) between the animal's internal fluids and the external environment.

A transport epithelium is usually a single sheet of cells facing the external environment or some channel that leads to the exterior through an opening on the body surface. The cells of the epithelium are joined by impermeable tight junctions (see Chapter 7), forming a continuous barrier at the tissue–environment boundary. This configuration, another example of how

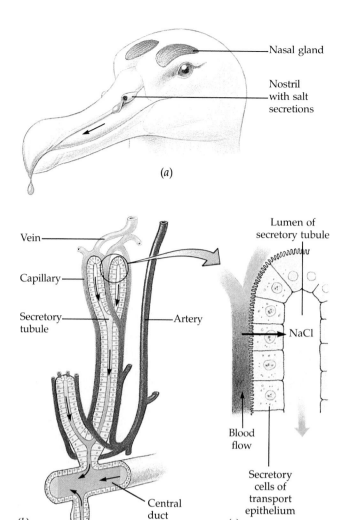

(a)

(b)

(c)

Figure 40.6
Salt glands in birds. (a) Many marine birds, such as this albatross, drink seawater and excrete excess salt through nasal glands. (b) A salt gland has many lobes, which contain several thousand tubules radiating from a central duct. Each tubule is surrounded by capillaries, where the blood flows counter to the flow of salt secretion (arrows). This countercurrent system facilitates salt transfer from the blood to the tubule (see Chapter 38). (c) Secretory cells of the transport epithelium lining the tubules pump salt from the blood into the tubules.

Labels in figure: Nasal gland; Nostril with salt secretions; Vein; Capillary; Secretory tubule; Artery; Central duct; Lumen of secretory tubule; NaCl; Blood flow; Secretory cells of transport epithelium

form fits function, ensures that any solute passing between the extracellular fluid and the environment must pass through the selectively permeable membranes of cells. It is the molecular composition of the epithelium's plasma membrane that determines the specific osmoregulatory functions. Transport epithelia vary in their passive permeabilities to water and salts, and in the number, type, and orientation of membrane proteins responsible for active transport. For example, differences in membrane structure and function account for the transport epithelia of the gills of marine fishes pumping salt outward, while the gills of freshwater fishes pump salt inward.

One of the most efficient transport epithelia is found in the nasal glands of marine birds, which drink seawater and excrete the excess salt via the nasal salt glands (Figure 40.6). As mentioned earlier, sharks also use salt glands, in this case located along the epithelium of the rectum, to expel salt.

The transport epithelia of salt glands are dedicated exclusively to osmoregulation—maintaining salt and water balance. In other cases, transport epithelia function in the excretion of nitrogenous wastes as well as osmoregulation. In the excretory systems of most animals, transport epithelia are arranged into tubular networks with extensive surface areas, as in the vertebrate kidney and a diversity of invertebrate excretory systems.

EXCRETORY SYSTEMS OF INVERTEBRATES

Protonephridia: Flame-Cell System of Flatworms

The simplest tubular excretory system is the **flame-cell system** of flatworms (Phylum Platyhelminthes). These animals have neither circulatory systems nor coeloms (see Chapter 29), and so the flame-cell system must regulate the contents of the interstitial fluid directly. The apparatus consists of a branched system of tubules ramifying throughout the body (Figure 40.7). Each of the smallest tubules at the tips of this excretory tree is capped by a bulbous cell called a flame cell. Interstitial fluid bathing the tissues of the animal passes through the flame cell and enters the tubular system. The flame cell has a tuft of cilia projecting into the tubule, and the beating of these cilia propels fluid along the tubule, away from the blind end where the flame cell is located. (The beating cilia look like a flickering flame, for which these cells are named.) In planaria, tributaries of the tubular system drain into excretory ducts that empty to the external environment through numerous openings called nephridiopores. The excreted fluid is very dilute in the case of freshwater flatworms; this helps to balance the osmotic uptake of water from the hypoosmotic environment. The cellular mechanisms for this osmoregulation are unknown, however. It is likely that the lining of the tubules is a transport epithelium specialized for reabsorbing certain salts before the fluid exits from the body. The flame-cell systems of freshwater flatworms function mainly in osmoregulation; most metabolic wastes are excreted into the gastrovascular cavity and eliminated through the mouth (see Chapter 36). However, some parasitic flatworms, which are isosmotic to the surrounding fluids of their host organisms, use their tubular excretory systems mainly to get rid of nitrogenous wastes.

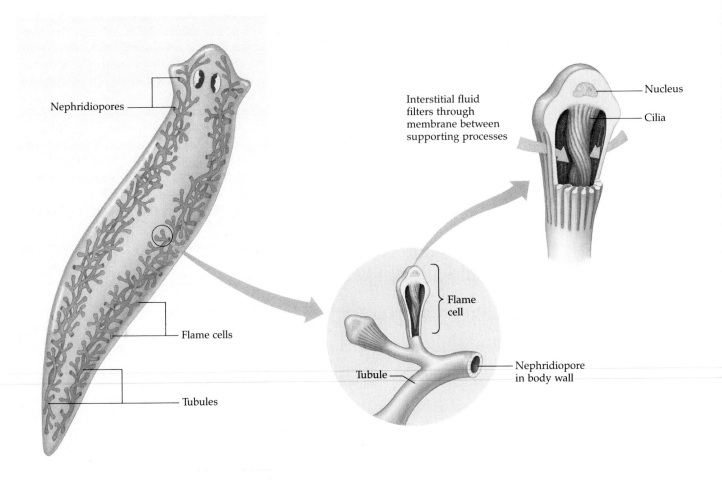

Figure 40.7
Protonephridia: the flame-cell system of planaria. Protonephridia are excretory tubules lacking internal openings. In planaria, interstitial fluid is filtered across the membranes of flame cells. Cilia projecting from the flame cells into the tubules keep the fluid moving along the tubular system, which opens to the exterior of the body through numerous pores. The transport epithelium lining the tubules functions in osmoregulation. For example, the epithelium of the flame-cell system of freshwater flatworms probably pumps salts from the tubular fluid back into the interstitial fluid, enabling the animals to excrete a dilute urine and balance the osmotic uptake of water from the hypoosmotic environment.

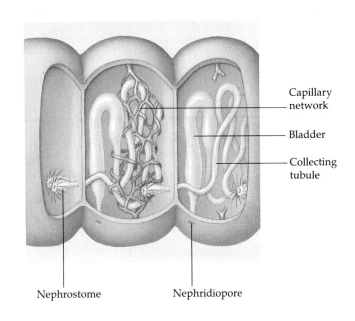

Figure 40.8
Metanephridia of annelids. Each segment of an earthworm contains a pair of metanephridia, which drain the adjacent anterior segment. Fluid enters the nephrostome, passes through the metanephridium, and empties into a storage bladder that opens to the outside through the nephridiopore. The urine becomes very dilute as the transport epithelium pumps certain salts back into the blood circulating in the capillary network enveloping the metanephridium.

The flame-cell system is one example of a simple type of tubular excretory system called a **protonephridium**, a network of closed tubules lacking internal openings. In addition to the flame-cell systems of flatworms, protonephridia are also found in rotifers, some annelids, the larvae of mollusks, and lancelets, which are invertebrate chordates (see Chapters 29 and 30 for a review of these animal phyla).

Metanephridia of Earthworms

In contrast to the closed protonephridia, another type of excretory tubule, the **metanephridium**, has internal openings that collect body fluids. Metanephridia are found in most annelids, including earthworms (Figure 40.8). Each segment of the worm has its own pair of metanephridia, which are serpentine tubules immersed in the coelomic fluid of that segment. As in most animals with closed circulatory systems, blood vessels are intimately associated with the excretory tubules of the earthworm; a network of capillaries envelops each metanephridium. The tubule drains to the outside of the body through a nephridiopore. At the opposite end of a metanephridium is the nephrostome, a ciliated funnel that collects coelomic fluid

from the body segment just anterior. (Notice again that the metanephridium, unlike the protonephridium of flatworms, is open at both ends.) As the fluid moves along the tubule, the transport epithelium bordering the lumen pumps essential salts out of the tubule, and the salts are reabsorbed into the blood circulating through the surrounding capillaries. The urine that exits through the nephridiopore is hypoosmotic to the body fluids of the earthworm. By excreting this dilute urine in amounts up to 60% of the body weight of the worm per day, the metanephridia offset the continuous osmosis taking place across the skin of the animal from the damp soil.

Malpighian Tubules of Insects

Insects and other terrestrial arthropods have open circulatory systems, with tissues bathed directly in hemolymph contained in sinuses (see Chapter 38). Excretory organs called **Malpighian tubules** remove nitrogenous wastes from the blood and function in osmoregulation (Figure 40.9). These organs open into the digestive tract at the juncture of the midgut and hindgut. The tubules, which dead-end at the tips away from the gut, dangle in the fluid of the body cavity.

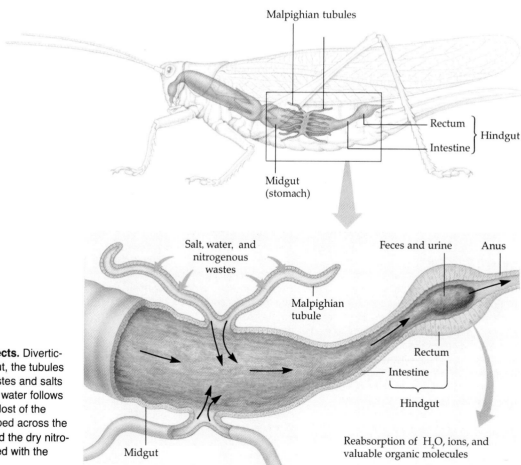

Figure 40.9
Malpighian tubules of insects. Diverticula (outpocketings) of the gut, the tubules accumulate nitrogenous wastes and salts from the coelomic fluid, and water follows these solutes by osmosis. Most of the salts and water are reabsorbed across the epithelium of the rectum, and the dry nitrogenous wastes are eliminated with the feces.

The transport epithelium that lines a tubule pumps certain solutes, including salts and nitrogenous wastes, from the blood into the lumen of the tubule. The fluid within the tubule then passes through the hindgut into the rectum. The epithelium of the rectum pumps most of the salt back into the blood, and water follows the salts by osmosis. The nitrogenous wastes are eliminated as nearly dry matter along with the feces. The insect excretory system is one adaptation that has contributed to the tremendous success of these animals on land, where conserving water is essential.

THE VERTEBRATE KIDNEY

Rather than being scattered throughout the body like the nephridia of earthworms, the excretory tubules of vertebrates, which are called **nephrons,** are collected into compact organs, the **kidneys.** Blood is cycled through the kidneys, which remove nitrogenous wastes and function in osmoregulation by adjusting the concentrations of various salts in the blood. The kidneys, the blood vessels that serve them, and the plumbing that carries urine formed in the kidneys out of the body are the components of the vertebrate excretory system. We shall focus first on the mammalian version of the system, and then compare the excretory systems of the various vertebrate classes.

Anatomy of the Excretory System

In humans, the kidneys are a pair of bean-shaped organs about 10 cm long (Figure 40.10). Blood enters the kidney via the **renal artery,** and leaves the kidney in the **renal vein**. Although the kidneys account for less than 1% of the weight of the human body, they receive about 20% of the blood pumped with each heartbeat. **Urine**, the waste fluid formed within the kidney, exits the organ through a duct called the **ureter**. The ureters of both kidneys drain into a common **urinary bladder**. The bladder is periodically emptied by micturition (urination); this final excretion of urine from the body is through a tube called the **urethra**, which empties near the vagina of females or through the penis of males. Sphincter muscles near the junction of the urethra and the bladder control micturition.

Structure of the Nephron

The functional unit of the kidney is the nephron, which consists of a **renal tubule** and its associated blood vessels (Figure 40.10b and c). Each kidney contains a large number of nephrons—about a million in a human kidney, representing a total of about 80 km of tubules. Water, urea, salts, and other small molecules present in blood flow from capillaries into the renal tubule, where the fluid is now called **filtrate**. The transport epithelium lining the renal tubule processes the filtrate to form the urine, which is eventually excreted from the kidney. From the 1100–2000 L of blood that flows through the human kidneys each day, the nephrons process about 180 L of filtrate, but excrete only about 1.5 L of urine. The rest of the filtrate, including about 99% of the water, is reabsorbed into the blood.

The blind end of the renal tubule, which receives filtrate from the blood, is expanded to form a cup-shaped receptacle called **Bowman's capsule,** which

---

Figure 40.10 ▶
Human excretory system. (a) The kidneys are compact excretory organs that regulate the composition of blood and form urine, which is transported from the kidneys to the urinary bladder in the ureters. The bladder empties to the outside of the body via the urethra. A renal artery supplies the kidney with up to 2000 L of blood per day, and a renal vein returns the blood to the general circulation. Two functional regions of the kidney are the outer cortex and the inner medulla. Urine formed in these regions drains into a central chamber, the renal pelvis, which leads to the ureter. **(b)** Arranged radially in the kidney are hundreds of thousands of nephrons, the functional units of the kidney. Juxtamedullary nephrons, found only in mammals and birds, have long loops of Henle, which extend into the medulla of the kidney. Cortical nephrons are restricted to the kidney cortex. **(c)** A nephron consists of a renal tubule and its associated blood vessels. The receiving end of a renal tubule is the Bowman's capsule, a cup-shaped swelling of the tubule. The major regions of the tubule are the proximal convoluted tubule, the descending and ascending limbs of the loop of Henle (only in juxtamedullary nephrons), and the distal convoluted tubule, which drains into a collecting duct that serves many other nephrons. Bowman's capsule envelops a ball of capillaries called the glomerulus. Blood pressure forces water and all small solutes from the blood plasma of these capillaries across the wall of the capsule and into the lumen of the renal tubule, forming a filtrate that is processed as it moves along the tubule. From the glomerulus, blood travels in an efferent arterial to a second bed of capillaries, the peritubular capillaries, which intertwine with the proximal and distal convoluted tubules. Extending downward is the vasa recta, the capillary network associated with the loop of Henle. The interstitial fluid bathing the nephron has a continuous traffic of substances passing between the various regions of the renal tubule and the blood plasma of the nephron capillaries. This chemical exchange consists of secretion, the selective addition of compounds from the blood to the filtrate within the the tubule, and reabsorption, the transport of substances back into the blood from the filtrate. Processing of the filtrate continues in the collecting duct, and the final product, now called urine, drains into the renal pelvis.

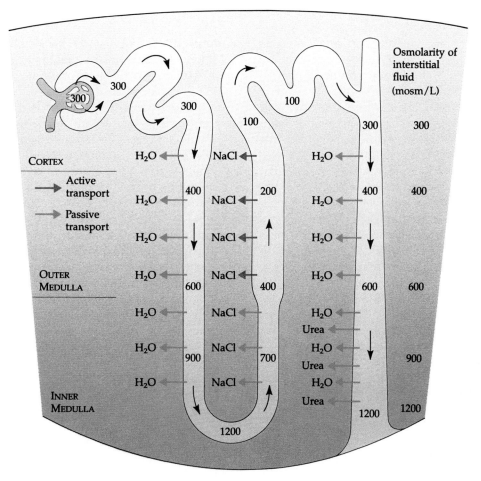

Figure 40.13

How the mammalian kidney concentrates urine: the two-solute model. Moving in a direction from outer cortex to inner medulla, the interstitial fluid of the kidney increases in osmolarity from about 300 to 1200 mosm/L. Two solutes contribute to this gradient of osmolarity: NaCl and urea. The loop of Henle maintains the interstitial gradient of NaCl. The filtrate concentration of this salt increases by the loss of water from the descending limb, and then the ascending limb leaks the salt into the interstitial fluid. Additional salt is actively transported out of the thick segment of the ascending limb. The second solute, urea, is added to the interstitial fluid of the kidney medulla by diffusion out of the collecting duct (urea remaining in the collecting duct is excreted). The filtrate makes a total of three trips between the cortex and medulla, first down, then up, and then down one more time in the collecting duct. As the filtrate flows in the collecting duct past interstitial fluid of increasing osmolarity, more and more water moves out of the duct by osmosis, thus concentrating the solutes, including urea, that are left behind in the filtrate. Under conditions when the kidney conserves as much water as possible, urine can reach an osmolarity of about 1200 mosm/L, considerably hyperosmotic to blood (about 300 mosm/L). This ability to excrete nitrogenous waste with a minimal loss of water is a key terrestrial adaptation of mammals.

imal loss of water from the body because osmosis causes the filtrate in the collecting duct to equal the osmolarity of the interstitial fluid, which can be as high as 1200 mosm/L in the inner medulla. Notice that urine, at its most concentrated, is actually *isosmotic* to the interstitial fluid of the inner medulla, but that makes it *hyperosmotic* to blood and interstitial fluid elsewhere in the body. The juxtamedullary nephron, with its urine-concentrating features, is a key adaptation to terrestrial life, enabling mammals to get rid of nitrogenous waste without squandering water.

Regulation of the Kidneys

While it is true that your kidneys *can* excrete a hyperosmotic urine, it is not always desirable to do so. Nevertheless, if you are dehydrated and water is unavailable, the kidneys can excrete a small volume of hyperosmotic urine as concentrated as 1200 mosm/L, making it possible to discharge wastes with a minimal water loss. But if you have consumed an excessive amount of fluid, the kidneys can actually excrete a large volume of *hypoosmotic* urine as dilute as 70 mosm/L, making it possible to eliminate a lot of water without losing essential salts. The kidney is a versatile osmoregulatory organ, where water and salt reabsorption are subject to a combination of nervous and hormonal controls. (Hormones, chemical signals between various organs of the body, are discussed in detail in the next chapter. Here, we are concerned only with the effects of a few hormones on the kidneys.)

One hormone important in osmoregulation is **antidiuretic hormone**, or **ADH** (Figure 40.14a). It is produced in a part of the brain called the hypothalamus and then stored and released from an organ called the pituitary gland, which is positioned just below the hypothalamus. **Osmoreceptor cells** located in the hypothalamus monitor the osmolarity of blood, triggering the release of ADH when an increase in the blood osmolarity is detected. The hormone is discharged into the bloodstream and reaches the kidney. The main targets of ADH are the distal convoluted tubules and the collecting ducts of the kidney, where the hormone increases the permeability of the epithelium to water. This amplifies water reabsorption, which reduces the osmolarity of the blood. By negative feedback, the subsiding osmolarity of the blood reduces the activity of osmoreceptor cells in the hypothalamus, and less ADH is secreted. When very little ADH is released, as would occur after consumption of a large volume of water has lowered the blood osmolarity, then the kidneys would absorb little water, resulting in copious excretion of dilute urine. (Voluminous urination is called diuresis, and it is because ADH opposes this state that it is called *anti*diuretic hormone.) Alcohol can perturb water balance by inhibiting the release of ADH, causing excessive loss of water in the urine and dehydrating the body—perhaps some of the symptoms of a hangover are due to this dehydration. Normally, however, blood osmolarity, ADH release, and water reabsorption in the kidney are all linked in a feedback loop that contributes to homeostasis.

A second system regulating kidney function involves a specialized tissue called the **juxtaglomerular apparatus (JGA)**, located in the vicinity of the afferent arteriole that supplies blood to the glomerulus (Figure 40.14b). When the blood pressure in the afferent arteriole drops, or when the Na^+ concentration of the blood is too low, the JGA releases an enzyme called **renin** to the bloodstream. Within the blood, renin activates a plasma protein called **angiotensin**. The active form of this protein, called angiotensin II, functions as a hormone, with multiple effects that increase the Na^+ concentration of the blood and raise blood pressure. For example, angiotensin II causes a generalized constriction of arterioles, which raises blood pressure. Increasing pressure within the afferent arterioles of the nephrons, in turn, increases filtration rate. Angiotensin II also acts remotely on the kidney by stimulating the adrenal glands, organs located on top of the kidneys, to release another hormone called **aldosterone**. This hormone acts on the distal convoluted tubules of the nephrons, stimulating the reabsorption of Na^+. Because water follows the Na^+ out of the renal tubule by osmosis, aldosterone also increases blood volume and blood pressure. It was a drop in blood pressure or a deficiency of Na^+ that triggered renin release from the JGA in the first place, and the various responses increase blood pressure and Na^+ concentration, thus reducing the release of renin—another example of a feedback circuit functioning in homeostasis.

It may seem that the functions of ADH and aldosterone are redundant, but this is not the case. It is true that both hormones increase water reabsorption, but they are enlisted to counter different osmoregulatory problems. The release of ADH is a response to an increase in the osmolarity of the blood, as occurs when the body is dehydrated—by inadequate intake of water, for instance. But imagine a situation that causes an excessive loss of salt and body fluids—an injury, for example, or severe diarrhea. This reduces the blood's

---

Figure 40.14 ▶
Hormonal regulation of the kidney.
(a) Antidiuretic hormone (ADH), produced in the hypothalamus of the brain and secreted from the pituitary gland, enhances fluid retention by increasing the permeability of the kidneys' collecting ducts to water. Release of ADH is triggered when osmoreceptor cells in the hypothalamus detect an increase in the osmolarity of the blood. The kidney responds to ADH with greater reabsorption of water, which reduces the osmolarity of the blood and inhibits the secretion of

ADH, completing the feedback circuit.
(b) The juxtaglomerular apparatus (JGA), a specialized tissue in the vicinity of the afferent arterioles leading to glomeruli of the kidneys, responds to a decrease in blood pressure or Na^+ concentration by releasing the enzyme renin into the bloodstream. In the blood, renin initiates the conversion of angiotensin to its active form, angiotensin II. A hormone, angiotensin II acts directly to increase blood pressure by causing arterioles to constrict. Angiotensin II also acts indirectly on the

kidney by signaling the adrenal glands to release aldosterone, a hormone that stimulates the active reabsorption of Na^+ across the epithelium of the distal convoluted tubules. Reabsorption of Na^+, in turn, results in the reabsorption of more water, which follows the solute by osmosis. Thus, the release of renin from the JGA leads to an increase in the concentration of Na^+ in the blood and an increase in blood volume and pressure, results that complete the feedback circuit by suppressing further release of renin.

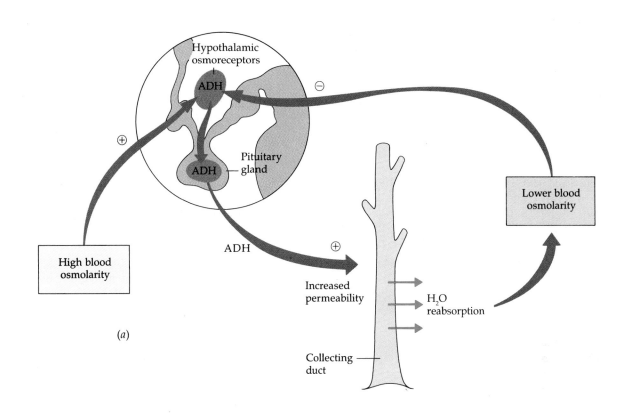

(a)

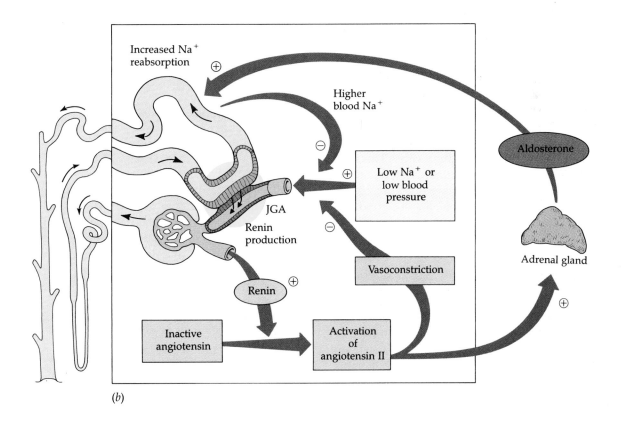

(b)

volume without increasing its osmolarity. Aldosterone would save the day by increasing water and a Na^+ reabsorption in response to the drop in blood volume caused by fluid loss. Normally, ADH and aldosterone are partners in homeostasis; ADH alone would lower blood Na^+ concentration by stimulating water reabsorption in the kidney, but aldosterone helps maintain balance by stimulating Na^+ reabsorption.

Still another hormone, **atrial natriuretic protein (ANP)**, opposes the renin-angiotensin-aldosterone system. The wall of the atrium of the heart releases ANP in response to an increase in blood volume and pressure, and the hormone counters by inhibiting the release of renin from the juxtaglomerular apparatus and also directly reducing aldosterone release from the adrenal glands. These actions decrease Na^+ reabsorption and lower blood volume and pressure. Thus, ADH, the renin-angiotensin-aldosterone trio, and ANP provide an elaborate system of checks and balances that regulate the kidney's ability to control the osmolarity, salt concentration, volume, and pressure of blood.

Having considered the mammalian kidney and its regulation in detail, we can now compare the structures and functions of kidneys in other vertebrate classes.

Comparative Physiology of the Kidney

Variations in nephron structure and physiology equip the kidneys of different vertebrates for osmoregulation in their various habitats. We have seen, for instance, that nephrons of the mammalian kidney can concentrate urine and conserve water. Among mammals, those able to excrete the most hyperosmotic urine, such as kangaroo rats and other mammals adapted to the desert, have exceptionally long loops of Henle that maintain steep osmotic gradients in the kidney. This results in urine becoming very concentrated as it passes from cortex to medulla in the collecting ducts. In contrast, beavers, which rarely face problems of dehydration, have nephrons with very short loops, resulting in a dilute urine.

Birds, like mammals, have kidneys with juxtamedullary nephrons that specialize in conserving water. However, the nephrons of birds have much shorter loops of Henle than those typical of mammalian nephrons, and birds are unable to concentrate urine to the osmolarities achieved by mammalian kidneys.

The kidneys of reptiles, having only cortical nephrons, produce urine that is, at best, isosmotic to body fluids. However, the epithelium of the cloaca (see Chapter 30) helps conserve fluid by reabsorbing some of the water present in urine and feces. Also, most terrestrial reptiles excrete nitrogenous wastes in an insoluble form known as uric acid, which helps to conserve water because it does not contribute to the osmolarity of the urine (this adaptation is discussed in more detail in the next section).

In contrast to mammals and birds, freshwater fishes face the problem of excreting excess water because the animal is hyperosmotic to its surroundings. Instead of conserving water, the nephrons use cilia to sweep a large volume of very dilute urine from the body. Freshwater fishes conserve salts by efficient reabsorption of ions from the filtrate in the nephrons.

Amphibians' kidneys function much like those of freshwater fishes. When in fresh water, the skin of the frog accumulates certain salts from the water by active transport, and the kidneys excrete a dilute urine. On land, where dehydration is the most pressing problem of osmoregulation, frogs conserve body fluid by reabsorbing water across the epithelium of the urinary bladder.

Bony fishes that live in seawater, being hypoosmotic to their surroundings, face the opposite problem to that of their freshwater relatives. In many species, nephrons lack glomeruli and capsules, and a concentrated urine is formed by secreting ions into the renal tubules. Thus, the kidneys of marine fishes excrete very little urine and function mainly to get rid of divalent ions such as Ca^{2+}, Mg^{2+}, and SO_4^{2-}, which the fish takes in by its incessant drinking of seawater. As mentioned previously, monovalent ions such as Na^+ and Cl^- are excreted mainly by the gills, as is most nitrogenous waste in the form of NH_4^+ (ammonium).

Osmoregulation—control of salt and water balance—was the original function of the kidney. In the course of evolutionary development, the excretion of nitrogenous wastes became a second function.

NITROGENOUS WASTES

Metabolism produces toxic by-products. Perhaps the most troublesome is the nitrogen-containing waste from the metabolism of proteins and nucleic acids. Nitrogen is removed from these nutrients when they are broken down for energy or when they are converted to carbohydrates or fats. The nitrogenous waste product is **ammonia**, a small and very toxic molecule. Some animals excrete their ammonia directly; others first convert it to less toxic wastes such as urea or uric acid (Figure 40.15). We shall see that the form of nitrogenous waste an animal excretes depends on both the animal's evolutionary history and its habitat.

Ammonia

Most aquatic animals excrete nitrogenous wastes as ammonia. Ammonia molecules are small and very

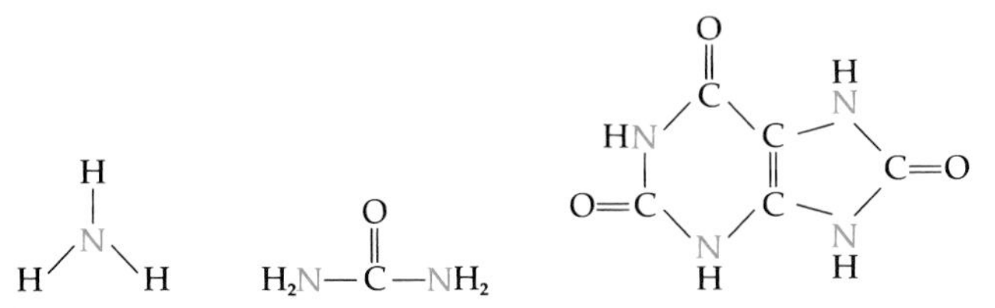

Figure 40.15

Nitrogenous wastes. Ammonia is a toxic by-product of the metabolic removal of nitrogen from proteins and nucleic acids. Most aquatic animals get rid of ammonia by excreting it in very dilute solutions. Most terrestrial animals convert the ammonia to urea or uric acid, which conserves water because these less toxic wastes can be transported in the body in more concentrated form.

AMMONIA   UREA   URIC ACID

soluble in water, so they easily permeate membranes. In soft-bodied invertebrates, ammonia diffuses across the whole body surface into the surrounding water. In freshwater fishes, most of the ammonia is lost as ammonium ions (NH_4^+) across the epithelium of the gills, with kidneys playing only a minor role in excretion of nitrogenous waste. The epithelium of the gills takes up Na^+ from the water in exchange for NH_4^+, which helps freshwater fishes maintain Na^+ concentrations much higher than that in the surrounding water.

Urea

Ammonia excretion, though it works in water, is unsuitable for disposing of nitrogenous waste on land. A terrestrial animal would have to urinate copiously to get rid of ammonia, because a compound so toxic could only be transported in the animal and excreted in a very dilute solution. Instead, mammals and most adult amphibians excrete urea. (Many marine fishes and turtles, which have the problem of conserving water in their hyperosmotic environment, also excrete urea.) This substance can be handled in much more concentrated form because it is about 100,000 times less toxic than ammonia. Urea excretion enables the animal to sacrifice less water to discard its nitrogenous waste, an important adaptation for living on land.

Urea is produced in the liver by a metabolic cycle that combines ammonia with carbon dioxide. The circulatory system carries the urea to the kidneys. As mentioned earlier, not all urea is excreted immediately by mammalian kidneys; some of it is retained in the kidneys, where it contributes to osmoregulation by helping to maintain the osmolarity gradient that functions in water reabsorption. Sharks, remember, also produce urea, which is retained at a relatively high concentration in the blood, which helps balance the osmolarity of body fluids with the surrounding seawater.

Amphibians that undergo metamorphosis generally switch from excreting ammonia to excreting urea during the transformation from an aquatic larva, the tadpole, to the terrestrial adult. This biochemical modification, however, is not inexorably coupled to metamorphosis. Frogs that remain aquatic, such as the South African clawed toad (*Xenopus*), continue excreting ammonia after metamorphosis. But if these animals are forced to stay out of water for several weeks, they begin to produce urea. Similarly, African lungfish switch from ammonia to urea excretion if their habitat dries up and they are forced to burrow in the mud and become inactive (see Chapter 30).

Uric Acid

Land snails, insects, birds, and some reptiles excrete uric acid as the major nitrogenous waste. Because it is thousands of times less soluble in water than either ammonia or urea, uric acid can be excreted as a precipitate after nearly all the water has been reabsorbed from the urine. In birds and reptiles, the pastelike urine is excreted into the cloaca and eliminated along with feces from the intestine.

Uric acid and urea represent two different adaptations that enable terrestrial animals to excrete nitrogenous wastes with a minimal loss of water. One factor that seems to have been important in determining which of these alternatives evolved in a particular group of animals is the mode of reproduction. Soluble wastes can diffuse out of a shell-less amphibian egg or be carried away by the mother's blood in the case of a mammalian embryo. The vertebrates that excrete uric acid, however, produce shelled eggs, which are permeable to gases but not to liquids. If an embryo released ammonia or urea within a shelled egg, the soluble waste would accumulate to toxic concentrations. Uric acid precipitates out of solution and can be stored within the egg as a solid that is left behind when the animal hatches.

In grouping the various vertebrates according to the nitrogenous wastes they excrete, the boundaries are not drawn strictly along phylogenetic lines but depend also on habitat. Among reptiles, for instance, lizards, snakes, and terrestrial turtles excrete mainly uric acid; crocodiles excrete ammonia in addition to uric acid; and aquatic turtles excrete both urea and ammonia. In fact, individual turtles modify their nitrogenous wastes when their environment changes. A tortoise that usually produces urea can shift to uric acid production when the temperature increases and water becomes less available. This is another example of how response to the environment occurs on two levels: Evolution determines the limits of physiological responses for a species, but individual organisms make adjustments within that range as their environment changes. This principle also applies to the regulation of body temperature.

REGULATION OF BODY TEMPERATURE

Metabolism is very sensitive to changes in the temperature of an animal's internal environment. For example, the rate of cellular respiration increases with temperature up to a certain point and then declines when temperatures are high enough to begin denaturing enzymes (see Chapter 6). The properties of membranes also change with temperature. Although different species of animals are adapted to different temperature ranges—some animals thrive in deserts where temperatures often reach 40°C, and others flourish in frigid polar climates—each animal has an optimal temperature range. Within that range, many animals can maintain a constant internal temperature as the external temperature fluctuates. To understand the problems and mechanisms of temperature regulation, it is important to first consider the exchange of heat between organisms and their environment.

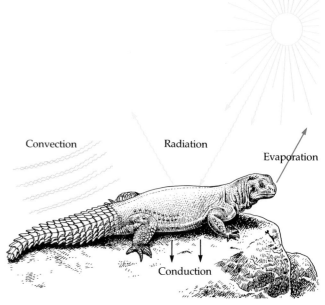

Figure 40.16
Heat transfer between an animal and its environment. Objects in direct contact can transfer heat by conduction, the transfer of molecular motion (heat). Convection is the mass movement of air or water over a body. Radiation transfers heat in the form of electromagnetic waves. Evaporation cools a surface (see Chapter 2).

Heat Production and Transfer Between Organisms and their Environment

An organism, like all objects, exchanges heat with its environment by four physical processes: conduction, convection, radiation, and evaporation (Figure 40.16).

Conduction is the direct transfer of thermal motion (heat) between molecules of the environment and those of the body surface, as when an animal sits in a pool of cold water, or on a hot rock. Heat will always be conducted from a body of higher temperature to one of lower temperature (see Chapter 3). Water is 50 to 100 times more effective than air in conducting heat. This is one reason you can cool your body on a hot day so rapidly by going for a swim.

Convection is the mass flow of air or liquid past the surface of a body, as when a breeze contributes to heat loss from the skin of an animal (or when a fan brings comfort to a human on a hot, still day). On the other hand, a "wind-chill factor" compounds the harshness of cold winter temperatures.

Radiation is the emission of electromagnetic waves produced by all objects warmer than absolute zero, including the body and the sun. Radiation can transfer heat between objects that are not in direct contact, as when an animal absorbs heat radiating from the sun. A unique adaptation for exploiting solar radiation has recently been discovered in polar bears and arctic seals. The fur of these animals is actually clear, not white. Each hair functions somewhat like an optical fiber that transmits ultraviolet radiation to the black skin, where the energy is absorbed and converted to body heat.

Evaporation is the loss of heat from the surface of a liquid that is losing some of its molecules as gas. Evaporation of water from an animal has a significant cooling effect on the animal's surface (see Chapter 3).

If you were to sit at rest in still air at a comfortable temperature cooler than your body (say, an air temperature of 23°C), conduction would account for only about 1% of your heat loss, convection for 40%, radiation for 50%, and evaporation for 9%. Convection and evaporation are the most variable causes of heat loss. A breeze of just 15 km/hr will increase total heat loss substantially by increasing convection fivefold. Evaporative cooling is increased greatly by the production of sweat. However, evaporation can only occur if the surrounding air is not saturated with water molecules (that is, the relative humidity is less than 100%). This is the biological basis for the common complaint, "The heat is not as bad as the humidity."

Ectotherms and Endotherms

One way to classify the thermal physiology of various animals is to emphasize the major source of body heat. Animals that warm their bodies mainly by absorbing heat from their surroundings are called **ectotherms**. Invertebrates, fishes, reptiles, and amphibians are generally ectotherms. In contrast, animals that derive most of their body heat from their own metabolism are called **endotherms**. Although endotherms usually maintain a consistent internal temperature even as the environmental temperature fluctuates, a constant body temperature does not necessarily distinguish endotherms from ectotherms; for example, many marine fishes and invertebrates inhabit water with such stable temperatures that these animals vary in body temperature less than humans and other endotherms. Also, the terms *cold-blooded* and *warm-blooded* are misleading. Many lizards, which are ectotherms, have active body temperatures higher than those of mammals. Note again that the terms *ectotherm* and *endotherm* are not based on the animals' body temperature, but rather on the main source of body heat. However, even this distinction of environmental versus metabolic heat sources is not absolute: Many ectotherms, including certain fishes and insects, obtain body heat from metabolism; and birds and mammals, which are endotherms, may add body heat by basking in the sun.

Endothermy solves certain problems of living on land. Endothermy enables terrestrial animals to maintain a constant body temperature in the face of environmental temperature fluctuations that are generally more severe than those an aquatic animal confronts. In general, the endothermic vertebrates—birds and mammals—are warmer than their surroundings, but these animals also have mechanisms for cooling the body in a hot environment. A consistently warm body temperature requires active metabolism, but, conversely, a warm body temperature contributes to the high levels of aerobic metabolism (cellular respiration) required for endurance of intense physical activity. This is one reason why endotherms can generally endure vigorous activity longer than ectotherms. These connections between body temperature, active aerobic metabolism, and mobility were important in the evolution of endothermy; moving on land requires considerably more effort than moving in water. The efficient circulatory and respiratory systems of birds and mammals can be thought of as adaptations accompanying the evolution of endothermy and a high metabolic rate. This is not to say that ectothermy is incompatible with terrestrial success. Among ectotherms are amphibians and reptiles, which have their own adaptations for coping with the temperature changes of terrestrial environments. In the next section, we compare the mechanisms that determine body temperature in various endotherms and ectotherms.

Thermoregulation in Terrestrial Mammals

Heat Production   Metabolism generates heat that can warm the body. Insulating layers of fat and fur help retain body heat. The rate of heat production can be increased in one of two ways: by the increased contraction of muscles (by moving around or by shivering) or by the action of certain hormones (especially epinephrine and thyroxine) that increase metabolic rate. The hormonal triggering of heat production is called **nonshivering thermogenesis**. The process takes place throughout the body, but some mammals have a tissue called **brown fat** in the neck and between the shoulders that is specialized for rapid heat production (Figure 40.17).

Being endothermic is liberating, but it is also energetically expensive, especially in a cold environment. For example, at 20°C, an endotherm such as a human male has a basal metabolic rate of 1800 kcal per day (see Chapter 37). In contrast, an ectotherm of similar weight, such as an American alligator, has a basal metabolic rate of only 60 kcal per day.

Mechanisms of Thermoregulation   A land mammal maintains a relatively constant body temperature by a combination of physiological and behavioral adjustments that fall into four general categories:

1. The rate of metabolic heat production can be changed. For example, when exposed to cold, many animals can double or triple their metabolic heat production through increased activity of skeletal muscles and nonshivering thermogenesis.

2. The rate of heat exchange between an animal and its environment can be adjusted. One such mechanism alters the amount of blood flowing to the skin. Increased blood flow usually results from **vasodilation**. In this process, certain nerves to the superficial blood vessels (those near the body surface) decrease their activity, relaxing the muscles of blood vessel walls and allowing blood flow through the vessels to increase. When this occurs, more heat is transferred to the environment by conduction, convection, and radiation. The reverse adjustment, **vasoconstriction** of superficial vessels, reduces blood flow and heat loss by decreasing the diameter of the vessels. In rabbits, for example, little blood flows to the large, poorly insulated ears when the rabbit sits in a cool environment. However, if the rabbit exercises, increasing its heat production and thus its body temperature, the blood vessels in its ears vasodilate, and warm blood carried to this region can lose heat to the environment. Vasodilation and vasoconstriction also contribute to regional temperature differences within the animal. For example, on a cool day, a human's temperature may be several degrees lower in the arms and legs than in the

(a)

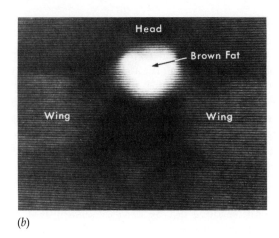

Head

Brown Fat

Wing   Wing

(b)

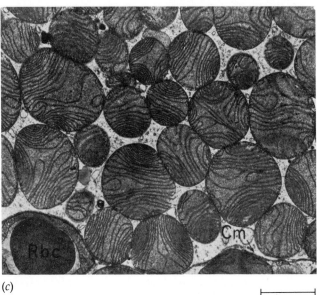

Rbc   Cm

(c)

├─ 1 μm ─┤

Figure 40.17
Brown fat. Many mammals, including those that hibernate, have adipose tissue called brown fat that is specialized for rapid heat production. Brown fat cells store lipids and contain large numbers of mitochondria. When brown cells oxidize the stored fat, the specialized mitochondria make little ATP and release most of the energy as heat instead. This quickly raises the body temperature to normal at the end of torpor, a daily period of inactivity when body temperature drops. (**a**) In bats, brown fat is located between the shoulder blades. (**b**) Heat-sensitive film shows that the area containing brown fat is much warmer than the rest of the body. (**c**) Of all the body tissues, brown fat contains the greatest number of mitochondria per cell (TEM).

trunk, where most vital organs are located. Hair is also important in regulating heat exchange between a mammal and its environment. Most land mammals react to cold by raising their fur, which increases insulation. (Humans just get goose bumps, a vestige from our furry ancestors. Although we have relatively little hair, the muscles that raise hair in the cold still contract.)

3. In mammals, evaporative loss of heat can cause substantial cooling. Terrestrial mammals lose water from their respiratory tract surfaces and across their skin. Evaporation from the respiratory tract can be increased by panting. Evaporation across the skin can be increased by sweating, which is under nervous control (Figure 40.18). If the humidity of the air is low enough, the

Figure 40.18
Skin as an organ of thermoregulation. Fat (adipose tissue) and hair help insulate mammals. Heat loss to the environment can be regulated by the constriction and dilation of superficial blood vessels and by erection and compaction of fur. Sweat glands, under nervous control, function in evaporative cooling.

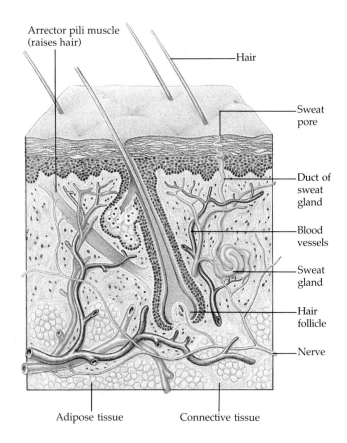

Arrector pili muscle (raises hair)

Hair

Sweat pore

Duct of sweat gland

Blood vessels

Sweat gland

Hair follicle

Nerve

Adipose tissue   Connective tissue

(a)

Figure 40.19
Behaviorial adaptations for thermoregulation. (a) Bathing in cool water brings immediate relief from the heat and continues to cool the surface for some time by evaporation. (b) Basking is an important mechanism for warming the body, especially in ectotherms such as these marine iguanas that inhabit the Galapagos Islands. (c) Dressing for the weather, illustrated by these Siberian children, is a thermoregulatory behavior unique to humans.

(b)

(c)

water will evaporate and cool the skin. Mammals lacking sweat glands have other mechanisms that promote evaporative cooling when heat stress is severe. For instance, some rodents spread saliva on their heads. Bats use both saliva and urine to cool themselves by evaporation.

4. Behavioral responses are also important to temperature control (Figure 40.19). Mammals and many other animals can increase or decrease body heat loss by relocating. They will bask in the sun in winter, find cool, damp areas or burrow in summer, or even migrate to a more suitable climate.

The Thermostat Nerve cells controlling thermoregulation, as well as those controlling other aspects of homeostasis, are concentrated in the hypothalamus of the brain (to be discussed in detail in Chapter 44). This small brain region has two thermoregulatory areas. One is called the **heating center** because it controls vasoconstriction of superficial vessels, erection of fur, shivering, and nonshivering thermogenesis. The other area is called the **cooling center** because it controls vasodilation and sweating (or panting). Nerve cells that sense temperature are located in the skin, the hypothalamus, and some other parts of the nervous system. Some of these temperature receptors, called

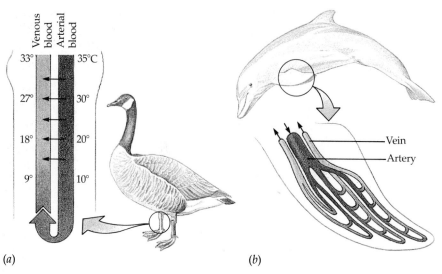

(a)   *(b)*

Figure 40.20
Countercurrent heat exchange.
(a) Some aquatic birds possess counter-current systems in their legs that reduce heat loss. The arteries carrying blood down the legs contact the veins that return blood to the body core. In a cold environment, the arterial blood transfers heat to the venous blood returning to the body (black arrows). The countercurrent flow facilitates heat exchange by setting up a thermal gradient between the arterial and venous blood over the entire length of the vessels (see Chapter 38). The body's core temperature is further protected by the constriction of surface veins. (b) In the flippers of marine mammals, where there is no blubber, each artery is surrounded by several veins in a countercurrent arrangement that allows efficient heat exchange between venous and arterial blood.

Ruffini organs (warm receptors), increase their nervous activity when the temperature increases. Others, called bulbs of Krause (cold receptors), increase their activity when the temperature decreases. The warm receptors excite the cooling center of the hypothalamus and inhibit the heating center, whereas the cold receptors have exactly the opposite effects on the two centers. Body temperature is thus controlled by feedback mechanisms, with the hypothalamus functioning as a thermostat that triggers heating and cooling responses (see Chapter 36).

Some Thermoregulatory Adaptations in Other Animals

Birds   The average body temperature of a bird is very high, around 40°C. Cooling mechanisms prevent body temperature from exceeding the optimal range. Birds have no sweat glands, but instead use panting to promote evaporative heat loss. Some species have a specialized, highly vascularized pouch in the floor of the mouth that they can flutter, increasing evaporation.

Birds also have mechanisms for reducing heat loss. Feathers provide excellent insulation. All species, especially those that walk or swim in water, face the problem of losing large amounts of heat from their legs and feet. Warm blood from the core of the body must flow to the cells of these extremities. A special arrangement of arteries and veins called a **countercurrent heat exchanger** reduces heat loss (Figure 40.20a). Arteries carrying warm blood down the legs are in close contact with veins conveying blood the opposite direction, back toward the trunk of the body. This countercurrent arrangement facilitates heat transfer from arteries to veins along the entire length of the blood vessels. Near the end of the leg, where the blood in an artery has been cooled to a temperature far below the animal's core temperature, the artery can still transfer heat to the even colder blood of a juxtaposed vein (remember, heat is conducted only from a warmer object to a cooler one). The venous blood can continue to absorb heat as it travels upward because it is passing warmer and warmer arterial blood flowing downward from the body core. By the time the venous blood reaches the top of a leg, it is almost as warm as the body core, minimizing the heat lost as a result of supplying blood to legs immersed in cold water. In some birds, blood can enter the limbs either through such an exchanger or by way of vessels that detour around the exchanger; the relative amount of blood that enters the limbs by way of the two different paths varies, controlling the amount of heat loss.

Marine Mammals   Seals and whales maintain a body temperature of about 36°–38°C, similar to that of other mammals. All marine mammals live in water that is colder than their body temperature, and many of them live at least part of the year in nearly freezing Arctic or Antarctic water. Although the loss of heat to water occurs 50 to 100 times more rapidly than heat loss to air, the metabolism of marine mammals is not much higher than that of land mammals of similar size, suggesting that marine mammals conserve heat more effectively.

The major adaptation for retaining body heat in a cold, wet environment is a very thick layer of insulating fat called blubber, just under the skin. In the tail and flippers, where there is no blubber, countercurrent heat exchange occurs between the arterial and venous blood, as in the legs of water-dwelling birds (Figure 40.20b).

Because many marine mammals move to warmer environments in their annual migrations, they also face the challenge of dissipating metabolic heat. This is often managed by dilating blood vessels that serve the skin and thus allowing greater amounts of blood to be cooled by the surrounding water, which even in these warmer seas is typically still well below body temperature.

Reptiles   Reptiles are generally ectotherms with relatively low metabolic rates that contribute little to the normal body temperature. Reptiles warm themselves mainly by behavioral adaptations. They seek warm places, orienting toward heat sources to increase heat uptake and expanding the body surface exposed to a heat source. Reptiles do not simply maximize heat uptake, however, but may behave in such a way as to truly regulate their temperature within a range. If a sunny spot is too warm, for instance, a lizard will alternately sit in the sun and in the shade, or will turn in another direction, which lowers the surface area exposed to the sun. By seeking favorable microclimates within the environment, many reptiles maintain body temperatures that are quite stable.

Indications are that some of the sophisticated regulatory mechanisms found in mammals are present in reptiles, at least in a rudimentary state. (Recall from Chapter 30 that mammals and birds evolved from reptiles.) For example, in diving reptiles, body heat is conserved by routing more blood to the body core during a dive. Reptiles can also increase thermogenesis, just as mammals do. For example, female pythons incubating their eggs generate considerable heat by shivering.

Amphibians   The optimal temperature range for amphibians varies substantially with the species. For example, closely related species of salamanders have average body temperatures ranging from 7°C to 25°C. Amphibians produce very little heat, and most lose heat rapidly by evaporation from their body surfaces, making it difficult to control body temperature. How-

ever, behavioral adaptations enable them to maintain body temperature within a satisfactory range most of the time—by moving to a location where solar heat is available or by moving into water, for instance. When the surroundings are too warm, the animals seek cooler microenvironments, such as shaded areas. Some amphibians, including bullfrogs, can vary the amount of mucus they secrete from their surface, a physiological response that regulates evaporative cooling.

Fishes   The body temperature of most fishes is usually within 1°–2°C of the surrounding water temperature. However, some large, active fishes maintain an elevated temperature at their body core. The heat generated by their swimming muscles is retained by specializations of the circulatory system. The bluefin tuna, for example, has its major blood vessels just under the skin. Branches deliver blood to the deep swimming muscles, where the small vessels are arranged into a countercurrent heat exchanger called the **rete mirabile**, which means "wonderful net." The apparatus enhances vigorous activity by keeping the swimming muscles several degrees warmer than tissues nearer the animal's surface, which is about the same temperature as the surrounding water (Figure 40.21). Such animals, called partial endotherms, illustrate how the distinction between ectothermy and endothermy can sometimes blur.

Invertebrates   Most invertebrates have very little control over their body temperature, but some do adjust temperature by behavioral or physiological mechanisms. The desert locust, for example, must reach a certain temperature to become active. It orients in a direction that maximizes absorption of sunlight.

Some species of large flying insects, such as bees and large moths, can generate internal heat. They are able to "warm up" before taking off by contracting all the flight muscles in synchrony, so that only slight movements of the wings occur but large amounts of heat are produced. (This is functionally analogous to shivering.) This higher temperature of the flight muscles enables the insects to sustain the intense activity required for flight. This mechanism has been elaborated to such an extent in bumblebees that the internal temperature of the abdomen is maintained above the environmental temperature at all times, not just in preparation for flight. Equally impressive are the endothermic adaptations of winter moths, which enable these insects to survive and fly during the cold winter months. For example, the thorax of a winter moth has a countercurrent heat exchanger that helps to maintain the flight muscles at a temperature of about 30°C, even on cold, snowy nights (Figure 40.22).

Honeybees use an additional mechanism that depends on social organization to increase body temperature. In cold weather, they increase their move-

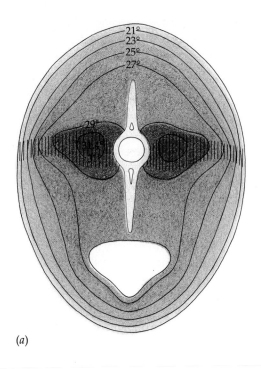

(a)

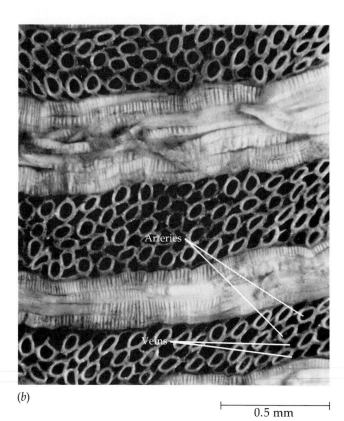

Arteries

Veins

(b)

0.5 mm

Figure 40.21
Thermoregulation in large, active fishes. (a) The bluefin tuna maintains an internal termperature much higher than the surrounding water. The fish is warmest around its dark swimming muscles (shad- ing). The temperature values were obtained for a tuna in 19°C water. (b) A shark, like the bluefin tuna, has a counter- current heat exchanger in its swimming muscles that reduces loss of metabolic heat. In this cross section of the counter- current system, arteries (thick walls) are surrounded by veins (thin walls).

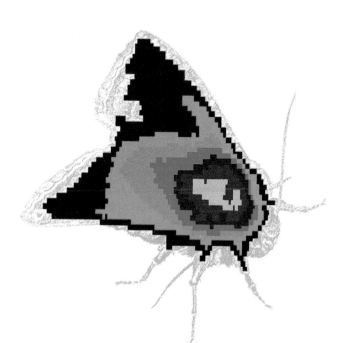

Figure 40.22
Thermoregulation in winter moths. Some species of the Noctuidae family of moths ("owlet moths") are active during cold winter months. Various endothermic adapta- tions, including a countercurrent heat exchanger in the thorax region, help keep the flight muscles warmed to a temperature of 30°C, even though the ambient tempera- ture may be subfreezing. An infrared map superimposed over an image of a winter moth portrays the distribution of heat immediately after a flight. Yellow, located here in the thorax region, indicates the highest temperature. Moving outward from the thorax, the variously colored zones correspond to regions of cooler and cooler body temperatures.

ments and huddle together, which retains their heat. They maintain a relatively constant temperature by changing the density of the huddling. Individuals move from the cooler outer edges of the cluster to the warmer center and back again, thereby circulating and distributing the heat. Honeybees also control the temperature of their hive by transporting water to it in hot weather and fanning with their wings, which promotes evaporation and convection. Hence, the whole colony of honeybees uses many of the mechanisms of temperature control also seen in single, larger organisms.

Temperature Acclimation

Many animals can adjust to a new range of environmental temperature over a period of many days or weeks, a physiological response called **acclimation.** Seasonal change is one context in which physiological adjustments to a new temperature range are important. For example, many frogs that inhabit temperate regions can tolerate winter temperatures that are low enough to be lethal to the same frogs if encountered during summer. The upper limits of temperature tolerance can also be adjusted. The bullhead catfish, for instance, can survive water temperatures up to 36°C in summer, but a temperature above 28°C is lethal during winter.

Physiological acclimation to a new temperature range is multifaceted. Cells may increase the production of certain enzymes, helping to compensate for the lowered activity of each enzyme molecule at temperatures that are not optimal. In other cases, cells produce variants of enzymes that have the same function, but different temperature optima (see Chapter 6). Membranes may also change in the proportions of saturated and unsaturated lipids they contain. This response helps keep membranes fluid at different temperatures (see Chapter 8). Other physiological responses to a new range of external temperatures involve adjustments in the mechanisms that control the internal temperature of the animal.

Torpor

Torpor is an alternative physiological state in which metabolism decreases and the heart and respiratory system slow down. One type of torpor is **hibernation,** during which body temperature is maintained at a lower level than normal (Figure 40.23a). Hibernation enables the animal to withstand long periods of cold temperatures and diminished food supplies. Another type of torpor characterized by slow metabolism and inactivity is **aestivation,** which allows an animal to survive long periods of elevated temperatures and diminished

(a)

(b)

Figure 40.23
Torpor. (a) During hibernation, metabolic rate is slowed substantially and body temperature is lowered, adaptations that enable the animal to survive long, cold periods when food is scarce. This hibernating animal is a golden-mantled ground squirrel. **(b)** Most bats, including these white tent bats in Costa Rica, exhibit diurnation, a daily torpor that alternates with a period of active feeding.

water supplies. **Diurnation** is a torpor physiologically similar to hibernation and aestivation, but lasting for much shorter times. Most bats, for instance, have a daily torpor during daylight hours and also hibernate for many weeks during the winter months (Figure 40.23b).

Many ectothermic animals enter a state of torpor when their food supply decreases. In addition to the other physiological changes, their optimal temperature range becomes lower. Environmental changes that tend to increase the food supply cause these animals to come out of their torpor in a matter of hours.

Some endothermic animals also exhibit torpor that appears to be adapted to their feeding patterns. All endotherms that show a daily torpor, such as hummingbirds, bats, and shrews, are relatively small and have a very high metabolic rate when active. The daily period of torpor allows the animals to survive the hours when they do not normally feed. However, this activity–torpor cycle may not be triggered by the availability of food. Even if food is made available to a shrew all day, it still goes through its daily torpor; the cycle is a built-in rhythm controlled by the biological clock (see Chapter 35). In fact, the need for sleep in humans may be a remnant of a more pronounced daily torpor in our early mammalian ancestors.

Hibernation and aestivation are often triggered by seasonal changes in the length of daylight. As the days shorten, some animals will eat huge quantities of food before hibernating. Ground squirrels, for instance, will more than double their weight in a month of gorging.

INTERACTION OF REGULATORY SYSTEMS

One of the major challenges in animal physiology is unraveling how the various regulatory systems are controlled and how they interact. For example, the regulation of body temperature involves mechanisms that also have impact on such parameters of the internal environment as osmolarity, metabolic rate, blood pressure, tissue oxygenation, and body weight. Under some conditions, usually at the physical extremes compatible with the organism's life, the demands of one system might come into conflict with those of other systems. For instance, in very warm and dry environments, the conservation of water takes precedence over evaporative heat loss. As a result, many desert animals tolerate occasional hyperthermia (abnormally high body temperature). Normally, however, the various regulatory systems act in concert to maintain homeostasis in the internal environment. The feedback circuits that integrate homeostasis involve nervous communication and hormones, which are chemical signals transmitted in the bloodstream. The next chapter addresses hormonal control in animals.

STUDY OUTLINE

1. Many animals can survive large fluctuations in their external environment by maintaining a relatively constant internal environment, a condition known as homeostasis.
2. Physiological adjustments, usually involving feedback mechanisms, regulate the hemolymph or interstitial fluid that bathes the body's cells.

Osmoregulation (pp. 875–879)

1. Water uptake must balance water loss, requiring various mechanisms of osmoregulation in different environments. Osmoconformers are isosmotic with their aqueous surroundings and do not regulate their osmolarity. Osmoregulators control water uptake and loss in a hyperosmotic or hypoosmotic environment.
2. Most marine invertebrates are osmoconformers, whereas most marine vertebrates osmoregulate. Sharks have an osmolarity slightly higher than seawater because they retain urea. Marine bony fish lose water to their hyperosmotic environment and must compensate by drinking large quantities of seawater. Marine vertebrates excrete excess salt through rectal glands, gill epithelia, or nasal salt glands.

3. Freshwater organisms constantly take in water from their hypoosmotic environment. Protozoa pump out excess water with contractile vacuoles, and freshwater animals excrete copious amounts of dilute urine. Salt loss is replaced by eating or ion uptake across gill epithelia.
4. Euryhaline animals are able to tolerate large osmotic changes in their environments, often by altering their osmoregulatory mechanisms. Some animals are capable of anhydrobiosis, surviving dehydration in a dormant, biochemically protected state.
5. Terrestrial animals combat desiccation by drinking and eating food with high water content, and through nervous and hormonal control of thirst, behavioral adaptations, and water-conserving excretory organs.
6. Osmoregulation is usually accomplished by the transport of salt across a transport epithelium, followed by the osmotic flow of water.

Excretory Systems of Invertebrates (pp. 879–882)

1. Extracellular fluid is filtered into the protonephridia of the flame-cell system in flatworms. These blind-ended tubules excrete a dilute fluid and function in osmoregulation.

2. Each segment of an earthworm has a pair of open-ended, tubular excretory organs closely associated with the capillaries. These metanephridia collect coelomic fluid, the transport epithelia pump out salts for reabsorption, and a dilute urine is excreted through nephridiopores.

3. In insects, Malpighian tubules function in osmoregulation and removal of nitrogenous wastes from the hemolymph, and produce a relatively dry waste matter, an important adaptation to terrestrial life.

The Vertebrate Kidney (pp. 882–892)

1. Nephrons, the excretory tubules of vertebrates, are arranged into compact organs, the kidneys, which, along with associated blood vessels and excretory ducts, comprise the excretory system. Urine exits the kidney through the ureter, is temporarily stored in the urinary bladder, and then eliminated through the urethra.

2. The major parts of the renal tubule are Bowman's capsule, proximal convoluted tubule, loop of Henle (with its descending and ascending limbs), distal convoluted tubule, and collecting duct, which passes urine into the renal pelvis. Most nephrons are radially arranged in the kidney cortex; mammals and birds also have juxtamedullary nephrons with loops of Henle that extent into the medulla. Blood supply to the nephron is through an afferent arteriole, which divides into the capillaries of the glomerulus. An efferent arteriole carries blood away from Bowman's capsule and subdivides into the peritubular capillaries embracing the convoluted tubules. The vasa recta is a system of capillaries that serves the loop of Henle.

3. Nephrons control the composition of the blood by filtration, secretion, and reabsorption. During filtration, blood pressure nonselectively filters water and small solutes from the glomerulus held within Bowman's capsule into the lumen of the nephron tubule. Additional substances destined for excretion are directly secreted from the interstitial fluid into the tubule by active and passive transport. Filtered substances that must be returned to the blood, such as vital nutrients and water, are reabsorbed at various points along the nephron.

4. Most of the salt and water filtered from the blood is reabsorbed by the proximal convoluted tubule. In addition, ammonia, drugs, and hydrogen ions (for the control of body pH), are selectively secreted into the filtrate, glucose and amino acids are actively transported out of the filtrate, and potassium is reabsorbed. The descending limb of the loop of Henle is permeable to water but not to salt; water moves by osmosis into the hyperosmotic interstitial fluid. Salt diffuses out of the concentrated filtrate as it moves through the salt-permeable ascending limb of the loop of Henle. The distal convoluted tubule is specialized for selective secretion and reabsorption, playing a key role in regulating potassium concentration and pH of the blood. The collecting duct, which is permeable to water but not to salt, carries the filtrate through the osmolarity gradient of the medulla, and more water exits by osmosis. Urea also diffuses out of the tubule, joining salt in forming the osmotic gradient that enables the kidney to produce urine that is hyperosmotic to the blood.

5. The osmolarity of the filtrate varies as it travels through a juxtamedullary nephron. The differing permeabilities to water, salt, and urea in regions of the tubule, combined with the active transport of salt, produce the increasing osmotic gradient in the medulla that results in the production of a hyperosmotic urine. The countercurrent flow of blood through the vasa recta supplies the medulla with oxygen and nutrients but without disrupting the osmotic gradient.

6. The osmolarity of the urine, which can vary widely depending on the hydration needs of the body, is regulated by nervous and hormonal control of water and salt reabsorption in the kidneys. ADH, released in response to a rise in blood osmolarity signaled by osmoreceptor cells in the hypothalamus, increases water reabsorption by the tubule. The juxtaglomerular apparatus responds to decreased blood pressure or Na^+ concentration by the release of renin, which activates angiotensin. This activated blood protein causes blood arterioles to constrict and the adrenal glands to release aldosterone, a hormone that stimulates reabsorption of Na^+ and the passive flow of water from the filtrate. Both the ADH and aldosterone control mechanisms result in the reabsorption of water and a more concentrated urine. ANP, released by the atrium in response to increased blood pressure, inhibits the release of renin and counters the renin-angiotensin-aldosterone system.

7. Adaptations of nephron structure and function in the vertebrates are related to the requirements for osmoregulation and excretion of nitrogenous wastes in various habitats.

Nitrogenous Wastes (pp. 892–894)

1. The metabolism of proteins and nucleic acids generates ammonia, a toxic waste product that is excreted in one of three forms, depending on the habitat and evolutionary history of the animal.

2. Most aquatic animals excrete ammonia, a highly toxic but very soluble molecule that easily passes across the body surface or gill epithelia into the surrounding water.

3. The liver of mammals and most adult amphibians converts ammonia into the less toxic urea, which is carried by the circulatory system to the kidneys and excreted in concentrated form with a minimal loss of water.

4. Uric acid is an insoluble precipitate excreted in the paste-like urine of land snails, insects, birds, and some reptiles.

5. The mode of reproduction of terrestrial animals is related to their form of nitrogenous waste product. Differences in waste product within phyla are related to habitat; some organisms can actually shift their nitrogenous waste form depending on environmental conditions.

Regulation of Body Temperature (pp. 894–902)

1. Heat transfer between the organism and the environment involves the physical processes of conduction, convection, radiation, and evaporation.

2. Ectotherms, such as invertebrates, fishes, reptiles, and amphibians, absorb their body heat from the environment. Mammals and birds are endotherms, which rely on the heat from their own metabolism. Endothermy allows terrestrial animals to maintain a constant body

temperature and a high level of aerobic metabolism that facilitates movement on land.

3. The physiological and behavioral processes that enable land mammals to regulate their body temperature include the adjustment of the rate of metabolic heat production by shivering or nonshivering thermogenesis, vasodilation or vasoconstriction of surface blood vessels, control of evaporative heat loss through panting or sweating, and various behavioral responses.

4. The heating center and cooling center are thermoregulatory areas of the hypothalamus that receive nervous impulses from warm and cold receptors and respond by initiating either cooling or warming physiological processes.

5. Birds may thermoregulate by panting, increasing evaporation from a vascularized pouch in the mouth, and by passing blood going to the legs through a countercurrent heat exchanger.

6. Marine mammals maintain their high body temperatures in cold water by a thick layer of insulating blubber and countercurrent heat exchange between arterial and venous blood. Vasodilation of skin vessels allows dissipation of heat in warm waters.

7. Reptiles and amphibians maintain internal temperatures within tolerable ranges by various behavioral adaptations.

8. Although the body temperature of most fishes matches the environment, some large, active species maintain a higher temperature in their swimming muscles with a countercurrent heat exchanger called a rete mirabile.

9. A few invertebrates use physiological or behavioral mechanisms, such as muscle contractions, countercurrent heat exchangers, or social huddling, to adjust their temperature.

10. Many animals can physiologically acclimate to a gradual change in temperature, such as a seasonal change.

11. Torpor, including hibernation, aestivation, and diurnation, is a physiological state characterized by a decrease in metabolic, heart, and respiratory rates. This state allows the animal to temporarily withstand varying periods of unfavorable temperatures or lack of food and water.

Interaction of Regulatory Systems (p. 902)
1. Homeostasis is a dynamic response to the environment, integrated by feedback circuits involving nervous communication and hormones.

SELF-QUIZ

1. Which of the following organisms is closest to being isosmotic to its environment?

   a. marine bony fish
   b. freshwater bony fish
   c. marine jellyfish
   d. freshwater protozoan
   e. marine reptile

2. *Unlike* an earthworm's metanephridia, a mammalian nephron

   a. is intimately associated with a capillary network
   b. forms urine by changing the composition of fluid inside the tubule
   c. functions both in osmoregulation and excretion of nitrogenous wastes
   d. filters blood instead of coelomic fluid
   e. has a transport epithelium

3. The majority of water and salt filtered into Bowman's capsule is reabsorbed by

   a. the brush-border of the transport epithelia of the proximal convoluted tubule
   b. diffusion from the descending limb of the loop of Henle into the hyperosmotic interstitial fluid of the medulla
   c. active transport across the transport epithelium of the thick upper segment of the ascending limb of the loop of Henle
   d. selective secretion and diffusion across the distal convoluted tubule
   e. diffusion from the collecting duct into the increasing osmotic gradient of the medulla

4. The high osmotic concentration of the kidney medulla is maintained by all of the following *except*

   a. diffusion of salt from the ascending limb of the loop of Henle
   b. active transport of salt from the upper region of the ascending limb
   c. spatial arrangement of juxtamedullary nephrons
   d. diffusion of urea from the collecting duct
   e. diffusion of salt from the collecting duct

5. Antidiuretic hormone affects the kidney by

   a. stimulating the release of renin
   b. constricting arterioles and thus raising blood pressure
   c. increasing the reabsorption of Na^+ in the distal convoluted tubules
   d. countering the renin-angiotensin-aldosterone mechanism
   e. increasing the water permeability of the distal convoluted tubule and collecting duct, thus increasing the osmolarity of the urine.

6. Ammonia (or ammonium) is excreted by most

   a. bony fishes
   b. organisms that produce shelled eggs
   c. adult amphibians
   d. land snails
   e. insects

7. Nonshivering thermogenesis is

   a. a behavioral adaptation for absorbing heat in ectotherms
   b. a hormone-triggered rise in metabolic rate, often associated with brown fat
   c. a countercurrent heat exchange of blood going to the limbs
   d. a heat-producing method of the rete mirabile of large, active fishes

e. isometric muscle contractions that have been identified in winter moths

8. Physiological adjustments or acclimation by an ectotherm to cooler seasonal temperatures may include

    a. an increase in metabolic rate

    b. aestivation

    c. changes in the lipid components of cell membranes

    d. increased vasodilation and countercurrent heat exchange

    e. hibernation

9. Malpighian tubules are excretory organs found in

    a. vertebrates

    b. insects

    c. flatworms

    d. annelids

    e. jellyfish

10. Which process in the nephron is *least* selective?

    a. secretion

    b. reabsorption

    c. transport across epithelium of collecting duct

    d. filtration

    e. salt pumping by loop of Henle

CHALLENGE QUESTIONS

1. A large part of the terrestrial success of arthropods and vertebrates is attributable to their osmoregulatory capabilities. Compare and contrast the Malpighian tubule with the nephron in regard to anatomy, relationship to circulation, and physiological mechanisms for conserving body water.

2. Humans are hyperosmotic to fresh water. Why don't we swell significantly and burst in a bathtub or swimming pool?

3. The longer a cat plays with a lizard or frog, the harder it is for the reptile or amphibian to escape. Explain this scenario in terms of the metabolic adaptations of the animals (assume that the cat has not injured its captive).

FURTHER READING

1. Heinrich, B. "Thermoregulation in Winter Moths." *Scientific American*, March 1987. Endothermy in insects.
2. Monastersky, R. "Dinosaurs in the Dark." *Science News*, March 19, 1988. How did dinosaurs living near the poles survive winter?
3. Pond, C. "Bearing Up in the Arctic." *New Scientist*, February 4, 1989. Thermoregulatory adaptations of polar bears may teach us some things about our own physiology.
4. Smith, H. W. *From Fish to Philosopher*. Boston: Little Brown and Co., 1953. Vertebrate evolution as revealed by kidney structure and function.
5. Tortora, G. J., and N. P. Anagnostakos. *Principles of Anatomy and Physiology*. 6th ed. New York: Harper & Row, 1990. An excellent chapter on the mammalian kidney, with colorful, lucid illustrations.

41

Chemical Coordination

P eople invoke hormones to explain the moodiness of teenagers and the howling of alley cats. One and a half million diabetics in the United States take the hormone insulin, and other hormones are used in cosmetics intended to keep the skin smooth or are added to livestock feed to fatten chickens and cows. Hormones affect metabolism, growth, development, emotion, and behavior and also contribute to homeostasis. But what are they?

Animals coordinate the activities of their specialized parts. The two major systems of internal communication that have evolved are the **nervous system** and the **endocrine system**. The nervous system, which we will discuss in Chapters 44 and 45, conveys high-speed signals along specialized cells called neurons. These rapid communications function in such activities as the movement of body parts in response to sudden environmental changes—jerking one's hand away from a flame, for instance. Other biological processes use slower means of communication. Different parts of the body, for example, must be informed how fast to grow and when to develop the characteristics that distinguish male and female or the juvenile and the adult of a species (Figure 41.1). In some cases, neighboring cells communicate by chemical messages that simply diffuse through the interstitial fluid from one cell to another. In many cases, however, messages must travel over greater distances to reach their targets. The endocrine system is a major source of such dispatches in the form of the chemicals called hormones. This chapter surveys the hormonal control of animal form and function.

Figure 41.1
Hormones and insect metamorphosis. Hormones are chemical signals that help control growth, development, reproduction, and homeostasis. For example, insect hormones are responsible for triggering the metamorphosis of a caterpillar into a butterfly. This monarch butterfly has just emerged from its cocoon.

CHEMICAL MESSENGERS OF THE BODY

All animals exhibit some form of coordination by chemical signals. The hormones of the endocrine system convey information between organs of the body, while other types of chemical messengers are used for other functions. Pheromones are like hormones in that they are chemical signals, but they are used to communicate between different individuals, as in mate attraction. Still other messengers act only between cells on a localized scale. The most familiar of these local regulators are the neurotransmitters that carry information between cells of the nervous system. In this major section of the chapter, we compare the different types of chemical messengers.

Hormones

Traditionally, an animal **hormone** (from the Greek *hormon,* "excite") has been defined by the following criteria: It is a specific molecule synthesized and secreted by a group of specialized cells usually referred to as an endocrine gland; it is released by these cells into the circulatory system; and it travels to another area of the body, where it elicits specific biological responses from selected **target cells.**

A gland is a secretory organ, and animal glands can be classified as exocrine or endocrine. Exocrine glands produce a variety of substances, such as sweat, mucus, and digestive enzymes, and convey their products to the appropriate locations by means of ducts. In contrast, **endocrine glands** are ductless glands. They produce hormones and secrete these chemical messengers into the bloodstream for distribution throughout the body. Many organs perform both endocrine and exocrine functions. The pancreas, for example, secretes at least two hormones directly into the circulatory system. However, endocrine cells make up only 1%–2% of the total weight of the pancreas; the rest of the organ is exocrine tissue that produces bicarbonate ions and digestive enzymes that are carried to the small intestine through a duct.

Hormones and the glands that produce them are two key components of the endocrine system. A third and equally important component is the target cells and their molecular receptors for hormones. These **receptors,** proteins within the target cell or on its surface, determine much of the specificity and action we ascribe to hormones. No amount of male sex hormone will produce masculine features if the target cells do not have the appropriate receptors.

The endocrine system, like other body systems, does not work in isolation; it often operates in close conjunction with the nervous system. Just as we have seen emergent properties arise from interactions among components beginning at the molecular level, the interaction of the nervous and endocrine systems creates a communication and control system that goes far beyond the capabilities of each system alone. This integration is vital for making various parts of the organism function as a whole. A failure in one part of the endocrine system can cause problems serious enough to threaten an animal's life.

Current research in **endocrinology,** the study of hormones, is moving so rapidly that any description of hormones and their actions must be kept flexible. There are more than 50 different hormones in the human body, and the endocrine systems of other animals are similarly elaborate. In terms of their chemical structure, these hormones can be grouped into three general classes. **Steroid hormones,** which include the sex hormones, are fat-soluble molecules that the body fashions from cholesterol (see Chapter 5). Another group consists of hormones derived from amino acids, most frequently from tyrosine. These water-soluble molecules, each smaller than a steroid molecule, include the "fight-or-flight" hormone epinephrine. The third group of hormones, the **peptides,** is the most diverse. Peptides are chains of amino acids. Some are only three amino acids long, but other hormones are full-fledged proteins with as many as 200 amino acids. Several peptide and amino-acid-derived hormones act as signal molecules in the nervous system and in the endocrine system. Notice that our classification of hormones as steroids, amino-acid derivatives, or peptides groups the hormones according to similarities in chemical structure without considering function. Two hormones belonging to the same chemical class may have completely unrelated functions.

Pheromones

Pheromones are chemical signals that function much like hormones, with one important exception: Instead of coordinating the parts of a single animal's body, pheromones are communication signals *between animals* of the same species. Pheromones are often classified according to their functions as mate attractants, territorial markers, or alarm substances, to name a few. One interesting example of a pheromone is the queen substance released by the queen of a honeybee colony. This pheromone causes the worker bees to settle down near the queen and prevents them from giving the young a special diet that makes new queens. Thus, the queen substance is essential to the stability of the hive.

All pheromones are small, volatile molecules that can disperse easily into the environment and, like hormones, are active in minute amounts. The sex attractants of some female insects can be detected by males as much as a mile away. The pheromone of the female gypsy moth elicits behavioral responses in males at concentrations as low as 1 molecule of pheromone in 10^{17} molecules of other gases in air.

Compared to most animals, humans do not have well-developed olfactory senses, but it is interesting to speculate whether we use pheromones to communicate. Some indirect evidence suggests we do. There have been cases of women living together for several months, as in dormitories, convents, or prisons, who begin to have synchronous menstrual cycles: They all menstruate at the same time of the month. The role of pheromones in social behavior is considered in more detail in Chapter 50.

Local Regulators

Chemical messengers that affect target cells adjacent to or near their point of secretion function in local regulation. The neurotransmitters are a familiar group of local regulators. These substances carry information from one neuron to another or from a neuron to a muscle, gland, or other target cell. Histamine and interleukins are among the local regulators that coordinate the immune response (see Chapter 39). Two additional groups of local regulators are growth factors and prostaglandins.

Growth Factors

Attempts to culture mammalian cells on artificial media led to the discovery of various **growth factors,** proteins that must be present in the extracellular environment in order for certain types of cells to grow and develop normally. For example, embryonic neurons will develop the long processes called axons only if a protein called nerve growth factor (NGF) is present in the culture medium. Epidermal growth factor (EGF) is required to grow epithelial cells in culture, but, in spite of its name, EGF also stimulates the growth of many other types of cells. Several other growth factors have also been discovered.

A variety of experiments indicate that growth factors not only work in the artificial environment of cultured cells, but also regulate development within the animal body. For instance, injecting epidermal growth factor into fetal mice accelerates epidermal (skin) development. Also, antibodies that destroy nerve growth factor obstruct normal development of the nervous system when injected into rodent embryos.

Experiments have yet to reveal how growth factors work. It is known that the target cells of a particular growth factor have proteins on their surfaces that function as receptors for that factor. It is the binding of the growth factor to the receptor that triggers the response of the target cell to the chemical stimulus. It is interesting that some oncogenes (genes that contribute to cancer—see Chapter 18) code for membrane proteins that mimic growth factor receptors. However, these aberrant receptors do not require the binding of a growth factor to stimulate growth and division of cells, and thus the tissue grows uncontrollably and abnormally.

Prostaglandins

The **prostaglandins (PGs)** are modified fatty acids, often derived from lipids of the plasma membrane. Released from most types of cells into the interstitial fluid, prostaglandins function as local regulators affecting nearby cells in various ways.

About 16 prostaglandins have been discovered, and very subtle differences in their molecular structures can result in profound differences in how these signals affect target cells (Figure 41.2). For example, two

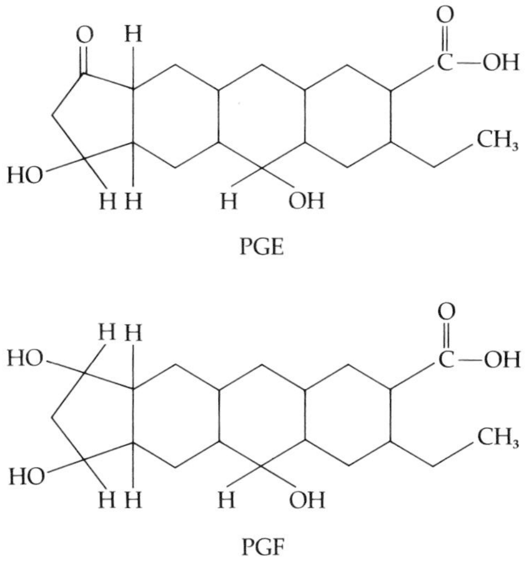

Figure 41.2
Prostaglandins. About 16 prostaglandins belonging to several families have been discovered so far. The differences in molecular structure are subtle; for example, the prostaglandin E family (top) differs from the prostaglandin F family (bottom) by one chemical group on the ring structure of molecules (color). These minor structural differences result in major functional differences in how these local chemical messengers affect target cells.

prostaglandins, called prostaglandin E and prostaglandin F, have opposite effects on the smooth muscle cells in the walls of blood vessels serving the lungs. PGE causes the muscles to relax, which dilates the blood vessels and promotes oxygenation of blood. PGF signals the muscles to contract, which constricts the vessels and reduces blood flow through the lungs. Thus, these two chemical signals are antagonistic, and shifts in their relative concentrations contribute to moment-to-moment adjustments an animal must make to cope with changing circumstances. The use of antagonistic signals as counterbalances is a common regulatory mechanism in both chemical and nervous coordination of the body.

Some of the best-known actions of prostaglandins are on the female reproductive system. For example, prostaglandins secreted by cells of the placenta cause the nearby muscles of the uterus to contract, helping to induce labor during childbirth.

Prostaglandins also function as local regulators in the defense mechanisms of vertebrates. Various prostaglandins help induce fever and inflammation and also intensify the sensation of pain (which can be thought of as contributing to the body's defense by sounding an alarm that something harmful is going on). Aspirin may reduce these symptoms of injury or infection by inhibiting the secretion of prostaglandins.

So far, biologists have probably recognized only a fraction of the multifarious functions of prostaglandins as local regulators of animal tissues. Scientists continue to discover new hormones and hormonelike substances and are learning more about their actions. In fact, a central question in endocrinology is: How do hormones actually trigger the changes they do?

MECHANISMS OF HORMONE ACTION

A set of general principles seems to govern the activity of those hormones we know about to date. First, hormones can act at very low concentrations. Second, a given hormone can affect different target cells in an animal differently, or vary in effects from one species of animal to another. A striking example is thyroxine, a hormone of the thyroid gland. In humans and other vertebrates, it is responsible for metabolic regulation. But thyroxine is also central in the development of an adult frog from the tadpole stage, triggering resorption of the tadpole's tail and other morphological changes during metamorphosis.

Although the diversity of responses to hormones is seemingly endless, a hormone triggers specific changes in a target cell by one of only two general mechanisms. Some hormones, especially the steroids, enter the nucleus of the target cell and influence the expression of the cell's genes. In contrast, most nonsteroid hormones, including peptides, attach to the cell surface and influence activity within the cell through cytoplasmic intermediaries called second messengers. Both mechanisms involve receptor molecules to which the hormones bind. The receptors are proteins, and it is the hormone–receptor complex that actually triggers the effects of the hormone.

Steroid Hormones and Gene Expression

Steroids pass through the target cell membrane and enter the nucleus, where they affect the transcription of specific genes (see Chapter 18). (Thyroxine, though not a steroid, probably works in a similar manner.) The general mechanism by which steroid hormones act was discovered by studying two vertebrate hormones, estrogen and progesterone. In most mammals, including humans, these hormones are necessary for the normal development and function of the female reproductive system. In the early 1960s, researchers demonstrated that cells in the reproductive tract of female rats accumulate estrogen. The hormone was found within the nuclei of these cells, but not in the cells of tissues such as the spleen, which do not respond to estrogen. Progesterone also enters the nuclei of target cells. Such observations led to the hypothesis that cells sensitive to a steroid hormone contain receptor molecules that bind specifically to that hormone. It is now known that the steroid receptors

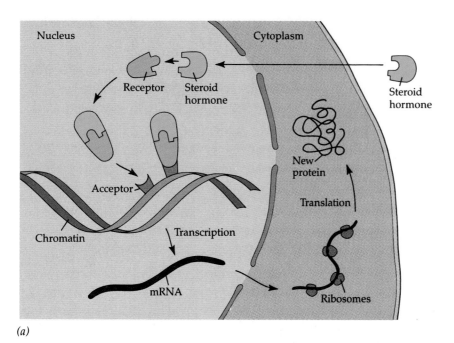

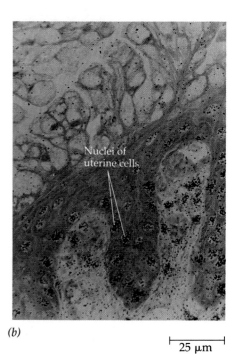

(a)

(b)

$\vdash\!\!\!-\!\!\!\dashv$
25 μm

Figure 41.3
Steroid hormone action. (a) In this model, the lipid-soluble steroid passes through the plasma membrane and binds to a receptor protein present only in target cells. Recent evidence suggests that the receptors reside in the nucleus rather than the cytoplasm. The hormone–receptor complex then binds to specific acceptor proteins located at certain sites along the chromatin, activating expression of certain genes. **(b)** In this autoradiogram (see Chapter 2), radioactively labeled proges-terone, a steroid hormone (black dots), is concentrated in the nuclei of specific target cells in the uterus of a guinea pig. Other cell types in the uterus do not bind the hormone (LM).

are specialized proteins within the cells, perhaps in the nucleus.

Specificity of receptor proteins is a key to understanding steroid hormone action. According to the model illustrated in Figure 41.3, a steroid hormone that crosses the membrane of its target cell binds to a receptor protein in the nucleus. This hormone–protein complex then has the proper conformation to bind to another specific protein, an acceptor protein located on certain regions of the chromatin. This association initiates the transcription of selected genes. The messenger RNA molecules thus produced are processed and transported to the cytoplasm, where they direct the synthesis of new proteins. For example, estrogen induces certain cells in the reproductive system of a female bird to synthesize large quantities of ovalbumin, the main storage protein stockpiled in egg white. Different proteins are synthesized in the liver cells of the same animal in response to estrogen.

How can two types of cells respond differently to a common chemical signal (estrogen, in this case)? It is likely that the estrogen receptor is the same protein in both target cells, but the chromosomal proteins that accept the hormone–receptor complex are associated with different genes in the two kinds of cells. This hypothesis implies that a key episode in cellular dif-ferentiation is the specific placement of chromosomal proteins that function as allies of steroid hormones in controlling gene expression.

Peptide Hormones and Second Messengers

A dramatic example of hormone action is illustrated when a frog's skin turns darker or lighter. This color change is controlled by a peptide hormone called melanocyte-stimulating hormone (MSH), secreted by the pituitary gland at the base of the brain. Melanocytes are specialized skin cells that contain the dark brown pigment melanin in cytoplasmic organelles called melanosomes. The frog's skin appears light when the melanosomes are clustered tightly around the cell nucleus and darker when the melanosomes spread throughout the cell. When MSH is added to the inter-stitial fluid surrounding the pigment-containing cells, the melanosomes disperse. However, direct microin-jection of MSH into individual melanocytes does not induce melanosome dispersion. Why won't some hor-mones act directly within a target cell?

Peptide hormones and most hormones derived from amino acids are unable to pass through the plasma membrane of their target cells, but they still influence cellular activity. Unlike steroids, these hormones bind to specific receptors on the cell's outer surface. The hormone receptors are proteins embedded in the plasma membrane. The surface of a target cell typically has about 10,000 receptors keyed to a particular hormone. (Based on this large number, one might think that the membrane is densely populated by proteins devoted to receiving a single type of chemical message. However, 10,000 receptors account for only about one ten-thousandth of the total number of proteins embedded in the plasma membrane of an animal cell.) On the extracellular surface of each receptor protein is a binding site that specifically fits a particular hormone molecule. This binding of hormone to the receptor sparks a burst of biochemical activity within the cell. But how can a chemical signal received on the *outside* of a cell affect metabolism *inside* the cell? The solution is a **second messenger,** an intracellular signal produced by the cytoplasmic surface of the plasma membrane in response to extracellular binding of the first messenger, the hormone. The most important secondary messengers are two compounds called cyclic AMP and inositol triphosphate.

Cyclic AMP Current understanding of how nonsteroid hormones act via second messengers has advanced from the pioneering work of Earl W. Sutherland, whose research led to a Nobel Prize in 1971. Sutherland and his colleagues at Vanderbilt University were investigating how the hormone epinephrine stimulates the hydrolysis of glycogen within liver cells and muscle cells. Glycogen is a storage polysaccharide, and its hydrolysis to release the sugar glucose-1-phosphate increases the energy supply for cells. Thus, one effect of epinephrine, secreted from the adrenal gland during times of physical or mental stress, is the mobilization of fuel reserves. Sutherland's research team discovered that epinephrine stimulates glycogen breakdown by activating a cytoplasmic enzyme called glycogen phosphorylase. However, if epinephrine was added to a test-tube mixture containing the phosphorylase and its substrate, glycogen, no hydrolysis occurred. Epinephrine could activate glycogen phosphorylase only if the hormone was added to the extracellular solution bathing intact cells with plasma membranes. Thus, the search began for a second messenger that transmits the signal from the plasma membrane to the metabolic machinery in the cytoplasm.

Sutherland found that the binding of epinephrine to the plasma membrane of a liver cell elevates the cytoplasmic concentration of a compound called cyclic adenosine monophosphate, abbreviated **cyclic AMP** or **cAMP** (Figure 41.4). An enzyme, **adenylate cyclase,** converts ATP to cAMP in response to a hormonal signal, in this case, epinephrine. Adenylate cyclase is a membrane protein with its active site facing the cytoplasm. The enzyme is idle until activated by the binding of epinephrine to the hormone's receptor protein. Thus, the first messenger, the hormone, triggers the synthesis of cAMP, which functions as a second messenger that relays the signal to the cytoplasm. It is not epinephrine itself, but its cytoplasmic emissary, cAMP, that activates glycogen hydrolysis within the cell. When the extracellular concentration of the hormone declines, this hushes the second message. Hormones bind reversibly to their receptors by weak bonds, and less epinephrine means that there are, at any particular instant, fewer hormone–receptor complexes to activate adenylate cyclase. Also, the second messenger does not persist for long in the absence of the first message, because another enzyme converts the cAMP to an inactive product. Another surge of epinephrine will again boost the cytoplasmic concentration of the second messenger. Epinephrine is only one of many hormones derived from amino acids and peptide hormones that use cAMP as a second messenger.

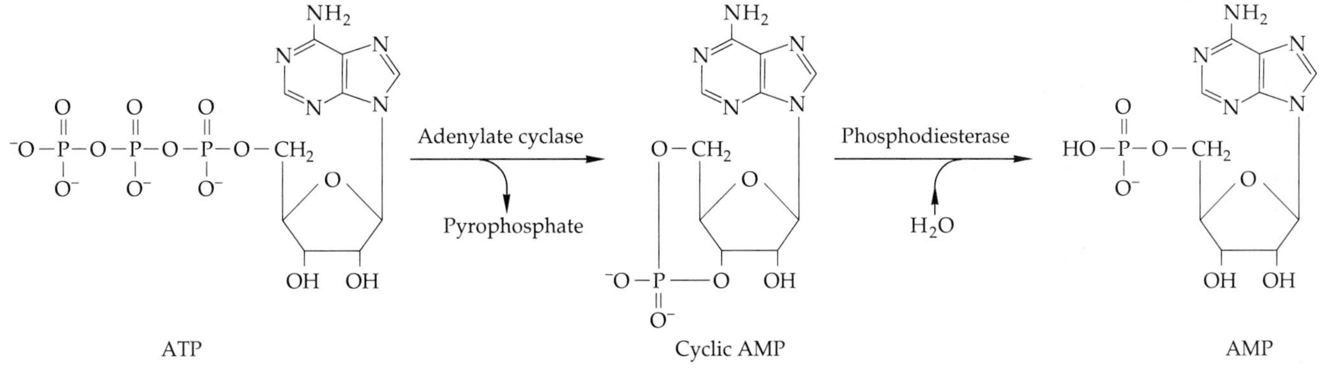

Figure 41.4
Cyclic AMP, a second messenger. Binding of a peptide hormone to its specific receptor on the surface of a target cell causes adenylate cyclase, an enzyme embedded in the plasma membrane, to convert ATP to cyclic AMP (cAMP). The cAMP functions as a second messenger that relays the signal from the membrane to the metabolic machinery of the cytoplasm. When the first message, the hormonal signal, ceases, the second message is also soon hushed by the action of phosphodiesterase, an enzyme that converts cAMP to inactive AMP.

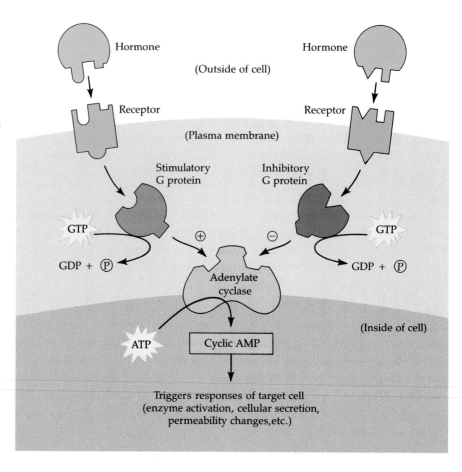

Figure 41.5
Control of adenylate cyclase. A hormone–receptor complex does not interact directly with adenylate cyclase, but uses intermediaries called G proteins (because they hydrolyze GTP). The cytoplasmic concentration of cAMP, the second messenger in responses to many hormones, is partly controlled by the effects of stimulatory and inhibitory G proteins on the activity of adenylate cyclase.

Subsequent research has revealed additional players in the second-messenger scenario of hormone action. Although the hormone receptor and adenylate cyclase are both built into the plasma membrane, they do not interact directly. How, then, does the binding of hormone to the receptor activate adenylate cyclase? A third membrane protein functions as an intermediary. It is called a **G protein** because it binds to GTP (guanosine triphosphate), an energy carrier closely related to ATP (see Chapter 9). Docking of the hormone with the receptor activates the G protein, which in turn activates adenylate cyclase to produce cAMP (Figure 41.5). The energy for converting the first message to the second message is provided by GTP, which the G protein hydrolyzes to GDP (guanosine diphosphate—this exergonic process is basically the same as the ATP → ADP + Ⓟ reaction).

The G protein provides a versatile link between a variety of hormone receptors and adenylate cyclase. The same G protein that stimulates cAMP production in response to epinephrine also transduces several other hormonal signals; for example, MSH, the hormone that triggers pigment dispersion in the melanocytes of a frog, works via the receptor-G protein-adenylate cyclase linkage.

A second type of G protein *down* regulates (inhibits) adenylate cyclase. Binding of an inhibitory hormone to its receptor causes this G protein to lower the activity of adenylate cyclase, thus reducing the cytoplasmic concentration of cAMP. These opposing effects of two G proteins in regulating adenylate cyclase enables the cell to fine-tune its metabolism in response to slight changes in the proportions of antagonistic hormones.

Having examined how membrane proteins convert a hormonal signal to the second message, let us next consider how cAMP is actually translated into a particular metabolic response by the cell. In the case of sugar mobilization in response to epinephrine, we can specifically ask how cAMP stimulates the activity of glycogen phosphorylase, the enzyme that catalyzes hydrolysis of glycogen. The stimulation is indirect, using two intermediaries (Figure 41.6). First, cAMP activates an enzyme called cAMP-dependent protein kinase. A **protein kinase** is an enzyme that catalyzes the transfer of a phosphate group from ATP to proteins, a process called phosphorylation. Depending on the type of protein that receives the phosphate group, phosphorylation either increases or decreases the activity of the protein. In the case of liver cells responding to epinephrine, the cAMP-dependent protein kinase stimulates *another* kinase enzyme, phosphorylase kinase. This second kinase adds a phosphate group to glycogen phosphorylase, the enzyme that actually hydrolyzes glycogen. Thus, the metabolic response to epinephrine involves an **enzyme cascade,** where each step activates an enzyme that in turn activates the next enzyme in the series: cAMP starts the cascade within the cytoplasm by activating

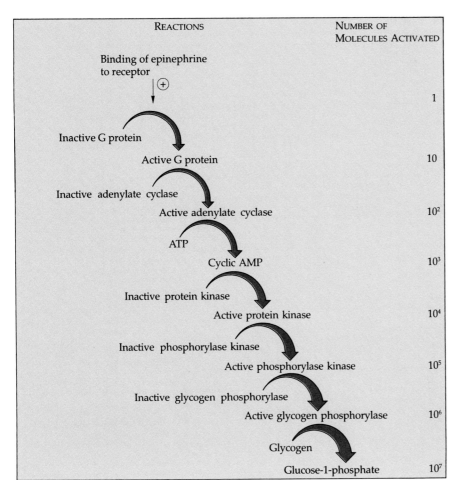

REACTIONS	NUMBER OF MOLECULES ACTIVATED
Binding of epinephrine to receptor	1
Inactive G protein → Active G protein	10
Inactive adenylate cyclase → Active adenylate cyclase	10^2
ATP → Cyclic AMP	10^3
Inactive protein kinase → Active protein kinase	10^4
Inactive phosphorylase kinase → Active phosphorylase kinase	10^5
Inactive glycogen phosphorylase → Active glycogen phosphorylase	10^6
Glycogen → Glucose-1-phosphate	10^7

Figure 41.6
How an enzyme cascade amplifies response to a hormone. In this system, the hormone epinephrine acts through cyclic AMP to activate a succession of enzymes, leading to extensive hydrolysis of glycogen. In reality, the number of activated molecules at each step is much larger than indicated. With each step, the hormonal signal is amplified.

protein kinase, which activates phosphorylase kinase, which activates glycogen phosphorylase, which mobilizes sugar by hydrolyzing glycogen.

One advantage of this elaborate enzyme cascade is that it amplifies response to the hormone. With each catalytic step in the cascade, the number of activated products is much greater than in the preceding step. A relatively small number of adenylate cyclase molecules can produce a large number of cAMP molecules, which activate even more molecules of protein kinase, which activate still more molecules of phosphorylase kinase, and so on. As a result of this amplification, a very few molecules of epinephrine binding to receptors on the surface of a liver or muscle cell can quickly result in the release of millions of sugar molecules from the storage polysaccharide glycogen.

Such an indirect mechanism of hormone action offers another advantage. An animal can use a single basic mechanism to mediate responses of diverse cell types to many different hormones. We have already seen that two different hormones can sound the same second message via a common linkage of G proteins to adenylate cyclase. The second messenger, cAMP, alters the activity of various cytoplasmic proteins by activating a cAMP-dependent protein kinase. Given this generalized action of hormones, how can we account for specificity in the responses of various target cells

to hormonal signals? First, a particular type of cell is only tuned-in to certain hormones—those that fit the receptors found on the surface of that cell type. Second, cAMP-dependent protein kinases vary somewhat in structure and function from tissue to tissue. Third, the steps in the enzyme cascade that follow the activation of protein kinase differ markedly from one cell type to another. Different kinds of specialized cells contain different proteins capable of being phosphorylated by protein kinase. In a liver cell responding to epinephrine, protein kinase activates phosphorylase kinase in the enzyme cascade that leads to glycogen breakdown. In a frog melanocyte responding to MSH, protein kinase apparently activates proteins of the cytoskeleton that function in pigment dispersion. In other cases, protein kinase phosphorylates membrane proteins and alters the transport of substances across the membrane. As it differentiates, each type of cell acquires its individual responses to peptide hormones by producing a unique set of proteins regulated by cAMP-dependent protein kinases. In this way, the same second messenger, cAMP, means different things to different cells.

Inositol Triphosphate   Many chemical messengers in animals, including neurotransmitters, growth factors, and some hormones, induce responses in their

Figure 41.7

Inositol triphosphate as a second messenger. The first messenger, the hormone, binds to its specific receptor, activating a G protein that in turn stimulates phospholipase C, an enzyme that resides in the plasma membrane. Phospholipase cleaves a membrane phospholipid into two products—inositol triphosphate (IP_3) and diacylglycerol—that function as second messengers. Diacylglycerol stimulates still another membrane enzyme, protein kinase C, which triggers various responses of the target cell by phosphorylated specific proteins. The IP_3 released from the membrane in response to the hormonal signal triggers the release of Ca^{2+} from the endoplasmic reticulum. This increases the cytoplasmic concentration of Ca^{2+}, an inorganic ion that regulates the activity of many cellular proteins, either by itself or by first binding to a protein called calmodulin.

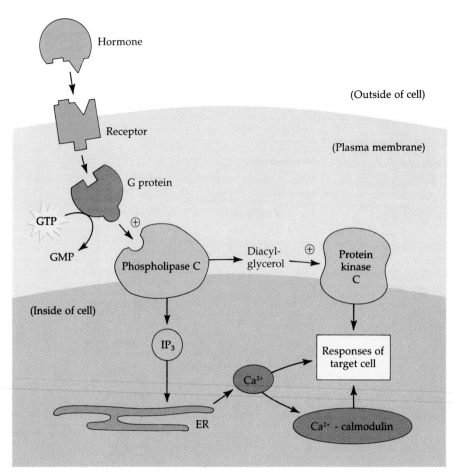

target cells by increasing the cytoplasmic concentration of calcium ions (Ca^{2+}). Binding of the messenger to a specific receptor on the cell surface results in the release of Ca^{2+} from a reservoir within the cell, usually the endoplasmic reticulum (see Chapter 7). The Ca^{2+} released to the cytoplasm alters the activities of specific enzymes, either directly or by first binding to a protein called **calmodulin,** which then binds to other proteins and changes their activities. Given these discoveries, calcium was once believed to be a second messenger in the responses of cells to extracellular signals. However, additional research has shown that calcium is actually a *third* messenger. The second messenger, which intermediates between the hormonal signal and the rise in cytoplasmic Ca^{2+} concentration, is a compound called **inositol triphosphate (IP_3).**

Inositol triphosphate is derived from a particular type of phospholipid present in the plasma membrane (Figure 41.7). Production of IP_3 involves a reaction sequence in the membrane analogous to the mechanism that transduces a hormonal signal into the production of cAMP. A hormone binds to its specific receptor, and the hormone–receptor complex activates a G protein that also resides in the membrane. This G protein, different from those that function in the cAMP system, stimulates a membrane enzyme called phospholipase C. This enzyme then cleaves a

phospholipid into two products, IP_3 and another compound called diacylglycerol. Both products may function as second messengers. The diacylglycerol activates another membrane enzyme called protein kinase C, which triggers some of the cytoplasmic responses to the hormone by phosphorylating specific proteins (similar to the action of cAMP-dependent protein kinase, but protein kinase C activates a different set of cytoplasmic proteins). The IP_3 released from the plasma membrane probably interacts with the membrane of the endoplasmic reticulum in some way that causes Ca^{2+} to leak from the ER. The signal is finally transmitted to various enzymes and other proteins by the increased concentration of cytoplasmic Ca^{2+}, and specific responses of the target cell to the original hormonal signal occur.

Let us review the two general modes of hormone action. Steroid hormones generally work by entering the nucleus of the cell and altering gene expression. Most other hormones, including peptides, bind to the cell surface and regulate cytoplasmic enzymes via second messengers, usually cAMP or IP_3. Notice that steroids affect the *synthesis* of proteins, while peptide hormones affect the *activity* of enzymes and other proteins already present in the cell. This is an important distinction. Target cells generally respond more slowly to steroids than to peptide hormones, but the changes

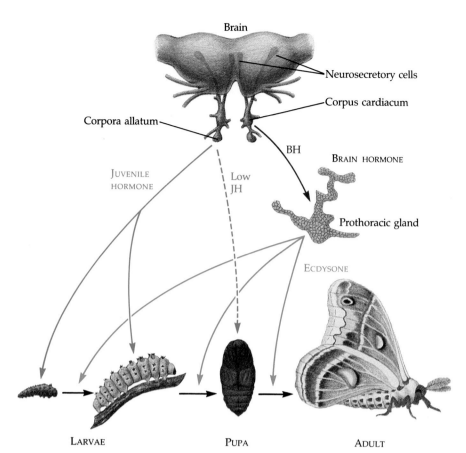

Brain

Neurosecretory cells

Corpus cardiacum

Corpora allatum

BH

BRAIN HORMONE

JUVENILE HORMONE

Low JH

Prothoracic gland

ECDYSONE

LARVAE   PUPA   ADULT

Figure 41.8
**Hormonal regulation of insect develop-
ment.** Juvenile hormone secreted by the
corpora allata inhibits formation of adult
characteristics and promotes the larval
status quo with each molt. Brain hormone
secreted by the corpora cardiaca stimulates production of ecdysone by the prothoracic gland. In the presence of juvenile
hormone, ecdysone triggers larval molts.
In the absence of juvenile hormone in the
pupa, ecdysone stimulates development of
adult structures.

induced by steroids usually last longer. Indeed, steroid hormones are commonly involved in the development of tissues. For example, steroid hormones
secreted by a woman's ovary stimulate gradual growth
of the wall of the uterus over a period of a few weeks
during each menstrual cycle. In contrast to this time
scale, it takes only minutes for MSH, a peptide hormone, to trigger pigment dispersion in the melanocytes of frog skin.

Now that we have considered the two general modes
of hormone action, let us examine the specific functions of some animal hormones.

INVERTEBRATE HORMONES

Although diverse invertebrate hormones function
in homeostasis—by regulating water balance, for
instance—the hormones that have been most extensively studied function in reproduction and development. For example, in *Hydra* one hormone stimulates
growth and budding (asexual reproduction) but prevents sexual reproduction. Annelids of the genus *Ner-
eis* produce a hormone that stimulates egg production
and inhibits body growth. One well-studied example
of nerve and hormone interaction in controlling both
reproductive physiology and behavior in mollusks

involves the egg-laying hormone of the sea slug *Aplysia*. This peptide hormone, when secreted from the
specialized neurons known as "bag cells," almost
immediately stimulates the laying of thousands of eggs.
At the same time, this egg-laying hormone inhibits
feeding and locomotion, activities that interfere with
reproduction.

All groups of arthropods have extensive endocrine
systems. The crustaceans, for example, have hormones for growth and reproduction, water balance,
movement of pigments in the integument and in the
eyes, and regulation of metabolism. The insects,
because they are the largest class of animals and are
so economically important, have been the most thoroughly studied of arthropods. Insects have exoskeletons that cannot stretch; thus, insects appear to grow
in spurts, shedding the old and secreting a new skeleton with each molt. Further, most insects acquire their
adult characteristics in a single, terminal molt. By
studying these events, scientists have identified how
three important hormones interact to control development (Figure 41.8).

The molt itself is triggered by the steroid hormone
ecdysone, secreted from a pair of prothoracic glands
just behind the head. Besides triggering the molt,
ecdysone also favors the development of adult characteristics, as in the change from a caterpillar to a butterfly, illustrated in Figure 41.1. Ecdysone has the same

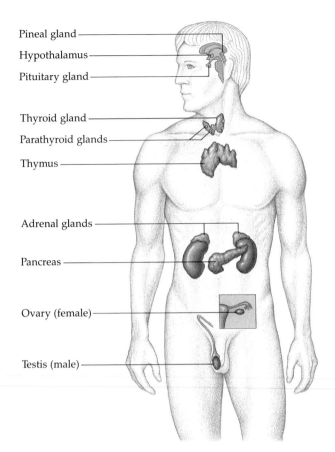

Pineal gland

Hypothalamus

Pituitary gland

Thyroid gland

Parathyroid glands

Thymus

Adrenal glands

Pancreas

Ovary (female)

Testis (male)

Figure 41.9
The major endocrine glands in humans.

mechanism of action as vertebrate steroid hormones, stimulating the transcription of specific genes. Ecdysone production is itself controlled by a second hormone, called **brain hormone.** A peptide from the brain, this hormone promotes development by stimulating the prothoracic glands to secrete ecdysone.

Brain hormone and ecdysone are balanced by **juvenile hormone (JH),** the third hormone in this system. JH is secreted by a pair of small glands just behind

the brain, the corpora allata. Juvenile hormone doesn't just prevent the action of ecdysone; it actively promotes retention of larval characteristics. Synthetic versions of JH are now being used as insecticides to prevent insects from maturing into reproducing adults.

In all these invertebrate examples, we see the importance of the nervous system to hormone activity. Often, the distinction between endocrine system and nervous system blurs. We will see these interactions repeatedly as we survey the vertebrate endocrine system.

THE VERTEBRATE ENDOCRINE SYSTEM

In vertebrates, more than a dozen tissues and organs secrete hormones. Some of these glands are endocrine specialists—secretion of one or more hormones is their major function. Figure 41.9 shows the location of the major endocrine glands in the human body and Table 41.1 summarizes the functions of the major vertebrate hormones.

Hormones vary in their range of targets. Some, like the sex hormones, which promote male and female characteristics, affect most of the tissues of the body. Others affect only one or a few tissues. Some hormones have other endocrine glands as their targets; these so-called **tropic hormones** are particularly important to our understanding of chemical coordination. In fact, we will begin our discussion of the vertebrate endocrine system by looking at the major control axis of the system.

The Hypothalamus and the Pituitary Gland

The **hypothalamus** plays an important role in integrating the vertebrate endocrine and nervous systems (Figure 41.10). This region of the lower brain receives

Figure 41.10
The pituitary gland and hypothalamus.
The pituitary gland, located on the bottom of the brain and surrounded by bone, consists of a posterior lobe (neurohypophysis) and an anterior lobe (adenohypophysis). The posterior lobe develops from the floor of the brain along with the hypothalamus, whereas the anterior lobe is derived in an embryo from the roof of the mouth.

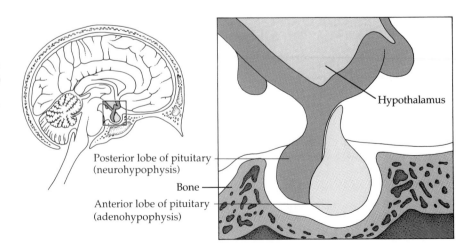

Posterior lobe of pituitary (neurohypophysis)

Bone

Anterior lobe of pituitary (adenohypophysis)

Hypothalamus

Table 41.1 Major endocrine glands of vertebrates and their hormones

Gland	Hormone	Chemical Structure	Representative Actions	Mechanism of Regulation
Pituitary				
Posterior lobe (releases hormones produced by hypothalamus)	Oxytocin	Peptide	Stimulates contraction of uterine muscles and mammary gland cells	Nervous system
	Antidiuretic hormone (ADH)	Peptide	Promotes reabsorption of water from collecting ducts of kidneys	Blood osmolarity
Anterior lobe	Growth hormone (GH)	Protein	Stimulates general growth, particularly skeletal; affects metabolic functions	Hypothalamic hormones
	Prolactin	Protein	Stimulates milk production and secretion (in humans)	Hypothalamic hormones
	Follicle-stimulating hormone (FSH)	Glycoprotein	Stimulates ovarian follicle and spermatogenesis	Estrogen in blood; hypothalamic hormones
	Luteinizing hormone (LH)	Glycoprotein	Stimulates corpus luteum and ovulation in females and interstitial cells in males	Progesterone or testosterone in blood; hypothalamic hormones
	Thyroid-stimulating hormone (TSH)	Glycoprotein	Stimulates thyroid gland to secrete hormones	Thyroxine in blood; hypothalamic hormones
	Adrenocorticotropin (ACTH)	Polypeptide	Stimulates adrenal cortex to secrete glucocorticoids (e.g., cortisol)	Cortisol in blood; hypothalamic hormones
Thyroid	Triiodothyronine (T_3) and thyroxine (T_4)	Iodinated amino acids	Increase oxygen consumption and heat production; stimulate and maintain metabolic processes	TSH
	Calcitonin	Peptide	Lowers blood calcium levels by inhibiting calcium release from bone	Ca^{2+} concentration in blood
Parathyroid	Parathyroid hormone (PTH)	Peptide	Raises plasma calcium levels by stimulating calcium release from bone and by promoting calcium uptake from gastrointestinal tract	Ca^{2+} concentration in blood
Pancreas	Insulin	Polypeptide	Lowers blood sugar, increases glycogen storage in liver, and stimulates protein synthesis	Glucose concentration in blood; somatostatin
	Glucagon	Polypeptide	Stimulates glycogen breakdown in liver	Glucose and amino acid concentration in blood
	Somatostatin	Peptide	Suppresses release of insulin and glucagon	Nervous control; feedback from growth hormone
Adrenal glands				
Medulla	Epinephrine	Catecholamine	Increases blood sugar; constricts blood vessels in skin, mucous membranes, and kidney	Nervous system
	Norepinephrine	Catecholamine	Increases heart rate and force of contraction of cardiac muscles; constricts blood vessels throughout the body	Nervous system

continued

Table 41.1 (continued) Major endocrine glands of vertebrates and their hormones

Gland	Hormone	Chemical Structure	Representative Actions	Mechanism of Regulation
Cortex	Glucocorticoids (e.g., cortisol)	Steroids	Increase blood sugar by affecting many aspects of carbohydrate metabolism	ACTH
	Mineralocorticoids (e.g., aldosterone)	Steroids	Promote reabsorption of sodium and excretion of potassium in the kidneys	K^+ in blood
Gonads				
Testis	Androgens (e.g., testosterone)	Steroids	Support spermatogenesis; develop and maintain secondary male sexual characteristics	FSH and LH
Ovary (follicle)	Estrogens	Steroids	Initiate building of uterine lining; develop and maintain female sexual characteristics	FSH and LH
Ovary (corpus liteum)	Progesterone and estrogens	Steroids	Promote continued growth of uterine lining	FSH and LH
Pineal	Melatonin	Catecholamine	Involved in circadian rhythms	Light/dark cycles
Thymus	Thymosin	Peptide	Stimulates T-lymphocyte development	Mechanism not established

information from peripheral nerves and from other parts of the brain and then initiates endocrine signals appropriate to the environmental conditions. In many vertebrates, for example, the brain passes sensory information about changes in season and the availability of a mate to the hypothalamus by means of nerve signals; the hypothalamus then triggers the release of reproductive hormones required for breeding.

The hormone-releasing cells in the hypothalamus are specialized neurons that differ from both the secretory cells of most endocrine glands and other neurons. Rather than being specialized just for hormone synthesis and release or just for conduction of nerve impulses, they do both. Called **neurosecretory cells,** these hypothalamic cells receive signals from other nerve cells, but instead of signaling to an adjacent nerve cell or muscle, they release hormones into the bloodstream. The hypothalamus contains two sets of neurosecretory cells. One set produces the hormones of the posterior pituitary. The other produces **releasing factors** that regulate the anterior pituitary.

The **pituitary gland,** a small appendage at the base of the hypothalamus, was formerly called the "master gland" because so many of its hormones regulate other endocrine functions. However, the pituitary itself obeys hormonal orders from the hypothalamus. The pituitary has two lobes, each with a different function. The **posterior lobe,** or **neurohypophysis,** is an extension of the brain that stores and secretes two peptide hor-

mones that are actually made by the hypothalamus. The two hormones released from the posterior lobe act directly on muscles and the kidneys, rather than affecting other endocrine glands. The **anterior lobe** of the pituitary, or **adenohypophysis,** produces its own hormones, several of which are tropic hormones that act on other endocrine glands. Figure 41.11 summarizes the names and actions of the pituitary hormones.

Posterior Pituitary Hormones   The posterior lobe of the pituitary stores and releases the hypothalamic hormones **oxytocin** and **antidiuretic hormone (ADH).** They are small peptides of nine amino acids, only two of which differ between oxytocin and ADH. The hormones are synthesized by the cell bodies of neurosecretory cells in the hypothalamus and move within extensions of these cells into the posterior pituitary. Oxytocin induces contraction of the uterine muscles during birth and causes the mammary glands to eject milk during nursing. ADH acts on the kidneys to increase water retention and thus decrease water volume.

Antidiuretic hormone is part of an elaborate feedback scheme that helps regulate the osmolarity of the blood. This mechanism was introduced in Chapter 40, but it is worth reviewing here to illustrate how hormones contribute to homeostasis and how negative feedback controls hormone levels. Blood osmolarity

Figure 41.11
Hormones of the pituitary gland.
(a) Posterior lobe. Neurosecretory cells in the hypothalamus synthesize oxytocin and antidiuretic hormone (ADH), peptide hormones that are transported down the axons to the posterior pituitary where they are stored. The pituitary gland releases the hormones into the blood, where they circulate and bind to target cells in the kidneys (ADH) and mammary glands and uterus (oxytocin). (b) Anterior lobe. Endocrine cells in the anterior pituitary manufacture a number of peptide hormones and secrete them into the circulation, but release of these hormones is controlled by the hypothalamus. Neurosecretory cells in the hypothalamus secrete releasing factors into a capillary network located in the median eminence, located above the stalk of the pituitary. Blood containing the releasing factors travels from the median eminence through short portal veins and into a second capillary network within the anterior pituitary. Various releasing factors stimulate or inhibit release of specific hormones by the pituitary cells.

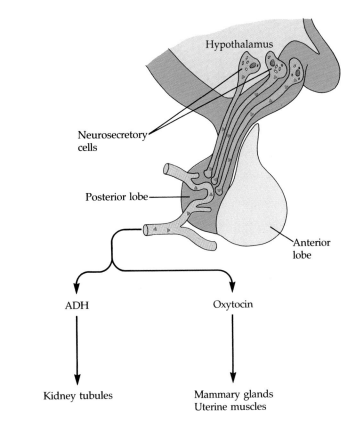

(a)

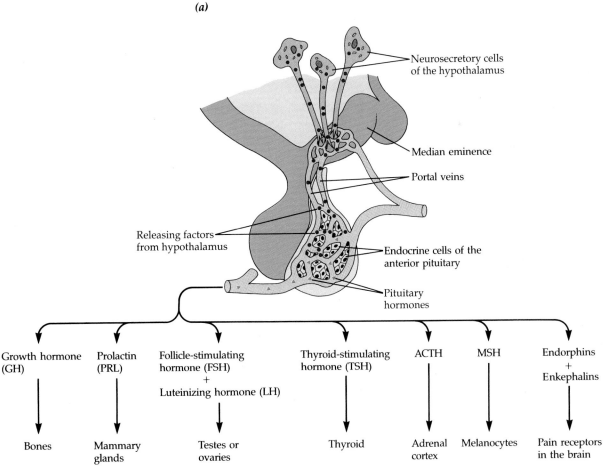

(b)

is monitored by a group of nerve cells that function as osmoreceptors in the hypothalamus. When the osmolarity of the plasma increases, these cells transmit nerve impulses to certain neurosecretory cells of the hypothalamus, which respond by releasing ADH from their tips (located in the posterior pituitary) into the general circulation. When ADH reaches the kidneys, it binds to receptors on the surface of the cells lining the collecting ducts. This binding increases the permeability of the epithelium of the collecting duct by acting through a cAMP second messenger system. Because of this increased permeability, water exits from the collecting ducts and enters nearby capillaries, reducing the blood's osmolarity. As the more dilute blood arrives at the brain, the hypothalamus responds to the reduction in osmolarity by slowing the release of ADH. Thus, the hormone's effect—water reabsorption by the kidneys—prevents overcompensation by shutting off further secretion of the hormone, which is why such a control circuit is referred to as a negative feedback loop (see Chapter 36).

Anterior Pituitary Hormones The anterior pituitary produces many different protein and peptide hormones. Four of these are tropic hormones that stimulate the synthesis and release of hormones from other endocrine glands: Thyroid-stimulating hormone (TSH) regulates release of thyroid hormones; adrenocorticotropin (ACTH) controls the adrenal cortex; and follicle-stimulating hormone (FSH) and luteinizing hormone (LH) govern reproduction by acting on the gonads. Other hormones produced by the anterior pituitary are growth hormone (GH), prolactin (PRL), melanocyte-stimulating hormone (MSH), and the endorphins and enkephalins.

Growth hormone (GH), a protein of almost 200 amino acids, affects a wide variety of target tissues. GH both promotes growth directly and stimulates production of other growth factors. For example, the ability of GH to stimulate growth of bones and cartilage is partly due to the hormone's signaling the liver to produce growth factors called **somatomedins,** which circulate in blood plasma and directly stimulate bone and cartilage growth. In the absence of GH, skeletal growth of an immature animal will stop. If GH is injected into an animal that has been deprived of its own GH, growth will be partially restored.

Several human growth disorders are related to abnormal GH production. Excessive production of GH during development can lead to gigantism, while excessive GH production during adulthood results in abnormal growth of bones in the hands, feet, and head—a condition known as acromegaly. Deficient GH production in childhood can lead to hypopituitary dwarfism. Children with GH deficiency have been treated successfully with human growth hormones isolated from cadaver pituitaries. But the supply falls short of demand, and growth hormones from most other animals are ineffective. One of the most dramatic achievements of genetic engineering has been the production of GH by bacteria with genes for human GH spliced into their genomes (see Chapter 19). The product is now being used to treat children with hypopituitary dwarfism, and some athletes are carelessly beginning to take GH (legally or illegally) because it stimulates protein synthesis.

Prolactin (PRL) is a protein so similar to GH that it is believed they are encoded in genes that evolved from the same ancestral gene. The physiological roles of these two hormones, however, are different. Prolactin's most remarkable characteristic is the great diversity of effects it produces in different vertebrate species. For example, PRL stimulates mammary gland growth and milk synthesis in mammals; regulates fat metabolism and reproduction in birds; delays metamorphosis in amphibians, where it may also function as a larval growth hormone; and regulates salt and water balance in freshwater fish. This list suggests that prolactin is an ancient hormone whose functions have diversified during the evolution of the various vertebrate classes.

Three of the tropic hormones secreted by the anterior pituitary are closely related chemically. **Follicle-stimulating hormone (FSH), luteinizing hormone (LH),** and **thyroid-stimulating hormone (TSH)** are all similar glycoproteins, protein molecules with carbohydrate attached to them. FSH and LH are also called **gonadotropins** because they stimulate the activities of both the male and female gonads, the testes and ovaries. TSH, as the name implies, stimulates the production of hormones by the thyroid gland.

The remaining hormones from the anterior pituitary all come from a single parent molecule called **pro-opiomelanocortin.** This large protein is cleaved into several short fragments inside the pituitary cells. At least four of these fragments are active peptide hormones. **Adrenocorticotropin (ACTH)** is a tropic hormone that stimulates the production and secretion of steroid hormones by the adrenal cortex. As described earlier, **melanocyte-stimulating hormone (MSH)** regulates the activity of pigment-containing cells in the skin of some vertebrates. Very small amounts of MSH are secreted by the human pituitary, but functions of this hormone in humans are uncertain. The other two derivatives of pro-opiomelanocortin are classes of hormones called **endorphins** and **enkephalins.** These molecules are also produced by certain neurons in the brain. They are sometimes called the body's "natural opiates" because they mimic the actions of morphine on the nervous system and appear to inhibit pain reception. It has been suggested that the so-called runner's high results partly from the release of endorphins when stress and pain in the body reach critical levels.

Hypothalamic Releasing Factors   Let us now return to the hypothalamus to see how the anterior pituitary is controlled. The hypothalamic hormones called releasing factors are produced by neurosecretory cells and released into capillaries in a region at the base of the hypothalamus called the **median eminence** (see Figure 41.11b). Unlike most veins that drain body organs, the veins exiting from the median eminence do not connect directly to the vena cava. Rather, they subdivide to form a second capillary bed within the anterior pituitary. (You may recall a similar portal circulation of the liver from Chapter 37.) In this way, the hypothalamic releasing factors have direct access to the gland they control.

Although they are called releasing factors, some of these signals actually inhibit the secretion of hormones from the anterior pituitary rather than stimulating secretion. Every anterior pituitary hormone is controlled by at least one releasing factor, and most have both a releasing factor and a release-inhibiting factor.

The Thyroid Gland

In humans and other mammals, the thyroid gland consists of two lobes located on the ventral surface of the trachea (see Figure 41.9). In other vertebrates, the halves of the gland may be more separated on the two sides of the pharynx. The thyroid gland produces two very similar hormones derived from the amino acid tyrosine, triiodothyronine (T_3), which contains three iodine atoms, and thyroxine (T_4), which contains four iodine atoms:

TRIIODOTHYRONINE (T_3)

THYROXINE (T_4)

In mammals, T_3 is usually the more active of the two hormones, although both have the same effects on their target cells.

The thyroid gland plays a crucial role in the development and maturation of vertebrates. A striking example is the thyroid's control of metamorphosis of a tadpole into a frog, which involves massive reorganization of many different tissues. The thyroid is

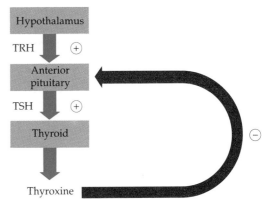

Figure 41.12
Feedback control loop regulating thyroxine secretion. The anterior pituitary gland produces thyroid-stimulating hormone, or TSH. When TSH binds to its receptors in the thyroid gland, it generates cAMP within the target cells, triggering the synthesis and release of the thyroid hormones. Secretion of TSH itself is controlled by TSH-releasing hormone, or TRH, from the hypothalamus. The system is balanced by negative feedback, with high levels of thyroxine inhibiting TSH secretion (probably by desensitizing pituitary cells to the hypothalamic releasing factor). The hypothalamus-pituitary-thyroid feedback system explains why iodine deficiencies lead to goiter. In the absence of sufficient iodine, the thyroid gland cannot synthesize adequate amounts of thyroid hormones. Consequently, the pituitary continues to secrete TSH, leading to an enlargement of the thyroid gland.

equally important in human development. An inherited condition of thyroid deficiency known as cretinism results in markedly retarded skeletal growth and poor mental development. These defects can often be overcome, at least partly, if treatment with thyroid hormones is begun early in life. Studies with animals have shown that thyroid hormones are required for the normal functioning of bone-forming cells and for branching of nerve cells during embryonic development of the brain.

The thyroid gland is also critical in regulating metabolism. Excessive secretion of thyroid hormones, known as hyperthyroidism, produces such symptoms as high body temperature, profuse sweating, weight loss, irritability, and high blood pressure. The opposite condition, hypothyroidism, can cause cretinism in infants and produce symptoms such as weight gain, lethargy, and intolerance to cold in adults. Another condition associated with a shortage of thyroid hormone is an enlargement of the thyroid called goiter, often caused by a deficiency of iodine in the diet (see Chapter 2).

The secretion of thyroid hormones is controlled by the hypothalamus and pituitary in a negative feedback loop (Figure 41.12).

The thyroid gland also contains endocrine cells that secrete **calcitonin.** This polypeptide lowers calcium levels in the blood as part of calcium homeostasis, described in the following section.

The Parathyroid Glands

The four parathyroid glands, embedded in the surface of the thyroid, function in homeostasis of calcium ions. They secrete **parathyroid hormone (PTH),** which raises blood levels of calcium and thus has an effect opposite to that of the thyroid hormone calcitonin. Parathyroid hormone elevates blood Ca^{2+} by stimulating Ca^{2+} absorption in the intestine and Ca^{2+} reabsorption in the kidney, and by inducing specialized bone cells called osteoclasts to decompose the mineralized matrix of bone and release Ca^{2+} to the blood. Calcitonin has just the opposite effects on the intestine, kidneys, and bone, thus decreasing blood Ca^{2+}. Vitamin D, synthesized in the skin and converted to its active form in many tissues, is essential to PTH function, so it is also required for complete calcium balance. A lack of PTH causes blood levels of calcium to drop dramatically, leading to convulsive contractions of the skeletal muscles. If unchecked, this condition, known as tetany, is fatal. Control of blood calcium is an example of how homeostasis is often maintained by the balancing of two antagonistic hormones—in this case, PTH and calcitonin (Figure 41.13).

The Pancreas

Insulin, the pancreatic hormone that regulates glucose uptake in the cells of the liver and other organs as well as protein and fat synthesis, is probably the most familiar hormone. Like many other hormones, it has been known largely by the condition that results from its deficiency in the body, diabetes mellitus. And as in many other cases, it was the search for a treatment that led to our understanding of the nature of the hormone.

Insulin is a protein produced by specialized clusters of pancreatic cells known as **islet cells.** Several distinct cell types are present within the islets, but only those known as **β cells** synthesize and secrete insulin. These cells first make proinsulin, an inactive form of the protein. Enzymatic removal of a portion of the polypeptide chain activates insulin (Figure 41.14). Insulin secretion from β cells is triggered by glucose receptors on the plasma membranes, which detect an increase in blood sugar levels. A different type of islet cells, called **α cells,** secretes another peptide hormone, **glucagon,** which counters insulin action by increasing blood sugar. Because insulin and glucagon work as antagonistic hormones to regulate carbohydrate metabolism, we will consider them together.

The principal effect of insulin is to lower blood sugar levels by facilitating the uptake of glucose by most cells, including adipose and muscle cells, and by promoting the formation and storage of glycogen in the liver. In addition, insulin stimulates the synthesis of

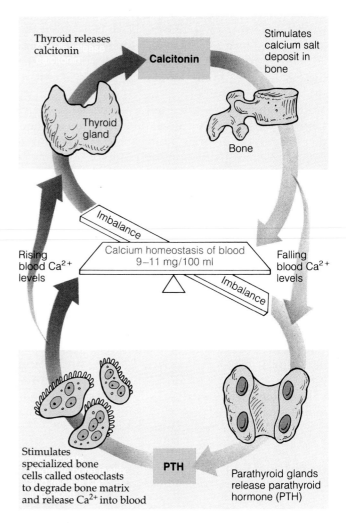

Figure 41.13
Hormonal control of calcium homeostasis. A negative feedback system involving two antagonistic hormones, calcitonin and parathyroid hormone (PTH), maintains the concentration of calcium in blood within a very narrow range of 9–11 mg per 100 ml. A rise in blood Ca^{2+} induces the thyroid gland to secrete calcitonin, which lowers the Ca^{2+} concentration by increasing bone deposition and by reducing Ca^{2+} absorption in the intestine and Ca^{2+} reabsorption in the kidneys (bone is the only target illustrated in this figure). These effects on the bone, intestine, and kidneys are reversed by PTH, which is secreted from the parathyroid glands when the concentration of blood Ca^{2+} falls below the set point of 10 mg/100 ml. Blood Ca^{2+} levels begin to increase, but are only allowed to rise so far before the thyroid counters by secreting more calcitonin. In classic feedback fashion, these two hormones balance one another to minimize fluctuations in the concentration of blood Ca^{2+}, an ion essential to the normal functioning of all cells.

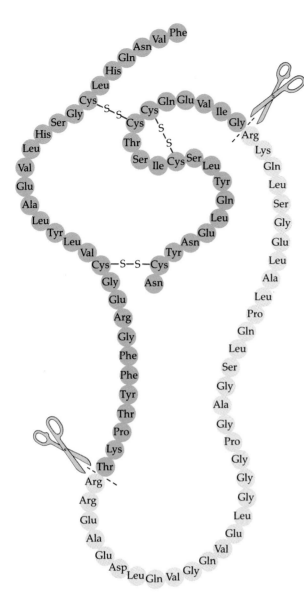

Figure 41.14
Structure of insulin. Beta cells in the pancreas first synthesize the protein as a single polypeptide chain with three disulfide bridges (see Chapter 5). This inactive molecule, called proinsulin, is converted to insulin by the excision of the middle portion of the polypeptide, leaving two small polypeptide chains (in color) joined by disulfide bridges.

protein and the storage of fat. Glucagon, conversely, increases the concentration of glucose in the blood by stimulating the conversion of glycogen to glucose in the liver. It also stimulates the breakdown of fats and proteins. In the homeostasis of blood sugar, we see another example of balance achieved by two hormones that oppose each other (Figure 41.15).

The secretion of both insulin and glucagon is controlled by the blood sugar level. When the glucose concentration is high, as it is shortly after eating, the β cells secrete more insulin. The insulin increases transport of glucose from the blood into cells, leading to a decrease in blood sugar and insulin secretion. Insulin is the only hormone that reduces blood sugar levels. Therefore, the absence of insulin in the bloodstream of diabetics means that glucose uptake by cells is severely restricted, leading to extremely high concentrations of glucose in the blood. The concentration is so high, in fact, that glucose is excreted into the urine—which explains why the presence of sugar in urine is used as one test for diabetes. As more glucose concentrates in the urine, more water is excreted along with it, resulting in serious dehydration. Because the cells of the diabetic cannot use glucose for fuel, the body begins to use up its stores of protein and fat, leaving the diabetic emaciated and weak.

Untreated, diabetes is eventually fatal. Mild cases often can be controlled by diet. Regular injections of insulin keep more serious cases of diabetes under control, although there may be long-term complications that can reduce life expectancy considerably. Until recently, the insulin used by diabetics had to be extracted from animal pancreases. However, genetic engineering has enabled researchers to insert the genes that code for human insulin into bacteria (see Chapter 19). Insulin from this source is now commercially available.

The Adrenal Glands

The **adrenal glands** are adjacent to the kidneys. In mammals, each adrenal gland is actually made up of two glands with different cell types, functions, and embryonic origins: the **cortex**, or outer portion, and the **medulla**, or central part of the gland. Nonmammalian vertebrates have quite different arrangements of the same tissues.

Adrenal Medulla   You have no doubt heard of the "fight-or-flight" syndrome. But just what makes your heart beat faster and your skin develop goose bumps when you sense danger or approach a stressful situation such as speaking in public? These reactions are stimulated by two hormones of the adrenal medulla, **epinephrine** (also known as adrenaline) and **norepinephrine** (noradrenaline). These compounds, called **catecholamines,** are synthesized from the amino acid tyrosine (Figure 41.16).

Epinephrine is secreted in response to positive or negative stress—everything from increased cold to life-threatening danger to extreme pleasure. Release of epinephrine into the blood produces rapid and dramatic effects involving several targets. Epinephrine, using cAMP as a second messenger, as described earlier in the chapter, mobilizes glucose from skeletal muscle and liver cells and stimulates the release of fatty acids from fat cells. The fatty acids may be used

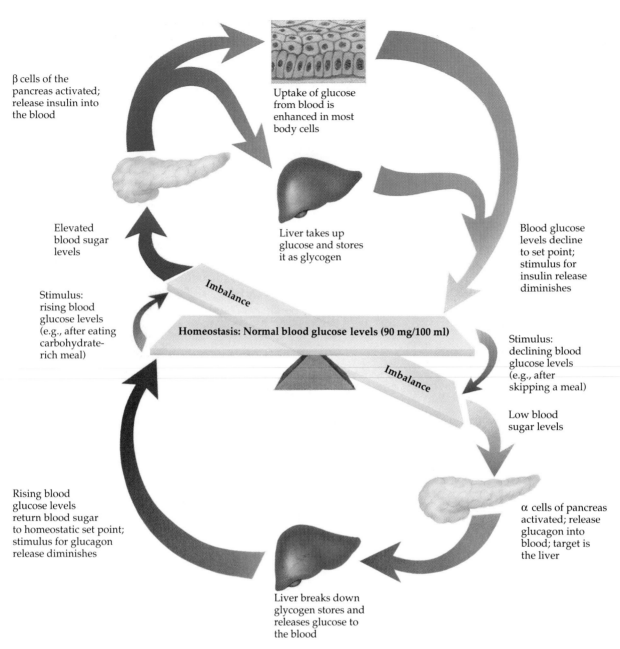

β cells of the pancreas activated; release insulin into the blood

Uptake of glucose from blood is enhanced in most body cells

Elevated blood sugar levels

Liver takes up glucose and stores it as glycogen

Blood glucose levels decline to set point; stimulus for insulin release diminishes

Stimulus: rising blood glucose levels (e.g., after eating carbohydrate-rich meal)

Imbalance

Homeostasis: Normal blood glucose levels (90 mg/100 ml)

Imbalance

Stimulus: declining blood glucose levels (e.g., after skipping a meal)

Low blood sugar levels

Rising blood glucose levels return blood sugar to homeostatic set point; stimulus for glucagon release diminishes

α cells of pancreas activated; release glucagon into blood; target is the liver

Liver breaks down glycogen stores and releases glucose to the blood

Figure 41.15
Glucose homeostasis maintained by insulin and glucagon. By countering one another in a feedback circuit, these antagonistic hormones keep the blood glucose concentration close to the set point of 90 mg/100 ml in humans. A rise in blood sugar stimulates the pancreas to secrete insulin, which acts on its target cells to lower blood sugar. When blood sugar concentration dips below the set point, the pancreas responds by secreting glucagon, which opposes insulin by raising blood sugar level.

directly by cells for energy or they may be converted to glucose by the liver. In addition to increasing the availability of energy sources, epinephrine and norepinephrine have profound effects on muscle contraction. For example, they increase both the rate and stroke volume of the heartbeat. They also cause smooth muscles of some blood vessels to contract and muscles of other vessels to relax, with an overall effect of shunting blood away from the skin, gut, and kidneys, while increasing the blood supply to the heart, brain, and skeletal muscles.

What causes the release of epinephrine during the response to stress? The adrenal medulla is under the control of nerve cells from the sympathetic division of the autonomic nervous system (see Chapter 44). In fact, nerve endings from the sympathetic nervous system are found in the adrenal medulla in close contact with individual cells known as chromaffin cells. When

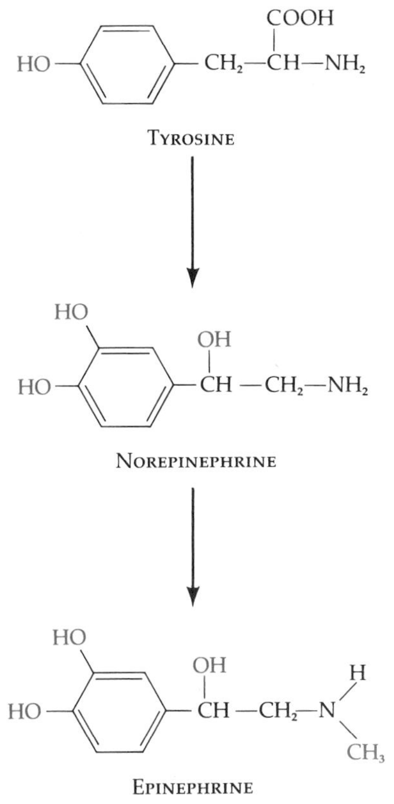

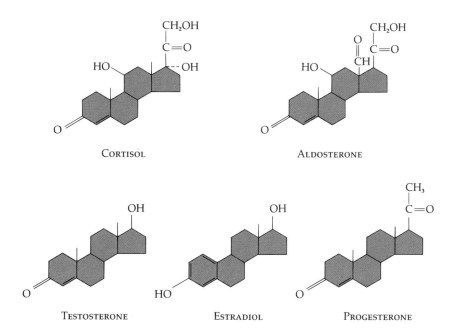

Figure 41.16
Synthesis of catecholamine hormones. Cells in the adrenal medulla called chromaffin cells synthesize the catecholamines norepinephrine and epinephrine from the amino acid tyrosine. Norepinephrine is made by removing a carboxyl group and adding hydroxyl groups. Epinephrine is made from norepinephrine by adding a methyl ($-CH_3$) group.

the nerve cells are excited by some form of stressful stimulus, they release the neurotransmitter acetylcholine. Acetylcholine combines with receptors on the chromaffin cells, stimulating the release of epinephrine. Norepinephrine is released independently of epinephrine. Its function is primarily that of sustaining blood pressure. Both epinephrine and norepinephrine also function as neurotransmitters in the nervous system, as we shall see in Chapter 44.

Adrenal Cortex   The adrenal cortex, like the adrenal medulla, reacts to stress. But it responds to endocrine signals rather than to nervous input. Stressful stimuli cause the hypothalamus to secrete a releasing factor that in turn stimulates the anterior pituitary to release the tropic hormone ACTH. When it reaches its target via the bloodstream, ACTH stimulates cells of the adrenal cortex to synthesize and secrete a family of steroids called **corticosteroids.** In another case of negative feedback, elevated levels of corticosteroids in the blood suppress secretion of ACTH.

Many corticosteroids have been isolated from the adrenal cortex; the two main types in humans are the **glucocorticoids,** such as cortisol, and the **mineralocorticoids,** such as aldosterone. The structures of these and several other important steroid hormones are shown in Figure 41.17.

The primary effect of glucocorticoids is on glucose metabolism. Glucocorticoids promote the synthesis of glucose from noncarbohydrate substrates, such as proteins, making more glucose available as fuel in response to a stressful situation. This effect is slower but of longer duration than the stress-induced action of epinephrine (Figure 41.18).

Figure 41.17
Structures of some steroid hormones. Cortisol (a glucocorticoid) and aldosterone (a mineralocorticoid) are made in the adrenal cortex. The gonads synthesize testosterone (an androgen), estradiol (an estrogen), and progesterone (a progestin). The precursor for the synthesis of steroid hormones is cholesterol (see Chapter 5).

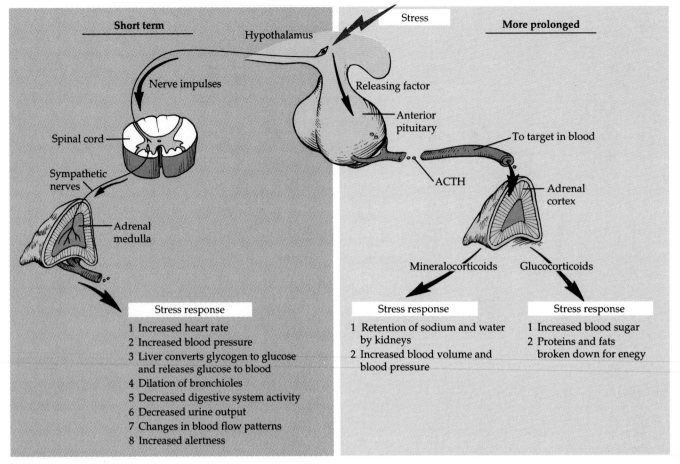

Figure 41.18
Stress and the adrenal gland. Stressful stimuli cause the hypothalamus to activate the adrenal medulla via nerve impulses and the adrenal cortex via hormonal signals. The medulla mediates short-term responses to stress by secreting epinephrine, while the cortex controls more prolonged responses by secreting its steroid hormones.

Very high doses of glucocorticoids administered as medication suppress certain components of the body's immune system—for example, the inflammatory reaction that occurs at the site of an infection. Glucocorticoids are used to treat diseases in which excessive inflammation is a problem. Cortisone, for instance, was once thought to be a miracle drug that could cure serious inflammatory conditions such as arthritis. It has become clear, however, that long-term use of the corticoid hormones can result in increased susceptibility to infection and disease because of their immunosuppressive effects.

The second major type of adrenal steroids, the mineralocorticoids, have their major effects on salt and water balance. Aldosterone, for example, stimulates cells in the kidney to reabsorb sodium ions from the filtrate, which also causes water reabsorption and raises blood pressure. Control of aldosterone secretion is largely independent of the pituitary and the hypothalamus. Instead, it is regulated by hormones produced in the liver and kidneys in response to changes in plasma ion concentration (see Chapter 40).

The Gonads

Steroids produced in the testes of males and ovaries of females affect growth and development and also regulate reproductive cycles and behaviors. There are three major categories of gonadal steroids: androgens, estrogens, and progestins. All three types are found in both males and females, but in different proportions (see Figure 41.17).

The testes primarily synthesize **androgens,** the main such hormone being **testosterone.** In general, androgens stimulate the development and maintenance of the male reproductive system. Androgens produced early in the development of an embryo determine that the fetus will develop as a male rather than a female. At puberty, high concentrations of androgens are responsible for the development of human male secondary sex characteristics, such as male patterns of hair growth and a low voice.

Estrogens, the most important of which is estradiol, have a parallel role in the maintenance of the female reproductive system and the development of female secondary sex characteristics. **Progestins,** at least in mammals, are primarily involved with preparing and maintaining the uterus, which supports the growth and development of an embryo.

Synthesis of both estrogens and androgens is controlled by gonadotropins from the anterior pituitary gland—follicle-stimulating hormone and luteinizing hormone. The secretion of FSH and LH is controlled by one hypothalamic releasing factor, GnRH (for gonadotropin-releasing hormone). The complex feedback relationships that regulate secretion of gonadal steroids will be described in detail in Chapter 42.

Other Endocrine Organs

Many organs with primarily nonendocrine functions also secrete hormones, many of which we have encountered in earlier chapters. The digestive tract, for example, is the source of at least eight hormones, including gastrin and secretin (see Chapter 37). Erythropoietin, from the kidney, stimulates red blood cell production (see Chapter 38). Atrial natriuretic hormone, secreted by the heart, helps to regulate salt and water balance and blood pressure (see Chapter 40). Two other endocrine organs that deserve some attention are the pineal and the thymus.

The **pineal gland** is a small mass of tissue near the center of the mammalian brain (closer to the brain surface in some other vertebrates). Although we know considerably more about the pineal than we did when Descartes described it as the seat of the soul, it still remains largely a mystery. The pineal secretes the hormone **melatonin,** a modified amino acid. The pineal contains light-sensitive cells or has nervous connections from the eyes, and melatonin regulates functions related to light and to seasons marked by changes in day length. For example, melatonin, like MSH, affects skin pigmentation in many vertebrates. Most of the pineal's functions, though, are related to biological rhythms associated with reproduction. Since melatonin is secreted at night, the amount secreted depends on the length of the night. In winter, for example, the days are short and the nights are long, so more melatonin is secreted. Thus, melatonin production is a link between a biological clock and daily or seasonal activities such as reproduction. However, the precise role of melatonin in mediating rhythms is not yet clear.

The **thymus** is another enigmatic gland. It was not until the 1960s that its role in the immune system was discovered. This gland lies across the front of the neck in humans and is quite large during childhood. At puberty, when the immune system is well established, the thymus begins to decline quickly and virtually disappears by adulthood. The thymus secretes several messengers, including **thymosin,** that stimulate the development and differentiation of T lymphocytes (see Chapter 39).

ENDOCRINE GLANDS AND THE NERVOUS SYSTEM

There have been many examples in this chapter of interactions between the endocrine system and the nervous system. Indeed, we have seen that the two systems are often inseparable and may function as a single unit. We can synthesize much of our understanding of the chemical communication and coordination in animals by examining three types of relationships between the endocrine system and the nervous system.

First, these two regulatory systems are *structurally* related. Many endocrine glands are made of nerve tissue. The vertebrate hypothalamus and posterior pituitary, the insect brain, and the "bag cells" of *Aplysia* are all examples of nerve tissues that secrete hormones into the blood. Other endocrine glands that are not nervous tissue in their present form have evolved from the nervous system. The adrenal medulla is derived from a modified ganglion (cluster of nerve cell bodies) that has become separated from the nervous system.

Second, the two systems are *chemically* related. Several vertebrate hormones are used as signals by the nervous system as well as by the endocrine system. Epinephrine, for example, functions in the body both as an adrenal hormone and as a neurotransmitter in the nervous system.

Third, the two systems are *functionally* related. We can see two types of functional relationships. In the first, the coordinating system controlling physiological processes involves both nervous and hormonal components arranged in series. For example, milk letdown, the release of milk by a mother during nursing, is controlled by a neuroendocrine reflex: Suckling stimulates sensory cells in the nipples, and nervous signals to the hypothalamus trigger release of oxytocin from the posterior pituitary. In the second type of functional relationship, each system affects the output of the other. We have seen several examples of how the nervous system controls the endocrine glands, including the stimulation of the adrenal medulla. But the endocrine system also affects both the development of the nervous system and its output—behavior.

Thus, we have seen that animals possess two coordinating systems. We have studied one, the endocrine system, in some detail, but that discussion has drawn us closer and closer to the other. In Chapters 44 and 45, we will turn to the other coordinating system, the nervous system. First, however, we will consider one of the most fundamental subjects in biology—one in which the role of the endocrine system is central not only to the survival of the individual but also to the propagation of the species. In Chapters 42 and 43, we will explore reproduction and development.

STUDY OUTLINE

1. The endocrine system utilizes chemical messengers, called hormones, to affect target organs at distant sites.

Chemical Messengers of the Body (pp. 907–909)

1. A hormone is a molecule that is secreted by an endocrine gland and that travels in blood to a target cell, where it binds with specific receptors and elicits a response.
2. Pheromones act as communication signals between different individuals of the same species.
3. Local regulators, such as neurotransmitters, growth factors, and prostaglandins, affect target cells in the immediate vicinity of their secretion.

Mechanisms of Hormone Action (pp. 909–915)

1. Steroid hormones penetrate the cell membrane and bind to specific protein receptors in the nucleus. Hormone–receptor complexes then bind to an acceptor protein on a chromosome and initiate transcription.
2. Peptide hormones, which cannot pass through the cell membrane, bind to specific receptors on the plasma membrane. Through second messengers such as cyclic AMP and inositol triphosphate, these hormones trigger a cascade of metabolic reactions within the cells.
3. Binding of a hormone to a surface receptor activates G protein, which in turn activates the membrane protein adenylate cyclase to produce cAMP. The cAMP then activates cAMP-dependent protein kinase, an enzyme that phosphorylates other proteins.
4. Inositol triphosphate (IP_3) serves as a second messenger for neurotransmitters, growth factors, and some hormones. Binding of a hormone to its receptor activates a G protein that stimulates a membrane enzyme to cleave a membrane phospholipid into IP_3 and diacylglycerol. Diacylglycerol activates a protein kinase, and IP_3 causes the release of Ca^{2+} from storage in the endoplasmic reticulum. Ca^{2+}, acting alone or bound to calmodulin, alters the activities of certain enzymes.

Invertebrate Hormones (pp. 915–916)

1. Invertebrate hormones control different aspects of development, reproduction, and homeostasis.
2. Arthropods have well-developed endocrine systems. In insects, molting and development are controlled by an interplay of ecdysone and juvenile hormone.

The Vertebrate Endocrine System (pp. 916–927)

1. The hypothalamus integrates endocrine and neural function by influencing the pituitary gland. Under the direction of releasing factors from the hypothalamus, the anterior pituitary produces several tropic hormones that act on other endocrine glands. The posterior pituitary is an extension of the brain that stores and releases two peptide hormones produced by the hypothalamus.
2. The posterior pituitary is a repository for oxytocin and antidiuretic hormone (ADH). Oxytocin induces uterine contractions and milk ejection, whereas ADH enhances water reabsorption in the kidneys.
3. The anterior pituitary produces an array of protein and peptide hormones, including thyroid-stimulating hormone (TSH), follicle-stimulating hormone (FSH), luteinizing hormone (LH), growth hormone (GH), prolactin (PRL), adrenocorticotropin (ACTH), melanocyte-stimulating hormone (MSH), and the endorphins.
4. The chemically related tropic hormones, TSH and the gonadotropins (FSH and LH), stimulate the thyroid gland and the gonads, respectively, to produce their hormones.
5. GH promotes growth either directly or by stimulating the production of other growth factors.
6. Prolactin, named for its stimulation of lactation in mammals, has diverse effects in different vertebrates.
7. The remaining four anterior pituitary hormones, all of which come from pro-opiomelanocortin molecules, include ACTH, which has a tropic effect on the adrenal cortex, and MSH, which influences skin pigmentation in some vertebrates. Endorphins and enkephalins mimic the action of morphine and appear to inhibit pain.
8. Hypothalamic releasing factors control secretion of specific hormones from the anterior pituitary.
9. The thyroid gland produces iodine-containing hormones that stimulate metabolism and influence development and maturation in vertebrates. The thyroid also secretes calcitonin, which lowers calcium levels in the blood.
10. The parathyroid glands raise plasma calcium levels by secreting parathyroid hormone (PTH). PTH works with calcitonin to effect calcium homeostasis by actions on bone, kidney, and intestine.
11. The endocrine portion of the pancreas consists of islet cells that secret insulin and glucagon. High plasma glucose stimulates the release of insulin, which increases cellular uptake of glucose, promotes formation and storage of glycogen in the liver, and stimulates protein synthesis and fat storage. Low plasma glucose triggers glucagon release, which increases blood glucose by stimulating the conversion of glycogen to glucose in the liver and increasing the breakdown of fat and protein.
12. The adrenal gland consists of an outer cortex and an inner medulla. The medulla releases epinephrine and norepinephrine in response to stress-activated impulses from the sympathetic nervous system. These hormones mediate a variety of "fight-or-flight" responses. The adrenal cortex releases two groups of corticosteroids, glucocorticoids and mineralocorticoids. Glucocorticoids influence glucose metabolism and the immune system; mineralocorticoids affect salt and water balance.
13. The gonads—testes and ovaries—produce varying proportions of androgens, estrogens, and progestins, steroids that affect growth, development, morphological differentiation, and reproductive cycles and behaviors.
14. The pineal gland secretes melatonin, which influences skin pigmentation, biological rhythms, and reproduction in various vertebrates.
15. The thymus gland functions during early life to stimulate the development of T lymphocytes by means of thymosin and other chemical messengers.

Endocrine Glands and the Nervous System (p. 927)

1. The endocrine and nervous systems function together in chemical communication and control in animals.

SELF-QUIZ

1. Which of the following is *not* an accurate statement about hormones?

    a. Hormones are chemical messengers that travel to target cells through the circulatory system.

    b. Hormones often regulate homeostasis through antagonistic functions.

    c. Hormones of the same chemical class usually have the same general function.

    d. Hormones are secreted by specialized cells usually located in endocrine glands.

    e. Hormones are often regulated through feedback loops.

2. A major difference in the mechanism of action between steroid and peptide hormones is that

    a. steroid hormones affect the synthesis of proteins, whereas peptide hormones affect the activity of proteins already in the cell

    b. target cells react more rapidly to steroid hormones than they do to peptide hormones

    c. steroid hormones enter the nucleus, whereas peptide hormones stay in the cytoplasm

    d. steroid hormones bind to a receptor protein, whereas peptide hormones bind to G protein

    e. steroid proteins affect metabolism, whereas peptide hormones affect membrane permeability.

3. Which of the following accurately represents the sequence of components in a cellular response to a peptide hormone?

    a. hormone binding to adenylate cyclase—G protein—protein kinase—phosphorylation of enzymes

    b. hormone binding to receptor—G protein—cAMP produced from GTP—protein kinase

    c. hormone binding to cAMP—G protein—cAMP-dependent protein kinase—adenylate cyclase

    d. hormone binding to G protein—adenylate cyclase—protein kinase—phosphorylation of proteins.

    e. hormone binding to receptor—G protein—adenylate cyclase—protein kinase

4. Growth factors are local regulators that

    a. are produced by the anterior pituitary

    b. are modified fatty acids that stimulate bone and cartilage growth

    c. are found on the surface of cancer cells and stimulate abnormal cell division

    d. are proteins that bind to surface receptors and stimulate target cells' growth and development

    e. include histamines and interleukins and are necessary for cellular differentiation

5. Which of the following hormones is *incorrectly* paired with its action?

    a. oxytocin—stimulates uterine contraction in childbirth

    b. thyroxine—stimulates metabolic processes

    c. insulin—stimulates glycogen breakdown in liver

    d. ACTH—stimulates release of glucocorticoids by adrenal cortex

    e. melatonin—affects biological rhythms, seasonal reproduction

6. An example of homeostasis control by antagonistic hormones is

    a. calmodulin and parathyroid hormone in calcium balance

    b. insulin and glucagon in glucose metabolism

    c. progestins and estrogens in sexual differentiation

    d. epinephrine and norepinephrine in "fight-or-flight" response

    e. oxytocin and prolactin in milk production

7. Which of the following human conditions is *incorrectly* paired with a hormone?

    a. acromegaly—growth hormone

    b. diabetes—insulin

    c. cretinism—thyroid hormones

    d. tetany—PTH

    e. hypopituitary dwarfism—ACTH

8. Which organ does *not synthesize* hormones?

    a. heart   d. posterior pituitary

    b. adrenal cortex   e. pineal gland

    c. anterior pituitary

9. A second messenger derived from membrane lipids is

    a. cyclic AMP   d. protein kinase

    b. calmodulin   e. calcium

    c. inositol triphosphate

10. The main target organs for tropic hormones are

    a. muscles   d. kidneys

    b. blood vessels   e. nerves

    c. endocrine glands

CHALLENGE QUESTIONS

1. A woman with a hypothyroid condition is treated with thyroxine. How is this medication likely to affect levels of TSH and TSH-releasing hormone (TRH)?

2. Describe three specific cases where the nervous and endocrine systems interact.

FURTHER READING

1. Berridge, M. J. "The Molecular Basis of Communication Within the Cell." *Scientific American*, October 1985. An illustrated guide to second messengers.
2. Cowen, R. "Speeding Up Wound Healing the EGF Way." *Science News*, July 15, 1989. Important function of a growth factor.
3. Snyder, S. H. "The Molecular Basis of Communication Between Cells." *Scientific American*, October 1985. An excellent article emphasizing the relationship between the endocrine and nervous systems.
4. Vaughn, C. "Second Word for Second Messenger Research." *Bioscience*, October 1987. Role of G proteins.

42

Animal Reproduction

The many aspects of animal form and function we have studied so far can be viewed, in the broadest context, as various adaptations contributing to reproductive success. Individuals are transient. A population transcends finite life spans only by reproduction, the creation of new individuals from existing ones. Animal reproduction, both sexual and asexual, is the subject of this chapter. We first compare the diverse reproductive mechanisms that have evolved and then examine the details of mammalian, particularly human, reproduction.

MODES OF REPRODUCTION

There are two principal modes of animal reproduction. **Asexual reproduction** occurs when a single individual produces offspring genetically identical to itself. All of these genetically identical individuals from one lineage form a clone. In **sexual reproduction**, two individuals produce offspring having a combination of genes inherited from both parents (see Chapters 12 and 34).

Asexual Reproduction

Many invertebrates can reproduce asexually by fission, splitting off new individuals from existing ones or separating into two or more individuals of equal size. Certain cnidarians and tunicates provide good examples of the former process, called **budding** (Fig-

Figure 42.1
Asexual reproduction. This sea anemone (*Epiactis prolitera*) is reproducing by budding. The small offspring, genetic clones of the parent from which they bud, will eventually detach from the parent.

ure 42.1). In this form of asexual reproduction, a new individual grows out from the body of the original. The offspring may detach from its parent, or they may remain joined, eventually forming extensive colonies. Stony corals, which may be more than a meter across, are colonies of several thousand individuals connected to one another. Another type of asexual reproduction is **fragmentation,** the breaking of the body into several pieces, each of which develops into a complete adult. When polychaetes, segmented worms of the Phylum Annelida, prepare to fragment, the anterior ends of what will become new individuals develop heads with sensory organs before the parent actually fragments. In another form of asexual reproduction, some invertebrates release specialized groups of cells that can grow into new individuals. For example, the **gemmules** of sponges are formed when cells of several types migrate together within the sponge and become surrounded by a protective coat. Finally, **regeneration** following an injury is also a form of asexual reproduction if it results in two or more individuals where there was only one before. Echinoderms provide one of the best examples: When one arm is removed from a sea star, the animal will grow a new arm. This in itself would not qualify as reproduction. However, if the free arm has even a small piece of the central disk attached, it too will regenerate into a complete new starfish (see Figure 18.14).

Because asexual reproduction involves only one parent, it is advantageous to sessile animals that cannot actively seek mates or to more active animals when population densities are low. The mechanisms of asexual reproduction often allow many offspring to be produced in a short amount of time, making this reproductive mode ideal for rapid population growth. Theoretically, asexual reproduction is most advanta-geous in stable, favorable environments because it perpetuates successful genotypes precisely.

Sexual Reproduction

In animals, sexual reproduction involves the fusion of two haploid **gametes** to form a diploid **zygote.** The female gamete, the **ovum** (unfertilized egg), is usually a relatively large and nonmotile cell. The male gamete, the **spermatozoon,** is generally a small, flagellated cell (though arthropods produce nonmotile sperm that must be placed near the female reproductive tract).

Sexual reproduction increases genetic variability among offspring by generating unique combinations of genes inherited from two parents (see Chapters 12 and 13). Many biologists believe that sexual reproduction, by producing offspring having varying phenotypes, may enhance the reproductive success of parents in certain situations—in a fluctuating environment, for instance. However, there is much new debate about the evolution of sex and about the relative advantages of sexual and asexual reproduction, as we shall see in Chapter 47.

Reproductive Cycles and Patterns

Most animals show definite cycles in reproductive activity, often related to changing seasons. The periodic nature of reproduction allows animals to conserve resources and reproduce when more energy is available than is needed for maintenance and when environmental conditions favor survival of offspring. Ewes (female sheep), for example, have 15-day reproductive cycles and ovulate at the midpoint of each cycle. But these cycles only occur during the fall and early winter, so lambs are born in the spring. Even animals that live in apparently stable habitats such as the tropics or the ocean generally reproduce only at certain times of the year. Reproductive cycles are controlled by a combination of hormonal and seasonal cues, examples of the latter including such factors as changes in temperature, rainfall, or day length.

Animals may use either asexual or sexual reproduction exclusively or alternate between the two. Aphids, rotifers, and the freshwater crustacean *Daphnia* all produce two types of eggs at different times of the year. One type is fertilized, but the other type develops by **parthenogenesis,** which means that the egg develops directly without being fertilized. The adults produced by parthenogenesis are often haploid, and their cells do not undergo meiosis in forming new eggs. In the case of *Daphnia,* the switch from sexual to asexual reproduction is often related to season. Asexual reproduction occurs under favorable conditions and sexual reproduction during times of environmental stress.

Parthenogenesis has a role in the social organization of certain species of bees, wasps, and ants. Male honeybees, or drones, are produced parthenogenetically, whereas females, both sterile workers and reproductive females (queens), develop from fertilized eggs.

Among vertebrates, several genera of fishes, amphibians, and lizards reproduce exclusively by a complex form of parthenogenesis requiring doubling of chromosomes after meiosis to create diploid "zygotes." For example, there are about 15 species of whiptail lizards (Genus *Cnemidophorus*) that reproduce exclusively by parthenogenesis. There are no males in these species, but the lizards imitate courtship and mating behavior typical of sexual species of the same genus. During the breeding season, one female of each "mating" pair impersonates a male. The roles change two or three times during the season, with female behavior occurring when the level of the female sex hormone estrogen is rising prior to ovulation (release of eggs) and male behavior occurring after ovulation when the level of estrogen drops (Figure 42.2). In fact, ovulation is more likely to occur if an individual is mounted by a pseudomale during the critical time of the hormone cycle; isolated lizards lay fewer eggs than individuals that are allowed to go through the motions of sex. Apparently, these parthenogenetic lizards, which evolved from species having two sexes, still require certain sexual stimuli for maximum reproductive success.

Sexual reproduction presents a special problem for sessile or burrowing animals such as barnacles and earthworms or for parasites such as tapeworms, which may have difficulty encountering a member of the opposite sex. One solution to this problem is **hermaphroditism.** Each individual has both functional male

(a)

(b)

Figure 42.2
Pseudosex in parthenogenetic lizards. The desert grassland whiptail lizard (*Cnemidophorus uniparens*) is an all-female species. These reptiles reproduce by parthenogenesis; eggs undergo a chromosome doubling after meiosis and develop into lizards without being fertilized. However, ovulation is enhanced by courtship and mating rituals that imitate the behavior of closely related species that reproduce sexually. (**a**) In this photo, the lizard on top is a female playing the role of a male. Every two or three weeks during the breeding season, individuals switch sexual roles. (**b**) The pseudosexual behavior of *C. uniparens* is correlated with a cycle of hormone secretion and ovulation. During the period when estrogen levels are increasing and the ovary is growing, a lizard is more likely to behave as a female. After ovulation estrogen levels drop, the concentration of a second steroid hormone, progesterone, increases, and the individual is more likely to behave as a male. Unisexual lizards lay fewer eggs if they are kept in isolation.

and female reproductive systems (the term *hermaphrodite* is contracted from "Hermes" and "Aphrodite"). Although some hermaphrodites fertilize themselves, most must mate with another member of the same species. When this occurs, each animal serves

Figure 42.3
Sex reversal in a sequential hermaphrodite. In many species of reef fishes called wrasses, gender is not fixed, but can change at some time in the animal's life. Sex reversal is often correlated with size. In this scene, a male Caribbean bluehead wrasse and two smaller females are feeding on a sea urchin. All wrasses of this species are born females, but the oldest, largest individuals change sex and complete their lives as males.

as both male and female, donating and receiving sperm. Each individual encountered is a potential mate, and twice as many offspring can be produced from such a union as could result if only one individual's eggs were fertilized.

Another remarkable reproductive pattern is **sequential hermaphroditism,** where an individual reverses its sex during its lifetime. In some species, the sequential hermaphrodite is **protogynous** (female first), while other species are **protandrous** (male first). In various species of reef fishes called wrasses, sex reversal is associated with age and size (Figure 42.3). For example, the Pacific cleaner wrasse is a protogynous species in which only the largest (usually the oldest) individuals change from female to male. These fishes live as harems consisting of a single male and several females. If the male dies or is removed in experiments, the largest female in the harem changes sex and becomes the new male. Within a week, the transformed individual is producing sperm instead of eggs. In this species, the male defends the harem against intruders, and thus larger size may confer a greater reproductive advantage to males than it does to females. In contrast, there are protandrous animals that change from male to female when size increases. In such cases, greater size may increase the reproductive success of females more than it does males. For example, production of huge numbers of gametes is an important asset for sedentary animals such as oysters that release their gametes into the surrounding water. Egg cells are generally much larger than sperm cells, and thus females produce fewer gametes than males. Of course, larger females produce more eggs than smaller ones, and species of oysters that are sequential hermaphrodites are generally protogynous.

The diverse reproductive cycles and patterns we observe in the animal kingdom are adaptations that

have evolved by natural selection. We shall see many other examples as we survey various mechanisms of sexual reproduction.

MECHANISMS OF SEXUAL REPRODUCTION

Patterns of Fertilization and Development

The two major mechanisms of fertilization that have evolved are external and internal fertilization, each having specific environmental and behavioral requirements.

Getting Gametes Together   The mechanics of fertilization play an important part in sexual reproduction. In **external fertilization,** the eggs are shed by the female and fertilized by the male in the environment. In contrast, **internal fertilization** occurs when the sperm are deposited in (or nearby) the female reproductive tract and egg and sperm unite within the body of the female. These patterns correlate more with the habitats of species than with their phylogenetic positions.

Because external fertilization requires an environment where an egg can develop without desiccation or heat stress, it occurs almost exclusively in moist habitats. Many aquatic invertebrates simply shed their eggs and sperm into the surroundings, and fertilization occurs without the parents actually making physical contact. Still, timing is crucial to ensure that mature sperm encounter ripe eggs. Often, environmental cues such as temperature or day length cause all the individuals of a population to release gametes at once, or pheromones from one individual releasing gametes trigger gamete release in others. In the South Pacific, the Palolo worm (a polychaete) lives in crevices of coral reefs but mates at the ocean surface. The entire population of worms inhabiting a Samoan reef reproduce on a single fall night during a certain phase of the moon. All the worms rise to the surface and shed their gametes simultaneously. People who live on nearby islands take advantage of this mating swarm to collect the worms, a favorite food delicacy.

Most fishes and amphibians that employ external fertilization show specific mating behaviors, resulting in one male fertilizing the eggs of one female. Courtship behavior acts as a mutual trigger for the release of gametes, with two effects: The probability of successful fertilization is increased, and the choice of mates may be somewhat selective.

Internal fertilization requires cooperative behavior that makes copulation possible. In some cases, uncharacteristic sexual behavior is eliminated by natural selection in a most direct manner; for example, female spiders will eat males if specific reproductive signals are not followed during mating. Many more examples of sexual selection and mating behavior are discussed in Chapter 50.

Internal fertilization also requires sophisticated reproductive systems. Copulatory organs for delivery of sperm and receptacles for its storage and transport to the eggs must be present.

Protection of the Embryo External fertilization usually results in enormous numbers of zygotes, but the proportion that survive and develop further is often quite small. In contrast, internal fertilization usually produces fewer zygotes, but this may be offset by greater protection of the embryos and parental care of the young. Major types of protection of the embryo include resistant eggs, development of the embryo within the reproductive tract of the female parent, and parental protection of eggs.

Many species of terrestrial animals have eggs that can withstand harsh environments. Among the vertebrates, we can compare the eggs of fishes and amphibians, which have only a gelatinous coat that allows free exchange of gases and water, with the amniote eggs of birds and reptiles, whose calcium and protein shells are resistant to water loss and physical damage (see Chapter 30).

Rather than secreting a protective shell around the egg, many animals retain the embryo, which develops within the female reproductive tract. Among mammals, the monotremes lay eggs reminiscent of their reptilian forebears, whereas marsupials such as kangaroos and opossums retain their embryos for a short period in the uterus; the embryos then crawl out and complete fetal development attached to a mammary gland in the mother's pouch. The embryos of placental mammals develop entirely within the uterus, being

Figure 42.4
Parental care in an invertebrate. Compared to most arthropods, a female scorpion produces a relatively small number of offspring, but parental protection enhances the survival of those offspring. Fertilization is internal and the eggs are retained within the reproductive tract of the mother, where they hatch. The young then cluster on the back of their mother, prolonging the parental protection.

nourished by the mother's blood supply through a special organ, the placenta (see Chapter 30).

When a bird hatches from an egg, or a kangaroo crawls out of its mother's pouch for the first time, or a human is born, it still is not capable of independent existence. We are familiar with adult birds feeding their young and mammals nursing their offspring, but parental care is much more widespread than we might suspect, and it takes a variety of unusual forms. In one species of tropical frog, for instance, the male carries the tadpoles in his stomach until they metamorphose and hop out as young frogs. There are also many cases of parental care among invertebrates (Figure 42.4).

Diversity in Reproductive Systems

To reproduce sexually, animals must have systems to produce and deliver gametes to the gametes of the opposite sex. These reproductive systems are varied. The least complex systems do not even contain distinct **gonads,** the organs that produce gametes in most animals. The most complex reproductive systems contain many sets of accessory tubes and glands to carry and protect the gametes and developing embryos. If we can make one generalization, it is that the complexity of the reproductive system is not related to the phylogenetic position of the animal; the reproductive systems of parasitic flatworms, for example, are among the most complex in the animal kingdom.

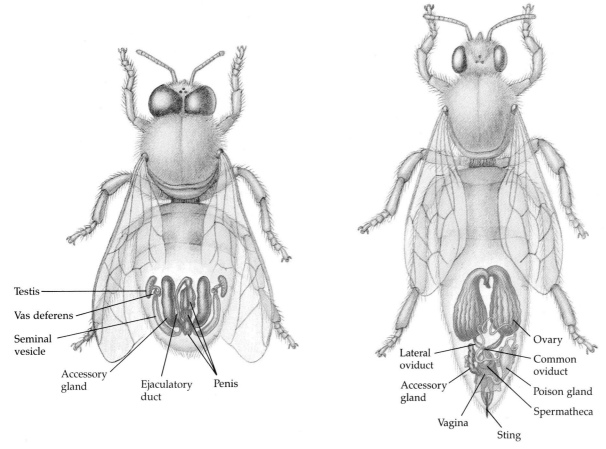

| Testis |
| Vas deferens |
| Seminal vesicle |
| Accessory gland | Ejaculatory duct | Penis |

(a) Male reproductive organs

Lateral oviduct	Ovary
Accessory gland	Common oviduct
	Poison gland
Vagina	Spermatheca
Sting

(b) Female reproductive organs

Figure 42.5
Insect reproductive anatomy. (a) Male. Sperm form in the testes, pass through the sperm duct, and are stored in the seminal vesicle. The male ejaculates sperm along with fluid from the accessory glands.

Some species of insects and other arthropods possess appendages called claspers to grasp the female during copulation. **(b)** Female. Eggs develop in the ovaries, pass through the oviducts, and are depos-

ited in the vagina. Some species have a spermatheca for the storage of sperm, which fertilize an egg as it passes through the vagina.

Invertebrate Reproductive Systems   Diverse reproductive systems have evolved among invertebrates. We will examine three examples.

Polychaete annelids are a group of mostly marine worms that have separate sexes with relatively simple reproductive systems. Most polychaetes do not have distinct gonads; rather, the eggs and sperm develop from undifferentiated cells lining the coelom. As the gametes mature, they are released from the body wall and fill the coelom. Depending on the species, mature gametes may be shed through the excretory openings, or the swelling mass of eggs may split the body open, killing the parent and spilling the eggs into the environment.

Insects have separate sexes with complex reproductive systems (Figure 42.5). In the male, sperm develop in the testes and pass through a duct to the seminal vesicles where they are stored. The seminal vesicles empty into an ejaculatory duct, which runs through

the penis. A pair of accessory glands that add fluid to the semen may also be present. Eggs in the female pass from the ovaries through the oviducts and are deposited in the vagina, which opens to the outside of the body. The female reproductive system may also include a **spermatheca,** a blind sac in which sperm may be stored for a year or more.

Flatworms (Phylum Platyhelminthes) are hermaphrodites with extremely complex reproductive systems (Figure 42.6). In addition to those structures described for insects, the female reproductive system includes yolk and shell glands, as well as a uterus where the eggs are fertilized and development begins in some species. The male system contains a complex copulatory apparatus sometimes called a penis. Copulation in flatworms is usually mutual, with each partner inseminating the other. The mechanisms of insemination range from insertion of the penis into a vagina to hypodermic impregnation, where the penis

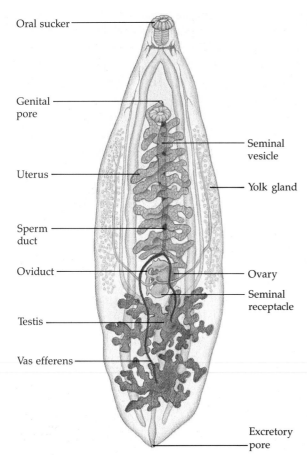

Figure 42.6
Reproductive anatomy in flatworms. Flatworms can reproduce either asexually by fission or sexually. During the breeding season, a flatworm develops both male and female reproductive organs that open through a common genital pore. Copulation usually results in mutual insemination. The flatworm depicted here is a fluke.

injects sperm into the body tissues and sperm must migrate on their own to the female reproductive tract.

Vertebrate Reproductive Systems   The basic plan of all vertebrate reproductive systems is quite similar, but there are some important variations on this common theme. Most mammals have separate openings for the digestive, excretory, and reproductive systems, but all other vertebrates have a common opening for all three systems, the **cloaca** (see Chapter 30). The uterus of most vertebrates is bicornate, having two separate branches. In humans and other mammals that develop only a few young in the uterus, as well as in birds and in snakes, the uterus has only one branch. Differences in the male reproductive systems center around the copulatory organs. Nonmammalian vertebrates do not have well-developed penes (plural of *penis*) and may just evert the cloaca to ejaculate.

MAMMALIAN REPRODUCTION

In describing mammalian reproduction, we will use humans as an example, but the reproductive anatomy is similar in other mammals.

Anatomy of Reproduction

Male Anatomy   The reproductive system is frequently described as two sets of organs, the internal reproductive organs and the external **genitalia** (Figure 42.7). The male genitalia are the **scrotum** and **penis.** The internal reproductive organs consist of gonads that produce gametes (sperm cells) and hormones, accessory glands that secrete products essential to sperm movement, and a set of ducts that carry the sperm and glandular secretions.

The male gonads, or **testes,** are a pair of tightly coiled tubes surrounded by several layers of connective tissue. These tubes are the **seminiferous tubules,** where sperm form. The **interstitial cells** scattered between the seminiferous tubules produce testosterone and other androgens, the male sex hormones.

Sperm production cannot occur at normal body temperatures in most mammals, but the testes of humans and many other mammals hang outside the abdominal cavity in the scrotum, a fold of skin. The temperature in a scrotum is about 2°C below that in the body cavity. The testes develop high in the abdominal cavity and descend into the scrotum just before birth. (In about 1%–2% of human males, the testes do not descend, but they can be brought down by hormone therapy or surgery.) In some mammals (not humans), the testes are drawn back into the body cavity between breeding seasons. Whales and bats are exceptional in retaining the testes within the body cavity permanently.

From the seminiferous tubules of the testes, the sperm pass into the coiled tubules of the **epididymis,** which stores sperm and is the site of their final maturation. During **ejaculation,** the sperm are propelled from the epididymis through the muscular **vas deferens.** These ducts run from the scrotum around and behind the urinary bladder where they join to form a short **ejaculatory duct.** The ejaculatory duct opens into the **urethra,** the tube draining both the excretory system and the reproductive system. Thus, the reproductive and excretory systems of the male are connected, but, as we shall see, this is not the case in females. The urethra runs through the penis and opens to the outside at the tip of the penis.

In addition to the testes and ducts, the male reproductive system contains three sets of glands that add their secretions to the **semen,** the fluid that is ejaculated. The **seminal vesicles** contribute about 60% of

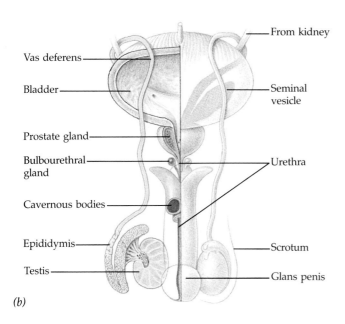

Seminal vesicle

Ejaculatory duct

Urinary bladder

Vas deferens

Pubis
(bone)

Prostate gland

Cavernous bodies

Urethra

Epididymis

Testis

Glans penis

Prepuce

Rectum

Bulbourethral
gland

Scrotum

(a)

Figure 42.7
**Reproductive anatomy of the human
male. (a)** Lateral view. **(b)** Frontal view.

From kidney

Vas deferens

Bladder

Seminal
vesicle

Prostate gland

Bulbourethral
gland

Urethra

Cavernous bodies

Epididymis

Scrotum

Testis

Glans penis

(b)

the total volume of the semen. This pair of glands lies below and behind the bladder and empties into the ejaculatory duct. The fluid from the seminal vesicles is thick and clear, containing mucus, amino acids, and large amounts of fructose, which provides energy for the sperm. The seminal vesicles also secrete prostaglandins (see Chapter 41). These hormonelike compounds, once in the female reproductive tract, stimulate contractions of the uterine muscles that help move the semen up the uterus. Proteins in the seminal fluid

cause the semen to coagulate after it is deposited in the female reproductive tract, making it easier for uterine contractions to move the semen.

The **prostate gland** is the largest of the accessory glands. It surrounds the initial segment of the urethra and secretes its products directly into the urethra through several small ducts. Prostatic fluid is thin, milky, and quite alkaline, which balances the acidity of any residual urine in the urethra and the natural acidity of the vagina. The prostate gland is the source of some of the most common medical problems of men over 40. Benign (noncancerous) enlargement of the prostate occurs in more than half of all men in this age group.

The **bulbourethral glands,** the final accessory structures, are a pair of small glands along the urethra below the prostate. Their function is still in question. They secrete a viscous fluid before emission of the sperm and semen. It has been suggested that this fluid lubricates the penis and vagina, but the volume (just one or two drops) seems insufficient to be very effective for this function. Bulbourethral fluid does carry some sperm released before ejaculation, which is one factor in the low success rate of the withdrawal method of birth control.

The human penis is composed of three cylinders of spongy tissue (cavernous bodies) derived from modified veins and capillaries. During sexual arousal, this erectile tissue fills with blood from the arteries. As it fills, the increasing pressure seals off the veins that drain the penis, causing it to engorge with blood. The resulting erection is essential to insertion of the penis

Figure 42.8
Reproductive anatomy of the human female. (a) Lateral view. **(b)** Frontal view.

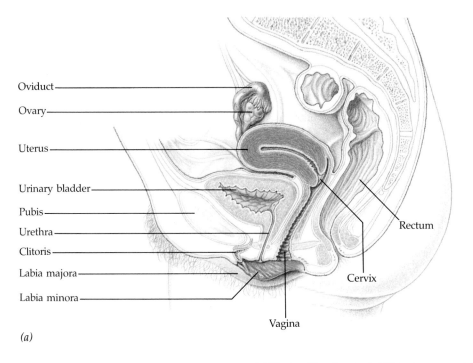

Oviduct

Ovary

Uterus

Urinary bladder

Pubis

Urethra

Clitoris

Labia majora

Labia minora

Rectum

Cervix

Vagina

(a)

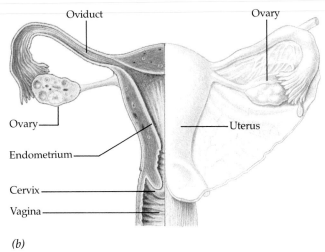

Oviduct

Ovary

Ovary

Endometrium

Uterus

Cervix

Vagina

(b)

into the vagina. Rodents, raccoons, walruses, and several other mammals also possess a **baculum,** a bone that stiffens the penis.

The main shaft of the penis is covered by relatively thick skin, whereas the head, or **glans penis,** has a much thinner covering and is consequently more sensitive to stimulation. The human glans is covered by a fold of skin called the foreskin, or **prepuce,** which may be removed by circumcision. Circumcision was almost routine in the middle part of this century, but now is less common. The practice arose from religious traditions and has no verifiable basis in health or hygiene.

Female Anatomy   The female contains a pair of gonads, a system of ducts and chambers to conduct the gametes as well as to house the embryo and fetus, and external genitalia that facilitate reproductive function (Figure 42.8). The female gonads, the **ovaries,** lie in the abdominal cavity below most of the digestive system. Each ovary is enclosed in a tough protective capsule and contains many **follicles.** A follicle consists of one egg cell surrounded by one or more layers of follicle cells, which nourish and protect the developing egg cell. All of the 400,000 follicles a woman will ever have are formed at birth. Of these, only several hundred will be released during the woman's reproductive years. After puberty, one (or rarely two or more) follicle matures and releases its egg during each menstrual cycle. The cells of the follicle also produce the primary female sex hormones, the estrogens. When **ovulation** occurs, the egg is expelled from the follicle (much like a small volcano), and the remaining

follicular tissue grows within the ovary to form a solid mass called the **corpus luteum.** The corpus luteum secretes progesterone, the hormone of pregnancy, and additional estrogen. If the egg is not fertilized, the corpus luteum degenerates and a new follicle matures during the next cycle.

The female reproductive system is not competely closed, and the egg cell is expelled into the abdominal cavity near the opening of the **oviduct,** or fallopian tube. The oviduct has a funnellike opening, and cilia on the inner epithelium lining the duct help collect the egg cell by drawing fluid from the body cavity into the duct. The cilia also convey the egg cell down the duct to the **uterus,** commonly called the womb. The uterus is a thick, muscular organ shaped much like an upside-down pear. It is remarkably small; the uterus of a woman who has never been pregnant is about 7 cm long and 4–5 cm wide at its widest point. The

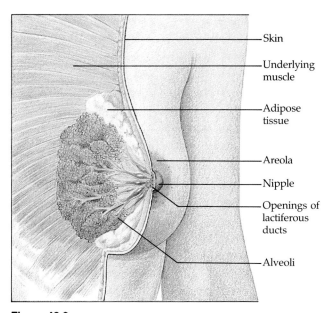

Figure 42.9
Human mammary gland.

Skin
Underlying muscle
Adipose tissue
Areola
Nipple
Openings of lactiferous ducts
Alveoli

unique arrangement of muscles that make up the bulk of the uterine wall allow it to expand to accommodate a 4-kg fetus. The inner lining of the uterus, the **endometrium,** is richly supplied with blood vessels.

The narrow neck of the uterus is the cervix, which opens into the vagina. The **vagina** is a thin-walled chamber that forms the birth canal through which the baby is expelled; it is also the repository for sperm during copulation. Although much thinner than the uterus, the walls of the vagina are sufficiently muscular that they can contract to hold any penis tightly or expand to allow passage of a baby's head.

The vagina is the terminal portion of the female reproductive system, but it is covered externally by two pairs of skin folds that form a **vestibule** containing the vaginal orifice and the opening of the urethra. (Notice that in contrast to the male reproductive tract, the female tract opens separately from the excretory system.) At birth and until intercourse or vigorous physical activity takes place, the human vaginal orifice is covered by a thin membrane called the **hymen,** which has no known function. The vestibule is bordered by the slender labia minora, which in turn are protected by a pair of thick, fatty ridges called the labia majora. Like the vagina, the labia minora are composed of erectile tissue and enlarge during arousal and intercourse. At the top of the vestibule a small bulb of erectile tissue peeks out from a hood of skin. This **glans clitoris,** or clitoris, is the female equivalent of the glans of the penis. Like that organ, it is composed of erectile tissue and is one of the most sensitive points of stimulation in sexual response.

The **mammary gland,** or breast, is another structure important to mammalian reproduction, although it is not part of the reproductive tract itself (Figure 42.9). The secretory apparatus consists of a series of **alveoli,** small sacs of epithelial tissue that secrete milk. The alveoli drain into a series of ducts that open at the nipple. Deposits of fatty tissue form the main mass of the mammary gland of a nonlactating mammal. The lack of estrogen in males prevents the development of both the secretory apparatus and the fat deposits, so the breasts remain small and the nipple is not connected to the ducts.

The external genitalia of both sexes arise from common **primordia,** or undifferentiated embryonic structures (Figure 42.10). Sex chromosomes determine the balance of male and female sex hormones in the embryo, and in turn the hormonal balance determines whether the primordial genitalia develop into female or male structures. If androgens are present, male structures form. Female structures form in the absence of androgens.

Hormonal Control of Mammalian Reproduction

The Male Pattern   The principal male sex hormones are the **androgens,** of which **testosterone** is the most important. Androgens, steroid hormones produced by the interstitial cells of the testes, are directly responsible for the primary and secondary sex characteristics of the male. The primary sex characteristics are those associated with the reproductive system—development of the vasa deferentia and other ducts, the external genitalia, and sperm production. The secondary sex characteristics are features we associate with maleness that are not directly related to the reproductive system. They include deepening of the voice; the male distributions of axillary (armpit), facial, and pubic hair; and muscle growth (androgens stimulate protein synthesis). Androgens are also potent determinants of behavior in mammals and other vertebrates. In addition to specific sexual behaviors and libido (sex drive), androgens increase general aggressiveness and are responsible for such actions as singing in birds and calling by frogs. Hormones from the anterior pituitary and hypothalamus control both androgen secretion and sperm production by the testes (Figure 42.11).

The Female Pattern   The pattern of hormone secretion controlling female reproduction differs strikingly from the male pattern, reflecting a cyclic nature of female reproduction. Whereas the male produces sperm continuously, females ovulate only one or a few eggs at one time during each cycle. The control of this cycle is considerably more complex than the control of male reproduction.

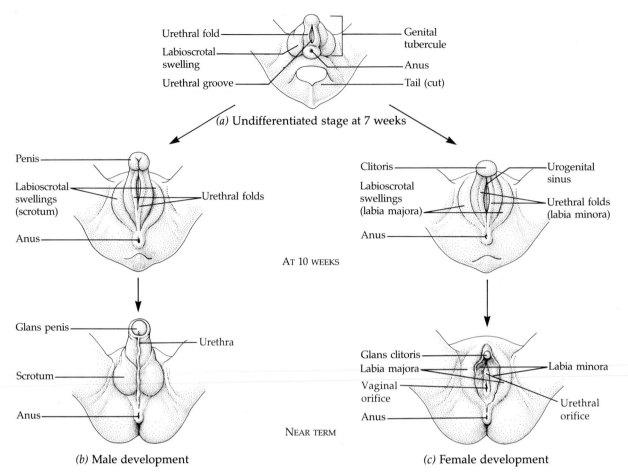

Figure 42.10

Development of external genitalia in humans. (a) Both external structures and internal reproductive organs are undifferentiated until about the eighth week of gestation. All embryos have a conical elevation called the genital tubercle, which has a shallow depression called the ure- thral groove. Urethral folds surround the groove, which in turn are bounded by labioscrotal swellings. Production of hor- mones, under genetic control, triggers sex- ual differentiation. **(b)** In the male, the ure- thral groove elongates and closes completely, and the urethral folds give rise to the shaft of the penis. **(c)** In the female, the elongated groove remains open and the same tissues become the labia minora. Similarly, the labioscrotal swelling develops into the scrotum of the male or the labia majora of the female.

Two different types of cycles occur in female mammals. Humans and many other primates have **menstrual cycles,** whereas other mammals have **estrous cycles.** In both cases, ovulation occurs at a time in the cycle after the endometrium has started to thicken and become more extensively vascularized, which prepares the uterus for the possible implantation of an embryo. One major difference between the two types of cycles involves the fate of the uterine lining if pregnancy does not occur. In menstrual cycles, the endometrium is shed from the uterus through the cervix and vagina in a bleeding called **menstruation.** In estrous cycles, the endometrium is reabsorbed by the uterus and no bleeding occurs.

Other distinctions include more pronounced behavioral changes during estrous cycles than during menstrual cycles and stronger effects of season and climate on estrous cycles. Whereas human females may be receptive to sexual activity throughout their cycles, most mammals will copulate only during the period surrounding ovulation. This period of sexual activity is called **estrus** (from the Latin *oestrus,* "frenzy" or "passion"), or heat because the body temperature increases slightly. The length and frequency of reproductive cycles vary widely among mammals. The human menstrual cycle averages 28 days, whereas the estrous cycle of the rat is only 5 days. Bears and dogs have one cycle per year, but elephants cycle several times per year.

Let us examine the menstrual cycle of the human female in more detail as a case study of how a complex function is coordinated by hormones. The cycle *averages* 28 days, but only about 30% of women have cycle lengths within a day or two of the statistical 28 days.

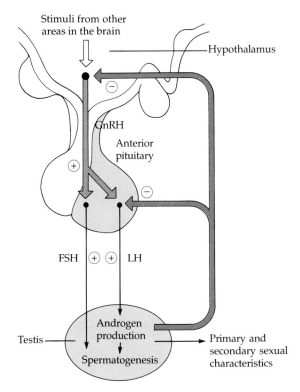

Figure 42.11
Hormonal control of the testes. The pituitary secretes two gonadotropic hormones with different effects on the testes. Luteinizing hormone (LH) stimulates androgen production by the interstitial cells. Follicle-stimulating hormone (FSH) acts on the seminiferous tubules to increase sperm production (spermatogenesis). Since androgens are also required for sperm production, LH stimulates spermatogenesis indirectly. LH and FSH are in turn regulated by a single hormone from the hypothalamus, gonadotropin-releasing hormone (GnRH). How GnRH controls the release of two different hormones at different times is still unknown. LH, FSH, and GnRH concentrations in the blood are regulated by negative feedback by androgens. GnRH is also controlled by negative feedback from the two pituitary gonadotropins. In human males, these feedback loops keep the hormones at relatively constant levels, but in many other mammalian species, there are seasonal cycles in hormone concentration associated with specific breeding seasons.

Cycles vary from one woman to another, ranging from about 20 to 40 days. In some women the cycles are usually very regular, but in other individuals the timing varies from cycle to cycle.

The term *menstrual cycle* refers specifically to the changes that occur in the uterus. By convention, the first day of a woman's "period"—that is, the first day of menstruation—is designated as day 1 of the cycle. Menstrual bleeding, sometimes called the **flow phase** of the cycle, usually persists for a few days. Then the endometrium begins to thicken for a week or two during what is called the **proliferative phase** of the men-

strual cycle. During the next phase, the **secretory phase,** usually about two weeks in duration, the endometrium continues to thicken, becomes more vascularized, and secretes a fluid rich in glycogen. If an embryo has not implanted in the uterine lining by the end of the secretory phase, a new menstrual flow commences, marking day 1 of the next cycle.

Paralleling the menstrual cycle is an **ovarian cycle.** It begins with the **follicular phase,** during which several follicles in the ovary begin to grow. The egg cell enlarges and the coat of follicle cells becomes multilayered. Of the several follicles that start to grow, only one usually continues to enlarge and mature, while the others degenerate. The maturing follicle develops an internal fluid-filled cavity and grows very large, forming a bulge near the surface of the ovary. The follicular phase ends with ovulation when the follicle and adjacent wall of the ovary rupture, releasing the egg cell. The follicular tissue that remains in the ovary after ovulation is transformed into the corpus luteum, an endocrine tissue that secretes female hormones during what is called the **luteal phase** of the ovarian cycle. The next cycle begins with a new growth of follicles.

Hormones coordinate the menstrual and ovarian cycles in such a way that growth of the follicle and ovulation are synchronized with preparation of the uterine lining for possible implantation of an embryo. Five hormones participate in an elaborate scheme involving both positive and negative feedback. These hormones are: gonadotropin-releasing hormone (GnRH) secreted by the hypothalamus; follicle-stimulating hormone (FSH) and luteinizing hormone (LH), the two gonadotropins secreted by the anterior pituitary gland; and estrogens and progesterone, the female sex hormones secreted by the ovary. The levels of the pituitary and ovarian hormones in blood plasma are traced in Figure 42.12, along with ovarian and menstrual cycles. Referring to the figure frequently as you read the following discussion will help you to understand how the female reproductive system is regulated.

During the follicular phase of the ovarian cycle, the pituitary secretes relatively small quantities of FSH and LH in response to stimulation by GnRH from the hypothalamus. At this time, the cells of immature follicles in the ovary have receptors for FSH, but not for LH. The FSH stimulates growth of follicles, and the cells of these growing follicles secrete estrogen. Notice in Figure 42.12 that the amount of estrogen secreted during this time is small, and because low levels of estrogen inhibit secretion of the pituitary hormones, the levels of FSH and LH are also relatively low during most of the follicular phase. These hormonal relationships change radically and rather abruptly when the rate of estrogen secretion by the growing follicle begins to rise steeply. Whereas a *low* concentration of estrogen kept secretion of pituitary gonadotropins at a low

Figure 42.12

Reproductive cycle of the human female. Hormones coordinate the ovarian cycle and the menstrual cycle, preparing the uterine lining (endometrium) for implantation of an embryo even before ovulation has occurred. The time scale used here assumes an idealized 28-day cycle, but actual reproductive cycles vary from about 20 to 40 days in length. The upper graph traces the levels of the two pituitary gonadotropins, FSH and LH, during the reproductive cycle. The second graph follows the levels of the two ovarian hormones, estrogens (actually a family of closely related hormones) and progesterone. Use these graphs as visual aids to support the text's discussion of the feedback circuits that regulate the levels of these hormones. The ovarian cycle consists of a follicular phase, during which follicles grow and secrete increasing amounts of estrogen, and a luteal phase, during which the corpus luteum formed from the follicular tissue after ovulation secretes estrogen and progesterone. The duration of the follicular phase varies extensively from woman to woman, and, in some cases, even in an individual woman from one cycle to the next. The luteal phase is more consistent in length, usually lasting about 13 to 15 days, regardless of the overall length of the cycle. The menstrual cycle consists of a flow phase, a proliferative phase, and a secretory phase. Menstruation, the bleeding associated with degeneration of the endometrium, occurs during the flow phase. The first day of flow marks day 1 of the menstrual cycle. During the proliferative phase, estrogen from the growing follicle stimulates the endometrium (uterine lining) to thicken and become more extensively vascularized. During the secretory phase, the endometrium continues to thicken, its arteries enlarge, and endometrial glands that secrete a fluid containing glycogen grow. These changes in the endometrium require estrogen and progesterone, which are secreted by the corpus

luteum after ovulation. Thus, the secretory phase of the menstrual cycle parallels the luteal phase of the ovarian cycle. Degeneration of the corpus luteum at the end of the luteal phase deprives the uterine lining of estrogen and progesterone, and the endometrium is sloughed off. The first day of menstruation, marking the beginning of the next cycle, usually occurs about 13–15 days after ovulation. However, because the follicular phase of the

ovary, and hence the proliferative phase of the uterus, is so variable in duration, it is not usually possible to predict how long after menstruation the next ovulation will occur (one of the problems with the rhythm method of contraception). In the event of pregnancy, additional mechanisms, to be discussed later in the chapter, maintain high levels of estrogen and progesterone and prevent loss of the endometrium.

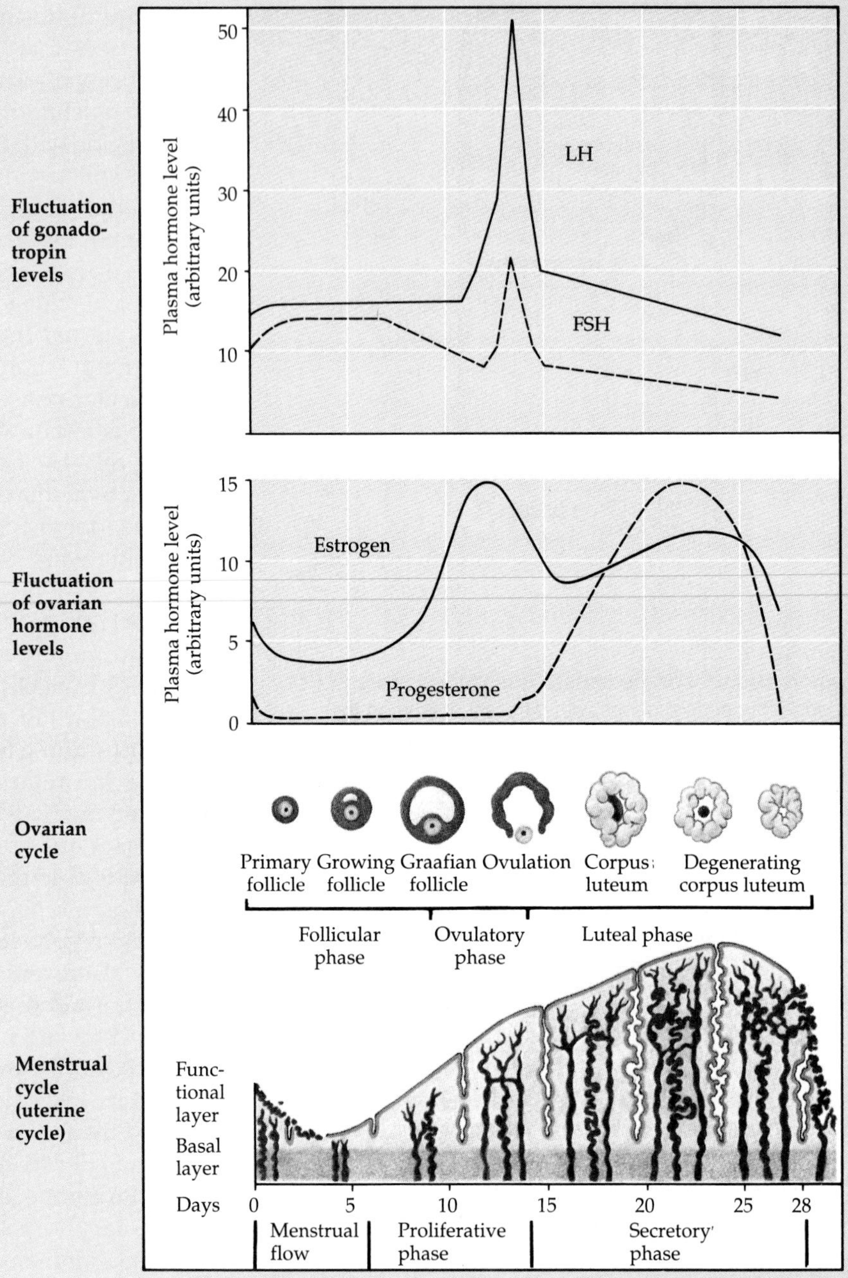

level, a *high* concentration of estrogen has the opposite effect and *stimulates* secretion of gonadotropins by acting on the hypothalamus to increase its output of GnRH. You can see this response in Figure 42.12 as a steep incline of FSH and LH levels that follows closely behind the increase in estrogen concentration. The effect is greater for LH because the high concentration of estrogen, in addition to stimulating GnRH secretion, also increases the sensitivity of LH-releasing mechanisms in the pituitary to the hypothalamic signal (GnRH). By now, the follicles have receptors for LH and can respond to this hormonal cue. In a case of positive feedback, the increase in LH concentration caused by increased estrogen secretion from the growing follicle induces final maturation of the follicle, and ovulation occurs about a day after the LH surge. Following ovulation, LH stimulates the transformation of the follicular tissue left behind in the ovary to form the corpus luteum, a glandular structure (it is for this "luteinizing" function that luteinizing hormone—LH—is named). Under continued stimulation by LH during the luteal phase of the ovarian cycle, the corpus luteum secretes estrogen and a second steroid hormone, progesterone. The corpus luteum usually reaches its maximum development about 8 to 10 days after ovulation. As the progesterone and estrogen levels rise, the combination of these hormones exerts negative feedback on the hypothalamus and pituitary, inhibiting the secretion of the gonadotropins LH and FSH. When the LH concentration plummets, the corpus luteum, which requires LH in order to function, begins to degenerate. Consequently, the concentrations of estrogen and progesterone decline sharply near the end of the luteal phase. The dropping levels of ovarian hormones liberate the hypothalamus and pituitary from the inhibitory effects of these hormones. The pituitary then begins to secrete enough FSH to stimulate growth of new follicles in the ovary, initiating the follicular phase of the next ovarian cycle.

How is the ovarian cycle synchronized with the menstrual cycle? Estrogen, secreted in increasing amounts by growing follicles, is a hormonal signal to the uterus, causing the endometrium to thicken. Thus, the follicular phase of the ovarian cycle is coordinated with the proliferative phase of the menstrual cycle. *Before* ovulation, the uterus is already being prepared for a possible embryo. After ovulation, estrogen and progesterone secreted by the corpus luteum stimulate continued development and maintenance of the endometrium, including an enlargement of arteries supplying blood to the uterine lining and growth of endometrial glands that secrete a fluid with nutrients that can sustain an early embryo before it actually implants in the uterine lining. Thus, the luteal phase of the ovarian cycle is coordinated with the secretory phase of the menstrual cycle. The rapid drop in the level of ovarian hormones when the corpus luteum

degenerates causes spasms of arteries in the uterine lining that deprives the endometrium of blood. Degeneration of the endometrium results in menstruation and the beginning of a new menstrual cycle. In the meantime, ovarian follicles that will stimulate renewed thickening of the endometrium are just beginning to grow. Cycle after cycle, the maturation and release of egg cells from the ovary is integrated with changes in the uterus, the organ that must accommodate an embryo should the egg cell be fertilized. In the absence of pregnancy, a new cycle begins. We shall soon see that there are "override" mechanisms that prevent degeneration of the endometrium in the event of pregnancy.

In addition to their role in coordinating reproductive cycles, estrogens are also responsible for the secondary sex characteristics of the female. The hormones induce deposition of fat in the breasts and hips, increase water retention, affect calcium metabolism, stimulate breast development, and mediate female sexual behavior.

Gamete Formation (Gametogenesis)

Spermatogenesis **Spermatogenesis,** the production of mature sperm cells, is a continuous and prolific process in the adult male. Each ejaculation of a human male contains about 400 million sperm cells, and males can ejaculate daily with no loss of fertilizing capacity. Spermatogenesis occurs in the seminiferous tubules of the testes (Figure 42.13).

The structure of a spermatozoon (sperm cell) fits its function (Figure 42.14). The thick head containing the haploid nucleus is tipped with a special body, the **acrosome,** which contains enzymes that help the sperm penetrate the egg. Behind the head, the neck contains large numbers of mitochondria (or a single large one in some species) that provide ATP for movement of the tail, a flagellum. Mammalian sperm morphology is quite variable, with head shapes ranging from a slender comma through the oval of the human sperm to nearly spherical.

Oogenesis Development of ova (mature, unfertilized egg cells), or **oogenesis,** differs from spermatogenesis in three important ways (Figure 42.15). First, during the meiotic divisions of oogenesis, cytokinesis is unequal, with almost all the cytoplasm monopolized by a single daughter cell. This large cell can go on to form the ovum, and the three smaller cells, called polar bodies, soon degenerate. This contrasts with spermatogenesis, when all four products of meiosis I and II develop into mature sperm (compare Figures 42.13a and 42.15a). Second, while the precursor cells of sperm continue to divide by mitosis throughout the reproductive years of a male, this is not the case for

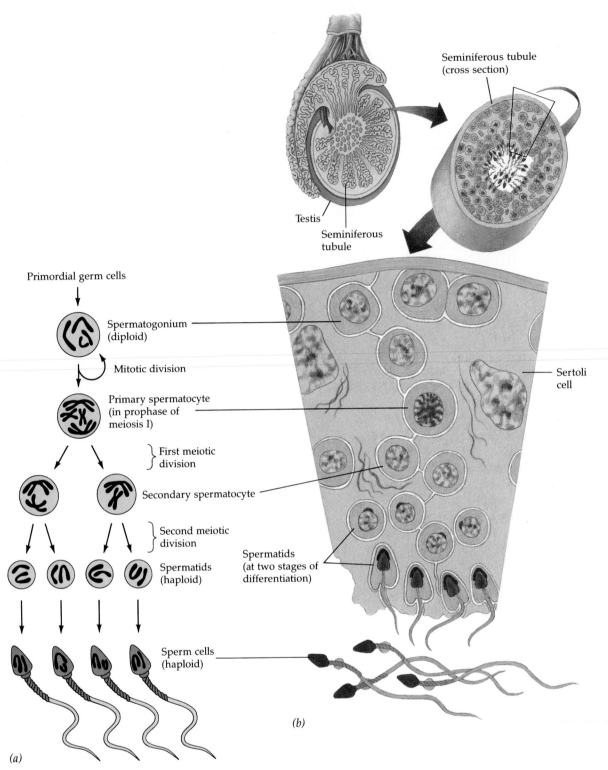

(a)

(b)

Figure 42.13
Spermatogenesis. Primordial germ cells of the embryonic testes differentiate into spermatogonia, the diploid cells that are the precursors of sperm. Located near the outer wall of the seminiferous tubules, spermatogonia undergo repeated mitoses to produce large, renewable populations of potential sperm. In a mature male, about three million spermatogonia per day differentiate into primary spermatocytes.

Meiosis, which generates haploid gametes, occurs in two steps. First, meiosis I produces two cells with haploid numbers of double-stranded chromosomes, the secondary spermatocytes. Then, the second meiotic division forms four spermatids with haploid numbers of single-stranded chromosomes. Spermatids do not much resemble mature sperm, and they must differentiate into mature sper-

matozoa, or sperm cells. This involves association of the developing sperm with Sertoli cells, which transfer nutrients to the spermatids. During spermatogenesis, the developing sperm are gradually pushed toward the center of the seminiferous tubule and make their way to the epididymis, where they acquire motility. This process, from spermatogonia to motile sperm, takes 65–75 days in the human male.

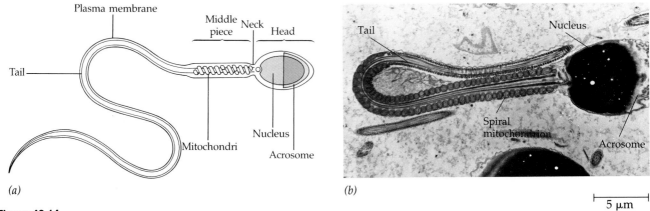

Figure 42.14
Sperm. (a) Human sperm structure. (b) Transmission electron micrograph of a sperm cell from a rhesus monkey, which is very similar to human sperm.

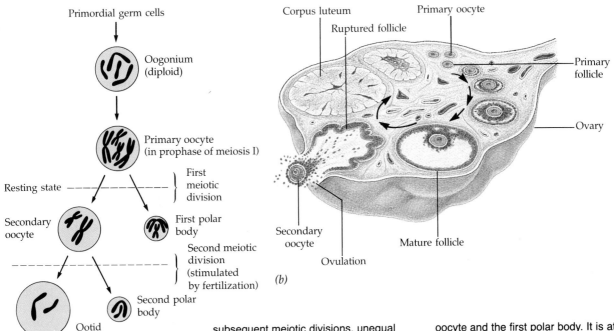

Figure 42.15
Oogenesis. (a) Oogenesis begins with mitosis of the primordial germ cells in the embryo to produce diploid oogonia, which develop into primary oocytes. During the subsequent meiotic divisions, unequal cytokinesis concentrates almost all of the cytoplasm in a single large ovum. The other products of meiosis are tiny cells called polar bodies that degenerate. This situation contrasts with spermatogenesis, which yields four sperm cells for each primary spermatocyte. (b) A human female is born with all her potential ova already present as primary oocytes. Between birth and puberty, the primary oocytes enlarge and the follicles around them grow. Under the influence of FSH in each reproductive cycle, a follicle enlarges and the first meiotic division produces a secondary oocyte and the first polar body. It is at this stage of oogenesis that LH triggers ovulation of the secondary oocyte. Only if a sperm cell penetrates the secondary oocyte is the second meiotic division initiated. After meiosis is completed and the second polar body divides from the ovum, the haploid nuclei of the sperm and the now-mature ovum fuse. In this diagram of an ovary, the growth of the follicle, ovulation, and formation and degeneration of the corpus luteum are schematically represented as a cycle. In reality, these stages are never present simultaneously because the ovarian cycle is temporal, not spatial.

oogenesis in the female. At birth, an ovary already contains all the presumptive egg cells it will ever have, though still in an immature state; eggs are a nonrenewable resource. Third, oogenesis has long "resting" periods before the process is complete, in contrast to spermatogenesis, which produces mature sperm from precursor cells in an uninterrupted sequence. After puberty in a female, a few potential ova undergo the first meiotic division within growing follicles during each ovarian cycle. The egg cell released during ovulation is not actually a mature ovum because it has not yet completed meiosis by undergoing the second division. In humans, penetration of the egg cell by the sperm triggers the second meiotic division, and only then is oogenesis actually complete.

Sexual Maturation

Mammals cannot reproduce until they have undergone substantial growth and development after birth. A human male can achieve erection at birth but has no sperm to ejaculate. In humans, the onset of reproductive ability is called puberty and is a gradual process that usually begins about two years earlier in females than in males. Between the ages of 8 and 14, depending on the individual, the hypothalamus begins secreting increasing amounts of GnRH, leading to higher levels of FSH followed by increased LH. These hormones trigger maturation of the reproductive system and development of the secondary sex characteristics (by initiating secretion of sex hormones from the gonads). The first indication of puberty is a growth spurt, followed by first menstruation at about age 12 or 13 in girls or first ejaculation of viable sperm at age 13 or 14 in boys. The age of puberty is quite variable, and recent analysis of historical data suggests that there has been little change in the average age throughout modern times.

Human Sexual Physiology

Many vertebrates and invertebrates have elaborate and complex mating behaviors, but these are usually stereotyped interactions that involve specific sequences of reciprocal behaviors (see Chapter 50). The hallmark of human sexuality is its diversity in stimuli and response. Behind this variable sexual behavior, however, is a common physiological pattern, often called the sexual response cycle. As is also true of reproductive anatomy, endocrinology, and gametogenesis, the sexual responses of males and females have similarities as well as differences.

Two types of physiological reactions predominate in both sexes. **Vasocongestion** is the filling of a tissue with blood caused by increased blood flow through the arteries of that tissue. Erection of the penis is the most obvious example of vasocongestion, but similar responses occur in the testes, labia, clitoris, vagina, and breasts. **Myotonia,** increased muscle tension, is also a widespread phenomenon in sexual response. Both skeletal and smooth muscle may show sustained or rhythmic contractions, including those associated with orgasm.

The sexual response cycle can be divided into four phases: excitement, plateau, orgasm, and resolution. During the **excitement phase,** vasocongestion is particularly evident in erection of the penis and clitoris; enlargement of the testes, labia, and breasts; and vaginal lubrication. Myotonia results in nipple erection and tension of the arms and legs.

The **plateau phase** continues these responses. In females, the outer third of the vagina becomes vasocongested, while the inner two-thirds becomes slightly expanded. This change, coupled with the elevation of the uterus, forms a depression that receives sperm at the back of the vagina. Reactions in nonreproductive organs continue as breathing increases and heart rate rises, sometimes to 150 beats per minute—not in response to the physical effort of sexual activity but as an involuntary response to stimulation of the autonomic nervous system (see Chapter 44).

Orgasm is characterized by rhythmic, involuntary contractions of the reproductive system in both sexes. Male orgasm has two phases. Emission is the contraction of the glands and ducts of the reproductive tract, which deposits semen in the urethra. Expulsion or ejaculation occurs when the urethra contracts and the semen is expelled. During female orgasm, the uterus and outer vagina contract, but the inner two-thirds of the vagina does not. Orgasm is the shortest phase of the sexual response cycle, usually lasting only a few seconds. In both sexes, contractions occur at about 0.8-second intervals and may involve the anal sphincter and several abdominal muscles.

The **resolution phase** completes the cycle and reverses the responses of the earlier stages. Vasocongested organs return to their normal size and color and muscles relax. Most of the changes during resolution are completed in 5 minutes. Loss of penile and clitoral erection, however, may take longer. An initial loss of erection, or detumescence, is rapid in both sexes, but a return of the organ to its nonaroused size may take as long as an hour.

Conception, Pregnancy, and Birth

Pregnancy, or **gestation,** is the condition of carrying developing embryos in the uterus. It begins at **conception,** the fertilization of the egg by a spermatozoon, and continues until the birth of the baby or babies. Human pregnancy averages 266 days (38 weeks) from conception or 40 weeks from the start of the last menstrual cycle. Duration of pregnancy in other spe-

Table 42.1 Milestones during pregnancy

	First Trimester	Second Trimester	Third Trimester
Mother	Estrogen and progesterone levels rise HCG detectable in urine 1 week after missed period Morning sickness, mood swings Slight enlargement of abdomen Weight gain 1 kg	Hormone levels stabilize HCG declines and corpus luteum degenerates; placenta secretes progesterone Nausea usually ends Pregnancy becomes evident by size of abdomen	Rapid weight gain (up to 0.5 kg per week) Pressure on viscera causes frequent urination, constipation Pubic symphysis (midline between left and right sides of pelvis) softens and cervix may dilate and efface Labor and delivery
Fetus	Fertilization, cleavage, and implantation in week 1 Heart begins beating in week 4 Limbs distinct and bones begin to form by week 8 Gender evident by week 10 All major organs formed by end; 5–7.5 cm; 28 g	Face appears human; eyelids and lashes distinct Fingers and toes elongate Body covered with fine hair (lanugo) May suck thumb Heartbeat 100–160, readily detected Fetal movements felt and seen through abdomen 30 cm, 600 g	Rapid growth Proportions change to reduce relative head size Activity decreases as fetus fills uterus May respond differentially to stimuli Birth 50 cm; 3.0–3.5 kg

cies correlates with body size and extent of development of the young at birth. Many rodents (mice and rats) have gestation periods of about 21 days, whereas those of dogs are closer to 60 days. In cows, gestation averages 270 days (almost the same as humans); in giraffes it is more than 420 days; and in elephants gestation is more than 600 days.

Human gestation can be divided for convenience of study into three **trimesters** of about three months each. Table 42.1 lists the principal changes in the mother and fetus during each trimester. The first trimester is the time of most radical change for both the mother and the baby. The egg is fertilized by the sperm in the oviduct, and the resulting **zygote** (fertilized egg) travels down the oviduct into the uterus, moved by the cilia of the oviduct (Figure 42.16). **Cleavage,** or cell division, begins about 24 hours after fertilization and continues more rapidly thereafter. This occurs while the embryo is still passing down the oviduct, a trip that usually takes 3–5 days. A few days after reaching the uterus, or about 1 week after fertilization, the zygote has developed into a hollow ball of cells called a **blastocyst,** which implants into the endometrium. Differentiation of body structures now begins in earnest (embryonic development is described in detail in Chapter 43.) During implantation, the blastocyst bores into the endometrium, which responds by growing over the blastocyst. Eventually, tissues grow out from the developing embryo and mingle with the endometrium to form the **placenta** (Figure 42.17). This disk-shaped organ containing embryonic and maternal blood

vessels grows to about the size of a dinner plate and weighs somewhat less than a kilogram. Diffusion of material between maternal and fetal circulations furnishes a means of respiratory gas exchange and nutrient transfer as well as waste removal for the embryo. Blood from the embryo travels to the placenta through arteries of the umbilical cord and returns via the umbilical vein, passing through the liver of the embryo.

The first trimester is also the main period of **organogenesis,** the development of the organs of the body (Figure 42.18). The heart begins beating by the fourth week and can be detected with a stethoscope by the end of the first trimester. By the end of the eighth week, all the major structures of the adult are present in rudimentary form. At this point, the embryo is called a **fetus.** Although well differentiated, the fetus is only 5 cm long by the end of the first trimester. Because of its rapid organogenesis the embryo is most sensitive during the first trimester to threats such as radiation and drugs that can cause birth defects.

The first trimester is also a time of rapid change for the mother. The embryo secretes hormones that signal its existence and control the mother's reproductive system. One embryonic hormone, **human chorionic gonadotropin (HCG),** acts like pituitary LH to maintain progesterone and estrogen secretion by the corpus luteum through the first trimester of pregnancy. In the absence of this hormonal override, the decline in maternal LH due to inhibition of the pituitary by progesterone would result in menstruation and spontaneous abortion of the embryo. Levels of HCG in the

Figure 42.16

Postfertilization events. (**a**) After fertilization occurs, the zygote travels down the oviduct toward the uterus. (**b**) The zygote begins to divide about 24 hours after fertilization and then continues to divide rapidly (cleavage). (**c**) Three to four days after ovulation, the zygote reaches the uterus and floats freely for several days, nourished by fluid secreted by endometrial glands. (**d**) At the blastocyst stage, the embryo implants itself into the endometrium about 7 days after ovulation.

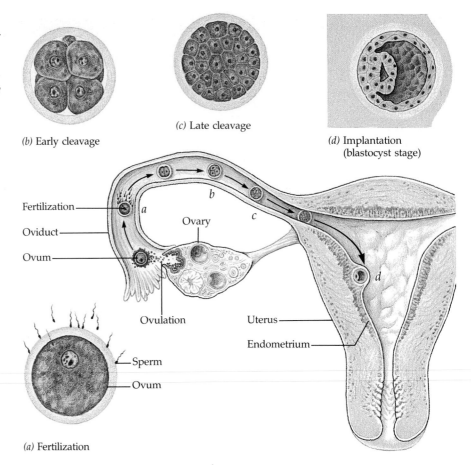

(b) Early cleavage

(c) Late cleavage

(d) Implantation (blastocyst stage)

(a) Fertilization

Figure 42.17

Fetal circulation. The embryo obtains nutrients directly from the endometrium during the first 2 months of development. From the third month until birth, the placenta transports nutrients and wastes to and from the embryo (now called a fetus). The placenta, a combination of maternal and fetal tissues, exchanges oxygen, carbon dioxide, glucose, and other substances between mother and fetus. The maternal blood enters the placenta in arterioles, flows through blood pools in the endometrium, and leaves via venules. The fetal blood, which remains in vessels, enters the placenta through arteries, passes through capillaries in fingerlike villi, where oxygen and nutrients are acquired, and exits in veins leading back to the fetus. Materials are exchanged via diffusion between the fetal capillary bed and the maternal blood pools.

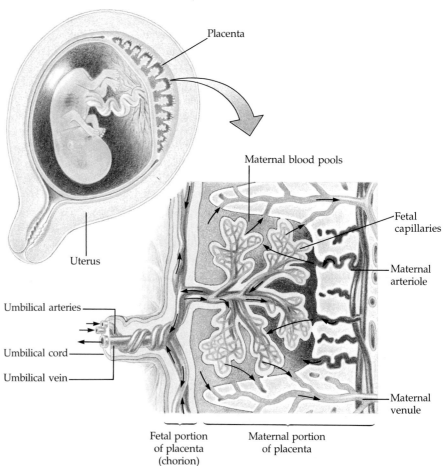

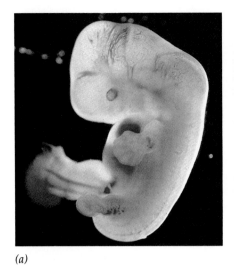
(a)

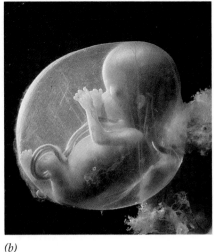
(b)

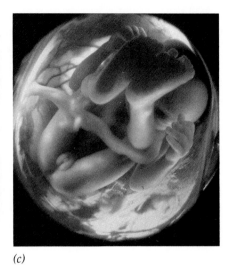
(c)

Figure 42.18
Human fetal development. (a) At 5 weeks, limb buds, eyes, heart, liver, and rudiments of all other organs have started to develop in the embryo, which is only 1 cm long. **(b)** Growth and development of the offspring, now called a fetus, continue during the second trimester. This fetus is 14 weeks old and about 6 cm long. **(c)** The fetus in this photo is 20 weeks old. By the end of the second trimester, the fetus grows to about 30 cm in length.

maternal blood are so high that some is excreted in the urine, where it can be detected in pregnancy tests. High levels of progesterone initiate changes in the pregnant woman's reproductive system that include increased mucus in the cervix to form a protective plug, growth of the maternal part of the placenta, enlargement of the uterus, and (by negative feedback on the hypothalamus and pituitary) cessation of ovulation and menstrual cycling. The breasts also enlarge rapidly and are often quite tender.

During the second trimester, the fetus grows rapidly to about 30 cm and is quite active; movements may be felt by the mother during the early part of the second trimester and be seen through the abdominal wall by the middle of pregnancy. Hormone levels stabilize as HCG declines, the corpus luteum degenerates, and the placenta secretes its own progesterone, which maintains the pregnancy. During the second trimester, the uterus will grow enough for the pregnancy to become obvious.

The third and final trimester is one of rapid growth of the fetus to about 3–3.5 kg and 50 cm. Fetal activity may decrease as the fetus fills the available space within the embryonic membranes. As the fetus grows and the uterus expands around it, the mother's abdominal organs become compressed and displaced, leading to frequent urination, digestive blockages, and strain in back muscles.

Birth, or **parturition,** occurs through a series of strong, rhythmic contractions of the uterus, commonly called **labor.** Prostaglandins from the uterus, oxytocin from the posterior pituitary (see Chapter 41), and nervous reflexes all play roles in regulating labor

contractions. During the first stage of labor, the opening of the cervix thins out (effaces) and dilates. Complete dilation of the cervix ends the first stage of labor. The second stage is the birth of the baby. The uterus is firmly attached to the floor of the abdomen, so the continued strong contractions force the fetus down and out of the uterus and vagina. The umbilical cord is commonly cut and clamped at this time. The final stage of labor is the expulsion of the placenta, which normally follows the baby.

Lactation is an aspect of postnatal care unique to mammals. After birth, decreasing levels of progesterone free the anterior pituitary from negative feedback and allow prolactin secretion. Prolactin stimulates milk production after a delay of 2 or 3 days. The release of milk from the mammary glands is controlled by oxytocin in the milk letdown response, described in Chapter 41.

Reproductive Immunology Pregnancy is an immunological enigma. Half of the embryo's genes are inherited from the father, and thus many of the chemical markers present on the surface of the embryo will be foreign to the mother. Why, then, does the mother not reject this foreign body, the embryo, much as she would repel a tissue or organ graft bearing antigens from another person? Reproductive immunologists are only beginning to solve this puzzle. Part of the answer is the presence of a physical barrier. A protective layer called the trophoblast prevents the embryo itself from actually contacting maternal tissue. But the trophoblast develops along with the embryo from the cells of the blastocyst, and this protective barrier, which penetrates the endometrium, may also be foreign to the mother. According to one hypothesis, the trophoblast does *not* develop paternal markers and thus does not trigger an immune response by the mother. However, several laboratories have recently discovered paternal antigens on portions of the trophoblast. There is evidence that the trophoblast produces a chemical signal that induces development of a special type of white blood cell in the uterus that prevents other white cells from mounting an attack on the foreign tissue. This suppressor cell may work by secreting a substance that blocks the action of interleukin II, the cytokine required for a normal immune response (see Chapter 39). One hypothesis suggests that this local dampening of the immune response occurs, paradoxically, only after white blood cells in the vicinity have first identified the trophoblast as foreign tissue and have taken the first steps of the immune response. If this immunological alarm is not intense enough—that is, if the father's cellular markers are too similar to the mother's—then no suppressor cells are produced.

Some researchers speculate that if the initial immune response is too weak to trigger suppression, then the persistent immunological attack on the foreign tissue, though weak, may lead to spontaneous abortion of the embryo. According to this view, failure to suppress the immune response in the uterus may account for many of the cases of women who have multiple miscarriages for no other apparent reason. There has been some success treating frequent miscarriers by sensitizing the woman's immune system to her mate's antigens through immunization—that is, by injecting appropriate chemical markers into the mother prior to pregnancy. The idea is to intensify the subsequent response to the foreign tissue of the embryo to a level that trips the suppressor mechanism. Some critics of this interpretation argue that the psychological support that women receive during this experimental treatment counts for more than the immunotherapy itself. Only more research will resolve the interesting and important questions about how a woman's immune system tolerates a 9-month parasitic relationship with a large foreign organism.

Contraception   **Contraception** literally means "against taking," in this case, the taking in of a child. The term has come to mean preventing a pregnancy through one of several methods. These methods fall into three main categories (Table 42.2): (1) preventing the egg and sperm from meeting in the female reproductive tract, (2) preventing implantation of a zygote, and (3) preventing the release of mature eggs and sperm from the gonads. The following brief introduction to the biology of these methods makes no pretense of being a contraception manual. For more complete information, consult a respected text on human sexuality or health (see the suggested readings at the end of this chapter).

Table 42.2 Contraceptive methods

Method	Failure Rate* Used Correctly	Failure Rate* Average U.S. Failure
Prevents Sperm and Egg from Meeting		
Rhythm	2–20	24
Withdrawal	16	23
Condom	2	10
Diaphragm and spermicide	2	19
Sponge and spermicide	9–11	10–20
Cervical cap	2	13
Spermicide (jelly, cream, or foam) alone	3–5	18
Prevents Implantation		
Intrauterine device (IUD)	1–2	5
Prevents Release of Gametes		
Birth control pill	0.5	2
Tubal ligation	0.4	0.4
Vasectomy	0.4	0.4

*Pregnancies per 100 women per year. Without contraception, about 90 pregnancies would occur.
(Data from R. Hatcher, et al., *Contraceptive Technology; 1986–1987.* New York: Irvington, 1986, p. 102).

Fertilization can be prevented by permanent or temporary abstinence or by any of several barriers that keep live sperm from contacting the egg. Temporary abstinence, often called the **rhythm method** of birth control, depends on refraining from intercourse when conception is most likely. Because the egg can survive in the oviduct for 24–48 hours and sperm for up to 72 hours, a couple practicing temporary abstinence should not engage in intercourse during the few days before and after ovulation, which is usually difficult to predict. The most effective methods for timing ovulation combine several indicators, including changes in cervical mucus and body temperature during the menstrual cycle. Because rhythm methods depend on regular menstrual cycles and changes that may not be apparent in all women, they are less effective than most other methods. A failure rate of 10%–20% is typical, even if the couple follows the rules and the woman's menstrual cycle is regular. (Expressing the failure rate of a contraceptive as a percentage gives the number of women who become pregnant during a year out of every hundred women using a particular contraceptive method.)

The several **barrier methods** of contraception that block the sperm from meeting the egg are more effective, having failure rates of less than 10%. The condom is a thin, natural membrane or latex rubber sheath that fits over the penis to collect the semen. The diaphragm is a dome-shaped rubber cap fitted into the upper portion of the vagina before intercourse. Both of these methods are more effective when used in conjunction with a spermicidal (sperm-killing) foam or jelly. More recently introduced barriers include the cervical cap (not yet approved by the Federal Drug Administration for use in the United States), which fits tightly around the opening of the cervix, held in place for a prolonged period by suction, and the contraceptive sponge, also inserted into the vagina.

The intrauterine device (IUD) probably prevents implantation of the blastocyst in the uterus by irritating the endometrium, but its precise mechanism of preventing pregnancy is unknown. IUDs are small, usually plastic devices, and they come in a variety of shapes that fit into the uterine cavity. IUDs have a low failure rate but carry a potential for serious side effects. Problems of persistent vaginal bleeding, uterine infection, perforation of the uterus, tubal pregnancy (implantation of the embryo in the oviduct), and spontaneous expulsion of the devices have been reported. Whether or not IUDs are safe for most women, they probably have little future as contraceptives in the United States because of the prohibitive legal expenses for manufacturers to defend against lawsuits.

As a method to pervent fertilization, *coitus interruptus*, or withdrawal (removal of the penis before ejaculation), is unreliable. Sperm may be present in secretions that precede ejaculation, and a lapse in timing or willpower can result in late withdrawal.

Besides complete abstinence, the methods that prevent release of gametes are the most effective means of birth control. Chemical contraception—birth control pills—have failure rates of less than 1%, and sterilization is nearly 100% effective. Birth control pills are combinations of a synthetic estrogen and a synthetic progestin (progesterone-like hormone). These two hormones act by negative feedback to stop the release of GnRH by the hypothalamus and FSH (an estrogen effect) and LH (a progestin effect) by the pituitary. By blocking LH release, the progestin prevents ovulation. As a backup measure, the estrogen inhibits FSH secretion so no follicles develop. Chemical contraception has been the center of much debate, particularly because of the long-term side effects of the estrogens. No solid evidence exists for cancers caused by the pill, but cardiovascular problems are a major concern. Birth control pills have been implicated in blood clotting, atherosclerosis, and heart attacks. Smoking while using chemical contraception increases the risk of mortality tenfold or more. Although the pill places women at risk for these diseases, it eliminates the dangers of pregnancy; women on birth control pills have mortality rates about one-half those of pregnant women.

Sterilization is the permanent prevention of gamete release. **Tubal ligation** in women involves cutting a short section out of the oviduct to prevent eggs from

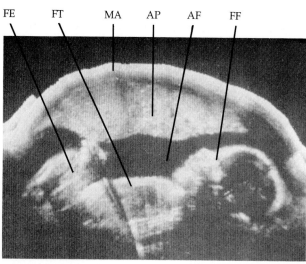

Figure 42.19
Ultrasound fetal monitoring. A small transducer emits pulses of sound waves that are reflected back by tissues of different densities. The echoes thus formed are viewed on a video monitor. Ultrasound can be used to determine gestational age of the fetus, detect physical abnormalities, and guide the physician performing amniocentesis. (FE = fetal extremities, FT = fetal thorax, MA = maternal abdomen, AP = anterior placenta, AF = amniotic fluid, FF = fetal face.)

traveling into the uterus. **Vasectomy** in men is the cutting of the vas deferens to prevent sperm from entering the urethra. Both male and female sterilization are relatively safe and free from side effects. Both are also difficult to reverse, so the procedures should be considered permanent.

Abortion is the termination of a pregnancy in progress, so it is not considered a form of contraception. Spontaneous abortion, or miscarriage, is very common, occurring in as many as one-third of all pregnancies, often before the woman is even aware of her pregnancy. Induced abortions, like contraception, receive much political, social, and religious attention.

Reproductive Technology

Recent scientific and technological advances have made it possible to deal with problems of reproduction in striking ways. For example, it is now possible to diagnose many genetic diseases and congenital (present at birth) defects while the fetus is in the uterus. Noninvasive procedures use high-frequency sound waves, or **ultrasound,** to detect fetal condition (Figure 42.19). Amniocentesis and chorionic biopsy are more invasive techniques (see Figure 13.18). In **amniocentesis,** a long needle is inserted into the amnion (the sac surrounding the fetus), and a sample of fluid is withdrawn. Fetal cells in the fluid are cultured for 2–4 weeks, and the cultured cells can then be analyzed for chromosomal and genetic defects (see Chapter 14). **Chorionic villi sampling** is a more recent technique wherein a small sample of tissue is removed for genetic and metabolic analysis from the chorion, the fetal part of the placenta. This test carries a greater risk than amniocentesis (5%–20% versus 1% spontaneous abortions following testing), but results may be obtained in days rather than weeks and chorionic sampling can be performed earlier in the pregnancy.

Ultrasonography, amniocentesis, and chorionic biopsies all pose important ethical questions. To date, essentially all detectable abnormalities remain untreatable in the uterus, and many cannot be corrected even after birth. Parents may be faced with difficult decisions about whether to terminate a pregnancy or cope with a child who may have profound defects and a short life expectancy. These are not easy questions; they demand careful, informed thought and competent counseling.

Another breakthrough in reproductive technology that has received considerable publicity is **in vitro fertilization.** First accomplished in 1978 in England, this procedure is now performed in major medical centers throughout the world. Women whose oviducts are blocked can have ova surgically removed from hormonally prepared follicles, and then the ova are fertilized in Petri dishes in the laboratory. After about 2½ days, when the embryo has reached the eight-cell stage, it is placed in the uterus and allowed to implant. It

sounds simple, but in vitro fertilization is a difficult and costly procedure. At this time, only about one out of six attempts is successful, at a cost of $4000 or more per attempt. This success rate may seem low, but in fact it is probably not different from the pregnancy rate resulting from insemination by intercourse. Multiple embryos are often placed in the uterus to increase the chances of a successful pregnancy. With several hundred children now conceived by in vitro fertilization, we have no evidence of any abnormalities associated with the procedure.

One area of reproductive research finally receiving much attention is male contraception. Male chemical contraceptives have proved quite elusive. Testosterone will block release of pituitary gonadotropins, but testosterone itself stimulates spermatogenesis. Estrogens are effective, but they inhibit libido and can be feminizing. The best prospects so far are for analogues of GnRH, which have been shown to be potent inhibitors of spermatogenesis. Other treatments under study include progestins and gossypol, a compound extracted from cotton seeds.

*   *   *

In this chapter, we have considered the structural and physiological bases of animal reproduction with little attention to the mechanics of development that transform a zygote into an animal form. The following chapter focuses on embryology and other topics of animal development.

STUDY OUTLINE

Modes of Reproduction (pp. 930–933)

1. Asexual reproduction produces a clone of genetically identical offspring from a single parent. Budding, fragmentation, gemmule formation, and regeneration are mechanisms of asexual reproduction used by various invertebrates.
2. Sexual reproduction requires the fusion of male and female gametes to form a diploid zygote. The production of offspring with varying genotypes and phenotypes may enhance reproductive success in fluctuating environments.
3. Animals may be exclusively sexual or asexual; or they may alternate between the two, depending on environmental conditions. Variations on these two modes are made possible through parthenogenesis, hermaphroditism, and sequential hermaphroditism.
4. Reproductive cycles are controlled by hormones and seasonal changes in temperature, rainfall, or day length.

Mechanisms of Sexual Reproduction (pp. 933–936)

1. External fertilization requires critical timing, mediated by environmental cues, pheromones, and/or courtship behavior. External fertilization is most common in aquatic or moist habitats, where the zygote can develop without desiccation and heat stress.

2. Internal fertilization requires important behavioral interactions between male and female animals, as well as compatible copulatory organs.
3. Insects have separate sexes and complex reproductive systems with various male and female specializations.
4. There are many similarities in reproductive anatomy among vertebrates. All but the mammals have a common opening, the cloaca, for reproductive, excretory, and digestive systems.

Mammalian Reproduction (pp. 936–952)

1. Male anatomy consists of internal reproductive organs and a scrotum and penis that make up the external genitalia. The gonads, or testes, reside in the cool environment of the scrotum. They possess endocrine interstitial cells surrounding sperm-forming seminiferous tubules that successively lead into the epididymis, vas deferens, ejaculatory duct, and urethra, which exits at the tip of the penis. Accessory glands add secretions to the semen.
2. Female anatomy consists of a pair of ovaries, oviducts, a uterus, a vagina, and external genitalia. The ovaries are stocked with gamete-containing follicles by the time of birth. Beginning at puberty, one or more follicles mature during each menstrual cycle. After ovulation, the

remaining tissue of the follicle forms a corpus luteum that secretes progesterone and estrogen for a variable duration, depending on whether or not pregnancy occurs.

3. The oviduct draws the gamete into its open end and transports it to the uterus by ciliary action. The uterus opens through the cervix into the muscular, distensible vagina, which serves as a sperm receptacle and birth canal.

4. Although separate from the reproductive system, the mammary gland, or breast, is a structure that evolved in association with parental care. Milk-secreting alveoli drain into ducts that open at the nipple.

5. In the embryo, genital primordia common to both sexes develop into male or female structures depending on the presence or absence of androgens.

6. Androgens from the testes cause the development of primary and secondary sex characteristics in the male. Androgen secretion and sperm production are both under the control of hypothalamic and anterior pituitary hormones.

7. Female hormones are secreted in a rhythmic fashion reflected in the menstrual or estrous cycle.

8. In both types of cycles, the uterine lining thickens in preparation for possible implantation. The menstrual cycle, however, is punctuated by endometrial bleeding and lacks the clear-cut period of sexual receptivity limited to the heat period of the estrous cycle.

9. The human menstrual cycle consists of the flow phase, proliferative phase, and secretory phase. The ovarian cycle includes the follicular and luteal phases.

10. The female reproductve cycle is orchestrated by cyclic secretion of GnRH from the hypothalamus and FSH and LH from the anterior pituitary. The developing follicle produces estrogen, and corpus luteum secretes progesterone and estrogen. Positive and negative feedback loops produce the changing levels of these five hormones, which coordinate the menstrual and ovarian cycles.

11. The production of sperm by spermatogenesis is a continuous and prolific process.

12. Unlike spermatogenesis, oogenesis occurs in discontinuous stages throughout life and involves unequal division of the cytoplasm during meiosis to form one large, functional gamete from each oogonium.

13. Puberty is a gradual process that involves the development of secondary sex characteristics and the onset of reproductive capability.

14. Common physiological patterns of the sexual response cycle underlie apparent differences in the rich diversity of human sexuality. Both males and females experience erection of certain body tissues due to vasocongestion and the increased muscle tension of myotonia, which culminates in orgasm.

15. Pregnancy, or gestation, begins at conception and continues until the birth of the young. Human pregnancy can be divided into three trimesters. Organogenesis is completed by eight weeks.

16. Birth, or parturition, results from strong, rhythmic uterine contractions that bring about the three stages of labor: dilation of the cervix, birth of the baby, and expulsion of the placenta.

17. The ability of a pregnant woman to accept her "foreign" fetus may be due to the suppression of the immune response in her uterus.

18. Prevention of pregnancy by a variety of contraceptive measures includes prevention of gamete union in the female tract, prevention of implantation of the zygote, or prevention of release of mature gametes from the gonads.

19. Current technological advances allow detection of fetal condition by ultrasound, amniocentesis, and chorionic villi sampling. Current technology has also given us *in vitro* fertilization and promises to yield new developments in contraception.

SELF-QUIZ

1. Which of the following characterizes parthenogenesis?
   a. An individual may change its sex during its lifetime.
   b. Specialized groups of cells may be released and grow into new individuals.
   c. An organism is first a male and then a female.
   d. An egg develops without being fertilized.
   e. Both members of a mating pair have male and female reproductive organs.

2. Which of the following structures is incorrectly paired with its function?
   a. gonads—gamete-producing organs
   b. spermatheca—sperm-transferring organ found in male insects
   c. cloaca—common opening for reproductive, excretory, and digestive systems
   d. baculum—bone that stiffens penis, found in some mammals
   e. endometrium—lining of the uterus, forms maternal part of placenta

3. Which of the following male and female structures are least alike in function?
   a. seminiferous tubules—vagina
   b. interstitial cells of testes—follicle cells
   c. testes—ovaries
   d. spermatogonia—oogonia
   e. vas deferens—oviduct

4. A difference between estrous and menstrual cycles is that
   a. nonmammalian vertebrates have estrous cycles, whereas mammals have menstrual cycles
   b. the endometrial lining is shed in menstrual cycles but reabsorbed in estrous cycles
   c. estrous cycles occur more frequently than menstrual cycles do
   d. estrous cycles are not controlled by hormones
   e. ovulation occurs before the endometrium thickens in estrous cycles

5. Peaks of LH and FSH production occur during
   a. the flow phase of the menstrual cycle
   b. the follicular phase of the ovarian cycle

c. the period surrounding ovulation

d. the end of the luteal phase of the ovarian cycle

e. the secretory phase of the ovarian cycle

6. The direct function of GnRH is to

   a. stimulate production of estrogen and progesterone

   b. initiate ovulation

   c. inhibit secretion of pituitary hormones

   d. stimulate secretion of LH and FSH

   e. initiate the flow phase of the menstrual cycle

7. During human gestation, organogenesis occurs

   a. in the first trimester

   b. in the second trimester

   c. in the third trimester

   d. while the embryo is in the fallopian tube

   e. during the blastocyst stage

9. An IUD probably functions in contraception by preventing

   a. fertilization

   b. implantation

   c. ovulation

   d. oogenesis

   e. conception

9. Fertilization of human eggs most often takes place in the

   a. vagina

   b. ovary

   c. uterus

d. oviduct (fallopian tube)

e. vas deferens

10. In mammalian males, the excretory and reproductive systems share the

   a. testes

   b. urethra

   c. ureter

   d. vas deferens

   e. prostate

CHALLENGE QUESTIONS

1. Describe how sexual and asexual reproduction differ in mechanism and result.

2. Explain why menstruation and ovulation do not occur during pregnancy.

3. What are some of the ethical issues raised by *in vitro* fertilization and test tube embryos?

FURTHER READING

1. Crews, D. "Courtship of Unisexual Lizards: A Model for Brain Evolution." *Scientific American*, December 1987. Research on all-female species.
2. Crooks, R., and K. Baur. *Our Sexuality.* 4th ed. Redwood City, Calif.: Benjamin/Cummings Publishing Company, Inc., 1990.
3. Daly, M., and M. Wilson. *Sex, Evolution and Behavior: Adapting for Reproduction.* Belmont, Calif.: Wadsworth Publishing Co., 1978. Compares diverse mechanisms of reproduction in vertebrates and arthropods.
4. Eberhard, W. G. "Runaway Sexual Selection." *Natural History,* December 1987. The evolution of male genitalia.
5. Weiss, R. "Test Screens Live Test Tube Embryos." *Science News,* March 4, 1989.

Animal Development

43

I t is difficult to imagine that each of us began life as a single cell about the size of the period at the end of this sentence. Less than a month after conception, our brains were taking form and our rudimentary hearts had already begun to pulsate. It took a total of only about nine months—the length of a school year—to be transfigured from zygote to human. The development of any animal is a marvel that biologists are only beginning to understand.

Form and function develop over the entire lifetime of an animal, from conception to death. Among the important episodes are the development of the embryo, growth and maturation after birth or hatching of the animal, metamorphosis in many invertebrates and amphibians that have larval stages, regeneration (the replacement of lost body parts), wound-healing, and aging. Although embryonic development is the main focus of this chapter, many of the basic cellular mechanisms that transform an embryo are also important in postembryonic development.

Three of the important processes in embryonic development are cell division, differentiation, and morphogenesis. Through a succession of mitotic divisions, the zygote gives rise to a large number of cells—trillions, in the case of a newborn human. Cell division alone, however, would produce a great ball of identical cells, which is nothing like an animal. During embryonic development, cells not only increase in number, but also undergo **differentiation** to become the specialized cells that are ordered in the tissues and organs of the animal. How two cells with the same genes can become as different as, say, a nerve cell and a muscle cell is one of the enduring puzzles of biology.

Figure 43.1
Fertilization. In this painting based on a scanning electron micrograph, sea urchin sperm surround an ovum, or unfertilized egg. Normally, only one of the sperm cells will successfully fertilize the ovum. Fertilization not only introduces a sperm cell's haploid genome into the ovum, but also activates the egg by triggering metabolic changes in the egg that mark the onset of embryonic development.

In Chapter 18, we discussed possible mechanisms for the control of gene expression during differentiation. The present chapter will emphasize **morphogenesis,** the development of an animal's shape and organization. The first half of the chapter describes *what* happens during the early stages of embryonic development, when the basic body plan of an animal takes form. The second half of the chapter discusses experiments designed to reveal how morphogenesis occurs.

FERTILIZATION

Embryonic development begins with **fertilization,** the union of a sperm and an egg (Figure 43.1). One function of fertilization is to combine haploid sets of chromosomes contained in the gametes from two individuals into a single diploid cell, the zygote. Fertilization also activates the egg, as contact of the sperm with the surface of the egg initiates a chain of metabolic reactions within the egg that triggers the onset of embryonic development.

Fertilization has been studied most extensively by combining the gametes of sea urchins in the laboratory. Although the details of fertilization vary with different animal groups, sea urchins (Phylum Echinodermata) provide a good general model for the important events of fertilization.

The Acrosomal Reaction

The eggs of sea urchins are fertilized externally after the animals release their gametes into the surrounding seawater. When a sperm cell contacts the jelly coat surrounding an egg, the acrosome, a vesicle at the tip of the sperm (see Chapter 42), discharges its contents by exocytosis (Figure 43.2). This acrosomal reaction releases hydrolytic enzymes that enable an extending **acrosomal process** to penetrate the jelly coat of the egg. The process is coated with a protein called **bindin** that adheres to specific receptor molecules located on the **vitelline layer** just external to the plasma membrane of the egg. "Lock-and-key" recognition of molecules ensures that eggs will be fertilized only by sperm of the same animal species. This is especially important when fertilization occurs externally in water, where gametes of other species are likely to be present.

Enzymes on the acrosomal process probably digest material in the vitelline layer. The tip of the process then contacts the plasma membrane of the egg, which forms a fertilization cone consisting of fingerlike projections that pull the sperm inward. The sperm's plasma membrane, which extends over the acrosomal process, fuses with the egg's plasma membrane, and the nucleus of the sperm enters the cytoplasm of the egg.

Even before the sperm nucleus slips into the egg, fusion of the gametes causes a nervelike electrical response by the plasma membrane of the egg. Union of the gametes' membranes opens ion channels that allow sodium ions to flow into the egg cell and change the membrane potential, the voltage across the membrane (see Chapter 8). This depolarization, as the electrical response is called, bars other sperm cells from fusing with the plasma membrane. A common phenomenon among animal species, this **fast block to polyspermy**—which occurs in much less than a second—prevents multiple fertilizations that would result in an aberrant chromosome count.

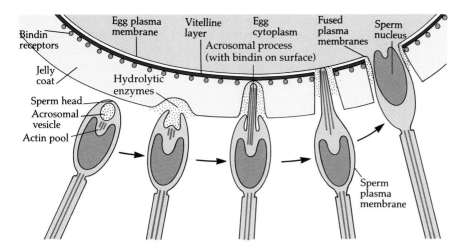

Figure 43.2

The acrosomal reaction during sea urchin fertilization. Upon contacting the jelly coat of the egg, the acrosomal vesicle in the head of the sperm releases proteins, some of them hydrolytic enzymes that excavate a hole in the jelly. An acrosomal process extends by the polymerization of actin, a protein. Bindin protein molecules on the surface of the acrosomal process attach to receptors on the vitelline layer of the egg, providing species specificity for fertilization. The plasma membranes of the gametes fuse, and the sperm nucleus enters the egg.

The Cortical Reaction

The acrosomal reaction of the sperm makes it possible for the plasma membranes of the gametes to fuse, and the egg responds to this stimulus with a cortical reaction (Figure 43.3). Depolarization of the egg cell membrane causes calcium to be released into the cytoplasm from some reservoir, probably the endoplasmic reticulum of the egg. The increased concentration of cytoplasmic Ca^{2+} then causes vesicles located in the cortex, the gelatinous outer zone of cytoplasm, to fuse with the plasma membrane. These **cortical granules** release their contents by exocytosis into the space between the plasma membrane and the vitelline layer. The secreted substances include enzymes that loosen the adhesive material between the two layers (plasma membrane and vitelline layer) and a high concentration of solutes that cause the space between the two

layers to swell by the osmotic uptake of water. This elevates the vitelline layer. Other enzymes polymerize molecules in the elevated vitelline layer to form a hardened **fertilization membrane** that resists entry of additional sperm. By this time, usually about a minute after sperm and egg fuse, the voltage across the plasma membrane has returned to normal, and thus the fast block to polyspermy no longer functions. But the fertilization membrane is a shield that functions, along with other changes of the egg's surface, as a **slow block to polyspermy**.

Activation of the Egg

Triggered by depolarization of the plasma membrane, the sharp rise in the egg's cytoplasmic concentration of Ca^{2+} not only stimulates the cortical reaction, but

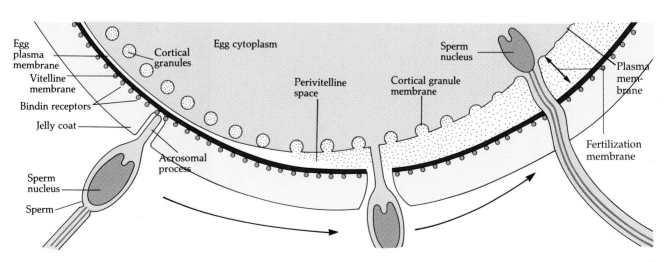

Figure 43.3

The cortical reaction. Cortical granules in the viscous cortex of the egg fuse with the plasma membrane and discharge their contents by exocytosis. Enzymes and other substances released from the cortical granules raise the vitelline layer and harden it to form a fertilization membrane that serves as a slow block to polyspermy.

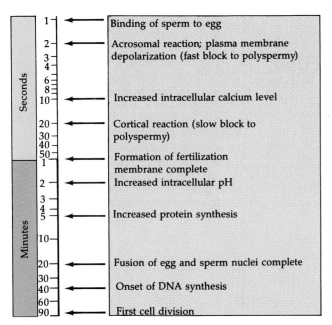

Seconds	1 — Binding of sperm to egg
	2 — Acrosomal reaction; plasma membrane depolarization (fast block to polyspermy)
	3
	4
	6
	8
	10 — Increased intracellular calcium level
	20 — Cortical reaction (slow block to polyspermy)
	30
	40
	50
Minutes	1 — Formation of fertilization membrane complete
	2 — Increased intracellular pH
	3
	4
	5 — Increased protein synthesis
	10
	20 — Fusion of egg and sperm nuclei complete
	30
	40 — Onset of DNA synthesis
	60
	90 — First cell division

Figure 43.4
Time line for fertilization and activation of sea urchin eggs. Notice that the scale is logarithmic.

also incites metabolic changes within the egg cell. The unfertilized egg has a very sluggish metabolism, but within a few minutes after fertilization, the rates of cellular respiration and protein synthesis increase substantially (Figure 43.4). With these rapid changes, the egg cell is said to be **activated.** In activating the egg, calcium acts indirectly by stimulating mechanisms that change the pH of the cytoplasm from 6.8 to 7.3; it is this pH change that seems to be directly responsible for many of the metabolic responses of the egg to fertilization.

Sperm entry and egg activation are distinguishable functions of fertilization. Eggs can be artificially activated by injection of calcium or by a variety of mildly injurious treatments, such as temperature shock. This artificial activation switches on the metabolic responses of the egg and causes it to begin developing by parthenogenesis (without fertilization by a sperm). It is even possible to artificially activate an egg that has had its own nucleus removed. (Of course, embryonic development of such an egg terminates at a very early stage.) The fact that an egg lacking a nucleus can begin making new kinds of proteins upon activation means that messenger RNA coding for these proteins must be stockpiled in an inactive form in the cytoplasm of the unfertilized egg. This is an example of controlling gene expression at the translational rather than the transcriptional level (see Chapter 18), but the mechanism by which the mRNA in the unfertilized egg is activated is unknown.

While the activated egg gears up its metabolism, the nucleus of the sperm cell that fertilized the egg begins to swell, and after about 20 minutes, it merges with the egg nucleus to produce the diploid nucleus of the zygote. Replication of DNA commences, and the first cell division occurs in about 90 minutes (in the case of sea urchins). Development of the embryo is under way.

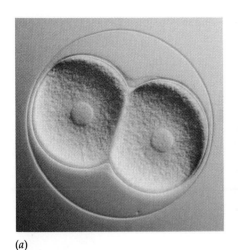

(a)

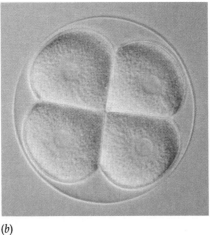

(b)

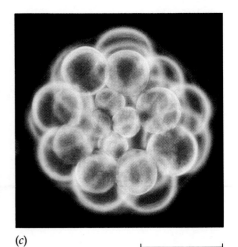

(c)

50 μm

Figure 43.5
Cleavage of an echinoderm zygote.
Cleavage divisions consist of relatively rapid mitosis and cytokinesis without growth of the cells between successive divisions. Following fertilization, cleavage transforms a single large cell, the zygote, into a ball of much smaller cells. These are phase contrast micrographs of living sea urchin embryos. **(a)** Two-cell stage following first cleavage division, which occurs about 45–90 minutes after fertilization (note that the fertilization membrane is still present). **(b)** Four-cell stage following second cleavage division. **(c)** In a few hours, repeated cleavage divisions have formed a multicellular ball. The embryo is still retained within the fertilization membrane, from which the motile larva that develops from the embryo will eventually "hatch."

EARLY STAGES OF EMBRYONIC DEVELOPMENT

Although mechanisms of morphogenesis sculpture an animal throughout its embryonic development, the basic body plan of the animal is established in the early stages of this process. For an overview of early development, we will feature the embryology of two animals, a sea urchin and amphioxus (*Branchiostoma*), a cephalochordate (see Chapter 30). The key stages are cleavage, gastrulation, and organogenesis.

Cleavage

Fertilization is followed by **cleavage,** a succession of rapid cell divisions that produces a ball of cells from the zygote (Figure 43.5). During cleavage, the embryos of sea urchins and amphioxus (and most other animals) nourish themselves on nutrients stored in the egg. The embryo does not grow during this period of development; the cytoplasm of one large cell is simply partitioned into many smaller cells, called **blastomeres.** The cells cycle between the S (DNA synthesis) and M (mitosis) phases of cell division, essentially skipping the G_1 and G_2 phases, when most cell growth normally occurs (see Chapter 11).

By dividing the zygote into many smaller blastomeres, cleavage increases the surface-to-volume ratio of each cell, which enhances oxygen uptake and other important exchanges with the surroundings. Also, cleavage substantially reduces the volume of cytoplasm that must be controlled by each nucleus.

The eggs of most animals have a definite polarity, and cleavage planes follow a specific pattern relative to the axis of the zygote (Figure 43.6). In the case of amphioxus, the axis is defined by the point, called the **animal pole,** where the polar body (superfluous haploid nucleus) is budded from the cell when meiosis reduces the chromosome count of the maternal nucleus. The opposite end of the zygote is known as the **vegetal pole.** The first two cleavage divisions are polar (vertical), dividing the zygote into four cells that each extend from animal to vegetal pole. The third division is equatorial (horizontal), producing an eight-celled embryo with a tier of four cells at the animal pole stacked on top of four cells at the vegetal pole. The pattern up to this point is the same in sea urchin embryos. In fact, echinoderms, chordates, and the other animal phyla grouped as deuterostomes share many features of early embryonic development (see Figure 29.4 on page 607). These similarities distinguish the deuterostomes from the protostomes, the evolutionary branch that includes the annelids, arthropods, and mollusks. For example, cleavage in deuterostomes is radial, meaning that the upper tier of four cells is aligned

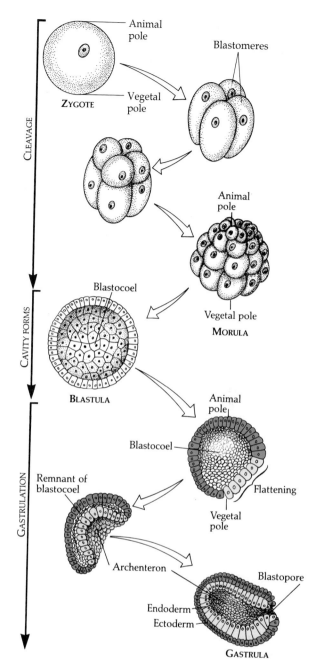

Figure 43.6
Cleavage and gastrulation of an amphioxus embryo. A series of cleavage divisions partitions the zygote into many smaller cells to form a solid ball, the morula. The embryo progresses to the blastula stage when a central cavity, the blastocoel, develops. Gastrulation forms a cup-shaped embryo with two layers of cells, an outer layer of ectoderm (blue) and an inner layer of endoderm (yellow), which lines the rudimentary digestive tract (archenteron).

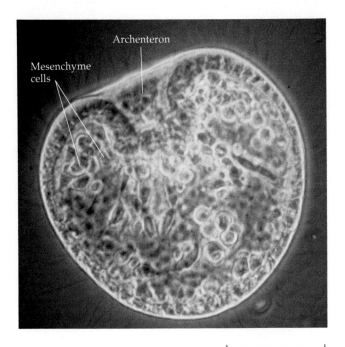

Figure 43.7
Gastrulation in an echinoderm (sea urchin). The blastula is transformed into a gastrula by two of the general mechanisms of morphogenesis: the folding of a cell layer and the migration of individual cells. Gastrulation begins with invagination, the infolding of a plate of cells at the vegetal pole of the embryo. The archenteron begins as this shallow depression. Then, cells at the leading edge of the invaginating region extend pseudopodia that attach to a specific region across the blastocoel near the animal pole of the embryo. These amoeboid cells, called mesenchyme cells, continue to pull the archenteron inward, and later detach to form mesoderm between the ectoderm and endoderm (LM).

with the lower tier at the eight-cell stage. In contrast, most protostomes exhibit spiral cleavage, in which the cells of the upper tier sit in the grooves between the cells of the lower tier.

Continued cell division produces a solid ball of cells known as the **morula** (from the Latin for "mulberry"). A fluid-filled cavity called the **blastocoel** then forms in the center of the morula, and soon the cells are arranged in a single epithelial layer surrounding the blastocoel. This hollow-ball stage of embryonic development is the **blastula.**

Gastrulation

The next stage of development transforms the blastula into a cup-shaped embryo, the **gastrula,** which has two layers of cells. In amphioxus, **gastrulation** (formation of the gastrula) begins when cells near the vegetal pole change their shape in such a way that the wall of the blastula in that location flattens and buckles inward, a process called **invagination** (see Figure 43.6).

In sea urchins and other echinoderms, migratory cells called mesenchyme cells also contribute to gastrulation (Figure 43.7).

The effect of gastrulation, as it occurs in amphioxus, is analogous to collapsing an uninflated basketball so that the wall on one side of the ball is pushed inward against the inside of the wall on the opposite side. Gastrulation obliterates the blastocoel and forms a new cavity called the **archenteron,** which will become the digestive tract of the animal. The archenteron opens to the outside through the **blastopore.** In chordates, echinoderms, and other deuterostomes, the blastopore becomes the anus, and the mouth forms later from a second opening that develops at the opposite end of the archenteron (see Chapter 29). The gastrula begins to elongate along what will become the anteroposterior axis of the animal, but this change in shape should not be confused with the increase in mass that accompanies real growth; the embryo is still no larger than the zygote was.

The early amphioxus gastrula has two layers of cells, one lining the archenteron and the other forming the outer wall of the embryo. In the space between these two layers, a third layer of cells develops from pouches that bud off from the lining of the archenteron. Thus, very early in development, the three-layered (triploblastic) body plan characteristic of chordates and most other animal phyla is established. Indeed, an echinoderm embryo hatches from the fertilization membrane as a motile, feeding larva, the pluteus, which is little more than a gastrula with a mouth, cilia, and a rudimentary skeleton (Figure 43.8).

The three epithelial layers of a late gastrula, **ectoderm** (outer), **mesoderm** (middle), and **endoderm** (inner), are known as the **primary germ layers,** and they will form the various organ systems of the animal as development continues.

Organogenesis

Various regions of the primary germ layers develop into the rudiments of organs, a process called **organogenesis.** The organs that begin to take shape first in amphioxus and other chordates are the neural tube and notochord, the skeletal rod characteristic of all chordate embryos (see Chapter 30). The **notochord** is formed from dorsal mesoderm just above the archenteron, and the neural tube originates from dorsal ectoderm above the developing notochord (Figure 43.9). Ectoderm in that location thickens to form a **neural plate.** In amphioxus, the neural plate sinks, surrounding ectoderm covers the plate, and the neural plate then rolls itself into a tube. This **neural tube** will become the central nervous system—the brain and spinal cord, which are hollow in chordates because of this mechanism of development.

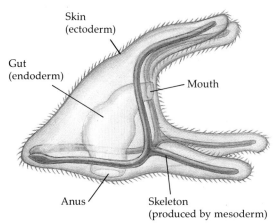

Figure 43.8
The sea urchin larva (pluteus). The basic body plan organized by gastrulation is still evident in this feeding larva. The archenteron is now a functional gut with two openings: an anus formed from the blastopore of the gastrula and a mouth formed from a secondary opening. Some of the mesenchyme cells of the mesoderm have secreted minerals that form a simple internal skeleton. The skin that develops from ectoderm is ciliated. The pluteus drifts near the sea surface as plankton, feeding on bacteria and unicellular algae.

As organogenesis progresses, many other rudimentary organs develop from the primary germ layers. In addition to forming the nervous system, ectoderm also gives rise to the animal's external covering (epidermis) and associated glands, as well as producing the inner ear and the lens of the eye. Developing from mesoderm are the notochord, lining of the coelom (body cavity), muscles, skeleton, gonads, kidneys, and most of the circulatory system. Endoderm forms the linings of the digestive tract and various organs that originate as outpocketings of the archenteron, such as the liver, pancreas, and lungs (in some vertebrates).

Morphogenesis and cellular differentiation continue to refine the organs that arise from the primary germ layers, but the fundamental plan of an animal with three concentric layers is already evident by the end of the gastrula stage.

COMPARATIVE EMBRYOLOGY OF VERTEBRATES

As chordates, vertebrates begin to develop along the same general lines as amphioxus, but with some important modifications. In this section, we compare early embryonic development of the frog, the chick, and the human—representatives of three vertebrate classes.

Amphibian Development

The frog egg is much larger than the egg of amphioxus (about 1 mm in diameter in many species), and it has more yolk. Stored as lipid and protein in **yolk platelets,** this food reserve must nourish the embryo until it hatches as a tadpole from the transparent jelly coat that encloses the egg.

The frog egg is also more noticeably polar than the amphioxus egg. Yolk platelets are concentrated in the lower, or vegetal, hemisphere of the egg. The animal hemisphere is distinguished by pigment granules, which are located in the viscous cortex of the cell. This asymmetric distribution of yolk and pigment gives frog eggs a two-tone appearance and, as we shall see, has profound effects on the mechanics of cleavage and gastrulation (Figure 43.10).

The first two cleavage divisions of a frog zygote are polar, as in amphioxus. The third division is horizon-

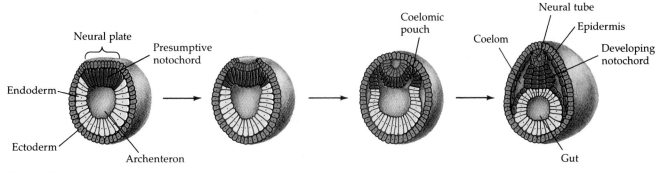

Figure 43.9
Early stages of organogenesis in amphioxus. As seen in these cross sections, mesoderm (red) arises from pouches that pinch off from the inner cell layer. The dorsal mesoderm forms the notochord. Ectoderm above the rudimentary notochord thickens to form a neural plate that sinks below the surface of the embryo and rolls itself into a neural tube, the rudiment of the brain and spinal chord.

Figure 43.10
Polarity of the frog egg. Cytoplasmic components are asymmetrically localized within the egg. The polarity of frog eggs is apparent even to the unaided eye, or when the eggs are viewed with a simple hand lens. Yolk granules are concentrated in the vegetal hemisphere, while pigment granules are located in the cortex of the animal hemisphere. Distribution of many other substances and organelles is also polarized. The heterogeneity of the cytoplasm has important effects on cleavage and gastrulation and, as we shall see, also influences differentiation in different parts of the embryo derived from specific regions of the egg.

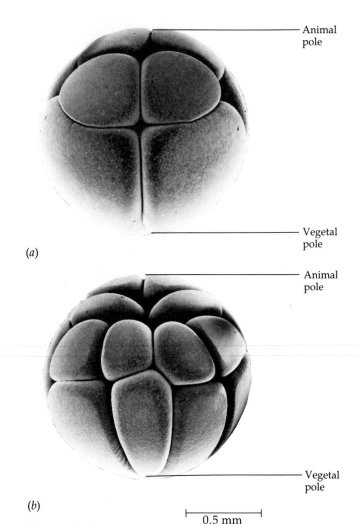

(a)

(b)

0.5 mm

Figure 43.11
Unequal cleavage of a frog egg. Yolk, concentrated near the vegetal pole of the egg, impedes the formation of cleavage furrows. **(a)** After two equal polar divisions, the third cleavage division is perpendicular to the polar axis but is displaced by yolk to the animal pole's side of the equator. Thus, the four blastomeres at the animal pole are smaller than the four blastomeres at the vegetal pole. **(b)** As cleavage continues, the cells at the animal pole divide more frequently than the yolk-laden cells near the vegetal pole. This accounts for the blastomeres at the animal pole being smaller and more numerous than those at the vegetal pole (SEMs).

tal; this cleavage plane, however, does not follow the equator of the embryo, but is displaced toward the animal hemisphere to form an eight-celled embryo with four small blastomeres on top of four larger ones (Figure 43.11). The yolk platelets of the vegetal hemisphere impede cleavage furrows, and thus the asymmetry of the third division results from the furrow following the path of least resistance. The yolk continues to affect cleavage, causing cells of the vegetal hemisphere to divide more slowly than cells of the animal hemisphere.

The frog blastula is not as simple as the ball-like blastula of amphioxus. In the frog blastula, the cells near the animal pole are smaller than those near the vegetal pole, where the yolky cells divided less frequently during cleavage (Figure 43.12). The blastocoel is restricted to the animal hemisphere, enclosed by a wall of cells several layers thick, unlike the single-layered wall of amphioxus.

Gastrulation during frog development results in a triple-layered embryo with an archenteron, as in amphioxus. However, the mechanics of gastrulation are much more complicated in the frog because of the large, yolk-laden cells of the vegetal hemisphere and because the wall of the blastula is more than one cell thick in most species. The first sign of gastrulation is

a small crease on one side of the blastula where the blastopore will eventually be located. This invagination is produced by "bottle cells," so named for the shape they assume when they lead the way into the interior of the embryo. As these cells burrow inward, they remain attached for a while to the surface by long necks, giving the cells their bottle shapes. The tuck produced by invagination will become the upper edge, or **dorsal lip,** of the blastopore. Then, in a process called **involution,** cells on the surface of the embryo move toward the lip, make a U-turn as they roll over the edge into the interior of the embryo, and then continue their migration by moving along the roof of

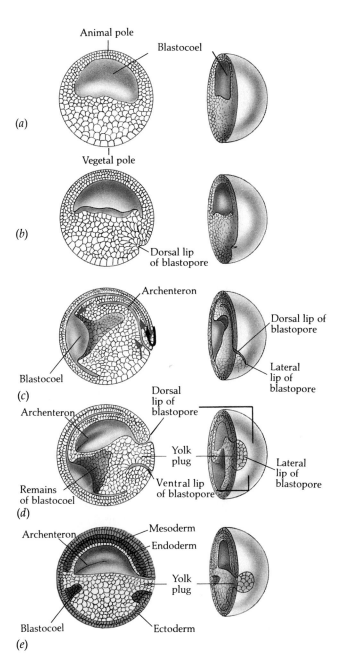

(a)

Animal pole

Blastocoel

Vegetal pole

(b)

Dorsal lip
of blastopore

(c)

Archenteron

Blastocoel

Dorsal lip of
blastopore

Lateral
lip of
blastopore

(d)

Archenteron

Dorsal
lip of
blastopore

Remains
of blastocoel

Yolk
plug

Ventral lip
of blastopore

Lateral
lip of
blastopore

(e)

Archenteron

Blastocoel

Mesoderm

Endoderm

Yolk
plug

Ectoderm

make up the ectoderm. The three primary germ layers
are in place, and organogenesis commences.

As in amphioxus, the first organs to take shape in
the frog embryo are the notochord, derived from dorsal mesoderm, and the neural tube, which originates
as a plate of ectoderm above the notochord (Figure
43.13). The notochord will elongate and stretch the
embryo along its anteroposterior axis. Later, the notochord will function as a core around which mesodermal cells gather in the formation of the vertebrae.
The strips of mesoderm lateral to the notochord separate into blocks, called **somites,** which are arranged
serially on both sides along the length of the notochord (Figure 43.14). Cells from the somites not only
give rise to the vertebrae of the backbone, but also
form the muscles associated with the axial skeleton.
This serial origin of the axial skeleton and muscles
reinforces the point made in Chapter 30 that chordates
are basically segmented animals, although the segmentation becomes less obvious later in development.
(There are, however, signs of segmentation even in
the adult, as in the series of vertebrae in a human or
the segments of chevron-shaped muscles in a fish.)
Lateral to the somites, the mesoderm separates into
two layers that form the lining of the body cavity, or
coelom.

Along the border where the neural tube pinches off
from the ectoderm, cells break loose to form the **neural
crest,** a band of cells unique to vertebrate embryos.
Cells of the neural crest later migrate to various parts
of the embryo, forming pigment cells in the skin, the
bones and muscles of the skull, the teeth, the medulla
of the adrenal glands, and peripheral components of
the nervous system, such as sensory and sympathetic
ganglia.

Embryonic development of the frog leads to a larval
stage, the tadpole, which hatches from the jelly coat
that originally cloaked the egg. Later, metamorphosis

the blastocoel. The epithelium near the animal pole,
which is more than one cell thick, spreads out as invo-
luting cells nearer the blastopore disappear from the
surface. As gastrulation progresses, the lip of the blastopore widens, first becoming frownlike in shape and
finally forming a complete circle. At this stage, the lip
of the blastopore surrounds a **yolk plug** consisting of
the large, food-laden cells from the vegetal pole of the
embryo. By now, the complex movements of gastrulation have eliminated the blastocoel and formed the
archenteron, lined by endoderm. Cells once on or near
the surface of the embryo now reside in the interior
as mesoderm, and the cells remaining on the surface

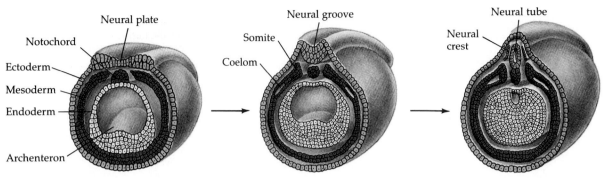

Figure 43.13
Early organogenesis in the frog embryo (cross-sectional views). Dorsal mesoderm forms the notochord. Blocks of mesoderm called somites flank the notochord and give rise to segmental structures, such as the vertebrae of the backbone and serially arranged skeletal muscles.

Lateral mesoderm separates into the two tissue layers that line the coelom. Dorsal ectoderm thickens to form the neural plate, which rolls to become the neural tube. Tissue at the meeting margins of the tube separate from the tube as the neural crest, a source of migrating cells that

eventually form many structures, including the bones and muscles of the skull, the pigment cells of the skin, the adrenal medulla glands, and peripheral ganglia of the nervous system. The accompanying scanning electron micrographs are dorsal views of the developing neural plate and tube.

will transform the frog from the aquatic, herbivorous tadpole to the terrestrial, carnivorous adult frog (see Figure 30.15 on page 652).

Avian Development

Birds and reptiles lay shelled eggs, an important adaptation to the terrestrial environment (see Chapter 30).

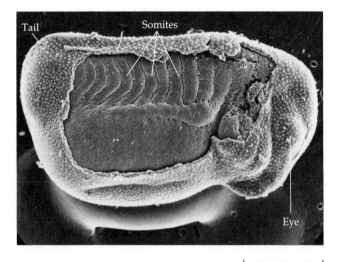

1 mm

Figure 43.14
Embryonic origin of segmentation in a frog embryo. At this stage of development, called the tailbud stage, serial somites are evident along each side of the embryo. The somites will give rise to such segmental structures as the vertebrae and axial muscles. Part of the ectoderm of this embryo has been removed to reveal the somites (SEM).

The part of an egg (such as a chicken egg) that we call the yolk is actually the egg cell (ovum), which is surrounded by a protein-rich solution (the egg white) that will provide nutrients for the growing embryo in addition to the food supplied by the yolk of the ovum itself.

Yolk is so abundant in the bird egg that cleavage is restricted to a small disc of cytoplasm at the animal pole of the egg cell. After fertilization, cell division partitions this yolk-free cytoplasm to form a cap of cells called the **blastodisc,** which rests on the large, yolky, undivided portion of the original egg cell (Figure 43.15). This incomplete division of a yolk-rich egg is known as **meroblastic cleavage.** It contrasts with **holoblastic cleavage,** the term used for the complete division of eggs having little yolk (amphioxus) or a moderate amount of yolk (frog).

Cleavage at the animal pole of the bird egg is followed by sorting of the blastodisc cells into an upper and lower layer. The cavity between these two layers is the blastocoel. This embryonic stage is the avian equivalent of the blastula, although its form is different from the hollow ball of a frog or amphioxus blastula.

Gastrulation of the bird egg begins with the formation of a straight invagination, the **primitive streak,** which extends in the upper layer of cells along what will become the anteroposterior axis of the embryo. The primitive streak is equivalent to the lip of the frog's blastopore, but it is arranged as a linear tuck rather than in a ring. Cells of the upper layer migrate toward the primitive streak, rolling over the edge and into the blastocoel. These cells form the mesoderm and also contribute to the endoderm. The ectoderm consists of cells remaining in the upper layer. The three primitive germ layers have now formed, but at this

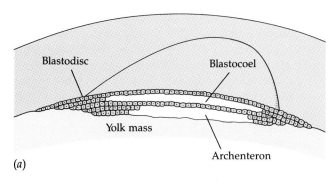

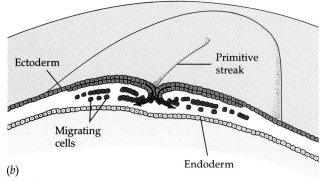

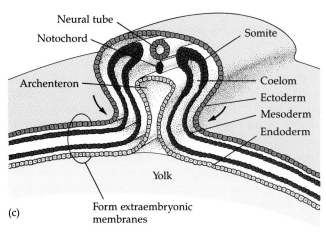

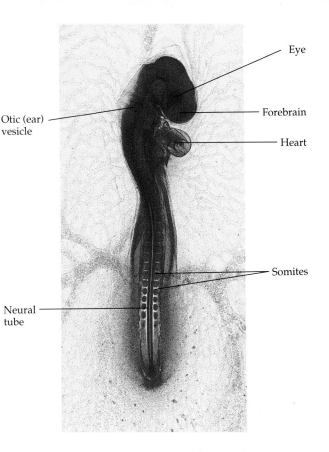

Figure 43.16
Organogenesis of the chick embryo. Rudiments of most major organs have already formed in this chick, which is about 56 hours old. (LM).

Figure 43.15
Cleavage and gastrulation of the chick embryo.
(**a**) For the eggs of birds and reptiles, which store a very large amount of yolk, cleavage is meroblastic, or incomplete. Cell division is restricted to a small cap of cytoplasm at the animal pole. Cleavage produces a blastodisc that rests on the large, undivided mass of yolk. The blastodisc becomes arranged into two layers that bound the blastocoel, forming the avian version of a blastula.
(**b**) During gastrulation, some cells of the outer layer migrate into the interior of the embryo through the primitive streak, a linear blastopore lip. The migrating cells form mesoderm and add to the endoderm. (**c**) The rudimentary gut (archenteron) is formed when lateral folds pinch the embryo from the yolk. About midway along its length, the embryo will remain attached to the yolk by a yolk stalk. Primary germ layers excluded from the embryo proper contribute to the system of extraembryonic membranes that support further development.

stage they are arranged as three stacked layers rather than concentric cylinders. This will change when the borders of the embryonic disc fold downward and then come together below the embryo, pinching the embryo into a triple-layered tube that remains attached to the yolk below by only a stalk at midbody. Neural tube formation, development of the notochord and somites, and other events in organogenesis occur much as in the frog embryo (Figure 43.16).

In addition to the embryo itself, the primary germ layers also give rise to four **extraembryonic membranes** that support further embryonic development (Figure 43.17). These four membranes, the **yolk sac,** the **amnion,** the **chorion,** and the **allantois,** are important avian and reptilian adaptations to the special problems of embryonic development in terrestrial habitats.

Mammalian Development

Mammalian eggs are fertilized in the oviduct, and the earliest stages of development occur while the embryo completes its journey down the oviduct to the uterus

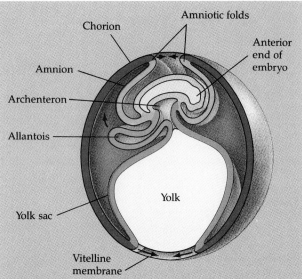

Figure 43.17
Development of extraembryonic membranes of the chick. Each of the four membranes, which provide support services for the embryo, develops from epithelial sheets that are external to the embryo proper. The *yolk sac* expands over the surface of the yolk mass. Cells of the yolk sac will digest yolk, and blood vessels that develop within the membrane will carry nutrients into the embryo. Lateral folds of extraembryonic tissue extend over the top of the embryo and fuse to form two more membranes, the *amnion* and the *chorion*, which are separated by an extraembryonic extension of the coelom. The amnion encloses the embryo in a fluid-filled amniotic sac, protecting the embryo from desiccation and buffering mechanical shocks. The fourth membrane, the *allantois*, originates as an outpocketing of the embryo's hindgut. The allantois is a sac that extends out into the extraembryonic coelom separating the chorion and amnion. The allantois functions as a disposal sac for uric acid, the insoluble nitrogenous waste of the embryo. As the allantois continues to expand, it presses the chorion against the vitelline membrane, the inner lining of the eggshell. Together, the allantois and chorion form a respiratory organ that serves the embryo. Blood vessels that form in the epithelium of the allantois transport oxygen to the embryonic chick. The extraembryonic membranes of reptiles and birds are adaptations associated with the special problems of development on land.

(see Chapter 42). In contrast to the large, yolky eggs of birds and reptiles, the egg of a placental mammal is quite small, storing little in the way of food reserves. Although cleavage of the mammalian zygote is holoblastic, gastrulation and early organogenesis follow a pattern similar to that of birds and reptiles (mammals, remember from Chapter 30, descended from reptilian stock during the early Mesozoic era).

Cleavage is relatively slow in mammals. In the case of the human, the first division is not complete until about 36 hours after fertilization, the second division

at 60 hours, and the third division at about 72 hours. The mammalian zygote has no apparent polarity. The cleavage planes seem to be randomly oriented, and the blastomeres are equal in size.

By 5 days after fertilization, the human embryo has over a hundred cells. At this stage, the cells form the tight junctions characteristic of a compact epithelium (see Chapter 7), which is arranged around a central cavity. This is the embryonic stage known as the blastocyst (see Chapter 42). Protruding into one end of the blastocyst cavity is a cluster of cells called the **inner cell mass,** which will subsequently develop into the embryo proper and some of the extraembryonic membranes. The outer epithelium surrounding the cavity is the **trophoblast,** which will form the fetal portion of the placenta (Figure 43.18).

The human embryo reaches the uterus by the blastocyst stage, implanting about a week after fertilization. The trophoblast secretes enzymes that enable the blastocyst to penetrate the uterine lining. Bathed in a pool of blood spilled from broken capillaries in the endometrium (uterine lining), the trophoblast thickens and extends fingerlike projections into the surrounding maternal tissue. The placenta forms from this proliferated trophoblast and the region of endometrium it invades (see Chapter 42). At about the time the blastocyst implants in the uterus, the inner cell mass forms a flat **embryonic disc** that develops into the embryo proper.

Four extraembryonic membranes homologous to those of reptiles and birds—chorion, amnion, yolk sac, and allantois—form during mammalian development (see Figure 43.18). The chorion, which develops from the trophoblast, completely surrounds the embryo and the other extraembryonic membranes. The amnion begins as a dome above the embryonic disc and encloses the embryo in a fluid-filled amniotic cavity. (The fluid from this cavity is the "water" expelled from the vagina of the mother when the amnion breaks just prior to childbirth.) Below the embryonic disc is another fluid-filled cavity enclosed by the yolk sac. Although this sac contains no yolk, it is given the same name as the homologous membrane in birds and reptiles. The yolk sac membrane of mammals is a site of early formation of blood cells, which later migrate into the embryo proper. The fourth extraembryonic membrane, the allantois, develops as an outpocketing of the embryo's rudimentary gut, as it does in the chick. The allantois is incorporated into the umbilical cord, where it forms blood vessels that transport oxygen and nutrients from the placenta to the embryo and rid the embryo of carbon dioxide and nitrogenous wastes.

Thus, the extraembryonic membranes that function in shelled eggs, where embryos are nourished with yolk, were conserved as mammals diverged from reptiles in the course of evolution, but with modifications suited for development within the reproductive tract of the mother.

Figure 43.18
Early development of human embryo and its extraembryonic membranes. (a) Cleavage produces a blastocyst, consisting of a trophoblast surrounding a cavity and an inner cell mass protruding into the cavity from one side. The blastocyst implants in the uterine lining. Although the placental mammal develops within the uterus of the mother rather than within a shelled and yolk-rich egg, the extraembryonic membranes found in birds and reptiles have persisted in the evolution of mammals. The mammalian embryo proper develops from a flat embryonic disc homologous to the blastodisc of the chick, and gastrulation occurs by involution of mesoderm and endoderm through a primitive streak. **(b)** Relationship of the extraembryonic membranes to the embryo and the uterus of the mother.

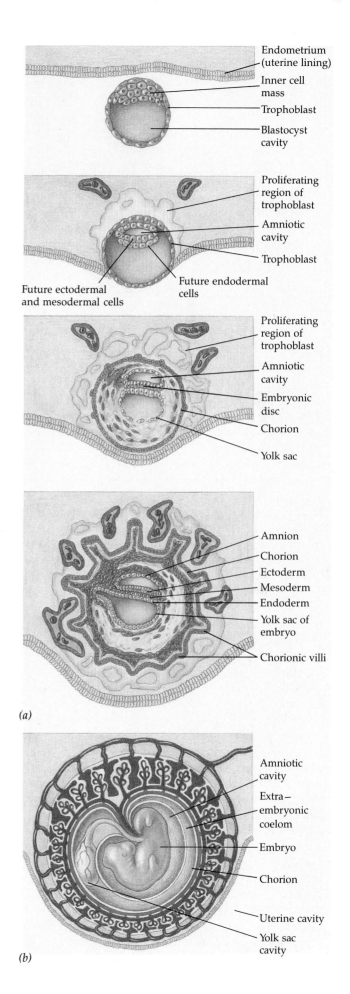

Early stages in the development of the embryo proper are also similar for reptiles and their two evolutionary offshoots, birds and mammals. As already mentioned, the mammalian embryo develops from the embryonic disc, the flat floor of the amniotic cavity. Homologous to the blastodisc of birds and reptiles, the embryonic disc has an upper and lower layer (see Figure 43.18). Gastrulation occurs by involution of cells from the upper layer through a primitive streak to form mesoderm and endoderm, just as gastrulation occurs in the chick. Organogenesis begins with the formation of the neural tube, notochord, and somites. By the end of the first trimester of human development, rudiments of all the major organs have developed from the three primary germ layers, as reviewed in Table 43.1.

MECHANISMS OF DEVELOPMENT

We have seen that the basic body plan of an animal is established quite early in embryonic development. Describing a sequence of changes, however, does not explain how those changes occur. By manipulating embryos in a variety of experiments, developmental biologists have discovered some of the morphogenetic mechanisms at work as an animal takes shape, although many questions remain. In this section, we will examine a few of the concepts that have emerged from experimental embryology.

Polarity of the Embryo

A bilaterally symmetric animal has an anterior-posterior axis, a dorsal-ventral axis, and left and right sides (see Chapter 29). In some cases, these three polarities are already established at the time of fertil-

Table 43.1 Derivatives of the primary germ layers in mammals

Ectoderm	Mesoderm	Endoderm
All nervous tissue	Skeletal, smooth, and cardiac muscle	Epithelium of digestive tract (except that of oral and anal cavities)
Epidermis of skin and epidermal derivatives (hairs, hair follicles, sebaceous and sweat glands, nails)	Cartilage, bone, and other connective tissues	Glandular derivatives of digestive tract (liver, pancreas)
Cornea and lens of eye	Blood, bone marrow, and lymphoid tissues	Epithelium of respiratory tract
Epithelium of oral and nasal cavities, of paranasal sinuses, and of anal canal	Endothelium of blood vessels and lymphatics	Thyroid, parathyroid, and thymus glands
Tooth enamel	Lining of body cavity	Epithelium of reproductive ducts and glands
Epithelium of pineal and pituitary glands and adrenal medulla	Organs of urogenital system (ureters, kidneys, gonads, and reproductive ducts)	Epithelium of urethra and bladder

SOURCE: Modified from Elaine N. Marieb, *Human Anatomy and Physiology.* Menlo Park: Benjamin/Cummings, 1989.

ization. In the frog, for example, the embryonic axes are set before the first cleavage division.

Recall that the eggs of most animals are already polarized along an animal-vegetal axis owing to the heterogeneous distribution of cytoplasmic components within the egg. The amphibian egg, for instance, has yolk platelets concentrated near the vegetal pole and cortical pigment granules concentrated near the animal pole (see Figure 43.10). A rearrangement of the amphibian egg cytoplasm occurs at the time of fertilization. The cortex of the egg is rotated toward the point of sperm entry, probably because the centriole brought into the egg by the sperm cell reorganizes the cytoskeleton. Contraction of the cortex pulls the edge of the pigmented layer that is opposite the point of sperm entry toward the animal pole, exposing lighter-colored cytoplasm that had been masked by the pigment (Figure 43.19). This contraction forms the **gray crescent,** a half-moon-shaped mark near the equator of the egg, on the side opposite where the sperm cell entered. The first cleavage division will bisect the gray crescent, and the dorsal lip of the blastopore will later form where the gray crescent was located in the zygote. Thus, all three axes of the embryo are defined before the onset of cleavage: The animal-vegetal axis of the egg becomes the anterior-posterior axis of the embryo, and the location of the gray crescent determines the dorsal-ventral axis and the left-right axis of the embryo. When the egg is experimentally manipulated to prevent the rotation of the cortex that forms the gray crescent, then gastrulation is abnormal and axial structures such as the neural tube fail to develop.

Although embryonic axes are fixed at the zygote stage or during early cleavage in most animals, there are important exceptions, including mammals. Mammalian eggs, remember, lack apparent polarity, and

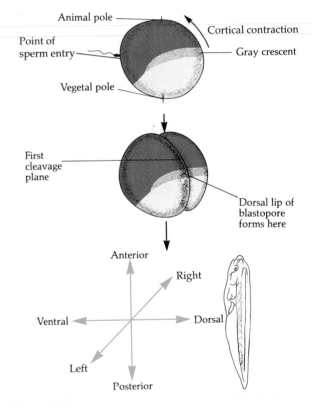

Figure 43.19
Establishment of embryonic axes in amphibians. At the time of fertilization, a gray crescent forms on the side of the zygote opposite the point where the sperm cell entered. Contraction of the pigmented cortex produces the crescent by exposing lighter-colored cytoplasm beneath the pigment granules. The first cleavage division bisects the gray crescent, and the dorsal lip of the blastopore will later form where the crescent was located. Thus, all three axes of the embryo are established before the zygote begins cleavage.

cell divisions are randomly oriented during early cleavage. We cannot distinguish one end of the embryo from another until the blastocyst stage.

Cytoplasmic Determinants

If cytoplasmic components are unevenly distributed within the egg, then the early cleavage divisions will partition the zygote into blastomeres that differ in cytoplasmic makeup. The localized substances sequestered within specific blastomeres may function as **cytoplasmic determinants** that fix the developmental fates of different regions of the embryo very early, presumably by controlling gene expression (see Chapter 18). In the development of many mollusks,

for example, the zygote forms a polar lobe that is restricted to one of the two blastomeres at the first cleavage division (Figure 43.20a). In experiments where the two blastomeres are pulled apart, only the cell with the polar lobe will develop into a normal embryo. Moreover, even that cell will fail to develop normally if the polar lobe is removed; the abnormal larva will lack a heart, eyes, and several other organs derived from mesoderm in mollusks. Apparently, the polar lobe contains cytoplasmic determinants required for the subsequent development of these mesodermal structures.

Even if cytoplasmic determinants are asymmetrically distributed in the zygote, the first cleavage may occur along an axis that produces two blastomeres of equal developmental potential. In amphibians, for

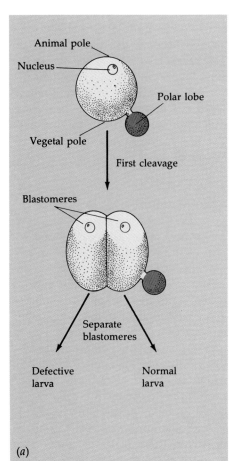

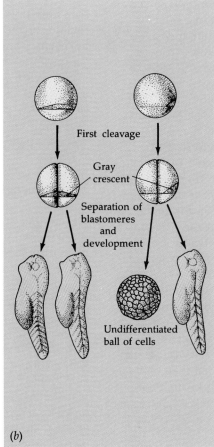

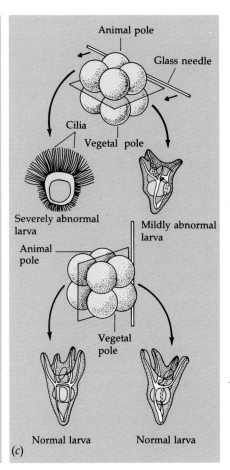

Figure 43.20
Cytoplasmic determinants. (a) In the development of the mollusk *Dentalium*, a polar lobe is restricted to one of the two blastomeres formed by the first cleavage. The lobe apparently contains determinants required for the development of certain structures. Thus, the cell lacking the lobe has already lost its totipotency after the

very first cell division. (b) The first cleavage division of an amphibian zygote normally divides the gray crescent between the two blastomeres, which retain their totipotency after the division. However, in experiments where pressure is used to divert the cleavage plane from the crescent, only the blastomere obtaining mate-

rial from the crescent develops normally. (c) Cytoplasmic determinants in the sea urchin egg apparently have a polar distribution. At the eight-cell stage, splitting the embryo vertically results in two normal larvae, but splitting the embryo horizontally produces a mildly defective larva and a severely defective one.

instance, blastomeres experimentally separated at the two-cell stage develop into normal tadpoles. The cells are said to be **totipotent,** meaning they retain the potential to form all parts of the animal. If, however, the zygote of an amphibian is experimentally manipulated so that the first cleavage plane misses the gray crescent instead of bisecting it, only the blastomere that gets the gray crescent develops into a normal tadpole (Figure 43.20b). Thus, a zygote's characteristic pattern of cleavage, along with its distribution of cytoplasmic determinants, affects the destiny of cells in the embryo.

Mosaic and Regulative Development

Development can be described as mosaic or regulative, depending on how early the blastomeres lose their totipotency. The development of a mollusk from an egg with a polar lobe is said to be **mosaic,** because each cell is rigidly set to give rise to specific parts of the embryo, much as each piece of a mosaic has a fixed place in the overall picture. In **regulative** development, cells are totipotent longer, and thus their fates can be experimentally altered. The distinction is relative, since at some point the developmental fate of most cells becomes narrowed in all animals. For example, sea urchin development is regulative up to the four-cell stage, but the third cleavage plane divides cytoplasmic determinants in such a way that some of the cells lose their totipotency (see Figure 43.20c).

Mammalian embryos remain regulative in development much longer than most other animals. The cells do not begin to lose their totipotency until they become arranged into the trophoblast and inner cell mass of the blastocyst. The mammalian egg, remember, has no distinct polarity, and the planes of early cleavage seem to be randomly oriented. Up to the eight-cell stage, the blastomeres of a mouse embryo all look alike, and indeed, each can form a complete embryo if isolated. Conversely, two mouse embryos at the eight-cell stage can be fused into an enlarged morula that, when transplanted to a uterus, will develop into a mouse combining traits of two sets of parents. In the regulative development of mammals, the early embryo can apparently adjust to such tampering, which is not the case for embryos with mosaic development.

Fate Maps

In embryos where the axes are defined early in development, it should be possible to determine which parts of the embryo will be derived from each region of the zygote or blastula. In the 1920s, German embryologist W. Vogt charted a **fate map** for the amphibian embryo (Figure 43.21). He used vital (nontoxic) dyes to label

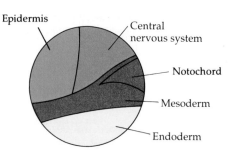

Figure 43.21
Fate map of an amphibian blastula. Fates were determined by marking different regions of the blastula surface with dyes of various colors and then determining the locations of dyed cells in the gastrula or at later stages of development. The embryo is viewed from the left side, with the future anterior end at the upper left.

cells of different regions of the blastula surface with different colors and subsequently sectioned the embryo at later stages to see where the colors turned up. Note by comparing Figures 43.21 and 43.12 that different regions of the embryo become rearranged between the blastula and gastrula stages. This is because of the complex movements that occur during gastrulation.

In the case of *C. elegans,* the tiny, transparent nematode worm discussed in Chapter 18, it has been possible to map the developmental fates of every cell, beginning with the first cleavage division of the zygote (see pages 395–396).

Morphogenetic Movements

The migration of cells and changes in cell shape are fundamental processes in the development of an animal's form. Three properties of cells—extension, contraction, and adhesion—are involved in the various morphogenetic movements that shape the embryo, including gastrulation and formation of the neural tube.

Changes in the shape of a cell usually involve reorganization of the cytoskeleton. Consider, for example, how the cells of the neural plate become distorted in shape during formation of the neural tube (Figure 43.22). First, microtubules lengthen the cells of the plate in the dorsal-ventral axis of the embryo. Then, ordered arrays of microfilaments contract the apical ends of the cells, giving them a wedge shape that deforms the epithelium inward. Similar shape changes are observed for other invaginations (inpocketings) and evaginations (outpocketings) of the epithelia throughout development.

Amoeboid movement is also important in morphogenesis. We have seen that in the gastrulation of some embryos, for example, invagination is initiated by a wedging of cells on the surface of the blastula, but the penetration of cells deeper into the embryo involves

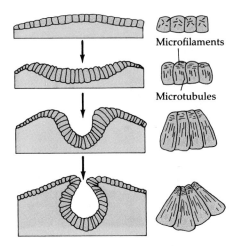

Figure 43.22
Change in cellular shape during morphogenesis.
Reorganization of the cytoskeleton is believed to function in invagination and evagination of epithelial layers. In the formation of the neural tube, microtubules elongate the cells of the neural plate, and then microfilaments at the apexes of the cells contract to deform the cells into wedge shapes.

the extension of pseudopodia by cells on the leading edge of the migrating tissue. These amoeboid cells, connected to their neighbors by intercellular junctions, drag the sheet of epithelium into the blastocoel to form the endoderm and mesoderm of the embryo. There are also many cases in morphogenesis where amoeboid cells migrate individually, as when the cells of the neural crest disperse to various parts of the embryo.

The extracellular matrix helps to guide cells in their morphogenetic movements. The matrix includes adhesive substances and fibers that may function as tracks, directing migrating cells along a particular route. One family of extracellular glycoproteins, the **fibronectins,** helps cells adhere to their substratum as they migrate. There is a close relationship between the orientation of fibronectin fibrils in the extracellular matrix and contractile microfilaments of the cytoskeleton within migrating cells (Figure 43.23). The extracellular fibrils are themselves ordered by the orientation of the cytoskeleton in the cells that secrete the extracellular materials. In this way, one group of cells can influence the path along which another group of cells migrates.

Equally important is the intercellular glue that holds certain cells together as tissues and organs are shaped. For example, if the cells of a gastrula are experimentally dissociated from one another and then allowed to reaggregate, they sort into three layers, with endoderm on the inside, ectoderm on the outside, and mesoderm in between. Contributing to selective association are substances on the surfaces of cells called cell adhesion molecules (CAMs), which vary either in amount or in chemical identity from one type of cell to another.

Induction

After the morphogenetic movements of gastrulation produce a three-layered embryo, interactions among the layers are important in the origin of most organ rudiments. The ability of one group of cells to influence the development of another is known as **induction.**

It is induction by the rudimentary notochord that causes the dorsal ectoderm of the gastrula to thicken into a neural plate. If chordamesoderm, the dorsal mesoderm that forms the notochord, is transplanted from its normal position to a position beneath the ectoderm in some other part of the embryo, a neural plate will form and give rise to a neural tube in an abnormal location. In a series of transplant experiments in the 1920s, Hilde Mangold and Hans Spemann discovered that the dorsal lip of the blastopore is responsible for setting up the interaction between chordamesoderm and the overlying ectoderm. For example, transplanting a small piece of the dorsal lip to some other part of the surface of an early amphibian gastrula causes tissue to involute and form chordamesoderm at two locations, and the dual notochords induce the formation of two neural tubes. Spemann referred to the dorsal lip of the blastopore as the primary **organizer** of the embryo because of its role in the early stages of organogenesis.

Induction of dorsal ectoderm to develop into the neural tube is just the first of many interactions that transform the primary germ layers into organ systems. For example, development of the vertebrate eye involves a series of reciprocal inductions between evaginations of the rudimentary brain and ectoderm (Figure 43.24). The mechanism for induction is not yet known.

Determination and Differentiation

The cells of the body differ in structure and function, not because they contain different genes, but because they express different portions of a common genome inherited from the zygote (see Chapter 18). We say that a cell is beginning to differentiate with the appearance of the trademarks—specific molecules or structures—of its specialized type. Before there is molecular or cytological evidence of differentiation, however, the cell may already be committed to follow a particular course of development. A cell is said to be **determined** when it is possible to predict its developmental fate.

One way to verify if cells have already been deter-

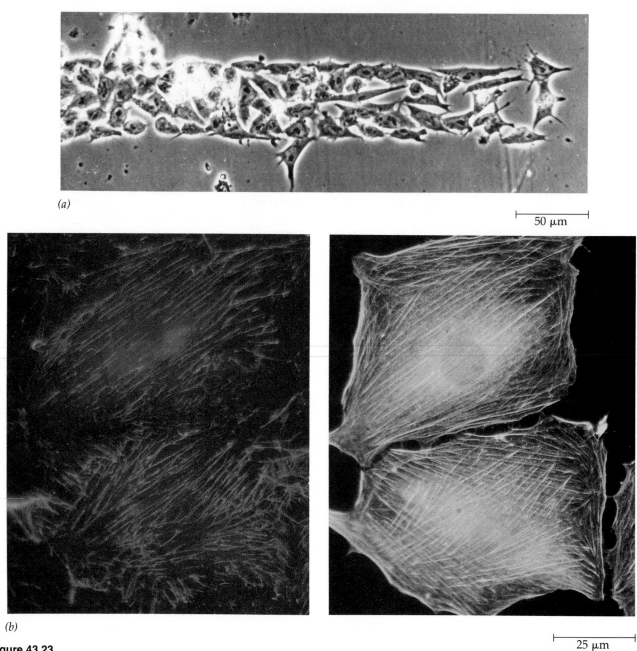

(a)

50 μm

(b)

25 μm

Figure 43.23
The extracellular matrix and cell migration. (a) Cells from the neural crest migrate along a strip of fibronectin fibrils placed on an artificial substratum (LM).

(b) Two different fluorescent dyes have been used to stain the fibronectin fibrils of the substratum (left) and the actin microfilaments within these two cells (right).

Notice that the orientation of the intracellular microfilaments and the extracellular fibrils correspond (LMs).

mined is to transplant them to some other part of the embryo to check if their developmental fate can be altered. For example, the larvae of *Drosophila* and other insects contain structures called **imaginal disks,** islands of cells destined to form various organs when the larva undergoes metamorphosis to form an adult insect (Figure 43.25). One imaginal disc may form a leg, another may give rise to an antenna, and still another may develop into an eye. In the larva, the cells of the imaginal disks are not yet differentiated, but they *are*

determined; if a disk that normally forms an antenna is experimentally removed and replaced with one determined to form a leg, then the adult fly will have a leg projecting from the head where an antenna is normally found. (Recall from Chapter 18 that genetic changes called homeotic mutations can produce similarly bizarre results, suggesting that determination involves the expression of specific regulatory genes.) Thus, determination has a "memory" aspect. A cell that has been determined passes that developmental

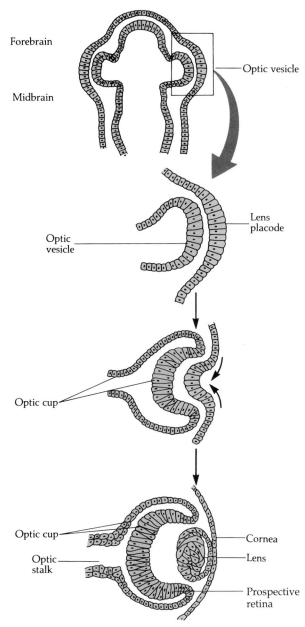

Figure 43.24
Induction during eye development. In a sequence of reciprocal inductions, the eye forms from two different layers of cells. The optic vesicle, an outgrowth of the rudimentary brain, induces ectoderm to form a thickened region called a lens placode, which in turn induces invagination of the optic vesicle to form a cup. The optic cup then induces the lens placode to invaginate, forming the lens, which in turn induces development of the cornea.

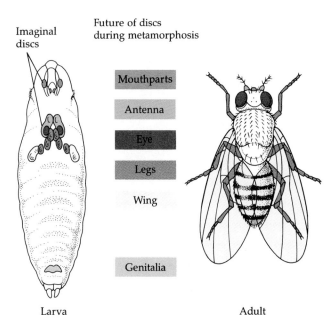

Figure 43.25
Imaginal disks and metamorphosis of *Drosophila*.
Embryonic development of a fly produces a feeding larva (maggot) that hatches from the egg. Within the larva are imaginal disks, bundles of cells that are as yet undifferentiated, but which are determined to develop into specific structures in the adult fly (the word *imago* means the adult form of an insect). Metamorphosis occurs after the larva ceases feeding and forms a pupa. Within the pupa, the imaginal disks develop into specialized structures such as wings, legs, and eyes, while much of the larval tissue is broken down and the resulting nutrients used by the developing disks.

commitment on to daughter cells, and they "remember" what they are supposed to become even if they are moved to a different environment.

Determination seems to be a serial process. Beginning with a totipotent zygote, the developmental options of cells become more and more narrow as development progresses. When differentiation finally occurs, it is the product of a cell's developmental history extending back to early cleavage. If, for example, dorsal ectoderm of an early amphibian gastrula is replaced with ectoderm from some other location, the transplanted epithelium will form a neural plate above the notochord; that ectoderm had not yet been determined to develop into epidermis and is still capable of being induced to form the neural plate. If, however, the transplanted ectoderm comes from a late-stage gastrula, it will not respond to induction and no neural plate will form. At some point during the gastrula stage, the ectoderm becomes more restricted in developmental potential. However, there is still some flexibility. For example, ectoderm no longer capable of forming a neural plate may still be able to differentiate into either epidermis or the lens of an eye, depending on its location. If an optic vesicle (the brain evagination that forms the cup of the eye) is moved to an abnormal location early enough, it will induce the overlying ectoderm to form a lens. If the manipulation is performed later, however, epidermis, rather than a lens, forms over the transplanted optic vesicle.

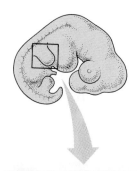

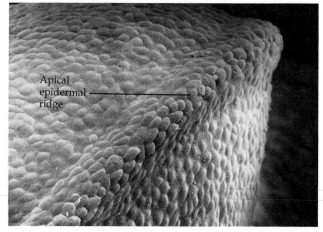

Apical epidermal ridge

(a)

50 μm

Figure 43.26
Abnormal pattern formation in a *Drosophila* mutant.
Homeotic mutations affect the development of the adult
body parts radically during metamorphosis of a fly larva
(see Chapter 18). In this case, a mutation alters the normal pattern of body segmentation and produces a fly
with an extra pair of wings (SEM).

1 mm

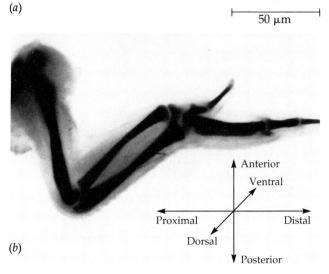

Anterior

Ventral

Proximal ———— Distal

Dorsal

(b)

Posterior

Figure 43.27
**Pattern formation during vertebrate limb develop-
ment.** **(a)** Vertebrate limbs develop from rudiments
called limb buds (SEM). **(b)** As the bud develops into a
limb, such as this wing of a chick embryo, a specific pattern of tissues emerges. This requires that each embryonic cell receives some kind of positional information that
indicates location along the three axes of the limb.

Determination apparently involves control of the
genome by the cytoplasmic environment of the cell.
Cytoplasmic cues can begin to determine cells as early
as the first cleavage division, as in the mosaic development of mollusks (see Figure 43.20). By partitioning
the heterogeneous cytoplasm of an egg, cleavage
exposes the nuclei of cells to different cytoplasmic
determinants that may preset which genes will be
expressed later when the cell begins to differentiate.
Morphogenetic movements also play a role in determination and differentiation by placing cells in embryonic neighborhoods of differing chemical and physical
environments.

Pattern Formation

There is more to building a specialized organ than the
differentiation of its cells. An arm and a leg have the
same mixture of tissues—muscle, connective tissue,
cartilage, and skin—but have different spatial
arrangements of these common tissues. The shaping
of an animal and its individual parts involves **pattern
formation,** the emergence of a body form with specialized organs and tissues all in their characteristic
places. The aberrant anatomy resulting from homeotic
mutations in *Drosophila* illustrates the importance of

genes that regulate the formation of body pattern
(homeotic mutations are discussed in Chapter 18). One
homeotic mutation, for example, affects the segmentation of a fly in such a way that an extra pair of wings
develops (Figure 43.26). Genes, however, are not the
whole story of pattern formation. Within each rudimentary organ, a specific pattern of specialized tis-

sues emerges even though all the cells have the same genes. Genes that regulate development must themselves respond to some kind of **positional information**—signals that indicate a cell's location relative to other cells in an embryonic structure.

Lewis Wolpert and his colleagues have been investigating the role of positional information in the development of chick limbs. The wings and legs begin as undifferentiated limb buds that go on to develop their unique forms (Figure 43.27). This requires that each bone, muscle, and other limb component have a precise location and orientation relative to three different axes: the proximal-distal axis, from the base of the limb to the tip of the digits; the anterior-posterior axis, from the front edge of the limb to the rear edge; and a dorsal-ventral axis, from the top of the limb to the bottom. Embryonic cells within a limb bud must receive some kind of positional information indicating location along these three axes. For example, a region of the limb bud called the "zone of polarizing activity" (ZPA), located where the posterior side of the bud is attached to the body, apparently assigns position along the anterior-posterior axis to developing tissues. Cells nearest the ZPA give rise to posterior structures, such as the digit homologous to our little finger, and cells farthest from the ZPA form anterior structures, such as the avian equivalent of our forefinger and thumb. Experiments support this hypothesis (Figure 43.28). Another region, consisting of mesoderm located just beneath an "apical epidermal ridge" at the tip of the wing bud, is believed to assign position to the embryonic tissue along the proximal-distal axis as the limb bud grows. Positional information along the dorsal-ventral axis probably reflects the relative distance of developing tissue from the dorsal and ventral ectoderm of the limb bud. In experiments where the ectoderm of a bud is separated from the mesoderm and then replaced with its orientation rotated by 180 degrees, the limb develops with its dorsal-ventral ori-entation reversed (equivalent to reversing the palm and back of our hand).

From experiments such as these, we can conclude that pattern formation requires that cells receive and interpret environmental cues that vary from one location to another. One possibility is a chemical signal that varies in concentration along a gradient, enabling cells to resolve their position along that gradient. Such a substance is called a **morphogen.** Gradients of morphogens along two other dimensions would provide the positional information a cell needs to determine its location in the three-dimensional realm of a developing organ. Until recently, morphogens were only hypothetical substances. In 1987, however, Christina Thaller and Gregor Eichele identified a possible morphogen, a derivative of vitamin A called **retinoic acid.** If this compound is dabbed onto the anterior margin of a limb bud, the effect is the same as transplanting an additional ZPA to this location; the result is two sets of digits forming a mirror image arrangement (see Figure 43.28). Furthermore, the researchers were able to demonstrate a gradient of retinoic acid in the normal limb buds of chick embryos, with the compound most concentrated near the ZPA.

Other experiments suggest that positional information functions in the development of the overall body form, not just in the patterning of individual parts, such as limbs. In one experiment, a small block of undifferentiated mesoderm from the part of the leg that would normally become the thigh was transplanted to a position just below the ectoderm at the tip of the wing bud. The transplanted tissue did not develop into wayward thigh tissue, nor did it form structures typical of wing tips, but instead gave rise to a toe. Apparently, the transplanted mesoderm had received its positional information in installments, first determining that it was in the posterior limb bud (leg) and subsequently sensing that it was near the tip of a developing limb.

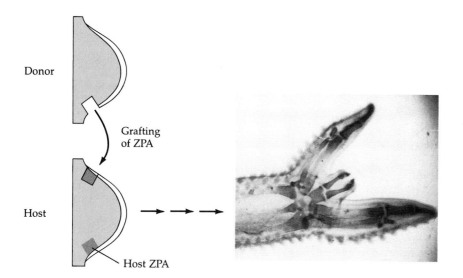

Donor

Grafting of ZPA

Host

Host ZPA

Figure 43.28
Altering positional information with grafting experiments. A region called the zone of polarizing activity (ZPA) is the reference point for indicating the position of cells along the anterior-posterior axis of a vertebrate limb bud. The ZPA is located where the posterior margin of the bud is attached to the body. In this experiment, a second ZPA is added to the anterior margin of a limb bud by transplanting the tissue from a donor. Cells near the grafted ZPA, like cells near the host bud's own ZPA, apparently receive positional information that indicates "posterior." The pattern that emerges in the developing limb is a mirror image, with an arrangement of digits equivalent to two human hands joined together at the thumbs.

Graded cues probably cannot operate over long distances in the embryo, but installments of positional information provided throughout development can first establish the gross form of the animal when the embryo is very small ("This will be the head end, this will be the tail end," and so on) and then refine its parts as development progresses.

Through a variety of experiments, developmental biologists are beginning to learn how the one-dimensional information encoded in the nucleotide sequence of a zygote's DNA is translated into the three-dimensional form of an animal. There are enough unanswered questions, however, to entertain many future generations of scientists.

* * *

STUDY OUTLINE

1. Development encompasses changes in form and function that occur from conception to death.
2. Developmental changes are most marked during embryonic life, in which cell division, differentiation, and morphogenesis transform a unicellular zygote into a multicellular organism.

Fertilization (pp. 956–958)

1. Fertilization both reinstates diploidy and activates the egg to begin a chain of metabolic reactions that triggers the onset of embryonic development.
2. The acrosomal reaction, which occurs when the sperm meets the egg, releases hydrolytic enzymes that digest through material surrounding the egg.
3. Gamete fusion depolarizes the egg cell membrane and sets up a fast block to polyspermy.
4. Egg cell depolarization also initiates the cortical reaction, in which calcium ions stimulate cortical granules to erect a fertilization membrane that functions as a slow block to polyspermy.
5. The calcium ions also trigger metabolic changes in the fertilized cell that greatly increase cellular respiration and protein synthesis, thus activating the egg.

Early Stages of Embryonic Development (pp. 959–961)

1. The basic body plan of an animal is determined in the early stages of development.
2. Fertilization is followed by cleavage, a period of rapid cell division without growth, which results in the production of a large number of small cells called blastomeres.
3. Cleavage planes usually follow a specific pattern relative to the animal and vegetal poles of the zygote.
4. Cleavage produces a solid ball of cells, called the morula, which develops a fluid-filled blastocoel, thereby becoming the blastula.
5. Gastrulation transforms the blastula by invagination into a cup-shaped, two-layered embryo called the gastrula. At this stage, the embryo has an archenteron that opens through the blastopore.
6. A mesoderm layer develops between the outer ectoderm and the inner endoderm, completing the three primary germ layers from which all body structures are derived during organogenesis.
7. The notochord forms from dorsal mesoderm, and the neural tube develops from the ectodermal neural plate above the notochord.

Comparative Embryology of Vertebrates (pp. 961–967).

1. The amphibian egg is noticeably polar, with yolk platelets concentrated in the vegetal hemisphere and pigment granules in the animal hemisphere. This asymmetry has profound effects on the mechanics of cleavage and gastrulation.
2. The shelled eggs of birds and reptiles have large yolks and undergo meroblastic cleavage restricted to a small disc of cytoplasm at the animal pole. A cap of cells called the blastodisc forms and begins gastrulation with the formation of the primitive streak. In addition to the embryo, primary germ layers give rise to the four extraembryonic membranes: yolk sac, amnion, chorion, and allantois.
3. The eggs of placental mammals are small and store little food, exhibiting a holoblastic cleavage with no apparent polarity. Gastrulation and organogenesis, however, resemble the patterns of birds and reptiles. After fertilization and early cleavage in the oviduct, the blastocyst implants in the uterus. The trophoblast initiates formation of the fetal portion of the placenta, and the embryo proper develops from an embryonic disc. Extraembryonic membranes homologous to those of birds and reptiles function in intrauterine development.

Mechanisms of Development (pp. 967–976)

1. In most animals except mammals, the polarity of the embryo is established during early cleavage.
2. The unequal relegation of cytoplasmic determinants into specific blastomeres can fix the developmental fates of different regions of the embryo very early.
3. In regulative development, cells remain totipotent longer than do the blastomeres in mosaic development.
4. Experimentally derived fate maps of polar embryos have shown that specific regions of the zygote or **blastula** develop into specific parts of the embryo.
5. Morphogenetic movements rely on the extensor, contractile, and adhesive properties of cells. Morphogenesis by reorganization of the cytoskeleton and amoeboid movement is aided by extracellular fibronectin fibrils and cell surface adhesion molecules.
6. Transplant experiments have demonstrated that certain groups of cells can influence the development of contiguous cells by induction. The dorsal lip of the blastopore is a primary organizer of the amphibian embryo.
7. Cell differentiation is orchestrated by the serial process of determination, which progressively narrows the developmental options of each cell.
8. Cells are ordered into specific three-dimensional positions to produce specific organs and body parts by pattern formation, which requires that cells receive and interpret positional information that varies with location.

SELF-QUIZ

1. The cortical reaction functions directly in the
   a. formation of a fertilization membrane
   b. production of a fast block to polyspermy
   c. release of hydrolytic enzymes from the sperm cell
   d. generation of a nerve-like impulse by the egg cell
   e. the fusion of egg and sperm nuclei

2. Which of the following is common to both avian and mammalian development?
   a. holoblastic cleavage
   b. primitive streak
   c. trophoblast
   d. yolk plug
   e. gray crescent

3. The archenteron develops into
   a. the mouth in protostomes
   b. the blastocoel
   c. the endoderm
   d. lumen of the digestive tract
   e. the placenta

4. In a frog embryo, the blastocoel is
   a. completely obliterated by yolk platelets
   b. lined with endoderm during gastrulation
   c. located primarily in the animal hemisphere
   d. restricted to the vegetal hemisphere
   e. the cavity that later forms the archerteron

5. Which embryonic membrane is closest to the shell in a bird egg?
   a. allantois
   b. amnion
   c. chorion
   d. yolk sac
   e. endoderm

6. In an amphibian embryo, a band of cells called the neural crest
   a. rolls up to form the neural tube
   b. develops into the main sections of the brain
   c. produces amoeboid cells that migrate to form teeth, skull bones, and other structures in the embryo
   d. has been shown by experiments to be the organizer region of the developing embryo
   e. induces the formation of the notochord

7. What is the difference between differentiated and determined cells?
   a. A determined cell induces the differentiation of neighboring cells.
   b. A determined cell has its developmental fate set, even though it may not yet be differentiated.
   c. A differentiated cell has its developmental fate set, even though it may not yet be determined.
   d. A differentiated cell is found in imaginal disks, whereas a determined cell is found only in adult structures.
   e. Cytoplasmic determinants have regulated some genes in a determined cell; morphogens have regulated genes in a differentiated cell.

8. The specific arrangement of tissues and organs in a body part, such as a leg, is called
   a. pattern formation
   b. induction
   c. differentiation
   d. determination
   e. organogenisis

9. In the early development of an amphibian embryo, the organizer is the
   a. neural tube
   b. notochord
   c. archenteron roof
   d. dorsal lip of blastopore
   e. dorsal ectoderm

10. The first cavity to form in an amphibian embryo is the
    a. coelom
    b. blastocoel
    c. archenteron
    d. gastrocoel
    e. optic vesicle

CHALLENGE QUESTIONS

1. Using the frog embryo as an example and being as specific as you can, describe the role of morphogenetic movements in gastrulation and the role of pattern formation in organogenesis.

2. Explain the role of fertilization in triggering development.

3. Describe the sequence of events that leads to induction of dorsal ectoderm and formation of the neural tube.

FURTHER READING

1. Edelman, G. M. "Topobiology." *Scientific American*, May 1989. How adhesion molecules at the cell surface contribute to embryonic form.
2. Fjose, A. "Spatial Expression of Homeotic Genes in *Drosophila*." *Bioscience*, September 1986. Genetic control of pattern formation in insects.
3. Gilbert, S. F. *Developmental Biology*. 2d ed. Sunderland, MA: Sinauer Associates, 1988. A widely used upper division text.
4. Hynes, R. O. "Fibronectins." *Scientific American*, June 1986. A beautifully illustrated article on adhesive proteins that serve as organizers in development.
5. Malacinski, G. M., A. W. Neff, J. A. Alberts, and K. A. Souza. "Developmental Biology in Outer Space." *Bioscience*, May 1989. Embryology in a weightless environment.
6. Wassaman, P. M. "Fertilization in Mammals." *Scientific American*, December 1988. The sperm receptor on the egg surface governs many events of fertilization.

44

Nervous Systems

An infielder's diving catch of a line drive is a spectacular display of animal coordination. The player sees a ball shot from the bat, and a living computer called the brain integrates information about the angle and speed of the ball and then signals multiple muscles to thrust a hand to meet the ball, all in a fraction of a second. Coordinated behavior, of course, is more than a game. To survive and reproduce, an animal must respond rapidly and appropriately to environmental stimuli, and it is mainly the animal's nervous system that controls these reactions.

The nervous system, as pointed out in Chapter 41, is one of two coordinating systems in animals, the other being the endocrine system. In many cases, these two systems cooperate to control physiology and behavior; for example, hormones and nerve impulses are both involved in the "fight-or-flight" responses of animals to threats and stress. However, there are three important differences in the roles of the nervous and endocrine systems. One is that the nervous system reacts much more quickly. Messages can travel along nerve cells as fast as 100 m/sec (over 300 mph), which means that nervous signals can be transmitted all the way from the brain of a human to the hands in just a few milliseconds. In contrast, there is usually a delay of several minutes between the stimulation of an endocrine gland and the responses of target organs to hormones secreted by that gland. A second distinguishing feature of the nervous system is its pinpoint control. Whereas the endocrine system dispatches its chemical messages to target cells via the bloodstream, nerve cells transmit signals directly to individual target cells. For example, the nervous system can control very fine movements of your hands by stimulating only one or a few muscles to contract at a particular instant. A third difference is that the nervous system has an incomparable structural complexity that enables it to integrate more kinds of information and to elicit a broader range of responses than can the endocrine system.

The nervous system is probably the most intricately organized aggregate of matter on Earth. Just 1 cm³ of the human brain may contain several million nerve cells, each of which may communicate with thousands of other nerve cells in data-processing networks that make the most elaborate circuit boards fashioned by computer engineers look primitive. These neural pathways not only control our every perception and movement, but, somehow, also enable us to learn and remember. A thought is an emergent property of a highly ordered mass of cells.

In general, the nervous system has three overlapping functions: *sensory input, integration,* and *motor ouput.* Input is the conduction of signals from sensory receptors, such as the light-detecting cells of the eye, to integration centers of the nervous system, such as the brain. Integrating the information requires that the sensations triggered by environmental stimulation of the receptors be interpreted and associated with appropriate responses of the body. Motor output is the conduction of signals from the brain or other processing center of the nervous system to **effector cells,** the muscle cells or gland cells that actually perform the body's responses to stimuli. From receptor to effector, the information is communicated by a combination of electrical and chemical transmissions along pathways of specialized nerve cells, or **neurons** (Figure 44.1).

Despite the complexity of the nervous system, neurobiologists are beginning to understand how the system processes and transmits signals, and the basic mechanisms turn out to be remarkably similar across a wide variety of animals. However, because so many important questions remain unanswered, neurobiology will continue to be one of the most active and exciting areas of scientific research for decades to come. This chapter focuses on how neurons work and compares how the nervous systems of diverse animals are organized. The next chapter connects the nervous system to its inputs and outputs by discussing sensory receptors and the physiology of movement. Throughout both chapters, the axiom that structure fits function applies.

CELLS OF THE NERVOUS SYSTEM

Two main classes of cells populate the nervous systems of animals: neurons and supporting cells. Neurons, already introduced, are the nerve cells that

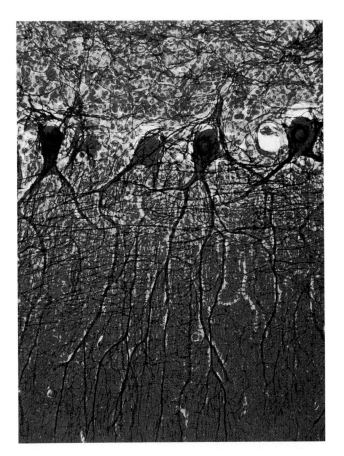

25 μm

Figure 44.1
Nerve cells in the human brain. In this neural network, the cell bodies of four neurons communicate with enormous numbers of additional neurons via neural processes (LM).

actually conduct messages along the pathways of the nervous system. More numerous are the **supporting cells,** which provide structural reinforcement in the nervous system and also protect, insulate, and generally assist neurons.

Neurons

Neurons, cells specialized for transmitting signals from one location in the body to another, are the functional units of the nervous system. Although there are many different types of neurons, which vary in details of their structure depending on their roles in the nervous system, most neurons share some common features. A neuron has a relatively large **cell body** containing the nucleus and most of the cytoplasm and organelles of the cell. The most striking features of the neuron are fiberlike extensions called *processes* that increase the distance over which the cells can conduct messages. The processes are of two types: **dendrites,** which convey signals toward the cell body, and **axons,**

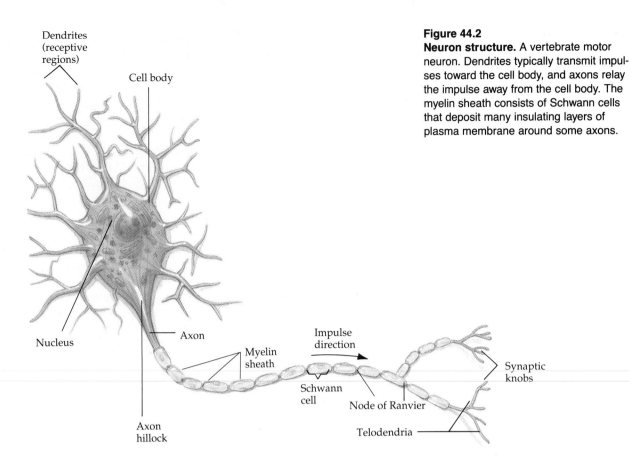

Dendrites
(receptive
regions)

Cell body

Nucleus

Axon

Axon
hillock

Myelin
sheath

Schwann
cell

Impulse
direction

Node of Ranvier

Telodendria

Synaptic
knobs

Figure 44.2
Neuron structure. A vertebrate motor neuron. Dendrites typically transmit impulses toward the cell body, and axons relay the impulse away from the cell body. The myelin sheath consists of Schwann cells that deposit many insulating layers of plasma membrane around some axons.

which conduct messages away from the cell body (Figure 44.2).

The dendrites of most types of neurons, such as the one illustrated in Figure 44.2, are short, numerous, and extensively branched (indeed, the name for these processes is derived from the Greek *dendron*, "tree"). Thus, the dendrites are structural adaptations that increase the surface area of the "receiving end" of the neuron, where the cell is most likely to be stimulated.

Most neurons have a single axon, which may be very long, as it is in the type of neuron represented in Figure 44.2. (In fact, the sciatic nerve in your leg has axons that carry impulses all the way from the lower part of your spinal cord to muscles of your toes, a distance of a meter or more.) The axon extends from a cone-shaped region of the cell body called the **axon hillock,** which, as we shall see, is the region where impulses that pass down the axon are usually generated. A series of supporting cells called **Schwann cells,** arranged along the length of the axon, encloses the axons of many neurons. Collectively, the chain of Schwann cells forms an insulating layer called the **myelin sheath.** The axon may be branched, and each branch usually terminates in hundreds or thousands

of branchlets called telodendria. The bulbous tips of the telodendria are the **synaptic knobs,** which relay nervous signals to other cells by releasing chemicals called **neurotransmitters.** These chemical messages diffuse across the **synapse,** a narrow gap between the synaptic knob and the signal-receiving portion (dendrite or cell body) of another neuron or an effector, such as a muscle cell. Thus, the synapse is where one neuron communicates with another neuron in a neural pathway.

The cell bodies of most neurons are located in the **central nervous system,** which consists of the brain and spinal cord. However, the cell bodies of certain types of neurons are located outside the central nervous system in clusters called ganglia. The **peripheral nervous system** consists mainly of nerves, bundles of neuron processes communicating between the central nervous system and the rest of the body. Based on the direction of transmission, we can distinguish two major types of neurons in the peripheral nervous system: sensory neurons and motor neurons. **Sensory neurons** communicate information about the external and internal environments from sensory receptors to the central nervous system. **Motor neurons** convey impulses from the central nervous system to effector cells. Sensory input and motor output of the nervous system are usually integrated by **interneurons,** or **association neurons,** which are located within the central nervous system. In most neural pathways, sensory neurons synapse with interneurons, which in turn synapse with motor neurons (Figure 44.3).

Supporting Cells

Supporting cells outnumber neurons tenfold to fiftyfold in the nervous system. Although the supporting cells do not actually conduct nerve impulses, they are essential for the structural integrity of the nervous system and for the normal functioning of neurons.

Supporting cells in the central nervous system are called **glial cells,** which means "glue" cells. There are several types of glial cells in the brain and spinal cord. **Astrocytes** line the capillaries in the brain and, along with the specialized walls of the capillaries themselves, contribute to the **blood–brain barrier.** This barrier restricts the passage of most substances into the brain, preventing dangerous fluctuations in the chemical environment of the central nervous system. (Medications used to treat diseases of the central nervous system must be cleverly chosen or designed so they will be able to cross the blood–brain barrier.) Astrocytes also help control the ionic environment around neurons, a function critical to the electrical properties of neurons. Glial cells called **oligodendrocytes** form myelin sheaths, the coatings that insulate the processes of some neurons. Most neurons in the peripheral nervous system are also myelinated, but there, Schwann cells are the supporting cells that form the sheath (see Figure 44.2).

Neurons become myelinated in a developing nervous system when Schwann cells or oligodendrocytes grow around axons in such a way that their plasma membranes form many concentric layers—somewhat like a jelly roll (Figure 44.4). The membranes are mostly lipid, which is a poor conductor of electric currents. Thus, the myelin sheaths provide electrical insulation between crowded neurons, analogous to the rubber insulation covering copper wires. We will soon see that myelin sheaths also increase the rate of nerve impulses. In the degenerative disease known as *multiple sclerosis,* myelin sheaths gradually deteriorate, and progressive loss of coordination results. Clearly, supporting cells are indispensable components of a working nervous system. That being said, let us return to the neuron and its ability to conduct an impulse.

TRANSMISSION ALONG NEURONS

The signal transmitted along the length of a neuron, from a dendrite or cell body to the tip of an axon, depends on electric currents in the form of ion fluxes across the plasma membrane of the cell. Understanding nerve impulses requires a detailed examination of how these ion currents occur, and how currents *across the membrane* can be converted to a signal that travels in a perpendicular direction *along the neuron.*

The Resting Potential

All living cells have a charge difference across their plasma membranes. Usually the cytoplasm just inside the membrane is negative in charge compared to the extracellular fluid just outside the membrane (see Chapter 8). The membrane is said to be *polarized.* Since opposite charges tend to move toward one another, a polarized membrane stores energy, much as a battery does, by holding opposite charges apart. The strength, or voltage, of this electrical potential energy can be measured with microelectrodes (see Methods Box, p. 983). The voltage across the plasma membrane of a cell is known as the *membrane potential* because of its potential to perform work. Specifically, the membrane potential of a nontransmitting neuron—one that is not presently conducting an impulse—is called the **resting potential.** Neurons typically have resting potentials of about -70 millivolts (mV); the minus sign indicates that the inside of the cell is negative compared to the outside. When a neuron is stimulated, the voltage across its membrane changes, and this change in membrane potential can give rise to a nerve impulse. Neurons, muscle cells, and certain other cells that can use changes in their membrane potentials to conduct signals are called **excitable cells.**

How is the resting potential of an excitable cell generated? The resting potential is due to differences in the ionic composition of the solutions on opposite sides

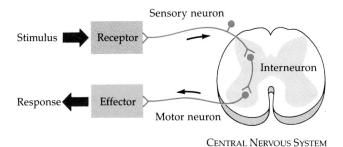

Figure 44.3
Nervous pathway between a receptor and an effector.

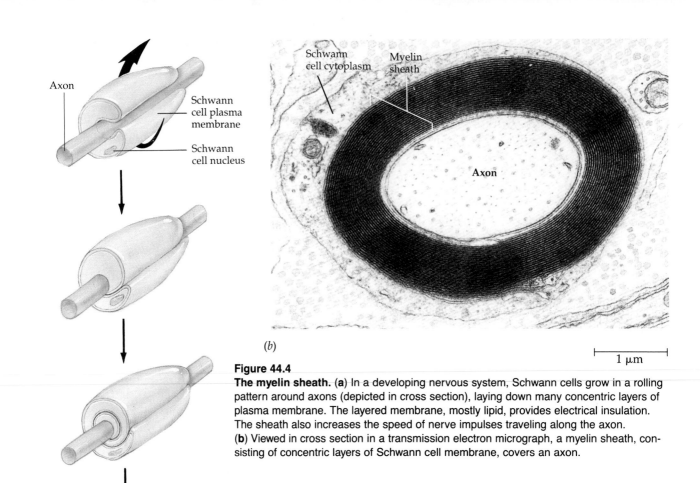

Axon

Schwann
cell plasma
membrane

Schwann
cell nucleus

Myelin
sheath

(a)

Schwann
cell cytoplasm

Myelin
sheath

Axon

(b)

1 μm

Figure 44.4

The myelin sheath. (a) In a developing nervous system, Schwann cells grow in a rolling pattern around axons (depicted in cross section), laying down many concentric layers of plasma membrane. The layered membrane, mostly lipid, provides electrical insulation. The sheath also increases the speed of nerve impulses traveling along the axon.
(b) Viewed in cross section in a transmission electron micrograph, a myelin sheath, consisting of concentric layers of Schwann cell membrane, covers an axon.

of the membrane and to the relative permeability of the membrane to these ions. The plasma membrane of a neuron has an abundance of sodium-postassium pumps, ATP-driven proteins that actively transport sodium ions (Na^+) out of the cell and potassium ions (K^+) into the cell (see Chapter 8). The pumps generate steep gradients of these two ions, with Na^+ much more concentrated outside the cell and K^+ much more concentrated inside the cell (Figure 44.5). However, there is a tendency for these ions to diffuse back across the membrane down their gradients. For K^+, the chemical gradient (concentration difference) is partially countered by the electrical attraction of the positive ions for the electrically negative interior of the cell. However, the force of the concentration gradient is greater than this attraction and tends to drive

K^+ out of the cell. In the case of Na^+, the two forces—the voltage and the concentration gradient—work in the same direction, intensifying the tendency for this ion to reenter the cell. The unstimulated neuron can maintain the steep gradients of Na^+ and K^+ generated by active transport because its plasma membrane is not very permeable to these ions. The membrane is *more* permeable to K^+ than it is to Na^+, but the driving force acting on Na^+ is much greater (because the chemical gradient and voltage are *both* driving Na^+ into the cell by diffusion). It is the combined effect of ion pumping (active transport) and leaking (diffusion) that accounts for the cytoplasmic and extracellular concentrations of Na^+, K^+, and other ions for the unstimulated neuron.

Stimulation and Graded Potentials

Changes in the local environment of a neuron can affect the permeability of the plasma membrane to ions and alter the membrane potential. Any environmental factor that induces such a change is called a stimulus. The stimulus may be physical pressure on the cell, an abrupt chemical change in the extracellular environment, an electric shock, a change in temper-

Due to a difference in the relative concentrations of cations and anions on opposite sides of the plasma membrane, the cytoplasmic side of the membrane is negative in charge compared to the extracellular side. This charge separation is called a membrane potential, or, for an unstimulated neuron, a resting potential.

Electrophysiologists can measure the membrane potential as a voltage by using microelectrodes connected to a sensitive voltmeter or oscilloscope. Precise mechanical devices called micromanipulators (moved with the large knobs being manipulated by this scientist) are used to position one electrode just inside the cell for comparison with a reference electrode located outside the cell. The voltmeter indicates the magnitude of the charge separation across the membrane, typically about -70 mV for an unstimulated neuron (the minus sign indicates that the inside of the cell is negative compared to the outside).

The giant neurons of certain invertebrates, including squid and lobsters, have been particularly useful model systems for studying the electrophysiology of nerve impulses, because the large cells are relatively easy to impale with electrodes. Some of these giant neurons have axons with diameters of about 1 mm. Once electrodes are in place, they can be used not only to measure the voltage of the resting potential, but also to record changes in voltage due to the ion currents that occur during transmission of a nerve impulse.

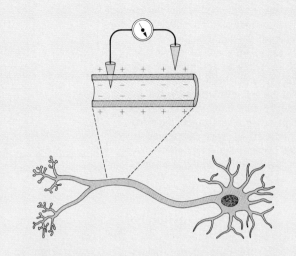

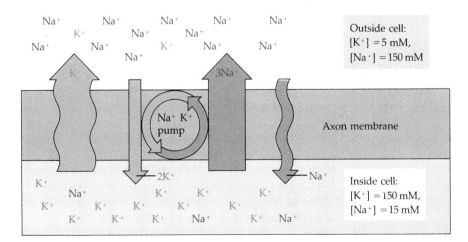

Figure 44.5
Generation of Na⁺ and K⁺ gradients.
The sodium-potassium pump, powered by ATP, actively transports 3 Na⁺ out of the neuron for every 2 K⁺ pumped into the neuron. In an unstimulated neuron, the sodium-potassium pump maintains steep gradients of Na⁺ and K⁺, which are possible only because the membrane of the unstimulated neuron is not very permeable to these ions. However, the diffusion gradients for Na⁺ and K⁺ are so strong that there is some leakage (wavy arrows) of the ions across the membrane, which is more permeable to K⁺ than it is to Na⁺. Thus, differential rates of both pumping and leaking of Na⁺ and K⁺ contribute to the membrane potential and to the differential ion concentration on opposite sides of the membrane.

Figure 44.6
Graded changes in membrane potential. Neurons are stimulated by environmental changes that alter the membrane potential of the cell. Depending on the type of neuron and the nature of the stimulus, the event may either depolarize or hyperpolarize the membrane. (**a**) During depolarization, the membrane potential decreases in voltage as it moves closer to zero. (**b**) Hyperpolarization makes the inside of the cell even more negative than the resting potential. In either case, depolarization or hyperpolarization, the initial effect of the stimulus is called a graded potential because the extent of the voltage change is proportional to the strength of the stimulus.

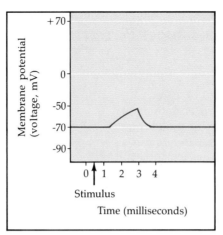
(*a*) Depolarization

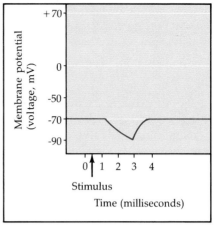
(*b*) Hyperpolarization

ature, or some other environmental episode. If the stimulus decreases the voltage of the neuron's membrane from the resting potential of −70 mV—that is, in the direction of zero voltage—then the membrane is said to be **depolarized.** For example, a stimulus that depolarizes the membrane may reduce the potential from −70 mV to −50 mV. In contrast, if a change in the local environment of a neuron pushes the membrane potential to a voltage even greater than the resting potential—say, from −70 mV to −90 mV—then the membrane has been **hyperpolarized.** These changes in the plasma membrane's electrical potential are the initial responses of neurons to environmental stimuli.

The local voltage changes induced by stimulation of a neuron are called **graded potentials,** because the magnitude of the change in membrane potential is proportional to the strength of the stimulus (Figure 44.6). Graded potentials last only about a millisecond. The magnitude of a graded potential, whether a depolarization or a hyperpolarization, is greatest at the point where the neuron has been stimulated, and decreases with distance away from that point.

As we shall see, hyperpolarization of a neuron reduces the probability that the cell will "fire," or transmit a nerve impulse, while depolarization increases the chances that a nerve impulse will be triggered. The actual impulse is generated when graded potentials that depolarize the membrane of a neuron are converted to an even larger depolarization called an action potential.

The Action Potential

A nerve impulse originates as an **action potential,** a rapid, reversible depolarization of the neuron's membrane near the point of stimulation. In fact, the action potential goes beyond depolarizing the membrane, actually making the inside of the cell positive compared to the outside; action potentials typically alter

the transmembrane voltage from the resting potential of −70 mV to about +35 mV, a change of more than 100 mV. This occurs within a millisecond or two of stimulation, and just as rapidly, the membrane voltage returns to the resting potential. If the event is recorded with microelectrodes that register the voltage across the membrane (see Methods Box), the graph of the action potential looks like a "spike." When neuroscientists speak of a neuron "firing," they are referring to the generation of an action potential (Figure 44.7).

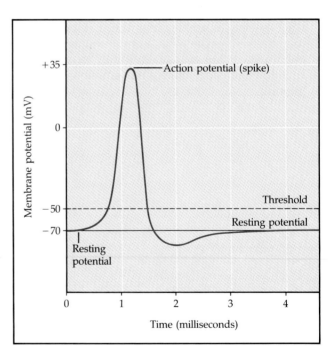

Figure 44.7
The action potential. A stimulus that depolarizes the neuron membrane from its resting potential to a reduced voltage called the threshold potential triggers an even larger depolarization called the action potential. The membrane potential "spikes" and quickly returns to the resting voltage.

The action potential is due to ion fluxes through specific channels in the membrane called **voltage-sensitive gates** (Figure 44.8). These ion channels, consisting of proteins embedded in the membrane, open and close in response to changes in the membrane potential—hence, the name voltage-sensitive gates. A stimulus depolarizes a neuron's plasma membrane by causing a local increase in the permeability of the membrane to Na^+. Sodium ions cross the membrane from the extracellular fluid and enter the cell, driven by the concentration gradient and by the attraction of the cations for the electrically negative interior of the cell. It is this permeability change to Na^+ that begins to depolarize the membrane, decreasing the voltage from the resting potential of -70 mV to a value closer to zero, say -50 mV. If the depolarization caused by the stimulus is sufficiently large, then the voltage change causes voltage-sensitive Na^+ gates to open, and permeability to Na^+ changes even more. Thus, a small depolarization is amplified by the opening of

Na^+ gates to generate the larger change in voltage that characterizes the action potential. This is one of the few examples of positive feedback known in a biological system.

The action potential is local and transient, lasting less than 2 milliseconds because the Na^+ gates close very soon after opening. In fact, until the membrane potential returns to the resting potential, the Na^+ gates are *inactivated*. This short time of insensitivity is called the **refractory period.**

How is the resting potential restored? The neuron membrane also has voltage-sensitive K^+ gates, which open slightly later than the Na^+ gates after a stimulus initiates depolarization (Figure 44.8). Remember, K^+ tends to diffuse out of the cell because of its concentration gradient, and thus the opening of the K^+ gates results in a rapid efflux of this ion from the cell. This increased permeability to K^+ helps to restore the resting potential and may even temporarily hyperpolarize the membrane to a voltage more negative than the

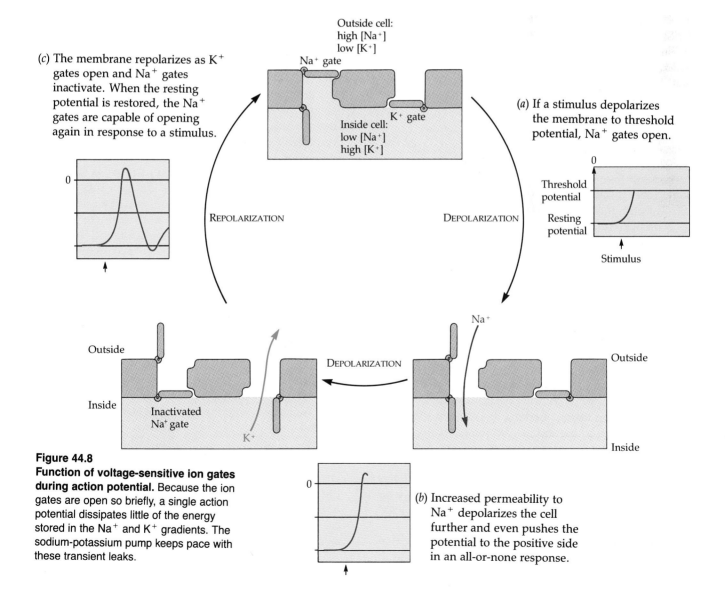

(c) The membrane repolarizes as K^+ gates open and Na^+ gates inactivate. When the resting potential is restored, the Na^+ gates are capable of opening again in response to a stimulus.

(a) If a stimulus depolarizes the membrane to threshold potential, Na^+ gates open.

Figure 44.8
Function of voltage-sensitive ion gates during action potential. Because the ion gates are open so briefly, a single action potential dissipates little of the energy stored in the Na^+ and K^+ gradients. The sodium-potassium pump keeps pace with these transient leaks.

(b) Increased permeability to Na^+ depolarizes the cell further and even pushes the potential to the positive side in an all-or-none response.

resting potential (see Figure 44.7). Now, you may think that these ion fluxes will dissipate the Na$^+$ and K$^+$ gradients that make the action potential possible in the first place, but the ion gates are open so briefly that a single action potential has only a small effect on the ion gradients. And between stimuli, the sodium-potassium pumps restore the gradients, storing energy that is tapped each time a neuron "fires" an action potential.

Unlike a graded potential, an action potential is an **all-or-none event;** a neuron either fires by generating an action potential or it does not. There is no *partial* action potential, and, for a given type of neuron, the voltage change corresponding to the action"spike" is of the same magnitude each time the neuron fires, regardless of the nature or strength of the stimulus. But not all stimuli are intense enough to trigger an action potential. A neuron has a **threshold potential,** a minimum depolarization that must occur before voltage-sensitive Na$^+$ gates open and an action potential is achieved. For example, many neurons have thresholds of about -50 mV. The stimulus, remember, induces a graded potential that is proportional to the intensity of the stimulus. If a stimulus is sufficiently strong to depolarize the membrane from the resting potential of -70 mV to the threshold of -50 mV, then Na$^+$ gates in the vicinity open, and there is an action potential. If the graded potential induced by the stimulus is too small to reach threshold, then the membrane potential simply returns to the resting voltage.

If the magnitude of an action potential is unaffected by the intensity of the stimulus, then how can the nervous system distinguish strong stimuli from weaker ones that are still sufficient to trigger action potentials? Strong stimuli result in a greater *frequency* of action potentials than weaker stimuli; if a stimulus is intense, then the neuron "fires" repeatedly after each refractory period until the stimulus is no longer strong enough to bring the membrane potential from the resting voltage to the threshold voltage. Thus, it is the number of action potentials per second, not their amplitude, that codes for stimulus intensity in the nervous system.

Action potentials are most commonly generated at the axon hillock, the point where the axon tapers away from the cell body of the neuron. However, the dendrites and cell body are the parts of the neuron that are usually stimulated. Because the graded depolarizations caused by a stimulus diminish with distance along the membrane, stimulation of a dendrite near the axon hillock is more likely to trigger an action potential than stimulation of the same intensity at a more distant site. However, even stimulation relatively remote from the axon hillock can trigger an action potential if the stimulus is intense enough.

Let us review what we have learned about the action potential up to this point: A stimulus increases the permeability of the neuron's plasma membrane to Na$^+$ and a certain amount of the cation enters the cell, causing a graded depolarization. If the stimulus is strong enough to depolarize the membrane at the axon hillock to at least the threshold potential, then Na$^+$ gates open and the membrane depolarizes all the way to the action potential of about $+35$ mV. Then, the Na$^+$ gates close and K$^+$ gates open, repolarizing the membrane by replacing positive charge outside the cell. How is this localized electrical event called an action potential propagated from its point of origin all the way to the very tip of the axon?

Propagation of the Nerve Impulse

Transmission of an impulse along an axon depends on local currents of ions associated with action potentials. The sodium ions that rush into a stimulated neuron at the point where the action potential is first triggered diffuse laterally from their point of entry. This positive charge depolarizes a neighboring region of the membrane to the threshold potential, causing voltage-sensitive Na$^+$ gates in this adjacent region to open and another action potential to occur, this one a little farther along the axon from the point of the original action potential (Figure 44.9). Lateral diffusion of the incoming Na$^+$ at this point incites an action potential still farther along the axon, and so on, all the way down to the telodendria at the tips of the axon. As depolarization spreads sequentially along the axon, repolarization trails closely behind due to the exit of K$^+$, which quickly follows the entry of Na$^+$. Thus, once an initial action potential occurs in the vicinity of the axon hillock, serial action potentials occur in turn, all along the length of the axon. It is like tipping the first member of a long row of standing dominoes; the first domino does not actually travel far, but its topple is relayed to the end of the row. Similarly, an action potential does not actually travel, but is regenerated anew at each patch along the membrane. In this way, local currents of ions *across* the plasma membrane give rise to a signal that travels in a perpendicular direction, *along* the neuron process. This signal, or nerve impulse, is the self-propagating wave of depolarization that spreads along the axon.

If an axon is stimulated somewhere along its axis, then the action potential is propagated in both directions. However, this rarely happens, since dendrites and the cell body are usually the only parts of a neuron located where stimuli from the internal or external environment of the organism are common. This is why the initial action potential usually occurs in the region of the axon hillock, the transitional zone between the cell body and axon. Once the action potential is propagating along the axon, what prevents Na$^+$ entry from

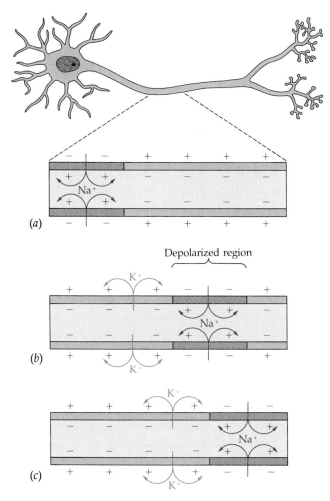

Figure 44.9
Propagation of the action potential. As sodium ions flow inward across the membrane, their lateral movement depolarizes the patch of membrane just ahead of the impulse. Depolarization of this patch triggers depolarization of the next region, and the action potential progresses along the axon or dendrite. As the action potential propagates, repolarization due to K$^+$ exit occurs in its wake.

reexciting the region *behind* the action potential, which would cause depolarization to spread back toward the axon hillock? Remember, the spike of an action potential is followed by a brief refractory period, when Na$^+$ gates are inactivated and cannot open. A wave of depolarization passing a point along the axon cannot induce another action potential *behind* it, but only in the *forward* direction. Thus, the axon is normally a one-way avenue for nerve impulses.

Rate of Transmission

One factor that affects the speed of nerve impulses is the diameter of the axon: The larger the diameter of the axon, the faster the rate of transmission. This is because the resistance to the flow of electrical current is inversely proportional to the cross-sectional area of the "wire" conducting the current. Because of this effect, an action potential is propagated more rapidly along a thick axon than a thinner one because the local ion currents can depolarize the next region of membrane sooner. Impulse rates vary from several centimeters per second for very thin axons to about 120 meters per second for the giant axons of certain invertebrates, including squid and lobsters (see Methods Box). These giant axons, which can be a millimeter in diameter, function in behavioral responses requiring great speed, such as in the backward thrust of a threatened lobster or crayfish.

A different means of speeding up nerve impulses has evolved in vertebrates. Recall that many axons in vertebrate nervous systems are myelinated, coated with insulating layers of membranes deposited by supporting cells such as Schwann cells (see Figures 44.2 and 44.4). The myelin sheath has small gaps called the **nodes of Ranvier** between the Schwann cells that are serially arranged along the neuron process. The voltage-sensitive ion channels that function in the action potential are concentrated in the node regions of the axon. Also, extracellular fluid is in contact with the neuron membrane only at these nodes, and thus the local currents of ions can complete their circuits between the outside and inside of the cell only in these regions. For these reasons, the action potential does not propagate in a continuum over the length of the axon, but rather "jumps" from node to node, skipping the insulated regions of membrane between the nodes (Figure 44.10). This mechanism, called **saltatory conduction** (from the Latin, *saltare*, "to leap"), results in faster transmission of the nerve impulse.

We have seen how stimulating the dendrite of a neuron can produce an action potential that is propagated along the axon to the very tip of the neuron. But how is the impulse transmitted from one neuron to the next in a neural pathway? Our next step toward understanding nervous communication is to examine the synapse, the tiny space between neurons.

THE SYNAPSE: TRANSMISSION BETWEEN CELLS

The synapse is a unique junction that controls communication between neurons. Synapses are also found between sensory receptors and sensory neurons, and between motor neurons and the muscle cells they control. Here, we shall focus on synapses between neurons, which usually conduct signals from the termini of axons to dendrites or cell bodies of the next cells in a neural pathway. The transmitting cell is called the **presynaptic cell** and the receiving cell is called the

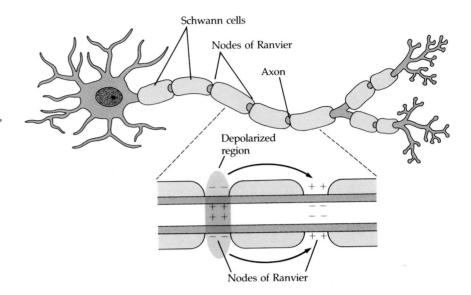

Figure 44.10
Saltatory conduction. Schwann cells form the myelin sheath that envelops the axons of many vertebrate neurons in the peripheral nervous system. The nodes of Ranvier are the bare patches of axonal membrane that lie between the myelinated sections. In saltatory conduction, the wave of depolarization jumps from node to node, speeding transmission of the impulse.

Schwann cells

Nodes of Ranvier

Axon

Depolarized region

Nodes of Ranvier

postsynaptic cell. Synapses are of two types: electrical synapses and chemical synapses.

Electrical Synapses

An **electrical synapse** allows action potentials to spread directly from the presynaptic to postsynaptic cell. The cells are connected by gap junctions (see Chapter 7), intercellular channels that allow the local ion currents of an action potential to flow between neurons. The giant neuron processes of lobsters and other crustaceans are actually composed of several short, thick neurons, connected end to end and coupled by electrical synapses. These make it possible for impulses to travel from neuron to neuron along the neural pathway without delay and with no loss of signal strength. Electrical synapses in the central nervous systems of vertebrates synchronize the activity of neurons responsible for some rapid, stereotyped movements. For example, electrical synapses in the brains of some fishes function in the tail-flapping reflex that helps the animal to escape from a predator. However, chemical synapses are much more common than electrical synapses in vertebrates and most invertebrates.

Chemical Synapses

At a **chemical synapse,** a narrow gap, or synaptic cleft, separates the presynaptic cell from the postsynaptic cell. Because of the cleft, the cells are not electrically coupled, and an action potential occurring in the presynaptic cell cannot be transmitted directly to the membrane of the postsynaptic cell. Instead, a series of events at the synapse converts the electrical signal of the action potential arriving at the terminus of an axon into a chemical signal that travels across the synapse and then converts it back into an electrical signal in the postsynaptic cell.

The key to understanding the function of a chemical synapse is to examine its structure. On the presynaptic side of the cleft is a synaptic knob, the swelling at the terminus of one of the axon's telodendria (Figure 44.11). Within the cytoplasm of the synaptic knob are numerous sacs called **synaptic vesicles.** Each of these sacs contains thousands of molecules of **neurotransmitter,** the substance that is released as an intercellular messenger into the synaptic cleft. Many different neurotransmitters have been discovered in the nervous systems of animals. Most neurons secrete only one kind of neurotransmitter. However, a single neuron may *receive* chemical signals from a variety of neurons that secrete different neurotransmitters from their synaptic knobs.

A neuron dispatches its neurotransmitter into the synapse when the action potential being propagated along the axon arrives at the synaptic knob and depolarizes the **presynaptic membrane,** the surface of the synaptic knob that fronts on the cleft. Calcium ions play a central role in this conversion of the electrical impulse into a chemical signal. Depolarization of the presynaptic membrane causes Ca^{2+} to rush into the cell through voltage-sensitive channels. The sudden rise in the cytoplasmic concentration of Ca^{2+} stimulates the synaptic vesicles to fuse with the presynaptic membrane and spill the neurotransmitter into the synaptic cleft by exocytosis (see Chapter 8). Thousands of vesicles may respond in unison to a single action potential. The neurotransmitter diffuses the short distance from the presynaptic membrane to the **postsynaptic membrane,** the plasma membrane of the cell body or dendrite on the other side of the synapse.

The postsynaptic membrane is specialized to receive

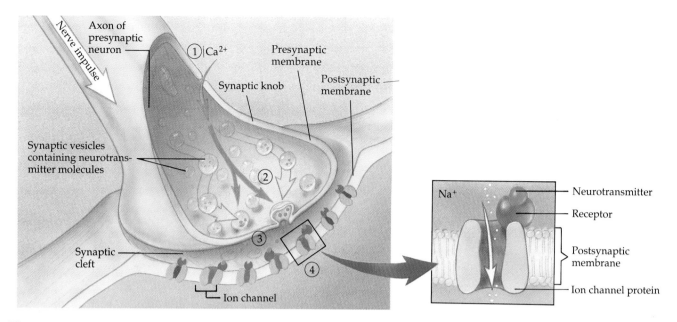

Figure 44.11

A chemical synapse. ① When an action potential (arrow) depolarizes a nerve terminal, it triggers an influx of Ca^{2+} that ② causes synaptic vesicles to fuse with the presynaptic membrane. ③ When the synaptic vesicles fuse with the membrane, they release neurotransmitter molecules into the synaptic cleft ④. These molecules diffuse across the cleft and bind to receptors on the postynaptic membrane. The receptors control selective ion channels; binding of a neurotransmitter to its specific receptors opens the ion channels. The resulting ion flux changes the voltage of the postsynaptic membrane. This either moves the membrane potential closer to the threshold required for an action potential (an excitatory synapse), or hyperpolarizes the membrane (an inhibitory synapse). In either case, the neurotransmitter molecules are quickly degraded by enzymes or are taken up by another neuron, terminating the synaptic response.

the chemical message. Projecting from the extracellular surface of the membrane are proteins that function as specific receptors for neurotransmitters. The receptors are associated with selective ion channels that open and close, controlling fluxes of ions across the postsynaptic membrane. A receptor is keyed to a particular type of neurotransmitter, and when it binds to this chemical, the ion gate opens, allowing a particular ion, such as Na^+, K^+, or Cl^-, to cross the membrane. Thus, the ion channels of the postsynaptic membrane are chemically sensitive, in contrast to the voltage-sensitive gates responsible for the action potential. The ion movements resulting from the binding of the neurotransmitter to its receptors alter the membrane potential of the postsynaptic cell. Depending on the type of receptors and the ion gates they control, neurotransmitters binding to postsynaptic membrane may either excite the membrane by bringing its voltage closer to the threshold potential or inhibit the postsynaptic cell by hyperpolarizing its membrane. In either case, the chemical signal is quickly muted by enzymes that break the neurotransmitter down into smaller chemical components that can be recycled back to the presynaptic cell. For example, the transmitter known as acetylcholine is rapidly degraded by cholinesterase, an enzyme present in the synaptic cleft and also on the postsynaptic membrane.

Notice that one important function of the synapse is that it allows nerve impulses to be transmitted in only a single direction over a neural pathway. Synaptic vesicles are found only in the knobs at the tips of axons, and thus only the presynaptic membrane can discharge neurotransmitters. And receptors are restricted to the postsynaptic membrane, ensuring that only this membrane can receive a chemical signal from another neuron. Another important function of the synapse is to integrate information consisting of excitatory and inhibitory signals.

Summation: Nervous Integration at the Cellular Level

A single neuron may receive information from numerous neighboring neurons via thousands of synapses, some of them excitatory and some of them inhibitory (Figure 44.12). Whether or not a neuron "fires" off an action potential at any particular instant depends on its ability to integrate these multiple positive and negative inputs.

Excitatory and inhibitory synapses have opposite effects on the membrane potential of the postsynaptic cell. At an excitatory synapse, the neurotransmitter receptors control a type of gated channel that allows

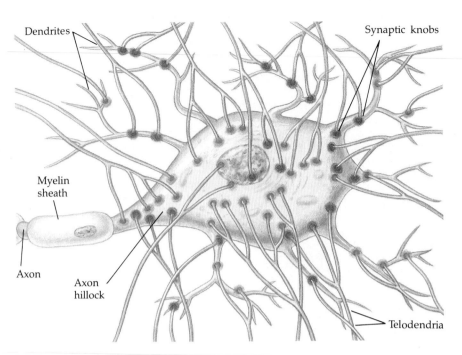

Figure 44.12
Integration of multiple synaptic inputs.
(a) Each neuron, especially in the central nervous system, is on the receiving end of thousands of synapses, some of them excitatory (green) and some of them inhibitory (red). At any instant, an action potential is generated at the axon hillock if the combined effect of ion currents induced by excitatory and inhibitory synapses depolarizes the membrane in the hillock region to the threshold potential. This mechanism of integrating multiple positive and negative inputs by adding their individual effects is called summation. (b) A scanning electron micrograph reveals numerous synaptic knobs that communicate with a single postsynaptic cell.

Dendrites

Synaptic knobs

Myelin sheath

Axon

Axon hillock

Telodendria

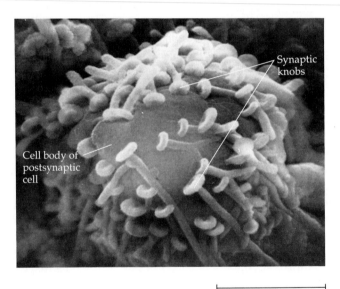

Synaptic knobs

Cell body of postsynaptic cell

5 μm

Na$^+$ to enter the cell and K$^+$ to leave the cell. Because the driving force is greater for Na$^+$ than for K$^+$ (remember, the voltage and concentration gradient both drive Na$^+$ into the cell), the effect of opening these gates is a net flow of positive charge into the cell. This depolarizes the cell, moving the membrane potential closer to the threshold voltage and making it more likely that the postsynaptic cell will generate an action potential. Therefore, the electrical change caused by binding of the transmitter to the receptor is called an **excitatory postsynaptic potential,** or **EPSP.**

At an inhibitory synapse, binding of the neurotransmitter molecules to the postsynaptic membrane hyperpolarizes the membrane by opening ion gates that make the membrane more permeable to K$^+$, which rushes out of the cell; or to Cl$^-$, which enters the cell

due to a large concentration gradient; or to both of these ions. These ion fluxes push the membrane potential to a voltage even more negative than the resting potential, making it more difficult for an action potential to be generated. Therefore, the voltage change associated with chemical signaling at an inhibitory synapse is called an **inhibitory postsynaptic potential,** or **IPSP.** Whether a particular neurotransmitter results in an EPSP or an IPSP depends on the type of receptors and gated ion channels on the postsynaptic membrane responding to that neurotransmitter.

Both EPSPs and IPSPs are graded potentials that vary in magnitude with the number of neurotransmitter molecules binding to receptors on the postsynaptic membrane. The change in voltage, either a local depolarization or hyperpolarization, lasts only a few

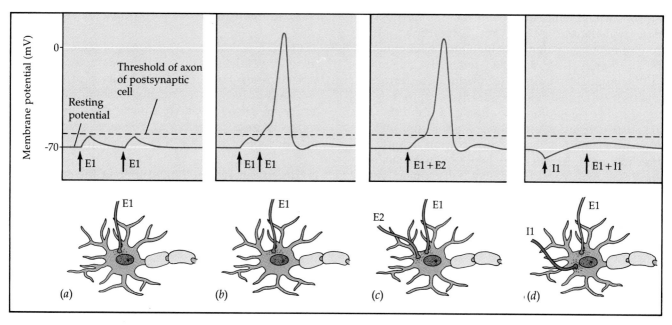

Figure 44.13

Summation of postsynaptic potentials.
(a) Generally, a single EPSP cannot depolarize the postsynaptic membrane all the way to the threshold level, and thus no action potential is generated at the axon hillock. Repeated EPSPs do not summate if they do not overlap in time. In the graph that traces changes in the membrane potential at the axon hillock, the arrow indicates the time of the signal at one excitatory synapse, labeled E1. (b) Temporal summation occurs when a second release of neurotransmitter at a synapse affects the postsynaptic membrane when it is still partially depolarized from a slightly earlier stimulation. Due to temporal summation, EPSPs that are individually subthreshold can generate an action potential if they overlap in time so that they reinforce one another. (c) Spatial summation occurs when two or more presynaptic cells release their neurotransmitters at the same time, causing a cumulative voltage change greater than the individual EPSPs. Here, two excitatory synapses, E1 and E2, team up to depolarize the postsynaptic cell to threshold voltage. (d) EPSPs and IPSPs also summate, but, of course, they move the membrane potential in opposite directions. In this case, an inhibitory synapse, I1, subtracts from the depolarization due to an excitatory synapse, E1.

milliseconds because the neurotransmitters are inactivated by enzymes soon after their release into the synapse. Also, the electrical impact on the postsynaptic cell decreases with distance away from the synapse. For the postsynaptic cell to fire, the local ion currents due to EPSPs must be strong enough to depolarize the membrane in the region of the axon hillock to the threshold potential, usually about -50 mV. The axon hillock, remember, is the zone of "spike" initiation, the region where voltage-sensitive Na^+ gates open to generate an action potential when some stimulus has depolarized the membrane to the threshold.

A single EPSP at one synapse, even one close to the axon hillock, is not usually strong enough to trigger an action potential. However, several synaptic knobs acting simultaneously on the same postsynaptic cell, or a smaller number of synaptic knobs discharging neurotransmitters repeatedly in rapid-fire succession, can have an accumulative impact on the membrane potential at the axon hillock. This additive effect of postsynaptic potentials is called **summation.**

There are two types of summation: temporal summation and spatial summation. In **temporal summation,** chemical transmissions from one or more synaptic knobs occur so close together in time that each postsynaptic potential affects the membrane before the voltage has returned to the resting potential after the previous stimulation (Figure 44.13b). In **spatial summation,** several different synaptic knobs, usually belonging to different presynaptic neurons, stimulate a postsynaptic cell at the same time and have an additive effect on the membrane potential (Figure 44.13c).

By reinforcing one another through temporal or spatial summation, the ion currents associated with several EPSPs can depolarize the axon hillock to the threshold voltage, causing the neuron to fire (generate an action potential). IPSPs also summate, teaming up to hyperpolarize the membrane to a voltage more negative than any single release of neurotransmitter at an inhibitory synapse can achieve. Furthermore, EPSPs summate with IPSPs, each countering the electrical effects of the other.

The axon hillock is the neuron's integrating center; its membrane potential at any instant is an average of the summated depolarization due to all EPSPs and the summated hyperpolarization due to all IPSPs. (Of course, synapses close to the axon hillock, compared to more remote synapses, have a disproportionate impact on the membrane potential at the integrating center.) Whenever the EPSPs overpower IPSPs enough for the membrane potential at the axon hillock to reach the threshold voltage, an action potential is generated and the impulse is transmitted along the axon to the next synapse. A few milliseconds later, after the refractory period, the neuron may fire again if the sum of all synaptic inputs at that moment is still sufficient to depolarize the membrane of the axon hillock to threshold level. On the other hand, by that time, the sum of all EPSPs and IPSPs may put the membrane potential at the axon hillock at a voltage more negative than the threshold, or even hyperpolarize the membrane to a potential more negative than the resting potential, desensitizing the neuron for the moment.

Action potentials, remember, are all-or-none events. But now we see that the occurrence of these nerve impulses depends on the ability of the neuron to integrate quantitative information in the form of multiple excitatory and inhibitory inputs, each involving the specific binding of a neurotransmitter to a receptor on the postsynaptic membrane.

Neurotransmitters and Receptors

Neuroscientists are only beginning to understand the chemistry of the nervous system. About ten different molecules are known to function as neurotransmitters, and dozens more are good candidates for addition to the growing list. For a particular compound to be recognized for certain as a neurotransmitter at a specific type of synapse, it must meet three criteria:

1. The presynaptic cell must contain the compound in synaptic vesicles and discharge the substance when the cell is appropriately stimulated; and the chemical must then affect the membrane potential of the postsynaptic membrane.

2. The compound should be able to cause an EPSP or IPSP when experimentally injected into the synapse with a micropipette.

3. The substance must be removed rapidly from the synapse, either by enzymatic degradation or uptake by a cell, allowing the postsynaptic membrane to return to resting potential.

(a) ACETYLCHOLINE

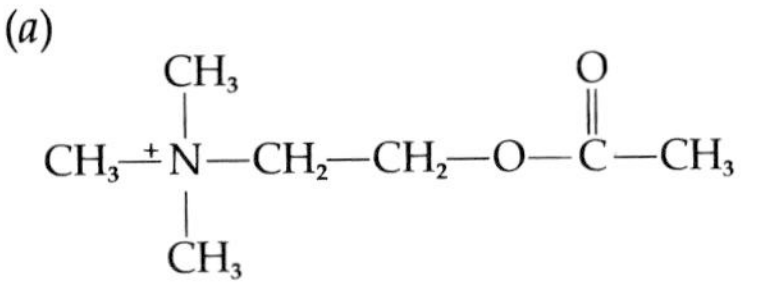

(b) EPINEPHRINE (ADRENALINE)

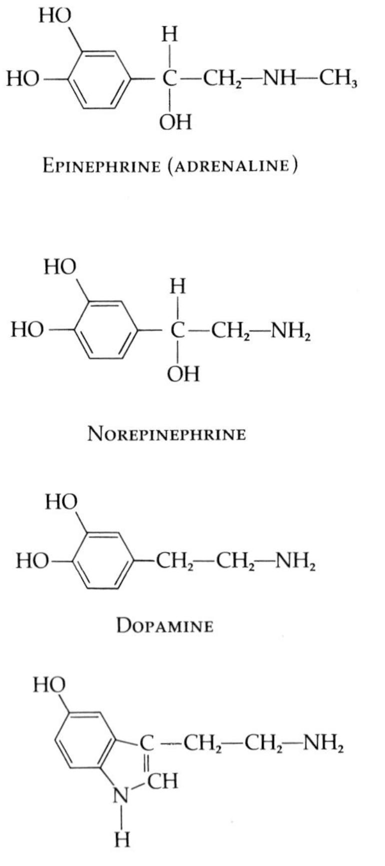

NOREPINEPHRINE

DOPAMINE

SEROTONIN

(c) GABA

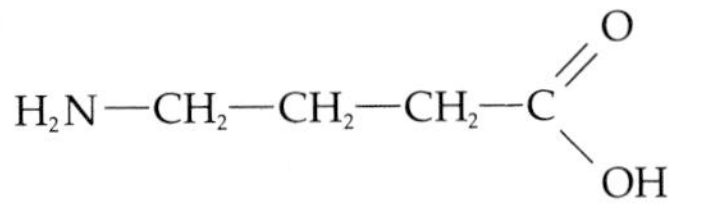

(d) Met—phe—gly—gly—tyr

MET-ENKEPHALIN

Figure 44.14
Survey of neurotransmitters. (a) Acetylcholine is the transmitter of the vertebrate neuromuscular junction, the synapse between a motor neuron and a skeletal muscle cell. It also functions in a branch of the nervous system called the autonomic system, and is an important transmitter in invertebrates. (b) The biogenic amines are derived from amino acids. The catecholamines—epinephrine, norepinephrine, and dopamine—are biogenic amines synthesized from the amino acid tyrosine; they share a common ring component called the catechol ring (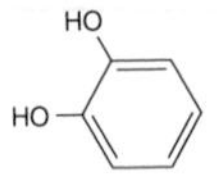). Serotonin is a biogenic amine derived from the amino acid tryptophan. (c) GABA is an example of an amino acid that functions as a neurotransmitter. It is the most common inhibitory transmitter in the brain. (d) The neuropeptides are short chains of amino acids. An example is met-enkephalin, a pentapeptide that functions as a natural pain reliever in the brain (the letters in this formula are the abbreviations for the amino acids that make up this neuropeptide).

As analytical methods in neurochemistry continue to improve, it is likely that many additional compounds will meet these criteria for neurotransmitters.

Acetylcholine is one of the most common neurotransmitters in both invertebrates and vertebrates (Figure 44.14a). At the vertebrate neuromuscular junction, the synapse between a motor neuron and a skeletal muscle cell, acetylcholine triggers contraction of the muscle cell, which is the postsynaptic cell of this specific synapse. In other cases, acetylcholine is inhibitory. For example, this transmitter slows down the hearts of vertebrates and mollusks. This versatility of a single transmitter, remember, depends on the diverse types of receptors found on different postsynaptic cells.

The **biogenic amines** are neurotransmitters derived from amino acids (Figure 44.14b). One family of biogenic amines, known more specifically as *catecholamines*, is produced from the amino acid tyrosine. This group includes **epinephrine, norepinephrine,** and **dopamine.** Another biogenic amine, **serotonin,** is synthesized from the amino acid tryptophan. The biogenic amines most commonly function as transmitters within the central nervous system. However, norepinephrine also functions in a branch of the peripheral nervous system called the autonomic system, which will be discussed shortly. Dopamine and serotonin are widespread in the brain and affect sleep, mood, attention, and learning. Imbalances of these transmitters are associated with mental illness; for example, an excess production of dopamine is one factor that has been tied to schizophrenia. Some psychoactive drugs, including LSD and mescaline, apparently induce hallucinations by binding to receptors for certain biogenic amines.

Three amino acids, **glycine, glutamate,** and **gamma aminobutyric acid (GABA)** are known to function as neurotransmitters in the central nervous system (Figure 44.14c). GABA, believed to be the transmitter at most inhibitory synapses in the brain, produces IPSPs by increasing the chloride permeability of the postsynaptic membrane. The brain has hundreds of times more GABA than it does any other neurotransmitter.

Several **neuropeptides,** relatively short chains of amino acids, probably serve as neurotransmitters (Figure 44.14d). The **endorphins** and **enkephalins** are two classes of neuropeptides that function as natural analgesics, decreasing perceptions of pain by the central nervous sysytem. They were first discovered in the 1970s by neurochemists studying the mechanism of opium addiction. These investigators found specific receptors for morphine and heroin on neurons in the brain. It seemed odd that humans would have receptors keyed to chemicals from a plant (the opium poppy). Further research found that, in fact, the drugs bind to these receptors by mimicking endorphins and enkephalins, the natural pain killers produced in the brain

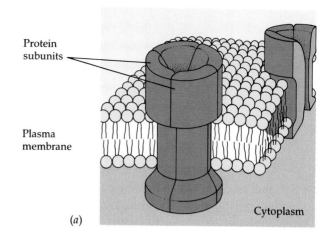

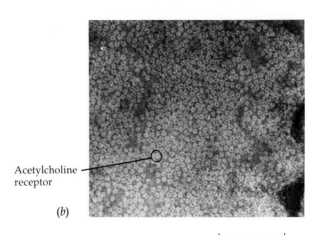

(b)

Figure 44.15
The acetylcholine receptor. (a) Consisting of four polypeptide subunits surrounding a central pit, the receptor spans the plasma membrane of the postsynaptic cell. **(b)** The donutlike shape of the receptor is evident in this transmission electron micrograph of a postsynaptic membrane. Ions move through the receptor when acetylcholine binds, but the control mechanism that opens and closes the ion channel is not yet understood.

during times of physical or emotional stress, such as during the labor of a pregnant woman. In addition to relieving pain, endorphins and enkephalins also decrease urine output (by affecting ADH secretion—see Chapter 41), depress respiration, produce euphoria, and have other emotional effects through specific pathways in the brain. An endorphin is also released from the anterior pituitary as a hormone that affects specific regions of the brain. Once again, we see the overlap between nervous and endocrine control.

A neurotransmitter affects the postsynaptic cell by one of two general mechanisms. Acetylcholine and the amino acid transmitters bind to receptors and alter the permeability of the postsynaptic membrane to specific ions, either depolarizing (EPSP) or hyperpolarizing (IPSP) the membrane (Figure 44.15). In contrast, the biogenic amines and the neuropeptides usually have a longer-lasting impact because they affect metabolism within the postsynaptic cell. Binding of these transmitters to their specific receptors on the postsynaptic membrane activates an enzyme that produces a second messenger within the cell, often cyclic AMP, the same molecule that functions as a second messenger in many responses of target cells to hormones (see Chapter 41).

It must be emphasized again that the action of a neurotransmitter depends on how the specific receptor on the postsynaptic membrane behaves when bound to the transmitter. Considerable research has focused on the acetylcholine receptor that functions at excitatory synapses, where acetylcholine depolarizes the postsynaptic membrane. The receptor is a protein consisting of four polypeptide subunits that span the plasma membrane. The protein is donutlike in shape, with a central pit that probably leads to a channel through which Na^+ and K^+ can pass (see Figure 44.15). The mechanism by which the binding of acetylcholine to the receptor protein opens the ion channel is not yet known.

Neural Circuits and Clusters

We have now seen how neurons, the individual units of the nervous system, function. But in a nervous system, the activity of millions or more of these units must be coordinated. Some organizational principles are common to all nervous systems.

Neurons are arranged in circuits, groups of neurons that feed into and away from one another to carry information along specific pathways. The three major patterns—**convergent circuits, divergent circuits,** and **reverberating circuits**—are described in Figure 44.16.

Nerve cell bodies are not uniformly distributed within the nervous system, but are arranged in functional groups called **ganglia** (singular, ganglion). When a ganglion occurs within the brain, it is often referred to as a **nucleus** (not to be confused with the nucleus of an individual cell). These clusters of cells allow parts of the nervous system to coordinate activities without involving the entire system. This centralization of nervous coordination is one of the key features of the evolution of nervous systems in both invertebrates and vertebrates.

INVERTEBRATE NERVOUS SYSTEMS

While there is remarkable uniformity in how nerve cells function throughout the animal kingdom, there is great diversity in how nervous systems are organized. The simplest type of nervous system occurs in *Hydra*, a cnidarian (Figure 44.17a). The cnidarian **nerve**

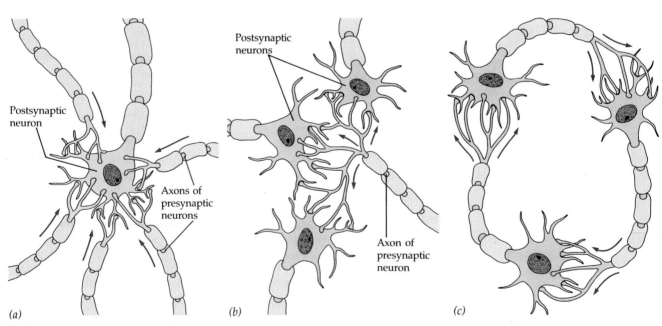

(a)   *(b)*   *(c)*

Figure 44.16
Neural circuits. Three types of circuits are typical in nervous systems. (**a**) In convergent circuits, information from several neurons (arrows) comes together along a pathway and feeds into one or a few cells. (**b**) In divergent circuits, the opposite occurs; information from one pathway spreads out in several directions. Whereas convergent circuits might bring information from several sources, such as vision, touch, and hearing, to identify an object in the environment, divergent circuits could take information from the optic nerve leading from the eye to the different parts of the brain involved in vision, coordinating movement, and memory. (**c**) Some neurons participate in circular reverberating circuits, in which the signal returns to its source. Reverberating circuits are thought to play a part in memory storage.

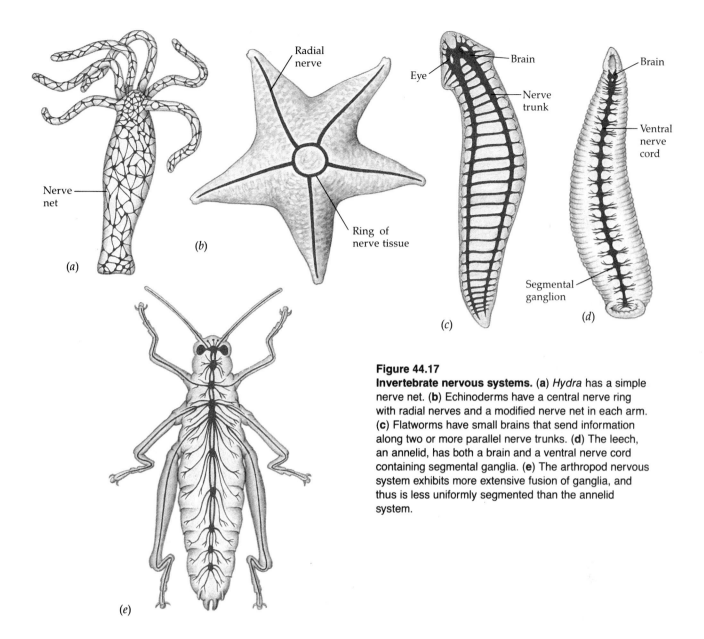

Radial
nerve

Ring of
nerve tissue

(b)

Nerve
net

(a)

Eye
Brain

Nerve
trunk

(c)

Brain

Ventral
nerve
cord

Segmental
ganglion

(d)

(e)

Figure 44.17
Invertebrate nervous systems. (a) *Hydra* has a simple nerve net. **(b)** Echinoderms have a central nerve ring with radial nerves and a modified nerve net in each arm. **(c)** Flatworms have small brains that send information along two or more parallel nerve trunks. **(d)** The leech, an annelid, has both a brain and a ventral nerve cord containing segmental ganglia. **(e)** The arthropod nervous system exhibits more extensive fusion of ganglia, and thus is less uniformly segmented than the annelid system.

net is a loosely organized system of nerves with no central control. Because most of the synapses in the nerve net are electrical, impulses are conducted in both directions, and stimulation at any point on the body of a hydra spreads from that site to cause movements of the entire body. Other cnidarians and ctenophores show the first signs of centralization. In jellyfish, clusters of nerve cells at the margin of the bell (associated with simple sensory structures) and pathways around the bell confer the ability to perform tasks such as swimming, which require more complex coordination of the entire body.

Such modified nerve nets are not restricted to Cnidaria and Ctenophora. The nervous systems of echinoderms are similar to those of jellyfish (Figure 44.17b). In the sea star, for example, radial nerves travel through each arm from a central nerve ring around the oral disc. Branches of the radial nerves form an interconnected network similar to the nerve net of *Hydra*. This

system coordinates movement regardless of which arm leads.

Cnidarians and echinoderms are radially symmetric and often sedentary. Bilateral animals with more active life styles tend to have sense organs and feeding structures localized at the anterior, or head, end. This concentration of sensory and feeding organs on the anterior end of a moving animal, the part of an animal most likely to make first contact with food or threatening stimuli, is an evolutionary trend called **cephalization.** Enlargement of anterior ganglia that receive sensory information and control feeding structures is a corresponding feature of cephalization that gave rise to the first brains.

Flatworms are the simplest animals to show two hallmarks of cephalization: a "brain" (cephalic ganglia) and one or more nerve trunks that serve as thoroughfares for information, constituting a central nervous system (Figure 44.17c). The relatively simple

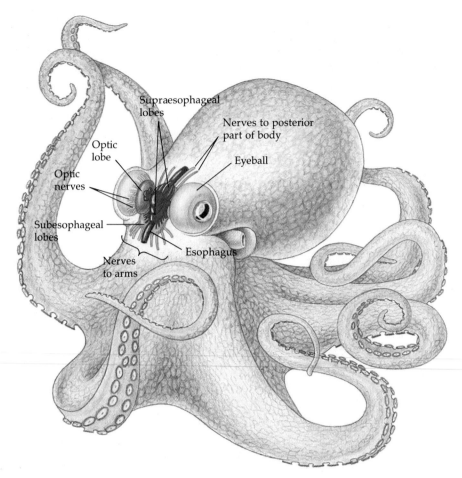

Figure 44.18
The octopus brain. An aggregate of ganglia surrounding the esophagus, the octopus brain has about a hundred million neurons. Proportional to body size, it is the largest brain of any invertebrate, and probably exceeds the integrative abilities of certain vertebrates, including lampreys and some fishes. The large optic lobes process visual information sensed by the well-developed eyes. Dorsal regions of the brain are involved in learning and memory, important in the animal's daily interactions with its environment.

Labels on figure:
Supraesophageal lobes
Nerves to posterior part of body
Optic lobe
Eyeball
Optic nerves
Subesophageal lobes
Esophagus
Nerves to arms

flatworm brain contains many large interneurons that coordinate most nervous functions, relaying signals from the sensory structures of the head to the muscles of the body. From two to several nerve trunks travel posteriorly from the brain, usually in a ladderlike system with transverse nerves connecting the main trunks. This simple nervous system controls such responses as the flatworm's ability to sense bright light and retreat to a darker place, such as the safety of a space beneath a rock.

Other invertebrates show increasing centralization of the nervous system. In contrast to the diffuse, ladderlike nervous system of flatworms, a well-defined ventral nerve cord with a prominent brain at the anterior end occurs in annelids and arthropods (Figure 44.17d and e). The nerve cords of these segmented animals often contain ganglia in each segment to coordinate the actions of that segment. The brains are much larger and more complex than those of flatworms. Some of the segmental ganglia, however, have only a few large cells.

The mollusks provide good examples of how nervous system complexity correlates with habitat and natural history as well as with phylogeny. Sessile or slow-moving mollusks such as clams have little or no cephalization and only simple sense organs. Their central nervous system is a chain of ganglia circling the body. In contrast, the cephalopods have the most sophisticated nervous systems of any invertebrates, rivaling even those of some vertebrates. The large brain of the octopus, accompanied by large, image-forming eyes and rapid conduction along giant axons, correlates well with the active predatory life of the animal (Figure 44.18). The octopus is capable of learning to discriminate between visual patterns and to perform specific tasks in laboratory experiments. Indeed, learning and memory are probably everyday functions of the octopus as it interacts with its natural environment.

THE VERTEBRATE NERVOUS SYSTEM

Because the vertebrate nervous system is so complex, it is convenient to divide it into components that differ in function (Figure 44.19). The primary division is between the central nervous system, which processes information, and the peripheral nervous system, which carries information to and from the central nervous system and the sensory, muscle, and gland cells.

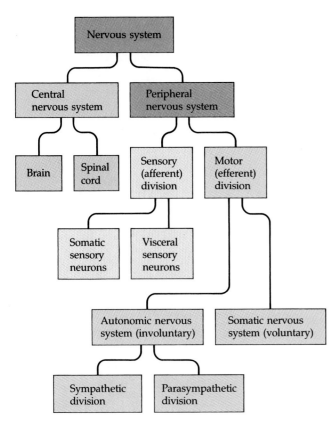

Figure 44.19
The divisions of the vertebrate nervous system.

Peripheral Nervous System

The peripheral nervous system really consists of two separate groups of cells. The **sensory,** or **afferent, nervous system** is made up of neurons that bring information *to* the central nervous system from sensory receptors, whereas the **motor,** or **efferent, nervous system** carries signals *from* the central nervous system to effector cells.

In the human peripheral nervous system, there are 12 pairs of cranial nerves that originate in the brain and innervate organs of the head and upper body, as well as 31 pairs of spinal nerves that innervate the entire body. Most of the cranial nerves and all the spinal nerves contain both sensory and motor neurons; a few of the cranial nerves are sensory only (the olfactory and optic nerves, for example).

Nervous systems have two basic functions: to control responses to the external environment and to coordinate the functions of internal organs (that is, to maintain homeostasis). The sensory nervous system contributes to both, bringing in stimuli from the external environment and monitoring the status of the internal environment. The motor nervous system has separate divisions associated with these two functions. The motor neurons of the **somatic nervous system** carry signals to skeletal muscles mainly in response

to external stimuli. The somatic nervous system is often considered "voluntary" because it is subject to conscious control, but a substantial proportion of skeletal muscle movement is actually determined by reflexes. A **reflex** is an automatic reaction to a stimulus, mediated by the spinal cord or lower brain. The **autonomic nervous system,** in contrast to the somatic system, regulates the internal environment by controlling the smooth and cardiac muscles and the organs of the gastrointestinal, cardiovascular, excretory, and endocrine systems. This control is generally involuntary.

The autonomic nervous system consists of two divisions that are anatomically, physiologically, and chemically distinguishable. These two divisions are the **sympathetic** and **parasympathetic nervous systems** (Figure 44.20). When sympathetic and parasympathetic nerves innervate the same organ, they often (but not always) have antagonistic (opposite) effects. In general, the parasympathetic division enhances activities that gain and conserve energy, such as stimulating digestion and slowing the heart. The sympathetic division increases energy expenditures and prepares an individual for action by accelerating the heart, increasing metabolic rate, and performing related actions.

The somatic and autonomic nervous systems often cooperate. In response to a drop in temperature, for example, the hypothalamus signals the autonomic nervous system to constrict surface blood vessels to reduce heat loss, at the same time signaling the somatic nervous system to cause shivering.

Central Nervous System

The central nervous system, or CNS, forms the bridge between the sensory and motor functions of the peripheral nervous system. The CNS consists of a bilaterally symmetric group of structures in two main organs. The **spinal cord,** which runs down the neck and back inside the vertebral column, or spine, receives information from the skin and muscles and sends out motor commands for movement. The **brain,** at the top of the spinal cord, contains centers for more complex integration of homeostasis, perception, movement, and (in humans, at least) intellect and emotions. The central nervous system is covered by a set of three protective layers of connective tissue, the **meninges.**

The axons and dendrites carrying signals in the CNS are located in well-defined bundles, or tracts, whose myelin sheaths give them a white appearance. In the brain, this **white matter** is located in the inner region, where there are pathways going to the cell bodies in the outer **gray matter.** The situation is reversed in the spinal cord, where the white matter is outside the gray matter.

All vertebrate nervous systems are hollow. Fluid-

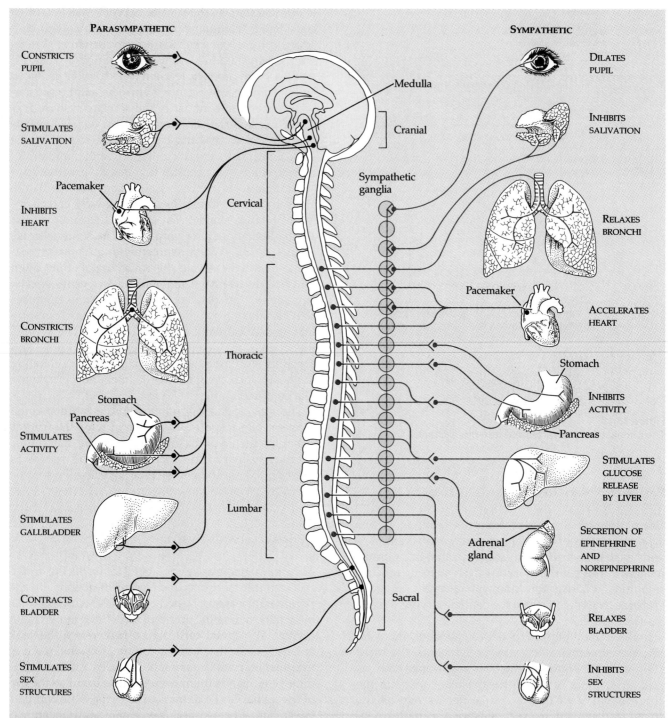

Figure 44.20
The autonomic nervous system. The symptathetic and parasympathetic systems differ anatomically and chemically. Anatomically, the divisions of the autonomic nervous system are distinguished by where their nerves originate. The nerves of the sympathetic nervous system emerge from the upper and central spinal cord (the thoracic and lumbar regions). Parasympathetic nerves originate at the top and bottom of the central nervous system, coming from some of the cranial nerves and from the sacral region of the spinal cord. All autonomic pathways contain two neurons in series, but the arrangements and positions of the junctions between the presynaptic and postsynaptic cells differ. Sympathetic nerves have short presynaptic axons that synapse with cell bodies of the second neurons in prominent ganglia near the spinal cord. The long axons of the postsynaptic cells leave these sympathetic ganglia and branch to travel to the target organs. The presynaptic axons of the parasympathetic nerves are much longer, and synapses with the second neurons usually occur at or near the target organ.

Both the sympathetic and parasympathetic divisions use acetylcholine at the synapse between neurons. The parasympathetic division also uses acetylcholine at the synapse with the effector organ. In the sympathetic division, however, the neurotransmitter at the effector organ is usually norepinephrine.

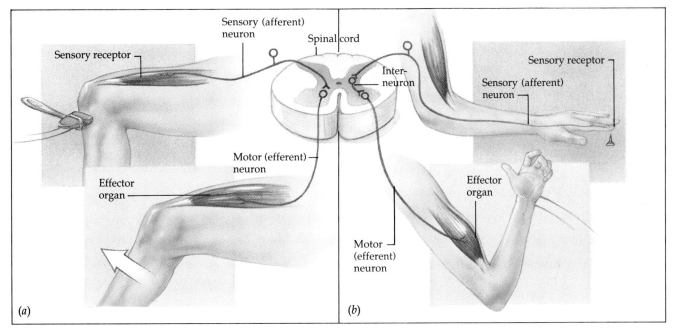

Figure 44.21
The spinal cord and spinal reflexes. The gray matter (butterfly-shaped region) in the center of the spinal cord contains the cell bodies of motor neurons and interneurons. The outer white matter consists of motor and sensory axons. The incoming sensory neurons have their cell bodies outside the spinal cord in dorsal ganglia. **(a)** The patellar knee-jerk reflex is a simple spinal reflex involving only two neurons. **(b)** Most reflexes involve at least one interneuron coordinating sensory input with motor output.

filled spaces called **ventricles** in the brain are continuous with the narrow **central canal** of the spinal cord. These spaces are filled with **cerebrospinal fluid,** which is formed in the brain by filtration of the blood. Among the most important functions of cerebrospinal fluid is absorption of shock to cushion the brain. Cerebrospinal fluid also carries out circulatory functions, bringing nutrients, hormones, and white blood cells to different parts of the brain. The cerebrospinal fluid normally circulates through the ventricles and central canal of the spinal cord and drains back into the veins.

The Spinal Cord   The spinal cord, shown in cross section in Figure 44.21, has two principal functions: It integrates simple responses to certain kinds of stimuli, and it carries information to and from the brain. Spinal integration usually takes the form of a reflex, an unconscious programmed response to a specific stimulus. The knee-jerk, or patellar, reflex is an example of the simplest type of reflex, involving only two neurons (Figure 44.21a). When a stretch receptor in the patellar tendon detects that the tendon has been stretched (as when struck by a physician's rubber hammer), a sensory neuron carries that information up the thigh to the spinal cord, where the sensory neuron synapses directly with a motor neuron. If the signal is strong enough, an action potential is generated in the motor neuron, causing contraction of the

quadriceps muscle and the forward knee jerk. Most reflexes are more complex, sometimes involving one or more interneurons between the sensory and motor neurons (Figure 44.21b). Also, branches in the pathway may carry signals to other segments of the spinal cord or to the brain to generate larger-scale or more complex responses.

Evolution of the Vertebrate Brain   The evolution of complex behaviors in vertebrates is largely due to increases in brain complexity. The vertebrate brain began as a set of three bulges at the anterior end of the spinal cord (Figure 44.22). These three regions, the **rhombencephalon** or **hindbrain,** the **mesencephalon** or **midbrain,** and the **prosencephalon** or **forebrain,** are present in all vertebrates. In more complex brains, they become further subdivided, providing additional capacity for integration of complex activities.

Three trends are evident in the evolution of the vertebrate brain (Figure 44.23). First, the relative size of the brain increases in certain evolutionary lineages. Brain size is a fairly constant function of body weight among fishes, amphibians, and reptiles, but increases dramatically relative to body size in birds and mammals. A rodent weighing 100 g would have a much larger brain than a 100 g lizard. But the brain of that lizard and the brain of a 100 g fish would be approximately the same size.

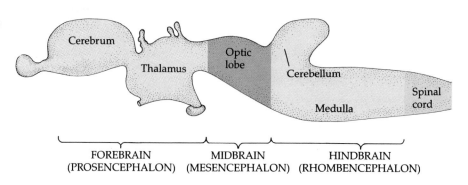

Figure 44.22
Regions of the generalized vertebrate brain.

Cerebrum

Thalamus

Optic lobe

Cerebellum

Medulla

Spinal cord

FOREBRAIN
(PROSENCEPHALON)

MIDBRAIN
(MESENCEPHALON)

HINDBRAIN
(RHOMBENCEPHALON)

A second evolutionary trend is increased compartmentalization of function. The three primitive divisions remain, but they become subdivided into areas assuming specific responsibilities. In the hindbrain, for example, the region known as the cerebellum becomes a prominent structure for coordinating movements. In the forebrain, one subdivision called the diencephalon contains the thalamus and hypothalamus and other groups of cells, while another subdivision called the telencephalon contains the cerebral cortex, the part of the brain most important in learning and memory. As these specific regions become more complex, the original divisions between the three bulges become blurred. The midbrain, hindbrain, and lower forebrain of mammals are not readily distinguishable in adults, although the three separate bulges are apparent in developing embryos.

The third trend in the evolution of the vertebrate brain is the increasing sophistication and complexity of the forebrain. As amphibians and reptiles made the transition from water to land, the vision and hearing functions of the midbrain and hindbrain became increasingly important, and natural selection favored enlargement of these regions. Beyond this, however, more complex behaviors parallel the growth of one important region of the forebrain—the **cerebrum.** Among mammals, in particular, more sophisticated behavior is associated with the relative size of the cerebrum and the presence of folds, or convolutions, which increase the surface area of the cerebrum. Because the cell bodies of the cerebrum are in the cortex, or outer layer, the surface area of the brain is more important in determining performance than is the volume. Although less than 5 mm thick, the human cerebral cortex occupies over 80% of the total brain mass. Marsupials such as the opossum have very little folding of the cerebral cortex, while cats and other placental mammals have substantially more. Primates and porpoises have dramatically larger and more complex cerebral cortices than any other vertebrates. In fact,

the cerebral cortex of the porpoise is second in surface area (relative to body size) only to that of the human.

The Human Brain   At 1.35 kg (about three pounds), the human brain is one of the largest organs in the body. Its soft, almost squishy texture belies the density of its cells and the complexity of its structure and function. As already noted, the brain develops from three primary bulges at the anterior end of the spinal cord. These, in turn, differentiate into several distinct structures with specific functions (Figure 44.24). The hindbrain and midbrain together make up the **brainstem,** and they form a cap on the spinal cord that extends to about the middle of the brain.

The hindbrain has three parts with functions in homeostasis, movement coordination, and signal conduction. The lowest parts of the brain, the **medulla oblongata** and the **pons** just above it, appear as swellings of the hindbrain at the top of the spinal cord. The medulla contains centers that control several visceral (autonomic, homeostatic) functions, including breathing, heart and blood vessel activity, swallowing, vomiting, and digestion. The pons also participates in some of these activities, having nuclei (ganglia) that regulate the breathing centers in the medulla, for example. All of the sensory and motor neurons passing to and from higher brain regions pass through the hindbrain, making conduction of information one of the most important functions of the medulla and pons. The hindbrain also helps coordinate large-scale body movements such as walking. Most of the motor neurons from the midbrain and forebrain that travel through the medulla cross from one side of the CNS to the other. As a result, the right side of the brain controls much of the movement of the left side of the body, and vice versa.

The primary function of the **cerebellum,** the third part of the hindbrain, is coordination of movement. This highly convoluted, semidome-shaped outgrowth on the dorsal surface of the hindbrain is tucked

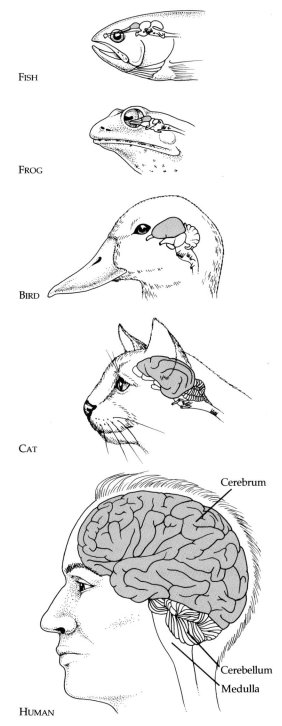

FISH

FROG

BIRD

CAT

Cerebrum

Cerebellum

Medulla

HUMAN

Figure 44.23
Evolution of the vertebrate brain. Three major trends are evident: an increase in overall brain size relative to body size, compartmentalization of function, and increased development of the forebrain (color), especially in mammals.

behind and partially beneath the cerebrum. The cerebellum receives sensory information about the position of the joints and the length of the muscles, as well as information from the auditory and visual systems. It also receives input from the motor pathways, telling it which actions are being commanded from the cerebrum. The cerebellum uses this information to provide unconscious coordination of movements and balance. If one part of the body is moved, the cerebellum will coordinate other parts to ensure smooth action and maintenance of equilibrium. Hand-eye coordination is one example of cerebellar function. If the cerebellum is damaged, the eyes can follow a moving object, but will not stop at the same place as the object.

The upper portion of the brainstem, or midbrain, contains centers for the receipt and integration of several types of sensory information. It also serves as a projection center, sending coded sensory information along neurons to specific regions of the forebrain. The most prominent areas of the midbrain are the **superior** and **inferior colliculi,** bumps on the dorsal surface of the brainstem that are part of the visual and auditory systems. All fibers involved in hearing either terminate in or pass through the inferior colliculi. The superior colliculi are important visual centers. In nonmammalian vertebrates, they may be the only visual centers, taking the form of prominent optic lobes. In mammals, vision is integrated in the forebrain, leaving the superior colliculi to coordinate visual reflexes and limited perceptual functions. The major nuclei in the midbrain are part of the **reticular formation,** a structure that regulates states of arousal (discussed shortly).

The most sophisticated neural processing occurs in the forebrain. The intricate networks of integrating centers and sensory and motor pathways allow pattern and image formation, as well as associative functions such as memory, learning, and emotions. Of the two major divisions of the forebrain, the lower **diencephalon** contains two integrating centers, the thalamus and the hypothalamus. The upper **telencephalon** contains the cerebrum, the most complex integrating center in the CNS.

Within the diencephalon, the **thalamus** is a major projection area for the cerebral cortex; most of the neural input to the cerebral cortex is relayed by neurons with cell bodies in the thalamus. The thalamus contains many different nuclei that project information to specific anatomical areas of the cerebrum. The thalamus also contains nuclei of the reticular formation, the network that helps to control which information gets sent to the cerebral cortex.

The **hypothalamus** has been called the "Casablanca of the nervous system" because so many mysterious messages are sorted out there. Although weighing only a few grams, this area of the brain is one of the most important sites for regulation of homeostasis.

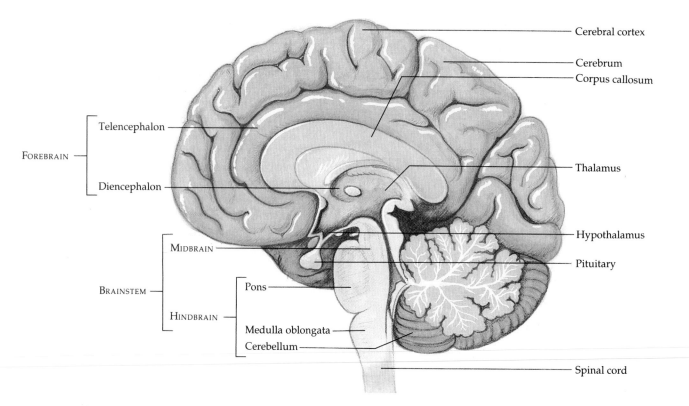

Cerebral cortex

Cerebrum

Corpus callosum

Telencephalon

FOREBRAIN

Diencephalon

Thalamus

MIDBRAIN

Hypothalamus

BRAINSTEM

Pituitary

Pons

HINDBRAIN

Medulla oblongata

Cerebellum

Spinal cord

Figure 44.24
The human brain. This midsagittal section shows the major structures of the brain. The hindbrain controls homeostasis and coordinates movement, while the midbrain functions as a relay center for sensory information. The forebrain is highly developed in mammals, with the diencephalon a relay and homeostasis center and the telencephalon the locus for information processing.

We have already seen that the hypothalamus is the source of two sets of hormones, the posterior pituitary hormones and the releasing factors for the anterior pituitary (see Chapter 41). It also contains the body's thermostat, centers for regulating hunger and eating, thirst and drinking, and many other basic survival functions. This region also plays a role in sexual response and mating behaviors, the "fight-or-flight" response, and pleasure.

The hypothalamic pleasure centers have been given that name because of the responses seen when they are stimulated in experimental animals, although we cannot really know whether a rat experiences what humans interpret as pleasurable sensations. Nevertheless, when electrodes are implanted in the pleasure centers, the animal will press a bar continually to receive electrical shocks in this region, to the exclusion of eating, drinking, and mating.

Below the cerebral cortex (the major structure of the telencephalon) lies a cluster of nuclei referred to as the basal ganglia. The basal ganglia are important centers for motor coordination, acting as switches for impulses from other motor systems. If the basal ganglia are damaged, a person may become passive and immobile because the ganglia no longer allow motor impulses to be sent to the muscles. Degeneration of cells entering the basal ganglia occurs in Parkinson's disease.

The **cerebral cortex** is the largest and most complex part of the human brain, and the part that has changed the most during vertebrate evolution. The highly folded human cortex has a surface area of about 0.5 m^2 and is divided into five lobes, four of which are visible from the surface (Figure 44.25). Like the rest of the brain, the cerebral cortex is bilaterally symmetric, and the two hemispheres are connected by a thick band of fibers known as the **corpus callosum.** The cortex contains both sensory and motor areas involved in direct processing of information, as well as association areas that integrate information from several sources. Some of the major functional areas of the cortex are indicated in Figure 44.25.

The primary sensory and motor areas are bilateral. The primary sensory area receives impulses generated by stimulation of tactile, pressure, and pain receptors scattered throughout the body. The primary motor area of the cortex initiates impulses to the various skeletal muscles. The sensory and motor cortex

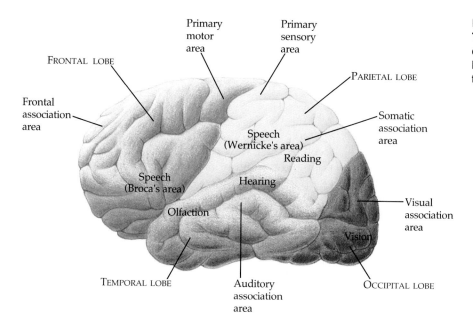

Figure 44.25
The cerebral cortex. The surface of the cerebral cortex can be divided into the four lobes visible here, with specialized functions localized in each one.

is a mosaic of regions corresponding to different parts of the body. The proportion of the sensory or motor cortex devoted to a particular part of the body is correlated with the importance of sensory or motor information for that part of the body. For example, more brain surface is committed to sensory and motor communication with the hands than with the entire torso of the body. Impulses transmitted from receptors to specific areas of the primary sensory region of the cerebral cortex enable us to associate pain, touch, pressure, heat, or cold with specific parts of the body subjected to those stimuli. However, the so-called *special senses*—vision, hearing, smell, and taste—are integrated by other regions of the cortex (see Figure 44.25).

Integration and Higher Brain Functions

Integration of nerve impulses occurs at all levels in the human nervous system. The simplest kinds of integration are the spinal reflexes illustrated in Figure 44.21; the most complex integration enables the cerebral cortex to generate works of art or scientific discovery. Four aspects of brain activity that are particularly interesting are arousal and sleep, emotions, lateralization in the brain (differential functions of the left and right hemispheres of the brain), and memory.

Arousal and Sleep   As anyone who has sat through a lecture on a warm spring day knows, attentiveness and mental alertness vary from moment to moment. Arousal is a state of awareness of the external world.

The counterpart of arousal is sleep, when an individual continues to receive external stimuli but is not conscious of them. The mechanisms of arousal and sleep are fairly well worked out, but the question of why we sleep remains a compelling problem. All birds and mammals sleep and show a characteristic sleep–wakefulness cycle, which may be mediated by the hypothalamus.

Sleep and wakefulness produce different patterns in the electrical activity of the brain, which can be recorded in an **electroencephalogram,** or **EEG** (Figure 44.26). As a general rule, the less mental activity taking place, the more synchronous the brain waves of the EEG. When a normal person is lying quietly with closed eyes, slow, synchronous *alpha waves* predominate. When the eyes are opened or the person solves a complex problem, faster *beta waves* take over, indicating desynchronization of the parts of the brain. Sleep produces a third type of pattern, *delta waves*, which are quite slow and highly synchronized.

Sleep, however, is a dynamic process, and the EEG of a sleeping person is far from constant. Sleep can be divided into two patterns. One is a pattern of slow, deep delta waves. At other times, a desynchronized EEG reminiscent of wakefulness occurs; during these periods, the eyes move actively across the visual field behind the lids, and this is therefore called REM (rapid eye movement) sleep. Most dreaming occurs during REM sleep. Like sleep, dreaming has been ascribed magical or prophetic importance, but its true function remains unknown. The two patterns of sleep typically alternate during a night, with a 90-minute cycle containing a 20- to 30-minute bout of REM sleep.

Sleep and arousal are controlled by several centers

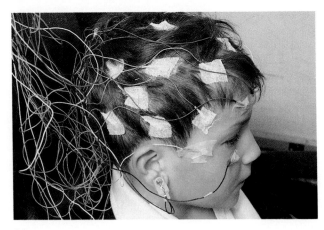

(a)

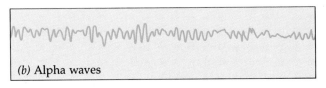

(b) Alpha waves

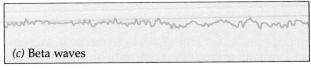

(c) Beta waves

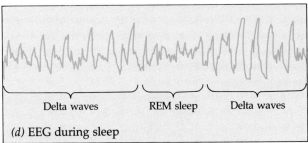

Delta waves   REM sleep   Delta waves

(d) EEG during sleep

Figure 44.26
The electroencephalogram. (a) Electrical contacts placed on the skin of the head detect electrical activity of the brain (brain waves). **(b)** When a person is awake but quiet, the electroencephalogram (EEG) is dominated by slow alpha waves. **(c)** During mental activity, the EEG consists of rapid, irregular beta waves. **(d)** A typical sleep cycle has periods of prominent delta waves, interspersed with bouts of beta waves during REM (rapid eye movement) sleep and dreaming.

in the cerebrum and the brainstem. The reticular formation is a vital link in determining states of arousal and consciousness. This group of over 90 separate nuclei (brain ganglia) extends from the medulla to the thalamus, through which almost all neuron processes reaching the cerebral cortex must pass. The reticular formation is essentially a sensory filter that selects which information reaches the cortex. The more input the cortex receives, the more alert and aware a person is. But arousal is not just a generalized phenomenon;

specific sets of information can be ignored ("turning out" stimuli) while the brain is actively processing other input. Also, specific centers regulate sleep and wakefulness. The pons and medulla contain nuclei that cause sleep when stimulated, but the midbrain has a center that causes arousal. It has been suggested that serotonin is the neurotransmitter of the sleep-producing centers. Drinking milk before bedtime might induce sleep because milk contains large amounts of tryptophan, the amino acid from which serotonin is synthesized.

Emotions   What causes us to laugh, cry, love, and fight has been the subject of much biological and philosophical speculation. Some say our emotions cause facial expressions; others suggest that contracting the facial muscles stimulates the emotional centers of the brain. Some theories suggest that emotions result from feedback from the body's organs and muscles to the central nervous system. Emotions are difficult to study experimentally, because even if an experimental animal seems to *show* emotion, we cannot say conclusively that the animal *feels* the emotion in the same sense that we do.

Despite these uncertainties, we know that much of human emotion depends on interactions of the cerebral cortex and a group of nuclei in the lower part of the forebrain called the **limbic system.** Much as the reticular system selects signals affecting arousal and sleep, the limbic system selects certain emotional and behavioral responses. Surgical destruction of the amygdala, part of the limbic system, has been used as a treatment for extreme aggression, but the docility produced is not necessarily a cure and may not be without serious effects on other brain functions.

Right Brain/Left Brain   The association areas of the cerebral cortex, unlike the primary motor and sensory areas (see Figure 44.25), are not bilaterally symmetric; each side of the brain controls different functions. Speech, language, and calculation, for example, are centered in the left hemisphere, while the right hemisphere controls artistic ability and spatial perception. Much of what we know about this **lateralization** of the brain comes from the work of Nobel Prize winner Roger Sperry and his colleagues, who study "split-brain" patients. Some forms of epilepsy involve reverberating circuits that send massive electrical discharges between hemispheres through the corpus callosum. These patients may be treated by surgical severing of the corpus callosum. Surprisingly, this drastic procedure does not seem to affect behavior overtly, but does have subtle effects on the patient's brain function.

A person with a severed corpus callosum may appear perfectly normal in most situations, but careful experiments reveal much about lateralization. A patient holding a key in the left hand, with both eyes open, will readily name it as a "key." If blindfolded, though, the subject will recognize the key and use it to open a lock (and may be able to describe it), but will be completely unable to name it. The center for speech is in the left hemisphere, but sensory information from the left hand crosses and enters the right side of the brain. Without the corpus callosum to function as a switchboard between the two sides of the brain, knowledge of the size, texture, and function of the object cannot be transferred from the right to the left hemisphere. Thus sensory input and spoken response are dissociated.

Language and Speech Two areas on the left hemisphere of the cerebral cortex are required for storing information related to speech (see Figure 44.25). **Wernicke's area** stores information required for speech content, arranging the words of a learned vocabulary into meaningful speech according to rules of grammar. **Broca's area** contains information required for speech production. It is told "what to say" by Wernicke's area, and then Broca's area programs the motor cortex to move the tongue, lips, and other speech muscles to articulate the words. Damage to either area causes very different kinds of aphasia, the inability to speak coherently. Wernicke's aphasia leads to the production of long strings of words and nonsense syllables. Broca's aphasia leads to loss of fluency in speech, but at least some content remains.

Memory The computer that this chapter was typed on weighs about 11 kg and can retain about 10^6 bits (300 pages) of information. Your brain, a much more powerful computer, weighs about 1.35 kg and stores literally hundreds of millions of bits of information dating back to the beginning of your life.

Memory, essential for learning, is the ability to store and retrieve information related to previous experiences. Human memory occurs in two stages. **Short-term memory** reflects immediate sensory perception of an object or idea and occurs before the image is stored. Short-term memory allows you to dial a phone number after looking it up but without looking at it directly. If you call the number frequently, it becomes stored in **long-term memory** and can be recalled several weeks after you originally looked it up. The transfer of information from short-term to long-term memory is enhanced by rehearsal ("practice makes perfect"), favorable emotional state (we learn best when we are alert and motivated), and association of new information with information previously learned and stored in long-term memory (it is much easier to learn a new card game if you already have a lot of "card sense" from playing other games).

In its ability to learn and remember, the human brain apparently distinguishes between facts and skills. When you acquire factual knowledge by memorizing dates, names, word definitions, the parts of a cell, and other information, this fact memory can be consciously and specifically retrieved from the data bank of your long-term memory. You can even recall visual images, such as the face of a friend. In contrast, skill memory usually involves motor activities that are learned by repetition without consciously remembering specific information. You perform learned motor skills, such as walking, tying your shoes, riding a bicycle, or handwriting without consciously recalling the individual steps required to do these tasks correctly. Once a skill memory is learned, it is difficult to "unlearn." For example, a "duffer" who has played golf for years with a self-taught, awkward swing has a much tougher time learning a smooth swing than does a novice just learning the game. Bad habits, as we know, are difficult to break.

By studying experimental animals and amnesia (memory loss) in humans, neuroscientists are beginning to map the major brain pathways involved in memory (Figure 44.27). In the pathway for fact memory, sensory information is transmitted from the sensory regions of the cerebral cortex to the **hippocampus** and **amygdala,** two components of the limbic system, which also function in emotions. In what is apparently a memory circuit, the hippocampus and amygdala then relay impulses to other regions of the forebrain. The circuit is completed when an integrating area called the basal forebrain, which has extensive neural connections with the sensory areas of the cortex, dispatches impulses back to the very region of the cortex where the sensory perception first occurred. Perhaps the returning impulses cause chemical or structural changes in the sensory cortex that store the event as a memory, such as a visual image, a particular sound, or a fragrance that we associate with a certain person or episode. A different neural pathway functions in skill memory.

Physiologist Karl Lashley spent several decades of this century searching for what he called the engram, the physical basis of a memory. But because partial damage to one area of the cerebral cortex does not destroy individual memories, he ultimately concluded that there is no highly localized memory trace in the nervous system. Rather, a memory seems to be stored within a certain association area of the cortex with some redundancy.

Many neuroscientists are investigating the cellular changes involved in memory or learning. According to one hypothesis, structural changes in dendrites have a function in learning. One version of this idea postulates that neural input causes a postsynaptic cell in the brain to take up calcium, which in turn activates enzymes that alter the cytoskeleton and change the shape of the dendrite in such a way that future transmission across that synapse is enhanced.

The human brain, with its billions of neurons, is too

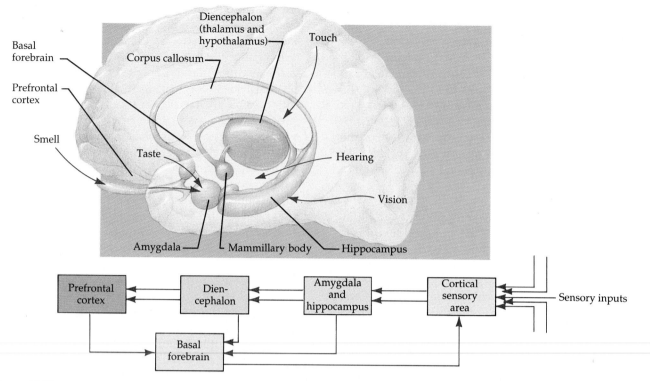

Figure 44.27
Possible memory pathway. According to this model, information received by the sensory regions of the cerebral cortex is relayed by the amygdala and hippocampus to the diencephalon (thalamus and hypothalamus), which in turn transmits impulses to forebrain regions called the prefrontal cortex and basal forebrain. Using multiple neural connections with the sensory cortex, the basal forebrain completes the memory circuit by transmitting impulses to the same sensory region that first perceived the sensory input. This may contribute to the storing of that particular experience in the cerebral cortex.

complex to serve as a model system to study the most basic mechanisms of memory and learning, which may, after all, be quite similar in diverse animals. Many neuroscientists are probing the much simpler nervous systems of certain invertebrates. For example, the sea hare *Aplysia* (Phylum Mollusca) has only about 20,000 neurons, yet it displays several forms of behavioral plasticity (learning). For instance, these animals eventually ignore mild touch stimuli that are presented repeatedly, a primitive type of learning called habituation (Figure 44.28). *Aplysia* can also be conditioned to evoke a stronger than normal withdrawal response to a soft touch that has been paired with another stimulus, such as a mild electrical stimulation of the tail.

This learning results from changes in the properties of ion channels at the synapses of interneurons in the central nervous system of *Aplysia*. Conditioning ("training") increases the amount of calcium that enters the synaptic knobs of certain neurons, resulting in more neurotransmitter being released with each action potential. This facilitates excitation of the next neuron in the circuit. Thus, a single synapse could "learn" from past experience.

Whether memory involves changes at individual synapses, or some still undiscovered mechanism, the sophistication and complexity of even relatively simple nervous systems remains one of the most exciting and fascinating aspects of modern biology.

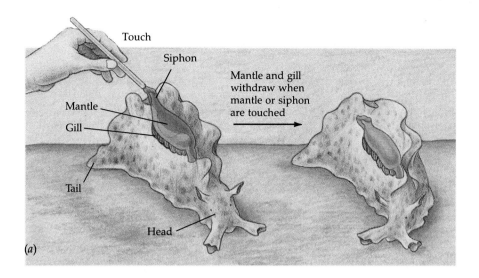

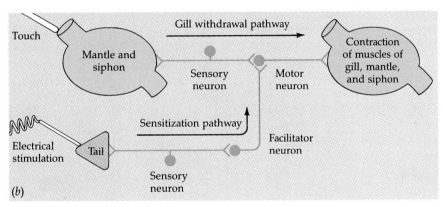

Figure 44.28

A learning pathway in the sea hare (*Aplysia*). (a) The mantle, with a siphon that allows water to flow over the gills, normally protrudes from the dorsal surface of this mollusk (see Chapter 29). Touching the mantle or siphon triggers a reflex that protects the mantle by withdrawing it. If the mantle is touched repeatedly, the withdrawal response becomes progressively weaker, a simple type of learning called habituation. (b) *Aplysia* can also be sensitized to stimuli so weak that they do not normally induce the withdrawal reflex. If the tail of the animal is stimulated prior to touching the mantle, the withdrawal reflex is stronger. If these treatments of dual stimulation are repeated several times, the enhanced withdrawal is still evident weeks later when the mantle is touched *without* stimulation of the tail. Apparently, *Aplysia* is capable of both short-term and long-term memory. The mechanism of this simple learning involves a facilitator neuron that fires when the tail is stimulated. This neuron intervenes in the withdrawal reflex by secreting a neurotransmitter (serotonin) that intensifies synaptic transmission between the sensory neuron and motor neuron of the reflex arc. In a series of chemical steps mediated by cyclic AMP as a second messenger within the synaptic knob of the sensory neuron, facilitation alters the calcium gates of the synaptic knob. When the tail is touched and an action potential reaches the terminus of the sensory neuron, more Ca^{2+} than usual enters the cell through the specific gates. This results in the release of more neurotransmitter from the sensory cell, strengthening the withdrawal response. By investigating learning pathways involving only a few neurons in invertebrates, neuroscientists may discover fundamental mechanisms that also work in the more complex nervous systems of humans and other vertebrates.

STUDY OUTLINE

1. Although functionally related to the endocrine system, the nervous system is characterized by more rapid communication, more precise control, and broader range of responses.

2. The nervous system's three main functions are sensory input, integration, and motor output to effector cells.

Cells of the Nervous System (pp. 979–981)

1. Cells of the nervous system include neurons, which transmit the signals, and supporting cells, which support, insulate, and protect the neurons.

2. A neuron's fiberlike dendrites and axons conduct impulses toward and away from the cell body, respectively. Axons,

which are often wrapped in a myelin sheath formed by Schwann cells, originate from the axon hillock and terminate in numerous branchlets, called telodendria. Synaptic knobs release neurotransmitters into the synapses and thus relay nervous signals to the dendrites or cell bodies of other neurons or effectors.

3. The central nervous system consists of the brain and spinal cord. The peripheral nervous system contains sensory neurons, which transmit information from internal and external environments to the central nervous system; and motor neurons, which carry information from the brain or spinal cord to effector organs. Interneurons, or association neurons, of the central nervous system integrate sensory input and motor output.

4. Glial cells—supporting cells in the central nervous system—include astrocytes, which line the capillaries in the brain and contribute to the blood–brain barrier; and oligodendrocytes, which wrap and insulate some neurons in a myelin sheath.

Transmission Along Neurons (pp. 981–987)

1. The membrane potential for a nontransmitting neuron is due to the unequal distribution of ions, particularly sodium and potassium, across the membrane; the cytoplasm is usually more negatively charged than the extracellular environment. This resting potential is maintained by differential ion permeabilities and the sodium-potassium pump.

2. A stimulus that affects the membrane's permeability to ions can either depolarize or hyperpolarize the membrane relative to the membrane's resting potential. This local voltage change is called a graded potential, and its magnitude is proportional to the strength of the stimulus.

3. An action potential is a rapid depolarization of the neuron's membrane that originates a nerve impulse. A local depolarization to the threshold potential opens voltage-sensitive gates, and the rapid influx of Na^+ brings the membrane potential to a positive value. A refractory period follows an action potential, while the movement of ions through voltage-sensitive potassium channels helps to restore the resting potential.

4. The all-or-none generation of an action potential always creates the same amplitude of voltage change for a given neuron. The frequency of action potentials varies with the intensity of the stimulus.

5. Once an action potential is initiated, local depolarizations spark a propagation of serial action potentials down the axon.

6. The rate of transmission of a nerve impulse is directly related to the diameter of the axon. Saltatory conduction, as action potentials jump between the nodes of Ranvier of myelinated axons, speeds nervous impulses in vertebrates.

The Synapse: Transmission Between Cells (pp. 987–994)

1. Synapses between neurons conduct impulses from the axon of a presynaptic cell to a dendrite or cell body of a postsynaptic cell.

2. Electrical synapses use gap junctions to directly pass an action potential between two neurons with little signal loss or delay.

3. In a chemical synapse, a depolarization stimulates the fusion of synaptic vesicles with the presynaptic membrane and the release of their neurotransmitter into the synaptic cleft. The diffusing transmitter binds to receptor proteins on the postsynaptic membrane that are associated with particular ion channels. The selective opening of these chemically sensitive gates either brings the membrane potential closer to the threshold potential (EPSP) or hyperpolarizes the membrane (IPSP). The neurotransmitter is rapidly broken down in the synaptic cleft by enzymes.

4. Whether an action potential is created in the postsynaptic cell depends on the temporal or spatial summation of EPSPs and IPSPs at the axon hillock.

5. One of the most common invertebrate and vertebrate neurotransmitters is acetylcholine. Other transmitters that have been identified include the biogenic amines (epinephrine, norepinephrine, and dopamine), several amino acids, and some neuropeptides, such as the opiumlike endorphins and enkephalins.

6. Groups of neurons may interact and carry information along specific pathways called circuits. Nerve cell bodies are arranged in functional groups called ganglia or nuclei.

Invertebrate Nervous Systems (pp. 994–996)

1. Invertebrate nervous systems range from the diffuse nerve nets of the cnidarians to the highly centralized nervous systems of the cephalopods, which possess large brains capable of fairly sophisticated learning.

The Vertebrate Nervous System (pp. 996–1007)

1. The peripheral nervous system consists of the sensory, or afferent, nervous system, which brings information from sensory receptors to the central nervous system; and the motor, or efferent, nervous system, which carries signals away from the central nervous system to effector muscles and glands.

2. The motor nervous system has a somatic portion, which carries signals to skeletal muscles, and an autonomic portion, which regulates the primarily automatic, visceral functions of smooth and cardiac muscles.

3. The autonomic system is subdivided into the parasympathetic and sympathetic nervous systems, which are anatomically, functionally, and chemically distinct and usually antagonistic in effect on target organs.

4. The central nervous system (CNS), composed of the brain and spinal cord, serves as the link between the sensory and motor subdivisions of the peripheral nervous system.

5. The spinal cord is the mediator of many reflexes that integrate sensory input with motor output. It also has tracts of neurons that carry information to and from the brain.

6. All vertebrate brains develop and diversify from three regions: the forebrain, the midbrain, and the hindbrain.

7. Three evolutionary changes in the vertebrate brain include increases in relative size, in compartmentalization of function, and in complexity of the forebrain, particularly the cerebral cortex.

8. The human brain develops from the three primary regions, which differentiate into specialized structures.

9. The medulla oblongata and pons of the hindbrain work together to control several homeostatic functions, such as respiration and digestion. The medulla and pons also

conduct sensory and motor information between the spinal cord and higher brain centers.

10. The cerebellum of the hindbrain coordinates movement and balance by integrating unconscious sensory and motor signals.

11. The midbrain receives, integrates, and projects sensory information to the forebrain. The reticular formation, a major group of nuclei in the midbrain, regulates states of arousal.

12. The forebrain is the site of the most sophisticated neural processing, with major integrative centers in the thalamus, hypothalamus, and cerebrum.

13. The thalamus routes neural input to specific areas of the cerebral cortex—the outer gray matter of the cerebrum. The functions of the hypothalamus range from hormone production to regulation of body temperature, hunger, thirst, sexual response, and the alarm response.

14. The cortex contains distinct sensory and motor areas, which directly process information; and association areas, which integrate information.

15. Sleep and arousal are controlled by several areas in the cerebrum and brainstem, the most important of which is the reticular formation, which filters the sensory input sent to the cortex. Characteristic changes in the patterns of electrical activity in the brain are produced by different states of arousal and sleep.

16. Human emotions are believed to originate from interactions between the cerebral cortex and the limbic system, a group of nuclei in the lower forebrain.

17. The two sides of the association areas of the cerebral cortex control different functions. Speech, language, and analytical ability are centered in the left hemisphere, whereas spatial perception and artistic ability predominate in the right. Nerve tracts of the corpus callosum link the two sides and allow the brain to function as an integrated whole.

18. Different aspects of language and speech are controlled by Wernicke's area and Broca's area.

19. Human memory consists of short-term and long-term memories. The learning and memory of facts appear to differ from that of skills. The hippocampus and amygdala, two components of the limbic system, participate in circular brain pathways involved in fact memory. Chemical or structural changes in the neurons of the sensory cortex may store memories.

20. Simpler organisms, such as *Aplysia,* are often used to study the mechanisms of memory and learning. Simple learning in this mollusk has been related to changes in the ion channels at the synapses of interneurons.

SELF-QUIZ

1. Which of the following occurs when a stimulus depolarizes a neuron's membrane?

   a. Voltage-sensitive gates that permit a rapid outflow of Na^+ open.

   b. The action potential of the membrane approaches zero.

   c. The membrane potential changes from the resting potential to a voltage closer to the threshold potential.

   d. The depolarization is all-or-none.

   e. The depolarization creates IPSP in a postsynaptic cell.

2. Action potentials are usually propagated in only one direction along an axon because

   a. the nodes of Ranvier only conduct in one direction

   b. the brief refractory period prevents a depolarization from occurring in the direction from which the impulse came

   c. the axon hillock has a higher membrane potential than the tips of the telodendria

   d. ions can only flow along the axon in one direction

   e. both sodium and potassium voltage-sensitive gates open in one direction

3. The depolarization of the presynaptic membrane of an axon *directly* causes

   a. voltage-sensitive calcium channels in the membrane to open

   b. synaptic vesicles to fuse with the membrane

   c. formation of an action potential in the postsynaptic cell

   d. chemical-sensitive gates that allow neurotransmitter to spill into the synaptic cleft to open

   e. an EPSP or IPSP in the postsynaptic cell

4. Gray matter is

   a. made up of three protective layers called meninges

   b. located on the outside of the spinal cord

   c. restricted to the brain

   d. populated by cell bodies of neurons

   e. found in the ventricles of the vertebrate brain

5. Which of the following structures or regions is incorrectly paired with its function?

   a. Broca's region—screening of information between spinal cord and the brain, regulates arousal and sleep

   b. medulla oblongata—homeostatic control centers

   c. cerebellum—unconscious coordination of movement and balance

   d. corpus callosum—band of fibers connecting left and right cerebral hemispheres

   e. hypothalamus—production of hormones and regulation of temperature, hunger, and thirst

6. Which of the following is a *true* statement about the mantle-withdrawal reflex of the sea hare *Aplysia*?

   a. The animal learns this behavior.

   b. Repeatedly stimulating the siphon over a short period of time intensifies the withdrawal reflex.

   c. *Aplysia* can learn to associate stimulation of the tail with stimulation of the siphon.

   d. There is evidence that the reflex involves conscious thought on the part of *Aplysia*.

   e. Even on the first try, stimulating the tail substitutes for stimulating the siphon in triggering the withdrawal reflex.

7. During the refractory period following an action potential, which of the following statements correctly describes the status of the neuron's membrane?

   a. Na^+ gates are open.

   b. Na^+ gates are inactivated.

   c. K^+ gates are closed.

   d. The membrane is completely impermeable to Na^+.

   e. The Na^+ and K^+ gradients across the membrane have been completely exhausted.

8. Receptor sites for neurotransmitters are located on the

   a. tips of telodendria

   b. axon membranes in the regions of the node of Ranvier

   c. postsynaptic membrane

   d. membranes of synaptic vesicles

   e. presynaptic membrane

9. Nerve nets are most characteristic of the nervous systems of

   a. annelids

   b. vertebrates

   c. insects

   d. cnidarians

   e. flatworms

10. All of the following electrical changes of neurons are *graded* events *except*

   a. EPSPs

   b. IPSPs

   c. action potentials

   d. depolarizations caused by stimuli

   e. hyperpolarizations caused by stimuli

CHALLENGE QUESTIONS

1. From what you know about the action potential, propose one feasible mechanism whereby anesthetics might prevent pain. (See Further Reading 8.)

2. Describe various ways in which drugs that are stimulants could increase activity of the nervous system by acting at the synapse.

3. Describe the role of calcium in nerve impulses.

FURTHER READINGS

1. Aoki, C., and P. Siekevitz. "Plasticity in Brain Development." *Scientific American,* December 1988. A brain protein plays a role in "rewiring" the visual cortex of young.

2. Kemp, M. "A Squid for All Seasons." *Discover,* June 1989. An important model organism in neurobiology.

3. Kimelberg, H. K. and M. D. Norenberg. "Astrocytes." *Scientific American,* April 1989.

4. Kuffler, S. W., J. G. Nicholls, and A. R. Martin. *From Neuron to Brain.* 2d ed. New York: Sinauer Press, 1984. General introduction to the nervous system.

5. Lent, C. M., and M. H. Dickinson. "The Neurobiology of Feeding in Leeches." *Scientific American,* June 1988.

6. Miskin, M., and T. Appenzeller. "The Anatomy of Memory." *Scientific American,* June 1987.

7. Rogers, L. "The Left and Right Brains at Work." *New Scientist,* February 11, 1989. Is lateral dominance inherited, or can it be developed?

8. Winter, P. M., and J. N. Miller. "Anesthesiology." *Scientific American,* April 1985. An interesting article on anesthetics, some history of their use, and how they affect the nervous system.

Sensory and Motor Mechanisms

45

Hearing a twig crack in the dry brush, a kangaroo rat freezes and then jumps, but it is too late. A rattlesnake executes a perfectly aimed strike and catches the rodent in midair. This display of agility and judgment is an example of processing sensory and motor information to produce a coordinated movement of the entire body. Perhaps its most impressive aspect is the rapid, unconscious calculation, based on subtle sensory cues, of just where the jumping rat will go and how the snake must move to meet it there. Figure 45.1 illustrates another remarkable example of sensory and motor virtuosity.

In Chapter 44, we saw how the nervous system transmits and integrates sensory and motor information. We will now examine the input and output of the coordinating system that controls so many aspects of animal behavior. We will first consider the sensory receptors that receive information from the environment, and then examine the structure and function of muscles, the motor effectors that bring about movement in response to that information. Skeletons will also be discussed in the context of body movement.

SENSORY RECEPTORS

Amputees often report pain or numbness in "phantom" limbs no longer present. To understand this phenomenon, we must consider the distinction between sensation and perception.

Figure 45.1
Salmon follow their noses home. Various species of Pacific salmon make a one-time round trip from small streams where the fish hatch, out to the sea via rivers such as the Columbia, and then back to the stream of their origin to spawn. While in the sea, salmon from many river systems school and feed together in the Gulf of Alaska. When they are a few years old, in a behavior apparently triggered by hormonal changes, sexually mature salmon segregate into groups of common geographic origin and migrate back toward the river from which they emerged as juveniles. During this first stage of the return, salmon may navigate by using the position of the sun. But once a salmon reaches the general region of the river leading to its home stream, a keen sense of smell takes over. The water that flows from each stream into the river carries a unique scent resulting from the types of plants, soil, and other environmental components of that stream. This scent is apparently imprinted in the memory of a young salmon before it migrates to the sea. Years later, on its return journey, a salmon follows these chemical cues at each fork in the river system. Battling white water, perhaps for hundreds of miles, the fish eventually arrives at its place of origin, spawns, and dies. The odyssey of the salmon highlights many of the sensory and motor mechanisms that are the subjects of this chapter.

Sensation and Perception

As we saw in Chapter 44, all information is transmitted in the nervous system in the form of action potentials. An action potential triggered by light striking the eye is no different from an action potential triggered by air vibrating in the ear, yet we readily distinguish sight from sound. The difference depends on the part of the brain that receives the signal. The air vibrations we call "sounds," for example, are converted by the ear into nerve impulses that are received by a particular region of the cerebral cortex. These nerve impulses, conveyed as action potentials along sensory neurons to the brain, are called **sensations.** Once the brain is aware of the sensations, it interprets them, giving us the **perception** of sounds. Other kinds of input are sent to other parts of the brain and trigger different perceptions. What matters, then, is where the impulse goes, not what triggers it.

It follows that if neurons from your eyes could be crossed with those from your ears, you might perceive a camera flash as a loud "boom," and a concert as bursts of light. In a less dramatic demonstration of this principle, you have probably experienced light flashes while rubbing your eyes. The rubbing provides enough energy to generate action potentials in the sensory neurons of the optic nerves. The stimulus is pressure, but the perception is a spot of light because this is the only perception that can be generated by the part of the brain receiving the sensations. To return to the example of pain in a phantom limb, severed nerves that carried impulses from the limb of an amputee may remain alive and respond to irritation. They can still transmit sensations to the brain and trigger perception. The resulting pain is just as real as that experienced by a person who irritates the nerves in an existing arm. But how does the sensory system actually generate sensations?

General Function of Sensory Receptors

Sensations, and the perceptions they evoke in the mind, begin with excitation of **sensory receptors,** structures that transmit information about changes in the external and internal environment of the animal. Receptors are usually modified neurons, which occur singly or in groups within sensory organs such as the eyes and ears. Receptors are specialized to respond to various stimuli, including heat, light, pressure, and chemicals. All of these stimuli represent forms of energy. The general function of receptor cells is to convert the energy of stimuli into the electrochemical energy of action potentials and carry those action potentials into the nervous system. This task can be broken down into five functions common to all receptor cells: reception, transduction, amplification, transmission, and integration.

Reception   The ability of a cell to absorb the energy of a stimulus is called reception. Each kind of receptor has a region specifically suited to absorbing a particular type of energy. As we will see, sensory cells in the human eye, for example, have membranes that contain a light-absorbing pigment molecule.

Transduction   The conversion of stimulus energy into electrochemical activity of nerve impulses is called transduction. Receipt of a stimulus changes the permeability of the receptor cell, causing a change in its membrane potential and, ultimately, changes in the number of action potentials transmitted along sensory pathways to the central nervous system. In some cases, a stimulus such as pressure can stretch the membrane and increase ion flow generally. In other cases, specific receptor molecules on the membrane of a receptor cell open or close gates to ion channels when the stimulus is present. We will discuss specific examples of sensory transduction later in this chapter.

Amplification   The stimulus energy is often too weak to be carried into the nervous system and must be amplified. Amplification of the signal may occur in accessory structures of a complex sense organ, as when sound waves are amplified by a factor of more than 20 before they reach the receptors of the inner ear. Amplification also may be a part of the transduction process itself. An action potential conducted from the eye to the brain has about 100,000 times as much energy as the few photons of light that triggered it.

Transmission   Once the energy in the stimulus has been transduced into changes in the membrane potential of the receptor cell, those changes must be transmitted to the nervous system. In some instances, such as in the case of "pain cells," the receptor itself is actually a sensory neuron that conducts action potentials to the central nervous system. Other receptors are separate cells that must transmit chemical signals across synapses to sensory neurons. In either case, the initial response of the receptor to the stimulus is a graded change in membrane potential called a **receptor potential.** (Recall from Chapter 44 that a graded potential is a change in the voltage across the membrane that is proportional to the strength of the stimulus.) If the receptor also functions as the sensory neuron, then the intensity of the receptor potential affects the frequency of action potentials that travel as sensations to the central nervous system. For separate receptor cells, the strength of the stimulus and receptor potential affect the amount of neurotransmitter released by the receptor into its synapse with a sensory neuron, which in turn determines the frequency of action potentials fired by the sensory neuron. Sensory neurons actually fire at a low rate spontaneously, and so a stimulus does not really switch the production of action potentials on or off, but rather modulates their frequency. In this way, the central nervous system is sensitive not only to the presence or absence of a stimulus, but also to changes in stimulus intensity.

Integration   The processing of information begins as soon as it starts to be received. Signals from receptors are integrated through summation of graded potentials, as are those within the nervous system (see Chapter 44).

One type of integration by the receptor cell is **sensory adaptation,** a decrease in sensitivity during continued stimulation (not to be confused with the term "adaptation" as used in an evolutionary context). Without sensory adaptation, you would feel every beat of your heart and every bit of clothing on your body. Receptors are selective in the information they send to the central nervous system, and adaptation reduces the likelihood that a continued stimulus will be transmitted.

Another important aspect of sensory integration is the sensitivity of the receptors. The threshold for firing in receptor cells varies with conditions. For example, the firing thresholds of glucose receptors in the human mouth and in the feet of flies can vary over several orders of magnitude of sugar concentration, as both the general state of nutrition and the amount of sugar in the diet change.

Integration of sensory information occurs at all levels within the nervous system, and the cellular actions just described are only the first steps. Complex receptors such as the eyes have higher levels of integration as signals converge on sensory nerves, and the central nervous system further processes all incoming signals.

Types of Receptors

The various types of sensory receptors can be divided on the basis of location into two broad groups: the **exteroreceptors** that receive information from the outside world (light, sound, touch) and the **interoreceptors** that provide information about the body's internal environment. Another way of categorizing receptors is in terms of the energy stimulus to which they respond. Based on this criterion the five types of receptors we will consider are mechanoreceptors, chemoreceptors, electromagnetic receptors, thermoreceptors, and pain receptors.

Mechanoreceptors   Mechanoreceptors are stimulated by their physical deformation caused by such stimuli as pressure, touch, stretch, motion, and sound, all forms of mechanical energy. Bending or stretching of the plasma membrane of a mechanoreceptor increases its permeability to both sodium and potassium ions, resulting in a depolarization (receptor potential).

The human sense of touch relies on mechanoreceptors that are actually modified dendrites of sensory neurons. **Pacinian corpuscles** are found in deep skin layers and respond to strong pressure. Closer to the surface, detecting light touch, are **Meissner's corpus-**

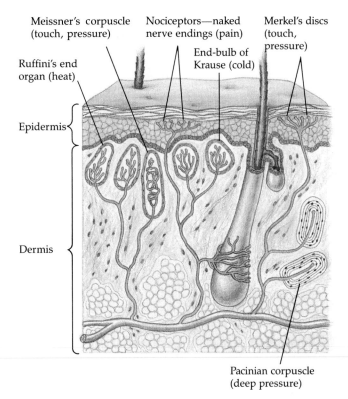

Meissner's corpuscle (touch, pressure)
Nociceptors—naked nerve endings (pain)
Merkel's discs (touch, pressure)
Ruffini's end organ (heat)
End-bulb of Krause (cold)
Epidermis
Dermis
Pacinian corpuscle (deep pressure)

Figure 45.2
Receptors in the skin. The sensory receptors illustrated here are described throughout this chapter. All of these tactile receptors are actually modified dendrites, encapsulated by connective tissue in some cases.

cles and **Merkel's discs.** Figure 45.2 illustrates the various sensory receptors in the human skin.

An example of an interoreceptor stimulated by mechanical distortion is the **muscle spindle,** or stretch receptor. This mechanoreceptor is used to monitor the position of body parts. The muscle spindle contains modified muscle fibers attached to sensory neurons and runs parallel to muscle. When the muscle is stretched, the fibers of the spindle are also stretched, depolarizing them and sending action potentials through the sensory neurons.

The **hair cell** is a common type of mechanoreceptor used to detect motion. Hair cells are found in the vertebrate ear; in the lateral line organs of fishes and amphibians, where they detect movement relative to the environment (see Chapter 30); and in the balance organs of arthropods. The "hairs" are either specialized cilia or microvilli (cellular projections supported by microfilaments—see Chapter 37). They project upward from the surface of the hair cell into either an internal compartment, such as the human inner ear, or an external environment, such as a pond. When the cilia bend in one direction, they stretch the hair

cell membrane and increase its permeability to sodium and potassium ions, thereby increasing the rate of impulse production in a sensory neuron. When the cilia bend in the opposite direction, ion permeability decreases, reducing the number of action potentials in the sensory neuron. This specificity allows hair cells to respond to the direction of motion as well as to its strength and speed. The role of hair cells in hearing and balance is explored later in this chapter.

Chemoreceptors **Chemoreceptors** include both general receptors that transmit information about the total solute concentration in a solution and specific receptors that respond to individual kinds of molecules. Osmoregulators in the mammalian brain, for example, are general receptors that detect changes in the total solute concentration of the blood and stimulate drinking behavior when osmolarity increases (see Chapter 40). Osmoreceptors in the feet of house flies respond to a dilute solution of virtually any substance. Most animals also have receptors specific to important molecules, including glucose, oxygen, carbon dioxide, and amino acids. In all these examples, the stimulus molecule binds to a specific site on the membrane of the receptor cell and initiates changes in membrane permeability. Two other groups of chemoreceptors show intermediate specificity. **Gustatory** (taste) and **olfactory** (smell) **receptors** respond to *categories* of related chemicals. Humans often classify such categories as "sweet," "sour," or "salty." (Taste and olfaction are discussed in detail later in this chapter.) One of the most sensitive and specific chemoreceptors known is found in the antennae of the male silkworm moth (Figure 45.3); it detects the female sex pheromone bombykol.

Electromagnetic Receptors Electromagnetic radiation consists of energy of different wavelengths, which takes such forms as visible light, electricity, and magnetism (see Chapter 10). **Photoreceptors,** which detect the radiation we know as visible light, are often organized into eyes. Snakes have extremely sensitive infrared receptors that detect the body heat of prey standing out against a colder background (Figure 45.4a). Some fishes discharge electric currents and use special electroreceptors to locate objects, such as prey, that disturb the electric currents. The platypus, a monotreme mammal (see Chapter 30), has electroreceptors on its bill that can probably detect electric fields generated by the muscles of prey, such as crustaceans, frogs, and small fishes. There is also evidence that some animals that home or migrate use the magnetic field lines of the Earth to help orient themselves (Figure 45.4b). Although the nature of the magnetoreceptors is unknown, the ferrous mineral magnetite has been found in the skulls of several animals.

Figure 45.3
Chemoreceptors in an insect. The antennae of the male silkworm moth *Bombyx mori* are covered with chemoreceptive hairs that are highly sensitive to the female sex pheromone bombykol. (Each filament of these feathery antennae actually bears hundreds of olfactory hairs, too small to be seen in this photograph.) Males show definite behavioral responses when as little as 50 of the more than 50,000 bombykol receptors on the antennae come into contact with one bombykol molecule per second.

(a)

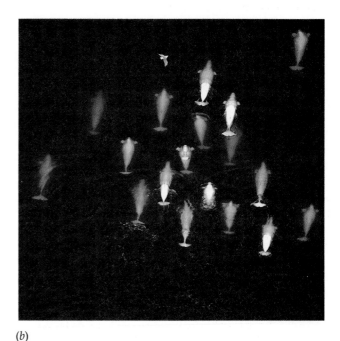

(b)

Figure 45.4
Specialized electromagnetic receptors.
(a) Rattlesnakes and other pit vipers, such as this Asian species, have a pair of infrared receptors, one between each eye and nostril. The organs are sensitive enough to detect the infrared radiation emitted by a warm mouse a meter away. The snake moves its head from side to side until the radiation is detected equally by the two receptors, indicating that the mouse is straight ahead. (b) Some migrating animals, such as these Beluga whales, can apparently sense the magnetic field of the Earth and use the information, along with other cues, for orientation. The mechanism of the magnetic sense is unknown.

Thermoreceptors   Responding to either heat or cold, **thermoreceptors** help regulate body temperature by signaling both surface and body core temperature. There is still debate about the identity of thermoreceptors in the mammalian skin. Possible candidates are two receptors consisting of encapsulated, branched dendrites: **Ruffini's end organs** may be heat receptors, and **end-bulbs of Krause** are possibly cold receptors (see Figure 45.2). Some researchers, however, believe that these structures are actually modified pressure receptors, and propose that naked dendrites of certain sensory neurons are the actual thermoreceptors of the skin. There is more agreement that interothermoreceptors in the hypothalamus of the brain function as the major thermostat of the mammalian body.

Pain Receptors   Virtually all animals experience pain, although we cannot say what perceptions they actually associate with stimulation of their pain receptors. Pain is one of the most important senses because the stimulus becomes translated into a negative reaction, such as withdrawal from danger. Rare individuals who are born without any pain sensation suffer large numbers of cuts and burns and may even die from such conditions as a ruptured appendix because they cannot feel the associated pain and so are unaware of the danger.

Pain is detected by a class of naked dendrites called **nociceptors** (see Figure 45.2). Different groups of pain receptors respond to excess heat, pressure, or specific classes of chemicals released from damaged or inflamed tissues. Some of the chemicals that trigger pain include histamine and acids. Prostaglandins increase pain by sensitizing the receptors, that is, lowering their threshold; aspirin reduces pain by inhibiting prostaglandin synthesis. Nociceptive neurons carry impulses to the spinal cord, synapsing with neural pathways leading to the brain.

We have seen that receptors can be classified according to the type of energy they receive. In many cases, large numbers of a particular type of receptor cell are collected into complex sense organs that detect specific environmental stimuli. We will now examine the structure and function of sense organs responsible for vision, hearing, equilibrium, taste, and olfaction.

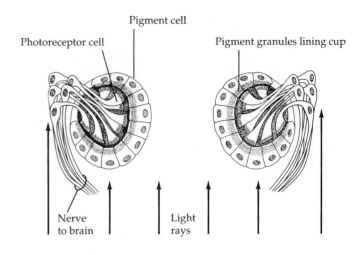

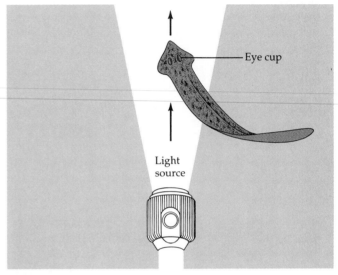

Figure 45.5
Eye cups and orientation behavior in planarians. The head of the planarian has two eye cups with photoreceptors that send nerve impulses to the brain. The wall of the cup consists of pigment cells that shade the photoreceptors. Light can reach the photoreceptors only through the mouth of the cup, which opens to the side of the head and slightly forward. The brain, comparing the frequency of nerve impulses from the two eye cups, directs the body to turn until the sensations from the two cups are equal and at a minimum. This orientation response causes the planarian to swim directly away from the light source until it reaches a dark pocket under a rock or some other shaded sanctuary where the animal is less obvious to predators.

VISION

Light Receptors and Vision of Invertebrates

Most invertebrates can detect light with receptors containing light-absorbing pigments. One of the simplest light receptors is the **eye cup** of planarians (Figure 45.5). These structures provide information about light intensity and direction without actually forming an image. The receptor cells are located within a cup formed by a layer of darkly pigmented cells that block light. Light can enter the cup and stimulate the photoreceptors only through an opening on one side of the cup where there are no pigment cells. The mouth of one eye cup faces left and slightly forward, and the mouth of the other cup faces right-forward. Thus, light shining from one side of the planarian can enter only

(a)

$\vdash\!\!\dashv$
1 mm

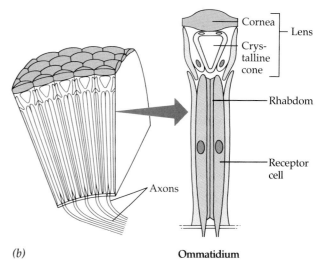

Cornea

Crys-
talline
cone

Lens

Rhabdom

Receptor
cell

Axons

(b)

Ommatidium

Figure 45.6
Compound eyes. (a) The faceted eyes of a horsefly,
photographed with a stereo microscope. **(b)** The cornea
and crystalline cone of each ommatidium function as lenses that focus light through a structure called a rhabdom,
a stack of pigmented plates formed on the inside of a circle of receptor cells. The image is a mosaic of dots
formed by the different intensities of light entering the
many ommatidia.

the eye cup on that side of the animal. The brain compares the rate of nerve impulses coming from the two
eye cups, and the animal turns until the sensations
from the two cups are equal and minimal. The result
is that the animal swims directly away from the light
source and reaches a shaded location beneath a rock
or some other object, a behavioral adaptation that helps
hide the planarian from predators.

True image-forming eyes of two major types have
evolved in invertebrates: the compound eye and the
single-lens eye. **Compound eyes** are found in insects
and crustaceans (Phylum Arthropoda) and some
polychaete worms (Phylum Annelida). A compound
eye consists of up to several thousand light detectors
called **ommatidia** (the "facets" of the eye), each with
its own cornea and lens. Each ommatidium registers
light from a tiny portion of the field of view. Differences in the intensity of light entering the many
ommatidia result in a mosaic image (Figure 45.6).
Although the image is not as sharp as that produced
by the human eye, the compound eye is more acute
at detecting movement, an important adaptation for
flying insects and small animals constantly threatened
with predation. This advantage of the compound eye
is partly due to the rapid recovery of the photoreceptors. The human eye can distinguish light flashes
occurring at 50 flashes per second, and thus a 60-cycle
light bulb is fused in time into a single image. The

compound eyes of some insects, however, recover from
excitation rapidly enough to sense a light flashing 330
times per second as separate flashes: An insect viewing a light bulb could easily resolve the 60 flashes per
second. Insects also have excellent color vision, and
some (including bees) can see into the ultraviolet range
of the spectrum, which is invisible to us. In studying
animal behavior, we cannot extrapolate our sensory
world to other species; different animals have different sensitivities.

Recent studies indicate that the compound eyes of
certain crabs and mayflies can actually form a single
unfragmented image under conditions of dim light.
The ommatidia have lenses that work like prisms and
parabolic mirrors, focusing the light entering several
ommatidia onto a single photoreceptor, thus increasing the sensitivity of the eye to light. When lighting
is brighter, these *superimposition eyes*, as they are called,
seem to work like typical compound eyes.

The second type of invertebrate eye is the **single-
lens eye,** found in some jellyfish, polychaetes, spiders, and many mollusks (Figure 45.7). It works on a
principle similar to that of a camera. The single lens
focuses light onto the retina, a bilayer of photosensitive receptor cells. The eyes of humans and other vertebrates are also of the camera type, but they evolved
independently and differ from the single-lens eyes of
invertebrates in several details.

Figure 45.7
Single-lens (camera-type) eye of a squid. The cephalopod eye and the vertebrate eye evolved independently, and their similarities are analogous, not homologous.

Vertebrate Vision

The human eye, shown in Figure 45.8, can detect an almost countless variety of colors, can form images of objects miles away, and can respond to as little as one photon of light. Remember, however, that it is actually the brain that "sees." Thus, to understand vision, we must begin by learning how the vertebrate eye generates sensations (action potentials), and then follow these signals to the visual centers of the brain, where sights are perceived.

Structure and Function of the Vertebrate Eye The globe of the vertebrate eye, or eyeball, consists of a tough, white outer layer of connective tissue called the **sclera** and a thin, pigmented inner layer called the **choroid.** At the front of the eye, the sclera becomes the transparent **cornea,** which lets light into the eye. The anterior choroid forms the donut-shaped **iris,** which gives the eye its color. By changing size, the iris regulates the amount of light entering the **pupil,** the hole in the center of the iris. Just inside the choroid, the **retina** forms the innermost layer of the eye. The retina contains the photoreceptor cells themselves. Information from the photoreceptors leaves the eye at the optic disc, where the optic nerve attaches to the eye. Because there are no photoreceptors in the optic disc, this spot on the lower outside of the retina is a blind spot: Light focused onto that part of the retina is not detected.

The **lens** and **ciliary body** divide the eye into two cavities, one between the lens and the cornea, and a much larger cavity behind the lens within the eyeball itself. The ciliary body constantly produces the clear, watery **aqueous humor** that fills the anterior cavity of the eye. Blockage of the ducts that drain the aqueous humor can produce glaucoma, increased pressure that leads to blindness by compressing the retina. The posterior cavity, filled with the jellylike **vitreous humor,** occupies most of the volume of the eye. The aqueous and vitreous humors function as liquid lenses that help focus light onto the retina. The lens itself is a transparent protein disc that focuses an image onto the retina. Many fishes focus by moving the lens forward or backward, camera-style. Humans and other mammals, however, focus by changing the *shape* of the lens. When viewing a distant object, the lens is flat. To focus on a close object, the lens becomes almost spherical, a change termed **accommodation** (Figure 45.9).

Signal Transduction in the Eye When the lens focuses a light image onto the retina, how do the cells of the retina transduce the stimuli into sensations—action potentials that transmit this information about the environment to the brain? Built into the human retina are about 125 million **rod cells** and 6 million **cone cells,** two types of photoreceptors named for their shapes. They account for 70% of all receptors in your body, a fact that underscores the importance of the eyes and visual information in how humans perceive their environment.

Rods and cones have different functions in vision, and the relative numbers of these two photoreceptors in the retina are partly correlated with whether an animal is most active during the day or at night. Rods are more sensitive to light, but do not distinguish colors; they enable us to see at night, but only in black and white. It takes more light to stimulate cones, and thus these photoreceptors do not function in night vision, but they can distinguish color during the day. Color vision is found in all vertebrate classes, though not in all species. Fishes, amphibians, reptiles, and birds have well-developed color vision, but humans and other primates are among the minority of mammals that can see color. Most mammals are nocturnal, and a maximum number of rods in the retina is an adaptation that gives these animals keen night vision. Cats, usually most active at night, do have limited color vision, and probably see a pastel world during the day. In the human eye, rods are found in greatest density at the lateral regions of the retina, and are completely absent from the **fovea,** the center of the visual field (see Figure 45.8). If you look directly at a dim star at night, it is harder to see than if you look at it on an angle, which focuses the starlight onto the regions of the retinas most populated by rods. You achieve your sharpest day vision, however, by looking straight at the object

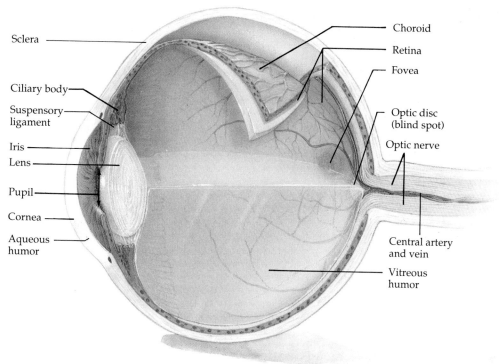

Figure 45.8
Structure of the vertebrate eye. In this longitudinal section of the eye, the jellylike vitreous humor is illustrated only in the lower half of the eyeball.

of interest. This is because cones are most dense at the fovea, where there are about 150,000 color receptors per square millimeter. Some birds have more than a million cones per square millimeter, which enables such species as hawks to spot field mice and other small prey from high in the sky. In the retina of the eye, as in all biological structures, we see variations representing evolutionary adaptations.

Each rod or cone cell has an outer segment with a stack of folded membranes in which visual pigments are embedded (Figure 45.10a). The actual light-absorbing molecule is **retinal,** which is synthesized

Figure 45.9
Focusing in the mammalian eye. The lens bends light and focuses it onto the retina. The thicker the lens, the more sharply the light is bent. The lens is nearly spherical when focusing on near objects and much flatter when focusing at a distance. The shape of the lens is controlled by the ciliary muscle. **(a)** When viewing a close object, the ciliary muscles contract, pulling the border of the choroid layer of the eye toward the lens and causing the suspensory ligaments to slacken. With this reduced tension, the elastic lens becomes thicker and rounder, bending light in such a way that near objects can be focused on the retina. This adjustment of the lens for close viewing is called accommodation. **(b)** When viewing a distant object, the ciliary muscle relaxes, allowing the choroid to expand and put tension on the suspensory ligament. The lens is pulled into a flatter shape, and the distant object is focused onto the retina. Notice that the focusing muscle is relaxed during distance viewing and is contracted for close work.

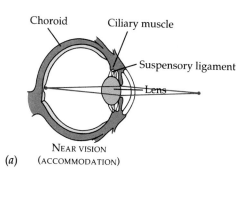

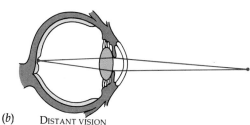

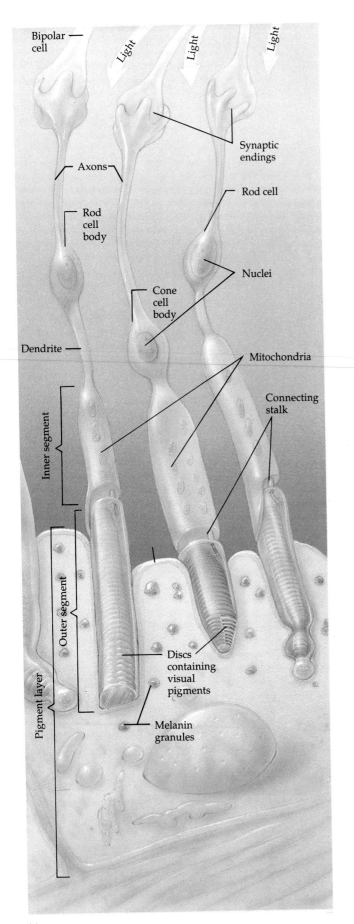

(a)

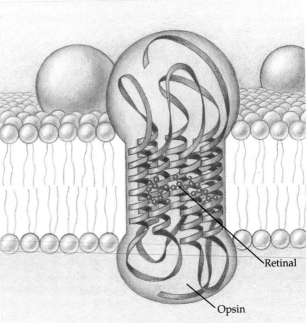

(b)

Figure 45.10

Photoreceptors of the retina. (a) The retina is populated by two types of photoreceptors: Rods are very sensitive to light and function in black-and-white vision at
night; cones are less sensitive to light and account for
color vision during the day. Both types of cells are modified neurons. Each rod and cone has an outer segment
partly embedded in a layer of darkly pigmented cells.
The outer segment is connected by a short stalk to an
inner segment, which is in turn connected to the cell
body. Axons of the rods and cones synapse with other
retinal neurons called bipolar cells. Within the outer segment of a rod or cone is a stack of folded membrane.
The visual pigments responsible for detecting light
focused onto the retina are built into these stacked membranes. **(b)** The visual pigments consist of a light-
absorbing molecule called retinal (derived from
vitamin A) bonded to a protein called an opsin. Each
type of photoreceptor has a characteristic kind of opsin,
which affects the absorption spectrum of the retinal. In
the case of rods, the whole pigment complex—retinal
plus the specific type of opsin—is called rhodopsin.
Notice that the opsin has several regions of alpha helix
(see Chapter 5) spanning the membrane. At the core of
the opsin is the light-absorbing retinal.

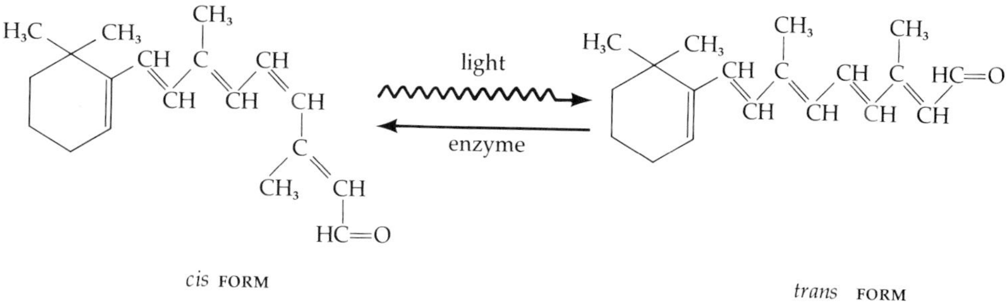

cis FORM   trans FORM

Figure 45.11

How light affects retinal. Retinal exists in two forms, isomers of one another. Absorption of light converts the pigment from the *cis* isomer to the *trans* isomer. When the photoreceptor is no longer stimulated by light, enzymes convert the reti- nal back to the *cis* form. The photochemical reaction changes the shape of retinal, causing it to dissociate from the opsin. This event triggers a chain of metabolic responses by the photoreceptor that ulti- mately changes the voltage across the plasma membrane of the cell, producing a receptor potential (which, in this case, is actually a *hyper*polarization, not a depolar- ization).

from vitamin A. The retinal is bonded to a membrane protein called an **opsin.** Opsins vary in structure from one type of photoreceptor to another, and the light-absorbing ability of the retinal is affected by the specific identity of its opsin partner. Rods have their own type of opsin, and, along with the retinal component, the molecule is called **rhodopsin** (Figure 45.10b). When rhodopsin absorbs light, the retinal changes shape and dissociates from the opsin. This photochemical reaction is referred to as "bleaching" of the rhodopsin. In the dark, enzymes convert the rhodopsin back to its original form (Figure 45.11). Bright light keeps the rhodopsin bleached and rods become unresponsive; cones take over. When we walk from a bright environment into a dark place, such as walking into a movie theater when a matinee is in progress, we are initially almost blind; there is not enough light to stimulate cones, and it takes at least a few minutes for the bleached rods to become functional again.

Color vision results from the presence of three subclasses of cones in the retina, each with its own type of opsin associated with retinal to form visual pigments collectively called **photopsins.** These photoreceptors are known as red cones, green cones, and blue cones, referring to the colors their brands of photopsin are best at absorbing. The absorption spectra for these pigments overlap, and the perception of intermediate hues depends on the differential stimulation of two or more types of cones. For example, when both red and green cones are stimulated, we may see yellow or orange, depending on which of these two populations of cones is most strongly stimulated. Color blindness, more common in males than females because it is generally inherited as a sex-linked trait (see Chapter 14), is due to a deficiency or absence of one or more types of cones.

Transduction of light energy into action potentials is not accomplished by the photoreceptors alone, but results from the cooperation of additional types of neurons in the retina. The chemical response of retinal to light triggers a complex chain of metabolic events that ultimately alters the membrane potential of the rod or cone cell. The photoreceptors themselves do not "fire" action potentials. The light-induced change in voltage across the membrane is a localized receptor potential, and like all receptor potentials, it is a graded response proportional to the strength of the stimulus. The membrane does not depolarize. Light actually *hyperpolarizes* the membrane by decreasing the permeability to sodium ions. This results in the receptor cell releasing less neurotransmitter in the light than in the dark. Thus, it is actually a *decrease* in the chemical signal to the cells with which rods and cones synapse that serves as a message that the photoreceptors have been stimulated by light. The axons of rods and cones synapse with neurons called **bipolar cells,** which in turn synapse with **ganglion cells** (Figure 45.12). Additional types of neurons in the retina, **horizontal cells** and **amacrine cells,** help to integrate the information before it is dispatched to the brain. The axons of ganglion cells then convey the resulting sensations to the brain as action potentials along the optic nerve.

Visual Integration   Processing of visual information begins in the retina itself. Signals from the rods and cones may follow either vertical or lateral pathways (see Figure 45.12). In the vertical pathway, information passes directly from the receptor cells to the bipolar cells to the ganglion cells. The horizontal and amacrine cells provide lateral integration of visual signals. Horizontal cells carry signals from one rod or cone to other receptor cells and to several bipolar cells; amacrine cells spread the information from one bipolar cell to several ganglion cells. When a rod or cone stimulates a horizontal cell, the horizontal cell stimulates nearby receptors, but inhibits more distant receptors and bipolar cells that are not illuminated, making the light spot appear lighter and the dark surroundings

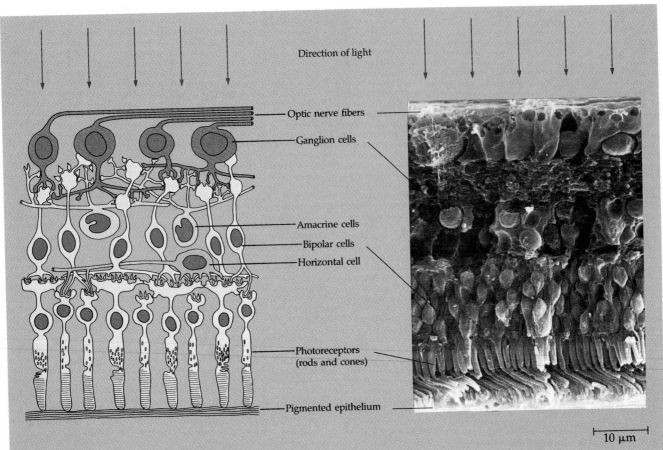

Direction of light

Optic nerve fibers

Ganglion cells

Amacrine cells

Bipolar cells

Horizontal cell

Photoreceptors (rods and cones)

Pigmented epithelium

10 μm

Figure 45.12
The human retina. Light must pass through several relatively transparent layers of cells before reaching the rods and cones. These photoreceptors communicate with ganglion cells via bipolar cells. The axons of the ganglion cells transmit the visual sensations (action potentials) to the brain. There is not a one-to-one relationship between the rods and cones, bipolar cells, and ganglion cells. Rather, each bipolar cell receives information from several rods or cones, and each ganglion cell from several bipolar cells. The horizontal and amacrine cells carry information across the retina to integrate the signals. All the rods or cones that feed information to one ganglion cell form the receptive field for that cell. The larger the receptive field (the more rods or cones that supply a ganglion cell), the less sharp the image, because it is less evident exactly where the light struck the retina. The ganglion cells of the fovea have very small receptive fields, so visual acuity is very sharp in this area. (SEM from *Tissues and Organs: A Text-Atlas of Scanning Electron Microscopy* by Richard G. Kessel and Randy H. Kardon. W. H. Freeman and Company. Copyright © 1979.)

even darker. This integration, called **lateral inhibition,** sharpens edges and enhances contrast in the image. One type of retinal integration based on lateral inhibition is described in Figure 45.13. Lateral inhibition is repeated by the interactions of the amacrine cells with the ganglion cells and occurs at all levels of visual processing.

Axons of ganglion cells form the optic nerves that transmit sensations from the eyes to the brain. The optic nerves from the two eyes meet at the **optic chiasma** near the center of the base of the cerebral cortex. The optic chiasma has its nerve tracts arranged in such a way that what is sensed in the left field of view by both eyes is transmitted to the right side of the brain, and what is detected in the right field of view goes to the left side of the brain (Figure 45.14). A few of the ganglion cell axons terminate in the midbrain, but most continue on to the **lateral geniculate nuclei** of the thalamus. Neurons of the lateral geniculate nuclei continue back to the **primary visual cortex** in the occipital lobe of the cerebrum. Additional interneurons carry the information to other, more sophisticated visual processing and integrating centers elsewhere in the cortex.

Point-by-point information in the visual field is projected along neurons onto the visual cortex according to its position in the retina, but the information the brain receives is highly distorted. How this coded set of spots, lines, and movement is converted into our perception and recognition of objects is still largely unknown; there is much left to learn about how we actually "see."

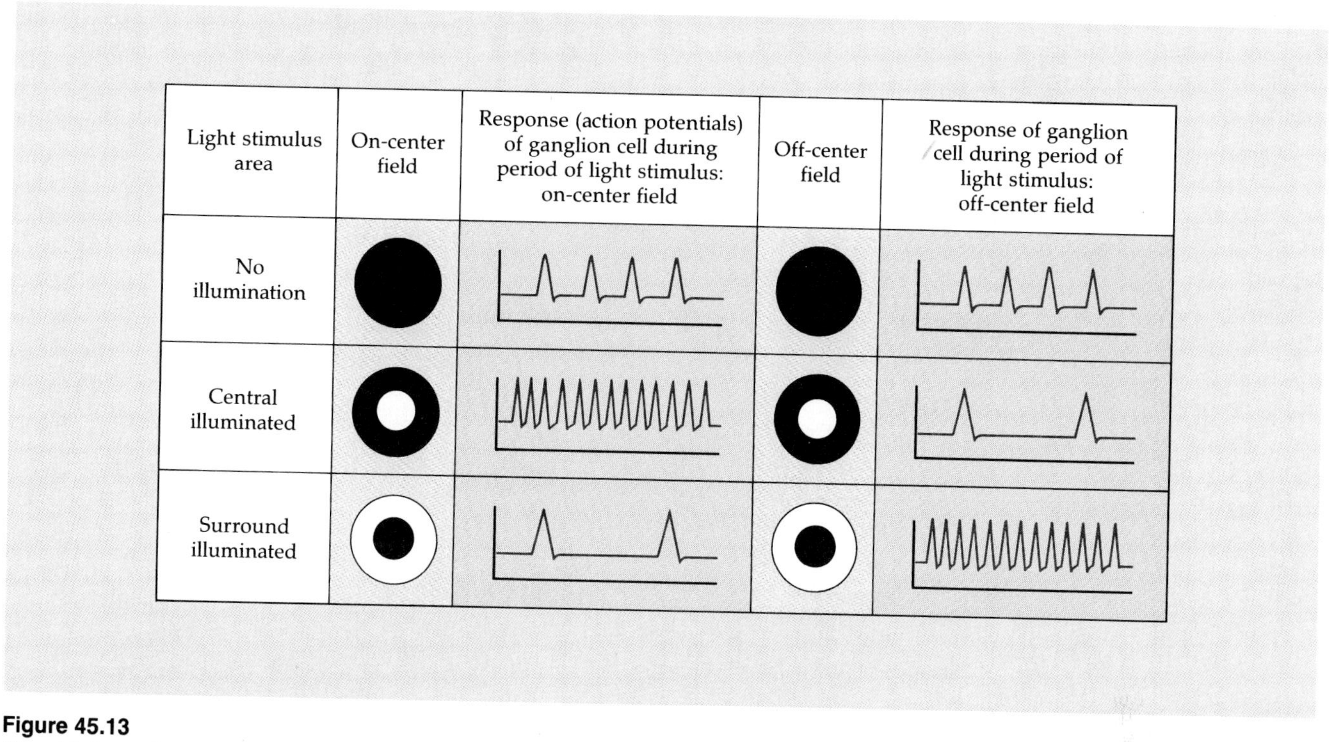

Light stimulus area	On-center field	Response (action potentials) of ganglion cell during period of light stimulus: on-center field	Off-center field	Response of ganglion cell during period of light stimulus: off-center field
No illumination	(black circle)	(low-frequency spikes)	(black circle)	(low-frequency spikes)
Central illuminated	(black circle, white center)	(high-frequency spikes)	(black circle, white center)	(low-frequency spikes)
Surround illuminated	(white circle, black center)	(low-frequency spikes)	(white circle, black center)	(high-frequency spikes)

Figure 45.13

Visual integration in the vertebrate retina. The retina is organized into visual fields consisting of the many photoreceptors that communicate with a common ganglion cell. Action potentials are propagated along the axons of the ganglion cell to the brain with a frequency based on integration of the inputs from the receptors making up that receptive field. Bipolar cells, horizontal cells, and amacrine cells function as the integrating network of the receptive field. Even in the dark, a ganglion cell "fires" action potentials with a constant frequency. If the entire receptive field is uniformly illuminated, there is no effect on this basal rate of discharge by the ganglion cell. It is when only a portion of the field is illuminated that the frequency of sensations sent from the ganglion cell to the brain changes. Two types of receptive fields are known as *on-center fields* and *off-center fields*. The ganglion cell of an on-center field is stimulated when a spot of light is focused on the center of the receptive field, and it is inhibited when only the peripheral region of the field is illuminated. The ganglion cell of an off-center field has just the opposite response: A spot of light focused onto the center of the field inhibits the ganglion cell, but illumination of the periphery of the field stimulates the cell. Thus, on-center fields increase the sharpness of bright spots against a dark background, and off-center fields make dark spots against a light background stand out more vividly. Much of the visual information transmitted from the retina to the brain is based on these enhanced patterns of bright and dark spots.

Figure 45.14

Neural pathways for vision. Objects in the left field of view are "seen" by the right side of the brain, while objects in the right field of view are projected into the left side of the brain. (We are seeing the brain from the bottom in this diagram.) The lateral geniculate nuclei, ganglia located in the thalamus, relay the sensations to the visual cortex.

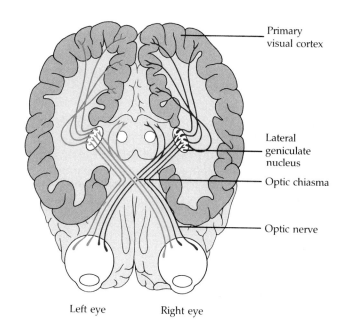

Primary visual cortex

Lateral geniculate nucleus

Optic chiasma

Optic nerve

Left eye   Right eye

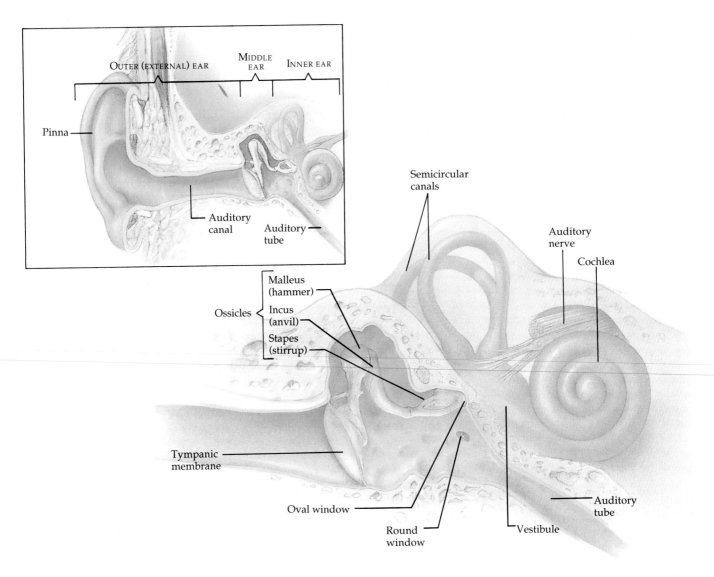

Figure 45.15
Structure of the human ear. The hair cells, the actual sensory receptors of the ear, are located within the labyrinth of canals of the inner ear. The cochlea contains hair cells that function in hearing. The vestibule and semicircular canals contain hair cells responsible for the sense of balance.

HEARING AND BALANCE

The senses of hearing and balance are related in most animals. Both senses involve mechanoreceptors containing hair cells that trigger action potentials when the hairs are bent by settling particles or moving fluid. In mammals and most other terrestrial vertebrates, the sensory organs for hearing and balance are located together within the ear.

The Mammalian Ear

The human ear is really two separate sense organs. It contains the organ of audition, or hearing, and that of equilibrium, or balance. Both of these senses are functions of hair cells in fluid-filled canals.

The ear itself can be divided into three regions (Figure 45.15). The outer ear consists of the external **pinna** and the **auditory canal,** which collect sound waves and channel them to the **tympanic membrane** (eardrum) of the **middle ear.** There the vibrations are conducted through three small bones—the **malleus** (hammer), **incus** (anvil), and **stapes** (stirrup)—to the **inner ear,** passing through the **oval window,** a membrane beneath the stapes. The middle ear also opens into the **auditory (Eustachian) tube,** which connects with the pharynx and equalizes pressure between the middle ear and the atmosphere—enabling you to "pop" your ears when changing altitude, for example. The inner ear consists of a labyrinth of channels within a

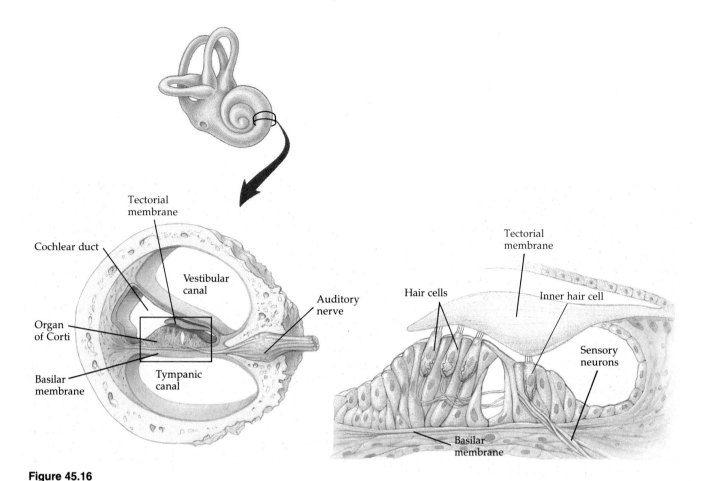

Figure 45.16
Structure and function of the cochlea.
The cochlea is a long coiled tube. A cross-sectional view reveals three canals. The vestibular canal and tympanic canal contain a fluid called perilymph. Between these two canals is a smaller cochlear duct, filled with endolymph. The receptor cells, hair cells, are part of the organ of Corti, which sits on the basilar membrane that forms the floor of the cochlear duct. Suspended over the organ of Corti is the tectorial membrane, to which many of the hairs of the receptor cells are attached. Vibrations of the tympanic membrane (eardrum) are transmitted to the oval window on the surface of the cochlea (see Figure 45.15), and this creates pressure waves in the cochlear fluid. As the basilar membrane vibrates, the hairs of the organ of Corti are repeatedly pushed against the tectorial membrane. This stimulus causes the hair cell to depolarize and release a neurotransmitter that triggers an action potential in a sensory neuron.

skull bone (the temporal bone). These channels are lined by a membrane and contain fluid that moves in response to sound or movement of the head.

The part of the inner ear involved in hearing is a complex coiled organ known as the **cochlea** (from the Latin for "snail"). It occupies part of the labyrinthine cavity within the temporal bone. The cochlea has two large chambers, an upper vestibular canal and a lower tympanic canal, separated by a smaller cochlear duct (Figure 45.16). The vestibular and tympanic canals contain a fluid called perilymph, and the cochlear canal is filled with a liquid named endolymph. The floor of the cochlear canal, the basilar membrane, bears the **organ of Corti.** Built into the organ of Corti are the actual receptor cells of the ear, hair cells with their hairs projecting into the cochlear canal. The tips of some of the hairs are attached to the tectorial mem-brane, which hangs over the organ of Corti like a shelf. Let us now see how this complex anatomy of the ear is correlated with the function of hearing.

The Process of Hearing The ear converts the energy of pressure waves traveling through air into nerve impulses that the brain perceives as sound. Vibrating objects, such as the reverberating strings of a guitar or the vocal cords of a speaking person, create percussion waves in the surrounding air. These waves cause the tympanic membrane to vibrate with the same frequency as the sound. Frequency is the number of vibrations per second, or hertz (Hz). The three bones of the middle ear amplify and transmit the mechanical movements to the oval window, a membrane on the surface of the cochlea. Vibrations of the oval window produce pressure waves in the fluid within the cochlea.

Figure 45.17

How the cochlea distinguishes pitch. (a) Vibrations of the stapes against the oval window agitates the fluid within the cochlea, setting up pressure ripples that have a frequency equivalent to the sound waves that entered the ear. (In this schematic representation, the cochlea is uncoiled so that it is easier to trace the fate of the pressure waves.) The waves pass through the vestibular canal to the apex of the cochlea, and then back toward the base of the cochlea via the tympanic canal. Much of the energy causes the cochlear duct, with its basilar membrane and organ of Corti (see Figure 45.16) to vibrate up and down. The bouncing of the basilar membrane stimulates the hair cells within the cochlear duct. (The remaining energy is dissipated by vibrations of the round window at the basal end of the tympanic canal.) **(b)** Spanning the width of the basilar membrane are fibers. Like harp strings, these fibers vary in length, being shorter near the basal end of the membrane and longer near the apex. And like harp strings, the length of the fibers "tunes" them to vibrate at specific frequencies. **(c)** Thus, pressure waves in the cochlea cause a specific region along the length of the basilar membrane to oscillate more vigorously than other regions not "tuned" to that particular frequency. The preferential stimulation of hair cells is perceived in the brain as sound of a certain pitch.

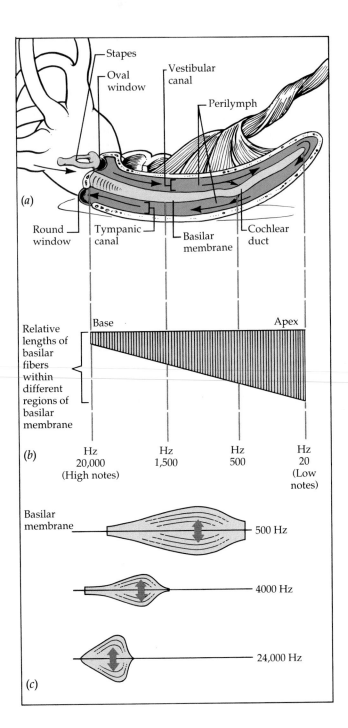

The cochlea transduces the energy of the vibrating fluid into action potentials. The stapes vibrating against the oval window creates a traveling pressure wave in the fluid of the cochlea that passes into the vestibular canal (see Figure 45.17). This wave continues around the tip of the cochlea and through the tympanic canal, dissipating as it strikes the **round window.** The pressure waves in the vestibular canal push downward on the central canal and basilar membrane. As the basilar membrane vibrates in response to the pressure waves, it alternately presses the hair cells into and draws them away from the tectorial membrane. This "tweaking" of the hairs distorts the plasma membranes of the receptor cell, making it more permeable to sodium. The resulting depolarization increases neurotransmitter release from the hair cell and the frequency of action potentials in the sensory neuron with which the hair cell synapses. This neuron carries the sensations to the brain through the auditory nerve.

Sound is detected by increases in the frequency of impulses in the auditory neuron, but how is the quality of that sound determined? Two important aspects of sound are volume and pitch. **Volume** (loudness) is determined by the **amplitude,** or height, of the sound wave. The greater the amplitude of a sound, the more vigorous the vibrations of fluid in the cochlea, the greater the bending of the hair cells, and the more action potentials generated in the sensory neurons. **Pitch** is a function of the **frequency** of sound waves and is usually expressed in hertz. Short, high-frequency waves produce high-pitched sound, while long, low-frequency waves generate low-pitched sound. Healthy young humans can hear sounds in the range of 40 to 20,000 Hz; dogs can hear sounds as high as 40,000 Hz; and bats can emit and hear sounds of even higher frequency, using this ability to locate objects by sonar (Figure 45.18).

Pitch can be distinguished by the cochlea because the basilar membrane is not uniform along its length. The proximal end near the oval window is relatively narrow and stiff, while the distal end near the tip is wider (see Figure 45.17) and more flexible. Each region of the basilar membrane is most affected by a partic-

Figure 45.18
Echolocation. Bats, which can hear sounds as high as 75,000 Hz, emit high-pitched clicking sounds and detect objects such as prey by the echos. In contrast, few humans can hear beyond 20,000 Hz.

ular frequency of vibrations, and the region vibrating most vigorously at any instant sends the most action potentials along the auditory nerve. But the actual perception of pitch depends on the mapping of the brain. Sensory neurons from the auditory pathway project onto specific auditory regions of the cerebral cortex according to the region of the basilar membrane in which the signal originated. When a particular site of the cortex is stimulated, we perceive a sound of a particular pitch.

Balance and Equilibrium in Mammals   The balance apparatus in humans and most other mammals is also centered in the inner ear. Behind the oval window is a vestibule (see Figure 45.15) that contains two chambers, the **utricle** and **saccule.** The utricle opens into three **semicircular canals** that complete the apparatus for equilibrium.

Sensations related to body position are generated much like sensations of sound in humans and most other mammals. Hair cells in the utricle and saccule respond to changes in head position with respect to gravity and movement in one direction. The hair cells are arranged in clusters, and all their hairs project into a gelatinous material containing many small calcium

carbonate particles called otoliths ("ear stones"). Because this material is heavier than the endolymph within the utricle and saccule, gravity is always pulling downward on the hairs of the receptor cells, sending a constant train of action potentials along the sensory neurons of the vestibular branch of the auditory nerve. When the position of the head changes with respect to gravity (as when the head bends forward), the force on the hair cell changes, and it increases (or decreases) its output of neurotransmitter. The brain interprets the resulting changes in impulse production by the sensory neurons to determine the position of the head. By a similar mechanism, the semicircular canals, arranged in the three planes of space, detect rotation of the head (Figure 45.19).

Hearing and Equilibrium in Other Vertebrates

Most fishes and aquatic amphibians have a **lateral line system** that runs along both sides of the body (see Chapter 30). The system contains mechanoreceptors that detect movement by a mechanism similar to the function of those in the inner ear. Water from the animal's surroundings enters the lateral line system through numerous pores and flows along a tube past the mechanoreceptors (Figure 45.20). The receptor units, called **neuromasts,** resemble structures called ampullae in the semicircular canals. Each neuromast has a cluster of hair cells, with the sensory hairs embedded in a gelatinous cap called the cupula. As pressure of the moving water bends a cupula, the hair cells transduce the energy into action potentials that are transmitted along a nerve to the brain. This information helps the fish perceive its movement through water or the direction and velocity of water currents flowing over its body. The lateral line system also detects water movements or vibrations generated by other moving objects, including prey and predators.

Like other vertebrates, fishes also have inner ears located near the brain. Along with the lateral line system, the inner ears enable a fish to hear. There is no cochlea, but there are a saccule, utricle, and semicircular canals, structures homologous to the equilibrium sensors of our own ears. Within these chambers in the inner ear of a fish, sensory hairs are stimulated by the movement of otoliths, tiny granules. Unlike the mammalian hearing apparatus, the inner ear of a fish has no eardrum and does not open to the outside of the body. Vibrations of the water caused by sound waves are conducted through the skeleton of the head to the inner ears, setting the otoliths in motion and stimulating the hair cells. The air-filled swim bladder (see Chapter 30) also vibrates in response to sound and may contribute to the transfer of sound to the inner ear. Some fishes, including catfishes and min-

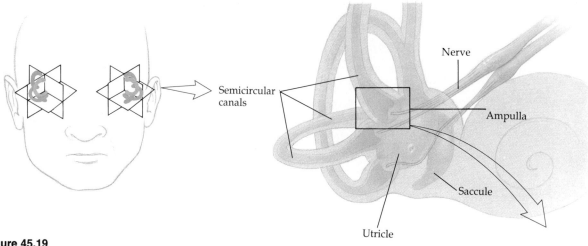

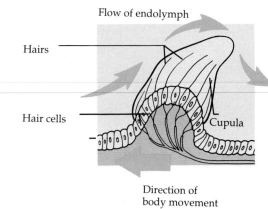

Figure 45.19
The semicircular canals and equilibrium. A swelling called an ampulla at the base of each semicircular canal contains a cluster of hair cells. The hairs from these cells project into a gelatinous mass called the cupula. When the head rotates, inertia prevents the endolymph in the canals from moving with it, so the fluid presses against the cupula, bending the hair cells. This bending increases the frequency of impulses in the sensory neurons in direct proportion to the speed of rotation. This receptor mechanism adapts quickly if rotation continues at a constant speed. Gradually, the endolymph begins to move with the head, and the pressure on the cupula is reduced. If rotation stops suddenly, however, the fluid continues to flow through the semicircular canals and again stimulates the hair cells. This new stimulus causes dizziness. The saccule and utricle contain hair cells that detect the pull of gravity on small particles called otoliths. This tells the brain which way is up and also informs the brain of any linear acceleration associated with movement.

nows, have a series of bones called the **Weberian apparatus,** which conducts vibrations from the swim bladder to the inner ear. Sound waves also stimulate the hair cells of the lateral line system, but only if the sound is of relatively low frequency. The inner ears extend the hearing of fishes to higher frequencies.

The lateral line system functions only in water. In terrestrial vertebrates, the inner ear has evolved as the sole organ of hearing and equilibrium. Some amphibians have a lateral line system as tadpoles, but not as adults living on land. In the ear of a terrestrial frog or toad, sound vibrations traveling in the air are conducted by a tympanic membrane and a single bone to the inner ear (the mammalian ear, remember, has three bones). There is recent evidence that the lungs of a frog also vibrate in response to sound and transmit their vibrations to the eardrum via the auditory tube. A small side pocket of the saccule functions as the main hearing organ of the frog, and it is this outgrowth of the saccule that gave rise to the more elaborate cochlea during the evolution of mammals. Birds also have a cochlea, but like amphibians and reptiles, sound is conducted from the tympanic membrane to the inner ear by a single bone, the stapes.

Sensory Organs for Hearing and Balance in Invertebrates

Many invertebrates sense sounds. Hearing structures have been most extensively studied in the arthropods. For example, the body hairs of many insects vibrate in response to sound waves of specific frequencies, depending on the stiffness and length of the hairs. The hairs are commonly tuned to frequencies of sounds produced by other organisms. Male mosquitos use fine hairs on their antennae to detect the hum produced by the beating wings of flying females, thus enabling them to locate a mate; a tuning fork that vibrates at the same frequency as a female mosquito's wings will also attract males. Some caterpillars, larvae of insects, use vibrating body hairs to hear the buzzing wings of predatory wasps, warning the caterpillars of danger. Many insects also have localized "ears," most commonly on their legs (Figure 45.21). A tympanic membrane (eardrum) is stretched over an internal air chamber. Sound waves vibrate the tympanic membrane, stimulating receptor cells attached to the inside of the membrane and resulting in nerve impulses that are transmitted to the brain. Some moths can hear

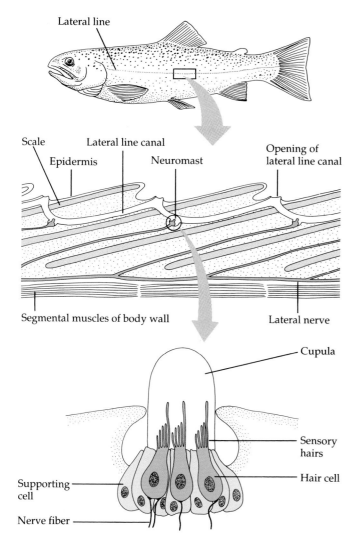

Figure 45.20
Lateral line system in fish. Water flowing through the system bends hair cells, which transduce the energy into action potentials conveyed to the brain. The lateral line system enables the fish to monitor water currents, pressure waves produced by moving objects, and low-frequency sounds conducted through water.

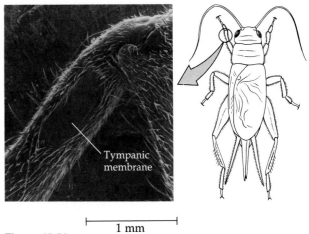

1 mm

Figure 45.21
An insect ear. The tympanic membrane, this one on the front leg of a cricket, vibrates in response to sound waves. This stimulates mechanoreceptor cells attached to the inside of the eardrum (SEM).

notes of such high pitch that they detect the sounds that bats produce for echolocation, and perception of these sounds triggers an escape maneuver in which the moths change directions abruptly.

Most invertebrates have mechanoreceptors called **statocysts** that function in their sense of equilibrium (Figure 45.22). A common type of statocyst consists of a layer of hair cells surrounding a chamber containing statoliths, which are grains of sand or other dense granules. Gravity causes the statoliths to settle to the low point within the chamber, stimulating hair cells in that location. (This is similar to how the saccule and utricle function in vertebrates, and indeed these structures in the vertebrate inner ear are considered to be specialized types of statocysts.) The statocysts of invertebrates have various locations. For example, many jellyfish have statocysts at the fringe of the "bell," giving the animals indications of body position. Lobsters and crayfish have statocysts near the bases of their antennules. Crayfish have been tricked into swimming upside down in experiments where the statoliths were replaced with metal shavings that could be pulled to the upper end of the statocysts with magnets.

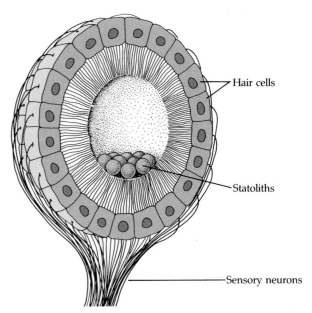

Figure 45.22
The statocyst of an invertebrate. The settling of statoliths to the low point within the chamber bends hair cells in that location, providing the brain with information about the position of the body.

TASTE AND SMELL

The senses of taste (gustation) and smell (olfaction) depend on chemoreceptors that detect specific chemicals in the environment. In the case of terrestrial animals, taste is the detection of certain chemicals that are present in a solution, and smell is the detection of airborne chemicals. However, these chemical senses are usually closely related, and there really is no distinction in aquatic environments.

Various animals use their chemical senses to find mates, to recognize territory that has been marked with some chemical substance, and to help navigate during migration (see Figure 45.1). Many animals, especially social insects, also use chemicals to communicate; in Chapter 50 you will learn how the social organization of a bee hive or ant hill is based on chemical "conversation." In all animals, taste and smell are important in feeding behavior. For example, the cnidarian *Hydra* begins to swallow when chemoreceptors detect a compound called glutathione, which is released from prey that has been captured by the tentacles of *Hydra*. Try adding some glutathione to the water around a *Hydra* in the laboratory, and you can observe this feeding behavior for yourself. (In fact, if there are two or more of the animals on your watch glass, they may resort to cannibalism.)

The taste receptors of insects are located within sensory hairs called **setae** on the feet and mouthparts. The animals use their sense of taste to select food. A tasting hair contains several chemoreceptor cells, each especially responsive to a particular class of chemical stimuli, such as sugar or salt (Figure 45.23). By integrating sensations (nerve impulses) from these different receptor cells, the brain of the insect can apparently distinguish a very large number of tastes. Insects can also smell airborne chemicals, using olfactory setae, usually located on the antennae (see Figure 45.3).

In humans and other mammals, the chemical senses of gustation and olfaction are functionally similar and interrelated. In both cases, a small molecule must dissolve in liquid to reach the receptor cell and stimulate the sensation. That molecule binds to a specific protein in the receptor cell membrane and triggers a depolarization of the membrane and release of neurotransmitter.

The receptor cells for taste are organized into **taste buds** scattered in several areas of the tongue and mouth. Most of the taste buds are on the surface of the tongue or are associated with raised papillae on the tongue. Although we cannot distinguish different types of taste receptors from their structures, we recognize four primary taste perceptions—sweet, sour, salty, and bitter—each detected in a distinct region of the tongue. These primary tastes are associated with specific molecular shapes or charges (the ring struc-

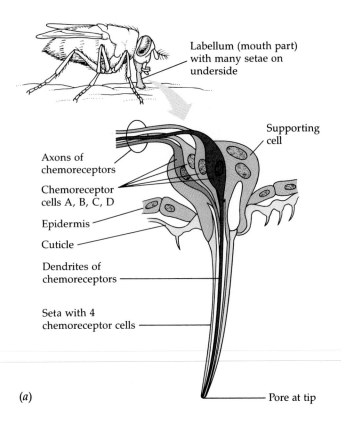

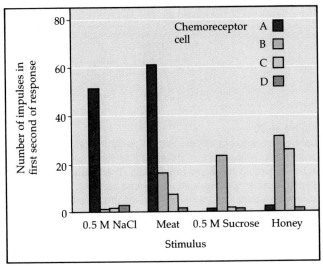

(b)

Figure 45.23
How a blowfly tastes. (a) Gustatory setae (hairs) on the feet and mouthparts each contain four chemoreceptor cells with dendrites that extend to the pore at the tip of the sensory hair. These receptor cells are labeled with letters. **(b)** Each taste cell is especially sensitive to a particular class of substance; for example, the receptor labeled B is most responsive to sugars. This specificity, however, is relative; each cell can respond to some extent to a broad range of chemical stimuli. Thus, any natural food probably stimulates two or more of the receptor cells. The brain apparently integrates the frequencies of impulses arriving along the axons of the four classes of receptor cells and distinguishes a great variety of tastes.

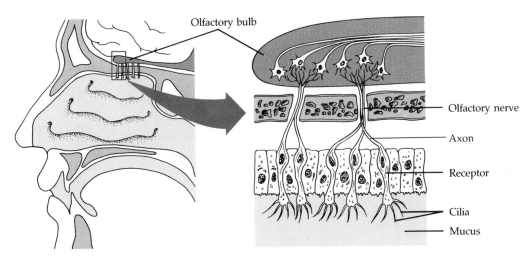

Figure 45.24
Olfaction in humans. Specific binding of molecules to the ciliary membrane of receptor cells triggers action potentials conveyed directly to the olfactory bulb of the brain by axons of the receptor cells.

ture of glucose for sweet, for instance, or the positive sodium ion for salty) that bind to separate receptor molecules. As with the taste receptors of insects, sensory data transmitted by sensory neurons from taste buds to the mammalian brain represent the differential stimulation of the various classes of receptors. Although each receptor cell is more reponsive to a particular type of substance, it can actually be stimulated by a broad range of chemicals. With each taste of food or sip of drink, the brain integrates the differential input from the taste buds, and a complex flavor is perceived.

The olfactory sense of mammals detects certain airborne chemicals. Olfactory receptor cells line the upper portion of the nasal cavity and send impulses along their axons directly to the olfactory bulb of the brain (Figure 45.24). The receptive ends of the cells contain cilia that extend into the layer of mucus coating the nasal cavity. When an odorous substance diffuses into this region, it binds to specific receptor molecules in the olfactory hairs and depolarizes the olfactory receptor cell. Humans can distinguish thousands of different odors, but these are probably based on a few primary odors, analogous to the primary tastes of the gustatory system.

Although the receptors and brain pathways for taste and olfaction are independent, the two senses do interact. Indeed, much of what we call taste is really smell. If the olfactory system is blocked, as by a head cold, the perception of taste is sharply reduced.

Throughout these discussions of sensory mechanisms, we have seen many examples of how these inputs to the nervous system result in the specific body movements that we observe as animal behavior.

The swimming of planarians away from light, the escape behavior of a moth that hears bat sonar, the feeding movements of *Hydra* when it tastes glutathione, and the homing of a salmon that can smell its breeding stream are just a few cases that have been mentioned so far in this chapter. The remainder of the chapter focuses on the motor mechanisms that make these animal responses possible—on how animals use their muscles and skeletons to move.

INTRODUCTION TO ANIMAL MOVEMENT

Movement is an animal hallmark. To catch food, an animal must either move through its environment or move the environmental milieu past itself. Sessile animals stay put, but they wave prehensile tentacles that capture prey or use beating cilia to generate water currents that draw and trap small food particles (see Chapter 37). Most animals, however, are mobile and spend a considerable portion of their time and energy actively searching for food, as well as escaping from danger and looking for mates.

The modes of animal locomotion are diverse. Several animal phyla include species that swim. On land and in the sediments on the floor of the sea and lakes, animals crawl, walk, run, or hop. Flight equipment has evolved in only a few animal classes—in the insects (Phylum Arthropoda) and in the reptiles, birds, and mammals (Phylum Chordata).

Whatever the mode of transportation, an animal must exert enough force against its environment to over-

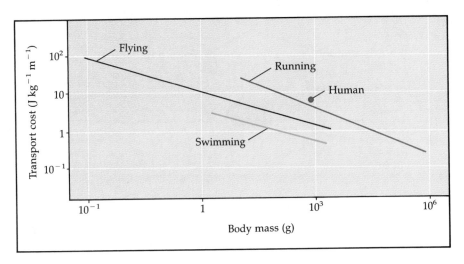

Figure 45.25
The cost of transport. This graph compares the transport cost, in joules per kilogram of body weight per meter traveled, for animals specialized for swimming, flying, and running (1 J = 0.24 calorie). Notice that both axes are plotted on logarithmic scales. Swimming is the most efficient mode of transport (assuming, of course, that an animal is specialized for swimming). If we were to compare energy consumption per *minute* rather than per meter, then we would find that flying animals use more energy than animals swimming or walking for the same amount of time. However, birds and flying insects are generally fast, and thus running requires more energy per *meter* than flying. To walk or run, an animal must not only work against gravity and friction, but must repeatedly accelerate limbs from standing starts, using considerable energy to overcome inertia. Notice also, however, that a larger animal travels more efficiently than a smaller species specialized for the same mode of transport. For example, a horse consumes less energy per kilogram of body weight than does a cat running the same distance (of course, *total* energy consumption is greater for the larger animal). If you compare the extremes of body mass on the x axis of this graph, you can see that a large, running animal actually moves at a lower energy cost than a small fish or bird.

come friction and gravity. The relative importance of these two impedances depends on the environment. Because most animals are reasonably buoyant in water, overcoming gravity is less of a problem for swimming animals than for species that move on land or through the air. On the other hand, water is a much denser medium than air, and thus the problem of resistance (friction) is a major one for aquatic animals. A sleek, fusiform (torpedo-like) shape is a common adaptation of fast swimmers. On land, a walking or running animal must be able to support itself and move against gravity, but, at least at moderate speeds, air poses relatively little resistance. To move in such an environment, strong skeletal support is more important than a streamlined shape. A walking or running animal must also overcome inertia with each step by accelerating a leg from a standing start. This is a major reason that running animals consume more energy per meter than do equivalently sized animals specialized for swimming or flying (Figure 45.25). A flying animal does not use its skeleton to support itself against the pull of gravity, but gravity poses a major problem in a different way: To be airborne, the wings must develop enough lift to enable the animal to completely overcome the downward force of gravity. In all of these situations, we observe body shapes, specialized appendages, and other adaptations of morphology representing evolutionary solutions to specific problems of movement.

Underlying these diverse forms of movement are fundamental mechanisms common to all animals. At the cellular level, all animal movement is based on one of two basic contractile systems, both of which use protein strands moving against one another. These two systems of cell motility, microtubules and microfilaments, were discussed in Chapter 7. Microtubules are responsible for the beating of cilia and the undulations of flagella. Microfilaments play a major role in amoeboid movement, and are also the contractile elements of muscle cells. It is the contraction of muscles that concerns us in this chapter, but the work of a muscle in itself cannot translate into movement of the animal. Swimming, crawling, running, and flying all result from muscles working against some type of skeleton.

SKELETONS AND THEIR ROLES IN MOVEMENT

Three functions of a skeleton are support, protection, and movement. Most land animals would sag from their own weight if they had no skeleton to support

(a)

(b)

|⎯⎯⎯⎯⎯⎯⎯⎯⎯⎯|
0.5 mm

Figure 45.26
Three types of skeletons. (a) Cnidarians, such as these jellyfishes, use a hydrostatic skeleton to aid in movement. Hydrostatic skeletons, consisting of pressurized fluid in a body compartment, are also typical of flatworms, nematodes, and annelids. **(b)** Many invertebrates, including arthropods, have exoskeletons that provide protection and support. In this scanning electron micrograph, an aphid is sitting next to its old exoskeleton, which was shed in a process called molting. **(c)** Vertebrates are among the animals that have endoskeletons. In this photo of a bat embryo, the red structures are bones that are in the process of ossifying (hardening by deposit of calcium salts), and the blue structures are cartilage.

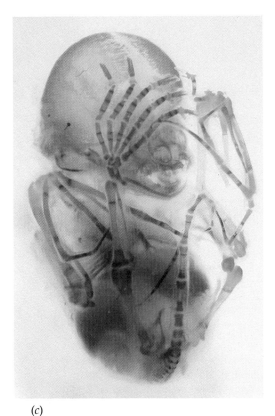

(c)

them. Even an animal living in water would be a formless mass with no framework to maintain its shape. Many animals have hard skeletons that protect soft tissues. For example, the vertebrate skull protects the brain, and the ribs form a cage around the heart, lungs, and other internal organs. And skeletons aid in movement by giving muscles something firm to work against. There are three main types of skeletons: hydrostatic skeletons, exoskeletons, and endoskeletons (Figure 45.26).

Hydrostatic Skeletons

A **hydrostatic skeleton** consists of fluid held under pressure in a closed body compartment. This is the main type of skeleton in most cnidarians, flatworms, nematodes, and annelids (see Chapter 29). These animals control their form and movement by using muscles to change the shape of the fluid-filled compartments. *Hydra,* for example, can elongate by closing its mouth and using contractile cells in the body wall to constrict the central gastrovascular cavity. Since water cannot be compressed very much, decreasing the diameter of the cavity forces it to increase in length. In planarians, the interstitial fluid is kept under pressure and functions as the main hydrostatic skeleton. Flatworms move by using muscles in the body wall to exert localized forces against this hydrostatic skeleton. Roundworms (nematodes) are able to hold the fluid in the body cavity (a pseudocoelom—see Chapter 29) at a high pressure, and contractions of longitudinal muscles result in thrashing movements. In earthworms, it is the coelomic fluid that functions as a hydrostatic skeleton. The coelomic cavity is divided by septa between the segments of the worm, and thus the animal can change the shape of each segment individually, using both circular and longitudinal muscles. Hydrostatic skeletons provide no protection, and they certainly could not support a large animal living on land.

Exoskeletons

An **exoskeleton** is a hard encasement deposited on the surface of an animal. For example, most mollusks are enclosed in calcareous (calcium carbonate) shells secreted by the mantle, a sheetlike extension of the body wall. As the animal grows, it enlarges the diameter of the shell by adding to its outer edge. Clams and other bivalves close their hinged shells using muscles attached to the inside of this exoskeleton.

The jointed exoskeleton typical of arthropods is a **cuticle,** a nonliving coat secreted by the epidermis. Muscles are attached to knobs and plates of the cuticle that extend into the interior of the body. About 30%–50% of the cuticle consists of **chitin,** a polysaccharide similar to cellulose (see Chapter 5). Fibrils of chitin are embedded in a matrix made of protein, forming a composite material, analogous to fiberglass, that combines strength and flexibility. Where protection is most important, the cuticle is hardened by the addition of organic compounds called quinones, which cross-link the proteins of the exoskeleton. Some crustaceans, such as crabs and lobsters, harden portions of their exoskeletons even more by adding calcium salts. In contrast, at the joints of the legs, where the cuticle must be thin and flexible, there is only a small amount of inorganic salts and little cross-linking of proteins. The exoskeleton of an arthropod cannot enlarge once it is deposited, and must periodically be shed (molted) and replaced by a larger case with each spurt of growth by the animal.

Endoskeletons

An **endoskeleton** consists of hard supporting elements, such as bones, buried within the soft tissues of an animal. Sponges are reinforced by hard spicules

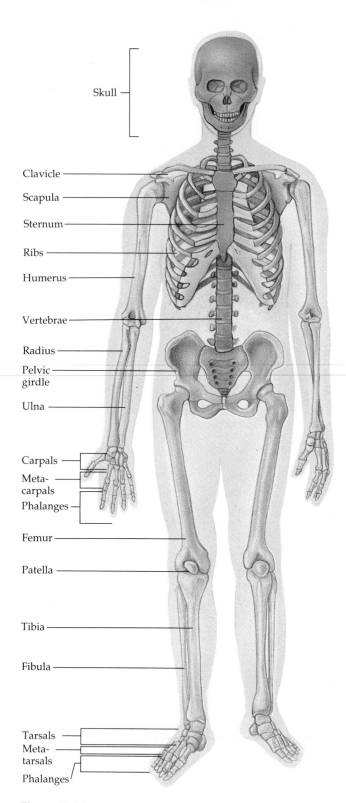

Skull
Clavicle
Scapula
Sternum
Ribs
Humerus
Vertebrae
Radius
Pelvic girdle
Ulna
Carpals
Meta-carpals
Phalanges
Femur
Patella
Tibia
Fibula
Tarsals
Meta-tarsals
Phalanges

Figure 45.27
The human skeleton. The axial skeleton is shaded green, and the appendicular skeleton is shaded gold.

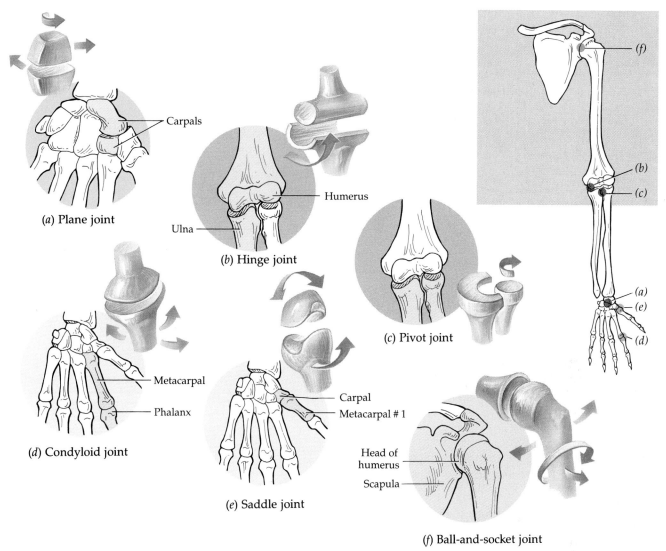

Figure 45.28
Types of joints in the vertebrate skeleton. With its many articulations, the vertebrate forelimb provides examples of six types of joints. The arrows indicate the movements facilitated by each joint.

(a) Plane joint

(b) Hinge joint

Carpals

Humerus

Ulna

(c) Pivot joint

(d) Condyloid joint

Metacarpal

Phalanx

(e) Saddle joint

Carpal

Metacarpal # 1

(f) Ball-and-socket joint

Head of humerus

Scapula

consisting of inorganic material or by softer fibers made of protein. Echinoderms have an endoskeleton of hard plates beneath the skin. These *ossicles* are composed of magnesium carbonate and calcium carbonate crystals, and the separate plates are usually bound together by protein fibers. Sea urchins have a skeleton of tightly bound ossicles, but the ossicles of sea stars are more loosely bound, allowing the animal to change the shape of its arms.

Chordates have endoskeletons consisting of cartilage, bone, or some combination of these materials (Figure 45.27). The mammalian skeleton is built from more than 200 bones, some fused together, and others connected at joints by ligaments that allow freedom of movement. Anatomists divide the vertebrate frame into an axial skeleton consisting of the skull, vertebral column (backbone), and rib cage; and an appendicular skeleton made up of limb bones and the pectoral and pelvic girdles that anchor the appendages to the axial skeleton. In just one appendage, the vertebrate forelimb, we can identify several types of joints (Figure 45.28).

In addition to functioning as buttresses that support the body, the bones of the vertebrate skeleton act as levers when the muscles attached to them contract.

MUSCLES

Animal movement, as we have seen, is based on the contraction of muscles working against some type of skeleton. The action of a muscle is *always* to contract; muscles can extend only passively. The ability to move a part of the body in opposite directions requires that muscles be attached to the skeleton in antagonistic

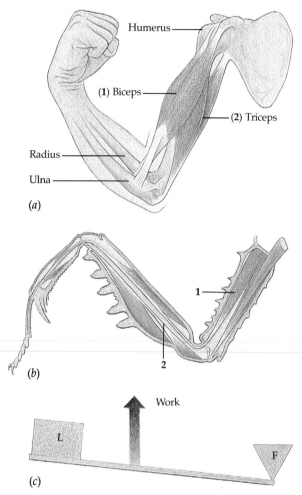

(a)

(b)

Work

(c)

Figure 45.29
Cooperation of muscles and skeletons in movement.
Muscles actively contract, but they elongate only when passively stretched. Back and forth movement is generally accomplished by antagonistic muscles, each working against the other. This arrangement works with either an endoskeleton or an exoskeleton. (**a**) In humans, contraction of the biceps (muscle 1) raises the forearm. Contraction of the triceps (muscle 2) lowers the forearm. (**b**) To raise the insect leg, muscle 2 contracts. When muscle 1 contracts, the leg lowers. Although contraction of muscle 1 produces an opposite effect in humans and insects, both muscle–skeletal systems work according to the same physical principle. (**c**) To raise the appendage, the skeleton acts as a lever, with the joint as a fulcrum (*F*) and the limb as the load (*L*). When the muscle applies force between the fulcrum and the load, the lever pivots upward about the fulcrum. Contraction of the muscles in the human and insect has different results because the muscles have different positions relative to the skeleton.

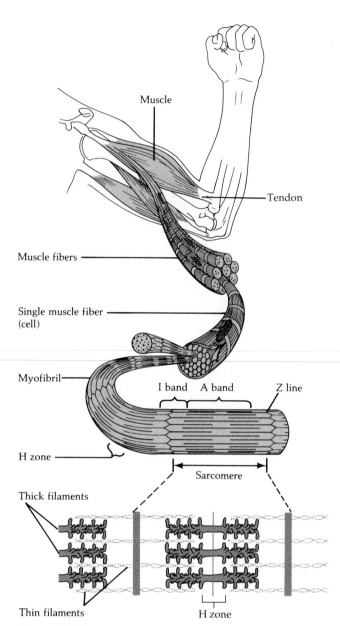

Figure 45.30
Structure of skeletal muscle. Skeletal muscle is attached by tendons to the bones that it moves. The muscle consists of bundles of multinucleated muscle fibers (cells), each consisting of a bundle of myofibrils. Each myofibril is made of thick and thin filaments aligned in a regular array of sarcomeres, units marked by Z lines at each end.

we will examine the structure and mechanism of contraction of vertebrate skeletal muscle and then compare this basic pattern with other types of muscle.

Structure and Physiology of Vertebrate Skeletal Muscle

Vertebrate **skeletal muscle,** which is attached to the bones and responsible for their movement, is characterized by a hierarchy of smaller and smaller parallel

pairs, each muscle working against the other (Figure 45.29). We flex our arm, for instance, by contracting the biceps, with the hinged joint of the elbow acting as the fulcrum of a lever. To extend the arm, we relax the biceps while the triceps on the opposite side contracts. But how does a muscle actually contract? The key to function, as usual, is structure. In this section,

units (Figure 45.30). A skeletal muscle is a bundle of long fibers running the length of the muscle. Each fiber is a single cell with many nuclei, reflecting its formation by the fusion of many embryonic cells. Each fiber is itself a bundle of smaller **myofibrils** arranged longitudinally. The myofibrils, in turn, are composed of two kinds of **myofilaments. Thin filaments** consist of two strands of actin and one strand of regulatory protein coiled around one another, while **thick filaments** are staggered arrays of myosin molecules.

Skeletal muscle is also called striated muscle because the regular arrangement of the myofilaments creates a repeating pattern of light and dark bands. Each repeating unit is a **sarcomere,** the fundamental unit of organization of the muscle. The borders of the sarcomere, the **Z lines,** are lined up in adjacent myofibrils to contribute to the striations visible with a light microscope. The thin filaments are attached to the Z lines and project toward the center of the sarcomere, while the thick filaments are centered in the sarcomere. At rest, the thick and thin filaments do not overlap completely, and the area near the edge of the sarcomere where there are only thin filaments is called the **I band.** The **A band** is the broad region where the thick and thin filaments do overlap. The thin filaments do not extend completely across the sarcomere, so the **H zone** in the center of the A band contains only thick filaments. This arrangement of thick and thin filaments is the key to how the sarcomere, and hence the whole muscle, contracts.

Molecular Mechanism of Muscle Contraction The mechanism of muscle contraction is explained by the **sliding filament model.** According to this model, the thin filaments slide, or more precisely ratchet, across the thick filaments to pull the Z lines together and shorten the sarcomere (Figure 45.31). The I bands shorten and the H zone disappears, but the A bands do not change in length. As all the sarcomeres in a myofibril shorten, they are pulled toward the center of the myofibril, allowing the muscle to contract to about half of its resting length. Note that the myofibril shortens and the individual filaments of the sarcomere slide across each other, but the myofilaments themselves do not contract.

The basis for the sliding filaments is the interaction of actin and myosin within the sarcomere. The heads of the myosin molecules on the thick filaments have an affinity for specific sites on the actin molecules and form **cross-bridges** between the thick and thin filaments (Figure 45.32). When a myosin head attaches to a thin filament, it bends inward on itself, exerting tension on the thin filament and pulling it toward the center of the sarcomere. These cross-bridges can form spontaneously because the myosin head is energetically more stable in the cross-bridge position. The head of the myosin molecule then cleaves a molecule of ATP and uses the energy to break the cross-bridge and return to its original position. The free head can now attach to another binding site farther along the thin filament, form another cross-bridge, and pull the thin

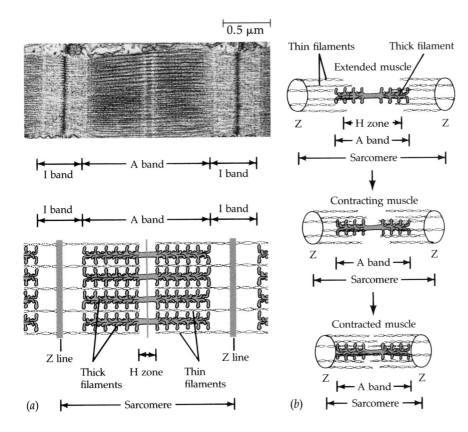

Figure 45.31
Sliding filament model of muscle contraction. (a) The alternating light and dark bands in the transmission electron micrograph of striated muscle from a frog show the regular organization of thick and thin filaments. As you can see in the diagram, the ends of the A bands include the regions where the thick and thin filaments overlap, while the H zone contains only the thick filaments. Only thin filaments occur in the I bands. The dark zone in the center of each I band, called the Z line, is attached to the thin filaments. A sarcomere is defined as the entire apparatus between successive Z lines. (b) Contraction of one sarcomere. The muscle contracts when the thick and thin filaments slide past each other. The sarcomere becomes shorter, although the lengths of the individual filaments do not change.

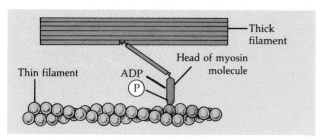

1. Myosin head of thick filament bound to actin monomer on thin filament, forming cross-bridge. Myosin head also attached to ADP and phosphate.

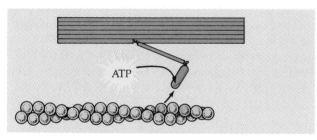

2. Power stroke: ADP + P released; bending of cross-bridge makes filaments slide past each other.

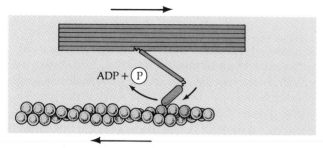

3. ATP energy releases myosin head from actin

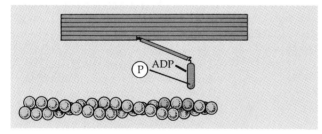

4. Hydrolysis of ATP returns myosin head to original position

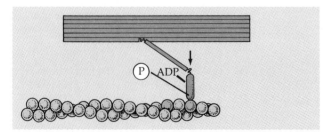

5. Myosin head attaches to new actin monomer

Figure 45.32
Mechanism of muscle contraction.

filament farther. This cycle—attach, bend, detach, straighten out—occurs repeatedly. In normal contraction, each of the appproximately 350 heads of a thick filament form and re-form about 5 cross-bridges per second.

Excitation–Contraction Coupling A muscle contracts only when stimulated by a motor neuron. When the muscle is at rest, the myosin-binding sites on the actin molecules are blocked by a strand of the regulatory protein **tropomyosin** in the thin filament (Figure 45.33). Another set of regulatory proteins, the **troponin complex,** is located at each binding site. For the muscle to contract, these obstacles must be displaced from the myosin-binding sites. This occurs when troponin binds calcium ions. The interactions between the calcium and the troponin alter the interaction of the troponin and the tropomyosin, causing the whole complex to change shape and expose the actin filament. When calcium is present, the muscle can contract, and when calcium concentrations fall, the binding sites of actin are covered and contraction stops.

Calcium concentration in the cytoplasm of the muscle-cell is regulated by the **sarcoplasmic reticulum,** a specialized endoplasmic reticulum (Figure 45.34). The membrane of the sarcoplasmic reticulum actively transports calcium from the cytoplasm into spaces within the sarcoplasmic reticulum, where it is stored. An action potential in the motor neuron innervating the muscle cell causes the axon to release the neurotransmitter acetylcholine into the neuromuscular junction. This induces a graded depolarization of the plasma membrane of the muscle cell, which is an excitable cell (see Chapter 44). A depolarization sufficiently large triggers an action potential that spreads across the muscle cell membrane. Infoldings of the plasma membrane called **transverse (T) tubules** carry the action potential deep into the muscle cell. Where the T tubules contact the sarcoplasmic reticulum, the action potential depolarizes the membrane of the sarcoplasmic reticulum, changing its permeability and causing it to release calcium ions. These calcium ions bind to troponin to allow the muscle to contract. As soon as the action potential passes, the sarcoplasmic reticulum pumps the calcium back out of the cytoplasm, and the tropomyosin–troponin complex again blocks the actin sites and prevents further contraction.

All this activity requires energy, but a muscle cell stores only enough ATP for a few contractions. The cells also store glycogen in columns between the myofibrils, but most of the energy is in substances called **phosphagens. Creatine phosphate,** the phosphagen of vertebrates, has a high-energy phosphate bond that it transfers to ADP to make ATP.

Graded Contractions of Whole Muscles Stimulation of a single muscle fiber by a motor neuron results in an "all-or-none" contraction. However, we know

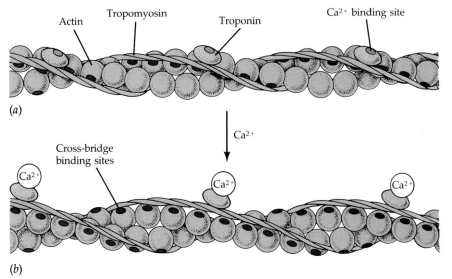

(a)

(b)

Figure 45.33
Control of muscle contraction. The thin filament has two strands of actin twisted to form a helix. (**a**) When the muscle is at rest, a long, rodlike tropomyosin molecule blocks the myosin-binding sites that are instrumental in forming cross-bridges. (**b**) When another protein complex, troponin, binds calcium ions, the actin binding sites are exposed, cross-bridges with myosin can form, and contraction occurs.

from experience that the action of a whole muscle, such as the biceps, is graded; we can vary the extent and strength of the contraction. The key to these graded contractions of muscles is the organization of the individual cells into motor units. Each muscle fiber is innervated by only one motor neuron, but one branched motor neuron may innervate many fibers. A **motor unit** consists of a single motor neuron and all the muscle fibers it controls (Figure 45.35). When the neuron fires, all of the muscle fibers of the motor unit contract. In muscles requiring fine control, such as those controlling eye movements, a motor neuron may innervate only one fiber. In larger muscles with less precise movements, like the biceps, a single neuron may innervate several hundred fibers. The muscle fibers of any one motor unit are scattered throughout the muscle, and thus stimulation of a single motor unit results in a weak contraction of the whole muscle. A more forceful contraction is achieved when additional motor units are recruited. This **multiple motor unit summation,** or **recruitment,** is one mechanism responsible for graded responses of skeletal muscles.

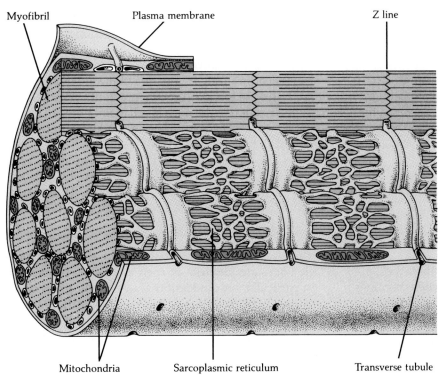

Figure 45.34
The sarcoplasmic reticulum and transverse tubules. The sarcoplasmic reticulum is a system of membranes that accumulates and releases the calcium ions that trigger muscle contraction. The transverse tubules are continuous with the plasma membrane of the muscle fiber and relay the contraction signal (action potential).

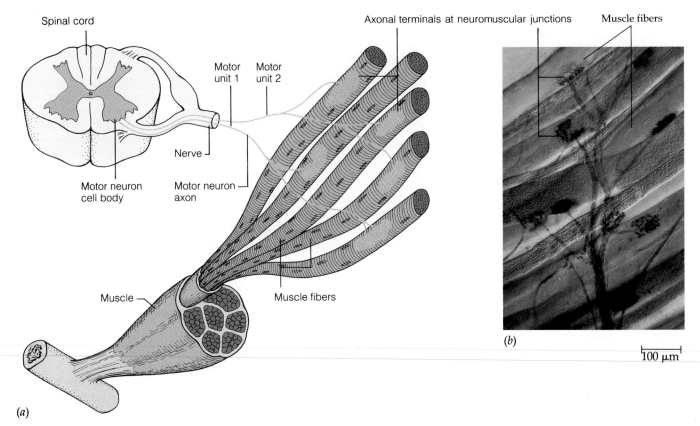

Spinal cord

Axonal terminals at neuromuscular junctions

Muscle fibers

Motor unit 1 Motor unit 2

Nerve

Motor neuron cell body

Motor neuron axon

Muscle

Muscle fibers

(b)

100 μm

(a)

Figure 45.35
Motor units. (a) A branched motor neuron typically innervates several muscle fibers scattered throughout the muscle. The entire contractile apparatus—the neuron and the fibers it controls—is called a motor unit. One factor that strengthens a muscular contraction is the recruitment of additional motor units. **(b)** In this light micrograph, you can see a single branched axon that synapses with several muscle fibers of a motor unit.

Depending on how many motor units your brain commands to contract, you can lift a fork, or something much heavier, such as your biology textbook.

A second mechanism responsible for graded contraction of muscles is **wave summation.** In laboratory experiments where an isolated muscle can be electrically stimulated, a single stimulus results in a **muscle twitch,** one rapid contraction followed by relaxation of the muscle. The stronger the stimulus, the more motor units recruited, and the more vigorous the twitch. If the muscle is given two electrical shocks of equal voltage in rapid-fire succession, the second twitch is stronger than the first. This wave summation occurs because the second contraction commences before the muscle is completely relaxed in the wake of the first contraction, and the muscle, still somewhat shortened, contracts further. If the muscle is stimulated repeatedly at a faster and faster rate, then the time allowed for relaxation between twitches becomes shorter and shorter. And if the rate of stimulation is fast enough, then the twitches blur into one smooth and sustained contraction called **tetanus** (not to be confused with the disease). Motor neurons usually deliver their stimuli in rapid-fire volleys, and wave summation results in the smooth contraction typical of tetanus rather than the jerky actions of muscle twitches.

Prolonged tetanus results in muscle fatigue. Three factors contributing to muscle fatigue are depletion of ATP, dissipation of the ion gradients required for depolarization of the muscle cell membrane, and accumulation of lactic acid, a waste product of fermentation in the muscle (see Chapter 9).

Some muscles, especially those that hold the body up and maintain posture, are almost always partially contracted, a condition known as **tonus.** The muscles do not fatigue because different motor units take turns maintaining the tonus.

Slow and Fast Muscle Fibers   Not all skeletal muscle fibers are identical in their contractile properties. In particular, we can identify slow and fast fibers based on the duration of their twitches. A twitch in a slow muscle fiber lasts about five times longer than one in a fast fiber because the slow fiber has less sarcoplasmic reticulum, so calcium remains in the cytoplasm longer.

Slow fibers are also specialized to make use of a steady supply of energy; they have many mitochondria, a rich blood supply, and an oxygen-storing protein called **myoglobin.** Myoglobin, the brownish-red pigment in the "dark meat" of poultry and fish, binds oxygen more tightly than hemoglobin, so it can effectively extract oxygen from the blood. Slow fibers are often found in muscles used to maintain posture, because they can sustain long contractions. Fast muscles are used for rapid, powerful contractions. Some, such as the flight muscles of birds, may be able to sustain long periods of contractile activity without fatiguing.

Other Types of Muscles

There are many types of muscles in the animal kingdom, but as noted before, they all share the same fundamental mechanism of contraction: the sliding of actin microfilaments and myosin microfilaments past one another. In Chapter 36, we saw that in addition to skeletal muscle, vertebrates also have cardiac and smooth muscle.

Vertebrate **cardiac muscle** is found in only one place— the heart. Like skeletal muscle, cardiac muscle is striated. The primary differences between skeletal and cardiac muscle are in their electrical and membrane properties. Cardiac muscle cells are branched, and the junctions between cells contain specialized regions called **intercalated discs** (see Figure 36.6). Intercalated discs are gap junctions that electrically couple all the muscle cells in the heart. If one cell is stimulated strongly enough, its action potentials will spread to all the cells in the heart, and the entire heart will contract (see Chapter 38). Cardiac muscle cells can also generate action potentials on their own, without any input from the nervous system. This ability results from a relatively high membrane permeability to sodium. The action potentials themselves last up to 20 times longer than those for skeletal muscle and have a decided plateau after the peak. The refractory period of these excitable cells is also quite long, so a contraction of the heart lasts a relatively long time, and the heart relaxes completely between beats.

Smooth muscles lack the striations of skeletal and cardiac muscle because the actin and myosin filaments are not all aligned along the length of the cell. Rather, the microfilaments may have a spiral arrangement within fibers of smooth muscle. Smooth muscle also appears to contain less myosin than striated muscle, and the myosin is not associated with specific actin strands. Because of this organization, smooth muscle cannot generate nearly as much tension as striated muscles, but it can contract over a much greater range of lengths. Further, smooth muscle does not have either a T-tubule system or a well-developed sarcoplasmic reticulum. Calcium ions must enter the cytoplasm via the plasma membrane during an action potential, and the amount reaching the filaments is rather small. Contractions are relatively slow, but there is a greater range of control than is possible in striated muscle.

Invertebrates possess muscles similar to vertebrate skeletal and smooth muscles. Arthropod skeletal muscles are nearly identical to vertebrate skeletal muscles. However, the flight muscles of insects are capable of independent, rhythmic contraction, so the wings of some insects can actually beat faster than action potentials can arrive from the central nervous system. Another interesting adaptation has been discovered in the muscles that hold clam shells closed. The thick filaments of these muscle fibers contain a unique protein called paramyosin that allows the muscles to remain in a fixed state of contraction for as long as a month.

*   *   *

Although we have talked about receptors and muscles separately in this chapter, they are the two ends of a single integrated system. An animal's behavior, so fundamental to how the animal interacts with its environment, is the product of a nervous system connecting certain sensations to certain responses. Behavior is discussed in the next unit within the broader context of ecology, the study of interactions between organisms and their environment.

STUDY OUTLINE

1. Animal behavior depends on the ability of the nervous system to integrate the input from sensory receptors about the internal and external environments and translate the information into appropriate responses in muscles and other effectors.

Sensory Receptors (pp. 1011–1016)

1. Sensations are action potentials traveling along sensory neurons that are interpreted by different parts of the brain as perceptions.

2. Sensory receptors are usually modified neurons that collect and transmit information from environmental stimuli.

3. Reception is the specialized ability of a receptor cell to absorb the particular energy of a stimulus.

4. Transduction is the conversion of the stimulus energy into the electrochemical energy of changes in membrane potentials and action potentials.

5. The stimulus energy may be amplified by accessory structures of sense organs or by the process of transduction.

6. Transmission of the receptor potential of the receptor cell to the nervous system occurs either as action potentials (when the receptor is a sensory neuron) or as release of neurotransmitter, which then initiates action potentials in the sensory neuron with which the receptor cell synapses. Changes in the spontaneous firing rate of sensory cells provide information on both the presence or absence of the stimulus and its intensity.

7. Integration of information begins at the receptor level by the processes of summation, adaptation, and variation in the sensitivity of the receptor. Integration and processing continue in the central nervous system.

8. Sensory receptors can be classified as exteroreceptors or interoreceptors. They can also be categorized on the basis of the type of stimulus energy they receive.

9. Mechanoreceptors respond to stimuli such as pressure, touch, stretch, motion, and sound. Bending or stretching of the cell membrane in response to such stimuli affects ion permeabilities and creates a receptor potential. Hair cells are common mechanoreceptors used in motion detection, in hearing, and in balance.

10. Chemoreceptors respond either to generalized solute concentrations or to specific molecules. Gustatory and olfactory receptors respond to groups of related chemicals.

11. Electromagnetic receptors detect energy in the form of different wavelengths or radiation, such as visible light, electricity, and magnetism.

12. Various types of thermoreceptors signal surface and core temperatures of the body.

13. Pain is detected by nociceptors, a group of diverse receptors that respond to excess temperature, pressure, or specific classes of chemicals.

Vision (pp. 1016–1023)

1. The light receptors and visual capabilities of invertebrates vary widely, ranging from the simple light-sensitive eye cup of planaria to the image-forming compound eye of insects and crustaceans and the single-lens eye found in some jellyfish, spiders, and many mollusks.

2. The vertebrate eye is composed of three concentric layers: an outer sclera with its transparent, anterior cornea; a middle pigmented choroid layer with the iris that surrounds the pupil; and the innermost retina, which contains the photoreceptor cells.

3. Transduction of the light signal occurs in specialized photoreceptors called rods and cones, which contain the light-absorbing molecule retinal bonded to a form of opsin. Rods, which are more light sensitive and are responsible for night vision, contain the molecule rhodopsin. The three types of cones contain photopsins and detect various colors in bright light.

4. When rods and cones absorb light, their membrane becomes hyperpolarized, and they release less neurotransmitter. This change in chemical message is transmitted to bipolar cells and then to ganglion cells, whose axons in the optic nerve convey action potentials to the brain. Horizontal cells and amacrine cells integrate information before it is sent to the brain—by contrast-enhancing lateral inhibition, for example.

5. The left and right optic nerves meet at the optic chiasma. Most axons of the ganglion cells go to the lateral genic-ulate nuclei of the thalamus, from which neurons lead to the primary visual cortex in the occipital lobe.

6. The process by which the cortex integrates visual sensations into our perception of sight is still largely unknown.

Hearing and Balance (pp. 1024–1029)

1. In mammals and most terrestrial vertebrates, the hair cell mechanoreceptors for hearing and balance are located within the ear.

2. The outer ear consists of the external pinna and auditory canal. The tympanic membrane transmits sound waves to the three small bones of the middle ear, which amplify and transmit the waves through the oval window to the fluid in the coiled cochlea of the inner ear. The resultant pressure waves vibrate the basilar membrane and the attached organ of Corti, which contains receptor hair cells. The bending of the hairs against the tectorial membrane depolarizes the hair cells, and the neurotransmitter they release initiates action potentials in the sensory neurons that then travel through the auditory nerve to the brain.

3. Volume is a function of the amplitude of the sound wave that results in greater bending of the hair cells and an increase in action potentials generated in sensory neurons. Pitch is related to frequency of sound waves. Regions of the basilar membrane vibrate more vigorously at different frequencies and transmit their impulses to specific regions of the cortex, where the sensations are perceived as particular pitches.

4. The inner ear vestibular apparatus, composed of the utricle, saccule, and three semicircular canals, functions in balance and equilibrium. The utricle and saccule detect static head position; the semicircular canals respond to rotation of the head.

5. The detection of movement in fishes and aquatic amphibians is accomplished by a lateral line system of clustered hair cell receptors. Fishes also have inner ears that sense vibrations of the water that are conducted through the skeleton of the head and the swim bladder.

6. Many arthropods use vibrating body hairs and localized "ears," consisting of a tympanic membrane and receptor cells, to sense sounds. Most invertebrates detect position in space by means of statocysts, chambers that contain hair cells and statoliths, or grains of sand.

Taste and Smell (pp. 1030–1031)

1. Taste and smell both depend on the stimulation of receptor cells by small, dissolved molecules that bind to proteins on a chemoreceptor membrane.

2. Setae containing taste receptors are found on the feet and mouthparts of insects. Insects also have olfactory setae, usually located on their antennae.

3. In mammals, taste receptors are organized into various kinds of taste buds that respond to distinct shapes of molecules. Olfactory receptor cells line the upper part of the nasal cavity and send their axons to the olfactory bulb of the brain. These chemoreceptors respond to certain airborne chemicals.

Introduction to Animal Movement (pp. 1031–1032)

1. Skeletal structure and body form reflect evolutionary solutions to the challenges of moving in a water, land, or air environment. Skeletons function in support, protection, and movement.

2. Animal movement is a function of protein strands moving past each other, either in the microtubules of cilia and flagella, or by microfilaments in amoeboid movement and muscle contraction.

Skeletons and Their Roles in Movement (pp. 1032–1035)

1. A hydrostatic skeleton, found in most cnidarians, flatworms, nematodes, and annelids, consists of fluid under pressure in a closed body compartment.
2. Exoskeletons, found in most mollusks and arthropods, are hard coverings deposited on the surface of an animal. Chitin is a primary component of the flexible cuticle of an arthropod.
3. Endoskeletons, found in sponges, echinoderms, and chordates, are hard supporting elements embedded within the animal's body. Cartilage and/or bone comprise the endoskeletons of chordates.

Muscles (pp. 1035–1041)

1. Muscles, often present in antagonistic pairs, contract and pull against elements of the skeleton to provide movement.
2. Vertebrate skeletal muscle consists of a bundle of muscle cells or fibers, each of which contains myofibrils composed of thin filaments of actin and thick filaments of myosin. The regular arrangement of these myofilaments into sarcomeres is the basis for contraction.
3. Contraction begins when impulses from a motor neuron are transmitted to the muscle cell membrane through release of acetylcholine at the neuromuscular junction. Action potentials travel to the interior of the cell along the transverse (T) tubules, stimulating the release of calcium from the sarcoplasmic reticulum. The calcium binds to the regulatory troponin–tropomyosin complex on the thin filaments, exposing the myosin-binding sites on the actin. Cross-bridges form, and bending of the myosin heads pulls the thin filaments toward the center of the sarcomere. Hydrolysis of ATP breaks the cross-bridges, which then form farther along the thin filament.
4. Energy for sustained contraction comes from glycogen and phosphagens, such as creatine phosphate, which regenerate ATP.
5. A motor unit consists of a branched motor neuron and the muscle fibers it innervates. Multiple motor unit summation, or recruitment, results in stronger contractions.
6. A muscle twitch results from a single stimulus. More rapidly delivered signals produce a graded contraction by wave summation. Tetanus is a state of smooth and sustained contraction, obtained when motor neurons deliver a volley of action potentials. Tonus is a state of partial contraction of the muscles involved in maintaining posture.
7. Slow muscle fibers, which have many mitochondria, a rich blood supply, and oxygen-storing myoglobin, are specialized for sustaining long contractions. Fast muscle fibers are used for rapid and powerful contractions.
8. Cardiac muscle, found only in the heart, consists of striated, branching cells that are electrically connected by intercalated discs. Cardiac muscle cells can generate action potentials without neural input.
9. Smooth muscle, which is unstriated owing to its spiral arrangement of actin and myosin, lacks T tubules and a well-developed sarcoplasmic reticulum. Contractions are slow, but can be sustained over long periods and over a greater range of lengths.

SELF-QUIZ

1. Which of the following receptors is *incorrectly* paired with its category?
   a. hair cell—mechanoreceptor
   b. muscle spindle—mechanoreceptor
   c. gustatory receptor—chemoreceptor
   d. rod—photoreceptor
   e. nociceptor—deep pressure receptor

2. The intensity of a receptor potential is
   a. dependent on the intensity of the action potentials produced by the receptor cell
   b. translated into the strength of the action potential when the receptor cell is a sensory neuron
   c. correlated with the frequency of action potentials fired by the sensory neuron
   d. referred to as sensation
   e. all-or-none of the above

3. Which of the following is an *incorrect* statement about the vertebrate eye?
   a. The vitreous humor regulates the amount of light entering the pupil.
   b. The transparent cornea is an extension of the sclera.
   c. The fovea is the center of the visual field and contains only cones.
   d. The ciliary muscle functions in accommodation.
   e. The retina lies just inside the choroid and contains the photoreceptor cells.

4. The utricle and saccule are
   a. involved in lateral integration of visual signals
   b. visual centers in the cortex
   c. part of the vestibular apparatus responsible for positional information about the head
   d. semicircular canals that respond to head rotation
   e. organs of balance found in insects and crustaceans

5. When you first walk from a brightly lit area into darkness, which of the following occurs?
   a. The photopsins in your cones become bleached.
   b. Your rod cells become hyperpolarized.
   c. The receptor cells release less neurotransmitter.
   d. Lateral inhibition caused by your horizontal cells ceases.
   e. Your rhodopsin is still dissociated into retinal and opsin, and your rods are nonfunctional.

6. The transduction of sound waves to action potentials takes place
   a. within the tectorial membrane as it is stimulated by the hair cells
   b. when hair cells are bent against the tectorial membrane, causing them to depolarize and release neurotransmitter molecules that stimulate sensory neurons
   c. as the basilar membrane becomes more permeable

to sodium and depolarizes, initiating an action potential in a sensory neuron

d. as the basilar membrane vibrates at different frequencies in response to the varying volume of sounds

e. within the middle ear as the vibrations are amplified by the malleus, incus, and stapes

7. The role of calcium in muscle contraction is

a. to break the cross-bridges as a cofactor in the hydrolysis of ATP

b. to bind with troponin, changing its shape so that the actin filament is exposed

c. to transmit the action potential across the neuromuscular junction

d. to spread the action potential through the T tubules

e. to reestablish the polarization of the plasma's membrane following an action potential

8. The term tetanus refers to

a. the partial sustained contraction of major supporting muscles

b. the all-or-none contraction of a single muscle fiber

c. a stronger contraction resulting from multiple motor unit summation

d. the result of wave summation, which produces a smooth and sustained contraction of a muscle

e. the state of muscle fatigue caused by depletion of ATP and accumulation of lactic acid

9. Which of the following is a *true* statement about cardiac muscle cells?

a. They lack an orderly arrangement of actin and myosin filaments.

b. They have less extensive sarcoplasmic reticulum and thus contract more slowly than do smooth muscle.

c. They are connected by intercalated discs, through which action potentials spread to all cells in the heart.

d. They have a relatively low sodium permeability and resting potentials more positive than an action potential threshold.

e. They only contract when stimulated by neurons.

10. Which of the following changes occurs when a skeletal muscle contracts?

a. The A bands shorten.

b. The I bands shorten.

c. The Z lines slide farther apart.

d. The actin filaments contract.

e. The thick filaments contract.

CHALLENGE QUESTIONS

1. Explain the difference between sensation and perception. Give some examples.

2. Compare vision and hearing with respect to receptor type, transduction, amplification, transmission, and integration characteristics.

3. Although skeletal muscles generally fatigue fairly rapidly, clam shell muscles, as mentioned in the chapter, have a unique protein called paramyosin that allows them to sustain contraction for up to a month. From your knowledge of the cellular mechanism of contraction, propose a hypothesis to explain how paramyosin might work. How would you investigate your hypothesis experimentally?

FURTHER READING

1. Alexander, R. "Muscles Fit for the Job." *New Scientist*, April 15, 1989. How are particular muscles adapted for diverse functions in movement?

2. Borg, E., and S. A. Counter. "The Middle-ear Muscles." *Scientific American*, August 1989. Muscles associated with the ear bones affect hearing.

3. Carafoli, E., and J. T. Penniston. "The Calcium Signal." *Scientific American*, November 1985. A close look at calcium and its regulation and function in processes like muscle contraction.

4. Masland, R. H. "The Functional Architecture of the Retina." *Scientific American*, December 1986. Encoding the visual world.

5. Nathans, J. "The Genes for Color Vision." *Scientific American*, February 1989. The genetics and physiology of color blindness.

6. Nilsson, D-E. "Vision Optics and Evolution." *Bioscience*, May 1989. Comparative anatomy and physiology of eyes.

7. Renouf, D. "Sensory Functions in the Harbor Seal." *Scientific American*, April 1989. Sensory adaptations of a mammal that must function on land and under water.

8. Schnapf, J. L., and D. A. Baylor. "How Photoreceptors Respond to Light." *Scientific American*, May 1987. How light is transduced into sensations.

Ecology

Interview: Jane Goodall

Inspired by her love of animals and her desire to write about them, an eighteen year old Jane Goodall traveled by boat from her home in England to the shores of Africa. There she met the famed anthropologist/paleontologist, Louis S. B. Leakey, who hired her as an assistant at Nairobi's Natural History Museum. Dr. Leakey later encouraged her to begin a field study observing chimpanzees at a remote site in East Africa. In 1960 Jane Goodall made her first trip to the Gombe Stream Reserve (now the Gombe National Park) in Tanzania. What she thought might be a three year study became a life-long mission to understand chimpanzee behavior and share that understanding with others.

Dr. Goodall's extraordinary discoveries have been closely followed by the world in her six published books, her numerous appearances on National Geographic Society television specials, and her frequent lecture tours.

In this interview, Dr. Goodall shares some fascinating stories and magical moments from her experience studying our closest relatives in their natural habitats.

Dr. Goodall, why did you make this long trip from Tanzania to New York?

This particular occasion is to receive one of the centennial awards of the National Geographic Society. I just happen to be one of the scientists who's been involved with them for a very long time.

You've just finished an interview with *Time* magazine, and you do a lot of public speaking. How do you view the responsibility of a scientist to public education?

I feel a sense of responsibility to the public. Then, too, I owe a great deal to the chimpanzees; they have given me so much. Chimpanzees need our help: They are endangered in the wild and often misused in captivity. To effect change it is often necessary to get the support of the public. Only if the public understands about the animals and the issues will they care enough to help and they can understand only if I share information with them.

When a scientist has celebrity status, are there special problems and benefits?

The main problem is needing to be in so many places at the same time—particularly now that I'm involved in trying to enforce more humane conditions for primates in medical research labs. When I'm asked to attend a conference "because it could make a difference"— well, I have to go, if I can. It all means I get less time to do field research at Gombe. And I spend days

and days in airplanes. Because a great deal of the cruelty inflicted on nonhuman animals is due to ignorance, I feel it is important to spend time with the media, to write books for children, and to talk at schools in addition to the college and public lectures. And this is not only in the United States but in Europe as well. However, so long as it really does make a difference, even a small one, it's worthwhile.

One of your current projects is the ChimpanZoo program. Can you tell us about that?

In 1984, I suddenly realized that chimp groups in zoos would make ideal subjects for study. Students, keepers, and volunteers could all become involved. The research would help zoo management improve their exhibits, which would, in turn, benefit the chimps. Finally, the project would create growing understanding of chimpanzees, their complex personalities, and their intelligence. The zoos that I approached were enthusiastic. Now, in 1989, fifteen zoos are committed to statistical data collection, and many others are contributing in a less formal way. Eventually, we shall be able to compare behavior in the different sites—a cross-cultural study.

What is the Jane Goodall Institute for Wildlife Research, Education and Conservation? And how is it related to the Gombe Research Center, the actual site of your research?

The Jane Goodall Institute was formed in 1976. We support research at Gombe and also contribute to chimpanzee research in other parts of Africa and to the ChimpanZoo program. We are working to enforce better living conditions for chimps in medical research labs. We are also launching a series of education programs for children, students, and the general public concerning our relationship with nonhuman animals in general and chimpanzees in particular.

What facilities exist at the Gombe center now, and what is the typical camp population like today?

I have a cement block house with a grass-covered, galvanized iron roof. It's simple; there's no running water or electricity. The windows are made of weld-mesh to keep the baboons and chimps out. The day-to-day research is carried on by Tanzanian field assistants; they have their own little "village," which, like my house, is on the shore of Lake Tanganyika. About half a mile inland is "chimp camp"—two aluminum huts—where we still feed bananas to the chimps (each chimp gets about six bananas once every ten days or so). And those really are all the facilities we have at Gombe.

Your childhood dream was to study animals in Africa. How did that interest develop at such a young age?

I was born that way. When I was two years old, I once took worms to bed with me. My mother was wonderful; when she found them, instead of saying "Yuck! How disgusting!" and throwing them out, she said, "Jane, if you leave them there they'll die, they need the damp earth." So I gathered them up as fast as I could and ran into the garden with them to save them. That early interest continued. I watched insects and birds in the garden and, as I got older, made notes about them. Then I began reading books about Africa. *Dr. Dolittle* and *Tarzan*—stories about African animals. By the time I was eight, I knew that I had to go to Africa.

How did you finance your first trip to Africa?

By giving up my very fascinating job with documentary film-making in London, and working as a waitress—it paid better.

How did Louis Leakey first learn about you?

I heard about him and went to see him. He asked me all kinds of questions and was clearly impressed with my knowledge of African animals and offered me a job as his secretary-assistant. Soon after that I accompanied him and his wife on their annual dig at Olduvai Gorge. That was in 1957, before the first of the hominid remains had been found there. It was the Africa of my dreams come true—utterly remote and wild with no car tracks but our own. There were lions and rhinos and giraffes all around us.

What was Leakey's vision in first sending you to study the chimpanzees of Gombe?

Louis told me he had been looking for ten years for the right person to go and study mankind's closest living relative in the wild, hoping the results would give him clues about the behavior of early humans. Today we take for granted his logic—that behaviors shared by modern man and modern chimpanzee are likely to have been present in the common ancestor and, therefore, in early man himself. But at that time it was a new approach, one of the signs of Leakey's genius. I was concerned that I had no college degree. In his eyes this was a plus: He wanted a person whose mind was uncluttered by scientific theory.

You've written that you do not personally find it necessary to justify the study of chimpanzees by suggesting

that the results will help us in our long search to comprehend human behavior. Would you please explain what you mean by that?

Well, first let me make it clear that I *do* believe that the results of the research help us to better understand some aspects of human behavior. But equally I believe that the study of creatures as complex and fascinating as chimpanzees is important in its own right. In fact, the most important spin-off of the chimp research is probably the humbling effect it has on us who do the research. We are not, after all, the only aware, reasoning beings on this planet.

What can ethology, the study of animal behavior in the field, contribute to general ecology?

You can't understand the ecology of a given area without knowing how the animal species behave, how much territory they need, and so on. It is particularly important to understand the needs of the various animal species if one is to effectively manage a reserve or national park. The sciences of ecology and ethology are interdependent.

Record-keeping at Gombe includes weighing chimpanzees. How do you weigh a wild chimpanzee?

At chimp camp we have a spring balance suspended on a chain between two trees. A rope hangs from this, near the top of which is attached a tin can. When we want to weigh a chimp, we put a banana in the can, the chimp then climbs the rope, and we read the weight off the scale.

Dr. Goodall, you once said that being accepted by the chimpanzees was one of the most momentous episodes of your life. How did you first earn that acceptance?

Well, when I first got to Gombe, the chimps would run off, even if I was 500 yards away. It was rather depressing. Then I discovered "the peak," a wonderful vantage point with a view over two valleys. I stopped trying to get close to the chimps. Instead, I climbed up to the peak day after day and sat there watching through my binoculars. As I gradually pieced together the daily behavior patterns of the chimps, they slowly got used to me and eventually lost their fear. Then I was able to move ever closer. Never shall I forget the day when I

approached to within 20 yards of David Greybeard and Goliath; they just glanced at me and went on grooming—I was accepted. Even now, 29 years later, I never take my relationship with the chimps for granted. When I sit among a group in the forest and a mother will allow her infant to sleep a few feet away from me, I am overwhelmed by the trust that the chimpanzees have in me. It's a terrific responsibility—I must never allow that trust to be broken.

You've also said that if you'd known the study would have continued for so long, you would not have established such close contact with the chimpanzees. Would you please explain?

First, let me say how the contact came about. Can you imagine how thrilled I was when, after struggling for so many months to get anywhere near the chimpanzees, I was actually able to touch some of them? There were moments I shall never forget—such as when David Greybeard first allowed me to groom him; when Flo let her infant reach out and touch me; when adolescent Figan joined me in a game, let me tickle him, and laughed. I would not have forgone those moments for anything. But when I realized that, with the help of students, the research could continue indefinitely, I knew I had to distance myself from the chimps. For one thing, I wanted to affect their natural behavior as little as possible. For another, chimps are much stronger than humans. Too close a relationship, I thought, might become dangerous, might destroy the inherent respect that most wild chimps have for humans. Indeed, Figan obviously learned when he played with me that he was stronger than I was. Thereafter he sometimes knocked observers over during his charging displays. We have had the same problem with some other particularly fearless chimps.

You have discovered that chimpanzees not only use tools, but *make* tools. What kind of tools?

They modify blades of grass, leafy twigs, strips of bark, and sticks to make them more suitable for a variety of purposes. Chimpanzees use more objects as tools in more contexts than any other creature except ourselves. At Gombe, they use grass stems, twigs and so on to extract termites from their mounds. Long, thin sticks are used to

fish for army ants and strong, thick ones to enlarge the entrances of bees' or birds' nests. Leaves are chewed, making them more absorbant, to sop up rain water from tree hollows. Leaves are also used to wipe dirt from the body. Sticks are flailed and rocks hurled in aggressive contexts. Most fascinating is the finding that in all the different areas where chimps have been studied across Africa, they have developed different tool-using traditions, or cultures, each one obviously having been originally invented by some chimpanzee genius in the past. Chimpanzees are able to learn new behaviors by watching the performance of others, then imitating and practicing. Each tool-using pattern can thus be handed down from one generation to the next. Unfortunately, we have not yet been able to document the appearance and spread of an innovative act, but I feel sure that we shall if the study goes on long enough.

Earlier you mentioned grooming behavior of chimpanzees—their picking flakes of skin off one another. What do you think is the social function of this behavior?

Social grooming is the single most important social activity in the chimp community. It improves bad relationships and maintains good ones. A few brief grooming movements serve to reassure, to appease a higher-ranking individual, or to calm a subordinate. A mother pacifies her nervous or hurt child by embracing and then grooming him or her. Adult males enjoy particularly long grooming sessions—this is important. Males do sometimes compete quite vigorously for dominance rank, and their relationships may then become tense. Yet it is crucial that they be able to cooperate in order to jointly protect the territory of their community. It's the long sessions of social grooming, enabling them to spend time in friendly physical contact, that permits them to relax after periods of social tension.

You mentioned dominance hierarchies among males in chimpanzee social groups. How is this hierarchy established and maintained?

Some male chimpanzees devote much time and effort to improving or maintaining their position in the hierarchy.

For the most part, the male uses the impressive *charging display*, during which he races across the ground, hurls rocks, drags branches, leaps up and shakes the vegetation—in other words, he makes himself look larger and more dangerous than he may actually be. In this way he can often intimidate a rival without having to risk an actual fight, which could be dangerous for him as well as for his rival. The more frequent, the more vigorous, and the more imaginative his charging display, the more likely it is that he will attain a high social position.

What are the benefits of a high position?

This is an interesting question. A high-ranking male has prior access to the best food, but this is not of great benefit since, if food is short, the chimps typically move about in ones and twos. He can usurp a choice resting place, but he almost never does. He can inhibit other males from copulating with a female in estrus when they are all in a group together, but there is a mechanism in chimp society that enables even a low-ranking male to appropriate a sexually attractive female. All he has to do is persuade her to follow him, away from the other males, to some peripheral area of the community range. Of course, this is not always easy. For one thing he must initiate this consortship at a stage of her estrus cycle when she is not interesting to higher-ranking males. She will not want to go with him, and he may have to use considerable force. Even when he has got her to his area of choice, he must keep her there, often against her will, until she ovulates. However, if he has the social skills to accomplish all of this, he has a good chance of impregnating her. This means that every male has the opportunity to pass on his genes—a fact that, without doubt, has contributed to the pronounced individual variation that we find among chimpanzees. But we are still left with the question: What is the advantage of high rank for a male chimpanzee? It's almost as though humans aren't the only creatures who value high rank for its own sake, and the power that it gives. We don't understand why the chimps devote so much time, effort, and risk. What are the evolutionary benefits, the reproductive benefits?

Females have a hierarchy too, varying according to whether or not the females are accompanied by their adult son or other offspring. Also, a female in estrus may be more assertive. The reproductive advantage to the high-ranking female is, however, clear. She can better appropriate choice food items, and thus make her milk richer. In addition, her offspring are likely to become high-ranked since she will support them. In the supportive family group situation, all have a better chance of survival.

During your time at Gombe, you observed the splitting of a social group into two factions. What were the consequences of that split?

Soon after the split, the males of the larger group, which remained in the north, began to make raids, in groups of three or more, into the area taken over by the southern subgroup. If they encountered a single individual, they chased him or her and attacked savagely. Fights between members of the same community may look ferocious, but they seldom last as long as half a minute, and they rarely result in wounding. By contrast, the assaults on members of this newly formed community were brutal, lasting between 10 and 20 minutes and resulting in the incapacitation of the victims, who were left to die. Within a four-year period all the members of the newly formed social group had been killed, or had disappeared and were presumed killed.

Laboratory studies indicate that chimpanzees have advanced cognitive abilities. Is there evidence from field studies that such cognitive abilities actually have adaptive value?

Yes, and this is most obvious in relation to social awareness. In chimpanzee society, individuals are continually separating and meeting again. This means, for example, that a young male may, at one moment, be the highest-ranking individual in a group of females, able to bully them at will, but then, if other males arrive, he may find himself the lowest-ranking male. He must be able to adjust to this change at once. If he can't, he may be chastised by one of the higher-ranking males for inappropriate behavior. Then too there is a constant need for decision making. Suppose a chimpanzee hears others calling on the far side of the valley. Because each chimp has a recognizable voice, he will know who is there. He must then make some decisions: *"Do I reply or do I stay quiet? Do I join those guys, do I stay where I am, or do I hasten in the opposite direction?"* In the morning when a chimpanzee wakes, he must decide: *"Do I go and eat figs with David Greybeard, or do I eat leaves by myself?"* Of course, other primates need to make some choices too, but usually to a lesser degree since they spend most of their time in stable troops. Due to their highly developed social intelligence, chimpanzees are often able to get their way, even when set against much higher ranked individuals. A subordinate can withhold information from a dominant. He can, for example, sit gazing directly away from a delectable fruit (if he looked at it, this would serve as a cue for the others) until the dominant departs—and then collect his prize. Or he can mate with a female in secret, without the dominant male's knowledge, by beckoning to her from behind some vegetation. I could give many other examples of this sort of behavior, all pointing to the adaptive value of well-developed cognitive abilities. Technical problem solving, involving the use of tools to obtain food that would not otherwise be available, is clearly adaptive as well.

Is there any evidence for altruistic behavior in chimpanzee populations?

Yes, indeed. Let me give an example. We had an epidemic of viral pneumonia in which one of the female chimpanzees died, leaving a three-year-old, sickly infant named Mel. Since chimps nurse for five years, Mel was still dependent on his mother's milk. Spindle, a nonrelated, 12-year-old male, formed an incredible bond with Mel. Spindle cared for the infant, carrying him and sharing his food with him, almost as though he were his mother. Sometimes Spindle risked being attacked himself as he tried to protect Mel, as when he seized him from the path of a displaying adult male. On several occasions, Spindle was actually bowled over when he interfered in this way. The relationship lasted more than a year, and there is little doubt but that Spindle saved Mel's life. Mel is still alive today.

What seems to be the fate of forest communities in Tanzania and other parts of Africa? Is chimpanzee habitat already threatened?

Chimpanzee habitat is dwindling rapidly as the great rain forests of Africa disappear to make way for farms and villages, and as timber merchants plunder the trees. But it's not only habitat destruction that is endangering the chimpanzees. In some countries they are hunted for food, or captured and sold to dealers for the pet trade, the entertainment industries, and for biomedical research. Only infants are wanted—adults are too big and dangerous. Infants are traditionally captured by shooting their mothers. The weapons are usually inefficient. Many mothers are wounded and die later, along with their infants. Other infants are too badly hurt at the time of capture to survive. Many of those that reach the holding stations alive fall sick because they are maintained in nonhygienic conditions with inappropriate food. Others die on the journey to their final destinations. We estimate that for every infant that survives, up to ten individuals perish. Because it was, at one time, believed that chimpanzees were the only suitable animal model for the study of AIDS, there was a sudden new demand for wild youngsters. Even in conservation-minded Tanzania, where chimpanzees are protected by law, there was an increase in poaching for awhile. Even so, Tanzanian chimps are relatively secure, for there are two national parks created specially for their protection, and they also live in a number of forest reserves.

How can your studies of chimpanzee life history and behavior help African governments make good decisions about habitat preservation?

Well, for one thing, our research shows clearly that it is necessary to protect sufficient habitat to support at least five chimpanzee communities—250 to 400 individuals. This would provide a reasonable gene pool. Our experience at Gombe also suggests that when smaller populations are conserved, great care must be taken to protect the chimpanzees from being infected with human diseases, which can occur as a result of uncontrolled tourism. It might be necessary, too, to protect nearby farmers from plantation raids by chimpanzees who have become habituated to humans. In 1986, a group of scientists established the Committee for the Conservation and Care of Chimpanzees (the Four C's). The Jane Goodall Institute is a major financial supporter of the Four C's. Under the Chairmanship of Dr. Geza Teleki, the committee is planning surveys of the chimpanzees throughout their range, with a view to creating protected areas across Africa.

Dr. Goodall, how long will you stay at Gombe?

As long as I can get there physically. And even if I can't get there myself, there is, of course, the team of Tanzanian field assistants following the chimpanzees daily, recording their behavior. They are using 8-millimeter video cameras now so that I can actually see what the chimps have been doing when I get back.

What advice would you give students who share your interest in animal behavior and would like to follow in your footsteps as field researchers?

I get hundreds of letters asking for advice. I tell these young people the truth—that this field is becoming increasingly competitive and funding for field research increasingly difficult to obtain. It's really important to work hard and get good grades at school, then go on and get a college education. Of course, I began my study without such qualifications—but it was different in those days, just after World War II. Today, academic credentials are all but essential. But having explained the difficulties, I always add, *"If you really and truly want to devote your life to the study of animals, somehow you will find a way to do it."* I'm convinced of that.

After all, I was told that it was not possible for a young girl to go out into the bush in Africa and study animals. But I found a way. And so I end up: *"Be prepared to take any opportunity to further your goal, even if it seems somewhat indirect. If you remain true to your underlying ambition, you will somehow find a way to get back on course. In the meantime, read books on the subject outside of school, try to get involved in any programs that are offered that relate to the study of animals."*

Of course, the ChimpanZoo program is a natural for anybody interested in nonhuman primates. Many of the students who have taken part have written me glowing letters about their experiences. Anyone who is seriously interested should write for advice to Dr. Virginia Landau of the Jane Goodall Institute. She is a primatologist and coordinates the ChimpanZoo program. She also has a list of schools that offer good graduate programs in primate behavior.

No young person should leave school feeling that to be a good scientist it's so important to be objective that it's absolutely impossible to become emotionally involved with the animals being studied. It is this reasoning that has led to the inhumanity we find in some animal research labs. It is perfectly possible to record data objectively despite feelings of personal involvement. It simply requires discipline. So often today young people get the message that in order to be a really good scientist you must be a scientist first and a human being second. That misconception must be eradicated from our society.

46

Diverse Environments of the Biosphere: An Introduction to Ecology

E cology (from the Greek *oikos*, "home") is the scientific study of the interaction of organisms with their environments. This seemingly straightforward definition masks an enormously complex and exciting area of biology that is also of critical practical importance. Let us first look more closely at the key terms in our definition.

As an area of *scientific* study, ecology incorporates the usual methods of careful observation and description, hypothesis development, and experimentation. As we shall see shortly, however, ecologists often encounter great difficulties in applying these methods because of the complexity of their questions and their subjects, and because of the large time and space scales over which studies must be conducted. Ecology is also challenging because of its multidisciplinary nature; ecological questions form a continuum with those from many other areas of biology, including genetics, evolution, physiology, and behavior.

The *environment*, as defined by ecologists, includes **abiotic** factors, such as temperature, light, water, and nutrients. Just as important in their effects on organisms are **biotic** factors—the other organisms that are part of any individual's environment. Other organisms may compete with an individual for food and other resources, prey upon it, or change its physical and chemical environment. As we shall see, questions about the relative importance of various environmental factors are frequently at the heart of ecological studies—and of the accompanying controversies.

Finally, in our dissection of the definition of ecology, we should highlight the *interaction* between organisms and their environment. Organisms are affected by their environment but, by their very presence and activi-

Figure 46.1
A sense of home. An external view of Earth, as in this photo of an earthrise taken by Apollo astronauts, gives a compelling sense of the planet as a finite home. This awareness helped propel the environmental movement, which began in the 1960s when *ecology* became a fashionable, but misunderstood, word. General principles of ecology are crucial in helping us understand the environmental impact of humans and their technology, but this application should not be confused with the objectives of ecology as a basic science that seeks to understand the interaction of factors determining the distribution and abundance of organisms. This distinction, however, does not require that ecologists sever their emotions and environmental concerns from their scientific work. As Jane Goodall put it in the preceding interview, "It is perfectly possible to record data objectively despite feelings of personal involvement."

ties, they also change it—often dramatically. Through their metabolism, microorganisms in a lake at night reduce the oxygen and lower the pH of the lake. Trees reduce light levels on the floor of a forest as they grow, sometimes making the environment unsuitable for their own offspring. Throughout our survey of ecology, we shall see many more examples of how organisms and their environments affect one another.

The basic science of "ecology," the study of how organisms interact with their environment, should be distinguished from the term "ecology" that is sometimes used informally in popular writing. You have probably read accounts of how a particular pollutant might destroy the ecology of an area. Although this is not what is meant by our definition (unless the pollutant also destroyed all the ecologists!), there is obviously a close link between ecology and the vast array of current environmental concerns, some of which extend to our very ability to survive as a species. Beginning primarily in the 1960s, when many people became alarmed by the adverse effects humans and their technology were having on the environment, "ecology" has worked its way into the public consciousness. Photos of the planet taken by the lunar astronauts contributed to the growing awareness of Earth as a finite home in the vastness of space (Figure 46.1), rather than an unlimited province for human activity. Acid rain, localized famine aggravated by land misuse and population growth, the growing list of species extinct or endangered because of loss of habitat, and the poisoning of soil and streams with toxic wastes are just a few of the problems that threaten the home we share with millions of other species. Ecology as a science helps us understand these problems and

their possible solutions. Most ecologists recognize a responsibility to educate legislators and the general public about decisions that affect the environment. Usually, however, any response must also involve the realms of economics, politics, and ethics. These are beyond our scope here, but the following discussions will point out many connections between basic ecology and environmental issues.

This chapter introduces ecology by defining its scope, discussing some of the physical factors that are important in the environments of organisms, and describing how these abiotic factors, especially climate, determine some of the larger-scale types of biological communities found in various parts of the world.

THE SCOPE AND DEVELOPMENT OF ECOLOGY

The Questions of Ecology

The questions of ecology are extremely wide-ranging. To help us keep a general focus, we can think of ecology as the *study of the distribution and abundance of organisms*. What factors determine where species are found, and what factors control their numbers in those locations? Specific aspects of these general questions apply to increasingly comprehensive levels of organization, from the environmental interactions of individual organisms to the dynamics of ecosystems.

Physiological ecology is concerned with how the individual organism meets the challenges of its physiochemical environment and how the organism's limits of tolerance for environmental stresses determine where it can live. (See Bartholomew interview on p. 777.)

The next level of organization in ecology is the **population,** a group of individuals in a particular geographic area that belong to the same species. Population ecology concentrates mainly on factors that affect population size, as described in Chapter 47.

A **community** consists of all the organisms that inhabit a particular area; it is an assemblage of populations of different species. Questions at this level involve the ways in which predation, competition, and other interactions between organisms affect distribution and abundance. Debates abound in community ecology, as we shall see in Chapter 48. The major types of communities typical of broad climatic regions, such as tropical rain forests and deserts, are called **biomes.** We will look more closely at biomes later in this chapter.

A level of ecological study even more inclusive than the community is the **ecosystem,** which includes all the abiotic factors as well as the community of species that exists in a certain area. Some critical questions at this level concern energy flow and the cycling of chemicals between the various biotic and abiotic components.

Ecology ultimately deals with the highest levels in the hierarchy of biological organization (Chapter 1). The web of interactions at the heart of ecological phenomena is what makes this branch of biology so engaging and challenging.

Ecology as an Experimental Science

By necessity, humans have always had an absorbing interest in other organisms and their environments. As hunters and gatherers, prehistoric people had to learn where game and edible plants could be found in greatest abundance. Observing and describing organisms in their natural habitat became an end in itself to natural historians from Aristotle to Darwin. Extraordinary insight can still be gained through this descriptive approach, as demonstrated by the contributions of Jane Goodall, who has spent 29 years observing chimpanzees in their natural habitat (see preceding interview).

Although ecology has a long history as a descriptive science, it is very young as an experimental science. The large scale on which much ecological research must be carried out has made it difficult to conduct experiments and control variables. Consider the simple-sounding question: How does the foraging of squirrels on acorns affect the distribution and abundance of oak trees? Think about what might be required to answer this experimentally: large areas of land from which squirrels could be removed, other large areas that could serve as controls, and a long period of time in which to observe results. In spite of the difficulty

Figure 46.2
Ecology as an experimental science. The foundations of ecology lie in natural history, the observation of organisms in their native, undisturbed environments. Over the past two decades, however, ecology has become more of an experimental science. Here, a team of researchers prepares an experiment to test hypotheses about the colonization of islands by insects and other animals. The scientists enclosed a number of small mangrove islands in the Florida Keys with scaffolding that was used to support a plastic tent over the island. The island was then fumigated to destroy its fauna of small invertebrates. When the tent was removed, the researchers could study recolonization (see Chapter 48 for further description of this work).

of conducting such experiments, an increasing number of creative ecologists are testing hypotheses in laboratory experiments and by manipulating communities in field experiments. For example, one researcher removed periwinkle snails from one tidepool and added them to another to investigate the effect of this key herbivore on algal populations. Another ecologist eliminated all the insects on small islands to study new colonization of the islands (Figure 46.2).

Many ecologists also devise mathematical models to help answer ecological questions. In this approach, important variables and their relationships are described through mathematical equations. The ways in which the variables interact can then be studied, usually with the help of a computer. This approach is very appealing because it allows ecologists to simulate large-scale experiments that would be impossible to actually conduct in the field. Of course, such simulations are only as good as the basic information on which the models are based, and obtaining that information still requires extensive field work. We will study a simple model describing population growth in Chapter 47.

Although ecology is advancing through continuing descriptive research, field and laboratory experiments, and mathematical modeling, most ecologists would probably agree that their discipline does not yet have a broad, universally applicable, and universally accepted set of principles. Questions such as "What determines the number of species of birds in a forest?" often must be answered by saying "It depends." Despite this uncertainty—or rather, because of it—ecology is currently an extraordinarily dynamic and exciting science. As ecologists debate a wide range of issues, ideas constantly ebb and flow, some finding favor for a while until displaced by another, perhaps only to return later in a modified form. This has been especially true in the study of communities, which is described in Chapter 48.

Ecology and Evolution

Though the field had not been defined at the time, Charles Darwin was an ecologist. He found evidence for evolution in the geographic distribution of organisms and in their exquisite adaptations to specific environments. And he realized how the short-term interactions of organisms with their environments could have long-term effects through natural selection. Put another way, the momentary episodes enacted in the frame of what is sometimes called *ecological time* translate into effects over the longer scale of *evolutionary time.* For instance, hawks feeding on field mice have an impact on the gene pool of the population of prey by curtailing the reproductive success of certain individuals. One long-term effect of such a predator–prey interaction may be a prevalence in the mouse population of fur coloration that camouflages the animals.

Another connection between ecology and evolution is apparent in the effects of past history on the current distribution of species. For example, the pattern of continental drift after the fragmentation of Pangaea helps explain the absence of native placental mammals in Australia (Chapter 23).

An evolutionary theme is inherent in these chapters on ecology: The distribution and abundance of organisms are products of long-term evolutionary changes as well as ongoing interactions with the environment.

ENVIRONMENTAL DIVERSITY OF THE BIOSPHERE

The entire portion of the Earth that is inhabited by life is called the **biosphere**; it is the sum of all the planet's ecosystems. The biosphere is a relatively thin layer consisting of the seas, lakes, and streams, the land to a soil depth of a few meters, and the atmosphere to an altitude of a few kilometers.

The biosphere is a mosaic of habitats differing in abiotic factors such as temperature, rainfall, and light, which in turn affect the distribution of biotic factors

Figure 46.3
Patchiness of the environment. If we were to move closer in to the different environments visible from this distance, we would find still more variation on a smaller scale.

such as the type of vegetation. We can see the patchiness of the biosphere on different levels, from the global to more localized environments such as a sunny meadow surrounded by forest (Figure 46.3).

Natural selection acting in the different environmental theaters of the biosphere has produced diverse adaptations to extremes of temperature, light, and other abiotic factors. Some organisms living in the Arctic and Antarctic tolerate air temperatures of –70°C and other organisms survive desert temperatures higher than 45°C. Aquatic life can be found in water with salt concentration near zero (snowmelt) and in salt marshes several times more saline than seawater. Few organisms survive, however, if they are transplanted from one extreme environment to another, as is evident to anyone who has tried to grow tropical plants outdoors in a temperate climate. The ranges of environmental variables that individual organisms can tolerate are important factors determining the distribution of species. Various structures and physiological mechanisms that have evolved as adaptations to the constraints of specific environments are discussed primarily in Units Six (for plants) and Seven (for animals). In the following section, we briefly survey some of the abiotic factors that are important in the environments of organisms.

Some Important Abiotic Factors

Temperature One reason temperature is such an important factor in the distribution of organisms is its effect on metabolism. Few organisms can maintain a sufficiently active metabolism at temperatures close to 0°C, and temperatures above 50°C denature the enzymes of most organisms. Extraordinary adaptations enable some organisms to live outside this temperature range. For example, some bacteria are restricted to hot springs because their enzymes function optimally at about 90° C.

The actual internal temperature of an organism is affected by the processes that transfer heat between the organism and its surroundings (see Chapter 40). Most organisms cannot maintain body temperatures more than a few degrees above or below the ambient temperature. Mammals and birds are the major exceptions, being endotherms that use metabolic processes to regulate their internal temperature. Even endotherms, however, function best within certain temperature ranges, which vary with the species.

Water Water, as you know, is essential to life. Its availability varies, as do the adaptations of organisms to that availability. Aquatic organisms have a seemingly unlimited supply of water, but they face problems of water balance if their osmolarity does not match that of the surrounding water. Most freshwater fishes, for instance, are hyperosmotic to their environment, and most marine fishes are hypoosmotic. In these animals, water balance is maintained partly by the pumping of salts between the animal and the external environment (see Chapter 40).

For terrestrial organisms, the main problem concerning water is the threat of desiccation. The varying abilities of plants and animals to obtain and retain water is an important factor in their distribution. Most desert plants, for instance, have very thick cuticles and other adaptations that help reduce water loss. Among the adaptations of desert mammals, such as the kangaroo rat, is the ability of the kidney to excrete very concentrated urine.

Light Solar energy drives nearly all ecosystems, though only plants and other photosynthetic organisms use this energy source directly. In many terrestrial environments, light intensity is not the most important factor limiting plant growth. In the understory of forests, however, shading by the canopy of trees makes competition for light intense. The natural environment of many house plants, which are adapted to relatively low light levels, is the floor of tropical forests.

In aquatic environments, the intensity and quality of light is an important abiotic factor limiting the distribution of photosynthetic organisms. Water selectively reflects and absorbs certain wavelengths of light. For example, about 45% of red light is absorbed for every meter of water it penetrates, whereas only about 2% of blue light is absorbed in the same distance. Additionally, the pigments of photosynthetic organ-

Figure 46.4
Effect of wind on plant morphology. The "flagging" of these alpine firs growing on a windy ridge on Mt. Hood, Oregon, is a result of the mechanical disturbance of the prevailing winds inhibiting limb growth on the windward sides of the trees.

isms often greatly increase the absorption of specific colors of light. As a result, most photosynthesis in aquatic environments occurs relatively near the surface of the body of water. Some photosynthetic organisms that live deeper, such as many species of red algae, have accessory pigments that absorb the wavelengths of light that selectively penetrate the water.

The physiology, development, and behavior of many plants and animals are sensitive to photoperiod, the relative lengths of daytime and nighttime. As we have seen, photoperiod is a more reliable indicator than temperature for cueing seasonal events, such as flowering or migration (see Chapters 35 and 50).

Soil The physical structure, pH, and mineral composition of soil are important environmental factors limiting the distribution of plants and hence of the animals that feed on those plants. For example, only plants with special adaptations such as salt glands can grow in the saline soils of dry lakes. Variations in soil contribute to the patchiness we see in landscapes.

(a)

(b)

(c)

Figure 46.5

Effects of fire on communities. Whereas some disturbances are too irregular and infrequent to be important in natural selection, fire is common enough in some communities to affect the course of evolution. (a) A brush fire seems to have devastated this chaparral community in southern Cali-fornia, but many of the plants have extensive storage roots that survive underground. (b) Within a few months after the fire, resprouting of plants from these root crowns initiates regeneration. Also, some of the plants have seeds that will germinate only after exposure to the high temperature of a fire. (c) In this photo, taken in 1989, only a short time after extensive fires in Yellowstone National Park, Wyoming, we see new growth taking advantage of the nutrients recycled by the fire.

Wind Wind amplifies the effects of environmental temperature on organisms by increasing heat loss due to evaporation and convection. You can experience this clearly on a cold, windy day, when what we call the wind chill factor increases the effects of cold temperatures. Wind also contributes to water loss in organisms by increasing the rate of evaporation in animals and transpiration in plants.

Wind can have a substantial effect on the morphology of plants. For example, the mechanical pressure of the wind may inhibit the growth of limbs on the windward side of trees, while limbs on the leeward side grow normally, resulting in a "flagged" appearance (Figure 46.4).

Fire and Other Disturbances Catastrophic events such as fires, hurricanes, typhoons, and volcanic eruptions can devastate biological communities (Figure 46.5). After the disturbance, the area is recolonized by organisms or repopulated by survivors, but the structure of the community may go through a succession of changes during this rebound (ecological succession is discussed in detail in Chapter 48).

Some disturbances, such as the eruptions of Mount St. Helens in Washington, are infrequent and highly unpredictable in space and time, so populations do not exhibit adaptations to them. By contrast, fire, although unpredictable over the short term, recurs frequently in some communities, such as the chaparral areas of coastal California. In such places, fire is a potent factor in natural selection, and adaptations to this periodic disturbance have evolved in many plants. Many of the shrub species in the chaparral store food reserves in fire-resistant root crowns, enabling these plants to resprout in the aftermath of a fire. The extensive fires in Yellowstone National Park, Wyoming, during 1988 attracted much public attention because of their apparently massive destruction. But many ecologists pointed out that most of the fires were natural and inevitable and were in fact overdue.

An Integrated Approach to the Physical Environment

Although we have dissected the physical environment to identify some of the major abiotic factors, it is important to understand that these factors are woven together in the integrated environments with which organisms interact. The success of an organism in surviving and reproducing reflects its overall tolerance to the entire set of environmental variables it confronts. Also, the ability to tolerate a particular factor may depend on another factor. For example, many aquatic organisms can survive reduced oxygen at low temperatures but not at high temperatures. Coping with a set of environmental problems usually involves evolutionary compromises, as, for instance, in sweating, which can cool the body on a hot day but can also lead to a water deficit in the animal. Such trade-offs are one reason perfect adaptations cannot evolve (see Chapter 21).

One concept physiological ecologists have found useful in assessing responses of organisms to their complex environments is the **principle of allocation**. This principle holds that each organism has a limited amount of total energy that can be allocated for growing, reproducing, obtaining nutrients, escaping from predators, and coping with such environmental fluctuations as changes in temperature. Energy budgets are like checking accounts with a restricted balance; overdrafts are not allowed. For example, in grasshoppers, which are moderately active ectotherms, about 30% of assimilated energy remains after the basic animal's maintenance needs are met. This energy can be channeled into growth or reproduction. By contrast, for a very active endotherm, such as a weasel, only 2.5% of assimilated energy remains, and for a wren, only 0.5% remains. For the latter organisms, there are apparently significant advantages to higher activity levels and endothermy that offset the disadvantage of lower growth rate.

Different priorities in energy allocation are related to the distribution of organisms. Species that live in very stable environments can channel more of their energy into growth and reproduction. However, the intolerance of such specialists to environmental change severely restricts their distribution. In contrast, organisms that allocate a larger fraction of their energy to coping with environmental changes are less efficient at growing and propagating, but such generalists are able to survive and reproduce over a wider range of variable environments.

RESPONSES OF ORGANISMS TO ENVIRONMENTAL CHANGE

Environments vary in time as well as space. Temperature, for example, fluctuates during the day in most terrestrial environments and changes even more with season in temperate regions. Individual organisms can respond to changing environments by mechanisms that are behavioral, physiological, or morphological.

Behavioral Responses

Behavioral response in the sense of muscular reaction to a stimulus is limited to animals. Behavioral mechanisms are so important to how animals interact with their environments that they are discussed in detail in Chapter 50. Let us consider them briefly here from an environmental perspective.

The quickest response of many animals to an unfavorable change in the environment is to move to a new location. Such movement may be fairly localized. For example, lake trout avoid the heat of the upper zone of a lake during summer by moving to deeper water. Many desert animals escape intense heat by burrowing. Some animals are capable of migrating great distances in response to such environmental cues as changes in temperature or the changes in photoperiod associated with seasonal transitions. Some ducks, geese, swallows, and many other migratory birds overwinter in Central and South America, returning to northern latitudes in the summer.

Some animals are able to modify their immediate environment by cooperative social behavior. Honeybees, for instance, can cool the inside of their hive on hot days by the collective beating of their wings. During cold periods, they seal the hive, helping to retain the heat generated by their activity inside. Many small mammals huddle within burrows during cold weather, a behavioral mechanism that reduces heat loss by minimizing the total amount of surface the animals expose to the cold air.

Physiological Responses

In general, physiological responses to environmental change are slower than behavioral reactions. If you moved from Boston, which is essentially at sea level, to the mile-high city of Denver, one physiological response to the lower oxygen pressure in your new environment would be an increase in the number of your red blood cells, but this reaction would require several days to a few weeks. Some physiological responses are much faster, however. When you venture outside on a very cold day, blood vessels in your skin may constrict within seconds, a physiological response that reduces loss of body heat.

In their physiological adjustments to a change in a particular environmental factor, organisms can be described as regulators or conformers. **Regulators** are able to maintain constant internal conditions when the external environment changes, whereas **conformers** are less able to regulate their internal environment, which varies with the external environment. Whether an animal is a regulator or a conformer is related to the stability of the environment in which it normally lives. For example, in brine shrimp, which live at high but variable salt concentrations, the ability to maintain a stable internal salt concentration by osmoregulation has evolved (see Chapter 40). Many other marine invertebrates live in environments where the salinity is very stable. These organisms have no

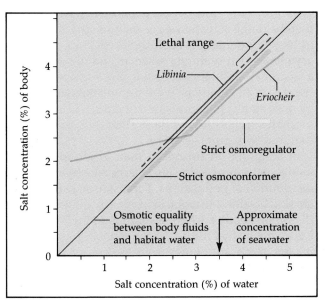

Figure 46.6
Two types of physiological response to changing salinity. Spider crabs *(Libinia)* are osmoconformers, with little ability to regulate internal osmolarity. Above or below a fairly narrow range, the crabs continue to conform and soon die. Chinese mitten crabs *(Eriocheir)* are also conformers at some salinities but are able to regulate at higher and lower levels. As a result, they are less restricted in their distribution and may migrate between freshwater and marine habitats. Generally, organisms are neither perfect conformers nor perfect regulators in their physiological responses to environmental changes in osmolarity, temperature, oxygen level, or other abiotic factors.

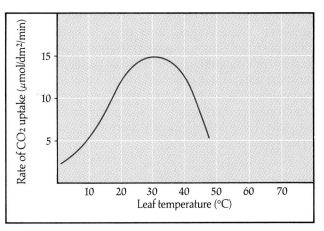

Figure 46.7
A tolerance curve. The rate of photosynthesis in salt bush *(Atriplex)* varies as a function of temperature.

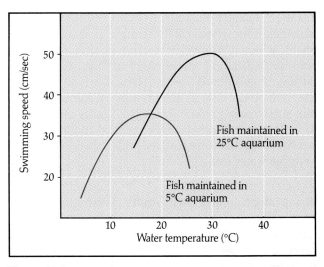

Figure 46.8
Acclimation. There is an optimal external (water) temperature for a goldfish and a range of temperatures the fish can tolerate. The range can be shifted somewhat by a gradual change in environment. In other words, the fish becomes acclimated to its new environment, tolerating higher or lower temperatures than it could if it had been transferred to a different environment abruptly. In this figure, swimming speed is used as a measure of the fish's general well-being.

ability to regulate, and if placed in water of varying salinity, they will lose or gain water to conform to the external environment. Many species are conformers over a certain range but can regulate to some extent at more extreme salinities (Figure 46.6). "Conforming" and "regulating" represent extremes of a spectrum of possible responses to an environmental change; few organisms are perfect conformers or regulators.

All organisms, whether regulators or conformers, function most efficiently under certain environmental conditions. We can study the optimal conditions for an organism by varying a single abiotic factor, such as temperature, and measuring some aspect of the organism's performance. The resulting curves are typically bell-shaped, with the tails of the curve representing the limits of the organism's tolerance to a particular environmental variable. Figure 46.7 illustrates such a curve for the effect of temperature on photosynthetic rate for salt bushes *(Atriplex lentiformis)* in a particular location. The rate is greatest at about 30°C, and declines at temperatures above or below this optimum. Tolerance limits are important in the spatial distribution of organisms, though bio-

logical interactions often override these physiochemical considerations.

Physiological responses to changing environment can shift the tolerance curves of organisms, as we can see in the response of goldfish to a colder temperature, shown in Figure 46.8. Such physiological adjustment to a change in an environmental factor is known as **acclimation.** The ability to acclimate is correlated to a large extent with the harshness of the environment within which a species has evolved. For example, the goldfish just described and other aquatic organisms generally require fairly long times without abrupt changes to acclimate to a different temperature without harm. This reflects the fact that water has a high heat capacity and therefore changes temperature only very slowly. Terrestrial organisms, however, live in an environment that has selected for individuals with the capacity for tolerating much more rapid changes in temperature.

Morphological Responses

Organisms may react to some change in the environment with developmental or growth responses that alter the form or internal anatomy of the body. Morphological responses are usually slower than behavioral or physiological responses.

In general, plants are more morphologically plastic than animals; this response helps them compensate for their inability to escape from an environment that changes. One remarkable example is the arrowleaf plant, which can grow on land or submerged in water and has a different morphology in each case. Leaves of submerged plants are flexible and lack a waxy cuticle, and so are able to absorb mineral nutrients from the surrounding water. Arrowleaf plants growing on land have more extensive root systems, and their leaves are more rigid and are covered with a thick cuticle that reduces water loss.

Adaptation over Evolutionary Time

The various behavioral, physiological, and morphological mechanisms we have been discussing are responses of individual organisms on the ecological time scale. These responses occur within the framework of adaptations fashioned by natural selection acting over evolutionary time. For example, plants are capable of changing the size of the stomata of their leaves, a physiological response that helps prevent desiccation under environmental conditions when transpiration would exceed delivery of water to the leaves. In plants living in the desert, the ability to adjust the size of the stomatal openings in response to water stress is superimposed on many other anatomical and physiological adaptations that have accumulated over evolutionary time as these plants have interacted with their arid environments. For instance, some desert plants have their stomata in pits, protected from the hot, dry winds that accelerate transpiration. Also common among desert plants is the CAM pathway of photosynthesis, which enables the plants to keep their stomata closed during the daytime (see Chapter 10).

The distinction between short-term adjustments on the scale of ecological time and adaptation on the scale of evolutionary time begins to blur when we consider that the range of responses of an individual to changes in the environment is itself the product of evolutionary history. For example, when an endotherm such as a mammal maintains constant body temperature in the face of fluctuations in environmental temperature by making physiological adjustments, it is employing mechanisms of homeostasis that are adaptations acquired by natural selection.

In adapting organisms to their localized environments, natural selection also places constraints on the distribution of populations. For instance, earthworms are skin breathers that obtain oxygen by diffusion across their moist body surface, a solution to the problem of gas exchange that restricts these animals to damp soils. If pine seeds are blown from the rim of the Grand Canyon to the canyon floor 2000 m below, where conditions are much hotter and drier, it is unlikely that the seeds can germinate and grow successfully in the new environment. Organisms locked by their adaptations into one type of environment may fail to survive if dispersed to some foreign environment, or they may face extinction if their local environment changes beyond what the organisms can tolerate. On the other hand, the absence of a species in a particular place does not necessarily imply that the species could not survive in that location. Pines would not live even on the rim of the Grand Canyon if they had not managed to disperse to that location at some time in their evolutionary history. Thus, the existence of a species in a particular place depends on two factors: The species must reach that location, and it must be able to survive in that location once it is there. We evaluate the importance of these factors in the geographic distribution of organisms in Chapter 48. For now, our focus shifts to the ways abiotic factors affect the general appearance of biological communities around the world.

TERRESTRIAL BIOMES

The general appearance of a community may be similar over a large continuous geographical area and may also be similar to a community in another, disjunct region of the Earth. For example, coniferous forest extends in a broad band across North America, Europe, and Asia. Extensive desert areas are scattered over many parts of the Earth. Earlier, we defined these general, large-scale communities as biomes. Although terrestrial biomes are often named for the predominant vegetation, each biome is also characterized by animals adapted to that particular environment—grasslands are more likely than forests to be populated by large grazing mammals, for instance.

The actual species composition throughout biomes varies from one location to another. In the North

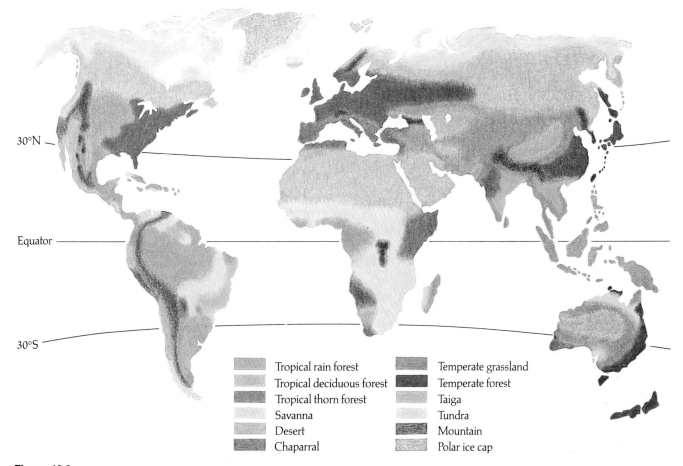

Figure 46.9
The major terrestrial biomes.

Legend:
- Tropical rain forest
- Tropical deciduous forest
- Tropical thorn forest
- Savanna
- Desert
- Chaparral
- Temperate grassland
- Temperate forest
- Taiga
- Tundra
- Mountain
- Polar ice cap

American coniferous forest, red spruce is common in the east but does not occur in most other areas, where black spruce and white spruce are abundant. Although the vegetation of African deserts resembles that of North American deserts, the plants are in completely different families. Such similarities can arise because of convergent evolution (see Chapter 23).

There is no strict way of defining specific biomes, and through the years different ecologists have recognized and organized biomes in different ways. Also, biomes usually grade into each other, without sharp boundaries. If the area of intergradation is itself large, it may be recognized as a separate biome (see the discussion on the savanna, pp. 1063–1064). Twelve major biomes are mapped in Figure 46.9.

Within a biome there may in fact be extensive patchiness, with several communities represented. Biomes are usually recognized on the basis of the communities that develop as the result of natural *succession* (changes in community structure through time), a topic discussed in Chapter 48. Disturbances often allow

representatives of earlier successional stages to occur. For example, snowfall may break branches and small trees and cause openings in the coniferous forest, allowing deciduous species such as aspen and birch to grow. Most of the eastern United States is classified as temperate forest, but human activity has eliminated all but a tiny percentage of the forest that would otherwise be present. In fact, ecologists recognize an "urban biome" scattered over much of the Earth, where the natural development of communities has been drastically altered.

We may now ask *why* a particular biome develops in a certain area. The general answer seems to be that the prevailing climate, particularly temperature and rainfall, is the most important factor in determining the kind of biome that will develop. Climate on both a global and local scale is determined by a complex interaction of factors, some of which are described in Figure 46.10.

We can see the great impact of climate on the distribution of biomes by constructing a *climograph* for

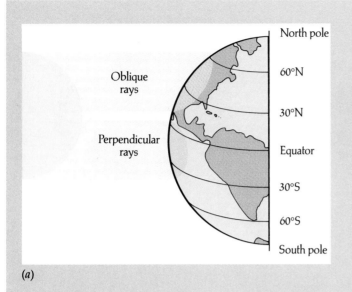

(a)

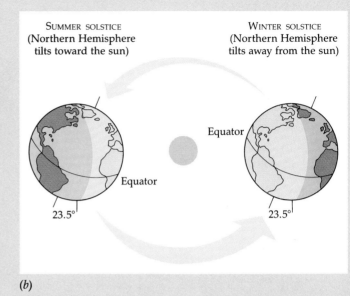

SUMMER SOLSTICE
(Northern Hemisphere
tilts toward the sun)

WINTER SOLSTICE
(Northern Hemisphere
tilts away from the sun)

(b)

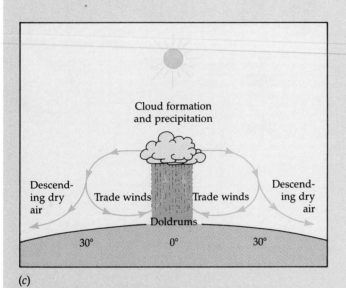

(c)

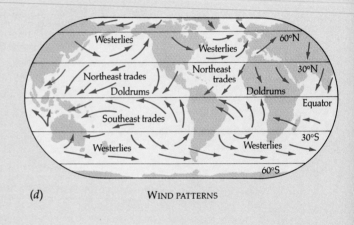

(d)   WIND PATTERNS

Figure 46.10
Some factors affecting climatic patterns on Earth. (a) Because sunlight strikes the equator perpendicularly, more heat is absorbed there per unit area than is absorbed in northern and southern latitudes, where the sunlight is oblique. The uneven heating of the Earth's surface is a major factor driving vertical air movements and surface currents, which result in circulation of air between the poles and equator. **(b)** Seasons occur because the Earth's axis is tilted by 23.5° relative to the plane of its orbit around the sun. December is winter in the Northern Hemisphere because the North Pole is tilted away from the

sun, reducing day length in the hemisphere and making sunlight more oblique during those shorter days. December is summer in the Southern Hemisphere because the South Pole is tilted toward the sun. The two hemispheres' seasons are reversed when the North Pole tilts toward the sun during June. **(c)** Warm air at the equator rises, creating an area of light, shifting winds known as the doldrums. As the air rises, it cools and drops rain. Tropical rain forests are concentrated in this zone of the Earth. The high-altitude air masses, now dry, then spread away from the equator until they cool enough to

descend again at latitudes of about 30° north and south. Many of the world's great deserts are located at these latitudes. Some of the air spreads back toward the equator, creating the familiar trade winds. Because it is warming, it tends to pick up and hold moisture until uplifted again. The air moving toward temperate areas also picks up moisture, but because the air is generally cooling, it tends to drop the water easily. Temperate areas are often moist, but local considerations greatly affect precipitation [see **(f)** on next page]. **(d)** Air movements over the Earth's surface—the prevailing winds—are

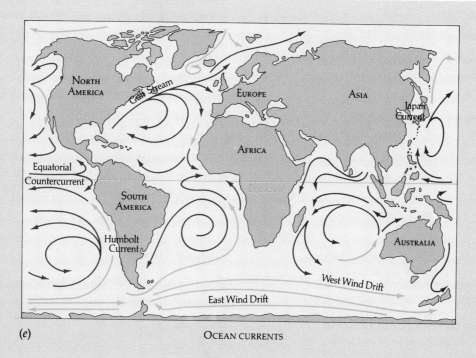

(e)   OCEAN CURRENTS

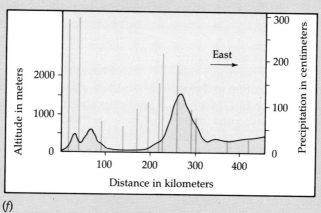

(f)

caused by the combined effects of the vertical movements of air masses, which cause pressure differentials, and by the rotation of the Earth. This rotation causes the trade winds to deflect so they blow from the northeast in the Northern Hemisphere and from the southeast in the Southern Hemisphere. In temperate areas, winds from the west tend to dominate. **(e)** Prevailing winds, the Earth's rotation, and the locations of the continents are the main factors determining the major ocean currents (red = warm currents; blue = cold currents). In turn, by the movement of either warm or cold water, the ocean cur-

rents affect regional climates. For instance, the Gulf Stream, a warm water current, tempers the climate on the west coast of the British Isles, making it warmer than the coast of New England, which is actually farther south. Similarly, the Japan Current, a cold water current, makes the western coast of California relatively cool. **(f)** On a more local scale, topography plays an important part in determining climate. Mountains may cause dramatic differences in precipitation over relatively short distances. As moist air approaches a mountain, it rises and cools, so precipitation may be heavy on the windward side.

On the back side, a "rain shadow" may occur where the dry air descends. In this graph, the tan area traces the topography across the state of Washington. The bars indicate rainfall. As moist air from the Pacific Ocean encounters lower coastal ranges, a large amount of precipitation is dropped. The biological community in this wet region is sometimes called a temperate rain forest. Precipitation peaks again as the air moves across higher mountains inland. On the back side of these mountains, however, there is little precipitation, so that near-desert conditions occur.

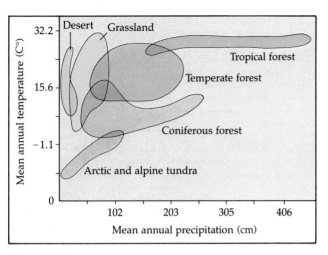

Figure 46.11
Climograph for some major North American biomes.
The areas plotted here encompass the annual mean temperatures and precipitation occurring in some major North American biomes. The climograph provides only circumstantial evidence, however, that these factors are important in explaining the distribution of the biomes. The areas of overlap, for example, show that these variables alone are not sufficient to explain the observed distribution.

some of the major North American biomes (Figure 46.11). A climograph is a plot of the temperatures and rainfall occurring in a particular region, often in terms of annual means. For example, in Figure 46.11, we see that the range of rainfall occurring in coniferous forest regions is similar to that of temperate forest areas, but temperatures are cooler. Grasslands, on the other hand, are generally drier than either kind of forest, and deserts are drier still.

Annual means for temperature and rainfall are reasonably well correlated with the biomes we find in different regions. However, we must always be careful to distinguish a *correlation* between variables from *causation*, a cause-and-effect relationship between the variables. Our climograph does not prove that annual mean temperature and precipitation cause biome patterns. It does, however, provide strong circumstantial evidence that these variables are important.

We can see in our climograph that factors other than mean temperature and precipitation also play a role in biome patterns; notice that there are regions of overlap on our climograph. For example, we can see a set of conditions where grassland, temperate forest, and coniferous forest all overlap. This means that there are areas in North America with a certain temperature and precipitation combination that support a temperate forest; other areas with the same values for these variables support a coniferous forest; still others, a grassland. How do we explain this variation? First, remember that our climograph is based on annual *means*. Often what is important is not the mean but the pattern of climatic factors. For example, some areas

may get regular precipitation throughout the year. Others may get the same overall amount, but with long periods of little or no rain. Inland areas may have the same mean temperature as coastal areas, but the annual extremes may be much greater inland, where the moderating influence of a large body of water is not present. Other factors such as the geology of an area may greatly affect mineral nutrient availability and soil structure, which in turn affect the kind of vegetation that will develop.

With these complex climatic considerations in mind, let us now survey the major terrestrial biomes, traveling generally in a direction from the equator to the poles.

Tropical Forest

Tropical forests are found near the equator, where the temperature varies little throughout the year, averaging around 25° C, and the length of daylight varies from 12 hours by less than one hour. Rainfall in equatorial areas, on the other hand, is quite variable, and the amount of precipitation, rather than temperature or photoperiod, is the prime determinant of the vegetation growing in an area. In lowlands relatively far from the equator, where rainfall is scarce, thorn forests occur. These areas have prolonged dry seasons, and the plants found there are a mixture of thorny shrubs and trees, and succulents. Nearer the equator, in areas that have a distinct wet and dry season, tropical deciduous forests dominate. Tropical deciduous trees and shrubs drop their leaves during the long dry season and releaf only during the following heavy rains or monsoons. The luxuriant **tropical rain forest** is found in areas near the equator where rainfall is abundant (greater than 250 cm per year) and the dry season lasts no more than a few months.

The tropical rain forest is the most complex of all communities, harboring more species of plants and animals than any other community in the world (Figure 46.12). Up to 300 species of trees, many of them 50–60 m tall, can be found in 1 hectare (10,000 m², or about 2.5 acres). Competition for light becomes a strong selective force in the plant communities of the tropical rain forest. Because of the density of trees, the rain forest is often a closed canopy, with little light reaching the ground below. When an opening does occur, perhaps by a tree fall, other trees and large woody vines known as lianas grow rapidly, competing as they fill the gap. Many of these giant trees are covered with epiphytes, such as orchids and bromeliads, and many have wide buttress bases that may provide support in the thin soil. The soils of tropical rain forests are typically poor because high temperatures and rainfall lead to rapid decomposition and recycling rather than buildup of organic material. Instead, at any given time, almost all the nutrients are tied up in living organisms.

Figure 46.12
Tropical rain forest. The density of this Costa Rican jungle is typical of rain forests.

Though the forest as a whole is extremely dense, individuals of each species are widely scattered, so many plants rely on mutualistic interactions with animals to deliver pollen. Animals are also important in dispersing fruits and seeds. The animals are typically tree dwellers; monkeys, birds, insects, snakes, bats,

and even frogs find food and shelter many meters above the ground. The warm environment allows the presence of numerous ectothermic animals—for example, there may be 30 species of frogs per hectare.

Human impact on the tropical rain forest is currently a source of great concern. It is a common practice to clear the forest for lumber, farm the land for a few years, and then abandon it. Mining has also devastated the land. Once stripped, the tropical rain forest recovers very slowly, if at all. The destruction of the tropical forest is proceeding at an alarming rate. More than one-half is already gone, and projections suggest that these communities will have disappeared entirely by the end of the century. The loss is more than aesthetic; destruction of tropical rain forests may cause large-scale changes in world climate, as well as large-scale destruction of species (see Chapter 49).

Savanna

Savanna is grassland with scattered individual trees (Figure 46.13). Extensive savanna covers wide areas of central South America, central and south Africa, and parts of Australia. There are generally three distinct seasons in these tropical and subtropical savannas: cool and dry, hot and dry, and warm and wet, in that sequence. Some savanna soils are fertile, but most are porous and have a thin humus layer.

Savanna is relatively simple in structure but often rich in number of species. With only scattered trees, there is little vertical structure. Frequent fires inhibit invasion by trees and maintain the small growth form of grasses and **forbs** (small broad-leaved plants that grow with grasses). Grasses are wind pollinated, but

Figure 46.13
Savanna. This Kenyan grassland is a showcase of large herbivores and their predators.

Figure 46.14
Desert. This saguaro "forest" is in the Sonoran desert of southern Arizona. Like all deserts, the Sonoran is a harsh environment, but a variety of plants provides modest structure, which in turn helps support diverse animals.

the forbs produce showy flowers that attract insect pollinators, often in great numbers in the summer.

The savanna is home to some of the world's large herbivores, including the giraffe, zebra, antelope, buffalo, and kangaroo. Burrowing animals are also common: mice, moles, gophers, snakes, ground squirrels, worms, and arthropods. Growth is rapid in grasslands, so animals have a good food source. But grasses do not provide much structure compared with a forest, so nest sites and shelters are primarily on or under the ground. Animals in the savanna are most active during the rainy season, and many species are nocturnal. In the winter or dry season, when the above-ground vegetation is dry, many animals are dormant or subsist on seeds and dead plant parts.

The term *savanna* is also applied to areas where forest and grassland biomes intergrade. For example, scattered savanna occurs in North America where the temperate forest and grasslands merge, in a band running roughly from Minnesota to east Texas. Here the climatic conditions and community features are intermediate between those of forest and grassland.

Desert

Deserts are among the harshest of all biomes; maximum temperatures may reach 54°C (Figure 46.14).

However, a desert is determined by low precipitation (less than 30 cm per year), not by temperature; both hot and cold deserts exist. The hot deserts are found in the southwestern United States, along the west coast of South America, in North Africa, the Middle East, and the center of Australia. The cold deserts are found west of the Rocky Mountains, in eastern Argentina, and in much of central Asia. The driest deserts, where average annual rainfall is less than 2 cm and some years have no rain at all, are the Atacama in Chile, the Sahara in Africa, and parts of central Australia.

The cycles of growth and reproduction in the desert are keyed to rainfall rather than temperature. The driest deserts have no perennial vegetation. In less arid regions, the dominant vegetation consists of widely scattered shrubs, often interspersed with cacti or succulents. The periods of rainfall (for example, late winter in the Sonoran desert of the southwestern United States) are marked by tremendous and spectacular blooms of annual plants.

Seed eaters such as ants, birds, and rodents are common in deserts, feeding on the enormous numbers of seeds many of the plants produce in this harsh environment. Reptiles such as lizards and snakes are important predators of these seed eaters. Like the desert plants, most of the desert animals have adapted to low amounts of water and extreme temperatures. Many live in burrows and are active only during the cooler nights. Others are light in color, thereby reflect-

Figure 46.15
Chaparral. Hot, dry summers and relatively frequent fires have shaped many of the adaptations of organisms inhabiting chaparral, such as this California scrubland.

ing the sunlight. And most desert animals have anatomical and physiological adaptations enabling them to conserve water.

Chaparral

Midlatitude areas along coasts where cool ocean currents circulate offshore are often characterized by mild, rainy winters and long, hot, dry summers. These areas are dominated by **chaparral** (sometimes called scrubland)—regions of dense, spiny shrubs with tough evergreen leaves (Figure 46.15). First described in the Mediterranean region, chaparral vegetation is also found along coastlines in California, Chile, southwestern Africa, and southwestern Australia. Plants from these various regions are unrelated, but resemble each other in form and function—for instance, the low-growing eucalyptus shrubs of Australia and the scrub live oak of California. Annual plants are also common in chaparral regions during the winter and early spring when rainfall is most abundant.

Chaparral is maintained by and adapted to periodic fires. As mentioned earlier, many of the shrubs have root systems adapted to fire; these permit quick regeneration and use of nutrients released by fires. In addition, many chaparral species produce seeds that will germinate only after a hot fire, whereas other species are clonal, reproducing asexually without complete reliance on seeds.

Animals characteristic of the chaparral are browsers such as deer, fruit-eating birds, and rodents that eat seeds of annual plants, as well as lizards and snakes.

Temperate Grassland

Temperate grasslands have some of the characteristics of tropical savanna, but they are found in regions of relatively cold temperatures. Temperate grasslands include the veldts of South Africa, the puszta of Hungary, the pampas of Argentina and Uruguay, the steppes of the Soviet Union, and the plains and prairies of central North America (Figure 46.16). The key to the persistence of all grasslands is occasional fires and drought, which prevent woody shrubs and trees from invading and dominating the landscape.

Grasslands expanded in range following the retreat of glaciers as hotter and drier climates prevailed worldwide after the last ice age. Coupled with this expansion was the proliferation of large grazing mammals. The bison of North America, the gazelles and zebras of the African veldt, and the wild horses and sheep of the Asian steppes are some examples. Large carnivores such as lions and wolves feed on the grazers.

Temperate Forest

Temperate forests grow throughout midlatitude regions where there is sufficient moisture to support the growth of large trees—most of the eastern United States, most of middle Europe, and part of eastern Asia. Temperate forests are characterized by broad-leaved, deciduous trees (Figure 46.17).

Temperatures range from very cold in the winter to hot in the summer ($-30°C$ to $+30°C$), with a five- to six-month growing season. Precipitation is relatively

Figure 46.16
Figure 46.16
Temperate grassland. This prairie is located in Wind Caves National Park, South Dakota.

high and generally fairly evenly distributed throughout the year. Temperate deciduous forests have a distinct annual rhythm in which trees drop leaves and become dormant in winter, then produce new leaves each spring. This cycle is thought to be an adaptation that conserves water. However, losing leaves is costly in terms of energy and nutrients, and trees shed leaves only in temperate areas where soils are relatively rich in nutrients.

More open than the tropical forest and not as tall, the temperate forest has several layers of vegetation, including herbs, shrubs, and one or two strata of trees. Species composition varies widely around the world; some of the dominant tree species are oak, birch, hickory, beech, and maple. Humans have dramatically altered temperate forests by logging for building materials and fuel and clearing for agriculture; only scattered remnants of the original worldwide forest remain today.

Because of the variety and abundance of food and habitats it offers, the temperate deciduous forest supports a rich diversity of animal life. A great variety of

Figure 46.17
Temperate forest. Dense stands of deciduous trees are trademarks of temperate forests, such as this woodland located in Marquette County, northern Michigan.

Figure 46.18
Taiga. In contrast to temperate forests, conifers prevail in taiga. This is a winter scene photographed in the Alaskan interior.

microorganisms, insects, and spiders, for instance, live in the soil or leaf litter or feed on the leaves and understory of trees and shrubs. The forest is also home to many species of birds and small mammals and, where human encroachment has not eliminated them, wolves, bobcats, foxes, and mountain lions.

Taiga

The **taiga,** also known as coniferous or boreal forest (Figure 46.18), extends in a broad band across North America, Europe, and Asia to the southern border of the arctic tundra. Taiga is also found at higher elevations in more temperate latitudes, as in much of the mountainous region of western North America. The taiga is characterized by harsh winters and short summers that can be warm occasionally. There may be considerable precipitation, mostly in the form of snow. (Coastal coniferous forests, such as those of the Pacific Northwest, are considerably warmer and moister than others because of their proximity to the ocean.) The soil is thin and acidic and forms slowly, owing to the low temperatures and the waxy covering of conifer needles, which decompose slowly.

The conifers typically consist of a few species of spruce, pine, fir, and hemlock, often in such dense stands that little undergrowth is present. Deciduous species include oak, birch, willow, alder, and aspen.

The heavy snowfall that may accumulate to several meters each winter has important ecological consequences. By insulating the soil before the coldest temperatures occur, snow helps reduce the permafrost

that greatly inhibits tree growth in the tundra to the north (see next section). Mice and other small mammals that would quickly freeze to death above the snow remain active all winter in snow tunnels at ground level, where they continue to forage on old vegetation. Heavy snows also break tree limbs and cause tree falls, allowing patches of increased light on the forest floor, which in turn allows more plant species to exist.

Because the taiga is dominated by a few coniferous tree species, animal species found there are those usually associated with the conifers. The animal communities consist mainly of seed eaters, such as squirrels, jays, and nutcrackers; herbivores such as insects that eat leaves and wood; and larger browsers such as deer, moose, elk, snowshoe hares, beavers, and porcupines. Predators of the taiga include grizzly bears, wolves, lynxes, and wolverines. These animals, like the plant species of the taiga, are well adapted for the cold winters.

Tundra

At the northernmost limits of plant growth and at high altitudes is the **tundra,** whose plant forms are limited to low shrubby or matlike vegetation (Figure 46.19). The arctic tundra encircles the North Pole southward to the coniferous forests. Alpine tundras are found above the trees on high mountains. The flora and fauna of the arctic and alpine tundra are generally similar, with perhaps 40% of the plant species in common. However, there are significant differences in the two environments.

In the arctic tundra, the climate is often very cold

Figure 46.19
Alaskan tundra in autumn. Permafrost is an important environmental factor that determines the kinds of plants that can populate the tundra.

with little light for long periods. During the brief warm summers, marked by nearly 24 hours of daylight, much of the productivity and plant reproduction occurs in a rapid burst. The arctic tundra is characterized by **permafrost,** continuously frozen ground, which prevents the roots of plants from penetrating very far into the soils. This likely accounts for the absence of taller plant forms such as trees. The tundra may receive as little precipitation as some deserts (see Figure 46.11), yet the combination of permafrost, low temperatures, and low evaporation leaves the soils continually saturated, further restricting the types of plants that can grow there. Dwarf perennial shrubs, sedges, grasses, mosses, and lichens are the dominant forms.

Alpine tundra occurs at all latitudes, even in the tropics, if the elevation is high enough. Tropical alpine tundra is confined to very high elevations on mountaintops, where nightly temperatures are usually below freezing. In contrast to the arctic tundra, daylight varies little from 12 hours per day throughout the year. Also, instead of a brief, intense period of productivity, vegetation in the tropical alpine tundra exhibits slow but steady rates of photosynthesis all year round.

Animals of the tundra withstand the cold by living in burrows or having good insulation that retains heat. Many animal species are migratory, and ectothermic animals are rare except for the abundant gnats and mosquitos in the summer. Many arthropods spend the winter at immature stages of growth, which are more resistant to cold than adult forms. Large animals of the tundra include the musk oxen and caribou in North America and the reindeer of Europe and Asia, all of which are herbivorous. Principal smaller animals are lemmings and a few predators such as the white fox and snowy owl.

AQUATIC COMMUNITIES

Life first arose in water and evolved there for almost 3 billion years before plants and animals moved into and diversified in terrestrial habitats. Today, the largest part of the biosphere is still occupied by aquatic habitats. The term *biome* is not commonly used to describe the major types of aquatic communities, but the same general concept applies: Where aquatic environments are similar, similar adaptations and similar communities are usually found.

Freshwater Communities

Lakes and Ponds   Standing bodies of water (Figure 46.20a) range from a few square meters to thousands of square kilometers. Except in the smallest, there is usually a significant stratification in the important physiochemical variables and in community structure. As we have seen, light is rapidly absorbed by both water itself and microorganisms in it, so that light intensity rapidly decreases with depth. Deeper water also tends to be colder; this is especially true in temperate areas during the summer. At these times, an upper layer, lower in density because it has been warmed by absorption of sunlight, may be separated from a much colder, deeper layer by only a narrow region where a sharp temperature change occurs.

A variety of phytoplankton, consisting of algae and cyanobacteria, are found in the upper area, or **photic zone,** where there is sufficient light for photosynthesis. Zooplankton, mostly rotifers and small crustaceans, graze on the phytoplankton.

(a)

(b)

Figure 46.20
Freshwater communities. (a) This freshwater pond is in Great Meadow
National Wildlife Refuge in Massachusetts. **(b)** Mountain stream—Georgia.

Nitrogen and phosphorus are often key "limiting nutrients" that determine the amount of phytoplankton growth. Many waters today are affected by large inputs of these nutrients, brought in by run-off from fertilized lawns and agricultural fields. The result is often a massive algal bloom, or population explosion, that interferes with many uses of the water and degrades its aesthetic value (see Chapter 49).

The smaller organisms produced in the upper waters are short-lived. After dying, they tend to sink into the deeper areas, so that there is a continuous "rain" of dead organic material, or **detritus.** The deeper areas are often an **aphotic zone,** where it is too dark for photosynthesis, but respiration by microbes and other organisms continues as they decompose the detritus from above. Of course, this removes oxygen, and because the deeper waters are cut off from the atmosphere, oxygen may eventually be totally depleted by the end of summer. The lack of oxygen then makes the deeper waters unsuitable for any organisms except anaerobic microbes. The decomposition also releases nutrients that were originally used in photosynthesis at the surface, but these are now trapped out of reach of the phytoplankton. In temperate areas, the surface waters eventually cool down as winter approaches. The denser, colder water then tends to mix with the deeper water, restoring oxygen to the depths and allowing mineral nutrients to return to the surface where phytoplankton can again use them.

The zooplankton are eaten by many small fish, which in turn become food for larger fish, semiaquatic snakes and turtles, and fish-eating birds such as the king-

fisher. In shallower areas, a diverse **benthic** (bottom-dwelling) community usually occurs. This community may include aquatic plants and many kinds of attached algae, especially diatoms. Animals include various clams and snails, crustaceans such as crayfish and isopods, and numerous insects. For many of the insects, such as dragonflies and midges, only the larval stage is strictly aquatic. The adults emerge from the water to become flying insects that may die on land, forming a link between the water and terrestrial environments. Other links include various mammals that may feed on this community, such as otters and raccoons.

Rivers and Streams Rivers and streams are bodies of water continuously moving in one direction (Figure 46.20b). The biological communities of rivers and streams are dynamic, changing significantly from the source, or headwaters, to the point at which rivers or streams empty into a lake or the ocean. At the source (perhaps a spring or snowmelt), the water is cold and clear with few nutrients. The channel is usually narrow, with a swift current that maintains a rocky substrate. Downstream, the water increases in turbidity and nutrients; the channel widens, current speed decreases, and the substrate becomes siltier.

Stream communities differ significantly from those in lakes because of the absence of a plankton community, which cannot be maintained in the flowing water. The food chain is supported by the photosynthesis of attached algae, but in many streams much of the food utilized comes from organic material carried into the stream. Leaf-fall from surrounding trees and

Figure 46.21
A marine community. This intertidal zone is located in Devon, England.

other organic material carried in by run-off from the land may be important. Upstream, fish such as trout may be present where their requirement for cool temperatures, high oxygen, and clear water are met. In the warmer, murkier waters downstream, catfish and carp may occur. Benthic communities also change. The larvae of caddisflies and mayflies, which usually prefer cool, fast-flowing water with a firm substrate, may be abundant upstream. Various midges and worms become abundant in lower reaches, where they burrow into the softer substrate.

Marine Communities

Nearly three-fourths of the world is covered by oceans. Their evaporation provides most of Earth's rainfall, and ocean temperatures have a major effect on world climate and wind patterns. Marine algae supply a substantial portion of the world's oxygen.

Marine communities (Figure 46.21) can be classified in several different ways. The distribution of marine organisms, like that of freshwater ones, depends on the penetration of light. As in fresh water, there is a photic zone where phytoplankton, zooplankton, and many fish species occur. Below is the aphotic zone. Because water absorbs light so well and because the ocean is so deep, most of the ocean is in fact devoid of light, except for tiny amounts produced by a few luminescent fish and invertebrates.

Marine communities are also classified on the basis of depth and tidal effects (Figure 46.22). The shallow zone where land meets water is called the **intertidal zone**. This area is inundated daily by tides and then left dry. The composition of intertidal communities varies depending on whether the substrate is rock, sand, or mud. The rocky intertidal is home to typically sedentary organisms such as barnacles, mussels, sponges, annelid worms, bryophytes, echinoderms, and algae, all of which compete for places to attach where they can avoid being washed away by tides and waves. In sandy substrates (beaches) or in mudflats, many of the organisms such as worms and clams bury themselves and filter food out of water brought in by wave action. Other surface-dwelling animals such as crabs are scavengers or predators of these organisms.

Beyond the intertidal zone is the **neritic zone,** the shallow regions over the continental shelves. Past the continental shelf is the **oceanic zone,** reaching very great depths. Open water is the **pelagic zone,** at the bottom of which is the seafloor or **benthic zone.** The benthic community consists of fungi, bacteria, sponges, burrowing worms, sea stars, crustaceans, fish, sea

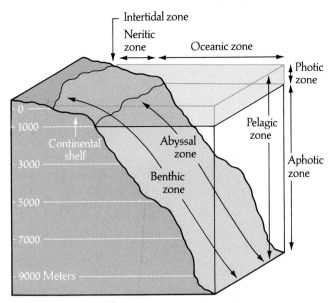

Figure 46.22
Oceanic zones. Marine areas can be classified according to light penetration (photic and aphotic), distance from shore and depth (intertidal, neritic, oceanic), and open water or bottom (pelagic and benthic). The abyssal zone is the deepest benthic area.

anemones, and clams. Deep benthic areas are called the **abyssal zone.** Here, there is continuous cold (about 3°C), extremely high water pressure, and, as already mentioned, almost no light. Unlike the situation we described for many freshwater lakes, however, the deep ocean generally has at least some oxygen, and a fairly diverse assemblage of fish and invertebrates is present. These live primarily on detritus from the photic zone.

Humans have harvested the ocean heavily and used it as a dump for waste, thinking that its vastness made it essentially invulnerable. We are now seeing the effects of our disregard for the ocean, however, as food fish are becoming scarcer, whales are in danger of extinction, and many coastal areas are polluted.

The area where a freshwater stream or river merges with the ocean is called an **estuary;** it is often bordered by intertidal mudflats or saltmarshes. Salinity ranges from nearly that of fresh water to that of the ocean. For reasons that are not fully understood, estuary communities are often not as diverse as those in either fresh water or the open ocean, and are dominated by marine forms. Nonetheless, estuaries are among the most productive environments on the Earth and support a wide variety of extremely valuable commercial species. Oysters, crabs, and many of the fish species that humans consume live continuously in estuaries, reproduce in them, or migrate through them. Estuaries are also crucial feeding areas for many semi-aquatic vertebrates, particularly waterfowl.

Unfortunately, areas around estuaries are also prime locations for commercial and residential developments and are at the receiving end for pollutants dumped upstream. Very little undisturbed estuary habitat remains, and a large percentage has been totally eliminated. Many states have now, rather belatedly, taken steps to preserve remaining estuaries.

* * *

The biome concept emphasizes the importance of physical factors in explaining the general distribution and appearance of communities. However, as we have already seen, physical factors alone are not sufficient to account for even the broad patterns we see at the biome level. There is even more variation to be explained on smaller scales. In the following chapters, we turn our attention to populations, communities, and ecosystems and begin addressing a much wider range of factors that affect the distribution and abundance of organisms.

STUDY OUTLINE

1. Ecology is the scientific study of the distribution and abundance of organisms, as determined by interactions with the biotic and abiotic parts of their environment.

The Scope and Development of Ecology (pp. 1051–1053)

1. Ecology spans increasingly comprehensive levels of organization, from the individual organism through populations and communities to the ecosystem.
2. Ecologists are increasingly using field manipulations and mathematical modeling to help understand complex interactions and complement descriptive studies.
3. Ecology is inextricably interwoven with evolution. The distribution and abundance of organisms depend not only on the immediate environment, but also on long-term evolutionary history.

Environmental Diversity of the Biosphere (pp. 1053–1056)

1. The biosphere is a variable and delicate envelope of life extending from a few meters below the surface of the Earth to a few kilometers into the atmosphere.
2. Although natural selection has produced a wide spectrum of adaptations to the environmental diversity of the biosphere, most species tolerate only a relatively narrow range of environmental variables. For example, all animals function best within certain temperature ranges, which vary with the species and its adaptations.

3. Among the other abiotic factors important in the distribution and abundance of organisms are water availability and quality, light intensity, soil characteristics, wind, and occasional disturbances such as fire.

4. The total amount of energy available for all of life's processes, including response to physical variables, is limited. Natural selection has resulted in different "budgeting" approaches, which are usually related in part to the stability of an organism's environment.

Responses of Organisms to Environmental Change
(pp. 1056–1058)

1. Environments vary in time as well as space, forcing organisms to make behavioral, physiological, or morphological adjustments.

2. Animals may respond to the environment by physiological changes, by moving or migrating to more favorable locations, or by modifying their surroundings by cooperative social behaviors.

3. Physiological responses range from regulation of internal conditions to conformity with the external environment, depending on the changing environmental variable and the normal environment of the organism. Tolerance limits can shift through physiological adjustments involved in acclimation.

4. Many plants exhibit morphological plasticity, which helps compensate for their inability to move to new locations.

5. Although behavioral, physiological, and morphological adjustments operate in the here-and-now of ecological time, they occur within the framework of adaptations resulting from natural selection over evolutionary time.

Terrestrial Biomes (pp. 1058–1068)

1. Biomes are major terrestrial communities of generally similar structure, usually classified according to the predominant vegetation.

2. Precipitation and temperature account for much, but not all, of the variation among biomes. A climograph shows a correlation between biome patterns and annual mean temperature and precipitation.

3. Tropical forests are found near the equator, where photoperiod and temperature are nearly constant but where rainfall varies with location and season. The tropical rain forest is the most complex and species-rich biome. Its rampant destruction by humans will likely have far-reaching ecological and climatic consequences.

4. Savanna is a tropical grassland with scattered trees. Precipitation varies greatly between wet and dry seasons, and fires are frequent. Large grazing herbivores and small burrowing animals are common.

5. Deserts are harsh biomes, with extremes in temperature and very low precipitation. Desert plants and animals have special physiological and morphological adaptations and generally exhibit striking spurts of growth and reproduction in the presence of rainfall.

6. Chaparral consists of fire-adapted scrublands of dense, spiny evergreen shrubs, usually found along coastlines characterized by mild, rainy winters and long, hot, dry summers.

7. Temperate grasslands occur in relatively cool climates where periodic fires and drought inhibit the growth of woody shrubs and trees. Large grazing animals and their predators are typical.

8. Temperate forests occur in midlatitudes where there is sufficient moisture to support the growth of large, broad-leaved deciduous trees, which show distinct seasonal rhythms of leaf drop and regrowth. The environment supports a rich diversity of animal life, owing to the variety and abundance of food and habitats.

9. Taiga constitutes the dense coniferous or boreal forests, characterized by harsh winters and short summers. Animal species are well adapted for cold winters and life among conifers.

10. Tundra occurs at the northernmost limits of plant growth and at high altitudes, where plant forms are limited by cold and winds to a low shrubby or matlike morphology. Particularly in the arctic, permafrost is important in preventing the growth of larger plants. Animals of the tundra may migrate, overwinter in resistant states, live in burrows, or have thick insulation.

Aquatic Communities (pp. 1068–1071)

1. Lakes and ponds are usually stratified vertically in regard to light, temperature, and community structure. Phytoplankton and zooplankton are the primary food source for the rest of the community.

2. Rivers and streams contain freshwater communities that change significantly from the source to the final destination in an ocean or lake. Upper areas contain organisms associated with clear water, cool temperatures, and rocky substrates. Downstream are organisms that can tolerate murkier water and warmer temperatures.

3. Marine communities comprise the nearly three-fourths of the Earth's surface covered by oceans. Oceanic zones can be classified according to degree of light penetration as photic or aphotic and according to depth as intertidal (shoreline), neritic, or oceanic (the last two being progressively deeper and more distant from the shoreline). Open water is called the pelagic zone. The bottom of the ocean in all regions is the benthic zone; in the deepest parts, this constitutes the abyssal zone, a region of constant darkness, cold temperatures, and high pressure.

4. An estuary is the transition zone between a river or stream and the ocean into which it empties. Such areas are extremely productive and support an abundance of both aquatic and terrestrial organisms.

SELF-QUIZ

1. Which level of ecology most considers the effects of predation and interspecific competition on the distribution and abundance of organisms?

   a. biotic

   b. physiological

   c. population

   d. community

   e. ecosystem

2. Which statement follows from the principle of allocation?

   a. The number of organisms an area can support is determined by its energy supply.

   b. Physiological adjustments to environmental changes can extend the tolerance limits of organisms.

   c. The total amount of energy available to an organism is partitioned into such processes as reproduction, obtaining nutrients, and coping with the environment.

d. Organisms that use more energy for growth and reproduction are able to survive in a wider range of variable environments.

e. Organisms allocate most of their energy for homeostasis.

3. Which statement about *tolerance limits* is *incorrect*?

a. They can often be tested experimentally and plotted as a tolerance curve.

b. They help determine whether organisms can live in particular environments.

c. They can be extended by acclimation.

d. They are likely to be greater in regulators than conformers.

e. They are generally greatest for organisms restricted to stable environments.

4. Which of the following biomes is *incorrectly* paired with the description of its climate?

a. savanna—cool temperature, precipitation uniform during year

b. tundra—extreme cold, permafrost, brief summer

c. chaparral—mild and wet winters, hot and dry summers

d. temperate grasslands—relatively cool climates, periodic drought

e. tropical forests—nearly constant photoperiod and temperature

5. A diagnostic difference between temperate forests and taiga involves

a. the type of dominant tree species

b. soil quality

c. the amount of precipitation

d. the latitude

e. species diversity

6. Which of the following is incorrectly paired with its description?

a. neritic zone—shallow area over continental shelf

b. abyssal zone—benthic region where light does not penetrate

c. pelagic zone—area of open water

d. aphotic zone—zone in which light penetrates

e. intertidal zone—shallow area at edge of water.

7. In which area are algal blooms and plankton most likely to be found?

a. headwaters of a stream

b. downstream area of a river

c. lake or pond

d. intertidal zone of ocean

e. benthic zone of ocean

8. In general, deserts are located at latitudes where

a. dry air is descending

b. moist air is descending

c. dry air is rising

d. rising air creates doldrums

e. air masses are stationary

9. The growing season would generally be shortest in which biome?

a. tropical rain forest

b. savanna

c. taiga

d. deciduous forest

e. temperate grassland

10. Imagine some cosmic catastrophe that jolts Earth so that its axis is perpendicular to the line between the sun and Earth. The most predictable effect of this change would be

a. no more night and day

b. big change in length of year

c. cooling of the equator

d. loss of seasonal variations at northern and southern latitudes

e. elimination of tides

CHALLENGE QUESTIONS

1. In which terrestrial biome is your college or university located? If you are in a city, would you still consider your location part of a biome? Why or why not?

2. The bald ibis is a foraging bird that feeds on beetles, grasshoppers, and small animals in natural grasslands and cultivated fields. The ibises thrive on annual burning of the habitat and show complex fire-adapted behaviors and breeding cycles. How might fire act to increase the overall foraging efficiency of these birds, and how might this affect their reproduction?

3. Some estuaries, such as the San Francisco Bay estuary, have experienced a reduced freshwater inflow, owing to diversion of water for reservoirs and agricultural irrigation. What consequences do you think this abiotic change in water availability may have on the biology of this aquatic community and on the ocean into which it empties? (See Further Reading 4.)

FURTHER READING

1. Abrahamson, W. G., T. G. Whitham, and P. W. Price. "Fads in Ecology." *Bioscience*, May 1989. Changing ideas in a dynamic field.

2. Begon, M., J. L. Harper, and C. R. Townsend. *Ecology: Individuals, Populations, and Communities.* Sunderland, MA: Sinauer, 1986. A thorough introductory text.

3. Bushbacher, R. J. "Tropical Deforestation and Pasture Development." *BioScience*, January 1986. A sobering look at the consequences of widespread destruction of tropical rain forest.

4. "Managing Planet Earth." *Scientific American*, September 1989. Special issue devoted to the environment.

5. Nichols, F. H., J. E. Cloern, S. N. Luoma, & D. H. Peterson. "The Modification of an Estuary." *Science* 231:567–573, 1986. A focus on the ecological impact of humans on the aquatic community of the San Francisco Bay estuary.

6. Ricklefs, R. E. *Ecology.* 3d ed. New York: Chiron Press, 1986.

47

Population Ecology

Trop Forest
Temp. Forest
Temp. Grassl.
Savanna
Desert
Chaparall
Tundra
Taiga

Every day the morning newspaper and the evening news on television report local and global problems that threaten our well-being or provoke conflicts between individuals and nations. The media give us updates on the greenhouse effect, acid rain, toxic wastes, and many other symptoms of an ailing biosphere. We hear of conflict between Western nations and those in the Middle East, aggravated in part by the demand for oil. News videos and wire photos of emaciated children starving to death in Africa and elsewhere depress the compassionate and remind everyone that the quality of life is dismal for millions of people. Closer to home, school boards argue about whether to include birth control as a topic in sex education—for students who may be jammed into trailers called temporary classrooms. Contributing to all of these seemingly unrelated problems is a common factor: the continued increase of the human population in the face of limited, and often dwindling, resources. The human population explosion is now Earth's most significant biological phenomenon. As a species of five billion individuals, we require vast amounts of materials and space, including places to live, ground to grow our food, and places to dump our waste. Pushing to make way for ourselves on Earth, we have devastated the environment for many other species, and now threaten to make it unfit for ourselves.

To understand human population growth on more than a superficial level, we must consider the problem within the context of the general principles of population ecology. It is obvious that no population can grow indefinitely. Species other than humans sometimes exhibit population explosions (Figure 47.1), but population crashes also occur. In contrast to these radical booms and busts, many populations are relatively

Figure 47.1

A population explosion. For years at a time, population densities of these tent caterpillars are moderate. But every decade or so, populations grow so explosively that the feeding caterpillars completely strip trees of their leaves; then the insect population returns to normal density. Population ecology is concerned with fluctuations in population size and the factors that regulate populations.

are regulated by common factors, whether biotic or abiotic. Population ecology therefore overlaps with both physiological ecology and community ecology: For example, the size of a population might be affected by the tolerance of individuals to cold temperature, as well as by predation on the individuals. Often, the dynamics of a population are determined by some complex interaction of factors.

Later in this chapter we will return to our discussion of the human population. For now, let us begin examining some of the ways we can describe and analyze populations in general.

DENSITY AND DISPERSION

Ecologists studying the dynamics of a population must begin by defining the boundaries of the groups they are interested in; researchers choose geographical boundaries appropriate to the questions they are asking. One ecologist may follow changes in the number of members of a particular barnacle species in a single New England tidepool; another researcher may monitor fluctuations in the number of caribou in the entire state of Alaska; and still another may be interested in growth of the human population on a global basis. Regardless of these differences in scale, two important characteristics of any population are its density and dispersion. Population **density** is the number of individuals per unit area or volume—the number of crayfish per 1 m^2 on the bottom of a lake, for example (Table 47.1). **Dispersion** refers to the pattern of spacing for individuals within the boundaries of the population.

stable over time, with only minor changes upward and downward. Population ecology, the subject of this chapter, is concerned with measuring changes in population size, identifying the causes of these fluctuations, and learning how populations are regulated. In our study of population dynamics, a familiar ecological theme will be evident: The characteristics of a population are shaped by interactions of individuals with their environment—both on the scale of ecological time and over a longer evolutionary span, during which natural selection can modify the organisms in a population.

We can think of populations in a variety of ways. In Chapter 21 our emphasis was on populations as interbreeding groups of individuals that are more or less isolated from other such groups. With a more ecological emphasis, we can also think of a population as a group of individuals who use common resources and

Table 47.1 Representative population densities

Organism	Density
Diatoms	5,000,000/m^3
Soil arthropods	5,000/m^2
Barnacles (adult)	20/100 cm^2
Trees	200/acre (494/ha)
Field mice	100/acre (247/ha)
Woodland mice	5/acre (12/ha)
Deer	4/km^2
Human beings	
Netherlands	346/km^2
Canada	2/km^2

SOURCE: From C. J. Krebs, *Ecology*, 3d ed. (New York: Harper & Row, 1985).

Figure 47.2
Indirect census of a bank swallow population. The number of birds nesting in this bank can be determined by counting the holes.

Measuring Density

In rare cases, it is possible to determine population size and density by actually counting all individuals within the boundaries of the population. We could count the number of trees of a particular species in a relatively small wooded area or tally the number of sea stars in a tidepool, for example. Herds of large mammals such as buffalo and giraffes can sometimes be counted accurately. In most cases, however, it is impractical or impossible to count all individuals in a population. Instead, ecologists often use a variety of sampling techniques to estimate densities and total population sizes. For example, they might base an estimate of the number of alligators in the Florida Everglades on a count of individuals in a few sample plots, or **quadrats**, of one square kilometer. The estimate becomes more accurate as the sample plots increase in number or size.

In some cases, population sizes are estimated not by counts of organisms but by indirect indicators such as the numbers of nests or burrows (Figure 47.2) or signs such as droppings or tracks. Another sampling technique commonly used to estimate wildlife popu-

lations is the **mark-recapture method**, described in the Methods Box on the next page.

Patterns of Dispersion

The geographic limits within which a population lives is that population's **range**. Within the general range, local densities may vary substantially because not all areas provide equally suitable habitat. Whatever the density, individuals also exhibit patterns of spacing in relation to other individuals (Figure 47.3). The possible patterns vary in a continuum, from **clumped** if the individuals are aggregated in patches; to **random,** where spacing varies in an unpredictable way; to **uniform,** where the spacing is even.

Clumping may result from the environment itself being heterogeneous, with resources concentrated in patches within the range. Plants may be clumped in certain sites because soil conditions and other environmental factors vary locally; although seeds may be randomly distributed, they germinate best where conditions are most suitable. For example, the eastern red cedar is often found clumped on limestone outcrops, where soil is less acidic than in nearby areas. Animals may move toward a particular microenvironment within the range that satisfies their requirements. For example, many forest insects are clumped under logs where the humidity remains high. Herbivorous animals of a particular species are likely to be most abundant where the plants they eat are concentrated. Clumping of animals may also be associated with mating or other social behavior. For example, crane flies often swarm in great numbers, a behavior that increases mating chances for these short-lived insects.

An evenly spaced, or uniform, pattern is often associated with antagonistic interactions of individuals of the population. For example, a tendency toward regular spacing of plants may result from shading and competition of the roots for water and minerals (Figure 47.4a). Some plants also secrete chemicals that inhibit germination and growth of nearby individuals. In animal populations, regular spacing may be caused by competition for some resource or by social interactions that set up individual territories for feeding, breeding, or nesting (Figure 47.4b). Territorial behavior is discussed along with other behavioral aspects of ecology in Chapter 50.

Figure 47.3
Patterns of dispersion within the range of a population. The same population may have different patterns on different scales; for example, the individuals *within* a clump will have their own pattern, as will the clumps themselves.

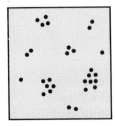
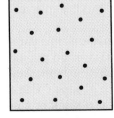
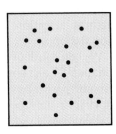

Clumped   Uniform   Random

METHODS: MARK-RECAPTURE ESTIMATE OF POPULATION SIZE

Traps are placed within the boundaries of the population being studied, and captured animals are marked with tags, collars, bands, or spots of dye and then released. After a certain amount of time has elapsed, usually a few days to a few weeks, traps are set again. The proportion of marked to unmarked animals that are captured during the second trapping gives an estimate of the size of the entire population. The following simple equation can be used to estimate the number of individuals in a population (N):

$$N = \frac{(\text{Number marked}) \times (\text{Total catch second time})}{\text{Number of recaptures}}$$

For example, suppose that 50 sanderlings are captured in mist net traps, marked with leg bands, and released. Two weeks later, 100 sanderlings are captured. If 10 of this second catch are marked birds that have been recaptured, we would estimate that 10% of the total sanderling population is marked. Since 50 birds were originally marked, we would then estimate that the entire population consists of about 500 birds. Note that this method assumes each marked individual has the same probability of being trapped as each unmarked individual. This is not always a safe assumption, however; an animal that has been trapped before, for instance, may be wary of the traps.

(a)

(b)

Figure 47.4
Examples of uniform dispersion. (a) The regular spacing of these creosote bushes in their desert habitat may be caused by competition for water. **(b)** The uniform spacing between individual territories is apparent in this housing development in Pennsylvania.

Random spacing occurs in the absence of strong attractions or repulsions among individuals of a population. For example, forest trees are sometimes randomly distributed. Overall, however, random patterns are not very common in nature; most populations show at least a tendency toward either clumping or uniformity.

DEMOGRAPHY

Changes in population size reflect the relative rates of processes that add individuals to the population versus processes that eliminate individuals from it. Additions occur through births (which we will define here to include all forms of reproduction) and immigration, the influx of new individuals from other areas (Figure 47.5). Opposing these additions are mortality (death) and emigration. In this chapter, we focus primarily on factors that influence birth and death rates.

The study of the vital statistics that affect population size is called **demography.** Birth and death rates usually vary among subgroups within a population, depending in particular on age and sex. It follows that future population size will be determined in part by the existing age and sex structures, two of the most important demographic factors.

Age and Sex Structure

When the average life span of individuals in a population is greater than the time it takes to mature and reproduce, generations overlap. The coexistence of generations gives a population its **age structure,** which is the relative number of individuals of each age.

Each age group has a characteristic birth and death rate. Often, juveniles and old individuals are more likely to die than individuals of intermediate age, who have the optimum combination of youthful vigor and the ability to find food or avoid predators that comes with maturity. Also, birth rate, the number of offspring produced during a certain amount of time, is often greatest for individuals of intermediate age. For example, as an age group, 10-year-old female fur seals produce about twice as many young as 5- or 18-year-old females do over the same amount of time. In humans, the death rate is highest in the first year and, of course, in old age; the birth rate is highest at about age 20. In general, a population with a large percentage of older, nonreproductive individuals will grow proportionately more slowly than a population that has an age structure heavily skewed toward individuals of prime reproductive age or groups slightly younger (the implications of this conclusion for human population growth will be discussed later in the chapter).

An important demographic feature related to age

Figure 47.5
A population built by immigration. Most populations of monarch butterflies that live during the summer in the central and eastern United States and Canada migrate south in the fall. The insects overwinter in remote mountains of central Mexico, creating seasonal populations such as the one shown here. Monarchs migrate north again during the spring. Each round trip may involve several generations: The individuals that migrate to specific sites in Mexico have never made the trip before.

structure is **generation time,** which is the average span between the birth of individuals and the birth of their offspring. Other factors being equal, a shorter generation time will result in faster population growth (assuming, of course, that the overall birth rate is greater than the death rate). This is simply because the increases over time are "compounded" more often.

The proportion of individuals of each sex is another important demographic statistic that affects population growth. The number of females is usually directly related to the number of births that can be expected, but the number of males may be less significant because in many species a single male can mate with several females. In elk, for example, individual males protect harems of females with whom they mate; the percentage of males of reproductive age is significantly

Table 47.2 Life table for a population of the barnacle *Balanus glandula*

Age in Years (x)	Observed Number Alive Each Year	Proportion Surviving at Start of Age Interval (x)	Number Dying in Interval (x)	Mortality Rate	Mean Remaining Life for Animals Alive at Start of Age x
0	142	1.000	80	0.563	1.58 years
1	62	0.437	28	0.452	1.97
2	34	0.239	14	0.412	2.18
3	20	0.141	4	0.225	2.35
4	16	0.109	5	0.290	1.89
5	11	0.077	4	0.409	1.45
6	7	0.046	5	0.692	1.12
7	2	0.014	0	0.000	1.50
8	2	0.014	2	1.000	0.50
9	0	0.0	—	—	—

SOURCE: Adapted from "A Predator-Prey System in thee Marine Intertidal Region. I. *Balanus glandula* and Several Predatory Species of *Thais*," by J. H. Connell, *Ecological Monographs*, 1970, 40:49-78. Copyright © 1970 by the Ecological Society of America. Reprinted by permission.

lower than that of females, but this has no significant effect on the number of births in the overall population. In Canada geese, by contrast, individuals form monogamous bonds for life; any significant reduction in males might be more likely to affect birth rate. Wildlife management often reflects these demographic considerations. For example, deer hunting regulations are usually more liberal regarding the killing of bucks than does (a buck typically mates with many does).

Life Tables and Survivorship Curves

About a century ago, when life insurance first became available, insurance companies needed to determine how much longer on average an individual of a given age could be expected to live. The result was the development of mortality summaries in what are called, ironically, **life tables.**

One way to construct a life table is to follow the fate of a group, or **cohort,** of new organisms throughout their lives until all are dead (Table 47.2). The essential information is simply the number of individuals remaining alive as a function of time; all other information in the life table follows from this. Obviously, this approach to constructing a life table (following the same cohort over time) can be used for only a very limited number of species. Fortunately, it is also possible to construct a life table for a hypothetical starting cohort using only the age at death of a sample of individuals.

The various columns in a life table are really just different ways of describing the same thing: how mortality varies with age over a time period correspond-

ing to the maximum life span. We can see from Table 47.2, for example, that the mortality rate during a given age interval follows directly from the number dying in that interval compared to the number alive at the start of it.

A graphic way of depicting some of the data in a life table is to draw a **survivorship curve,** which is simply a plot of the numbers in a cohort still alive at each age (Figure 47.6). Survivorship curves can be classified into three general types. A Type I curve is relatively flat during early and middle life, and then drops steeply as death rates increase among older age groups. Humans and many other large mammals

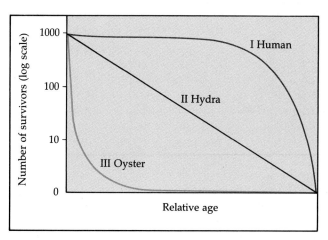

Figure 47.6
General types of survivorship curves. Notice that the y-axis is logarithmic and that the x-axis is on a relative scale so that species with widely varying life spans can be compared on the same graph.

exhibit this general kind of curve, which is usually associated with species that produce relatively few offspring but provide them with good care, resulting in most surviving into later ages. In contrast, a Type III curve involves very high death rates for the young, and then a period when death rates are much lower for those few individuals who survive to a certain age. This type of curve is usually associated with organisms that produce very large numbers of offspring but provide little or no care. An oyster, for example, may release millions of eggs, but most offspring die as larvae from predation or other causes. Those few that manage to survive long enough to find a suitable substrate and begin growing a hard shell, however, will probably survive for a relatively long time. Type II curves are intermediate, with mortality more constant over the lifespan. This kind of survivorship has been observed in various invertebrates, such as *Hydra*, and in rodents, such as the gray squirrel. Many species, of course, fall somewhere between these basic types of survivorship or show more complex patterns. In birds, for example, mortality is often high among the youngest individuals (as in a Type III curve), but fairly constant among adults (as in a Type II curve). Some invertebrates such as crabs may show a "stair-stepped" curve, with brief periods of increased mortality during molts (caused by physiological problems or greater vulnerability to predation), followed by periods of lower mortality (when the exoskeleton is hard).

Later in this chapter we will explore some of the evolutionary considerations related to survivorship and the timing of reproductive activities. For now, we turn our attention to a more detailed analysis of how populations grow and of the factors that regulate population growth.

THE LOGISTIC MODEL OF POPULATION GROWTH

To begin to understand the potential of populations for increase, consider a single bacterium that can reproduce by fission every 20 minutes under ideal laboratory conditions. At the end of this time there would be two bacteria, then four after 40 minutes, and so on. If this continued for only a day and a half—a mere 36 hours—there would be bacteria enough to form a layer a foot deep over the entire Earth. At the other extreme, elephants may produce only-six young in a hundred-year lifespan. Darwin calculated that it would take only 750 years for a population starting with a single mating pair to reach a size of 19 million elephants. Obviously, this kind of indefinitely sustained increase does not happen in the laboratory or in nature. A population that begins at a low level in a favorable environment may increase rapidly for a while, but eventually the numbers must, as a result of limited resources, stop increasing if other factors do not interfere sooner. In Figure 47.7 we see examples for three species of how populations actually grow.

As we discussed in Chapter 46, it is often difficult for ecologists to apply experimental methods to their questions and to predict the consequences of changes in ecological systems. Mathematical modeling, to the extent that it is based on accurate assumptions, provides an alternative approach to some of these problems. It allows an ecologist to study how variables interact or to make predictions about what would happen if some of the variables change. Particularly in population ecology, which deals mainly with variations in numbers and rates, mathematical modeling is

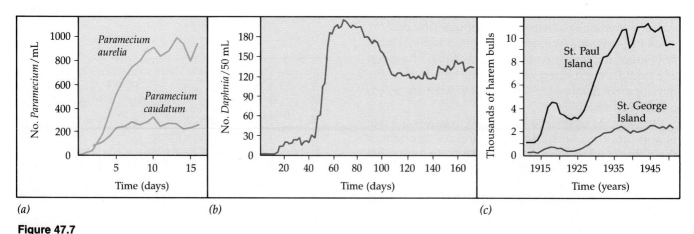

(a)   *(b)*   *(c)*

Figure 47.7
Growth curves for some populations.
(a) Growth of two species of *Paramecium* in laboratory cultures. *P. aurelia* is the smaller organism of the two, which probably accounts for its having a higher numerical population than *P. caudatum* in the face of similar resources. **(b)** Growth of a laboratory population of the small freshwater crustacean *Daphnia* in 50 mL of pond water. Notice that the population rose quickly over a short time to a high value, then settled back to a lower, more stable level. **(c)** Numbers of adult male fur seals with harems on two of the Pribilof Islands, Alaska. The numbers of these males reflect the general changes occurring in seal populations over time. Prior to 1911, seal populations were greatly depressed due to hunting. After hunting was stopped, populations increased dramatically, particularly on St. Paul Island. Both populations eventually reached an apparent general upper limit, but continued to oscillate.

a common method. This approach itself can be complex. Here, we will explore only a few basic considerations and one simple model as an example.

Imagine a hypothetical population consisting of only a few individuals in a very favorable environment. Under these conditions there is no restriction on the abilities of individuals to live, grow, and reproduce other than inherent physiological limitations. We would try to describe how numbers (N) in the population change as a function of time (t), which we could symbolize as $\Delta N / \Delta t$. Or, to denote the instantaneous changes occurring in the population, we can express this in terms of the differentials of calculus as dN/dt. And this net rate of instantaneous change in the population would be:

$$\frac{dN}{dt} = rN$$

Here, N is the existing number in the population, and r is the **intrinsic rate of increase.** This intrinsic rate is based on the net result of additions to a population from births minus removals by deaths. More specifically, r is the change in population size per individual per unit time—a per capita rate of change. Since births will exceed deaths under the maximally favorable conditions we have assumed, r is a positive number and therefore the population will grow (at the rate rN). Potential values for r vary over several orders of magnitude, depending on the kind of organisms (Table 47.3). Of course, the actual change in the population size also depends on N, the number of individuals in the population to which the intrinsic rate is applied. As our hypothetical population increases in size, N is getting bigger and bigger, and the same r value continuously applied would yield ever-greater increases—the same kind of exponential increase typical of any "compound interest" system.

In contrast to our hypothetical population living in an imaginary habitat with unlimited resources, a real population may grow exponentially for a while, but growth eventually slows and then stops. This deceleration of growth occurs as the population approaches **carrying capacity,** the population size that can be supported by available resources. Carrying capacity, symbolized by K, is the number of individuals the environment can just maintain ("carry") with no net increase or decrease. In other words, at carrying capacity, birth rate equals death rate. The value of K varies, depending on species and habitat.

Thus, a population in a limited environment grows at a slower and slower rate as it nears carrying capacity. To account for this, we can modify our original equation for population growth by adding a "nearness to carrying capacity" term:

$$\frac{dN}{dt} = rN\left(\frac{K - N}{K}\right)$$

Table 47.3 Maximum instantaneous rate of increase (r) and generation time for selected species

Taxon	Organism	Birth Rate (per Capita per day)	Generation Time (Days)
Bacterium	*Escherichia coli*	60.0*	0.014
Protist	*Paramecium aurelia*	1.24	0.33–0.50
Insect	Flour beetle	0.120	80*
Insect	Spider beetle	0.014	179
Insect	Gall fly	0.010	110
Mammal	Rat	0.015	150
Mammal	Mouse	0.013	171
Mammal	Dog	0.009	1,000*
Insect	17-year cicada	0.001	6,050
Mammal	Human	0.0003	7,000*

*Approximate
SOURCE: From E. R. Pianka, *Evolutionary Ecology*, 3d ed. (New York: Harper & Row, 1983).

This model is called the **logistic equation.** Let us examine how the new term, $(K - N)/K$, affects population growth as N increases. If the population is small, N will be small in comparison with K. The term $(K - N)/K$ will therefore be close to $K/K = 1$, and population growth will be described by something close to our simple exponential model (rN). If N is close to carrying capacity, however, $(K - N)/K$ will be a small fraction. In this case, our change in population size will be described not by (rN) alone, but by this value depreciated by the fraction $(K - N)/K$. At carrying capacity, $N = K$ and $(K - N)/K = 0$. At this point, therefore, the change in the population is zero. (Note that this is the *change*; the *number* may be large but has leveled off). Figure 47.8 graphically compares exponential growth versus growth described by the logistic equation. Notice that for logistic growth, the curve is S-shaped (sigmoidal); a period of increasing growth rate is followed by a leveling off as the population size approaches K, the carrying capacity. Although the actual populations graphed in Figure 47.7 deviate from perfect S-shaped growth curves, the effect of carrying capacity is evident.

What are the actual biological implications of the logistic model? The growth rate is small when the population size is *either* small or large, and highest when the population is at an intermediate level. In biological terms, at a low population level, resources are abundant and intraspecific competition is slight. The population is able to grow nearly exponentially,

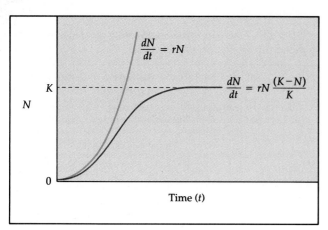

Figure 47.8
Exponential and logistic increase in number (N) with time. Exponential growth (blue curve) describes an idealized case, where resources are unlimited. The logistic equation (red curve) incorporates the concept of "carrying capacity," or K. This value is the maximum population size that can be maintained by the available resources of the environment. According to the logistic model, the rate of population growth decelerates as N approaches K—that is, as the size of the population nears what the environment can support.

and the changes are small mostly because N is small. In contrast, at a high population level, environmental pressures oppose the intrinsic potential for the population to continue increasing, and these pressures become more severe as N increases. There might be less food available per individual or fewer nest sites or shelters. Competition and other negative interactions become more intense. In a laboratory culture dish of small organisms, the buildup of wastes might become a factor that limits growth. As the population approaches carrying capacity, birth rates decrease, death rates increase, or both. A population growing as described by the logistic equation eventually stabilizes in size when birth rate equals death rate.

How well does the logistic equation describe growth of actual populations? The growth of laboratory populations of some small animals, such as beetles, and microorganisms, such as *Paramecium*, yeast, and bacteria, fit S-shaped curves fairly well (see Figure 47.7a). However, these experimental populations are grown in a constant environment lacking predators and other species that may compete for resources, idealized conditions that never occur in nature. Even under these laboratory conditions, not all populations stabilize at a clear carrying capacity, and most populations show unpredictable deviations from a smooth sigmoid curve.

Some of the basic assumptions built into the logistic model clearly do not apply to all populations. For example, the model incorporates the idea that even at low population levels each individual added to the population has a negative effect on population

growth rate; that is, *any* increase in N reduces the term (K − N)/K. Some populations, however, show an **Allee effect,** named after the researcher who first described it. This idea points out the *benefit* that individuals may experience as populations increase. In some species, individuals may have a more difficult time growing and reproducing if the population size is too low as well as too high. For example, a single plant standing alone may suffer from excessive wind but would be protected in a clump of individuals. A predator might be more likely to be spotted if a large number of prey are all together than it would be by a single individual. Some oceanic birds require large numbers at their breeding grounds to provide the necessary social stimulation for reproduction. In these cases, a greater number of individuals in the population has an overall positive effect, up to a point, on population growth. When a population is at a low level, even if this does not directly cause problems, there is a greater possibility that chance events will eliminate all individuals or that inbreeding will lead to a general reduction in fitness (see Chapter 21). As Jane Goodall mentioned in the interview at the beginning of this unit, epidemics and detrimental changes in genetic makeup can be problems for chimpanzees living in small, isolated groups.

The logistic model also incorporates the assumption that populations approach carrying capacity smoothly. In many populations, however, there is a lag time before the negative effects of an increasing population are realized. For example, as some important resources, say food or nest sites, become limiting for a bird population, mating activity may be reduced—but the birth rate is not immediately affected because the eggs resulting from *previous* matings do not hatch for a while. This may cause the population to overshoot the carrying capacity. Eventually deaths will exceed births, and the population may then drop below carrying capacity; even though reproduction begins again as numbers fall, there is a delay until new individuals actually appear. Apparently as a result of such time lags, many populations seem to oscillate about some general carrying capacity or to overshoot at least once before settling down to a more stable size (see Figure 47.7).

Finally, as we will discuss in the next section, populations do not necessarily remain at, or even reach, levels where population density is an important factor. In many insects and other small, quickly reproducing organisms that are sensitive to environmental fluctuations, physical variables such as temperature or moisture reduce the population well before resources become limiting. For describing changes in the numbers of these organisms, the idea of carrying capacity does not really apply.

Overall, the logistic model is a useful starting point for thinking about how populations grow and for con-

Figure 47.9
Density-dependent regulation of a population. These gannets nest in dense colonies on rocky islands in the sea. The number of such sites is limited, restricting populations from growing beyond a certain size. In some "gannetries," the number of breeding birds is known to have been almost constant for about 20 years. In other areas, however, populations were once greatly reduced because of predation by humans, who used the birds as fishbait. Now that this practice has largely stopped, gannet populations in these areas are again climbing steadily toward carrying capacity.

structing more complex models. Although it fits few if any real populations exactly, the logistic model incorporates basic ideas that, with modification, do apply to many populations. And like any good hypothesis, this model has stimulated many experiments and discussions that, whether they support the model or not, lead to a greater understanding of population ecology in general.

REGULATION OF POPULATIONS

Ecologists have long debated the most important factors regulating populations. At one time, population ecologists were somewhat divided into two camps on this subject: One emphasized the importance of density-dependent factors in population regulation and the other emphasized density-independent factors. The current consensus is that the relative importance of these factors varies with the kind of organism and its specific circumstances, and that often both density-dependent and density-independent factors interact to affect a population.

Density-Dependent Factors

In resisting population growth, a **density-dependent factor** is one that intensifies—that is, affects a greater percentage of the individuals in a population—as the number of individuals increases. Each individual has a dwindling share of resources, or perhaps a higher proportion of individuals succumbs to disease or predation. Population growth declines because the death

rate increases, the birth rate decreases, or both. As we have seen, the logistic model assumes this kind of density-dependent regulation.

It is obvious that resource limitation sets an ultimate cap on any population, and this factor significantly influences the actual size of many populations in nature. For example, oceanic birds known as gannets nest in colonies on rocky islands, where they are relatively safe from predators (Figure 47.9). Such areas are limited in number and size, and only a fraction of the total population can obtain suitable nest sites. Although the limited number of nest sites fits the general idea of a density-dependent factor, it probably does not operate with the kind of simple, gradually increasing severity that the logistic model predicts. That is, up to a certain population size, most birds could find a suitable nesting site. Few birds beyond that number, however, would be successful.

Density-dependent regulation is common in laboratory populations. For example, if several flour beetles are placed in a box with a constant amount of flour added each day for food, the population increases rapidly at first but then levels off or declines as population size outstrips the food supply. White-footed mice in a small enclosure will multiply from a few to a colony of 30 to 40 individuals, but eventually reproduction will decline until the population ceases to grow. Although this change is clearly associated with increasing density, it occurs even when food and shelter, the main resources needed by the mice, are provided in abundance. Through mechanisms that are not yet understood, high densities cause physiological changes that delay sexual maturation, cause reproductive organs to shrink, and generally inhibit reproduction.

Predation may be an important density-dependent regulator for some populations. A predator probably encounters and captures more prey as the population density of the prey increases. This is not considered density-dependent, however, because the same percentage of the prey population is still being taken. Many predators, however, exhibit switching behavior; they begin to concentrate on a particular kind of prey when it becomes energetically efficient to do so (see the discussion on optimal foraging in Chapter 50). As a prey population builds up, predators may feed preferentially on that species and take a higher percentage of individuals; this change is density-dependent. For example, trout may concentrate for a few days on a particular species of insect that is emerging from its aquatic larval stage, then switch as another insect species becomes more abundant. Many herbivorous animals immigrate into areas where there are abundant food plants, subsequently reducing plant density. Such interactions at the community level that are important in affecting population sizes are discussed in more detail in Chapter 48.

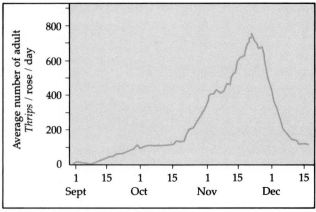

Figure 47.10
Density-independent regulation of a population. Populations of the insect *Thrips* build up quickly as the flowers they live in and feed on increase during spring. Before the populations reach carrying capacity, however, numbers are drastically reduced as the dry Australian summer causes most of the flowers to die. (Remember, in interpreting this graph, that seasons in the Southern Hemisphere occur at opposite times of the calendar year from those in the Northern Hemisphere.)

Density-Independent Factors

The occurrence and severity of density-independent factors are not correlated with population size; density-independent factors affect the same percentage of individuals regardless of population size. The most common and important density-independent factors are related to weather and climate. For example, a freeze in the fall may kill a certain percentage of the insects in a population. Obviously, the time when the first freeze occurs and just how cold it gets are not determined by the density of the insect population.

In some natural populations, density-independent effects routinely control population size before resources or other density-dependent factors become important. This seems to be the case, for instance, for population cycles of Australian insects known as *Thrips* (Figure 47.10). These insects feed on the pollen and flower tissues of plants belonging to the rose family, including apples, and *Thrips* can become severe pests. Population size of *Thrips* is closely linked to the number of flowers available, which in turn is correlated with the seasonal weather patterns. The insect populations are relatively small during the cool winters but rapidly increase during the spring as flowers bloom. Summers, however, are extremely dry, and flowers survive only in a few sheltered locations. During the favorable spring period, the *Thrips* population increases in a manner that may approximate logistic growth; with the onset of summer, however, population size suddenly crashes—well before density-dependent factors would become important and before the population shows any evidence of reaching carrying capacity. A few individuals remain in the surviving flowers, and these allow population growth to resume again when favorable conditions return. Many other species, particularly insect populations, are probably controlled at some point primarily by density-independent factors.

Environmental episodes more sporadic than seasonal changes in weather can also affect populations in a density-independent manner. For example, fires and hurricanes may strike often enough in some areas to have a significant impact on some populations. Volcanic eruptions, though extraordinarily dramatic and destructive, occur so infrequently that they are not very important as general mechanisms of population regulation.

Interaction of Regulating Factors

Over the long term, many populations remain fairly stable in size and are presumably close to a carrying capacity that is determined by density-dependent factors. Superimposed on this general stability, however, are short-term fluctuations due to density-independent factors. In a case study, researchers monitored the number of European herons, large birds, for 30 years in two different areas of England. The general pattern was the same for both areas: Populations were reasonably stable over the three-decade span, but major declines occurred following unusually cold winters.

In some cases, density-dependent and density-independent factors act together to regulate a population. For example, in very cold, snowy areas many deer may starve to death during the winter. The severity of this factor is proportional to the harshness of winter; colder temperatures increase energy requirements (and therefore the need for food), while deeper snow makes it harder to find food. But the effect is also density-dependent, since a larger population means less of whatever little food is available per individual.

The relative importance of density-dependent and density-independent regulation may also vary seasonally. This complex interaction of factors apparently regulates populations of the bobwhite quail. The range of this bird extends northward into southern Wisconsin, where survival is difficult during winter. The number of birds alive at the end of the winter is largely determined by the depth of snow cover, a density-independent factor. If there is little snow, as much as 80% of the population will survive, but with more snow this may decline to only 20%. Whatever happens in the winter, however, the population going into next fall is quite constant. This is because the population grows according to a density-dependent, logistic-like pattern during the spring and summer, increasing to the apparent carrying capacity. (Keep in mind an important aspect of this kind of growth: When numbers are reduced, the remaining individuals may exhibit a relatively high birth rate because of an abundance of resources per capita.) Even the *Thrips* populations described earlier as a classic example of density-independent regulation are probably regulated during part of the year by a density-dependent factor; that is, the number of flowers available in the winter is a limited resource that restricts the number of surviving insects. Most populations are probably regulated by some mix of density-dependent and density-independent factors.

Population Cycles

Some populations of birds, mammals, and insects fluctuate in density with remarkable regularity. For example, many species of arctic voles and lemmings have population cycles of three to four years, whereas snowshoe hares have a ten-year cycle. Several hypotheses have been proposed to explain population cycles.

One idea is that crowding regulates cyclical populations, perhaps by affecting the organisms' endocrine systems. Stress resulting from a high population density may alter hormonal balance and reduce fertility, as described earlier for mice under crowded laboratory conditions. However, it is not known if such changes are a common occurrence in the many species of animals that have population cycles.

Another hypothesis is that population cycles are

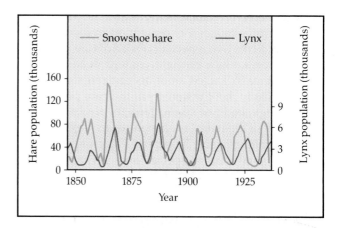

Figure 47.11

Population cycles of the snowshoe hare and its predator, the lynx. Oscillation in the population density of the hare followed by corresponding changes in lynx density was once interpreted as strong circumstantial evidence that these populations of prey and predator regulated each other. (Population counts are based on the number of pelts sold by trappers to the Hudson Bay Company.) However, snowshoe hare populations on islands where the lynx is absent show similar cycles. The periodic crashes in the hare population may be associated with changes in the food source brought about by overfeeding. Changing ideas on the snowshoe hare cycle emphasize the problem of inferring process from pattern. Most patterns of population growth are likely caused by multiple factors that are difficult to distinguish without careful experimentation.

caused by a time lag in the response to density-dependent factors; this may cause the overshooting and undershooting of carrying capacity mentioned earlier. Such a mechanism involving predation as the major density-dependent factor was once a popular explanation for a correlation in the cycles of the snowshoe hare and the Canadian lynx, which eats the hare (Figure 47.11). Now, however, the evidence is less clear that the lynx is the major factor regulating the hare population. An alternative explanation for population cycles of herbivores such as the snowshoe hare is that a high density causes a deterioration in food quality. Studies indicate that when certain plants are damaged by herbivores, their nutrient content decreases and they produce increased amounts of defensive chemicals. It is likely that the density of the hare population may be regulated not so much by its predators as by the quality of the hares' own food source—or perhaps by some other resource limitation. The causes of cycles probably vary among species and perhaps among populations of the same species; so far there is no satisfactory general explanation.

EVOLUTION OF LIFE HISTORIES

Birth, reproduction, and death—the personal episodes fundamental to the **life history** of an organism—affect the growth of populations over ecological time. Of course, a particular life history pattern, like most characteristics of an organism, is the result of natural selection operating over evolutionary time. Because of varying selection pressures, life histories are diverse. Pacific salmon, for example, require several years to mature in the ocean, after which they return to freshwater streams to spawn millions of small eggs in a single reproductive opportunity, and then they die. In contrast, a lizard may mature in about a year and produce only a few large eggs that year, but repeat the reproductive act annually for several years. The life histories of plants are just as variable. Some species of oaks do not reproduce until the tree is 20 years old, but then produce vast numbers of large seeds each year for a century or more. Annual desert wildflowers generally produce many small seeds in their lifetimes, which may last only a month. Important characteristics of life history may vary significantly among closely related species, or even among individuals in the same population.

What accounts for this variation in life histories? Although it is often impossible to disentangle the historical interrelationships of adaptations, we can see some obvious cases where features of life history are constrained by other evolutionary developments. For example, the mammalian mode of reproduction precludes a life history in which thousands of offspring are produced each year. Natural selection also operates more directly on life history, through mechanisms we will examine shortly.

Three major characteristics of life history affect the number of offspring an individual will produce; these in turn affect a population's intrinsic rate of increase (*r*, in the growth models introduced earlier):

1. *Number of reproductive episodes per lifetime.* Organisms that reproduce only once in a lifetime, such as

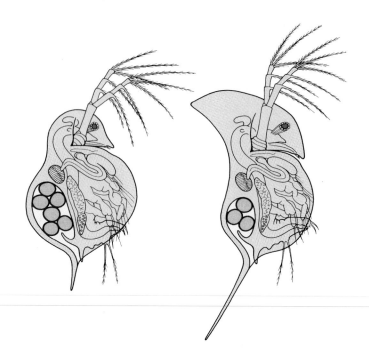

Figure 47.12
Variation in life history caused by predation. The freshwater crustacean *Daphnia retrocurva* differs markedly in morphology and in clutch size depending on the season. In the spring, the phytoplankton that serve as food for *Daphnia* are abundant and predators that eat *Daphnia* are scarce. Under these favorable conditions, individuals develop into a rounder form with a relatively large brood chamber that contains about six eggs, thus maximizing reproduction. In the summer, however, other plankton that feed on *Daphnia* become more abundant. *Daphnia* developing at this time have large "helmets" and long tail spines; these make it more difficult for the predators to "lock on" and hold the *Daphnia*. These morphological changes, however, cause the brood chamber to become more compressed and smaller overall so that only about half as many eggs are present. Presumably, of course, helmeted individuals survive longer under these conditions than would unhelmeted forms with more eggs. We see here that adaptations of life history can be interpreted only within the total environmental context affecting the organism.

the Pacific salmon or annual plants, are called **semelparous.** Those that reproduce more than once, as is the case with a wide variety of plants and animals including humans, are called **iteroparous.**

2. *Clutch size.* The number of offspring produced at each reproduction episode is called the clutch size. Generally, organisms that have a large clutch size tend to produce smaller offspring, and vice versa.

3. *Age at first reproduction.* Timing of reproduction in the overall life history can have a large effect on the numbers in the future generation. Consider two human populations where each female produces an average of three offspring, but reproduction starts at an average age of 18 years in one population and at age 28 in the other. The former population will grow at a much faster rate because "interest" (offspring) is compounded faster—remember, *r* is inversely related to generation time.

If we were to construct a hypothetical life history that would produce the greatest rate of increase, we might imagine a population of individuals that begin reproducing at an early age, have large clutch sizes, and reproduce many times in a lifetime. However, seldom if ever does natural selection maximize all of these variables simultaneously. This is partly because organisms have a finite energy budget that mandates trade-offs. For example, an individual might allocate some energy to egg production (reproduction) instead of general growth at an earlier than average age. This earlier reproduction may translate into more reproductive episodes, but because the parent is smaller than average, fewer eggs can be produced in each episode. Another individual that begins reproducing at a later age has fewer reproductive episodes, but each episode results in more eggs because the parent is larger.

Predation is among the other factors that can affect life history through natural selection. For example, selection may favor earlier reproduction for organisms that generally have short lifespans due to predators.

Figure 47.12 describes an interesting example of how predation pressure may cause characteristics of life history to vary seasonally even within the same population.

Darwinian fitness, of course, is measured ultimately not in terms of how many offspring are produced but by how many survive to produce their own offspring. For an individual in a population already at carrying capacity, producing a large number of offspring with little chance of survival may not be as effective as producing a few well-cared-for offspring that can compete vigorously for limited resources.

In the 1960s, population ecologists developed the concepts of ***r*-selection** and ***K*-selection** in an attempt to describe the evolutionary forces behind differences in life history. In an *r*-selected population, life history is centered around a high potential rate of increase (a high "*r*" value in the logistic equation). Individuals reproduce at an early age, are generally semelparous, and have large clutch sizes of relatively small offspring. This type of life history would be advantageous in variable environments, where populations are controlled mainly by density-independent factors such as fluctuations in the weather. Such populations seldom approach carrying capacity, where competitive interactions and other density-dependent controls would become more important. Also, factors causing death are relatively indiscriminate, killing individuals somewhat randomly. In an *r*-selected population, therefore, simple quantity of offspring is a better assurance of Darwinian success than "quality" of offspring. The desert annuals of Figure 47.13a generally fit this description.

(a)

(b)

Figure 47.13
Variations in life history. (a) Most desert annuals, such as these wildflowers at Organ Pipe Cactus National Monument, Arizona, exhibit the general characteristics of *r*-selected populations. They produce large quantities of offspring with a relatively low probability of individual survival. **(b)** Polar bears fit the general model of a *K*-selected population. They produce few offspring but give them good care, which enhances survival.

In contrast, populations that can tolerate the full range of environmental conditions they experience remain near carrying capacity, controlled in a density-dependent manner by limited resources. In this situation, fitness may depend more on the competitive abilities of individuals and their offspring than on sheer numbers. Here we would expect K-selected populations, with individuals producing young that are few in number but are likely to survive (the "K" here is for the symbol for carrying capacity used in the logistic equation). Many of the larger terrestrial vertebrates fit this general pattern (Figure 47.13b).

The concepts of r-selection and K-selection quickly became popular, not only because they integrated a number of important concepts but also because they generated testable hypotheses. Many field and laboratory studies generally support the view that features of life history are affected by environmental stability and the extent to which density-dependent factors control a population. However, it is important to understand that r-selection and K-selection represent extremes in what is actually a range of life histories; many populations have life histories and factors affecting them that fall somewhere between r- and K-selection. Some of the most enlightening studies compare closely related species, or different populations of the same species, that show different life history characteristics in different environments. For example, populations of the familiar dandelion may vary along the r- to K-selection continuum depending on how severely their habitats are disturbed. Plants of populations in frequently trampled areas tend to be smaller and produce more seed than dandelions in undisturbed areas (Table 47.4).

As with so many important ecological ideas developed in the past 20 to 30 years, the concepts of r-selection and K-selection are now seen as an oversimplification. Environmental fluctuations important to life history may operate at varying degrees of severity and over a wide range of frequencies. Recently, some ecologists have suggested *stress* as another important environmental characteristic that shapes life history. In this sense a stressful environment is not so much one that fluctuates, but rather one in which conditions remain generally difficult—for example, continuously cold or with dim light. Here we might expect to find organisms that grow slowly for a relatively long time and produce their few offspring infrequently—characteristics that are similar to those of K-selected species. In this case, however, the characteristics are not based on strong competitive interactions. We have seen too how interactions between species, such as predation, may affect life history in variable ways even within the same population. Increasingly, ecologists recognize that a population may show a mix of the traditional r- and K-selected characteristics; in most natural settings, some complex interplay of factors determines the most favorable life history pattern.

Now that we have examined some general concepts of population dynamics, let us apply these ideas to the specific case of human population growth.

HUMAN POPULATION GROWTH

The human population has been growing exponentially for centuries. Indeed, it is unlikely that any other population of large animals has ever sustained exponential growth for so long.

Human population increased relatively slowly until about 1650, when approximately 500 million people inhabited the Earth (Figure 47.14). The population doubled to 1 billion within the next two centuries, doubled again to 2 billion between 1850 and 1930, and doubled still again by 1975 to more than 4 billion. The population now numbers about 5.2 billion people and increases by about 80 million each year—it takes only about three years for world population to add the equivalent of another United States. If the present

Table 47.4 Percentage of each of four types of dandelions in three Michigan populations

Habitat	Number in Sample	Smaller Size, More Seeds		Larger Size, Fewer Seeds	
		A	B	C	D
Dry, full sun, frequently trampled	94	73	13	14	0
Dry, shade, occasional disturbance	96	53	32	14	1
Wet, semishade, undisturbed	94	17	8	11	64

SOURCE: Modified from Paul Colinvaux. *Ecology.* (New York: John Wiley & Sons, 1986), p. 251.

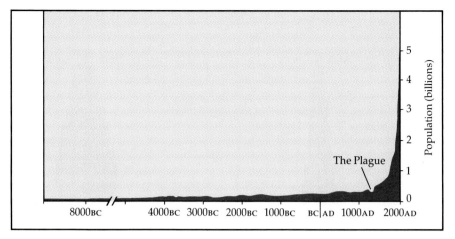

Figure 47.14
History of human population growth.
The human population has grown almost continuously throughout historical times, but has skyrocketed since the Industrial Revolution. No population can continue growing like this indefinitely; we know that increased death rates, decreased birth rates, or a combination of both will cause a leveling off or a decline. What we do not know are the exact mechanisms that will bring this about or the speed with which they will operate: a gradual movement toward an equilibrium primarily through decreased births or a rapid change due to mass death.

growth rate persists, there will be 8 billion people on Earth by the year 2017.

Human population growth is based on the same general parameters that affect other animal and plant populations: birth rates and death rates. Birth rates increased and death rates decreased when agricultural societies replaced a life style of hunting and gathering about 10,000 years ago. Since the Industrial Revolution, exponential growth has resulted mainly from a drop in death rates, especially infant mortality, even in the most underdeveloped countries (Figure 47.15). Improved nutrition, better medical care, and sanitation have all contributed to an increased percentage of newborns that survive long enough to leave offspring of their own. The effect of decreasing mortality on population growth is compounded in most underdeveloped countries, which tend to have relatively high birth rates.

A unique feature of human reproduction is that it can be consciously controlled by voluntary contraception or government sanctions. Social change also affects birth rates. For example, in most developed countries, many women are delaying marriage and reproduction, perhaps because of better opportunities for employment and advanced education.

Current worldwide population growth is a mosaic of various rates of growth in different countries. Some developed countries, such as Sweden, are near zero population growth because birth and death rates balance. The human population as a whole, however, continues to grow because birth rates greatly exceed death rates in most nations, particularly in underdeveloped and developing countries.

Age structure is an important demographic factor in present and future growth trends (Figure 47.16). The relatively uniform age distribution in Sweden, for instance, contributes to that country's stable population size; individuals of reproductive age or younger are not disproportionately represented in the population. In contrast, Mexico has an age structure that is bottom heavy—skewed toward young individuals who

will grow up and sustain the explosive growth by their own reproduction. Notice in Figure 47.16 that the age structure for the United States is roughly even except for a bulge that corresponds to the "baby boom" that lasted for about two decades after the end of World War II. Even though couples of that generation are now having an average of fewer than two children, the nation's overall birth rate still exceeds death rate because there are so many "boomers" of reproductive age. Immigration also contributes to the growth of the U.S. population. Nevertheless, the overall growth rate of the United States is relatively low—about 1% per year compared with 1.7% for the world as a whole.

Just as it is difficult to predict future sizes of other animal and plant populations, it is difficult to foresee the future for human population growth, though for

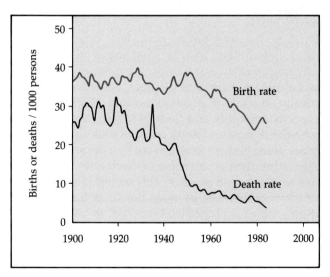

Figure 47.15
Changes in the birth rates and death rates in Sri Lanka (formerly Ceylon). Population growth has continued at a rapid pace despite a decline in birth rate, which has not been large enough to offset the drop in death rates.

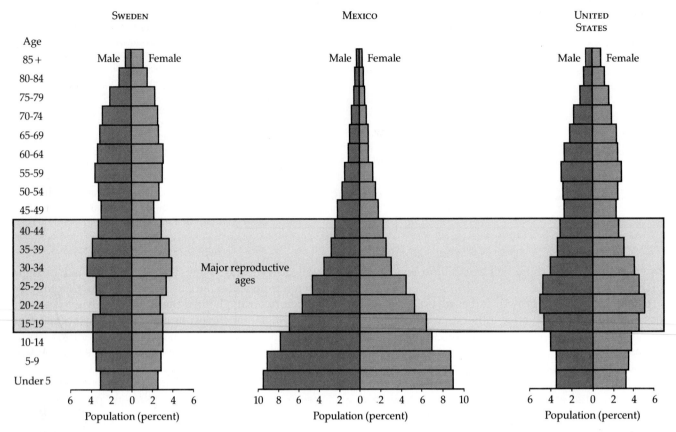

Figure 47.16
Age structures of three nations. The proportion of individuals in different age groups has a significant impact on the potential for future population growth. Mexico, for example, has a large fraction of individuals who are young and are likely to reproduce in the near future. In contrast, Sweden's population is distributed more evenly through all age classes, with a high percentage of individuals beyond prime reproductive years. The United States has a fairly even age distribution, except for the bulge corresponding to the post–World War II baby boom. (1985 data.)

somewhat different reasons. Technology has undoubtedly increased the carrying capacity of the Earth for humans, but no population can continue to grow indefinitely. Exactly what the world's human carrying capacity is and under what circumstances we will approach it are topics of great concern and debate. Ideally, human populations would reach carrying capacity smoothly and then level off. This will occur when birth rates and death rates are equal, and it seems more desirable for r to reach zero by a decrease in birth rate rather than an increase in death rate. If the population fluctuated about K, this would mean periods of increase followed by mass death, as has occurred during plagues or localized famines. In any case, the human population will eventually stop growing, whether by social changes, individual choice, government intervention, or increasing mortality due to the limitations of the environment.

* * *

We have seen throughout this chapter that population dynamics change during ecological time and evolutionary time as a result of interactions with the environment, which include not only physical factors but also other organisms living in the same place. In the next chapter, we examine in more detail the interaction of organisms within communities.

STUDY OUTLINE

1. Population ecology is the study of population fluctuations and factors influencing population dynamics.

Density and Dispersion (pp. 1075–1078)

1. A population is a group of individuals of a given species inhabiting a specified geographical range and exhibiting a characteristic density and dispersion.

2. Density is the number of individuals per unit area or volume. Density may be determined by actual counts or by various estimating techniques. Dispersion is the pattern of spacing for individuals and may range from clumped, to random, to uniform, as determined by various environmental or social factors.

Demography (pp. 1078–1080)

1. Demography is the study of vital statistics that affect population size and growth.
2. The age structure of a population influences population growth owing to the differential reproductive capabilities and probabilities of mortality of the different age groups.
3. Life tables describe the fate of a cohort of newborn organisms throughout their lives, tabulating mortality rates, the number of survivors from one age to the next, and life expectancies.
4. Survivorship curves, which plot the number of individuals of a cohort alive at each age, can be divided into three general types depending on whether mortality is greater among the young or old or constant over all ages.

The Logistic Model of Population Growth (pp. 1080–1083)

1. The exponential growth equation is a beginning way of thinking about the potential of a population for explosive growth. Defined as

$$\frac{dN}{dt} = rN$$

where N is the starting number, t the time, and r the intrinsic rate of increase, this model predicts very simply that the larger a population becomes, the faster it grows. Such exponential growth is never sustained for extended periods of time in any population.

2. A more realistic model limits growth by incorporating a term called the carrying capacity (K), which is the maximum population size that can be supported by the available resources. The logistic equation

$$\frac{dN}{dt} = rN\left(\frac{K - N}{K}\right)$$

fits an S-shaped curve in which population growth levels off as size approaches carrying capacity.

3. Few populations fit the logistic model exactly, but many show a generally similar pattern in which the growth rate is highest when the population size is intermediate.

Regulation of Populations (pp. 1083–1086)

1. Population size is regulated through density-dependent and density-independent factors.
2. Density-dependent factors have a greater effect as the population density increases, owing to more intense intraspecific competition for resources or increased predation and disease. The logistic equation reflects the effect of density-dependent factors, which ultimately stabilize populations around the carrying capacity.
3. Density-independent factors, such as climatic events, reduce population size by a given fraction, regardless of its density. Such factors are often superimposed on density-dependent factors.

4. Defining the importance of specific environmental factors in population regulation is complicated by overlap of density-dependent and density-independent factors and the difficulty in determining cause and effect relationships.
5. Some populations have remarkably regular population cycles that may be a result of the physiological effects of crowding or time lags in response to density-dependent factors.

Evolution of Life Histories (pp. 1086–1088)

1. Natural selection has led to the evolution of diverse life history strategies. Each population has characteristics of birth, reproduction, and death that operate under historical and anatomical constraints.
2. The number of reproductive episodes per lifetime, clutch size, and age at first reproduction are three important life history characteristics.
3. Some populations are said to be *r*-selected; these have a high intrinsic rate of increase, producing large clutch sizes of small offspring at an early parental age, usually in variable environments where density-independent factors are important regulators of population size. Such populations fluctuate widely with the environment and rarely approach a carrying capacity.
4. Stable environments with limited resources may favor *K*-selected populations controlled mostly by density-dependent factors. Reproductive efforts are concentrated on producing relatively few offspring with a good chance of survival and good competitive abilities.
5. Most populations have both *r*- and *K*-selected characteristics. The relative influence of the two may vary under different environmental conditions.

Human Population Growth (pp. 1088–1090)

1. Human population growth has been exponential for centuries, sustained by such factors as the Industrial Revolution and improved nutrition, medical care, and sanitation. Different countries have different rates of growth for various environmental, cultural, and historical reasons.
2. The importance of age structure as a demographic factor strongly affecting population growth can be seen in the post–World War II "baby boom" in the United States and in the "bottom-heavy" structure of some countries.
3. Although technology has increased our carrying capacity, the human population cannot grow indefinitely.

SELF-QUIZ

1. A uniform dispersion pattern for a population may indicate that
   a. the population is spreading out and increasing its range
   b. resources are heterogeneously distributed
   c. individuals of the population are competing for some resource, such as water and minerals for plants or nesting sites for animals
   d. there is an absence of strong attractions or repulsions among individuals
   e. the density of the population is low

2. In a mark-recapture study of a lake trout population, 40 fish were captured, marked, and released. In a second capture, 45 fish were captured; 9 of these were marked. What is the estimated number of individuals in the lake trout population?

a. 90 b. 200 c. 360 d. 800 e. 1800

3. The term $(K - N)/K$ influences dN/dt such that

a. the increase in actual population numbers is greatest when N is small

b. as N approaches K, r (the intrinsic rate of increase) becomes smaller

c. when N equals K, population growth is zero

d. when K is small, the influence of density-dependent factors is smaller

e. as N approaches K, birth rate approaches zero

4. The carrying capacity for a population is

a. the number of individuals in that population

b. reached when mortality exceeds natality

c. inversely related to r

d. the population size that can be supported by available resources for that species within the habitat

e. set at 8 billion for the human population

5. A Type III survivorship curve would be expected in a species in which

a. mortality occurs at a constant rate over the lifespan

b. parental care is extensive

c. a large number of offspring are produced, but parental care is minimal

d. mortality rate is quite low for the young

e. K selection prevails

6. A population that has a relatively low r value will most likely

a. have large clutch sizes with relatively small offspring

b. be found in environments that are highly variable

c. have an early age of first reproduction and a short generation time

d. produce fewer offspring with more competitive capabilities

e. be regulated by density-independent factors

7. The example of the population cycles of the snowshoe hare and its predator, the lynx, illustrates that

a. predators are the major factor in controlling the size of prey populations, and are, in turn, regulated in their numbers by the oscillating supply of prey

b. the two species must have evolved in close contact with each other since their life histories are intertwined

c. one should not conclude a cause and effect relationship when viewing population patterns without careful observation and experimentation

d. both populations are controlled by density-dependent factors

e. the hare population is r-selected whereas the lynx population is K-selected

8. The current size of the human population is closest to

a. 5 million

b. 3 billion

c. 4 billion

d. 5 billion

e. 10 billion

9. Consider five human populations that differ demographically *only* in their age structures. The population that will grow the most in the next 30 years is the one with the greatest fraction of people in which age group?

a. 10–20

b. 20–30

c. 30–40

d. 40–50

e. 50–60

10. All of these terms are characteristic of the human populations in developed countries *except*

a. relatively small "clutch" size

b. iteroparous

c. r-selected

d. type I survivorship curve

e. N<K

CHALLENGE QUESTIONS

1. East African chimpanzees eat fruit as a primary food source. Their 12 most important food species are relatively rare trees that show a clumped distribution and unpredictable fruiting patterns. Ripe fruit appears and disappears in a matter of days and varies greatly with the wet and dry seasons. How might such a distribution of fruit trees affect the density and social ecology of the chimpanzees?

2. From your reading of this chapter and Further Reading 1, would you consider the twentieth-century global human population as primarily r- or K-selected? Why?

FURTHER READING

1. Ackerman, L., et al. "The Successful Animal." *Science* 86, January/February 1986. Cultural and historical aspects of human population growth and control.
2. Ghiglieri, M. P. "The Social Ecology of Chimpanzees." *Scientific American*, June 1985. The availability of food in a territory affects the social structure and dispersion of apes.
3. Lewin, R. "Judging Paternity in the Hedge Sparrow's World." *Science*, March 31, 1989. Population biologists are using DNA fingerprinting to study the reproductive success of individual birds.
4. Pool, R. "Ecologists Flirt with Chaos." *Science*, January 20, 1989. Does chaos theory apply to population ecology?
5. Ricklefs, R.E. *The Economy of Nature.* 2d ed. New York: Chiron Press, 1983. Helpful chapters on populations.
6. Savonen, C. "One Salmon, Two Salmon . . . 10,000 Salmon: Counting the Fish in Alaska." *Oceans*, January/February 1985. A population density determination in action.

Communities

48

O n your next walk through a natural field or woodland, or even across campus or through a park, try to observe some of the interactions of species. You may see birds using trees as nesting sites, bees pollinating flowers, shelf fungi growing on trees, spiders trapping insects in their webs, ferns growing in shade provided by trees—any of the multifarious interactions that exist in any ecological theater. An organism's environment comprises, in addition to the physical and chemical factors discussed in Chapter 46, other individuals in the population as well as populations of various other species living in the same area. Such an assemblage of species living close enough together for potential interaction is called a **community** (Figure 48.1).

This chapter examines various kinds of biotic interactions in communities and addresses the central issue in community ecology: What factors are most significant in structuring a community—in determining the number of species and both the absolute and relative abundance of each species in a community? It will become apparent that community ecology is a field replete with unanswered questions and controversies.

TWO GENERAL VIEWS OF COMMUNITIES

Why are certain combinations of species found together as members of a community? Two divergent views on this question emerged among ecologists in the 1920s and 1930s, derived primarily from observations of plant distributions. One idea, first enunciated by H. S. Glea-

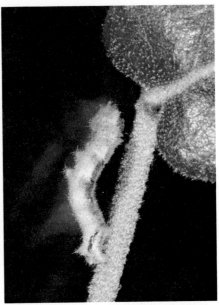

Figure 48.1
Community interactions shape adaptations. The larva (caterpillar) of *Nemoria arizonia*, a moth that lives in the southwestern United States, occurs in two seasonal forms. Caterpillars that hatch during the spring feed on catkins, the male flowers of oaks, and these caterpillars develop into forms that resemble the catkins in shape and coloration (left photo). A second brood of caterpillars hatches in summer. By then, catkins are gone and the new caterpillars feed on leaves. These "summer" caterpillars develop into forms that resemble oak stems rather than catkins (right photo). In laboratory experiments, caterpillars fed with catkins develop into the catkinlike form, and the caterpillars fed on leaves become stemlike individuals; diet determines the morphology of the caterpillar (in fact, the two caterpillars in these photos are full siblings raised on different diets). In the natural environment, each form of the caterpillar is camouflaged against the vegetation it feeds and rests on, an evolutionary adaptation that presumably makes the caterpillars less visible to predators such as birds. The case of *Nemoria* is a remarkable example of how interactions between species within a biological community affect organisms on both the evolutionary time scale and the ecological time scale.

son, depicted the community as a chance assemblage of species found in the same area because they have similar environmental requirements—for example, for temperature, rainfall, and soil type. The other view, advocated by F. E. Clements, saw the community as an assemblage of closely linked species, locked into association by mandatory interactions that cause the community to function as an integrated unit. Carried to the extreme, this view interpreted communities as a kind of "superorganism." Evidence was based on the observation that certain species of plants consistently cohabit. For example, deciduous forests in the northeastern United States include certain species of oaks, maples, birches, and beeches, along with specific shrubs and vines.

The two views of communities make contrasting predictions about how plant species should be distributed along a gradient of environmental variables, such as altitude or temperature. The Gleason hypothesis predicts that discrete boundaries should generally not exist between communities because each species will have an independent distribution along the environmental gradient. Along this continuum of chang-

ing vegetation, each species will be distributed according to its tolerance ranges for abiotic factors that vary along the gradient (Figure 48.2a). According to Clements' hypothesis, there should be more clustering of species, with relatively discrete boundaries between the communities (Figure 48.2b).

In most cases, the composition of plant communities does seem to change spatially on a continuum, with each species more or less independently distributed (Figure 48.3). This distribution supports the view of plant communities as relatively loose associations without discrete boundaries. Cases where neighboring communities are delineated by sharp boundaries can usually be explained by an abrupt change in key abiotic factors. For example, vegetation may be markedly different on the north and the south sides of a steep hill because of substantially different amounts of sunlight, which in turn may result in different temperature and moisture conditions.

We must be cautious in applying the same view to the animals in a community, which are often linked more rigidly to other organisms. For example, limpkins, long-beaked birds of Florida swamps, feed pri-

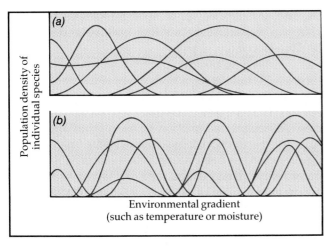

Figure 48.2
Two views of communities. (a) Is each species independently distributed, or **(b)** are species organized into discrete groupings? These two extreme possibilities are compared here through graphs of hypothetical distributions of plant species along a gradient of environmental factors, such as temperature or moisture. Each curve traces a single species.

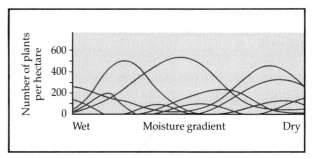

Figure 48.3
Actual distribution along a moisture gradient. At a particular elevation in the Santa Catalina Mountains of Arizona, each tree species has an independent distribution along the gradient, apparently conforming to the tolerance limits of that species for moisture. The species that cohabit at any location along the gradient have similar environmental requirements. Because the vegetation changes on a continuum, delineation of community boundaries is rather arbitrary.

marily on one species of snail. The birds' evolutionary adaptations make them extremely efficient in their specialized foraging, but their geographical distribution is restricted by the distribution of their prey. Gray squirrels of the eastern forests of the United States are not so strongly tied to a particular kind of organism, and may be found in a variety of habitats including areas where pine is abundant. However, these squirrels are most common in areas of mature hardwoods that include oak and hickory trees.

Thus, in trying to explain why certain species commonly occur together in communities, it is difficult to make simple generalizations that are accurate and widely applicable. The distribution of almost all organisms is probably affected to some extent by both abiotic gradients and interactions with other species.

PROPERTIES OF COMMUNITIES

Much as populations exhibit characteristics, such as density and spacing, that are not features of individual organisms, communities have several aggregate properties that are unique to that level of organization. Community properties include the following:

1. Species richness, equitability, and diversity. Species **richness** is the number of species in a community. Ecologists also refer to species **equitability,** a measure of the relative abundance of the different species. For example, imagine two communities, each with 100

individuals among four different species (A, B, C, and D). Assume, however, that the relative abundance of individuals for each species differs between the two communities as follows:

Community 1: 25A 25B 25C 25D
Community 2: 97A 1B 1C 1D

If these were, say, trees in a forest, and you were exploring Community 1, you would easily notice four different species. In Community 2, however, almost every tree you saw would be species A, and only rarely would you see anything else. Although these two communities are equally rich—they have the same number of species—Community 1 has a more equitable representation of the four species. Most people would intuitively think of Community 1 as more diverse. Indeed, the term species **diversity,** as used by ecologists, considers *both* components of diversity: species richness and equitability.

Researchers have proposed a variety of mathematical formulas, called diversity indexes, that combine data on species richness and equitability. As with many topics in this chapter, however, there is no general agreement on which of these indexes strikes the most appropriate balance of these two measures, or even whether variation in equitability provides much insight into the factors responsible for community structure. One reason for this controversy is that the relative abundance of a species in a community is not necessarily a measure of its importance in the overall structure of the community. The size of an organism or its activity may be the most important factor. For example, a large predator may play a dominant role in a community's structure, even though the predator is not among the most abundant species.

2. Prevalent form of vegetation. Our first impression of a terrestrial community is likely to be the overall form of the vegetation. What is the community's vertical profile—do trees, shrubs, or grasses dominate the landscape? Are the plants predominantly broadleaf species, or do needlelike leaves prevail? Is the vegetation stratified into canopy, subcanopy, and floor levels? (Grouping communities according to similarities in overall form, without regard to the actual species, is part of the basis for designating the biomes discussed in Chapter 46.) The general structure of the vegetation will in turn influence which other organisms are present.

3. Trophic structure. Who eats whom in the community? The various feeding relationships, or trophic structure (from the Greek *trophe*, meaning "nourishment"), of a community determine the flow of energy and cycling of nutrients from plants and other photosynthetic organisms to herbivores to carnivores. Since energy flow and chemical cycling also involve abiotic components, such as air and soil, these topics are more appropriately discussed at the ecosystem level (see Chapter 49) rather than the community level. Trophic structure is of interest in this chapter because specific plant–herbivore and predator–prey interactions may play a major role in determining community structure.

4. Stability. Community stability refers to the ability of a community to bounce back to its original composition in the wake of some disturbance, such as a fire or a disease that kills most individuals of a dominant species. Stability depends on both the type of community and the nature of the disturbance. For example, after a fire, a grassland may return to its original species composition much faster than a forest would. As we will see, the subject of stability provides one of the best examples of the pendulum-like swings in viewpoint that characterize ecology as a young, developing science.

COMMUNITY INTERACTIONS

Adaptations to Biotic Factors: An Overview

Just as the physical and chemical features of the environment are important factors in adaptation by natural selection, the biotic components of the environment—the other species with which an organism interacts—can also strongly influence the course of evolution. In some cases, the adaptation of one species to the presence of another has a relatively obvious evolutionary basis. For example, natural selection has apparently favored peppered moths that blend in with the living lichens on which the moths sometimes rest (see Figure 20.10, p. 433). For more elaborate relationships between species, evolutionary history and adaptive significance are difficult to ascertain. This is

the case, for instance, for complex interactions between chimpanzees and baboons that have been described by Jane Goodall.

A special type of interspecific adaptation is **coevolution.** The term is used to describe situations involving a series of reciprocal adaptations; a change in one species acts as a new selective force on another species, and counteradaptation of the second species in turn affects selection of individuals in the first species. Coevolution has been studied most extensively in predator–prey relationships and in symbiosis. One apparent general example of coevolution, that between species of flowering plants and their exclusive pollinators, was discussed in Chapter 27.

In many cases where two species apparently have adaptations that coevolved, it is difficult to establish that the selection factor resulting in a trait in one species was some evolutionary change in another species. As an example, let us examine a specific relationship between an herbivore and a group of plants that illustrates possible coevolution but also demonstrates the complexities that may confound our attempts to sort out evolutionary history. Passionflower vines of the genus *Passiflora* are protected against most herbivorous insects by the production of toxic compounds in young leaves and shoots. However, the larvae of the butterfly *Heliconius* can tolerate these defensive chemicals, apparently by using enzymes to destroy the compounds. This counteradaptation has enabled *Heliconius* larvae to become specialized feeders on plants few other insects can eat (Figure 48.4). Survival of the larvae is further enhanced by a behavioral adaptation of the butterflies. The eggs that female *Heliconius* butterflies lay on the leaves of passionflower vines are bright yellow, and other females generally avoid laying eggs on leaves marked by yellow dots. This presumably reduces competition among the larvae for food. An infestation of *Heliconius* larvae can devastate a passionflower vine, and we may assume that these poison-resistant insects are a strong selection force favoring the reproductive success of individual plants that have additional defenses. In some species of *Passiflora*, the leaves have conspicuous yellow spots that mimic *Heliconius* eggs, an adaptation that may divert the butterflies elsewhere in their search for sites to lay eggs. The story becomes more complicated, however. The yellow "spots" on the passionflower vine are actually nectaries, which also attract ants and wasps. The ants and wasps prey on *Heliconius* eggs or larvae. In fact, leaves may have many additional nectaries that do not closely resemble *Heliconius* eggs. There is evidence, too, that the presence of ants will discourage a *Heliconius* butterfly from laying eggs. Thus, adaptations that may seem, on superficial examination, to be evolutionary counterresponses between just two species may in fact result from interactions with many species, where it is difficult to sort out the importance

(a)

(b)

(c)

Figure 48.4
Evolutionary interactions of a plant and an insect. (a) Passionflower vines (*Passiflora*) produce toxic chemicals that help protect leaves from herbivorous insects. A counteradaptation has evolved in the butterfly *Heliconius:* The larvae can feed on the leaves because they have enzymes that break down the toxic compounds. **(b)** The females of some *Heliconius* species will avoid laying eggs where bright yellow egg clusters, such as the two seen here, have already been deposited by other females on passionflower vines. This behavior helps ensure an adequate food supply for offspring. **(c)** These vivid yellow nectaries are the egglike structures that some species of passionflowers have on their leaves and elsewhere. Experiments show that egg-laying *Heliconius* females avoid laying eggs on leaves with these yellow spots. These nectaries, as well as the smaller ones scattered over the leaf, also serve to attract ants and other insects that prey on the butterfly eggs and larvae.

of various selective forces. It may be that the simple idea of coevolution as adaptation–counteradaptation occurring exclusively between two species does not often apply.

In spite of the problems of assessing cause and effect in the specifics of natural selection, biologists agree that adaptation of organisms to other species in a community is a fundamental characteristic of life. Put another way, interactions of species in ecological time can translate into adaptations over evolutionary time. We shall encounter several examples as we now survey the interspecific interactions of greatest importance in community ecology: competition, predation, and symbiosis.

Competition Between Species

A major controversy in modern ecology centers on the role of competition in structuring communities. We saw in Chapter 47 that intraspecific competition—competition for resources among individuals of the same population—places limits on population growth. Similarly, if two species both require a resource that is present in limited quantities, the density of each species may have impact on the density of the other. This situation is called **interspecific competition.** For example, if several species of birds in a forest feed on

the same insects, which are a limiting resource, then the bird species may have negative effects on one another's population growth. Some species may even be eliminated from the community. This assumption is the basis for the hypothesis that competition for shared, limited resources is an important force determining community structure.

The Competitive Exclusion Principle   In 1934, G. F. Gause studied the effects of interspecific competition in laboratory experiments with two closely related species of the protozoan *Paramecium*, *P. aurelia* and *P. caudatum* (Figure 48.5). Grown in separate cultures under constant conditions and with a constant amount of bacteria added every day as food, each population of *Paramecium* grew until it leveled off at what was apparently the carrying capacity. When the two species were cultured together on a constant food supply, *P. caudatum*, seemingly unable to compete with *P. aurelia*, was driven to extinction in the microcosm of the culture dish. Gause concluded that two species so similar that they compete for the same limiting resources cannot coexist in the same place. One or the other will use the resources more efficiently and thus reproduce more rapidly. Even a slight reproductive advantage will eventually lead to the elimination of the inferior competitor. This concept was later termed the **competitive exclusion principle.** Subsequent lab-

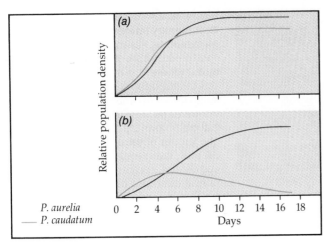

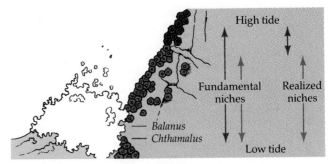

Figure 48.5
Competition in laboratory populations of _Parame-cium_ (a) In separate laboratory cultures with constant amounts of bacteria added every day for food, populations of the two species, _P. aurelia_ and _P. caudatum_, each grow to carrying capacity. **(b)** When the two species are grown together, _P. aurelia_ has a competitive edge in obtaining food, and _P. caudatum_ is driven to extinction in the culture.

Figure 48.6
Fundamental and realized niches. _Balanus balanoides_ and _Chthamalus stellatus_ are two species of barnacles that grow on the same rocks along the Scottish coast. These rocks are exposed during low tide. The barnacles have a stratified distribution, with _Balanus_ most concentrated on the lower portions of the shore. Although the swimming larvae of the barnacles may settle randomly on the rocks and begin to develop into sessile adults, _Balanus_ fails to survive high on the rocks, apparently because the species is unable to resist desiccation during low tides. Thus, its fundamental niche and its realized niche are the same. _Chthamalus_ is concentrated primarily on the upper parts of the rocks, which are out of water for a greater duration during low tides. However, in experiments where J. H. Connell removed _Balanus_ from tidepools, the _Chthamalus_ population spread lower on the rocks. Apparently _Chthamalus_ could survive lower on the rocks than it is generally found, if it were not for competition from _Balanus;_ thus, its realized niche is only a fraction of its fundamental niche. Connell actually documented _Balanus_ overgrowing, undercutting and crushing _Chthamalus_.

oratory experiments on several other species of animals and plants reinforced the principle.

Does competitive exclusion occur in natural communities as well as in laboratory populations? How important is competitive exclusion in determining how many species can coexist in a community? Are localized extinction or emigration the only fates for a species that is outcompeted, or can natural selection modify some interactions by favoring individuals that exploit a slightly different set of resources? These questions have not yet been thoroughly answered, and community ecologists are continuing research in this area. We can explore the questions first by considering theoretical aspects of competition and then by examining observations and experiments on competition in natural settings.

The Ecological Niche The way in which an organism "fits into" an ecosystem is sometimes described as its **niche.** The niche is not defined simply by habitat; it is the sum total of the organism's use of the biotic and abiotic resources of its environment. For example, the temperature range within which the individual lives, the type of substrate it lives on, the time of day it feeds, and the type and size of food it consumes are all part of the individual's niche. We most often think of a particular niche as being associated with an entire species or population because individuals of the same species are generally similar in the way they use resources. However, niche differences at the individual level may also occur, depending on sex, geographical location, and other factors.

The term **fundamental niche** refers to the resources an organism is theoretically capable of using. In reality, each population is embedded in a web of interactions with populations of other species, and thus there may be biological constraints such as competition or predation that restrict organisms within their fundamental niches to what is called the **realized niche.** Figure 48.6 illustrates the difference between fundamental and realized niches for two species of barnacle.

We can now restate the competitive exclusion principle to say that two species cannot coexist in a community if their niches are identical. However, ecologically similar species can coexist in a community if there are one or more significant differences in their niches.

The Role of Competition in Species Diversity If the competitive exclusion principle is correct, we can construct a more general theory of how competition might help organize communities and affect species richness. According to this theory, competition must be resolved either by localized extinction of the less successful species or by selection favoring individuals who do not use resources in exactly the same way as individuals of the other species. Interspecific compe-

Figure 48.7
Niche separation in a group of coexisting warbler species. These closely related warbler species all feed on insects in spruce trees of northeastern forests, but each species tends to forage in a different zone of the tree (indicated in these drawings with shading). This resource partitioning presumably allows coexistence by reducing direct competition.

tition would not only limit the number of species in a community but also determine which ecologically similar species could coexist.

What evidence exists that competition has been important to community structure? Although we cannot look directly at the evolutionary history of a community, many ecologists see much circumstantial evidence that they interpret as demonstrating the importance of competition. One of the first to provide this evidence was Robert MacArthur, whose study in the 1950s of small birds known as warblers has become a classic example documenting what ecologists call **resource partitioning.** If you explore the coniferous forest of eastern North America, you will find a variety of colorful warblers foraging throughout the trees. At first glance, it may seem to you that they are all doing pretty much the same thing, in terms of how and where they feed. But MacArthur made extensive observations on *exactly* where individuals of each species fed in the trees. Although there was some overlap, the species clearly tended to sort out, with each having its own feeding domain that varied by height in the tree and distance from the middle (Figure 48.7).

Many other studies have since documented a similar pattern among a wide variety of animals: Where ecologically similar species coexist, there is often at least one important niche difference among them. Put

another way, these species can apparently coexist because they do not compete directly.

Many ecologists are reluctant to embrace competition as the major factor structuring communities. Since the niche has so many facets, it is difficult to assess to what extent two species are actually competing in the present and even more difficult to know what has occurred in the evolutionary past. Furthermore, competition cannot be important unless populations are large enough so that a resource is significantly limiting. There is now considerable evidence that predation may be important in structuring many communities by limiting the size of prey populations, *preventing* competitive exclusion.

Predation

In a community interaction where one species eats another, the consumer is called a **predator** and the food species is known as the **prey.** We shall use these terms not only for cases of animals eating other animals, but also for plant–herbivore interactions, where the plant is prey.

One organism eating another is an individual interaction that has implications for the evolution of adaptations of both predator and prey. Many of the impor-

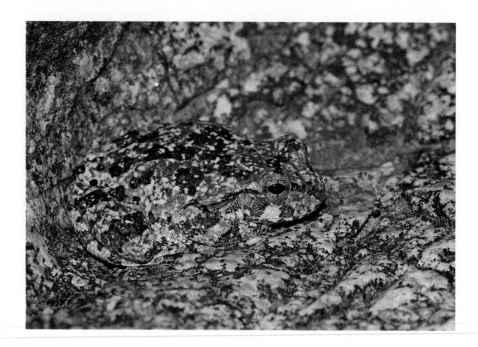

Figure 48.8
Camouflage. The canyon tree frog (*Hyla arenicolor*) disappears on a granite background.

tant adaptations of predators are fairly obvious and familiar. These include acute senses keyed to finding prey; speed and agility to prevent escape of prey; implements of destruction such as stingers, fangs, poisons, and claws; and camouflage for "ambushers." Equally important but sometimes less obvious are the defenses of prey, which we examine next.

Plant Defenses Against Herbivores   Plants cannot run away from herbivores. They do, however, have some mechanical and chemical defenses against the animals that would prey on them. Although some plants may escape the notice of animals by camouflage, mechanical and chemical defenses against herbivores are more common. Thorns may discourage large herbivores such as vertebrates, and some plants have microscopic hooks or spines that make feeding difficult for even small insects.

Many plants produce chemicals that function in defense by making the vegetation distasteful or harmful. These chemicals, classified as **secondary compounds,** are so named because they form through metabolic diversion of a product from a major pathway such as glycolysis or the Krebs cycle (see Chapter 9). For many years, plant physiologists were puzzled about the widespread occurrence and sometimes large amounts of these secondary compounds, and only recently has their importance as defensive materials become clear.

Secondary compounds believed to function as chemical weapons against herbivores include strychnine, produced by plants of the genus *Strychnos;* morphine, from the opium poppy; nicotine, produced by tobacco; and mescaline, a secondary product of the peyote cactus. Others include compounds responsible

for the familiar flavors of cinnamon, cloves, and peppermint. Tannins, used by humans in the preservation of leather, are widespread, especially in oaks. Some plants even produce secondary products that are analogues of insect hormones and cause abnormal development in the insects that eat these plants.

The specific defenses of a plant act as selective factors that may lead to counteradaptations enabling certain herbivores to nullify those defenses and eat the plant, as we saw earlier in the *Passiflora–Heliconius* interaction. A plant's defenses restrict the number of herbivorous species that can eat that plant, but in a world populated by evolving animals that must eat to reproduce, no defense measures are likely to be foolproof or eternal.

Figure 48.9
Deception. The wing markings of the io moth (*Automerisio*) resemble the eyes of a much larger animal. Potential predators may be momentarily startled, allowing the moth to escape.

(b)

Figure 48.10
Mechanical and chemical weapons against predators. (a) Sharp quills are a mechanical defense for the porcupine. (b) The bombardier beetle sprays attackers with a toxic chemical.

(a)

Animal Defenses Against Predators An animal can avoid being eaten by hiding, escaping, or defending against predators.

Camouflage is a particular type of hiding that can take various forms. One common form is **cryptic coloration,** which makes potential prey difficult to spot against its background (Figure 48.8). A classic example of how selection by predators can result in the evolution of cryptic coloration is the case of the English peppered moth, which was discussed in Chapter 21. The shape of the animal and its behavior may also help camouflage it (see Figure 48.1).

Another form of defense for an animal is deceptive markings. Large, fake eyes or false heads can startle predators momentarily, allowing the prey to escape, or they may cause the predator to strike a nonvital area (Figure 48.9).

Animals that are easy to find may be capable of rapid escape or have mechanical or chemical defenses against would-be predators (Figure 48.10). Most predators are strongly discouraged by the familiar defenses of porcupines and skunks. Some animals acquire chemical defense passively by accumulating toxins from the plants they eat. Monarch butterflies, for example, retain cardiac poisons they acquired by eating milkweed plants as larvae. The monarch is resistant to these poisons, but the compounds are distasteful and harmful to most other animals. Birds that try eating monarch butterflies quickly regurgitate their prey and

learn to avoid feeding on that particular species of insect.

Animals with effective physical or chemical defenses are often brightly colored, a warning to predators known as **aposematic coloration** (Figure 48.11).

Figure 48.11
Aposematic coloration. The conspicuous markings of the poison dart frog act as a warning to would-be predators, and probably train predators more quickly to avoid this particular animal.

(b)

Figure 48.12
Mimicry. (**a**) Batesian mimicry, in which a harmless species resembles an "armed" one. This nonstinging syrphid fly resembles a stinging honey bee. (**b**) Müllerian mimicry, in which two well-defended species resemble one another. This nomadidae bee resembles a yellowjacket; both species possess stingers.

(a)

This warning coloration is not in itself a defense, but seems to be adaptive because predators learn more quickly to avoid harmful prey that are conspicuous.

Mimicry   A predator or prey may gain a significant advantage through **mimicry** of another species. Defensive mimicry in prey often involves aposematic forms. For example, the bright orange viceroy butterfly mimics the monarch. Viceroys do not feed on toxic plants and are edible, but birds generally avoid them because of their similarity to the monarch. Many harmless snakes mimic the conspicuous red, white, and black markings of the poisonous coral snake. These are examples of **Batesian mimicry:** A palatable species mimics an unpalatable model (Figure 48.12). For the mimicry to be effective, the mimic must remain generally a great deal less abundant than the model; otherwise, predators would learn that, for example, orange butterflies are good rather than bad to eat.

In **Müllerian mimicry,** two or more unpalatable species resemble each other. Presumably such species gain additional advantage beyond their basic defenses because predators will learn more quickly to avoid any prey with a certain appearance.

Predators also use mimicry in a variety of ways. Snapping turtles have tongues that resemble a wriggling worm, thus attracting small fish; a fish approaching closely disappears in a flash as the turtle's strong jaws snap closed. The familiar fireflies of summer evenings provide a fascinating example of predatory mimicry. The flashes of light are actually species-specific signals between males and females that are used to locate a mate. Females of the genus *Photuris* have taken advantage of the system to become predators of the males of *Photinus*. The *Photuris* females flash the *Photinus* signal, luring in unsuspecting males seeking a mate. As one researcher noted, males home in on the signal expecting sex, only to be killed as food.

The Role of Predation in Structuring Communities
In simple laboratory experiments where a single predator species is kept with a single prey species having no refuge, the predator may devour all the prey and then itself perish from starvation (Figure 48.13). If this scenario were commonly enacted in nature, predators would always reduce the diversity of species in communities. In fact, the role of predators in structuring communities is more complex. One reason why predators rarely drive their prey and themselves to extinction is the defensive ability just described. Also important is the switching behavior of many predators. A predator may switch to an alternative food source when the population of one prey begins to dwindle due to overfeeding or some other factor. Moderate predation may even *help* maintain the presence of some prey species, by preventing them from overshooting their carrying capacity and then possibly oscillating downward to a level dangerously close to local extinction.

Probably the most important idea relating predation to community structure is that competitive interactions by prey species may be reduced if populations

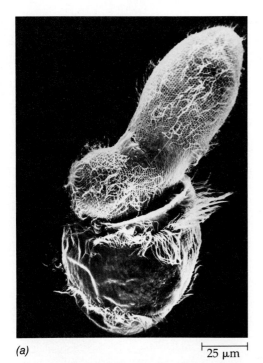

(a)

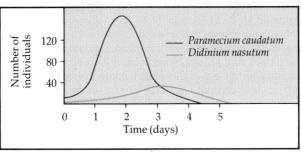

(b)

Figure 48.13
Predator–prey dynamics in a laboratory environment. (a) The ciliate *Didinium* (bottom) eats *Paramecium*, another ciliate (SEM). (b) When these two species are cultured together, *Didinium* eats all the *Paramecium*, and then becomes extinct as a result of starvation. In natural communities, however, which are much more complex than the simple two-species system of a culture flask, predator–prey relationships can actually help stabilize species diversity rather than reduce it.

are held down by predators. Experiments by Robert Paine in the 1960s were among the first to give strong impetus to this view. Paine removed the dominant predator, a sea star of the genus *Pisaster,* from experimental areas within the intertidal zone of the Washington coast. The favorite prey of *Pisaster,* the mollusk *Mytilus,* then outcompeted many of the other tidepool organisms for the important resource of space on the rocks. The number of species dropped from fifteen to eight. This work and a number of other field experiments have given rise to the concept of a **keystone predator.** Keystone predators help maintain higher

species diversity in a community by keeping the best competitors at low enough density, with the result that competitive exclusion of other species does not occur.

Other Interactions

Species commonly interact in ways other than competition and predation. Recall from Chapter 25 that in **parasitism,** one organism, the parasite, harms the host; in **commensalism,** one partner benefits without significantly affecting the other; and in **mutualism,** both partners benefit from the relationship. These interactions sometimes intergrade, blurring the distinctions among the categories. All three types of interactions are technically included under the general category of **symbiosis** ("living together"), but this term is sometimes used to refer exclusively to mutualism. Because parasitic, commensal, and mutualistic interactions, like competition and predation, affect population densities for better or worse, they too must be included as important determinants of community structure.

Parasitism   Parasitism can be treated as a special case of predation in the sense that one organism derives its nourishment from another organism, which is harmed in the process. A parasite absorbs organic nutrients from the body fluid of a living host. Generally, parasites are smaller than their host. Birds that feed on insects are predators, but insects such as mosquitos that suck blood from larger animals are parasites.

As in predator–prey relationships, natural selection favors parasites that are best able to find and feed on hosts, while also selecting for defensive capabilities on the part of the host populations. Some of the secondary plant products that are toxic to herbivores, for instance, are also toxic to parasites such as fungi and bacteria. In vertebrates, the immune system provides a multipronged defense against internal parasites (see Chapter 39). Many parasites, particularly microorganisms, have adapted to specific hosts, often a single species. In such specific interactions, coevolution generally results in a relatively stable relationship that does not kill the host, an excess that would eliminate the parasite as well.

An example will demonstrate how rapidly natural selection can temper a host–parasite relationship. In the 1940s, Australia was plagued by a population of hundreds of millions of rabbits, which had exploded from just 12 pairs of the animals imported a century earlier. In 1950, the myxoma virus, which parasitizes rabbits, was introduced in an effort to control the rabbit population. The virus spread rapidly and killed 99.8% of all rabbits infected. However, a second exposure to the virus killed only 90% of the remaining

Figure 48.14
Commensalism between a bird and a mammal. The cattle egret concentrates its feeding where grazing cattle and other large herbivores, such as this cape buffalo in Tanzania, flush insects from the vegetation, a benefit to the bird that occurs without known harm or benefit to the mammal. Of course, the relationship may help the buffalo in some way not yet discovered.

Figure 48.15
Mutualism between acacia trees and ants. Certain species of Central and South American acacia trees, called bull's horn acacias, have hollow thorns that house stinging ants of the genus *Pseudomyrmex*. The ants feed on sugar produced by nectaries on the tree and also on protein-rich swellings called Beltian bodies that grow at the tips of leaflets. The acacia benefits from its population of pugnacious ants, which attack anything that touches the tree. The ants sting other insects, remove fungal spores and other debris, and clip surrounding vegetation that happens to grow close to the foliage of the acacia. A series of experiments in the 1960s demonstrated the mutualistic nature of the relationship. When the ants on experimental trees were poisoned, the trees died, probably owing to damage from herbivores and failure to compete with the surrounding vegetation for light and growing space.

rabbits, and the third infection killed only about 50%. Today, the virus has only a mild effect on the host, and the rabbit population has rebounded. Apparently, viral infection selected for host genotypes that were better able to resist the parasite. At the same time, there was selection for less virulent strains of the virus, which had greater success at being propagated and transmitted by hosts that survived. Natural selection has stabilized this host–parasite relationship.

Commensalism　Few absolute examples of commensalism exist because of the unlikelihood that one partner will be completely unaffected by the other. "Hitchhiking" species, such as algae that grow on the shells of aquatic turtles or barnacles that attach to whales, are sometimes considered to be commensal. However, the hitchhikers may decrease the reproductive success of their hosts slightly by reducing the efficiency of the hosts' movement in their search of food or escape from predators. An association that may be truly commensal is the relationship between cattle egrets (heronlike birds) and grazing cattle (Figure 48.14). The egrets concentrate their predation near grazing cattle, which flush insects from the vegetation as they move. The birds clearly benefit from their association with cattle, and it is difficult to imagine that the cattle either benefit from, or are harmed by, the relationship in any way. Because one species is not affected, any evolutionary changes relevant to the relationship will usually be one-sided, involving only the beneficiary.

Mutualism　In contrast to commensalism, mutualism may involve two-way evolutionary adaptations because changes in one species are likely to affect the performance of the other. Adaptations that have co-evolved in numerous mutualistic relationships have been described in earlier chapters. Examples are legumes and the nitrogen-fixing bacteria of their root nodules; microorganisms that digest cellulose in the guts of termites and ruminants such as cows; single-celled algae and animal hosts such as the colonial cnidarians of coral reefs; mycorrhizae, the associations between fungi and the roots of plants; and the specific interactions of flowering plants and their pollinators. Figure 48.15 illustrates another interesting case of mutualism, the relationship between certain species of acacia trees and the ants that protect the trees from other insects.

Many mutualistic relationships may be derived from interactions that originally involved predation or parasitism. For example, reliance of certain angiosperms on animals for pollination and seed dispersal probably represents counteradaptation to the selective pressures that herbivores were placing on plant populations by eating pollen and seeds. Those plants that could derive some benefit from this sacrifice of organic matter would have greater reproductive success. In many cases, the pollen is spared when the pollinator is able to eat nectar instead.

Complex Effects of Species Interactions on Community Structure

The idea that competition is the most significant factor structuring communities is giving way to the more pluralistic view that a variety of interactions is important. For example, we have seen that predation, especially when it involves a keystone species, can affect the species diversity of a community.

Adding to the difficulty of determining which interactions are most important in structuring communities is the problem of sorting out the forces of natural selection that have given rise to present ecological relationships. This problem is illustrated by recent research on the nesting behavior of two species of birds living in the same trees in the western United States: MacGillivray's warbler nests low in trees; black-headed grosbeaks nest high. At first this might be interpreted as another case of resource partitioning that reduces competition, as described earlier. But Thomas Martin has recently demonstrated that predation may be the key factor in this segregation of nests; the pressure of predators that feed on the young of these birds is greater when a given number of nests is concentrated in the same general area rather than being spread more diffusely in the trees. In other words, the partitioning we see in nest sites reduces predation, and this, rather than a reduction in competition between the bird species, may be the key explanation.

Thus, careful studies lead to the conclusion that communities are structured by multiple interactions of organisms with their biotic environment and with abiotic factors as well. Which interactions are most important can vary from one type of community to another and even for different components of the same community. For example, in a particular community, the diversity of herbivores may be controlled mainly by predators with prey-switching behavior, whereas the diversity of these predators may depend mainly on competition for food. Other kinds of interactions may affect numerous species to varying degrees.

Because of the complexity of these community networks, there are few, if any, natural communities for which we have a good understanding of all the important connections and how these connections evolved.

Community ecology is made even more challenging by the fact that the structure of a community may change, sometimes in just a few years or less. Outside disturbances influence structure, and the actions of the organisms themselves may destabilize the existing structure and favor a new one. Our next section discusses some of these changes.

SUCCESSION

Transitions in community structure are most apparent after some disturbance—a flood, a fire, the advance and retreat of a glacier, volcanic eruption, or human activity—strips away the existing vegetation. The disturbed area may be colonized by a variety of species, which are subsequently replaced by yet other species. Such a transition in species composition is known as **succession.** (The term is used to describe changes in species composition in ecological time, not to species change associated with the origin of new species in evolutionary time.) The process is called **primary succession** if it begins in areas essentially barren of life because of the absence of soil—for example, on a new island formed by a volcano, on strip-mined land, or on the till left by a retreating glacier. In the case of glaciers, which are still actively retreating in such places as Glacier Bay, Alaska, the barren ground is first colonized by mosses and lichens, then by dwarf willows. After about 50 years, alders form dense stands. These eventually give way to Sitka spruce, which are later joined by hemlock to form a relatively stable spruce-hemlock forest. The entire process takes about 200 years (Figure 48.16).

Secondary succession occurs where an existing community has been cleared by some disturbance that leaves the soil intact and the area begins its return to a natural community. Old-field succession in the Piedmont region of the mid-Atlantic United States has been studied extensively. If an agricultural field is abandoned in this area, an herbaceous community develops first, followed by shrubby forms that give way to pine. The pines are eventually replaced by a more stable community dominated by oaks and hickories. As in glacial-till succession, the process takes about two centuries.

In some cases, succession may seem to have a final stage called the climax community. Many ecologists once viewed the climax as a common endpoint toward which communities in the same general area inevitably move—a condition that will then continue indef-

(a)

(b)

(c)

(d)

(e)

(f)

Figure 48.16
Succession after the retreat of glaciers.
Though these photographs are not all
from the same area, they illustrate the different stages of succession: (**a**) retreating
glacier; (**b**) barren landscape after the
retreat; (**c**) moss and lichen stage; (**d**) alders and cottonwoods covering the hillsides; (**e**) spruce coming into the alder
and cottonwood forest; (**f**) spruce and
hemlock forest. The entire sequence takes
about 200 years.

initely and in which diversity is maximized. As we will see, however, these ideas may not be as widely applicable as was once thought.

Causes of Succession

In most cases, a variety of interrelated factors determines the course of succession. Early species are typically good colonizers with excellent dispersal mechanisms. Many of these may be "fugitive" species that do not compete well in established communities, but maintain themselves by constantly colonizing new areas before better competitors reach the same places. Tolerance for the general abiotic conditions of the area also affects the species composition during early successional stages. A species may colonize but never become abundant if conditions are near the extremes of its tolerance limit for some factor. Important abiotic factors can differ on a fairly local level, as on opposite sides of a mountain. Variations in size and growth among colonizing species may also be important. For example, if two plant species, one an herbaceous plant and the other a tree, colonize a community at the same time, the herbaceous species may have earlier prominence because of faster growth. The ecological impact of the tree species may not be realized until the trees are relatively large.

Many of the changes in community structure during succession may be induced by the organisms themselves. Such changes are said to be **autogenic.** Direct biotic interactions may be involved, including **inhibition** of some species by others. Often the organisms also affect the abiotic environment. This may result in **facilitation,** by which the group of organisms representing one successional stage "paves the way" for species typical of the next stage. For example, the alders that are abundant in an intermediate stage of glacial-till succession lower soil pH as their leaves decompose. This facilitates the entry of spruce and hemlock, which require acidic soil. Sometimes the changes that facilitate the next stage make the environment unsuitable for the very species responsible for the changes.

Both inhibition and facilitation may be involved throughout the successional process. For example, horseweed is one of the earliest colonizers in the old-field succession described on p. 1105. For the first year or two, this plant may inhibit other species through shading. Then, the horseweed and other early species facilitate the entry of later species by adding organic matter to the soil, which aids in holding moisture. The pine trees that dominate a later successional stage require full sunlight; their growth becomes self-inhibitory as the trees shade the ground and prevent any offspring from growing. At the same time, the pines continue to add organic material to the soil, and the shade keeps the forest floor moist, conditions that facilitate germination and growth of the hardwoods that follow. At the climax stage, environmental conditions are such that the same species can continue to maintain themselves. For example, the oak-hickory forest that is the climax stage of the old-field succession maintains the moist, shaded environment that allows offspring of these species to grow, while inhibiting most of the species typical of earlier stages.

As noted earlier, some ecologists have challenged the idea of a climax community. For one thing, many communities are routinely disturbed by **allogenic** ("of outside origin") factors during the course of succession. For example, as we saw in Chapter 46, prairie grasslands are maintained by fire. Without fire, some grassland areas would become forest, at least in moister areas. We might therefore say that forest is the climax community for these areas, but that seems to make little sense if the forest community never develops. In this case, periodic fires stabilize the community at a stage that does not exactly fit the traditional idea of a climax community. There is also evidence that what appear to be climax communities may not be so stable over somewhat longer time frames. Studies of pollen, preserved in lake sediments, indicate that common tree species have nearly disappeared for hundreds of years from North American forests (for unknown reasons), only to eventually return.

Human Disturbance

Disturbance by human activities has had an impact on succession of communities all over the world. Logging and clearing for farmland have reduced large tracts of mature hardwood and pine forests to small patches of disconnected woodlots in many parts of the eastern and midwestern United States and throughout Europe. Similarly, agricultural development has disrupted what were once the vast grasslands of the North American prairie.

After a community is disturbed and then left alone, early stages of secondary succession, which are often dominated by weedy and shrubby vegetation, may persist for many years. This type of vegetation can be found extensively in forests that have been clear-cut, in agricultural fields no longer under cultivation, and in vacant lots and construction sites that are periodically cleared. Much of the United States is now a hodgepodge of early successional growth where mature communities once prevailed.

Human disturbance of communities is by no means limited to the United States, nor is it a recent problem. Tropical rain forests are quickly disappearing as a result of clear-cutting for lumber and pastureland. Centuries of overgrazing and agricultural disturbance have undoubtedly contributed to the current famine in parts of Africa by turning seasonal grasslands into great barren areas.

Equilibrium and Species Diversity

The traditional concept of a climax community includes the idea that species diversity reaches an equilibrium; there may be some turnover in species, but new colonists would be balanced by localized extinctions.

Species diversity generally increases during succession. As a community matures, species that are more durable competitors generally replace rapid colonizers that are not as successful when population densities begin to increase. Interactions—predation, symbiosis, and continuing competition—become more extensive and varied during succession, making increased diversity possible. According to the equilibrium model, succession reaches a climax when the web of interactions becomes so intricate that no additional species can "fit into" the community unless niches become available by localized extinction. Indeed, this pattern seems to occur on islands, for reasons that are discussed in the last two sections of this chapter.

An opposing view sees most communities as in continual nonequilibrium, with not only the identity but also the numbers of species changing during all stages of succession, even at the so-called climax stage. This nonequilibrium model accords more important roles in succession and species diversity to the less predictable aspects of dispersal and disturbance. In this view, the course of succession may vary depending on which species happen to colonize the area in the early stages. Disturbances such as fires, hurricanes, windstorms, landslides, and unseasonable temperatures can prevent a community from ever reaching a state of equilibrium, even when it is relatively mature.

According to the nonequilibrium model, disturbance also affects species diversity. When disturbance is severe and frequent, community membership may be restricted to good colonizers typical of early stages of succession. If disturbances are mild and rare in a particular location, then the late-successional species that are most competitive will make up the community. According to the **intermediate disturbance hypothesis,** species diversity is greatest where disturbances are moderate in frequency and severity, because organisms typical of different successional stages are likely to be present. Studies of species diversity in tropical rain forests provide some evidence for the intermediate disturbance hypothesis. Scattered all through these forests are small clearings formed where trees have fallen, taking along with them attached vines and other associated species. In these small disturbed areas, a rapid succession of immigrations and extinctions occurs, and species from various successional stages coexist within a relatively small area.

Diversity and Community Stability

In the 1960s many ecologists believed that increased diversity somehow causes communities to be more stable. However, there was disagreement about how to define stability. Some researchers considered stability in terms of *resistance* to change. Others seemed to mean *resiliency,* an ability to return to previous conditions after a disturbance had passed. For example, Robert Paine, whose work was described earlier, pointed out that intertidal areas are extremely diverse, but can be radically disturbed by a change in population size of a single species (the predator *Pisaster*). Then, in the 1970s, a number of mathematical models actually predicted decreasing stability with increasing diversity—just the reverse of earlier views. Some ecologists rejected these models as based on unrealistic assumptions.

Today, most ecologists at least agree that no simple relationship exists between diversity and stability. However, debates still rage, and some discussions become entangled in a "chicken and egg" situation. For example, do we think of the high diversity of the tropical rain forest as a *cause* of the general *stability* of this kind of community, or do we think of the high diversity as *resulting from* a certain amount of *instability* (as described previously)? After several decades of research, these questions are still unresolved.

BIOGEOGRAPHICAL ASPECTS OF DIVERSITY

Biogeography is the study of the past and present distribution of species. It is concerned not only with species diversity in a particular region but also with the identity of those species (species composition). For example, although the tropical rain forests of Africa and South America are both extremely diverse communities with similar growth forms, the species composition is radically different in each of these geographically remote forests.

Terrestrial life can be divided into biogeographical realms that have boundaries (diffuse, in some cases) associated with the patterns of continental drift after the breakup of the supercontinent Pangaea (Figure 48.17). Thus, the distribution of species reflects past history as well as the present interactions of organisms with their environments. We considered evolutionary aspects of biogeography in Chapters 20 and 23. Here, we address three general questions as examples of how biogeography connects to community ecology: What limits the ranges of species? How do we explain two of the major global clines (gradients) in diversity? And what determines the diversity of species on islands?

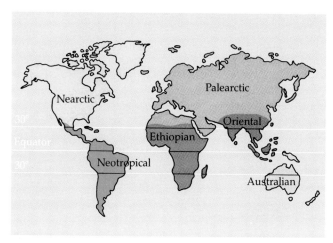

Figure 48.17
Biogeographical realms. Continental drift and barriers such as deserts and mountain ranges have contributed to each realm having its own suite of taxa. Except for Australia, the realms are not sharply divided but grade together in zones where taxa from adjacent realms coexist.

Limits of Species Ranges

We saw earlier in this chapter that the species composition of a community is determined in part by factors that limit the population distribution of the individual species. Two explanations can account for the limitation of a species to a particular range: (1) The species has never dispersed beyond that range, or (2) pioneers that have spread beyond the observed range failed to survive in those locations. One way to distinguish these two causes is by a transplant experiment, in which individuals of the plant or animal species in question are moved to areas of similar environment outside their normal range. Survival of the transplants suggests that the species had not managed to disperse to suitable locations outside the observed range. Nonsurvival at a particular test site is evidence that the species cannot spread into that site because of environmental resistance, such as inability to tolerate the physical environment in the new location or inability to compete with the natives.

Most transplant "experiments" are not the work of ecologists but are performed by travelers who unwittingly carry hitchhiking seeds or insects, or by people who intentionally introduce foreign plants or animals for agricultural or ornamental purposes. Many transplanted species, perhaps most, fail to survive outside their normal range. However, there are numerous examples of successful transplants. In the 1800s, starlings and English sparrows were imported to the northeastern United States from Europe; both species are now abundant in cities and the countryside throughout much of North America. Carp, freshwater fish from China, were introduced about the same time and have become common members of lake and stream communities.

Although any new species will probably have negative effects on some other species, some viable transplants have been especially noteworthy in this regard. The case of rabbits in Australia was mentioned earlier. Gypsy moths were imported to the United States from Europe by a scientist who was trying to crossbreed several silk-producing moths. In 1869, some of the moths escaped from the scientist's home near Medford, Massachusetts, and gypsy moths have been spreading southward ever since, defoliating large areas of forests. Another fiasco was importation of the Africanized honeybee by Brazilian honey producers in 1956. These bees, hybrids of the common European honeybee and an African variety, are very aggressive and have occasionally swarmed and attacked humans and livestock. In the past 30 years, the range of Africanized bees has spread across South America and through Central America and Mexico, and sightings have already been reported in the United States. The normal range of these bees was limited by barriers to dispersal, including the Atlantic Ocean. A similar problem is the northward march of fire ants, which were accidentally introduced to the southern United States from Brazil in the 1940s (Figure 48.18).

Species ranges sometimes change naturally, without an obvious explanation. For example, in the eastern United States, a number of species of birds and mammals have expanded generally northward over the past hundred years or so. For example, cardinals first nested in Michigan about 1892, and by 1970 had spread northward throughout the entire lower peninsula of the state. Mammals that have moved northward include the opossum and armadillo. A possible explanation for this expansion is the slight warming trend that has been occurring during this time, but there is no conclusive evidence that this is the cause.

Global Clines in Diversity

Ecologists have long been aware of major clines (gradual variation) in species diversity over large areas of the Earth. For example, there is a general increase in diversity from far northern and southern latitudes toward the equator (Table 48.1). In the ocean, diversity of the benthic community is relatively low in shallow coastal waters but increases dramatically in the dark, cold waters of the deep sea. How do we explain such gradations in species diversity?

For the latitudinal cline, we should first note that the general trend does not apply to all groups of organisms. Bears and rodents, for example, are more diverse in species number at northern latitudes. Overall, however, latitudes near the equator have greatly increased diversity compared to other areas. No sim-

Figure 48.18
March of the fire ants. (a) The fire ant (*Solenopsis invicta*) can inflict a beelike sting and tends to attack in large numbers. In the southern United States, more than 70,000 sting victims seek medical help each year. This is a winged female fire ant along with a much smaller worker. (b) Fire ants were accidentally imported from Brazil about 40 years ago, probably in the hold of a ship that docked in Mobile, Alabama. The range of the ant has been spreading ever since, as traced by this series of maps.

(a)

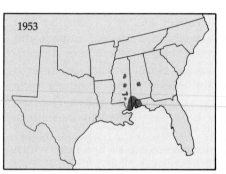 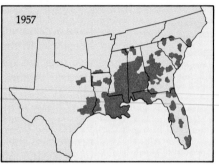 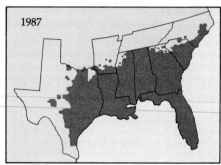

(b)

ple explanation accounts for this, and there is no general agreement on the cause. Undoubtedly many of the mechanisms we have already discussed, such as competition, predation, and periodic disturbance, interact in complex ways to contribute to this cline in diversity. Some ecologists suggest that an important factor is the great photosynthetic productivity near the equator, where light, temperature, and rainfall favor abundant plant growth. This provides a larger resource base that may support more species. More species may mean more complex predator–prey interactions, which, as we have seen, translate into even greater diversity. Other ecologists have noted, however, that in many communities, photosynthetic productivity is high but diversity remains rather low. Examples are estuaries, such as those discussed in Chapter 46. The shallow-to-deep water cline mentioned earlier provides another counterexample: The high diversity of the deep ocean is associated with a relatively unproductive habitat. Most ecologists now believe that although there is a

correlation between tropical diversity and photosynthetic productivity, there is no simple *causal* connection.

Island Biogeography

Islands provide ecologists an excellent opportunity to study some of the factors that affect the species diversity of communities. By "islands" we mean not only oceanic islands but also habitat islands on land, such as lakes or mountain peaks separated by lowlands—in other words, any area surrounded by an environment not suitable for the "island" species. In the 1960s, R. MacArthur and E. O. Wilson developed a general theory of island biogeography aimed primarily at explaining how and why species diversity should change on an island. The study of islands, which are relatively simple systems, may help us understand some of the interactions in more complex systems.

Imagine a newly formed oceanic island some dis-

Table 48.1 Diversity (by number of species) of beetles, ants, and birds arranged by latitude

Beetles		Ants		Birds	
Labrador	169	Alaska	7	Northern Alaska	26
Massachusetts	2000	Iowa	73	Southern Texas	153
Florida	4000	Trinidad	143	Central America	600

SOURCE: Modified from Paul Colinvaux. *Ecology.* (New York: John Wiley & Sons, 1986.)

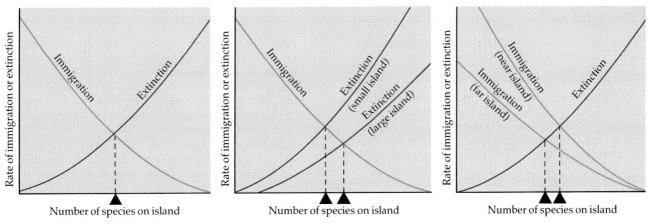

Figure 48.19
The theory of island biogeography. Equilibrium is indicated by the arrowheads.

tance from a "mainland" that will serve as a source of colonizing species. Two factors will determine the number of species that eventually inhabit the island: the rate at which new species manage to immigrate to the island, and the rate at which species become extinct on the island. Immigration and extinction rates are in turn affected by two important variables: the size of the island and its distance from the mainland. Small islands will generally have lower immigration rates, because it is less likely that potential colonizers will "find" the island. For example, birds blown out to sea by a storm are obviously less likely to land by chance on a small island than on a large one. Small islands will also have higher extinction rates. Fewer resources and less diverse habitat will limit the number of available niches, increasing the likelihood of competitive exclusion. Distance from the mainland is also important; for two islands of equal size, a closer island will have a higher immigration rate than one farther away.

The immigration and extinction rates are also affected at any given time by the number of species already present on the island. As the number of species increases, immigration rate decreases, since any individual reaching the island is less likely to represent a new species. At the same time, the more species an island has, the more there are to become extinct, whatever the reasons. These relationships are summarized in Figure 48.19, where immigration and extinction rates are plotted as a function of number of species present. The important point is that, eventually, an equilibrium will be reached where the rate of species immigration matches species extinction. The number of island species at this equilibrium point is correlated with the two variables already discussed. The theory of island biogeography predicts that species number on an island where immigration and extinction rates are balanced is directly proportional to the size of the island and inversely proportional to the distance of the island from the mainland. (Any equilibrium, of course, is

dynamic: Immigration and extinction continue, and the exact species composition may change over time.) This theory applies to a relatively short period of time, where colonization is the important process determining species composition; over a longer time, actual speciation on the island can affect the composition.

Observations and experiments furnish evidence that new islands do indeed reach an equilibrium in their species richness. For example, within 35 years after a volcanic eruption killed nearly all organisms on the island of Krakatoa in 1883, the diversity of birds that repopulated the island had reached an equilibrium of about 30 species. MacArthur and Wilson's studies of the diversity of amphibians and reptiles on many island chains, including the West Indies, support the prediction that species richness increases with island size. Species counts also fit the prediction that number decreases with remoteness of the island.

In the late 1960s, D. S. Simberloff and E. O. Wilson tested the theory of island biogeography in experiments on small islands of mangroves off the southern tip of Florida (see Chapter 46, Figure 46.2). Six islands, each about 12 m in diameter, were fumigated with a poison to kill all resident arthropods. The pesticide used, methyl bromide, decomposes rapidly, and thus the islands could be recolonized by arthropods from the mainland species pool. Within about a year, species numbers equilibrated on each island, with the fewest species on the island most distant from the coast. Although the equilibrium number for each island was about the same before fumigation and after recolonization, the species composition was different. Chance events—in this case, which species of arthropods happened to disperse to which islands—affected the species composition of the communities. This is another example of how ecological theory has become less deterministic in its predictions; in general, community structure and dynamics seem much less predictable than ecologists once believed.

1. A community is an assemblage of species living close enough together for potential interaction.

Two General Views of Communities (pp. 1093–1095)

1. Many plant species seem to be independently distributed; animals, however, are frequently linked to other species. For most species, presence in a given community probably reflects some combination of factors relating to tolerance range and biological interactions.

Properties of Communities (pp. 1095–1096)

1. Communities have several important properties: species richness and species diversity, the prevalent form of vegetation, trophic structure, and stability.

Community Interactions (pp. 1096–1105)

1. The term coevolution refers to a series of reciprocal interactions between two species. Strict coevolution is difficult to demonstrate because evolutionary interactions often involve a variety of species and varying selective forces that are difficult to disentangle.
2. Interspecific competition may be an important force determining species diversity in some communities.
3. The competitive exclusion principle states that two species competing for the same limiting resources cannot coexist in the same place. Although laboratory populations have demonstrated this principle, there is uncertainty about its applicability to natural communities.
4. The ecological niche is the sum total of the organism's use of the biotic and abiotic resources of its environment. Closely related species may be able to coexist if there are one or more significant differences in their niches.
5. Predation has important implications for the evolution of both predators and prey.
6. Plants defend against herbivores by mechanical defenses and the production of compounds that are irritating or even toxic to animals.
7. Animals resist predation by cryptic coloration, deceptive markings, rapid escape, and the possession of mechanical or chemical defenses that are sometimes advertised by aposematic coloration.
8. In Batesian mimicry, a palatable prey species resembles an unpalatable one; in Müllerian mimicry, a number of unpalatable prey species resemble one another.
9. Keystone predators may help maintain high diversity in a community by holding down populations of a prey species that could competitively exclude others.
10. Parasitism, commensalism, and mutualism are also important biotic interactions in communities.

Succession (pp. 1105–1108)

1. Succession involves changes in species composition of a community over ecological time. Primary succession occurs where no soil previously existed, whereas secondary succession begins in an area where soil remains after a disturbance.
2. Facilitation is the alteration of the environment by successive stages, each of which makes it more suitable for the next stage.

3. Some cases of succession involve inhibition, a phenomenon in which species inhibit the growth of newcomers or in some cases themselves.
4. Disturbances have a variable effect on communities, acting either to stabilize community structure or to initiate or alter succession, depending on the severity and duration of the disturbance.
5. Some ecologists believe that communities are in continual flux, with the identity and numbers of species changing at all stages of succession, even at climax. Dispersal and disturbance probably play key roles in this nonequilibrium.

Biogeographical Aspects of Diversity (pp. 1108–1111)

1. Biogeography is the study of the past and present distribution of species. The major biogeographic realms are associated with patterns of continental drift.
2. A species would be limited to a given range if it never dispersed beyond that range or if it dispersed but failed to survive in other locations. Both natural and artificially designed transplant experiments have been instructive in distinguishing between these two alternatives in specific situations.
3. A variety of interacting factors have probably contributed to biogeographic clines in species diversity.
4. A general theory of island biogeography maintains that species richness on an island levels off at some dynamic equilibrium point, where new immigrations are balanced by extinctions. Furthermore, the theory predicts that species richness is directly proportional to size and inversely proportional to distance of the island from the mainland.

SELF-QUIZ

1. The concept of "trophic structure" of a community emphasizes the
   a. prevalent form of vegetation
   b. keystone predator
   c. feeding relationships within a community
   d. effects of coevolution
   e. species richness of the community

2. According to the principle of competitive exclusion,
   a. two species cannot coexist in the same habitat
   b. extinction or emigration are the only possible results of competitive interactions
   c. intraspecific competition results in the success of the best-adapted individuals
   d. two species cannot share the same realized niche in a habitat
   e. resource partitioning will allow a species to utilize all the resources of its fundamental niche

3. The effect of a keystone predator within a community may be to
   a. competitively exclude other predators from the community

b. maintain species diversity by preying on the most abundant prey species, thus preventing competitive exclusion of other species

c. increase the relative abundance of the most competitive prey species

d. encourage the coevolution of predator and prey adaptations

e. create nonequilibrium in species diversity

4. All of the following statements are consistent with the nonequilibrium model of community structure *except*

a. Chance events such as dispersal and disturbance play major roles in determining species diversity.

b. Species diversity may be greatest when disturbances are intermediate in severity and frequency.

c. Even when a community represents a mature climax stage, species composition and the number of species may continue to change.

d. Succession reaches a climax when the intricate web of interactions allows for the addition of new species only when a niche is vacated by extinction.

e. The course of succession may vary depending on the chance arrival of early colonizers.

5. Transplant experiments provide evidence that

a. all transplants are ecological disasters

b. continental drift accounts for geographic distribution of species

c. the theory of island biogeography is valid

d. keystone predators maintain community structure

e. some species can live outside their normal ranges

6. An example of cryptic coloration is

a. the green color of a plant

b. the bright markings of the viceroy butterfly

c. the stripes of a skunk

d. the mottled coloring of moths that rest on lichens

e. the bright colors of an insect-pollinated flower

7. An example of Müllerian mimicry would be

a. a butterfly that resembles a leaf

b. two poisonous frogs that resemble one another in coloration

c. a minnow with spots that look like large eyes

d. a beetle that resembles a scorpion

e. a carnivorous fish with a worm-like tongue that lures prey

8. Which country is incorrectly matched with its biogeographical realm?

a. Canada-Palearctic

b. United States-Nearctic

c. Brazil-Neotropical

d. South Africa-Ethiopian

e. India-Oriental

9. To be certain that two species had coevolved, one would *ideally* need to establish that

a. the two species originated about the same time

b. local extinction of one species dooms the other species

c. each species affects the population density of the other species

d. one species has adaptations that *specifically* tracked evolutionary change in the other species, and vice versa

e. the two species are adapted to a common set of environmental conditions

10. According to the theory of island biogreography, species richness would be greatest on an island that is

a. small and remote

b. large and remote

c. large and close to a mainland

d. small and close to a mainland

CHALLENGE QUESTIONS

1. Tarantula hawk wasps and tarantulas engage in a nocturnal battle that often results in the paralysis of the tarantula by the wasp, which then lays its eggs on the spider. The eggs hatch and the young devour the living, immobilized tissues of the spider. Would you consider this interspecific relationship predation or parasitism? Give reasons for your answer. (See Further Reading 5.)

2. Compare and contrast some likely scenarios of primary succession on two new volcanic islands, a large one formed 100 miles off a large, species-rich mainland, and a smaller one formed 300 miles offshore.

FURTHER READING

1. Bergerud, A. T. "Prey Switching in a Simple Ecosystem." *Scientific American*, December 1983. A focus on an important predatory adaptation.
2. Lubchenco, J. "Relative Importance of Competition and Predation: Early Colonization by Seaweeds in New England." In J. Diamond and T. J. Case, *Community Ecology.* New York: Harper and Row, 1986. A quantitative, experimental assessment of the relative importance of competition and predation in community structure.
3. MacArthur, R. H., and E. O. Wilson. *The Theory of Island Biogeography.* Princeton, N.J.: Princeton University Press, 1967.
4. Miller, J. A. "Invasion of the Ecosystem." *Science News*, June 29, 1985. A brief highlight of the dangers involved in ecological transplant experiments.
5. "Of Sting and Fang." *Science 85*, January/February 1985. Interspecific interaction involving the egg-laying tarantula hawk wasp and its victim, a spider.

49

Ecosystems

An **ecosystem** consists of all the organisms in a given area along with the abiotic factors with which they interact; it is a community and its physical environment. As with populations and communities, the boundaries of an ecosystem are usually not discrete, the unit of study ranging from a microcosm such as a terrarium to lakes and forests to the entire biosphere, which can be regarded as an ecosystem of the grandest scale.

The most inclusive level in the hierarchy of biological organization, an ecosystem involves two processes that cannot be fully described at lower levels—energy flow and chemical cycling (Figure 49.1). Energy enters most ecosystems in the form of sunlight and is converted to chemical energy by autotrophic organisms, passed to heterotrophs as the organic compounds of food, and dissipated in the form of heat. Chemical elements such as carbon and nitrogen are cycled between abiotic and biotic components of the ecosystem. Photosynthetic organisms acquire these elements in inorganic form from the air and soil and assimilate them into organic molecules, some of which are consumed by animals. The elements are returned in inorganic form to the soil and air by the metabolism of plants and animals and by other organisms such as bacteria and fungi that break down organic wastes and dead organisms. Energy flow and chemical cycling are related, since both occur by the transfer of substances through the feeding levels of the ecosystem. However, because energy, unlike materials, cannot be recycled, an ecosystem must be powered by a continuous influx of new energy from an external source (the sun). This chapter describes the dynamics of energy flow

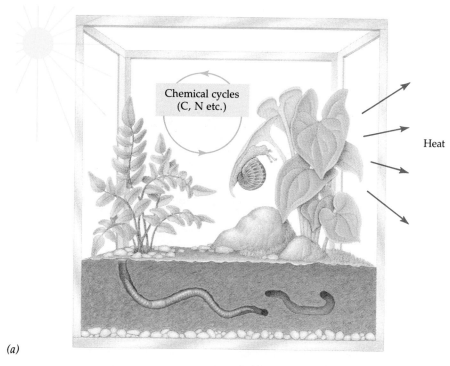

(a)

(b)

Figure 49.1
Ecosystem dynamics. (a) The two fundamental processes of any ecosystem are energy flow and chemical cycling. This terrarium is solar powered, with plants converting light energy to chemical energy by photosynthesis. Energy can flow in the form of organic fuels to animals that eat the plants and to bacteria and fungi that decompose organic refuse in the soil. Energy cannot be recycled, however, because every use of the chemical energy by the organisms dissipates energy to the surroundings in the form of heat; thus, there must be a continuous flux of energy through the ecosystem. In contrast, the terrarium is self-contained in terms of materials, because chemicals such as carbon and nitrogen are recycled between the biotic and abiotic (air, water, and soil) components of the ecosystem. **(b)** A much bigger terrarium, *Biosphere II,* will be an ambitious experiment on "closed" ecosystems (not truly closed, since there will be an influx of sunlight and a transfer of heat between the enclosure and the surroundings). Biosphere II, now under construction near Oracle, Arizona, will be about the size of two football fields in ground area and 26 meters high at its highest point. (This is a photograph of a scale model.) It is designed to be airtight and watertight. Sometime in 1990 or 1991, eight human volunteers will enter Biosphere II and remain for at least two years in the materially closed system. The "biospherians" will live in a five-story wing and grow their food in an agricultural "biome." The main wing of Biosphere II contains five natural biomes, ranging from tropical rain forest to desert. The experiment provides a unique opportunity to study ecosystem dynamics.

and chemical cycling in ecosystems, and considers some of the consequences of human intrusions into these processes.

TROPHIC LEVELS AND FOOD WEBS

Each ecosystem has a **trophic structure** of different feeding (trophic) levels that determines the route of energy flow and the pattern of chemical cycling. The trophic level that supports all others consists of autotrophs, which are referred to as the **primary producers** of the ecosystem. Most producers are photosynthetic organisms that use light energy to synthesize sugars and other organic compounds, which the producers then use as fuel for cellular respiration and as building material for growth. All other organisms in an ecosystem are consumers—heterotrophs directly or indirectly dependent on the photosynthetic output of producers. Herbivores, which eat plants or algae, are the **primary consumers** of photosynthetic products. The next trophic level includes all those species that are **secondary consumers,** carnivores that eat herbivores. These carnivores may in turn be eaten by other carnivores that are **tertiary consumers,** and some ecosystems have carnivores of even higher level. Some consumers, the **detritivores,** derive their energy from the organic wastes and dead organisms (detritus) from many trophic levels.

Figure 49.2

An ecosystem independent of light. In the late 1970s, marine ecologists exploring deep sea trenches in small submarines discovered rich biological communities on the dark seafloor, where life is generally scant. The so-called rift ecosystems, this one a few hundred kilometers northwest of the Galapagos Islands at a depth of about 2500 m, are located at spreading centers on the seafloor where hot magma super-heats the water. When it contacts the cold water outside the thermal vents, the super-heated water, with an initial temperature in excess of 250°C, cools rapidly. More than 200 species of bacteria have been identified living within or near the vents. About a dozen of these bacterial species are chemoautotrophs that obtain energy by oxidizing H_2S, formed by reaction of the hot water with dissolved sulfate (SO_4^{-2}). These chemosynthetic bacteria are the producers of the vent ecosystems. Among the animals in these ecosystems are giant tube-dwelling worms called beard worms, many more than a meter long. They are apparently nourished by chemosynthetic bacteria that live as symbionts within the worms. Many other invertebrates, including filter-feeding crabs, clams, mussels, and barnacles, consume bacteria around the vents.

The main producers in most terrestrial ecosystems are plants. In lakes and the open ocean, photosynthetic protists and cyanobacteria, which form the phytoplankton, are most important. In streams, much of the organic material used by consumers may be debris from terrestrial plants that fall or wash into the streams (see Chapter 46). Multicellular algae and aquatic plants often occur in the shallower benthic areas of both freshwater and marine habitats. In the aphotic zone on the deep seafloor, most life depends on photosynthetic production in the form of dead plankton and other detritus that rains down from the photic zone above. The most notable exceptions are communities of organisms that live around hot water vents deep in the seas, where the producers are not photosynthetic organisms but chemosynthetic bacteria that derive energy by oxidizing hydrogen sulfide. These ecosystems are driven by geothermal rather than solar energy (Figure 49.2).

The primary consumers, or herbivores, on land are mostly insects, snails, and certain vertebrates, including grazing mammals and birds that eat seeds and fruit. In aquatic ecosystems, phytoplankton is consumed mainly by zooplankton, which includes heterotrophic protists, various small invertebrates (especially crustaceans and, in the ocean, larval stages of many benthic forms), and some fish.

Examples of organisms functioning as secondary consumers in terrestrial ecosystems are spiders, frogs, and insect-eating birds, as well as lions and other carnivorous mammals that eat grazers such as antelope. In aquatic habitats, many fish feed on zooplankton, and are in turn fed upon by other fish. In the benthic zone of the seas, algae-eating invertebrates are prey to other invertebrates such as sea stars.

The organic material that forms the living organisms of an ecosystem is eventually recycled—that is, broken down and returned to the abiotic environment in forms that can be used by plants. Detritivores, which feed on nonliving organic material, are also called "decomposers," emphasizing their importance in this recycling process. The most important detritivores in most ecosystems are bacteria and fungi. These organisms secrete enzymes that digest organic material and then absorb the products; some bacteria and fungi can even break down cellulose. Earthworms and scavengers such as crayfish, cockroaches, and bald eagles are also detritivores, but these animals digest organic

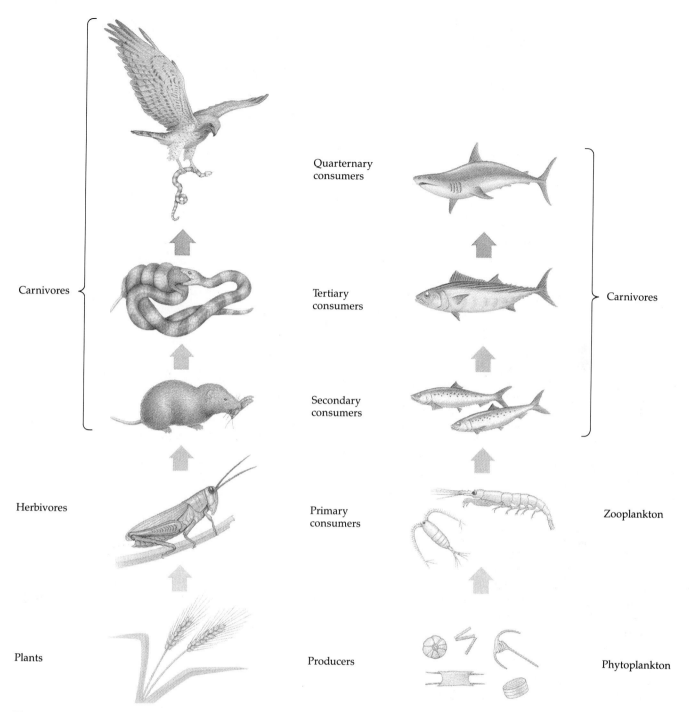

Figure 49.3
Examples of terrestrial and aquatic food chains.

material internally after ingesting their food. In fact, all heterotrophs, including humans, are decomposers in the sense that they break down organic material and release inorganic products, such as carbon dioxide and ammonia, to the environment. The main distinction is that detritivores feed on dead material. Detritivores often form a major link between primary producers and higher-level consumers. A crayfish, for example, might feed on detritus at the bottom of a lake and then be eaten by a bass. In a forest, birds might feed on earthworms that have been feeding on leaf litter in the soil.

The transfer of food from trophic level to trophic level, beginning with producers, is known as a **food chain** (Figure 49.3). However, few ecosystems are so simple as to have a single, unbranched food chain. Several different primary consumers may feed on the same plant, and one species of primary consumer may eat several different plants. Such branching of food chains occurs at the other trophic levels as well. For example, frogs, which are secondary consumers, eat several insect species, which may also be eaten by various birds. Some consumers feed at several different trophic levels. An owl, for instance, may eat primary consumers such as field mice and also feed on higher-level consumers such as snakes. Omnivores, including humans, eat producers as well as consumers of different levels. Thus, the feeding relationships of who eats whom in an ecosystem are usually woven into elaborate **food webs** (Figure 49.4).

ENERGY FLOW

All organisms require energy for growth, maintenance, reproduction, and, in some species, locomotion. Primary producers use light energy to synthesize energy-rich organic molecules, which can subsequently be broken down to make ATP (see Chapter 10). Consumers acquire their organic fuels secondhand (or even thirdhand or fourthhand) through food chains. Therefore, photosynthetic productivity sets the spending limit for the energy budget of the entire ecosystem.

Primary Productivity

Each day, the Earth is bombarded by 10^{22} joules (J) of solar energy, the equivalent of the energy of 100 million atomic bombs the size of the one dropped on Hiroshima. Most solar radiation is absorbed, scattered, or reflected by the atmosphere or the Earth's surface. Of the visible light that does reach leaves and algae, only about 1% is converted to chemical energy by photosynthesis (photosynthetic efficiency varies with the type of plant, light level, and other factors). Still, that is enough for the production of about 170 billion tons of organic material per year globally. Different types of ecosystems vary considerably in productivity.

The rate at which light energy is converted to chemical energy (organic compounds) by the autotrophs of an ecosystem is called **primary productivity**. The total of this productivity is known as **gross primary productivity** (see Methods Box on page 1121). Not all of this product is stored in the organic material of the growing plants, because the plants use some of the molecules as fuel in their own cellular respiration. The **net primary productivity** (NPP) is thus equal to the gross primary productivity (GPP) minus the energy used by the producers for respiration (Rs):

$$NPP = GPP - Rs$$

Or we may think in terms of the equations for photosynthesis and respiration:

$$6CO_2 + 6H_2O \underset{\text{Respiration}}{\overset{\text{Photosynthesis}}{\rightleftharpoons}} C_6H_{12}O_6 + 6O_2$$

Gross primary production results from the rightward reaction (photosynthetic reaction); net primary productivity is the difference between the yield of photosynthesis and the consumption of organic fuel symbolized by the leftward reaction. In less technical terms, net primary production accounts for the organic mass of the plants we see.

Net primary productivity is the measurement of interest to us because it represents the storage of chemical energy in an ecosystem available to consumers. For most primary producers, somewhere between 50% and 90% of the gross primary productivity remains as net primary productivity. The NPP to GPP ratio is generally smaller for large producers with elaborate nonphotosynthetic structures, such as trees, which have to support large root systems.

Primary productivity can be expressed in terms of energy per unit area per unit time (for example, $J/m^2/yr$), or as **biomass** added to the ecosystem per unit area per unit time ($g/m^2/yr$). Biomass is best expressed in terms of the dry weight of organic material because water molecules contain no usable energy. An ecosystem's primary productivity is *not* simply the total biomass of plants present at a given time, which is called the **standing crop**, but the *rate* at which the vegetation synthesizes new biomass. Although a forest has a very large standing crop, its productivity may actually be less than that of some grasslands, which do not accumulate vegetation because of grazing and because many of the plants are annuals.

Tropical rain forests are the most productive terrestrial ecosystems (Table 49.1). Estuaries and coral reefs also have very high productivity, but their total contribution to global productivity is relatively small because these ecosystems are not widespread. The open ocean contributes more primary production than any other ecosystem, but this is because of its very large size; productivity per unit area is relatively low. Deserts and tundra also have low productivity.

The factors most important in limiting productivity depend on the type of ecosystem and, for a single ecosystem, on temporal changes such as those involving the seasons. Productivity in terrestrial ecosystems is correlated with precipitation, heat, and light intensity. In general, terrestrial productivity increases at latitudes increasingly close to the equator.

In any ecosystem, inorganic nutrients may be important in limiting productivity. Recall from Chapter 33 that plants need a variety of nutrients, some in relatively large quantities and others only in trace amounts—but all are crucial. The exact proportions of mineral nutrients needed by different species will vary, as will the exact proportions available in the soil or water. As primary productivity continues in a system, nutrients are removed, sometimes at a rate faster

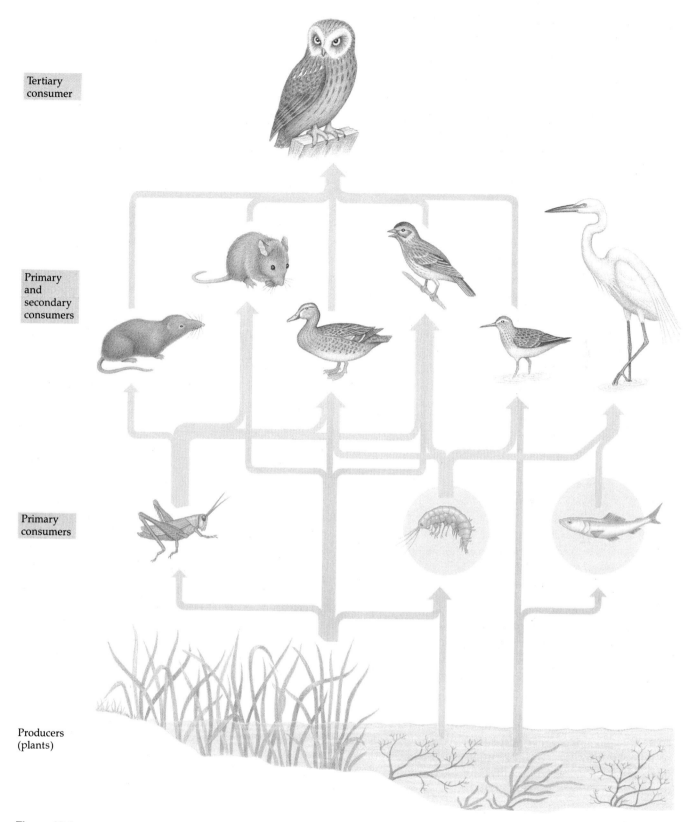

Tertiary
consumer

Primary
and
secondary
consumers

Primary
consumers

Producers
(plants)

Figure 49.4
A food web. This is a very simplified diagram of feeding relationships in a California salt marsh. The actual feeding relationships in any environment involve many more organisms at each trophic level as well as detritivores that feed at all levels. The arrows in this diagram are color-coded according to the trophic levels of the food and consumer.

Table 49.1 Net primary productivity of various ecosystems

Ecosystem	Area (million km²)	Average Net Primary Productivity per Unit Area (g*/m²/yr)	World Net Primary Productivity (billion tons*/yr)
Continental Ecosystems			
Tropical rain forest	17.0	2200	37.4
Tropical seasonal forest	7.5	1600	12.0
Temperate evergreen forest	5.0	1300	6.5
Temperate deciduous forest	7.0	1200	8.4
Boreal forest	12.0	800	9.6
Woodland and shrubland	8.5	700	6.0
Savanna	15.0	900	13.5
Temperate grassland	9.0	600	5.4
Tundra	8.0	140	1.1
Desert and semidesert scrub	18.0	90	1.6
Extreme desert, rock, sand, and ice	24.0	3	0.07
Cultivated land	14.0	650	9.1
Swamp and marsh	2.0	2000	4.0
Lake and stream	2.0	250	0.5
Total continental	149.0	773	115.0
Marine Ecosystems			
Open ocean	332.0	125	41.5
Upwelling zones	0.4	500	0.2
Continental shelf	26.6	360	9.6
Algal beds and reefs	0.6	2500	1.6
Estuaries	1.4	1500	2.1
Total marine	361.0	152	55.0
Total biosphere	510.0	333	170.0

*Units of measure are dry grams and dry metric tons of organic matter.
SOURCE: Adapted from Whittaker, *Communities and Ecosystems*, 2d ed. (New York: Macmillan, 1975).

than they are returned. At some point, further productivity slows or ceases because a specific nutrient is no longer present in sufficient quantity. It is unlikely that all nutrients will "run out" simultaneously, so the system is held back by the single nutrient—termed the **limiting nutrient**—that is no longer present in adequate amounts. Adding other nutrients to the system will have no effect because they are already present in sufficient quantity. Adding the limiting nutrient will stimulate the system to resume growth until some other nutrient or the same one becomes limiting at a higher level of productivity. In many situations, either nitrogen or phosphorus is the key limiting nutrient;

both are needed in large amounts, but often they are available in only moderate or small quantities. There is also evidence that CO_2 sometimes limits productivity. Increasing the amount of CO_2 around a plant may increase productivity, but sometimes only by a small amount because another nutrient then becomes limiting. We will see later how human intrusions affecting nutrient balance have disrupted aquatic ecosystems.

Productivity in the seas is generally greatest in the shallow waters near continents and along coral reefs. In the open oceans, light intensity and temperature affect the productivity of phytoplankton communi-

METHODS: MEASURING GROSS PRIMARY PRODUCTIVITY IN AQUATIC HABITATS

A transparent bottle and an opaque bottle are each filled with a water sample taken from the same depth. The bottles are then stoppered and returned to that depth for a specified time, often a day. A basic assumption is that the plankton communities inside each bottle are alike and are representative of the unconfined community in the area from which the water samples were taken. The bottles are removed from the pond after the set time, and their oxygen concentration is measured by a chemical method. These values are compared with the oxygen concentration of a separate sample of water from the same depth tested at the beginning of the study.

Since no photosynthesis can occur in the dark bottle, its oxygen concentration will decrease compared with the original amount, as a result of cellular respiration by the plankton. Assuming that respiration occurred at the same rate in the transparent bottle, any increase in the oxygen content of that bottle results from photosynthesis producing more O_2 than respiration consumes. Adding the amount of the oxygen decrease in the dark bottle to the amount of oxygen increase in the light bottle cancels respiration in the light bottle and gives the total amount of oxygen produced by photosynthesis. This method can be used to determine the gross primary productivity, based on the chemical equation for photosynthesis:

$$6CO_2 + 6H_2O \rightleftarrows C_6H_{12}O_6 + 6O_2$$

This "light bottle/dark bottle" method is relatively insensitive to small changes in oxygen concentration, and therefore may be difficult to use in unproductive waters such as the open ocean. The most common method today is a much more sensitive one that uses the radioactive tracer [14]C. Bicarbonate labeled with this form of carbon is introduced to a water sample in a bottle, which is then incubated at the depth from which it was taken, or under similar light/temperature conditions on board a ship. The label is incorporated as [14]CO_2 by the phytoplankton through photosynthesis. After the incubation period, which may be only an hour or so, the plankton are filtered out of the sample, and their radioactivity level is measured using an instrument called a scintillation counter. Since the radioactivity represents carbon retained by the algae, it can be used to estimate net primary production.

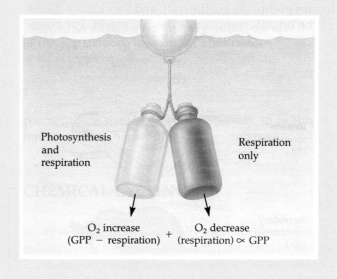

Photosynthesis and respiration

Respiration only

O_2 increase (GPP − respiration) + O_2 decrease (respiration) ∝ GPP

ties. Productivity is generally greatest near the surface and drops off steeply with depth, as light is rapidly absorbed by water and plankton. The primary productivity of the open ocean is relatively low overall because some inorganic nutrients, especially nitrogen and phosphorus, are in short supply near the surface; at depth, where nutrients are abundant, there is no light. Phytoplankton communities are most productive where upwelling currents bring nitrogen and phosphorus to the surface. We can see this, for instance, in Antarctic seas, which, in spite of the colder water and lower light intensity, are actually more productive than most tropical seas. The chemosynthetic ecosystems near hot-water vents are also very productive, as we saw in Figure 49.2, but these communities are not widespread and thus their overall contribution to marine productivity is small.

In freshwater ecosystems, light intensity seems to be the most important determinant of productivity on a day-to-day basis. Water temperature is also important, causing marked seasonal fluctuation in many areas. As in the ocean, the availability of inorganic nutrients may also limit productivity of freshwater ecosystems.

but cycling is more complicated because of the interaction of CO_2 with water and limestone. Carbon dioxide reacts with water to form carbonic acid (H_2CO_3). Carbonic acid in turn reacts with the limestone ($CaCO_3$) that is abundant in many waters, including the ocean, to form bicarbonates and carbonates:

$$H_2CO_3 + CaCO_3 \rightleftarrows Ca(HCO_3)_2 \rightleftarrows Ca^{2+} \ 2HCO_3^-$$
$$2HCO_3^- \rightleftarrows 2H^+ + \ 2CO_3^{2-}$$
$$\text{Bicarbonate} \qquad \qquad \text{Carbonate}$$

As CO_2 is used in photosynthesis, the equilibrium of this reaction series shifts toward the left, converting bicarbonates back to CO_2. Thus, bicarbonates serve as a CO_2 reservoir. Aquatic autotrophs may also use dissolved bicarbonate directly as their source of carbon. Overall, the amount of carbon tied up in the various inorganic forms in the ocean, not counting sediments, is about fifty times that available in the atmosphere. Because of these inorganic reactions of CO_2 in water, as well as uptake by marine phytoplankton, the ocean may be an important "buffer" that will absorb some of the CO_2 being added to the atmosphere by the burning of fossil fuels.

The Nitrogen Cycle

The atmosphere of the Earth is almost 80% N_2, but plants cannot assimilate nitrogen in this form into organic material. Only certain prokaryotes can fix nitrogen—that is, reduce atmospheric nitrogen to ammonia (NH_3), which can be used to synthesize nitrogenous organic compounds such as amino acids. Indeed, prokaryotes are vital links at several points in the nitrogen cycle (Figure 49.9).

Nitrogen is fixed in terrestrial ecosystems by free-living (nonsymbiotic) soil bacteria and the symbiotic bacteria (*Rhizobium*) of the root nodules of legumes and certain other plants (see Chapter 33). Some cyanobacteria fix nitrogen in aquatic ecosystems. Organisms that fix nitrogen, of course, are meeting their own metabolic requirements, but in the process they release excess amounts of ammonia that become available to other organisms. Small amounts of nitrogen are also fixed by lightning in the atmosphere. Far more important, however, is industrial fixation for the production of fertilizer, which now makes a significant contribution to the pool of nitrogenous materials in the soil and waters of agricultural regions.

Although plants can use ammonia directly, most of the ammonia in soil is used by certain aerobic bacteria as an energy source; this oxidizes ammonia to nitrite (NO_2^-) and then to nitrate (NO_3^-), a process called **nitrification.** Nitrate released from these bacteria can then be assimilated by plants and converted to organic forms, such as amino acids and proteins. Animals can assimilate only organic nitrogen, by eating plants or other animals.

Decomposition of organic nitrogen back to ammonia, a process called **ammonification,** is carried out by many bacteria and eukaryotes. Detritivores are especially important in recycling large amounts of nitrogen to the soil through this process. Some bacteria can obtain the oxygen they need for metabolism from nitrate rather than from O_2. As a result of this **denitrification** process, nitrogen from nitrate is converted back to N_2, returning to the atmosphere.

Overall, most of the nitrogen cycling in natural systems involves the nitrogenous compounds in soil and water, not atmospheric N_2. Although nitrogen fixation has been important in the buildup of a pool of available nitrogen and is essential for some plants with symbiotic bacteria, it contributes only a tiny fraction of the nitrogen assimilated annually by total vegetation. The amount of N_2 returned to the atmosphere by denitrification is also relatively small. The great bulk of assimilated nitrogen comes from nitrate, locally recycled from organic forms by ammonification and nitrification (Figure 49.9).

The Phosphorus Cycle

In some respects the phosphorus cycle is simpler than that of either carbon or nitrogen. There are no significant gaseous compounds, so cycling does not involve movement through the atmosphere. Also, there is essentially only one important inorganic form, which is phosphate (PO_4^{-3}). Weathering of rock gradually adds soil phosphate, which plants can absorb and use for organic synthesis (Figure 49.10). Phosphorus is transferred to consumers in organic form, and is added back to the soil by the excretion of phosphate by animals and by the action of decomposers on detritus. Humus and soil particles bind phosphate (see Chapter 33), and consequently the recycling of phosphorus tends to be quite localized in ecosystems. However, phosphorus does leach into the water table, gradually draining from terrestrial ecosystems to the sea. Severe erosion can hasten this drain, but in most natural ecosystems, weathering of rock can keep pace with the loss of phosphate. Phosphate that reaches the ocean gradually sediments and becomes incorporated into rocks that may turn up much later in terrestrial ecosystems as a result of geological processes that raise the seafloor or lower the sea level at that location. Thus, two phosphate cycles occur over very different time scales. Most phosphate recycles locally among soil, plants, and consumers on the scale of ecological time. A sedimentary cycle removes and restores terrestrial phosphorus over geological time. The same general pattern applies to other nutrients that lack atmospheric forms.

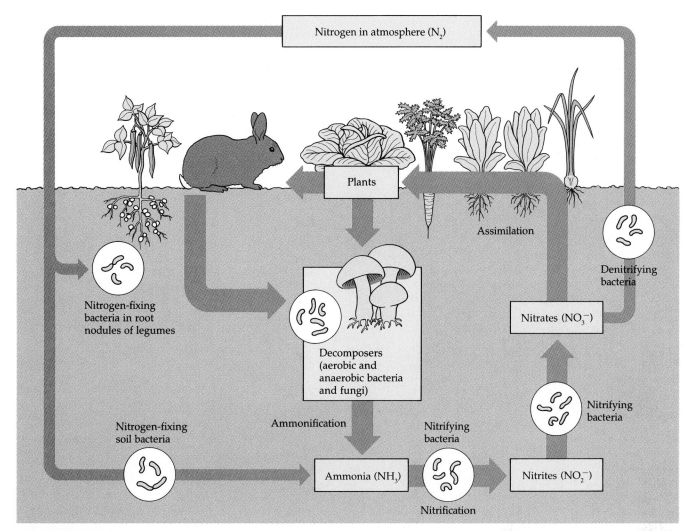

Figure 49.9
The nitrogen cycle. Most of the nitrogen cycling through food webs is taken up by plants in the form of nitrate. Most of this, in turn, comes from the nitrification of ammonia that results from the decomposition of organic material. Addition of nitrogen from the atmosphere through fixation and its return via denitrification involve relatively small amounts compared to the local recycling that occurs in the soil or water.

Variations in Nutrient-Cycling Time

In some parts of a tropical rain forest, key nutrients such as phosphorus occur in the soil at levels far below those typical of a temperate forest. At first this might seem paradoxical given the high productivity of these forests. In a tropical forest, the litter on the forest floor constitutes only about 1%–2% of the total biomass, whereas in temperate forests litter may make up 5%–20%. Because of the continuous warm temperature and abundant rainfall in many tropical areas, organic material decomposes rapidly. At the same time, demand for nutrients is very high because of the large biomass, so nutrients are taken up quickly as they are recycled. Overall, the tropical rain forest contains a large amount of nutrients, but a high percentage of these are in the living vegetation at any given time. In temperate forests, by contrast, a larger store of nutrients remains in the backlog of detritus yet to be decomposed, and the nutrients released may remain longer in the soil before being assimilated. Thus, the relatively low concentrations of some nutrients in the soil of tropical rain forests results from a fast cycling time, not an overall scantiness of these elements in the ecosystem. Among the factors that influence nutrient-cycling times in ecosystems are rate of decomposition, size of the standing crop, local soil chemistry, and frequency of fires.

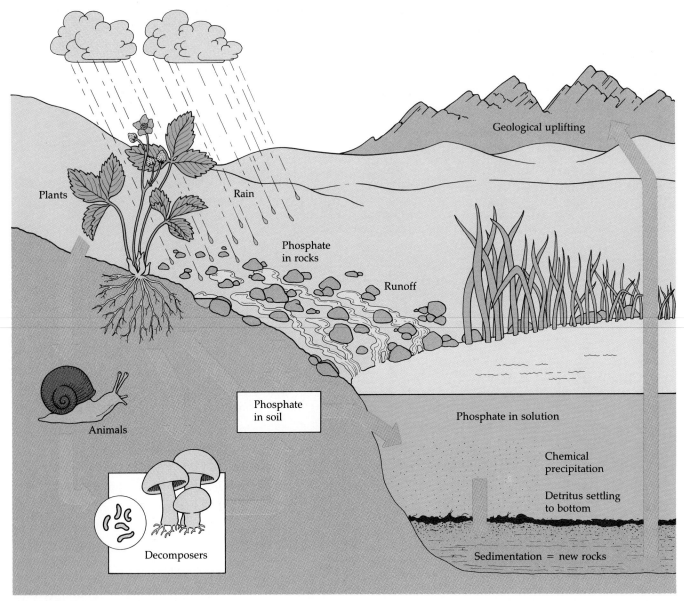

Figure 49.10
The phosphorus cycle. Phosphorus, which does not have an atmospheric component, tends to cycle locally. Exact rates vary depending on the system. Generally, small losses from terrestrial systems caused by leaching are balanced by inputs from weathering. In aquatic systems, as in terrestrial systems, phosphorus is cycled through food webs. Typically, however, some phosphorus is lost from the ecosystem because of chemical processes that cause precipitation or through settling of detritus to the bottom, where sedimentation may lock away some of the nutrient before biotic processes can reclaim it. On a much longer time scale, this phosphorus may become available to ecosystems again through geological processes such as uplifting. A general pattern similar to this applies to many other nutrients, including trace elements.

HUMAN INTRUSIONS IN ECOSYSTEM DYNAMICS

Human activities and technology have intruded in one way or another in the trophic structure, energy flow, and chemical cycling of ecosystems in most areas of the world. The effects are sometimes local or regional, but the ecological impact of humans can be widespread or even global. Acid rain, for example, may fall hundreds or thousands of miles from the smokestacks emitting the chemicals that produce it. One of the most critical environmental problems today, acid rain, was discussed in detail in Chapter 3. Here we address a few of the many other concerns ecologists have about human intrusions into the dynamics of ecosystems.

Agricultural Effects on Nutrient Cycling

Human activity generally intrudes in nutrient cycles by removing nutrients from one part of the biosphere and adding them to another. This may result in a depletion of key nutrients in one area and a disruption of the natural equilibrium in both areas. Consider, for example, the ecological impact of agriculture. After natural vegetation is cleared from an area, crops may be grown for some period without an additional supplement of nutrients because of the reserve of nutrients in both organic and inorganic forms in the soil. A substantial fraction of the nutrients is not recycled, but instead is sent out of the area in the form of crop biomass distributed to humans and animals, often long distances away. The "free" period for crop production—when there is no need to add nutrients to the soil—varies greatly, depending on the system. When some of the early North American prairie lands were first tilled, good crops could be produced for many years because of the large store of organic materials in the soil that continued to provide nutrients through decomposition. By contrast, in the tropics, as we have seen, soils may be poor in nutrients; cleared land can be farmed only one or two years before new areas must be opened up. Eventually, in any area, the natural store of nutrients becomes exhausted, and supplements must be added in the form of fertilizer. The industrially synthesized fertilizers used so extensively today are produced at considerable expense in terms of both money and energy (see Chapter 33).

Many of the nutrients in crops are quickly "short-circuited" out of terrestrial systems into aquatic ones. As we have seen, most nutrients, such as phosphate, are essentially lost from terrestrial systems once they exit to aquatic systems. Normally this loss is low, due only to gradual leaching. But nutrients in crops soon appear in the wastes of humans and livestock and then turn up in streams and lakes through runoff from fields and discharge as sewage. Someone eating a piece of broccoli on the east coast of the United States is consuming nutrients that only days before might have been in the soil in California; and a short time later, some of these nutrients will be in a nearby stream on their way to the sea, having passed through an individual's digestive system and the local sewage facilities. Once in aquatic systems, the nutrients may stimulate excessive growth of algae, a related problem to be discussed shortly. Researchers are attempting to find ways to reclaim nutrients at sewage treatment facilities and return them to the land to reduce the need for constant fertilizer supplements. There has been some public resistance to this idea because of an understandable concern about adding the products of human waste to the land that produces the food we eat. Technological advances in sewage treatment, however, make it possible to recycle these nutrients without any hazard to human health. Clearly, with nutrient cycling as well as many other ecological considerations, it is in our long-term interest to take new and realistic approaches to the way we interact with the rest of the biosphere.

Effects of Deforestation on Chemical Cycling: The Hubbard Brook Forest Study

For the past 30 years, a team of scientists led by Herbert Bormann of Yale University and Gene Likens of Cornell University has conducted an ambitious, long-term study of nutrient cycling in a forest ecosystem under natural conditions and after severe human intrusion. The study site is the Hubbard Brook Experimental Forest in the White Mountains of New Hampshire. It is a nearly mature deciduous forest with several valleys, each drained by a small creek that is a tributary of Hubbard Brook. Bedrock impenetrable to water is close to the surface of the soil, so each valley is a watershed that can drain only through its creek.

The research team first determined the mineral budget for each of six valleys by measuring influx and outflow of several key nutrients. They collected rainfall at several sites to measure the amount of water and dissolved minerals added to the ecosystem. To monitor the loss of water and minerals, they constructed V-shaped, concrete weirs across the creek at the bottom of each valley (Figure 49.11a). About 60% of the water added to the ecosystem as rainfall and snow exits through the stream, and the remaining 40% is lost by transpiration from plants and evaporation from the soil.

Preliminary studies confirmed that internal cycling within a mature terrestrial ecosystem conserves most of the mineral nutrients. Mineral influx and outflow were nearly balanced, and were relatively small compared with the quantity of minerals being recycled within the forest ecosystem. For example, only about 0.3% more Ca^{2+} left a valley via its creek than was added by rainwater, and this small net loss was probably made up by chemical decomposition of the bedrock. During most years, the forest actually registered small net gains of a few mineral nutrients, including nitrogenous ones.

In 1966, one of the valleys, with an area of 15.6 hectares, was completely logged in an experiment to test the effect of deforestation on nutrient cycling (Figure 49.11b). Influx and outflow of water and minerals in this experimentally altered watershed were compared with a control watershed for three years. Water runoff from the valley increased by 30%–40% after deforestation, apparently because there were no trees to absorb and transpire water from the soil. Net losses of minerals from the deforested watershed were huge. The concentration of Ca^{2+} in the creek increased fourfold, for example, and the concentration of K^+ increased by a factor of 15. Most remarkable was the loss of nitrate, which increased in concentration in the creek by sixtyfold (Figure 49.12). Not only was a vital mineral nutrient drained from the ecosystem, but also nitrate in the creek reached a level considered unsafe for drinking water.

Similar studies in other experimental forests have

(a)

(b)

Figure 49.11
Studying nutrient recycling in the Hubbard Brook Experimental Forest.
(a) Concrete weirs across streams at the bottom of watersheds enabled researchers to monitor the outflow of water and mineral nutrients from the ecosystem. Because these small concrete dams were anchored to the impervious bedrock, all water shed from the valleys had to pass through the weirs. The height of the water in the V-shaped spillway was correlated with the volume of flow. (b) Some watersheds were completely logged to study the effects of clear-cutting on drainage and nutrient cycling. All the logs and rubble from the clear-cutting operation were left in place, and care was taken not to disturb the soil any more than necessary. Herbicides were used to prevent regrowth.

corroborated the evidence from Hubbard Brook that deforestation severely alters both water drainage and nutrient cycling.

Accelerated Eutrophication of Lakes

Lakes change naturally. In a young lake, primary productivity is relatively low because nutrients required by phytoplankton are scarce. Gradually, runoff from the land brings in more nutrients that are captured by

the primary producers and then continuously recycled through the lake's food webs. Thus, the overall productivity increases, and the lake is said to be more **eutrophic** (from the Greek for "good food"). Under natural conditions, lakes eventually reach an equilibrium where input of new nutrients is balanced by the losses from the system due to outflow and sedimentation. Natural eutrophication is seldom a problem. A "rich" natural lake, one in which there is moderate growth of algae and plants, is usually aesthetically pleasing and valuable for human recreation.

Figure 49.12
Loss of nitrate from a deforested watershed. The graph shows the relative concentrations of nitrate in runoff from the deforested watershed and a control unlogged watershed, indicating that nitrate loss was 60 times greater in the deforested valley.

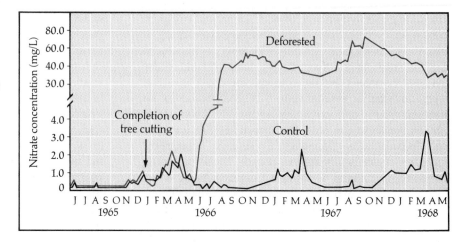

Unfortunately, human intrusion has disrupted almost all freshwater ecosystems by what may be termed "cultural eutrophication." Sewage and factory wastes, runoff of animal waste from pastures and stockyards, and leaching of fertilizer from both agricultural and urban areas has loaded many streams, rivers, and lakes with inorganic nutrients. This often results in an explosive increase in the density of photosynthetic organisms (Figure 49.13). Shallower areas become weed choked, making boating and fishing impossible. Large algal blooms become common, which may result in increased oxygen production during the day but greatly reduced oxygen levels at night, owing to respiration of the excessive biomass. As the algae die and the organic material accumulates, the metabolism of decomposers consumes all the oxygen. All of these effects may make it impossible for some organisms to survive. For example, eutrophication of Lake Erie resulting from pollution wiped out commercially important fish such as blue pike, whitefish, and lake trout by the 1960s. Since then, tighter regulations on the dumping of wastes into the lake have enabled some fish populations to rebound, but many of the native species of fishes and invertebrates have not recovered.

Figure 49.13
Experimental eutrophication of a lake. The far basin of this lake was separated from the near basin by a plastic curtain and fertilized with inorganic sources of carbon, nitrogen, and phosphorus. Within two months, the far basin was covered with an algal bloom, as can be seen in this photo. The near basin, which was treated with only carbon and nitrogen, remained unchanged. In this case, phosphorus is the key limiting nutrient, and its addition stimulates explosive growth of algal populations.

Poisons in Food Chains

An immense variety of toxic chemicals, including thousands of synthetics previously unknown in nature, have been dumped into ecosystems. Many of these poisons cannot be degraded by microorganisms and consequently persist in the environment for years or even decades. In other cases, the chemicals released into the environment may be relatively harmless, but they are converted to more toxic products by reaction with other substances or by the metabolism of microorganisms. For instance, mercury, a by-product of plastic production, was once expelled into rivers and the sea in a form that is insoluble. Bacteria in the bottom mud converted the waste to methyl mercury, an extremely toxic soluble compound that accumulates in the tissues of organisms.

One of the most serious environmental threats is radioactive fallout from nuclear accidents. The 1986 power plant disaster in Chernobyl, U.S.S.R., for example, belched great clouds of radioactive materials into the air, contaminating crops downwind.

Organisms acquire toxic substances from the environment along with nutrients or water. Some of the poisons are excreted, but others are accumulated in the tissues of the organism. Retained substances become more concentrated with each link in a food chain, a process called **biological magnification.** Magnification results from the biomass at each trophic level being produced from a much larger biomass ingested from the level below. Thus, top-level carnivores tend to be the organisms most severely affected by toxic compounds that have been released into the environment.

A well-known example of biological magnification involves the pesticide DDT, now banned in the United States. Because DDT persists in the environment and may be moved by water far from where the pesticide was sprayed to kill insects such as mosquitos and farm pests, it became a global problem. Traces of DDT have been found in nearly every organism tested, including penguins in the Antarctic. Because the compound is soluble in lipids, it collects in the fatty tissues of animals, each trophic level magnifying the concentration (Figure 49.14). The accumulation of DDT (and a product of its partial breakdown, called DDE) in the tissues of birds interferes with deposition of calcium in eggshells, reducing the number of offspring that hatch. The dwindling populations of several species of birds of prey were among the first signs that warned ecologists of a serious environmental problem.

Intrusions in the Carbon Cycle

Since the Industrial Revolution, the concentration of CO_2 in the atmosphere has been increasing as a result of combustion of fossil fuels. The burning of enormous amounts of wood removed by deforestation has interjected additional CO_2 into the carbon cycle (Figure 49.15). Various methods have estimated carbon dioxide concentration in the atmosphere before 1850

Figure 49.14
Biological magnification of DDT in a food chain. DDT concentration in a Long Island Sound food chain was magnified by a factor of about 10 million, from just 0.000003 parts per million (ppm) as a pollutant in the seawater to a concentration of 25 ppm in a fish-eating bird, the osprey.

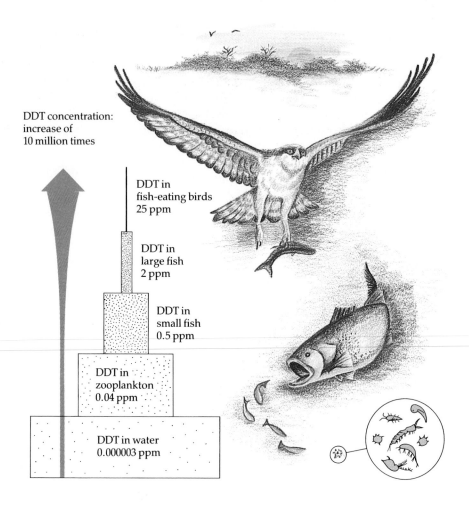

DDT concentration: increase of 10 million times

DDT in fish-eating birds 25 ppm

DDT in large fish 2 ppm

DDT in small fish 0.5 ppm

DDT in zooplankton 0.04 ppm

DDT in water 0.000003 ppm

as about 274 ppm. When a monitoring station on Hawaii's Mauna Loa began making very accurate measurements in 1958, the CO_2 concentration was 316 ppm (Figure 49.16). Today, the concentration of CO_2 in the atmosphere exceeds 340 ppm, an increase of more than 7% in just the past 30 years.

One effect we might expect increasing CO_2 levels to have on ecosystems is increased productivity by vegetation. In fact, when CO_2 concentrations are raised in experimental chambers, such as greenhouses, most plants respond with increased growth. Because C_3 plants are more limited than C_4 plants by CO_2 availability, one consequence of increasing CO_2 concentrations on a global scale may be the spread of C_3 species into terrestrial habitats previously favoring C_4 plants (see Chapter 10). This may have important agricul-

Figure 49.15
Interjecting carbon dioxide into the carbon cycle. One source of carbon from human activity is the burning of wood removed by deforestation, here taking place in Guatemala.

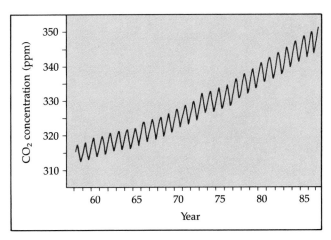

Figure 49.16
Increase in atmospheric CO₂ since 1958. Superimposed on normal seasonal pulses in CO₂ level is a steady increase in the total amount of CO₂ in the atmosphere. These measurements are being taken at a relatively remote site in Hawaii, where the air is "clean" and free from the variable short-term effects that occur near large urban areas.

tural implications. For example, corn, a C_4 plant and the most important grain crop in the United States, may be replaced on farms by wheat and soybeans, C_3 crops that will outproduce corn in a CO_2-enriched environment. No one can predict the gradual and complex effects that rising CO_2 levels will have on species composition in natural communities. One complicating factor is that temperature increases along with the CO_2 concentration, and this is the basis for the greatest concern about CO_2 levels.

Carbon dioxide in the atmosphere is an important factor in warming the Earth by what is known as the **greenhouse effect**. This gas (and also water vapor) is transparent to visible light, but absorbs infrared radiation and slows its escape from the irradiated Earth. If it were not for this effect, the average air temperature at the Earth's surface would be only $-18°C$.

Scientists continue to debate how increasing levels of CO_2 in the atmosphere will affect the temperature of the Earth. Several mathematical models discussed at a recent international meeting on this global issue all showed that a doubling of CO_2 concentration, which could occur by the end of the next century, would mean an average temperature increase of $3°-4°C$. (An increase of only $1.3°C$ would make the world warmer than at any time in the past 100,000 years.) The warming would be greatest near the poles, and the resultant melting of polar ice would raise sea levels by an estimated 100 m, flooding coastal areas 100 miles or more inland in some places. New York, Miami, Los Angeles and many other areas occupied by millions of humans will be underwater if this happens. Additionally, a warming trend will probably alter the geographical

distribution of precipitation. For instance, the grain belts of the central United States and the Soviet Union may become much drier. However, the various mathematical models disagree about how each region will be affected.

Coal, natural gas, gasoline, wood, and other organic fuels central to modern life cannot be burned without releasing CO_2. The warming of the Earth now under way as a result of CO_2 addition to the atmosphere is a problem of uncertain consequences and no simple solutions.

BIOTIC EFFECTS AT THE BIOSPHERE LEVEL

Some scientists postulate that life plays a major role in *regulating* the overall environment of the Earth and maintaining it within certain limits suitable to life. This idea is known as the Gaia hypothesis, named for a mythical goddess of Earth.

According to the Gaia hypothesis, first enunciated by the British scientist James Lovelock, life actually shapes the climate and atmosphere of Earth. For example, transpiration by tropical forests produces clouds that affect global weather patterns. Chemical cycling in ecosystems has generated an atmosphere with a far higher concentration of O_2 than would be found on any lifeless planet, and that concentration has been maintained within a range of 15%–25% of the atmosphere for the past 200 million years. Besides adjustments in the amount of photosynthesis and respiration occurring in the biosphere, another factor in controlling O_2 concentration is the release of methane (CH_4) by certain bacteria and by huge numbers of termites. In fact, termites probably account for about half of the methane in the atmosphere. Methane reacts with O_2 to form CO_2 and H_2O, and thus methane-producing organisms may help to control O_2 concentrations in the atmosphere. It is also possible that organisms moderate fluctuations in atmospheric CO_2 concentrations to some extent (Figure 49.17).

To many ecologists, the Gaia hypothesis, in its original form, erroneously viewed the Earth as a sort of "superorganism" with a self-regulated metabolism that helps counter fluctuations in the physical environment. This concept seemed to imply coordinated evolution of life on Earth, a phenomenon that cannot be accounted for by what we know about mechanisms of evolution. Recently, advocates of the Gaia hypothesis have been restating their ideas more in terms of regulation being a natural consequence of checks and balances inherent in the dynamics of ecosystems. If this means only that organisms have significant effects on the environment, with some feedback processes also occurring, then more scientists will probably agree.

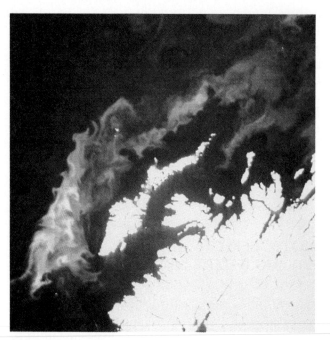

(b)

10 μm

(a)

Figure 49.17
Plankton as a CO₂ sink. (a) In this satellite photo (color enhanced) of the eastern Atlantic, a massive population of unicellular algae belonging to the phylum Haptophyta (yellow) drifts in the water off Scotland (white). (b) At one stage of the life cycle of these algae, plates of calcium carbonate cover the cell, which is then known as a coccolithophorid. This scanning electron micrograph shows one species. During great blooms of these algae, much carbonate is incorporated into the shells, which settle to the sea floor in vast numbers. (For example, a July 1988 bloom of the algae in the Gulf of Maine deposited 2 mm of sediment.) This may shift the complex equilibrium involving CO₂ and carbonate, causing more CO₂ to enter the sea from the atmosphere.

Figure 49.18
A remote ecosystem wounded by human culture. On March 24, 1989, a supertanker ran aground and spilled more than ten million gallons of crude oil into the pristine waters of Prince William Sound in Alaska. This ecological disaster was a shocking reminder that our technological tentacles reach far; we may be burning the gas in Los Angeles, Chicago, New York and other big cities, but the impact of our demand for oil is felt thousands of miles away.

In this more moderate view, Gaia is merely a metaphor for the interconnectedness of nature. It is clear from the study of ecosystems that processes occurring in one location can affect life far away. For this reason, it is impossible to predict how our own behavior, as the only technological species, will ripple throughout the biosphere. We do know that few if any ecosystems remain pristine (Figure 49.18). And we know that the rate of extinction of species is presently 50 times greater than at any time in the past 100,000 years, mostly as a result of our encroachment on habitat. We are treading rather clumsily on uncharted ecological territory. Perhaps our guide should be the words of the great American naturalist Aldo Leopold: "If the biota, in the course of aeons, has built something we like but do not understand, then who but a fool would discard seemingly useless parts? To keep every cog and wheel is the first precaution of intelligent tinkering."

STUDY OUTLINE

1. An ecosystem is the most inclusive level in the hierarchy of biological organization, consisting of the community of organisms in a defined area and the abiotic factors making up the physical environment.
2. Energy flow and chemical cycling are two interrelated processes that occur by transfer of substances through the feeding levels of ecosystems.

Trophic Levels and Food Webs (pp. 1115–1118)

1. The specific route of energy flow and the pattern of chemical cycling is dictated by trophic structure.
2. Trophic levels begin with primary producers, autotrophic organisms that support all other organisms of the community.
3. Heterotrophic organisms are consumers, beginning with herbivores, primary consumers that eat the autotrophs, and continuing with species that are secondary and tertiary consumers, carnivores that eat herbivores and carnivores that eat other carnivores, respectively. Detritivores live on organic wastes and dead organisms from all trophic levels.
4. The main primary producers in most ecosystems are photosynthetic autotrophs, although a few marine communities are powered by geothermal energy harnessed by chemosynthetic bacteria.
5. The sequential transfer of food from producers to the various levels of consumers is sometimes described as a food chain. However, natural feeding relationships usually are more like webs, since consumers feed at more than one trophic level or on a variety of items at the same level.

Energy Flow (pp. 1118–1123)

1. Primary productivity is the rate at which solar energy is converted to the chemical energy of organic compounds by autotrophs. It sets the spending limit for the energy budget of ecosystems.
2. Only net primary productivity, defined as gross primary productivity minus the energy used by the primary producers for respiration, is available to consumers.
3. Primary productivity is limited or affected by a variety of factors that depend on the specific ecosystem, as well as on temporal changes within ecosystems, such as a change in season.
4. The various levels of heterotrophs convert net primary productivity into new forms of biomass. Such secondary productivity declines with each trophic level, owing to the second law of thermodynamics, the inability of the consumers to eat or digest all of the available food, and the necessity for each level of organisms to use some of the energy for their own cellular respiration. Generally, only about 10% of the chemical energy available at one trophic level appears at the next.

Chemical Cycling (pp. 1123–1128)

1. Nutrient cycles involve both biotic and abiotic components of ecosystems and are termed biogeochemical cycles.
2. Biogeochemical cycles can be global or localized, depending on the mobility of the element in the environment, which in turn depends on its chemical nature and mechanism of routing through the trophic structure of the ecosystem.
3. The carbon cycle primarily reflects the reciprocal processes of photosynthesis and cellular respiration. Producers convert inorganic CO_2 into organic molecules that are eventually degraded by consumers into CO_2, which is reused by the autotrophs. In aquatic environments, CO_2 is also involved in a complex equilibrium with water and $CaCO_3$.
4. In the nitrogen cycle, certain prokaryotes fix abundant nitrogen in the atmosphere into ammonia, which other bacteria convert into nitrites and nitrates. Plants absorb ammonia and nitrates and convert them into protein that can be passed into the food chain. Detritivores decompose nitrogenous waste into ammonia, which can be reused by plants or reconverted into nitrites or nitrates; nitrogen in the soil is returned to the atmosphere in the form of free nitrogen by denitrifying bacteria.
5. The phosphorus cycle does not involve any atmospheric forms. Over ecological time and on a localized scale, phosphorus in the form of phosphate is absorbed by plants, transferred to consumers, and returned to the soil by animal excretion and the action of decomposers. A sedimentary cycle operates over geological time and involves weathering of rock, eventual leaching of the phosphorus into the sea, and incorporation into oceanic sediments that are later exposed to the air and weathered by geological processes.
6. The proportion of a nutrient in a particular form and its cycling time in that form vary among ecosystems. In the tropics, for example, nutrients pass quickly through the decomposition stage back into living biomass.

Human Intrusions in Ecosystem Dynamics (pp. 1128–1133)

1. Local and regional human impact on trophic structure, energy flow, and chemical cycling can have widespread or global ecological consequences.
2. Agricultural practices result in the constant removal of nutrients from ecosystems, so that large supplements are continually required. Distribution of the nutrients to humans and animals via crops often results in a "short-circuiting" to aquatic environments.
3. Controlled, long-term experiments at Hubbard Brook in New Hampshire showed that complete logging of a mature deciduous forest increased water runoff and caused huge losses of minerals, effectively destroying the balance that previously existed.
4. Dumping of nutrient-rich wastes into aquatic habitats accelerates eutrophication, and the increased biomass depletes O_2.
5. Release of toxic wastes has polluted the environment with harmful substances that often persist for long periods of time. Some of these substances become concentrated along the food chain by biological magnification, causing far-reaching deleterious effects.
6. Because of the burning of wood and fossil fuels, CO_2 in the atmosphere has been steadily increasing. The ultimate effects are uncertain but may include significant warming and other effects on climate.

Biotic Effects at the Biosphere Level (pp. 1133–1135)

1. The controversial Gaia hypothesis maintains that climate and the environment are regulated by the diverse actions of living organisms.
2. Whether or not the Gaia idea truly reflects the nature of the biosphere, the study of ecosystems has shown the complex interconnectedness of both the biotic and abiotic components.

SELF-QUIZ

1. Which of the following organisms is *mismatched* with its trophic level?
   a. cyanobacteria—primary producer
   b. grasshopper—primary consumer
   c. zooplankton—secondary consumer
   d. eagle—tertiary consumer
   e. fungi—detritivore

2. One of the lessons from an energy pyramid is that
   a. only one-half of the energy in one trophic level is passed on to the next level
   b. most of the energy from one trophic level is incorporated into the biomass of the next level
   c. the energy lost as heat or in cellular respiration is 10% of the available energy of each trophic level
   d. ecological efficiency is highest for primary consumers
   e. eating grain-fed beef is an inefficient means of obtaining the energy trapped by photosynthesis

3. The role of detritivores in the nitrogen cycle is to
   a. fix N_2 into ammonia

b. release ammonia from organic compounds, thus returning it to the soil
   c. denitrify ammonia to return N_2 to the atmosphere
   d. convert ammonia to nitrate, which then can be absorbed by plants
   e. incorporate nitrogen into amino acids and organic compounds

4. The Hubbard Brook Forest Study demonstrated all of the following *except*
   a. most minerals were recycled within a forest ecosystem
   b. mineral influx and loss within a natural watershed were nearly balanced
   c. deforestation resulted in an increase in water runoff
   d. the nitrate concentration in waters draining the deforested area became dangerously high
   e. regrowth of seedlings quickly restored normal nutrient cycling following deforestation

5. The increase in CO_2 concentration in the atmosphere is mainly a result of an increase in
   a. primary productivity
   b. methane production by termites and some bacteria
   c. the absorption of infrared radiation escaping from Earth
   d. the burning of fossil fuels and wood
   e. cellular respiration by the exploding human population

6. Which of the following is a result of biological magnification?
   a. Top predators may be most harmed by toxic environmental chemicals.
   b. DDT has spread throughout every ecosystem and is found in almost every organism.
   c. The greenhouse effect will be most significant at the poles.
   d. The biosphere is able to regulate the environment and adjust to disruptions.
   e. Many nutrients are being removed from agricultural lands and shunted into aquatic ecosystems.

7. Which of these ecosystems has the lowest primary productivity per square meter?
   a. salt marsh
   b. open ocean
   c. tidepool
   d. grassland
   e. tropical forest

8. Quantities of mineral nutrients in soils of tropical rain forests are relatively low because
   a. the standing crop is small
   b. microorganisms that recycle chemicals are not very abundant in tropical soils
   c. decomposition of organic refuse and reassimilation of chemicals by primary productivity occur rapidly
   d. nutrient cycles occur at a relatively slow rate in tropical soils
   e. the high temperatures destroy the nutrients

9. In the nitrogen cycle, the bacteria that restock the atmosphere with N_2 are
   a. nitrogen-fixing bacteria
   b. denitrifying bacteria
   c. nitrifying bacteria
   d. ammonifying bacteria
   e. *Rhizobium* bacteria of root nodules

10. Which of the following statements concerning the hydrologic cycle is *incorrect*?
    a. There is a net movement of water vapor from oceans to terrestrial environments.
    b. Precipitation exceeds evaporation on land.
    c. Most of the water that evaporates from oceans is returned by runoff from land.
    d. Transpiration makes a significant contribution to evaporative loss of water from terrestrial ecosystems.
    e. Evaporation exceeds precipitation over the seas.

CHALLENGE QUESTIONS

1. Do you agree or disagree with the Gaia hypothesis? If you decide in favor of the hypothesis, does this imply that regulatory mechanisms in the biosphere can eventually compensate for all of our intrusions in ecosystem dynamics? Justify your opinion.

2. Herbivores have generally been considered to have a negative impact on their plant prey. However, a controlled study of a natural community in which the crustacean herbivore *Daphnia pulex* feeds on algae showed that the *Daphnia* also had a stimulatory effect on the algal populations that approximately balanced *Daphnia's* impact on algal mortality. From your understanding of the nitrogen cycle, what do you think was the mechanism for the *Daphnia*-induced stimulation of algal growth?

3. Discuss some of the possible short-term and long-term consequences of the Chernobyl accident.

FURTHER READING

1. Aber, J. D., K. J. Nadelhoffer, P. Steudler, and J. M. Melillo. "Nitrogen Saturation in Northern Forest Ecosystems." *Bioscience*, June 1989. Nitrogenous exhaust from fossil fuel consumption may stress the biosphere.
2. Cohn, J. P. "Gauging the Biological Impacts of the Greenhouse Effect." *Bioscience*, March 1989. Projections for the fates of ecosystems as Earth warms.
3. Houghton, R. A., and G. M. Woodwell. "Global Climatic Change." *Scientific American*, April 1989. Evidence that the warming trend due to the greenhouse effect is already well under way.
4. Odum, H. T. *Systems Ecology: And Introduction.* New York: Wiley, 1983. A college-level text that emphasizes important aspects of ecosystems.
5. Turner, M. H. "Building an Ecosystem from Scratch." *Bioscience*, March 1989. Preview of the "Biosphere II" project.
6. "Toward a Holistic Management Perspective." *BioScience*, July/August, 1985. A special issue on doing something about stressed ecosystems.

50

Behavior

uppose you were peeling apples in the kitchen and one rolled off the counter and across the floor. What would you do? No doubt you would walk over, pick it up, and carry it back to the counter. If, while you were carrying it back, you inadvertently dropped it, you would certainly stop, pick it up again, and then resume your walk back to the counter. Now consider the graylag goose, a common European species whose nests consist of little more than shallow depressions in the ground. Occasionally, the goose accidentally bumps one of the eggs out of the nest, but then retrieves it, always in the same manner, with a side-to-side head motion (Figure 50.1). Why does the goose retrieve the egg? At first this may seem like a trivial question with an obvious answer. By extension of our own experience and motivations as humans, most people would probably simply assume that the goose wants the egg back because it is somehow valuable to the animal, in the same way that food or offspring are recognized as valuable and important by humans. Biologists might answer in more formal evolutionary terms: There has been selection for a tendency to retrieve eggs because individuals that did not retrieve eggs left fewer offspring.

The matter, however, is not so simple or obvious as it seems. If the egg slips away (or is experimentally pulled away) from the goose as it is being retrieved, the animal stops her side-to-side head motion but continues the normal retrieval movement with her head extended until she sits down, as though she were still pulling the egg. Only after she sits down does she "notice" that an egg is still outside the nest. Then she begins another retrieval. If the egg is again removed, the goose still goes through the retrieval motion as

Figure 50.1
Graylag goose retrieving an egg. The goose stands up, extends her neck, and retrieves the egg by pulling it slowly with the underside of her bill as she settles back into the nest. Because the egg is lopsided and tends to roll away, the goose uses a side-to-side motion of her head to keep the egg on course as it moves.

dent on nerve impulses, hormones, and other physiological mechanisms we have discussed in detail in Unit Seven. In this chapter, however, we focus not on the physiological basis of behavior but on **ethology** (from the Greek *ethos*, "custom," and *logos*, "speech"), the study of how animals behave in their natural environments.

Behavior is the most immediate means by which an animal interacts with its environment. Thus, the study of animal behavior is essential to a broad understanding of ecology. A lizard controlling its body temperature by moving between sun and shade is an example of the importance of behavior on the level of physiological ecology. Mating behavior, territorial defense, and social cooperation are types of behavior that affect populations (Figure 50.2). And foraging is one example of how animal behavior can influence community structure. This chapter discusses animal behavior in its many ecological contexts.

STUDYING BEHAVIOR

Humans, with their close relationship to other animals, have been describing animal behavior throughout recorded history. The literature documenting *what* animals do is vast and ever growing (see Methods Box, p. 1141). An understanding of *why* and *how* animals behave as they do has begun to develop only much more recently. That understanding is far from complete, partly because the mechanisms that account for behavior are so complex and variable. For decades, scientists have argued over the mechanisms behind animal behavior, and, as we shall see, the debates continue today.

The Problem of Anthropomorphism

The graylag goose probably does not have the ability to recognize that an unusual object near her nest is not an egg; in fact, she probably has no sense of "egg" at all. She is somehow programmed to retrieve objects near her nest, and even if she has some vague awareness that she is attracted to the object, it seems unlikely she has any sense of why. We are given here an important starting lesson in the example of the graylag goose: The ways in which other animals "think" and the cues that guide their behavior may be different from those of our own experience. It is important that any student of animal behavior recognize the pitfall of **anthropomorphism,** that is, the tendency to ascribe human feelings, reasoning, and motivation to other animals. This is not to say that what happens in us and in other animals is completely dissimilar; near the end of this chapter, we will examine some current

before. If an unusual object, such as a toy dog, is placed near the nest, the goose retrieves that also. Such an object is usually discarded, but only after the goose attempts to sit on it.

With this example, to which we will return, we are entering the complex and fascinating field of animal behavior. Feeding, courtship, mating, nest building, communication—essentially all the activities of animals fit under the rubric of **behavior,** which we define broadly as externally observable muscular activity, triggered by some stimulus. Any behavior is depen-

(a)

(b)

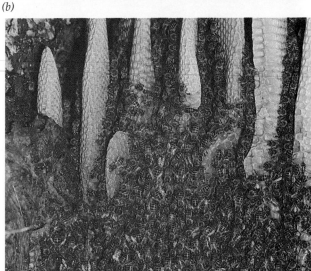

(c)

Figure 50.2
Individual behavior affects animal populations. (a)
Courtship among Arctic hares is accompanied by highly
ritualized "boxing matches," apparently among both
same-sex and opposite-sex individuals. **(b)** Territorial
contests between male elephant seals determine access
to breeding females, which surround the males. **(c)**
Thousands of worker bees cooperate in feeding young
and storing food, thereby contributing to survival of the
colony as a whole.

ideas and changing views on the subject of animal
mentality. But as yet we have no way of knowing for
sure what goes on in the mind of any other species
regarding matters we normally associate with conscious feelings or decision making—or even whether
the term *mind* has any meaning in relation to other
species.

In practice, it is very difficult for anyone to avoid
absolutely the appearance of anthropomorphism when
describing animal behavior. Familiar words such as
heard, noticed, recognized, or *painful* imply some kind of
conscious awareness. On the other hand, it is cumbersome to try to carefully couch all descriptions of
behavior in strictly mechanistic terms. This text uses
common terms like those just mentioned, but with
the understanding that they imply what may be only
analogies to human experience.

Proximate and Ultimate Cause

Asking why an animal does something may involve
more than one kind of answer, depending on the level
of causation we are considering. Behavior, like any
other feature of an organism, is subject to natural
selection. Biologists, as we have seen throughout this
text, view "why" questions from an evolutionary perspective. The evolutionary causes of behavior are
referred to as **ultimate causes. Proximate causes** explain
behavior in terms of more immediate interactions with
the environment. Often, the proximate cause is some
kind of cue that is not itself of prime importance to
the organism, but is correlated with something that *is*
important.

As an example of the two levels of causation, consider the question, "Why do bluegill sunfish repro-

Ethologists, like ecologists, are often confronted with great difficulties in making observations that reflect the natural situation and that can be rigorously quantified. For example, a researcher might be interested in how the overall amount of an animal's activity, as well as related considerations such as distance traveled or variation in the type of activity, changes with the time of day. For some species, of course, it is possible to obtain good data with equipment no more elaborate than binoculars, stopwatch, and notepad—combined with a great deal of patience and insight on the part of the observer. The extraordinary work of Jane Goodall, described at the beginning of this unit, attests to the value of this approach. However, there are a great many species for which the close physical presence of a human would be disruptive to the animal's natural behavior, and many others for which sustained direct observation is simply impossible. Turtles in the sea, eagles that forage over large distances, or a grizzly bear roaming about a rugged wilderness area cannot be monitored by any simple, direct means.

One of the most common methods for tracking the movements of wide-ranging animals is telemetry. A captured animal is fitted with a radio transmitter and released. A directional receiver that picks up the radio signals emitted by the transmitter can then be used to monitor the movements of the animal. The scientists in the top right photograph are following the activity of elk on Mount. St. Helens in Washington.

Computers have greatly simplified the collection of data related to animal activity. As an example, imagine the difficulty in trying to monitor the overall activity of a group of fish in a tank for an extended period, when at any given instant some individuals may be active and others inactive. The researcher in the bottom right photograph is testing a setup in which a television camera, connected to a computer, is trained on such a group of fish. The camera signal is converted to a high-contrast image in which the lighter fish stand out sharply on the computer monitor against the darker backgound. About 30 times

each second, the computer reads the image, keeping track of how much area has changed from dark to light or vice versa (an index correlated with the amount of activity). The computer stores these data, which can be analyzed in various ways. In an actual run, the fish shown here would be separated from the equipment and researcher by a partition. With modification, the same system can be set up to monitor activity ranging from that of microorganisms to that of large grazing animals in a field.

duce in the spring and early summer, instead of in the fall?" On the proximate level, we can say that increased light from the longer days acts on the pineal gland (see Chapter 41), triggering neural and hormonal effects that induce nest building and other reproductive behavior. But amounts of light *per se* are not of much

importance to the fish. The ultimate or evolutionary reason that this behavior has been selected is that conditions in the spring are the most favorable for reproduction: Warmer water allows faster growth of young, and food is more abundant. Fish that breed at other times would be at a selective disadvantage.

Figure 50.3

The genetic basis of behavior. Several species of brightly colored African parrots, commonly known as lovebirds, build cup-shaped nests inside tree cavities. Females typically make nests with thin strips of vegetation (or, in the laboratory, paper) that they cut with their beaks. (**a**) In one species, Fischer's lovebird (*Agapornis fischeri*), the bird cuts relatively long strips and carries them back to the nest one at a time in her beak. (**b**) The peach-faced lovebird (*A. roseicollis*), in contrast, cuts shorter strips and usually carries several at a time by tucking them into the feathers of the lower back. Tucking is a fairly complex behavior, since the strips must be held just right and pushed in firmly, with the feathers then smoothed over them.

These two species are closely related and have been experimentally interbred. The resultant hybrid females exhibited an intermediate kind of nest-building behavior. (**c**) The strips cut by such birds were intermediate in length; even more interesting was the birds' hybrid manner of handling the strips. They usually made some attempt to tuck them into the rear feathers, but in some cases they did not let go after turning and pushing them a short distance. In other cases, the strips were manipulated or inserted improperly or simply dropped. The result was almost total failure to transport strips by this method. Eventually, the birds learned to transport the strips in their beaks. Even so, they always made at least token tucking attempts. (**d**) After several years, the birds still turned their heads to the rear before flying off with a strip.

These observations demonstrate that the phenotypic differences in the behavior of the two species are based on different genotypes. We also see another important point: Behavior can be modified by experience, but some behaviors or components of behavior are firmly programmed into an animal and cannot be changed by experience or understood by the organism as inappropriate in certain circumstances.

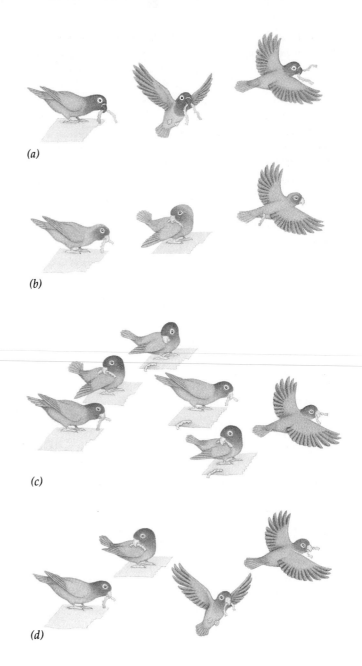

(a)

(b)

(c)

(d)

Nature Versus Nurture

Evolutionary explanations imply that patterns of behavior are to some extent inherited. The link between behavior and genetic makeup, however, is not simple. In a few cases, there is an apparent correspondence between a specific behavior and a single gene, in much the same way that a relatively simple trait such as eye color might correspond. In the fruit fly *Drosophila*, for example, a single gene causes affected individuals to exhibit an activity cycle based on 19 hours instead of the usual 24. But generally, behavior, particularly in vertebrates, is the result of a complex interaction of internal physiology and external stimuli. Reproductive behavior in an individual animal, for instance, typically depends on cues from the physical environment, the presence of certain hormones, and stimuli provided by the presence of another individual. It is unlikely that such behavior could be connected in any simple way to a single, or even a few, genes.

Unlike many phenotypic traits that have a relatively simple genetic basis, behavior can often be modified by experience; in other words, it may be phenotypically variable within the same individual. Still, even complex behavior, or at least the *capacity* for that behavior, must ultimately have a basis in an animal's genes. Figure 50.3 is an intriguing example of the role of genotype in behavior.

Debate about the degree to which genes program behavior is the contemporary version of a very old controversy over the relative importance of nature (genes) versus nurture (environment). We commonly speak of certain kinds of animal behavior, even our own, as "instinctive." At the same time, it is quite obvious that many behaviors, in many kinds of animals, are learned; that is, they develop or are modified as the result of experience. The question is the extent to which a particular behavior is innate or learned in an individual and, more broadly, the relative importance of each among animals in general. At one extreme, some scientists have argued that animals are essentially *tabulae rasae* (Latin for "blank tablets") whose behavior is determined and conditioned almost entirely by experience. In this view, genes may provide the capacity to behave in certain ways, but it is experience (or some kind of outside influence) that actually determines the behavior within the range of possibilities. At the other extreme is a view that strongly emphasizes the innate, genetic basis of behavior. Many behaviors probably fall somewhere on the continuum between these extremes. Only rarely can we identify specific components that are uninfluenced by experience and others that are totally directed by it. Genes and experience both contribute to behavioral development, especially in vertebrates.

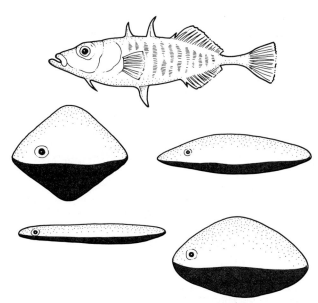

Figure 50.4
Sign stimuli used to demonstrate FAPs. Aggression in male three-spined stickleback fish is triggered by a simple cue. The realistic model at the top without a red underside produces no response. All the others produce strong responses because they have the required red underside.

INNATE COMPONENTS OF BEHAVIOR

Among the first to demonstrate clearly the importance of the innate components of behavior were Konrad Lorenz and Niko Tinbergen, European researchers whose studies of the graylag goose and other animals, first published in the 1930s, still form a major part of the foundation of ethology. The magnitude of their observations—many of them seemingly simple—was formally recognized in 1973 when, along with Karl von Frisch (who is discussed later), they received the Nobel Prize in physiology/medicine.

Fixed Action Patterns

The most fundamental concept associated with the "classical" ethology of Lorenz and Tinbergen is the **fixed action pattern (FAP)**, also called a fixed motor pattern. An FAP is a highly stereotyped behavior that is innate. Once an animal initiates an FAP, it usually carries the action pattern to completion even if, for some reason, other stimuli impinge on the animal or the activity becomes inappropriate. An FAP is triggered, or "released," by some kind of external sensory stimulus, known as a **sign stimulus.** In its simplest conceptual form, an FAP can be thought of as behavior that derives from an innate ability of an animal to detect a certain stimulus, combined with an innate "prewired" program that is turned on by the stimulus and directs some kind of motor activity. For the graylag

goose, the sight of an object near its nest is the stimulus that triggers the retrieval FAP. Some moths instantly fold their wings and drop to the ground in response to the ultrasonic signals sent out by predatory bats. Experiments strongly suggest that animals are not making any thoughtful or intelligent observations in the sense of integrating various kinds of information and then making a decision; rather, in the case of an FAP, they are behaving more like robots.

Any sign stimulus can be thought of as a signal that releases a specific behavior. However, the term **releaser** is usually restricted to sign stimuli that function as communication signals between individuals of the same species. A classic case of releasers in social behavior is seen in the male three-spined stickleback fish, which attacks other males who invade his territory. The releaser for the attack behavior is the red belly of the intruder. The stickleback will not attack an invading fish that lacks the red, but will readily attack distinctly nonfishlike models as long as some red is present (Figure 50.4). Tinbergen, who first reported these findings, was inspired to look into the matter by his casual observation that his fish responded aggressively when a red truck passed their tank.

Interactions of parents and young often involve FAPs. In many species of birds, when a parent returns to the nest with food, the newly hatched, blind young respond immediately with begging behavior, raising their heads, opening their mouths, and cheeping

Figure 50.5
Young larks gaping. The gape marks inside the birds' mouths are the releaser for feeding behavior, an FAP in adult birds.

loudly; the parent then stuffs the food into one of the gaping mouths. In thrushes, the response of the young is released specifically by the impact of the parent on the side of the nest—not by sound or any other potential stimulus associated with the return of the parent. (You can demonstrate this response yourself if you are near a nest by lightly tapping on the edge.) Later, when the birds' eyes are more developed, the sight of a parent releases gaping, but the specific stimuli required still are not complex; simple models of the parents work very well. Feeding by the parent also requires a releaser. Young birds typically have distinctive patterns inside the mouth that, combined with their gaping behavior, trigger the food-stuffing behavior by the parent (Figure 50.5).

FAPs occur in humans as well as other animals. Human infants grasp strongly with their hands in response to a tactile stimulus. An infant's smile is also an FAP; it is readily induced by simple stimuli such as a sound or a figure consisting of two dark spots on a white circle, a kind of rudimentary representation of a face.

The mechanistic component of behavior, as derived from FAPs, is demonstrated most clearly by some animals that seem to have an endless capacity to repeat stereotyped behavior. One example, first described by the nineteenth-century naturalist Jean-Henri Fabre, involves digger wasps. The female wasp builds a nest in the ground, places a paralyzed cricket in it, and then lays an egg that will eventually hatch into a larva that will consume the cricket. The wasp accomplishes all this with a highly stereotyped series of FAPs. The wasp places the cricket a short but rather precise distance (about 2.5 cm) from the nest. She enters the nest briefly, apparently for some kind of final inspection, and then returns to the outside to get the cricket and carry it down into the nest. If an observer moves the cricket a short distance while the wasp is inside, she searches for it after emerging and, after retrieving it, puts it back in its original spot near the entrance. The wasp then enters the nest for another inspection, even though she has just made one, and reemerges a short time later. If the cricket has been moved again, the wasp repeats the cycle—and will continue to repeat it at least 40 times, showing no signs of either tiring of the repetition or of circumventing the observer's game by changing behavior. The wasp can never get past the "inspect the nest" subroutine, which can be terminated only by immediately finding the cricket near the nest.

The Nature of Sign Stimuli

Ethologists have found from careful experiments that sign stimuli tend to be based on one or two simple characteristics that are associated with the relevant object or organism. In many cases, the stimulus is no doubt the most obvious or the only characteristic associated with a particular situation; for example, the ultrasonic signals of bats are the obvious cue that would trigger avoidance behavior in moths. In other cases, it appears that animals have settled on certain characteristics out of a potentially larger array of possible choices. When an adult herring gull arrives with food, it stands near its chick, bending its head downward and moving its beak, which has a red spot. The chick pecks the beak, in turn releasing actual feeding by the adult. The chick seemingly might be cued to peck its parent's bill by a variety of possible stimuli, including such obvious things as a lump at the end of a rectangle (simulating food at the end of a bill). Careful studies have demonstrated that the releaser specifically involves a red spot swung horizontally at the end of a relatively long, vertically oriented object. We would expect natural selection to have favored cues that have a high probability of association with the relevant object or activity, but in some cases there is probably a certain amount of randomness in the specific cues that have been settled on.

Close correlations exist between an animal's sensitivity to certain general stimuli and the specific sign stimuli that are important to it. For example, frogs have retinal cells that are especially good at detecting movement, and it is movement of an object that releases the frog's "tongue-shooting" behavior during feeding. A frog will starve to death if surrounded by dead or motionless flies, but will attack one readily if it moves.

Often, exaggeration of the relevant stimulus produces a stronger reponse. Wider gaping by young birds more readily releases the feeding behavior of the adult. Since a young bird gapes more widely in response to increased hunger, the hungriest young are most likely to be fed. This aspect of sign stimuli can be seen in simple experiments in which animals are presented with a **supernormal stimulus.** A graylag goose will attempt to roll a volleyball into her nest instead of her own egg, and an oystercatcher (a bird of seashore areas) will attempt to incubate a giant model of her egg instead of the real thing.

Some ethologists have suggested that the use of simple cues to release preprogrammed behavior prevents an animal from wasting time processing or integrating a wide variety of input. Perhaps a better way of interpreting the situation is in terms of the limitations of innate programming. The genes of a frog might feasibly encode for certain cells that merely detect movement, but it may be impossibly complex to encode specifically for cells or other components that have more detailed recognition capabilities. In any case, as we have noted earlier, natural selection does not necessarily result in the "best" imaginable system, but rather one that is "good enough" or "better than." Simple cues usually work quite well in an animal's normal sensory world.

We should note again that terms or expressions we use in our description of behavior such as *preprogrammed* or *subroutine*, are only general descriptions of what seems to happen. These are familiar terms in the current computer age, and the model they suggest is consistent with what we see, but we are not implying that the mechanisms of behavior are the same as what these terms specifically suggest.

Innate Releasing Mechanisms and Drive

So far, we have said little about how releasers and other sign stimuli actually work—that is, about the *link* between the stimulus and the resultant behavior. Early ethologists spoke of an **innate releasing mechanism (IRM)** that responded to a stimulus and initiated the associated behavior. They envisioned an IRM as a more or less distinct entity, perhaps a group or network of neurons or other cells. The idea has always been controversial because, for the most part, neurophysiologists have not been able to demonstrate any such centers designed for specific behaviors. Recently, however, some evidence has developed indicating that something like an IRM may exist in at least one invertebrate. The nudibranch *Tritonia* has a fixed swimming response that is released by the presence of a predator. Within the animal is a particular giant neuron that when artificially stimulated triggers the same escape behavior. For the most part, however—particularly in vertebrates—the concepts of releasers and IRMs are

vague models that are consistent with *what* happens but that tell us nothing of *how* it happens.

The concept of **drive,** or motivation, is also commonly used to describe or explain many kinds of behavior. The idea is that the stronger the drive, the more easily the associated behavior can be released. Like the concept of IRM, these terms, while more or less consistent with what happens, are not particularly enlightening as to how. Obviously, animals do not exhibit all possible behaviors at all times. Feeding may occur at only certain times of the day, or reproduction at only certain times of the year. Thus we may say that an animal has a variable drive to feed or mate. In some cases, a stronger tendency toward some behavior may be correlated with some physiological change. In chickens that have eaten recently (and so are not particularly hungry), stereotyped pecking is released by the sight of real food, but not by anything else. As the chickens become hungry, however, black pencil marks are sufficient to release the behavior, and eventually the animals will peck at a barren, white substrate. Again, this change in apparent motivation may result from internal physiological changes, in this case caused by a lack of food.

With other behaviors, it is not clear that the concept of drive is at all useful. Most animals flee from a predator, but they do not obviously have a "drive" to do so—these behaviors seem more like ad hoc responses to external stimuli.

LEARNING AND BEHAVIOR

So far, our discussion has concentrated on the innate components of behavior. Often, **learning**—the modification of behavior by experience—affects even strongly fixed actions. **Habituation,** a simple kind of learning involving loss of sensitivity to unimportant stimuli, prevents an animal from wasting time or energy. Examples are widespread: Hydra stop contracting if disturbed too often by water currents, gray squirrels stop responding to alarm calls in the absence of a visually apparent danger, and the dutiful graylag goose has a limit to the number of wayward eggs she will retrieve.

Some behaviors that are clearly innate may be performed more quickly or effectively as a result of practice, but this can be misleading. For example, pecking in young chickens is released by the sight of any small object on the ground. The chickens' aim is poor at first, but improves with time. However, this kind of change may not be the result of true learning, but rather of ongoing developmental changes that can more properly be called maturation. In some cases, this is clearly what is happening. We commonly speak of birds "learning" to fly, and you may have seen a hapless fledgling awkwardly fluttering about as though practicing. However, young birds have been experimentally reared in restrictive devices so that they could never flap their wings until an age when their normal kin were already flying. Such birds flew immediately and normally when released. Thus, the improvement must result from neuromuscular maturation, not from learning. But with the young chickens' pecking, missing the target might possibly modify the next attempt; these changes may still fit a broad definition of learning as refinement by experience. The distinction is often difficult to make.

Imprinting

Some of the most interesting cases where learning clearly interacts closely with innate behavior involve the phenomenon known as **imprinting.** You have probably seen young ducks or geese following their mother. This behavior makes good sense to us, since the adult bird is likely to have more experience than the young in finding food, avoiding predators, and otherwise getting along in the world. But how do the young know whom—or what—to follow? In perhaps his most famous study, Konrad Lorenz divided a clutch of graylag goose eggs, leaving some with the mother and putting the rest in an incubator. The young reared by the mother showed normal behavior, following her about as goslings and eventually growing up to mate and interact with other geese. When the artificially incubated eggs hatched, the geese spent their first few hours with the researcher, not with their real mother. From that day on, they steadfastly followed Lorenz and showed no recognition of their own mother or other adults of their own species (Figure 50.6). This early imprinting lingered into adulthood: The birds continued to prefer the company of Lorenz and other humans over that of their own species and sometimes even tried to mate with humans. In this species, the most important imprinting stimulus was movement of an object away from the young, although the effect was greater if the object emitted some sound. The sound did not have to be that of a goose, however; Lorenz found that a box with a ticking clock in it was readily and permanently accepted as a "mother."

Apparently, these animals have no innate sense of "mother" or "I am a goose, you are a goose." Instead, they simply respond to and identify with the first object they encounter that has certain simple characteristics. What is innate is the ability or tendency to respond; the outside world must fill in something to which a response will be directed.

Many other examples of imprinting are now known. Salmon are noteworthy for their long migrations in the open ocean, where they feed, grow, and mature after being hatched in freshwater streams (see Chapter 45). Some species remain at sea for several years.

Figure 50.6
Imprinting. Konrad Lorenz was "mother" to these imprinted geese.

Nonetheless, as demonstrated by the tagging of thousands of young fishes, each returning individual goes back to its specific home stream to spawn. This involves not only finding the proper major stream to enter at the coast but also making a number of correct choices among smaller and smaller tributaries as a fish winds its way upstream. Extensive research has demonstrated conclusively that this ability is based on olfactory imprinting. Young fishes artificially reared and exposed to a chemical called morpholine can be "directed" into a stream on their return if morpholine is introduced into it. Under natural conditions, the fishes apparently imprint on the complex bouquet of odors unique to their specific stream, and can recognize those odors and swim toward their source even after being removed long distances from the stream for long periods.

Closely associated with imprinting is the concept of a **critical period,** a limited time during which imprinting can occur. Lorenz found, for example, that geese totally isolated from any objects during their first two days failed to imprint afterward on anything. Imprinting has commonly been thought of as involving very young animals and rather short critical periods. But it

Figure 50.7
Imprinting on a brood parasite. Here a reed warbler feeds an absurdly outsized young—which happens to be a European cuckoo. Cuckoos and other brood parasites lay their eggs in the nests of other species. Often, because of the tendency of the parents to imprint on young by use of simple cues, the intruder is readily accepted.

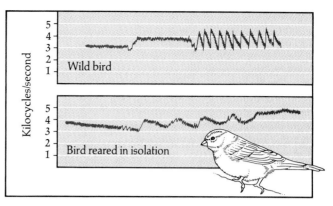

Figure 50.8
Songs of wild and hand-reared birds. These sonograms show the variation in sound patterns of the songs of male white-crowned sparrows. Birds reared by hand in isolation will not sing normally unless they are allowed to hear the usual song of an adult during the critical period of 10–50 days after hatching.

is now clear that imprinting-like behavior is not restricted to the young, and the critical period may be of various durations. For example, just as a young bird requires imprinting to "know" its parents, the adults must also imprint to recognize their young. For a day or two after their young hatch, adult herring gulls will accept, and even defend, a strange chick introduced to their nesting territory. After imprinting, which is probably based on a simple cue such as sound, the adults will kill and eat any strange chick. A more bizarre example involves some European cuckoos and other species that are "brood parasites." Such birds provide no parental care; instead, they simply lay their eggs in nests of other species. The other parents commonly imprint on the hatchling and feed it readily, even though it is usually absurdly larger than they are (Figure 50.7). It is interesting but perhaps predictable that parasitic young often have gape marks that are very similar to those of the host's young. Also, the parasite's mother sometimes pushes the other eggs or young out of the

nest, improving even more her offspring's chance of adoption.

In the case of Lorenz's geese, species identity and sexual orientation were imprinted early, at the same time as the resulting response. In many species, however, sexual imprinting occurs later, with a critical period lasting relatively long. For example, in one study involving two closely related species of finches, young males of one species were reared first with their own species, then with only members of the other species during their several-week-long critical period for sexual identity. Later, when exposed to females of their own species, they mated quite reluctantly. They readily mated with females of the other species, however, even when they had not seen any individuals for as long as eight years. Identification with the second species had been permanently imprinted.

In all these examples, we see a close interaction between learning and innate behavior. One further example is particularly instructive. As we will see in our discussion of aggressive behavior, the pleasant bird songs that brighten our backyards are often aggressive warnings by males to would-be competitors. Though somewhat variable among individuals of a population, songs are highly species-specific in their basic composition. How does a bird learn to sing a particular song? One study demonstrated that a male white-crowned sparrow reared in isolation will eventually sing, but quite abnormally (Figure 50.8). If it is allowed to hear the song of a normal male during its critical period of 10–50 days after hatching, it will sing normally. At first this may seem very similar to some of our other examples, where the animal has a very general innate tendency that is capable of being oriented in certain ways by any of a wide variety of external phenomena. However, if the young sparrow hears

Figure 50.9
Operant conditioning. Monarch butterflies contain poisonous, distasteful substances. Left: A native bluejay will feed readily on the butterfly. Right: But shortly afterward the bluejay regurgitates it. After just one or two such experiences, the bird will absolutely avoid this species. The bright orange color of the monarch probably helps make quick recognition easier. Some edible species, such as the viceroy butterfly, mimic the monarch and are also avoided by birds.

the song of some other species, even a closely related one, during its critical period, it does not learn that song. Even though learning is absolutely necessary for the bird, its innate programming to imprint on a song is somehow rather complexly and narrowly focused toward only one possibility. Genes determine what song a bird can sing, but it must nevertheless learn that song by hearing it from adults.

Classical Conditioning

Many animals can learn to associate one stimulus with another. One type of **associative learning,** called **classical conditioning,** is well known from the work of Ivan Pavlov, a Russian physiologist who worked around 1900. Pavlov sprayed powdered meat into dogs' mouths, causing them to salivate (primarily a physiological rather than a behavioral response). Just before the spraying, however, he exposed the dogs to a sound such as a ringing bell or clicking metronome. After a while, the dogs salivated readily in response to the sound alone, which they had learned to associate with the normal stimulus. Similar types of conditioning experiments have been carried out with various other animals, but this kind of association of a normally irrelevant stimulus with a fixed behavioral response is probably not particularly common or important in natural contexts. However, many kinds of perceptual sharpening probably do involve this same kind of conditioning. For example, the tendency of a young gull to respond to a red spot becomes more generalized to include the entire adult.

Operant Conditioning

A more likely kind of associative learning that directly affects behavior is **operant conditioning,** also called

trial-and-error learning. Here, an animal learns to associate one of its own behaviors with a reward or punishment; the animal then tends to repeat or avoid that behavior, depending on whether the reinforcement is positive or negative. The best known laboratory work involving operant conditioning is that of the American psychologist B. F. Skinner. A rat or other animal placed in a "Skinner box" has a choice of various levers it might push, some of which reward the animal by releasing food. The animal's choices may be random at first, but it quickly learns to choose those levers that provide food. Such learning is the basis for most of the animal training done by humans, in which the trainer typically induces a particular behavior at first by rewarding the animal.

Operant conditioning is undoubtedly very common in nature. For example, animals quickly learn to associate eating particular food items with good or bad tastes and modify their behavior accordingly (Figure 50.9). In some cases, animals may be able to skip some of their own trial and error and learn simply by watching the behavior of others. An interesting example is that of tits (chickadee-like birds) in England. Sometime in the early 1950s, one of these birds apparently learned to peck through the paper tops of milk bottles left on doorsteps and drink the cream on top. This was probably a case of operant conditioning, with the bird learning that its general pecking, probing behavior was rewarded if directed at the bottles. But the behavior quickly spread through the population and was "handed down" to succeeding generations that learned the behavior by watching adults.

Insight

Insight learning is the ability to perform a correct or appropriate behavior on the first attempt in a situation with which the animal has no prior direct experience

(some prefer to call this reasoning, rather than learning). If a chimpanzee is placed in an area with a banana hung too high above its head to be reached and several boxes on the floor, the chimp can "size up" the situation and then stack the boxes to allow it to reach the food. In general, insight is best developed in primates and other mammals, but even in these groups the amount of insight often varies substantially from one situation or species to another. Although the great majority of animals seem to have little or no ability to use insight (Figure 50.10), some recent work on bees suggests that an insightlike capability may be found even among invertebrates (as will be discussed in an upcoming section on animal cognition).

Very broadly, the capacity to learn can be thought of as another adaptation that enhances survival and reproductive success, and rests ultimately on some kind of genetic bases. However, the internal mechanisms of learning are very poorly known, and only recently has progress begun on linking some simple kinds of learning to internal biochemical or physiological changes. Although animals frequently appear to do complex things, most behavior can be understood as relatively fixed patterns, controlled by simple stimuli, that are often modified in their frequency and orientation by simple kinds of learning. Nonetheless, the behaviors of animals have been fine-tuned by natural selection, and the limited repertoire of abilities works very well in normal circumstances.

BEHAVIORAL RHYTHMS

What determines when a squirrel sleeps and when it gets up? As with most questions about behavior, the answer is neither as obvious nor as simple as it might seem. Animals exhibit all kinds of regularly repeated behaviors—feeding in the day and sleeping at night

(or vice versa), reproducing every spring, migrating both spring and fall, and so on. We have already discussed circadian (daily) rhythms in some detail in relation to plants (see Chapter 35). Animals too are subject to such rhythms and the various environmental cues that regulate them. But what would happen to rhythmic behavior if an animal were placed in an environment with no external cues about time? Or to put it another way: Is rhythmic behavior based only on *exogenous* (external) timers, or is there also an *endogenous* (internal) component?

For many years, researchers argued over the relative importance of endogenous versus exogenous components in rhythmic behavior. On the basis of numerous studies of many species, we can now safely say that there is usually a strong endogenous component—a biological clock—but because the rhythm does not exactly match that of the environment, an exogenous cue is necessary to keep the behavior properly timed with the real world. The exogenous cue is sometimes called a *zeitgeber* (German for "time-giver"). Light is undoubtedly the most common *zeitgeber* for circadian rhythms. For example, activity of the North American flying squirrel normally begins with the onset of darkness and ends at dawn, which suggests that light is an important exogenous regulator. However, if a squirrel is placed in constant light or constant darkness, the rhythmic activity is not abolished; in fact, it holds up quite well for at least a month or so. In other words, the cycle has a strong endogenous basis—a biological clock. The duration of each individual cycle (one period of activity plus one period of inactivity) deviates slightly from 24 hours, so that under constant (free-running) conditions, an animal gradually drifts out of phase with the outside world. Eventually, it will be active or inactive at the reverse times of animals in the forest (Figure 50.11). For some animals, however, light and dark are not important considerations for their activity cycle. Fiddler crabs living

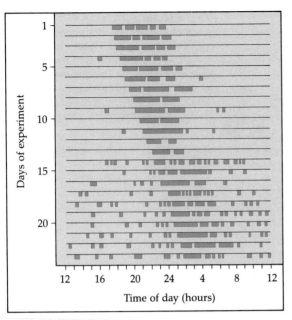

Figure 50.11
Rhythmic activity of a flying squirrel. The flying squirrel is a nocturnal animal that is most active during the first few hours after sunset. This graph traces the activity level of a squirrel kept in constant darkness for 25 days. The animal was placed in a rotating cage connected electronically to a chart recorder (which moves graph paper past a pen at a fixed speed). Whenever the animal ran, the rotating cage caused the pen to mark the graph paper—the thicker horizontal lines on this graph indicate these periods of activity. The pattern displays a definite rhythm, but the period of greatest activity is shifted slightly each day. Under free-running conditions—that is, in the absence of environmental cues about the time of day—the biological clock is not set to exactly 24 hours. The free-running cycle for this squirrel was 24 hours, 21 minutes. Thus, after 25 days, the activity cycle of this squirrel was about 8 hours out of synchronization with the actual timing of sunrise and sunset. In its normal environment, the daily cues of alternating light and dark periods would adjust the biological clock to a 24-hour cycle. Biological clocks are intrinsic to organisms, but the timing of rhythmic behavior must be adjusted to important events in the environment.

in the intertidal zone, for example, forage actively when the tide is out and retreat to their burrows when it comes in. If crabs are experimentally given a simulated continuous low-tide condition, they will show an activity rhythm based on the 12.4-hour tidal cycle.

Human circadian rhythms have also been studied by placing individuals in comfortable living quarters deep underground where they could make their own schedules with no external cues of any kind. Under these free-running conditions, the biological clock of humans seems to have a period of about 25 hours, but with much individual variation; as with other animals, humans use external cues to adjust their rhythms to 24 hours in the real world.

Whether endogenous timekeepers are important in rhythmic behavior that involves longer cycles is much less clear. In many species, **circannual** behaviors such as breeding and hibernation are known to be based, at least in part, on physiological and hormonal changes that are directly linked to exogenous factors such as day length. Little is known about possible endogenous factors, in large part because of practical difficulties in conducting experiments: Animals would have to be maintained for years (instead of days) under constant conditions. Some long-term studies on ground squirrels indicate that fat deposition, associated with hibernation, occurs regularly in a constant environment, but almost nothing is known about possible endogenous control of other kinds of circannual phenomena—particularly behaviors.

Even with short-term rhythms in which endogenous components are clearly present, the timing mechanism is unknown. Current hypotheses posit some kind of biochemical clock, perhaps derived from molecular interactions that occur regularly; as yet, however, no satisfactory model has been developed, much less experimentally verified.

ORIENTATION AND NAVIGATION

Animals can use a variety of environmental cues to orient and navigate as they move from one place to another. These behaviors are ecologically important because they affect the distribution of animals.

A **taxis** is a movement, more or less automatic, toward or away from some stimulus. We have seen taxis in bacteria (Chapter 25); an analogous phenomenon exists in animals. For example, housefly larvae are negatively phototactic after feeding, automatically moving away from light; this simple response presumably ensures that the flies remain in a relatively safe area. Trout are positively rheotactic (from Greek *rheos*, "current"); they automatically swim or orient in an upstream direction, which keeps them from being swept away. **Kinesis** involves a change in activity rate in response to a stimulus. Lice become more active in dry areas and less so in humid ones, a simple behavior that tends to keep the animals in moist environments. In contrast to a taxis, a kinesis is randomly directed; once animals reach a favorable environment, their random movement slows down and the animals tend to remain in that environment.

Animals that migrate long distances have sophisticated mechanisms of orientation and navigation. The most notable examples are migrations of birds and certain ocean-going fish. How is it, for example, that golden plovers find their way over 8000 miles from their arctic breeding grounds to southeastern South America? More remarkably, some populations return to the Hawaiian Islands, a small piece of land in a vast

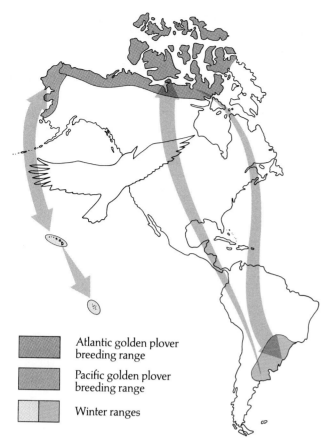

Figure 50.12
Migration routes of the golden plover. These birds are able to navigate across vast expanses of ocean to the relatively small Hawaiian and Marquesas islands.

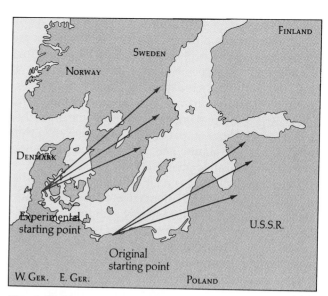

Figure 50.13
Solar navigation. Crows that were experimentally moved westward and then allowed to migrate maintained proper compass direction, probably by use of the sun, and ended up in the wrong location.

expanse of ocean (Figure 50.12). The answer, surprising at first but now firmly established, is that some species of birds and other animals commonly use the same means of navigation as early sailors: the heavens. The sun (for day-migrating species) and stars (for night migrants) provide excellent, albeit complex, cues to direction (Figure 50.13).

Navigating by the sun or stellar constellations requires an internal timing device to compensate for the continuous daily movement of celestial objects. Consider what would happen if you started walking one day, orienting yourself by keeping the sun on your left. In the morning you would be heading south, but by evening you would be heading back north, having made a circle and gotten nowhere. At night, the stars also shift their apparent position as the Earth rotates. One night-migrating bird called the indigo bunting seems to solve the problem by the same means as ancient navigators: fixing on the North Star, which moves little. Many migrants, however, have the ability to adjust to the apparent movement. For example, if an experimental sun is held in a constant position, starlings change their orientation steadily at a rate of about 15° per hour. How they do it is unknown. The problem is very complex, since the apparent location of the sun

shifts at a variable rate, being fastest at midday. Furthermore, the apparent position of celestial objects changes as the animal moves over its migration route. As with the various internal rhythms discussed earlier, almost nothing is known of the mechanisms that could account for this extraordinary ability.

With some bird species, experimentally obscuring the sky to simulate cloudy weather causes them to flutter about in all directions or to cease migratory behavior. Many species, however, are known to continue migrating quite accurately under clouds or even through fog. There is now good evidence that some birds have the ability to detect and orient to the Earth's magnetic field, which varies geographically. The migratory orientation of some species can be experimentally manipulated by changing the magnetic field around them. Very little is known about how birds detect magnetism, but it is intriguing that magnetite, the iron-containing ore once used by sailors as a primitive compass, has been found in the heads of some birds as well as in the abdomens of bees and in certain bacteria that orient with respect to magnetic field. Magnetic sensing may be a widespread, important orienting mechanism among animals, overlooked until recently because it is apparently completely out of our own sensory repertoire.

FORAGING BEHAVIOR

Feeding is obviously a behavior essential to survival and reproductive success, but what determines exactly what an animal eats? Animals feed in a great many ways, using many different kinds of behavior that are closely linked to morphological traits. Filter feeding, for example, requires behavior different from that required by active predation. Ecological and evolutionary considerations are also extremely important; food habits are a fundamental part of an animal's niche, and may be shaped in part by competition with other species. Let us examine here some aspects of feeding from a behavioral perspective.

Some feeding involves relatively simple behavior. A cuttlefish, for example, orients toward and then automatically strikes at a shrimp shortly after the prey comes near; this seems to be a fixed action pattern released by the sight of the shrimp.

Animals in many species can theoretically choose from a large array of potential foods. Some animals tend to be generalists, feeding on a wide variety of items. Gulls feed on material that may be living or dead, aquatic or terrestrial, plant or animal. In contrast, limpkins (wading birds of swamps in the southeastern United States) are specialists that feed only on one kind of snail. Specialists usually have morphological and behavioral adaptations that are highly specific to their food, and as a result they are extremely efficient at foraging. Generalists cannot be as efficient at securing any one type of food, but they have other options if any one food becomes unavailable.

Most generalists do not choose their food randomly. Often, an animal will concentrate on a particular item when it is abundant, sometimes to the exclusion of other foods. The animal is said to have a **search image** for the favored item. (We can understand search images from our own experience, for we often use them to help us find something more efficiently. For example, if you were looking for a particular package on a kitchen shelf, you would probably scan rapidly, looking for a package of particular size and color rather than reading labels.) Eventually, if the favored item becomes scarce in relation to others, the animal will develop a new search image. Search images therefore allow an animal to combine efficient short-term specialization with the flexibility of generalization. *Search image*, of course, is another term that describes what seems to happen but says little as to how.

The switching described here, and other aspects of the choices animals make while foraging, have generated considerable interest among behavioral ecologists in recent years. We would expect natural selection to favor animals that were most efficient at maximizing energy intake over expenditure. However, there are certain trade-offs when an animal is confronted with food choices. A given food item may be larger, and therefore contain more energy, but farther away, and so require spending more energy to get to it. In addition, because the item is larger it may

require more time to subdue or manipulate before swallowing, time lost when other prey could be pursued.

Consider, for example, a smallmouth bass, which readily consumes both minnows and crayfish. Minnows contain more usable energy per unit weight (the crayfish has a lot of hard-to-digest exoskeleton), but they may require more energy expenditure in pursuit. On the other hand, even though crayfish may be easier to catch, they require more handling, since their large claws and aggressive resistance make them hard to manipulate. These trade-offs are also modified by the relative abundance and size of each type of prey.

A basic question, then, has been whether and to what extent animals exhibit **optimal foraging.** Because of the many variables and the complexity of real-world situations, it has been difficult to demonstrate that any animal forages in an absolutely optimal manner. However, on the basis of numerous studies of many different species, we can say that animals have surprising abilities to modify their behavior in such a way that overall energy intake-to-expenditure is high. The smallmouth bass is somehow able to factor in all relevant variables and forage in a highly efficient manner, switching between minnows and crayfish as conditions change.

We know little about how animals do this. Presumably much of the ability is innate, although experience no doubt also plays a role. Bluegill sunfish provide an interesting example. These animals feed readily on small cladocerans (a type of crustacean) called *Daphnia*, generally selecting larger prey that supply the greatest energy intake in relation to expenditure. A smaller prey will be selected if a larger one is too far away (Figure 50.14a).

However, the exact proportions of small versus large prey eaten vary depending on the overall density of prey. Theoretical calculations of energy expenditure in relation to energy intake suggest that, at low prey density, bluegill sunfish should be less selective, eating small as well as large crustaceans as they are encountered. At higher prey densities, it is more efficient to concentrate on larger crustaceans. In actual experiments, bluegill sunfish did become more selective at higher prey densities, though not to the extent that would theoretically maximize efficiency (Figure 50.14b).

Young bluegill sunfish forage fairly efficiently, but not as close to the optimum as older individuals, who are apparently able to make more complex distinctions. Young fish may be prevented from doing so in part because their vision is not sufficiently developed to judge size and distance as accurately. It is unclear, however, whether the improvement in the older fish is merely a process of physical maturation or also involves learning.

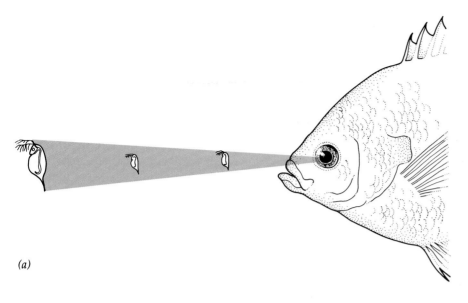

(a)

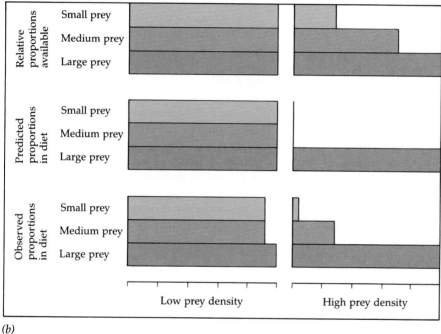

(b)

Figure 50.14
Feeding by young bluegill sunfish. (a) The fish do not feed randomly, but tend to select larger *Daphnia*. This seems to be accomplished by use of "apparent size": Confronted with various potential prey, the animal will pursue the one that looks largest. Thus, the middle prey in this example will be ignored. The closest one is in fact small (= low energy yield), but the animal would not have to go far to capture it (= low energy expenditure). The large one requires more energy expenditure to capture because it is farther away, but it may actually be a better choice because the amount of extra energy it provides is greater than the extra energy spent in capturing it. **(b)** When prey are at low density, calculations based on optimal foraging theory predict that bluegill will not be selective, but rather eat any size prey as it is encountered. At higher prey densities, the ratio of energy intake to energy expended can be maximized by concentrating only on larger prey. In the experiments described here, we see that, at lower prey density, bluegill did forage nonselectively. At higher prey density, they were more selective in favor of larger prey, though not to the extent predicted theoretically.

COMPETITIVE SOCIAL INTERACTION

Social behavior, broadly defined, is any kind of interaction of two or more animals, usually of the same species. Mating behavior is an example of social behavior that is mutually beneficial to the reproductive fitness of the interacting individuals. Many other interactions are competitive rather than cooperative, resulting from limited resources of some kind. In this section we discuss agonistic behavior, a form of behavior common in competitive interactions, and two systems of competitive interaction: dominance hierarchies and territories.

Agonistic Behavior

Since members of a population have a common niche, there is a strong potential for conflict, especially among members of species that normally maintain densities near carrying capacity.

In **agonistic behavior,** a contest of some kind, involving both threatening and submissive behavior, determines which competitor gains access to some resource such as food or mates. Sometimes the encounter involves tests of strength or, more commonly, threat displays that make the individuals look large or fierce, often involving exaggerated posturing

Figure 50.15
Ritual wrestling by rattlesnakes. Rattlesnakes attempt to pin each other to the ground, but never use their deadly fangs in such combat.

and vocalizations. Eventually one individual stops threatening and ends with a submissive or appeasement display, in effect surrendering. Much of this behavior includes **ritual,** the use of symbolic activity, with the result that usually no serious harm occurs to either combatant (Figure 50.15). Dogs and wolves show aggression by baring the teeth, erecting ears, tail, and fur, standing upright, and looking directly at their opponent—all of which make the animal look large and threatening. The eventual loser, on the other hand, sleeks its fur, tucks its tail, and looks away. This appeasement display inhibits any further aggressive activity. Although in some cases animals do inflict injury, natural selection favors a strong tendency in both individuals to end the contest as soon as the ultimate winner is clearly established because violent combat could reduce the reproductive fitness of the victor as well as the defeated. Once an encounter between two individuals has been decided, any future interactions of those same individuals are usually settled much more quickly in favor of the original victor. Examples of ritualized agonistic behavior are widespread.

Dominance Hierarchies

If several hens, unfamiliar with one another, are put together in a group, they respond by pecking one another and skirmishing about. Eventually, however, they establish a clear "peck order"—a more or less linear **dominance hierarchy.** The alpha (top-ranked) hen drives off all others, often by mere threats rather than actual pecking. The beta (second-ranked) hen similarly subdues all others except the alpha, and so on down the line to the omega, or lowest, animal. The

advantage to the top-ranked bird is obvious, since it is assured of access to resources such as food. Even for lower-ranked animals, the system ensures that they don't waste energy or risk harm in futile combat.

Wolves typically function in packs (cooperation is necessary in killing large prey) in which there is a dominance hierarchy among the females. The top female controls mating of the others. When food is generally abundant, she herself mates and allows others to do so also; when food is scarce, she allows less mating, sometimes monopolizing males for herself.

Territoriality

A **territory** is an area, usually fixed in location, that individuals defend and from which other members of the same species are usually excluded. Territories are typically used for feeding, mating, rearing young, or combinations of these activities. The size of a territory varies with the species, the function, and the amount of resources available (Figure 50.16). Song sparrow pairs, for example, may have territories of about 3000 m^2, in which they reside more or less permanently, feeding, mating, and nesting. Gulls and other seabirds, in contrast, mate and nest in territories of only a few square meters or less. They feed away from their territories in large flocks and display little of the vigorous agonistic activity typical of birds that defend territories. Bull sea lions defend small territories used only for mating, whereas red squirrels have rather large ones apparently based on feeding.

Note that a territory is distinguished from a home range, which is simply the area in which an animal roams about and which is often not defended. In some species, as with song sparrows, territory and home range may be the same; but for other species, a territory may be considerably smaller than the home range. The distinction cannot always be made clearly. Gray squirrels, for example, typically have home ranges that overlap extensively, but one individual may defend part of the range from competitors.

Territories are established and defended through agonistic behavior, and an individual that has gained a territory is often difficult to dislodge. Ownership is usually continually proclaimed; this is the primary function of most familiar bird songs, as well as the noisy bellowing of sea lions and the chatter of the red squirrel. Other animals may use scent marks or frequent patrols that warn potential invaders (Figure 50.17). Defense is usually directed only at conspecifics (animals of the same species); a white-throated sparrow may wander freely through a song sparrow's territory. This makes sense from an evolutionary perspective because, as we saw in Chaper 48, a different species usually has a different niche, and is less likely to be a direct competitor.

Although the evolution of dominance hierarchies

(a)

(b)

Figure 50.16
Territories. (**a**) Gannets nest virtually only a peck apart and defend their territories by calling and pecking at each other. This population is on Bonaventure Island, Quebec. (**b**) Langurs defend their territory largely by groups of females attacking other females for months at a time, so that territorial boundaries shift in favor of one group or another.

and territoriality can be explained in terms of advantage to individual organisms, such systems also have important effects at the population level because they tend to stabilize density. If resources were allocated evenly among all members of a population, the "fair share" that each individual received might not be enough to sustain anyone, particularly in the occasional bad year when, say, food supply is reduced.

Dominance and territoriality ensure that at least some individuals receive an adequate amount of a resource. Often, in fact, if a resource such as food becomes scarce, territories expand somewhat. In addition, there are usually individuals low in the hierarchy or lacking territories who are ready to move up or step in if one of the successful individuals dies. The result is relatively stable populations from year to year.

(a)

(b)

Figure 50.17
Staking out territory with chemical markers. (**a**) This male cheetah, a resident of Africa's Serengeti National Park, is spray-urinating on rocks. The odor will serve as a chemical "no trespassing" sign to other males. (**b**) Another male cheetah sniffs a marked rock. With their keen olfactory sense, cheetahs can distinguish their own marks from those left by others.

MATING BEHAVIOR

Courtship

Most animals probably do not have any conscious sense of reproduction as an important function in their life, nor do they have the kind of continuous attraction to the opposite sex familiar in humans. Often they are strongly programmed to view any organism of the same species as a threatening competitor, to be driven off if possible. Many have an extreme aversion to being touched, because any contact more often than not would be with a predator. How then is mating accomplished? In many animals, a complex courtship interaction, unique to the species, is required between the individuals before mating. Exact accomplishment of the programmed sequence of events assures each individual not only that the other is not a threat, but also that the other's species, sex, and physiological condition are all correct. In some species, courtship also plays an important role in allowing one or both sexes to choose a mate from a number of candidates. Such behavior may be complex in total, but (at least in part) it often consists of a series of fixed action patterns, each released by some action on the part of one partner and releasing in turn the next required behavior of the other. In many cases, the sequences probably evolved from agonistic interactions used in competitive situations; we can understand, for example, that an appeasement gesture might become part of a mating ritual, because it allows one individual to remain close to another.

One of the best-known examples of ritualized courtship is that of the three-spined stickleback fish mentioned earlier (Figure 50.18). Although the courtship sequence is based on releasers and fixed action patterns, it does not necessarily proceed smoothly or quickly. Often a female will begin to follow and then hesitate. She may cut off the sequence at any point, sometimes because she is simultaneously being courted by other nearby males. The final result is successful mating by some fish but not necessarily all.

The next time you are in a city park or a zoo with a duck pond, watch the animals carefully. Some of the repeated movements are intraspecific agonistic actions, but others are part of the duck's courtship ritual. Some of the specific behaviors may be common to several species, but each species typically has some unique behaviors or a unique sequence.

Ritualized acts, whether of courtship or agonistic behavior, probably evolved from actions whose meaning was more direct. We can see an interesting example of this process in a family of predaceous flies known as the Empididae (see Figure 22.4). As a male approaches a female for mating, he faces a serious risk

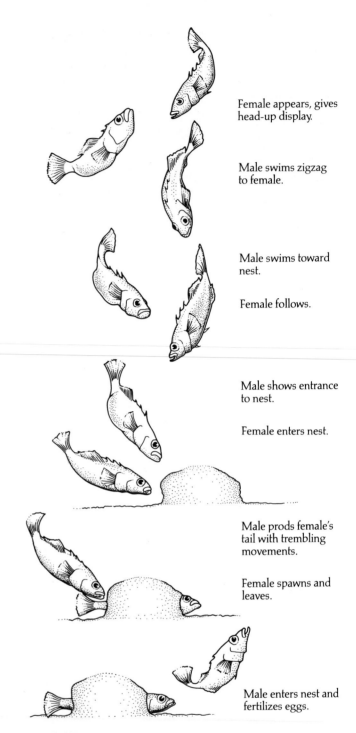

Female appears, gives head-up display.

Male swims zigzag to female.

Male swims toward nest.

Female follows.

Male shows entrance to nest.

Female enters nest.

Male prods female's tail with trembling movements.

Female spawns and leaves.

Male enters nest and fertilizes eggs.

Figure 50.18
Courtship behavior in the three-spined stickleback.
Males are strongly territorial, defending an area in which they have built a nest. If a gravid (egg-carrying) female approaches, her swollen belly releases zigzag swimming in the male. This entices her to swim closer, which in turn stimulates him to swim to the nest and stick his snout inside. This stimulates the female to wriggle into the tunnel. The male then nuzzles her tail, which stimulates her to spawn, after which she swims out the other end of the nest. The male then enters and deposits sperm on the eggs, after which he immediately and quite aggressively drives the female out of the area.

of being killed because the female's attack behavior is released by simple cues that do not differentiate a male of her own species. In some species, males spin an oval silk balloon, which they carry between their legs; the balloon contains nothing, but serves as an appeasement gesture, inhibiting the female's attack and also inducing a sexual response. A clue about how this ritual originated comes from examining the behavior of other species in the family. In some, the male directly inhibits attack by bringing a dead insect for the female to eat while he mates with her. In other species, the dead insect is carried inside a silk balloon like that just described, a variation that may have evolved because silk was helpful in subduing the insect or because it made the male's gift look larger. In some species, the appeasement gesture has apparently evolved into the ritual of merely bringing something formerly associated with food. As one entomologist remarked, it is as though over the course of an evolutionary time scale, a suitor wooed a lady first with diamonds, and then with a box containing diamonds, and finally with nothing but the box itself.

Another function of courtship behavior, mentioned earlier, may be selection of a specific mate from a number of potential ones. In species with sexual reproduction, an individual's own genes alone do not determine the success of its offspring; rather, it is the combination of that individual's genes with those of another individual. We might expect, then, that selection would propagate whatever genes enable individuals to identify those of the opposite sex who have high fitness. Of particular interest, and some controversy, is the role of female choice in the mating process. In some cases, there appears to be a clear relationship between male courtship "performance" and fitness, and females seem to use this in their mate selection. For example, male common terns (a gull-like bird of the seashore) carry fish and display them to potential mates as part of their mating ritual. Eventually, a male may begin feeding a female. The ability of the male to feed both the female and chicks is important to reproductive success, and presumably females select males who display a good potential for doing this. In other cases, the relationship between prominent male displays and fitness is less obvious. For example, the dramatic tail plumage of male peacocks is important in courting females, but it is not known what, if any, direct connection exists between a male's tail display and his fitness in matters other than attracting a mate. (Sexual selection is also discussed in Chapter 21.)

Mating Systems

The mating relationship between males and females varies a great deal among species. In many species,

mating is **promiscuous,** with no strong pair bonds or lasting relationships. In those species where the sexes remain together for a longer period, the relationship may be **monogamous** (one male mating with one female) or **polygamous** (an individual of one sex mating with several of the other). Many birds, for example, are monogamous, whereas mammals tend to be polygamous or promiscuous. Polygamous relationships most often involve a single male and many females (**polygyny**), though in some species this is reversed (**polyandry**).

Differences in mating relationships can be interpreted through evolutionary considerations. Bear in mind that males and females do not come together because of some magnanimous desire to share in the production and care of offspring. In the Darwinian sense, an individual is valuable to a member of the opposite sex only as a necessary vehicle to help his or her genes get into the next generation and succeed there. (The behavior of the male stickleback toward the female immediately after mating is a good example.) Even within the same species, the behavior by which one sex maximizes fitness may be different from—or even somewhat antagonistic to—that of the other sex. These differences are related to what is called **sexual investment.**

Generally, especially in vertebrates, females have a greater inherent investment in a particular offspring than do males. In a mammal, for example, females invest much time and energy in carrying the young before birth, and thus discriminant selection of a mate is important. For males, sperm is cheap, and therefore a different system may be best. Although it is to a male's advantage to choose mates carefully, it may also be advantageous to try to mate more often and with a larger number of partners. Even if some females are less fit, the energy cost is low, and any offspring at all contribute to the male's reproductive success. (By suggesting that mating systems are associated with maximizing Darwinian fitness, we do not mean to imply that animals are cognizant of this outcome.)

The needs of the young are another important consideration. Most newly hatched birds are unable to care for themselves, requiring a large, continuous food supply that one parent may not be able to provide. An individual parent can ultimately leave more viable offspring by staying and helping the other with the young than by going off and seeking more mates. In birds that are not monogamous, the young generally feed and care for themselves almost immediately after hatching; there is no need for parents to stay together, and the males in particular can maximize their success by seeking other mates. With mammals, often only the female is needed as a food source for the young. Males usually play no role or, if they protect females and young, typically take care of many at once in a harem.

COMMUNICATION

To be social, animals must communicate with one another. Vision or hearing are common methods, but many species have other means of communication through senses that in humans are not as well developed. Even in animals with excellent vision and hearing, olfaction is often a major source of information. We have already seen that salmon use odor for navigation and orientation. Another example is the familiar retrieval behavior of dogs. If a ball is thrown in an open field, the animal runs directly to it, probably using vision as its main cue. If the ball is thrown into cover of some kind, however, the dog immediately abandons vision and works only by its sensitive nose, weaving back and forth, sometimes in an indirect manner. You may notice that even as the dog approaches very close, to a point where the ball is easily visible ahead, the animal seems not to notice but continues a weaving, scent-oriented course until it contacts the object.

Odors are a means of relaying or obtaining information, and may be used in many social contexts. Many animals emit **pheromones,** chemical signals that are often used in reproductive behavior (see Chapter 41). Female silkworm moths, for example, emit a pheromone that can attract males from several kilometers away. Once the moths are together, pheromones are also important as releasers for specific courtship behaviors. Another example is the familiar trailing behavior of ants, which involves scents released by scouts, guiding others to the food.

One of the most complex communication systems—certainly among invertebrates—is that of honeybees. For maximum foraging efficiency, workers must communicate to one another the location of good food sources, which may change frequently as various flowers bloom or as new patches are located. How do honeybees talk to one another? The problem was studied in the 1940s by Austrian zoologist Karl von Frisch, who shared the Nobel Prize with Lorenz and Tinbergen. Von Frisch put out dishes of scented sugar water as food sources, varying their distance and direction from the hive. He then carefully watched the bees when they returned to the vertical face of an open hive, which he maintained under dimly lit conditions. (Later researchers have added to the story by use of photographic and auditory equipment.)

When a worker returns to the hive, others gather around it. The bee then goes through a "dance" of some kind that indicates the location of food. If the source is relatively close—less than 50 m or so—the bee moves rapidly sideways in tight circles (the "round dance"), causing the others to become excited (Figure 50.19). Often the dancer regurgitates some nectar that the others taste. The workers then leave the hive and begin foraging nearby. Thus, the round dance translates as "food is near" but does not indicate direction; however, tasting the nectar no doubt helps the bees quickly identify a scent to fly toward.

If the food is farther away, information about direction is needed. A worker returning from a longer distance does a "waggle dance" instead of the round dance. This involves a half-circle swing in one direction, followed by a straight run and then a half-circle swing in the other direction. During the straight run, the bee waggles its tail vigorously. This dance not only tells the others that food is farther away but, remarkably, also tells them the direction and distance. The method by which direction is communicated is quite intriguing. Remember that the bees are "dancing" on a vertical surface. The code is this: The angle of the straight run in relation to the vertical is the same as the horizontal angle of the food in relation to the sun. If the bee runs at a 30° angle to the left of vertical, the other workers will fly 30° to the left of the horizontal direction of the sun when they leave. If the bee runs directly downward, the others will fly directly away from the sun, and so forth. Bees also have some kind of innate timekeeping ability to adjust for the sun's movement. If a rainstorm prevents them from foraging for several hours, they will still fly in the proper direction to food even though the sun has moved.

Distance to the food is also indicated by variation in the speed at which cycles of the waggle dance are performed. A high rate (40 cycles/sec) translates as "about 100 m away," whereas a slow rate (18 cycles/ sec) means "about 1000 m away," with intermediate rates corresponding to intermediate distances. Also, as with the round dance, food is typically regurgitated so that the others have something to key on. Therefore, when a bee leaves to forage, it already "knows" the type of food to seek, its distance, and its direction.

It is not immediately obvious how bees could observe these dances of their fellow workers inside a dark hive. Apparently, swarming around a returner allows them to detect dance cues by physical contact. In addition, buzzing sounds the dancer makes during the straight run of the waggle dance are apparently closely correlated with the distance and direction to the food source. Although the honeybee dance, like much animal behavior, may be based on a limited number of cues and responses, these simple abilities may be combined to produce complex communications.

ALTRUISTIC BEHAVIOR

Throughout parts of the mountainous western United States are communities of Belding's ground squirrel. Although these squirrels hibernate nearly two-thirds of the year, they must spend much time above ground

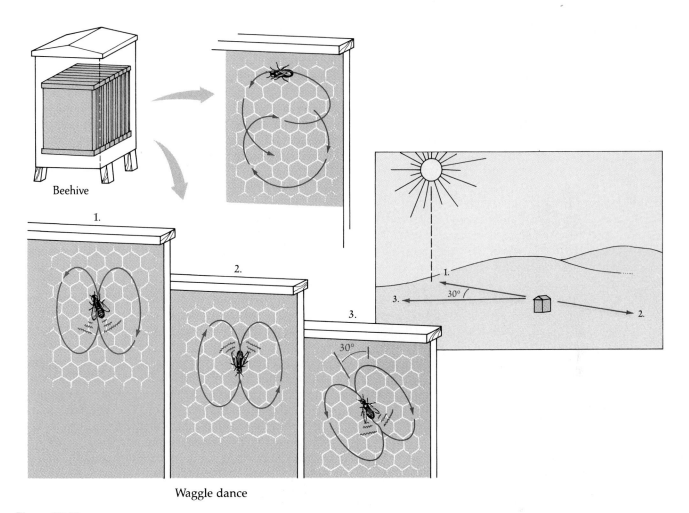

Beehive

Waggle dance

Figure 50.19
Communication in bees. The round dance indicates that food is nearby; the waggle dance is done when food is distant.

during the short summer season, foraging on seeds, mating, and so forth. During this time, they are vulnerable to predators such as coyotes and hawks. If a predator approaches, quite commonly one of the squirrels gives a high-pitched alarm call, alerting unaware individuals, who then retreat to their burrows (Figure 50.20).

This simple response to the presence of a predator, giving a signal to conspecifics, may seem reasonable at first. However, such "altruistic" behavior is in fact difficult to explain in Darwinian terms, and has generated a considerable amount of controversy. How can a squirrel enhance its individual fitness by aiding other members of the population, which, as we have seen, are apt to be its closest competitors? Careful observations have confirmed that the conspicuous alarm behavior increases the risk of being killed because it identifies the squirrel's location. Fitness would seemingly be maximized by quietly allowing a predator to take a competitor.

A very different but classic example of altruistic

behavior can be seen in bee societies, in which the workers are sterile. The workers themselves will never reproduce, but they labor on behalf of a single fertile queen. Furthermore, the workers will sting intruders, a behavior that helps defend the hive but results in the death of the individual.

Another interesting case involves red-cockaded woodpeckers, an endangered species of the southeastern United States. Within a territory, a single pair nests and produces young in a cavity high in an old pine tree. With them, however, are two to four younger birds. These other birds assist in practically all aspects of reproduction except for the actual mating and egg-laying; they help defend the territory, dig out the cavity, and even incubate the eggs and feed the young.

How can altruistic behavior evolve if it reduces the reproductive success of the self-sacrificing individuals? Natural selection will increase genes for altruism if individuals that benefit from the unselfish acts are themselves also carrying those genes for altruism. This is most likely to be the case if the altruists and their

Figure 50.20
Altruistic behavior. Ground squirrel giving an alarm call.

beneficiaries are related. Ordinary parental care of off-spring can be thought of, in a sense, as altruistic behavior that quite obviously helps perpetuate the very genotypes that produce that altruistic behavior. Similarly, siblings, nephews, and even more distantly related members of a population may also share genes for altruism, and individual acts of altruism may enhance propagation of those genes by increasing the overall reproductive success of kin. According to this idea, known as **kin selection**, altruistic behavior evolves because it increases the "inclusive fitness," which we can think of as a measure of the number of copies of a gene common to a group that appears in the next generation, regardless of which individual transmits them.

In fact, altruism commonly does involve relatives. For example, worker bees are genetic sisters of the queen, and their work ensures survival of a large number of their own genes, albeit vicariously. The red-cockaded woodpecker "helpers" are almost always previous offspring of the breeding pair.

Detailed studies of the Belding's ground squirrel demonstrate that the alarm behavior is consistent with a kin selection hypothesis. Alarm calls are usually given by females, only rarely by males. How is this discrepancy explained? Males tend to disperse to other col-onies after mating, and therefore are usually surrounded by nonrelatives. Females tend to stay in the same colony, surrounded by offspring as well as other animals of varying relatedness. Thus, survival of their genes is enhanced by dangerous altruistic behavior, whereas for males such behavior would be pointless.

In some cases, animals behave altruistically toward others who are not relatives. A baboon may help an unrelated companion in a fight, or a wolf may offer food to another even though they share no kinship. At the beginning of this unit, Jane Goodall described dramatic examples of a chimpanzee saving another's life and of another chimpanzee caring for an orphan. This is explained as **reciprocal altruism:** No immediate benefit accrues to the altruistic individual, but some benefit may be gained in the future when the current beneficiary (or another participant in the system) "returns the favor." Reciprocal altruism is also commonly invoked to explain altruism in humans.

At least among nonhuman animals, there is probably no such thing as true, genetically maintained unselfish behavior; any behavior that appears to benefit another animal is carried out simply because it directly or indirectly enhances survival of the individual's genes. If it did not, it would be selected against. For example, an individual that, in an extreme case, laid down its life to save another, unrelated individual would leave no genes to carry on that tendency. The great geneticist J. B. S. Haldane was once asked, long before ideas about kin selection had been developed, if he would lay down his life for his brother. He answered that he would lay down his life for *two* brothers or *four* cousins. Haldane's comment anticipated our understanding of relatedness and altruism, but the extent to which human behavior can be interpreted in such terms remains a problematical and controversial subject. Even with other animals, researchers have difficulty sorting out the various means that might be used to enhance fitness. Perhaps the most significant problem lies in demonstrating conclusively that seemingly altruistic behavior cannot be explained more simply in terms of the altruistic individual's own benefit. For example, the young red-cockaded woodpeckers may have their best chance of getting a suitable territory by staying as helpers and eventually "inheriting" the area from the parents. Some researchers would argue that most cases of apparent kin selection can be explained better by individual selection. Many ideas about kin selection and altruistic behavior are not yet fully worked out or accepted.

ANIMAL COGNITION

One dictionary defines "cognition" as "knowing, including both awareness and judgment." A simple but profound question is whether nonhuman animals are cognitive beings. Are other animals consciously *aware* of themselves and of the world around them? Do they "feel" pain or pleasure or sadness in the same way as we do? As yet we have no way of answering these questions or of really getting inside the mind of another creature. We have seen, of course, that other animals may sometimes behave more like programmed computers, and they certainly do not have the abilities to integrate information ("think") to the same extent that humans exhibit. But is this a matter of degree—of a continuum of abilities—or are humans fundamentally different in some behavioral respect?

Because of the impossibility of answering these questions, most researchers have taken a mechanistic approach referred to as **behaviorism.** They describe behavior in terms of stimulus/response, or more recently, in the same computer-derived terminology ("programmed," "subroutine") used earlier in this chapter, carefully avoiding any trace of the anthropomorphism we cautioned about. As an example, we might explain a dog's yelp as an unconscious response, involving muscular contractions resulting from nervous activity that was in turn stimulated by the tactile pressure of a foot on the dog's tail. But this narrow view seems unreasonable to many people, at least when applied to mammals. It would certainly be difficult to think of Dr. Goodall's chimpanzee subjects merely as highly sophisticated, unconscious robots.

Donald Griffin of Princeton University has been the foremost recent proponent of **cognitive ethology,** a view that sees conscious thinking as an inherent and routine part of the behavior of other animals. He argues that if other animals behave in ways that we associate with conscious processing in ourselves, perhaps it makes sense to assume that there is the same underlying awareness. (Note that this is the same argument we routinely apply to other humans, since we have no way of knowing what goes on in another human mind either.) Griffin also stresses the possible adaptive value of consciousness. In Jane Goodall's interview, she made the case that there is a clear advantage to a highly developed "intellect" and cognitive decision making in chimpanzees. Griffin suggests that such advantages may extend to many nonprimate branches of the phylogenetic tree. This view does not explain how cognition comes about (our understanding of the physiochemical basis for much of brain function is still too incomplete), but it sees cognitive ability arising through the normal process of natural selection—and, like many other major animal functions, forming a continuum that extends far back in evolutionary history. James Gould, also at Princeton, recently published evidence that bees can form and use cognitive maps of their foraging areas. As he notes, this kind of behavior is usually considered a basic form of thinking. Ultimately, answers to questions about animal

cognition may profoundly affect how we interact with other animals, and how we view ourselves as well.

HUMAN SOCIOBIOLOGY

In 1975, E. O. Wilson of Harvard University published a lengthy text entitled *Sociobiology* describing social systems in various animal species. Its thesis was that social behavior has an evolutionary basis—behavioral characteristics, like specific anatomical and physiological traits, exist because they are expressions of genes that have been perpetuated by natural selection. We have seen, for example, that altruistic behavior in honeybees is not truly unselfish, but occurs because, in bee societies, the overall reproductive fitness of a genotype shared by kin is enhanced by individual sacrifice. Much of Wilson's book applied evolutionary thinking to social behavior of insects. But in the last chapter of *Sociobiology,* the author speculated on the evolutionary basis of certain kinds of social behavior in humans. The book rekindled the nature-versus-nurture controversy, and the debate over sociobiology remains heated more than fifteen years later.

Let us examine the debate in the context of a specific example: avoidance of incest. Nearly all cultures have laws or taboos forbidding sexual relations or marriage between brother and sister. Is there an innate aversion to incest, or do we acquire this behavior as part of our socialization? Someone favoring the "nurture" side of the debate may argue that if this behavior were innate, cultural taboos would be superfluous. According to this argument, avoidance of incest is a learned behavior, and the social stigma attached to incest may be based on experience—people who break the taboo are more likely to have defective children. Some sociobiologists would begin by arguing that the occurrence of a specific behavior in diverse cultures is itself evidence that the behavior has an innate component. Taboos and laws have merely reinforced an avoidance of incest that has evolved by natural selection. Many animals lacking what we would call culture avoid incest. We should expect this behavior to evolve if incest decreases the reproductive fitness of individuals by reducing the average viability of offspring.

Some sociobiologists have cited the experience of the Israeli communal societies known as kibbutzim as evidence that innate repulsion of humans to incest is strong enough to counter cultural factors. From the time of birth, kibbutz children usually spend most of their time in large child-care centers, which gives them the equivalent of a sibling relationship with all other children in the kibbutz. Early in the kibbutz movement, parents encouraged their children to marry within their own kibbutz. However, according to a study of more than 5000 people who grew up in these societies, marriages between individuals raised together since early childhood were extremely rare, in spite of the wishes of parents and kibbutz leaders. Apparently, people who spend nearly all their time together as children have little sexual attraction for one another as adults. (In most situations, of course, individuals who cohabit throughout childhood are true siblings or other close relatives.) A sociobiologist might argue from this example that the human resistance to incest—in this case with "artificial" siblings—overrides cultural pressure encouraging it.

Some sociobiologists, including Wilson, envision cultural and genetic components of social behavior as linked in a cycle of reinforcement. If most members of a society share an innate avoidance to incest, the aversion is likely to become formalized in laws and taboos. The cultural stigma amplifies the evolutionary component of this behavior by acting as an environmental factor in natural selection. A society that shuns or imprisons members that practice incest lowers their reproductive fitness. According to this view, genes and culture are integrated in human nature.

Sociobiology does not reduce us to robots stamped out of a common, rigid genetic mold. Individuals vary extensively in anatomical features, and we should expect inherent variations in behavior as well. And even though we are locked into our genotypes, our nervous systems are not "hard wired." Indeed, our behavior is probably more plastic than that of any other animal. The spectrum of possible social behaviors may be circumscribed by our genetic potential, but that is very different from saying that genes are rigid determinants of behavior. Environment intervenes in the path from genotype to phenotype for physical traits, and environment is undoubtedly important for behavioral traits as well. To use a term introduced in Chapter 21, genes associated with complex behavior probably have very broad "norms of reaction." Thus, an evolutionary approach to social behavior has a place for environmental effects. Perhaps the nature-nurture debate may be a straw man after all (Figure 50.21).

Figure 50.21
Both genes and culture build human nature. Teaching of the younger generation by the older is one of the basic ways all cultures are transmitted. Sociobiologists would see tutoring as an innate tendency with adaptive value that has evolved in the human species.

*   *   *

The many facets of biology are reflected in the study of behavior; it is an intersection of biochemistry, genetics, physiology, evolutionary theory, and ecology. The study of behavior, especially social behavior, also connects biology to the social sciences and humanities. Through the study of life in general, it is inevitable that we will learn more about ourselves.

STUDY OUTLINE

1. Animal behavior can be broadly defined as any externally observable muscular activity, triggered by some stimulus.
2. The study of how animals behave in their natural environment is called ethology.

Studying Behavior (pp. 1139–1143)

1. Behavior has both immediate proximate causes and longer-term ultimate causes that fit into a more general evolutionary context.
2. Behaviors, or at least the *capacity* for behaviors, must ultimately have a genetic basis; however, most behaviors are not simple phenotypic characters, but are subject to environmental influence and modification by experience.

Innate Components of Behavior (pp. 1143–1145)

1. A fixed action pattern (FAP) is a highly stereotyped, innate behavior that continues to completion after initiation by an external sign stimulus. Releasers are specific sign stimuli that function as communication signals between individuals of the same species.
2. Sign stimuli tend to be based on simple cues associated with the relevant object or activity. Exaggeration of the

relevant stimulus in the form of a supernormal stimulus often produces a stronger response.

3. The concepts of innate releasing mechanisms and drive, which sometimes have been used to describe or explain many types of behaviors, shed little light on the actual mechanisms of behavior.

Learning and Behavior (pp. 1145–1149)

1. Behavior can be modified by learning in various ways. However, some apparent learning is due mostly to inherent maturation.
2. Habituation is a simple kind of learning involving loss of sensitivity to unimportant stimuli.
3. Imprinting, first described by Lorenz with newly hatched geese, is a type of learned behavior with a significant innate component, acquired during a limited critical period.
4. Associative learning involves linking one stimulus with another, as demonstrated by classical conditioning.
5. Operant conditioning, or trial-and-error learning, is a type of associative learning exhibited by many animals.
6. Insight learning involves the ability to reason by correctly performing a task on the first attempt in a situation in which the animal has had no earlier experience.
7. The internal mechanisms responsible for learning are so far largely undefined.

Behavioral Rhythms (pp. 1149–1150)

1. Various daily behaviors are dictated by endogenous clocks, which in turn require exogenous cues to keep the behavior properly timed with the real world.
2. Circannual behaviors such as breeding and hibernation seem to be controlled primarily by physiological and hormonal changes influenced by exogenous factors such as day length.
3. It is not known whether endogenous factors have a role in circannual rhythms, and even in circadian rhythms the precise mechanism of the biological clock is unknown.

Orientation and Navigation (pp. 1150–1151)

1. Taxis is an automatic movement toward or away from a specific stimulus, whereas kinesis is a random movement displaying a stimulus-specific change in activity rate.
2. The most sophisticated mechanisms of orientation and navigation are used by migrating animals. Some species of migratory birds are guided by the sun or stars. Others may be able to use the earth's magnetic field.

Foraging Behavior (pp. 1151–1153)

1. The large array of animal feeding patterns has generated a varied repertoire of foraging behaviors.
2. Species may be specialists or generalists. Even generalists, however, do not forage randomly, but form search images that can be changed if the food item becomes scarce.
3. Animals typically forage efficiently in nature by modifying their behavior in complex ways that favor a high energy intake to expenditure ratio.

Competitive Social Interaction (pp. 1153–1155)

1. Social behavior encompasses the spectrum of interactions between two or more animals, usually of the same species.

2. Agonistic behavior involves a contest in which one competitor gains an advantage in obtaining access to a limited resource such as food or mates. Natural selection has generally favored the evolution of symbolic, exaggerated ritual to resolve the conflict involving both aggressive and submissive behavior.
3. Some animals show dominance hierarchies, with differentially ranked individuals permitted options appropriate to their status in the "pecking order."
4. Territoriality is a behavior in which an animal defends a specific, fixed portion of its home range against intrusion by other animals of the same species through agonistic interactions.

Mating Behavior (pp. 1156–1157)

1. Courtship interactions are complex, species-specific behaviors that typically include a series of fixed action patterns and releasers, ensuring that the participating individuals are nonthreatening and of the proper species, sex, and physiological condition for mating.
2. Courtship may also involve selection of a specific mate from an array of potential candidates.
3. Some segments of courtship behavior likely evolved from agonistic interactions, and the ritualistic aspects of both courtship and agonistic behaviors probably evolved from actions whose meaning was more direct.
4. Mating relationships, which vary widely among different species, include promiscuity, monogamy, and polygamy, the latter involving polygyny more often than polyandry.
5. Evolution affects mating relationships depending on sexual investment in the next generation, which varies according to the degree of prenatal and postnatal input the parents are required to make.

Communication (p. 1158)

1. Animals communicate with one another through their various senses. Odors are particularly effective signals in many species, as attested by the functions of pheromones.
2. Honeybees have an impressive intraspecific communication system, which has been studied extensively by Karl von Frisch. Complex "round" and "waggle" dances executed by worker bees at different speeds signal the presence of food and its distance and direction from the hive. Further information is shared by providing samples of regurgitated nectar and by buzzing and physical contact.

Altruistic Behavior (pp. 1158–1160)

1. Altruistic behavior, which benefits a conspecific at the potential expense of the helpful individual, seems to be best explained by kin selection, a theory of inclusive fitness that maintains that genes enhance survival of copies of themselves by directing organisms to care for others who share those genes.
2. Altruism commonly involves close relatives, although not always. Reciprocal altruism between unrelated individuals may be explained as ultimately advantageous to the altruistic individual in the future when another individual "returns the favor."

Animal Cognition (pp. 1160–1161)

1. Because of the impossibility of knowing what goes on inside the minds of other animals, most researchers have

taken a mechanistic approach, called "behaviorism," to the study of animal behavior.

2. Recently, some workers have argued in favor of "cognitive ethology," which sees cognitive ability as a normal part of other animals that has arisen, like other traits, through natural selection.

Human Sociobiology (pp. 1161–1162)

1. The final chapter of E. O. Wilson's *Sociobiology*, published in 1975, suggested that human social behavior could be understood in evolutionary terms, igniting a heated and emotional debate in the scientific community.

2. Sociobiologists acknowledge human behavioral plasticity, which is manifested in an impressive spectrum of social behaviors that is circumscribed, but not rigidly determined, by our genetic potential.

SELF-QUIZ

1. Ethology is the study of how
   a. animals behave in their natural environments
   b. human emotions and thoughts apply to other animals
   c. ultimate causes of animal behavior evolved
   d. stimuli release the innate components of behavior
   e. animals behave in controlled laboratory experiments

2. The controversy between nature and nurture centers on the
   a. distinction between proximate and ultimate causes of behavior
   b. role of genes in learning
   c. whether animals have conscious feelings or thoughts
   d. the extent to which an animal's behavior is innate or learned
   e. importance of good parental care

3. Which of the following statements is *not* true of fixed action patterns?
   a. They are highly stereotyped, instinctive behaviors.
   b. They seem to lack adaptive significance.
   c. They are triggered by sign stimuli in the environment and, once begun, are continued to completion.
   d. They are often released by one or two simple cues associated with the relevant object or organism.
   e. A supernormal stimulus often produces a stronger response.

4. The return of salmon to their home streams to spawn is an example of
   a. olfactory imprinting
   b. insight
   c. associative learning
   d. operant conditioning
   e. habituation

5. Which of the following researchers is *incorrectly* paired with his work?
   a. Tinbergen—sign stimuli as releasers of FAPs
   b. Lorenz—imprinting with geese
   c. von Frisch—communication dances of honeybees
   d. Skinner—trial and error learning with rats in Skinner boxes
   e. Pavlov—critical learning period

6. Which of the following is *not* true of agonistic behavior?
   a. It is most common between members of the same species.
   b. It may be used to establish and defend territories.
   c. It often involves symbolic conflict and does not inflict serious harm to either the winner or loser of the encounter.
   d. It is a uniquely male behavior.
   e. It may be used to establish dominance hierarchies.

7. A sign stimulus that functions as a signal that triggers a certain behavior in another member of the same species is *specifically* called a (an)
   a. pheromone
   b. FAP stimulus
   c. releaser
   d. search image
   e. agonistic sign

8. The core idea of sociobiology is that
   a. human behavior is rigidly predetermined by inheritance
   b. humans cannot learn to alter their social behavior
   c. many aspects of social behavior have an evolutionary basis
   d. the social behavior of humans is comparable to that of honeybees
   e. environment outweighs genes in human behavior.

9. Learning to ignore insignificant stimuli is specifically called
   a. conditioning
   b. imprinting
   c. habituation
   d. insight
   e. critical learning

10. A honeybee returning to the hive from a food source performs a waggle dance with the "run" oriented straight to the left on the vertical surface. This means that the food is located
    a. 90° left of the hive
    b. 90° left of the line from the hive to the sun
    c. in the opposite direction—straight to the right of the hive
    d. very close to the hive

CHALLENGE QUESTIONS

1. Female crickets were tested for their ability to track synthesized male calling sounds. In one trial, trains of 5-Hz tone bursts were alternated with trains of four-syllable chirps simulating natural song. The female ignored the

burst but tracked the chirps avidly, changing direction approximately when the sound direction changed. Is this behavior a taxis or kinesis, and what might be its proximate and ultimate causes?

2. Experiments have demonstrated that tadpoles produce a chemical cue unique to their kin group, which they use to distinguish kin from nonkin. This chemical signal fosters the tadpoles' ability to form small, dense schools, which presumably increases their inclusive fitness by enhancing their success at finding and sharing food and escaping predators. Surprisingly, such kin recognition survived metamorphosis to appear in the adults, which are not known to form aggregations. Give at least one feasible explanation for the latter finding. (See Further Reading 2.)

3. In an attempt to demonstrate a circannual rhythm in bird migration, investigators separated migratory willow warblers into a group in their native Germany, where the birds were exposed to normal photoperiod; a group in equatorial Zaire, where there is little variation in photoperiod; and a group in the laboratory, where photoperiod was held constant. All three groups underwent migratory restlessness, a nocturnal behavior displayed by caged birds when they would otherwise be migrating, on the normal yearly schedule. Do these results support an internal timekeeping mechanism for a circannual rhythm? Why or why not? (See Further Reading 8.)

FURTHER READING

1. Alcock, J. *Animal Behavior: An Evolutionary Approach.* 4th ed. Sunderland, Mass.: Sinauer, 1989.
2. Blaustein, A. R., and R. K. O'Hara. "Kin Recognition in Tadpoles." *Scientific American,* January 1986. A focus on the genetic basis of kin selection.
3. Brooke, M. "Tricks of the Egg Trade." *Natural History,* April 1989. Behavior of a brood parasite.
4. Caro, T. "The Brotherhood of Cheetahs." *Natural History,* June 1989. Social behavior of cats.
5. Diamond, J. "Sexual Deception." *Discover,* August 1989. Speculations on the evolutionary basis of seduction.
6. Fackelmann, K.A. "Avian Altruism." *Science News,* June 10, 1989. Self-sacrifice in a species of African birds.
7. Funk, D.H. "The Mating of Tree Crickets." *Scientific American,* August 1989. The role of an insect song we have all heard.
8. Gwinner, E. "Internal Rhythms in Bird Migration." *Scientific American,* April 1986. Circannual rhythms may tell migratory birds when to begin and end their flight.
9. Jolly, A. "A New Science That Sees Animals as Conscious Beings." *Smithsonian,* March 1985. An excellent synopsis of current thought on animal thinking.
10. Lorenz, K. *On Aggression.* New York: Harcourt, Brace, and World, 1966. A classic book on competitive social interactions.
11. Miller, J.A. "Clockwork in the Brain." *Bioscience,* February 1989. How studies of a mutant hamster are helping biologists understand circadian rhythms.
12. Trivers, R. *Social Evolution.* Menlo Park, Ca.: Benjamin/Cummings, 1985.
13. Wilson, E. O. *Sociobiology: The New Synthesis.* Cambridge, Mass.: Harvard University Press, 1975.

Self-quiz Answers

Chapter 2

1. d	2. b	3. b	4. c	5. c
6. c	7. c	8. a	9. b	10. b

Chapter 3

1. d	2. c	3. b	4. c	5. b
6. c	7. d	8. c	9. c	10. e

Chapter 4

1. b	2. a	3. d	4. d	5. a
6. b	7. b	8. d	9. d	10. b

Chapter 5

1. d	2. c	3. d	4. c	5. b
6. b	7. b	8. b	9. c	10. b

Chapter 6

1. b	2. d	3. b	4. c	5. a
6. a	7. c	8. c	9. d	10. c

Chapter 7

1. d	2. c	3. b	4. d	5. d
6. b	7. c	8. c	9. c	

10. (a) centrioles—may help organize microtubules in animal cells
(b) mitochondrion—cellular respiration
(c) nucleus—genetic control of cell
(d) Golgi apparatus (dictyosome)—storage, refinement, sorting, and secretion of cellular products
(e) chloroplast—photosynthesis
(f) rough ER—membrane synthesis; synthesis of secretory proteins

Chapter 8

1. b	2. c	3. a	4. a	5. c
6. a	7. Fructose	8. Glucose	9. Into the cell	10. b

Chapter 9

1. d	2. b	3. c	4. d	5. a
6. a	7. b	8. b	9. b	10. b

Chapter 10

1. c	2. b	3. d	4. b	5. b
6. c	7. c	8. a	9. a	10. d

Chapter 11

1. c	2. b	3. b	4. b	5. c
6. a	7. a	8. c	9. c	

Chapter 12

1. d	2. b	3. d	4. d	5. c
6. a	7. b	8. d	9. a	10. c

Chapter 13

1. d	2. b	3. b	4. d	5. c
6. b	7. d	8. a	9. a	10. b

Genetics Problems

1. a. 1/64   b. 1/64   c 1/8   d 1/32

2. Albino is a recessive trait; black is dominant.

Parents	Gametes	Offspring
$BB \times bb$	B and b	All Bb
$bb \times Bb$	b and 1/2 B, 1/2 b	1/2 Bb, 1/2 bb

3.
Genotype	Gametes
$RrSs$	4; RS, Rs, rS, rs
$RRss$	1; Rs
$RrSS$	2; RS, rS
$rrss$	1; rs

4. F_1 cross is $AARR \times aarr$. Genotype of progeny is $AaRr$, phenotype is all axial pink.
F_2 cross is $AaRr \times AaRr$. Genotypes of progeny are 4 $AaRr$: 2 $AaRR$: 2 $AARr$: 2 $aaRr$: 2 $Aarr$: 1 $AARR$: 1 $aaRR$: 1 $AArr$: 1 $aarr$. Phenotypes of progeny are 6 axial-pink : 3 axial-red : 3 axial-white : 2 terminal-pink : 1 terminal-white : 1 terminal-red

5. a. $PPLl \times PPLl$, $PpLl$, or $ppLl$
b. $ppLl \times ppLl$
c. $PPLL \times$ any of the 9 possible genotypes
d. $PpLl \times Ppll$
e. $PpLl \times PpLl$

6. Man $I^A i$; woman $I^B i$; child ii. Other genotypes for children are 1/4 $I^A I^B$, 1/4 $I^A i$, 1/4 I^B_i

7. a. $3/4 \times 3/4 \times 3/4 = 27/64$
b. $1 - 27/64 = 37/64$
c. $1/4 \times 1/4 \times 1/4 = 1/64$
d. $1 - 1/64 = 63/64$

8. 750

9. a. 1/256   b. 1/16   c. 1/256   d. 1/64   e. 1/128

10. a. 1   b. 1/32   c. 1/8   d. 1/2

11. 1/9

12. 1/16, assuming both parents are heterozygous for the newly discovered disease gene

Chapter 14

1. b	2. b	3. d	4. a	5. b
6. a	7. d	8. b	9. a	10. d

Genetics Problems

1. 0; 1/2; 1/16

2. 17%

3. Between T and A, 12%; between A and S, 5%

4. Between T and S, 18%. Sequence of genes is $T - A - S$.

Chapter 15

1. c	2. c	3. d	4. a	5. d
6. c	7. c	8. b	9. a	10. d

Chapter 16

1. a	2. d	3. b	4. d	5. c
6. c	7. a	8. d	9. e	10. b

Chapter 17

1. a	2. c	3. d	4. c	5. a
6. a	7. a	8. c	9. b	10. b

Chapter 18

1. c	2. a	3. d	4. c	5. c
6. a	7. a	8. e	9. c	10. a

Chapter 19
1. b	2. d	3. b	4. c	5. e
6. c	7. c	8. d	9. b	10. a

Chapter 20
1. a	2. b	3. a	4. c	5. d
6. d	7. c	8. b	9. b	10. b

Chapter 21
1. b	2. d	3. c	4. a	5. a
6. a	7. c	8. b	9. e	10. b

Chapter 22
1. b	2. b	3. a	4. d	5. c
6. e	7. d	8. b	9. d	10. b

Chapter 23
1. b	2. b	3. a	4. e	5. c
6. b	7. a	8. d	9. c	10. b

Chapter 24
1. a	2. b	3. e	4. d	5. c
6. b	7. a	8. c	9. b	10. d

Chapter 25
1. c	2. a	3. d	4. b	5. b
6. a	7. c	8. c	9. a	10. b

Chapter 26
1. b	2. c	3. c	4. e	5. c
6. d	7. c	8. b	9. d	10. c

Chapter 27
1. c	2. d	3. a	4. b	5. a
6. d	7. a	8. b	9. c	10. a

Chapter 28
1. b	2. c	3. c	4. d	5. d
6. a	7. e	8. a	9. b	10. b

Chapter 29
1. e	2. e	3. b	4. c	5. a
6. d	7. b	8. a	9. e	10. c

Chapter 30
1. b	2. c	3. c	4. c	5. d
6. c	7. a	8. b	9. c	10. d

Chapter 31
1. e	2. b	3. a	4. c	5. b
6. c	7. d	8. a	9. d	10. b

Chapter 32
1. c	2. c	3. c	4. d	5. b
6. a	7. b	8. c	9. a	10. b

Chapter 33
1. b	2. b	3. e	4. d	5. d
6. c	7. b	8. b	9. b	10. b

Chapter 34
1. d	2. c	3. a	4. b	5. a
6. b	7. b	8. e	9. d	10. a

Chapter 35
1. b	2. a	3. d	4. c	5. b
6. c	7. a	8. b	9. b	10. c

Chapter 36
1. b	2. c	3. c	4. d	5. d
6. e	7. b	8. b	9. c	10. c

Challenge Question
1. (a) loose connective tissue; (b) nervous tissue; (c) stratified squamous epithelium; (d) striated (skeletal) muscle; (e) bone; (f) simple cuboidal epithelium

Chapter 37
1. b	2. b	3. c	4. c	5. d
6. d	7. c	8. d	9. b	10. b

Chapter 38
1. c	2. b	3. d	4. c	5. b
6. c	7. b	8. a	9. a	10. c

Chapter 39
1. a	2. b	3. c	4. d	5. c
6. a	7. d	8. b	9. b	10. b

Chapter 40
1. c	2. d	3. a	4. e	5. e
6. a	7. b	8. c	9. b	10. d

Chapter 41
1. c	2. a	3. e	4. d	5. c
6. b	7. e	8. d	9. c	10. c

Chapter 42
1. d	2. b	3. a	4. b	5. c
6. d	7. a	8. b	9. d	10. b

Chapter 43
1. a	2. b	3. d	4. c	5. c
6. c	7. b	8. a	9. d	10. b

Chapter 44
1. c	2. b	3. a	4. d	5. a
6. c	7. b	8. c	9. d	10. c

Chapter 45
1. e	2. c	3. a	4. c	5. e
6. b	7. b	8. d	9. c	10. b

Chapter 46
1. d	2. c	3. e	4. a	5. a
6. d	7. c	8. a	9. c	10. d

Chapter 47
1. c	2. b	3. c	4. d	5. c
6. d	7. c	8. d	9. a	10. c

Chapter 48
1. c	2. d	3. b	4. d	5. e
6. d	7. b	8. a	9. d	10. c

Chapter 49
1. c	2. e	3. b	4. e	5. d
6. a	7. b	8. c	9. b	10. c

Chapter 50
1. a	2. d	3. b	4. a	5. e
6. d	7. c	8. c	9. c	10. b

Classification of Life

This appendix presents the taxonomic classification used for the major groups of organisms discussed in this text; not all phyla are included. Plant and fungal divisions are the taxonomic equivalents of phyla.

Kingdom Monera*

Archaebacteria

   Methanogens

   Extreme halophiles

   Thermoacidophiles

Eubacteria

   Cyanobacteria

   Phototropic bacteria

   Pseudomonads

   Spirochetes

   Endospore-forming bacteria

   Enteric bacteria

   Rickettsias and chlamydias

   Mycoplasmas

   Actinomycetes

   Myxobacteria

Kingdom Protista

Phylum Rhizopoda (amoebas)

Phylum Actinopoda (heliozoans, radiolarians)

Phylum Foraminifera (forams)

Phylum Apicomplexa (apicomplexans)

Phylum Zoomastigina (zoomastigotes)

Phylum Ciliophora (ciliates)

Phylum Dinoflagellata (dinoflagellates)

Phylum Chrysophyta (golden algae)

Phylum Bacillariophyta (diatoms)

Phylum Euglenophyta (euglenoids)

Phylum Chlorophyta (green algae)

Phylum Phaeophyta (brown algae)

Phylum Rhodophyta (red algae)

Phylum Myxomycota (plasmodial slime molds)

Phylum Acrasiomycota (cellular slime molds)

Phylum Oomycota (water molds)

---

*Informal names are used here for the bacterial groups because there is not yet a concensus on how to divide Kingdom Monera into phyla.

Kingdom Plantae

Division Bryophyta (liverworts and mosses)

Division Psilophyta (whisk ferns)

Division Lycophyta (club mosses)

Division Sphenophyta (horsetails)

Division Pterophyta (ferns)

Division Coniferophyta (conifers)

Division Cycadophyta (cycads)

Division Ginkgophyta (ginkgos)

Division Gnetophyta (gnetae)

Division Anthophyta (flowering plants)

   Class Monocotyledones (monocots)

   Class Dicotyledones (dicots)

Kingdom Fungi

Division Zygomycota (zygomycetes)

Division Ascomycota (sac fungi)

Division Basidiomycota (club fungi)

Division Deuteromycota (imperfect fungi)

Lichens (symbiotic associations of algae and fungi)

Kingdom Animalia

Phylum Porifera (sponges)

Phylum Cnidaria

   Class Hydrozoa (hydrozoans)

   Class Scyphozoa (jellyfish)

   Class Anthozoa (sea anemones and coral animals)

Phylum Ctenophora (comb jellies)

Phylum Platyhelminthes (flatworms)

   Class Turbellaria (free-living flatworms)

   Class Trematoda (flukes)

   Class Cestoda (tapeworms)

Phylum Nemertina (proboscis worms)

Phylum Rotifera (rotifers)

Phylum Nematoda (roundworms)

Phylum Annelida (segmented worms)

   Class Oligochaeta (oligochaetes)

   Class Polychaeta (polychaetes)

   Class Hirudinea (leeches)

Phylum Mollusca (mollusks)

   Class Polyplacophora (chitons)

   Class Gastropoda (gastropods: snails and their relatives)

   Class Bivalvia (bivalves)

   Class Cephalopoda (cephalopods: squids and octopuses)

Phylum Arthropoda (arthropods)

    Class Arachnida (arachnids: spiders, ticks, scorpions)

    Class Crustacea (crustaceans)

    Class Diplopoda (millipedes)

    Class Chilopoda (centipedes)

    Class Insecta (insects)

Phylum Phoronida (phoronids)

Phylum Bryozoa (bryozoans)

Phylum Brachiopoda (brachiopods: lamp shells)

Phylum Echinodermata (echinoderms)

    Class Asteroidea (sea stars)

    Class Ophiuroidea (brittle stars)

    Class Echinoidea (sea urchins and sand dollars)

    Class Crinoidea (sea lilies)

    Class Holothuroidea (sea cucumbers)

Phylum Chordata (chordates)

    Subphylum Cephalochordata (cephalochordates: lancelets)

    Subphylum Urochordata (urochordates: tunicates)

    Subphylum Vertebrata (vertebrates)

        Class Agnatha (jawless vertebrates)

        Class Chondrichthyes (cartilaginous fishes)

        Class Osteichthyes (bony fishes)

        Class Amphibia (amphibians)

        Class Reptilia (reptiles)

        Class Aves (birds)

        Class Mammalia (mammals)

APPENDIX THREE

Concept Mapping: A Technique for Learning

Concept mapping is a process—a process of organizing your knowledge to both increase your understanding and help you learn. A concept map is a diagram that shows the organization of ideas and the relationship among concepts in a particular subject area. The *structure* of a concept map is a hierarchically organized cluster of concepts, enclosed in boxes and connected with lines that explicitly state the relationships among the concepts. The *function* of a concept map is to help you structure your understanding of a topic and create personal meaning. The *value* of a concept map arises from the process of thinking and evaluating what you must do in order to create the map.

Biology is a rich and diverse subject, filled with information and terminology that are necessary to understand the concepts and principles of the field. Beginning students often focus their attention on memorizing these bits of information, while missing how these pieces fit into the "big picture." A student must assimilate these details into broader concepts that are then related to other concepts, groups of concepts, and organizing principles. Neil Campbell's *BIOLOGY* helps the student in this process by regularly pointing out the themes that run through the study of biology: hierarchy of organization, emergent properties, correlation of structure and function, and evolution. Every chapter of this book has a set of key ideas or principles that are supported by examples and explanations. A student must recognize how these ideas are organized and develop a personal conceptual framework that can house the rich details of that chapter. Sketching a concept map will help to create such a structure.

To develop a concept map for a particular area, you must first identify the most important ideas or concepts. That process alone will help you sort out details from the organizing principles. Then you evaluate the relative importance of these key concepts (those which are most inclusive, those which are subordinate to other concepts); arrange the concepts in a meaningful cluster; and label the connections or relationships among the concepts.

A concept map is an individual picture of the understanding you had developed at the time you made the map. Meanings change and grow as you gain more knowledge and experience in an area. You develop a richer picture; you can make more connections among concepts and relate ideas in a more meaningful way. Knowledge is not static; it grows. As your understanding of an area develops, your concept map will evolve—sometimes becoming more simplified and streamlined, sometimes becoming more complex and interrelated.

Concept maps are context-dependent. The same group of concepts can be organized in several ways, depending on the focus of the map. The following examples show similar clusters of concepts from Neil Campbell's first chapter, on themes in the study of life. Notice how the hierarchy of concepts changes depending on whether you are looking at an overview of biology or at one of the specific themes. Read through these maps; read up and down and across the connections that are drawn. Recognize that these maps provide one way of structuring the ideas of this chapter. There is no one "right" or "wrong" concept map; they are individual representations of understanding. Maps may, however, be more or less accurate or valid, and sharing and talking through concept maps is a good way to assess your understanding.

Remember, the real value in concept mapping is in the process, in the weighing and relating and organizing of concepts that each individual must do to develop a personal and meaningful understanding of this fascinating subject of biology.

Many examples and concept mapping exercises are included in the *Student Study Guide* that accompanies Campbell's *Biology*. Additional information about concept maps may be found in *Learning How to Learn* by J. D. Novak and D. Gowin (New York: Cambridge University Press, 1984).

Martha Taylor
Cornell University

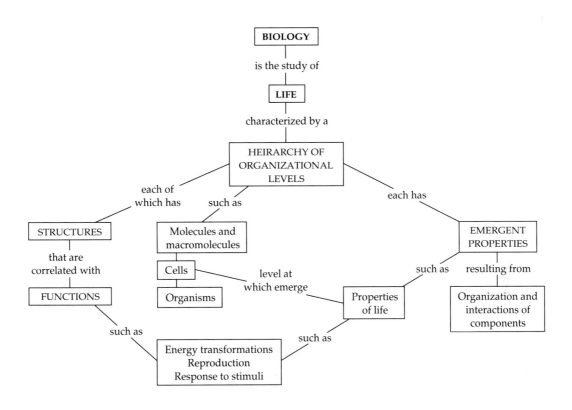

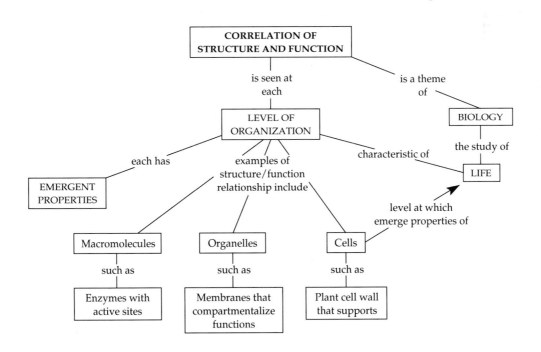

Glossary

abscisic acid *(ab-SIS-ik)* A plant hormone that generally acts to inhibit growth, promote dormancy, and help the plant withstand stressful conditions.

absorption spectrum The range of a pigment's ability to absorb various wavelengths of light.

abyssal zone *(uh-BISS-ul)* The portion of the ocean floor where light does not penetrate and where temperatures are cold and pressures intense.

acclimation *(AK-lim-AY-shun)* Physiological adjustment to a change in an environmental factor.

accommodation Automatic adjustment of an eye to focus on objects at different distances.

acetyl CoA The entry compound for the Krebs cycle in cellular respiration; formed from a fragment of pyruvic acid attached to a coenzyme.

acid A substance that increases the hydrogen ion concentration in a solution.

acoelomates *(ay-SEEL-oh-mates)* Solid-bodied animals lacking cavities between the gut and outer body wall.

actin *(AK-tin)* A globular protein that links into chains, two of which twist helically about each other, forming microfilaments in muscle and other contractile elements in cells.

acrosome *(AK-ruh-some)* An organelle at the tip of a sperm cell that helps the sperm penetrate the egg.

action potential The rapid change in the membrane potential of an excitable cell, caused by stimulus-triggered, selective opening and closing of voltage-sensitive ion channels.

activation energy The initial investment of energy in a reaction. Activation energy (E_a) is required to energize the bonds of the reactants to an unstable transition state that precedes the formation of the products.

active immunity The immunity achieved when antigens enter the body naturally or artificially. Active immunity is due to stimulation of antibody production.

active site That specific portion of an enzyme that attaches to the substrate by means of weak chemical bonds.

active transport The movement of a substance across a biological membrane against its concentration or electrochemical gradient with the help of energy input and specific transport proteins.

adaptive peak An equilibrium state in a population when the gene pool has allelic frequencies that maximize the average fitness of a population's members.

adaptive radiation The emergence of numerous species from a common ancestor introduced to an environment presenting a diversity of new opportunities and problems.

adhesion *(ad-HEE-zhun)* The clinging of one substance to another.

adrenal gland *(uh-DREEN-ul)* An endocrine gland located adjacent to the kidney in mammals; composed of two glandular portions: an outer cortex, which responds to endocrine signals in reacting to stress and effecting salt/water balance, and a central medulla, which responds to nervous inputs in reacting to stress.

age structure The relative number of individuals of each age in a population.

aggregate fruit A ripened ovary resulting from a single flower with several separate carpels.

agnathans *(AG-naa-thuns)* A jawless class of vertebrates represented today by the lampreys and hagfishes.

agonistic behavior *(ag-on-IS-tik)* A type of behavior involving a contest of some kind that determines which competitor gains access to some resource such as food or mates.

aldehyde *(AL-duh-hyde)* An organic molecule with a carbonyl group located at the end of the carbon skeleton.

all-or-none event An action that occurs either completely or not at all—e.g., the generation of an action potential by a neuron.

allantois *(AL-an-TOE-iss)* One of four extraembryonic membranes that serves as a repository for the embryo's nitrogenous waste.

alleles *(uh-LEELS)* Alternative forms of a gene.

allometric growth *(AL-low-MET-rik)* The variation in the relative rates of growth of various parts of the body, which helps to shape the organism.

allopatric speciation *(AL-oh-PAT-rik)* A mode of speciation induced when the ancestral population becomes segregated by a geographical barrier.

allopolyploid *(AL-oh-POL-ee-ploid)* A common type of polyploid species resulting from two different species interbreeding and combining their chromosomes.

allosteric site *(AL-oh-STEER-ik)* A specific receptor site on an enzyme molecule remote from the active site. Molecules bind to the allosteric site and change the shape of the active site, making it either more or less receptive to the substrate.

alpha (α) helix A spiral shape constituting one form of the secondary structure of proteins, arising from a specific hydrogen-bonding structure.

alternation of generations Occurrences of a multicellular diploid form, the sporophyte, with a multicellular haploid form, the gametophyte.

altruistic behavior *(AL-troo-ISS-tik)* The aiding of another individual at one's own risk or expense.

alveolus *(al-VEE-oh-lus;* pl. **alveoli** *al-VEE-oh-lie)* One of the dead-end, multilobed air sacs that constitute the gas exchange surface of the lungs; also, one of the milk-secreting sacs of epithelial tissue in the mammary glands.

amino acid–activating enzymes A family of enzymes, at least one for each amino acid, that catalyzes the attachment of an amino acid to its specific tRNA molecule; also known as aminoacyl-tRNA synthetases.

amino acids *(uh-MEE-noh)* Organic molecules possessing both carboxyl and amino groups and serving as the monomers of proteins.

amino group A functional group that consists of a nitrogen atom bonded to two hydrogen atoms. It can act as a base in solution, accepting a hydrogen ion and acquiring a charge of $+1$.

amniocentesis (*AM-nee-oh-sen-TEE-sis*) A technique for determining genetic abnormalities in a fetus by the presence of certain chemicals or defective fetal cells in the amniotic fluid, obtained by aspiration from a needle inserted into the amnion.

amnion (*AM-nee-on*) The innermost of four extraembryonic membranes that encloses a fluid-filled sac in which the embryo is suspended.

amniote egg A shelled egg with a self-contained reservoir of amniotic fluid, which enables some vertebrates to complete their life cycles on dry land.

amoeboid movement (*uh-MEE-boyd*) A streaming locomotion characteristic of *Amoeba* and other protists, as well as some individual cells, such as white blood cells, in animals.

anabolic pathways (*AN-uh-BOLL-ik*) Metabolic pathways that consume energy to build complicated molecules from simpler ones.

anagenesis (*AN-uh-JEN-eh-sis*) A pattern of evolutionary change involving the transformation of an entire population, sometimes to a state different enough from the ancestral population to justify renaming it as a separate species; also called phyletic evolution.

analogy (*uh-NAL-oh-jee*) The similarity of structure between two species that are not closely related; attributable to convergent evolution.

androgens (*ANN-droh-jens*) The principal male steroid hormones, such as testosterone, which are produced by the Leydig cells and which stimulate the development and maintenance of the male reproductive system and secondary sex characteristics.

aneuploidy (*AN-yoo-ploy-dee*) The chromosomal aberration in which certain chromosomes are present in extra copies or are deficient in number.

angiosperms (*ANN-jee-oh-spurms*) Flowering plants, which form seeds inside protective chambers called ovaries.

animal pole The point that defines the axis of the zygote of the invertebrate chordate amphioxus where the polar body is budded from the cell when meiosis reduces the chromosome count of the maternal nucleus. In amphibian, reptile, and bird blastulas, cells near the animal pole are smaller and poorer in yolk than those near the opposite end, the vegetal pole.

anion (*ANN-eye-un*) A negatively charged ion.

annuals Plants that complete their entire life cycles in a single year or growing season.

anterior The head end of a bilaterally symmetric animal.

anther (*ANN-thur*) The terminal pollen sac of a stamen, inside which pollen grains with male gametes form in the flower of an angiosperm.

anticodon (*AN-tee-COH-don*) A specialized base triplet on one end of a tRNA molecule that recognizes a particular complementary codon on an mRNA molecule.

antigen (*ANN-teh-jen*) A foreign macromolecule that does not belong to the host organism and that elicits an immune response.

antigenic determinants Localized regions on the surface of an antigen that are chemically recognized by antibodies.

aphotic zone (*ay-FOE-tik*) The part of the ocean beneath the photic zone where light does not penetrate enough for photosynthesis to take place.

apical dominance (*AY-pik-ul*) Concentration of growth at the tip of a plant shoot, where a terminal bud exerts partial inhibition of axillary bud growth.

apical meristems (*AY-pik-ul MARE-eh-stems*) Embryonic plant tissues in the tips of roots and in the buds of shoots that supply cells for the plant to grow in length.

apomixis (*Ap-oh-MIX-sis*) Asexual production of seeds.

apomorphic characters (*AP-oh-MORE-fik*) Derived phenotypic characters, or homologies, that evolved after a branch diverged from a phylogenetic tree.

apoplast (*AY-poe-plast*) In plants, the nonliving continuum formed by the extracellular pathway provided by the continuous matrix of cell walls.

aposematic coloration (*AP-oh-some-AT-ik*) The bright color of animals with effective physical or chemical defenses that acts as a warning to predators.

aqueous solution (*AY-kwee-us*) A solution in which water is the solvent.

archaebacteria (*AR-kuh-bak-TEER-ee-uh*) The more ancient of the two major lineages of prokaryotes, represented today by a few groups of bacteria inhabiting extreme environments.

archenteron (*ark-EN-ter-on*) The endoderm-lined cavity formed during the gastrulation process that develops into the digestive tract of the animal.

arteries Vessels that carry blood away from the heart to organs throughout the body.

arterioles (*ar-TEER-ee-oles*) Branches of the arteries present within organs.

artificial selection The selective breeding of domesticated plants and animals to encourage the occurrence of desirable traits.

asexual reproduction A type of reproduction involving only one parent that produces genetically identical offspring by budding or division of a single cell or the entire organism into two or more parts.

A-site The binding site on a ribosome that holds the tRNA carrying the next amino acid to be added to a growing polypeptide chain.

associative learning The acquired ability to associate one stimulus with another. Also called classical conditioning.

assortive mating A type of nonrandom mating in which mating partners resemble each other in certain phenotypic characters.

asymmetric carbon A carbon atom covalently bonded to four different atoms or groups of atoms.

atherosclerosis (*ATH-ur-oh-sklur-OH-sis*) A chronic cardiovascular disease in which plaques develop on the inner walls of arteries, narrowing their bore.

atomic number The number of protons in the nucleus of an atom, unique for each element and designated by a subscript to the left of the elemental symbol.

ATP (adenosine triphosphate) (*uh-DEN-oh-sin try-FOS-fate*) An adenine-containing nucleoside triphosphate that releases free energy when its phosphate bonds are hydrolyzed. This energy is used to drive endergonic reactions in cells.

ATPases A group of enzymes that function in producing or using ATP.

atria (*AY-tree-uh*) The chambers of the heart that receive blood returning to the heart.

autoimmune disease An immunological disorder in which the immune system goes awry and turns against itself.

autonomic nervous system (*AWT-uh-NAHM-ik*) A subdivision of the motor nervous system of vertebrates that regulates the internal environment; consists of the sympathetic and parasympathetic subdivisions.

autopolyploid (*AW-toe-POL-ee-ploid*) A type of polyploid species resulting from one species doubling its chromosome number to become tetraploids, which may self-fertilize or mate with other tetraploids.

autosomes (*AW-tuh-somes*) Chromosomes that are not directly involved in determining sex.

autotrophic nutrition (*AW-tuh-TROE-fik*) A mode of obtaining organic food molecules without eating other organisms. Autotrophs use energy from the sun or from the oxidation of inorganic substances to make organic molecules from inorganic ones.

auxin (*AWK-sin*) One of several hormone compounds in plants that have a variety of effects, such as phototropic response through stimulation of cell elongation, stimulation of secondary growth, and development of leaf traces and fruit.

auxotroph (*AWK-soh-trofe*) A nutritional mutant that is unable to synthesize and that cannot grow on media lacking certain essential molecules that are normally synthesized by wild-type strains of the same species.

axillary bud (*AKS-ill-air-ee*) An embryonic shoot present in the angle formed by a leaf and stem.

axon (*AKS-on*) A typically long outgrowth, or process, from a neuron that carries nerve impulses away from the cell body toward target cells.

axon hillock (*HILL-uk*) A cone-shaped region of the nerve cell body where impulses passing down the axon are usually generated.

bacteriophage (*bak-TEER-ee-oh-faj*) A virus that infects bacteria; also called a phage.

balanced polymorphism A type of polymorphism in which the frequencies of the coexisting forms do not change noticeably over many generations.

Barr body The dense object that lies along the inside of the nuclear envelope in cells of female mammals, representing the one inactivated *X* chromosome.

basal body A cell structure identical to a centriole that organizes and anchors the microtubule assembly of a cilium or flagellum.

basal metabolic rate (BMR) The minimal number of kilocalories a resting animal requires to fuel itself for a given time.

base A substance that reduces the hydrogen ion concentration in a solution.

basement membrane The floor of an epithelial membrane on which the basal cells rest.

Batesian mimicry (*BATESZ-ee-un MIM-ih-kree*) A type of mimicry in which a harmless species looks like a different species that is poisonous or otherwise harmful to predators.

B cells A type of lymphocyte that develops in the bone marrow and later produces antibodies, which mediate humoral immunity.

benign tumor (*bin-NINE*) An abnormal growth composed of cells that multiply excessively but remain at their place of origin in the body.

benthic (*BEN-thik*) Pertaining to the bottom surfaces of aquatic environments; bottom dwelling.

beta (β) pleated sheet A zigzag shape, constituting one form of the secondary structure of proteins, formed of hydrogen bonds between polypeptide segments running in opposite directions.

biennial (*by-EN-ee-ul*) A plant that requires two years to complete its life cycle.

bilateral symmetry A body form with a central longitudinal plane dividing the body into two equal but opposite halves.

Bilateria (*BY-leh-TEER-ee-uh*) A branch of eumetazoans possessing bilateral symmetry.

bile A mixture of substances containing bile salts, which emulsify fats and aid in their digestion and absorption.

binary fission The kind of cell division found in prokaryotes, in which dividing daughter cells each receive a copy of the single parental chromosome.

biogeochemical cycles The various nutrient circuits, which involve both biotic and abiotic components of ecosystems.

biogeography The study of the past and present distribution of species.

biological magnification A trophic process in which retained substances become more concentrated with each link in the food chain.

biological species A population or group of populations whose members have the potential to interbreed.

biomass The dry weight of organic matter comprising a group of organisms in a particular habitat.

biomes (*BY-omes*) The world's major communities, classified according to the predominant vegetation and characterized by adaptations of organisms to that particular environment.

biosphere (*BY-oh-sfeer*) The entire portion of the Earth that is inhabited by life. The sum of all the planet's ecosystems.

biotechnology The industrial use of living organisms or their components to improve human health and food production.

biotic (*by-OT-ik*) Pertaining to the living organisms in the environment.

blastocoel (*BLAS-toe-seel*) The fluid-filled cavity that forms in the center of the blastula embryo.

blastopore (*BLAS-toe-pore*) The opening of the archenteron in the gastrula that develops into the mouth in protostomes and the anus in deuterostomes.

blastula (*BLAS-tyoo-la*) The hollow ball of cells marking the end stage of cleavage during early embryonic development.

blood–brain barrier A specialized capillary arrangement in the brain that restricts the passage of most substances into the brain, thereby preventing dramatic fluctuations in the brain's environment.

blood pressure The hydrostatic force that blood exerts against the wall of a vessel.

bond energy The quantity of energy that must be absorbed to break a particular kind of chemical bond; equal to the quantity of energy the bond releases when it forms.

book lungs Organs of gas exchange in spiders, consisting of stacked plates contained in an internal chamber.

bottleneck effect Genetic drift resulting from reduction of a population, typically by a natural disaster, such that the surviving population is no longer genetically representative of the original population.

Bowman's capsule (*BOH-munz*) A cup-shaped receptacle that is the initial, expanded segment of the nephron where filtrate enters from the blood.

brainstem The hindbrain and midbrain of the vertebrate central nervous system. In the human, it forms a cap on the anterior end of the spinal cord, extending to about the middle of the brain.

bryophytes (*BRY-oh-fites*) The mosses, liverworts, and hornworts—a group of nonvascular plants that inhabit the land but lack many of the terrestrial adaptations of vascular plants.

budding An asexual means of propagation in which outgrowths from the parent form and pinch off to live independently or else remain attached to eventually form extensive colonies.

buffer A substance that consists of acid and base forms in solution and that minimizes changes in pH when extraneous acids or bases are added to the solution.

callus (*KAL-iss*) A mass of dividing, undifferentiated cells at the cut end of a shoot or in a tissue culture, from which adventitious roots develop for vegetative propagation.

calmodulin (*kal-MOD-yoo-lin*) An intracellular protein to which calcium binds in its function as a second messenger in hormone action.

calorie The amount of heat energy required to raise the temperature of 1 g of water 1°C; the amount of heat energy that 1 g of water releases when it cools by 1°C. The Calorie (with a capital C), usually used to indicate the energy content of food, is a kilocalorie.

Calvin cycle The second of two major stages in photosynthesis; involves atmospheric CO_2 fixation and reduction of the fixed carbon into carbohydrate.

CAM Crassulacean acid metabolism, an adaptation for photosynthesis in arid conditions, first discovered in the plant family Crassulaceae. Carbon dioxide entering open stomata during the night is converted into organic acids, which release CO_2 for the Calvin cycle during the day, when stomata are closed.

camera eye A type of eye found in cephalopods and vertebrates that has a single lens that focuses light onto the photosensitive retina.

capillaries (*KAP-ill-air-ees*) Microscopic blood vessels penetrating the tissues and consisting of a single layer of endothelial cells that allows exchange between the blood and interstitial fluid.

capsid The variously shaped protein shell that encloses the viral genome.

carbohydrate (*KAR-bo-HI-drate*) A sugar (monosaccharide) or one of its dimers (disaccharides) or polymers (polysaccharides).

carbonyl group (*KAR-bon-EEL*) A functional group found in aldehydes and ketones that consists of a carbon atom double-bonded to an oxygen atom.

carboxyl group (*kar-BOX-ul*) A functional group present in organic acids, consisting of a single carbon atom double-bonded to an oxygen atom and also bonded to a hydroxyl group.

carcinogen (*kar-SIN-oh-jen*) A chemical agent that causes cancer.

cardiac muscle (*KAR-dee-ak*) A type of muscle that forms the contractile wall of the heart. Its cells are joined by intercalated discs that relay each heartbeat.

carpels (*KAR-pels*) The female reproductive organs of a flower, consisting of the stigma, style, and ovary.

carrier protein A transport protein involved in facilitated diffusion, possessing a specific binding site for a specific substance.

carrying capacity The maximum population size that can be supported by the available resources; symbolized as K.

cartilage (*KAR-til-ij*) A type of flexible connective tissue with an abundance of collagenous fibers embedded in a rubbery ground substance called chondrin, which is secreted by chondrocytes.

Casparian strip (*kas-PAR-ee-un*) A water-impermeable ring of wax around endodermal cells that blocks the passive flow of water and solutes into the stele by way of cell walls.

catabolic pathways (*KAT-uh-BOL-ik*) Metabolic pathways that release energy by breaking down complex molecules to simpler compounds.

catabolite activator protein (CAP) (*ka-TAB-ul-LITE*) In *Escherichia coli*, a helper protein that stimulates gene expression by binding within the promoter region of an operon and enhancing the promoter's ability to associate with RNA polymerase.

cation (*KAT-eye-on*) An ion with a positive charge, produced by the loss of one or more electrons.

cell adhesion molecules (CAMS) A diverse group of molecules on the surface of cells that vary in amount or chemical identity from one cell type to another and that contribute to selective cell association during embryonic development.

cell center A region in the cytoplasm near the nucleus from which microtubules originate and radiate.

cell culture A technique of growing isolated cells outside the organism.

cell cycle An ordered sequence of events in the life of a dividing cell, composed of the M, G_1, S, and G_2 phases.

cell fractionation The disruption of a cell and separation of its organelles by centrifugation.

cell-mediated immune system The part of the immune system that functions in defense against fungi, protists, bacteria, and viruses inside host cells and against tissue transplants with highly specialized cells that circulate in the blood and lymphoid tissue.

cell plate A double membrane across the midline of a dividing plant cell, between which the new cell wall forms during cytokinesis.

cell wall A wall formed of cellulose fibers embedded in a polysaccharide/protein matrix, surrounding the plasma membrane of a plant cell.

cellular differentiation Divergence in structure and function of cells as they become specialized during a multicellular organism's development; depends on control of gene expression.

cellular respiration The most prevalent and efficient catabolic pathway for the production of ATP, in which oxygen is consumed as a reactant along with the organic fuel.

cellulose (*SELL-yoo-lose*) A structural polysaccharide of cell walls, consisting of glucose monomers joined by β-1, 4-glycosidic linkages.

Celsius scale (*SELL-see-us*) A temperature scale (°C) equal to ⁵⁄₉ (°F − 32) that measures the freezing point of water at 0°C and the boiling point of water at 100°C.

central nervous system In vertebrate animals, the brain and spinal cord.

central vacuole A membrane-enclosed sac taking up most of the interior of a mature plant cell and containing a variety of substances important in plant reproduction, growth, and development.

centrioles (*SEN-tree-oles*) Two structures in the center of animal cells, composed of cylinders of nine triplet microtubules in a ring. Centrioles help organize microtubule assembly during cell division.

centromere (*SEN-troh-mere*) The centralized region joining two sister chromatids.

cephalization (*SEF-uh-le-ZA-shun*) The evolution of a "head" end with sensory structures and a highly specialized brain to process sensory input; a feature of bilaterally symmetric animals, especially the vertebrates.

cerebellum (*SEH-reh-BELL-um*) Part of the vertebrate hindbrain (rhombencephalon) located dorsally; functions in unconscious coordination of movement and balance.

cerebral cortex (*seh-REEB-brul*) The surface of the cerebrum; the largest and most complex part of the mammalian brain, containing sensory and motor nerve cell bodies of the cerebrum; the part of the vertebrate brain most changed through evolution.

cerebrum (*seh-REEB-brum*) The dorsal portion, composed of right and left hemispheres, of the vertebrate telencephalon; the integrating center for memory, learning, emotions, and other highly complex functions of the central nervous system.

chaparral (*SHAP-uh-RAL*) A scrubland biome of dense, spiny evergreen shrubs found at midlatitudes along coasts where cold ocean currents circulate offshore; characterized by mild, rainy winters and long, hot, dry summers.

character displacement The divergence of overlapping characteristics in two species living in the same environment as a result of resource partitioning.

chemical equilibrium In a reversible chemical reaction, the point at which the rate of the forward reaction equals the rate of the reverse reaction.

chemical synapse (*SIN-aps*) A junction between two neurons or between a neuron and a muscle, receptor, or gland that uses chemicals to transmit information between the cells.

chemiosmosis (*KEE-mee-os-MOH-sis*) The production of ATP by coupling the transport of hydrogen ions across membranes with the phosphorylation of ADP.

chemoautotrophs (*KEE-moh-AW-tow-trohfs*) Organisms that need only carbon dioxide as a carbon source, but that obtain their energy by oxidizing inorganic substances.

chemoheterotrophs (*KEE-moh-HET-er-oh-trohfs*) Organisms that must consume organic molecules both for energy and as a source of carbon skeletons.

chemoreceptors (*KEE-moh-ree-SEP-turz*) A class of receptors that transmits information about the total solute concentration in a solution or about individual kinds of molecules.

chemosynthetic autotrophs Those autotrophs that use the energy from the oxidation of inorganic substances to drive the formation of organic molecules from inorganic substrates.

chiasmata (sing. **chiasma**) *(KYE-az-muh-tuh KYE-as-muh)* The X-shaped, microscopically visible figures representing homologous chromatids that have exchanged genetic material through crossing over during meiosis.

chitin *(KY-tin)* A structural polysaccharide of an amino sugar found in many fungi and in the exoskeletons of all arthropods.

chlorophyll *(KLOR-oh-fill)* Various green pigments located within chloroplasts.

chloroplasts *(KLOR-oh-plasts)* Organelles found only in plants and photosynthetic protists that absorb sunlight and use it to drive the synthesis of organic compounds from carbon dioxide and water.

cholesterol *(kol-ESS-teh-rol)* A steroid forming an essential component of animal cell membranes and acting as a precursor molecule for the synthesis of other biologically important steroids.

chordates *(KOR-dates)* A diverse phylum of animals that possess a notochord, a dorsal, hollow nerve cord, and pharyngeal gill slits as embryos.

chorion *(KOR-ee-on)* The outermost of four extraembryonic membranes; contributes to formation of the mammalian placenta.

chorionic villi sampling *(KOR-ee-on-ik VILL-eye)* A technique for diagnosing genetic and congenital defects while the fetus is in the uterus. A small sample of the fetal portion of the placenta is removed and analyzed.

chromatin *(KRO-muh-tin)* The aggregate mass of dispersed genetic material formed of DNA and protein and observed between periods of cell division in eukaryotic cells.

chromatography *(KRO-muh-TA-graf-ee)* A technique used to separate organic compounds from mixtures on the basis of their solubility in one or more solvents passed through a solid support medium.

chromosomes *(KRO-muh-somes)* Long, threadlike associations of genes found in the nucleus of all eukaryotic cells and most visible during mitosis and meiosis. Chromosomes consist of DNA and protein.

cilium *(SILL-ee-um; pl.* **cilia***)* One of many short cell appendages specialized for locomotion and formed from a core of nine outer doublet microtubules and two inner single microtubules ensheathed in an extension of plasma membrane.

circadian rhythms *(sur-KAY-dee-un)* Physiological cycles of about 24 hours, present in all eukaryotic organisms, that persist even in the absence of external cues.

citric acid cycle *(SIT-rik)* A cyclic metabolic pathway occurring in the matrix of the mitochondrion, which generates NADH, FADH$_2$, and some ATP by substrate-level phosphorylation; also called the tricarboxylic acid or Krebs cycle.

cladistics *(kluh-DIS-tiks)* A taxonomic approach that classifies organisms according to the order in time at which branches arise along a phylogenetic tree, without considering the degree of morphological divergence.

cladogenesis *(KLAY-doh-GEN-eh-sis)* A pattern of evolutionary change that produces biological diversity by budding one or more new species from a parent species that continues to exist; also called branching evolution.

classical conditioning A type of associative learning; association of a normally irrelevant stimulus with a fixed behavioral response.

classical evolutionary taxonomy *(TAX-on-um-ee)* An approach to biological classification and phylogeny that considers both overall homology and evolutionary branching sequences.

cleavage The process of cytokinesis in animal cells, characterized by pinching of the plasma membrane; also, the succession of rapid cell divisions without growth during early embryonic development that converts the zygote into a ball of cells.

cleavage furrow The first sign of cleavage in an animal cell; a shallow groove in the cell surface near the old metaphase plate.

cline *(KLYNE)* Variation in features of individuals in a population that parallels a gradient in the environment.

cloaca *(kloh-AY-kuh)* A common opening for the digestive, urinary, and reproductive tracts seen in all vertebrates except most mammals.

clonal selection *(KLON-ul)* The mechanism that determines specificity and accounts for memory of antigens in the immune system; occurs because an antigen introduced into the body selectively activates only a tiny fraction of inactive lymphocytes, which proliferate to form a clone of effector cells specific for the stimulating antigen.

clone *(KLONE)* A lineage of genetically identical individuals.

closed circulatory system A type of internal transport in which blood is confined to vessels.

coadapted gene complex Genes whose functions are interrelated.

cochlea *(KOH-klee-uh)* The complex, coiled organ of hearing that contains the organ of Corti.

codominant A phenotypic situation in which both alleles are expressed in the heterozygote.

codon *(KOH-don)* A three-nucleotide sequence of DNA or mRNA that specifies a particular amino acid or termination signal and that functions as the basic unit of the genetic code.

coelom *(SEE-loam)* A body cavity completely lined with mesoderm.

coelomates *(SEE-loh-mates)* Animals whose body cavities are completely lined by mesoderm, the layers of which connect dorsally and ventrally to form mesenteries.

coenocytic *(SEN-no-SIT-ik)* A multinucleated condition resulting from repeated division of nuclei without cytoplasmic division.

coenzyme *(ko-EN-zyme)* An organic molecule serving as a cofactor. Most of the vitamins function as coenzymes in important metabolic reactions.

coevolution The mutual influence on the evolution of two different species interacting with each other and reciprocally influencing each other's adaptations.

cofactor Any nonprotein molecule or ion that is required for the proper functioning of an enzyme. Cofactors can be permanently bound to the active site or may bind loosely with the substrate during catalysis.

cohesion *(ko-HEE-zhun)* The binding together of two or more substances, often by hydrogen bonds.

coleoptile *(KOAL-ee-OP-tyle)* The structure that covers the embryonic shoot in germinating monocot seeds.

collenchyma *(koal-EN-keh-muh)* A flexible cell type in plants that occurs in strands or cylinders that support young parts of the plant without restraining growth.

commensalism *(kuh-MEN-sul-iz-um)* A symbiotic relationship in which the symbiont benefits, but the host is neither helped nor harmed.

community All the organisms that inhabit a particular area; an assemblage of populations of different species living close enough together for potential interaction.

companion cell A type of plant cell that is connected to a sieve-tube member by many plasmodesmata and whose nucleus and ribosomes may serve one or more adjacent sieve-tube members.

competitive exclusion principle The concept that when the populations of two species compete for the same limiting resources, one population will use the resources more efficiently and have a reproductive advantage that will eventually lead to the elimination of the less efficient species population.

competitive inhibitor A substance that reduces the activity of an enzyme by entering the active site in place of the substrate whose structure it mimics.

complement A group of at least 20 blood proteins that cooperate with other defense mechanisms; may amplify the inflammatory response, enhance phagocytosis, or directly lyse pathogens; activated by onset of the immune response or by surface chemicals on microorganisms.

complementary DNA (cDNA) DNA that is identical to a native DNA containing a gene of interest except that the cDNA lacks noncoding regions (introns) because it is synthesized in the laboratory using mRNA templates.

complete digestive tract A digestive tube that runs between a mouth and an anus. An incomplete digestive tract has only one opening.

complete flower A flower that has sepals, petals, stamens, and carpels.

complete metamorphosis (MET-uh-MOR-fuh-sis) A type of development in certain animals, characterized by larval stages that look entirely different from the adult stage.

complex tissue A plant tissue composed of more than one type of cell.

compound A chemical combination, in a fixed ratio, of two or more elements.

compound eye A type of multifaceted eye present in insects and crustaceans that consists of up to several thousand light-detecting, focusing ommatidia and that is especially good at detecting movement.

cone cell One of two types of photoreceptors in the vertebrate eye; detects color during the day.

conjugation (KON-joo-GAY-shun) A sexual-like process in some organisms that results in exchange of genetic material between two cells that are temporarily joined.

contact inhibition The phenomenon observed in normal animal cells that causes them to stop dividing when they come into contact with one another.

contractile vacuole (kun-TRAK-tul VAK-you-ole) An organelle that pumps excess water out of many freshwater protozoan cells.

convection (kon-VEK-shun) The mass movement of warmed air or liquid to or from the surface of a body or object.

convergent evolution The independent development of similarity between species as a result of their having similar ecological roles and selection pressures.

cooperativity (koh-OP-ur-uh-TIV-eh-tee) The interaction of the constituent subunits of a protein that causes a conformational change in one subunit to be transmitted to all the others.

corepressor (KOH-re-PRESS-ur) A substance that, in complex with a repressor protein, attaches to an operator gene and blocks operon activity.

cork cambium (KAM-bee-um) A cylinder of meristematic tissue in plants that produces cork cells externally and phelloderm internally to replace the epidermis during secondary growth.

corpus callosum (KOR-pus KAHL-lose-um) A thick band of nerve fibers connecting the right and left cerebral hemispheres of the vertebrate brain.

corpus luteum (KOR-pus LOO-tee-um) Secreting tissue in the ovary that forms from the collapsed follicle after ovulation and produces progesterone.

cotransport The coupling of the downhill diffusion of one substance to the transport of another against its gradient.

cotyledons (KOT-eh-LEE-dons) The one (monocots) or two (dicots) seed leaves of an angiosperm embryo.

cotylosaurs (KOT-eh-luh-sores) The stem reptiles.

countercurrent exchange The opposite flow of adjacent fluids that maximizes transfer rates; for example, blood in the gills flows opposite to the direction in which water passes over the gills, maximizing oxygen uptake and carbon dioxide loss.

covalent bond (koh-VALE-ent) A type of strong chemical bond in which two atoms share one pair of electrons in a mutual valence shell.

C₃ plants Those plants using the Calvin-Benson cycle for the initial steps of the Calvin cycle; they form a three-carbon acid as the first stable intermediate.

C₄ plants Plants that preface the Calvin-Benson cycle with reactions that incorporate carbon dioxide into four-carbon compounds, the end product of which supplies CO_2 for the Calvin cycle.

cristae (KRIS-tee) The inner, infolded membrane of a mitochondrion that houses the electron transport chain and the enzyme catalyzing the synthesis of ATP.

critical period The limited time during which imprinting can occur.

crossing over The reciprocal exchange of genetic material between nonsister chromatids of a bivalent during synapsis of meiosis I.

cryptic coloration (KRIP-tik) A type of camouflage that makes potential prey difficult to spot against its background.

cuticle (KYOO-teh-kul) The exoskeleton of an arthoropod, consisting of layers of protein and chitin that are variously modified for different functions; also, a waxy covering on the surface of stems and leaves that acts as an adaptation to prevent desiccation in terrestrial plants.

cyanobacteria (sigh-AN-oh-bak-TEER-ee-uh) Photosynthetic, oxygen-generating prokaryotes belonging to the kingdom Monera.

cyclic adenosine monophosphate (cyclic AMP or cAMP) A small, ring-shaped molecule that acts as a chemical signal in slime molds, as an intracellular second messenger in vertebrate endocrine systems, and as a regulator of the *lac* operon.

cyclic electron flow A route of electron flow during the light reactions of photosynthesis that involves only photosystem I and produces ATP but no NADPH or oxygen.

cyclic photophosphorylation (FO-to-foss-FOR-i-lay-shun) Production of ATP during a cyclic pathway of electron flow.

cytochromes (SIGH-tuh-kromes) A group of iron-containing proteins that are components of electron transport chains in mitochondria and chloroplasts.

cytokinesis (SIGH-toh-kin-EE-sis) The division of the cytoplasm to form two separate daughter cells immediately after mitosis.

cytokinins (SIGH-toh-KY-nins) A class of related hormones in plants that retard aging and act in concert with auxins to stimulate cell division, influence the pathway of differentiation, and control apical dominance.

cytoplasm (SIGH-toh-plaz-um) The entire contents of the cell exclusive of the nucleus and bounded by the plasma membrane.

cytoplasmic determinants Localized cellular substances that become partitioned into specific blastomeres of an embryo, thereby fixing the developmental fates of different regions very early.

cytoskeleton (SIGH-toh-SKEL-eh-ton) A network of microtubules, microfilaments, and intermediate filaments that ramify throughout the cytoplasm and serve a variety of mechanical and transport functions.

cytosol (SIGH-toh-sol) The semifluid portion of the cytoplasm.

dalton (DAWL-ton) The atomic mass unit. A measure of mass for atoms and subatomic particles.

Darwinian fitness A measure of the relative contribution of an individual to the gene pool of the next generation.

day-neutral plants A group of plants whose flowering is not affected by photoperiod.

decomposers Saprophytic fungi and bacteria that absorb nutrients from nonliving organic material such as corpses, fallen plant material, and the wastes of live organisms, and convert them into inorganic forms.

dehydration synthesis A process of polymer formation in which monomers are linked together by removing a molecule of water; also called condensation.

deletion An aberration in chromosome structure resulting from an error in meiosis or from mutagens; a deficiency in a chromosome resulting from loss of a fragment through breakage.

demography *(de-MOG-ruf-ee)* The study of statistics relating to births and deaths in populations.

denaturation A process in which a protein unravels and loses its native conformation, thereby becoming biologically inactive. Denaturation occurs under suboptimal conditions of pH, salt concentration, and temperature.

dendrite *(DEN-dryt)* One of usually numerous, short, highly branched processes of a neuron that conveys nerve impulses toward the cell body.

density-dependent factor Any factor influencing population regulation that has a greater impact as population density increases.

density-independent factor Any factor influencing population regulation that acts to reduce population size by the same fraction whether the population is large or small.

deoxyribonucleic acid (DNA) *(DEE-ox-ee-rye-boh-noo-KLAY-ik)* A double-stranded, helical nucleic acid molecule capable of replicating and determining the inherited structure of a cell's proteins.

depolarization *(dee-POL-ur-iz-AY-shun)* An electrical state in an excitable cell whereby the inside of the cell is made less negative relative to the outside than was the case at resting potential. A nerve cell membrane is depolarized if a stimulus decreases its voltage from the resting potential of -70 mV in the direction of zero voltage.

deposit-feeders Heterotrophs, such as earthworms, that eat their way through detritus, salvaging bits and pieces of decaying organic matter.

desmosome *(DEZ-muh-soam)* A type of intercellular junction in animal cells that functions as a "spot weld."

determinate cleavage A type of embryonic development in protostomes that rigidly casts the developmental fate of each embryonic cell very early.

determinate growth A type of growth characteristic of animals, in which the organism stops growing after it reaches a certain size.

determination The progressive restriction of developmental potential, causing the possible fates of each cell to become more limited as the embryo develops.

detritivores *(deh-TRYT-eh-vores)* A special class of consumers that derives its energy from organic wastes and dead organisms representing all trophic levels.

detritus *(deh-TRY-tis)* Dead organic matter.

deuterostomes *(DOO-ter-oh-stomes)* One of two distinct evolutionary lines of coelomates, consisting of the echinoderms and chordates and characterized by radial, indeterminate cleavage, enterocoelous formation of the coelom, and development of the anus from the blastopore.

diaphragm A sheet of muscle that forms the bottom wall of the thoracic cavity in mammals; active in ventilating the lungs.

diastole *(die-ASS-tuh-lee)* The stage of the heart cycle in which the heart muscle is relaxed, allowing the chambers to fill with blood.

dicots *(DIE-kots)* A subdivision of flowering plants whose members possess two embryonic seed leaves, or cotyledons.

dictyosome *(DIK-tee-uh-soam)* One of several Golgi complexes in a plant cell.

differentiation *(DIF-ur-EN-shee-aye-shun)* The process of becoming specialized, as during embryonic development when cells take on a certain form and function and become ordered into tissues and organs that perform specific activities.

diffusion *(deh-FYU-shun)* The spontaneous tendency of a substance to move down its concentration gradient from a more concentrated to a less concentrated area.

digestion The process of breaking down food into molecules small enough for the body to absorb.

dihybrid cross *(DIE-HIGH-brid)* A breeding experiment in which parental varieties differing in two traits are mated.

dikaryons *(die-KAH-ree-ons)* Mycelia of certain septate fungi that possess two separate haploid nuclei per cell.

dioecious *(die-EE-shus)* A plant species that has staminate and carpellate flowers on separate plants.

diploid cell *(DIP-loid)* A cell containing two sets of chromosomes (2N), one set inherited from each parent.

directional selection Natural selection that favors relatively rare individuals on one end of the phenotypic range.

disaccharide *(die-SAK-ur-ide)* A double sugar, consisting of two monosaccharides joined by dehydration synthesis.

diversifying selection Natural selection that favors extreme over intermediate phenotypes.

DNA-DNA hybridization Comparison of whole genomes of two species by estimating the extent of hydrogen bonding that occurs between single-stranded DNA obtained from the two species.

DNA ligase *(LYE-gase)* A linking enzyme essential for DNA replication; catalyzes the covalent bonding of the 3' end of a new DNA fragment to the 5' end of a growing chain.

DNA methylation The addition of methyl groups ($-CH_3$) to bases of DNA after DNA synthesis; may serve as a long-term control of gene expression.

DNA polymerases *(pol-IM-ur-ases)* A family of enzymes that catalyze the synthesis of a new DNA strand once the old strands are separated and a primer is in place.

DNA probe A chemically synthesized, radioactively labeled segment of nucleic acid used to find a gene of interest by hydrogen-bonding to a complementary sequence.

domain A structural and functional portion of a polypeptide that may be coded for by a specific exon; a globular region of a protein with tertiary structure.

dominance hierarchy A linear "peck order" of animals, where position dictates characteristic social behaviors.

dominant allele In a heterozygote cell, the allele that is fully expressed in the phenotype.

dorsal The top (or back) half of a bilaterally symmetric animal.

dorsal lip The upper edge of the blastopore produced by invagination during gastrula formation in amphibian embryos; the site toward which surface cells of the gastrula converge and migrate inward along the roof of the blastocoel in the process of involution.

double circulation A circulation with separate pulmonary and systemic circuits, which ensures vigorous blood flow to all organs.

double fertilization A mechanism of fertilization found in angiosperms and unique in all of life, in which two sperm cells unite with two cells in the embryo sac, one containing the egg and the other two polar nuclei.

double helix A term for the form of native DNA, referring to its two adjacent polynucleotide strands wound into a spiral shape.

duplication An aberration in chromosome structure resulting from an error in meiosis or mutagens; duplication of a portion of a chromosome resulting from fusion with a fragment from a homologous chromosome.

Down syndrome (SIN-drome) A human genetic disease with characteristic symptoms, including severe mental retardation and heart and respiratory defects, resulting from having an extra chromosome 21.

dynein (DY-neen) A large contractile protein forming the sidearms of microtubule doublets in cilia and flagella.

ecdysone (EK-deh-sone) A steroid hormone that triggers molting in arthropods.

ecology (ee-KOL-uh-jee) The study of how organisms interact with their environments (from the Greek eikos, "home").

ecosystem (EE-koh-sis-tum) A level of ecological study that includes all the organisms in a given area along with the abiotic factors with which they interact; a community and its physical environment.

ectoderm (EK-tuh-durm) The outermost of the three primary germ layers in animal embryos; gives rise to the outer covering and, in some phyla, to the nervous system, inner ear, and lens of the eye.

ectotherms (EK-toh-thurms) Animals, such as reptiles, fishes, and amphibians, that must use environmental energy and behavioral adaptations to regulate their body temperature.

effector cells Muscle cells or gland cells that perform the body's responses to stimuli. They respond to signals from the brain or other processing center of the nervous system.

electrical synapse (SIN-aps) A junction between two neurons separated only by a gap junction, in which the local currents sparking the action potential pass directly between the cells.

electrochemical gradient The diffusion gradient of an ion, representing a type of potential energy that accounts for both the concentration difference of the ion across a membrane and its tendency to move relative to the membrane potential.

electron microscope A microscope that focuses an electron beam through a specimen, consequently having a thousandfold greater resolving power than a light microscope. A transmission electron microscope (TEM) is used to study the internal structure of thin sections of cells. A scanning electron microscope (SEM) is used to study the fine details of cell surfaces.

electron transport chain A group of molecules in the inner membrane of a mitochondrion that synthesize ATP molecules by means of an exergonic slide of electrons; the third major stage (oxidative phosphorylation) in cellular respiration.

electronegativity (eh-LEK-troh-neg-uh-TIV-eh-tee) The tendency for an atom to pull electrons toward itself.

element Any substance that cannot be broken down to any other substance by ordinary chemical means.

embryo sac The female gametophyte of angiosperms, formed from growth and division of the megaspore into a multicellular structure with eight haploid nuclei.

endergonic reaction (EN-dur-GON-ik) A nonspontaneous chemical reaction in which free energy is absorbed from the surroundings.

endocrine glands (EN-doh-krin) Ductless glands that secrete hormones directly into the bloodstream.

endocrine system The internal system of chemical communication involving hormones, the ductless glands that secrete hormones, and the molecular receptors on or in target cells that respond to hormones; functions in concert with the nervous system to effect internal regulation and maintain homeostasis.

endocytosis (EN-do-sigh-TOE-sis) The cellular uptake of macromolecules and particulate substances by localized regions of the plasma membrane that surround the substance and pinch off to form an intracellular vesicle.

endoderm (EN-doh-durm) The innermost of the three primary germ layers in animal embryos; lines the archenteron and gives rise to liver, pancreas, lungs, and the lining of the digestive tract.

endodermis (EN-doh-DER-mis) The innermost layer of the cortex in roots; a cylinder one cell thick that forms the boundary between the cortex and the stele.

endomembrane system The collection of membranes inside and around a eukaryotic cell, related either through direct physical contact or by transfer of membranous vesicles.

endometrium (EN-doh-MEE-tree-um) The inner lining of the uterus, which is richly supplied with blood vessels that provide the maternal part of the placenta and nourish the developing embryo.

endoplasmic reticulum (ER) (EN-doh-plaz-mik reh-TIK-yoo-lum) An extensive membranous network in eukaryotic cells, continuous with the outer nuclear membrane and composed of ribosome-studded (rough) and ribosome-free (smooth) regions.

endorphins (en-DOR-fins) Hormones produced in the brain and anterior pituitary that inhibit pain reception by mimicking the effects of morphine on the nervous system.

endoskeleton (EN-doh-SKEL-eh-ton) A hard skeleton buried within the soft tissues of an animal, such as the spicules of sponges, the plates of echinoderms, and the bony skeletons of vertebrates.

endosperm (EN-doh-spurm) A nutrient-rich structure formed by union of a sperm cell with two polar nuclei during double fertilization, which provides nourishment to the developing embryo in angiosperm seeds.

endosymbiotic hypothesis (EN-doh-SIM-by-OT-ik) A theory on the origin of the eukaryotic cell that maintains that the forerunners of eukaryotic cells were symbiotic associations of prokaryotic cells living inside larger prokaryotes.

endothelium (EN-doh-THEEL-ee-um) The innermost, simple squamous layer of lining cells in blood vessels; the only constituent structure of capillaries.

endotherms (EN-doh-thurms) Animals that use metabolic energy to maintain a constant body temperature, such as birds and mammals.

endotoxins (EN-doh-TOKS-ins) Components of the outer membranes of certain gram-negative bacteria that are responsible for generalized symptoms of fever and ache.

energy The capacity to do work by moving matter against an opposing force.

enkephalins (en-KEF-uh-lins) Hormones produced in the brain and anterior pituitary that inhibit pain reception by mimicking the effects of morphine on the nervous system.

entropy (EN-truh-pee) A quantitative measure of disorder or randomness, symbolized by S.

enzymes A class of proteins serving as catalysts, chemical agents that change the rate of a reaction without being consumed by the reaction.

epicotyl (EP-eh-KOT-ul) In angiosperm seeds, the portion of the embryonic axis above the cotyledons.

epidermis (EP-eh-DER-mis) The dermal tissue system in plants; also, the outer covering of animals.

epigenesis (EP-eh-JEN-eh-sis) The progressive development of form in an embryo.

epiphyte (EP-eh-fite) A plant that nourishes itself but grows on the surface of another plant for support, usually on the branches or trunks of tropical trees.

episomes (EP-eh-soams) Plasmids, such as the F factor, that are capable of integrating into the bacterial chromosome.

epistasis A phenomenon in which one gene alters the expression of another gene that is independently inherited.

epithelial tissue (EP-eh-THEEL-ee-ul) Sheets of tightly packed cells line organs and cavities.

equilibrium potential The membrane potential for a given ion at which the voltage exactly balances the chemical diffusion gradient for that ion.

erythrocyte *(er-RITH-roh-site)* A red blood cell; contains hemoglobin, which functions in transporting oxygen in the circulatory system.

essential nutrients Nutrient substances that an animal cannot make itself from any raw materials, but which must be obtained in food in prefabricated form.

estivation *(ES-teh-VAY-shun)* A physiological state characterized by slow metabolism and inactivity, which permits survival during long periods of elevated temperature and diminished water supplies.

estrogens *(ES-troh-jens)* The primary female steroid sex hormones, which are produced in the ovary by the developing follicle during the first half of the cycle and in smaller quantities by the corpus luteum during the second half. Estrogens maintain the female reproductive system and develop secondary female sex characteristics.

estrous cycle *(ES-trus)* A type of reproductive cycle in all female mammals except higher primates, in which the nonpregnant endometrium is reabsorbed rather than shed, and sexual response occurs only during midcycle at estrus.

estrus *(ES-trus)* The limited period of heat or sexual receptivity that occurs around ovulation in female mammals having estrous cycles.

ethology *(ee-THOL-uh-jee)* The study of how animals behave in their natural environments.

ethylene *(ETH-ul-een)* The only gaseous plant hormone, responsible for fruit ripening, growth inhibition, leaf abscission, and aging.

eubacteria *(YOO-bak-TEER-ee-uh)* The more recent of the two lineages of prokaryotes; includes the cyanobacteria and nearly all other contemporary bacteria.

euchromatin *(yoo-KROW-muh-tin)* The more open, unraveled form of eukaryotic chromatin, which is actively being transcribed.

eukaryotic cell *(YOO-kar-ee-OT-ik)* A type of cell with a membrane-enclosed nucleus and membrane-enclosed organelles, found in protists, plants, fungi, and animals.

Eumetazoa *(YOO-met-uh-ZOH-uh)* A subkingdom that includes all animals except sponges.

eutrophic *(yoo-TROFE-ik)* Highly productive; pertaining to a lake having a high rate of biological productivity supported by a high rate of nutrient cycling.

evaporative cooling The property of a liquid whereby the surface becomes cooler during evaporation, owing to loss of highly kinetic molecules to the gaseous state.

evolution All the changes that have transformed life on Earth from its earliest beginnings to the diversity that characterizes it today.

excitable cells Cells, such as neurons and muscle cells, that can use changes in their membrane potentials to conduct signals.

excitatory postsynaptic potential (EPSP) *(POSTE-sin-AP-tik)* An electrical change (depolarization) in the membrane of a postsynaptic neuron caused by the binding of an excitatory neurotransmitter from a presynaptic cell to a postsynaptic receptor; makes it more likely that a postsynaptic neuron will generate an action potential.

excretion *(ex-KREE-shun)* Disposal of nitrogen-containing waste products of metabolism.

exergonic reaction *(EX-ur-GON-ik)* A spontaneous chemical reaction in which there is a net release of free energy.

exocrine glands *(EX-oh-krin)* Glands, such as sweat and salivary glands, that excrete their products into tubes or ducts that empty onto an epithelial surface.

exocytosis *(EX-oh-sigh-TOE-sis)* The cellular secretion of macromolecules by the fusion of vesicles with the plasma membrane.

exons The coding regions of eukaryotic genes that are expressed. Exons are separated from each other by introns.

exoskeleton A hard encasement deposited on the surface of an animal, such as the shells of mollusks or the cuticles of arthropods, that provides protection and points of attachment for muscles.

exotoxins *(EX-oh-TOX-ins)* Toxic proteins secreted by a bacterial cell that produce specific symptoms even in the absence of the bacterium.

expressivity *(EX-pres-IV-eh-tee)* The degree to which a particular gene is expressed in individuals showing a particular trait.

extraembryonic membranes *(EX-truh-EM-bree-AHN-ik)* Four membranes (yolk sac, amnion, chorion, allantois) that support the developing embryo in reptiles, birds, and mammals. Important adaptations to life on land.

facilitated diffusion The spontaneous passage of molecules and ions, bound to specific carrier proteins, across a biological membrane down their concentration gradients.

facultative anaerobes *(FAK-ul-tay-tiv AN-uh-robes)* Organisms that make ATP by aerobic respiration if oxygen is present, but that switch to fermentation under anaerobic conditions.

fats (triacylglycerol) *(tri-A-sil-GLI-ser-all)* A biological compound consisting of three fatty acids linked to one glycerol molecule.

fauna *(FAWN-uh)* A general term for animal life.

feedback inhibition A method of metabolic control in which the end product of a metabolic pathway acts as an inhibitor of an enzyme within that pathway.

fermentation A catabolic process that makes a limited amount of ATP from glucose without an electron transport chain and that produces a characteristic end product, such as ethyl alcohol or lactic acid.

fertilization The union of haploid gametes to produce a diploid zygote.

F factor The fertility factor in bacteria, a plasmid that confers the ability to form pili for conjugation and associated functions required for transfer of DNA from donor to recipient.

F_1 generation The first filial or hybrid offspring in a genetic cross-fertilization.

F_2 generation Offspring resulting from interbreeding of the hybrid F_1 generation.

fiber A lignified cell type that reinforces the xylem of angiosperms and functions in mechanical support; a slender, tapered sclerenchyma cell that usually occurs in bundles.

fibrin *(FY-brin)* The activated form of the blood-clotting protein fibrinogen, which aggregates into threads that form the fabric of the clot.

fibroblasts *(FY-broh-blasts)* A type of cell in loose connective tissue that secretes the protein ingredients of the extracellular fibers.

fibronectins *(FY-broh-NEK-tins)* A family of extracellular glycoproteins that helps embryonic cells adhere to their substrate as they migrate.

filter-feeders Aquatic heterotrophs, such as clams and oysters, that sift small food particles from the water.

first law of thermodynamics *(THUR-moe-die-NAM-iks)* The principle of conservation of energy. Energy can be transferred and transformed, but it cannot be created or destroyed.

fixed action pattern (FAP) A highly stereotyped behavior that is innate and must be carried to completion once initiated.

flaccid *(FLAK-sed)* Limp. Walled cells are flaccid in isotonic surroundings, where there is no tendency for water to enter.

flagellum *(fluh-JEL-um; pl. **flagella**)* A long cellular appendage specialized for locomotion and formed from a core of nine outer doublet microtubules and two inner single microtubules, ensheathed in an extension of plasma membrane.

flame-cell system The simplest tubular excretory system, present in flatworms and acting to directly regulate the contents of the extracellular fluid.

flora (*FLOOR-uh*) A general term for plant life.

fluid mosaic model The currently accepted model of cell membrane structure, which envisions the membrane as a mosaic of individually inserted protein molecules drifting laterally in a fluid bilayer of phospholipids.

fluid feeders Animals that live by sucking nutrient-rich fluids from another living organism.

follicles (*FOL-eh-kuls*) Microscopic structures in the ovary that contain developing ova and secrete estrogens.

food chain The transfer of food from trophic level to trophic level, beginning with producers.

food web The elaborate, interconnected feeding relationships of who eats whom in an ecosystem.

fossil A relic or impression of an organism from the past, preserved in rock.

founder effect A cause of genetic drift attributable to colonization by a limited number of individuals from a parent population.

fragmentation A mechanism of asexual reproduction, in which the parent plant or animal separates into parts that re-form whole organisms.

frameshift mutation A mutation occurring when the number of nucleotides inserted or deleted is not a multiple of 3, thus resulting in improper grouping into codons.

free energy A quantity of energy that interrelates entropy (*S*) and enthalpy (*H*); symbolized by *G*. The change in free energy of a system is calculated by the equation $\Delta G = \Delta H - T\Delta S$, where *T* is the absolute temperature.

free energy of activation The initial investment of energy necessary to start a chemical reaction.

free-running periods Deviations in rhythmic responses, ranging from about 21 to 27 hours, from normal circadian rhythms of 24-hour duration.

frequency-dependent selection Decline in reproductive success of a morph resulting from the morph's phenotype becoming too common in a population; a cause of balanced polymorphism in populations.

fruit A ripened ovary of a flower that protects dormant seeds and aids in their dispersal.

functional groups Specific configurations of atoms that are commonly attached to the carbon skeletons of organic molecules and are usually involved in chemical reactions.

fundamental niche The total resources an organism is theoretically capable of utilizing.

gametangia (*GAM-eh-TANJ-ee-uh*) The reproductive organs of bryophytes, consisting of male antheridia and female archegonia; multichambered jackets of sterile cells in which gametes are formed.

gametes (*GAM-eets*) Haploid egg or sperm cells that unite during sexual reproduction to produce a diploid zygote.

gametophyte (*guh-ME-toh-fite*) The multicellular haploid form in organisms undergoing alternation of generations, which mitotically produces haploid gametes that unite and grow into the sporophyte generation.

ganglia (*GANG-lee-a*; sing. **ganglion** *GANG-lee-un*) Clusters (functional groups) of nerve cell bodies in a centralized nervous system.

gap junction A type of intercellular junction in animal cells that allows passage of material or current from cell to cell.

gastrovascular system A digestive pouch with a single opening; functions in digestion, circulation, and, often, gas exchange and excretion.

gastrula (*GAS-troo-la*) The two-layered, cup-shaped embryonic stage.

gastrulation (*GAS-truh-LAY-shun*) Formation of a gastrula from a blastula.

gel electrophoresis (*JELL eh-LEK-troh-for-EE-sis*) Separation of nucleic acids or proteins, on the basis of their size and electric charge, by measuring their rate of movement through an electric field in a gel.

gene One of many discrete units of hereditary information located on the chromosomes and consisting of DNA.

gene amplification The selective synthesis of DNA, which results in multiple copies of a single gene, thereby enhancing expression.

gene cloning Formation by a bacterium, carrying foreign genes in a recombinant plasmid, of a clone of identical cells containing the replicated foreign genes.

gene flow The loss or gain of alleles from a population due to the emigration or immigration of fertile individuals, or the transfer of gametes, between populations.

gene pool The total aggregate of genes in a population at any one time.

generation time The average age when females of a population begin reproducing, which greatly impacts on the intrinsic rate of increase.

genetic drift Changes in the gene pool of a small population due to chance.

genetic recombination The general term for the production of offspring that combine traits of the two parents.

genetics The science of heredity; the study of heritable information.

genome (*JEE-nome*) The complete complement of an organism's genes; an organism's genetic material.

genomic equivalence (*jen-OME-ik*) The presence of all of an organism's genes in all of its cells.

genomic library A set of thousands of DNA segments from a genome, each carried by a plasmid or phage.

genotype (*JEE-noh-type*) The genetic makeup of an organism.

genus (*JEE-nus*) The taxon above the species level, designated by the first word of a species's binomial Latin name.

geometric isomer (*EYE-so-mer*) A form of a chemical compound having the same covalent partnerships as other forms of the same compound, but having different spatial arrangements.

gibberellins (*JIB-ur-EL-ins*) A class of related hormones in plants that stimulate growth in the stem and leaves, trigger germination of seeds and breaking of bud dormancy, and stimulate fruit development with auxin.

gills Localized extensions of the body surfaces of many aquatic animals specialized for gas exchange.

glial cell (*GLEE-ul*) A nonconducting cell of the nervous system that provides support, insulation, and protection for the neurons.

glomerulus (*glum-AIR-yoo-lus*) A ball of capillaries surrounded by the Bowman's capsule in the nephron and serving as the site of filtration.

glycocalyx (*GLY-koh-KAY-liks*) A fuzzy coat on the outside of animal cells, made of sticky oligosaccharides.

glycogen (*GLY-koh-jen*) An extensively branched glucose storage polysaccharide found in the liver and muscle of animals; the animal equivalent of starch.

glycolysis (*gly-KOL-eh-sis*) The splitting of glucose into pyruvic acid. Glycolysis is the one metabolic pathway that occurs in all living cells, serving as the starting point for fermentation or aerobic respiration.

glyoxysomes (*gly-OX-eh-soams*) A family of microbodies commonly found in the fat-storing tissues of germinating seeds and containing enzymes that convert fats to sugar.

Golgi apparatus (*GOAL-jee*) An organelle in eukaryotic cells consisting of stacks of membranes that modify, store, and route products of the endoplasmic reticulum.

gonadotropins (*go-NAD-oh-TROH-pins*) Hormones that stimulate the activities of the testes and ovaries; a term for follicle-stimulating and luteinizing hormones.

gonads (*GONE-adz*) The male and female sex organs; the gamete-producing organs in most animals.

G proteins A group of membrane proteins that function as intermediaries between hormone receptors in cell membranes and the enzyme adenylate cyclase, which converts ATP to cAMP in the second messenger (cAMP) system in nonsteroid hormone action. Depending on the type of hormone, G proteins may increase or decrease the activity of adenylate cyclase in producing cAMP.

G_1 phase The "first gap" phase of the cell cycle, consisting of the portion of interphase before DNA synthesis begins.

G_2 phase The "second gap" phase of the cell cycle, consisting of the portion of interphase after DNA synthesis occurs.

graded potentials Local voltage changes in a neuron membrane induced by stimulation of a neuron, with strength proportional to the strength of the stimulus and lasting about one millisecond.

gradualism A view of Earth history that attributes profound change to the cumulative product of slow but continuous processes.

granum (*GRAN-um;* pl. **grana** *GRAN-uh*) Stacked portions of the thylakoid membrane in the chloroplast. Grana function in the light reactions of photosynthesis.

gravitropism (*GRAV-eh-TROH-piz-um*) A response of a plant or animal in relationship to gravity.

greenhouse effect The warming of the Earth due to atmospheric accumulation of carbon dioxide, which absorbs infrared radiation and slows its escape from the irradiated Earth.

gross primary productivity The total primary productivity of an ecosystem.

ground tissue system A tissue of mostly parenchyma cells that makes up the bulk of a young plant and fills the space between the dermal and vascular tissue systems.

growth factors Proteins that must be present in the extracellular environment (tissue culture medium or animal body) for the growth and normal development of certain types of cells.

guard cells Specialized epidermal cells in plants that form the boundaries of the stomata.

gymnosperms (*JIM-noh-spurms*) Vascular plants that bear naked seeds unenclosed in any specialized chambers.

habituation (*huh-BITCH-yoo-AY-shun*) A simple kind of learning involving loss of sensitivity to unimportant stimuli, allowing an animal to conserve time and energy.

hair cell A common type of mechanoreceptor found in the vertebrate ear, in the lateral line system of fishes and amphibians, and in arthropod statocysts, which detects motion of various kinds.

half-life The average amount of time it takes for one-half of a specified quantity of a substance to decay or disappear.

haploid cell (*HAP-loid*) A cell containing only one set of chromosomes (N).

Hardy-Weinberg equilibrium Stability in frequency of alleles and genotypes in a population generation after generation; a state of equilibrium in a population's gene pool.

Hardy-Weinberg theorem An axiom that maintains that the sexual shuffling of genes alone cannot alter the overall genetic makeup of a population.

Haversian system (*ha-VER-shun*) One of many structural units of vertebrate bone, consisting of concentric layers of mineralized bone matrix surrounding lacunae, which contain osteocytes, and a central canal, which contains blood vessels and nerves.

heat The total amount of kinetic energy due to molecular motion in a body of matter. Heat is energy in its most random form.

helper T cells A type of T cell that is required by some B cells to help them make antibodies or that helps other T cells to respond to antigens or secrete lymphokines or interleukins.

heme group (*heem*) A prosthetic (nonproteinaceous) chemical group consisting of a ring structure surrounding an iron atom.

hemoglobin (*HEE-moh-gloh-bin*) An iron-containing protein in red blood cells that reversibly binds oxygen.

hemophilia (*HEEM-uh-FILL-ee-uh*) A genetic disease characterized by excessive bleeding following injury; results from an abnormal sex-linked recessive gene.

herbivores Heterotrophic animals that eat plants.

hermaphrodite (*her-MAF-roh-dite*) An individual that functions as both male and female in sexual reproduction by producing both sperm and eggs.

heterochromatin (*HET-ur-oh-KROH-muh-tin*) Nontranscribed eukaryotic chromatin that is so highly compacted that it is visible with the light microscope during interphase.

heterocysts (*HET-ur-oh-sists*) Specialized cells that engage in nitrogen fixation on some filamentous cyanobacteria.

heterogametic sex (*HET-ur-oh-gam-ET-ik*) The sex that produces two different kinds of gametes and determines the sex of the offspring.

heteromorphic (*HET-ur-oh-MOR-fik*) A condition in the life cycle of all modern plants in which the sporophyte and gametophyte generations differ in morphology.

heterosporous (*HET-ur-OS-pur-us*) A condition in some plants in which the sporophyte produces two kinds of spores that develop into unisexual gametophytes, either female or male.

heterotrophic nutrition (*HET-ur-oh-TROH-fik*) A mode of obtaining organic food molecules by eating other organisms or their by-products.

heterozygote advantage (*HET-ur-oh-ZY-gote*) A mechanism that preserves variation in eukaryotic gene pools by conferring greater reproductive success to heterozygotes over individuals homozygous for any one of the associated alleles.

heterozygous (*HET-ur-oh-ZY-gus*) Having two different alleles for a given trait.

hibernation A physiological state, permitting survival during long periods of cold and diminished food, in which metabolism decreases, the heart and respiratory system slow down, and body temperature is maintained at a lower level than normal.

histamine (*HISS-tuh-meen*) A substance released by injured cells that causes blood vessels to dilate during an inflammatory response.

histones (*HISS-tones*) Abundant small proteins with a high proportion of positively charged amino acids that bind to the negatively charged DNA and play a key role in its folding into chromatin.

holoblastic cleavage (*HO-loh-BLAS-tik*) A type of cleavage in which there is complete division of the egg, as occurs in eggs with little or a moderate amount of yolk.

holotrophs (*HOLE-oh-trohfs*) Predators that use tentacles, pincers, claws, or jaws to kill their prey or tear off pieces of meat or vegetation.

homiobox Specific sequences of DNA that regulate patterns of differentiation during development of an organism.

homeostasis (*HOME-ee-oh-STAY-sis*) The steady-state physiological condition of the body; also called the "constant internal milieu."

homeotic genes (*HOME-ee-OT-ik jeens*) Genes controlling the overall body plan of animals by controlling the developmental fate of groups of cells.

homogametic sex (*HOME-oh-gam-ET-ik*) The sex that produces only one kind of gamete.

homologous chromosomes (*home-OL-uh-gus*) Chromosome pairs of the same length, centromere position, and staining pattern that possess genes for the same traits at corresponding loci. One homologous chromosome is inherited from the organism's father, the other from the mother.

homologous structures (*home-OL-uh-gus*) Structures in different species that are similar because of common ancestry.

homology (*home-OL-uh-gee*) Similarity in characteristics resulting from a shared ancestry.

homosporous plants (*home-OS-pur-us*) Plants in which a single type of spore develops into a bisexual gametophyte having both male and female sex organs.

homozygous (*HOME-oh-ZY-gus*) Having two identical alleles for a given trait.

hormone (*HOR-mone*) One of many types of circulating chemical signals found in all multicellular organisms that are formed in specialized cells, travel in body fluids, and coordinate the various parts of the organism by interacting with target cells.

host range The limited number of host species, tissues, or cells that a parasite (including viruses and bacteria) can infect.

humoral immune system (*HYOO-mur-al*) The part of the immune system that fights bacteria and viruses in body fluids with antibodies that circulate in blood plasma and lymph, fluids formerly called "humors."

hybrid vigor Increased vitality (compared to that of either parent stock) in the hybrid offspring of two different, inbred parents.

hybridoma (*HY-brid-OH-muh*) A hybrid cell that produces monoclonal antibodies in culture, formed by the fusion of a myeloma cell with a normal antibody-producing lymphocyte.

hydrocarbons (*HY-droh-kar-bons*) Organic molecules consisting only of carbon and hydrogen.

hydrogen bond A type of weak chemical bond formed when the partially positive hydrogen atom of a polar covalent bond in one molecule is attracted to the partially negative atom of a polar covalent bond in another molecule.

hydrolysis (*hy-DROL-eh-sis*) A chemical process in which water adds across a covalent bond and lyses or splits the molecule; an essential process in digestion.

hydrophilic (*HY-droh-FIL-ik*) Having an affinity for water.

hydrophobic (*HY-droh-FOBE-ik*) Having an aversion for water; tending to coalesce and form droplets in water.

hydrophobic interaction (*HY-droh-FOH-bik*) A type of weak chemical bond formed when molecules that do not mix with water coalesce to exclude the water. Once formed, these intermolecular associations are reinforced by van der Waals interactions.

hydrostatic skeleton (*HY-droh-STAT-ik*) A skeletal system composed of fluid held under pressure in a closed body compartment; the main skeleton of most cnidarians, flatworms, nematodes, and annelids.

hydroxyl group (*hy-DROKS-ul*) A functional group consisting of a hydrogen atom joined to an oxygen atom by a polar covalent bond. Molecules possessing this group are soluble in water and called alcohols.

hyperosmotic solution (*Hy-pur-os-MAH-tik*) A solution with a greater solute concentration than another, the hypoosmotic solution.

hyperpolarization (*HY-pur-POLE-ur-i-ZAY-shun*) An electrical state whereby the inside of the cell is made more negative relative to the outside than was the case at resting potential. A nerve cell membrane is hyperpolarized if a stimulus increases its voltage from the resting potential of −70 mV. Reduces the chance that a neuron will transmit a nerve impulse.

hypha (*HY-fa;* pl. **hyphae**) The filaments making up the body of a fungus.

hypocotyl (*HY-poh-kot-ul*) The embryonic axis of a plant below the point at which the cotyledons are attached that terminates in the radicle.

hypoosmotic solution (*HY-poh-oz-MAH-tik*) A solution with a lesser solute concentration than another, the hyperosmotic solution.

hypothalamus (*HYPE-oh-THAL-uh-mus*) The ventral part of the diencephalon of the vertebrate brain; functions in maintaining homeostasis, especially in coordinating the endocrine and nervous systems; secretes hormones of the posterior pituitary and releasing factors, which regulate the anterior pituitary.

imaginal disk (*im-OGG-in-ul*) An island of undifferentiated cells in an insect larva, which are committed (determined) to form a particular organ during metamorphosis to the adult.

imbibition (*IM-be-BE-shun*) The soaking of water into a porous material that is hydrophilic.

immunoglobulins (*IM-myoo-noh-GLOB-yoo-lins*) The class of proteins comprising the antibodies.

imperfect flower A flower possessing either stamens or carpels, but not both.

imprinting (*IM-print-ing*) A type of learned behavior with a significant innate component, acquired during a limited critical period.

incomplete dominance A type of inheritance in which F_1 hybrids have an appearance that is intermediate between the phenotypes of the parental varieties.

incomplete flower A flower missing sepals, petals, stamens, or carpels.

incomplete metamorphosis (*MET-uh-MOR-foh-sis*) A type of development in certain insects such as grasshoppers, in which the larvae resemble adults but are smaller and have different body proportions. The animal goes through a series of molts, each time looking more like an adult, until it reaches full size.

indeterminate cleavage (*IN-dee-TUR-min-it*) A type of embryonic development in deuterostomes, in which each cell produced by early cleavage divisions retains the capacity to develop into a complete embryo.

indeterminate growth A type of growth characteristic of plants, in which the organism continues to grow as long as it lives.

induced fit The change in shape of the active site of an enzyme so that it binds more snugly to the substrate; induced by entry of the substrate.

inducible enzymes Enzymes whose synthesis is stimulated by specific metabolites.

induction (*in-DUK-shun*) The ability of one group of embryonic cells to influence the development of another.

inflammation (*in-flum-MAY-shun*) Redness, heat, and swelling of tissue resulting from physical injury or infection; results from chemical signals released from injured cells and from outflow of fluid and cells from blood vessels.

inflammatory response A line of defense triggered by penetration of the skin or mucous membranes, in which small blood vessels in the vicinity of an injury dilate and become leakier, enhancing infiltration of leukocytes; may also be widespread in the body.

ingestion (*in-JEST-shun*) A heterotrophic mode of nutrition in which other organisms or detritus are eaten whole or in pieces.

inhibitory postsynaptic potential (IPSP) (*POSTE-sin-AP-tik*) An electrical charge (hyperpolarization) in the membrane of a postsynaptic neuron caused by the binding of an inhibitory neurotransmitter from a presynaptic cell to a postsynaptic receptor; makes it more difficult for the postsynaptic neuron to generate an action potential.

innate releasing mechanism (IRM) A hypothetical cellular entity in a nervous system that may respond to a stimulus and initiate a specific associated behavior.

inner cell mass A cluster of cells in a mammalian blastocyst that protrudes into one end of the cavity and subsequently develops into the embryo proper and some of the extraembryonic membranes.

inositol triphosphate (IP$_3$) *(in-NOS-i-tahl)* The second messenger, which functions as an intermediate between certain nonsteroid hormones and the third messenger, a rise in cytoplasmic Ca^{2+} concentration.

insertion A mutation involving the addition of one or more nucleotide pairs to a gene.

insight learning The ability of an animal to perform a correct or appropriate behavior on the first attempt in a situation with which it has had no prior experience.

integral proteins Those proteins of biological membranes that penetrate into or span the membrane.

insulin *(IN-sul-en)* The vertebrate hormone that lowers blood sugar levels by promoting the uptake of glucose by most body cells and promoting the synthesis and storage of glycogen in the liver; also stimulates protein and fat synthesis; secreted by islet cells of the pancreas.

interferon *(IN-tur-FEER-on)* A chemical messenger of the immune system, produced by virus-infected cells and capable of helping other cells resist the viruses.

interleukin I *(IN-tur-luke-in)* A chemical regulator (cytokine) secreted by macrophages that have ingested a pathogen or foreign molecule and bound with a helper T cell; stimulates T cells to grow and divide and elevates body temperature. Interleukin II, secreted by activated T cells, stimulates helper T cells to proliferate more rapidly.

intermediate filaments A component of the cytoskeleton that includes all filaments intermediate in size between microtubules and microfilaments.

interneurons *(IN-tur-NOOR-ahns)* Association neurons; nerve cells within the central nervous system that form synapses with sensory and motor neurons and integrate sensory input and motor output.

internode *(IN-tur-NODE)* The segment of a plant stem between the points where leaves are attached.

interstitial cells *(IN-tur-STISH-ul)* Cells scattered among the seminiferous tubules of the vertebrate testis that secrete testosterone and other androgens, the male sex hormones.

interstitial fluid The internal environment of vertebrates, consisting of the fluid filling the spaces between cells.

intertidal zone The shallow zone of the ocean where land meets water; also called the littoral zone.

intrinsic rate of increase The difference between natality and mortality, symbolized as *r*.

introgression *(IN-troh-GRES-shun)* Transplantation of genes between species resulting from fertile hybrids mating successfully with one of the parent species.

introns *(IN-trons)* Intervening sequences of noncoding regions in eukaryotic genes.

invagination *(in-VAJ-eh-NAY-shun)* The buckling inward of a cell layer, caused by rearrangements of microfilaments and microtubules; an important phenomenon in embryonic development.

inversion An aberration in chromosome structure resulting from an error in meiosis or from mutagens; reattachment in a reverse orientation of a chromosomal fragment to the chromosome from which the fragment originated.

in vitro **fertilization** *(VEE-troh)* Fertilization of ova in laboratory containers followed by artificial implantation of the early embryo in the mother's uterus.

ion *(EYE-on)* An atom that has gained or lost electrons, thus acquiring a charge.

ionic bond *(eye-ON-ik)* A chemical bond due to attraction between oppositely charged ions.

isogamy *(eye-SOG-uh-mee)* A condition in which male and female gametes are morphologically indistinguishable.

isomers *(EYE-sum-urs)* Organic compounds with the same molecular formula but different structures and hence different properties. There are three types: structural isomers, geometric isomers, and optical isomers.

isosmotic solutions *(EYE-soz-MAH-tik)* Solutions of equal solute concentration.

isotopes *(EYE-so-topes)* Atomic forms of an element, containing different numbers of neutrons and thus differing in atomic mass.

iteroparous *(IT-er-oh-PAIR-us)* Pertaining to organisms that reproduce more than once in a lifetime.

juvenile hormone (JH) A hormone in arthropods secreted by corpora allata, which promotes retention of larval characteristics.

karyotype *(KAR-ee-oh-type)* A method of organizing the chromosomes of a cell in relation to number, size, and type.

keystone predator A predatory species that helps maintain species richness in a community by reducing the density of populations of the best competitors so that populations of less competitive species are maintained.

kin selection A phenomenon of "inclusive fitness," used to explain altruistic behavior between related individuals.

kinesis *(kih-NEE-sis)* A change in activity rate in response to a stimulus.

kinetic energy *(kih-NET-ik)* The energy of motion, which is directly related to the speed of that motion. Moving matter does work by transferring some of its kinetic energy to other matter.

kinetochores *(kih-NET-oh-kores)* Specialized regions on the centromere that link each sister chromatid to the spindle.

Krebs cycle (citric acid cycle; tricarboxylic acid cycle) A chemical cycle involving nine steps that completes the metabolic breakdown of glucose molecules to carbon dioxide; occurs within the mitochondrion; the second major stage in cellular respiration.

K-**selection** The concept that in some (*K*-selected) populations, life history is centered around producing relatively few offspring that have a good chance of survival.

lacteal *(lak-TEEL)* A tiny lymph vessel extending into the core of the intestinal villus and serving as the destination for absorbed chylomicrons.

larva *(LAR-vuh)* A free-living, sexually immature form in some animal life cycles that may differ from the adult in morphology, nutrition, and habitat.

lateral line system A mechanoreceptor system consisting of a series of pores and receptor units (neuromasts) along the sides of the body of fishes and aquatic amphibians; detects water movements made by an animal itself and by other moving objects.

lateral meristems *(MARE-eh-stems)* The vascular and cork cambiums, cylinders of dividing cells that run most of the length of stems and roots and are responsible for secondary growth.

law of independent assortment Mendel's second law, stating that each allele pair segregates independently during gamete formation; applies when genes for two traits are located on different pairs of homologous chromosomes.

law of segregation Mendel's first law, stating that allele pairs separate during gamete formation, and then randomly re-form pairs during fusion of gametes at fertilization.

learning The modification of behavior by experience.

leukocyte (LUKE-oh-site) A white blood cell; typically functions in immunity, such as phagocytosis or antibody production.

life tables Tables of data summarizing mortality in populations.

ligament (LIG-uh-ment) A type of fibrous connective tissue that joins bones together at joints.

ligand (LIG-und) A molecule that binds specifically to a receptor site of another molecule.

light microscope An optical instrument with lenses that refract (bend) visible light to magnify images of specimens.

light reactions The steps in photosynthesis that occur on the thylakoid membranes of the chloroplast and convert solar energy into the chemical energy of ATP and NADPH, evolving oxygen in the process.

lignin (LIG-nin) A hard material embedded in the cellulose matrix of vascular plant cell walls that functions as an important adaptation for support in terrestrial species.

limbic system (LIM-bik) A group of nuclei (clusters of nerve cell bodies) in the lower part of the mammalian forebrain that interact with the cerebral cortex in determining emotions; includes the hippocampus and the amygdala.

linked genes Genes that are located on the same chromosome.

lipid (LI-pid) One of a family of compounds, including fats, phospholipids, and steroids, that are insoluble in water.

littoral zone (LIT-or-ul) Another term for the intertidal zone.

locus (LOH-kus) A particular place along the length of a certain chromosome where a given gene is located.

logistic equation (leh-JIS-tik) A mathematical model for population growth that is based on the assumption of potential exponential growth, but that includes a term that compares the actual size of the population to the carrying capacity.

long-day plant One of a group of plants that flower, usually in late spring or early summer, only when the light period is longer than a critical length.

loop of Henle (HEN-lee) The long hairpin turn of the renal tubule in the kidney of mammals and birds. Functions in water and salt reabsorption.

lungs The invaginated respiratory surfaces of terrestrial vertebrates, land snails, and spiders that connect to the atmosphere by narrow tubes.

lymph (LIMF) The colorless fluid, derived from interstitial fluid, in the lymphatic system of vertebrate animals.

lymphatic system (lim-FAT-ik) A system of vessels and lymph nodes separate from the circulatory system that returns fluid and protein to the blood.

Lyon hypothesis The explanation, proposed by British geneticist Mary F. Lyon, that one of the two X chromosomes in the cells of mammalian females is largely inactivated during early embryonic development. The inactive X chromosome contracts into a Barr body.

lysogenic cell (LYE-so-JEN-ik) A host cell carrying a prophage in its chromosome, which has the potential to lyse and release phages.

lysogenic cycle A type of viral replication cycle in which the viral genome becomes incorporated into the bacterial host chromosome as a prophage.

lysosome (LYE-so-soam) A membrane-enclosed bag of various hydrolytic enzymes found in the cytoplasm of eukaryotic cells.

lysozyme (LYE-so-zime) An enzyme in perspiration, tears, and saliva that attacks bacterial cell walls.

lytic cycle (LIT-ik) A type of viral replication cycle resulting in the release of new virions by death or lysis of the host cell.

macroevolution Evolutionary change on a grand scale, encompassing the origin of novel designs, evolutionary trends, adaptive radiation, and mass extinction.

macromolecule (MAK-roh-MOL-eh-kyool) A giant molecule of living matter formed by the joining of smaller molecules, usually by dehydration syntheses. Polysaccharides, proteins, and nucleic acids are examples.

macronutrients (MAK-roh-NOO-tree-ents) Nine chemical elements required by plants in relatively large amounts.

macrophage (MAK-roh-fahj) An amoeboid cell that moves through tissue fibers, engulfing bacteria and dead cells by phagocytosis.

major histocompatibility complex (MHC) (HIS-toe-kum-pat-uh-BIL-eh-tee) A large set of cell surface antigens in each individual, encoded by a family of genes serving as unique biochemical markers, that trigger T-cell responses that may lead to rejection of transplanted tissues and organs.

malignant tumor A cancerous growth; an abnormal growth whose cells multiply excessively, have altered surfaces, and may have unusual numbers of chromosomes and/or aberrant metabolic processes.

Malpighian tubules (mal-PIG-ee-un) The unique excretory organs of insects that empty into the digestive tract, remove nitrogenous wastes from the blood, and function in osmoregulation.

mantle A heavy fold of tissue in mollusks that drapes over the visceral mass and may secrete a shell.

marsupials (mar-SOOP-ee-uls) A group of mammals, such as koalas, kangaroos, and opossums, whose young complete their embryonic development inside a maternal pouch called the marsupium.

matrix The nonliving component of connective tissue, consisting of a web of fibers embedded in homogeneous ground substance that may be liquid, jellylike, or solid.

matter Anything that takes up space and has mass.

mechanoreceptors (meh-KAN-oh-ree-SEP-tursz) Sensory receptors that detect physical deformations in the body environment associated with pressure, touch, stretch motion, and sound.

medulla oblongata (meh-DULL-uh OBB-long-GAH-tuh) The lowest part of the vertebrate brain; a swelling of the hindbrain dorsal to the anterior spinal cord that controls autonomic, homeostatic functions, including breathing, heart and blood vessel activity, swallowing, digestion, and vomiting.

medusa (meh-DOO-suh) The floating, flattened, mouth-down version of the cnidarian body plan.

megapascal (MPa) (MEG-uh-pass-KAL) A unit of pressure equivalent to 10 bars.

megaspore (MEG-uh-SPORE) A haploid spore of plants that is produced meiotically by a megasporocyte, and that develops into an embryo sac (female gametophyte).

meiosis (my-OH-sis) A two-stage cell division found only in sexually reproducing organisms that results in gametes with half the chromosome number of the original cell.

membrane potential The charge difference between the cytoplasm and extracellular fluid in all cells, which is due to the differential distribution of ions. Membrane potential affects the activity of excitable cells and the transmembrane movement of all charged substances.

menstrual cycle (MEN-stroo-ul) A type of reproductive cycle in higher female primates, in which the nonpregnant endometrium is shed as a bloody discharge through the cervix into the vagina.

menstruation (MEN-stroo-AY-shun) Vaginal bleeding resulting from shedding of the uterine lining during a menstrual cycle.

meristems *(MARE-eh-stems)* Plant tissues that remain embryonic as long as the plant lives, allowing indeterminate growth.

meroblastic cleavage *(MARE-oh-BLAS-tik KLEE-vij)* A type of cleavage in which there is incomplete division of yolk-rich egg, characteristic of avian development.

mesenteries *(MEZ-en-ter-eez)* Membranes that suspend many of the organs of vertebrates inside fluid-filled body cavities.

mesoderm *(ME-zoh-durm)* The middle primary germ layer of an early embryo that develops into the notochord, the lining of the coelom, muscles, skeleton, gonads, kidneys, and most of the circulatory system.

mesoglea *(MEZ-oh-GLEE-uh)* The jellylike substance between the two layers of the body wall of a sponge or cnidarian.

mesophyll *(MEEZ-oh-fil)* The ground tissue of leaf, sandwiched between the upper and lower epidermis and specialized for photosynthesis.

mesophyll cell A cell in a leaf's mesophyll active in carbon fixation and transport of fixed carbon to the leaf's bundle sheath.

messenger RNA (mRNA) A type of RNA that is synthesized from DNA in the genetic material and that attaches to ribosomes in the cytoplasm and specifies the primary structure of a protein.

metabolism *(meh-TAB-oh-liz-um)* The totality of an organism's chemical processes, consisting of catabolic and anabolic pathways.

metamorphosis *(MET-uh-MOR-fuh-sis)* The resurgence of development in an animal larva that transforms it into a sexually mature adult.

metanephridium *(MET-uh-NEH-frid-ee-um)* A type of excretory tubule in annelid worms that has internal openings called nephrostomes that collect body fluids and external openings called nephridiopores.

metastasis *(meh-TAS-teh-sis)* The spread of cancer cells.

microbodies *(MY-crow-bod-ees)* A variety of membrane-enclosed compartments in eukaryotic cells, containing enzymes and substrates for specific metabolic pathways.

microevolution A change in the gene pool of a population over a succession of generations.

microfilaments Solid rods of actin protein found in the cytoplasm of almost all eukaryotic cells, making up part of the cytoskeleton and acting alone or with myosin to cause cell contraction.

micronutrients *(MIKE-roh-NOO-tree-ents)* Seven chemical elements required by plants in very small amounts; function mainly as cofactors of enzymes.

microtubule organizing center (MTOC) The material in a cell that organizes the spindle; associated with centrioles in animal cells; mode of action unknown.

microtubules Hollow rods of tubulin protein found in the cytoplasm of all eukaryotic cells and found in cilia, flagella, and as part of the cytoskeleton.

middle lamella *(luh-MEL-uh)* A thin layer of adhesive extracellular material, primarily pectins, found between primary walls of adjacent young plant cells.

missense mutation The most common type of mutation involving a base-pair substitution within a gene; alters a codon, but the new codon "makes sense" in that it still codes for an amino acid.

mitochondria *(MY-toe-KON-dree-uh)* Organelles in eukaryotic cells that serve as sites of cellular respiration.

mitochondrial matrix The compartment of the mitochondrion enclosed by the inner membrane and containing enzymes and substrates for the Krebs (citric acid) cycle.

mitosis *(my-TOE-sis)* A process of cell division in eukaryotic cells characterized by a sequence of distinctive stages that conserve chromosome number by equally allocating replicated chromosomes to each of two daughter cells.

modern synthesis Neo-Darwinism; a comprehensive theory of evolution emphasizing natural selection, gradualism, and populations as the fundamental units of evolutionary change.

molarity *(mul-AR-eh-tee)* A common measure of solute concentration, referring to the number of moles of solute in 1 L of solution.

mole The number of grams of a substance that equals its molecular weight in daltons and contains Avogadro's number of molecules.

molecular formula A type of molecular notation indicating only the quantity of the constituent atoms.

molecular weight The sum of the weights in daltons of all the atoms in a molecule of a substance.

molecule Two or more atoms of one or more elements held together by ionic or covalent chemical bonds.

molting A process in arthropods in which the exoskeleton is shed at intervals to allow growth by secretion of a larger exoskeleton.

monoclonal antibody *(MON-oh-KLONE-ul)* A defensive protein produced by cells descended from a single cell. An antibody that is secreted by a clone of cells and, consequently, is specific for a single antigenic determinant.

monocots *(MON-oh-kots)* A subdivision of flowering plants whose members possess one embryonic seed leaf, or cotyledon.

monoculture *(MON-oh-KUL-chur)* Cultivation of large land areas with a single plant variety.

monoecious *(mon-EE-shus)* In plants, possessing both staminate and carpellate flowers on the same individual.

monohybrid cross *(MON-oh-HY-brid)* A breeding experiment that employs parental varieties differing in a single trait.

monomers *(MON-uh-mers)* The subunits that serve as the building blocks of a polymer.

monophyletic *(MON-oh-fye-LEH-tik)* Pertaining to a taxon derived from a single ancestral species that gave rise to no species in any other taxa.

monosaccharides *(MON-oh-SAK-ur-ides)* The simplest carbohydrates active alone or serving as monomers for disaccharides and polysaccharides. Also known as simple sugars, their molecular formulas are generally some multiple of CH_2O.

monotremes *(MON-uh-treems)* A group of egg-laying mammals, represented by the platypus and echidnas.

morphogen *(MOR-fuh-jen)* An as yet hypothetical chemical signal in development that varies in concentration along a gradient, thus enabling cells to resolve their position along that gradient.

morphogenesis *(MOR-foe-jen-eh-sis)* The development of body shape and organization during ontogeny.

morphospecies *(MOR-foh-spec-shees)* Species defined by their anatomical features.

morphs *(morfs)* Two or more distinct forms of individuals in a population.

mortality Death; an essential factor in population dynamics.

mosaic development A pattern of development, such as that of a mollusk, in which the early blastomeres each give rise to a specific part of the embryo. In some animals the fate of the blastomeres is established in the zygote.

mosaic evolution The evolution of different features of an organism at different rates.

motor neurons *(NYOOR-onz)* Nerve cells that transmit signals from the brain or spinal cord to muscles or glands.

motor unit A single motor neuron and all the muscle fibers it controls.

M phase The mitotic phase of the cell cycle, which includes mitosis and cytokinesis.

Mullerian mimicry (*myoo-LER-ee-un*) A mutual mimicry by two unpalatable species.

multigene family A collection of genes with similar or identical sequences, presumably of common origin.

multiple fruit A fruit, such as a pineapple, that develops from an inflorescence.

mutagen (*MYOOT-uh-jen*) A chemical or physical agent that interacts with DNA and causes a mutation.

mutagenesis (*MYOOT-uh-JEN-uh-sis*) The production of mutations.

mutation (*myoo-TAY-shun*) A rare change in the DNA of genes that ultimately creates genetic diversity.

mutation pressure A factor causing populations to evolve through spontaneous alterations in alleles constituting the gene pool.

mutualism (*MYOO-chew-ul-iz-um*) A symbiotic relationship in which both the host and the symbiont benefit.

mycelium (*my-SEEL-ee-um*) The densely branched network of hyphae in a fungus.

mycorrhizae (*MY-koh-RYE-zee*) Mutualistic associations of plant roots and fungi.

myelin sheath (*MY-eh-lin*) An insulating coat of cell membrane from Schwann cells that is interrupted by nodes of Ranvier where saltatory impulse conduction occurs.

myofibrils (*MY-oh-FIBE-rills*) Fibrils arranged in longitudinal bundles in muscle cells (fibers); composed of thin filaments of actin and a regulatory protein and thick filaments of myosin.

myoglobin (*MY-uh-glow-bin*) An oxygen-storing, pigmented protein in muscle cells.

myosin (*MY-uh-sin*) A type of protein filament that interacts with actin filaments to cause cell contraction.

NAD⁺ Nicotinamide adenine dinucleotide (oxidized); a coenzyme present in all cells that assists enzymes in transferring electrons during the redox reactions of metabolism.

natality (*nuh-TAL-eh-tee*) Births; an essential factor in population dynamics.

natural selection Differential success in reproduction of different phenotypes resulting from interaction of organisms with their environment. Evolution occurs when natural selection causes changes in relative frequencies of alleles in the gene pool.

negative feedback loops A primary mechanism of homeostasis, whereby a change in a physiological variable that is being monitored triggers a response that counteracts the initial fluctuation.

nephridium (*neh-FRID-ee-um*; pl. **nephridia**) A serpentine excretory tube in earthworms immersed in the fluid of the coelom.

nephron (*NEF-ron*) The tubular excretory unit of the vertebrate kidney.

neritic zone (*neh-RIT-ik*) The shallow regions of the ocean overlying the continental shelves.

net primary productivity The gross primary productivity minus the energy used by the producers for cellular respiration; represents the storage of chemical energy in an ecosystem available to consumers.

net reproductive rate The expected number of female offspring that will be produced during the average lifetime of a female member of a population.

neural crest (*NYOOR-ul*) A band of cells along the border where the neural tube pinches off from the ectoderm; the cells migrate to various parts of the embryo and form the pigment cells in the skin, bones of the skull, the teeth, the adrenal glands, and parts of the peripheral nervous system.

neuron (*NYOOR-on*) A nerve cell. The fundamental unit of the nervous system, it has a structure and properties that allow it to conduct signals by taking advantage of the electric charge across its cell membrane.

neurosecretory cells Hypothalamus cells that receive signals from other nerve cells, but that instead of signaling to an adjacent nerve cell or muscle, release hormones into the bloodstream.

neurotransmitter The chemical messenger released from the synaptic knobs of a neuron at a chemical synapse that diffuses across the synaptic cleft and binds to and stimulates the postsynaptic cell.

neutral variation Genetic diversity that confers no apparent selective advantage.

niche (*nich*) The sum total of an organism's utilization of the biotic and abiotic resources of its environment.

nitrogen fixation The assimilation of atmospheric nitrogen by certain prokaryotes into nitrogenous compounds that can be directly used by plants.

nodes Points along the stem of a plant at which leaves are attached.

noncompetitive inhibitor A substance that reduces the activity of an enzyme by binding to a location remote from the active site, changing its conformation so that it no longer binds to the substrate.

noncyclic electron flow A route of electron flow during the light reactions of photosynthesis that involves both photosystems and produces ATP, NADPH, and oxygen. Net electron flow is from water to NADP⁺.

noncyclic photophosphorylation (*FO-toe-foss-FOR-i-lay-shun*) The production of ATP by noncyclic electron flow.

nondisjunction (*NON-dis-JUNK-shun*) An accident of meiosis or mitosis, in which both members of a pair of homologous chromosomes or both sister chromatids fail to move apart properly.

nonpolar covalent bond A type of covalent bond in which electrons are shared equally between two atoms of similar electronegativity.

nonsense mutations Mutations that alter an amino acid–encoding triplet to one of the three termination codons, resulting in a shorter and usually nonfunctional protein.

norm of reaction The range of phenotypic possibilities for a single genotype, as influenced by the environment.

notochord (*NO-toe-kord*) A longitudinal, flexible rod formed from dorsal mesoderm and located between the gut and the nerve cord in all chordate embryos.

nuclear envelope The membrane in eukaryotes that encloses the genetic material, separating it from the cytoplasm.

nucleic acid (polynucleotide) (*PAHL-ee-NOO-klee-o-tide*) Biological molecules (RNA and DNA) that allow organisms to reproduce; polymers composed of monomers called nucleotides that are joined by covalent bonds (phosphodiester linkages) between the phosphate of one nucleotide and the sugar of the next nucleotide.

nucleoid region (*NOO-klee-oid*) The region in a prokaryotic cell consisting of a concentrated mass of DNA.

nucleolus (*noo-KLEE-oh-lus*) A specialized structure in the nucleus, formed from various chromosomes and active in the synthesis of ribosomes.

nucleoside (*NOO-klee-oh-side*) An organic molecule consisting of a nitrogenous base joined to a five-carbon sugar.

nucleosome (*NOO-klee-oh-soam*) The basic, beadlike unit of DNA packaging in eukaryotes, consisting of a segment of DNA wound around a protein core composed of two copies of each of four types of histone.

nucleotide (*NOO-klee-oh-tide*) The building block of a nucleic acid, consisting of a five-carbon sugar covalently bonded to a nitrogenous base and a phosphate group.

obligate aerobes (*OB-lig-it AIR-obes*) Organisms that require oxygen for cellular respiration and cannot live without it.

obligate anaerobes (*AN-ur-obes*) Organisms that cannot use oxygen and are poisoned by it.

oceanic zone The region of ocean water lying over deep areas beyond the continental shelf.

Okazaki fragment A short segment of the lagging (3′ to 5′) strand of DNA, synthesized in the 5′ to 3′ direction. A series of Okazaki fragments unite to produce the lagging strand at a DNA replication fork.

ommatidium (*OM-uh-TID-ee-um*; pl. **ommatidia**) A light-detecting unit in a compound eye that has its own cornea and lens.

omnivores (*OM-neh-vores*) Heterotrophic animals that consume both meat and plant material.

oncogenes (*ON-koh-jeens*) Genes found in viruses or as part of the normal genome that are crucial for triggering cancerous characteristics.

ontogeny (*on-TOJ-en-ee*) The embryonic development of an organism.

oogamy (*oh-OG-um-ee*) A condition in which male and female gametes differ, such that a small, flagellated sperm fertilizes a large, non-motile egg.

oogenesis (*OH-uh-JEN-eh-sis*) The process in the ovary that results in the production of female gametes.

open circulatory system An arrangement of internal transport in which blood bathes the organs directly and there is no distinction between blood and interstitial fluid.

operant conditioning (*OP-ur-ent*) A type of associative learning that directly affects behavior in a natural context. Also called trial-and-error learning.

operator In *Escherichia coli*, a segment of DNA between a promoter and structural genes that functions as an on/off switch, controlling an operon.

operon (*OP-ur-on*) A unit of genetic function common in bacteria and phages and consisting of regulated clusters of genes with related functions.

optical isomer (*EYE-so-mer*) A molecule that is a mirror image of another molecule with the same molecular formula.

optimal foraging A type of foraging behavior that maximizes the ratio of energy intake to expenditure.

organ A specialized center of body function composed of several different types of tissues.

organ of Corti The actual hearing organ of the vertebrate ear, located in the floor of the cochlear canal in the inner ear; contains the receptor cells (hair cells) of the ear.

organelle (*OR-gan-EL*) One of several formed bodies with specialized function, suspended in the cytoplasm and found in eukaryotic cells.

organogenesis (*or-GAN-oh-JEN-eh-sis*) An early period of rapid embryonic development in which the organs are taking form from the primary germ layers.

osmoconformer (*OZ-moh-kun-FOR-mur*) An animal that does not actively adjust its internal osmolarity because it is isosmotic with its environment.

osmolarity (*OZ-muh-LAR-eh-tee*) Solute concentration expressed as molarity.

osmoregulation (*OZ-moh-reg-yoo-LAY-shun*) Adaptations to control water balance in organisms living in hyperosmotic, hypoosmotic, or terrestrial environments.

osmoregulator An animal whose body fluids have a different osmolarity than the environment, and that must either discharge excess water if they live in a hypoosmotic environment or take in water if they inhabit a hyperosmotic environment.

osmosis (*oz-MOH-sis*) The movement of water across a selectively permeable membrane.

osmotic pressure (*oz-MOT-ik*) A measure of the tendency of a solution to take up water when separated from pure water by a selectively permeable membrane.

ostracoderms (*os-TRAK-uh-durms*) Extinct agnathans that were fish-like creatures encased in an armor of bony plates.

ovarian cycle (*oh-VAIR-ee-un*) Cyclic recurrence of the follicular phase, ovulation, and the luteal phase in the mammalian ovary, regulated by hormones.

ovary (*OH-vur-ee*) In flowers, the portion of a carpel in which the egg-containing ovules develop; in animals, the structure producing female gametes and reproductive hormones.

oviduct (*OH-veh-dukt*) A tube passing from the ovary to the vagina in invertebrates or to the uterus in vertebrates.

oviparous (*oh-VIP-ur-us*) Referring to a type of development in which young hatch from eggs laid outside the mother's body.

ovoviviparous (*OH-vo-vy-VIP-ur-us*) Referring to a type of development in which young hatch from eggs that are retained in the mother's uterus.

ovules (*OV-yools*) Structures that develop in the plant ovary and contain the female gametophyte.

ovum (*OH-vum*) The female gamete; the haploid, unfertilized egg, which is usually a relatively large, nonmotile cell.

oxidation The loss of electrons from a substance involved in a redox reaction.

oxidative phosphorylation (*FOS-for-eh-LAY-shun*) The production of ATP using energy derived from the redox reactions of the electron transport chain.

oxidizing agent The electron acceptor in a redox reaction.

pacemaker A specialized region of the right atrium of the mammalian heart that sets the rate of contraction.

paedomorphosis (*PEE-doh-mor-FOE-sis*) The retention in an adult organism of the juvenile features of its evolutionary ancestors.

paleontology (*PAY-lee-un-TOL-uh-gee*) The scientific study of fossils.

Pangaea (*pan-JEE-uh*) The supercontinent formed near the end of the Paleozoic era when plate movements brought all the land masses of the Earth together.

parapatric speciation (*PAR-uh-PAT-rik SPEE-shee-AY-shun*) Formation of new species from two populations between which gene flow occurs at a rate too slow to overcome divergence of the gene pools of the two populations.

paraphyletic (*PAR-uh-FYE-leh-tik*) Pertaining to a taxon that excludes some species that share a common ancestor with species included in the taxon.

parasites (*PAR-uh-sites*) Organisms that absorb their nutrients from the body fluids of living hosts.

parasympathetic nervous system (*PAR-uh-SIM-peh-THET-ik*) One of two divisions of the autonomic nervous system; generally enhances body activities that gain and conserve energy, such as digestion and reduced heart rate.

Parazoa (*PAR-uh-ZOH-uh*) A subkingdom of animals consisting of the sponges.

parenchyma (*pur-EN-kim-uh*) In plants, relatively unspecialized cell types that carry on most of the metabolism, synthesize and store organic products, and develop into more differentiated cell types.

parental types Offspring whose phenotype is the same as one of their parents.

parthenogenesis (*PAR-then-oh-JEN-eh-sis*) A type of reproduction in which females produce offspring from unfertilized eggs.

passive immunity The immunity obtained by acquiring ready-made antibodies that lasts only a few weeks or months because the immune system has not been stimulated by the antigen.

passive transport The diffusion of a substance across a biological membrane.

pattern formation The ordering of cells into specific three-dimensional structures, which is an essential part of shaping an organism and its individual parts during development.

pedigree A family tree describing the occurrence of heritable traits in parents and offspring across as many generations as possible.

pelagic zone *(pel-AY-jik)* The area of the ocean past the continental shelf, with areas of open water often reaching to very great depths.

penetrance *(PEN-eh-trens)* The proportion of individuals who show the phenotype that is expected from their genotype.

PEP carboxylase *(KAR-box-i-lase)* The enzyme that adds CO_2 to phosphoenolpyruvic acid in the Calvin cycle of photosynthesis.

peptide bond The covalent bond between two amino acid units, formed by dehydration synthesis.

peptidoglycan *(PEP-tid-oh-GLY-kan)* A type of polymer found only in bacterial cell walls and consisting of modified sugars cross-linked by short polypeptides.

perception The interpretation of sensations by the brain.

perennials *(per-EN-ee-uls)* Plants that live for many years.

perfect flower A flower with both stamens and carpels.

pericarp *(PER-eh-karp)* The thickened wall of a ripened plant ovary, or fruit.

pericycle *(PER-eh-sigh-kul)* A layer of cells just inside the endodermis of a root, which may become meristematic and begin dividing again.

periderm *(PER-eh-durm)* The protective coat that replaces the epidermis in plants during secondary growth, formed of the cork cambium, cork, and phelloderm.

peripheral nervous system The sensory and motor neurons that connect to the central nervous system.

peripheral proteins Those proteins on the surface of biological membranes.

peristalsis *(PER-is-TAL-sis)* Rhythmic waves of contraction of digestive smooth muscle that push food along the tract.

peroxisomes *(per-OX-eh-soams)* A family of microbodies containing enzymes that transfer hydrogen from various substrates to oxygen, producing and then degrading hydrogen peroxide.

petiole *(PET-ee-ole)* The stalk of a leaf, which joins the leaf to a node of the stem.

P generation The parental organisms in a genetic cross-fertilization.

phagocytosis *(FAY-go-sigh-TOE-sis)* A type of endocytosis involving large, particulate substances.

pharynx *(FAH-rinks)* An area in the throat of vertebrates where the air and food passages cross; also, in flatworms, the muscular tube that protrudes from the ventral side of the worm and ends in the mouth.

phenetics *(feh-NEH-tiks)* An approach to taxonomy based entirely on measurable similarities and differences in phenotypic characters, without consideration of homology, analogy, or phylogeny.

phenotype *(FEE-nuh-type)* The expressed traits of an organism.

pheromones *(FAIR-uh-mones)* Small, volatile chemical signals functioning in communication between animals and acting much like hormones to influence physiology and behavior.

phloem *(FLOAM)* A portion of the vascular system in plants, consisting of living cells arranged into elongated tubes that transport sugar and other organic nutrients throughout the plant.

phosphate group *(FOS-fate)* A functional group important in energy transfer.

phospholipids *(FOS-foe-LIP-ids)* Molecules constituting the inner bilayer of biological membranes. Phospholipids have a polar, hydrophilic head and a nonpolar, hydrophobic tail.

photic zone *(FOE-tik)* The narrow top slice of the ocean where light penetrates sufficiently for photosynthesis to occur.

photoautotrophs *(FOE-toe-AW-to-trohfs)* Organisms that harness light energy to drive the synthesis of organic compounds from carbon dioxide.

photoheterotrophs *(FOE-toe-HET-ur-oh-trohfs)* Organisms that use light to generate ATP, but that must obtain their carbon in organic form.

photon *(FOE-tahn)* A quantum, or discrete amount, of light energy.

photoperiodism *(FOE-toe-PEER-ee-od-iz-um)* A physiological response to day length, such as flowering in plants.

photophosphorylation *(FOE-toe-fos-for-ul-AY-shun)* The process of generating ATP from ADP and phosphate by means of a proton-motive force generated by the thylakoid membrane of the chloroplast during the light reactions of photosynthesis.

photorespiration A metabolic pathway that consumes oxygen, evolves carbon dioxide, generates no ATP, and decreases photosynthetic output. Photorespiration generally occurs on hot, dry, bright days, when stomata close and the oxygen concentration in the leaf exceeds that of carbon dioxide.

photosystem The light-harvesting unit in photosynthesis, located on the thylakoid membrane of the chloroplast and consisting of the antenna complex, the reaction-center chlorophyll *a*, and the primary electron acceptor.

photosystem I A light-harvesting unit with a P700 chlorophyll *a* molecule at its center; absorbs light best at a wavelength of 700 nm.

photosystem II A light-harvesting unit with a P680 chlorophyll *a* molecule at its center; absorbs light best at a wavelength of 680 nm.

phototropism *(foh-TOH-treh-PIZ-um)* Growth of a plant shoot toward or away from light.

pH scale A measure of hydrogen ion concentration equal to $-\log[H^+]$ and ranging in value from 0 to 14.

phylogeny *(fye-LOJ-en-ee)* The evolutionary history of a species or group of related species.

phytochrome *(FYE-tuh-krome)* A pigment involved in many responses of plants to light.

pili *(PILL-eye)* Surface appendages in certain bacteria that function in adherence and transfer of DNA during conjugation.

pinocytosis *(PIN-no-sigh-TOE-sis)* A type of endocytosis in which the cell ingests extracellular fluid and its dissolved solutes.

pineal gland *(PIN-ee-ul)* A small endocrine gland on the dorsal surface of the diencephalon in vertebrates; secretes the hormone melatonin, which regulates body functions related to seasonal day length.

pituitary gland *(pi-TOO-i-tair-ee)* An endocrine gland at the base of the hypothalamus; consists of a posterior lobe (neurohypophysis), which stores and releases two hormones produced by the hypothalamus, and an anterior lobe (adenohypophysis), which produces and secretes many hormones regulating diverse body functions.

placenta *(pluh-SEN-tuh)* A structure in the pregnant uterus for nourishing a viviparous fetus with the mother's blood supply; formed from the uterine lining and embryonic membranes.

placental mammals A group of mammals, including humans, whose young complete their embryonic development in the uterus, joined to the mother by a placenta.

placoderms *(PLAK-oh-durms)* An extinct class of fishlike vertebrates that had jaws and were enclosed in a tough, outer armor.

plankton Communities of mostly microscopic organisms that drift passively or swim weakly near the surface of oceans, ponds, and lakes.

plasma *(PLAZ-muh)* The liquid matrix of blood in which the cells are suspended.

plasma cells Derivatives of B cells that secrete antibodies.

plasma membrane The membrane at the boundary of every cell that acts as a selective barrier to the passage of ions and molecules.

plasmid *(PLAZ-mid)* A small ring of DNA that carries accessory genes separate from those of a bacterial chromosome.

plasmodesmata *(PLAZ-moh-dez-MAH-tuh)* Open channels in the cell wall of plants through which strands of cytoplasm connect from adjacent cells.

plasmolysis *(plaz-MOL-eh-sis)* A phenomenon in walled cells in which the cytoplasm shrivels and the plasma membrane pulls away from the cell wall when the cell loses water to a hyperosmotic environment.

plastids A family of closely related plant organelles, including chloroplasts, chromoplasts, and leucoplasts.

pleiotropy *(PLY-eh-troh-pee)* The ability of a single gene to have multiple effects.

plesiomorphic character *(PLEEZ-ee-oh-MORE-fik)* A primitive phenotypic character possessed by a remote ancestor.

point mutation A change in the chromosome at a single nucleotide within a gene.

polar covalent bond A type of covalent bond between atoms that differ in electronegativity. The shared electrons are pulled closer to the more electronegative atom, making it partially negative and the other atom partially positive.

polar molecule A molecule (such as water) with opposite charges on opposite sides.

pollen grains Immature male gametophytes that develop within the anthers of stamens.

pollination *(POL-eh-NAY-shun)* The placement of pollen onto the stigma of a carpel by wind or animal vectors, a prerequisite to fertilization.

polyandry *(POL-ee-AN-dree)* A polygamous mating system involving one female and many males.

polygenic inheritance *(POL-ee-JEN-ik)* An additive effect of two or more gene loci on a single phenotypic characteristic.

polygyny *(pol-IJ-en-ee)* A polygamous mating system involving one male and many females.

polymer *(POL-eh-mur)* A large molecule consisting of many identical or similar monomers linked together.

polymorphic *(POL-ee-MOR-fik)* Said of a population in which two or more morphs are present in readily noticeable frequencies.

polymorphism *(POL-ee-MOR-fiz-um)* The coexistence of two or more distinct forms of individuals in the same population.

polyp *(POL-ip)* The sessile variant of the body plan observed in the phylum Cnidaria. The alternate form is the medusa.

polypeptide chain *(POL-ee-PEP-tide)* A polymer of many amino acids linked by peptide bonds.

polyphyletic *(POL-ee-fye-LET-ik)* Pertaining to a taxon whose members were derived from two or more ancestral forms not common to all members.

polyploidy *(POL-uh-ploid-ee)* A chromosomal alteration in which the organism possesses more than two complete chromosome sets.

polysaccharide *(POL-ee-SAK-ur-ide)* A polymer of up to over a thousand monosaccharides, formed by dehydration synthesis.

polysome *(POL-ee-some)* An aggregation of several ribosomes attached to one messenger RNA molecule.

polytene chromosome *(POL-ee-teen)* A many-stranded chromosome resulting from repeated chromosomal replication without separation of sister chromatids or cell division.

population A group of individuals of one species that live in a particular geographic area.

population genetics The scientific study of gene pools and genetic variation in biological populations.

portal system A circulatory pathway with two successive capillary beds in tandem.

positional information Signals, to which genes regulating development respond, indicating a cell's location relative to other cells in an embryonic structure.

positive feedback A physiological control mechanism in which a change in some variable triggers mechanisms that amplify the change.

postsynaptic membrane *(post-sin-AP-tik)* The surface of the cell on the opposite side of the synapse from the synaptic knob of the stimulating neuron that contains receptor proteins and degradative enzymes for the neurotransmitter.

postzygotic barrier *(POST-zig-OT-ik)* Any of several species-isolating mechanisms that prevent hybrids produced by two different species from developing into viable, fertile adults.

potential energy The energy stored by matter as a result of its location or spatial arrangement.

preaptation *(PREE-ap-TAY-shun)* A structure that evolves and functions in one environmental context, but can perform additional functions when placed in some new environment.

preformation The notion that a gamete contains a whole series of successively smaller embryos within embryos.

prezygotic barrier *(PREE-zye-GOT-ik)* A reproductive barrier that impedes mating between species or hinders fertilization of ova if interspecific mating is attempted.

primary consumers Herbivores, organisms in the trophic level that eat plants or algae.

primary germ layers The three layers (ectoderm, mesoderm, endoderm) of the late gastrula, which develop into all parts of an animal.

primary growth Growth initiated by the apical meristems of a plant root or shoot.

primary immune response The initial immune response to an antigen, which appears after a lag of several days.

primary producers Autotrophs; photosynthetic and chemosynthetic organisms.

primary productivity The rate at which light energy or inorganic chemical energy is converted to the chemical energy of organic compounds by autotrophs in an ecosystem.

primary structure The level of protein structure referring to the specific sequence of amino acids.

primary succession A type of ecological succession that occurs in an area where there were originally no organisms.

principle of allocation The concept that each organism has an energy budget, or a limited amount of total energy apportionable to all of its needs for maintenance and reproduction.

procambium *(pro-KAM-bee-um)* A primary meristem of roots and shoots that forms the vascular tissue.

progesterone *(pro-JES-ter-one)* The ovarian hormone produced by the corpus luteum after ovulation that acts to prepare the uterus for pregnancy and the breasts for lactation.

prokaryotic cell *(pro-KAR-ee-OT-ik)* A type of cell lacking a membrane-enclosed nucleus and membrane-enclosed organelles; found only in the kingdom Monera.

promoter A specific nucleotide sequence in DNA, flanking the start of a gene; instructs RNA polymerase where to start transcribing RNA.

prophage *(PRO-faj)* A phage genome that has become inserted into a specific site on the bacterial chromosome.

prostaglandins (*PROS-tuh-GLAN-dins*) A group of modified fatty acids secreted by virtually all tissues and serving a wide variety of functions as messengers.

protein (*PRO-teen*) A three-dimensional biological polymer constructed from a set of 20 different monomers called amino acids.

Protist (*PRO-tist*) The kingdom containing all eukaryotic organisms that do not fit the definition of plant, animal, or fungi. This kingdom is also referred to as Protista.

protoderm (*PRO-toe-durm*) The outermost primary meristem, which gives rise to the epidermis of roots and shoots.

proton-motive force The potential energy stored in the form of an electrochemical gradient, generated by the pumping of hydrogen ions across biological membranes in chemiosmosis.

proton pump (*PRO-tahn*) An active transport mechanism in cell membranes that consumes ATP to force hydrogen ions out of a cell and, in the process, generates a membrane potential.

protonephridium (*PRO-toe-NEF-rid-ee-um*) An excretory system, such as the flame-cell system of flatworms, consisting of a network of closed tubules having external openings called nephridiopores and lacking internal openings.

protooncogene (*PROE-toe-ONK-oh-jeen*) A normal cellular gene corresponding to an oncogene; a gene with a potential to cause cancer, but that requires some alteration to become an oncogene.

protostomes (*PRO-toh-stoams*) One of two distinct evolutionary lines of coelomates, consisting of the annelids, mollusks, and arthropods, and characterized by spiral, determinate cleavage, schizocoelous formation of the coelom, and development of the mouth from the blastopore.

provirus Viral DNA that inserts into a host genome.

pseudocoelomates (*SOO-doe-SEEL-oh-mates*) Animals, such as rotifers and roundworms, whose body cavity is not completely lined by mesoderm.

pseudogenes (*SOO-doe-JEENS*) Noncoding DNA segments with sequences similar to functional genes, but lacking signals necessary for gene expression.

pseudopodium (*SOO-doe-POE-dee-um*) Cellular extensions of amoeboid cells used in moving and feeding.

P-site A binding site on a ribosome that holds the tRNA carrying a growing polypeptide chain.

punctuated equilibrium A theory of evolution advocating spurts of relatively rapid change followed by long periods of stasis.

purine (*PYOOR-een*) One of two families of nitrogenous bases found in nucleotides consisting of two members: adenine (A) and guanine (G).

pyrimidine (*pir-IM-eh-deen*) One of two families of nitrogenous bases found in nucleotides consisting of three members: cytosine (C), thymine (T), and uracil (U).

quantitative traits Heritable traits in a population, each of which varies continuously as a result of environmental influences and the additive effect of two or more genes (polygenic inheritance).

quaternary structure (*KWAT-ur-nair-ee*) The particular shape of a complex, aggregate protein, defined by the characteristic three-dimensional arrangement of its constituent subunits, each a polypeptide.

radial cleavage Embryonic development characteristic of deuterostomes, in which the planes of cell division that transform the zygote into a ball of cells are either parallel or perpendicular to the polar axis, thereby aligning tiers of cells one above the other.

radial symmetry A body shaped like a pie or barrel, with many equal parts radiating outward like the spokes of a wheel. Present in cnidarians and echinoderms.

radicle An embryonic root of a plant.

radioactive Having a nucleus that gives off particles and energy; characteristic of unstable isotopes of a chemical element.

radioactive dating A method using half-lives of radioactive isotopes to determine the age of fossils and rocks.

range The geographic area in which a population lives.

reaction center The location of one or a pair of specialized chlorophyll *a* molecules in the pigment assembly system of the light reactions of photosynthesis.

realized niche The environmental resources that an organism actually uses; the niche that an organism actually occupies.

receptor-mediated endocytosis (*EN-do-si-TO-sis*) The movement of specific molecules into a cell by the inward budding of membranous vesicles containing proteins with receptor sites specific to the molecules being taken in; enables a cell to acquire bulk quantities of specific substances.

receptor potential An initial response of a receptor cell to a stimulus, consisting of a change in voltage across the receptor membrane proportional to the stimulus strength. The intensity of the receptor potential determines the frequency of action potentials traveling into the nervous system.

recessive allele In a heterozygous cell, the allele that is completely masked in the phenotype.

reciprocal altruism (*AL-troo-IZ-um*) Altruistic behavior between unrelated individuals; believed to produce some benefit to the altruistic individual in the future when the current beneficiary "returns the favor."

recognition concept of species Definition of species based on mate-recognition mechanisms; assumes that reproductive adaptations of a species consist of a set of features that maximize successful mating with members of the same population; an alternative to the biological species concept.

recombinant DNA technology A set of techniques for recombining genes from different sources *in vitro* and transferring this recombinant DNA into cells, where it may be expressed.

recombinants Offspring whose phenotype differs from that of the parents.

redox reactions (*REE-doks*) Chemical reactions involving the transfer of one or more electrons from one reactant to another; also called oxidation-reduction reactions.

reducing agent The electron donor in a redox reaction.

reduction The gaining of electrons by a substance involved in a redox reaction.

reflex An automatic reaction to a stimulus, mediated by the spinal cord or lower brain.

refractory period (*ree-FRAK-tor-ee*) The short time immediately after an action potential in which the neuron cannot respond to another stimulus, owing to an increase in potassium permeability.

regeneration The replacement of parts of an organism that are lost from injury.

regulative development A pattern of development, such as that of a mammal, in which the early blastomeres retain the potential to form the entire animal.

relative fitness The contribution of one genotype to the next generation compared to that of alternative genotypes for the same locus.

releaser A signal stimulus that functions as a communication signal between individuals of the same species.

releasing factors Hormones produced by neurosecretory cells in the hypothalamus of the vertebrate brain that stimulate or inhibit secretion of hormones by the anterior pituitary.

replication forks The Y-shaped points on a replicating DNA molecule where new strands are growing.

repressible enzymes Enzymes whose synthesis is inhibited by a specific metabolite.

repressor protein A protein that suppresses the expression of prophage or operon genes.

resolving power A measure of the clarity of an image; the minimum distance that two points can be separated and still be distinguished as two separate points.

resource partitioning Division of environmental resources by coexisting species populations such that the niche of each species differs by one or more significant factor from that of all coexisting species populations.

resting potential The membrane potential characteristic of a nonconducting, excitable cell, with the inside of the cell more negative than the outside.

restriction enzyme A degradative enzyme that recognizes and cuts up DNA (including that of certain phages) that is foreign to a cell.

restriction fragment length polymorphisms (RFLPs) Differences in DNA sequence on homologous chromosomes that result in different patterns of restriction fragment lengths (DNA segments resulting from treatment with restriction enzymes); useful as genetic markers for making linkage maps.

restriction mapping A method of determining similarity between the genomes or genes of two species by the electrophoretic comparison of restriction fragments of DNA of the two species.

retina *(REH-tin-uh)* The innermost layer of the eye, containing photoreceptor cells (rods and cones) and neurons; transmits images formed by the lens to the brain via the optic nerve.

retinal The light-absorbing pigment in rods and cones of the vertebrate eye.

retrovirus *(REH-troh-VIRE-us)* An RNA virus that reproduces by transcribing its RNA into DNA and then inserting the DNA into a cellular chromosome; an important class of cancer-causing viruses.

reverse transcriptase *(trans-KRIP-tase)* An enzyme encoded by some RNA viruses that uses RNA as a template for DNA synthesis.

rhizoids *(RYE-zoids)* Rootlike structures found in some fungi and nonvascular plants.

ribonucleic acid (RNA) *(RYE-bow-noo-KLAY-ik)* A single-stranded nucleic acid molecule involved in protein synthesis, the structure of which is specified by DNA.

ribosomal RNA (rRNA) The most abundant type of RNA. Together with proteins, it forms the structure of ribosomes that coordinate the sequential coupling of tRNAs to the series of mRNA codons.

ribosome A cell organelle constructed in the nucleolus, consisting of two subunits and functioning as the site of protein synthesis in the cytoplasm.

RNA polymerase *(pul-IM-ur-ase)* An enzyme that links together the growing chain of ribonucleotides during transcription.

rod cell One of two kinds of photoreceptors in the vertebrate retina; sensitive to black and white and allows night vision.

root hairs Thousands of minute projections that grow just behind the root tips of plants, increasing surface area for absorption of water and minerals.

root pressure Upward push of water within the stele of vascular plants, caused by active pumping of minerals into the xylem by root cells.

rough ER That portion of the endoplasmic reticulum studded with ribosomes.

r-selection The concept that in some (r-selected) populations a high reproductive rate is the chief determinant of life history.

rubisco (RuBP carboxylase) The enzyme that catalyzes the first step (the addition of CO_2 to ribulose bisphosphate) of the Calvin cycle.

ruminants Animals, such as cattle and sheep, with elaborate, multicompartmentalized stomachs specialized for a herbivorous diet.

SA (sinoatrial) node The pacemaker of the heart, located in the wall of the right atrium; sets the rate of contraction of the heart.

saltatory conduction *(SAHL-tuh-TORR-ee)* Rapid transmission of a nerve impulse along an axon resulting from the action potential jumping from one node of Ranvier to another, skipping the myelin-sheathed regions of membrane.

saprophytes *(SAP-ruh-fites)* Organisms that act as decomposers by absorbing nutrients from dead organic matter.

sarcomere *(SAR-koh-meer)* The fundamental, repeating unit of striated muscle, delimited by the Z lines.

sarcoplasmic reticulum *(SAR-koh-PLAZ-mik reh-TIK-yoo-lum)* A modified form of endoplasmic reticulum in striated muscle cells that stores calcium that is used to trigger contraction during stimulation.

satellite DNA DNA with sufficiently different base sequences and different natural density that it can be isolated from other DNA in a cell by ultracentifugation in a cesium chloride density gradient; apparently located at centromeres where it functions in DNA replication and nuclear division.

saturated fatty acid A fatty acid in which all carbons in the hydrocarbon tail are connected by single bonds, thus maximizing the number of hydrogen atoms that can attach to the carbon skeleton.

savanna *(suh-VAN-uh)* A tropical grassland biome with scattered individual trees, large herbivores, and three distinct seasons based primarily on rainfall, maintained by occasional fires and drought.

Schwann cells *(SHWAHN)* A chain of supporting cells enclosing the axons of many neurons and forming an insulating layer called the myelin sheath.

sclereids *(SKLER-ee-ids)* Short, irregular sclerenchyma cells found in nutshells and seed coats and scattered through the parenchyma of some plants.

sclerenchyma *(skler-EN-kim-uh)* A rigid, supportive cell type in plants usually lacking protoplasts and possessing thick secondary walls strengthened by lignin at maturity.

second law of thermodynamics The principle whereby every energy transfer or transformation increases the entropy of the universe. Ordered forms of energy are at least partly converted to heat, and in spontaneous reactions, the free energy of the system also decreases.

second messenger A chemical signal, such as cyclic AMP, that relays a hormonal message from the cell's surface to its interior.

secondary compounds Chemical compounds synthesized through diversion of products of major metabolic pathways that are used in defense by prey species.

secondary consumers The trophic level consisting of carnivores that eat herbivores.

secondary growth The increase in girth of the stems and roots of many plants, especially woody, perennial dicots.

secondary immune response The immune response elicited when an animal encounters the same antigen at some later time. The secondary immune response is more rapid, of greater magnitude, and of longer duration than the primary immune response.

secondary productivity The rate at which all the heterotrophs in an ecosystem incorporate organic material into new biomass, which can be equated to chemical energy.

secondary structure The localized, repetitive folding of the polypeptide backbone of a protein due to hydrogen-bond formation between peptide linkages.

secondary succession A type of succession that occurs where an existing community has been severely cleared by some disturbance.

sedimentary rocks *(SED-eh-MEN-tar-ee)* Rocks formed from sand and mud that once settled in layers on the bottom of seas, lakes, and marshes. Sedimentary rocks are often rich in fossils.

seed An adaptation for terrestrial plants consisting of an embryo packaged along with a store of food within a resistant coat.

selection coefficient The difference between two fitness values, representing a relative measure of selection against an inferior genotype.

selective permeability A property of biological membranes that allows some substances to cross more easily than others.

semen (*SEE-men*) The fluid that is ejaculated by the male during orgasm; contains sperm and secretions from several glands of the male reproductive tract.

semelparous (*SEM-el-PAIR-us*) Pertaining to organisms that reproduce only once in a lifetime.

semicircular canals A three-part chamber of the inner ear that detects angular acceleration.

semilunar valves Heart valves located at the exits of the heart, where the aorta leaves the left ventricle and the pulmonary artery leaves the right ventricle.

seminal vesicles (*SEM-in-ul VES-eh-kuls*) A part of the male reproductive tract that stores sperm in invertebrates and produces semen in vertebrates.

seminiferous tubules (*SEM-in-IF-er-us*) Highly coiled tubes in the testes in which sperm are produced.

semispecies (*SEM-ee-SPEE-sheesz*) Sympatric populations that were formerly allopatric and exhibit limited hybridization with hybrids being less fit than offspring of parents from one or the other parent population.

senescence (*seh-NESS-ents*) Aging; progression of irreversible change in a living organism, eventually leading to death.

sensations The impulses sent to the brain from activated receptors and sensory neurons.

sensory neurons Nerve cells that receive information from the internal and external environments and transmit the signals toward the brain or spinal cord.

sensory receptors Specialized structures that respond to specific stimuli from an animal's external or internal environment. They transmit the information of an environmental stimulus to the animal's nervous system by converting stimulus energy to the electrochemical energy of action potentials.

sepals (*SEE-puls*) A whorl of modified leaves in angiosperms that encloses and protects the flower bud before it opens.

sex chromosomes The pair of chromosomes that differentiate between and are responsible for determining the sexes.

sex-influenced trait The sex-dependent variation in penetrance or expressivity of autosomal genes.

sex-limited trait A characteristic appearing exclusively in one sex, but whose gene is often carried on an autosome.

sex-linked genes Genes located on one sex chromosome but not the other.

sexual dimorphism (*dye-MOR-fiz-um*) A special case of polymorphism based on the distinction between secondary sexual characteristics of males and females.

sexual reproduction A type of reproduction in which two parents give rise to offspring that have unique combinations of genes inherited from the gametes of the two parents.

sexual selection Selection based on variation in secondary sexual characteristics, leading to the enhancement of sexual dimorphism.

shoot system The aerial portion of a plant body, consisting of stems, leaves, and flowers.

short-day plant A plant that flowers, usually in late summer, fall, or winter, only when the light period is shorter than a critical length.

sieve-tube members Chains of living cells that form sieve tubes in phloem.

sign stimulus An external sensory stimulus that triggers a fixed action pattern.

simple fruit A fruit derived from a single ovary.

simple tissues Plant tissues composed of a single cell type.

sister chromatids (*KROH-muh-tids*) Replicated forms of a chromosome that are joined together by the centromere and eventually separated during mitosis or meiosis II.

skeletal muscle Striated muscle generally responsible for the voluntary movements of the body.

sliding filament model The theory explaining how muscle contracts, based on change within a sarcomere, the basic unit of organization of muscle; states that thin (actin) filaments slide (actually ratchet) across thick filaments, shortening the sarcomere; shortening of all sarcomeres in a myofibril shorten the entire myofibril.

small nuclear ribonucleoproteins (**snRNPs**) (*RIBE-oh-NUKE-lee-oh-pro-teens*) A variety of small particles in the cell nucleus, composed of RNA and protein molecules; functions not fully understood, but some form parts of spliceosomes, active in RNA splicing.

smooth ER That portion of the endoplasmic reticulum that is free of ribosomes.

sodium-potassium pump A special transport protein in the plasma membrane of animal cells that transports sodium out of and potassium into the cell against their concentration gradients.

solute (*SOL-yoot*) A substance that is dissolved in a solution.

solution A homogeneous, liquid mixture of two or more substances.

solvent The dissolving agent of a solution. Water is the most versatile solvent known.

somatic cell (*SOME-at-ik*) Any cell in a multicellular organism except a sperm or egg cell.

somatic nervous system The division of the motor nervous system of vertebrates composed of motor neurons that carry signals to skeletal muscles in response to external stimuli.

somites (*SO-mites*) Blocks of mesoderm along each side of a chordate embryo.

spatial summation An accumulative effect on the membrane potential of a postsynaptic cell, caused by several different synaptic knobs stimulating a postsynaptic cell simultaneously.

speciation (*SPEE-shee-AY-shun*) The origin of new species in evolution.

species A particular kind of organism; members possess similar anatomical characteristics and have the ability to interbreed.

species diversity The number and relative abundance of species in a biological community.

species equitability The relative abundance of individuals of a species in a biological community.

species richness The number of species in a biological community.

species selection A theory that maintains that species that live the longest and generate the greatest number of species determine the direction of major evolutionary trends.

specific heat The amount of heat that must be absorbed or lost for 1 g of a substance to change its temperature by 1°C.

spermatogenesis The continuous and prolific production of mature sperm cells in the testis.

spermatozoon (pl. **spermatozoa**) The male gamete; a haploid, usually small, flagellated cell.

S phase The synthesis phase of the cell cycle, constituting the portion of interphase during which DNA is replicated.

sphincter (*SFINK-ter*) A ringlike valve, consisting of modified muscles in a muscular tube, such as a digestive tract; closes off the tube like a purse string.

spindle An assemblage of microtubules that orchestrates chromosome movement during eukaryotic cell division.

spiracles Minute pores distributed over the surface of an insect's body that serve as entrances to the tracheal system.

spiral cleavage Embryonic development characteristic of protostomes, in which the planes of cell division that transform the zygote into a ball of cells occur obliquely to the polar axis, resulting in cells of each tier sitting in the grooves between cells of adjacent tiers.

spliceosome (*SPLISE-ee-oh-SOME*) A complex assembly that interacts with the ends of an RNA intron in splicing RNA; releases an intron and joins two adjacent exons.

spore In the life cycle of a plant or alga with alternation of generations, a meiotically produced haploid cell that divides mitotically, generating a multicellular individual, the gametophyte, without fusing with another cell.

sporophylls Spore-producing leaves specialized for reproduction.

sporophyte The multicellular diploid form in organisms undergoing alternation of generations that results from a union of gametes and that meiotically produces haploid spores that grow into the gametophyte generation.

stabilizing selection Natural selection that favors intermediate variants by acting against extreme phenotypes.

stamens The pollen-producing male reproductive organs of a flower, consisting of an anther and filament.

statocyst (*STAT-eh-SIST*) A type of mechanoreceptor that functions in equilibrium in invertebrates through the use of statoliths, which stimulate hair cells in relation to gravity.

statolith A starch grain in a plant cell that settles to the low point of the cells in relation to gravity. In animals, a granule of sand or limestone that, in responding to gravity, stimulates hair cells of a statocyst.

stele The central vascular cylinder in roots where xylem and phloem are located.

steroids A class of lipids characterized by a carbon skeleton consisting of four rings with various functional groups attached.

stigma The sticky tip of the carpel that receives the pollen in a flower.

stoma (pl. **stomata**) Microscopic pores surrounded by guard cells in the epidermis of leaves and stems that allow gas exchange between the environment and the interior of the plant.

striated muscle A type of muscle exhibiting a regular pattern of light and dark bands, caused by the regular arrangement of myofilaments.

strict aerobes (*AIR-obes*) Organisms that can survive only in an atmosphere of oxygen. The oxygen is used in aerobic respiration.

strict anaerobes Organisms that cannot survive in an atmosphere of oxygen. Other substances, such as sulfate or nitrate, are the terminal electron acceptors in the electron transport chains that generate their ATP.

stroma The ground substance of the chloroplast surrounding the thylakoid membrane and involved in the synthesis of organic molecules from carbon dioxide and water.

stromatolites Rocks made of banded domes of sediment in which are found the most ancient forms of life: prokaryotes dating back as far as 3.5 billion years.

structural formula A type of molecular notation in which the constituent atoms are joined by lines representing covalent bonds.

structural gene A gene that codes for a polypeptide.

structural isomers Organic molecules differing only in the covalent arrangements of their atoms and/or the location of double bonds.

style The slender, necklike portion of the carpel that leads to the basal ovary.

substrate The substance on which an enzyme works.

substrate-level phosphorylation The formation of ATP by directly transferring a phosphate group to ADP from an intermediate substrate in catabolism.

succession (*suk-SESH-un*) Transition in the species composition of a biological community, often following ecological disturbance of the community; establishment of a biological community in an area virtually barren of life.

summation A phenomenon of neural integration in which the membrane potential of the postsynaptic cell in a chemical synapse is determined by the total activity of all excitatory and inhibitory presynaptic impulses acting on it at any one time.

supergene A cluster of genes having a cooperative function and linked closely on a common chromosome.

suppressor T cells A type of T cell that causes B cells as well as other cells to ignore antigens.

surface tension A measure of how difficult it is to stretch or break the surface of a liquid. Water has a high surface tension because of hydrogen bonding of surface molecules.

survivorship curve A plot of the number of members of a cohort that are still alive at each age; one way to represent age-specific mortality.

symbiont (*SIM-bye-ont*) The smaller participant in a symbiotic relationship, which lives in or on the host.

symbiosis An ecological relationship between organisms of two different species that live together in direct contact.

sympathetic nervous system One of two divisions of the autonomic nervous system of vertebrates; generally increases energy expenditure and prepares the body for action.

sympatric speciation A mode of speciation occurring as a result of a radical change in the genome that produces a reproductively isolated subpopulation in the midst of its parent population.

symplast In plants, the continuum of cytoplasm connected by plasmodesmata between cells.

synapse (*SIN-aps*) The locus where one neuron communicates with another neuron in a nervous pathway; a narrow gap between a synaptic knob of an axon and a signal-receiving portion (dendrite or cell body) of another neuron or effector cell. Neurochemicals released by synaptic knobs diffuse across the synapse, relaying messages to the dendrite or effector.

synapsids (*SIN-ap-sids*) Predatory reptiles that evolved during the Permian period and gave rise to several lineages, including the mammal-like therapsids.

synapsis The pairing of replicated homologous chromosomes during prophase I of meiosis.

synaptic knob One of many swollen bulbs at the end of an axon, in which neurotransmitter is stored and released.

synaptic vesicle One of many membrane-enclosed sacs in a synaptic knob that contains neurotransmitter.

synaptonemal complex (*sin-APP-toe-NEEM-ul*) A protein structure, visible in electron micrographs of mid-prophase I of meiosis; function is unknown but correlated with, and essential for, chiasmata formation and crossing over; often zipperlike in appearance.

syngamy (*SIN-gam-ee*) The process of nuclear union during fertilization.

system (organ system) An organized group of organs that carries out one or more body functions.

systematics The branch of biology that studies the diversity of life. Systematics encompasses taxonomy and is involved in reconstructing phylogenetic history.

systole The stage of the heart cycle in which the heart muscle contracts and the chambers pump blood.

taiga (*TYE-guh*) The coniferous or boreal forest biome, characterized by considerable snow, harsh winters, short summers, and evergreen trees.

taxis (*TAKS-iss*) A movement toward or away from some stimulus.

taxon (pl. **taxa**) The named taxonomic unit at any given level.

taxonomy The branch of biology concerned with naming and classifying the diverse forms of life.

T cell A type of lymphocyte responsible for cellular immunity that differentiates under the influence of the thymus.

temperate forest A biome located throughout midlatitude regions where there is sufficient moisture to support the growth of large trees, which are mostly of the broad-leaved deciduous type.

temperate viruses Viruses that can reproduce without killing their hosts.

temperature A measure of the intensity of heat in degrees, reflecting the average kinetic energy of the molecules.

temporal summation An accumulative effect on the membrane potential of a postsynaptic cell; caused by chemical transmissions from one or more synaptic knobs occurring so close together in time that each postsynaptic potential affects the membrane before the voltage has returned to the resting potential after the previous stimulation.

tendon A type of fibrous connective tissue that attaches muscle to bone.

terminal bud Embryonic tissue at the tip of a shoot, consisting of developing leaves and a compact series of nodes and internodes.

tertiary structure (*TUR-she-air-ee*) Irregular contortions of a protein molecule due to interactions of side chains involved in hydrophobic interactions, ionic bonds, hydrogen bonds, and disulfide bridges.

testcross Breeding of an organism of unknown genotype with a homozygous recessive individual to determine the unknown genotype. The ratio of phenotypes in the offspring determines the unknown genotype.

testis (pl. **testes**) The male reproductive organ, or gonad, in which sperm and reproductive hormones are produced.

testosterone The most abundant androgen in the male body.

tetanus (*TET-un-us*) The maximal, sustained contraction of a skeletal muscle, caused by a very fast frequency of action potentials elicited by continual stimulation.

tetrapods Vertebrates possessing two pairs of limbs, such as amphibians, reptiles, birds, and mammals.

thalamus (*THAAL-uh-mus*) One of two integrating centers of the diencephalon of vertebrates. Neurons with cell bodies in the thalamus relay neural input to specific areas in the cerebral cortex and regulate what information goes to the cerebral cortex.

thecodonts A group that descended from the stem reptiles during the Permian period and that served as the ancestors of dinosaurs, crocodiles, and birds.

therapsids (*thur-AP-sids*) A lineage of reptiles that descended from synapsids during the Permian period and eventually gave rise to mammals.

thermodynamics The study of energy transformations that occur in a collection of matter.

thermoregulation Maintenance of internal temperature within a tolerable range.

thigmomorphogenesis (*THIG-moh-MORE-foh-JEN-eh-sis*) Decrease in plant stem length with concomitant thickening caused by mechanical disturbance.

thigmotropism (*THIG-moh-TROH-piz-um*) The directional growth of a plant in relation to touch.

threshold The potential an excitable cell membrane must reach for an action potential to be initiated.

thylakoids (*THIGH-luh-koids*) Flattened membrane sacs inside the chloroplast, used to convert light energy to chemical energy.

thymus (*THYE-mus*) An endocrine gland in the neck region of mammals that is active in establishing the immune system; secretes several messengers, including thymosin, that stimulate T lymphocytes.

tight junction A type of intercellular junction that prevents the leakage of material between animal cells.

tissue An integrated group of cells with a common structure and function.

tissue culture A technique of growing tissue fragments outside the organism.

tonoplast (*TAHN-oh-plast*) An endomembrane enclosing the large central vacuole usually present in mature plant cells.

totipotent Referring to embryonic cells that retain the potential to form all parts of the animal.

trace elements Those elements indispensable for life but required in extremely minute amounts.

trachea (*TRAY-kee-uh*) The windpipe. That portion of the respiratory tube that has C-shaped cartilagenous rings and passes from the larynx to two bronchi.

tracheae (*TRAY-kee-ee*) Tiny air tubes that ramify throughout the insect body for gas exchange.

tracheids (*TRAY-kee-ids*) A water-conducting and supportive element of xylem composed of long, thin cells with tapered ends and walls hardened with lignin.

transcription The transfer of information from a DNA molecule into an RNA molecule.

transduction A mode of bringing together genes of different bacteria using bacteriophages.

transfer RNA (tRNA) RNA molecules that function as interpreters between nucleic acid and protein language by picking up specific amino acids and recognizing the appropriate codons in the mRNA.

transformation A phenomenon in which external genetic material is assimilated by a cell.

translation The transfer of information from an RNA molecule into a polypeptide, involving a change of "language" from nucleic acids to amino acids.

translocation An aberration in chromosome structure resulting from an error in meiosis or from mutagens; attachment of a chromosomal fragment to a nonhomologous chromosome; transport via phloem of food in a plant.

transpiration The evaporative loss of water from a plant.

transposon (*trans-POE-son*) A transposable element or "jumping gene"; a mobile segment of DNA that serves as an agent of genetic change.

triglyceride A fat, composed of a glycerol molecule bonded to three fatty acids by ester linkages.

trisomic A type of aneuploidy in which one chromosome is present in triplicate in the fertilized egg.

trophic structure The different feeding levels of an ecosystem; they determine the route of energy flow and the pattern of chemical cycling.

trophoblast The outer epithelium of the blastocyst, which forms the fetal part of the placenta.

tropic hormones Hormones that have other endocrine glands as their targets.

tropical rain forest The most complex of all communities, located near the equator where rainfall is abundant; harbors more species of plants and animals than any other ecosystem in the world.

tropisms Growth responses that result in curvatures of whole plant organs toward or away from stimuli due to differential rates of cell elongation.

tumor A mass that forms within otherwise normal tissue, caused by uncontrolled growth of a transformed cell.

tundra A biome at the northernmost limits of plant growth and at high altitudes, where plant forms are limited to low shrubby or matlike vegetation.

turgid (*TUR-jid*) A firm condition; walled cells become turgid as a result of the entry of water from a hypoosmotic environment.

undulipodia (*UN-dyoo-luh-POE-dee-uh*) A recently proposed term for cilia or flagella in eukaryotes.

unsaturated fatty acid A fatty acid possessing one or more double bonds between the carbons in the hydrocarbon tail. Such bonding reduces the number of hydrogen atoms attached to the carbon skeleton.

urea Soluble form of nitrogenous waste excreted by mammals and most adult amphibians.

ureter A duct leading from the kidney to the urinary bladder.

urethra The tube that releases urine from the body near the vagina in females or through the penis in males; also serves in males as the exit tube for the reproductive system.

uric acid The insoluble precipitate of nitrogenous waste excreted by land snails, insects, birds, and some reptiles.

uterus A female reproductive organ where eggs are fertilized and/or development of the young occurs.

vaccination The presentation of an inactive or attenuated form of a pathogen to the body; this initiates a long-term capability of the immune system to respond quickly to the real, infective agent.

vaccine A harmless variant or derivative of a pathogen that stimulates a host's immune system to mount defenses against the pathogen.

vagina A thin-walled, cornified chamber that forms the birth canal and is the repository for sperm during copulation.

valence shell The outermost energy shell of an atom, containing the valence electrons involved in chemical reactions of that atom.

van der Waals interaction A type of weak, transient chemical bond formed between nonpolar molecules that are close together; a weak attraction between local, positively charged regions on one nonpolar molecule to local, negatively charged regions on an adjacent nonpolar molecule; an attraction between oppositely charged adjacent regions of nonpolar molecules.

vascular cambium A continuous cylinder of meristematic cells surrounding the xylem and pith that produces secondary xylem and phloem.

vascular plants Plants with vascular tissue, consisting of all modern species except the mosses and their relatives.

vas deferens The tube in the male reproductive system in which sperm travel from the epididymis to the urethra.

vector A means of biological transfer—for example, plasmid vectors that move recombinant DNA from test tubes back into the cell, and insect vectors that transmit viruses from plant to plant.

vegetal pole The end of the zygote that is opposite the animal pole.

vegetative reproduction Cloning of plants by asexual means.

veins Vessels that return blood to the heart.

ventilation Any method of increasing contact between the respiratory medium and the respiratory surface.

ventral The bottom (or belly) half of a bilaterally symmetrical animal.

ventricles (*VEN-treh-kuls*) The chambers of the heart that pump blood out of the heart.

venules Vessels into which capillaries converge and which empty into veins.

vertebrates Chordate animals with backbones, consisting of the mammals, birds, reptiles, amphibians, and various classes of fishes.

vessel elements Specialized short, wide cells in angiosperms, arranged end to end to form continuous tubes for water transport.

vestigial organ A type of homologous structure that is rudimentary and of marginal or no use to the organism.

villi Fingerlike projections of the small intestine that increase surface area.

virion (*VEER-ee-on*) A virus particle; a noncellular, obligatory parasite of cells, which in simplest form, consists of a nucleic acid in a protein shell.

viroids (*VY-roids*) An extremely tiny class of plant pathogens, composed of molecules of naked RNA only several hundred nucleotides long.

visceral muscle Smooth muscle found in the walls of the digestive tract, bladder, arteries, and other internal organs.

visible light That portion of the electromagnetic spectrum that is detected as various colors by the human eye, ranging in wavelength from about 400 nm to about 700 nm.

vitalism The belief that natural phenomena are governed by a life force outside the realm of physical and chemical laws.

vitamins Organic molecules required in the diet in very small amounts, serving primarily as coenzymes or parts of coenzymes.

viviparous (*vy-VIP-er-us*) A type of development in which the young are born alive after having been nourished in the uterus by blood from the placenta.

voltage-sensitive gate An ion channel in the membrane of excitable cells that is constructed of protein molecules and that allows only certain ions to pass under specific changes of electric potential.

water potential The physical property predicting the direction in which water will flow, governed by solute concentration and applied pressure.

water vascular system A network of hydraulic canals unique to echinoderms that branches into extensions called tube feet, which function in locomotion, feeding, and exchange.

wavelength The distance between crests of waves, such as those of the electromagnetic spectrum.

wild type Individuals with the normal phenotype.

wobble A violation of the base-pairing rules in that the third nucleotide (5' end) of a tRNA anticodon can form hydrogen bonds with more than one kind of base in the third position (3' end) of a codon.

X-ray crystallography A physical method useful in determining the overall shape of macromolecules; determination of the positions of atoms in a molecule by mapping the diffraction pattern of X-rays passed through the crystallized molecule.

xylem (*ZY-lum*) The tube-shaped, nonliving portion of the vascular system in plants that functions to carry water and minerals from the roots to the rest of the plant.

yolk sac One of four extraembryonic membranes that supports embryonic development. The first site of blood cells and circulatory system function.

zygote The diploid product of the union of haploid gametes in conception; a fertilized egg.

Photo Credits

Unit Three Interview Robin J. Williams.

Chapter 12 12.2: R.D. Campbell, University of California, Irvine/BPS. 12.4 (six photos) Carolina Biological Supply. 12.8: P.B. Moens, *Chromosoma*, 23(1968):418. Copyright 1968 by Springer-Verlag. Methods Box: © NYU Cytogenics Laboratory/Peter Arnold, Inc.

Chapter 13 13.1: Mary Evans Picture Library/Photo Researchers, Inc. 13.6 and 13.11a: Science Education Resources Pty. Ltd., Victoria, Australia. 13.17a and b: M. Murayama, Murayama Research Laboratory/BPS.

Chapter 14 14.3: © Darwin Dale. 14.5: Courtesy Garland Allen; photo by Dr. Tore Mohr, Fredrikstad, Norway. 14.10: P.J. Bryant, University of California, Irvine/BPS. 14.16: © Ron Kimball. 14.19a: © Science Photo Library/Photo Researchers, Inc. 14.19b: © Richard Hutchings/Photo Researchers, Inc. 14.21: J.N.A. Lott, McMaster University/BPS.

Chapter 15 15.1: Courtesy Nelson Max and Richard E. Dickerson. 15.5 and 15.8: From J.D. Watson, *The Double Helix*, Atheneum, New York, 1968, p. 215 © 1968 by J.D. Watson. 15.11: G.C. Fareed, C.F. Garon, and N.P. Salzman, "Origin and Direction of Simian Virus 40 Deoxyribonucleic Acid Replication," *J. Virology*, 10:484(1972). 15.16 a and b: Richard J. Feldman, National Institute of Health.

Chapter 16 16.5: O.L. Miller and B. R. Beatty, Oak Ridge National Laboratory. 16.6: Courtesy of D.M. Prescott, University of Colorado Medical School; reproduced from *Prog. Nucleic Acid Res. III*, 35(1964), with permission. 16.7c: Institut de Biologie Moléculaire et Cellulaire, Strasbourg. 16.9: Adapted from "The Ribosome," J.A. Lake, *Sci. Am.* 245(1981):86. 16.13b: Barbara Hamkalo, *Int. Rev. Cytology* 33(1972):7.

Chapter 17 17.1: A.J. Olsen, Scripps Clinic and Research Foundation. 3a and b: R.C. Williams, University of California, Berkeley. 17.3c: F.A. Murphy, Centers for Disease Control. 17.3d: R.C. Williams, University of California, Berkeley. 17.4: R.C. Williams, University of California, Berkeley. 17.5: © Lee D. Simon/Photo Researchers, Inc. 17.6: Photograph from G.S. Stent and R. Calendar. *Molecular Genetics*, Second Ed., W.H. Freeman, San Francisco, 1978. 17.8: © Karl Maramorosch. 17.12: O.L. Miller, Jr., B.A. Hamkalo, and C. A. Thomas, Jr., *Science* 169(24 July 1970):392. Copyright 1970 by the American Association for the Advancement of Science. 17.13: Christine L. Case, Skyline College. 17.14a: Courtesy of J. Cairns. 17.16: L.G. Caro and R. Curtiss.

Chapter 18 18.1(four photos): © Ed Reschke. 18.3 a and b: Courtesy of J.R. Paulsen and U.K.Laemmli, *Cell* 12(1977):817–828. 18.4a: S.C. Holt, University of Texas, Health Science Center, San Antonio/BPS. 18.4b: A.L. Olins, University of Tennessee/BPS. 18.4c: Courtesy Barbara Hamkalo. 18.4d: Courtesy of J.R. Paulsen and U.K. Laemmli, *Cell* 12(1977):817. Copyright MIT; published by the MIT Press. 18.4e and f: G.F. Bahr, Armed Forces Institute of Pathology. 18.5: Courtesy of J.M. Amabis. 18.8: Courtesy of O.L. Miller, Jr., Department of Biology, University of Virginia. 18.11: From *Tissues and Organs: A Text-Atlas of Scanning Electron Microscopy*, by Richard G. Kessel and Randy H. Kardon. W.H. Freeman and Company. Copyright 1979. 18.14: © Michael McCoy/Photo Researchers, Inc. 18.17 left: © David Scharf. 18.17 right: F.R. Turner, University of Indiana. Methods Box: Courtesy of Einhard Schierenberg.

Chapter 19 19.1: Keith Wood. 19.8: © Nita Winter/Cetus Corporation. 19.11: © Dan McCoy/Rainbow. Methods Box : Damon Biotech, Inc.

Unit Four Interview Robin J. Williams.

Chapter 20 20.1: © Charles H. Phillips. 20.3: © David Schwimer/Bruce Coleman, Inc. 20.4: © Tom Till. 20.5: Larry Burrows, *Life* Magazine © Time Inc. 20.7a: © Frans Lanting/Minden Pictures. 20.7b: © Alan Root/Bruce Coleman, Inc. 20.8a: © C.A. Morgan/Peter Arnold, Inc. 20.8b: Rudi Kuiter © *Discover* Magazine, Time Inc. 20.9: © John Colwell/Grant Heilman Photography. 20.10a and b: © Breck P. Kent/Animals Animals.

Chapter 21 21.1a: © David Cavagnaro. 21.1b: Produced from USAF DMSP (Defense Meteorological Satellite Program) film transparencies archived for NOAA/NESDIS at the University of Colorado, CIRES/National Snow and Ice Data Center. 21.4: John Alcock. 21.5: © John Colwell/Grant Heilman, Inc. 21.6: © R. Andrew Odum/Peter Arnold, Inc. 21.9: © Animals Animals/OSF. 21.10: © Animals Animals/OSF - J.A.L. Cooke. 21.12a: top, © Stan Goldblatt/Photo Researchers, Inc.; bottom, © Gilbert Grant/Photo Researchers, Inc. 21.12b: John Balik. 21.14a: © Dieter Blum/Peter Arnold, Inc. 21.14b: © Tom and Pat Leeson. 21.14c: © Stephen Kraseman/DRK Photo.

Chapter 22 22.2a: © Michael O'Neill. 22.2b: top, © Don and Pat Valenti, Tom Stack & Associates; bottom, © John Shaw, Tom Stack & Associates. 22.5: © Robert Lee/Photo Researchers, Inc. 22.7: © Jeff Foott. 22.10: © David Cavagnaro. 22.11b: Courtesy of Kenneth Kaneshiro.

Chapter 23 23.1: © Doug Lee. 23.2a and b: © Manfred Kage/Peter Arnold, Inc. 23.2c: Margo Crabtree. 23.2d: © Tom Till. 23.2e: W.H. Hodge/Peter Arnold, Inc. 23.5 top left: © Stephen Kraseman/Peter Arnold, Inc. 23.5 top right: © Gary Milburn/Tom Stack & Associates. 23.5 bottom left and right: © Tom McHugh/Photo Researchers, Inc. 23.10: © Jane Burton/Bruce Coleman, Inc. 23.15b: © Kevin Schafer. 23.15c: © Kal Muller/Woodfin Camp & Associates, Inc. 23.17: Courtesy Lawrence Berkeley Laboratory.

Unit Five Interview Robin J. Williams.

Chapter 24 24.1 © Sigurgier Jonasson. 24.3: S.M. Awramik, University of California/BPS. 24.4a: Courtesy John Stoltz. 24.4b and c: S.M. Awramik, University of California/BPS. 24.6a: Sidney W. Fox, University of Miami. 24.6b: D.W. Deamer and P.B. Armstrong, University of California, Davis.

Chapter 25 25.1: © Science Photo Library/Photo Researchers, Inc. 25.2a, b, and c: David M. Phillips/Taurus Photos, Inc. 25.3a: Courtesy S.W. Watson. 25.3b: N.J. Lang, University of California, Davis/BPS. 25.3c: S.C. Holt, University of Texas Health Science Center, San Antonio/BPS. 25.4: S. Abraham and E .H. Beachey, VA Medical Center, Memphis. 25.5: Courtesy J. Adler. 25.6a and b: R.M. Macnab and M.K. Ornston, *J. Mol. Biol.* 112:1–30(1977). 25.7: Photo by Richard P. Blakemore. Reproduced with permission, from the *Annual Review of Microbiology*, Volume 37, Copyright 1982 by Annual Reviews, Inc. 25.9: Christine L. Case, Skyline College. 25.10: Christine L. Case, Skyline College. 25.11a: J.R. Waaland, University of Washington/BPS. 25.11b: © T.E. Adams/ Peter Arnold, Inc. 25.11c: P.W. Johnson and J. Mcn. Sieburth, University of Rhode Island/BPS. 25.12: © M.I. Walker/Photo Researchers, Inc. 25.13: P.W. Johnson and J. Mcn. Sieburth, University of Rhode Island/BPS. 25.14: H.S. Pankratz, Michigan State University/BPS. 25.15: Willy Burgdorfer, Rocky Mtn. Lab. 25.16: Michael Gabridge, University of Illinois. 25.17: © David Scharf/Peter Arnold, Inc. 25.18: K. Stephens, Stanford University/BPS. Methods Box, top and bottom: Christine L. Case, Skyline College.

Chapter 26 26.1: J.R. Waaland, University of Washington/BPS. 26.2: © Peter Parks—OSF/Animals Animals. 26.3a and b: © Eric Grave/Photo Researchers, Inc. 26.4: © Manfred Kage/Peter Arnold, Inc. 26.6a: © Eric Grave/Photo Researchers, Inc. 26.6b: Courtesy of John Donelson and Steven Brentano, University of Iowa. 26.7a: © Eric Grave/Photo Researchers, Inc. 26.7b: © Roland Birke/Peter Arnold, Inc. 26.9a: © Biophoto Associates/Photo Researchers, Inc. 26.9b:© Eric Grave/Photo Researchers, Inc. 26.10: J.R. Waaland, University of Washington/BPS. 26.11: © Manfred Kage/Peter Arnold, Inc. 26.14a: © Manfred Kage/Peter Arnold, Inc. 26.14b: © Pat Lynch/Photo Researchers, Inc. 26.14c: Copyright E.R. Degginger. 26.14d: © Manfred Kage/Peter Arnold, Inc. 26.17a: © Chuck Davis. 26.17b: © Marty Snyderman. 26.19: © Gary R. Robinson/Visuals Unlimited. 26.20a: © P.W. Grace/Photo Researchers, Inc. 26.20b: © Karl H. Maslowski/Photo Researchers, Inc. 26.23: © David Scharf/Peter Arnold, Inc.

Chapter 27 27.1: © Jeff Gnass. 27.2a and b: J.R. Waaland, University of Washington/BPS. 27.5: © Stephen Kraseman/DRK Photo. 27.7: © John Shaw/Tom Stack & Associates. 27.9: © Rod Planck/Tom Stack and Associates. 27.11: © Dale and Marian Zimmerman/Bruce Coleman, Inc. 27.12: © Robert and Linda Mitchell. 27.13: © E.S. Ross. 27.14a: © Kevin Schafer.27.14b: Copyright Donald R. Kirk. 27.15: © Milton Rand/Tom Stack & Associates. 27.17: Field Museum of Natural History (Neg #75400C), Chicago. 27.18a: © James Castner. 27.18b: © Runk/Schoenberger/Grant Heilman Photography. 27.18c: © Michael Fogden/Bruce Coleman, Inc. 27.19: James W. Behnke. 27.21a: © John Shaw/Tom Stack & Associates. 27.21b: © W. H. Hodge/Peter Arnold, Inc. 27.24a: © Kevin Schafer. 27.24b: © Oxford Scientific Films, Animals Animals/Earth Scenes. 27.25: © John Colwell/Grant Heilman. 27.27a: © D. Wilder. 27.27b: © Thomas C. Boyden. 27.27c: Merlin D. Tuttle, Bat Conservation International.

Chapter 28 28.1a: © Kevin Schafer. 28.1b: © Thomas B. Boyden. 28.3: D.D. Kunkel, University of Washington. 28.4 G.L. Barron, University of Guelph/BPS. 28.6a: © Animals Animals/Earth Scenes—E.R. Degginger. 28.6b: ©

J.L. Lepore/Photo Researchers, Inc. 28.8a: © Kerry T. Givens/Tom Stack & Associates. 28.8b: © Frans Lanting. 28.8c: © A. Davies/Bruce Coleman, Inc. 28.8d: © Leonard Lee Rue III/Photo Researchers, Inc. 28.10: © Robert and Linda Mitchell. 28.11: N. Allin and G. L. Barron, University of Guelph/BPS. 28.12: © Robert Lee/Photo Researchers, Inc. 28.14: © Runk/Schoenberger/Grant Heilman, Inc.

Chapter 29 29.1: © Anne Wertheim. 29.6: © Jeff Rotman. 29.10a: © Gwen Fidler/Comstock, Inc. 29.10b: © Jeff Rotman. 29.10c: © Claudia Mills. 29.10d: © Chuck Davis. 29.10e: © Chris Huss/Ellis Wildlife Collection. 29.12: © Fred Bavendam/Peter Arnold, Inc. 29.13: © Bill Wood/Bruce Coleman, Inc. 29.16: © William H. Amos/Bruce Coleman, Inc. 29.17a and b: © L.S. Stepanowicz/Photo Researchers, Inc. 29.19: © Jeff Foott/Tom Stack & Associates. 29.21a: © Kevin Schafer. 29.21b: © Chris Huss. 29.22: © James M. Cribb/Photo Researchers, Inc. 29.24a: © Franklin Viola/Comstock, Inc. 29.24b: © Fred Bavendam/Peter Arnold, Inc. 29.24c: © Douglas Faulkner/Photo Researchers, Inc. 29.26a: © Sea Studios. 29.26b: © Kjell Sandved/Sandved and Coleman. 29.26c: © Robert and Linda Mitchell. 29.28: © Cliff B. Frith/Bruce Coleman, Inc. 29.29: © Ronald F. Thomas/Bruce Coleman, Inc. 29.30: © Zig Leszczynski/Animals Animals. 29.31a: © Robert and Linda Mitchell. 29.31b: © Paul Skelcher/Rainbow. 29.31c: © David Scharf. 29.33a: © Marty Snyderman. 29.33b: © Flip Nicklin. 29.33c: C.R. Wyttenbach, University of Kansas/BPS. 29.34a and b: © Robert and Linda Mitchell. 29.35: © Stephen Dalton/Animals Animals. 29.37a,b,d, and e: © John Shaw/Tom Stack & Associates. 29.37c: © Frans Lanting/Minden Pictures. 29.38a: © J.A.L. Cooke/Animals Animals. 29.38b: © Fred Bavendam/Peter Arnold, Inc. 29.39a and e: © Jeff Rotman/Peter Arnold, Inc. 29.39b: © Dave Woodward/Tom Stack & Associates. 29.39c: © Marty Snyderman. 29.39d: © Carl Roessler/Animals Animals. 29.43: Courtesy of Prof. K.G. Grell, Tubingen. 29.44a: © Carole Hickman/South Australian Museum. 29.44b: S. Conway Morris.

Chapter 30 30.1: © Pieter Folkens. 30.5: © Tom McHugh/Photo Researchers, Inc. 30.6: Neg. No. 124619 (photo by Rota). Courtesy Department of Library Services, American Museum of Natural History. 30.8a: © Tom McHugh/Photo Researchers, Inc. 30.8b: © Jeff Rotman. 30.9a: © E.R. Degginger/Bruce Coleman, Inc. 30.9b: © Fred Bavendam/Peter Arnold, Inc. 30.11: © Steve Martin/Tom Stack & Associates. 30.14a: © Zig Leszczynski/Animals Animals. 30.14b: © M.P.L. Fogden/Bruce Coleman, Inc. 30.14c: © R. Andrew Odum/Peter Arnold, Inc. 30.15a-c: © Hans Pfletschinger/Peter Arnold, Inc. 30.18a: © John Gerlach/Tom Stack & Associates. 30.18b: © R. Andrew Odum/Peter Arnold, Inc. 30.18c: © John R. MacGregor/Peter Arnold, Inc. 30.18d: © Patricia Caulfield/Photo Researchers, Inc. 30.19: © Stephen Dalton/Animals Animals. 30.20: © Kevin Schafer. 30.22: © Gerry Ellis. 30.23a: © John Henry Dick, Academy of Natural Sciences, Philadelphia/Vireo. 30.23b: © G. Nuechterlein, Academy of Natural Sciences, Philadelphia/Vireo. 30.23c: © Bob and Clara Calhoun/Bruce Coleman, Inc. 30.23d: © Art Wolfe. 30.24: © Gunter Ziesler/Peter Arnold, Inc. 30.25: © Mitch Reardon/Photo Researchers, Inc. 30.26: © E.H. Rao/Photo Researchers, Inc. 30.27a: © Tom McHugh/Photo Researchers, Inc. 30.27b: © Norman Owen Tomalin/Bruce Coleman, Inc. 30.28a: © E.R. Degginger/Animals Animals. 30.28b: © Ulrich Nebelsiek/Peter Arnold, Inc. 30.28c: © Nancy Adams E.P.I./Tom Stack & Associates. 30.28d: © Tom McHugh/Photo Researchers, Inc. 30.30a: © Cleveland Museum of Natural History, Cleveland, Ohio. 30.30b: Photograph by John Reader copyright 1982.

Unit Six Interview Robin J. Williams.

Chapter 31 31.1: © Larry Lefever/Grant Heilman Photography. 31.4: © Animals Animals/Earth Scenes—Doug Wechsler. 31.8a: © Larry Mellichamp/Visuals Unlimited. 31.8b and c: © Kevin Schafer. 31.8d: © E.S. Ross. 31.11a and b: Science Source © Biophoto Associates/Photo Researchers, Inc. 31.11c and d: Courtesy Nels Lersten. 31.11e: J.R. Waaland, University of Washington/BPS. 31.11f: Courtesy Nels Lersten. 31.14a: J.R. Waaland, University of Washington/BPS, 31.14b: Science Source © Biophoto Associates/Photo Researchers, Inc. 31.18: Courtesy F.A.L. Clowes. 31.19a: © Ed Reschke. 31.19b: © Omikron/Taurus Photos, Inc. 31.20: J.R. Waaland, University of Washington/BPS. 31.21, 31.22a and b, 31.24a and b: © Ed Reschke. 31.24c: © John Blaustein/Woodfin Camp & Associates. 31.25: © Nicholas Devore/Photographers Aspen. 31.27: © Lester Bergman. 31.28: © Russ Kinne/Photo Researchers, Inc.

Chapter 32 32.1: © Dennis Brokaw. 32.4: P. Gates, University of Durham/BPS. 32.6: © H. Reinhard/Bruce Coleman, Inc. 32.7: G. Montenegro. 32.12: © Alfred Pasieka/Taurus Photos, Inc. 32.13: © Biophoto Associates/Photo Researchers, Inc. 32.16a: © E.S. Ross. 32.16b: M.H. Zimmerman, Science 133(13 January 1961):76, Fig. 4. Copyright 1961 by the American Advancement of Science.

Chapter 33 33.1: © Grant Heilman. 33.3: © John Colwell/Grant Heilman. 33.4: © John H. Hoffman/Bruce Coleman, Inc. 33.5: © Grant Heilman. 33.8a: © Runk/Schoenberger/Grant Heilman, Inc. 33.8b: © Alan Pitcairn/Grant Heilman, Inc. 33.9: © John Colwell/Grant Heilman. 33.11a: © Breck P. Kent/Earth Scenes. 33.11b: E.H. Newcomb, University of Wisconsin, Madison/BPS. 33.13a: © W.H. Hodge/Peter Arnold, Inc. 33.13b: © Kevin Schafer. 33.13c: © Biophoto Associates/Photo Researchers, Inc. 33.14: © Thomas C. Boyden. 33.15a: © Jeff Lepore/Photo Researchers, Inc. 33.15b: © John Shaw/Tom Stack & Associates. 33.15c: © Robert and Linda Mitchell. 33.16: R.L. Peterson, University of Guelph/BPS.

Chapter 34 34.1: © Frans Lanting. 34.4a: © W.H. Hodge/Peter Arnold, Inc. 34.4b (left): © Michel Viard/Peter Arnold, Inc. 34.4b (right): From *Marijuana Grower's Guide*, copyright © 1976 by Mel Frank, published by AND/OR Press. 34.6a and b: © Alfred Owczarzak/Taurus Photos, Inc. 34.6c: © G. Cox/Bruce Coleman, Inc. 34.7: © Ed Reschke. 34.10a, b and c: W.H. Hodge/Peter Arnold, Inc. 34.10d: © Robert P. Carr/Bruce Coleman, Inc. 34.11a: © Phil Farnes/Photo Researchers, Inc. 34.11b: © John Colwell/Grant Heilman Photography. 34.11c: © Zig Leszczynski/Animals Animals. 34.11d: © Paolo Koch/Photo Researchers, Inc. 34.12a: © Runk/Schoenberger/Grant Heilman Photography. 34.12b: © David Cavagnaro/DRK Photo. 34.12c: © Dwight R. Kuhn. 34.15a: © Runk/Schoenberger/Grant Heilman, Inc. 34.15b: © Kevin Schafer. 34.15c: J.N.A. Lott, McMaster University/BPS. 34.15d: © Galen Rowell Photo. 34.15e: © David Cavagnaro/DRK Photo. 34.16: J.R. Waaland, University of Washington/BPS. 34.17: © Runk/Schoenberger/Grant Heilman Photography. 34.18: Courtesy Daniel Facciotti/Calgene Inc.

Chapter 35 35.1: © Runk/Schoenberger/Grant Heilman, Inc. 35.7: Courtesy of Sylvan Wittwer. 35.8: © Ed Reschke. 35.9: Courtesy of Michael Evans. 35.10: Courtesy of Randy Moore. 35.11: © Peter G. Aitken/Photo Researchers, Inc. 35.12a and b: © E.R. Degginger/Bruce Coleman, Inc. 35.13: Courtesy of Frank Salisbury.

Unit Seven Interview Robin J. Williams.

Chapter 36 36.1: © G.I. Bernard/Animals Animals. 36.4a: Courtesy of Dr. Jerome Gross, Developmental Biology Laboratory, Massachusetts General Hospital, Boston. 36.5 and 36.7: © Ed Reschke. 36.8: From *Tissues and Organs: A Text-Atlas of Scanning Electron Microscopy*, by Richard G. Kessel and Randy H. Kardon. W.H. Freeman and Company. Copyright 1979. 36.10b: © Biophoto Associates/Photo Researchers, Inc. Self-quiz: a: Marian Rice; b, c, and e: © Bruce Iverson; d: © Eric Grave/Photo Researchers, Inc. f: © Biophoto Associates/ Photo Researchers, Inc.

Chapter 37 37.1: © Gunter Ziesler/Peter Arnold, Inc. 37.2a: © C. Allan Morgan/Peter Arnold, Inc. 37.2b: Painting by Richard Schlecht, © National Geographic Society. 37.3: © Tom Eisner. 37.4: © Hans Pfletschinger/Peter Arnold, Inc. 37.15: © Manfred Kage/Peter Arnold, Inc. 37.20: © Fred Bavendam/Peter Arnold, Inc. Methods Box: left, © William Thompson; right, courtesy of Robert Full.

Chapter 38 38.1: Boehringer Ingelheim International GmbH © Lennart Nilsson, *The Incredible Machine*, National Geographic Society. 38.8b: From *Tissues and Organs: A Text-Atlas of Scanning Electron Microscopy*, by Richard G. Kessel and Randy H. Kardon. W.H. Freeman and Company. Copyright 1979. 38.12: © D.W. Fawcett/Photo Researchers, Inc. 38.15b: © Lennart Nilsson/Boehringer Ingelheim International GmbH. 38.16: © Lennart Nilsson, Boehringer Ingelheim International GmbH/*The Body Victorious*, Delacorte Press. 38.21c: © Tom Eisner. 38.22c: © David M. Phillips/Taurus Photo, Inc. 38.28: © Kevin Schafer.

Chapter 39 39.1: © Lennart Nilsson, Boehringer Ingelheim International GmbH. 39.2: © R. Dourmashkin/Photo Researchers, Inc. 39.3: © Lennart Nilsson, Boehringer Ingelheim International GmbH. 39.14b: A.J. Olson, Scripps Clinic and Research Foundation. 39.19a and b: Courtesy Ann Dvorak, from A.M. Dvorak et al. Fed. Proc. 42:2510–2515, 1983. 39.20: © Lennart Nilsson, Boehringer Ingelheim International GmbH.

Chapter 40 40.1: © Art Wolfe. 40.4a and b: John Crowe. 40.5: © Tom McHugh/Photo Researchers, Inc. 40.11c: From *Tissues and Organs: A Text-Atlas of Scanning Electron*

Microscopy, by Richard G. Kessel and Randy H. Kardon. W.H. Freeman and Company. Copyright 1979. 40.11d: Peter Andrews, reprinted with permission from *Biomed. Res* (Suppl.), 2:293–305. 40.17b and c: Hayward and Lyman, *Mammalian Hibernation III,* Kenneth C. Fisher, ed. (New York: Elsevier, 1967). 40.19a: © E. Lyons/Bruce Coleman, Inc. 40.19b: © Frans Lanting/Minden Pictures. 40.19c: Sovfoto. 40.21b: Courtesy Francis Carey. 40.23a: © Warren Garst/Tom Stack & Associates. 40.23b: © Michael Fogden/Bruce Coleman, Inc.

Chapter 41 41.1: © Frans Lanting/Minden Pictures. 41.3b: Dr. Walter Stumpf and Dr. Madhabananda Sar, University of North Carolina.

Chapter 42 42.1: © R.N. Mariscal/Bruce Coleman, Inc. 42.2a: David Crews. Photo by P. DeVries. 42.3: © Stephan Myers. 42.4: © K.H. Switak/Photo Researchers, Inc. 42.14b: © David M. Phillips/Taurus Photos, Inc. 42.18a–c: © Lennart Nilsson. 42.19: From *Ultrasonography in Obstetrics and Gynecology, ed 2.* by J.C. Hobbins, F. Winsberg, and R.L. Berkowitz, Baltimore, Williams and Wilkins, 1983, p. 123.

Chapter 43 43.1: © Science Photo Library/Photo Researchers, Inc. 43.5a–c and 43.7: George Watchmaker. 43.10: © Grant Heilman. 43.11a and b: Robert Waterman. 43.14: Courtesy of Thomas Poole. 43.16: Carolina Biological Supply. 43.23a: Dr. Jean-Paul Thiery. Reproduced from the *Journal of Cell Biology,* 1983, Vol. 96, pp. 462–473 by Copyright permission of The Rockefeller University Press. 43.23b: Richard Hynes, Scientific American (June 1986), p. 44. 43.26: © David Scharf. 43.27a: bottom, Katherine Tosney, University of Michigan. 43.27b and 43.28: Dennis Summerbell.

Chapter 44 44.1: © Manfred Kage/Peter Arnold, Inc. 44.4: Courtesy of G.L. Scott, J.A. Feilbach, and T.A. Duff. 44.12b: Dr. E.R. Lewis, ERL-EECS, University of California, Berkeley. 44.15b: Courtesy of F. Hucho. 44.26a: © Alexander Tsiaras/Photo Researchers, Inc. Methods Box: © William Thompson.

Chapter 45 45.1: © Joe Monroe/Photo Researchers, Inc. 45.3: Animals Animals/OSF. 45.4a: R. Andrew Odum/Peter Arnold, Inc. 45.4b: © Russ Kinne/Photo Researchers, Inc. 45.6a: Janice Sheldon. 45.7: © Roger J. Hanlon. 45.11: From *Tissues and Organs: A Text-Atlas of Scanning Electron Microscopy,* by Richard G. Kessel and Randy H. Kardon. W.H. Freeman and Company. Copyright 1979. 45.18: Andreas Feininger, *Life* Magazine © 1952 Time Inc. 45.21: From *Fundamentals of Entomology, Third Edition* by Richard J. Elzinga (1987). 45.26a: Animals Animals/OSF-P. Parks. 45.26b: © Biophoto Associates/Photo Researchers, Inc. 45.26c: © James Hanken/Photo Researchers, Inc. 45.31: Courtesy of Clara Franzini-Armstrong. 45.35: © Eric Grave/Photo Researchers, Inc.

Unit Eight Interview Ray Pfortner.

Chapter 46 46.1: Courtesy of NASA. 46.2: Courtesy of Dan Simberloff. 46.3: © Stephen J. Krasemann/Photo Researchers, Inc. 46.4: © David Cavagnaro. 46.5a: © John D. Cunnigham. 46.5b: © Tom McHugh/Photo Researchers, Inc. 46.5c: © Michael S. Quinton. 46.12: © Doug Wechsler, Animals Animals/Earth Scenes. 46.13: © Jonathan Scott/Planet Earth Pictures, Inc. 46.14: © Charlie Ott/Photo Researchers, Inc. 46.15: © Doug Wechsler/Animals Animals/Earth Scenes. 46.16: © Rod Planck/Photo Researchers, Inc. 46.17: © John Shaw. 46.18 and 19: © Charlie Ott/Photo Researchers, Inc. 46.20a: © Steve Selum/Bruce Coleman, Inc. 46.20b: © Wendell Metzen/Bruce Coleman, Inc. 46.21: © John Lythgoe/Seaphot, Ltd: Planet Earth Pictures.

Chapter 47 47.1: © David T. Overcash/Bruce Coleman, Inc. 47.2: © Victor H. Hutchinson/Visuals Unlimited. 47.4a: © Robert P. Carr/Bruce Coleman, Inc. 47.4b: © Breck P. Kent/Animals Animals/Earth Scenes. 47.5: © Jeff Foott/Bruce Coleman, Inc. 47.9: © A. Rainon/Jacana. 47.11: © Alan Carey/Photo Researchers, Inc. 47.13a: © Stewart M. Green/Tom Stack & Associates. 47.13b: © Leonard L. Rue III/Bruce Coleman, Inc. Methods Box: © Frans Lanting/Minden Pictures.

Chapter 48 48.1a and b: Erick Greene, *Science* 243(3 February 1989):644, Fig. 1. Copyright 1989 by the American Association for the Advancement of Science. 48.4a-c: © L.E. Gilbert, University of Texas, Austin/BPS. 48.8: © C. Allan Morgan/Peter Arnold, Inc. 48.9: © John L. Tveten. 48.10a: © Stephen Kraseman/Peter Arnold, Inc. 48.10b: © Tom Eisner and Daniel Aneshansley. 48.11: © E.S. Ross. 48.12a: © J. Alcock/Visuals Unlimited. 48.12b: © E.S. Ross. 48.13a: © Dr. Gregory A. Antipa. 48.14: © Stephen J. Krasemann/Photo Researchers, Inc. 48.15: © Robert and Linda Mitchell. 48.16a-f: © Tom Bean. 48.18a: Walter R. Tschinkel, © *Discover,* Time, Inc.

Chapter 49 49.1b: © Space Biospheres Ventures. 49.2: © Dudley Foster, Woods Hole Oceanographic Institution. 49.11a: © John D. Cunningham/Visuals Unlimited. 49.11b: Furnished by Dr. Robert Pierce/U.S. Department of Agriculture, Northeastern Forest Experiment Station, Durham, NH. 49.13: D.W. Schindler, *Science* 184 (24 May 1974):897, Fig. 1. Copyright 1974 by the American Association for the Advancement of Science. 49.15: © David Hiser/Photographers Aspen. 49.17a: Courtesy NASA and Patrick Holligan, Marine Biological Association, Plymouth, England. 49.17b: Courtesy of Dr. Susumu Honjo. 49.18: © Bill Nation/Sygma.

Chapter 50 50.2a: © Art Wolfe. 50.2b: © Jeff Foott. 50.2c: © Jane Burton/Bruce Coleman, Inc. 50.5: © Lloyd Beesley/Animals Animals, Inc. 50.6: Thomas McAvoy, *Life* Magazine © Time, Inc. 50.7: © John Markham/Bruce Coleman, Inc. 50.9: Lincoln P. Brower. 50.15: © Gordon Wiltsie. 50.16a: © Jean Michel Labat/Jacana. 50.16b: © Sarah Hrdy, Anthro-Photo. 50.17a: © Jonathan Scott/Seaphot Ltd:Planet Earth Pictures. 50.17b: © Michael Dick/Animals Animals. 50.20: © Stephen Kraseman/Peter Arnold, Inc. 50.21: © Ivan Polunin/Bruce Coleman, Inc. Methods Box: top, © 1989 Gary Braasch. bottom, Courtesy of Greg Capelli.

The artist and/or source of every illustration that appears in *BIOLOGY* is listed below. In those cases where two artists are credited, the first artist supplied a rough sketch and the second artist rendered the final illustration.

Interview Portraits: Jackie Osborn.

Interview Icons: Raychel Ciemma.

Chapter 1 1.2, 1.9c: Raychel Ciemma; 1.7, 1.11: Carla Simmons.

Chapter 2 2.4, 2.6, 2.7, 2.9, 2.11, 2.14: Georg Klatt/Carol Verbeeck. 2.8: Georg Klatt. 2.10: Georg Klatt/Chris Carothers. 2.12: Georg Klatt/Cecile Duray-Bito. 2.13: Linda McVay. Methods Box (p. 26): Carol Verbeeck.

Chapter 3 3.2: Georg Klatt/Chris Carothers. 3.3: Linda McVay. 3.6, 3.8, 3.10: Cecile Duray-Bito. 3.11: Georg Klatt/Carol Verbeeck. 3.12: Chris Carothers. Challenge Questions (p. 50): Georg Klatt/Cecile Duray-Bito.

Chapter 4 4.3, 4.6, Table 4.2: Cecile Duray-Bito. 4.4, 4.7–4.9, 4.10, 4.12: Cecile Duray-Bito/Janet Hayes. Methods Box (p. 54): Linda McVay.

Chapter 5 5.2, 5.3, 5.16, 5.17, 5.19, 5.23: Cecile Duray-Bito/Janet Hayes. 5.4–5.6, 5.15, 5.18, 5.22, 5.28: Cecile Duray-Bito. 5.8: Linda McVay. 5.9: From K.D. Johnson, D.L. Rayle, and H.L. Wedberg, *Biology: An Introduction.* (Redwood City, CA: Benjamin/Cummings, 1984), p. 21. 5.12, 5.21: Georg Klatt/Janet Hayes. 5.14, 5.20, 5.26, 5.29: Georg Klatt/Cecile Duray-Bito. 5.24: Georg Klatt/Carol Verbeeck; © Irving Geis. 5.25: Cecile Duray-Bito; (c) and (d) © Irving Geis. 5.30: Georg Klatt. Methods Box (p. 85): Chris Carothers.

Chapter 6 6.1: Cecile Duray-Bito; from B. Alberts, D. Bray, J. Lewis, M. Raff, K. Roberts, and J.D. Watson, *Molecular Biology of the Cell,* 2d ed. (New York: Garland, 1989). 6.4, 6.8, 6.18: Cecile Duray-Bito. 6.5, 6.11: Carol Verbeeck. 6.7, 6.9, 6.19: Cecile Duray-Bito/Janet Hayes. 6.10: Chris Carothers. 6.12, 6.14, 6.15, Challenge Questions (p. 112): Linda McVay. 6.17: Georg Klatt/Cecile Duray-Bito.

Chapter 7 7.2, 7.10, 7.11c, 7.17, 7.18, 7.20, 7.22, 7.23, 7.25, 7.27: Carla Simmons. 7.3, 7.7, 7.21: Fran Milner/Carla Simmons. 7.5, 7.6a, 7.8a, 7.9a, 7.12, 7.15, 7.29, 7.30, 7.31: Barbara Cousins. 7.14: Linda McVay. 7.16: Kenneth R. Miller. 7.32: From W.M. Becker, *The World of the Cell.* (Redwood City, CA: Benjamin/Cummings, 1986), p. 44. 7.37: Carla Simmons; from E.N. Marieb, *Human Anatomy and Physiology.* (Redwood City, CA: Benjamin/Cummings, 1989), p. 65.

Chapter 8 8.2, 8.3, 8.4–8.16, 8.18–8.22, 8.24, 8.26, Methods Box (p. 158), Self-Quiz (p. 177), Challenge Questions (p. 177): Barbara Cousins. 8.17: Carol Verbeeck.

Chapter 9 9.2–9.4, 9.6, 9.15, 9.16a, 9.21, 9.22: Carla Simmons. 9.5, 9.7, 9.9–9.11, 9.14: Cecile Duray-Bito/Janet Hayes. 9.12, 9.13, 9.17, 9.18, 9.20: Barbara Cousins

Chapter 10 10.2, 10.6, Methods Box (p. 212): Barbara Cousins. 10.4, 10.5, 10.9, 10.11–10.20: Carla Simmons. 10.7: Carol Verbeeck. 10.8: Cecile Duray-Bito/Janet Hayes.

Chapter 11 11.2, 11.12a: Chris Carothers. 11.6, 11.7, 11.8, 11.10, 11.14: Barbara Cousins. 11.9b: Linda McVay.

Chapter 12 12.3, 12.5–12.7, 12.10: Chris Carothers. 12.4: Barbara Cousins. 12.8b, 12.9, Methods Box (pp. 256–257): Linda McVay.

Chapter 13 13.2–13.5, 13.7–13.15, 13.20: Barbara Cousins. 13.16: Carol Verbeeck.

Chapter 14 14.1, 14.6, 14.7, 14.12, 14.14, 14.16, 14.17: Barbara Cousins. 14.2, 14.4, 14.11, 14.13, 14.19: From F.J. Ayala and J.A. Kiger, Jr., *Modern Genetics,* 2d ed. (Redwood City, CA: Benjamin/Cummings, 1984), pp. 60, 61, 694, 75, 690. 14.8: Linda McVay. 14.9, 14.15, 14.21: Carol Verbeeck.

Chapter 15 15.2: From W.M. Becker, *The World of the Cell.* (Redwood City, CA: Benjamin/Cummings, 1986), p. 402. 15.3, 15.6, 15.9–15.15: Cecile Duray-Bito. 15.4, 15.7: Linda McVay.

Chapter 16 16.1, 16.18, 16.20, Methods Box (p. 324): Linda McVay. 16.2, 16.3 16.7, 16.8, 16.10–16.17, 16.19, 16.21–16.23: Cecile Duray-Bito. 16.4: Kenneth R. Miller.

Chapter 17 17.2: Elizabeth Morales. 17.3, 17.5, 17.7, 17.12, 17.15, 17.18–17.23: Cecile Duray-Bito. 17.4: Chris Carothers. 17.11: Kenneth R. Miller. 17.10, 17.12: Linda McVay. 17.14: Adapted from G.J. Tortora, B.R Funke, and C.L. Case, *Microbiology: An Introduction,* 3d ed. (Redwood City, CA: Benjamin/Cummings, 1989), p. 200. 17.17: Cecile Duray-Bito/Georg Klatt.

Chapter 18 18.2, 18.15: Elizabeth Morales. 18.4, 18.6–18.10, 18.12, 18.18, 18.19: Cecile Duray-Bito/Georg Klatt. 18.13: From W.M. Becker, *The World of the Cell.* (Redwood City, CA: Benjamin/Cummings, 1986), p. 592. 18.16: Fran Milner/Elizabeth Morales. Methods Box (p. 396): From J.D. Watson, N.H. Hopkins, J.W. Roberts, J.A. Steitz, A.M. Weiner, *Molecular Biology of the Gene,* 4th ed., vol. 2. (Redwood City, CA: Benjamin/Cummings, 1987), pp. 804–805.

Chapter 19 19.2–19.4, 19.5, 19.7, 19.9, 19.10: Chris Carothers/Georg Klatt. 19.5, Methods Boxes (pp. 403, 404, 407): Chris Carothers. Methods Box (pp. 412–413): Linda McVay.

Chapter 20 20.2, 20.6: Chris Carothers. 20.11: Darwen and Vally Hennings.

Chapter 21 21.2, 21.3, 21.7: Darwen and Vally Hennings. 21.8, 21.13: Carol Verbeeck. 21.11: Carla Simmons.

Chapter 22 22.1, 22.3, 22.4, 22.11, 22.12, 22.14: Darwen and Vally Hennings. 22.6, 22.13: Carla Simmons. 22.8, 22.9: Chris Carothers. 22.11a: Darwen and Vally Hennings; data from H.L. Carson, D.E. Hardy, H.R. Speith, and W.J. Stone, "The Evolutionary Biology of the Hawaiian Drosophilidae." In M.K. Hecht and W.C. Steere, editors, *Essays in Evolution in Honor of Theodosius Dobzhansky.* (New York: Meredith, 1970), pp. 437–543.

Chapter 23 23.3: Carla Simmons; adapted from Biruta Hansen/ © 1986 Discover Publications. 23.4, 23.7–23.9, 23.11, 23.12: Darwen and Vally Hennings. 23.6: Carla Simmons. 23.13, Methods Box (p. 490): Carol Verbeeck. 23.14–23.16,: Chris Carothers.

Chapter 24 24.2: Carol Verbeeck. 24.5, 24.7–24.9: Barbara Cousins. 24.9: Adapted from B. Alberts, D. Bray, J. Lewis, M. Raff, K. Roberts, and J.D. Watson, *Molecular Biology of the Cell,* 2d ed. (New York: Garland, 1989), p. 10.

Chapter 25 25.5, 25.8, 25.19, Methods Box (p. 526): Barbara Cousins.

Chapter 26 26.5, 26.7, 26.8, 26.12, 26.13, 26.15, 26.16, 26.21, 26.22, 26.25, 26.26: Barbara Cousins. 26.24: Carla Simmons.

Chapter 27 27.3, 27.4, 27.6, 27.8, 27.10, 27.16, 27.20, 27.22, 27.23, 27.26: Barbara Cousins.

Chapter 28 28.2, 28.9, 28.13: Carla Simmons. 28.5, 28.7: Barbara Cousins.

Chapter 29 29.2a, 29.18, 29.42: Barbara Cousins. 292b-29.5, 29.7, 29.8, 29.9, 29.11, 29.14, 29.15, 29.19, 29.23. 29.25, 29.27, 29.32, 29.36, 29.40, Table 29.2: Carla Simmons. 29.41: Raychel Ciemma.

Chapter 30 30.2–30.4, 30.7, 30.10, 30.12, 30.13, 30.17, 30.21, Table 30.2: Carla Simmons. 30.29: Chris Carothers.

Chapter 31 31.2, 31.3, 31.9, 31.26: Carla Simmons. 31.5–31.7, 31.10, 31.12, 31.13, 31.15–31.17, 31.23: Cecile Duray-Bito/Linda McVay. 31.29: Carol Verbeeck.

Chapter 32 32.2: Barbara Cousins. 32.3: Fran Milner/Linda McVay. 32.5, 32.7a, 32.9, 32.10, 32.12, 32.15: Cecile Duray-Bito/Linda McVay. 32.8, 32.11, Methods Box (p. 712): Carla Simmons; from *Plant Physiology,* 4th ed., by Robert M. Devlin and Francis H. Witham © 1983 by PWS Publishers. Reprinted by permission of Wadsworth, Inc.

Chapter 33 33.2, 33.6, 33.7, 33.10, 33.12: Elizabeth Morales. Methods Box (p. 723): Carla Simmons.

Chapter 34 34.2, 34.3: Cecile Duray-Bito/ Elizabeth Morales. 34.5, 34.8, 34.9, 34.13, 34.14: Fran Milner/Elizabeth Morales.

Chapter 35 35.2–35.6, 35.14–35.17: Elizabeth Morales. 35.7: Elizabeth Morales; adapted from D. Rayle and L. Wedberg, *Botany: A Human Concern.* (Boston: W.B. Saunders Company, 1975). Methods Box (p. 764): Carla Simmons.

Chapter 36 36.2: Raychel Ciemma. 36.3, 36.6, 36.9, 36.10a: Darwen and Vally Hennings. 36.4: From W.M. Becker, *The World of the Cell.* (Menlo Park, CA: 1986), p. 308.

Chapter 37 37.5, 37.7–37.11, 37.13, 37.14, 37.16, 37.17: Darwen and Vally Hennings. 37.6: Raychel Ciemma. 37.12: Fran Milner; from L.G. Mitchell, J.A. Mutchmor, and W.D. Dolphin, *Zoology.* (Redwood City, CA: Benjamin/Cummings, 1988), p. 93. 37.15: Kenneth R. Miller; from E.N. Marieb, *Human Anatomy and Physiology.* (Redwood City, CA: Benjamin/Cummings, 1989), p. 771. 37.18: Carol Verbeeck. 37.19: Elizabeth Morales.

Chapter 38 38.2: Fran Milner; from L.G. Mitchell, J.A. Mutchmor, and W.D. Dolphin, *Zoology.* (Redwood City, CA: Benjamin/Cummings, 1988), p. 123. 38.3–38.6, 38.8a, 38.10, 38.11, 38.13–38.15, 38.17–38.19, 38.21, 38.22, 38.24, 38.27, Methods Box (p. 826): Darwen and Vally Hennings. 38.7: Barbara Cousins; from E.N. Marieb, *Human Anatomy and Physiology.* (Redwood City, CA: Benjamin/Cummings, 1989), p. 607. 38.9, 38.20: Carol Verbeeck. 38.23: From K.D. Johnson, D.L. Rayle, and H.L. Wedberg, *Biology: An Introduction.* (Menlo Park, CA: Benjamin/Cummings, 1984), p. 378. 38.25: From A.P. Spence and E.B. Mason, *Human Anatomy and Physiology,* 3d ed. (Menlo Park, CA: Benjamin/Cummings, 1987), p. 666. 38.26: Carla Simmons.

Chapter 39 39.4, 39.9, 39.11, 39.16, 39.17: Carla Simmons. 39.5, 39.6, 39.12, 39.13, 39.15, 39.18: Carla Simmons; from E.N. Marieb, *Human Anatomy and Physiology.* (Redwood City, CA: Benjamin/Cummings, 1989), pp. 679, 683, 698, 686, 691, 696. 39.7, 39.14a: Darwen and Vally Hennings. 39.8, 39.10, Table 39.1 (p. 862): Carol Verbeeck. Methods Box (p. 865): From W.M. Becker, *The World of the Cell.* (Menlo Park, CA: Benjamin/Cummings, 1986), p. 739.

Chapter 40 40.2, 40.7–40.10, 40.12–40.14: Raychel Ciemma. 40.3: Fran Milner/Martha Blake; from L.G. Mitchell, J.A. Mutchmor, and W.D. Dolphin, *Zoology.* (Redwood City, CA: Benjamin/Cummings, 1988), pp. 150, 151. 40.6:

Darwen and Vally Hennings: adapted from Schmidt and Nielsen, *Animal Physiology: Mechanisms and Adaptations.* (New York: W.H. Freeman, 1983). 40.16, 40.17a, 40.18, 40.20, 40.21a: Darwen and Vally Hennings. 40.11: Linda McVay. 40.15: Janet Hayes. 40.22: Kenneth R. Miller.

Chapter 41 41.2, 41.12, 41.16, 41.17: Janet Hayes. 41.3a, 41.5, 41.7, 41.9, 41.14: Raychel Ciemma. 41.4: Kenneth R. Miller. 41.6, 41.8, 41.10, 41.11: John and Judy Waller. 41.13: Barbara Cousins: from E.N. Marieb, *Human Anatomy and Physiology.* (Redwood City, CA: Benjamin/Cummings, 1989), p. 163. 41.15: Jeanne Koelling: from E.N. Marieb, *Human Anatomy and Physiology.* (Redwood City, CA: Benjamin/Cummings, 1989), p. 14. 41.18: Charles W. Hoffman; from E.N. Marieb, *Human Anatomy and Physiology.* (Redwood City, CA: Benjamin/Cummings, 1989), p. 554.

Chapter 42 42.2b, 42.5, 42.6, 42.13: Carla Simmons. 42.7–42.12, 42.14a, 42.15–42.17: Darwen and Vally Hennings

Chapter 43 43.2, 43.3, 43.6, 43.12, 43.19, 43.24: Linda McVay. 43.4, 43.20, 43.21: Chris Carothers. 43.22: Chris Carothers; from B.S. Guttman and J.W. Hopkins III, *Understanding Biology.* (New York: Harcourt Brace Jovanovich, 1983), p. 395. 43.8, 43.9, 43.13, 43.15, 43.17, 43.18, 43.25, 43.27, 43.28: Carla Simmons.

Chapter 44 44.2, 44.4a, 44.11, 44.21, 44.27: Charles W. Hoffman; from E.N. Marieb, *Human Anatomy and Physiology.* (Redwood City, CA: Benjamin/Cummings, 1989), pp. 336, 339, 358, 369, 482. 44.3, 44.6, 44.12, 44.13, 44.15, 44.17, 44.18, 44.28: Carla Simmons. 44.5, 44.7–44.10, 44.16, 44.19. 44.20, 44.25, 44.26, Methods Box (p. 983): John and Judy Waller. 44.14: Janet Hayes. 44.22: Linda McVay; based on A.S. Romer and P.S. Parsons, *The Vertebrate Body,* 6th ed., (Philadelphia: W.B. Saunders, 1986), p. 569. 44.23: John and Judy Waller/Carla Simmons. 44.24: Stephanie McCann; from E.N. Marieb, *Human Anatomy and Physiology.* (Redwood City, CA: Benjamin/Cummings, 1989), p. 389.

Chapter 45 45.2, 45.10b, 45.13, 45.23, 45.25: Carla Simmons. 45.5, 45.20: John and Judy Waller; from L.G. Mitchell, J.A. Mutchmor, and W.D. Dolphin, *Zoology.* (Redwood City, CA: Benjamin/Cummings, 1988), pp. 234, 283. 45.6b, 45.12, 45.22: John and Judy Waller. 45.8, 45.10a, 45.15, 45.16, 45.17, 45.19: Charles W. Hoffman: from E.N. Marieb, *Human Anatomy and Physiology.* (Redwood City, CA: Benjamin/ Cummings, 1989), pp. 496, 499, 513, 515, 523,

519. 45.9, 45.14, 45.24, 45.30–45.34: John and Judy Waller/Carla Simmons. 45.11: Janet Hayes. 45.27: Nadine Sokol; from E.N. Marieb, *Human Anatomy and Physiology.* (Redwood City, CA: Benjamin/Cummings, 1989), p. 174. 45.28: Barbara Cousins; from E.N. Marieb, *Human Anatomy and Physiology.* (Redwood City, CA: Benjamin/Cummings, 1989), p. 226. 45.29: Darwen and Vally Hennings. 45.35: Raychel Ciemma; from E.N. Marieb, *Human Anatomy and Physiology.* (Redwood City, CA: Benjamin/Cummings, 1989), p. 254.

Chapter 46 46.6, 46.10 (b, c, f), 46.11: Raychel Ciemma. 46.7: Elizabeth Morales; data from Mooney, Bjorkman, and Berry, in Hadley (editor), *Environmental Physiology of Desert Organisms.* (Stroudsberg, PA: Dowden, Hutchinson & Ross, 1975), p. 138. 46.9: Elizabeth Morales; based on data from Fry and Hart, 1948. 46.10(a, d, e), 46.22: Elizabeth Morales.

Chapter 47 47.3, 47.14, 47.16: Carol Verbeeck. 47.6–47.8, 47.10, 47.12, 47.15: Raychel Ciemma. 47.11: Carol Verbeeck; based on data from MacLulich, 1937.

Chapter 48 48.1, 48.2, 48.6, 48.17, 48.19: Janet Hayes. 48.5: Janet Hayes; based on data from G.F. Gause, *The Struggle for Existence,* 1934. Reprint. (New York: Macmillan, 1964). 48.7: Raychel Ciemma; from R. H. Ricklefs, *Ecology,* 3d ed. (New York: Chiron Press, 1988). 48.18b: Raychel Ciemma; from *Discover,* Nov. 1986, p. 31. © 1989 Discover Publications.

Chapter 49 49.1a, 49.3, 49.5, 49.9, 49.10, 49.16, Methods Box (p. 1121): Raychel Ciemma. 49.4: Raychel Ciemma; from R.L. Smith, *Ecology and Field Biology,* 3d ed. (New York: Harper & Row, 1980). 49.6, 49.14: Janet Hayes. 49.7: Raychel Ciemma; from R. H. Ricklefs, *Ecology,* 2d edition. (New York: Chiron Press, 1988). 49.8: Raychel Ciemma; from P. Colinvaux, *Ecology.* (New York: John Wiley & Sons, 1986). 49.12: Elizabeth Morales; adapted from G.E. Likens et al., "Effects of Forest Cutting and Herbicide Treatment on Nutrient Budgets in the Hubbard Brook Watershed-Ecosystem," *Ecological Monographs* 40, 1970.

Chapter 50 50.1, 50.10, 50.12, 50.13, 50.14a, 50.18, 50.19: Elizabeth Morales. 50.3, 50.11, 50.14b: Raychel Ciemma. 50.4: Elizabeth Morales; from N. Tinbergen, *The Study of Instinct.* (Oxford: Oxford University Press, 1951). 50.8: Elizabeth Morales; data from J. Alcock, *Animal Behavior,* 4th ed. (Sunderland, MA: Sinauer, 1989).

Index

Primitive streak, 964
Primordia, 939
Principle of allocation, 1056
Probes, 405, *406*
Proboscideans, characteristics of, *663*
Proboscis worms, 615
Procambium, 693
Prochlorothrix, 559
Productive cycle, 356
Products, 33
Progesterone, 909
  structure of, *925*
Progestins, 927
Proglottids, 615
Prokaryotes, 122, 522–37
  cell division in, 230
  cell surface of, 524–25
  chemical cycles and, 534
  diversity of, 529–34
  flagella of, *527*
  gene expression in, control of, 367–73
  genome of, 523–24
  importance of, 534–35
  membranes of, 524, *525*
  metabolic diversity of, 529
    origins of, 535–37
  morphology of, 523
  motility of, 525–26
  origin on Earth, 512
  photosynthesis in, 204–5
  photosynthetic, 206
  Precambrian, *512*
  reproduction and growth of, 526–29
Prokaryotic cells, 6, *6*, 122–23, *122*
  protein synthesis in, 334
  structures of, *150*
Prolactin (PRL), 920
Proliferative phase, 941
Proline, structure of, *76*
Promiscuous mating, 1157
Promoters, 327
Pro-opiomelanocortin, 920
Propanal, structure of, *58*, *59*
Propane, carbon skeleton of, *56*
Prophages, 354
Prophase, 233, *234*, *239*
Prophase I, *252*, *254*
Prophase II, *253*
Proplastids, 139–40
Prosencephalon, 999
Prosimians, 665, *665*
Prostaglandins, 908–9, *909*
Prostate gland, 937
Protandrous, 933
Protein conformation, 77–78, *78*
  determination of, 82–83
Protein deficiency, 810–11
Protein-folding problem, 83
Protein kinase, 912–13
Proteinoids, 515
Proteins, *52*, 74–83
  activator, 370
  antimicrobial, 851–53
  catabolism of, 198
  comparison of, molecular systematics
    and, 487–88
  denaturation of, 82–83, *83*
  functions of, *75*
  gel electrophoresis of, 403
  integral, 160
  membrane, functions of, *161*
  mutations and, 340–45
  peripheral, 160
  repressor, 354, 369
  single-strand binding, 316

structure of
  primary, 78–80, *79*
  quaternary, 81–82, *82*
  secondary, 80–81, *80*
  tertiary, 81, *81*
synthesis of, 328–35
  elongation cycle of, 332–33, *333*
  initiation of, 332, *332*
  nuclear control of, *129*
  rough endoplasmic reticulum and,
    131–32
  sites of, 334–35
  termination of, 333–34
  transport, 162–63
    classes of, *163*
Protista, 9, 520
  boundaries of, 541–42
  other kingdoms compared, *519*
Protists, 540–60
  algal, *542*, 546–54
  characteristics of, 541
  fungi-like, *542*, 554–58
  photosynthesis in, 204–5
Protobionts, 514
  catalysis in, *516*
  formation of, 515–16
  laboratory versions of, *515*
Protoderm, 693
Protogynous, 933
Protonephridia, 879–81, *880*
Proton gradient, generation of, 191–93
Proton-motive force, ATP synthesis and,
    193–94
Proton pump, 706
Protons, 23–24
Proto-oncogenes, 391
Protoplast fusion, 753–54, *754*
Protostomes, 606–7, 616–31
  early development in, *607*
Protozoa, 9, 542–46, *542*
Proviruses, 356
Proximal convoluted tubule, 883
  transport properties of, 886
Proximate cause, 1140
Pseudocoelom, 605
Pseudocoelomates, 606, 615–16
Pseudogenes, 367, 385–86
Pseudomonads, 532
Pseudopodia, 146, 543
Pseudosex, *932*
Psilophyta, 567, 572–73
Psilotum, 572–73
  sporophyte of, *572*
P site, 332
Pterophyta, 567, 574–75
Puffballs, *595*
Pulmonary circuit, 820
Pulse, 822–23
Punctuated equilibrium, 419–22, 474, *475*,
    496
Punnett, Reginald Crundall, 288
Punnett square, 265
Pupil, 1018
Purgatorius unio, 664
Purines, 86
Pyloric sphincter, 802
Pyridoxine, requirements for humans, *812*
Pyrimidines, 86
Pyrogens, 851
Pyruvic acid, catabolism and, 196–97, *197*

Quadrats, 1076
Quanta, 210
Quantitative traits, 274
Quaternary structure, 81–82, *82*

Radial symmetry, 605, *605*
Radiata, 605, 610–11
Radiation, 210, 894, *894*
Radiation damage, to forests, *25*
Radicle, 745
Radioactive dating, 482
Radioactive isotopes, 25
Radioactive tracers, use in biology, 26
Radioautography, 113–14
Radiolarians, *543*, 544
Radula, 617
Rain forests, tropical, 1062–63, *1063*
Ramapithecus, 668
Rana temporaria, life cycle of, *652*
Random dispersion, 1076, *1076*
Random fertilization, 258–59
Ratite birds, *659*
Rattlesnakes, ritual wrestling by, *1154*
Reactants, 33
Reaction center, 213–14, *215*
Reactions. *See* Chemical reactions
Reading frame, 332
Realized niche, 1098, *1098*
Reception, 1012–13
Receptor-mediated endocytosis, 173–74, *173*
Receptor potential, 1013
Receptors, 907
  electromagnetic, 1014, *1015*
  gustatory, 1014
  nervous pathway to, *981*
  neurotransmitter, 992–94
  olfactory, 1014
  pain, 1016
  sensory, 1011–16
    function of, 1012–13
    types of, 1013–16
Receptor sites, 352
Recessive allele, 265
Reciprocal altruism, 1160
Recognition concept of species, 473
Recombinant DNA, *402*
Recombinant DNA gene products,
  manufacture of, 406–8, *406*
Recombinant DNA technology, 399–416
  applications of, 408–16
  safety of, 416
Recombinants, 289
Recombination. *See* Genetic recombination
Recommended daily allowances (RDAs),
    811–12
Recruitment, 1039
Rectum, 806
Red algae, 553–54, *554*
  characteristics of, *548*
Red blood cells, 831
  sickle-cell allele and, *277*
Redox process, photosynthesis as, 208
Redox reactions, 183–85
  cellular respiration as, 194
  methane combustion as, *184*
Reducing agents, 183
Reduction, 183
Reflex, 997
  spinal, 999, *999*
Refractory period, 985
Regeneration, 389, *389*, 610, 931
Regulative development, 970
Regulators, 1056
Regulatory genes, 370
  macroevolution and, 493–94
Relative fitness, 452
Release factors, 333
Releaser, 1143
Releasing factors, 918
  hypothalamic, 921

eggs of, fertilization of, *957, 958*
gastrulation in, *960*
larva of, *961*
Seaweed, 553
Secondary cell wall, 147, *147*
Secondary compounds, 1100
Secondary consumers, 1115–16
Secondary immune response, 858
Secondary productivity, 1122
Secondary roots, 694
Secondary structure, 80–81, *80*
Secondary succession, 1105
Second law of thermodynamics, 94
Second messengers, peptide hormones
    and, 910–15
Secretin, 802
Secretory phase, 941
Sedimentary rock
    fossils in, 425–26, *426*
    stratification of, *426*
Seed, 566, 744–45
    structure of, *746*
Seed coat, 745
Seed dormancy, 748–49
Seed germination, 748–51, *749, 750*
    gibberellins and, 765
Seed plants, terrestrial adaptations of, 575
Segmented worms, 620–22, *621*
Segregation
    as chance event, *269*
    in dihybrid cross, *270*
    law of, 264–67, *266*
Seilacher, Adolf, 636
Selaginella, 573
Selection, 400–1. *See also* Natural selection
    against lethal allele, *452*
    directional, 455
    diversifying, 455
    sexual, 455–56, *455*
    species, 497
    stabilizing, 455
Selection coefficient, 452
Selective permeability, 154, 162–63, *162*
Self-fertilization, 445
Semelparous, 1086
Semen, 936–37
Semicircular canals, equilibrium and, *1028*
Semilunar valves, 821
Seminal vesicles, 936–37
Seminiferous tubules, 936
Semispecies, 472
Senescence, in plants, ethylene and, 766
Sensation, 1012
Sensory adaptation, 1013
Sensory nervous system, 997
Sensory neurons, 979
Sensory receptors, 1011–16
    function of, 1012–13
    types of, 1013–16
Sepals, 581, *582*, 739
Septa, 591
Serine, structure of, *76*
Serotonin, 993
    structure of, *992*
Setae, 1030
Set point, 789
Severe combined immunodeficiency
    (SCID), 869
Sex
    chromosomal basis of, 293–94
    heterogametic, 294
    homogametic, 294
Sex chromosomes, 250
    abnormalities in, 300–2, *301*
Sex determination, systems of, 294, *295*

Sex hormones, functional groups of, *58*
Sex-influenced traits, 297–98
Sex-limited traits, 297–98
Sex-linked genes, 287–88
Sex-linked inheritance, 294–98
Sex-linked traits, 294
Sex structure, 1078–79
Sexual dimorphism, 455–56, *455*
Sexual investment, 1157
Sexual life cycles, 250–51, *260*
    variety of, 259–60
Sexual maturation, 946
Sexual recombination, genetic variation
    and, *447*, 449
Sexual reproduction, 250, 930, 931
    mechanisms of, 933–36
Sexual response cycle, 946
Sexual selection, 455–56, *455*
Sharks, 647–48, *647*
Sheldon, Peter, 475
Shelf fungi, 595
Shoots, primary growth of, 695–701, *695*
Shoot system, 682–86
Short-day plants, 771
Short-term memory, 1005
Shrews, 664
Shrimp, 625, *626*
Sickle-cell anemia, 79, 272, 277–78, 340, 450
    molecular basis of, *341*
Sickle-cell trait, 278
Side chain, 75
Sieve plates, 691
Sieve-tube members, *689*, 690–91
Sieve tubes
    pressure flow in, 717, *717*
    translocation and, 715, *715*
Signal hypothesis, 132, *132*
Signal sequence, 132
Sign stimuli, 1143, *1143*, 1144–45
Silurian period
    agnathans and, 645
    green algae and, 566
    placoderms and, 646
Simberloff, D. S., 1111
Simple cuboidal epithelium, 783
Simple epithelium, 782
Simple fruit, 746–47
Simple tissue, 691, *691*
Simpson, George Gaylord, 506
Singer, S., 156
Single-lens eye, 1017, *1018*
Single-strand binding proteins, 316
Sink, 715
Sinoatrial (SA) node, 823
Siphonapterans, characteristics of, *628*
Sirenians, characteristics of, *663*
Sister chromatids, 230–31, *231*
Skeletal muscle, 786, 1036–41
    structure of, *1035*
Skeleton, 1032–35
    cooperation with muscles in movement,
    1035
    human, *1034*
    hydrostatic, 1033–34
    types of, *1033*
    vertebrate, *642*
        joints in, *1035*
Skin
    defense system and, 850
    pigmentation of, polygenic inheritance of,
    274, *275*
    thermoregulation and, *896*
    receptors in, 1013–14, *1014*
Skinner, B. F., 1148
Skoog, Folke, 762

Sleep, 1003–4
Sleep movements, 770, *770*
Slender loris, *665*
Sliding filament model, *1037*
Slime molds
    cellular, 555
        fruiting structure of, *557*
    life cycle of, *556*
    plasmodial, 555, *555*
Slow block to polyspermy, 957
Slugs, 616–17
Small intestine, 802–4
    activation of zymogens in, *804*
    digestion in, 802–4
    nutrient absorption in, 804
    structure of, *805*
Small nuclear ribonucleoproteins (snRNPs),
    337–38
    mRNA splicing and, *338*
Small nuclear RNA. *See* snRNA
Smell, 1030–31
Smooth endoplasmic reticulum (SER), 130
    functions of, 130
Smooth muscles, 1041
Smuts, 595
Snails, 616–17
Snakes, 655, *656*
    infrared receptors of, *1015*
Snapdragons, color of, incomplete
    dominance in, *272*
snRNA, 337
    function of, *340*
Social behavior, 1153–55
Sociobiology, human, 1161–62
Sodium, requirements for humans, *813*
Sodium chloride crystal, *32*
Sodium-potassium pump, 168, *169*, 983
Soil, 726–29
    cation exchange in, 728, *728*
    distribution of organisms and, 1054–55
    texture and composition of, 726–27
Soil horizons, 726, *727*
Soil management, 728–29
Soil particles, sizes of, *726*
Soil water, availability of, 727–28, *728*
Solar energy, light reactions and, 209–19
Solar navigation, *1151*
Solute concentration, 44–45
Solutes, 42
    diffusion of, *163*
    water mobility and, *165*
Solutions, 42
    aqueous, 43, 44–48
        pH of, *47*
    neutral, 45
Solvents, 42–44
Somatic cells, 250
    plant development from, 388–89, *388*
Somatic nervous system, 997
Somatomedins, 920
Somites, 643, *963*
Soredia, *597*
Sound, *1026*
Source, 715
Spatial summation, 991
Specialized transduction, 364
Speciation, 421–22
    allopatric, 465–68, *466*
    biogeography of, 465–72
    by divergence, 472–73
    genetic mechanisms of, 472–74
    interpretations of, gradual and
        punctuated, 474–76
    parapatric, 465, *466*, 468–70
    patterns of, *460*